ADLER'S
PHYSIOLOGY OF THE EYE

Commissioning Editor: *Russell Gabbedy*
Development Editor: *Ben Davie/Louise Cook*
Editorial Assistant: *Kirsten Lowson*
Project Manager: *Nayagi Athmanathan*
Design: *Charles Gray*
Illustration Manager: *Bruce Hogarth*
Illustrator: *Antbits*
Marketing Manager(s) (UK/USA): *Gaynor Jones/Helena Mutak*

ADLER'S
PHYSIOLOGY OF THE EYE

ELEVENTH EDITION

Editors:

Leonard A. Levin MD PhD
Canada Research Chair of Ophthalmology and Visual Sciences
University of Montreal
Montreal, Quebec, Canada
Professor of Ophthalmology and Visual Sciences
University of Wisconsin
Madison, WI, USA

Siv F. E. Nilsson PhD
Lecturer, Department of Medical and Health Sciences,
   Division of Drug Research
Linköping University
Linköping, Sweden

James Ver Hoeve PhD
Senior Scientist, Department of Ophthalmology and
   Visual Sciences
University of Wisconsin School of Medicine and Public Health
Madison, WI, USA

Samuel M. Wu PhD
Camille and Raymond Hankamer Chair in Ophthalmology
Professor of Ophthalmology, Neuroscience and Physiology
Baylor College of Medicine
Houston, TX, USA

Managing Editors:

Paul L. Kaufman MD
Peter A. Duehr Professor and Chairman
Department of Ophthalmology and Visual Sciences
University of Wisconsin School of Medicine and Public Health
Madison, WI, USA

Albert Alm MD
Professor
Department of Neuroscience & Ophthalmology
University Hospital
Uppsala, Sweden

SAUNDERS
ELSEVIER

Edinburgh, London, New York, Oxford, Philadelphia, St Louis, Sydney, Toronto

SAUNDERS
ELSEVIER

SAUNDERS is an imprint of Elsevier Inc.

Notices

Knowledge and best practice in this field are constantly changing. As new research and experience broaden our understanding, changes in research methods, professional practices, or medical treatment may become necessary. Practitioners and researchers must always rely on their own experience and knowledge in evaluating and using any information, methods, compounds, or experiments described herein. In using such information or methods they should be mindful of their own safety and the safety of others, including parties for whom they have a professional responsibility.

With respect to any drug or pharmaceutical products identified, readers are advised to check the most current information provided (i) on procedures featured or (ii) by the manufacturer of each product to be administered, to verify the recommended dose or formula, the method and duration of administration, and contraindications. It is the responsibility of practitioners, relying on their own experience and knowledge of their patients, to make diagnoses, to determine dosages and the best treatment for each individual patient, and to take all appropriate safety precautions.

To the fullest extent of the law, neither the Publisher nor the authors, contributors, or editors, assume any liability for any injury and/or damage to persons or property as a matter of products liability, negligence or otherwise, or from any use or operation of any methods, products, instructions, or ideas contained in the material herein.

ISBN: 978-0-323-05714-1

British Library Cataloguing in Publication Data
Adler's physiology of the eye.—11th ed.
  1. Eye–Physiology.
  I. Physiology of the eye II. Kaufman, Paul L. (Paul Leon), 1943- III. Alm, A. IV. Adler, Francis Heed, 1895-
  Physiology of the eye.
  612.8'4—dc22

Library of Congress Cataloging in Publication Data
A catalog record for this book is available from the Library of Congress

your source for books,
journals and multimedia
in the health sciences
www.elsevierhealth.com

Working together to grow
libraries in developing countries

www.elsevier.com | www.bookaid.org | www.sabre.org

ELSEVIER   BOOK AID International   Sabre Foundation

The publisher's policy is to use **paper manufactured from sustainable forests**

Printed in China
Last digit is the print number: 9 8 7 6 5 4 3 2

Contents

SECTION 7 Visual processing in the retina

SECTION 8 Non-perceptive vision

SECTION 9 Visual processing in the brain

SECTION 10 Visual perception

SECTION 11 Development and deprivation of vision

Preface

The 11th edition of Adler's sees a significant reorganization based on function and function/structure relationships within ocular cells, tissues and organs. Prior editions had been organized largely anatomically, but in the eight years since the 10th edition, basic, translational and clinical knowledge has increased exponentially. Capturing, synthesizing, organizing and conveying vast amounts of this new information in the context of a new organizational strategy has been both challenging and stimulating. We hope that the readers, especially our younger scientists and clinicians who represent the future of our field, will find this approach, material and presentation conducive to learning, retention and referral.

List of Contributors

Albert Alm MD
Professor
Department of Neuroscience &
Ophthalmology
University Hospital
Uppsala Sweden

David C Beebe PhD FARVO
Janet and Bernard Becker Professor of
Ophthalmology and Visual Science
Professor of Cell Biology and
Physiology
Department of Ophthalmology and
Vision Sciences
Washington University
Saint Louis MO USA

Carlos Belmonte MD PhD
Professor of Human Physiology
Medical School
Instituto de Neurociencias de Alicante
Universidad Miguel Hernandez
San Juan Alicante Spain

David M Berson PhD
Professor of Neuroscience
Department of Neuroscience
Brown University
Providence RI USA

Sai H S Boddu BPharm MS
PhD Candidate and Doctoral Fellow
School of Pharmacy
University of Missouri-Kansas City
Kansas City MO USA

Jamie D Boyd PhD
Research Assistant
Psychiatry, Faculty of Medicine
University of British Columbia
Vancouver BC Canada

Vivien Casagrande PhD
Professor, Cell & Developmental
Biology, Psychology, and
Ophthalmology & Visual Sciences
Departments of Cell & Developmental
Biology
Vanderbilt Medical School
Nashville TN USA

Yuzo M Chino PhD
Benedict-McFadden Professor
Professor of Vision Sciences
College of Optometry
University of Houston
Houston TX USA

Darlene A Dartt PhD
Senior Scientist
Harold F Johnson Research Scholar
Schepens Eye Research Institute
Associate Professor
Harvard Medical School
Boston MA USA

Chanukya R Dasari MS
MD Candidate
School of Medicine
University of Missouri-Kansas City
Kansas City MO USA

Daniel G Dawson MD
Ophthalmic Researcher
Emory University Eye Center
Atlanta GA USA

Henry F Edelhauser PhD
Professor of Ophthalmology
Department of Ophthalmology
Emory University Eye Center
Atlanta GA USA

Erika D Eggers PhD
Assistant Professor
Department of Physiology
University of Arizona
Tucson AZ USA

Ione Fine PhD
Assistant Professor of Psychology
Department of Psychology
University of Washington
Seattle WA USA

Laura J Frishman PhD
Professor of Vision Science, Optometry
and Biology
College of Optometry
University of Houston
Houston TX USA

B'Ann True Gabelt MS
Distinguished Scientist
Department of Ophthalmology and
Vision Sciences
University of Wisconsin
Madison WI USA

Juana Gallar MD PhD
Professor of Human Physiology
Instituto de Neurociencias and
Facultad de Medicina
Universidad Miguel Hernández-CSIC
San Juan de Alicante Spain

Adrian Glasser PhD
Professor of Optometry and Vision
Sciences and Biomedical Engineering
Benedict/Pitts Professor
College of Optometry
University of Houston
Houston TX USA

Jeffrey L Goldberg MD PhD
Assistant Professor of Ophthalmology,
and of Neurosciences
Bascom Palmer Eye Institute
University of Miami
Miami FL USA

Gregory J Griepentrog MD
Clinical Instructor
Department of Ophthalmology and
Visual Sciences
University of Wisconsin-Madison
Madison WI USA

Alecia K Gross PhD
Assistant Professor
Department of Vision Sciences,
Biochemistry and Molecular
Genetics, Cell Biology and
Neurobiology
The University of Alabama at
Birmingham
Birmingham AL USA

Ronald S Harwerth OD PhD
John and Rebecca Moores Professor of
Optometry
College of Optometry
University of Houston
Houston TX USA

Horst Helbig MD
Professor of Ophthalmology
Director of the University Eye Hospital
Klinik und Poliklinik fur
Augenheilkunde
Klinikum der Universitats Regensburg
Regensburg Germany

Robert F Hess PhD DSc
Professor of Ophthalmology
McGill Vision Research
McGill University
Montréal QC Canada

Jennifer Ichida PhD
Postdoctoral Fellow
Department of Ophthalmology &
Visual Science
Moran Eye Center
University of Utah
Salt Lake City UT USA

Chris A Johnson PhD DSc FARVO
Professor
Department of Ophthalmology
University of Iowa
Iowa City IA USA

Randy Kardon MD PhD
Professor and Director of
Neuro-ophthalmology
Pomerantz Family Chair of
Ophthalmology
Department of Ophthalmology and
Visual Sciences
University of Iowa and Veterans
Administration Hospitals
Iowa City IA USA

Pradeep K Karla PhD
Assistant Professor of Pharmaceutical
Sciences
Department of Pharmaceutical Sciences
School of Pharmacy
Howard University
Washington DC USA

Paul L Kaufman MD
Peter A Duehr Professor and
Chairman
Department of Ophthalmology &
Visual Science
University Wisconsin Madison
Madison WI USA

SM Koch MS
Graduate Student
Neuroscience Graduate Program
University of California, San Francisco
San Francisco CA USA

Ron Krueger MD MSE
Professor of Ophthalmology
Cleveland Clinic Lerner College of
Medicine
Medical Director
Department of Refractive Surgery
Cole Eye Institute
Cleveland Clinic Foundation
Cleveland OH USA

James A Kuchenbecker PhD
Senior Fellow
Department of Ophthalmology
University of Washington
WA USA

Trevor D Lamb BE ScD FRS FAA
Distinguished Professor, John Curtin
School of Medical Research
and Research Director, ARC Centre of
Excellence in Vision Science
The Australian National University
Canberra City ACT Australia

Dennis M Levi OD PhD
Professor of Optometry and Vision
Science
Professor, Helen Wills Neuroscience
Institute
Dean, School of Optometry
University of California, Berkeley
Berkeley CA USA

Lindsay B Lewis PhD
Postdoctoral Fellow
Department of Ophthalmology
McGill Vision Research
Montreal QC Canada

Mark J Lucarelli MD FACS
Professor
Director of Oculoplastics Service
Department of Ophthalmology and
Visual Sciences
University of Wisconsin at Madison
Madison WI USA

Peter D Lukasiewicz PhD
Professor of Ophthalmology &
Neurobiology
Department of Ophthalmology &
Visual Sciences
Washington University School of
Medicine
St Louis MO USA

Henrik Lund-Anderson MD DMSc
Professor of Ophthalmology
Department of Ophthalmology
University of Copenhagen
Glostrup Hospital
Copenhagen Denmark

Peter R MacLeish PhD
Professor of Neurobiology
Director
Neuroscience Institute
Morehouse School of Medicine
Atlanta GA USA

Clint L Makino PhD
Associate Professor of Ophthalmology
(Neuroscience)
Department of Ophthalmology
Massachusetts Eye and Ear Infirmary
& Harvard Medical School
Boston MA USA

Katherine Mancuso PhD
Postdoctoral Fellow
Department of Ophthalmology
University of Washington
Seattle WA USA

Robert E Marc PhD
Professor of Ophthalmology
John A Moran Eye Center
University of Utah
Salt Lake City UT USA

Roan Marion BS
Graduate Student
Neuroscience PhD Program
Casagrande Vision Research Lab
Vanderbilt University Medical School
Nashville TN USA

Joanne A Matsubara BA PhD
Professor and Director of Research
(Basic Sciences)
Eye Care Centre Department of
Ophthalmology and Visual Sciences
University of British Columbia
Vancouver BC Canada

Allison M McKendrick PhD
Senior Lecturer
Department of Optometry and Vision
Sciences
The University of Melbourne
Melbourne VIC Australia

Linda McLoon PhD
Professor of Ophthalmology and
Neuroscience
Departments of Ophthalmology and
Neuroscience
University of Minnesota
Minneapolis MN USA

David Miller MD
Associate Clinical Professor of
Ophthalmology
Department of Ophthalmology
Harvard Medical School
Boston MA USA

Ashim K Mitra PhD
Curators' Professor of Pharmacy
Vice-Provost for Interdisciplinary
  Research
Chairman, Division of Pharmaceutical
  Sciences
University of Missouri-Kansas City
Kansas City MO USA

Jay Neitz PhD
Bishop Professor
Department of Ophthalmology
University of Washington Medical
  School
Seattle WA USA

Maureen Neitz PhD
Ray H Hill Professor
Department of Ophthalmology
University of Washington Medical
  School
Seattle WA USA

Anthony M Norcia PhD
Professor
Department of Psychology
450 Serra Mall
Stanford University
Stanford, CA USA

Lance M Optican PhD
Chief, Section on Neural Modelling
Laboratory of Sensorimotor Research
National Eye Institute
National Institutes of Health
Bethesda MD USA

Carole Poitry-Yamate PhD
Senior physicist, Institute of Physics for
  Complex Matter, Centre d'Imagerie
  Biomédicale (CIBM), Laboratory for
  Functional and Metabolic Imaging,
  Ecole Polytechnique Fédérale de
  Lausanne
Lausanne Switzerland

Constantin J Pournaras MD
Professor in Ophthalmology
Department of Ophthalmology
Vitreo-Retinal Unit
Faculty of Medicine
University Hospitals of Geneva
Geneva Switzerland

Christian Quaia MSc PhD
Staff Scientist
Laboratory of Sensorimotor Research
National Eye Institute
National Institutes of Health
Bethesda MD USA

Charles E Riva DSc
Professor Emeritus, University of
  Lausanne
Professor a contratto
University of Bologna
Grimisuat Switzerland

Birgit Sander MSc PhD
Head of Laboratory
Department of Ophthalmology
Glostrup Hospital
Copenhagen Denmark

Clifton M Schor OD PhD
Professor of Optometry, Vision Science,
  Bioengineering
School of Optometry
University of California at Berkeley
Berkeley, CA, USA

Paulo Schor MD
Affiliated Professor of Ophthalmology
Department of Ophthalmology and
  Medical Informatics
Bioengineering Laboratory and
  Refractive Surgery Clinic
Federal University of Sao Paulo
São Paulo SP Brazil

Ricardo N Sepulveda MD
Refractive Surgery Fellow
Cole Eye Institute
The Cleveland Clinic Foundation
Cleveland OH USA

Olaf Strauss Prof. PhD
Professor of Experimental
  Ophthalmology
Research Director
Klinik und Poliklinik fur
  Augenheilkunde
Klinikum der Universität Regensburg
Regensburg Germany

Timo T Tervo MD PhD
Professor of Applied Clinical
  Ophthalmology
Division of Ophthalmology
Helsinki University Eye Hospital
Helsinki Finland

John L Ubels PhD FARVO
Professor of Biology, Calvin College,
  Grand Rapids MI
Adjunct Professor
Department of Ophthalmology
Wayne State University School of
  Medicine
Detroit MI USA

EM Ullian PhD
Assistant Professor
Department of Ophthalmology
University of California
San Francisco CA USA

Michael Wall MD
Professor of Neurology and
  Ophthalmology
Department of Ophthalmology and
  Visual Sciences
University of Iowa
Iowa City IA USA

Minhua H Wang BMed MS PhD
Research Associate
Department of Vision Science
College of Optometry
University of Houston
Houston TX USA

Theodore G Wensel PhD
Welch Professor
Departments of Biochemistry and
  Molecular Biology, Ophthalmology,
  and Neuroscience
Baylor College of Medicine
Houston TX USA

Kwoon Y Wong PhD
Assistant Professor
Department of Ophthalmology &
  Vision Sciences and
Department of Molecular, Cellular and
  Developmental Biology
University of Michigan
Ann Arbor MI USA

Samuel M Wu PhD
Camille and Raymond Hankamer
  Chair in Ophthalmology
Professor of Ophthalmology,
  Neuroscience and Physiology
Baylor College of Medicine
Houston TX USA

Acknowledgements

We are grateful most especially to our authors who took on our challenge to make a classic text into a new entity with a different organizational and contextual approach, to our editors (Leonard Levin, Siv Nilsson, James Ver Hoeve and Samuel Wu) who undertook the hard task of selecting, overseeing and coordinating these stellar authors and to our editorial executives at Elsevier (Ben Davie and Russell Gabbedy) who were supportive, tolerant, flexible and stern taskmasters all at once. All were essential to create a book that we hope will be useful to our readers and worthy of its tradition.

To our spouses who provided their never-failing support and sustenance,
to our readers who are the reason we undertook this,
and especially to our students, who are our hope for progress in our field.

Optics

Paulo Schor & David Miller

The young eye

Primate and human infants must normally pass head first through their mother's pelvis to accommodate the limited opening determined by the bony configuration. Therefore the size of the mother's pelvis limits the head and brain size of the infant. Specifically, the brain of an infant ape is 55 percent of its full size, and the brain of a present-day human infant is only 23 percent of the adult size.[1] The result is a human infant who is neurologically immature.[2] Notice that the baby monkey can immediately cling tightly to the fur on its mother's stomach, whereas the human infant has poor muscle strength, has little motor control, and is completely dependent on the mother for survival. While immature, the human infant lives in a restricted and artificial reality, interacting primarily with the mother. The human infant interacts little with the forces of life in the outside world.

It is possible that this early immaturity and restricted world contact are naturally beneficial. The infant's restricted curriculum concentrates on a few priorities necessary for survival. Without words, the infant must be able to announce all his or her needs and encourage a high level of motherly devotion. To communicate with the mother, the infant must be able to read facial expressions and respond with a non-verbal vocabulary. If this speculation is correct, what vision equipment does the infant have to perform these functions?

Relevant anatomy

Axial length

Larsen[3] noted that the axial length of the neonate's eye was 17 mm and that it increases 25 percent by the time the child reaches adolescence. The size of the normal infant's eye is about three-fourths that of the adult size. Geometric optics teaches us that the retinal image of the normal infant eye is therefore about three-fourths the size of the adult's image.* A smaller image also means that much less fine detail is

recorded. The small retinal image may be but one reason why an infant's visual acuity is poorer than that of the adult. In fact, experiments have shown that the neonate's visual appreciation for fine detail at birth is one-thirtieth, or approximately 3 percent, of the development of the adult,* yet the neonate appreciates large objects (e.g. nose, mouth, eyes of close faces) as does the adult.

Figure 1.1 shows that visual acuity swiftly improves, and by the age of 12 months, the infant's level of visual acuity is 25 percent (20/80) of optimal adult visual acuity. This improvement in acuity seems to parallel eyeball growth. By the age of 5 years, the child usually has 20/20 vision.[1,5–8] What other factors beyond eye size account for the young child's lowered visual acuity? As the eye grows, the optical power of the eye lens and cornea must weaken in a tightly coordinated fashion so that the world stays in sharp focus on the retina. Patients with myopia highlight the developmental process of balancing the growth of the eye while maintaining a sharp retinal image. Most cases of myopia have an elongated eye. The stretching and weakening of the sclera seems to depend on two major factors. First, that the intraocular pressure maintains a constant force on the sclera. Second, there is digestion of sclera architecture (collagen I fibers and extracellular matrix) by metalloproteinase enzymes.[9] This idea of strengthening the sclera to prevent myopic expansion is supported experimentally. In one series of experiments, 7-methylxanthine (a caffeine metabolite) was used to enhance concentration and thickness of collagen fibers in the posterior sclera of animals. In humans, scleral collagen fibers were cross-linked using riboflavin activated by ultraviolet light.

However, we do not know what activates the process of scleral weakening and stretching in myopia. Some human studies have shown that use of atropine drops in children partially inhibits the development of myopia. The reason is unclear. Some feel that atropine reduces the pull of the ciliary muscle on the sclera which otherwise allows the sclera to elongate. Another theory suggests that the atropine reduces vitreous pressure, thus reducing a stretching force. However, atropine has the same effect in chickens that have

---

*Specifically, the size of the retinal image depends on an entity known as the *nodal distance*, which averages 11.7 mm in the newborn and 16.7 mm in the adult emmetropic schematic eye, giving a ratio of adult to infant retinal images of 1.43.[4]

---

*The infant's visual acuity is about 20/600 versus the normal visual acuity of an adult, which is 20/20.

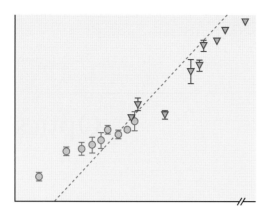

Figure 1.1 A graph showing the improvement of visual acuity in the infant as it ages. The method of preferential viewing was used to achieve these results. (From Teller DG: The development of visual function in infants. In: Cohen B, Bodis-Wollner I, eds. Vision and brain. New York: Raven Press, 1990.)

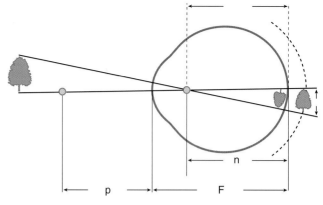

Figure 1.2 A diagram of a reduced human eye. *F*, Focal points; *n*, nodal point; *p*, principal point. The *dotted line* represents the retina of an enlarged eye.

aciliary muscle unresponsive to the effects of atropine. Another school of thought is that retinal receptors somehow activate the process of scleral weakening, e.g. M1 muscarinic receptors present in the neural retina but not in muscle. Animal experiments using specific M1 blockers such as pirenzepine show the same effects as atropine, without blockade of accommodation. The presence of pirenzepine-inhibited receptors within the choroid and retina could also explain why elongation in eyes with transected optic nerves is inhibited by pirenzepine. Human use of topical pirenzepine was reported in two studies which showed a small but statistically significant reduction in myopia and axial length.[10] These data suggest that during childhood the retina can record information concerning the sharpness of the retinal image and then use this information to control the eye's axial length via scleral stretching.

If this tight coordination of growth fails, the infant may become nearsighted or farsighted. Because the coordination of eye length growth and the focusing power of the cornea and eye lens may be imperfect, is there some compensation provided, early in life, guaranteeing that almost every child can process a sharply focused retinal image of the world? Accommodation is the safety valve that can help provide a sharp image, even if all the ocular components are not perfectly matched. In the young child the range of accommodation is greater than 20 diopters. This range, in addition to the farsightedness of almost all infants, means that most young eyes can focus almost any object by using part or all of this enormous focusing capacity.

Because of the infant's smaller pupil, a second factor that helps the infant achieve a sharper retinal image is an increased depth of focus.[5] Photographers use this device when they use larger F-stops (F32, F64) to keep objects at different distances all in focus.

Figure 1.2 shows the importance of the nodal point in determining the size of the retinal image in a typical human eye. To help us appreciate the basic optical principles operating within the human eye and avoid being confused by their many details (e.g. the many different radii of curvature, the different indices of refraction), an all-purpose, simplified eye was developed. Such model eyes have many names (e.g.

reduced, schematic, simplified eye) and were developed by some of the true giants of physiologic optics.*

Figure 1.2 depicts such an eye with its cardinal points (the principal points, focal points, and nodal points). Knowing the location of the cardinal points of a lens system, the optical designer can calculate all of the relationships between an object and an image. For example, to determine the image size of the reduced eye, one simply traces the ray, starting from the top of the object, which goes undeviated through the nodal point and lands on the top of the inverted retinal image. As the distance between the nodal point and the retina increases, the image size increases. The addition of a plus spectacle lens to the eye's optic system moves the nodal point of the new system forward (increasing the nodal point to retina distance), thus magnifying the retinal image. The reverse is true with a negative spectacle lens. Therefore a contact lens or a refractive cornea on which surgery has been performed enlarges the image size for a person with myopia who previously wore spectacles. The change in retinal image size should be taken into account when evaluating the visual acuity results after corneal refractive surgery. From an optical point of view, an 8 diopter hyperope exchanges a larger image produced by their +8 spectacle lens for a smaller image after corneal refractive surgery. Thus, they should theoretically lose a line of visual acuity after refractive surgery, but instead often gain a line. A possible reason is that the aberrations of the high plus spectacle lens cancel the effects of the larger retinal image, but this area requires further study.[11]

Emmetropization

The coordination of the power of the cornea, crystalline lens, and axial length to process a sharp retinal image of a distant object is known as *emmetropization*. In the United States, more than 70 percent of the population is either emmetropic or mildly hyperopic (easily corrected with a small accommodative effort).

---

*A partial list of giants of physiologic optics who have created schematic or reduced eyes includes Listing, Helmholtz, Wüllner, Tserning, Matthiessen, Gullstrand, Legrand, Ivanoff, and Emsley.

   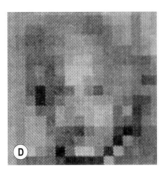

Figure 1.3 A computer display of the face of Albert Einstein, with pixels of different size. (From Lakshiminarayanan V et al. Human face recognition using wavelengths. In: Vision science and its application. Santa Fe: Optical Society of America, 1995.)

With age, the cornea, lens, and axial length undergo coordinated changes. Essentially, the optical components (cornea and lens) must lose refractive power as the axial length increases so that a sharp image remains focused on the retina.

The cornea, which averages 48 diopters of power at birth and has an increased elasticity, loses about 4 diopters by the time the child is 2 years of age.[12,13] One may assume that the spurt in growth of the sagittal diameter of the globe during this period pulls the cornea into a flatter curvature. The fact that the average corneal diameter is 8.5 mm at 34 weeks of gestation, 9 mm at 36 weeks, 9.5 mm at term, and about 11 mm in the adult eye supports this "pulling, flattening" hypothesis.[14]

On the other hand, other coordinated events also occur, such as the change of lens power and the coordinated increase in eye size (most important, an increase in axial length). The crystalline lens, which averages 45 diopters during infancy, loses about 20 diopters of power by age 6 years.[15,16] To compensate for this loss of lens power, the axial length increases by 5–6 mm in that same time frame.[3] (In general, 1 mm of change in axial length correlates with a 3-diopter change in refractive power of the eye.)

Now let us examine a possible mechanism that could account for most of the data.[15,17] As the cross-sectional area of the eye expands, there is an increased pull on the lens zonules and a subsequent flattening of the lens (the anterior lens surface is affected a bit more and the posterior lens surface a bit less), thus decreasing the overall lens power. There also may be a related decrease in the refractive index of the lens, which also contributes to the reduction in lens power. Because the incidence of myopia starts to accelerate significantly around the age of 10,[15] one may question whether there could be a decoupling of the previously described coordinated drop in lens power and increase in axial length. An increased amount of near work (e.g. schoolwork) is associated with a higher incidence of myopia.[18] It is also well known that genetic predisposition influences myopia incidence, as evidenced by the fact that more Asian children than Caucasian children are myopic.[18] Thus one might hypothesize that the long periods of accommodation that accompany schoolwork (ciliary body contraction) may tend to stretch and weaken the linkage between the enlarging scleral shell and the ciliary body. If this were to happen, the lens would flatten less during eye growth. Another way of looking at this phenomenon is to theorize that with the linkage weakened, the restraining effect of the lens-zonule

combination on eye growth is also weakened, which results in an increase in axial length in the myopic student. Many studies demonstrating that the myopic eye has a greater axial length than the emmetropic eye support this idea.[19]

Retinal receptors

The cone photoreceptors of the retina are responsible for sharp vision under daylight conditions. The denser the cones are packed, the more acute the vision.[1,7] To use a photographic analogy, film with the highest resolution has smaller photosensitive grains, packed tightly, whereas a film with large grains of silver halide yields a coarser picture.

The most sensitive part of the retina is the fovea.* Here the cones are even finer and are packed together even tighter. The fovea of the infant eye is packed less than one fourth the density of that of the adult. Furthermore, the synapse density in the neural portion of the retina, as well as the visual brain, is low at birth. The combination of these two anatomic configurations means that fewer fine details of the retinal image are recorded and sent to the brain.

Neural processing

Finally, the nerves that transmit the visual information to the brain, as well as the nerve fibers at the various levels within the brain, are poorly myelinated in the infant. Myelin is the insulating wrap around each nerve fiber. A normally myelinated nerve can transmit nerve impulses swiftly and without static or "cross talk" from adjacent nerves. To use a computer analogy, one might think of the infant brain as being connected with poorly insulated wires. Therefore sparks, short circuits, and static all slow or interfere with perfect transmission, and only the strongest messages get through. Figure 1.3 represents an appropriate analogy. The face of Albert Einstein is shown on a computer screen with larger and larger pixels. The infant's early vision might be comparable to the picture with the biggest block pixels. With maturation of the brain processing elements, the neurologic equivalent of pixel size gets smaller and more details can be registered. Thus the equivalent of the photographic film grain size in the retina and the equivalent of pixel size in the

---

*F. W. Campbell[20] quotes Stuart Ansti's clear analogy of how the fovea functions: "A retina with a fovea surrounded by a lower acuity periphery can be compared to a low magnification finder telescope with a large field of view which will find any interesting target and then steer on to it a high powered main telescope, with a very small field which could examine the target in detail."

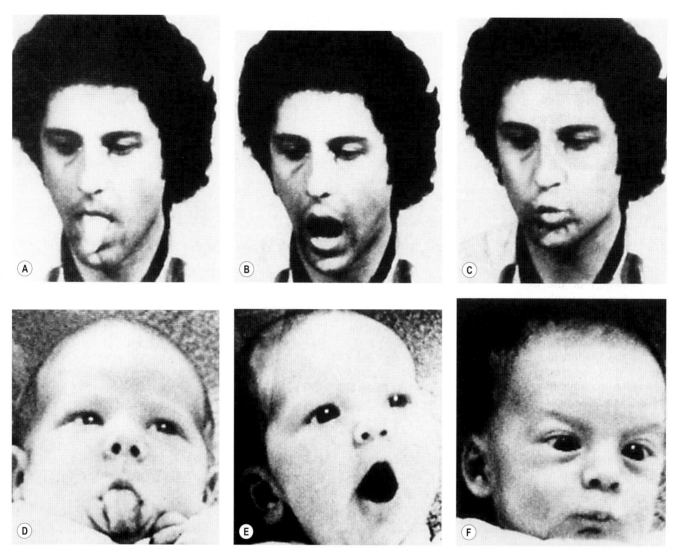

Figure 1.4 This photo shows a recently born infant mimicking the expression of psychologist Dr. A Meltzoff. The baby is obviously able to perceive the different expressions in order to mimic them. (From Klaus MH, Klaus PH. The amazing newborn. Reading, Mass: Addison Wesley, 1985.)

brain processor both get smaller as the child grows. Lakshiminarayanan, who created Figure 1.3, speculated that the immaturity of the infant's memory capacity may be one reason why its visual images have less detail. In other words, the coarseness of the infant visual system does not overtax the immature memory system.

Relevant early physiology

Experiments with infants reveal that good color vision does not appear for about 3 months. The infant also takes longer to "make sense" of the retinal image. Specifically, the infant must stare for relatively long periods (1–3 minutes), blinking rarely during this period.[1,7]

Recognizing faces

The remarkable thing about the infant's eyesight is that the relatively poor level of resolution just described still allows the infant to recognize different faces and different facial expressions. We know this to be true in some newborn infants, who can accurately imitate the expressions of an adult (Fig. 1.4), almost as if the baby uses his or her own face as a canvas to reproduce the facial expression of the onlooker.

Experiments with infants demonstrate that infants prefer to look at faces or pictures of faces rather than to look at other objects. By the age of 6 weeks, infants can discern specific features of the face. For example, they can lock in on the mother's gaze. By age 6 months, they can also recognize the same face in different poses. In fact, they are experts at recognizing a face, be it upside down or right side up, till the age of 6 years. After age 6, infants actually lose their skill at quickly recognizing upside-down faces.[7]

A closer examination of the eye at 6 months of age is worthwhile because an unusual change has started to take place in the optics of the eye at this time. Gwiazda et al[21] found that a significant amount of astigmatism develops in 56 percent of the infants studied. This condition remains for only 1–2 years.[18,19,21,22]

The transient astigmatism just described tends to elongate tiny dots of the retinal image into lines. In essence, these create the equivalent of a line drawing of the retinal image.

Figure 1.5 A series of computer simulations of the face of Albert Einstein with an extensive gray scale on the left (16 gray scale) and only a 2 gray scale on the right, the latter being similar to a line drawing. (From Lakshiminarayanan V et al. Human face recognition using wavelength. In: Vision science and its applications. Santa Fe: Optical Society of America, 1995.)

For example, think of a mime (i.e. a painted face with a few dark lines and spots for eyes and nose) as creating different line drawings of the face. What is astonishing is that, although made of only a few dark lines, the mime can recreate most human expressions. It seems reasonable to imagine that the mime presents faces similar to those seen by the young child or found in a child's drawing. The faces have no texture, no shadowing, and no creases – only a line for a mouth, circle for eyes, and occasionally, a dot for a nose. Is it not then possible that infant astigmatism helps represent faces as line drawings to the infant visual system? Line drawings also save memory storage space, which would be an advantage for the small infant brain.* Figure 1.5 illustrates this point in a different way. The face of Albert Einstein is shown with a complete gray scale on the left and is shown as a line drawing (only black and white) on the right. The line drawing requires much less computing power than a face with texture and would be more compatible with the child's immature processing system.

Line orientation receptors

As noted earlier, in many infants the amount of astigmatism can rise to a level of greater than 2 diopters in the first year of life. The orientation of the distortion is usually horizontal (180 degrees) initially. In the course of the next 2 years, the meridian of distortion rotates to the vertical and the amount of the astigmatism diminishes. This slow rotation of the axis of exaggeration can help activate different groups of brain cells, which become sensitive to features in the retinal image with different tilts. In fact, the discovery of these brain cells with orientation selectively leads to a ground-breaking understanding of the functional architecture in the higher brain. In 1958 Torsten Wiesel and David Hubel, working in their laboratory at the Harvard Medical School, implanted electrodes in the visual cortex of an anesthetized cat to record cortical cell responses to patterns of light, which they projected onto a screen in front of the cat. After 4 hours of intense work, the two scientists put the dark spot slide into the projector, where it jammed. As the edge of the glass slide cast an angled shadow on the retina, the implanted cell in the visual cortex fired a burst of action potentials. Torsten Wiesel described that moment as the "door to all secrets."

The pair went on to prove that cells in the cortex responded only to stimuli of a particular orientation. Similar responding cells were all located in the same part of the cortex. This work opened up the area of how and where the brain encodes specific features of the retinal image. Fittingly, Drs. Hubel and Wiesel were awarded the Nobel Prize for Medicine in 1981.[23]

Monitoring other's eye movements

Another function of great survival value to the infant is the ability to follow the eye movements of his or her caretakers. Consider this a near task involving the contrast discrimination of a 12-mm dark iris against a white sclera framed by the palpebral fissure; such a task can be accomplished with a visual acuity of 20/200.

The British psychologist Simon Baron-Cohen, in his book *Mindblindness,** suggests that a major evolutionary advance has been the human's ability to understand and then interact with others in a social group (e.g. playing social chess). He further suggests that we accomplish this social intelligence primarily by following the eye movements of others, which begins at a young age. For example, an infant of 2 months begins to concentrate on the eyes of adults. Infants have been shown to spend as much time on the eyes as all other features of an observer's face.

By 6 months of age, infants look at the face of an adult who is looking at them two to three times longer than an adult who is looking away. We also know that when infants achieve eye contact, a positive emotion is achieved (i.e. infants smile). By age 14 months, infants start to read the direction that an adult is looking. Infants turn in that direction and then continue to look back at the adult to check that both are looking at the same thing. By age 2, infants typically can read fear and joy from eye and facial expression.

Recognizing movement

Infants are capable of putting up their arms to block a threatening movement. This act tells us that infants appreciate both movement and the implied threat of this particular movement.[24,5] Admittedly, infants cannot respond if the threat

---

*This idea was suggested by David Marr in "Early Processing of Visual Information," published in Transactions of the Royal Society of London, Series B 1976; 275:483.

*In his book *Mindblindness: An Essay on Autism and Theory of Mind,* published by MIT Press in 1995, Baron-Cohen ferrets out the key features of "eye following" in the normal child by comparing them with those of the autistic child.

moves too quickly, probably because the immature myelinization of the nerves slows the neural circuits. Nevertheless, a definite appreciation of movement and threat exists.

For movement to be registered accurately, the infant retina probably records an object at point A. That image is then physiologically erased (in the brain and/or retina), and the object is now seen at point B. This physiologic erasure is important because, without it, movement would produce a smeared retinal image. Researchers think that the infant probably sees movement as a smoothed series of sharp images, not smears.[25] This hypothesis is supported by other experiments demonstrating that the infant can appreciate the on/off quality of a rapidly flickering light at an early age. It seems logical that the movement of an image across the retina (with the inherent erasures) is physiologically related to the rapid on/off registration of a flickering light.

A related reflex, the foveal reflex, is triggered by stimulation of the peripheral retina, activating the eye movement system so that the fovea is directed to the visual stimulus.

Summary – social seeing

Although the visual system of the infant is immature, some infants can recognize and respond to adult facial expressions on the first day of life and follow the mother's glances by 6 weeks of age. Clearly, the infant's top priority is to maintain a social relationship with his or her mother or other caring adults. This idea was described succinctly by the "language expert" Pinker:[26] "Most normally developing babies like to schmooze." As infants grow, they learn to see people and objects in the way that their culture demands and communicate in the expected manner; that is, "We don't see things as they are but as we are." Is it possible that the immature eye and visual system actually facilitate socially biased seeing? Perhaps the smaller, simpler retinal images, along with the less sophisticated brain processing, prevent the many other details of life from confusing the key message; that is, social interaction takes top priority.

Even in adulthood there is plasticity of the visual system, e.g. after the implantation of multifocal IOLs (intraocular lenses). These IOLs superimpose blurred near images on a sharply focused distant image. Thus, the patient's brain must filter and suppress the third, fourth and so on blurred images from the scene. This is a time-dependent phenomenon that may take days to months to occur.[27]

The image of the human adult eye

The image quality of the human adult eye is far superior to that of the human infant, although probably inferior to certain predator birds. Its wide focusing range is smaller than that of certain diving birds, and its fine sensitivity to low light levels is weaker than spiders or animals with a tapetum lucidum. Its ability to repair itself if probably not as efficient as some animals (e.g. newts, which can form a new lens if the original is damaged). Finally, the human eye has the ability to transmit emotional information[28] (e.g. excitement by means of pupil dilation, sadness by means of weeping), but with less forcefulness than some fish, who uncover a pigmented bar next to the eye when they are about to attack,[29] or the horned lizard, which squirts a jet of blood

from its eye when it is threatened.[30] Thus, in reading this chapter, one must appreciate the eye's level of performance in light of its large spectrum of functions.

Tuned to visible light waves

When our retinas receive an image of a spotted puppy in a room, what is really happening in terms of information transfer? Light waves from the ceiling light are cast onto the puppy. The puppy's body reflects and scatters the light waves into our eyes. In a sense, information about the puppy has been encoded into visible light waves. The optical elements of the eye focus the encoded light waves onto our retina as a map of bright and dim colored dots, known as the *retinal image*. Nerve signals report the retinal image to the brain. In the brain the nerve signals are recreated into the impression that a real puppy is in the room.

One might liken the function of the eye to that of a radio, which receives radio waves carrying a Beethoven symphony. The specific station broadcasting the symphony beams it out on a specific radio wavelength. The radio then receives the radio waves, and the speaker reconverts the waves to musical sounds. If the eye is to receive and process visible light, it must be constructed with the ability to be tuned to the wavelengths of visible light. The physicist would substitute the term *resonate* for *tune*. Literally, *resonate* means to resound another time. A simple example of resonance is an opera singer who can make a wine glass hum when the frequency (or wavelength) of the note is the same as the natural frequency of the glass. What is the natural frequency of anything? The answer has to do with composition and size. Organ pipes of different lengths resonate at different frequencies and wavelengths. Changing the length of the antenna on a car radio allows the frequencies of different stations to be received. The best receiver for a specific wavelength (frequency is the reciprocal of wavelengths) is physically the same size as the wavelength, a precise number of wavelengths, or a precise fraction (one-fourth, one-half) of the size of the wavelength. Therefore optical theory demands that the size of the key components of the eye be the size of a wavelength of visible light or some number (n) times the size of, or a fraction of the size of, the wavelengths of visible light; in addition, the key components must be made of a resonating material.

Role of the cornea

The human cornea is a unique tissue. First, it is the most powerful focusing element of the eye, roughly twice as powerful as the lens within the eye. It is mechanically strong and transparent. Its strength comes from its collagen fiber layers. Some 200 fiber layers crisscross the cornea in different directions. These fibers are set in a thick, watery jelly called *glycosaminoglycan*. The jelly gives the cornea pliability. For a long time, no one could explain convincingly the transparency of the cornea. No one could understand how nature combined tough, transparent collagen fibers (with their unique index of refraction) with the transparent glycosaminoglycan matrix, which had a different index of refraction, and still maintain clarity. Perhaps an everyday example of this phenomenon will help. When a glass is filled from the hot water tap, the solution looks cloudy. Looking closely, one can see many fine, clear expanded air bubbles (which have a unique index of refraction) within the water (which has different refractive properties). Conversely, cold water appears clear because its air bubbles are tiny. The normal corneal structure might be considered optically similar to the structure of the cold water (i.e. tiny components with different indices of refraction).

Miller & Benedek[31] were ultimately able to prove that if the spaces consisting of glycosaminoglycan and the size of the collagen fibers were smaller than one-half a wavelength of visible light, the cornea is clear, even if the fibers were arranged randomly. An orderly arrangement of the fibers also helps maintain corneal transparency.

Another way to explain it is to say that the cornea is basically transparent to visible light because its internal structures are tuned to the size of a fraction of the wavelengths of visible light. Figure 1.6 is an electron micrograph showing the fibers of the human cornea. The black dots are cross sections of collagen fibers imbedded in the glycosaminoglycan matrix. In this specimen (Fig. 1.6A), the fibers are spaced closer than half a wavelength of visible light apart, and the fibers in each of the major layers are arranged in an orderly manner. In an edematous, hazy cornea there are large spaces between collagen fibers (Fig. 1.6B).

This arrangement of corneal fibers serves a number of important functions. First, the arrangement offers maximal strength and resistance to injury from any direction. Second, the arrangement produces a transparent, stable optical element. Third, the spaces between the major layers act as potential highways for white blood cell migration if any injury or infection occurs. Horizontal arrangement of corneal lamellae, which can slide over each other during eye rubbing, facilitates the clinical development of pathological conditions such as keratoconus. The lack of interlamellar adhesion and corneal thinning facilitate a bulging under IOP and gravity. Enzymatic digestion seems to be the causative agent in the thinning, which eventually deforms the anterior surface of the cornea. This conical cornea is irregularly astigmatic and will produce multiple blurred images in the retina. Collagen cross-linking helps connect the adjacent lamellae, inducing greater corneal strength by resisting interlamellae sliding, and thus reduces the progression of keratoconus (Box 1.1).[32]

Role of the crystalline lens

Is it easier to see underwater with goggles?* Without goggles, the water practically cancels the focusing power of the cornea,† leaving objects blurred. The goggles ensure an envelope of air in front of the cornea, restoring its optical power. If water cancels optical power, how can we explain the focusing ability of the crystalline lens, which lies inside the eye and is surrounded by a fluid known as *aqueous humor*? The answer is that the focusing power resides in the unusually high protein content of the lens. The protein concentration

---

*Because the index of refraction of water is greater than air, objects underwater appear about one-third closer and thus one-third larger than they would in air (i.e. magnification = 1.33×).[33]

†The cornea is a focusing element for two reasons. First, it has a convex surface. Second, it has a refractive index greater than air. Actually, its refractive index is close to that of water. Thus, when one is underwater, the surrounding water on the outside and the aqueous humor inside the eye combine to neutralize the cornea's focusing power.

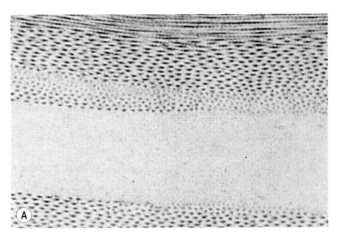

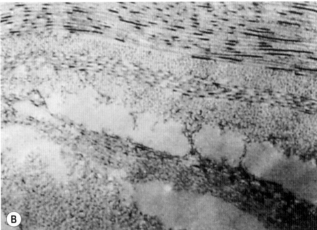

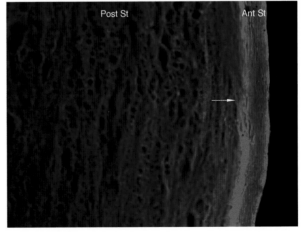

Figure 1.6 (**A**) Electron micrograph showing the neat pattern of corneal collagen fibers. The black dots are the fibers cut on end. In this photo the spacing between fibers is less than a wavelength of light apart. (**B**) Large spaces between collagen fibers, as seen in a waterlogged, hazy cornea. (From Miller D, Benedek G. Intraocular light scattering, Springfield, Ill: Charles C Thomas, 1973.)

Box 1.1

An example of immunofluorescent staining for confocal microscopy (40×) of a bovine cornea soaked with a riboflavin (0.01 percent) solution. Prior to staining the specimen was submitted to 30 minutes exposure to UV light (365 nm). Note the connection between lamellae in the anterior stroma (ant st) compared to the laxity of the posterior stroma (post s). (Courtesy Bottos, Schor, Chamon, Regatieri, Dreyfuss and Nader.)

may reach 50 percent or more in certain parts of the lens.* Such a high concentration increases the refractive index above that of water and allows the focusing of light. Now we are ready to appreciate the real secret of the lens.[31,34]

Normally, a 50 percent protein solution is cloudy, with precipitates floating about like curdled lumps of milk in a cup of coffee. However, the protein molecules of the normal lens do not precipitate. In a manner not fully understood, the large protein molecules known as *crystallins* (large protein molecules ranging in size from 45 to 2000 kD) seem to repel each other, or at least prevent aggregation, to maintain tiny spaces between each other. The protein size and the spaces between them are equivalent to a small fraction of a light wavelength. Spaced as they are, one might say that they are tuned to visible light and allow the rays to pass through unimpeded. On the other hand, if some pathologic process occurs, the protein molecules clump together and the lens loses its clarity. When this happens, light is scattered about as it passes through the lens. The result is a cataract.

Accommodation

If the emmetropic eye is in sharp focus for the distant world, it must refocus (accommodate) to see closer objects.[†] For example, the child's range of accommodation is large, as noted earlier. This allows the child to continue to keep objects in sharp focus from an infinite distance away to objects brought to the tip of the nose. The act of accommodation is fast, taking only about one-third of a second. Our range of accommodation decreases with each passing year, so by the age of 45, most of us are left with about 20 percent of the amplitude of accommodation we started with.

With age, the lens enlarges and becomes denser and more rigid. In so doing, it progressively loses the ability to accommodate. Parenthetically, the cornea of many birds, from pigeons to hawks, can change shape to accommodate.[35] The avian cornea does not change flexibility with age; therefore these birds do not become presbyopic. However, there is no "free lunch" in nature. The human lens, sitting within the eye surrounded by protective fluid is far less vulnerable to injury than the cornea.

Role of the retina

After light passes through the cornea, the aqueous humor, the lens, and the vitreous humor, it is focused onto the retinal photoreceptors. The light must pass through a number of retinal layers of nerve fibers, nerve cells, and blood vessels before striking the receptors. These retinal layers (aside from blood vessels) are transparent because of the small size of the elements and the tight packing arrangement.

The bird retina does not have blood vessels. The human retina has retinal blood vessels that cover some of the retinal receptors and produce fine angioscotomas. A bird's retina obtains much of its oxygen and nutritive supply from a tangle of blood vessels (the pectin), which is covered with black pigment and sits in the vitreous in front of the retina and above the macula (so as to function as a visor). The negative aspect of such a vascular system is vulnerability to a direct blunt or penetrating injury that can lead to a vitreous hemorrhage and sudden blindness. Obviously, this is less probable in the bird because of its lifestyle.

Rhodopsin

The rods and cones are made up of a biologic molecule that absorbs visible light and then traduces that event into an electrical nerve signal. The rhodopsin molecule is an example of Einstein's photoelectric effect.* In fact, only one quanta (the smallest possible amount of light) of visible light[†] is needed to trigger the molecule, that is, snap the molecule into a new shape.[‡]

The internal structure of the molecule allows the wavelengths of visible light to resonate within its electron cloud and within 20 million millionths of a second, inducing the change in the molecule that starts the reaction.

Probably the earliest chemical relative of rhodopsin is to be found in a primitive purple-colored bacteria called *Holobacterium halobium*. Koji Nakanishi, a biochemist at Columbia University, in an article titled "Why 11-cis-Retinal?"[37] (a type of rhodopsin) notes that this bacteria has been on the planet for the last 1.3 billion years.[38,39] Its preference for low oxygen and a salty environment places its origin at a time on earth where there was little or no oxygen in the atmosphere and a high salt concentration in the sea. Although found in primeval bacteria, bacteriorhodopsin is a rather complicated molecule, containing 248 amino acids. It is thought that this bacteria probably used rhodopsin for photosynthesis, rather than light sensing. Time-resolved spectroscopic measurements have determined that this molecule changes shape within one trillionth of a second after light stimulation.[40] This early form of rhodopsin absorbs light most efficiently in the blue-green part of the spectrum, although it does respond to all colors.[41]

To function as the transducer for vision, the photopigment must capture light and then signal the organism's brain that the light has registered. As noted earlier, one molecule needs only one quanta to start the reaction. Even more amazing is the molecule's stability. Although only one quanta of visible light is necessary to trigger it, the molecule will not trigger accidentally. In fact, it has been estimated that spontaneous isomerization of retinal (the light-sensing chromophore portion of rhodopsin) occurs once in a thousand years.[40] If this were not so, we would see light flashes every time there is a rise in body temperature (a fever). To better understand the rhodopsin mechanism, one may

---

*The chemical composition of a focusing element such as the crystalline lens determines the refractive index. Water has a refractive index of 1.33. As the protein concentration of the lens rises, the index of refraction approaches 1.42.

[†]The question "How does accommodation 'know' it has achieved the sharpest focus?" seems to be best answered by a sensing system in the brain. However, some have suggested that the system takes advantage of the naturally occurring chromatic aberration of the primate eye to fine-tune focusing.[36]

---

*Some wavelengths of light are powerful enough to knock electrons of certain molecules out of their orbits, thereby producing an electric current. Einstein was awarded the Nobel Prize for explaining the "photoelectric effect."

[†]In 1942, Selig Hecht and his co-workers in New York first proved that only one quanta of visible light could trigger rhodopsin to start a cascade of biochemical events eventuating in the sensing of light.

[‡]Photoactivation of one molecule of rhodopsin starts an impressive example of biologic amplification, in which hundreds of molecules of the protein transducer each activate a like number of phosphodiesterase molecules, which in turn, hydrolyze a similar number of cyclic guanine monophosphate (cGMP) molecules, which then trigger a neural signal to the brain.[42]

picture a hair trigger on a pistol that takes only the slightest vibration (but only a special type of vibration) to be activated. As noted, the activating quanta must be of the proper energy level to "kick in" the reaction. That is, the quanta of light must be made of wavelengths of visible light.

Receptor size and spacing

Retinal receptor factors that influence the optical limits of visual acuity occur in the foveal macular area. The fovea itself subtends an arc of about 0.3 degree. It is an elliptical area with a horizontal diameter of 100 μm. The area contains more than 2000 tightly packed cones. The distance between the centers of these tightly packed cones is about 2 μm. The cone diameters themselves measure about 1.5 μm (a dimension comparable to three wavelengths of green light) and are separated by about 0.5 μm.[43–46] Therefore the fine details of the retinal image occupy an elliptical area only about 0.1 mm in maximum width.

A discussion of the diffraction limits of resolution, in a theoretical emmetropic human eye, must involve the anatomic size of the photoreceptors and the pupil. A point or an object is focused on the retina as an Airy disc because of diffraction. The angular size of the Airy disc is determined as follows:

$$\text{Angular size (in radians)} = 1.22 \times \text{wavelength (mm)}/\text{pupil diameter (mm)}.$$

Let the wavelength be 0.00056 mm (560 nm; yellow green/light). Then

$$\text{Angular size} = 0.00068 \text{ radians}/\text{pupil diameter (mm)}$$

For a pupil of 2.4 mm (optimal balance between diffraction and spherical aberration in the human eye):

$$\text{Angular size} = 0.00028 \text{ radians, or about 1 minute of arc *}$$

Given this angle of 1 minute, the actual size of the Airy disc can be calculated if the distance from the nodal point to the retina is known. The optimal distance depends on the diameter of the photoreceptors. Because these act as light guides, the theoretical limit is 1–2 μm. To obtain the maximum visual information available, Kirschfield calculated that more than five receptors are required to scan the Airy disc.[31] Assume that each foveal cone is 1.5 μm in diameter and that there is an optimal space of 0.5 μm between receptors. The following equation describes the situation for three cones and two spaces (5.5 μm).

$$5.5 \, \mu m/\text{Tan 1 minute} = \text{Nodal point to retina distance}$$

Substituting 0.0003 from equation (2) into equation (3) gives equation (4):

$$5.5 \, \mu m/0.0003 = \text{Nodal point to retina distance}$$

From equation (4) the distance from the nodal point to the retina can be rounded off to 18.00 mm, which is close to the distance between the secondary nodal point and the retina in the schematic human eye.

How closely does optical theory agree with reality? The average visual acuity for healthy eyes in the age group younger than 50 was better than 20/16. In the distribution within the group younger than 40, the top 5 percent had an acuity of close to 20/10.[47]

Another related factor must be kept in mind. The fixating eye is in constant motion, as opposed to a camera on a tripod. Presumably, these movements prevent bleaching or fading within individual photoreceptors. These small movements, called either *tremors*, *drifts*, or *microsaccades*, range in amplitude from seconds to minutes of angular arc. Such movements tend to smear rather than enhance our traditional concept of visual resolution. It can only be presumed that to maintain high resolution within the context of this physiologic nystagmus, the visual system takes quick, short samples of the retinal image during these potentially smearing movements and then recreates an image of higher resolution.[48–50]

The unique essence of the vertebrate retina is that the structure of the transparent optical components, the rhodopsin molecule, and the size of the foveal cones are all tuned to interact optimally with wavelengths of visible light.[51,52] It is earth's unique atmosphere and its unique relationship to the sun that have allowed primarily visible light, a tiny band from the enormous electromagnetic spectrum of the sun, to rain down upon us at safe energy levels. Our eyes are a product of an evolutionary process that has tuned to these unique wavelengths at these levels of intensity.[53,54]

With this basic science background, we can discuss how functions such as visual acuity and contrast sensitivity are monitored in a clinical setting.

Visual acuity testing

The idea that the minimum separation between two point sources of light was a measure of vision dates back to Hooke in 1679, when he noted "tis hardly possible for any animal eye well to distinguish an angle much smaller than that of a minute: and where two objects are not farther distant than a minute, if they are bright objects, they coalesce and appear as one."[55] In the early nineteenth century, Purkinje and Young used letters of various sizes for judging the extent of the power of distinguishing objects. Finally, in 1863, Professor Hermann Snellen of Utrecht developed his classic test letters. He quantitated the lines by comparison of the visual acuity of a patient with that of his assistant, who had good vision. Thus 20/200 (6/60) vision meant that the patient could see at 20 feet (6 m) what Snellen's assistant could see at 200 feet (60 m).[55]

The essence of correct identification of the letters on the Snellen chart is to see the clear spaces between the black elements of the letter. The spacing between the bars of the "E" should be 1 minute for the 20/20 (6/6) letter. The entire letter is 5 minutes high. To calculate the height of "*x*" (i.e. a 20/20 or 6/6 letter), use the following equation:

$$\text{Tan 5 minutes} = x \text{ feet}/20$$

$$0.0015 = x/20 \text{ feet}$$

$$x = 0.36 \text{ inches (9.14 mm) *}$$

Chart luminance

In clinical visual acuity testing, the chart luminance should (1) represent typical real work photopic conditions and (2)

---

*One minute of arc is the spacing between the bars of a 20/20 symbol. Interestingly, the sizing of the symbol was originally determined empirically.

---

*The 20/200 (6/6) letter is 10 times larger, or 3.6 inches (9.14 cm).

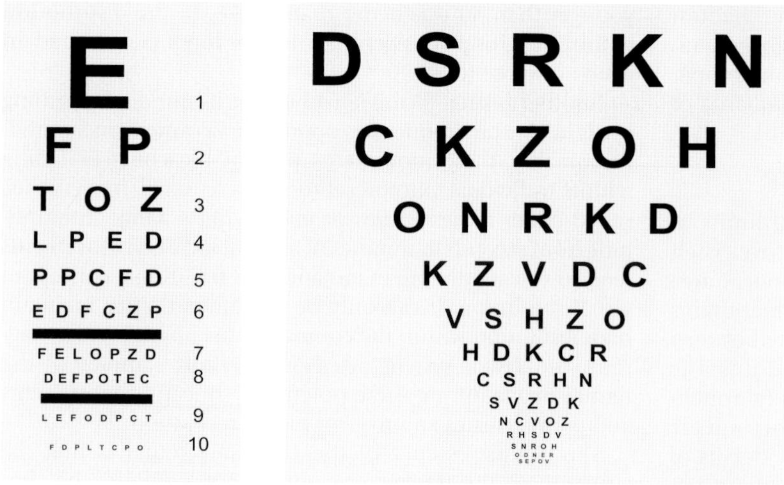

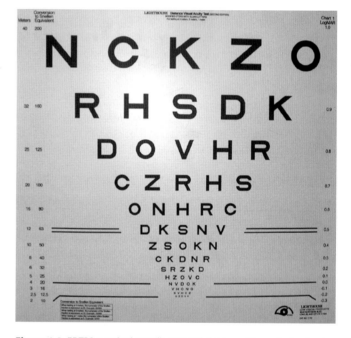

Figure 1.7 The standard Snellen chart and the Bailey–Lovie chart are examples of visual acuity charts.

be set at a level where variation produced by dust accumulation in the projector system, bulb decline, or normal variation in electrical current levels minimally affect visual performance. Thus chart luminance between 80 and 320 cd/m² meet such criteria (160 cd/m² is a favorite level of illumination).

Visual acuity as Log MAR

If one looks at a standard Snellen acuity chart (Fig. 1.7), the lines of symbols progress as follows: 20/400, 20/200, 20/150, 20/120, 20/100, 20/80, 20/70, 20/60, 20/50, 20/40, 20/30, 20/25, 20/20, 20/15, and 20/10. Thus the line-to-line decrease in symbol size varies from 25 percent (20/20 to 20/150) to 20 percent (20/120 to 20/100) to 16.7 percent (20/30 to 20/25).

Would it not be more logical to create a chart of uniform decrements, that is, a chart in which the line-to-line diminution in resolution angle were 0.1 steps? To create such a chart, one must first describe the spaces within a symbol (i.e. spaces between the bars of "E") in terms of "minutes of arc" (MAR) at 20 feet (6 m). Thus the 20/20 line represents a resolution of 1 MAR. If we take the log to the base of 10 of 1 (minute), we get 0. A spacing of 1.25 MAR (the equivalent of 20/30) yields a log value of 0.2, whereas a spacing of 1.99 MAR (the equivalent of 20/40) yields a log MAR of 0.3. See Table 1.1 for a complete listing of equivalents (courtesy of Prof. Dr. Wallace Chamon).[56,57]

The Bailey–Lovie acuity chart (Fig. 1.7) uses the log MAR sizing system. Figure 1.8 shows a standardized retroilluminated visual acuity chart commercially available, as used in the EDTRS protocol. Log MAR tests add precision to visual acuity testing. Thus, subtle individual variation may be identified under controlled conditions even in this high-contrast environment by counting the number of letters correctly identifiable by the subject.[58]

Visual acuity chart contrast

Clean printed charts using black characters on a white background usually have a character-to-background luminance contrast ratio between 1/20 and 1/33. For projected charts, the contrast ratio drops to a range of 1/5 to 1/10. Such a decrease in contrast is probably the result of the light scattering produced within the projector and the ambient

Figure 1.8 EDTRS standard retroilluminated chart used to evaluate visual acuity. (Produced by the Lighthouse Low Vision Products, Long Island City, NY)

light falling on the screen. Therefore the test should be performed in a dark room or using retroilluminated charts as shown in Figure 1.8.

Contrast sensitivity testing

Visual acuity testing is relatively inexpensive, takes little time to perform, and describes visual function with one notation, e.g. 20/40 (6/12 or 0.5). Best of all, for more than 150 years it has provided an end point for the correction of a patient's refractive error. However, contrast sensitivity testing, a time-consuming test born in the laboratory of the visual physiologist and described by a graph rather than a simple notation, has recently become a popular clinical test. It describes a number of subtle levels of vision, not accounted for by the visual acuity test; thus it more accurately quantifies the loss of vision in cataracts, corneal edema, neuro-ophthalmic diseases, and certain retinal diseases. These assets have been

Table 1.1

	Decimal	Numerator base 20 (20/x)	Angle (Minutes of Arc)	Spacial Frequency	Log Numerator Base 20	LogMAR	Jaeger	American Point Type
HM 60cm	0.001	20000	1000.00	0.03	4.30	3.00		
CF 60cm	0.01	2000	100.00	0.30	3.30	2.00		
	0.03	800	40.00	0.75	2.90	1.60		
	0.05	400	20.00	1.50	2.60	1.30		
	0.06	320	16.00	1.88	2.51	1.20		
	0.08	250	12.50	2.40	2.40	1.10		
	0.10	200	10.00	3.00	2.30	1.00	14	23
	0.13	160	8.00	3.75	2.20	0.90	13	21
	0.16	125	6.25	4.80	2.10	0.80	12	14
	0.18	114	5.70	5.26	2.06	0.76	11	13
	0.20	100	5.00	6.00	2.00	0.70	10	12
	0.25	80	4.00	7.50	1.90	0.60	9	11
	0.30	67	3.33	9.00	1.82	0.52		
	0.32	63	3.15	9.51	1.80	0.50	8	10
	0.33	60	3.00	10.00	1.78	0.48	7	9
	0.40	50	2.50	12.00	1.70	0.40	6	8
	0.50	40	2.00	15.00	1.60	0.30	5	7
	0.60	33	1.67	18.00	1.52	0.22		
	0.63	32	1.59	18.90	1.50	0.20	4	6
	0.67	30	1.50	20.00	1.48	0.18	3	5
	0.70	29	1.43	21.00	1.46	0.15		
	0.80	25	1.25	24.00	1.40	0.10	2	4
	0.90	22	1.11	27.00	1.35	0.05		
	1.00	20	1.00	30.00	1.30	0.00	1	3
	1.10	18	0.91	33.00	1.26	−0.04		
	1.20	17	0.83	36.00	1.22	−0.08		
	1.25	16	0.80	37.50	1.20	−0.10		
	1.33	15	0.75	40.00	1.18	−0.12		
	1.50	13	0.67	45.00	1.12	−0.18		
	1.60	13	0.63	48.00	1.10	−0.20		
	2.00	10	0.50	60.00	1.00	−0.30		

Shaded cells = LogMAR Standard progression.

CF = Count Finger; HM = Hand Motion.

Courtesy of Prof. Dr. Wallace Chamon.

known for a long time, but the recent enhanced popularity has arisen because of cataract patients. As life span increases, more patients who have cataracts request medical help. Often, their complaints of objects that appear faded or objects that are more difficult to see in bright light are not described accurately by their Snellen acuity scores. Contrast sensitivity tests and glare sensitivity tests can quantitate these complaints. Several validated quality of life assessment questionnaires are also available.[59] They offer the possibility of evaluating vision-related symptoms in a comparable manner, either over a given time frame (pre vs. post operative period) or among different interventions and patients (presbyopic LASIK vs. multifocal IOL). A patient happy with his or her vision but with visual acuity worse than 20/20 has been described as a "20/happy" patient. With physical evaluation alone it may not be possible to diagnose such "happiness," but subjective tests such as a quality of life questionnaire may help identify it. This is critically important when highly irregular astigmatism or highly aberrant human optical systems are to be evaluated, e.g. intracorneal ring placement for treating keratoconic patients.

Contrast sensitivity testing is related to visual acuity testing. Contrast sensitivity tests the equivalent of four to eight differently sized Snellen letters in six or more shades of gray.

Definition and units

Contrast

Whereas a black letter on a white background is a scene of high contrast, a child crossing the road at dusk and a car looming up in a fog are scenes of low contrast. Thus contrast may be considered as the difference in the luminance of a target against the background:

$$\text{Contrast} = (\text{Target luminance} - \text{Background luminance})/$$
$$(\text{Target luminance} + \text{Background luminance})$$

To compute contrast, one uses a photometer to measure the luminance of a target against the background. For example, a background of 100 units of light and a target of 50 units of light yields the following:

$$\text{Contrast} = (50 - 100)/(50 + 100) = 50/150 = 33\,\text{percent}$$

Contrast sensitivity

Suppose the contrast of a scene is 33 percent, or one-third, which also represents the patient's threshold (i.e. the patient cannot identify targets of lower contrast). The patient's contrast sensitivity is the reciprocal of the fraction (i.e. 3). A young, healthy subject may have a contrast threshold of 1 percent, or 1/100 (i.e. a contrast sensitivity of 100). Occasionally, subjects have even better contrast thresholds. A subject could have a threshold of 0.003 (0.03 percent, or 1/1000), which converts into a contrast sensitivity of 3000. In the visual psychology literature, the contrast threshold is described in logarithmic terms. Therefore a contrast sensitivity of 10 is 1, a contrast sensitivity of 100 is 2, and a contrast sensitivity of 1000 is 3.

However, the video engineer describes contrast by using a gray scale that may contain more than 100 different levels of gray. A newspaper printer may use the term *halftones* in place of gray scale and may need more than 100 different half tones (densities of black dots) to describe the contrast of a scene.

Targets

Both the visual scientist and the optical engineer use a series of alternating black and white bars as targets. The optical engineer describes the fineness of a target by the number of line pairs per millimeter (a line pair is a dark bar and the white space next to it); the higher the number of line pairs per millimeter, the finer the target. For example, about 100 line pairs per millimeter is equivalent to a space of 1 minute between two black lines, which is almost equivalent to the spacing of the 20/20 (6/6) letter. In experimental testing, 109 line pairs per millimeter is equivalent to 20/15 (6/4.5).

The vision scientist describes the alternating bar pattern in terms of spatial frequency; the units are cycles per degree (cpd). A cycle is a black bar and a white space. To convert Snellen units into cycles per degree, one must divide the Snellen denominator into 600, or 180 if meters are used, for example, 20/20 (6/6) converts to 30 cpd (600/20, or 180/6), and 20/200 (6/60) converts into 3 cpd (600/200, or 180/60).

Sine waves

So far, targets have been described as dark bars of different spatial frequency against a white background. These are also known as *square waves* or *Foucault gratings*. However, in optics, few images can be described as perfect square waves with perfectly sharp edges. Diffraction tends to make most edges slightly fuzzy, as do spherical aberration and oblique astigmatism. If the light intensity is plotted across a black bar with fuzzy edges against a light background, a sine wave pattern results. Sine wave patterns have great appeal because they can be considered the essential element from which any pattern can be constructed. The mathematician can break down any alternating pattern (be it an electrocardiogram or a trumpet's sound wave) into a unique sum of sine waves, known as a *Fourier transformation*. Joseph Fourier, a French mathematician, initially developed this waveform language to describe heat waves. Fourier's theorem states that a wave may be written as a sum of sine waves that have various spatial frequencies, amplitudes, and phases.

It also is thought that the visual system of the brain may operate by breaking down observed patterns and scenes into sine waves of different frequencies. The brain then adds them again to produce the mental impression of a complete picture. Fourier transformations may be the method the visual system uses to encode and record retinal images. In fact, it has been shown that different cells or "channels" occur in the retina, lateral geniculate body, and cortex and selectively carry different spatial frequencies. So far, six to eight channels have been identified. It also has been shown that all channels respond to contrast. Interestingly, the cortex shows a linear relationship between the amplitude of the neuronal discharge and the logarithm of the grating contrast.[60] As a result of the preceding reasoning, most contrast sensitivity tests are based on sine wave patterns rather than square wave patterns of different frequency.

Recording contrast sensitivity

Figure 1.9 shows a number of functions, including the contrast sensitivity testing function for a normal subject. The

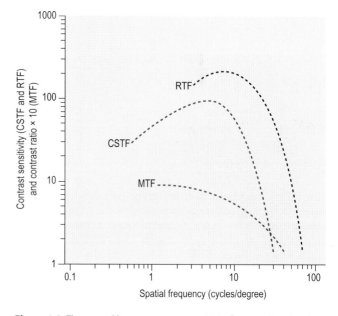

Figure 1.9 The normal human contrast sensitivity function (CSTF) is the sum of the contrast sensitivity of the purely optical contribution (MTF), and the neuroretinal enhancement system (RTF). (Modified from Mainster M. Surv Ophthalmol 1978; 23:135.)

shape of the human contrast function is different from that of almost all good optical systems, which have a high contrast sensitivity for low spatial frequency. The contrast sensitivity gradually diminishes at the higher spatial frequencies, as diffraction and other aberrations make discrimination of finer details more difficult. The contrast sensitivity function for the purely optical portion of the visual system (cornea and lens) is the modulation transfer function. The human contrast sensitivity function is different from the sum of its components because the retina–brain processing system is programmed to enhance the spatial frequencies in the range 2–6 Hz. Receptor fields, on/off systems, and lateral inhibition are the well-known physiologic mechanisms that influence the different spatial frequency channels and are responsible for such enhancement.

In Figure 1.9, the wave labeled "retinal testing function" represents the retinal neural system performance.[60–62] Normal variations are found in the contrast sensitivity function. For example, contrast sensitivity decreases with age. Two factors appear to be responsible. First, the normal crystalline lens scatters more light with increasing age, which thus blurs the edges of targets and degrades the contrast. Second, the retina–brain processing system itself loses some ability to enhance contrast with increasing age.

The contrast sensitivity also decreases as the illumination decreases. Thus contrast sensitivity for a spatial frequency of 3 cpd drops from 300 to 150 to 10 as the retinal luminance drops from 9 to 0.09 to 0.0009 trolands. (The *troland* is a psychophysical unit; 1 troland is the retinal luminance produced by the image of an object, the luminance of which is 1 lux, for an area of the entrance pupil of 1 mm^2.)

The contrast sensitivity function also is an accurate method by which to follow certain disease states. For example, the contrast sensitivity of a patient who has a cataract is diminished, as it is in another light-scattering lesion, corneal edema. Because the contrast sensitivity function depends on central nervous system processing, it is not surprising that conditions such as optic neuritis and pituitary tumors also characteristically have diminished contrast sensitivity functions.

Glare, tissue light scattering, and contrast sensitivity

When a transparent structure loses its clarity, the physicist describes it as a light *scatterer* rather than a light *transmitter.* This concept is foreign to the clinician whose textbooks talk about opaque lenses and corneas. The word *opaqueness* conjures up the image of a cement wall that stops light. Of all the experiments demonstrating that most cataracts scatter light rather than stop light, the most graphic involves the science of holography. If it is true that a cataract splashes or scatters oncoming light, resulting in a poor image focused on the retina, it should theoretically be possible to collect all the scattered light with a special optical element and recreate a sharp image. The essence of such an optical element, one that would take the scattered light of the cataract and rescatter it so that a proper image could be formed, would be a special inverse hologram of the cataract itself. Figure 1.10 shows how such a filter would work. Miller et al[63] were able to demonstrate how an extracted cataract (the patient's visual acuity was worse than 20/200) would be made relatively transparent by registering a special inverse hologram of that specific cataract in front of the cataract.

To follow the progress of conditions such as cataracts or corneal edema, a measure of tissue transparency or tissue backscattering is useful. Although photoelectric devices can be used to quantitate the amount of light scattered by various ocular tissues, a subjective discrimination system is needed to evaluate patient complaints. The Snellen visual acuity test was the traditional index, but it is not sensitive enough. Figure 1.11 shows a scene in a fog taken with a digital camera where the closer objects are sharp (good acuity through less cloudy media) and the more distant objects have poor contrast and resolution, as seen through a very cloudy media (i.e. cataract).

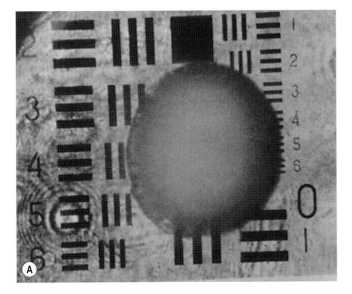

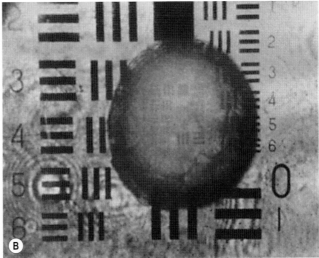

Figure 1.10 Note how the addition of the inverse conjugate hologram in registration with the cataract allows you to see the resolution chart. (From Miller D, Benedek G. Intraocular light scattering, Springfield, Ill: Charles C Thomas, 1973.)

Figure 1.11 A scene in a fog taken with a digital camera where the closer objects are sharp (good acuity through less cloudy media) and the more distant objects have poor contrast and resolution, as seen through a very cloudy media (i.e. cataract).

LeClaire et al[64] observed that many patients with cataracts showed good visual acuity but had poor contrast sensitivity in the face of a glare source. In fact, this should not come as a surprise because the essence of vision is the discrimination of the light intensity of one object as opposed to another, often with a natural glare source present. Thus a plane is seen against the sky because the retinal image of the plane does not stimulate the photoreceptors to the same degree that the sky does. Terms such as *contrast luminance* and *intensity discrimination* are used to describe differences in brightness between an object and its background.

How then can ocular light scattering, glare, and contrast sensitivity be linked together to give the clinician a useful index? An industrialist scientist named Holliday set the stage to solve this puzzle.[65] In 1926, Holliday developed the concept of glare and glare testing to measure the degrading effect of stray light. In the 1960s, Wolfe, a visual physiologist working in Boston, realized that glare testing could be a useful way to describe the increase in light scattering seen in different clinical conditions.[66,67] How does increased light scattering produce a decrease in the contrast of the retinal image in the presence of a glare source? Figure 1.12 shows how corneal edema splashes light from a naked light bulb onto the foveal image, reducing the contrast of the image of the target. Figure 1.13 illustrates the way a patient with a cataract or corneal edema would see a scene in the presence

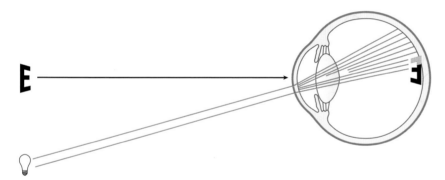

Figure 1.12 Corneal edema scatters the light from the peripheral light source onto the fovea, decreasing the contrast of the foveal image. (From Miller D, Benedek G. Intraocular light scattering, Springfield, Ill: Charles C Thomas, 1973.)

Figure 1.13 Photograph of the way the scene would appear to a normal patient (**A**) and a patient with corneal edema (**B**) in the face of glare. (From Miller D, Benedek G. Intraocular light scattering, Springfield, Ill: Charles C Thomas, 1973.)

of a glare source. In the mid-1970s, Nadler observed that many of his cataract patients complained of annoying glare. His observations rekindled interest in glare testing and led to the first clinical glare tester – the Miller–Nadler glare tester.[63]

Clinical conditions affecting glare and contrast sensitivity

Optical conditions

This section describes how contact lenses, cataracts, opacified posterior capsules, displaced intraocular lenses (IOLs), and multifocal IOLs affect glare sensitivity and contrast. With the exception of IOLs, these conditions primarily diminish contrast sensitivity because of increased light scattering.

Corneal conditions

Corneal edema

Studies tracing the progression of corneal decompensation have shown that the stroma increases in thickness before the epithelium changes.[68] The stroma may increase in thickness by up to 30 percent before the epithelium becomes edematous. Studies have shown that an increase in stromal thickness above 30 percent need not influence Snellen visual acuity results if there is no epithelial edema.[69] Unlike Snellen visual acuity, both contrast sensitivity and glare sensitivity are compromised as soon as the stroma thickens. Mild edema affects only the middle and high frequencies of a contrast sensitivity test, sparing the low frequencies. With further edema, the sparing of the low frequencies disappears and contrast sensitivity is decreased throughout the spatial frequency spectrum.[63] Glare sensitivity measurements also detect early epithelial edema. A mildly edematous epithelium is roughly equivalent to an increase of 10 percent in stromal thickness, whereas moderate to significant epithelial edema has a profound effect on glare and contrast sensitivity.

Contact lens wear

The wearing of contact lenses may reduce contrast sensitivity in a number of subtle ways. Patients with significant corneal astigmatism who wear thin soft contact lenses experience blur that affects their contrast sensitivity. Aging of the plastic material itself or surface-deposit accumulations can affect soft lens hydration and ultimately influence acuity, glare, and contrast sensitivity. Most important, contact-lens-induced epithelial edema produces increased glare disability and reduced contrast sensitivity.[63]

Keratoconus

Patients with keratoconus demonstrate attenuation of contrast sensitivity with relative sparing of low spatial frequencies despite normal Snellen visual acuity. However, once scarring develops in the keratoconic cornea, all frequencies become attenuated. In addition, glare sensitivity acutely increases as soon as scarring develops. Thus contrast sensitivity testing at a number of spatial frequencies, with or without a glare source, may be an excellent way of following the progression of keratoconus.[63]

Nephrotic cystinosis

In a study of patients with infantile-onset cystinosis, contrast sensitivities were reduced at all frequencies, although the loss at high frequencies was the greatest. Of 12 subjects, 10 showed glare disability, compared with a control population.[70]

Penetrating keratoplasty

Contrast sensitivity or glare testing may also be useful in detecting the earliest signs of graft rejection. In such cases, the earliest corneal damage is corneal edema. Although visual acuity may remain normal, contrast and glare performance start to slip. As the edema progresses to involve the epithelium, the degradation of these visual functions is accentuated. Similarly, reversal of graft rejection may be followed by an improvement in the contrast sensitivity function.[63]

Refractive surgery

Some patients who have undergone radial keratotomy or photorefractive keratoplasty with postoperative corneal haze have been reported to experience increased glare sensitivity.[71,72] The extent of the problem and the number of patients complaining of heightened glare sensitivity varies from study to study and depends on the time elapsed since the surgery and the method by which the glare was assessed. Modern refractive surgery approaches the haze problem decreasing the healing process intensity either by the use of lasik (using mechanical or laser microkeratome) or mitomycin-C and prk. Both techniques are effective and have its own indications. Glare due to haze is nowadays less frequent than glare due to postoperative spherical aberration.

Harper & Halliday[73] reported on four unilaterally aphakic patients and found significant contrast sensitivity losses in the eyes with epikeratoplasty when compared with the normal fellow eye.

Cataracts and opacified posterior capsules

Figure 1.12 demonstrates the way that an edematous cornea or cataract scatters stray light onto the fovea and degrades contrast sensitivity, thereby heightening glare disability. Thus measurements of contrast sensitivity are usually better correlated with patient complaints than with a visual acuity measurement. The addition of a glare source to a contrast sensitivity test causes a dramatic decrease in the contrast function. Of the various cataract types, the posterior subcapsular cataract degrades the glare and contrast function the most. It should be noted that the presence of a glare light diminishes both visual acuity and contrast sensitivity in cataract patients. In the presence of a glare light, the contrast sensitivity function gradually diminishes as a simulated cataract increases in severity, whereas the visual acuity function holds steady until an 80 percent simulated cataract produces a dramatic drop in visual acuity.

Progressive opacification of the posterior capsule after an extracapsular cataract extraction produces a progressive increase in glare disability.[74] A neodymium:yttrium-aluminum-garnet (Nd:YAG) laser capsulotomy in such cases improves the visual function. The improvement of contrast and glare sensitivity after Nd:YAG laser treatment depends on the ratio of the area of the clear opening to the area of the remaining opaque capsule. Thus a photopic pupil of 4 mm would require a 4-mm capsulotomy for best results in daylight. However, if the pupil dilates to 6 mm at night, an oncoming headlight would induce an annoying glare unless the capsulotomy were enlarged to 6 mm in diameter. Thus the smallest capsulotomy is not necessarily the best from an optical point of view.

Modulation transfer function

Optical engineers generally evaluate optical systems by means of a system similar to contrast sensitivity, called the *modulation transfer function* (MTF). The MTF is the ratio of image-to-object contrast as a function of spatial frequency, where the object is either a bar graph or a sinusoidal grating. It gives more information than the parameter of resolving power. For example, two systems may have the same resolving power, but one might be unable to form useful images of low-contrast objects, which the other could readily form. A smaller pupillary aperture introduces diffraction interference, which makes it difficult to resolve fine detail (higher spatial frequencies). Thus the spatial frequency cutoff occurs sooner with the small aperture system. In an MTF plot, the vertical axis is akin to contrast sensitivity (see Figure 1.9). Because it represents the ratio of contrast of image against contrast of object, its values decrease from 1.0 to 0. The MTF is conceptually similar to the manner in which electronic engineers evaluate an amplifier. The performance of an amplifier is described by the output/input ratio, or the gain, for different sound frequencies. The MTF concept also may be useful in the comparison of the performance of the eye with that of optical and electronic instruments. Campbell & Green[75] plotted the MTF of the human eye for various pupillary diameters and found that a smaller pupil system has a better contrast ratio than a larger pupil system. This probably reflects the opposing factors of a somewhat better performance with greater illumination and the degrading effect of spherical aberration with a larger pupil. As noted earlier, a pupil size in the range 2.0–2.8 mm gives the maximum MTF for high spatial frequencies.

Depth of focus

How can an insect or a small animal like a rat see objects clearly from 10 m to 10 cm without an accommodating mechanism? How does a pinhole allow a patient with presbyopia to read the newspaper without a reading correction? The answer is that both situations rely on an optical system with an increased depth of focus.

Recall that an image can be thought of as being made up of an array of points. Thus the finite size of the pixels, the photographic grains, or the photoreceptor clusters ultimately determine the fineness of details of the recorded image. This means that a blur of a focused point is tolerated if it is no bigger than the size of the receptor. Because a point of light is focused as an Airy disc, a cluster or two to five cones is considered the "limiting grain size." Let us review two important definitions:

> Depth of focus: *the amount of blur in diopters or millimeters from the retina that will be tolerated or go unnoticed*
> Depth of field: *the distance range, in object space, that an object can move toward or away from a fixed focus optical system and still be considered in focus*

Figure 1.14 is a schematic of the eye. For simplicity, we used a reduced eye, with a biconvex lens representing the cornea and lens. Let:

- () = the object that can move from infinity to a near point N
- ρ = pupillary diameter
- f = focal length of the model eye
- x = distance from the retina that near point remains in focus
- c = limiting photoreceptor cluster size (i.e. limiting grain size or pixel size)
- N = refractive index of the model eye
- $D_1 = n/f$ (i.e. dioptric power of eye when viewing a finite object)
- $D_2 = n/f$ (i.e. dioptric power of eye when near object at N)

(In this case, the eye can be considered to have lengthened (theoretically) by a distance x.)

Depth of focus is as follows:

$$D_1 - D_2 = n/\tilde{f} - n/f + x = nx(f + x) - nf/f(f + x) = nx/f(f + x)$$

(but by similar triangles). Thus

$$D_1 - D_2 = nc/f\rho$$

This equation tells us that the depth of focus in diopters $(D_1 - D_2)$ is proportional to the index of refraction (n) and the limiting photoreceptor of grain size c. The depth of focus is also inversely proportional to the pupil size (ρ) or the focal length of the system (f).

For example, determine the depth of focus for a reduced human eye under the following conditions. Let:

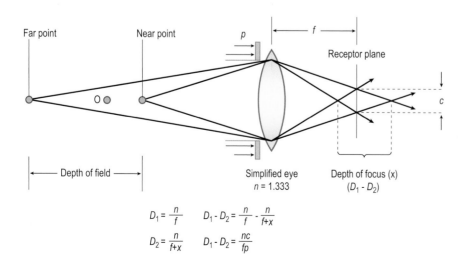

$$D_1 = \frac{n}{f} \qquad D_1 - D_2 = \frac{n}{f} - \frac{n}{f+x}$$

$$D_2 = \frac{n}{f+x} \qquad D_1 - D_2 = \frac{nc}{f\rho}$$

Figure 1.14 A schematic representation of a single refracting surface model eye where ρ = pupillary diameter, f = focal length distance from the retina where image of near object falls, c = limiting photoreceptor cluster size (i.e. similar to grain or pixel size), n = refractive index, D1 = n/f dioptric power of eye when viewing infinite object, D_2 = n/(f + x) (i.e. dioptric power of an eye when viewing near object). Thus depth of focus = $D_1 - D_2$ = n/f − n/(f + x). Ultimately, we find $D_1 - D_2$ = nc/fρ.

- Pupil (ρ) = 3 mm, or 0.003 m
- Focal length (f) = 22.2 mm, or 0.0222 m
- Limiting cone cluster (c) = 5 cones (Assume each cone is 1.5 μm in diameter and spacing between cones is 0.5 μm. The total cluster of 5 cones + 4 spaces = 9.5 μm, or 0.0000095 m.)
- Index of refraction (n) = 1.333

Depth of focus $(D_1 - D_2) = 1.33 \times 0.0000095/0.0222 \times 0.003$
$$= 0.189$$

Our calculation of 0.189 for the reduced (hypothetical) eye with a pupil of 3 mm is about half of the 0.40 diopter (Fig. 1.15A), which is the average value from four (real) human studies.[75–78] Thus we may conclude that the normal eye has a modest depth of focus.

Interestingly, Figure 1.15B represents a study in which artificial pupils (placed in front of the cornea) between 1 and 2 mm were used. Within this range of apertures, a depth of focus between 2 and 4 diopters was obtained.[31]

Do we ever see the equivalent of pupillary apertures of between 1 and 2 mm in human patients? Yes, we do in cases of trauma, disease, or use of strong miotics. Figure 1.16 demonstrates four examples of 1- to 2-mm clear areas within a corneal opacity. Actually, the patient shown in the upper left had an acuity of 20/20. Figure 1.17 shows four examples of 1- to 2-mm clear areas within traumatic cataracts or 1- to 2-mm spaces between traumatic cataracts and iris tears. In Figure 1.18, we can see the equivalent of a reduced pupillary aperture produced by a ptosis or a conscious reduction of the palpebral fissure as a result of squinting. It is comforting to know that in cases of trauma, disease, or significant refractive error, the human eye can call on a 2- to 3-diopter depth of focus mechanism.

The clinical diagnosis of "corneal visual acuity" has been proposed, e.g. the "Holladay Diagnostic Summary," first presented in the EyeSys corneal topography system. It predicts decreased visual acuity areas of the cornea that have irregular surfaces, assuming that there is a transparent media behind the corneal epithelium.

Optical aberrations

The famous nineteenth century German physiologist Herman von Helmholtz, in Volume 1 of his *Treatise on Physiologic Optics*,[79] wrote that the optical aberrations of the human eye are "of a kind that is not permissible in well constructed instruments." The implication is that the optical design of the human eye would receive low marks if evaluated by someone from the optical industry. If we are to simply compare the optical quality of the living human eye with that of our best cameras and telescopes under ordinary static daylight conditions, then Helmholtz was correct. The optical imperfections noted by Helmholtz are discussed in the following sections.

Light scattering

Fingerprints on spectacle lenses scatter light, making small letters difficult to read. Raindrops or windshield wiper smears on the car windshield make the reading of street signs difficult. In a similar way, small bubbles from the warm water tap give a haze to a glass of water and make it difficult to see the details at the bottom of the glass. These are all

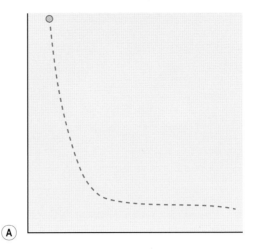

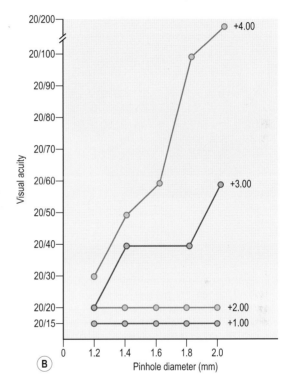

Figure 1.15 (A) An averaged depth of focus versus pupil size function representing four different studies.[24,62,80,81] **(B)** Plot of visual acuity versus pinhole diameters (between 1 and 2 mm) using different blurring lenses (i.e. +1 diopter, +2 diopters, +3 diopters, and +4 diopters). Thus, for example, the subject could achieve a visual acuity of 20/40 through a 1.8-mm pinhole, but vision would be blurred by a 3-diopter lens. (Modified from Miller D. Optics and refraction: a user friendly guide. St Louis: Gower Medical, 1991.)

examples of light scattering, which can obscure the details of any object.

As noted earlier, the cornea is clear, but technically speaking, it is not perfectly transparent. Its composition of fine collagen fibers, loaded into a watery matrix of glycosaminoglycans and populated by fine cells that swim in the matrix, scatter a small percentage (10 percent of incident light) and create a slight haze. This is unquestionably a flaw, as opposed to the glass optical systems of cameras and telescopes. However, the "imperfect" corneal structure allows for healing. Thus we can appreciate that a

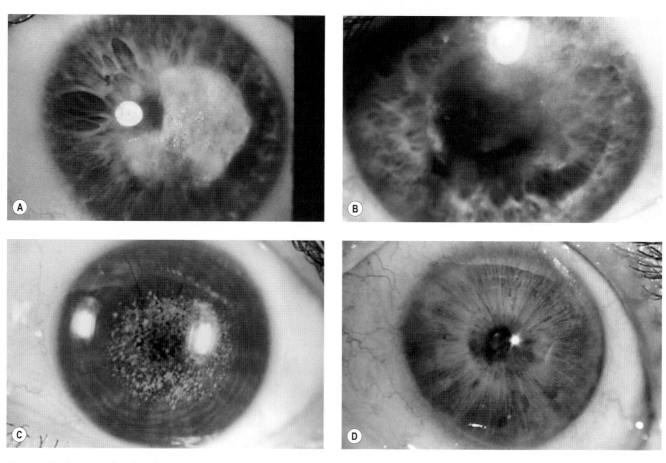

Figure 1.16 Four examples of small clear areas within corneal opacities that operated as pinholes.

Box 1.2

Holladay diagnostic summary maps (Holladay JT. J Cataract Refract Surg 1997; 23(2):209–21). The patient's refractive error was a –4.00 D, prior to a Lasik treatment 4 months earlier. Presently, the visual acuity is 20/15 in daylight (3mm pupil) and 20/25 in dim light (5 mm pupil). The upper maps include refractive power maps on standard and auto scales. The bottom left shows a profile difference map which demonstrates the patient's corneal shape compared to 'normal' cornea. The bottom right represents a distortion map which displays the optical quality of the cornea. Note that the 20/16 central (3 mm pupil) visual acuity predicted by the distortion map was achieved by the patient.

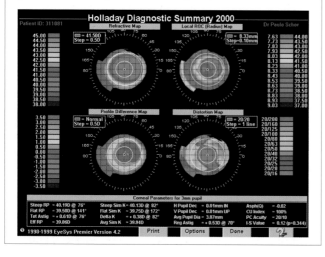

Box 1.3

The picture represents the cornea of a 51-year-old female patient who had an Acufocus Presbyopic inlay implanted 45 days earlier. Unaided near visual acuity changed from J9 to J2 after surgery. Distance vision changed from 20/20 to 20/25. (Courtesy of the Refractive Surgery Section – Department of Ophthalmology – Paulista School of Medicine, Sao Paulo, Brazil.)

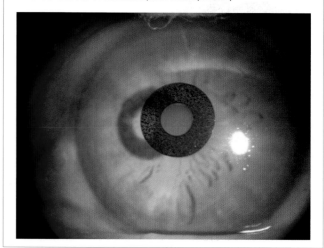

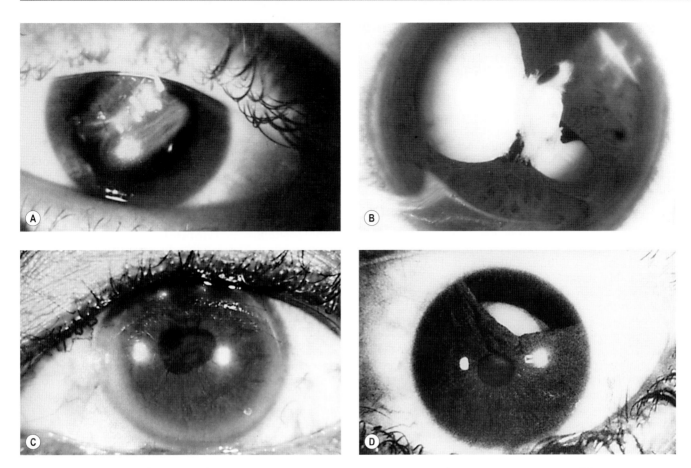

Figure 1.17 Four examples of 1- to 2-mm clear area within traumatic cataracts or 1- to 2-mm spaces between traumatic cataracts and iris tears.

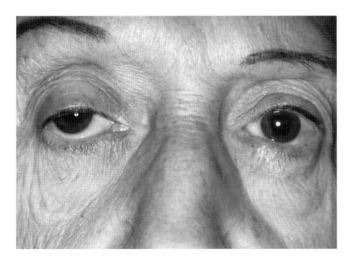

Figure 1.18 The equivalent of a reduced aperture produced by a ptosis.

10 percent level of light scatter is a fair price to pay for a self-healing system.

The lens of the eye is made up of tens of thousands of fine fibers (each a bag of clear protein solution) packed closely together. The living lens continually adds outside fibers and packs old cells in its nucleus throughout life, ultimately increasing its volume by a factor of 3 in the older eye. The refractive index of the fibers differs from that of the thin spaces between the fibers. Thus the young lens scatters about 20 percent of the incident light. Is there a practical advantage to the fiber structure of the lens? There are many ways that the lens may be injured, including inflammation from diseases inside the eye, blunt blows from fists or rocks, periods of malnutrition, poisoning, and osmotic upset from systemic disease. In all of these situations, the part of the lens being laid down at the time of the injury or disease loses transparency. This hazy section may also be lumpy in thickness. However, in a short time, new transparent layers cover and compress the hazy ones, smoothing and reducing the blurring effect of the injury. Again, we now understand that a small level of light scattering resulting from this fiber layering system is a fair price to pay for a self-repairing lens system.

Natural defenses against light scattering

The reader may have been led astray in thinking that the eye has no defense against normal light scattering. In fact, it does. For example, the birefringent capacity of the collagen fibers in the cornea, in combination with the birefringence of the fovea, may cancel out some annoying glare caused by light scattering through a process known as *destructive interference*, somewhat akin to the way polarizing sunglasses cancel annoying glare.

The retina has three defenses against the image-degrading effects of scattered light. To appreciate one of the defenses, one first must recognize that not all colors (wavelengths) are scattered equally. The fine components of eye tissue scatter blue light 16 times more than red light. This light produces lipofuscin-induced cell toxicity in in vitro experiments. It has also been proposed as a pathological mechanism for cataract formation and retinal damage.[82] Yellow filter spectacles reduce chromatic aberration and sharpen the retinal image by reducing the number of wavelengths striking the retina. However, a yellow filter makes the sky look gray rather than blue. This effect, although unnatural, enhances the contrast of objects seen against the sky. There is presently insufficient evidence to prove or disprove the claim that a yellow IOL will prevent or slow down macular degeneration.

Therefore a defense that can reduce blue scattered light would be disproportionately helpful. Sprinkled throughout the ultrasensitive fovea and its immediate surround is yellow pigment. The yellow pigment is efficient at absorbing the scattered blue light, thus preventing much of it from degrading the retinal image.[31]

The second defense used by the retina involves the positioning of the rods and cones. Each rod or cone functions as a light guide. To enter the guide, light must enter at a specific angle. Interestingly, normally focused light enters a photoreceptor at a different angle than does scattered light.[83] The photoreceptors of the retina are directed so that they primarily receive focused light, but not scattered light (Fig. 1.19). Stiles and Crawford established the fact that the human visual system has reduced sensitivity to light rays that enter near the edge of the pupil compared to those entering centrally. This effect, referred to as the psychophysical Stiles–Crawford effect, is the result of the orientation of photoreceptors, which preferentially guide the central rays to the retina more efficiently than the peripheral rays.[84]

The dark brown pigment of the retinal pigment epithelium and the choroid absorbs any stray light that has passed through the retina and prevents such light from backscatter, which would reverberate among the neighboring photoreceptors. None of these defenses is perfect, but all work to reduce the annoyance of scattered light.

The brow and eyelid may also be thought of as blocking annoying glare sources such as the overhead sun. Interestingly, the Asian lid has a double fold and serves as a more effective thicker visor than the Caucasian lid, thereby more effectively blocking the glaring effect of the sun overhead.

Chromatic aberrations

A rainbow is produced by millions of tiny round droplets of water vapor that hover over the earth during or after a rain. Each water droplet functions as a tiny, powerful round lens. Such a powerful lens bends each wavelength of color differently. A rainbow is caused by the chromatic aberration of the water droplets. Thus, like Newton's prism, the tiny droplets break up white light into the colors of the rainbow. The phenomenon of strong lenses producing colored fringes around a focused image is known as *chromatic aberration*.[85,86] The optical components of the eye (cornea and lens), like the fine water droplets, also produce chromatic aberration. The total chromatic aberration of the photopic human eye is about 3 diopters. However, we do not see colored fringes around objects because significant colored fringes of red and blue are less likely to be seen as a result of the cones' relative insensitivity to the colors at the ends of the spectrum. Also,

the visual processing in the retina and brain can sharpen the edges of the retinal image and "erase" the colored blur. Although we rarely consciously sense the chromatic aberration of the eye, some researchers believe that the retina may make use of the faint colored fringes around images to help accommodation reach a precise end point.[36,87]

Spherical aberration

A major distortion produced by many high-powered optical systems such as the cornea or lens is called *spherical aberration*. Figure 1.19 shows the results of this aberration. The rays at the edge of the lens are bent more than those going through the center of the lens, creating a smeared focus. The cornea (a strong optical element) is subject to spherical aberration. Figure 1.20 shows that the cornea sits at the front of the eye like a small, strongly curved dome. The steeper the dome (the shorter its radius of curvature), the more spherical aberration created. We have known since the time of the French mathematician Descartes that spherical aberration can be controlled by flattening the curvature of the edge of a lens, thereby weakening the focusing power of the lens periphery. Descartes described such a surface as *aspheric*. Most cameras today use lenses with aspheric surfaces.

The average cornea is also somewhat aspheric. It becomes slightly flatter at its periphery, allowing it to more smoothly merge with the sclera. However, there are some real-world considerations that make it advantageous to keep the cornea steeply curved. In the event of a direct blow to the eye by a blunt object, presumably the steep protruding cornea can

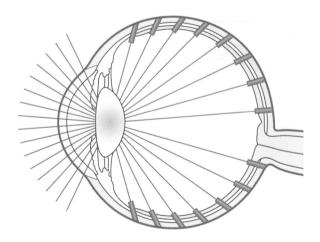

Figure 1.19 The orientation of the retinal photoreceptors in the normal human eye. Note that they are directed to the second nodal point of the eye.

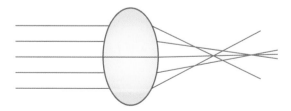

Figure 1.20 An illustration of the smearing of sharp focus created by spherical aberration. Note that the peripheral rays are bent more than the paraxial rays of light.

Figure 1.21 Three-dimensional drawing of the cornea sitting atop the eye as a small dome, vaulting over the internal structures. The viscoelastic material injected into the anterior chamber should remind the reader that the vitreous humor (similar properties) serves as a second damping element in the event of blunt trauma.

Box 1.4

A practical example is the comparison between the optical properties of the +20D trial frame lens and the aspheric indirect binocular ophthalmoscope (IBO) +20D lens. Both lenses focus light coming from infinity to a focal point of 5 centimeters behind the lens. However, when analyzing the quality of the entire image, one can observe a deformity in the periphery of the image formed by the trial frame lens, which is not present with the aspheric lens. The higher the power of a lens , the more the distortion produced by spherical aberration. The figure on the right demonstrates the spherical aberration (i.e. peripheral distortion) produced by a typical + 20D spectacle lens. The figure on the left demonstrates the reduced spherical aberration produced by an aspheric +20D indirect ophthalmoscopy lens.

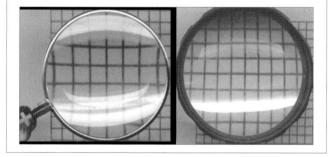

---

absorb the blow, much like a spring. The steeper the cornea, the more it vaults over the rest of the eye and the greater the spring effect (Fig. 1.21). Such a spring-dampening effect protects the deeper eye structures. How then does the eye satisfy both needs? In daylight, with the pupil being constricted, the iris tissue essentially blocks many of the light rays coming through the corneal periphery and effectively cancels most of the spherical aberration. Thus spherical aberration from the cornea has an important degrading effect only when illumination is dim and the pupil is large. Night myopia can develop in these situations and is more of a problem when the corneal asphericity is greater. Such greater asphericity is often found after corneal refractive surgery.

Happily, in dim light, the human eye switches to the rod retinal system in which seeing fine details takes a lower priority than simply seeing large shapes.

The crystalline lens is also a powerful optical element and therefore is also vulnerable to spherical aberration. Although an aspheric surface corrects the aberration in cameras and telescopes, nature has chosen a different approach for the crystalline lens. Recall that refraction, or the bending of light, can be controlled by either the curvature of the lens surface or the index of refraction of the composition of the lens. The high index of refraction of the crystalline lens is the result of a high concentration of protein. The lens has a lower refractive index near its edge than at its center. Therefore the lens periphery has a weaker focusing action and self-corrects spherical aberration, much as an aspheric surface.[29]

However, even with the aforementioned correction factors, the total spherical aberration of the human eye varies from 0.25 to 2.00 diopters. The corneal shape is the most important factor in the induction of spherical aberration.[88]

Abolition of spherical aberration not only sharpens focus but can be thought of as concentrating the light at the focus. Concentrating light energy at a focus makes it easier to see a dimly lit object. Therefore cameras, or creatures with an optical system of minimal spherical aberration, function well in low illumination.

Some intraocular lenses are aspheric. However, the degree of IOL asphericity must compensate for the patient's individual corneal asphericity in order to yield the sharpest retinal image of a distant object. On the other hand, the presence of some spherical aberration may blur a distant object but allow an object at an intermediate distance (1–2 meters) to be in focus. As yet, there is no IOL that produces completely sharp images for both distant and intermediate objects.

Light absorption

The lens of a typical 20-year-old absorbs about 30 percent of incident blue light. At age 60, the typical lens absorbs about 60 percent of incident blue light.[89] This increase in lens absorption of blue light results in both a decrease in subtle color discrimination and a decrease in chromatic aberration.

Summary – a compromise of eye function

The human eye (which is similar to the monkey eye) is a fairly good resolving optical instrument.[20,90,91] Admittedly, the eyes of certain birds are even better optical instruments, reaching the outer limits of the constraints of the laws of physics. One must appreciate that the evolution of better and better optics had to be balanced against other useful functions, such as glare prevention, injury prevention, injury repair, and use of the eyes for non-verbal communication.

The aging eye

The components of human eyes usually last a lifetime. This was reinforced to me years ago when our eye team did a visual screening of residents in a housing project for older adults. On average, most residents were older than 80 years. The number of residents with 20/40 vision in at least one eye was more than 90 percent.[92] This finding should not minimize the disability of people with cataracts (which affect 9.4 million people older than age 65 in the United States).[93] However, it is important to recognize that a high percentage of people live their entire lives with good vision

– thanks to a number of positive compensations that help the aging eye. For example, the human macula is particularly vulnerable to damage from ultraviolet and blue light. Fortunately, a yellow pigment effectively absorbs or scatters away most of these harmful wavelengths, thus diminishing the potential damage to the macula.[94]

One of the more remarkable aspects of the aging eye is that the eye lens continually acquires new layers of fibers, becoming both progressively thicker and steeper. These changes would normally lead to an increase in lens focusing power and a tendency toward nearsightedness in older eyes. In fact, this does not happen universally because the index of refraction of the cortex of the lens decreases in a perfectly compensatory fashion,[95] so the lens power often stays constant.

Evolution of ocular components

Human eyes are fascinating examples of a potpourri of components seen all along the evolutionary trail. Indeed, human eyes contain components previously developed for other uses. For example, our retinal rhodopsin may have come from an ancient bacterium, which may have used the rhodopsin for photosynthesis. Our rounded corneal curvature originally comes from primitive fish. As noted earlier, when a person is under water, the corneal refractive power is canceled by the surrounding water. However, rounded shape helps decrease water resistance because the eye is then more streamlined. Finally, and most difficult to explain, many of the special crystalline proteins that have been identified in animal eye lenses are similar to metabolic enzymes found elsewhere in the body. For example, lactic dehydrogenase-β is similar to E-crystalline found in the lens: arginosuccinate lyase is similar to γ-crystalline found in the lens.[96] Non-lenticular enzymes have been adapted to also function as special lens proteins, a process known as *gene sharing*. The term *gene sharing* simply names the phenomenon without giving us an idea of which function came first or what evolutionary pressures forced the new use of the molecules. In summary, one might think of this entire process of nature as reaching into its dusty attic to find new uses for old creations as the ultimate in recycling efficiency.

Non-optical brain mechanisms that enhance the retinal image

There is a group of visual phenomena in which the retinal image is enhanced or made complete by the brain.[97] They represent ways of improving the retinal image in non-optical ways. One might think of these brain-processing effects as examples of methods of going beyond the limits of the laws of optics to bring out visual information.[98]

Contrast enhancement

The visual brain has the ability to sharpen the contrast of elements in the retinal image. Figure 1.22 presents two faces with the same degree of grayness. In the right figure, the face is seen against a black background, whereas in the left figure, it is seen against a white background. The gray face on the black background looks lighter (enhancing its contrast), whereas the gray face looks darker on the white background. This effect is reduced when the edges of the faces are fuzzy.

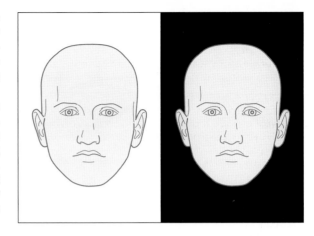

Figure 1.22 Right: Gray face is seen against a black background. Left: Same gray face is seen against a white background. A contrast enhancement function makes the gray face look lighter on the black and darker on the white side. (Illusion created by R. Miller, F. Miller, and D. Miller.)

One might speculate that a sharper edge on an object brings out a stronger contrast enhancement response.

Computer graphics professionals (the Joint Photographic Experts Group) have taken advantage of the knowledge of the physiology of human vision and released an image data compression format named after the group, i.e. JPEG. The JPEG compression process first separates the image into luminance and color signals. This step doesn't reduce the inherent quality of the image. The JPEG process then performs a variety of arithmetic procedures which separate out and remove small differences (in luminance and color) between adjacent pixels which would not be noticed by the human visual system. This process involves dividing the picture up into 8 × 8-pixel blocks, and analyzing details contained in each block. The high-frequency (fine details) information is than discarded. Such a process eliminates a great deal of data. The final process results in a very high compression ratio (i.e. often a 95 percent reduction in file size). As expected, such a trade-off results in a loss of high-frequency information (i.e. sharp transitions become blurred). Fortunately, our visual system accepts the slightly reduced image contrast.[99]

Edge sharpening

If an ametrope looks in the distance at a framed picture, the details within the picture will be fuzzy, yet the edges of the frame will be seen as a distinct boundary against the wall. The visual system places a priority on sharpening the edge of the retinal image, even though the details within remain hazy. One might suppose that in the case of the picture on the wall, recognizing the edges of the frame tells one that the fuzzy blob on the wall is a picture rather than a swarm of insects.

There is a second way that the brain processes the edge of a retinal image. If two objects of similar brightness are placed next to each other, they normally appear to merge into one object. However, if the connecting edge of one of these similar objects is slightly darker than the connecting edge of the other, the entire side with the darker edge appears darker than the lighter one. The greater contrast at the edge has spread across the whole panel.

Box 1.5

Figure A is an example of the "completing the picture" illusion. Because of our previous visual experience, we assume that the partially covered words represent "THE EYE." Wrong!

(A) Because of our previous visual experience, we assume the line covered words will spell "THE EYE." On removal of the cover, we see we were wrong. If there is a discontinuity in an object, we almost always assume that the object is partially covered. We simple create a full impression of the covered object in our mind. In Figure B, we are made to think that one geographic form covers the other, although the artist simply fitted the outline of one triangle next to the other.

(B) A group of triangles with some seeming to cover the others. In fact, the edges of the shapes simply abut each other. These illusions illustrate the ability of our visual processing system to fill in all sorts of blanks and incongruities within the retinal image to create a coherent story.

Suppose that the object of regard is not covered but that the observer has a brain lesion that has produced a small scotoma in the visual field. When such a patient is presented with a circle or square in which part of the figure resides inside the scotoma, after a brief period, the patient reports that the gap has filled in and the figure looks whole. The same phenomenon takes place if part of an image falls on the physiologic blind spot. It suggests that the visual system, faced with a gap in the information, hypothesizes (gambles) that the region surrounding the scotoma has the needed data and places that data within the scotoma to produce a complete scene.[100]

There is another series of illusions known as *gap figures*.[101] Figure C shows an example of a gap illusion. The defect (gap) is strongly highlighted by the radiating lines as a footprint might appear highlighted to a skilled guide.

(C) A gap figure (modified from an Ehrenstein illusion). Note that the radiating lines seem to create the outline of a foot. A related phenomenon is presented in Conan Doyle's story *The Adventure of Silver Blaze,* in which Sherlock Holmes identifies the murderer when he concludes that the watchdog must have known the murderer because no barking was heard on the night of the murder. As in the gap figure, the gap in the expected pattern of the dog's behavior takes on a heightened importance. Hearst appropriately called this phenomenon "getting something for nothing."[102]

Vision suppression is present in pediatric patients with strabismus in order to avoid diplopia. Vitreous floaters (mostly seen in bright environments) are another example of the nervous system suppressing annoying visual effects. The floaters are structures which have a different index of refraction than the surrounding vitreous and so can cast a shadow onto the retina. Floaters can be called a time domain phenomenon. They are often seen in post op refractive surgery patients, and usually disappear in 3 to 6 months.

Finally, mention should be made of a related temporal blocking of a visual scene, that is, our failure to notice the fleeting disappearance of an image during a blink, a twitch, a flicker, or a saccade.

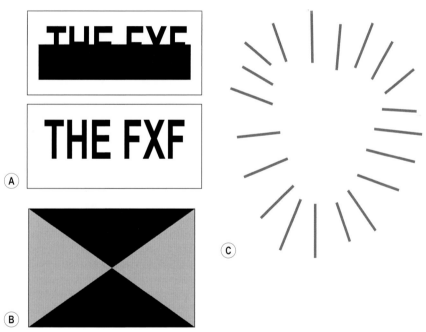

This phenomenon, known as the Craik–Cornsweet–O'Brien illusion, is seen in Figure 1.23. If a pencil is placed between the two vertical rectangles (occluding the boundary), it becomes clear that they are actually the same brightness.

Vernier acuity

Earlier in this chapter, it was noted that a normal-sighted human being (with 20/20 visual acuity) can detect a separation between two objects as small as 1 minute of angular subtense. Interestingly, it was said that Ted Williams, the great outfielder of the Boston Red Sox, had a visual acuity of 20/10 (could detect a half minute of angular separation). However, there is a visual task (Vernier acuity), which has a threshold of about 5 seconds of arc (1/12th of a minute of arc). Indeed, as seen in Fig. 1.24 most normal-sighted people can line up dots or notice a discrepancy in the alignment of dots or lines as small as 5 seconds of arc or less. We know that the brain is involved in the processing because the experiment can be redesigned so that one eye is

Figure 1.23 The Craik–Cornsweet–O'Brien illusion. The perceived difference in the darkness of the vertical panels disappears if one places a pencil or finger along the border between panels. The effect persists even if the panel details are blurred.

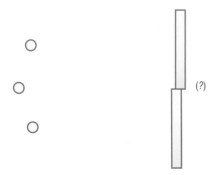

Figure 1.24 Two typical targets for measuring vernier acuity. The subject either notes misalignment of the vertical lines or of the three dots.

presented with the top and bottom dots (which are in alignment), while the other eye is allowed to see only the middle dot. Subjects show similar thresholds whether the experiment is done this way or done monocularly. Clearly, the brain must be where the ultimate processing occurs in these experiments. Because a threshold of less than 5 seconds is well beyond the optical diffraction limit of the eye, Vernier acuity represents a special example of sophisticated brain processing. The challenge is to try to conceptualize some important task necessary for the survival of our *Homo sapiens* ancestors that required such precision. Incidentally, we also know that the macaque monkey demonstrates a high level

Figure 1.25 Note the deer hidden in the bush. By almost closing one's eyes, the distraction of the high grass is canceled and the animal is easier to see. (From: Osborne C, Tanner O, eds. Animal defenses. New York: Time-Life Films, 1978.)

of Vernier acuity. In fact, the adult monkey averages a threshold of about 13 seconds of arc.[103] Thus it is possible that monkeys and early humans used Vernier acuity to detect the presence of an animal or an enemy hiding behind a stalk or a tree by noting a misalignment between the edge of the tree or stalk and the protruding body of the enemy. If that possible scenario was true, normal Vernier acuity might save a life. Once again, brain enhancement of the retinal image brings out details well beyond the limits of the best optical resolution.

Removing distractions

In Figure 1.25, a deer is hidden in the high grass. Almost completely closing the eyes creates a blurring of the overlying fine grass, making the animal more vividly seen. A similar effect can be achieved through a heavy rain. Again, almost closing the eyes allows figures to be seen more easily through the rain. This is an example of erasing the distractions that have a high spatial frequency.

Refractive errors

Thus far, when pertinent, this chapter has covered the physiologic optics of the average, emmetropic eye. However, the average refractive error in certain populations is not emmetropia. For example, in many Asian groups the incidence of myopia may be between 80 and 90 percent.

Therefore, for a broader review of physiologic optics, this last section covers some of the essential epidemiologic aspects of the refractive errors other than emmetropia.

Prevalence

Studies that tabulate the distribution of refractive errors often are taken from data on young army recruits,[81,104] which show the incidence of myopia to be about 10 percent. However, this group of healthy young men is not representative of the general population. Stenstrom's study in Uppsala, Sweden, consisted of clinic patients, colleagues, nurses, and cadet officers, which is a group more reflective of the general population.[81] His study showed that about 20 percent of the population has low myopia (less than

2 diopters), 7 percent has moderate myopia (2–6 diopters), and another 2.5 percent has high myopia (greater than 6 diopters). The great majority of this population (just less than 70 percent) clustered between emmetropia and 2 diopters of hyperopia; the rest were high hyperopes and aphakes.

The spectacle-wearing population of a typical Western country provides a different focus on emmetropes. Bennett[105] surveyed the distribution of spectacles dispensed in England. His study of refractions carried out by the average eye clinician showed that about 20 percent are myopic refractions and 75 percent of all refractions require prescriptions from −0.5 to +8.00 diopters. Subtraction of Stenstrom's estimate of the percentage of high hyperopes shows that about 65 percent of all refraction prescriptions are for presbyopia. Interestingly, Indians from the Amazon rainforest have an extremely low incidence of myopia. From an overall sample of 486 people, 259 indigenous people and 78 Brazilians between 12 and 59 years of age, only 2.7 percent of eyes showed myopia of −1.00 D or more and 1.6 percent had bilateral myopia of −1.00 D or more.[106]

Myopia

Pathologic myopia

Curtin[107] estimates that 2–3 percent of the population has pathologic myopia (a condition in which there is a significant enlargement of the eyeball with a lengthening of the posterior segment). This group falls into Stenstom's group of patients with myopia of greater than 6 diopters. The term *pathologic* is used because these patients show significant choroidal and retinal degenerative changes, a high incidence of retinal detachment, glaucoma, and increased occurrence of staphyloma development. At present, high myopia (greater than 6 diopters) is regarded as hereditary, either autosomal dominant, autosomal recessive, or X-linked recessive. Some high myopia might be inherited in a complex fashion. So far, eight loci for high myopia have been reported, including MYP1 (Xq28), MYP2 (18p11.31), MYP3 (12q21-q23), MYP4 (7q36), MYP5 (17q21-q22), MYP11 (4q22-q27), MYP12 (2q37.1), and MYP13 (Xq23-q25). Different genetic loci have been mapped in different populations, but none of the responsible genes has yet been identified.[108–110]

Physiologic, or school, myopia

As noted by Stenstom,[111] most patients have myopia that is less than 2 diopters; this type of myopia is called *physiologic*, or *school*, *myopia*. The word *physiologic* implies that this form of myopia is a normal, physiologic response to a stress. In fact, substantial evidence exists that increased time spent reading from early teenage to the mid-twenties is that stress.[4,108]

However, near work is not the sole cause of physiologic myopia. Racial and ethnicity studies show that myopia is more prevalent among Asian and Jewish persons and less prevalent among African–Americans.[50] The results of a study in Taiwan showed the incidence of myopia to be about 12 percent in children 6 years of age or younger, 55 percent in children 12 years of age or younger, 75 percent in children 15 years of age or younger, and 84 percent in those older than 18 years of age.[112] Thus it appears that many if not most cases of physiologic myopia result from a combination of an inherited predisposition and excessive close work during the student years.

Astigmatism

About 50 percent of term infants in their first years of life show astigmatism of more than 2 diopters.[113–115] This may arise from the influence of the recti muscles that pull on the delicate infant sclera, because the astigmatism seems to change in different gaze directions. Howland et al[114] suggested that the high astigmatism helps the infant bracket the position of best focus while learning to accommodate. By adulthood, this high incidence of astigmatism has disappeared. Studies show that about 15 percent of the adult population has astigmatism greater than 1 diopter and only 2 percent have astigmatism greater than 3 diopters. It is possible that much of the high astigmatism in this last group is related to some form of intraocular surgery (e.g. corneal transplants, cataract surgery, repair of corneal lacerations).

Presbyopia

Although presbyopia is age related, its age of onset varies around the world. For example, presbyopia develops earlier in people who live closer to the equator.[80,116] Specifically, the age of onset of presbyopia was noted to be 37 years in India, 39 years in Puerto Rico, 41 years in Israel, 42 years in Japan, 45 years in England, and 46 years in Norway. Further studies show the important variable to be ambient temperature rather than latitude. Thus the higher the ambient temperature, the earlier the onset of presbyopia.

On the other hand, life expectancy is lower in developing countries where the ambient temperatures are usually high. Thus, although presbyopia starts at a young age in the developing world, fewer persons with presbyopia are found in the general population. For example, in Haiti the prevalence rate of presbyopia is about 16 percent for the normal population, whereas in the United States it is 31 percent.[117] The lower rate of presbyopia in Haiti is paradoxical. The reason is probably that the average life span in Haiti is much shorter than in Western countries. Seen in perspective, presbyopia constitutes about 65 percent of all those who wear glasses in developed Western countries. Thus it is of little surprise that the first spectacles produced sometime in the fourteenth century were created for persons with presbyopia.

Components of ametropia

The overall refractive state of the eye is determined by four components:

- Corneal power (mean, 43 diopters)
- Anterior chamber depth (mean, 3.4 mm)
- Crystalline lens power (mean, 21 diopters)
- Axial length (mean, 24 mm)

Figure 1.26 shows the distribution of total refraction and the four components just mentioned for 194 eyes.[118] The most striking conclusion drawn from Figure 1.26 is that, although each of the individual optical components may be considered randomly distributed, the overall refractive status does not show a normal distribution of refractive errors, but rather a skew in the region of emmetropia. It seems that the

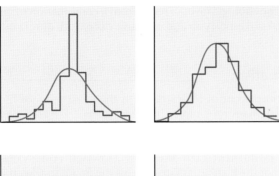

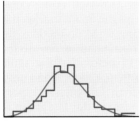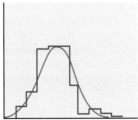

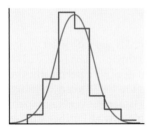

Figure 1.26 Curves of distribution of refraction and its components in 194 eyes. (Adapted from Sorsby A, Sheridan M, Leary GA: BMJ 1960; 1394; and Sorsby A et al: Med Res Counc Special Rep Serv Rep 1959; 293.)

various components cooperate to achieve a higher-than-expected incidence of refractive state between 0 and +2 diopters.

This cooperation of components to produce a higher-than-expected incidence of emmetropia and low hyperopia has been called *emmetropization*.[118] The process of emmetropization seems to be fully effective during the infantile growth of the eye. As noted earlier, the average sagittal diameter of the eye is approximately 18 mm at birth. By the age of 3 years, the axial length increases to about 23 mm. Such elongation of the eye theoretically yields a state of myopia of about 15 diopters. However, during this period, the data show that almost 75 percent of these young eyes are hyperopic.[119] Between 3 and 14 years of age, the elongation increases, on average, an additional millimeter. Again, this should theoretically produce another 3 diopters of myopia. However, at 14 years of age, the average refractive state shows a strong clustering in the emmetropic neighborhood. Because the cornea and anterior chamber depth change little during these periods of eye growth, it appears that the power of the crystalline lens must change to maintain emmetropia. It seems possible that the process is coordinated by the retina–brain complex, which might tune each component to ensure a sharp image. However, studies of infant monkeys who were raised in the dark or who have had their optic nerve sectioned suggest that emmetropization is largely programmed on a genetic basis.[120] The experiments further showed that procedures that result in significant degradation of the retinal image, such as the suturing of the lids together or induction of a corneal opacity during the early growth period, influence the axial growth process. Surprisingly, these types of opacification significantly increase the axial length and produce states of myopia of up to 12 diopters. Such excessive image degradation seems to override the emmetropization process and result in high levels of axial myopia.

In conclusion, it appears that a genetic bias underscores the refractive state of the eye. However, this genetic program can be tuned by environmental and intrinsic (i.e. intraocular) factors. Koretz et al,[121] Laties & Stone,[122] and Stone et al[8,111,121,123–126,] have provided further insights into the biophysical and biochemical controls of emmetropization and their failure.

References

1. Boothe RG, Dobson V, Teller DY. Post natal development of vision in human and non-human primates. Ann Rev Neurosci 1985; 8:495.
2. Collins D. The human revolution: from ape to artist. London: Phaidon, 1976.
3. Larsen JS. The sagittal growth of the eye: ultrasonic measurements of axial length of the eye from birth to puberty. Acta Ophthalmol 1971; 49:872.
4. Baldwin WR. Some relationships between ocular, anthropometric and refractive variables in myopia. Doctoral thesis. . Indianapolis: University of Indiana, 1964.
5. Frantz RL. Visual perception from birth as shown by pattern selectivity. Ann NY Acad Sci 1965; 118:793.
6. Green DG, Powers MK, Banks MS. Depth of focus, eye size, visual acuity. Vis Res 1980: 20:827.
7. Reynolds CR, Fletcher F, Janzen E. Handbook of clinical child neurophysiology. New York: Plenum Press, 1989.
8. Teller DY. First glances: the vision of infants. Invest Ophthalmol Vis Sci 1997; 38:2183.
9. Rada JAS, Shelton S, Norton TT. The sclera and myopia. Exp Eye Res 2006; 82: 185–200.
10. Siatkowski RM et al and U.S. Pirenzepine Study Group. Two-year multicenter, randomized, double-masked, placebo-controlled, parallel safety and efficacy study of 2% pirenzepine ophthalmic gel in children with myopia. J Am Assoc Pediatr Ophthalmol Strabismus. 2008; 4:332–339.
11. Applegate RA, Howland HC. Magnification and visual acuity in refractive surgery. Arch Ophthalmol 1993; 111:1335–1342.
12. Inagaki Y. The rapid change of corneal curvature in the neonatal period and infancy. Arch Ophthalmol 1986; 104:1026.
13. Insler MS et al. Analysis of corneal thickness and corneal curvature in infants. CLAO J 1987; 3:192.
14. Tucker SM et al. Corneal diameter, axial length and intraocular pressure in premature infants. Ophthalmology 1992; 99:1296.
15. Mutti DO et al. Optical and structural development of the crystalline lens in childhood. Invest Ophthalmol Vis Sci 1997; 39:120.
16. Wood ICJ, Mutti DO, Zadnik K. Crystalline lens parameters in infancy. Ophthalmol Physiol Opt 1996; 6:310.
17. Hofstetter HW. Emmetropization - biological process or mathematical artifact? Am J Optom Arch Am Acad Optom 1969; 46:447.
18. Zadnik K, Mutti DO. Development of ametropias. In: Benjamin WJ, ed. Borish's clinical refraction. Philadelphia: WB Saunders, 1998.
19. Goss DA. Development of the ametropias. In: Benjamin WJ, ed. Borish's clinical refraction. Philadelphia: WB Saunders, 1998.
20. Campbell FW, Gubisch RW. Optical quality of the human eye. J Physiol 1966; 186:558.
21. Gwiazda J et al. Astigmatism in children: changes in axis and amount from birth to six years. Invest Ophthalmol Vis Sci 1984; 25:99.
22. Mohindra I, Held R, Gwiazda J. Astigmatism in infants. Science 1978; 202:329.
23. Strickland C. Torsten Wiesel, winner of 1981 Nobel Prize for Vision Research. San Francisco: American Academy of Ophthalmology, 1995.
24. Bower TG. The perceptual world of the child. Cambridge, MA: Harvard University Press, 1977.
25. Tronick E. Simultaneous control and growth of the infant's effective visual field. Percep Psychophys 1972; 11:373.
26. Pinker S. The language instinct. New York: Penguin Books, 1994.
27. Wilson SE. Wave-front analysis: are we missing something? Am J Ophthalmol 2003; 136:340–342.
28. Kohda Y, Watanabe M. The aggression releasing effect of the eye-like spot of the Oyanirami Coreopera Kawamebari, a fresh water serranid fish. Ethology 1990; 84:162.
29. Fernald RD. Vision and behavior in an African cichlid fish. Am Sci 1984; 72:58.
30. Sherbrooke W. Personal communication, 1995.
31. Miller D, Benedek G. Intraocular light scattering. Springfield, IL: Charles C Thomas, 1973.
32. Wollensak G, Wilsch M, Spoerl E, Seiler T. Collagen fiber diameter in the rabbit cornea after collagen crosslinking by riboflavin/UVA. Cornea 2004; 23:503–507.
33. Mainster MD. Contemporary optics and ocular pathology. Surv Ophthalmol 1978; 23:135.

34. Miles S. Underwater medicine. Philadelphia: Heppesen Sandreson, 1966.

35. Pardue MT, Anderson ME, Sivak J. Accommodation in raptors. Invest Ophthalmol Vis Sci 1996; 37:725.

36. Aggurwala KR, Nowbotsing S, Kruger PB. Accommodation to monochromatic and white-light targets. Invest Ophthalmol Vis Sci 1995; 36:2695.

37. Nakanishi K. Why 11-cis-retinal? Am Zool 1991; 31:479.

38. Oesterhelt D, Stoekenius W. Rhodopsin-like protein from the membrane of Holobacterium halobium. Nature New Biol 1971; 233:149.

39. Spudich JL, Bogomolni RD. Sensory rhodopsins of Halobacteria. Annu Rev Biophys Chem 1988; 17:183.

40. Atkins GH et al. Picosecond time resoled fluorescence spectroscopy of K-590 in the bacteriorhodopsin photocycle. Biophys J 1989; 55:263.

41. Yokoyama S, Yokoyama R. Molecular evolution of human visual pigment genes. Mol Biol Evol 1989; 6:186.

42. Stryer L. Mini review: visual excitation and recovery. J Biol Chem 1991; 266:1071.

43. Campbell FW. The depth of field of the human eye. Optica Acta 1957; 4:157.

44. Fein A, Szutz EZ. Photoreceptors: their role in vision, Cambridge: Cambridge University Press, 1982.

45. Snyder AW, Bossomaier JR, Hughes A. Optical image quality and the cone mosaic. Science 1986; 231:499.

46. Snyder AW, Menzal R. Photoreceptor optics. Berlin: Springer-Verlag, 1975.

47. Elliott DB, Yang KGH, Whitaker D. Visual acuity changes throughout adulthood in normal healthy eyes seeing beyond 6/6. Optom Vis Sci 1995; 72:186.

48. Ratliff F. The role of physiologic nystagmus in monocular acuity. J Exp Psychol 1952; 43:163.

49. Riggs LA et al. The disappearance of steadily fixated visual test objects. J Opt Soc Am 1953; 43:495.

50. Riggs LA. Visual acuity. In: Graham CH, ed. Vision and visual perception. New York: John Wiley & Sons, 1965.

51. Eakin RM. Evolution of photoreceptors. In: Robzhansky T, Hetch MK, Steere WC, eds. Evolutionary biology, vol 2. New York: Appleton-Century-Crofts, 1968.

52. Williams DR. Topography of the foveal cone mosaic in the living human eye. Vis Res 1988; 28:433.

53. Von Ditfurther H. Children of the universe. New York: Athenaeum Press, 1976.

54. Zeilik M. Astronomy: the evolving universe, 3rd edn. New York: Harper & Row, 1982.

55. Levene JR. Clinical refraction and visual science. London: Butterworths, 1977.

56. Bailey H, Lovie JE. New design principles for visual acuity letter charts. Am J Optom Physiol Opt 1976; 53:740.

57. Ferris FL et al. New visual acuity charts for clinical research. Am J Ophthalmol 1982; 94:91.

58. Carkeet AD. Modeling logMAR visual acuity scores: effects of termination rules and alternative forced-choice options. Optom Vis Sci 2001; 78:529–538.

59. Nunes LM, Schor P. Evaluation of the impact of refractive surgery on quality of life using the NEI-RQL (National Eye Institute Refractive Error Quality of Life) instrument. Arq Bras Oftalmol 2005; 68(6):789–796.

60. Maffei L, Fiorentin A. The visual cortex as a spatial frequency analyzer. Vis Res 1973; 3:1255.

61. Campbell FW, Robson JG. Application of Fourier analysis to the visibility of gratings. J Physiol 1968; 197:551.

62. Campbell FW. The physics of visual perception. Phil Trans R Soc Lond B Biol Sci 1980; 290:5.

63. Miller D, Sanghvi S. Contrast sensitivity and glare testing in corneal disease. In: Nadler M, Miller D, Nadler DJ, eds. Glare and contrast sensitivity for clinicians. New York: Springer-Verlag, 1990.

64. LeClaire J et al. A new glare tester for clinical testing. Arch Ophthalmol 1982; 100:153.

65. Holliday LL. The fundamentals of glare and visibility. J Opt Soc Am 1926; 12A:492.

66. Wolfe E, Gardiner JS. Studies on the scatter of light in the dioptric median of the eye as a basis for visual glare. Arch Ophthalmol 1963; 37:450.

67. Wolfe E. Glare and age. Arch Ophthalmol 1960; 64:502.

68. Miller D, Dohlman CH. The effect of cataract surgery on the cornea. Trans Am Acad Ophthalmol Otolaryngol 1970; 74:369.

69. Lancon M, Miller D. Corneal hydration, visual acuity and glare sensitivity. Arch Ophthalmol 1973; 90:227.

70. Katz B, Melles RB, Schneider JA. Glare disability in nephrotic cystinosis. Arch Ophthalmol 1987; 105:1670.

71. Rowsey JJ, Balyeat HD. Preliminary results and complications of radial keratotomy. Am J Ophthalmol 1983; 93:347.

72. Waring GO et al. Results of the progressive evaluation of radial keratotomy (PERK) study one year after surgery. Ophthalmology 1985; 92:177.

73. Harper RA, Halliday BL. Glare and contrast sensitivity in contact lens corrected aphakia, epikeratophakia and pseudophakia. Eye 1989; 3:562.

74. Koch D et al. Glare following posterior chamber lens implantation. J Cataract Refract Surg 1986; 12:480.

75. Campbell FW, Green DG. Optical and retinal factors affecting visual resolution. J Physiol 1965; 181:576.

76. Charman WN, Whitefoot H. Pupil diameter and the depth of field of the human eye as measured by laser speckle. Optica Acta 1977; 24:1211.

77. Ogle KN, Schwartz JT. Depth of focus of the human eye. J Opt Soc Am 1959; 49:273.

78. Tucker J, Charman WN. The depth of focus of the human eye for Snellen letters. Am J Optom Physiol Opt 1975; 52:3.

79. Helmholtz H. Handbuch der Physiologische Optik. Leipzig: Hamburg University, 1909.

80. Miranda MH. The environmental factor in the onset of presbyopia. In: Stark L, Obrecht G, eds. Presbyopia. New York: Professional Press, 1987.

81. Sorsby A, Sheridan M, Leary GA. Vision, visual acuity and ocular refraction in young men. Br Med J 1960; 1394.

82. Fernandes BF, Marshall JCA, Burnier Jr MN. Blue light exposure and uveal melanoma. Ophthalmology 2006; 113:1062.

83. Enoch JM. Retinal receptor orientation and the role of fiber optics in vision. Am J Optometry 1972; 49:455.

84. Snyder AW, Pask C. The Stiles-Crawford effect explanation and consequences. Vision Res 1973; 13:1115–1137.

85. Thisbos LN, Zhang X, Ming Y. The chromatic eye: a new model of ocular chromatic aberration. Appl Opt 1992; 31:3594.

86. Wald G, Griffin DR. The change in refractive power of the human eye in dim and bright light. J Opt Soc Am 1947; 37:321.

87. Troelsta Z et al. Accommodative tracking: a trial and error function. Vis Res 1964; 4:585.

88. Bennett AG, Rabbetts RB. Clinical visual optics. London: Butterworths, 1984.

89. Said FS, Weale RA. The variation with age of the human spectral transmissivity of the living human crystalline lens. Gerontologia 1959; 3:213.

90. Barlow HB. Critical limiting factors in the design of the eye and visual cortex: the Ferrier lecture, 1980. Proc R Soc Lond 1981; 212B:1.

91. Katz M. The human eye as an optical system. In: Duane TD, ed. Clinical ophthalmology. New York: Harper & Row, 1990.

92. Miller D. Optics and refraction: a user friendly guide. St Louis: Gower Medical, 1991.

93. Pizzarellio LD. The dimensions of the problems of eye disease among the elderly. Ophthalmology 1987; 94:1191.

94. Weiter JJ. Phototoxic changes in the retina. In: Miller D, ed. Clinical light damage to the eye. New York: Springer-Verlag, 1987.

95. Hemenger RP, Garner LF, Ooi CS. Change with age of the refractive index gradient of the human ocular lens. Invest Ophthalmol Vis Sci 1995; 36:703.

96. Wistow GJ, Piatigorsky J. Recruitment of enzymes as lens structural proteins. Science 1987; 236:1554.

97. Frisby JP. Illusion, brain and mind. Oxford: Oxford University Press, 1980.

98. Lee DN. The optic flow field: the foundation of vision. Phil Trans R Soc London B Biol Sci 1980; 290:169.

99. Zeng W, Daly S, Lei S. An overview of the visual optimization tools in JPEG 2000. Signal Proc Image Commun 2002; 17:85–104.

100. Ramachandran VS. 2-D or not 2-D, that is the question. In: Gregory R et al, eds. The artful eye. Oxford: Oxford University Press, 1995.

101. Kanizsa G. Subjective contours. Sci Am 1976; 234:48.

102. Hearst E. Psychology of nothing. Am Sci 1991; 79:432.

103. Tang C, Kiorpes L, Morshon JA. Stereo acuity and vernier acuity in Macaque monkeys. Invest Ophthal 1995; 36:5365.

104. Stromberg E. Uber refraktion und Achsenlange des menchlicken Auges. Acta Ophthalmol 1936; 14:281.

105. Bennett AG, Rabbetts RB. Clinical visual optics, 2nd edn. London: Butterworths, 1989.

106. Thorn F, Cruz AA, Machado AJ, Carvalho RA. Refractive status of indigenous people in the northwestern Amazon region of Brazil. Optom Vis Sci 2005; 82:267–272.

107. Curtin BJ. The myopias: basic science and clinical management. Philadelphia: Harper & Row, 1985.

108. Angle J, Wissman DA. The epidemiology of myopia. Am J Epidemiol 1980; 111:220.

109. Wold KC. Hereditary myopia. Arch Ophthalmol 1949; 42:225.

110. Zhang Q, Li S, Xiao X, Jia X, Guo X. Confirmation of a genetic locus for X-linked recessive high myopia outside MYP1. J Hum Genet 2007; 52:469–472.

111. Stenstom S. Untersuchungen uber die variation fon kovariation des optishen elements des menschlichen auges. Acta Ophthalmol 1948; 26(suppl).

112. Luke LK et al. Epidemiological study of ocular refraction amount school children in Taiwan. Invest Ophthalmol Vis Sci 1996; 6:1002.

113. Bennett AG. Lens usage in the supplementary ophthalmic service. Optician 1965; 149:131.

114. Howland HC et al. Astigmatism measured by photorefraction. Science 1978; 202:331.

115. Mohundra I et al. Astigmatism in infants. Science 1978; 202:329.

116. Klemstein RN. Epidemiology of presbyopia. In: Start L, Obrecht G, eds. Presbyopia. New York: Professional Press, 1987.

117. Stark L, Obrecht G. Presbyopia: recent research and reviews from the Third International symposium. New York: Professional Press, 1987.

118. Sorsby A et al. Emmetropia and its aberrations. MRC Special Rep Serv Rep 293. London: Medical Research Council, 1959.

119. Cook RC, Glasscock RE. Refractive and ocular findings in the newborn. Am J Ophthalmol 1951; 34:1407.

120. Raviola E, Wiesel TN. Animal model for myopia. N Engl J Med 1985; 312:1609.

121. Koretz JF, Rogot A, Kaufman PL. Physiological strategies for emmetropia. Trans Am Ophthalmol Soc 1995; 93:105.

122. Laties AM, Stone RA. Some visual and neurochemical correlates of refractive development. Vis Neurosci 1991; 7:125.

123. Iuvone PM et al. Effects of apomorphine, a dopamine receptor agonist, on ocular refraction and axial elongation in a primate model of myopia. Invest Ophthalmol Vis Sci 1991; 32:1674.

124. Papastergiou GI et al. Induction of axial eye elongation and myopic refractive shift in one-year old chickens. Vis Res 1998; 38:1883.

125. Stark L, Obrecht G. Presbyopia: recent research and reviews from the Third International symposium. New York: Professional Press, 1987.

126. Stone RA et al. Postnatal control of ocular growth: dopaminergic mechanisms. Ciba Foundation Symposium 1990; 155:45.

CHAPTER 2

Optical Aberrations and Wavefront Sensing

Ricardo N. Sepulveda & Ron Krueger

Introduction

Myopia, hyperopia and cylinder are refractive errors known as second-order aberrations. These aberrations result in the inability of the eye to focus images appropriately on the retina. In myopia, light rays entering the eye focus anterior to the retina. This is most often seen in an elongated eye. In contrast, hyperopia occurs in a short eye where light rays tend to focus behind the retina. Astigmatism results from an irregular-shaped cornea or early cataractous lens, which causes the light rays to focus at multiple points along the pathway to the retina. With regular astigmatism, the rays focus into a line oriented in the same axis of the cylinder, and yet another oriented 90° away. These basic optical errors related to the eye are what we have been correcting for the past 200 years with the aid of spectacles, contact lenses and even refractive surgery (Fig. 2.1).

With the advent of wavefront technology, we have discovered a new way of conceptualizing how light rays behave when entering the eye. This technology allows us to visualize in two-dimensional images, the complex profile of refracted light as it passes through the cornea and the crystalline lens. We are able to now detect higher-order aberrations such as coma, trefoil and spherical aberrations. Laser technology for refractive surgery has evolved significantly over the past years. It now permits us to correct higher-order aberrations by performing a "customized" ablation of the cornea according to the data provided by wavefront sensors, improving the visual performance significantly. What is also exciting is how wavefront technology may be applied to customize contact lenses and even intraocular lenses.

This chapter intends to provide the reader with a basic understanding of optical aberrations, wavefront sensing technology and the benefit of correcting higher-order aberrations in the human eye.

Optical aberrations

In a perfect optical system, such as one free of aberrations, emerging image-forming rays are planar and converge onto a single point. In reality, however, we have learned that this is not true for our optical system, the human eye. In this section, we will review basic concepts of wavefront optics and explain the different types of optical aberrations.

Wavefront optics

In geometric optics, we study the relationships between refractive error and pupil size, which have an impact on the blur of an image. By reducing the pupil size of an eye with a given refractive error, the blur of an image improves by increasing the depth of focus. This can be understood by looking through a pin-hole, in which images appear to be sharper, but at the same time we decrease light and image resolution by inducing diffraction.

In physical optics, we describe light as energy which is transmitted in the form of a wave. The properties of a wave are wavelength, frequency, and velocity. In air, the speed of light remains relatively constant. When the light passes through a higher index of refraction, its properties change and aberrations are formed. This can be explained by the following equation:[1]

$$F = Vn/\lambda$$

where F = frequency, V = velocity, n = index of refraction, λ = wavelength.

The waves of light are joined at a single point in time by what is called a wavefront and always travel perpendicular to it. When the light waves emerge from a point source, the wavefront takes on a spherical shape. As the light waves move on, the wavefront becomes progressively more flat or planar. When light waves pass through an aberration-free optical system, they emerge from it perpendicular to the wavefront, forming a spherical shape which is either converging or diverging as if coming from a single point. When the wavefront is interrupted by an optical media with an irregular surface, the emerging wavefront is not planar, the light waves are irregular and unparallel to the wavefront. The distorted shape that a wavefront takes after emerging from an irregular optical media is called a wavefront aberration (Fig. 2.2).[1]

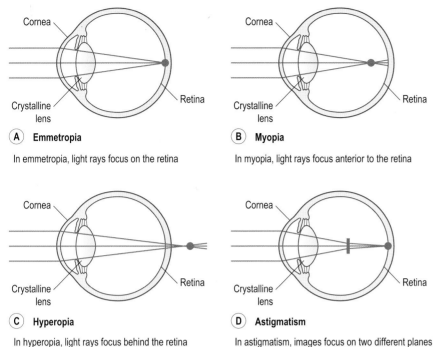

(A) Emmetropia

In emmetropia, light rays focus on the retina

(B) Myopia

In myopia, light rays focus anterior to the retina

(C) Hyperopia

In hyperopia, light rays focus behind the retina

(D) Astigmatism

In astigmatism, images focus on two different planes

Figure 2.1 Refractive states. (**A**) In emmetropia, light rays focus on the retina. (**B**) In myopia, light rays focus anterior to the retina. (**C**) In hyperopia, light rays focus behind the retina. (**D**) In astigmatism, variations in the surface of the cornea and lens cause light rays to focus at two different points. When one focal line is located anterior to the retina and the other on the retina, it is termed *myopic astigmatism*. When one focal line is located on the retina and the other behind it, it is termed *hyperopic astigmatism*.

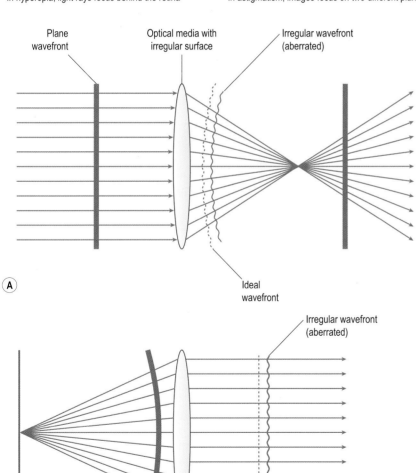

Figure 2.2 Wavefront aberrations. (**A**) The aberrated wavefront for light coming from an object at distance. The light emerging does not converge on a single point, but rather multiple different points. (**B**) Light diverging from a single point source forms an aberrated wavefront as it emerges from an irregular optical media unable to form a parallel beam of light.

Optical limitations to vision

We cannot discuss the limitations that a human eye's optics impose on the image quality without introducing the concepts of point spread function (PSF) and modulation transfer function (MTF). To further understand this, we must think of the eye as a camera. Where the cornea, crystalline lens and vitreous are the optical lenses of a camera, the pupil is the aperture, and the retina is the photographic film on which the images will be imprinted (Box 2.1).

PSF is the intensity with which an optical system distributes an image from a point source onto the retina. The point source is influenced by the pupil size. The larger the pupil, the more irregular the shape of the point source imaged on the retina (Fig. 2.3).[2–4]

MTF is the ability of the eye's optics to focus a sharp image on the retina with high contrast. As light passes through optical structures of the eye, it undergoes a process of "degradation" which can be measured by MTF. If we present an optical system with patterns of light and dark bars and measure their luminance, we are measuring the "modulation" or contrast of the light.[5–7]

$$M = \frac{(\text{Maximum Luminance} - \text{Minimum Luminance})}{(\text{Maximum Luminance} + \text{Minimum Luminance})}$$

MTF involves spatial frequency and measures the sine waves (Fourier transformation) of the light source in cycles per degree (c/deg), which is similar to sound frequency being measured in Hertz (cycles per second). MTF is defined as the modulation of the image, Mi, divided by the modulation of the stimulus (the object), Mo, giving rise to the following equation:[5–7]

$$\text{MTF}(v) = \text{Mi}/\text{Mo}$$

Low spatial frequency corresponds to large angular spacing between white bars (wide grating) and high spatial frequency corresponds to fine grating (Fig. 2.4). The spatial frequency of the images entering the eye can be influenced by the pupil size – the wider the pupil, the higher the spatial frequency of an object that can be perceived by the eye. However, the highest spatial frequency that can be detected by the visual system is also limited by the number of photoreceptors densely packed in the fovea also known as the Nyquist Sampling Limit. The Nyquist Sampling Limit states that spatial frequencies are only detected when they are less than one half the sampling frequency. The human eye cannot detect sampling frequencies higher than 60 c/deg, because the cones on the fovea provide a sampling rate of about 120 c/deg (Fig. 2.5).[8–11] Our brain compensates for much of this retinal undersampling, making us interpret images as being sharp.[11]

Diffraction is a phenomenon which occurs when light waves are bent as they enter an aperture – in the case of the

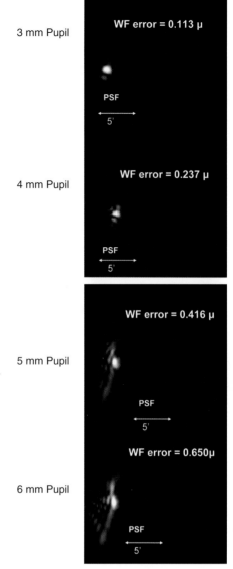

Figure 2.3 Point spread function and pupil size. Point spread function is the intensity with which an optical system focuses an image from a point source on the retina. Note how the blurring of the point source increases as the size of the pupil increases. (Modified from Azar D. *Refractive surgery*, 2nd edn. New York: Mosby. Copyright Elsevier 2006)

human eye, the pupil. In 1896, the German physicist, Arnold Sommerfeld, defined diffraction as "any deviation of light rays from a rectilinear path which cannot be interpreted as a reflection or refraction".[12] Diffraction is important in image quality because it sets limits to the resolution of an image.[13] When a wavefront propagates without interruption, the array of point sources combine and interfere to form a new wavefront of similar shape as the previous one. When the same wavefront is interrupted by an aperture, the waves from the array of point sources combine and form a different shape. As the aperture's diameter increases, light is diffracted less.[14,15] This principle was described by Huygens & Fresnel.[16]

The Stiles–Crawford effect[17] is another factor which influences image quality. This is the effect of light entering the cones transversely from the pupil margin, which is perceived

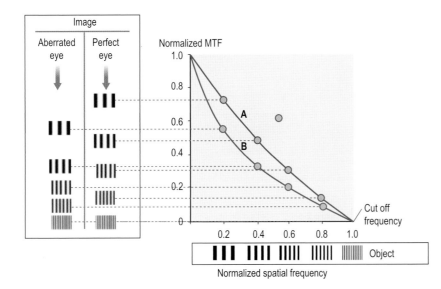

Figure 2.4 Modulation transfer function. MTF is the ratio of image to object contrast as determined by the spatial frequency of a sinusoidal grating. Curve A (blue) represents the MTF of an aberration-free optical system and curve B (purple) that of an aberrated optical system. Notice how image contrast is less at all spatial frequencies for the aberration-free optical system when compared to the aberrated optical system when looking at the images to the left. (From Azar D. *Refractive surgery*, 2nd edn. New York: Mosby. Copyright Elsevier 2006)

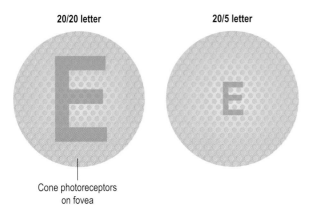

Figure 2.5 Nyquist sampling limit. The spatial frequency is limited by the number of cone photoreceptors densely packed in the fovea. A 20/20 letter is sampled by more photoreceptors than a 20/5 letter (>60 c/deg) where the image is undersampled and uncorrrectly detected.

Box 2.2 Monochromatic aberrations

- Piston
- Tilt
- Defocus
- Spherical aberration
- Coma
- Astigmatism

half as bright as the light entering the center of the pupil. In simpler terms, light that passes through the edge of the pupil contributes less to image quality than light entering the center of the pupil.[17]

Aberrations

Aberrations are dictated by the performance of any given optical system. They occur when light from one point of an object after transmission through the system does not converge into (or does not diverge from) a single point causing image blur. There are two classes of aberrations: chromatic and monochromatic. Theoretically, correcting both chromatic and monochromatic aberrations increases the contrast of images focused on the retina (contrast sensitivity).

Chromatic aberrations

Chromatic aberration is also known as "achromatism". It is defined as the inability of a lens to focus all colors of light on a single point. A rainbow is formed by droplets of water which bend color wavelengths differently producing chromatic aberration. Chromatic aberration arises because the index of refraction of the media is not the same for all wavelengths. It can be longitudinal or transverse. In longitudinal chromatic aberrations, the wavelengths focus at a different distance from the lens.[18] The index of refraction of a medium is higher for shorter wavelengths. This explains why the cornea focuses blue light (455–492 nm) in front of the retina when the eye is corrected for green light (492–577 nm). Transverse chromatic aberrations occur when wavelengths focus at different positions of a focal plane, such as seen when a pupil is de-centered.

Monochromatic aberrations

Monochromatic aberrations are defects of an image caused by the nature of a lens. An irregular-shaped cornea or an aging crystalline lens (e.g. cataract) produce monochromatic aberrations (Box 2.2).

Piston and tilt are not considered true optical aberrations, as they do not represent a curvature of wavefront. Piston and tilt are monochromatic aberrations which are not visually significant. Defocus and astigmatism are the lowest-order true optical aberrations.

Helmholtz[19] theorized the existence of these aberrations by observing a point source of light with his own eyes. He proposed that the human eye perceived aberrations which were not seen through conventional lenses. Liang & Williams[20] later would develop a technique to objectively measure the wavefront aberrations of the eye up to 10 radial orders using a Shack–Hartmann device.[21] The wavefront aberrations captured by the device are then transmitted to a computer which uses Zernike polynomials to calculate and represent them as a two-dimensional image. It is the correction of these monochromatic aberrations by spectacles, contact lenses, intraocular lenses and refractive surgery, which have been proved to result in improvement of visual performance.[22]

Measuring optical aberrations

The interest in measuring wavefront aberrations in ophthalmology grew as refractive surgery became more and more popular in the last 20 years. Coma and spherical aberrations are secondary effects of standard laser corneal ablations. In normal eyes, wavefront aberrations are not a significant source of image quality degradation. In post-refractive surgery eyes, the induction and worsening of pre-existing higher-order aberrations worsens as pupil size increases. The effect of "starbursts" and "glare" are the product of induced higher-order aberrations, which become more problematic for patients while functioning under poor light conditions. Modern technological advances in the field have enabled us to detect these aberrations and create "customized" corneal ablations to reduce them.

Aberrometry and wavefront sensing devices

The term "aberrometry" is used to describe the science of the detection and analysis of wavefront aberrations. To interpret optical aberrations, we must become familiar with the concept of optical path length (OPL). Earlier in this chapter, we learned how waves of light are slowed down as they travel through the higher index of refraction present in the ocular media compared to air for the same length. This results in higher oscillations of the waves within the eye. Thus, OPL is defined by the number of oscillations of these light waves from a point source of light, traveling through a media of x index of refraction at any given length.[23,24] To measure aberrations in any optical system, we compare the OPL of a light ray passing through any point (x,y) present at the plane of the pupil with the chief ray passing though the center of the pupil (0,0). This results in what is then called the optical path difference (OPD),[25] and its variation across the pupil is represented in a two-dimensional map.[24] Thus, optical aberrations represent differences in OPL which are influenced by the integrity of the optical media (i.e. tear film quality,[26] corneal defects and irregularities (i.e. keratoconus),[27] opacifications in the crystalline lens,[28] vitreous abnormalities, and decentering or tilting of any ocular component).

Christopher Scheiner, a German astronomer from the 1600s, set the platform for modern aberrometry and wavefront sensing devices with the invention of the Scheiner Disk. He reasoned that an optically imperfect eye would form two separate images on the retina when looking at a distant light source through a disk containing two pinholes (Fig. 2.6).

In 1895, Tscherning[29] developed an instrument to measure aberrations of the eye consisting of a grid with 1-mm squares on a +5 diopter lens. He called his instrument an "aberroscope", and it worked by placing the instrument in front of an individual's eye while viewing a light source. The individual would then interpret the shadows and draw them

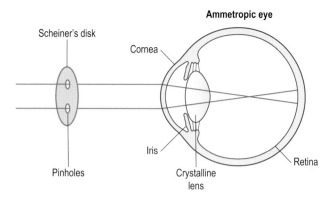

Figure 2.6 Scheiner's disk. An ammetropic eye forms two separate retinal images when viewing a distant light source through a disk with two pinholes.

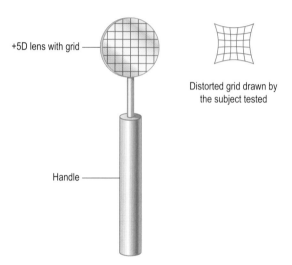

Figure 2.7 Tscherning aberroscope. This aberroscope used a +5 diopter lens with a grid composed of 1-mm squares. The subject views a distant light source through the grid and draws the distortions on paper.

(Fig. 2.7). This, of course, provided a subjective way of measuring aberrations, but is nonetheless a bright idea. In 1900, Hartmann developed a device, based on the principle of Scheiner's Disk, which constructed wavefront aberrations based on ray-tracing.[30] In the late 1970s Howland et al developed a different aberroscope and discovered that higher-order aberrations, specifically coma, were the principal aberrations present in normal eyes at every pupil diameter. They were also the first to introduce Zernike polynomials to measure wavefront aberrations.[30,31]

In the field of astrophysics, wavefront technology emerged from the need to perfect telescopic devices used to view distant stars, planets, and galaxies. The Hubble telescope uses the Shack–Hartmann wavefront sensor and adaptive optics in the form of deformable mirrors to compensate for monochromatic aberrations induced by the atmosphere as light enters the earth and increasing image resolution. In 1978, Josef Bille was the first to develop this type of wavefront technology for the eye at the University of Heidelberg, Germany. A Shack–Hartmann sensor device works by detecting aberrated light waves reflected by the retina emerging from the eye on an array of tiny lenses (lenslets) (Fig. 2.8).[32]

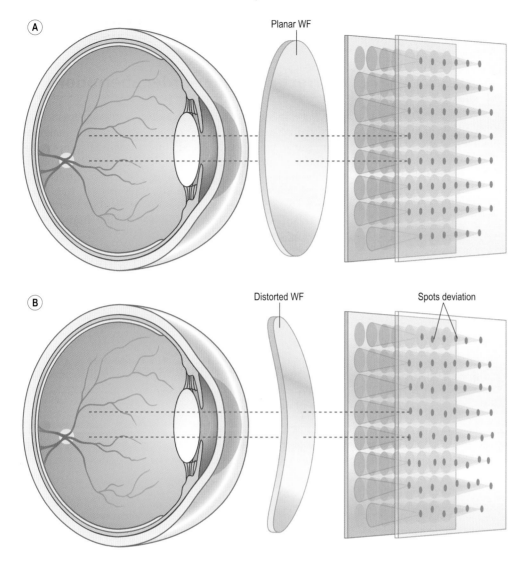

Figure 2.8 The Shack–Hartmann sensor. Light is reflected on the retina from a point source and the emerging wave from the eye is transmitted through an array of lenslets which focus the wavefront on a CCD device and processed in a computer to produce a two-dimensional image. In an aberration free eye (**A**) a planar wave emerges from the eye showing no distortion of the spots focused by the array of lenslets. In an aberrated eye (**B**), the wavefront emerging from the eye is distorted and the focused spots are deviated. (From Azar D. *Refractive surgery*, 2nd edn. New York: Mosby. Copyright Elsevier 2006)

This technology has also been employed by scientists from the University of Rochester to experimentally image the cone photoreceptors of the fovea in real time.[1]

In the eye, the cornea and the crystalline lens act much as the earth's atmosphere in inducing monochromatic aberrations. Because these structures have a circular form, the aberrations may be mathematically described by a series of polynomials called Zernike polynomials, named after Frits Zernike. Zernike polynomials are the preferred method of representing wavefront aberrations for various reasons.[33–36] They are mathematically stable, they are based on circular geometry, and they are continuously orthogonal over a unit circle.[33,37,38] They are reported as a single- or double-indexing scheme, or as a system of coordinates.[39] Figure 2.9 shows the first five radial orders of the Zernike pyramid used to describe the eye's wave aberrations.

Lower-order aberrations are simple refractive errors such as defocus and astigmatism, which can be corrected by conventional refraction. Third-order aberrations and higher are termed higher-order aberrations. Each Zernike mode has a value expressed as the root mean square in microns (RMS) or as a standard deviation across the pupil[40] and serves as a means of quantifying the severity of the aberrations present in an eye.

There are multiple wavefront sensing devices commercially available thanks to the increase in demand of customized laser refractive surgery in the past years. Companies like Alcon, Visx, Bausch and Lomb, Meditec, Schwind and Topcon use the Shack–Hartmann device as a basis for wavefront analyses. Only WaveLight utilizes a device named after Tscherning (Tscherning Aberrometer) in their systems. The Tscherning device detects wavefront patterns by shining a laser through a grid into the eye and analyzing the pattern imprinted on the retina (Fig. 2.10). A modification of this method using a rapidly placed, sequential pattern of rays is utilized by Tracey. These two methods are a form of "ray tracing" and the former was introduced by Theo Seiler (Fig. 2.11).[41,42] Emory University has developed an aberrometer

(Ingoing Adjustable Aberrometer) which measures light rays manually focused on the retina by the subject being tested. Nidek (Nidek OPD) uses a mechanism to measure aberrations called Slit Skiascopy,[43] in which a slit of light scans the eye along different axes. Photodetectors then determine the timing and scan rate of the light reflected to construct the wavefront.

Wavefront sensing devices not only measure aberrations but also function as refractometers, giving us an idea of the patient's spherocylindrical refractive error. The refraction provided by these devices is not entirely accurate and should be complemented by a proper manifest and cycloplegic refraction[44] before transferring these numbers to the laser device.

Wavefront sensing devices

See Box 2.3.

Correcting higher-order aberrations

The main focus of laser refractive surgery nowadays is to not only correct refractive errors such as defocus and astigmatism, but to create customized ablations which compensate for induced higher-order aberrations (HOA) such as spherical aberration (optimized ablation) and also correct pre-existing aberrations (customized ablation). It has been proven by experimental correction of ocular aberrations with adaptive optics that the image quality improves by increasing contrast sensitivity.[45]

Visual disturbances associated with HOA

Because an aberrated eye generally has multiple kinds of aberrations present, it is difficult to pinpoint exactly which aberration causes a certain kind of symptom. We know that

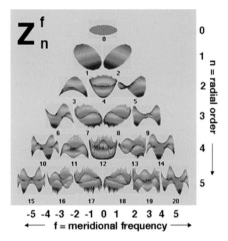

Figure 2.9 Zernike pyramid. The Zernike functions up to the fifth radial order are represented. n represents the radial order and f the meridional frequency. Lower-order aberrations are those of the 2nd radial order (astigmatism and defocus). 3rd radial order and higher are considered higher-order aberrations. Clinically significant HOAs such as coma and trefoil are of the 3rd radial order, and spherical aberrations are of the 4th radial order.

Box 2.3 Wavefront sensing devices	
Type of device	**Companies**
Shack–Hartmann	Alcon
	VisX
	Bausch & Lomb
	Meditec
	Schwind
	Topcon
Tscherning aberrometer	WaveLight
Laser ray tracing	Tracey Technologies
Slit skiascopy	Nidek OPD

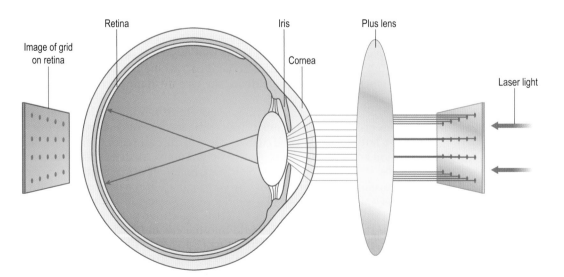

Figure 2.10 Tscherning aberrometer. Laser light is shined through a grid into the eye. The aberrated waves focused on the retina are then captured and analyzed.

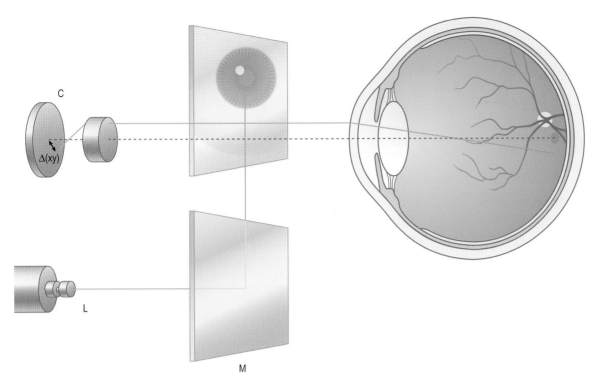

Figure 2.11 Wavefront analyses using the ray tracing method. Using the laser ray tracing method, a laser beam is reflected by a mirror (M) to enter the eye, and a camera (C) rapidly scans and registers the deviation of light emerging from the retina. (From Azar D. *Refractive surgery*, 2nd edn. New York: Mosby. Copyright Elsevier 2006)

Box 2.4 Visual effects of aberrations

- Glare
- Starbursts
- Haloes
- Ghost images
- Poor image contrast
- Poor night vision

---

in general, patients with significant higher-order aberrations, such as a side effect of conventional refractive surgery, complain of glare, starbursts, haloes, ghost-images, poor contrast and poor night vision (see Box 2.4). Fortunately, these symptoms do not happen frequently, and they tend to occur under conditions of dim illumination because of the resulting pupil dilatation. The extent to which these limit a person's ability to function normally, however, may vary from mild to severe.

Visual performance after correcting HOA

We are yet to develop an ideal system to clinically assess visual performance pre-operatively and post-operatively in patients undergoing refractive surgery. The Snellen acuity chart is the standard clinical form of assessing visual acuity and is far from perfect. A dim illuminated chart and "crowding" phenomenon limit the test's ability to accurately estimate visual acuity.[46] Furthermore, studies have shown poor reliability with up to 13% of subjects tested showing discrepancies of two lines or more.[47] The ideal method for testing visual acuity would be one that clearly defines the quality of

visual perception and the quality of the retinal image. The task of assessing visual perception is challenging because of the complex neural and cognitive processing involved. Retinal image quality can be assessed using aberrometers.

Most of the results highlighting the benefit of customized corneal ablations use the Snellen acuity chart, root mean square (RMS) wavefront values and contrast sensitivity. Data on the benefit of HOA correction varies between the different excimer laser platforms used for customized corneal ablations. Studies have shown stability of contrast sensitivity under mesopic conditions and less induction of HOA after customized LASIK when compared to conventional LASIK.[48,49] But wavefront guided treatments require greater corneal tissue ablation than conventional LASIK, thus putting the cornea at risk for biomechanical instability (i.e. keratectasia).[50] The studies suggest, however, that spherical aberrations are reduced almost 50% in custom treatments vs only 12% in conventional treatments. Thus, it is safe to conclude, that under the right conditions, customized treatments are superior to conventional treatments when using laser vision correction. Later in this chapter, the clinical application of HOA correction will be discussed in more detail.

Factors which limit the benefit of HOA correction

Customized corneal ablations are intended to correct the overall aberrations of the eye. The treatment not only reduces the aberrations induced by the cornea, but it is also designed to shape the cornea in a way to compensate for other aberrations that may be induced by the tear film quality, pupil or the crystalline lens. The tear film contributes to up to 70%

Optical system with NO optical aberrations

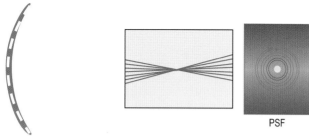

Optical system with high-order optical aberrations

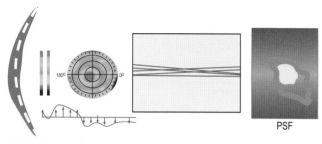

High order aberrations: advanced or retarded phases

Figure 2.12 (**A**) Point spread function in an aberration free optical system. (**B**) Point spread function in an optical system with higher-order aberrations. (From Azar D. *Refractive surgery*, 2nd edn. New York: Mosby. Copyright Elsevier 2006)

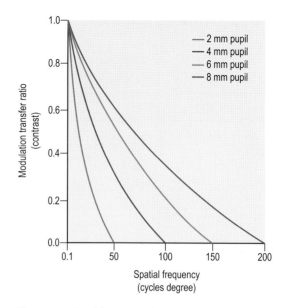

Figure 2.13 Pupil diameter and MTF. This graph represents the importance of pupil diameter and MTF in an aberration free optical system. As pupil diameter increases the MTF ratio (contrast) increases as a function of spatial frequency.

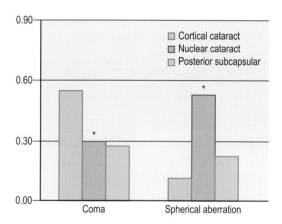

Figure 2.14 Higher-order aberrations in the aging crystalline lens. Wavefront analyses of patients with cortical, nuclear and posterior subcapsular cataracts measured at pupil diameters of 6.0 mm (Data from Rocha KM, Nose W, Bottos K, et al., Higher-order aberrations of age-related cataract. J Cataract Refract Surg 2007. 33(8): 1442-6.[28]).

to the corneal refractive power.[51] Dry eyes, for example, have greater coma and spherical aberrations compared to non-dry eyes.[52]

Diffraction caused by the pupil causes a great impact on the quality of the retinal image, and is directly proportional to the pupil size. These changes in image quality may be assessed by measuring PSF and optical transfer function (OTF). Figure 2.12 shows how pupil size affects the PSF at different pupil diameters in a typical eye compared to a diffraction-limited eye (free of aberrations). We can observe how at smaller pupil sizes, diffraction is the main source of poor image quality. But as the pupil diameter increases, aberrations degrade the image more than diffraction. As previously discussed in this chapter, pupil diameter also affects MTF. Figure 2.13 displays the ratio of image contrast to object contrast as a function of the spatial frequency of a sinusoidal grating.

Accommodation also plays a role in correcting HOAs. In young subjects accommodating on near targets, MTF decreases because of the miosis which occurs during accommodation,[53–57] and also because aberrations also increase during accommodation. As the patient ages, the correction of aberrations becomes less efficient because the initial correction plan is based on the patient's accommodative status at the time of examination.

It is well known and documented in the literature that ocular aberrations change as the eye ages.[57] Age-related corneal and crystalline lens changes alter the aberrated status of the eye.[58] In young subjects, the cornea usually has "with-the-rule" astigmatism, with the vertical meridian being steeper. In older subjects, the steepening of the cornea shifts horizontally, causing "against-the-rule" astigmatism. The

formation of cataracts in the crystalline lens and its change in radius of curvature as it ages, also produce aberrations in the eye, with coma being predominant.[28,59] In Figure 2.14, we can appreciate the aberrations caused by different types of cataracts. The group concluded that coma aberrations predominated in cortical cataracts, and spherical aberrations predominated in nuclear cataracts.[28] Interestingly, in a study by Lee et al[59] the amount of spherical aberrations present was found to be inversely proportional to the degree of nuclear sclerosis. This discordance may be caused by the different aberrometers used and difference in pupil sizes used to collect the data. Thus, this topic remains controversial. It is helpful to keep in mind that even though we correct HOAs in a patient at a certain age, these will change with time due to changes in the internal optics of the eye.

Clinical applications of wavefront aberration correction

In the previous sections of this chapter, we have attempted to provide the reader with a basic science knowledge of wavefront optics and how the correction of HOAs could potentially improve image quality overall. With an increasing market and demand for super sharp vision (i.e. "supervision"), scientists have worked hard at making this technology clinically available in many different forms. We will review the different types of wavefront guided corneal ablations and the future of HOA correction with spectacles, contact lenses and intraocular lenses.

Corneal ablations

There are several laser platforms utilized for corneal ablations. These range from the simple conventional treatments to the more sophisticated topographical guided and wavefront customized ablations. Differences in these treatment modalities lie in the type of beam delivery, spot size, treatment zone, and amount of tissue ablated for each diopter of refractive error treated. For the purpose of this chapter, we will only discuss the wavefront customized corneal ablations.

There are two types of wavefront customized ablations: wavefront guided ablations (WGA) and wavefront optimized ablations (WOA). Let us first discuss the differences between these two.

Both types of treatment aim to correct for HOAs in the eye. In WGA, the treatment is aimed to correct the preoperative HOAs, while in WOA, the treatment attempts to reduce HOAs generated during surgery. The WOA profile corrects expected HOAs for an average eye, and those that are anticipated as a result of the surgery. This means that an eye with higher than normal HOAs, will end up with near equally high HOAs after treatment.[60,61] Wavefront guided ablations may be better at correcting high pre-operative HOAs because new post-treatment HOAs may be less than HOAs present pre-treatment. Studies comparing the two profiles have concluded that WGA may be the preferred method in eyes with RMS errors >0.3 μm. Padmanabhan et al[61] found that WGA induced fewer HOAs and had better contrast sensitivity after surgery. Another study found higher total RMS errors in patients after WOA than in those after WGA.[62] After statistical analyses, however these differences were not statistically significant. With new programs available, there is the possibility of combining both treatment algorithms to make a better ablation profile (Allegretto Wave, WaveLight Laser Technologie AG).[60] Such programs are currently undergoing FDA trials and awaiting approval.

Excimer lasers used to treat wavefront-customized ablations typically use a small-diameter and circular flying spot laser. The cornea is ablated by delivering multiple pulses, and the energy used by each spot is generally Gaussian in order to obtain smooth surfaces. Because these ablation profiles can be used to treat very irregular corneas, many technical aspects must be addressed. In order to achieve adequate outcomes, the surgeon should obtain reliable wavefront analysis, compensate for cyclotorsion during treatment, ensure proper functioning of eye tracker, and calculate enough residual stromal bed to prevent ectasia.

Correcting HOA with spectacles, contact lenses and intraocular lenses

What happens to those individuals who are poor candidates for customized refractive surgery? These patients are obligated to correct their refractive errors with spectacles or contact lenses, or undergo corneal cross-linking with riboflavin and ultraviolet light prior to refractive surgery. Since there are inherent problems with how these devices are adequately aligned and synchronized to the eye, it has been challenging to apply wavefront technology to them. We will discuss HOA correction with intraocular lenses in the latter part of this section.

For wavefront-customized spectacles to work, they would have to be stable enough to not decenter with head movements, and they would have to be designed in such a way that the aberrations would be corrected in all fields of gaze and at different pupil sizes. Ophthonix, Inc. (Vista, CA), has designed a lens available for commercial use, which consists of a three-layer structure with a refractive index of 1.6. The middle layer consists of a patented photopolymer in between two coated lenses. The lens corrects HOAs from the 2nd to 6th order using wavefront guided technology. The company's aberrometer measures both lower- and higher-order aberrations using the "Talbot" effect.[63] Thibos & Miller, from the University of Indiana, are working on developing an electronic spectacle device which may potentially compensate for HOA. It consists of using transparent liquid-crystals embedded on glass providing it with the capability of varying its refractive index.[64] This would potentially provide patients with the advantage of electronically adjusting the spectacle according to their visual demands.

The use of customized contact lenses that would correct HOAs has also been proposed.[25] As we know, there are rigid gas permeable (RGP) and soft contact lenses. In a clinical setting, RGP lenses are helpful in correcting refractive errors in highly irregular corneas. The smooth contour of their anterior and posterior curvatures, coupled with the tear film formed within the interface, provide a new refracting surface for the eye. These lenses are designed to move freely, reaching up to 1mm of decentration from the referent visual axis of the eye with every blink.[65] Needless to say, a wavefront-customized RGP would have to sit on the eye in a stable manner such that its movement relative to the visual axis would be minimal. This property makes RGP lenses difficult for correcting HOAs such as coma and trefoil. These aberrations are present in higher magnitude in the general population[66] and are also the most sensitive to rotational misalignment. Soft contact lenses, on the other hand, are much more stable and less likely to undergo misalignment relative to the visual axis. This is especially true for prism-ballasted lenses, currently available for the correction of spherical and cylindrical errors (e.g. toric lenses). These lenses rotate less than 5° with every blink, which is ideal for HOA correction because coma, trefoil, and astigmatism are tolerant to mis-rotations of up to 60°.[67]

The most common surgical procedure in ophthalmology in patients over 60 years of age is cataract surgery with removal of the natural crystalline lens and implantation of an intraocular lens (IOL) in the remaining capsular bag. Since nearly 3 million cataract surgeries are performed in the US each year, one can infer that a significant amount of research is currently being focused on designing customized intraocular lenses. Studies have shown that implanted monofocal spherical IOLs induce positive spherical aberrations.[68,69] To compensate for positive spherical aberration, aspheric IOLs were designed and are commercially available, such as the Tecnis-Z9000 (Pfizer, New York, NY) and SN60WF (Alcon, Fort Worth, TX). These lenses are designed with a prolate anterior surface, hence the term *aspheric*, and have the same radius of curvature at every point on the surface. Clinical studies have shown that implanted aspheric IOLs markedly increase the MTF with a resultant improvement in image contrast in comparison to monofocal spherical IOLs.[70] Also commercially available are toric IOLs such as the STAAR toric IOL (STAAR Surgical, Monrovia, CA) and the Acrysof SA60TT (Alcon, Fort Worth, TX), which correct corneal astigmatism. The benefit of both aspheric and toric IOLs depends largely on their centration and stability within the capsular bag. In the future, we may encounter light adjustable IOLs, which once implanted in the eye, may be altered to compensate for residual refractive errors and even HOAs by irradiating the optic with blue light to polymerize monomers and set up a diffusion gradient to alter the lens shape (Calhoun Vision, Pasadena, CA).

In conclusion, the "Quest for Super-Vision", as described by leading investigators in the field of wavefront technology, holds a promising and exciting future. In the last decade, large contributions to the field have been made and continue to flourish. The clinician must be aware of these rapid changes to provide patients with state-of-the-art technology to improve their quality of vision and life. Currently there is no perfect way to treat HOAs. Customized corneal ablations, however, offer the most stable form of HOA correction, and the industry is working diligently at perfecting customized soft contact lenses and intraocular lenses.

References

1. Roorda A. Wavefront customized visual correction. In: Krueger RR, Applegate RA, MacRae SM, eds The quest for supervision II, Vol. 2. 2004, Thorofare, NJ: SLACK Incorporated, 2004 9–17.

2. Ligabue EA, Giordano C. Assessing visual quality with the point spread function using the NIDEK OPD-Scan II. J Refract Surg, 2009; 25(1 Suppl):S104–S109.

3. Logean E, Dalimier E, Dainty C. Measured double-pass intensity point-spread function after adaptive optics correction of ocular aberrations. Opt Express 2008; 16(22): 17348–17357.

4. Ijspeert JK, Van Den Berg TJTP, Spekreijse H. An improved mathematical description of the foveal visual point spread function with parameters for age, pupil size and pigmentation. Vision Res 1993; 33(1):15–20.

5. Deeley RJ, Drasdo N, Charman WN. A simple parametric model of the human ocular modulation transfer function. Ophthalmic Physiol Opt 1991; 11(1):91–93.

6. Charman WN. Wavefront aberration of the eye: a review. Optom Vis Sci 1991; 68(8):574–583.

7. Walsh G, Charman WN. Variation in ocular modulation and phase transfer functions with grating orientation. Ophthalmic Physiol Opt 1992; 12(3):365–369.

8. Anderson SJ, Mullen KT, Hess RF. Human peripheral spatial resolution for achromatic and chromatic stimuli: limits imposed by optical and retinal factors. J Physiol 1991; 442:47–64.

9. Williams DR, Coletta NJ. Cone spacing and the visual resolution limit. J Opt Soc Am A 1987; 4(8):1514–1523.

10. Hirsch J, Miller WH. Does cone positional disorder limit resolution? J Opt Soc Am A 1987; 4(8):1481–1492.

11. Artal P, Chen L, Fernandez EJ et al. Adaptive optics for vision: the eye's adaptation to point spread function. J Refract Surg 2003; 19(5):S585–S587.

12. Sommerfeld A. Mathematische Theorie der Diffraction. Mathematische Annalen 1896; 47(2–3):317–374.

13. McLellan JS, Prieto PM, Marcos S et al. Effects of interactions among wave aberrations on optical image quality. Vision Res 2006; 46(18):3009–3016.

14. Schwiegerling J. Theoretical limits to visual performance. Surv Ophthalmol 2000; 45(2):139–146.

15. Strang NC, Atchison DA, Woods RL. Effects of defocus and pupil size on human contrast sensitivity. Ophthalmic Physiol Opt 1999; 19(5):415–426.

16. Longhurst R. Geometrical and physical optics, 2nd edn. London: Longmans, 1968.

17. Stiles WS, Crawford B. The luminous efficiency of rays entering the pupil at different points. Proc R Soc Lond B Biol Sci 1933; 112:428–450.

18. Seong K, Greivenkamp JE. Chromatic aberration measurement for transmission interferometric testing. Appl Opt 2008; 47(35):6508–6511.

19. Hemltholtz H. Popular scientific lectures. New York: Dover Publications, Inc., 1962.

20. Liang J, Grimm B, Goelz S et al. Objective measurement of wave aberrations of the human eye with the use of a Hartmann-Shack wave-front sensor. J Opt Soc Am A Opt Image Sci Vis 1994; 11(7):1949–1957.

21. Platt B, Shack R. Lenticular Hartmann screen. Opt Sci Center Newsl 1971; 5:15–16.

22. Artal P, Guirao A, Berrio E et al. Compensation of corneal aberrations by the internal optics in the human eye. J Vis 2001; 1(1):1–8.

23. Preussner PR. The practicality of wavefront correction in ophthalmology. Klin Monatsbl Augenheilkd 2004; 221(6):456–463.

24. Hitzenberger C, Mengedoht K, Fercher AF. Laser optic measurements of the axial length of the eye. Fortschr Ophthalmol 1989; 86(2):159–161.

25. Thibos LN, Cheng X, Bradley A. Design principles and limitations of wave-front guided contact lenses. Eye Contact Lens 2003 29(1 Suppl):S167–S170; discussion S190–1, S192–4.

26. Koh S, Maeda N. Wavefront sensing and the dynamics of tear film. Cornea 2007; 26(9 Suppl 1):S41–S45.

27. Maeda N, Fujikado T, Kuroda T et al. Wavefront aberrations measured with Hartmann-Shack sensor in patients with keratoconus. Ophthalmology 2002; 109(11):1996–2003.

28. Rocha KM, Nose W, Bottos K et al. Higher-order aberrations of age-related cataract. J Cataract Refract Surg 2007; 33(8):1442–1446.

29. Tscherning M. Die monochromatischen aberrationen des menschlichen. Auges Z Psychol Physiol Sinn 1894; 6:456–471.

30. Howland B. Use of crossed cylinder lens in photographic lens evaluation. Appl Op 1960; 7:1587–1588.

31. Howland HC, Howland B. A subjective method for the measurement of monochromatic aberrations of the eye. J Opt Soc Am 1977; 67(11): 1508–1518.

32. Mirshahi A, Buhren J, Gerhardt D et al. In vivo and in vitro repeatability of Hartmann-Shack aberrometry. J Cataract Refract Surg 2003; 29(12):2295–2301.

33. Born M, Wolf E. Principles of optics, 7th edn. Cambridge, UK: Cambridge University Press, 1999.

34. Rocha KM, Vabre L, Harms F et al. Effects of Zernike wavefront aberrations on visual acuity measured using electromagnetic adaptive optics technology. J Refract Surg 2007; 23(9):953–959.

35. Dai GM. Comparison of wavefront reconstructions with Zernike polynomials and Fourier transforms. J Refract Surg 2006; 22(9):943–948.

36. Yoon G, Pantanelli S, MacRae S. Comparison of Zernike and Fourier wavefront reconstruction algorithms in representing corneal aberration of normal and abnormal eyes. J Refract Surg 2008; 24(6):582–590.

37. Dai GM, Mahajan VN. Zernike annular polynomials and atmospheric turbulence. J Opt Soc Am A Opt Image Sci Vis 2007; 24(1):139–155.

38. Dai GM. Wavefront expansion basis functions and their relationships. J Opt Soc Am A Opt Image Sci Vis 2006; 23(7):1657–1660.

39. Thibos LN, Applegate RA, Schwiegerling JT et al. Standards for reporting the optical aberrations of eyes. J Refract Surg 2002; 18(5):S652–S660.

40. Williams D, ed. How far can we extend the limits of vision? Vol. 2. Thorofare, NJ: SLACK Incorporated, 2004:19–38.

41. Sanchez MJ, Mannsfeld A, Borkensein AF et al. Wavefront analysis in ophthalmologic diagnostics]. Ophthalmologe 2008; 105(9):818–824.

42. Mrochen M, Bueeler M, Donitzky C et al. Optical ray tracing for the calculation of optimized corneal ablation profiles in refractive treatment planning. J Refract Surg 2008; 24(4):S446–S451.

43. MacRae S, Fujieda M. Slit skiascopic-guided ablation using the Nidek laser. J Refract Surg 2000; 16(5):S576–S580.

44. Perez-Straziota CE, Randleman JB, Stulting RD. Objective and subjective preoperative refraction techniques for wavefront-optimized and wavefront-guided laser in situ keratomileusis. J Cataract Refract Surg 2009; 35(2):256–259.

45. Yoon GY, Williams DR. Visual performance after correcting the monochromatic and chromatic aberrations of the eye. J Opt Soc Am A Opt Image Sci Vis 2002; 19(2):266–275.

46. McGraw P, Winn B, Whitaker D. Reliability of the Snellen chart. Br Med J 1995; 310(6993):1481–1482.

47. Gibson RA, Sanderson HF. Observer variation in ophthalmology. Br J Ophthalmol 1980; 64(6):457–460.

48. Netto MV, Dupps W, Jr., Wilson SE. Wavefront-guided ablation: evidence for efficacy compared to traditional ablation. Am J Ophthalmol 2006; 141(2):360–368.

49. Randleman JB, Perez-Straziota CE, Hu MH et al. Higher-order aberrations after wavefront-optimized photorefractive keratectomy and laser in situ keratomileusis. J Cataract Refract Surg 2009; 35(2):260–264.

50. Castanera J, Serra A, Rios C. Wavefront-guided ablation with Bausch and Lomb Zyoptix for retreatments after laser in situ keratomileusis for myopia. J Refract Surg 2004; 20(5):439–443.

51. Courville CB, Smolek MK, Klyce SD. Contribution of the ocular surface to visual optics. Exp Eye Res 2004; 78(5):417–425.

52. Montes-Mico R, Caliz A, Alio JL. Wavefront analysis of higher order aberrations in dry eye patients. J Refract Surg 2004; 20(3):243–247.

53. He JC, Burns SA, Marcos S. Monochromatic aberrations in the accommodated human eye. Vision Res 2000; 40(1):41–48.

54. Artal P, Fernandez EJ, Manzanera S. Are optical aberrations during accommodation a significant problem for refractive surgery? J Refract Surg 2002; 18(5):S563–S566.

55. Lopez-Gil N, Iglesias I, Artal P. Retinal image quality in the human eye as a function of the accommodation. Vision Res 1998; 38(19):2897–2907.

56. Atchison DA, Collins MJ, Wildsoet CF et al. Measurement of monochromatic ocular aberrations of human eyes as a function of accommodation by the Howland aberroscope technique. Vision Res 1995; 35(3):313–323.

57. Iida Y, Shimizu K, Ito M et al. Influence of age on ocular wavefront aberration changes with accommodation. J Refract Surg 2008; 24(7):696–701.

58. Radhakrishnan H, Charman WN. Age-related changes in ocular aberrations with accommodation. J Vis 2007; 7(7):111–121.

59. Lee J, Kim MJ, Tchah H. Higher-order aberrations induced by nuclear cataract. J Cataract Refract Surg 2008; 34(12):2104–2109.

60. Cheng AC. Wavefront-guided versus wavefront-optimized treatment. J Cataract Refract Surg 2008; 34(8):1229–1230.

61. Padmanabhan P, Mrochen M, Basuthkar S et al. Wavefront-guided versus wavefront-optimized laser in situ keratomileusis: contralateral comparative study. J Cataract Refract Surg 2008; 34(3):389–397.

62. Brint SF. Higher order aberrations after LASIK for myopia with alcon and wavelight lasers: a prospective randomized trial. J Refract Surg 2005; 21(6): S799–S803.

63. Seiple W, Szlyk JP. Clinical Investigation into the Visual performance provided by the i-Zon Spectacle Lens System. Rev Optom 2008; 2(Suppl):1–16.

64. Thibos LN, Bradley A. Use of liquid-crystal adaptive-optics to alter the refractive state of the eye. Optom Vis Sci 1997; 74(7):581–587.

65. Knoll HA, Conway HD. Analysis of blink-induced vertical motion of contact lenses. Am J Optom Physiol Opt 1987; 64(2):153–155.

66. Porter J, Guirao A, Cox IG et al. Monochromatic aberrations of the human eye in a large population. J Opt Soc Am A Opt Image Sci Vis 2001; 18(8):1793–1803.

67. Guirao A, Williams DR, Cox IG. Effect of rotation and translation on the expected benefit of an ideal method to correct the eye's higher-order aberrations. J Opt Soc Am A Opt Image Sci Vis 2001; 18(5):1003–1015.

68. Barbero S, Marcos S, Jimenez-Alfaro I. Optical aberrations of intraocular lenses measured in vivo and in vitro. J Opt Soc Am A Opt Image Sci Vis 2003; 20(10):1841–1851.

69. Atchison DA. Optical design of poly(methyl methacrylate) intraocular lenses. J Cataract Refract Surg 1990; 16(2):178–187.

70. Mester U, Dillinger P, Anterist N. Impact of a modified optic design on visual function: clinical comparative study. J Cataract Refract Surg 2003; 29(4):652–660.

CHAPTER 3

Accommodation

Adrian Glasser

Introduction

"There is no other portion of physiological optics where one finds so many differing and contradictory ideas as concerns the accommodation of the eye where only recently in the most recent time have we actually made observations where previously everything was left to the play of hypotheses"

H Von Helmholtz (1909)

It is primarily due to Helmholtz[1] that we owe our current understanding of the accommodative mechanism of the human eye (Fig. 3.1). His insight came from his own work and from pioneers before him. Thomas Young[2] was instrumental in demonstrating that accommodation occurs, not through changes in corneal curvature or axial length as those before him believed,[3] but through changes in the curvature of the lens. Young's painstaking anatomical investigations were insufficient for him to rule out the possibility that the crystalline lens received direct innervation from a branch of the ciliary nerves to allow it to contract as a muscle. It was only after the work of Crampton,[4] who first described the ciliary muscle from his investigation of bird eyes, that a mechanistic description of how the ciliary muscle might alter lens curvatures was proposed by Müller.[5] Understanding of human accommodation was mired by confusion from numerous investigations of the eyes of birds and other vertebrates, studied for their comparatively large size to gain insight into the human accommodative mechanism (Box 3.1). However, these species are now known to accommodate through mechanisms quite different from humans.[6-8] Current understanding of accommodation stems from the work of many early investigators including Brücke,[9] Cramer,[10] Hess,[11] Müller,[5] Helmholtz and Gullstrand.[1] This path was made tortuous by the diversity of accommodative mechanisms of the various vertebrates studied. Possibly the most ancient of accommodative mechanisms is that of the sauropsidae (lizards, birds and turtles). Although these eyes differ from the primate eye, these species share many unusual ocular characteristics among themselves including striated intraocular muscles, bony plates or ossicles in the sclera, attachment of the ciliary processes to the lens equator, the absence of a circumlental space, a lens annular pad, and in some species at least, corneal accommodation and iris-mediated lenticular accommodation.

The wide diversity of avian visual habitats (aerial, aquatic, terrestrial), eye shapes (tubular, globose and flattened), and feeding behaviors in all likelihood dictates their accommodative needs. Corneal accommodation, of considerable value to terrestrial birds, is of no value to aquatic birds where the corneal optical power is neutralized under water. The evolutionarily divergent accommodative mechanisms, or the absence of accommodation in other vertebrates is, by reasonable conjecture, determined by feeding behaviors. Herbivorous animals (sheep, horses, cows, etc.), those which forage and dig for food primarily using olfactory cues (pigs), or those with nocturnal eyes and relatively poor visual abilities (mice, rats, rabbits) have little need for accommodation. Carnivores have better-developed ciliary muscles than these other species, but still have relatively little accommodative ability; the raccoon is the only non-primate terrestrial mammal with substantial accommodative amplitude.[12] Cats are suggested[13-15] and raccoons[12] and fish shown to translate the lens forward without lenticular thickening.[16-18] Other adaptations in the lens, iris, or retina allow other lower vertebrates functional near and distance vision, although these cannot be classified as true accommodation since they rely on static optical adaptations.

Among the vertebrates that do accommodate, amplitudes vary considerably. Diving birds have among the largest amplitudes with cormorant having ~50 D[11] and diving ducks suggested to have 70–80D.[19] Among the mammals, vervet and cynomolgus monkeys have approximately 20 D,[20-22] young rhesus as much as 40 D[23] and raccoons about 20 D.[12] Humans, for only a few short childhood years, may have a maximum of about 10–15D measured subjectively[24] or about 7–8 D measured objectively,[25] but find much less accommodation adequate for most visual tasks. Although accommodative amplitude gradually declines until completely lost by about age 50 years, to most individuals the deficit appears to be of sudden onset when the accommodative amplitude is diminished to a few diopters as presbyopia develops. Although presbyopes may read at intermediate distances, this is almost certainly due to depth of field (see below) resulting from pupil constriction rather than active accommodation. The word presbyopia (Greek, *presbys* meaning an aged person and *opsis* meaning vision) possibly derives from Aristotle's use of the term *presbytas* to describe "those who see well at distance, but poorly at

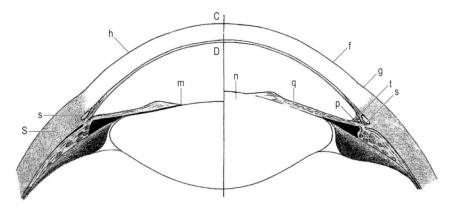

Figure 3.1 Diagram showing the mechanism of accommodation of the human eye as described by Helmholtz. The left half depicts the eye in the unaccommodated state and the right half depicts the eye in the accommodative state. Helmholtz described an increase in lens thickness, an increase in the anterior surface curvature, an anterior movement of the anterior lens surface, but no posterior movement of the posterior lens surface. Key: S, sclera; s, Schlemm's canal; h, cornea; F, side for far vision; m, unaccommodated lens; n, accommodated lens; q, iris; p, trabecular meshwork; f, clear cornea; g, limbus; N, side for near vision; C–D, optical axis). (From Helmholtz von HH. Helmholtz's Treatise on Physiological Optics. Translation edited by Southall JPC in 1924 (original German in 1909). New York: Dover, 1962: vol. 1, ch. 12.)

Box 3.1 Accommodative mechanism

- Accommodation is a dioptric change in optical power of the eye due to ciliary muscle contraction
- Accommodation occurs largely in accordance with the mechanism originally proposed by Helmholtz
- Ciliary muscle contraction moves the apex of the ciliary body towards the axis of the eye and releases resting zonular tension around the lens equator
- When zonular tension is released, the elastic lens capsule molds the young lens into a more spherical and accommodated form
- During accommodation, lens diameter decreases, lens thickness increases, the anterior lens surface moves anteriorly, the posterior lens surface moves posteriorly and the lens anterior and posterior surface curvatures increase, the thickness of the nucleus increases, but without a change in thickness of the cortex
- The increase in curvature of the lens anterior and posterior surfaces results in an increase in optical power of the lens
- The physical changes in the lens and eye result in an increase in optical power of the eye to focus on near objects

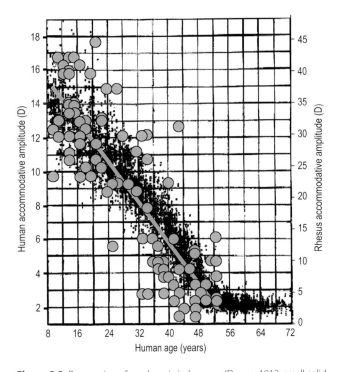

Figure 3.2 Progression of presbyopia in humans (Duane, 1912, small solid black symbols) as measured subjectively using a push-up test, and in rhesus monkeys (Bito et al, 1982, larger gray symbols and solid line) as measured objectively with a Hartinger coincidence refractometer following topical application of the cholinergic agonist pilocarpine. The horizontal axis is in human years and the rhesus data are scaled to human years such that 25 rhesus years is equivalent to 52 human years. The vertical axis is scaled such that the mean amplitudes of 14 D in humans is equivalent to 37 D in rhesus. In humans and rhesus monkeys, presbyopia progresses at the same rate relative to the absolute age span of each species. (From Bito LZ et al. Invest Ophthalmol Vis Sci 1982; 23:23; and Duane A. J Am Med Assoc 1912; 59:1010. Reproduced with permission from Association for Research in Vision and Ophthalmology.)

near".[26] Historically the term was used to describe the condition where the near point has receded too far from the eye due to a diminution in the range of accommodation.[27] Despite the wealth of studies of accommodation on vertebrates, only primates are shown to systematically lose the ability to accommodate with increasing age. It may be no coincidence that although absolute life spans differ considerably, the relative age course of the progression of presbyopia is similar in humans and monkeys (Fig. 3.2).

Accommodation

Accommodation is a dynamic, optical change in the dioptric power of the eye allowing the point of focus of the eye to be changed from distant to near objects. In primates this is mediated through a contraction of the ciliary muscle, release of resting zonular tension around the lens equator, a decrease in lens diameter and a "rounding up" of the crystalline lens through the force exerted on the lens by the lens capsule. The increased optical power of the lens is achieved through increased anterior and posterior surface curvatures and increased thickness. In an unaccommodated, emmetropic eye (an eye without refractive error) distant objects at or beyond what is considered optical infinity for the eye (6 m or 20 ft) are focused on the retina. When an object is brought closer to the eye, the eye must accommodate to maintain a clearly focused image on the retina. Myopic eyes, typically too long for the optical power of the lens and cornea combined, are unable to attain a sharply focused image for objects at optical infinity unless optical compensation is provided such as through negative powered spectacle lenses. Myopes can focus clearly on objects closer to the eye than optical infinity without accommodation (i.e. objects at their far point). Young hyperopes are only able to focus clearly on objects at optical infinity through an accommodative increase in the optical power of the eye provided their accommodative amplitude exceeds the amount of hyperopia.

Optics of the eye

Light from the environment enters the eye at the cornea and, in an emmetropic eye, is brought to a focus on the retina through the combined optical power of the cornea and the lens (see Chapter 1). When light from an object is focused on the retina, a clear, sharp image is perceived. This enables the performance of near visual tasks such as reading. If the image is not focused on the retina, these tasks become difficult or impossible to perform without optical compensation to bring the image to focus on the retina.

The optical elements of the eye, cornea, the aqueous humor, the crystalline lens and the vitreous humor all contribute to the optical power of the eye (see Chapters 1 & 2). Specific details for schematic eyes are given in Bennett & Rabbetts.[28] In the adult human eye an average, normal cornea has a radius of curvature of about +7.8 mm, a thickness of about 0.25 mm near the optical axis and the cornea provides about 70 percent of the optical refracting power of the eye. Light passes from an air environment, with a refractive index of approximately 1.00, through the tear film and into the cornea. The cornea is composed largely of fluid and proteins and therefore has a refractive index greater than air of about 1.376. The optical power of the cornea is due to a combination of the positive radius of curvature and the higher corneal refractive index than the surrounding air. Light then passes through the cornea and into the aqueous humor. Since the refractive index of the aqueous humor is close to that of the cornea (about 1.336) there is relatively little optical effect at the posterior cornea/aqueous interface. Light then enters the anterior surface of the crystalline lens. The surface of the crystalline lens has a refractive index slightly higher than that of the aqueous humor (about 1.386). The lens anterior surface has a radius of curvature of about +11.00 mm which adds to the optical power of the eye. The crystalline lens has a gradient refractive index that progressively increases from about 1.362 at the surface of the

cortex to about 1.406 at the center of the nucleus of the lens. The gradient refractive index of the lens adds additional optical power to the lens because the gradient results in refraction of light throughout the lens. This results in light taking a curved path rather than a straight path through the lens. For simplified optical calculations, the more complex gradient refractive index of the lens is often substituted with a single equivalent refractive index value.

The extent to which the gradient refractive index adds additional optical power to the lens is evident when it is realized that for an equivalent refractive index lens to have the same shape and optical power as a gradient refractive index lens, the equivalent refractive index value must be greater than the highest refractive index value at the center of the gradient refractive index lens. The posterior surface of the crystalline lens has a radius of curvature of about −6.50 mm. Although the posterior lens surface has a negative radius of curvature, it is still a convex surface which adds optical power to the eye, and relatively more so than does the anterior lens surface since the lens posterior surface is more steeply curved than the lens anterior surface. The lens anterior and posterior surface curvatures (as well as the lens gradient refractive index) are important to the optical power of the eye and it is these surfaces that become more steeply curved to allow the accommodative increase in optical power of the lens to occur. Historically it was suggested that the posterior lens surface does not move[1] and that the posterior lens surface curvature does not change appreciably with accommodation.[1,29,30] However, it is now known that the posterior lens surface does undergo an increase in curvature and moves posteriorly during accommodation as the lens thickness increases.[31-38] Gullstrand suggested that the lens equivalent refractive index must change during accommodation.[39] Since the lens shape, axial thickness and equatorial diameter change during accommodation, this dictates that the form of the gradient refractive index of the lens must also change during accommodation.[40,41] Although the form of the lens gradient refractive index changes as the lens changes shape during accommodation, this does not require a change in the equivalent refractive index of the lens during accommodation, at least to the extent that resolution limits of currently available technology permit this to be discerned.[42]

The optical requirements for accommodation

The optical power of the crystalline lens increases (i.e. the lens focal length decreases) during accommodation. As a consequence, the eye changes focus from distance to near so the image of a near object is brought to focus on the retina. The dioptric change in power of the eye defines accommodation and accommodation is measured in units of diopters (D). A diopter is a reciprocal meter and is a measure of the vergence of light. Light rays from a point object diverge and are by convention designated to have negative vergence. Light rays converging towards a point image are designated to have positive vergence (see Chapter 1). An object at optical infinity subtends zero vergence at the cornea. The optical interfaces of the eye (the cornea and lens) add positive vergence to draw light rays towards a focus on the retina (Fig. 3.3). When an object is moved from infinity to a point

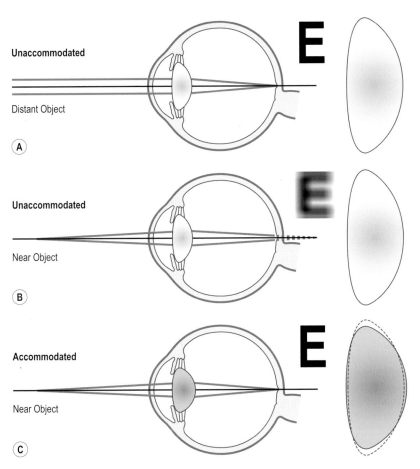

Unaccommodated

Distant Object

(A)

Unaccommodated

Near Object

(B)

Accommodated

Near Object

(C)

Figure 3.3 The accommodative optical changes in the eye occur through an increase in optical power of the crystalline lens. (**A**) The unaccommodated emmetropic eye is focused on a distant object with the lens in an unaccommodated state. (**B**) A near object subtends divergent rays and in the unaccommodated eye the image would be formed behind the retina and is therefore out of focus when the lens remains unaccommodated. (**C**) In the accommodated eye, the in focus image of the near object is formed on the retina when the lens is in an accommodated state.

closer to the eye, the near object subtends divergent rays on the cornea. To focus on the near object, the optical power of the eye must increase to add positive vergence to the now divergent rays to bring the refracted rays to a focus on the retina. When an emmetropic eye is focused on a distant object the eye is considered unaccommodated. If the eye accommodates from an object at optical infinity to an object 1.0 m in front of the eye, this represents 1.0 D of accommodation. If the eye accommodates from infinity to 0.5 m in front of the eye, this is 2 D of accommodation; from infinity to 0.1 m is 10 D, and so on. The accommodative response is therefore the increase in optical power that the eye undergoes to change focus from an object at optical infinity to the near object.

Depth of field

Clinically, the nearest point of clear vision is typically measured subjectively in an eye corrected for distance vision. This is done by moving a near reading chart towards the eyes while the subject is asked to report when they can no longer sustain clear vision on the near target or when the near target first becomes blurred. Although the reciprocal of this near reading distance expressed in meters is clinically referred to as the accommodative amplitude, this is technically inaccurate. The push-up test is a subjective measure of the difference between the far point and the near point expressed in units of diopters. However, this is not a measure of the true dioptric change in power of the eye because of the depth of field of the eye.

Depth of field is defined as the range over which an object can be moved towards or away from the eye in object space without a perceptible change in the blur or focus of the image. The depth of field of an eye depends on many factors. For example, depth of field is dependent on pupil size. A large pupil results in a wide and steep cone of light converging towards the retina. A small error in image focus with respect to the position of the retina therefore results in a large change in image blur. A large pupil results in a relatively small depth of field. A small pupil results in a narrow and flat cone of light converging towards the retina. Small errors in image focus with respect to the position of the retina result in relatively small changes in image focus. A small pupil therefore results in a relatively larger depth of field.

The depth of field of an eye is dependent on the level of illumination because of the effect that illumination has on pupil diameter. For a brightly illuminated object, pupil size will decrease resulting in an increase in depth of field. The presence of optical aberrations such as astigmatism, coma and spherical aberration also act to increase the depth of field of an eye. The presence of ocular aberrations results in an image that is not in sharp focus on the retina. Therefore, small movements of the object in object space would not perceptibly alter the focus of the image on the retina in an eye with aberrations. With accommodation and with increasing age the pupil size decreases. An effort to focus at near therefore increases the depth of field of the eye due to pupil constriction. When the nearest point of clear vision is assessed using subjective methods, such as the push-up or push-down method, the depth of field of the eye results in an overestimation of the dioptric change in optical power of the eye. When the near point of clear vision is measured using a subjective push-up method, this overestimates the objectively measured accommodative response amplitude by about 1–2 D.[43–46] Subjective testing of this nature in complete presbyopes might lead one to believe that about 1 D of accommodation is present, but this is not a true change in optical power of the eye and is therefore called pseudoaccommodation.

Visual acuity

In addition to the depth of field of the eye, acuity or contrast sensitivity of the eye also affect the subjective measurement of the near point of clear vision. The subjective push-up measurement relies heavily on the subject perceiving when the object can no longer be seen in sharp focus. As a near reading target is brought closer to the eye, the subject must decide at what point an object is no longer in acceptable focus. As mentioned, the level of illumination of the target can affect the depth of field of the eye, but illumination also affects the contrast and the brightness of the image. If the target is viewed in dim illumination, it is more difficult to detect when it is in clear and sharp focus. A brightly illuminated reading target will be seen more clearly. The increased illumination provides higher contrast on the target and so smaller changes in focus or blur of the target are more easily detected. While increasing the level of illumination will help to improve the contrast sensitivity and acuity, this will also decrease the pupil size and will thereby increase the depth of field of the eye and so result in a nearer point of perceived clear vision. Further, in cases of cataract or other opacities of the ocular optical media, the image of a near object is not seen clearly and so small changes in the focus of the image are less readily detected. With increasing age the optical clarity of the lens decreases and the prevalence of cataract increases. Retinal disease can also affect visual acuity. Elderly patients often have reduced visual acuity and/or reduced contrast sensitivity, although not solely due to decreased optical performance.[47] If the near point of clear vision is measured using the subjective push-up test in presbyopes or in patients with cataracts or retinal disease, this will overestimate the true objectively measured accommodative amplitude.[44,46,48]

The anatomy of the accommodative apparatus

The accommodative apparatus of the eye consists of the ciliary body, the ciliary muscle, the choroid, the anterior and posterior zonular fibers, the lens capsule and the crystalline lens (Fig. 3.4). Theoretical suggestions for a role for the vitreous in accommodation[49–52] and empirical evidence against a need for the vitreous in accommodation[53–55] exist. The ciliary muscle is located within the ciliary body beneath the anterior sclera. The ciliary muscle is comprised of three muscle fiber groups oriented longitudinally, radially (obliquely) and circularly. The anterior zonular fibers span the circumlental space extending from the ciliary processes to insert all around the lens equator. These zonular fibers constitute the suspensory elements of the crystalline lens. Posterior zonular fibers extend between the tips of the ciliary processes and the pars plana of the posterior ciliary body near the ora serrata. The crystalline lens consists of a central nucleus and a surrounding cortex. This lens is surrounded by the collagenous elastic lens capsule.

The ciliary body

The ciliary body is a triangular-shaped region bounded on its outer surface by the anterior sclera and on its inner surface by the pigmented epithelium. It lies between the scleral spur anteriorly and the retina posteriorly. The anterior ciliary body begins at the scleral spur at the angle of the anterior chamber. The base of the iris inserts into the anterior ciliary body. Posterior to the iris, the ciliary processes are found at the anterior-innermost point of the ciliary body and form the corrugated pars plicata of the ciliary body. Posterior to the pars plicata the smooth surface of the ciliary body is called the pars plana. The vitreal surface of the pars plana is spanned by longitudinally oriented posterior zonular fibers.[56–58] The most posterior aspect of the ciliary body joins to the ora serrata of the retina. The outer surface of the ciliary body beneath the anterior sclera is the suprachoroidal lamina or supraciliarus, formed by a thin layer of collagen fibers, fibroblasts and melanocytes.[59] Ultrastructural differences exist between the ciliary non-pigmented epithelial cells at the tips of the processes and those in the valleys, the former being adapted for fluid secretion and the latter for mechanical anchoring of the zonule.[60] The length of the ciliary body from the tips of the ciliary processes to the ora serrata is longest temporally and shortest nasally.[61]

The ciliary muscle

The ciliary muscle occupies a triangular-shaped region within the ciliary body beneath the anterior sclera (Fig. 3.5). It has an anterior origin at the scleral spur in close proximity to Schlemm's canal.[61,62] Anterior ciliary muscle tendons insert into the scleral spur and the trabecular meshwork, which serve as a fixed anterior anchor against which the ciliary muscle contracts.[59] Posterior to the scleral spur, the outer surface of the ciliary muscle is attached only loosely to the inner surface of the anterior sclera. The posterior attachment of the ciliary muscle is to the stroma of the choroid. The anterior and inner surfaces of the ciliary muscle are bounded anteriorly by the stroma of the pars plicata and posteriorly by the pars plana of the ciliary body. The ciliary muscle fiber bundles beneath the sclera are oriented such that a contraction of the ciliary muscle results in a forward and inward redistribution of the mass of the ciliary body and a narrowing of the ciliary ring diameter due to sliding ciliary muscle movement along the inner surface of the sphere formed by the anterior sclera. This causes the anterior choroid to be pulled forward. The ciliary muscle is a smooth muscle, with a dominant parasympathetic innervation causing contraction mediated by M3 muscarinic receptors and a sympathetic innervation causing relaxation mediated by β_2-adrenergic receptors. The ciliary muscle is atypical for smooth muscles, in its speed of contraction, the large size of its motor neurons, the distance between the muscle and the

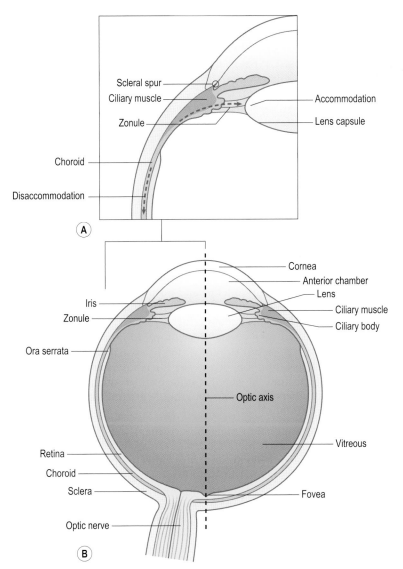

Figure 3.4 Schematic representation of a sagittal section of the anatomy of the accommodative apparatus at the ciliary region of the eye (**A**), in relation to a mid-sagittal section of the eye as a whole (**B**). The schematic shows that the lens capsule, zonule, ciliary muscle and choroid constitute a single elastic "sling" anchored at the posterior scleral canal where the optic nerve leaves the globe, and anteriorly at the scleral spur. The action of accommodation results in a movement of the ciliary body forward and towards the axis of the eye against the elasticity of the posterior attachment of the ciliary muscle and the posterior zonular fibers. With a cessation of an accommodative effort, the ciliary muscle is returned to its unaccommodated configuration through the elasticity of choroid and posterior zonular fibers.

motor neurons, and the unusual ultrastructure of the ciliary muscle cells which in some ways resemble skeletal muscles (indeed, in birds it is a striated skeletal muscle).

There are also regional differences in ultrastructure and histochemistry of the primate ciliary muscle, suggesting that the longitudinal portion may be acting like a fast skeletal muscle to "set" or "brace" the system rapidly, for the contraction of the inner portion to be most effective.[63] The ciliary muscle is comprised of three muscle fiber groups identified by their relative positions and orientations, forming a morphologically and functionally integrated three-dimensional structure.[59] The major group of muscle fibers is the peripheral meridional or longitudinal fibers or Brücke's muscle.[9] They extend longitudinally between the scleral spur and the choroid adjacent to the sclera. Located inward to the longitudinal fibers are the reticular or radial fibers. These constitute a relatively smaller proportion of the ciliary muscle. The radial fibers are branching V- or Y-shaped fibers. These radial fibers are attached anteriorly to the scleral spur and the peripheral wall of the anterior ciliary body at the insertion of the iris. They attach posteriorly to the elastic tendons of the choroid. Beneath the radial fibers and positioned more anteriorly in the ciliary body and closest to the lens are the equatorial or circular fibers or Müller's muscle. These constitute the smallest proportion of the ciliary muscle.

The division of the ciliary muscle into three muscle fiber groups is somewhat artificial. In reality, there is a gradual transition from the outermost longitudinal muscle fibers to the radial fibers to the innermost circular muscle fibers with some intermingling of the different fiber types. A contraction of the ciliary muscle results in a contraction of all three muscle fiber groups together. With a contraction of the ciliary muscle there is a gradual rearrangement of the muscle bundles, with an increase in thickness of the circular portion and a decrease in thickness of the radial and longitudinal portions.[59] A contraction of the entire ciliary muscle as a whole pulls the anterior choroid forward, moves that apex of the ciliary processes towards the lens equator and serves the primary function of releasing resting zonular tension at the lens equator to allow accommodation to occur.

The zonular fibers

The zonular fibers are a complex meshwork of fibrils. Fibrils 70–80 nm in diameter[64] are grouped into fiber bundles estimated to be between 4–6 to 40–50 micrometers in diameter.[64,65] The zonule is composed of the non-collagenous carbohydrate-protein mucopolysaccharide and glycoprotein complexes that are secreted by the ciliary epithelium. The zonular fibers are elastin-based elastic fibers and are thought to be much more elastic than the lens capsule. Their primary function is to stabilize the lens and allow accommodation to occur. Since the zonule is not a continuous tissue, but is composed of fibers, it also allows fluid flow from the posterior chamber behind the iris through to the vitreous chamber (see Chapter 11).

The attachment of the zonular fibers to the lens capsule is superficial with few fibers penetrating into the capsule to form a mechanical (possibly similar to Velcro) or chemical union.[68] From scanning electron microscopy this anterior zonule crossing the circumlental space and extending to the lens is alternatively described as:

(i)   consisting of three fiber strands running to the anterior, equatorial and posterior lens surfaces,[66] or

(ii) fibers that insert along a circular line on the anterior and posterior surface of the lens with some fibers inserting directly on the equator,[64,65,67] or

(iii) a zonular fork with two main fiber groups extending to the lens anterior and posterior surfaces with finer bundles seemingly of relative unimportance running to the lens equator,[57] or

(iv) successive sagittal lines of insertion from lens anterior to posterior surface and two coronal lines of insertion, one where the fibers insert onto the capsule around the anterior surface and another where the fibers insert onto the capsule around the posterior surface.[69]

Although no systematic crossing of anterior zonular fiber was observed by McCulloch,[69] crossing of anterior zonular fibers has been observed in other preparations[54,64,65] and was documented in early diagrams from histology of this tissue (Fig. 3.6). From histological preparations, when an appropriate plane of section is obtained, a continuous line of zonular insertion into the entire lens equator is seen.[69] Unfixed, dissected human eye specimens show a continuous meshwork of fibers uniformly covering the entire lens equator, and show crossing of zonular fibers.[54]

Observations of the ciliary region during accommodation show that the posterior ciliary body slides forward against the curvature of the anterior sclera, moving the posterior insertion of the posterior zonular fibers forward. However, contraction of the ciliary muscle stretches the posterior

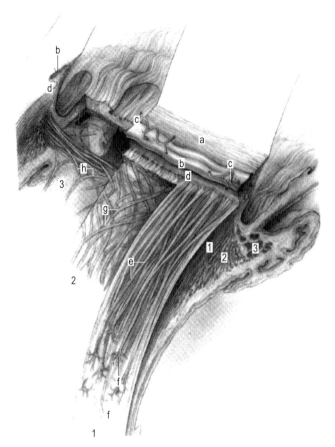

Figure 3.5 Drawing of the ciliary muscle showing a sequential dissection following removal of the outer layers of the globe to reveal the orientation of the underlying ciliary muscle fibers. After removal of the overlying sclera (right) first the meridional or longitudinal fibers, then the reticular or radial fibers and finally the equatorial or circular fibers (left) of the ciliary muscle are revealed. (Reproduced with permission from Hogan MJ, Alvarado JA, Weddell JE. Histology of the human eye. An atlas and textbook. Philadelphia: WB Saunders, 1971.)

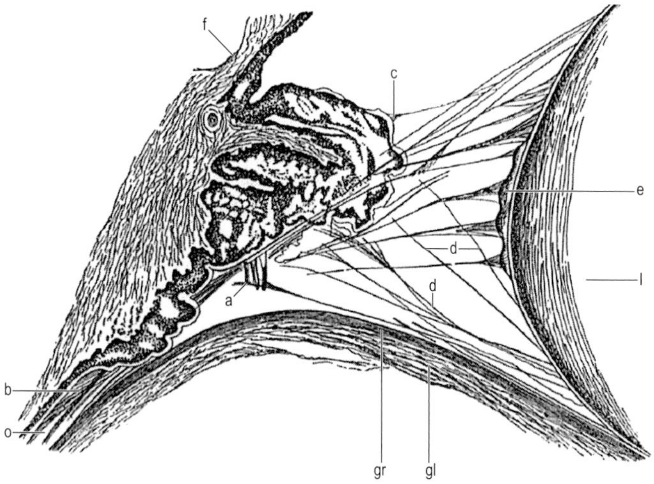

Figure 3.6 Due to the delicate nature of the zonule and the difficulties in observing it, descriptions of the insertion of the anterior zonule onto the lens equator differ. Early anatomists with relatively crude methods produced remarkably accurate diagrams of the structure of the anterior zonule showing crossing of zonular fibers, fiber bundles of varying thickness and insertion into a thickened region of the capsule at the lens equator. Some clumping of zonular bundles is evident in this depiction which is not seen in unfixed specimens. (From Helmholtz von HH. Treatise on Physiological Optics. Translation edited by Southall JPC in 1924 (original German in 1909). New York: Dover, 1962: vol. 1, ch. 12.)

attachment of the ciliary muscle due to a forward and inward movement of the tips of the ciliary muscle and ciliary processes. This suggests that the posterior zonular fibers may similarly assist in pulling the ciliary muscle back to the unaccommodated configuration after cessation of an accommodative effort.

The lens capsule

The crystalline lens is surrounded by the lens capsule (Fig. 3.6). This is a thin, transparent, elastic membrane secreted by the lens epithelial cells largely composed of collagen type IV.[70] Fincham was the first to attribute the accommodative change in shape of the lens to the forces exerted on the young lens by the lens capsule.[71] Fincham studied the capsule in histological section and found it to be of relatively uniform thickness in non-accommodating mammals. However, in primates Fincham found it to be thickest at the mid-peripheral anterior surface, thinner towards the lens equatorial region with a posterior peripheral thickening, but thinnest at the region of the posterior pole of the lens[31] (Fig. 3.7). Several aspects of Fincham's idealized description of

the capsule have been largely confirmed in a more recent study, although with some age-related changes in thickness.[70] The capsule is about 11–15 μm thick at the anterior pole. There is an anterior, mid-peripheral thickening of the capsule that is about 13.5–16 μm thick. This is located more central to the region of zonular insertion into the capsule around the lens equator. The equatorial region of the capsule, to which the anterior zonular fibers insert, is about 7 μm thick at the lens equator and does not appear to change systematically with age. The posterior capsule thickness decreases to a minimum at the posterior pole of about 4 μm, without a posterior mid-peripheral thickening[70] (Fig. 3.8).

The crystalline lens

The lens consists largely of lens fiber cells composing the nucleus and cortex. On the anterior lens surface beneath the capsule is a layer of lens epithelial cells. The embryonic nucleus remains present at the center of the lens throughout life as the cortex grows progressively around it by the addition of an increasing number of layers of lens fiber cells. The deeper layers of lens epithelial cells on the lens anterior surface differentiate to become lens fiber cells. The proliferation of lens epithelial cells and their differentiation into lens fiber cells continues throughout life. Because the lens is contained within the capsule, lens epithelial cells do not slough off as do epithelial cells in other organ systems such as those lining the skin and gut. Therefore, the lens continues to grow throughout life. After adolescence the human lens undergoes a linear increase in mass with increasing age.[72] In vivo, with increasing age, lens thickness increases[73,74] with a resulting increase in the anterior surface and posterior surface curvatures.[75] Although lens thickness and surface curvatures change systematically with increasing age, this occurs without a systematic age-related change in lens diameter.[74,76,77]

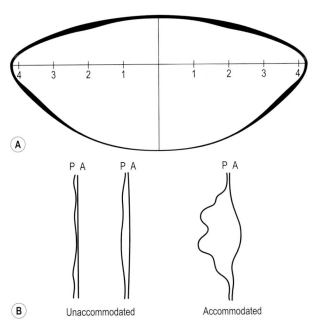

Figure 3.7 (**A**) Fincham's idealized depiction of the regional variations in thickness of the human lens capsule showing the anterior mid peripheral thickening. The equatorial region of the capsule, to which the zonular fibers insert show regional thinning. This idealized depiction is supported by more recent results, although with some age-related regional variations. (**B**) Appearance of the anterior (**A**) and posterior (**P**) capsule in the eye of a patient in whom the lens was displaced and lost from the eye after ocular injury. When the patient focused with the contralateral eye on a distant object (left) the capsule was relatively taught. When the patient focused on a near object, (middle) capsule became more flaccid. After contraction of the ciliary muscle (right) with eserine, the capsule became completely flaccid. Observation by Graves of the behavior of the empty capsule in this aphakic patient eye served as the basis for Fincham's recognition of the role of the capsular tension holding the lens in a flattened and unaccommodated state when the ciliary muscle is relaxed and the capsule rounding the lens into a more spherical and accommodated form when the ciliary muscle contracts. (From Fincham EF. The mechanism of accommodation. Br J Ophthalmol 1937; Monograph VIII:7–80.)

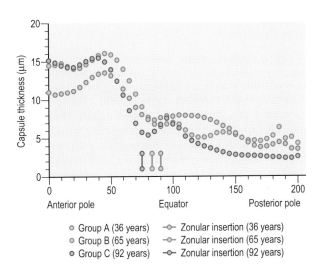

Figure 3.8 Average regional thickness of human capsules from lenses of three different age groups. Symbols connected by vertical lines depict the most anterior positions of the capsules to which the zonular fibers insert. Note regional thinning of the capsule at and posterior to this region is more pronounced in the older lenses. A pronounced anterior mid peripheral thickening of the capsule is evident in all age groups. (From Barraquer RI, Michael R, Abreu R, et al. Invest Ophthalmol Vis Sci 2006;47:2053–60. Reproduced with permission from The Association for Research in Vision and Ophthalmology.)

47

Unaccommodated

7.5 D Accommodation

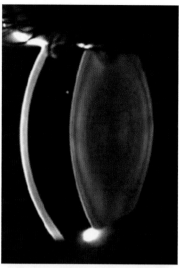

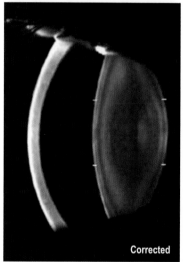
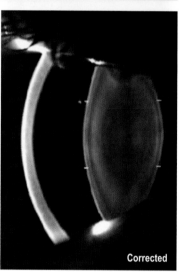

Corrected

Corrected

Figure 3.9 Scheimpflug slit-lamp images of a dilated human eye focused at distance (top left) and to a 7.5 D accommodative stimulus (top right). Each optical interface posterior of the anterior corneal surface is progressively distorted by the optical effects due to the preceding optical interfaces. Optical correction of this distortion applied to the images in the unaccommodated (lower left) and accommodated (lower right) eye shows the true dimensions and dimensional changes in anterior chamber depth, lens thickness and lens surface curvatures during accommodation. (From Dubbelman M, van der Heijde GL, Weeber HA. Vision Res 2005; 45:117–32. Reproduced with permission from Elsevier Science Ltd.)

The crystalline lens has a gradient refractive index, with a refractive index of 1.385 near the poles and a higher refractive index of 1.406 at the center of the nucleus. The lens is not optically homogeneous and when viewed through a slit-lamp, several optical zones of discontinuity are observed which allow visual differentiation of the lens nucleus from the surrounding lens cortex (Fig. 3.9). The unaccommodated young adult human lens is roughly 9.0 mm in diameter and 3.6 mm thick. The lens thickness increases by approximately 0.5 mm with 8 D of accommodation.[33]

The mechanism of accommodation

Current understanding of the accommodative mechanism is largely in accord with the description provided by Helmholtz in 1855[1,78] (Fig. 3.10). Although Fincham[71] and more recent investigations have added further to understanding the accommodative mechanism, the basic tenets are in accord with those originally described by Helmholtz. When the young eye is unaccommodated and focused for distance, the ciliary muscle is relaxed. Resting tension on the zonular fibers spanning the circumlental space and inserting around the lens equator (collectively called the *anterior* zonular fibers[57]) apply an outward directed tension around the lens equator through the lens capsule to hold the lens in a relatively flattened and unaccommodated state. For the eye to focus at near, the ciliary muscle contracts, the inner apex of the ciliary body moves forward and towards the axis of the eye[79] (Fig. 3.11). This inward movement of the apex of the ciliary muscle stretches the posterior attachment of the ciliary muscle and releases resting tension on all zonular fibers around the lens equator. The lens capsule then molds the lens into a more accommodated form.[54,71] The capsule provides the force to cause the lens to become accommodated.[54,71,72] A clear role for the lens capsule in accommodation stems from observations by Graves of the effect of accommodation on the empty lens capsule in an otherwise

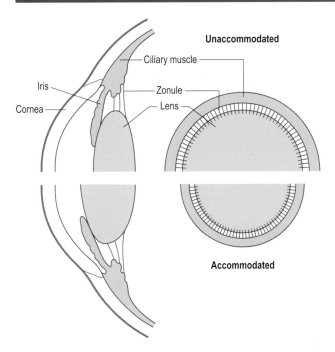

Figure 3.10 Diagram showing the Helmholtz accommodative mechanism. In the upper half of the diagram the eye is in the unaccommodated state. In the lower half the eye is in the accommodated state. The left side shows a sagittal section and the right side a frontal section through the anterior segment of the eye. In the unaccommodated state resting tension on the zonule at the lens equator holds the lens in a relatively flattened and unaccommodated state. When the ciliary muscle contracts, this resting zonular tension is released and the lens is allowed to round up through the force exerted on the lens substance by the lens capsule. Lens axial thickness increases, lens equatorial diameter decreases, anterior chamber depth decreases and vitreous chamber depth decreases with accommodation The lens anterior and posterior surface curvatures increase to increase the optical power of the lens. (Redrawn from Koretz JF, Handelman GH. Sci Am 1988; July:92–99.)

healthy aphakic eye in which the lens was absent.[80,81] When the patient looked at a distant object, the anterior and posterior capsule was taught and flat. When the patient focused on a near object, the capsule became mildly flaccid and the surfaces separated. An eserine stimulated contraction of the ciliary muscle resulted in a completely slack capsule (see Fig 3.7). Fincham concluded that resting zonular tension at the lens equator pulls outward on the capsule to hold the lens in the unaccommodated state and when the eye is accommodated, the resting tension on the zonular fibers is released to allow the capsule to mold the lens into the accommodated form.[71]

Further evidence of the role of the capsule comes from in vitro experiments. In dissecting a young human or monkey eye, when the zonular fibers around the lens equator are cut, the isolated young lens with the intact capsule assumes a maximally accommodated form. If the capsule is then cut and carefully removed from the isolated lens, the young decapsulated lens substance assumes a maximally unaccommodated form.[71,72,82] Further, mechanical stretching studies of partially dissected young human and monkey eyes show that applying an outward directed stretching force to the lens via the capsule and anterior zonular fibers will pull the lens into a flattened and unaccommodated state and releasing

the zonular tension will allow the lens to become accommodated through the forces exerted by the capsule on the lens (Fig. 3.12). Such in vitro mechanical stretching studies reliably and reproducibly produce accommodative optical changes in the lens that match the accommodative optical changes in vivo in the living eye.[54,55,83–85]

During accommodation, lens diameter decreases systematically in response to voluntary accommodation, brain stimulated accommodation or pharmacologically stimulated accommodation[77,86–89] (Figs 3.13, 3.14 & 3.15). Lens thickness increases[33–35,38,90] and the central anterior surface curvature and to a lesser extent the central posterior surface curvature increases.[36–38] These physical changes in the lens are relatively linearly correlated with the accommodative optical changes in the eye (Fig. 3.16). The increased lens surface curvature results in an increase in the optical power of the crystalline lens. Anterior chamber depth decreases due to the forward movement of the anterior lens surface and the vitreous chamber depth decreases due to the posterior movement of the posterior lens surface[33,34,38] (Fig. 3.16). About 75 percent of the increase in lens thickness is accounted for by the anterior movement of the anterior lens surface and about 25 percent of the increase in lens thickness is accounted for by a posterior movement of the posterior lens surface.[33,34]

When the accommodative effort ceases, the ciliary muscle relaxes and the elasticity of the posterior attachment of the choroid pulls the ciliary muscle back into its flattened and unaccommodated configuration. The outward movement of the apex of the ciliary body once again increases the tension on the anterior zonular fibers around the lens equator to pull the lens via the capsule into a flattened and unaccommodated form.

Variants on the Helmholtz accommodative mechanism have been suggested to include an essential role for the vitreous and differential pressure changes in the eye.[51,52,91] However, accommodation still occurs after vitrectomy[53] and mechanical stretching studies of dissected eyes in which the vitreous is absent and no pressure differential can exist, results in normal accommodative optical changes to the lens,[54,83,85] thereby obviating a role for the vitreous or for differential pressure changes in the eye. A revisionist theory of accommodation originally proposed by Tscherning[92] has been espoused.[93–95] This theory is opposite to the Helmholtz accommodative mechanism in that it paradoxically requires an *increase* in lens equatorial diameter during accommodation, a flattening of the peripheral lens surfaces and an increase in curvature of the lens central surfaces. However, there exists no experimental evidence in support of the supposed accommodative increase in lens diameter. In fact, numerous studies demonstrate that lens diameter decreases systematically during accommodation.[76,77,79,86,87,89]

Accommodative optical changes in the lens and eye

For the unaccommodated emmetropic eye to focus on a near object requires an increase in optical power of the eye. This occurs through an increase in optical power of the lens. The

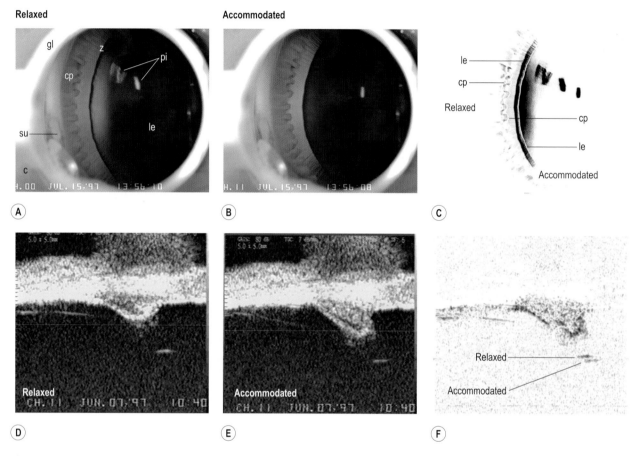

Figure 3.11 Gonioscopy images of an iridectomized rhesus monkey eye: (**A**) unaccommodated and (**B**) accommodated state. (**C**) The subtracted difference image shows the accommodative movements of the lens equator and ciliary processes as well as the relative stability of the eye. The lens equator and the ciliary processes move away from the sclera with accommodation to roughly the same extent. Key: c, cornea; gl, gonioscopy lens; su, suture; cp, ciliary processes; le, lens; z, zonule; pi, lens Purkinje images. Ultrasound biomicroscopy images of an iridectomized rhesus monkey eye in the (**D**) unaccommodated and (**E**) accommodated states. (**F**) The subtracted difference image shows the accommodative movements of the ciliary muscle and the lens equator. The apex of the ciliary muscle and the lens equator (short horizontal line and identified with arrows) move away from the sclera with accommodation. (Reprinted from Glasser A, Kaufman PL. Ophthalmology 1999: 106:863–72, with permission from Elsevier Science Ltd.)

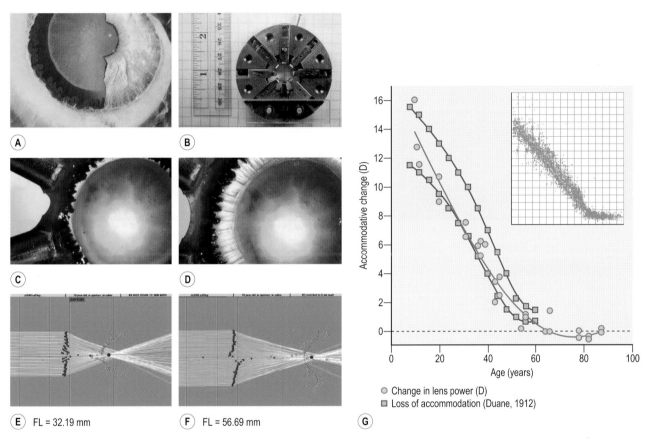

(A)

(B)

(C)

(D)

(E) FL = 32.19 mm

(F) FL = 56.69 mm

(G)

○ Change in lens power (D)
▢ Loss of accommodation (Duane, 1912)

Figure 3.12 (**A**) The anterior segment of a partially dissected 54-year-old human donor eye glued to (**B**), the arms of a mechanical stretching apparatus. The zonule can be completely relaxed (**C**) allow the lens to become maximally accommodated, or (**D**) the zonule stretched to disaccommodate the lens. (**E**) Scanning laser measurements of the focal length of a 10-year-old human lens measured in the unstretched, accommodated state (focal length = 34.39 mm) and (**F**) in the maximally stretched and unaccommodated state (focal length = 57.69 mm). Parallel laser beams enter the lens, are refracted by the lens (at red symbols, left), and cross the optical axis (dark horizontal line) at the position identified (yellow symbols, right). The distance from the lens (red symbols, left) to the average focus of all rays (blue symbol) represents the lens focal length. (**G**) The change in focal length converted to diopters (red line and circles) as a function of the applied stretch shows that young lenses undergo 12–16 D of change in power with stretching, but that by age 60, the same extent of applied stretch results in no change in lens power. The data from the human lenses are plotted in the inset in G, together with Duane's (1912) data (blue lines, diamonds) showing the range of accommodative amplitudes from some 1500 subjects as measured with a push-up technique. (Reprinted from Glasser A, Campbell MCW. Vision Res 1998; 38:209–29, with permission from Elsevier Science Ltd; and from Duane A. J Am Med Assoc 1912; 59:1010.)

Relaxed

Accommodated

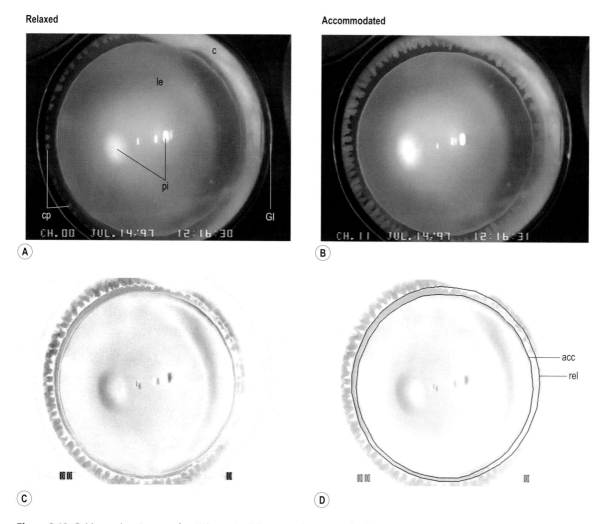

Figure 3.13 Goldmann lens images of an iridectomized rhesus monkey eye in the (**A**) unaccommodated and (**B**) accommodated states. Key: c, conjunctiva; cp, ciliary process; le, lens; pi, lens Purkinje images; Gl, Goldmann lens. (**C**) Subtracted difference image to show the accommodative movements of the lens. The lens undergoes a concentric decrease in diameter and the ciliary processes move concentrically inward with accommodation. There is a virtual absence of eye movements as evident from the absence of additional detail in the difference image. (**D**) The outlines of the accommodated and unaccommodated lens diameter are shown superimposed on the difference image demonstrating a concentric decrease in lens diameter with accommodation in accordance with the Helmholtz accommodative mechanism. Key: acc, accommodated; rel, relaxed. (From Glasser A, Kaufman PL. Ophthalmology 1999: 106:863–72, with permission from Elsevier Science Ltd.)

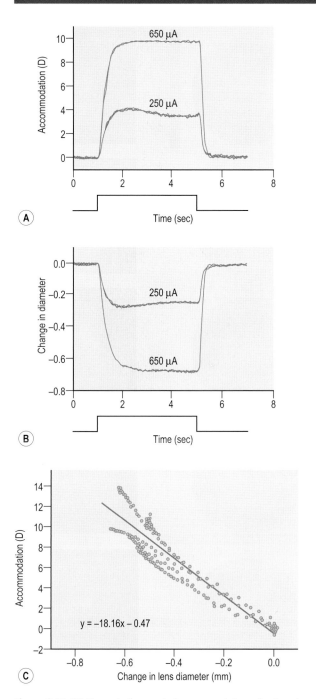

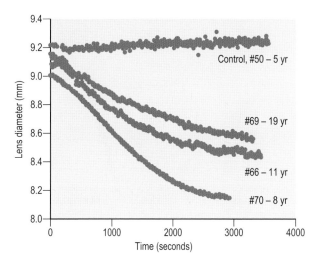

Figure 3.15 Time-course of topical pilocarpine stimulated accommodative changes in lens diameter in the eyes of anesthetized, iridectomized rhesus monkey and from a control experiment in which saline was applied to the eye. Pilocarpine stimulated accommodation causes a systematic and progressive decrease in lens diameter, the magnitude of which is larger for younger monkeys with higher accommodative response amplitudes. (From Wendt M, Croft MA, McDonald J, et al. Exp Eye Res 2008; 86:746–52. Reproduced with permission from Elsevier Science Ltd.)

Figure 3.14 (**A**) Dynamically recorded accommodative refractive changes in response to two different stimulus current amplitudes applied to the Edinger–Westphal nucleus of the brain in an anesthetized rhesus monkey. (**B**) Dynamically measured accommodative decreases in lens diameter in response to the same two current stimulus amplitudes for which accommodation was measured in panel A. (**C**) Correlation of the optical refractive changes and decreases in lens diameter during accommodation from panels A & B. Lens diameter decreases linearly by about 600 μm for about 12 D of accommodation. (From Glasser A, Wendt M, Ostrin L. Invest Ophthalmol Vis Sci 2006; 47:278–86. Reproduced with permission from The Association for Research in Vision and Ophthalmology.)

accommodative increase in optical power of the lens comes about from an increase in the lens anterior and, to a lesser extent, posterior surface curvatures. Several other physical changes in the eye and lens also occur during accommodation that have optical effects on the eye. Lens thickness increases, the lens anterior surface moves anteriorly to reduce anterior chamber depth and the lens posterior surface moves posteriorly to increase anterior segment length. The lens asphericity changes and the pupil constricts. In addition, because the lens has a gradient refractive index, the form of which is constrained by the shape and size of the lens, since the lens changes shape, so too must the form of the lens gradient refractive index. All of these accommodative physical changes in the eye and lens result not only in an increase in optical power, but also in other changes in the ocular aberrations of the eye.

Simple paraxial vergence calculations show that for parallel rays incident on a simple biconvex lens, if lens thickness alone is increased, lens power decreases. In an eye, however, if the lens thickness increases, this must occur in conjunction with either a decrease in anterior chamber depth or an increase in anterior segment length consequent to the increase in lens thickness. For a distant object, the lens inside the eye does not have parallel light incident on it, but rather convergent light due to refraction by the cornea. Simple paraxial schematic eye calculations show that if lens thickness alone is increased without a change in lens curvatures but with a resultant decrease in anterior chamber depth, the result is an overall increase in power of the eye. The accommodative increase in optical power of the lens however is primarily due to an increase in the lens surface curvatures.

Since the lens anterior surface is flatter than the lens posterior surface, for a given change in curvature the anterior surface will undergo a relatively greater increase in optical power than the posterior surface. The accommodative optical increase in power of the eye is therefore ultimately due to a complex combination of optical and physical changes in the lens and the eye. This results not only in an

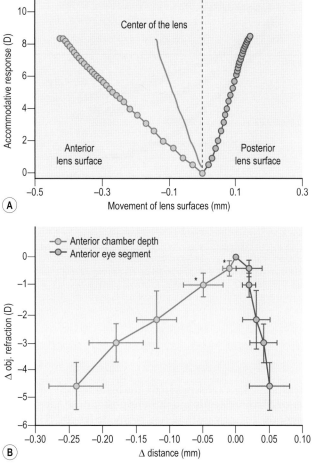

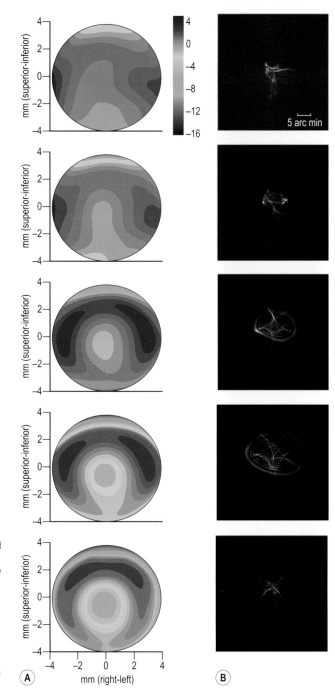

Figure 3.16 (**A**) Continuous a-scan ultrasound recordings of the changes in the anterior segment of the eye and the accommodative refractive change in a rhesus monkey eye in response to Edinger–Westphal stimulated accommodation. As the eye accommodates, the anterior lens surface (solid symbols) moves anteriorly towards the cornea and the posterior lens surface (open symbols) moves posteriorly towards the retina. The calculated geometric center of the lens (midway between the lens anterior pole and the lens posterior pole – thin line) shows a slight anterior movement towards the retina. (From Vilupuru AS, Glasser A. Exp Eye Res 2005; 80:349–60. Reproduced with permission from Elsevier Science Ltd.) (**B**) Objective accommodative refractive changes measured with infrared photorefraction and anterior segment biometric changes measured with partial coherence interferometry simultaneously during visual stimulus driven accommodation in humans show changes similar to those recorded in the monkey eye. In this case, the accommodative refraction is graphed as negative numbers. The anterior surface of the lens moves anteriorly and the posterior surface of the lens moves posteriorly during accommodation. (From Bolz M, Prinz A, Drexler W, Findl O. Br J Ophthalmol 2007; 91:360–5. Reproduced with permission from the BMJ Group.)

Figure 3.17 Accommodative changes in ocular wavefront aberrations in vivo in an iridectomized rhesus monkey eye. (**A**) Graphs show the wavefront maps with the defocus term removed calculated for an 8 mm entrance pupil diameter. (**B**) Images show the corresponding retinal point spread functions calculated with the defocus term removed. As accommodation increases progressively (top to bottom: 0 D, 1.41 D, 3.88 D, 5.93 D, and 10.91 D), there is a progressive increase in the ocular wavefront error resulting in an increase in the size of the retinal point spread function. There is a marked increase in negative spherical aberration with increasing accommodation with a relatively greater change in overall optical power of the eye towards the center than at the periphery. There is relatively little change in optical power near the periphery of the eye which is normally occluded by the iris. (From Vilupuru AS, Roorda A, Glasser A. J Vis 2004; 4:299–309. Reproduced with permission from The Association for Research in Vision and Ophthalmology.)

increase in optical power of the eye, but also accommodative changes in ocular aberrations of the eye. In particular, accommodation is accompanied by an increase in negative spherical aberration of the eye and lens[83,84,96–98] (Figs 3.17 & 3.18). In addition, the iris constricts during accommodation, to decrease the optical entrance pupil of the eye. This too has optical effects. Simply decreasing the entrance pupil diameter results in an overall reduction in optical aberrations of the eye.[84] Further, in an eye with negative spherical aberration, in which the paraxial rays are focused closer to

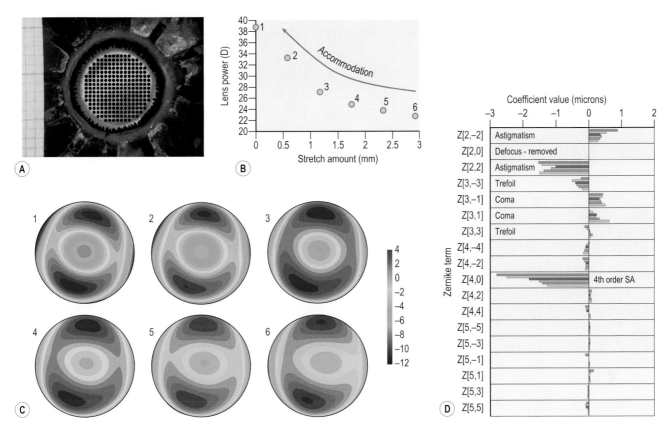

Figure 3.18 Accommodative optical changes during in vitro mechanical stretching of a rhesus monkey lens. (**A**) Optical scanning laser measurements of the rhesus monkey lens. A total of 241 laser beams are sequentially passed through the lens parallel to the optical axis. Reconstruction of the laser scans permits the optical wavefront aberrations of the lens to be calculated. (**B**) The optical accommodative change in lens power as a function of stretch applied to the lens. Position 1 corresponds to the unstretched, maximally accommodated lens. Position 6 corresponds to the maximally stretched, unaccommodated lens. (**C**) Wavefront aberration contour plots with the defocus term removed showing the decrease in wavefront error of the lens in going from the maximally accommodated, unstretched lens in position 1 to the maximally stretched and unaccommodated lens in position 6. (**D**) Graph showing the changes in the terms of the Zernike coefficients of the wavefront as the lenses are stretched. In particular, Zernike term Z[4, 0] shows a systematic increase in 4th order negative spherical aberration (SA) of the lens as the lens progresses from the unaccommodated to the maximally accommodated state. (From Roorda A, Glasser A. J Vis 2004; 4:250–61. Reproduced with permission from The Association for Research in Vision and Ophthalmology.)

the lens than the peripheral rays, a pupil constriction alone would result in an overall increase in optical power of the eye simply due to the constricted iris occluding the more weakly refracted peripheral rays.[84] In a young eye, when accommodation occurs, all these various optical changes occur in concert.

The stimulus to accommodate

At rest, the eyes have some residual or resting level of accommodation amounting to approximately 0.5–1.5 D. This is called tonic accommodation or a lead of accommodation. In a young eye, an effort to focus at near causes three physiological responses; the eyes accommodate, the pupils constrict and the eyes converge (Fig. 3.19). Together these three physiological functions are referred to as the accommodative triad or the near reflex. These three actions are neuronally coupled through the preganglionic parasympathetic innervation extending from the Edinger–Westphal (EW) nucleus in the brain. The intraocular muscles (iris and ciliary muscle) are innervated by the postganglionic ciliary nerves entering

the sclera. The extraocular muscles of the eyes are innervated by the oculomotor (III), the trochlear (IV) and abducent (VI) nerves, the axons of which originate from motor nerve nuclei in the brainstem, which receive impulses from the EW nucleus. Accommodation and convergence and the accompanying pupil constriction are neuronally coupled in the brain and therefore in the two eyes. An accommodative stimulus such as minus lens induced blur or a proximal stimulus presented monocularly to one eye results in binocular accommodation, convergence and pupil constriction. Similarly, a convergence stimulus presented monocularly to one eye results in pupil constriction, convergence and accommodation in both eyes.

Accommodation can be stimulated in a variety of ways. It can be driven by blur cues alone – if myopic blur is presented to one or both eyes by placing a negative powered lens in front of the eye(s), both eyes will accommodate to attempt to overcome the imposed defocus (Fig. 3.20). If convergence is stimulated in a young eye, such as by having the subject fixate a distant target and placing baseout prisms in front of the eyes, pupil constriction and accommodation will also occur. In an emmetropic eye,

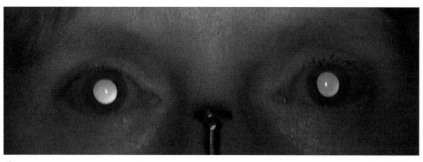

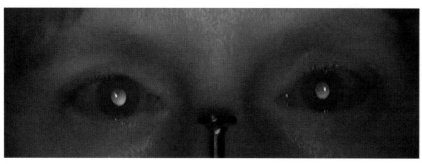

Figure 3.19 Infrared photorefraction of the eyes showing the accommodative triad of accommodation, pupil constriction and convergence as the subject changes fixation from **A,** a distant object to **B,** a near object held a few centimeters beyond the nose. The accommodative optical change in power of the eye is evident from the photorefraction images by virtue of the brighter crescents of light in the lower parts of the pupils when the eyes are focused on the near object.

blur and vergence driven accommodation can be induced simultaneously with a proximal stimulus. If a near object is presented, coupled accommodation and convergence occur. As the accommodative stimulus increases, the objectively measured accommodative response is typically less than the magnitude of the stimulus. This is called the lag of accommodation.

Studies in which the accommodative response is measured with a wavefront aberrometer show that calculated retinal image quality for the near object actually improves when the lag of accommodation is taken into account.[99] Therefore, although the overall refraction of the eye lags the accommodative stimulus the lag may serve to maximize retinal image quality for near objects due to the ocular aberrations of the eye. As the stimulus amplitude increases, there is a linear increase in the accommodative response, which due to the lag of accommodation has a slope less than 1. As the stimulus is increased further, lag increases as the maximum accommodative response amplitude is reached. The accommodative stimulus function is therefore "S" shaped with the initial lead, the intermediate linear region with some lag, and the final plateau region. Studies aimed at addressing how the eye detects defocus have shown that the longitudinal chromatic aberration (LCA) of the eye plays a role. The imperfect optics of the eye cause considerable LCA with the result that shorter wavelengths of light are focused closer to the lens than longer wavelengths. Removing the LCA by using monochromatic light or optically neutralizing or reversing the LCA disrupts the normal reflex accommodative response.[100–103]

Accommodation can also be pharmacologically stimulated. Topical application of muscarinic cholinergic agonists, such as pilocarpine, results in direct pharmacological stimulation of the ciliary muscle.[46,48,104] In rhesus monkeys pharmacologically stimulated accommodation is of higher amplitude than centrally stimulated accommodation.[105–107] This is attributed to a supramaximal pharmacological contraction of the ciliary muscle and iris which is greater than the contraction due to a parasympathetically driven stimulus from the brain.[108] In addition, drug stimulated accommodation ultimately produces a net forward movement of the natural lens that does not occur with centrally stimulated accommodation in monkeys[109] or voluntary accommodation in humans.[110] Rapid and strong pupil constriction also occurs with pharmacological stimulation, but convergence does not. Anticholinesterases, such as echothiophate iodide, when applied topically produce a resting tonus of accommodation.[22] This is due to the spontaneous release of acetylcholine at the neuromuscular junction, normally broken down by cholinesterases, inducing an accommodative tonus. Accommodative esotropia, often occurring in uncorrected hyperopes due to the need to accommodate to focus on distant objects, can be treated with topical echothiophate. By producing increased accommodative tonus without increased neuronal input, the stimulus for convergence is reduced, and the accommodation convergence /accommodation (AC/A) ratio is reduced, helping to alleviate the accommodative esotropia.[111] Anticholinesterases produce a long response to a single administration and so are more therapeutically useful than shorter-acting cholinomimetics like pilocarpine.

The pharmacology of accommodation

Accommodation occurs when the post-ganglionic parasympathetic innervation to the ciliary muscle releases the neurotransmitter acetylcholine at the neuromuscular junctions.

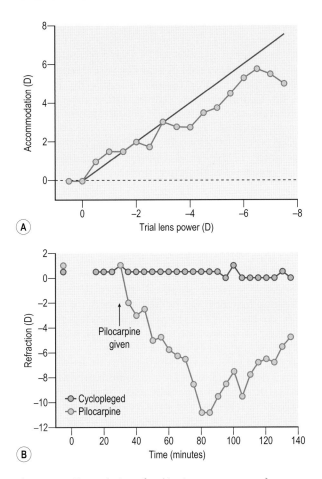

(A)

(B)

Figure 3.20 Two techniques for objective measurement of accommodation in humans. (**A**) The accommodative response of the right eye is measured with a Hartinger coincidence refractometer as increasingly powered negative trial lenses are placed in front of the left eye of a 35-year-old subject viewing a distant letter chart. As the letter chart is viewed through increasing powered negative trial lenses, the accommodative response increases towards the maximal accommodative amplitude of about 6 D. A further increase in lens power results in no further increase in accommodation. (**B**) The right eye is dilated with one drop of phenylephrine and the baseline resting refraction is measured 4 times in both eyes. The left eye (blue symbols) is then cyclopleged with 1 percent cyclopentolate and accommodation is stimulated in the right eye (red symbols) with one drop of 6 percent pilocarpine. Refraction is measured in each eye three times at the end of each 5-minute interval with a Hartinger coincidence refractometer. The pilocarpine stimulated accommodative response in the right eye of the same 35-year-old subject as in (A) reaches a maximum of 11 D approximately 30 minutes after instillation of pilocarpine. In this subject, the pilocarpine stimulated accommodative response is greater than the voluntary accommodative response elicited from negative lens induced defocus.

Acetylcholine is a muscarinic agonist which binds with the ciliary muscle muscarinic receptors to cause the muscle to contract. Topically applied muscarinic agonists, such as pilocarpine also bind to the muscarinic receptors and cause ciliary muscle contraction. Thus accommodation can be stimulated by topically applied pilocarpine.[46,48,104,110] This results in an involuntary monocular accommodative response which in some individuals can be of higher amplitude than voluntary accommodation and is greater in eyes with lighter colored irides than in eyes with darker

colored irides.[48] Individuals with dark irides (brown) are less sensitive to topically applied drugs because of increased pigment epithelium in the iris and ciliary muscle which binds topically applied agents and decreases their bioavailability. The effect of ocular pigmentation on the ocular hypotensive response to pilocarpine through the action of the drug on the ciliary muscle, is well known.[112] Accommodation can also be pharmacologically blocked. This is accomplished by temporary pharmacological paralysis of the ciliary muscle and is called cycloplegia. Cycloplegia can be induced by topical application of muscarinic antagonists such as atropine, cyclopentolate or tropicamide. These agents competitively bind to and block the muscarinic receptors, thereby preventing the agonists from causing accommodation.

Measurement of accommodation

Although objective methods are available for measurement of accommodation (Box 3.2), unfortunately, clinically the subjective measurement of the near point of clear vision is most often used. Subjective measurements overestimate the true accommodative optical change in power of the eye (Fig. 3.21). The subjective push-up method requires the patient to gradually move a near letter chart towards the eyes and report when a near letter chart can no longer be maintained in clear, sharp focus. The reciprocal of the distance from the eyes and the near reading chart is then used as a measure of accommodative amplitude in diopters. There are many reasons why subjective measurement of accommodation should be avoided. The endpoint of the subjective push-up test requires a subjective evaluation of best image focus by the subject and this endpoint varies between individuals. Subjective evaluation of the point of best focus can be influenced by depth of focus, visual acuity, contrast sensitivity of the eye, and contrast of the image, for example. A dimly illuminated reading chart may provide a poor stimulus to accommodate or may not allow an accurate recognition of defocus. Different levels of illumination alter pupil diameter and therefore depth of focus of the eye thus influencing the near point of clear vision. Subjective push-up measurements are also confounded by the increasing angular subtense of the object. As a reading chart is brought closer to the eye,

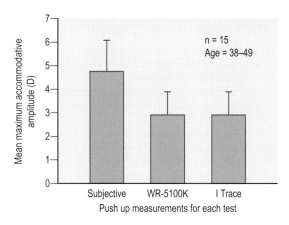

Push up measurements for each test

Figure 3.21 Comparison of subjectively measured and objectively measured accommodative amplitudes in 15 human subjects ranging in age from 38 to 49 years. Subjectively measured accommodation was assessed with the push-up test in which the subjects report the nearest point of clear vision. Accommodative amplitude is determined as the reciprocal of the distance between the eyes and the near reading chart in meters. This is compared with objective measurements of the dioptric change in power of the eye as measured with two objective instruments – the WR-5100K Grand-Seiko autorefractor and the iTrace aberrometer. For the objective tests, first distance refraction was measured and then the near stimulus was pushed-up towards the eyes as the objective instruments measured the refraction of the eyes. Objectively measured accommodative amplitude is determined as the maximum dioptric change in spherical refraction of the eyes. The subjective push-up test significantly overestimates the objectively measured accommodative response amplitudes. (From Win-Hall DM, Glasser A. J Cataract Refract Surg 2008; 34:774–84. Reproduced with permission from Elsevier Science Ltd.)

this results in an increased retinal image size and hence increased legibility of the letters as they are brought closer. Although this can be avoided by carefully controlling the image angular magnification with scaled letter sizes, this is not done with the subjective push-up test.

Subjective measurement of accommodative amplitude can also be done by placing negative powered trial lenses in front of one or both eyes to blur a distant letter chart. The optically induced blur stimulates accommodation in an attempt to maintain a sharp focused image on the retina. The negative lens power is progressively increased until the smallest legible letter line of a distance Snellen letter chart can no longer be maintained in clear focus.[113] Accommodative amplitude is determined by the strongest powered negative lens through which the smallest legible Snellen letter line can still be read clearly. This is still a subjective test and prone to the same sources of errors as the subjective push-up test. Subjective push-up tests are also an inaccurate measure of accommodative amplitude because of the lag of accommodation. It is well known that the accommodative optical response of the eye lags behind the stimulus and that this lag increases as the stimulus amplitude increases. Therefore the dioptric vergence of the stimulus is expected to be less than the accommodative response. Subjective methods traditionally used for evaluating accommodative amplitude are inherently inaccurate and overestimate true accommodative amplitude. Near vision can be improved through non-accommodative optical means. Multifocal intraocular lenses or multifocal contact lenses for example allow some degree

of functional near vision to presbyopes, but through static, non-accommodative, optical means. Similarly, astigmatism or ocular aberrations provide some degree of multifocality to the eye. The ability to read at near does not unequivocally imply that accommodation occurs, and subjective methods to measure accommodation cannot differentiate between true accommodation, depth of field or optical compensation such as with multifocal optics.

Since accommodation results in a change in the optical refractive power of the eye, accommodation can readily be measured objectively. Objective methods provide a true measure of accommodative amplitude of the eye. Accurate objective measurement of accommodation can be done statically[44,113,114] or dynamically.[115–119] Autorefractors, refractometers or aberrometers are suitable instruments for objective accommodation measurements. These instruments provide a measure of the refraction of the eye as the eye changes focus between a distant and a near target. The accommodative response amplitude is then determined as the difference between the refraction when looking at a distant target and the refraction when looking at a near target. Subjective measurements in presbyopes may suggest some accommodation is present, but it is only when objective methods are used that a complete loss of active accommodation is demonstrated at the endpoint of presbyopia.[45,46] The success of objective instruments to measure maximal accommodation relies on the accuracy of the instrument as well as on the ability to elicit the maximum accommodative response from the subject. If the subject does not produce an accommodative response, no accommodation can be measured.

Objective instruments differ in whether they measure statically or dynamically. If a single static measurement is made, this may miss the point of maximum accommodation. Dynamic measurements can provide an indication of how much the accommodative response varies over time. Dynamic optometers provide a real-time graphic display of the accommodative response and record data from which a reliable measure of true accommodative amplitude can be calculated. The success of these instruments at measuring maximal accommodation also depends on how well a distance and near target can be presented to the subject and whether the near target can be viewed monocularly or binocularly by the subject. To stimulate accommodation, the subject must be presented with a compelling accommodative stimulus and the subject must elicit an accommodative response.

Accommodation can be stimulated in a number of ways. If a negative powered trial lens is placed in front of one eye while viewing a distant letter chart, the consensual accommodative response can be measured in the contralateral eye.[46,48] This method of measuring the accommodative response suffers the disadvantage that the convergence response accompanying accommodation occurs entirely in the eye being measured since the eye being defocused with the trial lenses maintains primary gaze position as it fixates on the distant letter chart. Unless the instrument being used to measure accommodation is realigned with the optical axis of the converged eye, an off-axis refraction measurement will result which can introduce inaccuracies. Accommodation can also be stimulated by topically applied muscarinic agonists (pilocarpine, for example) and the resulting accommodative response measured periodically over 30–45 minutes using a refractometer or an autorefractor until the maximal accommodative response is attained[46,48] (see Fig. 3.20B). This is a slow time-course for an accommodative response, but if the refraction is measured frequently enough, the maximum accommodative response amplitude can be determined. The accommodative amplitude measured in this way is independent of a visual accommodative stimulus and of patient subjectivity since the application of the drug produces the accommodative response. However, the magnitude of the accommodative response does depend on drug concentration, intraocular pharmacokinetics, iris pigmentation and other non-accommodative factors that influence how much drug or how quickly the drug reaches the ciliary muscle.

Presbyopia

Presbyopia (Box 3.3) is the gradual age-related loss of accommodative amplitude which begins early in life and ultimately culminates in a complete loss of accommodation by about 50 years of age.[45,46,120] Subjective measurements of accommodation may suggest that about one diopter of accommodation remains after about 50 years of age.[24,45,46] However, this small remaining apparent response is due to depth of field effects that are inherent in the subjective measurement of accommodation. Objective measurement shows a linear decline of accommodation by about 2.5 D per decade to zero at about 50–55 years of age.[45,46,120]

Presbyopia results in the complete loss of the normal physiological function of accommodation roughly two-thirds of the way through the human life-span. Few other normal physiological functions undergo such a profound and systematic deterioration so soon and with such certainty in so many. Presbyopia is likely a consequence of age-related changes in the accommodative apparatus that begin early in life and continue beyond the point at which accommodation is ultimately completely lost, possibly continuing until death. Since two-thirds of human accommodative amplitude is lost between ages 15 and 45, this is the age group that may be of most interest in trying to understand the progression of presbyopia. However, while what happens after the age of 45–50 may not be of particular relevance as far as causes of presbyopia, these age-related changes may well represent a part of the continuum that earlier in life leads to presbyopia. The causes of presbyopia may be best understood by studying how and why accommodation is lost, but understanding the age-related changes that continue beyond 50 years of age may also provide important insights.

Factors contributing to presbyopia

Since the accommodative apparatus is composed of many different tissues and systems and accommodation is a complex interaction of these components, there are potentially many factors that contribute to the loss of accommodation. Aging affects many of these tissues and systems to differing extents and so the reasons why accommodation is lost as a consequence of aging are potentially many and complex. Although several fundamental changes occur, such as stiffening of the lens, which must have a profound impact on the ability of the eye to accommodate, other aspects of ocular aging may also impact accommodative amplitude. Further, many studies show age-related changes in the accommodative structures that progress well beyond that age at which accommodation is lost. Ultimately, at its endpoint, presbyopia is due to a loss of the fine balance of forces that permit the accommodative structures to cause a change in optical power of the lens in the young eye. In the following sections, age-related changes in the ciliary muscle, lens, lens capsule, zonule and associated tissues are considered in terms of their possible roles in presbyopia.

Age-related changes in rhesus ciliary muscle

Since accommodation is lost with increasing age, and accommodation is mediated by the ciliary muscle, the question arises whether presbyopia is due to a loss in ability of the ciliary muscle to contract. Since pupillary constriction and convergence are part of the near reflex but do not decrease with increasing age, this suggests that loss of muscle contractility is not a normal part of presbyopia.[121] The iris, like the ciliary muscle, is an intraocular muscle and the iris continues to contract with light stimulus and with an accommodative effort even in presbyopes, therefore it is likely that the ciliary muscle continues to contract with an accommodative effort in presbyopes. Accommodative excursion of the ciliary muscle is reported to be reduced in presbyopic rhesus monkeys, as seen from both direct observation in surgically aniridic animals in which accommodation is stimulated by an electrode placed in the Edinger–Westphal nucleus,[122] and from histologic study of ciliary muscle topography in eyes

---

Box 3.3 Presbyopia

- Presbyopia is the age-related loss of accommodative amplitude
- The accommodative optical change in power of the eye with an effort to focus at near is completely lost by about 55 years of age
- Many aspects of the accommodative apparatus of the eye change with increasing age
- Lens thickness increases, the lens anterior surface curvature increases, anterior segment length increases, the apex of the unaccommodated ciliary body progressively moves inward towards the axis of the eye, the elastic modulus of the capsule increases
- Lens stiffness increases exponentially with increasing age
- The stiffness gradient of the human lens increases with increasing age. In the young lens, the nucleus is softer than the cortex, but with increasing age the nucleus undergoes a greater increase in stiffness than the cortex such that in the older lens the nucleus becomes stiffer than the cortex
- Ultimately, the human lens completely loses the ability to undergo accommodative changes in optical power
- The ciliary muscle retains the ability to contract and to undergo accommodative movements in the presbyopic eye
- Presbyopia is, at its endpoint, due to a complete loss in accommodative ability of the lens

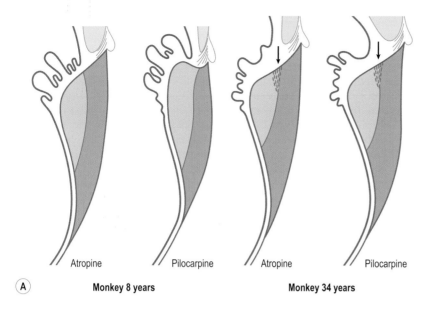

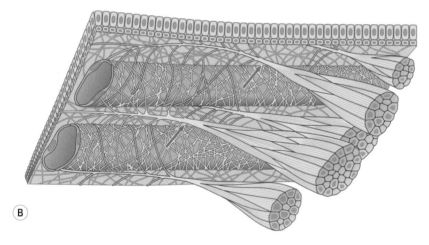

Figure 3.22 (**A**) Diagrams of the configuration of the ciliary muscle from an 8-year-old (left pair of images) and a 34-year-old (right pair of images) enucleated rhesus monkey eye after each globe was bisected with one half (left) placed in an atropine solution and the other half (right) placed in a pilocarpine solution. Representative sections based on histologic specimens. In the young, but not the old eye, the pilocarpine treated ciliary muscle showed a configurational accommodative change. The ciliary muscle from the older eye fails to undergo an accommodative configurational change due to the loss of elasticity of the posterior attachment of the ciliary muscle to the choroid. The 8-year-old rhesus monkey exhibits essentially no intramuscular connective tissue whereas the 34-year-old rhesus monkey exhibits connective tissue (arrows) only anteriorly between longitudinal and reticular zones of the ciliary muscle. (**B**) Diagram of the posterior attachment of the ciliary muscle (CM) in rhesus monkeys. The meridional muscle fiber bundles (arrows) are attached to the elastic layer of Bruch's membrane via the elastic tendons. Smaller elastic fibers connect (arrowheads) the tendons of different bundles to the elastic network that surrounds the vessels of the pars plana. (Panel B from Tamm, Lutjen-Drecoll, Jungkunz & Rohen. Invest Ophthalmol Vis Sci 1991; 32:1680, with permission from the Association for Research in Vision and Ophthalmology and from the author.)

fixed in the presence of pilocarpine or atropine[123] (Fig. 3.22). The posterior attachment of the rhesus ciliary muscle, comprising elastic tendons continuous with Bruch's membrane, shows structural changes with increasing age.[124] While the elastic tendons of the young monkey eye stain strongly for actin and desmin, in the aging eye this region exhibits increased collagen fibers that adhere to the elastic fibers, thickening of the elastic tendons and increased microfibrils.[124] These anatomical changes may lead to decreased compliance of the posterior attachment of the ciliary muscle and the choroid. This is supported by the observation that in aged rhesus monkey eyes in which the posterior attachment of the ciliary muscle is severed prior to pilocarpine stimulation, a configurational change occurs that is otherwise absent.[125] However, the contractile force of isolated rhesus ciliary muscle strips to pilocarpine stimulation is not reduced with increasing age[126] (Fig. 3.23). Although there is some loss of ciliary muscle mass, there is no loss of muscarinic receptor number or binding affinity, and no change in cholineacetyltransferase or acetylcholinesterase activity.[127]

These histological, histochemical and ultrastructural studies show that the reduced ciliary muscle accommodative movement in presbyopic monkey eyes is due to a loss of elasticity of the posterior attachment of the ciliary muscle and choroid. This tissue is normally elastic and is stretched during accommodation in the young eye. If the tissue becomes less extensible with advancing age, the ciliary muscle must work harder to move forward during accommodation. Studies comparing the accommodative movements of the ciliary processes and lens equator in iridectomized monkeys show that both ciliary process and lens edge movements are reduced with increasing age and the loss of accommodation. However, accommodative movements of the ciliary processes are always greater than accommodative movements of the lens edge, irrespective of age[128,129] (Fig. 3.24). Ciliary process movements are also greater than lens edge movements even in young monkeys.[128] Together this suggests that lens movement limits the accommodative amplitude at all ages, not ciliary body movement. Ultimately at the endpoint of presbyopia ciliary process movement still occurs even in the absence of accommodative movements of the lens.[129] Thus, although accommodative movement of the ciliary muscle/ciliary body may be systematically reduced with increasing age in rhesus monkey eyes, it is not the reduced ciliary body

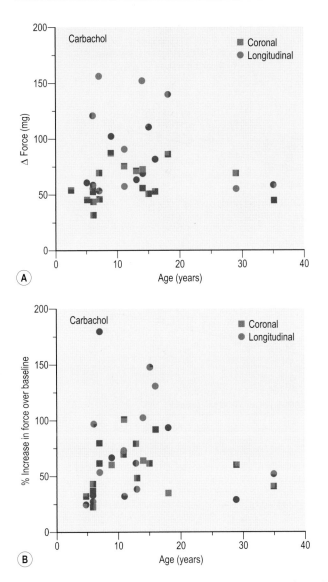

Figure 3.23 (A) Force of contraction and **(B)** percent increase in force of contraction over baseline of longitudinal (circles) and coronal (squares) strips of isolated rhesus monkey ciliary muscle strips stimulated with muscarinic agonists aceclidine (50 μM) or carbachol (1 μM). No age dependence in force of contraction of the isolated ciliary muscle was observed. (Graphs adapted with data from Poyer JF, Kaufman PL, Flügel C. Curr Eye Res 1993; 12:413–22, with permission from Swetz & Zeitlinger Publishers.)

movements that limits accommodative amplitude in rhesus monkeys.

Age-related changes in human ciliary muscle

With increasing age, the human ciliary muscle shows a loss of muscle fibers and an increase in connective tissue.[125,130,131] Despite this, studies using impedance cyclography[132,133] or modeling[134] to indirectly infer force of contraction of the human ciliary muscle, suggest that human ciliary muscle contractile force does not decrease, but indeed may increase and reach a maximum at the age at which presbyopia is manifest. These findings are consistent with observations of continued accommodative movement of the ciliary body with accommodative effort in human presbyopes and

pseudophakic eyes.[76,135,136] Histological study of the atropinized human ciliary muscle shows that total area, area of longitudinal and reticular portion, and length of the muscle decrease with age. In addition, there is a decrease in ciliary ring diameter[76] and the inner apex of the unaccommodated ciliary muscle resides further forward and inward towards the anterior–posterior axis in the aging eye so that the configuration of the older unaccommodated ciliary muscle appears more like that of the young accommodated ciliary muscle[125] (Fig. 3.25). Whether this is a cause or a consequence of the anterior zonule gradually pulling the ciliary muscle inward is not clear. Based on this result it has been suggested that, at rest, the aged human ciliary muscle may be less able to hold or pull the crystalline lens into its flattened and unaccommodated configuration.[121]

Age-related changes in the zonule

The anterior zonular attachment all around the equatorial region of the lens serves a fundamental role in the accommodative process. It is the outward force directed through these zonular fibers and the resulting tension on the lens capsule that maintains the unaccommodated lens in its flattened state. The release of this resting tension during accommodation allows the capsule to mold the lens into its more spherical and accommodated form.[71] Thus any age-related changes that may affect this zonular attachment are likely to impact the accommodative process and so may contribute to presbyopia. This is a very fine, delicate network of fibers and is especially difficult to study, and thus relatively few studies have been done on this tissue. Zonular spring constants determined indirectly by stretching human tissues show no correlation with age.[137] Scanning electron microscopic studies of human eyes over a range of ages show an anterior zonular/capsular shift on the lens with increasing age.[138] The distance from the zonular/capsule insertion to the lens equator increases, the distance from the zonular/capsule insertion to the ciliary processes is unchanged, the circumlental space decreases with age (see below), and the rate of increase in distance from zonular/capsular insertion to lens equator remains relatively constant until the 5th decade and then increases dramatically.[138] Based on the constancy of the distance between the zonular/capsular insertion and the ciliary body, it is suggested that there would be no change in zonular length or zonular tension with increasing age provided zonular elasticity remains unchanged.

It is therefore possible that the decreased circumlental space described above is due to centripetal pulling of the ciliary body by the zonule as the zonule/capsular shift occurs with increasing lens thickness or due to the inward expansion of the ciliary muscle.[125] The decreased circumlental space is not due to an age-related increase in equatorial diameter of the lens since lens diameter does not increase systematically with age (see below). Farnsworth & Shyne[138] theorized that the anterior zonular shift occurs because the capsule is thinner on the posterior lens surface and is stretched more than the anterior capsule as the lens continues to grow within the capsule. A zonular/capsular shift could reasonably occur since the posterior capsule is thinner[70,71] and therefore likely to stretch more than the anterior capsule as lens thickness increases. As a consequence of this zonular/capsular shift, in older eyes the attachment

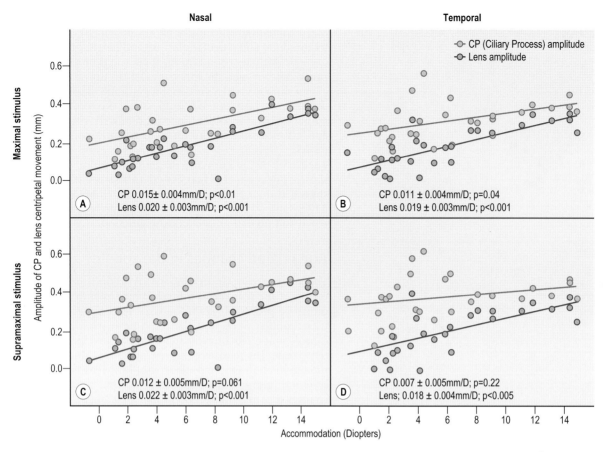

Figure 3.24 Graphs showing the accommodative excursions of the ciliary processes (solid symbols) and lens edge (open symbols) from the eyes of iridectomized rhesus monkeys. Quantitative goniovideographic analysis of the accommodative movements from the nasal (left) and temporal (right) sides of the monkey eyes for Edinger–Westphal stimulated accommodative responses to either maximal (top) or supramaximal (bottom) stimulus amplitudes. Accommodative movements of the lens and ciliary processes are reduced with the age-related loss of accommodative amplitude, but in all eyes, the magnitude of ciliary process movement exceeds the magnitude of the lens edge movements. With supramaximal stimulation there is a greater ciliary process movement than with the stimulus required to elicit maximum accommodation, and more so in the older eyes. This demonstrates that although increasing ciliary body movements can be produced even in the oldest eyes, this is without an increase in accommodative lens movements. (From Croft MA, Glasser A, Heatley G, et al. Invest Ophthalmol Vis Sci 2006; 47:1076–86. Reproduced with permission from The Association for Research in Vision and Ophthalmology.)

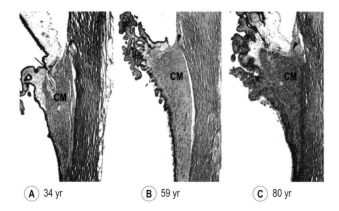

Figure 3.25 Age-related changes in the configuration of the atropinized human ciliary muscle. Histologic sections from (**A**) a 34-year-old, (**B**) a 59-year-old and (**C**) an 80-year-old human donor eye. The aging, atropinized human ciliary muscle looks more like a young accommodated ciliary muscle with the inner apex of the ciliary muscle moving forward and towards the axis of the eye. (Reprinted from Tamm S, Tamm E, Rohen JW. Mechanisms Ageing Development 1992; 62:209–21, with permission from Elsevier Science Ltd.)

of the anterior zonular fiber to the lens is anterior to the equator, the fine zonular fibers which reside at the lens equator in the young eye are found anterior to the equator, and there are fewer zonular fibers at the equator.[138] This would result in diminution of the outward directed force on the lens equator by the anterior zonular fibers as a whole and is suggested as a contributing factor in the age related loss of accommodation.[138] Measurements on unfixed human eyes from which the lens substance was removed by phaco-emulsification also show an age-dependent increase in the distance from the anterior zonular/capsular insertion to the equatorial edge of the capsular bag, a decrease in circumlental space, and an age-dependent increase in the distance from the anterior zonular/capsular insertion to the ciliary body.[139] However absence of the lens substance from the capsular bag complicates interpretation of these measurements.

Age-related changes in the capsule

The thickness of the anterior lens capsule has been reported to increase from about 11 microns at birth to approximately 20 microns at 60 years of age and then decreases slightly

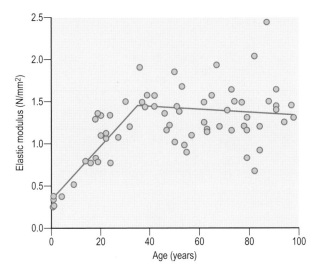

Figure 3.26 Age-related changes in the elastic modulus of the human anterior lens capsule in the 0–10 percent range of strain, the level of strain relevant for accommodation. Data are obtained from stretching isolated rings from the anterior capsular surface. (From Krag S, Andreassen TT. Prog Retin Eye Res 2003; 22:749–67. Reproduced with permission from Elsevier Science Ltd.)

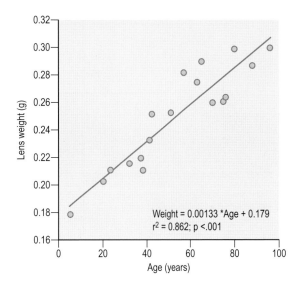

Figure 3.27 The human lens continues to grow throughout life as evident from an increase in mass of the isolated human lens. The wet weight of 18 human lenses was measured after isolating the lenses from human eye-bank eyes ranging in age from 5 to 96 years of age. (Reprinted from Glasser A, Campbell MCW. Vision Res 1999; 39:1991–2015, with permission from Elsevier Science Ltd.)

thereafter.[140] Krag[141] found an increase from 11 to 33 microns up to age 75 and then a slight decrease thereafter. More recent results show an increase in thickness of the anterior mid peripheral capsule with increasing age and a thinning of the mid peripheral posterior capsule.[70] Fisher[140] measured the extensibility of the human lens capsule by applying a fluid pressure behind the central part of the anterior lens capsule clamped between two rings and found it to be 29 percent and age independent. Despite the increased capsular thickness with increasing age, Fisher[140] showed a decrease in Young's modulus of elasticity of the capsule from 6×10^7 dyn/cm^2 in infancy to 2×10^7 dyn/cm^2 in old age. Fisher suggests that the force that can be transmitted per unit thickness of the capsule decreases by half by age 60, but that the increased thickness offers some compensation for the loss of elasticity. Age-related change in lens volume may account for the variations in thickness of the capsule at a specified region on the lens. Krag et al's[141] measurements of extensibility of a ring cut from the anterior capsule show that while the young capsule can be stretched to 108 percent of its unstretched length, there is a linear decline in strain to 40 percent at age 98. The force required to break the capsular ring remained constant until age 35 and decreased linearly thereafter. In the 0–10 percent strain level, the level of strain relevant for accommodation, there is a linear increase in elastic modulus of the anterior lens capsule up to about 35 years of age and a slight decrease after this age (Fig. 3.26). With increasing age the capsule gets thicker, less extensible and more brittle.[141,142]

Growth of the crystalline lens

The crystalline lens continues to grow throughout life. In humans, this results in a linear increase in mass of the isolated lens between age 5 and 96 years of age[72] (Fig. 3.27). The human lens also undergoes an increase in axial thickness as a consequence of the addition of lens fiber cells.[45,143–145]

However, lens equatorial diameter in the unaccommodated state does not increase systematically with increasing age[74,76,77,145] (Figs 3.28 & 3.29). Scheimpflug slit-lamp phakometry and MRI measurements show that the anterior lens surface curvatures increase systematically with increasing age, but posterior lens surface curvature is found to increase in some studies and not in others.[38,145,146] The increase in axial thickness of the lens results in a decrease in anterior chamber depth and an increase in anterior segment length[145] (Fig. 3.30). The distance between the cornea and the center of the lens does not change with increasing age.[147] Qualitatively similar age-related changes have been described in the anterior segment of the rhesus monkey eye.[148] Since the thickness and anterior and posterior surface curvatures of the lens increase with increasing age, but without an increase in lens equatorial diameter, the external shape of the aged lens begins to look more like that of an accommodated lens. However, the increased axial thickness with age is due to an increase in thickness of the anterior and posterior cortex whereas accommodation in a young lens is due to an increase in thickness of the nucleus.[29,146,147] In addition, with increasing age, the cortical layers of the lens increase to a greater extent than the nucleus.[147] Although the surface curvatures in the aging lens appear to be more like an accommodated lens, the presbyopic eye is clearly not focused for near since presbyopia results in a loss of near vision.

The incongruity between the increasing lens surface curvatures and the gradual loss of near vision has been termed the *lens paradox*.[149] The reason the presbyopic eye does not become nearsighted despite the increasing lens surface curvatures is due to a gradual age-related decrease in the equivalent refractive index of the lens.[75,145,150] The optical changes in the lens with accommodation and aging differ in several respects. While accommodation results in an increase in the extent of the negative spherical aberration of young lenses,

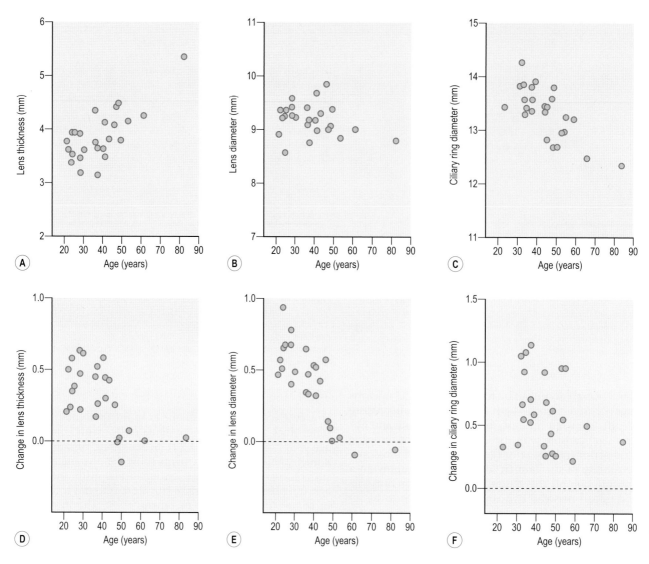

Figure 3.28 MRI measurements of resting, unaccommodated lens thickness (**A**), lens diameter (**B**) and ciliary ring diameter (**C**) in human eyes as a function of age. Subjects viewed a near target (8 diopters of accommodative demand) and MRI measurements were repeated to determine the change in lens thickness (**D**), the change in lens diameter (note: the change in lens diameter is graphed as positive values, but this actually represents a decrease in lens diameter with accommodation) (**E**) and the change in ciliary ring diameter (**F**) with accommodation. The accommodative changes in lens thickness and lens diameter reach zero (horizontal lines) by approximately 50 years of age, while the accommodative change in ciliary ring diameter, although reduced with increasing age, does not. (From Strenk SA, Semmlow JL, Strenk LM et al. Invest Ophthalmol Vis Sci, 1999; 40:1162–9. Reproduced with permission from The Association for Research in Vision and Ophthalmology.)

aging results in a systematic change in sign of spherical aberration from negative to positive.[54,72] The increased axial thickness of the lens would also tend to reduce the lens refractive power if no other changes occurred in the lens. Although the refractive index value at the center of the lens does not change systematically with increasing age, the shape of the lens gradient refractive index changes with increasing age – resulting in a larger plateau in the refractive index in the central regions of the lens and a steeper change in the gradient refractive index in the more peripheral lens cortical regions[41] (Fig. 3.31).

It has been suggested that lens equatorial diameter increases systematically with increasing age[94,151,152] and this has been suggested to be the primary cause of presbyopia.[153] The only evidence cited in support of an age-related increase in lens diameter is data from an original study by Priestly

Smith in 1883 in which lens equatorial diameter was measured in isolated human lenses.[154] Smith and others[54,71,72] recognized that when the zonular fibers are cut and the lens is isolated and removed from the eye and from external zonular traction forces, younger lenses tend to become accommodated and undergo a decrease in equatorial diameter while older presbyopic lenses do not undergo any change in shape upon removal from the eye.[54,71,72] Therefore, the diameter of isolated lenses is relatively smaller in young isolated lenses than in older isolated lenses. Measurements of lens equatorial diameter of isolated lenses, therefore compare young accommodated lenses with older unaccommodated lenses. Although such studies show an age-dependent increase in lens diameter, this trend is not due to age, but is due to the different accommodative states of isolated lenses.[154] Measurements from isolated adult lenses,

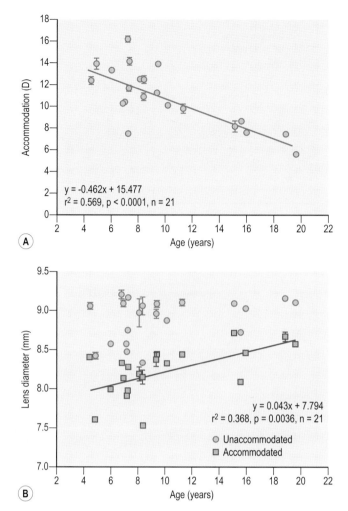

Figure 3.29 In rhesus monkeys in which (**A**), there is an age-related decrease in accommodative amplitude, (**B**), there is no systematic age-related change in unaccommodated lens equatorial diameter (circles) and accommodative decrease in lens diameter decreases with increasing age (squares). (Wendt M, Croft MA, McDonald J, et al. Exp Eye Res 2008; 86:746–52. Reproduced with permission from Elsevier Science Ltd.)

therefore, do not reflect the equatorial diameter of the lens in vivo. Recent studies have measured lens thickness and equatorial diameter in the living human[74,76,145] and monkey[77] eye. These studies show that although lens growth occurs as is evident from an increase in lens axial thickness, the progression of presbyopia occurs without a systematic increase in lens equatorial diameter.

Loss of ability of the human lens to accommodate

In vitro studies show that the human lens progressively loses the ability to undergo accommodative changes. Fisher subjected isolated human lenses to high-speed rotational forces designed to simulate the forces that act on the lens to hold the lens in an unaccommodated state in the living eye.[155] The results show an age-dependent decline in deformability of the human lens. For a given rotational stress, the equatorial and polar strain (change in lens diameter and thickness) decrease by about one-third between the ages of 15 and 65.

Fisher's calculated Young's modulus of polar and equatorial elasticity show a more than three-fold increase over this age range.[155] From these studies Fisher suggested that a decrease in elastic modulus of the capsule, an increase in elastic modulus of the lens substance and a flattening of the lens are sufficient to account for the loss of accommodation by the age of 61 years.[156] However Fisher's assumptions of no age-related change in lens shape or zonular insertion onto the lens are inaccurate[138,146] and the theoretical assumptions on which the calculations are based are questioned.[157] Although Fisher's calculated Young's modulus values may be inaccurate, his experiments do demonstrate reduced deformability of the aging human lens due to rotational forces.

Mechanical stretching experiments show that young human lenses undergo 12–16 D of changes in optical power from stretching forces applied through the intact zonular apparatus.[54] These mechanically induced changes in optical power of the lens correspond well with the accommodative dioptric change in power of the young eye in vivo. The change in optical power that the human lens undergoes with mechanical stretching gradually decreases with increasing age[54,134] (see Fig. 3.12). By about 55 years of age, human lenses are unable to undergo any change in optical power with the same degree of stretching that produces 12–16 D of change in optical power of young lenses.[54] Phakoemulsification and aspiration of the presbyopic lens and injection of a soft silicone polymer to refill the capsule restores the ability of the refilled lens to undergo accommodative changes in optical power with mechanical stretching.[85] These experiments show that regardless of what other age-related changes may occur in the human accommodative apparatus, the human lens ultimately completely loses the ability to undergo accommodative optical changes and that this is due to the increased stiffness of the lens. Presbyopic human lenses are unable to undergo any change in optical power from forces exerted on the lenses either through zonular traction or rotational forces.[54,155]

After the zonule is cut, and the young lens is removed from the eye, the young isolated lens is in a maximally accommodated form due to the forces exerted on the lens by the lens capsule.[71,72,154] When the lens capsule is removed from the isolated young lens, the decapsulated lens substance takes on an unaccommodated form[71,72] (Fig. 3.32). Removing the capsule from young lenses results in a decrease in optical power. However, removing the capsule from lenses over about 50 years of age results in no change in optical power.[72] This, together with the results from mechanical stretching experiments,[54] shows that the lens substance of older human lenses ultimately is incapable of undergoing capsule-induced optical alterations required for accommodation and disaccommodation.

High resolution magnetic resonance imaging (MRI) studies in living human eyes provide insights on aging of the accommodative apparatus.[76] When subjects are presented with an 8D accommodative stimulus, in the young subjects there is an accommodative increase in lens thickness and decrease in lens equatorial diameter, but with increasing age, these accommodative changes in the lens decline to zero by about age 50 (see Fig 3.28). However, an accommodative decrease in ciliary ring diameter still occurs even in the oldest subjects. This shows that in presbyopes, an effort to focus at near results in a ciliary muscle contraction

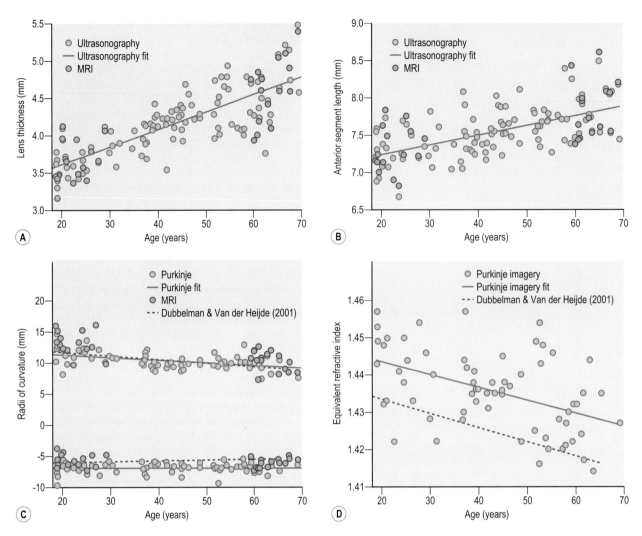

Figure 3.30 Age-related changes in the human lens. (**A**) Lens axial thickness as measured with A-scan ultrasound (red symbols) and MRI (blue symbols) increases with increasing age. (**B**) Anterior segment length as measured with A-scan ultrasound (red symbols) and MRI (blue symbols) increases with increasing age. (**C**) Lens anterior radius of curvature as measured from phakometry (red symbols) and MRI (blue symbols) decreases with increasing age, but lens posterior radius of curvature does not change systematically. (**D**) Calculated lens equivalent refractive index decreases with increasing age. (From Atchison DA, Markwell EL, Kasthurirangan S, et al. J Vis 2008; 8:29–30. Reproduced with permission from The Association for Research in Vision and Ophthalmology.)

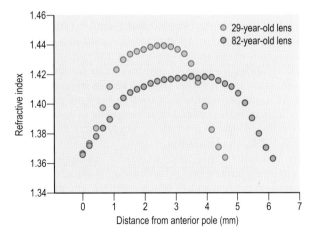

Figure 3.31 Since lens thickness increases systematically, but the refractive index at the center of the lens does not change systematically, the overall form of the gradient refractive index of the lens changes with increasing age. The refractive index profile becomes steeper near the lens surfaces and it becomes flatter near the lens center. (From Moffat BA, Atchison DA, Pope JM. Optom Visi Sci 2002; 79:148–50.)

and accommodative movements of the ciliary body, but without the required accommodative changes in the lens. Presbyopia, therefore results in a failure of the crystalline lens to undergo accommodative changes.

Age-related increase in stiffness of the human lens

The human lens undergoes an exponential age-related increase in stiffness.[72,158,159] In the young lens, the nucleus is softer than the cortex, but with increasing age, there is a relatively greater increase in the stiffness of the cortex[158,159] (Fig. 3.33). In the young eye, accommodation occurs through an increase in thickness of the lens nucleus, but not the cortex. In the young eye, it may be that the relatively stiffer cortical layers surrounding the nucleus mold the nucleus to change its shape during accommodation. With increasing age as the rate of stiffening of the nucleus exceeds that of the cortex, the cortical stiffness begins to exceed that of the nucleus by about 35 years of age.[158] There is a stiffness gradient to the lens (Fig. 3.34). In the young lens, stiffness progressively

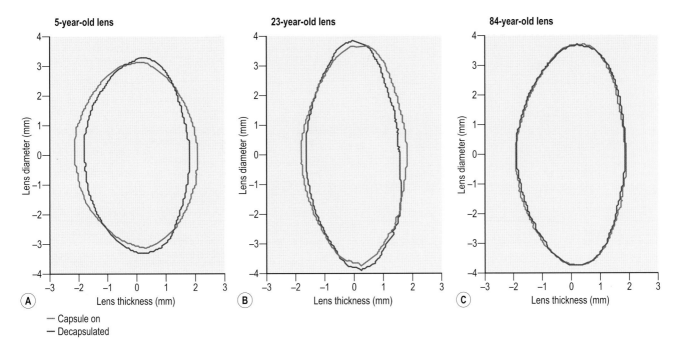

5-year-old lens

23-year-old lens

84-year-old lens

— Capsule on
— Decapsulated

Figure 3.32 Effect of removing the capsule (blue profile) on (**A**) 5-year-old, (**B**) 23-year-old, and (**C**) 84-year-old isolated human lenses. In the youngest lens, the capsule causes the isolated lens to become maximally accommodated. Removing the capsule results in the isolated lens substance taking on a more unaccommodated form. This effect of the capsule is reduced in the intermediate aged lens and removal of the lens capsule has no effect on the shape of the oldest lens. (From Glasser A, Croft MA, Kaufman PL. In: Friedlander MH, ed. International Ophthalmology Clinics. Philadelphia, PA: Lippincott Williams & Wilkins, 2001; vol. 41(2), ch. 1.)

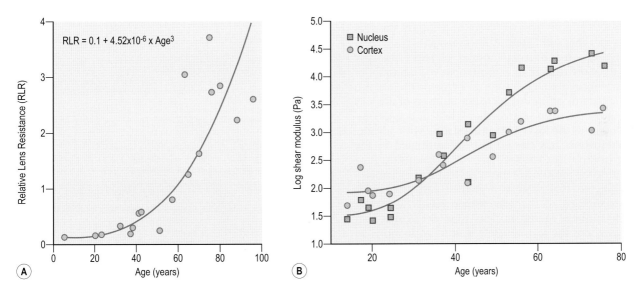

$RLR = 0.1 + 4.52 \times 10^{-6} \times Age^3$

Figure 3.33 Age-related increase in stiffness of human lenses. (**A**) Nineteen human lenses ranging in age from 5 to 96 years show an exponential increase in stiffness with an over four-fold increase in stiffness over the human life-span which continues well after the age at which accommodation is completely lost. (**B**) The nuclear regions of young lenses have a lower sheer modulus than the cortex, but there is an exponential increase in stiffness of both the nucleus and the cortex of the lens with a relatively greater increase in sheer modulus of the nucleus. (Panel A from Glasser A, Campbell MCW. Vision Res 1999: 39:1991–2015, with permission from Elsevier Science Ltd. Panel B from Heys KR, Cram SL, Truscott RJ. Mol Vis 2004; 10:956–63.)

decreases from the surface of the cortex to the center of the nucleus, whereas in older lenses the stiffness gradient increases from the surface of the cortex to the center of the nucleus. It may be this change in stiffness gradient that leads to an age-related change in the fine balance of forces that ultimately results in a loss of accommodation.[159,160] It is an essential requirement that if accommodation is to occur, the lens substance must remain sufficiently pliable for capsular and zonular forces to flatten the lens to hold it in an unaccommodated form and for capsular forces to increase the surface curvatures to mold the lens into an accommodated form. Since accommodation relies on forces exerted by the capsule on the lens, a small change in the fine balance of capsular and lens nuclear and cortical elastic forces would

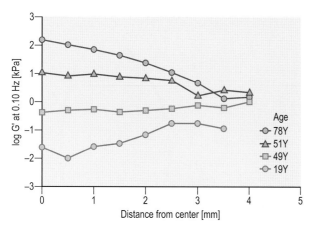

Figure 3.34 In younger human lenses, lens stiffness is lowest nearer the center of the nucleus and lens stiffness increases towards the periphery of the cortex. With increasing age, there is a progressive change in stiffness gradient of the lens. In the oldest lenses, the stiffness gradient has altered such that the center of the nucleus is stiffer than the periphery of the cortex. (From Weeber HA, Eckert G, Pechhold W, van der Heijde RGL. Graefe's Arch Clin Exp Ophthalmol 2007; 245:1357–66 with kind permission of Springer Science + Business Media.)

diminish the accommodative ability of the lens. Although presbyopia results in a complete loss of accommodation by the age of about 50 years, the stiffness of the human lens continues to increase beyond this age throughout the remaining years of the human life-span. The age-related stiffening of the lens may simply represent a continuation of the age-related changes that ultimately lead to cataract.

The progressive loss of compliance of the lens from early in life parallels the decline in accommodative amplitude. If no other age-related changes were to occur in the accommodative apparatus of the eye, increased stiffness of the lens and a change in the stiffness gradient of the lens could completely account for the loss of accommodation with advancing age. Presbyopia has classically been attributed to the hardening or "sclerosis" of the lens. Some confusion may exist regarding the meaning of the term "lenticular sclerosis". Evidence suggests that there is a gradual and progressive increase in stiffness of the lens with a relatively greater increase in stiffness of the nucleus than the cortex that occurs throughout life to ultimately lead to an inability of the lens to undergo the optical changes required for accommodation.

References

1. Helmholtz von HH. Handbuch der Physiologishen Optik, 3rd edn. Vol. 1. Menasha, Wisconsin: The Optical Society of America, 1909.
2. Young T. On the mechanism of the eye. Phil Trans R Soc Lond 1801; 91:23.
3. Home E. The Croonian lecture on muscular motion. Phil Trans R Soc Lond 1795; 85:1.
4. Crampton P. The description of an organ by which the eyes of birds are accommodated to the different distances of objects. Thompson's Annals of Philosophy 1813; 1:170.
5. Müller H. Ueber den accommodations-apparat im auge der vogel, besonders der falken. Archiv für Ophthalmologie 1857; 3:25.
6. Glasser A, Troilo D, Howland HC. The mechanism of corneal accommodation in chicks. Vision Res 1994; 34:1549.
7. Glasser A, Murphy CJ, Troilo D et al. The mechanism of lenticular accommodation in the chick eye. Vision Res 1995; 35:1525.
8. Glasser A, Howland HC. A history of studies of visual accommodation in birds. Q Rev Biol 1996; 71:475.
9. Brücke E. Ueber den musculus Cramptonianus und den spannmuskel der choroidea. Archiv für Anatomie, Physiologie und Wissenschaftliche Medicine 1846; 1:370.
10. Cramer A. Het accommodatievermogen der oogen, physiologisch toegelicht. Hollandsche Maatschappij der Wetenschappen te Haarlem 1:139. Haarlem: De Erven Loosjes, 1853.
11. Hess C. Vergleichende Untersuchungen über den Einfluss der Accommodation auf den Augendruck in der Wirbelthierreihe. Archiv fur Augenheilkunde 1909; 63:88.
12. Rohen JW, Kaufman PL, Eichhorn M et al. Functional morphology of accommodation in raccoon. Exp Eye Res 1989; 48:523.
13. Armaly MF. Studies on intraocular effects of the orbital parasympathetic pathway. I. Techniques and effects on morphology. AMA Arch Ophthalmol 1959; 61:14.
14. Vakkur GJ, Bishop PO. The schematic eye in the cat. Vision Res 1963; 3:357.
15. O'Neill WD, Brodkey JS. A nonlinear analysis of the mechanics of accommodation. Vision Res 1970; 10:375.
16. Sivak J, Howland HC. Accommodation in the northern rock bass (Ambloplites rupestris rupestris) in response to natural stimuli. Vision Res 1973; 13:2059.
17. Andison ME, Sivak JG. The naturally occurring accommodative response of the oscar, Astronotus ocellatus, to visual stimuli. Vision Res 1996; 36:3021.
18. Beer T. Die Accommodation des Fischauges. Pfluegers Archiv fuer die Gesamte Physiologie des Menschen und der Tiere 1894; 58:523.
19. Sivak JG, Hildebrand T, Lebert C. Magnitude and rate of accommodation in diving and nondiving birds. Vision Res 1985; 25:925.
20. Tornqvist G. Effect on refraction of intramuscular pilocarpine in two species of monkey (Cercopithecus aethiops and Macaca irus). Invest Ophthalmol Vis Sci 1965; 4(2):211.
21. Tornqvist G. Effect of topical carbachol on the pupil and refraction in young and presbyopic monkeys. Invest Ophthalmol Vis Sci 1966; 5(2):186.
22. Kaufman PL, Bárány EH. Subsensitivity to pilocarpine in primate ciliary muscle following topical anticholinesterase treatment. Invest Ophthalmol Vis Sci 1975; 14(4):302.
23. Bito LZ, DeRousseau CJ, Kaufman PL et al. Age-dependent loss of accommodative amplitude in rhesus monkeys: an animal model for presbyopia. Invest Ophthalmol Vis Sci 1982; 23:23.
24. Duane A. Normal values of the accommodation at all ages. J Am Med Assoc 1912; 59:1010.
25. Anderson HA, Hentz G, Glasser A et al. Minus-lens-stimulated accommodative amplitude decreases sigmoidally with age: a study of objectively measured accommodative amplitudes from age 3. Invest Ophthalmol Vis Sci 2008; 49:2919.
26. Hirschberg J. The history of ophthalmology, Ostend, Belgium: Wayenborgh Publishing, 1982.
27. Donders FC. On the anomalies of accommodation and refraction of the eye with a preliminary essay on physiological dioptrics. London: The New Sydenham Society, 1864.
28. Rabbetts RB. Bennett & Rabbetts' Clinical visual optics, 3rd edn. Boston: Butterworth Heinemann, 1998.
29. Koretz JF, Handelman GH, Brown NP. Analysis of human crystalline lens curvature as a function of accommodative state and age. Vision Res 1984; 24(10):1141.
30. Strenk SA, Strenk LM, Koretz JF. The mechanism of presbyopia. Prog Retin Eye Res 2005; 24:379.
31. Fincham EF. The changes in the form of the crystalline lens in accommodation. Trans Optical Soc 1825; 26:240.
32. Garner LF, Yap MKH. Changes in ocular dimensions and refraction with accommodation. Ophthalmic Physiol Opt 1997; 17:12.
33. Bolz M, Prinz A, Drexler W et al. Linear relationship of refractive and biometric lenticular changes during accommodation in emmetropic and myopic eyes. Br J Ophthalmol 2007; 91:360.
34. Vilupuru AS, Glasser A. The relationship between refractive and biometric changes during Edinger-Westphal stimulated accommodation in rhesus monkeys. Exp Eye Res 2005; 80:349.
35. Drexler W, Baumgartner A, Findl O et al. Biometric investigation of changes in the anterior eye segment during accommodation. Vision Res 1997; 37:2789.
36. Rosales P, Wendt M, Marcos S et al. Changes in crystalline lens radii of curvature and lens tilt and decentration during dynamic accommodation in rhesus monkeys. J Vis 2008; 8:18.
37. Rosales P, Dubbelman M, Marcos S et al. Crystalline lens radii of curvature from Purkinje and Scheimpflug imaging. J Vis 2006; 6:1057.
38. Dubbelman M, van der Heijde GL, Weeber HA. Change in shape of the aging human crystalline lens with accommodation. Vision Res 2005; 45:117.
39. Gullstrand A. Helmholtz's Treatise on physiological optics. New York: Dover, 1909.
40. Garner LF, Smith G. Changes in equivalent and gradient refractive index of the crystalline lens with accommodation. Optom Vis Sci 1997; 74:114.
41. Kasthurirangan S, Markwell EL, Atchison DA et al. In vivo study of changes in refractive index distribution in the human crystalline lens with age and accommodation. Invest Ophthalmol Vis Sci 2008; 49:2531.
42. Hermans EA, Dubbelman M, van der Heijde R et al. Equivalent refractive index of the human lens upon accommodative response. Optom Vis Sci 2008; 85:1179.
43. Duane A. Studies in monocular and binocular accommodation with their clinical applications. Am J Ophthalmol 1922; 5:865.
44. Win-Hall DM, Glasser A. Objective accommodation measurements in prepresbyopic eyes using an autorefractor and an aberrometer. J Cataract Refract Surg 2008; 34:774.
45. Koretz JF, Kaufman PL, Neider MW et al. Accommodation and presbyopia in the human eye – aging of the anterior segment. Vision Res 1989; 29:1685.
46. Ostrin LA, Glasser A. Accommodation measurements in a prepresbyopic and presbyopic population. J Cataract Refract Surg 2004; 30:1435.
47. Calver RI, Cox MJ, Elliott DB. Effect of aging on the monochromatic aberrations of the human eye. J Opt Soc Am A Opt Image Sci Vis 1999; 16:2069.
48. Wold JE, Hu A, Chen S et al. Subjective and objective measurement of human accommodative amplitude. J Cataract Refract Surg 2003; 29:1878.
49. Koretz JF, Handelman GH. A model for accommodation in the young human eye: the effects of lens elastic anisotropy on the mechanism. Vision Res 1983; 23:1679.
50. Coleman DJ, Rondeau MJ. Current aspects of human accommodation II. Heidelberg: Kaden Verlag, 2003.
51. Coleman DJ, Fish SK. Presbyopia, accommodation, and the mature catenary. Ophthalmology 2001; 108:1544.

52. Coleman DJ. On the hydrolic suspension theory of accommodation. Trans Am Ophthalmol Soc 1986; 84:846.

53. Fisher RF. Is the vitreous necessary for accommodation in man? Br J Ophthalmol 1983; 67:206.

54. Glasser A, Campbell MCW. Presbyopia and the optical changes in the human crystalline lens with age. Vision Res 1998; 38:209.

55. Manns F, Parel JM, Denham D et al. Optomechanical response of human and monkey lenses in a lens stretcher. Invest Ophthalmol Vis Sci 2007; 48:3260.

56. Glasser A, Croft MA, Brumback L et al. Ultrasound biomicroscopy of the aging rhesus monkey ciliary region. Optom Vis Sci 2001; 78:417.

57. Rohen JW. Scanning electron microscopic studies of the zonular apparatus in human and monkey eyes. Invest Ophthalmol Vis Sci 1979; 18:133.

58. Ludwig K, Wegscheider E, Hoops JP et al. In vivo imaging of the human zonular apparatus with high-resolution ultrasound biomicroscopy. Albrecht Von Graefes Arch Klin Exp Ophthalmol 1999; 237:361.

59. Tamm ER, Lütjen-Drecoll E. Ciliary body. Microsc Res Tech 1996; 33:390.

60. Hara K, Lutjen-Drecoll E, Prestele H et al. Structural differences between regions of the ciliary body in primates. Invest Ophthalmol Vis Sci 1977; 16:912.

61. Hogan MJ, Alvarado JA, Weddell JE. Histology of the human eye; an atlas and a textbook. Philadelphia: WB Saunders, 1971.

62. Lütjen-Drecoll E. Functional morphology of the trabecular meshwork in primate eyes. Prog Retin Eye Res 1998; 18:91.

63. Flügel C, Bárány EH, Lütjen-Drecoll E. Histochemical differences within the ciliary muscle and its function in accommodation. Exp Eye Res 1990; 50:219.

64. Farnsworth PN, Burke P. Three-dimensional architecture of the suspensory apparatus of the lens of the rhesus monkey. Exp Eye Res 1977; 25:563.

65. Davanger M. The suspensory apparatus of the lens. The surface of the ciliary body. A scanning electron microscopic study. Acta Ophthalmol 1975; 53:19.

66. Marshall J, Beaconsfield M, Rothery S. The anatomy and development of the human lens and zonules. Trans Opthal Soc UK 1982; 102(3):423.

67. Bernal A, Parel JM, Manns F. Evidence for posterior zonular fiber attachment on the anterior hyaloid membrane. Invest Ophthalmol Vis Sci 2006; 47:4708.

68. Farnsworth PN, Mauriello JA, Burke-Gadomski P et al. Surface ultrastructure of the human lens capsule and zonular attachments. Invest Ophthalmol Vis Sci 1976; 15(1):36.

69. McCulloch C. The zonule of Zinn: its origin, course, and insertion, and its relation to neighboring structures. Trans Am Ophthalmol Soc 1954; 52:525.

70. Barraquer RI, Michael R, Abreu R et al. Human lens capsule thickness as a function of age and location along the sagittal lens perimeter. Invest Ophthalmol Vis Sci 2006; 47:2053.

71. Fincham EF. The mechanism of accommodation. Br J Ophthalmol 1937; Monograph VIII:7.

72. Glasser A, Campbell MCW. Biometric, optical and physical changes in the isolated human crystalline lens with age in relation to presbyopia. Vision Res 1999; 39:1991.

73. Dubbelman M, van der Heijde GL, Weeber HA. The thickness of the aging human lens obtained from corrected Scheimpflug images. Optom Vis Sci 2001; 78:411.

74. Jones CE, Atchison DA, Pope JM. Changes in lens dimensions and refractive index with age and accommodation. Optom Vis Sci 2007; 84:990.

75. Dubbelman M, van der Heijde GL. The shape of the aging human lens: curvature, equivalent refractive index and the lens paradox. Vision Res 2001; 41:1867.

76. Strenk SA, Semmlow JL, Strenk LM et al. Age-related changes in human ciliary muscle and lens: A magnetic resonance imaging study. Invest Ophthalmol Vis Sci 1999; 40:1162.

77. Wendt M, Croft MA, McDonald J et al. Lens diameter and thickness as a function of age and pharmacologically stimulated accommodation in rhesus monkeys. Exp Eye Res 2008; 86:746.

78. Helmholtz von HH. Ueber die Accommodation des Auges. Archiv für Ophthalmologie 1855; 1:1.

79. Glasser A, Kaufman PL. The mechanism of accommodation in primates. Ophthalmology 1999; 106:863.

80. Graves B. Change of tension on the lens capsules during accommodation and under the influence of various drugs. Br Med J 1926; 1:46.

81. Graves B. The response of the lens capsules in the act of accommodation. Trans Am Ophthalmol Soc 1925; 23:184.

82. Glasser A, Croft MA, Kaufman PL. International ophthalmology clinics. Philadelphia, PA: Lippincott Williams & Wilkins, 2001.

83. Roorda A, Glasser A. Wave aberrations of the isolated crystalline lens. J Vis 2004; 4:250.

84. Vilupuru AS, Roorda A, Glasser A. Spatially variant changes in lens power during ocular accommodation in a rhesus monkey eye. J Vis 2004; 4:299.

85. Koopmans SA, Terwee T, Barkhof J et al. Polymer refilling of presbyopic human lenses in vitro restores the ability to undergo accommodative changes. Invest Ophthalmol Vis Sci 2003; 44:250.

86. Glasser A, Wendt M, Ostrin L. Accommodative changes in lens diameter in rhesus monkeys. Invest Ophthalmol Vis Sci 2006; 47:278.

87. Wilson RS. Does the lens diameter increase or decrease during accommodation? Human accommodation studies: a new technique using infrared retro-illumination video photography and pixel unit measurements. Trans Am Ophthalmol Soc 1997; 95:261.

88. Grossmann K. The mechanism of accommodation in man. Br Med J 1903; 2:726.

89. Grossmann K. The mechanism of accommodation in man. Ophthal Rev 1904; 23:1.

90. Beauchamp R, Mitchell B. Ultrasound measures of vitreous chamber depth during ocular accommodation. Am J Optom Physiol Opt 1985; 62:523.

91. Coleman DJ. Unified model for accommodative mechanism. Am J Ophthalmol 1970; 69:1063.

92. Tscherning M. Physiologic optics. Philadelphia: The Keystone, 1904.

93. Tscherning M. Physiologic optics. Philadelphia: The Keystone, 1920.

94. Schachar RA. Cause and treatment of presbyopia with a method for increasing the amplitude of accommodation. Ann Ophthalmol 1992; 24:445.

95. Schachar RA, Black TD, Kash RL et al. The mechanism of accommodation and presbyopia in the primate. Ann Ophthalmol 1995; 27:58.

96. Cheng H, Barnett JK, Vilupuru AS et al. A population study on changes in wave aberrations with accommodation. J Vis 2004; 4:272.

97. He JC, Burns SA, Marcos S. Monochromatic aberrations in the accommodated human eye. Vision Res 2000; 40:41.

98. Lopez-Gil N, Fernandez-Sanchez V, Legras R et al. Accommodation-related changes in monochromatic aberrations of the human eye as a function of age. Invest Ophthalmol Vis Sci 2008; 49:1736.

99. Buehren T, Collins MJ. Accommodation stimulus-response function and retinal image quality. Vision Res 2006; 46:1633.

100. Kruger PB, Nowbotsing S, Aggarwala KR et al. Small amounts of chromatic aberration influence dynamic accommodation. Optom Vis Sci 1995; 72:656.

101. Aggarwala KR, Nowbotsing S, Kruger PB. Accommodation to monochromatic and white-light targets. Invest Ophthalmol Vis Sci 1995; 36:2695.

102. Aggarwala KR, Kruger ES, Mathews S et al. Spectral bandwidth and ocular accommodation. J Opt Soc Am A 1995; 12:450.

103. Kruger PB, Mathews S, Aggarwala KR et al. Accommodation responds to changing contrast of long, middle and short spectral-waveband components of the retinal image. Vision Res 1995; 35:2415.

104. Croft MA, Oyen MJ, Gange SJ et al. Aging effects on accommodation and outflow facility responses to pilocarpine in humans. Arch Ophthalmol 1996; 114:586.

105. Koretz JF, Bertasso AM, Neider MW et al. Slit-lamp studies of the rhesus monkey eye. II Changes in crystalline lens shape, thickness and position during accommodation and aging. Exp Eye Res 1987; 45:317.

106. Vilupuru AS, Glasser A. Dynamic accommodation in rhesus monkeys. Vision Res 2002; 42:125.

107. Crawford K, Terasawa E, Kaufman PL. Reproducible stimulation of ciliary muscle contraction in the cynomolgus monkey via a permanent indwelling midbrain electrode. Brain Res 1989; 503:265.

108. Crawford KS, Kaufman PL, Bito LZ. The role of the iris in accommodation of rhesus monkeys. Invest Ophthalmol Vis Sci 1990; 31:2185.

109. Ostrin LA, Glasser A. Comparisons between pharmacologically and Edinger–Westphal-stimulated accommodation in rhesus monkeys. Invest Ophthalmol Vis Sci 2005; 46:609.

110. Koeppl C, Findl O, Kriechbaum K et al. Comparison of pilocarpine-induced and stimulus-driven accommodation in phakic eyes. Exp Eye Res 2005; 80:795.

111. Owens PL, Amos DM. Clinical ocular pharmacology. Boston: Butterworth-Heinemann, 1995.

112. Harris LS, Galin MA. Effect of ocular pigmentation on hypotensive response to pilocarpine. Am J Ophthalmol 1971; 72:923.

113. Win-Hall DM, Glasser A. Objective accommodation measurements in pseudophakic subjects using an autorefractor and an aberrometer. J Cataract Refract Surg 2009; 35:282.

114. Win-Hall DM, Ostrin LA, Kasthurirangan S et al. Objective accommodation measurement with the Grand Seiko and Hartinger coincidence refractometer. Optom Vis Sci 2007; 84:879.

115. Kasthurirangan S, Vilupuru AS, Glasser A. Amplitude dependent accommodative dynamics in humans. Vision Res 2003; 43:2945.

116. Seidemann A, Schaeffel F. An evaluation of the lag of accommodation using photorefraction. Vision Res 2003; 43:419.

117. Schaeffel F, Wilhelm H, Zrenner E. Inter-individual variability in the dynamics of natural accommodation in humans: relation to age and refractive errors. J Physiol 1993; 461:301.

118. Mathews S. Scleral expansion surgery does not restore accommodation in human presbyopia. Ophthalmology 1999; 106:873.

119. Bharadwaj SR, Schor CM. Acceleration characteristics of human ocular accommodation. Vision Res 2005; 45:17.

120. Hamasaki D, Ong J, Marg E. The amplitude of accommodation in presbyopia. Am J Optom Arch Am Acad Optom 1956; 33:3.

121. Bito LZ, Miranda OC. Accommodation and presbyopia. Ophthalmol Annu 1989; 103.

122. Neider MW, Crawford K, Kaufman PL et al. In vivo videography of the rhesus monkey accommodative apparatus: age-related loss of ciliary muscle response to central stimulation. Arch Ophthalmol 1990; 108:69.

123. Lütjen-Drecoll E, Tamm E, Kaufman PL. Age-related loss of morphologic responses to pilocarpine in rhesus monkey ciliary muscle. Arch Ophthalmol 1988; 106:1591.

124. Tamm E, Lütjen-Drecoll E, Jungkunz W et al. Posterior attachment of ciliary muscle in young, accommodating old, presbyopic monkeys. Invest Ophthalmol Vis Sci 1991; 32:1678.

125. Tamm S, Tamm E, Rohen JW. Age-related changes of the human ciliary muscle. A quantitative morphometric study. Mech Ageing Dev 1992; 62:209.

126. Poyer JF, Kaufman PL, Flügel C. Age does not affect contractile responses of the isolated rhesus monkey ciliary muscle to muscarinic agonists. Curr Eye Res 1993; 12:413.

127. Gabelt BT, Kaufman PL, Polansky JR. Ciliary muscle muscarinic binding sites, choline acetyltransferase, and acetylcholinesterase in aging rhesus monkeys. Invest Ophthalmol Vis Sci 1990; 31:2431.

128. Ostrin LA, Glasser A. Edinger–Westphal and pharmacologically stimulated accommodative refractive changes and lens and ciliary process movements in rhesus monkeys. Exp Eye Res 2007; 84:302.

129. Croft MA, Glasser A, Heatley G et al. Accommodative ciliary body and lens function in rhesus monkeys, I. normal lens, zonule and ciliary process configuration in the iridectomized eye. Invest Ophthalmol Vis Sci 2006; 47:1076.

130. Nishida S, Mizutani S. Quantitative and morphometric studies of age-related changes in human ciliary muscle. Jpn J Ophthalmol 1992; 36:380.

131. Pardue MT, Sivak JG. Age-related changes in human ciliary muscle. Optom Vis Sci 2000; 77:204.

132. Swegmark G. Studies with impedance cyclography on human ocular accommodation at different ages. Acta Ophthalmol 1969; 47:1186.

133. Saladin JJ, Stark L. Presbyopia: new evidence from impedance cyclography supporting the Hess–Gullstrand theory. Vision Res 1975; 15:537.

134. Fisher RF. The force of contraction of the human ciliary muscle during accommodation. J Physiol 1977; 270:51.

135. Strenk SA, Strenk LM, Guo S. Magnetic resonance imaging of aging, accommodating, phakic, and pseudophakic ciliary muscle diameters. J Cataract Refract Surg 2006; 32:1792.

136. Stachs O, Martin H, Kirchhoff A et al. Monitoring accommodative ciliary muscle function using three-dimensional ultrasound. Graefe's Arch Clin Exp Ophthalmol 2002; 240:906.

137. van Alphen GW, Graebel WP. Elasticity of tissues involved in accommodation. Vision Res 1991; 31:1417.

138. Farnsworth PN, Shyne SE. Anterior zonular shifts with age. Exp Eye Res 1979; 28:291.

139. Sakabe I, Oshika T, Lim SJ et al. Anterior shift of zonular insertion onto the anterior surface of human crystalline lens with age. Ophthalmology 1998; 105:295.

140. Fisher RF. Elastic constants of the human lens capsule. J Physiol 1969; 201:1.

141. Krag S, Olsen T, Andreassen TT. Biomechanical characteristics of the human anterior lens capsule in relation to age. Invest Ophthalmol Vis Sci 1997; 38:357.

142. Krag S, Andreassen TT. Mechanical properties of the human lens capsule. Prog Retin Eye Res 2003; 22:749.

143. Sorsby A, Leary GA, Richards M et al. Ultrasonographic measurements of the components of ocular refraction in life. Vision Res 1963; 3:499.

144. Weekers R, Delmarcelle Y, Luyckx-Bacus J, Collignon J. Morphological changes of the lens with age and cataract. Ciba Foundation Symposium 19 (new series). Elsevier, Excerpta Medica, North-Holland; Associated Scientific Publishers Amsterdam: London, New York. 1973; 25–43.

145. Atchison DA, Markwell EL, Kasthurirangan S et al. Age-related changes in optical and biometric characteristics of emmetropic eyes. J Vis 2008; 8:29.

146. Brown N. The change in lens curvature with age. Exp Eye Res 1974; 19:175.

147. Dubbelman M, van der Heijde GL, Weeber HA et al. Changes in the internal structure of the human crystalline lens with age and accommodation. Vision Res 2003; 43:2363.

148. Koretz JF, Neider MW, Kaufman PL et al. Slit-lamp studies of the rhesus monkey eye. I. Survey of the anterior segment. Exp Eye Res 1987; 44:307.

149. Koretz JF, Handelman GH. The "lens paradox" and image formation in accommodating human eyes. Topics Aging Res Europe 1986; 6:57.

150. Moffat BA, Atchison DA, Pope JM. Explanation of the lens paradox. Optom Vis Sci 2002; 79:148.

151. Weale RA. A biography of the eye: development, growth, age. London: H K Lewis, 1982.

152. Rafferty NS. The ocular lens: structure, function, and pathology. New York: Marcel Dekker, 1985.

153. Schachar RA. Presbyopia: a surgical textbook. Thorofare, NJ: SLACK Inc., 2002.

154. Smith P. Diseases of the crystalline lens and capsule: on the growth of the crystalline lens. Trans Opthal Soc UK 1883; 79.

155. Fisher RF. The elastic constants of the human lens. J Physiol 1971; 212:147.

156. Fisher RF. Presbyopia and the changes with age in the human crystalline lens. J Physiol 1973; 228:765.

157. Burd HJ, Wilde GS, Judge SJ. Can reliable values of Young's modulus be deduced from Fisher's (1971) spinning lens measurements? Vision Res 2006; 46:1346.

158. Heys KR, Cram SL, Truscott RJ. Massive increase in the stiffness of the human lens nucleus with age: the basis for presbyopia? Mol Vis 2004; 10:956.

159. Weeber HA, Eckert G, Pechhold W et al. Stiffness gradient in the crystalline lens. Graefe's Arch Clin Exp Ophthalmol 2007; 245:1357.

160. Weeber HA, van der Heijde RG. On the relationship between lens stiffness and accommodative amplitude. Exp Eye Res 2007; 85:602.

Cornea and Sclera

Daniel G. Dawson, John L. Ubels & Henry F. Edelhauser

Introduction

The outermost, fibrous tunic of the human eye is the cornea and the sclera (Fig. 4.1A,B).[1-5] Both are soft connective tissues designed to provide structural integrity of the globe and to protect the inner components of the eye from physical injury. The clear, transparent cornea (Fig. 4.1A,C) covers the anterior 1/6th of the total surface area of the globe, while the white, opaque sclera (Fig. 4.1A) covers the remaining 5/6ths. The cornea and the lens are the eye's primary refractive structures and both have two key optical properties to this end – refractive power (light refraction) and transparency (light transmission). The presence of a healthy cornea is essential for good vision as it is basically the window of the eye. The cornea is most analogous to the external lens of a compound lens camera. By comparison, the sclera predominantly serves more of a biomechanical function and is analogous to the housing of the camera and lens.

The cornea is 540 to 700 μm thick and is arranged in five basic layers – epithelium, Bowman's layer, stroma proper, Descemet's membrane, and endothelium (Fig. 4.1D); each having distinctly different structural and functional charactistics.[6] It also is composed of three major cell types – epithelial, stromal keratocyte, and endothelial cells. Two of these, epithelium and endothelium, form cellular barrier layers to the stroma. Thus, their resistance to diffusion of solutes and bulk fluid flow are of considerable importance to maintaining normal corneal function (resistance to diffusion of solutes and fluid flow: epithelium [2000] >> endothelium [10] > stroma [1]). All 3 can replicate through mitosis, but vary considerably in their in vivo proliferative capacity with epithelial cells having the highest rates of cell division and endothelial cells being the least renewable. This fact is seen clinically since epithelial cells can completely regenerate after injury (e.g. corneal abrasions), while endothelial cells, as a result of limited in vivo proliferation, are most commonly involved in age-related (e.g. Fuchs' dystrophy) or injury-related disease (e.g. pseudophakic bullous keratopathy – PBK) ultimately resulting in corneal edema and bullous keratopathy. Corneal stromal keratocytes are a middle ground compromise between these two extremes. One major disadvantage of the epithelium's high proliferative potential is that it can occasionally go unchecked, resulting in cancer (e.g. squamous cell cancers of the cornea), whereas keratocytes and endothelial cells do not have this risk.

The sclera is 0.3 to 1.35 mm thick and is arranged into three layers – episclera, scleral stroma proper, and lamina fusca; each having distinctly different structural and functional characteristics. The sclera is composed of only one major cell type, the sclera fibrocyte, which has moderate proliferative potential, like that of the corneal stromal keratoctye. As the sclera has no cellular barrier layers, its permeability properties are quite similar to that of the corneal stroma. The sclera is an excellent example of a tissue made for biomechanical stability as it is stiff, strong, and tough. As such, disease of the sclera commonly results in loss of tectonic support, abnormal size due to dysregulated growth as well as inflammatory conditions that commonly also affect the joints in the body. Many animals have a very rigid sclera that is often supported by bone or cartilage. Humans deviate from this extreme in that their sclera is a less rigid fibrous connective tissue, perhaps reflecting its need to maintain an even blood flow to the choroid and retina during large excursions in eye motility.

This chapter reviews and explains the structure and function of the cornea and sclera, providing a framework for understanding normal health and disease of each tissue with an emphasis on function.

Cornea

Embryology, growth, development, and aging

The foundations of contemporary corneal embryology and development stem from pioneering embryonic chick studies. Since few such detailed studies have been carried out in primates or other mammals, the information that follows has potential gaps in knowledge since species-related differences may exist. Following lens vesicle formation between 4 to 5 weeks' gestation (27–36 days), surface ectodermal cells cover the defect left by lens vesicle invagination and become the primitive, undifferentiated corneal epithelium, composed initially of two cell layers.[7] Similar to the avian cornea, the primitive corneal epithelium of primates and higher mammals immediately produces a primary acellular corneal stroma, or post-epithelial layer.[8,9] In primates, this is seen as the gradual subepithelial addition of diagonal and then randomly oriented fibrillar elements, which later thicken into collagen fibrils that are slightly smaller in diameter than

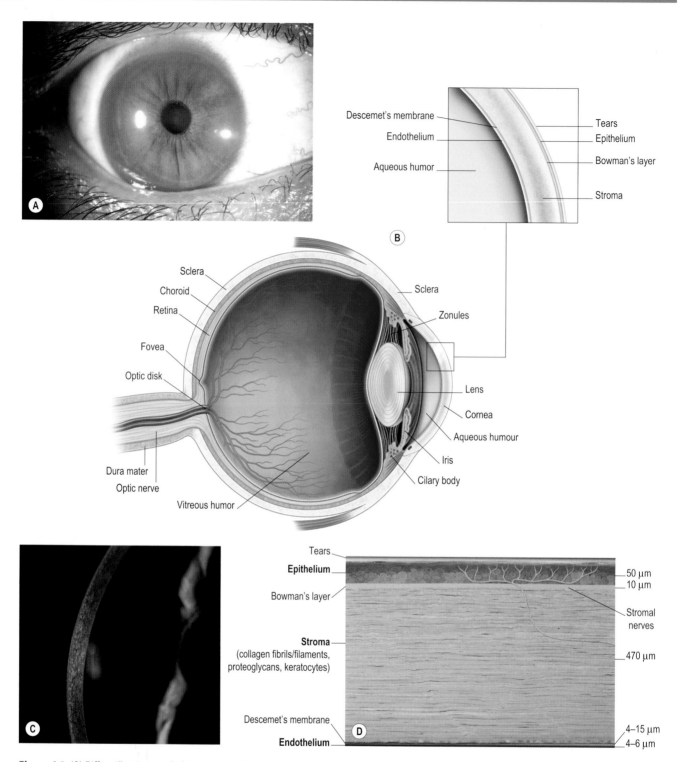

Figure 4.1 (**A**) Diffuse illumination slit-lamp image of the human cornea and sclera. (**B**) Main anatomical components of the globe with detailed emphasis on the corneal and scleral components. (**C**) Slit-beam illumination slit-lamp image of the human cornea shows an optical section of the tissue. Notice the slight light scattering that occurs in the tissue, mainly from cellular components in cornea. (**D**) Histologic section of the human cornea labeling the five main cellular and extracellular matrix layers (toluidine blue ×25).

stromal collagen fibrils.[9] The Bowman's layer is thought to be a distinct, dense anterior-most remnant of this embryologic layer, which is first detectable by light microscopy around 20 weeks' gestation.[9–11] In contrast to humans, lower mammals, such as rabbits and rodents, have an underdeveloped primary acellular corneal stroma and its residual

remnant, the Bowman's layer, is indistinct and rather thin.[12] Around 12 weeks' gestation, a time period between eyelid fusion at 8 weeks' gestation and eyelid opening at 26 weeks' gestation, the epithelium differentiates to become a stratified, squamous epithelium 4 cell layers thick, which then produces an epithelial basement membrane. It remains 4 cell layers thick until approximately 6 months after birth when it reaches adult levels of 4–6 cell layers thick. Early in gestation, the epithelial basement membrane and anchoring complexes on the basal surface of the epithelium are absent. Rudimentary epithelial basement membrane and anchoring complexes only become detectable by 17 weeks of gestation. With further development in utero, the thickness and number of these structures gradually increases.

A first wave of neural crest-derived mesenchymal cells begins to extend beneath the corneal epithelial cells from the limbus around 5 weeks' gestation (33 days); these cells form the primitive endothelium. The primitive endothelium is initially composed of two-cell layers. By 8 weeks' gestation, it becomes a monolayer that starts to produce Descemet's membrane, which becomes recognizable on light microscopy at 3–4 months' gestation. The epithelium and endothelium remain closely opposed until 7 weeks' gestation (49 days), when a second wave of mesenchymal cells begins to migrate centrally from the limbus between the epithelium and endothelium invading below and into most of the primary acellular stroma. The cells do not enter the anterior-most 10 μm of stroma, which lacks keratocan, a proteoglycan core protein signal thought to be required for cellular invasion.[10] This second wave of cells forms the stroma proper, or secondary cellular corneal stroma, as production of lamellar collagen begins within a few days in a posterior-to-anterior fashion. It is believed that the invading mesenchymal cells, destined to become keratocytes, use the primary acellular stroma as a scaffold, primarily in the anterior third of stroma proper. This concurs with the significant lamellar interweaving and oblique lamellar orientation in the anterior third of post-natal human corneal stroma as well as the fact that each successive lamellar layer is rotated 1–2 degrees clockwise. This directional rotation is the same in both right and left eyes. In the posterior two-thirds of the stroma, the corneal stroma is composed of essentially orthogonal lamellae. By 3 months' gestation, corneal nerves invade the stroma and eventually penetrate through the Bowman's layer so that nerve endings develop in the epithelium. Studies also suggest that by 5 months' gestation, tight junctions form around all the corneal endothelial cells and, by 5 to 7 months' gestation, the in utero cornea becomes transparent as the density of functioning endothelial Na^+/K^+-ATPase metabolic pump sites increase to adult levels.[13,14] By 7 months' gestation, the cornea resembles that of the adult in most structural characteristics other than size. At birth in the full-term infant, the horizontal corneal diameter is about 9.8 mm and the corneal surface area is around 102 mm². The cornea of the newborn infant is approximately 75–80 percent of the size of the adult human cornea (Fig. 4.2A,B,C), while the posterior segment is < 50 percent of adult size (Fig. 4.2D).[15] At birth, the cross-sectional thickness of the 4-cell layer epithelium averages 50 μm, the Bowman's layer averages 10 μm, the central stroma proper averages 500 μm, Descemet's membrane averages 4 μm, and the endothelium averages 6 μm thick (total mean central corneal thickness ~ 570 μm).[16–18]

During infancy, the cornea continues to grow, reaching adult size around 2 years of age with a horizontal diameter of 11.7 mm, surface area of 138 mm², anterior surface curvature of 44.1 D (Fig. 4.2A), and mean central corneal thickness of 544 μm (Fig. 4.2C).[15,17,18] Thereafter, it changes very little in size, shape, transparency, or curvature, although a

shift from with-the-rule to against-the-rule astigmatism has been associated with aging (Fig. 4.2B).[19–22] Post-natal aging is associated with several structural changes to the corneal tissue including: (1) epithelial basement membrane growth or thickening of an additional 100 to 300 nm or a rate of approximately 30 nm per decade of life; (2) decreased keratocyte, sub-basal nerve fiber, and endothelial cell density, presumably from stress-induced premature senescence; (3) increased stiffness, strength, and toughness of the stroma, from enzymatic maturation- and non-enzymatic age-related glycation-induced cross-linking of collagen fibrils; (4) Descemet's membrane thickening of an additional 6–11 μm or a rate of approximately 1 μm per decade of life; and (5) possible degeneration of extracellular matrix structures.[19,20,23] These structural and cellular changes, however, minimally affect the optical and barrier functions of the cornea and perhaps improve the mechanical function. For example, corneal ectasia from natural causes, like keratoconus, is rarely seen after 40 years of age. Only three documented negative functional consequences are associated with aging – impaired corneal wound healing, decreased corneal sensation, and decreased extensibility of its tissue.[19–31] In elderly individuals or younger individuals with lipid abnormalities, the cornea often becomes yellowish in the periphery of the cornea due to a fine deposition of lipid. This condition is called arcus senilis.

Major corneal reference points and measurements

When viewed anteriorly in the living eye, the adult human cornea appears elliptical (Fig. 4.3A) as the largest diameter is typically in the horizontal meridian (mean 11.7 mm) and the smallest is in the vertical meridian (mean 10.6 mm).[1] This elliptical configuration is brought about by anterior extension of the opaque sclera superiorly and inferiorly. When viewed from the posterior surface, the cornea is actually circular (Fig. 4.3A), with an average horizontal and vertical diameter of 11.7 mm. The average radius of curvature of the anterior and posterior corneal surface is 7.8 mm and 6.5 mm, respectively, which is significantly less than the 11.5 mm average radius of curvature of the sclera. This results in a small 1.5–2 mm transition zone that forms an external and internal surface groove, or scleral sulcus, where the steeper cornea meets the flatter sclera (Fig. 4.3A). This sulcus typically is not obvious clinically because it is filled in by overlying episclera and conjunctiva externally. The tissue in this transition zone is known as the limbus (Fig. 4.3B), which averages 1.5 mm wide in the horizontal meridian and 2 mm wide in the vertical meridian. It is important because it contains adult corneal stem cell populations, contains the trabecular meshwork, which is the conventional outflow pathway for the aqueous humor, and is the inciting site of pathology in a few immunologic diseases.[32] The limbus is also a major anatomic reference point for planning surgical entry into the anterior segment because it appears clinically as a blue transition zone. Therefore, an incision placed anterior to the blue zone is anatomically in the peripheral cornea, safely inside the trabecular meshwork and stem cells. The cornea is thinner in the center, measuring on average 544 ± 34 μm (range: 440–650 μm) with ultrasound pachymetry, and increases in thickness in the periphery to

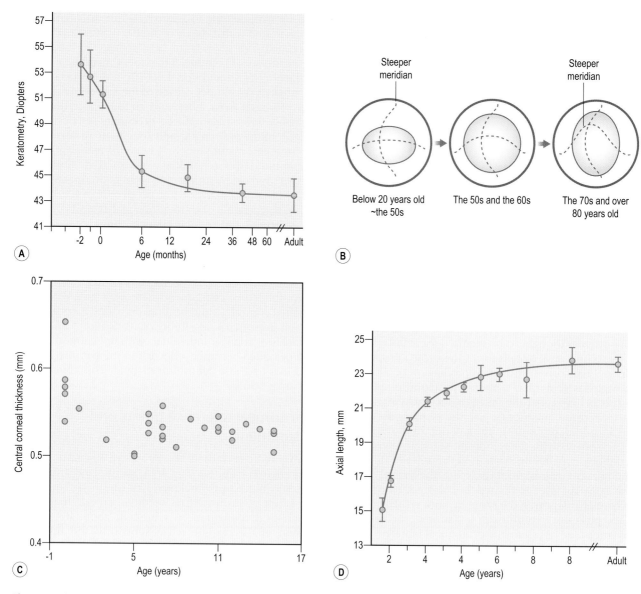

Figure 4.2 (**A**) Keratometry values plotted with respect to age on a logarithmic scale. Negative numbers represent months of prematurity; dot = mean value for each age group and bar = standard deviation. (From Gordon RA, Donzis PB. Arch Ophthalmol 1985; 103:785–9.) (**B**) Changes in corneal shape associated with aging showing the shift from with-the-rule astigmatism to against-the-rule astigmatism. (From Hayashi K, Hayashi H, Hayashi F. Cornea 1995; 14:527–32.) (**C**) Scatterplot showing the relationship between age and central corneal thickness from various studies in the published literature. (From Doughty MJ, Laiquzzaman M, Muller A, et al. Ophthalm Physiol Optics 2002; 22:491–504.) (**D**) Axial length plotted with respect to age. Dots = mean values for age group; bars = standard deviation. (From Gordon RA, Donzis PB. Arch Ophthalmol 1985; 103:785–9.)

Figure 4.3 (**A**) Coronal views show the elliptical shape of the right cornea when viewed anteriorly (upper left) and the circular shape when viewed posteriorly (lower left). Superior axial view (right) illustrates how the right globe deviates from a perfect sphere. Dashed lines = theoretical spherical globe; solid lines = actual contour of the globe. ES = external sulcus; TB = temporal bulge. (Modified from Bron AJ, Tripathi R, Tripathi B. In: Wolff's anatomy of the eye and orbit, 8th edn. London, UK: Chapman & Hall, 1997.) (**B**) Locations of the peripheral cornea, limbus, sclera, episclera, and conjunctiva (PAS ×20). (**C**) The cornea overlying the entrance pupil, known as the central or effective optical zone, directly impacts foveal vision, whereas the cornea peripheral to the entrance pupil, known as the peripheral optical zone, primarily impacts peripheral vision. (Modified from Uozato H, Guyton D. Am J Ophthalmol 1987; 103:264–75.) (**D**) The clinically definable and practically useful principal axes of the eye (left), line of sight and pupillary axis, and major corneal reference points of the cornea (right [right cornea shown]), corneal sightening center (CSC), corneal apex, and thinnest corneal point (TCP) are illustrated in reference to the theoretical, but not practically useful visual axis. The line of sight is the line from the fixation target to the corneal sighting center (CSC) that typically continues through the cornea into the eye, at or near the center of the entrance pupil, where it is refracted by the cornea and lens to finally reach the fovea. The light rays from the fixation target (shown by shaded areas) are usually centered on or near the entrance pupil, E, and are nearly symmetric around the line of sight. The line of sight is often confused with a theoretical principle axis of the schematic eye called the visual axis. Technically, there is no visual axis of the real human eye because the non-centered optics of the real eye do not allow a single straight line to describe this theoretical pathway of chief light rays, defined as an undeviated line from the fixation target that passes through the nodal points of the eye and ultimately onto the fovea. The visual axis is used for calculating the relation between object and image sizes using Gaussian optics and Gullstrand's schematic eye, but it has no meaning or usefulness in the real eye. The pupillary

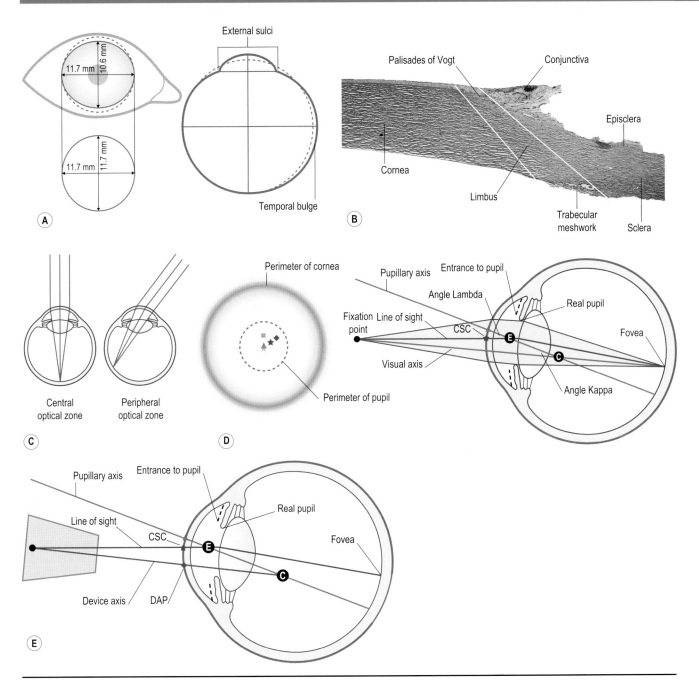

axis is an easily definable line from the center of the real pupil to the perpendicular or normal surface of the cornea, which is aligned with the center of curvature of the cornea, C_1. Angle kappa is the angle between real pupillary axis and theoretical visual axis; angle lambda is the angle between the real pupillary axis and real line of sight. In the clinical setting, it is the angle lambda that is measured by observing the displacement of the coaxially viewed corneal reflex from the pupil center of a fixating eye, even though it is erroneously called angle kappa. Usually angle lambda measures between 3° to 6°. The visual axis and line of sight are often assumed to be parallel, but this is only true for very distant objects near infinity. Pink dot = pupillary axis intercept with cornea; blue dot = line of sight intercept with cornea or CSC; yellow dot = corneal apex; violet dot = imaging device's axis point (DAP); gray dot = TCP. (**E**) The standard alignment position for all imaging devices occurs when a patient directly looks at a luminous fixation point centered in the circular rings of the imaging device and its reflected image is then aligned by the examiner so that it is centered in the operator's screen of the device. When in standard alignment position, the imaging device's optical axis is aimed perpendicular or vertex normal to the surface of cornea and directed toward the center of curvature of the cornea, C; it also is approximately twice the distance from the pupillary axis as the line of sight. The anterior corneal surface intercept with the device axis is the device axis point (DAP). The center of the circular rings of the imaging device's operator screen and the reflected first Purkinje image of the luminous fixation target are perfectly aligned in the standard alignment position. Only if angle lambda is zero will standard alignment position coincide with the line of sight and the corneal sighting center. E = center of entrance pupil. (Modified from Mandell R, Horner D. In: Gills J, Sanders D, Thornton S, et al, eds. Corneal topography. The state of the art. Thorofare, New Jersey: Slack, 1995.)

approximately 700 μm as it reaches the limbus.[1–4,6] A meta-analysis of all cross-sectional and longitudinal corneal thickness studies over a 30-year time period showed that no significant age-related change in central corneal thickness occurred beyond the infant years.[18]

The central cornea overlying the entrance pupil, which is a virtual image of the real pupil and is typically located 0.5 mm anterior to and 14 percent larger than the real pupil, contains the central or effective optical zone of the cornea (Fig. 4.3C). The central optical zone is the portion of the cornea that can successfully refract a cone-like bundle of light from a distant or near fixation target through the pupil of the eye and then directly refracts it onto the fovea. The central location and size of this central optical zone dynamically changes according to the location of the fixation target in relation to the cornea (e.g. more distant fixation target = larger-diameter central optical zone, off-center fixation target = off-center central optical zone) and the aperture of the real pupil in various lighting conditions since the bundle of light has the same cross-sectional shape (e.g. an oval pupil results in an oval central optical zone) as the pupil and its diameter are defined by the pupil's diameter as well as the location of fixation target. The central optical zone's diameter typically averages 3.6 ± 0.8 mm in photopic lighting and 5.8 ± 0.9 mm in scotopic light with a range between 1.5 to 9.0 mm depending on the lighting condition.[33–36] The remaining cornea peripheral to this central optical zone is the peripheral optical zone, which can refract light through the entrance pupil, but it does so at such an acute angle that it only affects the more peripheral aspects of the retina including the macula (Fig. 4.3C); it rarely directly impacts foveal vision.

Within the central optical zone are three major corneal reference points (Fig. 4.3D) that are extremely useful in determining the shape, refractive power, and biomechanical properties of the cornea, since they are statically fixed reference points.[37] The first is called the corneal apex; it is defined as the steepest point or area of the cornea. It typically is measured using a corneal keratographer, topographer, or tomographer and thus its exact location on the cornea is referenced in relation to the corneal intercept of the imaging device's optical axis (Fig. 4.3E).[38] The corneal intercept of the imaging device's optical axis is termed the device's axis point (DAP). On average, the corneal apex usually is located 0.8 mm temporal and 0.2 mm superior to the DAP or 0.5 mm temporal and 0.5 mm superior to the corneal intercept of the line of sight of the eye (Fig. 4.3D), but considerable inter-individual variability is found in the location of the corneal apex in comparison to the average.[39] The clinical utility of the corneal apex is that it is of paramount importance in the selection and fitting of contact lenses and in determining the geometric aspheric shape of the cornea.[37] Significant decentration of the corneal apex from the DAP may give a false interpretation of corneal shape (e.g. asymmetric shape). The easiest way to clarify whether this is real or due to artifact is to aim the imaging device's optical axis directly at the corneal apex, which is more cumbersome and difficult to do than standard alignment position.

The second major corneal reference point is the corneal intercept of the line of sight, also known as the corneal sighting center (CSC).[37,40] The line of sight is an actual principal axis of the real eye as opposed to a theoretical construct of a schematic eye (e.g. visual axis) and it is defined as the principal axis joining a distant fixation point to the fovea. It theoretically is always thought to cross the center of the entrance pupil, which may not be true in the real eye since it is known that the center of the entrance pupil dynamically and unpredictably shifts up to 0.7 mm in direction with changes in pupil diameter, whereas the line of sight is a statically fixed axis line. Thus, it perhaps is best thought of as a line connecting the fixation target to the CSC on the anterior surface of the cornea and then via an unknown non-linear pathway is refracted in the cornea and by the lens to focus on the fovea. The location of the line of sight and CSC are of utmost importance to know in certain clinical situations for getting the best visual results with keratorefractive surgery, particularly with retreatments and customized ablations, and for calculating the proper posterior chamber intraocular lens (PCIOL) power to put in during cataract extraction (CE), especially after previous refractive surgery or with anterior surface irregularities, since subclinical decentrations ≥ 0.5 mm or torsional misalignments ≥ 15° can yield unwanted visual symptoms (e.g. coma or other higher-order aberrations [HOAs]).[37,40–43] On average, the CSC is 0.4 mm nasal and 0.3 mm superior to the dynamically, unfixed pupillary axis or 0.5 mm inferior and 0.5 mm nasal to the static, fixed corneal apex reference point (Fig. 4.3D), but its location overall is highly variable between different individuals.[40] Using a keratographer, topographer, or tomographer in non-standard alignment position, the CSC can be directly determined by having the patient directly look at a luminous fixation point centered in the circular rings of the imaging device and then it is marked in reference to the center of the operator's screen. Because of practical difficulties in non-standard alignment, some clinicians approximate the CSC's position using the coaxially sighted corneal reflex, where the patient looks directly at a luminous fixation point centered in the circular rings of the imaging device and the anterior corneal surface's first Purkinje image is used to approximate the location of the CSC.[40] This approximation method reportedly locates a point on average 0.02 ± 0.17 mm (range: −0.43 mm to +0.68 mm) from the actual CSC in normal eyes; this, however, may not be the case in diseased or surgically altered corneas.[40]

The third major and newest corneal reference point is called the thinnest corneal point (TCP), defined as the thinnest point or zone of the entire cornea. It is measured using various tomography instruments, which enable a mathematical 3D reconstruction of the in vivo pachymetric distribution map, allowing one to evaluate the spatial variation of the thickness profile over the entire cornea. In the normal cornea, the average location and value of the TCP is 0.4 mm inferior and 0.4 mm temporal to CSC and 5 μm less thicker than the central corneal thickness at the CSC. The location and value of TCP is important because it allows clinical differentiation of corneas with normal biomechanical properties from those with current or previous keratectasia since no normal cornea was found in a group with a TCP more than 1 mm distance from the CSC and less than 500 μm of thickness.

If the central optical zone of the anterior corneal surface is regular, yet not uniformly spherical in each meridian, the condition of astigmatism usually results. With astigmatism, a distant fixation point is refracted by the cornea and lens to become two focal lines rather than a sharp image point. On the other hand, if the central optical zone is irregular, then irregular astigmatism usually results. In adult humans, the conjunctival surface area has been measured at approximately 17.65 cm² and the corneal surface area measures 1.38 cm², giving a conjunctival-to-corneal surface area ratio of 12.8, which is important for drug delivery calculations.[1]

Optical properties

Light refraction

The main optical measurements that determine the total refractive power of the eye are the anterior and posterior curvatures of both the central cornea and lens, the depth of the anterior chamber, and the axial length of the eye (Fig. 4.4A).[35] As this chapter is strictly about the cornea and sclera, we will focus only on the optical properties of the cornea. Refractive power and aberrations induced by the optics of the cornea are primarily due to corneal curvature and contour, respectively. Both are descriptors of corneal shape. The contour of the anterior corneal surface is basically of an aspheric geometry with the corneal apex defining the point of greatest refractive power or steepest curvature and then it gradually and variably flattens from the apex to the periphery.[44] Corneal asphericity has been known for over 100 years and has been modeled by various mathematical formulas in order to derive a quantitative approximation of contour.[45,46] The central optical zone of the anterior corneal surface best corresponds to that of a conic section using the following formula, which requires only knowing two conic fit parameters – Q and R:[47]

$$Q = p - 1 = \left(\frac{b}{a}\right)^2 - 1 = \frac{R}{a} - 1 = (1 - e^2) - 1$$

Q is a unitless asphericity factor or expression of the rate of curvature change from the apex of the cornea to the periphery (Fig. 4.4B); p is a geometric form factor; a and b are horizontal and vertical semi-meridian hemi-axes; R is the apical radius of curvature; and e is the eccentricity. Q averages −0.4 in early childhood, but then gets gradually slightly less negative with age such that the central optical zone has a mean Q of −0.2 in adulthood (range: −0.81 to +0.47).[47-49] Q < 0 describes a prolate contour where the rate of curvature change from the apex is less than that of a sphere (Fig. 4.4C); most normal corneas are prolate as it is advantageous in that it compensates for spherical aberrations induced by larger pupil sizes, which project misaligned peripheral rays of light on the fovea. Q = 0 describes a spherical contour where the rate of curvature change from the apex to the periphery is zero, while Q > 0 describes an oblate contour where the rate of curvature change from the apex is more than that of a sphere (Fig. 4.4C). Oblate contours are typically present in ≤ 20 percent of the normal population. Interestingly, asphericity can significantly change after surgery, especially excimer laser keratorefractive surgery, usually resulting in various oblate contours. Although the contour of the central optical zone of the anterior corneal surface is the most important for directly impacting foveal vision and on-axis aberrations (spherical aberrations, coma, and other HOAs), recent study has also shown that the peripheral optical zone is important for off-axis aberrations (e.g. glare, halos, starbursts). This peripheral optical zone does not fit a conic section well, but rather fits a ninth-order polynomial formula

best and has a measured Q of −0.4 in adulthood when the central 10 mm of cornea is best fit to this formula.[50] A few reports on the posterior corneal surface suggest it also has a prolate contour too with a Q of −0.4, but its contribution to the total optical aberrations of the eye are less well known.[46,48]

The actual total corneal dioptric power of the central 4.0 mm of cornea reportedly averages 42.4 ± 1.5 D (range: 38.4–46.3) of the eye's total dioptric power of 60 D.[35,45,51] The location of the corneal apex compared to the CSC (generally less than 1 mm from the CSC and on average 0.7 D steeper than that at the CSC), the degree of asphericity of the anterior corneal surface, and the degree of anterior corneal surface-to-posterior corneal surface ratios can vary widely from one individual to another or even change with age in an individual.[48,52-54] For these reasons, it is difficult to take these general population-averaged results as an empirically useful value. An individual's total corneal power along the line of sight of the eye should probably best be measured using the Gaussian optics formula:

$$P_{totalcornea} = \frac{n_c - n_{air}}{r_{ant}} + \frac{n_a - n_c}{r_{post}} - \left(\frac{d}{n_c} \times \left(\frac{n_c - n_{air}}{r_{ant}}\right) \times \left(\frac{n_a - n_c}{r_{post}}\right)\right)$$

Where $P_{totalcornea}$ equals diopters of optical power; n_{air}, n_c, and n_a are the indices of refraction in air (1.000), cornea (1.376), and aqueous humor (1.336), respectively; r_{ant} and r_{post} are the radii of curvature of the anterior (0.0078) and posterior (0.0065) corneal surface in meters, respectively; and d is the central corneal thickness (0.00054) in meters.

$$42.18\,D = \frac{1.376 - 1}{0.0078} + \frac{1.336 - 1.376}{0.0065} - \left(\frac{0.00054}{1.376} \times \left(\frac{1.376 - 1}{0.0078}\right) \times \left(\frac{1.336 - 1.376}{0.0065}\right)\right)$$

Therefore, the calculated total optical power of the cornea using known average major reference values is 48.21 − 6.15 + 0.12 = 42.18 D, which agrees closely with the actual mean of 42.4 D found in the study above.

Because the cornea is thinner in the center than in the periphery, it should act as a minus lens, but functions as a plus lens because the aqueous humor neutralizes most of the minus optical power on the posterior corneal surface. If we compute the power of the posterior corneal surface in air, we find the following:

$$P_{postcornea} = \frac{n_a - n_c}{r_{post}} = \frac{1 - 1.376}{0.0065} = -57.85$$

The resulting calculated total optical power of the cornea would then be 48.21 − 57.85 + 1.12 = −8.52 D, which is a minus lens.

From the foregoing calculations, it is obvious that the most important refractive surface for humans is the anterior corneal surface. However, if a large air bubble is placed in the anterior chamber so that it contacts the corneal endothelium or if the anterior surface of the cornea is submerged in water, tremendous changes in the refractive power of the eye occur. For example, when the eye is open underwater during swimming, the optical imagery is extremely blurred; the index of refraction of water (1.333) is quite similar to that of the tear film and cornea (1.376). Thus, most of the optical power of the anterior corneal surface is lost. If the air–tear

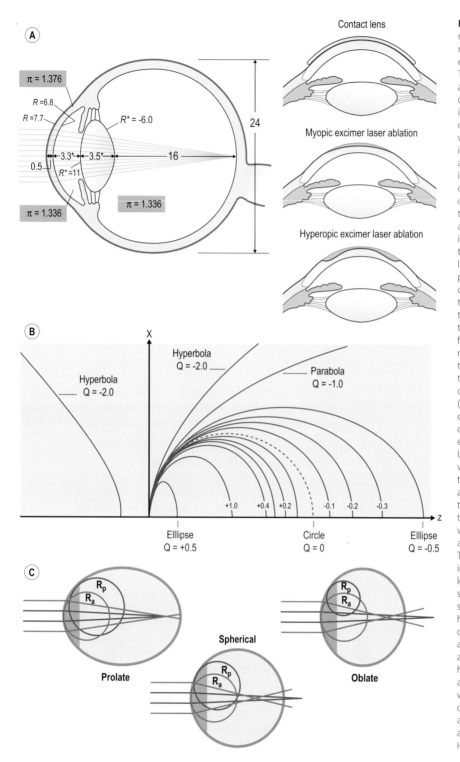

Figure 4.4 (**A**) Diagram of the schematic eye with representative average dimensions in millimeters and refractive indices of the relaxed, non-accommodating eye. The starred values change with accommodation. These dimensions are average values used to construct a representative or schematic eye – in real life, all dimensions of the eye can vary greatly between individuals and from these average values. Upper-right diagram shows that the principle of corneal contact lens wear is to make the anterior corneal surface ineffective as it is now bathed with aqueous tears and the new anterior refractive surface is the air–anterior contact lens interface. Middle-right diagram shows that the principle of myopic correction using the excimer laser is based on a graded removal of central anterior corneal tissue to decrease the central anterior corneal curvature, analogous to the removal of a biologic contact lens that is thicker in the center. Lower-right diagram shows that the principle of hyperopic correction using the excimer laser is based on a graded removal of a peripheral and paracentral concave lenticule of tissue to increase the central anterior corneal curvature (i.e. a donut-shaped trough) analogous to the removal of a biologic contact that is thicker in the periphery and of no or minimal thickness in the center. (**B**) Variations in conicoid shape for different asphericities, or Q-factors, with the same radius of curvature. (**C**) Schematic diagrams showing the three possible anterior corneal surface contours with ray tracings from a distant fixation point being refracted onto the retina. Upper left diagram is of a prolate cornea (Q factor < 0), which has a larger peripheral radius of curvature than at the apex ($R_p > R_a$). Lower middle diagram is of a spherical cornea (Q factor = 0), which has equal peripheral and apical radii of curvature ($R_a = R_p$). Upper right diagram is of an oblate cornea (Q factor > 0), which has a smaller peripheral radius of curvature than at the apex ($R_a > R_p$). As a prolate cornea reduces spherical aberrations, its image is more tightly or precisely focused than spherical or lastly oblate corneas. The mean value of the Q factor for normal healthy adult corneas is −0.2, which reduces natural spherical aberration by about half; a Q factor of −0.50 eliminates all spherical aberration. Thus, the human cornea typically is not designed to induce zero spherical aberrations. In fact, the human lens is optically coupled to cornea in such a way that spherical aberrations are reduced close to zero in youth since during childhood and young adulthood the lens has negative asphericity. Lens asphericity dynamically changes with aging, usually resulting in a Q of zero around age 40 and then has positive asphericity after age 40. The overall effect on the optical properties of the human eye is that it gains progressively more spherical and other optical aberrations with increasing age, which perhaps best explains the direct association of deterioration of visual performance and quality with aging. Most degradation of image quality is due to age-related changes in the lens. (Modified from Gatinel D, Haouat M, Hoang-Xuan T. J Fr Ophthalmol 2002; 25:81–90.)

film interface is maintained by the use of a mask or goggles, then underwater vision is as sharp and clear as normal terrestrial vision.

Light transmission

The cornea is an excellent example of the structural characteristics that a tissue needs to fulfill its dual role of transparency and mechanical support. Tissue transparency is rarely seen in the animal kingdom outside the eye.[2–3] In fact, the only structures in humans where this property is seen are in the eye (e.g. cornea, lens, and vitreous). Corneal transparency has occupied scientists for over half a century and initial transparency theories focused on the extracellular stromal matrix while ignoring the cells of the stroma.[66] Corneal transparency is now thought to be attributable to both the lattice-like arrangement of collagen fibrils in the

Box 4.1 Keratometry, corneal topography, and corneal tomography

Until recently, total corneal power was usually derived clinically from keratometric or topographical measurements of anterior corneal radius of curvature.[35,51] The cornea was regarded as a single refractive surface with an effective refractive index of 1.3375, also known as the keratometric index of the cornea, because no accurate measurement of an individual's posterior corneal radius of curvature could be measured clinically. Although not individualized for each patient, this mathematical approximation was used primarily for its convenience in the clinical setting; however, with the advent of clinical Scheimpflug tomography (e.g. Pentacam, Oculus Inc., Lynnwood, WA, USA) and other new tomography instruments soon to come on the market, the problems of accurately measuring an individual's actual posterior corneal surface curvature and corneal thickness along the line of sight in the clinical setting have been predominantly resolved, although finding the exact location of the line of sight is still sometimes cumbersome, particularly in post-refractive surgery patients.[55] Alternatively, total corneal power can still be approximated using these older devices, but at least now more accurately since the keratometric index has been calculated to be around 1.328.[51]

Box 4.2 Contact lenses

The principle of using contact lenses to correct refractive errors is that one essentially replaces the powerful anterior corneal refractive surface to that of the anterior contact lens surface (Fig. 4.4A, right top diagram). Upon application of a contact lens, the cornea's anterior surface is rendered ineffective as it is bathed with aqueous tears and the air–contact lens interface now becomes the predominate refractive surface of the eye.[56,57] Soft contact lenses typically are used to treat spherical and regular astigmatism and rarely cause any permanent pathologic changes or alterations to the cornea unless there is a complication, such as corneal infection, infiltrate, or toxic conjunctivitis.

The prevalence of complications in contact lens wearers is currently about 5 percent per year of wear.[56,57] Soft contact lenses, however, have been noted to cause some acute physiologic changes to the cornea, including epithelial thinning, hypoesthesia, superficial punctate keratitis, epithelial abrasions, stromal edema, and endothelial blebs. They also cause chronic changes including corneal neovascularization, stromal thinning, corneal shape alterations, and endothelial cell polymegathism and pleomorphism (signs of endothelial cell stress).[58] These are all thought to result from the contact lens-induced hypoxia and/or hypercapnia of the tissue.

Most of these physiologic alterations, particularly the chronic ones, are markedly less common now with the introduction of daily-wear high oxygen transmissibility lenses, like silicone hydrogel contact lenses. In contrast, hard contact lenses treat spherical, regular astigmatism, and even some cases of irregular astigmatism. Hard contact lenses cause the same acute and chronic physiologic changes to cornea as soft contact lenses, although they do more commonly cause corneal shape alternations to occur because they induce more mechanical pressure on the anterior corneal surface.[57] In fact, this is the basis for the practice of deliberately fitting tight, overly flat rigid gas-permeable contact lenses with the aim of flattening the central cornea to reduce myopia in a technique known as orthokeratology.[59] The fitting of contact lenses is highly empirical. Considerable trial and error can be involved in adjusting variables such as contact lens material, size, curvature, and other patient-related factors that are used to arrive at an appropriate correction and comfort level for each individual contact lens wearer.

corneal stroma and the transparency of cells that reside in the cornea.[66–71] In summary, all currently viable transparency theories agree with these major points:

1. Each corneal collagen fibril is an ineffective scatterer of light. Although inefficient, based on the large number of fibrils in the human corneal stroma, destructive interference of scattered light must occur due to the short range order of collagen fibrils in the stroma.

2. Each keratocyte nucleus mildly scatters some light, but since the cell body is an ineffective scatterer of light because of transparent intracellular cytoplasmic water-soluble corneal crystallins, its thinness, and because keratocytes are evenly distributed in the corneal stroma through a clock-wise circular arrangement, light transmission is hardly affected.

3. Scattering of light is minimal in the cornea because it is thin.

4. If an increased refractive index imbalance occurs between fibrils, keratocytes, or extrafibrillar matrix, light scattering can increase tremendously in the corneal stroma resulting in loss of transparency.

In order to understand these theories and generalized principles, one needs to start with the structure of the corneal stroma.

The corneal stroma accounts for 90 percent of the corneal thickness. It is predominantly composed of water (3.5 gram H_2O/gram dry weight) that is stabilized by an organized structural network of insoluble and soluble extracellular and cellular substances (Table 4.1).[72] The dry weight of the adult human corneal stroma is made up of collagen, keratocyte constituents, proteoglycans, corneal nerve constituents, glycoproteins, and salts (Table 4.1).[72,73] Overall, these corneal components work together to maintain and establish a transparent cornea. Although the cornea primarily absorbs most ultraviolet (UV) light, it transmits almost all visible

(400–700 nm) and infrared (IR) light up to a wavelength of 2500 nm, with its peak transmission rate of 85–99 percent in the visible spectrum (Fig. 4.5A).[74,75] The remaining portion (1–15 percent) is scattered in all directions by the cornea in a wavelength-dependent fashion with violet light being most affected.[76] Clinical slit-lamp examination and in vivo confocal microscopy suggest that most of the light scatter is due to cellular components in the cornea rather than extracellular matrix. Relative amounts of light scattering due to each stromal constituent are the following: endothelial cells > epithelial cells > nerve cells > keratocytes >> collagen fibrils or extracellular matrix (Fig. 4.5B).[77,78] In fact, the in vivo confocal microscope shows that the highest area of the light scatter occurs where differences in the indices of refraction are high, such as at the air–tear film interface of the epithelium.[71,77] Within the corneal stroma, light scatter predominantly comes from the stroma–plasma membrane interface of nerve cells and the cytoplasm–nuclear interface of keratocytes. With corneal edema or corneal scarring, loss of corneal transparency has been found to be primarily due to changes

Box 4.3 Refractive surgery

Several keratorefractive surgical procedures have been developed to permanently alter the curvature of the anterior corneal surface, thereby reducing refractive errors.[60,61] The procedures most commonly performed today use the 193-nm argon fluoride (ArF) excimer laser and include laser in situ keratomileusis (LASIK), a thin-flap variant of LASIK known as sub-Bowman's keratomileusis (SBK) as well as the surface ablation techniques, photorefractive keratectomy (PRK), laser-assisted subepithelial keratectomy (LASEK), and EpiLASIK.[60] The excimer laser reshapes the curvature and contour of the anterior corneal surface by removing anterior corneal stroma in a microscopically precise process known as ablative photodecomposition. This results in non-thermal, photochemical breakage of carbon–carbon covalent molecular bonds in the corneal tissue with submicron accuracy. Thus, excimer laser-based keratorefractive surgery is a very accurate, precise, and safe means to permanently change the curvature and contour of the anterior cornea surface. In fact, it has become the most commonly performed refractive procedure performed in the US since it was approved by the US Food and Drug Administration (FDA) in 1995.

The main reason that photoablation of stroma is effective is that the corneal stroma does not regenerate after it is ablated. It only undergoes reparative stromal scarring that at most replaces 5–20 percent of the ablated stromal tissue.[62] Excimer laser-based keratorefractive surgery has been used to successfully treat myopic and hyperopic refractive errors with mild to moderate degrees of astigmatism resulting in stable long-term (at least 12 years) uncorrected visual outcomes. However, it still is known to potentially deteriorate visual quality due to induction of on- or off-axis aberrations.[63] This occurs mainly because:

(1) the actual laser ablation profile is sometimes different from the intended profile

(2) the laser ablation profile is based on spherical geometry whereas the preoperative anterior corneal surface is an aspheric ellipse

(3) the myopic ablation profile makes the cornea more oblate in shape with a flatter curvature in the center and a steeper one in the periphery

(4) visually significant subclinical lateral decentrations or torsional misalignments occasionally occur

(5) the ablation zone is sometimes less than the diameter of the entrance pupil, particularly under low lighting conditions.[64,65]

The basic principle of myopic correction using the excimer laser is based on the graded removal of central tissue to flatten or increase the radius of curvature of the anterior corneal surface (Fig. 4.4A, right middle).[65] In contrast, the correction of spherical hyperopia involves the graded removal of peripheral and paracentral tissue to steepen or decrease the radius of curvature of the anterior corneal surface (Fig. 4.4A, right bottom).[65] Finally, the goal behind various astigmatic ablations is to reshape the anterior corneal surface to bring the two focal points of the eye to the same plane and then ultimately onto the retina with a subsequent second spherical treatment step, if needed.[65] The first step requires selective flattening of the steep meridian and/or steepening of the flat meridian, usually using plus or minus cylinder formats depending on which one removes the least amount of tissue.

Table 4.1 Composition of cornea and sclera

Component	Wet weight (percent)	Dry weight (percent)
Cornea		
Water	78	–
Matrix	66	
Cellular	12	
Collagen	15	71
Proteoglycans	1	9
Keratocytes	1	10
Other	5	10
Sclera		
Water	68	–
Collagen	27	77
Elastin	1	2
Proteoglycans	1	3
Fibrocytes	1	3
Other	2	6

in the light-scattering characteristics of keratocytes rather than alterations in the extracellular matrix.[71,77] In these conditions, the cell body of keratocytes scatters considerably more light than normal corneas, particularly their cell bodies and dendritic processes.

Collagen

Collagen is a structural protein organized into a relatively inextensible scaffold of water-insoluble fibrils that form the basic structural framework of a connective tissue. The corneal collagens are functionally important in establishing tissue transparency and in resisting tensile loads, ultimately defining the size and shape of the tissue. Collagen molecules measure 1.5 nm in width by 300 nm in length and are composed of a triple helix of three alpha chains.[79] Of the 28 different types of collagen, there currently are 13 known collagen types in the human cornea.[80,81] Upon secretion from the cell, the propeptide form of the collagen molecule is cleaved and the monomer of the collagen molecule is assembled into fibril-forming, non-fibril-forming, or fibril-associated collagens with interrupted terminals (FACIT) in a surface recess on the keratocyte or, sometimes, begins assembly inside the cell.

The most common collagen molecule in the cornea is type I (58 percent), which usually aggregates into structural, banded fibrils (Fig. 4.6B), as seen on transmission electron microscopy, by ordering themselves into a quarter-staggered parallel arrangement that is further stabilized in position by covalent intramolecular and head-to-tail intermolecular immature divalent cross-links in a post-translational enzymatic step using lysyl oxidase (Fig. 4.6A and 4.6C, top diagram).[82] With increasing maturity, a spontaneous conversion to mature cross-links occurs where intramolecular, intermolecular, and interfibrillar mature trivalent cross-links replace the immature divalent ones resulting in corneal collagen fibrils resulting in more optimal mechanical properties (Fig. 4.6C, middle diagram). Both immature and mature cross-links occur between lysine and hydroxylysine

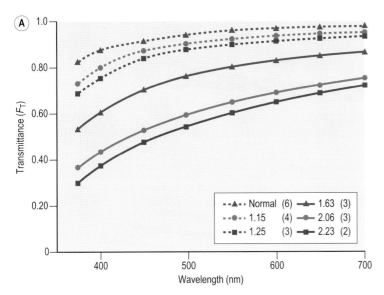

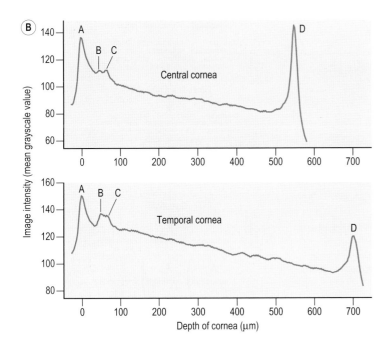

Figure 4.5 (**A**) Experimental values for the percentage of light transmitted through normal and edematous rabbit corneas as a function of wavelength. The ratio of the thickness of the edematous corneas to normal thickness values and the number of corneas used for each curve are given in the key. (From Farrell RA, McCally RL, Tatham PE. J Physiol 1973; 233:589–612). (**B**) In vivo confocal microscopy back-scattered light intensity profile from the central and temporal portions of a 25-year-old normal, healthy human cornea. Intensity peaks correspond to the (A) epithelium, (B) sub-basal nerve plexus, (C) most anterior keratocyte layer, and (D) endothelium. (From Patel S, McLaren J, Hodge D, Bourne W. Invest Ophthalmol Vis Sci 2001; 42:333–9.)

side-chains.[82] After maturation, the turnover or half-life of collagen molecules and fibrils becomes very slow; the concentration of mature cross-links, however, remains stable, whereas high levels of random intramolecular, intermolecular, and interfibrillar non-enzymatic glycation cross-links accumulate, predominantly between lysine and arginine residues (Fig. 4.6C, bottom diagram).[81,82] These non-enzymatic age- or diabetes-related glycation cross-links initially enhance the mechanical properties of fibrils resulting in stiffer, stronger, and tougher fibrils than normal, but occasionally they can go too far, making the tissue too brittle or inextensible to function normally. Type I fibrils are generally heterotypic (type I and type V [15 percent] collagen molecules) since they are composed of two or more types of collagen molecules (Fig. 4.6B). This may serve as a fine-tuning mechanism for controlling a fibril's structural characteristics, such as fibril size and interfibrillar connectivity.

Type I fibrils usually reach certain specific diameters based on their composition and ratio of heterotypic collagen molecular types, are restricted from further lateral accretion or growth, and are permitted to fuse and grow axially due to interactions with small leucine-rich proteoglycans covalently bound to its external surface, and through surface interactions connect to various other fibril-forming, non-fibril-forming, or FACIT collagens, which overall links different levels of structural organization (Fig. 4.6).[83] Thus, the surface properties of type I fibrils are a major determinant of intrafibrillar and interfibrillar biomechanical properties of the tissue. The other major determinants are the direction of the fibrils and the suprafibrillar architectures of the tissue, which overall defines the hierarchical structure of the tissue. They typically form uniform 25 nm diameter fibrils in the stroma proper with only slight variability (note: the diameters used in this chapter are based on transmission

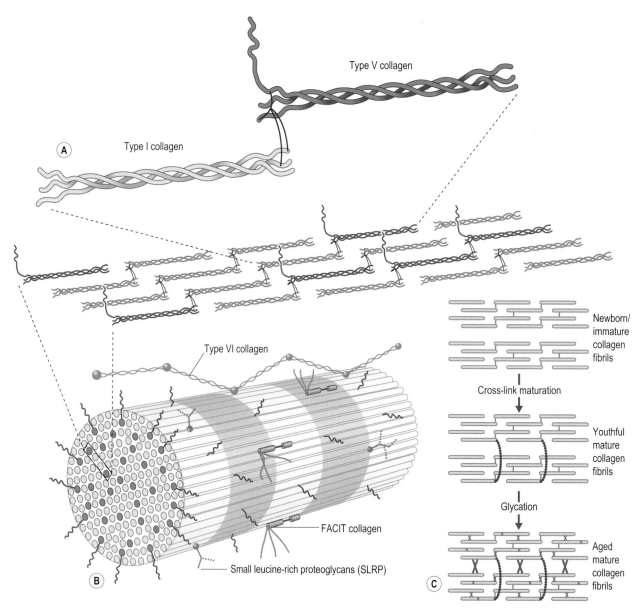

Figure 4.6 (**A,B**) Cross-sectional oblique view of a 25 nm diameter, heterotypic, banded (periodicity = 65 nm) corneal stromal collagen fibril composed of type I (white) and V (blue) collagen molecules (bottom). The amino-terminal domains on the type V collagen molecules appear to be important in regulating collagen fibril diameter as they project externally to the fibril surface and presumably block further accretion of collagen molecules through steric and/or electrostatic hindrance effects. The collagen molecules on longitudinal view are aligned in a parallel, quarter-staggered (68 nm) arrangement with 40 nm gaps between molecules (middle). The longitudinal view also clearly shows that the ends of the alpha chains in each collagen molecule form intermolecular cross-links with adjacent collagen molecules as well as intramolecular cross-links (top). (**C**) With maturity, these immature divalent cross-links become mature trivalent cross-links with the addition of interfibrillar cross-link branches. Finally, with aging, intramolecular, intermolecular, and interfibrillar non-enzymatic glycation cross-links form.

electron microscopic (TEM); x-ray scattering of ex vivo unprocessed human tissue suggests that a 24–36 percent shrinkage artifact occurs by fixing and processing tissue for TEM studies).[84–86] Bowman's layer is the main exception as it has uniform 22 nm diameter type I fibrils, which are epithelial in origin as opposed to keratocyte in origin. The diameter of type I fibrils remains constant across most of the central cornea (mean: 25 ± 2 nm; range: 18–32 nm), but then gradually thickens another 4 nm at about 5.5 mm from the center of the cornea, eventually increasing to up to 50 nm at the limbus.[86,87] Similarly, interfibrillar spacing between nearest neighbor type I fibrils remains constant in

the central cornea (mean: 20 ± 5 nm; range: 5–35 nm), then gradually increases another 5 nm at about 4.5 to 5 mm from the center of the cornea before increasing even more rapidly up to the limbus.[86] Type I fibril diameters and interfibrillar spaces do not seem to vary significantly with depth in the cornea.[86] Although the refractive index of type I fibrils (1.47) is different from that of the extrafibrillar matrix (1.35), the highly uniform small size and highly uniform small interfibrillar spaces along with the predominantly parallel directionality of these fibrils results in a highly ordered, lattice-like arrangement. This arrangement is not a true crystalline lattice, but more of a short-range order that allows transparency of the cornea due to destructive interference (Fig. 4.7).[66]

Type VI collagen (24 percent) is the second most common type of collagen in the corneal stroma. It is present in an unusually high amount compared to most other connective tissues in the body, but it is also unique in that it is only able to aggregate into repeating tetramers of type VI molecules that are stabilized by disulfide cross-links (Fig. 4.8A).[88,89] Thus, it forms 10–15 nm diameter, non-banded filaments with 20–30 nm diameter beaded ovals having a periodicity of 100 nm. Functionally, type VI filaments act as a bridging structure in the interlamellar space since it binds corneal lamellae together diffusely throughout the stroma where they directly cross one another (Fig. 4.8B,C). Along with type XII and XIV FACIT collagens, it also may bridge fibrils together in the interfibrillar space (Fig. 4.8D).[89] Overall, this suprafibrillar architecture results in a one-dimensionally-ordered Bowman's layer and a three-dimensionally ordered stroma proper.[6,90–92]

Although difficult to count precisely, the central corneal stroma reportedly consists of approximately 300 corneal lamellae, while the peripheral cornea consists of approximately 500.[90] Although most corneal stromal lamellae extend from limbus to limbus and cross adjacent lamellae at various angles, randomly in the anterior stroma and nearly orthogonal to one another in the posterior two-thirds, various regional differences in lamellar size, directionality, and amount of interweaving are also found (Figs 4.9 & 4.10). The anterior third of the stroma proper has thinner (0.2–1.2 μm thick), narrower (0.5–30 μm wide), and mostly obliquely oriented lamellae (mean 18° ± 11° [range 0–36°]) with extensive vertical and horizontal interweaving (Fig. 4.9A,B), while the posterior two-thirds has thicker (1–2.5 μm thick), wider (100–200 μm wide), and mostly parallel-oriented lamellae (mean 1° ± 2° [range 0–5°]) with only slight horizontal interweaving (Figs 4.9C & 4.10B,C).[90,93] In the most superficial layers of the stroma, the interwoven lamellae actually attach or possibly seem to originate from the posterior surface of Bowman's layer in a polygonal fashion creating an anterior corneal mosaic pattern (Fig. 4.9A, inset) that can be seen on the anterior corneal surface under certain circumstances (the attached fibrils to Bowman's layer seem to be homologous to sutural fibers in the shark cornea and embryologically may be remnants of the primary acellular stroma).[1] Those attaching or originating fibril bundles usually are oriented obliquely to Bowman's layer, but sometimes are noted to be almost perpendicular to it.[94] Finally, except those fibrils and lamellae in the anterior-most region of the stroma that attach to Bowman's layer, the remaining type I fibrils and corneal lamellae stretch across the cornea from limbus to limbus in a belt-like fashion where they turn and form a circumferential annulus approximately 1.0–2.5 mm wide around the cornea. This annulus is thought to maintain the curvature of the cornea, while blending with limbal collagen fibrils.[95,96]

Keratocytes

Keratocytes make up the second major component of the corneal stroma's dry weight. They are sandwiched between collagenous lamellae forming a closed, highly organized syncytium. They function as modified fibroblasts during neonatal life forming most of the extracellular matrix of the

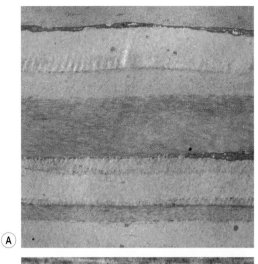

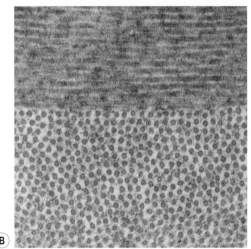

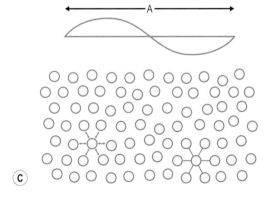

Figure 4.7 (**A**) Low-magnification (×4750) TEM of predominantly orthogonally stacked lamellae in the middle third of stroma proper. (**B**) Higher magnification (×72,500) TEM of two lamellae in the middle third of stroma proper. One lamella is in longitudinal view (top portion) and the other is in cross-sectional view (bottom portion). Notice the uniform 25 nm diameter type I collagen fibrils and uniform 20 nm diameter interfibrillar spaces, which demonstrates only a short-range order, but not a true crystalline lattice arrangement. (**C**) Cross-sectional diagram of collagen fibrils arranged in a true crystalline lattice arrangement. Size of a wavelength of light is shown above for comparison. (Modified from Maurice DM. J Physiol 1957; 136(2):263–86.)

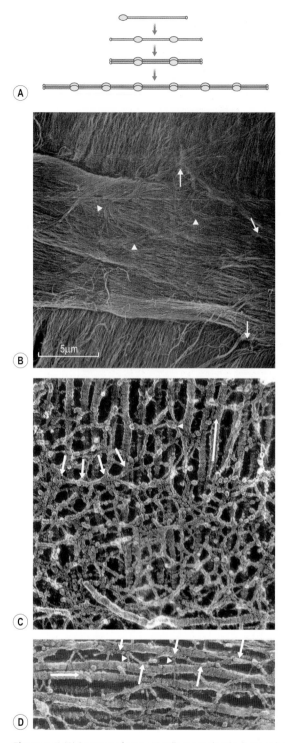

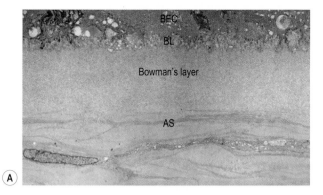

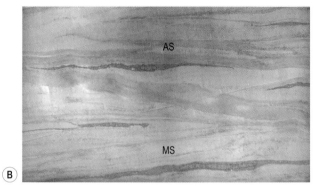

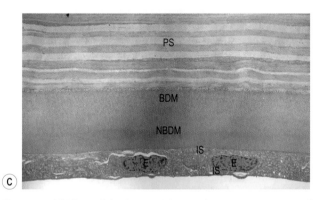

Figure 4.8 (**A**) Diagram of a type VI collagen molecule showing how it assembles into filaments by aggregating into repeating tetramers of type VI molecules with a periodicity of 100 nm. (**B**) Low-magnification (×6200) SEM showing bundles of (type VI) collagen filaments extending between lamellae (arrows) and a loose meshwork of collagen filaments on the posterior surface of a corneal lamella where an adjacent lamella crosses it (arrowheads). Bar = 5 μm. (From Komai Y, et al. Invest Ophthalmol Vis Sci 1991; 32: 2244–58.) (**C**) High-magnification (×115,000) quick-freeze, deep-etched electron micrograph showing a loose meshwork of interlamellar beaded (type VI) filaments with a periodicity of 100 nm (thick arrows) that appear to bind to collagen fibrils (long arrow) by their beads (arrowhead) and bridge fibrils from separate lamellae together. (**D**) Very-high-magnification (×185,000) quick-freeze, deep-etched electron micrograph of intralamellar collagen fibrils (long arrows) with beaded (type VI) filaments (thick arrows) criss-crossing between fibrils in the interfibrillar space and projecting three finger-like structures (FACIT collagens) (arrowheads), which both appear to function in joining neighboring fibrils together. (C and D are from Hirsch M et al. Exp Eye Res 2001; 72:123–35.)

Figure 4.9 (**A**) The acellular Bowman's layer and anterior-most portion of the stroma proper are shown. The type I fibrils and interweaving lamellae in this region branch extensively and some insert into the posterior surface of the Bowman's layer. This arrangement causes an anterior corneal mosaic pattern to occur on the anterior corneal surface after applying digital pressure through the eyelid and instilling fluorescein. (**B**) The anterior third of the stroma proper has predominantly oblique lamellar orientation and extensive vertical lamellar branching and interweaving. (**C**) The posterior third of the stroma proper along with Descemet's membrane and endothelium is shown. The parallel-oriented, orthogonal arrangement of lamellae in this region of corneal stroma is clearly apparent. Although some collagen fibrils randomly insert into Descemet's membrane, no interweaving or significant attachment occurs in this region of the cornea. BEC = basal epithelial cells. BL = basal lamina. AS = anterior stroma. MS = mid-stroma. PS = posterior stroma. BDM = banded portion of Descemet's membrane. NBDM = non-banded portion of Descemet's membrane. E = endothelial cells. IS = intercellular space of the endothelium. (TEM ×4750).

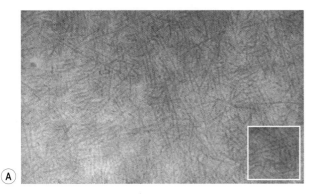

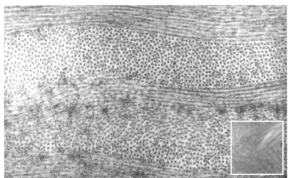

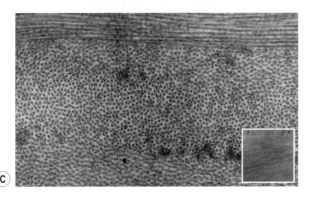

Figure 4.10 (**A**) The acellular Bowman's layer in cross-section (main photo) and tangential section (inset). Notice the random directionality of the 22 nm diameter collagen fibrils in Bowman's layer. (**B**) The transition zone between the anterior third and mid-third of the stroma proper in cross-section (main photomicrograph) and tangential section (inset). The thin and more obliquely oriented lamellae in this region of the corneal stroma and its lattice-like arrangement of 25 nm diameter type I collagen fibrils are shown. (**C**) The posterior third of the stroma proper in cross-section (main photomicrograph) and tangential section (inset). The thick and parallel-oriented lamellae in this region of the cornea stroma and its lattice-like arrangement of 25 nm diameter collagen fibrils are shown. Compare the orthogonal nature of crossing lamellae (inset) to that in the inset B. (TEM ×72,500).

stroma. Subsequently, they remain in the cellular stroma throughout life as modified fibrocytes, where they maintain the extracellular matrix of the corneal stroma. Keratocytes can become metabolically activated or fibroblastic again if the corneal stroma is wounded. The adult human corneal stroma has approximately 2.4 million keratocytes communicating with each other through gap junctions present on their long dendritic processes (Figs 4.11A & 4.12A,B).[97,98] In adulthood, keratocytes occupy 10 percent of the stromal volume, decreasing from 20 percent in infancy, and on

two-dimensional, cross-sectional views appear to be sparse, flattened, and quiescent (scant intracytoplasmic organelles) cells lying between corneal lamella (Fig. 4.11B).[73,99] In actuality, keratocytes are three-dimensional, stellate-shaped cells composed of a 15 to 20 μm diameter cell body with numerous dendritic processes that extend up to 50 μm from the cell body. Tangential sections of the normal cornea suggest that these cells more densely populate the stroma than originally thought and are more metabolically active in the resting state than initially presumed since in tangential section an abundance of cytoplasmic organelles are commonly seen (Fig. 4.11C,D).[100]

Tangential sections show that the anterior stromal keratocytes contain twice the number of mitochondria as the posterior two-thirds of the stroma, which correlates with the higher oxygen tension of the anterior stroma. It also has been demonstrated that a higher stromal cell density occurs in the anterior stroma than in the mid- or posterior stroma (Fig. 4.12A), whereas a higher cell volume-to-extracellular matrix ratio occurs in the posterior stroma compared to the anterior- or mid-stroma.[19,26,99] These views also show that in all levels the keratocytes are highly spatially ordered as they turn in a clock-wise direction like a cork-screw. In vivo confocal microscopy of normal human corneas has shown stromal cell densities averaging around 20,000 keratocytes/mm^3 with a focal zone of increased cell density directly under Bowman's layer, averaging 35,000 keratocytes/mm^3 in the anterior-most layer that gradually tapers to 20,000 keratocytes/mm^3 over the initial 60–100 μm in depth (Fig. 4.12C).[19,26] Confocal microscopy also has shown that stromal cell density decreases with age at a rate of approximately 0.5 percent per year of life with the anterior stroma declining 0.9 percent per year, mid-stroma 0.3 percent per year, and posterior stroma 0.3 percent per year.

Studies using immunohistochemistry or electron microscopy suggest that not all the cells in the corneal stroma are actually keratocytes, but some are one of three types of bone marrow-derived immune cells: "professional" dendritic cells, "non-professional" dendritic cells, and histiocytes (Fig. 4.13).[100–102] Recent studies also found evidence of a small resident subpopulation of adult stromal stem cells, also known as keratocyte progenitor cells, in the corneal stroma, primarily in the periphery of the corneal stroma near the limbus.[103,104] The immune cells appear to play a pivotal role in the induction of immune tolerance versus immune initiation in cell-mediated immunity, and the stromal histiocytes have a role in innate immunity as phagocytic effector cells. The presence of adult stromal stem cells helps explain how the slow replacement and renewal of keratocytes occurs after injury, surgery (e.g. epikeratophakia or penetrating keratoplasty [PK]), or toxicity to the central corneal stroma (e.g. mitomycin C).

Proteoglycans

Proteoglycans make up the third major component of the corneal stroma's dry weight. They are water-soluble glycoproteins made up of a core protein with a covalently attached anionic polysaccharide side chain called a glycosaminoglycan (GAG). The core proteins are non-covalently attached to collagen fibrils uniformly throughout the tissue, whereas the GAG side-chains extend into the interfibrillar space where it acts as a pressure-exerting polyelectrolyte gel.[105–107]

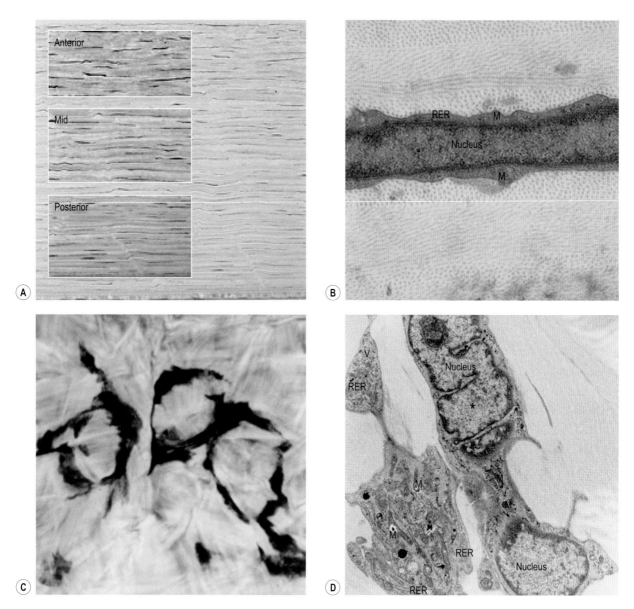

Figure 4.11 Light and TEM photomicrographs showing in cross-section and tangential-section the keratocytes in the stroma. (**A**) Cross-sectional light microscopy view shows that keratocytes are primarily obliquely aligned to corneal surface in the anterior-third of cellular corneal stroma or are aligned parallel to the corneal surface in the posterior two-thirds. (**B**) Cross-sectional TEM view shows that keratocyte nuclei occupy most of the area of the keratocyte seen in this perspective with only a thin rim of surrounding cytoplasm that contains only small numbers of cytoplasmic organelles. (**C**) Tangential-section light microscopy view shows that keratocytes are arranged in a circular fashion. (**D**) Tangential-section TEM view shows that the supposedly quiescent keratocytes may actually be more active in the base-line state than initially thought as an extensive amount of cytoplasmic organelles can be seen in this view. M = mitochondria, RER = rough endoplasmic reticulum, V = vacuoles, * = main portion of nucleus that contains nucleolus. (Modified from Muller LJ et al. Invest Ophthalmol Vis Sci 1995; 36:2557–67.)

The cornea collapses to approximately 20 percent of its original volume if the proteoglycans in the corneal stroma are precipitated out with cetylpyridinium.[105] It has become apparent that the primary functions of proteoglycans are to provide tissue volume, maintain spatial order of collagen fibrils, resist compressive forces, and give viscoelastic properties to the tissue as well as having a secondary role in regulating collagen fibril assembly.[105] Corneal proteoglycans previously were referred to as extrafibrillar amorphous ground substance since their water-soluble state made it difficult to fully delineate them with light and electron microscopy (Fig. 4.14A). It was not until an electron-dense,

cationic dye called cupromeronic blue and a critical electrolyte concentration of 0.1 M $MgCl_2$ were used in combination to specifically stain for the sulfate-ester groups on corneal proteoglycans that the shape, size, arrangement, and location of this material were observed with light and electron microscopy (Fig. 4.14B).[108]

Since that discovery, it has become apparent that corneal proteoglycans are not amorphous, but rather tadpole-shaped molecules composed of a 10–15 nm diameter globular core protein with a covalently attached 7 nm wide × 45–70 nm in length. The latter of which is where GAG sidechain attaches to. They are arranged in the corneal stroma perpendicular to

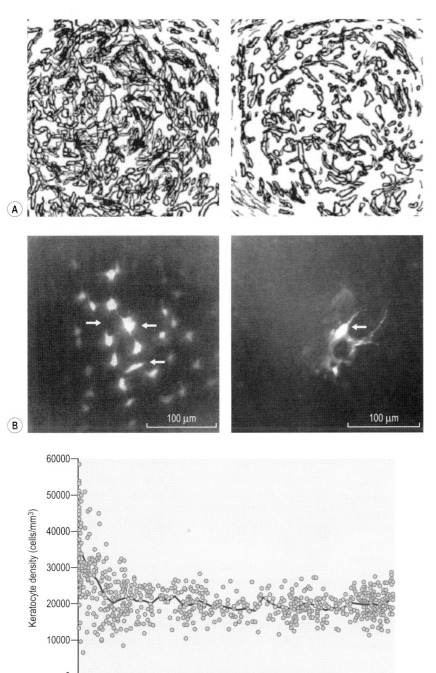

Figure 4.12 (**A**) Reconstruction of keratocyte outlines seen in tangential-section in the anterior and posterior third of the corneal stroma proper. (Modified from Muller LJ et al. Invest Ophthalmol Vis Sci 1995; 36:2557–67.) (**B**) Fluorescent dye spreading between many adjacent keratocytes in rabbit (center-left) and human corneas (center-right), which demonstrates the intimate importance of gap junctions in communication of keratocytes with one another. (From Watsky MA. Invest Ophthalmol Vis Sci 1995; 36:S22.) (**C**) Mean stromal cell density associated with depth in the corneal stroma is shown. The highest zone of increased cell density was closest to the Bowman's layer. Perhaps this is due to baseline, normal epithelial–stromal interactions. (From Patel SV et al. Invest Ophthalmol Vis Sci 2001; 42:333–9.)

collagen fibrils with a constant spacing of around 65 nm between each other along the collagen fibrils. Their core proteins non-covalently bind to collagen fibrils in specific gap zones along the peripheral portions of the collagen fibril. The core proteins with dermatan sulfate side-chains bind to 'd' and 'e' gap zones and those with keratan sulfate side chains bind to 'a' and 'c' gap zones.[109] GAGs are highly negatively charged, stiff polymers that extend into the interfibrillar space and form antiparallel duplexes with adjacent GAG side-chains (Fig. 4.14C), thereby, linking different next-nearest-neighbor collagen fibrils together by forming dumbbell-like structures. The genes that produce the core proteins have been cloned and four types of corneal stromal proteoglycan core proteins have been identified: decorin, lumican, keratocan, and mimecan.[110] Decorin contains a single dermatan sulfate GAG side-chain (Fig. 4.14D), while lumican and mimecan have a single keratan sulfate GAG side-chain and keratocan has three keratan sulfate GAG side-chains (Fig. 4.14D). Thus, there are four known types of proteoglycan core proteins and only two types of GAGs, keratan sulfate (60 percent) and dermatan sulfate (40 percent), found in the human corneal stroma. GAGs are polymers of repeating disaccharide units of galactose and N-acetylglucosamine or iduronic acid and N-acetylgalactosamine, respectively.[109]

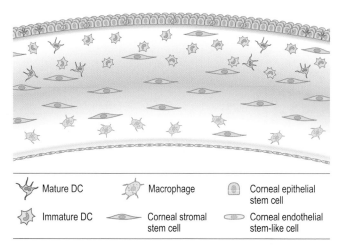

Mature DC
Immature DC
Macrophage
Corneal stromal stem cell
Corneal epithelial stem cell
Corneal endothelial stem-like cell

Figure 4.13 The mouse cornea suggests that 5–10 percent of the cells in the cellular corneal stroma are actually one of three types of bone marrow-derived immune cells. The anterior-third of the cellular corneal stroma contains professional (MHC-positive) dendritic antigen-presenting cells in periphery of the cornea, while non-professional (MHC-negative) dendritic cells are present in both the periphery and center of the anterior cornea. The posterior-most regions of the cellular corneal stroma appear to contain macrophages in the periphery and center of the cornea. (Modified from Hamrah P et al. Invest Ophthalmol Vis Sci 2003; 44:581–9.)

Because the core protein tail and their associated GAG side-chains are post-translationally added to core proteins in the Golgi apparatus, there seems to be some flexibility in how long or how sulfated they can become depending on the function of the connective tissue producing them. The human cornea is unique in that the core protein tail and their associated GAG sidechains are fibril-associated and small in length (KS ~45 nm and DS ~70 nm) with a higher amount being over-sulfated than in other connective tissues. A comparative study of corneas from 12 mammalian species suggests that dermatan sulfate is the preferred proteoglycan in oxygen-rich environments, such as in the thin cornea of mice, or is seen predominantly in the anterior portion of the thicker cornea of mammals, such as humans or rabbits. Keratan sulfate is a functional substitute produced through an alternate metabolic pathway in thicker corneas, especially in the posterior portion where oxygen levels may drop precipitously.[111] Functionally, this duality is quite useful because dermatan sulfate appears to be more efficient at holding water; it absorbs less water than keratin sulfate, but holds most of it in a tightly bound, non-freezable state.[112] This is consistent with the fact that dermatan sulfate is more abundant in the anterior corneal stroma in humans, which is the region of highest oxygen tension and most affected by evaporation. In contrast, keratan sulfate is more abundant in the posterior corneal stroma (Fig. 4.15). This is the region of lowest oxygen tension, least affected by evaporation, and the area where the need for loosely bound water is required for transport across the endothelium via the metabolic pumps.

Corneal nerves

The epithelium of the cornea is the most richly innervated tissue of the body with about 16,000 nerve terminals/mm² (~2.2 million nerve endings), about 300–400 times more dense than skin.[113–115] Most of the nerve fibers in the cornea are sensory in origin, responding to mechanical, chemical,

and temperature stimuli, and are derived from the ophthalmic branch of the trigeminal nerve (CN III₁) (Fig. 4.16A).[116] Refer to Chapter 16 (Sensory innervation of the eye) for full details of the innervation pathway of the cornea. Although all mammalian species have been found to receive variable proportions of nerve fibers in the cornea from the sympathetic and parasympathetic autonomic nervous system, human corneas appear to be on the extreme end of this spectrum as their corneas have a very small proportion of their nerve fibers derived from the autonomic nervous system.[116]

Electrophysiological studies have shown that the cornea's receptive nerve field primarily is composed of polymodal nociceptors (70 percent) followed by mechano-nociceptors (20 percent) and then cold-sensitive nociceptors (10 percent).[117] Using in vivo laser confocal microscopes, one can only evaluate morphologically the corneal nerves from the main nerve trunks up to sub-basal plexus (SBP) as the resolution and contrast of these images is not capable of visualizing the final nerve branches and free nerve terminals in the corneal epithelium.[118]

Since corneal nerve fibers ultimately terminate in the brainstem, it appears that interneuron intermediate pathways must relay the information to the sensation areas of cerebrum. Additionally, there must also be intermediate relays to efferent systems that trigger the reflex pathways of involuntary blinking via orbicularis motor innervation from CN VII and reflex tearing via parasympathetic innervation of lacrimal gland. The intricate central nervous system (CNS) details of these specific pathways are currently unknown.

Corneal sensitivity is a valuable clinical measure of corneal health since corneal nerves directly maintain the health of corneal epithelium through direct trophic factors as well as serve a protective role in warning the host of possible dangers to the normal healthy state and maintain an adequate basal tear secretion rate.[116,119] It is usually tested clinically in a semi-quantitative fashion with a Cochet and Bonnet esthesiometer, which is a thin (0.12 mm diameter), flexible, nylon filament of variable length (0–6 cm).[119–121] When the filament is long, it applies very little pressure to the corneal surface because it bends easily, while when short it applies a proportionally higher pressure before bending. The length is converted into pressure using a conversion table with a range of touch pressures between 11 and 200 mg/mm². Corneal sensitivity is defined as the reciprocal of corneal touch threshold and it can be evaluated subjectively by asking the patients when they feel touch upon the cornea or objectively when a reflexive blink response is triggered. Refer to Chapter 16 for full details on the subjective and objective threshold in normal healthy corneas and those with disease or after surgery.

Corneal sensation also variably decreases with corneal disease (e.g. herpes simplex keratitis, diabetes, corneal dystrophies, keratoconus), following surgical procedures on the anterior segment of the eye and sometimes the posterior segment (e.g. panretinal photocoagulation), after application of certain topical medications (e.g. anesthetics, non-steroidal anti-inflammatory drugs), and even after contact lens wear.[119,122] As nerve regeneration occurs at the rate of approximately 1 mm per month, it may take up to 3–12 months or longer for corneal re-innervation and sensation to maximally recover, depending on the type and degree of

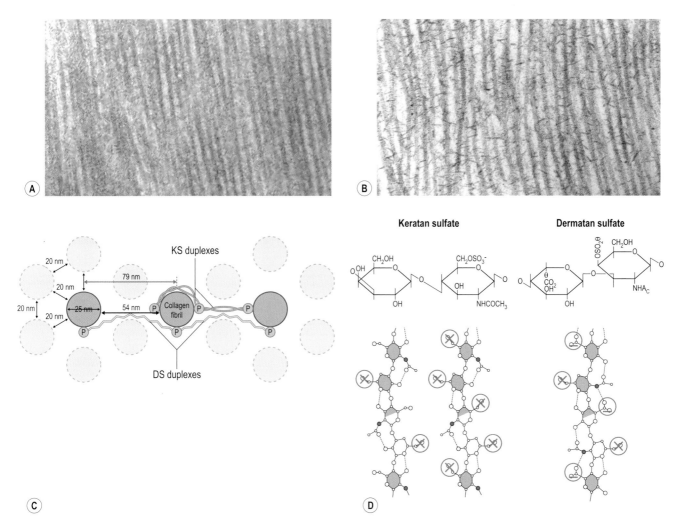

Figure 4.14 Tangential-section TEM views (×90,000) of longitudinally running type I corneal stromal collagen fibrils in a lamellae without (**A**) and with (**B**) cupromeronic blue staining. Notice the scattered "amorphous ground substance" in the interfibrillar spaces in (A), while in (B) duplexes of proteoglycans are clearly seen bridging next-nearest neighbor collagen fibrils. (**C**) Diagram of how proteoglycans attach along the periphery of type I fibrils via their core proteins (P) and how the core protein tail/GAGs duplex in an anti-parallel fashion in the interfibrillar space between next-nearest neighbor collagen fibrils. (**D**) Diagram of the polymer backbones of keratan sulfate and dermatan sulfate. The top portion shows the primary structure of the repeating disaccharide units of keratan sulfate (top left) and dermatan sulfate (top right). The bottom portion of the diagram shows the secondary structures of each proteoglycan. In the human cornea, ~50 percent of keratan sulfate is in the normal-sulfated state (left), while the other ~50 percent is in the over-sulfated state (center). All three displayed proteoglycan polymers have similar backbones and, therefore, form similar secondary structures of a two-fold helix (right). Anionic charges = sulfate esters (X) and carboxylic acid (−). (C and D are modified from Scott JE. Biochem Soc Trans 1991; 19:877–81.)

surgical injury.[119] After maximal recovery, the sensitivity in the portion of the cornea involved by the procedure often is variably less than that which was present prior to surgery, which means there is a potential long-term as well as short-term neurotrophic keratitis.

It has been recently discovered that after corneal nerve injury, microneuromas can develop during the regeneration process.[117,123] These microneuromas, as well as injured corneal nerves themselves, exhibit altered functional properties where responsiveness to normal stimuli is impaired, yet paradoxically display abnormal intrinsic electrical excitability (i.e. spontaneous impulses or even abnormal responsiveness to normally minimal stimuli).[117] Overall, this may produce hyperalgesia and/or dysesthesia, which may continue long-term after surgery despite some attenuation. These neuropathic pain impulses are often perceived by the patient as dryness, foreign body sensation, and irritation

after surgery, yet are not actually due to actual dryness or irritation of the cornea. These undesired, unpleasant sensations may best respond to ion channel antagonists rather than dry eye lubricating medications.[123]

The mechanisms by which corneal nerves maintain the ocular surface and promote healing after eye injuries are currently under active research in several laboratories. Corneal nerves secrete neuropeptides, such as substance P and calcitonin gene-related protein, and neurotransmitters, such as acetylcholine, vasoactive intestinal polypeptide, and neurotensin, which are believed to be important in corneal epithelial function and proliferation.[116,117]

Corneal stromal wound healing

The first published report specifically addressing the cellular reactions in the corneal stroma after injury appeared in 1958.[124] It described the morphologic changes of stromal

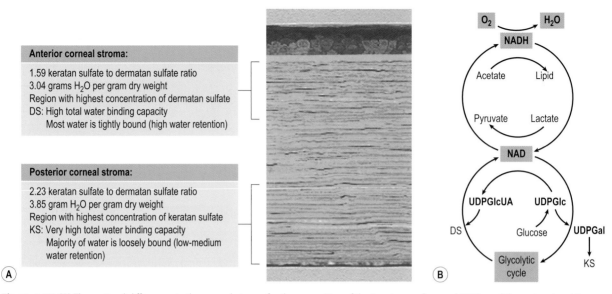

Figure 4.15 (**A**) The regional differences in the corneal stroma for the proportion of the two types of corneal GAGs and the water-absorbing properties in these regions are shown. (**B**) Diagram of the metabolic pathways for dermatan sulfate and keratan sulfate production. The supply of oxygen is the primary factor determining whether dermatan sulfate is made through an aerobic pathway or whether keratan sulfate is made through an anaerobic alternative pathway. (Modified from Scott JE. TIBS 1992; 17:340–3.)

cells after different types of trauma and found that stromal cells lose their interconnecting, dendritic processes immediately after injury with many cells subsequently developing signs of degeneration. That report also described the appearance of morphologically unique, spindle-shaped corneal fibroblastic cells invading into the wound region during later stages of stromal healing. Since that time, many excellent animal model corneal wound healing studies have further addressed the changes in the extracellular matrix and the stromal cells after stromal injury.[125–133] They suggest that corneal stromal injury is immediately followed by keratocyte apoptosis in the zone around the site of stromal injury with a subsequent influx of transient mixed acute and chronic inflammatory cells, proliferation and migration of surviving keratocytes, and finally differentiation of the keratocytes into transiently metabolically activated cell types called activated keratocytes. This latter cell type is functionally important because it synthesizes and deposits the extracellular matrix of the stromal scar, while also degrading and remodeling the damaged cellular and extracellular tissues around the wound. Epithelial injury alone can also cause transient cellular injury to the underlying stroma presumably from exposure of stroma to tear-related factors, resulting in apoptosis, proliferation, and differentiation into migratory keratocytes as well as resulting in some anterior stromal edema.[134–136] It, however, does not appear to cause differentiation into activated keratocytes, hypercellularity, differentiation into myofibroblasts, or stimulate extracellular matrix production – all of which are seen with corneal stromal injury whether by incision or excision.[134–136] Myofibroblasts are characterized by the intracellular cytoplasmic appearance of α-smooth muscle actin, which helps impart contractile properties to the cell.[131]

A number of studies have looked into the expression of cytokines and growth factors in normal and injured corneas.[137,138] These studies tried to assess the relative importance of each specific factor in cornea wound healing, since it was known clinically and experimentally that epithelial–stromal interactions increase the number of proliferatory and migratory keratocytes within the stromal wound compared to deeper corneal stromal injury, some of which differentiated beyond the activated keratocyte stage into myofibroblastic cell type.[133,139–142] Some studies focused even more specifically on strictly epithelial or tear-related cytokines or growth factors since the epithelium and aqueous tears were found to be a major source of cytokines and growth factors.[143–147] The major cytokines and growth factors studied to date include epithelial growth factor (EGF), fibroblast growth factor (FGF), interleukin-1, nerve growth factor (NGF), transforming growth factor-beta (TGF-β), insulin, retinol, and LPA. TGF-β is currently thought to be the most important growth factor of this group in regard to stimulating a fibrotic reparative stromal scar phenotype.[130,139,142,147,148] A recent study, however, has shown that local cytokine and growth factor-related influence on normal corneal stromal wound repair is predominantly an early wound healing phenomenon as cell–matrix interactions seem to take over in the later stages of wound healing.[147] For example, integrity or return of a complete epithelial cell basement membrane after stromal injury seems to regulate epithelial–stromal interactions long-term as it decreases the production and, more importantly, the release of TGF-β from epithelial cells into the stroma.[147] One gap in this area of research is how mechanotransduction pathways (mechanical load-induced intracellular signals) fit into this scheme, particularly in maintaining myofibroblast differentiation long-term and thus corneal haze. Other cytokines and growth factors of this group were also found to have some minor complementary or even competing roles with TGF-β, and in other cases they had no role at all in corneal wound healing. For example, NGF is complementary to TGF-β as it too is known to stimulate myofibroblastic

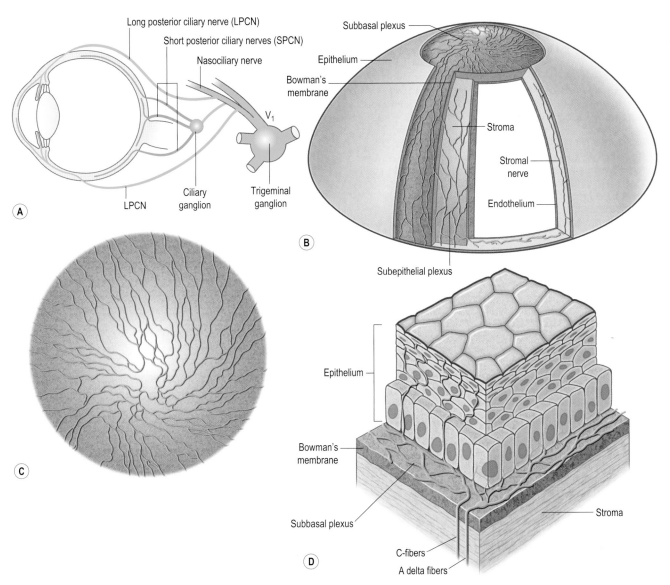

Figure 4.16 (**A**) The nerves of the cornea and sclera are mixed (motor, sensory, and autonomic) and come from the nasociliary branch of the 1st division of the Vth cranial nerve, which branch to form two long (LPCN) and several short posterior ciliary nerves (SPCN). The SPCN supply the posterior sclera, whereas the LPCN supply the cornea and equatorial and anterior sclera. The nerve supply of the cornea is the richest in the body and the sclera also has numerous nerves, commonly with many free nerve terminal endings found in the most vascular portions of the episclera, presumably to regulate blood flow in the anterior segment and to influence aqueous humor outflow. (Modified from Watson PG, Hazelman B, Pavesio C, Green WR. The sclera and systemic disorders, 2nd edn. Edinburgh, UK: Butterworth Heinemann, 2004.). (**B**) and (**C**) Distribution of corneal nerves in the stroma including the subepithelial plexus (SEP) and the sub-basal plexus (SBP). (Modified from Muller LJ et al. Exp Eye Res 2003;76: 521–42; Patel DV, McGhee CN, Patel DV, McGhee CNJ. Invest Ophthalmol Vis Sci 2005; 46(12):4485–8). (**D**) The architecture of the nerve bundles in the SBP (arrow) contain both straight and beaded fibers. The straight C fibers branch and turn upwards to extend into the epithelium. (Modified from Guthoff RF, Wienss H, Hahnel C et al. Cornea 2005; 24(5):608–13.)

cellular transformation independent of TGF-β, but it does so without enhancing keratocyte proliferation or extracellular matrix deposition.[146] In another example, all-*trans* retinol (vitamin A), which is reflexively secreted in the aqueous tears from its storage location in the lacrimal gland, has a completely different role than wound healing as it maintains the moist, mucosal ocular surface phenotype via gene transcription regulation; thus, it ultimately controls the rate of ocular surface cell proliferation and differentiation (i.e. prevents keratinization and squamous metaplasia).[149,150] Once activated, keratocytes exhibit a wide range of cellular responses,

including increased tritiated thymidine uptake (indicating increased proliferation); initiation of protease and collagenase activity; phagocytosis; interferon, prostaglandin and complement factor 1 production; and fibronectin, collagen, and proteoglycan secretion.[130,141,151–159]

Human studies of stromal wound healing concur with animal studies on most issues with the following notable differences: adult human corneas heal less aggressively, more slowly, and not as completely as animal corneas.[62,133,160–163] However, both animal and human studies show that corneal stromal wounds heal in two distinct

Box 4.4 Wound healing variability after common ophthalmic surgeries

The location of scar types along with the variable degree of wound healing responses in the human cornea is best exemplified when one reviews published human histopathology studies that describe findings in corneas that have had common ophthalmic procedures such as cataract extraction (CE), PK, radial keratotomy (RK), astigmatic keratomy (AK), PRK, or LASIK (Fig. 4.17). It is also apparent from these cases that acellular barrier layers, like Bowman's layer, are not reformed if damaged or excised, whereas acellular basement membranes, like the epithelial basement membrane and Descemet's membrane, can be regenerated.[168] Sutured and unsutured clear corneal CE wounds are corneal stromal incisions constructed at oblique angles to the corneal surface so that they self-seal. They usually heal with well-aligned external wound margins and wound edges, which results in a small (50–75 μm in depth) subepithelial zone of hypercellular fibrotic stromal scarring and a remaining deeper zone of hypocellular primitive scarring (Fig. 4.17A).[62]

Occasional small epithelial plugs are found in the external wound of unsutured clear corneal cataract wounds and sometimes Descemet's membrane is found partially detached or poorly re-aligned along the internal wound margin so that stromal ingrowth occurs. In marked contrast, limbal and scleral tunnel CE incisions heal because fibrovascular granulation tissue from the episclera completely grows into the wound by 15 days after surgery with remodeling up to 2.5 years after surgery (early wound repair mechanisms in vascularized tissue are controlled by bioactive substances released by platelets at the wound site, such as platelet-derived growth factor and TGF-β, as opposed to epithelial-derived factors, such as TGF-β, seen in the avascular cornea).[147,169–171] PK wounds heal similarly to sutured clear corneal cataract wounds with the notable differences of having significant wound compression because of the oversized nature of the donor button (usually 0.25–0.5 mm) and wound edge mismatch caused by the irregular, asymmetric nature of the trephine wound found between donor and host wound edges. Additionally, a high percentage of PK cases have overriding external or internal wound edges with Bowman's layer or Descemet's membrane incarceration, which serves to cause weak areas in the hypercellular fibrotic scar or regenerated Descemet's membrane.[166,167,172,173]

Although of partial thickness (70–95 percent depth) and being constructed perpendicular to the corneal surface, RK and AK incisions heal similarly to unsutured clear corneal cataract wounds.[174–178] The most notable difference from unsutured CE wounds is the more commonly present and more widely variable degree of external wound gaping found in these corneas, which commonly leads to epithelial ingrowth or plugging that rarely goes away long-term (Fig. 4.17B).[178] PRK heals entirely under the influence of epithelial–stromal interactions. Therefore, a disk-shaped hypercellular fibrotic stroma scar is produced (Fig. 4.17C), which usually is 12–20 percent in thickness of the amount initially ablated.[179,180] In contrast, LASIK heals similar to that of unsutured clear corneal cataract incisions (Fig. 4.17D). A sub-epithelial zone of hypercellular fibrotic scarring occurs at the flap wound margin and the remainder heals by producing a hypocellular primitive stromal scar usually with a thickness around 5–10 percent of the amount initially ablated.[181–183] However, a notable difference from unsutured CE wounds was that approximately 50 percent of the LASIK corneas were found to have at least some microscopic epithelial plugging present.

phases: (1) an active phase – results in the production of a stromal scar over the first six months after injury in humans, and (2) a remodeling phase – improves corneal transparency and increases wound strength. This 2nd phase occurs up to 3–4 years after injury in humans. Overall, the long-term result in human corneas is the production of a hypercellular fibrotic stromal scar in wound regions where epithelial–stromal interactions occur and a hypocellular primitive stromal scar in wound regions where keratocyte injury pathways occur. These two histological wound types have functional differences as the hypercellular fibrotic stromal scar is strong, but can look clinically hazy because of myofibroblastic cells populating this scar type.[69,131] In contrast, the hypocellular primitive stromal scar is transparent, but it is very weak in tensile and cohesive strength and serves as a potential space for fluid, inflammatory cells, and microbes.[164] An additional variable to consider in this scheme is the fact that more precisely re-aligned wounds, such as sutured and unsutured wounds with minimal gaping and no epithelial cell plugging, heal better than poorly aligned wounds, such as wounds with wide wound gaping, epithelial plugging, or incarceration of Bowman's layer, Descemet's membrane, or uvea.[164–167]

Barrier properties

Low-permeability barrier: the corneal epithelium

The external surface of the human cornea is covered by a stratified squamous epithelium (see Fig. 4.1D).[1,2] Unlike other epithelial surfaces, the corneal epithelium is specialized to exist over a moist, transparent, refractive, avascular surface and thus it is smooth and non-keratinized. The corneal epithelium is composed of 4–6 cell layers about 50 μm in total thickness across the entire anterior corneal surface (Fig. 4.18). It is continuous with the epithelium of the limbus (see Fig. 4.3B). The basal epithelial cells actively secrete a basement membrane, around 90 nm thick at birth, composed of type IV collagen fibrils, laminin, heparin sulfate, and fibronectin. The corneal epithelial basement membrane, or basal lamina, increases in thickness with age measuring around 300 nm thick in late adulthood.[23] By electron microscopy, the basal lamina appears to be composed of two distinct layers: a 20–30 nm thick more anterior lamina lucida and a 30–60 nm thick more posterior lamina densa. Its function is similar to most basement membranes in that it serves as a scaffold for epithelial cell movement and attachment.

The cytoplasm of all epithelial cells contains mainly cytoskeletal intermediate filaments and has sparse cytoplasmic organelles. This aids in maintaining transparency. The predominant cytoplasmic protein is keratin, while actin filaments and microtubules are the other major ones. The basal cell layer stores large glycogen granules as a source of metabolic energy for times of stress during hypoxia or wound healing. The epithelial cells are held together by desmosomes, while the basal surface of the epithelium adheres to the basal lamina and underlying Bowman's layer through an anchoring complex composed of hemidesmosomes, type VII collagen anchoring fibrils, and anchoring plaques (Fig. 4.19). The epithelial cells differentiate from the basal layer to form one to three mid-epithelial layers of wing cells and finally to form one to two superficial squamous cell layers

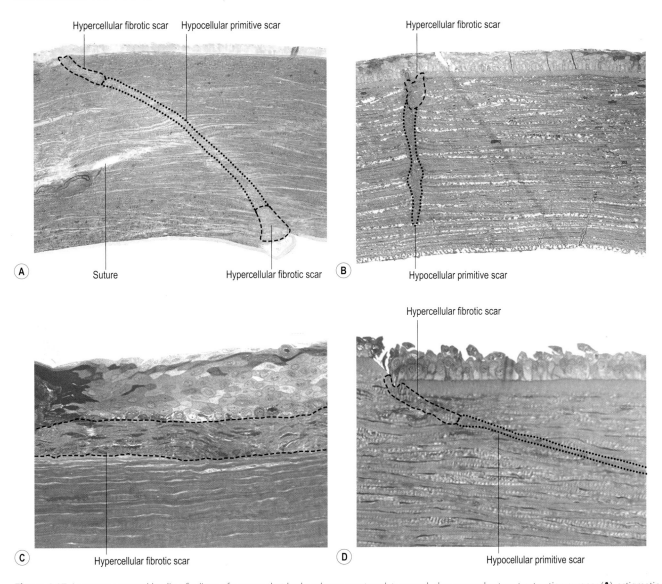

Figure 4.17 Long-term wound healing findings of corneas that had undergone sutured, temporal, clear-corneal cataract extraction surgery (**A**), astigmatic keratotomy (**B**), photorefractive keratectomy (**C**), and laser in situ keratomileusis (**D**). All corneas shown are greater than 4 years after surgery. (toluidine blue ×15 for A, ×25 for B, and ×100 for C and D). (Modified from reference 62.)

(Fig. 4.18). The most superficial squamous cells at the external surface form a high-resistance ($8–16$ k$\Omega \cdot$ cm^2) barrier to the external environment since they are all surrounded by a continuous band of apical zonula occludens tight junctions at their peripheral intercellular margins (Fig. 4.20).[184,185]

Zonula occludens tight junctions are characterized by fusion of the adjacent cell membranes resulting in obliteration of the intercellular space over variable distances and are made up of the tight junction proteins ZO-1, JAM-A, occludin, and claudin-1, as well as some other claudin subtypes.[186] This barrier prevents the movement of ions and thus fluid from the tears into the stroma, reduces some evaporation, and protects the cornea from infectious pathogens. A clinical test to determine if this barrier is disrupted uses the vital dye fluorescein, which is topically instilled on the external surface of the cornea and stains the corneal surface if there is a breakdown in the epithelial tight junctions. The apical surface of the corneal epithelium has microplicae and microvilli (Fig. 4.18) that are covered with a wettable, smooth glycocalyx layer, consisting of membrane-associated mucins (Fig. 4.20), MUC 1, MUC 4, and MUC 16.[187–189] These mucins are produced by surface epithelial cells and conjunctival goblet cells forming a 1.0 μm thick mucinous layer of the tear film.[2,190] The healthy total tear film typically measures 7 μm in thickness and contains three layers: mucinous, aqueous, and lipid layers (Fig. 4.20B). Recently, the aqueous layer of the tear film has also been found to contain other membrane-spanning, gel-forming mucins, MUC 5AC and 2, along with lysozyme, immunoglobulin A, transferrin, defensin, and trefoil factor.[189] Overall, the tear film functions in maintaining a healthy ocular surface by preventing evaporation, friction during eye blinking, and ocular infections. It also is of paramount importance in forming a moist, smooth optical surface on the cornea required for clear vision. Deficiencies in the components of tear film potentially can cause ocular surface disease. For example, loss of the lipid layer,

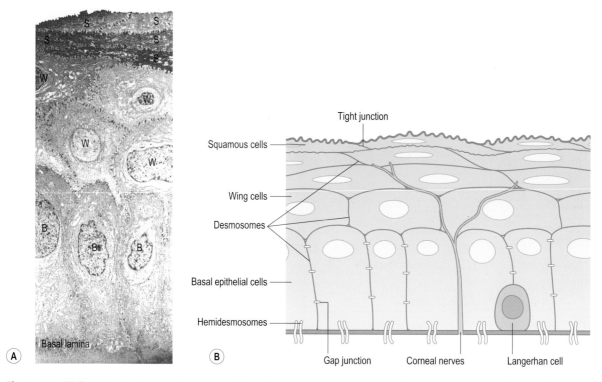

Figure 4.18 (**A**) Transmission electron micrograph (×3500) of the central corneal epithelium with a summary diagram (**B**). Microvilli project from the anterior corneal surface into the tear film. S = squamous cells. W = wing cells. B = basal epithelial cells. (B modified from Hogan MJ et al. Histology of the human eye. Philadelphia: WB Saunders, 1971.)

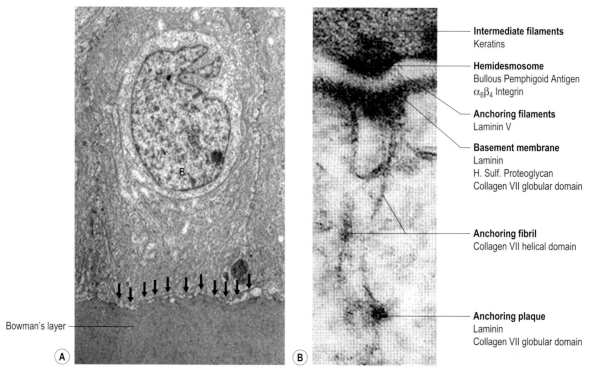

Figure 4.19 (**A**) Transmission electron micrograph (×10,000) of the basal epithelial cells showing the adhesion complexes (arrows) that anchor it in place into the Bowman's layer and a summary diagram (**B**). B = Basal epithelial cell. Bar = 1 μm. (B from Gipson I, Joyce N. In: Albert D, Miller J, eds. Principles and practice of ophthalmology, 3rd edn. Philadelphia: WB Saunders, 2008).

Mucinous layer of the tear film

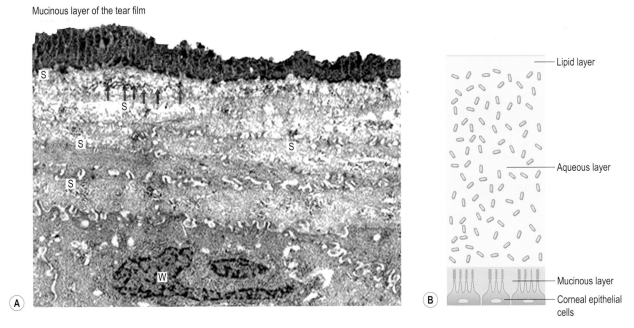

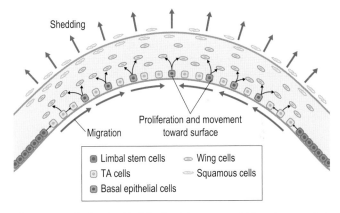

Figure 4.20 (**A**) Transmission electron micrograph (×10,000) of surface epithelium from a specimen specially preserved in glutaraldehyde and cetylpyridium chloride and specially stained with tannic acid to show the mucinous layer of the tear film (glycocalyx + membrane-bound mucins). (**B**) Summary diagram showing how the tear film layers interact with the microvillae of the surface squamous epithelial cells and mucinous layer. S = squamous cells. W = wing cells. Arrows = zonula occludens tight junctions.

which is produced by the meibomian glands, results in the condition of evaporative dry eye disease due to increased evaporative loss from the anterior corneal surface (from the normal baseline rate of 3 μl/hr cm^2 up to a maximum of 40 μl/hr cm^2).[191] In another example, loss of the aqueous layer, which is produced by the lacrimal glands, results in the condition of aqueous tear-deficiency dry eye disease. For more information about dry eye disease including the pathophysiology, classification, diagnosis, and management, see references or refer to Chapter 15.[192–194]

The corneal epithelium is in a state of constant renewal as the most superficial squamous cells are continuously shed into the tear film. It is estimated that the cell layers of the corneal epithelium turn over every 7–10 days. The epithelial surface is maintained by basal epithelial cells, which can usually undergo mitosis one time resulting in two daughter cells that are found anterior to the basal cell layer, resulting into two wing cells and eventually into two squamous cells (Fig. 4.21).[195–197] A delicate balance of shedding followed by proliferation is critical in maintaining a smooth and uniform epithelial surface. The shedding step is primarily induced by friction that occurs from involuntary or voluntary eyelid blinking, which happens on average every 7 seconds or 6–15 times/minute. The signal for basal epithelial cell proliferation probably comes via the gap junctions, especially in the basal cell layer since they express the gap junction protein connexin 43.[187,198] In addition to basal epithelial cell mitosis (vertical proliferation), the corneal epithelium is maintained by migration of new basal epithelial cells into the cornea from the limbus (horizontal proliferation). The cells migrate centripetally at a rate of approximately 120 μm/week and originate from a subpopulation of proliferative limbal progenitor epithelial cells (Fig. 4.21).[199–201] It appears that the

Figure 4.21 The basal epithelial cells of the cornea, which undergo vertical proliferation (both daughter cells move into the middle layers of the epithelium) are continually replenished by a stem cell population that resides in basal layer of the limbus. The transient amplifying (TA) basal epithelial cells, which are two horizontal progeny from the stem cells, migrate forward from the limbus to the periphery of the cornea and commonly to reach the center of the cornea. The TA undergo mixed (one daughter cell retained in the basal cell layer and the other moves into the middle layers of the epithelium) proliferation. The final or terminal cell cycle of mitosis is the vertical proliferation step, where the two daughters continually move toward the surface eventually being shed in the tear film. (Modified from Thoft RA et al. Invest Ophthalmol Vis Sci 1983; 24:1442)

corneal epithelium is maintained by a balance among the processes of limbal cell horizontal proliferation and migration, basal corneal epithelial cell horizontal migration and vertical terminal proliferation, and shedding of superficial squamous corneal epithelial cells.[201,202] This concept has been coined the X, Y, Z hypothesis by Thoft, where X (basal corneal epithelial cell proliferation) + Y (limbal cell proliferation) = Z (epithelial cell loss on the anterior surface from eyelid blinking).[203] When this equilibrium is disrupted, corneal epithelial cell wound healing typically begins. After injury, these processes typically return to equilibrium with a degree of flexibility. For instance, after epithelial and stromal injury, if a stromal deficit occurs (e.g. a healed corneal ulcer) then the epithelium can still maintain a smooth anterior corneal surface either by developing elongated hypertrophic basal epithelial cells in mild defects and/or by developing epithelial hyperplasia (> 6 cell layers) in moderate to severe defects.[204]

The subpopulation of limbal basal progenitor epithelial cells are called adult corneal epithelial stem cells.[201,205] Epithelial stem cells have the characteristics of being undifferentiated, slow cycling, and extremely long lived with a high proliferative potential. They are an excellent source for corneal epithelial cell reconstruction. The stem cells are located in a well-defined, protective niche microenvironment called the palisades of Vogt (see Fig. 4.3B), and are controlled via delicate regulatory mechanisms.[206–208] The progeny of these epithelial stem cells are referred to as transient amplifying (TA) cells, which are the basal epithelial cells of the limbus or peripheral cornea that migrate centripetally (Fig. 4.21). These TA cells divide more frequently than stem cells and undergo mixed (horizontal or vertical) proliferation. They do, however, have a finite proliferative potential, usually replicating at least twice, which is in marked distinction to the stem cells from which they are derived. Once TA cells reach the end of their proliferative capacity, usually when near the center of the cornea, they become basal epithelial cells, which terminally differentiate just once into two daughter wing cells (vertical proliferation). Limbal stem cell theory forms the basis for several surgical procedures using transplanted or cultured limbal stem cells to restore vision in patients with limbal stem cell deficiency.[209,210]

Finally, it has been shown in both animal and human specimens that the corneal epithelium is devoid of melanocytes, but does contain immune cells (see Fig. 4.13). The basal epithelial cell layer of the peripheral cornea along with the limbus and conjunctiva appear to have a subpopulation of cells that are bone-marrow-derived immune surveillance cells with high constitutive expression of major histocompatibility complex (MHC) type II antigen and co-stimulatory molecules.[211] This type of immune cell has previously been termed a Langerhans cell, which functions as a "professional" antigen-presenting cell with an extraordinary capacity to initiate T-cell lymphocyte-dependent responses.[212,213] The functional steps of this cell type include the up-take and processing of antigens and migration out of the cornea to lymph nodes where they stimulate naïve T-lymphocyte-mediated immune responses by presenting antigens and overexpressing co-stimulatory molecules. Recently, the central corneal epithelium also has been found to have a similar subpopulation of immune cells that are apparently all "immature" Langerhans cells because their constitutive expression of MHC type II antigen and costimulatory molecules is low.[214] These "immature" Langerhans cells or "non-professional" antigen-presenting cells under certain circumstances, such as inflammation or trauma, may develop the requisite signals for T-cell priming.[214]

High-permeability barrier: the corneal endothelium

The primary function of the corneal endothelium is to maintain corneal transparency by regulating corneal hydration and nutrition through a "leaky" barrier and metabolic pump function first described by David Maurice.[72] The "pump-leak" hypothesis basically suggests that an equilibrium is needed in the amount of passive fluid flow into the cornea and energy expended pumping out excess fluid for maintenance of corneal transparency and relative dehydration of the corneal stroma. Secondarily, it is also known to secrete an anteriorly located basement membrane called Descemet's membrane and a posterior-located glycocalyx layer.[1]

The endothelium of the infant cornea is composed of a monolayer of approximately 500,000 neural-crest-derived cells, each measuring around 6 μm in thickness by 20 μm in diameter, or covering a surface area of 250 μm² (Figs 4.22A & 4.23A).[6,215] The cells lie on the posterior surface of the cornea and form an irregular polygonal mosaic. The tangential apical or inner surface of each corneal endothelial cell is uniquely irregular, usually uniform-in-size to one another, and typically six-sided hexagons. The hexagon is the most energy-efficient geometric shape in order to cover a surface completely without leaving gaps; thus, minimizing intercellular boundary exposure to the aqueous humor.[215,216] They abut one another in an undulating, interdigitating fashion with a 20 nm wide intercellular space between each other, which serves to increase the internal surface area of the lateral cell membranes, making the length of the intercellular space sometimes at least ten times the width of the cell itself (Fig. 4.22A,B).[44,217] This is clearly seen on the endothelium's tangential basal or outer surface as corneal endothelial cells form an extremely complex jig-saw shape here as opposed to the six-sided hexagon on the inner surface.[218] The intercellular space is known to contain apical macula occludens tight junctions and lateral gap junctions (Fig. 4.22C,D); thereby forming an incomplete barrier with a preference to the diffusion of small molecules (Fig. 4.23B,C). The corneal endothelial cells have numerous cytoplasmic organelles, particularly mitochondria, and thus have been inferred to have the second highest aerobic metabolic rate of all the cells in the eye next to the retinal photoreceptors (Fig. 4.22B).[6]

At birth, the central endothelial cell density (ECD) of the cornea is around 5000 cells/mm².[215] Postnatally, the corneal endothelium of primates and higher mammals including humans, mature and lose most of their proliferative capacity, whereas lower mammals do not lose this proliferative capacity and thus can regenerate a normal endothelial cell monolayer after injury or disease as compared to humans. Because the human corneal endothelium has very limited innate in vivo proliferative capacity due to contact inhibition, high aqueous humor concentrations of TGF-β, and age-related gradual cellular senescence, particularly in the central regions of the cornea in part through the activity of the cyclin-dependent kinase inhibitor p21, there is a well-documented decline in central ECD with age that typically involves two phases: a rapid and a slow component.[24,25,219–221] In fact, recent studies suggest that an adult stem cell population of corneal endothelial cells exists near Schwalbe's line of the limbus, which is a transition zone between the

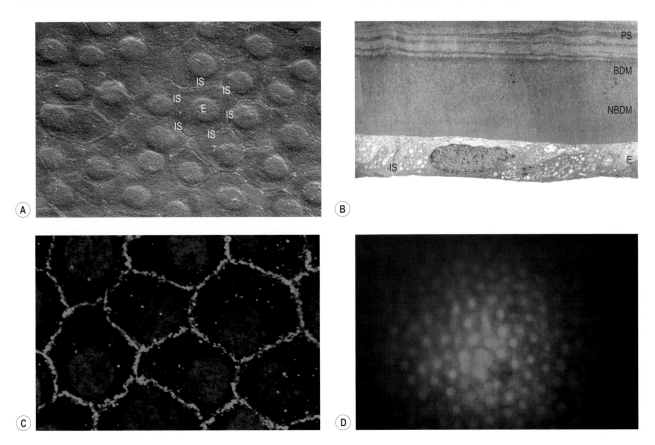

Figure 4.22 (**A**) Scanning electron micrograph (×1000) of the posterior surface of the corneal endothelium from a 65-year-old patient with healthy corneas. Note how the hexagonal endothelial cells form a uniform monolayer with small 20 nm intercellular spaces between adjacent endothelial cells. E = endothelial cells. IS = intercellular space. (**B**) Transmission electron micrograph (×4750) of the posterior corneal stroma, Descemet's membrane, and corneal endothelium from a 65-year-old patient with healthy corneas. PS = posterior stroma. BDM = banded portion of Descemet's membrane. NBDM = non-banded portion of Descemet's membrane. E = endothelial cells. IS = intercellular space. (**C**) Immunofluorescent laser confocal microscopic photomicrograph (×2000) of human corneal endothelial macula occludens tight junctional complexes stained with immunolabeled monoclonal antibodies to junctional adhesion molecule-A (green). Nuclei are counterstained with TO-PRO (blue). (Courtesy of Kenneth J. Mandell, MD, PhD.) (**D**) Photomicrograph (×400) of fluorescein dye spreading between many adjacent endothelial cells in a human cornea, which demonstrates the intimate importance of gap junctions in how endothelial cells communicate with one another. (Courtesy of Mitchell A. Watsky, PhD.)

trabecular meshwork and the corneal endothelial periphery, and proliferates in a limited fashion, particularly after injury.[222,223] Due to corneal growth and developmentally selective cell death, during the fast component, the central ECD decreases exponentially to about 3500 cells/mm² by age 5 and 3000 cells/mm² by age 14–20 (Fig. 4.24A).[25,221] Thereafter, a slow component occurs where central ECD decreases to a linear steady rate of 0.3–0.6 percent per year, resulting in central ECDs around 2500 cells/mm² in late adulthood (Fig. 4.24A).[24,25,221] Because the corneal endothelium maintains its continuity by migration and expansion or thinning of surviving cells to cover a larger surface area, it is not surprising that the percentage of hexagonal cells decreases (pleomorphism) and the coefficient of variation of cell area increases (polymegathism) with age.[221]

When reviewing normal physiologic variability of the corneal endothelium, realize that these average central ECDs are primarily from Caucasian US populations. Several studies reveal that important ethnic differences exist as Japanese, Filipino, and Chinese corneas have been found to have higher central ECDs than Caucasians at all ages, while Indian corneas have lower central ECDs (Table 4.2).[224–227] It is hypothesized that this ethnic variance of central ECDs may be predominantly due to population differences in mean corneal diameter and thus mean endothelial surface area between these groups (Japanese, Caucasian, and Indian horizontal corneal diameters averaged 11.2, 11.7, and 12.0 mm, respectively), but genetic and environmental factors cannot be excluded and thus need further study. Additionally, these data apply only to central ECD since a recent study has shown that higher ECDs can typically be found in the periphery of the cornea (Figs 4.23A & 4.24B).[228] This finding has been previously reported, but the recent study was the only one that was thorough enough in detail to document how the variance applies to the entire posterior corneal surface.[229–232] Therefore, it appears that total corneal endothelial cell numbers and ECDs decrease on average about 50 percent from birth to death in normal subjects without causing corneal disease or pathology. Because corneal decompensation typically doesn't occur until central ECDs reach 500 cells/mm² (Fig. 4.25A,B), which is a 90 percent decrease in central ECD from birth or an 80 percent decrease

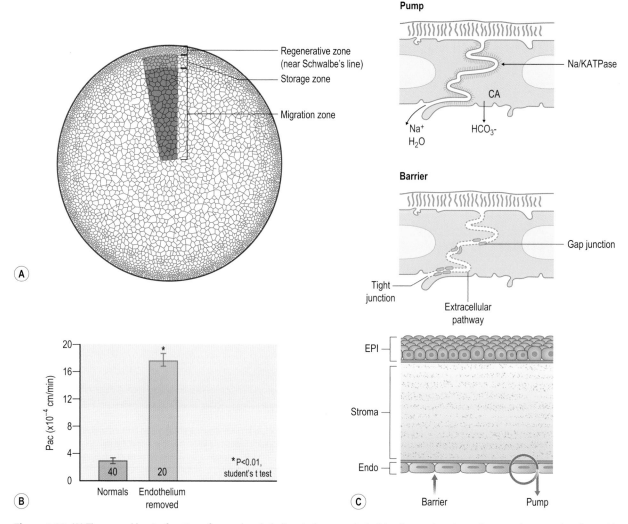

Figure 4.23 (**A**) The normal barrier function of corneal endothelium is due to endothelial cells covering the entire posterior corneal surface without gaps and the focal, discontinuous tight junctions in its apical intercellular space. (**B**) shows the normal permeability of the human endothelial cell monolayer to carboxyfluorescein (2.26×10^{-4} cm/min) compared to that without endothelium (12.85×10^{-4} cm/min), which resulted in a six-fold increase in permeability. (Modified from Watsky MA et al. Exp Eye Res 1989; 49:751–67.) (**C**) The opposing forces of the leaky corneal endothelial barrier and the metabolic pump sites are shown. When the leak rate equals the metabolic pump rate, the corneal stroma is 78 percent hydrated and the corneal thickness and transparency is maintained. (Modified from Waring GO et al. Ophthalmol 1982; 89:531).

from healthy adulthood levels, there appears to be plenty of cellular reserve remaining after an average human lifespan of 75–80 years of life.[25,221] Estimates suggest that healthy, normal human corneal endothelium should maintain corneal clarity up to a minimum of 224–277 years of life – if humans could live that long.[25]

Leaky barrier function

The barrier function of the corneal endothelium is dependent upon a sufficient number of endothelial cells to cover the posterior surface of the cornea (Fig. 4.23A) and intact macula occludens tight junctions (Fig. 4.23B,C) between the endothelial cells resulting in a low electrical resistance ($25 \, \Omega \cdot cm^2$) barrier to the aqueous humor flow. Macula occludens tight junctions are characterized by partial obliteration of the intercellular space and partial retention of a

10 nm wide intercellular space.[217] Clinically, the barrier function of the cornea can be assessed by using the specular microscope or the confocal microscope to measure ECD, and fluorophotometry to measure permeability.[233] In healthy human corneas, this barrier prevents the bulk flow of fluid from the aqueous humor to the corneal stroma, but it does still allow moderate diffusion of small nutrients, water, and other metabolites to cross into the stroma through the 10 nm wide intercellular spaces.[44]

The leakiness of the endothelial barrier may initially seem inefficient and counterproductive; however, most nutrients for all layers of the cornea come from the aqueous humor. Thus, leakiness of the endothelial cell monolayer is essential for corneal health due to the fact that the cornea is avascular and contains no lymph vessels or other channels for bulk fluid flow. Despite the loss of endothelial cells that occurs with aging, there appears to be no significant increase in the

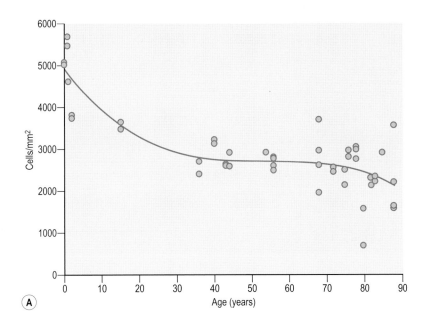

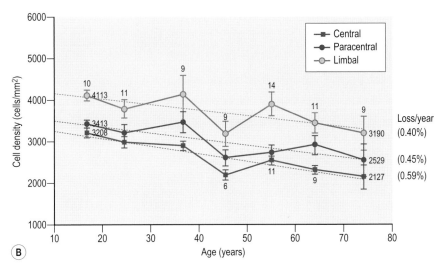

Figure 4.24 (**A**) Scatterplot with best fit curve showing the average central corneal endothelial cell density (ECD) for normal, healthy eyes of different ages. (From Williams KK et al. Arch Ophthalmol 1992; 110:1146–9.) (**B**) Graph illustrating the central, paracentral, and peripheral corneal ECDs for healthy, normal subjects of different ages. (From Edelhauser HF. Cornea 2000; 19:263–73; and Edelhauser HF. Invest Ophthalmol Vis Sci 2006; 47:1755–67.)

permeability of normal, healthy aged corneas.[234,235] Only when the endothelial cell junctions are disrupted does permeability increase, usually up to a maximum of 6-fold higher than baseline (Fig. 4.23B).[234] Corneal endothelial permeability does, however, gradually increase as central ECD decreases below 2000 cells/mm², but compensatory metabolic pump mechanisms keep the cornea at its normal dehydrated state until a central ECD of 500 cells/mm² is reached.

A number of factors are known to acutely affect the barrier function of the endothelium including reversible disruption of cell junctions during irrigation with calcium-free solutions or glutathione-restricted solutions, mechanical damage during intraocular surgery, or chemical injury due to introduction of non-physiologic or toxic solutions into the anterior chamber. Fortunately, the remaining viable cells are oftentimes able to migrate and re-cover the posterior corneal surface by spreading out over a larger surface area, re-establishing the intercellular cell junctions. Thus, the barrier function of the corneal endothelium is efficiently restored.

Metabolic pump function
In the early days of corneal transplantation, it was observed that when donor corneas were refrigerated, corneal thickness increased and transparency decreased.[236] When the cornea was brought back to its normal temperature of 35°C, a temperature reversal occurred during which the cornea returned to its normal thickness and regained clarity. In vitro perfusion studies of the cornea showed that temperature reversal can take place in the absence of the epithelium, leading to the conclusion that active metabolic processes in the endothelium are responsible for maintaining corneal dehydration.[236] Subsequent studies demonstrated that transporters, located primarily in the endothelial cell's basolateral cell membrane transport ions, principally sodium (Na^+) and bicarbonate (HCO_3^-), out of the stroma and into the aqueous humor (Fig. 4.23C).[236,237] An osmotic gradient is created and water is thus osmotically drawn from the stroma into the aqueous humor. This osmotic gradient can be maintained only if the endothelial barrier is intact. A major transport protein found to be essential for endothelial metabolic pump function is

Table 4.2 Comparison of endothelial cell density in Indian, American, Chinese, Filipino, and Japanese populations

Age groups (years)	Indian[A]		American[B]		Chinese[C]		Filipino[D]		Japanese[B]	
	# of eyes	Cell density (cells/mm^2)	# of eyes	Cell density (cells/mm^2)	# of eyes	Cell density (cells/mm^2)	# of eyes	Cell density (cells/mm^2)	# of eyes	Cell density (cells/mm^2)
20–30	104	2782 ± 250	11	2977 ± 324	100	2988 ± 243	114	2949 ± 270	18	3893 ± 259
31–40	96	2634 ± 288	6	2739 ± 208	100	2920 ± 325	112	2946 ± 296	10	3688 ± 245
41–50	97	2408 ± 274	11	2619 ± 321	97	2935 ± 285	112	2761 ± 333	10	3749 ± 407
51–60	98	2438 ± 309	13	2625 ± 172	97	2810 ± 321	102	2555 ± 178	10	3386 ± 455
61–70	88	2431 ± 357	8	2684 ± 384	90	2739 ± 316	114	2731 ± 299	6	3307 ± 330
>70	54	2360 ± 357	15	2431 ± 339	83	2778 ± 365	86	2846 ± 467	15	3289 ± 313

[A]Rao SK et al. Corneal endothelial cell density and morphology in normal Indian eyes. Cornea 2000; 19:820–3.

[B]Matsuda M et al. Comparison of the corneal endothelium in an American and a Japanese Population. Arch Ophthalmol 1985; 103:68–70.

[C]Yunliang S et al. Corneal endothelial cell density and morphology in healthy Chinese eyes. Cornea 2007; 26:130–2.

[D]Padilla MDB et al. Corneal endothelial cell density and morphology in normal Filipino eyes. Cornea 2004; 23:129–35.

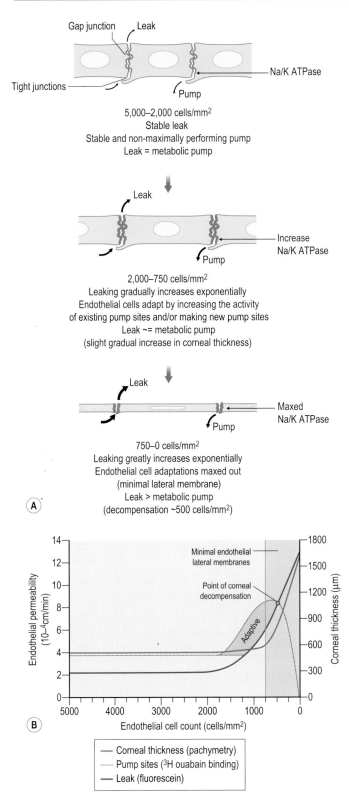

Figure 4.25 (**A**) and (**B**) The relationship between central ECD, barrier function, metabolic pump sites, and pachymetry is shown. Note that the pump sites are not all maximally used in the normal state (5000–2000 cells/mm²). With increased permeability (2000–750 cells/mm²), there is an adaptive phase in which the endothelial cells can maximally use all their pump sites or can make more pump sites to offset the leak up to a certain point. When the surface area of the corneal endothelial cell's lateral membranes becomes too small (750–0 cells/mm²), these adaptations max out and eventually decline. The point where endothelial cell pump site adaptations intersect with permeability (500 cells/mm²) is typically when corneal decompensation occurs. (Modified from Dawson DG et al. In: Duane's Foundation of Clinical Ophthalmology on CD-ROM. 2006;v. 2 c. 4:1–76.)

Na⁺/K⁺-ATPase.²³⁸,²³⁹ The number and density of Na⁺/K⁺-ATPase sites have been quantified using [3H]-ouabain, which binds specifically in a 1-to-1 ratio to Na⁺/K⁺-ATPase pump sites.²⁴⁰ These studies have shown that approximately 3 million Na⁺/K⁺-ATPase pump sites are present in the basolateral membrane of a single corneal endothelial cell. This corresponds to an average pump site density of 4.4 trillion sites/mm² along the lateral plasma membrane of an intact endothelial cell.²⁴⁰ Clinically, the metabolic pump function of the corneal endothelium can be assessed in vivo using pachymetry to measure how quickly the corneal thickness recovers after being purposefully swollen by wearing an oxygen-impermeable contact lens or, secondarily, by measuring the degree of diurnal fluctuation in corneal thickness.²³³,²⁴¹

A number of factors are known to alter endothelial pump function including pharmacologic inhibition of Na⁺/K⁺-ATPase, decreased temperature, lack of bicarbonate, carbonic anhydrase inhibitors, and a chronic reduction in ECD from mechanical injury, chemical injury, or disease states. Fortunately, with regard to the latter, compensatory metabolic pump mechanisms prevent corneal edema from occurring to a certain degree when central ECDs are between 2000 and 750 cells/mm² (Fig. 4.25A, B). This occurs by either increasing the metabolic activity of pump sites already present, which requires more ATP production by the cell, and/or by increasing the total number and density of pump sites on the lateral membranes of endothelial cells.²⁴⁰ A similar phenomenon occurs in the proximal tubule of the human kidney to adjust for an increased salt load. For example, in Fuchs' endothelial dystrophy, the cornea has been found to remain clear and of normal thickness despite having focal areas of very low ECD and increased endothelial monolayer permeability to fluorescein.²⁴² Apparently, this occurs because the metabolic activity and density of the Na⁺/K⁺ pump sites increase in adjacent healthier areas to compensate for the focal areas of increased permeability.²⁴² The point at which compensatory mechanisms ultimately fail is when the central ECD reaches approximately 500 cells/mm² or less.²¹⁵,²⁴³ At this low ECD, the permeability has increased to a point where the endothelial cells are spread so thin that they do not have enough room on their lateral cell membranes for more metabolic pump sites and all the existing pump sites are maximally active (Fig. 4.25A,B). At this point, the metabolic pump fails to balance the leak and corneal edema results.

A summary of the entire corneal endothelial cell transport system was recently reviewed by Bonanno.²⁴⁴ When the corneal endothelial barrier and metabolic pump are functioning normally, the corneal stroma has a total Na⁺ concentration of 179 mM (134.4 mM free and 44.6 mM bound to stromal PGs), while the aqueous humor has a total Na⁺ concentration of 142.9 mM (all free).²⁴⁵ Therefore, after accounting for chloride activity and stromal imbibition pressure, an osmotic gradient of +30.4 mmHg exists causing water to diffuse from the stroma into the aqueous humor.

Corneal edema

Corneal edema is a term often used loosely and non-specifically by clinicians, but it literally refers to a cornea that is more hydrated than its normal physiologic state of 78 percent water. The Donnan effect states that the swelling

Normal	Acute Glaucoma	Bullous Keratopathy	Phthisis
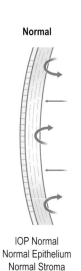	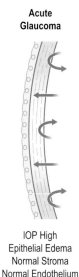	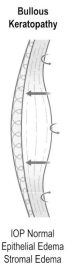	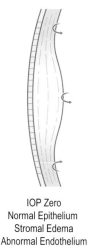
IOP Normal	IOP High	IOP Normal	IOP Zero
Normal Epithelium	Epithelial Edema	Epithelial Edema	Normal Epithelium
Normal Stroma	Normal Stroma	Stromal Edema	Stromal Edema
Normal Endothelium	Normal Endothelium	Abnormal Endothelium	Abnormal Endothelium

Figure 4.26 The delicate balance between stromal swelling pressure, endothelial barrier, metabolic pump function, and intraocular pressure are illustrated. Usually if endothelial cell pump function fails and IOP remains at normal physiologic levels, both stromal and epithelial edema occur (center right). Only when IOP increases above the swelling pressure of the stroma and the endothelium functions normally, does epithelial cell edema occur by itself (center left); and only when IOP is around zero and the endothelium functions abnormally, does stromal edema occur by itself (farther right). (Modified from Hatton MP et al. Exp Eye Res 2004; 78:549–52.)

pressure in a charged gel, like the corneal stroma, results from ionic imbalances. The fixed negative or anionic charges on the corneal stromal proteoglycan GAG side-chains have a central role in this effect. The anti-parallel GAG duplexes (tertiary structure) produce long-range electrostatic repulsive forces that induce an expansive force termed swelling pressure (SP). Because the corneal stroma has cohesive and tensile stiffness (elasticity) that resists expansion, the normal SP is around +55 mmHg.[246,247] If the stroma is compressed as occurs with increasing IOP or mechanical applanation or is expanded as occurs with corneal edema, the SP will correspondingly increase or decrease. Conversely, the negatively charged GAG side-chains also form a double-folded helix in aqueous solution (secondary structure) that attracts and binds Na^+ cations, resulting in an osmotic effect, leading to the diffusion and subsequent absorption of water. Thus, the central corneal thickness is maintained at an average value of 544 μm because the fixed negatively charged proteoglycans induce a constant swelling pressure through anionic repulsive forces, yet it still tends to imbibe more water via its cationic attractive forces.[105]

Under normal circumstances, the negative pressure drawing fluid into the cornea, called the imbibition pressure (IP) of the corneal stroma, is approximately −40 mmHg.[248] This implies that the negative charges on corneal proteoglycan GAG side-chains are just over one-quarter saturated or bound with Na^+, and that the remaining unbound proportion is still available to bind more Na^+ and absorb more water if given the opportunity. Normally, the highly impermeable epithelium and the mildly impermeable endothelium keep the diffusion of electrolytes and fluid flow in the stroma to such a low level that the endothelium's metabolic pumps can maintain stromal hydration in the normal range of 78 percent without significant effort. Although IP = SP when corneas are in the ex vivo state, IP is lower than SP in the in vivo state because the hydrostatic pressure induced by intraocular pressure (IOP) must now be accounted. This is best represented by the equation IP = IOP − SP and explains why the hydration level of a patient's cornea is not only dependent on having normal barrier function, but also on having a normal IOP.[248] Therefore, a loss

of corneal barrier function, an IOP ≥ 55 mmHg, or a combination of the two typically results in the clinical appearance of corneal edema (Fig. 4.26).[23,249,250]

The topic of corneal edema is important for clinicians to understand because it affects the function of the corneal stroma and the epithelium. With minor (< 5 percent) hydration changes, the corneal thickness changes with slight to no effect on the refractive, transparency, and mechanical functions of the cornea. For example, during sleep, a diurnal increase in hydration occurs causing on average a 6 ± 3 percent increase in corneal thickness of the cornea (range: 2–13 percent; stromal = 6 percent and epithelial = 8 percent), mainly because of reduced oxygen levels (from 155 to 55 mmHg) caused by eyelid closure and secondarily from decreased evaporative loss (from 3 μl/hr cm² to 0 μl/hr cm²) caused by eyelid closure.[191,251,252] Upon awakening and eyelid opening, the corneal hydration and thickness reverts back to normal within 1–2 hours.

If the cornea becomes hydrated by 5 percent or more above its normal physiologic level of 78 percent, it begins to scatter significant amounts of light and loses its transparency. Some loss of refractive function may also occur, particularly if the epithelial surface becomes too irregular. Epithelial edema clinically causes a hazy microcystic appearance to occur in the epithelium in mild to moderate cases, significantly decreasing vision and increasing glare. It also can cause the development of large painful, sub-epithelial bullae in severe cases. These changes correlate histopathologically with hydropic basal epithelial cell degenerative changes due to intraepithelial fluid accumulation and the development of interepithelial cellular fluid-filled spaces, known as cysts and bullae. If bullae are chronically present, a fibrocollagenous degenerative pannus will often times form in the subepithelial space decreasing vision further, but, paradoxically, this reduces the pain. In comparison, corneal stromal edema clinically appears as a painless, hazy, thickening of the corneal stroma resulting in a mild to moderate reduction in visual acuity and an increase in glare. At the same time, Descemet's membrane folds commonly appear on the posterior corneal surface. Histopathologically, these changes correlate with the light microscopic findings of thickening of the corneal stroma in the posterior cross-sectional direction with loss of the normally present artifacteous interlamellar stromal clefting.[253] Ultrastructural and biochemical studies have further shown that stromal edema results in an increase in the interfibrillar distance and disruption of the spatial order between collagen fibrils, a decrease in the refractive index of the extracellular matrix, loss of proteoglycans, and, perhaps most importantly, hydropic degenerative changes or cell lysis in the resident keratocyte population.[250,254–256]

Although depth-related differences in the concentration of the two types of negatively charged proteoglycans may account for the higher hydration levels in the posterior stroma compared to the anterior stroma, it appears that the directional orientation of the collagen fibrils and the degree of lamellar interweaving have the greatest influence on the amount of regional stromal thickening from edema-related swelling.[106,257] Because the collagen fibrils in the cornea run from limbus to limbus, the corneal stroma highly resists circumferential expansion; however, it can expand anterior-posteriorly, mostly in the posterior direction. The depth-related differences in lamellar interweaving explain why the anterior third of the cornea mildly swells and actually maintains the anterior corneal curvature even when the remaining posterior two-thirds swells up to three times its normal thickness.[250] Because fibrotic corneal scars have random directionally oriented collagen fibrils, they too have been found to resist swelling under edematous conditions.[258] Therefore, although it is commonly stated that corneal thickness and interfibrillar spacing increase linearly with increasing hydration of the corneal stroma, this relationship mainly applies to the posterior two-thirds of the corneal stroma.[247,250]

Finally, while epithelial and stromal edema commonly co-exist together, there are two notable exceptions. As the epithelium has much weaker cohesive and tensile strengths than the corneal stroma, its state of hydration is mainly dictated by IOP levels.[259] Conversely, because collagen fibrils in the corneal stroma are anchored at the limbus for 360°, they exert increasing or decreasing cohesive strength on the corneal stroma (i.e. compression or decompression of stromal tissue) as the IOP elevates above or decreases below normal, respectively. This results in the transmission of stromal edema to the epithelial surface in cases of high IOP or into the stroma in cases of low IOP. Therefore, if IOP is ≥ 55 mmHg with normal endothelial barrier and metabolic pump function, epithelial edema usually occurs by itself. In comparison, if endothelial cell dysfunction and hypotony (IOP ~0 mmHg) occur together, then stromal edema occurs alone (Fig. 4.26).

Box 4.5 Endothelial cell injury

There are many exogenous stresses that could potentially damage the corneal endothelium (Fig. 4.27). Although accidental trauma and infection are perhaps the most common, they usually are preventable or are difficult to prognosticate since so many variables need to be considered. Several common, more predictable intentional interventions that might stress a person's corneal endothelium are contact lens wear, excimer laser-based keratorefractive surgery, and intraocular surgery.

Contact lenses

Contact lens wear does not cause loss of ECD, but it can cause acute reversible corneal edema and, in the long-term, increased polymegathism and decreased pleomorphism.[215,260] The corneal endothelium utilizes the same carbohydrate metabolic pathways as the corneal epithelium. However, the transport function of the endothelial cell is higher than that of the epithelial cell due to its high baseline metabolic rate, which requires oxidative activity that is 5–6 times that of the epithelial cell.[261] Atmospheric oxygen is the primary source of oxygen to the endothelium. Interruption of this oxygen supply by low oxygen transmissibility contact lenses or a low oxygen environment (e.g. eyelid closing) will result in a shift to anaerobic metabolism, a concurrent increase in stromal lactic acid and CO_2 production, and a drop in stromal pH.[262] Hypoxia can also stimulate epithelial production of 12(R)HETE, a potent inhibitor of the endothelial Na^+/K^+-ATPase pump sites.[263,264] Acute reversible endothelial changes observed with hypoxia include stromal swelling, endothelial dysfunction, and endothelial blebbing. Chronic hypoxia can eventually lead to irreversible endothelial polymegathism and pleomorphism. In fact, Polse and associates have shown that chronic corneal hypoxia in humans alters the endothelium's ability to reverse induced swelling.[265]

Surgical injury

Excimer laser-based keratorefractive surgery has been found to induce acute loss of barrier function and reversible endothelial cell stress (increased polymegathism and decreased pleomorphism) only if performed on a cornea with a residual stromal bed thickness ≤ 200 μm due to the shockwave effect from photoablation. As refractive surgeons rarely or never ablate to this depth due to concerns for the risk of ectasia, there appear to be no short- or long-term endothelial effects from this type of surgery in the current clinical setting.[215]

By comparison, all intraocular surgeries have been found to cause varying degrees of both acute and, perhaps more importantly, chronic damage to the corneal endothelium. Modern small incision cataract surgery (≤ 1.5 mm incisions for microincision cataract surgery [MICS] ; ~3.5 mm incisions for Kelman phacoemulsification [KPE]) is the preferred technique in the USA, Europe, and other more developed countries as randomized studies have shown that it results in better clinical outcomes compared to larger incision techniques (~7–12 mm incisions for extracapsular cataract extraction [ECCE]).[266] However, KPE does cause significant endothelial cell injury due to a number of factors, such as corneal distortion, ricocheting of nuclear fragments, intraocular lens or intraocular instrument contact, and release of free radicals.[267] A recent randomized controlled trial showed that KPE caused an exponential reduction in central ECD up to 1 year after surgery with ECD loss averaging 10.5 percent.[267] However, no significant changes in polymegathism or pleomorphism were found.[267] This fast component period of cell loss is statistically similar to that observed following ECCE, another common surgical alternative technique used for performing cataract surgery, particularly in developing countries, which results in an average 9.1 percent reduction in ECD at 1 year after surgery.[267] Long-term slow component cell loss data are currently unknown for KPE as it has not been well studied beyond 1 year postoperatively, whereas ECCE data show that an annual cumulative cell loss rate of 2.5 percent exists from 1 to 10 years after surgery (four-fold higher than normal physiologic annual cell loss rates).[25,260,267,268] There

Box 4.5 Endothelial cell injury—cont'd

currently is no significant evidence to support MICS over KPE or to support one cataract technique over another (e.g. divide-and-conquer vs. phaco-chop) from an ECD loss standpoint as they all seem to be statistically similar.[269,270]

Recent data on the two phakic refractive IOL procedures commercially available in the USA, Verisyse (Ophthec, Gronigen, The Netherlands/AMO, Santa Ana, California) and the foldable Visian Implantable Collamer Lens (ICL, Visian ICL 4; STAAR Surgical Co., Monrovia, California), show less acute component damage than standard cataract surgery (Verisyse: ~7 percent ECD loss and Visian ICL: ~3 percent ECD at 1 year after surgery).[271,272] However, just as with ECCE, a chronic, slow component annual cumulative cell loss rate is found from 1 to 5 years postoperatively, which is approximately 4–5-fold higher than normal physiologic cell loss rates (Verisyse: ~2.7 percent ECD loss/year and Visian ICL: 2.5 percent ECD loss/year between 1 and 5 years after surgery).

Corneal transplantation surgery (e.g. PK) has been found to cause the greatest long-term decrease in central ECDs of all the commonly performed intraocular anterior segment surgeries, perhaps because of the peripheral loss of stem-like cells or the peripheral storage zone (Fig. 4.23A).[223,228] Long-term longitudinal studies up to 20 years after PK surgery show that ECD loss occurs in two phases: a fast and a slow component.[273] During the fast component, the central ECD decreases exponentially with 36.7 percent ECD loss at 1 year and 8.4 percent ECD loss/year up to 5 years after surgery.[266] Thereafter, a slow component occurs where central ECD loss decreases to a linear rate of 4.2 percent ECD loss/year.[273] Concurrently, polymegathism gradually increases and pleomorphism gradually decreases throughout the longitudinal follow-up period.

Over the last decade, surgical techniques for corneal transplantation have changed markedly as techniques such as posterior lamellar keratoplasty (PLK), deep lamellar endothelial keratoplasty (DLEK), Descemet's stripping endothelial keratoplasty (DSEK), Descemet's stripping automated endothelial keratoplasty (DSAEK), and Descemet's membrane endothelial keratoplasty (DMEK), have evolved to take the place of PK for endothelial cell disorders.[274,275] Currently, the simplest and most efficient technique is DSAEK surgery, which is preferred by patients and surgeons compared to the more traditional full-thickness PK surgery.[276] The preferred approach of DSAEK, however, may soon be supplanted by DMEK.[277,278] One of the major concerns with DSAEK surgery, second only to the high dislocation rate, is how the endothelium survives long-term in these grafts. Surgeons commonly fold the ~8 mm diameter DSAEK graft and insert them through small incisions (~3.5–5 mm incisions) using various grasping forceps, injectors, or insertion devices, and then inject air in the anterior chamber to keep the graft in place for 1 hour to up to 2 days in duration after the surgery. Long-term studies show similar short-term rates of ECD loss (35–39 percent central ECD cell loss at 1 year postoperatively) to those observed after PK.[279,280]

Finally, there are other surgical adjuvants or new procedures on the horizon that could affect the corneal endothelium. The surgical adjuvant that is currently of most interest is topical mitomycin C (MMC) application, which is applied after excimer laser-based keratorefractive surface surgery to prevent or treat corneal haze caused by subepithelial scarring. Currently, there appears to be no evidence of short- or long-term ECD loss or keratocyte toxicity, if a single local application of MMC is used at a concentration of 0.02 to 0.002 percent for an exposure duration of 12 s to 2 min.[281–285] The medical procedure that is currently of most interest is corneal collagen cross-linking with riboflavin and UVA light (CXL), which is a promising treatment option for the disease of corneal ectasia.[286,287] Currently, no evidence of endothelial cell damage has been found, but most of the studies to date are preliminary, short-term, small in number, and are under strict research protocols.[23,288,289]

Pharmacologic toxicity

In addition to surgical injury, the corneal endothelium can also be influenced by pharmacologic toxicity.[215,290,291] Past studies that helped guide the development of intraocular irrigating solutions found that the best were those most similar to aqueous humor in composition. BSS Plus (Alcon Laboratories, Ft Worth, TX, USA) is currently the most physiologically compatible intraocular solution (Table 4.3); BSS is probably the next best alternative (Table 4.3). The main ingredients required for intraocular solutions to be biologically compatible with the corneal endothelium are electrolyte levels matching those in the aqueous humor, glucose as an energy source, bicarbonate as a buffer, and glutathione as an antioxidant/free radical scavenger. Intraocular tissues, particularly the corneal endothelium, require a pH between 6.7 to 8.1 and an osmolality between 270 and 350 mOsm. Furthermore, any medication used intraocularly needs to be at a non-toxic concentration and contain no preservatives. Pharmacologic toxicity to the anterior segment tissues of the eye, including the corneal endothelium, has become better understood through the discovery of a condition known as toxic anterior segment syndrome (TASS).[290] TASS is a sterile postoperative inflammatory reaction caused by a non-infectious substance that enters the anterior segment, resulting in toxic damage to intraocular tissues. Most cases are severe, resulting in >50 percent loss of central ECD. The injury typically starts 12–48 hours after cataract or anterior segment surgery and is limited to the anterior segment of the eye. It is always Gram stain and culture negative and usually improves with steroid treatment. The primary differential diagnosis is infectious endophthalmitis.

The possible causes of TASS include intraocular solutions with an inappropriate chemical composition, drug concentration, pH, or osmolality. It can also be caused by preservatives, enzymatic detergents, bacterial endotoxin, oxidized metal deposits and residues, and factors related to intraocular lens processing, such as retained residues from polishing or sterilizing the lenses (Table 4.4).[290] Since TASS is an environmental and toxic control issue, it has made anterior segment surgeons and surgical staff more aware that maintaining the health of the corneal endothelium requires a thorough understanding of all medications and fluids used during surgery. Additionally, it has helped all involved in surgical eye care understand the importance of proper cleaning and sterilization of intraocular instruments since most cases of TASS appear to be directly caused by retained detergents or contaminated water sources.

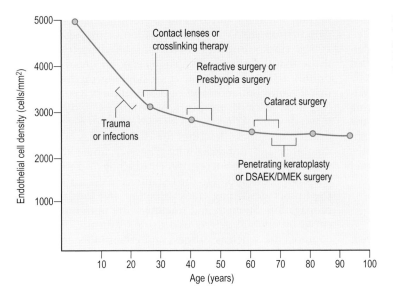

Table 4.3 Composition of aqueous humor compared to various intraocular irrigating solutions

Ingredient	Human aqueous humor	BSS Plus	BSS
Sodium	162.9	160.0	155.7
Potassium	2.2–3.9	5.0	10.1
Calcium	1.8	1.0	3.3
Magnesium	1.1	1.0	1.5
Chloride	131.6	130.0	128.9
Bicarbonate	20.15	25.0	–
Phosphate	0.62	3.0	–
Lactate	2.5–4.5	–	–
Glucose	2.7–3.7	5.0	–
Ascorbate	1.06	–	–
Glutathione	0.002–0.010	0.3	–
Citrate	–	–	5.8
Acetate	–	–	28.6
pH	7.38	7.6	7.4
Osmolarity (mOsm)	304	305	298
Protein	0.135–0.237	–	–

All concentrations are expressed in millimoles per liter or millequivalents per liter of solution.

Table 4.4 Causes of toxic anterior segment syndrome

1. *Irrigating solutions or viscoelastic devices*
   - Incomplete chemical composition
   - Incorrect pH (<6.7 or >8.1)
   - Incorrect osmolality (<270 mOsm or >350 mOsm)
   - Preservatives or additives (e.g. antibiotics, dilating medications)
2. *Ophthalmic instrument contaminants*
   - Detergent residues (ultrasonic, soaps, enzymatic cleaners)
   - Bacterial LPS or other endotoxin residues
   - Metal ion residues (copper and iron)
   - Denatured viscoelastics?
3. *Ocular medications*
   - Incorrect drug concentration
   - Incorrect pH (<6.7 or >8.1)
   - Incorrect osmolality (<270 mOsm or >350 mOsm)
   - Vehicle with wrong pH or osmolality
   - Preservatives in medication solution
4. *Contaminated water sources*
   - Water baths
   - Autoclave reservoirs
   - Non-sterile or non-pyrogen-free water
5. *Intraocular lenses*
   - Polishing compounds
   - Cleaning and sterilizing compounds

Basement membrane and glycocalyx

A secondary function of the corneal endothelium is its ability to secrete a basement membrane along its basal surface called Descemet's membrane, which is continuously deposited throughout life (Fig. 4.22B).[1,6] Although some collagen fibrils from the posterior stroma are embedded in the Descemet's membrane, it really has no major junctional or adhesional complexes to the posterior stroma other than a 0.5 µm thick layer of fibronectin.[1] Descemet's membrane is highly extensible and tough, but less strong and stiff than the posterior stroma since it is primarily composed of type IV and VIII collagen fibrils as well as the glycoproteins fibronectin, laminin, and thrombospondin. At birth, it averages 4 µm thick.[6] On electron microscopy, the fetal Descemet's membrane is composed of many wide-spaced, 110-nm banded collagen fibrils (see Fig. 4.22B).[292] Postnatally, collagen is gradually added posteriorly to this initial fetal layer throughout life, being notably different in morphology from the fetal layer as it is non-banded and contains small-diameter collagen fibrils that are arranged into a hexagonal lattice (see Fig. 4.22B).[292] Typically, at the end of a normal lifespan, Descemet's membrane measures around 10–15 µm thick (4 µm thick banded layer and a 6–11 µm thick non-banded layer). With disease (e.g. Fuchs' dystrophy) or injury (e.g. trauma or surgery), Descemet's membrane may become focally (corneal guttae) or diffusely thicker than normal from abnormal collagen deposition. This newly deposited abnormal basement membrane is called the posterior collagenous layer of the Descemet's membrane and is classified into one of three types: banded, fibrillar, or fibrocelluar.[293] Excess basement membrane deposition is a common strategy used by cells attempting to recover from injury, allowing them to remain attached to a tissue's scaffolding. Finally, the endothelium is also known to secrete a 0.7 μm thick glycocalyx layer on its apical or posterior surface.[215] Functionally, it is thought that the glycocalyx layer may protect the internal (or posterior) surface membrane of the endothelium.

Mechanical properties

As the cornea is a pressurized, thick-walled, partially woven, unidirectionally fibril-reinforced laminate biocomposite, it represents an excellent compromise between stiffness, strength, extensibility, and therefore toughness to withstand internal and external forces that may stress it, distort its shape, or threaten its integrity.[1,90,294,295] The biologic behavior of a soft fibrous connective tissue, like the human cornea, usually is much stronger in the direction of its collagen fibrils than perpendicular to it and these fibrils are assembled into various hierarchical structures, which give the tissue anisotropic mechanical properties.[296–298] Maturity and age-acquired covalent collagen cross-links serve to bolster the stiffness, strength, and toughness of these hierarchical structures, usually without significantly compromising extensibility (Fig. 4.6C). As the biomechanical properties of the cornea are dominated by the stroma, the macro-, micro-, and nano-mechanical behaviors of the cornea is primarily due to the hierarchical structure of essentially four composite-like regions: Bowman's layer, the anterior third of the stroma proper, the posterior two-thirds of the stroma proper, and Descemet's membrane (Fig. 4.28).[93,299–306] Like most composites, the cornea is biomechanically strong, light in weight (when dehydrated), and has an extraordinary capacity to absorb energy by material destruction due to its fluid-like matrix gel; however, it also has a composite's shortcomings in that its dynamic three-dimensional state of stress is difficult to definitively quantify and its biomechanical failure process is complex since it appears to break down at the nano- and micro-mechanical level.

When a biologic tissue is subjected to force (or load), it typically will deform (elongate, compress, or shear) in the direction of the applied force. The force divided by the cross-sectional area of the applied stress (N/m^2) is called the stress acting on the material, while the degree of deformation in the direction of the applied stress is called strain. The ratio of the stress and strain (i.e. the slope of the stress–strain curve) is called the elastic modulus or Young's modulus of the tissue, which basically describes the stiffness of the material. Stiff materials are needed to transmit forces and are especially crucial in resisting bending or buckling from compression. The strength of a material is defined as the maximum stress it can sustain before breaking. Strong materials are needed to carry a load. The extensibility of a tissue is defined as the maximum strain it can sustain before rupturing. Extensible materials are needed to reversibly change elastic shape instantaneously and then return to its original shape without damage. Toughness is defined as the ability of the material's micro- and nano-structure to dissipate deformation energy without initiation or propagation of a critical crack. Toughness is a common characteristic of a tissue that deforms to dissipate energy on impact (i.e. not brittle) and/or has natural crack-stopping mechanisms so that it avoids initiation and/or propagation of a critical crack (i.e. work of extension [work per volume or work per unit mass] or work of fracture [work per surface area created]). In essence, toughness describes the characteristics of a tissue's durability and unbreakability. The area under the stress–strain curve or the energy needed to elongate a pre-existing crack both correspond to the toughness of the material, with the former describing crack initiation and the latter crack propagation. Strength and toughness are both highly influenced by internal defects in a tissue's structure or material components, whereas stiffness and extensibility are not.

Corneal stress

The dome-like cornea encounters three types of loads or stresses: transmural pressure, piercing, and crushing. The static net internal pressure constantly stressing the cornea in an outward direction is the IOP (16 ± 3 [S.D.] mmHg above external atmospheric pressure [760 mmHg] and the external resting tension from the eyelids).[307–309] Additional dynamic stressors that can occur as a result of normal variability or external environmental stress include various causes of increased IOP, such as accommodation (4 mmHg), turning the eyes (10 mmHg), arterial pulsation (1–2 mmHg), diurnal changes (5 mmHg), respiration (5 mmHg), Valsalva maneuver (8 mmHg) and recumbent or inverted positions, external eyelid blinking (normal blinks = 5–10 mmHg; hard squeeze blinks = 50–110 mmHg), eye rubbing (increases IOP [light digital rubbing = 5–20 mmHg; hard knuckle grinding/rubbing, = 25–135 mmHg], induces shear and transverse compression on the corneal stroma, and may actually indent the cornea), or accidental eye impact.[307–311] An increase in internal pressure causes the cornea's stress behavior to be dominated by transmural longitudinal tension, which is usually very well tolerated since the net vector direction of stress is essentially parallel to the direction of most of the collagen fibrils in the structure. In contrast, external pressure, such as from eye rubbing (back and forth or circular pressure and friction), nocturnal external eye pressure, or direct blunt impact trauma, causes radial compressive stress, circumferential tensile stress, and/or deformation. Since the net vector direction of stress is oblique to the direction of most collagen fibrils in the cornea, external stress is more likely to initiate or propagate internal defects due to fatigue-related damage in the corneal stroma than internal stress. Moreover, eye rubbing's often repetitive or oscillatory ECM shearing and subsequent cellular damage results in both transmural pressure and crush-related injury.

Using a basic understanding of mechanical physics, the cornea should theoretically obey Pascal's Principle, which states that the pressure is the same everywhere inside an enclosed system or vessel at equilibrium in addition to contributions already caused by static fluid pressure. Despite this constant IOP, differences in wall tension exist because according to the Law of LaPlace – the tension in the walls of an enclosed vessel are directly dependent on the pressure in the system and its radius of curvature and inversely

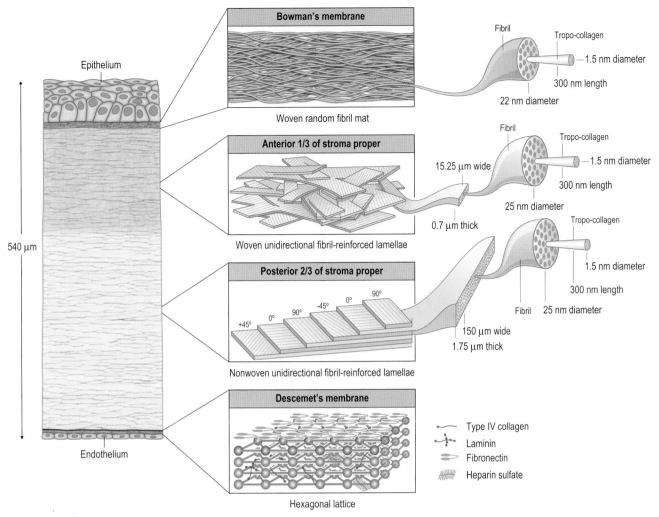

Figure 4.28 The hierarchical structure of the cornea showing that it is basically composed of three composite-like regions. A fourth composite-like region, the Descemet's membrane, is included for completion's sake. The macroscopic to microscopic to nanoscopic features are emphasized (from left to right) to help illustrate the various interactions between the tissue components. The Bowman's layer is essentially a random fibril, woven mat composite, which maximizes multiaxial stiffness and strength. The underlying anterior third of the stroma proper is a lamellar interwoven fabric composed of unidirectionally (UD) fibril-reinforced lamellae. This architectural hierarchy is much more reinforced against z-axis deformations compared to non-woven UD-laminates. In the human body, it is most similar to that of pericardium, which serves in mechanically preventing aneurysm formation of the heart. The posterior two-thirds of the stroma is essentially a non-woven, UD-fibril-reinforced lamellar composite, which maximizes longitudinal x- and y-axis stiffness and strength, but has weak transverse z-axis stiffness and strength. In the human body, it is most similar to that of the annulus fibrosis of intervertebral disk, which functions efficiently as a cushioning mechanism for the spine, but is prone to chronic biomechanical failure. The UD-orientation of collagen fibrils in each lamellae is important because this arrangement prevents fibril undulation and thus maximizes the initial axial tensile strength of each individual fibril. Descemet's membrane forms a hexagonal lattice. In toto, these composite-like regions characterize the overall stiffness, strength, extensibility, and toughness of the cornea. They also help explain how the cornea biomechanically behaves normally after surgery, disease, or injury.

dependent on the thickness of the vessel wall. For example, in an isotropic, thin-walled (wall thickness to cavity diameter ratio ≤ 0.1) elastic vessel, the Law of Laplace would be simplified to the following equation:

$$T = PR/2t$$

where T equals the radial wall tension or stress of a spherical vessel wall, P equals the internal pressure in the vessel (IOP), R equals the radius of curvature of the wall (the curvature of the posterior corneal surface), and t equals the thickness of the wall in meters (corneal thickness). Unfortunately, the Law of Laplace is not a perfect approximation of stress in the corneal eye wall in the real world since the cornea is actually anisotropic, inhomogeneous, asymmetric, and thick-walled

(Fig. 4.28). Thus, one can see that although the static stress can be approximated (Fig. 4.29A), the dynamic temporal stresses acting upon the corneal eye wall are quite complex and perhaps too difficult to definitively characterize, particularly since they overlap one another in real-time.[312,313]

Corneal stiffness, strength extensibility, and toughness

The two principal structural characteristics of the human cornea that help maintain its stiffness (rigidity or elasticity), strength (ultimate fracture stress), extensibility (ultimate fracture strain), and toughness (durability or unbreakability) are its thickness and its innate biomechanical properties (hierarchical structure and degree of collagen fibril cross-linking). While corneal thickness can be directly measured accurately in vivo, a major difficulty in the field of

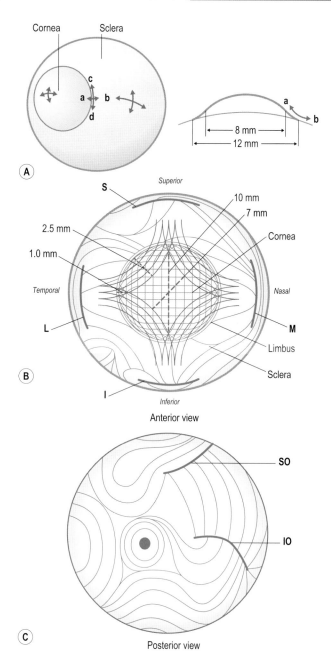

Figure 4.29 The stress on the coats of the eye wall (red lines with double-headed arrows are shown in (A)) and the preferential orientation of the collagen fibrils (solid gray and blue lines) are shown in (B) and (C). (**A**) The lowest radial stress is in the cornea due to its smaller radius of curvature; the largest radial stress is under the rectus muscle insertion sites or at the equator of the globe due to its large radius of curvature and the fact that these are the thinnest regions of the sclera. The largest circumferential stress is at the limbus due to the sudden change in curvature going from the cornea to the sclera, which doubles the stress in the circumferential direction compared to the radial direction. (**B**) The central 4–5 mm of the cornea has collagen fibrils that are relatively uniformly directed in either the superior–inferior or the nasal–temporal direction. These fibrils appear to bend in the periphery of the cornea to coalesce and form the circumferential annulus. The diamond-shaped appearance of the central orthogonal collagen fibrils suggests that additional anchoring collagen fibrils come in from the sclera and limbus at one of the principal meridians (inferior, nasal, etc.), then curve within the peripheral cornea to exit an adjacent principal meridian 90° away. Overall, this arrangement would aid in maintaining the peripheral flattening of the cornea and resisting the higher stress in that region of the cornea. It secondarily would not interfere with the central optical zone of the cornea. (**C**) The preferential direction of the sclera's collagen fibrils is radial in the anterior sclera, particularly in relation to the muscle insertion sites, whereas those at the equator are circumferential and those at the posterior sclera form circular whirls. The deep layers of sclera, which are not shown, form a net- or rhomboid-like layer diffusely throughout all regions of the sclera. (Modified from Meek KM. In: Frazl P, ed. Collagen: structure and mechanics. New York: Springer, 2008:359–96; Aghamohammadzadeh H, Newton RH, Meek KM et al. Structure 2004; 12(2):249–56; and Watson PG, Hazelman B, Pavesio C, Green WR. The sclera and systemic disorders, 2nd edn. Edinburgh, UK: Butterworth Heinemann, 2004.)

ophthalmology has been an adequate actual direct measurement or even an accurate approximation of its innate biomechanical properties.[314] The current understanding of the cornea's biomechanical properties still mainly relies on ex vivo experiments, primarily strip extensiometry and inflation testing.[315]

The cornea is an important biological mechanotransducer of stress due to its viscoelastic properties (i.e. solids with some characteristics of a fluid), which helps prevent premature biomechanical failure via energy dissipation. The property of viscoelasticity complicates our understanding of corneal biomechanical behavior since it contains both elastic and viscous component properties that overlap one another over time when forming the stress–strain curve. Elasticity is related to time-independent energy storage due to stretching or strain of the flexible domains of collagen molecules (potential energy) and deformation or transfer of energy to the sclera and other portions of the eye due to the rigid

domains of collagen molecules (kinetic energy). In essence, elasticity is a time-independent physical property of a solid that allows the cornea to return to its original shape after stress or, if significantly damaged, to be permanently altered in shape. Viscosity is related to time-dependent energy dissipation due to the sliding of interfibrillar collagen fibrils, filaments, and lamellae past one another in a hydrated proteoglycan matrix (mechanicothermal energy transduction). In essence, viscosity is a time-dependent physical property of a fluid that allows the cornea to deflect some input energy away from its elastic components and it allows the cornea the possibility to reversibly or permanently change its elastic shape with time.

Ex vivo experiments show that the human cornea has J-shaped, non-linear, stress–strain loading and unloading curves. The cornea's stress–strain curves have an immediate elastic response dominated initially by the low strain behavior of its non-collagenous matrix and the act of removing slack from its collagen fibrils (fibril uncrimping) as well as re-orienting its collagen fibrils to the direction of the stress (fibril recruitment) at the lower part of the J and then later by the high strain behavior of collagen fibrils that are maximally taught and oriented in the direction of stress in the upper part of the J.[316] Finally, a slower, time-dependent viscous response occurs in addition to the immediate J-shaped response due to interlamellar and interfibrillar slippage resulting in time-dependent strain (creep) or stress (stress relaxation) curves.[316] Creep is defined as a time-dependent elongation that occurs in three stages: (1) primary creep with its associated decelerating strain rate, (2) secondary creep with its associated constant strain rate, and (3) tertiary creep with its accelerating strain rate until rupture.[317]

Primary and secondary creep are reversible viscoelastic responses without accompanying damage, whereas tertiary creep is associated with an irreversible reduction in the strength and toughness of the structure due to damage by irreversible slippage or cracking. The cornea's unloading stress–strain curve does not overlap with the loading curve, being variably less dependent on the load or strain rate. The difference in area between the two curves represents the energy dissipated via viscous friction (i.e. hysteresis) during a single mechanical cycle. As the cornea has very small collagen fibril diameters and lengths compared to other connective tissues, like tendon and bone, and has biomechanical properties between skin and ligament, it is reported that the cornea converts approximately 35 percent of its total input energy into energy dissipation and the other 65 percent into energy storage, energy transmission, or deformation. It should be pointed out though that a single stress–strain curve has limited meaning when evaluating a material's or structure's total dynamic, temporal biomechanical properties since each curve is either load or strain rate dependent over a specific time interval and each represents only one of several different material components and orientations of the structure.

Based on the hierarchical structure of the corneal stroma (Fig. 4.28), it has been inferred that the anterior woven portion of the cornea is much stiffer and stronger than the posterior non-woven portion. In fact, this inference has been qualitatively supported by both clinical and experimental studies, which show that the posterior two-thirds of the corneal stroma is easier to bluntly dissect in a lamellar fashion than the anterior third and that the anterior corneal curvature remains relatively constant at various physiologic IOPs and stromal hydration levels compared to the posterior corneal curvature.[318–322] After induction of corneal edema in an ex vivo setting, ultrastructural studies by Muller and associates showed that the 10 µm thick Bowman's layer and the underlying anterior 100–120 µm of the stroma proper are the most stiff regions of the cornea since they did not swell significantly, even when stromal edema resulted in total central stromal thicknesses of up to 1200 µm.[255] Quantitative ex vivo direct measurements of human corneal stiffness are very consistent with these initial observations and inferences. Seiler and associates showed, using uniaxial strip extensiometry, that removing the Bowman's layer with the excimer laser reduced the longitudinal (x- and y-axis lengthwise dimension) Young's modulus on average by 4.75 percent.[302] This was similar to findings discussed in a 1989 National Eye Institute-sponsored corneal biophysics workshop that described the longitudinal Young's modulus of

Bowman's layer as about 50 MPa/mm^2 and that of the stroma proper as ranging between 2 to 10 MPa/mm^2.[323]

Uniaxial strip extensiometry further demonstrated that the anterior third of the corneal stroma is 2–3-fold stiffer and stronger than the posterior two-thirds.[303,324,325] Using inflation tests, Descemet's membrane was measured to have a longitudinal Young's modulus of 0.5–2.6 MPa/mm^2 depending on the physiologic stress levels used during testing, which overall is less than that of the corneal stroma.[304,326] Recent uniaxial strip extensiometry measurements show considerable regional and directional anisotropy in the longitudinal cornea as vertical and horizontal strips were stiffer (53 percent and 40 percent, respectively) and stronger (25 percent and 13 percent, respectively) than diagonal strips.[327] Cohesive or transverse (z-axis thickness-wise dimension) strength measurements, using modified strip extensiometry techniques, showed that Bowman's layer averages 50 g/mm, the anterior third of stroma proper averages 34 g/mm, the posterior two-thirds of stroma proper averages 20 g/mm, and Descemet's membrane averages 7 g/mm.[28,93] This same study importantly documented a depth-dependent decrease in the cohesive strength in the corneal stroma, with a maximum at Bowman's layer and then an exponential decrease until reaching a plateau at approximately 40–80 percent in depth into the stroma before decreasing again in the posterior-most 20 percent. An approximately 2-fold increase in cohesive strength was also found comparing the peripheral cornea to central cornea. Interestingly, these z-directional cohesive strength measurements were quantitatively 5–50-fold less than the longitudinal x- and y-direction tensile strengths.

Inflation studies have further shown that cutting or ablating Bowman's layer down to approximately 150 µm depth into the anterior corneal stroma results in anterior corneal flattening and peripheral corneal thickening; in comparison, deeper cuts or ablations into the stroma result in progressively more anterior corneal steepening rather than flattening.[328–330] According to Hjortdal, after taking into account load-induced volume change, shallow ablations resulted in anterior corneal flattening because of similar inward inner and outer corneal wall strains.[331] In contrast, deeper ablations result in anterior corneal steepening because outward outer corneal wall strains occur up to 2-fold higher than inward inner corneal wall strains.[331] Their conclusion was that the difference between outer and inner corneal wall strains was due to less shear resistance in the posterior two-thirds of the corneal stroma. Finally, using inflation testing, increasing age and increasing IOP have both been associated with increased stiffness of the human cornea, presumably due to age-related cross-linking of the collagen fibrils and the changes in the viscoelastic properties of the interfibrillar matrix with increasing IOP, respectively.[330] Clearly, further information about the innate biomechanical material properties of the human cornea are still needed since the current information is far from complete.

Chronic biomechanical failure of the cornea – ectasia
In classical mechanics, acute damage and rupture occur when the stress reaches a critical value, which in tension is called the ultimate tensile stress (UTS). In addition to overload-induced acute rupture, if you dissect a tendon and hang a lesser load (below the UTS) from it while keeping it moist,

in a chronic, time-dependent fashion the tendon will become damaged or fatigued and eventually ruptures. This occurs because of tertiary creep and its associated micro- and nano-structural pathogenesis of interfibrillar crack initiation and propagation (known as slippage for biologic tissues). In fact, susceptibility to fatigue is a universal phenomenon in biological tissues, which typically fail through constant static load (time-dependent damage due to magnitude of the constant load and time of application of this load). In comparison, man-made structural materials typically fail through dynamic oscillatory loads (cycle-dependent damage due to magnitude of the load and frequency of the cycling time). Cycling of the load may also accelerate the time-to-rupture of biological tissues.

The human cornea's innate biomechanical properties depend on the composition and organization of collagen fibrils and the matrix structures involved in force transmission between successive lamellar and fibrillar layers of collagenous tissue. This framework is essentially composed of 25-nm diameter, heterotypic type I collagen fibrils. The exact mechanical properties of these individual corneal collagen fibrils have not been evaluated or measured specifically. However, rat tendon studies found that individual 50-nm diameter, heterotypic type I collagen fibrils are moderately stiff, strong, and tough with only slight (~10 percent) extensibility, which is usually only evident when loaded under extreme physiologic stress.[332] Therefore, it can be assumed for practical purposes that corneal type I collagen fibrils do not stretch much under normal physiologic conditions or stressors. This seems to imply, at least from a theoretical perspective, that the matrix structures involved in force transmission between lamellar and fibrillar layers (analogous to mortar in brickwork masonry) are the primary structures of the corneal stroma that are vulnerable to fatigue (tertiary creep or irreversible slippage) under normal physiologic circumstances.[333] Ultrastructural and x-ray scattering studies of human corneas with natural ectasia, like keratoconus, or iatrogenic ectasia, like post-LASIK or post-PRK ectasia, show that a chronic two-phase biomechanical matrix failure process of the stroma proper probably is the initiating and evolving pathophysiology, resulting in irreversible interlamellar slippage and subsequent irreversible interfibrillar slippage as opposed to collagen fibril failure (Fig. 4.30).[334–341]

In the scientific field of composite sciences, man-made laminate polymer matrix composites that are most similar to that of the human cornea (e.g. woven and non-woven, UD fiber-reinforced laminate composites made of ductile matrix material) are known to undergo two similar chronic types of biomechanical matrix (or resin) failure processes known as delamination and interfiber fracture (IFF), whereby breakdown of the matrix between lamina or fibers occurs, respectively, causing redistribution of stresses within the laminate structure.[294,342–345] The quantitative amount of stress that causes this damage is the sum of the cycling rate of the stresses acting on the structure, the duration of time exposed to this stress, the temperature of the structure, the curvature and thickness of the structure, the laminate stacking sequence of the individual plies, and the innate biomechanical properties of the fibers and binding matrix. As a consequence of nano- and micro-mechanical damage, usually resulting from transverse compression or transverse shear stresses as opposed to transverse tension or

longitudinal shear, further damage to other plies can occur, eventually resulting in macroscopic interlaminar delamination between two plies and full-thickness intralaminar fractures of a ply. In essence, delamination and IFF cause internal stress concentration sites throughout the laminate composite and accumulation of too many of these damage sites can considerably reduce the strength and toughness of the material without detectable thinning or bulging.[339]

Eventually, once the biomechanical failure process progresses to at least a moderate stage, the structure usually thins to measurable or detectable levels. It typically only begins to bulge or undergo pathologic shape changes due to a significant reduction in stiffness or significant increase in extensibility in late moderate or severe stages before ultimately undergoing total structural failure during the end stage. Moreover, during the redistribution of stress phase, further damage to other adjacent plies may occur even without significant external or internal increases in wall stress. Thus, from a composite science perspective, corneal ectasia essentially is the biologic equivalent of both delamination and IFF in the stroma proper and is essentially a stereotypic, non-specific chronic biomechanical failure response to multiple different specific stimuli, such as eye rubbing, LASIK surgery, family history (genetic conditions associated with environmental or behavioral eye rubbing), or possibly external nocturnal eye pressure.[334,342] The irreversible slippage caused by delamination and IFF can either be

---

Box 4.6 Clinical biomechanical testing

Because reliable in vivo biomechanical tests have not yet been developed, the clinical setting currently relies on placido disk topography and tomography (2- and 3-dimensional reconstruction of thickness profiles) studies to indirectly infer information about the innate biomechanical properties of the cornea based on shape and thickness changes, respectively.[55,346] This is the unfortunate reality for clinical ophthalmology and presumably explains why we have trouble predicting the ectasia risk or susceptibility before keratorefractive surgery in some cases and also have trouble predicting the probability of ectasia progression in keratoconus patients (i.e. time-dependent thickness profile changes or shape changes are moderate to late stages in the chronic biomechanical failure process). Only with a proper understanding of the cornea's innate biomechanical properties can an accurate predictor of material behavior be developed, which would help predict ectasia risk and/or progression, optimize refractive surgery, improve accuracy of tonometry measurements, or improve the design of contact lenses. Recently, Reichert introduced an instrument, the Ocular Response Analyzer (ORA; Reichert Corporation; Depew, USA), that measures the corneal response to indentation with an air pulse (principally similar to air pulse non-contact tonometry).[347,348] This is the first instrument that purportedly directly measures the innate biomechanical properties of the cornea in vivo. The two biomechanical properties that it currently measures are corneal hysteresis (CH) and corneal resistance factor (CRF). CH is reported to be IOP-dependent and predominantly reflects the viscous properties of the cornea, whereas the CRF correlates with the elasticity of the cornea. However, discrepancies compared to expected results with these measures (e.g. a decrease in these properties with aging and no change in these properties after CXL treatment) suggest that further work is needed to determine precisely what these two properties measure or represent so that they are clinically meaningful to the clinician.[349–351]

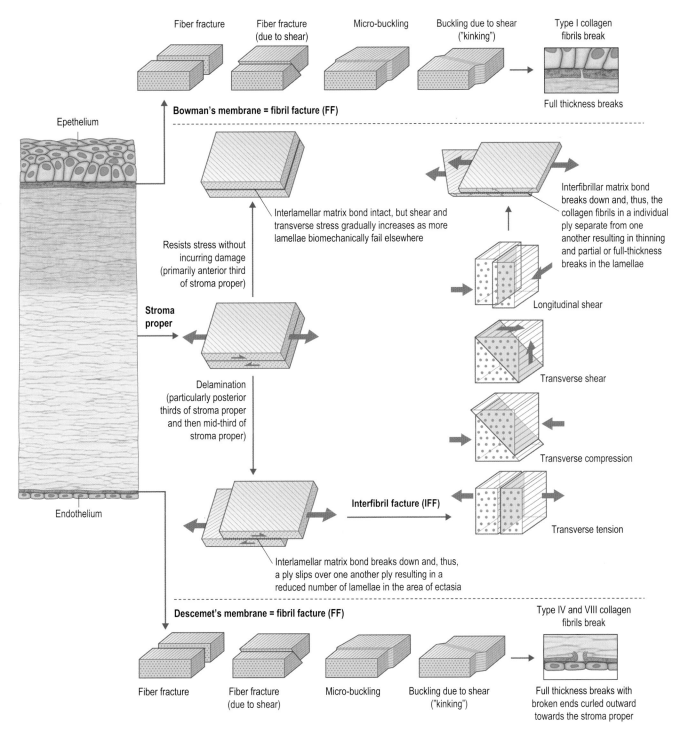

Figure 4.30 Diagram illustrating how the four composite-like regions of the cornea biomechanically fail when undergoing ectasia. In essence, ectasia appears to be a stereotypic chronic biomechanical failure response to multiple different stimuli. According to Alfred Puck's theory, laminated fiber polymer matrix composites, like the corneal stroma, biomechanically fail through three pathways: fiber fracture (FF), interfiber fracture (IFF), and/or delamination. Although FF is seen in Bowman's layer and Descemet's membrane in late stages of ectatic disease, both are not the cause for the ectasia, but rather are secondary events to the biomechanical failure process going on in the stroma proper. The initiating and evolving pathology causing the ectasia is due to delamination and subsequent IFF, which is seen in the stroma proper, particularly in the posterior-most regions of the cornea. This two-step biomechanical failure process leads to ectasia through gradual cumulative damage rather than an acute event. Interestingly, delamination and subsequent IFF can also occur in benign fashion, depending on type and degree of stress. Although delamination and IFF both take place and are clearly seen using ultrastructure studies in the posterior two-thirds of the corneal stroma, it does progress to the anterior third of the stroma proper. However, in the anterior third of the stroma proper, ultrastructurally IFF appears to dominate since the interlamellar slippage resulting from the delamination step is minimized by crack stopping mechanisms due to the interwoven nature of lamellae in this region of the stroma proper. FF usually is associated with acute ultimate failure of its individual composite-like layer, which results in acute loss of biomechanical and barrier function in that specific layer. For example, FF of Bowman's layer results in the clinical appearance of subepithelial fibrotic scarring, which is seen histopathologically as full-thickness breaks in Bowman's layer with outgrowth of hypercellular fibrotic stromal scar tissue. In another example, FF of Descemet's membrane results in the acute clinical appearance of corneal hydrops due to loss of endothelial barrier function, which is seen histopathologically as full-thickness Descemet's membrane breaks with curled outward fracture edges and endothelial cell discontinuity. Delamination may be seen clinically at the slit lamp as Vogt's striae, which are often found during early to moderate stages of ectasia in the posterior-most regions of the corneal stroma. (Adapted from Knops M. Analysis of failure in fiber polymer laminates: the theory of Alfred Puck. Heidelberg, Germany: Springer, 2008; and Reifsnider K. In: McNicol L, Strahlman E, eds. Corneal biophysics workshop. Corneal biomechanics and wound healing. Bethesda, MD: NEI, 1989.)

benign or harmful to overall cornea, with the latter causing a gradual, progressive biomechanical failure process eventually leading to the ultimate failure of the structure through loss of shape and loss of refractive function. Although both delamination and IFF are universally seen in the non-woven posterior two-thirds of the stroma proper, it can progress into the woven anterior third of the stroma proper. However, the woven anterior third of the stroma has a natural crack-stopping mechanism due to the highly interwoven nature of its lamellae. This perhaps also explains why keratoconus rarely results in spontaneous rupture of the eye.

Other functions

Drug delivery

Although there are several drug delivery routes into the anterior segment of the eye (Fig. 4.31A), topical instillation of ophthalmic drugs is the most common method used to administer treatments for ocular disease – evidenced by the fact that 90 percent of ophthalmic drug formulations are geared for topical drop use. The barriers to productive topical absorption of drugs into the anterior chamber are well documented and the two major pathways to permeate these barriers are the transcorneal and transconjunctival pathways. The primary absorption pathway for small, lipophilic drugs is the transcorneal route and the primary absorption pathway for large, hydrophilic drugs is the transconjunctival (conjunctiva–sclera–ciliary body) route. In order to understand these pathways, and therefore predict the drug's biological effects on targeted anterior segment tissues, knowledge of the physicochemical properties of the drug and the pharmacokinetics of the tissues the drug must traverse is essential. Physicochemical properties describe the relationship between the chemical structure of the drug and its biological effects. Various physicochemical properties of a drug – such as its molecular size (radii, weight, and shape), lipophilicity or hydrophilicity, degree of ionization – may optimize its biological effects in the targeted tissue and/or may improve its pharmacokinetic profile. Pharmacokinetics describes the process of absorption, distribution, metabolism, and elimination of a drug in a tissue, organ, or entire body, depending on the drug delivery route and its absorption pathways. It is a function of the intrinsic physicochemical properties of the drug being applied, the applied concentration and dosing frequency of the drug, the static permeability properties of the tissue, the dynamic metabolic and elimination mechanisms of the tissues it permeates, and the bioavailability of the administered drug at its target tissue site. Bioavailability describes how much of an instilled concentration of a drug actually reaches the target tissue.

The advantages of topical drug delivery are its convenience and non-invasiveness, its avoidance of first-pass metabolism in the liver, and its ability to locally target cornea and anterior segment tissues with high drug concentrations. The main disadvantages of this route are its high dynamic drug clearance rates and its static anatomic and physiologic permeability barriers to drug absorption, both of which contribute to low bioavailability. On average, usually less than 5–10 percent of instilled topical drugs reach the aqueous humor or even the corneal stroma (Fig. 4.31B), whereas the major portion (50–99 percent) goes to the systemic circulation, primarily via absorption in the conjunctiva vasculature and lymphatics and/or via absorption in the nasal mucosal vasculature due to drainage in the nasolacrimal duct. Moreover, on average, a drug resides in the conjunctival cul-de-sac for only about 5–6 minutes before being removed by the precorneal tear clearance mechanism (tear turnover rate of 0.5–2.2 μL/min). Even if the topical drug does get into the anterior or posterior chamber via the transcorneal or transconjunctival route, respectively, it usually only stays there briefly unless it is retained in depot form on the external surface or binds to proteins in the tissues it traverses since the aqueous humor removes it through the trabecular meshwork (aqueous humor turnover rate 2–3 μL/min) or venous blood flow removes it through the porous anterior uvea.

After topical drug instillation, the peak concentration in the aqueous humor, or even the corneal stroma, is usually reached 30 minutes to 3 hours after application, but the bioavailability or concentration of drug in the corneal stroma or aqueous humor is reduced to approximately 1/1500th for lipophilic drugs and 1/150,000th for hydrophilic drugs compared to the original concentration of the eye drop.[352,353] From the aqueous humor, the drug does have easy access to the uvea (iris and ciliary body). Here the drug may bind to melanin, which may form a reservoir source that gradually releases the drug to the surrounding cells, thereby prolonging the drug activity (e.g. beta-blockers).[353-355] Drug distribution to the lens is much slower than to the uvea since it is composed of tightly packed lens proteins without melanin; however, sequestration of lipophilic drugs in cell membranes of the lens may serve as another potential depot reservoir source.[353] Topical drugs also reportedly can sometimes reach the vitreous cavity, most likely via the conjunctival–scleral–choroid–RPE–retina or possibly the conjunctival–orbital–optic nerve head routes, where bioavailability is approximately one-millionth of the original concentration of the eye drop (Fig. 4.31B).[353] Various modifications can be made to topical drugs in order to increase their corneal absorption and bioavailability, and most fall into one of two categories: (1) drug formulations made to increase residence time in the conjunctival cul-de-sac (e.g. gels, suspensions, ointments, and inserts) or (2) compounds are added to topical drug solutions to increase corneal penetration (e.g. benzalkonium chloride [BAK], other preservatives, epithelial scraping [when treating fungal ulcers with antifungal drops]).

In addition to dynamic clearance mechanisms, topical drug delivery is also hampered by various anatomical and physiological barriers that affect the static permeability properties of the tissue.[354,356] Topical drug delivery via the transcorneal route (Fig. 4.31A) is challenging because the drug needs to be small (< 5 kDa) and lipophilic (e.g. prednisolone acetate) to bypass the intercellular zonula occludens tight junctions of the most superficial squamous corneal epithelial cells. In fact, on average, 90 percent of the corneal barrier to hydrophilic drug penetration is due to the resistance of the paracellular (intercellular) pathway. In contrast, on average, the corneal epithelium accounts for only 10 percent of the resistance to lipophilic drug permeation since these drugs can penetrate the epithelium easily via the transcellular (intracellular) pathway. In comparison to the epithelium, the corneal stroma is highly permeable to most lipophilic and hydrophilic drugs, although the main rate-limiting criterion is having a small molecular radius (size), reflecting the 20 nm wide interfibrillar spaces used for drug diffusion.[30,357] Although no significant changes in permeability have been

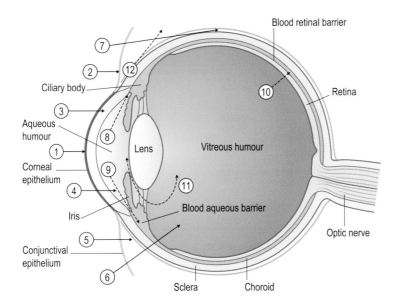

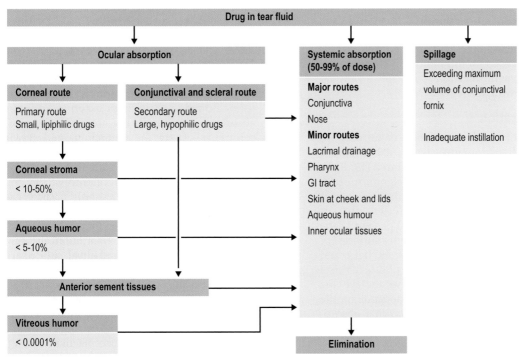

Figure 4.31 (**A**) Diagram of the eye with common drug delivery routes (solid arrows) and clearance pathways (dotted arrows) illustrated. The numbers refer to the following processes: (1) transcorneal route from the tear film across the cornea into the anterior chamber, (2) transconjunctival route across the conjunctiva, sclera, and anterior uvea into the posterior chamber, (3) intrastromal route directly into corneal stroma, (4) intracameral route directly into anterior chamber, (5) subconjunctival route from the anterior subconjunctival space across the sclera and anterior uvea into the posterior chamber or across the sclera, choroid, RPE, and retina into the anterior vitreous, (6) intravitreal drug injection directly into the vitreous, (7) sub-Tenon route from the posterior sub-Tenon space across the sclera, choroid, RPE, and retina into the posterior vitreous, (8) elimination of drug in the aqueous humor across the trabecular meshwork and Schlemm's canal into the systemic vascular circulation, (9) elimination of drug in the aqueous humor across the uvea into the systemic vascular circulation, (10) elimination of drug in the vitreous humor across the blood-retinal barrier to the systemic vascular circulation, (11) drug elimination from the vitreous across anterior hyaloid face to the posterior chamber or vice versa , (12) drug elimination from subconjunctival and/or episcleral space to systemic lymphatic or vascular circulation. (Modified from Urtti A, Urtti A. Advanced Drug Deliv Rev 2006; 58(11):1131–5.) (**B**) Pharmacokinetics of topical eye drop drug delivery. (Modified from Cruysberg L. Novel methods of ocular drug delivery. Ph.D thesis. University of Maastricht, 2008.)

reported with aging, the interfibrillar space of the corneal stroma does decrease 15 percent with aging.[30] Also, interfibrillar spacing in the corneal stroma dynamically changes depending on tissue hydration and IOP levels of the eye. Although it too shows a preference for lipophilic drugs, like the corneal epithelium, the corneal endothelium has static permeability properties for hydrophilic drugs that are only slightly less than that of the lipophilic drugs, reflecting the 10 nm wide intercellular spaces or discontinuities in the macula occludens tight junctions. The static permeability properties of the corneal endothelium and stroma are actually quite similar. Therefore, the corneal epithelium is the main anatomic and physiologic static permeability barrier for drug absorption using the transcorneal route. However, this can be overcome quite easily if drugs are small and lipophilic. Interestingly, ophthalmic topical drug formulations consist of drugs with low molecular radii that are relatively lipid soluble.

The transconjunctival route (Fig. 4.31A) allows passive diffusion of larger (< 20–40 kDa) or more hydrophilic molecules (e.g. beta-blockers, carbonic anhydrase inhibitors) than the transcorneal route because of its diffusely scattered goblet cell population, which contains leakier tight junctions, and a surface area is nearly 18 times greater than that of the cornea.[357,358] Thus, the static permeability properties of the conjunctival epithelium are 50–100-times greater than the corneal epithelium.[353]

Other possible anterior segment drug delivery options include intrastromal, intracameral, and subconjunctival injections (Fig. 4.31A). Although all three of these routes produce the highest peak concentrations of drugs in the cornea or anterior segment, the major disadvantage is that the peak is often followed by low and persistent trough concentrations, unless given repeatedly and frequently.[357] Thus, the overall biologic efficacy of the drug may be less than optimal using these routes compared to topical drug delivery's moderate, but sustained, drug concentrations in the targeted tissues. For example, sustained concentrations of drug are definitely needed for full biological effect against corneal or anterior segment infection. For a full review on this topic, see the following references.[354,356,357]

Ultraviolet light filtration

The external surface of the human body is exposed to terrestrial sunlight that contains UV (~295–400 nm wavelengths), visible (400–800 nm), and infrared (IR; 800–1200 nm) light.[359] Solar UV light undergoes significant scattering and absorption in the Earth's atmosphere such that most of the harmful, shortest wavelengths (< 290 nm), or all of the UVC rays (100–280 nm) and 70–90 percent of the UVB rays (280–315 nm) do not reach the Earth's surface. UV light contains more energy than visible or IR light and consequently has more potential to cause a photochemical injury or damage. When this radiation reaches the eye, the proportion absorbed by different structures depends on the wavelength and angle of incidence, with maximal absorption occurring when UV light rays are parallel to the pupillary axis.[75] Although the cornea is more sensitive to UV light injury than the skin since it has no melanin, individuals seldom experience acute UV sunburns unless direct high exposure occurs. Such injuries occur in "snow blindness" (this is common when skiing at mid-day without goggles as snow reflects 85 percent of incident UV light as opposed to only 1–2 percent from grass) or when welding without UV eye protection since welding arcs emit harmful UV-C and UV-B radiation.

There also is considerable variability in UV light exposure to the eye since the terrestrial UV-B exposure spectrum varies enormously with solar elevation in the sky and other exposure-related factors, such as the albedo ratio (percentage reflectance of UV light) from the ground surface, and the anatomy of the eyebrow and upper eyelid, which shields the eye from the expected terrestrial UV light dosage.[360] Since human photoreceptors and corneal nerves cannot detect UV light, suprathreshold and, more importantly, repeated subthreshold UV light injury can take place without the

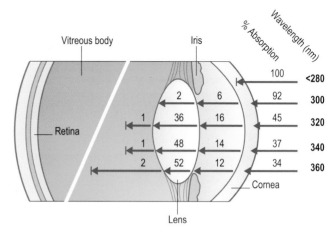

Figure 4.32 The UV light absorption spectra for different structures in the eye. (From Johnson GJ. Eye 2004; 18:1235–50.)

individual knowing it. This can cause acute photokeratitis of the epithelial surface or chronic irreversible keratopathies to the epithelium and anterior parts of the corneal stroma. The cornea absorbs 100 percent of incident UV-C, 90 percent of incident UV-B and 60 percent of incident UV-A radiation (Fig. 4.32).[359] It, therefore, is the major filter of UV light for wavelengths of 200–300 nm, protecting the lens and retina from damage.[361] All UV-C radiation is absorbed in the corneal epithelium due to its high ascorbic acid content. UV-B is absorbed primarily in the anterior 100 μm of the human cornea, especially the epithelium and Bowman's layer, due to high amounts of tryptophan residues in the proteins of the Bowman's layer and anterior stroma and the previously mentioned high ascorbic acid content in the epithelium.[362-366] UV-A (315–400 nm) is only partially attenuated by the cornea (Fig. 4.32), but the transmitted portion is nearly all absorbed by the lens so that only a minor percentage (<3 percent) reaches the retina. Short visible wavelengths, such as violet (~400 nm) and blue (~475 nm) light, pass through the cornea and lens predominantly unattenuated (85–90 percent transmittance) before being absorbed by the retina and RPE. These wavelengths are less harmful to biological tissues than UV light, but are still capable of photochemical injury and damage, particularly if the duration of exposure is long or cumulative. Evidence to support the detrimental health effects of UV light on the eye continues to expand.[367,368] The acute and chronic ocular conditions for which UV light has been suggested to be a causative agent are listed in Table 4.5 and includes several conjunctival, corneal, iris, lens, and possible retinal disorders.[359]

Acute overexposure to UV-B and UV-C damages the corneal epithelium by causing apoptosis, or programmed cell death.[369-371] UV-B and UV-C exposure directly activates signaling molecules and enzymes in corneal epithelia, including JNK, SEK, p53, caspase 9 and caspase 3, all of which lie in the pathway that leads to classic apoptotic events, such as DNA fragmentation and formation of apoptotic bodies.[372-374] This culminates in massive shedding of corneal epithelial cells within hours of exposure. Studies of corneal epithelial cells in culture show that a very early event leading to apoptosis is activation of K^+ channels by UV. This activation of channels occurs within minutes of exposure and apoptosis can be prevented by K^+ channel blockers. In

Table 4.5 Ocular diseases potentially due to ultraviolet radiation exposure

	Strength of evidence
EXTERNAL:	
Photokeratitis	High
Climatic droplet keratopathy	Medium
Pterygium	Medium
Pinguecula	Low
Carcinoma in situ/conjunctival squamous cell carcinoma	Low
IRIS:	
Melanoma	Low
LENS:	
Cataract	Medium
Exfoliation (more likely related to infrared light)	None
Anterior lens capsule changes	Medium
CHOROID/RETINA:	
Solar retinopathy	High
Photomaculopathy (more likely related to blue or visible violet light)	None
Uveal melanoma	Low
Age-related macular degeneration (ARMD)	Low

contrast, chronic overexposure to UV-B and UV-A best explains the high frequency and variable degree of acquired cytogenic DNA damage found in the keratocytes of normal adult corneas.[375]

A second injury mechanism by which UV, violet, and blue light works is through the generation of reactive oxygen species (ROS), such as hydrogen peroxide, singlet oxygen, and oxygen free radicals (superoxide anions and hydroxyl radicals), which cause cellular and extracellular damage by reacting with lipids, proteins, and DNA. Under normal physiological conditions, the cornea and anterior segment protect themselves from ROS by producing and maintaining sufficient antioxidant levels, including several low-molecular-weight (ascorbic acid, glutathione and alpha-tocopherol) and high-molecular-weight (catalase, superoxide dismutase, glutathione peroxidase, and reductase) antioxidants.[361] Ascorbic acid is thought to be the primary antioxidant in the cornea and anterior segment of the human eye since it is produced in such high concentrations in the aqueous humor.[376] The danger to the cornea from ROS appears to stem from excessive UV-B exposure as opposed to UV-A or UV-C exposure since experiments have shown that only UV-B excess leads to profound decreases in corneal antioxidants. Thus, an imbalance in the pro-oxidant to antioxidant ratio takes place in the anterior 100 μm of the cornea, which is very likely to lead to oxidative eye injury and inflammation since the tissue components of the cornea then become the predominant absorbers and detoxifiers of UV light.[361,366,377]

Sclera

The human sclera is a roughly spherical, relatively avascular, white, rigid, dense connective tissue that covers the globe posterior to the cornea (see Figs 4.1A,B, 4.3A, & 4.33A).[6,32,378–380] Little interest was shown in the sclera in the past primarily because infections and tumors did not penetrate it easily and foreign material was well tolerated in it, giving it the unjustified reputation for inertness. Evidence now demonstrates that although the sclera has low baseline metabolic requirements, it constantly remodels throughout life to maintain its functions and thus is far from inert.[381] In comparison to the cornea, major differences in the sclera are that it has variably larger collagen fibril diameters and interfibrillar spaces (compare Fig. 4.34B to 4.10A,B,C); is more opaque, interwoven (Fig. 4.34D), and rigid; has a regional zone of vascularity in the episclera (Fig. 4.35D); and does not have adjacent external or internal cellular barrier layers. The color of the sclera is opaquely white because it scatters all frequencies of visible light due to spatial fluctuations in the refractive index of the tissue, which have dimensions that are greater than a half-wavelength of visible light (Fig. 4.34B,C).[382,383] The opaqueness reduces internal light scattering, but some light actually does transmit through the sclera, evidenced when transilluminating the globe for locating intraocular tumors. The interwoven rigidity helps maintain a stable shape since deformation of the sclera could lead to poor vision or internal injury (Fig. 4.34D). It also is notable for containing a moderately rich nerve supply (Fig. 4.16A), predominantly around episcleral blood vessels, and for having no lymphatic channels, albeit the overlying conjunctiva has two well-formed lymphatic layers (Fig. 4.35C,D). The principal functions of the sclera are to provide a strong, tough external framework to protect the delicate intraocular structures and to maintain the shape of the globe so that the retinal image is undisturbed. Secondarily, it serves as a stable expansive-resistant semi-spherical structure to the forces generated by IOP, facilitates appropriate aqueous outflow, provides stable attachment sites for the extraocular muscles to rotate the globe and for the ciliary muscle to accommodate the lens, provides a conduit for vascular and neural pathways to go from inside-to-outside the eye and vice versa, and plays a critical role in determining the absolute size of the eye and thus the refractive error of the eye. For this latter function, the scleral wall must have a mechanism in place for controlling its growth.

Embryology, growth, development, and aging

The sclera is predominantly neural-crest-derived, except for a small temporal portion that comes from the mesoderm.[6] The development of the sclera begins by 6.5 weeks of gestation and proceeds in an anterior-to-posterior and inside-to-outside fashion as the anterior periocular mesenchyme, derived from the 2nd wave of neural crest cell invasion, condenses anteriorly on the optic cup. This anterior mesenchyme subsequently differentiates into an inner vascular layer that develops into the uvea (iris, ciliary body, and choroid) and an outer fibrous layer that develops into the sclera. The retinal pigment epithelium (RPE) and/or the

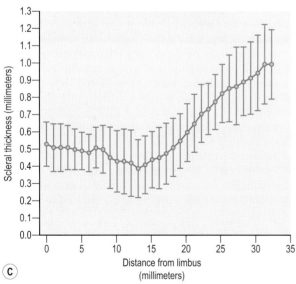

Figure 4.33 (**A**) Gross photo of the superior section of a horizontally bisected normal globe demonstrating the cross-sectional appearance of the sclera, limbus, and cornea. (**B**) Photomicrograph of a normal globe showing the cross-sectional appearance of the sclera, limbus, and cornea (PAS, ×2). (**C**) Line graph summarizing the average scleral thickness (± SD) vs. distance from the limbus in normal eyes (n = 55). (From Olsen TW et al. Am J Ophthalmol 1998; 125:237–41.)

choroid, is directly responsible for embryonic scleral development since if the RPE or choroid is absent or not in contact with the sclera, the sclera does not develop (e.g. chorioretinal colobomas) or will not grow, respectively.[32] The anterior sclera fully differentiates by 7 weeks' gestation followed by the equatorial sclera at 8 weeks' gestation and the posterior sclera by 11 weeks' gestation. It gradually increases in thickness and extracellular matrix denseness during the remaining months of gestation. It initially is primarily composed of scleral fibroblasts, collagen fibrils, and proteoglycans; elastin fibers are acquired mainly after birth, perhaps in response to IOP.[32] At birth, the sclera is relatively thin, highly distensible (in infancy, the sclera is on average one-quarter as stiff as it is in adulthood), and translucent (explaining why the blue color from the underlying uvea often times shows through the infant sclera).[384]

Post-natal growth and maturation continue in a similar anterior-to-posterior fashion. During the first 3 years of life, the sclera grows rapidly in diameter and thickness, gradually losing some of its high distensibility (perhaps due to increased thickness, decreased cellularity, increased extracellular matrix denseness, or proportionally increased type I collagen fibril deposition), but it still remains relatively translucent. This early loss of distensibility explains why the sclera can expand from increased IOP (e.g. infantile glaucoma) only from birth to around an age of 3 years old, resulting in a buphthalmic "ox" eye. Thereafter, the sclera distends only slightly from increased IOP, mainly in the lamina cribrosa. After age 3, the sclera thickens further, becomes more opaque, and grows in diameter at an exponentially slower rate than the first 3 years of life, reaching adult size by 13–16 years of age.[15,385] During this growth period, the anterior sclera develops and matures much more quickly (adult size by 2 years of age) than the equatorial (adult size by 13 years of age) and the posterior sclera (adult size by 13–16 years of age). The sclera continues to become less distensible and more rigid with advancing age, primarily due to maturation and age-related glycation-induced cross-linking of collagen fibrils (scleral stiffness reportedly increases 2–3-fold from age 3 to 20, and then another 2-fold from age 20 to 78), typically resulting in no further eye or scleral growth after age 16.[32,386–388]

Normal eye growth is thought to be controlled in part by a visual feedback mechanism that depends on the quality of the retinal image. This feedback influences the scleral fibrocytes and thus the extracellular matrix of the sclera to undergo constant remodeling during childhood eye growth, continuing to some extent into adult life – albeit to a lesser degree.[381] This visual feedback mechanism serves to guide childhood eye growth toward emmetropia and the attainment of adult eye size, resulting in the majority of the human adult population being free of significant refractive errors. Interestingly, this visual feedback mechanism is not dependent on the CNS. In animal studies, transection of the optic nerve or blocking ganglion cell action potentials does not prevent the development or the recovery of experimentally induced myopia.[381] Rather, it is directly dependent on paracrine cytokine or growth factor pathways (e.g. dopamine, acetylcholine) originating from the retina and/or RPE.[381] Animal experiments that pharmacologically damage the retinal photoreceptors or RPE prevent visual deprivation myopia.[32] Epidemiological studies suggest that environmental visual

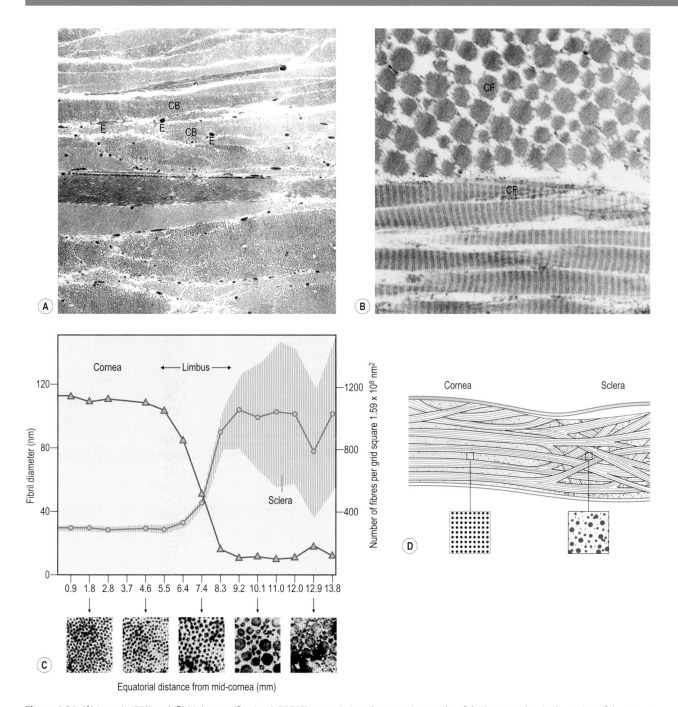

Figure 4.34 (**A**) Low- (×4750) and (**B**) high-magnification (×72,500) transmission electron micrographs of the human sclera in the region of the stroma proper. Compare these to Figs 4.9 and 4.10A,B&C to see how much more variably larger or irregular the collagen fibril diameters, interfibrillar spaces, and collagen bundles of the scleral stroma proper are to the corneal stroma. CB = collagen bundle. E = elastin fibers. CF = collagen fibril. (**C**) Diagram comparing the collagen fibril diameters (○) and densities (△) in the cornea, limbus, and sclera. (Modified from Borcherding MS et al. Exp Eye Res 1975; 21:59–70). (**D**) Diagram illustrating the greater degree of interweave of collagen bundles in the sclera than collagen lamellae in the corneal stroma. Additionally, the sclera has larger and more varied in collagen fibril diameters and interfibrillar spacing. (Modified from Bron AJ, Tripathi R, Tripathi B. In: Wolff's anatomy of the eye and orbit, 8th edn. London, UK: Chapman & Hall, 1997.)

stimuli (e.g. lengthy periods of near work or accommodation; visual deprivation – particularly early in life), developmental delay in the maturation of the posterior sclera, or genetic factors may alter this visual feedback mechanism resulting in physiologic or pathologic myopia. In contrast, genetic or developmental delay in overall globe growth may result in hyperopia.[32]

A minority of the population may actually have significant breakdown of this normal visually guided emmetropization process resulting in severe refractive errors or even refractive error progression outside the normal period of eye growth, such as adolescent-onset (16–20 years of age) or adult-onset (2nd to 4th decades of life) myopia, due to further axial length and posterior scleral elongation.[389,390] The exact pathogenesis

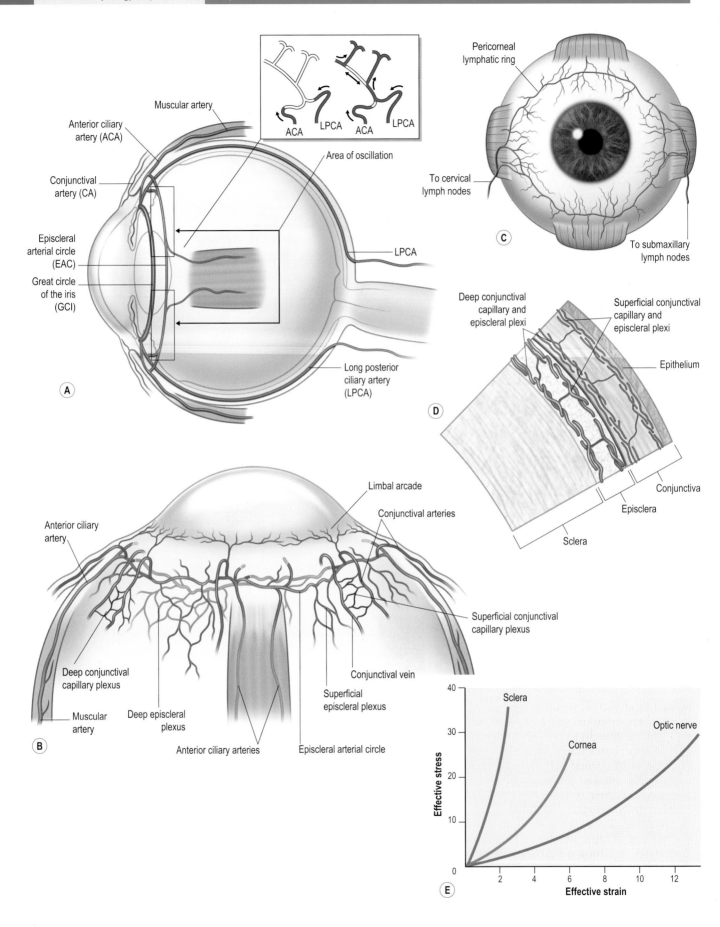

Figure 4.35 (**A**) The arterial supply of the anterior segment comes from the anterior ciliary arteries (ACA) and from the terminal branches of the long posterior ciliary arteries (LPCA). These vessels form two sagittal arterial circles (between superior or inferior ACA and LPCA arteries) and also anastomose together superficially and deeply to form two coronal arterial circles, called the episclera arterial circle (EAC) superficially and the greater circle of the iris (GCI) deeply. In the anterior episclera (inset), the deep perforating branches of the LPCA anastomose with the ACA, together forming the EAC. The blood flow in the EAC typically comes from LPCA (inside outwards) rather than from ACA. The EAC supplies both a superficial and a deep episcleral plexus (deep plexus not shown). The conjunctival artery derives from the EAC, passing posteriorly, while also giving off an anteriorly directed branch called the limbal artery, which subsequently forms the limbal arcade. Blood flow in the EAC (inset) is continuous near the rectus muscle insertion sites, but oscillates rather than flows between the rectus muscle insertion sites. (Modified from Watson PG, Hazelman B, Pavesio C, Green WR. The sclera and systemic disorders, 2nd edn. Edinburgh, UK: Butterworth Heinemann, 2004.). (**B**) Diagram shows a schematic of the arterial circulation of the episclera and conjunctival, which both have superficial and deep components that supply capillary plexi. (Modified from Meyer PAR et al. 1987; 71: 2–10.) (**C**) Schematic diagram showing the distribution of lymphatics (green) in relation to the blood vessels in the conjunctiva. Centripetal branches collect into a larger circular lymphatic ring, called the pericorneal lymphatic ring, which then drains medially or laterally into regional lymph nodes. (**D**) A cross-sectional representation of the conjunctiva, episclera, and sclera shows that episclera and sclera are devoid of lymphatic networks, while the conjunctiva has two lymphatic plexi – a superficial plexus (below the epithelium) and a deep plexus (within Tenon's fascia, but just above the episclera). Blood vessel plexi are found in the superficial and deep conjunctiva and episclera. (Modified from Robinson M, Lee S, Kim H, et al. Exp Eye Res 2006; 82:479–87.) (**E**) Exponential approximations of trilinear stress–strain curves for sclera, cornea (stroma), and lamina cribosa (optic nerve). (From Woo SL, Kobayashi AS, Schlegel WA, Lawrence C. Exp Eye Res 1972;14(1):29–39.)

of myopia is still not completely understood, but myopia induction in mammalian animal models using various visual deprivation stimuli suggests that the earliest biochemical and structural remodeling changes begin with reduced scleral fibrocyte proliferation and altered metabolism resulting in reduced type I collagen fibril synthesis, increased collagen degradation (increased matrix metalloproteinase-2 expression), reduced proteoglycan synthesis, scleral tissue volume loss (loss of both scleral wet and dry weight), and posterior scleral thinning.[381,391] Late or more chronic biochemical and structural remodeling changes include reduction of mean type I collagen fibril diameters, particularly in the outer layers of the sclera, and, in advanced end stages, localized areas of ectasia or staphyloma formation of the posterior sclera, especially at the optic nerve or the macula.[381,391]

All of these biochemical and structural changes appear to be guided by a retino–RPE–scleral visual signaling pathway, which dynamically controls scleral extensibility under the stress of normal IOP and thus controls the innate biomechanical properties of the sclera. Biomechanical testing of myopic human eyes by Avetisov and associates supports this theory since the thinned posterior sclera was 30–40 percent biomechanically weaker in tensile strength than normal eyes.[386] Animal studies by McBrien and associates have also shown that posterior scleral creep rates during acute periods of myopia induction are significantly higher in myopic eyes compared to normal control eyes.[381] Interestingly, using animal models, after removal of the myopia-inducing visual deprivation stimulus, recovery to normal eye size is quite rapid and is thought to be due to scleral keratocyte differentiation into contractile myofibroblasts and possibly due to resumption of normal proteoglycan synthesis.[381] In contrast to animals allowed to recover naturally, some myopic animals had their myopia corrected with optical devices (analogous to spectacles in humans). The myopic biomechanical scleral phenotype persisted and thus recovery from the induced myopic state was aborted – although the eyes did return to a more normal stable growth rate.[381] Such a finding may have important implications for the correction of myopia in humans since prevention or recovery from the aberrant remodeling process seems to be the key to any potentially successful medical or optical therapy for myopia. Another alternative promising medical treatment option is

that of cross-linking the sclera in a manner that might serve to stabilize or halt myopic progession.[392]

In elderly individuals, the sclera often becomes somewhat yellowish due to a fine deposition of lipid. Another common finding is a small rectangular area of grayish-blue translucency just anterior to the insertion of the medial and lateral rectus muscles. These changes are known as senile scleral plaques and are associated with deposition of calcium in scleral regions that are under strain and exposed to the environment.[6,378–380] They never are found adjacent to unexposed superior or inferior rectus muscles.

Major scleral reference points and measurements

The sclera is an incomplete sphere (Fig. 4.33A,B) resulting in an average outer surface area of 16.3 cm², an average outer diameter of 24 mm, and an average inner diameter of 23 mm (mean radius of curvature = 11.5 mm).[6] Knowing the topographic curvature of the sclera is important for fitting scleral contact lenses, which have certain advantages over corneal contact lenses including the ability to optically neutralize almost any corneal topographical shape since the lens is borne solely by the sclera and not the cornea; being more comfortable than some corneal contact lenses since the edge of the lens does not touch the eyelids; and being better for some forms of aqueous dry eye deficiency states since a large precorneal tear film reservoir is trapped between the lens and the entire corneal surface. On average, the sclera is thickest posteriorly near the optic nerve (1.0–1.35 mm), decreasing gradually as it approaches the equator of the globe (0.4–0.6 mm), before typically becoming thinnest under the rectus muscles just before it reaches its insertion sites (0.3 mm) (Fig. 4.33C).[378,393] It then gradually increases in thickness at the actual muscle insertion sites (0.6 mm) and continues getting thicker up to the limbus (0.8 mm), where it blends with the cornea (Fig. 4.33C).[378,393] These average scleral thickness measurements are subject to great variation. For example, the equatorial sclera is sometimes found to be < 0.1 mm thick.

There are two major openings in the sclera: the anterior scleral foramen (13.7 mm diameter; circumscribes the area of the cornea and limbus) and the postero-nasally located

119

fenestrated posterior scleral foramen or scleral canal (1.5–2.0 mm internal diameter and 3.5 mm external diameter). The inner third of the sclera forms a fenestrated scaffold in the scleral canal called the lamina cribosa that supports the optic nerve axons, whereas the outer posterior two-thirds of the sclera merge with the dura mater of the optic nerve leaving the outer posterior two-thirds of the scleral canal essentially free of any scleral support. Since the lamina cribosa is the least stiff and strong point of the adult human globe to expansile forces, diseases that cause chronically high IOP, like glaucoma, preferentially cause lamina cribosa ectasia and subsequent glaucomatous optic nerve cupping and optic neuropathy. There are also other numerous minor openings in the sclera, including 30–40 emissary channels for ciliary arteries, veins, and nerves, and 4–7 vortex vein channels. The outer surface of the sclera is smooth, except for where the tendons of the extraocular muscles insert (spiral of Tillaux and oblique muscle insertion sites) and where Tenon's capsule adheres (within 1 mm of limbus, over rectus muscle insertion sites, and around the optic nerve). The superficial layer of the sclera, called the episclera, is a thin, highly vascularized, dense connective tissue. It is around 15–20 μm thick near the limbus, progressively thinning as it extends into the posterior aspect of the eye. The scleral stroma proper is a white, avascular, dense connective tissue accounting for more than 95 percent of the total scleral thickness. Finally, the inner surface of the sclera is a brown, avascular, 5 μm thick layer called the lamina fusca, which contains a large number of elastin fibers and melanocytes. In fact, these melanocytes sometimes pass through an emissary channel with nerve loops producing dark spots on the surface of the sclera. This can be mistakenly confused with a melanoma.

Mechanical properties

The mechanical behavior of the sclera is most dependent on its thickness and innate biomechanical properties (hierarchical structure of collagen and its associated cross-links). Because of its toughness, cohesiveness, and tightly interwoven architecture, the sclera is not easily dissected in a blunt fashion. The sclera is predominantly composed of water (68 percent) that is stabilized in a disorganized structural network of insoluble and soluble extracellular proteins with fewer fibrocytes than the cornea (see Table 4.1).[378] The posterior sclera is more hydrated than the equatorial and anterior sclera (71 percent vs. 62 percent).[394] The dry weight of the adult human sclera (see Table 4.1) is comprised of collagen, proteoglycans, fibrocyte constituents, elastin, blood vessel constituents, and other substances (lipids, salts, glycoproteins, etc.).[379] The biomechanical properties of the sclera are dominated by the scleral stroma proper, which has a similar composite-like structure to that of the anterior third of the corneal stroma, albeit it is even more disorganized and highly irregular (Fig. 4.34D).[6]

Collagen is the major water-insoluble extracellular protein of the scleral stroma proper (80 percent type I, 5 percent type III, and minor amounts of types V, VI collagen); elastin is a minor component.[378] In contrast to the cornea, the sclera's heterotypic type I collagen fibrils are composed of types I, III, and V collagen molecules, and are larger and more variable in diameter (mean diameter: 100 ± 30 nm; range: 25–

300 nm), more irregularly spaced (mean center-to-center interfibrillar distances: 150 ± 40 nm; range: 30–375), and arranged in variably sized (0.5–6 μm thick; 1–50 μm wide), highly interwoven, irregularly directed lamellar collagen bundles.[6,87,379] The arrangement of collagen fibrils in the individual bundles is more random and they intermingle in a more wavy fashion than that of the cornea. Also, the bundles do not form a plywood-like stacking sequence, like that seen in the posterior two-thirds of the corneal stroma. Topographically, the sclera varies not only in thickness, but also in size, compactness, and angle of weave of its collagen bundles.[395] The anterior sclera consists of medium- to small-sized, moderately compact collagen bundles with wide-angle weave, while the equatorial sclera consists of small-sized, very compact collagen bundles with narrow-angle weave.[395] Finally, the posterior sclera consists of medium- to large-sized, loose collagen bundles with a wide-angle weave.[395]

No significant differences in the number of collagen bundles or elastin fibers have been reported when comparing these three topographic regions. There also are depth-related changes in the scleral stroma proper as the more superficial collagen fibrils are on average further apart from one another and are larger in diameter than the deeper layers where they are more compact and smaller in diameter. Additionally, the collagen bundles are thinner, narrower, and form whorl-like patterns superficially, whereas they are thicker, wider, and form a net- or rhomboid-like pattern in the deeper layers.[32] The scleral proteoglycans and their covalently linked GAG side chains differ markedly from the corneal stroma as they are about one-fourth the concentration and consist of the following: core proteins – decorin, biglycan, and aggrecan; GAGs – dermatan sulfate (36 percent), chondroitin sulfate (35 percent), hyaluronic acid (23 percent), and heparin sulfate (6 percent).[378,379] The scleral stroma proper also contains a syncytium of fibrocytes, like the corneal stroma; albeit at a much lower level of cellularity. Although the scleral stroma proper is traversed by ciliary blood vessels and nerves, it has no direct blood, lymph, or nerve supply.[378] It derives its nutrition solely by diffusion from the overlying episcleral and underlying choroidal vascular networks.

The most superficial layer of the sclera, called the episclera, differs from the scleral stroma proper in that its collagen bundles are more loosely arranged; it contains melanocytes and a few residential histiocytes. It also has a rich nerve supply with many unmyelinated and myelinated free nerve terminals (see Fig. 4.16A), which are most densely populated near the rich direct blood supply in the episclera (Fig. 4.35A) and lymph vasculature in the overlying conjunctiva (Fig. 4.35D), presumably to help regulate blood supply and to influence aqueous outflow.[6] The episcleral vasculature drains aqueous humor from the conventional outflow route via anastomosis with the aqueous collector channels to the episcleral venous system. Thus, outflow and IOP are dependent on the pressure gradient between IOP and episcleral venous pressure (EVP). Finally, the lamina fusca appears to merely serve as a transition zone from the sclera to the choroid.[6]

Although the sclera is constantly under stress from the IOP, which dynamically varies considerably as noted previously in this chapter, it acts in a similar manner to systemic muscular arteries in the body. It also typically displays a limited ability to distend in adulthood (range: 0.58–6.68 percent strain), termed scleral distensibility. Considering the different possible topographical thicknesses and internal curvatures of the sclera, marked regional differences in wall stress can occur depending on each individual globe (see Law of LaPlace, previously described in this chapter). Thus, accurately calculating regional wall stress of the sclera is very complex and perhaps even more difficult to do than for the cornea. In general, the anterior sclera under the rectus muscle insertion sites and the equatorial sclera should have the highest internal wall radial stress since they are the thinnest areas with comparable internal curvatures to the rest of the sclera, whereas the limbus should have the highest circumferential wall stress since a two-fold higher amount of stress is required to sustain the curvature change from the cornea to the sclera (see Fig. 4.29A).[86] Thus, when under acute extreme stress, such as direct blunt impact to the globe, there is a tendency for the globe to rupture in these areas. However, these regions also have the highest innate biomechanical properties, which were measured using ex vivo strip extensiometry or inflation tests, perhaps explaining the chronic lengthening of the posterior sclera in myopia or the chronic ectasia of the lamina cribosa seen in glaucoma that have no effect on these regions. In fact, structural alterations in the sclera caused by myopia or glaucoma may considerably change the expected normal stress distribution in the scleral eye wall (e.g. posterior scleral stress may increase up to four-fold with severe myopic globe elongation).

Using inflation testing and finite element modeling, the longitudinal Young's modulus of the adult human sclera averages 1.7×10^7 dyne/cm^2 (Fig. 4.35E).[396] This is approximately 3.7-fold stiffer than the cornea and 5.3-fold stiffer than the lamina cribosa (Fig. 4.35E).[396] Using uniaxial strip extensiometry, cohesive strength measurements of the anterior sclera averaged 55 g/mm, which was approximately two-fold stronger than the cornea.[93] Human uniaxial strip extensiometry testing has also shown topographical differences in the sclera's longitudinal Young's modulus as the equatorial sclera was the stiffest at 23 ± 8 g/mm^2, the posterior sclera was the least stiff at 3 ± 1 g/mm^2; the anterior sclera was between these at 4 ± 1 g/mm^2.[395] The greater extensibility of the posterior sclera had previously led to a "dual function" theory of the sclera where the anterior and equatorial sclera were thought to be essential for the rigid support and stability of the globe, whereas that of the posterior sclera served to act as a mechanical buffer or cushion against acute increases in IOP, which could be injurious to the delicate globe.[395]

Like most biological systems, the sclera's elastic stress–strain curve exponentially or non-linearly increases with higher elevations in IOP or stress and then, because of viscosity, shows a time-dependent decrease with time. The J-shaped stress–strain curve of the sclera involves an immediate elastic response dominated initially by elastin, non-collagenous matrix, and collagen fibrils at the lower part of the J and then later by collagen fibrils in the upper part of the J. Overall, the non-linearity arises from the gradual loading of highly extensible elastin fibers and the wavy collagen fibrils that take up slack initially, followed by gradually increasing in resistance to stress as maximal recruitment is attained. Finally, a slower, time-dependent viscous response results over time, where the resistance to stress gradually decreases with time due to creep. The viscoelastic properties of the sclera also mean the loading or unloading stress–strain curves do not overlap, with the unloading curve being less than the loading curve – depending on the loading rate. This allows energy dissipation via viscous friction during the mechanical cycle and, ultimately, the return of the original shape of the material in a slow, time-dependent fashion. Therefore, the amount of distension of the sclera is not fixed, but determined by the level of the pressure changes and how long they have acted on the sclera and on the innate biomechanical properties and thickness of the sclera.[397]

A good example of this is when measuring IOP by indentation methods. The pressure measurement recorded by indentation initially increases the IOP of the eye by indenting the cornea inward (immediate elastic response), but by the time the measurement is made several seconds later it is typically back to normal levels (slow viscous response). However, in some individuals (e.g. high myopes or severely thinned sclera from scleral malacia perforans), the sclera will distend much more than normal because of the lower innate biomechanical properties or thinner sclera – resulting in false low readings. In contrast, false high readings may occur in the cases of more rigid scleras (e.g. scleral buckles). Therefore, most people choose to use applanation (e.g. Goldman applanation tonometry) over indentation (e.g. Schiotz tonometry) to measure IOP since it is more accurate as it eliminates artifacts caused by the sclera.

Scleral dehydration and edema

Hydration of the sclera is most closely associated with the extracellular proteoglycan concentrations in the sclera. Since the normal, healthy sclera has one-fourth the concentration of proteoglycans (scleral swelling pressure of 20–30 g/cm) between collagen fibrils than the cornea, it is not surprising that it contains less water (68 percent) than the cornea (78 percent).[379] However, if the normal water content of sclera is reduced to < 40 percent, it becomes somewhat translucent. This is a common observation made during long surgical procedures on the sclera when scleral flaps are dissected and sometimes inadvertently exposed to air for prolonged time periods, which reduces the hydration level through evaporation. Similarly, if the water content is increased to > 80 percent, the sclera again becomes somewhat translucent due to the hydration of scleral proteoglycans.

Episcleral vasculature

The blood supply of the episclera is particularly prominent along a 4 mm zone anterior to the rectus muscle insertion sites, while being markedly less vascular in more posterior aspects. The former area is called the episcleral arterial circle (EAC) (see Fig. 4.35A), which is fed by seven anterior ciliary arteries (ACA) and several (≤ 12 per principal meridian) anastomotic terminal branches from the long posterior ciliary arteries (LPCA), which are more commonly found superiorly and inferiorly than medially or laterally.[378] This unusual dual artery-to-artery anastomosis typically flows inward to outward, but it can change direction if needed and thus ensures that the anterior segment of the eye is always supplied with adequate blood flow. The EAC has both superficial and deep branches (Fig. 4.35B,D) that directly nourish the episclera through corresponding superficial or deep episcleral capillary plexi. The EAC also has separate

The pre-capillary arterioles and post-capillary venuoles of the superficial and deep episcleral, limbus, and conjunctiva capillary plexi are notable for having continuous non-fenestrated endothelium without intercellular tight junctions (20 nm intercellular spaces).[398] Therefore, these vessels are leaky and are highly permeable to small molecules, but do at least resist bulk fluid flow. Since the pre-capillary arterioles of the anterior segment do not have a smooth muscle wall, they tend to be thrown into tortuous folds resulting in areas of turbulent blood flow. When you combine this with the fact that a notable disadvantage to having artery-to-artery anastomosis is that regions between the rectus muscles may not have continuous high arterial perfusion pressures, but rather oscillatory blood flow due to being further away from the feeding ACA and LPCA vessels, then it should not be surprising that extravasation of fluid or stagnation of cells commonly occurs in these regions. If the individuals have systemic infections or autoimmune diseases (such as hepatitis viral infections, systemic lupus erythematosus, rheumatoid arthritis, or Wegener's granulomatosis), deposition of antigens or immune complexes occurs in this region of stagnation, resulting in inflammatory microangiopathy (e.g. peripheral ulcerative keratitis, episcleritis, scleritis, Mooren's ulcers).[378] Moreover, since some of these regions (sclera and peripheral cornea) are far from lymphatic drainage sites, which are efficient mechanisms for removal of unwanted antigens or immune complexes, chronic problems often follow.[378]

conjunctival or limbal artery branches that have their own capillary plexi (Fig. 4.35B,D). For example, during internal eye infection or inflammation or with severe corneal infection or inflammation, the limbal plexus is notable for dilating and causing a limbal ciliary flush pattern to appear around the periphery of the cornea on slit lamp examination, usually indicating more serious disease is present than just a simple conjunctivitis.

Wound healing

Although injury converts scleral fibrocytes to metabolically active fibroblasts, similar to the cornea, superficial lacerations of the sclera heal more aggressively and completely than the cornea because episcleral fibrovascular granulation tissue migrates into and fills the wound.[6] Similarly, lacerations involving the inner portion of the sclera also heal predominantly through fibrovascular granulation tissue, which grows outward from the choroid.[6] Penetrating scleral lacerations typically heal because fibrovascular granulation tissue grows from both of these sites. Such healing is usually very strong since it causes fibrovascular scarring. As time passes, a gradual remodeling or reorganization occurs to this scar; however, it can always be identified histologically by the abrupt change in scleral collagen fibril orientation, the persistent vascularity, or the disorganization of the surrounding tissue architecture.

Drug delivery

Although most of the bulk transport of fluid out of the eye takes place through the anterior segment's trabecular meshwork/Schlemm's canal system or via uveoscleral outflow, an appreciable amount is also drained transretinally to the choroid where it can get into the systemic circulation by diffusion into the choroidal vasculature or can travel extra-ocularly via the trans-scleral pathway (see Fig. 4.31A). The trans-scleral pathway is very intriguing to researchers and clinicians since it may serve as a potential route for non-invasive delivery of medications into the eye.[399] Currently, the medical treatment for posterior segment disease is to a large extent limited by the difficulty in delivering therapeutically effective doses of drugs to the posterior segment tissues (vitreous and retina). Unfortunately, the topical drug delivery route does not consistently or even efficiently yield therapeutic drug levels in posterior segment tissues. Although systemic administration (oral or intravenous routes) can deliver drugs to the choroidal vasculature at therapeutic levels, the duration of choroidal drug delivery is too brief (usually less than 30 minutes) to result in meaningful intraocular drug levels and the large systemic doses necessary are often associated with significant systemic side effects or toxicities.

Intravitreal injection delivers agents directly into the vitreous and next to the retina (Fig. 4.31A). Thus, it has the advantages of achieving the highest intraocular peak drug levels, while minimizing systemic exposure. However, it also has the disadvantages of being the most invasive or traumatic technique available and can lead to persistent trough levels unless given repeatedly and frequently since drugs usually are rapidly eliminated via anterior segment and/or posterior segment bulk flow. In fact, the intravitreal route is often poorly tolerated and places patients at risk for high IOP, floaters, transient blurry vision, retinal hemorrhage, retinal tears, retinal detachment, endophthalmitis, and cataract. The acceptability of such an invasive direct approach is likely more a function of the poor visual state of the eye due to the disease and the change in prognosis due to intervention (e.g. endophthalmitis and intravitreal antibiotic injections, wet age-related macular degeneration [AMD] and intravitreal anti-angiogenic drugs).

Periocular drug delivery is an alternative, minimally invasive approach to drug delivery to the posterior segment as it commonly delivers moderately sustained concentrations of drugs.[357] The various periocular techniques include subconjunctival, retrobulbar, peribulbar, and sub-Tenon's injections (see Fig. 4.31A). They are far less invasive than the direct intravitreal route, but they do have notable shortcomings, mainly from having to permeate more static anatomical barriers and from encountering enhanced dynamic clearance mechanisms. In general, periocular drug delivery involves placing the drug, usually by needle injection, into the tissue surrounding or adjacent to the posterior segment of the eye. Corticosteroids are the most common drugs given by these techniques (e.g. sub-Tenon's triamcinolone injection). The feasibility of using these techniques depends, to a large extent, on the permeability of the sclera to the drugs and the accuracy of the clinician in injecting the drug to the desired location. In fact, although three possible absorption pathways exist (trans-scleral, systemic hematogenous circulation, and anterior routes) for the various periocular drug delivery techniques, the trans-scleral pathway has been proven to be the main route for delivering sufficient drug concentrations into the choroid, RPE, retina, and vitreous.[400] A recent in vivo dynamic periocular drug delivery study by Ghate and associates (clearance mechanisms were all fully functional and active as well as the static permeability properties of the tissue) using rabbits found that sub-Tenon's injection resulted in the highest and most persistent vitreous concentrations with the least systemic exposure of the various periocular techniques. Subconjunctival injection resulted in the highest and most persistent anterior segment concentrations with slightly less vitreous concentrations than that of sub-Tenon's injection – with the caveat of some higher risk for systemic exposure.[400]

Although preliminary studies several decades ago suggested that it might be possible to exploit the trans-scleral route for drug delivery to intraocular tissues in the posterior segment, it has only recently seen renewed interest and focused detailed attention – probably based more on the success of anti-vascular endothelial growth factor (VEGF) drug therapy for wet AMD and the risk involved with frequent repeated intravitreal injections required to keep the disease in check.[401] Of these initial studies, Barza was the first to clearly establish that drugs could permeate through various ocular tissues including the sclera when they were

administered by either subconjunctival or retrobulbar injection.[402,403] Anders Bill further demonstrated that albumin or dextran injected into the suprachoroidal space of the rabbit eye could diffuse across the sclera and accumulate in extraocular tissues.[404] In vitro permeability studies on human sclera by Olsen and associates further demonstrated that the tissue was permeable to drugs up to a molecular weight of 70 kDa.[405] Scleral permeability is expressed as a pharmacokinetic volume transfer coefficient known as its K^{trans} (cm/sec), which reflects the tissue's surface area permeability to a specific perfusion flow of the drug.[405] A number of permeability studies have been published, using essentially comparable methods, and have shown that the sclera is quite permeable to a wide range of drugs (Table 4.6).[352,399,406] Collectively, the results indicate that scleral permeability is 5–15 times better than that of the corneal stroma, depending on the molecular radius of the drug studied. The sclera shares a similar ultrastructure and composition to corneal stroma, albeit the sclera has variably larger interfibrillar spaces (50 nm [range:

Table 4.6 Known scleral permeability values for certain drugs or agents

Drug	Molecular weight (g/mol)	K_{trans} (cm/sec)	Source
Polymyxin B	1800	$3.90 \pm 0.59 \times 10^{-7}$	2
Doxil	580	$4.74 \pm 0.73 \times 10^{-7}$	8
Vancomycin BODIPY	1723	$6.66 \pm 1.46 \times 10^{-7}$	2
SS fluorescein-labeled oligonucleotide	7998.3	$7.67 \pm 1.8 \times 10^{-7}$	3
Dexamethasone-fluorescein	841	$1.64 \pm 0.17 \times 10^{-6}$	4
Rhodamine	479	$1.86 \pm 0.39 \times 10^{-6}$	4
Penicillin G	661.46	$1.89 \pm 0.21 \times 10^{-6}$	2
Methotrexate-fluorescein	979	$3.36 \pm 0.62 \times 10^{-6}$	4
Doxorubicin hydrochloride	580	$3.50 \pm 0.31 \times 10^{-6}$	8
Nanoparticle doxorubicin	580	$4.97 \pm 0.19 \times 10^{-6}$	8
Fluorescein	332	$5.21 \pm 0.71 \times 10^{-6}$	4
Vinblastine BODIPY FL	1043	$5.88 \pm 1.2 \times 10^{-6}$	-
Cisplatin in collagen matrix	300.05	$8.3 \pm 1.2 \times 10^{-6}$	5
Carboxyfluorescein	317	$9.93 \pm 3.5 \times 10^{-6}$	6
Carboplatin in fibrin sealant	371.25	$13.7 \pm 2.3 \times 10^{-6}$	7
Cisplatin in BSS	300.05	$20.1 \pm 1.8 \times 10^{-6}$	5
Carboplatin in BSS	371.25	$27.0 \pm 1.7 \times 10^{-6}$	7
Water (H_2O)	18	$51.8 \pm 18 \times 10^{-6}$	6

Sources:

1. Zhang L, Gu FX, Chan JM, et al. Nanoparticles in medicine: therapeutic applications and developments. Clin Pharmacol Ther Advance online publication 24 October 2007; doi: 10.1038/sj.clpt.6100400.

2. Kao JC, Geroski DH, and Edelhauser HF. Trans-scleral permeability of fluorescent-labeled antibiotics. J Ocul Pharmacol Ther 2005; 21:1–10.

3. Shuler RK Jr., Dioguardi PK, Henjy C, et al. Scleral permeability of a small single-stranded oligonucleotide. J Ocul Pharmacol Ther 2004; 20:159–68.

4. Cruysberg LPJ, Nuijts RMMA, Geroski DH, et al. In vitro human scleral permeability of fluorescein, dexamethasone-fluorescein, methotrexate-fluorescein, and rhodamine 6G and the use of a coated coil as a new drug delivery system. J Ocul Pharmacol Ther 2002; 18:559–69.

5. Gilbert JA, Simpson AE, Rudnick DE, et al. Transscleral permeability and intraocular concentrations of cisplatin from a collagen matrix. J Control Release 2003; 89:409–17.

6. Rudnick DE, Noonan JS, Geroski DH, et al. The effect of intraocular pressure on human and rabbit scleral permeability. Invest Ophthalmol Vis Sci 1999; 40:3054–8.

7. Simpson AE, Gilbert BS, Rudnick DE, et al. Transscleral diffusion of carboplatin: an in vitro and in vivo study. Arch Ophthalmol 2002; 120:1069–74.

8. Kim ES, Lee SJ, Zaffos JA et al. Transscleral delivery of doxorubicin: a comparison of hydrophilic and lipophilic nanoparticles. ARVO E-Abstract A590.

5–120 nm] vs. 20 nm [range: 5–35 nm]) one-fourth the ground substance, one-fifth the cellularity, a 10 percent less hydration level, and a much more variable thickness profile.[352]

Like the corneal stroma, the primary transportation route through the sclera is by passive diffusion through the interfibrillar spaces. Also, as in corneal stroma, sclera permeability was found to be primarily dependent on the molecular radius of the drug (scleral permeability declines roughly exponentially with increasing molecular radius; 8 nm molecular radii drugs or less have been successfully shown to permeate the sclera) and secondarily dependent on the shape and molecular weight of the drug.[405,407] As the sclera stroma proper is hypocellular and is essentially devoid of melanin, it has no intercellular barriers, few proteolytic enzymes, and few protein-binding sites to interfere with drug permeation and, thus, shows no preference for hydrophilic or lipophilic drugs. Although several studies have not found significant changes in scleral permeability to small molecules when associated with aging, tissue hydration, or various IOP levels, the hydration level and interfibrillar space of the scleral stroma have been found to decrease 35 percent and 20 percent with age, respectively, due to the effects of age-related collagen fibril cross-linking rather than an age-related change in the concentration of proteoglycans. The interfibrillar spaces also dynamically change based on the IOP of the eye (i.e. high IOP compresses the sclera and narrows the interfibrillar spaces, and vice versa).[30,394,405,406,408,409] Thus, the effect of aging, tissue hydration, or IOP may be more important for drugs in the size range of large macromolecules (e.g. monoclonal antibodies, vectors for gene-based therapy).[394]

Periocular drug delivery using the trans-scleral diffusion route is proven to consistently deliver drugs to choroid efficiently since this pathway is highly permeable to most ophthalmic drugs. However, this comes with one important caveat in that the clearance mechanisms in the subconjunctival space near the limbus appear to be superb (Fig. 4.31A), which can affect the amount and retention time of depot drug that is available for absorption trans-sclerally – especially when given via subconjunctival injection. Only recently has how the drug permeates through the inner tissue layers of the eye directly adjacent to the sclera to eventually reach the targeted tissues of the retina or vitreous been examined.[407,410] In fact, knowing a specific drug's scleral permeability properties is now known to be insufficient information to accurately predict the drug delivery rate to the retina or vitreous humor, particularly for large, hydrophilic drugs. The permeability properties of the choroid and Bruch's membrane have recently been studied and they are rather porous as their permeation properties are better than that of the sclera.[411] Nonetheless, the permeability properties of Bruch's membrane did gradually decline with age due to lipid accumulation. The major rate-limiting step for retinal or vitreous targeted drug delivery via the periocular route has recently been determined to be the RPE (10–100-fold less permeable than the human sclera and 14–16-fold less permeable than the choroid or Bruch's membrane for large, hydrophilic drugs; comparably similar to sclera, choroid, or Bruch's membrane for lipophilic drugs), which forms the outer part of the blood–retinal barrier due to the apical zonula occludens tight junctions found in its intercellular space (most similar to that of the corneal epithelium).[358,410]

While drug metabolism does not appear to be a major source of drug removal in the posterior segment, dynamic drug clearance mechanisms via orbital blood vessels and lymphatics, conjunctival blood vessels and lymphatics, episcleral blood vessels, and the choroidal vasculature are another confirmed inhibitory factor for successful periocular drug delivery.[412] Currently, animal studies show that the orbital (~25 percent drug removal rate using retrobulbar injection since drugs disperse throughout the orbit) and conjunctival (~5–80 percent drug removal rate using sub-Tenon's injection since drugs disperse circumferentially around the eye, but not into the orbit; the clearance rate for triamcinolone is estimated to be between 8 and 13 μg/hr) clearance mechanisms are of more importance than that of the choroid (~2–20 percent drug removal rate using sub-Tenon's injection).[400,412,413] Therefore, for large, hydrophilic drugs, the major barrier to drug entry into the retina or vitreous is the RPE, whereas small, lipophilic drugs are relatively unaffected since it transverses RPE quite easily using the intracellular (transcellular) route. Overall then, the rate-limiting factors for posterior segment drug delivery using this drug delivery route come down to avoiding or abrogating the dynamic physiologic clearance mechanisms and the individual physiochemical properties of the drug itself – predominantly the molecular radius (≤ 8 nm molecular radii drugs) and the lipophilicity of the drug, and, secondarily, the shape, MW, protein-binding properties, and ionic charge of the drug. Other issues on these various periocular drug delivery routes still need to be worked out and, as such, the information on this topic is far from complete.

In summary, there is now definitive in vitro and in vivo evidence to suggest that periocular drug delivery is the most efficient means to treat choroidal and RPE disease without significant systemic or intraocular risk since it diffuses through the static permeability barriers in the sclera and choroid quite easily. However, one caveat is that the drug depot site must have a drug release rate that exceeds the dynamic tissue clearance rates stated above. The feasibility of treating retinal or vitreous diseases is less certain, but is promising, likely depending more on the molecular size and lipophilicity of the drug being administered. Due to recently published data on static permeability properties of the sclera, choroid, Bruch's membrane, and RPE, we can at least now understand how the posterior segment is amenable to pharmacologic intervention other than direct, invasive intravitreal injection and should follow these periocular treatment modalities more closely as it may change the way retinal or posterior segment disease is treated since it basically is a crossroads between nanomedicine and ophthalmology. One promising offshoot drug delivery technology from the periocular drug delivery research efforts is that of minimally invasive drug-eluting microneedles (<1 mm diameter) that deliver drugs into the cornea, sclera, or suprachoroidal space.[414–416] Preliminary animal studies suggest that by circumventing the subconjunctival/episcleral clearance mechanism microneedle-based drug delivery markedly improves (80-fold) the intraocular bioavailability and moderately improves (3-fold) the duration of action over periocular drug delivery techniques. Because of our improved understanding of the pharmacokinetics involved in periocular drug delivery, attention should probably now shift more toward optimizing the physiochemical properties of existing drugs so that it prolongs residence time or enhances penetration into the retina and vitreous using these periocular drug delivery routes.[417]

Acknowledgments

Supported in part by NIH Grants P30 EY06360 (Departmental Core Grant), T32EY07092 (DGD), R01EY00933 (HFE), R01EY018100 (JLU), and Research to Prevent Blindness, Inc., New York.

References

1. Dawson DG, Watsky MA, Geroski DH, Edelhauser HF. Duane's Foundation of clinical ophthalmology on CD-ROM Vol. 2c. Philadelphia: Lippincott Williams & Wilkins, 2006:1–76.
2. Gipson I, Joyce N, Zieske J. The anatomy and cell biology of the human cornea, limbus, conjunctiva, and adnexa. In: Foster CS AD, Dohlman CH, eds. Smolin and Thoft's the cornea scientific foundations and clinical practice, 4th edn. Philadelphia: Lippincott Williams & Wilkins, 2005.
3. Klyce S. Corneal Physiology. In: Foster CS AD, Dohlman CH, eds. Smolin and Thoft's the cornea scientific foundations and clinical practice, 4th edn. Philadelphia: Lippincott Williams & Wilkins, 2005.
4. Nishida T. Cornea. In: Krachmer JH, Holland EJ, eds. Cornea. Fundamental, diagnosis, and management, 2nd edn. Philadelphia: Elsevier Mosby, 2005.
5. Rada J, Johnson J. Sclera. In: Krachmer JH, Holland EJ, eds. Cornea. Fundamental, diagnosis, and management, 2nd edn. Philadelphia: Elsevier Mosby, 2005.
6. Hogan M, Alvarado J, Wedell J. Histology of the human eye. Philadelphia: WB Saunders, 1971.
7. Barishak Y. Embryology of the eye and its adnexa. Switzerland: Karger, 2001.
8. Ozanics V, Rayborn M, Sagun D. Some aspects of corneal and scleral differentiation in the primate. Exp Eye Res 1976; 22:305–327.
9. Ozanics V, Rayborn M, Sagun D. Observations on the morphology of the developing primate cornea: epithelium, its innervation and anterior stroma. J Morphol 1977; 153:263–297.
10. Quantock AJ YR. Development of the corneal stroma, and the collagen–proteoglycan associations that help define its structure and function. Dev Dyn 2008:Epub ahead of print.
11. Hay ED. Development of the vertebrate cornea. Int Rev Cytol 1979; 63:263–322.
12. Hayashi S, Osawa T, Tohyama K. Comparative observations on corneas, with special reference to Bowman's layer and Descemet's membrane in mammals and amphibians. J Morphol 2002; 254:247–258.
13. Stiemke MM, McCartney MD, Cantu-Crouch D, Edelhauser HF. Maturation of the corneal endothelial tight junction. Invest Ophthalmol Vis Sci 1991; 32(10):2757–2765.
14. Stiemke MM, Edelhauser HF, Geroski DH. The developing corneal endothelium: correlation of morphology, hydration and Na/K ATPase pump site density. Curr Eye Res 1991; 10(2):145–156.
15. Gordon RA, Donzis PB. Refractive development of the human eye. Arch Ophthalmol 1985; 103(6):785–789.
16. Ehlers N, Sorensen T, Bramsen T, Poulsen EH. Central corneal thickness in newborns and children. Acta Ophthalmol 1976; 54(3):285–290.
17. Doughty MJ, Laiquzzaman M, Muller A et al. Central corneal thickness in European (white) individuals, especially children and the elderly, and assessment of its possible importance in clinical measures of intra-ocular pressure. Ophthal Physiol Optics 2002; 22(6):491–504.
18. Doughty MJ, Zaman ML. Human corneal thickness and its impact on intraocular pressure measures: a review and meta-analysis approach. Surv Ophthalmol 2000; 44(5):367–408.
19. Niederer RL, Perumal D, Sherwin T et al. Age-related differences in the normal human cornea: a laser scanning in vivo confocal microscopy study. Br J Ophthalmol 2007; 91(9):1165–1169.
20. Faragher RG, Mulholland B, Tuft SJ et al. Aging and the cornea. Br J Ophthalmol 1997; 81(10):814–817.
21. Hayashi K, Hayashi H, Hayashi F. Topographic analysis of the changes in corneal shape due to aging. Cornea 1995; 14(5):527–532.
22. Hayashi K, Masumoto M, Fujino S, Hayashi F. Changes in corneal astigmatism with aging. Nippon Ganka Gakkai Zasshi–Acta Societatis Ophthalmologicae Japonicae 1993; 97(10):1193–1196.
23. Alvarado J, Murphy C, Juster R. Age-related changes in the basement membrane of the human corneal epithelium. Invest Ophthalmol Vis Sci 1983; 24(8):1015–1028.
24. Edelhauser HF. The balance between corneal transparency and edema: the Proctor Lecture. Invest Ophthalmol Vis Sci 2006; 47(5):1754–1767.
25. Armitage WJ, Dick AD, Bourne WM et al. Predicting endothelial cell loss and long-term corneal graft survival. Invest Ophthalmol Vis Sci 2003; 44(8):3326–3331.
26. Patel S, McLaren J, Hodge D, Bourne W. Normal human keratocyte density and corneal thickness measurement by using confocal microscopy in vivo. Invest Ophthalmol Vis Sci 2001; 42(2):333–339.
27. Moller-Pedersen T. A comparative study of human corneal keratocyte and endothelial cell density during aging. Cornea 1997; 16(3):333–338.
28. Randleman JB, Dawson DG, Grossniklaus HE et al. Depth-dependent cohesive tensile strength in human donor corneas: implications for refractive surgery. J Refract Surg 2008; 24(1):S85–S89.
29. Daxer A, Misof K, Grabner B et al. Collagen fibrils in the human corneal stroma: structure and aging. Invest Ophthalmol Vis Sci 1998; 39(3):644–648.
30. Malik NS, Moss SJ, Ahmed N et al. Ageing of the human corneal stroma: structural and biochemical changes. Biochim Biophys Acta 1992; 1138(3):222–228.
31. Kanai A, Kaufman HE. Electron microscopic studies of corneal stroma: aging changes of collagen fibers. Ann Ophthalmol 1973; 5(3):285–287.
32. Watson PG, Hazelman B, Pavesio C, Green WR. The sclera and systemic disorders, 2nd edn. Edinburgh, UK: Butterworth Heinemann, 2004.
33. Cheng A, Rao S, Cheng L, Lam D. Assessment of pupil size under different light intensities using the Procyon pupillometer. J Cataract Refract Surg 2006; 32:1015–1017.
34. Kjesbu S, Moksnes K, Klepstad P et al. Application of pupillometry and pupillary reactions in medical research. Tidsskr Nor Laegeforen 2005; 125:29–32.
35. Atchison D SG. Optics of the human eye. Oxford: Butterworth-Heinemann, 2000.
36. Winn B, Whitaker D, Elliot D, Phillips N. Factors affecting light-adapted pupil size in normal human subjects. Invest Ophthalmol Vis Sci 1994; 35:1132–1137.
37. Mandell R, Chiang C, Klein S. Location of major corneal reference points. Optom Vis Sci 1995; 11:776–784.
38. Mandell R, Horner D. Alignment of videokeratographs. In: Gills J, Sanders D, Thornton S et al, eds. Corneal topography. The state of the art. Thorofare, New Jersey: Slack, 1995.
39. Edmund C. Location of the corneal apex and its influence on the stability of the central corneal curvature. A photokeratoscopy study. Am J Optom Physiol Optics 1987; 64:846–852.
40. Pande M, Hillman J. Optical zone centration in keratorefractive surgery. Entrance pupil center, visual axis, coaxially sighted corneal reflex, or geometric corneal center? Ophthalmology 1993; 100:1230–1237.
41. Uozato H, Guyton D. Centering corneal surgical procedures. Am J Ophthalmol 1987; 103:264–275.
42. Bueeler M, Mrochen M, Seiler T. Maximum permissible lateral decentration in aberration-sensing and wavefront-guided ablation. J Cataract Refract Surg 2003; 29:257–263.
43. Bueeler M, Mrochen M, Seiler T. Maximum permissible torsional misalignment in aberration-sensing and wavefront-guided corneal ablation. J Cataract Refract Surg 2004; 30:17–25.
44. Bron AJ, Tripathi R, Tripathi B. The cornea and sclera. In: Wolff's anatomy of the eye and orbit, 8th edn. London, UK: Chapman & Hall, 1997.
45. Klyce S, Martinez C. Cornea topography. In: Albert DM, ed. Principles and practice of ophthalmology, 2nd edn. Philadelphia: WB Saunders, 2000.
46. Gatinel D, Haouat M, Hoang-Xuan T. A review of mathematical descriptors of cornea asphericity. J Fr Ophthalmol 2002; 25:81–90.
47. Kiely P, Smith G, Carney L. The mean shape of the human cornea. Optica Acta 1982; 29:1027–1040.
48. Dubbelman M, Sicam VA, Van der Heijde GL, Sicam VADP. The shape of the anterior and posterior surface of the aging human cornea. Vision Res 2006; 46(6–7):993–1001.
49. Davis W, Raasch T, Mitchell G et al. Corneal asphericity and apical curvature in children: a cross-sectional and longitudinal evaluation. Invest Ophthalmol Vis Sci 2005; 46:1899–1906.
50. Read S, Collins M, Carney LG, Franklin R. The topography of the central and peripheral cornea. Invest Ophthalmol Vis Sci 2006; 47:1404–1415.
51. Ho JD, Tsai CY, Tsai RJ et al. Validity of the keratometric index: evaluation by the Pentacam rotating Scheimpflug camera. J Cataract Refract Surg 2008; 34(1):137–145.
52. Khoramnia R, Rabsilber TM, Auffarth GU et al. Central and peripheral pachymetry measurements according to age using the Pentacam rotating Scheimpflug camera. J Cataract Refract Surg 2007; 33(5):830–836.
53. Dingeldein SA, Klyce SD. The topography of normal corneas [erratum appears in Arch Ophthalmol 1989; 107(5):644]. Arch Ophthalmol 1989; 107(4):512–518.
54. Navarro R, Gonzalez L, Hernandez JL et al. Optics of the average normal cornea from general and canonical representations of its surface topography. J Optical Soc Am A, Optics, Image Sci Vis 2006; 23(2):219–232.
55. Konstantopoulos A, Hossain P, Anderson DF et al. Recent advances in ophthalmic anterior segment imaging: a new era for ophthalmic diagnosis? Br J Ophthalmol 2007; 91(4):551–557.
56. Bennett ES, Weissman BA. Clinical contact lens practice. Philadelphia: Lippincott Williams & Wilkins, 2004.
57. U.S. Food and Drug Administration, Department of health and human services, health Cfdar. Contact Lenses. Available at: http://www.fda.gov/cdrh/contactlenses/, 2008.
58. Liesegang TJ, Liesegang TJ. Physiologic changes of the cornea with contact lens wear. CLAO Journal 2002; 28(1):12–27.
59. Swarbrick HA, Swarbrick HA. Orthokeratology (corneal refractive therapy): what is it and how does it work? Eye Contact Lens Sci Clin Pract 2004; 30(4):181–185; discussion 205–6.
60. U.S. Food and Drug Administration, Department of health and human services, health Cfdar. LASIK. Available at: http://www.fda.gov/cdrh/lasik/, 2008.
61. Weiss JS. Basic and Clinical Science Course (BCSC): Refractive Surgery. San Francisco: American Academy of Ophthalmology, 2006.
62. Dawson DG, Edelhauser HF, Grossniklaus HE. et al. Long-term histopathologic findings in human corneal wounds after refractive surgical procedures. Am J Ophthalmol 2005; 139(1):168–178.
63. Kohnen T, Buhren J, Cichocki M et al. Optical quality after refractive corneal surgery. Ophthalmologe 2006; 103(3):184–191.
64. Mrochen M, Bueler M. Aspheric optics: physical fundamentals. Ophthalmologe 2008; 105(3):224–233.
65. Mrochen M, Hafezi F, Jankov M, Seiler T. Ablation profiles in corneal laser surgery. Current and future concepts. Ophthalmologe 2006; 103(3):175–183.
66. Farrell R, McCally R. Corneal transparency. In: Albert DM, ed. Principles and practice of ophthalmology, 2nd edn. Philadelphia: WB Saunders, 2000.
67. Meek KM, Leonard DW, Connon CJ et al. Transparency, swelling and scarring in the corneal stroma. Eye 2003; 17(8):927–936.
68. Maurice DM. The structure and transparency of the cornea. J Physiol 1957; 136(2):263–286.

69. Jester JV, Moller-Pedersen T, Huang J et al. The cellular basis of corneal transparency: evidence for "corneal crystallins". J Cell Sci 1999; 112(Pt 5):613–622.

70. Piatigorsky J. Enigma of the abundant water-soluble cytoplasmic proteins of the cornea: the "refraction" hypothesis. Cornea 2001; 20(8):853–858.

71. Jester JV, Jester JV. Corneal crystallins and the development of cellular transparency. Semin Cell Dev Biol 2008; 19(2):82–93.

72. Maurice DM. The cornea and sclera. In: Davson H, ed. The Eye, 3rd edn. New York: Academic Press, 1984.

73. Kaye GI. Stereologic measurement of cell volume fraction of rabbit corneal stroma. Arch Ophthalmol 1969; 82(6):792–794.

74. Lerman S. Biophysical aspects of corneal and lenticular transparency. Curr Eye Res 1984; 3(1):3–14.

75. Boettner E, Wolter J. Transmission of the ocular media. Invest Ophthalmol Vis Sci 1962; 6:776–783.

76. Farrell RA, McCally RL, Tatham PE. Wave-length dependencies of light scattering in normal and cold swollen rabbit corneas and their structural implications. J Physiol 1973; 233(3):589–612.

77. Moller-Pedersen T, Moller-Pedersen T. Keratocyte reflectivity and corneal haze. Exp Eye Res 2004; 78(3):553–560.

78. van den Berg TJ, Tan KE. Light transmittance of the human cornea from 320 to 700 nm for different ages. Vision Res 1994; 34(11):1453–1456.

79. Ihanamaki T, Pelliniemi LJ, Vuorio E et al. Collagens and collagen-related matrix components in the human and mouse eye. Progr Retin Eye Res 2004; 23(4):403–434.

80. Robert L, Legeais JM, Robert AM, Renard G. Corneal collagens. Pathologie Biologie 2001; 49(4):353–363.

81. Meek KM, Fullwood NJ. Corneal and scleral collagens–a microscopist's perspective. Micron 2001; 32(3):261–272.

82. Avery N, Bailey A. Restraining cross-links responsible for the mechanical properties of collagen fiber: natural and artificial. In: Fratzl P, ed. Collagen: structure and mechanics. New York: Springer, 2008.

83. Hulmes D. Collagen diversity, synthesis, and assembly. In: Fratzl P, ed. Collagen: structure and mechanics. New York: Springer, 2008.

84. Birk DE. Type V collagen: heterotypic type I/V collagen interactions in the regulation of fibril assembly. Micron 2001; 32(3):223–237.

85. White J, Werkmeister JA, Ramshaw JA, Birk DE. Organization of fibrillar collagen in the human and bovine cornea: collagen types V and III. Connect Tiss Res 1997; 36(3):165–174.

86. Meek KM. The cornea and sclera. In: Fratzl P, ed. Collagen: structure and mechanics. New York: Springer, 2008.

87. Borcherding MS, Blacik LJ, Sittig RA et al. Proteoglycans and collagen fibre organization in human corneoscleral tissue. Exp Eye Res 1975; 21(1):59–70.

88. Cho HI, Covington HI, Cintron C. Immunolocalization of type VI collagen in developing and healing rabbit cornea. Invest Ophthalmol Vis Sci 1990; 31(6):1096–1102.

89. Hirsch M, Prenant G, Renard G. Three-dimensional supramolecular organization of the extracellular matrix in human and rabbit corneal stroma, as revealed by ultrarapid-freezing and deep-etching methods. Exp Eye Res 2001; 72(2):123–135.

90. Komai Y, Ushiki T. The three-dimensional organization of collagen fibrils in the human cornea and sclera. Invest Ophthalmol Vis Sci 1991; 32(8):2244–2258.

91. Meek KM, Boote C, Meek KM, Boote C. The organization of collagen in the corneal stroma. Exp Eye Res 2004; 78(3):503–512.

92. Ojeda JL, Ventosa JA, Piedra S. The three-dimensional microanatomy of the rabbit and human cornea. A chemical and mechanical microdissection-SEM approach. J Anat 2001; 199(Pt 5):567–576.

93. Dawson DG, Grossniklaus HE, McCarey BE et al. Biomechanical and wound healing characteristics of corneas after excimer laser keratorefractive surgery: is there a difference between advanced surface ablation and sub-Bowman's keratomileusis? J Refract Surg 2008; 24(1):S90–S96.

94. Mathew J, Bergmanson J, Doughty MJ. Fine structure of the interface between the anterior limiting lamina and the anterior stromal fibrils of the human cornea. Invest Ophthalmol Vis Sci 2008; 49:3914–3918.

95. Newton RH, Meek KM. Circumcorneal annulus of collagen fibrils in the human limbus. Invest Ophthalmol Vis Sci 1998; 39(7):1125–1134.

96. Newton RH, Meek KM. The integration of the corneal and limbal fibrils in the human eye. Biophys J 1998; 75(5):2508–2512.

97. Moller-Pedersen T, Ledet T, Ehlers N. The keratocyte density of human donor corneas. Curr Eye Res 1994; 13(2):163–169.

98. Watsky MA. Keratocyte gap junctional communication in normal and wounded rabbit corneas and human corneas. Invest Ophthalmol Vis Sci 1995; 36(13):2568–2576.

99. Hahnel C, Somodi S, Weiss D, Guthoff R. The keratocyte network of the human cornea: a three-dimensional study using confocal laser scanning microscopy. Cornea 2000; 19:185–193.

100. Muller LJ, Pels L, Vrensen GF. Novel aspects of the ultrastructural organization of human corneal keratocytes. Invest Ophthalmol Vis Sci 1995; 36(13):2557–2567.

101. Hamrah P, Liu Y, Zhang Q et al. The corneal stroma is endowed with a significant number of resident dendritic cells. Invest Ophthalmol Vis Sci 2003; 44(2):581–589.

102. Poole CA, Brookes NH, Clover GM. Keratocyte networks visualised in the living cornea using vital dyes. J Cell Sci 1993; 106(Pt 2):685–691.

103. Du Y, Funderburgh ML, Mann MM et al. Multipotent stem cells in human corneal stroma. Stem Cells 2005; 23(9):1266–1275.

104. Du Y, Sundarraj N, Funderburgh ML et al. Secretion and organization of a cornea-like tissue in vitro by stem cells from human corneal stroma. Invest Ophthalmol Vis Sci 2007; 48(11):5038–5045.

105. Hedbys BO. The role of polysaccharides in corneal swelling. Exp Eye Res 1961; 1:81–91.

106. Bettelheim FA, Plessy B. The hydration of proteoglycans of bovine cornea. Biochim Biophys Acta 1975; 381(1):203–214.

107. Scott JE. Proteoglycan: collagen interactions and corneal ultrastructure. Biochem Soc Trans 1991; 19(4):877–881.

108. Scott JE. Proteoglycan histochemistry–a valuable tool for connective tissue biochemists. Collagen Rel Res 1985; 5(6):541–575.

109. Scott JE. Extracellular matrix, supramolecular organisation and shape. J Anat 1995; 187(Pt 2):259–269.

110. Hassell J, Blochberger T, Rada J et al. Proteoglycan gene families. Adv Mol Cell Biol 1993; 6:69–113.

111. Scott JE, Bosworth TR. The comparative chemical morphology of the mammalian cornea. Basic Appl Histochemi 1990; 34(1):35–42.

112. Castoro JA, Bettelheim AA, Bettelheim FA. Water gradients across bovine cornea. Invest Ophthalmol Vis Sci 1988; 29(6):963–968.

113. Guthoff RF, Wienss H, Hahnel C et al. Epithelial innervation of human cornea: a three-dimensional study using confocal laser scanning fluorescence microscopy. Cornea 2005; 24(5):608–613.

114. Rozsa AJ, Beuerman RW. Density and organization of free nerve endings in the corneal epithelium of the rabbit. Pain 1982; 14(2):105–120.

115. Schimmelpfennig B. Nerve structures in human central corneal epithelium. Graefes Arch Clin Exp Ophthalmol 1982; 218(1):14–20.

116. Muller LJ, Marfurt CF, Kruse F et al. Corneal nerves: structure, contents and function. [erratum appears in Exp Eye Res 2003; 77(2):253]. Exp Eye Res 2003; 76(5):521–542.

117. Belmonte C, Acosta MC, Gallar J et al. Neural basis of sensation in intact and injured corneas. Exp Eye Res 2004; 78(3):513–525.

118. Stachs O, Zhivov A, Kraak R et al. In vivo three-dimensional confocal laser scanning microscopy of the epithelial nerve structure in the human cornea. Graefes Arch Clin Exp Ophthalmol 2007; 245(4):569–575.

119. Lawrenson JG. Corneal sensitivity in health and disease. Ophthal Physiol Optics 1997; 17(Suppl 1):S17–S22.

120. Cochet P, Bonnet R. Corneal esthesiometry. Performance and practical importance. Bulletin des Societes d'Ophtalmologie de France 1961; 6:541–550.

121. Brennan NA, Bruce AS. Esthesiometry as an indicator of corneal health. Optom Vis Sci 1991; 68(9):699–702.

122. Kohlhaas M. Corneal sensation after cataract and refractive surgery. J Cataract Refract Surg 1998; 24(10):1399–1409.

123. Belmonte C, Belmonte C. Eye dryness sensations after refractive surgery: impaired tear secretion or "phantom" cornea? J Refract Surg 2007; 23(6):598–602.

124. Wolter JR. Reactions of the cellular elements of the corneal stroma; a report of experimental studies in the rabbit eye. AMA Arch Ophthalmol 1958; 59(6):873–881.

125. Cintron C, Hassinger LC, Kublin CL, Cannon DJ. Biochemical and ultrastructural changes in collagen during corneal wound healing. J Ultrastruct Res 1978; 65(1):13–22.

126. Hassell JR, Cintron C, Kublin C, Newsome DA. Proteoglycan changes during restoration of transparency in corneal scars. Arch Biochem Biophys 1983; 222(2):362–369.

127. Cintron C, Covington HI, Kublin CL. Morphologic analyses of proteoglycans in rabbit corneal scars. Invest Ophthalmol Vis Sci 1990; 31(9):1789–1798.

128. Funderburgh JL, Cintron C, Covington HI, Conrad GW. Immunoanalysis of keratan sulfate proteoglycan from corneal scars. Invest Ophthalmol Vis Sci 1988; 29(7):1116–1124.

129. Binder P, Wickham M, Zavala E, Akers P. Symposium on Medical and Surgical Diseases of the Cornea: Corneal anatomy and wound healing. New Orleans: American Academy of Ophthalmology, 1980.

130. Mohan RR, Hutcheon AE, Choi R et al. Apoptosis, necrosis, proliferation, and myofibroblast generation in the stroma following LASIK and PRK. Exp Eye Res 2003; 76(1):71–87.

131. Jester JV, Petroll WM, Cavanagh HD. Corneal stromal wound healing in refractive surgery: the role of myofibroblasts. Progr Retin Eye Res 1999; 18(3):311–356.

132. Wilson SE, Mohan RR, Ambrosio R, Jr. et al. The corneal wound healing response: cytokine-mediated interaction of the epithelium, stroma, and inflammatory cells. Progr Retin Eye Res 2001; 20(5):625–637.

133. Netto MV, Mohan RR, Ambrosio R, Jr. et al. Wound healing in the cornea: a review of refractive surgery complications and new prospects for therapy. Cornea 2005; 24(5):509–522.

134. Zhao J, Nagasaki T, Maurice DM. Role of tears in keratocyte loss after epithelial removal in mouse cornea. Invest Ophthalmol Vis Sci 2001; 42(8):1743–1749.

135. Wilson SE, Mohan RR, Hong J et al. Apoptosis in the cornea in response to epithelial injury: significance to wound healing and dry eye. Adv Exp Med Biol 2002; 506(Pt B):821–826.

136. Zieske JD, Guimaraes SR, Hutcheon AE. Kinetics of keratocyte proliferation in response to epithelial debridement. Exp Eye Res 2001; 72(1):33–39.

137. Fini ME. Keratocyte and fibroblast phenotypes in the repairing cornea. Progr Retin Eye Res 1999; 18(4):529–551.

138. Fini ME, Stramer BM, Fini ME, Stramer BM. How the cornea heals: cornea-specific repair mechanisms affecting surgical outcomes. Cornea 2005; 24(8 Suppl):S2–S11.

139. Jester JV, Barry-Lane PA, Cavanagh HD, Petroll WM. Induction of alpha-smooth muscle actin expression and myofibroblast transformation in cultured corneal keratocytes. Cornea 1996; 15(5):505–516.

140. Wilson SE, Liu JJ, Mohan RR. Stromal–epithelial interactions in the cornea. Progr Retin Eye Res 1999; 18(3):293–309.

141. Meltendorf C, Burbach GJ, Buhren J et al. Corneal femtosecond laser keratotomy results in isolated stromal injury and favorable wound-healing response. Invest Ophthalmol Vis Sci 2007; 48(5):2068–2075.

142. Meltendorf C, Burbach G, Ohrloff C et al. Intrastromal Keratotomy with Femtosecond Laser avoids profibrotic TGF-β1 Induction. Invest Ophthalmol Vis Sci 2009; 50(8):3688–3695.

143. Lim M, Goldstein M, Tuli S, Schultz G. Growth factor, cytokine, and protease interactions during corneal wound healing. The Ocular Surface 2003; 1:53–65.

144. Klenkler B, Sheardown H, Klenkler B, Sheardown H. Growth factors in the anterior segment: role in tissue maintenance, wound healing and ocular pathology. Exp Eye Res 2004; 79(5):677–688.

145. Klenkler B, Sheardown H, Jones L et al. Growth factors in the tear film: role in tissue maintenance, wound healing, and ocular pathology. The Ocular Surface 2007; 5(3):228–239.

146. Micera A, Lambiase A, Puxeddu I et al. Nerve growth factor effect on human primary fibroblastic-keratocytes: possible mechanism during corneal healing. Exp Eye Res 2006; 83(4):747–757.
147. Stramer BM, Zieske JD, Jung JC et al. Molecular mechanisms controlling the fibrotic repair phenotype in cornea: implications for surgical outcomes. Invest Ophthalmol Vis Sci 2003; 44(10):4237–4246.
148. Girard MT, Matsubara M, Fini ME. Transforming growth factor-beta and interleukin-1 modulate metalloproteinase expression by corneal stromal cells. Invest Ophthalmol Vis Sci 1991; 32(9):2441–2454.
149. Ubels J, Foley K, Rismando V. Retinol secretion by the lacrimal gland. Invest Ophthalmol Vis Sci 1986; 27:1261–1268.
150. Tsubota K. Tear dynamics and dry eye. Progr Retin Eye Res 1998; 17:565–596.
151. Woost PG, Brightwell J, Eiferman RA, Schultz GS. Effect of growth factors with dexamethasone on healing of rabbit corneal stromal incisions. Exp Eye Res 1985; 40(1):47–60.
152. Kirschner SE, Ciaccia A, Ubels JL. The effect of retinoic acid on thymidine incorporation and morphology of corneal stromal fibroblasts. Curr Eye Res 1990; 9(11):1121–1125.
153. Johnson MK, Gebhardt BM, Berman MB. Appearance of collagenase in pneumolysin-treated corneal fibroblast cultures. Curr Eye Res 1988; 7(9):951–953.
154. Kenney MC, Chwa M, Escobar M, Brown D. Altered gelatinolytic activity by keratoconus corneal cells. Biochem Biophys Res Commun 1989; 161(1):353–357.
155. Fujita H, Ueda A, Nishida T, Otori T. Uptake of india ink particles and latex beads by corneal fibroblasts. Cell Tissue Res 1987; 250(2):251–255.
156. Mishima H, Yasumoto K, Nishida T, Otori T. Fibronectin enhances the phagocytic activity of cultured rabbit keratocytes. Invest Ophthalmol Vis Sci 1987; 28(9):1521–1526.
157. Mondino BJ, Sundar-Raj CV, Brady KJ. Production of first component of complement by corneal fibroblasts in tissue culture. Arch Ophthalmol 1982; 100(3):478–480.
158. Taylor JL, O'Brien WJ. Interferon production and sensitivity of rabbit corneal epithelial and stromal cells. Invest Ophthalmol Vis Sci 1985; 26(11):1502–1508.
159. Taylor L, Menconi M, Leibowitz MH, Polgar P. The effect of ascorbate, hydroperoxides, and bradykinin on prostaglandin production by corneal and lens cells. Invest Ophthalmol Vis Sci 1982; 23(3):378–382.
160. Lemp MA. Cornea and sclera. Arch Ophthalmol 1976; 94(3):473–490.
161. Maurice DM. The biology of wound healing in the corneal stroma. Castroviejo lecture. Cornea 1987; 6(3):162–168.
162. Binder PS. Barraquer lecture. What we have learned about corneal wound healing from refractive surgery. Refract Corneal Surg 1989; 5(2):98–120.
163. Assil KK, Quantock AJ. Wound healing in response to keratorefractive surgery. Surv Ophthalmol 1993; 38(3):289–302.
164. Schmack I, Dawson DG, McCarey BE et al. Cohesive tensile strength of human LASIK wounds with histologic, ultrastructural, and clinical correlations. J Refract Surg 2005; 21(5):433–445.
165. Fournie PR, Gordon GM, Dawson DG et al. Correlations of long-term matrix metalloproteinase localization in human corneas after successful laser-assisted in situ keratomileusis with minor complications at the flap margin. Arch Ophthalmol 2008; 126(2):162–170.
166. Morrison JC, Swan KC. Bowman's layer in penetrating keratoplasties of the human eye. Arch Ophthalmol 1982; 100(11):1835–1838.
167. Morrison JC, Swan KC. Full-thickness lamellar keratoplasty. A histologic study in human eyes. Ophthalmology 1982; 89(6):715–719.
168. Obata H, Tsuru T, Obata H, Tsuru T. Corneal wound healing from the perspective of keratoplasty specimens with special reference to the function of the Bowman layer and Descemet membrane. Cornea 2007; 26(9 Suppl 1):S82–S89.
169. Christensen L. Cataract-wound closure and healing. Symposium on cataracts. St. Louis, MO: CV Mosby, 1965.
170. Flaxel JT, Swan KC. Limbal wound healing after cataract extraction. A histologic study. Arch Ophthalmol 1969; 81(5):653–659.
171. Flaxel JT. Histology of cataract extractions. Arch Ophthalmol 1970; 83(4):436–444.
172. Lang GK, Green WR, Maumenee AE. Clinicopathologic studies of keratoplasty eyes obtained post mortem. Am J Ophthalmol 1986; 101(1):28–40.
173. Morrison JC, Swan KC. Descemet's membrane in penetrating keratoplasties of the human eye. Arch Ophthalmol 1983; 101(12):1927–1929.
174. Deg JK, Binder PS. Wound healing after astigmatic keratotomy in human eyes. Ophthalmology 1987; 94(10):1290–1298.
175. Melles GR, Binder PS. A comparison of wound healing in sutured and unsutured corneal wounds. Arch Ophthalmol 1990; 108(10):1460–1469.
176. Melles GR, Binder PS, Anderson JA. Variation in healing throughout the depth of long-term, unsutured, corneal wounds in human autopsy specimens and monkeys. Arch Ophthalmol 1994; 112(1):100–109.
177. Melles GR, Binder PS, Moore MN, Anderson JA. Epithelial-stromal interactions in human keratotomy wound healing. Arch Ophthalmol 1995; 113(9):1124–1130.
178. Jester JV, Villasenor RA, Schanzlin DJ, Cavanagh HD. Variations in corneal wound healing after radial keratotomy: possible insights into mechanisms of clinical complications and refractive effects. Cornea 1992; 11(3):191–199.
179. Wu WC, Stark WJ, Green WR. Corneal wound healing after 193-nm excimer laser keratectomy. Arch Ophthalmol 1991; 109(10):1426–1432.
180. Taylor DM, L'Esperance FA, Jr., Del Pero RA et al. Human excimer laser lamellar keratectomy. A clinical study. Ophthalmology 1989; 96(5):654–664.
181. Anderson NJ, Edelhauser HF, Sharara N et al. Histologic and ultrastructural findings in human corneas after successful laser in situ keratomileusis. Arch Ophthalmol 2002; 120(3):288–293.
182. Kramer TR, Chuckpaiwong V, Dawson DG et al. Pathologic findings in postmortem corneas after successful laser in situ keratomileusis. Cornea 2005; 24(1):92–102.
183. Dawson DG, Kramer TR, Grossniklaus HE et al. Histologic, ultrastructural, and immunofluorescent evaluation of human laser-assisted in situ keratomileusis corneal wounds. [erratum appears in Arch Ophthalmol. 2005; 123(8):1087]. Arch Ophthalmol 2005; 123(6):741–756.
184. Klyce SD, Crosson CE. Transport processes across the rabbit corneal epithelium: a review. Curr Eye Res 1985; 4(4):323–331.
185. McLaughlin BJ, Caldwell RB, Sasaki Y, Wood TO. Freeze-fracture quantitative comparison of rabbit corneal epithelial and endothelial membranes. Curr Eye Res 1985; 4(9):951–961.
186. Ban Y, Dota A, Cooper LJ et al. Tight junction-related protein expression and distribution in human corneal epithelium. Exp Eye Res 2003; 76(6):663–669.
187. Gipson IK, Yankauckas M, Spurr-Michaud SJ et al. Characteristics of a glycoprotein in the ocular surface glycocalyx. Invest Ophthalmol Vis Sci 1992; 33(1):218–227.
188. Nichols BA, Chiappino ML, Dawson CR. Demonstration of the mucous layer of the tear film by electron microscopy. Invest Ophthalmol Vis Sci 1985; 26(4):464–473.
189. Argueso P, Gipson IK. Epithelial mucins of the ocular surface: structure, biosynthesis and function. Exp Eye Res 2001; 73(3):281–289.
190. Gipson I, Joyce N. Anatomy and cell biology of the cornea, superficial limbus, and conjunctiva. In: Albert D, Miller J, eds. Principles and practice of ophthalmology, 3rd edn. Philadelphia: WB Saunders, 2008.
191. Mishima S. Some physiological aspects of the precorneal tear film. Arch Ophthalmol 1965; 73:233–241.
192. Anonymous. The definition and classification of dry eye disease: report of the Definition and Classification Subcommittee of the International Dry Eye WorkShop (2007). The Ocular Surface 2007; 5(2):75–92.
193. Behrens A, Doyle JJ, Stern L et al. Dysfunctional tear syndrome: a Delphi approach to treatment recommendations [see comment]. Cornea 2006; 25(8):900–907.
194. Perry HD, Perry HD. Dry eye disease: pathophysiology, classification, and diagnosis. Am J Managed Care 2008; 14(3 Suppl):S79–S87.
195. Friedenwald J, Buscke W. Some factors concerned in the mitotic and wound healing activities of the corneal epithelium. Trans Am Ophthalmol Soc 1944; 42:371–383.
196. Hanna C, Bicknell DS, O'Brien JE. Cell turnover in the adult human eye. Arch Ophthalmol 1961; 65:695–698.
197. Hanna C, O'Brien JE. Cell production and migration in the epithelial layer of the cornea. Arch Ophthalmol 1960; 64:536–539.
198. Williams K, Watsky M, Williams K, Watsky M. Gap junctional communication in the human corneal endothelium and epithelium. Curr Eye Res 2002; 25(1):29–36.
199. Buck RC. Measurement of centripetal migration of normal corneal epithelial cells in the mouse. Invest Ophthalmol Vis Sci 1985; 26(9):1296–1299.
200. Kinoshita S, Friend J, Thoft RA. Sex chromatin of donor corneal epithelium in rabbits. Invest Ophthalmol Vis Sci 1981; 21(3):434–441.
201. Schermer A, Galvin S, Sun TT. Differentiation-related expression of a major 64K corneal keratin in vivo and in culture suggests limbal location of corneal epithelial stem cells. J Cell Biol 1986; 103(1):49–62.
202. Sharma A, Coles WH. Kinetics of corneal epithelial maintenance and graft loss. A population balance model. Invest Ophthalmol Vis Sci 1989; 30(9):1962–1971.
203. Thoft RA, Friend J. The X, Y, Z hypothesis of corneal epithelial maintenance. Invest Ophthalmol Vis Sci 1983; 24(10):1442–1443.
204. Dillon EC, Eagle RC, Jr., Laibson PR. Compensatory epithelial hyperplasia in human corneal disease. Ophthalm Surg 1992; 23(11):729–732.
205. Cotsarelis G, Cheng SZ, Dong G et al. Existence of slow-cycling limbal epithelial basal cells that can be preferentially stimulated to proliferate: implications on epithelial stem cells. Cell 1989; 57(2):201–209.
206. Tseng SC. Concept and application of limbal stem cells. Eye 1989; 3(Pt 2): 141–157.
207. Lavker RM, Tseng SC, Sun TT et al. Corneal epithelial stem cells at the limbus: looking at some old problems from a new angle. Exp Eye Res 2004; 78(3):433–446.
208. Boulton M, Albon J, Boulton M, Albon J. Stem cells in the eye. Int J Biochem Cell Biol 2004; 36(4):643–657.
209. Nakamura T, Kinoshita S, Nakamura T, Kinoshita S. Current regenerative therapy for the cornea. Nippon Rinsho – Jpn J Clin Med 2008; 66(5):955–960.
210. Koizumi N, Nishida K, Amano S et al. Progress in the development of tissue engineering of the cornea in Japan. Nippon Ganka Gakkai Zasshi – Acta Societatis Ophthalmologicae Japonicae 2007; 111(7):493–503.
211. Gillette TE, Chandler JW, Greiner JV. Langerhans cells of the ocular surface. Ophthalmology 1982; 89(6):700–711.
212. Langerhans P. Uber die: nerven der meschllichen haut. Virchows Arch Path Anat Physiol 1868; 44:325–337.
213. Steinman RM. The dendritic cell system and its role in immunogenicity. Ann Rev Immunol 1991; 9:271–296.
214. Hamrah P, Zhang Q, Liu Y et al. Novel characterization of MHC class II-negative population of resident corneal Langerhans cell-type dendritic cells. Invest Ophthalmol Vis Sci 2002; 43(3):639–646.
215. Edelhauser HF. The resiliency of the corneal endothelium to refractive and intraocular surgery. Cornea 2000; 19(3):263–273.
216. Hales T. The honeycomb conjecture. Discrete Comput Geom 2001; 25:1–22.
217. Iwanmoto T, Smelser G. Electron microscopy of the human corneal endothelium with reference to transport mechanisms. Invest Ophthalmol Vis Sci 1965; 4:270–284.
218. Hirsch M, Renard G, Faure J, Pouliquen Y. Study of the ultrastructure of the rabbit corneal endothelium by the freeze-fracture technique: apical and laterial junctions. Exp Eye Res 1977; 25:277–288.
219. Joyce NC, Joyce NC. Cell cycle status in human corneal endothelium. Exp Eye Res 2005; 81(6):629–638.
220. Mimura T, Joyce NC, Mimura T, Joyce NC. Replication competence and senescence in central and peripheral human corneal endothelium. Invest Ophthalmol Vis Sci 2006; 47(4):1387–1396.
221. Yee RW, Matsuda M, Schultz RO, Edelhauser HF. Changes in the normal corneal endothelial cellular pattern as a function of age. Curr Eye Res 1985; 4(6):671–678.
222. McGowan SL, Edelhauser HF, Pfister RR et al. Stem cell markers in the human posterior limbus and corneal endothelium of unwounded and wounded corneas. Molec Vis 2007; 13:1984–2000.
223. Whikehart DR, Parikh CH, Vaughn AV et al. Evidence suggesting the existence of stem cells for the human corneal endothelium. Molec Vis 2005; 11:816–824.
224. Matsuda M, Yee RW, Edelhauser HF. Comparison of the corneal endothelium in an American and a Japanese population. Arch Ophthalmol 1985; 103(1):68–70.
225. Rao SK, Ranjan Sen P, Fogla R et al. Corneal endothelial cell density and morphology in normal Indian eyes. Cornea 2000; 19(6):820–823.

226. Padilla MD, Sibayan SA, Gonzales CS et al. Corneal endothelial cell density and morphology in normal Filipino eyes. Cornea 2004; 23(2):129–135.

227. Yunliang S, Yuqiang H, Ying-Peng L et al. Corneal endothelial cell density and morphology in healthy Chinese eyes. Cornea 2007; 26(2):130–132.

228. Amann J, Holley GP, Lee SB et al. Increased endothelial cell density in the paracentral and peripheral regions of the human cornea [see comment]. Am J Ophthalmol 2003; 135(5):584–590.

229. Irvine AR, Irvine AR, Jr. Variations in normal human corneal endothelium; a preliminary report of pathologic human corneal endothelium. Am J Ophthalmol 1953; 36(9):1279–1285.

230. Schimmelpfennig BH. Direct and indirect determination of nonuniform cell density distribution in human corneal endothelium. Invest Ophthalmol Vis Sci 1984; 25(2):223–229.

231. Daus W, Volcker HE, Meysen H, Bundschuh W. Vital staining of the corneal endothelium – increased possibilities of diagnosis. Fortschritte der Ophthalmologie 1989; 86(4):259–264.

232. Daus W, Volcker HE, Meysen H. Clinical significance of age-related regional differences in distribution of human corneal endothelium. Klinische Monatsblatter fur Augenheilkunde 1990; 196(6):449–455.

233. Bourne WM, McLaren JW, Bourne WM, McLaren JW. Clinical responses of the corneal endothelium. Exp Eye Res 2004; 78(3):561–572.

234. Watsky MA, McDermott ML, Edelhauser HF. In vitro corneal endothelial permeability in rabbit and human: the effects of age, cataract surgery and diabetes. Exp Eye Res 1989; 49(5):751–767.

235. Carlson KH, Bourne WM, McLaren JW, Brubaker RF. Variations in human corneal endothelial cell morphology and permeability to fluorescein with age. Exp Eye Res 1988; 47(1):27–41.

236. Harris JE. Symposium on the cornea. Introduction: factors influencing corneal hydration. Investigative Ophthalmology 1962; 1:151–157.

237. Kaye G, Tice L. Studies on the cornea. V. Electron microscope localization of adenosine triphosphatase activity in the rabbit cornea in relation of transport. Invest Ophthalmol Vis Sci 1966; 5:22–32.

238. Lim JJ. Na⁺ transport across the rabbit corneal endothelium. Curr Eye Res 1981; 1(4):255–258.

239. Lim JJ, Ussing HH. Analysis of presteady-state Na⁺ fluxes across the rabbit corneal endothelium. J Membr Biol 1982; 65(3):197–204.

240. Geroski DH, Matsuda M, Yee RW, Edelhauser HF. Pump function of the human corneal endothelium. Effects of age and cornea guttata. Ophthalmology 1985; 92(6):759–763.

241. Bourne WM. Clinical estimation of corneal endothelial pump function. Trans Am Ophthalmol Soc 1998; 96:229–239; discussion 239–242.

242. Burns RR, Bourne WM, Brubaker RF. Endothelial function in patients with cornea guttata. Invest Ophthalmol Vis Sci 1981; 20(1):77–85.

243. Mishima S. Clinical investigations on the corneal endothelium – XXXVIII Edward Jackson Memorial Lecture. Am J Ophthalmol 1982; 93(1):1–29.

244. Bonanno JA, Bonanno JA. Identity and regulation of ion transport mechanisms in the corneal endothelium. Progr Retin Eye Res 2003; 22(1):69–94.

245. Stiemke MM, Roman RJ, Palmer ML, Edelhauser HF. Sodium activity in the aqueous humor and corneal stroma of the rabbit. Exp Eye Res 1992; 55(3):425–433.

246. Hedbys BO, Dohlman CH. A new method for the determination of the swelling pressure of the corneal stroma in vitro. Exp Eye Res 1963; 2:122–129.

247. Klyce SD, Dohlman CH, Tolpin DW. In vivo determination of corneal swelling pressure. Exp Eye Res 1971; 11(2):220–229.

248. Hedbys BO, Mishima S, Maurice DM. The inbibition pressure of the corneal stroma. Exp Eye Res 1963; 2:99–111.

249. Ytteborg J, Dohlman C. Corneal edema and intraocular pressure. 1. Animal experiments. Arch Ophthalmol 1965; 74:375–381.

250. Ytteborg J, Dohlman CH. Corneal edema and intraocular pressure. II. Clinical results. Arch Ophthalmol 1965; 74(4):477–484.

251. Harper CL, Boulton ME, Bennett D et al. Diurnal variations in human corneal thickness [erratum appears in Br J Ophthalmol 1997; 81(2):1175]. Br J Ophthalmol 1996; 80(12):1068–1072.

252. Feng Y, Varikooty J, Simpson TL. Diurnal variation of corneal and corneal epithelial thickness measured using optical coherence tomography. Cornea 2001; 20(5):480–483.

253. Van Horn DL, Doughman DJ, Harris JE et al. Ultrastructure of human organ-cultured cornea. II. Stroma and epithelium. Arch Ophthalmol 1975; 93(4):275–277.

254. Meek KM, Dennis S, Khan S et al. Changes in the refractive index of the stroma and its extrafibrillar matrix when the cornea swells. Biophys J 2003; 85(4):2205–2212.

255. Muller LJ, Pels E, Vrensen GF. The specific architecture of the anterior stroma accounts for maintenance of corneal curvature [see comment]. Br J Ophthalmol 2001; 85(4):437–443.

256. Kangas TA, Edelhauser HF, Twining SS, O'Brien WJ. Loss of stromal glycosaminoglycans during corneal edema. Invest Ophthalmol Vis Sci 1990; 31(10):1994–2002.

257. Cristol SM, Edelhauser HF, Lynn MJ. A comparison of corneal stromal edema induced from the anterior or the posterior surface. Refract Corneal Surg 1992; 8(3):224–229.

258. Connon CJ, Meek KM, Connon CJ, Meek KM. The structure and swelling of corneal scar tissue in penetrating full-thickness wounds. Cornea 2004; 23(2):165–171.

259. Hatton MP, Perez VL, Dohlman CH et al. Corneal oedema in ocular hypotony. Exp Eye Res 2004; 78(3):549–552.

260. Bourne WM. Biology of the corneal endothelium in health and disease. Eye 2003; 17(8):912–918.

261. Riley M. Transport of ions and metabolites across the corneal endothelium. In: Cell biology of the eye. New York: Academic Press, 1982.

262. Klyce SD. Stromal lactate accumulation can account for corneal oedema osmotically following epithelial hypoxia in the rabbit. J Physiol 1981; 321:49–64.

263. Edelhauser HF, Geroski DH, Woods WD et al. Swelling in the isolated perfused cornea induced by 12(R)hydroxyeicosatetraenoic acid. Invest Ophthalmol Vis Sci 1993; 34(10):2953–2961.

264. Davis KL, Conners MS, Dunn MW, Schwartzman ML. Induction of corneal epithelial cytochrome P-450 arachidonate metabolism by contact lens wear. Invest Ophthalmol Vis Sci 1992; 33(2):291–297.

265. Polse KA, Brand RJ, Cohen SR, Guillon M. Hypoxic effects on corneal morphology and function. Invest Ophthalmol Vis Sci 1990; 31(8):1542–1554.

266. Minassian DC, Rosen P, Dart JK et al. Extracapsular cataract extraction compared with small incision surgery by phacoemulsification: a randomised trial [see comment] [erratum appears in Br J Ophthalmol 2001; 85(12):1498]. Br J Ophthalmol 2001; 85(7):822–829.

267. Bourne RR, Minassian DC, Dart JK et al. Effect of cataract surgery on the corneal endothelium: modern phacoemulsification compared with extracapsular cataract surgery. Ophthalmology 2004; 111(4):679–685.

268. Bourne WM, Nelson LR, Hodge DO. Continued endothelial cell loss ten years after lens implantation. Ophthalmology 1994; 101(6):1014–1022; discussion 1022–1023.

269. Crema AS, Walsh A, Yamane Y et al. Comparative study of coaxial phacoemulsification and microincision cataract surgery. One-year follow-up. J Cataract Refract Surg 2007; 33(6):1014–1018.

270. Storr-Paulsen A, Norregaard JC, Ahmed S et al. Endothelial cell damage after cataract surgery: divide-and-conquer versus phaco-chop technique. J Cataract Refract Surg 2008; 34(6):996–1000.

271. Espandar L, Meyer JJ, Moshirfar M et al. Phakic intraocular lenses. Curr Opin Ophthalmol 2008; 19(4):349–356.

272. Lovisolo CF, Reinstein DZ, Lovisolo CF, Reinstein DZ. Phakic intraocular lenses. Sur Ophthalmol 2005; 50(6):549–587.

273. Bourne WM. Cellular changes in transplanted human corneas. Cornea 2001; 20(6):560–569.

274. Melles GR, Melles GRJ. Posterior lamellar keratoplasty: DLEK to DSEK to DMEK [comment]. Cornea 2006; 25(8):879–881.

275. Terry MA, Terry MA. Endothelial keratoplasty: history, current state, and future directions. Cornea 2006; 25(8):873–878.

276. Price MO, Price FW, Price MO, Price FW. Descemet's stripping endothelial keratoplasty. Curr Opin Ophthalmol 2007; 18(4):290–294.

277. Melles GR, Ong TS, Ververs B et al. Preliminary clinical results of Descemet membrane endothelial keratoplasty. Am J Ophthalmol 2008; 145(2):222–227.

278. Melles GR, Ong TS, Ververs B et al. Descemet membrane endothelial keratoplasty (DMEK). Cornea 2006; 25(8):987–990.

279. Price MO, Price FW, Jr., Price MO, Price FW, Jr. Endothelial cell loss after descemet stripping with endothelial keratoplasty influencing factors and 2-year trend. Ophthalmology 2008; 115(5):857–865.

280. Terry MA, Chen ES, Shamie N et al. Endothelial cell loss after Descemet's stripping endothelial keratoplasty in a large prospective series. Ophthalmology 2008; 115(3):488–496.e3.

281. Lacayo GO, 3rd, Majmudar PA, Lacayo GO, 3rd, Majmudar PA. How and when to use mitomycin-C in refractive surgery. Curr Opin Ophthalmol 2005; 16(4):256–259.

282. Mearza AA, Aslanides IM, Mearza AA, Aslanides IM. Uses and complications of mitomycin C in ophthalmology. Expert Opinion Drug Safety 2007; 6(1):27–32.

283. Diakonis VF, Pallikaris A, Kymionis GD et al. Alterations in endothelial cell density after photorefractive keratectomy with adjuvant mitomycin. Am J Ophthalmol 2007; 144(1):99–103.

284. Midena E, Gambato C, Miotto S et al. Long-term effects on corneal keratocytes of mitomycin C during photorefractive keratectomy: a randomized contralateral eye confocal microscopy study. J Refract Surg 2007; 23(9 Suppl):S1011–S1014.

285. Rajan MS, O'Brart DP, Patmore A et al. Cellular effects of mitomycin-C on human corneas after photorefractive keratectomy. J Cataract Refract Surg 2006; 32(10):1741–1747.

286. Kohlhaas M. Collagen crosslinking with riboflavin and UVA-light in keratoconus. Ophthalmologe 2008; 105:785–796.

287. Raiskup-Wolf F, Hoyer A, Spoerl E et al. Collagen crosslinking with riboflavin and ultraviolet-A light in keratoconus: long-term results. J Cataract Refract Surg 2008; 34(5):796–801.

288. Spoerl E, Mrochen M, Sliney D et al. Safety of UVA-riboflavin cross-linking of the cornea. Cornea 2007; 26(4):385–389.

289. Spoerl E, Raiskup-Wolf F, Pillunat LE. Biophysical principles of collagen cross-linking. Klinische Monatsblatter fur Augenheilkunde 2008; 225(2):131–137.

290. Mamalis N, Edelhauser HF, Dawson DG et al. Toxic anterior segment syndrome. J Cataract Refract Surg 2006; 32(2):324–333.

291. Holland SP, Morck DW, Lee TL et al. Update on toxic anterior segment syndrome. Curr Opin Ophthalmol 2007; 18(1):4–8.

292. Johnson DH, Bourne WM, Campbell RJ. The ultrastructure of Descemet's membrane. I. Changes with age in normal corneas. Arch Ophthalmol 1982; 100(12):1942–1947.

293. Waring GO, 3rd. Posterior collagenous layer of the cornea. Ultrastructural classification of abnormal collagenous tissue posterior to Descemet's membrane in 30 cases. Arch Ophthalmol 1982; 100(1):122–134.

294. U.S. Department of Defense. Polymer matrix composites. Volume 3. Materials usage, design, and analysis. In: Defense USDo, ed., 1997; v. 3.

295. Hayes S, Boote C, Lewis J et al. Comparative study of fibrillar collagen arrangement in the corneas of primates and other mammals. Anat Rec (Hoboken, NJ) 2007; 290(12):1542–1550.

296. Fung Y. Biomechanics. Mechanical properties of living tissues. New York: Springer-Verlag, 1981.

297. Baer E. Basic Properites and Responses. In: MicNicol L, Strahlman E, eds. Corneal biophysics workshop. Corneal biomechanics and wound healing. Bethesda, MD: NEI, 1989.

298. Silver FH, Kato YP, Ohno M, Wasserman AJ. Analysis of mammalian connective tissue: relationship between hierarchical structures and mechanical properties. J Long-Term Effects Med Implants 1992; 2(2–3):165–198.

299. Maurice D. Mechanics of the cornea. In: Cavanagh H, ed. The cornea: transactions of the world congress on the cornea. New York: Raven Press, 1988.

300. Boyce B, Grazier J, Jones R, Nguyen T. The mechanics of soft biological composites. Alburquerque, NM: Sandia National Laboratories, 2007.

301. Elsheikh A, Alhasso D, Rama P et al. Assessment of the epithelium's contribution to corneal biomechanics. Exp Eye Res 2008; 86(2):445–451.

302. Seiler T, Matallana M, Sendler S, Bende T. Does Bowman's layer determine the biomechanical properties of the cornea? Refract Corneal Surg 1992; 8(2):139–142.

303. Avetisov SE, Mamikonian VR, Zavalishin NN, Neniukov AK. Experimental study of mechanical characteristics of the cornea and the adjacent parts of the sclera. Oftalmologicheskii Zhurnal 1988(4):233–237.
304. Jue B, Maurice DM. The mechanical properties of the rabbit and human cornea [see comment]. J Biomech 1986; 19(10):847–853.
305. Meek KM, Newton RH. Organization of collagen fibrils in the corneal stroma in relation to mechanical properties and surgical practice. J Refract Surg 1999; 15(6):695–699.
306. Bron AJ. The architecture of the corneal stroma [comment]. Br J Ophthalmol 2001; 85(4):379–381.
307. Alward W. Glaucoma: the requisites. St. Louis: Mosby, 1999.
308. Chihara E, Chihara E. Assessment of true intraocular pressure: the gap between theory and practical data. Surv Ophthalmol 2008; 53(3):203–218.
309. Miller D. Pressure of the lid on the eye. Arch Ophthalmol 1967; 78(3):328–330.
310. Coleman DJ, Trokel S. Direct-recorded intraocular pressure variations in a human subject. Arch Ophthalmol 1969; 82(5):637–640.
311. McMonnies CW, Boneham GC, McMonnies CW, Boneham GC. Experimentally increased intraocular pressure using digital forces. Eye Contact Lens: Sci Clini Pract 2007; 33(3):124–129.
312. Reifsnider K. Assumptions and stress measurement. In: McNicol L, Strahlman E, eds. Corneal biophysics workshop. Corneal biomechanics and wound healing. Bethesda, MD: NEI, 1989.
313. McPhee TJ, Bourne WM, Brubaker RF. Location of the stress-bearing layers of the cornea. Invest Ophthalmol Vis Sci 1985; 26(6):869–872.
314. Ethier CR, Johnson M, Ruberti J et al. Ocular biomechanics and biotransport. Annu Rev Biomed Eng 2004; 6:249–273.
315. Elsheikh A, Anderson K, Elsheikh A, Anderson K. Comparative study of corneal strip extensometry and inflation tests. J Roy Soc Interface 2005; 2(3):177–185.
316. Wess T. Collagen fibrillar structure and hierachies. In: Fratzl P, ed. Collagen: structure and mechanics. New York: Springer, 2008.
317. Ker R. Damage and failure. In: Fratzl P, ed. Collagen: structure and mechanics. New York: Springer, 2008.
318. MacRae S, Rich L, Phillips D, Bedrossian R. Diurnal variation in vision after radial keratotomy. Am J Ophthalmol 1989; 107(3):262–267.
319. Maloney RK. Effect of corneal hydration and intraocular pressure on keratometric power after experimental radial keratotomy. Ophthalmology 1990; 97(7):927–933.
320. Simon G, Small RH, Ren Q, Parel JM. Effect of corneal hydration on Goldmann applanation tonometry and corneal topography. Refract Corneal Surg 1993; 9(2):110–117.
321. Simon G, Ren Q. Biomechanical behavior of the cornea and its response to radial keratotomy. Refract Corneal Surg 1994; 10(3):343–351; discussion 51–6.
322. Ousley PJ, Terry MA. Hydration effects on corneal topography. Arch Ophthalmol 1996; 114(2):181–185.
323. Maurice DM. Model Systems. In: McNicol L, Strahlman E, eds. Corneal biophysics workshop. Corneal biomechanics and wound healing. Bethesda, MD: NEI, 1989.
324. Mamikonian V, Krasnov M. Clinical effects of corneal surgery. In: McNicol L, Strahlman E, eds. Corneal biophysics workshop. Corneal biomechanics and wound healing. Bethesda, MD: NEI, 1989.
325. Kohlhaas M, Spoerl E, Schilde T et al. Biomechanical evidence of the distribution of cross-links in corneas treated with riboflavin and ultraviolet A light. J Cataract Refract Surg 2006; 32(2):279–283.
326. Danielsen CC. Tensile mechanical and creep properties of Descemet's membrane and lens capsule. Exp Eye Res 2004; 79(3):343–350.
327. Elsheikh A, Brown M, Alhasso D et al. Experimental assessment of corneal anisotropy. J Refract Surg 2008; 24(2):178–187.
328. Gilbert ML, Roth AS, Friedlander MH. Corneal flattening by shallow circular trephination in human eye bank eyes. Refract Corneal Surg 1990; 6(2):113–116.
329. Litwin KL, Moreira H, Ohadi C, McDonnell PJ. Changes in corneal curvature at different excimer laser ablative depths. Am J Ophthalmol 1991; 111(3):382–384.
330. Elsheikh A, Wang D, Brown M et al. Assessment of corneal biomechanical properties and their variation with age. Curr Eye Res 2007; 32(1):11–19.
331. Hjortdal JO, Ehlers N. Effect of excimer laser keratectomy on the mechanical performance of the human cornea. Acta Ophthalmol Scand 1995; 73(1):18–24.
332. Hiltner A, Cassidy J, Baer E. Mechanical properties of biological polymers. Annu Rev Mater Sci 1985; 15:455–482.
333. Provenzano PP, Vanderby R, Jr., Provenzano PP, Vanderby R, Jr. Collagen fibril morphology and organization: implications for force transmission in ligament and tendon. Matrix Biology 2006; 25(2):71–84.
334. Bron AJ. Keratoconus. Cornea 1988; 7(3):163–169.
335. Patey A, Savoldelli M, Pouliquen Y. Keratoconus and normal cornea: a comparative study of the collagenous fibers of the corneal stroma by image analysis. Cornea 1984; 3(2):119–124.
336. Pouliquen Y. 1984 Castroviejo Lecture. Fine structure of the corneal stroma. Cornea 1985; 3:168–176.
337. Pouliquen Y. Doyne lecture keratoconus. Eye 1987; 1(Pt 1):1–14.
338. Meek KM, Tuft SJ, Huang Y et al. Changes in collagen orientation and distribution in keratoconus corneas. Invest Ophthalmol Vis Sci 2005; 46(6):1948–1956.
339. Dawson D, Randleman J, Grossniklaus H et al. Corneal ectasia after excimer laser keratorefractive surgery: histopathology, ultrastructure, and pathophysiology. Ophthalmology 2008; 115(12): 2181–2191.
340. Andreassen TT, Simonsen AH, Oxlund H. Biomechanical properties of keratoconus and normal corneas. Exp Eye Res 1980; 31(4):435–441.
341. Edmund C. Corneal topography and elasticity in normal and keratoconic eyes. A methodological study concerning the pathogenesis of keratoconus. Acta Ophthalmol Suppl 1989; 193:1–36.
342. Knops M. Analysis of failure in fiber polymer laminates: the theory of Alfred Puck. Heidelberg: Springer, 2008.
343. Garg A. Delamination – a damage mode in composite structures. Eng Fract Mech 1988; 29:557–584.
344. Hinton M, Soden P, Kaddour A. Failure criteria in fibre reinforced polymer composites: the world-wide failure exercise. Manchester, UK: Elsevier, 2004.
345. Knops M, Bogle C. Gradual failure of fibre/polymer laminates. Composites Sci Technol 2006; 66:616–625.
346. Swartz T, Marten L, Wang M et al. Measuring the cornea: the latest developments in corneal topography. Curr Opin Ophthalmol 2007; 18(4):325–333.
347. Luce DA, Luce DA. Determining in vivo biomechanical properties of the cornea with an ocular response analyzer. J Cataract Refract Surg 2005; 31(1):156–162.
348. Kotecha A, Kotecha A. What biomechanical properties of the cornea are relevant for the clinician? Surv Ophthalmol 2007; 52(Suppl 2):S109–S114.
349. Moreno-Montanes J, Maldonado MJ, Garcia N et al. Reproducibility and clinical relevance of the ocular response analyzer in nonoperated eyes: corneal biomechanical and tonometric implications. Invest Ophthalmol Vis Sci 2008; 49(3):968–974.
350. Ortiz D, Pinero D, Shabayek MH et al. Corneal biomechanical properties in normal, post-laser in situ keratomileusis, and keratoconic eyes. J Cataract Refract Surg 2007; 33(8):1371–1375.
351. Kotecha A, Elsheikh A, Roberts CR et al. Corneal thickness- and age-related biomechanical properties of the cornea measured with the ocular response analyzer. Invest Ophthalmol Vis Sci 2006; 47(12):5337–5347.
352. Prausnitz MR, Noonan JS. Permeability of cornea, sclera, and conjunctiva: a literature analysis for drug delivery to the eye. J Pharmaceut Sci 1998; 87(12):1479–1488.
353. Maurice DM. Drug delivery to the posterior segment from drops. Surv Ophthalmol 2002; 47(Suppl 1):S41–S52.
354. Urtti A, Urtti A. Challenges and obstacles of ocular pharmacokinetics and drug delivery. Adv Drug Deliv Rev 2006; 58(11):1131–1135.
355. Salazar-Bookaman MM, Wainer I, Patil PN. Relevance of drug-melanin interactions to ocular pharmacology and toxicology. J Ocular Pharmacol 1994; 10(1):217–239.
356. Barar J, Javadzadeh AR, Omidi Y et al. Ocular novel drug delivery: impacts of membranes and barriers. Expert Opin Drug Deliv 2008; 5(5):567–581.
357. Ghate D, Edelhauser HF, Ghate D, Edelhauser HF. Ocular drug delivery. Expert Opin Drug Deliv 2006; 3(2):275–287.
358. Maurice DM, Mishima S. Ocular pharmacokinetics. In: Sear ML, ed. Handbook of experimental pharmacology. Berlin & Hieidelberg: Springer Verlag, 1984:69.
359. Johnson GJ. The environment and the eye. Eye 2004; 18(12):1235–1250.
360. Sliney DH, Sliney DH. Exposure geometry and spectral environment determine photobiological effects on the human eye. Photochem Photobiol 2005; 81(3):483–489.
361. Cejkova J, Stipek S, Crkovska J et al. UV Rays, the prooxidant/antioxidant imbalance in the cornea and oxidative eye damage. Physiol Res 2004; 53(1):1–10.
362. Ringvold A. Corneal epithelium and UV-protection of the eye. Acta Ophthalmol Scandinavica 1998; 76(2):149–153.
363. Ringvold A, Anderssen E, Kjonniksen I. Distribution of ascorbate in the anterior bovine eye. Invest Ophthalmol Vis Sci 2000; 41(1):20–23.
364. Brubaker RF, Bourne WM, Bachman LA, McLaren JW. Ascorbic acid content of human corneal epithelium. Invest Ophthalmol Vis Sci 2000; 41(7):1681–1683.
365. Mitchell J, Cenedella RJ. Quantitation of ultraviolet light-absorbing fractions of the cornea. Cornea 1995; 14(3):266–272.
366. Kolozsvari L, Nogradi A, Hopp B et al. UV absorbance of the human cornea in the 240- to 400-nm range. Invest Ophthalmol Vis Sci 2002; 43(7):2165–2168.
367. Report of an advisory group on non-ionizing radiation. Health effects from ultraviolet radiation. Documents of the NRPB. Chilton, UK: National Radiological Protection Board, 2002; v. 13.
368. Environmental Health Criteria 160. Ultraviolet radiation. Geneva: World Health Organization, 1994.
369. Doughty M, Cullen A. Long-term effects of a single dose of ultraviolet-B on albino rabbit cornea – I. in vivo analyses. Photochem Photobiol 1989; 49:185–196.
370. Ren H, Wilson G. The effect of ultraviolet-B irradiation on the cell shedding rate of the corneal epithelium. Acta Ophthalmol (Copenh) 1994; 72:447–452.
371. Podskochy A, Gan L, Fagerholm P. Apoptosis in UV-exposed rabbit corneas. Cornea 2000; 19:99–103.
372. Shimmura S, Tadano K, Tsubota K. UV dose-dependent caspase activation in a corneal epithelial cell line. Curr Eye Res 2004; 28:85–92.
373. Wang L, Li T, Lu L. UV-induced corneal epithelial cell death by activation of potassium channels. Invest Ophthalmol Vis Sci 2003; 44:5095–5101.
374. Lu L, Wang L, Shell B. UV-induced signaling pathways associated with corneal epithelial cell apoptosis. Invest Ophthalmol Vis Sci 2003; 44:5102–5109.
375. Pettenati M, Sweatt A, Lantz P et al. The human cornea has a high incidence of acquired chromosome abnormalities. Hum Genet 1997; 101:26–29.
376. Rose RC, Richer SP, Bode AM. Ocular oxidants and antioxidant protection. Proc Soc Exp Biol Med 1998; 217(4):397–407.
377. Kennedy M, Kim KH, Harten B et al. Ultraviolet irradiation induces the production of multiple cytokines by human corneal cells. Invest Ophthalmol Vis Sci 1997; 38(12):2483–2491.
378. Watson PG, Young RD, Watson PG, Young RD. Scleral structure, organisation and disease. A review. Exp Eye Res 2004; 78(3):609–623.
379. Foster C, de la Maza M. The sclera. New York, NY: Springer-Verlag, 1994.
380. Hogan M, Zimmerman L. Ophthalmic pathology. An atlas and textbook. Philadelphia: WB Saunders, 1962.
381. McBrien NA, Gentle A, McBrien NA, Gentle A. Role of the sclera in the development and pathological complications of myopia. Progr Retin Eye Res 2003; 22(3):307–338.
382. Vaezy S, Clark J. A quantitative analysis of transparency in the human sclera and cornea using Fourier methods. J Microsc 1991; 163:85–94.
383. Vaezy S, Clark J. Quantitative analysis of the microstructure of the human cornea and sclera using 2-D Fourier methods. J Microsc 1994; 175:93–99.
384. Girard LJ, Neely W, Sampson WG. The use of alpha chymotrypsin in infants and children. Am J Ophthalmol 1962; 54:95–101.
385. Jones LA, Mitchell GL, Mutti DO et al. Comparison of ocular component growth curves among refractive error groups in children. Invest Ophthalmol Vis Sci 2005; 46(7):2317–2327.
386. Avetisov S, Savitskaya N, Vinetskaya M, Imodina E. A study of biochemical and biomechanical qualities of normal and myopic eye sclera in humans of different age groups. Metab Pediatr Syst Ophthalmol 1983; 7(4):183–188.

387. Pallikaris IG, Kymionis GD, Ginis HS et al. Ocular rigidity in living human eyes. Invest Ophthalmol Vis Sci 2005; 46(2):409–414.

388. Schultz D, Lotz J, Lee S et al. Structural factors that mediate scleral stiffness. Invest Ophthalmol Vis Sci 2008; 49:4232–4236.

389. McBrien NA, Millodot M. A biometric investigation of late onset myopic eyes. Acta Ophthalmol 1987; 65(4):461–468.

390. McBrien NA, Adams DW. A longitudinal investigation of adult-onset and adult-progression of myopia in an occupational group. Refractive and biometric findings. Invest Ophthalmol Vis Sci 1997; 38(2):321–333.

391. Rada JA, Shelton S, Norton TT et al. The sclera and myopia. Exp Eye Res 2006; 82(2):185–200.

392. Wollensak G, Spoerl E. Collagen crosslinking of human and porcine sclera. J Cataract Refract Surg 2004; 30:689–695.

393. Olsen TW, Aaberg SY, Geroski DH, Edelhauser HF. Human sclera: thickness and surface area. Am J Ophthalmol 1998; 125(2):237–241.

394. Boubriak O, Urban J, Bron A. Differential effects of aging on transport properties of anterior and posterior human sclera. Exp Eye Res 2003; 76:701–713.

395. Curtin B. Physiopathologic aspects of scleral stress-strain. Trans Am Ophthalmol Soc 1969; 67:417–461.

396. Woo SL, Kobayashi AS, Schlegel WA, Lawrence C. Nonlinear material properties of intact cornea and sclera. Exp Eye Res 1972; 14(1):29–39.

397. St Helen R, Mc EW, McEwen WK. Rheology of the human sclera. 1. Anelastic behavior. Am J Ophthalmol 1961; 52:539–548.

398. Meyer PA, Watson PG. Low dose fluorescein angiography of the conjunctiva and episclera. Br J Ophthalmol 1987; 71(1):2–10.

399. Ambati J. Transscleral drug delivery to the retina and choroid. In: Jaffe G, Ashton P, Pearson P, eds. Intraocular drug delivery. New York: Taylor & Francis, 2006.

400. Ghate D, Brooks W, McCarey BE et al. Pharmacokinetics of intraocular drug delivery by periocular injections using ocular fluorophotometry. Invest Ophthalmol Vis Sci 2007; 48(5):2230–2237.

401. Ahmed I, Patton TF. Importance of the noncorneal absorption route in topical ophthalmic drug delivery. Invest Ophthalmol Vis Sci 1985; 26(4):584–587.

402. Barza M, Kane A, Baum JL. Regional differences in ocular concentration of gentamicin after subconjunctival and retrobulbar injection in the rabbit. Am J Ophthalmol 1977; 83(3):407–413.

403. Barza M, Kane A, Baum J. Intraocular penetration of gentamicin after subconjunctibal and retrobulbar injection. Am J Ophthalmol 1978; 85(4):541–547.

404. Bill A. Movement of albumin and dextran through the sclera. Arch Ophthalmol 1965; 74:248–252.

405. Olsen TW, Edelhauser HF, Lim JI, Geroski DH. Human scleral permeability. Effects of age, cryotherapy, transscleral diode laser, and surgical thinning. Invest Ophthalmol Vis Sci 1995; 36(9):1893–1903.

406. Ambati J, Canakis CS, Miller JW et al. Diffusion of high molecular weight compounds through sclera. Invest Ophthalmol Vis Sci 2000; 41(5):1181–1185.

407. Lawrence MS, Miller JW, Lawrence MS, Miller JW. Ocular tissue permeabilities. Intl Ophthalmol Clin 2004; 44(3):53–61.

408. Rudnick DE, Noonan JS, Geroski DH et al. The effect of intraocular pressure on human and rabbit scleral permeability. Invest Ophthalmol Vis Sci 1999; 40(12):3054–3058.

409. Anderson O, Jackson T, Singh J et al. Human transscleral albumin permeability and the effect of topographical location and donor age. Invest Ophthalmol Vis Sci 2008; 49:4041–4045.

410. Pitkanen L, Ranta VP, Moilanen H et al. Permeability of retinal pigment epithelium: effects of permeant molecular weight and lipophilicity. Invest Ophthalmol Vis Sci 2005; 46(2):641–646.

411. Moore DJ, Clover GM. The effect of age on the macromolecular permeability of human Bruch's membrane. Invest Ophthalmol Vis Sci 2001; 42(12):2970–2975.

412. Robinson M, Lee S, Kim H et al. A rabbit model for assessing the ocular barriers to the transscleral delivery of triamcinolone acetonide. Exp Eye Res 2006; 82: 479–487.

413. Kim SH, Lutz RJ, Wang NS, Robinson MR. Transport barriers in transscleral drug delivery for retinal diseases. Ophthalmic Res 2007; 39:244–254.

414. McAllister DV, Allen MG, Prausnitz MR. Microfabricated microneedles for gene and drug delivery. Annu Rev Biomed Eng 2000; 2:289–313.

415. Jiang J, Gill HS, Ghate D et al. Coated microneedles for drug delivery to the eye. Invest Ophthalmol Vis Sci 2007; 48:4038–4043.

416. Jiang J, Moore JS, Edelhauser HF, Prausnitz MR. Instrsclera drug delivery to the eye using hollow microneedles. Pharm Res 2009; 26:395–403

417. Simpson AE, Gilbert JA, Rudnick DE et al. Transscleral diffusion of carboplatin: an in vitro and in vivo study. Arch Ophthalmol 2002; 120(8):1069–1074.

CHAPTER 5

The Lens

David C. Beebe

The lens is a remarkably specialized epithelial tissue that is responsible for fine-tuning the image that is projected on the retina. To perform this function it must be transparent, have a higher refractive index than the medium in which it is suspended, and have refractive surfaces with the proper curvature. Because the refractive power of the lens is variable, it permits the diopteric apparatus to focus on objects that are near or far.

To maintain transparency and a high refractive index, lens fiber cells are precisely aligned with their neighbors, have minimal intercellular space, and accumulate high concentrations of cytoplasmic proteins, the crystallins. Disruption of the precise organization of the lens fiber cells or aggregation of the proteins within them can destroy the transparency of the lens – a process known as cataract formation. Cataracts are the leading cause of blindness worldwide and the removal of cataracts is the most common surgical procedure in the aged population.

This chapter provides an overview of the structure, development, biochemistry, physiology, and pathology of the lens. Information about the human lens is emphasized, although studies using animals are included when data are not available for humans. Information is provided on the causes of cataracts and the mechanisms that are thought to protect the lens from cataract. The unsolved issues in the biology and pathology of the lens are highlighted throughout. References to comprehensive reviews and original source material are included to assist the reader. The citations provided are not meant to be inclusive, but to highlight articles of special relevance.

The anatomy of the adult lens

The lens is formed from two populations of specialized epithelial cells (Fig. 5.1). A sheet of cuboidal cells, the lens epithelium, covers the surface of the lens closest to the cornea. The bulk of the lens consists of concentric layers of elongated fiber cells. The outer shells of fiber cells extend from just beneath the epithelium to the posterior lens surface, a distance of over 1cm in adults. An elastic extracellular matrix, the lens capsule, which is secreted by the epithelial and superficial fiber cells, surrounds the entire lens. In the adult lens, most epithelial cells and all fiber cells are not dividing. Cells near the equatorial margin of the lens epithelium, in a region called the germinative zone, proliferate slowly. Most of the cells produced by mitosis in this region migrate towards the posterior of the lens and differentiate into fiber cells at the lens equator.[1] These new fiber cells elongate and accumulate large amounts of crystallins. During elongation the posterior (basal) ends of the fiber cells move along the inner surface of the capsule and their anterior (apical) ends slide beneath the epithelium until they meet elongating cells from the other side of the lens near the posterior and anterior midlines (Fig. 5.1). The junctions between the ends of cells from the opposite sides of the lens are called the sutures. Once fiber cells reach the sutures they stop elongating and their basal ends detach from the capsule (Fig. 5.1, inset). Soon after reaching the sutures, fiber cells degrade all intracellular membrane-bound organelles, including their nuclei, mitochondria, and endoplasmic reticulum.[2-5] Mature fiber cells are gradually buried deeper in the lens as successive generations of fibers elongate and differentiate. In this way the lens continues to increase in size and cell number throughout life.[6] Because protein synthesis ceases just before organelle degradation,[7] the components of mature fiber cells must be much more stable than those in cells found in other parts of the body. In fact, since the fiber cells in the center of the lens are present at birth and persist until death (or cataract surgery), their constituent proteins and membranes may last for more than 100 years.

The lens is suspended in the anterior of the eye by a band of inelastic microfibrils, the zonules, which insert into the lens capsule near the equator (Figs 5.1 & 5.2). These fibrils originate in the non-pigmented layer of the ciliary epithelium, a tissue that is located just posterior to the iris. They are composed of the protein fibrillin, one of the components of elastic fibrils in many connective tissues throughout the body.[8,9] However, the zonules do not stretch appreciably. Changes in the tension applied to the zonules are responsible for the alterations in lens curvature that occurs during accommodation, the process by which the lens focuses on objects nearby or at a distance.

The basics of lens refraction and transparency

The refractive properties of the lens are the result of the high concentration and graded distribution of crystallin proteins

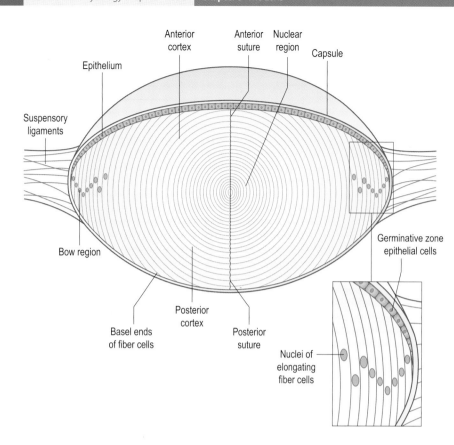

Figure 5.1 labels: Anterior cortex, Anterior suture, Nuclear region, Capsule, Epithelium, Suspensory ligaments, Bow region, Basel ends of fiber cells, Posterior cortex, Posterior suture, Nuclei of elongating fiber cells, Germinative zone epithelial cells

Figure 5.1 Diagram of the adult human lens. The expanded regions show the relationships between the elongating lens fiber cells and the posterior capsule as the basal ends of the fibers reach the posterior sutures and the changes in cell shape and orientation that occur as lens epithelial cells differentiate into lens fibers at the lens equator.

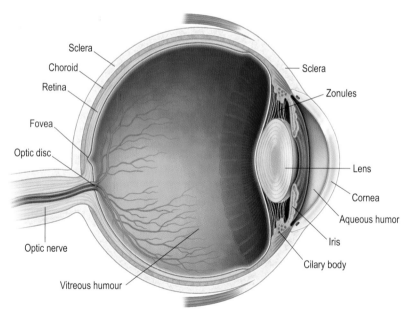

Figure 5.2 labels: Sclera, Choroid, Retina, Fovea, Optic disc, Optic nerve, Vitreous humour, Sclera, Zonules, Lens, Cornea, Aqueous humor, Iris, Cilary body

Figure 5.2 Diagram showing the relationship of the lens and zonules to the other structures in the adult eye.

in the cytoplasm of the lens fiber cells and the curvature of the lens refractive surfaces. Lens crystallins accumulate to concentrations that are up to three times higher than in typical cells.[10] This gives the fiber cells a significantly higher refractive index than the fluids around the lens. In the emmetropic eye the curvature of the refractive surfaces of the lens focuses light at the photoreceptors of the retina. The curvatures of the lens surfaces are the result of the tension on the zonules, the elasticity of the capsule, and the growth properties of the lens fibers and epithelial cells. In younger individuals refractive

error, myopia or hyperopia is often caused by abnormalities in corneal curvature or the length of the globe, but rarely by defects in the curvature or refractive index of the lens.

Transparency depends on minimizing light scattering and absorption. Light passes smoothly through the lens as a result of the regular structure of lens fibers (Fig. 5.3), the absence of membrane-bound organelles in the deeper fiber cells, and the small and uniform extracellular space between the fiber cells. Paradoxically, the high crystallin concentration of the lens fiber cell cytoplasm is an essential

Figure 5.3 Scanning electron micrograph showing the orderly arrangement of hexagonal lens fibers in the vertebrate lens. (Courtesy Dr. J. Kuszak.)

component of lens transparency. Reduced light scattering is due to short-range interactions between the highly concentrated crystallins.[11,12] Although it absorbs increasing amounts of the shorter wavelengths of visible light as it ages, giving the lens a pale yellow to brown color, the young human lens is nearly colorless.[13,14]

The early development of the lens

The cells that form the lens are originally part of the surface ectoderm covering the head of the embryo. Interactions between the future lens cells and nearby tissues during early development give these cells a "lens forming bias."[15] As a result of these interactions, patches of cells that lie on either side of the head are marked by expression of the transcription factor, Pax6.[16] At the same time, neural epithelial cells on either side of the diencephalon in the embryonic forebrain bulge laterally to form the optic vesicles, which eventually contact the surface ectoderm cells (Fig. 5.4A). The cells of the optic vesicle also express Pax6 and Pax6 function in these cells is essential for eye formation.[17,18] Some of the genes that regulate and are regulated by Pax6 at early stages of eye formation are now known, although the number of genes involved and knowledge about their interactions is likely to increase.[19–25]

Members of the bone morphogenetic protein (BMP) family of secreted factors play an essential role in lens formation. Deletion of the genes encoding BMP4 or BMP7 in the mouse embryo prevents lens formation in most cases (a minority of BMP7 null animals make a small, abnormal lens).[20,26–27] Deletion of BMP4 leads to decreased expression of the transcription factor Sox2 in the lens placode, while loss of BMP7 is associated with absence of Pax6 expression. BMP4 appears to be derived primarily by the optic vesicle; implantation of a bead containing BMP4 into the lumen of the vesicle can rescue lens formation in BMP4 knockout embryos.[27] BMP7 mRNA mainly localizes in the pre-lens surface ectoderm, suggesting that BMP7 acts as an autocrine or paracrine factor to promote lens formation.

After they make contact, the optic vesicles and the prospective lens ectoderm cells secrete an extracellular matrix that causes these cell layers to adhere tightly to each other (Fig. 5.4B).[28] The surface epithelial cells then elongate,

forming the thickened lens placode (Fig. 5.4B).[29] Soon afterward, the lens placode and the adjacent cells of the optic vesicle buckle inward to form the lens pit (Fig. 5.4C).[30] This morphological transformation is accompanied by the formation of the bi-layered optic cup. The invaginated lens placode soon separates from the surface ectoderm by a process involving the death of the cells in the connecting stalk between the cells of the surface ectoderm and the lens (Fig. 5.4D).[31] The Pax6-expressing ectodermal cells adjacent to the lens placode that remain on the surface of the eye become corneal and conjunctival epithelial cells.[32]

During lens invagination, the extracellular matrix between the optic vesicle and the lens begins to diminish and the two tissues separate (Fig. 5.4E).[28,33] The space that is formed between them is rapidly filled with a loose extracellular matrix, the primary vitreous body, that is secreted by the cells of inner layer of the optic cup (Fig. 5.4E).[34]

The epithelial cells that give rise to the lens vesicle originally lie on a thin basal lamina. In the process of invagination, this basal lamina comes to surround the lens vesicle. It gradually thickens by the deposition of successive layers of basal lamina material to form the lens capsule.[35–37]

Soon after the lens vesicle separates from the surface ectoderm, the cells in the portion of the vesicle that are closest to the retina begin to elongate. The elongation of these primary fiber cells soon obliterates the lumen of the vesicle as their apical ends contact the apical ends of the anterior epithelial cells (Fig. 5.4E,F). Primary fiber cell formation establishes the fundamental structure of the lens, with epithelial cells covering the anterior surface and elongated fiber cells filling the bulk of the lens.

At early stages of lens formation, most of the lens epithelial cells are actively proliferating. Cells at the margin of the epithelium are pushed or migrate towards the equator and are stimulated to differentiate into secondary fiber cells by factors present in the vitreous body.[38–41] As the secondary fibers elongate and their basal and apical ends curve towards the center of the lens, they displace the central primary fibers from their attachments with the capsule and the lens epithelium. This process buries the primary fiber cells in the center of the lens (Fig. 5.1). In the adult lens these cells comprise the "embryonic nucleus." Deposition of successive layers of secondary fiber cells continues throughout life. The rate of fiber cell formation, and therefore lens growth, is rapid in the embryo and slows greatly after birth (Fig. 5.5).[6]

In addition to Pax6, at least one additional transcription factor may be essential for lens formation.[21] In chicken embryos, prospective lens cells express L-maf, a member of the maf family of b-ZIP transcription factors, soon after they contact the optic vesicle.[42] When modified forms of this protein that bind to DNA but do not activate transcription are expressed in the prospective lens-forming region of chicken embryos, the formation of the lens placode and lens vesicle is suppressed. Expression of L-maf in regions of the head ectoderm outside of the lens placode (but within the area expressing Pax6) leads to the formation of extra lenses.[42] Although members of the maf family of transcription factors are expressed during lens formation in mammals, there is no factor comparable in function to chicken L-maf. Since deletion of individual members of the maf family does not prevent lens formation in mice, multiple maf family members may assume the function of L-maf in mammals.

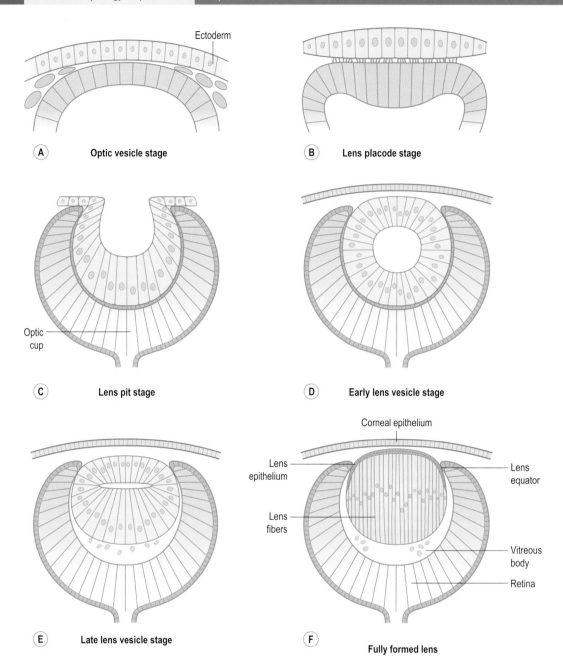

Figure 5.4 The early stages of lens formation. (**A**) The lens vesicle contacts the surface ectoderm. (**B**) The optic vesicle adheres to the surface ectoderm and the prospective lens cells elongate to form the lens placode. (**C**) The lens placode and the outer surface of the optic vesicle invaginate to form the lens pit and the optic cup, respectively. (**D**) The lens vesicle separates from the surface ectoderm. (**E**) The primary lens fibers elongate and begin to occlude the lumen of the vesicle. The posterior of the lens vesicle separates from the inner surface of the optic cup. Capillaries from the hyaloid artery invade the primary vitreous body. (**F**) The configuration of the lens as it begins to grow. Secondary fiber cells have not yet developed and organelles are still present in all fiber cells. (Modified from McAvoy J, Developmental biology of the lens. In Duncan G (Ed), Mechanism of cataract formation. Academic Press, pp 7-46. Copyright Elsevier 1981[480])

The proper separation of lens vesicle from the overlying presumptive corneal epithelium is defective in the mouse mutant, *dysgenetic lens* (*dyl*). A mutation was identified in the gene encoding the "forkhead" transcription factor, Foxe3, in the *dyl* strain. Foxe3 is selectively expressed in the lens placode and developing lens epithelium.[24,43] Lenses lacking Foxe3 show severe defects in lens epithelial cell proliferation and subsequently degenerate.[44] A similar phenotype is seen in lenses lacking the transcription factor, AP-2α. These also fail to separate from the surface ectoderm and show defects in epithelial cell differentiation.[45]

Targeted disruption of the genes encoding three additional transcription factors that are normally expressed during primary fiber cell differentiation, c-Maf, Sox1, and Prox1, causes failure of elongation of primary lens fiber cells.[46–49] In mice lacking any one of these genes, lens fiber cells do not elongate or express the high level of lens crystallins that are characteristic of primary fiber cells. The diffusible factors that trigger fiber cell formation and the mechanisms responsible for fiber cell elongation and the regulation of crystallin synthesis are discussed below (see Box 5.1).

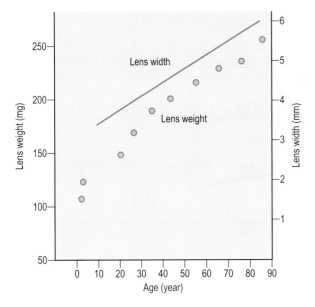

Figure 5.5 Changes in lens width and height from birth to 90 years of age. (Data from Scammon, R.E. and M.B. Hesdorfer, Growth in mass and volume of the human lens in postnatal life. Arch. Ophthalmol., 1937. 17: p. 104-112 and Brown, N., Slit-image photography and measurement of the eye. Med Biol Illus, 1973. 23(4): p. 192-203[481])

Figure 5.6 A montage of photomicrographs showing the organization of cells in the pre-germinative zone (curly brackets), germinative zone (square brackets) and transitional zone (curved brackets) of the equatorial lens epithelium. The image on the right is a continuation of the one on the left showing the change in cell morphology as the cells of the transitional zone elongate (asterisk) to form lens fiber cells. The images were obtained using a scanning electron microscope (magnification ×780). (Courtesy of Dr. J. Kuszak.)

Box 5.1 Mutations in genes involved in lens formation contribute to the formation of congenital cataracts and anterior segment dysgenesis

Several genes have been identified that are required for the formation of the normal lens. The proteins encoded by these genes, including Pax6, Pitx3, c-Maf and Foxe3, are transcription factors; they control the expression of batteries of genes required to make a lens. Most of these mutations are dominant and result from "haploinsufficiency;" loss of the function of one copy has a severe effect on lens development. Defects may range from mild cataracts to absence of the lens. Many of these mutations are associated with defects in the development of the anterior segment, including the corneal endothelium and stroma and the aqueous humor outflow pathway. Anterior segment defects can be secondary to Peters anomaly, the failure of the lens to separate from the surface ectoderm, or to less obvious interactions between the developing lens and the mesenchymal cells that form the structures of the outflow pathway.

Mice homozygous for the *aphakia* mutation form lens vesicles, but the lens cells degenerate soon after.[50] The *aphakia* mutation was mapped close to the gene encoding a homeodomain transcription factor, Pitx3, that is expressed early in lens development.[51,52] Two families were described in which affected individuals have hereditary congenital cataracts and point mutations in Pitx3.[53] Deletions were found in the region upstream and in the coding sequence of Pitx3 in the *aphakia* strain.[51,54] A microarray study indicated that Pitx3 mRNA expression is decreased in lenses lacking a single copy of Pax6.[25]

Lens fiber cell differentiation

Lens fiber cell differentiation is characterized by an exceptional degree of specialization. During their differentiation, fiber cells withdraw from the cell cycle, elongate greatly, express large amounts of proteins (the crystallins), acquire several specializations of their plasma membranes, and eventually degrade all membrane-bound organelles.

In mature lenses, the first evidence of fiber cell differentiation is the withdrawal from the cell cycle of epithelial cells near the lens equator. After their last mitosis, these cells move in a posterior direction into the transition zone (Fig. 5.6). This movement has been termed "migration," but it may simply result from displacement caused by the proliferation of cells in the germinative zone, a kind of "mitotic pressure." Cells in the transition zone are postmitotic, but have not yet begun to elongate. As they move posteriorly they are exposed to factors that stimulate their elongation. Studies in chicken embryos and newborn mice, in which the lens was rotated so that the epithelial cell faced the vitreous body, first demonstrated the existence of these factors.[38,55] Epithelial cells exposed to vitreous humor in this manner elongated like primary fiber cells. Furthermore, chicken embryo lens epithelial cells stopped progressing through the cell cycle within 9 hours after lens rotation, indicating that suppression of lens cell proliferation at the initiation of fiber cell differentiation is dependent on factors from outside the lens.[56]

Proteins that are known to be involved in regulating the cell cycle in many cell types appear to play an essential role in the cessation of lens fiber cell proliferation. Included in this group are the cyclin-dependent kinase inhibitor protein, p57 (KIP2) and the retinoblastoma protein (pRb). Progression through the cell cycle requires the phosphorylation of regulatory proteins, called cyclins, by cyclin-dependent kinases. The cyclin-dependent kinase inhibitor protein, p57, an inhibitor cell cycle progression, is expressed in the early stage of lens fiber cell differentiation, where it is found bound to cyclins.[57,58] In lenses that lack the gene for pRb, a well-known regulator of the cell cycle, p57 levels are low and lens fiber cells do not withdraw from the cell cycle.[59,60] This finding suggests that pRb is required for the expression of p57 and that at least one function of p57 is to maintain lens fiber cells in the non-proliferating state by complexing with cyclins. Absence of the two main cyclin-dependent kinase inhibitors expressed in the lens, p27 and p57, leads to extensive lens fiber cell proliferation, even in the presence of pRb, confirming that these molecules play an essential role in the withdrawal of fiber cells from the cell cycle.[61] Lenses lacking both p27 and p57 also do not make detectable levels of crystallins, suggesting that the function of these molecules is essential for normal gene expression during fiber cell differentiation.[61] In addition to the requirement for pRb, the transcription factor Prox1 is also required for high levels of p57 accumulation in the nuclei of lens fiber cells.[62] Prox1 is present in lens epithelial cells, but localizes to the cell nucleus at the onset of fiber cell differentiation.[63]

Some of the basic mechanisms responsible for the extensive cell elongation that characterizes lens fiber cell differentiation have been described, but this process is still not well understood. Early studies suggested that the microtubules of the lens fiber cytoskeleton are required for fiber cell elongation.[64,65] However, later experiments showed that fiber-like cells could elongate in the absence of microtubules.[66] Whether other components of the cytoskeleton play an essential role in fiber cell elongation in vivo has not been demonstrated.

In chicken embryo lenses, the early stages of lens fiber cell elongation may be driven by an increase in cell volume.[67] In these cells, initiation of fiber cell elongation is accompanied by changes in the ionic permeability of the plasma membrane. The resulting accumulation of potassium and chloride ions in the cytoplasm leads to an osmotically driven increase in cell volume.[68] Continued cell elongation depends on sustained protein synthesis.[69,70] In mouse lenses, secondary fiber cells appear to initiate elongation without increasing in volume.[71] The mechanism by which fiber cell elongation is regulated at later stages of their differentiation, especially when fiber cells reach the sutures and stop elongating, has not been studied.

Several growth factors are capable of initiating lens fiber cell differentiation when added to cultured lens epithelia or, in some cases, when over expressed in the lens in vivo. Among these are members of the fibroblast growth factor (FGF) and insulin-like growth factor (IGF) families. Studies in chicken embryos identified IGF-1 as a potent activator of fiber cell differentiation.[72] However, experiments in rats and mice showed that, although IGFs may play some role in fiber cell differentiation, members of the FGF family appear to be more important in mammals and also contribute to fiber cell differentiation in chicken embryos.[41,73–83] Studies using mouse lenses deficient in three of the four FGF receptors have conclusively demonstrated that FGF signaling is required for the formation of lens fiber cells.[84] Other soluble factors, especially members of the BMP family, appear to modulate or potentiate the effects of FGFs on lens fiber cell differentiation.[85–87]

Lens crystallins

The synthesis and accumulation of very large amounts of crystallin proteins is a major characteristic of lens fiber cell differentiation. As much as 40 percent of the wet weight of the lens fiber cell can be accounted for by crystallins,[10] a protein concentration that is about three times higher than in the cytoplasm of typical cells. Crystallin proteins can be classified as either "classical" or "taxon-specific". The classical crystallins are comprised of two α-crystallins and several members of the beta/gamma (β/γ)-crystallin superfamily. All vertebrate lenses accumulate large amounts of classical crystallins in their fiber cells. In many species, fiber cells also produce large amounts of taxon-specific crystallins (see reviews[88,89]). As the name implies, different taxon-specific crystallins are found in the lenses of different taxonomic groups. Taxon-specific crystallins are functional enzymes or proteins that are structurally very similar to enzymes, but with little or no enzymatic activity. The levels of taxon-specific crystallins sometimes exceed those of the classical crystallins. Adult human lenses do not produce taxon-specific crystallins, although the enzyme betaine-homocysteine methyltransferase is present at high levels in the embryonic nucleus of the rhesus macaque lens, suggesting that this enzyme may serve as a taxon-specific crystallin during the early development of the primate lens (see Box 5.2).[90]

Many of the transcription factors that promote the abundant expression of the crystallin genes in lens fiber cells have been identified. The details of these studies are described in comprehensive reviews and will not be considered further here.[21,23,91–96]

Human lenses express two α-crystallin genes, αA and αB. Examination of the protein structure of the α-crystallins revealed that they are members of the widely distributed family of small heat shock proteins.[97–99] An important function of small heat shock proteins is to stabilize proteins that are partially unfolded and prevent them from aggregating (this is termed their "chaperone" activity). The role of the α-crystallins in preventing protein aggregation and precipitation has been demonstrated in experiments performed in vitro.[100] A similar, in vivo role for αA crystallin has been inferred in mice in which αA crystallin gene was disrupted.[101] The function of the α-crystallins in preventing protein aggregation has obvious importance for the lens, because the proteins in lens fiber cells must persist for the life of the individual and excessive protein aggregation could lead to light scattering and cataract formation (see below).

Alpha-crystallins are also enzymes, since they possess serine-threonine auto-kinase activity.[102] It is not yet clear whether the α-crystallins phosphorylate other proteins in lens fiber cells. The α-crystallins are, themselves, normally phosphorylated in vivo and can be phosphorylated in vitro in a cyclic AMP-dependent manner.[103,104] The factors that regulate the phosphorylation of α-crystallins and the importance of this phosphorylation for the function of the α-crystallins in the living lens have not been determined.[105–107]

Alpha-crystallin proteins normally associate in the lens cell cytoplasm to make high-molecular-weight complexes containing approximately 30 subunits. The structure of these complexes has been revealed by cryoelectron microscopy.[108] These observations show that native α-crystallin complexes can be assembled in a number of configurations, indicating that there is substantial flexibility in the way the subunits associate. Alpha-crystallin monomers also readily exchange between high-molecular-weight complexes, further supporting the view that α-crystallin complexes are quite plastic.[109]

The phenotype of αA-crystallin knockout mice provides insight into the function of α-crystallins in vivo.[101] The lenses of these animals are slightly smaller than normal, but structurally quite similar to the normal lens. Mature fiber cells contain aggregates of proteins that lead to the formation of cataracts, beginning a few weeks after birth. Analysis of these aggregates shows that they contain large amounts of αB crystallin and smaller amounts of other proteins. These results suggest that, in the fiber cell cytoplasm, αA crystallin is partly responsible for preventing αB from aggregating. In addition, when lens epithelial cells from these animals were cultured in vitro, they grew more slowly than normal cells, were more sensitive to stress, and had a higher rate of apoptosis, accounting for the smaller size of the knockout lenses.[110] Therefore, αA crystallin appears to be important for the normal function of lens epithelial and fiber cells.

In addition to being expressed at very high levels in the lens, αB crystallin is present in a variety of cells throughout the body, especially in heart and skeletal muscle.[111] The protein is also found in damaged areas of the central nervous system in a variety of neurodegenerative diseases, including

in the "drusen" that accumulate at the basal surface of the retinal pigmented epithelial cells in age-related macular degeneration.[112–114] A naturally occurring mutation in the αB crystallin gene (CRYAB) leads to the formation of cataracts and "desmin-related" myopathy.[115] In vitro tests showed that the mutant form of the protein had no chaperone activity and even enhanced the aggregation of test proteins.[116] These studies suggest that αB crystallin has important chaperone functions in the lens and in other cells of the body. However, the fact that the mutant protein accelerated the denaturation of test proteins leaves open the possibility that the mutant is a gain-of-function. In this case, the cataracts and myopathy seen in individuals harboring this mutation could be due to the destabilizing function of the protein, rather than loss of its function as a chaperone. However, additional mutations in CRYAB, which do not have myopathy as part of their phenotype, support the view that loss of its chaperone function contributes to cataract.[117]

The β/γ crystallin superfamily is more diverse than the α-crystallins and the function of its members in the lens is less evident. The β- and γ-crystallins were originally thought to be two distinct protein families. However, once the protein sequences of representative members of these families were available, it was clear that they were closely related.[118] The major difference in these proteins is the tendency for most of the β-crystallins to form multimers, while the γ-crystallins exist as monomers. Solving the three-dimensional structure of these molecules confirmed their close structural relationship. It also confirmed that the N- and C-terminal extensions of the β-crystallins provided a structural rationale for why these family members form higher-order aggregates.[119,120] There are six beta crystallin polypeptides (βA1, βA3, βA4, βB1, βB2, βB3), and three gamma crystallins (γS, γC, γD) expressed in the human lens,[121] although the βA1 and βA3 polypeptides are derived from the same gene (βA3/A1).[122] Beta- and gamma-crystallins bind calcium in vitro and may buffer this important cation in the lens fiber cell cytoplasm.[123–125]

The lens fiber cell cytoskeleton

Microtubules are abundant beneath the plasma membranes of lens fiber cells where they probably play an important role in stabilizing the fiber cell membrane.[126] Microtubules may also be important for transporting vesicles to the apical and basal ends of elongating fiber cells, although neither function has been demonstrated in vivo. In addition to microtubules, there is an abundant network of actin-containing microfilaments beneath the plasma membrane of lens fiber cells. These microfilaments associate with the cytoplasmic surfaces of the adhesive junctions between lens fibers and are also likely to interact with the spectrin-containing sub-membrane meshwork.[127,128] This sub-membrane scaffold also contains tropomyosin and tropomodulin, proteins that may modulate the structure of the microfilaments.[129]

Lens fiber cells also contain an unusual complement of intermediate filaments, including those composed of vimentin.[130] This is unusual, because vimentin intermediate filaments are usually restricted to cells of mesodermal, not epithelial origin. The function of vimentin-containing intermediate filaments in the lens is not evident, because mice lacking vimentin appear to have normal lenses.[131]

Over-expression of vimentin in the lens leads to cataract formation and defects in fiber cell differentiation.[132] Whether this is a specific effect of excessive vimentin levels in the lens, or the response of the lens to protein over-expression in general, has not been demonstrated.

In addition to vimentin, the lens contains intermediate filaments composed of the proteins filensin and phakinin (see reviews[133,134]). These filaments have an unusual "knobby" structure, leading to the name "beaded filaments."[135] Beaded filament proteins have only been found in lens fiber cells, suggesting that they have a specialized role in these cells.[135] Mutations of both genes are associated with the formation of human cataracts, often involving the sutures.[136,137] Deletion of the phakinin gene (*Bfsp2*) in mice caused a mild, progressive cataract that was associated with decreased levels of its binding partner, filensin, but no obvious defects in fiber cell differentiation.[138]

Other cellular and biochemical specializations found in lens fiber cells

In fiber cells at early stages of elongation the lateral plasma membranes are smooth. However, as fiber cells reach the sutures, the membranes become progressively more interdigitated, forming interlocking "ball-and-socket" junctions (Fig. 5.7).[139-141] Ball-and-socket junctions may stabilize the lateral membranes of the fiber cells and assure that the cells remain tightly connected during accommodation. The mechanisms responsible for the formation of these unusual membrane specializations have not been identified.

The membranes of mature fiber cells have an unusual lipid composition. Human lens fiber cells have the highest proportion of cholesterol of any plasma membrane in the body and the amount of cholesterol increases as the fiber cells mature.[142,143] The cholesterol/phospholipid ratio of nuclear fiber cells is nearly three-fold greater than in cortical fiber cells. Partially purified fiber cell membranes have substantially lower cholesterol/phospholipid ratio than was found in the whole tissue, although membranes from nuclear fibers still have proportionally higher cholesterol content than membranes from the cortex.[142] This finding suggests that some of the cholesterol in mature lens fiber

cells may be associated with a complex that is not an integral component of the plasma membrane. There is also a high percentage of sphingomyelin in lens membrane phospholipids.[143] The presence of high concentrations of cholesterol and sphingomyelin is likely to cause lens fiber cell membranes to be quite rigid.[142] The functional significance of these biochemical specializations is not known.

In addition to its unusual lipid content, lens fiber cell plasma membranes contain several unique proteins. The most abundant of these is the "major intrinsic polypeptide" (MIP). MIP accounts for as much as 50 percent of the total protein of the lens fiber cell membrane and has not been detected in any other cells in the body. When the MIP cDNA was first cloned and sequenced it encoded a protein sequence that was unrelated to any known protein.[144] Later, when the aquaporin family of water channel proteins was identified, MIP was recognized to be its "founding" member.[145] In spite of its strong structural resemblance to the other aquaporins, MIP is a relatively inefficient water channel and may play a more important role in cell–cell adhesion.[146,147] Mutations in the MIP gene lead to cataracts in mice and humans.[148,149] These mutations could cause cataracts because they reduce MIP function, or because they interfere with the formation or function of lens fiber cell membranes.[148]

The gap junctions of the lens are assembled from a unique set of subunits, or connexins. The cell-to-cell transport of small molecules (< 1 kDa) mediated by these gap junctions is likely to be important for the function of the lens, since most of the fiber cells are far from the nutrients supplied by the aqueous and vitreous humors.[150] Not surprisingly, lens fiber cells have the highest concentration of gap junction plaques of any cells in the body (Fig. 5.8).

There are three connexins found in lens cells, α1, α3, and α8. Connexin α1 (also known as Cx43) is found in many tissues in the body. In the lens, it is present only in the epithelial cells.[151] Connexins α3 (Cx46) and α8 (Cx50) are abundant in fiber cells, but α8 is also expressed in lens

Figure 5.7 Visualization of the ball-and-socket interdigitations at the lateral surfaces of lens fiber cells. The tissue was fractured to show the surface morphology of the cells and viewed with a scanning electron microscope. (Courtesy of Dr. J. Kuszak.)

Figure 5.8 Scanning electron micrograph showing the abundant gap junction plaques on the surface of young lens fiber cells (magnification ×270,000). (Reproduced with permission from FitzGerald, P.G., D. Bok, and J. Horwitz, The distribution of the main intrinsic membrane polypeptide in ocular lens. Curr Eye Res, 1985. 4(11): p. 1203-18. p 1204[482])

epithelial cells.[152,153] In mice, epithelial α8-containing gap junctions are required for maximal proliferation of the epithelial cells in the first weeks of postnatal life.[154,155] Gap junction plaques containing both connexins α3 and α8 are present along the lateral membranes of lens fiber cells.[156] Mutations in α3 or α8 are responsible for dominant congenital cataracts in several families.[157]

While the presence of a large number of gap junctions would seem to be essential for the normal function of the lens, recent studies have shown that the lenses of mice lacking either the connexin α3 or α8 gene are relatively mildly affected. In the case of the α3 knockouts, fiber cell differentiation proceeds normally, although nuclear cataracts appear shortly after birth.[152] However, no cataracts form when the disrupted α3 connexin gene is bred onto a different genetic background, indicating that genetic modifiers can compensate for the absence of this connexin in maintaining lens transparency.[158] The lenses of the α8 knockouts develop diffuse nuclear cataracts.[154] Considering the abundance of specialized connexins in the lens, these phenotypes seem surprisingly mild. Mouse lenses lacking α3 and α8 have more severe cataracts, suggesting that the functions of these proteins partially overlap.[159]

A second pathway, which mediates the diffusion of larger molecules between lens cells has also been described. In chicken embryos, fiber cells fuse with their neighbors late in their differentiation, just before they degrade their organelles.[160] Fiber cell fusions have been described in previous morphological studies in the lenses of several species.[161,162] Examination of fiber cell fusions in postnatal mouse lenses suggests that they mainly mediate the diffusion of proteins between fiber cells at a similar depth in the lens, rather than providing a diffusion pathway between more superficial and deeper fiber cells.[163]

The second most abundant protein in the lens fiber cell membrane is MP20 (Lim2). The sequence of MP20 places in the tetraspanin family of proteins, with diverse functions in many cell types. Mutations in this gene lead to the formation of congenital cataracts in mice and humans. Knockout of Lim2 causes cataracts and defects in the lens refractive gradient.[164,165] The fiber cells in these lenses fail to fuse with their neighbors to establish the normal intercellular macromolecular diffusion pathway.[166]

Lens fiber cells are linked to their neighbors all along their lateral membranes by N-cadherin, a calcium-dependent, homophilic cell adhesion molecule.[167,168] N-cadherin is typically linked to the actin cytoskeleton by a complex of proteins that includes α- and β-catenin. As the lens changes shape during accommodation, the lateral membranes of the fiber cells are held together tightly by interlocking ball-and-socket junctions and N-cadherin-containing cell–cell adhesion complexes. These adhesion complexes, along with Lim2, which has a role in cell–cell adhesion in the lens, are likely to contribute to the close association between the lateral membranes of lens fiber cells. Minimizing the extracellular space is important for reducing light scattering to maximize transparency.

In addition to the lateral cell adhesion complexes, lens fiber cells are joined tightly to each other at their apical and basal ends.[168–170] These complexes contain abundant N-cadherin and vinculin, a protein that plays an important role in regulating the interaction between adhesion molecules and the actin cytoskeleton. Membrane complexes at the basal ends of the lens fiber cells, along the posterior capsule, probably attach the fiber cells to the capsule, since they contain a rich actin cytoskeleton, the contractile protein myosin, and integrin extracellular matrix receptors.[169,171] It is possible that the basal membrane complex also plays a role in the migration of the basal ends of the lens fiber cells towards the sutures.[169]

The lens grows by progressively adding new layers of fiber cells to its outer surface. These new fiber cells elongate, gradually extending their apical and basal ends towards the sutures. Once the tips of an elongating fiber cell reach the sutures they meet a fiber cell coming from the opposite side of the lens. The cells then form new junctions at the apical and basal ends and detach from the posterior capsule (see Fig. 5.1). After they detach from the capsule, the fiber cells are gradually buried deeper within the lens by successive layers of "younger" fiber cells.

Soon after they detach from the posterior capsule, fiber cells suddenly degrade all of their membrane-bound organelles, including their mitochondria, endoplasmic reticulum, and nuclei.[2,3,5,172] Organelle degradation is complete within a few hours.[3] The mechanism responsible for organelle degradation in lens fiber cells is not known, although it has been suggested that the enzyme 15-lipoxygenase mediates organelle loss in the lens, as it does during the maturation of erythrocytes.[173]

Numerous studies have pointed to the similarity between organelle degradation in the lens and the initiation of apoptosis (programmed cell death) in many other cell types. Although several biochemical similarities are striking, mice lacking the "executioner" caspases 3, 6 and 7 (which are critical mediators of apoptosis) showed normal organelle degradation.[174] Knockout of the lysosomal enzyme, DNase IIβ, revealed that it plays a role in the removal of DNA fragments remaining after nuclear membrane fragmentation, but is not required for the initiation of organelle breakdown.[175] Mice lacking the heat shock transcription factor, HSF4, have cataracts and failure of proper organelle loss, raising the possibility that HSF4 controls this process.[176,177]

In primates, the outer shell of fiber cells that contain organelles is only about 100 microns wide. These organelle-containing cells are located, for the most part, outside of the optical axis of the lens (see Fig. 5.1).[4] The absence of organelles in most lens fibers increases transparency by reducing light scattering.[175] However, organelle loss also makes the fiber cells in the center of the lens dependent on the superficial fiber cells, an arrangement that may contribute to age-related cataract formation (see below).

The control of lens growth

The human lens grows rapidly in the embryo and during the first postnatal year. The rate of lens growth slows between ages 1 and 10, then continues at a much slower, nearly linear rate throughout life (see Fig. 5.5).[6,178] The factors that regulate lens growth in humans are not known. Many growth factors have been shown to stimulate lens epithelial cell proliferation in vitro and the receptors for several families of growth factors have been identified in the human lens and in the lenses of other species. These include the fibroblast growth factor family,[179] the insulin-like growth factor family,[179] the epidermal growth factor family,[179,180] platelet-derived growth factor,[180,181] hepatocyte growth factor,[180,182] and vascular endothelial cell growth factor.[183] In addition to these growth factor receptors, there are a variety of other kinds of receptors, including muscarinic acetylcholine receptors and purinergic receptors.[184,185] It is not known whether any of these signaling systems is essential for lens growth in vivo. In this regard, deletion of EGF, PDGF, FGF or IGF receptors has not been associated with defects in lens growth or development.[84,186]

Recent studies revealed an endogenous system regulating the growth of rodent lenses in vivo. The environment around the lens is severely hypoxic. When mice or rats breathed increased amounts of oxygen, oxygen levels increased in the eye. In older animals, oxygen treatment was associated with a more than three-fold increase in lens epithelial cell proliferation and long-term treatment led to greater lens wet weight.[187] Younger lenses were not affected by oxygen levels and breathing lower oxygen did not alter lens growth at either age. These results suggest that the low oxygen levels that normally exist in the eye are required for the decline in lens growth that normally occurs with age. Subsequent studies showed that this decline probably required the oxygen-regulated transcription factor, HIF-1, since expressing oxygen-insensitive forms of HIF in lens epithelial cells suppressed the effect of oxygen on their rate of proliferation.[188] Since lens size is an important risk factor for age-related cataracts,[189-192] understanding how HIF-1 suppresses lens growth is likely to have clinical relevance.

Communication between lens epithelial and fiber cells

The apical ends of the lens epithelial cells abut the apical ends of elongating fiber cells. It has been suggested that nutrients are provided to the underlying fiber cells through gap junctions linking the apposed ends of these cells.[150,154,193] Other studies have questioned this interpretation, since gap junctions are rarely detected at the apical ends of central epithelial cells.[153,194] Since the peripheral fiber cells contain a full complement of organelles, it is unclear whether transport from the epithelium is necessary for lens viability. The possibility that the lens does not require gap junction communication between the epithelial and fiber cells is supported by studies in which the main lens epithelial connexin, Cx43, was deleted in the lens and ciliary body. Although loss of this connexin in the ciliary epithelium reduced the production of aqueous humor, the lenses appeared normal.[195] However, damage to lens epithelial cells compromises the viability of underlying lens fibers.[196] Whether this is due to the absence of metabolites provided by the epithelium, or to some other role of the epithelium has not been determined.

Vascular support during lens development

Soon after it is formed, the lens becomes covered with a meshwork of capillaries. In the posterior of the lens this network, the tunica vasculosa lentis, arises from the hyaloid artery. The capillaries at the anterior of the lens arise from blood vessels of the developing iris stroma to form the anterior pupillary membrane. These capillary networks join with each other near the lens equator. It has been assumed the fetal vasculature is important for normal lens development, although similar vessels are never present around the lenses of non-mammalian species. Deletion of vascular endothelial growth factor (VEGF-A) from the lens prevented the formation of the fetal vasculature, resulting in smaller lenses with transient nuclear cataracts.[183] Since mouse lens cells express functional VEGF receptors,[197] further studies are required to determine whether these cataracts were due to loss of VEGF signaling to lens cells, to the absences of the fetal vasculature, or both.

During the second trimester of human development, the capillaries of the tunica vasculosa lentis and the anterior pupillary membrane regress.[198] Decreasing levels of plasma-derived vascular endothelial cell growth factor may be one of the factors involved in the normal regression of these vessels.[199] Macrophages in the vitreous body also appear to play an essential role in capillary regression.[198,200] In mice, these macrophages cause the programmed death of the endothelial cells by secreting the morphogen, Wnt7b.[201]

A number of hereditary and acquired ocular diseases are accompanied by persistence of the fetal vasculature (see review[202]). At present, it is not clear why the fetal vasculature fails to regress in so many different syndromes and hereditary diseases. Better understanding of the factors that regulate vascular regression in normal ocular development is needed to address this question.

The lens as the organizer of the anterior segment

Studies performed in the 1960s and extended in the last decade show that the lens plays an important role in the development of the other tissues of the anterior segment.[51,203-206] Absence of the lens at an early stage of embryogenesis leads to the absence of the corneal endothelium, abnormal differentiation of the corneal stroma, and absence of the iris, ciliary body, and anterior chamber.[51,204-207] Therefore, the lens not only receives signals from its environment, but sends signals to nearby tissues that are essential for their normal development. The nature of these signals is unknown.

Special problems of lens cell metabolism

Overview

The lens, like all biological systems, is subjected to oxidative stress. Oxidation can be caused by molecular oxygen or free radicals. Oxygen free radicals are generated by the normal activity of mitochondria,[208] by other metabolic processes, and by the absorption of light. To counteract the effects of oxidation, all cells maintain a reducing environment in their cytoplasm. The generation of reducing equivalents requires the expenditure of energy and, therefore, presents an especially difficult problem for the deeper lens fiber cells, which lack mitochondria. Enzyme systems in these deeper cells are also less active or inactive, since, in older lenses, they would have been synthesized decades earlier. For this reason, central fiber cells maintain a precarious balance between the catastrophic damage that would be caused by the unchecked oxidation of membrane lipids and cytoplasmic proteins and the diffusion from the more superficial cells in the lens of molecules that protect against oxidative damage.[209,210]

The unique structure of the lens creates special problems for the majority of fiber cells that do not contact the lens epithelium or capsule. Nutrients must reach these cells by diffusion, either between cells or through specialized cell–cell junctions. To reduce light scattering and maintain transparency, lens fiber cells must maintain a very small extracellular space. Therefore, nutrient and metabolite transport is more likely to occur through cells, rather than between them. This predicts that metabolites will accumulate in the center of the lens and that diffusion will limit the availability of nutrients and essential metabolites to cells deeper in the lens. Mature fiber cells do not synthesize proteins and must deal with the consequences of molecular senescence without the capacity for repair.

The lens derives much of its energy from glycolysis.[211,212] The end product of glycolysis is lactic acid. As a result of lactate accumulation, intracellular pH drops significantly from peripheral to deeper fiber cells.[213,214] As a result, pH-sensitive processes will be differentially affected in different regions of the lens.[215] Gap junction conductance is one of the systems in the lens that could be affected, since gap junction permeability is generally decreased at low pH. Interestingly, lens connexins do not decrease their conductance in response to the decreased pH in the mature fiber cells of the mouse lens.[216,217] This adaptation preserves a diffusion pathway from the lens periphery to its center.[217] Other structural and metabolic adaptations to the standing pH gradient in the lens are likely to exist.

Another problem that the lens faces is the need to maintain protein stability for many decades. Once the lens is formed, proteins are synthesized only in the superficial fiber cells. Therefore, proteins made during embryogenesis in humans may have to last for more than 100 years. Accumulated damage leads to loss of enzymatic activity. Altering the structure of the crystallins, cytoskeletal proteins, and enzymes also increases their propensity for aggregation, a process that can lead to cataract formation.

Oxidants within and around the lens

All of the cells of the body exist in an oxidizing environment. Molecular oxygen is, directly or indirectly, the source of most oxidative damage. If cells could survive in an atmosphere free of oxygen, most oxidative damage could be avoided. For most cells this is not possible. However, human lenses can be maintained in an atmosphere of pure nitrogen for some time, as long as adequate amounts of glucose are provided.[218] This is possible because the lens obtains much of its energy from glycolysis.[211,212]

The oxygen tension around the lens in the living eye is quite low, <15 mmHg ($\sim2\%$ O_2) just anterior to the lens and <9 mmHg ($\sim1.3\%$ O_2) near its posterior surface.[219–224] Oxygen levels within the human lens are even lower (<2 mmHg).[225] The low oxygen tension around and within the lens helps to protect lens proteins and lipids from oxidative damage. Even with this low level of oxygen, the lens normally derives a proportion of its ATP from oxidative phosphorylation,[211] a process that, of necessity, generates free radicals.[208]

Hydrogen peroxide is another molecule that has been suggested to cause oxidative stress to the lens. Hydrogen peroxide is produced in mitochondria by the enzyme superoxide dismutase acting on superoxide anion, a byproduct of oxidative phosphorylation. Hydrogen peroxide can also be produced during the oxidation of ascorbic acid, which is present at high levels in the aqueous and vitreous humors (approximately 1.5–2.5 mM). Both processes may contribute to the hydrogen peroxide that has been reported within the intraocular fluids and lens.[226,227] The level of hydrogen peroxide in the aqueous humor has been reported to average over 30 μM and to exceed 200 μM in about one-third of cataract patients.[226,228] However, recent studies have suggested that, for methodological reasons, the level of hydrogen peroxide in aqueous humor was overestimated in these earlier studies.[229–231] Ascorbate interferes with some assays of hydrogen peroxide and aqueous humor produces hydrogen peroxide when exposed to air.[229,231] Careful re-examination of the levels of hydrogen peroxide around and within the human lens using methods not subject to these errors is needed.

The lens is exposed to solar irradiation throughout its life. Although the most energetic and potentially harmful ultraviolet light that reaches the eye is absorbed by the cornea, the remaining solar radiation could have harmful effects, especially on the metabolically vulnerable fiber cells.[232] If light is not absorbed, it produces no damage. However, ultraviolet light is readily absorbed by a number of cellular constituents, including DNA, proteins, nucleoside-containing metabolites, flavinoids and pigments. Flavinoids and pigments also absorb visible light, especially the shorter wavelengths of visible light. All of these interactions are potential sources of free radicals. Free radicals oxidize DNA, lipids, and proteins. Whether or not free radicals are produced by the absorption of light depends upon the chemical nature of the molecule that is interacting with light and the molecular environment.

In spite of its exposure to light throughout life and the presence of abundant targets for light damage, there is no evidence for the photo-oxidation of proteins in the central region of the lens, even in lenses from older individuals.[233,234] It is likely that lens constituents are protected against the harmful effects of light-generated free radicals by the high intracellular concentration of reducing substances (see below), and by the low concentration of oxygen around and within the lens.[221,222,224,225]

Protection against oxidative damage

Glutathione, a tripeptide of the amino acids glutamine, cysteine, and glycine, provides most of the protection against oxidative damage in the lens. It can prevent the oxidation of components of the lens cytoplasm because its concentration in the lens is very high, approximately 4–6 mM, and its sulfhydryl group is readily oxidized. When glutathione levels have been lowered in lens epithelial cells or whole lenses, cell damage and cataract formation follow rapidly.[235–237]

Lens epithelial and superficial fiber cells can synthesize glutathione and glutathione can be transported into lens from the aqueous humor.[237–239] Reduced glutathione (GSH) is regenerated from oxidized glutathione (GSSG) by glutathione reductase and NADPH (Fig. 5.9).[237] Much of the

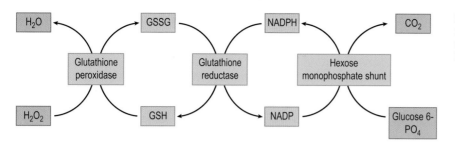

Figure 5.9 Diagram showing the major reactions responsible for the reduction of glutathione (right side) and the use of glutathione to reduce hydrogen peroxide (left side).

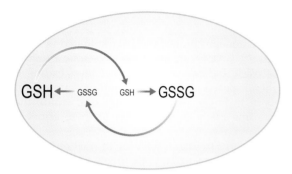

Figure 5.10 Diagrammatic representation of the distribution of reduced glutathione (GSH) and oxidized glutathione disulfide (GSSG) in the adult human lens. An increased fraction of glutathione is in the oxidized form in the center of the lens, a situation that is often increased in the aging lens.

NADPH in the lens is produced by the hexose monophosphate shunt, the activity of which is important for the continued production of reduced glutathione.[212]

However, fiber cells deeper in the lens have minimal capacity for the synthesis or reduction of glutathione. They must obtain most of their reduced glutathione by diffusion from more superficial fiber cells.[237,240] Glutathione can form disulfide bonds with the oxidized sulfhydryl groups of proteins. These glutathione-protein mixed disulfides can then be reduced by a second molecule of glutathione, a process that is facilitated by the enzyme thioltransferase.[241] This regenerates the protein sulfhydryl and forms GSSG (two molecules of glutathione linked by a disulfide bond). The GSSG that results from this process must then diffuse to more superficial layers of the lens where it can be reduced to regenerate GSH (Fig. 5.10). It is likely that this two-way diffusion is the rate-limiting step in maintaining a reducing environment in the center of the lens.[209,210] The rate of diffusion between the superficial and deeper layers of the lens diminishes with age.[240] Therefore, proteins and lipids in the nuclei of older lenses may be more susceptible to oxidative injury than those in younger lenses.

Ascorbic acid is also likely to protect the lens against oxidative damage. Ascorbate is actively transported from the blood to the aqueous humor by a sodium-dependent transporter located in the ciliary epithelium[242] and reaches concentrations in the aqueous humor that are 40–50 times higher than levels in the blood.[243] The ascorbate levels in the lens and other intraocular tissues are also substantial.[244,245] Dehydroascorbate, the oxidized form of ascorbic acid, enters lens cells by way of the glucose transporter, where it is reduced by glutathione-dependent processes.[243,246] Like glutathione, ascorbate is readily oxidized, forming dehydroascorbic acid in the process. Therefore, ascorbate can react with free radicals and other oxidants in the aqueous humor and the lens, preventing these molecules from damaging lens lipids, proteins, and nucleic acids. On the other hand, if dehydroascorbate accumulates in the lens, its metabolites can react with lens proteins, increasing lens color and decreasing protein stability.[245] The high GSH levels in the lens are likely to maintain most ascorbate in its reduced state, thereby avoiding much of this potential damage.

The lens has two enzyme systems to detoxify hydrogen peroxide. Lens epithelial cells have abundant levels of catalase, which converts hydrogen peroxide to water,[247] and glutathione peroxidase, an enzyme that couples the reduction of hydrogen peroxide to the oxidation of glutathione. Studies on cultured lenses and lens epithelial cells suggest that glutathione peroxidase provides most of the protection against the potential damaging effects of physiological levels of hydrogen peroxide, as catalase is only effective against relatively high concentrations of peroxide.[248]

Energy production in the lens

Due to the lack of blood supply, the oxygen concentration within and around the lens is much lower than in most other parts of the body.[219,221,222,224,225] The lens, therefore, depends on glycolytic metabolism to produce much of the ATP and the reducing equivalents required for its metabolic activities.[211,218] The glucose required for glycolytic metabolism is derived from the aqueous humor. Aqueous humor glucose levels are maintained by facilitated diffusion across the ciliary epithelium.

However, lens epithelial cells and superficial fiber cells also contain mitochondria. Therefore, cells near the surface of the lens use both glycolytic and oxidative pathways to derive energy from glucose. About 50 percent of the ATP produced by rabbit lens epithelial cells is derived from oxidative metabolism, while glycolysis accounts for nearly all the ATP produced in lens fiber cells.[211]

Water and electrolyte balance

Due to its high protein concentration and the lack of a blood supply, the lens faces special problems in regulating its water content and in providing nutrients and antioxidants to cells deeper in the lens.[249] Protein concentration increases as one goes from the more superficial fiber cells to fiber cells deeper in the lens.[10] However, this protein gradient is not associated with a reciprocal gradient in the osmotic activity of water, because there is no tendency of water to flow into the cells of the lens nucleus. The mechanism by which this large potential osmotic gradient is neutralized is not known. Decreased protein osmotic activity might result from the increase in the short-range interactions between protein molecules that contributes to lens transparency.[12] The mechanism by which proteins can be concentrated in osmotically neutral manner remains an important unsolved question in the biophysics of the lens.

Several studies provide evidence for an ionic circulation within and around the lens. Vibrating electrodes detect gradients of electrical potential around the lens.[250,251] Positive charges flow into the lens near the anterior and posterior sutures and out at the equator. Like most cells, the membranes of lens epithelial and superficial fiber cells contain sodium, potassium-activated ATPase activity. This Na^+, K^+-ATPase generates an electrochemical gradient across the surface membranes of the lens, with the interior of the lens more negative than the extracellular space. The electrochemical potential tends to drive positive charges, largely sodium ions, into lens cells. This appears to be the origin of the inward positive current at the sutures.[250-252] The positive current flowing out at the lens equator is likely to be carried by potassium ions.[252]

A model that takes into account the electrical and biophysical properties of the lens predicted that water will follow the flow of ions into and out of the lens, creating an internal circulation system.[253] While this is an appealing theory, direct evidence for water flow through the lens fiber cell cytoplasm has not been provided. Numerous studies in which fluorescent dyes were injected into fiber cells have, so far, failed to detect the directed, bulk movement of water that would be predicted by this model. Other studies detected inward flow of water across the lens epithelium and the whole lens, suggesting net flow of water from the anterior chamber to the vitreous body.[254] The importance to lens physiology of this potential flow has not been tested.

The transmembrane potential of the human lens decreases steadily with age.[255] This decrease is caused by an increase in the permeability of the fiber cell membranes to sodium and calcium ions through non-selective cation channels. It is not clear whether the increased cation permeability is due to increased numbers of these channels, to the appearance of the new type of channel, or to an increase in the activity (open probability) of pre-existing channels. The increase in cation permeability is balanced by an increase in the activity of membrane ATPases, which remove sodium and calcium from lens cells. In spite of the increased pump activity, free sodium and calcium levels increase in the cytoplasm of older lenses.[255] Since the transmembrane potential of all cells indirectly provides the driving force for the transport of many metabolites and nutrients, the age-related decrease in the transmembrane potential of the lens may have important consequences for lens metabolism and ionic homeostasis. Elevated calcium levels can also lead to metabolic disturbances and destruction of cell components through the activation of calcium-sensitive proteases.[256]

Like all cells, lens epithelial and fiber cells maintain a much lower concentration of free calcium ions in their cytoplasm than is found in the extracellular space. However, free calcium levels measured in fiber cell cytoplasm are substantially higher than the levels in epithelial cells.[255,257] In addition, there appears to be a gradient of free calcium that decreases from the posterior to the anterior ends of lens fiber cells.[257] These observations are consistent with the view that calcium slowly leaks into lens fiber cells and is removed by membrane pumps at the lens surface. The activity of these pumps in maintaining low cytoplasm calcium concentrations is important, because treatments that abruptly raise free calcium levels lead to rapid degradation of the lens fiber cell cytoskeleton, uncontrolled proteolysis, cell swelling, and opacification.[256,258-262]

Lens epithelial cells transport nutrients into their cytoplasm from the aqueous humor. Although the transport of small molecules from epithelial cells to fiber cells has been demonstrated,[154,193] the relative importance of this pathway in providing nutrients to the fiber cells, compared to transport directly across the surface membranes of superficial fiber cells, has not been determined. When metabolite transporters have been examined in lens fiber cells, these molecules have usually been found. The distribution of glucose transporters in the lens illustrates this issue. Lens epithelial cells express abundant levels of the glucose transporter, glut1, that is presumably used for the uptake of glucose from the aqueous humor.[263] Although fiber cells express little glut1, they express large amounts of the higher-affinity glucose transporter, glut3. Therefore, fiber cells can transport glucose into their cytoplasm from the extracellular milieu, a finding that raises questions about the relative importance of the epithelial cells in providing glucose to the fiber cells.

Lens transparency and refraction

Vertebrate lenses are remarkably effective optical devices. An efficient lens must be transparent to the wavelengths of light that can be detected by the photoreceptors, have a focal point that is appropriate for the optical system in which it functions, and have a minimum of spherical and chromatic aberration. Lens transparency depends on the organization of the cells of the lens and of the distribution of the proteins within them. The precise organization of the fiber cells, their high protein concentration, and the absence of organelles from the fiber cells that lie in the optical axis assures that a minimum of scattering occurs as light passes through the lens.

The cellular and molecular interactions responsible for establishing and maintaining the curvature of the lens surfaces are not known. Studies in chicken embryos demonstrated that influences from outside the lens normally regulate its shape and rate of growth.[39] Lenses in different species range from nearly spherical (rodents) to an axial ratio of more than 2:1 (humans). As the human lens grows, the radii of curvature of its anterior and posterior surfaces decrease significantly.[264] In spite of this, the focal point of the lens remains remarkably constant, suggesting that the refractive power of the lens cytoplasm changes in a way that compensates for the change in curvature of the refractive surfaces. Control of lens shape is one of the most fascinating unanswered (and little explored) aspects of lens biology.

The high protein concentration of the lens fiber cells causes the refractive index of the lens to be higher than that of the fluid around it. Fiber cells close to the surface of the lens have a lower protein concentration than fiber cells deeper in the lens, creating a gradient of refractive index that at least partially corrects for spherical aberration.[265–267]

Although the human lens is transparent to most wavelengths of visible light, it produces and accumulates chromophores that absorb the shortest wavelengths of the visible spectrum. At birth, the human lens is a pale yellow color. With increasing age the amount of yellow pigmentation in the lens increases. This pigmentation absorbs the shorter, more energetic wavelengths of light, preventing them from reaching the retina. The predominant yellow chromophores in the young human lens are metabolites of tryptophan, especially N-formyl kynurenine glucoside.[14] With aging, an increasing variety of soluble and protein-bound chromophores are found in the lens fiber cells.[268,269] When especially high concentrations of these chromophores accumulate, they can reduce visual acuity by increasing light absorbance, leading to the formation of what is termed a "brunescent" cataract. The factors responsible for the excessive accumulation of chromophores in some lenses are not known, although recent studies have suggested that the oxidative modification of proteins may play a significant role.[268] Brunescent cataract is more common in rural areas and in developing countries, a factor that suggests that nutritional or environmental factors, like cooking smoke, may also be important.[270]

Changes in the lens with aging

Fiber cells in the center of an adult lens were produced during early embryonic life, while those at the surface of the lens may be weeks or months old. Many have touted the lens as a valuable model system for studying aging. However, unique properties of the lens, especially the lens fibers, make it an inappropriate model for aging in most cell types. Gene expression and protein synthesis cease at about the time that the fiber cells degrade their organelles.[7] In nearly all other cell types, RNA and protein synthesis and turnover continue throughout life.

To follow changes in the composition of the lens that are related to age, one must correlate changes in fiber cells from lenses of different ages and from different layers of the same lens. Using this approach it was shown that, in the human lens, many of the soluble crystallins are sequentially truncated by proteolysis over a period of months to years.[271] Some of the β-crystallins are degraded relatively rapidly, while other crystallins much more slowly. However, by the end of the first 20 years of life, most of the crystallins appear to reach a steady state after which little additional degradation occurs. Therefore, in older lenses, crystallins in the center of the lens and crystallins a short distance from the surface of the lens show similar degrees of proteolysis.[271–273] These studies only examined changes in soluble crystallins. Further studies may show that crystallins in the insoluble fraction are more extensively degraded.

The proteins of the lens are often characterized by their relative solubility, typically being separated into fractions that are water-soluble, urea-soluble, and "insoluble". As one goes from more superficial to deeper fiber cells, an increasing percentage of the crystallins are found in the insoluble fraction. In addition, when one separates the water-soluble crystallins by size, an increasing proportion is found in high molecular aggregates. Therefore, there is a general tendency for proteins to become aggregated and less soluble in the older fiber cells.[274,275] Alpha-crystallin is a good example. In the lens nucleus from individuals of increasing age, one finds decreasing amounts of soluble α-crystallin. By age 45, no α-crystallin is detectable in the water-soluble fraction from the lens nucleus.[276,277]

The gradual insolubilization of α-crystallin is likely to be related to its function as a molecular chaperone.[100] When lens proteins are partially unfolded, hydrophobic domains are exposed. Alpha-crystallin binds to these hydrophobic regions, presumably preventing uncontrolled protein aggregation. It seems likely that accumulated damage to the crystallins and other proteins leads to increased association with α-crystallin. Interestingly, in spite of the loss of soluble α-crystallin, one does not see a precipitous increase in the rate of protein aggregation in the lens nucleus after age 45.

Another age-related change seen in crystallin structure is the increasing racemization of aspartic acid, methionine and tyrosine, and the deamidation of glutamine and asparagine.[272,273,278–280] Racemization and deamidation alter protein structure and/or charge. These changes correlate well with age, but do not differ significantly in clear and cataractous lenses of the same age.

Some components of the lens fiber cell cytoskeleton are disassembled in older fiber cells. Vimentin intermediate filaments are degraded in the deeper lens cortex, well after the loss of membrane-bound organelles.[281] Beaded filaments composed of phakinin and filensin persist into the lens nucleus and may last for the life of the lens.[133] Actin microfilaments also appear to survive in the oldest fiber cells, although the persistence and continued association of these filaments with the plasma membrane has been questioned.[282–284] Proteolysis and insolubilization of the components of the cytoskeleton appear to contribute to their disassembly.[133,281,284]

The structure and development of the lens sutures

Sutures form at the anterior and posterior poles of the lens where fiber cells growing from opposite sides of the lens abut at their apical and basal ends. In some species all fiber cells meet near the midline of the lens, forming an "umbilical" suture.[285] In most species, the sutures form along planes. In human embryos, elongated lens fiber cells meet at three planes, forming an upright "Y" at their anterior ends (with respect to the superior–inferior axis of the eye) and an inverted "Y" posteriorly (Fig. 5.11A). As the human lens grows, the suture planes formed by more superficial shells of fibers become increasingly complex. The first evidence of this typically occurs soon after birth, when two new suture planes form at the ends of each of the three branches of the Y sutures (Fig. 5.11B). As new fibers are added during lens growth, the branch points of the newly formed sutures gradually "migrate" towards the center, eventually forming a six-pointed "star" suture (Fig. 5.11C).[285] Branching again occurs at the tips of each of these six planes, eventually forming a total of 12 suture planes at the anterior and posterior

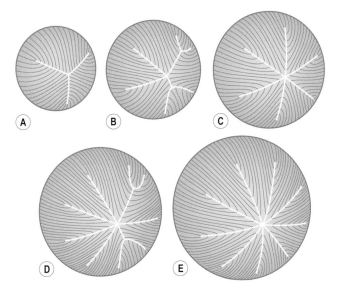

Figure 5.11 Diagram illustrating the increasing complexity of the sutures as the lens grows. The sizes of the lenses depicted are to scale. (**A**), (**B**), (**C**) The tripartite "Y" suture that forms during embryogenesis as a result of secondary fiber cell formation is converted into a six-pointed star suture by the continued deposition of new fiber cells. If one were to peel away the fibers from the lens depicted in (C), the initial tripartite suture pattern would be revealed. (**D**), (**E**) Further growth results in the formation of additional suture planes.

surfaces of the lens (Fig. 5.11D, E).[285] The increasing geometric complexity of the suture patterns in older human lenses results in lenses with better optical properties than in species that maintain a simple Y suture pattern throughout life.[286]

The formation and orientation of the sutures provides a challenging problem for developmental, cell, structural and theoretical biologists. The lens appears to be radially symmetrical about the optical axis. Yet, the sutures form in a pattern that breaks this symmetry. The spatial cues that lead to the precise alignment of the suture planes with respect to the axes of the eye and body are not known.

The lens capsule

When the lens placode invaginates from the surface ectoderm early in embryonic life, it is already supported by a thin basal lamina.[35] This extracellular matrix surrounds the lens vesicle after it detaches from the surface ectoderm. The lens epithelial and superficial fiber cells continue to secrete the components of the basal lamina, which thickens to become the lens capsule.[35,36] Examination of the lens capsule in the electron microscope shows that it is composed of multiple laminae, as if the basal lamina had been replicated many times.[287] Studies of the synthesis of the lens capsule in experimental animals showed that newly synthesized capsular materials are originally deposited close to the basal ends of epithelial and fiber cells.[288] With time, the labeled components of the capsule move farther away from the surface of the cells, as they are displaced by successive layers of newly synthesized capsular material.

Clinical observations have also supported the view that the human lens capsule is synthesized from the inside out.

In certain cases, foreign materials may be deposited in the capsule during its synthesis. These deposits can be viewed in slit lamp or Scheimpflug images.[289] Eventually, these deposits disappear from the lens. One interpretation of this observation is that, as new layers are laid down on the inner surface of the capsule, capsular material is lost from the surface of the lens. If this interpretation is correct, the rate of synthesis of components of the capsule at its inner surface and the rate of degradation at its outer surface regulate its thickness. The capsule also must be remodeled during embryonic life, when the surface area of the lens is increasing rapidly.[290] The enzymes that are responsible for the degradation or remodeling of the capsule have not been identified.

In keeping with its similarity to typical basal laminae, the lens capsule is composed predominantly of type IV collagen, laminin, entactin (nidogen), and heparan sulfate proteoglycans.[33,37,287,291] However, studies have shown that some components of the capsule differ in the anterior, posterior, and equatorial regions of the lens.[288,290,292] In the adult lens the capsule is significantly thicker over the epithelium and thinner over the basal ends of the superficial fiber cells. The factors that regulate the differential distribution and accumulation of the components of the lens capsule remain to be identified. Several diseases are associated with defects in the components of the capsule, including pseudoexfoliation syndrome (lysyl oxidase polymorphisms), Pierson syndrome (laminin-β2 mutations) and Alport syndrome (collagen IV mutations).[37]

The zonules

The zonules (suspensory ligament, zonules of Zinn) are composed of thin fibrils that suspend the lens in the anterior of the eye. Zonular fibers insert into the lens capsule near the equator and into the basal lamina of the non-pigmented layer of the ciliary epithelium (Fig. 5.12).[293] It is likely that the ciliary epithelial cells synthesize the components of these fibrils.[8] The primary structural protein in zonular fibers is fibrillin.[8,9,294] Mutations in the fibrillin gene are responsible for Marfan syndrome, in which dislocation of the lens is a common clinical finding.[295,296]

Examination of the insertion of zonular fibers into the lens capsule shows that these fibers are intimately interwoven with the components of the capsule.[8,293,297] If, as suggested above, the capsule is continuously degraded at its outer surface, it is unclear how the zonular fibers and capsular fibers maintain their connections.

This is only one of the topological paradoxes associated with the synthesis and maintenance of the zonules. For example, it is also not clear how the components of the fibrils are originally inserted into the lens capsule, especially when they are under tension, or how the zonular fibers are physically attached to the basal lamina of the ciliary epithelial cells while being synthesized by these same cells. Presumably, the zonular fibers establish their attachment between the lens capsule and ciliary epithelium early in development when these basal laminae are in direct contact. If this assumption is correct, it would predict that the number of zonular fibers should not change throughout life. If it is not correct, a mechanism must exist for these fibrils to be assembled across the space between the ciliary

Figure 5.12 Scanning electron micrograph showing tapering bundles of zonular fibers inserting into the lens capsule (magnification ×780). (Reproduced with permission from Streeten, B.W., The zonular insertion: a scanning electron microscopic study. Invest Ophthalmol Vis Sci, 1977. 16(4): p. 364-75. Reproduced with permission from Association for Research in Vision and Ophthalmology[293])

epithelium and the lens capsule, a remarkable feat of bioengineering.

As the lens grows, the position of insertion of the zonules shifts anteriorly.[297] Since the zonules are believed to maintain a fixed point of insertion into the lens capsule, this shift has been interpreted as evidence for a relative increase in capsule synthesis at the lens equator.[297] It is possible that the age-related anterior shift in the zonular insertion point alters the forces applied on the lens during accommodation and may contribute to presbyopia.[297]

Cataracts

A cataract is any opacification of the lens. Cataracts are considered to be clinically significant when opacification interferes with visual function. Loss of lens transparency can be due to an increase in light scattering or light absorption. Increased light scattering can be caused by disruption of the structure of lens fiber cells, increases in protein aggregation, phase separation in the lens cell cytoplasm, or by a combination of these processes.

Age-related cataracts are classified based on the region of the lens that is affected (Fig. 5.13). The most common types of age-related cataracts are nuclear, cortical, and posterior subcapsular. Nuclear cataracts occur in the oldest fiber cells near the center of the adult lens – those formed during embryonic and fetal life. Cortical cataracts occur in cells formed later in life. These cataracts typically begin in a sector of the lens cortex, affecting mature cells that have already degraded their organelles. Posterior subcapsular cataracts result from light scattering by a plaque of swollen cells at the posterior pole of the lens.

In addition to age-related cataracts, there is a wide variety of less common cataracts that are usually classified based on their etiology. These include the opacification of the posterior lens capsule that sometimes occurs following cataract surgery, referred to as secondary cataracts, after cataracts, or posterior capsular opacification (PCO).

Cataract epidemiology

The risk factors associated with different types of age-related cataracts provide clues to the environmental factors that promote cataracts or protect against them. Therefore, before considering the mechanisms that may cause age-related cataracts, it is useful to consider the factors that influence the risk of cataracts in human populations. Although epidemiologic factors are often difficult to identify with certainty, especially for a disease that occurs late in life, those factors that are frequently associated with cataracts in human populations are likely to be relevant to the disease mechanism.

General risk factors

Not surprisingly, age is the primary risk factor for the most common kinds of cataract. There is an exponentially increasing incidence of cataract after age 50. One risk factor common to most kinds of cataract is lower socioeconomic status and/or lower education. Since lower socioeconomic status may predispose to nutritional deficiencies, increased exposure to diseases, poor general health status, and increased occupational exposure to cataractogenic agents, it is difficult to determine the specific aspects of lower socioeconomic status that are important for cataract formation.

Sex is also an important influence on the incidence of cataract. Women are at increased risk for most kinds of cataract.[298-304] Conversely, studies have suggested that estrogen protects against cataract formation in humans and animals and that cataracts may be delayed by late menopause.[305-310] Lowering estrogen function with the anti-estrogen tamoxifen increased the risk of cataracts when used for long duration.[311,312] The protective effect of estrogen suggested by these studies makes the increased overall levels of cataract in women harder to understand. If estrogen is protective, other factors, as yet unknown, must strongly predispose women to cataract formation. Alternatively, higher levels of male sex steroids may be protective against cataract.

When the influences of age, sex and socioeconomic status are removed, specific risks for different types of cataracts are revealed. Smoking and high alcohol consumption have been identified in several studies as dose-dependent risk factors for nuclear and, in some cases, cortical cataracts.[304,313-315] Dark iris color is often associated with a higher incidence of all types of lens opacities,[304,316,317] a finding that may also be related to higher levels of cortical cataract in blacks than whites.[300] Exposure to anti-inflammatory steroids and

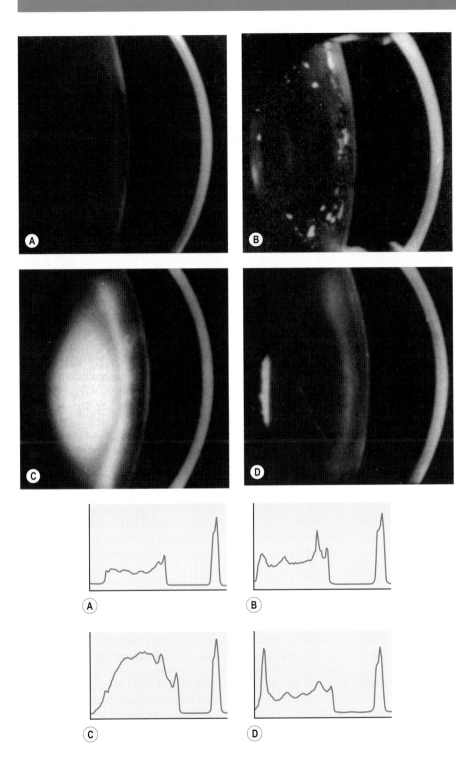

Figure 5.13 Photographs taken using a Scheimpflug camera of the anterior portion of human eyes. Light scattering from the cornea, the lens capsule-epithelium and any opacity in the lens can be seen in the photographs. Scattering intensity is shown in scans of the photographs made with a densitometer. (**A**) Normal lens. (**B**) Cortical cataract. (**C**) Nuclear cataract. (**D**) Posterior subcapsular cataract.

ionizing radiation are well-recognized risks for posterior subcapsular cataracts. Few additional risk factors have been consistently identified for PSCs. However, this could be the statistical consequence of the generally lower frequency of PSCs than of other types of cataract. Numerous epidemiological studies have identified additional risk factors for cataracts. These are not detailed here, either because their association with cataract was not especially strong, or because they have not been seen in the majority of epidemiological studies of cataract.

Several studies have found an interesting association between cataracts and increased mortality. When confounding variables, including smoking, age, race, and sex were removed, the presence of clinically significant cataracts remained as a strong, independent predictor of mortality, especially in younger patients with cataract.[318–321] These findings suggest a connection between cataracts and systemic phenomena influencing survival. Considered in another way, behavioral, nutritional and biochemical factors that prevent age-related cataract formation may lengthen life span. However, other studies failed to find an association between cataract and increased mortality.[320,322] A recent 5-year follow-up of patients undergoing cataract surgery in the UK in 2000–2001 revealed lower mortality than in the general population.[322] This paper also provides a thorough compilation of previous studies of cataract and mortality.

Several studies found an association between lens thickness and the most common types of age-related cataract. In the Beaver Dam Eye Study, a comprehensive evaluation of ocular disease in a US population of northern European ancestry, cortical cataracts correlated with thinner and nuclear cataracts with thicker lenses.[189] Five-year follow-up of subjects that initially had clear lenses revealed that having a thinner lens was a risk for developing cortical cataract. Thicker lenses at baseline were associated with increased risk of developing nuclear cataract.[190] Studies in Singapore and India subsequently found similar associations between lens thickness and age-related cataract.[191,192] Possible reasons for these associations are discussed below in the sections on nuclear and cortical cataracts.

Age-related nuclear cataracts

In the US and most Western countries, nuclear cataracts are most often associated with increased light scattering in the nuclear fiber cells. In many developing countries brunescent nuclear cataracts are more common. These opacities are associated with increased lens color and light absorption. Whether due to light scattering or absorption, nuclear cataract is the most common type in most countries or regions, typically accounting for more than 60 percent of cataract surgery. However, cortical cataracts predominate in some countries or regions, particularly in Asian populations living in temperate climates.

There is abundant evidence that nuclear cataracts are associated with increased oxidative damage to lens proteins and lipids.[228,237,323–325] The formation of disulfide bonds between protein subunits can lead to aggregation and increased light scattering,[237] although other forms of protein–protein interactions also occur. There is also evidence that crystallins may associate with lens fiber cell membranes in increased amount in nuclear cataracts.[326–329] The factors responsible for increased oxidation in nuclear cataracts are not fully understood. However the marked age dependence of nuclear cataracts and the concurrent increase in oxidized glutathione in the lens nucleus suggests that the balance between protein and lipid oxidation and glutathione-dependent reduction are likely to be involved (see Box 5.3).[237,240,330]

Persuasive evidence for an association between oxidation, age, and cataract formation was provided by studies of patients treated with hyperbaric oxygen to alleviate the complications of peripheral vascular disease.[331] These individuals were exposed to 2.0–2.5 atmospheres of pure oxygen for one hour each day for up to three years. Of the 25 patients treated in this study, all but one had increased light scatter in the lens nucleus and nearly half developed frank nuclear cataracts. All of the individuals who showed increased light scattering or opacification in the nucleus were over 50 years old. The individual who did not show increased light scatter was 24 years old. Patients of slightly older average age with the same presenting symptoms, but not treated with hyperbaric oxygen, served as the control group. None of the individuals in the control group developed cataracts over the study period. After the termination of treatment, some of the patients who had been treated with hyperbaric oxygen and had developed increased light scatter showed improvements

Box 5.3 Nuclear sclerotic cataracts; oxidation is the key

Nuclear cataract is associated with the oxidative damage to the proteins and lipids, leading to hardening of the lens nucleus and increased light scattering. Hardening of the lens nucleus increases its refractive power, causing a "myopic shift" which is later followed by cataract formation. The lens normally exists in an extremely hypoxic environment. Patients treated with long-term hyperbaric oxygen therapy develop a myopic shift and, eventually, nuclear cataracts. Removal of the vitreous body during retinal surgery increases oxygen levels at the posterior of the lens and leads to rapid-onset nuclear cataracts. Retina surgery performed without vitrectomy avoids cataract formation. The high level of ascorbate (vitamin C) that is present in the ocular fluids reacts with oxygen, reducing oxygen exposure to the lens. Age-related degeneration of the vitreous body is associated with lower ascorbate levels, slower oxygen consumption and increased risk of nuclear cataract. Protecting the lens from exposure to oxygen or preserving the gel structure of the vitreous body could prevent the formation of nuclear cataracts.

in their visual acuity and decreases in the amount of light scattering in their lenses. This result suggests that, even in older individuals, the lens has the capacity to reverse oxidative damage to the components of the nucleus.[331]

The onset of nuclear cataract formation is often associated with an increase in the refractive power of the lens.[332] For hyperopic patients this "myopic shift" temporarily causes an improvement in their near visual acuity, a phenomenon that is often called "second sight." This increase in refractive power is associated with hardening of the lens nucleus. It is not surprising that the term "nuclear sclerotic cataract" is used to describe the opacities that soon follow.

Patients treated with hyperbaric oxygen also experience significant myopic shift. As with the light scattering described above, these increases in lens refractive power usually reversed during treatment or soon thereafter.[331,333,334] This suggests that the hardening of the lens and its increased refractive power, as occurs in nuclear sclerotic cataracts, are caused by oxidative damage to the proteins and lipids of central lens fiber cells. In this case, reducing the source of oxidative damage (hyperbaric oxygen) permitted the lens to recover much of its normal refractive properties.

Isolated lenses and experimental animals have also been treated with hyperbaric oxygen. The results of these studies are in general agreement with the findings in patients receiving hyperbaric oxygen therapy, although frank nuclear opacities are rarely seen. These animal models provide useful tools to study the earlier biochemical effects of oxidative stress on cataract formation.[335–337]

Studies of patients treated with hyperbaric oxygen suggest that molecular oxygen or a metabolite derived from molecular oxygen contributes to nuclear sclerosis and nuclear cataract formation. Studies of the effect of hyperbaric oxygen on the lens also provide some of the strongest evidence that the cells of the lens nucleus are in a delicate balance between their ability to prevent or reverse oxidative damage and the tendency for oxidation to occur. The fact that hyperbaric oxygen leads to opacification of the lens nucleus, but not the more superficial lens fiber cells, shows that the central fibers are more susceptible to oxidative damage, even though they are farther away from the source of oxygen. Age-associated decrease in the rate of diffusion of glutathione to the lens nucleus is likely to be part of the reason for this increased susceptibility.[209,210,237,240,330,338]

Nuclear cataracts occur frequently in older patients within 6 months to 3 years after vitrectomy. Several studies have reported that the incidence of cataracts is as high as 95% 2 years after the removal of the vitreous body.[339–347] Evidence is accumulating which suggests that study of post-vitrectomy cataracts may help to understand the general mechanism responsible for the formation of age-related nuclear cataracts.

The appearance of the lens in post-vitrectomy nuclear cataracts is similar to typical age-related nuclear cataracts, the only difference being the rapidity of onset of cataracts after vitrectomy. If one assumes that the etiology of the cataracts in both cases is the same, there is likely to be an increase in oxidative damage in the lens following vitrectomy. The lens is usually not touched during vitrectomy and the time between surgery and cataract formation is at least several months. Therefore, the high incidence of nuclear cataracts after this procedure is probably due to changes in the environment around the lens, not to direct damage to the lens during surgery. Measurement of oxygen near the lens after vitrectomy showed that oxygen levels increased greatly during surgery and remained significantly elevated in the eyes of patients who returned for a second retinal surgery months or years later.[222] When retinal surgery could be performed without damaging the vitreous gel, cataract was prevented, even 5 years after the initial surgery.[348] In a related study, eye bank eyes with a greater degree of vitreous degeneration (liquefaction) were more likely to have nuclear opacities than eyes with a more intact vitreous gel. In eyes between 50 and 70, the extent of vitreous liquefaction was a stronger predictor of nuclear opacification than age. Diabetics, who have lower levels of oxygen in the vitreous,[224] also progress to cataract surgery more slowly after vitrectomy than non-diabetics.[349] These studies suggest that the vitreous gel protects against nuclear cataract, probably by protecting the lens from exposure to oxygen from the retinal vasculature.

The high concentration of ascorbate in the vitreous body also appears to protect the lens from exposure to oxygen.[350] There is usually ~2.0 mM ascorbate in the human vitreous.[350,351] Degeneration of the vitreous body or prior vitrectomy was associated with a ~40 percent decrease in the mean ascorbate concentration. An endogenous catalyst enabled the ascorbate in the vitreous to react with oxygen. Presumably, ascorbate-dependent oxygen consumption reduces the exposure of the lens to oxygen. When the level of ascorbate or the catalyst in the vitreous fluid decreased, the ability of the vitreous to consume oxygen also decreased.[350] Therefore, excessive vitreous degeneration or vitrectomy lowers the concentration of ascorbate in the vitreous, which would be expected to increase the exposure of the lens to oxygen (Fig. 5.14).

The increased light scattering associated with nuclear cataract formation could be due to protein aggregation or to changes in the organization of the fiber cell structure. Examination of the fiber cells in the lens nucleus with the electron microscope suggests that nuclear cataract formation is usually not associated with a gross disruption of cell membranes, obvious protein aggregation, phase separation, or

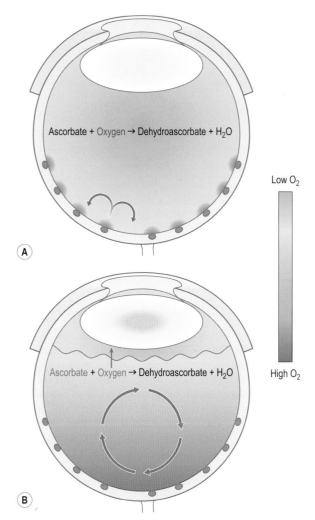

$$Ascorbate + Oxygen \rightarrow Dehydroascorbate + H_2O$$

Low O_2

$$Ascorbate + Oxygen \rightarrow Dehydroascorbate + H_2O$$

High O_2

Figure 5.14 Diagram illustrating the role of the gel vitreous body and ascorbate in the vitreous fluid play in protecting the lens from excessive exposure to oxygen from the retinal vasculature. The gel state of the vitreous body prevents stirring of the contents of the vitreous chamber, allowing the uptake of oxygen by adjacent retinal cells. Increased mixing of the vitreous fluids after vitreous degeneration or vitrectomy increases exposure of the fluids to oxygen, which increases the degradation of ascorbate, allowing more oxygen to reach the lens. The chemical reactions summarized in the figure show the initial reactants (ascorbate and oxygen) and the end products (dehydroascorbate and water). Hydrogen peroxide is an intermediate product in this reaction, which is degraded to water and oxygen by the enzyme, catalase. If not taken up by cells, dehydroascorbate is rapidly hydrolyzed to yield several additional degradation products.

with increased granularity of the fiber cell cytoplasm (Fig. 5.15).[352–354] Techniques aimed at detecting subtle alterations in the organization of the fiber cell cytoplasm did not detect significant differences between clear lenses and the majority of lenses with nuclear cataracts.[354] More subtle changes in the organization of the cytoplasm, whether they are due to protein aggregation or microscopic phase separation, cannot be ruled out. It is important to recognize that only a small fraction of the protein in the lens fiber cell cytoplasm needs to be aggregated in order to increase light scattering dramatically.[355] This makes it difficult to identify the changes that lead to opacification among the many alterations of lens proteins that occur with aging.

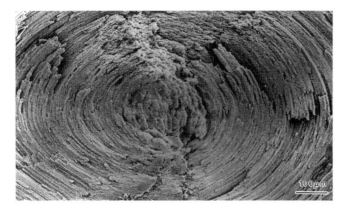

Figure 5.15 Scanning electron micrograph of the lens fiber cells in the embryonic nucleus of a lens from an 80-year-old. The fiber cells are still well organized and intact. (Courtesy of Dr. J. Kuszak.)

During life, the nuclei of clear human lenses become up to 1,000 times harder or stiffer than at the time of birth.[356,357] The increased hardness that typically accompanies nuclear cataract formation is superimposed upon this age-related increase.[356] As described above, since nuclear cataract involves increased oxidative damage, oxidation of the components of the nucleus is likely to contribute to the stiffening of the lens cytoplasm that characterizes these "sclerotic" cataracts. Whether oxidative modification of crystallins is also involved in the age-related stiffening of clear lenses is not known.

In summary, it is likely that the aging lens is susceptible to oxidative damage, due to the increased difficulty in maintaining the cytoplasm of the lens nucleus in the reduced state.[209] Any increase in oxidative load or further decrease in the ability of the lens nucleus to cope with the normal level of oxidation is likely to lead to nuclear cataract formation. In addition, experiments with animals, mentioned in the section on lens growth, showed that exposure of the lens to increased oxygen increased the rate of lens growth.[187] Since larger lenses are associated with increased risk of nuclear cataracts, exposure to oxygen may indirectly increase nuclear cataracts formation by increasing the size of the lens. Therefore, increased exposure of the lens to oxygen, as a result of hyperbaric oxygen therapy, after vitrectomy, or following age-related degeneration of the vitreous body, is likely to play a significant role in the formation of nuclear cataracts.

Age-related cortical cataracts

The cells that are affected in cortical cataracts are mature fiber cells that lie close to the surface of the lens. Cortical cataracts most often occur in the inferior half of the lens, with a tendency for these cataracts to occur in the inferior nasal quadrant.[358,359] Cortical opacities usually begin in small regions of the lens periphery and may spread around the circumference of the lens over a period of years. Early stage cortical cataracts are usually not clinically significant, since the opacity does not impinge on the visual axis. As the cataract progresses and the opacity spreads towards the visual axis, it may eventually interfere with vision. However, an individual may have a superficial cortical opacity for years without experiencing a diminution of visual function.

The superficial lens fiber cells that are involved in cortical cataracts, like all mature lens fiber cells, span the thickness of the lens from the posterior to the anterior suture. Therefore, the peripheral opacities that are characteristic of early-stage cortical cataracts occur only in the central part of the affected fiber cells, with the apical and basal ends of these cells remaining transparent.[360,361] The spread of the cortical cataract around the lens circumference involves progressive damage to an increasing number of lens fiber cells of similar age, while the extension of a cortical cataract into the visual axis involves progressive opacification of the extremities of the same cohort of fiber cells. It is remarkable that the central region of a group of fiber cells can be opaque while the tips of these same cells remain transparent. Extension of the opacity along the lengths of a small cluster of fiber cells leads to the formation of the "cortical spokes" that are frequently described in these cataracts.[362]

In contrast to the subtle morphological changes that occur during nuclear cataract formation, the damage found in cortical cataracts is catastrophic. Examination of the affected regions of lens fiber cells in a cortical cataract reveals almost complete disruption of cell structure.[361,363,364] Plasma membranes are ruptured and many whorls and vesicles of membrane-like material can be found in the cytoplasm. Cytoplasmic proteins are typically aggregated to such an extent that the cytoplasm takes on a chalky appearance when the cataractous region of the lens is disrupted during cataract surgery or during dissection of the lens postmortem.

Only a small group of cells is affected in the early stages of cortical cataract formation. One can imagine several mechanisms that could lead to the formation of this initial opacity, including physical or chemical damage to the fiber cell plasma membrane, inhibition of the ion pumps or transporters in these membranes, damage to the system responsible for calcium homeostasis or the local loss of protective molecules like glutathione. Any one of these damaging occurrences would be likely to lead to all the others, making it particularly difficult to identify the initiating event in cortical cataract formation.

No matter what the initiating event, loss of calcium homeostasis could underlie the radial and circumferential spread of opacification. Cytoplasmic calcium levels are abnormally high in the damaged cells of cortical cataracts.[364-366] As mentioned above, lens fiber cells typically maintain their intracellular calcium concentration in the low micromolar range. Experimental elevation of lens cytoplasmic calcium levels leads to widespread proteolysis.[365] If adjacent fiber cells are joined by membrane fusions, it may be difficult to prevent the eventual spread of high levels of calcium into adjacent cells.

The mechanism that initially prevents the spread of damage from the central portion of the lens fiber cell towards its basal and apical ends must involve the creation of a barrier between the central, cataractous region of the cell and the transparent, distal regions. Numerous membrane blebs and vesicles are seen in the opaque regions of cortical cataracts, a phenomenon that has been termed "globular degeneration."[367] If, during cataract formation, the cytoplasm were to fragment into such vesicles, it is likely that the proximal ends of the fiber cells, adjacent to the damage, would be sealed by fusion of the plasma membrane. This could "wall off" the damaged region of the cell, thereby slowing the spread of damage.[364] Since fiber cells are connected to their neighbors by gap junctions all along their lateral membranes, "walling off" the central region of the cell would not isolate the tips of the cell from the rest of the lens, permitting the passage of nutrients and antioxidants into and out of the transparent portions of the fiber cells.

The presence of a protective mechanism to prevent the spread of damage along the length of a fiber cell in typical age-related cortical cataracts is also evident in traumatic cataracts. Local physical damage to the lens, caused by traumatic injury to the eye or by touching the lens during surgery, usually leads to the formation of a local opacity. Such opacities may be restricted to the affected cells for many years, becoming buried deeper in the lens as new fiber cells are added, or the opacity may slowly spread to adjacent fiber cells in a manner that appears similar to the progression of non-traumatic cortical cataracts.[368]

The strong association between thinner lenses and increased risk of cortical cataract suggests a physical mechanism for the initiation of cortical cataracts.[189–191] Cortical cataracts typically occur after the lens has lost most or all of its ability to accommodate. However, even in older individuals accommodative stimuli cause contraction and relaxation of the ciliary muscle and, therefore, changes in the tension on the zonules, even though the lens may not change shape appreciably. Presbyopia is accompanied by continued hardening of the lens nucleus, while the cortex remains relatively soft. This results in lenses with increasingly steep gradients of stiffness between the nucleus and cortex.[357,369,370] It is likely that the forces generated during accommodation are focused at the interface between these layers, close to the location at which cortical cataracts occur.[362,371–373] In smaller lenses, there may be increased tension on the zonules during disaccommodation, due to the greater distance between the lens equator and the ciliary body. Influences that increase the stiffness of the lens nucleus, including higher levels of protein glycation in diabetics or increased sunlight exposure, could account for some of the increased risk of cortical cataract associated with these factors (see Box 5.4).

Posterior subcapsular cataracts

Posterior subcapsular cataracts (PSCs) are caused by light scattering in a cluster of swollen cells at the posterior pole of the lens, just beneath the capsule.[374–376] Because the opacity produced by these cells is in the optical axis, posterior subcapsular cataracts are particularly disabling. Careful examination has shown that, in lenses with PSCs, the most superficial fiber cells are disorganized in localized regions at the lens equator. A "stream" of epithelial-like cells leads from the affected region of the equator to the opacity at the posterior pole.[374] This suggests that PSCs result from the abnormal migration of lens epithelial cells or the aberrant differentiation of fiber cells.

Opacities can also occur at the posterior pole due to swelling of the posterior ends of the fibers along the suture planes. These "sutural" cataracts can sometimes resemble PSCs. However, sutural opacities are typically associated with inherited cataracts and are not common in age-related cataract.

"Pure" PSCs (PSCs not associated with another kind of cataract) are less common than either age-related nuclear or cortical cataracts, typically accounting for less than 10 percent of cataracts in humans. However, PSCs may occur in conjunction with nuclear or cortical opacities.

Although specific risk factors associated with PSC formation are known, we have less information about the cellular or molecular defects that lead to the formation of these cataracts in humans. An animal model of steroid-induced cataract provided some insight into this process, suggesting that defects in the adhesion of differentiating fiber cells may lead to PSC formation.[377] Experimentally induced radiation cataracts also provide a relevant model. Cortical and posterior subcapsular opacities develop over weeks to months after exposure to sufficient levels of ionizing radiation.[378–381] These opacities are preceded by increased death in the germinative zone epithelial cells and disorganization of the fiber cells that develop from those epithelial cells that survive the radiation damage.[382–384] Increasing use of intravitreal steroids has led to an increase in PSCs. These cataracts may provide useful material for further study of the mechanisms underlying this disease.

Mixed cataracts

Cataract patients sometimes have a combination of nuclear, cortical, and/or posterior subcapsular cataracts. It is not clear whether having one kind of cataract predisposes a person to developing an additional kind of cataract, although it seems likely that such an association would have been noticed. This is particularly likely in the case of a pre-existing cortical cataract, which may be detectable, but clinically insignificant for many years. However, the author is not aware of studies showing that the presence of a cortical cataract is a risk factor for the later formation of a nuclear cataract or PSC. In the case of mixed cataracts, it is possible that the factors that led to the formation of one kind of cataract also contributed to the formation of the second or third type.

In most developed countries cataracts are removed as soon as they become visually disabling. However, if a cataract is not removed, it may progress to a "total," "white" or "Morgagnian" cataract. There is no specific etiology for total cataracts. They are simply the result of the progression of a more localized cataract to the point at which it affects the entire lens.

Secondary cataracts

There are two general strategies for the removal of cataracts. The most common approach taken is to remove a portion of the anterior lens epithelium and capsule exposing the underlying fiber mass, extract the nuclear and cortical fiber cells, often using instruments for "phacoemulsification," and implant a plastic intraocular lens (IOL) in the capsular bag. This method of cataract removal is a form of "extracapsular" cataract extraction. Before intraocular lenses were available, the entire lens was removed to remove the cataract. This is called "intracapsular" cataract extraction.

A frequent complication of extracapsular cataract extraction is the formation of secondary cataracts, also called "after cataracts" or "posterior capsular opacification" (PCO). Removal of the cataractous fiber mass leaves the posterior capsule free of cells. Lens epithelial cells near the lens equator persist after cataract surgery and can migrate beneath the IOL onto the denuded posterior capsule. Epithelial cells that remain close to the equator may differentiate into a mass of fiber-like cells, forming a band called "Soemmering's ring" around the equator.[385] If the epithelial cells migrate further onto the posterior capsule they may differentiate into small "lentoid bodies," also called "Elschnig's pearls," or they may form fibrotic plaques. Both Elschnig's pearls and fibrotic plaques scatter light, degrade the visual image, and result in secondary cataract formation (see Box 5.5).

The cells found in Elschnig's pearls resemble the large, swollen cells seen in posterior subcapsular cataracts.[386] They are likely to result from lens epithelial cells that were stimulated to undergo aberrant fiber cell differentiation. Instead of forming precisely organized, elongated lens fibers, they form small, rounded clumps of cells that act as miniature lenses, thereby degrading the image produced by the implanted intraocular lens.

Fibrotic plaques are the most common cause of PCO. They contain contractile, myofibroblast-like cells in an abundant extracellular matrix, thereby resembling in many ways the fibrosis seen in anterior polar cataracts (see below).[387–390] Myofibroblasts are contractile cells found during wound healing in many parts of the body. They express the contractile protein α-smooth muscle actin and secrete an abundant collagenous extracellular matrix, properties that may be important in wound healing. In addition to the light scattered by the cells and matrix in these plaques, the contractile cells wrinkle the posterior capsule, further contributing to light scattering.

The cytokine TGF-β has been implicated as a stimulant of the capsular fibrosis seen in secondary cataracts.[391,392] Lenses and lens epithelial cells treated with activated TGF-β form plaques of myofibroblast-like cells similar to those seen in secondary cataracts.[393–395] However, there are no published studies showing that TGF-β is required for lens fibrosis and cataract formation.

Other studies have suggested that the migration or spreading of lens epithelial cells on the posterior capsule may, by itself, contribute to myofibroblast differentiation.[396] Since active TGF-β is likely to be found in the eye after cataract surgery and cataract surgery always provides a surface on the posterior capsule for the migration of lens epithelial cells, it is possible that both of these conditions contribute to the formation of myofibroblasts after cataract surgery.

Less common types of cataract

Congenital cataracts are cataracts that are present at birth or that appear soon after birth. They can include early-onset hereditary cataracts or cataracts caused by infectious agents. Most congenital cataracts are total, although some may affect specific regions of the lens. In most cases, the etiology of congenital cataracts is not known, although the presence of a family history of congenital cataracts suggests a genetic component.

One infectious cause of congenital cataracts is rubella infection in early pregnancy.[397] The rubella virus has been isolated from the lenses of affected individuals and experimental infection of early-stage human lens tissue with the rubella virus has been demonstrated. It is likely that the lens is susceptible to infection only early in gestation, prior to the separation of the lens vesicle from the overlying ectoderm. After this stage of lens development, the capsule appears to prevent the entry of the virus into the lens tissue.[398] Immunization against rubella in industrialized countries has reduced the incidence of rubella cataracts and the other types of birth defects associated with this disease. However rubella cataracts are still common in non-immunized populations.

One of the cytological characteristics of lenses infected with the rubella virus is the failure of organelle degradation in lens fiber cells.[399] The mechanism by which the rubella virus prevents organelle loss is not understood. However, several genetic and transgenic mouse models of cataract also show a failure of organelle degradation,[132,400] suggesting that inhibition of organelle loss may be a non-specific side effect of a variety of factors that perturb fiber cell differentiation.

Anterior polar cataract (APC) is another type of congenital or early-onset cataract. Anterior polar cataracts typically involve the formation of an opaque plaque near the center of the lens epithelium. Although they are usually rare, a recent hospital-based study found that prevalence of anterior polar cataracts in Koreans was 6 percent. The mean age in this study was 53 and 87 percent were male.[401] It is unclear whether the early onset, high risk in males and overall high prevalence is due to genetic or environmental effects.

Box 5.5 Posterior capsular opacification (PCO), a possible barrier to the introduction of advanced intraocular lenses (IOLs)

The opacities that sometimes form on the posterior capsule after cataract surgery are caused by the migration of lens epithelial cells across the newly denuded posterior capsule and their differentiation to "myofibroblast-like" cells. The migration and transdifferentiation of these cells is a kind of wound healing response. Opacification is caused by the excessive amount of extracellular matrix secreted by these cells and their tendency to cause wrinkling of the posterior capsule. PCO can be greatly reduced by preventing the migration of the cells onto the capsule by incorporating a square edge at the posterior of the IOL. However, newer types of "accommodating IOLs" do not effectively block the migration of lens epithelial cells and are subject to high rates of PCO. New biological or technical solutions to this problem are needed to permit advances in IOL technology.

Microscopic examination of the anterior plaques in APC shows spindle-shaped lens epithelial cells embedded in a large amount of extracellular matrix, the plaque taking on the appearance of a connective tissue. The fiber cells that lie beneath these plaques are often disrupted, a factor that may contribute to the opacity. The spindle-shaped lens cells in these plaques express several of the proteins characteristic of myofibroblast cells, including α-smooth muscle actin, fibronectin, type I collagen, transforming growth factor-β (TGF-β) and connective tissue growth factor (CTGF).[391,392,402] Experimental studies have shown that exposing lenses to TGF-β in vivo or in vitro can trigger the formation of myofibroblast-like cells and plaques that resemble those seen in anterior polar cataracts,[394,395] although it has not yet been shown that anterior polar cataracts in humans are associated with exposure of the lens to high levels of TGF-β.

Mutations in several lens-specific genes have been implicated in hereditary cataracts in humans. These include genes encoding αA- and αB crystallin,[115,403] members of the β/γ-crystallin superfamily,[404–406] the lens-specific connexins α3 and α8,[407,408] and the lens-specific intermediate filament proteins, phakinin and filensin.[136,137] To date, nearly all mutations identified in these genes have a dominant mode of inheritance.

The association of cataract and mutations in these genes does not necessarily demonstrate that the proteins they encode are essential to the function of the lens. For example, deletion of one copy of the genes encoding either of the two lens-specific connexins in mice does not result in cataract formation.[152,154] However, if one allele of these genes is mutated in humans, cataract can result.[407,408] Similarly, one mutant allele of any of several of the genes encoding members of the β-γ superfamily of crystallins is sufficient to cause cataracts.[400,404,406,409] In these cases, the mutant allele produces a protein that interferes with the function of the normal protein or other proteins in the lens, thereby disrupting their function and leading to cataract formation.

There are numerous hereditary syndromes that include cataracts as one of their characteristic features. The genes responsible for these syndromes are known in several cases, including the ocular-cerebral-renal syndrome of Lowe (a protein similar to inositol polyphosphate 5-phosphatase),[410] neurofibromatosis type 2 (a member of the ERM family of cell membrane-associated proteins),[411] galactokinase deficiency (galactokinase 1),[412] galactosemia (galactose-1-phosphate uridyltransferase),[413] hyperferritinemia (ferritin light chain),[414] Werner syndrome (recQ-related helicase),[415] and myotonic dystrophy (six-5 homeobox gene),[416,417] and several others. For most of these diseases, knowing the gene responsible has not provided an obvious explanation for why cataracts are part of the phenotype. Understanding the molecular and biochemical mechanisms underlying cataract formation in these syndromes may provide information about the mechanisms of age-related cataract formation and the importance of these genes in other tissues that are part of the disease syndrome.

Exceptions to this generalization include galactosemia and galactokinase deficiency, in which galactose accumulates to high levels in the body. High levels of galactose lead to the accumulation of the polyol galactitol in the lens fiber cell cytoplasm. This is due to the activity of the enzyme aldose reductase, which converts galactose to galactitol. Accumulation of galactitol leads to osmotic swelling of lens fiber cells, damage to fiber cell membranes, and cataract formation. The possible role of aldose reductase in the formation of cataracts seen in diabetics will be discussed further below.

Exposure of the eye to several kinds of electromagnetic radiation can lead to cataract formation. The best studied of these are cataracts caused by ionizing radiation.[418] In experimental animals, x- or γ-irradiation causes characteristic changes in the cells of the lens, leading to posterior subcapsular and cortical cataracts. The initial insult in the cataractogenic process may be damage to the proliferating cells in the germinative zone of the lens epithelium, which leads to extensive cell death in this region.[380,382,419–421] Exposure of only the central epithelial cells and mature fibers to very high levels of x-irradiation does not lead to cataract formation, as long as the germinative zone is protected from irradiation.[422] Furthermore, animals in which cell division in the lens can be stopped by environmental or experimental means are resistant to radiation cataracts.[423,424] Cell death in the germinative zone is followed by a delay in cell division and a wave of compensatory mitosis. When the epithelial cells that resulted from this increased proliferation begin to differentiate, the usually precise organization of the fiber cells is disrupted.[421] As they elongate, these fiber cells become distorted and may appear swollen, although their average cell volume is unchanged.[382] The appearance of fiber cell swelling may be due to the presence of distended and attenuated regions along the length of a fiber, instead of the uniform geometry of normal fiber cells. The nuclei of these elongating fibers, instead of remaining in a compact arc (the "lens bow"), move posteriorly. The posterior movement of the nuclei probably reflects the flow of the entire cytoplasm of these cells. Eventually, these abnormal fibers form "balloon cells" or "Wedl cells" at the posterior pole, resulting in the formation of a posterior subcapsular cataract. During the later stages of this process there is evidence of increased membrane permeability in the affected lens fibers. Glutathione levels decrease, the potassium concentration in the cytoplasm declines, the sodium concentration increases, and the synthesis of proteins slows.[425,426]

Exposure to x- or γ-radiation is a risk factor for cortical and PSC formation in humans. Radiologists routinely seek to minimize the exposure of the lens to ionizing radiation. In cases in which this is not possible or advisable, cataracts often follow and must be dealt with surgically.

Non-ionizing radiation can also contribute to cataract formation. The effects of UV light on the eye have been most intensively studied. Epidemiological studies have linked high lifetime exposure to UV light with the formation of cortical cataracts in humans.[359,427–430] The effects of UV light have been most clearly demonstrated in individuals who are exposed to high levels of UV as a consequence of their occupation.[427,431] The risk of developing cortical cataracts is only slightly increased in individuals who receive higher than normal levels of UV, but whose occupation does not routinely expose them to high levels of ultraviolet light (see Box 5.6).[429] The same studies have suggested that the shorter wavelengths of UV light reaching the eye, UV-B, and not the longer, less energetic UV-A, is likely to be responsible for increased risk of cortical cataract.[431] However, studies in

Box 5.6 Sunlight exposure and cataract; not as important as often believed

Sunlight exposure is widely thought to be a major contributor to cataract. Careful epidemiological studies have demonstrated increased risk of cataracts in individuals with higher lifetime sunlight exposure. However, most studies have found that only cortical cataracts are associated with increased exposure to sunlight, not nuclear or posterior subcapsular opacities. In a typical population, the increased risk of cortical cataracts attributable to sunlight is only about 10%. That added risk can be avoided by relatively simple means: wearing plastic glasses or a brimmed hat. Many other variables, like sex or a family history of cataracts, are more significant risk factors for cortical cataract than sunlight. The mechanism by which sunlight increases the risk of cortical cataracts is not understood. For example, the region of the lens most often affected in cortical cataracts is the part that is best protected from direct exposure to sunlight. Therefore, while exposure to excess sunlight is an avoidable risk factor, avoiding sunlight protects against only a small part of the total risk of age-related cataracts.

diurnal animals exposed to UV-A suggest that these longer wavelengths can also contribute to cataract formation (see below).

The mechanism by which UV exposure leads to cataract formation in humans has not been identified. Studies in experimental animals, on lenses cultured in vitro, and on isolated lens proteins, have been used to examine the potential damaging effects of UV light on the lens or lens components. While these studies have suggested that UV-generated free radicals can damage components of the lens in a similar manner to that seen in cataract formation, these studies often do not fully take into account the biology of the human lens. For example, most studies of the effects of UV light on the lens in experimental animals have been performed in rodents or rabbits. These are nocturnal animals that are not adapted to high levels of light exposure. Many potential protective mechanisms found in the human eye, like the high ascorbic acid content of the aqueous humor, are not present in these species. Furthermore, treatment of the isolated lenses or lens proteins with UV light has usually been performed at ambient oxygen concentrations (21 percent), not the low levels of oxygen that are typically found around the lens in vivo (less than 2 percent).[221] Experiments performed under these conditions may yield results that are substantially different from what is found as the result of light exposure in the eye. Finally, UV exposure in experimental animals is often acute, employing high levels of UV light. This differs significantly from the chronic, lower level of exposure over a lifetime that is typical in humans.

In a few cases, the effects of UV light on the lens were tested in diurnal species chronically exposed to more physiological levels of light over a substantial proportion of their lifetime. In contrast to epidemiological studies of human populations, these experiments indicated that UV-A light might also be a risk for cataract formation.[432] UV-A light can be absorbed by several chromophores into the lens, a process that may lead to the generation of free radicals.[268,433] These studies suggest that protection against free radical generation or the effects of free radicals might help reduce the incidence

of cataracts. They also suggested that the role of UV-A light in human cataract formation is worthy of further study.

One of the paradoxical findings concerning the role of UV light in cataract formation is that the cells that are affected are those that are best protected from light exposure. As indicated above, epidemiological studies have associated UV light with the formation of cortical, not nuclear cataracts. Cortical cataracts form in superficial fiber cells. Superficial fiber cells and the peripheral epithelial cells that give rise to them, because they lie behind the densely pigmented iris epithelium, are better protected from light exposure than any other part of the lens. This is especially true in bright light when the iris is maximally constricted and the size of the pupil is small. Conversely, epidemiological and experimental studies indicate that the fiber cells in the center of the lens, those directly exposed to light entering the eye through the pupil, are not damaged by exposure to light.[234,427,429]

Studies using model eyes have suggested that light entering the eye from the temporal side of the head may be focused by the curvature of the cornea onto the cortical fiber cells on the opposite (nasal) side of the lens.[358,434] While this hypothesis might provide an explanation for the higher frequency of cortical cataract formation on the naso-inferior side of the lens, it would not appear to account for the higher frequency of cortical cataracts in the remainder of the inferior half of the lens.[435] Studies are needed to explain the selective effects of light exposure on cortical fiber cells and to better understand the proposed role of ocular sunlight exposure in cortical cataract formation. As discussed above, it seems possible that sunlight exposure may increase the hardness of the lens nucleus, indirectly promoting cortical cataracts by focusing the stress of accommodation in the cortex. Modeling studies and non-invasive measures of lens hardness may be helpful in testing this possibility.[369,436,437]

Long-term exposure to infrared light and focused, high-energy microwaves can also cause cataracts. Evidence that infrared light can cause cataracts comes mostly from epidemiological studies in individuals exposed as a result of their occupation.[438,439] This type of cataract is also called "glass-blowers cataract." High-energy microwaves appear to cause cataracts by directly damaging lens fiber cell membranes.[440–442] It seems unlikely that either kind of cataract will provide significant insight into the etiology of more common types of cataracts, although lifetime exposure to increased ambient temperature has been suggested as a risk factor for nuclear cataracts.[443]

Diabetics are at increased risk for early-onset cataracts and the lens has provided a useful system with which to study diabetic complications. Like other organ systems affected in diabetes, damage to the lens seems to be due entirely to high levels of glucose. It is the mechanism by which elevated glucose leads to cellular damage that remains as the central question. Investigators studying diabetic cataracts have been leaders in illustrating the potential importance of the enzyme aldose reductase in diabetic complications.[444,445] Aldose reductase catalyzes the reduction of a wide variety of aldehydes, including glucose and galactose. These sugars are relatively poor substrates for the enzyme, but their metabolism by aldose reductase may become significant when concentrations exceed normal levels. Species that have high levels of aldose reductase in the lens seem to be particularly susceptible to diabetic cataract formation.[446] Inhibitors of aldose reductase protect these animals against diabetic cataracts.[444,446,447] Mice, which have low levels of aldose reductase in the lens, are not susceptible to diabetic cataracts. However, when aldose reductase is over expressed in the lenses of transgenic mice, these animals become very susceptible to hexose-induced cataract formation.[448] These studies convincingly show that a high level of aldose reductase in the lens is sufficient to cause diabetic cataracts in this species. Humans have relatively low levels of aldose reductase in the lens, a situation that has led to controversy about the relative role of aldose reductase in human diabetic cataracts.

It has been suggested that aldose reductase causes cataracts in diabetics because of the accumulation of the polyol sorbitol in lens fiber cells. Sorbitol accumulation could lead to osmotic damage. However, since aldose reductase uses reducing equivalents from NADH to produce polyols, it is possible the high glucose flux through the polyol pathway could also create oxidative stress in lens fiber cells.

There are three hypotheses that are currently favored to explain diabetic complications in affected organs. Increased flux through the polyol pathway mediated by aldose reductase is one of these. The second mechanism is the glucose-mediated activation of a specific isoform of protein kinase C.[449] A third mechanism is the generation of increased amounts of advanced glycation end products (AGEs).[450] AGEs are produced by the non-enzymatic reaction of aldehydes, like glucose, with a wide variety of chemical species. Diabetics and animals with experimental diabetes accumulate increased levels of AGEs in connective tissues and within cells. Treatments that block each of these three biochemical pathways have been shown to reduce or prevent diabetic complications in one or more experimental systems. A recent study suggested that the three major pathways thought to be responsible for diabetic complications might be explained by a single mechanism – the effect of high glucose on mitochondrial oxidative phosphorylation.[451]

In spite of the long association between diabetes and cataracts, more recent studies suggest that patients with reasonable control of blood glucose are not at increased risk for age-related cataract and that the types of cataracts that occur in these patients are similar to age-matched non-diabetics in the same population.[452] More effective management of this disease may reduce its impact on cataract prevalence and other systemic complications.

Overview of age-related cataract formation

As described above, excessive oxidative damage to lens crystallins and fiber cell membranes remains an attractive explanation for nuclear cataract formation. Increased oxidative damage in the lens nucleus must result from an imbalance between the factors that protect lens proteins from oxidative damage and increases in oxygen and free radicals that may occur with age. The age-related decrease in the diffusion of glutathione to the center of the lens is likely to be of central importance in this balance,[209,240] while degeneration of the vitreous body is an age-related change that appears to increase the delivery of oxidants to the lens.[453]

Cortical cataracts are associated with gross disorganization and disruption of fiber cells. This makes it even more difficult to identify the initiating insult for this type of cataract, since the wreckage that ensues obscures the early damage. While many possible inciting factors have been suggested, a comprehensive hypothesis that explains the subcellular and anatomical location of the damage in cortical cataracts (initial involvement of only the center of affected fiber cells, preferential location in inferior half of the lens) is lacking. While epidemiological studies suggest that light exposure contributes to cortical cataract formation, there is no evidence to indicate whether the damage to the lens results from the interaction of light with the affected cells or with other parts of the eye. For example, light could cause the release of substances from the iris that contribute to cortical cataract formation,[454] a possibility that seems consistent with the increase in cataract in persons with dark iris color.[304,316,317] It seems increasingly likely that physical forces acting on the aging lens, like those generated during accommodation, could rupture fiber cell membranes and initiate the formation of cortical cataracts (see Box 5.6).

Perspectives for preventing cataract blindness

One of the most unusual aspects of the lens is its continued growth throughout life. Since fiber cells are not replaced once they differentiate and epithelial cells continue to divide to produce new lens fibers, the number of fiber cells in the lens increases steadily with age, approximately doubling between birth and age 90. Although the rate of addition of new fiber cells is slow in postnatal life, most of the cells in the outer part of the lens, including the fiber cells that are involved in cortical cataract formation, are produced after childhood.

There is strong epidemiologic evidence supporting the possibility that the rate of growth of the lens plays a role in cataract formation. Individuals with larger lenses are more likely to have nuclear cataracts and those with smaller lenses are much more likely to have cortical cataracts.[189] Individuals with clear lenses at initial observation were more likely to develop nuclear or cortical cataracts over 5 years if they initially had larger or smaller lenses, respectively.[190] Could nuclear cataract formation be prevented or slowed by reducing the rate of lens cell proliferation? Recent studies support this view, since the rate of lens growth can be accelerated by supplying the adult lens with more oxygen and increased lens size and exposure to oxygen are likely to contribute to the formation of nuclear cataracts.[187,190,222,224,455] Would slowing the growth of larger lenses delay the onset of presbyopia? Why are smaller lenses at increased risk of cortical cataracts? These questions could be approached if there were an experimental means to slow or stop the formation of new lens fiber cells.

It is presently not possible to effectively slow lens growth in most mammals. However, in amphibians, removing the pituitary gland stops lens cell division. This approach was used to show that x-ray-induced cataracts could be avoided if cell division were halted, even if the pituitary gland were removed after the lens had been irradiated.[384,423,456] Restoration of lens cell growth by hormone supplementation led to cataract formation if the lenses had been irradiated prior to hypophysectomy.[384,457] These studies provide tantalizing support for the possibility that reducing the rate of lens growth could slow or prevent some kinds of cataract formation.

Another area that is likely to provide important information about cataract formation is the study of the genetics of human cataracts. Some of the genes responsible for congenital cataracts were described in a previous section. It is not surprising that the known genes that cause cataracts without affecting other parts of the body encode proteins that are expressed selectively and at high levels in the lens. Once a hereditary cataract is mapped close to one of these "lens-specific" genes it is relatively easy to connect the disease to a change in the sequence of that gene. However, several loci responsible for hereditary cataracts do not map close to the genes encoding known lens-specific proteins.[458–460] Identification of these genes and study of their function in the lens may give important clues about the factors required to maintain lens transparency.

Of even greater potential significance will be the identification of genes that predispose the lens to cataracts in middle or later life. Some of these adult-onset hereditary cataracts are likely to be caused by mutations in genes that protect the lens from cataract-causing damage over a lifetime. Studies of twins have suggested that heredity accounts for about 50 percent of the factors responsible for age-related cortical cataracts and 35 percent of the risk of nuclear cataracts.[461,462] Other statistical analyses of epidemiological data support these conclusions and suggest that a relatively small number of genes determine the probability of developing nuclear and cortical cataracts.[463,464] One recent study found a mutation in the cell-recognition receptor, EPHA2, to be associated with a rare, recessive form of juvenile cataracts. Further analysis showed that single nucleotide polymorphisms in the EPHA2 gene were significantly associated with age-related cortical cataracts, suggesting that variants of this protein may predispose individuals to cortical cataract formation.[465] Interestingly, deletion of the ephrin-A5, the ligand for Epha2 in the mouse, causes disruption of lens fiber cell organization.[466] These studies are among the first to identify candidate genes that may increase susceptibility to age-related cataracts.

Since degeneration of the vitreous body is a risk factor for nuclear cataracts,[453] it seems reasonable to broaden attention from lens-specific genes to the genetic variations that are responsible for maintaining the normal environment around the lens in older individuals. If one were to identify genes that alter stability of the vitreous gel, for instance, it might be possible to develop therapies to delay cataractogenesis by altering the pathways in which these genes work.

Cataract blindness is a serious and growing problem in the countries of the developing world. The early onset and higher incidence of cataracts in these populations suggests that basic nutritional, environmental, and hygienic factors contribute to cataract formation. Elimination of these predisposing conditions could preserve the vision of millions of individuals who would otherwise be deprived of their sight and productivity for a substantial proportion of their lifetime. The World Health Organization, through its program "Vision 2020: The Right to Sight," has developed plans to eliminate preventable blindness by the year 2020. Cataract blindness will be addressed in this program by providing increased access to affordable cataract surgery.[467] Although greatly increased numbers of cataract surgeries will be essential to the goals of this project, concomitant improvement of basic living conditions may reduce the rate of cataract formation, and along with it, a host of other diseases.

There are numerous epidemiological studies suggesting that nutrition plays an important role in cataract formation (see review[468]). There have been several interventional studies to test whether supplementation with specific nutrients can protect against cataract formation or progression. The AREDS study, a large-scale clinical trial of the effects of selected vitamin and nutrient supplementation on the progression of eye diseases, found that supplementation with known antioxidant vitamins did not provide measurable protection against cataract.[469] However, an interesting follow-up paper revealed that the multivitamin supplement offered to about two-thirds of the participants in this trial was associated with protection against "all cataracts" and, in subgroup analysis, nuclear cataracts.[470] This result is similar to that obtained in the Linxian Cataract Study, which provided supplements to a nutritionally deprived population in China.[471] In this case, supplementation was associated with a substantially decreased risk of nuclear, but not other types of cataract. However, supplementation with beta-carotene and vitamins C and E in a nutritionally deprived population in India demonstrated no protection against cataract.[472] Nuclear opacities were less common in a large group of older subjects receiving a multivitamin/mineral supplement in Italy.[473] However, this benefit was accompanied by increased incidence of posterior subcapsular opacities.

These and other clinical trials caution that consumption of higher levels of selected nutrients are likely to have marginal benefit and may even have adverse consequences.[474] Studying the importance of individual components of the diet on cataract incidence and progression may even obscure the larger picture. The message that seems to underlie most studies of the relationship between nutrition and cataract is one that is echoed in many studies of nutrition and disease; a moderate diet with abundant fruits and vegetables is likely to provide an effective foundation for good visual health and good health in general.

Secondary cataracts, the opacities that appear on the posterior capsule following extracapsular cataract surgery and IOL implantation, remain an important clinical problem. These opacities are routinely treated by ablating the posterior capsule with a laser – a procedure that is costly, depends on advanced technology, and is associated with increased risk of serious complications.[475] The frequency with which secondary cataracts develop depends on the age of the patient and the type of IOL that is implanted. Children have a high incidence of secondary cataract formation, a fact that complicates and limits the options for treating congenital or traumatic cataracts in children. Designs that limit the ability of lens epithelial cells to migrate behind the IOL can prevent this complication, but may require more complicated surgical procedures.[476] In older patients, the composition and design of the IOL can significantly influence the need for laser surgery after cataract extraction.[477–479] However, changes in IOL design to achieve other goals (smaller incision size, accommodative IOLs, for example) may increase the frequency of secondary cataracts. Further understanding of the biologic factors responsible for secondary cataract and its prevention could simplify pediatric cataract surgery and reduce the need to treat secondary cataracts in adults.

New technologies, such as accommodating IOLs, are likely to provide valuable advances in cataract surgery. Such improvements could increase the demand for replacement of the lens even before cataract formation. Although cataract surgery is now a relatively simple surgery, new technologies are likely to further simplify and speed this common procedure. Drugs to prevent or delay cataract formation have not been widely sought by the pharmaceutical industry. This may be due to the relatively low cost of cataract surgery coupled with the cost and potential risk of anti-cataract medications that may need to be taken for many years. Better understanding of the etiology of cataracts may point to therapies that are effective, minimally invasive, and inexpensive.

Cataract has been thought to be inevitable if one is fortunate enough to live a long life. However, better understanding of the cell biology, genetics, biophysics and physiology of the lens is beginning to suggest strategies to prevent or significantly delay this most common disease.

Acknowledgments

Many thanks to the colleagues, students, residents and fellows whose stimulating discussions and healthy skepticism helped me to formulate some of the ideas expressed in this chapter. I especially thank Drs. Steven Bassnett, Ying-Bo Shui, Nancy Holekamp, George Harocopos, Toshiyuki Nagamoto, Carmelann Zintz, Joram Piatigorsky, and Leo Chylack. Cheryl Armbrecht provided expert assistance with graphics and Dr. Jer Kuszak generously contributed figures. Support for preparing this manuscript was derived from Research to Prevent Blindness, NIH Grants EY04853, EY07528 and EY015863 and Core Grant EY02687 from the National Eye Institute to the Department of Ophthalmology and Visual Sciences at Washington University.

References

1. Rafferty NS, Rafferty KA. Cell population kinetics of the mouse lens epithelium. J Cell Physiol 1981; 107:309–315.
2. Kuwabara T. The maturation of the lens cell: a morphologic study. Exp Eye Res 1975; 20:427–443.
3. Bassnett S, Beebe DC. Coincident loss of mitochondria and nuclei during lens fiber cell differentiation. Dev Dyn 1992; 194(2):85–93.
4. Bassnett S. Mitochondrial dynamics in differentiating fiber cells of the mammalian lens. Curr Eye Res 1992; 11(12):1227–1232.
5. Bassnett S, Mataic D. Chromatin degradation in differentiating fiber cells of the eye lens. J Cell Biol 1997; 137(1):37–49.
6. Scammon RE, Hesdorfer MB. Growth in mass and volume of the human lens in postnatal life. Arch Ophthalmol 1937; 17:104–112.
7. Shestopalov VI, Bassnett S. Exogenous gene expression and protein targeting in lens fiber cells. Invest Ophthalmol Vis Sci 1999; 40(7):1435–1443.
8. Wheatley HM et al. Immunohistochemical localization of fibrillin in human ocular tissues. Relevance to the Marfan syndrome. Arch Ophthalmol 1995: 113(1):103–109.
9. Mecham RP et al. Development of immunoreagents to ciliary zonules that react with protein components of elastic fiber microfibrils and with elastin-producing cells. Biochem Biophys Res Commun 1988; 151(2):822–826.
10. Fagerholm PP, Philipson BT, Lindstrom B. Normal human lens – the distribution of protein. Exp Eye Res, 1981; 33(6):615–620.
11. Veretout F, Delaye M, Tardieu A. Molecular basis of eye lens transparency. Osmotic pressure and X-ray analysis of alpha-crystallin solutions. J Mol Biol 1989; 205(4):713–728.
12. Delaye M, Tardieu A. Short–range order of crystallin proteins accounts for eye lens transparency. Nature 1983; 302(5907):415–417.
13. Van Heyningen R. Fluorescent glucoside in the human lens. Nature 1971; 230(5293):393–394.
14. Hood BD, Garner B, Truscott RJ. Human lens coloration and aging. Evidence for crystallin modification by the major ultraviolet filter, 3-hydroxy-kynurenine O-beta-D- glucoside. J Biol Chem 1999; 274(46):32547–32550.
15. Henry J, Grainger R. Early tissue interactions leading to embryonic lens formation. Dev Biol 1990; 141(1):149–163.
16. Li HS et al. Pax-6 is first expressed in a region of ectoderm anterior to the early neural plate: implications for stepwise determination of the lens. Dev Biol 1994; 162(1):181–194.
17. Walther C, Gruss P. Pax-6, a murine paired box gene, is expressed in the developing CNS. Development 1991; 113(4):435–449.
18. Hill RE et al. Mouse Small eye results from mutations in a paired-like homeobox-containing gene. Nature 1992; 355(6362):750.
19. Xu PX et al. Mouse Eya homologues of the Drosophila eyes absent gene require Pax6 for expression in lens and nasal placode. Development 1997; 124(1):219–231.
20. Wawersik S et al. BMP7 acts in murine lens placode development. Dev Biol 1999; 207(1):176–188.
21. Ogino H, Yasuda K. Sequential activation of transcription factors in lens induction. Dev Growth Differ 2000; 42(5):437–448.
22. Yamada R et al. Cell-autonomous involvement of Mab21l1 is essential for lens placode development. Development 2003; 130(9):1759–1770.
23. Cvekl A, Duncan MK. Genetic and epigenetic mechanisms of gene regulation during lens development. Prog Retin Eye Res 2007; 26(6):555–597.
24. Brownell I, Dirksen M, Jamrich M. Forkhead Foxe3 maps to the dysgenetic lens locus and is critical in lens development and differentiation. Genesis 2000; 27(2):81–93.
25. Chauhan BK et al. Identification of genes downstream of Pax6 in the mouse lens using cDNA microarrays. J Biol Chem 2002; 277(13):11539–11548.
26. Jena N et al. BMP7 null mutation in mice: developmental defects in skeleton, kidney, and eye. Exp Cell Res 1997; 230(1):28–37.
27. Furuta Y, Hogan BLM. BMP4 is essential for lens induction in the mouse embryo. Genes Dev 1998; 12(23):3764–3775.
28. Hendrix RW, Zwaan J. The matrix of the optic vesicle–presumptive lens interface during induction of the lens in the chicken embryo. J Embryol Exp Morphol 1975; 33(4):1023–1049.
29. Hendrix RW, Zwaan J. Cell shape regulation and cell cycle in embryonic lens cells. Nature 1974; 247(437):145–147.
30. Schook P. A review of data on cell actions and cell interaction during the morphogenesis of the embryonic eye. Acta Morphol Neerl Scand 1978; 16(4):267–286.
31. Garcia-Porrero JA, Colvee E, Ojeda JL. The mechanisms of cell death and phagocytosis in the early chick lens morphogenesis: a scanning electron microscopy and cytochemical approach. Anat Rec 1984; 208(1):123–136.
32. Koroma BM, Yang JM, Sundin OH. The Pax-6 homeobox gene is expressed throughout the corneal and conjunctival epithelia. Invest Ophthalmol Vis Sci 1997; 38(1):108–120.
33. Parmigiani C, McAvoy J. Localisation of laminin and fibronectin during rat lens morphogenesis. Differentiation 1984; 28(1):53–61.
34. Smith GN Jr., Linsenmayer TF, Newsome DA. Synthesis of type II collagen in vitro by embryonic chick neural retina tissue. Proc Natl Acad Sci USA 1976; 73(12):4420–4423.
35. Silver PH, Wakely J The initial stage in the development of the lens capsule in chick and mouse embryos. Exp Eye Res 1974; 19(1):73–77.
36. Parmigiani CM, McAvoy JW. The roles of laminin and fibronectin in the development of the lens capsule. Curr Eye Res 1991; 10(6):501–511.
37. Danysh BP, Duncan MK. The lens capsule. Exp Eye Res 2010 In Press.
38. Coulombre JL, Coulombre AJ. Lens development: fiber elongation and lens orientation. Science 1963; 142:1489–1490.
39. Coulombre JL, Coulombre AJ. Lens development. IV Size, shape, and orientation. Invest Ophthalmol 1969; 8(3):251–257.
40. Beebe DC, Feagans DE, Jebens HA. Lentropin: a factor in vitreous humor which promotes lens fiber cell differentiation. Proc Natl Acad Sci USA 1980; 77(1):490–493.
41. Schulz MW et al. Acidic and basic FGF in ocular media and lens: implications for lens polarity and growth patterns. Development 1993; 118(1):117–126.
42. Ogino H, Yasuda K. Induction of lens differentiation by activation of a bZIP transcription factor, L-Maf. Science 1998; 280(5360):115–118.
43. Blixt A et al. A forkhead gene, FoxE3, is essential for lens epithelial proliferation and closure of the lens vesicle. Genes Dev 2000; 14(2):245–254.
44. Blixt A et al. Foxe3 is required for morphogenesis and differentiation of the anterior segment of the eye and is sensitive to Pax6 gene dosage. Dev Biol 2007; 302(1):218–229.
45. Pontoriero GF et al. Cell autonomous roles for AP-2alpha in lens vesicle separation and maintenance of the lens epithelial cell phenotype. Dev Dynam 2008; 237(3):602–617.
46. Nishiguchi S et al. Sox1 directly regulates the gamma-crystallin genes and is essential for lens development in mice. Genes Dev 1998; 12(6):776–781.
47. Ring BZ et al. Regulation of mouse lens fiber cell development and differentiation by the Maf gene. Development 2000; 127(2):307–317.
48. Kawauchi S et al. Regulation of lens fiber cell differentiation by transcription factor c--Maf. J Biol Chem 1999; 274(27):19254–19260.
49. Kim JI et al. Requirement for the c-Maf transcription factor in crystallin gene regulation and lens development. Proc Natl Acad Sci USA 1999; 96(7):3781–3785.
50. Varnum DS, Stevens LC. Aphakia, a new mutation in the mouse. J Hered 1968; 59(2):147–150.
51. Semina EV et al. Deletion in the promoter region and altered expression of Pitx3 homeobox gene in aphakia mice. Hum Mol Genet 2000; 9(11):1575–1585.
52. Semina EV, Reiter RS, Murray JC. Isolation of a new homeobox gene belonging to the Pitx/Rieg family: expression during lens development and mapping to the aphakia region on mouse chromosome 19. Hum Mol Genet 1997; 6(12):2109–2116.
53. Semina EV et al. A novel homeobox gene PITX3 is mutated in families with autosomal- dominant cataracts and ASMD. Nat Genet 1998; 19(2):167–170.
54. Rieger DK et al. A double-deletion mutation in the Pitx3 gene causes arrested lens development in aphakia mice. Genomics 2001; 72(1):61–72.
55. Yamamoto Y. Growth or lens and ocular environment: role of the neural retina in the growth of mouse lens as revealed by an implantation experiment. Dev Growth Diff 1976; 18:273–278.
56. Zwaan J, Kenyon RE. Cell replication and terminal differentiation in the embryonic chicken lens: normal and forced initiation of lens fibre formation. J Embryol Exp Morph 1984; 84:331–349.
57. Lovicu FJ, McAvoy JW. Spatial and temporal expression of p57(KIP2) during murine lens development. Mech Dev 1999; 86(1–2):165–169.

58. Gao CY et al. Changes in cyclin dependent kinase expression and activity accompanying lens fiber cell differentiation. Exp Eye Res 1999; 69(6):695–703.

59. Morgenbesser SD et al. p53-dependent apoptosis produced by Rb-deficiency in the developing mouse lens. Nature 1994; 371(6492):72–74.

60. Fromm L, Overbeek PA. Regulation of cyclin and cyclin–dependent kinase gene expression during lens differentiation requires the retinoblastoma protein. Oncogene 1996; 12(1):69–75.

61. Zhang P et al. Cooperation between the Cdk inhibitors p27(KIP1) and p57(KIP2) in the control of tissue growth and development. Genes Dev 1998; 12(20):3162–3167.

62. Wigle JT et al. Prox1 function is crucial for mouse lens-fibre elongation. Nat Genet 1999; 21(3):318–322.

63. Duncan MK et al. Prox1 is differentially localized during lens development. Mech Dev 2002; 112(1–2):195–198.

64. Piatigorsky J. Lens cell elongation in vitro and microtubules. Annals NY Acad Sci 1975; 253:333–347.

65. Piatigorsky J, Rothschild SS, Wollberg M. Stimulation by insulin of cell elongation and microtubule assembly in embryonic chick-lens epithelia. Proc Natl Acad Sci USA 1973; 70(4):1195–1198.

66. Beebe DC et al. Lens epithelial cell elongation in the absence of microtubules: evidence for a new effect of colchicine. Science 1979; 206(4420):836–838.

67. Beebe DC et al. The mechanism of cell elongation during lens fiber cell differentiation. Dev Biol 1982; 92(1):54–59.

68. Parmelee JT, Beebe DC. Decreased membrane permeability to potassium is responsible for the cell volume increase that drives lens fiber cell elongation. J Cell Physiol 1988; 134(3):491–496.

69. Milstone LM, Piatigorsky J. Rates of protein synthesis in explanted embryonic chick lens epithelia: differential stimulation of crystallin synthesis. Dev Biol 1975; 43:91–100.

70. Piatigorsky J, Webster H, Wollberg M. Cell elongation in the cultured embryonic chick lens epithelium with and without protein synthesis. J Cell Biol 1972: 55:82–92.

71. Bassnett S. Three-dimensional reconstruction of cells in the living lens: The relationship between cell length and volume. Exp Eye Res 2005; 81(6):716–723.

72. Beebe DC et al. Lentropin, a protein that controls lens fiber formation, is related functionally and immunologically to the insulin-like growth factors. Proc Natl Acad Sci USA 1987; 84(8):2327–2330.

73. Chamberlain CG, McAvoy JW. Evidence that fibroblast growth factor promotes lens fibre differentiation. Curr Eye Res 1987; 6(9):1165–1169.

74. McAvoy JW, Chamberlain CG. Fibroblast growth factor (FGF) induces different responses in lens epithelial cells depending on its concentration. Development 1989; 107(2):221–228.

75. Liu J, Chamberlain CG, McAvoy JW. IGF enhancement of FGF-induced fibre differentiation and DNA synthesis in lens explants. Exp Eye Res 1996; 63(6):621–629.

76. Chamberlain C, McAvoy J, Richardson N. The effects of insulin and basic fibroblast growth factor on fibre differentiation in rat lens epithelial explants. Growth Factors 1991; 4(3):183–188.

77. Leenders WP et al. Synergism between temporally distinct growth factors: bFGF, insulin and lens cell differentiation. Mech Dev 1997; 67(2):193–201.

78. Robinson M et al. Extracellular FGF01 acts as a lens differentiation factor in transgenic mice. Development 1995; 121(2):505–514.

79. Robinson ML et al. Disregulation of ocular morphogenesis by lens-specific expression of FGF-3/int-2 in transgenic mice. Dev Biol 1998; 198(1):13–31.

80. Chow RL et al. FGF suppresses apoptosis and induces differentiation of fibre cells in the mouse lens. Development 1995; 121(12):4383–4393.

81. Stolen CM, Griep AE. Disruption of lens fiber cell differentiation and survival at multiple stages by region-specific expression of truncated FGF receptors. Dev Biol 2000; 217(2):205–220.

82. Lang RA. Which factors stimulate lens fiber cell differentiation in vivo? Invest Ophthalmol Vis Sci 1999; 40(13):3075–3078.

83. Le AC, Musil LS. FGF signaling in chick lens development. Dev Biol 2001; 233(2):394–411.

84. Zhao H et al. Fibroblast growth factor receptor signaling is essential for lens fiber cell differentiation. Dev Biol 2008.

85. Beebe D et al. Contributions by members of the TGFbeta superfamily to lens development. Int J Dev Biol 2004; 48(8–9):845–856.

86. Rajagopal R et al. The functions of the type I BMP receptor, Acvr1 (Alk2), in lens development: cell proliferation, terminal differentiation and survival. Invest Ophthalmol Vis Sci 2008

87. Boswell BA, Lein PJ, Musil LS. Cross-talk between fibroblast growth factor and bone morphogenetic proteins regulates gap junction-mediated intercellular communication in lens cells. Mol Biol Cell 2008; 19(6):2631–2641.

88. Wistow G, Piatigorsky J. Recruitment of enzymes as lens structural proteins. Science 1987; 236(4808):1554–1556.

89. Piatigorsky J, Wistow GJ. Enzyme/crystallins: gene sharing as an evolutionary strategy. Cell 1989; 57(2):197–199.

90. Rao PV et al. Betaine-homocysteine methyltransferase is a developmentally regulated enzyme crystallin in rhesus monkey lens. J Biol Chem 1998; 273(46):30669–30674.

91. McDermott JB, Cvekl A, Piatigorsky J. A complex enhancer of the chicken beta A3/A1–crystallin gene depends on an AP–1–CRE element for activity. Invest Ophthalmol Vis Sci 1997; 38(5):951–959.

92. Cvekl A, Piatigorsky J. Lens development and crystallin gene expression: many roles for Pax-6. Bioessays 1996; 18(8):621–630.

93. Kodama R, Eguchi G. Gene regulation and differentiation in vertebrate ocular tissues. Curr Opin Genet Dev 1994; 4(5):703–708.

94. Piatigorsky J. Puzzle of crystallin diversity in eye lenses. Dev Dyn 1993; 196(4):267–272.

95. Piatigorsky J. Multifunctional lens crystallins and corneal enzymes. More than meets the eye. Ann NY Acad Sci 1998; 842:7–15.

96. Cvekl A et al. A complex array of positive and negative elements regulates the chicken alpha A-crystallin gene: involvement of Pax-6, USF, CREB and/or CREM, and AP-1 proteins. Mol Cell Biol 1994; 14(11):7363–7376.

97. Klemenz R et al. Alpha B-crystallin is a small heat shock protein. Proc Natl Acad Sci USA 1991; 88(9):3652–3656.

98. Ingolia TD, Craig EA. Four small Drosophila heat shock proteins are related to each other and to mammalian alpha-crystallin. Proc Natl Acad Sci USA 1982; 79(7):2360–2364.

99. Piatigorsky J. Molecular biology: recent studies on enzyme/crystallins and alpha-crystallin gene expression. Exp Eye Res 1990; 50(6):725–728.

100. Horwitz J. Alpha-crystallin can function as a molecular chaperone. Proc Natl Acad Sci USA 1992; 89(21):10449–10453.

101. Brady JP et al. Targeted disruption of the mouse alpha A-crystallin gene induces cataract and cytoplasmic inclusion bodies containing the small heat shock protein alpha B-crystallin. Proc Natl Acad Sci USA 1997; 94(3):884–889.

102. Kantorow, M and J Piatigorsky, Alpha-crystallin/small heat shock protein has autokinase activity. Proc Natl Acad Sci USA 1994; 91(8):3112–3116.

103. Chiesa R et al. Definition and comparison of the phosphorylation sites of the A and B chains of bovine alpha-crystallin. Exp Eye Res 1988; 46(2):199–208.

104. Spector A et al. cAMP-dependent phosphorylation of bovine lens alpha-crystallin. Proc Natl Acad Sci USA 1985; 82(14):4712–4716.

105. Wang K, Gawinowicz MA, Spector A. The effect of stress on the pattern of phosphorylation of alphaA and alphaB crystallin in the rat lens. Exp Eye Res 2000; 71(4):385–393.

106. Aquilina JA et al. Phosphorylation of αB-crystallin alters chaperone function through loss of dimeric substructure. J Biol Chem 2004; 279(27):28675–28680.

107. den Engelsman J et al. Nuclear import of αB-crystallin is phosphorylation-dependent and hampered by hyperphosphorylation of the myopathy-related mutant R120G. J Biol Chem 2005; 280(44):37139–37148.

108. Horwitz J. The function of alpha-crystallin in vision. Semin Cell Dev Biol 2000; 11(1):53–60.

109. Bova MP et al. Subunit exchange of alpha-crystallin. J Biol Chem 1997; 272(47):29511–29517.

110. Andley UP et al. The molecular chaperone alpha-crystallin enhances lens epithelial cell growth and resistance to UVA stress. J Biol Chem 1998; 273(47):31252–31261.

111. Dubin RA, Wawrousek EF, Piatigorsky J Expression of the murine alpha B-crystallin gene is not restricted to the lens. Mol Cell Biol 1989; 9(3):1083–1091.

112. van Rijk AF, Bloemendal H. Alpha-B-crystallin in neuropathology. Ophthalmologica 2000; 214(1):7–12.

113. Goldman JE, Corbin E. Rosenthal fibers contain ubiquitinated alpha B-crystallin. Am J Pathol 1991; 139(4):933–938.

114. Nakata K, Crabb JW, Hollyfield JG. Crystallin distribution in Bruch's membrane–choroid complex from AMD and age-matched donor eyes. Exp Eye Res 2005; 80(6):821–826.

115. Vicart P et al. A missense mutation in the alphaB-crystallin chaperone gene causes a desmin-related myopathy. Nat Genet 1998; 20(1):92–95.

116. Bova MP et al. Mutation R120G in alphaB-crystallin, which is linked to a desmin-related myopathy, results in an irregular structure and defective chaperone-like function. Proc Natl Acad Sci USA 1999; 96(11):6137–6142.

117. Berry V et al. Alpha-B crystallin gene (CRYAB) mutation causes dominant congenital posterior polar cataract in humans. Am J Hum Genet 2001; 69(5):1141–1145.

118. Driessen HP et al. Primary structure of the bovine beta-crystallin Bp chain. Internal duplication and homology with gamma-crystallin. Eur J Biochem 1981; 121(1):83–91.

119. Bax B et al. X-ray analysis of beta B2-crystallin and evolution of oligomeric lens proteins. Nature 1990; 347(6295):776–780.

120. Kroone RC et al. The role of the sequence extensions in beta-crystallin assembly. Protein Eng 1994; 7(11):1395–1399.

121. Lampi KJ et al. Sequence analysis of betaA3, betaB3, and betaA4 crystallins completes the identification of the major proteins in young human lens. J Biol Chem 1997; 272(4):2268–2275.

122. McDermott JB, Peterson CA, Piatigorsky J. Structure and lens expression of the gene encoding chicken beta A3/A1-crystallin. Gene 1992; 117(2):193–200.

123. Jobby MK, Sharma Y. Calcium-binding to lens betaB2- and betaA3-crystallins suggests that all beta-crystallins are calcium-binding proteins. FEBS 2007; 274(16):4135–4147.

124. Rajini B et al. Calcium binding properties of gamma-crystallin: calcium ion binds at the Greek key beta gamma-crystallin fold. J Biol Chem 2001; 276(42):38464–38471.

125. Bloemendal H et al. Ageing and vision: structure, stability and function of lens crystallins. Progr Biophys Mol Biol 2004; 86(3):407–485.

126. Kuwabara T. Microtubules in the lens. Arch Ophthalmol 1968; 79(2):189–195.

127. Lo WK. Adherens junctions in the ocular lens of various species: ultrastructural analysis with an improved fixation. Cell Tissue Res 1988; 254(1):31–40.

128. Lee A, Fischer RS, Fowler VM. Stabilization and remodeling of the membrane skeleton during lens fiber cell differentiation and maturation. Dev Dyn 2000; 217(3):257–270.

129. Fischer RS, Lee A, Fowler VM. Tropomodulin and tropomyosin mediate lens cell actin cytoskeleton reorganization in vitro. Invest Ophthalmol Vis Sci 2000; 41(1):166–174.

130. FitzGerald P. Methods for the circumvention of problems associated with the study of the ocular lens plasma membrane–cytoskeleton complex. Curr Eye Res 1990; 9(11):1083–1097.

131. Colucci-Guyon E et al. Mice lacking vimentin develop and reproduce without an obvious phenotype. Cell 1994; 79(4):679–694.

132. Capetanaki Y, Smith S, Heath J. Overexpression of the vimentin gene in transgenic mice inhibits normal lens cell differentiation. J Cell Biol 1989; 109(4, Pt 1):1653–1664.

133. Quinlan RA et al. The eye lens cytoskeleton. Eye 1999; 13(Pt 3b):409–416.

134. Georgatos SD et al. To bead or not to bead? Lens-specific intermediate filaments revisited. J Cell Sci 1997; 110(Pt 21):2629–2634.

135. Ireland M, Maisel H. A cytoskeletal protein unique to lens fiber cell differentiation. Exp Eye Res 1984; 38(6):637–645.

136. Jakobs PM et al. Autosomal-dominant congenital cataract associated with a deletion mutation in the human beaded filament protein gene BFSP2. Am J Hum Genet 2000; 66(4):1432–1436.

137. Conley YP et al. A juvenile-onset, progressive cataract locus on chromosome 3q21-q22 is associated with a missense mutation in the beaded filament structural protein-2. Am J Hum Genet 2000; 66(4):1426–1431.

138. Alizadeh A et al. Targeted deletion of the lens fiber cell-specific intermediate filament protein Filensin. Invest Ophthalmol Vis Sci 2003; 44(12):5252–5258.

139. Kuszak J, Alcala J, Maisel H. The surface morphology of embryonic adult chick lens-fiber cells. Am J Anat 1980; 159:395–410.
140. Willekens B, Vrensen G. The three-dimensional organization of lens fibers in the rhesus monkey. Graefe's Arch Clin Exp Ophthalmol 1982; 219:112–120.
141. Bassnett S, Winzenburger PA. Morphometric analysis of fibre cell growth in the developing chicken lens. Exp Eye Res 2003; 76(3):291–302.
142. Borchman D et al. Studies on the distribution of cholesterol, phospholipid, and protein in the human and bovine lens. Lens Eye Toxic Res 1989; 6(4):703–724.
143. Bloemendal H et al. The plasma membranes of eye lens fibres. Biochemical and structural characterization. Cell Diff 1972; 1(2):91–106.
144. Gorin MB et al. The major intrinsic protein (MIP) of the bovine lens fiber membrane: characterization and structure based on cDNA cloning. Cell 1984; 39(1):49–59.
145. Agre P et al. Aquaporin CHIP: the archetypal molecular water channel. Am J Physiol 1993; 265(4 Pt 2):F463–F476.
146. Chandy G et al. Comparison of the water transporting properties of MIP and AQP1. J Membr Biol 1997; 159(1):29–39.
147. Fotiadis D et al. Surface tongue-and-groove contours on lens MIP facilitate cell-to-cell adherence. J Mol Biol 2000; 300(4):779–789.
148. Shiels A, Bassnett S. Mutations in the founder of the MIP gene family underlie cataract development in the mouse. Nat Genet 1996; 12(2):212–215.
149. Berry V et al. Missense mutations in MIP underlie autosomal dominant 'polymorphic' and lamellar cataracts linked to 12q. Nat Genet 2000; 25(1):15–17.
150. Goodenough DA, Dick JS, Lyons JE. Lens metabolic cooperation: a study of mouse lens transport and permeability visualized with freeze-substitution autoradiography and electron microscopy. J Cell Biol 1980; 86(2):576–589.
151. Musil, L, Beyer E, Goodenough D. Expression of the gap junction protein connexin43 in embryonicchick lens: molecular cloning, ultrastructural localization, and post-translational phosphorylation. J Membr Biol 1990; 116(2):163–175.
152. Gong X et al. Disruption of alpha3 connexin gene leads to proteolysis and cataractogenesis in mice. Cell 1997; 91(6):833–843.
153. Dahm R et al. Gap junctions containing alpha8-connexin (MP70) in the adult mammalian lens epithelium suggests a re-evaluation of its role in the lens. Exp Eye Res 1999; 69(1):45–56.
154. White TW, Goodenough DA, Paul DL. Targeted ablation of connexin50 in mice results in microphthalmia and zonular pulverulent cataracts. J Cell Biol 1998; 143(3): 815–825.
155. White TW et al. Optimal lens epithelial cell proliferation is dependent on the connexin isoform providing gap junctional coupling. Invest Ophthalmol Vis Sci 2007; 48(12):5630–5637.
156. Kistler J et al. Ocular lens gap junctions: protein expression, assembly, and structure-function analysis. Microsc Res Tech 1995; 31(5):347–356.
157. Shiels A et al. A missense mutation in the human connexin50 gene (GJA8) underlies autosomal dominant "zonular pulverulent" cataract, on chromosome 1q. Am J Hum Genet 1998; 62(3):526–532.
158. Gong X et al. Genetic factors influence cataract formation in alpha 3 connexin knockout mice. Dev Genet 1999; 24(1–2):27–32.
159. Xia C-H et al. Absence of alpha3 (Cx46) and alpha8 (Cx50) connexins leads to cataracts by affecting lens inner fiber cells. Exp Eye Res 2006; 83(3):688–696.
160. Shestopalov VI, Bassnett S. Expression of autofluorescent proteins reveals a novel protein permeable pathway between cells in the lens core. J Cell Sci 2000; 113:1913–1921.
161. Kuszak JR et al. Cell-to-cell fusion of lens fiber cells in situ: correlative light, scanning electron microscopic, and freeze-fracture studies. J Ultrastruct Res 1985; 93(3):144–160.
162. Kuszak JR et al. The contribution of cell-to-cell fusion to the ordered structure of the crystalline lens. Lens Eye Toxic Res 1989; 6(4):639–673.
163. Shestopalov VI, Bassnett S. Development of a macromolecular diffusion pathway in the lens. J Cell Sci 2003; 116(20):4191–4199.
164. Steele EC Jr. et al. Identification of a mutation in the MP19 gene, Lim2, in the cataractous mouse mutant To3. Mol Vis 1997; 3:5.
165. Shiels, A et al. Refractive defects and cataracts in mice lacking lens intrinsic membrane protein-2. Invest Ophthalmol Vis Sci 2007; 48(2):500–508.
166. Shi Y et al. The stratified syncytium of the vertebrate lens. J Cell Sci 2009; 122(Pt 10):1607–1615.
167. Lagunowich LA, Grunwald GB. Expression of calcium-dependent cell adhesion during ocular development: a biochemical, histochemical and functional analysis. Dev Biol 1989; 135(1):158–171.
168. Beebe DC et al. Changes in adhesion complexes define stages in the differentiation of lens fiber cells. Invest Ophthalmol Vis Sci 2001; 42(3):727–734.
169. Bassnett S, Missey H, Vucemilo I. Molecular architecture of the lens fiber cell basal membrane complex. J Cell Sci 1999; 112:2155–2165.
170. Lo WK et al. Spatiotemporal distribution of zonulae adherens and associated actin bundles in both epithelium and fiber cells during chicken lens development. Exp Eye Res 2000; 71(1):45–55.
171. Walker JL, Menko AS. Alpha6 Integrin is regulated with lens cell differentiation by linkage to the cytoskeleton and isoform switching. Dev Biol 1999; 210(2):497–511.
172. Bassnett S. Lens organelle degradation. Exp Eye Res 2002; 74(1):1–6.
173. van Leyen K et al. A function for lipoxygenase in programmed organelle degradation. Nature 1998; 395(6700):392–395.
174. Zandy AJ et al. Role of the executioner caspases during lens development. J Biol Chem 2005.
175. Nishimoto S et al. Nuclear cataract caused by a lack of DNA degradation in the mouse eye lens. Nature 2003; 424(6952):1071–1074.
176. Min JN et al. Unique contribution of heat shock transcription factor 4 in ocular lens development and fiber cell differentiation. Genesis 2004; 40(4):205–217.
177. Fujimoto M et al. HSF4 is required for normal cell growth and differentiation during mouse lens development. EMBO J 2004
178. Spencer RP, Change in weight of the human lens with age. Ann Ophthalmol 1976; 8(4):440–441.
179. Bhuyan DK, Reddy PG, Bhuyan KC. Growth factor receptor gene and protein expressions in the human lens. Mech Ageing Dev 2000; 113(3):205–218.
180. Fleming TP, Song Z, Andley UP. Expression of growth control and differentiation genes in human lens epithelial cells with extended life span. Invest Ophthalmol Vis Sci 1998; 39(8):1387–1398.
181. Ray S et al. Platelet-derived growth factor D, tissue-specific expression in the eye, and a key role in control of lens epithelial cell proliferation. J Biol Chem 2005; 280(9):8494–8502.
182. Weng J et al. Hepatocyte growth factor, keratinocyte growth factor, and other growth factor-Receptor systems in the lens. Invest Ophthalmol Vis Sci 1997; 38(8):1543–1554.
183. Garcia CM et al. The function of VEGF-A in lens development: Formation of the hyaloid capillary network and protection against transient nuclear cataracts. Exp Eye Res 2008
184. Williams MR et al. Acetylcholine receptors are coupled to mobilization of intracellular calcium cultured human lens cells letter. Exp Eye Res 1993; 57(3):381–384.
185. Riach RA et al. Histamine and ATP mobilize calcium by activation of H1 and P2u receptors in human lens epithelial cells. J Physiol (Lond) 1995; 486(Pt 2):273–282.
186. Potts JD, Kornacker S, Beebe DC. Activation of the Jak-STAT-signaling pathway in embryonic lens cells. Dev Biol 1998; 204(1):277–292.
187. Shui, Y-B, Beebe DC. Age-dependent control of lens growth by hypoxia. Invest Ophthalmol Vis Sci 2008; 49(3):1023–1029.
188. Shui YB et al. HIF-1: an age-dependent regulator of lens cell proliferation. Invest Ophthalmol Vis Sci 2008; 49(11):4961–4970.
189. Klein B, Klein R, Moss S. Correlates of lens thickness: the Beaver Dam Eye Study. Invest Ophthalmol Vis Sci 1998; 39(8):1507–1510.
190. Klein BE, Klein R, Moss SE. Lens thickness and five-year cumulative incidence of cataracts: the Beaver Dam Eye Study. Ophthalmic Epidemiol 2000; 7(4):243–248.
191. Wong TY et al. Refractive errors, axial ocular dimensions, and age-related cataracts: the Tanjong Pagar Survey. Invest Ophthalmol Vis. Sci 2003; 44(4):1479–1485.
192. Praveen MR et al. Lens thickness of Indian eyes: impact of isolated lens opacity, age, axial length, and influence on anterior chamber depth. Eye 2008
193. Rae JL et al. Dye transfer between cells of the lens. J Membr Biol 1996; 150(1):89–103.
194. Bassnett S et al. Intercellular communication between epithelial and fiber cells of the eye lens. J Cell Sci 1994; 107(Pt 4):799–811.
195. Calera MR et al. Connexin43 is required for production of the aqueous humor in the murine eye. J Cell Sci 2006.
196. Hightower KR et al. Lens epithelium: a primary target of UVB irradiation. Exp Eye Res 1994; 59(5):557–564.
197. Shui Y-B et al. Vascular endothelial growth factor expression and signaling in the lens. Invest Ophthalmol Vis Sci 2003; 44(9):3911–3919.
198. Zhu M et al. The human hyaloid system: cell death and vascular regression. Exp Eye Res 2000; 70(6):767–776.
199. Meeson AP et al. VEGF deprivation-induced apoptosis is a component of programmed capillary regression. Development 1999; 126(7):1407–1415.
200. Diez-Roux G et al. Macrophages kill capillary cells in G1 phase of the cell cycle during programmed vascular regression. Development 1999; 126(10):2141–2147.
201. Lobov IB et al. WNT7b mediates macrophage-induced programmed cell death in patterning of the vasculature. Nature 2005; 437(7057):417–421.
202. Goldberg MF. Persistent fetal vasculature (PFV): an integrated interpretation of signs and symptoms associated with persistent hyperplastic primary vitreous (PHPV). LIV Edward Jackson Memorial Lecture. Am J Ophthalmol 1997; 124(5):587–626.
203. Genis-Galvez, JM, Role of the lens in the morphogenesis of the iris and cornea. Nature 1966; 210(32):209–210.
204. Genis-Galvez JM, Santos-Gutierrez L, Rios-Gonzalez A. Causal factors in corneal development: an experimental analysis in the chick embryo. Exp Eye Res 1967; 6(1):48–56.
205. Beebe DC, Coats JM. The lens organizes the anterior segment: specification of neural crest cell differentiation in the avian eye. Dev Biol 2000; 220(2):424–431.
206. Stroeva OG. Relation of proliferative and determinative processes in the morphogenesis of the iris and ciliary body of mammals. Zh Obshch Biol 1967; 28(6):684–696.
207. Thut CJ et al. A large-scale in situ screen provides molecular evidence for the induction of eye anterior segment structures by the developing lens. Dev Biol 2001; 231(1):63–76.
208. Turrens JF, Alexandre A, Lehninger AL. Ubisemiquinone is the electron donor for superoxide formation by complex III of heart mitochondria. Arch Biochem Biophys 1985; 237(2):408–414.
209. Truscott RJ. Age-related nuclear cataract: A lens transport problem. Ophthalmic Res 2000; 32(5):185–194.
210. Truscott RJW. Age-related nuclear cataract – oxidation is the key. Exp Eye Res 2004; 80(5):709–725.
211. Winkler BS, Riley MV. Relative contributions of epithelial cells and fibers to rabbit lens ATP content and glycolysis. Invest Ophthalmol Vis Sci 1991; 32(9):2593–2598.
212. Chylack LT Jr., Friend J. Intermediary metabolism of the lens: a historical perspective 1928–1989. Exp Eye Res 1990; 50(6):575–582.
213. Bassnett S, Duncan G. Direct measurement of ph in the rat lens by ion-sensitive microelectrodes. Exp Eye Res 1985; 40:585–590.
214. Mathias RT, Riquelme G, Rae JL. Cell-to-cell communication and pH in the frog lens. J Gen Physiol 1991; 98(6):1085–1103.
215. Bassnett S, Croghan P, Duncan G. Diffusion of lactate and its role in determining intracellular Ph. Exp Eye Res 1987; 44(1):143–147.
216. Lin JS et al. Spatial differences in gap junction gating in the lens are a consequence of connexin cleavage. Eur J Cell Biol 1998; 76(4):246–250.
217. Martinez-Wittinghan FJ et al. Lens gap junctional coupling is modulated by connexin identity and the locus of gene expression. Invest Ophthalmol Vis Sci 2004; 45(10):3629–3637.
218. Kinoshita JH, Kern HL, Merola OH. Factors affecting the cation transport of calf lens. Biochim Biophys Acta 1961; 47:458–466.
219. McLaren JW et al. Measuring oxygen tension in the anterior chamber of rabbits. Invest Ophthalmol Vis Sci 1998; 39(10):1899–1909.
220. Kwan M, Niinikoski J, Hunt TK. In vivo measurements of oxygen tension in the cornea, aqueous humor, and anterior lens of the open eye. Invest Ophthalmol 1972; 11(2):108–114.

221. Helbig H et al. Oxygen in the anterior chamber of the human eye. Ger J Ophthalmol 1993; 2(3):161–164.

222. Holekamp NM, Shui YB, Beebe DC. Vitrectomy surgery increases oxygen exposure to the lens: a possible mechanism for nuclear cataract formation. Am J Ophthalmol 2005; 139(2):302–310.

223. Shui Y-B et al. Oxygen distribution in the rabbit eye and oxygen consumption by the lens. Invest. Ophthalmol Vis Sci 2006; 47(4):1571–1580.

224. Holekamp NM, Shui Y-B, Beebe D. Lower intraocular oxygen tension in diabetic patients: possible contribution to decreased incidence of nuclear sclerotic cataract. Am J Ophthalmol 2006; 141(6):1027–1032.

225. McNulty R et al. Regulation of tissue oxygen levels in the mammalian lens. J Physiol (Lond) 2004; 559(3):883–898.

226. Spector A, Garner WH. Hydrogen peroxide and human cataract. Exp Eye Res 1981; 33(6):673–681.

227. Devamanoharan P, Ramachandran S, Varma S. Hydrogen peroxide in the eye lens: Radioisotopic determination. Curr Eye Res 1991; 10(9):831–838.

228. Spector A. Oxidative stress-induced cataract: mechanism of action. Faseb J 1995; 9(12):1173–1182.

229. Spector A, Ma W, Wang RR. The aqueous humor is capable of generating and degrading H_2O_2. Invest Ophthalmol Vis Sci 1998; 39(7):1188–1197.

230. Bleau G, Giasson C, Brunette I. Measurement of hydrogen peroxide in biological samples containing high levels of ascorbic acid. Anal Biochem 1998; 263(1):13–17.

231. Garcia-Castineiras S et al. Aqueous humor hydrogen peroxide analysis with dichlorophenol-indophenl. Exp Eye Res 1992; 55(1):9–19.

232. Dillon J et al. The optical properties of the anterior segment of the eye: implications for cortical cataract. Exp Eye Res 1999; 68(6):785–795.

233. Harding J. Cataract: biochemistry, epidemiology and pharmacology. London: Chapman & Hall, 1991.

234. Harding JJ. The untenability of the sunlight hypothesis of cataractogenesis. Doc Ophthalmol 1994; 88(3–4):345–349.

235. Reddan JR et al. Protection from oxidative insult in glutathione depleted lens epithelial cells. Exp Eye Res 1999; 68(1):117–127.

236. Calvin H et al. Rapid deterioration of lens fibers in GSH-depleted mouse pups. Invest Ophthalmol Vis Sci 1991; 32(6):1916–1924.

237. Reddy V. Glutathione and its function in the lens – An overview. Exp Eye Res 1990; 50(6):771–778.

238. Kannan R et al. Molecular characterization of a reduced glutathione transporter in the lens. Invest Ophthalmol Vis Sci 1995; 36(9):1785–1792.

239. Kannan R et al. Identification of a novel, sodium-dependent, reduced glutathione transporter in the rat lens epithelium. Invest Ophthalmol Vis Sci 1996; 37(11):2269–2275.

240. Sweeney MH, Truscott RJ. An impediment to glutathione diffusion in older normal human lenses: a possible precondition for nuclear cataract. Exp Eye Res 1998; 67(5):587–595.

241. Lou MF. Thiol regulation in the lens. J Ocul Pharmacol Ther 2000; 16(2):137–148.

242. Tsukaguchi H et al. A family of mammalian Na+-dependent L-ascorbic acid transporters. Nature 1999; 399(6731):70–75.

243. Rose RC, Bode AM. Ocular ascorbate transport and metabolism. Comp Biochem Physiol A 1991; 100(2):273–285.

244. Kern HL, Zolot SL. Transport of vitamin C in the lens. Curr Eye Res 1987; 6(7):885–896.

245. Fan X et al. Vitamin C mediates chemical aging of lens crystallins by the Maillard reaction in a humanized mouse model. Proc Natl Acad Sci USA 2006; 103(45):16912–16917.

246. Winkler BS, Orselli SM, TS Rex TS. The redox couple between glutathione and ascorbic acid: a chemical and physiological perspective. Free Radic Biol Med 1994; 17(4):333–349.

247. Reddan JR et al. Regional differences in the distribution of catalase in the epithelium of the ocular lens. Cell Mol Biol (Noisy-le-grand) 1996; 42(2):209–219.

248. Giblin FJ et al. The relative roles of the glutathione redox cycle and catalase in the detoxification of H_2O_2 by cultured rabbit lens epithelial cells. Exp Eye Res 1990; 50(6):795–804.

249. Beebe DC. Maintaining transparency: A review of the developmental physiology and pathophysiology of two avascular tissues. Semin Cell Dev Biol 2007

250. Robinson KR, Patterson JW. Localization of steady currents in the lens. Curr Eye Res 1982; 2(12):843–847.

251. Parmelee JT. Measurement of steady currents around the frog lens. Exp Eye Res 1986; 42(5):433–441.

252. Wind BE, Walsh S, Patterson JW. Equatorial potassium currents in lenses. Exp Eye Res 1988; 46(2):117–130.

253. Mathias RT, Rae JL, Baldo GJ. Physiological properties of the normal lens. Physiol Rev 1997; 77(1):21–50.

254. Fischbarg J et al. Transport of fluid by lens epithelium. Am J Physiol 1999; 276(3 Pt 1):C548–C557.

255. Duncan G et al. Human lens membrane cation permeability increases with age. Invest Ophthalmol Vis Sci 1989; 30(8):1855–1859.

256. Sanderson J, Marcantonio JM, Duncan G. A human lens model of cortical cataract: Ca^{2+}-induced protein loss, vimentin cleavage and opacification. Invest Ophthalmol Vis Sci 2000; 41(8):2255–2261.

257. Jacob TJ. A direct measurement of intracellular free calcium within the lens. Exp Eye Res 1983; 36(3):451–453.

258. Truscott RJ et al. Calcium-induced opacification and proteolysis in the intact rat lens. Invest Ophthalmol Vis Sci 1990; 31(11):2405–2411.

259. Marcantonio JM, Duncan G. Calcium-induced degradation of the lens cytoskeleton. Biochem Soc Trans 1991; 19(4):1148–1150.

260. Clark JI et al. Cortical opacity, calcium concentration and fiber membrane structure in the calf lens. Exp Eye Res 1980; 31(4):399–410.

261. Jacob TJ. Raised intracellular free calcium within the lens causes opacification and cellular uncoupling in the frog. J Physiol (Lond) 1983; 341:595–601.

262. Fagerholm PP. The influence of calcium on lens fibers. Exp Eye Res 1979; 28:211–222.

263. Merriman-Smith R, Donaldson P, Kistler J. Differential expression of facilitative glucose transporters GLUT1 and GLUT3 in the lens. Invest Ophthalmol Vis Sci 1999; 40(13):3224–3230.

264. Brown N. The change in lens curvature with age. Exp Eye Res 1974; 19(2):175–183.

265. Sivak JG, Luer CA. Optical development of the ocular lens of an elasmobranch, Raja eglanteria. Vision Res 1991; 31(3):373–382.

266. Kroger RH et al. Refractive index distribution and spherical aberration in the crystalline lens of the African cichlid fish Haplochromis burtoni. Vision Res 1994; 34(14):1815–1822.

267. Sivak JG, Kreuzer RO. Spherical aberration of the crystalline lens. Vision Res 1983; 23(1):59–70.

268. Fu S et al. The hydroxyl radical in lens nuclear cataractogenesis. J Biol Chem 1998; 273(44):28603–28609.

269. Pirie A. Color and solubility of the proteins of human cataracts. Invest Ophthalmol. 1968; 7(6):634–650.

270. Pokhrel AK et al. Case-control study of indoor cooking smoke exposure and cataract in Nepal and India. Int. J Epidemiol 2005.

271. Garland DL et al. The nucleus of the human lens: demonstration of a highly characteristic protein pattern by two-dimensional electrophoresis and introduction of a new method of lens dissection. Exp Eye Res 1996; 62(3):285–291.

272. Lampi KJ et al. Age-related changes in human lens crystallins identified by two-dimensional electrophoresis and mass spectrometry. Exp Eye Res 1998; 67(1):31–43.

273. Ma Z et al. Age-related changes in human lens crystallins identified by HPLC and mass spectrometry. Exp Eye Res 1998; 67(1):21–30.

274. Spector A, Li S, Sigelman J. Age-dependent changes in the molecular size of human lens proteins and their relationship to light scatter. Invest Ophthalmol 1974; 13(10):795–798.

275. Ringens PJ, Hoenders HJ, Bloemendal H. Effect of aging on the water-soluble and water-insoluble protein pattern in normal human lens. Exp Eye Res 1982; 34(2):201–207.

276. Roy D, Spector A. Absence of low-molecular-weight alpha crystallin in nuclear region of old human lenses. Proc Natl Acad Sci USA 1976; 73(10):3484–3487.

277. McFall-Ngai MJ et al. Spatial and temporal mapping of the age-related changes in human lens crystallins. Exp Eye Res 1985; 41(6):745–758.

278. Masters PM, Bada JL, Zigler JS Jr. Aspartic acid racemization in heavy molecular weight crystallins and water insoluble protein from normal human lenses and cataracts. Proc Natl Acad Sci USA 1978; 75(3):1204–1208.

279. Hoenders HJ, Bloemendal H. Lens proteins and aging. J Gerontol 1983; 38(3):278–286.

280. Garner WH, Spector A. Racemization in human lens: evidence of rapid insolubilization of specific polypeptides in cataract formation. Proc Natl Acad Sci USA 1978; 75(8):3618–3620.

281. Prescott AR et al. The intermediate filament cytoskeleton of the lens: an ever changing network through development and differentiation. A minireview. Ophthalmic Res 1996; 28(Suppl 1):58–61.

282. Kibbelaar MA et al. Is actin in eye lens a possible factor in visual accomodation? Nature 1980; 285(5765):506–508.

283. Maisel H, Ellis M. Cytoskeletal proteins of the aging human lens. Current Eye Res 1984; 3:369–381.

284. Clark JI et al. Lens cytoskeleton and transparency: a model. Eye 1999; 13(Pt 3b):417–424.

285. Kuszak JR. The development of lens sutures. Prog Retin Eye Res 1995; 14:567–592.

286. Kuszak JR et al. The interrelationship of lens anatomy and optical quality. II Primate lenses. Exp Eye Res 1994; 59(5):521–535.

287. Cammarata PR et al. Macromolecular organization of bovine lens capsule. Tissue Cell 1986; 18(1):83–97.

288. Young RW, Ocumpaugh DE. Autoradiographic studies on the growth and development of the lens capsule in the rat. Invest Ophthal 1966; 5:583–593.

289. Obara H et al. Usefulness of Scheimpflug photography to follow up Wilson's disease. Ophthalmic Res 1995; 27(Suppl 1):100–103.

290. Johnson MC, Beebe DC. Growth, synthesis and regional specialization of the embryonic chicken lens capsule. Exp Eye Res 1984; 38(6):579–592.

291. Onodera S. Presence of the basement membrane component – heparan sulfate proteoglycan – in bovine lens capsules. Chem Pharm Bull (Tokyo) 1991; 39(4):1059–1061.

292. Fitch JM, Linsenmayer TF. Monoclonal antibody analysis of ocular basement membranes during development. Dev Biol 1983; 95(1):137–153.

293. Streeten BW. The zonular insertion: a scanning electron microscopic study. Invest Ophthalmol Vis Sci 1977; 16(4):364–375.

294. Streeten BW, Licari PA. The zonules and the elastic microfibrillar system in the ciliary body. Invest Ophthalmol Vis Sci 1983; 24(6):667–681.

295. Lee B et al. Linkage of Marfan syndrome and a phenotypically related disorder to two different fibrillin genes. Nature 1991; 352(6333):330–334.

296. Dietz HC et al. Marfan syndrome caused by a recurrent de novo missense mutation in the fibrillin gene. Nature 1991; 352(6333):337–339.

297. Farnsworth TN et al. Surface ultrastructure of the human lens capsule and zonular attachments. Invest Ophthalmol 1976; 36–40.

298. Leske MC et al. Prevalence of lens opacities in the Barbados Eye Study. Arch Ophthalmol 1997; 115(1):105–111.

299. Hiller R, Sperduto RD, Ederer F. Epidemiologic associations with nuclear, cortical, and posterior subcapsular cataracts. Am J Epidemiol 1986; 124(6):916–925.

300. Leske MC et al. Incidence and progression of lens opacities in the Barbados Eye Studies. Ophthalmology 2000; 107(7):1267–1273.

301. Klein BE, Klein R, Lee KE. Incidence of age-related cataract: the Beaver Dam Eye Study. Arch Ophthalmol 1998; 116(2):219–225.

302. Klein BE, Klein R, Moss SE. Incident cataract surgery: the Beaver Dam eye study. Ophthalmology 1997; 104(4):573–580.

303. Carlsson B, Sjostrand J. Increased incidence of cataract extractions in women above 70 years of age. A population based study. Acta Ophthalmol Scand 1996; 74(1):64–68.

304. Delcourt C et al. Risk factors for cortical, nuclear, and posterior subcapsular cataracts: the POLA study. Pathologies Oculaires Liees a l'Age. Am J Epidemiol 2000; 151(5):497–504.
305. Klein BE. Lens opacities in women in Beaver Dam, Wisconsin: is there evidence of an effect of sex hormones? Trans Am Ophthalmol Soc 1993; 91:517–544.
306. Klein BE, Klein R, Ritter LL. Is there evidence of an estrogen effect on age-related lens opacities? The Beaver Dam Eye Study. Arch Ophthalmol 1994; 112(1):85–91.
307. Bigsby RM et al. Protective effects of estrogen in a rat model of age-related cataracts. Proc Natl Acad Sci USA 1999; 96(16):9328–9332.
308. Hales AM et al. Estrogen protects lenses against cataract induced by transforming growth factor-beta (TGFbeta). J Exp Med 1997; 185(2):273–280.
309. Cumming RG, Mitchell P. Hormone replacement therapy, reproductive factors, and cataract. The Blue Mountains Eye Study. Am J Epidemiol 1997; 145(3):242–249.
310. Benitez del Castillo JM, del Rio T, Garcia-Sanchez J. Effects of estrogen use on lens transmittance in postmenopausal women. Ophthalmology 1997; 104(6):970–973.
311. Paganini-Hill A, Clark LJ. Eye problems in breast cancer patients treated with tamoxifen. Breast Cancer Res Treat 2000; 60(2):167–172.
312. Gorin MB et al. Long-term tamoxifen citrate use and potential ocular toxicity. Am J Ophthalmol 1998; 125(4):493–501.
313. Phillips CI et al. Human cataract risk factors: significance of abstention from, and high consumption of, ethanol (U-curve) and non-significance of smoking. Ophthalmic Res 1996; 28(4):237–247.
314. Hodge WG, Whitcher JP, Satariano W. Risk factors for age-related cataracts. Epidemiol Rev 1995; 17(2):336–346.
315. West SK, Valmadrid CT. Epidemiology of risk factors for age-related cataract. Surv Ophthalmol 1995; 39(4):323–334.
316. Cumming RG, Mitchell P, Lim R. Iris color and cataract: the blue mountains eye study. Am J Ophthalmol 2000; 130(2):237–238.
317. Hammond BR Jr. et al. Iris color and age-related changes in lens optical density. Ophthalmic Physiol Opt 2000; 20(5):381–386.
318. Meddings DR et al. Mortality rates after cataract extraction. Epidemiology 1999; 10(3):288–293.
319. West SK et al. Mixed lens opacities and subsequent mortality. Arch Ophthalmol 2000; 118(3):393–397.
320. Street DA, Javitt JC. National five-year mortality after inpatient cataract extraction. Am J Ophthalmol 1992; 113(3):263–268.
321. Hirsch RP, Schwartz B. Increased mortality among elderly patients undergoing cataract extraction. Arch Ophthalmol 1983; 101(7):1034–1037.
322. Blundell MSJ et al. Reduced mortality compared to national averages following phacoemulsification cataract surgery: a retrospective observational study. Br J Ophthalmol 2008.
323. Truscott RJ, Augusteyn RC. Oxidative changes in human lens proteins during senile nuclear cataract formation. Biochim Biophys Acta 1977; 492(1):43–52.
324. Dische Z, Zil HA. Studies on the oxidation of cysteine to cystine in lens proteins during cataract formation. Am J Ophthalmol 1951; 34:104–113.
325. Takemoto LJ, Azari P. Isolation and characterizaion of covalently linked, high molecular weight proteins from human cataractous lens. Exp Eye Res 1977; 24(1):63–70.
326. Kodama T, Takemoto L. Characterization of disulfide-linked crystallins associated with human cataractous lens membranes. Invest Ophthalmol Vis Sci 1988; 29(1):145–149.
327. Takehana M, Takemoto L. Quantitation of membrane-associated crystallins from aging and cataractous human lenses. Invest Ophthalmol Vis Sci 1987; 28(5):780–784.
328. Ifeanyi F, Takemoto L. Differential binding of alpha-crystallins to bovine lens membrane. Exp Eye Res 1989; 49(1):143–147.
329. Ifeanyi F, Takemoto L. Specificity of alpha crystallin binding to the lens membrane. Curr Eye Res 1990; 9(3):259–265.
330. Bova LM et al. Major changes in human ocular UV protection with age. Invest Ophthalmol Vis Sci 2001; 42:200–205.
331. Palmquist BM, Philipson B, Barr PO. Nuclear cataract and myopia during hyperbaric oxygen therapy. Br J Ophthalmol 1984; 68(2):113–117.
332. Brown NA, Hill AR. Cataract: the relation between myopia and cataract morphology. Br J Ophthalmol 1987; 71(6):405–414.
333. Ross ME et al. Myopia associated with hyperbaric oxygen therapy. Optom Vis Sci 1996; 73(7):487–494.
334. Lyne AJ. Ocular effects of hyperbaric oxygen. Trans Ophthalmol Soc UK 1978; 98(1):66–68.
335. Giblin FJ et al. Exposure of rabbit lens to hyperbaric oxygen in vitro: regional effects on GSH level. Invest Ophthalmol Vis Sci 1988; 29(8):1312–1319.
336. Giblin FJ et al. Nuclear light scattering, disulfide formation and membrane damage in lenses of older guinea pigs treated with hyperbaric oxygen. Exp Eye Res 1995; 60(3):219–235.
337. Padgaonkar VA et al. Hyperbaric oxygen in vivo accelerates the loss of cytoskeletal proteins and MIP26 in guinea pig lens nucleus. Exp Eye Res 1999; 68(4):493–504.
338. McGinty SJ, Truscott RJ. Presbyopia: the first stage of nuclear cataract? Ophthalmic Res 2006; 38(3):137–148.
339. Melberg NS, Thomas MA. Nuclear sclerotic cataract after vitrectomy in patients younger than 50 years of age. Ophthalmology 1995; 102(10):1466–1471.
340. Van Effenterre G et al. Is vitrectomy cataractogenic? Study of changes of the crystalline lens after surgery of retinal detachment. J Fr Ophtalmol 1992; 15(8–9):449–454.
341. Thompson JT et al. Progression of nuclear sclerosis and long-term visual results of vitrectomy with transforming growth factor beta-2 for macular holes. Am J Ophthalmol 1995; 119(1):48–54.
342. Novak MA et al. The crystalline lens after vitrectomy for diabetic retinopathy. Ophthalmology 1984; 91(12):1480–1484.
343. Ogura Y, Kitagawa K, Ogino N. Prospective longitudinal studies on lens changes after vitrectomy – quantitative assessment by fluorophotometry and refractometry. Nippon Ganka Gakkai Zasshi 1993; 97(5):627–631.
344. de Bustros S et al. Nuclear sclerosis after vitrectomy for idiopathic epiretinal membranes. Am J Ophthalmol 1988; 105(2):160–164.
345. de Bustros S et al. Vitrectomy for idiopathic epiretinal membranes causing macular pucker. Br J Ophthalmol 1988; 72(9):692–695.
346. Cherfan GM et al. Nuclear sclerotic cataract after vitrectomy for idiopathic epiretinal membranes causing macular pucker. Am J Ophthalmol 1991; 111(4):434–438.
347. Cheng L et al. Duration of vitrectomy and postoperative cataract in the vitrectomy for macular hole study. Am J Ophthalmol 2001; 132(6):881–887.
348. Sawa M et al. Nonvitrectomizing vitreous surgery for epiretinal membrane: long-term follow-up. Ophthalmology 2005; 112(8):1402–1408.
349. Smiddy WE, Feuer W. Incidence of cataract extraction after diabetic vitrectomy. Retina 2004; 24(4):574–581.
350. Shui Y-B et al. The gel state of the vitreous and ascorbate-dependent oxygen consumption: relationship to the etiology of nuclear cataracts. Arch Ophthalmol 2009; 127(4):1–8.
351. Takano S et al. Determination of ascorbic acid in human vitreous humor by high-performance liquid chromatography with UV detection. Curr Eye Res 1997; 16(6):589–594.
352. al-Ghoul KJ et al. Distribution and type of morphological damage in human nuclear age-related cataracts. Exp Eye Res 1996; 62(3):237–251.
353. al-Ghoul KJ, Costello MJ. Fiber cell morphology and cytoplasmic texture in cataractous and normal human lens nuclei. Curr Eye Res 1996; 15(5):533–542.
354. Taylor VL, Costello MJ. Fourier analysis of textural variations in human normal and cataractous lens nuclear fiber cell cytoplasm. Exp Eye Res 1999; 69(2):163–174.
355. Velasco PT et al. Hierarchy of lens proteins requiring protection against heat-induced precipitation by the alpha crystallin chaperone. Exp Eye Res 1997; 65(4):497–505.
356. Tabandeh H et al. Water content, lens hardness and cataract appearance. Eye 1994; 8(Pt 1):125–129.
357. Heys KR, Cram SL, Truscott RJ. Massive increase in the stiffness of the human lens nucleus with age: the basis for presbyopia? Mol Vis 2004; 10:956–963.
358. Merriam JC. The concentration of light in the human lens. Trans Am Ophthalmol Soc 1996; 94:803–918.
359. Cruickshanks KJ, Klein BE, Klein R. Ultraviolet light exposure and lens opacities: the Beaver Dam Eye Study. Am J Public Health 1992; 82(12):1658–1662.
360. Brown NP et al. Is cortical spoke cataract due to lens fibre breaks? The relationship between fibre folds, fibre breaks, waterclefts and spoke cataract. Eye 1993; 7(Pt 5):672–679.
361. Vrensen G, Willekens B. Biomicroscopy and scanning electron microscopy of early opacities in the aging human lens. Invest Ophthalmol Vis Sci 1990; 31(8):1582–1591.
362. Michael R et al. Morphology of age-related cuneiform cortical cataracts: The case for mechanical stress. Vision Res 2008; 48(4):626–634.
363. Vrensen, GF, Aging of the human eye lens – a morphological point of view. Comp Biochem Physiol A Physiol 1995; 111(4):519–532.
364. Duindam JJ et al. Cholesterol, phospholipid, and protein changes in focal opacities in the human eye lens. Invest Ophthalmol Vis Sci 1998; 39(1):94–103.
365. Marcantonio JM, Duncan G, Rink H. Calcium-induced opacification and loss of protein in the organ-cultured bovine lens. Exp Eye Res 1986; 42(6):617–630.
366. Vrensen GF et al. Heterogeneity in ultrastructure and elemental composition of perinuclear lens retrodots. Invest Ophthalmol Vis Sci 1994; 35(1):199–206.
367. Creighton MO et al. Globular bodies: a primary cause of the opacity in senile and diabetic posterior cortical subcapsular cataracts? Can J Ophthalmol, 1978. 13(3):166–181.
368. Rafferty NS, Goossens W, March WF. Ultrastructure of human traumatic cataract. Am J Ophthalmol 1974; 78(6):985–995.
369. Kasthurirangan S et al. In vivo study of changes in refractive index distribution in the human crystalline lens with age and accommodation. Invest Ophthalmol Vis Sci 2008; 49(6):2531–2540.
370. Weeber HA et al. Dynamic mechanical properties of human lenses. Exp Eye Res 2005; 80(3):425–434.
371. Fisher RF. Senile cataract. A comparative study between lens fibre stress and cuneiform opacity formation. Trans Ophthalmol Soc UK 1970; 90:93–109.
372. Pau H. Cortical and subcapsular cataracts: significance of physical forces. Ophthalmologica 2006; 220(1):1–5.
373. Weeber HA, van der Heijde RGL. Internal deformation of the human crystalline lens during accommodation. Acta Ophthalmol 2008.
374. Streeten BW, Eshaghian J. Human posterior subcapsular cataract. A gross and flat preparation study. Arch Ophthalmol 1978; 96(9):1653–1658.
375. Eshaghian J, Streeten BW. Human posterior subcapsular cataract. An ultrastructural study of the posteriorly migrating cells. Arch Ophthalmol 1980; 98(1):134–143.
376. Eshagian, J, Human posterior subcapsular cataracts. Trans Ophthalmol Soc UK 1982; 102(Pt 3):364–368.
377. Lyu, J et al. Alteration of cadherin in dexamethasone-induced cataract organ-cultured rat lens. Invest. Ophthalmol Vis Sci 2003; 44(5):2034–2040.
378. Worgul BV et al. Radiation cataractogenesis in the amphibian lens. Ophthalmic Res 1982; 14:73–82.
379. Merriam GR, Jr., Worgul BV. Experimental radiation cataract – its clinical relevance. Bull NY Acad Med 1983; 59:372–392.
380. Palva, M, Palkama A. Ultrastructural lens changes in X-ray induced cataract of the rat. Acta Ophthalmol 1978; 56(4):587–598.
381. Shui YB et al. In vivo morphological changes in rat lenses induced by the administration of prednisolone after subliminal X-irradiation. A preliminary report. Ophthalmic Res 1995; 27(3):178–186.
382. Zintz C, Beebe DC. Morphological and cell volume changes in the rat lens during the formation of radiation cataracts. Exp Eye Res 1986; 42(1):43–54.
383. Worgul BV, Merriam GR Jr. The lens epithelium and radiation cataracts. II Interphase death in the meridional rows? Radiat Res 1980; 84(1):115–121.
384. Holsclaw DS et al. Modulating radiation cataractogenesis by hormonally manipulating lenticular growth kinetics. Exp Eye Res 1994; 59(3):291–296.
385. Kappelhof JP et al. The ring of Soemmerring in the rabbit: A scanning electron microscopic study. Graefes Arch Clin Exp Ophthalmol 1985; 223(3):111–120.
386. Kappelhof JP, Vrensen GF. The pathology of after-cataract. A minireview. Acta Ophthalmol Suppl 1992; 205:13–24.
387. McDonnell PJ, Stark WJ, Green WR Posterior capsule opacification: a specular microscopic study. Ophthalmology 1984; 91(7):853–856.

388. Marcantonio JM, Vrensen GF. Cell biology of posterior capsular opacification. Eye 1999; 13(Pt 3b):484–488.

389. Schmitt-Graff A et al. Appearance of alpha-smooth muscle actin in human eye lens cells of anterior capsular cataract and in cultured bovine lens-forming cells. Differentiation 1990; 43(2):115–122.

390. Novotny GE, Pau H. Myofibroblast-like cells in human anterior capsular cataract. Virchows Arch A Pathol Anat Histopathol 1984; 404(4):393–401.

391. Hales AM et al. TGF-beta 1 induces lens cells to accumulate alpha-smooth muscle actin, a marker for subcapsular cataracts. Curr Eye Res 1994; 13(12):885–890.

392. Liu J et al. Induction of cataract-like changes in rat lens epithelial explants by transforming growth factor beta. Invest Ophthalmol Vis Sci 1994; 35(2):388–401.

393. Kurosaka D et al. Growth factors influence contractility and alpha-smooth muscle actin expression in bovine lens epithelial cells. Invest Ophthalmol Vis Sci 1995; 36(8):1701–1708.

394. Hales AM et al. Intravitreal injection of TGFbeta induces cataract in rats. Invest Ophthalmol Vis Sci 1999; 40(13):3231–3236.

395. Hales AM, Chamberlain CG, McAvoy JW. Cataract induction in lenses cultured with transforming growth factor-beta. Invest Ophthalmol Vis Sci 1995; 36(8):1709–1713.

396. Nagamoto T, Eguchi G, Beebe DC. Alpha-smooth muscle actin expression in cultured lens epithelial cells. Invest Ophthalmol Vis Sci 2000; 41(5):1122–1129.

397. Webster WS. Teratogen update: congenital rubella. Teratology 1998; 58(1):13–23.

398. Karkinen-Jaaskelainen M et al. Rubella cataract in vitro: Sensitive period of the developing human lens. J Exp Med 1975; 141(6):1238–1248.

399. Zimmerman LE. Histopathologic basis for ocular manifestations of congenital rubella syndrome. Am J Ophthalmol 1968; 65(6):837–862.

400. Klopp N et al. Three murine cataract mutants (Cat2) are defective in different gamma– crystallin genes. Genomics 1998; 52(2):152–158.

401. Kim H, CK Joo CK. The prevalence and demographic characteristics of anterior polar cataract in a hospital-based study in Korea. Korean J Ophthalmol 2008; 22(2):77–80.

402. Lee EH, Joo CK. Role of transforming growth factor-beta in transdifferentiation and fibrosis of lens epithelial cells. Invest Ophthalmol Vis Sci 1999; 40(9):2025–2032.

403. Litt M et al. Autosomal dominant congenital cataract associated with a missense mutation in the human alpha crystallin gene CRYAA. Hum Mol Genet 1998; 7(3):471–474.

404. Stephan DA et al. Progressive juvenile-onset punctate cataracts caused by mutation of the gammaD-crystallin gene. Proc Natl Acad Sci USA 1999; 96(3):1008–1012.

405. Litt M et al. Autosomal dominant cerulean cataract is associated with a chain termination mutation in the human beta-crystallin gene CRYBB2. Hum Mol Genet 1997; 6(5):665–668.

406. Heon E et al. The gamma-crystallins and human cataracts: a puzzle made clearer. Am J Hum Genet 1999; 65(5):1261–1267.

407. Shiels A et al. A missense mutation in the human connexin50 gene (GJA8) underlies autosomal dominant "zonular pulverulent" cataract, on chromosome 1q. Am J Hum Genet 1998; 62(3):526–532.

408. Mackay D et al. Connexin46 mutations in autosomal dominant congenital cataract. Am J Hum Genet 1999; 64(5):1357–1364.

409. Graw J. Mouse models of congenital cataract. Eye 1999; 13(Pt 3b):438–444.

410. Attree O et al. The Lowe's oculocerebrorenal syndrome gene encodes a protein highly homologous to inositol polyphosphate-5-phosphatase. Nature 1992; 358(6383):239–242.

411. Rouleau GA et al. Alteration in a new gene encoding a putative membrane-organizing protein causes neuro-fibromatosis type 2. Nature 1993; 363(6429):515–521.

412. Gitzelmann R. Deficiency of erythrocyte galactokinase in a patient with galactose diabetes. Lancet 1965; 2(7414):670–671.

413. Goppert F. Galaktosurie nach Milchzuckergabe bei angeborenem, familiaerem chronischem Leberleiden. Klin Wschr 1917; 54:473–477.

414. Girelli D et al. A linkage between hereditary hyperferritinaemia not related to iron overload and autosomal dominant congenital cataract. Br J Haematol 1995; 90(4):931–934.

415. Yu CE et al. Positional cloning of the Werner's syndrome gene. Science 1996; 272(5259):258–262.

416. Sarkar PS et al. Heterozygous loss of Six5 in mice is sufficient to cause ocular cataracts. Nat Genet 2000; 25(1):110–114.

417. Klesert TR et al. Mice deficient in Six5 develop cataracts: implications for myotonic dystrophy. Nat Genet 2000; 25(1):105–109.

418. Cogan DG, Donaldson DD, Reese AB. Clinical and pathological characteristics of radiation cataract. AMA Arch Ophthalmol 1952:55–70.

419. Worgul BV et al. Lens epithelium and radiation cataract. Arch Ophthalmol 1976; 94:996–999.

420. Worgul BV, Rothstein H. Radiation cataract and mitosis. Ophthal Res 1975; 7:21–32.

421. Worgul BV, Rothstein H. On the mechanism of radiocataractogenesis. Medikon 1977; I:5–13.

422. Alter AJ, Leinfelder PJ. Roentgen-ray cataract. Effects of shielding the lens and ciliary body. Arch Ophthalmol 1953; 49:257–260.

423. Rothstein H et al. G0/G1 arrest of cell proliferation in the ocular lens prevents development of radiation cataract. Ophthalmic Res 1982; 14(3):215–220.

424. Leinfelder PJ, Dickerson J. Species variation of the lens epithelium to ionizing radiation. Am J Ophthalmol 1960; 50:175–176.

425. Matsuda H, Giblin FJ, Reddy VN. The effect of x-irradiation on cation transport in rabbit lens. Exp Eye Res 1981; 33:253–265.

426. Garadi R et al. Protein synthesis in x-irradiated rabbit lens. Invest Ophthalmol Vis Sci 1984; 25(2):147–152.

427. Taylor HR. Ultraviolet radiation and the eye: an epidemiologic study. Trans Am Ophthalmol Soc 1989; 87:802–853.

428. McCarty CA, Taylor HR. Recent developments in vision research: light damage in cataract. Invest Ophthalmol Vis Sci 1996; 37(9):1720–1723.

429. West SK et al. Sunlight exposure and risk of lens opacities in a population-based study: the Salisbury Eye Evaluation project. JAMA 1998; 280(8):714–718.

430. Cruickshanks, KJ, Sunlight exposure and risk of lens opacities in a population-based study. Arch Ophthalmol 1998; 116(12):1666.

431. Taylor HR et al. Effect of ultraviolet radiation on cataract formation. N Engl J Med 1988; 319(22):1429–1433.

432. Zigman S et al. Effect of chronic near-ultraviolet radiation on the gray squirrel lens in vivo. Invest Ophthalmol Vis Sci 1991; 32(6):1723–1732.

433. Dillon J et al. Electron paramagnetic resonance and spin trapping investigations of the photoreactivity of human lens proteins. Photochem Photobiol 1999; 69(2):259–264.

434. Coroneo MT, Muller-Stolzenburg NW, Ho A. Peripheral light focusing by the anterior eye and the ophthalmohelioses. Ophthalmic Surg 1991; 22(12):705–711.

435. Klein B, Klein R, Linton K. Prevalence of age-related lens opacities in a population: The Beaver Dam Eye Study. Ophthalmology 1992; 99(4):546–552.

436. Moffat BA, Atchison DA, Pope JM. Age-related changes in refractive index distribution and power of the human lens as measured by magnetic resonance micro-imaging in vitro. Vision Res 2002; 42(13):1683–1693.

437. Hermans EA et al. Change in the accommodative force on the lens of the human eye with age. Vision Res 2008; 48(1):119–126.

438. Lydahl E, Philipson B. Infrared radiation and cataract II Epidemiologic investigation of glass workers. Acta Ophthalmol (Copenh) 1984; 62(6):976–992.

439. Lydahl E. Infrared radiation and cataract. Acta Ophthalmol Suppl 1984; 166:1–63.

440. Appleton B, McCrossan GC. Microwave lens effects in humans. Arch Ophthalmol 1972; 88(3):259–262.

441. Milroy WC, Michaelson SM. Microwave cataractogenesis: a critical review of the literature. Aerospace Med 1972; 43(1):67–75.

442. Creighton MO et al. In vitro studies of microwave-induced cataract. II Comparison of damage observed for continuous wave and pulsed microwaves. Exp Eye Res 1987; 45(3):357–373.

443. Sasaki H et al. High prevalence of nuclear cataract in the population of tropical and subtropical areas. Dev Ophthalmol 2002; 35:60–69.

444. Kinoshita JH, A thirty year journey in the polyol pathway. Exp Eye Res 1990; 50(6):567–573.

445. Kinoshita JH, Nishimura C. The involvement of aldose reductase in diabetic complications. Diabetes Metab Rev 1988; 4(4):323–337.

446. Datiles MB, H Fukui H. Cataract prevention in diabetic Octodon degus with Pfizer's sorbinil. Curr Eye Res 1989; 8(3):233–237.

447. Varma SD, Mizuno A, Kinoshita JH. Diabetic cataracts and flavonoids. Science 1977; 195(4274):205–206.

448. Yamaoka T et al. Acute onset of diabetic pathological changes in transgenic mice with human aldose reductase Cdna. Diabetologia 1995; 38(3):255–261.

449. King GL et al. Biochemical and molecular mechanisms in the development of diabetic vascular complications. Diabetes 1996; 45(Suppl 3):S105–S108.

450. Brownlee M. Negative consequences of glycation. Metabolism 2000; 49(2 Suppl 1): 9–13.

451. Nishikawa T et al. Normalizing mitochondrial superoxide production blocks three pathways of hyperglycaemic damage. Nature 2000; 404(6779):787–790.

452. Kumamoto Y et al. Epithelial cell density in cataractous lenses of patients with diabetes: association with erythrocyte aldose reductase. Exp Eye Res 2007; 85(3):393–399.

453. Harocopos GJ et al. Importance of vitreous liquefaction in age-related cataract. Invest Ophthalmol Vis Sci 2004; 45(1):77–85.

454. Andley UP et al. The role of prostaglandins E2 and F2 alpha in ultraviolet radiation-induced cortical cataracts in vivo. Invest Ophthalmol Vis Sci 1996; 37(8):1539–1548.

455. Shui YB et al. HIF-1: an age-dependent regulator of lens cell proliferation. Invest Ophthalmol Vis Sci 2008

456. Hayden JH et al. Hypophysectomy exerts a radioprotective effect on frog lens. Experientia 1980; 36(1):116–118.

457. Rothstein H et al. Somatomedin C: restoration in vivo of cycle traverse in G0/G1 blocked cells of hypophysectomized animals. Science 1980; 208(4442):410–412.

458. Ionides A et al. Clinical and genetic heterogeneity in autosomal dominant cataract. Br J Ophthalmol 1999; 83(7):802–808.

459. Ionides A et al. The clinical and genetic heterogeneity of autosomal dominant cataract. Acta Ophthalmol Scand Suppl 1996; 219:40–41.

460. Berry V et al. A locus for autosomal dominant anterior polar cataract on chromosome 17p. Hum Mol Genet 1996; 5(3):415–419.

461. Hammond CJ et al. Genetic and environmental factors in age-related nuclear cataracts in monozygotic and dizygotic twins. N Engl J Med 2000; 342(24):1786–1790.

462. Hammond CJ et al. The heritability of age-related cortical cataract: the twin eye study. Invest Ophthalmol Vis Sci 2001; 42(3):601–605.

463. Heiba IM et al. Genetic etiology of nuclear cataract: evidence for a major gene. Am J Med Genet 1993; 47(8):1208–1214.

464. Heiba IM et al. Evidence for a major gene for cortical cataract. Invest Ophthalmol Vis Sci 1995; 36(1):227–235.

465. Shiels A et al. The EPHA2 gene is associated with cataracts linked to chromosome 1p. Mol Vis 2008; 14:2042–2055.

466. Cooper MA et al. Loss of ephrin-A5 function disrupts lens fiber cell packing and leads to cataract. Proc Natl Acad Sci USA 2008

467. Pararajasegaram R. VISION 2020 – the right to sight: from strategies to action. Am J Ophthalmol 1999; 128(3):359–360.

468. Chiu CJ, Taylor A. Nutritional antioxidants and age-related cataract and maculopathy. Exp Eye Res 2006

469. AREDS. A randomized, placebo-controlled, clinical trial of high-dose supplementation with vitamins C and E and beta carotene for age-related cataract and vision loss. AREDS Report no. 9. Arch Ophthalmol 2001; 119(10):1439–1452.

470. AREDS. Centrum use and progression of age-related cataract in the age-related eye disease study; a propensity score approach. AREDS Report No. 21. Ophthalmology 2006; 113(8):1264–1270.

471. Sperduto RD et al. The Linxian cataract studies. Two nutrition intervention trials. Arch Ophthalmol 1993; 111(9):1246–1253.

472. Gritz DC et al. The antioxidants in prevention of cataracts (APC) study: effects of antioxidant supplements on cataract progression in South India. Br J Ophthalmol 2006.

473. Maraini G et al. A randomized, double-masked, placebo-controlled clinical trial of multivitamin supplementation for age-related lens opacities. Clinical trial of nutritional supplements and age–related cataract report no. 3. Ophthalmology 2008; 115(4):599–607.

474. Meyer CH, Sekundo W. Nutritional supplementation to prevent cataract formation. Dev Ophthalmol 2005; 38:103–119.

475. Olsen G, Olson RJ. Update on a long-term, prospective study of capsulotomy and retinal detachment rates after cataract surgery. J Cataract Refract Surg 2000; 26(7):1017–1021.

476. Tassignon M-J et al. Bag-in-the-lens intraocular lens implantation in the pediatric eye. J Cataract Refract Surg 2007; 33(4):611–617.

477. Nagamoto T, Eguchi G. Effect of intraocular lens design on migration of lens epithelial cells onto the posterior capsule. J Cataract Refract Surg 1997; 23(6):866–872.

478. Nishi O, Nishi K, Sakanishi K. Inhibition of migrating lens epithelial cells at the capsular bend created by the rectangular optic edge of a posterior chamber intraocular lens. Ophthalmic Surg Lasers 1998; 29(7):587–594.

479. Nishi O, Nishi K, Wickstrom K. Preventing lens epithelial cell migration using intraocular lenses with sharp rectangular edges. J Cataract Refract Surg 2000; 26(10):1543–1549.

480. McAvoy J. Developmental biology of the lens. In: Duncan J, ed. Mechanism of cataract formation. Academic Press: London, 1981;7–46.

481. Brown N. Slit-image photography and measurement of the eye. Med Biol Illus 1973; 23(4):192–203.

482. FitzGerald PG, Bok D, Horwitz J. The distribution of the main intrinsic membrane polypeptide in ocular lens. Curr Eye Res 1985; 4(11):1203–1218.

CHAPTER 6

The Vitreous

Henrik Lund-Andersen & Birgit Sander

Introduction

The vitreous body makes up approximately 80% of the volume of the eye and thus is the largest single structure of the eye (Fig. 6.1). In the anterior segment of the eye, it is delineated by and adjoins the ciliary body, the zonules, and the lens. In the posterior segment of the eye, the vitreous body is delineated by and adjoins the retina.

The vitreous body has many normal physiological functions. This chapter focuses on the most important physiologic relationships, especially those that have a close clinical correlation. As background for the understanding of the physiology and the pathophysiology of the vitreous body, we focus on the main features of the anatomy, biochemistry, and biophysics.

The investigation of the vitreous body and its structure and function is hampered by two fundamental difficulties. Firstly any attempts to define vitreous morphology are in fact attempts to visualize a tissue, which by design is intended to be invisible. Secondly the various techniques that have previously been employed to define the structure of the vitreous body are combined with artifacts that make interpretations difficult in terms of the true in vivo physiological situation.

Anatomy

Embryology

Structural considerations of embryology

In the early stages the optic cup is mainly occupied by the lens vesicle. As the cup grows the space formed is filled by a system of fibrillar material, presumably secreted by the cells of the embryonic retina. Later, with the penetration of the hyaloid artery, more fibrillar material apparently originating from the cells of the wall of the artery and other vessels contribute to filling the space. The combined mass is known as *primary vitreous*.[22,23,39,110,127]

The *secondary vitreous* develops later, appearing at the end of the sixth week, and is associated with the increasing size of the vitreous cavity and the regression of the hyaloid vascular system. The main hyaloid artery remains for some time, but it eventually disappears and leaves in its place a tube of primary vitreous surrounded by the secondary vitreous, running from the retrolental space to the optic nerve (area of Martegiani). The tube is called *Cloquet's canal* (Fig. 6.1); this is not a liquid-filled canal, but simply a portion of differentiated gel devoid of collagen fibrils.

The term *tertiary vitreous* is related to the fibrillary material, which develops as the suspensor fibrils, the zonules, of the lens. During childhood the vitreous undergoes significant growth. The length of the vitreous body in the newborn eye is approximately 10.5 mm, and by the age of 13 years, the actual length of the vitreous increases to 16.1 mm in the male.[107] In the absence of refractive changes, the mean adult vitreous is 16.5 mm.[30,91]

Molecular and cellular considerations of embryology

The two main components of the vitreous, collagen and hyaluronic acid, are produced in the primary and secondary vitreous. In the primary vitreous, however, there is initial production of substances other than hyaluronic acid, such as galactosaminoglycans; later hyaluronic acid becomes the predominant constituent.[27,41,111,127]

The primary vitreous contains cells which in the secondary vitreous differentiate as hyalocytes and fibroblasts. The hyalocytes are believed to be involved in the production of glycosaminoglycans, especially hyaluronic acid, a non-sulfated glycosaminoglycan.[111]

Although the function of the fibroblasts is not known exactly, they are probably involved in the formation of collagen. The retina may also be a source of collagen synthesis.[88,108] The hyalocytes are found in the vitreous cortex, approximately 30 µm from the internal limiting membrane (ILM), with the highest density near the vitreous base and the posterior pole.[9]

Anatomy of the mature vitreous body

The mature vitreous body is a transparent gel which occupies the vitreous cavity. It has an almost spherical appearance, except for the anterior part, which is concave, corresponding to the presence of the crystallin lens. The vitreous body is a transparent gel; however, it is not completely homogeneous (Fig. 6.2). The outermost part of the vitreous, called the *cortex*, is divided into an anterior cortex and a posterior

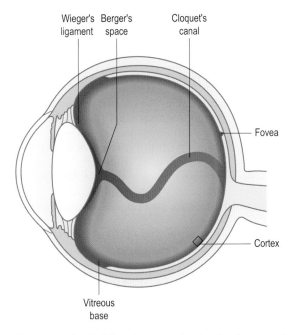

Figure 6.1 Sketch of the primate eye showing the vitreous body and its relations. Wieger's ligament is the attachment of the vitreous to the lens. Berger's space and the Cloquet's canal are the former sites of the hyaloid artery. (From Heegaard 1997.[45])

cortex, the latter being approximately 100 μm thick (Fig. 6.3). The cortex is also called the *anterior* and the *posterior hyaloid*. The cortex consists of densely packed collagen fibrils (Fig. 6.4). The vitreous base (see Fig. 6.1) is a three-dimensional zone. It extends approximately from 2 mm anterior to the ora serrate to 3 mm posterior to the ora serrata, and it is several millimeters thick. The collagen fibrils are especially densely packed in this region.

The vitreoretinal interface

The vitreoretinal interface can be defined from electron microscopy as the outer part of the vitreous cortex (posterior hyaloid), including anchoring fibrils of the vitreous body and the ILM of the retina (Fig. 6.5).[29,34,44-46,93] The ILM is a retinal structure between 1 and 3 μm thick, consisting mainly of type IV collagen and proteoglycans.[117] It contains several layers and can be considered the basal lamina of the Müller cells, the foot processes of which are in close contact with the membrane.

The vitreous cortex is firmly attached to the ILM in the vitreous base region, around the optic disc (Weiss ring), at the vessels, and in the area surrounding the foveola at a diameter of 500 μm.[48,106] Under normal conditions, the connection between the fibrils of vitreous cortex and the ILM is looser than in the rest of the vitreoretinal interface. The adhesion is strong in young individuals, and dissection of

Figure 6.2 Human vitreous dissection. (**A**) Vitreous of a 9-month-old child. The sclera, choroid and retina were dissected off the vitreous, which remains attached to the anterior segment. A band of gray tissue can be seen posterior to the ora serrata. This is peripheral retina that was firmly adherent to the posterior vitreous base and could not be dissected. The vitreous is solid and although situated on a surgical towel exposed to room air maintains its shape because at this age the vitreous is almost entirely gel. (**B**) Human vitreous dissected off the sclera, choroid and retina are still attached to the anterior segment. The specimen is mounted on a lucite frame using sutures through the limbus and is then immersed in a lucite chamber containing an isotonic, physiologic solution that maintains the turgescence of the vitreous and avoids collapse and artefactual distortion of vitreous structure. (Reprinted with permission from Sebag & Balazs 1984.[114])

the retina from the vitreous often leaves ILM tissue adherent to the vitreous cortex.[47,60,108,109] Under pathologic conditions, the tight connections between the vitreous cortex and the ILM play an important role, as is discussed later in this chapter.

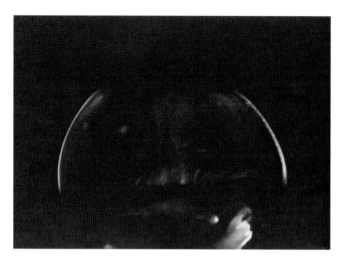

Figure 6.3 Human vitreous structure during childhood. This view of the central vitreous from an 11-year-old child demonstrates a dense vitreous cortex with hyalocytes. The posterior aspect of the lens is seen below, though dimly illuminated. No fibers are present in the vitreous.

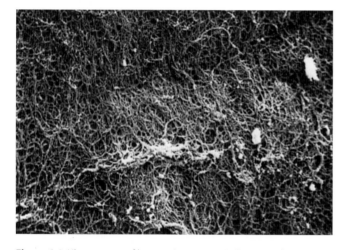

Figure 6.4 Ultrastructure of human vitreous cortex. Scanning electron microscopy demonstrates the dense packing of collagen fibrils in the vitreous cortex. To some extent this arrangement is exaggerated by the dehydration that occurs during specimen preparation for scanning electron microscopy (magnification = ×3750).

Ultrastructural, biochemical, and biophysical aspects

Ultrastructural and biochemical aspects

The vitreous contains more than 99% water; the rest is composed of solids. The vitreous acts as a gel (i.e. an interconnected meshwork) that surrounds and stabilizes a large amount of water compared with the amount of solids. The gel structure of the vitreous results from the arrangement of long, thick, non-branching, collagen fibrils suspended in a network of hyaluronic acid, which stabilize the gel structure and the conformation of the collagen fibrils (Figs 6.6 and 6.7).[8,10,107]

In the human eye the major part of the glycosaminoglycan is hyaluronic acid, with a molecular weight of $3-4.5 \times 10^6$.[111] The volume of non-hydrated hyaluronic acid is $0.66 \text{ cm}^3/\text{g}$, in contrast with the volume of the hydrated molecule, which is $2000-3000 \text{ cm}^3/\text{g}$.[8] The molecule forms into large, open coils, with the anionic sites spread apart. This arrangement of small-diameter fibers, separated by highly hydrated glycosaminoglycan chains, permits the transmission of light to the retina with minimal scattering.[11] The collagen fibrils in the vitreous are thin, with diameters of approximately 10–20 nm. Collagen fibrils are mostly of collagen type II. They are composed of three identical α-chains, which form a triple helix. The helix is stabilized by hydrogen bonds between opposing residues in different chains.[111] Collagen type IX is also present and may function as a bridge, linking type II collagen fibrils together.[10,32,120] Collagen V/XI is integrated with collagen II in the collagen fibers.[81] The collagen fibrils seem to interconnect with the hyaluronic acid, most likely via bridging glucoproteins.[111] The viscoelastic properties of the vitreous gel are neither due to hyaluronic acid or collagen alone but to the combination of the two molecules.[128]

Dissolved in the water of the vitreous gel are inorganic and organic substances as shown in Table 6.1, where plasma values are given for comparison.

According to Table 6.1, it appears that gradients exist in both directions between vitreous and plasma.[3,79,82,97,98] These gradients are a result of several mechanisms: presence of the blood–ocular barriers (i.e. active and passive passage across the barriers), metabolism in retina and ciliary body, and diffusion processes in the vitreous body (Box 6.1).

The values in Table 6.1 represent mean values for the whole vitreous. The methods used to quantitate vitreous concentrations are difficult and may differ between studies in absolute numbers. However, regional differences within the vitreous have been measured for some substances.[12]

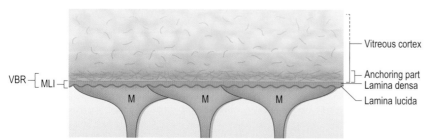

VBR—[MLI—[

Vitreous cortex

Anchoring part
Lamina densa
Lamina lucida

Figure 6.5 Sketch of the vitreoretinal interface / vitreoretinal border region (VBR). The VBR consists of two major components: the anchoring fibrils of the vitreous body and membrana limitans interna (MLI). The MLI is composed of three structures: the fusing point of the anchoring vitreous fibrils, lamina densa and lamina lucida. M = Müller cell. (From Heegaard 1997.[45])

Figures 6.8 to 6.10 show the regional difference for glucose (Fig. 6.8), lactate (Fig. 6.9), and oxygen (Fig. 6.10). The fall in vitreous oxygen tension towards the center, corresponding to the upper curve in Figure 6.10, was also found by Sakaue[100] and seems to result from an oxygen flux from the retina towards the vitreous corresponding to arterioles; the flux goes in the opposite direction corresponding to the venules (lower curve). Several studies have found an increase in preretinal oxygen after photocoagulation, indicating that the oxygen supply to the inner retina improves after destruction of the outer retina and a concomitant decrease in tissue metabolism and oxygen needs.[40,85,94,122–124]

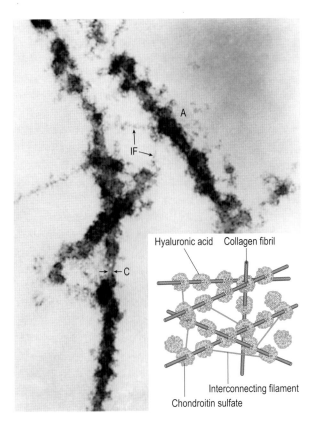

Figure 6.6 shows a schematic diagram with labels: Hyaluronic acid, Collagen fibril, Interconnecting filament, Chondroitin sulfate.

Figure 6.6 Ultrastructure of hyaluronic acid/collagen interaction in the vitreous. Specimen was fixed in glutaraldehyde/paraformaldehyde and stained with ruthenium red. Collagen fibrils (C) are coated with amorphous material (A) believed to be hyaluronic acid. The amorphous material may connect to the collagen fibril via another glycosaminoglycan, possibly chondroitin sulfate (see inset). Interconnecting filaments (IF) appear to bridge between collagen fibrils, inserting or attaching at sites of hyaluronic acid adhesion to the collagen fibrils (bar = 0.1 μm). (Reprinted with permission from: Asakura A. Histochemistry of hyaluronic acid of the bovine vitreous body as studied by electron microscopy. Acta Soc Ophthalmol J 1985; 89:179.)

Box 6.1 Vitreous – aging and ocular pathology

- The concentration of salts and organic substances of the vitreous differ substantially from plasma due to the blood–aqueous and blood–retinal barrier
- Small molecules move through the vitreous gel by diffusion
- Vitreous fluorometry is useful for evaluation of the vitreal morphology, the fluorescein profile is an indicator of physiologic aging such as vitreous liquefaction
- The aging process leads to posterior vitreous detachment, easily visualized by optical coherence tomography (OCT)
- Vitreoretinal traction may lead to formation of a macular hole and the traction can be conducted through the retinal layers
- Vitreoretinal traction is also implicated in some cases of macular edema
- In diabetes, the high glucose speeds up metabolism before visible retinopathy
- Increased demand for oxygen and capillary closure leads to retinal ischemia and an increased production of VEGF
- Increased leakage through the blood–retinal barrier leads to macular edema
- VEGF inhibition and steroids decrease macular edema

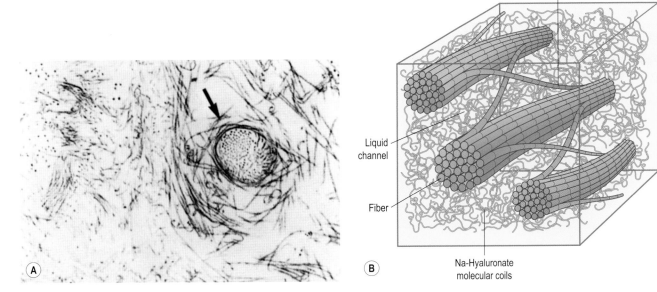

Figure 6.7 Ultrastructure of human vitreous. (**A**) Specimens were centrifuged to concentrate structural elements, but contained no membranes or membranous structures. Only collagen fibrils were detected. There were also bundles of parallel collagen fibrils such as the one shown here in cross-section (arrow). (**B**) Schematic diagram of vitreous ultrastructure, depicting the dissociation of hyaluronic acid (HA) molecules and collagen fibrils. The fibrils aggregate into bundles of packed parallel units. The HA molecules fill the spaces between the packed collagen fibrils and form "channels" of liquid vitreous. (Reprinted with permission from Sebag & Balazs 1989.[113] Reproduced with permission from Association for Research in Vision and Ophthalmology.)

Table 6.1 Concentration of various substances in the vitreous (weighted averages in mmol/kg H_2O)

	Sodium	Potassium	Calcium	Magnesium	Chloride	Phosphate	pH
Inorganic substances							
Vitreous	134	9.5	5.4*	2.3*	105	2	7.29**
Plasma	143	5.6	9.9*	2.2*	97	0.4	7.41**

Organic substances		
	Ascorbate	Glucose	Lactate
Vitreous	0.46	3.0	12.0
Plasma	0.04	5.7	10.3

*Human data from McNeil et al 1999.[82]

**Porcine data from Andersen 1991.[3]

All other values are rabbit data from Reddy & Kinsey 1960;[97] with modification from Kinsey 1967.[59]

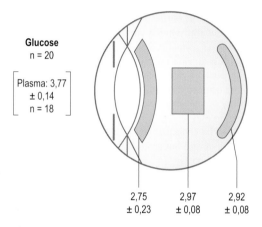

Glucose
n = 20

Plasma: 3,77
± 0,14
n = 18

2,75   2,97   2,92
± 0,23   ± 0,08   ± 0,08

Figure 6.8 Glucose concentration in different parts of the vitreous body and in plasma. All values are in μmol/g tissue weight (mean ± standard deviation, n = 20). (From Bourwieg et al 1974.[12] Reproduced with permission from Association for Research in Vision and Ophthalmology.)

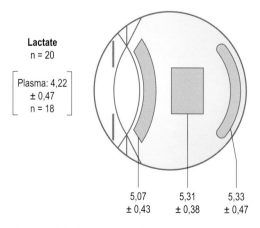

Lactate
n = 20

Plasma: 4,22
± 0,47
n = 18

5,07   5,31   5,33
± 0,43   ± 0,38   ± 0,47

Figure 6.9 Lactate concentration in different parts of the vitreous body and in plasma. All values are in μmol/g tissue weight (mean ± standard deviation, n = 20). (From Bourwieg et al 1974.[12] With kind permission of Springer Science + Business Media.)

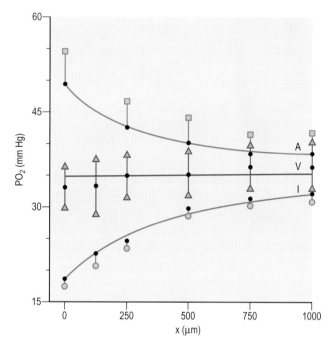

Figure 6.10 Heterogeneity of the PO_2 in the preretinal vitreous of a non-photocoagulated eye. Graphic representation of the PO_2 (± SEM) recorded when the O_2-sensitive microelectrode was withdrawn from the vitreal surface of the retina (x = 0) towards the vitreous. Curve A: opposite an arteriole: curve I: opposite an intervascular zone: curve V: opposite a vein. These results are averages of measurements made on one or both eyes of 11 miniature pigs. (From: Mohar I. Effect of laser photocoagulation on oxygenation of the retina in miniature pigs. Invest Ophthalmol Vis Sci 1985; 26:1410. Reproduced with permission from Association for Research in Vision and Ophthalmology.)

Biophysical aspects

The gel structure acts as a barrier against movement of solutes. Basically, substances may move by two different processes: diffusion or bulk flow. The diffusion process can be illustrated in humans by using fluorescein as a tracer substance for the biophysical behavior of the gel. The fluorescein concentration in the vitreous body can be estimated by vitreous fluorophotometry. After intravenous (IV) injection of fluorescein, a certain amount (in healthy humans only a very small amount) passes through the ocular barriers into the anterior chamber and into the vitreous body. The ILM, the vitreoretinal interface, and the vitreous cortex cannot be regarded as a diffusion restriction to smaller molecules (Box 6.1). In the vitreous the distribution versus time occurs according to the diffusion properties of a particular molecule in the vitreous gel.

An analysis of the fluorescein concentration gradient in the posterior part of the vitreous can be made with the aid of a simplified mathematical model of the relationship between the vitreous body and the blood–retinal barrier, as shown in Figures 6.11–6.14.[78]

In the model the vitreous body is considered as a globe with an outer delineation corresponding to the blood–retinal barrier (Fig. 6.11). Fluorescein passes the barrier passively with permeability P. Diffusion in the vitreous gel takes place with a diffusion coefficient D. The time-dependent plasma fluorescein concentration is given by $C_0(t)$ and the concentration in the vitreous body dependent on time (t) and distance (r) from the center of the eye is given by C(r,t).

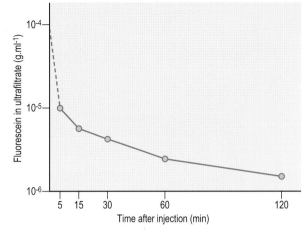

Figure 6.12 Concentration of free (non-protein bound) fluorescein in ultrafiltrate of plasma versus time after IV injection of the dye. The data were obtained in a normal subject. Fluorescein was injected at time 0. The first blood sample was obtained at 5 min, then 15, 30, 60 and 120 min after the injection. The concentration during the first min of injection was not directly measured, but calculated (dotted line) – see method. (From Lund-Andersen et al 1982.[75])

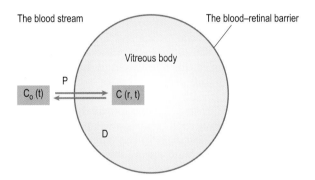

Figure 6.11 Simplified model of the eye used for the computerized calculation of a blood–retinal barrier permeability and vitreous body diffusion coefficient for the substance fluorescein. $C_0(t)$: concentration of free (not protein-bound) fluorescein in plasma at time (t): C (r,t): concentration of fluorescein in the vitreous body at time (t) and at the position (r) from the center of the eye. P: permeability of the blood–retinal barrier, symbolized by a single spherical shell: D: diffusion coefficient in the vitreous body. (From Lund-Andersen et al 1985.[78] Reproduced with permission from Association for Research in Vision and Ophthalmology.)

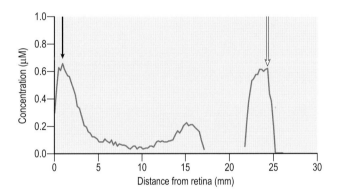

Figure 6.13 Vitreous fluorophotometry scan along the optical axis of the eye obtained 60 min after injection of fluorescein. The black arrow indicates the retina, the open arrow the fluorescein concentration in the anterior chamber. The autofluorescence signal from the lens has been removed. Note a small peak behind the lens (~15 mm from the retina) due to fluorescein leaking from the anterior chamber into the vitreous body. (From Sander et al 2001.[102] Reproduced with permission from Association for Research in Vision and Ophthalmology.)

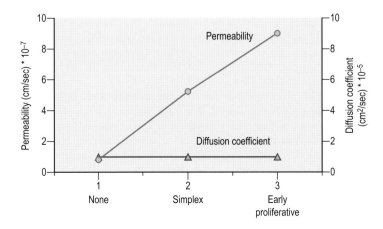

Figure 6.14 Diffusion coefficient for fluorescein in the vitreous body and fluorescein permeability of the blood–retinal barrier in diabetic patients with three different degrees of retinopathy. (Redrawn with permission from Lund-Andersen et al 1985.[77])

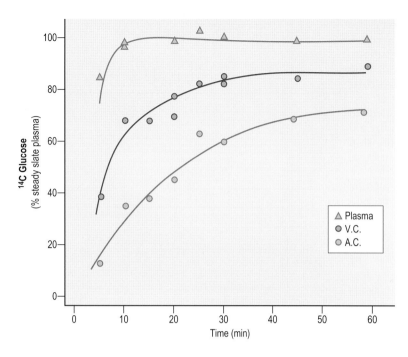

Figure 6.15 Entry of ^{14}C-glucose from the blood into the aqueous (A.C.) and vitreous (V.C.) humors. Blood and aqueous humor samples were fractionated on Sephadex columns before counting to separate glucose from its metabolic products. Curves are hand-fitted to the data; 6 animals were used. (From Riley 1972.[98] Reproduced with permission from Association for Research in Vision and Ophthalmology.)

The basic equations and the mathematical formalisms are as follows:[78]

$$C(r,t) = \int_0^t C_o(t-s) * F(r,s;\, a, D, P)\, ds \qquad (1)$$

where

$$F(r,s;a,D,P) = \frac{aP}{r\sqrt{D}} * \left[G\left(\frac{a-r}{2\sqrt{D}}, s; k\right) - G\left(\frac{a+r}{2\sqrt{D}}, s; k\right) \right] \qquad (2)$$

In equation (2), G is given by

$$G(x, s; k) = e^{-x2/4s} / \sqrt{\pi s} - k * e^{k(x+k*s)} \operatorname{erfc}\left(k\sqrt{s} + x/2\sqrt{s}\right) \qquad (3)$$

where $k = (P\sqrt{D}) - (\sqrt{D}/a)$ and erfc is the complementary error function. Radius [a] of the eye is determined experimentally. P and D are calculated from a set of experimental data by minimizing

$$S = \sum_{i=1}^{N} w_i \left[C_m(r_i, t) - C(r_i, t) \right]^2 \qquad (4)$$

where $C_m [r_i, t]$ is the measured value at $r = r_i$, and c is the corresponding value given by equation (1).

Equation (4) indicates that each point of the vitreous concentration profile is weighed equally during the fitting procedure. In the paper by Larsen et al[67] another weighing procedure was suggested. However, this procedure adds too much weight to the low values towards the center of the eye and, accordingly, the present equal weighting procedure is preferred.

Figure 6.12 shows the fluorescein concentration in the bloodstream versus time after IV injection.[75] Figure 6.13 shows the fluorescein concentration in the vitreous body and the anterior chamber 60 minutes after the injection as determined by vitreous fluorophotometry. A combination of plasma and vitreous values by aid of the simplified mathematical model results by curve fitting in a diffusion coefficient of approximately 6×10^{-6} cm^2/sec. This is close to a diffusion coefficient that would be expected in an unstirred

gel; experimentally, the diffusion coefficients for mannitol and inulin have been found to be 2.4 and 2.0×10^{-6} cm^2/sec respectively.[74,76]

The diffusion coefficient for fluorescein in the vitreous in diabetic patients with different degrees of retinopathy is shown in Figure 6.14. Although the permeability of the blood–retina barrier increases in relation to the degree of retinopathy, the diffusion coefficient is unchanged, indicating that the spread of fluorescein in the vitreous gel occurs with the same kinetics and rate during the earlier phases of diabetic retinopathy.

The permeability for the blood–retinal barrier relates to low-molecular-weight substances and ions; is low in the healthy eye, as shown here for fluorescein; and is close to the permeability of the blood–brain barrier.[13,131] The blood–aqueous barrier (i.e. the barrier in the ciliary body and the iris) is looser, although it is still tighter than capillaries elsewhere, such as in the muscles.[19]

The presence of the ocular barriers and the "slow" diffusion process have the consequence that transient changes in the bloodstream are reflected slowly in the total vitreous body (Fig. 6.13). The slow change of the vitreous body concentration can be used in some aspects of legal medicine regarding postmortem diagnosis.[82] The time constants for many substances (Figs 6.15 and 6.16) are of the same magnitude as those describing glucose transport between blood and brain; that is, half of the maximum is achieved in approximately 10 minutes.[73]

Bulk flow through the vitreous cavity as a result of a possible pressure gradient from the anterior part of the eye toward the posterior pole does not play any significant role in the distribution of low-molecular-weight substances in the intact vitreous; this aspect was not included in the mathematical model. However, high-molecular-weight substances or large particles are moving through the vitreous as a result of bulk flow (i.e. the flow of liquid that enters the vitreous body from the retrozonular space and leaves through the retina, as

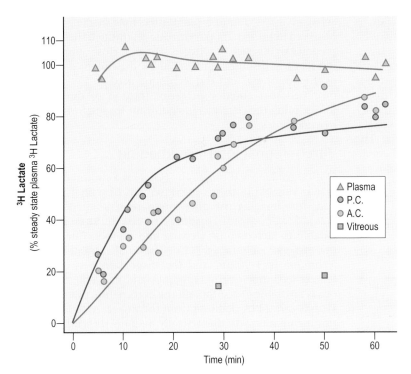

Figure 6.16 Entry of H-lactate from the blood into the aqueous (A.C., anterior chamber; P.C., posterior chamber) and vitreous humors. The two lower curves in this figure are best fit lines obtained by analog computer analysis in accordance with the theory of aqueous humor dynamics derived by Kinsey & Reddy. (From Riley 1972.[98] Reproduced with permission from Association for Research in Vision and Ophthalmology.)

described by Fatt[28]). If such a compound is placed in the anterior part of the vitreous, it moves slowly toward the retina; the diffusion process is virtually zero for large molecules. In contrast, diffusion is more rapid for the movement of small molecules. Low-molecular-weight substances move faster, diffusing in all directions, and are virtually unaffected by bulk flow; if placed in the vitreous, a low-molecular-weight substance will also be found in the anterior chamber.

Aging of the vitreous

The vitreous body goes through considerable physiological changes during life; changes that have great significance for its function. There is a sliding transition between the physiological aging changes and actual degenerative changes (retinitis pigmentosa, Wagner's disease).

The normal postnatal vitreous body is a homogeneous gel developed and biochemically composed as described above. The fundamental aging change is a disintegration of the gel structure, the so-called liquefaction or synchysis, especially notable in the center of the vitreous where the collagen concentration is lowest.[119,127,128] Liquefaction starts early in life, and a linear increase in the volume of vitreous liquid is found with age.[8,118]

Molecular mechanisms in aging

The mechanisms behind the liquefaction are not known exactly but could be linked to conformational changes in the collagen.[2] The apparent molecular weight of vitreous collagen increases with age because of the formation of new covalent cross links between the peptide chains equivalent to the aging process in collagen elsewhere in the body.[1] Bundles of collagen fibrils become biomicroscopically visible as coarse fibrous opacities.[113] The common aging

processes, such as the cumulative effect of light exposure and non-enzymatic glycosylation, seem to be important. Both hyaluronic acid and collagen may be affected by free radicals in the presence of a photosensitizer such as riboflavin (which is present in the eye) after irradiation with white light.[2] Enzymatic and non-enzymatic cross-linking have also been demonstrated.[115,116,134] Non-enzymatic glycosylation is well known from other tissues with a slow turnover of proteins, such as the lens.[58,125] Proteins are cross-linked due to the Maillard reaction with formation of a covalent binding between an amino group and glucose leading to insolubilized proteins (advanced glycation end-products – AGEs). The process is modulated by ultraviolet light and accelerated in persons with diabetes mellitus.[101] The vitreous glucose concentration is doubled in persons with diabetes compared with that of healthy subjects.[79] Sebag et al have found that collagens in the vitreous are cross-linked due to non-enzymatic glycolysation.[111,116,133]

Other mechanisms are probably involved. The network density of collagen decreases in childhood due to the growth of the eye, which could destabilize the gel. On the other hand, the hyaluronic acid concentration is increased and leads to gel stabilization.[111] The concentration of electrolytes, soluble protein and other substances such as metalloproteinase may change.[15,52] The soluble protein concentration increases with age due to an increase in the leakage through the blood–retina barrier, which may play a role both in the normal aging process and in pathologic conditions such as diabetic retinopathy.[127]

Structural changes

Regardless of the exact nature of the molecular changes, the structure of the gel is dissolved and replaced with aqueous lacunae, which melt together over time. The hyaluronic acid is redistributed from the gel to the liquid vitreous with concomitant conformational changes.[5] The liquefaction is qualitatively shown in Figures 6.17 and 6.18 and quantitatively in Figure 6.19.

Human vitreous structure can be observed using dark-field microscopy.[108,110,113,116] With this technique, the vitreous is optically empty in the young apart from the vitreous cortex. With time, fine parallel fibers appear, with anterior–posterior fibers attached to the vitreous base and ora serrata. Peripheral fibers are circumferential with the vitreous. The fibers probably correspond to aggregated collagen fibers no longer separated by hyaluronic acid and with increasing age the fibers become thickened and associated with pockets of liquid vitreous (lacunae).

In a series of investigations using India ink injection in the vitreous body, Worst argued that the adult vitreous body is composed of a number of cisterns (Fig. 6.20),[54,55,137] the lining of the cisterns corresponding to the fibers found by Sebag. A funnel-shaped cavity, the bursa premacularis, was found in front of the macula. Using fluorescein staining, Kishi has described a similar structure termed the posterior precortical vitreous pocket (PPVP).[61] In this study the anterior part was found to be lined by the vitreous gel proper and the posterior part by the posterior hyaloid membrane.

Sebag et al[113–116] found a premacular hole in the cortex and observed herniation of the vitreous gel through the hole. The presence or absence of vitreous cortex in close apposition to the fovea is disputed and might be dependent on the age of the donor eye and the technique used.[41,60,61,112,127] Worst proposed that anterior–posterior traction of the collagen fibrils lining the cisternae is significant for the formation of macular holes.[137]

Vitreo–retinal interface imaging

Optical coherence tomography (OCT) is a clinical technique that is suitable both for measurement of retinal thickness and for imaging of the vitreo–retinal interface (Fig. 6.21).[6,80,42,43,96] Measurements of retinal thickness are comparable to stereoscopic viewing and reproducible, even in the presence of cataract.[49,80,126] The axial resolution of the technique is 3–10 μm and the distinction between high-reflectivity intravitreal membranes and non-reflective vitreous is favorable,[18,136] although the exact nature of intravitreal strands cannot be deduced from the image. Anterior–posterior membranes will theoretically be more difficult to detect due to the minimal backscatter from the small cross-section of such fibers.

The aging process of the vitreo–retinal interface has been studied with the OCT in 209 healthy subjects (age 31–74 years).[132] Preretinal strands, presumably posterior cortex, were found in 60 percent of these non-symptomatic cases without biomicroscopic evidence of posterior vitreous detachment (PVD). Persistent attachment was found to the fovea, optic nerve head and mid-periphery. The study demonstrates that partial PVD demonstrated by OCT is frequent in healthy subjects.

Diffusion kinetics as an indicator of the biophysical status of the vitreous

Using fluorophotometry examination of the fluorescein diffusion profile in the vitreous reveals three different profiles. These profiles provide information on how the vitreous

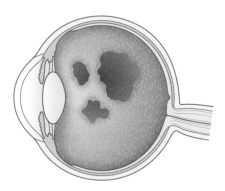

Figure 6.17 Schematic illustration of the vitreous body with laquena. (From Kanski J 1989.[57])

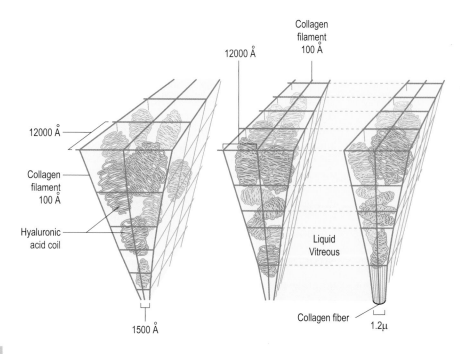

Figure 6.18 Schematic picture of the fine structure of the vitreous gel showing network reinforced with hyaluronic acid molecules. Left: Random distribution of the structural elements. Right: Formation of liquid pool and partial collapse of the network. From: Davson H, ed. The eye, vol. 1A. New York: Academic Press, 1984.)

Collagen filament 100 Å
12000 Å
Collagen filament 100 Å
12000 Å
Collagen filament 100 Å
Hyaluronic acid coil
Liquid Vitreous
Collagen fiber
1500 Å
1.2μ

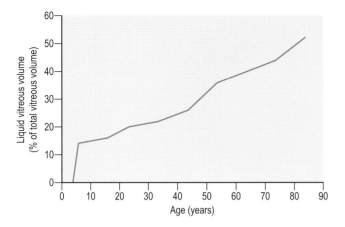

Figure 6.19 The formation of liquid vitreous (expressed as a percentage of the total vitreous space) during postnatal development and aging in the human eye. (From Seery 1994.[119])

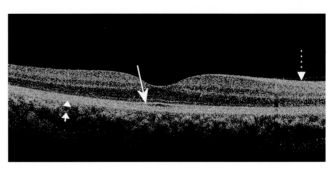

Figure 6.21 Optical coherence tomography scan from a healthy subject (6 mm scan through the fovea). The optically empty vitreous is in the top of the picture. The thin foveal center is seen in the middle of the scan and facing the vitreous, the retinal nerve fiber layer is visible (stippled arrow). The signal from the photoreceptors is relatively low (black or blue) while the reflectance from the retinal pigment epithelium is high and seen as red/white (repeated arrow). Also the junction of the inner and outer photoreceptor segments gives a highly reflecting line (full-drawn arrow). The vessels in choriocapillaris are partly visible.

Figure 6.20 Schematic diagram of Worst's interpretation of vitreous structure. "Cisterns" are visualized using white India ink to fill areas of the vitreous that take up this opaque dye. There are retrociliary, equatorial, and perimacular cisternal rings and a bursa premacularis. (From Jongebloed & Worst 1987.[55])

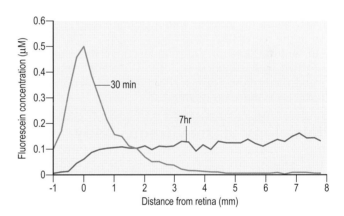

Figure 6.22 Pre-retinal vitreous concentration profiles for fluorescein in a patient with retinitis pigmentosa measured after 30 min and 7 h, which enables determination of passive permeability and outward, active transport. (From Moldow et al 1998.[83] With kind permission of Springer Science + Business Media.)

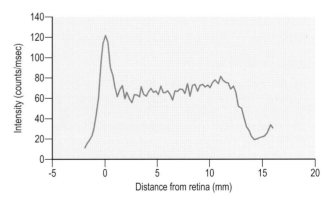

Figure 6.23 Fluorescein profile in the vitreous recorded at 30 min after fluorescein injection from a patient with retinitis pigmentosa. The absence of a preretinal gradient after 30 minutes indicates a significant vitreous detachment/liquefaction and the patient was excluded from the study. (From Moldow et al 1998.[83] With kind permission of Springer Science + Business Media.)

body behaves during aging from a diffusion-kinetic point of view.[77,83] The three different diffusion kinetic relationships are shown in Figures 6.22 to 6.24 from a study of patients with retinitis pigmentosa in different stages of the disease. The upper curve (see Fig. 6.22) shows a diffusion profile comparable to a normal vitreous body, in which the shape of the diffusion profile corresponds to the diffusion in an intact gel, and the diffusion coefficient corresponds to diffusion in a gel (see above). Figure 6.23 shows that there is no diffusion profile, as is the case when degeneration and liquefaction of the vitreous body are present, as observed in

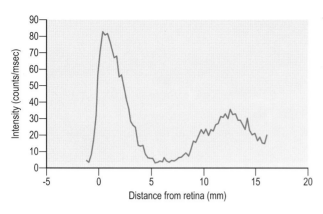

Figure 6.24 Vitreous profile recorded from a patient with retinitis pigmentosa 30 minutes after injection. The figure illustrates a typical "camel hump" which has no visual influence on the pre-retinal gradient. (From Moldow et al 1998.[83] With kind permission of Springer Science + Business Media.)

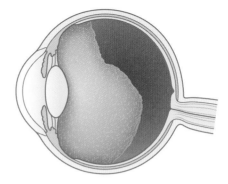

Figure 6.25 Schematic illustration of the vitreous body cavity with a posterior vitreous detachment. (From Kanski 1989.[57])

cases of retinitis pigmentosa. Figure 6.24 gives a diffusion profile interrupted by a convex hump, which can be related to a lacuna in the vitreous with high fluorescein concentration in the center (Box 6.1).

In studies where the diffusion kinetic relationships were compared with a clinical evaluation of the vitreous body, a good correlation between the diffusion kinetic and the clinical evaluation was reported.[95,99]

Physiology of the vitreous body

In the following the most essential physiological and pathophysiological relationships of the vitreous body will be illustrated. The normal physiology of the vitreous body can be divided into four main groups:

A. Support function for the retina and filling up function of the vitreous body cavity
B. Diffusion barrier between the anterior and the posterior segments of the eye
C. Metabolic buffer function
D. Establishment of an unhindered path of light

A. Support function for the retina and filling up function of the vitreous body cavity

Normal conditions

An intact vitreous body, which fills up the entire vitreous cavity, may retard or prevent the development of a larger retinal detachment. Presumably, the vitreous body can also absorb external forces and reduce mechanical deformation of the eye globe. The intact vitreous body supports the lens during trauma to the eye. However, this mechanical support is only of limited significance. Thus eyes in which the vitreous has been removed during vitrectomy can still have a normal function, and the retina is not detached.

Pathological/pathophysiological correlations

Posterior vitreous detachment

As previously mentioned, degeneration of the vitreous body is a normal physiologic aging phenomenon. When the central degeneration is sufficiently large, this leads to a collapse of the rest of the vitreous body, which causes the cortex to sink into the center of the vitreous body. This leads to the condition PVD (Figs 6.25 and 6.26).

PVD is the most common pathophysiologic condition of the vitreous body and is considered a normal physiologic aging phenomenon that might be related to a decrease in anchoring fibrils at the vitreoretinal interface.[46,47,132,139] Vitreous detachment leads to opacities in the vitreous body, but otherwise it typically has no clinical consequences. However, if there is a strong attachment between the posterior hyaloid and the ILM, a PVD can result in a retinal tear. This is the first step in a rhegmatogenous retinal detachment.

PVD can also induce traction on the retina, especially in the foveal region, if there is a strong attachment between the vitreous cortex and the ILM. This is seen especially in eyes with diabetic retinopathy. A shrinking of the posterior cortex when it is attached to the ILM can induce various forms of surface retinopathy, such as macular holes or macular edema.

Perifoveal vitreous detachment has been argued as a primary pathogenic event in idiopathic macular hole formation. However, this phenomenon may be secondary to the formation of the macular hole.

Gass[35,36] has postulated a centrifugal retraction of retinal receptors as an early event that might be caused by an early vitreoretinal degeneration of foveal Müller cells and the overlying vitreous cortex with subsequent split of the Müller cells. According to this hypothesis, the photoreceptors are still present in the cuff around the hole and the large improvement in visual acuity often found after vitrectomy is to be expected.

Other studies, using OCT and B-scan ultrasonography, indicate that vitreous change is the primary event.[37,53] Both techniques show a partial PVD, with attachment to the fovea, optic nerve and midperiphery in early stages of macular holes. As the vitreous shrinks, a traction develops in the anterior–posterior direction from the vitreous cortex to the fovea. An OCT image from the fellow-eye of a patient with a full-thickness macular hole in the other eye is shown in Figure 6.27A&B and clearly demonstrates the attachment sites of the posterior cortex. Discrete changes and a small break are seen in the outer retinal layers and the anterior–posterior traction might be conducted to the outer retinal layers through the thin, vertical fiber (probably Henle's fibers) in the middle of the central cyst. A similar picture, now with

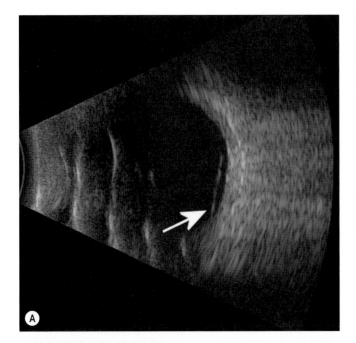

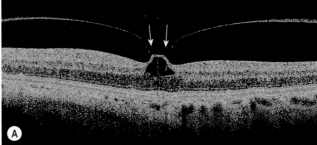

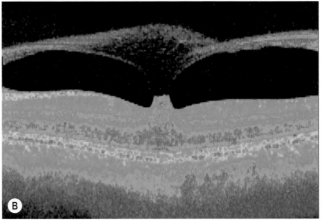

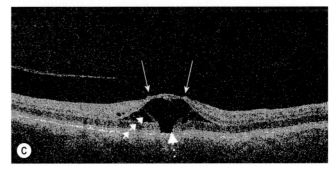

Figure 6.26 Posterior vitreous detachment as shown schematically in Fig. 6.25. (**A**) Ultrasound B-scan of a posterior vitreous detachment in a diabetic patient, in this case the partial detachment is located just anteriorly to the fovea (arrow). (**B**) OCT of the same case, the vitreous cortex is thick and reflects strongly (arrow). The retina is stretched out, giving a false impression of a strong curvature of the posterior cortex. Epiretinal fibrosis is seen on the right side of the foveal center.

Figure 6.27 (**A**) Optical coherence tomography scan from the fellow eye in a patient a with full-thickness macular hole in the other eye. The posterior vitreous cortex is clearly attached to the fovea (arrows), though the signal intensity from anterior–posterior structures as usual in OCT is weaker then for nearby transversal structures. A cystic formation is seen in the fovea, the vertical lines may originate from Henle's fibers or remaining Müller cells. Beneath the cyst, a discrete break seems to be present in the outer photoreceptor layer and the otherwise highly reflecting line from the junction of inner and outer segments. (**B**) A 3D picture of the same eye. The vitreo-retinal traction clearly forms a circular ring, attaching to the foveal center. (**C**) Prehole in another patient than in (A). The vitreo-retinal traction is very similar to the case shown in the previous figure, with vitreo-retinal traction (arrows) and a vertical line within the central cystic area (repeated arrow), indicating a conduction of the vitreo-retinal traction through the neuroretina. The cystic area expands into a split at the level of the outer plexiform layer. A dehiscence is seen of the photoreceptors (stippled arrow).

dehiscence of the photoreceptors, is seen in Figure 6.27C. Though clearly seen in Figure 6.27, PVD is not found in all cases and the etiology may not be the same in all patients (Box 6.1).[38,43,50,62,63,89]

Intravitreal string formation may also be implicated in macular hole or retinal edema formation and a vitreal schisis may represent a type of anomalous PVD as described by Sebag.[18,108,118]

Tangential forces cannot be excluded in the pathogenesis of macular holes.[6,86] Traction is well known from proliferative diabetic retinopathy, and the posterior vitreous is often attached to the optic nerve and to the midperiphery. A slack shape of the posterior cortex may present evidence for an argument in favor of an antero-posterior traction and against tangential traction.[17,37,53,86]

Development of macular edema

The basic principles behind the pathogenesis of macular edema are illustrated in Figure 6.28. The factors involved are related to the blood–retina barrier, that is, the passive permeability and the active transport correlated to blood flow and metabolic dysfunction with formation of osmotic

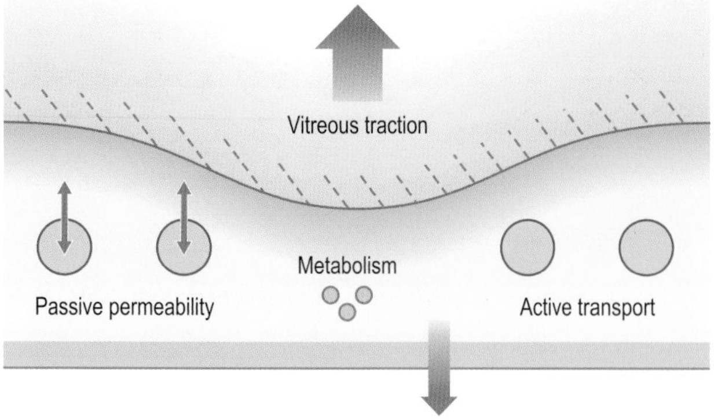

Figure 6.28 Factors related to the development of diabetic macular edema: vitreous traction, passive permeability, active transport (from the retina to the blood) and metabolism.

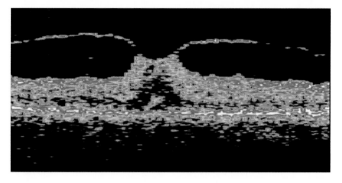

Figure 6.29 Optical coherence tomogram of a diabetic patient. The vitreous cortex is firmly attached to the fovea and the vascular arcades (not shown). After vitrectomy, normal anatomy and visual acuity were restored.

equivalents. Under normal conditions, the blood–retinal barrier is tight (low passive permeability). An increase in the passive permeability or a decrease in the outward active transport may lead to edema formation (see Table 6.2 below), and an increased passive permeability is predictive for later photocoagulation as well as substantial visual loss, though an increased leakage may not lead to an immediate increase in retinal thickening (Box 6.1).[104,105] Retinal ischemia seems to be a major cause of the pathophysiological changes leading to macular edema. As shown by Linsenmeyer, the oxygen tension in the inner retina of diabetic cats is nearly zero even before visible capillary dropout.[72] In diabetic patients without retinopathy, an acute increase in glucose leads to a faster impulse transduction (implicit time), as shown by electrophysiology, and the blood flow is increased, indicating an increased metabolism and oxygen demand.[16,64] In addition, in cases of capillary closure and drop out of the retinal vessels, the inner retina will be in a hypoxic state, which in turn induces production of vascular endothelial growth factor (VEGF) and an increased permeability. As a consequence, long-standing severe edema is now treated with VEGF inhibitors or steroid injection; both treatments decrease vascular permeability (Box 6.1).[7,31,135]

Vitreous traction is probably involved in some cases of macular edema and in diabetic patients OCT tomograms might have patterns similar to those seen in eyes with a macular hole (Fig. 6.29) (Box 6.1).[56] In several studies, eyes with otherwise untreatable edema have been shown to have partial PVD and vitrectomy seems to improve visual acuity in these eyes.[6,71,129,130,138] However, vitrectomy has also been helpful in cases without evidence of vitreomacular traction.[66]

As indicated above, the clinical significance of the tractional forces is complicated and difficult to quantitate, and the pathogenesis of these maculopathies is still a matter of debate.

B. Diffusion barrier between the anterior and the posterior segments of the eye

Normal conditions

The vitreous body is, as previously mentioned, a gel and in that context it has a considerable barrier function for bulk movements of substances between the anterior and posterior parts of the eye.

Substances that are liberated from the anterior segment of the eye will have difficulty reaching high concentrations in the posterior part of the eye when the vitreous body is intact because diffusion is slow and movement by bulk flow is very limited in a gel. An intact vitreous gel will also prevent topically administered substances reaching the retina and the optic nerve head in significant concentrations. Entrance of antibiotics from the bloodstream to the center of the vitreous will also be impeded by the normal vitreous.

Pathological/pathophysiological correlations

If the vitreous body is partly removed/degenerated/collapsed the exchange between the anterior and posterior part of the eye will be much faster and easier. This is the case when the lens is removed and anterior vitrectomy has been performed. Substances that are produced in the anterior segment of the eye or given topically enter the vitreous body through the less tight barrier systems in the anterior part of the eye (blood–aqueous barrier) and can be expected to reach the retina in higher concentrations than in eyes with an intact vitreous and an intact irido-lenticular barrier (Figs 6.30 and 6.31). The same is the case for topically administered pharmacologically active substances. Whether this has any clinical significance or not is not known, but the condition of the vitreous should always be taken into consideration in discussions on ocular pharmacokinetics, both for topically and systemically applied substances.

The preretinal oxygen tension is improved in diabetic patients after vitrectomy, indicating that oxygen transport increases with faster fluid currents.[122] This is of clinical relevance because retinal neovascularization and macular edema regress.

C. Metabolic buffer function

Normal conditions

The ILM and the posterior cortex do not act as diffusion barriers for smaller molecules. Because of the close anatomic relationship to the ciliary body and the retina, the vitreous body can act as a metabolic buffer and, to a certain degree, a reservoir for the metabolism of the ciliary body and especially the retina, as indicated in the oxygen profiles in Figure 9.10. Because of the tight blood–retina barrier, water-soluble

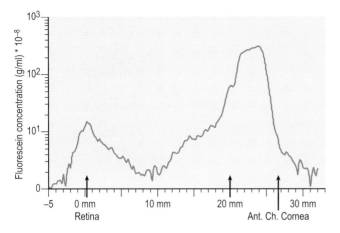

Figure 6.30 Fluorophotometric scan (30 minutes) in a patient after extracapsular cataract extraction with posterior chamber pseudophakos. Note the falling concentration of dye from the anterior and posterior towards the central vitreous. (From Ring et al 1987.[99])

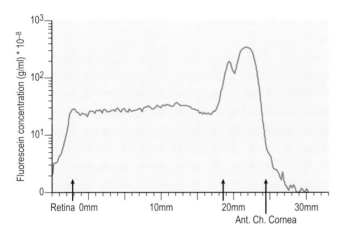

Figure 6.31 Fluorophotometric scan (30 minutes) in a patient after intracapsular cataract extraction with anterior chamber pseudophakos. There is a high concentration of dye uniformly throughout the vitreous cavity caused by gross vitreous detachment. (From Ring et al 1987.[99])

substances located in the retina have easier access to the vitreous cavity than to the bloodstream if the transport across the barrier is limited.

Substances present in or produced in the retina are thus diluted by diffusing into the vitreous body. Likewise glucose and glycogen in the vitreous body can supplement the metabolism of the retina, especially during anoxic conditions. The foot plates of the Müller cells also have close contact with the vitreous body. The vitreous body can thus act as a buffer in the physiological function of the Müller cells, for example in the potassium homeostasis of the retina.[10,14,87]

Vitamin C is also present in the vitreous body in relatively high concentrations, where it can act as a reservoir of antioxidants in stress situations, protecting the retina from metabolic- and light-induced free radicals (see Table 6.1).

Pathological/pathophysiological correlations

Because normal function of the retina can be obtained after total vitrectomy, the metabolic buffer functions of the vitreous do not seem to play an important role.

However, the condition of the vitreous may play a role for the effect of pathogenic factors produced in the retina. The concentration of vasoproliferative factors (produced in the retina and appearing in the vitreous body) near the retina depends on the condition of the vitreous body. If the vitreous acts as a diffusion barrier to these substances, they are retained in high concentrations close to the retina. Accordingly, one can speculate that vitrectomy can cure this condition, as can be seen in some aggressive cases of proliferative diabetic retinopathy. In other cases vitrectomy can lead to rapid movement of vasoproliferative factors from the posterior pole to the anterior pole, thus leading to neovascularization in the anterior segment of the eye (neovascular glaucoma).

D. Establishment of an unhindered path of light

Normal function

The normal physiological function of the vitreous body allows an unhindered passage of light to the retina, when visible light passes through the vitreous body. An important function for the vitreous body is to maintain this optimal transparency, which is primarily due to the low concentration of structural macromolecules (<0.2 % w/v)[118] and soluble proteins.[33] The transparency may also be maintained by the specific collagen/hyaluronic acid configuration, in analogy with the cornea. The scattering properties of the vitreous are anisotropic and the scatter decreases when the vitreous swells.[11]

Pathological/pathophysiological correlations

Degeneration of the vitreous body (as described in the section "Aging of the vitreous" above) with generation of opacities, so-called muches volantes, interferes with the path of light. This is also seen in connection with PVD as has already been mentioned. Synchysis scintillans, asteroid degeneration, hemorrhages, inflammatory material, fibrous tissue in the vitreous body, and lack of regression of the hyaloid artery are examples of pathological conditions which all interfere with the normal path of light through the vitreous body.[121]

The vitreous body as a sensor for the physiology of surrounding structures

As previously described neither the posterior cortex nor the internal limiting membrane can be regarded as diffusion barriers for smaller molecules. Due to the close anatomical relationship between the vitreous, the retina and the ciliary body, changes in concentrations of substances in the vitreous body will mirror processes in these tissues. Typically, fluorescein is used as a tracer of salt (and water) movement. In vitro studies have shown both passive and active outward transport through the blood–retina barrier (from the retina to the blood).[20,65,140,141] The passive movement (permeability) is bidirectional and probably due to fenestrations and/or leakage through tight junctions.[4,51] The outward transport from the retina to the blood, measured with fluorescein, is significantly reduced by metabolic and competitive inhibitors and is considered an active, carrier-mediated transport from the retina to the blood.[24,25,65,140] The active transport thus could be related to the metabolic activity of the pigment epithelium.

Determination of the blood–retinal barrier, passive permeability and active transport for fluorescein in humans, based upon concentration changes in the vitreous body

Mathematical models that are able to determine the blood–retinal barrier permeability in passive or active transport have been developed. The models are based on the assumption that the vitreous body is a globe surrounded by a homogeneous blood–retinal barrier. The following assumptions are made: The diffusion of fluorescein in the vitreous body takes place in accordance with conventional diffusion kinetics with a diffusion coefficient (D) the passive permeability of fluorescein through the blood–retinal barrier ($P_{passive}$) follows the usual passive permeability rules, and the active transport ($T_{out,active}$) is carrier mediated and non-saturable within the concentration range expected to be found in the vitreous body.

After IV injection of fluorescein, the concentration of fluorescein is determined in the bloodstream as the ultrafiltrated part in the plasma, and the concentration of fluorescein in the vitreous body is determined by vitreous fluorophotometry. The passive permeability and the diffusion coefficient are calculated as described earlier, based on diffusion profiles obtained up to 60 minutes after fluorescein injection.[21,67,68,69,70,74,77] The outward, active transport is calculated from the preretinal fluorescein concentration curve 8–10 hours after fluorescein injection (Fig. 6.22). The basic equation is as follows:

$$J = P_{passive} * Cp(t) - T_{out,active} * Cr(t)$$

where J is the flux, $P_{passive}$ the permeability, T is the outward, active transport and C is the fluorescein concentration in the plasma and vitreous.

Many hours after fluorescein injection, the plasma concentration is low and the first term small. Thus the outward (active) transport may be estimated from the preretinal gradient:

$$(D * dC/dx) * C_r^{-1}$$

where dC/dx is the pre-retinal fluorescein gradient and C_r^{-1} is the fluorescein concentration at the retina.[25,26,102]

The calculation of outward transport is not possible earlier than approximately 5–7 hours after fluorescein injection, when the preretinal gradient is reversed (Fig. 6.22). The time of reversal will vary according to the passive and active transport in the individual patient, and the equation above does not take into account the non-steady state of the whole diffusion process. With a mathematical model, the vitreous concentration curves can be simulated despite the non-steady state of the system.[102] In contrast to earlier studies,[90,92,140–143] the system is based on in vivo measurements of fluorescein isolated from the contribution of the fluorescent metabolite fluorescein-glucuronide. The model in Figure 6.32 is based on a spherical geometry of the eye; however the experimental and theoretical curves with this model did not fit in patients with high passive permeability.

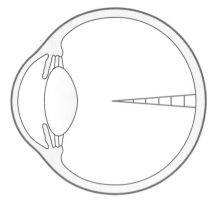

Figure 6.32 Simplified spherical model of the blood–retina barrier. The distance from the retina to the center of the eye is divided into a large number of conical cells. The fluorescein concentration in the cell next to the retina is estimated from the plasma fluorescein concentration curve and passive and active transport estimates. (From Sander et al 2001.[102] Reproduced with permission from Association for Research in Vision and Ophthalmology.)

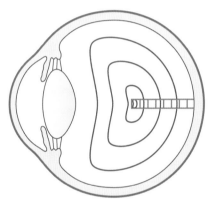

Figure 6.33 Simplified cylindrical model of the blood–retina barrier. Assuming diffusion from the anterior part of the eye into the vitreous and low leakage in the peripheral of the retina compared to the posterior pole, the concentration gradients will form ellipsoid curves in the vitreous (thin lines). The major direction of fluorescein flux is orthogonal to the concentration. (From Sander et al 2001.[102] Reproduced with permission from Association for Research in Vision and Ophthalmology.)

With a cylindrical model (Fig. 6.33) the match to experimental curves improved. In the final model a correction for diffusion of fluorescein from the anterior chamber was also included, as the concentration in anterior chamber often exceeds the posterior concentration up to several hours after injection and diffuses into the vitreous. The concentration profiles versus time with the final model are shown in Figure 6.34. By curve fitting it is possible to estimate the outward, active transport of fluorescein transport in the individual patient. The results are shown in Table 6.2.

It appears that the permeability for fluorescein in the normal person is very low and that the capacity of active transport process from vitreous to the blood is much larger than the passive permeability.[20,25,65,102,103,141] The passive permeability for fluorescein glucuronide equals that of fluorescein despite a large difference in lipid solubility between these two substances, indicating that the transport is related to water-filled pores.

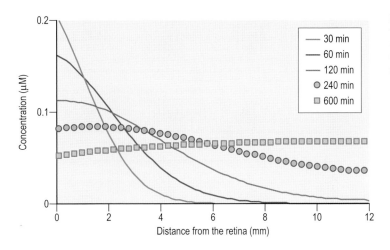

Figure 6.34 Vitreous curves calculated with the spherical model from 30 minutes to 8 hours after fluorescein injection. Input data ($P_{passive}$ 2,5 nm/sec: T_{active} 25 nm/sec) correspond to a healthy subject. (From Sander et al 2001.[102] Reproduced with permission from Association for Research in Vision and Ophthalmology.)

Table 6.2 Passive and active transport of fluorescein in patients with diabetes for eyes with and without clinically significant macular edema (CSME)

		Passive permeability (nm/sec)	Active transport (nm/sec)
CSME	mean	11.32*	62.20
	95% confidence interval	8.7–14.7	48.2–80.2
	range	2–72	2–265
No CSME	mean	3.57	71.95
	95% confidence interval	2.7–4.7	56.2–92.1
	range	1–13	23–195

From Sander et al.[103]

p < 0.05 compared to no CSME.

In healthy subjects, the passive permeability has been found to be 1.9 nm/sec, the active, outward transport is 43 nm/sec.[84,102] The same parameters in the condition of diabetic patients with and without macular edema are shown in Table 6.2. The passive permeability increases significantly with macular edema, whereas the active transport is unaffected by edema, although it appears to increase in patients with early diabetes compared with persons with healthy eyes.[102] The significance of this is unknown but could be related to an activation of the retinal pigment epithelium pump due to an increase of ions/non-ionic compounds in the retina of diabetic patients.

Concluding remarks

Although the vitreous body can be removed and almost normal function of the eye will still be maintained, the vitreous body plays an important role in the physiology and pathophysiology of the eye. The "silent" vitreous body exerts many important physiologic functions with close parallel to pathophysiology. An important implication for the future will be intravitreal drug application and release; a further understanding of the physiology of the vitreous body will be important for the development of rational applications of this new modality. The condition of the vitreous body is also important for movement of drugs within the eye and should be considered in discussions regarding the possibility that topically applied drugs will induce effects or side effects in the posterior segment of the eye.

Acknowledgments

We thank Drs. Ulrik Christensen, Kristian Krøyer and Mads Kofod for Optical Coherence Tomography images.

References

1. Akiba J, Ueno N, Chakrabarti B. Age-related changes in the molecular properties of vitreous collagen. Curr Eye Res 1993; 12:951.

2. Akiba J, Ueno N, Chakrabarti B. Mechanisms of photo-induced vitreous liquefaction. Curr Eye Res 1994; 13:505.

3. Andersen MN. Changes in the vitreous body pH of pigs after retinal xenon photocoagulation. Acta Ophthalmol 1991; 69:193.

4. Antoinetti DA, Barber AJ, Khin S et al. Vascular permeability in experimental diabetes is associated with reduced endothelial occludin content: vascular endothelial growth factor decreases occluding in endothelial cells. Diabetes 1998; 47:1953.

5. Armand G, Chakrabarti B. Conformational differences between hyaluronates of gel and liquid human vitreous: fractionation and circular dichroism studies. Curr Eye Res 1987; 6:445.

6. Asami T et al. Vitreoretinal traction maculopathy caused by retinal diseases. Am J Ophthalmol 2001; 131:134.

7. Audren F et al. Intravitreal triamcinolone acetonide for diffuse diabetic macular oedema: 6-months results of a prospective controlled trial. Acta Ophthalmol Scand 2006; 84:624.

8. Balazs EA, Denlinger JL. The vitreous. In Davson H, ed. The eye, vol 1A. New York: Academic Press, 1972.

9. Balazs EA, Toth LZ, Eckl EA. Studies on the structure of the vitreous body: XII. Cytological and histochemical studies on the cortical tissue layer. Exp Eye Res 1964; 3:57.

10. Berman E. Biochemistry of the eye. New York: Plenum Press, 1991.

11. Bettelheim FA, Balazs EA. Light scattering patterns of the vitreous humor. Biochem Biophysics Acta 1968; 158:309.
12. Bourwieg H, Hoffmann K, Riese K. Über Gehalt and Verteilung nieder- und hochmolekularer substanzen im glaskörper. Graefe's Arch Clin Exp Ophthalmol 1974; 191:53.
13. Bradbury MW, Lightman L. The blood-brain interface. Eye 1990; 4:249.
14. Brew H, Attwell D. Is the potassium channel distribution in glial cells optimal for spatial buffering of potassium? Biophys J 1985; 48:843.
15. Brown D et al. Cleavage of structural components of mammalian vitreous by endogenous matrix metalloproteinase-2. Curr Eye Res 1994; 13:639.
16. Bursell SE et al. Retinal blood flow changes in patients with insulin-dependent diabetes mellitus and no diabetic retinopathy. IOVS 1996; 37:886.
17. Chauhan DS, Antcliff RJ, Rai PA et al. Papillofoveal traction in macular hole formation: the role of optical coherence tomography. Arch Ophthalmol 2000; 118:32.
18. Chu TG, Lopez PF, Cano MR et al. Posterior vitreoschisis. An echographic finding in proliferative diabetic retinopathy. Ophthalmology 1996; 103:205.
19. Cunha-Vaz J. The blood-ocular barriers. Survey Ophthalmol 1979; 23:279.
20. Cunha-Vaz J, Maurice DM. The active transport of fluorescein by the retinal vessels and the retina. J Physiol 1967; 191:467.
21. Dalgaard P, Larsen M. Fitting numerical solutions of differential equations to experimental data: a case study and some general remarks. Biometrics 1990; 46:1097.
22. Davson D. Physiology of the eye, 3rd edn. New York: Churchill Livingstone, 1972.
23. Duke-Elder F. Textbook of ophthalmology. London: Henry Kimpton, 1938.
24. Eisner G. Zur anatomie des glaskörpers. Graefes Arch Clin Exp Ophthalmol 1975; 193:33.
25. Engler C et al. Fluorescein transport across the human blood-retina barrier in the direction vitreous to blood. Acta Ophthalmol 1994; 72:655.
26. Engler C, Sander B, Koefoed P, Larsen M et al. Probenecid inhibition of the outward transport of fluorescein across the human blood-retina barrier. Acta Ophthalmol 1994; 72:663.
27. Falbe-Hansen I, Ehlers N, Degn J. Development if the human foetal vitreous body. Biochemical changes. Acta Ophthalmol 1969; 47:39.
28. Fatt I. Flow and diffusion in the vitreous body of the eye. Bull Math Biol. 1975; 37:85.
29. Fine BS, Tousimis AJ. The structure of the vitreous body and the suspensory ligaments of the lens. Arch Ophthalmol 1961; 65:95.
30. Fledelius HC. Ophthalmic changes from age 10 to 18 years. A longitudinal study of sequels to low birth weight IV. Ultrasound oculometry of vitreous and axial length. Acta Ophthalmol 1983; 60:403.
31. Fraser-Bell S, Kaines A, Hykin PG. Update on treatments for diabetic macular edema. Curr Opin Ophthalmol 2008; 19:185.
32. Funderburgh J et al. Physical and biological properties of keratan sulphate proteoglycan. Biochem Soc Trans 1991; 19:871.
33. Gao Ben-Bo et al. Characterization of the vitreous proteome en diabetes without diabetic retinopathy and diabetes with proliferative retinopathy. J Proteome Res 2008; 7:2516.
34. Gartner J. Vitreous electron microscopic studies on the fine structure of the normal and pathologically changed vitreoretinal limiting membrane. Surv Ophthalmol 1964; 9:291.
35. Gass JDM. Müller cell cone, an overlook of the anatomy if the fovea centralis. Arch Ophthalmol 1999; 117:821.
36. Gass JDM. Reappraisal of biomicroscopic classification of stages of development of macular hole. Am J Ophthalmol 1995; 119:752.
37. Gaudric A, Haouchine B, Massin P et al. Macular hole formation. New data provided by optical coherence tomography. Arch Ophthalmol 1999; 117:744.
38. Giovannini A, Amato G, Mariotti C. Optical coherence tomography in diabetic macular edema before and after vitrectomy. Ophthalmic Surg Lasers 2000; 31:187.
39. Gloor BP. The vitreous. In Moses RA, Hart WM, eds. Adlers's physiology of the eye. St. Louis: Mosby, 1981.
40. Gottfredsdóttir M, Stefánsson E, Gíslason I. Retinal vasoconstriction after laser treatment for diabetic macular oedema. Am J Ophthalmol 1993; 115:64.
41. Grignoli A. Fibreous components of the vitreous body. Arch Ophthalmol 1951; 47:760.
42. Hee MR, Puliafito CA, Duker JS. Topography of diabetic macular edema with optical coherence tomography. Ophthalmology 1998; 105:360.
43. Hee MR, Puliafito CA, Wong C. Optical coherence tomography of macular holes. Ophthalmology 1995; 102:748.
44. Heegaard S. Structure of the human vitreoretinal border region. Ophthalmologia 1994; 208:82.
45. Heegaard S. Morphology of the vitreoretinal border region. Acta Ophthalmol Scand 1997; 75(suppl 222):1.
46. Heegaard S, Jensen OA, Prause JA. Structure and composition of the inner limiting membrane of the retina. Graefe's Arch Clin Exp Ophthalmol 1986; 224:355.
47. Hogan M. The vitreous, its structure, and relation to the ciliary body and retina. Invest Ophthalmol Vis Sci 1963; 2:418.
48. Hogan M. Histology of the human eye. Philadelphia: WB Saunders, 1971.
49. Hougaard J L, Wang M, Sander B, Larsen M. Effects of pseudophakic lens capsule opacification on optical coherence tomography of the macula. Curr Eye Res 2001; 23:415.
50. Ib M et al. Anatomical outcomes of surgery for idiopathic macular hole as determined by optical coherence tomography. Arch Ophthalmol 2002; 120:29.
51. Ishibashi T, Inomata H. Ultrastructure of retinal vessels in diabetic patients. Br J Ophthalmol 1993; 77:574.
52. Jin M, Kachiwagi K, Iizuka Y. Matrix metalloproteases in human diabetic and non-diabetic vitreous. Retina 2001; 21:28.
53. Johnson MW, van Newkirk MR, Meyer KA. Perifoveal vitreous detachment is the primary pathogenic event in idiopathic macular hole formation. Arch Ophthalmol 2001; 19:215.
54. Jongebloed WL, Humalda D, Worst JFG. A SEM-correlation of the anatomy of the vitreous body. Making visible the invisible. Doc Ophthalmologica 1986; 64:117.
55. Jongebloed WL, Worst JFG. The cisternal anatomy of the vitreous body. Doc J Ophthalmol 1987; 67:183.
56. Kaiser P, Riemann C, Sears J et al. Macular traction detachment and diabetic macular edema associated with posterior hyaloidal traction. Am J Ophthalmol 2001; 131:44.
57. Kanski J. Clinical ophthalmology, 2nd edn. London: Butterworths, 1989.
58. Kasai K, Nakamura T, Kase N et al. Increased glycosylation of proteins from cataractous lenses in diabetes. Diabetologia 1983; 25:36.
59. Kinsey VE. Further study of the distribution of chloride between plasma and the intraocular fluids in the rabbit eye. Invest Ophthalmol Vis Sci 1967; 6:395.
60. Kishi S, Demaria C, Shimizu K. Vitreous cortex remnants at the fovea after spontaneous vitreous detachment. Int Ophthalmol 1986; 9:253.
61. Kishi S, Hagimura N, Shimizu K. The role of premacular liquified pocket and premacular vitreous cortex in idiopathic macular hole development. Am J Ophthalmol 1996; 122:622.
62. Kishi S, Shimizu K. Posterior precortical vitreous pocket. Arc Ophthalmol 1990; 108:979.
63. Kishi S, Takahashi B. Three-dimensional observations of developing macular holes, Am J Ophthalmol 2000; 130:65.
64. Klemp K, Larsen M, Sander B, Vaag A et al. Effect of short-term hyperglycemia on multifocal electroretinogram in diabetic patients without retinopathy. Invest Ophthalmol Vis Sci. 2004; 45:3812.
65. Koyano S, Arai M, Eguchi S. Movement of fluorescein and fluorescein glucuronide across retinal pigment epithelium choroid. Invest Ophthalmol Vis Sci 1993; 34:531.
66. La Heij EC et al. Vitrectomy results in diabetic macular oedema without evident vitreomacular traction. Graefes Arch Clin Exp Ophthalmol. 2001; 239:264.
67. Larsen J, Lund-Andersen H, Krogsaa B. Transient transport across the blood-retina barrier. Bull Math Biol 1983; 45:749.
68. Larsen M, Dalgaard P, Lund-Andersen H. Determination of spatial coordinates in ocular fluorometry. Graefes Arch Clin Exp Ophthalmol 1991; 229:358.
69. Larsen M, Lund-Andersen H. Lens fluorometry: light attenuation effects and estimation of total lens transmittance. Graefes Arch Clin Exp Ophthalmol 1991; 229:363.
70. Larsen M, Loft S, Hommel E, Lund-Andersen H. Fluorescein and fluorescein glucuronide in plasma after intravenous injection of fluorescein. Acta Ophthalmol 1988; 66:427.
71. Lewis H. The role of vitrectomy in the treatment of diabetic macular edema. Am J Ophthalmol 2001; 131:123.
72. Linsenmeier RA, Braun RD, McRipley MA et al. Retinal hypoxia in long-term diabetic cats. IOVS 1998; 38:1647.
73. Lund-Andersen H. Transport of glucose from blood to brain. Physiol Rev 1979; 59:305.
74. Lund-Andersen H, Kjeldsen S. Uptake of glucose analogues by rat brain cortex slices: A kinetic analysis based upon a model. J Neurochem 1976; 27:361.
75. Lund-Andersen H, Krogsaa B, Jensen PK. Fluorescein in human plasma in vivo. Acta Ophthalmol.1982; 60:709.
76. Lund-Andersen H, Moller M. Uptake of inulin by cells in rat brain cortex. Exp Brain Res 1977; 23:37.
77. Lund-Andersen H, Krogsaa B, Larsen J et al. Fluorophotometric evaluation of the vitreous body in the development of diabetic retinopathy. In: Ryan S, ed. Retinal diseases. Orlando: Grune & Stratton and Harcourt Brace Jovanovich, 1985.
78. Lund-Andersen H et al. Quantitative vitreous fluorophotometry applying a mathematical model of the eye. Invest Ophthalmol Vis Sci 1985; 26:698.
79. Lundquist O, Österlin S. Glucose concentration in the vitreous of non-diabetic and diabetic human eyes. Grafe's Arch Clin Exp Ophthalmol 1994; 232:71.
80. Massin P, Vicaut E, Haouchine B, Erginay A et al. Reproducibility of retinal mapping using optical coherence tomography. Arch Ophthalmol 2001; 119:1135.
81. Mayne R, Brewton R, Mayne P et al. Isolation and characterization of type V/type XI collagen present in bovine vitreous. J Biol Chem 1993; 268:93.
82. McNeil A, Gardner A, Stables S. Simple method for improving the precision of electrolyte measurements in vitreous humor. Clinical Chemistry 1999; 45:135.
83. Moldow B et al. The effect of acetazolamide on passive and active transport of fluorescein across the blood-retinal barrier in retinitis pigmentosa complicated by macular edema. Graefes Arch Clin Exp Ophthalmol 1998; 236:881.
84. Moldow B et al. Effects of acetazolamide on passive and active transport of fluorescein across the normal BRB. Invest Ophthalmol Vis Sci 1999; 40:1770.
85. Molnar I, Poitry S, Tsacopoulos M et al. Effect of laser photocoagulation on oxygenation of the retina in minature pigs. Invest Ophthalmol Vis Sci 1985; 26:1410.
86. Mori K, Bae T, Yoneya S. Dome-shaped detachment of premacular cortex in macular hole development. Ophthalmic Surg Lasers 2000; 31:203.
87. Newman EA. Regulation of potassium levels by Muller cells in the vertebrate retina. Can J Physiol Pharmacol 1987; 65:1028.
88. Newsome D, Linsenmayer T, Trelstad R. Vitreous body collagen. J Cell Biol 1976; 71:59.
89. Nork TM, Giola V, Hobson R. Subhyaloid hemorrhage illustrating a mechanism of macular hole formation. Arch Ophthalmol 1991; 109:884.
90. Ogura Y, Tsukahara Y, Saito I, Kondo T. Estimation of the permeability of the blood-retinal barrier in normal individuals. Invest Ophthalmol Vis Sci 1985; 26:969.
91. Oksala A. Ultrasonic findings in the vitreous at various ages. Graefe's Arch Clin Exp Ophthalmol 1978; 207:275.
92. Palestine AG, Brubaker RF. Pharmacokinetics of fluorescein in the vitreous. Invest Ophthalmol Vis Sci 1981; 21:542.
93. Pedler C. The inner limiting membrane of the retina. Br J Ophthalmol 1961; 45:423.
94. Pournaras CJ, Ilic J, Gilodi N et al. Experimental venous thrombosis: preretinal pO_2 before and after photocoagulation. Klin Monatsbl Augenheilkd 1985; 186:500.
95. Prager T et al. The influence of vitreous change on vitreous fluorometry. Arch Ophthalmol 1982; 100:594.
96. Puliafito C et al. Imaging of macular diseases with optical coherence tomography. Ophthalmol 1995; 102:217.
97. Reddy DVN, Kinsey VE. Composition of the vitreous humor in relation to that of plasma and aqueous humors. Arch Ophthalmol 1960; 63:715.
98. Riley MV. Intraocular dynamics of lactic acid in the rabbit. Invest Ophthalmol Vis Sci 1972; 11:600.

99. Ring K, Larsen M, Dalgaard P. Fluorophotometric evaluation of ocular barriers and of the vitreous body in the aphakic eye. Acta Ophthalmol 1987; 65:160.

100. Sakaue H, Negi A, Honda Y. Comparative study of vitreous oxygen tension in human and rabbit eyes. Invest Ophthalmol Vis Sci 1989; 30:1933.

101. Sander B, Larsen M. Photochemical bleaching of fluorescent glycolysation products. Int Ophthalmol 1994; 18:195.

102. Sander B, Larsen M, Moldow B. Diabetic macular edema: passive and active transport of fluorescein through the blood-retinal barrier. Invest Ophthalmol Vis Sci 2001; 42:433.

103. Sander B, Larsen M, Engler C et al. Diabetic macular oedema: a comparison between vitreous fluorometry, angiography and retinopathy. Br J Ophthalmol 2002; 86:316.

104. Sander B, Thornit DN, Colmorn L et al. Progression of diabetic macular edema: correlation with blood-retinal barrier permeability, retinal thickness and retinal vessel diameter. Invest Ophthalmol Vis Sci 2007; 48: 3982.

105. Sander B, Hamann P, Larsen M. A 5-year follow-up of photocoagulation in diabetic macular edema: the prognostic value of vascular leakage for visual loss. Graefes Arch Clin Exp Ophthalmol 2008; 256:1535.

106. Schubert H. Cystoid macular edema: The apparent role of mechanical factors. Prog Clin Biol Res 1989; 312:299.

107. Sebag J. The vitreous. In: Hart W, ed. Adler's Physiology of the eye, 9th edn. St. Louis: Mosby, 1985.

108. Sebag J. Age-related differences in the vitreoretinal interface. Arch Ophthalmol 1991; 109:966.

109. Sebag J. Abnormalities of human vitreous structure. Graefe's Arch Clin Exp Ophthalmol 1993; 231:257.

110. Sebag J. Surgical anatomy of the vitreous and the vitreoretinal interface. In Tasman W, ed. Duane's clinical ophthalmology. Philadelphia: Lippincott, 1994.

111. Sebag J. Macromolecular structure of the corpus vitreum. Prog Polym Sci 1998; 23:415.

112. Sebag J. Letter to the editor. Arch Ophthalmol 1993; 109:1059.

113. Sebag J, Balazs EA. Morphology and ultrastructure of human vitreous fibers. Invest Ophthalmol Vis Sci 1989; 30:1867.

114. Sebag J, Balzac A. Pathogenesis of cystoid macular edema: an anatomical consideration of vitreoretinal adhesion. Surv Ophthalmol 1984; 28(suppl):493.

115. Sebag J, Buckingham B, Charles A et al. Biochemical changes in vitreous of humans with proliferative diabetic retinopathy. Arch Ophthalmol 1992; 110:1472.

116. Sebag J, Nie S, Reiser K et al. Raman spectroscopy of human vitreous in proliferative retinopathy. Invest Ophthalmol Vis Sci 1994; 35:2976.

117. Sebag J. Anomalous posterior vitreous detachment: a unifying concept in vitreo-retinal disease. Grafes Arch Clin Exp Ophthalmol 2004; 242:690.

118. Sebag J, Gupta P, Rosen R, Garcia P, Sadun A. Macular holes and macular pucker: The role of vitreoschisis as imaged by optical coherence tomography scanning laser ophthalmoscopy. Trans Am Ophthalmol Soc 2007; 105:121.

119. Seery M. Vitreous aging. In: Albert DM, Jakobiec FA, eds. Principles and practice of ophthalmology. Philadelphia: WB Saunders, 1994.

120. Snowden J. The stabilization of in vivo assembled collagen fibrils by proteoglycans/glycosaminoglycans. Biochim Biophys Acta 1982; 703:21.

121. Spraul CW, Grossniklaus HE. Vitreous hemorrhage. Survey Ophthalmol 1997; 42:3.

122. Stefánsson E. The therapeutic effects of retinal laser treatment and vitrectomy. A theory based on oxygen and vascular physiology. Acta Ophthalmol 2001; 79:435.

123. Stefánsson E, Peterson J, Wang Y. Intraocular oxygen tension measured with a fiber-optic sensor in normal and diabetic dogs. Am J Physiol 1989; 256: H1127.

124. Stefánsson E, Machemer R, McCueb BW et al. Retinal oxygenation and laser treatment in patients with diabetic retinopathy. Am J Ophthalmol 1992; 113:36.

125. Stitt A. Advanced glycation: an important pathological event in diabetic and age related ocular disease. Br J Ophthalmol 2001; 85:746.

126. Strom C, Sander B, Larsen N et al. Diabetic macular edema assessed with optical coherence tomography and stereo fundus photography. Invest Ophthalmol Vis Sci 2002; 43:241.

127. Swann D. Chemistry and biology of the vitreous body. Int Rev Exp Pathol 1989; 22:2.

128. Swann D, Constable I. Vitreous structure. I Distribution of hyaluronate and protein. Invest Ophthalmol Vis Sci 1972; 11:159.

129. Tachi N. Surgical management of macular edema. Semin Ophthalmol 1998; 13:20.

130. Tachi N, Ogino N. Vitrectomy for diffuse macular edema in cases of diabetic retinopathy. Am J Ophthalmol 1996; 122:258.

131. Törnquist P, Alm A, Bill A. Permeability of ocular vessels and transport across the blood-retinal barrier. Eye 1990; 4:303.

132. Uchino E, Uemura A, Ohba N. Initial stages of posterior vitreous detachment in healthy eyes of older persons evaluated by optical coherence tomography. Arch Ophthalmol 2001; 119:1474.

133. Ueno N, Sebag J, Hirokawa H et al. Effects of visible-light irradiation on vitreous structure in the presence of a photosensitizer. Exp Eye Res 1987; 44:863.

134. Vaughan-Thomas A, Gilbert S, Duance C. Elevated levels of proteolytic enzymes in the aging human vitreous. Invest Ophthalmol Vis Sci 2000; 41:3299.

135. Vinten M, Larsen M, Lund-Andersen H et al. Short-term effects of intravitreal triamcenolone on retinal vascular leakage and trunk vessel diameters in diabetic macular edema. Acta Ophthalmol Scand 2007; 85:21.

136. Wilkins JR, Puliafito CA, Hee MR. Characterization of epiretinal membranes using optical coherence tomography. Ophthalmology 1996; 103:2142.

137. Worst JGF. Cisternal systems of the fully developed vitreous in young adults. Trans Ophthalmol Soc UK 1977; 97:550.

138. Yamamoto T, Akabane N, Takeuchi S. Vitrectomy for diabetic macular edema: the role of posterior vitreous detachment and epimacular membrane. Am J Ophthalmol 2001; 132:369.

139. Yonomoto J, Ideta H, Sasaki K et al. The age of onset of posterior vitreous detachment. Grafes Arch Clin Ophthalmol 1994; 232:67.

140. Yoshida A, Ishiko S, Kojima M. Outward permeability of the blood-retina barrier. Grafes Arch Clin Exp Ophthalmol 1992; 230: 84.

141. Yoshida A, Ishiko S, Kojima M, Lipsky S. Blood-ocular barrier permeability in monkeys. Br J Ophthalmol 1992; 76:84.

142. Zeimer R, Blair NP, Cunha-Vz JG. Pharmacokinetic interpretation of vitreous fluorophotometry. Invest Ophthalmol Vis Sci 1983; 24:1374.

143. Zeimer R, Blair NP, Cunha-Vz JG. Vitreous fluorophotometry for clinical research. Arch Ophthalmol 1983; 101:1757.

The Extraocular Muscles

Linda McLoon

The extraocular muscles (EOM) are found within the bony orbit. They function in conjugate eye movements, maintenance of primary gaze position, and motor fusion – maintaining corresponding visual elements within the binocular field on corresponding retinal loci. In addition, the eyes must be able to follow moving objects (smooth pursuit) and accomplish rapid changes in fixation (saccades). This is accomplished by a very complex oculomotor control system, and the EOM are the final effector tissues. Understanding how the EOM adapt to changing visual demands is critical to the development of improved treatment strategies to realign the eyes when the system fails, as in strabismus or nystagmus.

The EOM have many distinct and complex properties that distinguish them from non-cranial skeletal muscles, many of which are normally associated with developing or regenerating muscle. This includes a population of multiply- and polyneuronally innervated myofibers, and retained expression of the immature subunit of the acetylcholine receptor, neural cell adhesion molecule, and "immature" myosin heavy chain isoforms. The EOM also have the capacity to continuously remodel throughout life. From a clinical perspective, the EOM have a distinct propensity for or sparing from a number of skeletal muscle diseases. EOM share some of their unusual characteristics with other craniofacial muscles, such as the laryngeal muscles, and the potential developmental basis for their unusual properties and disease profiles is presented.

The bony orbit

The eyes are protected in deep bony orbits that are roughly pyramidal in shape (Fig. 7.1). The orbit is largest just inside the orbital margin at its anterior extent and smallest at its posterior extent, the apex. The orbital margins are composed of the frontal bone superiorly, the zygomatic process of the frontal bone and the frontal process of the zygomatic bone laterally, the zygoma and the maxillary bone inferiorly, and the frontal process of the maxillary bone, the lacrimal bone and the maxillary process of the frontal bone medially (Fig. 7.1). However, direct impact to the bony margin can result in fracture (see Box 7.1). The orbit has a thick bony roof composed of frontal bone and a portion of the lesser wing of the sphenoid bone. The lateral wall is composed of the zygomatic bone and the greater wing of the sphenoid. A

relatively thin floor is composed of the maxillary bone, a small and variable part of the palatine bone, and the zygomatic bone. The thin medial wall is composed of the maxillary, the lacrimal, ethmoid, and sphenoid bones.[1] As a result of this bony configuration, the globe is somewhat protected from injury caused by direct impacts to the face, particularly if there are no bony fractures.

An understanding of the geometry of the bony anatomy relative to maintenance of eye position is critically important. The medial walls are parallel to each other, while the plane of the lateral wall in each orbit is 45° from the sagittal plane formed by the medial wall (Figs 7.1 and 7.2). Additionally, the geometry of the orbital bones requires that both globes be partially adducted in primary gaze. Maintenance of eye position in primary gaze requires a constant steady-state resting level of tension in all the EOM, referred to as tonus.[2]

The apex of the bony orbit has three major foramina: the optic foramen, and the superior and inferior orbital fissures. The nerves and blood vessels to the majority of structures within the orbit enter through these foramina. A number of small foramina also open into the orbit, allowing entry and exit of nerves and vasculature to a wide array of structures in the orbit and head.[3]

Normal extraocular muscles

Gross anatomy

There are six EOM in each orbit whose function is to move the eyes: four rectus muscles, superior, medial, inferior, and lateral; and two oblique muscles, inferior and superior (Fig. 7.3). In addition there is a seventh skeletal muscle in each orbit, the levator palpebrae superioris, which inserts into the upper eyelid and functions in elevating the palpebral fissure. While its cranial nerve innervation is similar to the EOM, functionally and metabolically it is distinct, and will not be discussed further in this chapter.

The four rectus muscles take their origin in part from the bones at the apex of the orbit, but also from the tendinous annulus. They course anteriorly to insert into the sclera anterior to the equator of the globe, a key factor when considering their functional effects on eye movements. Classically this insertion is described as external to the ora serrata; however, recent studies demonstrate that insertions of the

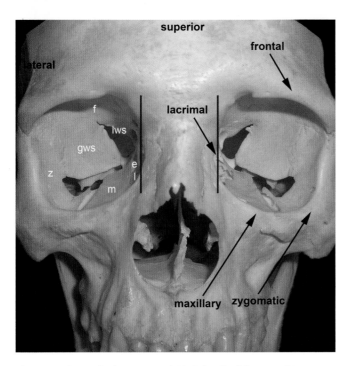

Figure 7.1 Bony orbit (anterior view). Medial walls of the two orbits are parallel to each other and in the sagittal plane (blue vertical lines). The walls form a pyramidal shape with the apex pointing posteriorly. f, frontal bone; lws, lesser wing of the sphenoid; gws, greater wing of the sphenoid; m, maxillary bone; l, lacrimal bone; e, ethmoid bone.

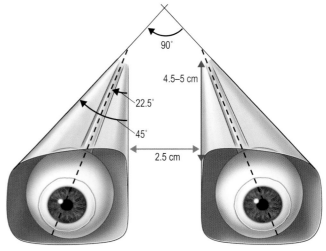

Figure 7.2 Geometry of the orbit. The orbit is a pyramidal-shaped structure with the base anterior and the apex posterior. The medial walls of the two orbits are parallel to each other and in the sagittal plane. The lateral walls are angled 45° relative to the medial wall. The lateral walls of both orbits form a 90° angle. The optic nerve emerges at an angle of 22.5° from the medial wall. The eyes in the primary gaze position results in adduction of the globe 22.5°. The orbital volume is 30 ml, with 6.5 mL filled by the globe.

Box 7.1 Blowout fractures

Bony fractures of the thin orbital walls can occur with blunt impact to the orbital margins. The increased force causes the bone to "blowout" into the sinuses, with the inferior and medial walls most susceptible to fracture. Sometimes the EOM become entrapped at the fracture site, as evidenced by restricted movements in the range of function of the entrapped muscle. These must be surgically repaired.[1]

---

rectus muscles range from 2.25 mm posterior to 2.25 mm anterior to the ora serrata, with 90 percent of the insertions within 1 mm.[4] Generally considered to be tendinous at the insertion site, the medial and lateral rectus muscles in humans may contain myofibers that extend directly to the sclera,[5] an important consideration for incisional strabismus surgery. The insertions of the four rectus muscles increase in distance from the corneal limbus circumferentially, with the medial rectus closest and the superior rectus furthest. The distances were originally determined on cadaveric material;[6] however, recent analyses on living adult patients during strabismus surgery show that the average distances from the corneal limbus to the rectus muscle insertions have large interindividual variations.[7-10] In part, this explains the disparate measurements seen in the literature. One typical study measured the distances from the limbus to muscle insertion as 6.2 ± 0.6 mm for the medial rectus, 7.0 ± 0.6 mm, for the inferior rectus, 7.7 ± 0.7 mm for the lateral rectus, and 8.5 ± 0.7 mm for the superior rectus.[10] Distances can vary up to 4 mm, even between the same muscles in both eyes of one patient, and does not correlate with primary position of the eye or surgical success for strabismus patients.[11] Thus, there is a significant amount of variation in rectus muscle insertions, and this variability has important consequences for incisional surgery of the EOM.

The superior and inferior oblique muscles have distinct paths compared to the rectus muscles. The superior oblique takes its origin from the dense connective tissue periosteum lining the orbit just superior and medial to the attachment of the tendinous annulus, and courses anteriorly along the border between the orbital roof and the medial orbital wall. Approximately 10–15 mm posterior to the orbital margin it becomes tendinous and enters the trochlea, a cartilaginous and dense connective tissue structure attached to the orbital periosteum. Emerging from the trochlea, the superior oblique muscle passes posteriorly at a 51° angle to the axis of the eye in primary position and inserts into the sclera. The trochlea thus serves as the "de facto" origin, creating the vector of force that moves the globe. The insertion of the superior oblique is on the superior pole deep to the superior rectus muscle, but in contrast to the rectus muscles, posterior to the equator of the globe (Fig. 7.3). The inferior oblique muscle is the only EOM that does not take its origin from the apex of the orbit; instead originating from the anteromedial orbital floor. The inferior oblique muscle courses posteriorly and inferior to the inferior rectus and inserts into the sclera posterior to the equator of the globe.

The shape, size, and orientation of the EOM from origin to insertion form the basis for the eye movements that result from their contraction (Table 7.1). While the effect of contraction of each EOM will be described separately, it is important to remember that they work in a coordinated fashion, maintaining significant tension or "tonus" even when the eye is in primary position and thus presumably "at rest".[2,12]

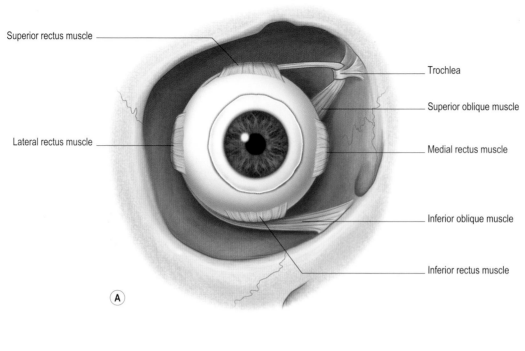

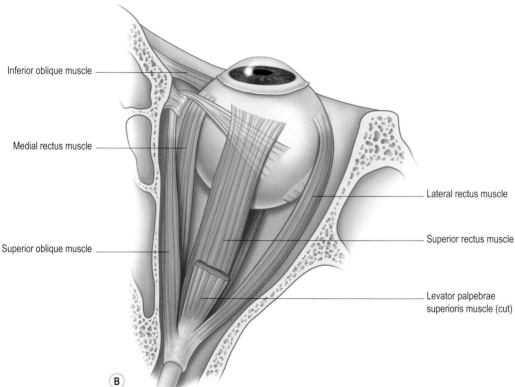

Figure 7.3 (**A**) Anterior view of the EOM in situ. Note that the origin for the inferior oblique is in the inferomedial aspect of the orbit and not the apex. (**B**) Superior view of the EOM in situ. Note the parallel arrangements of the horizontal muscles (medial and lateral rectus); the vertical muscles (superior and inferior rectus); and the insertional tendons of the superior and inferior obliques. (Modified from Clinical Orbital Anatomy, Marcos T. Doxanas and Richard L. Anderson eds, Baltimore, William and Wilkins, 1984.)

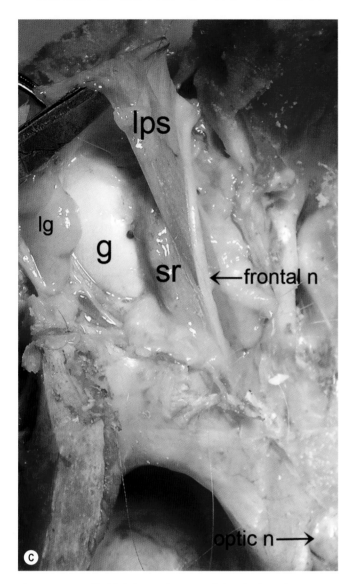

Figure 7.3, cont'd (**C**) Cadaveric dissection of the orbit from the superior approach. The levator palpebrae superioris (lps) is elevated. The superior rectus (sr) is directly inferior to it. The motor innervation cannot be seen from this view, as the oculomotor nerve (CNIII) runs within the muscle cone deep to the muscles. The frontal nerve is the most superior element and is a sensory branch of the ophthalmic nerve of CNV. The four rectus muscles insert into the anterior half of the globe, at approximately the level of the ora serrata. Insertions are found at increasing distances from the corneal limbus, medial rectus closest and superior rectus the furthest. The two oblique muscles insert into the posterior half of the globe. g, globe; l, lacrimal gland. Anterior is at the top of the photograph, and medial is to the right.

Horizontal movements are controlled by the medial and lateral rectus muscles, agonist–antagonist pairs with opposing primary functions; the medial rectus adducts the eye, while the lateral rectus abducts the eye. Vertical movements are more complex. The superior and inferior rectus muscles have a more complex effect on the direction of eye movements because the bony orbits are not parallel to each other (Fig. 7.2). In primary gaze, both the superior and inferior recti are angled laterally at approximately 22.5° from the sagittal plane. The primary action of the superior rectus

muscle is elevation, but it also adducts and intorts the eye (Table 7.1, Fig. 7.4). Intorsion is where the superior pole of the eye rotates medially. Thus, if the superior rectus muscle was acting alone, the direction of gaze would be superior and medial, that is, up and in towards the nose. The inferior rectus is parallel to the superior rectus, but inserts on the inferior surface of the globe. Thus, it primarily depresses the eye, but also adducts and extorts (Fig. 7.4); extorsion is rotation of the superior pole of the eye laterally.

Due to its insertion posterior to the equator of the globe, as well as the vector of force directed by the position of the trochlea in the superior and medial orbit, the superior oblique mainly intorts the eye (Table 7.1, Fig. 7.4). It also depresses and abducts. Thus, working unilaterally, gaze would be directed down and out. As the inferior oblique muscle parallels the superior oblique, but inserts on the inferior surface of the globe, its primary function is extorsion of the eye (Table 7.1, Fig. 7.4); it also elevates and abducts. In order for accurate positioning of the visual world on the fovea, activity of all the EOM must be tightly coordinated.

Cranial motor nerve innervation

The optic foramen is found within the lesser wing of the sphenoid bone at the orbital apex, through which runs the optic nerve (cranial nerve II, CNII) and ophthalmic artery. Between the greater and lesser wings of the sphenoid is the superior orbital fissure. The structures entering the orbit through this fissure are divided by the tendinous annulus (formerly the annulus of Zinn). Structures that enter the orbit superior to the annulus are the lacrimal and frontal nerves, both sensory branches of the ophthalmic division of the trigeminal nerve (CNV); the trochlear nerve (CNIV), motor nerve to the superior oblique muscle; and the superior ophthalmic veins (Fig. 7.5). Once through the annulus, structures enter what is referred to as the muscle cone, and are surrounded by the EOM and their connective tissue ensheathments. Within the annulus, the superior orbital fissure admits the superior and inferior divisions of the oculomotor nerve (CNIII) the motor nerve to the inferior rectus, inferior oblique, medial rectus, superior rectus and levator palpebrae superioris muscles; the nasociliary nerve which is a sensory branch of the ophthalmic division of the trigeminal nerve (CNV), and the abducens nerve (CNVI), the motor nerve to the lateral rectus muscle.[3] On the floor of the orbit is the inferior orbital fissure, which admits the zygomatic nerve, a sensory nerve innervating the lateral mid-face; communications of the inferior ophthalmic vein with the pterygoid plexus of veins inferiorly; and the lacrimal rami, carrying parasympathetic innervation from the facial nerve (CNVII) to the lacrimal gland.[13]

Once inside the bony orbit, the motor nerve branches of CNIII, CNIV and CNVI course anteriorly towards the muscles they innervate. The superior division of CNIII innervates the superior rectus muscle and continues superiorly to innervate the levator palpebrae superioris. The inferior division of CNIII innervates the medial and inferior rectus muscles, and the latter branch continues inferiorly to innervate the inferior oblique. All nerve branches enter the muscles on their deep surfaces within the muscle cone. CNVI also enters the lateral rectus muscle on its deep surface. Of the motor nerves, only the trochlear nerve, CNIV, enters the orbit outside the

Table 7.1 Extraocular Muscle Function: Primary and Secondary Actions

Muscle	Primary action	Secondary action	Motor innervation	Antagonists	Synergists
Lateral rectus	Abduction	None	Abducens n (CNVI)	Medial rectus	Superior and inferior oblique m
Medial rectus	Adduction	None	Oculomotor n (CNIII, inferior division)	Lateral rectus	Superior and inferior rectus m
Superior rectus	Elevation	Adduction Intorsion	Oculomotor n (CNIII, superior division)	Inferior rectus	Medial and inferior rectus m Superior oblique m
Inferior rectus	Depression	Adduction Extorsion	Oculomotor n (CNIII, inferior division)	Superior rectus	Medial and superior rectus m Inferior oblique and superior rectus m
Superior oblique	Intorsion	Depression Abduction	Trochlear n (CNIV)	Inferior oblique	Inferior rectus m Lateral rectus and inferior oblique m
Inferior oblique	Extorsion	Elevation Abduction	Oculomotor n (CNIII, inferior division)	Superior oblique	Superior rectus m Medial rectus and superior oblique m

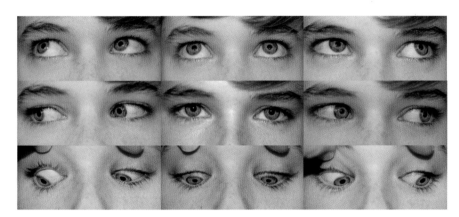

Figure 7.4 Composite photograph showing a subject looking in the nine cardinal positions of gaze. Center panel: primary gaze straight ahead. Top panels: up-gaze. Bottom panels: down-gaze. Left panels: gaze to the right. Right panels: gaze to the left. (From Christiansen and McLoon, Extraocular Muscles: Functional Assessment in the Clinic, In; Elsevier's Encyclopedia of the Eye, Ed. D. Dartt. 2010. Copyright Elsevier 2010)

tendinous annulus innervating the superior oblique muscle on its superior or lateral surface (Fig. 7.5). All motor nerves that innervate the EOM enter the body of the muscles at their posterior third (Fig. 7.5).

Neuromuscular junctions (NMJ) are the specialized sites of communication between a nerve and the myofibers it innervates. In non-cranial skeletal muscle, NMJs usually form in the middle one-third of each myofiber. The NMJs formed by the cranial motor nerves with individual EOM myofibers display some distinct differences compared to those in non-cranial skeletal muscle. Similar to body muscles, the EOM have singly innervated myofibers with NMJs referred to as "en plaque" endings (Fig. 7.6). However, the "en plaque" NMJs in EOM are smaller and less complicated structurally than those in non-cranial skeletal muscles.[14] In addition, the EOM have multiply innervated myofibers with neuromuscular junctions referred to as "en grappe" endings (Fig. 7.6). These are a linear array of small synaptic contacts often found towards the ends of individual myofibers, but can be continuous along the length of individual myofibers.[15] The "en grappe" NMJ contacts are structurally simple.[16] Thus, in EOM a single myofiber can have an "en plaque" NMJ somewhere along the middle one-third and also have multiple "en grappe" endings along the tapered ends (Fig. 7.6). Some myofibers in EOM that express the slow tonic myosin heavy chain isoform (MYH14) have "en grappe" endings along their entire myofiber length and do not have an "en plaque" ending.

The acetylcholine receptor found within NMJs is composed of 5 subunits, similar to skeletal muscle generally. In developing non-cranial muscle there are two alpha, 1 beta, 1 gamma, and 1 delta subunits ($\alpha 2\beta\gamma\delta$), and in the adult the γ subunit is replaced with the epsilon (ε) subunit.[17] In contrast, in adult EOM the majority of the "en grappe" endings express the "immature" gamma subunit, rather than the epsilon subunit of mature endings.[18,19] Both the "en plaque" and "en grappe" endings in the EOM can co-express both the epsilon and gamma subunits;[20] this appears to be unique to EOM. Due to the nature of EOM myofiber length, as discussed in a following section, NMJs can be seen throughout the origin-to-insertional length of EOM in most species where this has been examined.[21,22] This is in contrast to most limb skeletal muscles, which have a motor endplate zone, an NMJ band that is fairly contained within a defined area in the midbelly region of the muscle.

It has generally been assumed that the multiply innervated myofibers are innervated by a single motor neuron, but polyneuronally innervated myofibers are also present.[23,24] This means that more than one motor neuron can innervate a single myofiber. This has important implications for EOM physiology and will be discussed in that section.

Orbital connective tissue

A complex framework of connective tissue exists throughout the orbit, and this network has a clear structural organization and constant pattern (Fig. 7.7).[25] These connective tissue septa contain nerves, vessels and smooth muscle, and are postulated to play a role in supporting eye movements. Recent studies have confirmed and extended these initial detailed analyses of orbital connective tissue septa to include thickenings around individual EOM called orbital pulleys (Fig. 7.7).[26] These connective and smooth muscle septa and bands constrain the paths of the EOM, changing the vector of force as the EOM contract and stabilize muscle position during movement.[27]

Histological anatomy and physiologic implications

The EOM have a complex anatomy at the microscopic level. The overall cross-sectional areas of their myofibers are extremely small compared to non-cranial skeletal muscle. Each EOM is composed of two layers: an outer orbital layer composed of myofibers of extremely small cross-sectional area and an inner global layer with myofibers larger than in the orbital layer but still extremely small compared to non-cranial skeletal muscle (Fig. 7.8). Further descriptions will concentrate on human muscle, but the EOM of other mammals have the same general features despite some variations in detail.

In non-cranial skeletal muscles, two general fiber types are described, fast and slow, referring mainly to the myosin heavy chain isoforms (MyHC) they express, which in turn determines their shortening velocity.[28] Whether a fiber is "fast" or "slow" is due, in part, to their complement of contractile proteins. Non-cranial skeletal muscles that are largely fast MyHC-positive, such as the extensor digitorum longus, have an oxidative metabolism, rapid shortening velocities, and rapid fatigue with activation. Slow MyHC-positive myofibers in non-cranial skeletal muscle have a glycolytic metabolism, slower shortening velocities, larger force generation, and are fatigue-resistant. The EOM also have the two basic myofiber types. About 85 percent of the myofibers in both layers in adult EOM are fast MyHC-positive (Fig. 7.8).[29,30] The other 15 percent of the EOM myofibers immunostain for the slow MyHC. In contrast to non-cranial skeletal muscle, the EOM have extremely fast contractile characteristics yet are also extremely fatigue-resistant.[31] Several factors support these apparently contradictory characteristics. While the vast majority of non-cranial skeletal muscles express one of four MyHCs, fast fiber types IIa, IIx, or IIb or slow type 1, EOM myofibers can contain up to nine MyHCs.[32–35]

The MyHC expressed in EOM include: fast types IIa (MYH2), IIx (MYH1), or IIb (MYH4); MyHCs associated with immaturity – embryonic (developmental) (MYH3) and neonatal or fetal (MYH8); slow or type 1 (which is the same as beta-cardiac myosin) (MYH7); alpha-cardiac (MYH6); EOM-specific (MYH13); and the slow tonic myosin (MYH14).

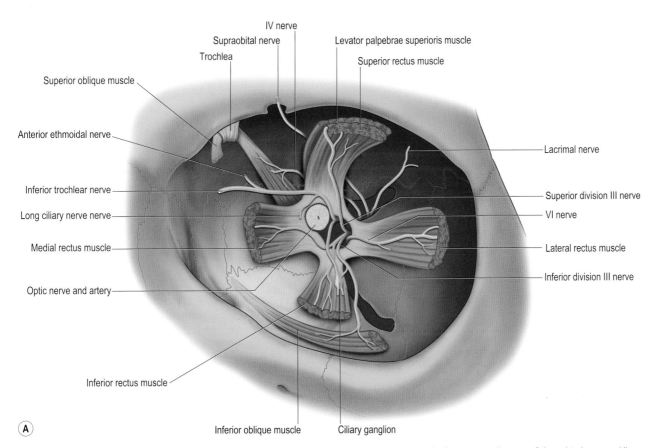

Figure 7.5 (**A**) Deep dissection of the orbit with the globe removed and the optic nerve sectioned, allowing visualization of the orbital nerves. All motor nerves except the trochlear nerve enter the muscles they innervate on their deep surfaces in the posterior third of their length. (Modified from Miller NR: Walsh and Hoyt's Clinical Neuro-Ophthalmology (ed 4), Vol. 1. Baltimore, Williams & Wilkins, 1982.)

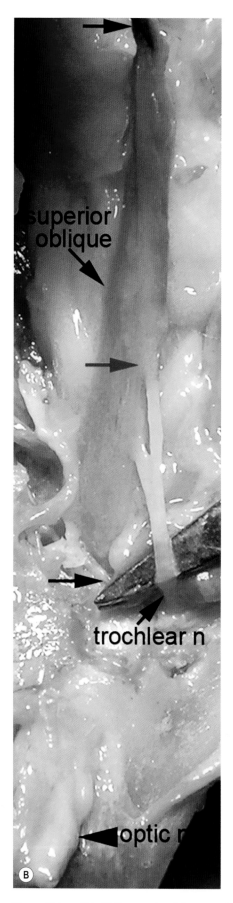

superior
oblique

trochlear n

optic n

B

Figure 7.5, cont'd (**B**) The superior oblique muscle is innervated by CNIV on its superior surface in the posterior third of the muscle length. Anterior is at the top and medial is to the left.

This large array of expressed MyHC within the EOM helps explain why its dynamic physiologic properties are significantly different than those in limb muscles.[36] It should be noted that patterns of MyHC composition vary between each EOM, with lateral rectus being the most different.[37] Support for the idea that the MyHC composition may be critical for understanding control of force generation is the physiological demonstration that there are significant contractile differences between muscles. For example, in the medial and lateral rectus muscles, motor units in the medial rectus generate faster twitch contractions and those in lateral rectus generate greater tetanic tensions.[38] This concept is critical in understanding the CNS control of eye movements as well as eye movement disorders like strabismus.

The expression patterns of EOM myofibers for these MyHC differ significantly between the global and orbital layers (Fig. 7.8); for example, the vast majority of the orbital layer myofibers are positive for developmental myosin (MYH3), yet the global layer only has scattered fibers that immunostain for this isoform (Fig. 7.8). Distinct patterns of differential staining are seen for all other MyHC present in EOM. From a physiological perspective, this issue becomes even more complex. Approximately 25 percent of cat single lateral rectus motor units examined had myofibers in both the orbital and global layers.[39] These bilayer motor units are stronger and faster than motor units contained within single EOM layers, and tend to be fatigable. Approximately 54 percent of the global motor units examined were stronger and faster than orbital layer units, in part a reflection of the properties imbued upon the individual myofibers by their contractile proteins.

The expression of individual MyHC also varies along the origin-to-insertion length of each muscle.[37,39,40] In part this is because many of the EOM myofibers do not run from origin to insertion in both the orbital and global layers;[22,23,37,41–43] they can be arranged in parallel or in series, and many branched and/or split myofibers exist (Fig. 7.9).[22] These short myofiber lengths have functional consequences. For example, in a series of studies examining summation of motor forces, individual cat and monkey motor units were stimulated, and the forces each unit produced in the lateral rectus muscle were determined. Then, using simultaneous stimulation, the evoked unit force of other motor units was added to the single motor unit force. In about 25 percent of the cases in cat and about 85 percent of the cases in monkey, the forces did not add linearly.[44,45] Thus, force is "lost" as multiple myofibers are activated, partly due to force dissipation laterally through myomyous junctions and myoconnective tissue connections formed by short myofibers.

In addition, the serial or parallel arrangement of myofibers with different myofibrillar isoforms would significantly affect contractile behavior. This was elegantly demonstrated using sets of single myofibers in vitro, one fast and one slow, that were tied together in series or in parallel.[46] In any given combination of one fast and one slow myofiber, the paired fibers showed a range of forces with greater or lesser fast or slow characteristics. In a fast/slow combination, the fast fiber begins to contract prior to the slow fiber, which is then slack and therefore not at its optimal length to generate full force. Serial or parallel arrangements of myofibers with disparate MyHC isoform composition would be expected, based on these results, to result in a

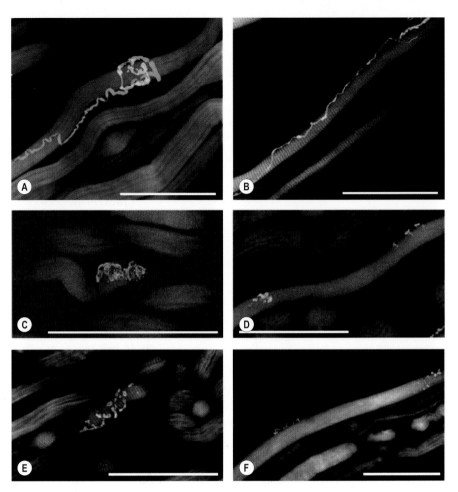

Figure 7.6 Confocal laser scanning microscopy images showing singly innervated myofibers with en plaque motor endplates and multiply innervated myofibers with en grappe motor terminals. Nerve fibers are labeled with anti-ChAT (green), motor terminals with α-bungarotoxin (red), and myofibers with phalloidin (white). (**A**) A ChAT-positive axon supplying an en plaque endplate that is positive for ChAT/α-bungarotoxin. (**B**) ChAT-positive axon supplying en grappe endings positive for ChAT/α-bungarotoxin. (**C**) En plaque endplate and (**D**) en grappe terminals showing ChAT/α-bungarotoxin reactivity. (**E, F**) Motor terminals are labeled with anti-VAChT (green) and α-bungarotoxin (red), myofibers with phalloidin (white): (E) en plaque endplate and (F) en grappe terminals showing VAChT/α-bungarotoxin-reactivity. Scale bars, 100 μm. (Reproduced from Blumer et al. IOVS 50:1176-1186, 2009, with permission from the Association of Research in Vision and Ophthalmology.)

range, or continuum, of forces produced by their co-activation. This suggests that a distributed model of motor unit recruitment at the CNS level is needed to address the non-linearity of the effector arm of the system, the EOM themselves.[47]

Individual EOM myofibers are polymorphic and also can express more than one MyHC.[31,40,41,48–50] This has been shown for a wide variety of species, including human muscle.[31] For example, in both singly and multiply innervated myofibers in the orbital layer in rats, the fiber ends express the neonatal MyHC (MYH8), but this isoform is completely eliminated at the position of the NMJ, where the fibers immunostained for fast MyHC.[40] In the orbital layer, individual myofibers express the embryonic MyHC (MYH3) at their fiber ends and the EOM-specific MyHC (MYH13) in the NMJ region. Orbital multiply innervated myofibers were found that expressed slow MyHC (MYH7) along their entire length, but also expressed embryonic MyHC (MYH3) at the fiber ends.[41] Single myofibers in the global layer were found that express EOM-specific MyHC (MYH13) at the NMJ region as well as fast myosins IIb and/or IIx.[49] In human EOM, orbital layer myofibers were found that expressed type 1 (MYH7), and of these (Fig. 7.10) some also expressed slow tonic (MYH14), alpha-cardiac (MYH6), and/or embryonic (MYH3) or EOM-specific (MYH13).[31,50] Physiologic examination of individual multiply innervated myofibers from the orbital layer of rats showed that electrical activity varied along the length of the fiber, with tonic characteristics at the fiber end and twitch characteristics in the central region near the endplate, the only location where fast MyHC was expressed.[24]

Thus, physiologic properties of the EOM motor units, when activated singly or collectively, reflect these MyHC isoform complexities, as well as differences in individual myofiber length and branching patterns.[45] It should be noted that the EOM show significant activity at all times, even when the eye is directed in what would be considered the off-direction for muscle action.[2] In addition, all motor units participate in all types of eye movements, and there appear to be no motor units that specialize in rapid saccades or slow vergence movements, for example.[51] Presumably, the MyHC composition is a reflection of demands placed on the EOM by the oculomotor control system, supported by the observation that altering the stimulation frequency of a muscle causes significant changes in MyHC expression.[52] Functional denervation also affects MyHC composition.[53] It is reasonable to suggest that a complex "conversation" is constantly occurring between EOM myofibers and the neurons innervating them, helping them to adapt to ever-changing physiologic demands.

Because of these complex MyHC expression patterns and physiological characteristics, past attempts to classify EOM myofibers into simple groups ultimately fail. Classically, myofibers in non-cranial skeletal muscles have been

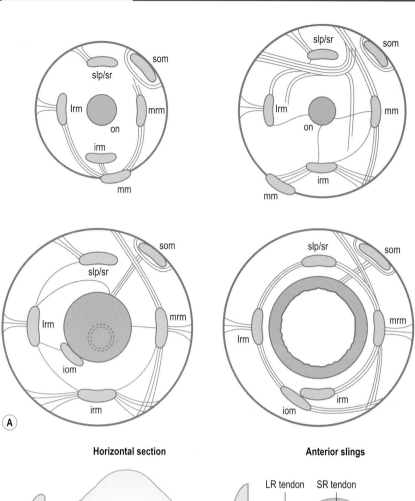

Figure 7.7 (**A**) Connective tissue septa at different levels in the orbit. Top left, near the orbital apex. Top right, halfway between apex and rear surface of globe. Bottom left, near the rear surface of the globe. Bottom right, area near the equator of the globe. slp/sr: superior levator palpebrae/superior rectus complex; lrm: lateral rectus muscle; ion: inferior oblique muscle; irm: inferior rectus muscles; mm: Muller's muscle; mrm: medial rectus muscles; som: superior oblique muscle; on: optic nerve. (Redrawn from Koornneef L: Spatial aspects of orbital musculo-fibrous tissue in man: a new anatomical and histological approach, Amsterdam, 1976, Swets and Zeitliinger.) (**B**) Diagrammatic representation of orbital connective tissues. IR, inferior rectus; SO, superior oblique; SR, superior rectus. The three coronal views are represented at the levels indicated by arrows in the horizontal section. (Modified from Demer et al. IOVS 36:1125–1136, 1995, with permission from the Association of Research in Vision and Ophthalmology.)

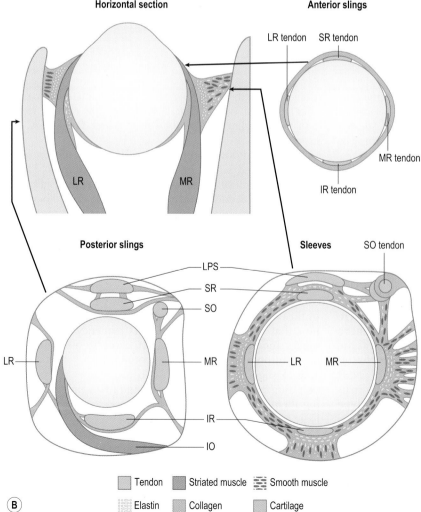

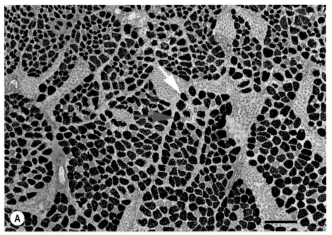

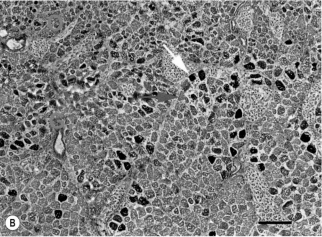

Figure 7.8 Photomicrographs of two serial sections from a normal monkey lateral rectus muscle immunostained for (**A**) fast MyHC and (**B**) neonatal MyHC. The orbital layer is at the top (O) and global layer on the bottom (G). Arrows point to two myofibers, one positive for both fast and neonatal MyHC and one positive for fast but negative for neonatal MyHC. Bar is 100 μm.

described by their MyHC expression profile (e.g. type IIa or type I). Even in non-cranial skeletal muscles, it is becoming increasingly clear that single myofiber MyHC polymorphism is more common than was previously believed.[54,55] Another fiber type, called "mismatched", has also been described in both non-cranial and cranial skeletal muscle.[54,56] Mismatched myofibers include those with "mixed" fast and slow characteristics, and can include fibers with mixtures of fast and slow MyHC or fast or slow MyHC with various regulatory proteins such as troponin[57] or myosin-binding protein C[58] that are not of the same "type". This complexity of protein expression and the heterogeneity of individual myofibers in EOM cannot be overstated. What these studies suggest is that rather than specific "types" of myofibers (Fig. 7.11), in EOM there is, in fact, a continuum of myofiber types. Each myosin heavy and light chain isoform results in a distinct shortening velocity,[29] which allows for an incredible plasticity in the control of muscle force generation. The modulation of the MyHC patterns with alterations in hormones or innervational changes also compound the heterogeneity of the EOM myofibers. These adaptive protein

changes are extremely rapid, and the control may be at the level of histone modifications[59] or at the translational level controlled by microRNA – known to be up-regulated in the EOM.[60] The EOM myofiber continuum hypothesis has been suggested previously for other skeletal muscles, including plantaris[61] and masseter.[62] This continuum of myofiber types in EOM combined with the non-linearity of eye muscle contractile properties[45] would allow CNS control of eye movement position and velocity to be finely tuned as the eyes are moved into an infinite number of positions.

Metabolism

The physiological properties of the EOM derive their dynamic and unusual characteristics from (1) their expression of specific contractile proteins including, but not limited to, isoforms of MyHC, myosin light chains, tropomyosin, and troponin, (2) the presence of myofibers shorter than the total origin-to-insertional length of each EOM, resulting in fibers connected in parallel or in series, (3) the presence of singly, multiply and polyneuronally innervated individual myofibers and (4) adaptations of their metabolic pathways.

The most studied metabolic property in EOM is calcium handling. Calcium plays a critical role in controlling the duration of muscle contraction. In part this is controlled by the sarcoplasmic reticulum Ca^{2+} ATPases (SERCA1 and SERCA2 in fast and slow myofibers respectively in non-cranial skeletal muscle). In the EOM, SERCA1 and 2 are co-expressed in the majority of individual myofibers.[63] The EOM contain an abundance of mitochondria, and EOM myofibers appear to use their mitochondria as fast calcium sinks to aid in regulation of calcium.[64] This effectively widens the dynamic range of EOM force production. Due to their efficient calcium handling, EOM myofibers are also resistant to pathological elevations of intracellular calcium levels.[65]

EOM are resistant to injury and oxidative stress, and contain higher levels of superoxide dismutases and glutathione peroxidase activity than limb skeletal muscle.[66] This is concordant with the highly aerobic nature of the EOM. Only cardiac muscle has a higher blood flow rate.[67] Despite their highly oxidative metabolism, they are also extremely fatigue-resistant.[68] Unlike non-cranial muscles, EOM do not depend on creatine kinase activity for their fatigue resistance;[69] instead, EOM can utilize lactate as a metabolic substrate during times of increased contractile activity.[70] This further illustrates how the EOM can maintain a highly oxidative metabolism and fatigue resistance simultaneously. These two opposite metabolic demands are also met by the expression in individual EOM myofibers of both succinate dehydrogenase and alpha-glycerophosphate dehydrogenase, enzymes in the oxidative and glycolytic pathways, respectively. In contrast, these two enzymes are fiber type specific in limb skeletal muscle (Fig. 7.12).[71] This supports the view that skeletal muscles consist of distinct allotypes,[72] the EOM representing the extreme end relative to its anatomy, innervation, metabolism, and physiologic function.

Another distinctive aspect of EOM function is their ability to undergo myonuclear addition and subtraction throughout life[73,74] while simultaneously maintaining their overall size, morphology and function. This is not seen in any of

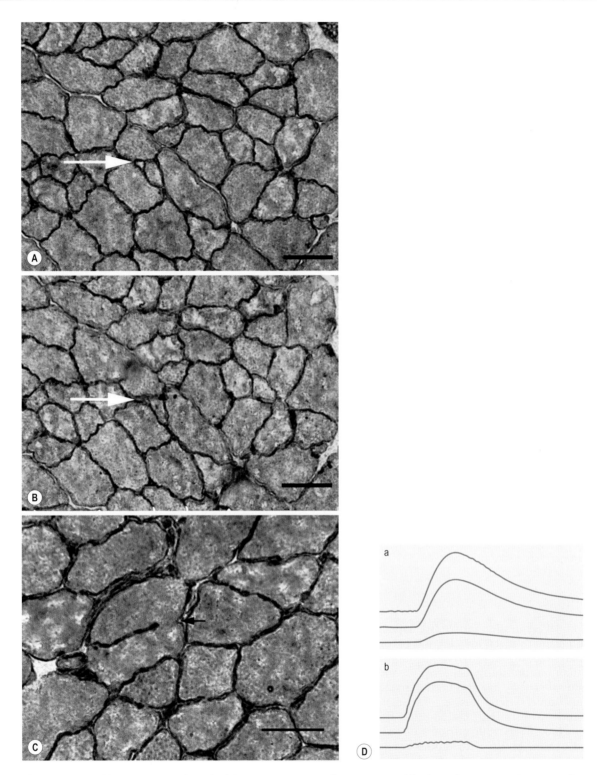

Figure 7.9 (**A**) and (**B**) Cross-sections through rabbit superior rectus muscle immunostained for dystrophin. Arrows indicate a myofiber present in one section that ended before the next section 24 μm distant. Bar is 20 μm. (**C**) Interconnected myofibers (arrow) in normal EOM immunostained for dystrophin. Bar is 20 μm. (**D**) Physiological demonstration of non-additive muscle forces in EOM using (a) twitch and (b) tetanic responses after stimulation. (a) Muscle responses to stimulation of one motor neuron (bottom trace, 45.9 mg), muscle responses to stimulation of several motor units (middle trace, 209.5 mg), and activation of a single motor unit plus nerve responses (upper trace, 257.5 mg). The twitch responses are additive; no force is "lost". (b) Muscle responses of the same motor unit to tetanic stimulation. Muscle responses from tetanic stimulation of a single motor unit (bottom trace, 398.7 mg), several motor units (middle trace, 4066 mg) and the single motor unit plus nerve responses (upper trace, 4226 mg). This unit's loses 40 percent of its force upon tetanic stimulation. Horizontal bar, 50 msec. Vertical bar, 917 mg. (From Goldberg et al., Muscle Nerve 20:1229–1235, 1997.)

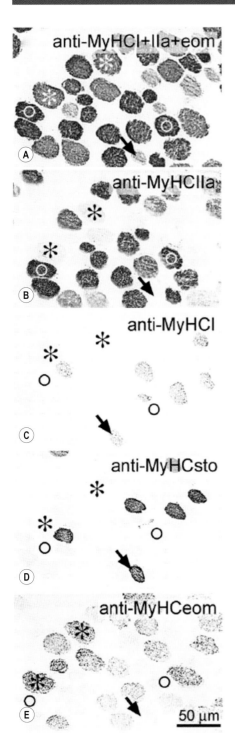

Figure 7.10 Photomicrographs of five serial sections from the global layer of an adult human superior oblique. Sections are immunostained with (**A**) anti-MyHCl+lla+eom, (**B**) anti-MyHClla, (**C**) anti-MyHCl, (**D**) anti-MyHCslowtonic, and (**E**) anti-MyHCeom. O represents two myofibers immunostained with anti-MyHClla; arrow represents a fiber stained with anti-MyHCl; * represents two MyHCeom-positive and MyHClla-negative fibers. (Reproduced from Kjellgren et al., IOVS 44:1419–1425, 2003, with permission from the Association of Research in Vision and Ophthalmology.)

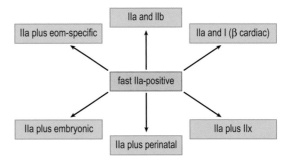

Figure 7.11 Schematic of human single myofiber co-expression of multiple MyHC isoforms. Myofiber can express MyHClla only, or it could co-express any one of the 6 MyHC isoforms shown, creating seven myofiber "types". However, single myofibers can co-express three, four, or more MyHC isoforms, increasing the possible number of "types", and this scheme only considers MyHC expression patterns. (Based on co-expression data found in Kjellgren et al, *Invest Ophthalmol Vis Sci* 44:1419–1425, 2003 and Stirn Kranjc et al, *Graefes Arch Clin Exp Ophthalmol* Epub July 17 2009.)

the non-cranial muscles examined thus far. Using labeling with the thymidine analog bromodeoxyuridine, activated satellite cells, the myogenic precursor cells of adult muscle, are seen; with sufficient post-labeling intervals, labeled myonuclei are present within existing myofibers in normal adult EOM of rabbits and mice. These new myonuclei are located peripherally, not centrally, indicating that this process of myonuclear addition is different than what occurs during muscle regeneration, when myotubes form followed by fibers with central nucleation (Fig. 7.13). The process of myofiber remodeling appears to continue throughout life, as the presence of activated satellite cells, identified by the myogenic lineage marker MyoD,[75] is seen in the EOM from elderly humans.[76] Ongoing myofiber remodeling occurs in other craniofacial muscles as well,[77] suggesting that the differences in genetic control of cranial muscle formation may play a role in retention of this dynamic process in adult EOM.[78] The control of this process in the adult EOM is unknown. In contrast to non-cranial skeletal muscles, EOM continue to express a number of myogenic growth factors and functional receptors for these growth promoting molecules (Fig. 7.13).[79,80] Pitx2, a mitogen and repressor of differentiation,[81,82] is a signaling factor expressed in adult EOM[81] and may play a role in maintaining a proliferative state in adult EOM. A conditional knock-out of Pitx2 in adult EOM results in alteration of the adult EOM phenotype to more resemble that of limb skeletal muscle.[201] This is further evidence of EOM's unique allotype.

EOM are also set apart from non-cranial skeletal muscles in their resiliency to injury and/or denervation. Botulinum toxin, a muscle paralytic agent used to treat strabismus and focal dystonias such as blepharospasm,[83] causes myofiber atrophy in limb skeletal muscle. However, botulinum toxin treatment in EOM actually causes some myofiber hypertrophy[84] (Fig. 7.14), short-term activation of satellite cell proliferation,[53] shifts in MyHC composition,[53] and few long-term effects except for loss of the EOM-specific MyHC isoform.[85] Recalcitrance to injury also is seen after injection of the local anesthetic bupivacaine. In limb skeletal muscles, bupivacaine is a potent myotoxin

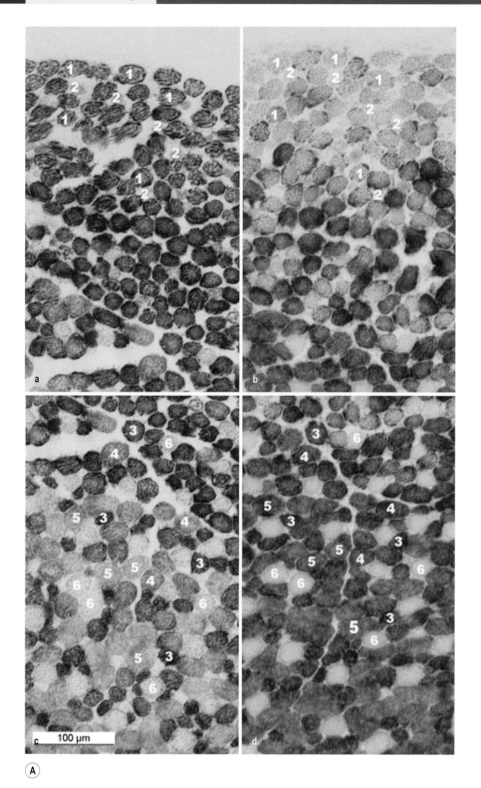

(A)

Figure 7.12 (**A**) Serial cross-sections through the orbital (a, b) and global (c, d) layers of rat superior rectus muscle stained histochemically for (a, c) succinic dehydrogenase (SDH) and (b, d) α-glycerophosphate dehydrogenase (GPDH). Individual myofibers are indicated by numbers 1–6. Different proportions of both enzymes are present in many myofibers (Reproduced with permission from Asmussen et al., Invest Opthalmol Vis Sci 49:4865–4871, 2008.).

and causes massive myofiber degeneration, but in the EOM of primates minimal lesions were seen after retrobulbar injection.[86,87] Surprisingly, denervation of adult EOM causes relatively little histological change;[88–90] the multiply innervated myofibers actually hypertrophy in the denervated EOM.[88]

Using a denervation/reinnervation animal model, previously denervated EOM actually has more myofibers than the non-operated control side.[91] These data illustrate two very important features of the EOM: they are remarkably resistant to various forms of injury, and they are incredibly adaptable.

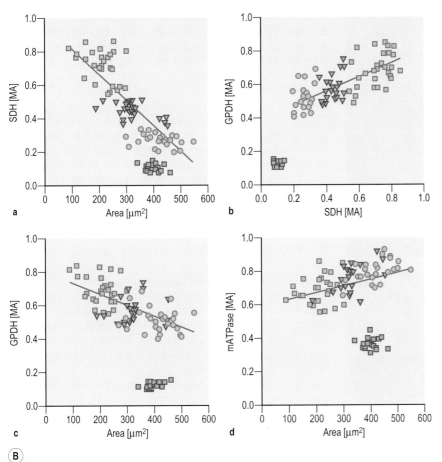

Figure 7.12, cont'd (**B**) Scatterplots of global layer data showing the relationship between cross-sectional area and SDH (a), GPDH (c) and mATPase (d) activities and between GPDH and SDH (b) expressed as the mean absorbance of the final reaction product in the cells of the fast twitch types 3 (■), 4 (▼), and 5 (◆) and the slow type 6 (□). The regression lines of the type 3 to 5 twitch fibers are indicated; they do not segregate into different "groups". (From Asmussen et al, *Invest Ophthalmol Vis Sci* 49:4865–4871, 2008, with permission from the Association of Research in Vision and Ophthalmology.)

Proprioception and proprioceptors

Proprioception in limb skeletal muscle is largely the job of muscle spindles, sensory receptors that detect changes in muscle length; this information is conveyed to the CNS which responds by regulating muscle tension to resist stretch. In the EOM, the role of proprioception is a matter of some controversy. Primate EOM does not appear to have a stretch reflex.[92] Deafferentation of the EOM bilaterally does not result in eye position or eye movement changes.[93] However, when the EOM of a cat is stretched, neurons within the brain respond, despite the fact that cat EOM lack muscle spindles.[94] The brain has access to eye position, because even in total darkness humans perceive passive changes in eye position.[94] Visual cortex responses are modulated by changes in eye position also,[95] and primary somatosensory cortex contains proprioceptive representation of eye position.[96] What are the sensory organs, then, that receive and transmit this proprioceptive input?

The presence of muscle spindles has been histologically demonstrated only in the EOM of man, monkey and even-toed ungulates. Muscle spindles are composed of two types of specialized intrafusal myofibers within a connective tissue sheath (Fig. 7.15): nuclear chain and nuclear bag fibers. Only nuclear chain fibers are common in human EOM. Intrafusal fibers receive sensory innervation at their equatorial region and motor innervation at their spindle poles, and are surrounded by an acidic mucopolysaccharide-containing fluid (Fig. 7.15).[97] In human EOM, the number and placement of muscle spindles varies quite significantly in the different muscles.[98] In general, they are higher in density in the more proximal and distal portions of the EOM and relatively devoid of spindles in the mid-belly region. The density varies between the individual EOM. Inferior rectus contains the most, with approximately 34 in the entire muscle. Superior oblique has two-fold more spindles than the other three rectus muscles. Inferior oblique has the fewest, averaging only four spindles which are located in the mid-belly region, rather than towards the origin and insertion.[98] Despite their well-described nature, the role muscle spindles play in the control of eye movements is unclear; one hypothesis suggests that they monitor static length and play no role in velocity monitoring, as no stretch reflex is seen in the EOM of monkey for example.[92]

A second organ associated with proprioception is palisade endings (formerly called myotendinous cylinders), present in all mammalian EOM examined thus far, including man (Fig. 7.16).[99] These endings appear to be unique to the EOM.[100] They are found in the distal myotendinous region, where nerve fibers exit the muscle and then turn back to form synaptic contacts on the distal muscle tips of multiply innervated myofibers (Fig. 7.16).[101] The synapses of palisade endings have been definitively shown to be motor (Fig. 7.16),[102,103] supported by experiments lesioning the oculomotor nucleus which results in loss of both myofiber NMJs and those in the palisade endings.[104] Thus, palisade endings have an effector function and would not appear to be able to provide sensory feedback on eye position.[102,103]

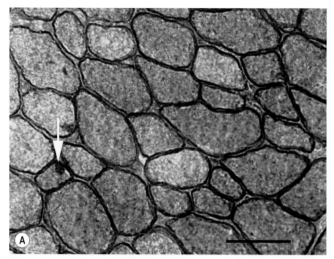

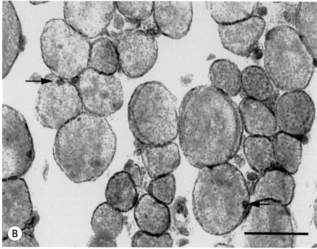

Figure 7.13 (A) Normal EOM incorporate the thymidine analog bromodeoxyuridine (brdU) into dividing satellite cells, and these are incorporated into apparently normal myofibers in adults. In this case, the rabbit was injected once daily with brdU for 7 days, followed by 14 days brdU-free. White arrow indicates brdU-positive nucleus. Myofibers are counterstained with an antibody to dystrophin for visualization of the sarcolemma. Bar is 20 μm. **(B)** Arrows indicate satellite cells positive for the insulin-like growth factor-1 receptor. The sections were counterstained for laminin for visualization of satellite cells. Bar is 20 μm.

Development

The distinct embryological origin of EOM may explain retention of its many unusual characteristics. While much more is understood about the early development of non-cranial skeletal muscles, recent studies have shed light on the genetic control of myogenesis in extraocular and other craniofacial muscles.[105-107] The early control of craniofacial myogenesis uses a distinct set of genes and proteins compared to non-cranial muscles (Fig. 7.17A),[108-110] and elucidating these differences will help lay the foundation for understanding clinical problems involving the EOM.

The EOM are derived from two sources of cranial mesoderm.[111] These early mesodermal precursor cells are not segmented, in contrast to the somites that give rise to non-cranial skeletal muscles. Additionally, cranial mesoderm does not separate into distinct dermamyotomes and sclerotomes.

Instead, EOM precursors arise from prechordal or paraxial mesoderm adjacent to the mes- and metencephalon, the future midbrain and hindbrain respectively, and migrate to their final locations.[112] When the transcription factor Pax3 is knocked out in mice, despite a complete absence of body and limb skeletal muscle, EOM develop normally.[108] Other myogenic regulatory factors such as Sonic Hedgehog (Shh), Wnt3a and bone morphogenetic protein 4 (BMP4), which turn on myogenesis in somitic mesoderm, have an inhibitory effect on myogenesis in cranial mesodermal cells both in vitro and in vivo.[105] These proteins are generated by the developing brain and surface ectoderm, delaying myogenesis onset. Soon after identifiable EOM anlagen form, neural crest-derived cells migrate into them,[112] and this migration depends on BMP signaling.[113]

Recent studies show that the homeodomain transcription factor Pitx2 is required for EOM development (Fig. 7.17A),[114,115] and the absence of this gene in development causes ocular defects such as Rieger syndrome.[116] EOM morphogenesis is tightly regulated by *Pitx2* gene dose, so if there is a small reduction in *Pitx2*, no superior and inferior oblique muscles form but the rectus muscles develop normally.[117] If *Pitx2* levels are reduced further, the rectus muscles become smaller and more disorganized, and a total deletion of *Pitx2* results in the complete absence of all the EOM (Fig. 7.17B). *Pitx2* is a retinoic acid (RA) responsive gene, and RA controls neural crest movement.[118] Neural crest infiltration into the EOM is required for their normal migration. In the absence of RA, the EOM form but do not successfully migrate.[118] Other signaling factors such as barx2 and Lbx1 are differentially expressed in early developing EOM, but not universally[110] (Fig. 7.17C), and their role in EOM formation is unclear.

After precursor migration, on a molecular level myogenesis proceeds similarly to that seen in non-cranial skeletal muscle, but it differs quite significantly on a temporal level.[109] Muscles derived from somites express myf5 and MyoD transcripts, early markers of muscle-specific differentiation, earlier in development than in craniofacial muscle precursors.[119] There is a prolonged period prior to expression of MyHC in craniofacial muscles compared to times between MyoD and MyHC expression in developing limb muscle. Neural crest cells express a number of signaling factors, including BMP, Wnt, and Shh, thought to be responsible for this early repression of myogenesis in developing craniofacial muscles,[105] whereas these same molecules promote muscle differentiation in somitic muscle precursors. The sequence and timing of MyHC expression also differs in cranial and non-cranial skeletal muscle.[120]

In the EOM from human fetuses, only primary myotubes are present at early gestational ages, and co-express two developmental isoforms of slow myosin (type I). In addition they express the two "immature" isoforms, embryonic (MYH3) and fetal (MYH8). These developmental stages consistently lag behind limb skeletal muscle by 2–4 weeks.[120] When secondary myotube formation begins, the first type IIa (MYH2) is expressed.[120,121] Even by 22 weeks of gestation in human fetuses, the alpha-cardiac (MYH6) and EOM-specific (MYH13) MyHC isoforms are not detected. This fiber type development and diversification in EOM occurs in the absence of morphologically mature endplates.[122] Studies in mouse embryos show that EOM position is dependent on cues from the local environment, but early EOM myogenesis

orbital layer

global layer

control

Botox treated

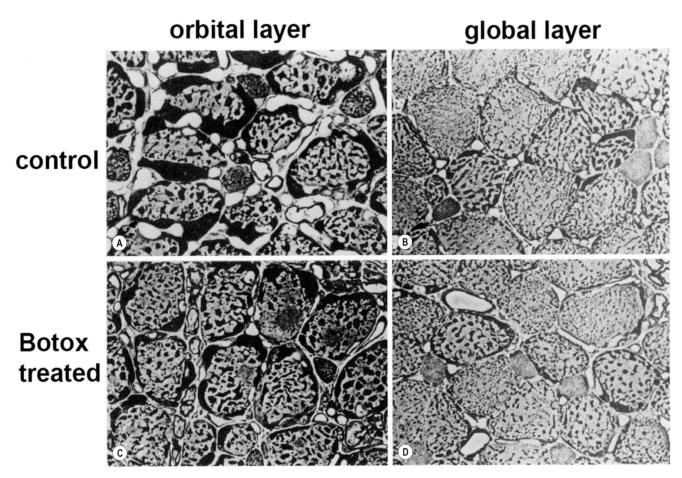

Figure 7.14 In (**A**, **B**) control or (**C**, **D**) botulinum toxin A (Botox)-treated (14 days post-injection) adult medial rectus muscle from primates, the orbital layer myofibers actually hypertrophy rather than atrophy. No change is seen in global layer fibers after Botox. Plastic sections stained with para-phenylenediamine. (From Spencer and McNeer Arch Ophthalmol 105:1703, 1987. Copyright 1987 American Medical Association. All rights reserved.)

is independent of neural crest cells, target tissues and nerves.[123] Modern molecular biological tools are helping to answer the question of how EOM becomes EOM and what controls its collective unusual properties at the molecular level, yet there is much still unknown. This becomes rapidly apparent when congenital conditions involving the EOM are examined.

Disease propensity

From a clinical perspective, the preferential involvement or sparing of the EOM in disease represent critical unsolved questions. Diseases that involve the EOM can be divided into three basic categories: (1) clinical disorders related to the specific motor functions of EOM, (2) diseases with preferential sparing of the EOM compared to limb skeletal muscle, and (3) diseases where the EOM are either primarily or preferentially involved compared to limb skeletal muscle. An exhaustive review of all these clinical entities is beyond the scope of this chapter; however, each of these general categories will be discussed.

Disorders of eye movements

Three examples of conditions that manifest with abnormal eye position and/or movements are discussed: strabismus,

nystagmus, and congenital cranial dysinnervation disorders (CCDD).

Strabismus

One of the most common motor disorders involving the EOM is strabismus, defined as misalignment of the eyes such that disparate images reach corresponding parts of each retina, disrupting binocular vision (see Chapter 36).

Many studies have examined the oculomotor control of eye position in strabismus, but much less is understood about the EOM themselves in these conditions. Studies measuring EOM forces during eye movements in strabismic patients yielded quite disparate results.[2,128,129] In one study of adults with concomitant strabismus, whether convergent (esotropia) or divergent (exotropia), no differences in mechanical or contractile properties were seen.[2] All patients with superior oblique palsy showed reduced values for mean peak velocity in most directions of gaze compared with normal levels, whereas mean peak and steady state tension did not vary between normal EOM and patients with either congenital or acquired superior oblique palsy.[129]

Studies of muscle structure in strabismic patients and animal models also yielded quite different results. Horizontal rectus muscle anatomy in naturally and artificially strabismic monkeys showed no differences in EOM size,

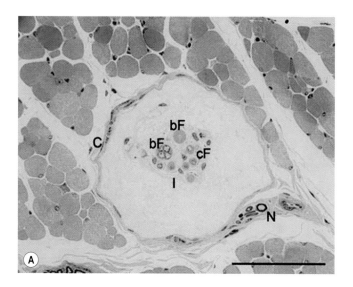

Figure 7.15 (**A**) Cross-section through the equatorial region of a cow muscle spindle. The muscle spindle contains two nuclear bag fibers (bF) and five nuclear chain fibers (cF). In one nuclear bag fiber the bag region containing several nuclei is visible. In some nuclear chain fibers their central nuclei are seen. Bag fibers have a larger diameter than chain fibers. An inner capsule (I) surrounds the intrafusal myofibers. Capsule (C), nerve (N). Scale bar: 100 mm. (From Blumer et al, Exp Eye Res 77:447–462, 2003. Copyright Elsevier 2003) (**B**) Reconstructions from camera lucida drawings illustrating the exact positions and relative lengths of spindles projected into one plane for all six EOM of the same orbit (72-year-old woman). MR = medial rectus; LR = lateral rectus; SR = superior rectus; IR = inferior rectus; SO = superior oblique; IO = inferior oblique; n = nasal; t = temporal; s = superior; i = inferior; a = anterior; p = posterior. Scale = 1:2.5. (Modified from Lukas et al. IOVS 35:4317–4327, 1994, with permission from the Association of Research in Vision and Ophthalmology.)

Distal

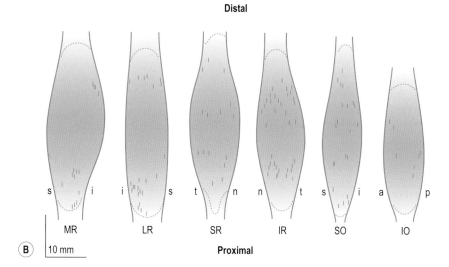

Proximal

structure, or innervation.[130] Interestingly, there was much less uniformity in cross-sectional area in the natural esotropes compared to those artificially induced. EOM obtained during strabismus surgery showed minimal changes at the light level, but an array of myofiber structural changes at the electron microscopic level, including fiber vacuolization and mitochondrial changes.[131,132] Others described minimal changes in the myofibers and their motor innervation, but significant pathology in the proprioceptive innervation at the myotendinous junctions.[133] However, after examination of EOM from a number of different ocular motility disorders, some of the unusual features in the strabismic muscles appeared in the control muscles also.[134] Increased satellite cell numbers were present in the inferior oblique muscles from patients with inferior oblique overaction[135] suggesting up-regulated myofiber remodeling. Similarly, underacting medial rectus muscles from exotropes had two-fold higher levels of activated satellite cells, while overacting medial rectus muscles from esotropes had fewer activated satellite cells compared to control.[136] These studies suggest that adaptive changes occur in strabismic muscles. Current treatments for strabismus can include patching and surgery (see Box 7.2). Future approaches to strabismus treatment may be able to tap into these processes.

Box 7.2 Treatment of strabismus

Eye deviations are first treated by patching the "good" eye, the one aligned in primary gaze. If this is ineffective, incisional surgery is performed on the muscles. Recession surgery is thought to weaken an "overacting" muscle, performed by moving the muscle's scleral insertion posteriorly. Resection surgery is thought to strengthen an "underacting" muscle, performed by removing a portion of the muscle and reattaching the muscle to its original site of insertion on the sclera.[127] If this misalignment is not corrected in early childhood, amblyopia may develop. Amblyopia is a condition of functional blindness in an otherwise normal eye.[124–126]

Nystagmus

Nystagmus is defined as a bilateral, involuntary, and conjugate oscillation of the eyes, and is primarily a problem of oculomotor control within the brain. Patterns of eye movements in nystagmus patients are complex, and many different forms exist including both congenital and acquired.[137,138] Nystagmus can be inherited genetically, and may have a

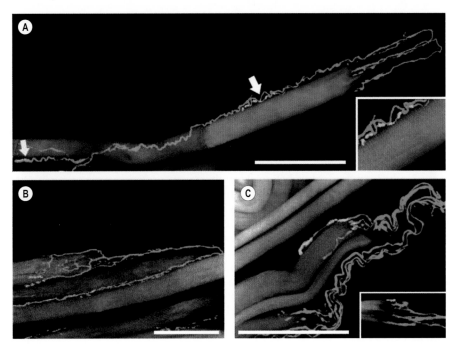

Figure 7.16 Confocal laser scanning microscopy images of palisade endings. Nerve fibers are labeled with anti-ChAT (green), nicotinic acetylcholine receptors with α-bungarotoxin (red), and myofibers with phalloidin (white). The tendon not labeled continues from the myofiber to the right. Neuromuscular contacts which exhibit ChAT/α-bungarotoxin reactivity appear yellow in the overlay. (**A**) ChAT-positive nerve fiber running alongside the myofiber forms palisade ending at the myofiber tip. The same nerve fiber established multiple, ChAT/α-bungarotoxin positive neuromuscular contacts (arrows) on the same myofiber. (Details of the neuromuscular contacts are shown in the insets.) Preterminal axons and nerve terminals of the palisade ending are also ChAT-positive. Palisade nerve terminals exhibit no α-bungarotoxin-binding, thus no neuromuscular contacts are present. (**B**) ChAT-positive nerve fiber forming a palisade ending at a myofiber tip and establishing ChAT/α-bungarotoxin-positive neuromuscular contacts on a neighboring myofiber outside the palisade complex. The palisade ending establishes neuromuscular contacts that are ChAT/α-bungarotoxin-positive. (**C**) Palisade endings with neuromuscular contacts exhibiting ChAT/α-bungarotoxin-reactivity. A detail of another palisade ending with a single ChAT/α-bungarotoxin-positive neuromuscular contact is shown in the inset. Scale bars, 100 μm. (Reproduced from Blumer et al. IOVS 50:1176–1186, 2009, with permission from the Association of Research in Vision and Ophthalmology.)

Box 7.3 Surgery for congenital nystagmus

It has been suggested that tenotomy in congenital nystagmus patients, removal of the muscle from its scleral insertion and reattachment, results in reduced amplitude and intensity.[144] Other investigators found tenotomy increased the velocity and intensity of the nystagmus in a monkey model.[145] Further studies are needed to clarify this issue; however, it is important to note that the myotendinous junctions thought to be disrupted by this procedure are motor, not sensory as originally proposed.[102]

---

sensory or motor origin.[139,140] The complex waveforms of the oscillatory eye movements have been extensively studied,[138] but little is known about adaptive changes at the EOM level. Ultrastructural studies on human EOM from patients with congenital nystagmus show that myofibrillar orientation as well as mitochondrial structure are altered.[141,142] While a variety of pharmacologic treatments have been suggested, controlled clinical trials are needed.[143] Surgery also is a common treatment option for these patients (Box 7.3).

Congenital cranial dysinnervation disorders (CCDD)

Several rare, non-progressive inherited strabismic disorders have been described and characterized by congenital fibrosis of one or more of the EOM resulting in a static eye position or directional impairment. The genetic bases for several of the congenital cranial dysinnervation disorders (CCDD), as they are now called, have been instructive in understanding EOM and cranial motor neuron development (Box 7.4).[146] Each identified genetic mutation causes distinct developmental defects of the oculomotor (CNIII), trochlear (CNIV) and/or abducens (CNVI) nerves.[147] While patients with these CCDDs are rare, they demonstrate the incredible complexity in the genetic control of the EOM and their innervation.

Diseases where EOM are preferentially spared

There are many skeletal muscle diseases that spare the EOM (Table 7.2). Duchenne muscular dystrophy (DMD) is a devastating X-linked genetic disorder characterized by the absence of dystrophin, repeated cycles of degeneration and regeneration, progressive muscle weakness, and ultimately death. The EOM are spared, both morphologically[153] and functionally.[154] The mechanism that allows for this sparing is unknown. None of the potential structural or metabolic characteristics examined thus far has proven to be mechanistic for this sparing including, but not limited to: calcium handling, antioxidative enzymes, nitric oxide location, cation handling, and utrophin up-regulation.[65,81,156–159]

The *constitutive* nature of EOM sparing is supported by the EOM being spared in other dystrophic diseases,[159] including Becker muscular dystrophy,[154] laminin alpha2-deficiency,[160] merosin-deficient muscular dystrophy,[161] sarcoglycan deficiency,[162] and congenital muscular dystrophy, where, despite some ultrastructural changes,[161] the EOM display full ocular motility.[163] EOM sparing has even been documented for inflammatory autoimmune diseases of skeletal muscle such as dermatomyositis.[164] This wide array of skeletal muscular dystrophies and degenerative disorders that spare the EOM suggests that the sparing mechanism must be intrinsic to EOM muscle. Two interesting hypotheses currently being tested as potential mechanisms for EOM sparing in dystrophies suggest that either the EOM are inherently able to resist the necrosis caused by the "hostile" tissue milieu within dystrophic muscle, or they have the capacity for increased regenerative potential, an idea supported by their ability to remodel and repair throughout life.[165] Their inherent ability to survive denervation supports the first hypothesis.[84,85,90] The ability of the EOM to remodel even in aging muscle[75] and the relative absence in the EOM of the sarcopenia and other defects seen in aging non-cranial skeletal muscle[166] suggest that the intrinsic regenerative capacity in adult EOM is sufficient to provide functional EOM throughout the lifetime of these patients. An understanding of the mechanism of sparing may lead to new treatments for these fatal dystrophic diseases.

The EOM, and the motor neurons which innervate them, have long been considered spared in amyotrophic lateral sclerosis (ALS), with histopathologic examination of nerve roots as supporting evidence.[167] However, many studies cast doubt on this hypothesis, suggesting instead that the overall time course of the disease plays a role in whether eye movement disorders are apparent. Analysis of ocular movements and muscle pathology is made difficult by the heterogeneity of the patient populations relative to onset and speed of decline. In a longitudinal study of eight patients, all but one showed changes in electro-oculography of the EOM; some were subclinical but most were progressive over time.[168] Differential vulnerability of cranial motor neurons innervating the EOM occurs in several mouse models of ALS, with differential loss or sparing of motor neurons in transgenic mice expressing the human mutation in the superoxide dismutase gene-1 compared to two spontaneous mouse models of ALS.[169] Direct examination of the EOM post-mortem from two subsets of ALS patients, those with either a short- or long-time course of neuron loss, showed there were degenerative changes in the cranial motor nerves within the EOM from these patients, and differential sparing of myofibers in the two groups.[170] The preferential sparing of the ocular motor neurons and the EOM in ALS patients may just be temporal; if an individual lives sufficiently long, the ocular motor neurons and associated EOM would show degenerative changes. Alternatively eye movements may not be

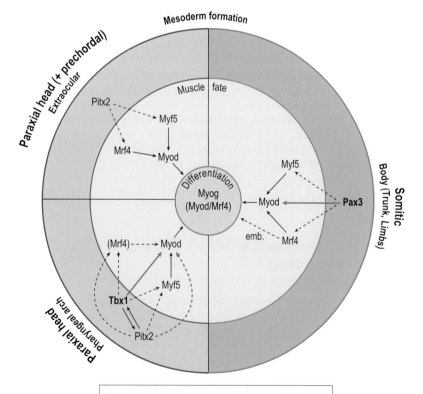

Figure 7.17 (**A**) EOMs do not have complementary regulatory pathways. MyoD expression is not rescued in the absence of Myf5 and Mrf4, and EOMs do not form. In pharyngeal (PA) muscles, Tbx1 cooperates with Myf5. In their combined absence, MyoD is not activated, and PA muscles are missing. In the body and EOMs, Mrf4 determines embryonic but not fetal MPC fate. Pitx2 may cooperate with Tbx1 at this regulatory step in the PA. Pax3 acts as the complementary pathway for body myogenesis to rescue MyoD expression. (Modified from Sambasivan et al, Dev Cell 16:810–821, 2009. Copyright Elsevier 2009.)

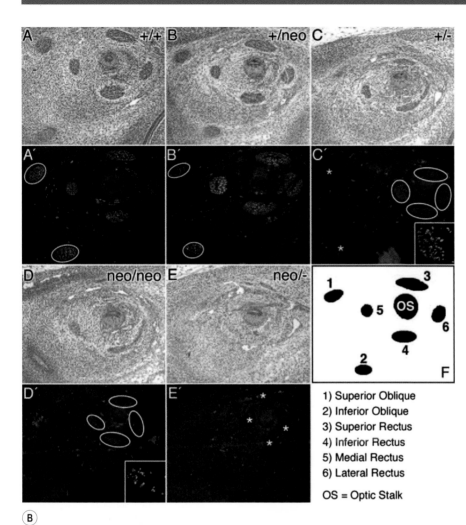

Figure 7.17, cont'd (**B**) EOM morphogenesis correlates strongly with *Pitx2* gene dose. Sagittal sections immediately medial to the optic cup taken from E14.5 mice embryos of the indicated genotypes. Sections were stained by hematoxylin and eosin (A–E) or developmental myosin heavy-chain (dMyh) immunofluorescence (A–E). *Insets*: medial rectus muscles. Superior and inferior oblique muscles (yellow ovals in A and B) are more sensitive to *Pitx2* gene dose than rectus muscles (white ovals in C and D). All EOM, including rectus muscles (E, *), are absent in *Pitx2neo-/-* embryos. (F) Schematic key for EOM position. Only rectus and oblique muscles are shown; the retractor bulbi muscle has been omitted. Magnification, ×20; *insets*, ×40. (Reproduced from Diehl et al., Invest Ophthalmol Vis Sci 47:1785–1793, 2006, with permission from the Association of Research in Vision and Ophthalmology.) (**C**) Genes expressed in all (*barx2, pax2*) or subsets of chick head muscles. *Lbx1* is, like *paraxis*, expressed in glossal and laryngeal myoblasts derived from occipital somites, and in the lateral rectus (and *lbx1* in the dorsal oblique). *Tbx1* is in all branchial myoblasts and also the lateral rectus, and is expressed in other epithelial and mesenchymal cells of the head. *Pitx2* is present in the 1st branchial arch muscle plate and also in several EOMs plus periocular neural crest cells. *HGF* is produced in the mandibular and dorsal oblique muscles, and serves as a neurotrophic factor for the mandibular and trochlear nerves. (From Noden and Francis-West. Dev Dyn 235:1194–1218, 2006.)

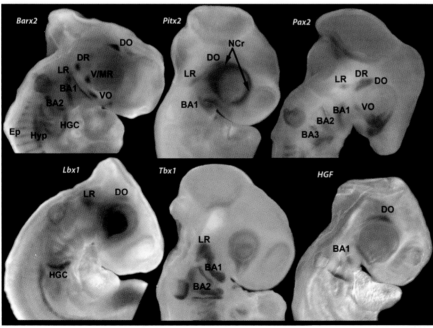

Box 7.4 Congenital fibrosis of the extraocular muscles

Patients with congenital fibrosis of the EOM type 1 (CFEOM1) have a defect in the development of the superior division of the CNIII resulting in atrophic superior rectus and levator palpebrae superioris muscles.[146] The mutation is in the KIF21A gene, a developmentally expressed kinesin which presumably delivers a molecule critical for early neurite survival.[148] CFEOM2 has a recessive inheritance, and all EOM are missing except the lateral rectus, which is innervated by CNVI (Fig. 7.18). This disorder is due to mutations in the PHOX2A gene,[149] a transcription factor restricted to several classes of differentiating neurons. Its role in development is unclear; however, in PHOX2A null mice, no

oculomotor or trochlear neurons develop along with other brain abnormalities.[150] Duane's retraction syndrome is a congenital EOM disorder characterized by the lack of CNVI, and aberrant innervation of the lateral rectus from CNIII is often seen (Fig. 7.18).[151] The genetic mutation in this disorder is the CHN1 gene which encodes alpha1-chimaerin, a Rac guanosine triphosphatase-activating protein signaling protein. This results in a gain-of-function mutation causing increased α2-chimaerin on cell membranes[152], and its over-expression in transgenic mice leads to pathfinding errors.

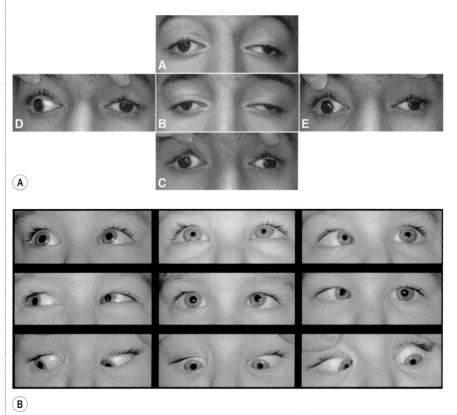

Figure 7.18 (**A**) Patient with congenital fibrosis of the EOM type 2 (CREOM2). In primary position, note eyelid ptosis and lack of fixation in either eye (center panel). Up-gaze (A) and down-gaze (C) show complete absence of vertical movement. Right gaze (D) and left gaze (E) show complete lack of adduction of the left eye with almost full abduction of the right eye but moderately reduced abduction of the left eye. (Reproduced with permission from Bosley et al., Brain 129:2363–2374, 2006.) (**B**) Child with Duane's retraction syndrome, the congenital absence of CNVI resulting in a paralytic lateral rectus muscle. In primary position (center panel), the child is esotropic. Note limited abduction (middle row, far right). (From Morad et al., J AAPOS 5:158–163, 2001. Copyright Elsevier 2001.)

abnormal in these patients because even in the absence of 25 percent of cranial motor neurons, eye movements are completely unchanged.[171] In addition, polyneuronal innervation, where more than one neuron can control the same myofiber, may protect the eye movement system against disease; if one motor neuron is lost, there are others that can take over functionally.[24]

Diseases where EOM are preferentially involved

There are many diseases where the EOM are exclusively or preferentially involved. One disorder with primary EOM involvement is thyroid eye disease (TED) or Graves' ophthalmopathy, an autoimmune disease with significant inflammatory cell invasion of the EOM (Fig. 7.19).[172] Despite many studies, the autoantigen responsible for this disease is still a

Box 7.5 Optic neuropathy of thyroid orbitopathy

Optic nerve compression in thyroid eye disease is a vision-threatening condition, and if the severity of the inflammation does not abate on its own and cannot be reduced medically with oral or intravenous pulse corticosteroids[173], surgical decompression of the orbit by surgically produced transpalpebral orbital floor removal or endoscopic decompression is required to prevent permanent vision loss.[174]

matter of debate.[175–180] Eye findings include exophthalmos, lid retraction, periorbital edema, pain, and diplopia.[181,182] Significant enlargement of the EOM can occur, and in severe cases causes compression of the optic nerve posteriorly, resulting in permanent vision loss (see Box 7.5). This is caused by the inability of the bony orbit to accommodate enlargement of soft tissues, resulting in compression injury.

Table 7.2 Extraocular Muscles Have Different Disease Profiles Compared to Limb Skeletal Muscle

Diseases where EOM are spared	Etiology	Limb muscle pathology
Duchenne muscular dystrophy	X-linked genetic mutation of dystrophin gene	Progressive. Muscle wasting and weakness
Becker muscular dystrophy	X-linked genetic mutation of dystrophin gene, less severe phenotype than Duchenne	Progressive. Muscle wasting and weakness
α, γ and δ-sarcoglycan deficiency (limb girdle muscular dystrophy)	Mutation of the sarcoglycan gene	Progressive. Muscle wasting and weakness
Laminin α2-congenital muscular dystrophy	Mutation of the laminin α2 gene	Progressive. Muscle wasting and weakness
Amyotrophic lateral sclerosis	Mutations of the superoxide dismutase gene; mitochondriopathy	Progressive. Muscle wasting and paralysis
Diseases where EOM are primarily or preferentially involved	**Etiology**	**EOM pathology and/or symptoms**
Graves' ophthalmopathy	Autoimmune disease of the EOM, resulting in enlargement; presumably due to one or more shared antigens with the thyroid gland	Inflammatory orbitopathy, myopathy
CPEO (chronic progressive external ophthalmoplegia)	Mitochondrial DNA deletion, mutation of DNA polymerase-gamma gene	Accumulation of mutant mitochondria leads to muscle paralysis
Kearns–Sayre syndrome	Longer mitochondrial DNA deletions than CPEO	Accumulation of mutant mitochondria leads to muscle paralysis
Ocular myasthenia gravis	Autoimmune disease to either the acetylcholine receptor or MuSK	EOM and levator palpebrae superioris muscle weakness
Myotonic dystrophy type 1	Expansion of a CTG repeat within the DMPK gene	Saccadic slowing, optokinetic nystagmus
Myotonic dystrophy type 2	Expansion of a CCTG repeat expansion of the CNBP gene	Rebound nystagmus
Childhood strabismus	Unknown. Complex genetic cause?	Under- or overactive EOM with loss of binocularity and eye alignment in primary gaze
Congenital nystagmus	Missense mutation in FRMD7 gene; function unknown. Clinically heterogeneous; multiple genes involved	Conjugate, horizontal eye oscillations, in primary or eccentric gaze
Miller-Fisher syndrome	Autoimmune disease against ganglioside GQ1b/GT1a	EOM paralysis
Congenital cranial dysinnervation disorders	Specific gene mutation for each type	EOM weakness or absence

Primary EOM involvement occurs with some mitochondrial myopathies, such as chronic progressive external ophthalmoplegia (CPEO). CPEO is not one specific disorder but a clinical set of symptoms associated with mitochondrial DNA deletions or mutations and primarily defined by paralysis or weakness of one or more of the EOM,[183,184] with no current treatments (see Box 7.6). CPEO is one of many diseases associated with clonal expansions of mitochondrial DNA mutations within myofiber segments.[184] Some mitochondrial myopathies affect both the EOM and other organs, including Kearns–Sayre syndrome (KSS) and Myoclonic Epilepsy associated with Ragged Red Fibers (MERRF).[185] In KSS, patients have a more severe phenotype than seen with CPEO; in addition to ophthalmoplegia, patients have retinopathy, proximal muscle weakness, cardiac arrhythmia, and ataxia. In MERRF, chronic neurodegenerative changes also occur. While it is still unclear why mitochondria with deletions would preferentially increase in the EOM and/or EOM and brain, both tissues are highly oxidative, with rates of

mitochondrial function three to four times greater than limb skeletal muscles.[186] Mitochondrial mutations reduce energy production, a problem for the oxidative EOM.[187]

Aging has been associated with the clonal expansion of defective mitochondria in this disorder,[188] as mutated mitochondria can completely replace normal ones over time.[189,190] Interestingly, magnetic resonance imaging (MRI) of the EOM in CPEO patients demonstrates that the EOM are

---

Box 7.6 Chronic progressive external ophthalmoplegia

Unfortunately there are no proven effective therapies for CPEO, KSS, and related mitochondrial cytopathies. In CPEO, for example, symptoms of strabismus, dry eye and ptosis can only be managed, as is true for those patients with concomitant cardiac defects.[192]

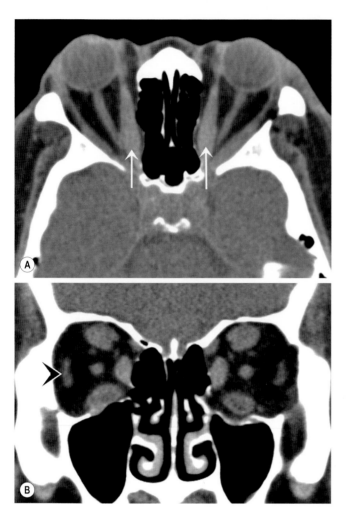

Figure 7.19 (**A**) Axial MRI of a patient with TED. Increased thicknesses of medial rectus muscles bilaterally (arrow) is prominent. (**B**) Coronal MRI of a patient with TED. Note all muscles are enlarged except the lateral rectus (arrowhead). (Courtesy of Dr. Michael Lee, University of Minnesota.)

Box 7.7 Treating ocular myasthenia

There are a number of drug treatments available for ocular myasthenia gravis. For example, prednisone can reduce symptoms in ocular myasthenic patients as well as significantly reduce the frequency of developing generalized myasthenia gravis both acutely and over the course of four or more years.[200]

alignment of the visual world on the retina causing diplopia, blur, vertigo, and other ocular symptoms (see Box 7.7).

The brain oculomotor control system cannot quickly adapt to the asymmetric or variable EOM weakness.[198] Therefore, EOM weakness is quickly discerned by patients. However, when patients with ocular myasthenia are examined, many have measurable, albeit subclinical, weakness in other muscle groups as well. Another hypothesis for EOM preferential involvement is the differences between their NMJ endplates compared to those in limb skeletal muscle.[20] Differences in both classical and alternative complement-mediated immune response pathways in EOM may also increase their susceptibility to ocular myasthenic symptoms.[199]

Conclusion

The EOM are a specialized collection of craniofacial muscles with many distinct characteristics compared to non-cranial skeletal muscle. These muscles have a complex anatomy and physiology, which should be carefully considered when determining treatment of the motor disorders that affect them. The EOM are also highly adaptable, altering their protein expression and physiological characteristics in response to hormones, denervation, and toxins. This adaptability should lend itself to pharmacologic manipulation of these muscles for the treatment of EOM motor disorders. Furthermore, understanding the mechanisms that result in EOM sparing or preferential involvement in diseases of skeletal muscle will hopefully suggest new treatments for EOM diseases and for skeletal muscle diseases in general.

significantly smaller than in controls.[191] This can aid in the differential diagnosis of this condition.

Myasthenia gravis is an example of a condition that preferentially involves the EOM, but can also involve multiple non-cranial skeletal muscles. Myasthenia gravis is an autoimmune disease of the NMJ resulting in skeletal muscle weakness and fatigability caused by functional NMJ dropout. While EOM weakness occurs in 90 percent of patients with myasthenia gravis, approximately 15 percent have only ocular myasthenia without other muscle involvement.[193] Interestingly, in generalized myasthenia patients, 90 percent have serum antibodies to the nicotinic acetylcholine receptor, while only 65 percent of the ocular myasthenics are positive for this autoantigen.[194] Other possible autoantigens that have been discussed include muscle-specific kinase (MuSK),[195] the ryanodine receptor,[196] and a number of eye muscle-specific proteins.[197] The frequent involvement of the levator palpebrae superioris and the EOM in myasthenia gravis may be due to the distinct NMJ characteristics of EOM compared to non-cranial skeletal muscles.[193] Early in the disease, slight weakness of the EOM results in lack of

References

1. Zaldivar RA, Lee MS, Harrison AR. Orbital anatomy and orbital fractures. In: Dartt D, ed. Elsevier's Encyclopedia of the eye. London: Elsevier, 2010; Vol 3:210–218.
2. Lennerstrand G, Schiavi C, Tian S, Benassi M, Campos EC. Isometric force measured in human horizontal eye muscles attached to or detached from the globe. Graefes Arch Clin Exp Ophthalmol 2006; 244:539–544.
3. Anderson BC, McLoon LK. Cranial nerve and autonomic innervation. In: Dartt D, ed. Elsevier's Encyclopedia of the eye. London: Elsevier, 2010; Vol 1:537–548.
4. White MH, Lambert HM, Kincaid MC, Dieckert JP, Lowd DK. The ora serrata and the spiral of Tillaux. Anatomic relationship and clinical correlation. Ophthalmology 1989; 96:508–511.
5. Jaggi GP, Laeng HR, Muntener M, Killer HE. The anatomy of the muscle insertion (scleromuscular junction) of the lateral and medial rectus muscle in humans. Invest Ophthalmol Vis Sci 2005; 46:2258–2263.
6. Fuchs E. Beitrage zur normulen Anatomie des Augapfels. Graefes Arch Clin Exp Ophthalmol 1884; 30:1–65.
7. Apt L, Call NB. An anatomical reevaluation of rectus muscle innervations. Ophthalmic Surg 1982; 13:108–112.
8. Souza-Dias C, Prieto-Diaz J, Uesugui CF. Topographical aspects of the insertions of the extraocular muscles. J Pediatr Ophthalmol Strabismus 1986; 23:183–189.
9. Stark N, Kuck H. Distance of muscle insertions in the corneal limbus. Klin Monatsbl Augenheilkd 1986; 189:148–153.
10. de Gottrau P, Gajisin S, Rother A. Ocular rectus muscle insertions revisited: an unusual anatomic approach. Acta Anat 1994; 151:268–272.

11. Otto J, Zimmermann E. Variations in the muscular insertion, the course and elasticity of the muscles in people suffering from squint. Klin Monatsbl Augenheilkd 1979; 175:418–427.

12. Robinson DA, O'Meara DM, Scott AB, Collins CC. Mechanical components of human eye movements. J Appl Physiol 1969; 26:548–553.

13. Ruskell GL. The fine structure of innervated myotendinous cylinders in extraocular muscles of rhesus monkeys. J Neurocytol 1978; 7:693–708.

14. Salpeter MM, McHenry FA, Feng HH. Myoneural junctions in the extraocular muscles of the mouse. Anat Rec 1974; 179:201–224.

15. Kupfer C. Motor innervation of extraocular muscle. J Physiol 1960; 153:522–526.

16. Pilar G, Hess A. Differences in internal structure and nerve terminals of the slow and twitch muscle fibers in the cat superior oblique. Anat Rec 1966; 154:243–252.

17. Mishina M, Takai T, Imoto K et al. Molecular distinction between fetal and adult forms of muscle acetylcholine receptor. Nature 1986; 321:406–411.

18. Oda K, Shibasaki H. Antigenic difference of acetylcholine receptor between single and multiple form endplates of human extraocular muscle. Brain Res 1988; 449:337–340.

19. Horton RM, Manfredi AA, Conti-Tronconi BM. The 'embryonic' gamma subunit of the nicotinic acetylcholine receptor is expressed in adult extraocular muscle. Neurology 1993; 43:983–986.

20. Kaminski HJ, Kusner LL, Block CH. Expression of acetylcholine receptor isoforms at extraocular muscle endplates. Invest Ophthalmol Vis Sci 1996; 37:345–351.

21. Mayr R, Gottschall J, Gruber H, Neuhuber W. Internal structure of cat extraocular muscle. Anat Embryol 1975; 148:23–34.

22. Harrison AR, Anderson BC, Thompson VL, McLoon LK. Myofiber length and three-dimensional localization of NMJs in normal and botulinum toxin treated adult extraocular muscles. Invest Ophthalmol Vis Sci 2007; 48:3594–3601.

23. Jacoby J, Chiarandini DJ, Stefani E. Electrical properties and innervation of fibers in the orbital layer of rat extraocular muscles. J Neurophysiol 1989; 61:116–125.

24. Dimitrova DM, Allman BL, Shall MS, Goldberg SJ. Polyneuronal innervation of single muscle fibers in cat eye muscles: inferior oblique. J Neurophysiol 2009; 101:2815–2821.

25. Koornneef L. New insights in the human orbital connective tissue. Arch Ophthalmol 1977; 95:1269–1273.

26. Miller JM, Demer JL, Rosenbaum AL. Effect of transposition surgery on rectus muscle paths by magnetic resonance imagining. Ophthalmology 1993; 100:475–487.

27. Demer JL, Miller JM, Poukens V, Vinters HV, Glasgow BJ. Evidence for fibromuscular pulleys of the recti extraocular muscles. Invest Ophthalmol Vis Sci 1995; 36:1125–1136.

28. Lowey S, Waller GS, Trybus KM. Function of skeletal muscle myosin heavy chain and light chain isoforms by an in vitro motility assay. J Biol Chem 1993; 268:20414–20418.

29. Barmack NH. Laminar organization of the extraocular muscles of the rabbit. Exp Neurol 1978; 59:304–321.

30. Kjellgren D, Thornell LE, Andersen J, Pedrosa-Domellöf F. Myosin heavy chain isoforms in human extraocular muscles. Invest Ophthalmol Vis Sci 2003; 44:1419–1425.

31. Robinson DA. Oculomotor unit behavior in the monkey. J Neurophysiol 1970; 33:393–403.

32. Wieczorek DF, Periasamy M, Butler-Browne GS, Whalen RG, Nadal-Ginard B. Co-expression of multiple myosin heavy chain genes, in addition to a tissue-specific one, in extraocular musculature. J Cell Biol 1985; 101:618–629.

33. Toniolo L, Maccatrozzo I, Patruno M, Caliaro F, Mascarello F, Reggiani C. Expression of eight distinct MHC isoforms in bovine striated muscles: evidence for MHC-2B presence only in extraocular muscles. J Exp Biol 2005; 208:4243–4253.

34. Kranjc BS, Smerdu V, Erzen I. Histochemical and immunohistochemical profile of human and rat ocular medial rectus muscles. Graefes Arch Clin Exp Ophthalmol 2009; 247:1505–1515.

35. Bicer S, Reiser PJ. Myosin isoform expression in dog rectus muscles: patterns in global and orbital layers and among single fibers. Invest Ophthalmol Vis Sci 2009; 50:157–167.

36. Close RI, Luff AR. Dynamic properties of inferior rectus muscle of the rat. J Physiol 1974; 236:259–270.

37. McLoon LK, Rios L, Wirtschafter JD. Complex three-dimensional patterns of myosin isoform expression: differences between and within specific extraocular muscles. J Muscle Res Cell Motil 1999; 30:771–783.

38. Meredith MA, Goldberg SJ. Contractile differences between muscle units in the medial rectus and lateral rectus muscles in the cat. J Neurophysiol 1986; 56:50–62.

39. Shall MS, Goldberg SJ. Lateral rectus EMG and contractile responses elicited by cat abducens motoneurons. Muscle Nerve 1995 ;18:948–955.

40. Jacoby J, Ko K, Weiss C, Rushbrook JI. Systematic variation in myosin expression along extraocular muscle fibers of the adult rat. J Muscle Res Cell Motil 1989; 11:25–40.

41. Rubinstein NA, Hoh JFY. The distribution of myosin heavy chain isoforms among rat extraocular muscle fiber types. Invest Ophthalmol Vis Sci 2000; 41:3391–3398.

42. Alvarado-Mallart RM, Pincon-Raymond M. Nerve endings on the intramuscular tendons of cat extraocular muscles. Neurosci Lett 1976; 2:121–125.

43. Davidowitz J, Philips G, Breinin GM. Organization of the orbital surface layer in rabbit superior rectus. Invest Ophthalmol Vis Sci 1977; 16:711–729.

44. Goldberg SJ, Wilson KE, Shall ME. Summation of extraocular motor unit tensions in the lateral rectus muscle of the cat. Muscle Nerve 1997; 20:1229–1235.

45. Shall MS, Dimitrova DM, Goldberg SJ. Extraocular motor unit and whole-muscle contractile properties in the squirrel monkey. Summation of forces and fiber morphology. Exp Brain Res 2003; 151:338–345.

46. Lynch GS, Stephenson DG, Williams DA. Analysis of Ca^{2+} and Sr^{2+} activation characteristics in skinned muscle fibre preparations with different proportions of myofibrillar isoforms. J Muscle Res Cell Motil 1995; 16:65–78.

47. Dean P. Motor unit recruitment in a distributed model of extraocular muscle. J Neurophysiol 1996; 76:727–742.

48. Kranjc BS, Sketelj J, Albis AD, Ambroz M, Erzen I. Fiber types and myosin heavy chain expression in the ocular medial rectus muscle of the adult rat. J Muscle Res Cell Motil 2000; 21:753–761.

49. Briggs MM, Schachat F. The superfast extraocular myosin (MYH13) is localized to the innervation zone in both the global and orbital layers of rabbit extraocular muscle. J Exp Biol 2002; 205:3133–3142.

50. McLoon LK. Christiansen SP. Orbital anatomy: the extraocular muscles. In: Dartt D, ed. Elsevier's Encyclopedia of the eye. London: Elsevier, 2010; Vol 2:89–98.

51. Keller EL, Robinson DA. Abducens unit behavior in the monkey during vergence movements. Vision Res 1972; 12:369–382.

52. Bacou F, Rouanet P, Barjot C, Janmot C, Vigneron P, d'Albis A. Expression of myosin isoforms in denervated, cross-innervated, and electrically stimulated rabbit muscles. Eur J Biochem 1996; 236:539–547.

53. Ugalde I, Christiansen SP, McLoon LK. Botulinum toxin treatment of extraocular muscles in rabbits results in increased myofiber remodeling. Invest Ophthalmol Vis Sci 2005; 46:4114–4120.

54. Stephenson GMM. Hybrid skeletal muscle fibres: a rare or common phenomenon? Clin Exp Pharmacol Physiol 2001; 28:692–702.

55. Caiozzo VJ, Baker MJ, Huang K, Chour H, Wu YZ, Baldwin KM. Single–fiber myosin heavy chain polymorphism: how many patterns and what proportions? Am J Physiol Regul Integr Comp Physiol 2003; 285:R570–R580.

56. Jacoby J, Ko K. Sarcoplasmic reticulum fast Ca^{2+} pump and myosin heavy chain expression in extraocular muscles. Invest Ophthalmol Vis Sci 1993; 34: 2848–2858.

57. Briggs MM, Jacoby J, Davidowitz J, Schachat FH. Expression of a novel combination of fast and slow troponin T isoforms in rabbit extraocular muscles. J Muscle Res Cell Motil 1988; 9:241–247.

58. Kjellgren D, Stal P, Larsson L, Furst D, Pedrosa-Domellöf F. Uncoordinated expression of myosin heavy chains and myosin-binding protein C isoforms in human extraocular muscles. Invest Ophthalmol Vis Sci 2006; 47:4188–4193.

59. Pandorf CE, Haddad F, Wright C, Bodell PW, Baldwin KM. Differential epigenetic modifications of histones at the myosin heavy chain genes in fast and slow skeletal muscle fibers in response to muscle unloading. Am J Physiol Cell Physiol 2009; 297:C6–C16.

60. Zeiger U, Khurana TS. Distinctive patterns of microRNA expression in extraocular muscles. Physiol Genomics 2010; Feb 9 (Epub ahead of print) PMID: 20145202.

61. Bottinelli R, Schiaffino S, Reggiani C. Force–velocity relations and myosin heavy chain isoform compositions of skinned fibres from rat skeletal muscle. J Physiol 1991; 437:655–672.

62. Morris TJ, Brandon CA, Horton MJ, Carlson DS, Sciotte JJ. Maximum shortening velocity and myosin heavy-chain isoform expression in human masseter muscle fibers. J Dent Res 2001; 80:1845–1848.

63. Kjellgren D, Ryan M, Ohlendieck K, Thornell LE, Pedrosa-Domellöf F. Sarco(endo) plasmic reticulum Ca^{2+} ATPases (SERCA1 and -2) in human extraocular muscles. Invest Ophthalmol Vis Sci 2003; 44:5057–5062.

64. Andrade FH, McMullen CA, Rumbaut RE. Mitochondria are fast Ca^{2+} sinks in rat extraocular muscles: a novel regulatory influence on contractile function and metabolism. Invest Ophthalmol Vis Sci 2005; 46:4541–4571.

65. Khurana TS, Prendergast RA, Alameddine HS et al. Absence of extraocular muscle pathology in Duchenne's muscular dystrophy: role for calcium homeostasis in extraocular muscle sparing. J Exp Med 1995; 182:467–475.

66. Ragusa RJ, Chow CK, St Clair DK, Porter JD. Extraocular, limb and diaphragm muscle group-specific antioxidant enzyme activity patterns in control and mdx mice. J Neurol Sci 1996; 139:180–188.

67. Wooten GF, Reis DJ. Blood flow in extraocular muscle of cat. Arch Neurol 1972; 26:350–352.

68. Fuchs AF, Binder MD. Fatigue resistance of human extraocular muscles. J Neurophysiol 1983; 49:28–34.

69. McMullen CA, Hayess K, Andrade FH. Fatigue resistance of rat extraocular muscles does not depend on creatine kinase activity. BMC Physiol 2005; 5:12.

70. Andrade FH, McMullen CA. Lactate is a metabolic substrate that sustains extraocular muscle function. Pflugers Arch – Eur J Physiol 2006; 452:102–108.

71. Asmussen G, Punkt K, Bartsch B, Soukup T. Specific metabolic properties of rat oculorotatory extraocular muscles can be linked to their low force requirements. Invest Ophthalmol Vis Sci 2008; 49:4865–4871.

72. Hoh JF, Hughes S. Myogenic and neurogenic regulation of myosin gene expression in cat jaw-closing muscles regenerating in fast and slow limb muscle beds. J Muscle Res Cell Motil 1988; 9:59–72.

73. McLoon LK, Wirtschafter JD. Continuous myonuclear addition to single extraocular myofibers in uninjured adult rabbits. Muscle Nerve 2002; 25:348–358.

74. McLoon LK, Rowe J, Wirtschafter JD, McCormick KM. Continuous myofiber remodeling in uninjured extraocular myofibers: Myonuclear turnover and evidence for apoptosis. Muscle Nerve 2004; 29:707–715.

75. Weintraub H, Tapscott SJ, Davis RL et al. Activation of muscle-specific genes in pigment, nerve, fat, liver and fibroblast cell lines by forced expression of MyoD. Proc Natl Acad Sci USA 1989; 86:5434–5438.

76. McLoon LK, Wirtschafter JD. Activated satellite cells in extraocular muscles of normal adult monkeys and humans. Invest Ophthalmol Vis Sci 2003; 44:1927–1932.

77. Goding GS, Al-Sharif KI, McLoon LK. Myonuclear addition to uninjured laryngeal myofibers in adult rabbits. Ann Otol Rhinol Laryngol 2005; 114:552–557.

78. Shih HP, Gross MK, Kioussi C. Muscle development: Forming the head and trunk muscles. Acta Histochem 2008; 110:97–108.

79. Fisher MD, Gorospe JR, Felder E et al. Expression profiling reveals metabolic and structural components of extraocular muscles. Physiol Genomics 2002; 9:71–84.

80. Anderson BC, Christiansen SP, Grandt S, Grange RW, McLoon LK. Increased extraocular muscle strength with direct injection of insulin-like growth factor-1. Invest Ophthalmol Vis Sci 2006; 47:2461–2467.

81. Zhou L, Porter JD, Cheng G et al. Temporal and spatial mRNA expression patterns of TGF-beta1, 2, 3 and Tbeta RI, II, III in skeletal muscles of mdx mice. Neuromusc Disord 2006; 16:32–38.

82. Martinez-Fernandez S, Hernandez-Torres F, Franco D, Lyons GE, Navarro F, Aranega AE. Pitx2C overexpression promotes cell proliferation and arrests differentiation in myoblasts. Dev Dyn 2006; 235:2930-2939.

83. Scott AB. Botulinum toxin injection into extraocular muscles as an alternative to strabismus surgery. J Pediatr Ophthalmol Strabismus 1980; 17:21–25.

84. Spencer RF, McNeer KW. Botulinum toxin paralysis of adult monkey extraocular muscle. Structural alterations in orbital, singly innervated muscle fibers. Arch Ophthalmol 1987; 105:1703–1711.

85. Kranjc BS, Sketelj J, D'Albis A, Erzen I. Long-term changes in myosin heavy chain composition after botulinum toxin A injection into rat medial rectus muscle. Invest Ophthalmol Vis Sci 2001; 42:3158–3164.
86. Porter JD, Edney DP, McMahon EJ, Burns LA. Extraocular myotoxicity of the retrobulbar anesthetic bupivacaine hydrochloride. Invest Ophthalmol Vis Sci 1988; 29:163–174.
87. Carlson BM, Emerick S, Komorowski T, Rainin E, Shepard B. Extraocular muscle regeneration in primates. Local anesthetic-induced lesions. Ophthalmology 1992; 99:582–589.
88. Asmussen G, Kiessling A. Hypertrophy and atrophy of mammalian extraocular muscle fibers following denervation. Experientia 1975; 31:1186–1188.
89. Ringel SP, Engel WK, Bender AN, Peters ND, Yee RD. Histochemistry and acetylcholine receptor distribution in normal and denervated monkey extraocular muscles. Neurology 1978; 28:55–63.
90. Porter JD, Burns LA, McMahon EJ. Denervation of primate extraocular muscle. A unique pattern of primate extraocular muscle. Invest Ophthalmol Vis Sci 1989; 30:1894–1908.
91. Baker RS, Christiansen SP, Madhat M. A quantitative assessment of extraocular muscle growth in peripheral nerve autografts. Invest Ophthalmol Vis Sci 1990; 31:766–770.
92. Keller EL, Robinson DA. Absence of a stretch reflex in extraocular muscles of the monkey. J Neurophysiol 1971; 34:908–919.
93. Lewis RF, Zee DS, Hayman MR, Tamargo RJ. Oculomotor function in the rhesus money after deafferentation of the extraocular muscles. Exp Brain Res 2001; 141:349–358.
94. Donaldson IM, Dixon RA. Excitation of units in the lateral geniculate and contiguous nuclei of the cat by stretch of extrinsic ocular muscles. Exp Brain Res 1980; 38: 245–255.
95. Wang X, Zhang M, Cohen IS, Goldberg ME. The proprioceptive representation of eye position in monkey primary somatosensory cortex. Nature Neurosci 2007; 10:640–646.
96. Zhang M, Wang X, Goldberg ME. Monkey primary somatosensory cortex has a proprioceptive representation of eye position. Prog Brain Res 2008; 171:37–45.
97. Blumer R, Konacki KZ, Streicher J, Hoetzenecker W, Blumer MJF, Lukas JR. Proprioception in the extraocular muscles of mammals and man. Strabismus 2006; 14:101–106.
98. Lukas JR, Aigner M, Blumer R, Heinzl H, Mayr R. Number and distribution of neuromuscular spindles in human extraocular muscles. Invest Ophthalmol Vis Sci 1994; 35:4317–4327.
99. Richmond FJR, Johnston WSW, Baker RS, Steinbach MJ. Palisade endings in human extraocular muscle. Invest Ophthalmol Vis Sci 1984; 25:471–476.
100. Ruskell GL. The fine structure of innervated myotendinous cylinders in extraocular muscles of rhesus monkeys. J Neurocytol 1978; 7:693–708.
101. Lukas JR, Blumer R, Denk M, Baumgartner I, Neuhuber W, Mayr R. Innervated myotendinous cylinders in human extraocular muscles. Invest Ophthalmol Vis Sci 2000; 41:2422–2431.
102. Konakci KZ, Streicher J, Hoetzenecker W et al. Palisade endings in extraocular muscles of the monkey are immunoreactive for choline acetyltransferase and vesicular acetylcholine transporter. Invest Ophthalmol Vis Sci 2005; 46:4548–4554.
103. Blumer R, Konakci KZ, Pomikal C, Wieczorek G, Lukas JR, Streicher J. Palisade endings: cholinergic sensory organs or effector organs? Invest Ophthalmol Vis Sci 2009; 50:1176–1186.
104. Sas J, Schab R. Die sogennanten "Palisaden-Endigungen" der Augenmusklen. Acta Morph Acad Sci Hung 1952; 2:259–266.
105. Tzahor E, Kempf H, Mootoosamy RC et al. Antagonists of Wnt and BMP signaling promote the formation of vertebrate head muscle. Genes Dev 2003; 17:3087–3099.
106. Gage PJ, Rhoades W, Prucka SK, Hjalt T. Fate maps of neural crest and mesoderm in the mammalian eye. Invest Ophthalmol Vis Sci 2005; 46:4200–4208.
107. Sambasivan R, Gayraud-Morel B, Dumas G et al. Distinct regulatory cascades govern extraocular and pharyngeal arch muscle progenitor cell fates. Dev Cell 2009; 16:810–821.
108. Tajbakhsh S, Rocancourt D, Cossu G, Buckingham M. Redefining the genetic hierarchies controlling skeletal myogenesis: Pax3 and Myf-5 act upstream of MyoD. Cell 1997; 89:127–138.
109. Hacker A, Guthrie S. A distinct developmental programme for the cranial paraxial mesoderm in the chick embryo. Development 1998; 125:3461–3472.
110. Noden DM, Francis-West P: The differentiation and morphogenesis of craniofacial muscles. Dev Dyn 2006; 235:1194–1218.
111. Noden DM. Interactions and fates of avian craniofacial mesenchyme. Development 1988; 103(suppl):121–140.
112. Noden DM. Patterning of avian craniofacial muscles. Dev Biol 1986; 116:347–356.
113. Kanzler B, Foreman RK, Labosky PA, Mallo M. BMP signaling is essential for development of skeletogenic and neurogenic cranial neural crest. Development 2000; 127:1095–1104.
114. Kitamura K, Miura H, Miyagawa-Tomita S et al. Mouse Pitx2 deficiency leads to anomalies of the ventral body wall, heart, extra- and periocular mesoderm and right pulmonary isomerism. Development 1999; 126:5746–5758.
115. Gage PJ, Suh H, Camper SA. Dosage requirement of Pitx2 for development of multiple organs. Development 1999; 126:4643–4651.
116. Gage PJ, Camper SA. Pituitary homeobox 2, a novel member of the bicoid-related family of homeobox genes, is a potential regulator of anterior structure formation. Hum Mol Genet 1997; 6:457–464.
117. Diehl AG, Zareparsi S, Qian M, Khanna R, Angeles R, Gage PJ. Extraocular muscle morphogenesis and gene expression are regulated by Pitx2 gene dose. Invest Ophthalmol Vis Sci 2006; 47:1785–1793.
118. Matt N, Ghyselinck NB, Pellerin I, Dupe V. Impairing retinoic acid signaling in the neural crest cells is sufficient to alter entire eye morphogenesis. Dev Biol 2008; 320:140–148.
119. Noden DM, Marcucio R, Borycki AG, Emerson CP. Differentiation of avian craniofacial muscles: I. Patterns of early regulatory gene expression and myosin heavy chain synthesis. Dev Dyn 1999; 216:96–112.
120. Pedrosa-Domellöf F, Holmgren Y, Lucas CA, Hoh JF, Thornell LE. Human extraocular muscles: unique pattern of myosin heavy chain expression during myotube formation. Invest Ophthalmol Vis Sci 2000; 41:1608–1616.
121. Marcucio RS, Noden DM. Myotube heterogeneity in developing chick craniofacial skeletal muscles. Dev Dyn 1999; 214:178–194.
122. Martinez AJ, McNeer KW, Hay SH, Watson A. Extraocular muscles: morphogenetic study in humans. Light microscopy and ultrastructural features. Acta Neuropath 1977; 38:87–93.
123. Von Scheven G, Alvares LE, Mootoosamy RC, Dietrich S. Neural tube derived signals and FGF8 act antagonistically to specify eye versus mandibular arch muscles. Development 2006; 133:2731–2745.
124. Sengpiel F, Blakemore C, Harrad R. Interocular suppression in the primary visual cortex: a possible neural basis of binocular rivalry. Vision Res 1995; 35:179–195.
125. Birch EE, Stager DR. Long-term motor and sensory outcomes after early surgery for infantile esotropia. J AAPOS 2006; 10:409–413.
126. Wong AM. Timing of surgery for infantile esotropia: sensory and motor outcomes. Can J Ophthalmol 2008; 43:643–651.
127. Kushner BJ. Perspective on strabismus, 2006. Arch Ophthalmol 2006; 124: 1321–1326.
128. Collins CC, O'Meara D, Scott AB. Muscle tension during unrestrained human eye movements. J Physiol Lond 1975; 245:351–369.
129. Tian S, Lennerstrand G. Vertical saccadic velocity and force development in superior oblique palsy. Vision Res 1994; 34:1785–1798.
130. Narasimhan A, Tychsen L, Poukens V, Demer JL. Horizontal rectus muscle anatomy in naturally and artificially strabismic monkeys. Invest Ophthalmol Vis Sci 2007; 48:2576–2588.
131. Martinez AJ, Biglan AW, Hiles DA. Structural features of extraocular muscles of children with strabismus. Arch Ophthalmol 1980; 98:533–539.
132. Spencer RF, McNeer KW. Structural alterations in overacting inferior oblique muscles. Arch Ophthalmol 1980; 98:128–133.
133. Domenici-Lombardo L, Corsi M, Mencucci R, Scrivanti M, Faussone-Pelligrini MS, Salvi G. Extraocular muscles in congenital strabismus: muscle fiber and nerve ending ultrastructure according to different regions. Ophthalmologica 1992; 205:29–39.
134. Berard-Badier M, Pellissier JF, Toga M, Mouillac N, Berard PV. Ultrastructural studies of extraocular muscles in ocular motility disorders. II. Morphological analysis of 38 biopsies. Albracht v Graefes Arch Klin Exp Ophthal 1978; 208:193–205.
135. Antunes-Foschini RM, Ramalho FS, Ramalho LN, Bicas HE. Increased frequency of activated satellite cells in overacting inferior oblique muscles from humans. Invest Ophthalmol Vis Sci 2006; 47:3360–3365.
136. Antunes-Foschini R, Miyashita D, Bicas HE, McLoon LK. Activated satellite cells in medial rectus muscles of patients with strabismus. Invest Ophthalmol Vis 2008; 49:215–220.
137. Stahl JS, Averbuch-Heller L, Leigh RJ. Acquired nystagmus. Arch Ophthalmol 2000; 118:544–549.
138. Abadi RV, Bjerre A. Motor and sensory characteristics of infantile nystagmus. Br J Ophthalmol 2002; 86:1152–1160.
139. Tarpey P, Thomas S, Sarvananthan N et al. Mutations in FRMD7, a newly identified member of the FERM family, cause X-lined idiopathic congenital nystagmus. Nat Genet 2006; 38:1242–1214.
140. Khanna S, Dell'Osso LF. The diagnosis and treatment of infantile nystagmus syndrome (INS). Sci World J 2006; 6:1385–1397.
141. Mencucci R, Domenici-Lombardo L, Cortesini L, Faussone-Pelligrini MS, Salvi G. Congenital nystagmus: fine structure of human extraocular muscles. Ophthalmologica 1995; 209:1–6.
142. Peng GH, Zhang C, Yang JC. Ultrastructural study of extraocular muscle in congenital nystagmus. Ophthalmologica 1998; 212:1–4.
143. McLean RJ, Gottlob I. The pharmacological treatment of nystagmus: a review. Expert Opin Pharmacother 2009; 10:1805–1816.
144. Hertle RW, Dell'Osso LF, FitzGibbon EJ, Thompson D, Yang D, Mellow SD. Horizontal rectus tenotomy in patients with congenital nystagmus: results in 10 adults. Ophthalmology 2003; 110:2097–2005.
145. Wong AM, Tychsen L. Effects of extraocular muscle tenotomy on congenital nystagmus in macaque monkeys. J AAPOS 2002; 6:100–107.
146. Engle EC, Goumnerov BC, McKeown CA et al. Oculomotor nerve and muscle abnormalities in congenital fibrosis of the extraocular muscles. Ann Neurol 1997;41: 314–325.
147. Engle EC. Genetic basis of congenital strabismus. Arch Ophthalmol 2007; 125:189–195.
148. Yamada K, Andrews C, Chan WM et al. Heterozygous mutations of the kinesin KIF21A in congenital fibrosis of the extraocular muscles type 1 (CFEOM1). Nat Genet 2003; 35:318–321.
149. Nakano M, Yamada K, Fain J et al. Homozygous mutations in ARIX (PHOX2A) result in congenital fibrosis of the extraocular muscles type 2 (CFEOM2). Nat Genet 2001; 29:315–320.
150. Pattyn A, Morin X, Cremer H, Goridis C, Brunet JF. Expression and interactions of the two closely related homeobox genes Phox2a and Phox2b during neurogenesis. Development 1997; 124:4065–4075.
151. Demer JL, Clark RA, Lim KH, Engle EC. Magnetic resonance imaging evidence for widespread orbital dysinnervation in dominant Duane's retraction syndrome linked to the DURS2 locus. Invest Ophthalmol Vis Sci 2007; 48:194–202.
152. Miyake N, Chilton J, Psatha M et al. Human CHN1 mutations hyperactivate alpha2-chimaerin and cause Duane's retraction syndrome. Science 2008; 321: 839–843.
153. Karpati G, Carpenter S, Prescott S. Small-caliber skeletal muscle fibers do not suffer necrosis in mdx mouse dystrophy. Muscle Nerve 1988; 11:795–803.
154. Kaminski HJ, Al-Hakim M, Leigh RJ, Katirji MB, Ruff RL. Extraocular muscles are spared in advanced Duchenne dystrophy. Ann Neurol 1992; 32:586–588.
155. Ragusa RJ, Chow CK, Porter JD. Oxidative stress as a potential pathogenic mechanism in an animal model of Duchenne muscular dystrophy. Neuromuscul Disord 1997; 7:379–386.
156. Wehling M, Stull JT, McCabe TJ, Tidball JG. Sparing of mdx extraocular muscles from dystrophic pathology is not attributable to normalized concentration or distribution of neuronal nitric oxide synthase. Neuromuscul Disord 1998; 8:22–29.
157. Porter JD, Karanthanasis P. Extraocular muscle in merosin-deficient muscular dystrophy: cation homeostasis is maintained but is not mechanistic in muscle sparing. Cell Tissue Res 1998; 292:495–501.

158. Porter JD, Merriam AP, Khanna S et al. Constitutive properties, not molecular adaptations, mediate extraocular muscle sparing in dystrophic mdx mice. FASEB J 2003; 17:893–895.

159. Andrade FH, Porter JD, Kaminski HJ. Eye muscle sparing by the muscular dystrophies: lessons to be learned? Microsc Res Tech 2000; 48:192–203.

160. Nyström A, Holmblad J, Pedrosa-Domellöf F, Sasaki T, Durbeej M. Extraocular muscle is spared upon complete laminin alpha2 chain deficiency: comparative expression of laminin and integrin isoforms. Matrix Biol 2006; 25:382–385.

161. Pachter BR, Davidowitz J, Breinin GM. A light and EM study in serial sections of dystrophic extraocular muscle fibers. Invest Ophthalmol 1973; 12:917–923.

162. Porter JD, Merriam AP, Hack AA, Andrade FH, McNally EM. Extraocular muscle is spared despite the absence of an intact sarcoglycan complex in sarcoglycan deficient mice. Neuromuscul Disord 2000; 11:197–207.

163. Mendell JR, Sahenk Z, Prior TW. The childhood muscular dystrophies: diseases sharing a common pathogenesis of membrane instability. J Child Neurol 1995; 10:150–159.

164. Scoppetta C, Morante M, Casali C, Vaccario ML, Mennuni G. Dermatomyositis spares extraocular muscles. Neurology 1985; 35:141.

165. Kallestad KM, McDonald AA, Hebert SL, Daniel ML, Cu SR, McLoon LK. Sparing of extraocular muscle in aging and dystrophic skeletal muscle: a myogenic precursor cell hypothesis. In press, 2010.

166. McMullen CA, Ferry AL, Gamboa JL, Andrade FH, Dupont-Versteegden EE. Age-related changes of cell death pathways in rat extraocular muscle. Exp Gerontol 2009; 44:420–425.

167. Sobue G, Matsuoka Y, Mukai E, Takayanagi T, Sobue I, Hashizume Y. Spinal and cranial motor nerve roots in amyotrophic lateral sclerosis and X-linked recessive bulbospinal atrophy: morphometric and teased-fiber study. Acta Neuropathol 1981; 55:227–235.

168. Polmowski A, Jost WH, Prudlo J et al. Eye movement in amyotrophic lateral sclerosis: a longitudinal study. Ger J Ophthalmol 1995; 4:355–362.

169. Haenggeli C, Kato AC. Differential vulnerability of cranial motoneurons in mouse models with motor neuron degeneration. Neurosci Lett 2002; 335:39–43.

170. Ahmadi M, Liu JX, Brännström T, Andersen PM, Stål P, Pedrosa-Domellöf F. Human extraocular muscles in ALS. Invest Ophthalmol Vis Sci 2010; 51:3494–3501.

171. McClung JR, Cullen KE, Shall MS, Dimitrova DM, Goldberg SJ. Effects of electrode penetrations into the abducens nucleus of the monkey: eye movement recordings and histopathological evaluation of the nuclei and lateral rectus muscles. Exp Brain Res 2004; 158:180–188.

172. Mizen TR. Thyroid eye disease. Semin Ophthalmol 2003; 18:243–247.

173. Stiebel-Kalish H, Robenshtok E, Hasanreysoglu M, Ezrachi D, Shimon I, Leibovici L. Treatment modalities for Graves' ophthalmology. Systemic review and meta-analysis. J Clin Endocrinol Metab 2009; 94:2708–2716.

174. Leong SC, Karbos PD, Macewen CJ, White PS. A systemic review of outcomes following surgical decompression for dysthyroid orbitopathy. Laryngoscope 2009; 119:1106–1115.

175. Molnar I, Szombathy Z, Kovacs I, Szentmiklosi AJ. Immunohistochemical studies using immunized Guinea pig sera with features of anti-human thyroid, eye and skeletal antibody and Graves' sera. J Clin Immunol 2007; 27:172–180.

176. Kloprogge SJ, Busuttil BE, Frauman AG. TSH receptor protein is selectively expressed in normal human extraocular muscle. Muscle Nerve 2005; 32:95–98.

177. Ohkura T, Taniguchi S, Yamada K et al. Detection of the novel autoantibody (anti-UACA antibody) in patients with Graves' disease. Biochem Biophys Res Commun. 2004; 321:432–440.

178. Conley CA, Fowler VM. Localization of the human 64kD autoantigen D1 to myofibrils in a subset of extraocular muscle fibers. Curr Eye Res 1999; 19:313–322.

179. Feldon SE, Park DJ, O'Louglin CW et al. Autologous T-lymphocytes stimulate proliferation of orbital fibroblasts derived from patients with Graves' ophthalmopathy. Invest Ophthalmol Vis Sci 2005; 46:3913–3921.

180. Khoo TK, Bahn RS. Pathogenesis of Graves' ophthalmopathy: the role of autoantibodies. Thyroid 2007; 17:1013–1018.

181. Nakase Y, Osanai T, Yoshikawa K, Inoue Y. Color Doppler imaging of orbital venous flow in dysthyroid optic neuropathy. Jpn J Ophthalmol 1994; 38:80–86.

182. Weber AL, Dallow RL, Sabates NR. Graves' disease of the orbit. Neuroimaging Clin N Am 1996; 6:61–72.

183. Hirano M, DiMauro S. ANT1, Twinkle, POLG, and TP. New genes open our eyes to ophthalmoplegia. Neurology 2001; 57:2163–2165.

184. Moslemi AR, Melberg A, Holme E, Oldfors A. Clonal expansion of mitochondrial DNA with multiple deletions in autosomal dominant progressive external ophthalmoplegia. Ann Neurol 1996; 40:707–713.

185. Schmiedel J, Jackson S, Schafer J, Reichmann H. Mitochondrial cytopathies. J Neurol 2003; 250:267–277.

186. Carry MR, Ringel SP, Starcevich JM. Mitochondrial morphometrics of histochemically identified human extraocular muscle fibers. Anat Rec 1986; 21:8–16.

187. Wallace DC. Mitochondrial DNA mutations and neuromuscular disease. Trends Genet 1989; 5:9–13.

188. Terman A, Brunk UT. Myocyte aging and mitochondrial turnover. Exp Gerontol 2004; 39:701–705.

189. Richter C. Oxidative damage to mitochondrial DNA and its relationship to ageing. Int J Biochem Cell Biol 1995; 27:647–653.

190. Cao Z, Wanagat J, McKiernan SH, Aiken JM. Mitochondrial DNA deletion mutations are concomitant with ragged red regions of individual, aged muscle fibers: analysis by laser-capture microdissection. Nucleic Acids Res 2001; 29:4502–4508.

191. Carlow TJ, Depper MH, Orrison WW. MR of extraocular muscles in chronic progressive external ophthalmoplegia. AJNR Am J Neuroradiol 1998; 19:95–99.

192. Lee AG, Brazis PW. Chronic progressive external ophthalmoplegia. Curr Neurol Neurosci Rep 2007; 2:413–417.

193. Kaminski HJ, Maas E, Spiegel P, Ruff RL. Why are eye muscles frequently involved in myasthenia gravis? Neurology 1990; 40:1663–1669.

194. Zimmermann CW, Eblen F. Repertoires of autoantibodies against homologous eye muscle in ocular and generalized myasthenia gravis differ. Clin Investig 1993; 71:445–451.

195. Sanders DB, El-Salem K, Masses JM, McConville J, Vincent A. Clinical aspects of MuSK antibody positive seronegative MG. Neurology 2003; 60:1978–1980.

196. Takamori M, Motomura M, Kawaguchi N et al. Anti-ryanodine receptor antibodies and FK506 in myasthenia gravis. Neurology 2004; 62:1894–1896.

197. Gunji K, Skolnick C, Bednarczuk T et al. Eye muscle antibodies in patients with ocular myasthenia gravis: possible mechanisms for eye muscle inflammation in acetylcholine receptor antibody-negative patients. Clin Immunol Immunopathol 1998; 87:276–281.

198. Schmidt D, Dell'Osso LF, Abel LA, Daroff RB. Myasthenia gravis: dynamic changes in saccadic waveform, gain, and velocity. Exp Neurol 1980; 68:365–377.

199. Soltys J, Gong B, Kaminski HJ, Zhou Y, Kusner LL. Extraocular muscle susceptibility to myasthenia gravis unique immunological environment? Ann NY Acad Sci 2008; 1132:220–224.

200. Kupersmith MJ. Ocular myasthenia gravis: treatment successes and failures in patients with long-term follow-up. J Neurol 2009; 256:1314–1320.

201. Zhou Y, Cheng G, Dieter L, Hjalt TA, Andrade FH, Stahl JS, Kaminski HJ. An altered phenotype in a conditional knockout of Pitx2 in extraocular muscle. Invest Ophthalmol Vis Sci 2009; 50:4531–4541.

CHAPTER 8

Three-Dimensional Rotations of the Eye

Christian Quaia & Lance M. Optican

Eye motility

In principle the eyeball, like any rigid object, has six degrees of freedom: three for rotation, and three for translation. The adult human eyeball is about 24–25mm in diameter, and can rotate about ±50° horizontally, 42° up and 48° down, and about ±30° torsionally. In contrast, the amount of translation possible for the eye is very limited: over the whole horizontal range the eyeball translates no more than 2 mm along the antero-posterior axis and 0.7 mm in the frontal plane.[1] Given the limited translational excursion, the globe can be treated as a spherical joint capable of rotating around its fixed center, and its motion has thus only three degrees of freedom.

Eye movements are easy to measure, in both the laboratory and the clinic. Technology for eye movement recording has been evolving for over 100 years, and although none are without their disadvantages, it is easy to find a commercial product to suit most needs (Box 8.1). Binocular recordings are especially important, because when both eyes fail to point in the same direction double vision (diplopia) results. Pathological conditions thus give rise to strabismus, the misalignment of the two gaze axes. This is commonly measured in the clinic with a Hess chart (Box 8.2), which gives a static measure of the misalignment. Binocular recordings can give both static and dynamic accounts of the misalignment as gaze angle changes.

Quantifying eye rotations

Describing translations is simple and intuitive. Once three orthogonal axes (e.g. Cartesian coordinates) are defined, a translation of a rigid object can be specified by simply providing the amount of translation of any one point of the object along each of the three axes. Importantly, the final position reached is independent of the order in which the three axes are considered (e.g., x-axis followed by y-axis movement yields the same position as y followed by x, i.e. translations are *commutative*). This happens because the familiar Euclidean space of translations is flat.

In contrast, rotations and their resulting orientations can not be described by any simple (i.e. intuitive) set of three coordinates. One of the fundamental reasons for this complexity is that the space of all rotations is curved. This can be easily noted by considering that if one keeps rotating an object around the same axis it will eventually (after 360°) get back to its initial orientation. One of the implications of this feature is that the final orientation, or attitude, reached after a sequence of rotations around different axes depends on their order. In the two panels of Figure 8.1 a camera, starting from the same initial orientation (left column), is rotated around the same pair of axes (arrows in the figure), but in different order. Clearly, the final orientation (right column) is different for the two sequences of rotations; unlike translations, rotations are *non-commutative*.

To address the inherent complexity of rotations, many mathematical tools have been developed over the last 150 years, such as quaternions, sequences of rotations, rotation matrices, rotation vectors, spinors, rotors, motors, etc. Although all these methods must be equivalent (they all describe the same rotations), each method has both advantages and disadvantages in different applications.[2-4] Accordingly, we can choose the mathematical formalism that best facilitates our thinking about eye rotations.

The first issue that must be addressed when describing rotations or orientations of the eye is the selection of three orthogonal axes. Obviously, they must pass through the center of the eye, but other than that there are few constraints. Of course, keeping the axes fixed in space, so that they point in the same spatial direction regardless of the movements of the eye or the head, would not be very useful, as the six extraocular muscles rotate the eye relative to the head. One reasonable arrangement is to fix them to the eye, so that their spatial orientation changes whenever either eye or the head move. Alternatively, they could be fixed to the head, so that their orientation in space would change only when the head moves. These two possibilities might sound radically different, but they are tightly connected if the head is fixed: starting from an initial orientation, a sequence of two eye rotations within one system leads to the same eye orientation as the reversed sequence of rotations in the other system. A final possibility is represented by the use of nested axes, in which one axis is fixed to the head, the second one rotates with it, and the third one is fixed to the eye and thus rotates with the other two. Unfortunately, these nested systems are often inappropriately called eye-fixed.

Box 8.1 Clinical eye movement recording methods

Axes	Range	Bandwidth	Resolution	Advantages	Disadvantages
Electro-oculography (EOG)					
H, V	~ ±25° H ~ ±15° V Not good for small movements (<5°)	~ 30 Hz	~0.5°	Easy to use, low cost	1, 2, 3, 4. EMG from scalp or jaw muscles can interfere
Infrared Reflection Device (IRD)					
H, V	~ ±30° H ~ ±20° V	~ 100 Hz	~ 0.02°	Easy to set up, moderate cost. Available for MRI, but limited range	1, 2, 5. Risk of ocular drying from IR source
Video-oculography (VOG)					
H, V, T	~ ±30° H, +30° up, −45 down, less for T	~ 25 Hz – 1 kHz H, V, but less for T	~ 0.05° H, V, ~ 0.1° T	Good for clinical use, good for children and infants, available for MEG & MRI	1, 2, 5, 6. Limited range, high cost
Magnetic Field Scleral Coil					
H, V, T	> ±45° in monkeys, but ~ ±30° in people because of eye coil limitations	>1 KHz	~ 0.01°	Coil can be permanently implanted in animals	6, 7. Contact lens causes discomfort, so inappropriate for children. High cost

1. Eye closure or eye lid artifacts.

2. Blink artifacts.

3. Poor linear range.

4. Signal drifts over time.

5. Sensitive to translation of head relative to device.

6. Torsion bandwidth and/or range reduced.

7. Limited recording time (~30 min) for people.

Box 8.2 3-D Effects of Muscle Weakness

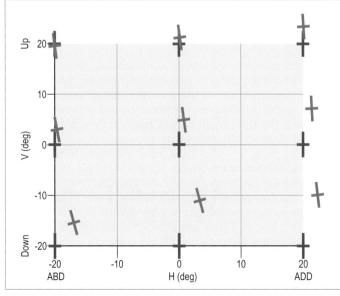

3-D Hess chart of ocular alignment following unilateral trochlear (IVth) nerve damage. The denervated superior oblique muscle can only develop a passive force. The graph shows the effects of a superior oblique palsy (SOP) in a monkey. These measurements were made with 3-D eye coils in each eye. The normal eye views the target, but the palsied eye is covered. The monkey follows a light as it jumps to each of the nine target positions. The blue cross shows the attitude of the normal eye. The red cross shows the attitude of the palsied eye. Note that as the animal looks down in adduction (the main effect of the SO muscle), the deficit worsens. When the deviation between the two eyes is not constant with gaze angle, it is called an *incomitant strabismus*.

Note that the error is not only a displacement of the cross (up and out), but also includes a CCW twist. Similar recordings can be made in human patients with eye coils on a contact lens. In a lower technology method subjects wear (red or blue) colored filters over each eye. A blue cross is projected onto the screen, and is seen only by the normal eye. A red cross is also projected, which can only be seen by the affected eye. The subject moves and twists the red cross until it is aligned, perceptually, with the blue cross.

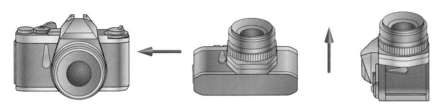

Figure 8.1 Non-commutativity of rotations. The image on the right of each arrow is obtained by rotating the image on its left around an axis collinear with the arrow. (**A**) The camera first rotates 90° around a vertical axis, and then 90° around a horizontal axis. (**B**) The order of rotations is reversed. The final orientation of the camera is clearly different in the two cases. (Redrawn from Quaia C, Optican LM. Commutative saccadic generator is sufficient to control a 3-D ocular plant with pulleys. J Neurophysiol 1998 Jun; 79(6):3197–3215. Used with permission.)

In oculomotor research only head-fixed and nested-axes systems have been extensively used. The decision of which one to use is affected by personal preference, and by the specific oculomotor task under study.

Nested-axes coordinates

Nested-axes coordinate systems are inspired by the classical mechanical method for mounting a rotating object, such as a camera. The simplest way to mount a camera is to have one axis for panning the camera left or right (yaw or Z-axis), one for tilting it up or down (pitch or Y-axis), and one for twisting it clockwise or counter-clockwise about the lens's optical axis (roll or X-axis). These axes are nested, one within the other, in a system of gimbals (since there are three rotation axes, there are six possible nesting sequences). As noted above, final orientation depends crucially on the order of rotations. However, in a gimbal system the mathematical order of the rotations is determined by the nesting order of the gimbals, and not by the order in which the mounted object is moved within the gimbals. Note that this mechanical coupling of the axes does not render rotations commutative; the space of rotations is still curved.

Two such nested-axes systems have been extensively used in the field of oculomotor research.[5] The Fick system starts with a *horizontal* rotation around the vertical axis, followed by a *vertical* rotation around the new horizontal axis, and finally a *torsional* rotation about the new line of sight. The Helmholtz system starts with a *vertical* rotation around the horizontal axis, followed by a *horizontal* rotation around the new vertical axis, and finally a *torsional* rotation about the new line of sight. The leftmost column of Figure 8.2 shows a Fick gimbal system, and the rightmost column shows a Helmholtz gimbal system (torsional axes not shown). Initially (top row) the eye looks straight ahead, in *primary* position. When the eye rotates from primary position around the head-fixed horizontal or vertical axis, it is said to move into a *secondary* position (note: all secondary positions lie along the horizontal or vertical meridian of the globe; cf. Fig. 8.5 below). This is shown in

he middle row for Fick and Helmholtz gimbals: the first Fick rotation turns the eye to the left, whereas the first Helmholtz rotation turns the eye upward. The bottom row shows the eye rotated away from the horizontal or vertical meridian, into what is called a *tertiary* position. Note that, in spite of the magnitude of the two rotations being the same, the final orientation of the eye is different in the two cases.

Head-fixed coordinates

Over the years several different mathematical formalisms have been used to quantify eye rotations using head-fixed coordinate systems. The formalism that we prefer (because we consider it the most intuitive), and the only one that we describe here, is the so-called axis-angle form (Fig. 8.2, middle column), which follows from Euler's theorem. This theorem states that *any orientation of a rigid body with one point fixed can be achieved, starting from a reference orientation, by a single rotation around an axis (through the fixed point) along a unit-length vector $\hat{n}$ by an angle Φ.*[6] Euler's theorem highlights an aspect common to all the methods that can be used to represent rotary motion: the need to define a *reference*, or *primary*, orientation. Although its choice is totally arbitrary, the one most commonly adopted in eye movement research is the orientation with the head upright and the eye looking straight ahead. (Note: when eye orientations are discussed in the context of Listing's law, a somewhat different reference orientation is chosen for convenience; this will be discussed below.) The three main axes of rotation then point straight ahead (X-axis, roll or torsional rotations), straight to the left (Y-axis, pitch or vertical rotations) and straight up (Z-axis, yaw or horizontal rotations). The X-, Y-, and Z-axes define a right-handed system of head-fixed coordinates (x, y, z), that describe, for each eye orientation, Euler's axis of rotation. In a right-handed coordinate system, positive rotations are in the direction that the fingers of the right hand curl when the thumb points along the axis $\hat{n}$.

With this convention, for example, after a 45° rotation to the left, the orientation is described by {(0, 0, 1), 45}, as that orientation is achieved, starting from the reference

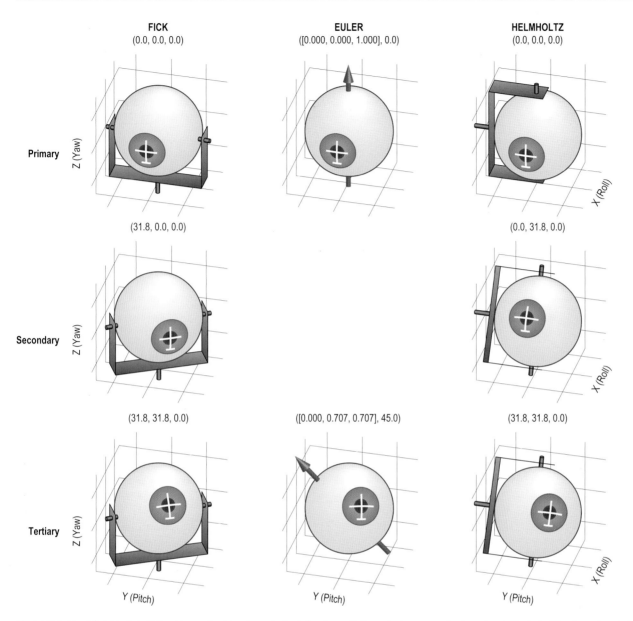

FICK
(0.0, 0.0, 0.0)

EULER
([0.000, 0.000, 1.000], 0.0)

HELMHOLTZ
(0.0, 0.0, 0.0)

Primary

Z (Yaw)

X (Roll)

(31.8, 0.0, 0.0)

(0.0, 31.8, 0.0)

Secondary

Z (Yaw)

X (Roll)

(31.8, 31.8, 0.0)

([0.000, 0.707, 0.707], 45.0)

(31.8, 31.8, 0.0)

Tertiary

Z (Yaw)

Y (Pitch)

Y (Pitch)

Y (Pitch)

X (Roll)

Figure 8.2 Head-fixed and nested-axes coordinate systems. In the left column Fick's nested-axes coordinate system is illustrated (torsion axis not shown). The axes can be represented by a gimbal system. The nesting order is: vertical axis, horizontal axis, torsional axis. In the middle column rotations are described in a head-fixed coordinate system, and Euler's axis is shown. Any orientation is represented by a single axis, tilted appropriately. In the right column another nested-axes system is shown (Helmholtz). In the Helmholtz system, the nesting order is: horizontal axis, vertical axis, torsional axis. Orientations are referred to as primary (looking straight ahead, top row), secondary (on the horizontal or vertical meridian, middle row), or tertiary (off both the horizontal and vertical meridians, bottom row). In all three cases we applied a rotation of 31.8° around the vertical and horizontal axes. In Fick coordinates, that corresponds to (31.8, 31.8, 0), in Helmholtz coordinates to (31.8, 31.8, 0), and in head-fixed coordinates to {(0, 0.707, 0.707), 45}. NB: even when the coordinates in two systems are the same numerically, the final orientations are different because rotations do not commute. In Euler coordinates, these Fick and Helmholtz rotations would be {(−0.198, 0.693, 0.693), 44.7} and {(0.198, 0.693, 0.693), 44.7}, respectively (note the opposite sign for the torsional component).

orientation, by rotating 45° around the vertical axis (0,0,1) (Fig. 8.3A; note that we are looking at the camera from in front, so the X, Y, and Z axes point out of the page, to the right, and up, respectively). Similarly, after a rotation 45° up and to the left, the orientation is {(0, 0.707, 0.707), 45} (Fig. 8.3B, and Fig. 8.2 middle column, bottom row).

Listing's law

The purpose of voluntary eye movements is to point the region of highest visual acuity (the fovea) at the object of

interest. Because the eye can rotate about the line of sight without changing the direction of gaze, the latter has only two degrees of freedom, whereas eye orientation has three. This situation, called *kinematic redundancy*,[7] implies that an infinite number of different eye orientations correspond to each direction of gaze. Despite this redundancy, observation of voluntary eye movements reveals that the brain constrains the torsion to be a function of the horizontal and vertical gaze direction. This reduces the number of degrees of freedom of eye orientation from three to two, so that each gaze direction (achieved with saccadic or smooth pursuit movements) corresponds to a unique eye orientation,

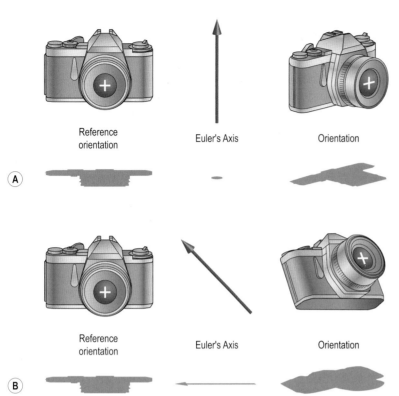

Reference orientation   Euler's Axis   Orientation

Reference orientation   Euler's Axis   Orientation

Figure 8.3 Representing orientation using the axis-angle form. For each panel, the reference orientation is shown on the left. (**A**) The camera is rotated 45° to the left, and the corresponding Euler's axis points straight up, (0, 0, 1) in the XYZ coordinate system (see text). (**B**) The camera is rotated 45° around the tilted axis (0, 0.707, 0.707). Note that the central cross on the camera's lens appears twisted with respect to the vertical axis, even though the Euler axis has no torsional component. (Redrawn from Quaia C, Optican LM. Commutative saccadic generator is sufficient to control a 3-D ocular plant with pulleys. J Neurophysiol 1998 Jun;79(6):3197–3215. Used with permission.)

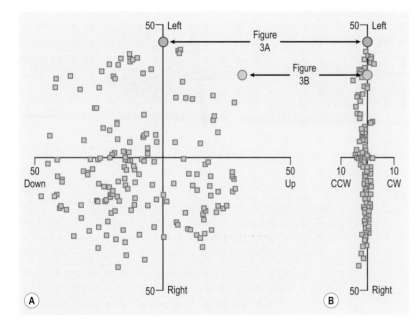

Figure 8.4 Example of Listing's plane in a human subject. Each small square represents the 3-D components of the tip of the Euler axis from one eye orientation. (**A**) Front view, vertical and horizontal components. (**B**) Side view, vertical and torsional components. The blue disks represent the orientation shown in Fig. 8.3A, while the green disks represent the orientation shown in Fig. 8.3B. Note that the horizontal component is reversed, as in Fig. 8.3 we were looking at the vectors from the front, whereas here they are plotted from the camera's point of view. (Adapted, with the permission of Cambridge University Press, from Crawford,[34] 1998.)

regardless of previous movements and orientations. This observation is known as *Donder's law*.[8]

Donder's law states that the torsional component of the eye's attitude is a function of the horizontal and vertical components, but does not specify the relationship. Listing extended Donder's law, by specifying the torsional angle for any line of sight.[8] *Listing's law* states that if the Euler vectors describing the eye orientations attained by a subject looking around with his head fixed in space are plotted, they lie in or near a plane (the so-called *Listing's plane*). In Figure 8.4 orientations measured in a human subject are shown; each

point indicates the orientation of the eye during a period of fixation. Figure 8.4A shows the vertical and horizontal components of the Euler axis (from the subject's point of view), whereas Figure 8.4B shows the vertical and torsional components (the components of the unit-length axis are multiplied by the eccentricity). As an example, on top of the human data we have added two symbols indicating where the Euler axes for the camera orientations shown in Figure 8.3A (blue disks) and Figure 8.3B (green disks) would be in this graph (note that the horizontal component is reversed, as in Figure 8.3 we were looking at the vectors from the front, and not from the camera's point of view). Obviously the points in Figure 8.4 form a thin cloud; Listing's plane is defined as the plane that best fits this cloud.

With the head upright, Listing's plane is normally tilted backwards (about 20°), i.e. it is not aligned with the vertical plane. However, the data are usually transformed to a new coordinate system so that the torsional axis is perpendicular to Listing's plane (Tweed et al.[9] show how to compute a unique reference position, called *primary position*, which is tilted away from straight ahead, and is perpendicular to Listing's plane). This was done, for example, in Figure 8.4. The major advantage of using this transformation is that Listing's law then simply states that *only eye orientations with zero torsion are allowed*.

It must be stressed that Listing's law is enforced only when the head is fixed; it breaks down when the head is moving. When the head turns, the vestibulo-ocular reflex (VOR) counter-rotates the eye, so that the visual world is kept (approximately) stable on the retina (cf. Chapter 9). When this involves head motions with the eyes in an elevated or depressed position, accumulated VOR slow phases can carry the eye out of Listing's plane by as much as 30°; this deviation is usually pre-compensated by adding a predictive torsional component to the innervation of the preceding VOR quick phase.[10] Furthermore, when binocular orientations (cf. Chapter 9) are considered, Listing's plane varies as a function of the distance to the target, so that Listing's plane for each eye rotates outward as the eyes converge.[11] This implies that Listing's law cannot arise from a mechanical property of the peripheral oculomotor complex (e.g. by gimbals), but must be enforced by generating an appropriate set of innervation signals. (Note: Obviously the lack of a torsional component for movements in Listing's plane does not imply that when the eye orientation is in Listing's plane the brain sends no innervation to the oblique muscles; Table 8.1).

Table 8.1 Classic description of muscle actions with the eye in primary position

Muscle	Primary	Secondary	Tertiary
Lateral rectus	Abduction	None	None
Medial rectus	Adduction	None	None
Superior rectus	Elevation	Intorsion	Adduction
Inferior oblique	Extorsion	Elevation	Abduction
Inferior rectus	Depression	Extorsion	Adduction
Superior oblique	Intorsion	Depression	Abduction

The axis of action is perpendicular to the plane formed by the center of the eye and the (functional) origin and insertion of the muscle. Each axis has some orientation, which can be described in head-fixed coordinates. The head-fixed axis where most of the force is generated by the muscle gives rise to its primary action. The axis with the next most force gives rise to the secondary action, and the axis with the least force projection gives rise to the tertiary action. Note that muscles work in pairs. From primary position, the lateral and medial recti move the eye horizontally, the superior rectus and the inferior oblique elevate the eye, the inferior rectus and the superior oblique depress the eye, and the superior oblique and the inferior oblique twist the eye. This classification is difficult to maintain once the eye moves away from primary position.

False torsion

Listing's law clearly puts a constraint on the normal orientation of the eyes, but the quantification of this constraint is closely intertwined with the coordinate system used. For example, with the head-fixed system we can simply say (with the above-mentioned caveats) that the Euler's axis describing orientation should have no torsional component. However, it would be incorrect to say that in the Fick or Helmholtz systems Listing's law means no rotation around the torsional axis. To demonstrate this fact, in Figure 8.5 we show a set of eye orientations. In this figure the orientation of the eye is indicated by a short line. Think of this line as a small piece of tape glued to the eye. With the eye in primary position this piece of tape, centered on the pupil, lies along the horizontal meridian. In red we show orientations described by an Euler axis with zero torsional component. In green we show orientations that are reached through rotations about the horizontal and/or vertical, but not torsional, axes in the Fick system. The orientations shown in blue are reached through rotations about the horizontal and/or

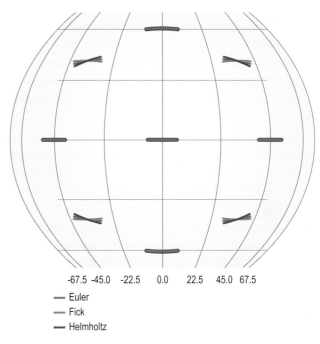

-67.5 -45.0 -22.5 0.0 22.5 45.0 67.5

— Euler
— Fick
— Helmholtz

Figure 8.5 False torsion demonstrated by rotations in nested-axes and head-fixed axes coordinates (orthographic projection). Gray lines show the longitude and latitude lines of the globe (every 22.5°). A colored line (Euler: red, Fick: green, Helmholtz: blue) is drawn on an invisible eyeball to show its orientation. The eye is then rotated in the three coordinate systems so that the same gaze directions are achieved. If Listing's plane were orthogonal to straight ahead, the eye would assume the orientation given by the Euler rotation (red line). In primary position ($\theta = 0°$ and $\varphi = 0°$) and in secondary positions (along the horizontal and vertical meridians, $\varphi = 0°$ or $\theta = 0°$) all three orientations are aligned. In tertiary positions (rotated ±45° around an axis tipped ±45°) the Fick lines (green) remain horizontal, but the others are twisted. This twist is called false torsion. Fick lines have always zero false torsion (unless a rotation around the torsional axis is imposed); for the eccentricity tested here Euler lines have ≈9.7° and Helmholtz lines have ≈19.4° of false torsion. Rotations are not commutative, so the Euler axis-angle rotation {(0, 0.707, 0.707), 45} corresponds to Fick and Helmholtz rotations of about (35.3, 30, 0) and (30.0, 35.3, 0), respectively.

vertical, but not torsional, axes in the Helmholtz system. The rotations applied are such that the direction of the line of gaze is the same regardless of the coordinate system used.

Obviously, the lines are identical in primary and secondary orientations, but they differ in tertiary orientations. More precisely, in tertiary positions the line is horizontal in the Fick coordinate system, but in the other two it is twisted. This last observation has led to a lot of confusion in the study of eye rotations. Remember that only the red lines represent orientations that obey Listing's law. Saying that orientations in Listing's plane have zero torsion is jargon for saying that the Euler vector describing an orientation in Listing's plane has a torsional component equal to zero. But that obviously does not mean that the horizontal meridian is not twisted relative to the gravitational vertical. To distinguish these two measures of torsion, the twist shown in Figure 8.5 is called *false torsion*. Three simple rules apply to false torsion. First, tertiary gaze directions reached using the two outermost gimbals in the Fick system always have zero false torsion. Second, gaze directions reached using the two outmost gimbals in Helmholtz systems always have values of false torsion larger (in magnitude) than if those same gaze directions were reached obeying Listing's law. Third, orientations in Listing's law are *not* characterized by zero false torsion. In Figure 8.5, the false torsion for the Euler lines (red) is about 9.7°, and for the Helmholtz lines (blue) it is about 19.4°.

Neural control of ocular orientation

The study of eye rotations is fundamental to identifying the patterns of innervation to the extraocular muscles that the brain must generate to accurately control the orientation of the eyes, suppress ocular drift, and (when necessary) enforce Listing's law. Using as an approximation to plant mechanics the biomechanical model of the eye plant developed by Robinson[12,13] it is possible to show that, because of the viscoelastic properties of the orbital tissues, the *torque* applied to the eyeball to generate an eye movement can be interpreted as the sum of three components: a *Step* (i.e. a signal proportional to the current eye eccentricity), a *Slide* (i.e. a low-pass filtered version of the velocity profile), and a *Pulse* (i.e. a signal proportional to the velocity of the eyes). This decomposition of the torque always holds, as it simply mirrors (a simplified representation of) the mechanical make-up of the orbital tissues.

The brain does not have direct control over this torque; instead, it indirectly controls it by generating the innervation signals that are delivered to the extraocular muscles. Unfortunately, the muscles are not very good actuators, especially at high speeds of shortening or lengthening, and the brain must take their properties into account when generating the innervation signals. More precisely, an analysis of Robinson's linear model of the eye plant reveals that, in the process of transferring force to its tendon, each muscle absorbs both a *Step* and a *Pulse* of force. The former is a function of the length of the muscle, while the latter is proportional to its speed of shortening. The muscles behave so poorly that approximately 90 percent of the energy consumed during a saccadic eye movement is dissipated by the muscles, and only the remaining 10 percent is actually used to rotate the eyeball (Fig. 8.6C&D, Pulse). Even during periods of fixation only 20 percent of the energy is transferred to the tendons, while the remaining 80 percent is used to maintain the length of the muscles (Fig. 8.6C&D, Step). In other words, to deliver the appropriate torque to the eyeball (Fig. 8.6D, note ten-fold gain increase for this graph), an extra, and much larger, innervation (Fig. 8.6B) must be supplied to account for the loss in the extraocular muscles (Fig. 8.6C).

The decomposition of the innervation signal into the force absorbed by the muscles and the force delivered to the tendons, as well as the decomposition of these forces into their basic components, follows from the properties of the eye plant, and thus holds for any movement, regardless of its dynamics or of the innervation pattern. However, it cannot be stressed enough that the only signal that must exist in the brain is the overall innervation command, which is carried by the motoneurons (Fig. 8.6B, bottom row). The various force/torque components described above (Pulse, Slide, and Step for the orbital tissues, and Pulse and Step for the muscles) are the result of an objective, but artificial, decomposition; these components do not need to have a corresponding neural signal in the brain. Nevertheless, each of these components is associated with different tissue properties, and thus their adaptation requirements are quite different. Accordingly, it would make sense to generate the innervational signals by computing separate neural signals for each of the physical signals described above, and then sum them together, with adaptable weights, at the level of the motoneurons (Fig. 8.6B, top row). The neural signals would thus have to be *matched* to the physical components. This does not however imply that these signals would have to be computed independently from each other. In fact, because they would be associated with physical signals that are related to each other (e.g. eye velocity and orientation are related, as the latter is a function of the history of the former), these neural signals would also have to be related to each other.

Robinson[14] recognized these issues and proposed an elegant solution for the one-dimensional case, i.e. for rotations around a single axis. In this simplified case, there is a direct proportionality between muscle length and eye orientation, and between eye velocity and rate of change of muscle length. Robinson noted that the Step of innervation could then be computed by simply integrating (in the mathematical sense) the Pulse. Similarly, the Slide can be computed by low-pass filtering the Pulse (i.e. passing it through a leaky integrator). This works well because, for rotations around one axis, the orientation (associated, when matching is perfect, with the Step component) is equal to the integral of angular velocity (associated, also under matching conditions, with the Pulse component).

If the three components are appropriately matched, the movement is fast and the eyes stop abruptly (Fig. 8.7A). However, if the neural Pulse, Slide and Step are not matched to the corresponding force/torque components, a fast movement is still generated, but a slow post-saccadic ocular drift follows (Fig. 8.7B). Since the purpose of eye movements is to serve vision, and any drift after a movement degrades vision,[15] perfect matching is very important.

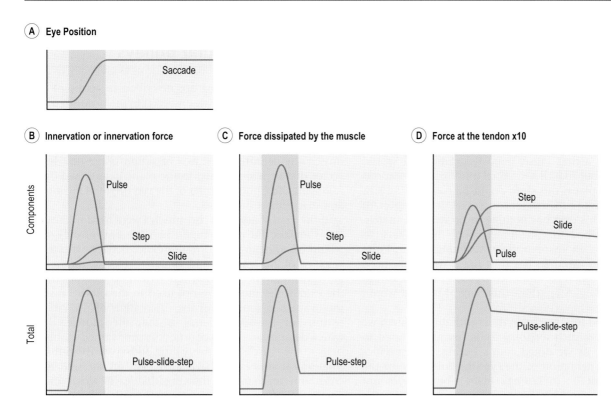

Figure 8.6 Distribution of forces between extraocular muscles and orbital tissues. (**A**) Schematic saccadic eye movement. (**B**) The saccadic innervation consists of a Pulse, Slide, and Step, which generate corresponding forces in the muscles. (**C**) Most of the Pulse and the Step are used to change muscle length. Thus the muscle itself "eats up" most of the force. (**D**) The remaining force delivered to the tendon consists of a small piece of the Pulse, the Slide, and part of the Step (Note: the forces in D are magnified about 10 times). The ratio of the Pulse dissipated by the muscle and the Pulse delivered to the tendon is about 20. The ratio of the Step dissipated to that delivered is about 3.3 (Second row shows components of innervation and force; third row shows total innervation and force; the gray area indicates saccade duration).

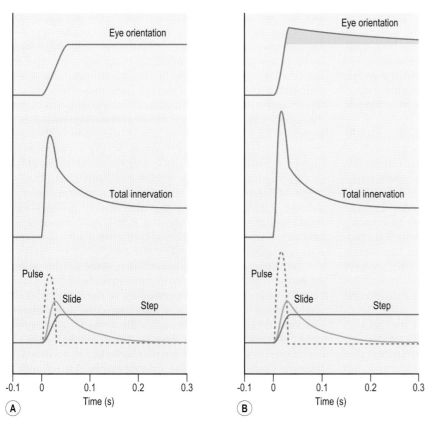

Figure 8.7 Components of the innervation signal. The innervation signal is the sum of three components: a Pulse, a Slide and a Step. (**A**) When the three signals are appropriately matched, the eyes move quickly to the target and stop abruptly. In this case the eye orientation follows the Step of innervation. (**B**) If the components are mismatched (in this case the Pulse is too large), the eyes drift back toward the target. The Step is no longer a faithful representation of eye orientation.

Extending this line of reasoning to three dimensions, when rotations around arbitrary axes must be considered, is not trivial, because in general *the derivative of eye orientation is not equal to eye angular velocity*.[6] This inequality, which is true for any rigid body rotating around a point fixed in space, is due to the non-commutativity of rotations (cf. Fig. 8.1). Thus, if the neural Pulse were proportional to the angular velocity, its integral would not be proportional to the orientation; if the neural Step were computed by integrating the neural Pulse, a post-saccadic drift would ensue. It turns out that, if the initial and final orientations are in Listing's plane, this drift would be confined mostly, but not exclusively, to the torsional axis. It is important to realize that this drift has nothing to do with the matching issue described for one-dimensional movements, as it is not due to an inappropriate weighting of the Pulse and the Step. To avoid such a drift the brain would need to perform a different, more complex, computation to produce the neural Step: the mathematical integral would not suffice.

To understand how the brain deals with this problem we must see whether such torsional drifts (usually referred to as blips) occur. As we noted above, Listing's law specifies only the final orientation of the eye (the Euler axis for the orientation must lie in Listing's plane), but it does not say anything about what happens during or around the movement. However, it was found that, for rotations in which the initial and/or final orientation is tertiary, the orientation of the eye remains in Listing's plane throughout saccadic and pursuit movements, i.e. torsional blips are very small.[16,17] For this to occur, the angular velocity vector about which the eye rotates must tilt *out* of Listing's plane.[18] This tilt of the angular velocity axis is given by the so-called half-angle rule,[16] i.e. if the eyes are elevated 45° up, and you want to rotate the eyes from, say, left 10° to right 10°, the axis of angular velocity must be tipped back from the vertical by 22.5° for the orientation to stay in Listing's plane (Fig. 8.8B, dashed line). In other words, to ensure that the orientation vector has no torsional component throughout the movement, the angular velocity vector must have a torsional component whose amplitude is a function of the orientation of the eyes.

How does all this influence the task of generating the innervation command appropriate to rotate the eyes in 3-D? Obviously whenever a rotation to/from a tertiary orientation is required, if the Pulse of innervation determines the angular velocity of the eyes, it will need to have a torsional component, or the eye orientation will not be confined to Listing's plane. Furthermore, the neural Step would have to be computed by passing the neural Pulse through a rotational operator: simply integrating it would not produce the correct signal. There is one caveat, though. Nothing of what we have said so far about the plant or the controller implies that the angular velocity of the eye is proportional to the neural Pulse. The innervation signal delivered to a muscle determines only the length and speed of length change of that muscle, not the orientation or angular velocity of the eye. It is the geometrical arrangement of the muscles that determines how the former set of signals is converted into the latter. Without a deeper understanding of the mechanics of the eye orbit it is not possible to make further inferences about the neural processing required to control eye rotations.

Orbital mechanics can simplify neural control: extraocular pulleys

Clinically, muscle actions are described in terms of how much they rotate the eye around each of the head-fixed axes, when the eye is in primary position (cf. Table 8.1). A more accurate analysis considers that each muscle tends to turn the eye around a specific axis, called the *axis of action* of the muscle. By definition, the axis of action is the unit length vector that is perpendicular to the plane determined by the following three points: center of the eye, origin of the muscle, and insertion of the muscle.

If the axes of action of the muscles were fixed in the orbit (i.e. the axes did not change when the eye moved), there would be a one-to-one correspondence between shortening velocity of the muscle and angular velocity of the eye. However, it has been demonstrated that the axes of action of the extraocular muscle are not fixed in the orbit.[19] Instead, they vary as a function of the orientation of the eye. The reason for this dependency is that the muscle's path, which determines its axis of action, is constrained so that the belly of the muscle moves very little during eye rotations.[19-21] Anatomical studies[22-24] have shed light on the underlying constraint mechanism, showing that each rectus muscle passes through a ring or sleeve just posterior to the equator of the globe (Box 8.3). The ring is made of collagen[25] and is linked to Tenon's fascia, adjacent muscles, and the wall of the orbit by bands consisting of collagen, elastin and smooth muscle. This anatomical structure thus forms a functional pulley.[26,27]

How does the presence of these pulleys affect oculomotor control? The mechanical effect of the pulleys is to make the axes of action of the extraocular muscles vary dramatically as a function of the orientation of the eye. Before the discovery of the pulleys it was assumed that the axis of action of each rectus muscle was perpendicular to the plane formed by its origin on the annulus of Zinn, its insertion, and the center of the globe. Under these conditions (Fig. 8.8A), changing the orientation of the eye (e.g. from straight ahead to 45° up) would only minimally affect the axis of action of the muscles (horizontal recti shown). However, with orbital pulleys (Fig. 8.8B), the axis of action of each muscle is now perpendicular to the plane formed by the location of its pulley, its insertion on the globe, and the center of the globe. In other words, the pulley acts as the functional point of origin of the muscle. If we now consider what happens when the eye is elevated, we see that the axes of action of the muscles change considerably (Fig. 8.8B). As a consequence, heterotopy of the pulleys (being in the wrong location; Box 8.4) can cause misalignment of the eyes.[28]

Quantitatively it can be shown[29] that, if the orbital pulleys are properly located, the velocity of shortening of the muscles (which, as we pointed out previously, is associated with the viscous force that must be compensated by the Pulse of innervation) closely approximates the *derivative of eye orientation*, and not eye angular velocity. The derivative of the eye orientation signal, unlike angular velocity, is confined to Listing's plane whenever the orientation is. Accordingly, it becomes much easier to implement Listing's law: all that is needed to keep the orientation of the eye in Listing's plane is to generate the Pulse of innervation in that plane. If the Step is then computed by integrating the Pulse, the resulting

Box 8.3 Scale Model of Orbital Geometry

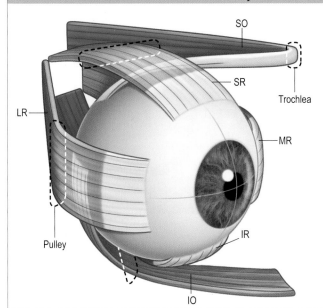

Scale model of the human orbit (globe radius = 12.5 mm). The eye moves as if its center is (approximately) stationary in the orbit. The orientation of each eye is controlled by six muscles: lateral rectus (LR, blue), medial rectus (MR, purple), superior rectus (SR, orange), inferior rectus (IR, green), superior oblique (SO, greenish blue), inferior oblique (IO, magenta). The muscles act in push–pull, i.e. as agonist/antagonist pairs (LR/MR, SR/IR, SO/IO). The force generated by a muscle is a function of both its length and its innervation. For a given length, the higher the innervation the stronger the force. For a given innervation, the longer the muscle the stronger the force. Each muscle (dark color) is made up of two layers. The *global layer* is continuous with the tendon (light color), which inserts on the globe. The *orbital layer* inserts on fibromuscular connective tissue pulleys, which constrain the path of the muscle. These are analogous to the cartilaginous trochlea of the SO. Great circles drawn on the eye show the horizontal (red) and vertical (green) meridians of the eye. Primary position is where they cross. All secondary positions have the gaze axis on one of those meridians; all tertiary positions are not on them. NB: Pulleys are drawn as compact objects at the point of maximum muscle deflection, but they are actually more distributed in front and behind that point.

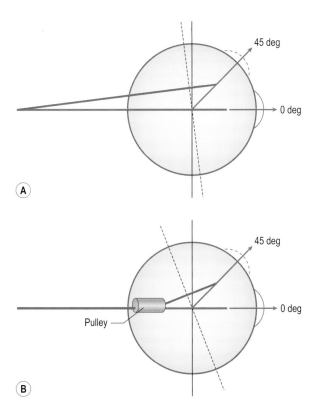

Figure 8.8 Axis of action of the horizontal recti for two different models of orbital mechanics. The schematics are a scaled version of an actual human orbit.[12] (**A**) If the muscles can move freely in the orbit, the muscular path does not change much whether the eye is in primary position (red solid line) or elevated by 45° (blue solid line). Correspondingly, the axis of action (black solid and dashed lines) is approximately fixed in the orbit. (**B**) If the path of the muscles through the orbit is constrained by pulleys, the muscular path from the origin to the pulleys is essentially constant in the orbit, regardless of the orientation of the eye. However, the axis of action of the muscles changes dramatically with orientation; the magnitude of this change is clearly a function of the position of the pulleys. (Note: the axis of action is collinear with the angular velocity vector about which the eye would spin if moved by that pair of muscles.) (Redrawn from Quaia C, Optican LM. Commutative saccadic generator is sufficient to control a 3-D ocular plant with pulleys. J Neurophysiol 1998 Jun;79(6):3197–3215. Used with permission.[29])

movement will have very small post-saccadic drifts, certainly small enough not to impede vision.[30]

This simplification *must not be confused with a mechanical implementation of Listing's law.* If the brain generates a Pulse of innervation that has a non-zero torsional component, the eye will rotate out of Listing's plane. Indeed, during head-free gaze shifts, the vestibulo-ocular reflex (VOR) rotates the eyes opposite to the head's rotation. If the head rotates out of Listing's plane, the VOR will be constantly violating Listing's law, and the saccade generator will need to compute torsional components to compensate for this.[10] Thus, Listing's law must be enforced by the brain, although the pulleys make the required computations much simpler than they would be if the muscle axes were fixed in the orbit.

It is important to realize that the mechanical configuration in the orbit only simplifies the neural machinery because the orbital tissues are viscoelastic. If the orbital tissues had no viscous components (i.e. if they were purely elastic), pulleys would be unnecessary. Without viscosity there would be no Pulse of torque, and the eye orientation would instantly track the torque generated by the muscles; because torques are vectors (i.e. they are commutative) the non-commutativity of rotations would become irrelevant.

Any confusion regarding the difference between a neural and a mechanical implementation of Listing's law can be avoided by analyzing the sequence of events that lead to a refixation movement (Fig. 8.9). First, a sensory-to-motor transformation combines 2-D retinal information with ocular orientation to determine what 3-D rotation is required to foveate the target.[31] Second, the brain generates the corresponding 3-D Pulse of innervation, which it integrates to get the Step. Third, the geometrical arrangement of muscles and pulleys converts the torque generated by the muscles into an angular velocity vector that will not induce post-movement drift. It is in the second step that Listing's law is enforced, whereas the third step takes care of the non-commutativity of rotations. So, for example, the control law for the Pulse could be changed, e.g. so that the eye uses Fick's coordinates. The pulleys would still guarantee that there

Box 8.4 3-D Effects of Mislocated Pulleys

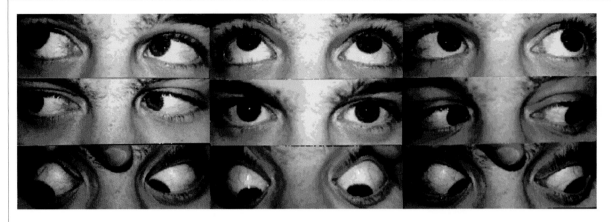

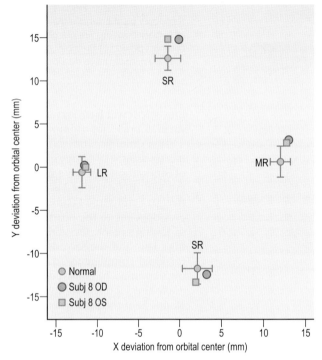

Clinical photographs show the eyes of a patient with a "V" pattern of esotropia in nine diagnostic positions. Note under-elevation bilaterally when the eye is adducted (top row, eye towards the nose), but normal depression in adduction (bottom row). Graph shows average positions of the eye muscle paths in the coronal plane. Red spots show the average muscle positions of normal subjects (error bars show 95 percent confidence intervals). SR, LR, MR, and IR are superior, lateral, medial, and inferior rectus muscles.

Blue spots show the muscle position for the right eye, and green squares show the muscle position for the subject's left eye. This indicates that the pulley positions in this patient are deviated by from 0 to 2.5 mm. A "V" pattern of esotropia with under-elevation in adduction could be caused by an under action of the inferior oblique muscle. However, in this patient the misplaced pulleys caused the direction of action of the muscles to shift, giving rise to the strabismus. (Adapted, with permission, from Clark et al.[28])

would be no post-saccadic drift, even though Listing's law would clearly not hold.

Needless to say, the pulleys can play this simplifying role only if they are appropriately placed. More precisely, they have to be located between the equator and the posterior pole of the globe, in the position that causes the angular velocity vector to tilt half as much as the change in elevation, i.e., to enforce the half-angle rule. For this action to remain unaltered as the eye moves around, the pulleys would also have to move,[32] but simulations show that this would not be particularly critical, as ensuing drifts would be within the physiological

range. However, if the pulleys were static, for large rotations the agonist muscles' insertions would move behind the pulleys, and the muscle would stop being effective. Indeed, just 30° of adduction would cause the medial rectus muscle to lose tangency; it would then translate, rather than rotate, the eye.[30] To avoid this situation, as a muscle contracts its pulley must be moved backwards. This behavior was indeed observed in high-resolution MRI studies of the human orbit.[33]

The mechanism proposed to achieve such a dynamic relocation of the pulleys turns out to be quite simple, albeit surprising. It has been known for a long time that the fibers

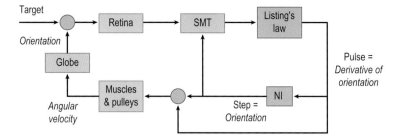

Figure 8.9 Cartoon of transformations in visually guided 3-D eye movements. Retinal target information is combined with eye orientation to obtain the desired eye displacement (a sensory-to-motor transformation, or SMT). A control law is imposed to reduce the motor system's degrees of freedom from three to two by generating the Pulse in Listing's plane. The Pulse, which is the derivative of orientation, is integrated (neural integrator, NI) to obtain the Step, which is the orientation. The total innervation (Pulse + Step) specifies the forces applied by the muscles. The geometry of the muscles and pulleys mechanically shift the axes of action of the muscles according to ocular orientation (i.e., by the half-angle rule), which tilts this force so that the globe has the correct angular velocity vector.

that make up each extraocular muscle can be histologically differentiated into two groups: the global fibers and the orbital fibers.[8] Whereas the global fibers of the rectus muscles go through the pulley and insert anterior to the globe's equator, the orbital fibers insert directly on the pulley.[33] Thus, when the whole muscle contracts, part of its tension will be delivered to the globe, and part will be delivered to the pulley itself, moving it as required.

Summary

Our vision is poor when our eyes move relative to a subject of interest; consequently, movements of the eyes must be as fast as possible, and prolonged drifts must be avoided. Once a target of interest has been chosen, the innervation signal to the extraocular muscles required to point the foveae toward the target must be produced. Unfortunately, the viscoelastic nature of the orbital contents, and some basic properties of rotations, makes generating this signal a far from trivial proposition.

The first challenge that the brain faces is that gaze direction has only two degrees of freedom whereas the eye can move around three axes. The reduction in dimensionality is achieved by generating innervation signals that keep the eyes in Listing's plane. There are many possible alternatives for reducing this kinematic redundancy, but Listing's law is the evolutionary choice.

The non-commutativity of rotations, coupled with the viscoelasticity of the orbital tissues, adds further complexity to the process of producing drift-free movements, regardless of whether Listing's law is obeyed or not. Fortunately, the brain does not need to take into account such complexity in its computations, because that problem is solved by the orbital pulley rig. It appears to be simpler, and perhaps more reliable, to control the location of the pulleys rather than to implement neural circuitry to perform non-commutative operations. This probably reduces the overall complexity of the system.

References

1. Carpenter RHS. Movements of the eyes, 2nd edn. London: Psion; 1988.
2. Bayro-Corrochano E. Modeling the 3D kinematics of the eye in the geometric algebra framework. Pattern Recognition 2003; 36(12):2993–3012.
3. Schreiber KM, Schor CM. A virtual ophthalmotrope illustrating oculomotor coordinate systems and retinal projection geometry. J Vis 2007; 7(10):4–14.
4. Tweed D. Kinematic principles of three-dimensional gaze control. In: Fetter M, Haslwanter T, Misslisch H, Tweed D, eds. Three-dimensional kinematics of eye, head and limb movements. Amsterdam: Harwood, 1997:17–31.
5. Haslwanter T. Mathematics of 3-dimensional eye rotations. Vision Res 1995; 35(12):1727–1739.
6. Goldstein H. Classical mechanics, 2nd edn. Reading: Addison-Wesley, 1980.
7. Crawford JD, Vilis T. How do motor systems deal with the problems of controlling 3-dimensional rotations? J Motor Behav 1995; 27(1):89–99.
8. Leigh RJ, Zee DS. The neurology of eye movements, 4th edn. New York: OUP, 2006.
9. Tweed D, Cadera W, Vilis T. Computing three-dimensional eye position quaternions and eye velocity from search coil signals. Vision Res 1990, 30(1):97–110.
10. Crawford JD, Ceylan MZ, Klier EM, Guitton D. Three-dimensional eye-head coordination during gaze saccades in the primate. J Neurophysiol 1999; 81(4):1760–1782.
11. Mok D, Ro A, Cadera W, Crawford JD, Vilis T. Rotation of Listing's plane during vergence. Vision Res 1992; 32(11):2055–2064.
12. Miller JM, Robinson DA. A model of the mechanics of binocular alignment. Comput Biomed Res 1984; 17(5):436–470.
13. Robinson DA. The mechanics of human saccadic eye movement. J Physiol 1964; 174:245–264.
14. Robinson DA. Models of the saccadic eye movement control system. Kybernetik 1973; 14(2):71–83.
15. Westheimer G, McKee SP. Visual acuity in the presence of retinal-image motion. J Opt Soc Am 1975; 65(7):847–850.
16. Tweed D, Vilis T. Geometric relations of eye position and velocity vectors during saccades. Vision Res 1990; 30(1):111–127.
17. Tweed D, Fetter M, Andreadaki S, Koenig E, Dichgans J. Three-dimensional properties of human pursuit eye movements. Vision Res 1992; 32(7):1225–1238.
18. Tweed D, Vilis T. Implications of rotational kinematics for the oculomotor system in three dimensions. J Neurophysiol 1987; 58(4):832–849.
19. Miller JM. Functional anatomy of normal human rectus muscles. Vision Res 1989; 29(2):223.
20. Miller JM, Robins D. Extraocular muscle sideslip and orbital geometry in monkeys. Vision Res 1987; 27(3):381–392.
21. Simonsz HJ, Harting F, de Waal BJ, Verbeeten BW. Sideways displacement and curved path of recti eye muscles. Arch Ophthalmol 1985; 103(1):124–128.
22. Clark RA, Miller JM, Demer JL. Three-dimensional location of human rectus pulleys by path inflections in secondary gaze positions. Invest Ophthalmol Vis Sci 2000; 41(12):3787–3797.
23. Demer JL, Miller JM, Poukens V, Vinters HV, Glasgow BJ. Evidence for fibromuscular pulleys of the recti extraocular-muscles. Invest Ophthalmol Vis Sci 1995; 36(6):1125–1136.
24. Demer JL, Poukens V, Miller JM, Micevych P. Innervation of extraocular pulley smooth muscle in monkeys and humans. Invest Ophthalmol Vis Sci 1997; 38(9):1774–1785.
25. Porter JD, Poukens V, Baker RS, Demer JL. Structure-function correlations in the human medial rectus extraocular muscle pulleys. Invest Ophthalmol Vis Sci 1996; 37(2):468–472.
26. Demer JL, Miller JM, Poukens V. Surgical implications of the rectus extraocular muscle pulleys. J Pediatr Ophthalmol Strabismus 1996; 33(4):208–218.
27. Demer JL. Current concepts of mechanical and neural factors in ocular motility. Curr Opin Neurol 2006; 19(1):4–13.
28. Clark RA, Miller JM, Rosenbaum AL, Demer JL. Heterotopic muscle pulleys or oblique muscle dysfunction? J AAPOS 1998; 2:17–25.
29. Quaia C, Optican LM. Commutative saccadic generator is sufficient to control a 3-D ocular plant with pulleys. J Neurophysiol 1998; 79(6):3197–3215.
30. Quaia C, Optican LM. Dynamic eye plant models and the control of eye movements. Strabismus 2003; 11(1):17–31.
31. Crawford JD, Guitton D. Visual-motor transformations required for accurate and kinematically correct saccades. J Neurophysiol 1997; 78(3):1447–1467.
32. Demer JL. Orbital connective tissues in binocular alignment and strabismus. In: Lennerstrand GYY, ed. Advances in strabismus research: basic and clinical aspects. London: Portland Press, 2000:17–32.
33. Demer JL, Oh SY, Poukens V. Evidence for active control of rectus extraocular muscle pulleys. Invest Ophthalmol Vis Sci 2000; 41(6):1280–1290.
34. Crawford JD. Listing's law: what's all the hubbub? In: Vision and Action (Harris LR, Jenkin M, eds.), 1998; pp 139–162. Cambridge: Cambridge University.

CHAPTER 9

Neural Control of Eye Movements

Clifton M. Schor

Introduction

Three fundamental visual sensory-motor tasks

The neural control of eye movements is organized to optimize performance of three general perceptual tasks. One task is to resolve the visual field while we move either by translation or rotation through space (self motion). Our body motion causes the image of the visual field to flow across the retina and reflexive eye movements reduce or stabilize this image motion to improve visual performance. The second task is to resolve objects whose position or motion is independent of the background field (object motion). Eye movements improve visual resolution of individual objects by maintaining alignment of the two foveas with both stationary and moving targets over a broad range of directions and distances of gaze. The third task is to explore space and shift attention from one target location to another. Rapid eye movements place corresponding images on the two foveas as we shift gaze between targets lying in different directions and distances of gaze.

Three components of eye rotation

All three perceptual tasks require three-dimensional control of eye position. These dimensions are controlled by separate neural systems. As described Chapter 8, three pairs of extraocular muscles provide control of horizontal, vertical, and torsional position of each eye. Eye movements are described as rotations about three principal axes as illustrated in Figure 9.1. Horizontal rotation occurs about the vertical Z-axis, vertical rotation about the horizontal X-axis, and torsion about the line of sight or Y-axis. As described in Chapter 8, the amount of rotation about each of the three principal axes that is needed to describe a certain direction of gaze and torsional orientation of the eye depends upon the order of sequential rotations (e.g. horizontal, followed by vertical and then torsional).[1] Some oculomotor tasks, such as retinal image stabilization, utilize all three degrees of freedom whereas other tasks, such as voluntary gaze shifts, only require two degrees of freedom, i.e. gaze direction and eccentricity from primary position. As described by Donder's law, torsional orientation of the eye is determined by horizontal and vertical components of eye position. Ocular torsion is independent of the path taken by the eye to reach a given eye position and is constrained by the gaze direction. Listing's law quantifies the amount of ocular torsion at any given eye position, relative to the torsion of the eye in primary position of gaze.

Binocular constraints on eye position control

Binocular alignment of retinal images with corresponding retinal points places additional constraints on the oculomotor system. Because the two eyes view the world from slightly different vantage points, the retinal image locations of points subtended by near objects differ slightly in the two eyes. This disparity can be described with three degrees of freedom (horizontal, vertical, and torsional components) that are analogous to the angular rotations of the eye shown in Figure 9.1. The main task of binocular eye alignment is to minimize horizontal, vertical, and cyclodisparities subtended by near targets on the two foveas. This requires a conjugate system that rotates the two eyes in the same direction and amount, and a disconjugate system that rotates the visual axes in opposite directions. As described by Hering,[2] a common gaze direction for the two eyes is achieved by a combination of conjugate and disconjugate movements that are controlled by separate systems. The *version* system controls conjugate movements and the *vergence* system controls disconjugate movements.

Pure version and vergence movements are described respectively by the isovergence and isoversion contours shown in Figure 9.2. The isovergence circle describes the locus of points that stimulate the same vergence angle in all directions of gaze.[3] A different isovergence circle exists at each viewing distance. The isoversion lines describe the locus of points that stimulate the same version angle over a range of viewing distances in a common direction of gaze relative to the head. Pure vergence movements occur along any of the isoversion lines and not just along the central or midsagittal plane. Fixation changes along any other contour result from a combination of version and vergence movements. Both version and vergence movements are described as combinations of horizontal, vertical, and torsional rotations. For example there can be horizontal and vertical version and vergence movements. Torsional rotations are usually referred to as cyclorotations (e.g. cycloversion or cyclovergence). Hering's law implies that there is equal innervation of yoked muscle pairs: "one and the same

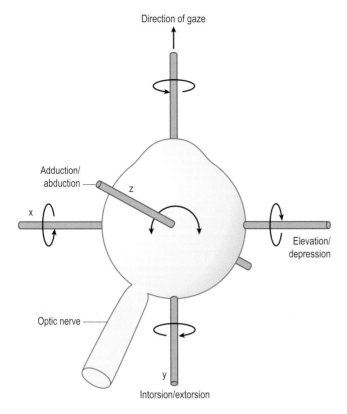

Figure 9.1 The three principal axes of eye rotation. Horizontal rotation occurs about the vertical axis (Z), vertical rotation about the transverse axis (X), and torsion about the anterior–posterior axis (Y). (From Goldberg ME, Eggers HM, Gouras D: The ocular motor system. In: Kandel ER, Schwartz JH, Jessell TM (eds): Principles of Neural Science, 3rd ed, Appleton and Lange 1991.)

impulse of will directs both eyes simultaneously as one can direct a pair of horses with single reins." The law should not be taken literally, because common gaze commands from higher levels are eventually parceled into separate innervation sources in the brainstem that control individual muscles in the two eyes.

Feedback and feedforward control systems

The oculomotor system requires feedback to optimize sensory stimuli for vision with a sufficiently high degree of precision. Feedback provides information about motor response errors based upon their sensory consequences, such as unwanted retinal image motion or displacement. This visual error information usually arrives too late to affect the current movement, because the time delays in the visual system are about 50–100 msec. Instead, it is used to adaptively adjust motor responses to minimize subsequent errors. Oculomotor systems use sensory information to guide eye movements in two different ways. Motor responses can be guided in a closed-loop mode with an ongoing feedback signal that indicates the difference between the desired and actual motor response, or they can operate without concurrent feedback in an open-loop mode. The closed-loop feedback mode is used to reduce internal system errors or external perturbations. A physical example of a closed-loop system is the thermostatic regulation of room temperature, e.g. if the outside temperature drops, the furnace will turn on so that the room temperature stays constant. Motor responses can also be controlled in an open-loop mode, without a concurrent feedback signal. A physical example of an open-loop system is a water faucet, e.g. if the pressure drops, the flow of water will also drop, because the valve does not compensate for the pressure drop.

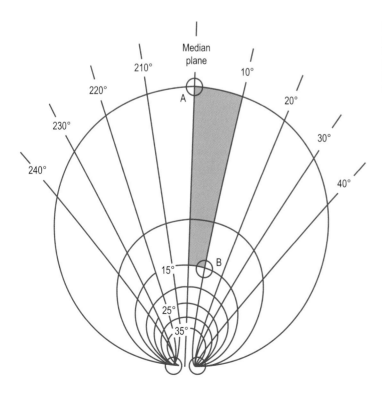

Figure 9.2 Geometric representation of the two components of eye movements by locations of the intersections of the two visual axes. The value in degrees marked along each isovergence circle denotes the convergence angle and the value in degrees marked on the hyperbola isoversion lines denotes the visual direction when a line is not very close to the eyes.

The mode of the response depends on the latency of the response, its duration and velocity. In most examples, visual feedback in a closed-loop system is used to maintain or regulate a fixed position or slow movements of the eyes when there is adequate time to process the error signal. Errors in eye posture or movement are sensed from displacement of the object of regard from the fovea or slippage of the retinal image and negative feedback control mechanisms attempt to reduce the errors to zero during the response.

Feed-forward control systems do not utilize concurrent visual feedback and are described as open-loop. These systems can respond to non-visual (extra-retinal) stimuli, or they respond to advanced visual information with short latencies and brief durations. For example, brief rapid head movements stimulate vestibular signals that evoke compensatory eye movements to stabilize the retinal image. These head movements can produce retinal image velocities of 300–400 deg/sec, yet the eyes respond with a counter rotation within 14 msec of the movement.[4] The oculomotor response to head motion must rely on vestibular signals since retinal image velocities produced by head rotation exceed the upper velocity limit for sensing motion by the human eye. Retinal image velocities that exceed this upper limit appear as blurred streaks rather than as moving images. Visual feedback is not available when the response to head rotation begins because the latency is too short to utilize concurrent visual feedback. A minimum of 50 msec is needed to activate cortical areas that initiate ocular following,[5] such that any motor response with a shorter latency must occur without concurrent visual feedback. Some open-loop systems, such as brief rapid gaze shifts (saccades), respond to visual information sensed prior to the movement rather than during the movement. Their response is too brief to be guided by negative visual feedback. Accuracy of a feed-forward system is evaluated after the response is completed. Visually sensed post-task errors are used by feedforward systems to improve the accuracy of subsequent open-loop responses in an adaptive process that calibrates motor responses. Calibration minimizes motor errors in systems that do not use visual feedback during their response. All feed-forward oculomotor systems are calibrated by adaptation and this plasticity persists throughout life.

Hierarchy of oculomotor control

The following sections present a functional classification of eye position and movement control systems used to facilitate three general perceptual tasks and a hierarchial description of their neuro-anatomical organization (Box 9.1). A hierarchy of neural control exists within each of the functional categories of eye movements that plans, coordinates,

and executes motor activity. Three pairs of extraocular muscles that rotate each eye about its center of rotation are at the bottom of this hierarchy. The forces applied by these muscle pairs to the eye are controlled at the level above by the motor nuclei of cranial nerves III, IV, and VI. Motor neurons in these nuclei make up the final common pathway for all classes of eye movements. Axon projections from these neurons convey information to the extraocular muscles for executing both slow and fast eye movements. Above this level, premotor nuclei in the brainstem coordinate the combined actions of several muscles to execute horizontal, vertical, and torsional eye rotations. These gaze centers orchestrate the direction, amplitude, velocity, and duration of eye movements. Interneurons from the premotor nuclei all converge on motor nuclei in the final common pathway. Premotor neurons receive instructions from supranuclear regions including the superior colliculus, the substantia nigra, the cerebellum, frontal cortical regions including the frontal eye fields (FEF) and supplementary eye fields (SEF), and extrastriate regions including the medial temporal visual area (MT), the medial superior temporal visual area (MST), the lateral intraparietal area (LIP), and the posterior parietal area (PP). These higher centers plan the desired direction and distance of binocular gaze in 3-D space. Cortical–spatial maps of visual stimuli are transformed into temporal codes for motor commands between cerebral cortex and ocular motoneurons, to which the superior colliculus and cerebellum contribute.[6] They determine when and how fast to move the eyes to fixate selected targets in a natural complex scene or to return them to a remembered gaze location. The following sections will discuss the hierarchical control for each of three functional classes of eye movements. The next section describes the final common pathway that conveys innervation for all classes of eye movements.

Final common pathway

Cranial nerves: III, IV, & VI and motor nuclei

Cranial nerves III, IV, and VI represent the final common pathway as defined by Sherrington,[7] for all classes of eye movements. All axon projections from these cranial nuclei carry information for voluntary and reflex fast and slow categories of eye movements.[8] The oculomotor (III), trochlear (IV), and abducens (VI) nuclei innervate the six extraocular muscles, iris and ciliary body. The abducens nucleus innervates the ipsilateral lateral rectus. Premotor interneurons also project from VI to the contralateral oculomotor nucleus for control of the contralateral medial rectus, to produce yoked movements on lateral gaze that are consistent with Hering's law. The trochlear nucleus innervates the contralateral superior oblique. The oculomotor nucleus innervates the ipsilateral medial rectus, inferior rectus, and inferior oblique, and the contralateral superior rectus. The anterior portion of the oculomotor nucleus also contains motor neurons that control pupil size and accommodation in a specialized region called the Edinger–Westphal nucleus.[9] Afferents from this nucleus synapse in the ciliary ganglion prior to innervating their target muscles.[10] The regions of the oculomotor nucleus that control various eye muscles are illustrated in Figure 9.3.

Box 9.1 Hierarchy of motor control

Subcortical oculomotor disorders are classified by lesion sites in the hierarchy of motor control:

- Peripheral (cranial nerves and muscles)
- Nuclear (cranial motor nuclei making up the final common pathway)
- Premotor (coordinates combined actions of several muscles)
- Internuclear (connections between nuclear and premotor sites)
- Supranuclear (motor planning stage)

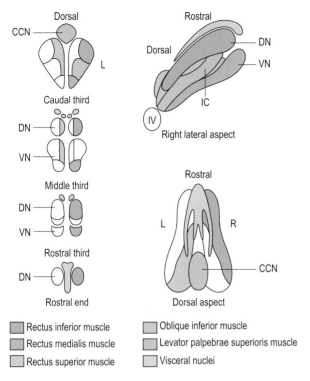

Figure 9.3 Representation of motor neurons for right extraocular muscles in the oculomotor nucleus of monkey. Transverse sections at levels as indicated in the complex. DN, dorsal nucleus; VN ventral nucleus, CCN caudal central nucleus; IC, intermediate column, IV trochlear nucleus. Lateral and dorsal views are shown to the right. (Modified from Warwick R, Representation of the extra-ocular muscles in the oculomotor nuclei of the monkey: J Comp Neurol 1953; 98: 449.)

Legend:
- Rectus inferior muscle
- Rectus medialis muscle
- Rectus superior muscle
- Oblique inferior muscle
- Levator palpebrae superioris muscle
- Visceral nuclei

Motor neuron response

The motor neurons control both the position and velocity of the eye. They receive inputs from burst and tonic cells in premotor nuclei. The tonic inputs are responsible for holding the eyes steady, and the more phasic or burst-like inputs are responsible for initiating all eye movements to overcome orbital viscosity and for controlling eye movements. All motor neurons have the following characteristics as illustrated in Figure 9.4.[11]

1. They have on–off directions (they increase their firing rate in the direction of agonist activity).
2. All cells participate in all classes of eye movements including steady fixation.
3. Each cell (especially tonic) has an eye position threshold at which it begins to fire. Motor neurons have thresholds that range from low to high. Cells with low thresholds begin firing when the eye is in the off field of the muscle that it innervates. Cells with higher thresholds can begin to fire after the eye has moved past the primary position by as much as 10 degrees into the on field of the muscle. The graded thresholds of motor neurons are responsible for the recruitment of active cells as the eyes move into the field of action for the muscle.
4. Increasing the frequency of spike potentials for a given neuron increases contractile force. Once their threshold is exceeded all cells increase their firing rate as the eye moves further along in the on direction of the muscle until they saturate. Cells increase their firing rate linearly as the eye moves into their on field.

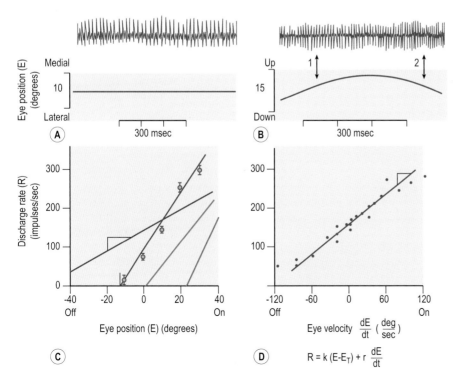

Figure 9.4 Discharge rate of oculomotor neurons in relation to eye movement. On the left, the steady firing rate is shown when the eye is stationary; and below, the rate–position curve illustrates an increase of firing rate with eye position for four neurons. On the right, the firing rate is shown during a slow voluntary eye movement. The arrows indicate points where the eye passes through the same position with velocity of opposite signs, and the associated firing rate is different. Below, firing rate is plotted for a single unit and a particular deviation of the eye as a function of eye velocity.[11] (From Robinson DA, Keller EL: Bibl Ophthalmol 82:7, 1972.)

$$R = k(E - E_T) + r\frac{dE}{dt}$$

Functional classification into three general categories

I. Stabilization of gaze relative to the external world

Movements of the head during locomotion tasks such as walking are described by a combination of angular rotations and linear translations. The oculomotor system keeps gaze fixed in space during these head movements by using extra-retinal and retinal velocity information about head motion. The primary extra-retinal signal comes from accelerometers in the vestibular apparatus.

A. Extra-retinal signals

The vestibular system contains two types of organs that transduce angular and linear acceleration of the head into velocity signals (Fig. 9.5).[12,13] Three semicircular canals lie on each side of the head in three orthogonal planes that are approximately parallel to a mirror image set of planes on the contralateral side of the head. These canals are stimulated by brief angular rotations of the head and the resulting reflexive ocular rotation is referred to as the vestibulo-ocular reflex (angular VOR). In addition, two otoliths (the utricle and sacculus) transduce linear acceleration caused by head translation as well as head pitch (tilt about the interaural axis) and roll (tilt about the nasal–occipital axis) into translation velocity and head orientation signals (linear VOR). Angular acceleration signals stimulate the semicircular canals and result in eye rotations that are approximately equal and opposite to the motion of the head. This stabilization reflex has a short 7–15-msec latency because it is mediated by only three synapses[4] and is accurate for head turns at velocities in excess of 300 deg/s.[14] Hair cells in the canals can be stimulated by irrigation of one ear with cold water. This produces a caloric-vestibular nystagmus that causes the eyes to rotate slowly to the side of the irrigated ear.[15] These slow-phase

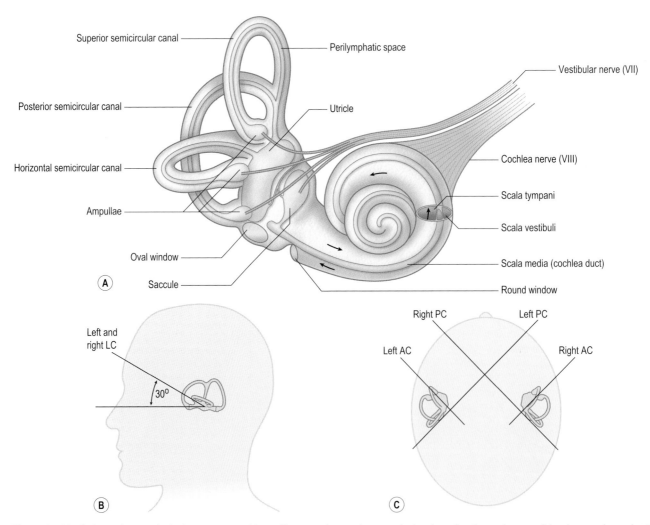

Figure 9.5 Vestibular end-organs in the human temporal bone. Three canals transduce angular head acceleration and two otoliths, the sacculus and utricle, transduce linear acceleration and head orientation. Right labyrinth and cochlea are viewed from horizontal aspect. (Drawings by Ernest W. Beck: courtesy Beltone Electronics Corp., Chicago, Ill.) The canals are in three orthogonal planes that are approximately parallel to a mirror image set of planes on the contralateral side of the head that lie roughly in the pulling direction of the three muscle planes. **A,** Lateral canals. **B,C** Anterior and posterior vertical canals. LC, lateral canal; AC anterior vertical canal: PC, posterior vertical canal. (Modified from Barber HO, Stickwell CW: Manual of electronystagmography, St Louis, CV Mosby, 1976).

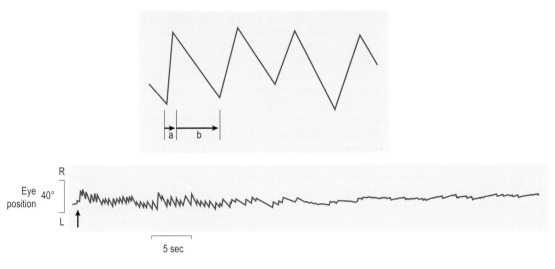

Figure 9.6 OKN and VOR are composed of a slow phase (beta) that rotates the eye in a direction that stabilizes the retinal image and a fast phase (alpha) that resets the eye's position. The figure illustrates the vestibulo-ocular response to sustained rotation. The slow phase is in the direction opposite to head rotation. Horizontal position is plotted against time. The reflex gradually habituates and has disappeared by about 30 seconds. (Redrawn with permission from Miller, NR (ed) Walsh and Hoyt's Clinical Neuro-Ophthalmology 4th Edition. 1982. Page 611, Figure 30.02. Williams & Wilkins, Baltimore.)

movements are interrupted by fast saccadic eye movements that reset eye position in the reverse direction (fast phase). A sequence of slow and fast phases is referred to as jerk nystagmus (Fig. 9.6). Head rotations about horizontal, vertical, and nasal–occipital axes produce VOR responses with horizontal, vertical, and torsional counter-rotations of the slow phase of the nystagmus.[16]

To be effective, these reflex eye rotations must stabilize the retinal image. If the axis of angular head rotation coincided with the center of eye rotation, perfect compensation would occur if angular eye velocity equaled angular head velocity. However, the axis of head rotation is the neck and not the center of eye rotation such that when the head rotates, the eyes both rotate and translate with respect to the visual field. This is exacerbated during near viewing conditions. To stabilize the retinal image motion of a nearby target caused by angular head movements, the eye must rotate more than the head. Indeed, the gain of the VOR increases with convergence.[17] Mismatches between eye and head velocity also occur with prescription spectacles that magnify or minify retinal image motion. Because the VOR responds directly to vestibular and not visual stimuli, its response is classified as open loop. The VOR compensates for visual errors by adapting its gain in response to retinal image slip to produce a stabilized retinal image.[12] Perfect compensation would occur if angular eye velocity were equal and opposite to angular head velocity while viewing a distant scene. However empirical measures show that at high oscillation frequencies (2 Hz) compensation by the VOR is far from perfect and yet the world appears stable and single during rapid head shaking without any perceptual instability or oscillopsis.[18] Thus, to perceive a stable world, the visual system must be aware of both the amount of head rotation and the inaccuracy of compensatory eye movements so that it can anticipate any residual retinal image motion during the head rotation.

Linear acceleration signals and gravity stimulate the otoliths and result in ocular rotations that are approximately 10 percent of the static head tilt caused by pitch and roll from vertical. Head pitch causes vertical eye rotations and head roll causes ocular torsion in the direction opposite to the head roll, "the ocular counterrolling reflex".[19,20] Head roll can also elicit skew movements. When the otolith membranes are displaced on their maculae by inertial forces during rapid linear acceleration, such as during takeoff of an aircraft, a false sense of body pitch called the somatogravic illusion can occur. This illusion results from perceiving vertical as the non-vertical combination of gravitational and acceleration-inertial force vectors. The non-vertical gravitoinertial force causes pilots to perceive the aircraft as pitched upward during takeoff, and if the pilot corrects the false climb this can cause the aircraft nose to pitch downward with dangerous consequences.[21]

Disorders of the vestibular system can produce imbalanced otolith inputs that can result in large amounts of vertical divergence of the eyes (skew deviation) by as much as 7 degrees, bilateral conjugate ocular torsion by as much as 25 degrees, and paradoxical head tilt known as the ocular tilt response (OTR).[22] For example, if the head is tilted to the left, there is paradoxical torsion of both eyes in the direction of head roll and the left eye is rotated downward with respect to the right eye.[23] Manifestations of the OTR will be discussed in the section on nystagmus.

B. Retinal signals

Head motion also produces whole-field retinal image motion of the visual field (optic flow).[24] These retinal signals stimulate reflexive compensatory eye rotations that stabilize the retinal image during slow or long-lasting head movements. The eyes follow the moving field with a slow phase that is interrupted by resetting saccades (fast phase) 1 to 3 times per second.[25] This jerk nystagmus is referred to as optokinetic nystagmus (OKN) and it complements the VOR by responding to low-velocity sustained head movements such as those that occur during walking and posture instability. Like the VOR, OKN also responds with horizontal, vertical and cyclo eye rotations to optic flow about vertical, horizontal, and nasal–occipital axes.[26]

The optokinetic response to large fields has two components, including an early and delayed segment (OKNe and OKNd). OKNe is a short-latency ocular following response (<50 msec) that constitutes the rapid component of OKN,[5] and OKNd builds up slowly after 7 seconds of stimulation.[27] OKNe is likely to be mediated by the pursuit pathway.[28] The delayed component is revealed by the continuation of OKNd in darkness (optokinetic after nystagmus, OKAN). OKNd results from a velocity memory or storage mechanism.[27,29] The time constant of the development of the OKAN matches the time constant of decay of the cupula in the semicircular canals.[30] Thus OKAN builds up so that vision can compensate for loss of the vestibular inputs during prolonged angular rotation that might occur in a circular flight path. OKN can be used clinically to evaluate visual acuity objectively by measuring the smallest texture size and separation in a moving field that elicits the reflex.

C. Neuro-control of stabilization reflexes

1. Vestibulo-ocular reflex

The transducer that converts head rotation into a neural code for driving the VOR consists of a set of three semicircular canals paired on each side of the head.[12] The horizontal canals are paired and the anterior canal on one side is paired with the posterior canal on the contralateral side (Fig. 9.5). These are opponent pairs so that when one canal is stimulated by a given head rotation its paired member on the contralateral side is inhibited. For example, downward and forward head motion to the left causes increased firing of the vestibular nerve for the left anterior canal and decreased firing of the vestibular nerve projections from the right posterior canal. The three canals lie roughly in the pulling directions of the three muscle planes.[12] Thus the left anterior canal and right posterior canal are parallel to the muscle planes of the left eye vertical recti and the right eye obliques. Pathways for the horizontal VOR are illustrated in Figure 9.7 for a leftward head rotation.[8] Excitatory innervation projects from the left medial vestibular nucleus to the right abducens nucleus to activate the right lateral rectus, and an interneuron from the right abducens nucleus projects to the left oculomotor nucleus to activate the left medial rectus. The abducens serves as a premotor nucleus to coordinate conjugate horizontal movements to the ipsilateral side in accordance with Hering's law.

The cerebellar flocculus is essential for adaptation of the VOR to optical distortions such as magnification. The flocculus receives excitatory inputs from retinal image motion (retinal slip) and head velocity information (canal signals) and inhibitory inputs from neural correlates of eye

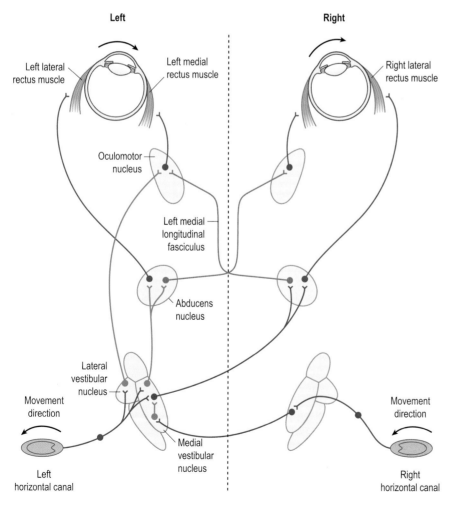

Figure 9.7 Pathways of the horizontal VOR in the brain stem for leftward head rotation. Inhibitory connections are shown as filled neurons, excitatory connections as unfilled neurons. Leftward head rotation stimulates the left horizontal canal and inhibits the right horizontal canal. This results in an increased discharge rate in the right lateral and left medial rectus and decreased discharge rate in the left lateral and right medial rectus.[31] (From Goldberg ME, Eggers HM, Gouras D: The ocular motor system. In: Kandel ER, Schwartz JH, Jessell TM (eds): Principles of Neural Science, 3rd ed, Appleton and Lange 1991.)

movements that provide a negative feedback signal.[32] Adaptation only occurs if retinal image motion and head turns occur together.[33] The gain of the VOR will be adapted to decrease whenever retinal slip and head turns are in the same direction, and to increase whenever they are in opposite directions. Following adaptation, an error correction signal is projected from the flocculus by Purkinje cells to floccular target neurons (FTN) in the vestibular nucleus to make appropriate changes in VOR gain.[32]

2. Optokinetic nystagmus

The visual stimulus for OKN is derived from optic flow of the retinal image.[34–36] The retina contains ganglion cells that respond exclusively to motion in certain directions or orientations. This information passes along the optic nerve, decussates at the chiasm and projects to the cortex via the geniculate body (LGN) or to the midbrain via the accessory optic tract (Fig. 9.8).[37] This tract has several nuclei in the pretectal area.[38] One pair of these nuclei, the nucleus of the optic tract (NOT), is tuned to horizontal target motion to the ipsilateral side (i.e. nasal to temporal motion). The lateral and medial terminal nuclei (LTN and MTN) are tuned for vertical target motion.[39] Neurons in these nuclei only receive subcortical inputs from the contralateral eye. They have large receptive fields and respond to large textured stimuli moving in specific directions. Stimulation of the right NOT with rightward motion causes following movements of both eyes to the right or ipsilateral side, and similarly stimulation of the left NOT with leftward motion causes leftward conjugate following movements. Each NOT projects

signals via the inferior olive to the vestibular nuclei and possibly to the flocculus via the climbing fibers of the cerebellum.[35] The NOT provides a visual signal to the vestibular nucleus, and the motor response is the same as for velocity signals originating from the semicircular canals.

The cortical region that organizes motion signals is the medial superior temporal lobe (MST). This region is important for generating motion signals for both pursuit and OKN.[34] Binocular cortical cells receive projections from both eyes and code ipsilateral motion from the contralateral visual field at higher velocities than the subcortical system.[36] The cortical cells project to the ipsilateral NOT.

Until the age of 3–4 months, the monocular subcortical projections predominate because the cortical projection has not yet developed.[40] As a result, OKN in infancy is driven mainly by the crossed subcortical input. The consequence is that monocular stimulation only evokes OKN with temporal-to-nasal motion but not with nasal-to-temporal motion. After 3–4 months of development, the infant's cortical projections predominate and horizontal OKN responds to both monocular temporal-ward and nasal-ward image motion. The cortical projections to the NOT fail to develop in infantile esotropia and as adults these patients exhibit the same asymmetric OKN pattern as observed in immature infants[15] This anomalous projection is responsible for a disorder known as latent nystagmus in which a jerk nystagmus occurs when one eye is occluded with the slow phase directed toward the side of the covered eye. During monocular fixation, the stimulated retina increases the activity of neurons in the contralateral NOT via subcortical crossed projections, but it is unable to innervate the ipsilateral NOT via the ineffective cortical-tectal projection. The result is that both eyes' positions are drawn to the side of the stimulated NOT (i.e. the side of the covered eye). The fixation error is corrected with a saccade and a repeated sequence is described as latent or occlusion nystagmus.[41]

II. Foveal gaze lock (maintenance of foveal alignment with stationary and slowly moving targets)

A. Static control of eye alignment (fixation)

The oculomotor system enhances visual resolution by maintaining alignment of the fovea with attended stationary and moving targets.[42,43] During fixation of stationary targets, the eyes sustain foveal alignment over a wide range of target locations in the visual field. Gaze direction is controlled by a combination of eye position in the orbit and head position.[44] Gaze is mainly controlled by eye position for targets lying at eccentricities of less than 15 degrees from primary position.[45] Steady fixation at larger gaze eccentricities is accomplished with a combination of head and eye position. Holding eye fixation at large eccentricities (>30 degrees) without head movements is difficult to sustain and the eye drifts intermittently toward primary position in gaze evoked nystagmus.[46] This drift is exacerbated by alcohol.[47] Even within the 15-degree range, the fixating eye is not completely stationary. It exhibits physiological nystagmus that is composed of slow horizontal, vertical, and torsional drifts (0.1 deg/sec), micro-saccades (<0.25 deg), and a small amplitude (<0.01 deg) high frequency tremor (40–80 Hz).[48]

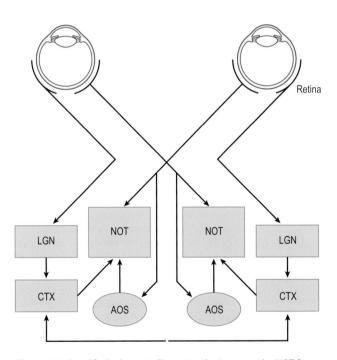

Figure 9.8 Simplified schematic illustrating the inputs to the NOT from subcortical crossed retinal projections and cortical-tectal projections. Each NOT gets direct retinal input from the contralateral nasal retina, which is excited by temporal-to-nasal motion. It also receives indirect cortical input (CTX) from the temporal retina of the ipsilateral eye, which tends to be excited more by nasal-to-temporal motion as well as input via the accessory optic system (AOS), and input from the contralateral visual cortex via the corpus callosum.[37] (Modified from Hoffmann KP, Distler C, Ilg U: J Comp Neurol 321:150, 1992.)

Some of the drifts and saccades are error producing, while others are error correcting when they serve to continually minimize motion and adjust the alignment of the target of regard with the fovea.[49,50] The high frequency tremor reflects the noise possibly originating from asynchronous firing of individual motoneurons that is filtered by the mechanical properties of the eye. Gaze evoked nystagmus can be associated with impaired smooth pursuit and paramedian pontine lesions.[51]

Stereoscopic depth perception is enhanced by accurate binocular fixation.[40,52] Precise bifoveal alignment requires that the eyes maintain accurate convergence at the distance of the attended target. Small constant errors of convergence during attempted binocular fixation (<15 arc min) are referred to as fixation disparity,[53] and these can impair stereo performance.[54] Fixation disparity is a closed-loop error because it occurs in the presence of retinal image feedback from binocular disparity. Fixation disparity results from incomplete nullification of an open-loop error of convergence known as heterophoria. Heterophoria equals the difference between the convergence stimulus and the convergence response measured under open-loop conditions (e.g. monocular occlusion). The magnitude of fixation disparity increases monotonically with the disparity stimulus to convergence when it is varied with horizontal prisms.[53] The slope of this function is referred to as the forced-duction fixation disparity curve.[53] Shallow slopes of these curves indicate that the heterophoria is reduced by adaptation during binocular or closed-loop conditions.[40] Vergence adaptation is very rapid. For example, when convergence is stimulated for only one minute with a convergent disparity and then one eye is occluded, the convergence response persists in the open-loop state.[40] Typically the convergent disparity is produced by prisms that deflect the perceived direction of both eye's images inward, or in the nasal-ward direction (i.e. the base of the prism before each eye is temporal-ward or "base-out"). This adapted change in heterophoria is constant in all directions of gaze and is referred to as concomitant adaptation. Prism adaptation is in response to efforts by closed-loop disparity vergence to nullify the open-loop vergence error (heterophoria). Prism adaptation improves the accuracy of binocular alignment and stereo-depth performance.

Horizontal vergence equals the horizontal component of the angle formed by the intersection of the two visual axes and it is described with three different units of measurement or scales that include the degree, prism diopter (PD), and meter angle (MA). *Prism diopters* equal 100 times the tangent of the angle. *Meter angles* equal the reciprocal of the viewing distance (specified in meters) at which the visual axes intersect as measured from the center of eye rotation. A target at one meter stimulates one meter angle of convergence (approximately 3.4 degrees) and one diopter of accommodation. Prism diopters can be computed from meter angles by the product of MA and the interpupillary distance (IPD), measured in cm. For example, convergence at a viewing distance of 50 cm by two eyes with a 6 cm IPD equals either 2 MA or 12 PD. The advantage to units in MA is that the magnitude of the stimulus to accommodation in diopters is approximately equal to the magnitude of the stimulus to convergence, assuming that both are measured from a common point such as the center of eye rotation. This

assumption produces large errors for viewing distances less than 20 cm because accommodation is usually measured in reference to the corneal apex that is 13 mm anterior to the center of eye rotation. The advantage to units in prism diopters is that they are easily computed from the product of MA and IPD. The advantage to units in degrees is that they accurately quantify asymmetric convergence where targets lie at different distances from the two eyes.

The Maddox classification[55] describes open- and closed-loop components that make up the horizontal vergence response. The classification includes three open-loop components that influence heterophoria. These include an adaptable intrinsic bias (tonic vergence), a horizontal vergence response to monocular depth cues (proximal vergence), and a horizontal vergence response that is coupled with accommodation by a neurological cross-link known as accommodative-convergence. Tonic vergence has an innate divergence posture (5 degrees) known as the anatomical position of rest.[56] Tonic vergence adapts rapidly to compensate for vergence errors during the first 6 weeks of life. At birth the eyes diverge during sleep, but after 6 weeks the visual axes are nearly parallel during sleep.[57] The adapted vergence, measured in an alert state, in the absence of binocular stimulation and accommodative effort, is referred to as the physiological position of rest. The physiological position of rest equals the sum of tonic vergence and the anatomical position of rest. Tonic vergence adaptation ability continues throughout life to compensate for trauma, disease, and optical distortions from spectacles and aging factors.

Proximal vergence describes the open-loop vergence response to distance percepts stimulated by monocular depth cues such as size, overlap, linear perspective, texture gradients, and motion parallax.[58,59] This proximal response accounts for a large portion of the convergence response to changing distance.[60,61] The open-loop convergence response is also increased by efforts of accommodation.[62,63] The eyes maintain approximately 2/3 of a meter angle of convergence (4 prism diopters) for every diopter of accommodation.[53] A target at one meter stimulates one meter angle of convergence (approximately 3.4 degrees) and one diopter of accommodation. Thus when the eyes accommodate 1 diopter the accommodative-convergence response increases by approximately 2.3 degrees or 4 prism diopters which is only 68 percent of the convergence stimulus. The sum of the three open-loop components of convergence typically lags behind the convergence stimulus and produces an open-loop divergence error (exophoria) that is usually less than 2 degrees under far viewing conditions and 4 degrees at near viewing distances.[64] Esophoria describes open-loop vergence errors caused by excessive convergence that lead the stimulus. The distribution of heterophoria in the general population is not Gaussian or normally distributed. It is narrowly peaked about a mean close to zero,[65] indicating that binocular errors of eye alignment are minimized by an adaptive calibration process. During closed-loop stimulation, vergence error is reduced to less than 0.1 degree by the component of the Maddox classification controlled by visual feedback, disparity vergence, which is stimulated by retinal image disparity. Variation in any of the three open-loop components of convergence will influence the magnitude of heterophoria and the resulting closed-loop fixation disparity. Fixation disparity acts as a stimulus to maintain activity of the disparity vergence so that it continues to sustain its nulling response of the underlying heterophoria during attempted steady binocular fixation.[66] Other dimensions of vergence also adapt to disparity stimuli. Vertical vergence adapts to vertical prism before one eye,[67] and cyclovergence adapts to optically produced cyclodisparity stimuli.[68–70]

B. Dynamic control of eye alignment (smooth tracking responses to open- and closed-loop stimuli)

1. Conjugate smooth pursuit tracking

Smooth following pursuit movements allow the eyes to maintain foveal alignment with a moving target that is voluntarily selected.[71] Pursuit is defined as the conjugate component of smooth following eye movement responses to target motion. As described below, disconjugate following responses to changes in target distance are referred to as smooth vergence. Conjugate pursuit differs from the delay portion of OKN, which is a reflex response to optic flow of the entire visual field. A conflict occurs between the pursuit and optokinetic systems when the eyes pursue an object that is moving across a stationary background. Pursuit stabilizes the moving target on or near the fovea but it causes optic flow or retinal slip of the stationary background scene. Conflicts can also occur between pursuit and the VOR if gaze is controlled with head tracking movements. Stabilizing the retinal image of a moving object with head movements produces a vestibular signal from the canals. Consequently, pursuit of a small moving target against a stationary background with eye or head movements requires that OKN and the VOR be ignored or suppressed. This is accomplished most effectively when the background lies at a different distance than the pursuit target.[72]

Pursuit responds to target velocities ranging from several minutes of arc/sec to over 175 deg/sec.[71] The gain or accuracy of pursuit is reduced as target velocity increases above 100 deg/sec.[73] The VOR has a much higher velocity range than pursuit. This can be demonstrated by comparing two views of your index finger. Either keep your finger stationary while you shake your head rapidly from side to side at 2–3 Hz, or shake your finger at the same frequency without moving your head. The eye cannot follow the moving finger, but it can follow the stationary finger while shaking your head even though the head-relative motion is identical in these two examples. The pursuit response is more accurate when the target motion is predictable such as with pendular motion.[74] Pursuit errors are reduced by modifying pursuit velocity and with small catch-up saccades. The combination of pursuit and catch-up saccades that appear at low stimulus velocities in patients with pursuit deficits is referred to as cogwheel pursuit. The ratio of eye velocity over target velocity (gain) or accuracy of pursuit is normally affected by target visibility (contrast) as well as drugs and fatigue.[75,76]

The pursuit response to sudden changes in target velocity has a short latency (80–130 msec),[77] and is composed of two general phases referred to as open-loop and closed-loop (Fig 9.9).[77] Pursuit is initiated during the open-loop phase and it is maintained during the closed-loop phase. The

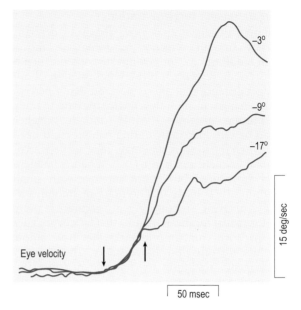

Figure 9.9 Eye velocity during the onset of pursuit to a 15 deg/sec ramp target motion, the ramp of motion beginning at different eccentricities as indicated at the right portion of each trace. The velocity of the early component (indicated by arrows) was the same for all starting positions, but the velocity of the late component varied.[77] (From Lisberger SG, Westbrook LE: J Neurosci 5:1662, 1985.)

open-loop response is divided into an early and late component. The early component is a feed forward phase that lasts for only 20 msec. During this early phase, there is a rapid acceleration of the eye (40–100 deg/sec/sec) that is in the correct direction but is independent of the stimulus velocity and initial retinal image position.[77] During the late open-loop component that lasts 80 msec, the initiation of pursuit depends strongly on target velocity and retinal image position.[78] Eye acceleration is highest in response to targets imaged near the fovea and decreases sharply with increasing eccentricity up to 21 degrees. These open-loop components are calibrated by adaptation.[78]

Pursuit is maintained during the closed-loop phase in response to negative feedback from retinal image velocity (retinal slip) and position, as well as an internal estimate of target velocity relative to the head. If the eye lags behind the stimulus, the retinal image velocity is not nulled leaving a residual retinal image slip and position error away from the fovea. The pursuit system accelerates to correct both retinal position and velocity errors.[76] When pursuit is very accurate and there is no retinal error, the eye continues to pursue the target without the eye-referenced signals. Pursuit is maintained by an internal estimate of target velocity or a head referenced motion signal that is computed from a combination of retinal slip and an internal representation of eye velocity.[71] This can be demonstrated by attempting to fixate an ocular floater or a retinal afterimage that is located near the fovea. Attempts to foveate the stabilized retinal image lead to smooth following eye movements even though the retinal image is always motionless. The eye is tracking an internal correlate of its own motion that causes the target to appear to move with respect to the head. Cognitive factors including attention, prediction, and learning are able to influence the execution of smooth pursuit.[79]

2. Disconjugate smooth vergence tracking

Foveal alignment of targets that move slowly in depth is maintained by smooth vergence following eye movements.[60,80] In addition to improving stereoscopic depth perception, smooth vergence provides information about changing target distance that affects size and depth perception.[81–83] Slow changes in smooth vergence respond to body sway and posture instability. While smooth vergence responses can be very inaccurate during natural rapid head movements, when the head is stationary they are very accurate at temporal frequencies up to 1 Hz.[84] At higher frequencies accuracy is reduced but it can improve with small disparity stimuli.[80] It is likely that accuracy of smooth vergence is task dependent. Smooth vergence accuracy is more demanding for spatial localization tasks that lack depth cues other than disparity, compared to tasks that have ample monocular cues for direction and distance. The accuracy of smooth vergence responses to changing disparity improves with predictable target motion.[84] Stimuli for smooth vergence tracking include magnitude and velocity of retinal image disparity[84] and perceptual cues to motion in depth, including size looming.[58,80] Smooth vergence tracking is very susceptible to fatigue and central nervous system suppressants.[84]

3. Adaptable interactions between smooth pursuit and smooth vergence

In natural scenes, motion of a target in the fronto-parallel plane is tracked binocularly with conjugate smooth pursuit. However in conditions of anisometropia corrected with spectacle lenses, the image motion of the two eyes is magnified unequally and this produces variations of binocular disparity that increase with target eccentricity from the optical centers of the lenses. Tracking motion of targets in the fronto-parallel plane then requires both smooth pursuit and smooth vergence eye movements such that one eye moves more than the other does. The oculomotor system can adapt to the binocular disparity that changes predictably with eye position. Adaptation produces open-loop non-conjugate variations of heterophoria that compensate for the horizontal and vertical disparities produced by the unequal magnifiers during smooth vergence tracking responses.[85] After only one hour of pursuit tracking experience with anisometropic spectacles, one eye can be occluded and the two eyes continue to move unequally during monocular tracking.[85] The adapted heterophoria is coupled to vary with both eye position and direction of eye movement.[86]

C. Neuro-control of smooth foveal tracking

1. Smooth pursuit tracking system

Smooth following eye movements result from cortical motion signals in extrastriate cortex in areas MT and MST that lie in the superior temporal sulcus.[87,88] Area MT encodes speed and direction of visual stimuli in three dimensions relative to the eye. MT receives inputs from the primary visual cortex and projects visual inputs to area MST and the frontal eye fields (FEF). Cells in MST fire in concert with head-centric target movement; i.e. they combine retinal and efference copy signals.[89] Each hemisphere of the MST codes motion to the ipsilateral side. Cells have two types of visual motion sensitivity; they respond to motion of large-field patterns and small spots but the direction preferences for the two stimulus types are in opposite directions. The anti-directional large-field responses could facilitate pursuit of small targets moving against a far stationary field, and motion parallax stimuli. Efferents from MST and FEF project ipsilaterally to the NOT to generate OKN and to the dorsal lateral pontine premotor nuclei (DLPN) for pursuit tracking.[35]

The DLPN plays a large role in maintaining steady-state smooth pursuit eye velocity, and the nucleus reticularis tegmenti pontis (NRTP) contributes to both the initiation and maintenance of smooth pursuit. Neurons in these areas are primarily encoding aspects of eye motion with secondary contributions from retinal signals such as those coded in the NOT.[90] Velocity signals are projected from DLPN to the floccular region and to the vermis lobules VI and VII of the cerebellum.[35] The DPN serves as a precerebellar relay for both pursuit and saccade-related information,[91] and these two types of eye movements may represent different outcomes from a shared cascade of sensory-motor functions.[92] The flocculus is thought to maintain pursuit eye movements during steady constant tracking while the vermis is important when the target velocity changes or when initiating pursuit. The role of the cerebellum is to sort out eye and head rotations in the tracking process and to sort out the ocular pursuit signal from visual and eye–head motor inputs.[93] From here, activity passes via parts of the vestibular nuclei, which perform the necessary neural integration of the velocity signal to a position signal that is sent to the eye muscle motor neurons.

2. Smooth vergence tracking system

Vergence results from the combined activity of intrinsic tonic activity, accommodative vergence, and responses to binocular disparity and perceived distance.[8] The sensory afferent signals for vergence (binocular disparity and blur) are coded in the primary visual cortex (area V1).[94] Some cells in V1 incorporate vergence to code egocentric (head-referenced) distance.[1] Cells in area MT and MST respond to retinal disparity and changing size.[95–97] Cells in the parietal cortex respond to motion in depth.[98] Efferent commands for vergence appear in cells in the frontal eye fields.[99] In the midbrain the premotor NRTP, located just ventral to the rostral portion of the paramedian pontine reticular formation (PPRF), receives projections from the frontal eye fields and the superior colliculus.[100] Lesions in this region result in deficits of slow-continuous and fast-step vergence control.[101,102] The NRTP projects to the cerebellum, and appears to be associated with vergence and accommodation.[103] The dorsal vermis is involved in the conversion of 3D pursuit signals to control signals for vergence eye movements.[104,105] The posterior interposed nucleus (PIN) of the cerebellum projects to supraoculomotor regions that contain near response cells in the mesencephalic reticular formation (MRF).[106,107] The supraoculomotor nucleus contains both burst and tonic neurons.[108] The burst cells code velocity signals for smooth vergence, and the tonic neurons code position signals to maintain static vergence angle. Excitatory connections of the supraoculomotor nucleus project to the oculomotor nucleus, driving the medial recti.[107,109,110] Inhibitory connections project to the abducens nucleus to inhibit the lateral rectus.[108]

The supraoculomotor nucleus relays control of both accommodative and disparity vergence.[109]

III. Foveal gaze shifts: target selection and foveal acquisition

A. Rapid conjugate shifts of gaze direction (saccadic eye movements)

Saccades are very fast, yoked eye movements that have a variety of functions.[111,112] They produce the quick phase of the VOR and OKN to avoid turning the eyes to their mechanical limits. They reflexively shift gaze in response to novel stimuli that appear unexpectedly away from the point of fixation. Saccades shift gaze during reading from one group of words to another. Saccades search novel scenes to assist us in acquiring information. They also return gaze to remembered spatial locations. Two primary functions in all of these tasks are to move the eye rapidly from one position to another and then maintain the new eye position. The rapid movement is controlled by a pulse and slide innervation pattern and the position is maintained by a step innervation.

The separate components of innervation for the saccade match the characteristics of the plant (i.e. the globe, muscles, fat, and suspensory tissues). The rapid changes in orbital position are made by the saccade at a cost of considerable energy. Saccade velocities can approach 1000 deg/sec.[111] In order to achieve these high velocities a phasic level of torque is needed to overcome the viscosity of orbital tissues, most of which is in the muscles.[56] The phasic-torque is generated by a large brief force resulting from a pulse or burst of innervation. The torque is dissipated or absorbed by the muscles, so that the force developed by the pulse of innervation does not reach the tendon (see Chapter 8 – Three-Dimensional Rotations of the Eye). At the end of the saccade, a lower constant force resulting from a step innervation generates a tonic level of torque that is needed to hold the eye still against the elastic restoring forces of the orbital tissues.[112] The eye positions resulting from the pulse and step forces must be equal to produce rapid gaze shifts. Pulse-step mismatches will result in rapid and slow components of gaze shifts. For example, if the pulse is too small, the saccade will slide (post-saccadic drift, called a glissade) to the new eye position at the end of the rapid phase of the saccade. The slide component is adaptable as has been shown by long-term exposure to artificially imposed retinal image slip immediately after each saccade.[113] The adapted post-saccadic drift cannot be explained by an adjustment of the pulse-step ratio, suggesting that the slide innervation is an independent third component of saccadic control (pulse-slide-step). Slide innervation produces a phasic-torque in addition to the pulse component that adjusts the duration and velocity of the saccade so that its amplitude matches the position maintained by the step component.[113] The pulse, slide and step are all under independent, cerebellar control, with the primary goal of protecting vision by preventing retinal slip, and a secondary goal of making accurate saccades.

The amplitude of the saccade determines its dynamic properties (e.g. its peak velocity and duration). The main sequence diagram plots these two dynamic parameters as a function of amplitude (Fig. 9.10).[95] As saccade amplitude

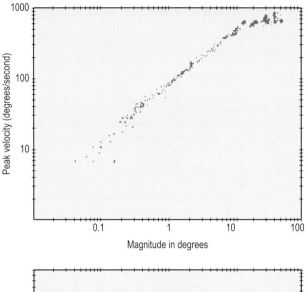

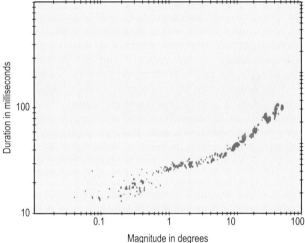

Figure 9.10 Main sequence diagram (Box 9.2). Peak velocity and saccade duration are plotted against magnitude of human saccadic eye movements.[114] (Redrawn with permission from Bahill AT, Clark MR and Stark, L: Math Biosci 24:191, copyright 1975 by Elsevier Science Publishing Co, Inc.)

Box 9.2 Main sequence for saccades

The main sequence (peak velocity and duration of saccades) illustrates abnormalities of the spike frequency and duration of the pulse innervation for saccades

increases from 0.1 to 10 degrees, duration increases from 20 to 40 msec and peak velocity increases from 10 to 400 deg/sec. Peak velocity saturates for saccades larger than 20 degrees, such that the amplitude of larger saccades increases primarily with duration. Abnormal saccade amplitudes (dysmetria) can either be too small (hypometric) or too large (hypermetric). Large gaze shifts are normally accomplished with a sequence of hypometric saccades that are composed of a series of short-latency corrective saccades in the same direction.[111] The normal latency of a saccade to an unpredictable stimulus is 180–200 msec.[48] However, corrective saccades occur with shorter latencies (100–150 msec). Saccade latency can be reduced by a blank or gap interval before the saccade, resulting in an express saccade with latencies less than 100 msec.[115] Saccade latencies to predictable target changes, such as occur when watching a tennis match, can be reduced to zero (Box 9.3).[74]

Although saccades are too brief to utilize visual feedback during their response, they do use fast internal feedback based upon an internal representation of eye position (efference copy signal) that helps control the position of the eye on a moment-to-moment basis.[116] Thus saccadic eye movements are not ballistic in that they are guided by extra-retinal information during their flight. The goal of the saccade is to reach a specified direction in head-centric space. Normally the perceived head-centric direction of a target does not change when the eye changes position. However, spectacle refractive corrections that magnify or minify the retinal image produce changes in perceived direction with eye position. Because the entrance pupil of the eye translates when the eye rotates, prior to the saccade the eye views a non-foveal eccentric target through a different part of the lens than after the saccade when the target is viewed directly along the line of sight. The prismatic power of the lens increases with distance from the optical center of the lens such that when viewing a target through a magnifier, the saccade amplitude needed to fixate an eccentric target is larger than the gaze eccentricity sensed prior to the saccade. Saccades are controlled by a feedforward system that does not utilize visual feedback during the motor response. Consequently, initial saccadic responses to visual distortion produced by the magnifier are hypometric. However, using position errors after the saccade, the system adapts rapidly (within 70 trials) to minimize its errors.[117] In cases of anisometropia in which the retinal images are magnified unequally by the spectacle refractive correction, the saccadic system adapts to produce unequal or disconjugate saccades that align both visual axes with common fixation targets.[118-120] The same adaptive process is likely to calibrate the conjugate saccades and maintain that calibration throughout life in spite of developmental growth factors and injury.

B. Disconjugate shifts of gaze distance (the near response in symmetrical convergence)

Large abrupt shifts in viewing distance stimulate adjustments of several motor systems including accommodation, convergence, and pupil constriction.[60] Separate control systems initiate and complete these responses.[121] Initially the abrupt adjustments are controlled by feedforward systems that do not use visual feedback until the responses are nearly completed. However they are guided by fast feedback from efference copy signals to monitor both the starting point of the near response and the accuracy of its end point.[61,122] Visual feedback is unavailable during the response because the blur and disparity cues are too large at the beginning of the gaze shift to be sensed accurately. The motor responses are initiated by high-level cues for perceived distance and by

voluntary shifts of attention. Retinal cues from blur and disparity are only used as feedback to refine the responses once the stimuli are reduced to amplitudes that lie within the range of visual sensitivity, i.e. as the eyes approach alignment at their new destination. The three motor systems are synchronized or coordinated with one another by cross-links. When they approach their new target destination, accommodation and convergence use visual feedback to refine their response.

The cross-couplings are demonstrated by opening the feedback loop for one motor system while stimulating a coupled motor system that is under closed-loop control. For example, accommodative convergence, measured during monocular occlusion, increases linearly with changes in accommodation stimulated by blur.[62,63] Similarly, convergence accommodation, measured during binocular viewing through pinhole pupils, increases linearly with changes in convergence stimulated by disparity.[123] The pupil constricts with changes in either accommodation or convergence to improve the clarity of near objects.[124] These interactions are greater during the dynamic changes in the near response than during the static endpoint of the response.[125]

The cross-couplings between accommodation and convergence are illustrated with a heuristic model (Fig. 9.11).[125] Three of the Maddox components are represented in the model by an adaptable slow tonic component, the cross-links between accommodation and convergence, disparity driven vergence and blur driven accommodation (fast phasic components). The enhanced gain of the cross-link interactions associated with dynamic stimulation of vergence and accommodation results from the stimulation of the cross-links by the phasic but not the tonic components. When accommodation or convergence is stimulated, the faster transient-phasic system responds first, but it does not sustain its response. The slower and more sustained adaptable tonic system gradually takes over the load of keeping the eyes aligned and focused by resetting the level of tonic activity. Because the cross-links are mainly stimulated by the phasic component, accommodative vergence and vergence accommodation are stimulated more during the dynamic response than during steady fixation when the tonic components control eye alignment and focus.

Traditionally, only the horizontal component of vergence has been considered as part of the near response. However, vertical and cyclovergence must also be adjusted during the near response to optimize the sensory stimulus for binocular vision.[126-129] The primary goal of the near response is to minimize large changes in horizontal, vertical, and cyclodisparities at the fovea that normally accompany large shifts in viewing distance. Horizontal disparities arise from targets that are nearer or farther than the convergence distance, assuming that the horopter equals the Veith–Muller circle. The isovergence circle describes the locus of points that stimulates a constant vergence angle in all directions of gaze,[3] and it is equivalent to the Veith–Muller circle described in Chapter 36 (see Fig. 9.2). Convergence and divergence stimuli lie closer or farther, respectively, from the isovergence circle. Vertical disparities arise from targets in tertiary gaze directions at finite viewing distances, because these targets lie closer to one eye than the other and their retinal images have unequal size and vertical eccentricity (vertical disparity) (Fig. 9.12). Torsional disparities arise from elevated targets

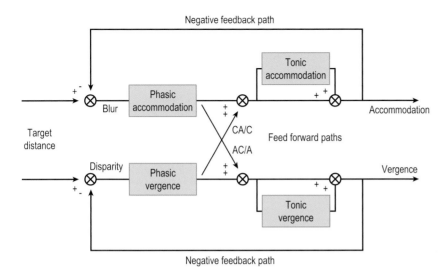

Negative feedback path

Figure 9.11 Model of cross-link interactions between vergence and accommodation. A fast phasic system drives the cross-links from accommodation to convergence (AC/A) and from convergence to accommodation (CA/C). The slow-tonic system adapts to the faster phasic system and gradually replaces it. Cross-link innervation is reduced when the tonic system reduces the load on the fast phasic system. (Modified from Kotulak J, Schor CM: The dissociability of accommodation from vergence in the dark. Invest Ophthal Vis Sci 1986; 27: 544. Reproduced from Association for Research in Vision and Ophthalmology.)

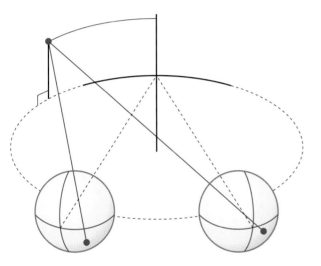

Figure 9.12 For convergence at a finite viewing distance, points in tertiary directions subtend unequal vertical visual angles at the two eyes, which produce vertical disparities.

at finite viewing distances, because during convergence, Listing's law predicts that the horizontal meridians of the two eyes will be extorted in upward gaze and intorted in downward gaze. These torsional eye postures would produce incyclodisparity in upward gaze and excyclodisparity in downward gaze.[126] With large shifts of viewing distance, horizontal, vertical, and cyclodisparities can exceed the stimulus operating range for continuous feedback control of disparity vergence. Initially the near response is stimulated by perceived distance and voluntary changes in horizontal vergence respond without disparity feedback in feedforward control. Unlike horizontal vergence, neither vertical vergence nor cyclovergence is normally under voluntary control. They participate in the near response through cross-couplings with motor responses that are under voluntary control. This allows potential vertical and torsional eye alignment errors to be reduced during the near response without feedback from retinal image disparity.[130]

Empirical measures demonstrate that the gains for the three coupling relationships for horizontal, vertical, and

cyclovergence are optimal for reducing disparity at the fovea to zero during the open-loop phase of the near response.[125,128,129,131] The tuned coupling gains for all three vergence components of the near response are the product of neural plasticity that adapts each of these cross-couplings to optimize binocular sensory functions.[69,132,133] Plasticity also exists for other couplings. For example, vertical vergence and cyclovergence can both be adapted to vary with head roll.[68,134,135] The coupling of pupil constriction with accommodation and convergence may also be under adaptive control. The pupil constriction component of the near response does not appear until the end of the second decade of life,[124] suggesting that it responds to accommodative errors resulting from the aging loss of accommodative amplitude. The pupil constriction component of the near response is an attempt to restore clarity of the near retinal image.

C. Interactions between conjugate and disconjugate eye movements (asymmetric vergence)

In natural viewing conditions it is rare for the eyes to converge symmetrically from one distance to another. Usually, gaze is shifted between targets located at different directions and distances from the head. Rapid gaze shifts to these targets require a combination of conjugate saccades and disjunctive vergence. In asymmetric convergence, the velocity of disparity vergence, accommodative vergence and accommodation are enhanced when accompanied by gaze shifting saccades.[136-138] Without the saccade, symmetrical vergence responses are sluggish, reaching velocities of only 10 deg/sec.[84] Symmetrical vergence has a latency of 160 msec and response time of approximately 1 second. Similarly accommodation that is not accompanied by a saccade has a peak velocity of only 4 deg/sec, a latency of approximately 300–400 msec and response time of approximately 1 second. However when accompanied by a saccade, vergence velocity approaches 50 deg/sec,[137] and accommodation velocity approaches 8–9 deg/sec.[138] Latency of accommodation accompanied by saccades is also reduced by 50 percent so that the accommodative response is triggered in synchrony with the saccade that has a latency of only 200 msec. As shown in Figure 9.13, response times for both accommodation and accommodative vergence are reduced dramatically when accompanied by saccades. Figure 9.14

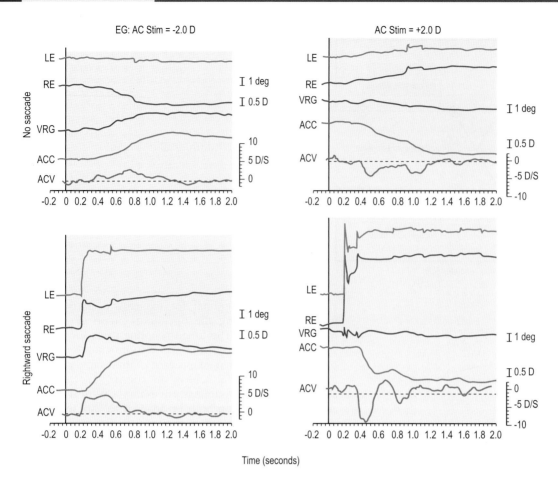

Figure 9.13 Examples of eye movement and accommodation traces during 6 deg rightward Saccade (bottom panels) and No-Saccade (top panels) conditions (left panel = trials requiring *increased* accommodation; right panel = trials requiring *decreased* accommodation). Time 0 corresponds to ACStim onset. The following conventions apply: LE = Left (viewing) eye position; RE = Right (non-viewing) eye position; VRG = Vergence position (LE-RE); ACC = Accommodation (D); ACV = Accommodation velocity (D/sec) = derivative of ACC. (Modified from Schor CM, Lott L, Pope D, et al.: Saccades reduce latency and increase velocity of ocular accommodation. Vision Res 1999; 39: 3769. Reproduced from Association for Research in Vision and Ophthalmology.)

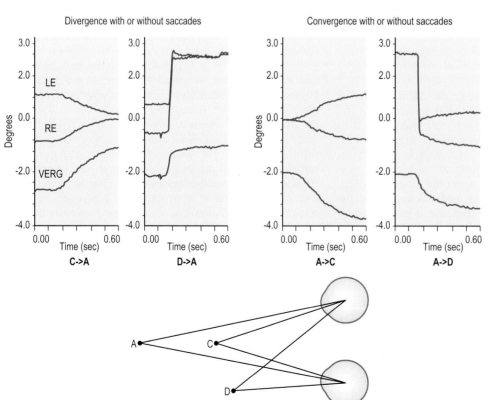

Figure 9.14 Vergence changes with or without an accompanying saccade, shown for a rhesus monkey. LE, Left eye; RE, right eye VERG vergence change. Vergence traces (right–left eye position) are offset for clarity. Convergence is negative. Note the increase in vergence velocity when a saccade is conjoined with vergence. The facilitation is greater for divergence because of the inherent divergence associated with horizontal saccades.[8] (From Leigh JR, Zee DS: The neurology of eye movements, 3rd edition, Oxford University Press 1999.)

compares symmetric and asymmetric disparity vergence. The high velocity asymmetric vergence appears to result in part from yoked saccades of unequal amplitude. These unequal saccades result from an asynchronous onset of binocular saccades. The abducting saccade begins before the adducting saccade, causing a brief divergence.[139,140] There also appears to be additional accelerated vergence and accommodation responses that are triggered with the saccade.[140–142] Both the vergence and accommodation responses continue after the completion of the saccade but overall the responses are completed in less time than when not accompanied by a saccade.

D. Neuro-control of foveal gaze shifts

1. Saccadic gaze shifting system

The frontal eye fields (FEF) mediate voluntary control of contralateral saccades. The FEF is active whether saccades occur or not.[143] The activity is related to visual attention, and when saccades occur the related activity in the FEF precedes them by 50 msec. The surface of the FEF has a coarse retinotopic organization. Stimulation of a particular area causes a saccade to change eye position in a specific direction and amplitude. These cells are active before saccades to certain regions of visual space. These regions are called the movement field of the cell, and they are analogous to the receptive fields of sensory neurons in the visual cortex. Stimulation of FEF cells in one hemisphere causes conjugate saccades to the contralateral side. Vertical saccades require stimulation of both hemispheres of the FEF. Modalities that can stimulate movement fields include vision, audition, and touch. The FEF project two main efferent pathways for the control of saccades. One projection is to the superior colliculus (SC). The other projection is to the midbrain, to the paramedian reticular formation (PPRF) and the rostral interstitial medial longitudinal fasciculus (riMLF) for the control of horizontal and vertical saccades, respectively.[144,145] The fibers from the frontal eye fields descend to the ipsilateral superior colliculus and cross the midline to the contralateral PPRF. Neither the superior colliculus nor FEF are required exclusively to generate saccades. Either one of them can be ablated without abolishing saccades, however if both are ablated, saccades are no longer possible. The function of the SC is to represent intended gaze direction resulting from combinations of head and eye position. Stimulation of a specific region in the intermediate layers of the colliculus can result in several combinations of head and eye position that achieve the same gaze direction relative to the body.[146] Cells in the superior colliculus respond to all sensory modalities including vision, audition, and touch. The spatial locations of all of these sensory stimuli are mapped in the colliculus relative to the fovea. Like the FEF, stimulation of one SC causes a conjugate saccade to the contralateral side; stimulation of both sides is necessary to evoke purely vertical saccades.

The output of the SC and FEF project to two premotor nuclei, the PPRF and the riMLF, that shape the velocity and amplitude of horizontal and vertical components of saccades, respectively.[88,144] The PPRF projects to the ipsilateral abducens nucleus, which contains motor neurons that innervate the ipsilateral lateral rectus and interneurons that project to the contralateral oculomotor nucleus to innervate the medial rectus. The PPRF also projects inhibitory connections to the contralateral PPRF and vestibular nucleus to reduce innervation of the antagonist during a saccade. The riMLF projects to the ipsilateral trochlear nucleus (IV) and to both oculomotor nuclei (III). Four types of neurons including burst cells, tonic cells, burst-tonic cells, and pause cells control saccades in several premotor sites. The pulse component of the saccade is controlled by medium-lead burst neurons. Long-lead burst neurons discharge up to 200 msec before the saccade and receive input from the SC and FEF. They drive medium-lead burst neurons (MLB) that begin discharging at a high frequency (300–400 spikes/sec) immediately at the beginning of the saccade and throughout its duration. Duration of MLB activity ranges from 10 to 80 msec. They project to the motor nuclei and control pulse duration and firing frequency, which determine saccade duration and velocity. Inhibitory burst neurons inhibit antagonist muscles by suppressing neurons in the contralateral abducens nucleus.

Initiation of the pulse is gated by the omnipause neuron (OPN), which is located in the nucleus of the dorsal raphe, below the abducens nucleus (Fig. 9.15). Normally the OPNs prevent saccades by constantly inhibiting burst cells. The OPN discharges continuously except immediately prior to and during saccades, when they pause. Omnipause neurons engage the saccade by releasing their inhibition of the burst cells. The same OPN inhibit saccades in all directions.

Upon completion of the saccade, the new eye position is held by the discharge step of the tonic cell. Integrating the pulse derives the discharge rate of the premotor tonic cell. At least two sites are known to integrate horizontal pulses; these are the medial vestibular nuclei and the nucleus prepositus hypoglossi (NPH). Pulses for vertical saccades are integrated in the interstitial nucleus of Cajal (INC). The flocculus of the cerebellum is also involved in integrating the velocity signals to position signals controlling eye movements. Some anomalies occur that appear to result from lesions of the integrator. In these cases the eyes make a saccade and then drift back to primary position. Affected patients are unable to hold fixation away from primary position and a jerk gaze nystagmus develops in which the slow-phase drift of the eyes is toward primary position and the fast phase is toward the desired eccentric gaze direction. Combined eye position and velocity signals are carried by burst-tonic neurons. They are active during ipsilateral saccades and inhibited during contralateral saccades.

2. Vergence gaze shifting system: the near triad and interactions with saccades

The supraoculomotor nucleus in the mesencephalic reticular formation contains near response cells. This is a heterogeneous population made up of cells that respond to accommodative stimuli, or vergence stimuli or a combination of accommodation and vergence stimuli.[107,109,110,141] This nucleus contains burst, tonic, and burst-tonic cells that are characteristic of premotor nuclei for saccades. Velocity signals related to disparity stimuli activate burst cells, and a position signal from tonic innervation is derived by integration of the burst cell activity. These cells are believed to provide velocity and position signals to the medial rectus motoneurons in the control of vergence as well as commands to the Edinger–Westphal nucleus to stimulate accommodation.[9] The Edinger–Westphal nucleus, located at the rostral portion of the midbrain at the oculomotor nucleus, contains

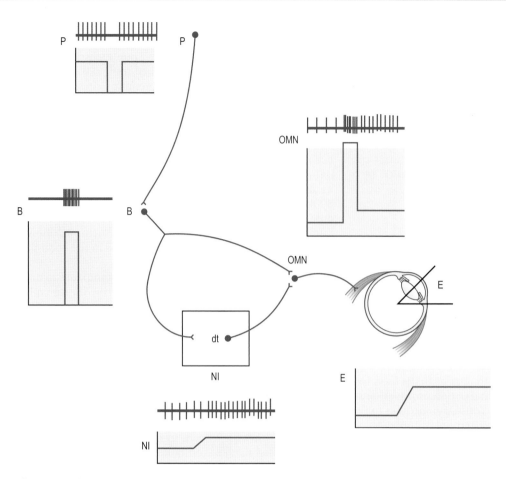

Figure 9.15 The relationship among omnipause cells (P), burst cells (B) and cells of the neural integrator (NI), in the generation of the saccadic pulse and step. Omnipause cells cease discharging just before each saccade, allowing the burst cells to generate the pulse. The pulse is integrated by the neural integrator (NI) to produce the step. The pulse and step combine to produce the innervational change on the ocular motoneurons (OMN) that produces the saccadic eye movement (E). Vertical lines represent individual discharges of neurons. Underneath the schematized neural (spike) discharge is a plot of discharge rate versus time.[8] (From Leigh JR, Zee DS: The neurology of eye movements, 3rd edition, Oxford University Press 1999.)

parasympathetic motor neurons that project to the ciliary muscle, and drive accommodation.[9] Each hemisphere of the Edinger–Westphal nucleus projects to its ipsilateral eye. Parasympathetic outflow of the Edinger–Westphal nucleus also results in miosis (pupillary contraction). Inhibition of the Edinger–Westphal nucleus causes pupil dilation.

Saccades enhance the velocity of both vergence and accommodation. Current models of saccade–vergence interactions suggest that OPNs in the midbrain gate the activity of both saccade bursters and vergence bursters.[140] Accelerated vergence caused by saccadic facilitation results from a release from inhibition of vergence bursters by reduced firing of the OPNs, which share their inhibition with saccadic and near response bursters. This facilitation is correlated with an augmentation of firing rate of a sub-set of convergence burst neurons of the near response cells.[141,142] This model has been developed further to include the potential for saccadic facilitation of accommodation.[141,142] Because near response cells provide innervation for both accommodation and vergence, release of inhibition from OPNs augments activity of both accommodation and vergence when associated with saccades.[138] This augmentation would not only enhance velocity of the near response, but it would also gate or synchronize innervation for vergence and accommodation with saccades.

Neurological disorders of the oculomotor system

Neurological disorders have greater variability across individuals than is found for the normal oculomotor system. The sources of variability come from the location and extent of lesions, associated anomalies or syndromes, different sources or causes of the problem, history of prior treatment interventions, and adaptive responses that attempt to compensate for the disorder. A multi-perspective classification can reflect this individual variation. It views the anomaly in different ways and highlights very different types of information about the condition. Its versatility describes the range of anomalous behaviors that may develop as a function of various combinations of causal factors. The pattern of associated conditions or syndromes facilitates estimating the prognosis for cosmetic and functional correction. Oculomotor disorders are classified in terms of behavioral descriptions, etiology, and neuro-anatomical sites of involvement. Behavioral categories can be descriptive such as the magnitude and direction of a strabismus (an eye turn), gaze restrictions, saccadic disorders, and nystagmus. They can also be described by an associated cluster of anomalies that collectively characterize a disease (syndrome).

Categories of etiology include congenital, developmental, and acquired. Congenital disorders appear in early infancy (<18 months). Developmental anomalies can result from interference of normal sensory–motor interactions during the first 6 years of life that constitute the critical period for visual development. Both congenital and developmental disorders are associated with anomalies throughout the oculomotor pathways, including sensory or afferent components of the visual system and they are classified in terms of syndromes. Acquired disorders can result from trauma or disease and they are classified in terms of specific anatomical sites of involvement and by syndromes.[8]

I. Strabismus

Non-paralytic strabismus falls into both congenital and developmental categories (Box 9.4). A strabismus or tropia describes a misalignment of the two eyes that is not corrected by disparity vergence during binocular fixation. In congenital and developmental forms of strabismus, the eye turn is usually concomitant, meaning that its amplitude does not vary with direction of gaze with either eye fixating. Usually non-paralytic strabismus is a horizontal deviation, where convergent and divergent eye turns are referred to as esotropia and exotropia respectively. Esotropia is more common than exotropia in childhood. Infantile or early-onset esotropia appears within the first 18 months of life and is associated with two forms of nystagmus that include latent or occlusion nystagmus (LN) and asymmetric optokinetic nystagmus, as well as cog-wheel temporal-ward pursuits (asymmetric pursuit) that were described earlier in this chapter. It is also associated with a vertical misalignment of the eyes when one eye is occluded, that is referred to as dissociated vertical deviation (DVD). When either the right or left eye is occluded, the covered eye rotates upward. This distinguishes DVD from a vertical strabismus in which one of the two eyes remains elevated with respect to the position of the other eye, independent of which eye is fixating. The combination of LN, asymmetric OKN, asymmetric pursuit and DVD are referred to as the infantile strabismus syndrome (Box 9.5).[147]

The presence of this syndrome in an adult is a retrospective indicator of an early age of onset for a strabismus. The early onset of eye misalignment prevents sensory fusion and causes one eye to be suppressed. This disrupts the development of both monocular and binocular sensory functions and can lead to the development of amblyopia, anomalous correspondence and stereoblindness that are described in Chapter 36. Accommodative esotropia is a developmental form of non-paralytic strabismus (Box 9.6) that results from either large uncorrected hyperopic refractive errors or abnormally large interactions between the cross coupling between accommodation and convergence. Uncorrected hyperopia produces a mismatch between the stimulus between accommodation and convergence. Excessive convergence is caused by accommodative attempts to clear the retinal image and the cross coupling between accommodation and convergence. The disparity-divergence system is unable to align the eyes and compensate for the accommodative esotropia.

Acquired strabismus usually results from lesions in the brainstem produced by trauma or disease that affect the integrity of the cranial nerves of the final common pathway. The lesions result in muscle palsies that cause deviations between the two visual axes that increase as gaze is directed into the field of action of the affected muscle (paralytic strabismus). The deviation of the paretic eye when the normal eye fixates (primary deviation) is usually smaller than the deviation of the normal eye when the paretic eye fixates (secondary deviation). This non-concomitant variation of eye turn is a diagnostic feature of paralytic strabismus. Lesions of the third, fourth, and sixth nerve are referred to as oculomotor palsy, trochlear palsy, and abducens palsy, respectively. Paresis of the muscles innervated by the III nucleus, i.e. the medial rectus, vertical recti, inferior oblique, levator of the lid and the pupilloconstrictor muscle, result in a fixed-dilated pupil and ptosis, with the eye remaining in a downward and abducted position. Trochlear palsy is characterized by a hyper deviation of the affected eye that increases during adduction and depression of the affected eye and head tilt to the side of the affected eye. Abducens palsy is characterized by an esotropia that increases during abduction of the affected eye.

II. Gaze restrictions

Lesions in premotor nuclei, supranuclear and cortical sites restrict movements of both eyes (Box 9.7). The medial longitudinal fasciculus (MLF) is the fiber bundle that interconnects premotor regions with the III, IV, and VI cranial nuclei. Any lesion that disconnects these fibers from the premotor

---

Box 9.4 Etiological classification of motor disorders

Motor anomalies are classified as congenital, developmental, and acquired

---

Box 9.5 Infantile strabismus syndrome

The infantile strabismus syndrome consists of asymmetric horizontal OKN and pursuits, dissociated vertical deviation (DVD), and latent nystagmus (LN)

Box 9.6 Accommodative esotropia

Accommodative esotropia results from the combination of uncorrected hyperopia and a high AC/A ratio

---

Box 9.7 Functional classification of motor disorders

- Peripheral and nuclear lesions are categorized as paresis and paralysis
- Premotor, internuclear, and supranuclear lesions are categorized as gaze restrictions or palsies

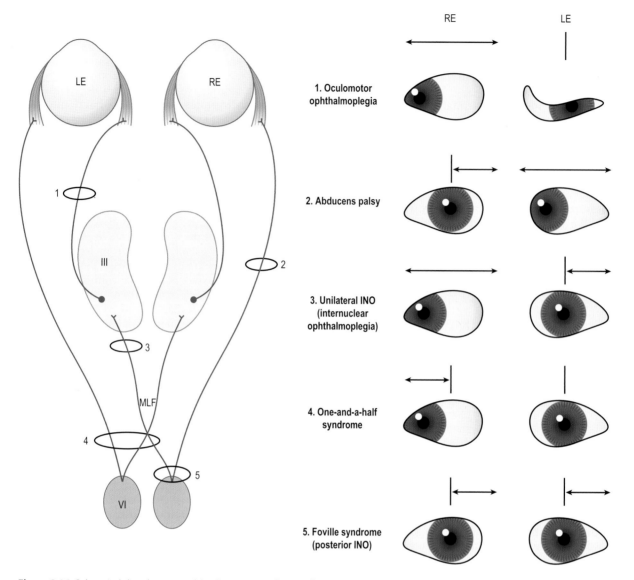

Figure 9.16 Subcortical disorder: gaze palsies. Eye positions shown reflect attempted right gaze in each case, but the arrows show the full range of horizontal gaze for each eye. (Original drawing by Scott B Stevenson, courtesy University of Houston.)

to the motor nuclei is referred to as ophthalmoplegia (Fig. 9.16). Lesions caudal to the oculomotor nucleus cause exotropia and failure of adduction, however convergence of the two eyes is spared. Internuclear ophthalmoplegia (INO) refers to an adduction failure caused by disruption of interneuron projections from the abducens nucleus to the contralateral III nerve nucleus. The affected eye is unable to adduct to the contralateral side and the eye drifts in the temporal-ward direction. Sparing of convergence distinguishes this lesion from oculomotor palsy. Patients with INO can develop a convergence nystagmus in an attempt to bring the exotropic eye into primary gaze. One-and-a-half syndrome is a combination of horizontal gaze palsy and INO that is caused by a lesion of the abducens (affecting the ipsilateral lateral rectus) and interneurons projecting from both abducens nuclei (affecting the ipsilateral and contralateral medial recti).

Foville's syndrome is a unilateral lesion at or near the abducens nucleus which causes conjugate gaze palsy,

contralateral limb paralysis, and ipsilateral facial paralysis. Lesions of the abducens nucleus block horizontal movement of both eyes to the side of the lesion because interneurons that project from the abducens nucleus to the contralateral oculomotor nucleus are also affected. Because the abducens is the final common pathway for all lateral conjugate eye movements, lesions there affect saccades, pursuit and the VOR. Lesions in the PPRF only limit horizontal saccades of both eyes to the ipsilateral side and cause drift of the eyes to the contralateral side. Lesions of the DLPN only affect horizontal pursuit toward the side of the lesion.

Lesions rostral to the III nucleus cause paralysis of vertical gaze (Parinaud's syndrome) and failure of convergence, but retention of normal horizontal gaze ability. Parinaud's syndrome occurs with lesions in the vicinity of the riMLF and INC and affects all vertical eye movements including saccades. It often results from tumors of the pineal gland that compress the superior colliculus and pretectal structures. Unilateral lesions in this area may also cause skew deviations in which there is a vertical deviation of one eye. Unilateral lesions can also cause unilateral nystagmus where the affected eye has a slow upward drift and fast downward saccade (down beat nystagmus).

Cortical lesions produce gaze restrictions of both saccades and pursuits. Lesions of the primary visual cortex produce blindness in the corresponding parts of the visual field contralateral to the lesion (scotoma). Pursuit and saccade targets presented in the scotoma are invisible to the patient. If the entire visual cortex in one hemisphere is destroyed, the vision near the fovea may be intact (macular sparing), allowing the patient to track targets over the full range of eye movements. Like the primary visual cortex, lesions in MT produce contralateral scotomas of the visual field. Saccades to fixed targets may be accurate but pursuit responses to moving targets presented in the affected field will be absent or deficient. Lesions in MST produce visual effects similar to MT, but they also produce a unidirectional pursuit deficit for targets in both visual hemifields moving toward the side of the lesion. Lesions in the posterior parietal cortex (PP) produce attentional deficits, which make pursuit and saccades to small targets more difficult than larger ones. Lesions of the FEF produce a deficit for horizontal pursuit and OKN toward the side of the lesion and saccades to the contralateral side. Lesions to the SEF impair memory-guided saccades.

III. Saccade disorders

Saccade disorders consist of abnormal metrics (velocity and amplitude), and inappropriate, spontaneous saccades that take the eye away from the target during attempted fixation. Saccades are classified as too fast or slow if their velocity does not fall within the main sequence plot of velocity versus amplitude (see Fig. 9.10). However, saccade velocity may be normal while its amplitude is in error. If saccades are too small, the eye could start to make a large saccade and would accelerate to an appropriate high velocity but then be stopped short of the goal by a physical restriction or by rapid fatigue, such as in ocular myasthenia. Saccades can also be interrupted by other saccades in the opposite direction (back-to-back saccades) such as is observed in voluntary nystagmus. These are truncated saccades. Slow saccades can result from muscle palsies and a variety of anomalies in premotor neurons.[8,148]

Cerebellar disorders can cause saccade *dysmetria* (inaccuracy). Saccades can be too large (hypermetric) or too small (hypometric) relative to the target displacement. Inaccuracy can lead to macrosaccadic oscillations when the eye repeatedly attempts to correct its fixation errors with inaccurate saccades. Either the pulse or the step component of the saccade can be inaccurate (Fig. 9.17). If the pulse is too small the saccade will be slow, and if the step is not constant but decays, eye position will drift toward primary position; repeated attempts to fixate eccentrically will result in gaze-evoked nystagmus. If the pulse and step are mismatched, there will be post-saccadic drift or glissade to the final eye position.

Cerebellar disorders and progressive supranuclear palsy can cause saccadic intrusions. These are conditions in which spontaneous saccades occur at the wrong time and move the eye away from the target during attempted fixation. Ocular flutter is characterized by rapid back-and-forth horizontal

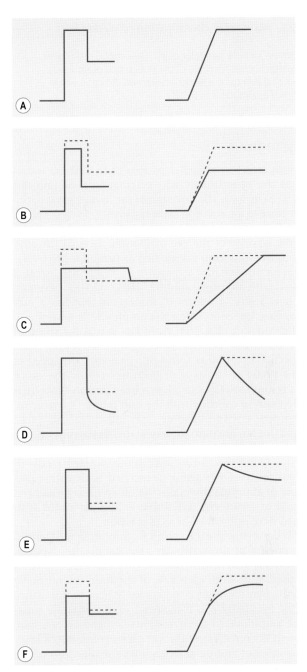

Figure 9.17 Disorders of the saccadic pulse and step. Innervation patterns are shown on the left, and eye movements on the right. Dashed lines indicate the normal response. (**A**) Normal saccade. (**B**) Hypometric saccade: pulse amplitude (width X height) is too small but pulse and step are matched appropriately. (**C**) Slow saccade: decreased pulse height with normal pulse amplitude and normal pulse-step match. (**D**) Gaze-evoked nystagmus: normal pulse, poorly sustained step. (**E**) Pulse-step mismatch (glissade); Step is relatively smaller than pulse. (**F**) Pulse-step mismatch due to internuclear ophthalmoplegia (INO): the step is larger than the pulse, and so the eye drifts onward after the initial rapid movements. (From Leigh JR, Zee DS: The neurology of eye movements, 3rd edition, Oxford University Press 1999.)

saccades without normal saccade latency or intersaccadic interval. Opsoclonus is ocular flutter in all directions. There are also square wave jerks that move the eye away from a point of fixation and then back again. These movements have a normal intersaccadic interval.

Box 9.8 Waveform classification of nystagmus

Nystagmus can have pendular or jerk (saw-tooth) waveforms. Jerk nystagmus disorders are of the slow phase of unsteady fixation. Null point is the gaze distance and direction where nystagmus amplitude is minimal.

IV. Nystagmus

Nystagmus refers to a regular pattern of to and fro movements of the eyes, usually with alternating slow and fast phases (see Fig. 9.6). The direction of the nystagmus is specified according to the direction of the fast phase (e.g. "jerk right"), because the fast phase is more visible than the slow phase. When the velocity of the eye oscillations is equal in both directions the waveform is classified as pendular (Box 9.8).

Congenital nystagmus appears early in life and is often associated with albinism, aniridia, and congenital acromatopsia. It is a jerk waveform of nystagmus that has a null point or gaze direction where the amplitude is minimized. The differentiating characteristic of CN is that the speed of the slow phase *increases* exponentially, until a resetting quick phase occurs. In all other forms of nystagmus, the speed of the slow phase is constant or decreases before the quick phase. Persons with CN can adopt a head turn to allow their gaze direction to coincide with the null point. Convergence also dampens CN and in some cases individuals adopt an esotropia in an attempt to block the nystagmus at the expense of binocularity (nystagmus blocking syndrome). Latent nystagmus described earlier in this chapter is a developmental form of nystagmus associated with early-onset esotropia or abnormal binocular vision and it is thought to be related to asymmetric OKN that is also associated with disrupted development of binocular vision.[40]

Vestibular nystagmus can result from central and peripheral anomalies of the vestibular gaze stabilization system. The vestibular system is organized in a push–pull fashion where the inputs from each side of the head are normally balanced when the head is stationary. When an imbalance exists, the eyes behave as if the head were constantly rotating. The amplitude of the nystagmus is highest when gaze deviates in the direction of the fast phase (Alexander's law). In some cases of vestibular nystagmus, the direction of the nystagmus reverses every 2 minutes (periodic alternating nystagmus, PAN). The reversal reflects the action of the normal cerebellar adaptive control mechanism to correct an imbalance in the vestibular system. Rebound nystagmus is a related condition associated with gaze-evoked nystagmus that was described earlier in this chapter. In gaze-evoked nystagmus, when fixation is held eccentrically, the slow phase is toward primary position. In rebound nystagmus, the amplitude of the nystagmus is reduced after 30 or more seconds of sustained fixation. However, when the eyes return to primary position, the nystagmus reverses direction, i.e. rebounds. The attenuation and reversal of slow phase direction demonstrates an attempt by the cerebellum to reduce the slow drifts during attempted steady fixation caused by a leaky tonic innervation for eye position.

Nystagmus can also have vertical and torsional components. For example, see-saw nystagmus is an acquired pendular form of nystagmus in which there is a combination of vertical and torsional oscillation of eye orientation. The eyes bob up and down by several degrees in opposite directions (skew movements). As one eye elevates and intorts, the other eye depresses and extorts. The exact cause is unknown, but see-saw nystagmus is associated with visual loss in bitemporal hemianopia and optic chiasm disorders. The nystagmus is thought to result from an inappropriate ocular counter-roll during head tilt orchestrated by the cerebellum in association with the otoliths.

All of the oculomotor anomalies described above show some evidence of attempts by the oculomotor system to correct them using the same adaptive mechanisms that normally calibrate the various classes of eye movement systems. However, the anomalies are beyond the corrective range of the adaptive processes. If it were not for these adaptive processes, these anomalies would be far more prevalent and the oculomotor system would be extremely susceptible to permanent injury resulting from disease and trauma. The consequences would be very dramatic. For example, about 50 years ago a physician named John Crawford lost function of his vestibular apparatus as a result of an overdose of streptomycin used to treat tuberculosis located in his knee. He reported that every movement of his head caused vertigo and nausea, even when his eyes were open. If his eyes were shut, the symptoms intensified. He attempted to steady his head by lying on his back and gripping the bars at the head of the bed. However even in this position the pulse beat in his head became a perceptible motion, disturbing his equilibrium. It is difficult to imagine how we would survive without the adaptive processes that continually calibrate our oculomotor system and allow us to distinguish between motion of our head and eyes from motion of objects in the world.

Acknowledgments

Comments by David Zee, James Maxwell, and Lance Optican are appreciated.

References

1. Trotter Y, Celebrini S, Stricanne B et al. Neural processing of stereopsis as a function of viewing distance in primate visual cortical area V1. J Neurophysiol 1996; 76:2872.
2. Hering E. The theory of binocular vision. In: Bridgeman B, Stark L, eds. New York: Plenum Press, 1977.
3. Ono H. The combination of version and vergence. In: Schor CM, Ciuffreda K, eds. Vergence eye movements: basic and clinical aspects. Boston: Butterworth, 1983:373–400.
4. Lisberger SG. The latency of pathways containing the site of motor learning in the monkey vestibulo-ocular reflex. Science 1984; 225:74.
5. Miles FA, Kawano K, Optican LM. Short-latency ocular following responses of monkey. I. Dependence on temporospatial properties of visual input. J Neurophysiol 1986; 56: 1321.
6. Leigh RJ, Kennard C. Using saccades as a research tool in the clinical neurosciences. Brain 2004; 127:460–477.
7. Sherrington C. The integrative action of the nervous system, 2nd ed. New Haven: Yale University Press, 1947.
8. Leigh JR, Zee DS. The neurology of eye movements, 3rd ed. Contemporary Neurology Series. Oxford: Oxford University Press, 1999.
9. Gamlin PD, Zhang H, Clendaniel RA et al. Behavior of identified Edinger–Westphal neurons during ocular accommodation. J Neurophysiol 1994; 72:2368.
10. Westheimer G, Blair S. The parasympathetic pathways to internal eye muscles. Invest Ophthalmol 1973; 12:193.
11. Robinson DA, Keller EL. The behavior of eye movement motoneurons in the alert monkey. Bibl Ophthalmol 1972; 82:7.
12. Melville Jones G. The vestibular contribution. In: Carpenter RHS, ed. Vision and visual dysfunction, vol 8, Eye movements. Boca Raton, Florida: CRC Press, 1991: 13–44.
13. Simpson JI, Graf W. The selection of reference frames by nature and its investigators. In: Berthoz A, Melville Jones G, eds. Adaptive mechanisms in gaze control. Amsterdam: Elsevier, 1985:3–20.

14. Keller EL. Gain of the vestibulo-ocular reflex in monkey at high rotational frequencies. Vision Res 1978; 20:535.

15. Cogan DG. Neurology of the ocular muscles, 2nd ed. Springfield, Illinois: Charles C Thomas, 1956.

16. Seidman SH, Leigh RJ. The human torsional vestibulo-ocular reflex during rotation about an earth-vertical axis. Brain Res 1989; 504:264.

17. Snyder LH, King WM. Effect of viewing distance and the location of the axis of rotation on the monkey's vestibulo-ocular reflex (VOR). I. Eye movement responses. J Neurophysiol 1992; 67:861.

18. Collewijn H, Steinman RM, Erkelens CJ. Binocular fusion, stereopsis and stereoacuity with a moving head. In: Regan D, ed. Binocular vision. London: Macmillan, 1991:121–136.

19. Bockisch CJ, Haslwanter T Three-dimensional eye position during static roll and pitch in humans. Vision Res 2001: 41:2127–2137.

20. Collewijn H, Van der Steen J, Ferman L, Jansen TC. Human ocular counter-roll: assessment of static and dynamic properties from electromagnetic search coil recordings. Exp Brain Res 1985; 59:185–196.

21. Clement G, Moore ST, Raphan T, Cohen B. Perception of tilt (somatogravic illusion) in response to sustained linear acceleration during space flight. Exp Brain Res 2001; 138(4):410–418.

22. Zee DS. Considerations on the mechanisms of alternating skew deviation in patients with cerebellar lesions. J Vestib Res 1996; 395–401.

23. Brandt T, Dieterich M. Pathological eye-head coordination in roll: tonic ocular tilt reaction in mesencephalic and medullary lesions. J Neurol Neurosurg Psychiat 1987; 54:549–666.

24. Collewijn H. The optokinetic contribution. In: Carpenter RHS, ed. Vision and visual dysfunction, vol 8, eye movements. Boca Raton, Florida: CRC Press, 1989:45–70.

25. Cheng M, Outerbridge JS. Optokinetic nystagmus during selective retinal stimulation. Exp Brain Res 1975; 23:129.

26. Cheung BSK, Howard IP. Optokinetic torsion: Dynamics and relation to circular vection. Vision Res 1991; 31:1327.

27. Raphan T, Cohen B. Velocity storage and the ocular response to multidimensional vestibular stimuli. In: Berthoz A, Melville Jones G, eds. Adaptive mechanisms in gaze control. Amsterdam: Elsevier, 1985:123–144.

28. Miles FA. The sensing of rotational and translational optic flow by the primate optokinetic system. In: Miles FA, Wallman J, eds. Visual motion and its role in the stabilization of gaze. Amsterdam: Elsevier, 1993:393–403.

29. Collewijn H. Integration of adaptive changes of the optokinetic reflex, pursuit and the vestibulo-ocular reflex. In: Berthoz A, Melville Jones G, eds. Adaptive mechanisms in gaze control. Amsterdam: Elsevier, 1985:51–75.

30. Robinson DA. Control of eye movements. In: Brooks V, ed. Handbook of physiology. Section I: The nervous system. Bethesda, MD: William and Wilkins, 1981: 1275–1320.

31. Goldberg ME, Eggers HM, Gouras P. The ocular motor system. In: Kandel ER, Schwartz JH, Jessell TM, eds. Principles of neural science, 3rd ed. Norwalk, Connecticut: Appleton & Lange, 1991:660–677.

32. Lisberger SG. The neural basis for learning of simple motor skills. Science 1988; 242:728.

33. Ito M. Synaptic plasticity in the cerebellar cortex that may underlie the vestibulo-ocular adaptation. In: Berthoz A, Melville Jones G, eds. Adaptive mechanisms in gaze control. Amsterdam: Elsevier, 1985:213–221.

34. Albright TD. Cortical processing of visual motion. In: Miles FA, Wallman J, eds. Visual motion and its role in the stabilization of gaze. Amsterdam: Elsevier, 1993:177–201.

35. Fuchs AF, Mustari MJ. The optokinetic response in primates and its possible neuronal substrate. In: Miles FA, Wallman J, eds. Visual motion and its role in the stabilization of gaze. Amsterdam: Elsevier, 1993:343–369.

36. Wallman J. Subcortical optokinetic mechanisms. In: Miles FA, Wallman J, eds. Visual motion and its role in the stabilization of gaze. Amsterdam: Elsevier, 1993:321–342.

37. Hoffmann KP, Distler C, Ilg U. Collosal and superior temporal sulcus contributions to receptive field properties in the macaque monkey's nucleus of the optic tract and dorsal terminal nucleus of the accessory optic tract. J Comp Neurol 1992; 321:150.

38. Gamlin PDR. The pretectum: connections and oculomotor-related roles. Prog Brain Res 2006; 151:379–405.

39. Pasik T, Pasik P. Optokinetic nystagmus: an unlearned response altered by section of chiasma and corpus callosum in monkeys. Nature 1964; 203:609.

40. Schor CM. Development of OKN. In: Miles FA, Wallman J, eds. Visual motion and its role in the stabilization of gaze. Amsterdam: Elsevier, 1993:301–320.

41. Dell'Osso LF, Schmidt D, Daroff RB. Latent, manifest latent and congenital nystagmus. Arch Ophthalmol 1979; 97:1877.

42. Kowler E, van der Steen J, Tamminga EP et al. Voluntary selection of the target for smooth eye movement in the presence of superimposed, full-field stationary and moving stimuli. Vision Res 1984; 12:1789.

43. Westheimer G, McKee SP. Visual acuity in the presence of retinal image motion. J Opt Soc Am 1975; 65:847.

44. Land MF. Predictable eye-head coordination during driving. Nature 1992; 359:318.

45. Bahill AT, Adler D, Stark L. Most naturally occurring human saccades have magnitudes of 15 degrees or less. Invest Ophthalmol 1975; 14:468.

46. Able LA, Parker L, Daroff RB et al. Endpoint nystagmus. Invest Ophthalmol Vis Sci 1977; 17:539.

47. Baloh RW, Sharma S, Moskowitz H et al. Effect of alcohol and marijuana on eye movements. Aviat Space Environ Med 1979; 50:18.

48. Carpenter RHS. Movements of the eyes, 2nd ed. London: Psion, 1988.

49. Kowler E. The stability of gaze and its implications for vision. In: Carpenter RHS, ed. Vision and visual dysfunction, vol 8, Eye movements. Boca Raton, Florida: CRC Press, 1991:71–94.

50. Van Rijn LJ, Van der Steen J, Collewijn H. Instability of ocular torsion during fixation: cyclovergence is more stable than cycloversion. Vision Res 1994; 34:1077.

51. Ahn BY, Choi D, Kim SJ, Park KP, Bae JH, Lee TH. Impaired ipsilateral smooth pursuit and gaze-evoked nystagmus in paramedian pontine lesions. Neurology 2007; 68: 1436.

52. Schor CM. Binocular vision. In: De Valois K, ed. Seeing. New York: Academic Press, 2000:177–257.

53. Ogle KN, Martens TG, Dyer JA. Oculomotor imbalance in binocular vision and fixation disparity. Philadelphia: Lea & Febiger, 1967.

54. Badcock DR, Schor CM. Depth-increment detection function for individual spatial channels. J Opt Soc Am A 1985; 1211.

55. Maddox EC. The clinical use of prism, 2nd ed. Bristol, UK: John Wright & Sons, 1893:83.

56. Robinson DA. A quantitative analysis of extraocular muscle cooperation and squint. Invest Ophthalmol 1975b; 14:801.

57. Rethy I. Development of the simultaneous fixation from the divergent anatomic eye-position of the neonate. J Ped Ophthalmol 1969; 6:92.

58. McLin L, Schor CM, Kruger P. Changing size (looming) as a stimulus to accommodation and vergence. Vision Res 1988; 28:883.

59. Wick B, Bedell HE. Magnitude and velocity of proximal vergence. Invest Ophthalmol Vis Sci 1989; 30:755.

60. Judge SJ. Vergence. In: Carpenter RHS, ed. Vision and visual dysfunction, vol 8, Eye movements. Boca Raton, Florida: CRC Press, 1991:157–170.

61. Schor, CM, Alexander J, Cormack L et al. Negative feedback control model of proximal convergence and accommodation. Ophthal Physiol Optics 1992; 12:307.

62. Alpern M, Ellen P. A quantitative analysis of the horizontal movements of the eyes in the experiments of Johannes Muller. I. Methods and results. Am J Ophthalmol 1956; 42:289.

63. Muller J. Elements of physiology, Vol. 2. Trans: W. Baly W. London: Taylor & Walton, 1826.

64. Borish IM. Clinical refraction, 3rd ed. Chicago: Professional Press, 1970.

65. Tait EF. A report on the results of the experimental variation of the stimulus conditions in the responses of the accommodative convergence reflex. Am J Optom 1933; 10:428.

66. Schor CM. Fixation disparity and vergence adaptation. In: Schor CM, Ciuffreda K, eds. Vergence eye movements: basic and clinical aspects. Boston: Butterworth, 1983:465–516.

67. Schor CM, Gleason G, Maxwell J et al. Spatial aspects of vertical phoria adaptation. Vision Res 1993; 33:73.

68. Maxwell JS, Schor CM. Adaptation of torsional eye alignment in relation to head roll. Vision Res 1999; 39:4192.

69. Schor CM, Maxwell JS, Graf E. Plasticity of convergence-dependent variations of cyclovergence with vertical gaze. Vision Res 2001; 41:3353.

70. Taylor MJ, Roberts DC, Zee DS. Effect of sustained cyclovergence on eye alignment: rapid torsional phoria adaptation. Invest Ophthalmol Vis Sci 2000; 41:1076.

71. Pola J, Wyatt HJ. Smooth pursuit response characteristics, stimuli and mechanisms. In: Carpenter RHS, ed. Vision and visual dysfunction, vol 8, Eye movements. Boca Raton, Florida: CRC Press, 1991:138–156.

72. Howard IP, Marton C. Visual pursuit over textured backgrounds in different depth planes. Exp Brain Res 1997; 90:625.

73. Myer CH, Lasker AG, Robinson DA. The upper limit for smooth pursuit velocity. Vision Res 1985: 25:561.

74. Stark L, Vossius G, Young LR. Predictive control of eye tracking movements. IRE Trans. Hum Factors Electron 1962; 3:52.

75. O'Mullane G, Knox PC. Modification of smooth pursuit initiation by target contrast. Vision Res 1999; 39:3459.

76. Rashbass C. The relationship between saccadic and smooth tracking eye movements. J Physiol (Lond) 1961; 159:326.

77. Lisberger SG, Westbrook LE. Properties of visual inputs that initiate horizontal smooth pursuit eye movements in monkeys. J Neurosci 1985; 5:1662.

78. Carl JR, Gellman RS. Adaptive responses in human smooth pursuit. In: Keller EL, Zee DS, eds. Adaptive processes in visual and oculomotor systems. Advances in the biosciences, volume 57. Oxford: Pergamon Press, 1986:335–339.

79. Ilg UJ. Commentary: Smooth pursuit eye movements: from low-level to high-level vision. In: Hyona J, Munoz DP, Heide W, Radach R, eds. Progress in brain research, vol 140. Amsterdam: Elsevier, 2002:279–299.

80. Collewijn H, Erkelens CJ. Binocular eye movements and the perception of depth. In: Kowler E, ed. Eye movements and their role in visual and cognitive processes. Amsterdam: Elsevier, 1990:213–261.

81. Bradshaw MF, Glennerster A, Rogers BJ. The effect of display size on disparity scaling from differential perspective and vergence cues. Vision Res 1996; 36:1255.

82. Foley JM. Primary distance perception. In: Held R, Leibowitz HW, Teuber H-L, eds. Handbook of sensory physiology: Vol VIII Perception. Berlin: Springer Verlag, 1978:181–214.

83. Hollins M, Bunn KW. The relation between convergence micropsia and retinal eccentricity. Vision Res 1997; 17:403.

84. Rashbass C, Westheimer G. Disjunctive eye movements. J Physiol (Lond) 1961; 159:339.

85. Schor CM, Gleason J, Horner D. Selective nonconjugate binocular adaptation of vertical saccades and pursuits. Vision Res 1990; 30:1827.

86. Gleason G, Schor CM, Lunn R et al. Directionally selective short-term non-conjugate adaptation of vertical pursuits. Vision Res 1993; 33:65.

87. Albright TD. Centrifugal directional bias in the middle temporal visual area (MT) of the macaque. Vis Neurosci 1989; 2:177.

88. Keller EL. The brainstem. In: Carpenter RHS, ed. Vision and visual dysfunction, vol 8, Eye movements. Boca Raton, Florida: CRC Press, 1991:200–223.

89. Newsome WT, Wurtz RH, Komatsu H. Relation of cortical areas MT and MST to pursuit eye movements. II. Differentiation of retinal from extraretinal inputs. J Neurophysiol 1988; 60:604.

90. Ono S, Das VE, Economides JR, Mustari MJ. Modeling of smooth pursuit-related neuronal responses in the DLPN and NRTP of the rhesus macaque. J Neurophysiology 2005; 93:108–116.

91. Dicke PW, Barash S, Ilg UJ, Thier P. Single-neuron evidence for a contribution of the dorsal pontine nuclei to both types of target-directed eye movements, saccades and smooth pursuit. Eur J Neurosci 2004; 19:609–624.

92. Krauzlis RJ. Recasting the smooth pursuit eye movement system. J Neurophysiol 2004; 91:591–603.

93. Noda H, Warabi T. Responses of Purkinje cells and mossy fibers in the flocculus of the monkey during sinusoidal movements of a moving pattern. J Physiol (Lond) 1987; 387:611.

94. Poggio G. Mechanisms of stereopsis in monkey visual cortex. Cereb Cortex 1995; 3:195.

95. DeAngelis GC, Newsome WT. Organization of disparity-selective neurons in macaque area MT. J Neurosci 1999; 19:1398.

96. Maunsell JHR, Van Essen DC. Functional properties of neurons in middle temporal visual area of the macaque monkey. II. Binocular interactions and sensitivity to binocular disparity. J Neurophysiol 1983; 49:1148.

97. Roy J-P, Wurtz RH. The role of disparity-sensitive cortical neurons in signaling the direction of self-motion. Nature 1990; 348:160.

98. Colby CL, Duhamel JR, Goldberg ME. Ventral intraparietal area of the macaque: anatomic location and visual response properties. J Neurophysiol 1993; 69:902.

99. Gamlin PD, Yoon, K. An area for vergence eye movement in primate frontal cortex. Nature 2000; 407:1003.

100. Gamlin PD, Clarke RJ. Single-unit activity in the primate nucleus reticularis tegmenti pontis related to vergence and ocular accommodation. J Neurophysiol 1995; 73:2115.

101. Rambold H, Neumann G, Helmchen CI. Vergence deficits in pontine lesions. Neurology 2004; 62:1850–1853.

102. Rambold H, Sander T, Neumann G, Helmchen CI. Palsy of fast and slow vergence by pontine lesions. Neurology 2005; 62:338–340.

103. Ohtsuka K, Maekawa H, Sawa, W. Convergence paralysis after lesions of the cerebellar peduncles. Ophthalmologica 1993; 206:143.

104. Takagi M, Tamargo R, Zee DS. Effects of lesions of the cerebellar oculomotor vermis on eye movements in primate binocular control. In: Progress in Brain Research, Vol 142. Amsterdam: Elsevier, 2004:19–33.

105. Nitta T, Akao T, Kurkin S, Fukushima K. Involvement of cerebellar dorsal vermis in vergence eye movements in monkeys. Cerebral Cortex 2008; 18:1042–1057.

106. May PJ, Porter JD, Gamlin PD. Interconnections between the primate cerebellum and midbrain near-response regions. J Comp Neurol 1992; 315:98.

107. Zhang H, Gamlin PD. Neurons in the posterior interposed nucleus of the cerebellum related to vergence and accommodation. I. Steady-state characteristics. J Neurophysiol 1998; 79:1255.

108. Mays LE, Porter JD. Neural control of vergence eye movements: activity of abducens and oculomotoneurons. J Neurophysiol 1984; 52:743.

109. Judge S, Cumming, B. Neurons in the monkey midbrain with activity related to vergence eye movements and accommodation. J Neurophysiol 1986; 55:915.

110. Zhang Y, Mays LE, Gamlin PD. Characteristics of near response cells projecting to the oculomotor nucleus. J Neurophysiol 1992; 67:944.

111. Becker W. Chapter 2. In: Wurtz RH, Goldberg ME, eds. The neurobiology of saccadic eye movements. Amsterdam: Elsevier, 1989:13–68.

112. Van Gisbergen JAM, Van Opstal AJ. Models. In: Wurtz RH, Goldberg ME, eds. The neurobiology of saccadic eye movements. Amsterdam: Elsevier, 1989:69–101.

113. Optican LM, Miles FA. Adaptive properties of the saccadic system. Visually induced adaptive changes in primate saccadic oculomotor control signals. J Neurophysiol 1985; 54:940.

114. Bahill AT, Clark MR, Stark L. The main sequence, a tool for studying human eye movements. Mathemat Biosci 1975; 24:191.

115. Fischer B, Ramsperger E. Human express saccades: extremely short reaction times of goal directed eye movements. Exp Brain Res 1984; 57:191.

116. Robinson DA. Oculomotor control signals. In: Lennerstrand G, Bach-Y-Rita P, eds. Basic Mechanisms of ocular motility and their clinical implications. Oxford: Pergamon Press, 1975:337–374.

117. Deuble H. Separate adaptive mechanism for the control of reactive and volitional saccadic eye movements. Vision Res 1995; 35:3520.

118. Erkelens, CJ, Collewijn H, Steinman RM. Asymmetrical adaptation of human saccades to anisometropia spectacles. Invest Ophthalmol Vis Sci 1989; 30:1132.

119. Lemij HG, Collewijn H. Nonconjugate adaptation of human saccades to anisometropic spectacles: meridian-specificity. Vision Res 1992; 32:453.

120. Oohira A, Zee DS, Guyton DL. Disconjugate adaptation to long-standing, large-amplitude, spectacle-corrected anisometropia. Invest Ophthalmol Vis Sci 1991; 32:1693.

121. Semmlow JL, Hung GK, Horng JL et al. Disparity vergence eye movements exhibit preprogrammed motor control. Vision Res 1994; 34:335.

122. Zee DS, Levi L. Neurological aspects of vergence eye movements. Rev Neurol (Paris) 1989;145:613.

123. Fincham EF, Walton J. The reciprocal actions of accommodation and convergence. J Physiol (Lond) 1957; 137:488.

124. Wilhelm H, Schaeffel F, Wilhelm B. Age relation of the pupillary near reflex. Klin Monatsbl Augenheilkd 1993; 203:110.

125. Schor CM, Kotulak J. Dynamic interactions between accommodation and convergence are velocity sensitive. Vision Res 1986; 26:927.

126. Tweed D. Visual-motor optimization in binocular control. Vision Res 1997; 37:1939.

127. Allen MJ, Carter JH. The torsion component of the near reflex. Am J Optom 1967; 44:343.

128. Schor CM, Maxwell J, Stevenson SB. Isovergence surfaces: the conjugacy of vertical eye movements in tertiary positions of gaze. Ophthal Physiol Optics 1994; 14:279.

129. Ygge J, Zee DS. Control of vertical eye alignment in three-dimensional space. Vision Res 1995; 35:3169.

130. Schor CM. Plasticity of the near response. In: Harris L, ed. Levels of perception. New York: Springer, 2003:231–255.

131. Somani RAB, Desouza JFX, Tweed D et al. Visual test of listing's law during vergence. Vision Res 1998; 38:911.

132. McCandless JW, Schor CM. An association matrix model of context-specific vertical vergence adaptation. Network Comput Neural Syst 1997; 8:239.

133. Miles FA, Judge SJ, Optican LM. Optically induced changes in the couplings between vergence and accommodation. J Neurosci 1987; 7:2576.

134. Maxwell JSM, Schor CM. Head-position-dependent adaptation of non concomitant vertical skew. Vision Res 1997; 37:441.

135. Maxwell JS, Schor CM. The coordination of binocular eye movements: vertical and torsional alignment. Vision Res 2006; 46:3537–3548.

136. Collewijn H, Erkelens CJ, Steinman RM. Trajectories of the human binocular fixation point during conjugate and non-conjugate gaze-shifts. Vision Res 1997; 37:1049.

137. Enright JT. Changes in vergence mediated by saccades. J Physiol (Lond) 1984; 350:9.

138. Schor CM, Lott L, Pope D et al. Saccades reduce latency and increase velocity of ocular accommodation. Vision Res 1999; 39:3769.

139. Maxwell JS, King WM. Dynamics and efficacy of saccade-facilitated vergence eye movements in monkeys. J Neurophysiol 1992; 68:1248.

140. Zee DS, FitzGibbon EJ, Optican LM. Saccade-vergence interactions in humans. J Neurophysiol 1992; 68:1624.

141. Mays LE, Gamlin PD. Neuronal circuitry controlling the near response. Curr Opin Neurobiol 1995; 5:763.

142. Mays LE, Gamlin PD. A neural mechanism subserving saccade-vergence interactions. In: Findlay JM, Walker R, Kentridge RW, eds. Eye movement research: mechanisms, processes and applications. New York: Elsevier, 1995.

143. Goldberg ME, Segraves MA. The visual and frontal cortices. In: Wurtz RH, Goldberg ME, eds. The neurobiology of saccadic eye movements. Amsterdam: Elsevier, 1989: 283–313.

144. Hepp K, Henn V, Vilis T et al. Brainstem regions related to saccade generation. In: Wurtz RH, Goldberg ME, eds. The neurobiology of saccadic eye movements. Amsterdam: Elsevier, 1989:105–212.

145. Sparks DL, Hartwich-Young R. The deep layers of the superior colliculus. In: Wurtz RH, Goldberg ME, eds. The neurobiology of saccadic eye movements. Amsterdam: Elsevier, 1989:213–256.

146. Freedman EG, Sparks DL. Activity of cells in the deeper layers of the superior colliculus of the rhesus monkey: evidence for a gaze displacement command. J Neurophysiol 1997; 78:1669.

147. Schor CM, Wilson N, Fusaro R et al. Prediction of early onset esotropia from the components of the infantile squint syndrome. Invest Ophthalmol Vis Sci 1997; 38:719.

148. Ramat S, Leigh RJ, Zee DS, Optican L. What clinical disorders tell us about the neural control of saccadic eye movements. Brain 2007; 130:10–35.

Ocular Circulation

Charles E. Riva, Albert Alm & Constantin J. Pournaras

Introduction

The ocular circulation is unique and complex due to the presence of two distinct vascular systems, namely the retinal and uveal systems. The part of this circulation, the one that supplies the fundus of the eye, has the useful property that it can be observed using the ophthalmoscope, an optical instrument introduced by Helmholz in the middle of the 19th century. In the early 1960s, the fundus camera, a device based on the principle of indirect ophthalmoscopy, allowed the recording of the passage of fluorescein through the human retinal vascular system, from which Hickam and Frayser derived the first quantitative measurements of retinal hemodynamics.[1] Technological developments in the field of optics and lasers have since led to a variety of non-invasive techniques, which have permitted the investigation of various parameters pertaining to human ocular hemodynamics and the response of these parameters to a number of physiologic and pharmacologic stimuli. These techniques have provided information on human ocular circulatory physiology and may help in understanding the role of ocular blood flow in the pathogenesis of eye diseases of vascular origin. Yet much of our basic knowledge of ocular blood flow regulation is based on data from animal experiments; thus, in the following, basic principles from animal experiments will be presented together with the human data available.

Anatomy of the ocular circulation

Knowledge of the anatomy and the regulatory mechanisms of the various ocular vascular beds is crucial for understanding the pathophysiologic changes occurring during the evolution of several systemic and local diseases threatening vision. The delivery of metabolic substrates and oxygen to the retina in higher mammals, including humans and other primates, is accomplished by two separate vascular systems, the retinal and the choroidal systems. In lower mammals, such as rabbit and guinea pig, the retinal vessels are present in only a small area of the retina or are lacking. In this case, the retinal metabolism depends almost completely upon the choroidal circulation.[2]

The ocular vessels are all derived from the ophthalmic artery (OA), a branch of the internal carotid artery. The retinal and choroidal vessels differ morphologically and functionally from each other. The OA gives off two to three main ciliary arteries (CAs), i.e. the nasal and the temporal, which supply the corresponding hemispheres of the choroid via branches of the posterior CAs (PCAs)[3,4] and recurrent branches of the anterior ciliary arteries (ACAs). The ACAs arise from the extraocular muscular arterial branching from the OA.[5] The central retinal artery (CRA) may branch from the same segment of the OA as the PCAs or from one of the main PCAs.[6,7] It enters the optic nerve from below, approximately 10–15 mm behind the globe,[8] from where it assumes its central position in the optic nerve up to the optic disk (Fig. 10.1).

Vascular supply of the retina

The retinal circulation is an end-arterial system without anastomoses. The CRA emerges at the optic disk where it divides into two major branches. These in turn divide into arterioles extending outward from the optic disk, each supplying one quadrant of the retina, although multiple branchings of the retinal arterioles towards the peripheral retina may occur. Retinal arteries and veins divide by dichotomous and side-arm branchings. The terminal arterioles branch off at almost right angles from the main vessel. In approximately 25 percent of human eyes, a cilioretinal artery emerging from the temporal margin of the optic disk supplies the macular region,[9] exceptionally feeding the foveal region (Fig. 10.2).

The larger vessels lay in the innermost portion of the retina, close to the inner limiting membrane. Their walls are in close spatial relationship with glial cells, mainly astrocytes (Fig. 10.3).[10] Astrocytes play important roles in constraining retinal vessels to the retina and in maintaining their integrity.[11] At the arterio-venous crossing sites the deeper vessels may indent the retina as far as the outer plexiform layer.[2] The retinal arterioles give rise to a plexus of capillaries (~5 μm in diameter). These capillaries form an interconnecting two-layer network. The first layer is located in the nerve fiber and ganglion cell layer and the second lies deeper, in the inner nuclear layer. In the peripapillary area an additional capillary network lies in the superficial portion of the nerve fiber layer, constituting the radial peripapillary capillaries distributed around the optic disk and along the temporal superior and inferior retinal vessels.[12]

Figure 10.1 Ophthalmic artery (1) branching to the central retinal artery (2) and a main lateral ciliary artery (3). (From Ducasse et al 1986.[529])

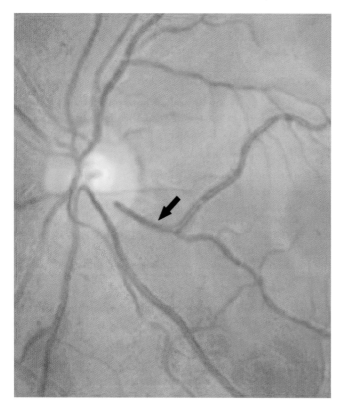

Figure 10.2 Cilioretinal artery branching around the foveal area.

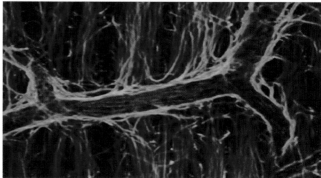

Figure 10.3 Spatial relationship between astrocytes and endothelial cells. Double immunostaining showing GAFP-positive astrocytes (green) in close relation to endothelial cells delimitated by occludin positive interendothelial junctions (red). (From Pournaras et al 2008.[530])

an approximately 1.5 mm wide area is avascular. A capillary free zone also surrounds the arterioles, probably resulting from the local high oxygen tension affecting the capillary remodelling during maturation of the retinal vascular system.[12]

The terminal vessels, namely the precapillary arterioles and the postcapillary venules, are linked through the capillary bed, the venous system presenting a similar arrangement as the arterioles' distribution. The central retinal vein (CRV) leaves the eye through the optic nerve to drain venous blood into the cavernous sinus.

Vascular supply of the choroid

The vasculature of the choroid derives from the OA via branches of the 2–3 nasal and temporal main CAs and the ACAs, which supply the corresponding hemisphere of the choroid.[3,4] The main CAs branch into 10–20 short PCAs, which enter into the globe at the posterior pole and assume a paraoptic and perimacular pattern[13] before branching peripherally in a wheel-shaped arrangement, and two long PCAs. Secondary and tertiary branches of the short PCAs are subsequently divided into the major choroidal arteries.[3,4] Some branches of these PCAs are selectively directed to the macular region vessels, namely the very short CAs.[14]

Paraoptic pattern

Medial and lateral paraoptic short PCAs converge towards the ONH and form an elliptical anastomotic circle, the so-called circle of Haller and Zinn, through the formation of superior and inferior perioptic optic nerve arteriolar anastomoses[13] (Fig. 10.4) The circle of Haller and Zinn is incomplete in 23 percent of cases and complete with narrowed sections in 33 percent of cases.[15] These variations could possibly make this part of the optic nerve head (ONH) circulation particularly vulnerable to ischemic attacks. Paraoptic short PCAs also supply the peripapillary choroid, whereas distal short PCAs supply only the peripheral choroid.[15]

Perimacular pattern

The PCAs follow long, oblique intrascleral courses and travel in the virtual suprachoroidal space, giving off recurrent branches to the macula and to the anterior choroid at the

Towards the periphery, the deep capillary net disappears leaving a single layer of wide-mesh capillaries. Similar anastomotic capillaries connect the perifoveal terminal arterioles with the venules, leaving a capillary free zone of 400–500 μm in diameter. At the extreme retinal periphery

ora serrata. Recurrent branches supply choroidal areas at 3 and 9 o'clock meridians.[16] The ACAs perforate the sclera at the insertion of the tendons of the muscles and pass through the suprachoroidal space to enter the ciliary body. They primarily supply the ciliary body and iris; in addition, the ACAs are divided into recurrent branches that supply the anterior choroid.[17] The choroid is composed of the choriocapillaris layer, the medium vessels layer (Sattler's layer) and the outer layer of large vessels (Haller's layer). In primates, the medium layer contains large arteries measuring 40–90 µm, large veins measuring 20–100 µm and nerves.[18] The choriocapillaris and the medium-sized choroidal vessels lie between the apical retinal pigment epithelium (RPE) of neuroepithelial origin and the outer choroidal pigment of neural crest origin. Two collagenous structures, one on the inside (Bruch's membrane) and one on the outside (subcapillary fibrous tissue), are connected by the intercapillary pillars. Most medium-sized vessels are external to the subcapillary fibrous tissue.[19] Some branches of the short PCAs are selectively directed to the macular region. They have a spiral-shaped configuration, consistent with the vascular pattern of the arterial phase in indocyanine angiography. This pattern differs from that of those short PCAs which are not directed to the macular area, in that it expands in a typical chevron (V-shaped) configuration (Fig. 10.5A).[20]

In vivo studies show that the PCAs and all their branches right down to their terminal arterioles and the choriocapillaries have segmental distribution with no functional anastomoses between them and thus behave as end-arteries.[4,21] The borderline area between the territories of distribution of any two end-arteries is defined as watershed zone, considered as an area of comparatively poor blood flow supply, more vulnerable to hypoxia–ischemia.[22] In exudative age-related macular degeneration (AMD), a stellate pattern of the watershed zone is often observed.[23] Choroidal neovascularization arises within this watershed zone in 88 percent of AMD patients.[23] The relation between the site of choroidal neovascularization and the macular watershed zones suggests that these zones may be vulnerable to choroidal neovascularization induced by hypoxia–ischemia in AMD.

The choriocapillaris are distributed as a dense network of one layer of freely connected capillaries in the peripapillary and submacular area. This pattern changes to a lobule-like arrangement in the posterior pole and to a palm-like organization more peripherally.[24] Each lobule, 0.6–1.0 mm diameter, consists of a capillary meshwork with radial and circumferential arrangement with an arteriole in the middle and a venule at its periphery (Fig. 10.5B).[24,25] Lobules in the equatorial part of the choriocapillaris are larger (200 µm) than those located both at the posterior pole (100 µm), and in the submacular area (30–50 µm).[14] The lobules subdivide the choroid into several functional islands. The choroidal capillaries are connected by a discontinuous row of "zonulae occludens". The part of the choriocapillaris that faces the RPE has large fenestrations, larger and more numerous in the submacular area,[26] covered by a thin membrane with a central thickening.[27,28] Blood is discharged from the lobules of the choriocapillaris by collecting venules that join the afferent veins. At the posterior pole, a venule is located at the periphery of the lobule, stays on the same plane of the lobule, and possibly also drains adjacent lobules.[16] Intervenular channels between choriocapillaries collecting venules and larger veins, direct arteriovenous anastomosis, and interdigitation between the choriocapillaris and venules are present in humans.[16,24] The meshwork of the venous plexus

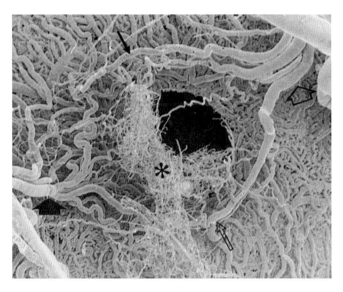

Figure 10.4 Scanning electron microphotograph of circle of Haller and Zinn formed by branches of lateral paraoptic short PCAs (empty arrowhead) and a medial paraoptic short PCA (empty arrowhead) forming a superior (long solid arrow) and inferior anastomosis (long empty arrow). Star: Retrolaminal capillaries plexus. (Reprinted from Olver et al[531] by permission from Macmillan Publishers Ltd, copyright 1990.)

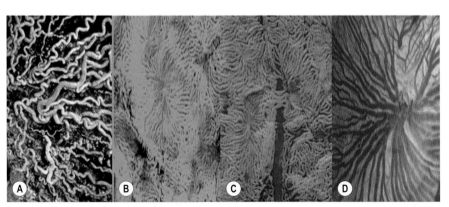

Figure 10.5 (**A**) V-shaped configuration of the posterior sort ciliary arteries. (**B**) Subfoveal choriocapillaris lobules. (**C**) Vertical venous trunk seen through the dense net of the choriocapillaris. (**D**) Net of a vortex vein: efferent veins centripetally towards the ampulla. (From Cerulli et al 2008.[532])

becomes less dense with increasing distance from the macula and the vessels become straighter, losing the tortuous aspect, which is characteristic of the macular region (Fig. 10.5C,D). Vessels of larger lumen form the subcapillaris plexus and eventually flow into the vortex veins.[16] Four to six vortex veins are located 2.5–3 mm posterior to the equator, closer to the vertical meridian than to the horizontal one and drain into the superior and inferior orbital veins.[29] Some drainage also occurs through the anterior ciliary veins of the ciliary body. Segmental venous drainage with watershed areas oriented horizontally through the disk and fovea and vertically through the papillomacular area is observed.[30,31]

Fine structure and innervation of retinal and choroidal vessels

Retinal arteries differ from arteries of the same size in other organs in that they have an unusually well-developed smooth muscle layer and lack an internal elastic lamina (Fig. 10.6). The smooth muscle cells are oriented both circularly and longitudinally, each being surrounded by a basal lamina that contains an increasing amount of collagen toward the adventitia.[32] Retinal arteriolar precapillary annuli are observed in a number of animals at the arterial side-arm branches, but not in humans.[33]

The capillary wall is composed of three distinct elements: endothelial cells, intramural pericytes, and a basement lamina. At their thickest point the endothelial cells present a nucleus bulging into the vessel lumen and express several cytoplasmic processes. Tight junctional complexes are found along the opposing surfaces of adjacent cells.[34] The continuous endothelial cell layer is surrounded by a basal lamina within which there is a discontinuous layer of intramural pericytes in almost a one-to-one ratio with the endothelial cells. The recognition that pericytes are highly contractile cells, coupled with their uniquely high representation in the retinal microvasculature has led to the hypothesis that these cells play an important role in the regulation of retinal blood flow.[35,36,37] However, direct in vivo evidence for this role is still lacking.

Histologic studies and autonomic nerve stimulation experiments reveal a rich supply of autonomic vasoactive nerves to the choroid but not to retinal vessels. Sympathetic nerves derived from the superior cervical sympathetic ganglion innervate the choroidal vascular bed as well as the CRA up to the lamina cribrosa but not further into the retina.[38,39] Nevertheless, along the ONH and within the retina, α- and β-adrenergic receptors[40] and receptors for angiotensin II are present.[41] As sympathetic nerve fibers to other tissues, the choroidal sympathetic nerve fibers are immunoreactive to neuropeptide Y.[42]

In the human choroid there are not only numerous nerve fibers but also choroidal ganglion cells (CGCs).[43,44] Histochemical and immunohistochemical studies performed in whole mount preparations allowed precise classification and quantification of these CGCs. A comparative study of different mammalian eyes demonstrated that larger numbers of CGCs (up to 2000/eye in humans) are present only in eyes of primates with a well-developed fovea centralis, namely humans and higher monkeys.[45] These CGCs stain for neural nitric oxide (NO) synthase (nNOS) and vasoactive intestinal polypeptide (VIP).[45,46] Targets innervated by these CGCs

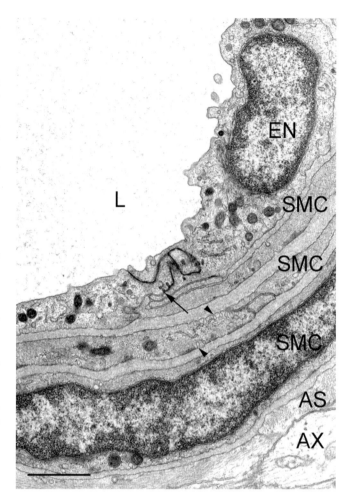

Figure 10.6 Electron microscopy image of the wall of an artery in the inner pig retina. The vessel wall is composed of the endothelium (EN) and three layers of smooth muscle cells (SMC), with numerous caveolae (arrowheads). Note high contrast in intercellular cleft and vesicular pits up to the site where a tight junctional element is located (arrow). L = lumen, AS = astrocytic processes, AX = axon. (From Pournaras et al 2008.[59])

include arteries and non-vascular smooth muscle cells attached to the choroidal elastic network and the walls of the choroidal arteries. Thus, these CGCs seem to be involved in control of choroidal thickness and blood flow. In addition, the postganglionic nerve fibers of the CGCs presumably support the vasodilative function of the facial nerve.[47]

The parasympathetic system provides vasodilating fibers to the choroid through the facial nerve,[48] as confirmed by stimulation experiments in rabbit, cat, and monkey.[49,50] VIP immunoreactive nerve fibers, originating in the pterygopalatine ganglion, have been observed around choroidal blood vessels in several species, including man.[51] Furthermore, the parasympathetic perivascular nerves are immunoreactive for both nNOS and VIP.[45,52] Additional neuropeptides like peptide histidine isoleucine (PHI) and pituitary adenylate cyclase polypeptide (PACAP) could be present in some of the parasympathetic nerve fibers as well.[53]

Vascular supply of the anterior segment

The vascularization of the anterior segment originates from the ACAs and the long PCAs. These PCAs have an intrascleral

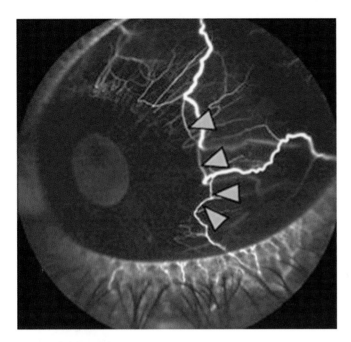

Figure 10.7 ICG angiography indicating the episcleral arteries running forward in the episcleral space assuming a circumferential course near the limbus and forming the episcleral arteriolar circle (arrowheads). (From Pournaras et al 2008.[530])

course beneath the medial and lateral rectus muscles.[54] They run forward to the ciliary body and the iris, where they give rise to their terminal branches. The ACAs have an extraocular course within the rectus muscles.[54] Close to muscle tendons, they exit the muscles and run radially towards the limbus within Tenon's capsule. Close to the limbus they give rise to superficial (episcleral) and deep (scleral) branches.[54,55] The episcleral arteries run forwards in the episcleral space; near the limbus they assume a circumferential course and anastomose each other to form a fragmentary episcleral arterial circle (Fig. 10.7).

Penetrating vessels through the sclera give rise to branches that vascularize the iris (major arterial circle of the iris) and the ciliary body (intramuscular arterial circle). In addition, species-dependent intramuscular arterial circle located within the ciliary body has been described.[56] The penetrating vessels give rise, through recurrent branches, to a communication between the ACAs and PCAs.[17]

The vascular supply of the ciliary processes is complex and species-related.[56] Three territories of supply have been described in humans.[17] The first territory, situated mainly at the broadened base of the anterior edge of the major ciliary process, is irrigated by anterior arterioles leading to a capillary network. The capillaries drain into venules located deep in the ciliary processes with little connection to the other vascular territories. The second territory is also derived from the anterior portion of the major ciliary processes and drains into a broad marginal vein at the inner edge of the ciliary process. The third territory provides blood irrigation to the posterior of the major ciliary processes and the minor ciliary ones. A schematic representation of the vascular supply of the human ciliary body is shown in Figure 11.2B.

The arterial supply of the iris originates mainly from the long PCAs and the ACAs. The long PCAs divide themselves in terminal branches to form the big arterial circle of the iris circumscribing the pupil. This circle receives also the contribution of the ACAs. Thus, the iris is supplied by the posterior and anterior ciliary net.[54,57] Vessels with radial trajectory and a corkscrew arrangement to take care of pupillary movement arise from the iris big arterial circle to reach the rim of the pupil. They divide to form the minor arterial circle of the iris.[58] The capillaries within the iris are difficult to see; they are localized mainly in the pars pupillaris. The venules originate from the pars pupillaris of the iris and run in parallel, but in more depth than the arteries to the root of the iris, increasing their diameter.[58] At this level, they receive also a contribution from the ciliary body and head for the suprachoroidal space before reaching the vortex veins.

Transport through blood–retinal barriers

Optimal cell function requires an appropriate, tightly regulated environment. This regulation is determined by cellular barriers, which separate functional compartments, maintain their homeostasis, and control transport between them. The close inter-relationship of epithelia and vascular endothelium with extracellular structures, namely extracellular matrix and glycocalyx, may modulate the dynamic responsiveness of barrier cells. As reviewed elsewhere,[59] two major pathways control the passage through barriers, namely the transcellular pathway involving vesicles, specific carriers, pumps, and channels and the paracellular pathway through the intercellular cleft.

Transcellular pathway (transcytosis)

The transcellular pathway actively and passively transports water, ions, non-electrolytes, small nutrients, and macromolecules in an energy-dependent manner. Most proteins are transported non-selectively within vesicles, either in their fluid phase or adsorbed at the vesicular membrane (vesiculovacuolar transport).[60] Constitutive transcytosis (the process by which macromolecules are transported across the interior of a cell) of albumin across the vascular endothelium is particular in that each transcytotic routing, whether receptor-mediated, by adsorption or as bulk fluid, is implicated in regulating the transvascular oncotic pressure gradient of albumin.[61,62] Fenestrations of the endothelial cell membrane are characterized by high permeability.[63] They are found in choroid plexi of the brain, in the choroid and in the neovessels formed during the angiogenic processes (Fig. 10.8).

Paracellular pathway

Tight junctions confer firm intercellular adhesion and regulate paracellular permeability through the intercellular cleft. Indeed, barrier properties depend on the specific molecular architecture of the tight junctions. Tight junctions are polymeric adhesion complexes with a transmembrane component composed of occludin (Fig. 10.9), claudins and junction adhesion molecules, linked to a cytoplasmic plaque containing, among other components, zonula occludens protein and cingulin.

The plaque itself is anchored to the actin cytoskeleton and to various signaling molecules that participate in the control of cell proliferation and differentiation.[64–67] In general, water,

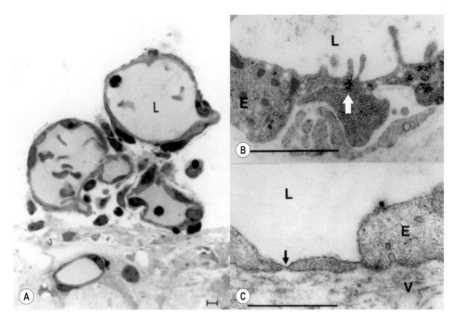

Figure 10.8 Neovessels at the vitreoretinal interface in an experimentally induced vasoproliferative microangiopathy of the minipig. (**A**) Semi-thin section showing new vessels lying over an ischemic retinal area. The vitreoretinal interface is severely altered. (**B**) Electron micrograph of a non-fenestrated endothelium of a neovessel. The intercellular cleft is connected by an adherens junction (white arrow). (**C**) Electron micrograph of a fenestration (black arrow) in a neovascular endothelial cell. V = vitreous; L = vessel lumen; E = endothelial cell. Bars: 1 μm. (From Pournaras et al 2008.[530])

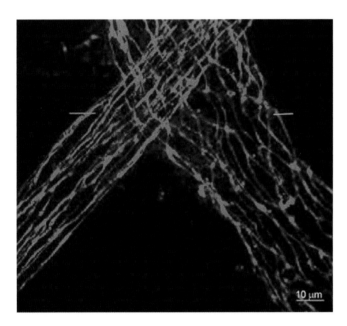

Figure 10.9 Immunolocalization of the tight junctional protein occludin in a retinal artery (left) and retinal vein (right). Confocal microscopy of flatmounted minipig retina. Regular reticulate pattern outlining the endothelial cells. Bar: 10 μm. (From Rungger-Brändle & Leuenberger 2008.[533])

ions and small non-charged solutes employ the paracellular pathway by passive diffusion and with low selectivity along electrochemical and osmotic gradients that are built up by the activity of transcellular transporters or, by external gradients of solutes.[68] The characteristics of both paracellular and transcellular pathways determine the preference for small solutes for either pathway and, therefore, are important parameters for drug absorption. Drugs with high lipid solubility readily cross the lipid cell membrane bilayer, but could still be poorly absorbed into the eye due to efflux pumps in the ocular barriers (see Chapter 17).

Extracellular structures

Glycocalyx

The apical and luminal surfaces of epithelial and endothelial cells are endowed with macromolecules that are bound to the plasma membrane and build up the glycocalyx.[69] This coat, composed of glycoproteins, proteoglycans, and glycosaminoglycans, is polyanionic and thus negatively charged. Molecules like fibrinogen and albumin, enzymes, cytokines and growth factors, associate with the glycocalyx thanks to their cationic charge. By these particular electrostatic features, the glycocalyx acts as a charge-selective barrier that determines the accessibility of certain molecules to the plasma membrane.[62]

A major function of the glycocalyx of the vascular endothelium is to sense mechanical forces generated by the blood flow and modulate the shear stress-induced release of NO.[70] Long-standing degradation of distinct components of the glycocalyx contributes to increased vulnerability of the endothelial cells, barrier function, block mechanotransduction and lead to inhibition of NO production that eventually abolishes vasodilation.

Extracellular matrix

This matrix (ECM) consists of higher-order assemblies of numerous components that are secreted in a cell-type-specific manner.[71,72] The major barrier tissues, epithelium and vascular endothelium, both synthesize a specialized ECM scaffold, the basal lamina, at their basal and albuminal side, respectively. Basal laminas serve as cell attachment sites by means of a variety of transmembrane receptors[73] that are able to transmit signals between cell and ECM in either direction. The best known among these receptors are the integrins, the specificity of which is determined by the molecular characteristics of the underlying ECM. Disruption of integrin-matrix binding causes the detachment of the cells from their substrate and also causes an increase in the permeability of tight junctions.[74,75] The interactions of cell to cell-matrix are of great importance to insure intercellular permeability in microvascular systems. Physical, oxidative stress, and inflammatory events may affect the permeability of the intercellular junctions.[76,77]

Inner blood–retinal barrier

The endothelium of intraretinal blood vessels is considered as the main component of the inner blood–retinal barrier (BRB) resembling the blood–brain barrier in that both separate blood from neural parenchyma. The presence of a complex network of tight junctions, the absence of fenestrations and a relative paucity of caveolae make up the tightness of the iBRB. Numerous transport systems account for the selectivity of the barrier,[78–81] such as the transport system for glucose across the retinal capillary endothelial cells of the inner BRB, which is mediated by the sodium-independent glucose transporter GLUT1.[82]

Pericytes

Both pericytes and smooth muscle cells provide structural support to the vasculature. The increased density of pericytes in the retinal microvasculature as compared to other organs[83,84] has led to the presumption that pericytes play a role in the regulation of retinal perfusion by contraction and relaxation.[37,85] Circumstantial evidence for contractility of pericytes is the expression of a number of contractile proteins, in particular α-smooth muscle actin, desmin, and non-muscle myosin.[86–88] Moreover, pericytes express receptors for vasoactive substances, enabling them to respond by contraction or relaxation. Recent work suggests that pericytes, rather than endothelial cells, control capillary diameter, because diameter changes do not occur in pericyte-free regions. Thus, pericytes would play a major role in redirecting blood flow at the capillary level.[89] However, direct in vivo evidence for this role in the retina is still lacking.

Pericytes modulate barrier permeability since they express a high number of caveolae that appear to be involved in transcellular transport.[90] Moreover, pericytes and smooth muscle cells participate in the permeability properties of the basal lamina by secretion of matrix components.[86–88] Deficient contractility of pericytes and smooth muscle cells may thus have a negative impact on the regulation of blood flow that, in turn, may impair pericyte–endothelial interactions, critical for the maintenance of the BRB.

Glial cells

In the retina, both astrocytes (Fig. 10.3) and Müller cells participate in vessel ensheathment, whereby astrocytes are confined to the optic nerve fiber and the ganglion cell layer.[91] Glial cells play an important role in vessel integrity and barrier properties, both via direct contact[92] and through release of humoral factors.[93] Glial cell line-derived neurotrophic factor (GDNF) has been recognized to enhance barrier tightness,[94] whereas transforming growth factor-β (TGF-β) appears to decrease it.[95] Several other growth factors and cytokines, such as tumor necrosis factor-α (TNF-α), interleukin-6 (IL-6), and vascular endothelial growth factor (VEGF) are produced by retinal glial cells[96] and affect the tightness of the iBRB. VEGF directly downregulates tight junctional proteins, evoking a decrease in transendothelial resistance[97] (Box 10.1).

> **Box 10.1 Anti-VEGF treatment of age-related macular degeneration (AMD)**
>
> VEGF is the predominant vasoproliferative factor involved in the development of retinal or subretinal neovascularization, occurring either during the evolution of retinal ischemic microangiopathies, or exudative neovascular AMD. Ocular neovascularization regresses following intravitreal injections of monoclonal antibodies inhibiting isoforms of VEGF-A. Anti-VEGF treatment is actually the predominant approach for the treatment of neovascular AMD.

Outer blood–retinal barrier

The outer blood–retina barrier (BRB) is composed of three structural entities, the fenestrated endothelium of the choriocapillaris, Bruch's membrane, and the retinal pigment epithelium (RPE).

Fenestrated endothelium

The fenestrations of the choriocapillaris are covered by a thin membrane with a central thickening.[27,28] They have high permeability,[98] similar to that of an isoporous membrane with a pore diameter of 4 Å,[99] which probably accounts for the maintenance of an adequate concentration of glucose and other nutrients at the RPE level.[27]

Retinal pigment epithelium

The RPE cells are equipped with an elaborate transcellular pathway system and an apico-lateral seal formed by intercellular junctions, the adherens junction and the tight and gap junctions, as well as, in some species, desmosomes.[100] The RPE is considered as a relatively tight epithelium with a 10 times higher paracellular than transcellular resistance (Fig. 10.10).[101,102]

Nutrients and vitamin A are transported in basolateral to apical direction from the blood to the photoreceptors. In the opposite direction, transcellular transport from the subretinal space towards the choriocapillary bed is required for removal of metabolites, water, and ions. The transepithelial ion transport is linked to the transport of lactic acid, the major metabolic end product of neuronal function.

Bruch's membrane

Using electron microscopy, five distinct layers of Bruch's membrane have been defined: the two basal laminas of the RPE and the endothelium of the choriocapillaris, the inner and outer collagenous layers, and a central discontinuous elastic layer.[103,104] Bruch's membrane provides tensile strength and, thanks to the presence of proteoglycans, constitutes a reservoir of growth factors. Its overall negative charge is due to the high content of glycoconjugates and it is responsible for charge-selective restriction to the passage of ions and solutes.

Blood–aqueous barriers

Secretion of the aqueous humor and its transport towards the posterior chamber is controlled by the blood–aqueous barrier (BAB). These barriers, which possess active transport

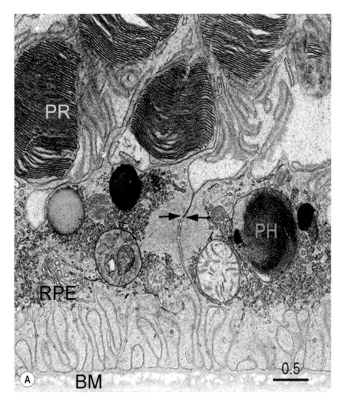

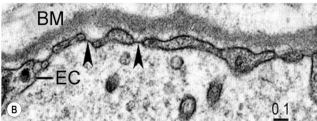

Figure 10.10 (**A**) Electronic micrography of the outer blood–retinal barrier. The retinal pigment epithelium (RPE) encloses the extremities of the external segments of the photoreceptors (PR) by means of long apical digitations, which are indicated by the two opposed arrows. At the basal pole, the plasma membrane forms numerous invaginations, the extremities of which rest on the basal lamina and form a part of Bruch's membrane (BM). PH = phagosome resulting from the internalization and lysosomal digestion of external segments. (Bar: 0.5 μm). (**B**) Detail of the fenestrated endothelium of the choriocapillaris. The fenestrations of the endothelium cell (EC) are indicated by arrowheads. The contrast of the membranes and basal lamina was increased by tannic acid. (Bar: 0.1 μm). (From Rungger-Brändle & Leuenberger 2008.[533])

mechanisms, include the endothelium of the iris capillaries, the iris posterior epithelium, and the non-pigmented posterior epithelium of the iris. The passive permeability of the BAB depends on the ionic concentration gradients.

Composed of iris capillaries and pigmentary epithelium, the anterior BAB allows the transcellular transport by means of vesicles. The paracellular transport is controlled by the extension of the tight junctions. Associated with the ciliary and retinal pigment epithelium, the iris pigment epithelium seems to constitute an obstacle to the passage of type T activated lymphocytes.[105] The anterior surface of the iris, formed by only one layer of fibroblasts, does not constitute a barrier.

It allows free access of the aqueous humor to the stroma and the iris muscles, thus causing the rapid resorption of drugs present in the aqueous humor.

The posterior BAB is formed by tight junctions, which are present on the lateral pole of the cells of the non-pigmented ciliary epithelium (Fig. 10.11). These tight junctions are permeable to small, non-ionic molecules, such as sucrose.[106-108] The capillary endothelium of the ciliary stroma possesses fenestrations that make its permeability relatively high. In contrast, the capillaries of the ciliary muscle appear relatively tight, similar to those of the iris.

Techniques for measuring ocular blood flow

Decades of technical developments have led to a number of methods to obtain quantitative information on ocular blood flow (BF) in animals and humans.

Techniques used in experimental animals

These techniques have been valuable for quantitative determinations of BF in experimental animals. They include direct measurements of BF through cannulation of uveal veins in rabbits and cats,[109] and radioactively labeled microspheres.[110] Calorimetry,[111] or heated thermocouple,[112] radioactive krypton desaturation,[113] hydrogen clearance,[114] iodoantipyrine I It should be I 125, 5,[115] laser speckle phenomenon-/flowgraphy,[116] targeted dye delivery,[117] leukocyte dynamics by fluography,[118] and invasive laser Doppler flowmetry[119] have been used to determine changes in flow in the various ocular tissues.

Intracardiac injections of radioactively labeled, non-labeled colored or fluorescent microspheres permit determination of regional BF to the different tissue beds within the eye by analyzing the distribution of the microspheres.[120-124] Ideally, the relative quantity of particles recovered in each organ, compared to the total number of injected particles, equals the fraction of the cardiac output to each organ. The precision of the various microsphere techniques depends on several factors, including the number of particles trapped in the tissue sample. The accuracy of BF measurements within small tissue samples, such as the retina, the iris, or the optic nerve, can be improved if the number of samples is increased.[125]

With direct cannulation of the uveal veins, the BF within the uveal tract has been measured in rabbits[109] and cats.[112] In cats, there is a large intrascleral venous plexus that permits sampling of venous blood from either the anterior uvea or the posterior choroid. This plexus has been used to determine arteriovenous differences of oxygen[126] and glucose.[98] A similar plexus exists in the retrobular space of the rabbit eye. Retinal arteriovenous differences can be studied in a variety of animals, including the pig, in which the retinal veins form a ring-shaped plexus around the optic nerve within the retrobulbar space.[78]

Studies of the oxygen partial pressure profile within the retina with microelectrodes have provided information on the relative importance of the choroidal and retinal circulation in providing oxygen to the retina.[127,128] Glucose consumption in vivo can be studied by determining the tissue

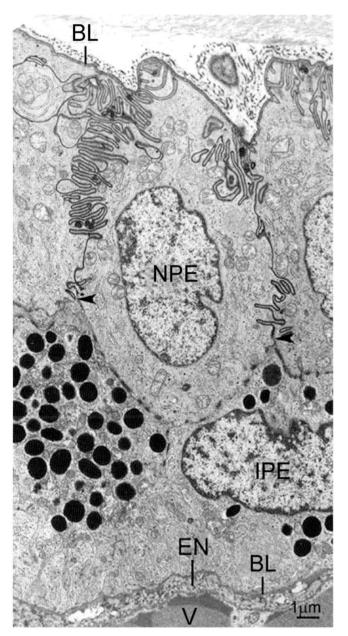

Figure 10.11 Pars plicata of the ciliary body of the rabbit eye. Electronic micrograph showing the internal retinal pigment epithelium (IPE) and the external, non-pigmented epithelium (NPE). The limitans lamina separating this layer of the vitreous body is a basal lamina (BL), which is tightly associated with the plasma membrane of the non-pigmented cells. Collagen fibers (shown as transversal or sometimes oblique cut) are present between the basal lamina of the pigmented epithelium of the perivascular space. Deep interdigitations are visible on the lateral faces of the non-pigmented cells, which adhere to the pigmented cells through numerous junctional complexes (arrowheads) including tight junctions. The contrast of the membranes and intercellular spaces is accentuated by the addition of tannic acid during tissue fixation. EN = choroidal endothelium; V = endovascular space. (Bar: 1 μm) (From Rungger-Brändle & Leuenberger 2008.[533])

uptake of labeled 2-deoxy-D-glucose. Although it is not a direct measure of BF, this technique gives an estimate of alterations in metabolic demands and has proved very useful for studying the effects of increased intraocular pressure (IOP) and light on the retina and the optic nerve.[129,130]

Non-invasive techniques used in physiological and clinical research

A number of techniques have been developed to obtain quantitative information on the physiology, pharmacology, and pathology of the blood circulation in the different ocular vascular beds.

The diameter (D) of retinal vessels has been measured in the past from magnified fundus photographs using a caliper or by scanning across the vessels.[131] In recent years, the Retinal Vessel Analyzer (RVA)[132] has simplified markedly this measurement, allowing also the quasi-continuous recording of D changes evoked by various physiological maneuvers (dynamic measurements) (Fig. 10.12).

The mean circulation or transit time (MCT or MTT) of fluorescein injected intravenously through a retinal segment (Fig. 10.13A) has been assessed from digitized fluorescein angiograms. The gray scale level is used as an estimate of the relative dye concentration (dilution curves) in the artery feeding and the vein draining the segment and the MCT/MTT has been determined from these dilution curves (Fig. 10.13B). Specific conditions related to the retinal segment, the properties of the dye, the mode of its injection, the method of recording the dilution curves and the correction for dye recirculation must be fulfilled to obtain precise data on MCT/MTT.[133–137] In the application of this technique to the eye, some of these conditions cannot be strictly satisfied, particularly in cases of retinal vascular pathology (for instance, in proliferative retinopathy, diffusion of fluorescein from the retinal vessels may distort the dilution curves), so that the measured MCT/MTT must be regarded as an approximation of the true MCT/MTT. Other parameters have been derived from the time course of the fluorescence intensity: the mean velocity of the dye (MVD)[138] in a retinal artery, the arterio-venous passage time (AVP),[139] which is the time difference between the first appearance of the dye at a reference point at the temporal retinal artery and that at the adjacent vein. Contrary to the MCT, which gives more weight to the peripheral retinal circulation, the AVP represents mainly the passage time of dye through the shortest segments close to the papilla and thus does not reflect the flow of fluorescein through the periphery of the retina. A study in monkeys, undertaken to determine the correlations between MCT and AVP and retinal BF, the latter measured by labeled microspheres, showed that these correlations were not statistically significant ($p > 0.05$),[140] even in the case of calculation of the MCT using the more sophisticated impulse response technique.[137] Clearly, MCT and AVP data should not be interpreted in terms of BF.

For studies on the choroidal circulation, indocyanine green (ICG) is a more suitable dye.[141] ICG has two significant advantages over fluorescein for choroidal angiography: the fluorescence of ICG is in the near infrared and, therefore, not blocked by the pigment epithelium and ICG is almost completely bound to proteins, which means that it does not easily pass the walls of the choriocapillaris. Single choroidal vessels can be observed even late in the angiogram. The need for a special camera and frequent difficulty in interpretation of ICG angiograms have somewhat limited widespread acceptance. An obvious use for ICG is to examine subretinal neovascular membranes, where it is a valuable complement to, but not a substitute for, fluorescein angiography.[142]

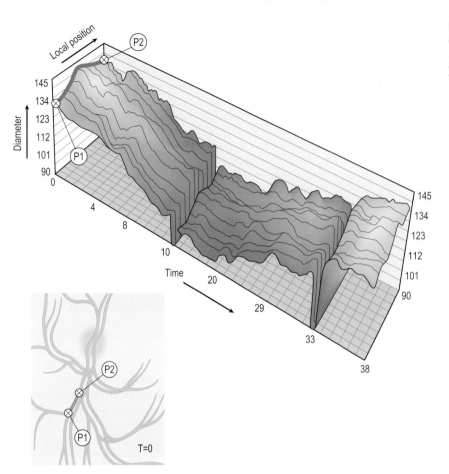

Figure 10.12 Diameter, D, of a retinal vessel along a segment P1–P2 and variation of D with time measured using the Retinal Vessel Analyzer (RVA). At time t = 0, the subject was given 100 percent O_2 to breathe. (From Seifert & Vilser 2002.[534])

Using recently developed techniques allowing simultaneous optical coherence tomography and ICG fluorescence[143] may offer a new approach to clinical studies of choroidal blood flow.

Bidirectional laser Doppler velocimetry (BLDV) allows the measurement of *absolute* blood velocity. BLDV is based on the Doppler effect (Fig. 10.14). Retinal BF (in μl/min) in the main retinal vessels is calculated from the centerline velocity (V_{max}) of the red blood cells (RBCs) combined with D measurements of these vessels and calculating the instantaneous

$$BF_{mean} = \pi \frac{D^2}{4} V_{mean}$$

V_{mean} represents an average over the vessel cross-section. The average RBC velocity during the heart cycle is obtained by integrating V_{mean} over this cycle. For a parabolic velocity profile, $V_{mean} = V_{max}/2$. The assumption of this type of profile is justified when measurements are performed at straight portions of a vessel, a few vessel diameters away from an arterial branching or a venous junction (Fig. 10.15),[144-146] as recently confirmed by Doppler Optical Coherence Tomography.[147,148] In the human and primate monkey eyes, the velocity in the main retinal arteries shows a strong pulsatile component and V_{max} increases linearly with D, as found in most vascular beds.[149,150]

Laser Doppler flowmetry (LDF) allows the measurement of the change in flux of RBCs in the superficial layer of the

ONH, the subfoveal choroid, and the iris. Calculation of this flux is based on the model of Bonner and Nossal (Fig. 10.16A).[151] The change in flux is proportional to the change in BF if this change is not accompanied by a change in hematocrit. LDF can operate in a dynamic or scanning mode. The dynamic mode, which records the flux in a continuous way at discrete sites of the capillary bed, is particularly appropriate for recording changes of flux over time.[152] The scanning mode provides a two-dimensional image of the flux in the capillaries of the optic disk and peri-papillary retina, as well as an intensity image of the retinal vessels perfused (Fig. 10.16B).[153]

The temporal variations of laser speckle resulting from the interference of laser waves scattered by the tissue can be used to determine the velocity of the RBCs in the ONH, retinal and choroidal circulations.[154,155] This laser speckle flowgraphy and the LDF approaches are different ways of looking at the same phenomenon.[156] Both techniques measure at a single point in the tissue. In both cases, adding scanning provides a map of the spatial velocity and flux.[157,158] One important point when applying these laser-based techniques is that the measured flux depends on the scattering and optical absorption properties of the tissue. Therefore, direct comparison between flux values from different eyes may not be valid due to variations in the scattering properties resulting from differences in tissue structure and composition. Similarly for a valid comparison between flux values obtained at different times in the same eye by dynamic and scanning

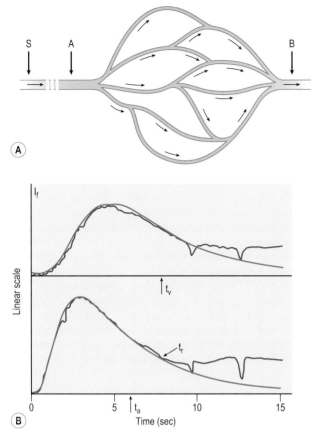

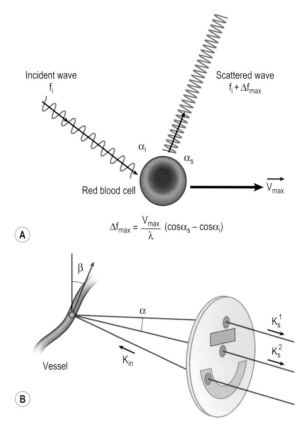

$$\Delta f_{max} = \frac{V_{max}}{\lambda} (\cos\alpha_s - \cos\alpha_i)$$

Figure 10.13 (**A**) Diagram of a retinal vascular segment extending from artery to a vein with paths of different lengths. S is the point of injection of the fluorescein dye, A the site of measurement of the fluorescence along the artery (inflow), and B the site of measurement of the fluorescence along the vein (outflow). (**B**) Fluorescein dilution curves obtained from an artery and vein of a retinal vascular segment in a human eye. t_a and t_v are the mean transit times of fluorescein measured from the first appearance of the dye in the artery. $t_a - t_v = MCT$, the mean segmental circulation time of fluorescein. I_f = fluorescence intensity. The smooth curves are the best log-normal fits to the recordings. These recordings were obtained using a two-point fluorophotometer.[535] (From Riva et al 1978.[136])

Figure 10.14 (**A**) The Doppler effect. The frequency of laser light scattered by a red blood cell (speed V_{max}) is shifted by an amount Δf_{max} compared to the frequency of the light incident (f_i) on the red blood cell. α_i and α_s are the angles between V_{max} and the directions of the incident and scattered light, respectively. λ is the wavelength of the incident light. (With permission from Riva & Petrig 2003.[159]) (**B**) Principle of bidirectional LDV (BLDV). Laser light scattered from blood in a retinal vessel is detected along two directions K_s^1 and K_s^2 (angle α between them). β is the angle between the plane of the vessel at the site of measurement and the direction of the blood velocity. V_{max}, the centerline velocity of the RBCs is determined from the maximum shifts in the Doppler shift power spectrum obtained in each scattering direction and the geometrical scattering parameters. The measurement is independent from the direction of the incident light defined by K_{in}.[536,537] (From Riva et al 1981.[537])

LDF and laser speckle flowgraphy, the tissue must be assumed to maintain the same scattering properties over time.[159]

The velocity, number and velocity pulsatility of leukocytes moving in the retinal capillaries of the macular region has been quantified using the blue field simulation technique, which is based on the entoptic perception of one's own leukocytes.[160] Visual acuity must be better than 20/50 for a reliable measurement.[161] The ability of a subject to do the blue field simulation test can be assessed by having subjects match the speed and number of two simulated leukocyte's motions displayed on computer screens. Blue field data have been confirmed by the objective SLO-adaptive optics imaging technique.[162]

Color Doppler imaging combines B-scan imaging of tissue structure, color representation of blood velocity based on ultrasound Doppler-shifted frequencies, and pulse-Doppler measurements.[163,164] At the level of the OA, CRA, and PCAs, the maximum and minimum values of the velocity are identified to determine the peak systolic (PSV)

and end diastolic (EDV) values of this velocity from which a resistivity index (RI) is obtained:

$$RI = \frac{(PSV - EDV)}{PSV}$$

RI is often considered as a measure of the downstream vascular resistance. However, experimental data show that RI in CRA is not a satisfactory measure of this resistance.[165] Studies on cerebral vessels have not confirmed a useful correlation between RI and cerebrovascular resistance.[166]

Pneumatonometric measurement of ocular pulse amplitude estimates the pulsatile component of choroidal BF (POBF).[167,168] Since the ratio of pulsatile/non-pulsatile ocular BF cannot be assumed to be constant, especially in the event of changes in systemic blood pressure, pulsatile BF is not a good estimate of total ocular BF. Another technique that makes use of the pulsatile nature of the IOP is the interferometric measurement of the IOP-induced variation of the distance between cornea and retina to

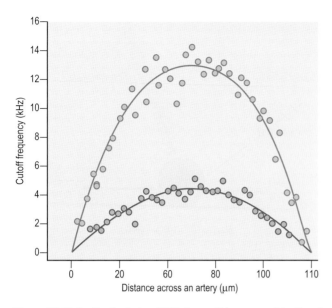

Figure 10.15 Profile of velocity of RBCs (∞ cutoff frequency of the Doppler power spectrum) in a main retinal vessel of a normal eye during the peak systolic and end diastolic phases of the cardiac cycle. The continuous lines are the best fits based on function of the type:

$$V(d) = V_{max} \left[1 - \frac{|d - d_o|^K}{R_i} \right]$$

where V(d) represents the velocity at a distance d from the internal wall. d_o is the position of the center of the vessel. R_i is the internal radius of the vessel. V_{max} is the centerline (maximum) velocity. K = 2.38 in the systolic phase and 1.94 in the diastolic phase. (From Logean et al 2003.[144])

estimate changes in choroidal blood volume during the cardiac cycle.[169]

The information given here on ocular BF techniques is far from exhaustive. Additional details can be found in a number of excellent reviews.[164,170–172]

Ocular circulatory physiology

General hemodynamic considerations

BF through a blood vessel depends upon the perfusion pressure (PP), the pressure that drives blood through the vessel, and the resistance (R) generated by the vessels. For an incompressible uniform viscous liquid (dynamic viscosity η) flowing through a cylindrical tube (length L) with radius (r), BF is given by the Hagen–Poiseuille law: BF = PP/R, where $R = \eta L / 2\pi r^4$. Many factors make it difficult to directly apply this law to a microvascular bed. These include the η-dependence on local hematocrit, the changes in the velocity profile of the RBCs and shear rate at branchings and junctions and others. Another approach at characterizing BF through a system of vessel is based on Murray's law,[173] which says that through each vessel of a circulatory system with optimal design (blood flowing with minimal loss of energy) $BF = k(r^3 / \sqrt{\eta})$. The constant k depends upon the lengths and the radius of the vessels.[174]

The mean ocular PP driving blood through the eye is the mean blood pressure in the ophthalmic artery (OA) minus

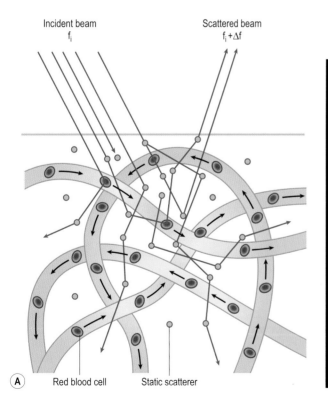

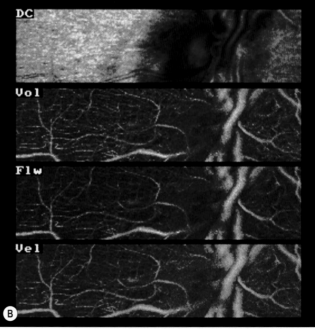

Figure 10.16 (A) Tissue scattering model of Bonner & Nossal[151] provides the basis for Laser Doppler Flowmetry. The incident laser beam is scattered by the moving RBCs and non-moving tissue components. A spectrum of Doppler shifts, Δf, is detected in the scattered light reaching the detector due to the multitude of RBC velocities and scattering directions before the laser light reaches the detector. (Adapted with permission from Riva & Petrig 1997.[159]) **(B)** Intensity and perfusion images of the optic nerve and peripapillary temporal retina obtained by laser scanning Doppler flowmetry. Vol, Flw and Vel and DC are the two-dimensional blood volume, flow, velocity and intensity maps.[153] (From Zinser 1999.[158])

the pressure in the veins leaving the eye. The venous pressure is close to the IOP.[175] With the subject in the sitting or standing position, mean ocular PP is about 2/3 of the mean brachial artery blood pressure (ABP), i.e.

$$2/3\,[ABP_{diast} + 1/3\,(ABP_{syst} - ABP_{diast})] - IOP$$

The factor 2/3 stands for the drop in pressure between the heart and the OA. ABP_{diast} and ABP_{syst} are the brachial ABP during diastole and systole, respectively. It is understood that this expression for PP is based on a group average and therefore provides only an approximate value for a single individual. The value of η has been experimentally related to hematocrit at defined shear rates. It diminishes with increasing shear rate to become almost constant.[176] An increase in viscosity (e.g. in cases of hyperglobulinemia, high hematocrit, leukemia, sickle cell anemia) substantially alters retinal BF, which may induce a stasis in the veins and ultimately to their occlusion.[177]

The main resistance to BF is located in the arterioles with half of the resistance in vessels with a radius of 10–25 μm. As R is proportional to $1/r^4$ even a small change in r will have a considerable effect on R. The vessel radius/diameter is modulated by the interaction of multiple systemic and local control mechanisms affecting the tone of the smooth muscle cells and perhaps the pericytes.

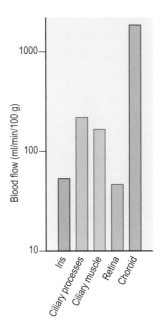

Figure 10.17 BF through tissues of monkey eye (values for BF through ocular tissues are from Alm et al 1973[198]). (From Alm 1992.[178])

Ocular hemodynamic data under basal physiological conditions

Retina

Retinal BF represents only about 4 percent of total ocular BF.[178] In primates, total retinal BF obtained by the microsphere technique was reported to be between 25–50 ml/min/100g.[110,179,180] In humans, recent measurements of total retinal BF (40.8–52.9 μl/min) using Fourier-Domain OCT[181] are in the range of values obtained by BLDV.[182] Retinal BF calculated from BLDV is larger in the temporal human retina than in the nasal region, supposedly due to the larger size (by 20–25 percent) and higher metabolic rate of the former. Retinal BF in the superior and inferior hemispheres are similar. This pertains also to BF in the superior and inferior parts of the macular region.[183]

Choroid

In animals, choroidal BF is higher than in most tissues with estimates ranging from 500 to 2000 ml/min/100 g tissue (Fig. 10.17).[110,112,114,126,184,185] There is no technique allowing reasonably precise measurements of choroidal BF in humans. The anatomical organization and dense vascularization of the choroid, as well as this high choroidal BF are very important. These properties optimize the partial pressure and concentration gradients for efficient diffusive exchange across the relatively long distance between the choroid and the retina for efficient delivery of oxygen and nutrients and the removal of carbon dioxide and metabolic waste.[98,186–189] Thus about 65 percent of the oxygen and 75 percent of the glucose consumed by the monkey retina is delivered by the choroidal vessels.[121,121a] Furthermore, in both primates and non-primates, the foveola is devoid of retinal vessels, presumably to enhance visual acuity and therefore is nourished solely by the choroid.[185,190]

A possible function of the choroid is that of thermoregulation.[180,191,192] In this case, the high BF would tend to maintain ocular temperature at or near body temperature during environmental cooling or heating, and also help protect the eye from thermal damage even under extreme conditions of temperatures and brightness. It also would prevent damaging increments in tissue temperature when light is focused on the fovea.[188,193] In support of this role, scleral temperature falls several degrees when choroidal BF is reduced significantly.[191]

Ciliary circulation

Present knowledge of ciliary hemodynamics is limited due mostly to the inaccessible location and complexity of the ciliary vascular bed.[189] Direct measurements of ciliary arterial pressure have not been made, but a reasonable estimate for this pressure just outside the eye is 67 mmHg for humans in an upright position with an arterial blood pressure (ABP) of 100 mmHg. Episcleral venous pressure is approximately 9 mmHg in humans,[194,195] which is a reasonable approximation for vortex vein pressure. However, the venous pressure inside the eye is determined by the IOP. In animals, it is 1–2 mmHg higher than the IOP.[175,196,197] Capillary pressure is similarly IOP-dependent and approximately 8 mmHg higher than the IOP in the rabbit choriocapillaris.[197] Assuming that ciliary hemodynamics is similar in humans, the ciliary capillary and venous pressures should be approximately 25 and 17 mmHg, respectively, at a normal IOP of 15 mmHg.

Measurements by labeled microspheres in anesthetized monkeys indicate a ciliary BF of 81 μl/min.[198] Plasma clearance of ascorbate provides a rough estimate of ciliary plasma flow of 73 μl/min, or a ciliary BF of 133 μl/min in the human eye, assuming a normal hematocrit.[199,200]

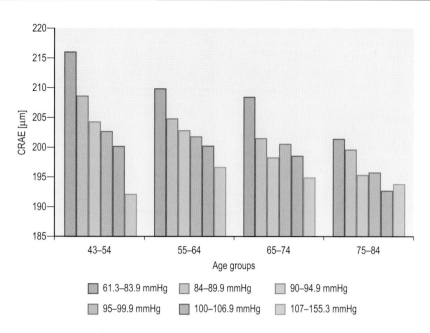

Figure 10.18 Central retinal artery equivalent (CRAE) by mean of ABP in persons of different age groups. (From Wong et al 2003.[209] Reproduced with permission from Association for Research in Vision and Ophthalmology.)

Vasomotion

Direct observation of the retinal and choroidal vasculature[113] reveals no precapillary sphincters and BF appeared to be uninterrupted in all capillaries. In the retina of young cats some spontaneous contractions and dilatations in the small arterioles and precapillary sphincters have been observed, but there was much less vasomotion in the retinal vessels of adult cats.[201] Temporal fluctuations of ONH and subfoveal choroidal BF, with frequencies < 10 cycles/min and distinct peaks at the breathing and heart beat frequencies, have been measured in the eyes of cats,[202,203] minipigs,[204] and humans.[205,206]

Effects of age on ocular blood flow

D of the main retinal arterioles and venules depends on age and blood pressure (Fig. 10.18). It is narrower in older than in younger people, independent of ABP and other factors, the decrease in D amounting to about 2 μm for each decade of increased age. Furthermore, independent of age and other risk factors, arteriolar D is narrower in subjects with higher ABP.[207–209]

Studies of the effect of aging on blood velocity in the OA and CRA in women and men have failed to reveal uniform findings. However, clinical studies of the microcirculation of the retina and the ONH indicate that aging leads to a decrease in blood volume,[210] RBC velocity,[210–212] and velocity and number of leukocytes in the macular area,[213] due presumably to the age-related morphologic decreases of retinal cells and nerve fibers.[214,215] An age-dependent decrease in subfoveal choroidal BF has also been reported,[216] which is mainly due to a decrease in subfoveal blood volume reflecting, presumably, a reduction in the diameter and number of choriocapillaries in the subfoveal region.[217,218]

Regulation of ocular BF

The regulation of ocular BF in the human eye is a complex process due to the presence of two vascular systems which differ anatomically and physiologically: the retinal vessels, which supply the neural region of the retina and the prelaminar portion of the ONH, and the uveal blood vessels, which supply the rest of the eye. Contrary to the autonomically innervated extraocular and choroidal vessels, the retinal and ONH vessels have no neural innervation so that total reliance for matching metabolism must be placed on local vascular control mechanisms.[39]

The mechanisms controlling BF in the different ocular vascular beds are of different types: systemic, local, neural, endothelial, endocrine, and paracrine. Not all operate in the various beds. Since the retinal and choroidal arterioles do not possess sphincters,[33] BF in these tissues is only a function of the muscular tonus of the arterioles and, possibly, the state of contraction of the pericytes, although the latter still remains to be demonstrated. Vessel tonus is modulated by the interaction of multiple control mechanisms: myogenic, metabolic, neurogenic, and humoral, which are mediated by the release of vasoactive molecules by the vascular endothelium or by the glial cells surrounding the vessels.[219,220]

Autoregulation of ocular blood flow

Autoregulation of BF in a tissue is the intrinsic ability of this tissue to maintain BF relatively constant despite variations in the PP.[221] Investigation of this local process, which requires isolation of tissue from neural and hormonal influences, can be ideally carried out in the retina, a tissue that lacks sympathetic innervations and is largely uninfluenced by circulating hormones due to the presence of a tight blood–retinal barrier.[222]

Retina and ONH

A moderate reduction in ocular PP has only a minor or no effect on retinal and ONH BF in experimental animals[110,115,121,122,223,224] In humans such a reduction also induces an autoregulatory response, as evidenced by the resulting increase in D of arteries[225] and the maintenance of constant retinal and ONH BF if PP is not reduced by more than 50 percent.[226–228] An autoregulatory response is also

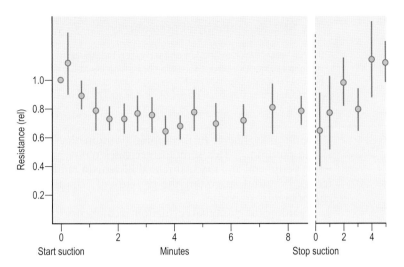

evoked by an increase in PP above normal induced by a decrease in the IOP below normal.[227,229,230] The autoregulatory response of the retinal vascular resistance and BF to these rapid changes in PP is shown in Figure 10.19. Thanks to the autoregulatory capability of the retina and anterior part of the ONH, the oxygen tension (PO_2) in the inner half of the retina and anterior part of the ONH remains largely unaffected by moderate increases in IOP (decreases in PP).[231-233]

Choroid

In animals, measurements of choroidal BF by various techniques have demonstrated a linear relationship between choroidal BF and ocular PP, indicating a lack of autoregulatory capacity of the choroid.[110,114,121,184,234] The microsphere measurements obtained simultaneously from the retina and choroidal have demonstrated a marked difference in the autoregulatory ability of the two tissues.[110,121] On the other hand, LDF data under manipulation of the ABP by occluders on the aorta and inferior vena cava in rabbits[119,235] and bleeding in pigeons[236] suggest that peripheral choroidal BF compensates for declines in ABP (decrease in PP) so as to remain relatively stable within a physiological range of PPs. This difference in response, which could be due to the different techniques used, still needs to be clarified.

In humans, measurements of subfoveal choroidal BF by LDF reveal a non-linear decrease in BF in response to a reduction in PP induced by increasing the IOP by suction cup,[237,238] which suggests some autoregulatory capacity of the subfoveal choroid. A neural local mechanism involving the dense vasodilatory innervations of the human choroid, specifically localized in the temporal-central portion of the choroid adjacent to the fovea,[45] could be responsible for the observed behavior. The foveal avascular zone and most of the outer retina receive nutrition exclusively from the choroidal circulation. They may therefore be more susceptible to ischemia than other retinal regions. The observed subfoveal choroidal BF response may represent a protective mechanism against moderate increases in IOP above normal.

Anterior uvea

Animal studies with labeled microspheres have shown autoregulation of BF in the iris and ciliary body.[110,121]

However, in humans, iris BF by LDF showed no evidence of being autoregulated when the ocular PP was decreased by raising the IOP by suction cup.[239] Methodological or species differences and anesthesia have been invoked to explain the discrepancy between the results in animals and humans.[239]

Mechanisms underlying retinal and ONH autoregulation

It is likely that the mechanisms underlying autoregulation rely on a balanced contribution of myogenic and metabolic components (see below)[222] involving the interaction of factors released by the retinal metabolism and the vascular endothelium.[240] It has been suggested that, when PP is acutely reduced by increasing the IOP, the vasodilatory response is induced by an increase in the concentration of lactate.[226,230] Ionic, molecular, or gas modifications either of the blood or the surrounding tissue could also be involved in the vasomotor response.[240,241] In the ONH, the autoregulatory response to a small decrease in PP was found to take place during a time of the order of a second,[227] similar to the brain response.[242] The rapidity of this response points towards a metabolic mechanism. Studies on isolated cerebral and pial arteries of rats[243] and monkeys,[244] indeed suggests that a myogenic process develops much slower, on a time scale of 1–10 minutes. Recent evidence suggests a major role for NO in ONH autoregulation.[245]

Regulation of blood flow in response to increase in arterial blood pressure (ABP)

A number of studies have investigated the effect of an increase in systemic ABP on ocular BF. Most frequently, this increase is induced by static and dynamic exercises and posture changes, which both produce marked effects on sympathetic tone.

Static exercises

Isometrics increases heart rate, ABP, and sympathetic nerve activity.[246] In humans, during isometrics, retinal, subfoveal choroidal, and ONH BF remain unchanged until the mean ocular PP is elevated by an average of 34–60 percent above

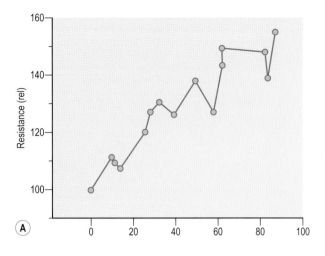

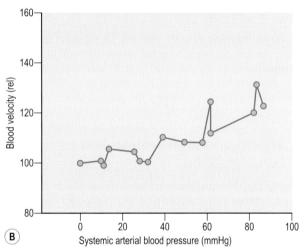

Figure 10.20 Relative average change of retinal vascular resistance (**A**) and velocity of RBCs (**B**) in retinal veins, as a function of the change of the relative brachial ABP. Each data point represents the average of the mean value obtained from three subjects. (From Robinson et al 1986.[247] Reproduced with permission from Association for Research in Vision and Ophthalmology.)

baseline (Fig. 10.20).[247-251] Since the vessels of the retina and ONH have no neural innervations, the regulation must be achieved, at least in part, through a local increase in vascular resistance,[247] as evidenced by the constriction of the retinal arterioles.[252] In the choroid, the sympathetic nervous system constricts the vessels in response to increased ABP.[253,254] Both ET-1[255] and NO[256] appear to play an important role in the regulation of subfoveal choroidal BF during isometrics.

Dynamic exercises

In response to such exercises the IOP decreases,[257] the ABP increases[258] but peripheral and macular retinal, as well as choroidal, BF remain rather unaffected.[259-262] The site of choroidal regulation (local or due to sympathetic activity) is still unclear. A sympathetic mechanism protecting the choroid from overperfusion has been postulated.[262]

Change in posture

The change from upright (or sitting) to supine induces a 16 percent decrease in heart rate, a small decrease in brachial diastolic ABP, no significant change in brachial systolic ABP,[263-266] and an increase in IOP.[267] The OA blood pressure

is significantly higher in the supine position than in the sitting position.[266] This stimulus decreases retinal vessel D while peripheral and macular retinal BF remain rather unaffected,[268,269] thus suggesting the presence of a local retinal compensatory response.

Regarding the choroidal circulation, recumbency increases subfoveal choroidal BF by approximately 11 percent,[265] a value corresponding to the increase in ocular PP calculated from experimental data.[266,270] This result and the linear relationship between subfoveal choroidal BF and ocular PP found in a subsequent study[271] are indicative of a passive response of this vascular bed to the increase in PP.

Regulation of blood flow in response to changes in blood gases

The role of arterial blood partial pressures of O_2 (P_aO_2) and CO_2 (P_aCO_2) on ocular BF regulation has been studied in animals and human subjects through the inhalation of various gas mixtures.

Hyperoxia

Increasing the P_aO_2 through the breathing of 100 percent O_2 results in a marked vasoconstriction of the retinal arterioles in both anesthetized animals[272-274] and healthy human subjects,[275-277] as well as a marked decrease in retinal BF (Fig. 10.21) in both the peripheral and macular region.[275,276,278,279] This BF response is 3–4 times greater than that of the cerebral circulation.[280] In contrast, choroidal BF does not change measurably when the P_aO_2 is increased.[281-283] In humans, the selective endothelin (ET) receptor antagonist BQ-123 (see below) dose-dependently blunts the retinal response to hyperoxia, indicating that ET-1 contributes to the hyperoxia-induced retinal vasoconstriction.[284]

The response of the retinal PO_2 to hyperoxia is species-dependent. In cats, with 100 percent O_2 inspiration, the vitreal and inner retinal PO_2 increases by 30–40 mmHg.[233,285,286] It reaches 53–88 mmHg in rats,[287,288] and remains almost unchanged in monkeys (average 20 mmHg in the inferior retina and about 32 mmHg in the parafovea and foveal area) and in minipigs.[273,289,290] The increased inner retinal PO_2 is not solely due to the O_2 derived from the retinal circulation, but also depends on the choroid, because during hyperoxia, O_2 diffuses all the way into the inner retina.[286,288,291,292] In the optic disc of minipigs, the PO_2 increases moderately by about 5 mmHg during hyperoxia and increases much more (about 13 mmHg) after injection of acetazolamide under systemic hyperoxia, due to the vasodilatory effect of elevated systemic P_aCO_2 (Box 10.2).[293]

Hypoxia

In humans and anesthetized animals, a decrease in P_aO_2 induces in the retina a vasodilatation of the arterioles and an increase in BF,[272,294-296] which is similar to the increase in cerebral BF.[280,297] The vasodilatation, which contributes to

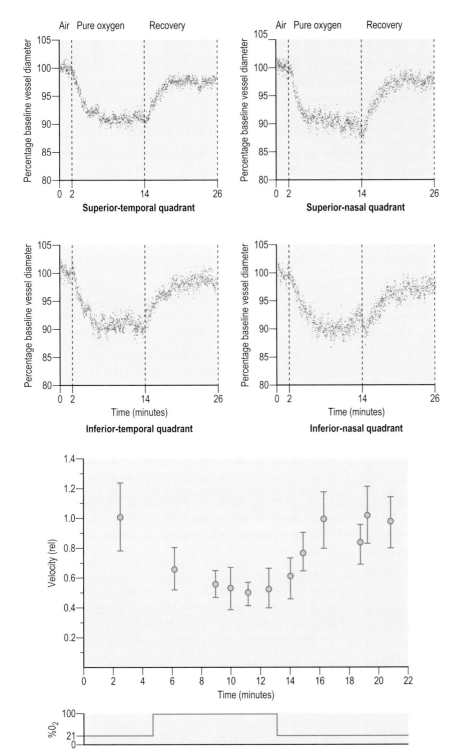

Figure 10.21 Effect of hyperoxia on (top) the diameter or retinal arteries in the four quadrants of the fundus and (bottom) the diastolic arterial velocity of a human retinal artery. (From Jean-Louis et al 2005[277] and Riva et al 1983.[276] Reproduced with permission from Association for Research in Vision and Ophthalmology.)

the maintenance of a rather stable P_{O_2} in the inner retina, could be due to the release of retinal lactate,[294,298,299] a process presumably mediated by endothelium-derived NO.[300] The increase in retinal BF allows inner retinal and vitreal P_{O_2} to remain relatively constant when the P_aO_2 is above 40 mmHg in cats and minipigs.[121,301–303] However, below this point, inner retinal P_{O_2} falls. There is a similar protection of inner retinal P_{O_2} during modest elevation of the IOP.[231,304] In contrast, the P_{O_2} measured close to the choroid and in the outer retina decreases linearly with the P_aO_2.[302,303,305]

Hypercapnia

Ocular blood vessels, like the cerebral vessels, are very sensitive to variations in the P_aCO_2. In animals, hypercapnia dilates retinal vessels and shortens the MCT.[306] Quantitatively, a rise in P_aCO_2 of 1 mmHg induces a 3 percent rise in retinal BF.[306] In conscious humans, the achievable changes in P_aCO_2 are smaller. Some studies report a dilatation of retinal arteries,[295,307] or shortening of the MCT,[295] while others observe no such changes.[308] Furthermore, hypercapnia increases the mean fluorescein (plasma) velocity (MVD) in retinal arteries,[309] macular leukocyte velocity,[308,310] RBCs velocity in the peripapillary,[311] perifoveal capillaries, and retinal and ONH BF.[307,312,313]

Raising the P_aCO_2 increases choroidal BF in cats[121,281] and humans,[282,283] but not in rats.[314] The increase in the P_aCO_2 increases also retinal PO_2.[233] There is a tight parallelism between the P_aCO_2 and interstitial pH in the inner retina. Thus, an increase of the P_aCO_2 of 38 mmHg induces a decrease in pH of $\approx$ 0.16 units.[315] The fact that acidification of the blood by infusing HCl or by injecting lactate does not affect interstitial pH nor retinal BF[299] suggests that interstitial rather than systemic acidosis contributes to the vasomotor response during hypercapnia.[315]

The hypercapnia-induced increase in retinal BF occurs through a mechanism involving NOS-1,[316] the release of PGE_2[294] or PGE_2-mediated endothelial NO.[317] Endothelial interaction between NO and PGs,[318,319] likely controls the arteriolar tone during changes of P_aCO_2.[320] Perfusion of the eye through the sublingual artery with indometacin, an inhibitor of PG synthase, induces a transient, reversible vasoconstriction of the retinal arterioles in normoxia/normocapnia.[299] In rabbits intravenous administration of indometacin reduces retinal BF by about 20 percent without any effect on the uveal circulation.[321] An increase in P_aCO_2[322] following metabolic acidosis induced by intravenous injection of acetazolamide (a carbonic anhydrase inhibitor) increases preretinal and ONH PO_2 and dilates the retinal vessels.[323–326]

Carbogen is a gas mixture containing 5 percent CO_2 and 95 percent O_2. The amount of CO_2 added to the O_2 was assumed to prevent the O_2-induced vasoconstriction and therefore maintain or even increase BF while providing the retina with increased oxygenation.[327] This gas decreases retinal vessel D and retinal BF.[328,329] It increases ONH and choroidal BF by LDF.[283,330,331] Carbogen inspiration leads to elevations of inner and optic disk PO_2 that are greater than those observed with hyperoxia alone.[233,288,323,324] The clinical value of this mixture is doubtful. It has no effect in fully developed retinal venous occlusions, whereas some improvement has been reported for selected cases of pre-occlusion of retinal veins.[332]

Metabolic control of retinal blood flow

In the retina, interactions between substances released either by endothelial or glial cells or by neurons affect the arteriolar tone, thus regulating the vasomotor responses. They may be either ionic or molecular or related to arterial blood gas modifications.[36,219,220,240,333] These factors can be tone relaxing or tone contracting. The former include NO and prostacyclin (PGI_2), the latter endothelin-1 (ET-1), angiotensin II, and cyclo-oxygenase (COX) products, such as thromboxane-A_2 (TXA_2) and prostaglandin H_2 (PGH_2). Extracellular lactate leads to contraction or relaxation of the vessel wall, according to the metabolic needs of the tissue.[334] In humans, lactate-induced retinal arteriolar dilatation occurs predominantly via stimulation of NO synthase and subsequent activation of guanylyl cyclase. Lactate likely mediates either the release of endothelial vasoactive substances or interferes with the metabolism of cells

surrounding the arterioles (i.e. astrocytes, neuronal cells), leading to the release of vasoactive substances, such as NO and PGs.[335] In addition, retinal arterial vasodilatation is mediated by an as yet unknown substance released by the retinal tissue and involving the activation of plasma membrane Ca^{2+}-ATPase.[336–338]

Retinal metabolism and vasoreactivity

The mammalian retina sustains an unusually high rate of glycolysis in that about 90 percent of the total glucose utilized is converted to lactate.[298] In addition, 70 percent of O_2 consumption in the retina is due to oxidation of glucose to CO_2.[339] Under normoglycemic conditions, neurons such as photoreceptor and ganglion cells, Müller glia, and pigment epithelial cells produce lactate aerobically and anaerobically at linear rates, even in the presence of a high starting ambient concentration of lactate. The anaerobic rate is 2–3-fold higher than the aerobic.[340] Thus, the retinal cells produce lactate and the retinal neurons utilize glucose as their major energy substrate. An alternative hypothesis suggests that the retinal neurons utilize preferentially lactate derived from Müller cells.[341] In rats, mediated by the endothelium, extracellular lactate leads to a rise in calcium in pericytes and, in turn, to contraction of microvessels in conditions of rich energy supply. By contrast, hypoxic microvessels relax during exposure to lactate.[334] This dual vasoactive capability may provide an efficient mechanism to match microvascular function to local metabolic needs.

In minipigs under normoxic conditions, however, preretinal microinjections (30–100 nl) of L-lactate (0.5 mol/l, pH 7.4) close to the retinal arterioles locally dilate the arteriolar wall.[299] In addition, intravenous administration of sodium lactate increases retinal BF at steady state[342] or during flicker stimulation.[343,344] High-dose lactate reduces the flicker-induced retinal D-response, whereas low-dose lactate increases it,[342] suggesting that the ratio of cytosolic free NADH to NAD$^+$ plays a critical role in the maintenance of retinal vascular tone. Uptake of intravenous or intravitreal lactate by vascular endothelial cells or astrocytes via monocarboxylate transporters[345–348] causes retinal arteriolar dilation predominantly via stimulation of NO synthase[335] and subsequent activation of guanylyl cyclase. In turn, guanylyl cyclase/cGMP signaling triggers opening of KATP channels for vasodilation. This process suggests that lactate either mediates the release of endothelial vasoactive substances (i.e. NO) or interferes with the metabolism and release of vasoactive substances by cells surrounding the arterioles (i.e. glial cells). Recent data from minipigs suggest that neuronal-derived NO is indeed an important mediator of lactate-induced vasodilatation.[349]

Blood flow response to visual stimulation

Light/dark transition

Studies have examined the effect of such a stimulus on glucose uptake in the outer and inner retinal layers,[130] retinal vessel D[350] and blood velocity[351–353] and retinal and ONH tissue PO_2.[128,354] The emerging picture is still rather unclear. Thus, D of retinal vessels measured after a transition from a period of darkness to ambient light was found to be respectively 2–3 percent and 5–8 percent, larger than during the light adaptation period prior to darkness. Blood velocity measured by LDV with a *visible* helium-neon laser was also found to be larger (40–70 percent) after the transition.[351,352] These findings were attributed to a presumed increase in retinal metabolism occurring during darkness.[351] However, similar studies performed with near-infrared light[350,353,355] suggest rather that these increases are induced by the transition itself.

Subfoveal choroidal BF decreases by 15 percent after a transition from light to darkness,[192] an effect observed in both eyes, even if only one eye is subjected to the transition.[356] This change was attributed to a mechanism centrally located and mediated by NO either from endothelial or neural source.[356,357]

Flicker

Luminance flicker (illumination with alternating brightness) increases glucose uptake in the ganglion cell layer (Fig. 10.22),[130] retinal vessel D (Fig. 10.23),[358–361] as well as retinal and ONH BF (Fig. 10.24).[362–366] The characteristics of this functional hyperemic response, namely the temporal dynamics and magnitude dependency upon flicker frequency, luminance, modulation depth, color ratio for chromatic flicker, dark adaptation, area of the retina stimulated, site of measurement at the retina (macula versus periphery) have been described in great detail for the ONH.[364,367–369] By comparison, little is known about these characteristics for the retina. Simultaneous measurements of ONH BF and the flicker electroretinogram indicate the presence, under certain conditions of flicker stimulation, of a coupling between the changes in ONH BF and retinal neural activity.[369] The effects of various physiological and pathological conditions on the flicker-induced retinal and ONH hyperemic responses have been studied. For example, increases in ABP, IOP, hyperglycemia, untreated arterial hypertension, diabetes, and glaucoma all reduce this response.[343,367,370–372]

Hypotheses on the function of this hyperemia include:[368] (a) increased glucose metabolism of the retinal tissue;[130,373] (b) increased delivery of glucose required by the astrocytes which play a crucial role in neurotransmitter recycling;[374] (c) increased delivery of oxygen required by the increased

activity in neurons and axons;[375] (d) NO production due to a signaling cascade fueled by cytosolic free NADH.[344] Studies suggest that potassium[376] and NO[377–379] act as putative mediators of the hyperemia, and demonstrate that dopamine,[380] adenosine,[381] lactate,[343] as well as arachidonic acid metabolites released by glial cells[220] modulate the response.

Control of arterial tone by endothelium or neuro-glial activity

The importance of endothelium-derived relaxing factors (EDRF) in regulating ocular BF[36,219] was recognized more than 20 years ago following the pioneering work of Furchgott.[382] These vascular tone-controlling substances have been reviewed in a number of previous publications.[59,383,384]

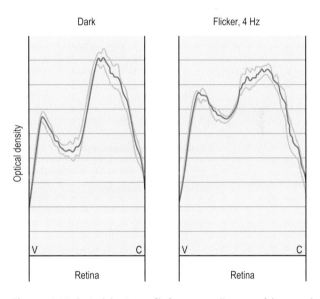

Figure 10.22 Optical density profile from autoradiograms of the central retina at a position 1–4 mm temporal to the optic disc at different conditions after an intravenous injection of labeled 2-deoxy-D-glucose. High density indicates high glucose uptake. V = vitreous, C = choroid. Flickering light increases glucose uptake in ganglion cell layer (red lines) compared to glucose uptake in the dark. (From Bill & Sperber 1990.[130] Reproduced by kind permission of Springer Science + Business Media.)

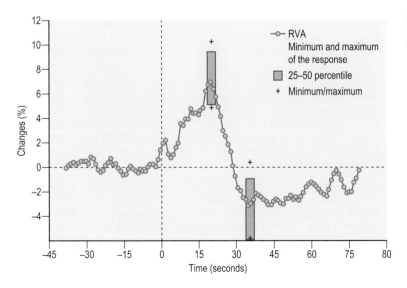

Figure 10.23 Flicker-induced change of diameter of retinal vessel, measured by RVA in 5 subjects. Stimulus started at time 0 sec and lasted 20 sec. (From Nagel & Vilser 2004.[360])

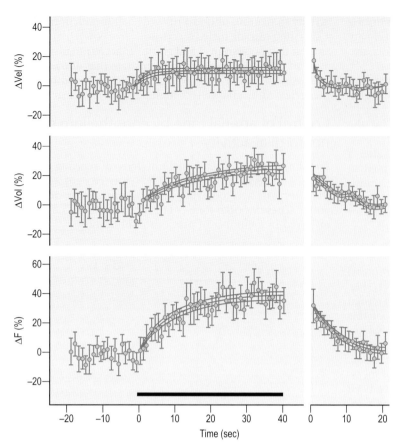

Figure 10.24 Mean (± 95 percent confidence interval, 15 healthy subjects) time course of changes (in percent) in velocity (ΔVel), volume (ΔVolh) and flux (ΔF) of red blood cells measured at a temporal site of the human optic disc in response to a diffuse luminance flicker (15 Hertz, 25-degree field). Horizontal bar: flicker period. (From Riva et al 2004.[371])

Nitric oxide

One of these EDRFs is NO produced by nitric oxide synthase (NOS), which has three isoforms, classified as neuronal NOS (NOS-1) and endothelial NOS (NOS-3), both activated via a calcium/calmodulin complex, and NOS-2 that is independent of calcium and capable of generating large amounts of NO in the presence of inflammatory and immunological stimuli. In general, NOS-1 is found in neurons of the central and peripheral nervous system. In the mammalian retina, it is present in ganglion, amacrine, horizontal and photoreceptor cells, as well as in Müller glial cells.[52,385–388] NOS-3 is mainly expressed by the vascular endothelial cells including the retinal[389,390] and choroidal vessels[390] and pericytes of the retinal capillaries.[391]

Analogs of L-arginine such as NG-monomethyl-L-arginine (L-NMMA), NG-nitro-L-arginine methyl ester (L-NAME), and NG-nitro-L-arginine (L-NA) can be used as specific inhibitors of NO production.[392] These competitive inhibitors of NO production[383] block all three isoforms of NOS.

The role of NO in the control of ocular BF has been discussed in a previous review.[383] In summary, in isolated porcine ophthalmic and ciliary artery rings, L-NMMA induces dose-dependent contractions.[393,394] NO is also capable of relaxing the contractile tone of retinal bovine pericytes[395] and Müller cells,[396] which may both be involved in the control of retinal BF. In addition, endothelium-derived NO, released under basal conditions or stimulated by bradykinin, regulates the ophthalmic circulation of the perfused porcine eye.[397]

In various animal species, a constant release of NO maintains the basal vascular diameter and BF in the various ocular tissues.[377,378,398–402] In humans, systemic inhibition of NOS with intravenous administration of L-NMMA reduces ONH BF.[403] The effect of L-NMMA can be reversed with high-dose of L-arginine, indicating that NO contributes to the maintenance of basal vascular tone in the choroid and the ONH. Furthermore, investigations in minipigs show that a continuous production of NO is necessary to maintain the arteriolar tone, at least in the inner retinal vessels.[378] In addition, L-NMMA decreases D of retinal arterioles and venules[404] and reduces retinal BF,[405] indicating that NO is also continuously released in retinal vessels under physiological conditions.

NO also appears to be involved in a variety of agonist-induced vasodilatations.[219] In humans this has been shown for histamine,[406] insulin,[407] hypercapnia,[408] and adrenomedullin.[409] In addition, NO plays a role in flicker-induced ONH hyperemic response in cats[377] and in the human retinal vasculature.[404] The role of NO in BF in the human eye is demonstrated by the ability of sodium nitroprusside and nitrates to increase BF in the various ocular tissues.[410,411]

Endothelins

Three isoforms of endothelin, ET-1, ET-2, ET-3, have been identified in the vascular endothelium.[412] ET-1 is also expressed in retinal neurons and astrocytes[413] and in the brain.[414] ET-1, the most potent vasoconstricting factor presently known, affects both pericytes and smooth muscle cells.[395] It induces potent contractions in isolated ophthalmic, ciliary, and retinal arterioles[393,415–417] and, in vivo, in rabbit[418] and cat retinal vessels.[419] Moreover, ET-1 contributes to the vasoconstriction of human retinal arterioles.[420] Furthermore i.v. administration of ET-1 strongly reduces cat ONH[421] and rabbit choroidal BFs.[422]

Two types of ET receptors, ET_A and ET_B, with different sensibilities for the three ligands have been identified in the retina and ONH from human and porcine eyes.[413] The ET_A receptor is expressed on pericytes and vascular smooth muscle cells[423] and presents a high affinity to ET-1. The ET_B receptor has been detected in cultured bovine retinal pericytes[423] and endothelial cells in culture.[424] Systemic administration of the selective ET_A receptor antagonist BQ-123 prevents the decrease in retinal BF induced by exogenous ET-1.[420]

ET_B has two subtypes of receptors. The ET_{B1} receptor is expressed in endothelial cells and has equal affinity for each ET isoform and mediates vasorelaxation through the release of NO by endothelial cells in culture,[424] in the pulmonary circulation of lambs,[425] and in rabbit kidneys[426] under physiological and hypoxic conditions.[427] Furthermore, ET_{B1} receptors appear to influence ABP homeostasis by reducing plasma ET-1 levels and thus minimizing ET_A activation.[428] In contrast to the vasodilating effect of ET_{B1} signaling, ET_{B2} receptors have a high affinity for ET-3 and directly mediate vasoconstriction.[429] Dual block of ET_A and ET_B receptors by bosentan, a drug used in the treatment of pulmonary hypertension, increased retinal BF in healthy and glaucomatous eyes.[430]

NO and ET appear to be major players in choroidal BF regulation.[383,431,432] In anesthetized rabbits, a dynamic interplay between NO-induced vasodilation and ET-induced vasoconstriction has been demonstrated as ocular PP is varied by mechanical manipulation of arterial pressure.[431] Thus, administration of L-NAME eliminates the NO-induced vasodilation and the unopposed ET vasoconstriction causes a dramatic downward shift of the ocular PP-choroidal BF relationship over a wide range of PPs. However, subsequent administration of the non-selective ET antagonist A-182086 reverses the vasoconstriction and returns the relationship back towards control.

Prostaglandins (PGs)

Major metabolites synthesized from arachidonic acid, namely prostacyclin (PGI_2) and contracting factors such as thromboxane A_2 (TXA_2) and prostaglandin H_2 (PGH_2), or superoxide anions are mainly produced in the cerebral circulation, under physiological conditions.[219,433] Isolated pericytes of retinal capillaries[434] and isolated rabbit retinas[435] also release some subclasses of PGs (PGE_2, $PGF_{2\alpha}$, PGI_2).

PGI_2 has been shown to exert a vasodilating action on isolated bovine retinal arterioles[436] and in the rabbit eye.[437] Microinjection of PGE_2 induces a segmental vasodilation of retinal arterioles in minipigs[294] suggesting that, during normocapnia, the release of vasodilating PGs sets the basal arteriolar tone. PGE_2 and $PGF_{2\alpha}$ are the predominant PGs produced by the retina and choroid and may play a role during physiological regulation in response to hypercapnia[294,438] and changes in ocular PP.[438,439] The regulation of retinal BF may also involve glial-evoked dilation, mediated by the arachidonic acid metabolites epoxyeicosatrienoic acid (EET), whereas glial-evoked constriction is mediated by 20-hydroxy-eicosatetraenoic acid (20-HETE).[220]

In healthy subjects, intravenous injection of PGE_1 does not alter the parameters of retinal and choroidal circulations, suggesting that efficient autoregulation mechanisms operate in response to the injection.[440] In the cynomolgus monkey eye, $PGF_{2\alpha}$-IE has been shown[441] to cause a dramatic increase in anterior uvea BF, but only weak effect was detected with selective FP prostanoid receptor agonists and an EP_1 receptor agonist after topical administration. Intravenous infusion of latanoprost at a dose range of 0.6 ± 6 mg/kg had little effect on BF in most ocular tissues, and the same was true for 17-phenyl-PGE_2-IE, a relatively selective EP_1 receptor agonist, after intracardiac infusion at about the same dose range. Intravenous infusion of the EP_2 receptor agonist 19R-OH-PGE_2 markedly reduced the vascular resistance in the eye. No significant effect was seen on the blood volume in the ocular tissues with any of the FP receptor agonists after topical administration. $PGF_{2\alpha}$-IE increased the capillary permeability to albumin in the anterior segment and possibly the retina, but 17-phenyl-$PGF_{2\alpha}$ IE and latanoprost/PhXA34 had no effect on capillary permeability in any of the ocular tissues. Based on the results of previous studies and recent experiments,[441] it is evident that $PGF_{2\alpha}$ has significant microvascular effects in the rabbit, cat and monkey eye, causing vasodilation and/or increased capillary permeability, whereas selective FP receptor agonists such as latanoprost exert no or minimal effects in the primate eye, and markedly reduced microvascular effects in the rabbit eye. However, little difference between $PGF_{2\alpha}$ and the selective FP receptor agonists was seen in the cat eyes. It also appears that the EP1 receptor like the FP receptor is not involved in the regulation of vascular tone in the primate eye, whereas stimulation of the EP_2 receptor reduces the vascular resistance in the monkey eye.

$PGF_{2\alpha}$ is often used to induce vasoconstriction in vitro, an effect seen at very high doses. Analogs that are more selective for the prostanoid FP receptor such as the analogs used in clinical practice today, have no effect on vessel diameter,[442] and cannot be expected to affect blood flow in the posterior pole.

Neural, endocrine, and paracrine control

A number of endocrine and paracrine factors are involved in choroidal BF regulation, as described in more detail in a previous publication.[384]

Effects of vasoactive nerves

Sympathetic stimulation reduces choroidal BF in a variety of species,[253,254,443–446] an effect which in rabbits is mediated partially via vasoconstrictive α-receptors and partially via non-adrenergic neuropeptide Y.[447] Furthermore, NPY is a potent vasoconstrictor in the rabbit uvea, when infused intravenously.[448] Contrary to this, NPY has only minor effects in the cat uvea, and the response to sympathetic nerve stimulation is completely blocked by α-adrenoreceptor blockade.[449] In rats, the decrease in anterior choroidal BF is mainly mediated via the α_1-receptor subtype. In addition, α-adrenoceptor blockade unmasks a sympathetically evoked choroidal vasodilator response. This residual vasodilatation appears to be mediated via the β_1-adrenoceptor subtype.[450] Experimental studies in vivo as well as in vitro have shown that norepinephrine is released already at low stimulation frequencies, whereas NPY is released at high frequencies.[451] In humans, NPY is only released during high levels of stress, e.g. exhaustive exercise, vaginal delivery, panic attacks, and cold exposure[452] Thus, the role of NPY in normal blood flow regulation in the choroid is unclear. The sympathetic system presumably prevents hyper-perfusion during increased ocular PP[443] induced, for example, during isometric exercises,[453] a situation where NPY may play a role. NPY has also been implicated in ocular pathology (Box 10.3).

There is unequivocal evidence that parasympathetic (facial nerve) stimulation increases choroidal BF in a variety of species.[49,50] Choroidal vasodilatation is blocked by ganglion blockade,[50] an effect indicative of a nicotinic synapse between the nervous system and the eye. A variety of neurotransmitters have been implicated in the choroidal vasodilator response to parasympathetic stimulation. These include acetylcholine, vasoactive intestinal peptide (VIP),[454] pituitary adenylate cyclase activating polypeptide (PACAP),[455] and NO;[45,432,456] all of them are present in nerve cell bodies in pterygopalatine ganglion. VIP and neuronal NOS have also been localized to intrinsic choroidal neurons,[45,46] that may be a part of the same nervous pathway. Physiological experiments indicate that the significance of the different neurotransmitters vary between species. In rabbits, VIP and PACAP are potent vasodilators when given intravenously,[457,458] and the vasodilatation caused by facial nerve stimulation is only slightly affected by muscarinic blockade and NOS-inhibition.[459] In the cat, the responses are reversed, that is the response to nerve stimulation is significantly reduced by combined muscarinic blockade and NOS-inhibition,[432] whereas intravenous infusion of VIP[460] or PACAP (Nilsson, SFE, personal communication) has no effect on uveal blood flows. Thus, acetylcholine and NO seem to be the most important transmitters in the cat, whereas in the rabbit, peptides seem to play a larger role.

Adenosine

Adenosine, a breakdown product of cellular adenosine triphosphate, is a modulator of synaptic transmission and a potent vasodilator in many vascular beds. In newborn piglets adenosine induces retinal vasodilatation via the adenosine A_2-receptor subtype[461] and plays a role in hypoxia-induced vasodilatation and retinal autoregulation.[462] Adenosine increases choroidal BF in cats and both choroidal and ONH BF in humans.[463,464] It plays a role in hypoxia-induced vasodilatation and retinal autoregulation by potentiating the action of endogenous extracellular adenosine.[462] It also enhances the flicker-induced functional hyperemic response in the cat.[381]

Endogenous and pharmacological substances

Role of administration route

The effect of exo- or endogenous vasoactive substances will to a large extent depend on the route of administration and the tissue. Close-arterial injections can provide high local concentration with little effect on the systemic circulation, and thus on ABP. A vasodilator may increase choroidal BF when given as a close-arterial injection, or reduce it due to a general vasodilatation and reduced ABP.[465] For the retina the blood–retinal barrier will prevent most systemically administered drugs from reaching the smooth muscles. Only lipid-soluble drugs, such as papaverine, will pass the barrier.[466] Circumvention of the blood–retinal barrier can also be achieved by injecting the drug into the vitreous body and it has been shown that adrenergic agonists, such as epinephrine, norepinephrine, and phenylepineprhine contract retinal vessels in vitro.[467,468]

Also, drugs applied topically as eye drops are unlikely to have any clinically beneficial effect on ocular BF. There are little data on ocular pharmacokinetics in the human eye[469] and data from animal studies cannot be extrapolated to the human eye due to the large difference in body weight. In most experimental animals the systemic dose will be about 20 times that in humans, and much of the drug that reaches the posterior pole will reach it by the systemic route. In monkeys one drop of nipradilol in one eye caused a local concentration in the posterior pole of the other eye that was about 82 percent of that in the treated eye 6 hours after administration.[470]

Thus there are no effective tools to improve BF in the posterior segment of the eye with today's drugs, but as many drugs used in the treatment of glaucoma have marked effects on blood vessels, significant changes can be expected in BF through the anterior uvea. They are, in fact, dosed in order to have an effect in the anterior segment of the eye. Further back in the eye, the tissue concentration will be lower, and in most studies on ocular pharmacokinetics in experimental animals the concentration found in the posterior pole is less than 10 percent of that in the anterior pole.[471]

Vasoconstrictors

In the anterior segment both the unselective adrenergic agonist epinephrine[472] and the alpha$_2$ adrenergic agonist brimonidine[473] decrease BF when applied topically as a single dose.

The role of adrenergic drugs in the control of BF in the back of the eye has attracted much interest because topical α-receptor agonists and β-receptor antagonists are used in glaucoma treatment. Exogenous norepinephrine does not affect retinal BF in experimental animals or healthy humans.[466,474] Both increases and decreases in choroidal BF have been reported after systemic administration of norepinephrine in experimental animals,[475,476] but the data are difficult to interpret due to the pronounced concomitant increase in ABP. β-receptors have been located in the retina and the choroid,[477,478] and have been shown to be involved in vasodilation in the choroid during sympathetic stimulation after blocking the adrenergic α-receptors.[479] Still, stimulation of the β-receptor has no effect on retinal vessels in vitro[468,480] and no consistent effect on ocular BF has been seen in clinical studies.[471]

Alpha-receptor subtypes can be found in the ocular tissues of pigs and rabbits.[481] In porcine ciliary arteries the α_2-receptor agonist brimonidine is a potent vasoconstrictor.[482] The majority of animal and human studies suggest a lack of influence of brimonidine on retinal, ONH, and choroidal BF.[471] The role of the dopaminergic system in the control of ocular BF appears to be complex. It has been reported that dopamine antagonists increase BF in all rabbit eye tissues.[483] On the other hand, dopamine itself induces vasodilatation in the rabbit choroid via the D_1/D_5 receptor subtype.[484] In humans, dopamine i.v. causes a slight but significant increase in retinal D.[380] These results may reflect the complex interplay of the different types of dopamine receptors in the control of ocular BF with the D_1 and D_5 receptors mediating vasodilatation and the D_2, D_3, and D_4 receptors tending to mediate vasoconstriction.

Vasodilators

While intracranial stimulation of the oculomotor nerve induces a vasoconstriction in the rabbit iris,[485] topical application of the cholinergic agonist pilocarpine in monkeys markedly increases anterior segment BF while having no effect in the back of the eye.[198]

Potential effects on ocular circulation for two other commonly used glaucoma drugs, topical carbonic anhydrase inhibitors and prostaglandin $F_{2\alpha}$-analogs, have also been examined. However, in clinical use none of them can be expected to have an effect in the back of the eye. Systemic administration of carbonic anhydrase inhibitors, such as acetazolamide, have marked effects on cerebral BF,[486] most likely secondary to increased local carbon dioxide tension. A similar effect is seen in the eye where systemic administration of acetazolamide or dorzolamide increases oxygen tension close to the optic disc in pigs.[487] A tendency to increased oxygen tension could also be seen when dorzolamide was applied as eye drops applied every 2.5 min for 3 hours for a total dose of about 40 mg,[488] but a single dose cannot be expected to have any significant effect.

Prostaglandin $F_{2\alpha}$ is often used to induce vasoconstriction in vitro, an effect seen at very high doses. Analogs that are more selective for the prostanoid FP receptor such as the analogs used in clinical practice today, have no effect on vessel diameter,[442] and in monkeys topical application of the prostaglandin $F_{2\alpha}$-analog latanoprost in a dose 10 times the clinically used dose induced conjunctival hyperemia and a tendency to increased BF in the anterior sclera, but no effect in the back of the eye.[441]

Ciliary blood flow regulation

Ciliary BF demonstrates some autoregulatory capacity in animals.[189,489] In humans, data on this capacity are lacking due to the difficulties of developing a measuring technique in this inaccessible tissue. Several systems play a role in the regulation of ciliary BF, such as the cholinergic system,[198] NO and the adrenergic system.[490–492] Dopamine increases ciliary BF in rabbits, presumably by acting via D_1/D_5 dopamine receptor activation, as in the choroid.[484]

Ciliary BF plays an important role by supplying the necessary nutrients for the production of aqueous humor. It has been found that in anesthetized rabbits and monkeys aqueous production is independent of moderate reductions in ABP or ciliary BF.[489,493,494] In rabbits aqueous flow is unaffected until ciliary BF declines below 74 percent of control; when BF is below this critical value, aqueous production becomes flow-dependent, decreasing with further BF decreases.[494]

Ocular blood flow and its regulation in diseases

Diabetes

The hypothesis that altered retinal BF plays a role in the development of diabetic retinopathy (DR) was made more than 25 years ago.[495] It has since led to numerous investigations of BF in the various tissues of the eye in this disease aimed at getting some insight into the progression of DR. At present this knowledge is rather controversial since some of the observations appear to be contradictory.[59,496] Discrepancies in the findings may be due to differences in the type of patients, i.e. type I versus type II diabetes, controlled versus poorly controlled diabetes, improvement in the degree of glucose control over the years, short- versus long-duration of the disease, and presence or absence of other concomitant diseases, such as systemic hypertension.[497]

Increased D of the retinal arterioles and veins is seen early in the disease.[497,498] On the other hand, retinal BF seems to be unaffected in eyes with well-regulated diabetes until more severe retinopathy develops.[498–501] In patients with proliferative DR, retinal hemodynamics seems to depend on the specific pathologic features. For example, reduced retinal BF and vessel staining seem to be associated with severe capillary non-perfusion.[502]

Retinal vessels in eyes with minimal or no DR show a reduced vasoconstriction in response to isocapnic hyperoxia,[503] and an altered retinal BF response, which is correlated with the degree of DR (Fig. 10.25).[504] This altered reactivity occurs in the absence of any difference in baseline hemodynamic values[499] and is related to the objectively defined magnitude of retinal edema.[499]

The ability of retinal BF to respond to changes in ocular PP is altered in diabetes, whether the PP is decreased through an increase of the IOP,[505] increased by treatment with tyramine[506] or by isometric exercises.[507] This alteration is further accentuated by hyperglycemia[506] and more prevalent in patients with autonomic dysfunction than in those with an intact autonomic nervous system.[508] Presumably, the hyperglycemia prevents a normal autoregulatory response to the additional stress induced by the increase in PP.[506]

The response of D of retinal vessels to diffuse luminance flicker is blunted in insulin-dependent diabetic patients in comparison to healthy controls, due either to a vascular abnormality (endothelial dysfunction or loss of pericytes) and/or decreased neural activity response[509] resulting from selective abnormalities of Müller glial cell function.[510] These cells probably play an important role in the coupling between retinal neural activity and BF.[368] The blunting occurs

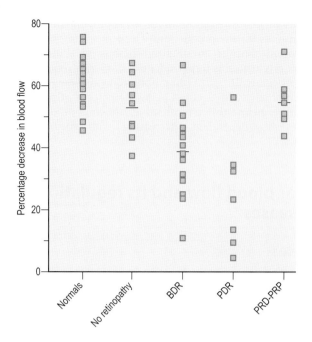

Figure 10.25 Decrease in retinal BF by BLDV at 5 min of 100 percent O₂ breathing in diabetic patients with various degrees of diabetic retinopathy. BDR = background retinopathy, PDR = proliferative retinopathy, PDR-PRP = proliferative retinopathy after panretinal photocoagulation. Horizontal bars: average value in each group. (From Grunwald et al 1984.[504])

already in patients without DR and increases with the stage of DR.[511]

Excessive choroidal basement membrane thickening and degeneration,[512] as well as capillary dropout, has been reported for eyes with DR.[513] Reduced subfoveal choroidal BF has been reported for patients with diabetes and particularly in those with macular edema.[514,515] Altered regulation of this BF in response to increased PP by isometrics has been demonstrated in type 1 diabetic patients with DR, but not in those without DR.[453] Contrary to the response in healthy subjects and diabetics with no DR, subfoveal choroidal BF increases linearly with PP in DR patients. For detailed information on BF and BF regulation in diabetes, see recent reviews.[59,384]

Glaucoma

The role of vascular disturbances in the pathogenesis of glaucoma remains controversial, but most experts agree that elevated IOP alone does not explain the whole spectrum of individuals with open-angle glaucoma. Apart from IOP, a variety of other systemic and ocular risk factors for glaucoma have been identified, such as heart disease,[516] low systolic blood pressure, and low ocular perfusion pressure.[517]

A large number of studies, using different clinical techniques, have demonstrated reduced ONH BF values in eyes with glaucoma when compared to normal eyes. Whether this is secondary to loss of neural tissue or a causative factor is, however, not known. As BF through the retina and the ONH is efficiently autoregulated, moderate increments in IOP would have no or little effect on BF through these tissues. The situation may be different in the presence of deficient autoregulation and ocular BF would then be affected even by small increments in IOP. An impaired autoregulation of

the retinal circulation was indeed found in eyes with glaucoma with the blue field simulation technique.[518] This investigation determined the maximum, acutely increased IOP above resting IOP at which retinal macular leukocyte speed is maintained constant by autoregulation. This maximum IOP was 25 ± 1.5 mmHg (± 1 SD) for the POAG patients and 30 ± 3.6 mmHg in the normal subjects. Studies with other techniques have also suggested that autoregulation of retinal and/or ONH BF is reduced in eyes with glaucoma. Thus retinal venous D response to short-term elevations in IOP was found to be altered[519] and changes in rim perfusion after a therapeutic IOP reduction[520] suggest that autoregulation may be defective in eyes with glaucoma while intact in ocular hypertension. Also the increase of D in retinal veins and of ONH BF in response to flicker were found to be significantly diminished in glaucoma patients as compared with healthy volunteers,[367,521] both indicative of an impairment of neurally mediated vasoreactivity. A reduced ONH BF-response was also observed in patients with ocular hypertension.[367] Whether altered ONH BF regulation represents a risk factor for the development and progression in glaucoma remains to be shown.

Age-related macular degeneration

The pathophysiology of this disease is largely unknown.[522] In recent years, based on the assumption that early in the development of AMD there is thickening and loss of elasticity of ocular vessel walls due to atherosclerotic processes, it was proposed that reduced choroidal BF could play an important role in the pathogenesis of AMD.[523-525] The hemodynamic model of AMD is supported by a number of in vitro studies showing an association between morphological changes in Bruch's membrane, loss of RPE, and capillary dropout.[526] To investigate the possibility that a vascular component is involved in AMD subfoveal choroidal BF was measured, demonstrating lower LDF flow rates in eyes with AMD compared to healthy eyes.[527] However, one cannot exclude that morphological changes in AMD eyes may well contribute to the difference. On the other hand, measurements of subfoveal choroidal BF changes in response to isometric exercises, have revealed an altered BF regulation.[528] Thus, whereas these in vitro and in vivo studies have established a relation between AMD and choroidal BF alterations, it still remains to be demonstrated that these alterations represent a risk factor in the pathogenesis of AMD and choroidal neovascularization.

References

1. Hickam JB, Frayser R. A photographic method for measuring the mean retinal circulation time using fluorescein. Invest Ophthalmol 1965; 4:876–884.
2. Wise GN, Dollery CT, Henkind P. The retinal circulation. New York: Harper and Row, 1975:20–82.
3. Weiter JJ, Ernest JT. Anatomy of the choroidal vasculature. Am J Ophthalmol 1974; 78:583–590.
4. Hayreh SS. Segmental nature of the choroidal vasculature. Br J Ophthalmol 1975; 59:631–648.
5. Hayreh SS. The ophthalmic artery: Iii. Branches. Br J Ophthalmol 1962; 46:212–247.
6. Singh S, Dass R. The central artery of the retina. I. Origin and course. Br J Ophthalmol 1960; 44:193–212.
7. Singh S, Dass R. The central artery of the retina. II. A study of its distribution and anastomoses. Br J Ophthalmol 1960; 44:280–299.
8. Onda E, Cioffi GA, Bacon DR, Van Buskirk EM. Microvasculature of the human optic nerve. Am J Ophthalmol 1995; 120:92–102.
9. Hayreh SS. The cilio-retinal arteries. Br J Ophthalmol 1963; 47:71–89.

10. Rungger-Brändle E, Kolb H, Niemeyer G. Histochemical demonstration of glycogen in neurons of the cat retina. Invest Ophthalmol Vis Sci 1996; 37:702–715.
11. Zhang Y, Stone J. Role of astrocytes in the control of developing retinal vessels. Invest Ophthalmol Vis Sci 1997; 38:1653–1666.
12. Henkind P. Radial peripapillary capillaries of the retina. I. Anatomy: human and comparative. Br J Ophthalmol 1967; 51:115–123.
13. Olver JM. Functional anatomy of the choroidal circulation: methyl methacrylate casting of human choroid. Eye 1990; 4(Pt 2):262–272.
14. Carella E, Carella G. Microangioarchitecture of the choroidal circulation using latex casts. In: Yannuzzi LA, Flower RW, Slatker JS, eds. Indocyanine green angiography. St. Louis: Mosby, 1997:24–28.
15. Olver JM, Spalton DJ, McCartney AC. Quantitative morphology of human retrolaminar optic nerve vasculature. Invest Ophthalmol Vis Sci 1994; 35:3858–3866.
16. Hayreh SS. The long posterior ciliary arteries. An experimental study. Albrecht Von Graefes Arch Klin Exp Ophthalmol 1974; 192:197–213.
17. Funk R, Rohen JW. Scanning electron microscopic study on the vasculature of the human anterior eye segment, especially with respect to the ciliary processes. Exp Eye Res 1990; 51:651–661.
18. Sattler H. Über den feineren Bau der Choroidea des Menschen. Beitragen zur pathologischen und vergleichenden Anatomie der Aderhaut. Von Graefe's Arch 1976; 428–440.
19. Nuel IP. De la vascularisation de la choroide et de la nutrition de la rétine principalement au niveau de la fovea centralis. Arch Ophthalmol 1992; 70–87.
20. Hayreh SS. Submacular choroidal vascular pattern. Experimental fluorescein fundus angiographic studies. Albrecht Von Graefes Arch Klin Exp Ophthalmol 1974; 192:181–196.
21. Hayreh SS. In vivo choroidal circulation and its watershed zones. Eye 1990; 4:273–289.
22. Hayreh SS. Posterior ciliary artery circulation in health and disease: the Weisenfeld lecture. Invest Ophthalmol Vis Sci 2004; 45:749–757.
23. Mendrinos E, Pournaras CJ. Topographic variation of the choroidal watershed zone and its relationship to neovascularization in patients with age-related macular degeneration. Acta Ophthalmol 2009; 87:290–296.
24. Hayreh SS. The choriocapillaris. Albrecht Von Graefes Arch Klin Exp Ophthalmol 1974; 192:165–179.
25. Fryczkowski AW. Topographic anatomy of the central retina and segmental choroidal circulation. In: Yannuzzi LA, Flower RW, Slatker JS, eds. Indocyanine green angiography. St. Louis: Mosby, 1997:29–34.
26. Sugita A, Hamasaki M, Higashi R. Regional difference in fenestration of choroidal capillaries in Japanese monkey eye. Jpn J Ophthalmol 1982; 26:47–52.
27. Bill A, Tornquist P, Alm A. Permeability of the intraocular blood vessels. Trans Ophthalmol Soc UK 1980; 100:332–336.
28. Spitznas M, Reale E. Fracture faces of fenestrations and junctions of endothelial cells in human choroidal vessels. Invest Ophthalmol 1975; 14:98–107.
29. Torczynski E, ed. Choroid and suprachoroid. Philadelphia: JB Lippincott, 1987.
30. Fryczkowski AW, Sherman MD, Walker J. Observations on the lobular organization of the human choriocapillaris. Int Ophthalmol 1991; 15:109–120.
31. Hayreh SS, Baines JA. Occlusion of the vortex veins. An experimental study. Br J Ophthalmol 1973; 57:217–238.
32. Hogan MJ, Feeney L. The ultrastructure of the retinal blood vessels. I. The large vessels. J Ultrastruct Res 1963; 39:10–28.
33. Henkind P, De Oliveira LF. Retinal arteriolar annuli. Invest Ophthalmol 1968; 7:584–591.
34. Shakib M, Cunha-Vaz JG. Studies on the permeability of the blood–retinal barrier. IV. Junctional complexes of the retinal vessels and their role in the permeability of the blood–retinal barrier. Exp Eye Res 1966; 5:229–234.
35. Hirschi KK, D'Amore PA. Pericytes in the microvasculature. Cardiovasc Res 1996; 32;687–698.
36. Haefliger IO, Meyer P, Flammer J, Luscher TF. The vascular endothelium as a regulator of the ocular circulation: a new concept in ophthalmology? Surv Ophthalmol 1994; 39:123–132.
37. Chakravarthy U, Gardiner TA. Endothelium-derived agents in pericyte function/dysfunction. Prog Retin Eye Res 1999; 18:511–527.
38. Ehinger B. Distribution of adrenergic nerves in the eye and some related structures in the cat. Acta Physiol Scand 1966; 66:123–128.
39. Laties AM. Central retinal artery innervation. Absence of adrenergic innervation to the intraocular branches. Arch Ophthalmol 1967; 77:405–409.
40. Denis P, Elena PP. Retinal vascular beta-adrenergic receptors in man. Ophtalmologie 1989; 3:62–64.
41. Ferrari-Dileo G, Davis EB, Anderson DR. Angiotensin binding sites in bovine and human retinal blood vessels. Invest Ophthalmol Vis Sci 1987; 28:1747–1751.
42. Stone RA. Neuropeptide Y and the innervation of the human eye. Exp Eye Res 1986; 42:349–355.
43. Feeney L, Hogan MJ. Electron microscopy of the human choroid. II. The choroidal nerves. Am J Ophthalmol 1961; 51:1072–1083.
44. Wolter JR. Nerves of the normal human choroid. Arch Ophthalmol 1960; 64: 120–124.
45. Flügel C, Tamm ER, Mayer B, Lütjen-Drecoll E. Species differences in choroidal vasodilative innervation: evidence for specific intrinsic nitrergic and VIP-positive neurons in the human eye. Invest Ophthalmol Vis Sci 1994; 35:592–599.
46. Flügel-Koch C, Kaufman P, Lütjen-Drecoll E. Association of a choroidal ganglion cell plexus with the fovea centralis. Invest Ophthalmol Vis Sci 1994; 35:4268–4272.
47. Lütjen-Drecoll E, Neuhuber W. Innervation choroiidenne: Focalisation sur les neurons intranséques. In: Pournaras CJ, ed. Pathologies vasculaires oculaires. Paris: Elsevier, 2008:29–32.
48. Ruskell GL. Facial parasympathetic innervation of the choroidal blood vessels in monkeys. Exp Eye Res 1971; 12:166–172.
49. Stjernschantz J, Bill A. Vasomotor effects of facial nerve stimulation: non-cholinergic vasodilation in the eye. Acta Physiol Scand 1980; 109:45–50.
50. Nilsson SF, Linder J, Bill A. Characteristics of uveal vasodilation produced by facial nerve stimulation in monkeys, cats and rabbits. Exp Eye Res 1985; 40:841–852.
51. Stone RA, Tervo T, Tervo K, Tarkkanen A. Vasoactive intestinal polypeptide-like immunoreactive nerves to the human eye. Acta Ophthalmol (Copenh) 1986; 64:12–18.
52. Yamamoto R, Bredt DS, Snyder SH, Stone RA. The localization of nitric oxide synthase in the rat eye and related cranial ganglia. Neuroscience 1993; 54:189–200.
53. Nilsson SFE. Neuropeptides in the autonomic nervous system influencing uveal blood flow and aqueous humor dynamics. In: Troger J, Kieselbach G, eds. Neuropeptides in the eye. Kerala: Research Signpost, 2009:147–167.
54. Meyer PA. Patterns of blood flow in episcleral vessels studied by low-dose fluorescein videoangiography. Eye 1988; 2(Pt 5):533–546.
55. Meyer PA, Watson PG. Low dose fluorescein angiography of the conjunctiva and episclera. Br J Ophthalmol 1987; 71:2–10.
56. Morrison JC, DeFrank MP, Van Buskirk EM. Regional microvascular anatomy of the rabbit ciliary body. Invest Ophthalmol Vis Sci 1987; 28:1314–1324.
57. Saunders RA, Bluestein EC, Wilson ME, Berland JE. Anterior segment ischemia after strabismus surgery. Surv Ophthalmol 1994; 38:456–466.
58. Van Nerom PR, Rosenthal AR, Jacobson DR, Pieper I, Schwartz H, Greider BW. Iris angiography and aqueous photofluorometry in normal subjects. Arch Ophthalmol 1981; 99:489–493.
59. Pournaras CJ, Rungger-Brandle E, Riva CE, Hardarson SH, Stefansson E. Regulation of retinal blood flow in health and disease. Prog Retin Eye Res 2008; 27:284–330.
60. Feng D, Nagy JA, Hipp J, Dvorak HF, Dvorak AM. Vesiculo-vacuolar organelles and the regulation of venule permeability to macromolecules by vascular permeability factor, histamine, and serotonin. J Exp Med 1996; 183:1981–1986.
61. Minshall RD, Sessa WC, Stan RV, Anderson RG, Malik AB. Caveolin regulation of endothelial function. Am J Physiol Lung Cell Mol Physiol 2003; 285:L1179–L1183.
62. Mehta D, Malik AB. Signaling mechanisms regulating endothelial permeability. Physiol Rev 2006; 86:279–367.
63. Palade GE, Simionescu M, Simionescu N. Structural aspects of the permeability of the microvascular endothelium. Acta Physiol Scand Suppl 1979; 463:11–32.
64. Tsukita S, Furuse M, Itoh M. Multifunctional strands in tight junctions. Nat Rev Mol Cell Biol 2001; 2:285–293.
65. D'Atri F, Citi S. Molecular complexity of vertebrate tight junctions (Review). Mol Membr Biol 2002; 19:103–112.
66. Schneeberger EE, Lynch RD. The tight junction: a multifunctional complex. Am J Physiol Cell Physiol 2004; 286:C1213–C1228.
67. Aijaz S, Balda MS, Matter K. Tight junctions: molecular architecture and function. Int Rev Cytol 2006; 248:261–298.
68. Van Itallie CM, Anderson JM. The molecular physiology of tight junction pores. Physiology (Bethesda) 2004; 19:331–338.
69. Nieuwdorp M, Meuwese MC, Vink H, Hoekstra JB, Kastelein JJ, Stroes ES. The endothelial glycocalyx: a potential barrier between health and vascular disease. Curr Opin Lipidol 2005; 16:507–511.
70. Gouverneur M, Berg B, Nieuwdorp M, Stroes E, Vink H. Vasculoprotective properties of the endothelial glycocalyx: effects of fluid shear stress. J Intern Med 2006; 259:393–400.
71. Kalluri R. Basement membranes: structure, assembly and role in tumour angiogenesis. Nat Rev Cancer 2003; 3:422–433.
72. Brown B, Lindberg K, Reing J, Stolz DB, Badylak SF. The basement membrane component of biologic scaffolds derived from extracellular matrix. Tissue Eng 2006; 12:519–526.
73. Macarak EJ, Howard PS. Adhesion of endothelial cells to extracellular matrix proteins. J Cell Physiol 1983; 116:76–86.
74. Alexander JS, Elrod JW. Extracellular matrix, junctional integrity and matrix metalloproteinase interactions in endothelial permeability regulation. J Anat 2002; 200:561–574.
75. Wu MH. Endothelial focal adhesions and barrier function. J Physiol 2005; 569:359–366.
76. Tilling T, Korte D, Hoheisel D, Galla HJ. Basement membrane proteins influence brain capillary endothelial barrier function in vitro. J Neurochem 1998; 71:1151–1157.
77. Savettieri G, Di Liegro I, Catania C, et al. Neurons and ECM regulate occludin localization in brain endothelial cells. Neuroreport 2000; 11:1081–1084.
78. Tornquist P, Alm A, Bill A. Studies on ocular blood flow and retinal capillary permeability to sodium in pigs. Acta Physiol Scand 1979; 106:343–350.
79. Takata K, Hirano H, Kasahara M. Transport of glucose across the blood–tissue barriers. Int Rev Cytol 1997; 172:1–53.
80. Nilius B, Droogmans G. Ion channels and their functional role in vascular endothelium. Physiol Rev 2001; 81:1415–1419.
81. Mann GE, Yudilevich DL, Sobrevia L. Regulation of amino acid and glucose transporters in endothelial and smooth muscle cells. Physiol Rev 2003; 83:183–252.
82. Kumagai AK. Glucose transport in brain and retina: implications in the management and complications of diabetes. Diabetes Metab Res Rev 1999; 15:261–273.
83. Cogan DG, Kuwabara T. Comparison of retinal and cerebral vasculature in trypsin digest preparations. Br J Ophthalmol 1984; 68:10–12.
84. Frank RN, Turczyn TJ, Das A. Pericyte coverage of retinal and cerebral capillaries. Invest Ophthalmol Vis Sci 1990; 31:999–1007.
85. Wu DM, Kawamura H, Sakagami K, Kobayashi M, Puro DG. Cholinergic regulation of pericyte-containing retinal microvessels. Am J Physiol Heart Circ Physiol 2003; 284:H2083–H2090.
86. Allt G, Lawrenson JG. Pericytes: cell biology and pathology. Cells Tissues Organs 2001; 169:1–11.
87. Bandopadhyay R, Orte C, Lawrenson JG, Reid AR, De Silva S, Allt G. Contractile proteins in pericytes at the blood–brain and blood–retinal barriers. J Neurocytol 2001; 30:35–44.
88. Tomasek JJ, Haaksma CJ, Schwartz RJ, et al. Deletion of smooth muscle alpha-actin alters blood–retina barrier permeability and retinal function. Invest Ophthalmol Vis Sci 2006; 47:2693–2700.
89. Peppiatt CM, Howarth C, Mobbs P, Attwell D. Bidirectional control of CNS capillary diameter by pericytes. Nature 2006; 443:700–704.
90. Frey A, Meckelein B, Weiler-Guttler H, Mockel B, Flach R, Gassen HG. Pericytes of the brain microvasculature express gamma-glutamyl transpeptidase. Eur J Biochem 1991; 202:421–429.

91. Rungger-Brändle E, Messerli JM, Niemeyer G, Eppenberger HM. Confocal microscopy and computer-assisted image reconstruction of astrocytes in the mammalian retina. Eur J Neurosci 1993; 5:1093–1106.

92. Tao-Cheng JH, Nagy Z, Brightman MW. Tight junctions of brain endothelium in vitro are enhanced by astroglia. J Neurosci 1987; 7:3293–3299.

93. Wolburg H, Lippoldt A. Tight junctions of the blood–brain barrier: development, composition and regulation. Vascul Pharmacol 2002; 38:323–337.

94. Igarashi Y, Chiba H, Utsumi H, et al. Expression of receptors for glial cell line-derived neurotrophic factor (GDNF) and neurturin in the inner blood–retinal barrier of rats. Cell Struct Funct 2000; 25:237–241.

95. Behzadian MA, Wang XL, Windsor LJ, Ghaly N, Caldwell RB. TGF-beta increases retinal endothelial cell permeability by increasing MMP-9: possible role of glial cells in endothelial barrier function. Invest Ophthalmol Vis Sci 2001; 42:853–859.

96. Drescher KM, Whittum-Hudson JA. Herpes simplex virus type 1 alters transcript levels of tumor necrosis factor-alpha and interleukin-6 in retinal glial cells. Invest Ophthalmol Vis Sci 1996; 37:2302–2312.

97. Ghassemifar R, Lai CM, Rakoczy PE. VEGF differentially regulates transcription and translation of ZO-1alpha+ and ZO-1alpha- and mediates trans-epithelial resistance in cultured endothelial and epithelial cells. Cell Tissue Res 2006; 323:117–125.

98. Tornquist P. Capillary permeability in cat choroid, studied with the single injection technique (II). Acta Physiol Scand 1979; 106:425–430.

99. Chylack LT, Jr., Bellows AR. Molecular sieving in suprachoroidal fluid formation in man. Invest Ophthalmol Vis Sci 1978; 17:420–427.

100. Rizzolo LJ. Polarity and the development of the outer blood–retinal barrier. Histol Histopathol 1997; 12:1057–1067.

101. Miller SS, Steinberg RH. Active transport of ions across frog retinal pigment epithelium. Exp Eye Res 1977; 25:235–248.

102. Miller SS, Steinberg RH. Passive ionic properties of frog retinal pigment epithelium. J Membr Biol 1977; 36:337–372.

103. Nguyen-Legros J. Fine structure of the pigment epithelium in the vertebrate retina. Int Rev Cytol Suppl 1978; 287–328.

104. Lin WL, Essner E, McCarthy KJ, Couchman JR. Ultrastructural immunocytochemical localization of chondroitin sulfate proteoglycan in Bruch's membrane of the rat. Invest Ophthalmol Vis Sci 1992; 33:2072–2075.

105. Streilein JW. Ocular immune privilege: the eye takes a dim but practical view of immunity and inflammation. J Leukoc Biol 2003; 74:179–185.

106. Pederson JE, Green K. Aqueous humor dynamics: experimental studies. Exp Eye Res 1973; 15:277–297.

107. Pederson JE, Green K. Aqueous humor dynamics: a mathematical approach to measurement of facility, pseudofacility, capillary pressure, active secretion and X c. Exp Eye Res 1973; 15:265–276.

108. Lightman SL, Palestine AG, Rapoport SI, Rechthand E. Quantitative assessment of the permeability of the rat blood–retinal barrier to small water-soluble non-electrolytes. J Physiol 1987; 389:483–490.

109. Bill A. Quantitative determination of uveal blood flow in rabbits. Acta Physiol Scand 1962; 55:101–110.

110. Alm A, Bill A. Ocular and optic nerve blood flow at normal and increased intraocular pressures in monkeys (Macaca irus): a study with radioactively labelled microspheres including flow determinations in brain and some other tissues. Exp Eye Res 1973; 15:15–29.

111. Armaly MF, Araki M. Effect of ocular pressure on choroidal circulation in the cat and Rhesus monkey. Invest Ophthalmol 1975; 14:584–591.

112. Bill A. A method for quantitative determination of the blood flow through the cat uvea. Arch Ophthalmol 1962; 67:156–162.

113. Friedman E, Kopald HH, Smith TR. Retinal and choroidal blood flow determined with Krypton-85 anesthetized animals. Invest Ophthalmol 1964; 3:539–547.

114. Yu DY, Alder V, Cringle SJ, Brown MJ. Choroidal blood flow measured in the dog eye in vivo and in vitro by local hydrogen clearance polarography: validation of a technique and response to raised intraocular pressure. Exp Eye Res 1988; 46:289–303.

115. Sossi N, Anderson DR. Effect of elevated intraocular pressure on blood flow. Occurrence in cat optic nerve head studied with iodoantipyrine I 125. Arch Ophthalmol 1983; 101:98–101.

116. Suzuki Y, Masuda K, Ogino K, Sugita T, Aizu Y, Asakura T. Measurement of blood flow velocity in retinal vessels utilizing laser speckle phenomenon-. Jpn J Ophthalmol 1991; 35:4–15.

117. Guran T, Zeimer RC, Shahidi M, Mori MT. Quantitative analysis of retinal hemodynamics using targeted dye delivery. Invest Ophthalmol Vis Sci 1990; 31:2300–2306.

118. Nishiwaki H, Ogura Y, Kimura H, Kiryu J, Honda Y. Quantitative evaluation of leukocyte dynamics in retinal microcirculation. Invest Ophthalmol Vis Sci 1995; 36:123–130.

119. Kiel JW, Shepherd AP. Autoregulation of choroidal blood flow in the rabbit. Invest Ophthalmol Vis Sci 1992; 33:2399–2410.

120. O'Day DM, Fish MB, Aronson SB, Coon A, Pollycove M. Ocular blood flow measurement by nuclide labeled microspheres. Arch Ophthalmol 1971; 86:205–209.

121. Alm A, Bill A. The oxygen supply to the retina. II. Effects of high intraocular pressure and of increased arterial carbon dioxide tension on uveal and retinal blood flow in cats. A study with radioactively labelled microspheres including flow determinations in brain and some other tissues. Acta Physiol Scand 1972; 84:306–319.

121a. Törnquist P, Alm A. Retinal and choroidal contribution to retinal metabolism in vivo: a study in pigs. Acta Physiol Scand 1979;106:351-357.

122. Geijer C, Bill A. Effects of raised intraocular pressure on retinal, prelaminar, laminar, and retrolaminar optic nerve blood flow in monkeys. Invest Ophthalmol Vis Sci 1979; 18:1030–1042.

123. Chiou GC, Zhao F, Shen ZF, Li BH. Effects of D-timolol and L-timolol on ocular blood flow and intraocular pressure. J Ocul Pharmacol 1990; 6:23–30.

124. Nork TM, Kim CB, Shanmuganayagam D, Van Lysel MS, Ver Hoeve JN, Folts JD. Measurement of regional choroidal blood flow in rabbits and monkeys using fluorescent microspheres. Arch Ophthalmol 2006; 124:860–868.

125. Hillerdal M, Sperber GO, Bill A. The microsphere method for measuring low blood flows: theory and computer simulations applied to findings in the rat cochlea. Acta Physiol Scand 1987; 130:229–235.

126. Alm A, Bill A. Blood flow and oxygen extraction in the cat uvea at normal and high intraocular pressures. Acta Physiol Scand 1970; 80:19–28.

127. Alder VA, Cringle SJ, Constable IJ. The retinal oxygen profile in cats. Invest Ophthalmol Vis Sci 1983; 24:30–36.

128. Linsenmeier RA. Effects of light and darkness on oxygen distribution and consumption in the cat retina. J Gen Physiol 1986; 88:521–542.

129. Sperber GO, Bill A. Blood flow and glucose consumption in the optic nerve, retina and brain: effects of high intraocular pressure. Exp Eye Res 1985; 41:639–653.

130. Bill A, Sperber GO. Aspects of oxygen and glucose consumption in the retina: effects of high intraocular pressure and light. Graefes Arch Clin Exp Ophthalmol 1990; 228:124–127.

131. Delori FC, Fitch KA, Feke GT, Deupree DM, Weiter JJ. Evaluation of micrometric and microdensitometric methods for measuring the width of retinal vessel images on fundus photographs. Graefes Arch Clin Exp Ophthalmol 1988; 226:393–399.

132. Nagel E, Munch K, Vilser W. Measurement of the diameter of segments of retinal branch vessels in digital fundus images – an experimental study of the method and reproducibility. Klin Monatsbl Augenheilkd 2001; 218:616–620.

133. Meier P, Zierler KL. On the theory of the indicator-dilution method for measurement of blood flow and volume. J Appl Physiol 1954; 6:731–744.

134. Gonzalez-Fernandez JM. Theory of the measurement of the dispersion of an indicator in indicator-dilution studies. Circ Res 1962; 10:409–428.

135. Riva CE, Ben-Sira I. Injection method for ocular hemodynamic studies in man. Invest Ophthalmol 1974; 13:77–79.

136. Riva CE, Feke GT, Ben-Sira I. Fluorescein dye-dilution technique and retinal circulation. Am J Physiol 1978; 234:H315–H322.

137. Sperber GO, Alm A. Retinal mean transit time determined with an impulse-response analysis from video fluorescein angiograms. Acta Ophthalmol Scand 1997; 75:532–536.

138. Wolf S, Arend O, Reim M. Measurement of retinal hemodynamics with scanning laser ophthalmoscopy: reference values and variation. Surv Ophthalmol 1994; 38 Suppl:S95–100.

139. Wolf S, Jung F, Kiesewetter H, Korber N, Reim M. Video fluorescein angiography: method and clinical application. Graefes Arch Clin Exp Ophthalmol 1989; 227:145–151.

140. Tomic L, Maepea O, Sperber GO, Alm A. Comparison of retinal transit times and retinal blood flow: a study in monkeys. Invest Ophthalmol Vis Sci 2001; 42:752–755.

141. Bischoff PM, Flower RW. Ten years experience with choroidal angiography using indocyanine green dye: a new routine examination or an epilogue? Doc Ophthalmol 1985; 60:235–291.

142. Hasegawa Y, Hayashi K, Tokoro T, De Laey JJ. Clinical use of indocyanine green angiography in the diagnosis of choroidal neovascular diseases. Fortschr Ophthalmol 1988; 85:410–412.

143. Podoleanu AG, Dobre GM, Cernat R, et al. Investigations of the eye fundus using a simultaneous optical coherence tomography/indocyanine green fluorescence imaging system. J Biomed Opt 2007; 12:14–19.

144. Logean E, Schmetterer L, Riva CE. Velocity profile of red blood cells in human retinal vessels using confocal scanning laser Doppler velocimetry. Laser Phys 2003; 13:45–51.

145. Leitgeb RA, Schmetterer L, Drexler W, Fercher AF. Real-time assessment of retinal blood flow with ultrafast acquisition by color Doppler Fourier domain optical coherence tomography. Opt Express 2003; 11:3116–3121.

146. Yazdanfar S, Rollins AM, Izatt JA. In vivo imaging of human retinal flow dynamics by color Doppler optical coherence tomography. Arch Ophthalmol 2003; 121:235–239.

147. Wang Y, Bower BA, Izatt JA, Tan O, Huang D. In vivo total retinal blood flow measurement by Fourier domain Doppler optical coherence tomography. J Biomed Opt 2007; 12:412–415.

148. Szkulmowska A, Szkulmowski M, Szlag D, Kowalczyk A, Wojtkowski M. Three-dimensional quantitative imaging of retinal and choroidal blood flow velocity using joint spectral and time domain optical coherence tomography. Opt Express 2009; 17:10584–10598.

149. Le-Cong P, Zweifach BW. In vivo and in vitro velocity measurements in microvasculature with a laser. Microvasc Res 1979; 17:131–141.

150. Zweifach BW, Lipowsky HH. Quantitative studies of microcirculatory structure and function. III. Microvascular hemodynamics of cat mesentery and rabbit omentum. Circ Res 1977; 41:380–390.

151. Bonner RF, Nossal R. Principles of Laser Doppler Flowmetry. In: Shepherd AP, Öberg PA, eds. Laser-Doppler blood flowmetry. Boston: Kluwer, 1990:57–72.

152. Riva CE, Petrig BL. Laser Doppler techniques in ophthamology – principles and applications. In: Fankhauser F, Kwasniewska S, eds. Lasers in ophthalmology – basic, diagnostic and surgical aspects. The Hague, The Netherlands: Kugler, 2003:51–59.

153. Michelson G, Schmauss B. Two dimensional mapping of the perfusion of the retina and optic nerve head. Br J Ophthalmol 1995; 79:1126–1132.

154. Fercher AF, Briers JD. Flow visualization by means of single-exposure speckle photography. Optical Comm 1981; 37:326–330.

155. Briers JD, Fercher AF. Retinal blood–flow visualization by means of laser speckle photography. Invest Ophthalmol Vis Sci 1982; 22:255–259.

156. Briers JD. Laser Doppler and time-varying speckle: a reconciliation. J Opt Soc Am A 1996; 13:345–350.

157. Tamaki Y, Araie M, Kawamoto E, Eguchi S, Fujii H. Non-contact, two-dimensional measurement of tissue circulation in choroid and optic nerve head using laser speckle phenomenon-. Exp Eye Res 1995; 60:373–383.

158. Zinser G. Scanning Laser Doppler flowmetry. In: Pillunat LE, Harris A, Anderson DR, Greve EL eds. Current concepts on ocular blood flow in glaucoma. The Hague, The Netherlands: Kugler, 1999:197–204.

159. Riva CE, Petrig BL. Laser doppler flowmetry in the optic nerve head. In: Drance SM, ed. Vascular risk factors and neuroprotection in glaucoma – update 1996. Amsterdam / New York: Kugler, 1997:43–55.

160. Riva CE, Petrig B. Blue field entoptic phenomenon- and blood velocity in the retinal capillaries. J Opt Soc Am 1980; 70:1234–1238.

161. Loebl M, Riva CE. Macular circulation and the flying corpuscles phenomenon-. Ophthalmology 1978; 85:911–917.

162. Martin JA, Roorda A. Direct and non-invasive assessment of parafoveal capillary leukocyte velocity. Ophthalmology 2005; 112:2219–2224.

163. Lieb WE, Cohen SM, Merton DA, Shields JA, Mitchell DG, Goldberg BB. Color Doppler imaging of the eye and orbit. Technique and normal vascular anatomy. Arch Ophthalmol 1991; 109:527–531.

164. Harris A, Kagemann L, Cioffi GA. Assessment of human ocular hemodynamics. Surv Ophthalmol 1998; 42:509–533.

165. Polska E, Kircher K, Ehrlich P, Vecsei PV, Schmetterer L. RI in central retinal artery as assessed by CDI does not correspond to retinal vascular resistance. Am J Physiol Heart Circ Physiol 2001; 280:H1442–H1447.

166. Taylor GA, Short BL, Walker LK, Traystman RJ. Intracranial blood flow: quantification with duplex Doppler and color Doppler flow US. Radiology 1990; 176:231–236.

167. Langham ME, Farrell RA, O'Brien V, Silver DM, Schilder P. Blood flow in the human eye. Acta Ophthalmol Suppl 1989; 191:9–13.

168. Langham ME, Farrell RA, O'Brien V, Silver DM, Schilder P. Non-invasive measurement of pulsatile blood flow in the human eye. In: Lambrou GN, Greve EL, eds. Ocular blood flow in glaucoma means, methods and measurements. Amstelveen, The Netherlands: Kugler & Ghedini, 1989:93–99.

169. Schmetterer L. Measurement of in vivo fundus pulsations on the eye by laser interferometry. Opt Eng 1997; 34:711–716.

170. Flammer J. The concept of vascular dysregulation in glaucoma. In: Haefliger IO, Flammer J, eds. Nitric oxide and endothelin in the pathogenesis of glaucoma. Philadelphia: Lippincott-Raven, 1998:14–19.

171. Rechtman E, Harris A, Kumar R, et al. An update on retinal circulation assessment technologies. Curr Eye Res 2003; 27:329–343.

172. Riva CE. Débit vasculaire oculaire. In: Pournaras CJ, ed. Pathologies vasculaires. Issy-les-Molineaux: Masson, 2008:53–65.

173. Murray CD. The physiological principle of minimum work: I. The vascular system and the cost of blood volume. Proc Natl Acad Sci USA 1926; 12:207–214.

174. LaBarbera M. The design of fluid transport systems: a comparative perspective. In: Bevan JA, Kaley G, Rubanyi GM, eds. Flow-dependent regulation of vascular function. New York: Oxford University Press, 1995:3–27.

175. Glucksberg MR, Dunn R. Direct measurement of retinal microvascular pressures in the live, anesthetized cat. Microvasc Res 1993; 45:158–165.

176. Stoltz JF, Donner M. New trends in clinical hemorheology: an introduction to the concept of the hemorheological profile. Schweiz Med Wochenschr Suppl 1991; 43:41–49.

177. Knabben H, Wolf S, Remky A, Schulte K, Arend O, Reim M. Retinal hemodynamics in patients with hyperviscosity syndrome. Klin Monatsbl Augenheilkd 1995; 206:152–156.

178. Alm A. Ocular circulation. In: Hart W, ed. Adler's Physiology of the eye. St Louis: Mosby-Year Book, 1992:198–227.

179. Alm A, Tornquist P, Stjernschantz J. Radioactively labelled microspheres in regional ocular blood flow determinations. Bibl Anat 1977; 24–29.

180. Bill A. Circulation in the Eye. In: Renkin EM, Michel CC, eds. Handbook of physiology. Baltimore: Waverly Press, 1984:1001–1034.

181. Wang Y, Lu A, Gil-Flamer JH, Tan O, Izatt JA, Huang D. Measurement of total blood flow in the normal human retina using Doppler Fourier-domain optical coherence tomography. Br J Ophthalmol 2009; 93:634–637.

182. Riva CE, Grunwald JE, Sinclair SH, Petrig BL. Blood velocity and volumetric flow rate in human retinal vessels. Invest Ophthalmol Vis Sci 1985; 26:1124–1132.

183. Kimura I, Shinoda K, Tanino T, Ohtake Y, Mashima Y, Oguchi Y. Scanning laser Doppler flowmeter study of retinal blood flow in macular area of healthy volunteers. Br J Ophthalmol 2003; 87:1469–1473.

184. Friedman E. Choroidal blood flow. Pressure-flow relationships. Arch Ophthalmol 1970; 83:95–99.

185. Bill A. Blood circulation and fluid dynamics in the eye. Physiol Rev 1975; 55:383–417.

186. Ruskell G. Blood vessels of the orbit and globe. In: Prince J, ed. The rabbit in eye research. Springfield, IL: Charles C Thomas, 1964:514–553.

187. Alder VA, Cringle SJ. Vitreal and retinal oxygenation. Graefes Arch Clin Exp Ophthalmol 1990; 228:151–157.

188. Bill A, Sperber GO. Control of retinal and choroidal blood flow. Eye 1990; 4(Pt 2):319–325.

189. Reitsamer HA, Kiel JW. Circulation choroïdienne. In: Pournaras CJ, ed. Pathologies vasculaires. Issy-les-Molineaux: Masson, 2008:75–86.

190. Alm A, Bill A. Ocular circulation. In: Moses RA, Adler HW, eds. Adler's Physiology of the eye. St. Louis: Mosby, 1987:183–203.

191. Bill A, Sperber G, Ujiie K. Physiology of the choroidal vascular bed. Int Ophthalmol 1983; 6:101–107.

192. Longo A, Geiser M, Riva CE. Subfoveal choroidal blood flow in response to light-dark exposure. Invest Ophthalmol Vis Sci 2000; 41:2678–2683.

193. Parver LM. Temperature modulating action of choroidal blood flow. Eye 1991; 5(Pt 2):181–185.

194. Brubaker RF. Determination of episcleral venous pressure in the eye. A comparison of three methods. Arch Ophthalmol 1967; 77:110–114.

195. Podos SM, Minas TF, Macri FJ. A new instrument to measure episcleral venous pressure. Comparison of normal eyes and eyes with primary open-angle glaucoma. Arch Ophthalmol 1968; 80:209–213.

196. Bill A. Blood pressure in the ciliary arteries of rabbits. Exp Eye Res 1963; 2:20–24.

197. Maepea O. Pressures in the anterior ciliary arteries, choroidal veins and choriocapillaris. Exp Eye Res 1992; 54:731–736.

198. Alm A, Bill A, Young FA. The effects of pilocarpine and neostigmine on the blood flow through the anterior uvea in monkeys. A study with radioactively labelled microspheres. Exp Eye Res 1973; 15:31–63.

199. Linner E. A method for determining the rate of plasma flow through the secretory part of the ciliary body. Acta Physiol Scand 1951; 22:83–86.

200. Linner E. Ascorbic acid as a test substance for measuring relative changes in the rate of plasma flow through the ciliary processes. I. The effect of unilateral ligation of the common carotid artery in rabbits on the ascorbic acid content of the aqueous humour at varying plasma levels. Acta Physiol Scand 1952; 26:57–69.

201. Lemmingson W. The occurrence of vasomotion in the retinal circulation. Albrecht Von Graefes Arch Klin Exp Ophthalmol 1968; 176:368–377.

202. Braun RD, Linsenmeier RA, Yancey CM. Spontaneous fluctuations in oxygen tension in the cat retina. Microvasc Res 1992; 44:73–84.

203. Buerk DG, Riva CE. Vasomotion and spontaneous low-frequency oscillations in blood flow and nitric oxide in cat optic nerve head. Microvasc Res 1998; 55:103–112.

204. Riva CE, Pournaras CJ, Poitry-Yamate CL, Petrig BL. Rhythmic changes in velocity, volume, and flow of blood in the optic nerve head tissue. Microvasc Res 1990; 40:36–45.

205. Osusky R, Schoetzau A, Flammer J. Variations in the blood flow of the human optic nerve head. Eur J Ophthalmol 1997; 7:364–369.

206. Riva CE, Maret Y, Polak K, Logean E. Wavelet transform (WT) of temporal fluctuations in optic nerve and choroidal blood flow and retinal vessel diameter. Invest Ophthalmol Vis Sci 2000; 41:516.

207. Leung H, Wang JJ, Rochtchina E, et al. Relationships between age, blood pressure, and retinal vessel diameters in an older population. Invest Ophthalmol Vis Sci 2003; 44:2900–2904.

208. Tien Yin W, Ronald K, Sharrett AR, et al. The prevalence and risk factors of retinal microvascular abnormalities in older persons: the Cardiovascular Health Study. Ophthalmology 2003; 110:658–666.

209. Wong TY, Klein R, Klein BE, Meuer SM, Hubbard LD. Retinal vessel diameters and their associations with age and blood pressure. Invest Ophthalmol Vis Sci 2003; 44:4644–4650.

210. Embleton SJ, Hosking SL, Roff Hilton EJ, Cunliffe IA. Effect of senescence on ocular blood flow in the retina, neuroretinal rim and lamina cribrosa, using scanning laser Doppler flowmetry. Eye 2002; 16:156–162.

211. Groh MJ, Michelson G, Langhans MJ, Harazny J. Influence of age on retinal and optic nerve head blood circulation. Ophthalmology 1996; 103:529–534.

212. Rizzo JF, 3rd, Feke GT, Goger DG, Ogasawara H, Weiter JJ. Optic nerve head blood speed as a function of age in normal human subjects. Invest Ophthalmol Vis Sci 1991; 32:3263–3272.

213. Grunwald JE, Piltz J, Patel N, Bose S, Riva CE. Effect of aging on retinal macular microcirculation: a blue field simulation study. Invest Ophthalmol Vis Sci 1993; 34:3609–3613.

214. Gao H, Hollyfield JG. Aging of the human retina. Differential loss of neurons and retinal pigment epithelial cells. Invest Ophthalmol Vis Sci 1992; 33:1–17.

215. Jonas JB, Nguyen NX, Naumann GO. The retinal nerve fiber layer in normal eyes. Ophthalmology 1989; 96:627–632.

216. Grunwald JE, Hariprasad SM, DuPont J. Effect of aging on foveolar choroidal circulation. Arch Ophthalmol 1998; 116:150–154.

217. Ito YN, Mori K, Young-Duvall J, Yoneya S. Aging changes of the choroidal dye filling pattern in indocyanine green angiography of normal subjects. Retina 2001; 21:237–242.

218. Ramrattan RS, van der Schaft TL, Mooy CM, de Bruijn WC, Mulder PG, de Jong PT. Morphometric analysis of Bruch's membrane, the choriocapillaris, and the choroid in aging. Invest Ophthalmol Vis Sci 1994; 35:2857–2864.

219. Haefliger IO, Beny JL, Luscher TF. Endothelium-dependent vasoactive modulation in the ophthalmic circulation. Prog Retin Eye Res 2001; 20:209–225.

220. Metea MR, Newman EA. Glial cells dilate and constrict blood vessels: a mechanism of neurovascular coupling. J Neurosci 2006; 26:2862–2870.

221. Guyton AC, Jones CJ, Coleman TJ. Circulatory physiology: cardiac output and its regulation, 2nd edn. Philadelphia: WB Saunders, 1973.

222. Delaey C, Van De Voorde J. Regulatory mechanisms in the retinal and choroidal circulation. Ophthalmic Res 2000; 32:249–256.

223. Weinstein JM, Funsch D, Page RB, Brennan RW. Optic nerve blood flow and its regulation. Invest Ophthalmol Vis Sci 1982; 23:640–645.

224. Riva CE, Cranstoun SD, Petrig BL. Effect of decreased ocular perfusion pressure on blood flow and the flicker-induced flow response in the cat optic nerve head. Microvasc Res 1996; 52:258–269.

225. Nagel E, Vilser W. Autoregulative behavior of retinal arteries and veins during changes of perfusion pressure: a clinical study. Graefes Arch Clin Exp Ophthalmol 2004; 242:13–17.

226. Riva CE, Loebl M. Autoregulation of blood flow in the capillaries of the human macula. Invest Ophthalmol Vis Sci 1977; 16:568–571.

227. Riva CE, Hero M, Titze P, Petrig B. Autoregulation of human optic nerve head blood flow in response to acute changes in ocular perfusion pressure. Graefes Arch Clin Exp Ophthalmol 1997; 235:618–626.

228. Pillunat LE, Anderson DR, Knighton RW, Joos KM, Feuer WJ. Autoregulation of human optic nerve head circulation in response to increased intraocular pressure. Exp Eye Res 1997; 64:737–744.

229. Grunwald JE, Sinclair SH, Riva CE. Autoregulation of the retinal circulation in response to decrease of intraocular pressure below normal. Invest Ophthalmol Vis Sci 1982; 23:124–127.

230. Riva CE, Grunwald JE, Petrig BL. Autoregulation of human retinal blood flow. An investigation with laser Doppler velocimetry. Invest Ophthalmol Vis Sci 1986; 27:1706–1712.

231. Yancey CM, Linsenmeier RA. Oxygen distribution and consumption in the cat retina at increased intraocular pressure. Invest Ophthalmol Vis Sci 1989; 30:600–611.

232. Shonat RD, Wilson DF, Riva CE, Cranstoun SD. Effect of acute increases in intraocular pressure on intravascular optic nerve head oxygen tension in cats. Invest Ophthalmol Vis Sci 1992; 33:3174–3180.

233. Alm A, Bill A. The oxygen supply to the retina. I. Effects of changes in intraocular and arterial blood pressures, and in arterial PO_2 and PCO_2 on the oxygen tension in the vitreous body of the cat. Acta Physiol Scand 1972; 84:261–274.

234. Bill A. Intraocular pressure and blood flow through the uvea. Arch Ophthalmol 1962; 67:336–348.

235. Kiel JW, van Heuven WA. Ocular perfusion pressure and choroidal blood flow in the rabbit. Invest Ophthalmol Vis Sci 1995; 36:579–585.

236. Reiner A, Zagvazdin Y, Fitzgerald ME. Choroidal blood flow in pigeons compensates for decreases in arterial blood pressure. Exp Eye Res 2003; 76:273–282.

237. Riva CE, Titze P, Hero M, Petrig BL. Effect of acute decreases of perfusion pressure on choroidal blood flow in humans. Invest Ophthalmol Vis Sci 1997; 38:1752–1760.

238. Simader C, Lung S, Weigert G, et al. Role of NO in the control of choroidal blood flow during a decrease in ocular perfusion pressure. Invest Ophthalmol Vis Sci 2009; 50:372–377.

239. Chamot SR, Movaffaghy A, Petrig BL, Riva CE. Iris blood flow response to acute decreases in ocular perfusion pressure: a laser Doppler flowmetry study in humans. Exp Eye Res 2000; 70:107–112.

240. Pournaras CJ. Autoregulation of ocular blood flow. In: Kaiser HJ, Flammer J, Hendrickson P, eds. Ocular blood flow: new insights into the pathogenesis of ocular diseases. Basel: Karger, 1996:40–50.

241. Anderson DR. Introductory comments on blood flow autoregulation in the optic nerve head and vascular risk factors in glaucoma. Surv Ophthalmol 1999; 43(Suppl 1):S5–S9.
242. Aaslid R, Lindegaard KF, Sorteberg W, Nornes H. Cerebral autoregulation dynamics in humans. Stroke 1989; 20:45–52.
243. Halpern W, Osol G. Influence of transmural pressure of myogenic responses of isolated cerebral arteries of the rat. Ann Biomed Eng 1985; 13:287–293.
244. Bevan JA, Hwa JJ. Myogenic tone and cerebral vascular autoregulation: the role of a stretch-dependent mechanism. Ann Biomed Eng 1985; 13:281–286.
245. Okuno T, Oku H, Sugiyama T, Yang Y, Ikeda T. Evidence that nitric oxide is involved in autoregulation in optic nerve head of rabbits. Invest Ophthalmol Vis Sci 2002; 43:784–789.
246. Lind AR, Taylor SH, Humphreys PW, Kennelly BM, Donald KW. The circulatory effects of sustained voluntary muscle contraction. Clin Sci 1964; 27:229–244.
247. Robinson F, Riva CE, Grunwald JE, Petrig BL, Sinclair SH. Retinal blood flow autoregulation in response to an acute increase in blood pressure. Invest Ophthalmol Vis Sci 1986; 27:722–726.
248. Dumskyj MJ, Eriksen JE, Dore CJ, Kohner EM. Autoregulation in the human retinal circulation: assessment using isometric exercise, laser Doppler velocimetry, and computer-assisted image analysis. Microvasc Res 1996; 51:378–392.
249. Riva CE, Titze P, Hero M, Movaffaghy A, Petrig BL. Choroidal blood flow during isometric exercises. Invest Ophthalmol Vis Sci 1997; 38:2338–2343.
250. Movaffaghy A, Chamot SR, Petrig BL, Riva CE. Blood flow in the human optic nerve head during isometric exercise. Exp Eye Res 1998; 67:561–568.
251. Kiss B, Dallinger S, Polak K, Findl O, Eichler HG, Schmetterer L. Ocular hemodynamics during isometric exercise. Microvasc Res 2001; 61:1–13.
252. Blum M, Bachmann K, Wintzer D, Riemer T, Vilser W, Strobel J. Non-invasive measurement of the Bayliss effect in retinal autoregulation. Graefes Arch Clin Exp Ophthalmol 1999; 237:296–300.
253. Alm A. The effect of sympathetic stimulation on blood flow through the uvea, retina and optic nerve in monkeys (Macacca irus). Exp Eye Res 1977; 25:19–24.
254. Alm A, Bill A. The effect of stimulation of the cervical sympathetic chain on retinal oxygen tension and on uveal, retinal and cerebral blood flow in cats. Acta Physiol Scand 1973; 88:84–94.
255. Fuchsjager-Mayrl G, Luksch A, Malec M, Polska E, Wolzt M, Schmetterer L. Role of endothelin-1 in choroidal blood flow regulation during isometric exercise in healthy humans. Invest Ophthalmol Vis Sci 2003; 44:728–733.
256. Luksch A, Polska E, Imhof A, et al. Role of NO in choroidal blood flow regulation during isometric exercise in healthy humans. Invest Ophthalmol Vis Sci 2003; 44:734–739.
257. Marcus DF, Edelhauser HF, Maksud MG, Wiley RL. Effects of a sustained muscular contraction on human intraocular pressure. Clin Sci Mol Med 1974; 47:249–257.
258. McArdle WD, Katch FL, Katch VL. The cardiovascular system. Exercise physiology energy, nutrition, and human performance. Baltimore: Williams & Wilkins, 1996:267–283.
259. Michelson G, Groh M, Grundler A. Regulation of ocular blood flow during increases of arterial blood pressure. Br J Ophthalmol 1994; 78:461–465.
260. Harris A, Arend O, Bohnke K, Kroepfl E, Danis R, Martin B. Retinal blood flow during dynamic exercise. Graefes Arch Clin Exp Ophthalmol 1996; 234:440–444.
261. Forcier P, Kergoat H, Lovasik JV. Macular hemodynamic responses to short-term acute exercise in young healthy adults. Vision Res 1998; 38:181–186.
262. Lovasik JV, Kergoat H, Riva CE, Petrig BL, Geiser M. Choroidal blood flow during exercise-induced changes in the ocular perfusion pressure. Invest Ophthalmol Vis Sci 2003; 44:2126–2132.
263. Evans DW, Harris A, Garrett M, Chung HS, Kagemann L. Glaucoma patients demonstrate faulty autoregulation of ocular blood flow during posture change. Br J Ophthalmol 1999; 83:809–813.
264. James CB, Smith SE. The effect of posture on the intraocular pressure and pulsatile ocular blood flow in patients with non-arteritic anterior ischaemic optic neuropathy. Eye 1991; 5(Pt 3):309–314.
265. Longo A, Geiser MH, Riva CE. Posture changes and subfoveal choroidal blood flow. Invest Ophthalmol Vis Sci 2004; 45:546–551.
266. Sayegh FN, Weigelin E. Functional ophthalmodynamometry. Comparison between brachial and ophthalmic blood pressure in sitting and supine position. Angiology 1983; 34:176–182.
267. Kothe AC. The effect of posture on intraocular pressure and pulsatile ocular blood flow in normal and glaucomatous eyes. Surv Ophthalmol 1994; 38(Suppl):S191–S197.
268. Lovasik JV, Kergoat H. Gravity-induced homeostatic reactions in the macular and choroidal vasculature of the human eye. Aviat Space Environ Med 1994; 65:1010–1014.
269. Feke GT, Pasquale LR. Retinal blood flow response to posture change in glaucoma subjects compared with healthy subjects. Ophthalmology 2008; 115:246–252.
270. Sayegh FN, Weigelin EF. Functional ophthalmodynamometry: comparison between dynamometry findings of healthy subjects in sitting and supine positions. Ophthalmologica 1983; 187:196–201.
271. Kaeser P, Orgul S, Zawinka C, Reinhard G, Flammer J. Influence of change in body position on choroidal blood flow in normal subjects. Br J Ophthalmol 2005; 89:1302–1305.
272. Eperon G, Johnson M, David NJ. The effect of arterial PO_2 on relative retinal blood flow in monkeys. Invest Ophthalmol 1975; 14:342–352.
273. Riva CE, Pournaras CJ, Tsacopoulos M. Regulation of local oxygen tension and blood flow in the inner retina during hyperoxia. J Appl Physiol 1986; 61:592–598.
274. Stefansson E, Wagner HG, Seida M. Retinal blood flow and its autoregulation measured by intraocular hydrogen clearance. Exp Eye Res 1988; 47:669–678.
275. Hickam JB, Frayser R. Observation on vessels diameter, arteriovenous oxygen difference and mean circulation time. Circulation 1966; 33:302–316.
276. Riva CE, Grunwald JE, Sinclair SH. Laser Doppler velocimetry study of the effect of pure oxygen breathing on retinal blood flow. Invest Ophthalmol Vis Sci 1983; 24:47–51.
277. Jean-Louis S, Lovasik JV, Kergoat H. Systemic hyperoxia and retinal vasomotor responses. Invest Ophthalmol Vis Sci 2005; 46:1714–1720.
278. Fallon TJ, Maxwell D, Kohner EM. Retinal vascular autoregulation in conditions of hyperoxia and hypoxia using the blue field entoptic phenomenon-. Ophthalmology 1985; 92:701–705.
279. Kiss B, Polska E, Dorner G, et al. Retinal blood flow during hyperoxia in humans revisited: concerted results using different measurement techniques. Microvasc Res 2002; 64:75–85.
280. Kety SS, Schmidt CE. Effect of altered arterial tensions of carbon dioxide and oxygen on cerebral blood flow in normal young man. J Clin Invest 1948; 27:484–492.
281. Friedman E, Chandra SR. Choroidal blood flow. 3. Effects of oxygen and carbon dioxide. Arch Ophthalmol 1972; 87:70–71.
282. Riva CE, Cranstoun SD, Grunwald JE, Petrig BL. Choroidal blood flow in the foveal region of the human ocular fundus. Invest Ophthalmol Vis Sci 1994; 35:4273–4281.
283. Geiser MH, Riva CE, Dorner GT, Diermann U, Luksch A, Schmetterer L. Response of choroidal blood flow in the foveal region to hyperoxia and hyperoxia-hypercapnia. Curr Eye Res 2000; 21:669–676.
284. Dallinger S, Dorner GT, Wenzel R, et al. Endothelin-1 contributes to hyperoxia-induced vasoconstriction in the human retina. Invest Ophthalmol Vis Sci 2000; 41:864–869.
285. Linsenmeier RA, Yancey CM. Effects of hyperoxia on the oxygen distribution in the intact cat retina. Invest Ophthalmol Vis Sci 1989; 30:612–618.
286. Braun RD, Linsenmeier RA. Retinal oxygen tension and the electroretinogram during arterial occlusion in the cat. Invest Ophthalmol Vis Sci 1995; 36:523–541.
287. Berkowitz BA. Adult and newborn rat inner retinal oxygenation during carbogen and 100 percent oxygen breathing. Comparison using magnetic resonance imaging delta PO_2 mapping. Invest Ophthalmol Vis Sci 1996; 37:2089–2098.
288. Yu DY, Cringle SJ, Alder V, Su EN. Intraretinal oxygen distribution in the rat with graded systemic hyperoxia and hypercapnia. Invest Ophthalmol Vis Sci 1999; 40:2082–2087.
289. Yu DY, Cringle SJ, Su EN. Intraretinal oxygen distribution in the monkey retina and the response to systemic hyperoxia. Invest Ophthalmol Vis Sci 2005; 46:4728–4733.
290. Pournaras CJ, Riva CE, Tsacopoulos M, Strommer K. Diffusion of O_2 in the retina of anesthetized miniature pigs in normoxia and hyperoxia. Exp Eye Res 1989; 49:347–360.
291. Alder VA, Ben-Nun J, Cringle SJ. PO_2 profiles and oxygen consumption in cat retina with an occluded retinal circulation. Invest Ophthalmol Vis Sci 1990; 31:1029–1034.
292. Pournaras CJ, Tsacopoulos M, Bovet J, Roth A. Diffusion of O_2 in the normal and the ischemic retina of miniature pigs. Ophtalmologie 1990; 4:17–19.
293. Petropoulos IK, Pournaras JA, Stangos AN, Pournaras CJ. Effect of systemic nitric oxide synthase inhibition on optic disc oxygen partial pressure in normoxia and in hypercapnia. Invest Ophthalmol Vis Sci 2009; 50:378–384.
294. Pournaras C, Tsacopoulos M, Chapuis P. Studies on the role of prostaglandins in the regulation of retinal blood flow. Exp Eye Res 1978; 26:687–697.
295. Hickam JB, Frayser R, Ross JC. A study of retinal venous blood oxygen saturation in human subjects by photographic means. Circulation 1963; 27:375–385.
296. Ahmed J, Pulfer MK, Linsenmeier RA. Measurement of blood flow through the retinal circulation of the cat during normoxia and hypoxemia using fluorescent microspheres. Microvasc Res 2001; 62:143–153.
297. Kogure K, Scheinberg P, Reinmuth OM, Fujishima M, Busto R. Mechanisms of cerebral vasodilatation in hypoxia. J Appl Physiol 1970; 29:223–229.
298. Winkler BS. A quantitative assessment of glucose metabolism in the isolated rat retina. In: Doly CY, Droy-Lefaix MT, eds. Les seminaires ophthalmologiques d'IPSEN: vision et adaptation. Amsterdam: Elsevier, 1995:78–96.
299. Brazitikos PD, Pournaras CJ, Munoz J-L, Tsacopoulos M. Microinjection of L-lactate in the preretinal vitreous induces segmental vasodilation in the inner retina of miniature pigs. Invest Ophthalmol Vis Sci 1993; 34:1744–1752.
300. Nagaoka T, Sato E, Yoshida A. The effect of nitric oxide on retinal blood flow during hypoxia in cats. Invest Ophthalmol Vis Sci 2002; 43:3037–3044.
301. Enroth-Cugell C, Goldstick TK, Linsenmeier RA. The contrast sensitivity of cat retinal ganglion cells at reduced oxygen tensions. J Physiol 1980; 304:59–81.
302. Linsenmeier RA, Braun RD. Oxygen distribution and consumption in the cat retina during normoxia and hypoxemia. J Gen Physiol 1992; 99:177–197.
303. Pournaras CJ. Retinal oxygen distribution. Its role in the physiopathology of vasoproliferative microangiopathies. Retina 1995; 15:332–347.
304. Alder VA, Cringle SJ. Intraretinal and preretinal PO_2 response to acutely raised intraocular pressure in cats. Am J Physiol 1989; 256:H1627–H1634.
305. Moret P, Pournaras CJ, Munoz JL, Brazitikos P, Tsacopoulos M. Profile of pO_2. I. Profile of transretinal pO_2 in hypoxia. Klin Monatsbl Augenheilkd 1992; 200:498–499.
306. Tsacopoulos M, David NJ. The effect of arterial PCO_2 on relative retinal blood flow in monkeys. Invest Ophthalmol 1973; 12:335–347.
307. Dorner GT, Garhoefer G, Zawinka C, Kiss B, Schmetterer L. Response of retinal blood flow to CO_2-breathing in humans. Eur J Ophthalmol 2002; 12:459–466.
308. Tomic L, Bjarnhall G, Maepea O, Sperber GO, Alm A. Effects of oxygen and carbon dioxide on human retinal circulation: an investigation using blue field simulation and scanning laser ophthalmoscopy. Acta Ophthalmol Scand 2005; 83:705–710.
309. Harris A, Arend O, Wolf S, Cantor LB, Martin BJ. CO_2 dependence of retinal arterial and capillary blood velocity. Acta Ophthalmol Scand 1995; 73:421–424.
310. Sponsel WE, DePaul KL, Zetlan SR. Retinal hemodynamic effects of carbon dioxide, hyperoxia, and mild hypoxia. Invest Ophthalmol Vis Sci 1992; 33:1864–1869.
311. Chung HS, Harris A, Halter PJ, et al. Regional differences in retinal vascular reactivity. Invest Ophthalmol Vis Sci 1999; 40:2448–2453.
312. Venkataraman ST, Hudson C, Fisher JA, Flanagan JG. The impact of hypercapnia on retinal capillary blood flow assessed by scanning laser Doppler flowmetry. Microvasc Res 2005; 69:149–155.
313. Harris A, Anderson DR, Pillunat L, et al. Laser Doppler flowmetry measurement of changes in human optic nerve head blood flow in response to blood gas perturbations. J Glaucoma 1996; 5:258–265.
314. Wang L, Grant C, Fortune B, Cioffi GA. Retinal and choroidal vasoreactivity to altered $PaCO_2$ in rat measured with a modified microsphere technique. Exp Eye Res 2008; 86:908–913.
315. Tsacopoulos M, Levy S. Intraretinal acid-base studies using pH glass microelectrodes: effect of respiratory and metabolic acidosis and alkalosis on inner-retinal pH. Exp Eye Res 1976; 23:495–504.
316. Sato E, Sakamoto T, Nagaoka T, Mori F, Takakusaki K, Yoshida A. Role of nitric oxide in regulation of retinal blood flow during hypercapnia in cats. Invest Ophthalmol Vis Sci 2003; 44:4947–4953.

317. Checchin D, Hou X, Hardy P, et al. PGE(2)-mediated eNOS induction in prolonged hypercapnia. Invest Ophthalmol Vis Sci 2002; 43:1558–1566.
318. Doni MG, Whittle BJ, Palmer RM, Moncada S. Actions of nitric oxide on the release of prostacyclin from bovine endothelial cells in culture. Eur J Pharmacol 1988; 151:19–25.
319. Shimokawa H, Flavahan NA, Lorenz RR, Vanhoutte PM. Prostacyclin releases endothelium-derived relaxing factor and potentiates its action in coronary arteries of the pig. Br J Pharmacol 1988; 95:1197–1203.
320. Petropoulos IK, Munoz J-L, Pournaras CJ. Metabolic regulation of the hypercapnia-associated vasodilation of the optic nerve head vessels. Invest Ophthalmol Vis Sci 2005; 46:E-Abstract 3908.
321. Bill A. Effects of indomethacin on regional blood flow in conscious rabbits–a microsphere study. Acta Physiol Scand 1979; 105:437–442.
322. Taki K, Kato H, Endo S, Inada K, Totsuka K. Cascade of acetazolamide-induced vasodilatation. Res Commun Mol Pathol Pharmacol 1999; 103:240–248.
323. Pournaras JA, Petropoulos IK, Munoz JL, Pournaras CJ. Experimental retinal vein occlusion: effect of acetazolamide and carbogen (95 percent O_2/5 percent CO_2) on preretinal PO_2. Invest Ophthalmol Vis Sci 2004; 45:3669–3677.
324. Petropoulos IK, Pournaras JA, Munoz J-L, Pournaras CJ. Effect of carbogen breathing and acetazolamide on optic disc PO_2. Invest Ophthalmol Vis Sci 2005; 46:4139–4146.
325. Pedersen DB, Koch Jensen P, la Cour M, et al. Carbonic anhydrase inhibition increases retinal oxygen tension and dilates retinal vessels. Graefes Arch Clin Exp Ophthalmol 2005; 243:163–168.
326. Pedersen DB, Stefansson E, Kiilgaard JF, et al. Optic nerve pH and PO_2: the effects of carbonic anhydrase inhibition, and metabolic and respiratory acidosis. Acta Ophthalmol Scand 2006; 84:475–480.
327. Nielsen NV. Treatment of acute occlusion of the retinal arteries. Acta Ophthalmol (Copenh) 1979; 57:1078–1813.
328. Pakola SJ, Grunwald JE. Effects of oxygen and carbon dioxide on human retinal circulation. Invest Ophthalmol Vis Sci 1993; 34:2866–2870.
329. Luksch A, Garhofer G, Imhof A, et al. Effect of inhalation of different mixtures of O_2 and CO_2 on retinal blood flow. Br J Ophthalmol 2002; 86:1143–1147.
330. Haefliger IO, Lietz A, Griesser SM, et al. Modulation of Heidelberg Retinal Flowmeter parameter flow at the papilla of healthy subjects: effect of carbogen, oxygen, high intraocular pressure, and beta-blockers. Surv Ophthalmol 1999; 43(Suppl 1):S59–S65.
331. Wimpissinger B, Resch H, Berisha F, Weigert G, Schmetterer L, Polak K. Response of choroidal blood flow to carbogen breathing in smokers and non-smokers. Br J Ophthalmol 2004; 88:776–781.
332. Sedney SC. Photocoagulation in retinal vein occlusion. Doc Ophthalmol 1976; 40:1–241.
333. Brown SM, Jampol LM. New concepts of regulation of retinal vessel tone. Arch Ophthalmol 1996; 114:199–204.
334. Yamanishi S, Katsumura K, Kobayashi T, Puro DG. Extracellular lactate as a dynamic vasoactive signal in the rat retinal microvasculature. Am J Physiol Heart Circ Physiol 2006; 290:H925–H934.
335. Hein TW, Xu W, Kuo L. Dilation of retinal arterioles in response to lactate: role of nitric oxide, guanylyl cyclase, and ATP-sensitive potassium channels. Invest Ophthalmol Vis Sci 2006; 47:693–699.
336. Delaey C, Van de Voorde J. Retinal arterial tone is controlled by a retinal-derived relaxing factor. Circ Res 1998; 83:714–720.
337. Kaley G. Novel vasodilator released by retinal tissue. Circ Res 1998; 83:772–773.
338. Delaey C. Retinal tissue modulates retinal arterial tone through the release of a potent vasodilating factor. Verh K Acad Geneeskd Belg 2001; 63:335–357.
339. Winkler BS. Glycolytic and oxidative metabolism in relation to retinal function. J Gen Physiol 1981; 77:667–692.
340. Winkler BS, Starnes CA, Sauer MW, Firouzgan Z, Chen SC. Cultured retinal neuronal cells and Muller cells both show net production of lactate. Neurochem Int 2004; 45:311–320.
341. Poitry-Yamate CL, Poitry S, Tsacopoulos M. Lactate released by Muller glial cells is metabolized by photoreceptors from mammalian retina. J Neurosci 1995; 15:5179–5191.
342. Garhofer G, Zawinka C, Resch H, Menke M, Schmetterer L, Dorner GT. Effect of intravenous administration of sodium-lactate on retinal blood flow in healthy subjects. Invest Ophthalmol Vis Sci 2003; 44:3972–3976.
343. Garhofer G, Zawinka C, Huemer KH, Schmetterer L, Dorner GT. Flicker light-induced vasodilatation in the human retina: effect of lactate and changes in mean arterial pressure. Invest Ophthalmol Vis Sci 2003; 44:5309–5314.
344. Ido Y, Chang K, Williamson JR. NADH augments blood flow in physiologically activated retina and visual cortex. Proc Natl Acad Sci USA 2004; 101:653–658.
345. Oldendorf WH. Carrier-mediated blood–brain barrier transport of short-chain monocarboxylic organic acids. Am J Physiol 1973; 224:1450–1453.
346. Poole RC, Halestrap AP. Transport of lactate and other monocarboxylates across mammalian plasma membranes. Am J Physiol 1993; 264:C761–C782.
347. Gerhart DZ, Leino RL, Drewes LR. Distribution of monocarboxylate transporters MCT1 and MCT2 in rat retina. Neuroscience 1999; 92:367–375.
348. Pierre K, Pellerin L. Monocarboxylate transporters in the central nervous system: distribution, regulation and function. J Neurochem 2005; 94:1–14.
349. Mendrinos E, Petropoulos IK, Mangioris G, Papadopoulou DN, Stangos AN, Pournaras CJ. Lactate-induced retinal arteriolar vasodilation implicates neuronal nitric oxide synthesis in minipigs. Invest Ophthalmol Vis Sci 2008; 49:5060–5066.
350. Barcsay G, Seres A, Nemeth J. The diameters of the human retinal branch vessels do not change in darkness. Invest Ophthalmol Vis Sci 2003; 44:3115–3118.
351. Feke GT, Zuckerman R, Green GJ, Weiter JJ. Response of human retinal blood flow to light and dark. Invest Ophthalmol Vis Sci 1983; 24:136–141.
352. Riva CE, Grunwald JE, Petrig BL. Reactivity of the human retinal circulation to darkness: a laser Doppler velocimetry study. Invest Ophthalmol Vis Sci 1983; 24:737–740.
353. Riva CE, Petrig BL, Grunwald JE. Near-infrared retinal laser Doppler velocimetry. Lasers Ophthalmol 1987; 1:211–215.
354. Stefansson E, Wolbarsht ML, Landers MB, 3rd. In vivo O_2 consumption in rhesus monkeys in light and dark. Exp Eye Res 1983; 37:251–256.
355. Riva CE, Logean E, Petrig BL, Falsini B. Effect of dark adaptation on retinal blood flow. Klin Monatsbl Augenheilkd 2000; 216:309–310.
356. Fuchsjager-Mayrl G, Polska E, Malec M, Schmetterer L. Unilateral light-dark transitions affect choroidal blood flow in both eyes. Vision Res 2001; 41:2919–2924.
357. Huemer KH, Garhofer G, Aggermann T, Kolodjaschna J, Schmetterer L, Fuchsjager-Mayrl G. Role of nitric oxide in choroidal blood flow regulation during light/dark transitions. Invest Ophthalmol Vis Sci 2007; 48:4215–4219.
358. Formaz F, Riva CE, Geiser M. Diffuse luminance flicker increases retinal vessel diameter in humans. Curr Eye Res 1997; 16:1252–1257.
359. Polak K, Schmetterer L, Riva CE. Influence of flicker frequency on flicker-induced changes of retinal vessel diameter. Invest Ophthalmol Vis Sci 2002; 43:2721–2726.
360. Nagel E, Vilser W. Flicker observation light induces diameter response in retinal arterioles: a clinical methodological study. Br J Ophthalmol 2004; 88:54–56.
361. Kotliar KE, Vilser W, Nagel E, Lanzl IM. Retinal vessel reaction in response to chromatic flickering light. Graefes Arch Clin Exp Ophthalmol 2004; 242:377–392.
362. Riva CE, Petrig BL. The regulation of retinal and optic nerve blood flow: effect of diffuse luminance flicker determined by the laser Doppler and the blue field simulation techniques. Les séminaires ophtalmologiques d'IPSEN Vision et adaptation. Paris: Elsevier, 1995:61–71.
363. Scheiner AJ, Riva CE, Kazahaya K, Petrig BL. Effect of flicker on macular blood flow assessed by the blue field simulation technique. Invest Ophthalmol Vis Sci 1994; 35:3436–3441.
364. Vo Van T, Riva CE. Variations of blood flow at optic nerve head induced by sinusoidal flicker stimulation in cats. J Physiol 1995; 482(Pt 1):189–202.
365. Michelson G, Patzelt A, Harazny J. Flickering light increases retinal blood flow. Retina 2002; 22:336–343.
366. Garhofer G, Zawinka C, Resch H, Huemer KH, Dorner GT, Schmetterer L. Diffuse luminance flicker increases blood flow in major retinal arteries and veins. Vision Res 2004; 44:833–838.
367. Riva CE, Salgarello T, Logean E, Colotto A, Galan EM, Falsini B. Flicker-evoked response measured at the optic disc rim is reduced in ocular hypertension and early glaucoma. Invest Ophthalmol Vis Sci 2004; 45:3662–3668.
368. Riva CE, Logean E, Falsini B. Visually evoked hemodynamical response and assessment of neurovascular coupling in the optic nerve and retina. Prog Retin Eye Res 2005; 24:183–215.
369. Falsini B, Riva CE, Logean E. Flicker-evoked changes in human optic nerve blood flow: relationship with retinal neural activity. Invest Ophthalmol Vis Sci 2002; 43:2309–2316.
370. Dorner GT, Garhofer G, Huemer KH, Riva CE, Wolzt M, Schmetterer L. Hyperglycemia affects flicker-induced vasodilation in the retina of healthy subjects. Vision Res 2003; 43:1495–1500.
371. Riva CE, Logean E, Falsini B. Temporal dynamics and magnitude of the blood flow response at the optic disk in normal subjects during functional retinal flicker-stimulation. Neurosci Lett 2004; 356:75–78.
372. Pemp B, Garhofer G, Weigert G, et al. Reduced retinal vessel response to flicker stimulation but not to exogenous nitric oxide in type 1 diabetes. Invest Ophthalmol Vis Sci 2009; 50:4029–4032.
373. Wang L, Bill A. Effects of constant and flickering light on retinal metabolism in rabbits. Acta Ophthalmol Scand 1997; 75:227–231.
374. Heeger DJ, Ress D. What does fMRI tell us about neuronal activity? Nat Rev Neurosci 2002; 3:142–151.
375. Attwell D, Laughlin SB. An energy budget for signaling in the grey matter of the brain. J Cereb Blood Flow Metab 2001; 21:1133–1145.
376. Buerk DG, Riva CE, Cranstoun SD. Frequency and luminance-dependent blood flow and K^+ ion changes during flicker stimuli in cat optic nerve head. Invest Ophthalmol Vis Sci 1995; 36:2216–2227.
377. Buerk DG, Riva CE, Cranstoun SD. Nitric oxide has a vasodilatory role in cat optic nerve head during flicker stimuli. Microvasc Res 1996; 52:13–26.
378. Donati G, Pournaras CJ, Munoz JL, Poitry S, Poitry-Yamate CL, Tsacopoulos M. Nitric oxide controls arteriolar tone in the retina of the miniature pig. Invest Ophthalmol Vis Sci 1995; 36:2228–2237.
379. Kondo M, Wang L, Bill A. The role of nitric oxide in hyperaemic response to flicker in the retina and optic nerve in cats. Acta Ophthalmol Scand 1997; 75:232–235.
380. Huemer KH, Garhofer G, Zawinka C, et al. Effects of dopamine on human retinal vessel diameter and its modulation during flicker stimulation. Am J Physiol Heart Circ Physiol 2003; 284:H358–H363.
381. Buerk DG, Riva CE. Adenosine enhances functional activation of blood flow in cat optic nerve head during photic stimulation independently from nitric oxide. Microvasc Res 2002; 64:254–264.
382. Furchgott RF, Zawadzki JV. The obligatory role of endothelial cells in the relaxation of arterial smooth muscle by acetylcholine. Nature 1980; 288:373–376.
383. Schmetterer L, Polak K. Role of nitric oxide in the control of ocular blood flow. Prog Retin Eye Res 2001; 20:823–847.
384. Riva CE, Schmetterer L. Microcirculation of the ocular fundus. In: Tuma RF, Duran WN, Ley K, eds. Handbook of physiology: microcirculation. Amsterdam, The Netherlands: Academic Press, 2008:735–765.
385. Venturini CM, Knowles RG, Palmer RM, Moncada S. Synthesis of nitric oxide in the bovine retina. Biochem Biophys Res Commun 1991; 180:920–925.
386. Goureau O, Hicks D, Courtois Y, De Kozak Y. Induction and regulation of nitric oxide synthase in retinal Muller glial cells. J Neurochem 1994; 63:310–317.
387. Koch KW, Lambrecht HG, Haberecht M, Redburn D, Schmidt HH. Functional coupling of a Ca2+/calmodulin-dependent nitric oxide synthase and a soluble guanylyl cyclase in vertebrate photoreceptor cells. Embo J 1994; 13:3312–3320.
388. Osborne NN, Barnett NL, Herrera AJ. NADPH diaphorase localization and nitric oxide synthetase activity in the retina and anterior uvea of the rabbit eye. Brain Res 1993; 610:194–198.
389. Knowles RG, Moncada S. Nitric oxide synthases in mammals. Biochem J 1994; 298(Pt 2):249–258.
390. Meyer P, Champion C, Schlotzer-Schrehardt U, Flammer J, Haefliger IO. Localization of nitric oxide synthase isoforms in porcine ocular tissues. Curr Eye Res 1999; 18:375–380.
391. Chakravarthy U, Stitt AW, McNally J, Bailie JR, Hoey EM, Duprex P. Nitric oxide synthase activity and expression in retinal capillary endothelial cells and pericytes. Curr Eye Res 1995; 14:285–294.

392. Rees DD, Palmer AM, Moncada S. Role of endothelium-derived nitric oxide in the regulation of blood pressure. Proc Natl Acad Sci USA 1989; 86:3375–3378.
393. Haefliger IO, Flammer J, Luscher TF. Nitric oxide and endothelin-1 are important regulators of human ophthalmic artery. Invest Ophthalmol Vis Sci 1992; 33:2340–2343.
394. Yao K, Tschudi M, Flammer J, Luscher TF. Endothelium-dependent regulation of vascular tone of the porcine ophthalmic artery. Invest Ophthalmol Vis Sci 1991; 32:1791–1798.
395. Haefliger IO, Zschauer A, Anderson DR. Relaxation of retinal pericyte contractile tone through the nitric oxide-cyclic guanosine monophosphate pathway. Invest Ophthalmol Vis Sci 1994; 35:991–997.
396. Kawasaki Y, Fujikado T, Hosohata J, Tano Y, Tanaka Y. The effect of nitric oxide on the contractile tone of Muller cells. Ophthalmic Res 1999; 31:387–391.
397. Meyer P, Flammer J, Luscher TF. Endothelium-dependent regulation of the ophthalmic microcirculation in the perfused porcine eye: role of nitric oxide and endothelins. Invest Ophthalmol Vis Sci 1993; 34:3614–3621.
398. Granstam E, Wang L, Bill A. Vascular effects of endothelin-1 in the cat; modification by indomethacin and L-NAME. Acta Physiol Scand 1993; 148:165–176.
399. Seligsohn EE, Bill A. Effects of NG-nitro-L-arginine methyl ester on the cardiovascular system of the anaesthetized rabbit and on the cardiovascular response to thyrotropin-releasing hormone. Br J Pharmacol 1993; 109:1219–1225.
400. Deussen A, Sonntag M, Vogel R. L-arginine-derived nitric oxide: a major determinant of uveal blood flow. Exp Eye Res 1993; 57:129–134.
401. Mann RM, Riva CE, Stone RA, Barnes GE, Cranstoun SD. Nitric oxide and choroidal blood flow regulation. Invest Ophthalmol Vis Sci 1995; 36:925–930.
402. Granstam E, Granstam SO, Fellstrom B, Lind L. Endothelium-dependent vasodilation in the uvea of hypertensive and normotensive rats. Curr Eye Res 1998; 17:189–196.
403. Luksch A, Polak K, Beier C, et al. Effects of systemic NO synthase inhibition on choroidal and optic nerve head blood flow in healthy subjects. Invest Ophthalmol Vis Sci 2000; 41:3080–3084.
404. Dorner GT, Garhofer G, Kiss B, et al. Nitric oxide regulates retinal vascular tone in humans. Am J Physiol Heart Circ Physiol 2003; 285:H631–H636.
405. Delles C, Michelson G, Harazny J, Oehmer S, Hilgers KF, Schmieder RE. Impaired endothelial function of the retinal vasculature in hypertensive patients. Stroke 2004; 35:1289–1293.
406. Schmetterer L, Wolzt M, Graselli U, et al. Nitric oxide synthase inhibition in the histamine headache model. Cephalalgia 1997; 17:175–182.
407. Schmetterer L, Muller M, Fasching P, et al. Renal and ocular hemodynamic effects of insulin. Diabetes 1997; 46:1868–1874.
408. Schmetterer L, Findl O, Strenn K, et al. Role of NO in the O_2 and CO_2 responsiveness of cerebral and ocular circulation in humans. Am J Physiol 1997; 273:R2005–R2012.
409. Dorner GT, Garhofer G, Huemer KH, et al. Effects of adrenomedullin on ocular hemodynamic parameters in the choroid and the ophthalmic artery. Invest Ophthalmol Vis Sci 2003; 44:3947–3951.
410. Grunwald JE, Iannaccone A, DuPont J. Effect of isosorbide monon-itrate on the human optic nerve and choroidal circulations. Br J Ophthalmol 1999; 83:162–167.
411. Iannaccone AE, DuPont J, Grunwald JE. Human retinal hemodynamics following administration of 5-isosorbide monon-itrate. Curr Eye Res 2000; 20:205–210.
412. Inoue A, Yanagisawa M, Kimura S, et al. The human endothelin family: three structurally and pharmacologically distinct isopeptides predicted by three separate genes. Proc Natl Acad Sci USA 1989; 86:2863–2867.
413. Ripodas A, de Juan JA, Roldan-Pallares M, et al. Localisation of endothelin-1 mRNA expression and immunoreactivity in the retina and optic nerve from human and porcine eye. Evidence for endothelin-1 expression in astrocytes. Brain Res 2001; 912:137–143.
414. Hosli E, Hosli L. Autoradiographic evidence for endothelin receptors on astrocytes in cultures of rat cerebellum, brainstem and spinal cord. Neurosci Lett 1991; 129:55–58.
415. Nyborg NC, Prieto D, Benedito S, Nielsen PJ. Endothelin-1-induced contraction of bovine retinal small arteries is reversible and abolished by nitrendipine. Invest Ophthalmol Vis Sci 1991; 32:27–31.
416. Haefliger IO, Flammer J, Luscher TF. Heterogeneity of endothelium-dependent regulation in ophthalmic and ciliary arteries. Invest Ophthalmol Vis Sci 1993; 34:1722–1730.
417. White LR, Bakken IJ, Sjaavaag I, Elsas T, Vincent MB, Edvinsson L. Vasoactivity mediated by endothelin ETA and ETB receptors in isolated porcine ophthalmic artery. Acta Physiol Scand 1996; 157:245–252.
418. Takei K, Sato T, Non-oyama T, Miyauchi T, Goto K, Hommura S. A new model of transient complete obstruction of retinal vessels induced by endothelin-1 injection into the posterior vitreous body in rabbits. Graefes Arch Clin Exp Ophthalmol 1993; 231:476–481.
419. Granstam E, Wang L, Bill A. Ocular effects of endothelin-1 in the cat. Curr Eye Res 1992; 11:325–332.
420. Polak K, Luksch A, Frank B, Jandrasits K, Polska E, Schmetterer L. Regulation of human retinal blood flow by endothelin-1. Exp Eye Res 2003; 76:633–640.
421. Nishimura K, Riva CE, Harino S, Reinach P, Cranstoun SD, Mita S. Effects of endothelin-1 on optic nerve head blood flow in cats. J Ocul Pharmacol Ther 1996; 12:75–83.
422. Kiel JW. Endothelin modulation of choroidal blood flow in the rabbit. Exp Eye Res 2000; 71:543–550.
423. McDonald DM, Bailie JR, Archer DB, Chakravarthy U. Characterization of endothelin A (ETA) and endothelin B (ETB) receptors in cultured bovine retinal pericytes. Invest Ophthalmol Vis Sci 1995; 36:1088–1094.
424. Hirata Y, Emori T, Eguchi S, et al. Endothelin receptor subtype B mediates synthesis of nitric oxide by cultured bovine endothelial cells. J Clin Invest 1993; 91:1367–1373.
425. Wong J, Vanderford PA, Fineman JR, Chang R, Soifer SJ. Endothelin-1 produces pulmonary vasodilation in the intact newborn lamb. Am J Physiol 1993; 265:H1318–H1325.
426. D'Orleans-Juste P, Claing A, Telemaque S, Maurice MC, Yano M, Gratton JP. Block of endothelin-1-induced release of thromboxane A2 from the guinea pig lung and nitric oxide from the rabbit kidney by a selective ETB receptor antagonist, BQ-788. Br J Pharmacol 1994; 113:1257–1262.
427. Carville C, Raffestin B, Eddahibi S, Blouquit Y, Adnot S. Loss of endothelium-dependent relaxation in proximal pulmonary arteries from rats exposed to chronic hypoxia: effects of in vivo and in vitro supplementation with L-arginine. J Cardiovasc Pharmacol 1993; 22:889–896.
428. Reinhart GA, Preusser LC, Burke SE, et al. Hypertension induced by blockade of ET(B) receptors in conscious non-human primates: role of ET(A) receptors. Am J Physiol Heart Circ Physiol 2002; 283:H1555–H1561.
429. Sokolovsky M, Ambar I, Galron R. A novel subtype of endothelin receptors. J Biol Chem 1992; 267:20551–20554.
430. Resch H, Karl K, Weigert G, et al. Effect of dual endothelin receptor blockade on ocular blood flow in patients with glaucoma and healthy subjects. Invest Ophthalmol Vis Sci 2009; 50:358–363.
431. Kiel JW. Modulation of choroidal autoregulation in the rabbit. Exp Eye Res 1999; 69:413–429.
432. Nilsson SF. The significance of nitric oxide for parasympathetic vasodilation in the eye and other orbital tissues in the cat. Exp Eye Res 2000; 70:61–72.
433. Hagen AA, White RP, Robertson JT. Synthesis of prostaglandins and thromboxane B2 by cerebral arteries. Stroke 1979; 10:306–309.
434. Hudes GR, Li WY, Rockey JH, White P. Prostacyclin is the major prostaglandin synthesized by bovine retinal capillary pericytes in culture. Invest Ophthalmol Vis Sci 1988; 29:1511–1516.
435. Preud'homme Y, Demolle D, Boeynaems JM. Metabolism of arachidonic acid in rabbit iris and retina. Invest Ophthalmol Vis Sci 1985; 26:1336–1342.
436. Nielsen PJ, Nyborg NC. Contractile and relaxing effects of arachidonic acid derivatives on isolated bovine retinal resistance arteries. Exp Eye Res 1990; 50:305–311.
437. Starr MS. Effects of prostaglandin on blood flow in the rabbit eye. Exp Eye Res 1971; 11:161–169.
438. Stiris T, Suguihara C, Hehre D, Goldberg RN, Flynn J, Bancalari E. Effect of cyclooxygenase inhibition on retinal and choroidal blood flow during hypercapnia in newborn piglets. Pediatr Res 1992; 31:127–130.
439. Chemtob S, Beharry K, Rex J, Chatterjee T, Varma DR, Aranda JV. Ibuprofen enhances retinal and choroidal blood flow autoregulation in newborn piglets. Invest Ophthalmol Vis Sci 1991; 32:1799–1807.
440. Dorner GT, Zawinka C, Resch H, Wolzt M, Schmetterer L, Garhofer G. Effects of pentoxifylline and alprostadil on ocular hemodynamics in healthy humans. Invest Ophthalmol Vis Sci 2007; 48:815–819.
441. Stjernschantz J, Selen G, Astin M, Resul B. Microvascular effects of selective prostaglandin analogues in the eye with special reference to latanoprost and glaucoma treatment. Prog Retin Eye Res 2000; 19:459–496.
442. Spada CS, Nieves AL, Woodward DF. Vascular activities of prostaglandins and selective prostanoid receptor agonists in human retinal microvessels. Exp Eye Res 2002; 75:155–163.
443. Bill A, Linder M, Linder J. The protective role of ocular sympathetic vasomotor nerves in acute arterial hypertension. Bibl Anat 1977; 30–35.
444. Linder J. Cerebral and ocular blood flow during alpha 2-blockade: evidence for a modulated sympathetic response. Acta Physiol Scand 1981; 113:511–517.
445. Linder J. Effects of cervical sympathetic stimulation on cerebral and ocular blood flows during hemorrhagic hypotension and moderate hypoxia. Acta Physiol Scand 1982; 114:379–386.
446. Steinle JJ, Krizsan-Agbas D, Smith PG. Regional regulation of choroidal blood flow by autonomic innervation in the rat. Am J Physiol Regul Integr Comp Physiol 2000; 279:R202–R209.
447. Granstam E, Nilsson SF. Non-adrenergic sympathetic vasoconstriction in the eye and some other facial tissues in the rabbit. Eur J Pharmacol 1990; 175:175–186.
448. Nilsson SF. Neuropeptide Y (NPY): a vasoconstrictor in the eye, brain and other tissues in the rabbit. Acta Physiol Scand 1991; 141:455–467.
449. Granstam E, Nilsson SF. Effects of cervical sympathetic nerve stimulation and neuropeptide Y (NPY) on cranial blood flow in the cat. Acta Physiol Scand 1991; 142:21–32.
450. Kawarai M, Koss MC. Sympathetic vasodilation in the rat anterior choroid mediated by beta(1)-adrenoceptors. Eur J Pharmacol 1999; 386:227–233.
451. Lundberg JM. Pharmacology of cotransmission in the autonomic nervous system: integrative aspects on amines, neuropeptides, adenosine triphosphate, amino acids and nitric oxide. Pharmacol Rev 1996; 48:113–178.
452. Kuo LE, Abe K, Zukowska Z. Stress, NPY and vascular remodeling: Implications for stress-related diseases. Peptides 2007; 28:435–440.
453. Movaffaghy A, Chamot SR, Dosso A, Pournaras CJ, Sommerhalder JR, Riva CE. Effect of isometric exercise on choroidal blood flow in type I diabetic patients. Klin Monatsbl Augenheilkd 2002; 219:299–301.
454. Stone RA, Kuwayama Y, Laties AM. Regulatory peptides in the eye. Experientia 1987; 43:791–800.
455. Elsas T, Uddman R, Sundler F. Pituitary adenylate cyclase-activating peptide-immunoreactive nerve fibers in the cat eye. Graefes Arch Clin Exp Ophthalmol 1996; 234:573–580.
456. Yamamoto R, Bredt DS, Dawson TM, Snyder SH, Stone RA. Enhanced expression of nitric oxide synthase by rat retina following pterygopalatine parasympathetic denervation. Brain Res 1993; 631:83–88.
457. Nilsson SF, Bill A. Vasoactive intestinal polypeptide (VIP): effects in the eye and on regional blood flows. Acta Physiol Scand 1984; 121:385–392.
458. Nilsson SF. PACAP-27 and PACAP-38: vascular effects in the eye and some other tissues in the rabbit. Eur J Pharmacol 1994; 253:17–25.
459. Nilsson SF. Nitric oxide as a mediator of parasympathetic vasodilation in ocular and extraocular tissues in the rabbit. Invest Ophthalmol Vis Sci 1996; 37:2110–2119.
460. Nilsson SF, Maepea O. Comparison of the vasodilatory effects of vasoactive intestinal polypeptide (VIP) and peptide-HI (PHI) in the rabbit and the cat. Acta Physiol Scand 1987; 129:17–26.
461. Gidday JM, Park TS. Microcirculatory responses to adenosine in the newborn pig retina. Pediatr Res 1993; 33:620–627.
462. Gidday JM, Park TS. Adenosine-mediated autoregulation of retinal arteriolar tone in the piglet. Invest Ophthalmol Vis Sci 1993; 34:2713–2719.
463. Portellos M, Riva CE, Cranstoun SD, Petrig BL, Brucker AJ. Effects of adenosine on ocular blood flow. Invest Ophthalmol Vis Sci 1995; 36:1904–1909.
464. Polska E, Ehrlich P, Luksch A, Fuchsjager-Mayrl G, Schmetterer L. Effects of adenosine on intraocular pressure, optic nerve head blood flow, and choroidal blood flow in healthy humans. Invest Ophthalmol Vis Sci 2003; 44:3110–3114.

465. Chandra SR, Friedman E. Choroidal blood flow. II. The effects of autonomic agents. Arch Ophthalmol 1972; 87:67–69.

466. Alm A. Effects of norepinephrine, angiotensin, dihydroergotamine, papaverine, isoproterenol, histamine, nicotinic acid, and xanthinol nicotinate on retinal oxygen tension in cats. Acta Ophthalmol (Copenh) 1972; 50:707–719.

467. Hoste AM, Boels PJ, Brutsaert DL, De Laey JJ. Effect of alpha-1 and beta agonists on contraction of bovine retinal resistance arteries in vitro. Invest Ophthalmol Vis Sci 1989; 30:44–50.

468. Nielsen PJ, Nyborg NC. Adrenergic responses in isolated bovine retinal resistance arteries. Int Ophthalmol 1989; 13:103–107.

469. Maurice DM. Drug delivery to the posterior segment from drops. Surv Ophthalmol 2002; 47:Suppl 1:S41–S52.

470. Mizuno K, Koide T, Saito N, et al. Topical nipradilol: effects on optic nerve head circulation in humans and periocular distribution in monkeys. Invest Ophthalmol Vis Sci 2002; 43:3243–3250.

471. Costa VP, Harris A, Stefansson E, et al. The effects of antiglaucoma and systemic medications on ocular blood flow. Prog Retin Eye Res 2003; 22:769–805.

472. Alm A. The effect of topical l-epinephrine on regional ocular blood flow in monkeys. Invest Ophthalmol Vis Sci 1980; 19:487–491.

473. Reitsamer HA, Posey M, Kiel JW. Effects of a topical alpha2 adrenergic agonist on ciliary blood flow and aqueous production in rabbits. Exp Eye Res 2006; 82:405–415.

474. Jandrasits K, Luksch A, Soregi G, Dorner GT, Polak K, Schmetterer L. Effect of noradrenaline on retinal blood flow in healthy subjects. Ophthalmology 2002; 109:291–295.

475. Gherezghiher T, Okubo H, Koss MC. Choroidal and ciliary body blood flow analysis: application of laser Doppler flowmetry in experimental animals. Exp Eye Res 1991; 53:151–156.

476. Kitanishi K, Harino S, Okamoto N, Tani Y, Nishimura K. Optic nerve head and choroidal circulation measured by laser Doppler flowmetry in response to intravenous administration of noradrenaline. Nippon Ganka Gakkai Zasshi 1997; 101:215–219.

477. Bruinink A, Dawis S, Niemeyer G, Lichtensteiger W. Catecholaminergic binding sites in cat retina, pigment epithelium and choroid. Exp Eye Res 1986; 43:147–151.

478. Grajewski AL, Ferrari-Dileo G, Feuer WJ, Anderson DR. Beta-adrenergic responsiveness of choroidal vasculature. Ophthalmology 1991; 98:989–995.

479. Kawarai M, Koss MC. Sympathetic vasodilation in the rat anterior choroid mediated by beta(1)-adrenoceptors. Eur J Pharmacol 1999; 386:227–233.

480. Hoste AM, Boels PJ, Andries LJ, Brutsaert DL, De Laey JJ. Effects of beta-antagonists on contraction of bovine retinal microarteries in vitro. Invest Ophthalmol Vis Sci 1990; 31:1231–1237.

481. Wikberg-Matsson A, Uhlen S, Wikberg JE. Characterization of alpha(1)-adrenoceptor subtypes in the eye. Exp Eye Res 2000; 70:51–60.

482. Wikberg-Matsson A, Simonsen U. Potent alpha(2A)-adrenoceptor-mediated vasoconstriction by brimonidine in porcine ciliary arteries. Invest Ophthalmol Vis Sci 2001; 42:2049–2055.

483. Chiou GC, Chen YJ. Improvement of ocular blood flow with dopamine antagonists on ocular-hypertensive rabbit eyes. Zhongguo Yao Li Xue Bao 1992; 13:481–484.

484. Reitsamer HA, Zawinka C, Branka M. Dopaminergic vasodilation in the choroidal circulation by d1/d5 receptor activation. Invest Ophthalmol Vis Sci 2004; 45:900–905.

485. Stjernschantz J, Alm A, Bill A. Effects of intracranial oculomotor nerve stimulation on ocular blood flow in rabbits: modification by indomethacin. Exp Eye Res 1976; 23:461–469.

486. Vorstrup S. Tomographic cerebral blood flow measurements in patients with ischemic cerebrovascular disease and evaluation of the vasodilatory capacity by the acetazolamide test. Acta Neurol Scand Suppl 1988; 114:1–48.

487. Stefansson E, Jensen PK, Eysteinsson T, et al. Optic nerve oxygen tension in pigs and the effect of carbonic anhydrase inhibitors. Invest Ophthalmol Vis Sci 1999; 40:2756–2761.

488. Stefansson E, Pedersen DB, Jensen PK, et al. Optic nerve oxygenation. Prog Retin Eye Res 2005; 24:307–332.

489. Bill A. The role of ciliary blood flow and ultrafiltration in aqueous humor formation. Exp Eye Res 1973; 16:287–298.

490. Kiel JW, Reitsamer HA, Walker JS, Kiel FW. Effects of nitric oxide synthase inhibition on ciliary blood flow, aqueous production and intraocular pressure. Exp Eye Res 2001; 73:355–364.

491. Reitsamer HA, Kiel JW. Effects of dopamine on ciliary blood flow, aqueous production, and intraocular pressure in rabbits. Invest Ophthalmol Vis Sci 2002; 43:2697–2703.

492. Reitsamer H, Posey M, Kiel J. Brimonidine vasoconstriction: effects on ciliary blood flow, ciliary oxygen tension, aqueous production and episcleral venous pressure in rabbits. Invest Ophthalmol Vis Sci 2003; 82:405–415.

493. Bill A. The effect of changes in arterial blood pressure on the rate of aqueous humour formation in a primate (Cercopithecus ethiops). Ophthal Res 1970; 1:193–200.

494. Reitsamer HA, Kiel JW. Relationship between ciliary blood flow and aqueous production in rabbits. Invest Ophthalmol Vis Sci 2003; 44:3967–3971.

495. Kohner EM. Dynamic changes in the microcirculation of diabetics as related to diabetic microangiopathy. Acta Med Scand Suppl 1975; 578:41–47.

496. Schmetterer L, Wolzt M. Ocular blood flow and associated functional deviations in diabetic retinopathy. Diabetologia 1999; 42:387–405.

497. Grunwald JE, DuPont J, Riva CE. Retinal haemodynamics in patients with early diabetes mellitus. Br J Ophthalmol 1996; 80:327–331.

498. Feke GT, Buzney SM, Ogasawara H, et al. Retinal circulatory abnormalities in type 1 diabetes. Invest Ophthalmol Vis Sci 1994; 35:2968–2975.

499. Gilmore ED, Hudson C, Nrusimhadevara RK, et al. Retinal arteriolar diameter, blood velocity, and blood flow response to an isocapnic hyperoxic provocation in early sight-threatening diabetic retinopathy. Invest Ophthalmol Vis Sci 2007; 48:1744–1750.

500. Gilmore ED, Hudson C, Nrusimhadevara RK, et al. Retinal arteriolar hemodynamic response to an acute hyperglycemic provocation in early and sight-threatening diabetic retinopathy. Microvasc Res 2007; 73:191–197.

501. Lorenzi M, Feke GT, Cagliero E, et al. Retinal haemodynamics in individuals with well-controlled type 1 diabetes. Diabetologia 2008; 51:361–364.

502. Grunwald JE, Brucker AJ, Grunwald SE, Riva CE. Retinal hemodynamics in proliferative diabetic retinopathy. A laser Doppler velocimetry study. Invest Ophthalmol Vis Sci 1993; 34:66–71.

503. Sieker HO, Hickam JB. Normal and impaired retinal vascular reactivity. Circulation 1953; 7:79–83.

504. Grunwald JE, Riva CE, Brucker AJ, Sinclair SH, Petrig BL. Altered retinal vascular response to 100 percent oxygen breathing in diabetes mellitus. Ophthalmology 1984; 91:1447–1452.

505. Sinclair SH, Grunwald JE, Riva CE, Braunstein SN, Nichols CW, Schwartz SS. Retinal vascular autoregulation in diabetes mellitus. Ophthalmology 1982; 89:748–750.

506. Rassam SM, Patel V, Kohner EM. The effect of experimental hypertension on retinal vascular autoregulation in humans: a mechanism for the progression of diabetic retinopathy. Exp Physiol 1995; 80:53–68.

507. Dumskyj MJ, Kohner EM. Retinal blood flow regulation in diabetes mellitus: impaired autoregulation and no detectable effect of autonomic neuropathy using laser doppler velocimetry, computer assisted image analysis, and isometric exercise. Microvasc Res 1999; 57:353–356.

508. Lanigan LP, Clark CV, Allawi J, Hill DW, Keen H. Responses of the retinal circulation to systemic autonomic stimulation in diabetes mellitus. Eye 1989; 3(Pt 1):39–47.

509. Garhofer G, Zawinka C, Resch H, Kothy P, Schmetterer L, Dorner GT. Reduced response of retinal vessel diameters to flicker stimulation in patients with diabetes. Br J Ophthalmol 2004; 88:887–891.

510. Mizutani M, Gerhardinger C, Lorenzi M. Muller cell changes in human diabetic retinopathy. Diabetes 1998; 47:445–449.

511. Dawczynski J, Mandecka A, Blum M, Muller UA, Ach T, Strobel J. Endothelial dysfunction of central retinal vessels: a prognostic parameter for diabetic retinopathy? Klin Monatsbl Augenheilkd 2007; 224:827–831.

512. Caldwell RB, Fitzgerald ME. The choriocapillaris in spontaneously diabetic rats. Microvasc Res 1991; 42:229–244.

513. McLeod DS, Lutty GA. High-resolution histologic analysis of the human choroidal vasculature. Invest Ophthalmol Vis Sci 1994; 35:3799–3811.

514. Nagaoka T, Kitaya N, Sugawara R, et al. Alteration of choroidal circulation in the foveal region in patients with type 2 diabetes. Br J Ophthalmol 2004; 88:1060–1063.

515. Schocket LS, Brucker AJ, Niknam RM, Grunwald JE, DuPont J. Foveolar choroidal hemodynamics in proliferative diabetic retinopathy. Int Ophthalmol 2004; 25:89–94.

516. Gordon MO, Beiser JA, Brandt JD, et al. The Ocular Hypertension Treatment Study: baseline factors that predict the onset of primary open-angle glaucoma. Arch Ophthalmol 2002; 120:714–720; discussion 729–30.

517. Leske MC, Wu SY, Hennis A, Honkanen R, Nemesure B. Risk factors for incident open-angle glaucoma: the Barbados Eye Studies. Ophthalmology 2008; 115:85–93.

518. Grunwald JE, Riva CE, Stone RA, Keates EU, Petrig BL. Retinal autoregulation in open-angle glaucoma. Ophthalmology 1984; 91:1690–1694.

519. Nagel E, Vilser W, Lanzl IM. Retinal vessel reaction to short-term IOP elevation in ocular hypertensive and glaucoma patients. Eur J Ophthalmol 2001; 11:338–344.

520. Hafez AS, Bizzarro RL, Rivard M, Lesk MR. Changes in optic nerve head blood flow after therapeutic intraocular pressure reduction in glaucoma patients and ocular hypertensives. Ophthalmology 2003; 110:201–210.

521. Garhofer G, Zawinka C, Resch H, Huemer KH, Schmetterer L, Dorner GT. Response of retinal vessel diameters to flicker stimulation in patients with early open angle glaucoma. J Glaucoma 2004; 13:340–344.

522. la Cour M, Kiilgaard JF, Nissen MH. Age-related macular degeneration: epidemiology and optimal treatment. Drugs Aging 2002; 19:101–133.

523. Friedman E. A hemodynamic model of the pathogenesis of age-related macular degeneration. Am J Ophthalmol 1997; 124:677–682.

524. Friedman E. The role of the atherosclerotic process in the pathogenesis of age-related macular degeneration. Am J Ophthalmol 2000; 130:658–663.

525. Friedman E. Update of the vascular model of AMD. Br J Ophthalmol 2004; 88:161–163.

526. Lutty G, Grunwald J, Majji AB, Uyama M, Yoneya S. Changes in choriocapillaris and retinal pigment epithelium in age-related macular degeneration. Mol Vis 1999; 5:35.

527. Grunwald JE, Hariprasad SM, DuPont J, et al. Foveolar choroidal blood flow in age-related macular degeneration. Invest Ophthalmol Vis Sci 1998; 39:385–390.

528. Pournaras CJ, Logean E, Riva CE, et al. Regulation of subfoveal choroidal blood flow in age-related macular degeneration. Invest Ophthalmol Vis Sci 2006; 47:1581–1586.

529. Ducasse A, Segal A, Delattre JF. Macroscopic aspects of the long posterior ciliary arteries. Bull Soc Ophtalmol Fr 1986; 86:845–848.

530. Pournaras CJ, Bek T, Rungger-Brandle E. Circulation rétinienne. In: Pournaras CJ, ed. Pathologies vasculaires. Issy-les-Molineaux: Masson, 2008:33–35.

531. Olver JM, Spalton DJ, McCartney AC. Microvascular study of the retrolaminar optic nerve in man: the possible significance in anterior ischaemic optic neuropathy. Eye 1990; 4(Pt 1):7–24.

532. Cerulli A, Carella E, Iuliano L, Nucci C, Carella G. Afférents et éfférents choroïdiens. In: Pournaras CJ, ed. Pathologies vasculaires. Issy-les-Molineaux: Masson, 2008:26–28.

533. Rungger-Brändle E, Leuenberger MP. Barrières hémato-oculaires. In: Pournaras CJ, ed. Pathologies vasculaires. Issy-les-Molineaux: Masson, 2008:44–51.

534. Seifert BU, Vilser W. Retinal vessel analyzer (RVA) – design and function. Biomed Tech 2002; 47(Suppl 1).

535. Riva CE, Ben-Sira I. Two-point fluorophotometer for the human ocular fundus. Applied Optics 1975; 14: 2691–2693.

536. Riva CE, Feke GT, Eberli B, Benary V. Bidirectional LDV system for absolute measurement of blood speed in retinal vessels. Appl Opt 1979; 18:2301–2306.

537. Riva CE, Grunwald JE, Sinclair SH, O'Keefe K. Fundus camera based retinal LDV. Appl Optics 1981; 20:117–120.

538. Zukowska Z. Atherosclerosis and angiogenesis: what do nerves have to do with it? Pharmacol Rep 2005; 57(Suppl):229–234.

539. Niskanen L, Voutilainen-Kaunisto R, Terasvirta M, et al. Leucine 7 to proline 7 polymorphism in the neuropeptide y gene is associated with retinopathy in type 2 diabetes. Exp Clin Endocrinol Diabetes 2000; 108:235–236.

540. Koulu M, Movafagh S, Tuohimaa J, et al. Neuropeptide Y and Y2-receptor are involved in development of diabetic retinopathy and retinal neovascularization. Ann Med 2004; 36:232–240.

541. Yoon HZ, Yan Y, Geng Y, Higgins RD. Neuropeptide Y expression in a mouse model of oxygen-induced retinopathy. Clin Experiment Ophthalmol 2002; 30:424–429.

CHAPTER 11

Production and Flow of Aqueous Humor

B'Ann True Gabelt & Paul L. Kaufman

In the healthy eye, flow of aqueous humor against resistance generates an intraocular pressure (IOP) of approximately 15 mmHg, necessary for the proper shape and optical properties of the globe.[1] The circulating aqueous nourishes the cornea and lens, structures which must be transparent and therefore devoid of blood vessels[1] as well as the trabecular meshwork (TM). The aqueous also provides a transparent and colorless medium of refractive index 1.33332 between the cornea and lens, thus constituting an important component of the eye's optical system.[1] The basic anatomy of the primate anterior ocular segment and the normal pathways of aqueous humor flow are illustrated schematically in Figures 11.1 and 11.2.

Aqueous humor is secreted by the ciliary epithelium lining the ciliary processes (as a consequence of active ionic transport across the ciliary epithelium and hydrostatic and osmotic gradients between the posterior chamber and the ciliary process vasculature and stroma), and enters the posterior chamber. It then flows around the lens and through the pupil into the anterior chamber. There is convection flow of aqueous in the anterior chamber – downward close to the cornea where the temperature is cooler, upward near the lens where the temperature is warmer (seen clinically when there is particulate matter in the anterior chamber – e.g. cells). Aqueous leaves the eye by passive bulk flow mainly via two pathways at the anterior chamber angle:

(1) through the TM, across the inner wall of Schlemm's canal into its lumen, and thence into collector channels, aqueous veins, and the episcleral venous circulation – the trabecular or conventional route;

(2) across the iris root, uveal meshwork, and the anterior face of the ciliary muscle, through the connective tissue between the muscle bundles, the suprachoroidal space, and thence out through the sclera – the uveoscleral, posterior, or unconventional route.

Whether or not there is a net osmotic resorption of some of the aqueous passing through the uvea by the uveal venous circulation has not been verified.[2,3] In young individuals of various monkey species, total aqueous drainage is relatively evenly divided between the two pathways.[4–8] In the human eye, the uveoscleral pathway has more recently been shown to play a more prominent role in aqueous outflow than was originally suggested by earlier studies.[9,10] By using indirect methods, such as fluorophotometry, uveoscleral outflow accounts for approximately 35–55 percent of total aqueous drainage in younger humans.[10,11] In normal healthy subjects 60 years of age or older, uveoscleral outflow is significantly reduced compared to that of 20–30-year-olds (mean ± SD = 1.10 ± 0.81 vs 1.52 ± 0.81 μl/min; p<0.009).[11] Similarly, in rhesus monkeys >25 years of age (human equivalent >62 years of age) uveoscleral outflow is also dramatically reduced (0.33 ± 0.08 vs 0.63 ± 0.07 μl/min, mean ± SEM).[8] There is no significant net fluid movement across the cornea, iris vasculature, or vitreoretinal interface, although ion fluxes exist.[12,13]

Due to the autonomic innervation and receptors of the relevant structures, adrenergic and cholinergic mechanisms play major roles in aqueous humor formation and drainage in terms of both normal physiology and glaucoma therapeutics,[14–17] while evidence for other mechanisms, including serotonergic,[18] dopaminergic,[19–22] adenosinergic,[23–27] nitrergic,[28–31] cannabinergic,[32–35] prostaglandin(PG)ergic,[36,37] and cytoskeletal (see review[38]), has been established.

Aqueous humor formation

Physiology of aqueous humor formation

Until the early twentieth century, aqueous humor was regarded as a stagnant fluid.[39] Since that time, however, it has been shown to be continuously formed and drained[39] and the associated anatomic drainage portals (Schlemm's canal, collector channels, aqueous veins, ciliary muscle interstices) have been described.[13,40]

Three physiologic processes contribute to the formation and chemical composition of the aqueous humor: diffusion, ultrafiltration (and the related dialysis), and active secretion.[1] Diffusion and ultrafiltration are responsible for the formation of the "reservoir" of the plasma ultrafiltrate in the stroma, from which the posterior chamber aqueous is derived, via active secretion across the ciliary epithelium. Energy-dependent active transport of sodium into the posterior chamber by the non-pigmented ciliary epithelium (Fig. 11.3) results in water movement from the stromal pool into the posterior chamber. It seems fairly certain that under normal conditions active secretion accounts for perhaps

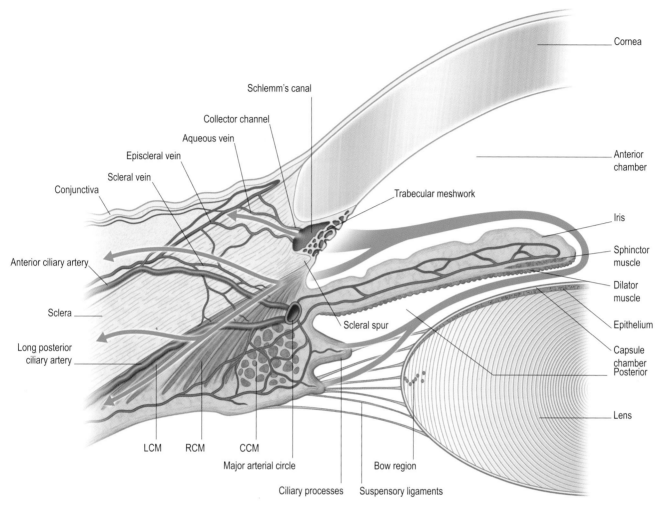

Figure 11.1 Schematic representation of the primate anterior ocular segment. *Arrows* indicate aqueous humor flow pathways. Aqueous humor is formed by the ciliary processes, enters the posterior chamber, flows through the pupil into the anterior chamber, and exits at the chamber angle via the trabecular and uveoscleral routes. (From Kaufman PL, Wiedman T, Robinson JR: Cholinergics. In Sears ML [ed]: Pharmacology of the eye: handbook of experimental pharmacology, Berlin, 1984, Springer-Verlag. Reproduced with kind permission of Springer Science + Business Media.)

80–90 percent of total aqueous humor formation.[12,41–43] The observation that moderate alterations in systemic blood pressure and ciliary process blood flow have little effect on aqueous formation rate supports this notion.[12,44]

Furthermore, Bill[45] noted that the hydrostatic and oncotic forces that exist across the ciliary epithelium–posterior aqueous interface favor resorption, not ultrafiltration, of aqueous humor. Active secretion is essentially pressure-insensitive at near-physiologic IOP. However, the ultrafiltration component of aqueous humor formation is sensitive to changes in IOP, decreasing with increasing IOP. This phenomenon is quantifiable and is termed facility of inflow, or pseudofacility, the latter because a pressure-induced decrease in inflow will appear as an increase in outflow when techniques such as tonography and constant-pressure perfusion are used to measure outflow facility.[46–50] Although measurements vary, pseudofacility in the non-inflamed monkey and human eye constitute a very small percentage of total facility.[50–54] Recently it has been argued that fluid transport rates exhibited by the ciliary epithelia are insufficient to account for the rate of aqueous humor formation and that there may

be some contribution from fluid directly entering the anterior chamber across the anterior surface of the iris.[55,56]

In most mammalian species, the turnover constant of the anterior chamber aqueous humor is ~0.01 to 0.015 × min⁻¹, that is, the rate of aqueous humor formation and drainage is ~1.0 to 1.5 percent of the anterior chamber volume per minute.[50,55,57] This is true also in the normal human eye, in which the aqueous formation rate is ~2.5 μl × min⁻¹.[11,58]

More comprehensive theoretic analyses of the fluid mechanics of aqueous production can be found elsewhere.[1,56,59–62]

Biochemistry of aqueous humor formation

The structural basis for aqueous humor secretion is the bilayered ciliary epithelium.[63] This consists of the pigmented epithelium (PE) facing the ciliary stroma and the non-pigmented epithelium (NPE) facing the aqueous humor. The PE and NPE are connected to each other at their apical membranes through gap junctions, thereby forming a complex syncytium. Tight junctions are present between the apical borders

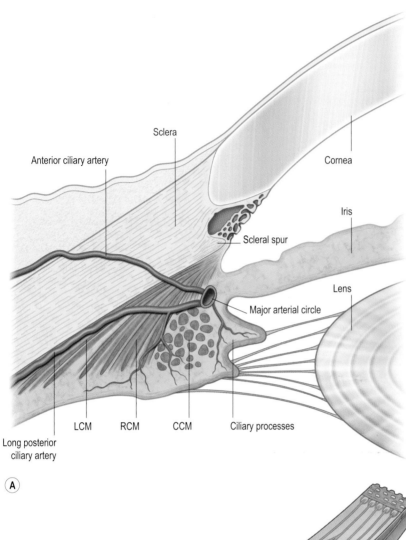

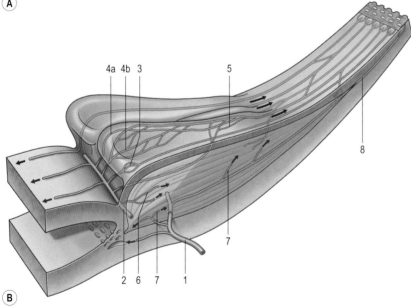

Figure 11.2 (**A**) Blood supply to the ciliary processes. *CCM,* Circular ciliary muscle; *LCM,* longitudinal ciliary muscle; *RCM,* radial ciliary muscle. (**B**) Vascular architecture in the human ciliary body. *1,* Perforating branches of the anterior ciliary arteries; *2,* major arterial circle of iris; *3,* first vascular territory. The second vascular territory is depicted in *4a,* marginal route, and *4b,* capillary network in the center of this territory. *5,* Third vascular territory; *6* and *7,* arterioles to the ciliary muscle; *8,* recurrent choroidal arteries. (Modified from Caprioli J: The ciliary epithelia and aqueous humor. In Hart WM [ed]: Adler's physiology of the eye: clinical application, ed 9, St Louis, 1992, Mosby; and Funk R, Rohen JW: Exp Eye Res 51:651, 1990.)

of NPE cells where they form a barrier that restricts paracellular diffusion.

The active process of aqueous secretion is mediated via selective transport of certain ions and substances across the basolateral membrane of the non-pigmented ciliary epithelium against a concentration gradient. Two enzymes abundantly present in the non-pigmented ciliary epithelium are intimately involved in this process: sodium-potassium-activated adenosine triphosphatase (Na^+-K^+-ATPase) and carbonic anhydrase.[64–71] Na^+-K^+-ATPase is found predominantly bound to the plasma membrane of the basolateral infoldings of the non-pigmented ciliary epithelium.[66,69–71]

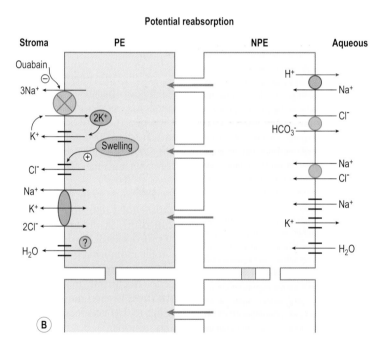

Unidirectional secretion

Potential reabsorption

Figure 11.3 Pathways for unidirectional secretion (**A**) and possible reabsorption (**B**) across the ciliary epithelium. PE, pigmented ciliary epithelial cells; NPE, non-pigmented ciliary epithelial cells. ((A) From McLaughlin CW et al: Am J Physiol Cell Physiol 293:C1455, 2007. Used with permission.)

As a result of the primary active transport of Na^+, other ions and molecules are transported over the epithelium by secondary active transport. Thus, aqueous humor in humans exhibits increased levels of ascorbate, some amino acids, and certain ions such as Cl^- as compared to plasma.[72] There is also a passive transporter for HCO_3^-.[1,73] A summary of the biochemistry of aqueous secretion is shown in Figure 11.3.

Inhibition of the ciliary process Na^+-K^+-ATPase by cardiac glycosides (e.g. ouabain) or vanadate (VO_3^-, VO_4^{3-}) significantly reduces the rate of aqueous humor formation and IOP in experimental animals[66,74–81] and humans,[79] showing that the primary active transport of Na^+ is the primary driving force for the secretion of aqueous humor. To maintain electroneutrality, anions must accompany the actively secreted Na^+, chloride can pass through chloride channels in the basolateral membrane and HCO_3^- can enter aqueous via exchange with chloride. The active transport of Na^+ and the accompanying anions create high osmolarity on the basolateral side of the NPE cells, which causes diffusion of water out of the cells. The movement of water is facilitated by aquaporins in NPE cells (aquaporins 1 and 4).[82–84] AQP1 null mice have decreased IOP and aqueous humor production compared to normals.[82] Sodium and chloride must continuously enter the pigmented epithelial cells for the continuous secretion of aqueous humor. This is achieved by Na^+/H^+ and Cl^-/HCO_3^- antiports and by the Na-K-2Cl cotransporter. In vivo as well as in vitro studies support the involvement of these transporters and channels in the formation of aqueous humor.[85–87]

Transport of Cl^- ions plays an equally important role as the active transport of Na^+ ions as the driving force for aqueous humor formation.[88–90] In isolated bovine eyes, aqueous humor is formed mostly by processes involving active secretion and chloride transport.[88] In rabbit preparations, the release of Cl^- ions is enhanced by agonists of A3 adenosine receptors.[85] A3 adenosine receptor-activated agonists and antagonists, respectively, increase and lower IOP in mice.[91] Bestrophin-2, a protein which regulates Ca^{2+}-activated Cl^- conductance, is localized to the non-pigmented ciliary epithelium, suggesting that it may play a role in regulating aqueous humor formation. Mice deficient in bestrophin-2 have significantly diminished IOP.[92]

In bovine eyes in vitro, 4,4'-diisothiocyanatostilbene-2,2'-disulfonic acid (DIDS, a probable inhibitor of the Cl^-/HCO_3^- exchanger, the Na-HCO3 cotransporter, and chloride channel) and 5-nitro-2-(3-phenylpropylamino)-benzoic acid (NPPB, a chloride channel blocker in non-pigmented cells) reduce aqueous humor formation by 55 percent and 25 percent respectively.[88] In mice, topical inhibitors of Na^+/H^+ antiports lower IOP.[93] Bumetanide (a specific inhibitor of Na-K-2Cl cotransport) and furosemide (a nonspecific anion transport inhibitor) reduce aqueous humor formation by 35 and 45 percent in bovine eyes in vitro.[88] Also, catecholamines such as epinephrine, norepinephrine, isoproterenol and dopamine, that stimulate aqueous humor formation in humans also stimulate Na-K-Cl cotransport.[89] However, topical application of bumetanide to mice[93] or monkeys[94] has no effect on IOP.

Carbonic anhydrase (CA) is abundantly present in the basal and lateral membranes and cytoplasm of the pigmented epithelium and non-pigmented epithelium of the ciliary processes.[95–99] Isoenzymes of CA (II, IV and XII) are present in the ciliary processes.[100,101]

The conversion of CO_2 and H_2O to carbonic acid and its subsequent dissociation to H^+ and HCO_3^- provides the HCO_3^-, which is essential for the active secretion of aqueous humor. Inhibition of the production of HCO_3^- also leads to an inhibition of the active transport of Na^+ across the non-pigmented ciliary epithelium into the first-formed aqueous, thereby reducing active aqueous humor formation. There are several hypotheses to explain the relationship between reduction in non-pigmented ciliary epithelium intracellular HCO_3^- and inhibition of Na^+ transport including:

(i) inhibition of CA causing a decrease in HCO_3^- available for transport with Na^+ from the cytosol of the non-pigmented ciliary epithelium to the aqueous, required to maintain electroneutrality;

(ii) reduction in intracellular pH inhibiting Na^+-K^+-ATPase, and

(iii) decreased availability of H^+ decreasing H^+/Na^+ exchange and reducing the availability of intracellular Na^+ for transport into the intercellular channel.

In addition, inhibition of renal and erythrocyte CA leads to a systemic acidosis which promotes inhibition of aqueous humor formation.[102]

CA inhibitors given systemically can reduce secretion by up to 50 percent[95,98,103–111] and have been in use for clinical glaucoma therapy for over 50 years.[102,104,107] It was once thought that the drug concentration at the ciliary epithelium required to produce nearly continuous, total ciliary CA inhibition necessary to achieve adequate and sustained reduction in aqueous humor formation (over 99 percent of the ciliary enzyme must be inhibited to achieve significant secretory suppression) might never be attainable via the eyedrop route. Fortunately, there are now available topically effective CA inhibitors such as dorzolamide[112–116] and brinzolamide[116–118] which may achieve substantial (albeit usually not quite as much) reduction in IOP as the earlier oral CA inhibitors but without their systemic side-effects.[102,114,119]

Aqueous humor composition

The composition of aqueous humor differs from that of plasma as a result of two important physiological characteristics of the anterior segment: a mechanical epithelial/endothelial blood–aqueous barrier, and active transport of various organic and inorganic substances by the ciliary epithelium. The greatest differences are the low protein and high ascorbate concentrations in the aqueous relative to plasma (200 times less and 20 times greater, respectively).[120–123]

The protein concentration in the peripheral portion of the anterior chamber, close to the meshwork, may be much higher than in the more central region because of protein entry directly from the peripheral iris, as demonstrated in monkey and human eyes.[124–127] When the aqueous protein concentration rises much above its normal 20 mg/100 mL,[128] as in uveitis, the resultant light scattering (Tyndall effect) makes the slit-lamp beam visible as it traverses the anterior chamber (a phenomenon known as "flare").

The high ascorbate concentration may help protect the anterior ocular structures from ultraviolet light-induced oxidative damage.[129] Ascorbate functions as an antioxidant, regulates the sol–gel balance of mucopolysaccharides in the TM, or partially absorbs ultraviolet radiation,[130] as diurnal mammals have approximately 35 times the concentration of aqueous ascorbate compared to nocturnal mammals.[131] Extensive and repeated oxidative stress in vivo may result in reduced TM cell adhesion, leading to TM cell loss which is associated with glaucomatous conditions.[132,133]

Lactate is also normally in excess in the aqueous, presumably as a result of glycolytic activity of the lens, cornea, and other ocular structures.[107,128] Other compounds or ions in excess in the aqueous relative to plasma are Cl^- and certain amino acids.[72] Glucose, urea, and non-protein nitrogen concentrations are slightly less than in plasma.[72,134] Oxygen is also present in the aqueous humor, at a tension determined to lie between 13 to 80 mmHg, depending upon the method of measurement.[135,136]

Blood–aqueous barrier

The blood–aqueous barrier is a functional concept, rather than a discrete structure, invoked to explain the degree to which various solutes are relatively restricted in travel from the ocular vasculature into the aqueous humor. The capillaries of the ciliary processes and choroid are fenestrated, but the interdigitating surfaces of the retinal pigment epithelia and the ciliary process non-pigmented epithelia respectively are joined to each other by tight junctions (zonulae occludens) and constitute an effective barrier to intermediate- and high-molecular-weight substances, such as proteins.[128,137–147] The endothelia of the inner wall of Schlemm's canal are similarly joined,[140,148] preventing retrograde movement of solutes and fluid from the canal lumen into the TM and anterior chamber. Tight junctions are also present in the iridial vascular endothelium[149] as well as between the iris epithelia.[150] For present purposes, one may say that the blood–aqueous barrier comprises the tight junctions of the ciliary process non-pigmented epithelium, the inner wall endothelium of Schlemm's canal, and the iris vasculature, and the outward-directed active transport systems of the ciliary processes (see below). A more universal concept of the blood–aqueous barrier must deal with the movement of smaller molecules, lipid-soluble substances, and water into the eye.[128]

With disease-, drug-, or trauma-induced breakdown of the blood–aqueous barrier (Table 11.1), plasma components enter the aqueous humor. Net fluid movement from blood to aqueous increases, but so does its IOP dependence (pseudofacility).[151] Total facility, as measured by IOP-altering techniques cannot distinguish pseudofacility from total outflow facility (Ctot) and therefore erroneously record the pseudofacility component as increased Ctot (hence the term, "pseudofacility") and therefore underestimate the extent to which the outflow pathways have been compromised by the insult. Under these circumstances, increased pseudofacility provides some protection against a precipitous rise in IOP; as IOP rises, aqueous inflow by ultrafiltration is partly suppressed, blunting (but not completely suppressing)[138,152] further IOP elevation. Additionally, the inflammatory process which occurs during blood–aqueous barrier breakdown leads to a reduction in active secretion of aqueous humor, possibly via interference with active transport mechanisms.[6,153] This in turn may actually produce ocular hypotony, despite compromised conventional outflow pathways

(due to plasma protein blockage of the TM). Prostaglandin release during inflammation may contribute to the hypotony by increasing aqueous outflow via the uveoscleral route.[154,155] When the noxious stimulus is removed, however, the ciliary body may recover before the TM, and the resulting normalization of the aqueous humor formation and uveoscleral outflow rates in the face of still compromised conventional outflow pathways leads to elevated IOP, as seen from the modified Goldmann equation:

$$IOP = [(AHF - F_u)/C_{trab}] + P_e$$

where IOP = intraocular pressure, AHF = aqueous humor formation, F_u = uveoscleral outflow, C_{trab} = facility of outflow from the anterior chamber via the TM and Schlemm's canal, P_e = episcleral venous pressure (the pressure against which fluid leaving the anterior chamber via the trabecular–canalicular route must drain).[156]

Active transport

The ciliary processes possess the ability to actively transport a variety of organic and inorganic compounds and ions out of, or to exclude them from, the eye, that is, to move them from the aqueous or vitreous to the blood against a concentration gradient. Para-aminohippurate (PAH), diodrast, and penicillin are examples of large anions that are actively transported out of the eye. These systems are similar to those in the renal tubules and satisfy all the criteria for active transport: saturability, energy and temperature dependence, Michaelis Menten kinetics, inhibition by ouabain and probenecid, etc.[72,157–166] In addition, another system can actively excrete injected iodide from the aqueous, resembling iodide transport in the thyroid and salivary glands.[159] The physiologic role of these outward-directed systems is unknown. With the discovery that prostaglandins may be actively transported out of the eye,[167,168] some workers have suggested that such outward-directed mechanisms may rid the eye of biologically active substances which are no longer needed or may even be detrimental.[157,161,169,170] The various

Table 11.1 Factors interrupting the blood–aqueous barrier

I. Traumatic	II. Pathophysiologic	III. Pharmacologic
A. Mechanical	A. Vasodilation	A. Melanoycte-stimulating hormone
1. Paracentesis	1. Histamine	B. Nitrogen mustard
2. Corneal abrasion	2. Sympathectomy	C. Cholinergic drugs, especially cholinesterase inhibitors
3. Blunt trauma	B. Corneal and intraocular infections	D. Plasma hyperosmolality
4. Intraocular surgery	C. Intraocular inflammation	
5. Stroking of the iris	D. Prostaglandins (varies with type, dose and species)	
B. Physical	E. Anterior segment ischemia	
1. Radiotherapy		
2. Nuclear radiation		
C. Chemical		
1. Alkali		
2. Irritants (e.g. nitrogen mustard)		

Modified from Stamper RL: Aqueous humor: secretion and dynamics. In Tasman W, Jaeger EA (eds): Clinical ophthalmology, vol 2, Philadelphia, 1979, Lippincott.

efflux transporters in ocular tissues are discussed in more detail in Chapter 17.

Pharmacology and regulation of aqueous humor formation and composition (Box 11.1)

IOP displays a circadian rhythm in animals and humans which is under the control of endogenous pacemaker structures including the suprachiasmic nucleus, which control the activity of sympathetic and parasympathetic ocular innervation.[171] Sympathetic and parasympathetic nerve terminals are present in the ciliary body[172–174] arising from branches of the long and short posterior ciliary nerves. These nerve fibers are of both the myelinated and non-myelinated variety. Parasympathetic fibers originate in the Edinger–Westphal nucleus of the third cranial nerve, run with the inferior division of this nerve in the orbit, and synapse in the ciliary ganglion.[175] Vasodilatory parasympathetic nerve fibers originating in the pterygopalatine ganglion are likely to release nitric oxide and vasoactive intestinal peptide (VIP) in addition to acetylcholine (see Chapter 10). Nerves displaying VIP immunoreactivity are also detected in the ciliary processes, posterior third of the ciliary muscle and around small to medium-sized blood vessels in the posterior uvea of the cat.[176] Only a few VIP-positive fibers are found in the ciliary processes of humans and monkeys.[177] In the cat eye, pituitary adenylate cyclase-activating peptide (PACAP)-containing nerve fibers are detected in the iris, ciliary body and conjunctiva. PACAP immunoreactivity co-localizes with VIP in the sphenopalatine ganglion and with calcitonin gene-related peptide (CGRP) in the trigeminal ganglion.[178] Sympathetic fibers synapse in the superior cervical ganglion and distribute to the muscles and blood vessels of the ciliary body. Stimulation of the cervical sympathetic nerves in vervet monkeys significantly increases the rate of aqueous humor formation.[179] Numerous unmyelinated nerve fibers surround the stromal vessels of the ciliary processes; these are most likely noradrenergic and subserve vasomotion.[128]

Sympathetic denervation in monkeys does not alter resting aqueous humor formation and marginally affects the aqueous humor formation response to timolol or epinephrine.[180] In humans with unilateral Horner's syndrome, IOP, daytime aqueous humor formation, tonographic outflow facility and the flow and IOP response to timolol are similar in both eyes.[181] However, eyes with Horner's syndrome show a decrease in aqueous humor formation in response to epinephrine instead of the normal increase. Isoproterenol, on the other hand, increases flow during sleep in both normal and Horner's syndrome eyes but has no significant effect on flow in either during the day.[182] Chemical sympathectomy with guanethidine sulfate in glaucomatous humans lowers IOP, presumably by reducing aqueous humor production measured indirectly by tonography.[183] VIP stimulated aqueous formation following intravenous administration in monkeys is secondary to activation of the sympathetic nervous system while the effect of intracameral administration of VIP is a direct effect on the ciliary epithelium.[184] PACAP appear to be slightly more potent than VIP as a stimulator of aqueous humor flow in the monkey.[177] The role of sympathetic innervation in mediating aqueous inflow and outflow responses to pharmacologic agents remains largely uncertain. Also the role of sympathetic innervation in mediating inflow responses can vary depending on the species used, i.e. rabbits vs monkeys, as is found, for example, with opioid agonists.[185,186]

Cholinergic mechanisms

The effects of cholinergic drugs on aqueous humor formation and composition and on the blood–aqueous barrier are unclear. Overall, cholinomimetics have little effect on the volumetric rate of aqueous humor formation. In general, cholinergic drugs cause vasodilation in the anterior segment resulting in increased blood flow to the choroid, iris, ciliary processes, and ciliary muscle (see Chapter 10). However cholinergic drugs may also promote vasoconstriction in the rabbit eye.[187] Congestion in the iris and ciliary body is a well-recognized clinical side-effect of topical cholinomimetics, especially the anticholinesterases.[188] The presence of flare and cells in the aqueous humor by biomicroscopy indicates that these agents can cause breakdown of the blood–aqueous barrier and perhaps frank inflammation.[188] Pilocarpine increases blood–aqueous barrier permeability to iodide[189] and inulin.[190] Cholinergic drugs may alter the aqueous humor concentration of inorganic ions[191] and the movement of certain amino acids from the blood into the aqueous humor and may also influence the outward-directed transport systems of the ciliary processes.[192,193] Cholinergic agents can disrupt the coupling and Na^+ currents between pigmented and non-pigmented epithelial cells in vitro, suggesting a possible inhibitory effect on aqueous secretion[194,195] although this is likely to be minimal as stated below.

Under certain conditions, pilocarpine may increase pseudofacility.[59,196] Using a variety of species, conditions, and experimental techniques, cholinergic agents or parasympathetic nerve stimulation is reported to increase, decrease, or

---

Box 11.1 Aqueous Humor Formation

- β-adrenergic antagonists such as timolol, that lower IOP by decreasing aqueous humor formation, continue to be a mainstay in clinical glaucoma therapy
- Nocturnal IOP elevation in glaucomatous eyes may not be effectively eliminated with adrenergic-based glaucoma therapies
- α2-adrenergic agonists such as apraclonidine and brimonidine, are powerful ocular hypotensive agents, lowering IOP primarily by decreasing aqueous humor formation
- Long-term suppression of aqueous humor formation and redirection of aqueous drainage from the trabecular to the uveoscleral pathway could lead to underperfusion of the trabecular meshwork with detrimental morphologic and functional effects
- Brinzolamide or timolol added to latanoprost have similar ocular hypotensive effects and safety in POAG, NTG, or OHT.[678] The small difference in efficacy between the secretory suppressants is buried beneath the bigger IOP effect of latanoprost. This has to do with the physics of IOP; at a lower pressure, it is harder to get the same magnitude of effect

not alter the aqueous humor formation rate and to increase slightly the episcleral venous pressure.[197-208] These apparently confusing results may indicate that cholinergic drug effects on these parameters are extremely dependent on species and technique-related factors and on the ambient neurovascular milieu. In any event, the effects on the rate of aqueous humor formation and episcleral venous pressure are surely minor in most instances, and not responsible for the drug-induced decrease in IOP that forms the basis of the therapeutic efficacy of pilocarpine in chronic glaucoma; the latter resides in its ability to decrease outflow resistance via its effect on the ciliary muscle and perhaps, to a much lesser extent, directly on the TM[209] (see Aqueous Humor Drainage).

Adrenergic mechanisms

The precise role and receptor specificity of adrenergic mechanisms in regulating the rate of aqueous humor formation are unclear. At one time it was generally believed that long-term topical administration of epinephrine, a combined α_1, α_2, β_1, β_2-adrenergic agonist, would decrease the rate of aqueous humor formation.[107] This effect was thought to be mediated by β-adrenergic receptors in the non-pigmented ciliary epithelium, via activation of a membrane adenylate cyclase.[68,210,211] Further, forskolin, which directly and irreversibly activates intracellular adenylate cyclase, in some,[212,213] albeit not all,[214] studies decreases the rate of aqueous humor formation when given topically or intravitreally.

Fluorophotometric studies show that short-term topical administration of epinephrine increases aqueous humor formation;[10,180,215] studies with other adrenergic agonists, including salbutamol,[216] isoproterenol (isoprenaline),[182] and terbutaline,[217] support this finding and are consistent with many studies showing that β-adrenergic antagonists unequivocally decrease aqueous humor formation.[15,218,219]

The ocular hypotensive action of β-antagonists led to their becoming mainstays of clinical glaucoma therapy: the non-selective β_1, β_2 antagonists timolol,[215,220-222] levobunolol,[223] and metipranolol,[224] the non-selective β_1, β_2 partial agonist carteolol,[225,226] and the relatively β_1 selective antagonist betaxolol.[227] Adrenergic receptors in the ciliary epithelium are of the β_2 subtype,[228-230] but antagonists that are relatively selective for β_1 receptors (e.g. betaxolol) are effective (although less potent) in suppressing aqueous humor formation.[58,227,231-233] However, the apparent β_1 efficacy may be related to a sufficiently high concentration reaching the ciliary epithelium so that non-selective blockade of β_2 receptors may occur.[234]

It is questioned whether β-adrenergic antagonists suppress aqueous humor formation via their effect on ciliary epithelial β-adrenergic receptors.[77,181,211,214,235-237] There is evidence that classical β-adrenergic receptor blockade may not be involved, and that other receptor types such as 5HT$_{1A}$, may be relevant (see Other Agents).[238] Furthermore, Na/K/2Cl cotransport stimulated by epinephrine and isoproterenol can be inhibited by β_2-adrenergic receptor antagonists in human fetal non-pigmented ciliary epithelial cells.[239]

Aqueous humor formation is reduced by nearly 50 percent during sleep,[58,240,241] due to the β-arrestin/cAMP cascade regulation of signal transduction from the β-adrenergic receptor

to the ciliary epithelium. This results in a circadian fluctuation of IOP.[242] However it is suggested that IOP may actually increase in humans at night and that technical limitations may be responsible for previously reported IOP decreases.[243,244] Nocturnal IOP elevation may play a more important role in glaucomatous progression than previously appreciated.[241] β-Antagonists produce little additional reduction in aqueous humor formation during sleep[245] or in pentobarbital-anesthetized monkeys.[246] Therefore adrenergic-based glaucoma therapies which target aqueous humor formation reduction may be less efficacious for combating the progression of glaucomatous damage than therapies targeting other mechanisms.

Sympathetic fibers synapse in the superior cervical ganglion and distribute to the muscles and blood vessels of the ciliary body, regulating blood flow to the ciliary process vasculature (see Chapter 10). Catecholaminergic and NPY-ergic nerve fibers preferentially supply the vasculature and epithelium of the monkey anterior ciliary processes, suggesting that they assist in the precise regulation of aqueous humor formation.[247] Stimulation of the cervical sympathetic nerves in vervet monkeys significantly increases the rate of aqueous humor formation.[179] However adrenergic tone to the ciliary epithelium may be humoral catecholaminergic rather than sympathetic innervational. Two major hormones of the adrenal gland, epinephrine and cortisol can regulate aqueous formation.[248] Aqueous flow in human subjects correlates with circulating catecholamine levels during sleep and wakefulness.[249] Adrenergic drugs may exert their effect by causing localized constriction in the arterioles that supply the ciliary processes.[250]

Topically applied α_1-adrenergic agonists and antagonists appear to have little effect on fluorophotometrically determined aqueous humor formation in the normal intact human eye.[251,252] Clonidine, which has both α_1-antagonist and α_2-agonist properties, decreases aqueous humor formation and ocular blood flow.[217,253-255] Therefore epinephrine may have a dual effect on aqueous humor formation: stimulation via β-adrenoceptors, and inhibition via α_2-adrenoceptors.[256-258] α_2-Adrenergic agonists, such as apraclonidine and brimonidine, are powerful ocular hypotensive agents when applied topically, and lower IOP primarily by decreasing aqueous humor formation.[16,259] However, one study in humans suggests that brimonidine suppresses aqueous humor formation early on, but after 8–29 days of treatment the predominant effect is on uveoscleral outflow.[260]

Other agents

There are many other ways in which aqueous humor formation can be reduced pharmacologically (Table 11.2). Effective compounds include: the guanylate cyclase activators atrial natriuretic factor[261,262] and the nitrovasodilators sodium nitroprusside,[263,264] sodium azide,[265] and nitroglycerin.[263] 8-Bromo cyclic GMP also reduces the aqueous humor formation rate by 15–20 percent in the monkey.[266] Atrial natriuretic factor injected intravitreally reduces IOP and aqueous humor formation in rabbits and monkeys.[262] However, intracameral and intravenous administration of atrial natriuretic factor to monkeys increases aqueous humor formation.[267] Atrial natriuretic factor levels are elevated in aqueous humor of rabbits following topical treatment with kappa

Table 11.2 Factors causing reduced aqueous humor secretion

I. General	II. Systemic	III. Local	IV. Pharmacologic	V. Surgical
A. Age	A. Artificial reduction in internal carotid arterial blood flow	A. Increased IOP (pseudofacility)	A. β-Adrenoceptor antagonists (e.g. timolol, betaxolol, levobunolol , carteolol, metipranolol)	A. Cyclodialysis
B. Diurnal cycle	B. Diencephalic stimulation	B. Uveitis (especially iridocyclitis)	B. Carbonic anhydrase inhibitors	B. Cyclocryothermy
C. Exercise	C. Hypothermia	C. Retinal detachment	C. Nitrovasodilators; atrial natriuretic factor (route and species dependent)	C. Cyclodiathermy
	D. Acidosis	D. Retrobulbar anesthesia	D. 5-HT$_{1A}$ antagonists (e.g. ketanserin)	D. Cyclophotocoagulation
	E. General anesthesia	E. Chroroidal detachment	E. DA$_2$ agonists (e.g. pergolide, lergotrile, bromocriptine)	
			F. α$_2$-Adrenoceptor agonists (e.g. apraclonidine, brimonidine)	
			G. Opioid agonists	
			H. Δ9-Tetrahydrocannabinol (Δ9 - THC)	
			I. Metabolic inhibitors (e.g. DNP, fluorocetamide)	
			J. Cardiac glycosides (e.g. ouabain, digoxin)	
			K. Spironolactone	
			L. Plasma hyperosmolality	
			M. cGMP	

cGMP, cyclic guanosine monophosphate; DA$_2$, dopamine; DNP, dinitrophenol; IOP, intraocular pressure.

Modified from Stamper RL: Aqueous humor: secretion and dynamics. In Tasman W, Jaeger EA (eds): Clinical ophthalmology, vol 2, Philadelphia 1979, Lippincott.

opioid agonists, which results in IOP reduction and aqueous humor formation suppression.[268] However, non-kappa opioid mechanisms are responsible for aqueous humor formation and IOP reductions in monkeys.[185] Nitric oxide may be involved in mediating the IOP lowering response to mu3 opioid agonists in rabbits.[269-271]

The serotonergic antagonist ketanserin reduces the aqueous humor formation rate in rabbits, cats, and monkeys.[272] Serotonergic receptors of a 5-HT$_{1A}$-like subtype are reported to exist in the iris-ciliary body of rabbits and humans.[238,273,274] These receptors may be antagonized by timolol and other β-blockers.[238] The 5-HT$_{1A}$ agonist 8-OH-DPAT dose-dependently decreases IOP in normotensive rabbits during light and dark cycles.[275,276] The more selective 5-HT$_{1A}$ agonist flesinoxan also decreases IOP in rabbits.[277] However, the precise nature of the putative 5-HT$_{1A}$-like receptor subtype in the ciliary epithelium is still in question. Also, IOP and aqueous flow suppression of 5-HT$_{1A}$ receptor agonists in non-human primates is variable.[278] A selective 5-HT2 agonist increases aqueous humor formation following topical treatment in monkeys but lowers IOP by increasing uveoscleral outflow.[279]

Metabolic inhibitors such as dinitrophenol decrease aqueous humor formation[280] as do the cardiac glycosides ouabain and digoxin, which inhibit the ciliary epithelial Na$^+$-K$^+$-ATPase enzyme.[88,281] None of these has yet been shown to be useful clinically as glaucoma therapeutics agents. Ibopamine exhibits activity at dopaminergic and adrenergic receptors. In humans, it can increase aqueous humor formation but can decrease[282] or increase IOP.[283,284]

A component of marijuana (cannabis), Δ^9-tetrahydro-cannabinol, reduces secretion of aqueous in human volunteers[285] when injected intravenously or when inhaled via marijuana smoking. In contrast, topical Δ^9-tetrahydro-cannabinol has no effect on the human eye.[286,287] Studies demonstrate the presence of functional CB1 receptors in the ciliary processes and TM of human and animal tissue.[288-290] Topical application of the cannabinoid receptor agonist, WIN-55-212-2, significantly reduces aqueous humor formation in normal and glaucomatous cynomolgus monkeys, but without increasing outflow facility. The decrease in aqueous humor formation after a single application is insufficient to explain the reduction in IOP, suggesting that other mechanisms are involved.[32] The endocannabinoid, 2-arachidonylglyceral, increases outflow facility in a porcine organ culture system by an effect mediated through both CB1 and CB2 cannabinoid receptors and changes in the actin cytoskeleton in porcine TM cells.[291]

Other local, systemic, and surgical factors can reduce aqueous humor formation rate, as can age and exercise (Table 11.2). In a study of 300 normal volunteers, ages 5–83 years, there is an average decline in aqueous flow of 25 percent between the ages of 10–80 years.[58] A smaller decline occurs in humans over 65 years of age.[11] There is no age-related decline in aqueous humor formation in rhesus monkeys 25–29 years of age (human equivalent 62–73 years) compared to those 3–10 years or 19–23 years of age.[8]

Long-term use of drugs that decrease IOP by decreasing aqueous humor formation could adversely affect the eye.[292] During the course of several weeks of oral acetazolamide therapy in humans, IOP gradually reverts toward pretreatment values but tonographic outflow facility is reduced.[292] Similarly, monkeys receiving prolonged treatment with timolol show a reduction in outflow facility.[219] Underperfusion of the TM can lead to detrimental morphologic effects.[293]

Aqueous humor drainage

Fluid mechanics

The tissues of the anterior chamber angle normally offer a certain resistance to fluid outflow. IOP builds up, in response to the inflow of aqueous humor, to the level sufficient to drive fluid across that resistance at the same rate it is produced by the ciliary body; this is the steady-state IOP. In the glaucomatous eye, this resistance is often unusually high, causing elevated IOP. Results from three major clinical trials confirm the value of reducing IOP in patients with ocular hypertension or glaucoma to prevent the onset of glaucoma in the former case and the progression of disease in the latter.[294–297] Understanding the factors governing normal and abnormal aqueous humor formation, aqueous humor outflow, IOP, and their interrelationships and manipulation is vital in understanding and treating glaucoma.

Briefly, let:

- F = flow (μl/min)
- F_{in} = total aqueous humor inflow: human = ~2.5 μl $\times$ min^{-1}.[11,58,221,298]
- F_S = inflow from active secretion
- F_f = inflow from ultrafiltration
- F_{out} = total aqueous humor outflow
- F_{trab} = outflow via trabecular pathway
- F_u = outflow via uveoscleral pathway, 0.3 μl/min, measured invasively by radioactive tracer, in primarily elderly human eyes with malignant melanoma, tumor or glaucoma;[9] 1.64 μl/min in normal subjects age 20–30 yrs and 1.16 μl/min in subjects >60 years, measured by non-invasive fluorophotometry[11]
- P = pressure (mmHg)
- P_i = IOP: humans = 16 mmHg[299]
- P_e = episcleral venous pressure: human = 9 mmHg[300,301]
- R = resistance to flow (mmHg $\times$ min/μl)
- C = facility or conductance of flow (μl/min/mmHg) = 1/R
- C_{tot} = total aqueous humor outflow facility (OF): humans age 20–30 years, 0.25 μl/min/mmHg; >60 years, 0.19 μl/min/mmHg;[11] OF = 0.54 – (0.0042 $\times$ age in years) μl/min/mmHg;[302] 0.24 μl/min/mmHg;[303] <40 years, 0.33; >60 years, 0.23 μl/min/mmHg;[304] 0.28 μl/min/mmHg,[305] all measured by indentation tonography
- C_{trab} = facility of outflow via trabecular pathway: humans 20–30 years, 0.21 μl/min/mmHg; >60 years, 0.25 μl/min/mmHg (fluorophotometry);[306] 0.22 μl/min/mmHg (tonography)[305]
- C_u = facility of outflow via uveoscleral pathway: humans = 0.02 μl/min/mmHg[307] this was probably based on reported monkey values[152,308]
- C_{ps} = facility of inflow: human = 0.06 μl/min/mmHg;[305] 0.08 μl/min/mmHg.[53] These values for C_{ps} are most likely overestimates: under normal circumstances in a non-inflamed eye the phenomenon is negligible;[54] values for the normal monkey measured by a more precise tracer technique are <0.02 μl/min/mmHg[152,308]

Then:

$$F_{in} = F_S + F_t$$

$$F_{out} = F_{trab} + F_u$$

$$C_{tot} = C_{trab} + C_u + C_{ps}$$

At steady-state:

$$F = F_{in} = F_{out}$$

The simplest hydraulic model, represented by the classic Goldmann equation, views aqueous flow as passive non-energy-dependent bulk fluid movement down a pressure gradient, with aqueous leaving the eye only via the trabecular route, where $\Delta P = P_i - P_e$, so that $F = C_{trab}(P_i - P_e)$. This relationship is correct as far as it goes, but it is vastly oversimplified. Since there is no complete endothelial layer covering the anterior surface of the ciliary body and no delimitation of the spaces between the trabecular beams and the spaces between the ciliary muscle bundles,[140] fluid can pass from the chamber angle into the tissue spaces within the ciliary muscle. These spaces in turn open into the suprachoroid, from which fluid can pass through the scleral substance or the perivascular/perineural scleral spaces into the episcleral tissues. Some investigators believe that some fluid may also be drawn osmotically into the vortex veins by the high protein content in the blood of these vessels.[2] Along these uveal routes, the fluid mixes with tissue fluid from the ciliary muscle, ciliary processes, and choroid. Thus, this flow pathway may be analogous to lymphatic drainage (also see Prostaglandin Mechanisms) of tissue fluid in other organs and provide an important means of ridding the eye of potentially toxic tissue metabolites.[153,154] The eye has long been considered to be devoid of lymphatics, but expression of lymphatic markers in the ciliary body and muscle have recently been reported.[308a–308d] Whether there is a true lymphatic flow system in the eye remains to be established, however.

Flow from the anterior chamber across the TM into Schlemm's canal is pressure-dependent, but drainage via the uveoscleral pathway is virtually independent of pressure at IOP levels greater than 7–10 mmHg.[12,48,309] While the actual drainage rates (μl/min) via the trabecular and uveoscleral routes in the monkey may be approximately equal, measured facility of uveoscleral outflow (C_u, determined by measuring uveoscleral outflow at two different IOP levels) is only about 0.02 μl/min/mmHg, or less than one-twentieth the facility of trabecular outflow.[309] The reasons for the pressure-independence of the uveoscleral pathway are not entirely clear but might be consequent to the complex nature of the pressure and resistance relationships between the various fluid compartments within the soft intraocular tissues along the route.[12] For instance, pressure in the potential suprachoroidal space (Ps) is directly dependent on IOP, such that at any IOP level, Ps is considerably but constantly less than IOP.[12] Since the pressure gradient between the anterior chamber and suprachoroid is independent of IOP, bulk fluid flow between these compartments will also be IOP-independent. Intraorbital pressure is such that under normal circumstances there is always a positive pressure gradient between the suprachoroidal and intraorbital spaces.[12] Fluid and solutes, including large protein molecules, can thus easily exit the eye by passing through the spaces surrounding the neural or vascular scleral emissaria or through the scleral substance itself.[4,50,310,311] At very low IOP levels, the net pressure gradient across the uveoscleral pathways is apparently so low that uveoscleral drainage decreases.[48]

The absence of an outflow gradient from the suprachoroid may contribute to the development of choroidal detachments seen during the ocular hypotony that sometimes follows intraocular surgery.[12] However, other investigators find unconventional outflow to be more pressure-sensitive than described by Bill.[3,312,313] Uveoscleral outflow may become more sensitive to pressure following prostaglandin treatment.[314] There is no need for fluid flow to carry tracer across the sclera since it can diffuse across on its own, based on calculations of diffusional transport porperties.[2] Direct evidence for a uveovortex pathway[3] is demonstrated after perfusion of the anterior chamber with fluorescein, and finding the fluorescein concentration in the vortex veins is higher that in the general circulation. Also, flow across the sclera is pressure-dependent.[313] Uveovortex flow explains the relative insensitivity of the flow to pressure because most of the driving force is the colloidal osmotic pressure of the blood that draws the fluid into these vessels.[2] Clearly, more needs to be learned about this flow pathway. Since, under normal steady-state conditions, C_{ps} and C_u are so low compared with C_{trab}, the hydraulics of aqueous dynamics may be reasonably approximated for clinical purposes by:

$$F_{in} = F_{out} = C_{trab}(P_i - P_e) + F_u$$

or rearranged:

$$P_i = P_e + (F_{in} - F_u)/C_{trab}$$

Clinically significant increases in inflow occur only in situations involving breakdown of the blood–aqueous barrier. The pressure sensitivity of the ultrafiltration component of aqueous secretion blunts the tendency for IOP to rise under such conditions, that is, C_{ps} is increased. Elevated episcleral venous pressure, such as may occur with arterio-venous communications resulting from congenital malformations or trauma[315] and perhaps, as shown for the first time in one study, in POAG and normal tension glaucoma,[300] causes a nearly mmHg for mmHg increase in IOP. Clinically relevant reductions in IOP are produced by decreasing F_{in} or by increasing C_{trab} and F_u. Pharmacologic agents may exert small and probably clinically insignificant effects on P_e.[59,179,196,197,316–321]

Structural components

Approximately one-half to three-quarters of the aqueous leaves the eye through the TM and Schlemm's canal.[11,306] This outflow is pressure dependent. Schlemm's canal and the TM lie within the internal scleral groove between the scleral spur and the ring of Schwalbe's line at the termination of Descemet's membrane. An anterior non-filtering portion, which presents minimal resistance to fluid outflow, can be distinguished from a posterior filtering portion of the meshwork. The TM itself consists of three functionally and structurally different parts: the iridic and uveal part, which represents the innermost portion of the meshwork; the corneoscleral part, which extends between the scleral spur and the cornea; and the juxtacanalicular part or cribriform layer, which lies adjacent to the inner wall of Schlemm's canal (Fig. 11.4).[322–326]

The major resistance site within the trabecular structures likely resides in the cribriform portion of the meshwork,[322–324,326–328] the outermost part of the mesh consisting of several layers of endothelial cells embedded in a ground substance comprising a wide variety of macromolecules, including hyaluronic acid, other glycosaminoglycans, collagen, fibronectin, and other glycoproteins presumably produced by meshwork endothelial cells.[329–331] The cribriform layer is supported by an elastic-like fiber network and fine collagen fiber bundles (Fig. 11.5),[332] having the same orientation as the elastic-like network in the central core of the trabecular lamellae. This network is connected to the inner wall endothelium of Schlemm's canal by fine, bent, connecting fibrils. This cribriform plexus is far more extensive in the juxtacanalicular tissue of human eyes as compared to other species, perhaps contributing to the lack of washout in human eyes in vitro.[333] The washout-associated increase in outflow facility in nonhuman species[334–337] correlates with the extent of physical separation between the juxtacanalicular tissue and the inner wall endothelium lining Schlemm's canal or, in some species, the aqueous plexus.[338]

However some investigators consider the main resistance to lie slightly proximal to the cribriform meshwork tissue,[339–341] i.e. in the inner wall itself and the very flimsy basement membrane of the inner wall endothelial cells. Separation of the basal lamina cells lining of the inner wall of Schlemm's canal from the underlying juxtacanalicular tissue as seen using quick-freeze/deep-etch electron microscopy may represent flow pathways.[342] Under normal conditions, the inner wall of Schlemm's canal and juxtacanalicular cells may be in a contracted state, limiting the routes available for fluid flow as demonstrated with gold particle infusion studies in non-human primates (Fig. 11.6).[343,344] Expansion of the area available for fluid drainage can increase the rate of fluid outflow.[343,344] The accompanying loss of extracellular material may not be responsible for the decrease in resistance to fluid outflow.[342–345]

Fluid movement across the inner canal wall endothelium itself appears to be predominantly via passive pressure-dependent transcellular pathways,[323] including giant vacuole formation (Fig. 11.7) especially near collector channels.[346–350] Pore formation is often associated with the giant vacuoles, but may also be found in thin, flat regions in the inner wall.[323] Calculations of the number and size of pores and openings in the inner wall endothelium of Schlemm's canal are too large to account for most of the outflow resistance.[351,352] These and other findings lead to the assumption that the main resistance to aqueous outflow is located internal to the endothelial lining, i.e. within the subendothelial or cribriform layer.[353]

Although the pores themselves contribute negligible flow resistance, because they force the fluid to "funnel" through those regions of the cribriform meshwork tissue nearest the pores, their number and size can greatly influence the

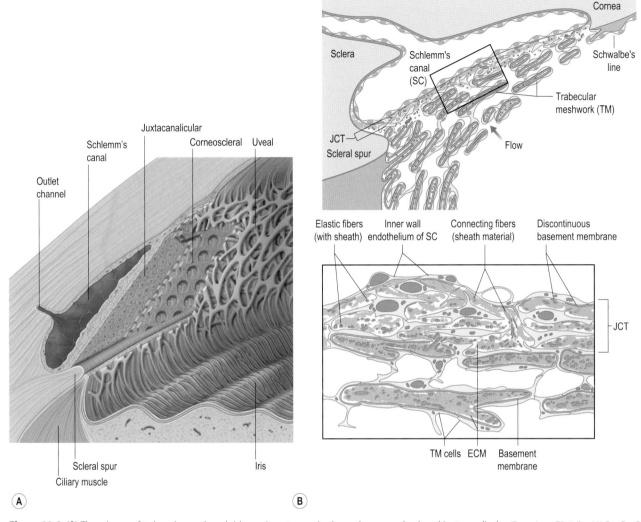

Figure 11.4 (**A**) Three layers of trabecular meshwork (shown in cutaway view): uveal, corneoscleral, and juxtacanalicular. (From Acott TS, Kelley MJ. Exp Eye Res 86:543, 2008. Copyright Elsevier 2008.) (**B**) Diagram of the outflow pathway and juxtacanalicular (JCT) or cribriform region. The lower portion of the figure shows a stylized view of the TM and the upper inset shows an expanded view of the JCT region.

effective outflow resistance of the cribriform meshwork tissue.[354] However, two studies[351,352] failed to find a correlation between outflow facility and inner wall pore density as would be expected if the funneling effect contributed to aqueous outflow resistance.[2] Studies with gold particles suggest that exclusion of large segments of the trabecular outflow pathway may help maintain high resistance to flow. Expanding the areas available for fluid drainage, by relaxation of the inner wall cells and the adjacent cribriform meshwork region may be a possible approach to glaucoma therapy (Fig. 11.6).[343]

Experimental studies of the transendothelial passage of ferritin particles in monkeys indicate that ferritin also traverses tortuous paracellular routes that lie between the endothelial cells of Schlemm's canal.[355] The functional significance of these paracellular routes for aqueous outflow is not known. Tight junctions between endothelial cells of Schlemm's canal become less complex with increasing pressure, suggesting that the paracellular pathway into Schlemm's canal in the normal eye may be sensitive to modulation within a range of physiologically relevant pressures.[356] However, gold tracer does not pass across junctions

between the cells at a pressure of 25 mmHg; perhaps a higher pressure gradient is required to disrupt normal cellular junctions.[343]

Arguments about trans- vs paracellular routes and inner wall vs juxtacanalicular resistance seem to be relatively minor questions from a clinical/translational viewpoint.

Pumping model for trabecular outflow

An alternative to the bulk flow model of aqueous outflow is one that characterizes it as a pumping phenomenon that depends on tissue compliance rather than static resistance (see reviews[357,358]). Research supports the hypothesis that the aqueous outflow pump receives power from transient increases in IOP such as occur in systole of the cardiac cycle, during blinking and during eye movement. In the aqueous outflow pump model, all aqueous flow is through Schlemm's canal valves spanning the inner and outer walls of Schlemm's canal (Fig. 11.8). Clinical evidence of pulsatile flow into Schlemm's canal, from Schlemm's canal into the collector channels, and from aqueous veins into episcleral veins supports the model (visit *youtube.com/majohnstone* to see a video of this courtesy of Robert Stegmann). Tissue

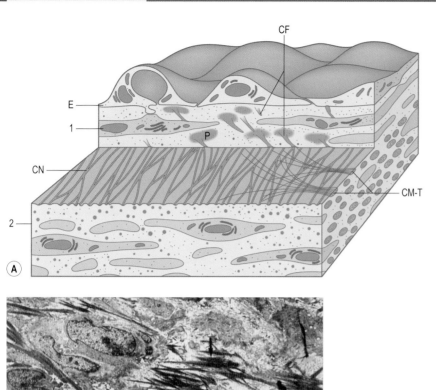

Figure 11.5 (**A**) Schematic drawing of the cribriform meshwork and the endothelial lining of Schlemm's canal (*E*). Note the connection between the ciliary muscle tendons (*CM-T*) and the elastic-like fiber plexus, or "cribriform plexus" (*CN*), located mainly in the region between the first and second subendothelial cell layers (*1*. and *2*.). The cribriform plexus is connected to the inner wall endothelium and the plaque material (*P*) by a system of fine fibrils or "connecting fibrils" (*CF*). (**B**) Electron micrograph of a tangential section through the cribriform region almost at the level between the second subendothelial cell layer and the first corneoscleral trabecular lamellas (normal eye). The cells seen in the *upper left* are subendothelial. The elastic-like fibers of the cribriform region (*arrows*) form a plexus that shows the same equatorial orientation as the network of the elastic-like fibers of the trabecular lamellas. (From Rohen JW, Futa R, Lütjen-Drecoll E: Invest Ophthalmol Vis Sci 21:574, 1981, reproduced with permission from the Association for Research in Vision and Ophthalmology.)

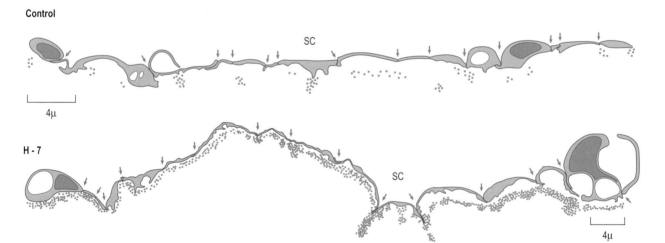

Figure 11.6 Distribution of perfused gold particles in the juxtacanalicular area of eyes treated with H-7 (1-[5-isoquinoline sulfonyl]-2-methyl piperazine) and in control eyes. Schematic drawings depicting 15-cell stretches (cell–cell junctions marked by *arrows*) along the Schlemm's canal (*SC*) of control and H-7-treated eyes. The location of individual gold particles is represented by *red dots*. (Bars = 4 μm.) (From Sabany I et al: Arch Ophthalmol 118:955, 2000. Copyright 2000 American Medical Association. All rights reserved.)

biomechanics allows the aqueous outflow pump to undergo short-term pressure-dependent stroke volume changes to maintain short-term homeostasis. Failure of the pumping mechanism occurs in glaucoma as a result of TM stiffening and reduced trabecular movement and persistent closure of Schlemm's canal.

Active involvement of the TM in regulating outflow

Another emerging concept portrays the TM as a living, active and reactive organ, rather than just a passive mélange of tissue components. Long-term homeostasis in response to physical stress requires modulation by cellular constituents

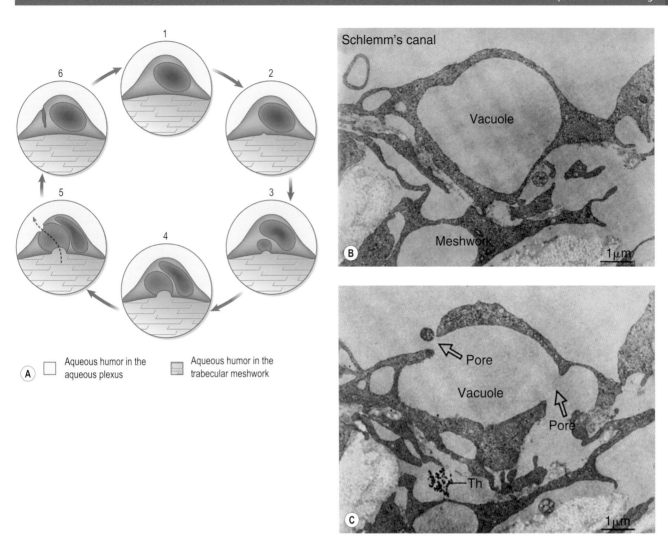

Figure 11.7 (**A**) Theory of transcellular aqueous transport in which a series of pores and giant vacuoles opens (probably in response to transendothelial hydrostatic pressure) on the connective tissue side of the juxtacanalicular meshwork (2–4). Fusion of basal and apical cell plasmalemma creates a temporary transcellular channel (5) that allows bulk flow of aqueous into Schlemm's canal. (From Tripathi RC: Exp Eye Res 11:116, 1971. Copyright Elsevier 1971.) (**B**) Transmission electron micrograph of the inner wall of Schlemm's canal and the adjacent subendothelial tissue showing empty spaces or "giant vacuoles" within the endothelial cells. (From Inomata H et al: Am J Ophthalmol 73:760, 1972. Copyright Elsevier 1972.) (**C**) Serial sections of the inner wall of Schlemm's canal indicate that the "giant vacuoles" of the endothelial cells have openings toward the trabecular side, indicating that they are invaginations from the trabecular side. Some of the invaginations also have openings (pores) into Schlemm's canal. Aqueous humor can pass through the cells via the invaginations and the pores. Th = thorotrast. (From Inomata H et al: Am J Ophthalmol 73:760, 1972. Copyright Elsevier 1972.)

of the conventional outflow pathway: Schlemm's canal endothelia, juxtacanalicular cells, endothelium lining the lumen of Schlemm's canal valves, and endothelium lining trabecular lamellae. Extracellular load-bearing constituents include the ECM of the trabecular lamellae and the juxta-canalicular meshwork. How might all this work? The TM is in essence suspended between two fluid compartments (anterior chamber and Schlemm's canal) at different pressures. The TM can likely "sense" this pressure differential by "monitoring" stretch, deformation, shear stress, etc. and strives to maintain these parameters within a homeostatic range. Schlemm's canal endothelial cells may transmit sheer stress or other deformational stimuli through cell processes to the endothelium of the juxtacanalicular region and the trabecular lamellae and to their corresponding basement membrane via integrins which in turn mediate regulation of ECM deposition and relaxation/contractility of the cells themselves. ECM rigidity in turn, modulates cytoskeletal

structures, protein expression patterns, signal transduction, and fibronectin deposition in HTM cells.[359] Thus the tissue senses the physical as well as the biochemical environment in which it resides, and makes modifications to its physical and conformational properties to affect its overall hydraulic conductivity and thereby allow the eye to reach a specific target pressure (see also Cytoskeletal and Cell-Junctional Mechanisms). One possibility for derangement in glaucoma is that the set points become altered.

Outflow obstruction

Extracellular matrix accumulation and POAG

In glaucomatous eyes there is an increase in ECM beneath the inner wall of Schlemm's canal and in the cribriform region of the TM and thickening of the trabecular lamellae compared with age-matched normal controls.[40,360] In advanced cases of primary open-angle glaucoma (POAG) there is

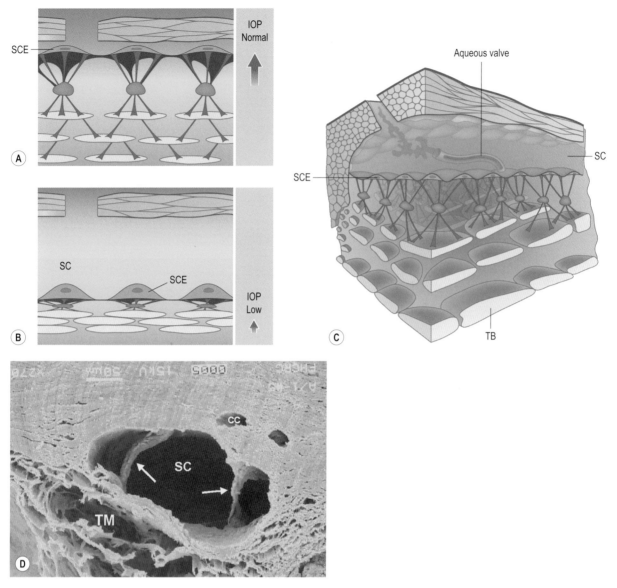

Figure 11.8 (**A**), (**B**), (**C**) Schematic representation of the biomechanical pump model. The pump is powered by transient increases in IOP, primarily caused by the cardiac cycle, blinking and eye movements. As the pressure increases, fluid is forced into one-way collector valves (C) that span across Schlemm's canal. At the same time, the increase in IOP pushes the endothelium of the inner wall of Schlemm's canal (A, B) outward and forces aqueous in the canal to move circumferentially into collector channels and aqueous veins. As the pressure drops, the tissues rebound, causing a pressure drop inside Schlemm's canal, moving fluid from the one-way valves (C) into the canal. SC, Schlemm's canal; SCE, Schlemm's canal endothelium; TB, trabecular beams. (From Johnstone MA: Rev Ophthalmol 14:9, 2007. Reproduced with permission from Jobson Publishing LLC and Murray A. Johnstone, MD.) (**D**) Valve-like structures in Schlemm's canal shown by scanning electron microscopy. Monkey eye – *Macaca nemistrina*. Viscoelastic dilation of Schlemm's canal (SC). Two valve-like structures (white arrows) span between trabecular meshwork (TM) and corneoscleral wall of SC. (cc) collector channel ostia. (From Smit BA, Johnstone MA: Ophthalmology 109:786, 2002. Copyright Elsevier 2002.)

additional loss of trabecular cells beyond that associated with normal aging.[132,361] In this condition, the inner uveal and corneoscleral lamellae can be "glued" together,[361] and Schlemm's canal can be partly obliterated. Areas with more ECM are less perfused, presumably because of the higher resistance of the area.[362] The origin of the increased amounts of ECM in glaucomatous eyes is still unknown. Additionally, most specimens of glaucomatous eyes investigated morphologically are derived from patients who have been treated for many years with antiglaucoma drugs which can themselves induce changes in the biology of the trabecular cells.[363]

Transforming growth factor (TGF) β2, a component of normal aqueous humor in many mammalian eyes,[364–367] influences ECM production in the TM and has been implicated in IOP elevation.[368,369] Increased levels of total and active TGFβ2 are found in the aqueous humor of POAG patients compared to age-matched controls.[367,370] TGFβ2 decreases the activity of matrix metalloproteinases (MMP), and thus possibly contributes to increased ECM in the TM of glaucomatous eyes.[371] TGFβs have an inhibitory effect on the rate of cell proliferation and motility of TM cells in vitro, which could contribute to the decreased cellularity of the TM.[372] Perfusion of human anterior segments in vitro with TGFβ2 results in decreased outflow facility and increased focal accumulation of ECM under the inner wall of Schlemm's canal.[369] Similarly, TGFβ2 also enhances production of cochlin (an ECM protein found only in the TM of POAG eyes[373]), increases IOP and decreases outflow facility in monkey and pig organ-cultured anterior segments.[374] Over-expression of cochlin alone can also increase IOP in monkey organ-cultured anterior segments.[682] In vitro treatment of cultured TM cells with TGFβ2 results in elevated production of ECM proteins such as fibronectin and plasminogen activator inhibitor (PAI-1).[368] In addition, interactions between bone morphogenic protein (BMP) and Wnt signaling, together with those between BMP and TGF-β, as well as TGF-β and Wnt, represent potential checkpoints to regulate TM cell function, IOP, and glaucoma pathogenesis. Increased expression of secreted frizzled-related protein-1 (sFRP-1) in glaucomatous TM may be responsible for the glaucoma phenotype of elevated IOP in humans. Studies are being conducted to determine the prevalence of defects in the Wnt pathway in the glaucoma population, the primary cause of altered sFRP-1 expression, and the potential deficiency of other members of the WNT pathway involved in glaucoma.[683]

How important "extra" ECM is in increasing resistance to aqueous outflow (and thus increasing IOP) in open-angle glaucoma is uncertain. The endothelial cells of the TM have phagocytic capabilities.[375–377] It has been proposed that the TM is in effect a self-cleaning filter, and that in most of the open-angle glaucomas, the self-cleaning (i.e. phagocytic) function is deficient or at least inadequate to cope with the amount of material present.[12,378] Phagocytosis, especially of particulate matter and red blood cells, is also carried out by trabecular macrophages. Although this process may be important in clearing the anterior chamber of some inflammatory debris or as a normal housekeeping function, it is probably not significant for bulk outflow of aqueous. However, this is another example of the TM as a reactive rather than a passive tissue, sensing its environment and modifying it. Artificial perfusion of the anterior chamber in non-human primates is associated with a progressive time-dependent decrease in outflow resistance as described above (Structural components – washout). Resistance washout does not change with age in monkeys.[379]

Combining the clogged-filter concept of glaucoma and the washout concept of perfusion-induced resistance decrease inevitably led to interest in compounds that might disrupt or remodel the structure of the meshwork and canal inner wall so as to enhance flow through the tissue and/or promote washout of normal and pathologic resistance-producing ECM (see Cytoskeletal Mechanisms). Such compounds might provide insights into cellular and extracellular mechanisms governing outflow resistance in normal and glaucomatous states. Additionally, if normal or pathologic ECM required many years to accumulate to the extent that IOP became elevated, perhaps a one-time washout would provide years of normalized outflow resistance and IOP.[380,381]

Cell and other particulates

Normal erythrocytes are deformable and pass easily from the anterior chamber through the tortuous pathways of the TM and the inner wall of Schlemm's canal.[323] However, non-deformable erythrocytes such as sickled or clastic (ghost) cells may become trapped within and obstruct the TM, elevating outflow resistance and IOP.[382–384] Similarly, macrophages that are swollen after ingesting lens proteins leaking from a hypermature cataract[385] or breakdown products from intraocular erythrocytes[386] or pigmented tumors (or the tumor cells themselves)[387] may produce meshwork obstruction. Pigment liberated from the iris spontaneously

(pigmentary dispersion syndrome) or iatrogenically (following laser iridotomy) may clog or otherwise alter the meshwork function, presumably without prior ingestion by wandering macrophages,[388,389] as may zonular fragments following iatrogenic enzymatic zonulolysis[390,391] or lens capsular fragments following laser posterior capsulotomy.[392,393] Ocular amyloidosis can lead to elevated IOP as a result of blockage of the TM with amyloid particles.[394]

Protein and other macromolecules

Glaucoma secondary to hypermature cataract (phacolytic glaucoma) or uveitis has long been ascribed to trabecular obstruction. In the former entity, the presence of protein-laden macrophages lining the chamber angle seems adequate to account for increased outflow resistance.[385] The uveitis-related glaucomas comprise many different entities, and the etiology of the increased outflow resistance seems less clear; postulated mechanisms include trabecular involvement by the primary inflammatory process, trabecular obstruction by inflammatory cells, or secondary alteration of trabecular cellular physiology by inflammatory mediators or by-products released elsewhere in the eye.

Small amounts of purified high-molecular-weight, soluble lens proteins[395,396] or serum itself,[397] when perfused through the anterior chamber of freshly enucleated human eyes, causes an acute and marked increase in outflow resistance. Thus, it may be that specific proteins, protein subfragments, or other macromolecules are themselves capable of obstructing or altering the meshwork so as to increase outflow resistance, perhaps contributing to the elevated IOP in entities such as the phacolytic, uveitic, exfoliation,[398] and hemolytic[386] glaucomas. Experimental perfusion of the enucleated calf eye[399,400] or monkey eye in vivo[401] with medium containing higher protein concentrations than found in normal aqueous humor indeed reduces or eliminates resistance washout. Perhaps protein in the TM is essential for maintenance of normal resistance either by providing resistance itself, or by signaling or modifying some property of the meshwork such as stimulation of focal adhesion and stress fiber formation, to enhance adhesion of TM cells to the ECM.[401–403]

Hyaluronate- and chondrotin sulfate-based agents used as tissue spacers during intraocular surgery ("viscoelastic agents") may raise IOP in human eyes if not completely removed from the eye by irrigation/aspiration at the conclusion of a procedure, presumably by obstructing trabecular outflow.[404,405]

Pharmacology and regulation of outflow

Cholinergic mechanisms

Conventional (trabecular) outflow

In primates the iris root inserts into the ciliary muscle and the uveal meshwork just posterior to the scleral spur, while the ciliary muscle inserts at the scleral spur and the posterior inner aspect of the TM.[406] The anterior tendons of the longitudinal portion of the ciliary muscle insert into the outer lamellated portion of the TM and into the cribriform meshwork, and via elastic elements into specialized cell surface adaptations on the inner wall endothelial cells (see Fig. 11.5). Ciliary muscle contraction results not only in spreading of the lamellated portion of the meshwork but also in an inward pulling of the cribriform elastic fiber plexus and straightening of the connecting fibrils and perhaps dilation of Schlemm's canal. Movement of the inner wall region might affect the area and configuration of the outflow pathways and thereby outflow resistance.[407] Voluntary accommodation (human),[408] electrical stimulation of the third cranial nerve (cat),[409] topical, intracameral, or systemically administered cholinergic agonists (monkey and human),[410,411] and, in enucleated eyes (monkey and human), pushing the lens posteriorly with a plunger through a corneal fitting[412] all decrease outflow resistance, while ganglionic blocking agents and cholinergic antagonists increase resistance.[410,413-415]

Furthermore, the resistance-decreasing effect of intravenous pilocarpine in monkeys is virtually instantaneous, implying that the effect is mediated by an arterially perfused structure or structures.[416] However, not all the experimental evidence supports this strictly mechanical view of cholinergic and anticholinergic effects on meshwork function. For example, in monkeys, intravenous atropine rapidly reverses some but not all of the pilocarpine-induced resistance decrease,[417,418] and topical pilocarpine causes a much greater resistance decrease per diopter of induced accommodation than does systemic pilocarpine (monkey)[418] or voluntary accommodation (human).[419] The inability of atropine to rapidly and completely reverse the pilocarpine-induced facility increase in normal eyes could be due to mechanical hysteresis of the meshwork.[420] The variation in the relative magnitude of pilocarpine-induced accommodation and resistance decrease when the drug is administered by different routes might reflect differences in bioavailability of the drug to different regions of the muscle.[418]

However, following ciliary muscle disinsertion and total iris removal (but not iris removal alone), there is virtually no acute resistance response to intravenous or intracameral pilocarpine and no response to topical pilocarpine.[421] Thus it seems virtually certain that the acute resistance-decreasing action of pilocarpine, and presumably other cholinomimetics, is mediated entirely by drug-induced ciliary muscle contraction, with no direct pharmacologic effect on the meshwork itself.

Cholinergic and nitrenergic nerve terminals that could induce contraction and relaxation of TM and scleral spur cells are present in primate TM and scleral spur. Terminals in contact with the elastic-like network of the TM and containing substance P immunoreactive fibers resemble afferent mechanoreceptor-like terminals.[422] Afferent mechanoreceptors that measure stress or strain in the connective tissue elements of the scleral spur have also been identified.[423] These findings raise the possibility that the TM may have some ability to self-regulate aqueous humor outflow.

Muscarinic receptors and contractile elements are present in the TM. The m3 mRNA muscarinic receptor transcript is detected in human TM of cadaver eyes.[424] Carbachol (CARB)-induced mobilization of Ca^{2+} and phosphoinositide production in human TM cells in culture is associated with the M3 muscarinic receptor.[425] Smooth muscle specific contractile proteins are present in cells within the human TM and adjacent to the outer wall of Schlemm's canal and the collector channels.[426-428] Isolated bovine TM strips contract

isometrically in response to CARB, pilocarpine, aceclidine (ACEC) and acetylcholine and endothelin-1.[429-431] However, in the organ cultured perfused bovine anterior segment, endothelin-1 and CARB-induced contractions result in an increase in resistance and a reduction of the outflow rate.[432] Also, low (10^{-8} to 10^{-6}M) but not high (10^{-4} to 10^{-2}M) doses of pilocarpine, ACEC, or CARB induce increased outflow facility in human perfused anterior ocular segments, devoid of ciliary muscle.[433,434] However, low doses of pilocarpine have no effect on outflow facility in living monkeys, eyes;[435] facility increases occur only at doses which also produced miosis and accommodation.

Direct effects of cholinergics on the TM and outflow facility could be mediated by cAMP[436] and/or increased intracellular calcium and phospholipase C activity.[425] Since receptor subtype specificity is concentration-dependent, low pilocarpine doses theoretically could stimulate M2 receptor-mediated cAMP elevation in the TM, and consequent relaxation of the meshwork.[209] These potential direct actions of cholinomimetics notwithstanding, the major effect of cholinomimetics on outflow facility is almost surely related to ciliary muscle contraction.

Pilocarpine-induced ciliary muscle contraction spreads the TM, perhaps with the same physiological consequences as direct relaxation of the meshwork – decreased tissue density and thus decreased flow resistance due to opening of new flow pathways. Combination of threshold facility-effective doses of pilocarpine with maximal outflow facility-effective doses of H-7 (see Cytoskeletal Mechanisms) can further enhance the outflow facility response with minimal effects on accommodation.[437]

At least two different subtypes of muscarinic receptors, M2 and M3, are present in the ciliary muscle.[424,438,439] The M2 receptor shows preferential localization to the longitudinal,[438] putatively more facility relevant portion of the ciliary muscle, but to date no functional role for this subtype has been elucidated. mRNA from the M2, M3, and M5 subtypes is strongly expressed in the longitudinal and circular portions of human ciliary muscle cells and tissue.[439] The M3 subtype appears to mediate the outflow facility and accommodative responses to pilocarpine and aceclidine in monkeys.[440,441] In monkeys, the outer longitudinal region of the ciliary muscle differs ultrastructurally and histochemically from the inner reticular and circular portions.[442] Although differential distribution of muscarinic subtypes is probably not responsible for the outflow facility/accommodation dissociation occurring under certain conditions, functional dissociation might be produced by combinations of drugs from different classes.

Alterations in cholinergic sensitivity of the outflow apparatus

Given the long-term use of cholinergic agonists in glaucoma therapy and the vital role of ciliary muscle tone in regulating outflow resistance, it is important to note that in the monkey, topical administration of the cholinesterase inhibitor echothiophate or the direct-acting agonist pilocarpine can induce subsensitivity of the outflow facility and accommodative responses to pilocarpine, accompanied by decreased numbers of muscarinic receptors in the ciliary muscle.[443-449] Even a single dose of pilocarpine or carbachol reduces receptor number.[448] Since cholinergic drug therapy is not the

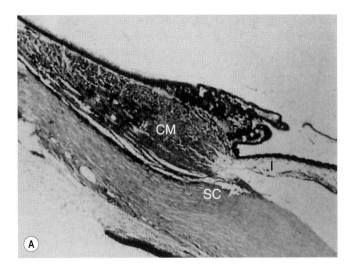

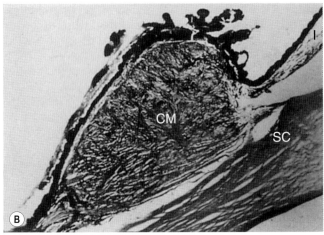

Figure 11.9 Histologic sagittal sections through the chamber angle region of monkey eyes treated acutely with atropine (**A**) or pilocarpine (**B**) topically. After treatment with pilocarpine, the muscle moves anteriorly and inwardly, thereby expanding the trabecular lamellas and widening Schlemm's canal. This contraction also obliterates the spaces between the ciliary muscle bundles, obstructing uveoscleral outflow. Conversely, atropine relaxes the ciliary muscle bundles. *CM*, Ciliary muscle; *I*, iris; *SC*, Schlemm's canal (Vervet monkey, Azan stain; original magnification ×25). (From Lütjen-Drecoll E, Kaufman PL: J Glaucoma 2:316, 1993.)

mainstay of glaucoma therapy that it once was, the issues related to refractoriness to long-term therapy versus disease progression are no longer prominent areas of research.

Unconventional (uveoscleral) outflow

When the ciliary muscle contracts in response to exogenous pilocarpine (Fig. 11.9), the spaces between the muscle bundles are essentially obliterated.[363,406,410] Conversely, during atropine-induced muscle relaxation, the spaces are widened.[410] If mock aqueous humor containing albumin labeled with iodine-125 or iodine-131 (which under resting conditions leave the anterior chamber essentially by bulk flow via the trabecular and uveoscleral drainage routes) is perfused through the anterior chamber, autoradiographs may be made to show qualitatively the distribution of the flow.[9,317] In the pilocarpinized eye, radioactivity is present in the iris stroma, the iris root, the region of Schlemm's canal and the surrounding sclera, and the most anterior portion

of the ciliary muscle. In the atropinized eye, radioactivity is found in all these tissues, as well as throughout the entire ciliary muscle and even further posteriorly in the suprachoroid/sclera.[9,317] In other perfusion experiments quantifying uveoscleral drainage, pilocarpinized eyes demonstrate but a fraction of the uveoscleral flow in atropinized eyes.[197,317,318] Thus, to generalize in the primate eye, the magnitude of the pilocarpine (and presumably all cholinergic agonists) enhancement of aqueous humor drainage via the trabecular route is greater than the reduction of drainage via the uveoscleral route thus resulting in a reduction in IOP.

Adrenergic mechanisms

Conventional (trabecular) outflow

Topical and intracameral epinephrine increase outflow facility in primate eyes.[179,316,450–453] Much work has been done attempting to define the time course, type of receptors (e.g. α, β, adenosine), and biochemical pathways (e.g., prostaglandins, cyclic adenosine monophosphate[454]) involved in these responses. Adrenergic receptor stimulation could alter intraocular, intrascleral, and extrascleral vascular tone, as well as have possible direct effects on the outflow pathways, all of which might alter IOP. These potential sites of action are not mutually exclusive, and indeed that might account for much of the variability and confusion in the literature.

In surgically untouched, aniridic, and ciliary muscle disinserted monkey eyes with widely varying starting facilities, epinephrine and norepinephrine increase facility by a constant percentage of the starting facility, indicating that neither the iris nor the ciliary muscle is involved in the responses and that the drugs exert their effects on whatever is responsible for the major part of the variation in starting facility (see Cytoskeletal Mechanisms). One possibility is an increase in the hydraulic conductivity per unit filtering area.[455,456] Trabecular outflow facility increases in response to β-adrenergic agonists.[10,179,457] Human TM or Schlemm's canal endothelial cells grown on porous filter supports separate and shrink when exposed to isoproterenol or epinephrine, resulting in increased transendothelial fluid flow.[458] Biochemical evidence also points to the meshwork as the target tissue. It appears that the facility-increasing effect of epinephrine and norepinephrine is mediated by β$_2$-adrenergic receptors on the trabecular endothelial cells, and the subsequent G-protein adenylate cyclase-cyclic AMP cascade.[61,459] The facility-increasing effect of epinephrine is blocked by timolol[460,461] but not betaxolol,[462–464] in both humans and monkeys, consistent with the hypothesis that there are no β$_1$ receptors present in the primate TM.

Trabecular cells synthesize cAMP in response to stimulation with β-adrenoceptor-selective agonists.[14] The increase in cAMP synthesis by TM cells in response to epinephrine can be blocked by timolol,[14] although not by betaxolol (a β$_1$-receptor antagonist), consistent with the hypothesis that there are only β$_2$-receptors present in the TM. Topically applied adrenergic agonists (epinephrine, norepinephrine, isoproterenol) elevate aqueous humor cAMP levels, and intracameral injection of cAMP or its analogs, but not the inactive metabolite 5'AMP, lowers IOP and increases outflow facility.[465–467] The cAMP-induced facility increase is not additive to that induced by adrenergic agonists, and vice-versa.[468] Epinephrine increases facility and perfusate cAMP levels in the organ-cultured perfused human anterior segment, effects that are blocked by timolol and the selective β_2-antagonist ICI 118,551.[469]

In monkey ciliary muscle, ciliary processes, TM and iris tissue in vitro, epinephrine stimulates cAMP production. Part of the facility response in vivo and cAMP production in vitro is inhibited by indometacin, suggesting that part of the mechanism of action of epinephrine may be via prostanoid production or release.[470] However, the epinephrine-induced facility increase in human anterior segments in vitro is attributed to protein synthesis and not to prostaglandin production.[471] Topical pretreatment of rabbits with an adenosine A1 antagonist inhibits both the epinephrine induced hypotensive response and the epinephrine-induced outflow facility increase, suggesting activation of ocular adenosine receptors as part of the ocular hypotensive action of epinephrine (see Other Mechanisms).[472]

Adrenergic innervation of the primate TM is sparse and concentrated mainly in the region of the meshwork near the ciliary muscle tendons. No functional significance can as yet be ascribed to these terminals.[473-475] Studies suggest an epinephrine-induced disruption of actin filaments within the TM cells, consequent alteration in cell shape and cell–cell and cell–extracellular matrix (C–ECM) adhesions within the meshwork, resulting in altered meshwork geometry, and increased hydraulic conductivity across the meshwork. Relaxation of the TM also could play a role in the outflow facility response to epinephrine.[476] Thus cytochalasin B (a disruptor of actin filament formation) potentiates the facility-increasing effect of epinephrine,[477] while phalloidin (a stabilizer of actin filaments) inhibits it.[478] Continuous exposure to epinephrine at a concentration of 10 μM produces arrest of normal cytokinetic cell movements, inhibits mitotic and phagocytic activity, marked cell retraction, and separation from the substrate and cellular degeneration after 4–5 days in cultured human trabecular cells.[479] Similarly, the hydraulic conductivity of trabecular cell monolayer cultures grown on filters increases in response to epinephrine and is associated with changes in cell shape and with separation between cells.[480] These actions of epinephrine were all partially blocked by pretreatment with timolol. β-Adrenergic agonists (epinephrine or isoproterenol) and analogs of cAMP increase phosphorylation of intermediate filaments of vimentin-type in human or rabbit ciliary processes or cultured ciliary epithelial-derived cells. Phosphorylation of vimentin increased by β-adrenergic agonists can be blocked by the β-adrenergic antagonist timolol[481] (see Cytoskeletal Mechanisms).

Unconventional (uveoscleral) outflow

β-Adrenergic receptors are present in the primate ciliary muscle, and their physiological or pharmacological stimulation relaxes the muscle.[482-485] Epinephrine, in addition to increasing trabecular outflow, also increases uveoscleral outflow in monkeys[453] and humans.[10,215] The mechanism is unknown, but may be in part due to the mildly relaxant effect of epinephrine on the ciliary muscle, presumably acting via its β-adrenergic receptors.[172,482-484] Pretreatment with the cyclo-oxygenase inhibitor indometacin inhibits the ocular hypotensive effect of topically applied epinephrine in humans[486] suggesting that the IOP-lowering action of epinephrine may be mediated at least in part by prostaglandins or other cyclo-oxygenase products.[457,486]

Numerous clinical studies in humans claim that topical application of timolol, a non-selective β1-, β2-adrenergic receptor antagonist, induces no change in distance refraction.[487] However, a single topical application of 0.5 percent timolol may increase myopia by nearly 1 diopter, presumably by blocking the effect of endogenous ciliary muscle-relaxing sympathetic neuronal tone.[488] Indirect fluorophotometric estimates fail to demonstrate any effect of timolol on uveoscleral outflow per se.[215] However, timolol may reduce epinephrine-induced increases in uveoscleral outflow when the two drugs are applied concurrently.[215] These findings are consistent with the data concerning adrenergic influences on ciliary muscle contractile tone and also illustrate the importance of the ambient neuronal and pharmacologic adrenergic tone in determining the response of a target tissue to an exogenous adrenergic agent.

In addition to suppressing aqueous formation (see above), α_2-adrenergic agonists may enhance uveoscleral outflow. In humans apraclonidine- and brimonidine-induced IOP reductions are associated with decreased aqueous humor formation and decreased (apraclonidine) or increased (brimonidine) uveoscleral outflow.[489,490] Treatment of ocular hypertensive patients with brimonidine for one month results in a suppression of aqueous formation early on with a later increase in uveoscleral outflow.[260] A single topical application of either brimonidine or apraclonidine decreases IOP and aqueous flow by similar amounts in timolol-treated normal human eyes,[259] suggesting that both α_2-agonists act by a similar mechanism.

Bunazosin, a selective α_1-adrenoceptor antagonist, can further enhance IOP lowering when used as an adjunctive therapy with latanoprost in monkeys (normotensive)[491] and humans (glaucoma)[492] or with timolol[491] by a mechanism partly due to ciliary muscle (bovine) relaxation but independent of an effect on matrix metalloproteinase activities (cultured monkey ciliary muscle cells).[492]

Cytoskeletal and cell junctional mechanisms (Box 11.2)

Agents that alter the cytoskeleton, cell junctions, contractile proteins, or ECM are capable of producing a "pharmacologic trabeculotomy". The adhesion of cells to their neighbors or to the ECM has multiple effects on cell shape and dynamics. Cell–cell and cell–extracellular matrix adherens junctions are complex and dynamic in nature, comprising a myriad of proteins, and are modulated by the ambient physical

---

Box 11.2 Cytoskeletal, cell contractility/relaxation and cell junctional mechanisms

- Agents that alter the actin cytoskeleton and cellular contractility are in development to lower IOP by directly targeting the trabecular meshwork
- Relaxation of the trabecular meshwork and Schlemm's canal increases the area available for fluid drainage
- The first clinical trial using a selective ROCK inhibitor SNJ-1656 in healthy human eyes shows a dose-dependent decrease in IOP after single and multiple topical applications, but with some ocular surface toxicity[523]

(pressure, shear stress) and chemical (endogenous hormonal and biochemical, exogenous pharmacological) milieu. They play a role in signaling to the cell the state of its external environment. Actin filaments play a central role as the "backbone" of the submembrane plaque in both types of junctions (Fig. 11.10), with the coupling of actin and myosin being essential for cell contractility.[493,494]

Changes in TM and/or Schlemm's canal cells, which could affect trabecular outflow resistance by altering the dimensions or direction of flow pathways and the amount and composition of the ECM, can be modulated directly by actin-disrupting agents or indirectly by inhibition of specific protein kinase(s) or cellular contractility through administration of protein kinase inhibitors or gene therapy.[38,495]

Early studies involving anterior chamber perfusions of living monkey eyes with cytochalasins, fungal metabolites that block actin filament assembly, resulted in marked increases in total outflow facility within minutes.[477,496,497] Tracer studies demonstrate that the increase in total outflow facility represents increased facility across the meshwork and inner canal wall[498] and is not related to contraction of the ciliary muscle since the effect is similar in eyes with and without surgically disinserted ciliary muscles.[308,496] More recently, studies show that potent actin-disrupting agents such as marine macrolides like latrunculins, which sequester monomeric G-actin, lead to massive disassembly of filamentous actin (Fig. 11.10). Treatment of HTM cells with latrunculin A or B (LAT-A, LAT-B) results in cell rounding and retraction of the lamellipodium, which is accompanied by an apparent "arborization" of the cells.[499,500] LAT-A or -B administered intracamerally or topically to living monkey eyes induces 2–4-fold increases in outflow facility.[501,502] In organ-cultured postmortem porcine eyes or human eyes, LAT-B significantly increases outflow facility by 60–70 percent.[500,503] Single or multiple topical treatments with LAT-A and/or -B also significantly decrease IOP in monkeys.[501,502,504]

Morphologically, the LAT-B-induced decrease in outflow resistance in monkey eyes in vivo is associated with massive "ballooning" of the juxtacanalicular region, leading to expansion of the space between the inner wall of Schlemm's canal and the trabecular collagen beams without observable separations between inner wall cells.[505] However, in postmortem human eyes, the facility increase is accompanied by increased openings between inner wall cells with only very modest rarefaction of the juxtacanalicular tissue and separation of the inner wall of Schlemm's canal from juxtacanalicular tissue.[503]

Perturbation of the cellular actomyosin system by inhibition of myosin light chain kinase (MLCK) and/or Rho kinase is demonstrated with the nonselective serine-threonine kinase inhibitor H-7.[500,506–509] H-7, administered intracamerally or topically to living monkey eyes, doubles outflow facility and decreases IOP[508] by directly affecting the TM.[510] In porcine, human or monkey eye cultured anterior segments, H-7 also significantly increases outflow facility.[500,511,512] The H-7-induced increase in outflow facility in the live monkey eye is associated with cellular relaxation and drainage-surface expansion of the TM and Schlemm's canal (see Fig. 11.6), accompanied by loss of ECM. The inner wall cells of Schlemm's canal become highly extended, yet cell–cell junctions are maintained.[343,344] A specific Rho kinase inhibitor, Y-27632, induces reversible changes in

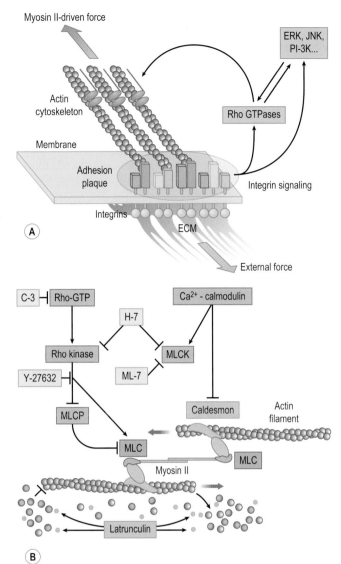

Figure 11.10 (**A**) Focal adhesion (FA) as a mechanosensor. Focal adhesions are multi-molecular complexes connecting the extracellular matrix with the actin cytoskeleton. Heterodimeric transmembrane integrin receptors (red) bind matrix proteins via their extracellular domains, while their cytoplasmic domains are associated with a dense submembrane plaque containing more than 50 different proteins ("boxes" enclosed in the oval area) including structural elements as well as signal transduction proteins such as FAK, Src, ILK, etc. The plaque, in turn, is connected to the termini of actin filament bundles. The assembly and maintenance of FA depend on local mechanical forces. This force may be generated by myosin II-driven isometric contraction of the actin cytoskeleton, or by extracellular perturbations such as matrix stretching or fluid shear stress. Force-induced assembly of the adhesion plaque leads to the activation of a variety of signaling pathways that control cell proliferation, differentiation, and survival (e.g., MAP kinase and PI 3-kinase pathways) as well as the organization of the cytoskeleton (e.g., Rho family GTPase pathways). Rho, in particular, is an indispensable regulator of FA assembly affecting, via its immediate targets Dia1 and ROCK, actin polymerization and myosin II-driven contractility. (From Geiger B, Bershadsky A: Cell 110:129, 2002. Copyright Elsevier 2002.) (**B**) Schematic drawing illustrating targets for agents known to disrupt the actin cytoskeleton to enhance outflow facility. C-3, Y-27632 and H-7 block the Rho cascade, inhibiting actomyosin contraction and disrupting actin stress fibers; H-7 and ML-7 block myosin light chain kinase phosphorylation of the myosin light chain to interfere with actin-myosin interactions; latrunculin sequesters monomeric G actin leading to microfilament disassembly; caldesmon negatively regulates actin-myosin interactions. (Modified with permission from the original by Alexander Bershadsky.)

cell shape and decreases in actin stress fibers, focal adhesions, and protein phosphotyrosine staining in human TM cells and Schlemm's canal cells.[513,514] In isolated bovine TM strips, Y-27632 completely blocks Ca^{2+}-independent phorbol myristate acetate or endothelin-1-induced contraction.[515–517] Rho kinase inhibitors increase outflow facility and/or decrease IOP in enucleated porcine eyes and/or living monkeys.[513,514,518–521] Morphological studies in bovine eyes indicate that, with Y-27632, the structural correlate to the increase in outflow facility is physical separation between the juxtacanalicular connective tissue and inner wall of Schlemm's canal.[522] A clinical study demonstrates that topical administration of a selective ROCK (Rho-associated coiled coil-forming) protein kinase inhibitor (SNJ-1656, an ophthalmic solution of Y-39983 which is 30 times more potent at ROCK than Y-27632[520]) can lower IOP in humans[523] although ocular surface toxicity occurs.

TM relaxation is also induced by modulating proteins, such as caldesmon, that negatively regulate actin–myosin interactions. When caldesmon is overexpressed, actin becomes uncoupled from myosin. Additionally, the exoenzyme C3 transferase disrupts actin–myosin interactions by inhibiting Rho-GTP thereby blocking the whole Rho cascade (Fig. 11.10). Recently, adenovirus-delivered exoenzyme C3 transferase (C3-toxin) cDNA and non-muscle caldesmon cDNA were successfully expressed in cultured human TM cells.[524,525] Outflow facility in organ-cultured human or monkey eyes is dramatically increased following overexpression of these genes.[525,526] Specific inhibition of Rho-kinase activity in the TM by dominant negative Rho expression also increases outflow facility in organ-cultured human anterior segments.[527]

Uncoupling of cellular adhesions from the ECM is another approach to disrupt the actin cytoskeleton resulting in increased outflow facility. It is well established that signaling events mediated by the ECM play a critical role in maintaining tissue architecture by regulating the organization of the actin cytoskeleton and cell contacts. Hence, these signaling events could potentially regulate outflow facility. Recent studies support this idea and show that the Heparin II (HepII) domain of fibronectin, an extracellular matrix protein found in the TM, increases outflow facility when perfused through cultured human anterior segments.[528] Presumably this domain increases outflow facility by mediating the disassembly of the actin cytoskeleton in TM cells.[529] The HepII domain plays an important role in regulating the organization of the actin cytoskeleton by acting as a ligand for members of the syndecan and integrin family of receptors.[530]

Microtubules comprise 25-nm-diameter hollow polar fibers, densely packed near the nucleus, and extending toward the cell periphery. They are not intrinsically contractile but are important for directional cell motility and, driven by specific microtubule motor proteins such as kinesins and dyneins, for cytoplasmic trafficking of vesicles and organelles. Associated proteins bind to microtubules and can affect their stability and potentially attach them to other cellular structures, including other cytoskeletal filaments. Microtubule function could affect outflow pathway events through direct cellular mechanical effects, influences on extracellular or cell membrane turnover, or secondary signaling leading to activation of the actin cytoskeleton.[531] Microtubule disrupting agents such as ethacrynic acid, colchicine, and vinblastine, increase outflow facility and cause cellular contraction in HTM cells if the actin cytoskeleton is intact.[532,533]

Ethacrynic acid (ECA), a sulfhydryl-reactive diuretic drug, inhibits microtubule assembly in vitro, and induces a rapid decrease in phosphotyrosine levels of focal adhesion kinase and a more subtle decrease in paxillin phosphorylation. Dephosphorylation of these proteins disrupts signaling pathways that normally maintain the stability of the actin microfilaments and cellular adhesions, indicating a close relationship between the microtubule system and the actomyosin system. This action leads both to cell shape change in culture[533,534] and to facility changes in vivo.[535–538] In enucleated human eyes, lower resistance-effective ethacrynic acid doses do not produce morphologic changes in the TM, whereas higher doses induce separations between TM and Schlemm's canal cells.[538,539] Recent evidence shows that several new derivatives of ECA significantly decrease IOP in cats and monkeys.[540,541] These ECA derivatives are more potent than ECA in terms of inducing cell shape alterations and decreasing actin stress fiber content in human TM cells,[542] suggesting that microtubule disruption may reduce outflow resistance at least partially through perturbation of the actomyosin system.

Perfusion of the monkey anterior chamber with calcium- and magnesium-free mock aqueous humor containing 4–6 mM Na_2EDTA or with calcium-free mock aqueous containing 4 mM ethylene glycol bis (aminoethylether) tetra-acetate (EGTA) also causes large facility increases, accompanied by ultrastructural changes in which the junctions are clearly fractured;[543] the latter is not the case with H-7 (see above). Since EDTA chelates both calcium and magnesium, while EGTA is much more specific for calcium,[544] calcium would appear to be a critical cation in maintaining the structural and functional integrity of the conventional outflow pathway. The calcium channel antagonist, verapamil, increases outflow facility in organ-cultured human eyes.[545]

Thus, agents that alter the cytoskeleton, cell junctions, contractile proteins, or ECM are capable of, in effect, producing a "pharmacologic trabeculotomy". Derivatives of some of these compounds (latrunculins, Rho kinase inhibitors) are currently in clinical trials to lower IOP in humans.

Corticosteroid mechanisms

Studies suggest that glucocorticoids may play a major role in the normal physiologic regulation of outflow facility and IOP. Glucocorticoid receptors are identified in the cells of the outflow pathways.[546] Corticosteroid regulation of IOP is proposed to occur via 11β-hydroxysteroid dehydrogenase (HSD)-1 expression which are localized in the non-pigmented ciliary epithelium of human eyes.[547–549] This enzyme catalyzes the conversion of cortisone to cortisol which, in turn, induces Na^+-K^+-ATPase,[550] leading to sodium and concomitant water transport into the posterior chamber, through epithelial sodium channels, resulting in aqueous production. Levels of cortisol compared to cortisone in the aqueous humor are normally much greater than in the systemic circulation.[548,551]

However, long-term interactions of cortisol in the aqueous with glucocorticoid receptors in the TM, could contribute to increasing outflow resistance in susceptible individuals.[551]

In the normal population, approximately 40 percent of patients treated with topical or systemic corticosteroids are termed "steroid responders," and develop markedly elevated IOP after several weeks.[552,553] This contrasts to patients with POAG, 90 percent of which are considered steroid responders.[553,554] One hundred percent of cow eyes treated in vivo with prednisolone acetate develop elevated IOP after 3 weeks of treatment.[555] Glaucoma patients also have increased plasma levels of cortisol compared to normal individuals[556,557] and increased vascular sensitivity to glucocorticoids.[558] Oral administration of the glucocorticoid biosynthesis inhibitor metyrapone to glaucoma patients[559] or the 11β-hydroxysteroid dehydrogenase inhibitor carbenoxolone to ocular hypertensive patients[547] elicits a small, transient decline in IOP. A recent pilot study shows that a single anterior juxtascleral depot of anecortave acetate (an angiostatic cortisene used in age-related macular degeneration) is effective in lowering IOP in POAG eyes for up to 12 months.[560] The mechanism of action is currently unknown. IOP elevation in a sheep model of steroid induced glaucoma could be reversed or prevented by a single dose of a gene therapy vector carrying an inducible metalloproteinase human gene.[681]

The glucocorticoid dexamethasone (DEX) alters complex carbohydrate, hyaluronic acid, protein, and collagen synthesis and distribution in cells and tissues of human aqueous humor outflow systems.[377,561–563] In TM cells, DEX inhibits prostaglandin synthesis (90 percent), reduces phagocytic[564,565] and extracellular protease activity, and changes gene expression.[566–568] Cortisol metabolism may be altered in cultured TM cells from patients with primary open-angle glaucoma. These cells accumulate 5β-dihydrocortisol and, to a lesser extent, 5α-dihydrocortisol, metabolites which potentiate the facility-decreasing and IOP-increasing effects of DEX; these cells produce relatively little 3α,5β-tetrahydrocortisol from cortisol.[569] Normal human TM cells show no accumulation of active dihydrocortisol intermediates; all cortisol is rapidly metabolized to the inactive tetrahydrocortisols. 3α,5β-Tetrahydrocortisol applied topically decreases IOP and increases outflow facility in glaucomatous human eyes,[569] and antagonizes DEX-induced cytoskeletal reorganization (see below) in normal HTM cells[569,570] but has no effect on outflow facility in normotensive monkeys following intracameral injection or 10 days of topical administration.[571]

Glucocorticoid glaucoma and POAG eyes both exhibit increased amounts of ECM in the meshwork.[572] However, the ECM that accumulates in eyes with corticosteroid-induced glaucoma differs from that seen in eyes with POAG.[573] The human anterior segment in organ culture exposed to DEX exhibits morphological changes similar to those reported for corticosteroid glaucoma and a resulting increase in IOP and decrease in outflow facility.[574] Studies of DEX effects in cultured human TM cells, combined with studies of changes in laminin and fibronectin and alterations in phagocytosis and surface binding properties, may provide clues to the changes in pressure-dependent outflow seen with corticosteroid use.[575] Inhibition of DEX-induced overexpression of laminin or collagen using antisense oligonucleotides increases permeability of DEX-treated HTM cells.[576] Prolonged treatment of HTM cells with DEX decreases hyaluronan synthesis. Hyaluronan is an inert molecule which may be necessary to prevent adherence of larger molecules to the cribriform meshwork.[577] These and other DEX-induced alterations in several other molecules involved in the regulation of the ECM have recently been reviewed.[578] Some glucocorticoid effects on the TM could also be mediated by alterations in growth factor/receptor expression.[578]

Human TM and Schlemm's canal cell monolayers grown on filters exhibit enhanced tight junction formation and decreased hydraulic conductivity in the presence of DEX.[579,580] There is a 2-fold increase in the number of tight junctions, a 10–30-fold reduction in the mean area occupied by inter-endothelial "gaps" or preferential flow channels, and a 3–5-fold increase in the expression of the junction-associated protein ZO-1. Inhibition of ZO-1 expression abolishes the DEX-induced increase in resistance and the accompanying alteration in cell junctions and gaps. These results support the hypothesis that intercellular junctions are necessary for the development and maintenance of transendothelial flow resistance in cultured TM and Schlemm's canal cells and may be involved in the mechanism of increased resistance associated with glucocorticoid exposure.

Glucocorticoids are shown to reorganize the TM cytoskeleton. Most intriguing in terms of outflow resistance, are the unusual geodesic dome-like cross-linked actin networks (CLANs) formed in response to DEX treatment of TM cells and perfusion cultured anterior segments (Fig. 11.11).[581–583] Higher basal levels of CLANs are present in glaucomatous TM cells and tissues and a greater CLAN response to glucocorticoids occurs in glaucomatous TM cells.[584] Recent studies have shown that CLANs exist within TM cells in situ in both normal and glaucoma donor eyes. There may be a CLAN in all cells in glaucomatous TM in situ and in two-thirds of cells in normal TM.[585] CLAN formation can be regulated by beta1 and beta3 integrin signaling pathways.[586]

Cultured human TM cells exposed to DEX also exhibit increased cell and nuclear size, an unusual stacked arrangement of smooth and rough endoplasmic reticulum, proliferation of the Golgi apparatus, and pleomorphic nuclei.[581] TM cell volume can be altered by an early DEX-induced enhancement of Na-K-Cl cotransport.[587]

One of the first TM cell genes/proteins induced by glucocorticoids is myocilin (MYOC) whose progressive induction over time matches the time course of clinical steroid effects on IOP and outflow facility, and is hypothesized to play a role in glaucoma pathogenesis (see review[588]).

The MYOC gene is directly linked to patients with open-angle glaucoma.[589] Mutations in the MYOC gene are present only in a minor percentage (~3–4 percent) of patients with adult forms of POAG[590] but are prevalent in juvenile open-angle glaucoma (see reviews[588,591]). Disease-causing MYOC mutants are misfolded, leading to accumulation in the endoplasmic reticulum which can result in cell death.[592] Recombinant MYOC increases outflow resistance in the organ-cultured human anterior segment.[593] In contrast, overexpression of mutant MYOC in vivo in cats does not elevate IOP.[594] It may be that IOP elevation is dependent on other factors in addition to a mutant MYOC protein or that longer duration of exposure may be required in vivo.

MYOC may play a role in maintaining normal outflow pathways. Maintaining elevated pressure in human anterior segment organ culture for 7 days causes a marked increase in MYOC mRNA that could be part of a protective response

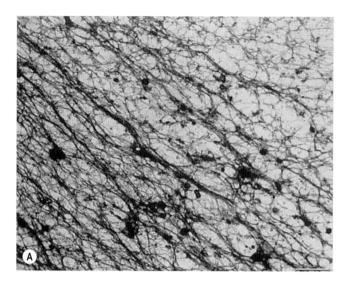
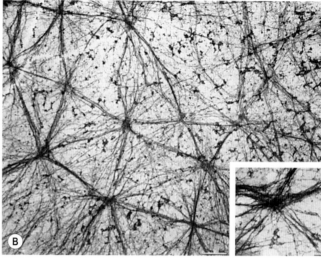

Figure 11.11 Whole mount transmission electron micrographs of the cytoskeleton of control trabecular meshwork cells (**A**) and trabecular meshwork cells exposed to 10^{-7} M dexamethasone for 14 days (**B**). The stress fibers in the control cells are arranged in normal linear arrays, whereas dexamethasone-treated microfilaments are grouped into 90- to 120-nm bundles radiating from electron-dense vertices. (From Clark AF et al: Invest Ophthalmol Vis Sci 35:281, 1994, reproduced with permission from the Association for Research in Vision and Ophthalmology.)

mechanism.[595] Over expression of the amino terminal domain or full-length MYOC protein in the human anterior segment organ culture produces an increase in outflow facility.[596] The regulation of human trabecular cell MYOC mRNA appears to be distinct in HTM cells compared with other cells types.[597] MYOC is localized both intracellularly and extracellularly (in a glycosylated form, 66 kDa) at multiple sites (associated with mitochondria, cytoplasmic filaments, elastic-like fibers in trabecular beams, extracellular matrices in the juxtacanalicular region) where it may exert diverse biological functions.[567,598,599] It may serve a structural function within the cytoplasm, or it may associate with other molecules within the cell, perhaps as a molecular chaperone. Extracellularly, it may be involved in creating resistance to aqueous outflow by binding to other extracellular molecules, such as the heparin II domain of fibronectin,[600] or to the cell membrane of trabecular cells.[601] Dynamic mechanical stimuli maintain MYOC expression in TM in situ.[602] Thus, the role of normal MYOC in glaucoma has yet to be established.

Prostaglandin mechanisms (Box 11.3)

There is a 60 percent increase in aqueous outflow via the uveoscleral pathway in monkeys after a single submaximal dose of $PGF_{2\alpha}$-1-isopropyl ester ($PGF_{2\alpha}$-IE).[7] Following multiple submaximal doses (Table 11.3), there is more than a 100 percent increase in uveoscleral outflow.[314] In both instances, aqueous outflow is substantially redirected from the trabecular to the uveoscleral pathway. In vitro, $PGF_{2\alpha}$ produces a weak dose-dependent relaxation of carbachol-precontracted rhesus monkey ciliary muscle strips.[603] Such relaxation may contribute to widening the intermuscular spaces in vivo.[604] However, the majority of the outflow effect resulting from $PGF_{2\alpha}$ treatment is likely due to ECM remodeling in the anterior segment characterized by an increase in MMP-1, -2, and -3 and reduction in collagen types I, II and IV, within the ciliary muscle, iris root and periciliary body sclera[605–607] possibly associated with activation of the proto-oncogene c-fos.[608] $PGF_{2\alpha}$ and latanoprost

(13,14-dihydro-17-phenyl-18,19,20-trinor-$PGF_{2\alpha}$-isopropyl ester) cause reductions in collagen types I, III, and IV, fibronectin, laminin and hyaluronan immunoreactivity in the ciliary muscle and adjacent sclera, while MMP-2 and -3 are increased.

Plasmin generation, an activator of MMPs, is enhanced.[605,609] $PGF_{2\alpha}$ can stimulate the formation of endogenous prostaglandins by stimulation of phospholipase A2 and release of arachidonic acid for prostaglandin synthesis.[610] Human ciliary muscle cells exposed to PGF_2 ethanolamide or latanoprost for 9 days show a downregulation of the FP receptor. In the same study, downregulation of the aquaporin-1 and versican genes are proposed to increase flow through the ciliary muscle and decrease IOP.[611] Long-term treatment with various subtype selective prostaglandins

Table 11.3 Uveoscleral outflow on day 5 of twice-daily unilateral treatment with PGF$_{2\alpha}$

	Treated	Control	Treated/Control
(a) Spontaneous IOP; Approximately 235–325 min (n = 6)			
Albumin	0.78 ± 0.12*	0.46 ± 0.03	1.66 ± 0.20†
(b) IOP = 17–18 mmHg; Approximately 240–335 min (n = 2)			
Albumin	1.45 ± 0.01§	0.62 ± 0.11	2.41 ± 0.42
(c) IOP = 17–18 mmHg; Approximately 135–195 min (n = 1)			
Albumin	2.03	0.63	3.21
(d) All IOP Time Group Combined (n = 9)			
Albumin	1.07 ± 0.17§	0.52 ± 0.04	2.00 ± 0.24‡

IOP, intraocular pressure; PGF$_{2\alpha}$, prostaglandin F$_{2\alpha}$.

On day 5 anterior chambers were exchanged with 2 mL of either ^{125}I- or ^{131}I-albumin. Infusion was continued at a lower rate for the balance of the indicated times. Pressures other than spontaneous were maintained by tracer flow from an elevated reservoir. Animals were then sacrificed and the equivalent anterior chamber fluid recovered in the ocular and periocular tissues was determined. Overall, PGP$_{2\alpha}$ increased uveoscleral outflow approximately two-fold compared with control eyes. Data are mean ± standard error of mean uveoscleral outflow (μl/min) for n animals, each contributing one treated and one control eye, following the ninth unilateral dose of PGF$_{2\alpha}$ on day 5; min indicates time window following PGF$_{2\alpha}$ encompassed by the measurement. Significantly different from 1.0 by the two-tailed paired t-test: †$p < .05$; §$p < .01$. Significantly different from contralateral control by the two-tailed two sample t-test: *$p < .05$; §$p < .01$.

(From Gabelt BT, Kaufman PL: Exp Eye Res 49:389, 1989, Copyright Elsevier 1989.)

results in intermuscular spaces became more organized and lined with an incomplete layer of endothelial cells resembling lymphatic pathways.[612]

Prostaglandin-induced changes in the sclera also are important in the regulation of uveoscleral outflow and may be used to enhance trans-scleral delivery of peptides and other high-molecular-weight substances to the posterior segment of the eye. Five days of topical treatment with PGF$_{2\alpha}$-isopropyl ester increases MMP-1, -2, and -3 in the sclera of monkeys.[613] Immunocytochemistry studies and mRNA analysis of human sclera and cultured human scleral fibroblasts shows the presence of EP$_1$, EP$_2$ and FP receptor subtypes but not EP$_3$ and EP$_4$ subtypes.[614] Human scleral permeability to dextrans, measured in an Ussing chamber following exposure to PGF$_{2\alpha}$ and latanoprost acid for 1–3 days, increases in a dose- and time-dependent manner. This is accompanied by an increase in MMP concentration in the media with the greatest increases in MMP-2 and -3 compared to MMP-1.[615] PGF$_{2\alpha}$ and latanoprost acid also induce increases in mRNA for MMPs and tissue inhibitors of matrix metalloproteinases in human scleral organ cultures.[616] MMPs alone are shown to directly increase scleral permeability of mouse eyes.[617]

Mice deficient in various prostaglandin receptors are used to determine the role of prostanoid receptor subtypes in mediating the IOP-lowering response to clinical prostaglandin analogs. Studies in FP receptor-deficient mice show that the FP receptor is essential for the early IOP-lowering response to topical latanoprost, travoprost, bimatoprost and unoprostone.[618] The involvement of the FP receptor in the IOP reduction with long-term dosing is unknown. Upregulation of MMP-2, -3, -9 and FP mRNA in the sclera following

7 days of topical treatment with latanoprost is also dependent on an intact FP receptor gene.[618a,618b] EP receptor-deficient mice are studied in similar ways. When EP$_1$, EP$_2$, and EP$_3$ receptor-deficient mice and their wild-type background strain are treated topically with latanoprost, travoprost, bimatoprost, or unoprostone, EP$_3$ receptors are involved in the IOP – lowering response to latanoprost, travoprost, and bimatoprost at 3 hours after drug administration – but EP$_1$ and EP$_2$ receptors are not.[619] This is in contrast to what is expected from in vitro receptor binding[620,621] and functional assays[620] in which FP receptor binding and functional responses (phosphoinositide turnover) are at least two orders of magnitude greater than for EP3 receptors. Also, immunohistochemistry studies show that EP2 receptors are the most abundantly expressed EP subtype in human ocular tissues.[622] EP2 selective agonists are effective in lowering IOP in monkeys.[623,624] Topical treatment with a combination of EP1, EP2, EP3 and FP receptor agonists increases the magnitude of the IOP-lowering response in monkeys when compared to FP agonist therapy alone.[624]

The uveoscleral outflow system likely evolved to protect the eye in several ways during inflammation. The TM becomes compromised by inflammation or obstructed by inflammatory debris, and the choroid is overloaded with debris and extravasated proteins that must be removed from the eye.[154] In this situation, prostaglandins are released and, as autacoids or hormones that are synthesized, released, and locally acting, induce the changes described. Since the eye has no true lymphatics, uveoscleral outflow may serve as an analog to an intraocular lymphatic drainage system.[12] However, as mentioned above, long-term exposure to prostaglandins, may lead to lymphatization of the uveoscleral pathway.[612] The normal low flow rate that is sufficient to remove normal levels of extravascular protein may be inadequate when protein levels are increased as in uveitis. Redirection of aqueous outflow from the trabecular to the uveoscleral pathway would both rid the eye of excess proteins and maintain physiologic IOP. This could also explain the very low IOP that often accompanies uveitis; during experimental iridocyclitis in monkeys, uveoscleral outflow increases approximately four-fold (Table 11.4).[153]

The increase in uveoscleral outflow in response to these compounds is so great that a larger reduction in IOP is possible than with any other known substance. PGF$_{2\alpha}$ analogs and metabolites are clinically useful ocular hypotensive agents, despite some undesirable side effects (conjunctival foreign body sensation, conjunctival hyperemia, stinging pain, photophobia),[625,626] and increased iris pigmentation in some instances.[627] In fact, they are the most commonly used glaucoma medications in the developed world. Gene therapy approaches to overexpress prostaglandin pathway genes in the anterior segment to lower IOP are being investigated so that problems patient compliance for drug administration may be reduced. Success has been demonstrated in a cat model where IOP was decreased for 5 months following lentiviral vector delivery of prostaglandin pathway genes.[684]

Cell volume related mechanisms

Na-K-Cl cotransporter is a plasma membrane protein that participates in vectorial transport of Na and Cl across epithelia and also regulates intracellular volume in a variety of epithelial and non-epithelial cell types.[628–631] Changes in TM

Table 11.4 Distribution of dextran tracer, uveoscleral outflow, protein and intraocular pressure in control and inflamed cynomolgus monkey eyes

	Control Eye (μl)	Inflamed Eye (μl)	Probability *
Iris	1.2 ± 0.3	2.4 ± 0.3	0.07
Anterior uvea	8.5 ± 2.4	18.6 ± 4.3	0.43
Posterior uvea	0.3 ± 0.2	4.2 ± 0.8	0.006
Anterior sclera	8.4 ± 1.6	28.4 ± 4.4	0.01
Posterior sclera	1.9 ± 0.9	21.5 ± 3.9	0.006
Retina	0.1 ± 0.1	2.1 ± 0.9	0.08
Fluid[†]	0.7 ± 0.4	11.2 ± 3.4	0.03
Total	**21.0 ± 4.7**	**88.4 ± 14.7**	**0.009**
Uveoscleral outflow (μl/min)	0.7 ± 0.2	2.9 ± 0.5	0.009
AC protein (mg/ml)	0.20 ± 0.02	7.8 ± 4.0	0.006
Pre-IOP	16.7 ± 0.8	15.2 ± 1.3	0.43
Post IOP	14.2 ± 1.6	3.0 ± 1.1	<0.001

AC, Anterior chamber; *IOP* intraocular pressure.

Inflammation was induced by intravitreal injection of bovine serum albumin. Two days later, tracers were perfused through the anterior chamber and F$_u$ was determined after 30 minutes at 15 mm Hg. Uveoscleral outflow was increased in inflamed eyes up to fourfold with the 70,000 molecular weight (MW) fluoresceinated dextran. Values are mean standard error of mean; *n* = 6.

*Paired t-test analysis value.

[†]Includes vitreous, posterior chamber fluid, suprachorodial fluid.

From Toris CB, Pederson JE: Invest Ophthalmol Vis Sci 28:477, 1987, reproduces with permission from the Association for Research in Vision and Ophthalmology.

cell volume by agents which modulate Na-K-Cl cotransport activity affect outflow facility in human and calf eyes in vitro.[632] However, no change in outflow facility is observed in monkeys in vivo following administration of the Na-K-Cl inhibitor bumetanide.[94] Although this was an intense area of interest for a time, no recent advances have been made.

Water channels (aquaporins), are also found in TM and Schlemm's canal cells.[82,633,634] They may be involved in transcellular water movement in these tissues.[635] However, no role in aquaporin 1 in regulating outflow facility is demonstrated after over-expression in human organ-cultured anterior segments.[636]

Nitric oxide donor-induced increases in outflow facility may be due partly to decreases in TM cell volume (also see Other Agents).

Although chloride channels are thought to be involved in the regulation of cellular volume and intracellular chloride concentration, they do not appear to contribute significantly to the regulation of outflow facility.[637]

Hyaluronidase and protease-induced facility increases

Glycosaminoglycans (GAGs) contribute to the filtration barrier of aqueous outflow through the TM. A quantitative biochemical profile of GAGs from normal and POAG TM

suggests that there is depletion of hyaluronic acid and an accumulation of chondroitin sulfates in the POAG TM.[638,639] Substantial hyaluronan is present in the non-glaucomatous outflow pathway associated with the endothelial cells lining the trabecular beams. This finding supports potential roles for this glycosaminoglycan in the regulation of the physiological aqueous outflow resistance or in the maintenance of the outflow channels, or both.[640] Hyaluronic acid covering the surfaces of the outflow pathways might prevent adherence of molecules to ECM components within the cribriform region and thereby prevent clogging of the outflow pathways.[407] It is hypothesized that POAG is characterized by a decreased concentration of hyaluronic acid and increased turnover and downregulation of the hyaluronic acid receptor CD44 in the eye, which, in turn, may influence cell survival of TM and retinal ganglion cells.[641]

Intracameral infusion of hyaluronidase markedly increases facility in the bovine eye, presumably as the result of washout of acid mucopolysaccharide-rich ECM in the chamber angle tissues.[642] Effects in primates are much more variable.[643–647] The variations are attributed to interspecies differences, the type and source of hyaluronidase, and the conditions used for the enzymatic digestion that may have contributed to a variable and incomplete degradation of hyaluronic acid. In the enucleated human eye perfused at room temperature, α-chymotrypsin has little effect on facility.[645] However, effects of trypsin may be masked at low temperatures[648] and a combination of trypsin and ethylenediaminetetra-acetate (EDTA) may have a marked effect in dissociating cultured cells not easily dissociated by either agent alone.[649] Perfusion of the anterior chamber of living monkeys with 50 U/mL α-chymotrypsin gives a large facility increase that persists for several hours even after the enzyme is removed from the infusate.[650] The facility increase induced by intracameral 0.5 mM Na$_2$EDTA is augmented and prolonged by α-chymotrypsin.[650]

Exposure of porcine TM cells to growth factors such as TGFβ induces increases in matrix metalloproteinases (MMPs) such as stromelysin, gelatinase B and collagenase, suggesting a role in the regulation of ECM turnover by TM cells.[651] However, prolonged elevation of TGFβ2, as occurs in POAG, can have the opposite effects and contribute to outflow obstruction (see Outflow Obstruction). Purified MMPs increase outflow facility in organ-cultured human anterior segments by 160 percent for at least 125 hours.[652]

The composition of the ECM in the TM has been recently reviewed.[653] Presumably any of the enzymes involved in the biosynthesis or degradation of the ECM are potential targets to be manipulated for enhancing trabecular outflow.

Other agents

Nitric oxide mimicking nitrovasodilators can decrease IOP in monkeys by altering outflow resistance. In human eyes, the TM and ciliary muscle are enriched sites of nitric oxide systhesis.[28] One role for nitric oxide in the anterior segment may be to modulate outflow resistance either directly at the level of the TM, Schlemm's canal and collecting channels, or indirectly through alteration in the tone of the longitudinal ciliary muscle. Nitrovasodilators are shown to relax TM[654] and ciliary muscle[654] strips precontracted with carbachol in vitro. Nitric oxide synthase-immunoreactive nerve fibers are abundant in the primate TM, especially an the cribriform region adjacent to the inner wall of Schlemm's canal.[422] Nitric oxide synthase released from nerve terminals could cause relaxation of TM cells and allow an increase in outflow facility independent from the ciliary muscle. In human glaucoma eyes there are dramatic reductions in staining indicative of nitric oxide synthase activity in ciliary muscle, TM and Schlemm's canal compared to control eyes,[655,656] that are unrelated to the use of multiple glaucoma therapies, or the severity of the disease.[655] Nitric oxide may also alter TM cell volume (see Cell Volume Related Mechanisms).

Purinergic P2 receptors are present in TM cells. Activation of these receptors in TM cells leads to intracellular calcium mobilization and/or extracellular signal-regulated kinase 1 and 2 (ERK1/2) pathway activation.[657] IOP is decreased and outflow facility increased in monkeys following topical application of adenosine A1 agonists.[23] In bovine organ-cultured anterior segments, the outflow facility increase produced by adenosine A1 agonist cyclohexyladenosine is also associated with MMP activation.[25] Aqueous adenosine levels are positively correlated with IOP in ocular hypertensive individuals and could possibly serve as an endogenous modulator of IOP.[658]

There is evidence that the eye contains a renin-angiotensin system and that it may be involved in the regulation of IOP. The presence of angiotensin-converting enzyme activities, the concentration of angiotensinogen and angiotensin II, and the density of angiotensin-II AT1 receptors in ocular tissues and fluids are demonstrated in several species, including humans.[659–663] Oral administration of an angiotensin II receptor type 1 antagonist[664,665] or an angiotensin-converting enzyme inhibitor[666,667] reduces IOP in normotensive and in POAG patients. Topical application of a renin inhibitor decreases IOP in rabbits and monkeys without affecting systemic blood pressure or heart rate.[668]

Multiple topical doses of an angiotensin AT1 receptor antagonist, CS-088, decreases IOP in monkey eyes with unilateral laser-induced glaucoma.[669] In humans, the oral AT1 receptor antagonist, Losartan, increases outflow facility.[665] The angiotensin-converting enzyme inhibitor, captopril[666] decreases IOP and increases total outflow facility without affecting blood pressure, heart rate or pupil diameter. Another study suggests the IOP-lowering responses to ACE inhibition is due to prostaglandin biosynthesis and increased uveoscleral outflow.[670] Angiotensin itself slightly decreases outflow facility in monkeys following intracameral injection.[455]

Physical enhancement of outflow

Aqueous outflow via the conventional drainage pathway is a physical process that can be altered pharmacologically as reviewed above. Ultimately, physical methods to alter trabecular outflow are accomplished by fistulation to allow fluid movement from one compartment to another.

Spacing ~50–100 small (50 μm), low-intensity argon, krypton or diode laser burns evenly around the circumference of the meshwork of the glaucomatous human eye can result in a significant and long-term increase in outflow facility and decrease in IOP, apparently without actually producing a "hole" in the meshwork or inner canal wall.[671,672] It may be that contracture of the laser-produced scars tightens and narrows the trabecular ring, and the distortion somehow expands the meshwork, opens aqueous channels, and improve hydraulic conductance.[672,673] Other data suggest that laser energy-induced alterations in trabecular cell biosynthetic, biodegradative, or phagocytic functions result in less hydraulic resistance.[674,675] Definitive proof of the facility-increasing mechanism of laser trabeculoplasty remains to be established.

Canaloplasty is a recently developed non-penetrating technique utilizing a flexible microcatheter to access the entire length of Schlemm's canal. The entire circumference of the canal or a portion of it can be dilated with viscoelastic. A trabecular tensioning suture can be inserted for applying additional tension. IOP can be effectively lowered with a relatively low surgical complication rate, although the pressure may not fall to the extent needed by some patients.[676,677]

References

1. Millar C, Kaufman PL. Aqueous humor: secretion and dynamics. In: Tasman W, Jaeger EA, eds. Duane's Foundations of clinical ophthalmology. Philadelphia: Lippincott-Raven, 1995.
2. Johnson M, Erickson K. Mechanisms and routes of aqueous humor drainage. In: Albert DM, Jakobiec FA, eds. Principles and practice of ophthalmology. Philadelphia: WB Saunders Co, 2000:2577.
3. Pederson JE et al. Uveoscleral aqueous outflow in the rhesus monkey: importance of uveal reabsorption. Invest Ophthalmol Vis Sci 1977; 16:1008.
4. Bill A. The aqueous humor drainage mechanism in the cynomolgus monkey (Macaca irius) with evidence for unconventional routes. Invest Ophthalmol 1965; 4:911.
5. Sperber GO, Bill A. A method for near-continuous determination of aqueous humor flow: effects of anaesthetics, temperature and indomethacin. Exp Eye Res 1984; 39:435.
6. Kaufman PL, Crawford K. Aqueous humor dynamics: how PGF$_{2\alpha}$ lowers intraocular pressure. In: Bito LZ, Stjernschantz J, eds. The ocular effects of prostaglandins and other eicosanoids. New York: Alan R. Liss, 1989:387.
7. Nilsson SFE et al. Increased uveoscleral outflow as a possible mechanism of ocular hypotension caused by prostaglandin F$_{2\alpha}$-1-isopropylester in the cynomolgus monkey. Exp Eye Res 1989; 48:707.
8. Gabelt BT et al. Aqueous humor dynamics and trabecular meshwork and anterior ciliary muscle morphologic changes with age in rhesus monkeys. Invest Ophthalmol Vis Sci 2003; 44:2118.
9. Bill A, Phillips I. Uveoscleral drainage of aqueous humor in human eyes. Exp Eye Res 1971; 21:275.
10. Townsend DJ, Brubaker RF. Immediate effect of epinephrine on aqueous formation in the normal human eye as measured by fluorophotometry. Invest Ophthalmol Vis Sci 1980; 19:256.
11. Toris CB et al. Aqueous humor dynamics in the aging human eye. Am J Ophthalmol 1999; 127:407.
12. Bill A. Blood circulation and fluid dynamics in the eye. Physiol Rev 1975; 55:383.
13. Bill A. Basic physiology of the drainage of aqueous humor. In: Bito LZ et al, eds. The ocular and cerebrospinal fluids. Fogarty International Center Symposium. London: Academic Press, 1977:291.
14. Sears ML, Neufeld AH. Adrenergic modulation of the outflow of aqueous humor. Invest Ophthalmol 1975; 14:83.
15. Zimmerman TJ. Topical ophthalmic beta blockers. A comparative review. J Ocular Pharmacol 1993; 9:373.
16. Kaufman PL, Gabelt B. α2-Adrenergic agonist effects on aqueous humor dynamics. J Glaucoma 1995; 4(Suppl 1):S8.
17. Gabelt BT, Kaufman PL. Cholinergic drugs. In: Netland PA, ed. Glaucoma medical therapy, principles and management, 2nd edn. New York: Oxford University Press in cooperation with the American Academy of Ophthalmology, 2008:103.
18. May JC et al. Evaluation of the ocular hypotensive response of serotonin 5-HT1A and 5-HT-2 receptor ligands in conscious ocular hypertensive cynomolgus monkeys. J Pharmacol Exp Ther 2003; 306:301.
19. Mekki QA et al. Bromocriptine lowers intraocular pressure without affecting blood pressure. Lancet 1983; 1:1250.
20. Chiou GCY. Treatment of ocular hypertension and glaucoma with dopamine antagonists. Ophthalmic Res 1984; 16:129.
21. De Vries GW et al. Stimulation of endogenous cyclic AMP levels in ciliary body by SK&F28526, a novel dopamine receptor agonist. Curr Eye Res 1986; 5:449.
22. Geyer O et al. Hypotensive effect of bromocriptine in normal eyes. J Ocul Pharmacol 1987; 3:291.
23. Tian B et al. Effects of adenosine agonists on intraocular pressure and aqueous humor dynamics in cynomolgus monkeys. Exp Eye Res 1997; 64:979.
24. Husain S et al. Mechanisms linking adenosine A1 receptors and extracellular signal-regulated kinase1/2 activation in human trabecular meshwork cells. J Pharmacol Exp Ther 2007; 320:258.
25. Crosson CE et al. Modulation of conventional outflow facility by the adenosine A1 agonist N6-cyclohexyladenosine. Invest Ophthalmol Vis Sci 2005; 46:3795.

26. Polska E et al. Effects of adenosine on intraocular pressure, optic nerve head blood flow, and choroidal blood flow in healthy humans. Invest Ophthalmol Vis Sci 2003; 44:3110.
27. Shearer TW, Crosson CE. Adenosine A1 receptor modulation of MMP-2 secretion by trabecular meshwork cells. Invest Ophthalmol Vis Sci 2002;43:3016.
28. Nathanson JA, McKee M. Identification of an extensive system of nitric oxide-producing cells in the ciliary muscle and outflow pathway of the human eye. Invest Ophthalmol Vis Sci 1995; 36:1765.
29. Shahidullah M et al. Cyclic GMP, sodium nitroprusside and sodium azide reduce aqueous humour formation in the arterially perfused pig eye. Br J Pharmacol 2005; 145:84.
30. Schneemaan A et al. Elevation of nitric oxide production in human trabecular meshwork by increased pressure. Graefe's Arch Clin Exp Ophthalmol 2003; 241:321.
31. Schneemann A et al. Nitric oxide/guanylate cyclase pathways and flow in anterior segment perfusion. Graefes Arch Clin Exp Ophthalmol 2002; 240:936.
32. Chien FY et al. Effect of WIN 55212-2, a cannabinoid receptor agonist, on aqueous humor dynamics in monkeys. Arch Ophthalmol 2003; 121:87.
33. McIntosh BT et al. Agonist-dependent cammabinoid receptor signalling in human trabecular meshwork cells. Br J Pharmacol 2007; 152:1111.
34. Oltmanns MH et al. Topical WIN 55 212-2 alleviates intraocular hypertension in rats through a CB1 receptor mediated mechanism of action. J Ocul Pharmacol Ther 2008; 24:104.
35. Chen J et al. Finding of endocannabinoids in human eye tissues: implications for glaucoma. Biochem Biophys Res Commun 2005; 330:1062.
36. Stjernschantz J. Studies on ocular inflammation and development of a prostaglandin analogue for glaucoma treatment. Exp Eye Res 2004; 78:759.
37. Weinreb RN et al. Effects of prostaglandins on the aqueous humor pathways. Surv Ophthalmol 2002; 47:S53.
38. Tian B et al. The role of the actomyosin system in regulating trabecular fluid outflow (review). Exp Eye Res 2008; 713–717
39. Seidël E. Weitre experimentelle Untersuchungen über die Quelle und den Verlauf der introkulären Saftsrömung. XII. Metteilung. Uber den manometrischen Nachweis des physiologischen Druckgefalles zwishen Vorderkammer und Schlemmshen Kanal. Graefe's Arch Clin Exp Ophthalmol 2008; 107:101.
40. Lütjen-Drecoll E, Rohen JW. Morphology of aqueous outflow pathways in normal and glaucomatous eyes. In: Ritch R et al, eds. The glaucomas. St Louis: CV Mosby, 1989:41.
41. Green K, Pederson JE. Contribution of secretion and filtration to aqueous humor formation. Am J Physiol 1972; 222:1218.
42. Pederson JE. Fluid permeability of monkey ciliary epithelium in vivo. Invest Ophthalmol Vis Sci 1982; 23:176.
43. Cole DF. Secretion of the aqueous humor. Exp Eye Res 1977; 25(Suppl):161.
44. Wilson WS et al. The bovine arterially-perfused eye and in vitro method for the study of drug mechanisms on IOP, aqueous humour formation and uveal vasculature. Curr Eye Res 1993; 12:609.
45. Bill A. The role of ciliary blood flow and ultrafiltration in aqueous humor formation. Exp Eye Res 1973; 16:287.
46. Bárány EH. Pseudofacility and uveoscleral outflow routes: some nontechnical difficulties in the determination of outflow facility rate and rate of formation of aqueous humor. In: Leydhecker W, ed. Glaucoma Symposium, Tutzing Castle. Basel: Karger, 1966:27.
47. Bill A, Bárány EH. Gross facility, facility of conventional routes, and pseudofacility of aqueous humor outflow in the cynomolgus monkey. Arch Ophthal 1966; 75:665.
48. Bill A. Further studies on the influence of the intraocular presssure on aqueous humor dynamics in cynomolgus monkeys. Invest Ophthamol 1967; 6:364.
49. Goldmann H. On pseudofacility. Bibl Ophthalmol 1968; 76:1.
50. Bill A. Aqueous humor dynamics in monkeys (*Macaca irus* and *Cercopithecus ethiops*). Exp Eye Res 1971; 11:195.
51. Brubaker RF. The measurement of pseudofacility and true facility by constant pressure perfusion in the normal rhesus monkey eye. Invest Ophthalmol 1970; 9:42.
52. Kupfer C, Sanderson P. Determination of pseudofacility in the eye of man. Arch Ophthal 1968; 80:194.
53. Beneyto MP et al. Determination of the pseudofacility by fluorophotometry in the human eye. Internat Ophthalmol 1995–1996; 19:219.
54. Moses RA et al. Pseudofacility. Arch Ophthalmol 1985; 103:1653.
55. Freddo TF. Shifting the paradigm of the blood-aqueous barrier. Exp Eye Res 2001; 3:581.
56. Candia OA, Alvarez LJ. Fluid transport phenomena in ocular epithelia. Prog Retin Eye Res 2008; 27:197.
57. Erickson KA et al. The cynomolgus monkey as a model for orbital research. III. Effects on ocular physiology of lateral orbitotomy and isolation of the ciliary ganglion. Curr Eye Res 1984; 3:557.
58. Brubaker RF. Flow of aqueous humor in humans [The Friedenwald Lecture]. Invest Ophthalmol Vis Sci 1991; 32:3145.
59. Bárány EH. A mathematical formulation of intraocular pressure as dependent on secretion, ultrafiltration, bulk outflow, and osmotic reabsorption of fluid. Invest Ophthalmol 1963; 2:584.
60. Pederson JE, Green K. Aqueous humor dynamics: a mathematical approach to measurement of facility, pseudofacility, capillary pressure, active secretion and Xc. Exp Eye Res 1973; 15:265.
61. Nilsson SFE, Bill A. Physiology and neurophysiology of aqueous humor inflow and outflow. In: Kaufman PL, Mittag TW, eds. Glaucoma. London: Mosby-Year Book Europe Ltd, 1994:1.17.
62. Krupin T, Civan MM. Physiologic basis of aqueous humor formation. In: Ritch R et al, eds. The glaucomas, 2nd edn. St Louis: CV Mosby, 1996.
63. Raviola G, Raviola E. Intercellular junctions in the ciliary epithelium. Invest Ophthalmol Vis Sci 1978; 17:958.
64. Wistrand PJ. Carbonic anhydrase in the anterior uvea of the rabbit. Acta Physiol Scand 1951; 24:144.
65. Cole DF. Effects of some metabolic inhibitors upon the formation of the aqueous humor in rabbits. Br J Ophthalmol 1960; 44:739.
66. Bonting SL, Becker B. Studies on sodium-potassium activated adenosinetriphosphatase. XIV. Inhibition of enzyme activity and aqueous humor flow in the rabbit eye after intravitreal injection of ouabain. Invest Ophthalmol 1964; 3:523.
67. Bhattacherjee P. Distribution of carbonic anhydrase in the rabbit eye as demonstrated histochemically. Exp Eye Res 1971; 12:356.
68. Tsukahara S, Maezara N. Cytochemical localization of adenyl cyclase in the rabbit ciliary body. Exp Eye Res 1978; 26:99.
69. Riley MV, Kishida K. ATPases of ciliary epithelium: cellular and subcellular distribution and probable role in secretion of aqueous humor. Exp Eye Res 1986; 42:559.
70. Flügel C, Lütjen-Drecoll E. Presence and distribution of Na^+/K^+-ATPase in the ciliary epithelium of the rabbit. Histochemistry 1988; 88:613.
71. Usukura J et al. [3H]Ouabain localization of Na-K ATPase in the epithelium of the rabbit ciliary body pars plicata. Invest Ophthalmol Vis Sci 1988; 29:606.
72. Davson H. The aqueous humor and the intraocular pressure. In: Davson H, ed. Physiology of the eye. New York: Pergamon Press, 1990:3.
73. Maren TH. Biochemistry of aqueous humor inflow. In: Kaufman PL, Mittag TW, eds. Glaucoma. London: Mosby-Year Book Europe Ltd, 1994:1:35.
74. Simon KA, Bonting SL. Possible usefulness of cardiac glycosides in treatment of glaucoma. Arch Ophthalmol 1962; 68:227.
75. Becker B. Ouabain and aqueous humor dynamics in the rabbit eye. Invest Ophthalmol 1963; 2:325.
76. Riley MV. The sodium-potassium-stimulated adensosine triphosphatase of rabbit ciliary epithelium. Exp Eye Res 1964; 3:76.
77. Becker B. Vanadate and aqueous humor dynamics. Proctor Lecture. Invest Ophthalmol Vis Sci 1980; 19:1156.
78. Krupin T et al. Topical vanadate lowers intraocular pressure in rabbits. Invest Ophthalmol Vis Sci 1980; 19:1360.
79. Podos SM et al. The effect of vanadate on aqueous humor dynamics in cynomolgus monkeys. Invest Ophthalmol Vis Sci 1984; 25:359.
80. Lee P-Y et al. Intraocular pressure effects of multiple doses of drugs applied to glaucomatous monkey eyes. Arch Ophthalmol 1987; 105:249.
81. Lee P-Y et al. Pharmacological testing in the laser-induced monkey glaucoma model. Curr Eye Res 1985; 4:775.
82. Verkman AS. Role of aquaporin water channels in eye function. Exp Eye Res 2003; 76:137.
83. Patil RV et al. Fluid transport by human nonpigmented ciliary epithelial layers in culture: a homeostatic role for aquaporin-1. Am J Physiol 2001; 281:1139.
84. Frigeri A et al. Immunolocalization of the mercurial-insensitive water channel and glycerol intrinsic protein in epithelial cell plasma membranes. Proc Natl Acad Sci USA 1995; 92:4328.
85. Civan MM, Macknight ADC. The ins and outs of aqueous humour secretion (review). Exp Eye Res 2004; 78:625.
86. McLaughlin CW et al. Electron microprobe analysis of rabbit ciliary epithelium indicates enhances secretion posteriorly and enhanced absorption anteriorly. Am J Physiol Cell Physiol 2007; 293:C1455.
87. Do CW, Civan MM. Species variation in biology and physiology of the ciliary epithelium. Similarities and differences. Exp Eye Res 2009; 88:631.
88. Shahidullah M et al. Effects of ion transport and channel-blocking drugs on aqueous humor formation in isolated bovine eye. Invest Ophthalmol Vis Sci 2003; 44:1185.
89. Hochgesand DH et al. Catecholaminergic regulation of Na-K-Cl contransport in pigmented ciliary epithelium: difference between PE, NPE. Exp Eye Res 2001; 72:1.
90. Do CW, To CH. Chloride secretion by bovine ciliary epithelium: a model of aqueous humor formation. Invest Ophthalmol Vis Sci 2000; 41:1853.
91. Do CW, Civan MM. Swelling-activated chloride channels in aqueous humour formation: on the one side and the other. Acta Physiol (Oxf) 2006; 187:345.
92. Bakall B et al. Bestrophin-2 is involved in the generation of intraocular pressure. Invest Ophthalmol Vis Sci 2008; 49:1563.
93. Avila MY et al. Inhibitors of NHE-1 Na^+/H^+ exchange reduce mouse intraocular pressure. Invest Ophthalmol Vis Sci 2002; 43:1897.
94. Gabelt BT et al. Anterior segment physiology following bumetanide inhibition of Na-K-Cl cotransport. Invest Ophthalmol Vis Sci 1997; 8:1700.
95. Maren TH. Carbonic anhydrase. Chemistry, physiology, and inhibition. Physiol Rev 1967; 47:595.
96. Maren TH. HCO_3-formation in aqueous humor: mechanism and relation to the treatment of glaucoma. Invest Ophthalmol 1974; 13:479.
97. Muther TF, Friedland BR. Autoradiographic localization of carbonic anhydrase in the rabbit ciliary body. J Histochem Cytochem 1980; 28:1119.
98. Lütjen-Drecoll E et al. Carbonic anhydrase distribution in the human and monkey eye by light and electron microscopy. Graefe's Arch Clin Exp Ophthalmol 1983; 220:285.
99. Mudge GH, Weiner IM. Agents affecting volume and composition of body fluids. In: Gilman AG et al, eds. The pharmacological basis of therapeutics, 8th edn. New York: McGraw-Hill, 1990:682.
100. Brechue WF, Maren TH. A comparison between the effect of topical and systemic carbonic anhydrase inhibitors on aqueous humor secretion. Exp Eye Res 1993; 57:67.
101. Mincione F et al. The development of topically acting carbonic anhydrase inhibitors as anti-glaucoma agents (review). Curr Top Med Chem 2007; 7:849.
102. Kaufman PL, Mittag TW. Medical therapy of glaucoma. In: Kaufman PL, Mittag TW, eds. Glaucoma. London: Mosby-Year Book Europe Ltd, 1994:9.7.
103. Becker B. Decrease in intraocular pressure in man by a carbonic anhydrase inhibitor, Diamox. Am J Ophthalmol 1954; 37:13.
104. Maren TH. The development of ideas concerning the role of carbonic anhydrase in the secretion of aqueous humor. Relations to the treatment of glaucoma. In: Drance SM, Neufeld AH, eds. Glaucoma. Applied pharmacology in medical treatment. Orlando: Grune & Stratton, 1984:325.
105. Maren TH et al. The transcorneal permeability of sulfonamide carbonic anhydrase inhibitors and their effect on aqueous humor secretion. Exp Eye Res 1983; 36:457.
106. Schoenwald RD et al. Topical carbonic anhydrase inhibitors. J Med Chem 1984; 27:810.
107. Hoskins HD Jr., Kass MA. Aqueous humor formation. In: Klein EA, ed. Becker-Shaffers's Diagnosis and therapy of the glaucomas, 6th edn. St. Louis: CV Mosby, 1989:18.
108. Eller MG et al. Topical carbonic anhydrase inhibitors, III. Optimization model for corneal penetration of ethoxzolamide analogues. J Pharm Sci 1985; 74:155.
109. Sugrue MF et al. A comparison of L-671,152 and MK927, two topically effective ocular hypotensive carbonic anhydrase inhibitors, in experimental animals. Curr Eye Res 1990; 9:607.
110. Sugrue MF et al. MK-927. A topically active ocular hypotensive carbonic anhydrase inhibitor. J Ocul Pharmacol 1990; 6:9.

111. Pierce WMJ et al. Topically active ocular carbonic anhydrase inhibitors-novel biscarbonylamidothiadiazole sulfonamides as ocular hypotensive agents. Proc Soc Exp Biol Med 1993; 203:360.
112. Lippa EA et al. Dose response and duration of action of dorzolamide, a topical carbonic anhydrase inhibitor. Arch Ophthalmol 1992; 110:495.
113. Gunning FP et al. Two topical carbonic anhydrase inhibitors sezolamide and dorzolamide in Gelrite vehicle: a multiple-dose efficacy study. Graefes Arch Clin Exp Ophthalmol 1993; 231:384.
114. Wilkerson M et al. Four-week safety and efficacy study of dorzolamide, a novel, active topical carbonic anhydrase inhibitor. Arch Ophthalmol 1993; 111:1343.
115. Vanlandingham BD et al. The effect of dorzolamide on aqueous humor dynamics in normal human subjects during sleep. Ophthalmology 1998; 105:1537.
116. Herkel U, Pfeiffer N. Update on topical carbonic anhydrase inhibitors. Curr Opin Ophthalmol 2001; 12:88.
117. Ingram CJ, Brubaker RF. Effect of brinzolamide and dorzolamide on aqueous humor flow in human eyes. Am J Ophthalmol 1999; 128:292.
118. Silver LH. Clinical efficacy and safety of brinzolamide (Azopt), a new topical carbonic anhydrase inhibitor for primary open-angle glaucoma and ocular hypertension. Brinzolamide Primary Therapy Study Group. Am J Ophthalmol 1998; 126:400.
119. Podos SM, Serle JB. Topicaly active carbonic anhydrase inhibitors for glaucoma. Arch Ophthalmol 1982; 109:38.
120. Krause U, Raunio V. Proteins of the normal human aqueous humor. Ophthalmologica 1969; 159:178.
121. Stjernschantz J et al. The aqueous proteins of the rat in the normal eye and after aqueous withdrawal. Exp Eye Res 1973; 16:215.
122. Fielder AR, Rahi AHS. Immunoglobulins of normal aqueous humor. Trans Ophthalmol Soc UK 1979; 99:120.
123. DiMatteo J. Active transport of ascorbic acid into lens epithelium of the rat. Exp Eye Res 1989; 49:873.
124. Kolodny NH et al. Contrast-enhanced MRI confirmation of an anterior protein pathway in normal rabbit eyes. Invest Ophthalmol Vis Sci 1996; 37:1602.
125. Bert R et al. Confirmation of anterior large-molecule diffusion pathway in the normal human eye. Invest Ophthalmol Vis Sci 1999;40(Suppl):S198.
126. Freddo TF et al. The source of proteins in the aqueous humor of the normal rabbit. Invest Ophthalmol Vis Sci 1990; 31:125.
127. Barsotti M et al. The source of proteins in the aqueous humor of the normal monkey eye. Invest Ophthalmol Vis Sci 1992; 33:581.
128. Caprioli J. The ciliary epithelia and aqueous humor. In: Hart WM, ed. Adler's Physiology of the eye. Clinical application, 9th edn. St Louis: CV Mosby, 1992:228.
129. Reddy VN et al. The effect of aqueous humor ascorbate on ultraviolet-B-induced DNA damage in lens epithelium. Invest Ophthalmol Vis Sci 1998; 39:344.
130. Ringvold A. The significance of ascorbate in the aqueous humour protection against UV-A and UV-B. Exp Eye Res 1996; 62:261.
131. Koskela TK et al. Is the high concentration of ascorbic acid in the eye an adaptation to intense solar radiation? Invest Ophthalmol Vis Sci 1989; 31:2265.
132. Alvarado J et al. Trabecular meshwork cellularity in primary open-angle glaucoma and nonglaucomatous normals. Ophthalmology 1984; 91:564.
133. Grierson I, Howes RC. Age-related depletion of the cell population in the human trabecular meshwork. Eye 1987; 1:204.
134. Duke-Elder S. The aqueous humor. In: Duke-Elder S, ed. The physiology of the eye and of vision. System of ophthalmology, vol 4. St Louis: CV Mosby, 1968:104.
135. Kleinstein RN et al. In vivo aqueous humor oxygen tension-as estimated from measurements on bare stroma. Invest Ophthalmol Vis Sci 1981; 21:415.
136. McLaren JW et al. Measuring oxygen tension in the anterior chamber of rabbits. Invest Ophthalmol Vis Sci 1998; 39:1899.
137. Bill A. The drainage of albumin from the uvea. Exp Eye Res 1964; 3:179.
138. Bill A. Capillary permeability to and extravascular dynamics of myoglobin, albumin, and gammaglobulin in the uvea. Acta Physiol Scand 1968; 73:204.
139. Shiose Y. Electron microscopic studies on blood-retinal and blood-aqueous barriers. Nippon Ganka Gakkai Zasshi- Acta Societatis Ophthalmologicae Japonicae 1969; 73:1606.
140. Hogan MJ et al. Ciliary body and posterior chamber. In: Hogan MJ et al, eds. Histology of the human eye. Philadelphia: WB Saunders, 1971:260.
141. Smith RS. Ultrastructural studies of the blood-aqueous barrier. 1. Transport of an electron dense tracer in the iris and ciliary body of the mouse. Am J Ophthalmol 1971; 71:1066.
142. Vegge T. An epithelial blood-aqueous barrier to horseradish peroxidase in the ciliary processes of the vervet monkey (*Cercopithecus aethiops*). Zeitschr Zellforsch Mikrosk Anat 1971; 114:309.
143. Uusitalo R et al. An electron microscopic study of the blood-aqueous barrier in the ciliary body and iris of the rabbit. Exp Eye Res 1973; 17:49.
144. Rodriguez-Peralta L. The blood-aqueous barrier in five species. Am J Ophthalmol 1975; 80:713.
145. Uusitalo R et al. Studies on the ultrastructure of the blood-aqueous barrier in the rabbit. Acta Ophthalmol 1974; 123:61.
146. Alm A. Ocular Circulation, Ch 6. In: Hart WM, ed. Adler's Physiology of the eye. Clinical application, 9th edn. St Louis: CV Mosby, 1992:198.
147. Vinores SA et al. Electron microscopic immunocytochemical demonstration of blood-retinal barrier breakdown in human diabetics and its association with aldose reductase in retinal vascular endothelium and retinal pigment epithelium. Histochem J 1993; 25:648.
148. Raviola G. Effects of paracentesis on the blood-aqueous barrier. An electron microscopic study on *Macaca mullata* using horseradish peroxidase as a tracer. Invest Ophthalmol 1974; 13:828.
149. Sonsino J et al. Co-localization of junction-associated proteins of thhe human glood-aqueous barrier: occludin, ZO-1 and F-actin. Exp Eye Res 2002; 74:123.
150. Hogan MJ et al. Iris and anterior chamber. In: Histology of the human eye: an atlas and textbook. Philadelphia: WB Saunders Co, 1971:202.
151. Masuda M, Mishima Y. Effects of prostaglandins on inflow and outflow of the aqueous in rabbits. Japn J Ophthamol 1973; 17:300.
152. Bill A. Effects of longstanding stepwise increments in eye pressure on the rate of aqueous humor formation in a primate (*Cercopithecus ethiops*). Exp Eye Res 1971; 12:184.
153. Toris CB, Pederson JE. Aqueous humor dynamics in experimental iridocyclitis. Invest Ophthalmol Vis Sci 1987; 28:477.
154. Kaufman PL et al. The effects of prostaglandins on aqueous humor dynamics. In: Kooner KS, Zimmerman TJ, eds. New ophthalmic drugs. Ophthalmological Clinics of North America. Philadelphia: WB Saunders, 1989:141.
155. Kaufman PL, Gabelt BT. Presbyopia, prostaglandins and primary open angle glaucoma. In: Krieglstein GK, ed. Glaucoma Update V. Proceedings of the Symposium of the Glaucoma Society of the International Congress of Ophthalmology in Quebec City, June 1994. New York: Springer-Verlag, 1995:224.
156. Kaufman PL. Pressure-dependent outflow. In: Ritch R et al, eds. The glaucomas, 2nd edn. St Louis: CV Mosby, 1996:307.
157. Becker B. The transport of organic anions by the rabbit eye. I. In vitro iodopyracet (Diodrast) accumulation by ciliary body-iris preparations. Am J Ophthalmol 1960; 50:862.
158. Forbes M, Becker B. The transport of organic anions by the rabbit eye. II. In vivo transport of iodopyracet (Diodrast). Am J Ophthalmol 1960; 50:867.
159. Becker B. Iodide transport by the rabbit eye. Am J Physiol 1961; 200:804.
160. Bárány EH. Inhibition by hippurate and probenecid of in vitro uptake of iodipamide and o-iodohippurate-composite uptake system for iodipamide in choroid plexus, kidney cortex, and anterior uvea of several species. Acta Physiol Scand 1972; 86:12.
161. Bito LZ. Accumulation and apparent active transport of prostaglandins by some rabbit tissues in vitro. J Physiol 1972; 221:371.
162. Bárány EH. The liver-like anion transport system in rabbit kidney, uvea, and choroid plexus. II. Efficiency of acidic drugs and other anions as inhibitors. Acta Physiol Scand 1973; 88:491.
163. Bárány EH. Bile acids as inhibitors of the liver-like anion transport system in the rabbit kidney, uvea, and choroid plexus. Acta Physiol Scand 1974; 92:195.
164. Bárány EH. In vitro uptake of bile acids by choroid plexus, kidney cortex, and anterior uvea. I. The iodipamide sensitive transprot systems in the rabbit. Acta Physiol Scand 1975; 93:250.
165. Stone RA. Cholic acid accumulation by the ciliary body and by the iris of the primate eye. Invest Ophthalmol Vis Sci 1979; 18:819.
166. Stone RA. The transport of para-aminohippuric acid by the ciliary body and by the iris of the primate eye. Invest Ophthalmol Vis Sci 1979; 18:807.
167. Bito LZ. Species differences in the response of the eye to irritation and trauma: a hypothesis of divergence in ocular defense mechanisms, and the choice of experimental animals for eye research. Exp Eye Res 1984; 39:807.
168. Bito LZ. Prostaglandins: old concepts and new perspectives. Arch Ophthalmol 1987; 105:1036.
169. Bito LZ, Salvador EV. Intraocular fluid dynamics. 3. The site and mechanism of prostaglandin transfer across the blood intraocular fluid barriers. Exp Eye Res 1972; 14:233.
170. Bito L et al. Inhibition of in vitro concentrative prostaglandin accumulation by prostaglandins, prostaglandin analogues, and by some inhibitors of organic anion transport. J Physiol 1976; 256:257.
171. Chiquet C, Denis P. The neuroanatomical and physiological bases of variations in intraocular pressure. J Fr Ophthalmol 2004; 27:2S11.
172. Ehinger B. Adrenergic nerves to the eye and to related structures in man and in the cynomolgus monkey (*Macaca irus*). Invest Ophthalmol 1966; 5:42.
173. Laties AMD, Jacobowitz D. A comparative study of the autonomic innervation of the eye in monkey, cat and rabbit. Anat Record 1966; 156:383.
174. Ruskell GL. Innervation of the anterior segment of the eye. In: Lütjen-Drecoll E, ed. Basic aspects of glaucoma research,. Stuttgart: Schattauer, 1982:49.
175. Bryson JM et al. Ganglion cells in the human ciliary body. Arch Ophthalmol 1966; 75:57.
176. Uddman R et al. Vasoactive intestinal peptide nerves in ocular and orbital structures of the cat. Invest Ophthalmol Vis Sci 1980; 19:878.
177. Nilsson SFE. Neuropeptides in the autonomic nervous system influencing uveal blood flow and aqueous humour dynamics. In: Troger J, Kieselbach G, eds. Neuropeptides in the eye. Kerala: Research Signpost, 2009:1.
178. Elsas T et al. Pituitary adenylate cyclase-activating peptide-immunoreactive nerve fibers in the cat eye. Graefes Arch Clin Exp Ophthalmol 1996; 234:573.
179. Bill A. Effects of norepinephrine, isoproterenol and sympathetic stimulation on aqueous humour dynamics in vervet monkeys. Exp Eye Res 1970; 10:31.
180. Gabelt BT et al. Superior cervical ganglionectomy in monkeys: aqueous humor dynamics and their responses to drugs. Exp Eye Res 1995; 60:575.
181. Wentworth WO, Brubaker RF. Aqueous humor dynamics in a series of patients with third neuron Horner's syndrome. Am J Ophthalmol 1981; 92:407.
182. Larson RS, Brubaker RF. Isoproterenol stimulates aqueous flow in humans with Horner's syndrome. Invest Ophthalmol Vis Sci 1988; 29:621.
183. Bonomi L, Di Comite P. Outflow facility after guanethidine sulfate administration. Arch Ophthalmol 1967; 78:337.
184. Nilsson SFE et al. Effects of timolol on terbutaline- and VIP-stimulated aqueous humor flow in the cynomolgus monkey. Curr Eye Res 1990; 9:863.
185. Rasmussen CA et al. Aqueous humor dynamics in monkeys in response to the kappa opioid agonist bremazocine. Trans Am Ophthalmol Soc 2007; 105:225.
186. Russell KR et al. Modulation of ocular hydrodynamics and iris function by bremeazocine, a kappa opioid receptor agonist. Exp Eye Res 2000; 70:675.
187. Bill A, Stjernschantz J. Cholinergic vasoconstrictor effects in the rabbit eye: vasomotor effects of pentobarbital anesthesia. Acta Physiol Scand 1980; 108:419.
188. Hoskins HD, Kass MA. Cholinergic drugs. In: Hoskins HD, Kass MA, eds. Becker-Shaffer's Diagnosis and therapy of the glaucomas, 6th edn. St. Louis: CV Mosby, 1989:420.
189. Becker B. The measurement of rate of aqueous flow with iodide. Invest Ophthalmol 1962; 1:52.
190. Swan K, Hart W. A comparative study of the effects of mecholyl, doryl, pilocarpine, atropine, and epinephrine on the blood-aqueous barrier. Am J Ophthalmol 1940; 23:1311.
191. Bito LZ et al. The relationship between the concentrations of amino acids in the ocular fluids and blood plasma of dogs. Exp Eye Res 1965; 4:374.
192. Wålinder P-E. Influence of pilocarpine on iodopyracet and iodide accumulation by rabbit ciliary body-iris preparations. Invest Ophthalmol 1966; 5:378.

193. Wålinder P-E, Bill A. Aqueous flow and entry of cycloleucine into the aqueous humor of vervet monkeys (*Cercopithecus ethiops*). Invest Ophthalmol 1969; 8:434.
194. Shi XP et al. Adreno-cholinergic modulation of junctional communications between the pigmented and nonpigmented layers of the ciliary body epithelium. Invest Ophthalmol Vis Sci 1996; 37:1037.
195. Stelling JW, Jacob TJ. Functional coupling in bovine ciliary epithelial cells is modulated by carbachol. Am J Physiol 1997; 273:C1876.
196. Gaasterland D et al. Studies of aqueous humor dynamics in man. IV. Effects of pilocarpine upon measurements in young normal volunteers. Invest Ophthalmol 1975; 14:848.
197. Bill A, Walinder P-E. The effects of pilocarpine on the dynamics of aqueous humor in a primate (*Macaca irus*). Invest Ophthalmol 1966; 5:170.
198. Berggren L. Further studies on the effect of autonomic drugs on in vivo secretory activity of the rabbit eye ciliary processes. A. Inhibition of the pilocarpine effect by isopilocarpine, arecoline, and atropine. B. Influence of isoproterenol and norepinephrine. Acta Ophthalmol 1970; 48:293.
199. Uusitalo R. Effect of sympathetic and parasympathetic stimulation on the secretion and outflow of aqueous humor in the rabbit eye. Acta Physiol Scand 1972; 86:315.
200. Kupfer C. Clinical significance of pseudofacility. Sanford R. Gifford memorial lecture. Am J Ophthalmol 1973; 75:193.
201. Macri FJ, Cevario SJ. The induction of aqueous humor formation by the use of Ach+ eserine. Invest Opthalmol 1973; 12:910.
202. Macri FJ, Cevario SJ. The dual nature of pilocarpine to stimulate or inhibit the formation of aqueous humor. Invest Ophthalmol 1974; 13:617.
203. Macri FJ, Cevario SJ. A possible vascular mechanism for the inhibition of aqueous humor formation by ouabain and acetazolamide. Exp Eye Res 1975; 20:563.
204. Stjernschantz J. Effect of parasympathetic stimulation on intraocular pressure, formation of aqueous humor and outflow facility in rabbits. Exp Eye Res 1976; 22:639.
205. Green K, Padgett D. Effects of various drugs on pseudofacility and aqueous homor formation in rabbit eye. Exp Eye Res 1979; 28:239.
206. Chiou GC et al. Studies of action mechanism of antiglaucoma drugs with a newly developed cat model. Life Sci 1980; 27:2445.
207. Liu HK, Chiou GCY. Continuous, simultaneous, and instant display of aqueous humor dynamics with a microspectrophotometer and a sensitive drop counter. Exp Eye Res 1981; 32:583.
208. Nagataki S, Brubaker RF. The effect of pilocarpine on aqueous humor formation in human beings. Arch Ophthalmol 1982; 100:818.
209. Erickson KA, Schreoder A. Direct effects of muscarinic agents on the outflow pathways in human eyes. Invest Ophthalmol Vis Sci 2000; 41:1743.
210. Neufeld AH, Sears ML. Cyclic-AMP in ocular tissues of the rabbit, monkey, and human. Invest Ophthalmol 1974; 13:475.
211. Gregory D et al. Intraocular pressure and aqueous flow are decreased by cholera toxin. Invest Ophthalmol Vis Sci 1981; 20:371.
212. Smith BR et al. Forskolin, a potent adenylate cyclase activator, lowers rabbit intraocular pressure. Arch Ophthalmol 1984; 102:146.
213. Shibata T et al. Ocular pigmentation and intraocular pressure response to forskolin. Curr Eye Res 1988; 7:667.
214. Shahidullah M et al. Effects of timolol, terbutaline and forskolin on IOP, aqueous humour formation and ciliary cyclic AMP levels in the bovine eye. Curr Eye Res 1995; 14:519.
215. Schenker JI et al. Fluorophotometric study of epinephrine and timolol in human subjects. Arch Ophthalmol 1981; 99:1212.
216. Coakes RL, Siah PB. Effects of adrenergic drugs on aqueous humor dynamics in the normal human eye. I. Salbutamol. Br J Ophthalmol 1984; 68:393.
217. Gharagozloo NZ et al. Terbutaline stimulates aqueous humor flow in humans during sleep. Arch Ophthalmol 1988; 106:1218.
218. Novack GD. Ophthalmic b-blockers since timolol. Surv Ophthalmol 1987; 31:307.
219. Kiland JA et al. Studies on the mechanism of action of timolol and on the effects of suppression and redirection of aqueous flow on outflow facility. Exp Eye Res 2004; 78(Special Issue):639.
220. Coakes RL, Brubaker RF. The mechanism of timolol in lowering intraocular pressure. Arch Ophthalmol 1978; 96:2045.
221. Yablonski ME et al. A fluorophotometric study of the effect of topical timolol on aqueous humor dynamics. Exp Eye Res 1978; 27:135.
222. Dailey RA et al. The effects of timolol maleate and acetazolamide on the rate of aqueous formation in normal human subjects. Am J Ophthalmol 1982; 93:232.
223. Yablonski ME et al. The effect of levobunolol on aqueous humor dynamics. Exp Eye Res 1987; 44:49.
224. Mills KB, Wright G. A blind randomized cross-over trial comparing metipranolol 0.3 percent with timolol 0.25 percent in open-angle glaucoma. A pilot study. Br J Ophthalmol 1986; 70:39.
225. Henness S et al. Ocular carteolol: a review of its use in the management of glaucoma and ocular hypertension. Drugs Aging 2007; 24:509.
226. Maruyama K, Shirato S. Additive effect of dorzolamide or carteolol to latanoprost in primary open-angle glaucoma: a prospective randomized crossover trial. J Glauc 2006; 15:341.
227. Stewart RH et al. Betaxolol vs. timolol. A six-month double-blind comparison. Arch Ophthalmol 1986; 104:46.
228. Nathanson JA. Adrenergic regulation of intraocular pressure. Identification of beta 2-adrenergic-stimulated adenylate cyclase in the ciliary process epithelium. Proc Natl Acad Sci USA 1980; 77:7420.
229. Nathanson JA. Human ciliary process adrenergic receptor. Pharmacological characterization. Invest Ophthalmol Vis Sci 1981; 21:798.
230. Sears ML. Autonomic nervous system. Adrenergic agonists. In: Sears ML, ed. Pharmacology of the eye. Handbook of experimental pharmacology. Berlin: Springer-Verlag, 1984:193.
231. Berrospi AR, Leibowitz HM. Betaxolol. A new b-adrenergic blocking agent for the treatment of glaucoma. Arch Ophthalmol 1982; 100:943.
232. Berry DPJ et al. Betaxolol and timolol. A comparison of efficacy and side-effects. Arch Ophthalmol 102:42, 1984;
233. Levy NS et al. A controlled comparison of betaxolol and timolol with long-term evaluation of safety and efficacy. Glaucoma 1985; 7:54.
234. Vuori ML et al. Concentrations and antagonist activity of topically applied betaxolol in aqueous humour. Acta Ophthalmol 1993; 71:677.
235. Schmitt CJ et al. Beta-adrenergic blockers: lack of relationship between antagonism of isoproterenol and lowering of intraocular pressure in rabbits. In: Sears M, ed. New directions in ophthalmic research. New Haven CT: Yale University Press, 1981:147.
236. Caprioli J et al. Forskolin lowers intraocular pressure by reducing aqueous inflow. Invest Ophthalmol Vis Sci 1984; 25:268.
237. Chiou GCY et al. Are β-adrenergic mechanisms involved in ocular hypotensive actions of adrenergic drugs? Ophthalmic Res 1985; 17:49.
238. Osborne NN, Chidlow G. Do beta-adrenoceptors and serotonin 5-HT1A receptors have similar functions in the control of intraocular pressure (IOP) in the rabbit? Ophthalmologica 1996; 210:308.
239. Crook RB, Riese K. Beta-adrenergic stimulation of Na^+, K^+, Cl^- cotransport in fetal nonpigmented ciliary epithelial cells. Invest Ophthalmol Vis Sci 1996; 37:1047.
240. Reiss GR et al. Aqueous humor flow during sleep. Invest Ophthalmol Vis Sci 1984; 25:776.
241. Sit AJ et al. Circadian variation of aqueous dynamics in young healthy adults. Invest Ophthalmol Vis Sci 2008; 49:1473.
242. Wan XL et al. Circadian aqueous flow medicated by beta-arrestin induced homologous desensitization. Exp Eye Res 1997; 64:1005.
243. Liu JH et al. Twenty-four-hour pattern of intraocular pressure in the aging population. Invest Ophthalmol Vis Sci 1999; 40:2912.
244. Liu JH et al. Elevation of human intraocular pressure at night under moderate illumination. Invest Ophthalmol Vis Sci 1999; 40:2439.
245. Topper JE, Brubaker RF. Effects of timolol, epinephrine, and acetazolamide on aqueous flow during sleep. Invest Ophthalmol Vis Sci 1985; 26:1315.
246. Robinson JC, Kaufman PL. Dose-dependent suppression of aqueous humor formation by timolol in the cynomolgus monkey. J Glaucoma 1993; 2:251.
247. Rittig MG et al. Innervation of the ciliary process vasculature and epithelium by nerve fibers containing catecholamines and neuropeptide Y. Ophthalmic Res 1993; 25:108.
248. Jacob E et al. Combined corticosteroid and catecholamine stimulation of aqueous humor flow. Ophthalmology 1996; 103:1303.
249. MacCumber MW et al. Endothelin mRNA's visualized by in situ hybridization provides evidence for local action. Proc Natl Acad Sci USA 1989; 86:7285.
250. Van Buskirk EM et al. Ciliary vasoconstriction after topical adrenergic drugs. Am J Ophthalmol 1990; 109:511.
251. Lee DA, Brubaker RF. Effect of phenylephrine on aqueous humor flow. Curr Eye Res 1982; 83:89.
252. Lee DA et al. Acute effect of thymoxamine on aqueous humor formation in the epinephrine-treated normal eye as measured by fluorophotometry. Invest Ophthalmol Vis Sci 1983; 24:165.
253. Bill A, Heilmann K. Ocular effects of clonidine in cats and monkeys (*Macaca irus*). Exp Eye Res 1975; 21:481.
254. Krieglstein GK et al. The peripheral and central neural actions of clonidine in normal and glaucomatous eyes. Invest Ophthalmol Vis Sci 1978; 17:149.
255. Lee DA et al. Effect of clonidine on aqueous humor flow in normal human eyes. Exp Eye Res 1984; 38:239.
256. Jin Y et al. Characterization of alpha 2-adrenoceptor binding sites in rabbit ciliary body membranes. Invest Ophthalmol Vis Sci 1994; 35:2500.
257. Liu JH, Gallar J. In vivo cAMP level in rabbit iris-ciliary body after topical epinephrine treatment. Curr Eye Res 1996; 15:1025.
258. Schutte M et al. Comparative adrenocholinergic control of intracellular Ca^{2+} in the layers of the ciliary body epithelium. Invest Ophthalmol Vis Sci 1996; 37:212.
259. Maus TL et al. Comparison of the early effects of brimonidine and apraclonidine as topical ocular hypotensive agents. Arch Ophthalmol 1999; 117:586.
260. Toris CB et al. Acute versus chronic effects of brimonidine on aqueous humor dynamics in ocular hypertensive patients. Am J Ophthalmol 1999; 128:8.
261. Mittag TW et al. Atrial natriuretic peptide (ANP), guanylate cyclase, and intraocular pressure in the rabbit eye. Curr Eye Res 1987; 6:1189.
262. Korenfeld MS, Becker B. Atrial natriuretic peptides. Effects on intraocular pressure, cGMP, and aqueous flow. Invest Ophthalmol Vis Sci 1989; 30:2385.
263. Nathanson JA. Nitrovasodilators as a new class of ocular hypotensive agents. J Pharmacol Exp Ther 1992; 260:956.
264. Nathanson JA. Nitric oxide and nitrovasodilators in the eye: implications for ocular physiology and glaucoma. J Glaucoma 1993; 2:206.
265. Shahidullah M, Wilson WS. Atriopeptin, sodium azide and cyclic GMP reduce secretion of aqueous humour and inhibit intracellular calcium release in bovine cultured ciliary epithelium. Br J Pharmacol 1999; 127:1438.
266. Kee C et al. Effect of 8-Br cGMP on aqueous humor dynamics in monkeys. Invest Ophthalmol Vis Sci 1994; 35:2769.
267. Samuelsson-Almén M et al. Effects of atrial natriuretic factor (ANF) on intraocular pressure and aqueous humor flow in the cynomolgus monkey. Exp Eye Res 1991; 53:253.
268. Russell KR, Potter DE. Dynorphin modulates ocular hydrodynamics and releases atrial natriuretic peptide via activation of kappa-Opioid receptors. Exp Eye Res 2002; 75:259.
269. Dortch-Carnes J, Russell K. Morphine-stimulated nitric oxide release in rabbit aqueous humor. Exp Eye Res 2007; 84:185.
270. Dortch-Carnes J, Russell KR. Morphine-induced reduction of intraocular pressure and pupil diameter: role of mitric oxide. Pharmacology 2006; 77:17.
271. Bonfiglio V et al. Possible involvement of nitric oxide in morphine-induced miosis and reduction of intraocular pressure in rabbits. Eur J Pharm 2006; 534:227.
272. Chang FW et al. Mechanism of the ocular hypotensive action of ketanserin. J Ocular Pharmacol 1985; 1:137.
273. Barnett NL, Osborne NN. The presence of serotonin (5-HT1) receptors negatively coupled to adenylate cyclase in rabbit and human ciliary processes. Exp Eye Res 1993; 57:209.
274. Chidlow G et al. Localization of 5-hydroxytryptamine$_{1A}$ and 5-hydroxytryptamine$_7$ receptors in rabbit ocular and brain tissues. Neuroscience 1998; 87:675.
275. Chidlow G et al. The 5-HT1A receptor agonist 8-OH-DPAT lowers intraocular pressure in normotensive NZW rabbits. Exp Eye Res 1999; 69:587.
276. Chu T-C et al. 8OH-DPAT-induced ocular hypotension: sites and mechanisms of action. Exp Eye Res 1999; 69:227.
277. Chidlow G et al. Flesinoxan, a 5-HT1A receptor agonist/alpha 1-adrenoceptor antagonist, lowers intraocular pressure in NZW rabbits. Curr Eye Res 2001; 23:144.

278. Gabelt BT et al. Effects of serotonergic compounds on aqueous humor dynamics in monkeys. Curr Eye Res 2001; 23:120.
279. Gabelt BT et al. Aqueous humor dynamics in monkeys after topical R-DOI. Invest Ophthalmol Vis Sci 2005; 46:4691.
280. Kodama T et al. Pharmacological study on the effects of some ocular hypotensive drugs on aqueous humor formation in the arterially perfused enucleated rabbit eye. Ophthalmic Res 1985; 17:120.
281. Hoffman BF, Bigger JTJ. Digitalis and allied cardiac glyosides. In: Gilman AG et al, eds. The pharmacological basis of therapeutics, 8th edn. New York: McGraw-Hill, 1990:814.
282. McLaren JW et al. Effect of ibopamine on aqueous humor production in normotensive humans. Invest Ophthalmol Vis Sci 2003; 44:4853.
283. Azevedo H et al. Effects of ibopamine eye drops on intraocular pressure and aqueous humor flow in healthy volunteers and patients with open-angle glaucoma. Eur J Ophthalmol 2003; 13:370.
284. Giuffrè I et al. The effects of 2 percent ibopamine eye drops on the intraocular pressure and pupil motility of patients with open-angle glaucoma. Eur J Ophthalmol 2004; 14:508.
285. Purnell WD, Gregg JM. Δ^9-tetrahydrocannabinol, euphoria and intraocular pressure in man. Ann Ophthalmol 1975; 7:921.
286. Green K, Roth M. Ocular effects of topical administration of Δ9-tetrahydrocannabinol in man. Arch Ophthalmol 1982; 100:265.
287. Jay WM, Green K. Multiple-drop study of topically applied 1 percent Δ9-tetrahydrocannabinol in human eyes. Arch Ophthalmol 1983; 101:591.
288. Porcella A et al. The synthetic cannabinoid WIN55212–2 decreases the intraocular pressure in human glaucoma resistant to conventional therapies. Eur J Neurosci 2001; 13:409.
289. Stamer WD et al. Cannabinoid CB(1) receptor expression, activation and detection of endogenous ligand in trabecular meshwork and ciliary process tissues. Eur J Pharm 2001; 431:277.
290. Straiker AJ et al. Localization of cannabinoid CB1 receptors in the human anterior eye and retina. Invest Ophthalmol Vis Sci 1999; 40:2442.
291. Njie YF et al. Aqueous humor outflow effects of 2-arachidonylglycerol. Exp Eye Res 2008; 87:106.
292. Becker B. Does hyposecretion of aqueous humor damage the trabecular meshwork? J Glaucoma 1995; 4:303.
293. Lütjen-Drecoll E, Kaufman PL. Long-term timolol and epinephrine in monkeys. II. Morphological alterations in trabecular meshwork and ciliary muscle. Trans Ophthalmol Soc UK 1986; 105:196.
294. Institute NE. Glaucoma and optic neuropathies program. Natl Plan Eye Vis Res 2006
295. Kass MA et al. A randomized trial determines that topical ocular hypotensive medication delays or prevents the onset of primary open-angle glaucoma. Arch Ophthalmol 2002; 120:701.
296. Leske MC et al. Factors for glaucoma progression and the effect of treatment: the early manifest glaucoma trial. Arch Ophthalmol 2003; 121:48.
297. Wahl J. Results of the collaborative initial glaucoma treatment study (CIGTS). Ophthalmologe 2005; 102:222.
298. Jones RF, Maurice DM. New methods of measuring the rate of aqueous flow in man with fluorescein. Exp Eye Res 1966; 5:208.
299. Armaly MF. On the distribution of applanation pressure. Arch Ophthalmol 1965; 73:11.
300. Selbach JM et al. Episcleral venous pressure in untreated primary open-angle and normal-tension glaucoma. Ophthalmologica 2005; 219:357.
301. Sultan M, Blondeau P. Episcleral venous pressure in younger and older subjects in the sitting and supine positions. J Glaucoma 2003; 12:370.
302. Croft MA et al. Aging effects on accommodation and outflow facility responses to pilocarpine in humans. Arch Ophthalmol 1996; 114:586.
303. Grant WM. Tonographic method for measuring the facility and rate of aqueous flow in human eyes. Arch Ophthalmol 1950; 44:204.
304. Becker B. The decline in aqueous secretion and outflow facility with age. Am J Ophthalmol 1958; 46:731.
305. Kupfer C, Ross K. Studies of aqueous humour dynamics in man. I. Measurements in young normal subjects. Invest Ophthalmol 1971;10:518.
306. Toris CB et al. Aqueous humor dynamics in ocular hypertensive patients. J Glaucoma 2002; 11:253.
307. Hart WM. Intraocular pressure. In: Hart WM, ed. Adler's Physiology of the eye. Clinical application, 9th edn. St Louis: CV Mosby, 1992:248.
308. Kaufman PL et al. Formation and drainage of aqueous humor following total iris removal and ciliary muscle disinsertion in the cynomolgus monkey. Invest Ophthalmol Vis Sci 1977; 16:226.
308a N Gupta et al. 2008 missing – author, please supply
308b K Birke et al. 2009 missing – author, please supply
308c ARVO EAbstract 2879 missing – author, please supply
308d ARVO EAbstract 4861 missing – author, please supply
309. Bill A. Conventional and uveoscleral drainage of aqueous humor in the cynomolgus monkey (*Macaca irus*) at normal and high intraocular pressures. Exp Eye Res 1966; 5:45.
310. Bill A. Movement of albumin and dextran through the sclera. Arch Ophthalmol 1965; 74:248.
311. Jackson TL et al. Human scleral hydraulic conductivity: age-related changes, topographical variation, and potential scleral outflow facility. Invest Ophthalmol Vis Sci 2006; 47:4942.
312. Toris CB, Pederson JE. Effect of intraocular pressure on uveoscleral outflow following cyclodialysis in the monkey eye. Invest Ophthalmol Vis Sci 1985; 26:1745.
313. Kleinstein RN, Fatt I. Pressure dependency of trans-scleral flow. Exp Eye Res 1977; 24:335.
314. Gabelt BT, Kaufman PL. Prostaglandin $F_{2\alpha}$ increases uveoscleral outflow in the cynomolgus monkey. Exp Eye Res 1989; 49:389.
315. Hoskins HD, Kass MA. Secondary open-angle glaucoma. In: Hoskins HD, Kass MA, eds. Becker-Shaffer's Diagnosis and therapy of the glaucomas, 6th edn. St. Louis: CV Mosby, 1989:308.
316. Bárány EH. Topical epinephrine effects on true outflow resistance and pseudofacility in vervet monkeys studied by a new anterior chamber perfusion technique. Invest Ophthalmol 1968; 7:88.
317. Bill A. Effects of atropine and pilocarpine on aqueous humour dynamics in cynomolgus monkeys (macaca irus). Exp Eye Res 1967; 6:120.
318. Bill A. Effects of atropine on aqueous humor dynamics in the vervet monkey (*Cercopithecus ethiops*). Exp Eye Res 1969; 8:284.
319. Gaasterland D et al. Studies of aqueous humor dynamics in man. III. Measurements in young normal subjects using norepinephrine and isoproterenol. Invest Ophthalmol Vis Sci 1973; 12:267.
320. Kupfer C et al. Studies of aqueous humor dynamics in man. II. Measurements in young normal subjects using acetazolamide and l-epinephrine. Invest Ophthalmol 1971; 10:523.
321. Wilke K. Early effects of epinephrine and pilocarpine on the intraocular pressure and the episcleral venous pressure in the normal human eye. Acta Ophthalmol 1974; 52:231.
322. Bill A, Svedbergh B. Scanning electron microscopic studies of the trabecular meshwork and the canal of Schlemm: an attempt to localize the main resistance to outflow of aqueous humor in man. Acta Ophthalmol 1972; 50:295.
323. Inomata H et al. Aqueous humor pathways through the trabecular meshwork and into Schlemm's canal in the cynomolgus monkey (*Macaca irus*). An electron microscopic study. Am J Ophthalmol 1972; 73:760.
324. Lütjen-Drecoll E. Structural factors influencing outflow facility and its changeability under drugs: a study of *Macaca arctoides*. Invest Ophthalmol 1973; 12:280.
325. Lütjen-Drecoll E et al. Ultrahistochemical studies on tangential sections of the trabecular meshwork in normal and glaucomatous eyes. Invest Ophthalmol Vis Sci 1981; 21:563.
326. Lütjen-Drecoll E et al. Acute and chronic structural effects of pilocarpine on monkey outflow tissues. Trans Am Ophthalmol Soc 1998; 96:171.
327. Ellingsen BA, Grant WM. Influence of intraocular pressure and trabeculotomy on aqueous outflow in enucleated monkey eyes. Invest Ophthalmol 1971; 10:705.
328. Ellingsen BA, Grant WM. Trabeculotomy and sinusotomy in enucleated human eyes. Invest Ophthalmol 1972; 11:21.
329. Schachtschabel DO et al. Production of glycosminoglycans by cell cultures of the trabecular meshwork of the primate eye. Exp Eye Res 1977; 24:71.
330. Hassel JR et al. Isolation and characterization of the proteoglycans and collagens synthesized by cells in culture. Vision Res 1981; 21:49.
331. Rohen JW et al. Structural changes in human and monkey trabecular meshwork following in vitro cultivation. Albrecht von Graefes Arch Klin Exp Ophthalmol 1982; 218:225.
332. Rohen JW et al. The fine structure of the cribriform meshwork in normal and glaucomatous eyes as seen in tangential sections. Invest Ophthalmol Vis Sci 1981; 21:574.
333. Erickson-Lamy K et al. Absence of time-dependent facility increase ('wash-out') in the perfused enucleated human eye. Invest Ophthalmol Vis Sci 1990; 31:2384.
334. Bárány EH. Simultaneous measurement of changing intraocular pressure and outflow facility in the vervet monkey by constant pressure infusion. Invest Ophthalmol 1964; 3:135.
335. Gaasterland DE et al. Rhesus monkey aqueous humor composition and a primate ocular perfusate. Invest Ophthalmol Vis Sci 1979; 18:1139.
336. Erickson KA, Kaufman PL. Comparative effects of three ocular perfusates on outflow facility in the cynomolgus monkey. Curr Eye Res 1981; 1:211.
337. Kaufman PL et al. Time-dependence of perfusion outflow facility in the cynomolgus monkey. Curr Eye Res 1988; 7:721.
338. Scott PA et al. Comparative studies between species that do and do not exhibit the washout effect. Exp Eye Res 2007; 84:435.
339. Alvarado JA et al. Juxtacanalicular tissue in primary open-angle glaucoma and in nonglaucomatous normals. Arch Ophthalmol 1986; 104:1517.
340. Murphy CG et al. Juxtacanalicular tissue in pigmentary and primary open-angle glaucoma. The hydrodynamic role of pigment and other constituents. Arch Ophthalmol 1992; 110:1779.
341. Hamard P et al. Confocal microscopic examination of trabecular meshwork removed during ab externo trabeculectomy. Br J Ophthalmol 2002; 86:1046.
342. Gong H et al. A new view of the human trabecular meshwork using quick-freeze, deep-etch electron microscopy. Exp Eye Res 2002; 75:347.
343. Sabanay I et al. H-7 effects on structure and fluid conductance of monkey trabecular meshwork. Arch Ophthalmol 2000; 118:955.
344. Sabanay I et al. Functional and structural reversibility of H-7 effects on the conventional aqueous outflow pathway in monkeys. Exp Eye Res 2004; 78:137.
345. Overby D et al. The mechanism of increasing outflow facility during washout in the bovine eye. Invest Ophthalmol Vis Sci 2002; 42:3455.
346. Johnstone MA, Grant WM. Pressure-dependent changes in structure of the aqueous outflow system of human and monkey eye. Am J Ophthalmol 1973; 75:365.
347. Grierson I, Lee WR. The fine structure of the trabecular meshwork at graded levels of intraocular pressure. I. Pressure effects within the near physiological range (8–30mmHg). Exp Eye Res 1975; 20:505.
348. Grierson I, Lee WR. The fine structure of the trabecular meshwork at graded levels of intraocular pressure. 2. Pressures outside the physiological range (0 and 50 mmHg). Exp Eye Res 1975; 20:523.
349. Grierson I, Lee WR. Light microscopic quantitation of the endothelial vacuoles in Schlemm's canal. Am J Ophthalmol 1977; 84:234.
350. Parc C et al. Giant macuoles are found preferentially near collector channels. Invest Ophthalmol Vis Sci 2000; 41:2924.
351. Sit AJ et al. Factors affecting the pores of the inner wall endothelium of Schlemm's canal. Invest Ophthalmol Vis Sci 1997; 38:1517.
352. Ethier CR et al. Two pore types in the inner-wall endothelium of Schlemm's canal. Invest Ophthalmol Vis Sci 1998; 39:2041.
353. Johnson M. What controls aqueous humour outflow resistance? Exp Eye Res 2006; 82:545.
354. Johnson M et al. Modulation of outflow resistance by the pores of the inner wall endothelium. Invest Ophthalmol Vis Sci 1992; 33:1670.
355. Epstein DL, Rohen JW. Morphology of the trabecular meshwork and inner wall endothelium after cationized ferritin perfusion in the monkey eye. Invest Ophthalmol Vis Sci 1991; 32:160.

356. Ye W et al. Interendothelial junctions in normal human Schlemm's canal respond to changes in pressure. Invest Ophthalmol Vis Sci 1997; 38:2460.

357. Johnstone MA. The aqueous outflow system as a mechanical pump. Evidence from examination of tissue and aqueous movement in human and non-human primates. J Glauc 2004; 13:421.

358. Johnstone MA. A new model describes an aqueous outflow pump and explores causes of pump failure in glaucoma. In: Grehn F, Stamper R, eds. Essentials in ophthalmology. Berlin: Springer, 2006:3.

359. Schlunck G et al. Substrate riggidity modulates cell matrix interactions and protein expression in human trabecular meshwork cells. Invest Ophthalmol Vis Sci 2008; 49:262.

360. Lütjen-Drecoll E et al. Quantitative analysis of 'plaque material' in the inner and outer wall of Schlemm's canal in normal and glaucomatous eyes. Exp Eye Res 1986; 42:443.

361. Grierson I et al. The effects of age and antiglaucoma drugs on the meshwork cell population. Res Clin Forums 1982; 4:69.

362. de Kater AW et al. Patterns of aqueous humor outflow in glaucomatous and nonglaucomatous human eyes. Arch Ophthalmol 1989; 107:572.

363. Lütjen-Drecoll E, Kaufman PL. Morphological changes in primate aqueous humor formation and drainage tissues after long-term treatment with antiglaucomatous drugs. J Glaucoma 1993; 2:316.

364. Cousins SW et al. Identification of transforming growth factor-beta as an immunosuppressive factor in aqueous humor. Invest Ophthalmol Vis Sci 1991; 32:2201.

365. Granstein RD et al. Aqueous humor contains transforming growth factor b and a small (<3500 dalton) inhibitor of thymocyte proliferation. J Immunol 1990; 144:3021.

366. Jampel HD et al. Transforming growth factor-beta in human aqueous humor. Curr Eye Res 1990; 9:963.

367. Tripathi RC et al. Aqueous humor in glaucomatous eyes contains an increased level of TGF-beta 2. Exp Eye Res 1994; 59:723.

368. Fleenor DL et al. TGFβ2-induced changes in human trabecular meshwork: implications for intraocular pressure. Invest Ophthalmol Vis Sci 2006; 47:226.

369. Gottanka J et al. Effects of TGF-beta2 in perfused human eyes. Invest Ophthalmol Vis Sci 2004; 45:153.

370. Picht G et al. Transforming growth factor beta 2 levels in the aqeuous humor in different types of glaucoma and the relation to filtering bleb development. Graefes Arch Clin Exp Ophthalmol 2001; 239:199.

371. Fuchshofer R et al. The effect of TGF-β2 on human trabecular meshwork extracellular proteolytic system. Exp Eye Res 2003; 77:757.

372. Borisuth NS et al. Identification and partial characterization of TGF-beta 1 receptors on trabecular cells. Invest Ophthalmol Vis Sci 1992; 33:596.

373. Bhattacharya SK et al. Proteomics reveal cochlin deposits associated with glaucomatous trabecular meshwork. J Biol Chem 2005; 280:6080.

374. Bhattacharya SK et al. Cochlin expression in anterior segment organ culture models after TGF(beta)2 treatment. Invest Ophthalmol Vis Sci 2009; 50:551.

375. Rohen JW, Van der Zypen E. The phagocytic activity of the trabecular meshwork endothelium. An electron microscopic study of the vervet (Cercopithecus ethiops). Albrecht von Graefes Arch Klin Exp Ophthalmol 1968; 175:143.

376. Grierson I, Lee WR. Erythrocyte phagocytosis in the human trabecular meshwork. Br J Ophthalmol 1973; 57:400.

377. Polansky JR et al. Trabecular meshwork cell culture in glaucoma research. Evaluation of biological activity and structural properties of human trabecular cells in vitro. Ophthalmology 1984; 91:580.

378. Alvarado JA, Murphy CG. Outflow obstruction in pigmentary and primary open angle glaucoma. Arch Ophthalmol 1992; 110:1769.

379. Kiland JA et al. Effect of age on outflow resistance washout during anterior chamber perfusion in rhesus and cynomolgus monkeys. Exp Eye Res 2005; 81:724.

380. Kaufman PL et al. Medical trabeculocanalotomy in monkeys with cytochalasin B or EDTA. Ann Ophthalmol 1979; 11:795.

381. Epstein DL. Open angle glaucoma. Why not a cure? [editorial]. Arch Ophthalmol 1987; 105:1187.

382. Campbell DG et al. Ghost cells as a cause of glaucoma. Am J Ophthalmol 1976; 81:441.

383. Goldberg MF. The diagnosis and treatment of sickled erythrocytes in human hyphemias. Trans Am Ophthalmol Soc 1978; 76:481.

384. Campbell DG, Essingman EM. Hemolytic ghost cell glaucoma. Further studies. Arch Ophthalmol 1979; 97:2141.

385. Flocks M et al. Phacolytic glaucoma. Clinicopathologic study of 138 cases of glaucoma associated with hypermature cataract. Arch Ophthalmol 1955; 54:37.

386. Fenton RH, Zimmerman LE. Hemolytic glaucoma. An unusual cause of acute open-angle secondary glaucoma. Arch Opthalmol 1963; 70:236.

387. Yanoff M. Glaucoma mechanisms in ocular malignant melanomas. Am J Ophthalmol 1970; 70:898.

388. Peterson HP. Can pigmentary deposits on the trabecular meshwork increase the resistance of the aqueous outflow? Acta Ophthalmol 1969; 47:743.

389. Quigley HA. Long-term follow-up of laser iridotomy. Ophthalmology 1981; 88:218.

390. Anderson DR. Experimental alpha chymotrypsin glaucoma studied by scanning electron microscopy. Am J Ophthalmol 1971; 71:470.

391. Worthen DM. Scanning electron microscopy after alpha chymotrypsin perfusion in man. Am J Ophthalmol 1972; 73:637.

392. Channell MM. Intraocular pressure changes after Neodynium-YAG laser posterior capsulotomy. Arch Ophthalmol 1984; 102:1024.

393. Ge J et al. Long-term effect of Nd:YAG laser posterior capsulotomy on intraocular pressure. Arch Ophthalmol 2000; 118:1334.

394. Nelson GA et al. Ocular amyloidosis and secondary glaucoma. Ophthalmol 1999; 106:1363.

395. Epstein DL et al. Obstruction of aqueous outflow by lens particles and by heavy-molecular weight soluble lens proteins. Invest Ophthalmol Vis Sci 1978; 17:272.

396. Epstein DL et al. Identification of heavy molecular weight soluble lens protein in aqueous humor in phakolytic glaucoma. Invest Ophthalmol Vis Sci 1978; 17:398.

397. Epstein DL et al. Serum obstruction of aqueous outflow in enucleated eyes. Am J Ophthalmol 1978; 86:101.

398. Davanger M. On the molecular composition and physiochemical properties of the pseudoexfoliation material. Acta Ophthalmol 1977; 55:621.

399. Johnson M et al. Serum proteins and aqueous outflow resistance in bovine eyes. Invest Ophthalmol Vis Sci 1993; 34:3549.

400. Sit AJ et al. The role of soluble proteins in generating aqueous outflow resistance in the bovine and human eye. Exp Eye Res 1997; 64:813.

401. Kee C et al. Serum effects on aqueous outflow during anterior chamber perfusion in monkeys. Invest Ophthalmol Vis Sci 1996; 37:1840.

402. Chrzanowska-Wodnicka M, Burridge K. Tyrosine phosphorylation is involved in reorganization of the actin cytoskeleton in response to serum or LPA stimulation. J Cell Sci 1994; 107:3643.

403. Seufferlein T, Rozengurt E. Lysophosphatidic acid stimulates tyrosine phosphorylation of focal adhesion kinase, paxillin, and p130. J Biol Chem 1994; 269:9345.

404. Dada VK et al. Postoperative intraocular pressure changes with use of different viscoelastics. Ophthal Surg 1994; 25:540.

405. Shibasaki H et al. Viscoelastic substance in the anterior chamber elevates intraocular pressure. Ann Ophthalmol 1994; 26:10.

406. Rohen JW et al. The relation between the ciliary muscle and the trabecular meshwork and its importance for the effect of miotics on aqueous outflow resistance. Albrecht von Graefes Arch Klin Exp Ophthalmol 1967; 172:23.

407. Lütjen-Drecoll E. Functional morphology of the trabecular meshwork in primate eyes. Prog Retinal Eye Res 1998; 18:91.

408. Armaly MF, Burian HM. Changes in the tonogram during accommodation. Arch Ophthalmol 1958; 60:60.

409. Armaly MF. Studies on intraocular effects of the orbital parasympathetics. II. Effects on intraocular pressure. Arch Ophthalmol 1959; 62:117.

410. Bárány EH, Rohen JW. Localized contraction and relaxation within the ciliary muscle of the vervet monkey (Cercopithecus ethiops). In: Rohen JW, ed. The structure of the eye. Second symposium. Stuttgart: Schattauer, 1965:287.

411. Kaufman PL, Gabelt BT. Cholinergic mechanisms and aqueous humor dynamics. In: Drance SM et al, eds. Pharmacology of glaucoma. Baltimore: Williams & Wilkins, 1992:64.

412. van Buskirk EM, Grant WM. Lens depression and aqueous outflow in enucleated primate eyes. Am J Ophthalmol 1973; 76:632.

413. Schimek RA, Lieberman WJ. The influence of Cyclogyl and Neosynephrine on tonographic studies of miotic control in open angle glaucoma. Am J Ophthalmol 1961; 51:781.

414. Bárány EH, Christensen RE. Cycloplegia and outflow resistance. Arch Ophthalmol 1967; 77:757.

415. Harris LS. Cycloplegic-induced intraocular pressure elevations. Arch Ophthalmol 1968; 79:242.

416. Bárány EH. The immediate effect on outflow resistance of intravenous pilocarpine in the vervet monkey. Invest Ophthalmol 1967; 6:373.

417. Bárány EH. The mode of action of pilocarpine on outflow resistance in the eye of a primate (Cercopithecus ethiops). Invest Ophthalmol 1962; 1:712.

418. Bárány EH. The mode of action of miotics on outflow resistance. A study of pilocarpine in the vervet monkey (Cercopithecus ethiops). Trans Ophthalmol Soc UK 1966; 86:539.

419. Shaffer RN. In: Newell FW, ed. Glaucoma: transactions of the fifth conference. New York: Josiah Macy Jr. Foundation, 1960:234.

420. Kaufman PL, Bárány EH. Residual pilocarpine effects on outflow facility after ciliary muscle disinsertion in the cynomolgus monkey. Invest Ophthalmol 1976; 15:558.

421. Kaufman PL, Bárány EH. Loss of acute pilocarpine effect on outflow facility following surgical disinsertion and retrodisplacement of the ciliary muscle from the scleral spur in the cynomolgus monkey. Invest Ophthalmol 1976; 15:793.

422. Selbach JM et al. Efferent and afferent innervation of primate trabecular meshwork and scleral spur. Invest Ophthalmol Vis Sci 2000; 41:2184.

423. Tamm ER et al. Nerve endings with structural characteristics of mechanoreceptors in the human scleral spur. Invest Ophthalmol Vis Sci 1994; 35:1157.

424. Gupta N et al. Localization of M3 muscarinic receptor subtype and mRNA in the human eye. Ophthalmic Res 1994; 26:207.

425. Shade DL et al. Effects of muscarinic agents on cultured human trabecular meshwork cells. Exp Eye Res 1996; 62:201.

426. de Kater AW et al. Localization of smooth muscle myosin-containing cells in the aqueous outflow pathway. Invest Ophthalmol Vis Sci 1990; 31:347.

427. de Kater AW et al. Localization of smooth muscle and nonmuscle actin isoforms in the human aqueous outflow pathway. Invest Ophthalmol Vis Sci 1992; 33:424.

428. Flügel C et al. Age-related loss of a-smooth muscle actin in normal and glaucomatous human trabecular meshwork of different age groups. J Glaucoma 1992; 1:165.

429. Lepple-Wienhues A et al. Differential smooth muscle-like contractile properties of trabecular meshwork and ciliary muscle. Exp Eye Res 1991; 53:33.

430. Lepple-Wienhues A et al. Endothelin-evoked contractions in bovine ciliary muscle and trabecular meshwork: interaction with calcium, nifedipine and nickel. Curr Eye Res 1991; 10:983.

431. Wiederholt M et al. Contractile response of the isolated trabecular meshwork and ciliary muscle to cholinergic and adrenergic agents. German J Ophthalmol 1996; 5:146.

432. Wiederholt M et al. Regulation of outflow rate and resistance in the perfused anterior segment of the bovine eye. Exp Eye Res 1995; 61:223.

433. Schroeder A, Erickson K. Low dose cholinergic agonists increase trabecular outflow facility in the human eye in vitro. Invest Ophthalmol Vis Sci 1995; 36(Suppl):S722.

434. Schroeder A, Erickson K. Cholinergic agonists do not increase trabecular outflow facility in the human eye. Invest Ophthalmol Vis Sci 1994; 34(Suppl):2054.

435. Kiland JA et al. Low doses of pilocarpine do not significantly increase outflow facility in the cynomolgus monkey. Exp Eye Res 2000; 70:603.

436. Zhang X et al. Expression of adenylate cyclase subtypes II and IV in the human outflow pathway. Invest Ophthalmol Vis Sci 2000; 41:998.

437. Tian B, Kaufman PL. Combined effects of H7 and pilocarpine on anterior segment physiology in monkey eyes. Curr Eye Res 2007; 32:491.

438. Gupta N et al. Muscarinic receptor M1 and M2 subtypes in the human eye. QNB, pirenzipine, oxotremorine, and AFDX-116 in vitro autoradiography. Br J Ophthalmol 1994; 78:555.

439. Zhang X et al. Expression of muscarinic receptor subtype mRNA in the human ciliary muscle. Invest Ophthalmol Vis Sci 1995; 36:1645.

440. Gabelt BT, Kaufman PL. Inhibition of outflow facility, accommodative, and miotic responses to pilocarpine in rhesus monkeys by muscarinic receptor subtype antagonists. J Pharmacol Exp Ther 1992; 263:1133.
441. Gabelt BT, Kaufman PL. Inhibition of aceclidine-stimulated outflow facility, accommodation and miosis by muscarinic receptor subtype antagonists in rhesus monkeys. Exp Eye Res 1994; 58:623.
442. Flügel C et al. Histochemical differences within the ciliary muscle and its function in accommodation. Exp Eye Res 1990; 50:219.
443. Kaufman PL, Bárány EH. Subsensitivity to pilocarpine in primate ciliary muscle following topical anticholinesterase treatment. Invest Ophthalmol 1975; 14:302.
444. Kaufman PL, Bárány EH. Subsensitivity to pilocarpine of the aqueous outflow system in monkey eyes after topical anticholinesterase treatment. Am J Ophthalmol 1976; 82:883.
445. Kaufman PL. Anticholinesterase-induced cholinergic subsensitivity in primate accommodative mechanism. Am J Ophthalmol 1978; 85:622.
446. Croft MA et al. Accommodation and ciliary muscle muscarinic receptors after echothiophate. Invest Ophthalmol Vis Sci 1991; 32:3288.
447. Bárány E. Pilocarpine-induced subsensitivity to carbachol and pilocarpine of ciliary muscle in vervet and cynomolgus monkeys. Acta Ophthalmol 1977; 55:141.
448. Bárány E et al. The binding properties of the muscarinic receptors of the cynomolgus monkey ciliary body and the response to the induction of agonist subsensitivity. Br J Pharmacol 1982; 77:731.
449. Bárány EH. Muscarinic subsensitivity without receptor change in monkey ciliary muscle. Br J Pharmacol 1985; 84:193.
450. Ballintine EJ, Garner LL. Improvement of the coefficient of outflow in glaucomatous eyes. Prolonged local treatment with epinephrine. Arch. Ophthalmol 1961; 66:314.
451. Krill AE et al. Early and long-term effects of levo-epinephrine on ocular tension and outflow. Am J Ophthalmol 1965; 59:833.
452. Sears ML. The mechanism of action of adrenergic drugs in glaucoma. Invest Ophthalmol 1966; 5:115.
453. Bill A. Early effects of epinephrine on aqueous humor dynamics in vervet monkeys (*Cercopithecus ethiops*). Exp Eye Res 1969; 8:35.
454. Camp JJ et al. Three-dimensional reconstruction of aqueous channels in human trabecular meshwork using light microscopy and confocal microscopy. Scanning 1997; 19:258.
455. Kaufman PL, Rentzhog L. Effect of total iridectomy on outflow facility responses to adrenergic drugs in cynomolgus monkeys. Exp Eye Res 1981; 33:65.
456. Kaufman PL, Bárány EH. Adrenergic drug effects on aqueous outflow facility following ciliary muscle retrodisplacement in the cynomolgus monkey. Invest Ophthalmol Vis Sci 1981; 20:644.
457. Anderson L, Wilson WS. Inhibition by indomethacin of the increased facility of outflow induced by adrenaline. Exp Eye Res 1990; 50:119.
458. Alvarado JA et al. Effect of b-adrenergic agonists on paracellular width and fluid flow across outflow pathway cells. Invest Ophthalmol Vis Sci 1998; 39:1813.
459. Wax MB et al. Characterization of beta-adrenergic receptors in cultured human trabecular cells and in human trabecular meshwork. Invest Ophthalmol Vis Sci 1989; 30:51.
460. Thomas JV, Epstein DL. Timolol and epinephrine in primary open angle glaucoma. Transient additive effect. Arch Ophthalmol 1981; 99:91.
461. Cyrlin MN et al. Additive effect of epinephrine to timolol therapy in primary open-angle glaucoma. Arch Ophthalmol 1982; 100:414.
462. Allen RC, Epstein DL. Additive effects of betaxolol and epinephrine in primary open angle glaucoma. Arch Ophthalmol 1986; 104:1178.
463. Allen RC et al. A double-masked comparison of betaxolol vs timolol in the treatment of open-angle glaucoma. Am J Ophthalmol 1986; 101:535.
464. Robinson JC, Kaufman PL. Effects and interactions of epinephrine, norepinephrine, timolol and betaxolol on outflow facility in the cynomolgus monkey. Am J Ophthalmol 1990; 109:189.
465. Kaufman PL. Adenosine 3′,5′ cyclic monophosphate and outflow facility in monkey eyes with intact and retrodisplaced ciliary muscle. Exp Eye Res 1987; 44:415.
466. Neufeld AH, Sears ML. Adenosine 3′,5′-monophosphate analogue increases the outflow facility of the primate eye. Invest Ophthalmol 1975; 14:688.
467. Neufeld AH et al. Cyclic-AMP in the aqueous humor: the effects of adrenergic agonists. Exp Eye Res 1972; 14:242.
468. Neufeld AH. Influences of cyclic nucleotides on outflow facility in the vervet monkey. Exp Eye Res 1978; 27:387.
469. Erickson-Lamy KA, Nathanson JA. Epinephrine increases facility of outflow and cyclic AMP content in the human eye in vitro. Invest Ophthalmol Vis Sci 1992; 33:2672.
470. Crawford KS et al. Indomethacin and epinephrine effects on outflow facility and cAMP formation in monkeys. Invest Ophthalmol Vis Sci 1996; 37:1348.
471. Erickson K et al. Adrenergic regulation of aqueous outflow. J Ocular Pharmacol 1994; 10:241.
472. Crosson CE, Petrovich M. Contributions of adenosine receptor activation to the ocular actions of epinephrine. Invest Ophthalmol Vis Sci 1999; 40:2054.
473. Ehinger B. A comparative study of the adrenergic nerves to the anterior segment of some primates. Zeitschr Zellforschung 1971; 116:157.
474. Nomura T, Smelser GK. The identification of adrenergic and cholinergic nerve endings in the trabecular meshwork. Invest Ophthalmol 1974; 13:525.
475. Ruskell GL. The source of nerve fibres of the trabeculae and adjacent structures in monkey eyes. Exp Eye Res 1976; 23:449.
476. Wiederholt M. Direct involvement of trabecular meshwork in the regulation of aqueous humor outflow. Curr Opin Opthalmol 1998; 9:46.
477. Robinson JC, Kaufman PL. Cytochalasin B potentiates epinephrine's outflow facility increasing effect. Invest Ophthalmol Vis Sci 1991; 32:1614.
478. Robinson JC, Kaufman PL. Phalloidin inhibits epinephrine's and cytochalsin B's facilitation of aqueous outflow. Arch Ophthalmol 1994; 112:1610.
479. Tripathi BJ, Tripathi RC. Effect of epinephrine in vitro on the morphology, phagocytosis, and mitotic activity of human trabecular endothelium. Exp Eye Res 1984; 39:731.
480. Alvarado JA et al. Epinephrine effects on major cell types of the aqueous outflow pathway. In vitro studies/clinical implications. Trans Am Ophth Soc 1990; 88:267.
481. Coca-Prados M. Regulation of protein phosphorylation of the intermediate-sized filament vimentin in the ciliary epithelium of the mammalian eye. J Biol Chem 1985; 260:10332.
482. Törnqvist G. Effect of cervical sympathetic stimulation on accommodation in monkeys. An example of a beta-adrenergic inhibitory effect. Acta Physiol Scand 1966; 67:363.
483. van Alphen GWHM et al. Drug effects on ciliary muscle and choroid preparations in vitro. Arch Ophthalmol 1962; 68:111.
484. van Alphen GWHM et al. Adrenergic receptors of the intraocular muscles. Comparison to cat rabbit, and monkey. Arch Ophthalmol 1965; 74:253.
485. Casey WJ. Cervical sympathetic stimulation in monkeys and the effects on outflow facility and intraocular volume. A study in the East African vervet (*Cercopithecus aethiops*). Invest Ophthalmol 1966; 5:33.
486. Camras CB et al. Inhibition of the epinephrine-induced reduction of intraocular pressure by systemic indomethacin in humans. Am J Ophthalmol 1985; 100:169.
487. Hoskins HD, Kass MA. Adrenergic agonists. In: Hoskins HD, Kass MA, eds. Becker-Shaffer's Diagnosis and therapy of the glaucomas, 6th edn. St. Louis: CV Mosby, 1989:435.
488. Gilmartin B et al. The effect of timolol maleate on tonic accommodation, tonic vergence, and pupil diameter. Invest Ophthalmol Vis Sci 1984; 25:763.
489. Toris CB et al. Effects of brimonidine on aqueous humor dynamics in human eyes. Arch Ophthalmol 1995; 113:1514.
490. Toris CB et al. Effects of apraclonidine on aqueous humor dynamics in human eyes. Ophthalmology 1995;102:456.
491. Kobayashi H et al. Effficacy of bunazosin hydrochloride 0.01 percent as adjunctive therapy of latanoprost or timolol. J Glauc 2004; 13:73.
492. Akaishi T et al. Effects of bunazosin hydrochloride on ciliary muscle constriction and matrix metalloproeinase activities. J Glauc 2004; 13:312.
493. Geiger B et al. Molecular interactions in the submembrane plaque of cell-cell and cell-matrix adhesions. Acta Anat 1995; 154:46.
494. Yamada KM, Geiger B. Molecular interactions in cell adhesion complexes. Curr Opin Cell Biol 1997; 9:75.
495. Liu X et al. Gene therapy targeting glaucoma: where are we? Surv Ophthalmol 2009
496. Kaufman PL, Bárány EH. Cytochalasin B reversibly increases outflow facility in the eye of the cynomolgus monkey. Invest. Ophthalmol Vis Sci 1977; 16:47.
497. Johnstone M et al. Concentration-dependent morphologic effects of cytochalasin B in the aqueous outflow system. Invest Ophthalmol Vis Sci 1980; 19:835.
498. Kaufman PL et al. Effect of cytochalasin B on conventional drainage of aqueous humor in the cynomolgus monkey. In: Bito LZ et al, eds. The ocular and cerebrospinal fluids. Fogarty International Center Symposium. Exp Eye Res 1977; 25(Suppl):411.
499. Cai S et al. Effect of latrunculin-A on morphology and actin-associated adhesions of cultured human trabecular meshwork cells. Mol Vis 2000; 6:132.
500. Epstein DL et al. Acto-myosin drug effects and aqueous outflow function. Invest Ophthalmol Vis Sci 1999; 40:74.
501. Peterson JA et al. Latrunculin-A increases outflow facility in the monkey. Invest Ophthalmol Vis Sci 1999; 40:931.
502. Peterson JA et al. Effect of latrunculin-B on outflow facility in monkeys. Exp Eye Res 2000; 70:307.
503. Ethier CR et al. Effects of latrunculin-B on outflow facility and trabecular meshwork structure in human eyes. Invest Ophthalmol Vis Sci 2006; 47:1991.
504. Okka M et al. Effect of low-dose latrunculin B on anterior segment physiologic features in the monkey eye. Arch Ophthalmol 2004; 122:1482.
505. Sabanay I et al. Latrunculin B effects on trabecular meshwork and corneal endothelial morphology in monkeys. Exp Eye Res 2006; 82:236.
506. Bershadsky A et al. Involvement of microtubules in the control of adhesion-dependent signal transduction. Curr Biol 1996; 6:1279.
507. Liu X et al. Effect of H-7 on cultured human trabecular meshwork cells. Mol Vis 2001; 7:145.
508. Tian B et al. H-7 disrupts the actin cytoskeleton and increases outflow facility. Arch Opthalmol 1998; 116:633.
509. Volberg T et al. Effect of protein kinase inhibitor H-7 on the contractility, integrity and membrane anchorage of the microfilament system. Cell Motility Cytoskel 1994; 29:321.
510. Tian B et al. H-7 increases trabecular facility and facility after ciliary muscle disinsertion in monkeys. Invest Ophthalmol Vis Sci 1999; 40:239.
511. Bahler CK et al. Pharmacologic disruption of Schlemm's canal cells and outflow facility in anterior segments of human eyes. Invest Ophthalmol Vis Sci 2004; 45:2246.
512. Hu Y et al. Monkey organ-cultured anterior segments; technique and response to H-7. Exp Eye Res 2006; 82:1100.
513. Honjo M et al. Effects of rho-associated protein kinase inhibitor Y-27632 on intraocular pressure and outflow facility. Invest Ophthalmol Vis Sci 2001; 42:137.
514. Rao PV et al. Modulation of aqueous humor outflow facility by the Rho kinase-specific inhibitor Y-27632. Invest Ophthalmol Vis Sci 2001; 42:1029.
515. Renieri G et al. Effects of endothelin-1 on calcium-independent contraction of bovine trabecular meshwork. Graefe's Arch Clin Exp Ophthalmol 2008; 246:1107.
516. Rosenthal R et al. Effects of ML-7 and Y27632 on carbachol- and endothelin-1-induced contraction of bovine trabecular meshwork. Exp Eye Res 2005; 80:837.
517. Thieme H et al. Mediation of calcium-independent contraction in trabecular meshwork through protein kinase C and rho-A. Invest Ophthalmol Vis Sci 2000; 41:4240.
518. Honjo M et al. A myosin light chain kinase inhibitor, ML-9, lowers the intraocular pressure in rabbit eyes. Exp Eye Res 2002; 75:135.
519. Rao PV et al. Regulation of myosin light chain phosphorylation in the trabecular meshwork: role in aqueous humour outflow facility. Exp Eye Res 2005; 80:197.
520. Tokushige H et al. Effects of topical administration of Y-39983, a selective rho-associated protein kinase inhibitor, on ocular tissues in rabbits and monkeys. Invest Ophthalmol Vis Sci 2007; 48:3216.
521. Waki M et al. Reduction of intraocular pressure by topical administration of an inhibitor of the rho-associated protein kinase. Curr Eye Res 2001; 22:470.
522. Lu Z et al. The mechanism of Rho-kinase inhibitor, Y27632, on outflow facility in monkey vs human eyes. Invest Ophthalmol Vis Sci 2007; Abnr 1146.
523. Tanihara H et al. Intraocular pressure-lowering effects and safety of topical administration of a selective ROCK inhibitor, SNJ-1656, in healthy volunteers. Arch Ophthalmol 2008; 126:309.

524. Grosheva I et al. Caldesmon effects on the actin cytoskeleton and cell adhesion in cultured HTM cells. Exp Eye Res 2006; 82:945.
525. Liu X et al. The effects of C3 transgene expression on actin and cellular adhesions in cultured human trabecular meshwork cells and on outflow facility in organ cultured monkey eyes. Mol Vis 2005; 11:1112.
526. Gabelt BT et al. Caldesmon transgene expression disrupts focal adhesions in HTM cells and increases outflow facility in organ-cultured human and monkey anterior segments. Exp Eye Res 2006; 82:935.
527. Rao PV et al. Expression of dominant negative Rho-binding domain of Rho-kinase in organ cultured human eye anterior segments increases aqueous humor outflow. Mol Vis 2005; 11:288.
528. Santas AJ et al. Effect of heparin II domain of fibronectin on aqueous outflow in cultured anterior segments of human eyes. Invest Ophthalmol Vis Sci 2003; 44:4796.
529. Gonzalez JM, Jr. et al. Effect of heparin II domain of fibronectin on actin cytoskeleton and adherens juctions in human trabecular meshwork cells. Invest Ophthalmol Vis Sci 2006; 47:2924.
530. Gonzalez JM, Jr. et al. Identification of the active site in the heparin II domain of fibronectin that increases outflow facility in cultured monkey anterior segments. Invest Ophthalmol Vis Sci 2009; 50:235.
531. Tian B et al. Cytoskeletal involvement in the regulation of aqueous humor outflow. Invest Ophthalmol Vis Sci 2000; 41:619.
532. Pitzer Gills J et al. Microtubulte disruption leads to cellular contraction in human trabecular meshwork cells. Invest Ophthalmol Vis Sci 1998; 39:653.
533. Erickson-Lamy KA et al. Ethacrynic acid induces reversible shape and cytoskeletal changes in cultured cells. Invest Ophthalmol Vis Sci 1992; 33:2631.
534. O'Brien ET et al. A mechanism for trabecular meshwork cell retraction. Ethacrynic acid initiates the dephosphorylation of focal adhesion proteins. Exp Eye Res 1997; 65:471.
535. Croft MA et al. Effect of ethacrynic acid on aqueous outflow dynamics in monkeys. Invest Ophthalmol Vis Sci 1994; 35:1167.
536. Melamed S et al. The effect of intracamerally injected ethacrynic acid on intraocular pressure in patients with glaucoma. Am J Ophthalmol 1992; 113:508.
537. Epstein DL et al. Influence of ethacrynic acid on outflow facility in the monkey and calf eye. Invest Ophthalmol Vis Sci 1987; 28:2067.
538. Liang L-L et al. Ethacrynic acid increases facility of outflow in the human eye in vitro. Arch Ophthalmol 1992; 110:106.
539. Johnson DH, Tschumper RC. Ethacrynic acid: outflow effects and toxicity in human trabecular meshwork in perfusion organ culture. Curr Eye Res 1993; 12:385.
540. Shimazaki A et al. Effects of the new ethacrynic acid derivative SA9000 on intraocular pressure in cats and monkeys. Biol Pharm Bull 2004; 27:1019.
541. Shimazaki A et al. Effects of the new ethacrynic acid oxime derivative SA12590 on intraocular pressure in cats and monkeys. Biol Pharm Bull 2007; 30:1445.
542. Rao PV et al. Effects of novel ethacrynic acid derivatives on human trabecular meshwork cell shape, actin cytoskeletal organization, and transcelular fluid flow. Biol Pharm Bull 2005; 28:2189.
543. Bill A et al. Effects of intracameral Na_2EDTA and EGTA on aqueous outflow routes in the monkey eye. Invest Ophthalmol Vis Sci 1980; 19:492.
544. Rodewald R et al. Contraction of isolated brush borders from the intestinal epithelium. J Cell Biol 1976; 70:541.
545. Erickson KA et al. Verapamil increases outflow facility in the human eye. Exp Eye Res 1995; 61:565.
546. Weinreb RN et al. Detection of glucocorticoid receptors in cultured human trabecular cells. Invest Ophthalmol Vis Sci 1981; 21:403.
547. Rauz S et al. Inhibition of 11b-hydroxysteroid dehydrogenase type 1 lowers intraocular pressure in patients with ocular hypertension. Q J Med 2003;96:481.
548. Rauz S et al. Expression and putative role of 11b-hydroxysteroid dehydrogenase isozymes within the human eye. Invest Ophthalmol Vis Sci 2001; 42:2037.
549. Stokes J et al. Distribution of glucocorticoid and mineralocorticoid receptors and 11b-hydroxysteroid dehydrogenases in human and rat ocular tissues. Invest Ophthalmol Vis Sci 2000; 41:1629.
550. Whorwood CB et al. Regulation of sodium-potassium adenosine triphosphate subunit gene expression by corticosteroids and 11b-hydroxysteroid dehydrogenase. Endocrinology 1994; 135:901.
551. Walker EA, Stewart PM. 11b-Hydroxysteroid dehydrogenase: unexpected connections. Trans Endocrin Metab 2003; 14:334.
552. Armaly MF. Effect of corticosteroids on intraocular pressure and fluid dynamics. I. The effect of dexamethasone in the normal eye. Arch Ophthalmol 1963; 70:482.
553. Becker B. Intraocular pressure response to topical corticosteroids. Invest Ophthalmol Vis Sci 1965; 4:198.
554. Armaly MF. Effect of corticosteroids on intraocular pressure and fluid dynamics. II. The effect of dexamethasone in the glaucomatous eye. Arch Ophthalmol 1963; 70:492.
555. Gerometta R et al. Steroid-induced ocular hypertension in normal cattle. Arch Ophthalmol 2004; 122:1492.
556. Schwartz B, Levene RZ. Plasma cortisol differences between normal and glaucomatous patients before and after dexamethasone suppression. Arch Ophthalmol 1972; 87:369.
557. Schwartz B et al. Increased plasma free cortisol in ocular hypertension and open-angle glaucoma. Arch Ophthalmol 1987; 105:1060.
558. Stokes J et al. Altered peripheral sensitivity to glucocorticoids in primary open-angle glacuoma. Invest Ophthalmol Vis Sci 2003; 44:5163.
559. Levi L, Schwartz B. Decrease of ocular pressure with oral metyrapone. A double masked crossover trial. Arch Ophthalmol 1987; 105:777.
560. Robin AL et al. Anterior juxtascleral delivery of anecortave acetate in eyes with primary open-angle glaucoma: a pilot investigation. Am J Ophthalmol 2009; 147:45.
561. Weinreb RN et al. Acute effects of dexamethasone on intraocular pressure in glaucoma. Invest Ophthalmol Vis Sci 1985; 26:170.
562. Tripathi BJ et al. Corticosteroids induce a sialated glycoprotein (Cort-GP) in trabecular cells in vitro. Exp Eye Res 1990; 51:735.
563. Steely HT et al. The effects of dexamethasone on fibronectin expression in cultured trabecular meshwork cells. Invest Ophthalmol Vis Sci 1992; 33:2242.
564. Shirato S et al. Kinetics of phagocytosis in trabecular meshwork cells: flow cytometry and morphometry. Invest Ophthalmol Vis Sci 1989; 30:2499.
565. Polansky JR et al. In vitro correlates of glucocorticoid effects on intraocular pressure. In: Krieglestein GK, ed. Glaucoma update IV. Berlin: Springer-Verlag, 1991:20.
566. Patridge CA et al. Dexamethasone induces specific proteins in human trabecular meshwork cells. Invest Ophthalmol Vis Sci 1989; 30:1843.
567. Polansky JR et al. Eicosanoid production and glucocorticoid regulatory mechanisms in cultured human trabecular meshwork cells. In: Bito LZ, Stjernschantz J, eds. The ocular effects of prostaglandins and other eicosanoids. New York: Alan R. Liss, 1989:113.
568. Yun AJ et al. Proteins secreted by human trabecular cells. Glucocorticoid and other effects. Invest Ophthalmol Vis Sci 1989; 30:2012.
569. Southren AL et al. Treatment of glaucoma with 3a,5b-tetrahydrocortisol. A new therapeutic modality. J Ocular Pharmacol 1994; 10:385.
570. Clark AF et al. Inhibition of dexamethasone-induced cytoskeletal changes in cultured human trabecular meshwork cells by tetrahydrocortisol. Invest Ophthalmol Vis Sci 1996; 37:805.
571. Seeman J et al. 3 alpha, 5 beta-tetrahydrocortisol effect on outflow facility. J Ocular Pharmacol Ther 2002; 18:35.
572. Rohen JW et al. Electron microscopic studies on the trabecular meshwork in two cases of corticosteroid glaucoma. Exp Eye Res 1973; 17:19.
573. Johnson D et al. Ultrastructural changes in the trabecular meshwork of human eyes treated with corticosteroids. Arch Ophthalmol 1997; 115:375.
574. Clark AF et al. Dexamethasone-induced ocular hypertension in perfusion-cultured human eyes. Invest Ophthalmol Vis Sci 1995; 36:478.
575. Dickerson JEJ et al. The effect of dexamethasone on integrin and laminin expression in cultured human trabecular meshwork cells. Exp Eye Res 1998; 66:731.
576. Tane N et al. Effect of excess synthesis of extracellular matrix components by trabecular meshwork cells: possible consequence on aqueous outflow. Exp Eye Res 2007; 84:832.
577. Engelbrecht-Schnur S et al. Dexamethasone treatment decreases hyaluronan formation by primate trabecular meshwork cells in vitro. Exp Eye Res 1997; 64:539.
578. Clark AF, Wordinger RJ. The role of steroids in outflow resistance. Exp Eye Res [Epub ahead of print] 2008
579. Underwood JL et al. Glucocorticoids regulate transendothelial fluid flow resistance and formation of intercellular junctions. Am J Physiol 1999; 277:C330.
580. O'Brien ET et al. Dexamethasone inhibits trabecular cell retraction. Exp Eye Res 1996; 62:675.
581. Wilson K et al. Dexamethasone induced ultrastructural changes in cultured human trabecular meshwork cells. Exp Eye Res 1993; 12:783.
582. Clark AF et al. Glucocorticoid-induced formation of cross-linked actin networks in cultured human trabecular meshwork cells. Invest Ophthalmol Vis Sci 1994; 35:281.
583. Clark AF et al. Dexamethasone alters F-actin architecture and promotes cross-linked actin network formation in human trabecular meshwork tissue. Cell Motil Cytoskel 2005; 60:83.
584. Clark AF et al. Cytoskeletal changes in cultured human glaucoma trabecular meshwork cells. J Glauc 1995; 4:183.
585. Hoare MJ et al. Cross-linked actin networks (CLANs) in the trabecular meshowrk of the normal and glaucomatous human eye in situ. Invest Ophthalmol Vis Sci 2009; 50:1255.
586. Filla MS et al. Beta1 and beta3 integrins cooperate to inducesyndecan-4-containing cross-linked actin networks in human trabecular meshwork cells. Invest Ophthalmol Vis Sci 2006; 47:1956.
587. Putney LK et al. Effects of dexamethasone on sodium-potassium-chloride cotransport in trabeuclar meshwork cells. Invest Ophthalmol Vis Sci 1997; 38:1229.
588. Fingert JH et al. Myocilin glaucoma. Surv Ophthalmol 2002; 47:547.
589. Stone EM et al. Identification of a gene that causes primary open angle glaucoma. Science 1997; 275:668.
590. Alward WL et al. Clinical features associated with mutations in the chromosome 1 open-angle glaucoma gene. N Engl J Med 1998; 338:1022.
591. Turalba AV, Chen TC. Clinical and genetic characteristics of primary juvenile-onset open-angle glaucoma (JOAG). Semin Ophthalmol 2008; 23:19.
592. Liu Y, Vollrath D. Reversal of mutant myocilin non-secretion and cell killing: implications for glaucoma. Hum Mol Genet 2004; 13:1193.
593. Fautsch MP et al. Recombinant TIGR/MYOC increases outflow resistance in the human anterior segment. Invest Ophthmol Vis Sci 2000; 41:4163.
594. Khare PD et al. Durable, safe, multi-gene lentiviral vector expression in feline trabecular meshwork. Mol Ther 2008; 16:97.
595. Borrás T et al. Effects of elevated intraocular pressure on outflow facility and TIGR/MYOC expression in perfused human anterior segments. Invest Ophthalmol Vis Sci 2002; 43:33.
596. Caballero M et al. Altered secretion of a TIGR/MYOC mutant lacking the olfactomedin domain. Biochim Biophys Acta 2000; 1502:447.
597. Polansky JR et al. Regulation of TIGR/MYOC gene expression in human trabecular meshwork cells. Eye 2000; 14(Pt 38):503.
598. Ueda J et al. Ultrastructural localization of myocilin in human trabecular meshwork cells and tissues. J Histochem Cytochem 48:1321.
599. Nguyen TD et al. Glucocorticoid (GC) effects on HTM cells. Molecular biology approaches. In: Lütjen-Drecoll E, ed. Basic aspects of glaucoma research III. Stuttgart: Schattauer, 1993:331.
600. Filla MS et al. In vitro localization of TIGR/MYOC in trabecular meshwork extracellular matrix and binding to fibronectin. Invest Ophthalmol Vis Sci 2002 2000; 43:151.
601. Johnson DH. Myocilin and glaucoma. A TIGR by the tail. Arch Ophthalmol 2000; 118:974.
602. Tamm ER et al. Modulation of myocilin/TIGR expression in human trabecular meshwork. Invest Ophthalmol Vis Sci 1999; 40:2577.
603. Poyer JF et al. Prostaglandin F2a effects on isolated rhesus monkey ciliary muscle. Invest Ophthalmol Vis Sci 1995; 36:2461.
604. Tamm E et al. Elektronenmikroskopische und immunhistochemische Untersuchungen zur augendrucksenkenden Wirkung von Prostaglandin F_{2a}. Fortschr Ophthalmol 1990; 87:623.
605. Sagara T et al. Topical prostaglandin F_{2a} treatment reduces collagen types I, III, and IV in the monkey uveoscleral outflow pathway. Arch Ophthalmol 1999; 117:794.
606. Lindsey JD et al. Prostaglandins increase proMMP-1 and proMMP-3 secretion by human ciliary smooth muscle cells. Curr Eye Res 1996; 15:869.
607. Lindsey JD et al. Prostaglandin action on ciliary smooth muscle extracellular matrix metabolism - implications for uveoscleral outflow. Surv Ophthalmol 1997; 41(Suppl 2):S53.

608. Lindsey JD et al. Induction of c-fos by prostaglandin F2 alpha in human ciliary smooth muscle cells. Invest Ophthalmol Vis Sci 1994; 35:242.
609. Ocklind A. Effect of latanoprost on the extracellular matrix of the ciliary muscle. A study on cultured cells and tissue sections. Exp Eye Res 1998; 67:179.
610. Yousufzai SYK, Abdel-Latif AA. Prostaglandin F2α and its analogs induce release of endogenous prostaglandins in iris and ciliary muscles isolated from cat and other mammalian species. Exp Eye Res 1996; 63:305.
611. Zhao X et al. Effects of prostaglandin analogues on human ciliary muscle and trabecular meshwork cells. Invest Ophthalmol Vis Sci 2003; 44:1945.
612. Richter M et al. Morphological changes in the anterior eye segment after long-term treatment with different receptor selective prostaglandin agonists and a prostamide. Invest Ophthalmol Vis Sci 2003; 44:4419.
613. Weinreb RN. Enhancement of scleral macromolecular permeability with prostaglandins. Tr Am Ophth Soc 2001; 99:319.
614. Anthony TL et al. Prostaglandin F2α receptors in the human trabecular meshwork. Invest Ophthalmol Vis Sci 1998; 39:315.
615. Kim J-W et al. Increased human scleral permeability with prostaglandin exposure. Invest Ophthalmol Vis Sci 2001; 42:1514.
616. Weinreb RN et al. Prostaglandin FP agonists alter metalloproteinase gene expression in sclera. Invest Ophthalmol Vis Sci 2004; 45:4368.
617. Lindsey JD et al. Direct matrix metalloproteinase enhancement of transscleral permeability. Invest Ophthalmol Vis Sci 2007; 48:752.
618. Ota T et al. The effects of prostaglandin analogues on IOP in prostanoid FP-receptor-deficient mice. Invest Ophthalmol Vis Sci 2005; 46:4159.
618a JG Crowston et al. 2008 missing – author, please supply
618b ARVO EAbstract 1551 missing – author, please supply
619. Ota T et al. The effects of prostaglandin analoques on prostaoid EP1, EP2, and EP3 receptor-deficient mice. Invest Ophthalmol Vis Sci 2006; 47:3395.
620. Sharif NA et al. Ocular hypotensive FP prostaglandin (PG) analogs. PG receptor subtype binding affinities and selectivities, and agonist potencies at FP and other PG receptors in cultured cells. J Ocul Pharmacol Ther 2003; 19:501.
621. Abramovitz M et al. The utilization of recombinant prostanoid receptors to determine the affinities annd selectivities of prostaglandins and related analogs. Biochim Biophys Acta 2000; 1483:285.
622. Biswas S et al. Prostaglandin E2 receptor subtypes, EP1, EP2, EP3 and EP4 in human and mouse ocular tissues - a comparative immunohistochemical study. Prostagland Leukot Essent Fatty Acids 2004; 71:277.
623. Nilsson SF et al. The prostanoid EP2 receptor agonist butaprost increases uveoscleral outflow in the cynomolgus monkey. Invest Ophthalmol Vis Sci 2006; 47:4042.
624. Gabelt BT et al. Prostaglandin subtype-selective and non-selective IOP lowering comparison in monkeys. J Ocul Pharmacol 2009; 25:1.
625. Giuffré G. The effects of prostaglandin F_{2a} in the human eye. Graefe's Arch Clin Exp Ophthalmol 1985; 222:139.
626. Villumsen J, Alm A. Prostaglandin F_{2a}-isopropylester eye drops. Effects in normal human eyes. Br J Ophthalmol 1989; 73:419.
627. Camras CB et al. Latanoprost, a prostaglandin analog, for glaucoma therapy: efficacy and safety after 1 year of treatment in 198 patients. Ophthalmol 1996; 103:1916.
628. O'Donnell ME. Role of Na-K-Cl contransport in vascular endothelial cell volume regulation. Am J Physiol 1993; 264:C1316.
629. Haas M. The Na-K-Cl cotransporters. Am J Physiol 1994; 267:C869.
630. O'Grady SM et al. Characteristics and functions of Na-K-Cl cotransport in epithelial tissues. Am J Physiol 1987; 253:C177.
631. O'Donnell ME et al. Na-K-Cl cotransport regulates intracellular volume and monolayer permeablility of trabecular meshwork cells. Am J Physiol 1995; 268:C1067.
632. Al-Aswad LA et al. Effects of Na-K-2Cl cotransport regulators on outflow facility in calf and human eyes in vitro. Invest Ophthalmol Vis Sci 1999; 40:1695.
633. Stamer WD et al. Localization of aquaporin CHIP in the human eye: implications in the pathogenesis of glaucoma and other cisorders of ocular fluid balance. Invest Ophthalmol Vis Sci 1994; 35:3867.
634. Stamer WD et al. Cultured human trabecular meshwork cells express aquaporin-1 water channels. Curr Eye Res 1995; 14:1095.
635. Stamer WD et al. Expression of aquaporin-1 in human trabecular meshwork cells: role in resting cell volume. Invest Ophthalmol Vis Sci 2001; 42:1803.
636. Stamer WD et al. Aquaporin-1 expression and conventional aqueous outflow in human eyes. Exp Eye Res 2008; 87:349.
637. Comes N et al. Identification and functional characterization of ClC-2 chloride channels in trabecular meshwork cells. Exp Eye Res 2006; 83:877.
638. Knepper PA et al. Glycosaminoglycans of the human trabecular meshwork in primary open-angle gluacoma. Invest Ophthalmol Vis Sci 1996; 37:1360.
639. Knepper PA et al. Glycosaminoglycan stratification of the juxtacanalicular tissue in normal and primary open-angle glaucoma. Invest Ophthalmol Vis Sci 1996; 37:2414.
640. Lerner LE et al. Hyaluronan in the human trabecular meshwork. Invest Ophthalmol Vis Sci 1997; 38:1222.
641. Knepper PA et al. Aqueous humor in primary open-angle glaucoma contains an increased level of CD44S. Invest Ophthalmol Vis Sci 2002; 43:133.
642. Bárány EH, Scotchbrook S. Influence of testicular hyaluronidase on the resistance to flow through the angle of the anterior chamber. Acta Physiol Scand 1954; 30:240.
643. Francois J et al. Perfusion studies on the outflow of aqueous humor in human eyes. Arch Ophthalmol 1956; 55:193.
644. Pedler WS. The relationship of hyaluronidase to aqueous outflow resistance. Trans Ophthalmol Soc UK 1956; 76:51.
645. Grant WM. Experimental aqueous perfusion in enucleated human eyes. Arch Ophthalmol 1963; 69:783.
646. Peterson WS, Jocson VL. Hyaluronidase effects on aqueous outflow resistance. Quantitative and localizing studies in the rhesus monkey eye. Am J Ophthalmol 1974; 77:573.
647. Hubbard WC et al. Intraocular pressure and outflow facility are unchanged following acute and chronic intracameral chondroitinase ABC and hyaluronidase in monkeys. Exp Eye Res 1997; 65:177.
648. Rees D et al. Control of grip and stick in cell adhesion through lateral relationships of membrane glycoproteins. Nature 1977; 267:124.
649. Tokiwa T et al. Mechanism of cell dissociation with trypsin and EDTA. Acta Med Okayama 1979; 33:1.
650. Bill A. Effects of Na_2EDTA and alpha-chymotrypsin on aqueous humor outflow conductance in monkey eyes. Uppsala J Med Sci 1980; 85:311.
651. Alexander JP et al. Growth factor and cytokine modulation of trabecular meshwork matrix metalloproteinase and TIMP expression. Curr Eye Res 1998; 17:276.
652. Bradley JM et al. Effect of matrix metalloproteinases activity on outflow in perfused human organ culture. Invest Ophthalmol Vis Sci 1998; 39:2649.
653. Acott TS, Kelley MJ. Extracellular matrix in the trabecular meshwork. Exp Eye Res 2008; 86:543.
654. Wiederholt M et al. Relaxation of trabecular meshwork and ciliary muscle by release of nitric oxide. Invest Ophthalmol Vis Sci 1994; 35:2515.
655. Nathanson JA, McKee M. Alterations of ocular nitric oxide synthase in human glaucoma. Invest Ophthalmol Vis Sci 1995; 36:1774.
656. Chen Z et al. Histochemical mapping of NADPH-diaphorase in monkey and human eyes. Curr Eye Res 1998; 17:370.
657. Crosson CE et al. Evidence for multiple P2Y receptors in trabecular meshwork cells. J Pharmacol Exp Ther 2004; 309:484.
658. Daines BS et al. Intraocular adenosine levels in normal and ocular-hypertensive patients. J Ocul Pharmacol Ther 2003;19:113.
659. Danser AH et al. Vitreous level of antiotensin-II in patients with diabetic retinopathy. Invest Ophthalmol Vis Sci 1994; 35:1008.
660. Wallow IH et al. Ocular renin angiotensin. EM immunocytochemical localization of prorenin. Curr Eye Res 1993; 12:945.
661. Meyer P et al. Local action of the rennin angiotensin system in the porcine opthalmic circulation: effects of ACE-inhibitors and angiotensin receptor antagonists. Invest Ophthalmol Vis Sci 1995; 36:555.
662. Sramek SJ et al. An ocular renin-angiotensin system. Immunohistochemistry of angiotensinogen. Invest Ophthalmol Vis Sci 1992; 33:1627.
663. Cullinane AB et al. Renin-angiotensin system expression and secretory function in cultured human ciliary body nonpigmented epithelium. Br J Ophthalmol 2002; 86:676.
664. Costagliola C et al. Effect of Losartan potassium oral administration on intraocular pressure in humans. Clin Drug Invest 1999; 29:329.
665. Costagliola C et al. Effect of oral losartan potassium administration on intraocular pressure in normotensive and glaucomatous human subjects. Exp Eye Res 2000; 71:167.
666. Costagliola C et al. Effect of oral captopril (SQ 14225) on intraocular pressure in man. Eur J Ophthalmol 1995; 5:19.
667. Constad WH et al. Use of an angiotensin converting enzyme inhibitor in ocular hypertension and primary open-angle glaucoma. Am J Ophthalmol 1988; 105:674.
668. Giardina WJ et al. Intraocular presuure lowering effects of the renin inhibitor ABBOTT-64662 diacetate in animals. J Ocular Pharmacology 1990; 6:75.
669. Wang R-F et al. Effect of CS-088, an angiotensin AT1 receptor antagonist, on intraocular pressure in glaucomatous monkey eyes. Exp Eye Res 2005; 80:629.
670. Lotti VJ, Pawlowski N. Prostaglandins mediate the ocular hypotensive action of the angiotensin converting enzyme inhibitor MK-422 (enalaprilat) in African green monkeys. J Ocul Pharmacol 1990; 6:1.
671. Wilensky JT, Jampol LM. Laser therapy for open-angle glaucoma. Ophthalmology 1981; 88:213.
672. Wise JB. Long-term control of adult open-angle glaucoma by argon laser treatment. Ophthalmology 1981; 88:197.
673. Wickham MG, Worthen DM. Argon laser trabeculotomy. Long-term follow-up. Ophthalmology 1979; 86:495.
674. Parshley DE et al. Laser trabeculoplasty induces stromelysin expression by trabecular juxtacanalicular cells. Invest Ophthalmol Vis Sci 1996; 37:795.
675. Parshley DE et al. Early changes in matrix metalloproteinases and inhibitors after in vitro laser treatment to the trabecular meshwork. Curr Eye Res 1995; 14:537.
676. Lewis RA et al. Canaloplasty: circumferential viscodilation and tensioning of Schlemm's canal using a flexible microcatheter for the treatment of open-angle glaucoma in adults: interim clinical study analysis. J Cataract Refract Surg 2007; 33:1217.
677. Shingleton B et al. Circumferential viscodilation and tensioning of Schlemm canal (canaloplasty) with temporal clear corneal phacoemulsification cataract surgery for open-angle glaucoma and visually significant cataract: one-year results. J Cataract Refract Surg 2008; 34:433.
678. Miura K et al. Comparison of ocular hypotensive effect and safety of brinzolamide and timolol added to latanoprost. J Glauc 2008; 17:233.
679. Lim KS et al. Mechanism of action of bimatoprost, latanoprost, and travoprost in healthy subjects. A crossover study. Ophthalmology 2008; 115:790.
680. Mansouri K et al. Quality of diurnal intraocular pressure control in primary open-angle patients treated with latanoprost compared with surgically treated glaucoma patients: a prospective trial. Br J Ophthalmol 2008; 92:332.
681. Gerometta R, Spiga MG, Borrás T, Candia OA. Treatment of sheep steroid-induced ocular hypertension with a glucocorticoid-inducible MMP1 gene therapy virus. Invest Ophthalmol Vis Sci 2010; 51:3042–3048.
682. Lee E, Gabelt BT, Faralli JA, Peters DM, Brandt CR, Kaufman PL, Bhattacharya SK: COCH transgene expression in cultured human trabecular meshwork cells and its effect on outflow facility in monkey organ-cultured anterior segments. Invest Ophthalmol Vis Sci 2010; 51:2060–2066.
683. Wang W-H, McNatt LG, Pang I-H, Millar JC, Hellberg PE, Hellberg MH, Steely HT, Rubin JS, Fingert JH, Sheffield VC, Stone EM, Clark AF. Increased expression of the WNT antagonist sFRP-1 in glaucoma elevates intraocular pressure. J Clin Invest 2008; 118:1056–1064.
684. Barraza RA, McLaren JW, Poeschla EM. Prostaglandin pathway gene therapy for sustained reduction of intraocular pressure. Mol Ther 2010; 18:491–501.

CHAPTER 12

Metabolic Interactions between Neurons and Glial Cells

Carole Poitry-Yamate & Constantin J. Pournaras

Introduction

The vasculature of both the retina and brain can autoregulate, meaning that blood flow is altered in response to neuronal activity. This tight coupling between neuronal activity and blood flow, or neurovascular coupling, was first described in the brain more than a century ago by Roy & Sherrington.[1] Retinal glia are not just passive supportive cells but rather they play an active role in directly changing neuronal activity. Glial cells sense neuronal activity and alter blood flow accordingly by directly communicating with the inner retinal vasculature. Thus, glia play a pivotal role not only in maintaining normal neuronal function but also in ensuring adequate retinal blood flow.[2,3] Even though glia are essential for normal retinal function, for the most part their roles are in response to neuronal function. Stimulation of retinal whole-mounts by light or direct glial stimulation led to either vasoconstriction or vasodilation of inner retinal blood vessels.[3] In particular, these vascular caliber changes were linked to increases in intracellular calcium within retinal glia. Therefore, like their counterparts in the brain, retinal glia induce caliber changes in capillaries in response to neuronal activity.

Recently, with the advent of ion-sensitive fluorescent indicators and calcium imaging, it has emerged that glial cells can generate active responses and play a profound role in directly modulating both the activity of neurons and the vasculature.[4,5] Glia communicate with one another via increases in intracellular calcium in the form of a calcium wave that propagates from one astrocyte to another via gap junctions or by the release of bursts of extracellular ATP.[5,6]

Studies in cell culture, brain slices and more recently, retinal whole-mounts have demonstrated that transmitters released from neurons induce transient elevations in intracellular calcium in glia.[4,6] Mechanical, chemical and light stimulation can evoke increases in intracellular calcium in both astrocytes and Müller cells that propagate to neighboring glia as a "wave".[6] It is well established that extracellular ATP evokes large increases in intracellular calcium within both astrocytes and Müller cells and that glia release ATP extracellularly in response to stimulation.[6] The source of the calcium elevations in retinal glia is thought to be primarily from intracellular stores, although there is also a range of calcium-permeable channels, pumps and exchangers that could mediate calcium influx into glia from the environment.

The functional significance of the elevation in intracellular calcium within retinal glia is two-fold: direct modulation of neurons and alteration of vessel caliber.[5] Studies in cortical glial cultures have shown that glia can release chemical transmitters such as ATP, glutamate and the NMDA receptor co-agonist D-serine in a calcium-dependent manner.[4,7] Moreover, these released "gliotransmitters" can directly alter neuronal function.[8] For example, many types of retinal neurons are known to express receptors to ATP (called P2 receptors), including photoreceptors, amacrine cells and ganglion cells.[9–12] Moreover, photoreceptor function is modulated by extracellular ATP.[11] Therefore, it is possible that the release of ATP extracellularly by glia could in turn modulate a variety of neuron cell types from photoreceptors to ganglion cells. In regards to a glial-dependent modulation of vessel caliber, increases in intracellular calcium within cortical astrocyte endfeet are linked to marked vasodilation in the adjacent arteriole, suggesting a relationship between astrocytes and the vasculature within different regions of the central nervous system[13,14] in response to an increase in neural activity.

Energy for excitatory glutamatergic synaptic transmission in the mammalian retina, as elsewhere in the central nervous system (CNS), is provided by the metabolism of blood-borne glucose. Indeed, retinal preparations become synaptically silent as a result of glucose depletion.[15] From the ultrastructural point of view, the remarkable concentration of mitochondria in synaptic terminals of axons[16] indicates that glutamatergic synapses have high capacity for oxygen consumption and are major users of metabolic energy.[17]

Although electrophysiological evidence shows that neurotransmission through the inner retina is supported by glycolysis[15] there is presently no experimental evidence showing that synaptic activity of pre- and post-synaptic retinal neurons is directly sustained by glucose. Indirect evidence, however, is suggested from the classical work of Lowry and co-workers[18] on the distribution of enzymes of glucose metabolism determined from pure samples of each retinal layer from monkey and rabbit. A brief introductory overview of this particular paper is given below as it offers invaluable insight to the contribution of Müller glia to overall retinal function and energy metabolism.

All enzymes of glycolysis are in the cytoplasm rather than in the mitochondria. To initiate glycolysis, hexokinase irreversibly phosphorylates glucose to glucose-6-phosphate (G6P). The distribution of hexokinase was confined to the layer containing inner segments of photoreceptors, the inner synaptic layer, and to the inner-most retinal layer bordering the vitreous. The second step of glycolysis is the conversion of G6P to fructose-6P by glucose-phosphate isomerase. This enzyme's distribution was largely confined to the inner- and outer-synaptic layers. Phosphofructokinase, the third enzyme in glycolysis, irreversibly phosphorylates fructose-6P to fructose-1,6diP and its distribution was confined to both synaptic layers and the innermost retinal layer. The ninth enzyme in glycolysis, phosphoglyceromutase converts 3-phosphoglycerate to 2-phosphoglycerate and its distribution was similar to the distribution of phosphofructokinase.

In tissues with adequate oxygen supply, pyruvate formation is the 11th and last step of glycolysis. The metabolism of pyruvate for energy production consumes oxygen and completes the breakdown of glucose to CO_2 and water through the process of oxidative metabolism. Upon careful study, the distribution of the aforementioned glycolytic enzymes corresponds to the morphological position of Müller glial cells in situ. Müller cells extend radially through all of the retinal layers from the photoreceptor inner segments to the inner limiting membrane bordering the vitreous; and they extend fine filaments laterally in both synaptic layers. They also form an additional physical and functional cell layer to the diffusion of substances from the blood to neurons. Indeed, Kuwabara & Cogan[19,20] undertook the first comprehensive histochemical study to identify Müller cells as the primary glucose utilizing cells in the retina.

Together, these landmark studies paved the way to our present day understanding of cellular coupling in maintaining retinal function and metabolism, and which forms the major theme of this chapter. As a concluding introductory note, one is reminded that regardless of how the energy consumption of the retina (or other parts of the CNS) is altered locally to meet changing demands, it is of major importance to know from a neurophysiological viewpoint just which cell types and what cellular events are associated with local changes in blood flow, metabolism and tissue oxygenation.

1. Retinal oxygen distribution and consumption

Under normal conditions, O_2 is the limiting factor in retinal metabolism. Oxygen is used by mitochondria, and their distribution is important in understanding locations of high O_2 demand. The O_2 consumption (QO_2) of the retinal pigment epithelial (RPE) cells is about 20% of that of the retina per mg protein.[21] Mitochondria are densely observed in the inner segments (IS) of photoreceptors. Cones have more mitochondria than rods.[22,23] Mitochondria are also found in each rod spherule and cone pedicle. In the inner retina, the inner plexiform layer has a larger amount of mitochondria than the nuclear layers.[24]

Inner retina The distribution of oxygen tension (PO_2) close to the vitreo–retinal interface is heterogeneous, being higher close to the arteriolar wall.[25] Preretinal and transretinal PO_2

profiles indicate that O_2 diffusion from the arterioles affects the PO_2 in the juxta-arteriolar areas (Fig 12.1).[25,26] O_2 reaches the vitreous by diffusion from the retinal circulation. In contrast, far from the vessels, the preretinal PO_2 remains constant and the average preretinal PO_2 from the vitreal side is similar to that measured in the inner retina.[25,27,28] In the inner retina, PO_2 averages about 20 mmHg, but up to 60 mmHg close to the arteriolar wall.[25,29,30]

Dark and light O_2 consumption Inner retinal oxygen consumption was the same in light and darkness,[31] indicating no influence of light adaptation as there is in the outer retina. In the inner retina, ganglion cells have much higher firing rates if a stimulus is presented repeatedly than if the same amount of light is delivered as a steady background. Consequently, one would expect the inner retina to use more energy when a stimulus is flickering. Indeed, there is, in response to a flickering stimulus, a higher lactate production in inner retina of rabbit than during darkness or steady illumination[15] and a higher deoxy-D-glucose uptake in monkey retina.[32]

Outer retina Trans-retinal PO_2 measurements have provided data about the O_2 supply to the photoreceptors and their QO_2. PO_2 profiles made in cat,[33,34] pig[25] and monkey[35] indicate that oxygen diffuses from the inner retina and from the choroid towards the middle of the retina, i.e. the outer plexiform layer (OPL). The choroidal circulation supplies about 90% of the photoreceptor's O_2 use.[29]

Photoreceptor QO_2 in darkness In the dark-adapted retina, photoreceptor QO_2 (Q_{OR}) depends strongly on choriocapillary PO_2 in cat and monkey.[29,36] In the inner segments of photoreceptors (IS) the local value of QO_2 is about five times higher than in the outer segments (OS). A similar range for QO_2 in the outer retina was obtained in rat, rabbit and pig retinas.[15,37,38] In both cat and monkey, the average value of PO_2 in the choriocapillaries is about 50 mmHg, and the corresponding average value of QO_2 in the outer retina is 4–5 mL $O_2/100g^{-1}/min^{-1}$.[29,36]

The average PO_2 value in cat was 5 mmHg, the minimum was frequently indistinguishable from zero in the dark[29] as predicted by Dollery et al.[39] The amount of oxygen consumed is an indirect measure of ATP synthesis and thus of ATP utilization. The ATP produced in the dark fuels many cellular processes,[15] mainly the Na^+/K^+-ATPase in the IS, which extrudes a large amount of sodium that enters through the light-dependent channels in the OS.[1,15,40] An additional process is the turnover of cGMP that holds these channels open.[15,41]

As noted above, there is no evidence about whether individual rods and cones use different amounts of O_2. Cones in the primate fovea appear to use slightly less O_2 than the parafoveal photoreceptors.[35,36]

Photoreceptor QO_2 in light QO_2 in the outer retina is lower in steady light than in darkness in various animals investigated.[15,29,35,42] The activity of the Na^+/K^+-ATPase decreases in light, but the turnover of cGMP increases,[43,44] so the decrease in Q_{OR} is not as great as the decrease in the pump rate. The maximum size of the overall change appears to be species-dependent.

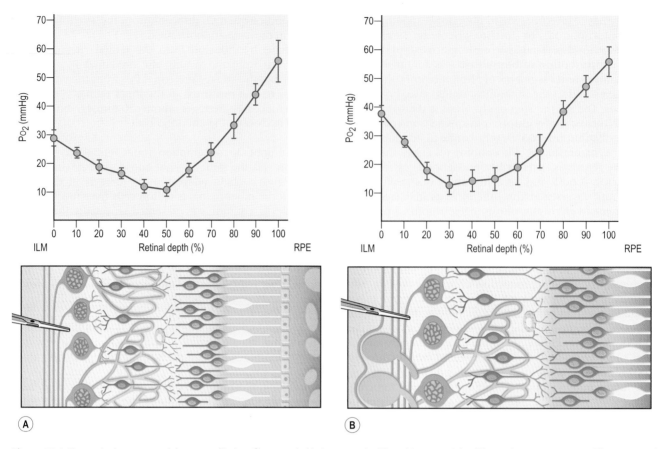

Figure 12.1 Transretinal oxygen partial pressure (PO₂) profiles recorded in intravascular (**A**) and juxta-arteriolar (**B**) retinal areas in mini pigs. The intraretinal values indicate a progressive decrease of the tissue PO₂ from both the vitreoretinal interface (ILM) and the pigment epithelium (PE) toward the middle of the retina, with the minimum mean value recorded at 40% and 50% retinal depth. At the vitreoretinal interface, the higher PO₂ suggests that oxygen diffusing from the larger vessels reaches the inner retina. Each point is the mean ± SE of 13 measurements. The drawings indicate the pathway of the microelectrode through the retina. ILM: internal limiting membrane, RPE: retinal pigment epithelium. (Modified from Pournaras CJ, *Retina*. 1995;15(4):332–47.)

2. Role of glycolysis underlying retinal function: from whole retina to its parts

Visualization of dynamic functional activity in the retina is an *indirect* measurement of neuronal activity. It is important to appreciate that any dynamic "functional imaging" of the central nervous system, e.g. fMRI or PET imaging, measures local changes in brain metabolism and physiology that are associated with neuronal activity. Therefore, examining and evaluating retinal function necessarily involves understanding the energy metabolism of this tissue.

The significance of glucose and its metabolism down the glycolytic pathway in mammalian retina is attested to by the measured high rate of aerobic glycolysis in vitro (high capacity for oxygen consumption),[22,45] its susceptibility to iodoacetate,[46] a strong Pasteur effect (inhibition of glucose utilization)[22,47–49] and the aerobic and anaerobic production of lactate.[47,50,51]

The adult retina, as is the case for every CNS region, depends on an uninterrupted supply of blood-borne glucose. Under normal conditions, glucose is virtually the sole substrate supporting the intense energy metabolism required to maintain retinal function, e.g. normal electrical responsiveness to light and neurotransmission.[15,51,52] Lactate generated from either anaerobic glycolysis or glycogenolysis within the retina has recently been proposed as another important

energy source during synaptic transmission,[53] but its uptake and metabolism by retinal cells has yet to be demonstrated in vivo. A critical evaluation of lactate use in the CNS has been presented.[54–59]

The distribution of key enzymes in glucose metabolism through the individual neuronal and synaptic retinal layers and the dehydrogenases for several of the intermediate stages of glucose degradation in retina have been documented in a series of pioneering histochemical work[18,19,60] and set the stage early for later metabolic studies of the glycolytic and oxidative capacity of this tissue. Kuwabara and Cogan's studies suggested for the first time that the Müller glial cells may serve a significant metabolic role, in stark contrast to the general view at that time as having only a structural function. Indeed, of the two major cell types in retina – the Müller glia and photoreceptors – metabolic studies have demonstrated that Müller cells both in situ and when acutely isolated, preferentially and massively take up and phosphorylate glucose, part of which is stored as glycogen.[61,62] Further evidence to confirm this finding has been performed in vivo.[63]

More recent supporting evidence comes from the use of iodoacetate (IAA), a well-known glycolytic poison that exerts its effect when the transformation of glucose is committed to proceed through glycolysis. In the 1950s, it was shown that intravenous delivery of sodium iodoacetate in rabbit, monkey and cat abolished the electrical response of the visual pathway to illumination within minutes, with a resulting histological picture similar to that presented in human retinitis pigmentosa.[64] This led to the speculation that the initial effect must be on the visual cells,[65] although Warburg suggested that the different cell types in the retina may not contribute equally to the general biochemical picture.[64] Indeed, this suggestion has been unambiguously supported by the differential suppression of the component waves of the extracellular electroretinogram (ERG).[66,67] However, beyond this evidence, the identity of the retinal cell types taking up IAA remains unknown. Recently this was explored using synchrotron-based x-ray fluorescence of IAA at the cellular level in situ (Fig. 12.2). The fluorescence map (Fig. 12.A) generated from the dark-adapted retina (Fig. 12.2B) showed that IAA was taken up specifically by Müller glia and not by retinal neurons, including photoreceptors, indicating that the effect of IAA on neurons is not direct but secondary to inhibiting glycolysis in glia (Poitry-Yamate 2009, unpublished results). Together, these results suggest a key role played by Müller glia in transporting glucose from the blood into the retina.

Acutely isolated mammalian photoreceptors produce $^{14}CO_2$ from $^{14}C(U)$-glucose[53] while photoreceptor outer segments produce both lactate from glucose and $^{14}CO_2$ from $^{14}C(U)$-glucose.[68] These authors interpreted their results as indicating that both glycolysis and the pentose phosphate pathway contribute to maintaining photoreceptor function. Considering that only one in six carbons from $^{14}C(U)$-glucose is converted to $^{14}CO_2$ through the pentose phosphate pathway, and the abundance of photoreceptor mitochondria, it is likely that the $^{14}CO_2$ reflects the rate of mitochondrial respiration.[53] Photoreceptors of some species do not express Gpi 1, the enzyme catalyzing the isomerization of glucose-6-phosphate to fructose-6-phosphate,[69] leaving the phosphate pentose pathway as the only possible downstream path, with a gain of 2 NADPH molecules potentially serving as reducing agent for the reduction of retinaldehyde.

A number of studies of glucose metabolism have been undertaken in intact retinal tissue or with acutely isolated cell models in vertebrate/mammalian retina.[53,61–63,68,70,71] Of these, one is unique for studying not only glucose metabolism but also metabolic compartmentation: the cell model of acutely isolated Müller cells still attached to photoreceptors (termed "the cell complex") shown in Figure 12.3 and further discussed in sections 4, 6 and 8. This study confirmed not only the previous work by Kuwabara and Cogan, but showed for the first time in a mammalian preparation of the CNS tissue that glial cells *transform rather than simply transfer* the primary energy substrate glucose and supply neurons with a glucose-derived metabolite.[53,72,73]

Cell culture models of transformed rat Müller cells, human RPE and transformed mouse photoreceptor cells and ganglion cells were all found to produce lactate, aerobically and anaerobically in the presence of 5 mM glucose.[74] This may not be surprising as culturing techniques influence cell metabolism and function. The composition of culturing medium may be a key underlying factor to explain why cells are predominantly glycolytic, irrespective of cell type.[75] However, the production of lactate by these cells did not significantly differ with the addition of 10 mM lactate.[74] In general, lactate production and its release into the extracellular space as a lactate anion plus a proton (H⁺) creates an extracellular pH gradient, i.e. increased proton concentration

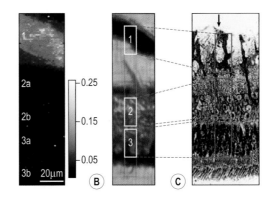

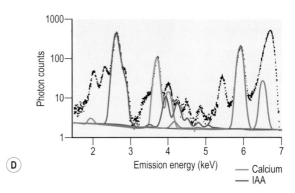

Figure 12.2 The glycolytic poison iodoacetate is localized to Müller glial cells and not to retinal neurons: indirect evidence of which cell type depends directly on glucose metabolism. (A) Synchrotron radiation-based hard x-ray fluorescence map (90 μm height × 20 μm width) of the period table element iodine from iodoacetate (orange) subsequent to scanning the retina from the inner (top) to outer (bottom) layers. Iodoacetate was localized in the retina to Region 1 comprising the endfeet of Muller glia. The metabolic experiment consisted of dark-adapting the tissue prior to a 50 min exposure ex vivo to bicarbonate buffered Ringer's physiological solution carrying iodoaceate and D-glucose. Note that iodoacetate inactivates the glyceraldehyde-3-phosphate dehydrogenase reaction and therefore inhibits the 6th step in glycolysis. **(B)** Phase contrast image of the retina from which the fluorescence map shown in A was obtained. Region 1, Müller glial endfeet; Region 2a, inner plexiform layer; Region 2b, inner nuclear layer; Region 3a, photoreceptor layer; Region 3b, photoreceptor segments. **(C)** Reference, methyl-blue stained coronal section used for identifying individual retinal layers. **(D)** X-ray fluorescence energy emission spectra from Region 1. An La energy emission at 3.9 keV is specific to iodine from iodoacetate (blue trace). The y-axis expresses the number of fluorescence photons emitted from iodine subsequent to exposing the preparation to synchrotron ligtt.

and consequently pH values become less than 7.4. Depending on the magnitude, direction and time course of the extracellular pH gradient, lactate may accumulate extracellularly, or alternatively, lactate may be taken up on a proton-linked monocarboxylate transporter.[76] In this context, transformed, cultured neurons did not transport or metabolize exogenous lactate,[74] consistent with a lactate-containing solution adjusted to a pH of 7.4 rather than a pH of <7.4.

3. Biochemical specialization of glial cells

The major glial cell type in vertebrate retina is the radial Müller glial cell (also termed radial fibers or sustentacular cells of Heinrich Müller). Structurally, they are elongated, possess a prominent specialized region called endfeet at the

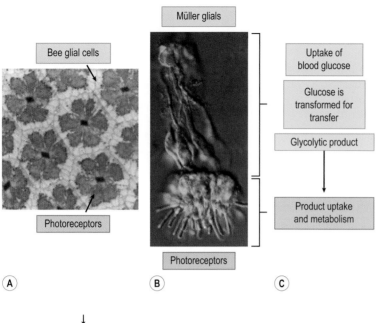

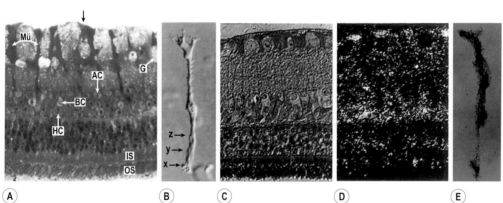

Figure 12.3 Acutely isolated cell models for exploring retinal function, metabolism and the trafficking of metabolites between glial cells and photoreceptor-neurons. (**A**) Honeybee drone, illustrating its crystalline-like structure of 6 photoreceptors that form a rosette surrounded by glia. The outline of the extracellular space between glia is seen as web-like. The compartmentation of glycolysis to glia and of oxidative metabolism to photoreceptors renders this CNS model unique for studying neuron-glial interactions (modified from Tsacopoulos et al, *Proc Natl Acad Sci USA* 1998; 85(22): 8727–31). (**B**) Mammalian Müller glial cells still attached to photoreceptors (cell complex) after their acute isolation and purification from guinea pig. Prominent glial endfeet are oriented at top. The distal ends of Müller cells are hidden by photoreceptor cell bodies (modified from Poitry-Yamate et al, *J Neurosci* 1995;15(7 Pt 2): 5179–91). (**C**) In both models, glia take up exogenous glucose and transform it to a glycolytic product (green), which in turn is released extracellularly, then taken up and metabolized by the photoreceptors (blue).

Figure 12.4 Müller glia in vertebrate retina. (**A**) Methyl-blue stained retinal section highlighting the large endfeet and radial structure of Müller glia (Mü) through the thickness of the retina. Laminar organization of this tissue allows for clear identification of nuclear and synaptic layers. G, ganglion cell; AC, amacrine cell; BC, bipolar cell; HC, horizontal cell; IS and OS, inner and outer segments of photoreceptors. Arrow at top indicates direction of light hitting the retina; *, synaptic layers. Müller endfeet form the vitreoretinal interface. (**B**) Müller cell after acute isolation and purification; approximately to scale with Müller cells shown in (A). The descending radial process (Z), radial strands (y) and terminal angular buttons (x) of the Müller cell are landmarks of that part of the Müller cell in contact with the outer synaptic layer, and the cell body and inner segments of photoreceptor, respectively (modified from Poitry-Yamate et al, *J Neurosci* 1995;15(7 Pt 2): 5179–91). (**C**) Phase contrast image of retinal cells from guinea pig in situ, oriented as in a. Note that individual cell types can be distinguished using the section in (A) (modified from Poitry-Yamate and Tsacopoulos, *Neurosci Lett* 1991;122(2): 241–4). (**D**) High resolution ³H-DG-6P autoradiogram of retinal preparation shown in (C). The silver grains shown in white were determined by high pressure liquid chromatography and correspond to ³H-DG-6P. (**E**) High resolution ³H-DG-6P autoradiogram of a single Müller cell similar to that shown in (B). illustrating homogeneity of phosphorylated DG intracellulary extending about 120 μm starting from the endfoot (at top) to the distal end of the cell (at bottom). Autoradiograms in (D) and (E) provide evidence that Müller glia, in situ and when acutely isolated, take up and phosphorylate the sugar analog deoxy-D-glucose.

inner limiting membrane, and are vertically oriented with respect to the retinal layers (Fig. 12.4). Since Müller cells extend through the synaptic and nuclear layers of the retina from the inner to outer limiting membranes they are in intimate apposition to every neuron cell type.[77] Müller glia also serve as an additional physical and functional cell layer to the diffusion of substances into and out of the extracellular space, the vitreous, the subretinal space and retinal vascular supply. Histochemical evidence has shown that glycogen synthesis, glycogenolysis and anaerobic glycolysis are localized to Müller glial cells in situ.[19] This was confirmed and quantitated in living intact, dark-adapted retina using biochemical and autoradiographic methodologies[61,62] and provided strong experimental evidence for the working

hypothesis of all retinal cell types, Müller glia play a major role beyond the blood–retinal barriers in transporting glucose from the blood into the neural retina. However, once in the neural retina, it remains to be shown along the Müller cell's entire radial length whether the distribution of transporters related to energy substrate uptake and release are tailored to this cell's own metabolic needs, but yet adapted to the function and metabolic needs of their immediate neuronal environment.

Two functional and biochemical specializations unique to Müller glia are their capacity to inactivate the excitatory neurotransmitter glutamate[78,79] and inhibitory neurotransmitters GABA and glycine.[80–82] They are the exclusive and/or predominant cellular site of:

(1) glutamine synthetase activity for the synthesis and release of glutamine, a precursor for photoreceptor neurotransmitter resynthesis;[83] and

(2) carbonic anhydrase for the conversion of water and CO_2 of neuron origin to bicarbonate, an enzymatic activity implicated in the regulation of intracellular and extracellular pH and volume.[84–86]

4. Role of glycogen

As an endogenous source of glucose-6-phosphate, glycogen, as well as glycogen phosphorylase, are exclusively localized to the cytoplasm of Müller glial cells in situ in a variety of mammalian species.[87] This important study suggests that Müller cells can effectively mobilize this energy store but it remains unclear whether it is for the purpose of meeting their own energy needs,[62] or alternatively partly those of the surrounding neurons in the form of glycogenolytically generated lactate[53] one isoform of glycogen phosphorylase is expressed in cone photoreceptors (brain type), while another isoform (muscle type) is expressed in the inner plexiform synaptic layers of primate retina.[71] So, although glycogen metabolism in Müller glia is established (e.g. see Ripps & Witkovsky[88] for a review), the function of Müller cell glycogen is still far from clear. The prevailing view is that glycogen plays the role of an *emergency* retinal carbohydrate reservoir supporting neuron function when glucose delivery is compromised such as during retinal ischemia.[89] This is in contrast to another view that Müller cell glycogen is mobilized as an immediate and accessible energy store under normal physiological conditions (Fig. 12.5), such as changes in illumination, i.e. light and darkness, and which are linked to the direct effects of altering photoreceptor function.[53,90] In addition, glycogenolysis can be stimulated in the retina by neurotransmitters increasing cAMP, such as vasoactive intestinal peptide, which are contained in and released from amacrine cells.[91] Glycogen is also localized in neurons of the cat retina, particularly of the rod-driven pathway, whose rod-driven components are selectively sensitive to prolonged hypoglycemia,[92,93] but whether this glycogen is mobilized is not presently known.

5. Functional neuronal activity and division of metabolic labor

A continuous supply of blood-borne glucose is vital to maintaining retinal function, but does not necessarily dictate that all retinal cells (both neurons and glia) take up and metabolize this energy substrate as is generally believed. Indeed, the conventional view of glucose metabolism in brain is that glucose is the principal substrate for oxidative metabolism in both cell types, i.e. neurons and glia.[94] The experimental evidence described in the previous sections is a major departure from this view and raises three highly controversial and still unresolved questions about the brain and about the generality of the findings in retina to other parts of the central nervous system:

(1) is glucose, the major brain energy substrate, taken up in a cell-type specific manner?

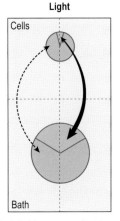
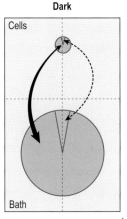

Light   **Dark**

Labeled lactate $_{bath}$ > lactate $_{cell}$ Nonlabeled lactate in the bath ↑

▨ Radiolabeled lactate from ^{14}C-glucose

▨ Nonlabeled lactate from glycogen

Figure 12.5 Lactate released by Müller glial cells is formed from exogenous radiolabeled glucose in the light-adapted cell complex but from glycogen in the dark-adapted cell complex. The preparation shown in Fig. 12.3B was maintained either in darkness or light before exposure to uniformly radiolabeled glucose (^{14}C(U) glucose) with the aim to determine the contribution of glucose and glycogen to the production of lactate when modulating photoreceptor metabolism and neurotransmitter release. It is known that: (1) photoreceptors release neurotransmitter glutamate and that their metabolism is increased in darkness; and (2) glutamate release stimulates glycolysis and production of lactate. The results shown are simplified and drawn as a pie to reflect changes both inside the cells (top) and in the surrounding bath (bottom). Pie size represents the total pool size of lactate, i.e. radiolabeled + non-radiolabeled lactate. The size of the wedge represents the specific activity of lactate, so the larger the wedge, the larger is the contribution of exogenous glucose to the formation of lactate. The direction of the solid arrow in the left panel indicates that the amount of radiolabeled lactate (pink) in the bathing solution was much greater than that inside the preparation. This is only possible when lactate is produced from ^{14}C-glucose and is released earlier than non-radiolabeled lactate formed from glycogen. In the right panel, the direction of the solid arrow indicates that the amount of non-radiolabeled lactate (blue) in the surrounding bath greatly increased. This is only possible when glycogen (gly) in glia is mobilized to produce additional, unlabeled lactate.

(2) does glycolysis predominate in one particular cell type, physiological condition or brain region and oxidative metabolism of glucose in another? and

(3) does coupling of metabolic and physiological changes to changes in neuronal activity involve the transformation of blood glucose by glial cells and do they supply neurons with glucose-derived metabolite(s)?

Underlying these issues is the idea of a predominant division of metabolic labor between neurons and glia. In other words, energy substrate production and substrate use is partitioned in a relatively cell-specific manner. This working hypothesis, summarized in Figure 12.6, was developed and tested in retina in the mid-1990s, in insect and mammal, and has seen a revival in recent years, particularly with regards to the recognition that the coordinated action of glial cells and neurons extends to energy metabolism throughout the central nervous system.

As will be developed in the next sections, the major theme is that glial cells *transform rather than simply transfer* the primary energy substrate glucose, and supply activated

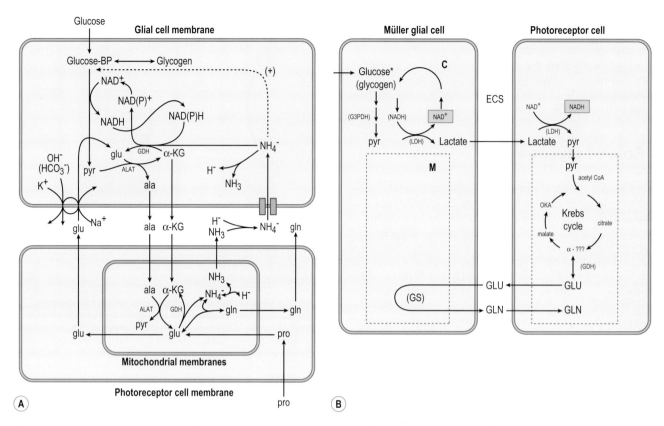

Figure 12.6 Proposed scheme of retinal function, from bee to mammals, highlighting metabolite trafficking between glial cells and photoreceptors. The honeybee drone model is shown in (**A**) and the mammalian model is shown in (**B**). Both models display relative metabolic compartmentation: glycolysis to glia and oxidative metabolism to photoreceptor-neurons. Commonalities of the models include production of a glycolytic product, i.e. lactate or alanine, their release extracellularly and uptake by photoreceptors; maintenance of the redox potential in glia, i.e. NADH/NAD ratio; and photoreceptor release of glutamate that acts to stimulate glial glycolysis. Glutamate is thus a chemical signal that turns the nourishing of neurons by glia into a function, instead of a passive process. In (B), the cytosol is indicated as **C**; mitochondria are indicated as **M**.

neurons with a glucose-derived metabolite, i.e. lactate and/or alanine. In this regards, the vertebrate retina has proven to be a CNS model of choice because its laminar organization of metabolism and blood flow lends itself to the study of compartmentation by virtue of its structure. It should, however, be recognized that the high specialization of this nervous tissue, e.g. phototransduction and high energy metabolism, makes it different from other regions of the CNS and any comparisons must be made with caution. Moreover, the retina has two barriers, the retinal capillary endothelial cells of the inner blood–retinal barrier (BRB), and the retinal pigment epithelium (RPE) comprising the outer BRB which is located between the photoreceptors and choroid. Thus, their expression and distribution of glucose transporters[95] and glycolysis will inevitably determine whether one, or alternatively both, of these barriers is the rate-limiting step for glucose delivery to the neuroretina.

6. Cellular compartmentation of energy substrates other than glucose

One distinctive property of the mammalian retina is its large production of endogenous lactate in the presence and absence of oxygen.[48] The production of lactate is a normal function of retinal tissue and can effectively replace glucose

in oxygenated Ringer's solution in maintaining retinal oxidative metabolism and photoreceptor function.[51] It was only in the mid-1990s that this phenomenon was explored at the cellular level[62] and when the idea was launched that energy metabolism underlying retinal function in mammals was relatively compartmentalized, and moreover orchestrated between glial cells and neurons.[53]

The use of high-resolution light microscopic autoradiography of ^{3}H-2DG in the dark-adapted retina in situ, coupled to HPLC for the identification of silver grains in autoradiograms[61] showed that glucose is not taken up by the majority of retinal cells but preferentially taken up and phosphorylated by the Müller glia (Fig. 12.4C,D), leaving the question as to the identification of the energy substrate used by neurons. Using a cell model of acutely isolated Müller cells, the metabolic fate of glucose-6-phosphate and the identity of metabolites released by these cells was first assessed by Poitry-Yamate and Tsacopoulos.[62] The experimental results of that study raised the hypothesis that lactate, synthesized and released in situ by a retinal cell type having high glycolytic capacity, may be transferred to and metabolized by photoreceptors that possess a high respiratory capacity. This working hypothesis was tested in the acutely isolated retinal model of the cell complex (see Fig. 12.2B) comprising Müller cells still attached to photoreceptors, and where the effect of altering illumination on glucose metabolism was quantitated.[53] A major finding of that study (see section 8 below) showed that lactate *formation* versus lactate *use* was cell-type dependent. As already discussed in section 4, sources of this glial lactate were from exogenous glucose, or alternatively, from endogenous glycogen (Fig. 12.5). The functional role for the release into the extracellular space of lactate and H^+ by Müller glial cells during either glycolysis or glycogenolysis is expected to prevent intracellular acidification and maintain the regeneration of NAD^+ for glycolysis to proceed. Lactate and other glucose-derived metabolites released from the Müller glia would participate in the regeneration of the excitatory neurotransmitter glutamate, released by photoreceptors in darkness.[53]

Within a larger framework, the current debate in the neuroscientific community is whether in other parts of the CNS, a net production and extracellular release by glial cells of glucose/glycogen-derived lactate provides sufficient fuel for activated neurons in vivo. NMR studies on brain function and energy metabolism have in recent years provided indirect support of a lactate flux from glia to neurons as providing the coupling mechanism between increased fluxes of the glucose carbon into the glycolytic pathway in glia, and into the oxidative pathway in neurons, in the form of lactate, during glutamatergic/excitatory synaptic transmission.[96] Unraveling the current debate surrounding the glial-neuron lactate shuttle hypothesis (ANLSH)[97,98] in the CNS at large is a key step towards: (1) understanding the regulation of retinal and cerebral energy metabolism during neuronal activity; and (2) interpreting [18]FDG PET and fMRI images in both clinical medicine and fundamental neuroscience, and for which there is accumulating evidence for the central importance of glial cells in brain imaging signals.[99]

7. Experimental models used to study the interaction between photoreceptors and glial Müller cells

In vitro studies of the retina of the honeybee drone

The most comprehensive and quantitative experimental tests in the literature demonstrating that glial cells supply neurons with a metabolic substrate to sustain neuron function were first formulated and established in the retina of the honeybee drone.[72,100–105] This elegant crystalline-like CNS preparation is characterized by mitochondrial-rich, glycogen-poor photoreceptors grouped in rosette-shaped clusters and their surrounding mitochondrial-poor and glycogen-rich glial cells (see Fig. 12.3), indicating that glycolysis/glycogenolysis is largely confined to glia and that oxidative metabolism is largely confined to photoreceptors. What follows is a historical and chronological overview of the major experimental questions, experimental results and interpretations.

If there are no conventional synapses in drone retina and only the photoreceptors are directly excitable by light, what is the evidence that photoreceptors depend on surrounding glia for their metabolic needs?
In retinal drone slices, three lines of evidence argue that photoreceptors depend on the surrounding glial cells for their metabolic needs:

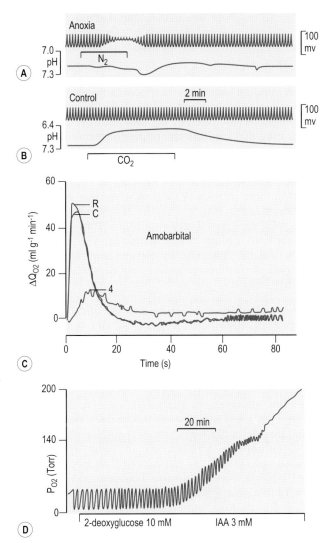

Figure 12.7 Characterization of photoreceptor metabolism and function in honeybee drone: evidence that photoreceptors metabolically depend on surrounding glia. (A) The effect of anoxia on abolishing receptor potential in photoreceptors but as shown in (**B**). The absence of an extracellular acidification as would be expected if anaerobic glycolysis is activated, indicates that photoreceptor energy metabolism is obligatorily aerobic. (**C**) The effect of blocking mitochondrial respiration (noted as 4) that uses a carbohydrate substrate reversibly suppressed the light-induced consumption of oxygen by photoreceptors. C and R are control and reversibility responses, respectively. This result indicated that the substrate of photoreceptor metabolism is obligatorily a carbohydrate. (**D**) The glucose analog, 2-deoxy-D-glucose (2DG) leads to a decrease in the partial pressure of oxygen (PO_2), so had no effect on the light-induced increase in oxygen consumption in photoreceptors. This result indicated that the continued function of photoreceptors does not depend on their direct uptake of glucose. The effect of the glycolytic poison, iodoacetate (IAA) on oxygen consumption and thereby of the partial pressure of oxygen in bee photoreceptors is not direct as glycolysis takes place only in the glia in this preparation (see Fig. 12.7A).

(1) Anoxia rapidly abolishes the receptor potential of photoreceptors stimulated every 5s by a light flash (Fig. 12.7A). However, simultaneous monitoring of the pH in the extracellular space changed by less than 0.02 units (Fig. 12.7B), in contrast with that of many other tissues where hypoxia leads to anaerobic glycolysis and

acidification. These results indicated that anoxia does not lead to anaerobic glycolysis in drone retinal slices and that *photoreceptor* energy metabolism is obligatorily aerobic.

(2) Amobarbital (or amytal) a specific blocker of mitochondrial respiration that uses carbohydrate substrate reversibly suppresses the light-induced change of oxygen consumption by photoreceptors (Fig. 12.7C), indicating that the substrate of photoreceptor metabolism is a carbohydrate.

(3) Superfusing slices of drone retina with the glucose analog 2-deoxy-D-glucose (2DG) had no effect on the light-induced increase of QO_2, as monitored by decreases in PO_2 (Fig. 12.7D). As 2DG is transported into cells, phosphorylated, and little of it is metabolized further, 2DG6P accumulates intracellularly. This indicates that the continued function of photoreceptors does not depend directly on their uptake of glucose.

When bee retinal slices are exposed to the glycolytic poison IAA, the light-induced change in QO_2 is gradually abolished. Is this modulation of QO_2 a direct effect of IAA in photoreceptors?

IAA exerts its effect in cells where glucose uptake and metabolism are committed to the glycolytic pathway. Identifying whether the cellular site of glucose phosphorylation is cell-type specific is a first step towards this answer. The distribution and density of black grains biochemically identified as ^{3}H-2DG-6-phosphate (^{3}H-2DG6P) in light microscopic autoradiographs (Fig. 12.8) of bee retinal slices exposed to tritiated 2DG provides this answer. Transport and phosphorylation of this sugar analog are localized in situ to the glia and absent in photoreceptors. Hence, IAA has obligatorily exerted its direct effect on the metabolism of glucose in glial cells, which in turn lead to downstream effects on the oxygen consumption in photoreceptors.

What is the evidence that photostimulation of drone retinal slices increases the carbohydrate metabolism in the glia to sustain photoreceptor respiration?

In the absence of metabolic substrate, i.e. glucose, a drone retinal slice consumes up to 70 µl O_2 per mL tissue per minute with light stimulation for > 2 hours. After 2 hours of repetitive light flashes (Fig. 12.9), the concentration of glycogen decreased by 30% (or by 15 mg per mL tissue), a value comparable with the carbohydrate requirements of photoreceptors, i.e. equivalent to 10 mg of glycogen per mL tissue. This led to the working hypothesis that glycogen in the glia is mobilized, and that a carbohydrate substrate downstream to glucose-6-phosphate is transferred from the glia to photoreceptors to maintain mitochondrial respiration and responsiveness to light.

Glucose is not the principal energy substrate used by photoreceptors, so what is the identity of the energy metabolite maintaining photoreceptor function and respiration?

This involved determining the metabolic fate of glucose-6-phosphate in glia, the identity of glial-derived metabolites released into the extracellular space, and their uptake by photoreceptors. More than 50% of uniformly labeled glucose (^{14}C(U)glucose) transported into drone glia was transformed to alanine by transamination of pyruvate with glutamate (Fig. 12.10A). This transformation was inhibited by IAA which is expected if alanine is formed from glucose by transamination of pyruvate with glutamate through the enzymatic action of alanine aminotransferase (ALAT) and glutamate dehydrogenase (GDH) schematized in Figure 12.6A. This was confirmed when the transformation was preserved in the presence of IAA and when the labeled carbohydrate in the perfusate was ^{14}C-pyruvate (Fig. 12.10B). The glial cells released into the extracellular space the equivalent of 216 µmol of radiolabeled alanine / liter of retina (Fig. 12.11A) and this alanine was taken up

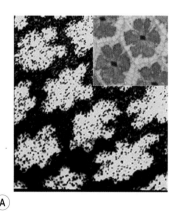

(A)

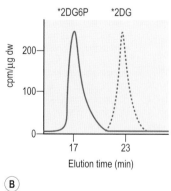

(B)

Figure 12.8 Autoradiographic and biochemical evidence that glucose transport and phosphorylation take place in glial cells and not in photoreceptors in retina. (**A**) The high resolution, light microscopic ^{3}H-2DG autoradiogram obtained after exposure of honeybee retina to the glucose analog ^{3}H-2-deoxy-D-glucose (DG) indicates that DG has been transported and phosphorylated specifically in the glia. The accumulation of black grains in glia corresponds to the intracellular accumulation of phosphorylated ^{3}H-2DG, as DG is not further metabolized. The sparse distribution of black grains within the photoreceptor rosettes (inset, see also Fig. 12.2A) is probably the fin-like glial structures that separate individual photoreceptors within the rosette. (**B**) The identity of the black grains in (A) was determined biochemically using high-pressure liquid chromatography (HPLC). The chromatograph shows that the black grains correspond to the phosphorylated form of the radiolabeled substrate (*2DG6P), eluted at 17 min. Note that the expected eluted time of the non-phosphorylated substrate *2DG is at 23 min. (Modified from Tsacopoulos et al., *Proc Natl Acad Sci U S A* 1998;85(22): 8727–31.)

by the tissue against a concentration gradient and transported in a Na^+-dependent fashion (Fig. 12.11B) and which is only effective in the photoreceptors. Moreover, alanine was consumed by the photoreceptors: repetitive light stimulation lead to >60% increase in $^{14}CO_2$ production from $^{14}C(U)$-alanine.

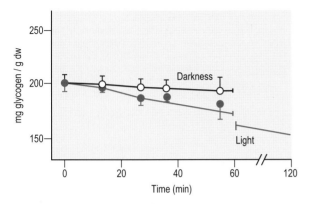

Figure 12.9 In the absence of glucose, photostimulation increases carbohydrate metabolism in the form of glycogen mobilization in glia to sustain photoreceptor respiration. The effect of darkness and light on the concentration of glycogen, expressed as mg glycogen per gram dry weight of tissue, was determined at different time points from bee retina deprived of exogenous glucose. After 2 hours of repetitive light flashes glycogen concentration decreased by 30% or by 15 mg per mL of tissue. This value is comparable to the measured carbohydrate needs of photoreceptors: 10 mg of glycogen per mL tissue. This result suggested that glial glycogen is mobilized and transformed downstream to phosphorylated glucose to a metabolite made available to photoreceptors. This metabolite would serve to maintain mitochondrial respiration and the photoreceptor's response to light.

What is the biochemical evidence for the metabolic effects of NH_4^+ and glutamate in glia?

Photoreceptors also consume proline from pollen and this consumption leads to the production and release of photoreceptor ammonia and glutamate in a stimulus-dependent manner. The production of alanine from glucose by the glial cells is coupled to the resynthesis of NAD^+ via the activity of glutamate dehydrogenase (GDH) (see Fig. 12.6A). Since NH_3 feeds into this reaction, consuming NADH to produce glutamate, it was hypothesized that NH_3/NH_4^+ and glutamate act on alanine production in the glial cells. In an elegant series of experiments[105] using bundles of freshly isolated bee glial cells, it was shown that:

(1) $(NH_4^+ +$ glutamate) are signals that photoreceptors send to glial cells to activate glycolysis and/or glycogenolysis, leading to an increased synthesis of alanine by glia and its release into the extracellular space.

(2) glycogen is the likely substrate for alanine *formation*, but glucose is the likely substrate for alanine *release*.

Overall scheme for metabolic compartmentation and metabolic trafficking in honeybee drone retina

Honeybee drone retinal glia transform glucose to a fuel substrate taken up and used by photoreceptors, which in turn consume both alanine supplied by glia and exogenous proline (see Fig 12.5A). Ammonium (NH_4^+) and glutamate, produced and released in a stimulus-dependent manner by the photoreceptors, contribute to the biosynthesis of alanine in the glia. NH_4^+ and glutamate are subsequently transported into glia and a transient rise in the intraglial concentration

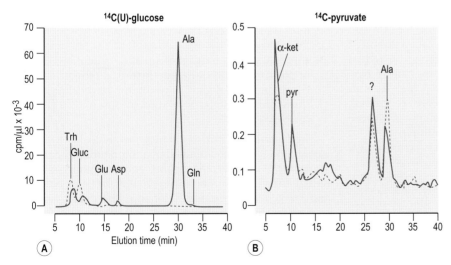

Figure 12.10 The transport of $^{14}C(U)$-glucose into bee glia is transformed to alanine by transamination of pyruvate with glutamate. The identity of the metabolite maintaining photoreceptor function and respiration was explored by first determining the metabolic fate of uniformly radiolabeled glucose in glia. The HPLC chromatogram in (**A**) shows that over 50% of $^{14}C(U)$-glucose was transformed to alanine (Ala). As summarized in Fig. 12.5A, alanine was formed from glucose by transamination of pyruvate with glutamate through the combined enzymatic action of alanine aminotransferase (ALAT) and glutamate dehydrogenase (GDH). This hypothesis was tested by exposing the preparation to IAA with either $^{14}C(U)$-glucose or with $^{14}C(U)$-pyruvate. Alanine formation was inhibited by IAA + $^{14}C(U)$-glucose (not shown), indicating the requirement for pyruvate. However, as shown by the dashed-line chromatograph in (**B**) alanine formation was preserved in the presence of IAA + $^{14}C(U)$-pyruvate since IAA exerts its effect upstream to the production of pyruvate. Thr, trehalose; Gluc, glucose; Glu, glutamate; Asp, aspartate; Gln, glutamine; αket, α-ketoglutarate; pyr, pyruvate. (Modified from Tsacopoulos et al., *J Neurosci* 1994; 14(3 Pt 1): 1339–51.)

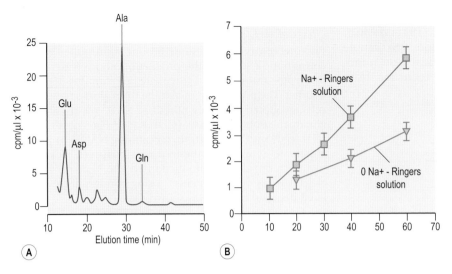

Figure 12.11 Glucose-derived alanine in glia is released extracellularly and taken up by photoreceptors. Following the determination of radiolabeled metabolites derived from $^{14}C(U)$ glucose in glia, the second step was to determine which of these metabolites were released extracellularly by performing HPLC analysis of the bathing solution. The chromatogram shown in (**A**) shows that the major radiolabeled metabolite released by glia was alanine (modified from Poitry-Yamate et al 1995, Neurosci Lett 122(2): 241–244. Copyright Elsevier 1995). This radiolabeled extracellular alanine was taken up principally on a sodium-dependent transport system shown in (**B**). Alanine was subsequently shown to be taken up by photoreceptors since repetitive light stimulation lead to >60% increase in $^{14}CO_2$ production from $^{14}C(U)$-alanine.

of NH_4^+ or of glutamate causes a net increase in the level of reduced nicotinamide adenine dinucleotides (NAD(P)H). Biochemical measurements indicated that this was due to the activation of glycolysis in the glia by the direct action of NH_4^+ and glutamate on two enzymatic reactions: those catalyzed by phosphofructokinase and glutamate dehydrogenase. The overall effect of NH_4^+ on glial metabolism is to increase the production and release of alanine, and that of glutamate is to adjust this production to the metabolic needs of photoreceptors. In this model (NH_4^+) and glutamate released by photoreceptors return to glial cells not simply for nitrogen conservation but as information signals. This tight regulation of metabolic coupling between neurons and glial cells by means of chemical signals turns the nourishing of neurons by glia into a function, instead of a passive process.

Experimental models in vertebrates

The most comprehensive model in vivo of metabolism and function in the inner and outer retina is the cat eye followed by that in rabbit. It is unique in the literature on retinal metabolism that only one group, headed by Günter Niemeyer, has fully exploited both the in vivo and in vitro models. This group has quantitated the effects of controlled, stepwise and transient changes in arterial supply of glucose on retinal and optic nerve function at multiple levels of retinal information processing (e.g. light-evoked electrical signals of the ERG b-wave and light-independent retinal field potentials of negative polarity) in the arterially perfused enucleated eye of the cat[106,107] and verified directly these in vitro findings in anesthetized cats under insulin-induced glucose clamping.[93,108] Although the entire body of work by this group cannot be justly summarized here, Niemeyer and colleagues made a number of important observations which have current clinical and fundamental relevance in understanding retinal function as related to metabolism:

(1) Electrophysiological measurements of the ERG b-wave, the STR (scotopic threshold response) and optic nerve action potentials indicate an exquisite, immediate and reversible sensitivity of the inner retina to changes in the supply of glucose between ~1 and 10 mmol/l glucose.

(2) In contrast to the cone-driven light-evoked signals, the rod-driven light-evoked signals increased or decreased in parallel to changes in glucose.

(3) Blood-borne insulin enhanced the reduction in b-wave amplitude at low (1–2 mM) glucose concentrations.

(4) Glycogen is distributed in Müller glia, and central perivascular astrocytes. It is present in ganglion cells, on–off amacrine cells and rod bipolar cells of the rod pathway, but entirely absent in cone and rod photoreceptors[92] under conditions where the ERG is affected by changes in plasma glucose concentrations.

These studies raise a number of interesting questions regarding retinal energy metabolism in regards to insulin-induced hypoglycemia and diabetes:

(1) Can glycogen reserves in the retina in vivo serve as glucose equivalents during hypoglycemia, as has been recently shown for brain?[109]

(2) Does impaired glucose sensing, i.e. hypoglycemia unawareness in the brain, also affect the retina?

(3) If retinal glycogen metabolism and content are insulin or glucose-sensitive, then do increased retinal concentrations of glucose result in a lowered glycemic threshold of counter-regulation as observed in chronic hypoglycemia?

The experiments carried out by Ames and colleagues using isolated cat retinas maintained in a miniaturized heart–lung apparatus[15,45] characterized the energy requirements of retinal function by monitoring retinal lactate production and oxygen consumption.[15] Finally, Lisenmeier and colleagues[29,110] have devoted a great deal of effort to studying the distribution and consumption of oxygen throughout the inner and outer retina in anesthetized cats during systemic normoxia and hypoxia and more recently retinal pH changes during hypoxemia, hyperglycemia and during anaerobic glycolysis. Other important contributions to energy metabolism of the retina have been carried out in monkey and miniature pigs in vivo,[111-113] and in rabbit[22] and rat in vitro.[51,114]

Complementary to laser Doppler flow measurements[115-117] and oxygen electrode measurements in anesthetized pig,[25,118] recent, exciting developments using BOLD fMRI (blood-oxygen-level-dependent functional magnetic resonance imaging) and intravascular contrast agents in the intact eye of the cat[119–121] provide significant promise for non-invasive, real-time visualization of retinal function, structure, blood flow and oxygenation in health and disease, for example in diabetic retinopathy. Multiple MRI contrast mechanisms presently permit laminar spatial visualization of the retina, corresponding to the inner and outer retina and choroid microvasculature.[122]

8. Metabolic interactions between vertebrate photoreceptors and Müller glial cells

Can the experimental findings in insect retina be generalized to mammals? This question was explored in the retina of the guinea pig[53] (see Figs 12.3 and 12.4) where analogous experimental findings supported the concept of metabolic compartmentation (see Fig. 12.6B). Four conclusions were drawn from that study and are summarized in Figure 12.12:

(1) Müller glial cells massively produced lactate from radiolabeled glucose, most of which was found extracellularly in the corresponding cell bath (Fig. 12.12A and B, blue bars).

(2) The effects of altering illumination of the cell complex (thereby specifically altering photoreceptor function and metabolism) affect the production, release and/or uptake of glucose/glycogen-derived lactate in Müller glia (Fig. 12.12A and B, a purple and green bars).

(3) From numerical calculations, the "missing lactate" in the bath of the cell complex (vertical arrow in Fig. 12.12B) matched the expected production of $^{14}CO_2$ by photoreceptor oxidative metabolism (see Box 12.1) and was antagonized by the addition of unlabeled lactate (Fig 12.12C).

(4) Glial-derived lactate is preferentially consumed by purified populations of acutely isolated photoreceptors (Fig 12.12D), even in the presence of millimolar concentrations of glucose (Fig 12.12E).

These results show that a *relative* compartmentation of substrate *formation* to the glial cells and substrate *use* to the photoreceptor neurons previously established in bee retina can be generalized to mammals (see Fig. 12.6). Such compartmentation may underlie the need for a continuous trafficking of lactate/alanine from glia to neurons (see also section 6). It should be noted that guinea pig photoreceptors do in fact take up and phosphorylate glucose,[53] but when challenged with either one alone or a combination of both, lactate is a preferred substrate as evaluated by $^{14}CO_2$ produced (Fig. 12.12E). Lastly the reader is reminded of the finding that the presence of photoreceptors, thus the model of Müller cells still attached to photoreceptors, modifies in a light/dark-dependent manner either the synthesis, release or metabolic fate of glucose-derived metabolites, i.e. lactate and glutamine, in the Müller glia.[53]

To summarize, the neuroretina can serve as an important CNS model in which to assess the use of energy fuel as related to function. Neurons and glia divide metabolic labor in a way that is vital to function. The general finding that substances released by neurons, i.e. excitatory neurotransmitter

---

Box 12.1 Appendix: Numerical calculations of missing lactate and its consumption by photoreceptors (PH)

- To a first approx, the amount of CO_2 produced by PH = amount of oxygen consumed
- PH ~50% of tissue; major contributors to CO_2 production; consume 40 μl O_2/gram min
- 30 min incubation: ~16,000 cpm μg protein lactate in Muller cell bath
- ~3000 in cell complex bath (dark conditions)
- 13,000 cpm/μg prot • (45 dpm 40 cpm) • [2.22 × 10^6 dpm μCi • 146 μCi/μmol glucose]^{-1}
- => 4.6 • 10^{-6} μmol glucose/μg prot or 4.6 • 10^{-6} μmol glucose/10^{-8} liters or 4.6 mM glucose
- Therefore, 4.6 mM glucose 30 min ≈ 0.15 mM glucose min
- 1 glucose →1 6O_2, so 0.9 mM O_2 min; 0.9 mM O_2 l • min × 22.4 mL O_2 /mM
- => 20 mL O_2/l • min is 20 μl O_2 cm^3 • min; => 40 μl O_2/cm^3 photoreceptors • min
- Therefore, 40 μl O_2 40 μl O_2 or (value from literature for PH in darkness/cm^3 • PH • min):
- (measured value for PH in cell complex/cm^3 • PH • min) × 13,000 cpm/μg prot
- Yields 13,000 cpm/μg protein
- => amount of CO_2 that would be produced from missing lactate ~ amount of O_2 consumed by PH

---

glutamate, constitute signals to glia as well as being materials to be removed or regulated underscores the emerging paradigm that neurons and glia act in concert as metabolic partners, with information in the form of metabolic intermediates and signaling molecules continuously flowing between them. Lastly, the similarity of metabolic shuttling between insect and mammalian retina strongly suggests that this mechanism might be universally operative in the CNS, although direct experimental evidence remains difficult to come by.

9. Metabolic interaction between photoreceptors and retinal pigment epithelia

The retinal pigment cell epithelium (RPE) forms part of the choroidal blood /outer retinal barrier, facing the outer photoreceptor segments on its apical membrane side, Bruch's membrane on its basolateral membrane side and plays an essential role in maintaining photoreceptor excitability and sensitivity and the reisomerization of all-*trans*-retinal after photon absorption.[123] In rod outer segments (ROS), all-*trans*-retinal is transformed to all-*trans*-retinol by retinol dehydrogenase using NADPH. NADPH is restored in ROS by the pentose phosphate pathway utilizing high amounts of glucose supplied by choriocapillaries. The retinal formed is transported to PE cells where regeneration of 11-*cis*-retinal occurs (see Chapter 13).

The RPE apical membrane is separated from the neuroretina by the subretinal space which serves to mediate the

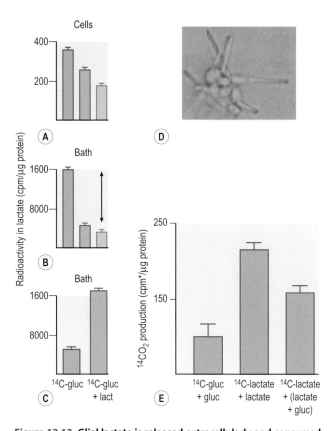

Figure 12.12 Glial lactate is released extracellularly and consumed by photoreceptors in mammalian retina. Metabolic trafficking between neurons and glia was explored using three cell models: purified populations of acutely isolated Müller glia (Fig. 12.4B), acutely isolated photoreceptors (**D**) and the acutely isolated cell complex (Fig. 12.3B) from guinea pig retina. The effect of altering illumination of the cell complex (purple: maintained in light vs green: maintained in darkness) was the strategy used to specifically alter photoreceptor function and metabolism. Darkness increases photoreceptor metabolism and the release of neurotransmitter glutamate. Retinal glia are not directly affected by light. Müller cells alone massively produced lactate from uniformly radiolabeled glucose, most of which was released into the extracellular space, as shown by the blue columns in (**A**) and (**B**). The effects of altering illumination of the cell complex, particularly in darkness, lowered significantly the amount of radiolabeled lactate in the corresponding bath (B, second and third columns) compared to the bath of Müller cells alone (B, first column). This "missing lactate" matched the expected production of $^{14}CO_2$ by photoreceptor oxidative metabolism. As expected, this "missing lactate" was also antagonized by the addition of unlabeled lactate to the cell complex bath (**C**), indicating that lactate released extracellularly by glia was consumed. Finally, (**E**) shows the amount of $^{14}CO_2$ produced by the cells in (D) as a function of 3 different substrate mixtures. The conclusion drawn was that glial lactate was preferentially consumed by photoreceptors, even in the presence of millimolar concentrations of glucose. Overall, the effects of altering illumination of the cell complex affected the production, release and or uptake of glucose-derived glial lactate. (Modified from Poitry-Yamate et al 1995, *Neurosci Lett* 122(2): 241–4. Copyright Elsevier 1995.)

transport of metabolic intermediates, ions and water between the choroid and the retina. Considering the vascular sources of fuel substrate, the choroidal microcirculation is one source of glucose for photoreceptor outer segments. However, it is presently unclear whether glucose crosses the RPE on glucose transporters GLUT1 and GLUT3[124] or if it is consumed and transformed through the glycolytic pathway to lactate for transfer and to a lesser extent oxidized through

the Kreb's cycle.[21] This is a relevant point since RPE cells depend on glucose as their primary energy substrate.[125] The expression of the monocarboxylate transporter MCT3 on the RPE basolateral side for the efflux of lactate from RPE to the choroidal microcirculation and the expression of MCT1 on the apical side for the transport of lactate and water from the retina to choroid[76] (see Chapter 13) suggests that lactate is not used by RPE cells, at least when glucose is non-limiting.[125] Thus, if RPE cells synthesize and release glucose-derived lactate in vivo, as has been demonstrated in acutely isolated RPE[21] and in cultured RPE,[74] then any lactate leaving the eye as determined by either choroidal venous drainage or femoral artery-choroidal vein differences[38] is a two-fold contribution (neuroretina and RPE). Presently the contributions from these two sources have not been quantitated nor their contribution to the amount of effluxed lactate to the choroidal vein. These unknowns need to be taken into account when critically evaluating the ANLSH hypothesis that aerobic glycolysis in glia increases in response to activation of neurotransmitter glutamate systems and the neuronal oxidation of glial lactate.[55] In the eye, *endogenous* lactate production does not appear to be linked only to energy metabolism: the cotransport of H+ and lactate with water in the direction from retina to choroid plays a role in volume regulation and thereby retinal adhesion.[126]

10. Metabolic factors in the regulation of retinal blood flow

In a series of original papers on the metabolic factors involved in the regulation of retinal blood flow in monkey and miniature pigs, Tsacopoulos proposed that an extravascular, intraretinal acidosis was a key step in the process leading to retinal arteriolar vasodilation, either by hypercapnia, hypoxia or ischemia.[111,112,127–129] Moreover, he proposed that this acidosis would provoke the release of a substance from the periarteriolar glial cells which in turn would act on the smooth muscle cells of the arteriolar wall. This idea that retinal glia plays a central role in the mechanisms linking neuron function to blood flow was highly original at that time and recognized the anatomical disposition of the glia as cells interpositioning themselves between the neurons and retinal vasculature. Indeed, his working hypothesis that the retinal arteriolar tone might be controlled by factors, such as K+, lactate and NO released from the retinal tissue surrounding the arterioles was tested in the miniature pig.[130–132] Of these factors, intraretinal lactate and NO, both of which can be synthesized from glucose in Müller glia, were found to either mediate acute hypoxia-induced vasodilation or basal retinal arteriolar tone (e.g. for review, see Pournaras et al[133]).

Further experimental evidence supporting the role of extracellular lactate as a dynamic vasoactive signal was provided by studying freshly isolated rat retinal microvessels using a combination of patch-clamp, fluorescence imaging and time-lapse photography.[134] In that study, lactate exposure was associated with an intracellular rise of calcium in pericytes, cell contraction and lumen constriction when energy supplies were adequate. Hypoxia, on the other hand, switched lactate's effect from vasocontraction to

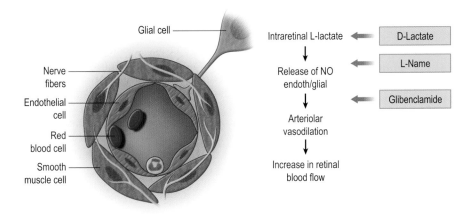

Figure 12.13 Components of the blood vessel wall hypothesized to play a role in arteriolar vasodilation and retinal blood flow. Endothelial cells and smooth muscle cells are two major components of retinal arteries and arterioles. The retinal glial cells send their endfoot processes out and over the vascular wall. The putative regulatory role of lactate is shown in the schema. Retinal arteriolar tone is controlled by the dynamic vasoactive signals lactate and NO. Intraretinal L-lactate, but not D-lactate, causes a pH-independent dilation of retinal arterioles. Juxta-arteriolar application, but not intravenous delivery, of L-NAME, an inhibitor of NO synthase, suppresses the vasodilatory effect of L-lactate, suggesting a glial rather than an endothelial source of NO. The L-lactate-induced vasodilatory response is mediated by stimulation of NO synthase and subsequent activation of smooth muscle guanylyl cyclase, leading to the opening of glibenclamide-sensitive K_{ATP} channels in smooth muscle. The regulation of arterial smooth muscle membrane potential through activation or inhibition of K^+ channel activity provides an important mechanism to dilate or constrict arteries or arterioles. Local changes in retinal arteriolar diameter in response to intraretinal vasoactive metabolites links the regulation of retinal blood flow to the metabolic needs of the surrounding retinal tissue.

vasorelaxation. This dual vasoactive capability may provide an efficient mechanism to match microvascular function to local metabolic needs.

In mini-pigs preretinal microinjections of L-lactate (0.5 mol/l, pH 2 and 7.4) close to the retinal arterioles produced a local, segmental and reversible arteriolar dilation while D-lactate had no effect. This effect was not pH-dependent nor mediated by prostaglandins. On the other hand, both systemic and sublingual administration of L-lactate (0.1 mol/l, pH 2) had no effect on retinal arteriolar diameter,[131] suggesting that exposure to endothelial cells of the arterioles to lactate was not sufficient to trigger a vasomotor response. These authors concluded that to induce vasodilation, lactate had to reach the vascular smooth muscle cells from the vitreoretinal side, raising the hypothesis that intraretinal lactate is a mediator of acute hypoxia-induced vasodilation.

Intravenous administration of sodium lactate has been reported to increase retinal blood flow at steady state[135] or during flicker stimulation.[135,136] High-dose lactate reduced flicker response, whereas low-dose lactate increased it. This is most likely related to the fact that high-dose lactate leads to pronounced pre-vasodilation and a reduced flicker response, whereas low-dose lactate shifts the ratio of cytosolic free NADH to NAD⁺, resulting in increased flicker response. Both observations indicate that the ratio of cytosolic-free NADH to NAD⁺ plays a critical role in the maintenance of retinal vascular tone (also known as the "pseudohypoxia hypothesis").[135]

Uptake of intravenous or intravitreal lactate by vascular endothelial cells or glial cells via monocarboxylate transporters[137–140] causes retinal arteriolar dilation, predominantly via stimulation of NO synthase[141] and subsequent activation of guanylate cyclase. In turn, guanylate cyclase/cGMP signaling triggers opening of K_{ATP} channels for vasodilation. This process suggests that lactate either mediates the release of endothelial vasoactive substances (i.e. NO) or interferes with the metabolism and release of vasoactive substances by cells surrounding the arterioles (i.e. glial cells).

Juxta-arteriolar administration of L-lactate induced vasodilation of retinal arterioles in minipigs was also observed during continuous intravenous infusion of L-NAME, inhibiting endothelial-derived NO. In contrast, juxta-arteriolar L-NAME microinjection inhibiting NOS within the neuronal/glial cells significantly suppressed the vasodilatory effect of L-lactate. Those findings suggested that, in vivo, neuronal/glial cells are the major source of NO mediating the retinal arteriolar vasodilatation response to lactate in minipigs (Fig. 12.13).[142]

11. Metabolic pathway leading to nitric oxide release

The diffusible and non-polar gas, nitrogen monoxide (NO) is synthesized from L-arginine by nitric oxide synthase (NOS). Until recently, NO has been generally recognized as an endothelium-derived releasing factor that controls vascular tone. Acutely isolated mammalian retinal Müller cells were shown to synthesize the NO precursor L-arginine from glucose as well as the key requisite amino acids – fumarate and aspartate – through which arginine is synthesized.[132,143] Moreover, NOS has been shown to be localized to Müller cells using NADH diaphorase histochemistry and NOS immunocytochemistry.[144,145] Hence, the conversion of L-arginine and oxygen into NO and L-citrulline through the enzymatic action of NOS highlights the role of glia in the generation and release of extravascular, intraretinal and periarteriolar NO, for either target neurons or in promoting arteriolar vasodilation and/or controlling the basal dilating arteriolar tone.

Support for this hypothesis in vivo in the inner retina of the miniature pig is the finding that:

(1) an increasing NO gradient from the vitreous towards the inner retina is measurable using an NO microprobe; this gradient is reversibly modulated by light; and light flicker stimulation of the dark-adapted retina induces a reversible increase of NO;[132]

(2) focal microinjection, but not intravenous infusion, of the NO-synthase inhibitor nitro-L-arginine leads to a segmental and reversible arteriolar vasoconstriction of 45%.[132] Intravenous[135] and intravitreal administration of L-arginine also increases retinal blood flow, but it remains to be shown that this effect is related to an increase of NO production; and

(3) in vivo, neuronal/glial cells are the major source of NO mediating the retinal arteriolar vasodilatation response to lactate in minipigs.[142]

Local factors released by neurons and perivascular glial cells appear to be involved in the regulation of the vascular tone of the retinal resistance vessels through the interaction of myogenic and metabolic mechanisms. These factors may be ionic, molecular or related to arterial blood gas modifications. They affect the arteriolar tone and thus regulate the retinal vasomotor responses, relaxing or contracting the tone. Nitric oxide (NO) and prostacyclin (PGI_2) are probable relaxing factors, whereas the contracting factors include endothelin-1 (ET-1), angiotensin II, and cyclo-oxygenase (COX) products, such as thromboxane-A2 (TXA_2) and prostaglandin H_2 (PGH_2).[133]

What may be the role of extracellular lactate in the contraction and relaxation of the vessel wall according to the metabolic needs of the tissue? Lactate could either mediate the release of endothelial vasoactive substances (i.e. NO), or alternatively, lactate could interfere with the metabolism of cells surrounding the arterioles namely glial cells and neurons, leading either to release of vasoactive substances (NO, PGs) or activation of retinal endothelin receptors.

The close interaction between PG and NO metabolic pathways could insure that, when one process is inhibited, the other may rapidly compensate for the deficiency and maintain blood flow constant. However, whether this interaction is a major component of the control of vasomotion and autoregulation in the inner retina remains to be shown.

Impairment of structure and function of the neural tissue and endothelium is the cause of abnormal retinal blood flow regulation observed during the evolution of ischemic microangiopathies.[133] However, impaired metabolism may underlie alterations of both neural structure and function. This raises the question whether impaired glycolysis leads to a decrease in preretinal lactate, which in turn affects the NO pathway. Equally as important is the transport of glucose to neural cells which is compromised in ischemia and/or by alterations of the retinal–blood barrier, both of which are functional triggers leading to neovascularization and formation of macular edema.

In this regard, the expression of hypoxia-related vascular endothelial growth factor (VEGF), known as the major inducer of angiogenesis, correlates in a temporal and spatial manner with the formation of neovessels and retinal–blood barrier alterations. It would be of future clinical importance to determine whether a restoration of blood flow in retinas with chronic ischemic microangiopathies is correlated with a decrease in VEGF expression. Insights into the multiple pathways controlling retinal blood flow, tissue oxygenation and metabolism should offer new therapeutic strategies to restore retinal blood flow and prevent damage observed during the evolution of ischemic microangiopathies.

References

1. Kimble EA, Svoboda RA, Ostroy SE. Oxygen consumption and ATP changes of the vertebrate photoreceptor. Exp Eye Res 1980; 31(3):271–288.
2. Mulligan SJ, MacVicar BA. Calcium transients in astrocyte endfeet cause cerebrovascular constrictions. Nature 2004; 431(7005):195–199.
3. Metea MR, Newman EA. Glial cells dilate and constrict blood vessels: a mechanism of neurovascular coupling. J Neurosci 2006; 26(11):2862–2870.
4. Araque A, Carmignoto G, Haydon PG. Dynamic signaling between astrocytes and neurons. Annu Rev Physiol 2001; 63:795–813.
5. Haydon PG, Carmignoto G. Astrocyte control of synaptic transmission and neurovascular coupling. Physiol Rev 2006; 86(3):1009–1031.
6. Newman EA, Zahs KR. Calcium waves in retinal glial cells. Science 1997; 275(5301):844–847.
7. Guthrie PB, Knappenberger J, Segal M, Bennett MV, Charles AC, Kater SB. ATP released from astrocytes mediates glial calcium waves. J Neurosci 1999; 19(2):520–528.
8. Fiacco TA, McCarthy KD. Intracellular astrocyte calcium waves in situ increase the frequency of spontaneous AMPA receptor currents in CA1 pyramidal neurons. J Neurosci 2004; 24(3):722–732.
9. Puthussery T, Fletcher EL. Synaptic localization of P2X7 receptors in the rat retina. J Comp Neurol 2004; 472(1):13–23.
10. Puthussery T, Fletcher EL. P2X2 receptors on ganglion and amacrine cells in cone pathways of the rat retina. J Comp Neurol 2006; 496(5):595–609.
11. Puthussery T, Fletcher EL. Neuronal expression of P2X3 purinoceptors in the rat retina. Neuroscience 2007; 146(1):403–414.
12. Puthussery T, Yee P, Vingrys AJ, Fletcher EL. Evidence for the involvement of purinergic P2X receptors in outer retinal processing. Eur J Neurosci 2006; 24(1):7–19.
13. Takano T, Tian GF, Peng W et al. Astrocyte-mediated control of cerebral blood flow. Nat Neurosci 2006; 9(2):260–267.
14. Zonta M, Angulo MC, Gobbo S et al. Neuron-to-astrocyte signaling is central to the dynamic control of brain microcirculation. Nat Neurosci 2003; 6(1):43–50.
15. Ames A, 3rd, Li YY, Heher EC, Kimble CR. Energy metabolism of rabbit retina as related to function: high cost of Na+ transport. J Neurosci 1992; 12(3):840–853.
16. Palay SL. Synapses in the central nervous system. J Biophys Biochem Cytol 1956; 2(4,suppl):193–207.
17. Attwell D, Laughlin SB. An energy budget for signaling in the grey matter of the brain. J Cerebral Blood Flow Metab 2001; 21:1133–1145.
18. Lowry OH, Roberts NR, Schulz DW, Clow JE, Clark JR. Quantitative histochemistry of retina. II. Enzymes of glucose metabolism. J Biol Chem 1961; 236:2813–2820.
19. Kuwabara T, Cogan DG. Retinal glycogen. Arch Ophthalmol 1961; 66:680–688.
20. Kuwabara T, Kogan DG. Tetrazolium studies on the retina: III. Activity of metabolic intermediates and miscellaneous substrates. J Histochem Cytochem 1960; 8:214–224.
21. Glocklin VC, Potts AM. The metabolism of retinal pigment cell epithelium. II. respiration and glycolysis. Invest Ophthalmol 1965; 4:226–234.
22. Cohen LH, Noell WK. Glucose catabolism of rabbit retina before and after development of visual function. J Neurochem 1960; 5:253–276.
23. Hoang QV, Linsenmeier RA, Chung CK, Curcio CA. Photoreceptor inner segments in monkey and human retina: mitochondrial density, optics, and regional variation. Vis Neurosci 2002; 19(4):395–407.
24. Kageyama GH, Wong-Riley MT. The histochemical localization of cytochrome oxidase in the retina and lateral geniculate nucleus of the ferret, cat, and monkey, with particular reference to retinal mosaics and ON/OFF-center visual channels. J Neurosci 1984; 4(10):2445–2459.
25. Pournaras CJ, Riva CE, Tsacopoulos M, Strommer K. Diffusion of O_2 in the retina of anesthetized miniature pigs in normoxia and hyperoxia. Exp Eye Res 1989; 49(3):347–360.
26. Pournaras CJ. Retinal oxygen distribution. Its role in the physiopathology of vasoproliferative microangiopathies. Retina 1995; 15(4):332–347.
27. Alder VA, Cringle SJ. The effect of the retinal circulation on vitreal oxygen tension. Curr Eye Res 1985; 4(2):121–129.
28. Pournaras CJ, Miller JW, Gragoudas ES et al. Systemic hyperoxia decreases vascular endothelial growth factor gene expression in ischemic primate retina. Arch Ophthalmol 1997; 115(12):1553–1558.
29. Linsenmeier RA, Braun RD. Oxygen distribution and consumption in the cat retina during normoxia and hypoxemia. J Gen Physiol 1992; 99(2):177–197.
30. Riva CE, Pournaras CJ, Tsacopoulos M. Regulation of local oxygen tension and blood flow in the inner retina during hyperoxia. J Appl Physiol 1986; 61(2):592–598.
31. Braun RD, Linsenmeier RA. Retinal oxygen tension and the electroretinogram during arterial occlusion in the cat. Invest Ophthalmol Vis Sci 1995; 36(3):523–541.
32. Bill A, Sperber GO. Aspects of oxygen and glucose consumption in the retina: effects of high intraocular pressure and light. Graefes Arch Clin Exp Ophthalmol 1990; 228:124–127.
33. Alder VA, Cringle SJ, Constable IJ. The retinal oxygen profile in cats. Invest Ophthalmol Vis Sci 1983; 24(1):30–36.

34. Linsenmeier RA. Effects of light and darkness on oxygen distribution and consumption in the cat retina. J Gen Physiol 1986; 88(4):521–542.
35. Yu DY, Cringle SJ, Su EN. Intraretinal oxygen distribution in the monkey retina and the response to systemic hyperoxia. Invest Ophthalmol Vis Sci 2005; 46(12):4728–4733.
36. Birol G, Wang S, Budzynski E, Wangsa-Wirawan ND, Linsenmeier RA. Oxygen distribution and consumption in the macaque retina. Am J Physiol Heart Circ Physiol 2007; 293(3):H1696–H1704.
37. Medrano CJ, Fox DA. Oxygen consumption in the rat outer and inner retina: light- and pharmacologically-induced inhibition. Exp Eye Res 1995; 61(3):273–284.
38. Wang L, Kondo M, Bill A. Glucose metabolism in cat outer retina. Effects of light and hyperoxia. Invest Ophthalmol Vis Sci 1997; 38(1):48–55.
39. Dollery CT, Bulpitt CJ, Kohner EM. Oxygen supply to the retina from the retinal and choroidal circulations at normal and increased arterial oxygen tensions. Invest Ophthalmol 1969; 8(6):588–594.
40. Haugh-Scheidt LM, Griff ER, Linsenmeier RA. Light-evoked oxygen responses in the isolated toad retina. Exp Eye Res 1995; 61(1):73–81.
41. Haugh-Scheidt LM, Linsenmeier RA, Griff ER. Oxygen consumption in the isolated toad retina. Exp Eye Res 1995; 61(1):63–72.
42. Ahmed J, Braun RD, Dunn RJ, Linsenmeier RA. Oxygen distribution in the macaque retina. Invest Ophthalmol Vis Sci 1993; 34(3):516–521.
43. Ames A, 3rd, Walseth TF, Heyman RA, Barad M, Graeff RM, Goldberg ND. Light-induced increases in cGMP metabolic flux correspond with electrical responses of photoreceptors. J Biol Chem 1986; 261(28):13034–13042.
44. Goldberg ND, Ames AA, 3rd, Gander JE, Walseth TF. Magnitude of increase in retinal cGMP metabolic flux determined by 18O incorporation into nucleotide alpha-phosphoryls corresponds with intensity of photic stimulation. J Biol Chem 1983; 258(15):9213–9219.
45. Ames A, 3rd, Nesbett FB. In vitro retina as an experimental model of the central nervous system. J Neurochem 1981; 37(4):867–877.
46. Babel J, Stangos N. Essai de corrélation entre l'ERG et les ultrastructures de la rétine. I. Action du monoiodoacétate. Arch Opht (Paris) 1973; 33:297–312.
47. Futterman S, Kinoshita JH. Metabolism of the retina. I. Respiration of cattle retina. J Biol Chem 1959; 234(4):723–726.
48. Krebs HA. The Pasteur effect and the relations between respiration and fermentation. Essays Biochem 1972; 8:1–34.
49. Racker E. History of the Pasteur effect and its pathobiology. Mol Cell Biochem 1974; 5(1–2):17–23.
50. Matchinsky FM. Energy metabolism of the microscopic structures of the cochlea, the retina, and the cerebellum. In: Costa E, Giacobini E, eds. Biochemistry of simple neuronal models. Advances in Biochemical Psychopharmacology. New York: Raven Press; 1970:217–243.
51. Winkler BS. Glycolytic and oxidative metabolism in relation to retinal function. J Gen Physiol 1981; 77(6):667–692.
52. Niemeyer G. The function of the retina in the perfused eye. Doc Ophthalmol 1975; 39(1):53–116.
53. Poitry-Yamate CL, Poitry S, Tsacopoulos M. Lactate released by Müller glial cells is metabolized by photoreceptors from mammalian retina. J Neurosci 1995; 15(7 Pt 2):5179–5191.
54. Chih CP, Lipton P, Roberts EL, Jr. Do active cerebral neurons really use lactate rather than glucose? Trends Neurosci 2001; 24(10):573–578.
55. Dienel GA, Cruz NF. Nutrition during brain activation: does cell-to-cell lactate shuttling contribute significantly to sweet and sour food for thought? Neurochem Int 2004; 45(2–3):321–351.
56. Pellerin L, Magistretti PJ. Food for thought: challenging the dogmas. J Cereb Blood Flow Metab 2003; 23(11):1282–1286.
57. Gladden LB. Lactate metabolism: a new paradigm for the third millennium. J Physiol 2004; 558(Pt 1):5–30.
58. Hyder F, Patel AB, Gjedde A, Rothman DL, Behar KL, Shulman RG. Neuronal-glial glucose oxidation and glutamatergic-GABAergic function. J Cereb Blood Flow Metab 2006; 26(7):865–877.
59. Schurr A. Lactate: the ultimate cerebral oxidative energy substrate? J Cereb Blood Flow Metab 2006; 26(1):142–152.
60. Kuwabara T, Cogan DG, Futterman S, Kinoshita JH. Dehydrogenases in the retina and Müller's fibers. J Histochem Cytochem. 1959 Jan;7(1):67–68.
61. Poitry-Yamate C, Tsacopoulos M. Glial (Müller) cells take up and phosphorylate [3H]2-deoxy-D-glucose in mammalian retina. Neurosci Lett 1991; 122(2):241–244.
62. Poitry-Yamate CL, Tsacopoulos M. Glucose metabolism in freshly isolated Müller glial cells from a mammalian retina. J Comp Neurol 1992; 320(2):257–266.
63. Wilson DJ. 2-deoxy-d-glucose uptake in the inner retina: an in vivo study in the normal rat and following photoreceptor degeneration. Trans Am Ophthalmol Soc 2002; 100:353–364.
64. Graymore C, Tansley K. Iodoacetate poisoning of the rat retina. I. Production of retinal degeneration. Br J Ophthalmol 1959; 43(3):177–185.
65. Graymore C. Biochemistry of the retina. In: Graymore CN, ed. Biochemistry of the eye. London: Academic Press, 1970:645–735.
66. Dick E, Miller RF. Extracellular K^+ activity changes related to electroretinogram components. I. Amphibian (I-type) retinas. J Gen Physiol 1985; 85(6):885–909.
67. Lachapelle P, Benoit J, Guite P, Tran CN, Molotchnikoff S. The effect of iodoacetic acid on the electroretinogram and oscillatory potentials in rabbits. Doc Ophthalmol 1990; 75(1):7–14.
68. Hsu SC, Molday RS. Glucose metabolism in photoreceptor outer segments. Its role in phototransduction and in NADPH-requiring reactions. J Biol Chem 1994; 269(27):17954–17959.
69. Archer SN, Ahuja P, Caffe R et al. Absence of phosphoglucose isomerase-1 in retinal photoreceptor, pigment epithelium and Müller cells. Eur J Neurosci 2004; 19(11):2923–2930.
70. Sperling HG, Harcombe ES, Johnson C. Stimulus controlled labeling of cones in the macaque monkey with 3H-2-D-deoxyglucose. In: Hollyfield JG, Vidrio EA, eds. The structure of the eye. New York: Elsevier, 1982:56–60.
71. Nihira M, Anderson K, Gorin FA, Burns MS. Primate rod and cone photoreceptors may differ in glucose accessibility. Invest Ophthalmol Vis Sci 1995; 36(7):1259–1270.
72. Tsacopoulos M, Magistretti PJ. Metabolic coupling between glia and neurons. J Neurosci 1996; 16(3):877–885.
73. Tsacopoulos M, Poitry-Yamate CL, MacLeish PR, Poitry S. Trafficking of molecules and metabolic signals in the retina. Prog Retin Eye Res 1998; 17(3):429–442.
74. Winkler BS, Starnes CA, Sauer MW, Firouzgan Z, Chen SC. Cultured retinal neuronal cells and Müller cells both show net production of lactate. Neurochem Int 2004; 45(2–3):311–320.
75. Hertz L. The astrocyte-neuron lactate shuttle: a challenge of a challenge. J Cereb Blood Flow Metab 2004; 24(11):1241–1248.
76. Halestrap AP, Price NT. The proton-linked monocarboxylate transporter (MCT) family: structure, function and regulation. Biochem J 1999; 343(2):281–299.
77. Polyak SL. Neuroglia of the retina. The retina. Chicago: University of Chicago Press; 1941:343–365.
78. Sarthy VP, Pignataro L, Pannicke T et al. Glutamate transport by retinal Müller cells in glutamate/aspartate transporter-knockout mice. Glia 2005; 49(2):184–196.
79. Brew H, Attwell D. Electrogenic glutamate uptake is a major current carrier in the membrane of axolotl retinal glial cells. Nature 1987; 327(6124):707–709.
80. Biedermann B, Bringmann A, Reichenbach A. High-affinity GABA uptake in retinal glial (Müller) cells of the guinea pig: electrophysiological characterization, immunohistochemical localization, and modeling of efficiency. Glia 2002; 39(3):217–228.
81. Gadea A, Lopez E, Hernandez-Cruz A, Lopez-Colome AM. Role of Ca^{2+} and calmodulin-dependent enzymes in the regulation of glycine transport in Müller glia. J Neurochem 2002; 80(4):634–645.
82. Sarthy PV. The uptake of [3H]gamma-aminobutyric acid by isolated glial (Müller) cells from the mouse retina. J Neurosci Methods 1982; 5(1–2):77–82.
83. Derouiche A, Rauen T. Coincidence of L-glutamate/L-aspartate transporter (GLAST) and glutamine synthetase (GS) immunoreactions in retinal glia: evidence for coupling of GLAST and GS in transmitter clearance. J Neurosci Res 1995; 42(1):131–143.
84. Riepe RE, Norenburg MD. Müller cell localization of glutamine synthetase in rat retina. Nature 1977; 268(5621):654–655.
85. Linser PJ, Sorrentino M, Moscona AA. Cellular compartmentalization of carbonic anhydrase-C and glutamine synthetase in developing and mature mouse neural retina. Brain Res 1984; 315(1):65–71.
86. Nagelhus EA, Mathiisen TM, Bateman AC et al. Carbonic anhydrase XIV is enriched in specific membrane domains of retinal pigment epithelium, Müller cells, and astrocytes. Proc Natl Acad Sci USA 2005; 102(22):8030–8035.
87. Pfeiffer-Guglielmi B, Francke M, Reichenbach A, Fleckenstein B, Jung G. Glycogen phosphorylase isozyme pattern in mammalian retinal Müller (glial) cells and in astrocytes of retina and optic nerve. Glia 2004; 49:84–95.
88. Ripps H, Witkovsky P. Neuron-glial interaction in the brain and retina. In: Progress in retinal research 1985:181–219.
89. Wasilewa P, Hockwin O, Korte I. Glycogen concentration changes in retina, vitreous body and other eye tissues caused by disturbances of blood circulation. Albrecht Von Graefes Arch Klin Exp Ophthalmol 1976; 199(2):115–120.
90. Coffe V, Carbajal RC, Salceda R. Glycogen metabolism in the rat retina. J Neurochem 2004; 88(4):885–890.
91. Schorderet M, Hof P, Magistretti PJ. The effects of VIP on cyclic AMP and glycogen levels in vertebrate retina. Peptides 1984; 5(2):295–298.
92. Rungger-Brandle E, Kolb H, Niemeyer G. Histochemical demonstration of glycogen in neurons of the cat retina. Invest Ophthalmol Vis Sci 1996; 37(5):702–715.
93. Hirsch-Hoffmann CH, Niemeyer G. Changes in plasma glucose level affect rod-, but not cone-ERG in the anesthetized cat. Clin Vis Sci 1993; 8:489–501.
94. Lajtha A. Brain energetics. Integration of molecular and cellular processes. In: Gibson GE, Dienel GA, eds. Handbook of neurochemistry and molecular neurobiology, 3rd edn, ed: Berlin: Springer-Verlag, 2007.
95. Kumugai AK. Glucose transport in brain and retina: implications in the management and complications of diabetes. Diabetes Metab Res Rev 1999; 15:261–273.
96. Frackowiak RSJ, Magistretti PJ, Shulman RG, Altman JS, Adams M. Neuroenergetics: relevance in functional brain imaging. Strasbourg: Human Frontiers Science Program, 2001: 1–214.
97. Magistretti PJ, Pellerin L. Astrocytes couple synaptic activity to glucose utilization in the brain. News Physiol Sci 1999; 14:177–182.
98. Magistretti PJ, Pellerin L. Functional brain imaging: role metabolic coupling between astrocytes and neurons. Rev Med Suisse Romande 2000; 120(9):739–742.
99. Raichle ME. Functional brain imaging and human brain function. J Neurosci 2003; 23(10):3959–3962.
100. Dimitracos SA, Tsacopoulos M. The recovery from a transient inhibition of the oxidative metabolism of the photoreceptors of the drone (*Apis mellifera O*). J Exp Biol 1985; 119:165–181.
101. Evêquoz V, Stadelmann A, Tsacopoulos M. The effect of light on glycogen turnover in the retina of the intact honeybee drone (*Apis mellifera*). J Comp Physiol 1983 150:69–75.
102. Tsacopoulos M, Coles JA, Van de Werve G. The supply of metabolic substrate from glia to photoreceptors in the retina of the honeybee drone. J Physiol (Paris) 1987; 82(4):279–287.
103. Tsacopoulos M, Evequoz-Mercier V, Perrottet P, Buchner E. Honeybee retinal glial cells transform glucose and supply the neurons with metabolic substrate. Proc Natl Acad Sci USA 1988; 85(22):8727–8731.
104. Tsacopoulos M, Veuthey AL, Saravelos SG, Perrottet P, Tsoupras G. Glial cells transform glucose to alanine, which fuels the neurons in the honeybee retina. J Neurosci 1994; 14(3 Pt 1):1339–1351.
105. Tsacopoulos M, Poitry-Yamate CL, Poitry S. Ammonium and glutamate released by neurons are signals regulating the nutritive function of a glial cell. J Neurosci 1997; 17(7):2383–2390.
106. Niemeyer G, Steinberg RH. Differential effects of pCO_2 and pH on the ERG and light peak of the perfused cat eye. Vision Res 1984; 24(3):275–280.
107. Macaluso C, Onoe S, Niemeyer G. Changes in glucose level affect rod function more than cone function in the isolated, perfused cat eye. Invest Ophthalmol Vis Sci 1992; 33(10):2798–2808.

108. Niemeyer G. Glucose concentration and retinal function. Clin Neurosci 1997; 4(6):327–335.

109. Gruetter R. Glycogen: the forgotten cerebral energy store. J Neurosci Res 2003; 74(2):179–183.

110. Padnick-Silver L, Linsenmeier RA. Quantification of in vivo anaerobic metabolism in the normal cat retina through intraretinal pH measurements. Vis Neurosci 2002; 19(6):793–806.

111. Tsacopoulos M, Baker R, Levy S. Studies on retinal oxygenation. Adv Exp Med Biol 1976; 75:413–416.

112. Tsacopoulos M, Levy S. Intraretinal acid-base studies using pH glass microelectrodes: effect of respiratory and metabolic acidosis and alkalosis on inner-retinal pH. Exp Eye Res 1976; 23(5):495–504.

113. Tsacopoulos M, Beauchemin ML, Baker R, Babel J. Studies of experimental retinal focal ischaemia in miniature pigs. In: Vision and circulation. Proceedings of the Third W Mackenzie Memorial Symposium. London: Kimpton, 1976:93.

114. Bui BV, Kalloniatis M, Vingrys AJ. The contribution of glycolytic and oxidative pathways to retinal photoreceptor function. Invest Ophthalmol Vis Sci 2003; 44(6):2708–2715.

115. Riva CE, Pournaras CJ, Poitry-Yamate CL, Petrig BL. Rhythmic changes in velocity, volume, and flow of blood in the optic nerve head tissue. Microvasc Res 1990; 40(1):36–45.

116. Riva CE, Harino S, Shonat RD, Petrig BL. Flicker evoked increase in optic nerve head blood flow in anesthetized cats. Neurosci Lett 1991; 128(2):291–296.

117. Riva CE, Grunwald JE, Petrig BL. Autoregulation of human retinal blood flow. An investigation with laser Doppler velocimetry. Invest Ophthalmol Vis Sci 1986; 27(12):1706–1712.

118. Petropoulos IK, Pournaras CJ. Effect of indomethacin on the hypercapnia-associated vasodilation of the optic nerve head vessels: an experimental study in miniature pigs. Ophthalmic Res 2005; 37(2):59–66.

119. Duong TQ, Ngan SC, Ugurbil K, Kim SG. Functional magnetic resonance imaging of the retina. Invest Ophthalmol Vis Sci 2002; 43(4):1176–1181.

120. Shen Q, Cheng H, Pardue MT et al. Magnetic resonance imaging of tissue and vascular layers in the cat retina. J Magn Res Imaging 2006; 23(4):465–472.

121. Duong TQ, Pardue MT, Thule PM et al. Layer-specific anatomical, physiological and functional MRI of the retina. NMR Biomed 2008; 21(9):978–996.

122. Cheng H, Nair G, Walker TA et al. Structural and functional MRI reveals multiple retinal layers. Proc Natl Acad Sci USA 2006; 103(46):17525–17530.

123. Strauss O. The retinal pigment epithelium in visual function. Physiol Rev 2005; 85(3):845–881.

124. Bergersen L, Johannsson E, Veruki ML et al. Cellular and subcellular expression of monocarboxylate transporters in the pigment epithelium and retina of the rat. Neuroscience 1999; 90(1):319–331.

125. Wood JP, Chidlow G, Graham M, Osborne NN. Energy substrate requirements of rat retinal pigmented epithelial cells in culture: relative importance of glucose, amino acids, and monocarboxylates. Invest Ophthalmol Vis Sci 2004; 45(4):1272–1280.

126. Zeuthen T, Hamann S, la Cour M. Cotransport of H+, lactate and H2O by membrane proteins in retinal pigment epithelium of bullfrog. J Physiol 1996; 497(Pt 1): 3–17.

127. Tsacopoulos M, Baker R, David NJ, Strauss J. The effect of arterial PCO_2 on inner-retinal oxygen consumption rate in monkeys. Invest Ophthalmol 1973; 12(6):456–460.

128. Bardy M, Tsacopoulos M. Metabolic changes in the retina after experimental microembolism in the miniature pig. Klin Monatsbl Augenheilkd 1978; 172(4):451–460.

129. Tsacopoulos M. Role of metabolic factors in the regulation of retinal blood minute volume. Adv Ophthalmol 1979; 39:233–273.

130. Poitry-Yamate CL, Tsacopoulos M, Pournaras CJ. Changes in the concentration of K+ induced by a flash of light in the preretinal vitreous of the anesthetized minipig. Klin Monatsbl Augenheilkd 1990; 196(5):351–353.

131. Brazitikos PD, Pournaras CJ, Munoz JL, Tsacopoulos M. Microinjection of L-lactate in the preretinal vitreous induces segmental vasodilation in the inner retina of miniature pigs. Invest Ophthalmol Vis Sci 1993; 34(5):1744–1752.

132. Donati G, Pournaras CJ, Munoz JL, Poitry S, Poitry-Yamate CL, Tsacopoulos M. Nitric oxide controls arteriolar tone in the retina of the miniature pig. Invest Ophthalmol Vis Sci 1995; 36(11):2228–2237.

133. Pournaras CJ, Rungger-Brandle E, Riva CE, Hardarson SH, Stefansson E. Regulation of retinal blood flow in health and disease. Prog Retin Eye Res 2008; 27(3):284–330.

134. Yamanishi S, Katsumura K, Kobayashi T, Puro DG. Extracellular lactate as a dynamic vasoactive signal in the rat retinal microvasculature. Am J Physiol Heart Circ Physiol 2006; 290(3):H925–H934.

135. Garhofer G, Resch H, Lung S, Weigert G, Schmetterer L. Intravenous administration of L-arginine increases retinal and choroidal blood flow. Am J Ophthalmol 2005; 140(1):69–76.

136. Ido Y, Chang K, Williamson JR. NADH augments blood flow in physiologically activated retina and visual cortex. Proc Natl Acad Sci USA 2004; 101(2):653–658.

137. Poole RC, Halestrap AP. Transport of lactate and other monocarboxylates across mammalian plasma membranes. Am J Physiol 1993; 264(4 Pt 1):C761–82.

138. Oldendorf WH. Carrier-mediated blood-brain barrier transport of short-chain monocarboxylic organic acids. Am J Physiol 1973; 224(6):1450–1453.

139. Gerhart DZ, Leino RL, Drewes LR. Distribution of monocarboxylate transporters MCT1 and MCT2 in rat retina. Neuroscience 1999; 92(1):367–375.

140. Pierre K, Pellerin L. Monocarboxylate transporters in the central nervous system: distribution, regulation and function. J Neurochem 2005; 94(1):1–14.

141. Hein TW, Xu W, Kuo L. Dilation of retinal arterioles in response to lactate: role of nitric oxide, guanylyl cyclase, and ATP-sensitive potassium channels. Invest Ophthalmol Vis Sci. 2006; 47(2):693–699.

142. Mendrinos E, Petropoulos IK, Mangioris G, Papadopoulou DN, Stangos AN, Pournaras CJ. Lactate-induced retinal arteriolar vasodilation implicates neuronal nitric oxide synthesis in minipigs. Invest Ophthalmol Vis Sci 2008; 49(11):5060–5066.

143. Poitry-Yamate CL. Biosynthesis, release and possible transfer of glucose-derived carbohydrate intermediates and amino acids from mammalian glial cells to photoreceptor-neurons. Geneva: Edition Médecine et Hygiène (University of Geneva), 1994.

144. Liepe BA, Stone C, Koistinaho J, Copenhagen DR. Nitric oxide synthase in Müller cells and neurons of salamander and fish retina. J Neurosci 1994; 14(12):7641–7654.

145. Kim IB, Lee EJ, Kim KY et al. Immunocytochemical localization of nitric oxide synthase in the mammalian retina. Neurosci Lett 1999; 267(3):193–196.

CHAPTER 13

The Function of the Retinal Pigment Epithelium

Olaf Strauss & Horst Helbig

Introduction

The retinal pigment epithelium (RPE) is a monolayer of pigmented cells located between the light-sensitive photoreceptor outer segments and the fenestrated endothelium of the choriocapillaris. On both sides, specialized extracellular matrices enable a close interaction of the RPE with its adjacent tissues. On the basolateral side the multilayered Bruch's membrane combines barrier function and selective transport matrix as part of the blood–retina barrier. On the apical side of the RPE, the interphotoreceptor matrix provides an interface for interaction between the photoreceptor outer segments and the RPE. In this interaction, the RPE forms a functional unit with the photoreceptors (Fig. 13.1).[1–3] In humans, every RPE cell is in interaction with a mean of 23 photoreceptors.[4] The formation of that functional unit is already of importance during the embryonic development.[5] The RPE and the neuronal retina depend on each other in the process of differentiation and maturation. Mutations in genes expressed in the RPE can lead to primary photoreceptor degeneration and mutations in genes expressed in photoreceptors can lead to primary RPE degenerations. In this functional unit the RPE fulfills a multitude of tasks which are essential for visual function. The failure of one of these functions leads to retinal degeneration.

Absorption of light

As a pigmented layer of cells the RPE covers the inner wall of the bulbus and absorbs scattered light to improve the optical quality. In the human eye light is focused by a lens onto the macula. This represents a high density of energy which is absorbed by the melanin granula of the RPE leading to an increase in the temperature of the RPE choroid complex.[6] The heat is transported away by the bloodstream in the choriocapillaris. However, this functional arrangement bears the danger of photo-oxidative damage.[7,8] The relative blood perfusion of the choriocapillaris is higher than that of the kidney.[9,10] However, only small amounts of oxygen are extracted by the adjacent tissues and venous blood from the choroid still shows an oxygen saturation of more than 90 percent. Thus, there is an overflow of oxygen in combination with a large density of light energy. This leads to a large production of reactive oxygen species. The RPE is protected by various lines of defense against this intoxication:[11] melanin of the melanosomes and the carotenoids lutein and zeaxanthin absorb light energy, ascorbate, α-tocopherol and β-carotene, and glutathione are non-enzymatic antioxidants and melanin itself can function as an anti-oxidant. This is supplemented by the cell's natural ability to repair damaged DNA, lipids and proteins.

Transepithelial transport

The RPE cells form a part of the blood–retina barrier. The RPE is by its electrophysiological parameters a tight epithelium: the paracellular resistance is at least ten times larger than that of the transcellular resistance.[12,13] This barrier function is of great importance for the immune privilege of the eye (see below). Due to this tight barrier function, the complete exchange of molecules and ions between the bloodstream and the photoreceptor side relies on transepithelial transport through the RPE.

Transport from the blood side to the photoreceptor side

The transport from the blood to the photoreceptor side consists mainly of transport of nutrients such as glucose, ω3 fatty acids and retinal.

For glucose transport, the RPE contains an abundance of the glucose transporters GLUT1 and GLUT3 in both the apical and the basolateral membrane.[14,15] GLUT3 mediates the basic transport whereas GLUT1 is responsible for an inducible glucose transport to adopt the glucose transport to different metabolic demands.

All-*trans* retinol is taken up from the bloodstream via a receptor-mediated process involving a serum-retinol-binding protein/transthyretin complex. Retinol is formed to all-*trans* retinyl ester and then enters directly the visual cycle (see below).

Docosahexaenoic acid (a polyunsaturated ω-3 fatty acid; 22:6ω3) is essential for the renewal process of the photoreceptor outer segments and cannot be synthesized by the photoreceptors.[16–19] The compound is synthesized in the liver from the precursor linolenic acid and transported to

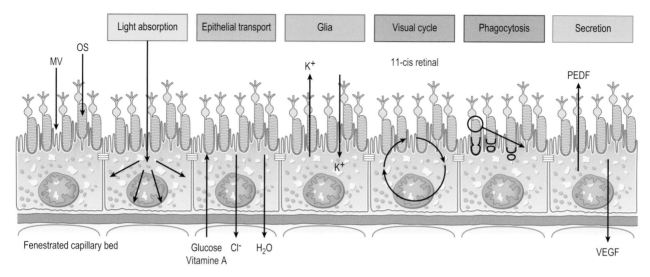

Figure 13.1 Summary of the different functions of the RPE by which a functional unit between photoreceptors is formed. (Modified from Strauss O: The retinal pigment epithelium in visual function, Physiol Rev 85:845–81, 2005. Used with permission.[3])

the eye in the blood bound to plasma protein. The RPE preferentially takes up docosahexaenoic acid in concentration-dependent manner. Docosahexaenoic acid is incorporated into glycerolipids for synthesis and storage.

Transport from the retinal side to the blood side

The retina is the tissue with the highest density of cells. Neuronal cells show a high metabolic activity. This results in the production of large amounts of water and accumulation of lactic acid. Additional amounts of water are moved towards the retina by intraocular pressure from the vitreous. Since the RPE is a tight epithelium water cannot pass in the paracellular route. Water and lactic acid are eliminated from subretinal space by active transcellular transport by the RPE (see Box 13.1).[20]

The transport of water is driven by an active transport of Cl^- from the retina to blood side (Fig. 13.2).[12,21,22] It is energized by the apically located Na^+/K^+-ATPase which uses the energy of ATP to transport Na^+ out of the cell in exchange with a transport of K^+ to the cytosol of the RPE. The K^+ ions recycle across the apical membrane through inward rectifier potassium channels which provide the large K^+-conductance of the apical membrane. This recycling keeps the K^+ gradient across the apical membrane small to support the activity of the Na^+/K^+-ATPase. In the apical membrane a $Na^+/2Cl^-/K^+$-co-transporter uses the Na^+ gradient across the apical membrane to transport K^+ and Cl^- into the cytosol of the RPE. Since K^+ moves back through inward rectifier channels to the subretinal space the $Na^+/2Cl^-/K^+$-cotransporter accumulates Cl^- in the intracellular space of the RPE cells (40–60 mM).[23] This high Cl^- concentration provides a driving force for Cl^- to leave the cell. The basolateral membrane displays a large conductance for Cl^- through which Cl^- leaves the cell from intracellular space to the blood side. This is the last step of transepithelial Cl^- transport from subretinal space to the blood side resulting in a basolateral negative transepithelial potential between −6 to −15 mV.[20] ClC2 Cl^- channels seem to provide the major part of the

Box 13.1 Retinal Detachment

Retinal detachment is characterized by a separation of the retinal pigment epithelium from the photoreceptor layer of the neuroretina. This separation prevents close interaction between the two layers leading to reduction of visual function. After successful closure of the retinal break, preventing access of fluid from the vitreous to the subretinal space, remaining subretinal fluid reabsorbs spontaneously. Subretinal fluid is eliminated by the electrolyte and fluid pump of the retinal pigment epithelium.[71]

RPE cells usually do not divide, but under certain circumstances they may become activated to play a role in wound repair. These wound repair mechanisms can also have devastating consequences by creating so called "proliferative vitreoretinopathy" (PVR). RPE cells migrating through retinal tears in the vitreous cavity and on the retinal surface may metaplastically transform to myofibroblasts, creating contracting membranes which cause complicated tractive forms of retinal detachment that are difficult to repair.[72]

basolateral Cl^- conductance.[24] Other channels providing basolateral Cl^- conductance are Cl^- channels linked to intracellular second-messenger systems. Cl^- transport can be increased by rises in intracellular free Ca^{2+}[25] or by increases in cytosolic cAMP concentration[26] resulting from either activation of Ca^{2+}-dependent Cl^- channels or cAMP-dependently regulated Cl^- channels.

The metabolic activity in the retina results in large subretinal concentrations of lactic acid which is transported away by the RPE.[14,27] This transport requires a tight regulation of the intracellular pH.[28,29] Lactic acid is removed from subretinal space by the monocarboxylate transporter 1 (MCT1)[30] which transports H^+ in cotransport with lactic acid (can be as high as 19mM). This transport is driven by the activity of the Na^+/H^+-exchanger which eliminates H^+ from cytosol of the RPE cells using the gradient for Na^+ provided by the activity of the Na^+/K^+-ATPase.[31] Across the basolateral membrane lactic acid leaves the cell via the activity of the monocarboxylate transporter 3 (MCT3) which is

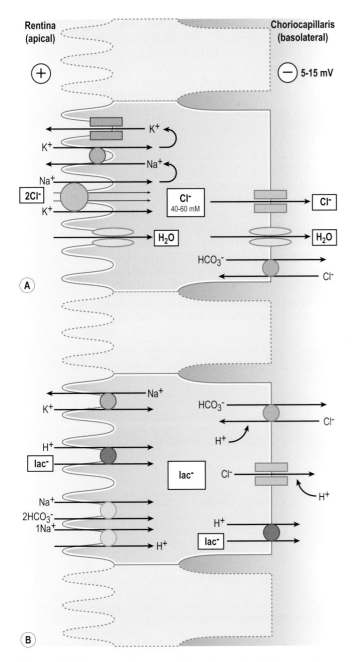

Rentina
(apical)

Choriocapillaris
(basolateral)

$-$ 5-15 mV

$+$

(A)

(B)

Figure 13.2 Transport of water and ions from subretinal space to blood side. (A) Using the energy of ATP hydrolysis the Na^+/K^+-ATPase establishes a gradient for Na^+ which drives the uptake of Na^+, K^+ and Cl^- by the $Na^+/K^+/2Cl^-$ cotransporter. K^+ recycles across the apical membrane through inward rectifier K^+ channels. The concerted transport activity of the three transport proteins results in the accumulation of Cl^- in the cytosol of the RPE. Cl^- leaves the cell through a variety of basolateral Cl^- channels. The net transport of Cl^- results in a basolateral negative transepithelial potential. The transport of Cl^- as osmolytes drives a transcellular transport of water through aquaporine water channels located in both the apical and the basolateral membrane. **(B)** Lactate is taken up from subretinal space via the activity of the MCT1 (monocarboxylate transporter-1) and accumulated in the cell. The driving force for this transport is large because the activity of retinal neurons produces large amounts of lactic acid (concentration in the subretinal space can be as high as 19 mM). Lactate leaves the cell through the basolateral membrane via the activity of the MCT3 transporter. The required pH regulation occurs at the apical membrane by the activity of sodium-dependent transporters: Na^+/H^+ exchanger and the Na^+/HCO_3^- cotransporter which use the sodium gradient of the Na^+/K^+ATPase. At the basolateral membrane pH regulation occurs by the activity of Cl^--dependent transport proteins which use the large intracellular Cl^- activity. In this part of the pH regulation a Cl^-/HCO_3^- exchanger and the Cl^- channel ClC2 are involved.

an H^+/lactic acid cotransporter too. Intracellular pH is stabilized by the transport of bicarbonate.[28,29] HCO_3^- is transported into the cell across the apical membrane by the $Na^+/2HCO_3^-$ cotransporter which uses the Na^+ gradient established by the Na^+/K^+-ATPase. HCO_3^- leaves the RPE cell across the basolateral membrane by the activity of the HCO_3^- exchanger.[32]

Production of metabolic water and accumulation of lactic acid are linked to each other. Therefore, transport of water and pH regulation are also coupled.[27] An increase in the transport of lactic acid leads to intracellular acidification which inhibits the transport activity of the Cl^-/HCO_3^- exchanger. Since the Cl^-/HCO_3^- exchanger transports Cl^- in the opposite direction to the Cl^- channels the exchanger decreases the Cl^- transport efficiency in resting conditions. The inhibition of the HCO_3^- exchanger, thus, increases the transepithelial transport of Cl^- and water. This is further increased by the ClC2 channels which are activated by a decrease in extracellular pH. Extracellular acidification would result from the increased transport of lactic acid by the MCT3 in the basolateral membrane.

Capacitative compensation of fast changes in the ion composition in the subretinal space

In the dark cGMP-dependent cation channels in the photoreceptor outer segments generate an inward current of Na^+ and Ca^{2+} which is counterbalanced by a K^+ outward current at the inner segments. In the light, cGMP-dependent cation channels close and the K^+ current is decreased resulting in a decrease of the K^+-concentration in the subretinal space from 5 to 2 mM. The decrease of the Na^+ conductance in the outer segments leads to an increase in the subretinal Na^+ concentration. Furthermore, illumination of the retina results in an increase in the subretinal volume. Both changes are compensated for in the moment when they occur by modulation of the ion transport by the RPE.[2,33–36]

Both, compensation for changes in the K^+ concentration and in extracellular volume are linked to the potassium transport.[20] The apical membrane of the RPE displays a large K^+ conductance. Decrease in the subretinal K^+ concentration leads to hyperpolarization of the apical membrane which corresponds to the c-wave in the electroretinogram. The decrease in the potassium concentration decreases the activity of the $Na^+/K^+/2Cl^-$ cotransporter which subsequently decreases the intracellular Cl^- activity and hyperpolarizes the basolateral membrane.[37] In the electroretinogram this can be seen as the delayed hyperpolarization. The hyperpolarization of the apical membrane activates inward rectifier K^+ channels which generate an efflux of K^+ into the subretinal space to compensate for the light-induced decrease in the subretinal K^+ concentration. At the same time the apical hyperpolarization decreases the activity of the Na^+/HCO_3^- cotransporter which is an electrogenic transporter by its stoichiometry. The subsequent intracellular acidification increases the transepithelial Cl^- and water transport as described above.

The light-induced increase in subretinal Na^+ concentration is compensated for by the activity of the $Na^+/K^+/2Cl^-$ cotransporter and by the Na^+/H^+ exchanger.[38,39] The Na^+/K^+-ATPase seem to uptake Na^+ to compensate for the increase in the subretinal Na⁺ concentration during the transition from light to dark. It is believed that this task is the reason for its localization in the apical membrane of the RPE.

Visual cycle

Vision starts in photoreceptor outer segments with absorption of a photon by the chromophore of rhodopsin, 11-*cis* retinal, which undergoes a conformational change from 11-*cis* retinal to all-*trans* retinal. For the absorption of the next photon, rhodopsin replaces all-*trans* retinal to a new 11-*cis* retinal. To maintain visual function, a re-isomerization of all-*trans* to 11-*cis* retinal takes place in the RPE (Fig. 13.3).[1,40–43] All-*trans* retinal leaves the disc membrane as N-retinylidine-phosphatidylethanolamine (N-retinylidine-PE) by the transport activity of the ABC (ATP-binding cassette protein) transporter ABCR4. Released from N-retinylidine-PE, all-*trans* retinal is reduced to all-*trans* retinol by RDH (all-*trans* retinol dehydrogenase), leaves the photoreceptor outer segments into subretinal space, is loaded onto the IRPB

(interphotoreceptor matrix retinal binding protein) and transported to the RPE.[44] In the cytosol of the RPE it binds to the CRBP (cellular retinol binding protein) as entry into a reaction cascade catalyzed by a protein complex composed of LRAT (lecithin:retinol transferase), RPE65 (RPE specific protein 65 kDa, function as isomerase[45]) and 11cRDH (11-*cis*-retinol dehydrogenase or RDH5) which performs the re-isomerization. After re-isomerization 11-*cis* retinal loaded to CRALBP (cellular retinaldahyde binding protein), leaves the cell and is transported back to photoreceptors by IRBP. This pathway accounts for the re-isomerization of retinal from rods. Cones might have an additional secondary pathway involving Müller cells.

Since the visual cycle maintains the excitability of the photoreceptors (see Box 13.2) it needs to be adapted to the different photoreceptor activities between light and darkness.[46] In the light there is a fast turnover of retinal whereas in the dark the turnover occurs much slower. Retinal which is not needed in the state of a slower turnover must be quickly available for the transition from darkness to light and vice versa. The different retinal binding proteins represent

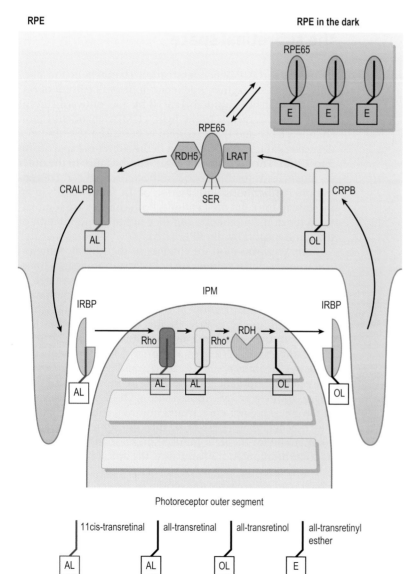

Figure 13.3 Cycle of retinal (visual cycle). The process of vision is started by the conformational change of the chromophore of rhodopsin from 11-*cis* retinal to all-*trans* retinal. All-*trans* retinal is re-isomerized to 11-*cis* in the RPE. The first step into this cycle is the reduction of all-*trans* retinal into all-*trans* retinol. all-*trans* retinol leaves the photoreceptor and binds to IRPB and is transported to the RPE. Inside the RPE all-*trans* retinol binds to CRBP and enters with this step the re-isomerization pathway. The reaction is catalyzed by a protein complex consisting of LRAT, RPE65 and RDH5. The reaction product, 11-*cis* retinal binds to CRALBP, leaves the cell where it binds to IRPB again to be transported to the photoreceptors. RPE65 modulation adapts this cycle to different light intensities. In the light it is palmotylated and membrane bound with isomerase function. In the dark it is freely diffusible in the cytosol and serves as retinal pool when the visual cycle has a lower turn-over rate (CRBP = cellular retinol binding protein; CRALBP = cellular retinaldehyde binding protein; IPM = interphotoreceptor matrix; IRBP = interstitial retinal binding protein; LRAT = lecithin retinol acyltransferase; RDH = retinol dehydrogenase; RPE 65 = retinal pigment epithelium specific protein 65 kDa; RDH5 = 11-*cis* retinol dehydrogenase; Rho = rhodopsin; SER = smooth endoplasmic reticulum).

connected pools for retinal. In the transition from darkness to light IRPB represent the first pool which delivers retinal.[44] This pool cannot be depleted because it refills retinal from the pool of CRALBP proteins. CRALBP in turn can recruit retinal from RPE65. Thus, with increasing light intensity retinal can be recruited from the different pools and in the transition from light to darkness retinal can be stored away. Of importance are IRBP which can carry in the dark larger amounts of retinal than in the light. Another key function is the modulation of the RPE65 function.[47] In the dark RPE65 is water soluble and freely diffusible in the cytosol. With this configuration RPE65 functions as a retinal store. In the light RPE65 is palmotylated and bound to intracellular membranes. In this configuration, RPE65 predominantly catalyze the re-isomerization of all-*trans* retinol to 11-*cis* retinol.

Phagocytosis of photoreceptor outer segments

The focus of a large density of light energy onto the retina and increased formation of reactive oxygen species by photo-oxidation constantly exposes photoreceptor outer segments to toxic compounds. This leads to destruction of the photoreceptor outer segments (POS). In order to maintain the excitability of the photoreceptors, outer segments are newly built at the connecting cilium from the inner segments.[48,49] The destroyed tips of the POS are shed and phagocytosed by the RPE.[50–52] The process of shedding and phagocytosis must be tightly coordinated to maintain the proper length of the outer segments. On the other hand the tight coordination ensures the efficient degradation of the internalized POS (see Box 13.3). The process of regeneration is regulated in a circadian manner.[53–56] With the onset of light in the morning the phagocytic activity is triggered and maintains a peak for 1–2 hours. Every 11 days a whole length of a photoreceptor outer segment is renewed.

Three receptors are involved in the process of phagocytosis (Fig. 13.4). Binding occurs via α5-integrins.[56] This binding

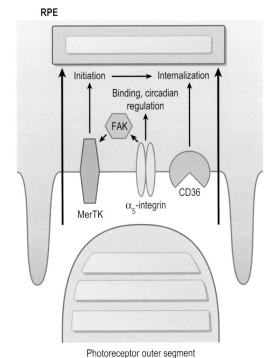

Figure 13.4 Phagocytosis of the photoreceptor outer membranes.
With onset of light destroyed tips of the photoreceptor outer segments are shed from the photoreceptors and phagocytosed by the RPE. The binding and circadian regulation occurs via αv-integrin which enables the initiation of the cascade which regulates the phagocytosis. This cascade is initiated by the activation of the MerTK receptor. The internalization of the shed photoreceptor outer segments is permitted by the activity CD36 (αv-integrin = vitronectin binding integrin; CD36 = macrophage phagocytosis receptor; FAK = focal adhesion kinase; MerTK = c-mer tyrosinekinase receptor).

initiates a signal cascade in the RPE involving the activation of focal adhesion kinase (FAK).[57] This process serves two functions. One leads to phosphorylation of another receptor which is involved in the regulation of phagocytosis, MerTK (c-mer tyrosine kinase). The second function leads to the circadian regulation of phagocytosis. MerTK is a receptor tyrosine kinase which triggers the process of internalization of POS[58-60] after binding to α5-integrin. The downstream cascade initiated after activation of MerTK involves the generation of inositol-1,4,5-trisphosphate and an increase of cytosolic free Ca^{2+} as second-messenger. The third receptor, CD36, is important for the process of internalization.[61]

Secretion

The RPE is capable of secreting a large variety of growth factors, cytokines or immune modulators.[62,63] This function serves to maintain the structural integrity of photoreceptors,[64] to maintain the fenestrated structure of the endothelium of the choriocapillaris and to actively interact with the immune system[3] (Fig. 13.5). The intracellular regulation of the secretion involves voltage-dependent L-type Ca^{2+} channels of the neuroendocrine subtype (Cav1.3) which are regulated by tyrosine kinase. With this regulation the secretion is linked to the influence of growth-factor receptors which mainly act via activation of tyrosine kinase.

Structural integrity of neighboring tissues

For the maintenance of structural integrity of photoreceptors the RPE secretes PEDF (pigment epithelium-derived factor; protects against hypoxia- and glutamate induced apoptosis), CNTF (ciliary neurotrophic factor; protects against photoreceptor cell death) and members of the fibroblast growth factor family. To stabilize the choroidal endothelium the RPE secretes and VEGF (vascular endothelial growth factor) and tissue inhibitors of matrix metalloproteases (TIMP) (see Box 13.4). The latter protein stabilizes the extracellular

matrix and prevents neovascularization. Basically, the regulation of secretion is embedded into a network of paracrine and autocrine stimulation.[65,66] To coordinate the secretion with the neighboring tissues, the RPE is equipped with receptors which are activated by factors as such IGF-1 (insulin-like growth factor-1), TNFα (tumor necrosis factor-α), VEGF or glutamate. The secretion of ATP by the RPE seems to be of importance for an autocrine regulation of phagocytosis.[67,68]

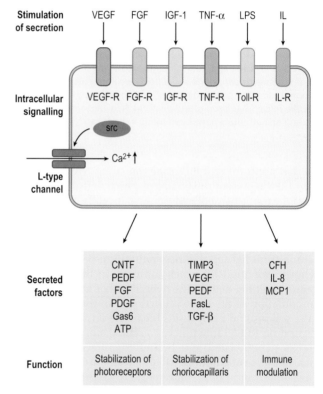

Figure 13.5 Secretion by the RPE. The secretion of different factors serves to maintain the structural integrity of the neighboring tissues and to modulate the immune system which helps to establish the immune privilege of the eye. To establish communication between the RPE and the according tissues the RPE is equipped with a large variety of receptors. The secretion itself is mainly regulated by voltage-dependent Ca^{2+} channels of the neuroendocrine subtype.[63] This Ca^{2+} channel is regulated by src-kinase which activated by different receptors of mediators leading to higher secretion rates. This illustration shows the major receptor types. From each different group several different receptors are expressed (e.g. VEGF-R includes VEGFR-1, VEGFR-2, VEGFR-3). The factors indicated represent the major family (e.g. VEGF includes VEGF-A, VEGF-C). Furthermore, as can be concluded from the figure, the secretion itself is not only regulated in a paracrine but also in an autocrine manner (Factors: ATP = adenosine triphosphate; CFH = complement factor H; CNTF = ciliary neurotrophic factor; FasL = CD95 apoptosis receptor ligand; FGF = fibroblast growth factor; Gas6 = growth arrest protein 6; IL = interleukin; IGF-1 = insulin-like growth factor-1; LPS = lipopolysaccharides; PEDF = pigment epithelium-derived factor; PDGF = platelet-derived growth factor; TIMP3 = tissue inhibitor of metallo matrix proteases; VEGF = vascular endothelial growth factor; TGF-β = transforming growth factor-β; TNF-α = tumor necrosis factor alpha. Receptors: FGF-R = fibroblast growth factor receptor; IGF-R = insulin-like growth factor receptor; IL-R = interleukin receptor; MCP-1 = monocyte chemotactic protein 1; TNF-R = tumor necrosis factor receptor; Toll-R = toll-like receptors; VEGF-R = vascular endothelial growth factor receptor. Intracellular signaling: L-type channel = voltage-dependent L-type Ca^{2+} channel; src = src-type intracellular tyrosine kinase)

Box 13.4 Advanced Age-related Macular Degeneration (AMD)

In the development of age-related maculopathy (ARM), degenerative changes in RPE-cells, extracellular matrix and possibly choriocapillaries probably lead to malnutrition of photoreceptors and RPE-cells. Two different types of responses to this malnutrition lead to the two forms of advanced AMD: neovascular and atrophic.

The atrophic form ("geographic atrophy") is characterized by insufficient production of survival factors by damaged RPE cells, leading to apoptosis of the functional complex formed by choriocapillaries, RPE and photoreceptors.[77]

In the neovascular type of AMD RPE-cells and possibly photoreceptors produce excess angiogenic factors, especially vascular endothelial growth factor (VEGF). This stimulates outgrowth of new capillaries from the choroid (choroidal neovascularization, CNV) through irregularities in Bruch's membrane under the RPE (occult CNV) or through gaps in the RPE under the neuroretina (classic CNV).[77] Clinically antagonists of VEGF are injected into the eye in exudative AMD, often stabilizing or even improving vision.[78]

Box 13.5 Clinical Role of Immune Privilege

The immune privilege of the inner eye is believed to represent a specific modulation of the cellular immune response, down-regulating defence mechanisms, which would lead to destruction of surrounding cells ("innocent bystander destruction").[70] This phenomenon can be used clinically for transplantation of RPE cells in cases of macular degenerations. Functional results have been disappointing so far, however, significant inflammation or graft rejection has not been observed.[83] For retinal gene therapy viral vectors with the new gene are injected in the subretinal space. For this approach the immune privilege is helpful, since there must be no significant immune reaction after injecting viral antigen into the subretinal space.[84]

Immune privilege of the eye

The inner eye represents an immune privileged space.[69,70] This is enabled on one hand by the barrier function of the RPE. On the other hand the RPE is able to actively participate in the modulation of the immune system by secretion.[66] The RPE expresses a large variety of membrane receptors which coordinate the regulation of the immune system such as MHC receptors, toll-like receptors and receptors for cytokines which are involved in immune signaling such as receptors for TNFα or interleukins. Furthermore, the RPE is able to secrete a variety of immune modulatory factors such as TGF-β transforming growth receptor-β, members of the interleukin family or factors of the complement systems, for example factor H (see Box 13.3 and 13.5).

References

1. Bok D. The retinal pigment epithelium: a versatile partner in vision. J Cell Sci Suppl 1993; 17:189–195.
2. Steinberg RH. Interactions between the retinal pigment epithelium and the neural retina. Doc Ophthalmol 1985; 60:327–346.
3. Strauss O. The retinal pigment epithelium in visual function. Physiol Rev 2005; 85:845–881.
4. Gao H, Hollyfield JG. Aging of the human retina. Differential loss of neurons and retinal pigment epithelial cells. Invest Ophthalmol Vis Sci 1992; 33:1–17.
5. Marmorstein AD et al. Morphogenesis of the retinal pigment epithelium: toward understanding retinal degenerative diseases. Ann NY Acad Sci 1998; 857:1–12.
6. Parver LM, Auker C, Carpenter DO. Choroidal blood flow as a heat dissipating mechanism in the macula. Am J Ophthalmol 1980; 89:641–646.
7. Boulton M. Ageing of the retinal pigment epithelium. Prog Ret Eye Res 1991; 125–151.
8. Boulton M, Dayhaw-Barker P. The role of the retinal pigment epithelium: topographical variation and ageing changes. Eye 2001; 15:384–389.
9. Alm A, Bill A. Blood flow and oxygen extraction in the cat uvea at normal and high intraocular pressures. Acta Physiol Scand 1970; 80:19–28.
10. Alm A, Bill A. The oxygen supply to the retina. I. Effects of changes in intraocular and arterial blood pressures, and in arterial PO_2 and PCO_2 on the oxygen tension in the vitreous body of the cat. Acta Physiol Scand 1972; 84:261–274.
11. Boulton M et al. Regional variation and age-related changes of lysosomal enzymes in the human retinal pigment epithelium. Br J Ophthalmol 1994; 78:125–129.
12. Miller SS, Steinberg RH. Active transport of ions across frog retinal pigment epithelium. Exp Eye Res 1977; 25:235–248.
13. Miller SS, Steinberg RH. Passive ionic properties of frog retinal pigment epithelium. J Membr Biol 1977; 36:337–372.
14. Adler AJ, Southwick RE. Distribution of glucose and lactate in the interphotoreceptor matrix. Ophthalmic Res 1992; 24:243–252.
15. Ban Y, Rizzolo LJ. Regulation of glucose transporters during development of the retinal pigment epithelium. Brain Res Dev Brain Res 2000; 121:89–95.
16. Bibb C, Young RW. Renewal of fatty acids in the membranes of visual cell outer segments. J Cell Biol 1974; 61:327–343.
17. Anderson RE et al. Conservation of docosahexaenoic acid in the retina. Adv Exp Med Biol 1992; 318:285–294.
18. Bazan NG, Gordon WC, Rodriguez de Turco EB. Docosahexaenoic acid uptake and metabolism in photoreceptors: retinal conservation by an efficient retinal pigment epithelial cell-mediated recycling process. Neurobiology of Essential Fatty Acids 1992; 295–306.
19. Bazan NG, Rodriguez de Turco EB, Gordon WC. Pathways for the uptake and conservation of docosahexaenoic acid in photoreceptors and synapses: biochemical and autoradiographic studies. Can J Physiol Pharmacol 1993; 71:690–698.
20. Dornonville de la Cour M. Ion transport in the retinal pigment epithelium. A study with double barrelled ion-selective microelectrodes. Acta Ophthalmol 1993; Suppl 1–32.
21. La Cour M. Cl- transport in frog retinal pigment epithelium. Exp Eye Res 1992; 54:921–931.
22. Miller SS, Edelman JL. Active ion transport pathways in the bovine retinal pigment epithelium. J Physiol 1990; 424:283–300.
23. Wiederholt M, Zadunaisky JA. Decrease of intracellular chloride activity by furosemide in frog retinal pigment epithelium. Curr Eye Res 1984; 3:673–675.
24. Bosl MR et al. Male germ cells and photoreceptors, both dependent on close cell-cell interactions, degenerate upon ClC-2 Cl(-) channel disruption. Embo J 2001; 20:1289–1299.
25. Rymer J, Miller SS, Edelman JL. Epinephrine-induced increases in [Ca^{2+}](in) and KCl⁻coupled fluid absorption in bovine RPE. Invest Ophthalmol Vis Sci 2001; 42:1921–1929.
26. Hughes BA, Segawa Y. cAMP-activated chloride currents in amphibian retinal pigment epithelial cells. J Physiol 1993; 466:749–766.
27. Edelman JL, Lin H, Miller SS. Acidification stimulates chloride and fluid absorption across frog retinal pigment epithelium. Am J Physiol 1994; 266:C946–56.
28. la Cour M. Kinetic properties and Na⁺ dependence of rheogenic Na(⁺)-HCO_3^- co-transport in frog retinal pigment epithelium. J Physiol 1991; 439:59–72.
29. la Cour M et al. Lactate transport in freshly isolated human fetal retinal pigment epithelium. Invest Ophthalmol Vis Sci 1994; 35:434–442.
30. Bergersen L et al. Cellular and subcellular expression of monocarboxylate transporters in the pigment epithelium and retina of the rat. Neuroscience 1999; 90:319–331.
31. Keller SK et al. Regulation of intracellular pH in cultured bovine retinal pigment epithelial cells. Pflugers Arch 1988; 411:47–52.
32. Keller SK et al. Interactions of pH and K⁺ conductance in cultured bovine retinal pigment epithelial cells. Am J Physiol 1986; 250:C124–37.
33. Bialek S, Miller SS. K⁺ and Cl⁻ transport mechanisms in bovine pigment epithelium that could modulate subretinal space volume and composition. J Physiol 1994; 475:401–417.
34. Edelman JL, Lin H, Miller SS. Potassium-induced chloride secretion across the frog retinal pigment epithelium. Am J Physiol 1994; 266:C957–66.
35. Joseph DP, Miller SS. Apical and basal membrane ion transport mechanisms in bovine retinal pigment epithelium. J Physiol 1991; 435:439–463.
36. Li JD, Govardovskii VI, Steinberg RH. Light-dependent hydration of the space surrounding photoreceptors in the cat retina. Vis Neurosci 1994; 11:743–752.
37. Adorante JS, Miller SS. Potassium-dependent volume regulation in retinal pigment epithelium is mediated by Na,K,Cl cotransport. J Gen Physiol 1990; 96:1153–1176.
38. Ames A, 3rd et al. Energy metabolism of rabbit retina as related to function: high cost of Na+ transport. J Neurosci 1992; 12:840–853.
39. Hodson S, Armstrong I, Wigham C. Regulation of the retinal interphotoreceptor matrix Na by the retinal pigment epithelium during the light response. Experientia 1994; 50:438–441.
40. McBee JK et al. Confronting complexity: the interlink of phototransduction and retinoid metabolism in the vertebrate retina. Prog Retin Eye Res 2001; 20:469–529.
41. Thompson DA, Gal A. Genetic defects in vitamin A metabolism of the retinal pigment epithelium. Dev Ophthalmol 2003; 37:141–154.
42. Baehr W, Wu SM, Bird AC, Palczewski K. The retinoid cycle and retina disease. Vis Res 2003; 43:2957–2958.
43. Besch D et al. Inherited multifocal RPE-diseases: mechanisms for local dysfunction in global retinoid cycle defects. Vis Res 2003; 43:3095–3108.
44. Gonzalez-Fernandez F. Evolution of the visual cycle: the role of retinoid-binding proteins. J Endocrinol 2002; v175:75–88.
45. Jin M et al. Rpe65 is the retinoid isomerase in bovine retinal pigment epithelium. Cell 2005; 122:449–459.
46. Lamb TD, Pugh EN, Jr. Phototransduction, dark adaptation, and rhodopsin regeneration the proctor lecture. Invest Ophthalmol Vis Sci 2006; 47:5137–5152.
47. Xue L et al. A palmitoylation switch mechanism in the regulation of the visual cycle. Cell 2004; 117:761–771.
48. Young RW. The renewal of photoreceptor cell outer segments. J Cell Biol 1967; 33:61–72.
49. Young RW, Bok D. Participation of the retinal pigment epithelium in the rod outer segment renewal process. J Cell Biol 1969; 42:392–403.
50. Bok D, Hall MO. The role of the pigment epithelium in the etiology of inherited retinal dystrophy in the rat. J Cell Biol 1971; 49:664–682.
51. Custer NV, Bok D. Pigment epithelium-photoreceptor interactions in the normal and dystrophic rat retina. Exp Eye Res 1975; 21:153–166.
52. LaVail MM. Rod outer segment disk shedding in rat retina: relationship to cyclic lighting. Science 1976; 194:1071–1074.
53. LaVail MM. Rod outer segment disc shedding in relation to cyclic lighting. Exp Eye Res 1976; 23:277–280.
54. Young RW. The daily rhythm of shedding and degradation of rod and cone outer segment membranes in the chick retina. Invest Ophthalmol Vis Sci 1978; 17:105–116.
55. Besharse JC, Hollyfield JG. Turnover of mouse photoreceptor outer segments in constant light and darkness. Invest Ophthalmol Vis Sci 1979; 18:1019–1024.
56. Nandrot EF et al. Loss of synchronized retinal phagocytosis and age-related blindness in mice lacking αvβ5 integrin. J Exp Med 2004; 200:1539–1545.
57. Finnemann SC. Focal adhesion kinase signaling promotes phagocytosis of integrin-bound photoreceptors. Embo J 2003; 22:4143–4154.
58. D'Cruz PM et al. Mutation of the receptor tyrosine kinase gene Mertk in the retinal dystrophic RCS rat. Hum Mol Genet 2000; 9:645–651.
59. Feng W et al. Mertk triggers uptake of photoreceptor outer segments during phagocytosis by cultured retinal pigment epithelial cells. J Biol Chem 2002; 277:17016–17022.
60. Gal A et al. Mutations in MERTK, the human orthologue of the RCS rat retinal dystrophy gene, cause retinitis pigmentosa. Nat Genet 2000; 26:270–271.
61. Finnemann SC, Silverstein RL. Differential roles of CD36 and alphavbeta5 integrin in photoreceptor phagocytosis by the retinal pigment epithelium. J Exp Med 2001; 194:1289–1298.

62. Wimmers S, Karl MO, Strauss O. Ion channels in the RPE. Prog Retin Eye Res 2007; 26:263–301.

63. Rosenthal R, Strauss O. Ca²⁺-channels in the RPE. Adv Exp Med Biol 2002; 514:225–235.

64. Bazan NG. Neurotrophins induce neuroprotective signaling in the retinal pigment epithelial cell by activating the synthesis of the anti-inflammatory and anti-apoptotic neuroprotectin D1. Adv Exp Med Biol 2008; 613:39–44.

65. Schlingemann RO. Role of growth factors and the wound healing response in age-related macular degeneration. Graefes Arch Clin Exp Ophthalmol 2004; 242:91–101.

66. Holtkamp GM et al. Retinal pigment epithelium-immune system interactions: cytokine production and cytokine-induced changes. Prog Retin Eye Res 2001; 20:29–48.

67. Besharse JC, Spratt G. Excitatory amino acids and rod photoreceptor disc shedding: analysis using specific agonists. Exp Eye Res 1988; 47:609–620.

68. Mitchell CH. Release of ATP by a human retinal pigment epithelial cell line: potential for autocrine stimulation through subretinal space. J Physiol 2001; 534:193–202.

69. Streilein JW. Ocular immune privilege: therapeutic opportunities from an experiment of nature. Nat Rev Immunol 2003; 3:879–889.

70. Zamiri P, Sugita S, Streilein JW. Immunosuppressive properties of the pigmented epithelial cells and the subretinal space. Chem Immunol Allergy 2007; 92:86–93.

71. Marmor MF, Maack T. Enhancement of retinal adhesion and subretinal fluid resorption by acetazolamide. Invest Ophthalmol Vis Sci 1982; 23:121–124.

72. Machemer R. Proliferative vitreoretinopathy (PVR): a personal account of its pathogenesis and treatment. Proctor lecture. Invest Ophthalmol Vis Sci 1988; 29:1771–1783.

73. Zarbin MA. Current concepts in the pathogenesis of age-related macular degeneration. Arch Ophthalmol 2004; 122:598–614.

74. Fritsche LG et al. Age-related macular degeneration is associated with an unstable ARMS2 (LOC387715) mRNA. Nat Genet 2008; 40:892–896.

75. Scholl HP et al. An update on the genetics of age-related macular degeneration. Mol Vis 2007; 13:196–205.

76. Yang Z et al. Toll-like receptor 3 and geographic atrophy in age-related macular degeneration. N Engl J Med 2008

77. Holz FG et al. Pathogenesis of lesions in late age-related macular disease. Am J Ophthalmol 2004; 137:504–510.

78. Rosenfeld PJ et al. Ranibizumab for neovascular age-related macular degeneration. N Engl J Med 2006; 355:1419–1431.

79. Goodwin P. Hereditary retinal disease. Curr Opin Ophthalmol 2008; 19:255–262.

80. Thompson DA, Gal A. Vitamin A metabolism in the retinal pigment epithelium: genes, mutations, and diseases. Prog Retin Eye Res 2003; 22:683–703.

81. Travis GH et al. Diseases caused by defects in the visual cycle: retinoids as potential therapeutic agents. Annu Rev Pharmacol Toxicol 2007; 47:469–512.

82. Bainbridge JW et al. Effect of gene therapy on visual function in Leber's congenital amaurosis. N Engl J Med 2008; 358:2231–2239.

83. Binder S et al. Transplantation of the RPE in AMD. Prog Retin Eye Res 2007; 26:516–554.

84. Allocca M et al. AAV-mediated gene transfer for retinal diseases. Expert Opin Biol Ther 2006; 6:1279–1294.

Functions of the Orbit and Eyelids

Gregory J. Griepentrog & Mark J. Lucarelli

Introduction

The cranium protects the brain and provides scaffolding for facial structures. During primate evolution, the orbits were enlarged and reoriented towards the front of the face.[1] This, along with the gradual flattening of the face, allowed for improved binocular vision due to overlapping visual fields. Along with morphologic skull changes, facial mimetic musculature evolved in primates as a means of close-proximity non-vocal communication.[2] It is in this context that the function of the eyelids to protect, lubricate, and cleanse the ocular surface, is most fully appreciated. This chapter will examine anatomy of the eyelids, orbits, and related facial structures.

Orbital anatomy and function

Orbit osteology

Derived from cranial neural crest cells, the bony orbit consists of seven individual bones combining into four walls that surround the globe, extraocular muscles, nerves, fat, and blood vessels. The orbital bones include the sphenoid (greater and lesser wings), frontal, ethmoid, maxillary, zygomatic, palatine, and lacrimal bones. Posteriorly approaching the apex, this four-sided pyramid becomes three-sided with the gradual merger of the medial wall and orbital floor after the floor is cut off by the inferior orbital fissure. The adult orbit has a volume of 25–30 mL, with the globe filling approximately 7 mL or 25% of the space.[3] The orbit depth as measured from the center of the orbital margin to the apex is approximately 45 mm. The widest diameter of the orbit occurs 1 cm behind the orbital rim. The lateral walls of the orbit are oriented 90 degrees from one another and run approximately 40–45 mm, while the medial walls are parallel to one another (Fig. 14.1). Due to this bony orientation, the eyes tend to diverge, and thus are tonically held in adduction by the medial rectus muscles to achieve ocular alignment.

The orbital rim provides a hard tissue shield to surround and protect the eye. It is a thick discontinuous spiral that begins medially at the anterior lacrimal (maxillary bone) crest and coils to end at the posterior lacrimal (lacrimal bone) crest (Fig. 14.2). The lacrimal fossa, which contains the lacrimal excretory sac, lies between these two crests. Formed by the zygomatic bone and the zygomatic process of the frontal bone, the lateral orbital rim is the strongest and thickest portion of the rim. While the posteriorly directed concavity of the lateral orbital rim allows for a wide visual field, it also makes the eye prone to injury from objects approaching from a lateral direction. The superior orbital rim is formed by the frontal bone. Medially, the superior rim contains the supraorbital notch through which the supraorbital nerve and artery pass. In some individuals, a supraorbital foramen replaces the notch.[4] The infraorbital nerve and artery exit through the infraorbital foramen approximately 4–6 mm below the inferior orbital rim medially.

The orbital roof is comprised of the orbital plate of the frontal bone along with a minor contribution from the lesser wing of the sphenoid posteriorly. The frontal bone is the strongest component of the craniofacial skeleton, withstanding between 800 and 2200 pounds of force before fracturing.[5] This is equivalent to the force achieved in a frontal collision at 30 mph for an unrestrained adult passenger. Orbital roof "blow-in" fractures may be associated with a number of ocular and neurologic injuries including proptosis, ptosis, optic nerve contusion, orbital hematoma, as well as contusive and hemorrhagic injuries to the ipsilateral frontal and parietal lobes.[6] Important bony landmarks of the orbital roof include the lacrimal gland fossa temporally and another small depression 3–5 mm behind the orbital rim anteromedially known as the trochlear fossa. Here resides the fibrocartilaginous trochlea through which the superior oblique tendon passes.

The frontal sinus begins to develop between 1 and 2 years of age through invagination of the frontal bone by frontoethmoidal air cells. The filling of air or pneumatization of the sinus begins at about 5–8 years old and continues into adulthood.[7] Therefore, there is a relative absence of frontal sinusitis in very young children. Yet, orbital complications due to frontal sinusitis such as orbital cellulitis and abscess most often occur between the ages of 5 and 10 years.[8] It has been hypothesized that direct extension through the relatively thin orbital roof in children or congenital bony dehiscences (gaps) posterior to the trochlear and supraorbital notch may account for these findings.[9]

The lateral wall is comprised of the zygomatic bone anteriorly and the greater wing of the sphenoid posteriorly. The vertically oriented zygomaticosphenoidal suture represents the thinnest portion of the lateral wall, and is a convenient breaking point during lateral orbitotomy. The posterior boundaries of the wall are marked by the superior and inferior orbital fissures. The frontosphenoidal suture forms the boundary between the lateral wall and orbital roof. About one-third of individuals have a meningeal foramen just superior to this suture which transmits the recurrent meningeal artery (a branch of the external carotid system) to form

an anastomosis with the lacrimal artery (a branch of the internal carotid system). This collateral system has potential importance if the primary internal carotid vascular supply becomes compromised. An important bony prominence of the lateral wall is Whitnall's lateral tubercle, a small rounded protuberance of the zygomatic bone 3–4 mm inside of the lateral orbital wall and approximately 11 mm below the frontozygomatic suture.[10,11]

Fractures to the lateral wall may result from blunt trauma to the zygomatic bone and lateral orbital rim. Due to its multiple bony articulations, fractures of the zygoma may disrupt the wider anatomy and are often referred to as zygomaticomaxillary complex fractures or ZMC fractures (Fig. 14.3). Clinical signs include lateral canthal dystopia (inferior displacement of the lateral canthal angle), cheek depression and trismus (difficulty opening the mouth owing to spasm of the masticatory muscles or impingement of the coronoid process of the mandible). Hypoesthesia may be noted at the lateral midface from disrupted branches of the zygomaticotemporal and zygomaticofacial neurovascular bundles (V1 branches traveling through the zygomatic bone), as well as in the distribution of the infraorbital nerve (V2 branch).

The orbital floor consists of the maxillary bone, the zygoma anterolaterally, and the palatine bone posteriorly. It is triangular in shape and extends from the maxillary-ethmoid buttress (the relatively dense bone strut at the union of the orbital floor and medial wall) to the inferior orbital fissure. The shortest of the walls, it travels posteriorly from the rim 35–40 mm and ends before the orbital apex at the pterygopalatine fossa. The infraorbital neurovascular bundle passes in the infraorbital groove and infraorbital canal of the orbital floor as it travels towards infraorbital foramen. The floor remains strong laterally, but thins posteromedially with expansion of the maxillary sinus. This thinned maxillary bone is where the floor usually fractures during trauma and is also a convenient site to initiate an inferior wall decompression. The infraorbital nerve is often contused during an orbital floor fracture but is rarely severed,

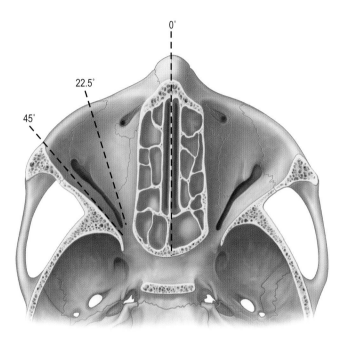

Figure 14.1 Horizontal section through orbits. Medial walls are nearly parallel, and lateral walls diverge 45 degrees from midline (© Virginia Cantarella).

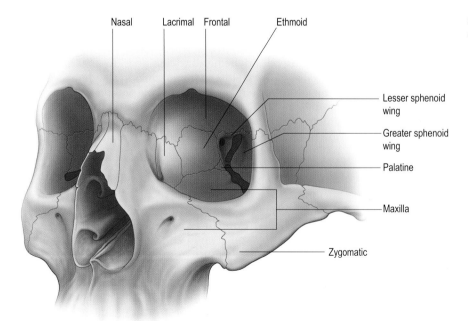

Nasal   Lacrimal   Frontal   Ethmoid

Lesser sphenoid wing

Greater sphenoid wing

Palatine

Maxilla

Zygomatic

Figure 14.2 Osteology of the orbital bones and orbital apex (© Virginia Cantarella).

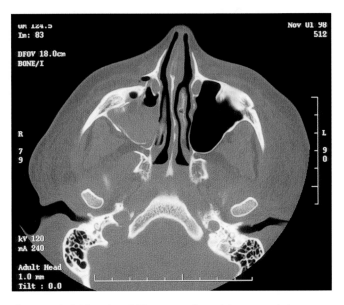

Figure 14.3 Axial section of CT scan revealing a right zygomatic fracture as a part a zygomaticomaxillary complex (ZMC) fracture.

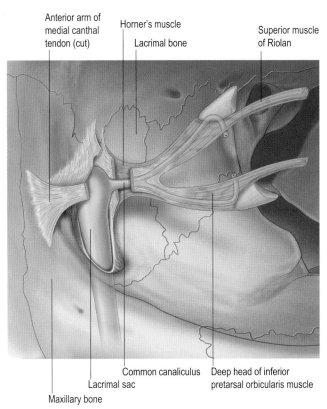

Figure 14.4 Lacrimal drainage system, superficial anatomy. (Modified from Dutton JJ. Atlas of clinical and surgical orbital anatomy. Philadelphia: WB Saunders, 1994.)

thus, initial hypoesthesia in the V2 distribution will commonly resolve over several months' time.

The medial wall consists of the maxillary, lacrimal and ethmoid bones, as well as the lesser wing of the sphenoid bone. The two medial walls are parallel to one another and extend 45–50 mm from the anterior lacrimal crest to the orbital apex. The lamina papyracea of the ethmoid bone is an extremely thin portion of bone that has a honeycombed structure. Giving support to the medial wall, the multiple bullae that form this structure develop secondarily to pneumatization of the ethmoid bone. This helps to explain why the medial wall fractures less often than the thicker orbital floor. Landmarks important to the orbital surgeon include the anterior and posterior ethmoidal foramina that reside at the frontoethmoidal suture and convey respective branches of the ophthalmic artery and nasociliary nerve. The anterior ethmoidal foramen is located approximately 24 mm posterior to the anterior lacrimal crest and the posterior ethmoidal foramen 36 mm posterior to the rim. The orbital foramen is located approximately 6 mm behind the posterior ethmoidal foramen. The frontoethmoidal suture may also be used surgically to approximate the level of the floor of the anterior cranial fossa.

Providing support for the inferomedial wall and helping to maintain globe position is the thickened region of bone known as the inferomedial strut. The strut is constructed using elements of multiple orbital and facial bones including the maxillary bone, ethmoid bone, and palatine bone posteriorly.[12,13] Preservation of the anterior portion of the inferomedial strut during orbital decompression has been shown to reduce the incidence of post-operative globe dystopia.[14]

As previously mentioned, the lacrimal sac fossa lies between the anterior (maxillary bone) crest and posterior (lacrimal bone) crest of the orbital rim (Fig. 14.4). The relative contribution of these two bones may vary. The lacrimal bone at the lacrimal sac fossa has a mean thickness of 106 μm, which allows easy penetration during

dacryocystorhinostomy surgery (surgical procedure to restore the normal flow of tears into the nose from the lacrimal sac by bypassing the nasolacrimal duct).[15] The maxillary bone is considerably denser than the lacrimal bone, and may require the surgeon to create a more posterior osteotomy during dacryocystorhinostomy. The nasolacrimal canal directs the nasolacrimal duct to the inferior meatus of the nose under the inferior turbinate.

The orbital apex

Many important neural and vascular structures pass through the orbit apex including cranial nerves II through VI, the origins of all the extraocular muscles except the inferior oblique, and arterial and venous blood supplies (Fig. 14.5 and Box 14.1). Pathology in this crucial location of the orbit may lead to the orbital apex syndrome with characteristic hallmarks of visual loss from optic neuropathy and ophthalmoplegia.[16]

The optic foramen conveys the optic nerve and ophthalmic artery from the optic canal into the orbit. The canal is housed in the lesser wing of the sphenoid bone with a contribution from the inferomedial optic strut. It runs 8–10 mm in length and is 5–6 mm in diameter. The optic canal and foramen attain adult dimensions by 3 years and are symmetric in most persons. A canal/foramen that is larger in diameter by at least 1 mm than the contralateral side may be considered abnormal.

The superior orbital fissure is located just lateral to the optic canal. Approximately 20–22 mm in length, the fissure

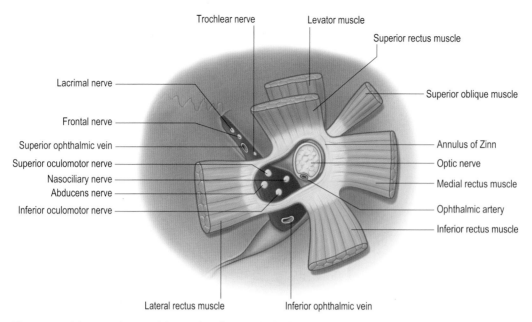

Figure 14.5 Schematic drawing of the annulus of Zinn and orbital apex. (Adapted from: Lemke BN, Lucarelli MJ: Anatomy of the ocular adnexa, orbit, and related facial structures. In Nesi FA, Lisman RD, Levine MR [eds]: Smith's ophthalmic plastic and reconstructive surgery, ed 2, St. Louis, 1998, Mosby.)

Box 14.1 Orbital apex

- Passing through the optic foramen: optic nerve, ophthalmic artery, and sympathetic fibers
- Passing through the superior orbital fissure: oculomotor nerve, trochlear nerve, abducens nerve, the ophthalmic division of the trigeminal nerve (frontalis, lacrimal, and nasociliary nerves), sympathetic fibers, and the superior ophthalmic vein
- Passing through the annulus of Zinn: optic nerve, ophthalmic artery, oculomotor nerve, abducens nerve, nasociliary nerve, and sympathetic fibers
- Passing through the inferior orbital fissure: maxillary division of the trigeminal nerve, branches from the sphenopalatine ganglion, and the inferior ophthalmic vein

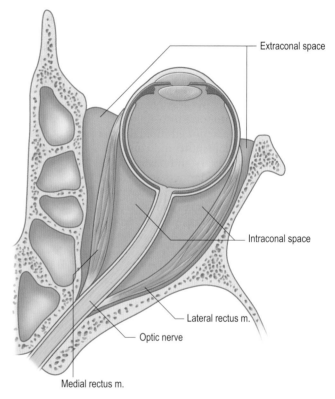

Figure 14.6 Schematic drawing of the compartments of the orbital space. (Adapted from: Imaging of Orbital and Visual Pathway Pathology, W.S. Muller-Forell, Springer-Verlag, 2002. Reproduced by kind permission of Springer Science + Business Media.)

divides the lesser and greater wings of the sphenoid bone. The superior ophthalmic vein, as well as the lacrimal, trochlear, and frontal nerves, passes through the superolateral portion of the fissure outside the annulus of Zinn, a fibrous ring formed by the common origins of the rectus muscles. The annulus is further subdivided by the oculomotor foramen through which the superior and inferior divisions of the oculomotor nerve, the abducens nerve, the nasociliary nerve (a terminal sensory branch of the ophthalmic division of the trigeminal nerve), and sympathetic fibers all pass.[17,18] Also passing through the annulus of Zinn are the optic nerve and ophthalmic artery. As the extraocular rectus muscles proceed forward to their insertions on the globe, they form a conoid enclosure known as the intraconal space. This compartment is useful radiologically (Fig. 14.6). Intraconal pathology of the fat, vessels, and optic

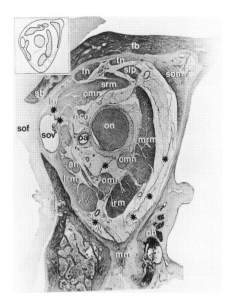

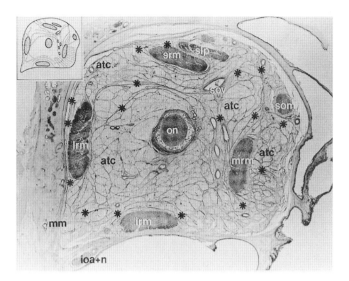

Figure 14.7 A view 18.4 mm from back of globe. Vertical diameter 1.5 cm; transversal diameter 1.1 cm; magnification approximately 11×. The following artifacts are present in this section. Inside muscle cone several holes in adipose tissue are seen. Outside the cone adipose tissue is torn off from frontal and trochlear nerves, superior levator palpebrae/superior rectus complex, medial and inferior recti muscles and medial orbital wall (*fb*, frontal bone; *sb*, sphenoid bone; *sof*, superior orbital fissure; *mm*, Müller's muscle; *pb*, palatine bone; *on*, optic nerve; *fn*, frontal nerve; *ln*, lacrimal nerve; *ncn*; nasociliary nerve; *tn*, trochlear nerve; *an*, abducens nerve; *omn*, oculomotor nerve; *oa*, ophthalmic artery; *sov*, superior ophthalmic vein; *iov*, inferior ophthalmic vein; *slp*, superior levator palpebrae muscle; *srm*, superior rectus muscle; *lrm*, lateral rectus muscle; *irm*, inferior rectus muscle; *mrm*, medial rectus muscle; *som*, superior oblique muscle; *asterisks* (*), connective tissue septa). (Reproduced from Koorneef L: Spatial aspect of orbital musculo-fibrous tissue in man: a new anatomical and histological approach, Amsterdam, 1976. Swets en Zeitlinger B.V.)

Figure 14.8 A view 1.4 mm from hind surface of eye. Vertical diameter, 2.4 cm; transversal diameter, 2.7 cm; magnification approximately 3.5×. Note artifacts in superior, medial, and inferolateral areas (*on*, optic nerve; *sov*, superior ophthalmic vein; *slp*, superior levator palpebrae muscle; *srm*, superior rectus muscle; *lrm*, lateral rectus muscle; *irm*, inferior rectus muscle; *mrm*, medial rectus muscle; *som*, superior oblique muscle; *asterisks* (*), connective tissue septa; *atc*, adipose tissue compartment; *ioa + n*, infraorbital artery and nerve; *mm*, Müller's muscle. (Reproduced from Koorneef L: Spatial aspect of orbital musculo-fibrous tissue in man: a new anatomical and histological approach, Amsterdam, 1976. Swets en Zeitlinger B.V.)

nerve-sheath complex may be distinguished from extraconal pathology of the lacrimal gland, bony orbit, and remaining extraconal fat.[19,20]

The inferior orbital fissure divides the greater wing of the sphenoid laterally from the maxillary bone of the orbital floor inferomedially. The fissure measures approximately 20 mm in length. It communicates with the pterygopalatine fossa which rests behind the maxillary sinus. Traveling through the fissure are the maxillary division of the trigeminal nerve, branches from the sphenopalatine ganglion, and the inferior ophthalmic vein. Branching away from the fissure and entering the infraorbital groove and canal are the infraorbital (V2) and the terminal branch of the internal maxillary artery.

Orbital soft tissues

Periorbital fascia

The periorbital fascia is a single interconnecting network emanating from the periosteal lining of the orbital walls, globe (Tenon's capsule), and extraocular muscles (Figs 14.7 & 14.8).[21] There are also check ligament extensions from the extraocular muscle fascia that attach to the bony orbit as well as sheaths that extend between the rectus muscles. The periorbita is firmly attached at the suture lines, foramina, fissures, arcus marginalis, and the posterior lacrimal crest.

Elsewhere, it is loosely attached to the bone which creates a potential space for accumulation of blood, pus, or tumor growth. It is most firmly attached along the arcus marginalis. The orbital septum is a thin, multilayered extension of the periorbita and is the anterior soft tissue boundary of the orbit. It functions as a physical barrier to pathogens and contributes to the normal posterior position of the orbital fat pads. In the upper eyelid, the fibrous lamellae of the orbital septum gradually blends with those of the levator aponeurosis on average 3.4 mm above the superior tarsal border (with a range of 2–5 mm).[22] In the lower eyelid, the septum inserts onto the inferior border of the tarsus after joining with the inferior retractors 4–5 mm below the tarsus.[23] Laterally the septum fuses with the lateral canthal tendon and attaches 2–3 mm posterior to the rim at the lateral orbital tubercle (Whitnall's tubercle). Medially the septum splits and inserts onto the anterior and posterior lacrimal crest. Finally, it is anchored anteriorly to the orbicularis muscle by multiple fibrous attachments.[24]

Tenon's capsule, or fascia bulbi, is a fibroelastic membrane that extends anteriorly from the dural sheath to encircle the globe and fuse with the conjunctiva just behind the limbus. Posteriorly it separates orbital fat from the globe and muscles. Openings within the fascia allow passage of the extraocular muscles which are in turn surrounded by their own muscular fascia. Anteriorly the muscular fascia thickens and connects with the orbital wall to form the check ligament of the extraocular muscles, preventing overaction of the muscle from which they extend.[25] The strongest of these is the lateral check ligament that inserts on the lateral orbital tubercle, while the medial rectus check ligament inserts behind the posterior lacrimal crest.[26]

Orbital fat

Orbital fat provides cushioned support for the globe and other intraocular structures. Eyelid fat pads are the anterior projections of the orbital fat. Changes in orbital fat appear to have a significant role in the development of thyroid-associated orbitopathy (TAO). Orbital fat and extraocular muscle volume expansion may result in soft tissue swelling, proptosis, lid retraction (with characteristic temporal flare), strabismus (usually due to muscle fibrosis), and compressive optic neuropathy.[27,28] Fibroblasts residing within the orbital connective/adipose tissue are believed to be the targets of an autoimmune attack consisting of primarily T lymphocytes,[29] and there is ongoing research to isolate key autoantibodies.[30] Patients with a hyperthyroid state are not the only patients affected with TAO. Euthyroid patients may also develop TAO. There appears to be a shift in the phenotypes of peripheral blood lymphocytes in euthyroid patients with TAO when compared to normal controls which may permit an orbital immune reaction.[31]

Orbital nerves

Entering the orbit are the optic (cranial nerve II), oculomotor (cranial nerve III), trochlear (cranial nerve IV), abducens (cranial nerve VI), the first and second divisions of the trigeminal (cranial nerve V), parasympathetic, and sympathetic nerves. The trigeminal nerve will be discussed in the eyelid portion of this chapter while the extraocular muscles are discussed in Chapter 7.

Composed of approximately 1.2 million nerve fibers, the optic nerve is an extension of the central nervous system. It has supportive neuroglial cells and is surrounded by the three layers of the meninges: the dura mater, arachnoid layer, and pia mater. Cerebrospinal fluid flows freely within the space between the arachnoid and pia. The nerve arises from the ganglion cell layer of the retina. These axons converge at the optic nerve head which is approximately 1.5 mm in diameter. As the nerve exits the globe at the lamina cribrosa and enters the orbit, its diameter increases to 3.5 mm due to the addition of the myelin sheath. The intraorbital portion of the nerve runs approximately 25 mm, although the distance from the globe to the optic foramen is 18 mm.

This extra slack allows freedom of movement of the globe and a certain degree of proptosis. When excessively stretched, the nerve may exert tension on the globe – seen radiologically as "globe tenting" – which may result in vision loss due to compromise of the nerves' vascular supply.[32] The intracanalicular portion of the nerve is tethered within the bone, making it susceptible to damage from blunt trauma.[33,34] The intracranial portion of the optic nerve runs approximately 15 mm just medial to the internal carotid artery ending at the optic chiasm.

Motor innervation of the orbit involves the oculomotor (cranial nerve III), trochlear (cranial nerve IV), and abducens nerves (cranial nerve VI) (Fig. 14.9). The oculomotor nerve innervates the superior, inferior, and medial recti, the inferior oblique, and the levator palpebrae superioris muscles. Additionally, parasympathetic fibers travel within the inferior division of the nerve. The nerve exits the brainstem medially and travels through the superolateral portion of the cavernous sinus while dividing into a superior and inferior division. Both branches enter the orbit through the oculomotor foramen within the annulus of Zinn. The superior branch enters the underside of the superior rectus, on its way to innervate the overlying levator palpebrae superioris muscle. The inferior branch of the oculomotor nerve travels beneath the optic nerve and supplies innervation to the medial and inferior rectus, as well as the inferior oblique.

Studies in both monkeys and man have revealed that parasympathetic fibers, issued from cells of the pterygopalatine ganglion, pass into the orbit.[35] Nilsson et al showed that parasympathetic fibers responsible for uveal vasodilation travel to the eye from the facial nerve via the pterygopalatine ganglion.[36] As previously mentioned, a small bundle of parasympathetic fibers destined for the ciliary ganglion travel within the inferior division of the oculomotor nerve. Included within this parasympathetic bundle are pupillomotor fibers. During orbital floor repair, damage to these fibers may result in post-operative mydriasis.[37]

The trochlear nerve is the only cranial nerve to exit dorsally from the midbrain; crossing the midline to emerge adjacent to the superior cerebellar peduncle. It travels within the wall of the cavernous sinus and enters the orbit via the superior orbital fissure *outside* of the annulus of Zinn. It is

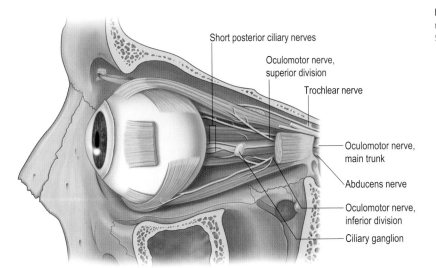

Figure 14.9 Lateral orbital view and major orbital motor nerves. (Adapted from: Dutton JJ. Atlas of Clinical and Surgical Orbital Anatomy. Philadelphia: WB Saunders; 1994.)

Short posterior ciliary nerves

Oculomotor nerve, superior division

Trochlear nerve

Oculomotor nerve, main trunk

Abducens nerve

Oculomotor nerve, inferior division

Ciliary ganglion

the only extraocular cranial nerve that travels extraconally within the orbit. It crosses medially over the origin of the superior rectus and levator palpebrae superioris, and enters the superior oblique muscle. The long intracranial course of the trochlear nerve makes it particularly vulnerable to closed head trauma.[38]

The abducens nerve may exit between the pons and the medulla in one or multiple trunks.[39] The nerve ascends through the subarachnoid space, then climbs along the bony clivus before passing through Dorello's canal into the posterior cavernous sinus. Unlike the oculomotor and trochlear nerves, the abducens nerve is not located in the lateral wall of the cavernous sinus. Instead it travels within the body of the sinus with the internal carotid artery. It enters the orbit through the oculomotor foramen and inserts onto the inner surface of the lateral rectus at approximately one-third the distance from the muscle origin to its insertion.

Sympathetic fibers, originating in the superior cervical ganglion, travel along the internal carotid artery through the cavernous sinus, and into the orbit along the ophthalmic artery. Fibers have also been shown to travel with the abducens and ophthalmic (V1) division of the trigeminal nerve.[17]

While in the orbit, some of the sympathetic fibers pass through the ciliary ganglion without synapsing. Orbital sympathetics allow for pupillary dilation of the iris dilator muscle, function of the smooth muscle of the eyelid (Müller's muscle), and vasoconstriction.

Vascular anatomy

Arterial supply

The arterial supply to the orbit is an anastomotic network with contributions from the internal and external carotid circulation. Most of the arterial supply of the eye and orbit is provided through the ophthalmic artery; the first intracranial branch of the internal carotid artery (Fig. 14.10). The most important branch of the external carotid artery in regards to orbital supply is the internal maxillary artery.

The ophthalmic artery branches off from the internal carotid artery just before the latter artery exits the cavernous sinus. It travels with the optic nerve within the inferolateral portion of a common dural sheath, and gradually bends medially at which point branching of the artery occurs.[40,41] There is a great deal of variability in the order of branching

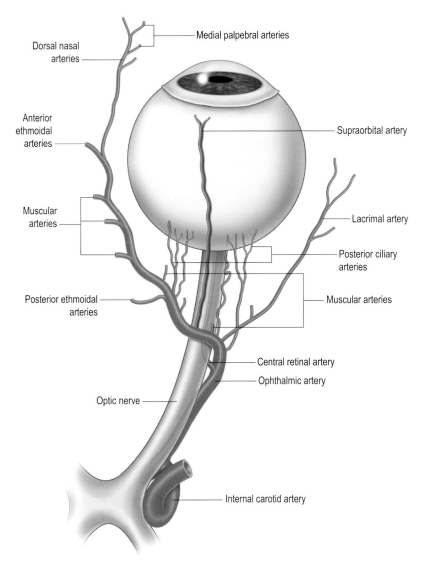

Figure 14.10 Typical branching pattern of the ophthalmic artery. (Adapted from: Lemke BN, Lucarelli MJ: Anatomy of the ocular adnexa, orbit, and related facial structures. In Nesi FA, Lisman RD, Levine MR [eds]: Smith's ophthalmic plastic and reconstructive surgery, ed 2, St. Louis, 1998, Mosby.)

Medial palpebral arteries

Dorsal nasal arteries

Anterior ethmoidal arteries

Muscular arteries

Posterior ethmoidal arteries

Optic nerve

Supraorbital artery

Lacrimal artery

Posterior ciliary arteries

Muscular arteries

Central retinal artery

Ophthalmic artery

Internal carotid artery

vessels originating from the ophthalmic artery. In general, the first branch is the central retinal artery. Originating at the level of the orbital apex and eventually supplying the outer layers of the retina, the central retinal artery travels lateral or inferior to the optic nerve and penetrates the nerve at 5–16 mm posterior to the globe (average 10 mm).[42,43] Also arising near the medial bend of the ophthalmic artery are the lateral and medial posterior ciliary arteries. Constituted of typically two or three branches, these arteries give rise to 15–20 short posterior ciliary arteries as well as the two long posterior ciliary arteries. The short posterior ciliary arteries supply the optic nerve head and choroid, while the long posterior ciliary arteries travel within scleral canals anteriorly to supply the ciliary muscle, anterior choroid, and iris.

Other intraorbital structures supplied by the ophthalmic artery include the extraocular muscles and lacrimal gland. There are two main muscular branches which supply the extraocular muscles: the medial and lateral branches. The medial muscular branch supplies the medial rectus, inferior rectus, and inferior oblique muscles, while the lateral branch supplies the levator palpebrae superioris, superior rectus, superior oblique, and lateral rectus muscles. The vessels run along the muscle belly or medial surface of the muscle. Each rectus muscle artery branches into two anterior ciliary arteries, except for the lateral rectus muscular artery, which terminates into one anterior ciliary artery. The anterior ciliary arteries anastomose with the long posterior ciliary arteries to supply the globe. Simultaneous surgery on three rectus muscles or two muscles in patients with poor vascular supply may result in anterior segment ischemia, characterized by corneal edema and a mild anterior uveitis.[44] The lacrimal artery travels superolaterally in the orbit to reach the posterior surface of the lacrimal gland. It also gives off the zygomaticotemporal, zygomaticofacial, and the lateral palpebral arterial branches. The latter branch supplies the lateral portion of the arcades of the upper and lower eyelids. Portions of the lacrimal artery anastomose with the middle meningeal artery through branches of the recurrent meningeal artery and temporal arteries.

The posterior and anterior ethmoidal arteries course medially from the ophthalmic artery. While traveling at the level of the frontoethmoidal suture, they exit the orbit through the posterior ethmoid foramen and anterior ethmoid foramen respectively. They supply the ethmoid and frontal sinus mucosa as well as the frontal dura.

The supraorbital artery comes off from a more distal portion of the ophthalmic, exits through the supraorbital foramen or notch, and supplies mainly the eyebrow, forehead, and the medial portion of the eyelids. The anterior continuation of the ophthalmic artery is the nasofrontal artery. Just posterior to the trochlea, the nasofrontal artery divides into the supratrochlear artery which supplies the medial forehead and scalp, and the dorsal nasal artery. The dorsal nasal artery and its branches, the medial palpebral arteries, supply the medial eyelids and nose.

The two terminal branches of the external carotid artery are the maxillary and superficial temporal arteries. The maxillary artery branches off from the external carotid artery just behind the neck of the mandible an average of 25.7 mm (range, 24.86–27.47 mm) below the condyle.[45] It passes deep to the mandible and enters the pterygopalatine fossa through the infratemporal fossa.[46] Among its branches are the middle meningeal and infraorbital arteries. The middle meningeal artery, which has been previously discussed, may anastomose with the lacrimal artery via the recurrent meningeal artery. Rarely, it may provide the majority of orbital blood in patients with ipsilateral carotid stenosis or congenital atresia (lack of patency) of the ophthalmic artery.[47] Branches from the infraorbital artery supply the inferior rectus and inferior oblique muscles, the lacrimal sac, orbital fat, upper cheek, and lower eyelid.[42] The superficial temporal artery supplies the forehead and may anastomose with the supraorbital and supratrochlear arteries of the internal carotid circulation.

Venous drainage

Venous drainage of the orbit consists of a valveless anastomotic system with multiple tributaries that drain anteriorly to the facial veins or posteriorly into the cavernous sinus or pterygoid plexus. The major venous branches of the orbit are the superior and inferior ophthalmic veins. The superior ophthalmic vein is formed at the confluence of the angular, supraorbital, and supratrochlear veins. It travels posteriorly along the roof of the orbit, before diving into the muscle cone where it picks up venous drainage from the globe through the superior vortex and ciliary veins. The superior ophthalmic vein then drains into the cavernous sinus after passing through the superior orbital fissure. The inferior orbital vein is responsible for venous drainage of the inferior orbit. It collects venous blood from the inferior extraocular muscles and inferior vortex veins, and drains into both the cavernous sinus and pterygoid plexus. Some of the venous blood from the surrounding periorbita drains through a system of anterior facial veins. These typically drain directly into the external jugular vein.

Orbital lymphatic drainage

The goal to identify the orbital lymphatic drainage system has been pursued in both human and animal studies.[48–50] Traditionally thought to be absent in the orbit, the more recent use of selective lymphatic markers has identified lymphatics in at least some areas of the human orbit.[51,52] Using CD-34 and D2-40 monoclonal antibodies, specific for blood vessel endothelium and a protein on lymphatic endothelium respectively, lymphatics were confirmed in human lacrimal gland and optic nerve dura.[52] Further studies should help to clarify and expand these findings.

Facial and eyelid anatomy and function

The eyebrow and forehead

The eyebrow plays an important role in facial expressions. Eyebrow position and contour are maintained by a group of muscles innervated by the temporal branch of the facial nerve (CN VII). The muscles of the eyebrow and forehead may be divided into superficial (frontalis, procerus, orbicularis oculi), intermediate (depressor supercilii) and deep (corrugator supercilii) groups.[53] (Fig. 14.11). The skin of the brow and accompanying subcutaneous fibroadipose are the thickest layers on the face.

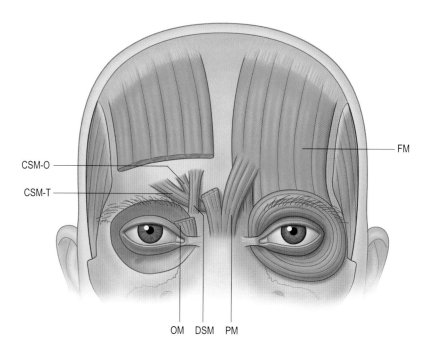

Figure 14.11 Forehead muscles. *FM*, frontalis muscle; *PM*, procerus muscle, *DSM*, depressor supercilii muscle, *CSM-O*, corrugator supercilii muscle – oblique head; *CSM-T*, corrugator supercilii muscle – transverse head. (Adapted from: David M. Knize, Mel Drisko. The Forehead and Temporal Fossa: Anatomy and Technique. Lippincott Williams & Wilkins, 2001.)

The main elevators of the eyebrows and forehead are the paired frontalis muscles. Recruitment of the vertically oriented frontalis may be used to compensate in cases of eyelid ptosis. The frontalis muscle interdigitates with the remaining superficial forehead muscles including the orbicularis oculi, which surrounds the anterior orbit. The orbicularis may be conceptually divided into three topographic regions: the orbital, preseptal, and pretarsal regions. The orbital portion of the muscle is responsible for forced eyelid closure and is an important depressor of the eyebrow and forehead. The pretarsal portion of the muscle is responsible for spontaneous blinking. The procerus muscle arises from the nasal bones and upper lateral cartilages and inserts into eyebrow skin medially. It depresses the glabella to form horizontal skin folds.

The depressor supercilii muscle originates on the frontal process of the maxillary bone and inserts into the dermis of the medial eyebrow.[54] During this course, it travels over the origin of the corrugator muscle. The bilateral corrugator supercilii muscles (transverse head and deep head) lie deep to the frontalis muscle and travel from their insertion on the superomedial orbital margin to insert into the muscle and skin of the respective medial eyebrow. Contraction of the corrugator draws the eyebrow medially and inferiorly resulting in vertical glabellar rhytids (wrinkles).

The midface

Our understanding of the anatomy of the midface and its dynamic relationship with other periocular structures has advanced considerably during the past 30 years. The midface extends from the upper lip to the inferior orbital rim. The facial soft tissue layers include skin, subcutaneous fat, superficial fascia (superficial musculoaponeurotic system), mimetic muscles, deep facial fascia, and a plane containing the parotid duct, buccal fat pad, and facial nerve.

The superficial musculoaponeurotic system (SMAS) was first described in the parotid and cheek area by Mitz &

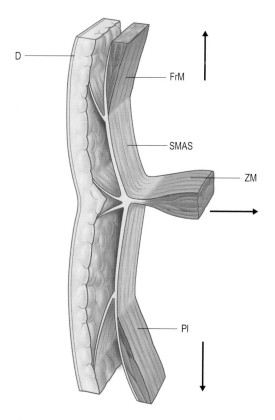

Figure 14.12 Schematic view of the superficial musculoaponeurotic system (SMAS) showing relation to frontalis muscle (*FrM*), zygomatic muscle (*Zm*), dermis (*D*), and platysma (*Pl*). (From Mitz V, Peyronie M: Plast Reconstr Surg 58:80, 1976.)

Peyronie (Fig. 14.12 and Box 14.2).[55] Multiple fibrous extensions from the SMAS to the skin are thought to distribute muscular contractions that aid in facial expression. Initially described as dividing the parotid and cheek fat into two layers, it has subsequently been shown to lie 11–13 mm

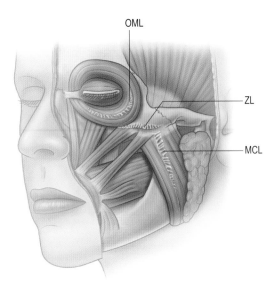

Figure 14.13 Ligamentous support of the midface. *MCL*, Masseteric cutaneous ligaments; *OL*, orbitomalar ligament; *ZL*, zygomatic ligaments. (From Lucarelli MJ et al: Ophthal Plast Reconstr Surg 16(1):7–22, 2000.)

Box 14.2 Key midfacial support components

- The SMAS is an aponeurotic system that is contiguous with the occipitalis, frontalis, orbicularis, and platysma
- Fibrous extensions from SMAS to the skin may help to transmit facial muscular contractions
- Midface bony attachments: zygomatic ligaments and orbitomalar ligaments; attenuation of these ligaments contributes to mid-facial ptosis.

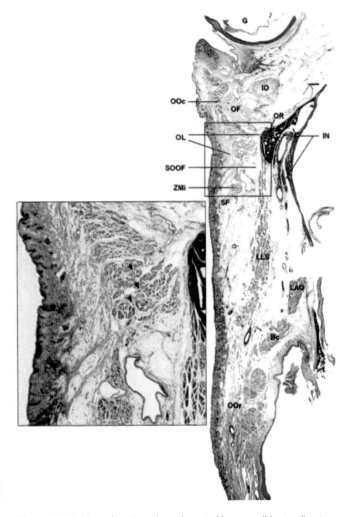

Figure 14.14 Merged sections through central lower eyelid extending to the oral commissure from specimen with midfacial ptosis (Masson trichrome staining). The globe (G), inferior oblique (IO), and orbital fat (OF) are visible superiorly. The orbitomalar ligament (OL) extends off the orbital rim (OR) and passes through the orbicularis oculi muscle (OOc) to the dermis. The nasolabial fold (NLF) is seen inferiorly. The subcutaneous fat (SF) can be seen anterior to the orbicularis oculi and zygomatic minor (ZMi) muscles. The suborbicularis oculi fat (SOOF) rests posterior to the orbicularis oculi muscle and surrounds the proximal portion of the levator labii superioris muscle (LLS). Note the continuity of the inferior SOOF and the subcutaneous fat. The infraorbital nerve (IN) and accompanying vessel can be seen traversing the SOOF deep to the levator labi superioris muscle. The levator labi superioris muscle extends into the upper lip. The levator anguli oris (LAO) arises from the maxilla inferior to the infraorbital foramen. The orbicularis oris (OOr) and buccinator (Bc) are visible in the inferior portion of the specimen. The inset details the orbitomalar ligament (arrowheads) traversing the orbicularis oculi muscle. (From Lucarelli MJ et al: Ophthal Plast Reconstr Surg 16(1):7–22, 2000.)

---

deep to the skin at the mid-cheek and invest the zygomaticus major, zygomaticus minor, and levator labii superioris.[56] It is continuous with the platysma in the lower face and the frontalis in the upper face as well as the anterior and posterior fascia of the orbicularis oculi muscle.[57,58] The facial nerve courses beneath the SMAS.

The SMAS has a number of bony attachments. Shortly after its initial description, the SMAS was recognized as an important structure in facial rhytidectomy (face-lift).[59] In the midface, the zygomatic ligaments and orbitomalar ligaments have been characterized (Fig. 14.13).[60,61] The mandibular ligaments have been described in the lower face. The orbitomalar ligament has been characterized to provide major osseocutaneous midfacial support due to its steadfast lateral component found approximately 5 mm lateral to the lateral orbital rim.[62] Work by Lucarelli and colleagues showed that attenuation of the zygomatic, masseteric cutaneous, and orbitomalar ligament is associated with age-related midfacial ptosis (Fig. 14.14).[62] Others have recently referred to the orbitomalar ligament as the orbital retaining ligament.[63]

The facial mimetic muscles are arranged into lamella. Freilinger et al have categorized in detail the mimetic muscle into four distinct layers.[64] A superficial layer consists of the orbicularis oculi, zygomatic minor, and depressor anguli oris. The second layer contains the platysma, depressor labii inferioris, the levator labii superioris alaeque nasi, and zygomatic major muscles. The third layer consists of the levator labii superioris and orbicularis oris. Finally, the deepest layer consists of the mentalis, the levator anguli oris, and the buccinator. Branches of the facial nerve supply the first three layers of muscles from the underside, while the deepest layer is innervated from the outer surface.

Paralysis of the facial nerve may have profound effects on facial symmetry and corneal protection. Idiopathic facial

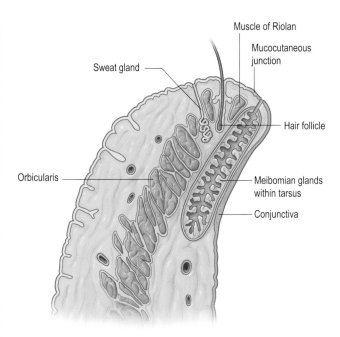

Figure 14.15 Sagittal section of the eyelid margin. (From Lemke BN, Lucarelli MJ: Anatomy of the ocular adnexa, orbit, and related facial structures. In Nesi FA, Lisman RD, Levine MR [eds]: Smith's ophthalmic plastic and reconstructive surgery, ed 2, St. Louis, 1998, Mosby.)

nerve palsy or Bell's palsy is estimated to have an incidence of 25 cases per 100,000 (Box 14.3).[65] The differential diagnosis of facial palsy includes infectious (Ramsey Hunt, herpes zoster/Lyme disease), neoplastic (facial nerve neuroma or hemangioma, acoustic neuroma, meningioma, metastatic), inflammatory (Wegener's granulomatosis, sarcoidosis), traumatic (basilar skull fracture, temporal bone trauma, parotid injury), and congenital etiologies. Lagophthalmos (inability to close the eyelid fully) and a poor blink may lead to corneal exposure. Intensive lubrication, patching, lid taping and use of a moisture chamber may be adequate therapies to prevent complications of corneal exposure. More severe cases may require surgery (upper lid weight or eyelid closure from a tarsorrhaphy).[66]

The eyelid

The eyelid margin

The eyelid margin may be divided into anterior and posterior lamella. The anterior lamella is composed of skin, muscle, and associated glands, while the posterior lamella consists of the tarsal plate, conjunctiva, and associated glands (Fig. 14.15). The two lamella may be separated from one another along the gray line; a terminal extension of the orbicularis known as the muscle of Riolan. This portion of the muscle has been hypothesized to keep the eyelid edges in close approximation to the surface of the globe as well as help expel glandular contents during blinking.[67,68] The cilia serve to protect the globe from large airborne particles. They are also highly sensitive to touch, and elicit the blink reflex when stimulated.

The anterior lamella functions as a single anatomic unit. The eyelid skin is the thinnest in the body, due in part to a relative thinning of dermis and absence of a subcutaneous fat layer. In adults, the palpebral fissure measures 9–12 mm vertically while the horizontal distance between the medial and lateral commissures measures approximately 30 mm. The lower eyelid rests at or up to 1.5 mm above the inferior limbus while the upper eyelid rests at approximately 1.5–2.0 mm below the superior limbus.[42] The highest point of the upper eyelid rests just nasal to the central pupillary axis.

The orbicularis muscle is located beneath the eyelid skin. The pretarsal portion of the muscle originates from the anterior and posterior arms of the medial canthal ligament. The deep head of the muscle (Horner's muscle) inserts onto the posterior lacrimal crest and lacrimal fascia while the anterior head attaches to the anterior lacrimal crest (see Fig. 14.4). The muscle therefore surrounds the lacrimal sac, and during

contraction assists the lacrimal pump mechanism.[69] Laterally, the pretarsal orbicularis muscle inserts in to the lateral canthal tendon.

Glands of the anterior lamella include the glands of Zeis and Moll. The Zeis glands are sebaceous glands while the tubular glands of Moll are apocrine sweat glands. The Zeis glands empty into the cilial follicular infundibulum while the tubular glands of Moll are located in between the lash follicles.

The posterior lamella of the eyelid consists of the tarsal plate, meibomian glands, and conjunctiva. The tarsal plates are composed of collagens (types I and III collagen) as well as components more typical of cartilage (aggrecan, chondroitins 4 and 6 sulfate).[70] They provide both rigidity and allow for dynamic movement of the eyelids. The height of the upper tarsus measures approximately 10 mm centrally, while the height of the lower tarsus measures on average 3–4 mm.[71] Medially and laterally, both the upper and lower tarsal plates taper to become the medial and lateral canthal tendons. As previously described, the medial canthal tendon splits into an anterior and posterior head which attach to the anterior lacrimal crest and posterior lacrimal crest respectively. The lateral canthal tendon inserts at Whitnall's lateral tubercle, a small rounded protuberance of the zygomatic bone a few millimeters inside of the lateral orbital wall. Laxity of the canthal tendon may lead to ectropion (turning outward of the eyelid) and may require an eyelid-tightening procedure by shortening and reattaching the lateral canthal tendon. Whitnall's lateral tubercle is also an insertion site for the lateral horn of the levator aponeurosis, the inferior (Lockwood's) and superior (Whitnall's) transverse ligament, the deep pretarsal orbicularis, and the expansion of the superior rectus muscle sheath.[11,72]

The tarsal plates of both the upper and lower eyelids contain the sebaceous meibomian glands. There are approximately 25 in the upper eyelid and 20 in the lower eyelid.[22] The glands secrete the lipid layer (outer layer) of the tear film which helps prevents evaporation of the tears. In the setting of meibomian gland dysfunction, the relative absence of the lipid layer may result in dry eyes.[73] Chronic inflammation of the glands or meibomitis, may lead to dysplastic hair follicles that curve towards the globe, known as distichiasis.[74] Treatment for distichiasis (misdirected lashes from the meibomian gland orifices) or trichiasis (misdirection of normal eyelash follicles) may include cryotherapy, argon laser, electrolysis, trephination and other lid-modifying modalities.[75-77]

Tightly bound to the posterior aspect of the tarsal plates is the palpebral conjunctiva. It consists of a non-keratinized stratified columnar epithelium overlying loose connective tissue known as substantia propria. The accessory glands of Wolfring, located in the tarsal conjunctiva, and Krause, located in conjunctival fornix, help to produce the aqueous layer of the tear film. This middle layer of the tear film is tightly bound to the hydrophilic mucin layer (inner layer) produced from conjunctival goblet cells. Mucocutaneous disorders such as ocular cicatricial pemphigoid, Stevens–Johnson syndrome, and bullous pemphigoid, may lead to destruction of goblet cells and their progenitor stem cells leading to severe chronic inflammation with scarring and conjunctival contraction.[78]

The inferior and superior puncta provide entrance to the lacrimal excretory system. The lower eyelid punctum is approximately 10 mm from the lacrimal tear sac with the upper punctum 8 mm from the sac. The puncta are normally oriented towards the tear lake.

Eyelid musculature

The major protractor of the eyelids is the orbicularis muscle, which may be divided into the orbital, preseptal, and pretarsal components. The orbicularis myofibers have the smallest diameter of any skeletal muscle, and although they vary somewhat in length, are overall relatively short.[79] The variability in myofiber length may help to explain the variety of complex eyelid movements including blinking, winking, and spasm.

The main retractor of the upper eyelid is the levator palpebrae superioris (Fig. 14.16). Assistance is also provided by the superior tarsal muscle (Müller's muscle) which is supplied by sympathetic innervation. The levator is divided into muscular and aponeurotic portions. The muscular portion of the levator measures approximately 36 mm in length and lies just above the superior rectus.[80] The two muscles can be separated easily except along the medial borders, where they are bound by a fascial sheath. As the muscle moves

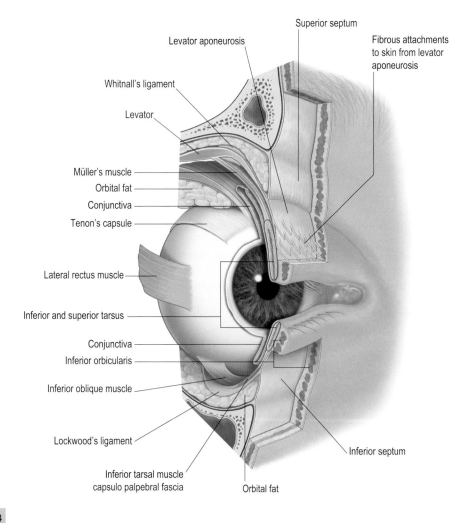

Figure 14.16 Oblique section through the upper and lower eyelids. (Modified from Lemke BN, Lucarelli MJ: Anatomy of the ocular adnexa, orbit, and related facial structures. In Nesi FA, Lisman RD, Levine MR [eds]: Smith's ophthalmic plastic and reconstructive surgery, ed 2, St. Louis, 1998, Mosby.)

Superior septum

Levator aponeurosis

Fibrous attachments to skin from levator aponeurosis

Whitnall's ligament

Levator

Müller's muscle

Orbital fat

Conjunctiva

Tenon's capsule

Lateral rectus muscle

Inferior and superior tarsus

Conjunctiva

Inferior orbicularis

Inferior oblique muscle

Lockwood's ligament

Inferior tarsal muscle capsulo palpebral fascia

Orbital fat

Inferior septum

anteriorly in the orbit it begins to widen, eventually joining Whitnall's ligament; a transverse fibrous condensation just posterior to the superior orbital rim.[10] Whitnall's ligament provides structural support to the upper eyelids. It has been suggested that it acts as either a fulcrum or swinging suspender for the levator.[13,81] After passing Whitnall's ligament, the levator muscle transitions into the fibrous levator aponeurosis. The aponeurosis courses another 14–20 mm to insert onto the inferior third of the anterior surface of the tarsus. An anterior portion of the aponeurosis sends fine attachments to the upper eyelid skin as well as multiple elastic fiber insertions of the levator muscle complex that includes the levator aponeurosis, the conjoined fascia, the lid crease area, and Muller's muscle tendon.[82] The upper eyelid crease is formed by attachments to the skin.

The smooth superior tarsal muscle of Müller arises from the underside of the levator muscle approximately 15 mm above the superior tarsal border and inserts at the upper border of the tarsus.[83] This sympathetically innervated muscle provides approximately 2 mm of upper eyelid lift. Interruption of this innervation may result in Horner's blepharoptosis (oculosympathetic paresis). Approximately 2 mm of ptosis along with miosis and occasional unilateral anhidrosis, is highly suggestive of Horner's syndrome due to an interruption of the sympathetic system. The vascular arterial arcade is found between the levator aponeurosis and Müller's muscle. This serves as a useful surgical landmark to identify the underlying Müller's muscle.

The lower eyelid retractors are palpebral extensions from the inferior rectus muscle and are utilized, along with the rectus muscle, during down-gaze. This muscular extension splits to surround the inferior oblique muscle as it travels anterior in the orbit. The outer division or capsulopalpebral fascia is analogous to the levator palpebrae superioris, while the inner division, which is composed of smooth muscle fibers, is analogous to Müller's muscle.[23] The two layers are typically not distinct during surgical dissection. The lower eyelid retractors have three distinct insertions. Anteriorly, the capsulopalpebral fascia fuses with the orbital septum 4 mm inferior to the tarsus. A middle layer of inferior tarsal muscles fibers terminates a few millimeters inferior to the tarsus. Finally, a posterior layer of retractors inserts on Tenon's fascia.

Blinking

Spontaneous blinking occurs every 3–8 seconds with an average of 12 blinks per minute. During the normal blink cycle, eyelid closure is a result of activity *and* co-inhibition of two groups of muscles: the protractors of the eyelids (the orbicularis oculi, corrugator, and procerus muscles), and the voluntary retractors of the eyelids (the levator palpebra superioris and frontalis muscles), respectively. The pretarsal portion of the orbicularis muscle is thought to be responsible for spontaneous blinking.

Blinking assists in the maintenance of a normal ocular tear surface. A blink initiates a cycle of secretion, dispersal, evaporation, and drainage of tears.[84] Closure of the eyelids occludes the puncta, thus preventing fluid regurgitation. At the same time the canaliculi and sac are compressed, which forces fluid down the nasolacrimal duct. When the eyelid opens, both the puncta and canalicular system open, creating a partial vacuum which sucks in tears from the ocular

---

Box 14.4 Benign essential blepharospasm

- Bilateral, involuntary, spasmodic forced eyelid closure
- Typically presents in 5th–7th decade of life
- Absent during sleep
- May be due to an abnormality of the basal ganglia
- Repeated administration of botulinum toxin type A (Botox) injections is the treatment of choice

---

surface. Closure of the eyelids also stimulates the expression of the aqueous tear layer from the lacrimal and accessory lacrimal glands.

During delayed blinking the normal tear cycle is altered. This may result in dry eye or secondary reflexive tearing. Seen in a variety of clinical scenarios including reading, older age, and in patients with Parkinson's disease, delayed blinking results in decreased tear expression, increased tear evaporation, and potentially decreased tear drainage.[85]

Benign essential blepharospasm (BEB) is a bilateral, involuntary, spasmodic forced eyelid closure with accompanying brow depression in the absence of any other ocular or adnexal cause (Box 14.4). It may initially begin with simply an increased blink rate. Blepharospasm may also be secondary to ocular surface disease or neurodegenerative diseases such as Parkinson's.[86] BEB affects women more often than men, and typically presents during the fifth to seventh decades of life.[87]

During BEB, the normal activation/inhibition pathways (possibly basal ganglia in origin) are disrupted, resulting in sustained activation of the protractors with sustained inhibition of the retractors. Repeated botulinum toxin type A (Botox) approximately every 3 months is the treatment of choice for many patients with BEB. The neurotoxin, which is produced from *Clostridium botulinum*, inhibits the release of acetylcholine (ACh) and may be used to temporarily paralyze the eyelid protractors.[88]

Eyelid fat

Bounded by the orbital septum anteriorly and the levator muscle/aponeurosis posteriorly, there are two upper eyelid fat pads separated by the trochlea and superior oblique tendon. The most medial fat pad takes on a fibrous white appearance while the central preaponeurotic fat pad is yellow, due to a high level of free-radical scavenging carotenoids.[89] The fatty acid levels found in the fat are similar in samples throughout the orbit, despite these color variations. Orbital fats are typically unsaturated. Some of these fatty acids include oleic acid (18:1) and linoleic acid (18:2), among others. The preponderance of unsaturated fats may provide a lubricating surface with advantages for ocular motility. Laterally, the pink and firm lacrimal gland should not be confused with eyelid fat during surgery. The lower eyelid contains three fat pads that are enclosed in fibrous compartments: medial, central, and lateral.

Eyelid vasculature

The eyelids are highly vascularized with contributions from both the external and internal carotid systems. The external carotid system supplies the eyelid through branches from the

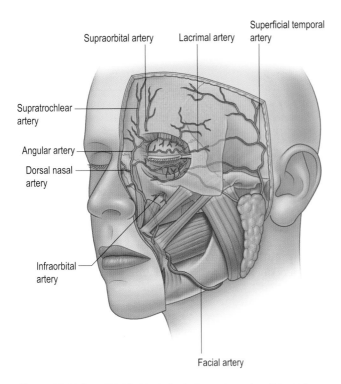

Figure 14.17 Superficial facial arteries in the ocular region. (Adapted from: Lemke BN, Lucarelli MJ: Anatomy of the ocular adnexa, orbit, and related facial structures. In Nesi FA, Lisman RD, Levine MR [eds]: Smith's ophthalmic plastic and reconstructive surgery, ed 2, St. Louis, 1998, Mosby.)

facial artery, superficial temporal artery, infraorbital artery (via the maxillary artery), and superficial temporal artery (Fig. 14.17). The facial artery arises from the external carotid artery just below the angle of the jaw and travels superomedially along the nasolabial fold. As it approaches the medial canthus it becomes the angular artery. The artery runs approximately 5–6 mm medial to the medial canthus. External dacryocystorhinostomy incisions should be positioned to avoid damage to the artery. The artery perforates the orbital septum to anastomose with branches of the ophthalmic artery. The superficial temporal artery supplies the forehead and may anastomose with the supraorbital and supratrochlear arteries of the internal carotid circulation. The maxillary artery and its branches were discussed in the orbital section.

The main arterial supply of the upper eyelids is supplied by four arterial arcades: the marginal, peripheral, superficial orbital, and deep orbital arcades.[90] In the upper and lower eyelids, the marginal arcades lie anterior to the tarsus approximately 2–4 mm from the eyelid margin. Anastomoses between the supratrochlear, lacrimal, and angular arteries compose the superior arcade, while anastomoses from the infraorbital, angular, and zygomaticofacial arteries form the inferior arcade. During tarsal sharing procedures such as free tarsoconjunctival grafts or tarsoconjunctival flaps (e.g. Hughes procedure), proper care should be taken to maintain vascular supply to both eyelids.

Arising from the confluence of the frontal and supraorbital veins, the facial vein is the main venous drainage pathway of the eyelids. It starts as the angular vein 6–8 mm medial to the medial canthus and travels roughly with the facial artery, ending at the external jugular vein. A minor proportion of the facial venous drainage passes posteriorly through veins that empty into cavernous sinus and pterygoid plexus, as discussed earlier.

Eyelid lymphatics

Lymphoscintigraphy of Cynomolgus monkey eyelids has revealed discrete lymphatic drainage pathways for the upper and lower eyelids and a dual pathway for the central upper eyelid.[91,92] The lateral and medial portions of the upper and lower eyelids drain to the superficial preauricular and submandibular lymph nodes, respectively. This drainage continues through the deep cervical nodes.

Eyelid innervation

The eyelids and periocular structures are innervated by the oculomotor nerve (CN III), facial nerve (CN VII), trigeminal nerve (CN V), and sympathetic fibers.

The facial nerve supplies most of the motor innervation to the face. After passing through the parotid gland the nerve divides into five divisions: temporal, zygomatic, buccal, mandibular, and cervical nerves. There is a great deal of overlap of regions innervated by the nerve. In general, the temporal branch innervates the frontalis muscle while the zygomatic, temporal, and buccal divisions innervate the orbicularis oculi. Motor innervation to the levator palpebrae is provided by the superior division of the oculomotor nerve. Sympathetic fibers innervate the underlying Müller's muscle. The exact course of these sympathetic fibers is controversial.

Sensory innervation to the face is supplied by the trigeminal nerve. The nerve is composed of the ophthalmic (V1), maxillary (V2), and mandibular divisions (V3), of which the ophthalmic and maxillary divisions provide sensory innervation to the periorbita (Fig. 14.18).

The ophthalmic (V1) division enters the orbit through the superior orbital fissure after dividing into three branches: the lacrimal, frontal, and nasociliary nerves. The frontal and lacrimal nerves enter the orbit in the superolateral portion of the superior ophthalmic fissure outside of the annulus of Zinn. A superior division of the lacrimal nerve supplies sensation to the nearby conjunctiva, the lateral upper eyelid, and the lacrimal gland. An inferior division of the nerve anastomoses with the zygomaticotemporal branch of the maxillary trigeminal nerve and is joined by parasympathetic secretory fibers destined for the lacrimal gland.

While traveling just below the periorbita in the anterior orbit, the frontal nerve divides into the supraorbital and supratrochlear divisions. The supratrochlear nerve innervates the glabellar skin, medial upper eyelid, and lower forehead. The supraorbital nerve typically divides into a superficial (medial) division that supplies the anterior margin of the scalp and forehead skin, and a deep (lateral) division that provides sensation to the frontoparietal scalp (Fig. 14.19).[93] Damage to this deep (lateral) division during a forehead lift may result in a large area of scalp numbness and paresthesia.[93]

The nasociliary branch of the ophthalmic trigeminal nerve passes through the annulus of Zinn and travels within the intraconal space. It gives off a small sensory branch to the ciliary ganglion and long ciliary nerves to the globe,

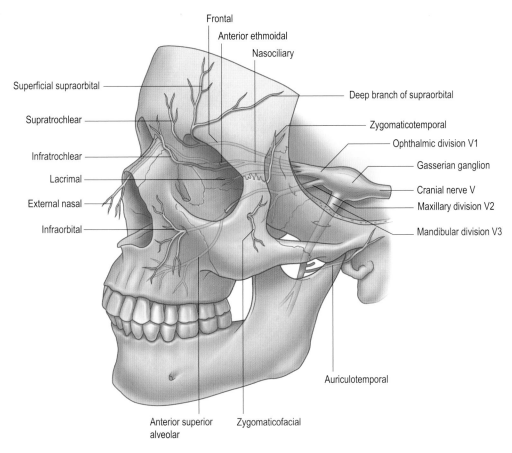

Figure 14.18 Trigeminal nerve pathways. (Adapted from: Lemke BN, Lucarelli MJ: Anatomy of the ocular adnexa, orbit, and related facial structures. In Nesi FA, Lisman RD, Levine MR [eds]: Smith's ophthalmic plastic and reconstructive surgery, ed 2, St. Louis, 1998, Mosby.)

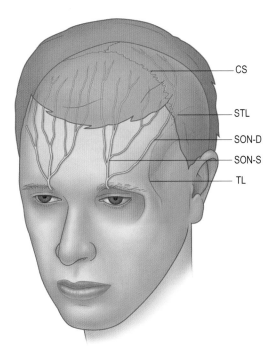

Figure 14.19 Diagram illustrating the courses of the deep (SON-D) and the superficial (SON-S) divisions of the supraorbital nerve trunk. (Adapted from Knize, DM. A Study of the Supraorbital nerve. Plast Reconstr Surg 1995;96:564–569.)

before dividing into the ethmoid and infratrochlear end branches. The long ciliary nerves innervate the iris, cornea, and ciliary muscle. The ethmoid branch provides sensation to the nasal mucosa and skin on the anterior tip of the nose. A small unilateral vesicular rupture of herpes zoster on the tip of the nose is known as Hutchinson's sign, and is associated with an increased risk of herpes zoster ophthalmicus.

The maxillary (V2) division of the trigeminal nerve exits the middle cranial fossa at the foramen rotundum, and passes through the lateral wall of the cavernous sinus and pterygopalatine fossa to enter the inferior orbital fissure. Within the pterygopalatine fossa, it also gives rise to the zygomatic, sphenopalatine, and posterior superior alveolar branches. The infraorbital nerve is the terminal branch of the maxillary nerve which enters the infraorbital groove approximately 30 mm posterior to the orbital rim. The zygomatic nerve divides into the zygomaticofacial and zygomaticotemporal nerves, which provide innervation to the lateral cheek and lateral brow respectively. Parasympathetic secretory fibers join the zygomaticotemporal nerve prior to joining the lacrimal nerve en route to the lacrimal gland.

References

1. Langdon JH. The human strategy: an evolutionary perspective on human anatomy. Oxford: Oxford University Press, 2005.

2. Burrows AM. The facial expression musculature in primates and its evolutionary significance. Bioessays 2008; 30(3):212–225.

3. Zide BM, Jelks GW. Surgical anatomy of the orbit. New York: Lippincott Williams & Wilkins, 1985.

4. Webster RC. Supraorbital and supratrochlear notches and foramina: anatomical variations and surgical relevance. Laryngoscope 1986; 96:311–315.

5. Nahum AM. The biomechanics of maxillofacial trauma. Clin Plast Surg 1975; 2:59–64.

6. Karesh JW, Kelman SE, Chirico PA et al. Orbital roof "blow-in" fractures. Ophthal Plast Reconstr Surg 1991; 7(2):77–83.

7. Lee D, Brody R, Har-El G. Frontal sinus outflow anatomy. Am J Rhinol 1997; 11(4):283–285.

8. Goldberg AN, Oroszlan G, Anderson TD. Complications of frontal sinusitis and their management. Otolaryngol Clin North Am 2001; 34(1):211–225.

9. Garcia CE, Cunningham MJ, Clary RA et al. The etiologic role of frontal sinusitis in pediatric orbital abscesses. Am J Otolaryngol 1993; 14:449–452.

10. Whitnall SE. On a tubercle on the malar bone, and on the lateral attachments of the tarsal plates. J Anat Physiol 1911; 45:426–432.

11. Whitnall SE. On a ligament acting as a check to the action of the levator palpebrae superioris. J Anat Phys 1910; 14:131.

12. Kim JW, Goldberg RA, Shorr N. The inferomedial orbital strut: an anatomic and radiographic study. Ophthal Plast Reconstr Surg 2002; 18(5):355–364.

13. Goldberg RA, Wu JC, Jesmanowicz A, Hyde JS. Eyelid anatomy revisited. Dynamic high-resolution magnetic resonance images of Whitnall's ligament and upper eyelid structures with the use of a surface coil. Arch Ophthalmol 1992; 110(11):1598–1600.

14. Goldberg RA, Shorr N, Cohen MS. The medial orbital strut in the prevention of postdecompression dystopia in dysthyroid ophthalmopathy. Ophthal Plast Reconstr Surg 1992; 8(1):32–34.

15. Hartikainen J, Aho HJ, Seppa H et al. Lacrimal bone thickness at the lacrimal sac fossa. Ophthalmic Surg Lasers 1996; 27:679–684.

16. Yeh S, Foroozan R. Orbital apex syndrome. Curr Opin Ophthalmol 2004; 15(6):490–498.

17. Lyon DB, Lemke BN, Wallow I, Dortzbach RK. Sympathetic nerve anatomy in the cavernous sinus and retrobulbar orbit of the cynomolgous monkey. Ophthal Plast Reconstr Surg 1992; 8:1–12.

18. Monard M, Tcherekayev V, deTribolet N. The superior orbital fissure: a microanatomical study. Neurosurgery 1994; 35:1087–1093.

19. Aviv RI, Casselman J. Orbital imaging: Part 1. Normal anatomy. Clin Radiol 2005; 60(3):279–287.

20. Aviv RI, Casselman J. Orbital imaging: Part 2. Intraorbital pathology. Clin Radiol 2005; 60(3):288–307.

21. Koornneef L. Details of the orbital connective tissue system in the adult. In: Spatial aspect of orbital musculo-fibrous tissue in man. New York: Swets & Zweitlinger, 1977.

22. Meyer DR, Linberg JV, Wobig JL et al. Anatomy of the orbital septum and associated eyelid connective tissues. Ophthal Plast Reconstr Surg 1991; 7:104.

23. Hawes MJ, Dortzbach RK. The microscopic anatomy of the lower eyelid retractors. Arch Ophthalmol 1982; 100:1313–1318.

24. Anderson RL, Beard C. The levator aponeurosis attachments and their clinical significance. Arch Ophthalmol 1977; 95:1437–1441.

25. Hargiss JL. Surgical anatomy of the eyelids. Trans Pacif Coast Otolaryngol Soc. 1963; 44:193–202.

26. Fink WH. An anatomical study of the check mechanism of the vertical muscles of the eye. Trans Am Ophthalmol Soc 1956; 54:193–213.

27. Garrity JA, Bahn RS. Pathogenesis of graves ophthalmopathy: implications for prediction, prevention, and treatment. Am J Ophthalmol 2006; 142:147–153.

28. Kazim M, Goldberg RA, Smith TJ. Insights into the pathogenesis of thyroid-associated orbitopathy: evolving rationale for therapy. Arch Ophthalmol 2002; 120(3):380–386.

29. Weetman AP, Cohen S, Gatter KC et al. Immunohistochemical analysis of the retrobulbar tissue in Graves' ophthalmopathy. Clin Exp Immunol 1989; 75:222–227.

30. Khoo TK, Bahn RS. Pathogenesis of Graves' ophthalmopathy: the role of autoantibodies. Thyroid 2007; 17:1013–1018.

31. Douglas RS, Gianoukakis AG, Goldberg RA et al. Circulating mononuclear cells from euthyroid patients with thyroid-associated ophthalmopathy exhibit characteristic phenotypes. Clin Exp Immunol 2007; 148(1):64–71.

32. Dalley RW, Robertson WD, Rootman J. Globe tenting: a sign of increased orbital tension. Am J Neuroradiol 1989; 10:181–186.

33. Sarkies N. Traumatic optic neuropathy. Eye 2004; 18:1122–1125.

34. Anderson RL, Panje WR, Gross CE. Optic nerve blindness following blunt forehead trauma. Ophthalmology 1982; 89:445–455.

35. Ruskell GL. The orbital branches of the pterygopalatine ganglion and their relationship with the internal carotid nerve branches in primates. J Anat 1970; 106:323–339.

36. Nilsson SF, Linder J, Bill A. Characteristics of uveal vasodilation produced by facial nerve stimulation in monkeys, cats, and rabbits. Exp Eye Res 1985; 40(6):841–852.

37. Hornblass A. Pupillary dilation in fractures of the floor of the orbit. Ophthalmic Surg 1979; 10:44–46.

38. Mansour AM, Reinecke RD. Central trochlear palsy. Surv Ophthalmol 1986; 30:279–297.

39. Lyon DB, Lemke BN, Wallow I, Dortzbach RK. Sympathetic nerve anatomy in the cavernous sinus and retrobulbar orbit of the cynomolgous monkey. Ophthal Plast Reconstr Surg 1992; 8:1–12.

40. Lang H, Kageyama I. The ophthalmic artery and its branches, measurements, and clinical importance. Surg Radiol Anat 1990; 12:83–90.

41. Bergen MP. A literature review of the vascular system in the human orbit. Acta Morphol Neerl Scan 1981; 19:273–305.

42. Dutton JJ. Atlas of clinical and surgical orbital anatomy. Philadelphia: WB Saunders, 1994.

43. Singh S, Dass R. The central artery of the retina. Br J Ophthalmol 1960; 44:193–212.

44. Saunders RA, Bluestein EC, Wilson ME et al. Anterior segment ischemia after strabismus surgery. Surv Ophthalmol 1994; 38:456–466.

45. Orbay H, Kerem M, Ünlü RE et al. Maxillary artery: anatomical landmarks and relationship with the mandibular subcondyle. Plast Reconst Surg 2007; 120:1865–1870.

46. Pearson BW, Mackenzie RG, Goodman WS. The anatomical basis of transantral ligation of the maxillary artery in severe epistaxis. Laryngoscope 1969; 79:969–984.

47. Lemke BN, Lucarelli MJ. Anatomy of the ocular adnexa, orbit, and related facial structures. In: Nesi FA, Lisman RD, eds. Smith's ophthalmic plastic and reconstructive surgery. St Louis CV Mosby, 1998.

48. McGetrick JJ, Wilson DG, Dortzbach RK, Kaufman PL, Lemke BN. A search for lymphatic drainage of the monkey orbit. Arch Ophthalmol 1989; 107:255–260.

49. Gausas RE, Gonnering RS, Lemke BN, Dortzbach RK, Sherman DD. Identification of human orbital lymphatics. Ophthal Plast Reconst Surg 1999; 15(4):252–259.

50. Sherman DD, Gonnering RS, Wallow IHL et al. Identification of orbital lymphatics: enzyme histochemical light microscopic and electron microscopic studies. Ophthal Plast Reconstr Surg 1993; 9(3):153–169.

51. Dickinson AJ, Gausas RE. Orbital lymphatics: do they exist? Eye 2006; 20:1145–1148.

52. Gausas RE. Advances in applied anatomy of the eyelid and orbit. Curr Opin Ophthalmol 2004; 15:422–425.

53. Daniel RK, Landon B. Endoscopic forehead lift: anatomic basis. Aesthetic Surg J 1997; 17:97–104.

54. Cook BE, Lucarelli MJ, Lemke BN. Depressor supercilii muscle: anatomy, histology, and cosmetic implications. Ophthal Plast Reconstr Surg 2001; 17(6):411–414.

55. Mitz V, Peyronie M. The superficial musculo-aponeurotic system (SMAS) in the parotid and cheek area. Plast Reconstr Surg 1976; 58:80–88.

56. Rose JG, Lucarelli MJ, Lemke BN. Radiological measurement of the subcutaneous depth of the SMAS in the midface. Orlando, FL (Oct 18–19,2002). In Proceedings of American Society of Ophthalmic Plastic and Reconstructive Surgery, 2002.

57. Wassef M. Superficial fascial and muscular layers in the face and neck: a histologic study. Aesthetic Plast Surg 1987; 11:171–176.

58. Thaller SR, Kim S, Patterson H et al. The submuscular aponeurotic system (SMAS): a histologic and comparative anatomy evaluation. Plast Reconstr Surg 1990; 86(4):690–696.

59. Jost G, Lamouche G. SMAS in rhytidectomy. Aesthetic Plast Surg 1982; 6:69–74.

60. Furnas DW. The retaining ligaments of the cheek. Past Reconstr Surg 1989; 83(1):11–16.

61. Kikkawa DO, Lemke BN, Dortzbach RK. Relations of the superficial musculoaponeurotic system to the orbit and characterization of the orbitomalar ligament. Ophthal Plast Reconstr Surg 1996; 12:77–88.

62. Lucarelli MJ, Khwarg SI, Lemke BN et al. The anatomy of midfacial ptosis. Ophthal Plast Reconstr Surg 2000; 16(1):7–22.

63. Muzaffar AR, Mendelson BC, Adams WP Jr. Surgical anatomy of the ligamentous attachments of the lower lid and lateral canthus. Plast Reconstr Surg 2002; 110(3):873–884.

64. Freilinger G, Gruber H, Happak W et al. Surgical anatomy of the mimic muscle system and the facial nerve: importance for reconstructive and aesthetic surgery. Plast Reconst Surg 1987; 80(5):686–690.

65. Katusic SK, Beard CM, Wiederholt WC et al. Incidence, clinical features, and prognosis in Bell's palsy, Rochester, Minnesota, 1968–1982.

66. Rahman I, Sadiq SA. Ophthalmic management of facial nerve palsy: a review. Surv Ophthalmol 2007; 52(2):121–144.

67. Whitnall SE. The anatomy of the human orbit and accessory organs of vision, 2nd edn. New York: Oxford University Press, 1932.

68. Lipham WJ, Tawfik HA, Dutton JJ. A histologic analysis and three-dimensional reconstruction of the muscle of Riolan. Ophthal Plast Reconstr Surg 2002; 18(2):93–98.

69. Becker BB. Tricompartment model of the lacrimal pump mechanism. Ophthalmology 1992; 99:1139–1145.

70. Milz S, Neufang J, Higashiyama I, et al. An immunohistochemical study of the extracellular matrix of the tarsal plate in the upper eyelid in human beings. J Anat 2005; 206:37–45.

71. Wesley RE, McCord CD, Jones NA. Height of the tarsus of the lower eyelid. Am J Ophthalmol 1980; 90(1):102–105.

72. Jones LT. A new concept of the orbital fascia and rectus muscle sheaths and its surgical implications. Trans Am Acad Ophthalmol Otolaryngol 1968; 72:755–764.

73. Mathers WD. Ocular evaporation in meibomian gland dysfunction and dry eye. Ophthalmology 1993; 100(3):347–351.

74. Scheie HG, Albert DM. Distichiasis and trichiasis: origin and management. Am J Ophthalmol 1966; 61:718–720.

75. Frueh BR. Treatment of distichiasis with cryotherapy. Ophthalmic Surg 1981; 12(2):100–103.

76. Bartley GB, Lowry JC. Argon laser treatment of trichiasis. Am J Ophthalmol 1992; 113(1):71–74.

77. McCracken MS, Kikkawa DO, Vasani SN. Treatment of trichiasis and distichiasis by eyelash trephination. Ophthal Plast Reconstr Surg 2006; 22(5):349–351.

78. Foster CS. Cicatricial pemphigoid. Trans Am Ophthalmol Soc 1986; 84:527–563.

79. Lander T, Wirtschafter JD, McLoon LK. Orbicularis oculi muscle fibers are relatively short and heterogeneous in length. Invest Ophthalmol Vis Sci 1996; 37(9):1732–1739.

80. Lemke BN, Stasior OG, Rosenberg PN. The surgical relations of the levator palpebrae superioris muscle. Ophthal Plast Reconstr Surg 1988; 4(1):25–30.

81. Anderson RL, Dixon RS. The role of Whitnall's ligament in ptosis surgery. Arch Ophthalmol 1979; 97:705.

82. Stasior GO, Lemke BN, Wallow III et al. Levator aponeurosis elastic fiber network. Ophthal Plast Reconstr Surg 1993; 9(1):1–10.

83. Kuwabara T, Cogan DG, Johnson CC. Structure of the muscles of the upper eyelid. Arch Ophthalmol 1975; 93:1189–1197.

84. Doane MG. Blinking and the mechanics of the lacrimal drainage system. Ophthalmology 1981; 88(8):844–851.

85. Sahlin S, Laurell CG, Chen E et al. Lacrimal drainage capacity, age and blink rate. Orbit 1998; 17:155–159.

86. Mauriello JA, Carbonaro P, Dhillon S et al. Drug-associated facial dyskinesias – a study of 238 patients. J Neuroophthalmol 1998; 18(2):153–157.

87. Frueh B, Callahan A, Dortzbach RK et al. A profile of patients with intractable blepharospasm. Trans Sect Ophthalmol Am Acad Ophthalmol Otolaryngol 1976; 81:591–594.

88. Dutton JJ, Buckley EG. Long-term results and complications of botulinum A toxin in the treatment of blepharospasm. Ophthalmology 1988; 95:1529–1534.

89. Sires BS, Saari JC, Garwin GG et al. The color difference in orbital fat. Arch Ophthalmol 2001; 119(6):868–871.

90. Kawai K, Imanishi N, Nakajima Y et al. Arterial anatomical features of the upper palpebra. Plast Reconstr Surg 2004; 113(2):479–484.

91. Cook BE Jr, Lucarelli MJ, Lemke BN et al. Eyelid Lymphatics I: Histochemical comparisons between the monkey and human. Ophthal Plast Reconstr Surg 2002; 18(1):18–23.

92. Cook BE Jr, Lucarelli MJ, Lemke BN et al. Eyelid Lymphatics II: A search for drainage patterns in the monkey and correlations with human lymphatics. Ophthal Plast Reconstr Surg 2002; 18(2):99–106.

93. Knize DM. A study of the supraorbital nerve. Plast Reconst Surg 1995; 96(3):564–569.

CHAPTER 15

Formation and Function of the Tear Film

Darlene A. Dartt

1. Tear film overview

The tear film overlays the ocular surface, which is comprised of the corneal and conjunctival epithelia, and provides the interface between these epithelia and the external environment. The tear film is essential for the health and protection of the ocular surface and for clear vision as the tear film is the first refractive surface of the eye.[1]

Tears produced by the ocular surface epithelia and adnexa are distributed throughout the cul-de-sac. Using ocular coherence tomography the thickness of the precorneal tear film was $3.4 \pm 2.6\ \mu m$,[2] agreeing with previous measurements using less accurate techniques.[3–7]

The tear film is an exceedingly complex mixture of secretions from multiple tissues and epithelia (Fig. 15.1) and consists of four layers (Fig. 15.2). The innermost layer is a glycocalyx that extends from the superficial layer of the ocular surface epithelium. The second is a mucous layer that covers the glycocalyx and may mix with the third aqueous layer. The outermost layer contains lipids. Similarly to mucous and aqueous layers, aqueous and lipid layers may mix. Production and function of tear film layers are distinct and will be presented separately.

Tear secretion by all ocular adnexa and ocular surface epithelia must be coordinated. For mucous and aqueous layers, secretion is regulated by neural reflexes. Sensory nerves in cornea and conjunctiva are activated by mechanical, chemical, and thermal stimuli that in turn activate the efferent parasympathetic and sympathetic nerves, which innervate the lacrimal gland and the conjunctival goblet cells, and cause mucous and fluid secretion. For the lipid layer, the blink itself regulates release of pre-secreted meibomian gland lipids stored in the meibomian gland duct. When the eyelids retract a thin film of lipid overspreads the underlying aqueous and mucous layers.

Tear secretion is balanced by drainage and evaporation. Tears on the ocular surface are drained through lacrimal puncta into the lacrimal drainage system. Drainage of tears can be regulated by neural reflexes from the ocular surface that cause vasodilation and vasoconstriction of the cavernous sinus blood supply of the drainage duct (Fig. 15.3). Both vasoconstriction and vasodilation cause a change in geometry of the lumen that decreases drainage.[8] Evaporation depends on the amount of time the tear film is exposed between blinks and temperature, humidity, and wind speed. The remainder of the chapter will focus on regulation of tear secretion.

2. Glycocalyx

A. Structure

The glycocalyx is a network of polysaccharides that project from cellular surfaces. In corneal and conjunctival epithelia, the glycocalyx can be found on the apical portion of the microvilli that project from the apical plasma membrane of the superficial cell layer (Fig. 15.2). Mucins are a critical component of the glycocalyx. Mucins consist of a protein core of amino acids linked by O-glycosylation to carbohydrate side chains of varying length and complexity. Mucins are classified by the nomenclature MUC1–21 and are divided into secreted and membrane-spanning categories (Fig. 15.4). Membrane-spanning mucins consist of a short intracellular tail, membrane-spanning domain, and large, extended extracellular domain that forms the glycocalyx. Secreted mucins are either gel-forming or small soluble. Gel-forming mucins are large molecules (20–40 million Da) secreted by exocytosis from goblet cells. Small soluble mucins are secreted by the lacrimal gland.

The ocular surface contains the membrane-spanning mucins, MUC1, MUC4, and MUC16.[9] These mucins are produced by stratified squamous cells of the cornea and conjunctiva and are stored in small clear secretory vesicles in the cytoplasm (Fig. 15.1). Fusion of secretory vesicles with plasma membrane inserts these molecules into the plasma membrane. Mucin molecules are localized to the tips of the squamous cell microplicae.

There is limited information on regulation of membrane-spanning mucin synthesis and secretion. Regulation of secretion would be by regulation of insertion of mucins into plasma membranes or by control of ectodomain shedding whereby matrix metalloproteinases (MMPs) cleave the mucin releasing the extracellular, active domain of the protein into the extracellular space. In immortalized stratified corneal-limbal cells, tumor necrosis factor induced the ectodomain shedding of MUC1, MUC4, and MUC16,

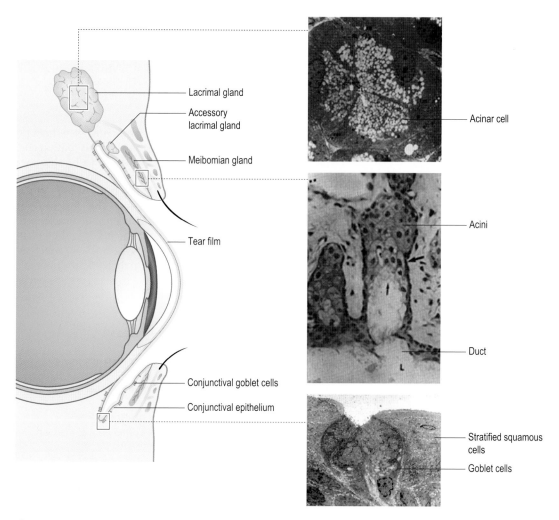

Figure 15.1 Schematic of the glands and epithelia that secrete tears. The ocular surface epithelia are in beige and lacrimal glands in pink and their contribution to the tear film is in blue. An electron micrograph of conjunctival goblet cells is in the bottom inset. An electron micrograph of a lacrimal gland acinus, is shown in the top inset. An electron micrograph of a meibomian gland acinus with its attached duct is shown in the middle inset. (Modified from Dartt DA. The Lacrimal Gland and Dry Eye Diseases. In: Ocular Disease: Mechanism and Management. Ed. Levin and Albert. Elsevier Ltd. Amsterdam. Copyright Elsevier 2010.)

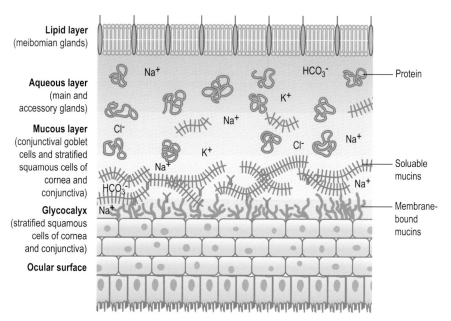

Figure 15.2 Schematic of the tear film. The outer lipid layer is secreted by meibomian glands. The middle layer contains electrolytes, water, proteins, and small soluble mucins produced by lacrimal glands and conjunctival epithelium. The inner mucous layer consists of MUC5AC, proteins, electrolytes, and water secreted by conjunctival goblet cells. The glycocalyx is comprised of membrane-bound mucins produced by stratified squamous cells of the corneal and conjunctival epithelial cells. The tear film overlies the ocular surface. (Modified from Hodges RR, Dartt DA. Regulatory pathways in lacrimal gland epithelium. Int Rev Cytol. 2003;231:129–196. Copyright Elsevier 2003.)

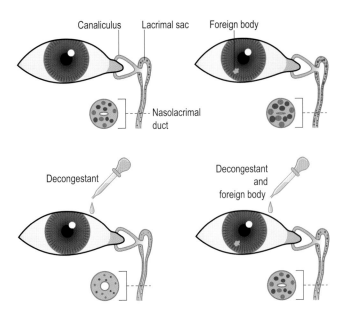

Figure 15.3 Schematic anatomic model of the state of the cavernous body and lacrimal passage. The top left illustration shows the lumen of the nasolacrimal duct under resting conditions. A foreign body in the eye causes activation of nerves in the cavernous body causing vasodilation of the blood vessels and a narrowing of the lumen that results in decreased tear drainage. The third panel shows the placement of decongestant of the ocular surface causes vasoconstriction of the blood vessels of the cavernous body, and opening of the lumen of the nasolacrimal duct, and surprisingly a decrease in drainage. The final panel shows that the vasoconstriction caused by the foreign body and vasodilation by the topical congestant prevent a change in the shape of the lumen and in the drainage of tears. (Modified from Ayub, M., A. B. Thale, J. Hedderich, B. N. Tillmann and F. P. Paulsen. 2003. Invest Ophthalmol Vis Sci. 44(11), 4900–7, reproduced from the Association for Research in Vision and Ophthalmology.)

whereas MMP-7 and neutrophil elastase induced the shedding of MUC16 only.[10] These compounds that caused shedding are elevated in the tears of dry eye patients and thus may cause the increase in Rose-Bengal staining found in dry eye and associated with loss of MUC16 (Box 15.1).

B. Function

The membrane-spanning mucins function to hydrate the ocular surface and serve as a barrier to pathogens. Membrane-spanning mucins are considered to be disadhesive, allowing the mucous layer to move over the ocular surface. The carbohydrate side chains hold water at the surface of the apical cell membranes. The function of individual membrane-spanning mucins remains unclear.

Membrane-spanning mucins appear to be altered in dry eye. MUC16 protein levels decreased in conjunctival epithelium and increased in tears of patients with Sjogren's

---

Box 15.1 Definition of Dry Eye Disease

- Dry eye is a multifactorial disease of the tears and ocular surface
- Dry eye results in symptoms of discomfort, visual disturbance, and tear film instability
- Dry eye can potentially damage the ocular surface
- Dry eye is accompanied by increased osmolarity of the tear film and inflammation of the ocular surface

Modified from 2007 Report of the International Dry Eye Workshop, The Ocular Surface, vol 5, number 2, page 27, 2007

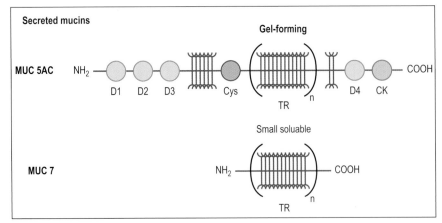

Figure 15.4 Structural motifs of the secreted and membrane-spanning mucins. MUC5AC consists of four cysteine-rich D domains for disulfide cross-linking and flank a region of variable number tandem repeats (TR). MUC7 is monomeric, has a variable number of tandem repeats. The membrane-spanning mucins have an SIG domain, a variable number of tandem repeats, a site for cleavage of the extracellular domain, a transmembrane domain (TM) for insertion of the molecule into membranes, and a carboxyterminus (CT) that is the intracellular domain. SIG: peptide signal sequence; CK: cysteine knot. (Modified from Gipson IK and Argueso P. 2003. Role of mucins in the function of the corneal and conjunctival epithelia. Int Rev Cytol. 231, 1–49. Copyright Elsevier 2003.)

syndrome.[11,12] MUC1 splice variants also play a role in dry eye.[13] Human cornea and conjunctiva contain five previously identified MUC1 splice variants and a new splice variant. These splice variants have unique changes that could affect their ectodomain shedding, signaling properties of the intracellular domain, and water retention, lubrication, and barrier properties of the extracellular domains. When the type of MUC1 splice variant was determined in dry eye patients (both evaporative dry eye and Sjogren's syndrome) compared to control patients there was a reduced frequency of MUC1/A variant and an increase in MUC1/B variant in dry eye patients.[13,14]

3. Mucous layer

A. Structure

The mucous layer backbone is the gel-forming mucin MUC5AC, synthesized and secreted by conjunctival goblet cells. MUC5AC is encoded by one of the largest genes known, producing a protein of about 600 kDa. The protein backbone of MUC5AC consists of four D domains (cysteine-rich domains) (Fig. 15.4) that flank a tandem repeat sequence in which amino acids in the protein backbone are O-glycosylated and linked to carbohydrate side chains. The D domains provide sites for disulfide bonds cross-linking multiple MUC5AC molecules that forms the framework of the mucous layer. Also contained in the mucous layer are shed ectodomains of membrane-spanning mucins, membrane-spanning mucins secreted by a soluble pathway, other proteins synthesized and secreted by goblet cells, and electrolytes and water secreted by goblet and stratified squamous cells.

B. Conjunctival goblet cells

Goblet cells are interspersed among stratified squamous cells of the conjunctiva (see Fig. 15.1). Goblet cells occur in clusters in rat and mouse, but singly in rabbit and humans.[15–17] In all species studied, goblet cells are unevenly distributed over the conjunctiva.

Goblet cells are identified by the large accumulation of mucin granules in the apex (see Fig. 15.1). Secretory product can be visualized using Alcian blue-periodic acid stain, the lectins *Ulex europeaus* agglutinin I (UEA-I) or *Helix pomatia* agglutinin (HPA), or antibodies to MUC5AC. MUC5AC is synthesized in the endoplasmic reticulum and carbohydrate side chains added in the Golgi apparatus. The mature proteins are stored in secretory granules. Upon stimulation secretory granules fuse with each other and apical membrane releasing secretory product into the tear film. Upon cell stimulation the entire complement of granules is released, known as apocrine secretion. The amount of secretion is controlled by regulating the number of cells that are activated by a given stimulus.

C. Regulation of goblet cell mucin production

Overview

Mucin production is regulated by controlling the rate of mucin synthesis, rate of mucin secretion, and number of goblet cells present in the conjunctiva. The rate of mucin synthesis has yet to be studied in conjunctival goblet cells. Thus the remainder of this section will focus on regulation of secretion and proliferation.

Regulation of goblet cell secretion

Nerves are the primary regulators of conjunctival goblet cell secretion. The conjunctiva is innervated by afferent sensory nerves and efferent sympathetic and parasympathetic nerves. Sensory nerves end as free nerve endings between the stratified squamous cells. The parasympathetic and sympathetic nerve endings also surround the goblet cells at the level of the secretory granules (Fig. 15.5). Stimulation of the sensory nerves in the cornea by a neural reflex induces goblet cell secretion via the efferent nerves. Goblet cells have receptors for neurotransmitters from both parasympathetic and sympathetic nerves. Parasympathetic nerves release both acetylcholine (Ach) and vasoactive intestinal peptide (VIP). Muscarinic receptors of the M_3AchR and M_2AchR subtypes are located similarly to the location of efferent nerves (Fig. 15.5).[18,19] VIP receptors of the VIPAC2 receptor subtype are located in the same area as M_3AchR.[19] Sympathetic nerves release norepinephrine and NPY. Several subtypes of both α_1- and β-adrenergic receptors are present on goblet cells.[18] It appears that parasympathetic nerves using Ach and VIP are the primary stimuli of goblet cell secretion. The function of the sympathetic nerves remains unknown.

Components of the signaling pathways used by Ach, but not VIP, have been delineated.[19] Cholinergic agonists use M_3AchR and M_2AchR to stimulate goblet cell secretion.[20] Cholinergic agonists presumably use the $G_{\alpha q/11}$ subtype of G protein that activate phospholipase C to breakdown phosphatidylinositol 4,5 bisphosphate to produce inositol 1,4,5 trisphosphate (IP_3) and diacylglycerol. IP_3 would then release intracellular Ca^{2+} by binding to its receptors on the endoplasmic reticulum. Based on a variety of experiments cholinergic agonists are known to increase the intracellular $[Ca^{2+}]$ to stimulate secretion.[21,22]

The diacylglycerol (DAG) released with IP_3 activates protein kinase C (PKC). Nine PKC isoforms are present in

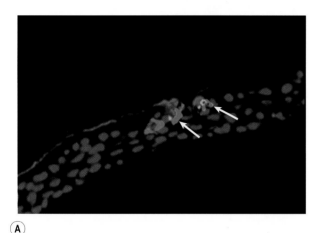

(A)

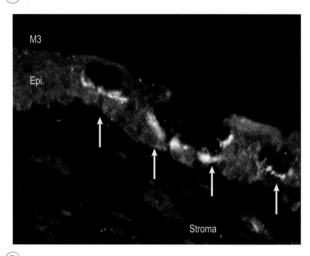

M3

Epi.

Stroma

(B)

Figure 15.5 Immunofluorescence micrographs of parasympathetic nerves and M₃Ach receptors in rat conjunctiva. (**A**) The parasympathetic neurotransmitter VIP is shown in green. The lectin *Ulex europeaus* is shown in red and indicates the location of goblet cells. The goblet cell body is subjacent to the secretory granules. The cell nuclei are shown in blue. (Reprinted with permission from Diebold, Y., J. D. Rios, R. R. Hodges, I. Rawe and D. A. Dartt. 2001. Invest Ophthalmol Vis Sci. 42(10), 2270–82.) (**B**) The M₃AchR is shown. The dark areas indicate the goblet cell secretory granules. Arrows indicate cell bodies of goblet cells. (Reprinted with permission from Rios, J. D., D. Zoukhri, I. M. Rawe, R. R. Hodges, J. D. Zieske and D. A. Dartt. 1999. Invest Ophthalmol Vis Sci. 40(6), 1102–11, reproduced from the Association for Research in Vision and Ophthalology.)

conjunctival goblet cells and phorbol esters, activators of PKC isoforms, stimulate goblet cell secretion.[21] Although a role for PKC isoforms in secretion could not be substantiated,[21] it is likely that cholinergic agonists use PKC isoforms to stimulate secretion as PKC inhibitors block a distal step in the signaling pathway, activation of extracellular regulated kinase (ERK1/2)[23] (Fig. 15.6).

Cholinergic agonists are known to activate the epidermal growth factor (EGF) signaling pathway. In goblet cells, cholinergic agonists activate the non-receptor tyrosine kinases Pyk2 and p60Src.[20] These kinases transactivate (phosphorylate) the EGF receptor (EGFR). This transactivation is usually mediated by ectodomain shedding by MMP causing the release of the extracellular, active domain of one member of the EGF family of growth factors. This has yet to

be tested in goblet cells. The released growth factor would bind to the EGF receptor inducing two receptors to associate and be autophosphorylated. This attracts the adapter proteins Shc and Grb2 that are phosphorylated, activating the guanine nucleotide exchange factor SOS to increase Ras activity. Ras activates MAPK kinase (Raf) that phosphorylates MAPK kinase (MEK) that phosphorylates ERK1/2. In both rat and human goblet cells, cholinergic agonists increase the intracellular [Ca²⁺] and activate PKC to stimulate goblet cell secretion by activating Pyk2 and p60Src to transactivate the EGF receptor, inducing the signaling cascade that activates ERK1/2 (Fig. 15.6).[20,23,24]

Other factors that stimulate goblet cell secretion include P2Y₂ purinergic receptors[25] that are activated by ATP and the neurotrophin family of growth factors. Of this family nerve growth factor and brain-derived neurotrophic factor stimulate secretion.[26]

Regulation of goblet cell proliferation

The amount of goblet cell mucin on the ocular surface can also be modulated by controlling goblet cell proliferation. Serum, which contains a variety of growth factors, stimulates proliferation of cultured conjunctival goblet cells, both human and rat.[27] EGF, transforming growth factor (TGFα), and heparin binding-EGF (HB-EGF), but not heregulin stimulate proliferation.[27] EGF, TGFα, and HB-EGF bind to the EGF receptor (EGFR or erb-B1), HB-EGF binds to erb-B4 and heregulin binds to erb-B3 and erb-B4. That EGF, TGFα, and HB-EGF are equipotent in stimulating conjunctival goblet cell suggests that the EGFR and erb-B4 are the primary receptors used.[27]

EGF was used as the prototype of the EGF family. EGF increases the activation of the EGFR in rat conjunctival goblet cells and activates ERK1/2 (also known as p44/p42 mitogen activated protein kinase (MAPK)) (Fig. 15.7) causing its biphasic translocation to the nucleus.[28] The slower sustained second peak response is responsible for cell proliferation, but the role of the rapid transient first peak is not known.

Activation of PKC isoforms also mediates EGF-stimulated goblet cell proliferation (Fig. 15.7). In rat and human goblet cells, EGF uses PKCα and PKCε to stimulate conjunctival goblet cell proliferation.

D. Regulation of conjunctival electrolyte and water secretion

The electrolyte and water composition of the external environment has an important effect on release of mucins from secretory granules. Both the stratified squamous and goblet cells of the conjunctiva express ion and water transport proteins and appear to secrete electrolytes and water. Mucin and electrolyte and water secretion may not be coordinated as stimuli of mucin secretion appear to differ from those of electrolyte and water secretion.

The conjunctival epithelium secretes Cl⁻ and absorbs Na⁺ with the ratio of 1.5 to 1 causing a net secretion of fluid into the tear film.[29] Interestingly, both absorption and secretion occur within the same cell in the conjunctiva. The ion transport proteins are distributed evenly in all areas of the conjunctiva, suggesting that the entire conjunctiva participates in fluid secretion.

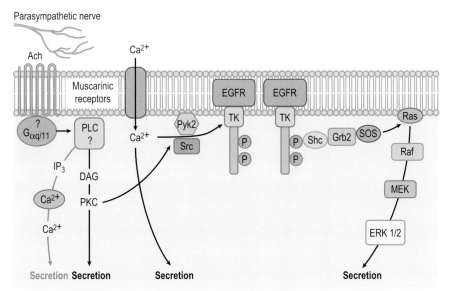

Figure 15.6 Signaling pathways used by cholinergic agonists to stimulate conjunctival goblet cell secretion – the pathway is described in the text. TK-tyrosine kinase, P-phosphorylation site. (Modified from Dartt DA. Prog Ret Eye Res 2002; 21:555–76.)

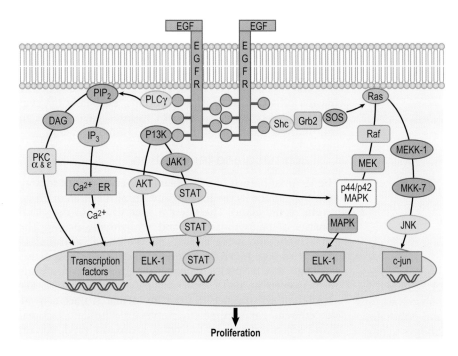

Figure 15.7 Schematic of EGF-dependent signaling pathways.

The driving force for conjunctival fluid secretion is the basolaterally located Na^+,K^+-ATPase that extrudes 3 Na^+ ions for the influx of 2 K^+ ions and generates a negative intracellular voltage (Fig. 15.8). Also on the basolateral side is the Na^+-K^+-2Cl^- cotransporter (NKCC1). The Cl^-/HCO_3^- exchanger takes up Cl^- into the cell. Cl^- is secreted from the apical side of the cell through a variety of Cl^- channels. K^+ is also secreted from the apical side with Na^+ moving through the paracellular pathway to the apical side. The net effect is isotonic secretion of Na^+, K^+, and Cl^- into the tears.[30]

Na^+ is absorbed across the apical membrane into the cell by the Na^+-dependent glucose cotransporter. The Na^+ that enters the cell on the apical side is extruded on the basolateral side by the Na^+,K^+-ATPase. The Na^+ influx mechanism can also be used to transport amino acids (AA) into the cell.[31]

Fluid secretion by the conjunctiva can be stimulated and contribute a substantial amount to the tear film. Increasing the intracellular $[Ca^{2+}]$ or the cellular level of cAMP causes Cl^- secretion.[32,33] The β-adrenergic agonist epinephrine stimulates secretion via cAMP. Increasing cellular cAMP levels activate protein kinase A (PKA) that stimulates both Cl^- secretion and Na^+ absorption by activating K^+ channels in the basolateral membrane to hyperpolarize the cell. Hyperpolarization stimulates Na^+ influx (into the cell) and Cl^- efflux (into tears) across the apical membrane. PKA also stimulates apical Cl^- channels. Because there is a higher rate of Cl^- secretion relative to Na^+ absorption, water moves into tears in the basolateral to apical direction. In addition Na^+ moves paracellularly to produce an isotonic NaCl secretion into tears. Water moves through water channels known as

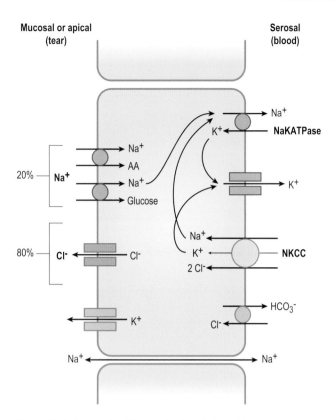

Mucosal or apical
(tear)

Serosal
(blood)

Figure 15.8 Schematic of the mechanism of electrolyte and water secretion by conjunctival epithelial cells – a description of the process is found in the text. (Modified from Dartt DA. Prog Ret Eye Res 2002; 21:555–76.)

aquaporins (AQP). AQP 5 is expressed in the apical membrane and AQP 3 in the lateral membrane.

Activation of the purinergic receptor $P2Y_2$ by UTP and ATP also stimulates electrolyte and water secretion from the conjunctiva. $P2Y_2$ receptors increase intracellular $[Ca^{2+}]$.[34]

E. Mucous layer function

The major roles of the conjunctiva are to contribute to tear production by secreting electrolytes and water, modify composition of the tear film by secreting mucins, absorb various organic compounds found in tears (including ophthalmic therapeutics), and contribute to resistance of the eye to infection by providing protection against microorganisms.[35] Mucins serve as wetting agents that keep the apical epithelia hydrated. Finally, conjunctival cells release paracrine signaling agents such as growth factors that affect ocular surface properties.

4. Aqueous layer

A. Overview

The main lacrimal gland is the major producer of the aqueous layer. Other ocular surface epithelia also contribute to the aqueous layer, most notably the conjunctiva, accessory lacrimal glands, and to a small extent corneal epithelium. Accessory lacrimal glands, which are similar to the main gland, are embedded in the conjunctiva. Function of the accessory lacrimal glands has not been determined as they

are difficult to study in isolation. Thus, this section will focus on the main lacrimal gland, which will be referred to as the lacrimal gland.

B. Lacrimal gland structure

The lacrimal gland is comprised of acini that drain into progressively larger tubules or ducts that coalesce into one or more excretory ducts. Acini are made up of basal myoepithelial cells that surround an inner layer of columnar secretory cells known as acinar cells (see Fig. 15.1). Stellate-shaped myoepithelial cells are known to contract, but their role in lacrimal gland secretion is still unknown. Acinar cells comprise about 80 percent of the gland and are organized in clusters. In cross-section an acinus contains a ring of pyramidal-shaped acinar cells that are joined at the junction of the apical and lateral membranes by tight junctions. The tight junctions polarize the cells and ensure unidirectional secretion of electrolytes, water, and protein, the secretory product of the lacrimal gland.

The lacrimal gland ducts are lined with one to two layers of cuboidal duct cells that make up 10–12 percent of the gland. Similarly to acinar cells, duct cells are joined at the apical side by tight junctions and secrete water, electrolytes, and protein modifying the primary lacrimal gland fluid secreted by the acinar cells.

The lacrimal gland also contains a small population of lymphocytes, plasma cells, mast cells, and macrophages. The plasma cells express immunoglobulins, especially IgA that along with secretory component forms secretory IgA, an important component of the mucosal immune system.

C. Lacrimal gland innervation

The lacrimal branch of the trigeminal nerves carries sensory stimuli from the lacrimal gland and forms the afferent innervation of the gland. The efferent innervation of the gland consists of parasympathetic and sympathetic nerves. The parasympathetic nerve endings predominate, surrounding most acini (Fig. 15.9). Sympathetic innervation is less dense, although this is species-dependent. Sensory nerves are the most sparsely located.

Afferent sensory nerves from the ocular surface activate the efferent parasympathetic nerves to stimulate lacrimal gland secretion. This neural reflex forms the predominant stimulus of lacrimal gland secretion and allows the gland to protect the ocular surface by responding to environmental stimuli.

Parasympathetic nerves release the neurotransmitters Ach and VIP that activate receptors on the myoepithelial, acinar, and duct cells. The M_3AchR is the only cholinergic, muscarinic receptor in the lacrimal gland. The VIP receptor VIPAC1 is on acinar cells and VIPAC2 on myoepithelial cells. Sympathetic nerves release the neurotransmitters norepinephrine and neuropeptide Y (NPY). Norepinephrine activates α_{1D}-adrenergic and β-adrenergic receptors located on acinar cells.

D. Protein secretion regulation

Types of protein secretion

The majority of lacrimal gland secretory proteins are secreted by exocytosis. Secretory proteins are stored in secretory granules. Upon the appropriate stimulus secretory granule

Parasympathetic nerves

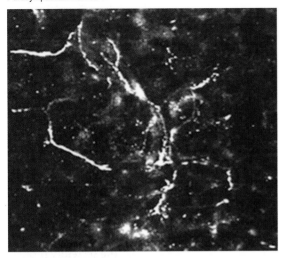

Sympathetic nerves

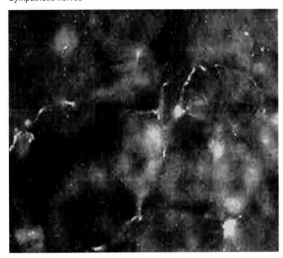

Figure 15.9 Immunofluorescence micrographs of the distribution of parasympathetic and sympathetic nerves in the lacrimal gland. Top panel: anti-VIP demonstrates location of parasympathetic nerves. Bottom panel: anti-tyrosine hydroxylase (TH) indicates sympathetic nerves.

membranes fuse with apical plasma membranes, releasing the contents into the lumen. This secretion is known as merocrine as only a small percentage of granules are released. The exocytotic process is controlled by trafficking effectors that include targeting and specificity factors such as SNAREs (soluble N-ethylmaleimide (NEM) sensitive factor (NSF) attachment protein receptors (SNAPs) that function in bringing the secretory granules and apical membrane together for release of cargo secretory proteins) and Rabs (small Ras-like GTPases that regulate membrane trafficking and are active when bound to GTP and recruit proteins necessary for membrane fusion and vesicle maturity) as well as transport factors such as microtubules, actin filaments, and motor proteins.[36]

Proteins such as secretory IgA are secreted by a combination of transcytosis and ectodomain shedding.[37] Transcytosis is a complex trafficking process in which the compound is endocytosed at the basolateral membrane, transported across the cell in a series of endomembrane components, and

secreted at the apical membrane. Secretory component and the EGF family of growth factors are secreted by ectodomain shedding.

The signaling pathways used by neurotransmitters to stimulate lacrimal gland secretion were correlated to exocytosis, not to transcytosis or ectodomain shedding. Thus the next section on signaling pathways will focus on exocytosis.

Cholinergic agonists

The cholinergic agonist Ach binds to M_3AchR on lacrimal gland acinar cells stimulating the $G_{\alpha q/11}$ subtype of G protein that activates phospholipase Cβ (Fig. 15.10). Phospholipase Cβ breaks down the membrane phospholipid phosphatidylinositol bisphosphate into IP_3 and diacyglycerol. IP_3 binds to its receptor on the endoplasmic reticulum and releases Ca^{2+}, increasing the intracellular $[Ca^{2+}]$. The depletion of intracellular Ca^{2+} stores causes Ca^{2+} influx across the plasma membrane, known as capacitative Ca^{2+} influx, to sustain the Ca^{2+} increase and replenish the stores. Intracellular Ca^{2+} stimulates exocytosis by activating target proteins involved in the release process.

The diacylglycerol produced with IP_3 activates the PKC isoforms PKCα, PKCδ, and PKCϵ to stimulate protein secretion. PKC stimulates secretion by phosphorylating unknown substrate proteins involved in exocytosis.[38]

Cholinergic agonists also activate inhibitory pathways that attenuate secretion. Cholinergic agonists activate Pyk2 and p60Src that cause a Ca^{2+}-dependent activation of ERK1/2 (also known as MAPK) (Fig. 15.10). The step in the pathway that activates Ras is unknown. However, the Ras, Raf, MEK pathway leads to activation of ERK1/2. ERK1/2 decreases cholinergic agonist stimulated secretion.

A second pathway activated by cholinergic agonists is the phospholipase D pathway that breaks down phosphatidylcholine to phosphatidic acid and choline. Phosphatidic acid can be metabolized to diacylglycerol that through PKC stimulates ERK1/2 to attenuate secretion. Even though cholinergic agonists activate inhibitory pathways, their ultimate effect is to stimulate secretion.[39]

VIP

VIP released from parasympathetic nerves activates a separate pathway from cholinergic agonists. VIP binds to VPAC1 and VPAC2 to activate the G protein $G_{\alpha s}$ that stimulates adenylyl cyclase (Fig. 15.11).[40] Adenylyl cyclase produces cAMP from ATP and cAMP activates PKA that phosphorylates target proteins in the exocytotic machinery. VIP also increases the intracellular $[Ca^{2+}]$.[40] Together cAMP and Ca^{2+} induce protein secretion.

VIP also interacts with ERK1/2. Unlike cholinergic agonists, VIP inhibits ERK1/2. VIP inhibition of ERK1/2 plays a critical role in the interaction of the VIP pathway with those stimulated by cholinergic and α_1-adrenergic agonists.[41]

α_1-Adrenergic agonists

The sympathetic neurotransmitter norepinephrine binds to α_{1D}-AR on the acinar cells (Fig. 15.12). To stimulate protein secretion α_1-adrenergic agonists activate endothelial nitric oxide synthase (eNOS) co-localized with caveolin on the basolateral membranes of acinar cells. eNOS produces nitric oxide (NO) that activates soluble guanylate cyclase.[42]

357

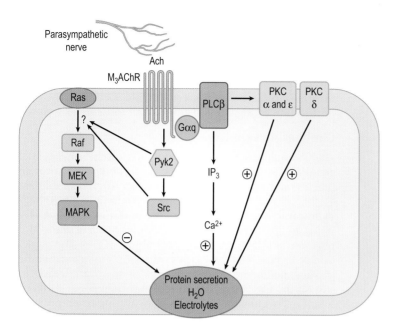

Figure 15.10 Schematic of parasympathetic signaling pathway – the pathway is described in the text.

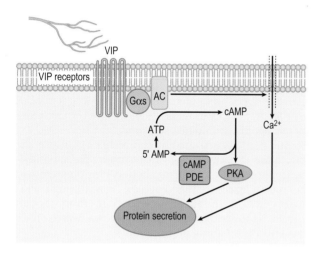

Figure 15.11 Schematic of VIP signaling pathway – the pathway is described in the text.

Guanylate cyclase (GC) produces cGMP from GTP. cGMP stimulates secretion by phosphorylating target proteins in the exocytotic machinery.

α_1-Adrenergic agonists also give a small increase in intracellular Ca^{2+} that could interact with cGMP to stimulate secretion.[43] The mechanism by which α_1-adrenergic agonists activate eNOS in the lacrimal gland is not known.

α_1-Adrenergic agonists also activate PKC isoforms. The effector enzyme that produces diacylglycerol is not known, but is not phospholipase C or D. Diacylglycerol activates PKCε to stimulate secretion.[43] Unlike cholinergic agonists that activate PKCα and PKCδ to stimulate secretion, α_1-adrenergic agonists activate these PKC isoforms to inhibit secretion.[44]

α_1-Adrenergic agonists also activate the ERK1/2 inhibitory pathway (Fig. 15.12).[45] The mechanism by which α_1-adrenergic agonists activate ERK1/2 (also known as MAPK) is different from that used by cholinergic agonists.

α_1-Adrenergic agonists activate a matrix metalloproteinase (MMP) ADAM17, which cleaves the membrane-spanning growth factor EGF, releasing its extracellular domain.[46] The extracellular domain contains the active EGF, binding domain that binds to the EGFR causing it to dimerize. The EGFR dimer autophosphorylates through the signaling pathway described in section 3.C and activates ERK1/2 to attenuate secretion. Even though α_1-adrenergic agonists activate inhibitory pathways, the overall action of these agonists is to stimulate secretion.

EGF

Addition of exogenous EGF (the 6 kDa EGF) gives a small stimulation of protein secretion. EGF uses IP_3, PKCα, and PKCδ to stimulate lacrimal gland secretion.[47]

Interaction of pathways

Activation of parasympathetic nerves alone or both parasympathetic and sympathetic nerves releases two or more neurotransmitters that cause secretion different than one agonist alone. Release of cholinergic and α_1-adrenergic agonists stimulates secretion that is additive as these agonists activate separate and different signaling pathways.[43] Release of VIP with cholinergic or α_1-adrenergic agonists causes secretion that is synergistic or potentiated.[48] The potentiation is caused by VIP inhibition of ERK1/2 that prevents activation of ERK1/2 by cholinergic and α_1-adrenergic agonists (Fig. 15.13). When ERK1/2 activation is blocked, it cannot inhibit secretion and the secretory response is increased.[41]

E. Regulation of electrolyte and water secretion

Mechanism of acinar electrolyte and water secretion

Electrolyte and water secretion is driven by the basolateral Na^+,K^+-ATPase (NKA) that produces the energy to drive the

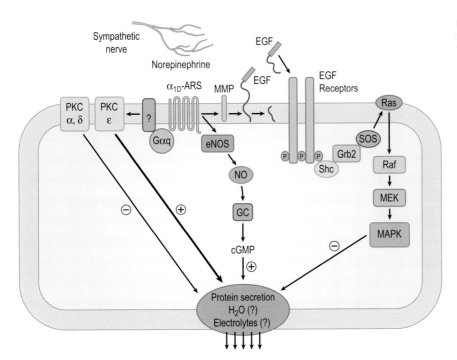

Figure 15.12 Schematic of α_{1D}-adrenergic signaling pathway – the pathway is described in the text.

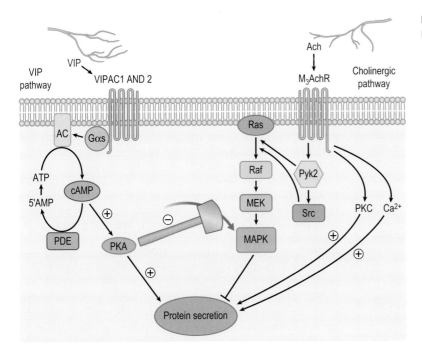

Figure 15.13 Schematic of the interaction between Ca^{2+}/protein kinase C and cAMP-dependent signaling pathways.

efflux of Na^+ and influx of K^+ across the basolateral membrane into the cell against their electrochemical gradients (Fig. 15.14). The Na^+/H^+ (NHE) and Cl^-/HCO_3^- (AE) exchangers and the Na^+-K^+-2Cl^- symporter (NKCC1) use the energy of the favorable inward Na^+ gradient to drive Cl^- influx across the basolateral membrane. This Cl^- influx establishes a favorable gradient for Cl^- efflux across the apical membrane through Cl^- selective channels into the acinar lumen generating a lumen negative transepithelial potential difference. This potential difference drives the flux of Na^+ from the basal to apical side of the cell through the paracellular pathway. K^+ selective channels allow the K^+ to recycle

across the basolateral membrane so that acini can produce a Na^+-Cl^- rich fluid.

Mechanism of ductal electrolyte and water secretion

Duct cells use mechanisms similar to those used by acini to secrete fluid with a few exceptions. First the K^+ channels are placed in the apical rather than basolateral membranes with the Cl^- channels producing a K^+-Cl^- rich fluid. Those channels include the K^+-Cl^- symporter KCC1, intermediate conductance calcium-activated K^+ channel ($IK_{Ca}1$), CFTR channel, and ClC3 Cl^- channel.

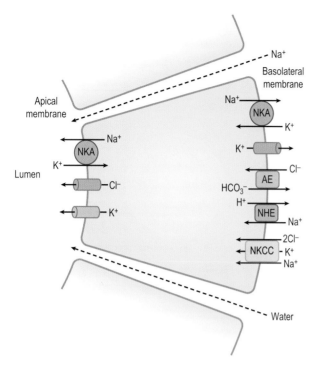

Figure 15.14 Mechanism of electrolyte and water secretion by lacrimal gland acinar cells – see text for description. (Modified from Selvam et al. Am J Cell Physiol 2007; 293:C1412–9. Used with permission.)

Neural activation of electrolyte and water secretion

The parasympathetic agonists Ach and VIP each stimulate electrolyte and water secretion.[49] Cholinergic agonists cause fluid secretion by activating apical Cl^- and K^+ channels, basolateral Na^+/H^+ exchangers, and increase the translocation of the Na^+,K^+-ATPase from the cytoplasm to the basolateral membranes. VIP stimulates fluid secretion by cAMP activating the Ca^{2+}-dependent large conductance (BK) K^+ channel in the basolateral membrane. The mechanisms by which cholinergic agonists and VIP stimulate fluid secretion are not yet completely investigated.

F. Lacrimal gland fluid composition

Lacrimal gland fluid is isotonic, and in human and rabbit it is 300 mOsm. The electrolyte composition is similar to plasma except that it has a decreased $[Na^+]$ and increased $[Cl^-]$ and $[K^+]$. The lacrimal gland also secretes a myriad of proteins.[50] The most extensively studied are lysozyme, lactoferrin, secretory IgA, and lipocalin. Other minor proteins secreted include EGF and other growth factors, lacritin a protein unique to the lacrimal gland, and surfactant proteins A–D.[51] Many proteins are antibacterial and function in the defense of the ocular surface.

G. Aqueous layer function

The aqueous layer has multiple functions critical to the health and defense of the ocular surface from the environment. The electrolyte composition of the aqueous layer is critical to the health of the ocular surface as small changes in osmolarity or electrolyte composition leads to aqueous

deficiency dry eye. Multiple causes can alter lacrimal gland secretion leading to aqueous deficiency dry eye and damage to the ocular surface (see Box 15.2). The proteins in the aqueous layer protect the ocular surface from bacterial infection and alter corneal and conjunctival function. Reflex secretion of the aqueous layers helps to wash away noxious substances and particulates into the drainage duct. The buffer system of the fluid helps to protect against changes in pH.

5. Lipid layer

A. Structure of meibomian glands and mechanism of lipid production

The lipid layer is secreted by the meibomian glands, modified sebaceous glands, that line the upper and lower eyelids in a single row.[52,53] A single meibomian gland consists of multiple acini that secrete into ducts that converge to form a common duct that exits onto the eyelid near the mucocutaneous junction (see Fig. 15.1). The acini are surrounded by nerves and blood vessels. The cells in the acini are arranged in a specific order that reflects their function such that the outer cell layer consists of undifferentiated, flattened basal cells. As these cells mature they move inward to the center of the acinus and concomitant with this inward migration is continuing synthesis of lipids. Meibomian gland lipids are stored in vesicles. As the cells move closer to the center they contain increasingly more lipid secretory vesicles reflecting continued lipid synthesis. Upon the appropriate stimulus (unknown), the cells in the acinus center burst, releasing the entire cell contents in their secretory granule lipids and the other components of the cell into the duct system. This releases the entire cell contents and is known as holocrine secretion. The secreted cells are replaced by proliferation of the basal cells.[54] Thus the secretory product contains a complex mixture of lipids and proteins and is termed meibum. The synthesized lipids are the major component of meibum, with the minor lipid classes probably reflecting the cellular membrane lipids. Meibomian fluid is a mixture of non-polar lipids (wax esters, cholesterol, and cholesterol esters) and polar lipids (mainly phospholipids). Meibum is liquid at lid temperature.

The secreted lipid is stored in the duct system that terminates in orifices, with a muscular cuff, that open onto the lid.[53] Meibum is released on to the ocular surface in small amounts with each blink forming a casual reservoir with about 30 times more lipid than needed for each blink. With the up-phase of each blink the upper lid draws oil from the lid reservoir and spreads it over the anterior tear film surface. With the down phase the lipid film is returned to the marginal reservoir as the lid closes.[53] The lipids mix only slightly with the lipid reservoir as it folds up as an intact sheet providing a gradual turnover of the lipids in the tear film.

Butovich et al[55] have proposed that proteins and mucins contribute to the lipid layer so that the proteins, which contain hydrophobic, hydrophilic, and charged portions, can unfold and form a variety of shapes depending upon the local milieu (Fig. 15.15). The proteins could extend across the lipid layer suggesting a complex mixture of

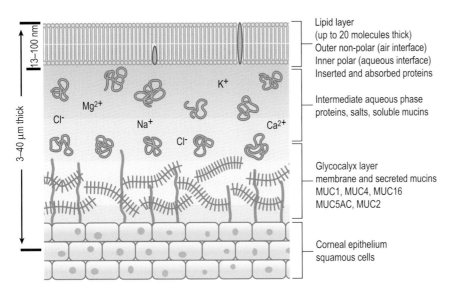

Lipid layer
(up to 20 molecules thick)
Outer non-polar (air interface)
Inner polar (aqueous interface)
Inserted and absorbed proteins

Intermediate aqueous phase
proteins, salts, soluble mucins

Glycocalyx layer
membrane and secreted mucins
MUC1, MUC4, MUC16
MUC5AC, MUC2

Corneal epithelium
squamous cells

Figure 15.15 Schematic diagram of the tear film showing a new concept for the structure of the lipid layer. (Modified from Butovich IA et al. Curr Eye Res 2008; 33(5):405–20.)

islands of proteins, mucins, and lipids, similar to that of lung surfactant. This model would be a non-collapsible, viscoelastic gel and allow for the lowest free energy states of the proteins in contact with lipids. Changes in the osmolarity of tears would affect the ability of the proteins to unfold and interact with the lipids. This model also accounts for the lack of change of interference patterns after each blink.

B. Regulation of meibum secretion

Neural regulation

Meibomian glands are richly innervated by sensory, sympathetic, and parasympathetic nerves.[56] The sensory nerves contain substance P and calcitonin gene-related peptide (CGRP). The sympathetic fibers use catecholamines and NPY as neurotransmitters. Parasympathetic nerves are immunopositive for cholinergic neurotransmitters, VIP and NO. However, the role of nerves in meibomian gland function is unknown.

Hormonal regulation

The meibomian glands have both androgen and estrogen receptors. Androgens regulate lipid secretion by controlling lipid synthesis. Meibomian acinar cells have nuclear androgen receptors[57] and the ratio of androgens and estrogens is critical for controlling lipid synthesis.[58] Stimulating androgen receptors increases gene transcription for enzymes associated with fatty acid and cholesterol synthetic pathways and hence stimulate lipid production. Androgen deficiency is associated with meibomian gland dysfunction.

C. Function

The meibomian gland secretion forms a hydrophobic barrier to prevent tear overflow onto the lids and sebum from the skin from entering the tear film.[59] The mucocutaneous junction itself separates the tear-wettable conjunctiva from the oil-wettable eyelid skin preventing overgrowth in both directions.[60] The meibom forms a water-tight seal of the apposed lid margins during sleep.[52,53] The tear lipids also reduce tear evaporation during eyelid opening and aid in lubrication for

Box 15.3 Meibomian Gland Diseases Causing Evaporative Dry Eye

- Reduced number of glands in congenital deficiency and acquired disease
- Replacement of glands in distichiasis and distichiasis lymphedema syndrome
- Hypersecretory gland dysfunction in meibomian seborrhea
- Hyposecretory gland dysfunction in retinoid therapy
- Obstructive disease from primary or secondary causes, of focal or diffuse nature, that is simple or cicatricial or can be atrophic or inflammatory
- Primary or secondary due to the local diseases anterior blepharitis
- Primary or secondary due to systemic disease such as acne rosacea, seborrhoeic dermatitis, and atopy
- Primary or secondary due to systemic toxicity from 13-cis retinoic acid, polychlorinated biphenyls, and epinephrine
- Cicatricial disease from chemical burns, trachoma, pemphigoid, erythema multiforme, acne rosacea, vernal keratoconjunctivitis, or atopic keratoconjunctivitis

Modified from 2007 Report of the International Dry Eye Workshop, The Ocular Surface, vol 5, number 2, page 82, 2007

the eyelids during blinking. Lipids enhance the stability of the tear film and provide a smooth optical surface for the cornea at the air/lipid interface. A decrease or an alteration in meibomian gland secretion leads to evaporative dry eye. There are multiple ways in which meibomian gland function can be altered to cause evaporative dry eye (Box 15.3) and damage the ocular surface.

References

1. Cotlier E. The lens. In: Moses R, ed. Adler's Physiology of the eye. St Louis: CV Mosby, 1975;275–297.
2. Wang J, Aquavella J, Palakuru J, Chung S, Feng C. Relationships between central tear film thickness and tear menisci of the upper and lower eyelids. Invest Ophthalmol Vis Sci 2006; 47(10):4349–4355.
3. Chen HB, Yamabayashi S, Ou B, Tanaka Y, Ohno S, Tsukahara S. Structure and composition of rat precorneal tear film. A study by an in vivo cryofixation. Invest Ophthalmol Vis Sci 1997; 38(2):381–387.

4. Ehlers N. The precorneal film. Biomicroscopical, histological and chemical investigations. Acta Ophthalmol 1965; 81(suppl):1–134.

5. King-Smith PE, Fink BA, Fogt N, Nichols KK, Hill RM, Wilson GS. The thickness of the human precorneal tear film: evidence from reflection spectra. Invest Ophthalmol Vis Sci 2000; 41(11):3348–3359.

6. Mishima S. Some physiological aspects of the precorneal tear film. Arch Ophthalmol 1965; 73:233–241.

7. Wang J, Fonn D, Simpson TL, Jones L. Precorneal and pre- and postlens tear film thickness measured indirectly with optical coherence tomography. Invest Ophthalmol Vis Sci 2003; 44(6):2524–2528.

8. Ayub M, Thale AB, Hedderich J, Tillmann BN, Paulsen FP. The cavernous body of the human efferent tear ducts contributes to regulation of tear outflow. Invest Ophthalmol Vis Sci 2003; 44(11):4900–4907.

9. Gipson IK, Argueso P. Role of mucins in the function of the corneal and conjunctival epithelia. Int Rev Cytol 2003; 231:1–49.

10. Blalock TD, Spurr-Michaud SJ, Tisdale AS, Gipson IK. Release of membrane-associated mucins from ocular surface epithelia. Invest Ophthalmol Vis Sci 2008; 49(5):1864–1871.

11. Danjo Y, Watanabe H, Tisdale AS et al. Alteration of mucin in human conjunctival epithelia in dry eye. Invest Ophthalmol Vis Sci 1998; 39(13):2602–2609.

12. Caffery B, Joyce E, Heynen M et al. MUC16 Expression in Sjogren's syndrome, KCS, and control subjects. Mol Vis 2008; 14:2547–2555.

13. Imbert Y, Darling DS, Jumblatt MM et al. MUC1 splice variants in human ocular surface tissues: possible differences between dry eye patients and normal controls. Exp Eye Res 2006; 83(3):493–501.

14. Imbert Y, Foulks GN, Brennan MD et al. MUC1 and estrogen receptor alpha gene polymorphisms in dry eye patients. Exp Eye Res 2009; 88(3):334–338.

15. Tseng SC, Hirst LW, Farazdaghi M, Green WR. Goblet cell density and vascularization during conjunctival transdifferentiation. Invest Ophthalmol Vis Sci 1984; 25(10):1168–1176.

16. Kessing SV. Investigations of the conjunctival mucin. (Quantitative studies of the goblet cells of conjunctiva; preliminary report). Acta Ophthalmol (Copenh) 1966; 44(3):439–453.

17. Srinivasan B, Worgul B, Iwamoto T, Merriam G. The conjunctiva epithelium:II. Histochemical and ultrastructural studies on human and rat conjunctiva. Ophthalmic Res 1977; 9:65–79.

18. Diebold Y, Rios JD, Hodges RR, Rawe I, Dartt DA. Presence of nerves and their receptors in mouse and human conjunctival goblet cells. Invest Ophthalmol Vis Sci 2001; 42(10):2270–2282.

19. Rios JD, Zoukhri D, Rawe IM, Hodges RR, Zieske JD, Dartt DA. Immunolocalization of muscarinic and VIP receptor subtypes and their role in stimulating goblet cell secretion. Invest Ophthalmol Vis Sci 1999; 40(6):1102–1111.

20. Kanno H, Horikawa Y, Hodges RR et al. Cholinergic agonists transactivate EGFR and stimulate MAPK to induce goblet cell secretion. Am J Physiol Cell Physiol 2003; 284(4):C988–C998.

21. Dartt DA, Rios JD, Kanno H et al. Regulation of conjunctival goblet cell secretion by Ca^{2+} and protein kinase C. Exp Eye Res 2000; 71(6):619–628.

22. Jumblatt J, Jumblatt M. Detection and quantification of conjunctival mucins. Adv Exp Med Biol 1998; 438:239–246.

23. Hodges RR, Horikawa Y, Rios JD, Shatos MA, Dartt DA. Effect of protein kinase C and Ca^{2+} on p42/p44 MAPK, Pyk2, and Src activation in rat conjunctival goblet cells. Exp Eye Res 2007; 85(6):836–844.

24. Horikawa Y, Shatos MA, Hodges RR et al. Activation of mitogen-activated protein kinase by cholinergic agonists and EGF in human compared with rat cultured conjunctival goblet cells. Invest Ophthalmol Vis Sci 2003; 44(6):2535–2544.

25. Jumblatt JE, Jumblatt MM. Regulation of ocular mucin secretion by P2Y2 nucleotide receptors in rabbit and human conjunctiva. Exp Eye Res 1998; 67(3):341–346.

26. Rios JD, Ghinelli E, Gu J, Hodges RR, Dartt DA. Role of neurotrophins and neurotrophin receptors in rat conjunctival goblet cell secretion and proliferation. Invest Ophthalmol Vis Sci 2007; 48(4):1543–1551.

27. Gu J, Chen L, Shatos MA et al. Presence of EGF growth factor ligands and their effects on cultured rat conjunctival goblet cell proliferation. Exp Eye Res 2008; 86(2):322–334.

28. Shatos MA, Gu J, Hodges RR, Lashkari K, Dartt DA. ERK/p44p42 mitogen-activated protein kinase mediates EGF-stimulated proliferation of conjunctival goblet cells in culture. Invest Ophthalmol Vis Sci 2008; 49(8):3351–3359.

29. Hosoya K, Horibe Y, Kim KJ, Lee VH. Na$^+$-dependent L-arginine transport in the pigmented rabbit conjunctiva. Exp Eye Res 1997; 65(4):547–553.

30. Turner HC, Alvarez LJ, Candia OA. Identification and localization of acid-base transporters in the conjunctival epithelium. Exp Eye Res 2001; 72(5):519–531.

31. Turner HC, Alvarez LJ, Bildin V,N, Candia OA. Immunolocalization of Na-K-ATPase, Na-K-Cl and Na-glucose cotransporters in the conjunctival epithelium. Curr Eye Res 2000; 21(5):843–850.

32. Kompella UB, Kim KJ, Shiue MH, Lee VH. Cyclic AMP modulation of active ion transport in the pigmented rabbit conjunctiva. J Ocul Pharmacol Ther 1996; 12(3):281–287.

33. Shiue MH, Kim KJ, Lee VH. Modulation of chloride secretion across the pigmented rabbit conjunctiva. Exp Eye Res 1998; 66(3):275–282.

34. Li Y, Kuang K, Yerxa B, Wen Q, Rosskothen H, Fischbarg J. Rabbit conjunctival epithelium transports fluid, and P2Y2(2) receptor agonists stimulate Cl$^-$ and fluid secretion. Am J Physiol Cell Physiol 2001; 281(2):C595–C602.

35. Candia O, Alvarea L. Overview of electrolyte and fluid transport across the conjunctiva, In: Dartt D, ed. Encyclopedia of the eye. London: Elsevier, 2010.

36. Wu K, Jerdeva GV, da Costa SR, Sou E, Schechter JE, Hamm-Alvarez SF. Molecular mechanisms of lacrimal acinar secretory vesicle exocytosis. Exp Eye Res 2006; 83(1):84–96.

37. Evans E, Zhang W, Jerdeva G et al. Direct interaction between Rab3D and the polymeric immunoglobulin receptor and trafficking through regulated secretory vesicles in lacrimal gland acinar cells. Am J Physiol Cell Physiol 2008; 294(3):C662–C674.

38. Zoukhri D, Hodges RR, Sergheraert C, Dartt DA. Cholinergic-induced Ca^{2+} elevation in rat lacrimal gland acini is negatively modulated by PKCdelta and PKCepsilon. Invest Ophthalmol Vis Sci 2000; 41(2):386–392.

39. Zoukhri D, Dartt DA. Cholinergic activation of phospholipase D in lacrimal gland acini is independent of protein kinase C and calcium. Am J Physiol 1995; 268(3 Pt 1):C713–C720.

40. Hodges RR, Zoukhri D, Sergheraert C, Zieske JD, Dartt DA. Identification of vasoactive intestinal peptide receptor subtypes in the lacrimal gland and their signal-transducing components. Invest Ophthalmol Vis Sci 1997; 38(3):610–619.

41. Funaki C, Hodges, RR, Dartt DA. Role of cAMP inhibition of p44/p42 mitogen-activated protein kinase in potentiation of protein secretion in rat lacrimal gland. Am J Physiol Cell Physiol 2007; 293(5):C1551–C1560.

42. Hodges RR, Shatos MA, Tarko RS, Vrouvlianis J, Gu J, Dartt DA. Nitric oxide and cGMP mediate alpha1D-adrenergic receptor-Stimulated protein secretion and p42/p44 MAPK activation in rat lacrimal gland. Invest Ophthalmol Vis Sci 2005; 46(8):2781–2789.

43. Hodges RR, Dicker DM, Rose PE, Dartt DA. Alpha 1-adrenergic and cholinergic agonists use separate signal transduction pathways in lacrimal gland. Am J Physiol 1992; 262(6 Pt 1):C1087–G1096.

44. Zoukhri D, Hodges RR, Sergheraert C, Toker A, Dartt DA. Lacrimal gland PKC isoforms are differentially involved in agonist-induced protein secretion. Am J Physiol 1997; 272(1 Pt 1):C263–C269.

45. Ota I, Zoukhri D, Hodges RR et al. Alpha 1-adrenergic and cholinergic agonists activate MAPK by separate mechanisms to inhibit secretion in lacrimal gland. Am J Physiol Cell Physiol 2003; 284(1):C168–C178.

46. Chen L, Hodges RR, Funaki C et al. Effects of alpha1D-adrenergic receptors on shedding of biologically active EGF in freshly isolated lacrimal gland epithelial cells. Am J Physiol Cell Physiol 2006; 291(5):C946–C956.

47. Tepavcevic V, Hodges RR, Zoukhri D, Dartt DA. Signal transduction pathways used by EGF to stimulate protein secretion in rat lacrimal gland. Invest Ophthalmol Vis Sci 2003; 44(3):1075–1081.

48. Dartt DA, Baker AK, Vaillant C, Rose PE. Vasoactive intestinal polypeptide stimulation of protein secretion from rat lacrimal gland acini. Am J Physiol 1984; 247(5 Pt 1):G502–G509.

49. Dartt DA, Moller M, Poulsen JH. Lacrimal gland electrolyte and water secretion in the rabbit: localization and role of (Na$^+$-K$^+$)-activated ATPase. J Physiol 1981; 321:557–569.

50. Dartt DA, Hodges RR, Zoukhri D. Tears and their secretion. In: Fischbarg J, ed. Advances in organ biology, vol. 10: the biology of the eye. London: Elsevier, 2006;21–82.

51. McKown RL, Wang N, Raab RW et al. Lacritin and other new proteins of the lacrimal functional unit. Exp Eye Res 2009; 88(5):848–858.

52. McCulley JP, Shine WE. Meibomian gland function and the tear lipid layer. Ocul Surf 2003; 1(3):97–106.

53. Foulks GN, Bron AJ. Meibomian gland dysfunction: a clinical scheme for description, diagnosis, classification, and grading. Ocul Surf 2003; 1(3):107–126.

54. Olami Y, Zajicek G, Cogan M, Gnessin H, Pe'er J. Turnover and migration of meibomian gland cells in rats' eyelids. Ophthalmic Res 2001; 33(3):170–175.

55. Butovich IA, Millar TJ, Ham BM. Understanding and analyzing meibomian lipids – a review. Curr Eye Res 2008; 33(5):405–420.

56. Seifert P, Spitznas M. Immunocytochemical and ultrastructural evaluation of the distribution of nervous tissue and neuropeptides in the meibomian gland. Graefes Arch Clin Exp Ophthalmol 1996; 234(10):648–656.

57. Sullivan DA, Sullivan BD, Ullman MD et al. Androgen influence on the meibomian gland. Invest Ophthalmol Vis Sci 2000; 41(12):3732–3742.

58. Suzuki T, Schirra F, Richards SM, Jensen RV, Sullivan DA. Estrogen and progesterone control of gene expression in the mouse meibomian gland. Invest Ophthalmol Vis Sci 2008; 49(5):1797–1808.

59. Nicolaides N, Kaitaranta JK, Rawdah TN, Macy JI, Boswell FM 3rd, Smith RE. Meibomian gland studies: comparison of steer and human lipids. Invest Ophthalmol Vis Sci 1981; 20(4):522–536.

60. Norn M. Meibomian orifices and Marx's line. Studied by triple vital staining. Acta Ophthalmol (Copenh) 1985; 63(6):698–700.

CHAPTER 16

Sensory Innervation of the Eye

Carlos Belmonte, Timo T. Tervo & Juana Gallar

Introduction

The sensory innervation of the eye is provided by the peripheral axons of primary sensory neurons located in the trigeminal ganglion. The sensory nerves enter the eyeball mainly through the ciliary nerves and reach all ocular tissues with the exception of the lens and the retina. Ocular innervation is particularly rich in the cornea but all tissues of the anterior segment of the eye have an abundant sensory supply. Ocular sensory fibers are functionally heterogeneous and include mechanoreceptor, polymodal nociceptor, and cold receptor types of sensory units. Consequently, the nerves respond to a variety of physical and chemical stimuli. Impulse activity originated at peripheral sensory nerve fibers of the eye reaches second-order neurons located at the trigeminal nucleus caudalis in the lower brainstem. Sensory information travels subsequently to relay stations in the contralateral posterior thalamic nucleus and finally reaches the cortex.

Pain is the main sensation elicited by stimulation of ocular sensory nerves although tactile sensations are evoked at the conjunctiva and feelings of cold are also evoked by mild cooling of the corneal surface. After repeated noxious stimulation or tissue injury, ocular polymodal nociceptors become sensitized and cause sustained pain and hyperalgesia. They also contribute to local inflammation through the antidromic release of peptide transmitters stored in their peripheral endings ("neurogenic inflammation"). Intact sensory nerve endings, particularly in the corneal epithelium, are subjected to a continuous remodeling. Ocular sensory nerves damaged by accidental or surgical injury or following pathological processes, may also experience long-lasting changes in excitability, giving rise to ocular dysesthesia or neuropathic pain referred to the eye.

Sensations originated at the ocular surface can be explored by esthesiometry. The cornea has an overall higher sensitivity than the conjunctiva. Sensitivity of the ocular surface is reduced by age and by a variety of pathological processes that affect sensory innervation such as herpetic keratitis, some types of corneal trauma and post-infectious conditions, certain hereditary corneal dystrophies and anterior segment or vitreo-retinal surgery, especially scleral buckling operations. Although the principal eye diseases leading to impaired vision – such as retinal pathologies, chronic open-angle glaucoma or cataract – course without pain, pain is a cardinal symptom of inflammatory and traumatic disturbances of the anterior segment of the eye. Pain is also evoked by various ocular diseases affecting other structures within the globe or orbital tissues, and may become an undesired consequence of ocular surgery. Dry eye, a multietiological condition that leads to alteration of the ocular surface epithelium, is the most common cause of superficial ocular discomfort and pain. Management of ocular pain includes the use of topical anesthetics, cycloplegic agents and anti-inflammatory drugs or ocular patching to reduce peripheral nociceptive input, and systemic analgesics to act on nociceptive pathways at the central nervous system.

1. Anatomy of ocular sensory nerves

1.1 Origin of the ocular sensory nerves

Trigeminal ganglion neurons

The sensory innervation of the eye is provided by primary sensory neurons mainly of small or medium size, clustered in the ophthalmic (medial) region of the ipsilateral trigeminal ganglion (TG). Most ocular sensory neurons are devoted to the cornea and represent about 1.5 percent of the total number of neurons of the TG.[1] The axons of the pseudounipolar trigeminal neurons divide into a peripheral branch that projects to the peripheral target tissues and a central branch that enters the brainstem to reach the trigeminal sensory complex. Based on the size and the presence of a myelin sheath in the peripheral axons, corneal trigeminal neurons can be classified as myelinated (20 percent in the mouse) and unmyelinated (80 percent in the mouse).[1,2] Myelin content is reflected in the conduction velocity of the peripheral axons (see below).

TG neurons innervating the eyeball contain several neuropeptides, including CGRP (calcitonin gene-related peptide, present in about 50 percent of corneal neurons), the tachykinin substance P (present in about 20 percent), cholecystokinin, somatostatin, opioid peptides, pituitary adenylate cyclase activating peptide (PACAP), vasoactive intestinal peptide (VIP), galanin and neuropeptide Y (NPY), as well as neuronal nitric oxide synthase (nNOS).[1,3–5]

The ophthalmic nerve and its branches

The peripheral axonal branch of most ocular sensory neurons in humans exits the TG running with the ophthalmic division of the ganglion, which traverses the superior orbital fissure and then branches into the nasociliary, the frontal and the lachrymal nerves. The nasociliary nerve gives several sensory branches: (1) two long ciliary nerves that reach the eyeball and pierce the sclera, constituting the major sensory output of the eye; (2) the infratrochlear nerve that innervates the medial aspect of lids, nose, and lachrymal sac; (3) the external nasal nerves; (4) a communicating branch to the ciliary ganglion. The ciliary ganglion is a parasympathetic ganglion located within the orbit that sends 5–10 short ciliary nerves carrying parasympathetic postganglionic fibers, postganglionic sympathetic axons originating at the superior cervical ganglion, and the trigeminal sensory nerve fibers arriving through the communicating branch of the nasociliary nerve to the eye. These mixed, short ciliary nerves enter the eyeball around the optic nerve. The second branch of the ophthalmic nerve is the frontal nerve, which bifurcates into the supraorbital nerve to innervate the upper eye lid and the frontal sinus, and the supratrochlear nerve that provides innervation to the forehead and the upper eyelid. Finally, the third branch of the ophthalmic nerve, the lachrymal nerve, innervates the lachrymal gland and some areas of the conjunctiva and the skin of the upper lid. It also incorporates postganglionic parasympathetic nerve fibers from the pterygopalatine ganglion (Fig. 16.1).

A minor part of the sensory fibers from the eye is conveyed, together with the innervation of the conjunctiva and the skin of the lower eye lid, by the infraorbital nerve, a branch of the maxillary nerve, which in turn constitutes the second major branch of the trigeminal ganglion (Fig. 16.1).

Trigeminal sensory fibers end in the epithelia, connective tissue and blood vessels of the lids, orbit, extraocular muscles, ciliary body, choroid, scleral spur, sclera, cornea, and conjunctiva – the retina and lens being the only ocular tissues not receiving direct trigeminal sensory innervation.

1.2 Distribution of sensory nerve fibers within the eye

After penetrating the sclera around the optic nerve at the posterior eye pole, long and short ciliary nerves containing a mixture of sensory, sympathetic, and parasympathetic nerve fibers form a ring around the optic nerve, keeping the lamina cribosa devoid of innervation. Ciliary nerve fascicles travel anteriorly towards the cornea within the suprachoroidal space, giving a few collaterals that provide the sparse innervation of the posterior part of the eye. Along their trajectory to the eye front, these nerve fiber bundles undergo repetitive branching. Some of the nerve fibers innervate the sclera itself, being more numerous in the episcleral tissue; others leave the sclera to enter the choroid, while most nerve filaments continue to innervate the ciliary body, the iris, and the cornea. Most nerve fibers finally form a series of ring meshworks of nerve fibers at the limbus around the cornea – the so-called limbal or pericorneal plexuses.

Nerve filaments entering the choroid are composed of sensory, sympathetic, and parasympathetic axons, that

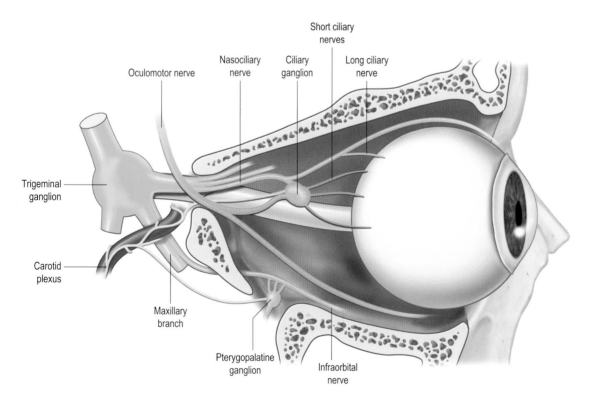

Figure 16.1 Innervation of the eye. Medial view of the orbit showing the sensory and autonomic nerves directed to the eye. The ophthalmic branch of the trigeminal ganglion gives the nasociliary nerve that sends long and short ciliary nerves to the eyeball, the last through the ciliary ganglion. Frontal and lacrimal nerves are not shown in this picture. The trigeminal maxillary branch gives the infraorbital nerve that innervates part of the eye and the lower lid. Sympathetic fibers from the superior cervical ganglion, travelling within the carotid plexus and parasympathetic branches of the ciliary and the pterygopalatine ganglia join short ciliary nerves. (Adapted from Rubin M, Safdieh JE, Netter's Concise Neuroanatomy, Elsevier, 2007.)

branch extensively forming a dense network in which individual parasympathetic ganglion cells are also found.[6]

Limbal plexus nerves supply the sensory and autonomic innervation of the limbal vessels, the trabecular meshwork and the scleral spur. Nerve bundles also form at the root of the iris, a circumferential plexus that gives origin to the iridal innervation. Finally, a variable number of nerve trunks arising from the limbal plexus enter the corneal stroma (corneal stromal nerves). Despite its comparatively small size, the cornea receives a significant portion of the ocular innervation, presenting a nerve density that is estimated to be 300–400 times higher than that of the human fingers or the teeth.[4,7]

The majority of the nerves directed to the eye are composed by thin, unmyelinated nerve fibers often incompletely surrounded by Schwann cells, while less than 30 percent present a myelin sheath. All branch extensively and terminate as free nerve endings showing small enlargements (varicosities) along their terminal course. In addition, some encapsulated nerve endings similar to Krause and Meissner bodies have been observed in the choroid, in particular near the ciliary body[8,9] and the iris,[10] and in the episclera and chamber angle but not in the cornea.[11]

1.3 Architecture of corneal sensory nerves

The corneal innervation is anatomically organized in four levels from the penetrating stromal nerve bundles up to the intraepithelial nerve terminals (Fig. 16.2A).

Corneal stromal nerves

The anatomy of nerve bundles radially entering the corneal stroma is quite similar among mammals varying only in their number (6–8 in rat, 15–40 in cat or dog, about 60 in human).[12–14] Stromal nerves branch immediately after entering the cornea and run within the stroma as ribbon-like fascicles enclosed by a basal lamina and Schwann cells. Myelinated axons included in stromal nerves (about 20 percent of the nerve fibers) lose their myelin sheath within a millimeter after penetrating the stroma (Fig. 16.2B). The distal branches of this arborization anastomose extensively, forming the anterior stromal nerve plexus, a dense and complex network of intersecting small and medium-size nerve bundles and individual axons with no preferred orientation laying in the anterior 25–50 percent of the corneal stroma, depending on the species, being most dense in the anterior layers. In contrast, the posterior half of the human stroma and the corneal endothelium are devoid of sensory nerve fibers.

In addition to stromal nerve innervation, a small number of small nerve fascicles originating at the limbal plexus and at the conjunctiva superficially enter the peripheral cornea to innervate the perilimbal and peripheral corneal zones.[13]

Subepithelial nerve plexus

In humans and higher mammals, the most superficial layer of the anterior stromal nerve plexus, located in a narrow strip of stroma immediately beneath Bowman's membrane, is especially dense and is referred to as the corneal subepithelial nerve plexus, its nerve density generally higher in the peripheral than in the central cornea. Two anatomically distinct types of nerve bundles are distinguished in the subepithelial plexus. One forms a highly anastomotic meshwork made of single axons and thin nerve fascicles located immediately beneath Bowman's membrane without penetrating it towards the corneal epithelium. The second type consists of about 400–500 medium-sized, curvilinear bundles that penetrate Bowman's membrane mainly in the peripheral and intermediate cornea, shed the Schwann cell coating, bend at a 90° angle and divide, each into 2–20 thinner nerve fascicles that continue into the corneal epithelium as the sub-basal nerve plexus. A relatively low number of stromal nerves penetrate Bowman's membrane in the central cornea, which receives most of its innervation from long sub-basal nerves that enter the peripheral cornea directly from the limbal plexus.

Sub-basal nerve plexus

The sub-basal nerve plexus in humans is formed by 5,000–7,000 nerve fascicles in an area of about 90 mm². A sub-basal nerve bundle gives several side branches, each containing 3–7 individual axons. Thus, the total number of axons in the sub-basal plexus is estimated to vary between 20,000 and 44,000.[4]

The sub-basal nerve axons may travel up to 6 mm between the basal epithelial cells and their basal lamina, roughly parallel to one another. The arrangement of a stromal nerve bundle branching into multiple, parallel daughter nerve fascicles, constitutes a unique neuroanatomical structure characteristic of the cornea and is termed an epithelial leash. Epithelial leashes are constituted by up to 40 individual unmyelinated straight and beaded nerve fibers of variable diameter (0.05–2.5 microns). Nerve fibers in adjacent leashes interconnect repeatedly such that they are no longer recognizable as individual leashes, forming finally a relatively homogeneous nerve plexus. In vivo confocal microscopy has shown that the sub-basal nerve plexus forms a whorl-like, spiral pattern of nerve fibers, the center of which is called the "vortex" (Fig. 16.2C). In humans it is located 2–3 millimeters inferior and nasal to the corneal apex. The mechanisms that govern the formation and maintenance of this spiral pattern remain unknown. However, as basal corneal epithelial cells and sub-basal nerves appear to migrate centripetally in tandem,[15] it is possible that basal epithelial cells derived from limbal stem cells migrate centripetally in a whorl-like fashion towards the corneal apex in response to a chemotropic guidance, to electromagnetic cues, and/or to population pressures; and the sub-basal nerves that occupy the narrow intercellular spaces between these migrating cells would be pulled along their trajectory. An alternate possibility is that sub-basal nerves would develop their whorl-like orientation independently of epithelial cell dynamics, providing in turn a structural scaffolding that directs epithelial cell migration.[16–19]

Despite their considerable branching, most stromal and subepithelial nerve bundles pass uninterrupted through the stroma to reach the corneal epithelium and apparently do not provide functional innervation to stromal tissue. However, a small proportion of corneal nerve fibers appear to travel downwards to terminate in the stroma as expanded structures resembling free nerve endings.

Intraepithelial nerve terminals

From the sub-basal nerves running horizontally through the basal epithelium, single fibers split off and turn 90°

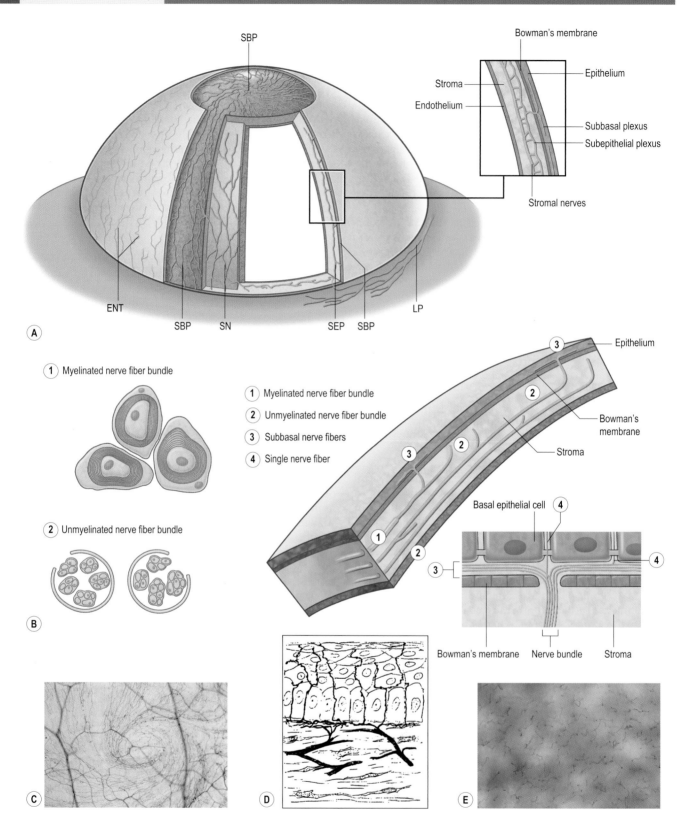

SBP

Bowman's membrane

Stroma

Endothelium

Epithelium

Subbasal plexus

Subepithelial plexus

Stromal nerves

ENT

SBP   SN   SEP   SBP

LP

A

1 Myelinated nerve fiber bundle

1 Myelinated nerve fiber bundle
2 Unmyelinated nerve fiber bundle
3 Subbasal nerve fibers
4 Single nerve fiber

2 Unmyelinated nerve fiber bundle

B

Epithelium

Bowman's membrane

Stroma

Basal epithelial cell

Bowman's membrane   Nerve bundle   Stroma

C

D

E

Figure 16.2 Corneal innervation. (**A**) Schematic distribution of nerves in the cornea. From the limbal plexus (LP), stromal nerve trunks (SN) penetrate the stroma radially and divide dichotomously to form the subepithelial plexus (SEP). Branches of this plexus ascend towards the epithelium, traverse Bowman's layer and form the sub-basal plexus (SBP) between the epithelium basal layer and its basal lamina, where nerve branches run horizontally as families of long parallel nerves (leashes) which in turn give rise to intraepithelial nerve terminals (ENT). (**B**) Characteristics of nerve fibers in the peripheral and central cornea. Myelinated nerve bundles (1) penetrate the stroma and immediately lose their myelin cover. Thus, only unmyelinated nerves (2) are found throughout the cornea. Subepithelial nerves ascend to form the sub-basal nerve plexus (3) from which individual axons (4) ascend vertically towards the corneal surface, with axons terminating at various levels of the corneal epithelium. (**C**) Architecture of the sub-basal nerve plexus in the infero-central region of the cornea. Whole mount of a stained mouse cornea depicting the sub-basal whorl area. (**D**) Drawing of epithelial nerve endings. Varicose intraepithelial nerve endings stained with the gold chloride impregnation method run between the epithelial elements up to the outer layers. (**E**) Single intraepithelial nerve ending. Whole mount of the mouse cornea stained with the gold chloride impregnation method, showing epithelial nerve endings at the corneal surface. (A: Modified from Müller et al 2003;[4] B: Adapted from Müller et al 1996;[14] Reproduced from Association for Research in Vision and Ophthalmology. C: From C. Belmonte and E. Raviola, unpublished results; D: From Ramon y Cajal S, Textura del sistema nervioso del hombre y de los vertebrados, Vol. I. Imprenta y Librería de Nicolas Moya, Madrid, 1899; E: From F. De Castro and C. Belmonte, unpublished results.)

vertically as a profusion of thin, short, and beaded terminal axons ascending between the epithelial cells, often with a modest amount of additional branching, up to the more superficial layers of the corneal epithelium (Fig. 16.2D,E).[4,20] Intraepithelial fibers end as free nerve endings, appearing as prominent, bulbous terminal expansions that appear morphologically homogeneous when visualized using optical or electron microscopy, though immunocytochemical staining reveals differences in the expression of neuropeptides and other neurotransmitters, suggesting a functional heterogeneity.[4] At the ultrastructural level, corneal nerve endings contain abundant small clear vesicles, perhaps filled with excitatory amino acids, and large, dense-cored vesicles containing neuropeptides such as calcitonin gene-related peptide (CGRP), substance P, and/or other peptides. They also contain mitochondria, glycogen particles, neurotubules, and neurofilaments.[3,4,21] The nerve terminals are located throughout all layers of the corneal epithelium, extending up to a few microns of the corneal surface, but being especially numerous in the wing and basal cell layers. Occasionally, epithelial cell membranes facing the nerve terminals show invaginations that may eventually completely surround the nerve ending.[4,19,20] This intimate relationship raises the possibility of bidirectional exchange of diffusible substances between both structures. The intimate contacts also allow nerve endings to detect changes in epithelial cell shape or volume, such as those produced by ocular surface desiccation or swelling.

The innervation density of the corneal epithelium is probably the highest of any surface epithelium. Although the actual number of corneal nerve endings remains a matter of speculation, considering that each sub-basal nerve fiber gives at least 10–20 intraepithelial nerve terminals, it is reasonable to speculate that the human central cornea contains approximately 3,500–7,000 nerve terminals/mm^2. This rich innervation provides the cornea with a highly sensitive detection system, and it has been hypothesized that injury of a single epithelial cell may be sufficient to trigger pain perception.[19] Nerve terminal density, and thus, corneal sensitivity, are higher in the central cornea and decrease progressively when moving towards the periphery. Similarly, corneal sensitivity and nerve density decrease progressively as a function of age and in several ocular pathologies.

An individual stromal axon entering at the corneoscleral limbus undergoes repetitive branching and eventually travels across as much as three-quarters of the cornea before terminating.[20] As a result, individual receptive fields of corneal

sensory fibers range in size from less than 1 mm^2 to as much as 50 mm^2 and may cover up to 25 percent of the corneal surface. The extensive branching also explains the significant overlapping of receptive fields found in electrophysiological studies of single corneal nerve fibers.

1.4 Central sensory pathways

Sensory information from the eye is carried by trigeminal ganglion neurons to the ventral portion of the ipsilateral trigeminal brainstem nuclear complex (TBNC), to activate ocular sensory second-order neurons located mainly in the intermediate zone between interpolaris and caudalis subnuclei (Vi/Vc), in laminae I-II of the subnucleus caudalis/upper cervical spinal cord (Vc/C1), and in the adjacent bulbar lateral reticular formation (Fig. 16.3).[22–24] Additionally, few trigeminal neurons innervating ocular and periocular tissues project to the principal nucleus of the TBNC, and a very sparse number are confined to a few locations along the ventral border of the pars oralis and interpolaris of the spinal trigeminal nucleus. Second-order ocular neurons located in the Vi/Vc area project to different places in the central nervous system including brainstem regions as the superior salivatory/facial motor nucleus and the contralateral thalamus.[25–27] Neurons receiving corneal information have been identified in the thalamic posterior nucleus and the zona incerta. These thalamic neurons project in turn to primary (SI) and secondary (SII) somatosensory cortical areas responsible for pain sensations and reactions. Non-visual sensory input from the eye is represented in contralateral cortical SI areas 3b and 1 (Fig. 16.3), located together with those that represent nose, ear, and scalp. Projections to the insula and to the cingulate and prefrontal cortex provide the neural substrate for the affective and cognitive components of sensations evoked by eye stimulation.

2. Development and remodeling of corneal innervation

2.1 Development of corneal nerves

The cornea lacks sensory innervation until the fifth gestational month, when corneal nerve endings appear. During development, sensory axons first form the nerve ring around the cornea (the limbal nerve plexus) and subsequently axons

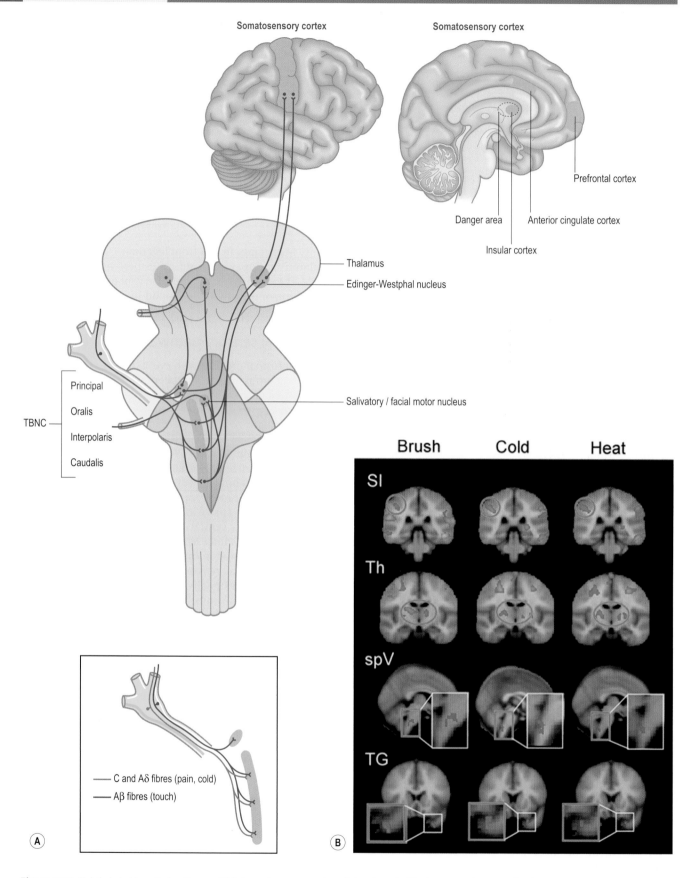

Figure 16.3 Ophthalmic trigeminal pathways. (**A**) Schematic representation of the trigeminal brainstem nuclear complex (TBNC), composed of the principal nucleus, located in the mesencephalon, and the spinal nucleus, subdivided into subnucleus oralis, interpolaris, and caudalis. TBNC projects to the contralateral thalamus, and from there to the primary somatosensory areas cortex (insular, cingulate, and prefrontal cortex). Thalamic projections also reach other areas of the cortex involved in the processing of affective and cognitive aspects of ocular pain. Trigeminal projections to the salivatory/facial motor nucleus, and to the Edinger–Westphal nucleus are also represented. Inset: Distribution of thick myelinated mechanosensory fibers and thin (thermosensitive and nociceptive) fibers within the TBNC. (**B**) MRI images showing the activation by noxious stimulation of the trigeminal sensory pathway. TG: trigeminal ganglion. spV: spinal trigeminal nucleus. Th: thalamus. SI: Primary somatosensory cortex. (From Becerra et al, 2006 J Neurosci. 26:10646–10657.)

grow radially into the corneal tissue (Fig. 16.4A).[20,28] The molecular signals controlling ocular nerve growth remain unknown, though several guidance molecules may regulate the process. Corneal and lens-derived semaphorins appear to mediate the initial repulsion of trigeminal sensory axons from the cornea,[29] thus inducing the proper formation of the pericorneal nerve ring meshwork, and also contribute to the positioning of a subset of axonal projections in the choroid fissure, that form a ventral plexus from which the iris innervation is later supplied.[30] Upon completion of the pericorneal nerve ring, sensory nerves enter the cornea as bundles, at uniformly spaced sites around its entire circumference and begin to extend radially inside the cornea, first innervating its periphery and then the entire stromal surface. The depth at which nerves enter the corneal stroma correlates with the area of the cornea that they will innervate, so that the deepest stromal nerves innervate almost the entire corneal surface, whereas the more superficial nerves which enter the stroma nearest to the epithelium mainly supply the periphery.[17]

The uniform separation between corneal nerve bundles during embryogenesis seems to be due to the release of neuro-repulsive factors from pioneering growth cones as they leave the limbal nerve ring and enter the corneal stroma, keeping nerve bundles growing relatively straight towards the center of the cornea as centripetal radii and separated one from another. The extension or radial innervation within the cornea is produced by bifurcation of nerve fascicles in successive, concentric zones, thus suggesting that the position and orientation of intracorneal nerves is determined by the intra-corneal milieu[16] and the migration of epithelial cells.[17] Stromal composition and neurotrophic factors released from corneal epithelium also participate in the regulation of nerve density and orientation of corneal nerves during development.[31,32]

2.2 Dynamic remodeling of adult corneal innervation

Corneal nerves are subjected to a remodeling in the adult cornea throughout life. Deep stromal nerve fiber bundles maintain a relatively constant position and configuration within the cornea while the corneal sub-basal nerve plexus, and especially intraepithelial nerve terminals, experience an extensive rearrangement (Fig. 16.4B; Box 16.1). Time-lapse, in vivo confocal microscopic examination of living human eyes reveals that the human sub-basal corneal nerve plexus is a dynamic structure with a slow but continuous centripetal movement (5–15 microns/day) that changes the whorl-centered plexus in a 6-week period.[33] Intraepithelial nerve endings are subjected to faster morphofunctional changes; their continuous remodeling follows the long-term, sub-basal nerve reconfiguration. The continuous shedding of corneal epithelial cells also strongly influences the rearrangement and dynamics of epithelial nerve terminals.

Reorganization of differentiating corneal epithelial cells migrating up to the most superficial layers of the corneal epithelium induces changes in intraepithelial nerve ending architecture that occur in less than 24 hours.[34] The molecular mechanisms responsible for this continuous rearrangement of corneal nerves are unknown. Adult corneal sensory nerves retain the ability to respond to semaphorins,

---

Box 16.1 Corneal nerve remodeling and ocular pathologies

The continuous remodeling of corneal nerves may be disturbed by an altered corneal architecture or prolonged wound healing. Abnormal patterns in sub-basal nerve organization have been observed following corneal epithelial pathologies such as basement membrane dystrophy, recurrent erosions, or dry eye.

---

which inhibit nerve sprouting both in transected and intact nerves.[35] In several mammalian species, corneal nerve terminal density decreases with age,[36,37] and this may be the basis of the decreased corneal sensitivity reported in the elderly.[38–41]

2.3 Regeneration of injured corneal nerves

Adult corneal nerves retain the ability to regenerate following injury. Nevertheless, after damage or transection, the morphology and functional properties of corneal nerves change substantially. Nerve processes distal to the site of lesion degenerate, whereas central stumps start to regenerate, finally producing a nerve pattern rather different from the original corneal nerve architecture.[42,43] The regenerative process takes place in several stages (Fig. 16.4C). When corneal nerves are cut, the denervated area is first invaded by sprouts of adjacent, intact nerve fibers. Later, the central stump of injured axons starts to regenerate, forming microneuromas from which newly formed sprouts begin to develop, while the early branches from intact fibers begin to degenerate.[4,42] It may be that interruption in the injured nerve terminals of the uptake of signal molecules such as nerve growth factor (NGF) produced by corneal cells, which are centripetally transported along the parent axon to the neuron's soma to regulate gene expression, causes morphological and functional changes in regenerating neurons.[44]

3. Functional characteristics of ocular sensory innervation

3.1 Trigeminal ganglion neurons

Electrophysiological studies of ocular sensory neurons have been centered mainly on extracellular recordings of the propagated impulse activity in the peripheral axons of trigeminal ganglion neurons, in most cases nerve fibers innervating the cornea and bulbar conjunctiva. Ocular sensory neurons innervating the cornea and uveal tissues possess in general thin myelinated (Aδ) axons (conducting action potentials at velocities between 3–15 m/s) or unmyelinated (C) axons, which conduct at less than 2 m/s. A reduced number of neurons with thick myelinated axons (Aβ) provide innervation to the limbus, lids, and conjunctiva.

3.1.1 Sensory fibers of the cornea and conjunctiva

As in other parts of the body, the response levels of the peripheral endings of sensory axons to different modalities of physical and chemical stimuli have been used to

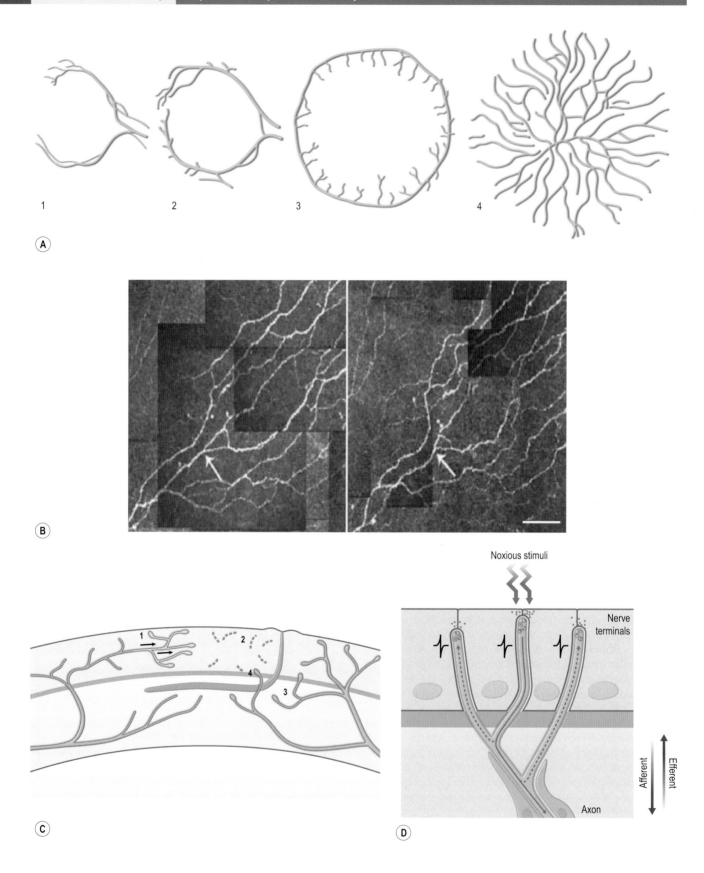

Figure 16.4 Development, remodeling and regeneration of corneal sensory nerves. (**A**) Schematic representation of corneal sensory nerve growth during development, in the chick cornea. Growing trigeminal sensory axons avoid the cornea and extend dorsally and ventrally (A1), forming a pericorneal nerve ring at E8 (A2). Upon completion of this ring, sensory nerves begin to enter the cornea radially, first innervating its periphery at E9–E10 (A3), and finally its entire surface at E15 (A4). (**B**) Continuous remodeling of nerves in the healthy cornea. In-vivo confocal microscopy images showing changes in human sub-basal nerve morphology occurring within a 2-week period. (**C**) Schematic representation of the changes taking place in corneal sensory axons following injury. Sprouts of intact nerves (1) invade the injured area while the distal end of axotomized fibers degenerate (2). Central stumps of lesioned nerves form neuroma endings (3) from which sprouts begin to invade the denervated area (4) initiating the recovery of corneal sensitivity. (**D**) Schematic representation of afferent and efferent functions of polymodal nociceptors. Stimulation of a section of the peripheral branches of a polymodal nociceptor fiber by a noxious substance directly depolarizes the nerve endings affected by the stimulus. This depolarization generates action potentials that travel centripetally (continuous arrow), and also cause the local release of neuropeptides (dots) contained in sensory nerve terminals, which contribute to local inflammation. Action potentials traveling centrally also antidromically invade (interrupted arrows) unstimulated neighboring branches. These then become depolarized and release neuropeptides in an area of the tissue that was not directly affected by the stimulus, thus extending inflammation ("neurogenic inflammation"). (B: From Patel & McGhee 2008;[33] Reproduced from Association for Research in Vision and Ophthalmology. C and D: Modified from Belmonte et al 2004b.[86])

distinguish various functional classes of sensory fibers innervating the various eye tissues (Fig. 16.5).

Polymodal nociceptors

About two-thirds of the sensory fibers innervating the cornea and the bulbar conjunctiva are activated by physical and chemical stimuli of intensity within the noxious or near noxious range, including mechanical forces, heat, intense cold, exogenous chemical irritants and a large variety of endogenous molecules released by tissue injury. Accordingly, they were named polymodal nociceptors (Figs 16.5 & 16.6).[45-47] Most of them are unmyelinated while a small portion, depending on species, belongs to the group of thin myelinated (Aδ) nerve fibers. Their receptive fields are round or oval, are usually large, often covering up to one-quarter or more of the cornea, and may extend several millimeters beyond the cornea onto the adjacent limbus and bulbar conjunctiva (Fig. 16.5). The large size and extensive overlapping of adjacent receptive fields of polymodal nociceptors, coupled with convergent mechanisms in the central nervous system, explain why stimuli of the corneal surface are poorly localized.

Corneal polymodal nociceptors respond to their natural stimuli with a continuous, irregular discharge of nerve impulses at a frequency roughly proportional to the magnitude of the stimulus, which persist as long as the stimulus is maintained, thus codifying its intensity and duration. Occasionally, the nerve impulse discharge overpasses the duration of the stimulus (post-discharge). All polymodal nociceptors respond to mechanical stimuli and to temperatures over 39–40°C (Fig. 16.6A:ii,iii). About half of the polymodal fibers in the cat also develop a low-frequency response with corneal temperature reductions below 29°C.[45,47,48] Polymodal nociceptors are additionally excited by many chemical agents. Acidic solutions (pH < 6.5) (Fig. 16.6A:iv) or gas jets containing increasing concentrations of CO_2 (the carbonic acid formed at the corneal surface drops the local pH) evoke robust impulse discharges in corneal polymodal nociceptors.[46-49] The sensitivity of polymodal nociceptors to acidic stimulation with CO_2 has been used for corneal esthesiometry. A large number of endogenous chemicals (inflammatory mediators) also activate polymodal nociceptors (see below). Expression of different membrane transduction molecules provides polymodal nociceptors with the ability to respond to stimuli of different quality. These include various members of the TRP ion channels superfamily: (a) TRPV1, gated by heat, protons, endogenous mediators and external chemical irritants; (b) TRPA1, that responds to a large number of volatile irritant chemicals; (c) TRPV2 and TRPV3, which are excited by high temperatures. Additionally, polymodal neurons express ASIC channels, which respond to acid and endogenous mediators, while various types of stretch-activated, mechanosensory channels mediate the mechanical responsiveness (Fig. 16.5C).

Mechano-nociceptors

Around 15–20 percent of the nerve fibers innervating the cornea respond only to mechanical forces of near-noxious intensity. Accordingly, they were classified as mechano-nociceptors. All the axons of this class of receptor are thinly myelinated (Aδ). The receptive fields are generally round and of medium size, covering about 10 percent of the corneal surface (Fig. 16.5). The force required to activate corneal mechano- and polymodal nociceptors (that is, their mechanical threshold) is around 0.6 mN – about 10 times lower than the force required to activate the equivalent fibers in the skin. This is possibly due to the proximity of corneal nerve terminals to the surface and to the absence of a keratinized surface epithelium in the cornea. Mechano-nociceptors fire only one or a few nerve impulses in response to brief or to sustained mechanical stimuli (Fig. 16.6A:i). Therefore, corneal mechano-nociceptors have a limited capacity to codify the intensity and duration of the stimulus and probably serve mainly to signal the presence of noxious mechanical stimuli (touching of the corneal surface, foreign bodies, etc.), presumably being responsible for the acute, sharp sensation of pain produced by sudden mechanical insults.

Cold thermal receptors

About 10–15 percent of corneal nerve fibers discharge spontaneously at the resting temperature of the corneal surface (around 33°C), and increase their firing rate when this temperature is decreased while they are transiently silenced upon warming.[47,50,51] Cold receptor fibers have small receptive fields (around 1 mm diameter) located all over the corneal surface but more abundant in the peripheral areas (Fig. 16.5). Corneal cold-sensitive thermal receptors increase their firing rate as soon as the temperature of the cornea drops, which occurs, for example, during evaporation of the corneal surface tear film, application of cold solutions or blowing of cold air on the cornea. In mammals, corneal cold receptor fibers are able to detect small falls in temperature (0.1°C or less) and encode the final temperature in its static impulse frequency.[47] Small corneal temperature reductions are perceived as a sensation of innocuous cooling[48] with a variable

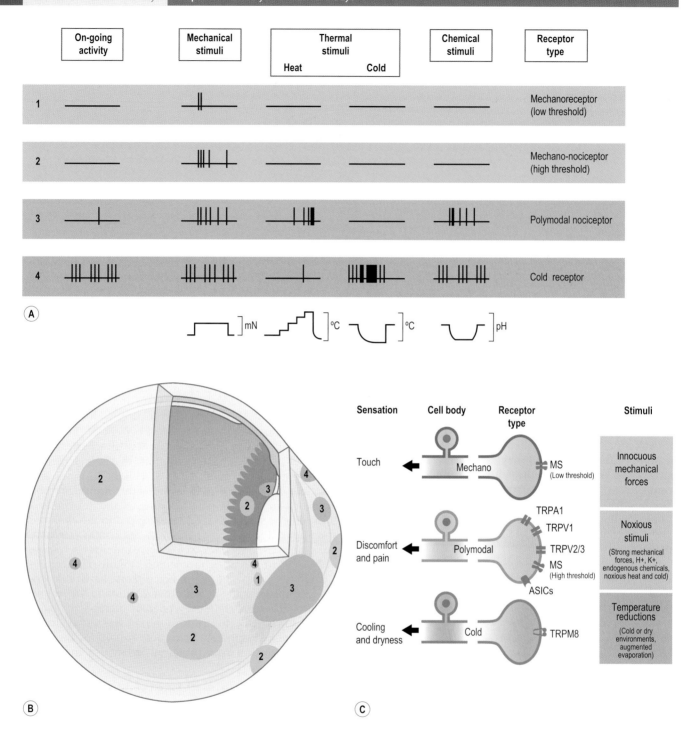

Figure 16.5 Functional characteristics of ocular sensory fibers. (A) The presence of ongoing activity at rest and impulse discharges in response to different stimuli are represented for each functional type of sensory receptor. **(B)** Diagram of the eyeball showing the distribution, size and location on the ocular surface, the ciliary body and the iris of the receptive field of mechanosensory, polymodal and cold sensory fibers innervating the eye. **(C)** Schematic representation of various modality-specific primary sensory neuron peripheral endings, and the putative membrane ion channels involved in the detection and transduction of the different stimuli. (A, B: Adapted from Belmonte et al 1997,[2] 2004a.[44], C: Adapted from Belmonte & Viana, 2008.[145])

irritative component. It is possible that corneal cold thermal receptors contribute to ocular dryness sensations when excessive evaporation reduces corneal temperature. Cold sensitivity of TG ocular neurons depends critically on the expression of TRP8 ion channels. Corneal cold receptors of TRPM8 null mice remain silent and do not respond to cooling. Basal tearing flow in these animals is reduced to half, while the tearing response to ocular surface irritation, which is mediated by polymodal nociceptor stimulation, remains normal.[52] This support the interpretation that this type of fiber is involved in signaling dryness of the ocular surface and contribute to regulate basal tearing flow.

"Silent" nociceptors

The existence in the cornea of nociceptor fibers that are insensitive to all types of stimulus when the tissue is intact

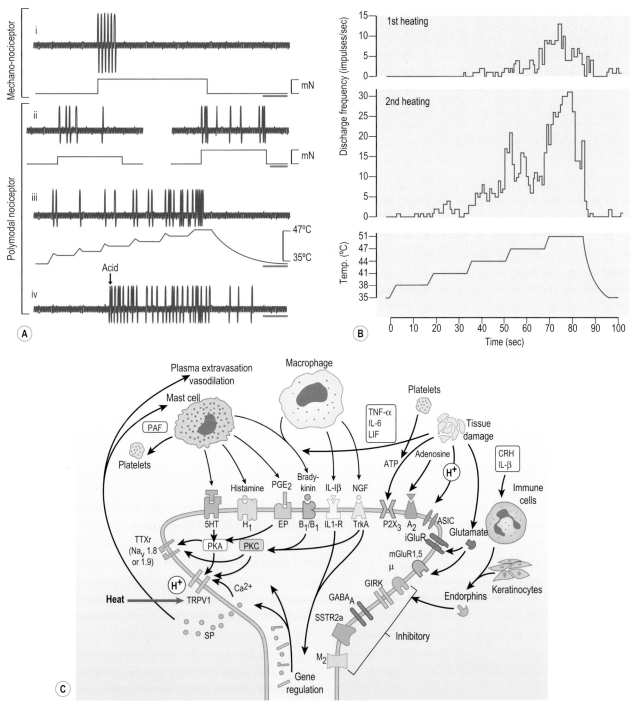

Figure 16.6 Response characteristics of corneal sensory fibers. (**A**) Effect of different types of stimuli on mechano- and polymodal nociceptors. (i) Transient (phasic) discharge evoked by a sustained mechanical indentation in a corneal mechano-nociceptor fiber. (ii) Response of a polymodal nociceptor fiber to mechanical indentations of increasing amplitude (80 and 150 μm). (iii) Activation of a polymodal nociceptor fiber by stepwise heating (from 35°C to 47°C) of the corneal surface. (iv) Response of a polymodal nociceptor fiber to application of a drop of 10 mM acetic acid (arrow) on the corneal receptive field. Upper traces depict nerve impulse recordings and lower traces the stimulus waveform. (**B**) Sensitization of corneal polymodal nociceptors. Frequency histograms of the impulse discharge evoked by two identical stepwise heating cycles of the corneal surface separated by a 3-min interval, showing the lowered threshold and enhanced impulse firing in response to the second heating cycle. (**C**) Receptor mechanisms underlying activation and sensitization of polymodal nociceptive afferents by inflammatory mediators released after tissue injury. In turn, neuropeptides released by activated nociceptors contribute to local inflammation (neurogenic inflammation). (A: From Belmonte & Gallar 2000;[78] B: Adapted from Belmonte & Giraldez 1981[45], C: Adapted from Black et al. 2004.[146])

but become excitable to mechanical, chemical, or thermal stimuli after a local inflammation develops ("silent" nociceptors)[53] was suggested by McIver & Tanelian.[54] Although experimental evidence for their presence in the cornea is indirect, silent nociceptors have been identified in many somatic tissues, particularly in humans – where they seem to play an important role in inflammatory pain. Therefore it is conceivable that they also exist in the cornea.

3.1.2 Sensory fibers of the sclera, iris, and ciliary body

Although the functional properties of sensory nerve fibers innervating the conjunctiva and the sclera have not been studied in great detail, the same main functional classes of sensory afferents as in the cornea, i.e. polymodal nociceptors, mechano-nociceptors, and cold receptors, have been identified in the sclera,[47,55,56] iris and ciliary body,[10,56,57] and the bulbar conjunctiva[45,58] (Fig. 16.5). Mechanosensitive and polymodal nociceptor fibers of the sclera, uvea, and cornea are readily activated by externally applied pressure but only transiently by sudden increases of intraocular pressure up to 100 mmHg,[56] which explains the absence of pain in most forms of glaucoma. However, combination of high intraocular pressure with inflammation, as occurs in congestive glaucoma, possibly causes nociceptor sensitization and a more sustained nociceptive inflow from ocular polymodal nociceptors to the brain, producing the intense pain characteristic for this disease (see below). On the other hand, a small number of low threshold, myelinated mechanosensory nerve fibers responding to moderate changes of intraocular pressure have been identified functionally. They may correspond to axons of the encapsulated nerve endings found in the chamber angle and scleral spur and may be associated with neural regulation of the intraocular pressure.[59,60]

Cold-sensitive nerve fibers with response properties similar to thermal receptors present in the cornea and conjunctiva are also found in the iris and in the posterior sclera. Cold receptor endings in these locations are not exposed to environmental temperature changes but are close to blood vessels and may serve to detect choroidal and retinal blood flow changes, thus contributing to reflex blood flow regulation rather than to the production of conscious thermal sensations or tearing regulation.[55]

3.1.3 Ocular trigeminal ganglion neurons

The cell bodies of TG neurons innervating the eye are heterogeneous not only in their expression of different transducing channels and their conduction velocity but also in their passive and active membrane properties, thus presenting different impulse firing characteristics. When recorded intracellularly, corneal mechanosensory neurons associated to thin myelinated fibers exhibit fast and short-lasting action potentials, possibly due to the expression of tetrodotoxin-sensitive Na^+ channels, while Aδ and C polymodal nociceptive neurons are rich in tetrodotoxin-resistant Na^+ channels and present slower and longer action potentials. Finally, a small population of neurons with a high excitability and very high input resistance has been identified as cold-sensitive neurons.[61]

3.2 Central pathways

Most ocular neurons found in the trigeminal subnucleus caudalis are mechanoreceptive neurons responding only to

noxious ocular stimulation or to stimulation of the cornea and periocular skin.[62] A modality-specific distribution of corneal neurons within the trigeminal complex has been proposed. Neurons of the interpolaris and caudalis subnuclei responded to all modalities of stimuli while those within the superficial laminae of the subnucleus caudalis and the cervical cord transition responded only to heat and chemical irritation, suggesting that the input to this region is restricted to TG polymodal nociceptor neurons.[24,63]

Second-order ocular neurons project to the thalamus and brainstem regions as the superior salivatory or facial motor nucleus – depending on their peripheral input.[25-27] They participate in several trigeminal-evoked blink reflexes, like the corneal blink reflex, in which corneal afferent information projects to orbicularis oculi motor neurons responsible for lid closing.[64] Likewise, a population of neurons specifically inhibited by wetting and excited by drying of the ocular surface may be involved in reflex tearing and fluid homeostasis of the ocular surface.[27] We can speculate that these neurons receive at least part of their input from peripheral cold fibers innervating eye surface tissues.

Ocular neurons with large receptive fields have been identified in SII areas, where a systematic representation of the face including the eye, and the rest of the body has also been described.[65,66] Cortical processing of ocular sensory information is sustained by reciprocal interactions between cortical and thalamic areas, and also by a direct modulation of the pre-thalamic relays. So, processing of nociceptive input at the ocular neurons of the TBNC is modified by activation of descending pathways from the pontine parabrachial area and the nucleus raphe magnus mediated by GABA(A) receptors.[24,67]

4. Inflammation and injury effects on ocular sensory neurons

4.1 Local inflammation

A key role of ocular sensory nerves is protection of the eye against injuring or potentially injuring forces acting on the eye. Accordingly, impulse firing in ocular nociceptors evoked by the direct action of such forces leads to immediate pain and aversive motor responses. As well as this immediate effect, tissue damage also produces a slower increase in the excitability of corneal, scleral, and uveal nociceptor endings. This phenomenon is called "sensitization" and is a general property of polymodal nociceptors in all superficial and deep tissues of the body, including the eye.[68,69] It is manifested by a long-lasting, irregular, low-frequency firing that reappears seconds after the impulse discharge directly caused by the stimulus has subsided, and persist for minutes in the absence of new stimulation.[45] This ongoing discharge is accompanied by an enhanced responsiveness and a lowered threshold to new non-noxious and noxious stimuli (Fig. 16.6B).[10,45,47]

Sensitization of ocular nociceptor fibers explains the presence of spontaneous pain after injury or inflammation of the eye (due to the development of ongoing activity) and the "tenderness" of damaged tissues to non-noxious (allodynia) and noxious stimuli (hyperalgesia). Sensitization of ocular nociceptors is the consequence of the action on nerve endings of a large variety of inflammatory mediators (including H^+ and K^+ ions, adenosine and ATP, serotonin, histamine, platelet-activating factor, bradykinin, prostaglandins, leukotrienes, thromboxanes, interleukins, tumornecrosis factor, NGF, among others) that are released locally by damaged cells and by resident and invasive cells of the immune system attracted to the injured area.[70] These mediators act on a variety of membrane receptor proteins at the ocular nociceptor nerve endings. Several ion channels present in the membrane of polymodal nociceptor endings as TRPV1 and TRPA1 are activated by many of these inflammatory mediators, thus enhancing ion flow, which causes the increased membrane excitability, augmented impulse firing frequency and reduced threshold that characterize sensitization of injured ocular nerve terminals (Fig. 16.6C).

These effects on membrane ion channel activity are exerted either directly or through different signaling pathways. Membrane receptors for prostaglandin E2, ATP, adenosine, and 5-HT ultimately activate the protein kinase A (PKA) signaling pathway, whereas receptors for other mediators such as bradykinin act on protein kinase C (PKC) to induce sensitization. Inflammatory mediators are also responsible for the local inflammatory reaction (vasodilatation, plasma extravasation, and cell migration). Drugs known to interfere with the formation and/or release of inflammatory agents, such as the non-steroidal antiinflammatory drugs (NSAIDs) that inhibit the production of prostaglandins, not only decrease local inflammation but also reduce sensitization of ocular nociceptors.[71,72] This explains the analgesic effect exerted by these drugs on post-surgical ocular pain.[73,74]

Sensitization of corneal nociceptors develops within minutes of acute corneal injury and normally persists as long as the inflammation remains. During long-lasting inflammatory processes, as in chronic dry eye, more permanent changes take place in the cell bodies and peripheral terminals of nociceptor neurons, including a modified expression of membrane receptors and ion channels present in both structures, possibly induced by growth factors, in particular NGF (Box 16.2).[75,76]

In addition to signaling damage, ocular nerves also contribute to the local vascular and cellular response to injury. Nerve impulses evoked by noxious stimuli in nociceptors not only travel centripetally through the axon to the central nervous system but also antidromically invade the non-stimulated peripheral branches of the axon[51,77] (see Fig. 16.4D). This evokes the release of several neuropeptides contained in its peripheral endings.[44,78] Neuropeptides CGRP and SP are found in most trigeminal neurons and their terminals. When released by depolarized nociceptor endings, CGRP and SP contribute to the local inflammatory response inducing vasodilatation, plasma extravasation, and cytokine release, thus amplifying the inflammatory effect of the endogenous mediators. This "neurogenic inflammation" caused by sensory nerves affects both the injured and the non-injured areas innervated by branches of the affected nerves, which may explain the extension of inflammation to distant, intact tissues (conjunctiva, iris, ciliary body, etc.) following a limited corneal lesion.[44]

Second-order neurons responsive to stimulation of the ocular surface at the trigeminal subnucleus interpolaris/caudalis transition and subnucleus caudalis/cervical cord junction also exhibit increased responses to chemical stimulation of the ocular surface days after development of an experimental uveitis, reflecting an alteration of the properties of second-order neurons following sustained and recurrent peripheral inflammation.[79] This "central sensitization" contributes to the hyperalgesia shown by inflamed tissues.

4.2 Nerve injury

Several circumstances – such as ocular surgery, accidental trauma and certain ocular or systemic diseases – cause damage of ocular sensory nerves at different points in their trajectory. As previously noted, lesion of the peripheral axons of ocular neurons substantially changes the morphology and functional properties of nerve fibers projecting to the eye, in particular to the cornea, where denervated areas are invaded by outgrowths of adjacent, non-injured nerve fibers and by sprouts of the injured axons, which are sometimes entrapped by scar tissue and form microneuromas.[80] Morphological alterations in peripheral axons are accompanied by functional disturbances in the cell body of primary sensory neurons. Expression of transcription factors and neuropeptides and of genes that encode ion channel and receptor proteins is altered to a variable degree depending on the location and extension of nerve damage.[81] In particular, Na and K channels are overexpressed.[82,83] As a consequence, during axonal regeneration nerve excitability of the neuroma endings formed by damaged neurons is disturbed, leading to ectopic discharges[84] and abnormal responsiveness to natural stimuli.[85] This impaired activity is the origin of peripheral neuropathic pain.[82,86] Such functional alterations have been reported in axotomized corneal neurons at different times after lesion. Polymodal nociceptors[87,88] and corneal cold-sensitive nerve terminals exhibit an increase in spontaneous and stimulus-evoked activity weeks after injury, presumably due to an enhanced expression of tetrodotoxin-resistant Na channels.[89]

5. Trophic effects of ocular sensory nerves

As well as their role in signaling external or internal stimuli, ocular sensory nerves play an important role in the trophic maintenance of their innervated tissues. This effect is particularly prominent in the cornea.

Impairment of corneal sensory innervation causes a number of functional disturbances and development of epithelial defects and recurrent ulcers (neurotrophic keratitis).

Box 16.2 Ocular nociceptive nerve fibers

Sensitization of nociceptive nerve fibers innervating the eye is the basis of persistent pain developed after ocular injury or inflammation. It also accounts for the development of allodynia (pain sensations evoked by non-noxious stimuli) and primary hyperalgesia (enhanced pain evoked by noxious stimulation) in the inflamed eye. It explains for instance the development of sustained pain after the iris has been touched accidentally during ocular surgery or after argon laser pulse application to the posterior uvea, and is possibly also the reason for the intense sustained ocular pain experienced in congestive glaucoma, in which uveal inflammation developed during congestion probably induces sensitization of nociceptors, increasing their excitability.

Table 16.1 Causes of reduced or enhanced corneal sensitivity

Reduced sensitivity	
Genetic diseases	• Familial corneal hypoesthesia • Congenital corneal anesthesia (CCA) • CCA associated with trigeminal hypoesthesia and neurological disorders (Moebius syndrome; hereditary sensory and autonomic neuropathies type III – Riley-Day syndrome, IV, or V – congenital insensitivity to pain, cerebellar ataxia, etc.) • CCA associated with somatic disorders (Goldenhar–Gorlin syndrome or oculo-auriculo-vertebral dysplasia, ectodermal dysplasia, etc.) • CCA associated with multiple endocrine neoplasia 2b (prominent corneal nerves) • CCA associated with other genetic ocular alterations (corneal dystrophies, epithelial basement dystrophy, microsomia, etc.)
Other systemic diseases	• Diabetes • CNS disturbances (infarctions, tumors, and other space-occupying lesions) • Leprosy • Neurosarcoidosis • Orbital tumors and inflammations (occasionally)
Eye surgery	• Long cataract incisions • Refractive surgery • Keratoplasty • Retinal surgery (especially circular buckling operations) • Extensive cyclophotocoagulation • Orbital surgery damaging sensory pathways to the eye • Lid speculum damage (occasionally)
Infections	• Ocular zoster • Other herpetic infections • Large bacterial ulcers • Acanthamoeba keratitis (initially hypersensitive)
Drugs	• Topical anesthetics (abuse) • Topical beta-blockers • Topical non-steroidal anti-inflammatory drugs (NSAIDs) • Cocaine abuse
Other conditions	• Chemical or thermal corneal burns • Ocular surface disease/dry eye in some cases • Contact lens wear • Neurosurgical procedures (acoustic neuroma, trigeminal neuralgia, etc.) • Aging
Enhanced sensitivity or irritation sensations	
Diseases of the ocular surface	• Dry eye with conjunctival or corneal epithelial lesions (sensitivity to air conditioning, contact lenses, cigarette smoke, etc.) or unstable tear fluid (lid function, pinguecula, stitches) • Recurrent erosion or map-dot-fingerprint corneal dystrophy, exposure to toxic chemicals
Acute or chronic inflammation of the ocular surface	• Inflammatory dry eye in Sjögren's syndrome, rheumatoid arthritis, graft-versus-host disease • Conjunctivalization of the cornea, pterygium • Post-refractive surgery pain (occasionally) • Acute Acanthamoeba infection • Corneal ulcers and inflammations (acute stage)

Adapted from Belmonte & Tervo 2006.[122]

Corneas with reduced sensitivity resulting from ocular or systemic conditions affecting its nerves (Table 16.1) show altered healing capabilities,[4,90–94] increased epithelial permeability,[95] punctate staining following instillation of fluorescein or Rose–Bengal, poor healing of corneal ulcers and reduced transparency. Epithelial micro-defects present in patients with decreased corneal sensitivity expressive of an altered innervation may be seen in the cornea or the conjunctiva, usually in the areas not covered by the lids and therefore continuously exposed to environmental changes.

As mentioned above, corneal sensory nerves release neuropeptides, in particular SP and CGRP, that either alone or in combination with other growth factors present in the cornea (e.g. insulin-like growth factor, epidermal growth factor, NGF, etc.) appear to be important agents in maintaining corneal epithelial integrity and promoting wound healing in the normal cornea.[91,96–100] On the other hand, ocular tissues produce several growth factors necessary for the development, remodeling, and regeneration of corneal nerves. Tissue-derived growth factors, such as glial cell line-derived neurotrophic factors (GDNF), brain-derived neurotrophic factor (BDNF), opioid growth factor (OGF), ciliary neurotrophic factor (CNTF), pigment epithelium-derived factor (PEDF), and neurotrophins such as nerve growth

Box 16.3 Symbiosis between ocular tissues and sensory neurons

Ocular tissues and their sensory nerves maintain a symbiotic, dynamic interdependence. Sensory neurons innervating the eye exert a well-known trophic effect on target ocular structures. Long-standing clinical observations show that accidental or surgical lesion of the trigeminal nerve leads to the appearance of severe lesions in the cornea (keratitis neuroparalytica). Conversely, ocular tissues strongly contribute to the development, remodeling, and survival of their sensory and autonomic innervations.

factor (NGF), neurotrophin-3 and -4 (NT-3, NT-4), have been found in the cornea and possibly participate in the maintenance of corneal innervation (Box 16.3).[4,91,101-103]

6. Sensations arising from the eye

6.1 Techniques for testing ocular surface sensitivity

Traditionally, clinical exploration of corneal or conjunctival sensitivity to mechanical stimulation has been performed by gentle touching of the ocular surface with a wisp of cotton, asking the subject about the sensation experienced or observing the evoked blink reflex. A more quantitative approach was obtained using modified von Frey filaments which are employed to explore cutaneous mechanosensitivity. These are nylon threads of fixed diameter and variable length, whose tip is pressed against the skin until they start to bend. At this moment, the force per surface unit exerted by the filament is constant and proportional to its length. The Cochet–Bonnet esthesiometer, composed of a filament of adjustable length,[104] is the best-known instrument based on this principle and has been widely used to explore mechanical sensitivity of the cornea in normal and pathological conditions.

To prevent the risk of tissue injury associated with contact esthesiometry, particularly in hypoesthesic corneas, non-contact esthesiometers that use an air jet to apply controlled mechanical forces on the cornea have been developed.[105,106] The Belmonte esthesiometer[107] is a non-contact gas esthesiometer that allows evaluation of the mechanical, and also thermal and chemical, sensitivity of the eye surface. It uses a gas jet of adjustable flow to apply a controlled pressure (over or below mechanical sensitivity threshold) on any point of the ocular surface. The temperature of the air jet can be adjusted, thus allowing cold or hot thermal stimulation. Finally, when air in the gas jet is mixed with CO_2 at variable concentrations, CO_2 combines with water at the ocular surface and forms carbonic acid, thus decreasing local pH proportionally to the CO_2 concentration and causing a chemical stimulation.[40,107] Gas esthesiometry has been used in recent years to define normal values of corneal and conjunctival mechanical, thermal, and chemical sensitivity in relation to age, sex, pregnancy, iris color, and use of contact lenses, and also to assess sensitivity changes associated with surgery, ocular surface pathologies such as herpes virus infections, dry eye, keratitis, iritis, uveitis and glaucoma, or systemic diseases like fibromyalgia or diabetes.[40,41,108-110]

6.2. Psychophysics of corneal and conjunctival sensations

Classical psychophysical studies applying crude mechanical and thermal stimulation to the cornea concluded that pain is the only sensory modality that can be evoked from this tissue, although some authors proposed that specific sensations of touch, heat, cold, and pain were evoked when more selective corneal stimulation was used.[111-115] During the past decade, application of controlled stimuli of different types to the cornea with the Belmonte esthesiometer evidenced that mechanical, acidic, and thermal stimulation each evoked sensations of different quality, allowing the subject to identify the mode of stimulus applied.[48] Moreover, electrophysiological recordings of impulse activity in nerve fibers of the cornea and conjunctiva of the cat stimulated with the Belmonte esthesiometer showed that different types of stimuli evoked variable degrees of response from the different functional types of sensory receptors. Therefore, the distinct quality of the conscious sensations evoked by stimulus of the cornea or conjunctiva appears to be primarily determined by variable recruitment of the mechano-nociceptors, polymodal nociceptors, and cold receptors innervating these structures.[48] Sensations produced in healthy human subjects by corneal stimulation with mechanical forces, heat, or acid in all cases include a component of irritation, possibly because polymodal nociceptors were always evoked by these stimuli.[48,116] In contrast, application of moderate cold stimuli evoked almost exclusively a non-noxious cooling sensation that became irritating only when more pronounced temperature reductions were applied (Fig. 16.7).[48,114,116] In the conjunctiva, sensitivity is comparatively lower than in the cornea; also, low-intensity mechanical stimuli are felt as non-irritating.[41,116-118]

6.3 Sensitivity of the injured cornea

In a number of pathological conditions affecting the ocular surface (Table 16.1) or after surgical procedures for retinal detachment, cataract removal and glaucoma, and particularly following corneal surgery (radial keratotomy, PRK, LASIK, keratoplasty, etc.), sensory nerves directed to the cornea become damaged at a level determined by the location and extent of wounding. Injured corneal nerves start to regenerate soon after injury, possibly stimulated by an enhanced local production of NGF, but successful recovery is only partially attained and corneal nerves remain disorganized and significantly reduced in number for months and even years after surgery (Fig. 16.8).[90] In parallel with morphological disturbances, severed nerves in healing corneas experience functional alterations evidenced by an altered threshold and an abnormal responsiveness to mechanical and chemical stimuli applied to the wounded area (Box 16.4).[89]

As a result of a nerve lesion during ocular surgery, corneal sensitivity to mechanical stimulation is altered, with a remarkable increase in threshold of the denervated areas that takes months to recover and may never return to normal values.[119-122] Transplanted corneas or implanted lenticules in epikeratophakia remain totally anesthetic for years, or recover at best a very limited mechanical sensitivity usually restricted to the periphery of the implant. After LASIK surgery, corneal sensitivity to mechanical and chemical stimulation measured

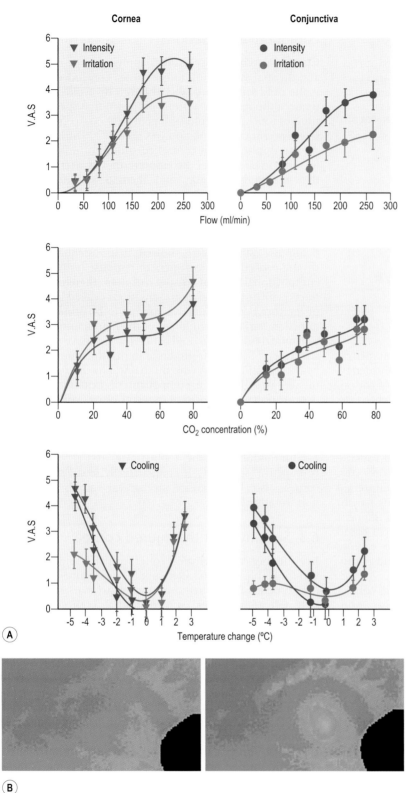

Figure 16.7 Sensations evoked by selective stimulation of the cornea and conjunctiva. (**A**) Subjective score given to the intensity and the irritation components of the sensation evoked by selective mechanical, chemical, and thermal stimuli on the cornea (triangles) and the conjunctiva (circles) with the Belmonte esthesiometer. Sensation parameters are expressed in arbitrary units marked on a visual analog scale (VAS) ranging from 0 (no sensation) to 10 (maximal value of the parameter). For thermal stimulation, VAS scores for the cooling component of the sensation have also been represented. (**B**) Thermographic images recorded during application of a gas pulse at neutral temperature for selective mechanical stimulation (left) and at 29°C (right) for cold stimulation of the cornea. (A: From Acosta et al 2001b;[116] Reproduced from Association for Research in Vision and Ophthalmology. B: From Acosta, MC 1999, Ph.D. Dissertation, Universidad Miguel Hernández de Elche, Spain.)

with the Belmonte esthesiometer is markedly impaired and normal sensitivity is recovered only 2 years later.[121] In addition to the reduced sensitivity (hypoesthesia), spontaneous pain sensations and abnormal sensations (dysesthesias), in particular dryness, appear in a significant number of patients. It has been proposed that these are the result of the activation of non-injured corneal nerve endings by the modest reduction of tear production, stability and clearance rate reported after photorefractive surgery.[123,124] Alternatively, abnormal firing originating from injured corneal nerve fibers may be read erroneously as ocular dryness by the brain, similar to the ectopic activity in the sensory nerve stump of amputated body extremities that is interpreted as aberrant sensations from the lost limb ("phantom limb") (Fig. 16.9).[125]

Unoperated control   PRK patient

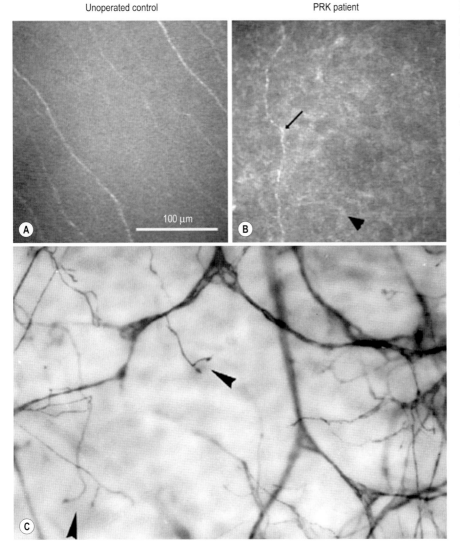

100 µm

Figure 16.8 Disturbances of corneal innervation following refractive surgery. (**A**), (**B**) In-vivo confocal microscopy images of human corneal nerves. (A) Sub-basal leashes in an intact cornea (B). Appearance of a sub-basal nerve fiber (arrow) 3 years after PRK, to illustrate incomplete nerve recovery and the persistence of haze signs (arrowhead). (**C**) Regenerating stromal nerve fibers in the rabbit cornea after PRK. Acethycholinesterase-stained corneal nerves 1 month after surgery, showing abundant regenerating leashes, sometimes with "growth cone-like" endings (arrowheads). (A: Adapted from Müller et al 2003;[4] B: From Tervo & Moilanen, 2003.[90])

Box 16.4 Alternations of ocular sensitivity

Until recently, most alterations of ocular sensitivity were associated with ocular inflammatory processes (that cause hypersensitivity, sustained pain, itching, etc.) or were the consequence of surgical procedures directed to preserve or recover vision (such as keratoplasty, cataract removal, vitreo-retinal surgery, etc., usually producing ocular hypoesthesia). Nowadays, reduced sensitivity and in particular dysesthesias (abnormal sensations) are increasingly associated with lifestyle (pollution, air conditioning, etc.) and the growing number of ocular interventions, in particular corneal refractive surgery.

---

7. Ocular pain

Several ocular and neurological conditions produce pain in or around the eye.[126] Based upon the source, ocular pain can be classified as: superficial pain, deep ocular or orbital pain, and pain sensed in the eye but originated at processes in other tissues (referred pain). Ocular pain is due to conditions of diverse etiology and is often an extension of head and face pain, although complete ocular examination (Table 16.2)

Table 16.2 Complete ocular examination in ocular pain

- Visual acuity (best-corrected)
- Slit lamp examination (conjunctiva, episclera, sclera and cornea: dry eye, punctuate keratitis, abrasions, etc.; anterior chamber and iris)
- Assessment of dry eye (Schirmer test, lissamine green, osmolality, tear film evaluation)
- Pupil analysis (size, reactivity, reflexes)
- External exam of the lids (erythema, ptosis, lag, retraction, etc.), periorbital skin (rashes, sores, etc.) and lacrimal glands, including palpation (focal tenderness, presence of masses, etc.), checking of preauricular lymphatic ganglia, and transillumination of maxillar and frontal sinuses
- Ocular asymmetries
- Relative position of the lids, cornea, and pupil
- Exophthalmometry (proptosis, enophthalmos)
- Ocular motility examination
- Functional examination of trigeminal system, including face (response to periorbital cutaneous pinprick) and corneal sensitivity (esthesiometry)
- Intraocular pressure measurement
- Ophthalmoscopy examination
- In-vivo confocal microscopy

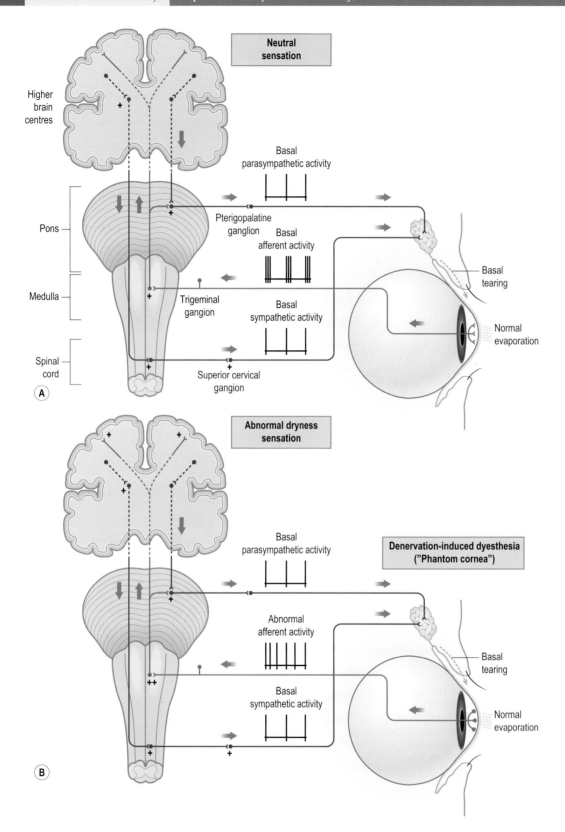

Figure 16.9 Corneal sensation and tear secretion before and after photorefractive surgery. Schematic representation of the mechanisms involved in ocular sensation and lacrimal secretion. (**A**) Basal tearing is maintained by low-frequency impulse activity of corneal sensory nerves. This activates efferent parasympathetic, and perhaps also sympathetic, pathways in the central nervous system stimulating basal tear secretion. However, basal sensory activity is too low to produce conscious sensations of ocular dryness. (**B**) When corneal nerves are surgically injured, severed axons generate ectopic impulse discharges that reach the brain and evoke abnormal dryness sensations despite the normal tear production. (Adapted from Belmonte 2007.[125])

should always be performed to define or exclude the real ocular origin of the eye pain experienced by the patient.[122]

7.1 Superficial ocular pain

Superficial pain, i.e. pain originated at the ocular surface, is a complex sensation usually caused by infectious or inflammatory conditions such as corneal ulcers and conjunctivitis, and also to systemic diseases affecting connective tissue, such as rheumatoid arthritis, systemic lupus erythematosus (LED), Sjögren's syndrome, or sarcoidosis, and some heavy metal intoxications.[127,128] Superficial ocular pain is often felt as foreign body sensation, and presents a prominent itching component in allergic inflammation and chronic blepharitis. This pain results presumably from the direct effect of local inflammation on peripheral endings in episclera or sclera (episcleritis or scleritis, respectively) causing an ipsilateral intense pain that can be triggered by pressing the area involved. Very intense pain is experienced during bacterial infections producing corneal ulcers (Acanthamoeba keratitis), in which corneal nerves are thickened probably due to inflammation, as well as 12–24 hours after photorefractive keratectomy (PRK).[90] As corneal nerve endings become easily stimulated at the areas not covered by an intact epithelium, the intense pain present in these conditions diminishes or stops when the epithelial defect is healed or covered with a therapeutic contact lens. Nevertheless, during some corneal stromal inflammations, such as stromal herpes virus keratitis, the eye may remain painful even in the presence of an intact epithelium, probably due to the abnormal activity in corneal sensory neurons affected by the virus infection (Box 16.5).

The section of a part of the ocular sensory nerves during orbital or ocular surgery causes hypoesthesia but spontaneous acute and chronic pain sensations may also appear. This problem has become more prominent in recent years with the extended use of surgical procedures on the cornea for correction of refractive errors (radial keratotomy, PRK, LASIK). In PRK, pain usually begins 15–60 min after surgery (when the local anesthetic drops have lost their effect) and becomes maximal within 3–6 hours. It is relieved after the first night but may persist as a moderate or low discomfort sensation for at least one week.[129] Such acute pain is presumably due to sensitization of nerves innervating the injured tissue. Pain of lower intensity may also be present for 1–3 days following LASIK.[130] It is worth noting that long-lasting symptoms after LASIK or PRK, including sharp pain, eye soreness, or dryness sensations, are found in 10–40 percent of these patients.[130,131] This is significantly more common, severe and prolonged after PRK. However, anti-inflammatory medication with NSAIDs, steroids, and cyclosporine normally reduces these symptoms significantly in clinical practice (see below). As mentioned above, these abnormal sensations may result from an aberrant spontaneous and stimulus-evoked nerve impulse firing of the damaged corneal nerves during their regeneration, evoking a perceptual disturbance known as "phantom cornea" (Fig. 16.9).[125]

7.2 Deep ocular pain

Deep ocular pain arises from structures inside the eye globe, generally as a consequence of inflammation of uveal structures, such as iritis/uveitis, which is typically associated not only with pain but also with photophobia. Pain after inflammation of the uveal tract is attributable to enhanced activity in the abundant polymodal nociceptor fibers innervating these structures. The mechanism for the additional discomfort evoked by light exposure is uncertain but reflex contraction of ciliary and/or iris muscles triggered by light or accommodation may become painful if nociceptor sensory nerve endings in these structures were also sensitized. Consequently, besides anti-inflammatory and analgesic treatment, cycloplegic drugs and dark glasses are recommended for these patients.

Although the retina lacks sensory nerves, retinitis and endophthalmitis are usually painful, probably because of the parallel involvement of uveal tract inflammation. Intraocular tumors are normally painless, unless the tumor causes necrosis.[125,132] Unlike chronic forms of glaucoma, acute congestive glaucoma, producing a sharp elevation of the intraocular pressure, is usually very painful, thus provoking a vegetative reaction that includes a decrease in blood pressure, headache, nausea, and vomiting.

There are many inflammatory processes and infections in the orbit that elicit pain. Most non-infectious (sterile) orbital inflammations imply accumulation of leukocytes within the orbital structures, with local release of inflammatory mediators and cytokines that are the likely cause of the signs of inflammation, including pain and edema.[122] Some of the symptoms of these orbital inflammatory processes are related not only to the action of inflammatory mediators on sensory nerve endings but also to the pressure exerted by the inflammatory mass within the orbit and to vascular, neural or extraocular muscle involvement. Orbital tumors are usually painless unless they become infiltrated by blood vessels or cause orbital hemorrhage. Also, lachrymal gland tumors are often painful. Inflammations of the optic nerve (optic nerve neuritis) are associated with decreased vision, orbital and ocular tenderness and pain provoked by eye movements, probably due to the involvement of adjacent orbital structures. Eye movement-evoked pain can also be observed after fractures of the orbital floor or in the presence of a foreign body (e.g. after scleral buckling operations performed for retinal detachment).

Refractive errors resulting in eye strain or over-accommodation may cause ocular and eye-referred pain of variable intensity. Anisometropia may not only produce ocular discomfort and "tired eyes" but also aggravate headache and evoke migraine symptoms. Pronounced astigmatism and uncorrected presbyopia, which may go in parallel with dry eye in middle age, often causes ocular discomfort. As mentioned above, muscles responsible for accommodation contain sensory terminals that may become sensitized during continuous, ineffective accommodation,

---

Box 16.5 Dry eye

The most common reason for superficial eye discomfort is possibly "dry eye" also termed "ocular surface disease". This is a common multietiological condition in which impaired tear production or/and altered tear film stability compromise the ocular surface, producing dot-like micro lesions in conjunctival and corneal epithelia, suggestive of neurotrophic defects.

giving rise to pain sensations. The phantom eye syndrome develops after eye enucleation in about 26 percent of patients,[133] and is characterized by chronic pain referred to the eye as well as non-painful visual sensations (photopsia). Neuropathic pain may also result from lesion of intraocular sensory nerves due to compression after retinal surgery.

7.3 Pain referred to the eye

As noted above, pain referred to the eye and periorbital region is often attributable to a systemic condition. Eye-referred pain is often due to a pathological process of the trigeminal ganglion and nerves innervating the orbit and adjacent tissues that leads to an abnormal activation of second- and higher-order ocular neurons caused by neural input from other structures. Augmented nociceptive input from the damaged structures elsewhere may induce sprouting of trigeminal sensory projections to ocular neurons, inducing central sensitization and giving rise to pain referred to the eye. Sharp, short-lasting, severe, unilateral, repeated painful sensations referred to the eye are typical of trigeminal neuralgia, related to a general, abnormal activation of trigeminal pathways. Referred pain can also be observed in a variety of conditions such as orbital surgery, cluster headache,[134] or SUNCT syndrome.[135] Migraine hemicrania continua and tension headache also lead to facial and ocular pain.[136] On the other hand, certain forms of headache may be due to eye disease.[137]

8. Drugs acting on ocular sensory nerves

Drugs affecting ocular sensory nerve activity are widely used in ophthalmology – in part for their therapeutic effects but mainly because common ophthalmological explorations as tonometry or gonioscopy, and manipulations such as foreign body extraction, paracentesis, and most surgical ophthalmological operations such as cataract and vitreo-retinal surgery are performed under local, regional anesthesia. General anesthesia is only rarely used in elective surgery.

Topical anesthetics

Nearly all local anesthetics are highly lipid-soluble amine compounds that cross cell membranes and nerve sheaths and block propagated action potentials by blocking Na channels.[138] Instillation into the eye of a drop of a topical anesthetic solution (e.g. proparacaine, bupivacaine, oxybuprocaine, amethocaine, lidocaine) anesthetizes the cornea in about one minute due to blockade of nerve impulse propagation in corneal and conjunctival sensory nerve fibers. Thus, pain and thermal sensations arising from these structures are immediately suppressed, although some sensitivity may remain in subconjunctival structures and lids. Topical anesthetics are also used for ophthalmological explorations, including differentiation between superficial and deep ocular pain and ocular surgery (see below). Nevertheless, repeated application of local anesthetics for the management of ocular surface discomfort and pain is contraindicated because prolonged administration of topical anesthetics leads to development of corneal epithelial defects, serious corneal healing problems, and eventually corneal melting.[139]

Anti-inflammatory drugs

Both steroidal (e.g. fluorometholone, dexamethasone, prednisolone) and non-steroidal anti-inflammatory drugs (NSAIDs),[140] and cyclosporine A or tacrolimus,[141,142] are used as topical or systemic drugs to prevent or treat ocular discomfort and pain associated with inflammatory processes. NSAIDs are potent inhibitors of the two cyclo-oxygenases (COX-1, constitutive and COX-2, inducible) that form active prostaglandins and thromboxane from arachidonic acid released by damaged cell membranes. NSAIDs include different compounds with variable potency and speed of elimination, including for example diclofenac, flurbiprofen, ketorolac, nepafenac, indometacin, or acetyl salicylic acid. When applied topically to the cornea, some of the NSAIDs reduce pain not only through inhibition of the formation of prostaglandins and of their sensitizing effects on nociceptors but also through a direct action on membrane excitability of polymodal nociceptor endings.[71,72] This direct analgesic action explains the usefulness of NSAIDs in treating not only ocular and orbital inflammatory processes, but pain associated with hyphema and post-surgical pain. Topical cyclosporine A is mainly used in the USA and has been reported to diminish inflammation and discomfort related to various dry eye syndromes.[141]

Cycloplegic agents

As indicated above, prevention of movement in the dilator and sphincter muscles of the iris, and in the ciliary muscle, using cycloplegic drugs (e.g. atropine, cyclopentolate, homatropine, tropicamide) often relieves pain associated with inflammatory conditions such as uveitis, and has become a standard therapeutic procedure. Furthermore, the iris usually becomes constricted during ocular inflammation and surgery due to local release of prostaglandins, which facilitates the formation of adhesions with the lens. Cycloplegic drugs also help to prevent this complication.

Analgesics

Analgesic drugs acting on the central nervous system such as non-acidic antipyretic analgesics (paracetamol) and weak (codeine) or strong (morphine) opioids are occasionally used to treat ocular pain. Pain of neuropathic origin such as phantom eye, post-refractive surgery eye pain, or postherpetic neuralgia and pain caused by acute congestive glaucoma often requires strong, systemic pain medication not only to abolish the acute pain sensations but also to prevent pain-related systemic symptoms such as nausea, vomiting, or even cardiac arrhythmia and shock. Drugs known generically as antidepressant analgesics, such as amitriptyline, butriptyline and some anticonvulsants such as carbamazepine or gabapentin, that antagonize neural transmission in the central nervous system and block neuronal Na channels, are being used to treat neuropathic pain.[143]

Prevention of surgical pain

General anesthesia is seldom used today in eye surgery, although it is still occasionally needed in severe trauma, invasive tumor surgery, poor patient cooperation, or extensive plastic surgery.[144] When corneal pain must be prevented but blockade of the extraocular muscles (akinesia) is not needed, topical anesthetics applied either as drops or gels are sufficient when a small corneal incision is performed, e.g. in cataract removal through a corneal tunnel using phacoemulsification. Surgery directed towards deeper ocular structures may require subtenon, periocular, or retrobulbar injection of the anesthetic drugs, performed alone or in combination with topical anesthesia.[144] For long-duration surgical procedures, compounds with a long-lasting effect such as bupivacaine are combined with compounds of shorter duration but faster onset such as lidocaine. Postoperative pain is prevented or treated with NSAIDs, alone or combined with codeine or oxycodone, in the latter case following posterior segment surgery.

References

1. De Felipe C, González GG, Gallar J, Belmonte C. Quantification and immunocytochemical characteristics of trigeminal ganglion neurons projecting to the cornea: effect of corneal wounding. Eur J Pain 1999; 3:31–39.
2. Belmonte C, Garcia-Hirschfeld J, Gallar J. Neurobiology of ocular pain. Progr Ret Eye Res 1997; 16:117–156.
3. Tervo T, Tervo K, Eränkö L. Ocular neuropeptides. Med Biol 1982; 60:53–60.
4. Müller LJ, Marfurt CF, Kruse F, Tervo TMT. Corneal nerves: structure, contents and function. Exp Eye Res 2003; 76:521–542.
5. Troger J, Kieselbach G, Teuchner B et al. Peptidergic nerves in the eye, their source and potential pathophysiological relevance. Brain Res Rev 2007; 53:39–62.
6. Lütjen-Drecoll E. Choroidal innervation in primate eyes. Exp Eye Res 2006; 82(3):357–361.
7. Rozsa AJ, Beuerman RW. Density and organization of nerve endings in the corneal epithelium of the rabbit. Pain 1982; 14:105–120.
8. Kurus E. Über ein Ganglienzellsystem der menschlichen Aderhaut. Klin Mobl Augenheilk 1955;127:198–206.
9. Castro-Correia J. Inervacao de coroideia. Anales del Instituto Barraquer 1961; 2:487–518.
10. Mintenig GM, Sanchez-Vives MV, Martin C, Gual A, Belmonte C. Sensory receptors in the anterior uvea of the cat's eye. An in vitro study. Invest Ophthalmol Vis Sci 1995; 36:1615–1624.
11. Lawrenson JG, Ruskell GL. The structure of corpuscular nerve endings in the limbal conjunctiva of the human eye. J Anat 1991; 177:75–84.
12. Sasaoka A, Ishimoto I, Kuwayama Y et al. Overall distribution of substance P nerves in the rat cornea and their three-dimensional profiles. Invest Ophthalmol Vis Sci 1984; 25(3):351–356.
13. Zander E, Weddell G. Observations on the innervation of the cornea. J Anat (London) 1951; 85:68–99.
14. Müller LJ, Pels L, Vrensen GFJM. Ultrastructural organization of human corneal nerves. Invest Ophthalmol Vis Sci 1996; 37:476–488.
15. Patel DV, Sherwin T, McGhee CN. Laser scanning in vivo confocal microscopy of the normal human corneoscleral limbus. Invest Ophthalmol Vis Sci 2006; 47:2823–2827.
16. Bee JA, Hay RA, Lamb EM, Devore JJ, Conrad GW. Positional specificity of corneal nerves during development. Invest Ophthalmol Vis Sci 1986; 27:38–43.
17. Riley NC, Lwigale PY, Conrad GW. Specificity of corneal nerve positions during embryogenesis. Mol Vis 2001; 7:297–304.
18. Nagasaki T, Zhao J. Centripetal movement of corneal epithelial cells in the normal adult mouse. Invest Ophthalmol Vis Sci 2003; 44:558–566.
19. Marfurt CF. Nervous control of the cornea. In: Burnstock G, Sillito AM, eds. Nervous control of the eye. Amsterdam: Harwood Academic, 2000:41–92.
20. Marfurt CF. Corneal nerves anatomy. In: Dartt D, ed. Encyclopedia of the eye. Oxford: Elsevier, 2009.
21. Tervo T, Joo F, Huikuri KT, Toth I, Palkama A. Fine structure of sensory nerves in the rat cornea: an experimental nerve degeneration study. Pain 1979; 6:57–70.
22. Marfurt CF. The somatotopic organization of the cat trigeminal ganglion as determined by the horseradish peroxidase technique. Anat Rec 1981; 201(1):105–118.
23. Martinez S, Belmonte C. C-Fos expression in trigeminal nucleus neurons after chemical irritation of the cornea: reduction by selective blockade of nociceptor chemosensitivity. Exp Brain Res 1996; 109:56–62.
24. Meng ID, Hu JW, Bereiter DA. Parabrachial area and nucleus raphe magnus inhibition of corneal units in rostral and caudal portions of trigeminal subnucleus caudalis in the rat. Pain 2000; 87:241–251.
25. Pellegrini JJ, Horn AKE, Evinger C. The trigeminally evoked blink reflex. I. Neuronal circuits. Exp Brain Res 1995; 107:166–180.
26. Toth IE, Boldogkoi Z, Medveczky I, Palkovits M. Lacrimal preganglionic neurons form a subdivision of the superior salivatory nucleus of rat: transneuronal labelling by pseudorabies virus. J Auton Nerv Syst 1999; 77:45–54.
27. Hirata H, Okamoto K, Tashiro A, Bereiter DA. A novel class of neurons at the trigeminal subnucleus interpolaris/caudalis transition region monitors ocular surface fluid status and modulates tear production. J Neurosci 2004; 24:4224–4232.
28. Bee JA. The development and pattern of innervation of the avian cornea. Dev Biol 1982; 92:5–15.
29. Lwigale PY, Bronner-Fraser M. Lens-derived Semaphorin3A regulates sensory innervation of the cornea. Dev Biol 2007; 306(2):750–759.
30. Kirby M, Diab I, Mattio T. Development of adrenergic innervation of the iris and fluorescent ganglion cells in the choroid of the chick eye. Anat Rec 1978; 191:311–320.
31. Chan KY, Järveläinen M, Chang JH, Edenfield MJ. A cryodamage model for studying corneal nerve regeneration. Invest Ophthalmol Vis Sci 1990; 31:2008–2021.
32. Leiper LJ, Ou J, Walczysko P et al. Control of patterns of corneal innervation by Pax6. Invest Ophthalmol Vis Sci 2009; 50(3):1122–1128.
33. Patel D, McGhee CNJ. In vivo laser scanning confocal microscopy confirms that the human corneal sub-basal nerve plexus is a highly dynamic structure. Invest Ophthalmol Vis Sci 2008; 49:3409–3412.
34. Harris LW, Purves D. Rapid remodelling of sensory endings in the cornea of living mice. J Neurosci 1989; 9:2210–2214.
35. Tanelian DL, Barry MA, Johnston SA, Le T, Smith GM. Semaphorin III can repulse and inhibit adult sensory afferents in vivo. Nat Med 1997; 3:1398–1401.
36. Erie JC, McLaren JW, Hodge DO, Bourne WM. The effect of age on the corneal subbasal nerve plexus. Cornea 2005; 24:705–709.
37. Dvorscak L, Marfurt CF. Age-related changes in rat corneal epithelial nerve density. Invest Ophthalmol Vis Sci 2008; 49:910–916.
38. Boberg-Ans J. On the corneal sensibility. Acta Ophthalmol (Copenh) 1956; 34:149–162.
39. Millodot M. The influence of age on the sensitivity of the cornea. Invest Ophthalmol Vis Sci 1977; 16:240–242.
40. Bourcier T, Acosta MC, Borderie V et al. Decreased corneal sensitivity in patients with dry eye. Invest Ophthalmol Vis Sci 2005; 46:2341–2345.
41. Acosta MC, Alfaro ML, Borrás F, Belmonte C, Gallar J. Influence of age, gender and iris color on mechanical and chemical sensitivity of the cornea and conjunctiva. Exp Eye Res 2006; 83:932–938.
42. Rózsa AJ, Guss RB, Beuerman RW. Neural remodelling following experimental surgery of the rabbit cornea. Invest Ophthalmol Vis Sci 1983; 24:1033–1051.
43. Yu CQ, Rosenblatt MI. Transgenic corneal neurofluorescence in mice: a new model for in vivo investigation of nerve structure and regeneration. Invest Ophthalmol Vis Sci 2007; 48:1535–1542.
44. Belmonte C, Acosta MC, Gallar J. Neural basis of sensation in intact and injured corneas. Exp Eye Res 2004; 78:513–525.
45. Belmonte C, Giraldez F. Responses of cat corneal sensory receptors to mechanical and thermal stimulation. J Physiol 1981; 321:355–368.
46. Belmonte C, Gallar J, Pozo MA, Rebollo I. Excitation by irritant chemical substances of sensory afferent units in the cat's cornea. J Physiol 1991; 437:709–725.
47. Gallar J, Pozo MA, Tuckett RP, Belmonte C. Response of sensory units with unmyelinated fibres to mechanical, thermal and chemical stimulation of the cat's cornea. J Physiol 1993; 468:609–622.
48. Acosta MC, Belmonte C, Gallar J. Sensory experiences in humans and single-unit activity in cats evoked by polymodal stimulation of the cornea. J Physiol 2001; 534:511–525.
49. Chen X, Gallar J, Pozo MA, Baeza M, Belmonte C. CO_2 stimulation of the cornea: A comparison between human sensation and nerve activity in polymodal nociceptive afferents of the cot. Eur J Neurosci 1995; 7:1154–1163.
50. Tanelian DL, Beuerman RW. Responses of rabbit corneal nociceptors to mechanical and thermal stimulation. Exp Neurol 1984; 84:165–178.
51. Brock JA, McLachlan EM, Belmonte C. Tetrodotoxin-resistant impulses in single nociceptor nerve terminals in guinea-pig cornea. J Physiol 1998; 512:211–217.
52. Parra A, Madrid R, Echevarria D, del Olmo S, Morenilla-Palao C, Acosta MC, Gallar J, Dhaka A, Viana F, Belmonte C. Ocular surface wetness is regulated by TRPM8-dependent cold thermoreceptors of the cornea. Nat Med (in press).
53. Schaible HG, Schmidt RF. Effects of an experimental arthritis on the sensory properties of fine articular afferent units. J Neurophysiol 1985; 54(5):1109–1122.
54. MacIver MB, Tanelian DL. Structural and functional specialization of A-δ and C fibre free nerve endings innervating rabbit corneal epithelium. J Neurosci 1993; 13:4511–4524.
55. Gallar J, Acosta MC, Belmonte C. Activation of scleral cold thermoreceptors by temperature and blood flow changes. Invest Ophthalmol Vis Sci 2003; 44:697–705.
56. Zuazo A, Ibañez J, Belmonte C. Sensory nerve responses elicited by experimental ocular hypertension. Exp Eye Res 1986; 43:759–769.
57. Tower SS. Units for sensory reception in the cornea. With notes on nerve impulses from sclera. iris and lens. J Neurophysiol 1940; 3:486–500.
58. Aracil A, Belmonte C, Gallar J. Functional types of conjunctival primary sensory neurons. Invest Ophthalmol Vis Sci 2001; 42: S662, ARVO abstract.
59. Belmonte C, Simón J, Gallego A. Effects of intraocular pressure changes on the afferent activity of the ciliary nerves. Exp Eye Res 1971; 12:342–355.
60. Tamm ER, Flügel C, Stefani FH, Lütjen-Drecoll E. nerve endings with structural characteristics of mechanoreceptors in the human scleral spur. Invest Ophthalmol Vis Sci 1994; 35:1157–1166.
61. Lopez de Armentia M, Cabanes C, Belmonte C. Electrophysiological properties of identified trigeminal ganglion neurons innervating the cornea of the mouse. Neurosci 2000; 101:1109–1115.
62. Pozo MA, Cervero F. Neurons in the rat spinal trigeminal complex driven by corneal nociceptors: receptive-field properties and effects of noxious stimulation of the cornea. J Neurophysiol 1993; 70:2370–2378.
63. Hirata H, Takeshita S, Hu JW, Bereiter DA. Cornea-responsive medullary dorsal horn neurons: modulation by local opioids and projections to thalamus and brain stem. J Neurophysiol 2000; 84:1050–1061.
64. Henriquez VM, Evinger C. The three-neuron corneal reflex circuit and modulation of second-order corneal responsive neurons. Exp Brain Res 2007; 179(4):691–702.
65. DaSilva AF, Becerra L, Makris N et al. Somatotopic activation in the human trigeminal pain pathway. J Neurosci 2002; 22:8183–8192.
66. Jain N, Qi HX, Catania KC, Kaas JH. Anatomic correlates of the face and oral cavity representations in the somatosensory cortical area 3b of monkeys. J Comp Neurol 2001; 429:455–468.
67. Hirata H, Okamoto K, Bereiter DA. GABA(A) receptor activation modulates corneal unit activity in rostral and caudal portions of trigeminal subnucleus caudalis. J Neurophysiol 2003; 90:2837–2849.
68. Bessou P, Perl ER. Response of cutaneous sensory units with unmyelinated fibers to noxious stimuli. J Neurophysiol 1969; 32(6):1025–1043.
69. Reeh PW, Bayer J, Kocher L, Handwerker HO. Sensitization of nociceptive cutaneous nerve fibers from the rat's tail by noxious mechanical stimulation. Exp Brain Res 1987; 65(3):505–512.
70. Handwerker HO. Electrophysiological mechanisms in inflammatory pain. Agents Actions Suppl 1991; 32:91–99.
71. Chen X, Gallar J, Belmonte C. Reduction by anti-inflammatory drugs of the response of corneal sensory nerve fibers to chemical irritation. Invest Ophthalmol Vis Sci 1997; 38:1944–1953.

72. Acosta MC, Luna CL, Graff G et al. Comparative effects of the nonsteroidal anti-inflammatory drug nepafenac on corneal sensory nerve fibers responding to chemical irritation. Invest Ophthalmol Vis Sci 2007; 48:182–188.

73. Stein R, Stein HA, Cheskes A, Symons S. Photorefractive keratectomy and postoperative pain. Am J Ophthalmol 1994; 117:403–405.

74. Acosta MC, Berenguer-Ruiz L, Garcia-Galvez A, Perea-Tortosa D, Gallar J, Belmonte C. Changes in mechanical, chemical, and thermal sensitivity of the cornea after topical application of nonsteroidal anti-inflammatory drugs. Invest Ophthalmol Vis Sci 2005; 46:282–286.

75. Senba E, Kashiba H. Sensory afferent processing in multi-responsive DRG neurons. Prog Brain Res 1996; 113:387–410.

76. Diogenes A, Akopian AN, Hargreaves KM. NGF up-regulates TRPA1: implications for orofacial pain. J Dent Res 2007; 86:550–555.

77. Weidner C, Schmidt R, Schmelz M, Torebjork HE, Handwerker HO. Action potential conduction in the terminal arborisation of nociceptive C-fibre afferents. J Physiol 2003; 547(Pt 3):931–940.

78. Belmonte C, Gallar J. The primary nociceptive neuron: A nerve cell with many functions. In: Rowe MJ, ed. Somatosensory processing: from single neuron to brain imaging. Sydney: Gordon & Breach Science Publisher, 2000:27–49.

79. Bereiter DA, Okamoto K, Tashiro A, Hirata H. Endotoxin-induced uveitis causes long-term changes in trigeminal subnucleus caudalis neurons. J Neurophysiol 2005; 94(6):3815–3825.

80. Beuerman RW, Rozsa AJ. Collateral sprouts are replaced by regenerating neurites in the wounded corneal epithelium. Neursoci Lett 1984; 44:99–104.

81. De Felipe C, Belmonte C. c-Jun expression after axotomy of corneal trigeminal ganglion neurons is dependent on the site of injury. Eur J Neurosci 1999; 11:899–906.

82. Black JA, Waxman SG. Molecular identities of two tetrodotoxin-resistant sodium channels in corneal axons. Exp Eye Res 2002; 75:193–199.

83. Waxman SG, Dib-Hajj S, Cummins TR, Black JA. Sodium channels and pain. Proc Natl Acad Sci USA 1999; 96:7635–7639.

84. Matzner O, Devor M. Hyperexcitability at sites of nerve injury depends on voltage-sensitive Na' channels. J Neurophysiol 1994; 72:349–359.

85. Rivera L, Gallar J, Pozo MA, Belmonte C. Responses of nerve fibres of the rat saphenous nerve neuroma to mechanical and chemical stimulation: an in vitro study. J Physiol 2000; 527:305–313.

86. Belmonte C, Aracil A, Acosta MC, Luna C, Gallar J. Nerves and sensations from the eye surface. Ocul Surf 2004; 2:248–253.

87. Charco P, Chen X, Gallar J, Belmonte C. Nociceptive activity in the wounded cornea of the cat. Proc. 8th World Congress on Pain, International Society for the Study of Pain, 1996.

88. Gallar J, Acosta MC, Gutierrez AR, Belmonte C. Impulse activity in corneal sensory nerve fibers after photorefractive keratectomy. Invest Ophthalmol Vis Sci 2007; 48:4033–4037.

89. Belmonte C, Donovan-Rodriguez T, Luna C et al. Sodium channel blockers modulate abnormal activity of regenerating corneal sensory nerves. Invest Ophthalmol Vis Sci 2009; 50: Arvo abstract. No. 908.

90. Tervo T, Moilanen JAO. In vivo confocal microscopy for evaluation of wound healing following corneal refractive surgery. Proc Ret Eye Res 2003; 22:339–358.

91. Bonini S, Rama P, Olzi D, Lambiase A. Neurotrophic keratitis. Eye 2003; 17:989–995.

92. Clarke MP, Sullivan TJ, Koyabashi J, Rootman DS, Cherry P. Familial congenital corneal anaesthesia. Aust NZ J Ophthalmol 1992; 20:207–210.

93. Liesegang TJ. Varicella zoster virus eye disease. Cornea 1999;18:511–531.

94. Klintworth GK. The molecular genetics of the corneal dystrophies – current status. Front Biosci 2003; 8:687–713.

95. Beuerman RW, Schimmelpfennig B. Sensory denervation of the rabbit cornea affects epithelial properties. Exp Neurol 1980; 69:196–201.

96. Gallar J, Pozo MA, Rebollo I, Belmonte C. Effects of capsaicin on corneal wound healing. Invest Ophthalmol Vis Sci 1990; 31:1968–1974.

97. Garcia-Hirschfeld J, Lopez-Briones LG, Belmonte C. Neurotrophic influences on corneal epithelial cells. Exp Eye Res 1994; 59:597–605.

98. Tan MH, Bryars J, Moore, J. Use of nerve growth factor to treat congenital neurotrophic corneal ulceration, Cornea 2006; 25:352–355.

99. Lambiase A, Manni L, Bonini S, Rama P, Micera A, Aloe L. Nerve growth factor promotes corneal healing: structural, biochemical, and molecular analyses of rat and human corneas. Invest Ophthalmol Vis Sci 2000; 41:1063–1069.

100. Imanishi J, Kamiyama K, Iguchi I, Kita M, Sotozono C, Kinoshita S. Growth factors: importance in wound healing and maintenance of transparency of the cornea. Prog Retin Eye Res 2000; 19:113–129.

101. Kruse FE, Tseng SC. Growth factors modulate clonal growth and differentiation of cultured rabbit limbal and corneal epithelium. Invest Ophthalmol Vis Sci 1993; 34:1963–1976.

102. de Castro F, Silos-Santiago I, Lopez de Armentia M, Barbacid M, Belmonte C. Corneal innervation and sensitivity to noxious stimuli in trkA knockout mice. Eur J Neurosci 1998; 10:146–152.

103. You L, Kruse FE, Volcker HE. Neurotrophic factors in the human cornea. Invest Ophthalmol Vis Sci 2000; 41:692–702.

104. Cochet P, Bonnet R. L'esthésie cornéenne. Clin Ophthalmol 1960; 4:3–27.

105. Zaidman GW, Weinstein C, Weinstein S, Drozdenko R. A new corneal microaesthesiometer. Invest Ophthalmol Vis Sci 1988; 29(4):454, ARVO Abstract.

106. Murphy PJ, Patel S, Marshall J. A new non-contact corneal aesthesiometer (NCCA). Ophthalmic Physiol Opt 1996; 16:101–107.

107. Belmonte C, Acosta MC, Schmelz M, Gallar J. Measurement of corneal sensitivity to mechanical and chemical stimulation with a CO_2 esthesiometer. Invest Ophthalmol Vis Sci 1999; 40:513–519.

108. du Toit R, Vega JA, Fonn D, Simpson T. Diurnal variation of corneal sensitivity and thickness. Cornea 2003; 22:205–209.

109. Murphy PJ, Patel S, Kong N, Ryder RE, Marshall J. Noninvasive assessment of corneal sensitivity in young and elderly diabetic and nondiabetic subjects. Invest Ophthalmol Vis Sci 2004; 45:1737–1742.

110. Gallar J, Morales C, Freire V, Acosta MC, Belmonte C. Corneal sensitivity and tear production are decreased in fibromyalgia patients. Invest Ophthalmol Vis Sci 2009; 50.

111. Von Frey, M. Beiträge sur Sinnesphysiologie der Aut. Sächsischen Akademie der Wissenschaften zu Leipzig. Math-Phys Cl 1895; 47:166–184.

112. Lele PP, Weddell G. The relationship between neurohistology and corneal sensibility. Brain 1956; 79(1):119–154.

113. Kenshalo DR. Comparison of thermal sensitivity of the forehead, lip, conjunctiva and cornea. J Appl Physiol 1960; 15:987–991.

114. Beuerman RW, Maurice DM, Tanelian DL. Thermal stimulation of the cornea. In: Anderson D, Matthews B, eds. Pain in the trigeminal region. Amsterdam: Elsevier, 1977:422–423.

115. Beuerman RW, Tanelian DL. Corneal pain evoked by thermal stimulation. Pain 1979; 7(1):1–14.

116. Acosta MC, Tan ME, Belmonte C, Gallar J. Sensations evoked by selective mechanical, chemical, and thermal stimulation of the conjunctiva and cornea. Invest Ophthalmol Vis Sci 2001; 42:2063–2067.

117. Feng Y, Simpson TL. Nociceptive sensation and sensitivity evoked from human cornea and conjunctiva stimulated by CO_2. Invest Ophthalmol Vis Sci 2003; 44:529–532.

118. Situ P, Simpson TL, Fonn D. Eccentric variation of corneal sensitivity to pneumatic stimulation at different temperatures and with CO_2. Exp Eye Res 2007; 85:400–405.

119. Rao GN, John T, Ishida N, Aquavella JV. Recovery of corneal sensitivity in grafts following penetrating keratoplasty. Ophthalmology 1985; 92:1408–1411.

120. Campos M, Hertzog L, Garbus JJ, McDonnell PJ. Corneal sensitivity after photorefractive keratectomy. Am J Ophthalmol 1992; 114:51–54.

121. Gallar J, Acosta MC, Moilanen JA, Holopainen JM, Belmonte C, Tervo TM. Recovery of corneal sensitivity to mechanical and chemical stimulation after laser in situ keratomileusis (LASIK). J Refract Corneal Surg 2004; 20:229–235.

122. Belmonte C, Tervo T. Pain in and around the eye. In: McMahon SB, Kotzenburg M eds. Wall and Melzack's Textbook of Pain. London: Elsevier, 2006:887–901.

123. Benitez-del-Castillo JM, del Rio T, Iradier T, Hernandez JL, Castillo A, Garcia-Sanchez J. Decrease in tear secretion and corneal sensitivity after laser in situ keratomileusis. Cornea 2001; 20:30–32.

124. Toda I, Asano-Kato N, Komai-Hori Y, Tsubota K. Dry eye after laser in situ keratomileusis. Am J Ophthalmol 2001; 132:1–7.

125. Belmonte C. Eye dryness sensations after refractive surgery: impaired tear secretion or "phantom" cornea? J Refract Surg 2007; 23(6):598–602.

126. Yanowsky NN. The acute painful eye. Emerg Med Clin North Am 1988; 6:21–42.

127. Brazis PW, Lee AG, Stewart M, Capobianco D. Clinical review: the differential diagnosis of pain in the quiet eye. Neurolog 2002; 8:82–100.

128. Tsubota K. Understanding dry eye syndrome. Adv Exp Med Biol 2002; 506:3–16.

129. Epstein RL, Laurence EP. Effect of topical diclofenac solution on discomfort after radial keratotomy. J Cataract Refract Surg 1994; 20:378–380.

130. Autrata R, Rehurek J. Laser-assisted subepithelial keratectomy for myopia: two-year follow-up. J Cataract Refract Surg 2003; 29:661–668.

131. Hovanesian JA, Shah SS, Maloney RK. Symptoms of dry eye and recurrent erosion syndrome after refractive surgery. J Cataract Refract Surg 2001; 27:577–584.

132. Kalina RA, Orcutt JC. Ocular and periocular pain. In: Bonica JJ Ed. The management of pain. Vol I. Philadelphia: Lea & Febiger, 1990:759–768.

133. Soros P, Vo O, Husstedt IW, Evers S, Gerding, H. Phantom eye syndrome: its prevalence, phenomenology, and putative mechanisms. Neurology 2003; 60:1542–1543.

134. Sjaastad O, Bakketeig LS. Cluster headache prevalence. Vaga study of headache epidemiology. Cephalalgia 2003; 23:528–533.

135. Pareja JA, Caminero AB, Sjaastad O. SUNCT Syndrome: diagnosis and treatment. CNS Drugs 2002; 16:373–383.

136. Goadsby PJ, Lipton RB, Ferrari MD. Migraine – current understanding and treatment. N Engl J Med 2002; 346:257–270.

137. Tomsak RL. Ophthalmologic aspects of headache. Med Clin North Am 1991; 75:693–706.

138. Hille B. Ion channels of excitable membranes, 3rd edn. Sunderland: Sinauer, 2001.

139. Henkes HE, Waubke TN. Keratitis from abuse of corneal anesthetics. Br J Ophthal 1978; 62:62–65.

140. Colin J. The role of NSAIDs in the management of postoperative ophthalmic inflammation. Drugs 2007; 67(9):1291–1308.

141. Donnenfeld E, Pflugfelder SC. Topical ophthalmic cyclosporine: pharmacology and clinical uses. Surv Ophthalmol 2009; 54(3):321–338.

142. Jap A, Chee SP. Immunosuppressive therapy for ocular diseases. Curr Opin Ophthalmol 2008; 19(6):535–540.

143. Zin CS, Nissen LM, Smith MT, O'Callaghan JP, Moore BJ. An update on the pharmacological management of post-herpetic neuralgia and painful diabetic neuropathy. CNS Drugs 2008; 22(5):417–442.

144. Gills JP, Hestead RF, Sanders DR. Ophthalmic anaesthesia, Throvare: SLACK Inc, 1993.

145. Belmonte C, Viana F. Molecular and cellular limits to somatosensory specificity. Mol Pain 2008; 4:14.

146. Meyer RA, Ringkamp M, Campbell JN, Raja SN. Peripheral mechanisms of cutaneous nociception. In: McMahon SB, Koltzenbrug M, editors. Wall and Melzack's Textbook of Pain, 5th ed. London: Elsevier, 2006. p. 3–34.

CHAPTER 17

Outward-Directed Transport

Pradeep K. Karla, Sai H.S. Boddu, Chanukya R. Dasari & Ashim K. Mitra

Introduction

In ocular therapeutics, the current literature describes drug delivery limitations posed by outward directed transport mechanisms of the eye. Drug delivery to ocular tissues is restricted by various factors such as limited tissue penetration, rapid clearance rate, and dose-limiting toxicity. Though diffusion plays a role, efflux transport proteins constitute a major driving force behind selective removal of therapeutic agents from ocular tissues. Therapeutic agents must penetrate the globe through the anterior or posterior route, depending on the mode of administration. The cornea forms a major barrier for anterior segment drug delivery. Presence of epithelial tight junctions, along with a wide array of efflux transporters causes poor permeation of drugs across the cornea. Apart from these, the drug molecules should possess optimum physicochemical properties to penetrate the different corneal layers.[1,2] The blood–retinal barrier (BRB) forms a major barrier to posterior segment ocular drug delivery. The BRB is composed of retinal vascular endothelium and retinal-pigmented epithelium and is categorized as inner BRB and outer BRB.[3,4] Inner BRB is composed of a monolayer of endothelial cells, which are non-fenestrated, and outer BRB is composed of retinal-pigmented epithelial cells.[4,5]

Multi-drug resistance mediated by the efflux transporter (pump) is a widely explored field in the area of cancer research. Efflux pumps belong to the ATP binding cassette (ABC) transporter family.[6] Family members such as P-glycoprotein (P-gp), multidrug resistance associated protein (MRP), breast cancer resistance protein (BCRP) and lung resistance protein (LRP) are known to play a vital role in imparting drug resistance to cancer cells. Efflux transporters are known to be protective in preventing the entry of toxic substances into cells which results in outward directed transport of toxins.[6]

The presence of efflux transporters on various physiological barriers protecting the eye was only recently discovered. Their existence in various ocular tissues such as cornea, retina, and conjunctiva has been established at molecular and functional level by various researchers. To a large extent the role of these efflux transporters in ocular drug resistance remains to be explored.

Various drug delivery strategies are employed to bypass the efflux transporters. The prodrug strategy, which involves linking of the active drug molecule to a promoiety, is known to reduce or eliminate the substrate specificity of drugs to the efflux transporters. Our recent findings revealed that amino acid and peptide prodrugs of quinidine, val-quinidine and val-val-quinidine were able to bypass the MRP class efflux transporters on the corneal epithelial cells.[7] Earlier it was shown that val-quinidine and val-val-quinidine can effectively bypass P-gp mediated efflux on the cornea.[8] This chapter comprises a discussion on the presence of efflux transporters in various ocular tissues and their possible role in outward directed transport. Various strategies employed to modulate or bypass efflux transporters are presented.

Efflux transporters – brief history

Though outward directed transport mediated by efflux transporters is a new field in the area of ocular drug delivery research, it has been widely explored in the area of cancer therapeutics. The role of ABC transporters in treatment failure has been attributed to overexpression of efflux transporters.[9]

First evidence of outward directed transport resulting in drug resistance was recognized in 1973.[10] Ehrlich astrocytes accumulated lower amounts of daunorubicin intracellularly primarily due to outward transport.[10] Later a higher-molecular-weight glycoprotein (P-gp) was identified in multidrug-resistant cells.[11] It took a considerable amount of time and effort to confirm the hypothesis that a protein binds to the therapeutic agents or toxins intracellularly and transports them out of the cells utilizing energy from ATP hydrolysis. P-gp was confirmed as a primary drug efflux transporter leading to drug resistance.[12] This protein was later found to be a major efflux pump involved in resistance to various cancer drugs including paclitaxel and doxorubicin. P-gp is also demonstrated as ATP-binding cassette subfamily B member 1 (ABCB1), MDR1 and PGY1. It is also designated as cluster of differentiation 243 (CD243). It is a ~170 kDa transmembrane glycoprotein with 10–15 kDa of N-terminal glycosylation.[13] P-gp has wide substrate specificity and is involved in the transport of lipids, steroids, bilirubin, xenobiotics, steroids, cardiac glycosides (digoxin), glucocorticoids (dexamethasone), HIV drugs (non-nucleoside reverse transcriptase inhibitors and protease inhibitors) and immunosuppressive agents.[14] Apart from being overexpressed in cancerous tissues, P-gp is expressed on various tissues including capillary endothelial cells of the blood–brain barrier (BBB), liver, intestine and renal proximal tubules.[13,15]

Another drug efflux transporter resulting in outward directed transport is multidrug resistance associated protein (MRP). It was identified in 1992 in a human lung cancer cell line (H69AR) demonstrating multidrug resistance. The cell line did not express P-gp despite conferring significant resistance to the absorption of doxorubicin. Later, the efflux transporter responsible for outward directed transport was found to be MRP.[16]

Apart from MRP and P-gp, various other human ATP binding cassette (ABC) transporters have been identified. A new efflux protein which is currently known as breast cancer resistance protein (BCRP) has been identified at protein level and cloned by three research groups.[17] It is also known as the mitoxantrone-resistance protein (MXR) or placenta-specific ABC protein (ABCP). This protein is an ~70 kDa efflux transporter belonging to the ABC family and proved to play a significant role in drug resistance.[18–20]

Another new efflux transporter belonging to the ATP binding cassette family is lung resistance protein (LRP).[21] It is also known as major vault protein (MVP) and can cause multidrug resistance. LRP was initially identified in P-gp negative MDR lung cancer cells[20,22] and is composed of complex ribonucleoprotein particles.[23,24] Vaults are present mainly in the cytoplasm and partially in the nuclear membrane. These structures are assumed to cause nucleocytoplasmic and intracellular transport.[25] The presence of LRP in cancer cells was found to result in resistance development to cisplatin, doxorubicin, carboplatin and melphalan.[25,26] Other than the eye it is present in various tissues such as colon, lung, macrophages, adrenal cortex and renal proximal tubules.[27] However, contradictory reports have appeared in the literature questioning the role of LRP or MVP. A few studies revealed that MVP does not play a significant role in drug resistance. The primary physiological function of LRP is not completely understood and is being further evaluated.[28]

Multidrug resistance conferred by efflux transporters is a relatively new area of interest in ocular drug delivery. Most of the transporters are ubiquitously expressed in various tissues of the body and have a protective role. The presence of efflux transporters in various ocular tissues and their role in ocular drug efflux resulting in drug resistance is discussed in the following sections.

Efflux transporters in ocular tissues

P-gp

P-gp in isolated primary culture of rabbit corneal epithelial cells was reported by Kawazu et al.[29] The presence of P-gp at protein level and a series of uptake experiments demonstrated that the tear-side-facing membrane (apical) of rabbit corneal epithelial cells expressed P-gp. This article further concluded that the presence of P-gp on corneal cells protects them from toxic agents.[29] Later the presence of P-gp in human cornea was demonstrated by Dey et al.[30] This report further confirmed the expression of P-gp in rabbit cornea and rabbit corneal epithelial cells. Uptake of Rho 123, a P-gp substrate, was significantly higher in the presence of P-gp inhibitors like cyclosporine A (CsA) indicating the role of P-gp in corneal drug efflux. Ocular pharmacokinetic studies involving male New Zealand Albino rabbits revealed a significant elevation in aqueous humor drug concentrations in the presence of various P-gp inhibitors, i.e. testosterone, quinidine, verapamil and CsA.[30] Moreover dose-dependent suppression in P-gp efflux was evident at various inhibitor concentrations. The presence of P-gp in the cornea was corroborated by a recent report on the quantitative expression of P-gp in human cornea.[31] The relative role of P-gp in the outward directed efflux of drugs used in glaucoma, antibacterial and anti-viral agents has yet to be established.

An ocular microdialysis technique was applied to determine the role of efflux transporters on the cornea.[30] Figure 17.1 shows a schematic diagram of the surgical implantation of the microdialysis probe. A well is placed on the top of cornea filled with drug solution. The sample is collected from the anterior end of the probe which protrudes from the cornea. Drug diffuses across the dializing membrane placed in the aqueous humor. This technique helps to quantitatively determine the drug concentrations in the aqueous humor at different time points. This study reported that P-gp inhibitors significantly raised drug concentration in aqueous humor (both C_{max} and AUC) (Table 17.1). The results demonstrated that P-gp can play a vital role in conferring resistance to drug absorption across the cornea. This article also suggested the use of efflux pump inhibitors along with therapeutic agents – a strategy that helps to increase the overall ocular bioavailability of drugs which are substrates of efflux pumps.

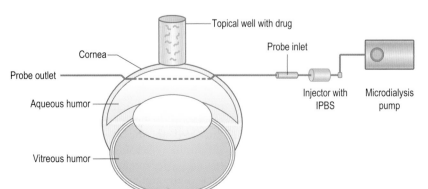

Figure 17.1 Schematic representation of ocular microdialysis technique that shows surgical implantation of linear probe in rabbit eye and adhering topical well on the surface of cornea as described by Dey S et al.[1] A microdialysis pump infuses isotonic phosphate buffer (IPBS), pH 7.4 at constant rate and sample is collected at the other end. Reservoir of drug loaded in the well is also shown. The extent of outward directed transport by an efflux pump can be estimated with the use of specific inhibitors.

Along with cornea, the BRB limits the permeability of various therapeutic agents into the eye.[32] Retinal pigmented epithelium (RPE), situated between the retina and its choroidal blood supply, forms a diffusional barrier controlling access to subretinal space. Transport proteins present on the RPE regulate ionic composition and hydration of the subretinal space.[33] Immunohistochemistry revealed that P-gp is localized both on the apical and basolateral sides of human RPE. Basolateral P-gp is attributed to the function of neural retina in clearing toxic and unwanted molecules from the subretinal space. Apical expression of P-gp appears to assist in the delivery of lipophilic substrates to subretinal space and engulfment by photoreceptors.[33] Three RPE cell lines, ARPE-19, D-407 and h1RPE, are usually employed for in vitro studies. Screening of these three cell lines for P-gp expression revealed that D-407 is suitable for functional expression and drug transport screening. ARPE-19 cells lacked both molecular and functional expression of P-gp, whereas h1RPE lacked functional expression of P-gp. D-407 demonstrated both molecular and functional expression of P-gp. Also, P-gp was found to be expressed on the choroidal blood vessels.[34] Further work is in progress to understand the role of P-gp in the choroid.

P-gp is also expressed in rabbit conjunctival epithelial cells. It is present on the apical side of the conjunctiva, conferring resistance to the absorption of cyclosporine A and other lipophilic drugs.[34] Use of various P-gp inhibitors – such as verapamil, progesterone, cyclosporine A, rhodamine 123 and metabolic inhibitor 2,4-DNP – enhanced the uptake of the model drug propranolol in rabbit conjunctival epithelial cells.[34] Yang et al[35] concluded that P-gp can play a vital role in decreased absorption of various lipophilic drugs across the conjunctiva.

P-gp appears to play an important role in the outward transport of cytotoxic drugs and in the modulation of chloride channels which regulate and initiate osmotic cataract. Further, findings by Miyazawa et al[36] revealed that P-gp might be induced by aldose reductase overexpression and osmotic stress. P-gp was found to maintain lenticular osmotic balance in sugar cataract. The role of P-gp in the drug disposition of various ocular agents is being further evaluated.

Mode of action and structure of P-gp

P-gp requires ATP for its function. There are two ATP binding sites, also known as nucleotide binding domains (NBD). Apart from NBD, there are two transmembrane domains incorporating six transmembrane domains at each site (a total of 12). A pictorial representation of P-gp is provided in Figure 17.2.[37]

MRP

Various classes of MRPs are expressed in the retina and other ocular tissues; the outer BRB expresses MRP. Functional characterization showed that the BRB is similar to the BBB. Functional studies using P-gp and MRP modulators revealed that outer BRB epithelium demonstrates similar pharmacological properties to endothelial BBB and expresses the same efflux transport systems.[38] Also, Wilson et al[39] demonstrated that retinoblastoma tumors express MRP1, 2 and 4. Immunohistochemistry revealed intense localization of these efflux pumps along with P-gp/BCRP in tumor tissue.[33,39] The possible role of these transporters, either individually or collectively, in tumor therapy needs to be evaluated.

Table 17.1 Effect of various P-gp inhibitors on the $AUC_{0-\infty}$ and $C_{aq,max}$ of [^{14}C]erythromycin in rabbit cornea

Drug and/or Inhibitor	$AUC_{0-\infty}$	C_{max}
Erythromycin	31.67 ± 4.99	0.136 ± 0.009
Erythromycin + 100 μM test	40.21 ± 5.29	0.151 ± 0.021
Erythromycin + 150 μM test	43.51 ± 7.75	0.183 ± 0.033
Erythromycin + 250 μM test	88.58 ± 6.34*	0.489 ± 0.095**
Erythromycin + 500 μM test	118.76 ± 10.72**	0.672 ± 0.095**
Erythromycin + 20 μM CsA	65.18 ± 7.09*	0.356 ± 0.060*
Erythromycin + 200 μM Quin	52.63 ± 7.79*	0.286 ± 0.052*
Erythromycin + 500 μM Vera	43.56 ± 6.99	0.205 ± 0.039

As evident from the data, 250 and 500 μM testosterone inhibited P-gp-mediated efflux to a significant extent. Among the other inhibitors studied, CsA and quinidine also inhibited P-gp-mediated efflux. Each point represents mean ± SD ($n = 4$). (Adapted with permission from Dey et al.[30])

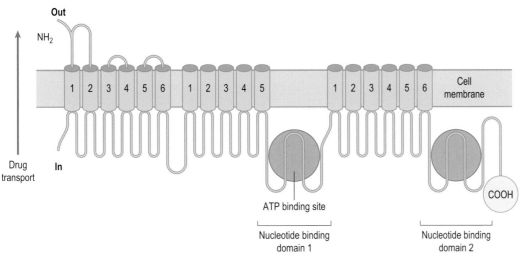

Figure 17.2 P-glycoprotein (P-gp) structure comprising 12 transmembrane domains, two nucleotide binding domains and two ATP binding sites. Drugs and toxins are effluxed out of the cell. (Drawn based on inspiration from Perez et al.[38] Reproduced from Association for Research in Vision and Ophthalmology.)

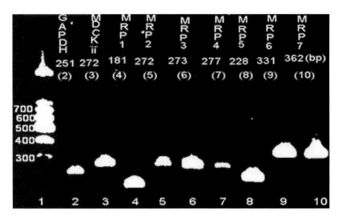

Figure 17.3 RT-PCR amplification of various MRPs using total RNA from human corneal epithelium (lanes 4–10); lanes 2 and 3 represent GAPDH and MDCKII-MDR-1 as controls.

Table 17.2 Primer sequences designed from Pubmed data base using OligoDT primer design software provided by Invitrogen (Carlsbad, CA)

MRP1 (NM_004996)	AGGTGGACCTGTTTCGTGAC CCTGTGATCCACCAGAAGGT
MRP2 (NM_000392)	GCAGCGATTTCTGAAACACA CAACAGCCACAATGTTGGTC
MRP3 (AF085691)	GCTTCTATGCAGCCACATCA ACAAGGACCCTGAACACCAG
MRP4 (AY081219)	GGTTCCCCTTGGAATCATTT GACAAACATGGCACAGATGG
MRP5 (AB209454)	ACCCGTTGTTGCCATCTTAG GCTTTGACCCAGGCATACAT
MRP6 (AF076622)	CACCTGCTAAACCGCTTCTC GAGCATTGTTCTGAGCCACA
MRP7 (NM_033450)	GAACGGCTGCTTAACTTTGC AGGTATACACAGGCCGCATC

MRP is another major class of ABCC efflux transporter imparting drug resistance. The MRP family has nine members (MRP 1–9) and substrate specificity varies significantly among them.[40] As shown in Figure 17.3, RT-PCR performed on primary human corneal epithelial cells (based on protocol established in our laboratory[1]) revealed the expression of MRP1–7. Primary culture of human corneal epithelial cells was obtained from Cascade Biologics (Portland, OR) and RT-PCR performed based on an established protocol[2] – primers used for the RT-PCR are presented in Table 17.2. MRP is regarded as an organic anion transporter, but some MRP (MRP 4,5) are known to transport nucleoside and nucleotide analogs.[41]

The presence of MRP2 at the gene and protein level, as well as its functional expression in rabbit corneal epithelial cells, was established in 2007 in our laboratory.[2] A possible role of MRP2 in conferring resistance to the absorption of erythromycin across rabbit corneal epithelial cells was reported. In an article by Karla et al,[2] MRP2 expression on rabbit corneal epithelial cells at the gene and protein level was reported. [^{14}C]-erythromycin was selected as a model

substrate and affinity to the transporter was measured employing a specific inhibitor, MK571. This work demonstrated apically polarized localization of the transporter in rabbit corneal epithelial cells. Also, uptake and transport of [^{14}C]-erythromycin was significantly elevated in the presence of MK571, suggesting that the transporters might act in conjunction towards outward directed transport.[2]

Later studies by Karla et al revealed significant expression of MRP2 conferring drug resistance in human corneal epithelial cells.[42] (As stated in our earlier publication:[2]

"Significant elevation in uptake of model drugs [14C]-erythromycin and [3H]-cyclosporine-A in the presence of MK-571 indicated the functional activity of MRP2 in human corneal epithelium. Expression of ABCC2 may play a critical role in reducing the bioavailability of topically applied drugs. Cyclosporine-A is widely used as an immunosuppressive agent in corneal transplants and it is also indicated in the treatment of various corneal opacities. Erythromycin in various topical formulations has been used a drug of choice for treating minor corneal opacities. Though these compounds are substrates for P-gp, use of MK-571 reveals contribution of only MRP. A significant elevation in uptake of [14C]-erythromycin in the presence of unlabeled cyclosporine-A was also observed. The results indicate a dose-dependent inhibition of ABCC2 in both rPCEC and SV40-HCEC cell lines. Further studies indicate similar MDR expression profiles in human and rabbit corneal epithelium, which validate rabbit as a suitable animal model for in-vivo MDR-mediated drug absorption studies. Transport studies across SV40-HCEC and rabbit cornea revealed a significant increase in drug concentration in receiver chamber in the presence of inhibitor. Enhanced permeability of [14C]-erythromycin and [3H]-cyclosporine-A from apical to basolateral side across SV40-HCEC in the presence of MK-571 reveals the important role of ABCC2 as a barrier in corneal drug influx. In human corneal epithelial cells apical to basolateral side transport of [3H]-cyclosporine-A was elevated almost ∼4-fold as compared to basal to apical transport, which increased ∼2-fold in the presence of an inhibitor. Similar trend is observed for the passage of [14C]-erythromycin where elevation in apical to basolateral transport was almost ∼4-fold as compared to control whereas basal to apical transport was only ∼1.5-fold higher than control. Acceleration in the basal to apical transport might be due to the presence of a homolog in MRP family.")

These results indicate that the MRP class of transporters can confer significant resistance to the absorption of drugs across human corneal epithelium. Literature review indicates that many ocular drugs are substrates for MRP2 – these include sulfated steroids, macrolides such as azithromycin, erythromycin, antibiotics and quinolones such as ciprofloxacin and grepafloxacin.[2,43–49] Fluoroquinolones are applied topically for treating conjunctivitis and keratitis[50] in a large number of cases.[42,51] As demonstrated in our earlier studies, erythromycin is a substrate for the MRP class of efflux transporters. Fluoroquinolones such as sparfloxacin, levofloxacin and grepafloxacin interacted with erythromycin. The use of MRP inhibitors in ophthalmic formulations or use of combination drugs which interact with the transporters might be an efficient strategy to increase drug delivery to the anterior segment.

Preliminary results from our laboratory indicate the expression of the nucleoside and nucleotide efflux transporters MRP4 and MRP5 in human corneal epithelial cells (Fig. 17.4). However, the functional role of these efflux pumps in ocular drug resistance has yet to be established. Preliminary results from our laboratory indicate that these transporters play a vital role in the outward directed transport of adefovir and acyclovir (unpublished results). Also, additional experimentation revealed that some drugs used in the treatment of glaucoma (bimatoprost and latanoprost) are effluxed by MRP on the corneal epithelium (unpublished data). Ocular microdialysis in an in vivo rabbit animal model revealed the functional role of the MRP4/MRP5 system in conferring resistance to the absorption of acyclovir.[52] Use of the specific inhibitor MK571 increased aqueous humor concentrations of acyclovir by ~2-fold.[37,53] Expression of various MRPs in the cornea and their relative role resistance to various ophthalmic drugs needs to be further explored.

Mode of action and structure of MRP

It is well known that MRP utilizes ATP for its functional activity. Similar to P-gp there are two ATP binding sites, also known as NBD. Apart from NBD, the protein is made up of three transmembrane sites (two sites with six transmembrane domains and one site with five transmembrane domains – a total of 17).[37,54] Also, the N-terminal loop is present outside the cell membrane – unlike P-gp, where it is present in the cytoplasm.[54] A pictorial representation of MRP structure is shown in Figure 17.5.

BCRP

BCRP is expressed on the retina and choroid and may be involved in outward directed transport. It is also expressed on retinal capillary endothelial cells which form the barrier at the inner BRB. Dual immunolabeling of BCRP and glucose transporter 1 in mouse retina by Asashima et al[55] revealed that BCRP is localized on the luminal side of retinal capillary endothelial cells. Using a BCRP inhibitor, Ko143, Asashima et al also established that BCRP acts as an efflux transporter of photosensitive toxins such as protoporphyrin IX and pheophorbide. BCRP is also expressed on the retinoblastoma cells. Its role in drug resistance still needs to be elucidated.

BCRP is another efflux transporter expressed on human corneal epithelial cells.[56,57] Studies in our laboratory have demonstrated that BCRP plays a significant role in conferring resistance to the absorption of drugs (mitoxantrone; MTX) across transfected human corneal epithelial cells (SV40-HCEC). We also reported the functional expression of BCRP in human corneal epithelial cells and proved it can play a role in drug efflux.[57] Significant efflux of [³H]-MTX in the presence of FTC and GF120918 (inhibitors of BCRP) established the functional activity of BCRP in SV40-HCEC and its possible role in limiting the ocular bioavailability of topically applied drugs. Dose-dependent inhibition of efflux was noticed with both fumtremorgic-C (FTC) and GF120918 in MDCKII-BCRP (+ve control) and SV40-HCEC. FTC and GF120918 are well-known inhibitors of BCRP and their use

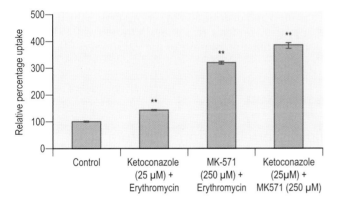

Figure 17.4 Collective role of P-gp and MRP in corneal drug efflux. Ketoconazole, a specific P-gp inhibitor (25 μM) along with MK-571 (250 μM) was used to demonstrate the dual efflux. Each data point represents the mean ± SD of three determinations. Statistical significance was tested by two-factor ANOVA. *$P \leq 0.05$; **$P \leq 0.01$. (Adapted with permission from Karla et al.[2])

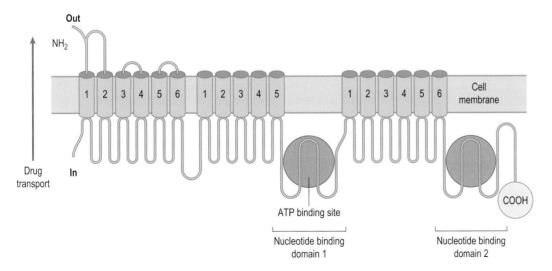

Figure 17.5 Multidrug resistance associated protein (MRP) structure comprising 17 transmembrane domains, two nucleotide binding domains and two ATP binding sites. Drugs and toxins are effluxed out of the cell. (Drawn based on inspiration from Perez et al[38] and Gottesman et al.[55])

in SV40-HCEC resulted in diminished efflux and increased uptake.[57]

An interesting study by Umemoto et al[58] revealed that basal limbal epithelial cells of the rat eye express BCRP significantly. Also, side population stem cells which help in the regeneration of corneal epithelium express BCRP – but at a relatively very low level compared to limbal side population cells. BCRP is expressed at a relatively low level in stem cells which help in regenerating corneal epithelium. Hence, it is regarded as a putative stem cell marker used to identify the corneal stem cells responsible for regeneration of epithelial cells. This is of high clinical significance as constructs generated from limbal epithelial cells were successfully used in clinical transplantation.[57,58] The fact that BCRP is expressed on the stem cells and on human corneal epithelial cells provides a new insight into the mechanism of ocular drug resistance development.[57] Relative expression of BCRP in stem cells and epithelial cells should be investigated.

Mode of action and structure of BCRP

BCRP requires ATP for its function. Unlike P-gp and MRP there is one ATP binding site, also known as the nucleotide binding domain (NBD). There is also a single transmembrane site incorporating six domains – hence it is regarded as half transporter.[54] A pictorial representation of BCRP is shown in Figure 17.6.

LRP

The presence of LRP in human corneal tissue was first indicated by Becker et al.[56] As described earlier there are contradicting reports with respect to the role of LRP in drug resistance. It is located inside the cell (intracellular) and is regarded as a major vault protein which aids in the outward directed transport of toxins.[21,59] Figure 17.7 demonstrates the intracellular localization of LRP.

Localization of transporters

Efflux transporters are localized on either apical/basal sides or are present intracellularly. Localization of transporters in biological membranes is represented in Figure 17.7. However, the localization may vary in some tissues. The BBB shows differential localization of transporters. MRP1, MRP3 and MRP5 are expressed on both apical and basal sides of the BBB.[60] Preliminary results from our laboratory indicated that MRP5 is localized on both apical and basolateral sides of the cornea. However, expression is predominantly on the basal side (unpublished data). Based on the localization and distribution in various tissues the transporters appear to confer a concerted protective role. The exact nature and function of most of the transporters still need to be explored.

Discussion

Clinical correlates from literature

Depending on the route of drug delivery, different structural barriers must be overcome to improve ocular bioavailability. Traditionally drug delivery mechanisms are divided into two categories: anterior segment and posterior segment. A topically administrated dose is cleared rapidly by tear turnover and blinking reflex before encountering the conjunctiva, which absorbs a substantial portion. Often the cornea is the target tissue, but it is composed of a low permeability cell structure with efflux pumps which effectively inhibit therapeutic agents. Even drugs delivered directly into aqueous humor either by intracameral injection or ocular inserts, are prone to removal by Schlemm's canal. Drug delivery to the posterior segment is similarly hindered. Delivery to the retina can be achieved via intravitreal, periorbital, or systemic routes. Intravitreal injection must allow for a long residence time in the vitreous humor, in order for the drug to traverse multiple neuronal layers before reaching the intended segment of the retina. Periorbital applications intended for the retina must traverse the thick sclera, choroid

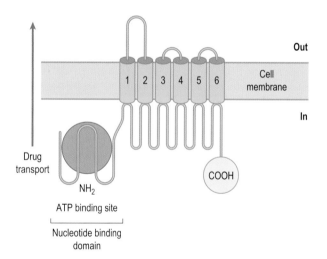

Figure 17.6 Breast cancer resistance protein (BCRP) structure comprising six transmembrane domains and one nucleotide binding domain. Drugs and toxins are effluxed out of the cell. (Drawn based on inspiration from Perez et al[38] and Gottesman et al.[55])

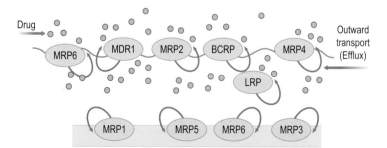

Figure 17.7 Schematic representation of apical and basal distribution of efflux transporters on cells. In general, apical transporters comprise of MDR1, MRP2, 4, 6 and BCRP. MRP1, 3, 5 and 6 are present basolaterally. LRP is present in the cytoplasm. This scheme represents the common insight on localization of transporters, but various sites like blood–brain barrier and cornea were found to show differential localization.

Box 17.1 ABCA and Macular Degeneration

ABCA mutations have been directly related to the pathogenesis of Stargardt and age-related macular degeneration, retinitis pigmentosa, and recessive cone rod dystrophy. In normal retinal pigment epithelial cells, ABCA-4 transports photobleached rhodopsin out of cells to prevent damage caused by toxic end products. A deficiency in the efflux protein leads to accumulations of N-retinylidene-PE, all-*trans* retinal, and A2E in disc membranes, eventually leading to photoreceptor death by accumulation of photo-oxidation products.[59]

Box 17.2 ABCC and Ocular Bleeding in PXE

Ocular bleeding observed in pseudoxanthoma elasticum (PXE) syndrome is caused by a recessive mutation in ABCC6 (*MRP6*) on chr 16p, next to the ABCC1 gene. ABCC6 is highly present in the liver and kidney, and is responsible for the efflux of glutathione conjugates such as LTC$_4$, and perhaps even triglycerides and HDL. Transport may be inhibited by organic anions or mis-sense mutations in pseudoxanthoma elasticum, which presents with dermal connective tissue calcification and cardiovascular disease.[59]

Box 17.3 ABCD1 and Adrenoleukodystrophy

ABCD1 is a peroxisome transporter that mediates the efflux of very-long-chain, unbranched, saturated fatty acids from cells and cholesterol esters of the brain and adrenal cortex. Several hundred mutations in the ABCD1 gene have been identified and implicated in adrenoleukodystrophy – a variable onset, X-linked, recessive disease characterized by adrenal deficiency and neuronal deficits. Ocular manifestations include progressive vision loss, optic nerve hypoplasia, exotropia, esotropia, retinal ganglionic atrophy; cataracts; and altered macular pigment.[70]

VLCFA levels in the blood can be normalized by conventional treatments; however cerebral fatty acid levels remain elevated leading to disease progression over time. Experimental studies have been conducted using fibrates, a class of cholesterol-lowering drugs, to induce ABCD2, a transport protein closely related to ABCD1 and found principally in the brain and adrenals. The hope is to upregulate fatty acid efflux from peroxisomes in neuronal and ocular tissue.[59]

Table 17.3 Tabulation of various classes of transporters that can result in outward directed transport in ocular tissues and their function based on literature review

Family	Members identified in ocular tissue	Function
ABC-A	ABCA-4	Transports lipids and cholesterol
ABC-B	P-gp	Some are located in the blood-brain barrier, liver, mitochondria and transports peptides and bile, for example
ABC-C	MRP, ABCC6	Significant role in ion transport, and toxin efflux
ABC-D		Transporter on peroxisome membrane
ABC-E / ABC-F		These proteins are not thought to have a role in transmembrane transport
ABC-G	BCRP	Transports bile, lipids, cholesterol, and steroids

research is focusing on the development of additional inhibitors which will lead to more effective chemotherapy.

Outward directed transport mediated by various transporters is a relatively new and exciting field of research. A better understanding of these transporters might result in the development of efficient drug delivery strategies to improve therapy for both anterior and posterior segment diseases.

Strategies to evade efflux transporters

Expression of efflux transporters on the protective barriers of the eye prevents the entry of many drug molecules. Drug delivery strategies which can aid in the evasion of efflux transporters would be ideal for enhancing drug delivery to the eye. Various options are discussed in the literature on modulating or bypassing efflux at various sites. One promising approach is a prodrug strategy where active drug molecules are modified by chemical synthesis to attach a peptide, amino acid, vitamin or other substrate to target a specific transporter/receptor. Targeting the active drug molecules to various influx transporters present on the cornea and blood–ocular barrier (BOB) can be an efficient strategy to bypass the efflux transporters.[8] Peptide transporters are the major influx transporters and are widely selected as a target. The oligopeptide transport system is also expressed on the BOB.[7,67]

The BOB is the major barrier that protects the eye from toxins in internal blood circulation. Atluri et al[67] proposed that this transport system can be employed to bypass the efflux transporters present on the BOB. The presence of the oligopeptide transporter on the cornea, which is a major barrier to anterior segment drug delivery, was proposed by Anand et al in 2002.[7] Val-acyclovir, an amino acid prodrug of acyclovir, is currently available on the market under the trade name Valtrex. It is employed in the treatment of

and RPE, which also contain efflux pumps. Systemic drug delivery to the retina must bypass the BRB to reach target tissue. Given the unique characteristics of multistep barriers and target tissues in ophthalmic drug delivery, the characterization of specific tissue types has been a recent focus. Each of the various ocular tissues, such as the conjunctiva, sclera, cornea, and retina, has a different milieu of membrane proteins which confer unique functional characteristics with respect to drug clearance (Boxes 17.1–17.3).

Table 17.3 summarizes various classes of transporters that can result in outward directed transport in ocular tissues and their function based on literature review. Highly specific ABCB1 inhibitors demonstrate the ability to decrease the extent of drug resistance, and perpetuate the therapeutic attributes of existing pharmacological regimens. Current

Box 17.4 ABCG and Sitosterolemia

Xanthelasma and corneal arcus are occasionally seen in sitosterolemia, a rare autosomal recessive lipid metabolism disorder caused by mutation in ABCG5 (Asian sitosterolemia) or ABCG8 (Caucasian). Unique to this class of efflux proteins is the mechanism of forming heterodimers to transport non-cholesterol plant and shellfish sterols which accumulate to form xanthomas.[59] Coincidentally the presence of a related protein ABCG2 (BRCP), has been demonstrated in retinal blood vessels but its function is not apparent.[56] Current treatment for this sitosterolemia involves predominantly dietary management and lipid-lowering therapy.

Box 17.5 Transporters and Multidrug Resistance to Cancer Chemotherapy

Normal cells which are exposed to toxic and xenobiotic compounds overexpress efflux pathways that ultimately lead to drug resistance. In cancer cells the process of drug resistance is accelerated by rapid growth processes. The predominant mode by which some cancers develop resistance is through overexpression of efflux protiens.[60] With regards to ocular neoplasms, P-gp, MRP and LRP exist on primary and metastatic melanomas and confer resistance to chemotherapy for these neoplasms.[61] LRP has a larger influence in choroidal melanomas, but in uveal melanomas demonstrating drug resistance P-gp expression is higher.[62] Also in enucleated retinoblastomas which failed to respond to chemotherapy, high P-gp, MRP-1, and MRP-2 expression has been histologically demonstrated.[63]

Box 17.6 Prodrug Strategy for Increased Ocular Bioavailability and Evasion of Efflux Transporter

A prodrug of acyclovir, Valtrex is currently in the market and is administered as an oral dose.[70] The presence of an oligopeptide transporter system on the blood–ocular barrier is a recent finding.[66] Valtrex was shown to permeate the tissues using the oligopeptide transporter, hence this might be the reason for increased permeability of Valtrex across the blood–ocular barrier. Preliminary results from our laboratory suggest that acyclovir is a substrate for the MRP4/MRP5 class of transporters present on the cornea (unpublished data). Targeting of acyclovir, using a valine promoiety might be aiding efficient bypass of the efflux transporters present on the blood–ocular barrier.

keratitis of the cornea caused by the herpes simplex virus. As stated earlier, prodrug derivatization can aid in bypassing various efflux transporters present on the cornea. Both the amino acid prodrug (val-acyclovir) and the peptide prodrug (val-val-acyclovir) are transported across the cornea by the oligopeptide transporter, leading to increased ocular penetration.[7,68] The peptide prodrug of acyclovir showed significantly improved anti-viral properties (Boxes 17.4–17.6).[68]

Administering specific inhibitors of efflux transporters might be another strategy to improve drug delivery across the cornea. Dey et al[30] demonstrated increased absorption of erythromycin across the cornea in an in vivo rabbit animal model in the presence of various inhibitors. Also, modulation of the MRP class of efflux transporters with the use of specific inhibitors elevated the absorption of erythromycin significantly across the cornea.

Outward directed transport mediated by efflux transporters at various ocular tissues is a relatively new field of research. Recent work has revealed the presence of various efflux transporters on the cornea and other protective barriers of the eye. However, extensive research work is needed to delineate their role in ocular drug delivery.

Acknowledgment

This is supported by NIH grants NIH R01- EY09171-14, R01-EY10659-12, and Howard University faculty seed grant. Some of the information and diagrams were obtained from a thesis published by Pradeep K. Karla (Copyright Holder) at the University of Missouri-Kansas City in 2008 and primary/corresponding author for this chapter.

References

1. Dey D, Patel J, Anand BS et al. Molecular evidence and functional expression of P-glycoprotein (MDR1) in human and rabbit cornea and corneal epithelial cell lines. Invest Ophthalmol Vis Sci 2003; 44:2909–2918.
2. Karla PK, Pal D, Mitra AK. Molecular evidence and functional expression of multidrug resistance associated protein (MRP) in rabbit corneal epithelial cells. Exp Eye Res 2007; 84:53–60 .
3. Mannermaa E, Vellonen KS, Urtti A. Drug transport in corneal epithelium and blood-retina barrier: emerging role of transporters in ocular pharmacokinetics. Adv Drug Deliv Rev 2006; 58:1136–1163.
4. Vinores SA. Assessment of blood-retinal barrier integrity. Histol Histopathol 1995; 10:141–154.
5. Xu Q, Qaum T, Adamis AP. Sensitive blood-retinal barrier breakdown quantitation using Evans blue. Invest Ophthalmol Vis Sci 2001; 42:789–794.
6. Schinkeland AH, Jonker JW. Mammalian drug efflux transporters of the ATP binding cassette (ABC) family: an overview. Adv Drug Deliv Rev 2003; 55:3–29.
7. Anand BS, Mitra AK. Mechanism of corneal permeation of L-valyl ester of acyclovir: targeting the oligopeptide transporter on the rabbit cornea. Pharm Res 2002; 19:1194–1202.
8. Katragadda S, Talluri RS, Mitra AK. Modulation of P-glycoprotein-mediated efflux by prodrug derivatization: an approach involving peptide transporter-mediated influx across rabbit cornea. J Ocul Pharmacol Ther 2006; 22:110–120.
9. Xu D, Fang L, Zhu Q, Hu Y, He Q, Yang B. Antimultidrug-resistant effect and mechanism of a novel CA-4 analogue MZ3 on leukemia cells. Pharmazie 2008; 63:528–533.
10. Dano K. Active outward transport of daunomycin in resistant Ehrlich ascites tumor cells. Biochim Biophys Acta 1973; 323:466–483.
11. Juliano RL, Ling V. A surface glycoprotein modulating drug permeability in Chinese hamster ovary cell mutants. Biochim Biophys Acta 1976; 455:152–162.
12. Germann UA. P-glycoprotein – a mediator of multidrug resistance in tumour cells. Eur J Cancer 1996; 32A:927–944.
13. Borstand P, Schinkel AH. Genetic dissection of the function of mammalian P-glycoproteins. Trends Genet 1997; 13:217–222.
14. Dean M, Rzhetsky A, Allikmets R. The human ATP-binding cassette (ABC) transporter superfamily. Genome Res 2001; 11:1156–1166.
15. Ueda K, Clark DP, Chen CJ, Roninson IB, Gottesman MM, Pastan I. The human multidrug resistance (mdr1) gene. cDNA cloning and transcription initiation. J Biol Chem 1987; 262:505–508.
16. Cole SP, Bhardwaj G, Gerlach JH et al. Overexpression of a transporter gene in a multidrug-resistant human lung cancer cell line. Science 1992; 258:1650–1654.
17. Doyle LA, Yang W, Abruzzo LV et al. A multidrug resistance transporter from human MCF-7 breast cancer cells. Proc Natl Acad Sci USA 1998; 95:15665–15670.
18. Miyake K, Mickley L, Litman T et al. Molecular cloning of cDNAs which are highly overexpressed in mitoxantrone-resistant cells: demonstration of homology to ABC transport genes. Cancer Res 1999; 59:8–13.
19. Allikmets R, Schriml LM, Hutchinson A, Romano-Spica V, Dean M. A human placenta-specific ATP-binding cassette gene (ABCP) on chromosome 4q22 that is involved in multidrug resistance. Cancer Res 1998; 58:5337–5339.
20. Mao Q, Unadkat JD. Role of the breast cancer resistance protein (ABCG2) in drug transport. AAPS J 2005; 7:E118–33.
21. Zhu Y, Kong C, Zeng Y, Sun Z, Gao H. Expression of lung resistance-related protein in transitional cell carcinoma of bladder. Urology 2004; 63:694–698.
22. Scheper RJ, Broxterman HJ, Scheffer GL et al. Overexpression of a M(r) 110,000 vesicular protein in non-P-glycoprotein-mediated multidrug resistance. Cancer Res 1993; 53:1475–1479.
23. Moffat AS. Making sense of antisense. Science 1991; 253:510–511.
24. Scheffer GL, Wijngaard PL, Flens MJ et al. The drug resistance-related protein LRP is the human major vault protein. Nature Med 1995; 1:578–582.
25. Kedershaand NL, Rome LH. Isolation and characterization of a novel ribonucleoprotein particle: large structures contain a single species of small RNA. J Cell Biol 1986; 103:699–709.

26. Izquierdo MA, Scheffer GL, Flens MJ et al. Relationship of LRP-human major vault protein to in vitro and clinical resistance to anticancer drugs. Cytotechnology 1996; 19:191–197.

27. Izquierdo MA, Shoemaker RH, Flens MJ et al. Overlapping phenotypes of multidrug resistance among panels of human cancer-cell lines. Int J Cancer 1996; 65:230–237.

28. Huffman KE, Corey DR. Major vault protein does not play a role in chemoresistance or drug localization in a non-small cell lung cancer cell line. Biochemistry 2005; 44:2253–2261.

29. Kawazu K, Yamada K, Nakamura N, Ota A. Characterization of cyclosporin A transport in cultured rabbit corneal epithelial cells: P-glycoprotein transport activity and binding to cyclophilin. Invest Ophthalmol Vis Sci 1999; 40:1738–1744.

30. Dey S, Gunda S, Mitra AK. Pharmacokinetics of erythromycin in rabbit corneas after single-dose infusion: role of P-glycoprotein as a barrier to in vivo ocular drug absorption. J Pharmacol Exp Ther 2004; 311:246–255.

31. Zhang T, Xiang CD, Gale D, Carreiro S, Wu EY, Zhang EY. Drug transporter and cytochrome P450 mRNA expression in human ocular barriers: implications for ocular drug disposition. Drug Metab Dispos 2008; 36:1300–1307.

32. Senthilkumari S, Velpandian T, Biswas NR, Sonali N, Ghose S. Evaluation of the impact of P-glycoprotein (P-gp) drug efflux transporter blockade on the systemic and ocular disposition of P-gp substrate. J Ocul Pharmacol Ther 2008; 24:290–300.

33. Kennedy BG, Mangini NJ. P-glycoprotein expression in human retinal pigment epithelium. Mol Vis 2002; 8:422–430.

34. Saha P, Yang JJ, Lee VH. Existence of a p-glycoprotein drug efflux pump in cultured rabbit conjunctival epithelial cells. Invest Ophthalmol Vis Sci 1998; 39:1221–1226.

35. Yang JJ, Kim KJ, Lee VH. Role of P-glycoprotein in restricting propranolol transport in cultured rabbit conjunctival epithelial cell layers. Pharm Res 2000; 17:533–538.

36. Miyazawa T, Kubo E, Takamura Y, Akagi Y. Up-regulation of P-glycoprotein expression by osmotic stress in rat sugar cataract. Exp Eye Res 2007; 84:246–253.

37. Perez-Plasencia C, Duenas-Gonzalez A. Can the state of cancer chemotherapy resistance be reverted by epigenetic therapy? Mol Cancer 2006; 5:27.

38. Steuer H, Jaworski A, Elger B et al. Functional characterization and comparison of the outer blood-retina barrier and the blood-brain barrier. Invest Ophthalmol Vis Sci 2005; 46:1047–1053.

39. Wilson MW, Fraga CH, Fuller CE et al. Immunohistochemical detection of multidrug-resistant protein expression in retinoblastoma treated by primary enucleation. Invest Ophthalmol Vis Sci 2006; 47:1269–1273.

40. Borst P, Evers R, Kool M, Wijnholds J. A family of drug transporters: the multidrug resistance-associated proteins. J Natl Cancer Inst 2000; 92:1295–1302.

41. Dallas S, Schlichter L, Bendayan R. Multidrug resistance protein (MRP) 4- and MRP 5-mediated efflux of 9-(2-phosphonylmethoxyethyl)adenine by microglia. J Pharmacol Exp Ther 2004; 309:1221–1229.

42. Karla PK, Pal D, Quinn T, Mitra AK. Molecular evidence and functional expression of a novel drug efflux pump (ABCC2) in human corneal epithelium and rabbit cornea and its role in ocular drug efflux. Int J Pharm 2007; 336:12–21.

43. Terashi K, Oka M, Soda H et al. Interactions of ofloxacin and erythromycin with the multidrug resistance protein (MRP) in MRP-overexpressing human leukemia cells. Antimicrob Agents Chemother 2000; 44:1697–1700.

44. Seral C, Carryn S, Tulkens PM, Van Bambeke F. Influence of P-glycoprotein and MRP efflux pump inhibitors on the intracellular activity of azithromycin and ciprofloxacin in macrophages infected by Listeria monocytogenes or Staphylococcus aureus. J Antimicrob Chemother 2003; 51:1167–1173.

45. Zelcer N, Reid G, Wielinga P et al. Steroid and bile acid conjugates are substrates of human multidrug-resistance protein (MRP) 4 (ATP-binding cassette C4). Biochem J 2003; 371:361–367.

46. Michot JM, Van Bambeke F, Mingeot-Leclercq MP, Tulkens PM. Active efflux of ciprofloxacin from J774 macrophages through an MRP-like transporter. Antimicrob Agents Chemother 2004; 48:2673–2682.

47. Chu XY, Huskey SE, Braun MP, Sarkadi B, Evans DC, Evers R. Transport of ethinylestradiol glucuronide and ethinylestradiol sulfate by the multidrug resistance proteins MRP1, MRP2, and MRP3. J Pharmacol Exp Ther 2004; 309:156–164.

48. Chen ZS, Guo Y, Belinsky MG, Kotova E, Kruh GD. Transport of bile acids, sulfated steroids, estradiol 17-beta-D-glucuronide, and leukotriene C4 by human multidrug resistance protein 8 (ABCC11). Mol Pharmacol 2005; 67:545–557.

49. Sugie M, Asakura E, Zhao YL et al. Possible involvement of the drug transporters P glycoprotein and multidrug resistance-associated protein Mrp2 in disposition of azithromycin. Antimicrob Agents Chemother 2004; 48:809–814.

50. Cochereau-Massin I, Bauchet J, Marrakchi-Benjaafar S et al. Efficacy and ocular penetration in experimental streptococcal endophthalmitis. Antimicrob Agents Chemother 1993; 37:633–636.

51. Block SL, Hedrick J, Tyler R et al. Increasing bacterial resistance in pediatric acute conjunctivitis (1997–1998). Antimicrob Agents Chemother 2000; 44:1650–1654.

52. Sikri V, Pal D, Jain R, Kalyani D, Mitra AK. Cotransport of macrolide and fluoroquinolones, a beneficial interaction reversing P-glycoprotein efflux. Am J Ther 2004; 11:433–442.

53. Karla PK, Quinn TL, Herndon BL, Thomas P, Pal D, Mitra A. Expression of multidrug resistance associated protein 5 (MRP5) on cornea and its role in drug efflux. J Ocul Pharmacol Ther 2009; 25:121–132.

54. Gottesman MM, Fojo T, Bates SE. Multidrug resistance in cancer: role of ATP-dependent transporters. Nat Rev Cancer 2002; 2:48–58.

55. Asashima T, Hori S, Ohtsuki S et al. ATP-binding cassette transporter G2 mediates the efflux of phototoxins on the luminal membrane of retinal capillary endothelial cells. Pharm Res 2006; 23:1235–1242.

56. Becker U, Ehrhardt C, Daum N et al. Expression of ABC-transporters in human corneal tissue and the transformed cell line, HCE-T. J Ocul Pharmacol Ther 2007; 23:172–181.

57. Karla PK, Earla R, Boddu SH, Johnston TP, Pal D, Mitra A. Molecular expression and functional evidence of a drug efflux pump (BCRP) in human corneal epithelial cells. Curr Eye Res 2009; 34:1–9.

58. Umemoto T, Yamato M, Nishida K et al. Rat limbal epithelial side population cells exhibit a distinct expression of stem cell markers that are lacking in side population cells from the central cornea. FEBS Lett 2005; 579:6569–6574.

59. Izquierdo MA, Scheffer GL, Flens MJ et al. Broad distribution of the multidrug resistance-related vault lung resistance protein in normal human tissues and tumors. Am J Pathol 1996; 148:877–887.

60. Zhang Y, Schuetz JD, Elmquist WF, Miller DW. Plasma membrane localization of multidrug resistance-associated protein homologs in brain capillary endothelial cells. J Pharmacol Exp Ther 2004; 311:449–455.

61. Strauss O. The retinal pigment epithelium in visual function. Physiol Rev 2005; 85:845–881.

62. Diekmann U, Zarbock R, Hendig D, Szliska C, Kleesiek K, Gotting C. Elevated circulating levels of matrix metalloproteinases MMP-2 and MMP-9 in pseudoxanthoma elasticum patients. J Mol Med 2009

63. Pulkkinen L, Nakano A, Ringpfeil F, Uitto J. Identification of ABCC6 pseudogenes on human chromosome 16p: implications for mutation detection in pseudoxanthoma elasticum. Hum Genet 2001; 109:356–365.

64. McNamara M, Clynes M, Dunne B et al. Multidrug resistance in ocular melanoma. Br J Ophthalmol 1996; 80:1009–1012.

65. van der Pol JP, Blom DJ, Flens MJ et al. Multidrug resistance-related proteins in primary choroidal melanomas and in vitro cell lines. Invest Ophthalmol Vis Sci 1997; 38:2523–2530.

66. Filho JP, Correa ZM, Odashiro AN et al. Histopathological features and P-glycoprotein expression in retinoblastoma. Invest Ophthalmol Vis Sci 2005; 46:3478–3483.

67. Atluri H, Anand BS, Patel J, Mitra AK. Mechanism of a model dipeptide transport across blood-ocular barriers following systemic administration. Exp Eye Res 2004; 78:815–822.

68. Anand BS, Katragadda S, Gunda S, AK. In vivo ocular pharmacokinetics of acyclovir dipeptide ester prodrugs by microdialysis in rabbits. Mol Pharm 2006; 3:431–440.

Biochemical Cascade of Phototransduction

Alecia K. Gross & Theodore G. Wensel

Overview

Phototransduction is the series of biochemical events that lead from photon capture by a photoreceptor cell to its hyperpolarization and slowing of neurotransmitter release at the synapse.[1-10] This overall process includes an activation phase and a recovery phase. While the fundamental mechanisms of phototransduction seem to be constant over a wide range of light intensities and very similar in rods and cones, the quantitative features differ strikingly depending on the level of ambient light (see Chapter 20) and the cell type. While we know more about the biochemistry of phototransduction than we do about any other neuronal signaling pathway, there are numerous remaining mysteries.

Location and compartmentalization of rods and cones

Rod and cone photoreceptors are densely packed in the outermost layer of the neural retina, stretching from their synapses in the outer plexiform layer to their most distal tips, which are embedded in the membrane processes of the retinal pigmented epithelium (RPE) (Fig. 18.1). They are long, thin cells with their long axes aligned along the radii of the eye to maximize light collection. Efficient phototransduction depends upon the organization of photoreceptors into distinct membrane compartments (Figs 18.1–18.3). The biochemical cascade of phototransduction are restricted to the outer segment, an elongated membrane compartment separated from the rest of the cell, by a thin connecting cilium or ciliary transition zone, consisting of a 9 + 0 bundle of microtubules surrounded by a very thin layer of cytoplasm and a plasma membrane. Immediately adjacent is the inner segment, the site of active biosynthesis and oxidative metabolism. Although small molecules and proteins can diffuse along the connecting cilium between the inner and outer segments, this process occurs in general on a slower time scale than that of a light response.[11] This inter-compartment transport plays an important role in producing the molecules necessary for phototransduction and modulating their levels, but it does not contribute directly to the phototransduction cascade. While cytoplasmic diffusion is limited by the constriction at the connecting cilium, this constriction does not pose a major barrier to the propagation of electrical changes along the plasma membrane. Therefore, changes in membrane potential can be passively communicated from the outer segment to the synaptic terminal at the other end of the cell, where they control neurotransmitter release and transmit the signal to higher order neurons.

The phototransduction cascade takes place in the outer segments. There are two membrane systems in rods (Figs 18.3, 18.4). The disk membranes, which can be thought of as flattened sealed vesicles, are stacked up along the long axis of the outer segment and make up the vast majority of membrane surface in the cells. The plasma membrane surrounds the disks and the cytoplasm (Fig. 18.4), and forms the interface between the cell and the extracellular space. Although there are proteins linking these two membranes, there does not appear to be any flow of lipid or trans-membrane proteins between them in the outer segment. In contrast, cones have a single membrane which forms multiple invaginations that resemble disks but are not sealed and are continuous with the plasma membrane. A cone-like relationship between plasma membrane and nascent disks appears to be present at the base of rod outer segments (ROS),[12] although there is some controversy as to whether even those nascent disks may be distinct from, and surrounded by, plasma membrane.

The components of phototransduction, discussed in detail below, are primarily localized in the outer segments, and are mostly peripheral or integral membrane proteins. The disk and plasma membranes have distinct protein compositions.

Dark-adapted rods

The most information at the molecular level is available for the dim light responses of dark-adapted rods (see Chapter 19). These cells in this state can respond with amazingly high efficiency to single photon capture events.[13] It is useful conceptually to begin with the resting dark state and then consider the chain of events set in motion when a photon is absorbed.

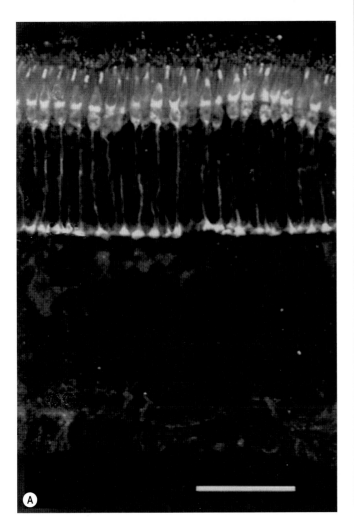

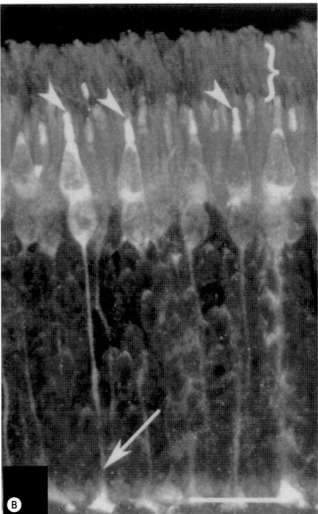

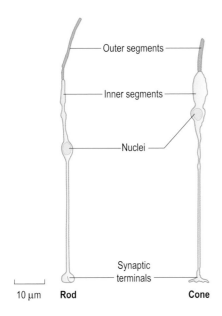

Figure 18.1 Immunofluorescence image of mammalian rods and cones. (**A**) Low magnification (scale bar, 50 μm), showing photoreceptor positions with respect to other retinal layers: OS, photoreceptor outer segments; IS, photoreceptor inner segments; ONL, outer nuclear layer (photoreceptor nuclei); OPL, outer plexiform layer (photoreceptor synapses and processes of bipolar and horizontal cells); INL, inner nuclear layer (nuclei of inner retinal neurons); IPL, inner plexiform layer (inner retinal synapses and processes); GCL, ganglion cell layers (cell bodies of retinal ganglion cells). (**B**) Higher magnification showing photoreceptors. Arrow, cone axon; arrowheads, cone outer segments; bracket, rod outer segments. The antibody used for immunofluorescence staining is specific for RGS9-1, whose concentration is 10-fold higher in cones than in rods. (Reproduced from Cowan et al 1998.[91]) (**C**) Schematic diagram of rods and cones and their compartments, with sizes and shapes based on a fluorescence micrograph of human retina.

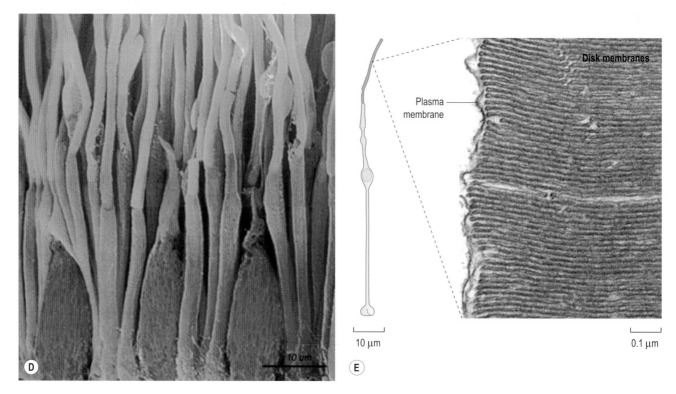

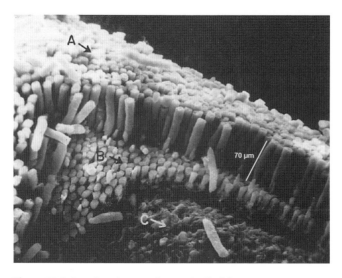

Plasma membrane

Disk membranes

10 µm

0.1 µm

Figure 18.1, cont'd (**D**) Scanning electron micrograph of rods and cones in primate retina. (Reproduced from webvision.med.utah.edu/imageswv/scanEMphoto.jpg with permission). (**E**) Magnified region of rod outer segment from transmission electron micrograph, showing the tightly packed disk membranes, and their relationship to the plasma membrane.

Figure 18.2 Scanning electron micrograph of bull frog outer segments. (Reproduced with permission from Bownds & Brodie 1975.[113])

The resting dark-adapted state

The membrane potential

In the dark, rods have a resting membrane potential of about −40 mV.[14] The negative value means that there is more positive charge on the outside of the plasma membrane than on the inside. In most resting neurons, the potential is closer to −70 mV, so rods are considered to be relatively depolarized as compared to other cells or, as we shall see, as compared to fully light-activated rods. In rods, as in other neurons, the source of membrane potential is the inequality of ion concentrations on either side of the plasma membrane generated by the Na^+/K^+-ATPase. This protein uses the energy released upon ATP hydrolysis to pump sodium ions out of the cell and potassium ions into the cell, and in rods it is found in the plasma membrane of the inner segment (see Chapter 19). In most neurons, potassium channels are the predominant conductance at resting potentials, and these allow potassium ions to flow out of the cell down their concentration gradient. Because these channels pass a specific cation, but no anions, a net positive charge accumulates on the outside of the membrane and a net negative charge accumulates on the inside of the membrane. This occurs until the electrical energy stored in the charge separation equals the energy stored in the potassium gradient, the relationship described by the Nernst equation, described in detail in many textbooks. Strictly speaking, this equation only works when the membrane in question contains a single type of channel permeable to only a single type of ion. In reality there are always additional minor conductances (other channels and transporters) so that the resting potential is modified by these to give the typical resting value of about −70 mV to −90 mV for excitable cells.

The dark current and the cGMP-gated channel

Rods also have a negative resting membrane potential, but its magnitude is less than in many other neurons because the effects of potassium channels in the inner segment are balanced by a major additional channel current in the outer segment. The less negative resting potential, or partially depolarized state, of rods is due to the cyclic-GMP-gated cation channel (see Chapter 19), also known as CNG channel (for cyclic nucleotide-gated) or light-sensitive channel or

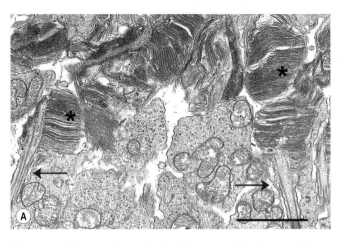
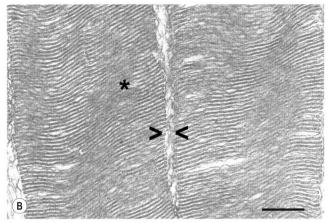

Figure 18.3 Electron micrographs showing postnatal development of outer retina in mouse. (**A**) postnatal day 10; (**B**) 5 weeks postnatal. *, rod outer segment disks; arrows, connecting cilia; arrowheads, plasma membrane separating individual rods.

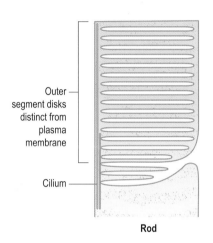
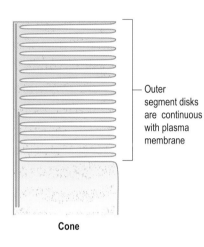

Outer segment disks distinct from plasma membrane

Cilium

Outer segment disks are continuous with plasma membrane

Rod   Cone

Figure 18.4 Schematic representation of base of outer segments. Both rod and cone outer segments are connected to the inner segments via the connecting cilium, or ciliary transition zone. Membrane topologies of rods and cones are similar near the base, with both rods and cones showing disk-like membranes as invaginations of the plasma membrane. Along the outer segment axis moving from the inner segment toward the distal tips, the rod disks become sealed compartments, completely surrounded by, but distinct from, the plasma membrane, whereas the cone disk and plasma membranes are continuous with the plasma membranes, and their interior is continuous with the extracellular space.

phototransduction channel (see Chapter 19). This multi-subunit channel only passes cations, but does not discriminate strictly among different cations, allowing Na^+, K^+, and Ca^{2+} to pass. Among physiologically important cations it has the highest permeability to Ca^{2+}, but because of its much higher concentration outside the cell (over 140 mM, as compared to ~1.5 mM for Ca^{2+}), Na^+ is the primary carrier of the current through the CNG channel in the dark. Because it tends to dissipate the gradient of charge, by allowing positive charge to enter the cell, the open CNG channel causes the cells to be partially depolarized. Put another way, because there are substantial inward sodium currents in the outer segment to balance outward potassium currents in the inner segment, the resting potential is moved away from the potassium "equilibrium" or Nernst potential of ~−90 mV, and toward the sodium "equilibrium" or Nernst potential of ~+60 mV. Because there is net flux of cations out of the inner segment plasma membrane, and a net flux of cations into the outer segment plasma membrane, as well as electrical conductance between the inner and outer segments, a complete circuit is made which is known as the circulating current or dark current.

Ca^{2+} and the exchanger

The CNG channel also has an influence on the concentration of Ca^{2+}, which in rods and cones, as in other cells, plays an important role in signaling and regulation. A Na^+/Ca^{2+}, K^+ exchanger protein, NCKX2,[15] uses the energy stored in the Na^+ gradient across the plasma membrane to push Ca^{2+} ions outside the cell, against their concentration gradient (see Chapter 19). In the absence of any inward flux of Ca^{2+}, this mechanism would lower the Ca^{2+} concentration in the outer segment cytoplasm to about 10 nM. In the dark, there is an inward leak of Ca^{2+} through the CNG channel, so the resting level of Ca^{2+} is a few hundred nanomolar.

Even though in the dark more channels are open than at any other time, because cGMP levels are at their highest, the percentage of total channels that are open is fairly small; most of the channels are closed even in the dark (see Chapter 19). One reason for this situation is that binding of more than one cGMP is necessary to give each channel a high probability of opening, i.e. the response of the channel to cGMP is non-linear and shows positive cooperativity. The other reason is that the dark concentration of cGMP is well below the concentration of cGMP at which channel opening probability is 50 percent, a number that reflects the affinity of cGMP for the channel subunits as well as the positive cooperativity. Because the dark cGMP concentration is at the low end of the dose response curve for the channel (therefore very far from saturation), even a small change in cytoplasmic concentration of cGMP can be immediately sensed as a change in the number of open channels and therefore of the dark current.

Control of [cGMP] by guanylate cyclase and PDE6

The resting, or dark, level of cGMP is determined by the balance between the activity of the enzyme that synthesizes it, guanylate cyclase (GC),[16–18] and the enzyme that degrades it, the cGMP phosphodiesterase, PDE6.[19–21] Both of these have much lower activities in the dark than they do in the light. When fully activated PDE6 is one of the most efficient enzymes known, and the most active of any of the PDE superfamily of cyclic nucleotide phosphodiesterases. The maximal turnover at saturating cGMP concentrations, k_{cat}, has been reported to be between 1000 s^{-1} and 7000 s^{-1}, and the Michaelis constant, the cGMP concentration at which activity is half-maximal is about 40 μM. Thus the enzymatic efficiency, k_{cat}/K_m, is thus in the range of 2.5×10^7 M^{-1}s^{-1} to 1.75×10^8 M^{-1}s^{-1}, implying that it operates near the diffusion limit. PDE6 requires two kinds of metal ions for activity. Mg^{2+} binds reversibly along with cGMP, and Zn^{2+} is permanently bound to high affinity sites essential for catalysis.[22] The other key players in photoactivation are also in quiescent states under dark conditions. The mechanisms for regulation of these enzymes by light are discussed below.

Guanylate cyclase uses GTP as a substrate, and produces cGMP and inorganic pyrophosphate as products. Inorganic pyrophosphate is rapidly broken down by the enzyme inorganic pyrophosphatase,[23] which is found at higher levels in rod outer segments than in any other cell type in which it has been measured. Guanylate cyclase is a transmembrane protein that is thought to exist in photoreceptor membranes as a dimer, with each dimer bound to the calcium binding protein, GCAP (guanylate cyclase activating protein).[24–26] PDE6 is a heterotetrameric peripheral membrane protein consisting of two large catalytic subunits, PDE6α and PDE6β, as well as two smaller identical inhibitory PDE6γ subunits. The catalytic subunits are each subject to the set of post-translational modifications associated with isoprenylation, as described below, so the enzyme is bound in a peripheral way to disk membranes. The PDE6α and PDE6β subunits bound in rods are similar in structure to one another, whereas in cones, two identical PDE6α subunits are bound. PDE6 catalytic subunits are related in sequence to a large family of cyclic nucleotide phosphodiesterases, the PDE superfamily.[27] These share sequence similarity in their catalytic domains, where the PDE6 isoforms most closely resemble in structure the PDE5 isoforms, which are the targets of sildenafil citrate (Viagra) and other drugs used to treat erectile dysfunction. In addition to the catalytic domains, PDE6 catalytic subunits each contain two GAF domains, one of which contains a non-catalytic binding site for cGMP. Because the total concentration of cGMP-binding GAF domains in the outer segments is about 50 μM, most of the cGMP in the cell is bound to PDE6. The physiological function of these non-catalytic sites is unclear, but their occupancy is coupled to binding of the catalytic subunits by the inhibitory PDE6γ subunit.[28–30] It is the PDE6γ subunit which interacts most extensively with the activated form of the G-protein, as described below.

Rhodopsin

In the dark, the photon receptor rhodopsin[3,31–33] is in its inactive state[34–36] with an inverse agonist, 11-*cis* retinal, covalently attached to its active site (Figs 18.5–18.7). Ligands that

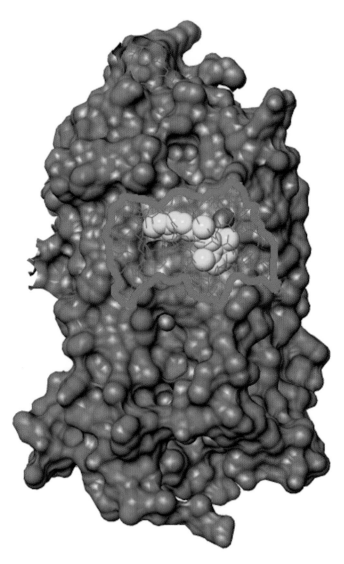

Figure 18.5 Surface rendering of rhodopsin crystal structure with cut-out of retinal binding pocket (red) to show 11-*cis* retinal chromophore (yellow spacefill). PDB coordinates, 1U19.[114]

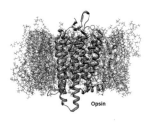  

Figure 18.6 Crystal structures highlighting helical arrangement in lipid bilayer of opsin (left, PDB structure 3CAP[115]), dark rhodopsin (middle, PDB structure 1U19[114]), and active opsin with peptide derived from the alpha subunit of transducin (right, PDB structure 3DQB[116]).

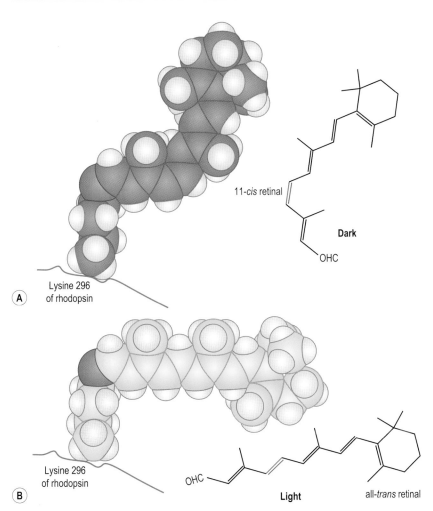

Figure 18.7 (**A**) Molecular structure, in space-filling representation, of 11-*cis* retinal in rhodopsin. The lower left depicts the sidechain of lysine 296, with the Schiff's base nitrogen shown in blue. Carbon atoms are shown in red, and hydrogens in white. (**B**) Molecular structure, in space-filling representation, of all *trans* retinal in rhodopsin. The lower left depicts the sidechain of lysine 296, with the Schiff's base nitrogen shown in blue. Carbon atoms are shown in yellow, and hydrogens in white.

activate receptors are called agonists, those that block the action of agonists are called antagonists, and those that actually reduce receptor activity below the level it has in the absence of any ligand are called inverse agonists. Rhodopsin has seven membrane-spanning helices and is a part of the G protein-coupled receptor (GPCR) super-family of signal transducing receptors. Like other GPCRs, its activity consists of its ability to catalyze the activation of a heterotrimeric G-protein. This is the activity of rhodopsin which is inhibited by the chromophore and inverse agonist 11-*cis*-retinal. The apo-form of rhodopsin without a ligand, known as opsin, has much higher G protein-activating activity than rhodopsin does, a fact that becomes important at high light levels but is relatively unimportant in the dark-adapted state. Rhodopsin is present at very high concentrations in the disk membrane (and somewhat lower concentrations in the plasma membrane), so that about one-third the surface area of the disks is occupied by rhodopsin, and the other two-thirds is occupied by phospholipids, cholesterol, and other minor lipids.

Rhodopsin is subject to post-translational modifications that are important for its function. N-linked carbohydrates are found on the domain of rhodopsin that faces the inside of the disks and the extracellular solution, and they appear to be important for proper transport of rhodopsin

from its site of synthesis in the inner segment to the outer segment membranes. Two palmitoyl groups are attached via thio-ester linkages to two adjacent cysteine residues near its carboxyl terminus, which tether the polypeptide chain to the cytoplasmic surface of the disk membrane forming a fourth cytoplasmic loop. In addition, the aldehyde moiety of 11-*cis* retinal and all-*trans* retinal is not bound to rhodopsin as a free aldehyde, but rather a covalent Schiff's base linkage is formed between ll-*cis* retinal and lysine residue 296 in the transmembrane domain of rhodopsin (Fig. 18.7).

There are 348 amino acids that comprise rhodopsin. Whereas mutations at only four different positions are found in patients with the disease congenital stationary night blindness (CSNB; Box 18.1), there are over 100 amino acid positions that are found mutated in patients with the blinding disease autosomal dominant retinitis pigmentosa (ADRP; Box 18.2). Therefore, the molecule is not only sensitive in that it can become activated with only one photon of light, it is also remarkably sensitive to its amino acid composition.

G-protein, G_t

Like rhodopsin, the G-protein, known as transducin[37–41] or G_t (Fig. 18.9), is also found in an inactive state in the dark.

Control

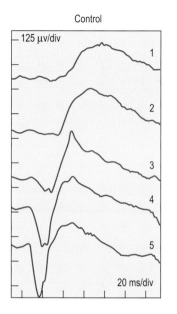

Patient V-2

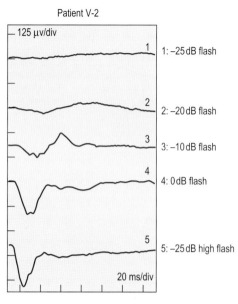

1: −25 dB flash

2: −20 dB flash

3: −10 dB flash

4: 0 dB flash

5: −25 dB high flash

Figure 18.8 Electroretinograms (ERGs) of a patient with CNSB due to rhodopsin A295V show altered signaling. The ERG represents the massed potential changes across the retina in response to light. The early negative deflection is the a-wave, derived from hyperpolarization of the photoreceptors. The later positive deflection is the b-wave, largely derived from depolarization of ON bipolar cells, which receive inputs from the photoreceptors. Scotopic ERGs of a representative control subject (left) and a patient (right). Intensities represented are 1, 0.01; 2, 0.03; 3, 0.03; 4, 1.0; 5, 3.0 cds s m^{-2}. The rod response is severely diminished in the patient. (Reproduced from Zeitz et al 2008.[117] Reproduced with permission from the Association for Research in Vision and Ophthalmology.)

Box 18.1 Congenital stationary night blindness

Congenital stationary night blindness (CSNB) comprises a group of genetically and clinically heterogeneous non-progressive retinal disorders, mainly due to defects in rod photoreceptor signal transduction and transmission. Mutations in genes coding for proteins of the phototransduction cascade (*RHO*, *GNAT1*, *PDE6B*, *RHOK*, *SAG* and *RDH5*) or genes associated with the transmission of the signals from the photoreceptors to bipolar cells (*NYX*, *GRM6*, *CANCA1F* and *CABP4*) can lead to this disease.

Night blindness, reduced or absent dark adaptation, is a typical and early sign of various forms of retinal dystrophies. Forms differ in inheritance pattern (autosomal dominant, autosomal recessive or X-linked), electroretinograms (presence or absence of a-wave), refractive error (presence or absence of myopia) and fundus appearance (Fig. 18.8). All patients with stationary night blindness have severely reduced rod ERG amplitudes and many may have modestly reduced cone ERG amplitudes. Rod sensitivity in patients is decreased by 100× to 1000× compared with normals. Almost all have cone responses with a normal peak implicit time (time interval between the light flash and subsequent peak of b-wave).

Box 18.2 Stargardt's disease

Stargardt's disease is an autosomal recessive form of juvenile macular degeneration with variable progression and severity. It is caused by mutations in the *ABCR* (*ABCA4*) gene on chromosome 1 which encodes a retina-specific ATP-binding cassette transporter protein, in the rims of rod and cone outer segment disks, proposed to serve as a flippase protein for phospholipids conjugated to all-*trans* retinal. Mutations in this gene have also been attributed to some cases of cone–rod dystrophy, retinitis pigmentosa, and age-related macular degeneration. Pathologic features of Stargardt's include accumulation of fluorescent lipofuscin pigments in the retinal pigmented epithelium (RPE) cells and retinal degeneration.

An important fluorophore in lipofuscin is the bis-retinoid pyridinium salt N-retinylidene-N-retinylethanolamine (A2E) formed by the condensation of all-*trans* retinaldehyde with phosphatidylethanolamine. Significant accumulation of A2E is seen in the RPE of patients with Stargardt's disease.[80,94] A2E has several potential cytotoxic effects on RPE cells, including destabilization of membranes, release of apoptotic proteins from mitochondria, sensitization of cells to blue light damage, impaired degradation of phospholipids from phagocytosed outer segments.

For a G-protein, this means that it has GDP, rather than GTP, bound to its α subunit, $G_{t\alpha}$, and exists in a heterotrimeric form, $G_{t\alpha\beta\gamma}$. The G-protein associates with the disk membranes by virtue of its covalently attached lipids.[42,43] A heterogeneous mixture of saturated and unsaturated 12- and 14-carbon fatty acids are found attached to the amino terminal glycine residue of $G_{t\alpha}$[44–46] linked through an amide bond. This reaction occurs during translation and is important for G-protein localization in the outer segment.[47] There is a 15-carbon isoprenyl group, farnesyl, attached to the carboxyl terminus of the γ subunit[48,49] through a thioether linkage. This sort of isoprenylation is found in other phototransduction proteins; in each case it occurs at a cysteine residue four residues from the carboxyl terminus of the initial translation product. The final three residues are cleaved by the action of a protease,[50] and the hydrophobicity conferred by the isoprenyl group is enhanced by the methyl

esterification of the carboxyl group on what has been converted into a C-terminal cysteine residue.

Importance of lipid milieu

The lipids of the disk membrane play an important role in assembling the phototransduction machinery[51] and providing the proteins a stage on which to act. Rhodopsin and guanylate cyclase are transmembrane proteins, and G_t and PDE6 are peripheral membrane proteins, attached to the membrane by covalently attached lipids (Figs 18.9, 18.10). GCAPs are also covalently lipidated by fatty acids on their amino termini, and another phototransduction protein, rhodopsin kinase, is isoprenylated (discussed below). The transmembrane CNG channel is found only in the plasma membrane, not in the disks. Experiments with purified proteins have shown that the proper assembly of these proteins on the disk membrane is essential for efficient interactions among them. Rhodopsin diffuses rather rapidly within the lipids of the disk membrane,[52-56] so that many encounters between it and the G-protein occur every second even in the dark. ROS membranes have an unusual lipid composition, with a high content of polyunsaturated fatty acids. Close to 40 percent of the fatty acyl groups of ROS phospholipids are the ω-3 fatty acid, docosahexaenoic acid (DHA or 22:6), which has 22 carbons and 6 double bonds.[57] DHA can only be synthesized in the body if essential ω-3 fatty acids are consumed in the diet. Cholesterol, while making up only 10–15 percent of total outer segment lipids, plays an important role in establishing the lipid environment in which phototransduction reactions occur.[56]

The activation phase of a light response

Photoisomerization of rhodopsin

Phototransduction begins with absorption of light by rhodopsin (Figs 18.9, 18.11). Rhodopsin has a broad absorption spectrum with a wavelength of peak absorbance close to 500 nm in the green portion of the visible spectrum (see Fig. 18.14 below). The probability of capture by rhodopsin of a photon of green light passing through the retina is fairly

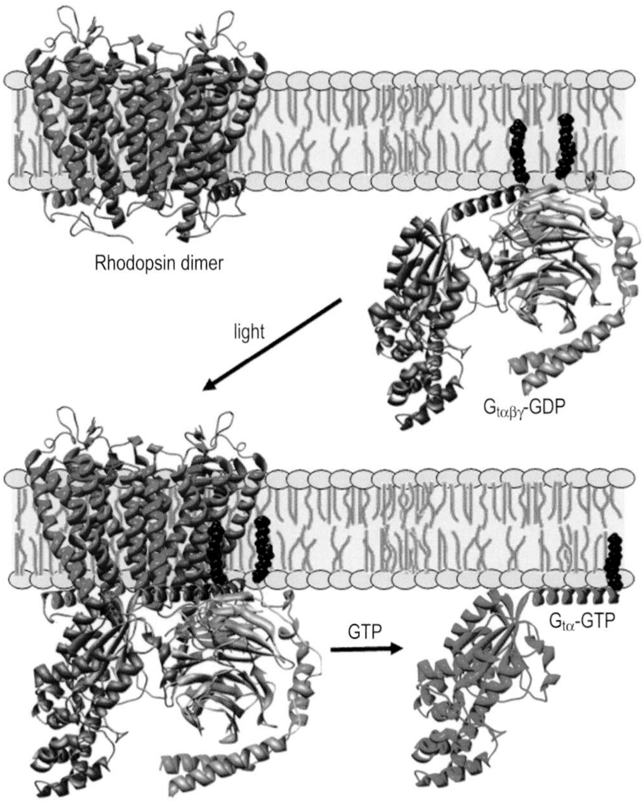

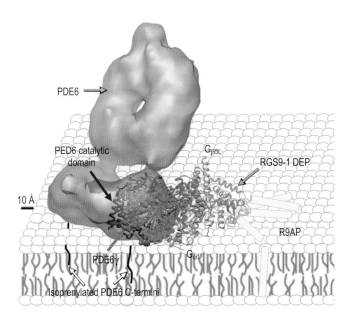

Figure 18.9 Structural models for interactions among transducin subunits, metarhodopsin II, and the disk membrane during light activation. (Modifed from Wang et al 2008[56].)

Figure 18.10 A multi-subunit complex essential for normal photoresponse recovery kinetics. PDE6γ C-terminal fragment is shown in red space-filling representation. (Modified from Wensel 2008.[51])

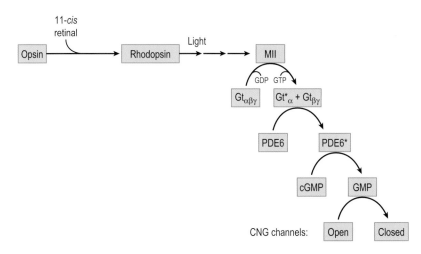

Figure 18.11 Schematic representation of phototransduction activation cascade.

high (~50 percent for 20 μm long human rods, and ~89 percent for 60 μm long toad rods), due to the large number of disks (~1000) through which it must pass, and the high concentration of rhodopsin in each one. When rhodopsin absorbs light, the probability that it will undergo a structural transition, or photoisomerization, is extraordinarily high: 65 percent.[58] This is one of the most efficient photochemical reactions known, and this efficiency helps explain the exquisite sensitivity of rod cells to a single photon of light. The bound 11-*cis* retinal chromophore is responsible for the light absorption (polypeptides do not absorb visible light unless they are bound to some chromophoric ligand), and in its excited state induced by the light absorption, 11-*cis* retinal undergoes a photoisomerization at the bond between carbon 11 and carbon 12 in its structure from 11-*cis* to all-*trans* retinal (Fig. 18.7). The rearrangement of bonds causes a dramatic change in the structure of retinal, with the new all-*trans* form representing a straightening out of a kinked molecule. This molecule no longer acts as an inverse agonist, but now as an agonist, or activator.

The retinal chromophore, covalently attached via a Schiff's base linkage to a lysine residue at amino acid position 296 on rhodopsin (see above), induces structural changes in the protein upon photoisomerization. After photoisomerization of the chromophore, the protein undergoes a series of conformational changes in the arrangement of its helices to accommodate this event. In other words, when retinal stretches out into the all-*trans* shape, it pushes part of the protein out of the way. These conformational changes ultimately lead to formation of Metarhodopsin II, a significantly different structure to dark rhodopsin. The intermediate conformations of photolyzed rhodopsin can be monitored by high speed and/or low temperature spectroscopy , including UV-visible absorbance, infrared, Raman, and fluorescence spectroscopies.[59-61] These and other studies have shown that upon light activation, rhodopsin first changes to bathorhodopsin, lumirhodopsin, metarhodopsin I (MI) and metarhodopsin II (MII or R*, the form that activates transducin), and ultimately metarhodopsin III (MIII) before the all-*trans* chromophore becomes hydrolyzed, leaving the apoprotein opsin. At physiological temperatures, almost immediately upon photoactivation, MI and MII establish a dynamic equilibrium which can be monitored by the differences in their absorbance spectra. It is MII rather than MI that preferentially binds to and activates the G protein.

G-protein activation

The key functional difference between rhodopsin and metarhodopsin II is in their interactions with the G-protein transducin (G_t). MII binds to the G-protein much more tightly than does rhodopsin and serves as an efficient catalyst for the activation of G_t. G-protein activation occurs about 10 million times faster at high MII concentrations than it does in the dark.[62] The G-protein binds to MII with GDP bound, but rapidly releases the bound GDP (Fig. 18.9). This GDP release is the slow step (~once per 10,000 s) for G-protein activation in the dark, but occurs very rapidly when G_t is bound to MII. Once GDP is released, the millimolar concentrations of GTP in the cell ensure rapid binding of GTP. Once GTP is bound to the $G_{t\alpha}$ subunit, the G-protein is in its active state. The activated $G_{t\alpha}$-GTP complex has much lower affinity for both the $G_{\beta\gamma}$ subunits and for MII than do either the

nucleotide-free state or the GDP-bound state of $G_{t\alpha}$. One MII molecule can activate many G-proteins in a catalytic fashion, at a rate exceeding 150 per second in mammalian rods.

PDE6 activation

When $G_{t\alpha}$ loses its affinity for MII and $G_{t\beta\gamma}$ upon binding GTP, it also gains greatly enhanced affinity for the cGMP phosphodiesterase, PDE6.[21,63] When $G_{t\alpha}$-GTP is bound, the catalytic activity of PDE6 goes up about one-thousand fold. This dramatic increase in enzymatic activity means that cytoplasmic levels of cGMP begin to fall rapidly in the vicinity of activated PDE6. Fully activated PDE6 is one of the most efficient enzymes known, with a catalytic efficiency, k_{cat}/K_m, of about 4×10^8 $M^{-1} \cdot s^{-1}$. This high catalytic efficiency means that nearly every time a cGMP molecule collides by diffusion with an activated PDE6 molecule, it will be hydrolyzed to form 5'-GMP. If 350 PDE6 molecules are fully activated in a single primate rod cell, they can hydrolyze most of its cGMP in one-tenth of a second.

Channel closing

Individual cGMP molecules are constantly dissociating from and binding to their sites on the CNG channel, even in the dark (see Chapter 19). Thus each CNG channel samples the cGMP concentration in the adjacent cytoplasm once every few milliseconds. For practical purposes, the channels respond immediately to the decline in cGMP in the cytoplasm, and they do this with a very steep concentration dependence, because of the cooperativity noted above. Closing of channels reduces the flow of Na^+ ions into the cell, reducing the dark current and making the membrane potential more negative (i.e. hyperpolarized). Simultaneously, the flux of Ca^{2+} into the cell is reduced as well[64] by the action of the Na^+/Ca^{2+} exchanger discussed above.

Slowing of neurotransmitter release

Communication of rods with downstream bipolar cells is by the release of the neurotransmitter, glutamate. High levels of glutamate release by rods signal total darkness to bipolar cells, and reductions in the levels of glutamate release signal absorption of light. This change results from the hyperpolarization of the cell membrane, which is passively propagated along the plasma membrane from outer segment to synaptic terminal. Because rod cells are relatively short as compared to neurons with long axons, active propagation of the signal by action potentials is not needed to communicate potential changes at the outer segment to the synapse.

The recovery phase

The biochemical changes that occur during activation contain within them the seeds of destruction for the activated state; they set the stage for the recovery phase. It is simplest to consider this process, which is rather more complex than activation, in terms of recovery from activation by a single dim flash of light after which no more MII formation occurs.

Rhodopsin phosphorylation, retinoid recycling and regeneration

Because of the catalytic activity of MII, as long as MII survives intact downstream signaling will continue, and cGMP will

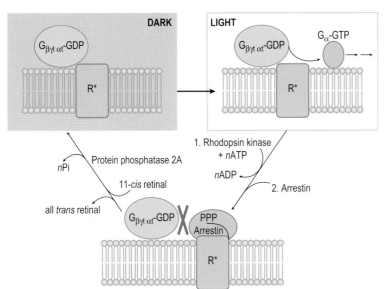

Figure 18.12 The role of phosphorylation by rhodopsin kinase in the inactivation of photoexcited rhodopsin, MII or R*. In the dark-adapted rod, rhodopsin is not the preferred substrate for rhodopsin kinase (RK), so on photoactivation and conversion to R* it catalyzes rapid guanosine diphosphate-guanosine triphosphate (GDP-GTP) exchange on $G_{\alpha t}$. Activated $G_{\alpha t}$-GTP activates cyclic guanosine monophosphate (cGMP) phosphodiesterase (PDE), leading to a closure of cGMP-gated cation channels and plasma membrane hyperpolarization. R* binds to and activates rhodopsin kinase, which uses adenosine triphosphate (ATP) to add multiple phosphates to the carboxyl-terminal region of R*, inducing a state that binds with high affinity to arrestin. Arrestin binding to phosphorylated rhodopsin effectively quenches the activity of R* until decay of metarhodopsin II and regeneration of rhodopsin lead to dephosphorylation by protein phosphatase 2A (PP2A). (Modifed from Gross et al 2008.[118])

be subject to rapid degradation. There are multiple mechanisms for inactivating MII, but the fastest-acting and most important one appears to be phosphorylation by rhodopsin kinase,[65] RK or RK1 (Fig. 18.12). Like the G-protein, RK preferentially recognizes the activated form of rhodopsin, MII. RK uses ATP to attach phosphate groups to serine residues in the carboxyl-terminal tail of rhodopsin. Up to eight phosphates may be added, and if any of the key residues are missing, MII inactivation kinetics are slowed.[66,67] Phosphorylation of MII reduces its affinity for the G-protein, and slows the activation of the G-protein. Rhodopsin kinase is a member of a family of kinases specific for activated forms of G-protein-coupled receptors, which includes a kinase found only in cones (but not found in all species with cones), RK7.[68,69] RK-1 associates with the disk membranes via a covalently attached farnesyl group,[70,71] and RK7 via a geranylgeranyl group. The greater hydrophobicity of the 20 carbon geranylgeranyl group as compared to 15 carbon farnesyl moiety may enhance the efficiency of interactions between RK7 and photoexcited cone pigments.

On a slower time scale than that of phosphorylation, photo-excited rhodopsin also undergoes loss of covalently attached all-trans retinal. The conformational change accompanying cis–trans isomerizations and light activation renders the Schiff's base between lysine 296 and retinal accessible to water, so that unlike the dark form of rhodopsin, MII is subject to Schiff's base hydrolysis and release of all-trans retinal from its binding pocket. All trans-retinal, like other aldehydes, is a chemically reactive species that can associate with proteins in the disk membrane, including rhodopsin, and form new Schiff's base linkages to amino groups from protein lysine residues or from phosphatidylethanolamine in the lipid phase into which it segregates. The conjugate between phosphatidylethanolamine and all-trans retinal, sometimes referred to as APE, is the precursor of a compound known as A2E (Fig. 18.13) which accumulates in animals and human patients deficient in the product of the Stargardt disease gene ABCA4 (or ABCR, see Box 18.2). ABCA4[72,73] belongs to a family of ATP-binding cassette transmembrane transporters which includes multidrug resistance proteins and lipid flippases. A proposed function for ABCA4

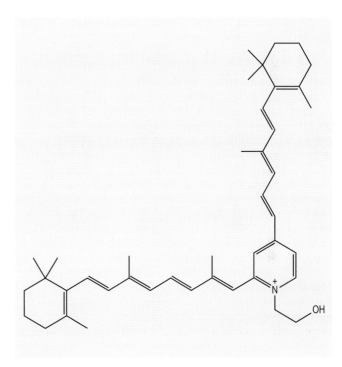

Figure 18.13 Chemical structure of N-retinylidene-N-retinylethanolamine (A2E).

is the trans-bilayer transport of APE. The aldehyde all trans-retinal must be converted to its alcohol form before being transported to the retinal pigmented epithelium (RPE) for recycling into 11-cis-retinal (see Chapter 13). This reduction reaction is catalyzed by one or more dehydrogenase enzymes, known as retinol dehydrogenases, all of which use NADPH as the reducing agent.

Opsin, whether phosphorylated or not, can bind 11-cis-retinal, supplied by the retinoid cycle in the RPE, to regenerate rhodopsin in its dark state. Phosphorylated rhodopsin does not bind arrestin (see below) with high affinity, and can be de-phosphorylated by the action of protein phosphatase 2A (PP2A). Because rhodopsin with 11-*cis*-retinal bound does not activate RK, rhodopsin in the dark is predominantly in a non-phosphorylated state.

Arrestin binding

Phosphorylation of rhodopsin by RK is not, in itself, sufficient for shutting off the activity of MII completely. A key additional step is binding of the capping protein, arrestin.[74,75] When MII is multiply phosphorylated, arrestin binds very tightly and blocks further binding and activation of G_t (Fig. 18.12). This MII is unable to propagate phototransduction due to steric hindrance of the G_t binding site. Eventually, all-*trans*-retinal will be removed by hydrolysis of the Schiff's base linkage and dissociation from the protein, yielding opsin, from which arrestin will dissociate and phosphates will be removed by a phosphoprotein phosphatase. This latter series of reactions occurs on a much longer time scale than the recovery from a dim flash excitation, as does regeneration of rhodopsin by binding of 11-*cis*-retinal to opsin (see discussion of dark adaptation below). Arrestin is a 48 kDa protein also known as 48 K or S-antigen, the latter a reference to its involvement in experimentally induced autoimmune uveitis.

cGMP restoration by guanylate cyclase activation

On an even faster time-scale than MII activation,[76,77] changes in guanylate cyclase activity occur via a calcium-feedback mechanism that serve to halt the lowering of cytoplasmic [cGMP] and eventually restore it to dark levels. The calcium feedback consists of a lowering of cytoplasmic [Ca^{2+}] that in turn activates guanylate cyclase.[78] The mechanism by which [Ca^{2+}] is lowered is simply that the Na^+/Ca^{2+} exchanger continues to pump Ca^{2+} out of the cell while the leak of Ca^{2+} into the cell through the CNG channels is reduced or blocked when the channels close (see Chapter 19). Changes in [Ca^{2+}] work through calcium binding proteins[79] known as GCAPs (guanylate cyclase binding proteins), members of the calmodulin superfamily of calcium-binding proteins. They bind to GC and keep its activity low when they have calcium bound, but dramatically enhance its activity when [Ca^{2+}] becomes low. Ca^{2+} inhibition of the GCAP-GC complex is also co-operative, so that the dependence of GC activity is very steeply dependent on Ca^{2+} concentration. This cooperativity allows very rapid changes in GC activity in response to changes in CNG channel current. GC activation through Ca^{2+} feedback begins to balance the activation of PDE6 in less than 100 ms after a dim flash (in mice), and eventually restores [cGMP] to the resting, dark, level.

The retinal photoreceptor guanylate cyclase (GC) isoforms[26,80–83] are members of a family of transmembrane or particulate guanylate cyclases, some of which, unlike the photoreceptor isoforms are receptors for peptides or other extracellular ligands. Transmembrane GC enzymes are found in membranes as dimers. Each GC polypeptide has a glycosylated extracellular domain, which participates in dimer formation, a single transmembrane alpha-helix, and intracellular domains. The intracellular domains include a kinase homology domain, a dimerization domain, the catalytic domain, which is responsible for the production of cGMP, and a C-terminal domain which may mediate interactions with proteins at the cytoplasmic surface of the membrane. The GCAP proteins, which have fatty acids attached to their amino termini, remain bound to GC even at low concentrations of intracellular Ca^{2+}. Binding of Ca^{2+} changes their conformation in such a way as to allow dramatic stimulation of the activity of GC.

There are two isoforms each of GC and GCAPs in human photoreceptors and most mammals. The importance of GC1 (also known as GCE or retGC1 or ROS-GC1) in human phototransduction is underscored by mutations in the gene (GUCY2D) encoding this protein which causes Leber's congenital amaurosis (LCA) in human patients.[84,85] The importance of GCAP1 is underscored by the association of mutations in GCAP1 with autosomal dominant cone dystrophy.[86] In mice, a double-knockout of the GCAP1 and GCAP2 genes, which are in close proximity on the chromosome, leads to dark-adapted rod flash responses that are much slower and larger in amplitude than those in wildtype rods. Transgenic expression of GCAP2 only partially rescues this phenotype, whereas GCAP1 expression appears to provide complete rescue if expression levels are sufficient.[87]

G-protein and PDE6 inactivation by RGS9-1

Even after MII has been quenched by phosphorylation and arrestin binding, signaling continues as long as GTP remains bound to the G-protein so that PDE6 remains in an activated state. The dark GDP state of $G_{t\alpha}$ is restored through hydrolysis of the bound GTP. Once phosphate is released, causing a conformational transformation in $G_{t\alpha}$ that restores its affinity for $G_{t\beta\gamma}$ and dramatically lowers its affinity for PDE6, PDE6 returns to its dark inactive state. This GTP hydrolysis step is the slowest step[88,89] in rod vision (time constant of about 200 ms in mammals and about 2 s in amphibians), and is regulated by a complex of proteins that act as GTPase accelerating proteins (GAPs). The core of this complex is RGS9-1,[90,91] a member of the regulator of G-protein signaling (RGS) family of GAPs found only in photoreceptor cells (Box 18.3). It is always bound to two obligate subunits $G_{\beta5L}$, and a trans-membrane anchor protein R9AP. $G_{\beta5L}$[92,93] is a member of the G-protein β subunit family, but its function is distinct from the other four members which bind very tightly to G-protein γ subunits and help G-protein α subunits couple to receptors. Rather, $G_{\beta5L}$ binds tightly to a G-γ-like domain,[94] or GGL domain, found in the R7 family of RGS proteins. $G_{\beta5L}$ binding is important for stability and specificity of RGS9-1 and other R7 RGS proteins, which tethers the complex to the disk membrane. This complex binds to the $G_{t\alpha}$-GTP-PDE6 complex and stimulates rapid hydrolysis of GTP by $G_{t\alpha}$ (Fig. 18.10), with the result that PDE6 activity falls to its dark levels and cGMP levels recover to what they were before the flash. As the slowest step in the recovery phase concludes, CNG channels are once again open to the same extent as before the flash, and both the membrane potential and [Ca^{2+}] have recovered as well (see Chapter 19).

Amplification

One of the most remarkable features of rod vision is that all the events described above can be triggered in a robust way by an individual photon activating a single molecule of MII within a rod cell. Rod cells thus achieve the quantum limit of sensitivity, the ability to sense and communicate the presence of a single photon of light. One reason for this ability

Patients with bradyopsia, or slow vision, show a prolonged response suppression in electroretinogram recordings. Bradyopsia is a recently discovered rare genetic disorder of the retina characterized by an inability to rapidly shut off the phototransduction cascade following the stimulation of the photoreceptors by a photon of light. Patients present with symptoms of photophobia, problems adjusting to bright light, and difficulties seeing moving objects; they often complain that they cannot play ball sports such as soccer, as they cannot see the ball in motion.[119] The mutation in these patients is in the *RGS9* (regulator of G-protein signaling 9) gene, the product of which is involved in the deactivation of photoreceptor responses (Fig. 18.12), or in the gene encoding R9AP, the RGS9-1 anchor protein, whose presence is essential to prevent degradation of RGS9-1. The discovery of bradyopsia and its molecular basis provides a fascinating insight into the ever-expanding role of medical genetics and animal models of human disease.

When the RGS9 gene was first discovered as one encoding a key player in phototransduction, it was recognized as a candidate gene for human retinal disease. However, sequencing of DNA from numerous patients with hereditary retinal degenerations uncovered polymorphisms in the RGS9 gene, but none with any disease association.[120,121] Results from mice engineered to have a genetic defect in their RGS9 gene revealed that homozygotes suffered not from retinal degeneration, but from slowed recovery from light responses in rod and cone photoreceptors. R9AP knockout mice had retinal phenotypes almost identical to those of RGS9 knockouts. These findings led researchers at the Massachusetts Eye and Ear Infirmary to search through patient records and the ophthalmological literature for patients whose ERG results resembled those of the knockout mice. This work, combined with DNA sequencing led to the discovery of this previously uncharacterized disease.[122]

is the multiple stages of amplification achieved by the catalytic activity of phototransduction components. A single MII molecule can produce many activated $G_{t\alpha}$-GTP molecules. Each of these, when bound to PDE6, can destroy thousands of cGMP molecules per second. Each cGMP molecule removed from the cytoplasm by hydrolysis decreases the number of open channels, each one of which passes nearly 2000 positively charged ions per second. Moreover, the cooperativity of the channel means that a 10 percent change in [cGMP] yields a nearly 30 percent change in the number of open channels. Depending on the species and their rod geometry, the overall gain in terms of numbers of ions blocked per rhodopsin photoisomerized is between 5×10^5 and 10^7.[95] These staggering numbers would seem to make it obvious why a single photon can be detected, but there is one problem that always accompanies amplification: noise is amplified along with the signal. Ultimately, reliable detection depends not just on a strong signal, but on a high signal-to-noise ratio. There are many sources of noise in the phototransduction cascade, and detailed studies suggest that many of the features of the phototransduction cascade and the molecules that mediate it have evolved to minimize noise. Individual rods produce false single-photon-like signals with a frequency of about once every 160 seconds, making it difficult for us to discriminate real light from darkness at intensity levels below that needed to induce single photon responses with this frequency.

Responses to saturating light levels

When light levels exceed that needed to photoisomerize one rhodopsin in every 10,000 (0.01 percent), nearly all the cGMP in the cell is hydrolyzed, pushing the concentration near zero, and all the channels close, completely shutting off the dark current. Beyond this intensity, increasing the light intensity only accelerates the kinetics of the rising phase, and prolongs the time in saturation. As described in Chapter 19, following a flash of saturating intensity, further rod responses cannot be elicited until several seconds later when the cell has recovered from the initial response. Over a fairly wide range of intensities, the time from the flash to recovery of the dark current to a criterion level (e.g. 90 percent or 50 percent) is linear with the logarithm of the flash intensity,[96] indicating a single exponential process, characterized by a "dominant time constant,"[97] which represents the slowest step in the transduction cascade over that intensity range. Results from transgenic mice with increased levels of the RGS9-1 GAP complex[88] have revealed that the levels of this complex, and therefore, the kinetics of GTP hydrolysis by $G_{t\alpha}$, determine the timing of the slow step in photoresponse recovery, and therefore the time resolution of rod vision.

Adaptation to changing levels of ambient lighting (see Chapter 20)

In the presence of even a fairly dim background light, the responses of rods to increments in intensity are attenuated and accelerated relative to those obtained in completely dark-adapted conditions. This desensitizing effect becomes more pronounced as the intensity of the background light increases. These changes are referred to as light adaptation. The changes are reversible upon return to darkness, with the return of sensitivity occurring over a time scale of minutes, proceeding through both fast and slow phases. This restoration of sensitivity is referred to as dark adaptation. The slow phase of dark adaptation is usually attributed to the regeneration of rhodopsin from opsin that has been formed by prolonged exposure to bright light, i.e. the rebinding of 11-*cis*-retinal to the pool of apoprotein formed in bright background light. Light adaptation can be observed on a time scale of seconds, and while it is not completely understood, appears to result from changes at several steps in the phototransduction cascade. The most important of these is simply an acceleration of cGMP hydrolysis and synthesis caused by PDE6 activation and low-[Ca^{2+}]-induced guanylate cyclase activation. Because these reactions are already proceeding at a fast clip due to the background light, further increases in light elicit lower amplitude responses, but these occur more rapidly. In addition, the low [Ca^{2+}] present under these conditions may modulate the activity of rhodopsin kinase, the CNG channel, and the RGS9-1 GAP complex.

Turnover of guanine nucleotides

While light greatly accelerates the turnover of guanine nucleotides in photoreceptors, even in the dark there is a very high rate of flux through cycles of synthesis and hydrolysis.[98] Although each PDE6 molecule may have only 0.1 percent or so of its maximal enzymatic activity in the dark, the high concentration of PDE6 (on the order of 20 µM) and its high enzymatic efficiency means that the entire cellular pool of cGMP in a mammalian rod is hydrolyzed at a rate of about once every 2.5 seconds. This hydrolytic flux is balanced by continual cGMP synthesis catalyzed by guanylate cyclase to achieve a constant steady-state level of cGMP in the dark. This dark cycling through cGMP consumes high amounts of GTP, which must be continually replenished by energy metabolism. To that end photoreceptors can use both aerobic respiration by dense concentrations of mitochondria in the region of the inner segment proximal to the outer segment, and glycolysis, whose enzymes are found in both the outer and inner segments. GTP is produced from GDP and inorganic phosphate, Pi, in the citric acid cycle by the action of succinate dehydrogenase, from GDP and ATP by the action nucleoside diphosphate kinase, and from two molecules of GDP by the action of a GDP phosphotransferase.[99] This apparent futile cycling and squandering of cellular energy sources may seem wasteful. However, it likely contributes to the speed of phototransduction, and its cost in terms of ATP use is dwarfed by the needs of the Na^+/K^+ ATPase pump in the inner segment.

Comparison of cones and rods

Similarities and differences of phototransduction molecules

Many of the phototransduction molecules of rods have homologs that perform similar functions in cones. The cone pigments, which are very similar in structure to rhodopsin have different absorption spectra, and different photophysical properties (Fig. 18.14). The thermal isomerizations rates of cone pigments are higher than those for rhodopsin,[100] as are the rates of spontaneous dissociation of 11-cis-retinal.[101] The evolution of different absorption spectra, known as spectral tuning, for the color pigments and rhodopsin arises due to varying amino acid side chains in and near the chromophore binding pocket. Pioneering work by George Wald[102] (for which he shared the Nobel Prize in Physiology or Medicine in 1967;[103]), Ruth Hubbard[35] and others showed that rod cells are most sensitive to green light (absorbance

maximum around 498 nm), and are not sensitive in the red (wavelengths longer than 640 nm). Many studies have addressed the molecular mechanism of spectral tuning, and it has been determined that the protonation state of the Schiff's base, as well as the polarity of the amino acids adjacent to it, dictate whether the absorption spectrum of the photoreceptor will shift to the blue (short wavelengths) or to the red (long wavelengths) (Fig. 18.14).[104–109]

The G-protein subunits of rods and cones are distinct but closely related. Cone $G_{\alpha t}$ is known as $G_{\alpha t2}$ or cone transducin, and is 81 percent identical to rod transducin. Rod transducin's beta subunit is $G_{\beta 1}$, whereas cone transducin uses $G_{\beta 3}$ which has 83 percent sequence identity. The rod $G_{\gamma ?}$ and cone $G_{\gamma 8}$ subunits are less than 40 percent identical; both undergo proteolytic cleavage of three residues, farnesylation, and methyl esterification at their C-termini.

Unlike rods, in which there are two related but distinct PDE6 catalytic subunits, $PDE6_{\beta}$ and $PDE6_{\alpha}$, cones contain a single $PDE6_{\alpha}$ subunit, which forms a homodimer that together with two copies of cone-specific $PDE6_{\gamma C}$ forms the holoenzyme heterotetramer. The CNG channel has cone-specific α and β subunits. Both rod rhodopsin kinase, GRK-1 and a cone-specific rhodopsin kinase, GRK-7 are found in cones, and there is a cone-specific form of arrestin. The RGS9-1-$G_{\beta 5}$-R9AP complex is identical in composition in cones (except that some of the $G_{\beta 5}$ is a shorter splice variant not found in rods), but is present at ten-fold higher levels in cones than in rods,[91,110] to allow the fast kinetics described below.

Physiological differences

Cones are much less sensitive than rods, they have much faster kinetics, and they are nearly impossible to saturate even in bright sunlight. The reasons for these differences are currently the subject of research in numerous laboratories and remain controversial. However, most current data point away from the reactions of the activation phase. The quantum sensitivity of the cone pigments is very similar to that of rhodopsin, as is their efficiency at activating the G-protein. The cone PDE6 has similar kinetic properties to those of the rod isoform, and the CNG channels have similar responsiveness to cGMP. Rather, most research points toward faster mechanisms for shutting off the response as the major sources for differences in kinetics and sensitivity between rods and cones. These include, possibly, faster phosphorylation of the activated cone pigments as compared to MII phosphorylation,[111,112] faster dissociation of all-trans-retinal, and faster G-protein GTP hydrolysis, driven by ten-fold higher concentrations of the RGS9-1/Gβ5/R9AP GAP complex in cones than in rods. Low activity (poor G-protein activation) of bleached pigment (opsin-like apo-protein or phosphorylated and arrestin-capped forms) may also play a role in preventing saturation of cones.

Phototransduction and disease

The major classes of disease resulting from defects in phototransduction include retinal degeneration, night blindness, color blindness, and achromatopsia (Box 18.5). Of these, each of the heritable diseases, except for X-linked color

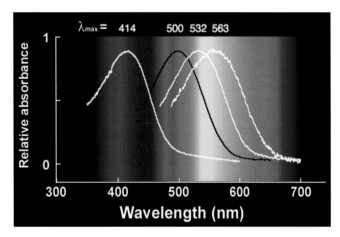

Figure 18.14 Absorption spectra of rhodopsin and the red-sensitive (L), green-sensitive (M) and blue-sensitive (S) cone pigments. (Courtesy of Dr. Masahiro Kono.)

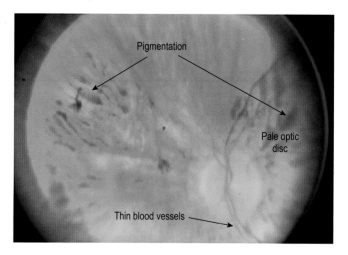

Figure 18.15 Fundus photograph taken of patient with autosomal dominant retinitis pigmentosa. Note the pale optic disc, presence of pigmentation and thinning of blood vessels (where indicated). (Courtesy of Dr. Leo P. Semes.)

blindness, is quite rare, but in total these conditions affect many thousands of patients who suffer various degrees of visual impairment.

Retinal degeneration and night blindness

Retinal degeneration is a progressive loss of visual function, usually accompanied by death of both rod and cone photoreceptor cells, which can lead to severe visual impairment and complete blindness. Such conditions are classified under a variety of names according to the appearance of the retina and pathology of the disease. Hereditary degenerative disorders, including retinitis pigmentosa (Fig 18.15), can arise from mutations in genes encoding photoreceptor proteins, as well as from mutations in many other non-phototransduction genes (see Box 18.4). Many such mutations have been identified in humans and animals, and lists of these can be accessed on the web site for the Retinal Information Network: http://www.sph.uth.tmc.edu/Retnet/ or the site for the Ocular Molecular Genetics Institute: http://eyegene.meei.harvard.edu/. Among the phototransduction proteins discussed above, retinal degeneration has been linked to defects in the genes encoding rhodopsin, PDE6, guanylate cyclase, CNG channel, guanylate cyclase, and guanylate cyclase activating proteins (GCAPs). Congenital stationary night blindness, an absence or severe deficiency in rod vision (referred to as "stationary" because it does not progress) has been linked to genes encoding rhodopsin, the transducin (G-protein) alpha subunit, rhodopsin kinase and arrestin.

What we don't know

Although the basic biochemistry of the major events in phototransduction is fairly well understood, most of the structural and mechanistic details remain to be determined. Although some hints are available, we do not know exactly how MII activates the G-protein or how the G-protein activates PDE6. Many of the biochemical events are mediated

Box 18.4 Retinitis pigmentosa

Retinitis pigmentosa (RP) is a progressive rod–cone dystrophy that presents with progressive field loss and eventual visual activity decline. The worldwide prevalence of RP is 1 in 4000 individuals.[123]

Typical symptoms include night blindness followed by decreasing visual fields, leading to tunnel vision and eventually blindness. Clinical hallmarks include gradually increasing bone-spiculed pigmentation, attenuation of retinal vasculature, waxy disc pallor, as well as diminished, abnormal or absent electroretinograph responses. Typically, symptoms start in the early teenage years, and severe visual impairment occurs by the ages of 40 to 50. There are early-onset forms of RP, with the underlying genetic cause proving to be a useful predictor of the severity of the disease.

The earliest of symptoms, measurable before visible onset of the disease, are abnormal light-evoked ERGs.[124] The ADRP form is less prevalent and less severe than the recessive form and has been linked to a variety of genes involved in phototransduction or rod cell function. The most prevalent mutation in human ADRP patients, in which Pro23 is replaced by His (P34H) in the rhodopsin gene, has been identified in ~15 percent of ADRP families in the United States. A comprehensive list of genes causing RP and other retinopathies can be found at the RetNet website (http://www.sph.uth.tmc.edu/Retnet/). In addition to simple forms of RP, there are syndromes involving pleiotropic effects (FIX). The most frequent form of syndromic RP is Usher syndrome. This disease presents as early-onset hearing loss followed by development of RP in the teenage years.[125]

by large membrane-attached multi-protein complexes, but we do not know how these are organized. For example, during the rate limiting step of vision, G-protein GTP hydrolysis, a minimum of eight different polypeptides ($G_{t\alpha}$, PDEα, PDEβ, 2 PDEγ, RGS9-1, Gβ5, R9AP) are present, and we have no idea how they are arranged on the surface of the disk membrane. The multisubunit CNG channel, and Na/Ca exchanger are associated with one another in the plasma membrane, and tightly attached to proteins in the disk rims, but the structures of these complexes are also unknown. Rhodopsin can form ordered rows of dimers in isolated disk membranes, but the physiological roles of these arrays and of the dimers themselves remain controversial. These structural questions remain an active area of current research.

There are also many remaining questions about the inactivation phase of the light response and about the events of light and dark adaptation. For example, the active lifetime of MII molecules seems to be under much more stringent control than would be expected for an excited species whose inactivation occurs by way of a randomly timed event such as phosphorylation or any other chemical reaction. Modeling studies suggest that multiple sequential chemical changes are needed to explain what is known as the reproducibility of the single photon responses. Studies of the kinetics of rhodopsin phosphorylation have mostly been carried out at very high levels of MII formation, under conditions very far from single photon responses; so how MII phosphorylation proceeds at very low light intensities and how it is influenced by [Ca^{2+}] remain controversial topics for further research. Certain features of long-term light adaptation are still without a molecular explanation.

Box 18.5 Cone-specific disease

Whereas retinitis pigmentosa and CSNB are thought of as diseases that primarily or initially affect the rods, there are additional hereditary visual disorders that primarily target cones. These include color-blindness (see Chapter 19), achromatopsia (see Chapter 19), cone- or cone/rod dystrophies, and inherited forms of macular dystrophy. Numerous genes have been implicated in these diseases, and these include some cone-specific phototransduction genes, including cone pigment genes, cyclic nucleotide gated channel subunit genes, a gene encoding cone-specific G protein, GCAP genes, and genes encoding potassium and calcium channels. In addition to *ABCA4*, discussed above, a gene encoding a fatty acid elongase, *ELOV4*, has been implicated in Stargardt-like macular dystrophy.

There is only a partial understanding of how the levels of the phototransduction proteins are regulated, even though it has been known for years that the levels of many of them are remarkably constant under a given set of conditions, but can vary in response to genetic defects or metabolic stress. Our understanding of how the phototransduction proteins are selectively transported to the outer segments from their sites of synthesis in the inner segment, and how they are kept there, is still at a rudimentary stage. In recent years there has been much interest in the migration during prolonged bright illumination of the G-protein transducin from the outer to the inner segment, and the movement of arrestin, under the same conditions, from the inner to the outer segment. The mechanisms by which these dramatic changes are achieved and their physiological significance are still under study.

One of the most important remaining areas of inquiry has to do with the connection between phototransduction and disease. Although defects in phototransduction components are known to lead to retinal degeneration, the molecular mechanisms by which cell death occurs are very poorly understood. It will be very important to understand these mechanisms in order to develop new therapeutic interventions, and to discover common pathways that might some day lead to therapies that could be applied to a number of different degenerative disorders.

Where the field is headed

The study of phototransduction is quite satisfying, because each new piece of information can be fitted into an overall scheme that is very comprehensive. It is like working on a jigsaw puzzle that is one-third to one-half finished. Each new piece of the puzzle can often be fit into the emerging picture and immediately reveals something new about the scene. The field at present is shifting: from the activation phase, whose mechanisms are fairly well understood, to the recovery phase, which holds more mysteries; from rods, whose properties make them easier to study, to cones, which present greater challenges; from the functions of proteins in the outer segments to understanding how they are transported and maintained there and how they are organized; from functioning of normal photoreceptors to how their malfunctioning leads to disease. There will be many exciting opportunities for discovery in these areas in the coming years.

References

1. Chabre M. Trigger and amplification mechanisms in visual phototransduction. Annu Rev Biophys Chem 1985; 14:331–360.
2. Schwartz EA. Phototransduction in vertebrate rods. Annu Rev Neurosci 1985; 8:339–367.
3. Hargrave PA, McDowell JH. Rhodopsin and phototransduction: a model system for G protein-linked receptors. Faseb J 1992; 6(6):2323–2331.
4. Arshavsky VY, Lamb TD, Pugh EN, Jr. G proteins and phototransduction. Annu Rev Physiol 2002; 64:153–187.
5. Burns ME, Arshavsky VY. Beyond counting photons: trials and trends in vertebrate visual transduction. Neuron 2005; 48(3):387–401.
6. Chen CK. The vertebrate phototransduction cascade: amplification and termination mechanisms. Rev Physiol Biochem Pharmacol 2005; 154:101–121.
7. Stryer L. Cyclic GMP cascade of vision. Annu Rev Neurosci 1986; 9:87–119.
8. Baylor DA. Photoreceptor signals and vision. Proctor lecture. Invest Ophthalmol Vis Sci 1987; 28(1):34–49.
9. Schnapf JL, Baylor DA. How photoreceptor cells respond to light. Sci Am 1987; 256(4):40–47.
10. Stryer L. Molecular mechanism of visual excitation. Harvey Lect 1991; 87:129–143.
11. Young RW. Passage of newly formed protein through the connecting cilium of retina rods in the frog. J Ultrastruct Res 1968; 23(5):462–473.
12. Peters KR, Palade GE, Schneider BG, Papermaster DS. Fine structure of a periciliary ridge complex of frog retinal rod cells revealed by ultrahigh resolution scanning electron microscopy. J Cell Biol 1983; 96(1):265–276.
13. Baylor DA, Lamb TD, Yau KW. Responses of retinal rods to single photons. J Physiol (Lond) 1979; 288:613–634.
14. Schneeweis DM, Schnapf JL. Photovoltage of rods and cones in the macaque retina. Science 1995; 268(5213):1053–1056.
15. Schnetkamp PP. The SLC24 Na$^+$/Ca^{2+}-K$^+$ exchanger family: vision and beyond. Pflugers Arch 2004 Feb; 447(5):683–688.
16. Koch KW. Purification and identification of photoreceptor guanylate cyclase. J Biol Chem 1991; 266(13):8634–8637.
17. Shyjan AW, de Sauvage FJ, Gillett NA, Goeddel DV, Lowe DG. Molecular cloning of a retina-specific membrane guanylyl cyclase. Neuron 1992; 9(4):727–737.
18. Dizhoor AM, Lowe DG, Olshevskaya EV, Laura RP, Hurley JB. The human photoreceptor membrane guanylyl cyclase, RetGC, is present in outer segments and is regulated by calcium and a soluble activator. Neuron 1994; 12(6):1345–1352.
19. Miki N, Baraban JM, Keirns JJ, Boyce JJ, Bitensky MW. Purification and properties of the light-activated cyclic nucleotide phosphodiesterase of rod outer segments. J Biol Chem 1975; 250(16):6320–6327.
20. Yee R, Liebman PA. Light-activated phosphodiesterase of the rod outer segment. Kinetics and parameters of activation and deactivation. J Biol Chem 1978; 253(24):8902–8909.
21. Wensel TG. The light-regulated cGMP phosphodiesterase of vertebrate photoreceptors: Structure and mechanism of activation by Gta. In: Dickey BF, Birnbaumer L, eds. GTPases in Biology II. Berlin: Springer-Verlag, 1993:213–223.
22. He F, Seryshev AB, Cowan CW, Wensel TG. Multiple zinc binding sites in retinal rod cGMP phosphodiesterase, PDE6alpha beta. J Biol Chem 2000; 275(27):20572–20577.
23. Yang Z, Wensel TG. Inorganic pyrophosphatase from bovine retinal rod outer segments. J Biol Chem 1992; 267(34):24634–24640.
24. Baehr W, Palczewski K. Guanylate cyclase-activating proteins and retina disease. Subcell Biochem 2007; 45:71–91.
25. Palczewski K, Sokal I, Baehr W. Guanylate cyclase-activating proteins: structure, function, and diversity. Biochem Biophys Res Commun 2004; 322(4):1123–1130.
26. Koch KW, Duda T, Sharma RK. Photoreceptor specific guanylate cyclases in vertebrate phototransduction. Mol Cell Biochem 2002; 230(1–2):97–106.
27. Conti M, Beavo J. Biochemistry and physiology of cyclic nucleotide phosphodiesterases: essential components in cyclic nucleotide signaling. Annu Rev Biochem 2007; 76:481–511.
28. Mou H, Cote RH. The catalytic and GAF domains of the rod cGMP phosphodiesterase (PDE6) heterodimer are regulated by distinct regions of its inhibitory gamma subunit. J Biol Chem 2001; 276(29):27527–27534.
29. Yamazaki M, Li N, Bondarenko VA, Yamazaki RK, Baehr W, Yamazaki A. Binding of cGMP to GAF domains in amphibian rod photoreceptor cGMP phosphodiesterase (PDE). Identification of GAF domains in PDE alphabeta subunits and distinct domains in the PDE gamma subunit involved in stimulation of cGMP binding to GAF domains. J Biol Chem 2002; 277(43):40675–40686.
30. Cote RH. Cyclic guanosine 5′-monophosphate binding to regulatory GAF domains of photoreceptor phosphodiesterase. Methods Mol Biol 2005; 307:141–154.
31. Palczewski K, Kumasaka T, Hori T, et al. Crystal structure of rhodopsin: A G protein-coupled receptor. Science 2000; 289(5480):739–745.
32. Palczewski K, Hofmann KP, Baehr W. Rhodopsin – advances and perspectives. Vision Res 2006; 46(27):4425–4426.
33. Nathans J. Rhodopsin: structure, function, and genetics. Biochemistry 1992; 31(21):4923–4931.
34. Wald G, Brown PK. Human rhodopsin. Science 1958; 127(3292):222–226.
35. Hubbard R, Wald G. Cis-trans isomers of vitamin A and retinene in vision. Science 1952; 115:2977.
36. Wald G. The chemistry of rod vision. Science 1951; 113(2933):287–291.
37. Fung BK, Hurley JB, Stryer L. Flow of information in the light-triggered cyclic nucleotide cascade of vision. Proc Natl Acad Sci USA 1981; 78(1):152–156.
38. Stryer L. Transducin and the cyclic GMP phosphodiesterase: amplifier proteins in vision. Cold Spring Harb Symp Quant Biol 1983; 48(Pt 2):841–852.

39. Baehr W, Morita EA, Swanson RJ, Applebury ML. Characterization of bovine rod outer segment G protein. J Biol Chem 1982; 257:6452–6460.

40. Kuhn H, Bennett N, Michel-Villaz M, Chabre M. Interactions between photoexcited rhodopsin and GTP-binding protein: kinetic and stoichiometric analyses from light-scattering changes. Proc Natl Acad Sci USA 1981; 78(11):6873–6877.

41. Liebman PA, Pugh EN, Jr. Gain, speed and sensitivity of GTP binding vs PDE activation in visual excitation. Vision Res 1982; 22:1475–1480.

42. Bigay J, Faurobert E, Franco M, Chabre M. Roles of lipid modifications of transducin subunits in their GDP-dependent association and membrane binding. Biochemistry 1994; 33(47):14081–14090.

43. Zhang Z, Melia TJ, He F, et al. How a G protein binds a membrane. J Biol Chem 2004 Aug 6; 279(32):33937–33945.

44. Neubert TA, Johnson RS, Hurley JB, Walsh KA. The rod transducin alpha subunit amino terminus is heterogeneously fatty acylated. J Biol Chem 1992; 267(26):18274–18277.

45. Yang Z, Wensel TG. N-myristoylation of the rod outer segment G protein, transducin, in cultured retinas. J Biol Chem 1992; 267(32):23197–23201.

46. Kokame K, Fukada Y, Yoshizawa T, Takao T, Shimonishi Y. Lipid modification at the N terminus of photoreceptor G-protein alpha-subunit. Nature 1992; 359(6397):749–752.

47. Kerov V, Rubin WW, Natochin M, Melling NA, Burns ME, Artemyev NO. N-terminal fatty acylation of transducin profoundly influences its localization and the kinetics of photoresponse in rods. J Neurosci 2007; 27(38):10270–10277.

48. Fukada Y, Takao T, Ohguro H, Yoshizawa T, Akino T, Shimonishi Y. Farnesylated gamma-subunit of photoreceptor G protein indispensable for GTP-binding. Nature 1990; 346:658–660.

49. Lai RK, Perez-Sala D, Canada FJ, Rando RR. The gamma subunit of transducin is farnesylated. Proc Natl Acad Sci USA 1990; 87(19):7673–7677.

50. Cheng H, Parish CA, Gilbert BA, Rando RR. A novel endoprotease responsible for the specific cleavage of transducin gamma subunit. Biochemistry 1995; 34(51):16662–16671.

51. Wensel TG. Signal transducing membrane complexes of photoreceptor outer segments. Vision Res 2008; 48(20):2052–2061.

52. Cone RA. Rotational diffusion of rhodopsin in the visual receptor membrane. Nature New Biol 1972; 236(63):39–43.

53. Poo M, Cone RA. Lateral diffusion of rhodopsin in the photoreceptor membrane. Nature 1974; 247(441):438–441.

54. Wey CL, Cone RA, Edidin MA. Lateral diffusion of rhodopsin in photoreceptor cells measured by fluorescence photobleaching and recovery. Biophys J 1981; 33(2):225–232.

55. Liebman PA, Weiner HL, Drzymala RE. Lateral diffusion of visual pigment in rod disk membranes. Methods Enzymol 1982; 81:660–668.

56. Wang Q, Zhang X, Zhang L, et al. Activation-dependent hindrance of photoreceptor G protein diffusion by lipid microdomains. J Biol Chem 2008; 283(44):30015–30024.

57. Nielsen JC, Maude MB, Hughes H, Anderson RE. Rabbit photoreceptor outer segments contain high levels of docosapentaenoic acid. Invest Ophthalmol Vis Sci 1986; 27(2):261–264.

58. Kim JE, Tauber MJ, Mathies RA. Wavelength dependent cis-trans isomerization in vision. Biochemistry 2001; 40(46):13774–13778.

59. Applebury ML. Dynamic processes of visual transduction. Vision Res 1984; 24(11):1445–1454.

60. Imai H, Mizukami T, Imamoto Y, Shichida Y. Direct observation of the thermal equilibria among lumirhodopsin, metarhodopsin I, and metarhodopsin II in chicken rhodopsin. Biochemistry 1994; 33(47):14351–14358.

61. Lewis JW, van Kuijk FJ, Thorgeirsson TE, Kliger DS. Photolysis intermediates of human rhodopsin. Biochemistry 1991; 30(48):11372–11376.

62. Ramdas L, Disher RM, Wensel TG. Nucleotide exchange and cGMP phosphodiesterase activation by pertussis toxin inactivated transducin. Biochemistry 1991; 30(50):11637–11645.

63. Malinski JA, Wensel TG. Membrane stimulation of cGMP phosphodiesterase activation by transducin: comparison of phospholipid bilayers to rod outer segment membranes. Biochemistry 1992; 31(39):9502–9512.

64. Yau KW, Nakatani K. Light-induced reduction of cytoplasmic free calcium in retinal rod outer segment. Nature 1985; 313(6003):579–582.

65. Maeda T, Imanishi Y, Palczewski K. Rhodopsin phosphorylation: 30 years later. Prog Retin Eye Res 2003; 22(4):417–434.

66. Doan T, Mendez A, Detwiler PB, Chen J, Rieke F. Multiple phosphorylation sites confer reproducibility of the rod's single-photon responses. Science 2006; 313(5786):530–533.

67. Mendez A, Burns ME, Roca A, et al. Rapid and reproducible deactivation of rhodopsin requires multiple phosphorylation sites. Neuron 2000; 28(1):153–164.

68. Zhao X, Yokoyama K, Whitten ME, Huang J, Gelb MH, Palczewski K. A novel form of rhodopsin kinase from chicken retina and pineal gland. FEBS Lett 1999; 454(1–2):115–121.

69. Hisatomi O, Matsuda S, Satoh T, Kotaka S, Imanishi Y, Tokunaga F. A novel subtype of G-protein-coupled receptor kinase, GRK7, in teleost cone photoreceptors. FEBS Lett 1998; 424(3):159–164.

70. Anant JS, Fung BK. In vivo farnesylation of rat rhodopsin kinase. Biochem Biophys Res Commun 1992; 183(2):468–473.

71. Inglese J, Glickman JF, Lorenz W, Caron MG, Lefkowitz RJ. Isoprenylation of a protein kinase. Requirement of farnesylation/alpha-carboxyl methylation for full enzymatic activity of rhodopsin kinase. J Biol Chem 1992; 267(3):1422–1425.

72. Molday RS. ATP-binding cassette transporter ABCA4: molecular properties and role in vision and macular degeneration. J Bioenerg Biomembr 2007; 39(5–6):507–517.

73. Ahn J, Beharry S, Molday LL, Molday RS. Functional interaction between the two halves of the photoreceptor-specific ATP binding cassette protein ABCR (ABCA4). Evidence for a non-exchangeable ADP in the first nucleotide binding domain. J Biol Chem 2003; 278(41):39600–39608.

74. Gurevich VV, Gurevich EV, Cleghorn WM. Arrestins as multi-functional signaling adaptors. Handb Exp Pharmacol 2008; 186:15–37.

75. Palczewski K. Structure and functions of arrestins. Protein Sci 1994; 3(9):1355–1361.

76. Mendez A, Burns ME, Sokal I, et al. Role of guanylate cyclase-activating proteins (GCAPs) in setting the flash sensitivity of rod photoreceptors. Proc Natl Acad Sci USA 2001; 98(17):9948–9953.

77. Burns ME, Mendez A, Chen J, Baylor DA. Dynamics of cyclic GMP synthesis in retinal rods. Neuron 2002; 36(1):81–91.

78. Koch KW, Stryer L. Highly cooperative feedback control of retinal rod guanylate cyclase by calcium ions. Nature 1988; 334(6177):64–66.

79. Gorczyca WA, Gray-Keller MP, Detwiler PB, Palczewski K. Purification and physiological evaluation of a guanylate cyclase activating protein from retinal rods. Proc Natl Acad Sci USA 1994; 91(9):4014–4018.

80. Sokal I, Alekseev A, Palczewski K. Photoreceptor guanylate cyclase variants: cGMP production under control. Acta Biochim Pol 2003; 50(4):1075–1095.

81. Dizhoor AM, Hurley JB. Regulation of photoreceptor membrane guanylyl cyclases by guanylyl cyclase activator proteins. Methods 1999; 19(4):521–531.

82. Pugh EN, Jr., Duda T, Sitaramayya A, Sharma RK. Photoreceptor guanylate cyclases: a review. Biosci Rep 1997; 17(5):429–473.

83. Hurley JH. The adenylyl and guanylyl cyclase superfamily. Curr Opin Struct Biol 1998; 8(6):770–777.

84. Perrault I, Rozet JM, Calvas P, et al. Retinal-specific guanylate cyclase gene mutations in Leber's congenital amaurosis. Nat Genet 1996; 14(4):461–464.

85. Duda T, Koch KW. Retinal diseases linked with photoreceptor guanylate cyclase. Mol Cell Biochem 2002; 230(1–2):129–138.

86. Payne AM, Downes SM, Bessant DA, et al. A mutation in guanylate cyclase activator 1A (GUCA1A) in an autosomal dominant cone dystrophy pedigree mapping to a new locus on chromosome 6p21.1. Hum Mol Genet 1998; 7(2):273–277.

87. Howes KA, Pennesi ME, Sokal I, et al. GCAP1 rescues rod photoreceptor response in GCAP1/GCAP2 knockout mice. Embo J 2002; 21(7):1545–1554.

88. Krispel CM, Chen D, Melling N, et al. RGS expression rate-limits recovery of rod photoresponses. Neuron 2006; 51(4):409–416.

89. Pugh EN, Jr. RGS expression level precisely regulates the duration of rod photoresponses. Neuron 2006; 51(4):391–393.

90. He W, Cowan CW, Wensel TG. RGS9, a GTPase accelerator for phototransduction. Neuron 1998; 20(1):95–102.

91. Cowan CW, Fariss RN, Sokal I, Palczewski K, Wensel TG. High expression levels in cones of RGS9, the predominant GTPase accelerating protein of rods. Proc Natl Acad Sci USA 1998; 95(9):5351–5356.

92. Watson AJ, Aragay AM, Slepak VZ, Simon MI. A novel form of the G protein beta subunit Gbeta5 is specifically expressed in the vertebrate retina. J Biol Chem 1996; 271:28154–28160.

93. Makino ER, Handy JW, Li T, Arshavsky VY. The GTPase activating factor for transducin in rod photoreceptors is the complex between RGS9 and type 5 G protein beta subunit. Proc Natl Acad Sci USA 1999; 96(5):1947–1952.

94. Snow BE, Krumins AM, Brothers GM, et al. A G protein gamma subunit-like domain shared between RGS11 and other RGS proteins specifies binding to Gbeta5 subunits. Proc Natl Acad Sci USA 1998; 95(22):13307–13312.

95. Rodiek RW. The first steps in seeing. Sunderland, MA: Sinauer Associates, Inc., 1998:539.

96. Pepperberg DR, Cornwall MC, Kahlert M, et al. Light-dependent delay in the falling phase of the retinal rod photoresponse. Vis Neurosci 1992; 8(1):9–18.

97. Nikonov S, Engheta N, Pugh EN, Jr. Kinetics of recovery of the dark-adapted salamander rod photoresponse. J Gen Physiol 1998; 111(1):7–37.

98. Ames A, 3rd, Walseth TF, Heyman RA, Barad M, Graeff RM, Goldberg ND. Light-induced increases in cGMP metabolic flux correspond with electrical responses of photoreceptors. J Biol Chem 1986; 261(28):13034–13042.

99. Panico J, Parkes JH, Liebman PA. The effect of GDP on rod outer segment G-protein interactions. J Biol Chem 1990; 265(31):18922–18927.

100. Sampath AP, Baylor DA. Molecular mechanism of spontaneous pigment activation in retinal cones. Biophys J 2002; 83(1):184–193.

101. Kefalov VJ, Estevez ME, Kono M, et al. Breaking the covalent bond – a pigment property that contributes to desensitization in cones. Neuron 2005; 46(6):879–890.

102. Wald G, Brown PK. The molar extinction of rhodopsin. J Gen Physiol 1953; 37:189–200.

103. Wald G. Molecular basis of visual excitation. Science 1968; 162:230–239.

104. Fasick JI, Lee N, Oprian DD. Spectral tuning in the human blue cone pigment. Biochemistry 1999; 38(36):11593–11596.

105. Kochendoerfer GG, Lin SW, Sakmar TP, Mathies RA. How color visual pigments are tuned. Trends Biochem Sci 1999; 24(8):300–305.

106. Lin SW, Kochendoerfer GG, Carroll KS, Wang D, Mathies RA, Sakmar TP. Mechanisms of spectral tuning in blue cone visual pigments. Visible and raman spectroscopy of blue-shifted rhodopsin mutants. J Biol Chem 1998; 273(38):24583–24591.

107. Fasick JI, Applebury ML, Oprian DD. Spectral tuning in the mammalian short-wavelength sensitive cone pigments. Biochemistry 2002; 41(21):6860–6865.

108. Merbs SL, Nathans J. Role of hydroxyl-bearing amino acids in differentially tuning the absorption spectra of the human red and green cone pigments. Photochem Photobiol 1993; 58(5):706–710.

109. Fasick JI, Robinson PR. Spectral-tuning mechanisms of marine mammal rhodopsins and correlations with foraging depth. Vis Neurosci 2000; 17(5):781–788.

110. Zhang X, Wensel TG, Kraft TW. GTPase regulators and photoresponses in cones of the eastern chipmunk. J Neurosci 2003; 23(4):1287–1297.

111. Tachibanaki S, Tsushima S, Kawamura S. Low amplification and fast visual pigment phosphorylation as mechanisms characterizing cone photoresponses. Proc Natl Acad Sci USA 2001; 98(24):14044–14049.

112. Tachibanaki S, Arinobu D, Shimauchi-Matsukawa Y, Tsushima S, Kawamura S. Highly effective phosphorylation by G protein-coupled receptor kinase 7 of light-activated visual pigment in cones. Proc Natl Acad Sci USA 2005; 102(26):9329–9334.

113. Bownds D, Brodie AE. Light-sensitive swelling of isolated frog rod outer segments as an in vitro assay for visual transduction and dark adaptation. J Gen Physiol 1975; 66(4):407–425.

114. Okada T, Sugihara M, Bondar AN, Elstner M, Entel P, Buss V. The retinal conformation and its environment in rhodopsin in light of a new 2.2 A crystal structure. J Mol Biol 2004; 342(2):571–583.

115. Park JH, Scheerer P, Hofmann KP, Choe HW, Ernst OP. Crystal structure of the ligand-free G-protein-coupled receptor opsin. Nature 2008; 454(7201):183–187.

116. Scheerer P, Park JH, Hildebrand PW, et al. Crystal structure of opsin in its G-protein-interacting conformation. Nature 2008; 455(7212):497–502.

117. Zeitz C, Gross AK, Leifert D, et al. Identification and functional characterization of a novel rhodopsin mutation associated with autosomal dominant CSNB. Invest Ophthalmol Vis Sci 2008; 49(9):4105–4114.

118. Gross AK, Wang Q, Wensel TG. Regulation of photoresponses by phosphorylation. In: Tombran-Tink J, ed. Visual transduction and non-visual light perception. Totowa, NJ: Humana Press, 2008:125–140.

119. Hartong DT, Dange M, McGee TL, Berson EL, Colman RG, Dryja TP. A defect in Krebs cycle in retinitis pigmentosa. IOVS 2008; E-3284.

120. Hagstrom SA, Zhang K, Howes K, Baehr W, He W, Wensel TG, Berson EL, Dryja TP. Screen for mutations in the RGS9 gene in patients with retinitis pigmentosa or an allied disease. Invest Ophthal Vis Sci 1999; 40:S600.

121. Hagstrom SA, Zhang K, Baehr W, He W, Wensel TG, Berson EL, Dryja TP. Comprehensive mutation screen in the RGS9 gene in 558 patients with inherited retinal degenerations. Invest Ohthal Vis Sci 2001; 45:S646.

122. Nishiguchi K, Sandberg MA, Kooijman AC, et al. Defects in RGS9 or its anchor protein R9AP in patients with bradyopsia, a novel form of retinal dysfunction. IOVS 2004; E-1020.

123. Weleber RG, Majewski J, Schultz DW, Matise TC, Ott J, Acott TS, Klein ML. Age-related macular degeneration: A genome-wide scan in extended families. IOVS 2003; E-1503.

124. Dryja et al., Nature (London) (90) 343:364–369

125. Keats, Savas Am J Med Genet A (04) 130:13–16

CHAPTER 19

Photoresponses of Rods and Cones

Peter R. MacLeish & Clint L. Makino

Retinal rods and cones are highly specialized neurons that respond to light with an electrical signal (see Chapter 18) and provide the sensory input for vision. In contrast to most other neurons, rods and cones maintain a relatively depolarized membrane potential at rest (in darkness) and when stimulated (by light) decrease Na^+ entry by closing ion channels. In turn, the resultant hyperpolarization closes Ca^{2+} channels at the synapse and the ensuing fall in intracellular Ca^{2+} reduces an ongoing vesicular release of neurotransmitter onto second-order neurons (see review[1]). This chapter describes the signaling properties of rods and cones, which subserve vision in dim and bright light, respectively. An overview of the voltage changes induced by light will be followed by a description of the properties of the photocurrent from which the voltage response originates and then an explanation of how voltage-dependent, inner segment conductances shape the final voltage response. For readers interested in the underlying molecular designs and mechanisms, summaries on ion channels and exchangers are included. Even though the retinal photoreceptors of all vertebrates operate similarly, there are some qualitative as well as quantitative differences. So wherever possible, we will focus on primate photoreceptors.

Photovoltage response to flashes

In darkness, rods and cones maintain a membrane potential near −40 mV.[2] When stimulated by flashes, rods and cones do not "fire" action potentials but instead respond with graded hyperpolarizations (Fig. 19.1)[3] that in some cells, exceed 25 mV.[4] The hyperpolarization then spreads passively to the synapse. In comparing rod and cone responses, the latter are considerably faster and require many more photons. Photoisomerization of 75 rhodopsins gives rise to a half maximal response in rods whereas in cones, it takes close to a thousand photoisomerizations (estimated from[2,5]).

The amplitude of the single photon response in rods is about a millivolt, a few percent of the maximal response.[2,4,6] Its time to peak of 200 ms or more is relatively slow considering that Olympic sprinter Usain Bolt covered more than two meters in that span of time. The response recovers even more slowly and may take over a second to disappear. The slow time course of the single-photon response provided compelling evidence for an internal, second messenger(s) and for amplification steps in the signaling pathway. The duration can be specified as an integration time, calculated by dividing the response integral by the peak amplitude. For rods, the integration time of the quantal response is several hundred ms. Although cones are less sensitive than rods, they too "respond" to single photons, but the amplitude of their quantal response is much smaller and does not emerge from the baseline noise.[2] The cone response is faster, with a time to peak of 30 ms and an integration time of only 25 ms.[5,7] Response kinetics and sensitivity are intimately linked and the relationship will be discussed below (see also Chapter 18). The waveforms of rod and cone flash responses change with flash strength. Responses peak sooner and the initial recovery is faster as the flash strength increases but with responses greater than half maximal, recovery stalls to a plateau before the final recovery. In cones, the recovery overshoots the baseline before the final recovery, while in rods, the plateau slowly returns to baseline without an overshoot.

Background light attenuates the response to a flash as the cells adapt (for discussions of the underlying mechanisms, see Chapters 18 and 20). The dim flash response reduces in size by two-fold with a background producing ~150 photoisomerizations per second in rods and ~8700 photoisomerizations per second in cones.[7] The latter value obtains for an isolated cone. In reality, light has a greater effect on the cone response because of electrical coupling between rods and cones in the retina (see Chapter 22). Background light quickens flash responses in rods, but has little effect on response kinetics in cones (ignoring the rod component of the response in cones communicated electrotonically).

Photocurrent response to flashes

The photon response derives from a decrease in the number of open cyclic nucleotide gated (CNG) channels that perturbs the flow of ions across the plasma membrane and shifts the transmembrane potential to more negative voltages. CNG channels are located exclusively in the outer segments of rods and cones. In darkness, a small fraction of the

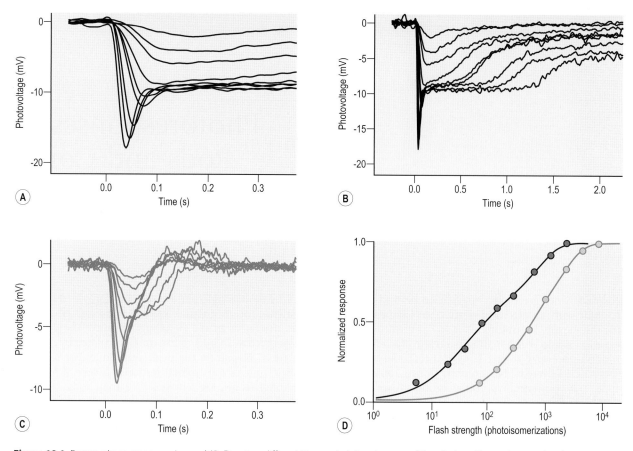

Figure 19.1 Faster voltage responses in a rod (**A**, **B** on two different time scales) than in a cone (**C**) to flashes. Photovoltage is the change in membrane potential induced by a flash, given at time zero. (**D**) The stimulus–response relation of the cone (green circles) is to the right of that of the rod (black circles), reflecting the lower sensitivity of the cone. Each response has been divided by the maximal, saturating response for that cell. (Panels A and B from Schneeweis DM and Schnapf JL: Photovoltage of rods and cones in the macaque retina, Science 268: 1053–1056, 1995. Reprinted with permission from AAAS;[2] reprinted with permission from AAAS. Panels C and D courtesy of D. Schneeweis, University of Illinois at Chicago and J. Schnapf, University of California at San Francisco, unpublished results.)

channels are in the open state, allowing an influx of cations. The intracellular concentration of Na^+ is low while that of K^+ is high relative to the extracellular space due to the action of Na^+,K^+-ATPases in the inner segment. Thus Na^+ and also some Ca^{2+} ions flow in through the outer segment channels. K^+ ions flow out through leak and voltage-gated ion channels in the inner segment to complete the electrical circuit, which is referred to as the "dark" or circulating current (Fig. 19.2). Light absorbed by a visual pigment molecule in the outer segment closes CNG channels in an annulus of plasma membrane surrounding the site of photon absorption on the disk membrane.[8-10]

When two photons are absorbed, it is unlikely for both absorptions to occur in close proximity. The response to each photon is generated independently and the overall response is twice as big. In other words, dim flashes lie within a linear operating range wherein the response simply scales with the flash strength. However, with increasing numbers of photon absorptions, local effects begin to overlap so fewer channels are closed per photon absorbed. With enough photons, all of the channels are closed and the rod response is maximal (Fig. 19.3). Hence the normalized response grows with flash strength according to a saturating exponential function:[9]

$$r/r_{max} = 1 - \exp(-k_f i) \qquad (1)$$

where r is the photocurrent response amplitude, r_{max} is the amplitude of the maximal saturating response, i is flash strength, k_f is a constant equal to $\ln(2)/i_{0.5}$ and $i_{0.5}$ is the flash strength giving rise to a half maximal response. The $i_{0.5}$ for rods produces 30–70 photoisomerizations on average,[11-13] while that for cones produces $\sim$10 times as many photoisomerizations.[14]

As is the case with photovoltage, photocurrent responses speed up with flash strength though the effect is less pronounced in cones.[11,14] Responses are faster in cones than in rods (note the difference in time scales in Fig. 19.3A and C) and have a marked undershoot (although cf. refs[15,16]). Responses of different cone types are indistinguishable in waveform. However, the photovoltage and photocurrent responses of rods and cones are not mirror images of each other; photocurrent responses lack the "nose" in the photovoltage response to brighter flashes.

Once the flash is bright enough to close all of the CNG channels, more light cannot produce a larger response, so instead, the duration of the photocurrent response increments with the natural logarithm of the flash strength (Fig. 19.3D). The basis for this behavior is that a single molecular process, namely the shutoff of transducin[17] (see also Chapter 18), that has a stochastic, exponential time course[18,19] dominates the recovery from phototransduction activation in

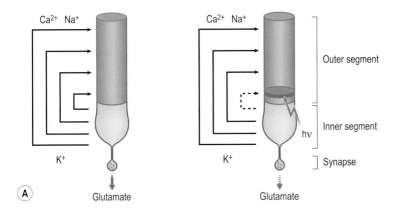

Figure 19.2 The dark current. (**A**) Light suppresses the entry of Na⁺ and to a lesser extent Ca²⁺ into the outer segment in a ring of plasma membrane around the photoexcited rhodopsin (darkened red band). The rod hyperpolarizes, which closes Ca²⁺ channels at the synapse and attenuates the release of neurotransmitter. (**B**) The inward Na⁺ current in this rod, plotted downward by convention, was −13.3 pA. A saturating flash, given at time zero, closes all the CNG channels and completely shuts down the inward Na⁺ current. After a while, channels reopen and the membrane current is restored. For simplicity, the change in membrane current is often measured as photocurrent, shown on the right.

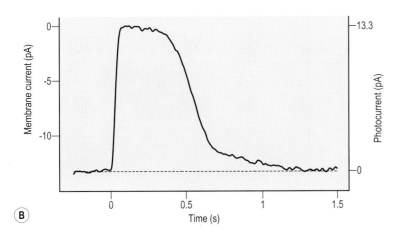

rods. The slope of the saturation function or "Pepperberg Plot" yields the dominant time constant of about ∼0.2 s[20] (a similar value has been reported by many others for murine rods). The identity of the rate-limiting step in the recovery of the cone response has not been firmly established but its time constant is an order of magnitude faster (T. Kraft, personal communication).[15,16,21]

Detecting single photons

Rods count single photons.[22] In order to do so, they must satisfy several conditions: noise must be minimal and quantal responses must be reproducible and highly amplified. Amplification is achieved by cascading several enzymatic reactions (see Chapter 18). Each chemical reaction takes time, hence the decrement in the circulating current after photon absorption is delayed by a few ms. Then the response rises to a peak amplitude of 0.1 to 0.7 pA approximately 200 ms later reflecting the suppression of the circulating current by a few per cent.[11–13,23] With the cascade working as fast as it can, the size of the response is determined by when shutoff and recovery processes kick in. In rods, hundreds of CNG channels are closed during the single photon response. Na⁺ ions traverse the channel at a rate of ∼10⁴ s⁻¹ (see review[25]), so one photon blocks the entry of a million Na⁺ ions. Responses to brighter flashes in rods and in cones appear with a shorter delay and rise more steeply because many more photoexcited rhodopsins are activating the phototransduction machinery. But if the single photon

response in cones is considered, its early rise is actually pretty similar to that in rods.[24] However, the cone response starts to shut off in 50 ms or less at a time when it has only reached an amplitude of a few tens of fA.[14,16,21]

It is imperative that single photon responses do not vary in size or duration because if they did, the rod would be unable to distinguish one large, slow quantal response from two or more small, brief quantal responses occurring close together in time. The coefficient of variation of the rod response, defined as the quotient of the standard deviation divided by the mean amplitude, has a low value of 0.2.[22,26–28] When stimulated by dim flashes, the rod does not respond the same way every time, but the unpredictability arises from the Poisson distribution of photon absorptions, rather than from single photon response variability. Whenever the rod does respond, the amplitude is quantized (Fig. 19.4A).

In darkness, there are two physiological kinds of electrical fluctuations in the current baseline: discrete noise and continuous noise (Fig. 19.4B).[11,29] Discrete noise arises from thermal isomerizations of rhodopsin. Even though rhodopsin has a half life of ∼300 years at body temperature, a rod contains so many copies (around a hundred million) that one spontaneously activates every minute or two. Cone visual pigments are far less stable than rod pigments,[14,30,31] making single photon detection impossible. Continuous noise springs from the spontaneous activity of PDE[32] and is much more prevalent than discrete events in rods but typically has a lower amplitude. The continuous noise amplitude distribution is approximately Gaussian so, occasionally, there is a single-photon-response-like event. However, such events

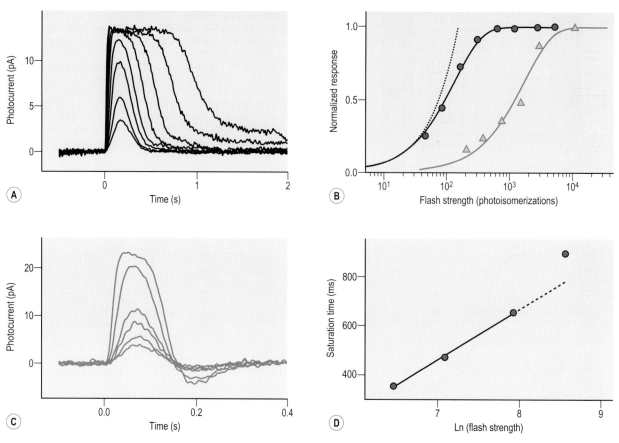

Figure 19.3 Current responses of a human rod (**A**) and cone (**C**) to flashes. (**B**) The stimulus–response relations of the rod (black circles) and cone (green triangles) follow eqn (1) (black and green lines). The dotted line shows a linear relation on semilogarithmic coordinates for comparison. (**D**) Saturation times for the rod in A, measured from midflash to 25 percent recovery. The line passing through the initial points has a slope of 205 ms and has been extended with a dashed line. (Panel A from T.W. Kraft, D.M. Schneeweis and J.L. Schnapf, 1993: Visual transduction in human rod photoreceptors, Journal of Physiology 464: 747–765.[12] reproduced with permission from Blackwell Publishing. Panel B contains results from Kraft TW, Schneeweis DM, Schnapf JL. Visual transduction in human rod photoreceptors. J Physiol. 1993; 464: 747–65[12] and unpublished results from T. Kraft, University of Alabama. Panel C from Kraft TW, Neitz J, Neitz M. Spectra of human L cones. Vision Research 1998; 38: 3663–70; reproduced by permission of Elsevier Science Copyright Elsevier 1998. Panel D courtesy of T. Kraft, University of Alabama; D. Schneeweis, University of Illinois at Chicago; and J. Schnapf, University of California at San Francisco.)

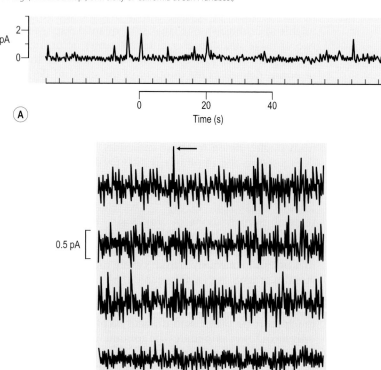

Figure 19.4 Responses of a primate rod to single, rhodopsin isomerizations. (**A**) For flashes that photoisomerize one rhodopsin on average, many trials fail to elicit a response because no photons were absorbed. Other trials produce responses whose amplitudes are multiples of the unitary response, dependent upon the number of photons absorbed. (**B**) Discrete noise events (e.g. arrow in top trace), resembling single photon responses, appear sporadically in darkness after spontaneous activations of individual rhodopsins. In addition, a continuous, lower amplitude noise component arises from basal PDE activity (top three traces). Both noise components disappear in bright light after closure of all of the CNG channels (bottom trace), leaving only the instrumental noise in the recording. (From D.A. Baylor, B.J. Nunn and J.L. Schnapf, 1984: The photocurrent, noise and spectral sensitivity of rods of the monkey Macaca fascicularis, Journal of Physiology 357: 575–607. Reproduced by permission of Blackwell Publishing.)

are inconsequential because they take place ten times less frequently than the thermal isomerizations of rhodopsin.

Thermal isomerizations of rhodopsin limit the detection of very dim light, because real photons cannot be distinguished from virtual ones. To guard against false alarms, the visual system relies on coincidence detection. A single photon response in one rod does not suffice for vision; a few rods in a small cluster of the retina must each generate a single photon response within a certain period of time before a flash is "seen".[33] The temporal requirement is specified by the integration time of the response. Thus shortening the integration time detracts from sensitivity of the system. This consideration also applies below, when rods and cones respond to steps of light. The integration time for rods is ~300 ms,[11,12,23] while for cones, it is ~20 ms (segment preceding undershoot[14]).

Thus rods are well suited for single photon detection. Oddly though, rods sometimes generate single photon responses that are truly aberrant.[11,34] For reasons unknown, one photoexcited rhodopsin out of several hundred fails to shut off properly[35,36] and gives rise to a two-fold larger than average response that can last for a very, very long time (Fig. 19.5). Aberrant response durations are exponentially distributed with a mean value of about 4 s, but some last for tens of seconds. Aberrant responses have little impact on photon counting due to their rarity, yet they prolong the recovery after exposure to bright light (flashes and steps), leaving rods vulnerable to saturation (Box 19.1).

Photocurrent response to steady light

The rod response to dim, steady light summates individual single photon responses.[11] But with brighter light, adaptation within the outer segment causes the step responses to fall short of the amplitude expected from a simple saturation behavior[23,37,38] (see Chapter 20). Midrange responses droop as additional adaptational mechanisms with a slower time course reduce the cascade gain and cause the stimulus–response relation to rise even more gently (Fig. 19.6). Light producing 300–600 photoisomerizations per second decreases the circulating current in darkness by 50 percent.[23,39] In contrast, the peak of the cone response to steps nearly does obey simple saturation.[14] A hundred ms in the light later, however, cone responses begin to droop as they too adapt. For dim steps, the droop reduces the amplitude to less than half that at the peak. The initial droop is typically larger and faster in cones than in rods.

In steady light, incremental flashes give rise to responses in rods that are smaller (Fig. 19.7) and recover more rapidly than in darkness, another manifestation of light adaptation[11–13,23,37,38] (see Chapter 20). Flash response kinetics change very little with background light in primate cones, differing in this respect from cones of cold-blooded vertebrates.[14] Light producing one to a few hundred photoisomerizations per second reduces flash sensitivity to half its value in darkness in rods,[11,12,23,39] where the value may increase slowly over tens of seconds in the light.[40,41] Adaptation is more powerful in cones, for which 10,000 or more photoisomerizations per second reduce sensitivity two-fold.[7,14] Mammalian rods adapt over several log units of background light intensity before saturating.[23,37,38] Amphibian rods adapt

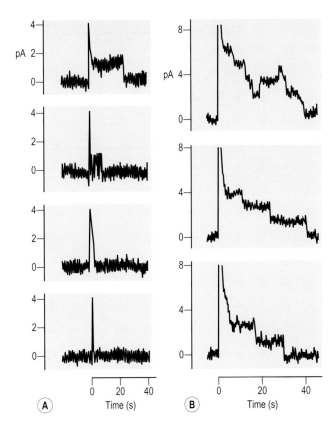

Figure 19.5 Aberrant responses in rods. The peaks of the responses are not shown. (**A**) In these four responses to a flash eliciting about 250 photoisomerizations, an aberrant response component can be seen at late times in the top three traces. The appearance of an aberrant response after a flash, as well as its duration is random. The aberrant response in the top trace persists for more than 20 s. The second trace does not initially contain an aberrant response. Then one appears and recovers only to reappear before shutting off for good. (**B**) In these three responses to a flash over ten times brighter than that in A, multiple individual aberrant responses add together to create protracted, stair-cased recoveries. (From D.A. Baylor, B.J. Nunn and J.L. Schnapf, 1984: The photocurrent, noise and spectral sensitivity of rods of the monkey Macaca fascicularis, Journal of Physiology 357: 575-607. Modified with permission from Blackwell Publishing.)

over an even greater range of intensities at least in part because they lack aberrant responses.

Light that is bright enough to bleach a significant fraction of visual pigment causes rods to behave for a period as if they were being exposed to a virtual, "equivalent light" that lingers after the light is removed (see review[42]). Rods adapt to the equivalent light with a reduction in circulating current and accelerated flash response kinetics as well as a profound loss of flash sensitivity exceeding that expected from the decrease in photon capture.[43] It turns out that bleached rhodopsin constitutively activates the phototransduction cascade. Although the activity of a single bleached rhodopsin is miniscule, the summated activity over hundreds of millions of bleached rhodopsin molecules is substantial. The equivalent light fades as visual pigment regenerates (see Chapter 13). Circulating current after full bleach recovers halfway in 15 min and is fully restored in ~25 min.[39] Cones are not impaired in this way by bleaching;[14,44] their bleached pigment seems to be inactive (although the situation is not yet clear for bleached blue cone pigment[45]). That means that when a very bright light is turned on, cones may saturate, but they

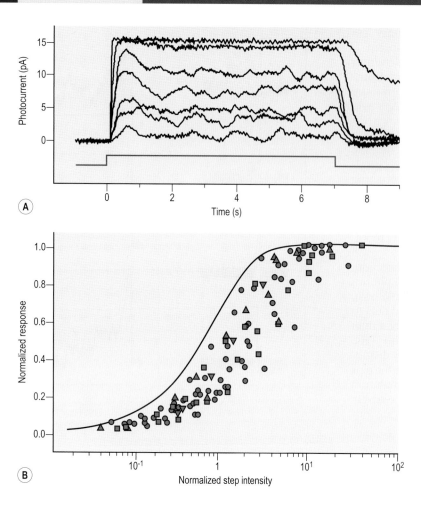

Figure 19.6 Adaptation of a rod to steady light. (**A**) The Poisson nature of light absorption causes the bumpiness in the responses to dim steps despite averaging. Midrange responses droop as the rod adapts. At the two highest intensities, both the droop and the bumpiness are compressed as the rod saturates so the traces become smoother. (**B**) In the stimulus–response relations for the rod in A and 14 other rods measured after several seconds of light exposure, light intensity on the abscissa has been divided by $I_{0.5}$, the intensity giving rise to a half-maximal response measured at the peak, for each rod. The continuous line plots the saturating exponential that would obtain in the absence of adaptation: $r/r_{max}=1 - \exp(-k_s\,I)$, where $k_s = k_f t_i$, k_f is from eqn (1), t_i is the integration time (see above) and I is the intensity of the step. Thus lacking adaptation, rods would saturate at intensities that are lower by as much as an order of magnitude. (From T. Tamura, K. Nakatani and K.-W. Yau, 1991. Originally published in The Journal of General Physiology 98: 95–130. Reproduced with permission.)

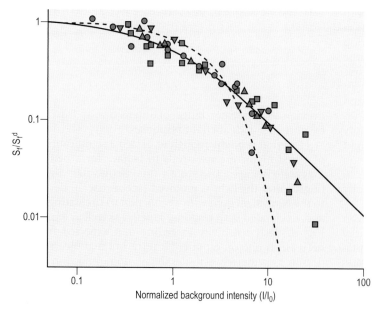

Figure 19.7 Loss in incremental flash sensitivity with background light. In the absence of adaptation, relative flash sensitivity, S_f/S_f^d, defined as response amplitude divided by flash strength for responses in the linear range normalized by its value in darkness, would drop off along a saturating exponential function (dashed line). Instead, rods exhibit Weber–Fechner behavior: $S_f/S_f^d = (1 + I/I_0)^{-1}$ where I_0 is the background intensity that reduces S_f/S_f^d to 0.5. Rods sacrifice some sensitivity at lower background intensities (symbols below dashed line) in order to maintain sensitivity at intensities that would otherwise be saturating (symbols above the dashed line). (From T. Tamura, K. Nakatani and K.-W. Yau, 1991. Originally published in The Journal of General Physiology 98: 95–130. Reproduced with permission.)

actually manage to recover circulating current during the exposure as loss of some of their visual pigment by bleaching lowers the ongoing rate of photon capture. Even after extensive bleaching, cones regain their full circulating current and the reduction in sensitivity scales in proportion to the loss in photon capture.[14,44,46] Cone pigment regeneration is faster than in rods, with full sensitivity returning after complete bleach within minutes. So while cones lack the absolute sensitivity of rods, they are better equipped to operate over a wide range of brighter light intensities without saturating.

Action spectra of rods and cones

Rods and cones respond to a wide range of wavelengths (Fig. 19.10), their action spectra being determined by the spectral absorptions of the visual pigment they express. For a given number of photons, the response amplitude varies with wavelength because the probability of absorption by the visual pigment varies with wavelength. The response itself to a photoisomerization is independent of wavelength. Rods respond maximally to ~493 nm (blue-green light).[11,12] There are three types of cones, commonly referred to as blue, green and red with maxima near 430 nm (violet light), 530 nm (green light) and 560 nm (greenish-orange light), respectively.[55,56] Since the names do not correspond to the colors of the maxima in every case, many prefer the designations: short-wavelength sensitive or S, middle-wavelength sensitive or M and long-wavelength sensitive or L (Box 19.2).

Box 19.1 Diseases caused by hyperactive signaling

Excessive phototransduction cascade activity in photoreceptors can interfere with vision and lead to retinal disease (see review[47]). For example, mutations that target rhodopsin kinase (see Chapter 18), can cause a form of night blindness known as Oguchi disease. Mutant mouse rods lacking rhodopsin kinase generate aberrant single photon responses like those shown in Fig. 19.5 for every rhodopsin isomerization. Cones are only mildly affected in the disease,[48] because they express a second type of rhodopsin kinase.[49,50] Patients with defects in the nuclear receptor NR2E3 develop enhanced S cone syndrome in which there is a higher prevalence of blue cones in the retina at the expense of red and green cones.[51] Very interestingly, the blue cones in these patients fail to express both types of rhodopsin kinases. Hence flash responses of their blue cones (but not red or green cones) do take an abnormally long time to recover.[15]

Mutations in arrestin, the protein normally responsible for quenching photoexcited rhodopsin's activity (see Chapter 18), can also cause Oguchi disease. Flash responses from rods of mutant, arrestin knockout mice are very prolonged (Fig. 19.8). Since longer-lasting responses enhance absolute sensitivity, it might at first seem surprising that the mutations in either rhodopsin kinase or arrestin should lead to night blindness. The problem is that the mutant rods saturate at very low light levels and then take an inordinately long period of time and much dimmer conditions to recover from saturation. In theory, Oguchi patients with rhodopsin kinase or arrestin mutations might actually see better than normal persons under the dimmest conditions when given adequate time for dark adaptation. Photopic vision is largely spared because cones express a unique arrestin.[52,53]

Mutations in either RGS9 or R9AP can sabotage the timely shutoff of transducin (see Chapter 18). Because these two proteins are used in rods and cones, there is night blindness and a problem in daytime vision. The ability of cones to light adapt enables them to escape saturation, so photopic vision is still possible. But long lasting photoresponses translate into a disturbing persistence in sensation and the person has difficulty following moving objects and adjusting to luminance changes, a condition termed bradyopsia.[54]

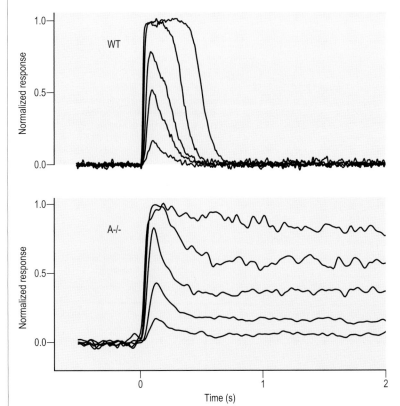

Figure 19.8 Murine model for Oguchi disease. Flash responses from a mutant rod lacking arrestin (A-/-, bottom) recover a hundred times more slowly than wild-type (WT, top) rod responses. (From Makino CL, Flannery JG, Chen J, Dodd RL. Effects of photoresponse prolongation on retinal rods of transgenic mice. In: Williams TP, Thistle AB, eds. Photostasis and related phenomena. New York: Plenum Press, 1998. Modified with permission from Springer Science and Business Media.)

Box 19.1 Diseases caused by hyperactive signaling—cont'd

Genetic defects in RPE65, the isomerase that converts all-*trans* to 11-*cis* retinal (see Chapter 13) can prevent the de novo synthesis of rhodopsin. The persistent presence of catalytically active apo-opsin (rhodopsin lacking 11-*cis* retinal – essentially equivalent to bleached rhodopsin)

gives rise to Leber congenital amaurosis, an especially severe form of retinal degeneration. RPE65 knockout mouse rods exist in an inescapable state of light adaptation, with greatly attenuated circulating current and smaller, faster flash responses (Fig. 19.9).

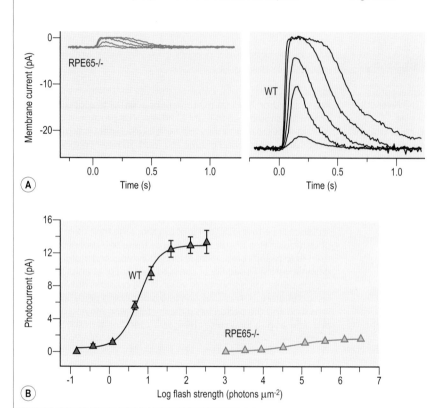

Figure 19.9 Murine model for Leber congenital amaurosis. (**A**) The outer segment membrane current and averaged flash responses of mutant rods lacking RPE65 (*n* = 3) are smaller than those of wild-type rods (*n* = 4). Mutant rods also generate flash responses with faster kinetics. (**B**) On average, the mutant rods (*n* = 16) are thousands of times less sensitive than WT rods (*n* = 32). (From Woodruff ML, Wang Z, Chung HY et al. Spontaneous activity of opsin apoprotein is a cause of Leber congenital amaurosis. Nature Genetics 2003; 35:158–64. Reprinted with permission from Macmillan Publishers Ltd.)

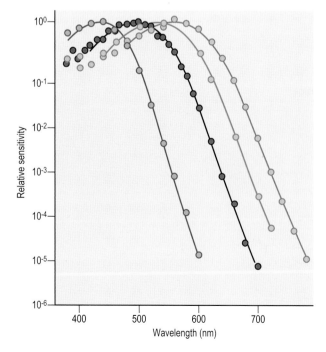

Figure 19.10 Spectral sensitivities of primate rod and cones. Rod (black symbols), green cone (green symbols) and red cone (red symbols) spectra are from human.[12,56] The blue cone spectrum (blue symbols) is from a macaque monkey[55] whose color vision is very similar to that in humans.

Box 19.2 Population differences in the spectral responses of cones

Normal color vision requires the proper electrical signaling of light by all three cone types. Evolutionarily, the red and green cone pigment genes arose from a duplication event and they lie together on the X chromosome. Their high sequence similarity has caused "confusion" during crossing over in gametogenesis creating within the population, hybrid genes, fragmented genes and variable numbers of copies of pigment genes on the chromosome (Fig. 19.11). In persons with many genes in the array, expression of the first two predominates. Yet some of these individuals and also women with different sets of pigment gene variants on their X chromosomes actually possess more than three cone types in their retinas. The genetics help to explain the substantial individual variation in color vision in "normal" persons, anomalous color vision as well as some forms of color "blindness" (see reviews[57,58]). The last category includes extreme cases where failure to express one or both types of pigment genes on the X chromosome gives rise to dichromacy or blue cone monochromacy, respectively.

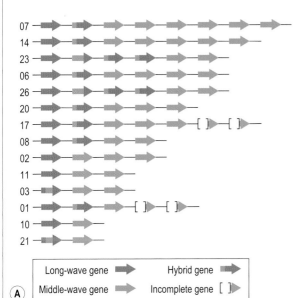

Figure 19.11 Genetic basis for heterogeneity in human color vision. (**A**) X-chromosome arrays of "red" (long-wave) and "green" (middle-wave) pigment genes are diverse, as shown by this small sample of male subjects with normal color vision. (Adapted from Neitz M, Neitz J, Grishok A. Polymorphism in the number of genes encoding long-wavelength sensitive cone pigments among males with normal color vision, Vision Research 35: 2395-2407. Copyright Elsevier 1995.) (**B**) Two common red cone pigment variants in the human population have maxima near 555 nm and 560 nm, as exemplified by the red cone spectral sensitivities of these two groups of individuals. (Adapted from Kraft TW, Neitz J, Neitz M. 1998: Spectra of human L cones, Vision Research 38: 3663-3670. Copyright Elsevier 1998.)

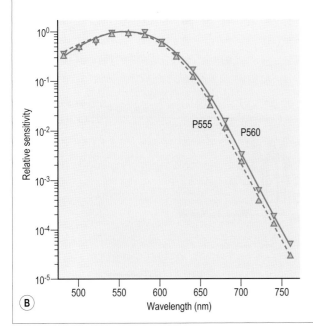

CNG channel and Na⁺/K⁺,Ca²⁺ exchanger

Two types of molecules are the principal mediators of ionic movement across the outer segment plasma membrane: the CNG channel and the Na⁺/K⁺,Ca²⁺ exchanger. CNG channels are heterotetramers consisting of CNGA1 and CNGB1 subunits (Fig. 19.12A) in a ratio of 3:1, respectively in rods[59-61] and CNGA2 and CNGB3 subunits in a B3-B3-A3-A3 arrangement in cones.[62] (Box 19.3) Although CNG channels are members of the same superfamily as voltage-gated K⁺ channels,[63] they are only mildly voltage-dependent. Instead, CNG channels in rods and cones are directly gated by cGMP.[64,65] Cyclic AMP also works, but with a 50-fold lower $K_{0.5}$ reflecting the reduced affinity of the channel for cAMP and the

reduced efficacy in opening once cAMP has bound.[66] While other ligand-gated ion channels in the body desensitize in the continued presence of their ligand, the CNG channel does not have any intrinsic mechanisms of desensitization and thus steadily monitors the [cGMP] in darkness.[67,68]

The rod CNG channel allows many different cations to pass: Ca²⁺, Mg²⁺, Na⁺, even K⁺. Permeability is greatest for Ca²⁺, but under normal conditions, Na⁺ carries ~80 percent of the current while Ca²⁺ carries 15–20 percent[69,70] because the extracellular concentration of Na⁺ is ten times higher than that of Ca²⁺. Mg²⁺ influx comprises a few per cent of the circulating current. The proportion of current carried by Ca²⁺ is twice greater through the cone channel because of its two-fold higher Ca²⁺ to Na⁺ permeability ratio.[71,72] Nevertheless, divalents are critical to the physiology of both rods and cones (see below and Chapters 18 and 20).

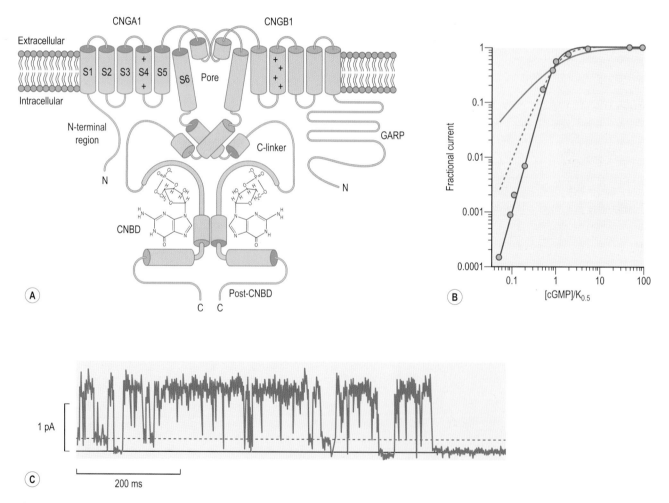

Figure 19.12 CNG channel. (**A**) The channel consists of 4 subunits, 2 of which are shown in this schematic diagram. Each subunit has six transmembrane segments, a pore-forming domain between helices 5 and 6 and a cyclic nucleotide binding domain (CNBD) on the carboxy terminus. Helix 4 contains lysine residues and functions as a voltage sensor. The carboxy and amino termini are located intracellularly. The amino terminus of the CNGB1 subunit contains a large glutamic acid-rich protein (GARP) segment. (Courtesy of W. Zagotta, University of Washington.) (**B**) CNG channel opening is steeply dependent on [cGMP]. Lines show Hill functions: fractional current = $[cGMP]^n/([cGMP]^n + K_{0.5}^n)$, where $K_{0.5}$ is the concentration of cGMP that opens half of the channels and n is the Hill coefficient taking values of 1 (red, continuous line), 2 (red, dashed line) and 3 (blue line). Hill coefficients greater than 1 indicate cooperativity. (Courtesy of A. Zimmerman, Brown University and Denis Baylor, Stanford University.) (**C**) A single CNG channel in a patch of membrane excised from a salamander rod flickers between an open state, a second open state of lower conductance (dashed line) and a closed state (horizontal line). The "fuzziness" of the trace above the dashed line represents extremely brief transitions out of the open state that were unresolved by the recording. The activity was recorded in symmetric salt solutions containing low concentrations of divalent cations and 5 μM cGMP on the intracellular face with a voltage of +70 mV applied to the membrane. (From W.R. Taylor and D.A. Baylor: Conductance and kinetics of single cGMP-activated channels in salamander rod outer segments, Journal of Physiology 483: 567–582. Adapted with permission from Blackwell Publishing, copyright 1995.)

For fast responsivity to increases in [cGMP], it would be advantageous for the CNG channel to avidly bind cGMP. On the other hand, the channel would release cGMP too slowly and fail to detect decreases in cGMP rapidly. As a compromise, the channel has a fairly high $K_{0.5}$ for cGMP (i.e. moderately low sensitivity), on the order of 10 μM.[64,65] The channel subunits work together in such a way that cGMP binding affinity increases with each cGMP bound, making channel opening steeply dependent on ligand concentration (Fig. 19.12B). This cooperativity is one of Nature's ways of building a biological switch. Cyclic GMP binding stabilizes the open state of the channel so by the law of microscopic reversibility, cGMP unbinds best from the closed state. Therefore, to be constantly vigilant for a decrease in [cGMP], the channel reverts incessantly to the closed state (see review[25]). The channel can then respond to changes in [cGMP] on a time-scale that is considerably faster than the photoresponse kinetics. Flickering makes the channel extremely noisy though (Fig. 19.12C). To minimize the noise, the channel detains Ca^{2+} and Mg^{2+} during their passage through the pore.[73] As long as divalents reside in the pore, no other ions can pass and the conductance of the CNG channel drops by two orders of magnitude from about 25 pS to 100 fS. Given low channel conductance and relative insensitivity to cGMP, the cell must express a lot of channels (a few hundred per μm^2) in order to maintain a reasonable circulating current. The arrangement averages channel noise across many channels, but it also puts the cell at risk. Normally, the free [cGMP] in darkness is only a few micromolar and a small percentage of channels are open. Should the cell synthesize too much cGMP the number of open channels would increase by more than an order of magnitude and annihilate the transmembrane ion gradients.

For the purposes of photon counting, it would be desirable for the light-induced decrease in circulating current to be proportional to the number of channels closed. But CNG channel closure hyperpolarizes the rod and Ohm's law dictates an increase in current flowing through each of the remaining open channels. Two mechanisms conspire against a linear Ohmic relation between current and voltage. First, hyperpolarization enhances divalent block by increasing the driving force on Ca^{2+} and Mg^{2+} to enter the channel pore.[74,75] Second, the CNG channel is slightly voltage dependent so hyperpolarization tends to close the channel. The net effect is that over the physiological range of membrane potentials, the inward current is nearly constant (Fig. 19.13). The CNG channel in cones is also outwardly rectifying over the physiological range of voltages, but to a lesser extent than the rod channel.[65] So in brighter light, as the cone becomes more hyperpolarized, the driving force sends more ions through the remaining open channels. Cones are not concerned with single photons and more interested in dynamic range. The reduced effect of closing channels at hyperpolarized potentials may contribute to their remarkable capacity to operate over a wide range of intensities. Another consequence is that the photocurrent in cones will not isolate events in the outer segment and can be influenced by voltage changes induced by external sources (see Chapters 21 and 22).

Diltiazem is a potent and reversible inhibitor of the CNG channel. Diltiazem blocks Ca^{2+} channels and is often prescribed as a vasodilator to treat hypertension. Fortunately, Ca^{2+} channels and CNG channels are sensitive to different

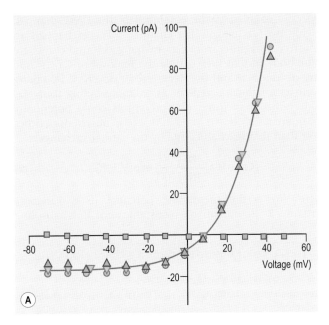

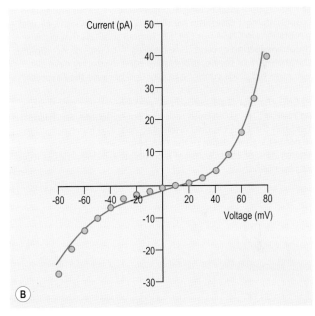

Figure 19.13 Outward rectification of the current-to-voltage relations of CNG channels in rods (**A**) and cones (**B**). Physiologically relevant voltages range from about –40 to –65 mV. The purple squares in A represent measurements made during exposure to light that closed all of the CNG channels. (Panel A from Baylor DA, Nunn BJ. Electrical properties of the light-sensitive conductance of rods of the salamander *Ambystoma tigrinum*. Journal of Physiology 1986; 371: 115–145. Adapted with permission from Blackwell Publishing. Panel B from L.W. Haynes and K.-W. Yau: Cyclic GMP-sensitive conductance in outer segment membrane of catfish cones, Nature 317: 61-64,[65] adapted with permission from Macmillan Publishers Ltd.)

diastereoisomers. Other CNG channel inhibitors include the Na$^+$ channel-blocking amilorides, the Ca^{2+}-activated K$^+$ channel-blocker dequalinium and the local anesthetic tetracaine (see review[76]).

Ca^{2+} is removed from the outer segment by a special exchanger, consisting of a dimer[79] of NCKX1 in rods[80] and of NCKX2 in cones[81] (Fig. 19.14A). Although there may be some exchangers at the cone synapse, the exchangers in rods are located exclusively in the outer segment[82] where they are physically coupled to the α-subunit of the CNG channel, roughly two exchangers per channel.[83] The exchangers in photoreceptors differ from Ca^{2+} exchangers found elsewhere in the body in that they couple the removal of Ca^{2+} to the entry of four Na^+ and the departure of one K^+.[84] By taking advantage of the K^+ gradient as well as sending four Na^+ inside rather than three, photoreceptors could potentially take internal Ca^{2+} down to 0.2 nM, a level hundreds of times lower than that reachable, e.g., by the cardiac exchanger:

$$[Ca]_i = [Ca]_o \frac{[Na]_i^4 [K]_o}{[Na]_o^4 [K]_i} \exp(V_m F/RT) \tag{2}$$

where Vm is the membrane potential, F is Faraday's constant, R is Boltzman's constant and T is absolute temperature. In reality, internal Ca^{2+} never gets that low, possibly because the exchanger inactivates. Since there is a net movement of charge, the action of the exchanger comprises a small percentage of the dark current. To a first approximation, Ca^{2+} falls exponentially in the rod outer segment with a time constant between 40 and 90 ms (Fig. 19.14B).[23,28] In cones, the combined properties of exchanger and CNG channel support a higher Ca^{2+} flux and faster light-induced changes in Ca^{2+} concentration.

Role of inner segment conductances

Closure of CNG channels in response to light hyperpolarizes rods and cones towards the equilibrium potential for K^+, however, the transmembrane voltage is subject to modification by a variety of conductances located on the inner segment, cell body and synaptic terminal. A net inward current develops that increases in magnitude as the transmembrane potential becomes more negative. The effect of these conductances, referred to collectively as inner segment conductances, is readily apparent from a comparison of the photocurrents to the photovoltages for either a rod or a cone (Fig. 19.15). For dim flashes, the inner segment conductances are little affected so the time course of the voltage response is similar to that of the current response.[4,7,11,14] But for bright flashes, the time courses differ; there is a prominent "nose" in the voltage response while there is no such nose in the current response.

Overall, activation of the inner segment conductances tends to return the membrane potential to the resting value. Such an action helps ensure that the membrane potential remains within the dynamic range of neurotransmitter release and prevents regenerative spiking.[87,88]

Box 19.3 Color blindness due to a channelopathy

Rod monochromacy is an autosomal recessive condition presenting with achromatopsia, poor visual acuity, photophobia and nystagmus. It is fairly rare, affecting fewer than 1 in 30,000 persons, although the prevalence is as high as 1 in 10 amongst the native population of Pingelap (see review[77] for an interesting account). In most cases, the condition is a channelopathy, caused by mutations in CNGA3 or CNGB3 channel subunits that disable phototransduction in cones (see reviews[76,78]). Cases of incomplete achromatopsia or severe, progressive cone dystrophy have also been reported. Rods continue to function because they express distinct CNG channel subunits. In contrast, null mutations in the rod CNG channel cause autosomal recessive retinitis pigmentosa (see review[76]), in which both rods and cones degenerate. Failure to express the CNG channel is lethal to rods and for reasons that are not yet known, death of the rods results in a secondary degeneration of cones so there is complete blindness rather than just night blindness.

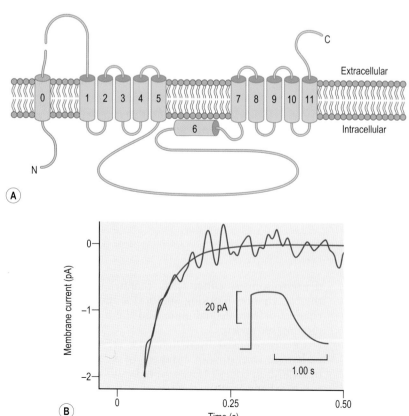

Figure 19.14 Calcium exchange. (**A**) The NCKX2 exchanger subunit in cones has an extracellular C terminus in contrast to the cardiac NCX, where it is intracellular.[85,86] A leader sequence (red) is cleaved off prior to localization of the exchanger to the outer segment plasma membrane. (**B**) A bright flash given at time zero, that closes all of the CNG channels very suddenly, reveals a residual current due to the electrogenic removal of Ca^{2+} by the $Na^+/K^+, Ca^{2+}$ exchanger (black trace; inset: flash response in its entirety). The exchange current declines (exponential fit, blue trace) in this rod with a time constant of 59 ms, as the internal Ca^{2+} drops to a minimum. (From G.D. Field and F. Rieke: Mechanisms regulating variability of the single photon responses of mammalian rod photoreceptors, Neuron 35: 733–747. Reproduced by permission of Cell Press, copyright 2002.)

Large depolarizations that might arise from activation of voltage-dependent Ca^{2+} currents, particularly in small compartments, are thwarted by voltage- and Ca^{2+}-activated K^+ currents. Large hyperpolarizations that might arise for example, from the electrogenic Na^+K^+-ATPase, are counteracted by a non-specific cationic current activated at voltages more negative than the resting potential. Changes in various of these currents at different voltages within the normal operating range of rods and cones occur slowly with respect to the photocurrent and therefore act to high pass filter the response to bright light (Fig. 19.15). A third role is to control the entry of Ca^{2+}, which in turn regulates transmitter release and other cellular activities. The intracellular concentrations of Ca^{2+} and other ions may also mediate effects in the cell beyond vesicular release. Five currents, three voltage-gated and two Ca^{2+}-activated, are described below.

Delayed rectifier potassium current, I_{KV}

The channels underlying I_{KV} activate at $V_m = -70$ mV in rods[89] and at $V_m = -50$ mV for monkey cones.[90] The outward flow of K^+ ions through I_{KV} together with a leak current constitute the inner segment leg of the circulating current and help to set the membrane potential of the photoreceptor in darkness. I_{KV} channels close slowly when the photoreceptor hyperpolarizes (Fig. 19.16). In so doing, they limit the light-induced hyperpolarization and may contribute to the nose in the response to brighter flashes and steps. This current is blocked by extracellular tetraethyl ammonium ions.

Hyperpolarization-activated current, I_H

A cationic conductance appears at potentials negative to -50 mV[90] (Fig. 19.17). It has a reversal potential of -30 to -40 mV, reflecting its permeability to K^+, Na^+ and even Ca^{2+}. I_H turns on slowly and tends to counteract closure of the CNG channels in the outer segment by allowing the entry of Na^+ and depolarizing the photoreceptor. It therefore has an effect similar to closure of I_{KV} channels, but comes into play with brighter flashes since it has a more negative activation potential. The gradual building of I_H upon hyperpolarization precludes it from greatly affecting the initial rising phase of the photoresponse, though. It is primarily responsible for quickening the photovoltage response at its peak and creating the "nose." The channels are formed from HCN1 subunits[91] that share structural similarities with CNG channel subunits (see review[92]). As is the case with CNG channels,

HCN1 channels are gated by cyclic nucleotides that shift slightly the I_H current-to-voltage relation to the right, i.e. to voltages a few mV less negative.

Voltage-activated calcium current, I_{Ca}

The major calcium conductance is the "long-lasting" or L-type. The channel is composed of α_1, $\alpha_2\delta$, β and γ subunits.

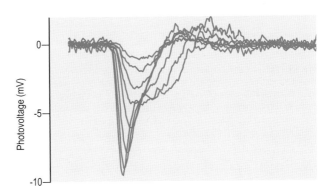

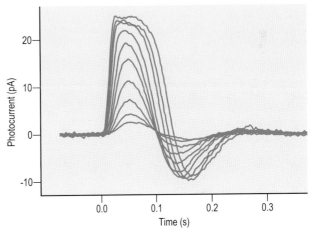

Figure 19.15 Differences in the current and voltage response waveforms. The photovoltages and photocurrents were recorded from different cones at 37°C. In the top panel (from Fig. 19.1C), flashes deflect the transmembrane potential from its resting level of -43 mV whereas in the lower panel, flashes decrease the circulating current from a value in darkness of -24.9 pA. (Lower panel from D.A. Baylor, B.J. Nunn and J.L. Schnapf: Spectral sensitivity of cones of the monkey Macaca fascicularis, Journal of Physiology 390: 145-160. Adapted with permission from Blackwell Publishing, copyright 1987.)

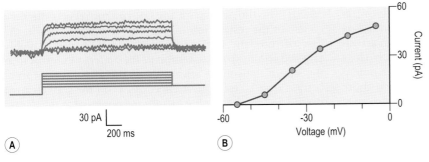

Figure 19.16 Voltage-dependent K^+ current. (**A**) I_{KV} turns on slowly in response to voltage steps from a holding potential of -105 mV to command voltages ranging from -55 mV to -5 mV at 33–36°C. The I_{KV} was isolated pharmacologically and by careful selection of the ionic composition of the bath and pipette solutions in voltage clamp experiments on monkey cone inner segments. (**B**) At steady state the current-to-voltage relations of I_{KV} indicate an outward current at potentials less negative than -60 mV in this cell. (From T. Yagi and P.R. MacLeish, 1994: Ionic conductances of monkey solitary cone inner segments, Journal of Neurophysiology 71: 656–665,[90] with permission from the American Physiological Society.)

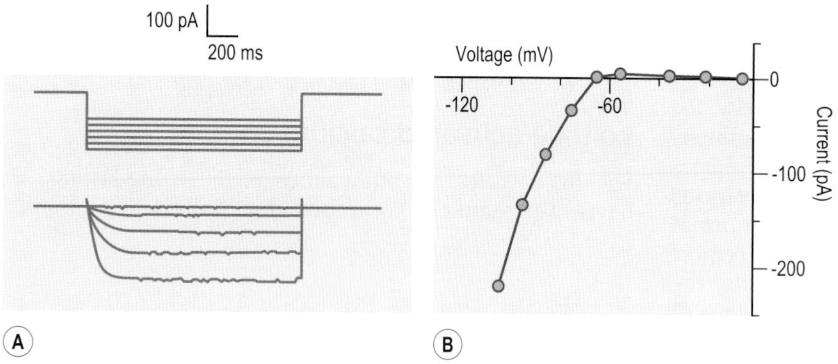

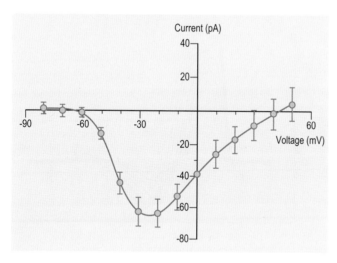

Figure 19.17 Slow development of an inwardly rectifying current (I_H) with hyperpolarization. (**A**) Stepping the membrane potential of an isolated, salamander rod under voltage-clamp from −5 mV to values between −55 and −105 mV turns on an inward current over a period of ~100 ms at 33–36°C. Currents other than I_H were blocked pharmacologically and by manipulating the ionic composition of the bathing solutions or were subtracted from the records. (**B**) The current-to-voltage relations of I_H at steady state indicate an inward current at negative potentials. Under physiological conditions with proper ionic solutions and cyclic nucleotide concentrations, I_H activates at slightly less negative potentials. Nevertheless, I_H remains negligible at the dark resting potential. (From T. Yagi and P.R. MacLeish, 1994: Ionic conductances of monkey solitary cone inner segments, Journal of Neurophysiology 71: 656–665,[90] with permission from the American Physiological Society.)

Figure 19.18 Current-to-voltage relations for I_{Ca}. Measurements were made in porcine rods with Ca^{2+} to minimize interference from other currents. The physiological voltage range controls nearly the full span of I_{Ca}. (From Cia D, Bordais A, Varela C, et al. 2005: Voltage-gated channels and calcium homeostasis in mammalian rod photoreceptors, Journal of Neurophysiology 93: 1468–1475,[96] with permission from the American Physiological Society.)

Three isoforms of the pore-forming α_1 subunit are expressed in rods and cones: α_{1C}, α_{1D} and α_{1F}.[93-95] Ca^{2+} channels are found on somata and synaptic terminals, the latter localization mediating synaptic transmission in rods and cones (Box 19.4). Studies on isolated salamander rods report an activation potential around −45 mV which is paradoxically close to the membrane potential in the dark. The activation potential for Ca^{2+} channels in primate rods and cones is more reasonable, close to −60 mV, leaving a wide operating range of voltage for modifying Ca^{2+} entry in response to light (Fig. 19.18). The Ca^{2+} channels in rods and cones react promptly to voltage changes and are non-inactivating, a property that is well suited for a high rate of sustained vesicular neurotransmitter release in darkness. The channels are blocked by diltiazem, verapamil and dihydropyridines such as nifedipine and nimodipine and are enhanced by Bay K 8644.[88,96]

Calcium-activated potassium current, $I_{K(Ca)}$

Two currents in rods and cones are activated by a rise in intracellular Ca^{2+}: the Ca^{2+}-activated anion current and the Ca^{2+}-activated K^+ current.[89] The latter shows the N-shaped current-to-voltage relationship characteristic of other calcium-activated currents. The magnitude of the outward current is reduced by charybdotoxin[99,100] indicating the involvement of maxi-K or BK channels. Although $I_{K(Ca)}$ is sizable in rods at positive voltages, the current is minimal within the physiological range and absent in cones. For that matter, $I_{K(Ca)}$ does not appear to be expressed in primate cones at all.[90] $I_{K(Ca)}$ may therefore serve as a safety net, capable of moving V_m towards the equilibrium potential for K^+ to guard against large depolarizations that would flood the rod interior with Ca^{2+}. Cones rely on other currents to fulfill that function.

Calcium-activated anion current, $I_{Cl(Ca)}$

Both rods and cones express a robust Ca^{2+}-activated anion current.[88-90,100] The main permeant anion is Cl^- and given an equilibrium potential for Cl^- near −20 in rods[101] and −45 mV in cones,[102] $I_{Cl(Ca)}$ helps to set and stabilize the resting potential. Experimentally, the current is observed most prominently as a tail current following a step of voltage that opens Ca^{2+} channels and allows Ca^{2+} to enter the cytoplasm (Fig. 19.20). The time-course of the tail current reflects the time course of the restoration of free Ca^{2+} to basal levels. In the case of rods, the Ca^{2+}-activated Cl^- conductance is highly targeted to the synaptic terminal. Its strategic location near the site of transmitter release raises the possibility of local control over I_{Ca} by a direct effect of Cl^- on the Ca^{2+} channel itself[103] and by altering the membrane potential at the release site, indirectly affecting Ca^{2+} channel gating. Given the sizes of the pedicle and spherule and their separation from the soma by a thin fiber, the membrane potential in the terminal region could differ from that in the soma. After hyperpolarizations lasting hundreds of milliseconds, the absence of Cl^- efflux through $I_{Cl(Ca)}$ would prime Ca^{2+} channels and shift their activation potential to lower voltages.[103] This

Box 19.4 Night blindness resulting from a defective rod synapse

Incomplete Schubert–Bornstein congenital stationary night blindness is an X-linked, recessive condition that can result from genetic defects in *CACNA1F*.[97,98] Over 57 disease-causing mutations have been documented (http://www.hgmd.cf.ac.uk/ac/index.php). Since the gene encodes the pore-forming α_{1F} subunit of the calcium channel in rods and cones, disruptive mutations

(Fig. 19.19) preclude rod signals from reaching second-order neurons. Transmission at the cone synapse is affected, but maintained nonetheless because cones also express the α_{1D} isoforms.[95] Night blindness is incomplete because gap junctions afford parallel albeit less-efficient rod signaling through the cone neural circuitry.

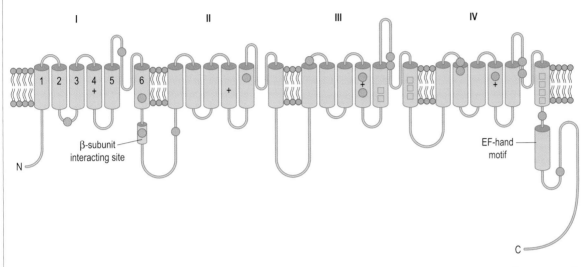

Figure 19.19 Partial spectrum of mutations in *CACNA1F* that cause night blindness. There are four repeating domains of six transmembrane segments in the α_{1F}-subunit. The S4 segment in each domain is a voltage sensor (pluses). Residues comprising the dihydropyridine binding site are present within repeats III and IV (green squares). Disease-causing mutations (red circles) are found in each of the four repeats. (From M. Nakamura , S. Ito, H. Terasaki and Y. Miyake: Novel CACNA1F mutations in Japanese patients with incomplete congenital stationary night blindness, Investigative Ophthalmology and Visual Science 42:1610–1616. Adapted with permission from the Association for Research in Vision and Ophthalmology, copyright 2001.)

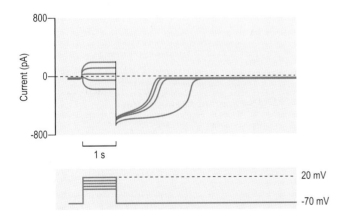

Figure 19.20 Ca^{2+}-activated Cl^- conductance. Depolarizations of salamander rods from −70 mV to voltages between −20 and +20 mV at 10 mV intervals evoke an anionic current with a reversal potential near 0 mV, the equilibrium potential for Cl^- under the nonphysiological conditions of these experiments. The depolarization increases intracellular Ca^{2+} so upon return of the membrane potential to −70 mV, $I_{Cl (Ca)}$ gives rise to an enormous tail current. Ca^{2+} currents have been subtracted and other currents blocked to isolate $I_{Cl (Ca)}$. Tail currents do not appear if the cell is bathed in a Ca^{2+}-free solution. Recordings were made near 10°C. (From MacLeish PR and Nurse CA, 2007: Ion channel compartments in photoreceptors: evidence from salamander rods with intact and ablated terminals, Journal of Neurophysiology 98: 86-95,[100] adapted with permission from the American Physiological Society.)

mechanism could hasten the recovery after bright flashes and sensitize the synapse to bright steady light being turned off.

The approximate contribution of each inner segment conductance to the photovoltage response is summarized in Figure 19.21. Rapid closure of CNG channels hyperpolarizes the photoreceptor and slowly shuts down an outward current (I_{KV}) at voltages slightly more negative than the resting potential and activates an inward current at even more negative voltages (I_H) to carve out the nose in the response to a bright flash. Ca^{2+}-activated Cl^- currents may sharpen the voltage response at the synapse. Ultimately the light-induced cessation of Ca^{2+}-mediated synaptic release gets dampened and proceeds over a faster time course.

Electrotonic coupling

In addition to inner segment conductances, the membrane potentials of rods and cones are influenced by neighboring cells via cell–cell coupling and synaptic feedback.[5-7,104] Single photon responses from rods induce a voltage response in cones but with an amplitude that is eight times smaller. Time to peak is about twice as fast because the network behaves as a high pass filter. A consequence of coupling between rods and cones and between cones of different spectral types is that if the spot of light falling on the retina is large, the size and shape of the response in a given cone

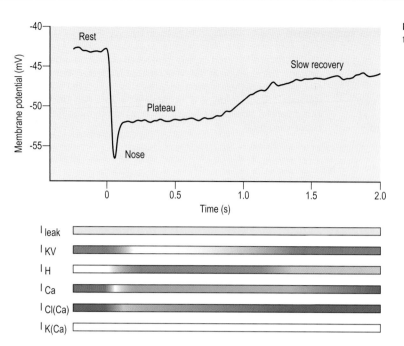

Figure 19.21 Influence of inner segment conductances on the photovoltage response to a bright flash in a rod.

will depend on wavelength. In addition, modulation by horizontal cells confers an antagonistic center-surround receptive field. These effects are described in greater detail in Chapter 22.

Summary

Rods and cones divide the intensity range for vision. Rods attain the ultimate in sensitivity by adeptly counting single photons. Their sensitivity is limited only by photon capture. To achieve high sensitivity, they sacrifice rapidity in their response kinetics and dynamic range. The photocurrent response generated in the outer segment is delayed, peaks in 200 ms and has an integration time of 300 ms. Since cones are not concerned with dim light, their photocurrent response to single photons is an order of magnitude smaller, time to peak is several times shorter and integration time is reduced over ten-fold. It is important that cones do not saturate, so their operating range not only shifts to the right, it also stretches out compared to rods and accommodations have been made to prevent bleaching from doing more than reducing photon capture. Higher intensities allow a point in visual space to be encoded by multiple receptors so that information about color may be extracted. Fast cone responses improve on temporal resolution. Inner segment conductances in rods and cones set the resting membrane potential, speed up the kinetics of the photovoltage response near its peak, and modulate intracellular Ca^{2+}. Certain heritable diseases may target rods, cones or both, depending upon their particular patterns of gene expression.

Acknowledgment

We thank D. Baylor for comments on the text.

References

1. Roof DJ, Makino CL. The structure and function of retinal photoreceptors. In: Albert DM, Jakobiec FA, eds. Principles and practice of ophthalmology, 2nd edn. Philadelphia: WB Saunders, 2000:1624–1673.
2. Schneeweis DM, Schnapf JL. Photovoltage of rods and cones in the macaque retina. Science 1995; 268:1053–1056.
3. Tomita T. Electrophysiological study of the mechanisms subserving color coding in the fish retina. Cold Spring Harb Symp Quant Biol 1965; 30:559–566.
4. Schneeweis DM, Schnapf JL. Noise and light adaptation in rods of the macaque monkey. Vis Neurosci 2000; 17:659–666.
5. Hornstein EP, Verweij J, Schnapf JL. Electrical coupling between red and green cones in primate retina. Nature Neurosci 2004; 7:745–750.
6. Hornstein EP, Verweij J, Li PH, Schnapf JL. Gap-junctional coupling and absolute sensitivity of photoreceptors in macaque retina. J Neurosci 2005; 25: 11201–11209.
7. Schneeweis DM, Schnapf JL. The photovoltage of macaque cone photoreceptors: adaptation, noise, and kinetics. J Neurosci 1999; 19:1203–1216.
8. McNaughton PA, Yau K-W, Lamb TD. Spread of activation and desensitisation in rod outer segments. Nature 1980; 283:85–87.
9. Lamb TD, McNaughton PA, Yau K-W. Spatial spread of activation and background desensitization in toad rod outer segments. J Physiol 1981; 319:463–496.
10. Matthews G. Spread of the light response along the rod outer segment: an estimate from patch-clamp recordings. Vision Res 1986; 26:535–541.
11. Baylor DA, Nunn BJ, Schnapf JL. The photocurrent, noise and spectral sensitivity of rods of the monkey *Macaca fascicularis*. J Physiol 1984; 357:575–607.
12. Kraft TW, Schneeweis DM, Schnapf JL. Visual transduction in human rod photoreceptors. J Physiol 1993; 464:747–765.
13. Pepperberg DR, Birch DG, Hood DC. Photoresponses of human rods in vivo derived from paired-flash electroretinograms. Vis Neurosci 1997; 14:73–82.
14. Schnapf JL, Nunn BJ, Meister M, Baylor DA. Visual transduction in cones of the monkey *Macaca fascicularis*. J Physiol 1990; 427:681–713.
15. Cideciyan AV, Jacobson SG, Gupta N et al. Cone deactivation kinetics and GRK1/GRK7 expression in enhanced S cone syndrome caused by mutations in NR2E3. Invest Ophthalmol Vis Sci 2003; 44:1268–1274.
16. Friedburg C, Allen CP, Mason PJ, Lamb TD. Contribution of cone photoreceptors and post-receptoral mechanisms to the human photopic electroretinogram. J Physiol 2004; 556:819–834.
17. Krispel CM, Chen D, Melling N et al. RGS expression rate-limits recovery of rod photoresponses. Neuron 2006; 51:409–416.
18. Pepperberg DR, Cornwall MC, Kahlert M et al. Light-dependent delay in the falling phase of the retinal photoresponse. Vis Neurosci 1992; 8:9–18.
19. Nikonov S, Engheta N, Pugh EN Jr. Kinetics of recovery of the dark-adapted salamander rod photoresponse. J Gen Physiol 1998; 111:7–37.
20. Birch DG, Hood DC, Nusinowitz S, Pepperberg DR. Abnormal activation and inactivation mechanisms of rod transduction in patients with autosomal dominant retinitis pigmentosa and the Pro-23-His mutation. Invest Ophthalmol Vis Sci 1995; 36:1603–1614.
21. van Hateren JH, Lamb TD. The photocurrent response of human cones is fast and monophasic. BMC Neurosci 2006; 7:34.
22. Baylor DA, Lamb TD, Yau K-W. Responses of retinal rods to single photons. J Physiol 1979; 288:613–634.

23. Tamura T, Nakatani K, Yau K-W. Calcium feedback and sensitivity regulation in primate rods. J Gen Physiol 1991; 98:95–130.

24. Pugh EN Jr, Lamb TD. Phototransduction in vertebrate rods and cones: molecular mechanisms of amplification, recovery and light adaptation. In: Stavenga DG, DeGrip WJ, Pugh EN Jr, eds. Handbook of biological physics molecular mechanisms in visual transduction. Amsterdam: Elsevier Science BV, 2000:183–255.

25. Yau K-W, Baylor DA. Cyclic GMP-activated conductance of retinal photoreceptor cells. Annu Rev Neurosci 1989; 12:289–327.

26. Rieke F, Baylor DA. Origin of reproducibility in the responses of retinal rods to single photons. Biophys J 1998; 75:1836–1857.

27. Whitlock GC, Lamb TD. Variability in the time course of single photon responses from toad rods: termination of rhodopsin's activity. Neuron 1999; 23:337–351.

28. Field GD, Rieke F. Mechanisms regulating variability of the single photon responses of mammalian rod photoreceptors. Neuron 2002; 35:733–747.

29. Baylor DA, Matthews G, Yau K-W. Two components of electrical dark noise in toad retinal rod outer segments. J Physiol 1980; 309:591–621.

30. Lamb TD, Simon EJ. Analysis of electrical noise in turtle cones. J Physiol 1977; 272:435–468.

31. Rieke F, Baylor DA. Origin and functional impact of dark noise in retinal cones. Neuron 2000; 26:181–186.

32. Rieke F, Baylor DA. Molecular origin of continuous dark noise in rod photoreceptors. Biophys J 1996; 71:2553–2572.

33. Hecht S, Shlaer S, Pirenne MH. Energy, quanta and vision. J Gen Physiol 1942; 25:819–840.

34. Kraft TW, Schnapf JL. Aberrant photon responses in rods of the macaque monkey. Vis Neurosci 1998; 15:153–159.

35. Chen J, Makino CL, Peachey NS, Baylor DA, Simon MI. Mechanisms of rhodopsin inactivation in vivo as revealed by a COOH-terminal truncation mutant. Science 1995; 267:374–377.

36. Chen C-K, Burns ME, Spencer M et al. Abnormal photoresponses and light-induced apoptosis in rods lacking rhodopsin kinase. Proc Natl Acad Sci USA 1999; 96:3718–3722.

37. Tamura T, Nakatani K, Yau K-W. Light adaptation in cat retinal rods. Science 1989; 245:755–758.

38. Nakatani K, Tamura T, Yau K-W. Light adaptation in retinal rods of the rabbit and two other nonprimate mammals. J Gen Physiol 1991; 97:413–435.

39. Thomas MM, Lamb TD. Light adaptation and dark adaptation of human rod photoreceptors measured from the a-wave of the electroretinogram. J Physiol 1999; 518:479–496.

40. Calvert PD, Govardovskii VI, Arshavsky VY, Makino CL. Two temporal phases of light adaptation in retinal rods. J Gen Physiol 2002; 119:129–145.

41. Krispel CM, Chen C-K, Simon MI, Burns ME. Novel form of adaptation in mouse retinal rods speeds recovery of phototransduction. J Gen Physiol 2003; 122:703–712.

42. Lamb TD, Pugh EN Jr. Dark adaptation and the retinoid cycle of vision. Prog Retin Eye Res 2004 23:307–380.

43. Jones GJ, Cornwall MC, Fain GL. Equivalence of background and bleaching desensitization in isolated rod photoreceptors of the larval tiger salamander. J Gen Physiol 1996; 108:333–340.

44. Paupoo AAV, Mahroo OAR, Friedburg C, Lamb TD. Human cone photoreceptor responses measured by the electroretinogram a-wave during and after exposure to intense illumination. J Physiol 2000; 529:469–482.

45. Nikonov SS, Kholodenko R, Lem J, Pugh EN Jr. Physiological features of the S- and M-cone photoreceptors of wild-type mice from single-cell recordings. J Gen Physiol 2006; 127:359–374.

46. Kenkre JS, Moran NA, Lamb TD, Mahroo OAR. Extremely rapid recovery of human cone circulating current at the extinction of bleaching exposures. J Physiol 2005; 567:95–112.

47. Paskowitz DM, LaVail MM, Duncan JL. Light and inherited retinal degeneration. Br J Ophthalmol 2006; 90:1060–1066.

48. Cideciyan AV, Zhao X, Nielsen L, Khani SC, Jacobson SG, Palczewski K. Null mutation in the rhodopsin kinase gene slows recovery kinetics of rod and cone phototransduction in man. Proc Natl Acad Sci USA 1998; 95:328–333.

49. Weiss ER, Ducceschi MH, Horner TJ, Li A, Craft CM, Osawa S. Species-specific differences in expression of G-protein-coupled receptor kinase (GRK) 7 and GRK1 in mammalian cone photoreceptor cells: implications for cone cell phototransduction. J Neurosci 2001; 21:9175–9184.

50. Chen C-K, Zhang K, Church-Kopish J et al. Characterization of human GRK7 as a potential cone opsin kinase. Mol Vis 2001; 7:305–313.

51. Haider NB, Jacobson SG, Cideciyan AV et al. Mutation of a nuclear receptor gene, NR2E3, causes enhanced S cone syndrome, a disorder of retinal cell fate. Nat Genet 2000; 24:127–131.

52. Craft CM, Whitmore DH, Wiechmann AF. Cone arrestin identified by targeting expression of a functional family. J Biol Chem 1994; 269:4613–4619.

53. Sakuma H, Inana G, Murakami A, Higashide T, McLaren MJ. Immunolocalization of x-arrestin in human cone photoreceptors. FEBS Lett 1996; 382:105–110.

54. Nishiguchi KM, Sandberg MA, Kooijman AC et al. Defects in RGS9 or its anchor protein R9AP in patients with slow photoreceptor deactivation. Nature 2004; 427:75–78.

55. Baylor DA, Nunn BJ, Schnapf JL. Spectral sensitivity of cones of the monkey Macaca fascicularis. J Physiol 1987; 390:145–160.

56. Schnapf JL, Kraft TW, Baylor DA. Spectral sensitivity of human cone photoreceptors. Nature 1987; 325:439–441.

57. Neitz M, Neitz J. Molecular genetics of human color vision and color vision defects. In: Chalupa LM, Werner JS, eds. The visual neurosciences. Cambridge: MIT Press, 2004:974–988.

58. Deeb SS. Genetics of variation in human color vision and the retinal cone mosaic. Curr Opin Genet Dev 2006; 16:301–307.

59. Weitz D, Ficek N, Kremmer E, Bauer PJ, Kaupp UB. Subunit stoichiometry of the CNG channel of rod photoreceptors. Neuron 2002; 36:881–889.

60. Zheng J, Trudeau MC, Zagotta WN. Rod cyclic nucleotide-gated channels have a stoichiometry of three CNGA1 subunits and one CNGB1 subunit. Neuron 2002; 36:891–896.

61. Zhong H, Molday LL, Molday RS, Yau K-W. The heteromeric cyclic nucleotide-gated channel adopts a 3A:1B stoichiometry. Nature 2002; 420:193–198.

62. Peng C, Rich ED, Varnum MD. Subunit configuration of heteromeric cone cyclic nucleotide-gated channels. Neuron 2004; 42:401–410.

63. Jan LY, Jan YN. A superfamily of ion channels. Nature 1990; 345:672.

64. Fesenko EE, Kolesnikov SS, Lyubarsky AL. Induction by cyclic GMP of cationic conductance in plasma membrane of retinal rod outer segment. Nature 1985; 313:310–313.

65. Haynes L, Yau K-W. Cyclic GMP-sensitive conductance in outer segment membrane of catfish cones. Nature 1985; 317:61–64.

66. Tanaka JC, Eccleston JF, Furman RE. Photoreceptor channel activation by nucleotide derivatives. Biochemistry 1989; 28:2776–2784.

67. Karpen JW, Zimmerman AL, Stryer L, Baylor DA. Molecular mechanics of the cyclic-GMP-activated channel of retinal rods. Cold Spring Harb Symp Quant Biol 1988; 53:325–332.

68. Watanabe S-I, Matthews G. Cyclic GMP-activated channels of rod photoreceptors show neither fast nor slow desensitization. Vis Neurosci 1990; 4:481–487.

69. Nakatani K, Yau K-W. Calcium and magnesium fluxes across the plasma membrane of the toad rod outer segment. J Physiol 1988; 395:695–729.

70. Ohyama T, Hackos DH, Frings S, Hagen V, Kaupp UB, Korenbrot JI. Fraction of the dark current carried by Ca^{2+} through cGMP-gated ion channels of intact rod and cone photoreceptors. J Gen Physiol 2000; 116:735–753.

71. Frings S, Seifert R, Godde M, Kaupp UB. Profoundly different calcium permeation and blockage determine the specific function of distinct cyclic nucleotide-gated channels. Neuron 1995; 15:169–179.

72. Picones A, Korenbrot JI. Permeability and interaction of Ca^{2+} with cGMP-gated ion channels differ in retinal rod and cone photoreceptors. Biophys J 1995; 69:120–127.

73. Bodoia RD, Detwiler PB. Patch-clamp recordings of the light-sensitive dark noise in retinal rods from the lizard and frog. J Physiol 1985; 367:183–216.

74. Zimmerman AL, Baylor DA. Cyclic GMP-sensitive conductance of retinal rods consists of aqueous pores. Nature 1986; 321:70–72.

75. Matthews G. Comparison of the light-sensitive and cyclic GMP-sensitive conductances of the rod photoreceptor: noise characteristics. J Neurosci 1986; 6:2521–2526.

76. Brown RL, Strassmaier T, Brady JD, Karpen JW. The pharmacology of cyclic nucleotide-gated channels: emerging from the darkness. Curr Pharmaceut Des 2006; 12:3597–3613.

77. Sachs OW. The Island of the Colorblind. New York: Vintage Books, 1997.

78. Michaelides M, Hunt DM, Moore AT. The cone dysfunction syndromes. Br J Ophthalmol 2004; 88:291–297.

79. Schwarzer A, Kim TSY, Hagen V, Molday RS, Bauer PJ. The Na/Ca-K exchanger of rod photoreceptor exists as dimer in the plasma membrane. Biochemistry 1997; 36:13667–13676.

80. Cook NJ, Kaupp UB. Solubilization, purification, and reconstitution of the sodium-calcium exchanger from bovine rod outer segments. J Biol Chem 1988; 263:11382–11388.

81. Prinsen CFM, Szerencsei RT, Schnetkamp PPM. Molecular cloning and functional expression of the potassium-dependent sodium-calcium exchanger from human and chicken retinal cone photoreceptors. J Neurosci 2000; 20:1424–1434.

82. Kim TSY, Reid DM, Molday RS. Structure-function relationships and localization of the Na/Ca-K exchanger in rod photoreceptors. J Biol Chem 1998; 273:16561–16567.

83. Schwarzer A, Schauf H, Bauer PJ. Binding of the cGMP-gated channel to the Na/Ca-K exchanger in rod photoreceptors. J Biol Chem 2000; 275:13448–13454.

84. Cervetto L, Lagnado L, Perry RJ, Robinson DW, McNaughton PA. Extrusion of calcium from rod outer segments is driven by both sodium and potassium gradients. Nature 1989; 337:740–743.

85. Cai X, Zhang K, Lytton J. A novel topology and redox regulation of the rat brain K^+-dependent Na^+/Ca^{2+} exchanger, NCKX2. J Biol Chem 2002; 277:48923–48930.

86. Kinjo TG, Szerencsei RT, Winkfein RJ, Kang KJ, Schnetkamp PPM. Topology of the retinal cone NCKX2 Na/Ca-K exchanger. Biochemistry 2003; 42:2485–2491.

87. Fain GL, Quandt FN, Gerschenfeld HM. Calcium-dependent regenerative responses in rods. Nature 1977; 269:707–710.

88. Maricq AV, Korenbrot JI. Calcium and calcium-dependent chloride currents generate action potentials in solitary cone photoreceptors. Neuron 1988; 1:503–515.

89. Bader CR, Bertrand D, Schwartz EA. Voltage-activated and calcium-activated currents studied in solitary rod inner segments from the salamander retina. J Physiol 1982; 331:253–284.

90. Yagi T, MacLeish PR. Ionic conductances of monkey solitary cone inner segments. J Neurophysiol 1994; 71:656–665.

91. Demontis GC, Moroni A, Gravante B et al. Functional characterisation and subcellular localisation of HCN1 channels in rabbit retinal rod photoreceptors. J Physiol 2002; 542:89–97.

92. Craven KB, Zagotta WN. CNG and HCN channels: two peas, one pod. Annu Rev Physiol 2006; 68:375–401.

93. Nachman-Clewner M, St Jules R, Townes-Anderson E. L-type calcium channels in the photoreceptor ribbon synapse: localization and role in plasticity. J Comp Neurol 1999; 415:1–16.

94. Morgans CW, Gaughwin P, Maleszka R. Expression of the α_{1F} calcium channel subunit by photoreceptors in the rat retina. Mol Vision 2001; 7:202–209.

95. Morgans CW, Bayley PR, Oesch NW, Ren G, Akileswaran L, Taylor WR. Photoreceptor calcium channels: insight from night blindness. Vis Neurosci 2005; 22:561–568.

96. Cia D, Bordais A, Varela C, Forster V, Sahel JA, Rendon A, Picaud S. Voltage-gated channels and calcium homeostasis in mammalian rod photoreceptors. J Neurophysiol 2005; 93:1468–1475.

97. Bech-Hansen NT, Naylor MJ, Maybaum TA et al. Loss-of-function mutations in a calcium-channel α_1-subunit gene in Xp11.23 cause incomplete X-linked congenital stationary night blindness. Nat Genet 1998; 19:264–267.

98. Strom TM, Nyakatura G, Apfelstedt-Sylla E et al. An L-type calcium-channel gene mutated in incomplete X-linked congenital stationary night blindness. Nat Genet 1998; 19:260–263.

99. Xu JW, Slaughter MM. Large-conductance calcium-activated potassium channels facilitate transmitter release in salamander rod synapse. J Neurosci 2005; 25:7660–7668.

100. MacLeish PR, Nurse CA. Ion channel compartments in photoreceptors: evidence from salamander rods with intact and ablated terminals. J Neurophysiol 2007; 98:86–95.

101. Thoreson WB, Stella SL Jr, Bryson EJ, Clements J, Witkovsky P. D2-like dopamine receptors promote interactions between calcium and chloride channels that diminish rod synaptic transfer in the salamander retina. Vis Neurosci 2002; 19:235–247.

102. Thoreson WB, Bryson EJ. Chloride equilibrium potential in salamander cones. BMC Neurosci 2004; 5:53.

103. Thoreson WB, Bryson EJ, Rabl K. Reciprocal interactions between calcium and chloride in rod photoreceptors. J Neurophysiol 2003; 90:1747–1753.

104. Verweij J, Hornstein EP, Schnapf JL. Surround antagonism in macaque cone photoreceptors. J Neurosci 2003; 23:10249–10257.

CHAPTER 20

Light Adaptation in Photoreceptors

Trevor D. Lamb

1. Vision from starlight to sunlight

The human visual system operates effectively over an enormously wide range of intensities, of at least a billion-fold, from around 10^{-4} cd m^{-2} under starlight conditions to around 10^5 cd m^{-2} under intense sunlight. Changes in pupil area account for only about 1 log unit of this 9 log unit range, since the pupil diameter changes from a maximum of 8 mm to a minimum of 2.5 mm, corresponding to about a 10-fold reduction in area. Instead, the great bulk of the operational range is achieved by the combination of, firstly, a switch between the scotopic (rod-based) and photopic (cone-based) pathways in our duplex visual system and, secondly, the ability of each of these photoreceptor systems to operate over a range of 5 log units (100,000-fold) or more.

Light adaptation versus dark adaptation

This ability of the visual system (or of any of its component parts, such as a photoreceptor) to adjust its performance to the ambient level of illumination is known as "light adaptation". This adjustment typically occurs very rapidly (within seconds), whether the light intensity is increasing or decreasing. The term "dark adaptation" is reserved for the special case of recovery in darkness, following exposure of the eye to extremely bright and/or prolonged illumination that activates (and thereby bleaches) a substantial fraction of the visual pigment, rhodopsin or its cone equivalent. Dark adaptation occurs slowly, and full recovery of the scotopic visual system after a very large bleach can take as much as an hour.

Light adaptation and the changes that accompany it are beneficial to the possessor of the eye. At very low intensities, visual sensitivity is increased to the utmost that is possible, so that the rod photoreceptors reliably signal the arrival of individual photons and the scotopic visual system operates in a photon-counting mode. This ability of the scotopic system to operate at incredibly low intensities is enhanced by two deliberate trade-offs that permit more reliable detection of small signals in the presence of noise: firstly, reduced spatial resolution (i.e. increased spatial summation), and secondly reduced time resolution (i.e. increased temporal integration). Similar trade-offs are used in the photopic system, so that as the ambient illumination decreases from daylight levels towards twilight levels, one's spatial and temporal resolution deteriorate, making it very difficult to play fast ball-games when the light fades.

In contrast, the changes that characterize dark adaptation are disadvantageous. To be effectively blind to dim stimuli, for some considerable time after intense light exposure, cannot in any way be useful to an organism. Indeed, for our ancestors (as well as ourselves) entering a cave from bright sunshine is likely to have been quite dangerous, given that visual sensitivity is greatly reduced for tens of minutes. Why should such an apparently unsatisfactory situation have persisted? One possibility is that this slowness of recovery represents an inevitable "cost" of the evolutionary changes that were needed to enable the scotopic system to signal individual photon hits.

Purposes of light adaptation

A photoreceptor that did not adapt to the ambient light intensity would have a very narrow operating range: low intensities would provide negligible response, whereas high intensities would saturate the cell. Accordingly, photoreceptor light adaptation can be viewed as a means for extending the operating range of the cell.

The purpose of light adaptation can be viewed more generally as being to permit the visual system (or any neuron within it) to provide the best possible performance at that particular level of illumination; however, in this context it is not always clear what constitutes "best". For example, it is difficult for us to specify the time-course of response that is best at any given level of illumination. A brief response is likely to provide a better reaction time for the organism, but it may result in very poor sensitivity; conversely, allowing the response to integrate for a longer period will provide greater sensitivity, but may result in a very poor reaction time.

In setting the best performance of a photoreceptor, one crucial task is to endow the cell with very high sensitivity at low light levels yet prevent it from saturating at higher light levels. Light adaptation accomplishes this by reducing the cell's sensitivity to light as the background light level increases; however, this task needs to be performed in a manner that avoids excessive reduction in sensitivity. Cone photoreceptors excel in this capacity, and are able to avoid saturation no matter how intense the steady light becomes, and to continue to signal contrast effectively. Rods, on the other hand, are only able to adjust their sensitivity over a relatively narrow range of intensities, before they are driven into saturation, thereby becoming completely unresponsive. One advantage of this saturation of the rods, though, is that it substantially reduces the metabolic demand of the rods, at intensities where the cones are already functional, and where the metabolic demand for chromophore re-synthesis begins increasing.

In addition to the very important function of optimizing the photoreceptor's sensitivity over a broad range of light intensities, there are two other ways in which photoreceptor light adaptation optimizes the cell's response. Firstly, as will be explained below when eqn (1) is presented, Weber Law light adaptation permits automatic extraction of the contrast in the visual scene, independent of the absolute level of illumination. Secondly, light adaptation provides real-time adjustment of the time-course of the response to an incremental flash of light, in a manner that is presumed to be optimal for the visual system, though, as explained above, we are not yet able to quantify optimality in this regard.

2. Performance of the photopic and scotopic divisions of the visual system

Photopic vision: the cone system is the workhorse of vision

For humans, the photopic cone system can be considered the "workhorse" of vision, because it is operational under almost all of the conditions that we experience (in the 21st century). Thus, it is the photopic system that underlies our sense of vision at all light levels apart from the exceptionally low intensities experienced in moonlit and starlit conditions. Under moonlight levels of illumination, our scotopic and photopic systems are both functional, over an intensity range that is termed "mesopic". In order to determine whether you are using your photopic system under twilight or nighttime conditions, there is a simple test: if you are able to detect any color in the scene, then your cones are active; if you cannot, then it is likely that only your rods are active. As the ambient intensity increases at dawn, the photopic system remains functional, up to the brightest sunlit conditions that we ever experience.

Despite their enormous importance to our vision, cones make up only about 5 percent of the population of photoreceptors. The low density of cone photoreceptors in the peripheral retina is quite adequate for our peripheral vision, even in daytime, as we require only relatively low spatial acuity in the periphery. Even though the great majority of peripheral photoreceptors are rods, they are not in fact used under most of the circumstances that we think of as vision – instead, they are only used at exceedingly low ambient lighting levels. The reason for having a very high density of rods is to be able to capture every available photon under starlight conditions.

The responses of cones are rapid and moderately sensitive

One of the greatest advantages of cones over rods is their much faster speed of response. Our rods, even when they are light-adapted, have responses that are much too slow to allow us to function visually at the speeds that are required to escape predators and to capture prey. Cones, instead, are specialized so as to permit extremely rapid signaling of visual stimuli to the brain.

Cones are often described as being far less sensitive than rods, but this view is misleading especially when considered in terms of the rapidly changing visual stimuli that the cones are specialized for signaling. Although a cone may exhibit a *peak* sensitivity to a brief flash of light that is perhaps 30-fold lower than in a rod, the sensitivity to rapidly fluctuating stimuli is considerably *higher* in cones than in rods; thus, the slow response of the rod makes it very insensitive to rapidly changing stimuli. When calculated in terms of the efficacy of activation in the G-protein cascade of phototransduction, the "amplification" in cones and rods appears to be essentially indistinguishable. The observed difference in sensitivity measured at the peak of the response to a flash instead stems from a difference in the speed of response inactivation.

Comparison of photopic and scotopic light adaptation

A classical result comparing light adaptation in the photopic and scotopic divisions of the visual system is illustrated in Figure 20.1A, from the work of Stiles.[1] The threshold of a human subject for the detection of a flash is plotted against background intensity in double logarithmic coordinates. In this panel, the stimulus conditions provided a rod-based scotopic sensitivity that was only about 30× better than the cone-based photopic sensitivity – thus, the parafoveal region was tested with a small diameter yellow/green test flash (1° dia., 60 ms, 580 nm) on a green background. The measurements are well described by Weber Law curves, in both the scotopic (blue trace) and photopic (red trace) regions; see eqn (1) below. However, the "dark light" that will be described shortly is over 10,000× (4.1 log units) higher in the cone system compared with the rod system.

Scotopic vision: the rod system provides specialization for night vision

Light adaptation is examined under conditions that optimized detection by the scotopic (rather than photopic) system using the blue symbols of Figure 20.1B. To achieve scotopic dominance, the test stimulus was presented in the peripheral retina, and comprised a large area, long duration green flash (9° dia., 200 ms, 520 nm) on a red background. The measurements are from Fig. 3 of [2] and their troland values have been converted using a factor of $K = 8.6$ photoisomerizations s^{-1} per troland.

For comparison, the red symbols plot the desensitization of primate rod photoreceptors, obtained from Fig. 9A and Table III in [3]. Their measurements of sensitivity have been converted to desensitization by taking the reciprocal; in addition the symbols have been shifted vertically to provide

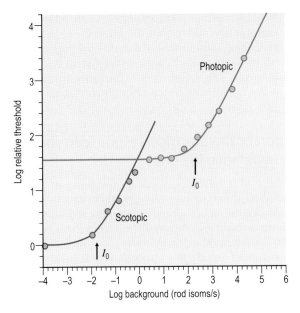

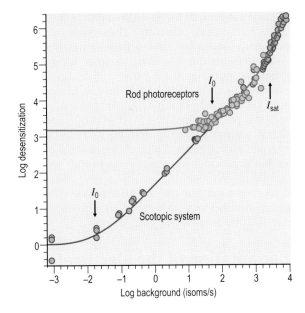

Figure 20.1 Light adaptation of the photopic (cone) and scotopic (rod) divisions of the human visual system. Both panels plot psychophysical threshold as a function of background intensity in double logarithmic coordinates. Left: Scotopic and photopic thresholds for a test flash that was only about 30× more sensitive to the scotopic system. Data are from [1], converted to isomerizations s^{-1} in the rods, horizontally, and to threshold relative to the dark-adapted scotopic value, vertically. Right: Blue symbols are for conditions that optimized scotopic detection; data from [2]. Red symbols are suction pipette measurements from isolated rod photoreceptors of monkeys (*Macaca fascicularis*), shifted vertically to align with the blue symbols in the upper intensity range; data from [3]. All curves plot Weber's Law; see text. (A: Data from Stiles WS. P Natl Acad Sci USA. 1959;45(1):100–114. B: Data from Aguilar M, Stiles WS. Optica Acta 1954;1:59–65 and Tamura T, Nakatani K, Yau K-W. J Gen Physiol 1991;98(1):95–130.)

a good fit to the psychophysical data in the high-intensity range.

The crucial difference between the blue and red symbols is that the overall scotopic visual system begins desensitizing at intensities around 1000 times lower than those required to begin desensitizing the rod photoreceptors. Thus, a considerable region at the lowest end of scotopic light adaptation is *post*-receptoral, rather than being internal to the rods. Because the post-receptoral scotopic system is able to integrate photon signals from large numbers of rod photoreceptors, and thereby gain increased sensitivity, it needs to begin desensitizing at much lower background intensities in order to avoid saturation. As a result the rod photoreceptors maintain their maximal sensitivity for several log units of the lowest-intensity regimen (up to ~10 isomerizations s^{-1}) over which the visual system exhibits gradual desensitization.

When the background intensity is reduced from relatively high scotopic intensities (moving from right to left along the x-axis in Fig. 20.1B) the desensitization of the rods, and of the scotopic visual system, steadily falls. However, below the intensities indicated by the red and blue arrows, the desensitization of first the rods, and second the visual system, fails to continue falling, as if the respective mechanism were experiencing a phenomenon equivalent to light. Accordingly, the arrowed intensities for the rods and for the scotopic visual system have been referred to as "equivalent background intensities", or "dark light". Clearly, the equivalent background for the scotopic system (around 0.016 photoisomerizations s^{-1}) is more than 1000× lower than the equivalent background intensity for the rods (around 50 isomerizations s^{-1}).

Each of the curves in Figure 20.1 plots desensitization according to Weber's Law, though the curves in Figure 20.1B additionally include saturation at high intensities, as described by the equation:

$$\frac{Desens}{Desens_D} = \{1 + (I/I_0)\} \exp(I/I_{sat}) \qquad (1)$$

where *Desens* is desensitization, $Desens_D$ is its dark-adapted value, and I is the background intensity. The first term on the right expresses Weber's Law, where I_0 is the equivalent background intensity mentioned above. This first term indicates that at low background intensities (when $I \ll I_0$) the desensitization approaches a constant level (its dark-adapted value, $Desens_D$), while for brighter backgrounds (when $I \gg I_0$) the desensitization increases linearly with background intensity.

At higher scotopic intensities, both the rods and the overall scotopic system exhibit saturation, characterized by a steep rise in desensitization with increasing background intensity. This behavior is described by the second term on the right in eqn (1), where I_{sat} is termed the saturation intensity. It is almost certain that saturation of the overall scotopic system results directly from saturation of the rods. Thus, for the blue and red curves in Figure 20.1B, the same value of saturation intensity, I_{sat} = 2500 photoisomerizations s^{-1} (black arrow), has been used. In contrast, the dark light is enormously different between the two curves, with I_0 = 0.016 and 50 photoisomerizations s^{-1}, respectively, for the psychophysical data (blue arrow) and the rod photoreceptor data (red arrow).

The span of intensities from I_0 (the dark light) to I_{sat} (the saturation intensity) represents the Weber region, and in this range of background intensities the desensitization is directly proportional to background intensity. The reciprocal of desensitization is the sensitivity, $S = 1/Desens$, and hence within the Weber region the sensitivity declines inversely with background intensity; i.e. $S \propto 1 / I$.

Since the contrast in a visual stimulus is likewise inversely proportional to background intensity (i.e. contrast = $\Delta I / I$), this Weber region is characterized by a fixed level of contrast sensitivity; that is, a given level of contrast elicits a fixed size of response. Hence, an important feature of Weber Law light adaptation is that it provides automatic extraction of visual contrast.

Figure 20.1B shows that for the overall scotopic system the Weber region covers a very wide range, of at least 5 log units (i.e. over 100,000-fold).

3. Light adaptation of the electrical responses of cones and rods

In the presence of background illumination, it is not only the overall visual system that adapts – the rod and cone photoreceptors themselves display light adaptation, characterized by desensitization and acceleration of their electrical response to an incremental flash.

For rods, though, such adaptation occurs over only a restricted range of intensities before saturation sets in. Thus, the red data in Figure 20.1B show that the Weber region for mammalian rods encompasses only 1–2 log units of intensity before saturation; for the larger rods of lower vertebrates the Weber region may encompass a slightly wider range of about 3 log units.

In contrast, cone photoreceptors undergo light adaptation over an enormously wide range of intensities, and they manage to escape saturation no matter how intense the steady light. In the overall photopic visual system it is likely that the observed adaptation results primarily from these changes at the level of the cones, rather than through post-receptoral processing.

Saturation of the electrical response in rods and its avoidance in cones

The response of a salamander rod to the onset of steady illumination at different intensities is illustrated in Figure 20.2. At the beginning of the step of light, the rod's response begins rising according to the prediction from the time integral of the flash response (noisy traces). But under normal conditions (upper panel) the response very soon deviates, falling well below the linear prediction. Characteristically, the response to such a step of light typically exhibits an early peak followed by a sag. This deviation from the simplest linear prediction is a crucial aspect of light adaptation – if this deviation did not occur, then the rod would be driven into saturation by lights of very low intensity. Saturation can readily be induced by exposing the rod to a solution that clamps the cytoplasmic calcium concentration (lower panel); the measured responses then follow the predictions of the theoretical curves, and a very low intensity (labeled 2) saturates the rod. This result shows that at least a part of the rod's ability to continue operating in backgrounds of moderate intensity is a consequence of changes in cytoplasmic calcium concentration; the molecular mechanisms that contribute to this will be discussed in Section 4.

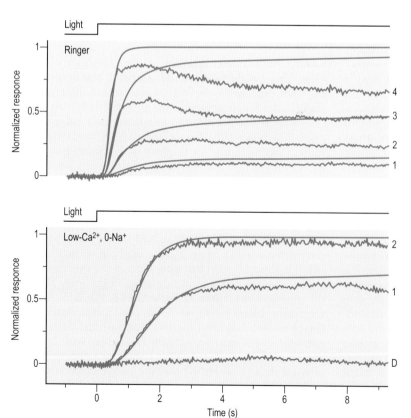

Figure 20.2 Responses of a salamander rod to onset of steps of light of different intensity. Upper panel: under control conditions (Ringer solution). Lower panel: in the presence of Ca^{2+}-clamping solution. The step intensities increased by factors of ~4 for traces labeled 1–4; D, darkness. The smooth curves are predictions obtained by integrating the measured dim flash response (not shown), and represent the step responses that are predicted in the absence of any adaptation. (From Fain GL, Lamb TD, Matthews HR, Murphy RLW. Cytoplasmic calcium as the messenger for light adaptation in salamander rods. J Physiol 416:215–243.[24])

At higher background intensities, though, the circulating current in the normal rod becomes completely suppressed. Thus, for the upper panel of Figure 20.2, intensities higher than that labeled 4 (not shown) causes the response simply to rise to its maximum level, corresponding to the closure of all cGMP-gated channels in the outer segment, with the consequence that incremental stimuli are unable to elicit any incremental response, and the cell's response is "saturated". Typically, such saturation sets in exponentially with increasing background intensity, as described earlier by the second term on the right of eqn (1).

Cone photoreceptors exhibit even more pronounced "sag" from an initial peak, when steady background illumination is turned on. An example is illustrated for modest steady intensities in the early region of the traces in Figure 20.3. The phenomenon is extremely robust in cones and, no matter how high the light intensity, the cone photoresponse always recovers to a reasonable operating level, even if it is transiently driven into saturation at the onset of intense illumination.

Desensitization and acceleration of the photoreceptor's electrical response

In addition to showing the transient nature of the responses elicited by the onset of backgrounds in cones, the main purpose of Figure 20.3 is to illustrate the desensitization of the responses to incremental flashes that is elicited by these backgrounds.

The traces illustrate the responses of a cone photoreceptor to an identical set of flashes presented under three different conditions. In panel (A) the flashes (a–f, of progressively greater intensity) were presented in darkness. In panel (B) the same flashes were presented on a dim steady background, and in panel (C) they were presented on a brighter background. In the presence of background illumination, the responses to dim flashes were smaller. For example, for the second flash intensity, b, the response amplitude became markedly smaller from panel (A) to panel (B) to panel (C). In other words, backgrounds of increasing intensity progressively desensitized the cone's incremental response. Such behavior is very characteristic of photoreceptors (both cones and rods).

The changes in the dim flash response elicited by backgrounds of different intensity are shown in Figure 20.4A. The largest trace is the response to a dim flash presented under fully dark-adapted conditions, while the other traces are for the same flash presented on backgrounds of progressively brighter intensity. (In fact, in order to maintain responses of measurable amplitudes, the flash intensities were increased in the presence of backgrounds, and the traces actually plot response divided by flash intensity; i.e. response sensitivity.)

The traces in Figure 20.4A demonstrate that the effect of increasing background intensity is both to desensitize and to accelerate the response to an incremental dim flash. This behavior of cones is very similar to that exhibited by rods, as illustrated in Figure 20.4B.

Unaltered rising phase but accelerated recovery

For the incremental flash responses of the rod illustrated in Figure 20.4B, the vertical scale has been adjusted to take

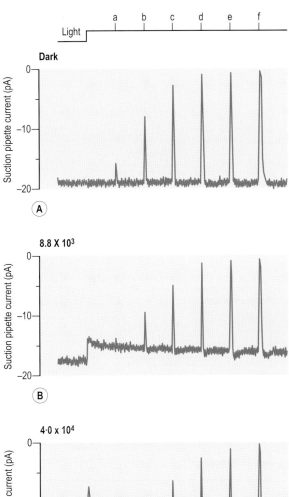

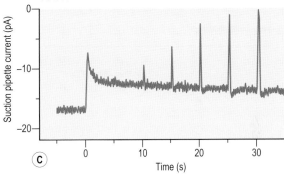

Figure 20.3 Circulating current of a salamander cone in response to flashes and steps of illumination. Timing of illumination is indicated by the marker trace at the top; flashes a–f increased in intensity by factors of ~4, and were the same intensity in each of panels (A)–(C). In panel (**A**) these flashes were presented in darkness; in panels (**B**) and (**C**) the same flashes were presented on steady backgrounds that had been switched on at time zero; the background in (C) was ~4 times brighter than in (B). (From Matthews HR, Fain GL, Murply RLW, Lamb TD. Light adaptation in cone photoreceptors of the salamander: a role for cytoplasmic calcium concentration. J Physiol 420:447–469.[25])

account of changes in the level of circulating current remaining in the presence of the different background intensities. Thus, rather than plotting raw sensitivity (response per photoisomerization), as was done in panel A for the cone, panel B for the rod instead plots the fractional response (i.e. the incremental response as a fraction of the circulating current at that background) per photoisomerization. This has been done in order to provide a direct measure of the level of activation of the guanine nucleotide binding protein (G-protein) cascade of phototransduction; thus, it can be shown

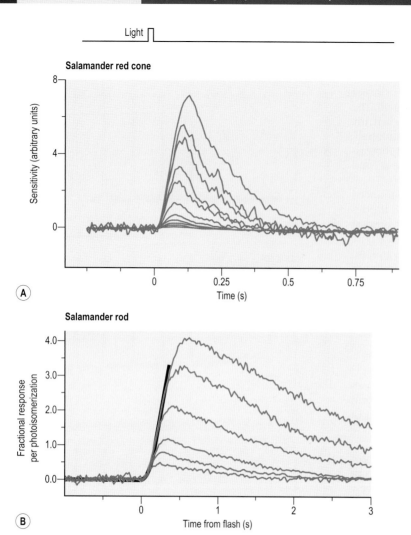

Light ⊓

Salamander red cone

(A)

Salamander rod

(B)

Figure 20.4 Incremental responses of cone and rod photoreceptors to test flashes presented on backgrounds of progressively higher intensity. In each panel, the largest trace is for a dim flash presented in darkness, while the other traces correspond to the same test flash presented on backgrounds. (**A**) Salamander red-sensitive cone. (**B**) Salamander rod. Smooth black curve indicates common rising phase. (A: From Matthews HR, Fain GL, Murply RLW, Lamb TD. Light adaptation in cone photoreceptors of the salamander: a role for cytoplasmic calcium concentration. J Physiol 420:447–469. B: From Pugh EN, Nikonov S, Lamb TD. Molecular mechanisms of vertebrate photoreceptor light adaptation. Curr Opin Neurobiol 9(4):410–418, 1999.[25,26])

that the level of cascade activation is best measured by the fractional channel opening, which in turn is measured by the incremental response expressed as a fraction of the existing circulating current.

When plotted in this manner, the incremental responses in Figure 20.4B demonstrate the remarkable property that the onset phase of the response is invariant; i.e. the traces for different background intensities exhibit a common rise at early times, indicated by the smooth black trace. This behavior indicates that the amplification parameter describing the activation steps in phototransduction is unaltered during light adaptation; in other words, light adaptation causes no change in the efficacy of the activation steps in phototransduction. Instead, what light adaptation does is to cause a marked speeding-up of the shut-off steps in the transduction cascade. The molecular identity of the steps that are accelerated is considered in Section 4.

Dependence of sensitivity on background intensity: Weber's Law

By plotting the peak amplitude of each of the traces in Figure 20.4A as a function of the background intensity on which it was measured, one obtains a sensitivity versus background plot of the kind illustrated in Figure 20.5.

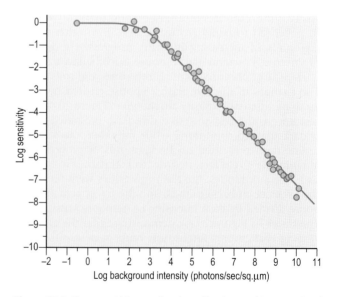

Figure 20.5 Cone sensitivity as a function of background intensity. Results are plotted in double logarithmic coordinates, and were obtained from intracellular measurements averaged from 15 cones in the turtle eye-cup preparation. These experiments monitored step sensitivity rather than the more conventional flash sensitivity. The smooth curve plots Weber's Law, given by eqn (2). (Reproduced with permission from [4].)

The results plotted in Figure 20.5 were obtained over an extremely wide range of background intensities by Burkhardt,[4] using a laser source of illumination. Importantly, the preparation was the intact eye-cup (of the turtle), so that the photoreceptors remained in contact with the retinal pigment epithelium and thereby experienced normal regeneration of visual pigment, so that physiological results could be obtained even at very high background intensities.

For background intensities from 10^3 to 10^{11} photons $\mu m^{-2} s^{-1}$, the relationship between log sensitivity and log background intensity is a straight line with a slope of -1; in other words, over roughly 8 log units of background, the turtle cone's sensitivity declines inversely with background intensity. The curve plotted near the points in Figure 20.5 represents Weber's Law, described by:

$$\frac{S}{S_D} = \frac{1}{1+(I/I_0)} \qquad (2)$$

where S is flash sensitivity, S_D is its dark-adapted value, I is the background intensity, and I_0 is the half-desensitizing intensity, also known as the dark-adapted equivalent background intensity. This is the same equation as used in Figure 20.1B for light adaptation of human rods, though without the saturation term (and in coordinates of sensitivity rather than its reciprocal). The good fit of the Weber Law expression shows, very importantly, that cone photoreceptors in the intact eye-cup are able to completely avoid saturation, even at enormously high intensities of steady illumination. This feature represents a crucial distinction between the properties of cones and rods. As illustrated earlier in Figure 20.2, the circulating current of rods is shut off at quite low background intensities so that the rods become unresponsive to superimposed stimuli in the presence of background illumination of moderate intensity.

In the same set of experiments, Burkhardt[4] measured the intensity at which 90 percent of the pigment was in the bleached state, and found this to be around 10^6 photons $\mu m^{-2} s^{-1}$. Hence, for all intensities above that level, the observed Weber Law behavior can be accounted for in terms of pigment bleaching. For each additional 10-fold increase in intensity there will be a 10-fold reduction in the amount of pigment remaining available to absorb light, and hence there will necessarily be a 10-fold reduction in sensitivity in the absence of any other change of parameters of transduction in the outer segment. In other words, if the photoreceptor is able to avoid saturation up to intensities that cause substantial bleaches, then it will be able to exhibit Weber Law desensitization at all higher intensities purely by means of pigment bleaching. Cones are able to function up to this critical intensity, whereas rods saturate at much lower intensities.

Extremely rapid recovery of human cone photocurrent

Experiments on photoreceptors at high intensities must be undertaken in a preparation in which the retina is in contact with the retinal pigment epithelium (as was the case in the experiments above on turtle cones), so that visual pigment is regenerated. Thus, it is not possible to investigate cone responses at high intensities using single-cell or isolated retina preparations. Instead, to assess the performance of human cones at high background intensities, one is essentially restricted to experiments measuring the electroretinogram (ERG) in the intact eye.

Figure 20.6 illustrates results from an experiment designed to measure the kinetics of recovery of the circulating current of human cone photoreceptors upon extinction of steady illumination that bleached 90 percent of the visual pigment. The traces show recordings of the human photopic ERG a-wave, which monitors the massed response primarily of the cone photoreceptors.

The four colored traces plot ERG responses from two subjects, at two different flash intensities, while the light stimulus is monitored by the black trace at the top. The left panel is for an intense flash presented on the intense steady background, while the right panel is for extinction of that background. The "ON" a-wave and b-wave elicited by the bright flash are indicated by the red arrows, while the "OFF"

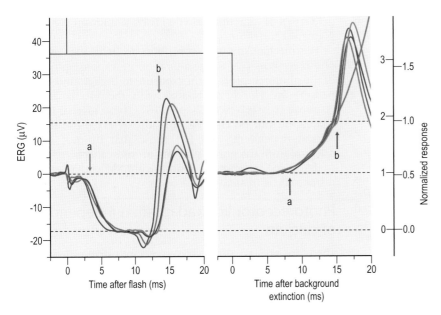

Figure 20.6 Extremely rapid recovery of human cone photocurrent upon extinction of intense illumination, measured with the ERG. The left panel shows the response to a bright flash superimposed on the intense steady background, while the right panel shows the response obtained at the extinction of that background. Dashed horizontal lines represent the following levels of cone circulating current (from bottom): zero level, steady level during intense background, dark level, as indicated by the two normalized scales on the right. See text for details. (From Kenkre JS, Moran NA, Lamb TD, Mahroo OAR. Extremely rapid recovery of human cone circulating current at the extinction of bleaching exposures. J Physiol 567(1):95–112.[21])

a-wave and *b*-wave elicited by extinction of the background are indicated by the blue arrows. The *b*-wave is roughly similar in the two cases, and arises from post-receptoral activity. The OFF *a*-wave represents recovery of the cone circulating current, and begins around 7 ms after the intense background is turned off.

Figure 20.6 shows two very important results relating to human cone responses in the presence of extremely bright illumination. Firstly, the amplitude of the ON *a*-wave in the left panel shows that the steady cone circulating current remained at roughly 50 percent of its original level in darkness. Secondly, the early time at which the OFF *a*-wave in the right panel begins rising (indicated by the blue arrow at ~7 ms) shows that the cone circulating current begins recovering extremely rapidly.

Thus, although little change occurs for the first 7 ms, thereafter a substantial upward response occurs until about 15 ms after extinction of the background; at this point the *a*-wave is obscured by spike-like activity of the *b*-wave. There is compelling evidence that the *a*-wave traces for these subjects monitor the recovery of the cone circulating current. On this basis, the cone circulating current is essentially fully recovered within about 15 ms after extinction of illumination so intense that it bleaches 90 percent of the cone pigment. This is extremely rapid recovery. Section 4 considers the speed of the phototransduction shut-off reactions responsible for generating this rapid recovery, and the smooth curve near the measured traces in the right-hand panel of Figure 20.6 was calculated from the model presented there, using the short time constants listed in Table 20.1 below.

Extremely rapid recovery of cone circulating current, as inferred from the results of Figure 20.6, is required in order to account for classical experiments on the flicker fusion frequency of human subjects. Even at quite low photopic intensities, human subjects are able to detect square-wave flicker at a frequency of around 50 Hz using peripheral vision. But at higher intensities the flicker fusion frequency increases, to 100 Hz or more; at this frequency the illumination is being switched on and off for durations of 5 ms each. Thus, in order for the flicker to be detectable, some degree of recovery of cone circulating current must have occurred within 5 ms. Hence, human flicker sensitivity is broadly consistent with the time-course inferred from Figure 20.6.

4. Molecular basis of photoreceptor light adaptation

The phototransduction cascade

For a full description of the biochemistry of phototransduction, see Chapter 18. Figure 20.7 presents a simplified schematic of these steps, emphasizing the shut-off reactions using red arrows. The lifetimes (or turnover times) of the important shut-off steps are: τ_R, the lifetime of activated rhodopsin; τ_E, the lifetime of activated transducin/PDE; $\tau_{cG} = 1/\beta$, the turnover time of cyclic GMP; and τ_{Ca}, the turnover time of cytoplasmic free Ca^{2+}.

In summary, activation and recovery occur as follows. Light (of intensity I) activates the visual pigment, and the activated pigment (R^*) is inactivated with a time constant

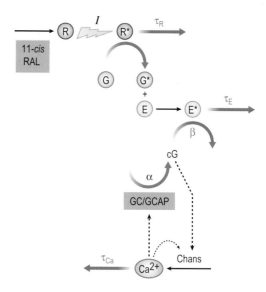

Figure 20.7 Simplified schematic of reaction steps underlying the recovery steps in phototransduction. See text for details. (From Lamb TD & Pugh EN. Avoidance of saturation in human cones is explained by very rapid inactivation reactions and pigment bleaching. Invest Ophthalmol Vis Sci 47;e3714.[10] Reproduced from Association for Research in Vision and Ophthalmology.)

τ_R. R^* catalyzes activation of the guanine nucleotide binding G-protein transducin to G^*, which then binds to phosphodiesterase (E) to form the activated G^*–E^* complex (indicated E^*), which has a lifetime τ_E. The activated E^* catalyzes hydrolysis of cGMP (cG) with a rate constant β. The increase in β is proportional to steady intensity I, with a scaling factor set by the product of the R^* and E^* lifetimes; i.e. $\beta = \beta_{Dark} + (A/n_{cG}) \, \tau_R \, \tau_E \, I$ (see [5]). cGMP formation is catalyzed by guanylyl cyclase (GC), under the regulation of Ca^{2+}-sensitive guanylyl cyclase activating proteins (GCAPs). When present, cG causes the opening of ion channels (chans) in the plasma membrane, admitting Ca^{2+} ions. Cytoplasmic Ca^{2+} has a powerful negative-feedback action via GCAPs onto the rate α of cGMP synthesis. Ca^{2+} ions are removed from the cytoplasm with a turnover time for free Ca^{2+} concentration of τ_{Ca}. Finally, and very slowly, visual pigment (R) is re-formed by delivery of 11-*cis* retinaldehyde (11-*cis* RAL) from the retinal pigment epithelium.

The mechanisms that contribute to light adaptation in photoreceptors (i.e. to the alteration in response properties of the photoreceptors upon exposure to background illumination) are closely associated with the mechanisms of response recovery (see Chapter 18). These mechanisms of adaptation can be classified broadly as (1) those that are calcium-dependent, and (2) those that do not involve calcium. Both categories play major roles, and here the less well-known category of non-calcium-dependent mechanisms will be treated first.

Photoreceptor light adaptation independent of calcium

There are at least three classes of non-calcium-dependent phenomena that represent mechanisms of light adaptation in photoreceptors, where light adaptation is interpreted as an alteration of response properties in comparison with the dark-adapted state. First, there is response compression, whereby the reduced level of circulating current in the presence of steady background illumination reduces the size of the flash response. This phenomenon will not be discussed here, in part because it is both very well known and very simple, but also because (in philosophical terms) it can be viewed as a failure of light adaptation. Cones substantially avoid response compression through powerful feedback mechanisms that maintain the circulating current. Second, there is pigment depletion, which is relevant to cones at high intensity; however, it is never relevant to rods, because they are driven into saturation by even very low levels of bleached pigment (see Section 6). Third, there is a direct effect of PDE activation, which is now considered.

Accelerated turnover of cGMP

In darkness the activity of the PDE is relatively low, so that the turnover rate of cGMP hydrolysis (denoted β) is low and the turnover time for cGMP ($\tau_{cGMP} = 1/\beta$) is long. Under dark-adapted conditions, τ_{cGMP} is around 1 s in amphibian rods, and around 200 ms in mammalian rods. The magnitude of this parameter has a major effect on both the sensitivity and the kinetics of the photoreceptor's response to a flash. Thus, when the PDE activity increases in steady illumination, the shorter turnover time for cGMP contributes both to desensitization and to acceleration of the photoresponse.

To provide an intuitive understanding of this mechanism, it is helpful to consider what we have referred to previously as the "bathtub analogy".[5] Imagine a container of water, such as a tall cylinder, and let the height of water in the cylinder represent the level (concentration) of cGMP in the outer segment. The rate at which water runs out of the cylinder, through a drain-hole at the base, is proportional both to the height of water and to the size of the opening, representing the cGMP level and the PDE activity, β, respectively. Likewise, the rate at which water flows in to the cylinder through a tap at the top represents the activity of guanylyl cyclase, α. When a steady state is reached, the height of water will equal the rate of influx divided by the size of the drain-hole; i.e. cGMP = α/β. Importantly, whenever the water level is perturbed from this steady-state level (e.g. upon a brief opening of an additional drain-hole), the level will re-equilibrate with a time constant $\tau_{cGMP} = 1/\beta$ (provided that the rate of influx through the tap remains constant). Hence, if the drain-hole is small (and the inflow via the tap correspondingly small) then any perturbation in water level elicited by a transient additional outflow will be corrected only slowly; if the drain-hole is large (and the influx correspondingly large) then any perturbation will be rapidly corrected. Furthermore, though perhaps less intuitively, it can be shown that for a non-instantaneous perturbation, corresponding to the normal flash response, not only will the kinetics of recovery be faster but also the peak will be smaller.

Hence the effect of the increased PDE activity during steady illumination is both to accelerate the response kinetics and to reduce the peak amplitude to an incremental flash. Calculations show that, in amphibian rods, the 20-fold increase in β during steady illumination provides the primary mechanism underlying the measured shortening of time-to-peak and the decrease in flash sensitivity.[5]

Calcium-dependent mechanisms of rapid light adaptation: re-sensitization through prevention of saturation

When cGMP-gated ion channels in the outer segment are closed in response to light, the cytoplasmic concentration of calcium drops (see Chapter 19). This drop in Ca^{2+} concentration is vitally important to light adaptation, though it is crucial to emphasize that it does not cause the desensitization that characterizes photoreceptor light adaptation. Quite the contrary: the drop in Ca^{2+} concentration actually rescues the rod from the saturation that would otherwise be induced by light, and thereby prevents the onset of massive desensitization at relatively low intensities of background illumination. Thus, the light-induced drop in Ca^{2+} acts to increase the rod's sensitivity above the drastically reduced level that would occur, either if the Ca^{2+} concentration did not alter, or if the rod's calcium-dependent mechanisms were inoperative.

Powerful negative feedback loop mediated by calcium

Calcium is the cytoplasmic messenger for a very powerful negative feedback loop that tends to stabilize the photoreceptor's circulating current. If ever the Ca^{2+} concentration drops (e.g. in response to light, or as a result of some other perturbation) then, as described in Chapters 18 & 19 and below, a number of changes occur very rapidly. These changes are stimulated by the unbinding of Ca^{2+} from at least three classes of calcium-sensitive protein: (1) guanylyl cyclase activating proteins (GCAPs 1 and 2), which activate guanylyl cyclase; (2) recoverin, which regulates the lifetime of activated R*; and (3) a protein that modulates the opening of cGMP-gated channels – in rods this is calmodulin. Calcium's action via each of these pathways leads to the opening of cGMP-gated channels, thereby increasing the circulating current and admitting Ca^{2+} ions from the extracellular medium. This influx of Ca^{2+} ions tends to counteract the initial reduction in Ca^{2+} concentration, thereby completing a negative feedback loop.

Each of these molecular mechanisms involved in the calcium negative feedback loop contributes towards extending the cell's operational range of light intensities, by helping to prevent saturation of the circulating current. Thus, each of these three molecular mechanisms assists in *rescuing* the photoreceptor from saturation and hence *increasing*, rather than decreasing, the sensitivity compared with the case that would exist if the mechanism were absent. Each of the three mechanisms is most effective over a particular range of calcium levels, and hence over a particular range of light intensities. The most powerful of the three (at least in rods) is the GCAP activation of guanylyl cyclase.

Because the various components of the calcium negative feedback loop act quite rapidly, they contribute not only to determining the photoreceptor's sensitivity in the presence of background illumination, but also to determining the kinetics of its response to an incremental flash presented on the background. The importance of altered Ca^{2+} concentration in setting the incremental flash response kinetics can be demonstrated by incorporating a calcium buffer (such as BAPTA) into the outer segment. Figure 20.8A shows that, although the flash response begins rising exactly as in control

Toad rods

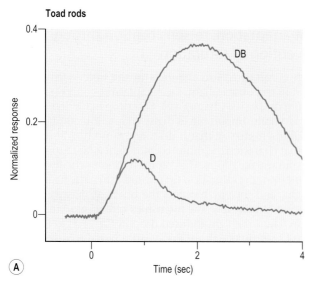

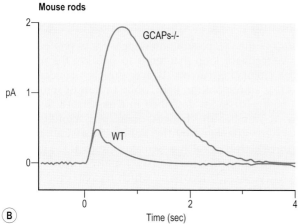

Figure 20.8 Comparison of calcium buffering and GCAPs knockout on the rod's dim-flash response. Both panels plot suction pipette recordings from single rods. (**A**) Salamander rod response to a flash in darkness (D) and to the same flash after incorporation of the calcium buffer BAPTA (DB). (Reproduced with permission from [27].) (**B**) Mouse rod single-photon responses, for wild type (WT) and GCAPs knock-out (GCAPs –/–) animals. (A: From Torre V, Matthews HR, Lamb TD. Role of calcium in regulating the cyclic GMP cascade of phototransduction in retinal rods. Proc Natl Acad Sci USA. 83:7109–7113.[27] B: From Burns ME, Mendez A, Chen J, Baylor DA. Dynamics of cyclic GMP synthesis in retinal rods. Neuron 36(1):81–91.[28])

conditions, it does not recover as rapidly, and instead rises to a substantially larger and later peak with slower final recovery. The three calcium-sensitive molecular pathways will now be described.

Guanylyl cyclase activation

In response to a drop in calcium concentration, Ca^{2+} will unbind from the GCAP proteins (GCAP1 and GCAP2), thereby activating guanylyl cyclase and stimulating the production of cGMP at a greatly increased rate, leading to the opening of cGMP-gated channels.

The cyclase activity increases roughly as the fourth power of the drop in Ca^{2+} concentration, and furthermore (as for the other two routes considered below) the number of channels open increases approximately as the cube of the cGMP concentration. Because of the cascading of two such steep dependencies, any small fractional change in Ca^{2+}

concentration stimulates a large and opposite fractional change in channel opening; that is, the fractional change in channel opening is opposite in sign to, and up to 12× the magnitude of, the originating fractional change in Ca^{2+} concentration. As a result, this molecular mechanism is the most potent of the three that contribute to the calcium negative feedback loop and hence to setting the adaptational state in rods. It is especially dominant at relatively bright background intensities, corresponding to low Ca^{2+} concentrations, and is therefore the most important in extending the rod's operating range to high intensities.

The role of the GCAP/guanylyl cyclase component of the Ca^{2+} feedback loop in setting the waveform of the incremental flash response is illustrated in Figure 20.8B, where averaged responses are shown for rods from WT and GCAPs knockout mice. In a manner very similar to that seen in amphibian rods containing the calcium buffer BAPTA, the response in the GCAP knockout case begins rising exactly as for the control (WT) case, but it does not recover as soon, so that the response continues rising and reaches a larger and later peak.

Shortened R* lifetime

Activated rhodopsin (R*) is inactivated by multiple phosphorylation steps mediated by rhodopsin kinase (GRK1) followed by binding of arrestin (for details, see Chapter 18). It is generally assumed that the decline in R* activity follows exponential kinetics, and can therefore be described by a characteristic lifetime, τ_R; however, it is worth bearing in mind that there is no direct evidence for this assumption. It was established by Kawamura[6] that GRK1's phosphorylation of R* is calcium-dependent and that the effect is mediated by the calcium-binding protein recoverin (termed S-modulin in frog). The molecular mechanism of this dependence is not entirely clear, but some evidence suggests that the calcium-bound form of recoverin binds to GRK1, thereby preventing it from interacting with R*. In any case, it is proposed that a reduction in Ca^{2+} concentration leads to a shortened R* lifetime, τ_R.

The slowest time constant in the phototransduction cascade (the so-called dominant time constant, τ_{dom}) can be estimated from the steepness of the relationship between the duration that the rod is held in saturation by a bright flash and the flash intensity (see [7]). Over the years there has been considerable debate as to whether this dominant time constant is set by the R* lifetime, τ_R, or instead by the lifetime τ_E of the transducin-phosphodiesterase (PDE) complex (the effector). The situation may differ between rods and cones, and may also be species dependent. But in mouse rods Burns and colleagues have now clearly established that under dark resting conditions the dominant time constant is that of transducin-PDE, with $\tau_E \approx 200$ ms, while the R* lifetime is shorter, with $\tau_R \leq 80$ ms[8] (see Table 20.1). In this scenario, where the R* lifetime is shorter than the transducin-PDE lifetime, then further light-induced shortening of τ_R is likely to have very little effect on the response kinetics, but it will instead cause a reduction in sensitivity, because fewer molecules of transducin will be activated during the R* lifetime.

Although it remains difficult to establish the effectiveness of any individual mechanism in an intact photoreceptor with a functional calcium feedback loop, it appears that in rods the recoverin-mediated reduction in R* lifetime plays a moderate role, especially at relatively low background intensities.

Channel reactivation

In response to a drop in calcium concentration, Ca^{2+} unbinds from calmodulin (in the case of the rods), leading to a lowered dissociation constant ($K_{1/2}$) for the binding of cGMP to the channels. The effect of the lowered $K_{1/2}$ is that any given concentration of cGMP will cause the opening of a larger fraction of the cGMP-gated channels, leading to an increase in circulating current and an influx of more calcium. However, the potency of this effect is low in rods, and the mechanism contributes only weakly to rod adaptation. In contrast, cone channels possess a more powerful regulatory mechanism, mediated by a different calcium-sensitive protein[9] (see Chapter 19).

Molecular basis of the cone's incredibly rapid recovery from light exposure

The speed of recovery of human cones from prolonged intense illumination is extremely rapid (Section 4), and the time constants of the various shut-off reactions shown in the schematic of Figure 20.7 have been estimated in a number of recent studies of primate cones using intact preparations. In the case of monkey cones, the parameters were extracted via theoretical modeling of results obtained from intracellular recordings of horizontal cells in the retina-RPE-choroid preparation. In the case of human cones, the parameters were extracted via theoretical modeling of ERG results, including those of the kind illustrated in Figure 20.6. The shut-off reactions have been found to be extremely rapid, and the parameters that have been reported are summarized in Table 20.1.

The collected estimates in Table 20.1 are consistent with the notion that all four of the shut-off time constants in human cones are extremely short, with three of the time constants around 5 ms or less, and one around 10–15 ms. Hence the shut-off time constants for both the activated visual pigment (R*) and the activated G-protein/PDE (E*), τ_R and τ_E, are around 20-fold shorter in human cones than in mouse rods, where values of ~70 and ~200 ms have been reported. It has recently been discovered that disruption of a molecular step involved in shut-off of phototransduction can lead to slowed light adaptation in cones (see Box 20.1).

Cone avoidance of saturation

A theoretical analysis[10] has shown that the ability of mammalian cones to avoid saturation is explicable in terms of the combination of these two 20-fold shorter time constants, together with the bleaching of cone visual pigment.

In human rod photoreceptors in vivo, the circulating current is halved at a steady intensity of ~70 scotopic trolands or 600 R* s^{-1},[11] with complete saturation occurring at ~1000 scotopic trolands ($\sim 10^4$ R* s^{-1}) (see also Fig. 20.1B). On the assumption that the gain of activation in transduction is the same in human cones as in human rods, then the two very short cone time constants would elevate the intensities required for half- and full-saturation by some ~400×, to

---

Box 20.1 Altered light adaptation when phototransduction shut-off is disrupted

Changes in light adaptation performance as a result of retinal disease are uncommon – any disturbance of the phototransduction cascade tends to cause major disruption of signaling, rather than a subtle change in adaptation. For example, Oguchi disease, which impairs the shut-off of the transduction cascade (through mutations in either rhodopsin kinase or arrestin), leads to symptoms resembling stationary night blindness.

Recently, it has been discovered that altered light-adaptation behavior can result from impairment of the reactions that inactivate the light-activated phosphodiesterase (PDE6). Thus, mutations in the *RGS9-1* or *R9AP* genes have been shown to affect light adaptation,[29,30] though in a manner that may appear counter-intuitive.[31]

Patients with mutations in the *RGS9-1* or *R9AP* report difficulty in adapting to sudden changes in photopic luminance, and experience difficulty seeing moving objects, especially at low contrast. This stationary condition has been termed bradyopsia (slow vision).[29,30] This slowness is expected, because disruption of the PDE shut-off process should lead to an extended time-course of the photoreceptor response to illumination.

Paradoxically, at low light levels the responses of cones lacking *RGS9-1* are actually faster than those of normal cones.[31] When a second effect of PDE is taken into consideration, this result turns out to be exactly as predicted. Not only is the PDE involved in shutting off the light response, by hydrolyzing cyclic GMP, but it is also involved in light adaptation and in response acceleration by shortening the "turnover time" for cyclic GMP.[5] Thus, with RGS9-1 absent, there is more PDE active at low light levels, and hence a shorter turnover time for cyclic GMP and thereby a "more adapted" state than in the normal case. Accordingly, although these patients will have slower than normal vision at high lighting levels, they will have faster than normal vision (i.e. apparent light adaptation) at dim lighting levels.

Table 20.1 Shut-off time constants estimated for primate cones and mouse rods

	τ_R ms	τ_E ms	τ_{Ca} ms	$\tau_{cG} = 1/\beta$ (Intense) ms	Preparation	Reference
Cone		18			Human ERG	[20]
	5	13		4	Human ERG	[21]
	3	9	3	4	Monkey retina	[22]
	3	10	3	6	Human ERG	[23]
Rod	70	200			Mouse suction pipette	[8]

Estimates for the shut-off time constants, τ_R, τ_E, τ_{Ca}, and τ_{cG}, obtained from recent experiments. It is not usually possible to determine which of the two time constants τ_R and τ_E is which; however, in the case of the mouse rod results, this identification was made using other experiments. Note that the turnover time for cyclic GMP, τ_{cG}, represents the value applicable during very intense illumination; under dark-adapted conditions, when the light-stimulated activity of the PDE is much lower, this time constant will be much longer.

levels of ~240,000 and ~4×10^6 R* s^{-1} in human cones. An additional factor is that the cGMP-gated channels of mammalian cones show increased cGMP binding affinity when Ca^{2+} falls,[9] thus further increasing the R* rate required for saturation.

How do these estimated rates of isomerization compare with the maximum rate at which the cone visual pigment can be bleached during steady illumination? At steady-state, the rate of photoisomerization equals the rate of pigment regeneration, which is set by the delivery of 11-*cis* retinal to opsin. For human L/M cones, the maximal rate of regeneration has been measured as ~45 percent min^{-1}, or 0.75 percent s^{-1}. If the outer segment contains ~40 million pigment molecules, then the maximal rate of photoisomerization during intense steady light will be ~300,000 R* s^{-1}. This rate simply cannot be exceeded in the steady-state, because a higher initial rate of isomerization would deplete the available pigment thereby lowering the rate.

Hence, from the numbers in the preceding paragraphs, the rate of photoisomerization required to saturate the human cone current exceeds the highest rate of isomerization of cone pigment molecules (~300,000 R* s^{-1}) that can be elicited by a steady light of arbitrarily high intensity. Therefore, the human cone photoreceptor cannot be saturated by steady lights, no matter how bright they are. This does not mean that the cone can never be saturated; if an intense light is presented from dark-adapted conditions (when the cone initially has a full complement of visual pigment), then it will transiently be driven into saturation until bleaching reduces the amount of visual pigment to a suitably low level.

Modeling of human cone light adaptation

A computational model of human cone light adaptation has been developed,[12] that puts factors of the kind described in the preceding section into a comprehensive theoretical/numerical model. The molecular description that they use is closely similar to that illustrated in Figure 20.7, including pigment bleaching, and they express the system as a set of differential equations. Their simulations confirm that human cones will indeed not saturate with steady intensities of any level, and predict that the sensitivity will conform to Weber's Law over a very wide range of background intensities.

5. Slow changes in rods: light adaptation or dark adaptation?

In addition to the conventional features of rod photoreceptor light adaptation that occur extremely rapidly (on a subsecond time-scale), other changes have been reported to occur over a time frame of minutes of exposure, in response to lights that saturate the cell's response. As the effects of these changes are very slow, and can only be observed in darkness when the adapting exposure is extinguished, there is a semantic issue as to whether these phenomena should be thought of as light adaptation or as dark adaptation.

Light-induced change in the dominant time constant

Exposure of mouse rods to a just-saturating intensity of around 1000 photoisomerizations s^{-1}, for 1 min or more,

leads to a persistent speeding of the bright-flash response upon extinction of the background.[13] The change does not involve any reduction in the activation phase of transduction, but instead involves a reduction in the dominant time constant of response recovery; typically, the dominant time constant τ_{dom} drops from around 200 ms under dark-adapted conditions to around 100 ms immediately after extinction of the saturating light. This adaptational effect develops relatively slowly, building up over 60 s or so, and it requires a rhodopsin bleach level of around 2 percent for full effect. The effect is relatively long-lasting, declining with a time constant of around 80 s.

The molecular mechanism giving rise to this adaptational effect is not known, though some evidence suggests that it corresponds to a reduction in lifetime of the activated transducin-PDE complex. If so, it represents a phenomenon distinct from the actions of dimmer adapting lights.

Light-induced translocation of proteins

Light-induced translocation of transducin, recoverin, and arrestin in photoreceptors is considered in Chapter 18 and will briefly be mentioned here. Movements of protein are elicited only at quite high intensities (generally in the saturating range), and they occur over a time-scale of many minutes. In mouse rods, intensities above 3000 photoisomerizations s^{-1} for 30 min (which bleach a substantial fraction of the rhodopsin) trigger the movement of transducin from the outer segment to the inner segment, while slightly lower intensities of 1000 photoisomerizations s^{-1} or more trigger the movement of arrestin in the opposite direction; recoverin also leaves the outer segment in bright light.

Protein movements of these kinds may well affect the adaptational state of the rod, though this has not yet been definitively established. Because the movements are triggered only by saturating light intensities, the electrical effects cannot readily be observed during the illumination, because the circulating current is completely suppressed. One possibility is that the changes help prepare the rod for its return to lower intensities, as occurs around dusk. Interestingly, in one attempt that was made to detect any change in the amplification constant of human rods (using the ERG) following exposures to intensities that elicit transducin translocation in mouse rods, no change in amplification was detectable.[14]

6. Dark adaptation of the rods: very slow recovery from bleaching

After our eye has been exposed to very intense illumination, our visual threshold is greatly elevated and may take tens of minutes to recover fully.[15] Comparable effects can be also measured at the level of the rod bipolar cells[16] or the rod photoreceptors.[11] The slow recovery of sensitivity is referred to as "dark adaptation" or "bleaching adaptation", but this use of the term "adaptation" is misleading. Adaptation normally refers to beneficial adjustments, yet the changes that persist after extinction of intense illumination are distinctly disadvantageous – thus, there can be no advantage in having one's vision greatly compromised following exposure to intense light.

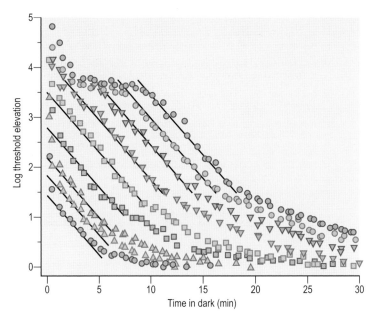

Figure 20.9 Human psychophysical dark adaptation. Recovery of log threshold elevation in a normal human observer is plotted as a function of time in darkness, after a wide range of bleaching exposures (from 0.5 percent to 98 percent). Parallel black lines represent component S2, with a slope of −0.24 decades min⁻¹ (see text). The lateral shift between the lines is consistent with the rate-limited delivery of 11-*cis* retinal from the RPE to opsin in the outer segments. (From Lamb TD & Pugh EN. Phototransduction, dark adaptation, and rhodopsin regeneration. The Proctor Lecture. Invest Ophthalmol Vis Sci 47(12):5138–5152. Reproduced from Association for Research in Vision and Ophthalmology.[18])

The recovery of visual threshold for a human subject is plotted in Figure 20.9, following the cessation of nine light exposures that bleached from 0.5 percent to 98 percent of the rhodopsin.[17,18] For a bleach of 20 percent, the visual threshold was initially elevated by 3.5 log units. This indicates that the elevation of threshold is out of all proportion to the fraction of pigment remaining unbleached; even though 80 percent of the rhodopsin remained functional, the threshold was raised 3000-fold. Instead, there is now overwhelming evidence that the phenomenon arises from the presence within the outer segment of unregenerated opsin (i.e. the presence of the protein part of the visual pigment, prior to its recombination with the regenerated 11-*cis* retinal).

Remarkably, the recovery of scotopic (rod-mediated) threshold exhibits a region of common slope across all the bleach levels, as indicated by the parallel black lines in Figure 20.9. This region is termed the S2 component of recovery, and has a slope $\Psi_{S2} = -0.24$ log unit min⁻¹ that is characteristic of dark adaptation recovery in normal (young adult) human eyes. Also characteristic is the nature of the rightward shift of the recovery traces as a function of increasing bleach level – the form of this shift is as expected for a rate-limited (zero-order) recovery process, as distinct from an exponential (first-order) recovery process.

From a detailed analysis of results of this kind, in combination with knowledge of the retinoid cycle, Lamb & Pugh[19] developed a cellular model that can account for human dark adaptation behavior. They postulated (1) that the presence of opsin (without chromophore) gives rise to a phenomenon closely equivalent to light, through activation of the G-protein cascade of transduction, and (2) that the elimination of opsin via its reconversion to rhodopsin follows rate-limited kinetics because of a limitation in the supply of 11-*cis* retinal that results from the diffusion of this substance from a pool in the retinal pigment epithelium.

Application of this cellular model has provided an accurate account of: (i) the regeneration of visual pigment in humans and other mammals, measured by retinal densitometry; (ii) normal human dark adaptation behavior (as in Fig. 20.9); and (iii) the slowed regeneration of pigment and the slowed dark adaptation that is characteristic of a number of diseases that affect the photoreceptors and/or RPE (see Boxes 20.2 and 20.3).[19]

Box 20.2 Diseases exhibiting slowed dark adaptation

A variety of diseases lead to slowed dark adaptation, and the great majority of these do so by affecting reactions of the retinoid cycle, and thereby reducing the effective concentration of 11-*cis* retinal in the retinal pigment epithelium.[18]

Mild cases of systemic vitamin A deficiency (VAD) lead to a slowing of the S2 component of dark adaptation, yet without any alteration in the final fully dark-adapted visual threshold.[32] This occurs because all the bleached opsin is able (eventually) to recombine with retinoid – the recombination is simply slowed. In these mild cases, supplementation with vitamin A can lead to a rapid and complete recovery of dark adaptation, showing that no other pathology was present. With more pronounced or extended deficiency, the final visual threshold becomes elevated, and eventually the photoreceptors may degenerate.

Fundus albipunctatus typically results from mutation of the 11-*cis* retinol dehydrogenase *RDH5* gene. The S2 component of rod dark adaptation is greatly slowed, but (as in vitamin A deficiency) the final visual thresholds are entirely normal.[33]

Sorsby fundus dystrophy is caused by a mutation in *TIMP3*, a metalloproteinase of the matrix in Bruch's membrane. All the features of this disease are consistent with the slowing of dark adaptation being caused by ocular VAD.[32]

Bothnia dystrophy is one of a group of rare diseases that involve mutations in CRALBP, the chaperone protein for 11-*cis* retinal. In these patients, loss of CRALBP leads to enormously slowed dark adaptation.[34]

Box 20.3 Slowed dark adaptation as an early indicator of age-related maculopathy (ARM)

The speed of dark adaptation is slowed in age-related maculopathy (ARM).[35-37] Thus, the slope of the S2 component of recovery is noticeably lower in patients with early ARM than in healthy subjects of similar age. Characteristic changes in ARM include a thickening of Bruch's membrane and the deposition of neutral lipids. Hence, the likely explanation for the slowing of dark adaptation in ARM is that the transport of vitamin A from the choroidal circulation is hindered, so that the RPE/retina becomes vitamin A deficient; i.e. that there is ocular VAD.

As a result, measurement of the slowing of dark adaptation is likely to provide a valuable and non-invasive bioassay, both as a diagnostic tool for predicting the likelihood of onset of macular degeneration, and as a means of assessing the efficacy of treatments of the disease.

References

1. Stiles WS. Color vision: the approach through increment-threshold sensitivity. Proc Natl Acad Sci USA 1959; 45(1):100–114.
2. Aguilar M, Stiles WS. Saturation of the rod mechanism of the retina at high levels of stimulation. Optica Acta 1954; 1:59–65.
3. Tamura T, Nakatani K, Yau K-W. Calcium feedback and sensitivity regulation in primate rods. J Gen Physiol 1991; 98(1):95–130.
4. Burkhardt DA. Light adaptation and photopigment bleaching in cone photoreceptors in situ in the retina of the turtle. J Neurosci 1994; 14(3):1091–1105.
5. Nikonov S, Lamb TD, Pugh EN. The role of steady phosphodiesterase activity in the kinetics and sensitivity of the light-adapted salamander rod photoresponse. J Gen Physiol 2000; 116(6):795–824.
6. Kawamura S. Rhodopsin phosphorylation as a mechanism of cyclic GMP phosphodiesterase regulation by S-modulin. Nature 1993; 362(6423):855–857.
7. Pepperberg DR, Cornwall MC, Kahlert M et al. Light-dependent delay in the falling phase of the retinal rod photoresponse. Visual Neurosci 1992; 8(1):9–18.
8. Krispel CM, Chen D, Melling N et al. RGS expression rate-limits recovery of rod photoresponses. Neuron 2006; 51(4):409–416.
9. Rebrik TI, Korenbrot JI. In intact mammalian photoreceptors, Ca²⁺-dependent modulation of cGMP-gated ion channels is detectable in cones but not in rods. J Gen Physiol 2004; 123(1):63–75.
10. Lamb TD, Pugh EN. Avoidance of saturation in human cones is explained by very rapid inactivation reactions and pigment bleaching. Invest Ophth Vis Sci 2006; 47:E-Abstract 3714.
11. Thomas MM, Lamb TD. Light adaptation and dark adaptation of human rod photoreceptors measured from the a-wave of the electroretinogram. J Physiol Lond 1999; 518(2):479–496.
12. van Hateren JH, Snippe HP. Simulating human cones from mid-mesopic up to high-photopic luminances. J Vision 2007; 7(4):1.
13. Krispel CM, Chen CK, Simon MI, Burns ME. Prolonged photoresponses and defective adaptation in rods of G beta 5(-/-) mice. J Neurosci 2003; 23(18):6965–6971.
14. Lamb TD, Mahroo OAR, Pugh EN. Absence of a change in amplification constant of transduction in human rod photoreceptors following bright illumination. J Physiol Lond 2002; 543:110P.
15. Stiles WS, Crawford BH. Equivalent adaptational levels in localized retinal areas. Report of a Joint Discussion on Vision. London: Cambridge University Press 1932:194–211. (Reprinted in Stiles WS. Mechanisms of colour vision. London: Academic Press, 1978.)
16. Cameron AM, Mahroo OAR, Lamb TD. Dark adaptation of human rod bipolar cells measured from the b-wave of the scotopic electroretinogram. J Physiol Lond 2006; 575(2):507–526.
17. Pugh EN. Rushton's paradox: rod dark adaptation after flash photolysis. J Physiol Lond 1975; 248(2):413–431.
18. Lamb TD, Pugh EN. Phototransduction, dark adaptation, and rhodopsin regeneration. The Proctor Lecture. Invest Ophth Vis Sci 2006; 47(12):5138–5152.
19. Lamb TD, Pugh EN. Dark adaptation and the retinoid cycle of vision. Prog Retin Eye Res 2004; 23(3):307–380.
20. Friedburg C, Allen CP, Mason PJ, Lamb TD. Contribution of cone photoreceptors and post-receptoral mechanisms to the human photopic electroretinogram. J Physiol Lond 2004; 556(3):819–834.
21. Kenkre JS, Moran NA, Lamb TD, Mahroo OAR. Extremely rapid recovery of human cone circulating current at the extinction of bleaching exposures. J Physiol Lond 2005; 567(1):95–112.
22. van Hateren H. A cellular and molecular model of response kinetics and adaptation in primate cones and horizontal cells. J Vision 2005; 5(4):331–347.
23. van Hateren JH, Lamb TD. The photocurrent response of human cones is fast and monophasic. BMC Neurosci 2006; 7:34.
24. Fain GL, Lamb TD, Matthews HR, Murphy RLW. Cytoplasmic calcium as the messenger for light adaptation in salamander rods. J Physiol Lond 1989;416:215–243.
25. Matthews HR, Fain GL, Murphy RLW, Lamb TD. Light adaptation in cone photoreceptors of the salamander: a role for cytoplasmic calcium. J Physiol Lond 1990; 420:447–469.
26. Pugh EN, Nikonov S, Lamb TD. Molecular mechanisms of vertebrate photoreceptor light adaptation. Curr Opin Neurobiol 1999; 9(4):410–418.
27. Torre V, Matthews HR, Lamb TD. Role of calcium in regulating the cyclic GMP cascade of phototransduction in retinal rods. Proc Natl Acad Sci USA 1986; 83(18):7109–7113.
28. Burns ME, Mendez A, Chen J, Baylor DA. Dynamics of cyclic GMP synthesis in retinal rods. Neuron 2002; 36(1):81–91.
29. Nishiguchi KM, Sandberg MA, Kooijman AC et al. Defects in RGS9 or its anchor protein R9AP in patients with slow photoreceptor deactivation. Nature 2004; 427(6969):75–78.
30. Hartong DT, Pott JWR, Kooijman AC. Six patients with bradyopsia (slow vision) – clinical features and course of the disease. Ophthalmology 2007; 114(12):2323–2331.
31. Stockman A, Smithson HE, Webster AR et al. The loss of the PDE6 deactivating enzyme, RGS9, results in precocious light adaptation at low light levels. J Vision 2008;8(1).
32. Cideciyan AV, Pugh EN, Lamb TD, Huang YJ, Jacobson SG. Plateaux during dark adaptation in Sorsby's fundus dystrophy and vitamin A deficiency. Invest Ophth Vis Sci 1997; 38(9):1786–1794.
33. Cideciyan AV, Haeseleer F, Fariss RN et al. Rod and cone visual cycle consequences of a null mutation in the 11-cis-retinol dehydrogenase gene in man. Visual Neurosci 2000; 17(5):667–678.
34. Burstedt MSI, Sandgren O, Golovleva I, Wachtmeister L. Retinal function in Bothnia dystrophy. An electrophysiological study. Vision Res 2003; 43(24):2559–2571.
35. Steinmetz RL, Haimovici R, Jubb C, Fitzke FW, Bird AC. Symptomatic abnormalities of dark adaptation in patients with age-related Bruch's membrane change. Br J Ophthalmol 1993; 77(9):549–554.
36. Owsley C, Jackson GR, White M, Feist R, Edwards D. Delays in rod-mediated dark adaptation in early age-related maculopathy. Ophthalmology 2001; 108(7):1196–1202.
37. Owsley C, McGwin G, Jackson GR, Kallies K, Clark M. Cone- and rod-mediated dark adaptation impairment in age-related maculopathy. Ophthalmology 2007; 114(9):1728–1735.

The Synaptic Organization
of the Retina

Robert E. Marc

The basic architecture, signal flow, and neurochemistry of signaling through the vertebrate retina is well-understood: photoreceptors, bipolar cells (BCs), and ganglion cells (GCs) are all thought to be glutamatergic neurons[1] and the fundamental synaptic chain that serves vision is photoreceptor → BC → GC.[2] But, our understanding of detailed signaling is far from adequate and a complete description of synaptic interactions or signaling mechanisms is lacking for any retinal network.[3] For example, GCs express different mixtures of ionotropic glutamate receptors (iGluRs) and each receptor can be composed of many different subunits leading to a vast array of possible functional varieties.[1] At a larger scale, network topologies are too numerous to resolve with current physiological or pharmacologic data.[4] Each GC contacts many different amacrine cells (ACs) and a full description of the inputs to any given GC does not yet exist.[5] Physiology can screen only a limited parameter space for any cell. Pharmacology is still an emergent field with many incomplete tools and an immense diversity of neurotransmitter receptor subunit combinations, modulators, and downstream effectors remains to be screened for any cell type. Molecular genetics, despite its power to modulate signaling elements, remains an ambiguous tool for analyzing retinal networks. Morphology, augmented by immunochemistry and physiology, remains the core tool in discovering new details of retinal organization.

Nothing has been as powerful as transmission electron microscopy for discovering retinal networks. Mammalian night (scotopic) vision is a prime example. Its unique pathways were described by Kolb & Famiglietti using electron microscopy.[6] Subsequent physiological analyses[7,8] provided clarification of how the network functions but would not have yielded the correct network architecture. Further complexities have been discovered by anatomical studies,[9–11] including the fact that the network rewires in retinal degenerations (Box 21.1).[12] But, electron microscopy has not kept pace with the demands for high-throughput imaging until recently. We are now on the verge of a new era in imaging that will provide a deluge of new information about retinal circuitry.[4] Finally, the basics of retinal development and new findings in neuroplasticity are beyond the scope of this chapter,[13,14] but the implications should be held in mind

throughout: the connections we have long considered as static or hard-wired in retina display many of the same molecular attributes as plastic pathways in brain.

The basic signal flow in retina is overlaid on a well-studied cell architecture (Fig. 21.1). Retinal ON and OFF BC polarities are generated in the outer plexiform layer and mapped onto the inner plexiform layer into largely separated zones. The distal sublamina *a* receives inputs from OFF BCs and therein the dendrites of OFF GCs collect signals via BC synapses. The proximal sublamina *b* receives inputs from ON BCs and therein the dendrites of OFF GCs collect signals via BC synapses. ON-OFF GCs thus collect inputs from both sublayers.

Kinds of neurons

The retina is a thin, multilayered tissue sheet … an image screen … containing three developmentally distinct, interconnected cell groups that form signal processing networks:

- Class 1 :: sensory neuroepithelium (SNE) :: photoreceptors and BCs
- Class 2 :: multipolar neurons :: GCs, ACs, and axonal cells (AxCs)
- Class 3 :: gliaform neurons :: horizontal cells (HCs)

These three cell groups comprise over 60–70 distinct classes of cells in mammals[2,3,15–17] and well over 100–120 in most non-mammalian retinas.[18]

The SNE phenotype includes photoreceptors and BCs. These cells are polarized neuroepithelia with apical ciliary-dendritic and basal axonal-exocytotic poles.[19] They form the first stage of synaptic gain in the glutamatergic photoreceptor → BC → GC → CNS vertical chain. This aggregates photoreceptor signals into BC receptive fields and amplifies their signals. The basal ends of the BCs form the inner plexiform layer. There are at least 12 kinds of BCs in mammals[16,20] and BCs delimit different functional zones in the IPL, suggesting nearly 1 micron precision in lamination. Both photoreceptors and BCs use high fusion-rate synaptic ribbons as their output elements, fueled by hundreds to thousands of

Box 21.1 Retinal Remodeling in Retinal Degenerations

- Primary photoreceptor or RPE degenerations leave the neural "inner" retina deafferented
- The neural retina responds by remodeling in phases, first by subtle changes in neuronal structure and gene expression and later by large-scale reorganization
- In *Phase 1*, expression of a primary insult activates photoreceptor and glial stress signals
- In *Phase 2*, ablation of the sensory retina via complete photoreceptor loss or cone-sparing rod loss triggers revision in downstream neurons
- Total photoreceptor loss triggers wholesale bipolar cell remodeling
- Cone-sparing degenerations trigger BCs reprogramming, down-regulating mGluR6 expression and up-regulating iGluR expression
- Loss of cone triggers *Phase 3*: a protracted period of global remodeling, including:
  - neuronal cell death
  - neuronal and glial migration
  - elaboration of new neurites and synapses
  - rewiring of retinal circuits
  - glial hypertrophy and the evolution of a fibrotic glial seal
- In advanced disease, glia and neurons may enter the choroid and emigrate from the retina
- Retinal remodeling represents the pathologic invocation of plasticity mechanisms
- Remodeling likely abrogates or attenuates many cellular and bionic rescue strategies
- However, survivor neurons are stable, healthy, active cells
- It may be possible to influence their emergent rewiring and migration habits

nearby vesicles. The retina is the only known tissue where SNE cells are arrayed in a serial chain.

As summarized in Figure 21.2, most mammals possess three classes of photoreceptors: rods expressing RH1 visual pigments, blue cones expressing SWS1 visual pigments, and green cones expressing Long-Wave System green (LWSG) visual pigments.[21] Conversely, the most visually advanced and diverse vertebrate classes (teleost fish, avians, reptiles) possess up to seven known classes of photoreceptors (RH1 rods, SWS1 UV/violet cones, SWS2 blue cones, LWSR and RH2 green members of double cones, LWSR and RH2 green single cones).[22]

Similarly, the diversity of BCs in mammalians is lower (10–13) than non-mammalians (>20). This reduced diversity is a result of the Jurassic collapse of the mammalian visual system, where over half of the visual pigment genes, half of the neuronal classes and almost two-thirds of the photoreceptor classes were abandoned to exploit nocturnal niches. In addition, the disproportionate proliferation of rods in the mammalian retina was accompanied by the loss of mixed rod–cone BCs in mammals and their replacement with pure rod BCs. How this occurred is unknown, but it cannot be due to an absolute selectivity of rod BCs for rods, as they will readily make contacts with cones when rods are lost in retinal degenerations. As we will see, the mammalian

retina has exploited a re-entrant use of synapses to enhance scotopic vision. The relationship between BCs and photoreceptors is still unclear, but there is both anatomical and molecular evidence that BCs were initially photoreceptors. For example, many non-mammalians possess BCs Landolt clubs, which are apical extensions extending from a BC primary cilium, extending past the outer plexiform layer into the outer nuclear layer, and containing packets of outer-segment-like membranes. Whether they are photosensitive has never been determined. Further, SWS1 blue cones and cone BCs share some SWS1 *cis*-regulatory sequences.[23]

The multipolar neuron phenotype

The multipolar neuron phenotype[2] includes ACs, AxCs, and GCs. Multipolar neurons can be further divided into axon-bearing (GCs, AxCs) and amacrine cells (ACs). Mammals display ≈ 30 kinds of ACs.[15] The 15–20 kinds of mammalian GCs[3,17] are classical projection neurons. GCs are postsynaptic at their dendrites and presynaptic at their axon terminals in CNS projections. So far, all are presumed to be glutamatergic. ACs are local circuit neurons similar to periglomerular cells in the olfactory bulb. ACs lack classical axons and often have mixed pre- and postsynaptic contacts on their dendrites, though some ACs partition inputs and outputs into different parts of their dendritic arbors. Most ACs are GABAergic and the remainder are glycinergic.[24] Several classes of ACs are dual transmitter cells, expressing both acetylcholine and GABA, serotonin and GABA (in non-mammalians) or peptides and GABA or glycine.[1] In between are the AxCs, also known as polyaxonal cells and intraretinal GCs, which have distinct axons that project within the retina.[25–27] One dramatic example of the AxC phenotype is the TH1 dopaminergic AxC.[28] This cell releases dopamine at unknown but probably axonal sites and likely glutamate at others,[29] similar to nigrostriatal neurons.[30] Some polyaxonal cells are GABAergic. There is no evidence for a glycinergic AxC. Multipolar neurons are characterized by numerous neurites branching in the plane of the retina, most collecting signals from BCs. Multipolar neurons are among the earliest to develop in the retina and quickly define the borders of the IPL and its stratifications. Multipolar neurons all manifest somewhat classical "Gray"-like synapses, generally with small clusters of less than 200 vesicles.

The gliaform cell phenotype

This phenotype contains the horizontal cells (HCs), whose somas and processes are restricted to the outer plexiform layer.[31] Though HCs are multipolar, neuron-like, and may display axons, they do not spike. Further, they express many glial features such as intermediate filament expression and very slow voltage responses. Further, HCs produce high levels of glutathione and make direct contact with capillary endothelial cells in some species, suggesting they play homeostatic roles similar to glia. Even so, HCs clearly mediate a powerful network function, collecting large patches of photoreceptor input via AMPA receptors and providing a wide-field, slow signal antagonistic to the vertical channel. The mechanism of HC antagonism remains a matter of uncertainty and debate. HCs do make conventional-appearing synapses onto neuronal processes in the outer plexiform layer, and in fishes these synapses are made onto

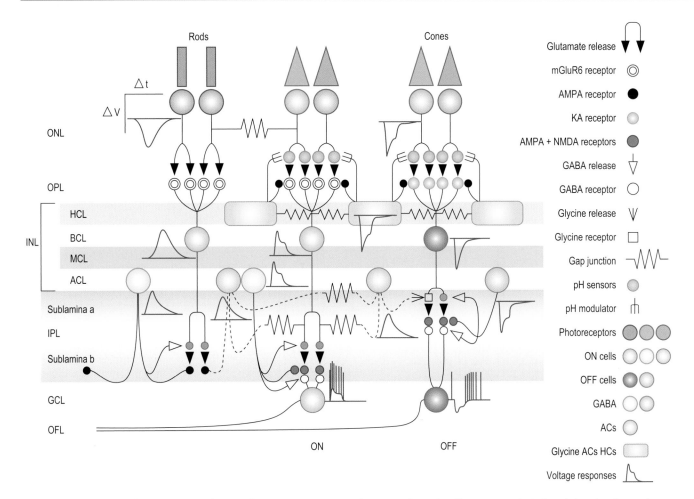

Figure 21.1 A summary of major cell superclasses and synaptic connections in the mammalian retina. Photoreceptors include rods (cyan) and cones (green, blue) that hyperpolarize in response to light. All photoreceptors are glutamatergic and drive HC AMPA receptors on HCs, ON BC mGluR6 receptors, and OFF BC KA or AMPA receptors. All BCs are glutamatergic and drive either predominantly AMPA receptors on rod pathway interneurons or various mixtures of AMPA and NMDA receptors on cone pathway ACs and GCs. Homocellular gap junctions are formed between like pairs of cells (HCs, certain ACs) and heterocellular gap junctions are formed between different cell pairs (rods and green LWS cones; glycinergic rod ACs and ON cone BCs; some ACs and certain GCs). Two classes of feedback pathways exist. There is a putative HC → cone feedback path mediated by a pH-sensitive process. AC → BC feedback is primarily GABAergic, as is AC → GC feedforward. Mammalian rod pathways are unique and not shared by any other vertebrate class. Rod BC signals are collected by a glycinergic rod AC that mediates a reentrant bifurcation into cone ON BC channels via gap junctions and cone OFF BC channels via glycinergic synapses. The outflow from the retina is largely split into ON GC channels that spike in response to light increments and OFF GC channels that spike in response to light decrements. The retina is precisely laminated into cellular and synaptic zones distal-to-proximal starting with the outer nuclear layer (ONL), the outer plexiform layer (OPL), the inner nuclear layer (INL), the inner plexiform layer (IPL), ganglion cell layer (GCL), and optic fiber layer (OFL). The INL is subdivided into the horizontal cell layer (HCL), bipolar cell layer (BCL), Müller cell layer (MCL) and amacrine cell layer (ACL). The IPL is subdivided into sublamina a that receives the output of OFF BCs and sublamina b that receives the output of ON BCs.

dendrites of glycinergic interplexiform cells, a form of AxC.[32] However, these are so sparse in all species and contain so few vesicles that they *cannot* be the source of the large sustained opponent surrounds of retinal neurons that HC generate. HCs must use some other mechanism.

The phylogenetics of HCs has been thoroughly reviewed.[33] HCs in mammals are postsynaptic to cones at their somatic dendrites. One class of HCs common in mammals (foveal type I in primates, type A in rabbits and cats, and absent in rodents) contacts cones alone. A second class of HCs (extrafoveal type I in primates, type B in rabbit and cats, and the only known HC in rodents) displays axons several hundred microns long that branch profusely and form massive arborizations contacting hundreds to thousands of rods. Another class of primate HC (type II) has axon terminals contacting

cones and rods. Importantly, the axon of HCs appears to be electrically inactive and these somatic and terminal regions are believed to act independently. HCs also appear to be early-developing pioneer cells that define the outer plexiform layer. After the GCs and HCs define the layout of the inner and outer plexiform layers respectively, photoreceptors and BCs mature and search for connections.

True glia and vasculature

The neurons of the retina are embedded in an array of vertical Müller glia that span the entire neural retina, forming one-third to one-half of the retinal mass and generating high-resistance seals at the distal and proximal limits of the retina. Most mammalian retinas are vascularized in three

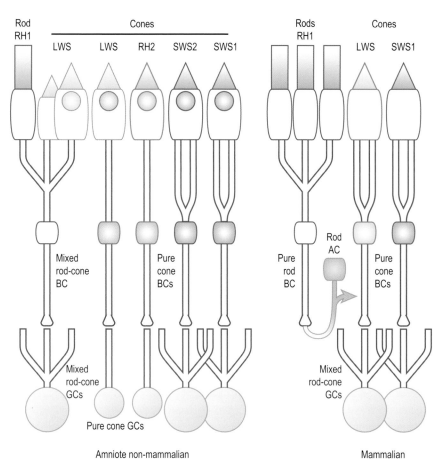

Figure 21.2 Photoreceptor cohorts and connections in vertebrates. Non-mammalians display multiple pigment classes and cone types. There are five pigment classes and seven photoreceptor types for a fresh water turtle, including rods (comprising less than 10 percent of the photoreceptors) expressing class RH1 rhodopsins, three kinds of LWS cones (short members of double cones, long members of double cones with orange oil droplets, single cones with rod oil droplets), single cones expressing RH2 green cone pigments and a yellow oil droplet, single cones expressing SWS2 blue cone pigments and a UV-opaque clear oil droplet, and single cones expressing SWS1 UV cone pigments and a UV-transparent clear oil droplet. The connection patterns for non-mammalians are mixed rod–cone BCs and pure cone BCs, leading to mixed rod–cone GCs and pure cone GCs. Mammals display three pigment classes (one rod and two cone), two cone color types in non-primates and three color types in primates, including RH1 rods, SWS1 cones, and LWS cones. The LWS cone class forms one green type in most mammals, and red (LWSR) and green (LWSG) chromatypes in primates. The main connection rules for mammalians are pure rod BCs and pure cone BCs, with only cone BCs driving GCs, with rod ACs (cyan) providing the re-entrant crossover.

capillary beds: at the GC-inner plexiform layer border, the AC-inner plexiform layer border, and the outer plexiform layer. Squirrels (Sciurids) display two beds (at the GC-inner plexiform layer and AC-inner plexiform layer borders; and rabbits (Lagomorphs) have none at all, similar to all other non-mammalian vertebrates. The GC layer of many species also displays classical astrocytes, though their role remains unclear. In brain, astrocytes carry out some of the operations attributed to retinal Müller glia, including transport of spillover K^+ and glutamate, and glucose supply via vascular > glial cell > neuron transcellular transport. Why and how most vertebrate retinas function without vasculature remains uncertain, but it is likely that Müller glia act as a surrogate vascular system with the added ability to accumulate large glycogen stores (like hepatocytes) as part of a glucose-skeleton homeostasis. The segregation of retinal astrocytes away from the inner plexiform layer remains a mystery.

Basic synaptic communication

With the discovery of the signaling mechanisms of the neuromuscular junction decades ago,[34] one might have thought that the archetypal synaptic format had been discovered. Yet it has become clear, especially in retina, that every kind of synapse is subtly different, with diverse physics, topologies, and molecular mechanisms leading to very different forms of synapses, most of which do not follow the single presynaptic "bouton" → single postsynaptic target pattern of brain. Further, the arrangement of these systems into synaptic chains in retina is unlike any other known network, including olfactory bulb. In retina, the first stage of synaptic signaling is a direct SNE → SNE synapse (Fig. 21.3): photoreceptor → BC. No other instance of this topology has been discovered in any organism. There are at least six modes of presynaptic–postsynaptic pairing in retina.

1. Photoreceptor ribbon synapses: small-volume multi-target signaling

It is thought that all photoreceptor signaling is glutamatergic, but sporadic indications of cholinergic physiology and molecular markers have been found in many non-mammalians.[1] Glutamate release from photoreceptors is effected by high rates of vesicle fusion at active sites on either side of a large synaptic ribbon[35] positioned close to the pre-synaptic membrane. The presynaptic zone is a protrusion or ridge with vesicle fusion sites positioned on the slopes of the ridge (Fig. 21.4). The releasable vesicle pool is so large that photoreceptors and BCs are capable of maintaining continuous glutamate release in response to steady depolarizations. This, among other things distinguishes photoreceptors and BCs from ACs, which have very small presynaptic vesicle clusters.

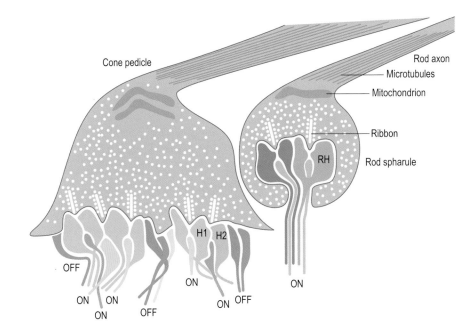

Figure 21.3 Basic organization of mammalian photoreceptor synaptic terminals. Primate cone terminals contain many ribbons, mitochondria clustered at the head of the pedicle and thousands of synaptic vesicles (white dots), some of which form organized zones near the plasma membrane opposite an array of cone-specific postsynaptic processes including type H1 and H2 HCs as lateral invaginating elements (primate H1 cells tend to avoid SWS1 cones, while H2 cells contact all cones). Cone ON BCs tend to center their dendrites between the lateral HC processes at varying distances from the synaptic ribbon, forming so-called invaginating and semi-invaginating contacts. Most cone OFF BCs position their dendrites outside the HC processes at so-called flat contacts. It is thought that most of these express KA receptors. Some occasional OFF BCs processes invaginate, and they may express AMPA receptors.

Various vertebrate rods and cones differ greatly in the number of ribbons and postsynaptic targets arrayed within the presynaptic terminals. For example, most mammalian and teleost fish rods have small grape-like presynaptic spherules $\approx$ 3 µm in diameter with a small entrance aperture leading to an enclosed extracellular invagination or vestibule in which thin postsynaptic dendrites are contained (Fig. 21.3). Importantly, glial processes are excluded from the interior of the spherule and any glutamate release must diffuse out of the spherule to reach the Müller glia. However, mammalian rods express the EAAT5 glutamate transporter[1] and likely regulate their own intrasynaptic glutamate levels. Each spherule contains one or two synaptic ribbons and a few postsynaptic targets.[36] In fishes, the postsynaptic targets are the dendrites of roughly five kinds of mixed rod–cone bipolar cells[37] and one kind of rod horizontal cell.[33] Thus each ribbon serves no less than six different types of postsynaptic targets. In mammals, only two targets are common: the dendrites of one kind of rod BC and the axon terminals of HCs. There are some instances of sparse OFF BC contact in mammals, but this seems to vary with species and may be an evolutionary relict with variable expression rather than a major signaling pathway.[38–40] In sum, rod spherules form a sparse-ribbon $\rightarrow$ small volume, sparse-target architecture.

Cones and rod terminals in some non-mammalians (e.g. urodele amphibians) adopt a different topology, with the presynaptic ending expanding to form a foot-piece or pedicle some 3–5 µm wide shaped either like a cupola (fishes) whose broadly concave interior admits some 50–100 or more fine dendrites served by roughly 12 synaptic ribbon sites;[41] or like a true pediment (e.g. primate cones) whose shallow concavity is studded with up to 50 ribbon sites (Fig. 21.3).[42] Cone pedicles in primates target at least ten different kinds of BCs and at least two kinds of HCs. Mouse cone pedicles are smaller but still target 11 kinds of BCs[20] and one kind of HC. In sum, *cone pedicles form a multi-ribbon $\rightarrow$ small volume, multi-target architecture*.

2. BC ribbon synapses: semi-precise target signaling

Like photoreceptors, BC signaling is generally considered glutamatergic.[1] Sporadic evidence of exceptions exists. In mammals (especially primates) and amphibians, some BCs contain biomarkers of GABA-related metabolism.[24] In contrast to photoreceptors, BC synaptic endings are topological spheroids, usually multiple (depending on BC type), with dozens to hundreds of ribbons abutting the surface. BCs form no invaginations, so there is no restricted volume into which glutamate is injected by vesicle fusion. In most cases each ribbon is directly apposed to a pair of postsynaptic targets, usually ACs. This is termed a dyad and, while monads, triads and tetrads do occur, dyads dominate. Large BC terminals such as those found in teleost fishes can drive up to 200 distinct processes. Mammalian BCs drive many fewer targets and most BCs have elaborate, branched terminals with connecting neurites often as small as 100 nm. In contrast to photoreceptors, the targets of BCs are focal. BC terminals are largely fully encapsulated by neuronal processes at their release sites to which they are presynaptic or postsynaptic, with rarely direct contact between the terminal and Müller glia near the synaptic release zone. This means that any glutamate that escapes from the synaptic cleft may travel some distance before glial glutamate transporters can clear it. Thus the potential for glutamate overflow at BC synapses is substantial. This may be particularly important for the activation of NMDA receptors, as they are suspected to be displaced from primary AMPA receptors. Thus, *BCs form multi-ribbon $\rightarrow$ semi-precise target architectures*.

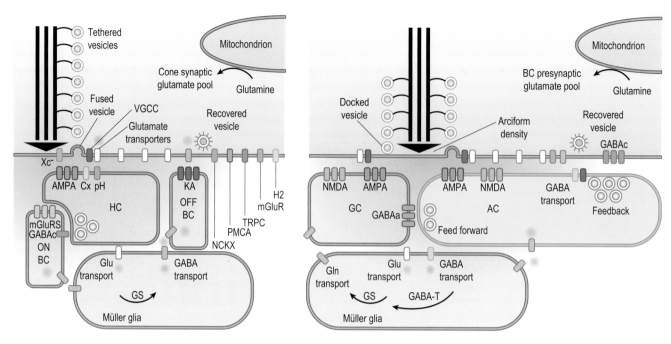

Figure 21.4 A detailed schematic of synaptic organization at cone (left) and BC (right) ribbon synapses. Each synaptic ribbon is a pentalaminar structure in cross-section, in reality a disc or lozenge-shaped solid with its two broad faces serving as attachment sites for tethered vesicles and its small paramembrane face attached to a dense structure known as the arciform density (composition unknown). Ribbons serve as the major site for the "readily releasable" pools of synaptic vesicles for continuous glutamate transmission, and facilitate high-speed formation of docked vesicles. Upon depolarization of the presynaptic membrane, voltage-gated calcium channels (VGCCs, black barrels) open, allowing docked vesicles to fuse and release glutamate into the synaptic cleft. At some distanced from the ribbon, endocytosis mediates vesicle recovery. BC ribbons tend to be shorter than photoreceptor ribbons. Cytoplasmic glutamate (orange) is formed glutamine via mitochondrial phosphate-activated glutaminase and loaded into vesicles via vGlut vesicular transporters. Glutamate release by vesicle fusion diffuses away from the release site (shaded orange) and is cleared by both presynaptic and distant Müller glia glutamate transporters (white barrels). Müller glia synthesise glutamine from glutamate via glutamine synthetase (GS) and exports glutamine via transporters (grey barrels). Similarly, neurons import glutamine via transporters. Vertebrate photoreceptors also express presynaptic cystine-glutamate (Xc-) exchangers (orange barrels) whose function is unknown. Glutamate released by cones activates ON BCs via mGluR6 receptors (light blue barrels), HCs via AMPA receptors (dark blue barrels) and OFF BCs via either AMPA or KA receptors (bright blue barrels). HCs are positioned at the highest glutamate concentration zone and OFF BCs at the lowest. Glutamate released by BCs activates ACs and GCs via AMPA receptors (dark blue barrels) and NMDA receptors (gold barrels). Feedback at photoreceptors appears to be mediated by either focal connexin (yellow barrel) or a pH modulator (magenta barrel) close to the photoreceptor VGCC. Feedforward from HCs to BCs may be GABAergic in some species and mediated by GABAC receptors (dark red barrels). Feedback at BCs is mediated by vesicular GABA (red shading) release from ACs targeting largely GABAC receptors. Feedforward from ACs to BCs is mediated by largely by GABAA receptors (bright red barrels). ACs and mammalian Müller glia also have GABA transporters that clear the synaptic space. GABA is converted via the GABA-transaminase (GABA-T) complex to glutamate in glia. Cone synaptic terminals also have a number of other proteins including Na-Ca exchangers (NCKX), plasma membrane Ca transporters (PMCA), transient receptor potential channels (TRPC), metabotropic glutamate receptors (mGluR) and possibly histamine receptors (H2). BC terminals may share some of these.

3. AC and AxC conventional fast synapses: precise presynaptic → postsynaptic signaling

ACs and AxCs are the only retinal cells that make synaptic contacts resembling CNS "Gray"-like, non-ribbon conventional synapses. ACs target BCs, GCs, or other ACs. The targets of most AxCs are not well known but appear likely to be ACs and GCs. Though each AC may make many hundreds of synapses, each synapse contacts only and only one postsynaptic target, similar to classical multipolar neurons in CNS and spinal cord.[43] The dominant fast transmitters of AC systems are GABA and glycine, with GABAergic neurons making up half to two-thirds of the AC population depending on species.[1] Additional transmitters such as acetylcholine, peptides, or serotonin (in non-mammalians) are also associated with GABAergic (in most cases) or glycinergic systems.[1,44] Acetylcholine (ACh) is a fast excitatory transmitter and is found in paramorphic starburst ACs in mammals and also uses conventional synapses.[1] However, we know of no distinguishing anatomical differences between GABA- and ACh-utilizing synapses in retina.

4. AC, AxC, and efferent slow transmitter synapses: large volume signaling

Dopamine (and possibly norepinephrine/epinephrine) as well as peptides in retina appear to be released by a non-focal, Ca^{2+}-dependent vesicular system,[45] but without any clear postsynaptic associations. Dopamine and the other slow transmitters likely act via volume conduction[46] and modulate a range of cellular responses largely via G-protein coupled receptors (GPCRs). In non-mammalians, efferent systems from CNS target ACs with fast neurotransmitter synapses, especially GABA.[5] In mammals, all known efferents appear to release either histamine or serotonin, likely as volume signaling systems.[47]

5. HC non-canonical signaling

HCs generate potent, large-field, slow surround signals in retinal GCs, BCs, and even in non-mammalian cone photoreceptors.[48-50] There is evidence for both feedforward signaling via the cone → HC → BC path[51] and feedback signaling via the cone → HC → cone → BC path[52,53] and, now, the rod → HC → rod path.[54] The efficacy and sustained nature of the feedback signal is such that no known vesicular mechanism could maintain it (other than a ribbon-style synapse). Vesicular HC synapses are very rare and small. Several models of non-canonical signaling have been proposed including synaptic pH regulation,[55] hemi-junction mediated ephaptic signaling,[56] and even transporter-mediated signaling. Some of these will be discussed in detail below, but this unusual functionality is further evidence that HCs are not classical neurons.

6. Coupling types and coupling patterns

While gap junctional coupling was first discovered between HCs, only in the past decade has it become clear how powerful and pervasive intercellular coupling is in retina.[57,58] There are two simple classes of coupling: homocellular and heterocellular (between like and different classes of cells, respectively). The participant connexins in each case are respectively likely to be homo- or heterotypic (similar or dissimilar connexin types).[37] The strength of coupling is associated with the size of the junctions, as they represent summed parallel conductances, and with functional modulation by various signaling pathways. Activated dopamine D1 receptors decrease conductances between coupled HCs[45,59] and coupled ACs,[60] and dopamine D2 receptors modulate rod–cone coupling.[61] The significance of coupling is clear in certain cases, such as the ability of HCs to spatially integrate signals over large fields (> 1 mm diameter) or the crossover of rod signals into cones via heterocellular rod–cone and rod AC–cone BC coupling. However, such knowledge does not readily extrapolate, and other coupling patterns are poorly understood, such as heterocellular AC–GC coupling and even HC–BC coupling.

Fast, focal neurochemistry, synaptic currents, and amplification

One of the most powerful discoveries of the last two decades has been the diversity of the primary fast neurotransmitter receptors of the vertebrate nervous system. Again, the primary signaling channel of retina is the vertical glutamatergic chain from photoreceptors to brain.[1] Rods, cones, and BCs encode their voltage responses as time-varying glutamate release. The targets of photoreceptors and BCs, in turn, decode time-varying extracellular glutamate levels as time-varying currents with glutamate receptors. There are two major classes of glutamate receptors: ionotropic and metabotropic (iGluRs and mGluRs, respectively). The iGluRs are separable into two distinct families: the AMPA/KA receptors and NMDA receptors. AMPA and KA receptors are related but pharmacologically and compositionally distinct. Four basic classes of glutamate receptor subunits (GluR1, 2, 3, 4) can be recruited to form a tetrameric AMPA receptor. Similarly, five basic classes of KA receptors (GluR5, 6, 7 and KA1, 2) can be assembled into tetrameric KA receptors. With some exceptions, these receptor assemblies can have nearly any stoichiometry. NMDA receptors are a distinct group of iGluRs in several ways. First, they have an obligate tetrameric subunit composition. Second, they are coincidence detectors that require dual activations by glutamate and by a glycine-like endogenous agonist. There is substantial evidence that this co-ligand may be D-serine released from Müller glia.[62] Finally, the mGluRs represent a complex collection of GPCRs whose functions are far from clear.

Different classes of neurons express different types or different combinations of receptors and. in the end, the glutamate receptor profile of a cell is diagnostic for its class. Mammalian BCs are unique in expressing either mGluR6, KA, or AMPA receptors as their glutamate decoding system. BCs seem functionally monolithic in having their signals dominated by one of these three receptor systems. But immunochemical and mRNA expression analysis suggest that these associations are not so precise, and iGluR subunit expression occurs in nominally mGluR6-driven cells.[63] HCs predominantly express AMPA receptors, but show no NMDA-mediated responses. Finally, ACs and GCs resemble CNS neurons in expressing AMPA receptors augmented by varying amounts of NMDA receptors.

The key glutamate receptor systems of retina operate on the principle of cation permeation.[1,64] When activated, iGluRs generate increased channel conductances and carry inward currents, carried mostly by Na^+ and Ca^{2+}. Thus the canonical iGluR AMPA, KA, and NMDA families of receptors are nominally *sign-conserving* (>) depolarizing systems that "copy" the polarity of the presynaptic source voltage input in the postsynaptic target. The facts that many inputs converge on one postsynaptic cell; that small presynaptic voltages can modulate the release of many vesicles (in SNE cells); and that glutamate gates large postsynaptic conductance changes to cations with a positive reversal potential means that such synapses have high gain. Signals from photoreceptor to brain are successively amplified by a chain of glutamate synapses.

The group III mGluR6 system is unique and, in retina, is expressed by ON BCs (Box 21.2). No known multipolar neuron in the CNS uses this receptor as its primary signaling modality. As a classical GPCR, with Go_α as its cognate G-protein,[65] the binding of glutamate triggers a cascade of signals that ultimately leads to the *closure* of cation channels on BC dendrites, thus moving the BC membrane potential closer to the K^+ equilibrium potential. Thus mGluR6 receptors are nominally *sign-inverting* (>i) hyperpolarizing systems that invert the polarity of the input in the postsynaptic target. The modulation of a strong cation current renders the mGluR6 mechanism high-gain in spite of its inverted polarity.

The differential expression of iGluRs and mGluR6 in BCs creates the two fundamental signal-processing channels of the retina: OFF and ON BCs respectively.[1] Unknown mechanisms regulate the expression of glutamate receptors in BCs. In general, BCs that express iGluRs such as KA or AMPA receptors do not express *functional* mGluR6 receptor display, and vice versa. However, there is evidence that BCs expressing mGluR6 also express low levels of iGluR protein, but there is as yet no evidence that such iGluR

Box 21.2 The mGluR6 receptor and vision

- The retina expresses mGluR6 receptors in both rod and cone ON BCs
- The human mGluR6 gene GRM6 is expressed predominantly in retina
- The mGluR6 receptor converts rod/cone hyperpolarizations into depolarizing BC signals
- The mGluR6 receptor dominates scotopic, but may not be essential for photopic vision
- Some human night-blind patients have GRM6 mutations,[109] with no photopic deficit
- The mGluR6 drives blue ON center cone BCs and GCs
- fMRI in rodents[110] shows that blockade of mGluR6 receptors attenuates blue-driven signals in the LGN
- Simultaneously, blue-driven cortical signals are enhanced, suggesting that blue percepts are effectively or even primarily carried by OFF channels
- OFF center BCs can support photopic vision
- The perceptual world under pure OFF vision is yet unknown

Box 21.3 Glutamate excitotoxicity

- Glutamate excitotoxicity has often been invoked as a mechanism in retinal diseases
- The evidence is mixed and controversial. On balance:
  - Elevations of vitreal glutamate in glaucoma[111] have not been validated
  - Glaucoma-mediated loss of GCs does not match known excitotoxic patterns in retina
  - Starburst ACs are the most glutamate sensitive cells in the retina[64]
  - There are no established AC losses in primate glaucoma
  - NMDA receptor antagonists appear neuroprotective in models of glaucoma[112]
  - The mechanism of NMDA neuroprotection may be indirect
    - retinal GC death in glaucoma likely involves Ca^{2+}-mediated apoptosis
    - NMDA antagonists (like many drugs) will decrease Ca^{2+} loads in neurons
    - weak NMDA antagonists likely have no lasting role in glaucoma therapeutics
- Glutamate is likely a major player in retinal damage in diabetes and ischemic insults
- Neuroprotection is difficult to achieve in those cases as it involves AMPA receptors
- Excitoxicity in hypoxic retina is likely initiated by reverse transport of glutamate by BCs
- Competitive, non-translocated transporter ligands may be safer ocular neuroprotectants

subunits contribute to an electrically detectable signaling event.[63]

There are additional mGluRs expressed in retina, including both the group I mGluR1 and mGluR5 and group III mGluR 4, 7, and 8, all largely expressed in varying patterns in the inner plexiform layer.[66] Their roles are thought to be associated with presynaptic glutamate feedback.

Global neurochemistry and modulation

There are a number of alternative neurochemical mechanisms that appear to operate on a larger spatial scale than conventional synapses. The most thoroughly described, though incompletely understood, is the dopamine.[45] In mammalian retinas, dopaminergic neurons are largely AxCs that arborize primarily in the distal-most layer of the IPL, appear to receive direct inputs from BCs, and have predominantly ON-type responses.[67] The exact sites of dopamine release are not known, although these cells clearly have small accumulations of synaptic vesicles distributed sparsely in their processes. Data from many species suggest that dopamine acts largely via volume conduction:[46] i.e. dopamine diffuses throughout the retina and targets high-affinity type D1 and D2 receptors distributed on almost every class of cell, including Müller cells. Actions triggered by D1 receptors include the uncoupling of HCs,[68] the uncoupling of rod-driven glycinergic ACs,[60] and enhanced spike speeds[69] (faster waveforms). All of these are consistent with the process of converting the retina from a scotopic to a photopic state. Dopaminergic neurons are also thought to contain a fast transmitter. Previous evidence in the mouse suggested that they were GABAergic, but other studies support the possibility that they are glutamatergic.

A similar role is posited for the rapidly diffusing neuroactive gas NO. A variety of retinal cells are posited to produce NO via the neuronal NO synthase (nNOS) pathway, activated by binding of Ca-activated calmodulin.[70] Thus synaptic currents with high Ca-permeability (e.g. those mediated by NMDA receptors) can turn on nNOS, which catalyzes the oxidation of one of the guanido nitrogens of arginine to NO. NO appears to have the capacity to diffuse through transcellular regimes (how far is in debate) and activate soluble guanyl cyclases to produce cGMP. Acting via cGMP-dependent protein kinase G or directly on cyclic-nucleotide gated cation channels, cGMP can effect a number of modulatory actions. One site of action is thought to be the heterocellular coupling of ON cone BCs and glycinergic rod ACs,[60] and increased cGMP appears to uncouple this network, again a light-adaptive action. While many cells appear to be able to produce NO, the best-known architecture for NOS-containing neurons is a wide-field GABAergic AC.[71]

Many retinas contain a number of neuropeptides as cotransmitters, largely in wide-field GABAergic ACs with cone BC inputs. These peptides include somatostatin, substance P, and neurotensin, with many more in non-mammalian, such as enkephalins.[44] This is probably the least understood neurochemical aspect of the retina. In general, it is thought that peptides act via specific peptide receptor GPCRs to modulate various ion channels. However, very little research on retinal peptides has been carried out in recent years.

Modulation by transporters

An important part of every signaling process is termination. Given the ability of glutamate to activate excitotoxic processes, the expression of high-affinity sodium-glutamate transporters by Müller glia likely represents the last line of defense.[1] A molecule of glutamate must have already left the synaptic cleft by diffusion to encounter glial glutamate transporters. However, both photoreceptors and BCs express glutamate transporters, presumably near the sites of synaptic release[1] and these likely play a major role in determining the temporal dynamics of extracellular glutamate levels, though this has been difficult to quantify. Even so, every fast transmitter has a corresponding presynaptic transporter mechanism. Though numerous studies support the roles of transporters in signal termination, the precise localization of transporters has not been achieved.

Signal processing

The roles of synaptic networks are to convert graded sensory photoreceptor potentials into patterns of action potentials for long-range transmission to the CNS, and perform spatial, temporal, and spectral signal processing on the input signals of the photoreceptors. This latter action converts the retinal image into the parallel signaling behaviors of 15–20 different classes of retinal GCs in mammals. The concept of signal processing, as derived from electrical engineering, is particularly relevant.[49] Each kind of synapse, each kind of cell, and each topology of network is invoked in various ways to generate the kinds of "filters" through which the visual scene must be encoded. The physiological analysis of retina in the 1970s (especially as carried out by Naka) represented a sea-change in thinking; a move away from Sherringtonian concepts of spinal excitation, inhibition, and circuits (loops) and towards engineering notions of polarity, inversion, networks, and filters.

Sign-conserving (>) and sign-inverting (>i, >m) transfers

The behavior of a photoreceptor is neither excitatory nor inhibitory. Photoreceptors encode time-varying changes in light intensity with fairly faithful (though nonlinear) time-varying changes in voltage. As glutamatergic neurons, one would normally think of them as "excitatory" in brain, but a more robust concept is derived by looking at the behaviors of target neurons. HCs (driven by AMPA receptors) and OFF BCs (driven by AMPA or KA receptors) merely copy the polarity of presynaptic photoreceptors (see Fig. 21.1). When light hyperpolarizes photoreceptors, this decreases the rate of synaptic glutamate release and (in conjunction with glutamate transport) leads to a decrease in synaptic glutamate levels. Since AMPA and KA receptors are iGluRs, decreased synaptic glutamate means that AMPA and KA receptor-gated currents will decrease and the HCs and OFF BCs will hyperpolarize. Conversely, when a fly navigates across the visual field, local darkening will depolarize some photoreceptors and the HCs and OFF BCs will follow. Thus photoreceptor → HC and OFF BC signaling is termed sign-conserving (>). In addition, iGluRs typically mediate high-gain responses (i.e. strong amplification) and, over modest voltage ranges this amplification is symmetric and polarity-invariant. ON BCs behave in a totally different manner. In mammals, all ON BCs express functional mGluR6 receptors that activate a cation channel when unbound and close it upon binding of glutamate. Thus, when photoreceptors decrease their glutamate release, this leads to decreased mGluR6 receptor binding and the opening of cation channels and depolarization of ON BCs (see Box 21.3). This is an explicit, high-gain, metabotropic sign-inverting (>m) synaptic transfer.

Non-mammalians display a twist on this mechanism that perhaps reveals the evolutionary history of mammalian ON BCs. The apparent homolog of the ancestral mammalian rod BC exists in the retinas of modern fishes as the mixed rod–cone BC. This cell has an unusual behavior in that it has different reversal potentials and conductance changes for different stimuli. Scotopic lights that activate rods generate ON responses that display a positive reversal potential (like a cation) and an increase in conductance (like a channel opening). This is very like mammalian rod and cone ON BCs, and indeed it appears to have the same pharmacology: 2-amino-4-phosphonobutyrate (AP4) is an agonist at mGluR6 receptors[72] and blocks rod ON BCs responses in fishes. However, upon light adaptation, fish ON BCs change their behaviors. In response to photopic lights that activate cones, the "ON" reversal potential moves to very negative values (like an anion) and the cells display a decrease in conductance (like a channel closure). In fact, the cone-driven ON responses of fish BCs are mediated by an anion channel coupled to a glutamate transporter.[73,74] Thus, in photopic "dark", glutamate release activates the transporter and its coupled chloride current, leading to hyperpolarization of the BC. Thus the fish cone → ON BC synapse is sign-inverting, but not metabotropic. The degree of its amplification is also unknown.

In the inner plexiform layer, BC → AC and GC signaling is all mediated by AMPA or AMPA+NMDA receptors.[3,64,75] Thus all BC output synapses are sign-conserving (Fig. 21.1). The bulk of AC → BC, AC or GC signaling is either GABAergic or glycinergic via increased anion conductances.[76] Thus these synapses are characteristically sign-inverting (>i). GABAergic and glycinergic transmission is also usually very low gain, often because the reversal potential is very close to the membrane potential and/or the total conductance change experienced by a target cell leads to a tremendous decrease in total cell input resistance, thus decreasing signal efficacy. In any case, it takes a significant amount of inhibition to control glutamate synapses. Some inhibitory mechanisms involve the metabotropic GABAB receptor, which is a GPCR that can lead to a tremendous increase in potassium currents, but can also show paradoxical excitation.[73] Since potassium currents are usually outward (positive current flowing outward), GABAB can produce a strong and long-lasting inhibition near threshold. Different GABA receptors tend to be expressed at different sites (ionotropic GABAA on ACs and BCs, ionotropic GABAC on BCs), but the distribution of GABAB receptors is less well understood.

Synaptic chains and polarity

The effect of cascading synapses through various pathways can be estimated in terms of polarity and gain. For example, though the mechanism of HC action is very poorly understood, its efficacy is not in doubt. As first established in fish retinas, currents injected into HCs have stereotyped actions on different GCs. The net pathway from HCs to ON GCs is sign-conserving. The net pathway from HCs to OFF GCs is sign-inverting.[48,49] From a signal processing perspective, this means that the polysynaptic chain from a given HC, somehow reaching a BC and thence to an ON GC must contain either no sign-inverting elements or an even number of them. The path to an OFF GC must contain an odd number of sign-inverting elements. This poses fundamental constraints on where and how signals flow in the retina and can be used as a model for network investigations. Similarly, we know that there are chains of two and three serial ACs in the inner plexiform layer of most retinas.[5] While there are no network models that use such chains, the minimum architecture for such a chain is BC → AC → AC → BC or GC. If the AC outputs are sign-inverting, the net transfers of the chain are > >i >i > (sign-conserving). Importantly, the low gain of GABAergic and glycinergic synapses prevents such chains from being runaway excitations.[5,77] What roles might such networks play? In non-mammalians, the bulk of GABAergic ACs also receive some form of GABAergic input as evidenced by their pharmacology. It was presumed that similar networks existed in mammals, but recently Hsueh et al[78] argued that in rabbit, the only synaptic crossover networks are glycine → GABA.

Feedback, feedforward, and nested feedback/feedforward

Designing analog operational amplifier networks is very similar to evolving a retina: every stage of forward amplification needs feedback control.[79] In retina, sign-inverting GABAergic mechanisms are used as feedback and feedforward control systems. Feedback is the most powerful way to set synaptic gain, improve signal-to-noise ratios, and improve synaptic bandpass. We will skip the mathematic demonstration of this, but note that it has been widely discussed.[76,80] On the other hand, feedforward is an effective way to generate strong antagonistic mechanisms in target cells. These architectures are clearly at play in retina but we have only a hazy idea of their importance. For example, blocking GABAergic inhibition converts directionally selective GCs into non-selective cells, but has little effect on the center-surround organization of other GCs, despite the abundance of GABAergic synapses in the inner plexiform layer. Thus is it hard to generalize function from anatomy. Conversely, it is impossible to understand function without anatomy.

Caveats

Three major problems have emerged in understanding how GABAergic (or any classical inhibitory transmitter) works in retina. The first is the chloride reversal potential. We have a very poor idea of which way GABA receptor-gated anion currents will flow: inward or outward. Small retinal cells may have the ability to adjust intracellular chloride levels with various ion transport systems. The KCC2 system tends to export Cl while the NKCC system tends to import it.[51] If intracellular Cl is locally high, opening an anion channel may evoke an outward negative current and depolarize the cell. However, most studies of GABAC receptors at the synaptic terminals of retinal BCs support the view that it is inhibitory.[1] A second problem is temporal delay. Imagine a cell responding to a light input with a sinusoidal voltage. Then imagine a surrounding cell giving a similar response and providing feedback with a sign-inverting polarity. If the

feedback was delayed so that the phase is shifted by 180 degrees, the "inhibitory" local surround would sum with the center response. There is much evidence to suggest that simple AC → BC inhibitory feedback cannot explain all BC responses.[81] Finally, it has been long assumed that BCs were effectively isopotential, and that simple lumped-parameter calculations would suffice to model their network functions. But reconstructions of BCs in mammals[4] suggest that the isopotentiality assumption is not correct and that complex local information processing can be effected at the synaptic terminals without any evidence of that filtering appearing at the BC soma.

Networks

The synaptology of center-surround organization

The long-standing view of any BC or GC receptive field is that it has antagonistic center-surround organization (Fig. 21.5). Where does the surround come from? Anatomically, the vast number of AC synapses at the BC synaptic terminal, as well as evidence of GABAC receptor function, suggested that ACs should have a powerful surround effect.[5] Conversely, direct current injection into non-mammalian HCs clearly shows an effective, low-frequency dominated, sustained path from HCs to GCs. Reconciling the mechanism has been problematic but is likely simple. HCs and ACs function on very different time and space scales. HCs are slow, sustained (beyond the capacity of any normal neuron) and have immense receptive fields due to strong coupling by gap junctions. Thus the presence of very large antagonistic surrounds in GCs is likely driven through HCs. Experiments using fast pH buffers such as HEPES block these surrounds.[55] Conversely, GABAergic drugs have no effect on these large surrounds (in mammals). So what about ACs and all those synapses? Why don't they create the large surrounds of GCs? First, ACs have much smaller receptive fields than HCs and their range of action will thus be smaller. Second, many ACs themselves show antagonistic center-surround organization, likely due to AC >i AC chains.[82,83] Third, ACs are very fast and their actions at the BC terminal likely have more to do with feedback stabilization of synaptic gain than creating large, slow antagonistic surrounds. ACs work in a highly time-and-space-restricted domain.

The synaptology of mammalian rod pathways – evolution of a new amplification scheme

As we have described earlier, the synaptic chains that drive GCs in all retinas are grouped into ON and OFF pathways. In non-mammalians, rod and cone pathways both use this direct chain to target the CNS. Thus rod signals undergo two-stage amplification before being encoded as a spike train: rods >m ON BCs, rods > OFF BCs and BCs > GCs. In mammals, a new amplification scheme evolved using cone BCs as the output stages, with rod BCs and glycinergic (gly) rod ACs as interneurons. Mammalian rod BCs are homologous to non-mammalian mixed rod–cone BCs, but have lost both cone inputs and the ability to target GC dendrites.

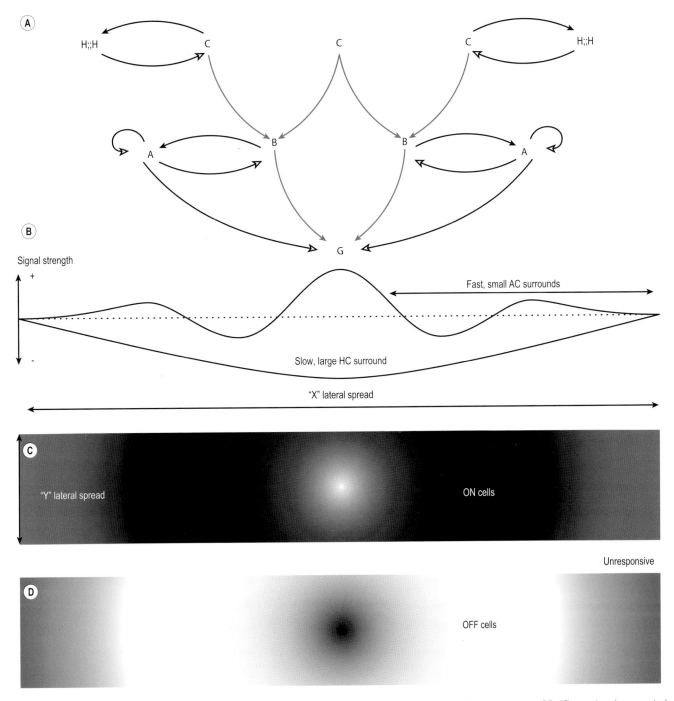

Figure 21.5 The synaptic flow that forms GC receptive fields. Cone signals (C) converge on BCs (B) which then converge on GCs (G), creating the canonical receptive field center, represented by a peak in the signal strength form the GC. The coupled HC layer (H:H) forms the large, slow antagonistic surround, while narrower ACs (A) form fast, small surrounds with damp oscillatory wings. The HCs dominate sustained signaling, so typical receptive field maps of the light required to excite cells represent BC+HC contributions. For ON cells, a spot of bright light will excite, while flanking regions of darkness will excite. For OFF cells, a spot of darkness will excite, while flanking regions of light excite. The red zones indicate regions outside the field where neither excites.

Nevertheless, six possible rod networks arising from three primary pathways (Fig. 21.6) exist in mammals, here grouped by amplification.

Three-stage amplification:

(1) rods >m ON rod BCs > gly rod ACs :: ON cone BCs > ON GCs

(2) rods >m ON rod BCs > gly rod ACs >i OFF cone BCs > OFF GCs.

Two-stage amplification:

(3) rods :: cones >m ON cone BCs > ON GCs

(4) rods :: cones > OFF cone BCs > OFF GCs

(5) rods > OFF cone BCs > OFF GCs (sparse and species variable)

(6) rods >m ON cone BCs[84] > ON GCs (sparse).

Thus, rod vision is parsed into ranges served by different networks: (i) the gly rod AC network with two arms of

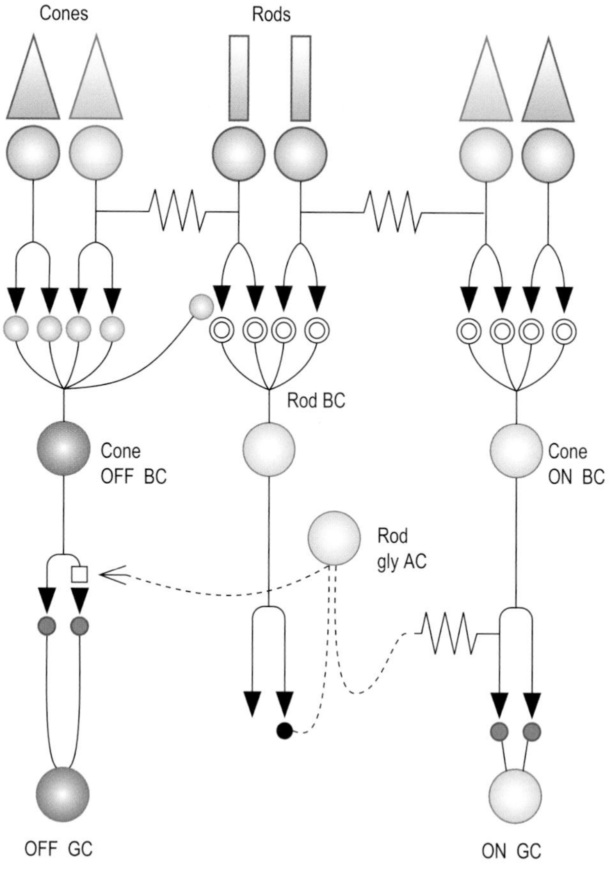

Figure 21.6 The synaptology of the mammalian scotopic network. Rod signals predominantly reach GCs by three pathways. The main dark-adapted pathway flows from rods → rod BCs → glycinergic rod ACs which then redistributes the signal back into the cone BC channels via gap junctions (ON BCs) or glycinergic synapses (OFF BCs), and thence to GCs. The second, less-sensitive path is via rod:cone gap junctions. A rarer path, not found in all mammals, is the occasional sampling of rod signals by OFF BCs. (Symbols as in Fig. 21.1.)

three-stage amplification for threshold scotopic vision and (ii) the rod::cone → cone BC → GC two-stage amplification for high brightness (moonlit) scotopic vision. Additional rod > cone BC contacts have been shown in some mammals,[38,39] but whether these additional pathways are structural errors in evolution or functional is not certain, as their incidences vary across mammalian species.[84]

The rod circuit is all the more complex for the involvement of GABAergic ACs, also known as S1 and S2 classes.[11] These γ rod ACs have dendritic arbors 1 mm in diameter and contact over 1000 rod BCs with reciprocal feedback synapses, with S2 cells providing twice the number of feedback synapses as S1.[85] This feedback likely further speeds the initially sluggish rod threshold response.

The synaptology of motion – AC surrounds from afar

While the roles of ACs in forming the center-surround features of sustained GCs are cryptic, their primacy in encoding motion is established. Directionally selective (DS) GCs respond to targets moving in a preferred direction, but

remaining silent when targets move in the opposite, "null" direction.[86–88] DS GCs come in two classes: ON-OFF and ON GCs. These networks engage BCs and perhaps several different AC inputs, including the ON and OFF subtypes of[88] starburst GABAergic/cholinergic ACs and other GABAergic ACs.[89,90] OFF starburst ACs hyperpolarize to light and are driven by OFF cone BCs. ON starburst cells have somas displaced to the GCL, depolarize to light and are driven by ON cone BC inputs. Each class stratifies with and synapses on the dendrites of DS GCs. The precise classes of other γ ACs in DS GC networks are not known, but the functional roles of GABAergic inhibition are emerging. At least one GABAergic AC inhibits the starburst ACs, and others inhibit DS GCs. Thus, as stimuli come from the preferred side, a combination of excitatory glutamatergic BC and cholinergic starburst AC signals converge on the GC in advance of GABAergic inhibition. The excitatory gain is likely enhanced by the BC > starburst AC > GC chain, which should have greater gain than a direct BC > GC transfer. In the null direction, a strong GABA signal reaches the DS GC in advance of the excitatory input and prevents it from reaching spike threshold (Fig. 21.7). GABAA receptor antagonists block this strong inhibition and convert DS GCs into non-directional cells.[86] The inhibition seems so strong (almost like veto synapses in cerebellum) that the BC > starburst AC > GC circuit can't break through. In fact, that may be the raison d'etre for starburst ACs: to break through any residual inhibition in the preferred direction as GABA inhibition is strong *even in the preferred direction*. This is likely an archetype for all AC circuits, where spatial properties, timing, and convergence of multiple cell classes select for fine grain features such as edges, texture, or flicker.

The synaptology of color – HC surrounds again?

Humans and old-world primates have cone mosaics with sparse blue (B, SWS1) cone arrays[91] surrounded by randomly distributed red cones and green cones (R, LWSR; G, LWSG). Most mammals possess dichromatic vision via B and G cone opponencies. Complete trichromatic vision has two opponent processes:[92] (1) blue/yellow (B/Y) opponency (where the Y signal is the sum of R and G cones signals);[93,94] and red/green (R/G) opponency. Both pose conceptual problems. The R and G pigment genes are tandem head-to-tail LSW arrays on the X chromosome.[21,92] LSW cones can express only one pigment, either LWSR or LWSG, creating either R or G cones[95,96] and this may be the *only* gene product that discriminates R and G cones. This suggests that the connectivity of R and G systems is probabilistic. Even so, R/G opponency is robust in trichromatic primates.

R/G opponency

In the foveola, each midget BC contacts only one cone and each midget GC contacts only one midget BC. Thus four types of center/surround R/G color opponency emerge (Fig. 21.8).[97] If the behavior of a single foveal R cone is not confounded by R::G coupling, a midget BC > midget GC chain should manifest a pure R or G center. Thus all midget GCs will be color opponent (Fig. 21.8) since their surrounds, whether derived from HCs or ACs, should be "yellower" than both: always greener than R cones or redder than G

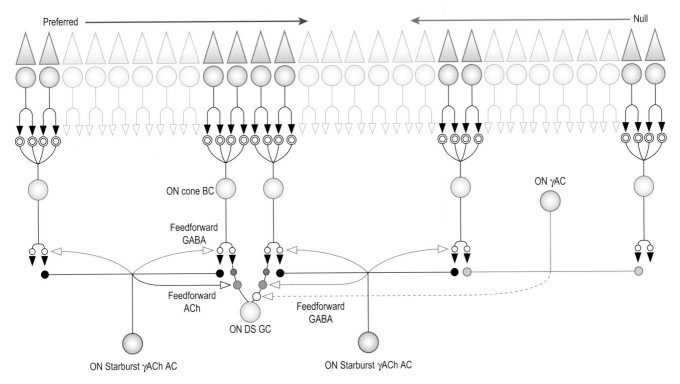

Figure 21.7 One possible synaptology for the mammalian ON-center DS GC network. ON DS GCs collect glutamatergic excitatory signals from ON cone BCs and cholinergic excitatory signals from ON starburst ACs. This amplifies the center response. ON starburst ACs also provide GABAergic feedback onto BCs. However, additional GABAergic inputs exist in the DS strata of the IPL. Such cells may receive inputs in the surround on one side and send axons (dotted) to target distant DS GC dendrites. Thus stimuli approaching form the left will excite and those from the right will inhibit. (Symbols as in Fig. 21.1.)

cones. HCs do not show any spectral selectivity for R or G cones and sum their inputs.[98] It was once posited that the GC surrounds were pure (pure R versus pure G) via selective contact of opponent BCs by ACs, but electron microscopy shows this is not so;[99] that the AC driven surrounds of midget GCs encode mixtures of R and G cones. Even so, some midget GCs show nearly pure opponent surrounds,[100–102] perhaps because of the patchiness of R and G cone distributions[103] and the small size of midget GC surrounds. Thus there is much more to be understood about midget GC networks. For example, why don't broad yellow-sensitive (Y) HC surrounds dominate midget BCs, as they do for B/Y opponent GCs?

B/Y opponency

GCs that convey blue signals are thought to be of two varieties: large and small bistratified B+/Y- GCs[93,97] receive B cone ON BCs synapses in sublamina *b* and Y inputs from diffuse OFF BCs in sublamina *a*. Though it was thought that B/Y centers and surrounds overlapped tremendously, recent data suggest that the Y surround comes from HCs via HC >i cone feedback and is much larger.[53] Recent anatomical evidence suggests the existence of a midget B cone OFF BC pathway in monkey[104] and a diffuse B cone OFF BC in rabbit.[105] Further, melanopsin GCs are putative B-/Y+ GCs with large receptive fields.[106] Which cells carry "the" blue signal remains uncertain as human patients lacking the mGluR6 receptor (and thus lacking ON BC signaling) apparently have quite excellent photopic sensitivity and apparently no color deficits.

Revising the retinal synaptic networks with disease

It was once thought that retinal networks were laid down once and for all in development by a process independent of sensory experience, but that is clearly incorrect. No less correct is the idea that, during the process of photoreceptor deconstruction and death, the neural retina remains normal. Many studies show that retina behaves much like CNS in response to challenges such as oxidative stress, denervation and trauma by remodeling its synaptic connectivity and reprogramming neural signaling rules. For example, the loss of photoreceptors in retinitis pigmentosa leads to the retraction of BC dendrites and the evolution of new axon-like structures; the generation of abundant new processes from retinal neurons of all kinds; the formation of new synaptic zones in the form of microneuromas; the switch from mGluR6 to iGluR expression in former ON rod BCs; and the ultimate death of many neurons.[12,107,108] These changes challenge many strategies to restore vision by genetic, molecular, cellular, and bionic schemes. But beyond that, they demonstrate two very important concepts. First, synaptic communication is likely never static and that signaling mechanisms are stabilized by active mechanisms. Second, the rules used by any neuron to decide which glutamate (or other) receptors to express are not known. We do not understand which transcriptional regulators make the decision to choose mGluR6 initially, much less choose AMPA or KA receptors in response to reprogramming.

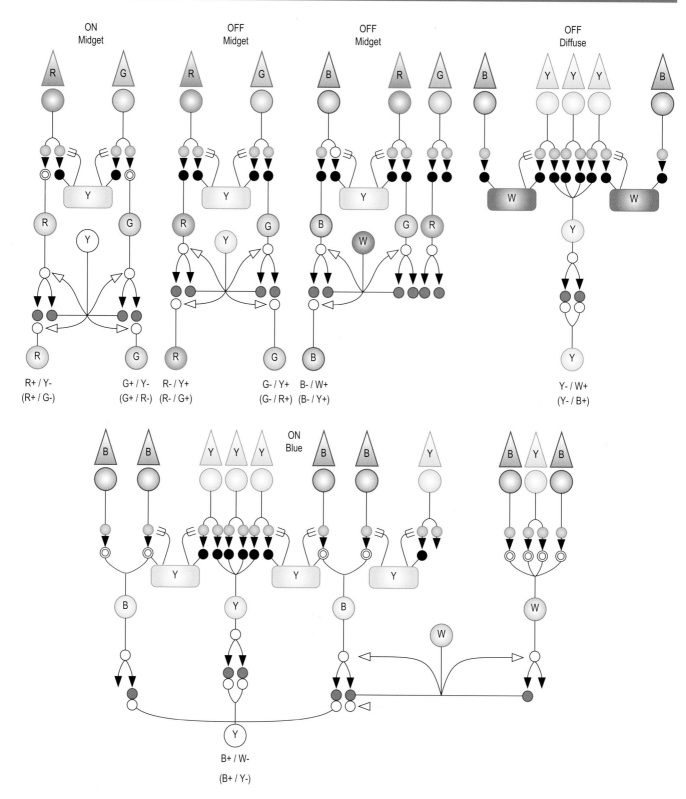

Figure 21.8 Possible color channel synaptologies for the mammalian (primate) retina. Midget pathways arise from midget BCs that contact only one cone and midget GCs that contact only one BC. This creates pure red (R) or green (G) center GCs in both ON and OFF channels. However the surround information from HCs and ACs collect from both R and G channels, generated "yellow" (Y) surrounds. GCs that are R/Y will generate net R/G percepts (shown in parenthesis), with similar outcomes for the three other possible R and G channels. OFF midget blue channels are likely antagonists by "yellow" H1 HCs and either Y or "white" (W) ACs that collect from all classes of cone BCs. GCs that are B/W will generate net B/Y percepts (shown in parenthesis), with similar outcomes for the three other possible R and G channels. It is also possible that diffuse BCs that contact all LWS cones (summarized as Y cones) may have surrounds from "white" H2 HCs (either by feedback as shown or feedforward) that generate Y-/W+ fields leading to Y-/B+ percepts. Finally, the best known B pathway involves selective for SWS1 B cones. The GCs that collect these signals are small bistratified cells with ON B BC inputs in sublamina b and Y selective OFF BC inputs in sublamina a. There are at least three possible -Y channels: H1 HC feedforward to B cones, direct OFF BC inputs from LWS-selective BCs, and W opponent inputs from non-selective ACs. The inability of GABA antagonists to block Y opponency in these cells suggest that the HC network dominates. (Symbols as in Fig. 21.1.)

References

1. Marc RE. Retinal neurotransmitters. In: Chalupa LM, Werner J, editors. The visual neurosciences. Cambridge, MA: MIT Press, 2004:315–330.
2. Marc RE. Functional neuroanatomy of the retina. In: Albert D, Miller J, eds. Albert and Jakobiec's principles and practice of ophthalmology, 3rd edn. New York: Elsevier, 2008:1565–1592.
3. Marc RE, Jones BW. Molecular phenotyping of retinal ganglion cells. J Neurosci 2002; 22:413–427.
4. Anderson JR, Jones BW, Yang J-H et al. A computational framework for ultrastructural mapping of neural circuitry. PLoS Biol 2009; 7:e1000074.
5. Marc RE, Liu W. Fundamental GABAergic amacrine cell circuitries in the retina: nested feedback, concatenated inhibition, and axosomatic synapses. J Comp Neurol 2000; 425:560–582.
6. Kolb H, Famiglietti EV. Rod and cone pathways in the inner plexiform layer of cat retina. Science 1975; 186:47–49.
7. Deans MR, Volgyi B, Goodenough DA, Bloomfield SA, Paul DL. Connexin36 is essential for transmission of rod-mediated visual signals in the mammalian retina. Neuron 2002; 36:703–712.
8. Muller F, Wassle H, Voigt T. Pharmacological modulation of the rod pathway in the cat retina. J Neurophysiol 1988; 59:1657–1672.
9. Zhang J, Li W, Trexler EB, Massey SC. Confocal analysis of reciprocal feedback at rod bipolar terminals in the rabbit retina. J Neurosci 2002; 22:10871–10882.
10. Mills SL, O'Brien JJ, Li W, O'Brien J, Massey SC. Rod pathways in the mammalian retina use connexin 36. J Comp Neurol 2001; 436:336–350.
11. Vaney DI. Morphological identification of serotonin-accumulating neurons in the living retina. Science 1986; 233:444–446.
12. Marc RE, Jones BW, Watt CB, Strettoi E. Neural remodeling in retinal degeneration. Progress in Retinal and Eye Research 2003; 22:607–655.
13. Tian N. Visual experience and maturation of retinal synaptic pathways. Vision Res 2004; 44:3307–3316.
14. Tian N, Copenhagen D. Plasticity of retinal circuitry. In: Masland RH, Albright T, editors. The senses: a comprehensive reference – vision. Amsterdam: Elsevier, 2008:473–490.
15. MacNeil MA, Heussy JK, Dacheux RF, Raviola E, Masland RH. The shapes and numbers of amacrine cells: matching of photofilled with Golgi-stained cells in the rabbit retina and comparison with other mammalian species. J Comp Neurol 1999; 413:305–326.
16. MacNeil MA et al. The population of bipolar cells in the rabbit retina. J Comp Neurol 2004; 472:73–86.
17. Rockhill RL, Daly FJ, MacNeil MA, Brown SP, Masland RH. The diversity of ganglion cells in a mammalian retina. J Neurosci 2002; 22:3831–3843.
18. Marc RE, Cameron DA. A molecular phenotype atlas of the zebrafish retina. J Neurocytol 2002; 30:593–654.
19. Morgan JL, Dhingra A, Vardi N, Wong RO. Axons and dendrites originate from neuroepithelial-like processes of retinal bipolar cells. Nat Neurosci 2006; 9:85–92.
20. Wässle H, Puller C, Muller F, Haverkamp S. Cone contacts, mosaics, and territories of bipolar cells in the mouse retina. J Neurosci 2009; 29:106–117.
21. Nathans J. The evolution and physiology of human color vision: insights from molecular genetic studies of visual pigments Neuron 1999; 24:299–312.
22. Loew ER, Govardovskii VI. Photoreceptors and visual pigments in the red-eared turtle, Trachemys scripta elegans. Vis Neurosci 2001; 18:753–757.
23. Chiu M, Nathans J. Blue cones and cone bipolar cells share transcriptional specificity as determined by expression of human blue visual pigment-derived transgenes. J Neurosci 1994; 14:3426–3436.
24. Kalloniatis M, Marc RE, Murry RF. Amino acid signatures in the primate retina. J Neurosci 1996; 16:6807–6829.
25. Famiglietti EV. Polyaxonal amacrine cells of rabbit retina: size and distribution of PA1 cells. J Comp Neurol 1991; 316:406–421.
26. Famiglietti EV. Polyaxonal amacrine cells of rabbit retina: PA2, PA3, and PA4 cells. Light and electron microscopic studies with a functional interpretation. J Comp Neurol 1992; 316:422–446.
27. Völgyi B, Xin D, Amarillo Y, Bloomfield SA. Morphology and physiology of the polyaxonal amacrine cells in the rabbit retina. J Comp Neurol 2001; 440:109–125.
28. Dacey DM. The dopaminergic amacrine cell. J Comp Neurol 1990; 301:461–489.
29. Marc RE, Jones BW, Pandit P, Anderson JR, Raleigh TM. Excitatory drive patterns of TH1 dopaminergic polyaxonal cells in rabbit retina. Invest Ophthalmol Vis Sci 2008; 49:24–32.
30. Descarries L, Bérubé-Carrière N, Riada M et al. Glutamate in dopamine neurons: Synaptic versus diffuse transmission. Brain Res Rev 2007; 58:290–302.
31. Marc RE. Functional anatomy of the retina. In: Jaeger E, ed. Duane's foundations of clinical ophthalmology, 2009.
32. Marc RE, Liu WL. Horizontal cell synapses onto glycine-accumulating interplexiform cells. Nature 1984; 312:266–269.
33. Perlman I, Kolb H, Nelson R. Anatomy, circuitry, and physiology of vertebrate horizontal cells. In: Chalupa LM, Werner J, eds. The visual neurosciences. Cambridge, MA: MIT Press, 2004:369–394.
34. Katz B, Miledi R. A study of synaptic transmission in the absence of nerve impulses. J Physiol 1967; 192:407–436.
35. Matthews G. Synaptic mechanisms of bipolar cell terminals. Vision Res 1999; 39:2469–2476.
36. Migdale K, Herr S, Klug K et al. Two ribbon synaptic units in rod photoreceptors of macaque, human, and cat. J Comp Neurol 2003; 455:100–112.
37. Ishida A, Stell W, Lightfoot D. Rod and cone inputs to bipolar cells in goldfish retina. J Comp Neurol 1980; 191:315–335.
38. Tsukamoto Y, Morigiwa K, Ishii M et al. A novel connection between rods and ON cone bipolar cells revealed by ectopic metabotropic glutamate receptor 7 (mGluR7) in mGluR6-deficient mouse retinas. J Neurosci 2007; 27:6261–6267.
39. Li W, Keung JW, Massey SC. Direct synaptic connections between rods and OFF cone bipolar cells in the rabbit retina. J Comp Neurol 2004; 474:1–12.
40. Protti DA, Flores-Herr N, Li W, Massey SC, Wässle H. Light signaling in scotopic conditions in the rabbit, mouse and rat retina: A physiological and anatomical Study. J Neurophysiol 2005; 93:3479–3488.
41. Stell WK, Lightfoot DO, Wheeler TG, Leeper HL. Goldfish retina functional polarization of cone horizontal cell dendrites and synapses. Science (Washington DC) 1975; 190:989–990.
42. Wässle H. Decomposing a cone's output (parallel processing). In: Masland RH, Albright T, eds. The senses: a comprehensive reference – vision. Amsterdam: Elsevier, 2008.
43. Dowling JE, Boycott BB. Organization of the primate retina: electron microscopy. Proc R Soc Lond B Biol Sci 1966; 166:80–111.
44. Brecha NC. Peptide and peptide receptor expression in function in the vertebrate retina. In: Chalupa LM, Werner J, eds. The visual neurosciences. Cambridge, MA: MIT Press, 2004:334–354.
45. Witkovsky P. Dopamine and retinal function. Doc Ophthalmol 2004; 108:17–39.
46. Witkovsky P, Nicholson C, Rice ME, Bohmaker K, Meller E. Extracellular dopamine concentration in the retina of the clawed frog, Xenopus laevis. Proc Natl Acad Sci USA 1993; 90:5667–5671.
47. Gastinger MJ, Tian N, Horvath T, Marshak DW. Retinopetal axons in mammals: emphasis on histamine and serotonin. Curr Eye Res 2006; 31:655–667.
48. Naka K-I, Witkovsky P. Dogfish ganglion cell discharge resulting from extrinsic polarization of the horizontal cells. J Physiol 1972; 223:449–460.
49. Naka K. Functional organization of catfish retina. J Neurophysiol 1977; 40:26–43.
50. Baylor DA, Fuortes MGF, O'Bryan PM. Receptive fields of cones in the retina of the turtle. J Physiol 1971; 214:265–294.
51. Vardi N, Zhang L-L, Payne JA, Sterling P. Evidence that different cation chloride cotransporters in retinal neurons allow opposite responses to GABA. J Neurosci 2000; 20:7657–7663.
52. McMahon MJ, Packer OS, Dacey DM. The classical receptive field surround of primate parasol ganglion cells is mediated primarily by a non-GABAergic pathway. J Neurosci 2004; 24:3736–3745.
53. Field GD, Sher A, Gauthier JL et al. Spatial properties and functional organization of small bistratified ganglion cells in primate retina. J Neurosci 2007; 27:13261–13272.
54. Thoreson WB, Babai N, Bartoletti TM. Feedback from horizontal cells to rod photoreceptors in vertebrate retina. J Neurosci 2008; 28:5691–5695.
55. Davenport CM, Detwiler PB, Dacey DM. Effects of pH buffering on horizontal and ganglion cell light responses in primate retina: Evidence for the proton hypothesis of surround formation. J Neurosci 2008; 28:456–464.
56. Kamermans M, Fahrenfort I. Ephaptic interactions within a chemical synapse: hemichannel-mediated ephaptic inhibition in the retina. Cur Opin Neurobiol 2004; 14:531–541.
57. Vaney DI, Weiler R. Gap junctions in the eye: evidence for heteromeric, heterotypic and mixed-homotypic interactions. Brain Res – Brain Res Revs 2000; 32:115–120.
58. Massey SC. Circuit functions of gap junctions in the mammalian retina. In: Masland RH, Albright T, eds. The senses: a comprehensive reference – vision. Amsterdam: Elsevier, 2008:457–471.
59. Hampson ECGM, Weiler R, Vaney DI. pH-Gated dopaminergic modulation of horizontal cell gap junctions in mammalian retina. Proc Roy Soc B: Biol Sci 1994; 255:67–72.
60. Mills SL, Massey SC. Differential properties of two gap junctional pathways made by AII amacrine cells. Nature 1995; 377:734–737.
61. Krizaj D, Gabriel R, Owen WG, Witkovsky P. Dopamine D2 receptor-mediated modulation of rod-cone coupling in the Xenopus retina. J Comp Neurol 1998; 398:529.
62. Gustafson EC, Stevens ER, Wolosker H, Miller RF. Endogenous D-serine contributes to NMDA-receptor-mediated light-evoked responses in the vertebrate retina. J Neurophysiol 2007; 98:122–130.
63. Hanna MC, Calkins DJ. Expression of genes encoding glutamate receptors and transporters in rod and cone bipolar cells of the primate retina determined by single-cell polymerase chain reaction. Mol Vis 2007; 13:2194–2208.
64. Marc RE. Mapping glutamatergic drive in the vertebrate retina with a channel-permeant organic cation. J Comp Neurol 1999; 407:47–64.
65. Dhingra A, Lyubarsky A, Jiang M et al. The light response of ON bipolar neurons requires Gao. J Neurosci 2000; 20:9053–9058.
66. Quraishi S, Gayet J, Morgans CW, Duvoisin R. Distribution of group-III metabotropic glutamate receptors in the retina. J Comp Neurol 2007; 501:931–943.
67. Zhang D-Q, Zhou T-R, McMahon DG. Functional heterogeneity of retinal dopaminergic neurons underlying their multiple roles in vision. J Neurosci 2007; 27:692–699.
68. Piccolino M, Neyton J, Gerschenfeld HM. Decrease of gap junction permeability induced by dopamine and cyclic adenosine 3′:5′ monophosphate in horizontal cells of turtle retina. J Neurosci 1984; 4:2477.
69. Vaquero CF, Pignatelli A, Partida GJ, Ishida AT. A dopamine- and protein kinase A-dependent mechanism for network adaptation in retinal ganglion cells. J Neurosci 2001; 21:8624–8635.
70. Eldred WD, Blute TA. Imaging of nitric oxide in the retina. Vision Res 2005; 45:3469–3486.
71. Vaney DI. Retinal amacrine cells. In: Chalupa LM, Werner J, eds. The visual neurosciences. Cambridge, MA: MIT Press, 2004:395–409.
72. Slaughter MM, Miller RF. 2-amino-4-phosphonobutyric acid: a new pharmacological tool for retina research. Science 1981; 211:182–185.
73. Grant GB, Dowling JE. A glutamate-activated chloride current in cone-driven ON bipolar cells of the white perch retina. J Neurosci 1995; 15:3852–3862.
74. Grant GB, Dowling JE. On bipolar cell responses in the teleost retina are generated by two distinct mechanisms. J Neurophysiol 1996; 76:3842–3849.
75. Marc RE. Kainate activation of horizontal, bipolar, amacrine, and ganglion cells in the rabbit retina. J Comp Neurol 1999; 407:65–76.
76. Slaughter MM. Inhibition in the retina. In: Chalupa LM, Werner J, eds. The visual neurosciences. Cambridge, MA: MIT Press, 2004:355–368.
77. Zhang J, Jung CS, Slaughter MM. Serial inhibitory synapses in retina. Vis Neurosci 1997; 14:553–563.
78. Hsueh HA, Molnar A, Werblin FS. Amacrine-to-amacrine cell inhibition in the rabbit retina. J Neurophysiol 2008; 100:2077–2088.

79. Marmarelis PZ, Marmarelis VZ. Analysis of physiological systems. New York: Plenum Press, 1978.

80. Marc RE. Structural organization of GABAergic circuitry in ectotherm retinas. Progr Brain Res 1992; 90:61–92.

81. Zhang A-J, Wu SM. Receptive fields of retinal bipolar cells are mediated by heterogeneous synaptic circuitry. J Neurosci 2009; 29:789–797.

82. Bloomfield SA, Xin D. Surround inhibition of mammalian AII amacrine cells is generated in the proximal retina. J Physiol Lond 2000; 15:771–783.

83. Wilson M, Vaney DI. Amacrine cells. In: Masland RH, Albright T, eds. The senses: a comprehensive reference – vision. Amsterdam: Elsevier, 2008:361–367.

84. Protti DA, Flores-Herr N, Li W, Massey SC, Wassle H. Light signaling in scotopic conditions in the rabbit, mouse and rat retina: A physiological and anatomical study. J Neurophysiol 2005; 93:3479–3488.

85. Zhang J, Li W, Trexler EB, Massey SC. Confocal analysis of reciprocal feedback at rod bipolar terminals in the rabbit retina. J Neurosci 2002; 22:10871–10882.

86. Wyatt HJ, Daw NW. Directionally sensitive ganglion cells in the rabbit retina: specificity for stimulus direction, size, and speed. J Neurophysiol 1975; 38:613–626.

87. Kittila CA, Massey SC. Pharmacology of directionally selective ganglion cells in the rabbit retina. J Neurophysiol 1997; 77:675–689.

88. Dacheux RF, Chimento MF, Amthor FR. Synaptic input to the on-off directionally selective ganglion cell in the rabbit retina. J Comp Neurol 2003; 456:267–278.

89. Famiglietti EVJ. Dendritic co-stratification of ON and ON-OFF directionally selective ganglion cells with starburst amacrine cells in rabbit retina. J Comp Neurol 1992; 324:322–335.

90. Grzywacz NM, Amthor FR, Merwine DK. Necessity of acetylcholine for retinal directionally selective responses to drifting gratings in rabbit. [see comments]. J Physiol 1998; 512:575–581.

91. Marc RE, Sperling HG. Chromatic organization of primate cones. Science 1977; 196:454–456.

92. Carroll J, Jacobs GH. Mammalian photopigments. In: Masland RH, Albright T, eds. The senses: a comprehensive reference – vision. Amsterdam: Elsevier, 2008:247–268.

93. Lee BB. Blue-ON Cells. In: Masland RH, Albright T, eds. The senses: a comprehensive reference – vision. Amsterdam: Elsevier, 2008:433–438.

94. Kaplan E. The P, M and K streams of the primate visual system: What do they do for vision? In: Masland RH, Albright T, eds. The senses: a comprehensive reference – vision. Amsterdam: Elsevier, 2008:369–381.

95. Wang Y, Smallwood PM, Cowan M et al. Mutually exclusive expression of human red and green visual pigment-reporter transgenes occurs at high frequency in murine cone photoreceptors. Proc Natl Acad Sci USA 1999; 96:5251–5256.

96. Smallwood PM, Wang Y, Nathans J. Role of a locus control region in the mutually exclusive expression of human red and green cone pigment genes. Proc Natl Acad Sci USA 2002; 99:1008–1011.

97. Dacey DM. Parallel pathways for spectral coding in primate retina. Annu Rev Neurosci 2000; 23:743–775.

98. Dacey DM, Lee BB, Stafford DK, Pokorny J, Smith VC. Horizontal cells of the primate retina: cone specificity without spectral opponency. Science 1996; 271:656–659.

99. Calkins DJ, Sterling P. Absence of spectrally specific lateral inputs to midget ganglion cells in primate retina. Nature 1996; 381:613–615.

100. Reid RC, Shapley RM. Spatial structure of cone inputs to receptive fields in primate lateral geniculate nucleus. Nature 1992; 356:716–718.

101. Reid RC, Shapley RM. Space and time maps of cone photoreceptor signals in macaque lateral geniculate nucleus. J Neurosci 2002; 22:6158–6175.

102. Sun H, Smithson HE, Zaidi Q, Lee BB. Specificity of cone inputs to macaque retinal ganglion cells. J Neurophysiol 2006; 95:837–849.

103. Hofer H, Carroll J, Neitz J, Neitz M, Williams DR. Organization of the human trichromatic cone mosaic. J Neurosci 2005; 25:9669–9679.

104. Klug K, Herr S, Ngo IT, Sterling P, Schein S. Macaque retina contains an S-cone OFF midget pathway. J Neurosci 2003; 23:9881–9887.

105. Liu P-C, Chiao C-C. Morphologic identification of the OFF-type blue cone bipolar cell in the rabbit retina. Invest Ophthalmol Vis Sci 2007; 48:3388–3395.

106. Dacey DM, Liao HW, Peterson BB et al. Melanopsin-expressing ganglion cells in primate retina signal colour and irradiance and project to the LGN. Nature 2005; 433:749–754.

107. Marc RE, Jones BW, Anderson JR et al. Neural reprogramming in retinal degenerations. Invest Ophthalmol Vis Sci 2007; 48:3364–3371.

108. Marc RE, Jones BW, Watt CB et al. Extreme retinal remodeling triggered by light damage: Implications for AMD. Molec Vis 2008; 14:782–806.

109. Zeitz C, Forster U, Neidhardt J et al. Night blindness-associated mutations in the ligand-binding, cysteine-rich, and intracellular domains of the metabotropic glutamate receptor 6 abolish protein trafficking. Hum Mutat 2007; 28:771–780.

110. Mauck MC, Salzwedel A, Kuchenbecker J et al. Probing neural circuitry for blue-yellow color vision using high resolution functional magnetic resonance imaging. Invest Ophthalmol Vis Sci 2008; 49:3251.

111. Dreyer EB, Zurakowski D, Schumer RA, Podos SM, Lipton SA. Elevated glutamate levels in the vitreous body of humans and monkeys with glaucoma. Arch Ophthalmol 1996; 114:299–305.

112. Seki M. Targeting excitotoxic/free radical signaling pathways for therapeutic intervention in glaucoma. Progr Brain Res 2008; 173:495–510.

CHAPTER 22

Signal Processing in
the Outer Retina

Samuel M. Wu

Light-evoked hyperpolarizing signals in rods and cones are transmitted to and processed by neurons in the rest of the retina through a complex, but highly organized network of electrical and chemical synapses.[14] In the outer retina, photoreceptors are electrically coupled to one another,[5] and rods and cones send output signals to second-order retinal cells, the horizontal cells (HCs) and bipolar cells (BCs), via chemical synapses. HCs make feedback synapses on cones, and send feedforward synaptic signals to BCs.[5] BCs can be classified into two major types, the on-center (or depolarizing bipolar cells, DBCs) and the off-center (or hyperpolarizing bipolar cells, HBCs).[78] Additionally, they can be further classified according to their relative rod/cone inputs, color opponency, axonal outputs and other parameters. BCs are the first neurons along the visual pathway that exhibit clear center-surround antagonistic receptive field (CSARF) organization, the basic code for spatial information processing in the visual system,[25] and the synaptic circuitry underlying bipolar cell receptive fields will be discussed in this chapter.

Electrical synapses (coupling) between photoreceptors

Electrical coupling between photoreceptors was first discovered in the turtle retina by Baylor et al in 1971.[5] Electron microscopic studies revealed that gap junctions exist between photoreceptors in many vertebrate species. In the primate retina, anatomical data have shown large gap junctions between cones and smaller junctions between rods and cones, suggesting that cone–cone coupling is stronger than rod–cone coupling.[59,60] In the turtle and ground squirrel retina, couplings are found between cones of the same spectral sensitivity and between rods;[5,9,11,63] whereas in amphibians, rods are strongly coupled to each other, but cones are not.[3,4] Coupling between rods and cones has been observed in turtles, tiger salamander and mammals, but the coupling strength is weaker than that between cones (in turtles) and between rods (in tiger salamander).[4,62] Recent evidence has shown that gap junctions between photoreceptors are mediated by connexin 35/36 gap junction proteins (Fig. 22.1)[10,53,94] and double patch recordings have revealed detailed properties of electrical synapses between photoreceptors. For

example, junctional conductance between salamander rods is linear, with an average value of 500 pS (Fig. 22.2A).[95] In ground-squirrel, green–green cone pairs are coupled with an average conductance of 220 pS, whereas coupling is undetectable in blue–green cone pairs (Fig. 22.2B).[39] In the monkey retina, coupling has been found between red-cones, between green-cones and between red- and green-cones (Fig. 22.2C).[22]

Although most rods are weakly coupled with cones in the tiger salamander retina, a small fraction of rods (10–15%) are strongly coupled with adjacent cones. Voltage responses of these rods (named rod$_C$s) to current injection into a next-neighbor cone are about three to four times larger than those of the other rods.[87] Rod$_C$s behave like hybrids of rods and cones, and one of their functions is to generate off overshoot responses in higher-order retinal neurons under light-adapted conditions.[82]

Photoreceptor coupling decreases resolution (visual acuity) of the visual system by spatially averaging photoreceptor signals over some lateral distance in the retina. However, the advantage of this arrangement is that it improves the signal-to-noise ratio of the photoreceptor output when the retina is uniformly illuminated. This is especially important for the rods to detect dim images under dark-adapted conditions when the voltage noise is high. It can be shown mathematically, however, that photoreceptor coupling decreases the signal-to-noise ratio of the photoreceptor network if the number of illuminated photoreceptors is less than the square root of the number of photoreceptors that are effectively coupled (Box 22.1).

Rod and cone photoreceptors operate in different ranges of illumination, and they exhibit different spectral sensitivities. Electrical coupling between rods and cones broadens the operating ranges and spectral sensitivity spans of both types of photoreceptor by mixing their light-evoked signals. Additionally, under conditions when the signal in one type of photoreceptor is suppressed (e.g. rod response in the presence of background light), its output synapse can be used to transmit signal from the other type of photoreceptor (e.g. cones). It has been shown that rod–cone coupling in the tiger salamander retina is enhanced by background light.[88] This allows transmission of cone signals through the rod output synapses. Such "synapse sharing" minimizes the

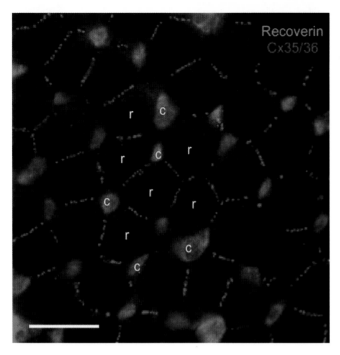

Figure 22.1 Connexin36 in the photoreceptor network. Confocal image of a salamander flatmount retina (with focal plane at the level of the distal region of rod cell bodies) double labeled with anti-Cx35/36 (red) and recoverin (green). Recoverin differentially labeled rods (r, weak green) and cone inner segments (c, strong green). The strong Cx35/36 punctate labeling on membrane contacts outlined the mosaic of the rod network in the field. Scale bar = 20 μm.

Box 22.1 Photoreceptor Coupling Alters the Signal-to-Noise Ratio of the Photoreceptor Network

If N photoreceptors are coupled and k photoreceptors are uniformly illuminated, then the average (sample) signal is $S = kS_1/N$ (where S_1 is the light-evoked signal in one photoreceptor and σ_1 is the noise, or the standard deviation of S_1 from the mean), and the average (sample) variance:

$$\sigma^2 = \text{var}(S) = \text{var}((1/N)(S_1 + S_2 + \ldots)) = (1/(N^2))\text{var}(S_1 + S_2 + \ldots)$$
$$= (1/(N^2))(\sigma_1^2 + \sigma_2^2 + \ldots) = ((\sigma_1^2)/N), \text{ and thus}$$
$$\sigma = ((\sigma_1)/(\sqrt{N}))$$

The average (sample) signal-to-noise ratio:

$$S/\sigma = (kS_1/N)/(\sigma_1/\sqrt{N}) = (k/\sqrt{N})(S_1/\sigma_1).$$

Therefore when $k > \sqrt{N}$, $S/\sigma > S_1/\sigma_1$, the average signal-to-noise ratio improves, and when $k < \sqrt{N}$, $S/\sigma < S_1/\sigma_1$, the average signal-to-noise ratio worsens.

---

amount of neural hardware and facilitates cone inputs in second-order retinal cells.[84]

Glutamatergic synapses between photoreceptors and second-order retinal neurons

Vertebrate photoreceptors use glutamate as their neurotransmitter, which is released continuously in darkness as the photoreceptors are depolarized.[8] Calcium enters photoreceptor synaptic terminals through voltage-gated calcium channels and triggers exocytosis of glutamatergic vesicles,[73] releasing packages of glutamate, each of these packages activates a number of glutamate receptors that initiate transient postsynaptic currents termed spontaneous excitatory postsynaptic currents, or sEPSCs.[41,55] sEPSCs are seen in most HBCs, but analogous (sign inverted) discrete postsynaptic currents are generally not observed in the DBCs.[57,85] This probably reflects a difference in the kinetics of the glutamate receptors in DBCs and HBCs.[2] Postsynaptic responses of the HBCs and horizontal cells (HCs) are mediated primarily by ionotropic glutamate receptors (kainate (KA) or α-amino-3-hydroxy-5-methylisoxazole-4-propionic acid (AMPA) receptors),[13,40,45,55] and the responses of the DBCs are mediated mainly by metabotropic (mGluR6, or L-α-amino-4-phosphonobutyrate (L-AP4)-sensitive) receptors (Fig. 22.3).[50,51,64,66] Glutamate released from photoreceptors in darkness binds to the KA/AMPA receptors in HBCs and HCs and opens postsynaptic cation channels. It also binds to the mGluR6 receptors in DBCs, resulting in a G protein-mediated process, which closes cation channels. Light suppresses glutamate release from photoreceptors, closes AMPA/KA-mediated cation channels and results in hyperpolarization in HBCs and HCs, and it opens mGluR-mediated cation channels and results in depolarization in DBCs. In fish, cone transmission to DBCs has been shown to be mediated by a glutamate-activated chloride current (TBOA-sensitive) that is shut down in the light, resulting in membrane depolarization.[18,80]

It has been shown that the AMPA/KA receptors in HBCs are largely desensitized in darkness, leading to small postsynaptic currents and reduced response dynamic ranges.[13] In the salamander retina, rod-dominated HBCs have been found to evade postsynaptic receptor desensitization by using synchronized multiquantal release at multiple invaginating ribbon junctions and GluR-4 AMPA receptors as the postsynaptic receptors.[55] The large multiquantal events have fast decay time so that self-desensitization is avoided and they allow long intervals between events so that mutual desensitization is minimized. Consequently HBC_Rs are not desensitized in darkness, allowing light signals to be encoded by the full operating range of the glutamate-gated postsynaptic currents.

The exact gating mechanism of the mGluR6 receptor-mediated cation channels is unclear.[67] The mGluR6 receptors generate substantially slower conductance changes compared with the AMPA/KA receptors in HBCs, and thus the sEPSC events in DBCs are heavily filtered.

Detailed descriptions of ionotropic and metabotropic glutamate receptors and glutamate transporters in the retina are also given in Chapter 21.

Horizontal cell responses

Intracellular recordings from the horizontal cells were first made by Svaetichin in the bream retina,[72] and thus the light responses of these cells were termed S-potentials. In the fish retina, two major types of horizontal cells which receive cone inputs may be distinguished: the L-type (luminosity), which hyperpolarizes to light stimuli of all wavelengths; and the C-type (chromaticity or color), which hyperpolarizes to

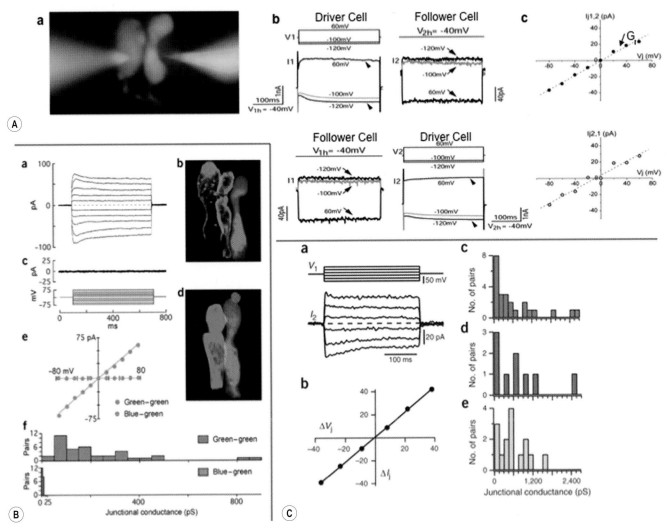

Figure 22.2 Rod–rod coupling in the tiger salamander retina (A), cone–cone coupling in the ground squirrel retina (B), and cone–cone coupling in the monkey retina (C). (A) (a) A pair of rods simultaneously patch clamped and filled with Lucifer yellow through two recoding pipettes. (b) Simultaneous dual whole-cell voltage clamp recordings from a pair of next-neighboring rods. The membrane potential of two rods was held at −40 mV. Upper panel: when a series of voltage step commands (V1) (from −120mV to 60mV with an increment of 20mV) were applied to cell 1 (driver cell), the voltage activated current responses (I1) (arrowheads) were recorded in cell 1 (left panel) and the junctional currents of the opposite polarity (I2) (arrows) were recorded in cell 2 (follower cell, right panel). Lower panel: switching the position of driver/follower cells. (c) Relations of transjunctional current (Ij) and transjunctional voltage (Vj) obtained in upper and lower panels in b. The junctional conductance (Gj) measured in either direction is 500 pS. (From Zhang & Wu 2005.[95]) **(B)** (a) Junctional currents from a pair of green cones during a series of steps in transjunctional voltage (lower traces in c). (b) The green cones in (a) were labeled with Alexa Fluor 568 (red) and 488 (green). A blue cone (blue) is adjacent to the green cone pair. (c) Junctional currents (upper traces) from a pair of blue and green cones during a series of transjunctional voltage steps (lower traces). (d) The blue cone in (c) was labeled with Alexa Fluor 568 (red) and the green cone was labeled with Alexa Fluor 488 (green) (e) Plot of the steady transjunctional current versus transjunctional voltage for the green-green pair (green) and blue-green pair (blue). (f) Histogram of junctional conductance for 38 green-green and 20 blue-green cone pairs. Bin widths, 50 pS for green-green pairs; 5 pS, blue-green pairs. (From Li & DeVries 2004.[39]) **(C)** (a) Change in membrane current in a red cone (I_2) in response to change in membrane potential (V_1) in an adjacent green cone. The red cone was voltage clamped to −55 mV and the green cone voltage was stepped from an initial value of −55 mV. (b) The slope of the relationship between the transjunctional current I_j and the transjunctional voltage V_j gives a coupling conductance of 1,086 pS. (c–e) Histograms of cone junctional conductances for 24 green-green pairs (c), 9 red-red pairs (d) and 15 red-green pairs (e). (From Hornstein et al 2004.[22]).

light stimuli within certain ranges of wavelength but depolarizes to light in other wavelength ranges. The C-type horizontal cells may be divided into the biphasic and the triphasic cells. Two kinds of biphasic C-cells have been reported: the green/blue cells, which depolarize to red and green flashes and hyperpolarize to blue light; and the red/green cells which depolarize to red light and hyperpolarize to green and blue lights. The triphasic C-cells, on the other hand, hyperpolarize to both blue and red lights and depolarize to green light.[48] Horizontal cells which receive input only

from rods are found in many species. The light responses of these cells differ from those of the cone horizontal cells in two ways: Rod S-potentials in the carp are found to have peak spectral sensitivity around 520 nm while the L-type cone S-potentials peak around 600 nm. The sensitivity to light is much higher in rod S-potentials than the L-type S-potentials.[34] The receptive fields of horizontal cells are much larger than those of the photoreceptors. Naka & Rushton showed that the amplitude of the L-type horizontal cell response continued to increase up to a light spot

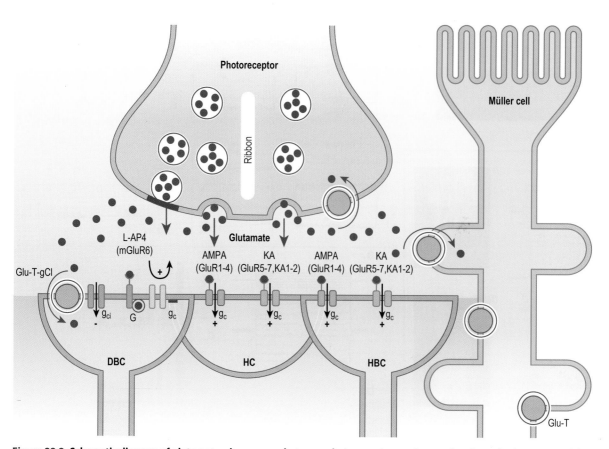

Figure 22.3 Schematic diagram of glutamatergic synapses between photoreceptors and second-order retinal neurons. HC: horizontal cell, DBC: depolarizing bipolar cell, HBC: hyperpolarizing, bipolar cell, Glu-T: glutamate transporters, mGluR6: metabotropic glutamate receptor 6, AMPA: α-amino-3-hydroxy-5-methylisoxazole-4-propionic acid receptor, KA: kainate receptor, L-AP4: L-α-amino-4-phosphonobutyrate. g_c: cation channel, g_{cl}: chloride channel, G: G-protein, Glu-T-g_{cl}: glutamate transporter coupled chloride channel.

diameter of 1.2 mm.[49] Very similar results have been obtained in the amphibian and cat horizontal cells.[69,93]

Horizontal cells in the vertebrate retina are electrically coupled to one another. In fish retina, the electrical (gap) junctions between HCs are very extensive. Gap junctions are observed between HC perikarya or between HC processes, but the couplings are only made between HCs of homologous types. For example, the H1 cone HCs in the fish couple only with other H1 cells, H2s couple only with other H2 cells, and HC axon terminals couple only with other HC axon terminals.[31,79] In the salamander retina and most mammals, there are two types of HC somas, one not bearing axons (axonless, A-type) and the other has axons (B-type).[16,38] B-type axon terminals are functionally isolated from the B-type somas and form their own network. All three types of HCs (A-type soma, B-type soma and B-type axon terminals) are electrically coupled with their own kinds. A-type somas are weakly coupled and thus they have narrow receptive fields, whereas B-type soma and B-type axon terminals are strongly coupled (with their own type) and thus they have wide receptive fields (Fig. 22.4).[93]

Horizontal cell output synapses

Horizontal cells mediate bipolar cell surround responses through a feedback synaptic pathway (HC → cone → bipolar cell) and through feedforward synapses (HC → bipolar cell). The HC → cone sign-inverting feedback synapse was discovered by Baylor et al (the same paper that discovered photoreceptor coupling),[5] and the HC → bipolar cell feedforward synapses were suggested by anatomical evidence.[15,37] Injection of hyperpolarizing current into horizontal cells elicits depolarizing responses in cones (Fig. 22.5A–D)[5] and HBCs and hyperpolarizing responses in DBCs,[44,47] consistent with the notion that horizontal cells drive the surround responses in these cells. Figure 22.5E shows the depolarizing surround responses in a cone, mediated by the sign-inverting feedback synapse from horizontal cells to the cone.[65] Three synaptic mechanisms have been proposed for the HC feedback actions on cones. The first is that HCs release an inhibitory neurotransmitter (GABA in several species) in darkness that opens chloride channels in cones, and surround light hyperpolarizes the HCs, suppresses feedback transmitter release, depolarizes the cones, and results in depolarization in HBCs and hyperpolarization in DBCs.[46,83]

The second mechanism is that surround light hyperpolarizes HCs, resulting in an outward current through hemichannels in their dendrites near the cones, charging the cone membrane and modulating calcium currents in cones, thereby increasing their calcium-dependent glutamate release which depolarizes the HBCs and hyperpolarizes the DBCs.[28,29]

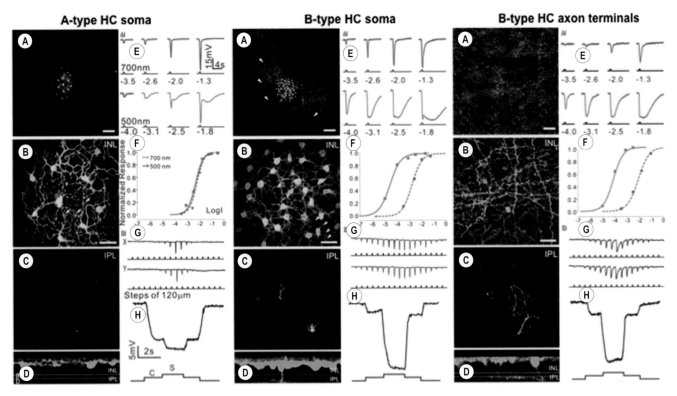

A-type HC soma   **B-type HC soma**   **B-type HC axon terminals**

Figure 22.4 Morphology, dye coupling, light responses and receptive field properties of three types of horizontal cells. Confocal fluorescent micrographs (**A**: low magnification at distal INL, **B**: high magnification at distal INL, **C**: at IPL and **D**: z-axis rotation) stained by injection of Lucifer yellow (red) and neurobiotin (green), voltage responses to 700 nm and 500 nm lights of various intensities (**E**), the response-intensity relations (**F**), voltage responses to a stepwise (120-μm steps) moving light bar (50 μm wide) in the x- and y-directions (**G**), and voltage responses to center and surround light stimuli (**H**) of A-type horizontal cell somas (left panels), B type horizontal cell somas (middle panels) and B type horizontal cell axon terminals (right panels). Scale bar = 200 μm (A), 50 μm (B). The spectral difference, ΔS (see definition in Fig. 22.6 legend), was determined by the separation of the two V-Log I curves for a criterion response of 5 mV. (From Zhang et al 2006.[93])

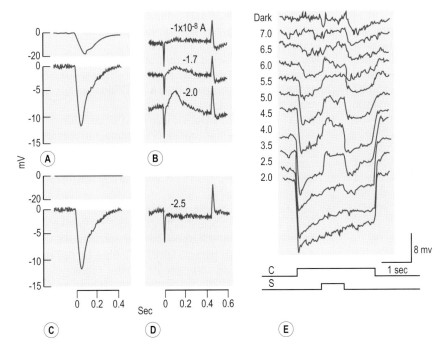

Figure 22.5 Feedback synapse from horizontal cells to cones. (**A**) Voltage responses of a horizontal cell (H) and a cone (C) to a light flash (600 μm radius) in the turtle retina. (**B**) of hyperpolarizing current (−1 to −2 × 10⁸A) in the horizontal cell produced graded depolarization in the cone. (**C**) light responses after withdrawal of the electrode from the horizontal cell. (**D**) No depolarization was observed in the cone when hyperpolarizing current was passed through the extracellular electrode (withdrawn from the horizontal cell). (**E**) Voltage responses of a cone in the tiger salamander retina to a center light spot (300 μm diameter) of various intensities (numbers on the left indicate log units of light attenuation) and a light annulus (650 μm inner diameter, 2.5 log unit attenuation). (A–D from Baylor et al 1971;[5] E from Skrzypek & Werblin 1983.[65])

The third mechanism is that surround-induced HC hyperpolarization elevates the pH in the HC-cone synaptic cleft, leading to an increase of calcium current in cones and a higher rate of glutamate release which depolarizes the HBCs and hyperpolarizes the DBCs.[21,75] It is possible that different species under different conditions favor different feedback synaptic mechanisms, and/or different types of HC-cone synapses in the same animal may use one or more of these three HC-cone feedback mechanisms. This idea is supported by a recent study demonstrating that the responses of salamander GCs to dim surround stimuli are sensitive to GABA blockers and those to bright surround stimuli are sensitive to carbenoxolone, a gap junction/hemichannel blocker.[27]

Properties of the feedforward synapses are less well understood. Results from salamander retina suggest that HC → DBC feedforward synapses may be functional, as application of L-AP4 (which blocks the cone → DBC synapse) converts the DBC depolarizing response to a sustained hyperpolarizing response.[20,89] However, application of GABA on BC dendrites in Co^{2+} Ringer's does not elicit any response, indicating that if feedforward synapses are chemical, the neurotransmitter is not GABA.[42] Histochemical evidence suggests that only about 50–60% of the HCs in the salamander retina are GABAergic, and the identity of neurotransmitter(s) in the remaining HCs is unknown (but unlikely to be glycine).[84,96] Recent studies on the rabbit and salamander retinas have shown that HCs and subpopulations of DBC are dye coupled (see Fig. 22.7 below).[74,92] This raises the possibility that HC–DBC electrical coupling may mediate the HC–DBC feedforward signal for the DBC surround response.

Rod and cone pathways and bipolar cell output synapses

Rods are responsible for vision under conditions of dim illumination when cones are not responsive. Cones operate under conditions of bright illumination, and they mediate color vision and provide high spatial resolution, features that rods do not handle.[61] In addition to dividing into DBCs and HBCs, BCs are further classified according to their rod/cone inputs. In mammals, rods and cones make synaptic connections with rod and cone BC separately, and one type of rod BC and 9–10 types of cone BCs have been identified.[7,17] In fish small BCs make synaptic contacts exclusively with cones, and large BCs make contacts with both rods and cones.[32,70] BCs in amphibian retinas contact both rods and cones, but some are clearly rod-dominated where others are cone-dominated.[90] Despite these differences between species, BCs in most vertebrate retinas have many common functional and morphological features.

It has been established in many vertebrate species, for example, that axon terminals of DBCs ramify in the proximal half (sublamina B) of the IPL whereas those of the HBCs ramify in the distal half (sublamina A) of the IPL.[1,52] Axon terminals of rod BCs in mammalian retinas end near the proximal margin of the IPL, and axon terminals of cone BCs ramify in more central regions of the IPL.[7] This agrees with the patterns of BC axon terminal stratification in the salamander: axon terminals of rod-dominated BCs ramify at the two margins of the IPL whereas those of cone-dominated

BCs ramify at central regions of the IPL.[86] Additionally, despite the fact that mammalian rod and cone BCs make segregated synaptic contacts with rods and cones, their light responses are somewhat mixed with rod/cone signals like most BCs in lower vertebrates, partially because rods and cones are electrically coupled (see above), and partially because rod BC signals can be transmitted to cone BCs through AII amacrine cells.[54,56,71] Furthermore, although rod HBCs have not been identified anatomically in mammalian retinas, recent physiological evidence suggests that rod signals may be transmitted by HBC-mediated channels in at least some mammals.[12,68]

Based on the properties of BCs across vertebrate species, it is reasonable to propose that most vertebrate retinas have 6 major functional types of BCs: the rod (or rod-dominated), cone (or cone-dominated) and mixed (rod/cone) depolarizing and hyperpolarizing bipolar cells (DBC_R, DBC_C, DBC_M, HBC_R, HBC_C and HBC_M). Each carries a characteristic set of light response attributes and projects them to the inner retina through axons that terminate at segregated regions (strata) of the IPL. Such stratum-by-stratum projection of light response attributes is exemplified by a large-scale patch clamp study of the salamander BC responses and morphology (Fig. 22.6).[57] This study reveals several rules for the function–morphology relationships of retinal bipolar cells:

(1) Cells with axon terminals in strata 1–5 (sublamina A) are HBCs (generated by outward light-evoked cation currents (ΔI_C)) and those in strata 6–10 (sublamina B) are DBCs (generated by inward ΔI_C). This agrees with the sublamina A/B rule observed in many vertebrate species.

(2) Cells with axon terminals in strata 1, 2 and 10 are rod-dominated, those in strata 4–8 are cone-dominated, and those in strata 3 and 9 exhibit mixed rod/cone input.

(3) ΔI_C at light onset in rod-dominated HBCs and DBCs are sustained, that of the cone-dominated HBCs exhibits a smaller sustained outward current followed by a transient inward current at the light offset, and that of the cone-dominated DBCs exhibit a sustained inward current followed by a small transient off outward current.

(4) ΔI_{Cl} (light-evoked chloride currents) in rod-dominated bipolar cells are sustained ON currents whereas those in cone-dominated bipolar cells are transient ON–OFF currents. ΔI_{Cl} in all BCs are outward, and thus they are synergistic to ΔI_C in HBCs and antagonistic to ΔI_C in DBCs.

(5) Bipolar cells with axon terminals stratified in multiple strata exhibit combined light response properties of the narrowly monostratified cells in the same strata.

(6) Bipolar cells with pyramidally branching or globular axons exhibit light response properties very similar to those of narrowly monostratified cells whose axon terminals stratified in the same stratum as the axon terminal endings of the pyramidally branching or globular cells.[56,57]

Electrical coupling between bipolar cells has been reported in several vertebrate retinas. Electron microscopic studies have revealed that gap junctions are made between bipolar

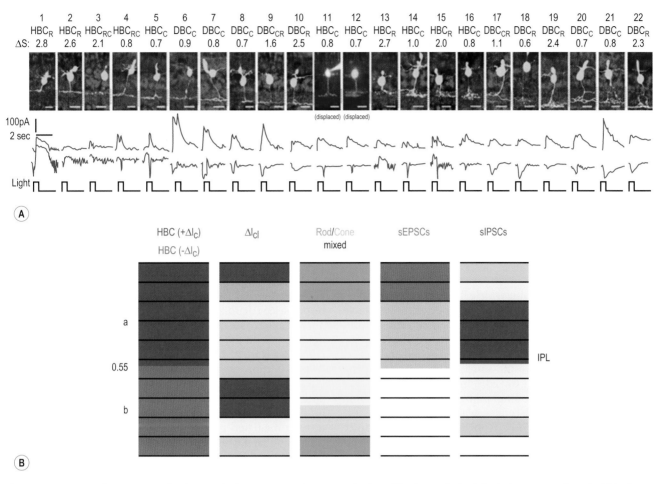

Figure 22.6 **Stratum-by-stratum rules for correlating patterns of axon terminal ramification and physiological responses in retinal bipolar cells.** (**A**) 22 morphologically identified (by Lucifer yellow filling) BCs and their light-evoked excitatory cation current (ΔI_C) and inhibitory chloride current (ΔI_{Cl}) recorded from dark-adapted salamander retinal slices. Each cell is named according to their spectral difference (ΔS) and ΔI_C polarity as rod-dominated, cone-dominated, mixed rod/cone hyperpolarizing or depolarizing bipolar cells (HBC$_R$, HBC$_C$, HBC$_{R/C}$, DBC$_R$, DBC$_C$ or DBC$_{R/C}$). BCs with inward ΔI_C are DBCs and with outward ΔI_C are HBCs. The spectral difference, ΔS, is defined as $S_{700} - S_{500}$ (where S_{700} and S_{500} are intensities (in log units) of 700 nm and 500 nm light eliciting responses of the same amplitude). Since ΔS for the salamander rods is about 3.4 and that for the cones is about 0.1, BCs with $\Delta S > 2.0$ are rod-dominated (HBC$_R$ or DBC$_R$), with $\Delta S < 1.0$ are cone-dominated (HBC$_C$ or DBC$_C$), and with $1.0 < \Delta S < 2.0$ are mixed rod/cone cells (HBC$_{R/C}$ or DBC$_{R/C}$) (also named HBC$_M$ or DBC$_M$, see Figs 22.7 and 22.8). displaced HBC$_C$s: HBC$_C$s with somas displaced in the outer nuclear layer. (Modified from Pang et al 2004;[56] and Maple et al 1994.[41]) (**B**) Stratum-by-stratum correlation of BC axon terminal ramification in the IPL with physiological function. ΔI_C: light-evoked excitatory cation current, $+\Delta I_C$: outward ΔI_C, $-\Delta I_C$: inward ΔI_C, ΔI_{Cl}: light-evoked excitatory chloride current, sEPSC: spontaneous excitatory postsynaptic current, sIPSC: spontaneous inhibitory postsynaptic current. More saturated hues indicate stronger response parameters.

cell axon terminals in the inner plexiform layer.[43,81] Current injection into fish bipolar cells elicits sign-preserving responses in adjacent bipolar cells.[36] In the tiger salamander retina, the diameter of the bipolar cell receptive field center is about 10–20 times larger than the diameter of the dendritic arbors of these cells.[6,19] These results suggest that bipolar cells of the same type are extensively coupled to one another at least in certain species or certain areas of the retina. Recent evidence has shown that bipolar cells of the same types are dye coupled (see Fig. 22.7),[92] consistent with the notion that bipolar cell receptive field centers are mediated by dendrites from more than one bipolar cell. The primary function of bipolar cell coupling is to increase the receptive field center of these cells. Additionally, bipolar cell coupling also improves the signal-to-noise ratio of the bipolar cell output signals. The disadvantage of bipolar cell coupling, however, is that it reduces the spatial resolution of bipolar cell outputs.

Bipolar cell responses and center-surround antagonistic receptive field (CSARF) organization

Center-surround antagonistic receptive fields (CSARF) are the basic alphabet for encoding spatial information in the visual system.[25] CSARF was first discovered by Kuffler in 1953, who demonstrated that light falling on the central region of the receptive field of cat retinal ganglion cells elicited a response of opposite sign as that elicited by light falling on the surround region of the receptive field.[35] Later studies showed that such CSARF also exist upstream in retinal bipolar cells[78] and downstream in cells in the lateral geniculate nucleus and the primary visual cortex.[24,26] Furthermore, complex receptive fields of neurons in higher visual centers are probably mediated by arrays of upstream neurons that exhibit CSARF organization.

The first neurons along the visual pathway that exhibit ON-center and OFF-center CSARFs are the retinal bipolar cells, first reported by Werblin & Dowling in 1969[78] and Kaneko in 1970.[30] The center input of a bipolar cell is mediated by direct photoreceptor output synapses made on its dendrites (and the dendrites of adjacent bipolar cells electrically coupled). The surround input is mediated by the horizontal cells in the outer plexiform layer and by amacrine cells in the inner plexiform layer.[76-78] As mentioned earlier, two major types of bipolar cells are found in all vertebrates, the depolarizing bipolar cells (DBCs) depolarize to central illumination and hyperpolarize to concentric surround illumination (ON-center cells); the hyperpolarizing bipolar cells (HBCs) hyperpolarize to central illumination and depolarize to concentric surround illumination (OFF-center cells). In the mudpuppy and salamander retinas annular illumination does not polarize the bipolar cells unless the centers of their receptive fields are simultaneously illuminated.[65,78]

The CSARF is also a major vehicle for carrying color information in the visual system. In some species the receptive field of certain bipolar cells and ganglion cells exhibit "double-opponent" organization. For instance, the receptive field center of a bipolar cell is depolarized by red and hyperpolarized by green, while the surround is depolarized by green and hyperpolarized by red.[33] In other species bipolar cells have other color coding configurations that include opponent receptive field centers with non-opponent surround, or vice versa.[91] These types of color-coded CSARF organizations help the visual system to detect stationary and moving color edges.

In addition to projecting signals to various strata of the IPL (described in Fig. 22.6), BCs with various light response attributes have different CSARF organizations. Figure 22.7 shows the morphology, patterns of dye coupling, light responses, CSARF properties and membrane resistance changes associated with the center and surround voltage responses of the six functional types of BCs (HBC$_R$, HBC$_M$, HBC$_C$, DBC$_C$, DBC$_M$ and DBC$_R$) in the tiger salamander retina.[92] The BC receptive field center diameters (RFCD) vary with the relative rod/cone input: RFCD is larger in DBCs with stronger cone input, and it is larger in HBCs with stronger rod input. RFCD also correlates with the degree of dye coupling: BCs with larger RFCD are more strongly dye coupled with neighboring cells of the same type, suggesting that BC–BC coupling significantly contributes to the BC receptive field center.

Membrane resistance measurements in Figure 22.7E demonstrate that the center responses of all HBCs are associated with a resistance increase and the center responses of all DBCs are accompanied with a resistance decrease. This is consistent with the notion that glutamate released from rods and cones in darkness opens AMPA/kainate receptor-mediated cation channels in HBCs and closes mGluR6 receptor-mediated cation channels in DBCs. Center light stimuli hyperpolarize rods and cones, suppress glutamate release, and result in a resistance increase (close ion channels) in HBCs and a resistance decrease (open ion channels) in DBCs.

Figure 22.8 is a schematic representation of the possible and unlikely synaptic pathways underlying surround inputs of various types of BCs, based on the surround response polarity and accompanying resistance changes shown in Figure 22.7.[92] These results suggest that the HC–cone–BC feedback synapses may contribute to the surround responses of all six types of BCs. The negative HC–cone feedback synapses (pathway I in Fig. 22.8) partially "turn off" the center responses by depolarizing the cones, as the membrane resistance changes associated with surround responses of all BCs are opposite to the resistance changes associated with center responses (Fig. 22.6E).

In addition to the common HC–cone–BC feedback pathway, various types of BCs use different HC and AC synaptic pathways to mediate their surround responses. It is unlikely, for example, that HBC surround responses are directly mediated by chemical synaptic inputs from hyperpolarizing lateral neurons such as HCs and AC$_{OFF}$s because of resistance change mismatch, and thus HBCs may only receive surround inputs from HC-cone-HBC and AC$_{ON}$–HBC synapses. On the other hand, resistance analysis suggests that DBC surround responses can be mediated by HC–cone–DBC, HC–DBC and AC$_{OFF}$–DBC chemical synapses, but not the AC$_{ON}$–DBC synapses. Moreover, dye coupling (Fig. 22.7A) results indicate that DBC$_C$s receive additional surround inputs from wide-field HCs through electrical synapses. Despite the heterogeneity, it is interesting to point out that an ON/OFF crossover inhibition rule applies here: cells with OFF (hyperpolarizing) responses (HCs and AC$_{OFF}$s) mediate surround inhibitory inputs to ON cells (DBCs); and cells with ON (depolarizing) responses (AC$_{ON}$s) mediate surround inhibitory inputs to OFF cells (HBCs). ON/OFF crossover inhibition from amacrine cells to ganglion cells have been reported in the salamander and mammalian retinas,[23,58] and the data shown in Figure 22.7 suggest that it

Figure 22.7 Morphology, light responses and receptive fields of six types of bipolar cells in the tiger salamander retina. (**A**): Fluorescent micrographs of a neurobiotin-filled HBC$_R$ (column 1), a HBC$_M$ (column 2), HBC$_C$ (column 3), a DBC$_C$ (column 4), a DBC$_M$ (column 5), and a DBC$_R$ (column 6) viewed with a confocal microscope at the outer INL/OPL level (**ai**), the IPL level (**aii**) and with z-axis rotation (**aiii**). Calibration bar: 100 μm. (**B**) (**bi**): BC voltage responses to 500nm and 700nm light steps of various intensities. (**bii**): response-intensity (V-Log I) curves of the responses to 500 nm and 700 nm lights. ΔS (spectral difference, defined as S$_{700}$ − S$_{500}$, where S$_{700}$ and S$_{500}$ are intensities of 700 nm and 500 nm light eliciting responses of the same amplitude) of the 6 BCs are 2.13, 1.51, 0.30, 0.57, 1.45 and 2.25. Since ΔS for the rods is about 3.4 and that for the cones is about 0.1 in the salamander retina, BCs with ΔS >2.0 are rod-dominated BCs, BCs with ΔS<1.0 are cone-dominated BCs, and BC with 1.0 <ΔS<2.0 are mixed rod/cone BCs. (**C**): Measurements of BC receptive field center diameters (RFCD) by recording voltage responses to a 100-μm wide light bar moving stepwise (with 120 μm step increments) across the receptive field. (**D**): Voltage responses of the 6 types of BCs elicited by a center light spot (300 μm) and a surround light annulus (700 μm inner diameter, 2000 μm outer diameter). The surround light annulus was of the same intensity (700 nm, −2) for all 6 cells whereas the intensity of the center light spot was adjusted so that it allowed the annulus to produce the maximum response. (**E**): Voltage responses of the 6 types of BCs elicited by a center light spot and a surround light annulus (same as in **D**), and by a train of −0.1-nA/200-msec current pulses passed into the cell by the recording microelectrode through a bridge circuit. (From Zhang & Wu 2009.[92])

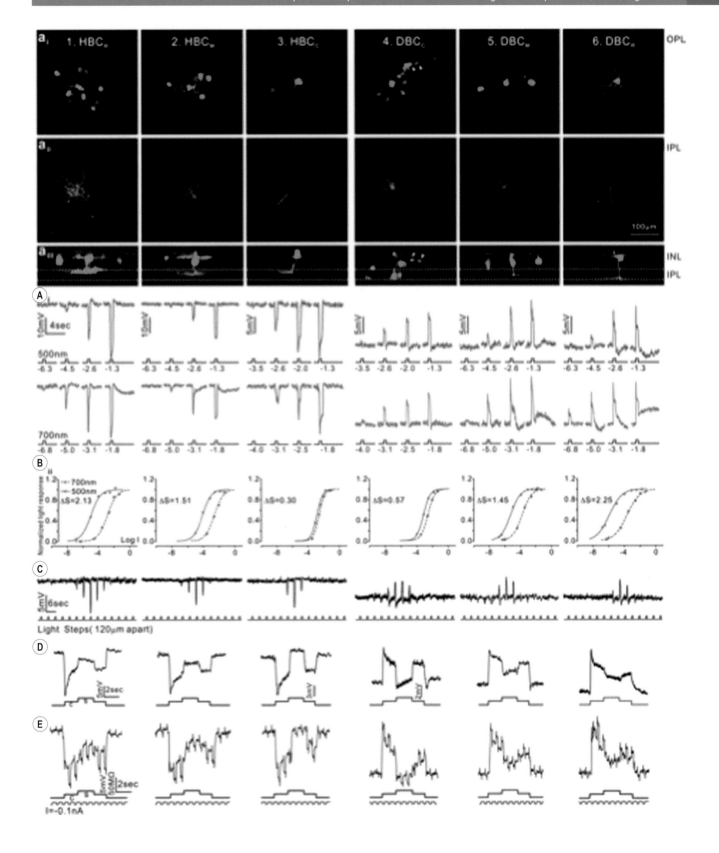

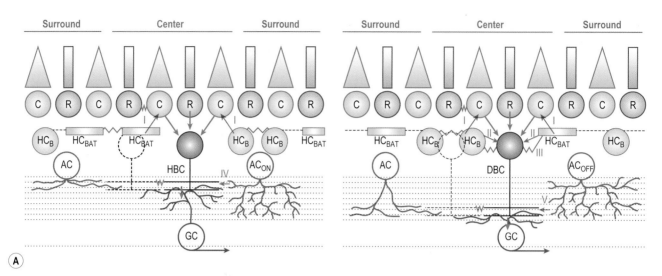

(A)

	Center synaptic organisation			Surround synaptic pathways			
	Rod/Cone inputs	BC-BC coupling	HC feedback (I) HC→Cone→BC	HC feedbackward HC→BC (II) chemical (III) electrical		AC feedback (chemical) (IV)AC$_{ON}$→BC (V)AC$_{OFF}$→BC	
HBC$_R$	Rod: +++ Cone: +	+++	+ $\underset{hyp \to dep \to dep}{(-) \quad (+)}$	no	no	yes $\underset{dep \to dep}{(+)}$	no
HBC$_M$	Rod: ++ Cone: +++	++	++ $\underset{hyp \to dep \to dep}{(-) \quad (+)}$	no	no	yes $\underset{dep \to dep}{(+)}$	no
HBC$_C$	Rod: + Cone: ++++	++	+++ $\underset{hyp \to dep \to dep}{(-) \quad (+)}$	no	no	yes $\underset{dep \to dep}{(+)}$	no
DBC$_C$	Rod: + Cone: +++	+ HCs	+++ $\underset{hyp \to dep \to hyp}{(-) \quad (-)}$	yes $\underset{hyp \to hyp}{(+)}$	yes $\underset{hyp \to hyp}{(+)}$	no	yes $\underset{hyp \to hyp}{(+)}$
DBC$_M$	Rod: ++ Cone: ++	+	++ $\underset{hyp \to dep \to hyp}{(-) \quad (-)}$	yes $\underset{hyp \to hyp}{(+)}$	no	no	yes $\underset{hyp \to hyp}{(+)}$
DBC$_R$	Rod: +++ Cone: +	0	+ $\underset{hyp \to dep \to hyp}{(-) \quad (-)}$	yes $\underset{hyp \to hyp}{(+)}$	no	no	yes $\underset{hyp \to hyp}{(+)}$

(B)

Figure 22.8 Center-surround antagonistic receptive field organization of bipolar cells. (A) Schematic diagrams of center (green) and surround (red) synaptic pathways of HBCs (left) and DBCs (right). R: rod, C: cone, HC$_B$: B-type HC somas, HC$_{BAT}$: B-type HC axon terminals, HBC: hyperpolarizing bipolar cell, DBC: depolarizing bipolar cell, AC: amacrine cell, AC$_{ON}$: ON amacrine cell, AC$_{OFF}$: OFF amacrine cell, GC: ganglion cell, arrows: chemical synapses, zig-zags: electrical synapses. I-V: five surround synaptic pathways list in B. **(B)** Variations in synaptic pathways mediating center (green) and surround (red) responses of the HBC$_R$, HBC$_M$, HBC$_C$, DBC$_C$, DBC$_M$ and DBC$_R$s. +++: strong, ++: intermediate, +: moderate, yes: possible, no: unlikely. For the possible pathways, response polarities (hyp: hyperpolarization, dep: depolarization) in each neuron and the synaptic sign ((+): sign-preserving or (–): sign-inverting) in each synapse in the pathways are indicated [e.g. in the HBC$_R$ HC → cone → BC pathway: (hyperpolarization in HC) via a sign-inverting synapse → (depolarization in cone) via a sign-preserving synapse → (depolarization in BC)]. (From Zhang & Wu 2009.[92])

may be a general rule for lateral inhibition in the visual system.

The AC–BC contribution to BC surround responses is mediated by GABAergic or glycinergic synapses, and detailed studies on these synapses are discussed in the next chapter. As illustrated in Figures 22.7 and 22.8, surround responses of different functional types of BCs in the salamander retina are mediated by different combinations of synaptic circuitries: HC → cone → BC (pathway I in Fig. 22.8), HC→BC (chemical/gap junction, pathways II and III in Fig. 22.8) and

AC→BC (GABA/glycine, pathways IV and V in Fig. 22.8). It is conceivable that the surround responses of different BCs/GCs from different animals under different conditions are mediated by different combinations of surround synaptic pathways, and thus they are sensitive to different synaptic blockers. The wide variation of synaptic circuitries underlying surround responses of various functional types of BCs allows for flexibility in function-specific modulation of BC/GC receptive fields. Hence different features of spatial and contrast information, such as rod/cone and ON/OFF signals, can be differentially modulated by different lighting and adaptation conditions.

Acknowledgments

This work was supported by grants from NIH (EY 04446), NIH Vision Core (EY 02520), the Retina Research Foundation (Houston), and the Research to Prevent Blindness, Inc.

References

1. Ammermuller J, Kolb H. The organization of the turtle inner retina. I. ON- and OFF-center pathways. J Comp Neurol 1995; 358:1–34.
2. Ashmore JA, Copenhagen DR. Different postsynaptic events in two types of retinal bipolar cell. Nature 1980; 288:84–86.
3. Attwell D, Wilson M. Behaviour of the rod network in the tiger salamander retina mediated by membrane properties of individual rods. J Physiol 1980; 309:287–315.
4. Attwell D, Wilson M, Wu SM. A quantitative analysis of interactions between photoreceptors in the salamander (Ambystoma) retina. J Physiol 1984; 352:703–737.
5. Baylor DA, Fuortes MG, O'Bryan PM. Receptive fields of cones in the retina of the turtle. J Physiol 1971; 214:265–294.
6. Borges S, Wilson M. Structure of the receptive fields of bipolar cells in the salamander retina. J Neurophysiol 1987; 58:1275–1291.
7. Boycott B, Wassle H. Parallel processing in the mammalian retina: the Proctor Lecture. Invest Ophthalmol Vis Sci 1999; 40: 1313–1327.
8. Copenhagen DR, Jahr CE. Release of endogenous excitatory amino acids from turtle photoreceptors. Nature 1989; 341:536–539.
9. Copenhagen DR, Owen WG. Coupling between rod photoreceptors in a vertebrate retina. Nature 1976; 260:57–59.
10. Deans MR, Volgyi B, Goodenough DA, Bloomfield SA, Paul DL. Connexin36 is essential for transmission of rod-mediated visual signals in the mammalian retina. Neuron 2002; 36:703–712.
11. Detwiler PB, Hodgkin AL. Electrical coupling between cones in turtle retina. J Physiol 1979;291:75–100.
12. Devries SH, Baylor DA. An alternative pathway for signal flow from rod photoreceptors to ganglion cells in mammalian retina. Proc Natl Acad Sci USA 1995;92:10658–10662.
13. Devries SH, Schwartz EA. Kainate receptors mediate synaptic transmission between cones and 'Off' bipolar cells in a mammalian retina. Nature 1999; 397:157–160.
14. Dowling JE. The retina, an approachable part of the brain. Harvard: Harvard University Press, 1987.
15. Dowling JE, Werblin FS. Organization of retina of the mudpuppy, Necturus maculosus. I. Synaptic structure. J Neurophysiol 1969; 32:315–338.
16. Fisher SK, Boycott BB. Synaptic connections made by horizontal cells within the outer plexiform layer of the retina of the cat and the rabbit. Proc R Soc Lond B Biol Sci 1974; 186:317–331.
17. Ghosh KK, Bujan S, Haverkamp S, Feigenspan A, Wassle W. Types of bipolar cells in the mouse retina. J Comp Neurol 2004; 469:70–82.
18. Grant GB, Dowling JE. A glutamate-activated chloride current in cone-driven ON bipolar cells of the white perch retina. J Neurosci 1995; 15:3852–3862.
19. Hare WA, Owen WG. Spatial organization of the bipolar cell's receptive field in the retina of the tiger salamander. J Physiol 1990;421:223–245.
20. Hare WA, Owen WG. Effects of 2-amino-4-phosphonobutyric acid on cells in the distal layers of the tiger salamander's retina. J Physiol 1992;445:741–757.
21. Hirasawa H, Kaneko A. pH changes in the invaginating synaptic cleft mediate feedback from horizontal cells to cone photoreceptors by modulating Ca^{2+} channels. J Gen Physiol 2003;122:657–671.
22. Hornstein EP, Verweij J, Schnapf JL. Electrical coupling between red and green cones in primate retina. Nature Neurosci 2004; 7:745–750.
23. Hsueh HA, Molnar A, Werblin FS. Amacrine to amacrine cell inhibition in the rabbit retina. J Neurophysiol 2008; 100:2077–2088.
24. Hubel DH, Wiesel TN. Integrative action in the cat's lateral geniculate body. J Physiol 1961; 155:385–398.
25. Hubel DH, Wiesel TN. Receptive field, binocular interaction and functional architecture in the cat's visual cortex. J Physiol 1962; 160:106–154.
26. Hubel DH, Wiesel TN. Receptive fields and functional architecture of monkey striate cortex. J Physiol 1968; 195:215–243.
27. Ichinose T, Lukasiewicz PD. Inner and outer retinal pathways both contribute to surround inhibition of salamander ganglion cells. J Physiol 2005; 565:517–535.
28. Kamermans M, Fahrenfort I. Ephaptic interactions within a chemical synapse: hemichannel-mediated ephaptic inhibition in the retina. Curr Opin Neurobiol 2004; 14:531–541.
29. Kamermans M, Fahrenfort I, Schultz K, Janssen-Bienhold U, Sjoerdsma T, Weiler R. Hemichannel-mediated inhibition in outer retina. Science 2001; 292:1178–1180.
30. Kaneko A. Physiological and morphological identification of horizontal, bipolar and amacrine cells in goldfish retina. J Physiol 1970; 207:623–633.
31. Kaneko A. Electrical connexions between horizontal cells in the dogfish retina. J Physiol 1971; 213:95–105.
32. Kaneko A. Receptive field organization of bipolar and amacrine cells in the goldfish retina. J Physiol 1973; 235:133–153.
33. Kaneko A, Tachibana M. Retinal bipolar cells with double colour-opponent receptive fields. Nature 1981; 293:220–222.
34. Kaneko A, Yamada M. S-potentials in the dark-adapted retina of the carp. J Physiol 1972; 227:261–273.
35. Kuffler SW. Discharge patterns and functional organization of the mammalian retina. J Neurophysiol 1953; 16:37–68.
36. Kujiraoka T, Saito T. Electrical coupling between bipolar cells in carp retina. Proc Natl Acad Sci USA 1986; 83:4063–4066.
37. Lasansky A. Organization of outer synaptic layer in the retina of larval tiger salamander. Phil Trans R Soc Lond B Biol Sci 1973; 265:471–489.
38. Lasansky A, Vallerga S. Horizontal cell responses in the retina of the larval tiger salamander. J Physiol 1975; 251:145–165.
39. Li L, Devries SH. Separate blue and green cone network in the mammalian retina. Nature Neurosci 2004; 7:751–756.
40. Maple BR, Gao F, Wu SM. Glutamate receptors differ in rod- and cone-dominated off-center bipolar cells. Neuroreport 1999; 10:3605–3610.
41. Maple BR, Werblin FS, Wu SM, Miniature excitatory postsynaptic currents in bipolar cells of the tiger salamander retina. Vision Res 1994; 34:2357–2362.
42. Maple BR, Wu SM. Synaptic inputs mediating bipolar cell responses in the tiger salamander retina. Vision Res 1996; 36:4015–4023.
43. Marc RE, Liu WL, Muller JF. Gap junctions in the inner plexiform layer of the goldfish retina. Vision Res 1988; 28:9–24.
44. Marchiafava PL. Horizontal cells influence membrane potential of bipolar cells in the retina of the turtle. Nature 1978; 275:141–142.
45. Massey SC, Miller RF. Excitatory amino acid receptors of rod- and cone-driven horizontal cells in the rabbit retina. J Neurophysiol 1987; 57:645–659.
46. Murakami M, Shimoda Y, Nakatani K, Miyachi E, Watanabe S. GABA-mediated negative feedback from horizontal cells to cones in the carp retina. Jpn J Physiol 1982; 32:911–926.
47. Naka KI. The horizontal cells. Vision Res 1972; 12:573–588.
48. Naka KI, Rushton WA. S-potentials from colour units in the retina of fish (Cyprinidae). J Physiol 1966; 185:536–555.
49. Naka KI, Rushton WA. S-potentials from luminosity units in the retina of fish (Cyprinidae). J Physiol 1966; 185:587–599.
50. Nawy S. The metabotropic receptor mGluR6 may signal through G(o), but not phosphodiesterase, in retinal bipolar cells. J Neurosci 1999; 19:2938–2944.
51. Nawy S, Jahr CE. Suppression by glutamate of cGMP-activated conductance in retinal bipolar cells. Nature 1990; 346:269–271.
52. Nelson R, Kolb H. Synaptic patterns and response properties of bipolar and ganglion cells in the cat retina. Vision Res 1983; 23:1183–1195.
53. O'Brien J, Nguyen HB, Mills SL. Cone photoreceptors in bass retina use two connexins to mediate electrical coupling. J. Neurosci 2004; 24:5632–5642.
54. Pang JJ, bd-El-Barr MM, Gao F, Bramblett DE, Paul DL, Wu SM. Relative contributions of rod and cone bipolar cell inputs to AII amacrine cell light responses in the mouse retina. J Physiol 2007;580:397–410.
55. Pang JJ, Gao F, Barrow A, Jacoby RA, Wu SM. How do tonic glutamatergic synapses evade receptor desensitization? J Physiol 2008; 586:2889–2902.
56. Pang JJ, Gao F, Wu SM. Light-evoked current responses in rod bipolar cells, cone depolarizing bipolar cells and AII amacrine cells in dark-adapted mouse retina. J Physiol 2004; 559:123–135.
57. Pang JJ, Gao F, Wu SM. Stratum-by-stratum projection of light response attributes by retinal bipolar cells of Ambystoma. J Physiol 2004; 558:249–262.
58. Pang JJ, Gao F, Wu SM. Cross-talk between ON and OFF channels in the salamander retina: indirect bipolar cell inputs to ON-OFF ganglion cells. Vision Res 2007; 47:384–392.
59. Raviola E, Gilula NB. Gap junctions between photoreceptor cells in the vertebrate retina. Proc Natl Acad Sci USA 1973; 70:1677–1681.
60. Raviola E, Gilula NB. Intramembrane organization of specialized contacts in the outer plexiform layer of the retina. A freeze-fracture study in monkeys and rabbits. J Cell Biol 1975; 65:192–222.
61. Rodieck RW. The first steps in seeing. New York: Sinauer Associates Inc., 1998.
62. Schwartz EA. Cones excite rods in the retina of the turtle. J Physiol 1975; 246:639–651.
63. Schwartz EA. Rod-rod interaction in the retina of the turtle. J Physiol 1975; 246:617–638.
64. Shiells RA, Falk G. Potentiation of 'on' bipolar cell flash responses by dim background light and cGMP in dogfish retinal slices. J Physiol 2002; 542:211–220.
65. Skrzypek J, Werblin F. Lateral interactions in absence of feedback to cones. J Neurophysiol 1983; 49:1007–1016.
66. Slaughter MM, Miller RF 2-amino-4-phosphonobutyric acid: a new pharmacological tool for retina research. Science 1981; 211:182–185.
67. Snellman J, Kaur T, Shen Y, Nawy S. Regulation of ON bipolar cell activity. Prog Retin Eye Res 2008; 27:450–463.
68. Soucy E, Wang Y, Nirenberg S, Nathans J, Meister M. A novel signaling pathway from rod photoreceptors to ganglion cells in mammalian retina. Neuron 1998; 21:481–493.
69. Steinberg RN. Rod-cone interaction in S-potentials from the cat retina. Vision Res 1969; 9:1331–1344.
70. Stell WK. The structure relationships of horizontal cells and photoreceptor-bipolar synaptic complexes in goldfish retina. Am J Anat 1967; 121:401–424.
71. Strettoi E, Dacheux RF, Raviola E. Cone bipolar cells as interneurons in the rod pathway of the rabbit retina. J Comp Neurol 1994; 347:139–149.
72. Svaetichin G. The cone action potential. Acta Physiol Scand 1953; 29:565–572.
73. Thoreson WB, Rabi K, Townes-Anderson E, Heidelberger R. A highly Ca^{2+}-sensitive pool of vesicles contributes to linearity at the rod photoreceptor ribbon synapse. Neuron 2004; 42:595–605.
74. Trexler EB, Li W, Mills SL, Massey SC. Coupling from AII amacrine cells to ON cone bipolar cells is bidirectional. J Comp Neurol 2001; 437:408–422.
75. Vessey JP, Stratis AK, Daniels BA, et al. Proton-mediated feedback inhibition of presynaptic calcium channels at the cone photoreceptor synapse. J Neurosci 2005; 25:4108–4117.
76. Werblin FS. Lateral interactions at inner plexiform layer of vertebrate retina: antagonistic responses to change. Science 1972; 175:1008–1010.
77. Werblin FS, Copenhagen DR. Control of retinal sensitivity. 3. Lateral interactions at the inner plexiform layer. J Gen Physiol 1974; 63:88–110.

78. Werblin FS, Dowling JE. Organization of the retina of the mudpuppy, Necturus maculosus. II. Intracellular recording. J Neurophysiol 1969;32:339–355.

79. Witkovsky P, Owen WG, Woodworth W. Gap junctions among the perikarya, dendrites, and axon terminals of the luminosity-type horizontal cell of the turtle retina. J Comp Neurol 1983; 216:359–368.

80. Wong KY, Cohen ED, Dowling JE. Retinal bipolar cell input mechanisms in giant danio. II. Patch-clamp analysis of on bipolar cells. J Neurophysiol 2005; 93:94–107.

81. Wong-Riley MTT. Synaptic organization of the inner plexiform layer in the retina of the tiger salamander. J Neurocytol 1974; 3:1–33.

82. Wu SM. The off-overshoot responses of photoreceptors and horizontal cells in the light-adapted retinas of the tiger salamander. Exp Eye Res 1988; 47:261–268.

83. Wu SM. Input-output relations of the feedback synapse between horizontal cells and cones in the tiger salamander retina. J Neurophysiol 1991; 65:1197–1206.

84. Wu SM. Signal transmission and adaptation-induced modulation of photoreceptor synapses in the retina. Progr Retin Res 1991; 10:27–44.

85. Wu SM, Gao F, Maple BR. Functional architecture of synapses in the inner retina: segregation of visual signals by stratification of bipolar cell axon terminals. J Neurosci 2000; 20:4462–4470.

86. Wu SM, Gao F, Maple BR. Integration and segregation of visual signals by bipolar cells in the tiger salamander retina. Progr Brain Res 2001; 131:125–143.

87. Wu SM, Yang XL. Electrical coupling between rods and cones in the tiger salamander retina. Proc Natl Acad Sci USA 1988; 85:275–278.

88. Yang XL, Wu SM. Modulation of rod-cone coupling by light. Science 1989; 244: 352–354.

89. Yang XL, Wu SM. Feedforward lateral inhibition in retinal bipolar cells: input-output relation of the horizontal cell-depolarizing bipolar cell synapse. Proc Natl Acad Sci USA 1991; 88:3310–3313.

90. Yang XL, Wu SM. Response sensitivity and voltage gain of the rod- and cone-bipolar cell synapses in dark-adapted tiger salamander retina. J Neurophysiol 1997; 78:2662–2673.

91. Yazulla S. Cone input to bipolar cells in the turtle retina. Vision Res 1976; 16:737–744.

92. Zhang AJ, Wu SM. Receptive fields of retinal bipolar cells are mediated by heterogeneous synaptic circuitry. J Neurosci 2009; 29:789–797.

93. Zhang AJ, Zhang J, Wu SM. Electrical coupling, receptive fields, and relative rod/cone inputs of horizontal cells in the tiger salamander retina. J Comp Neurol 2006; 499:422–431.

94. Zhang J, Wu SM. Connexin35/36 gap junction proteins are expressed in photoreceptors of the tiger salamander retina. J Comp Neurol 2004; 470:1–12.

95. Zhang J, Wu SM. Physiological properties of rod photoreceptor electrical coupling in the tiger salamander retina. J Physiol 2005; 564:849–862.

96. Zhang J, Zhang AJ, Wu SM. Immunocytochemical analysis of GABA-positive and calretinin-positive horizontal cells in the tiger salamander retina. J Comp Neurol 2006; 499:432–441.

CHAPTER 23

Signal Processing in the Inner Retina

Peter D. Lukasiewicz & Erika D. Eggers

The inner plexiform layer (IPL) is the second synaptic layer of the retina (Fig. 23.1) and the final stage for processing visual information before it leaves the eye. Visual signals from rod and cone photoreceptors are first processed in the outer plexiform layer (OPL; Fig. 23.1), where horizontal cells modulate their signaling to bipolar cells. Bipolar cells then transmit these signals to the inner plexiform layer (IPL), where amacrine cells shape bipolar cells signaling to ganglion cells. The visual signals are first separated into separate signaling streams at the bipolar cells. Distinct classes of bipolar cells form parallel pathways that convey information about different aspects of the visual world. Synaptic connections between bipolar cells, amacrine cells and ganglion cells in the IPL produce complex processing of the visual signals. These synaptic interactions in the IPL process spatial, motion, and directional information within the visual scene. The outputs of the IPL are transmitted to distinct ganglion cell types that convey visual signals to different brain regions.

Bipolar cells form parallel pathways and provide excitatory input to the IPL

A variety of bipolar cell types form parallel signaling pathways that relay the processed photoreceptor output to the IPL. Excitatory inputs to the IPL originate from ON bipolar cells and OFF bipolar cells that depolarize in response to increasing or decreasing illumination within their receptive field centers, respectively. These distinct responses are determined by separate classes of dendritic glutamate receptors.[1,2] ON and OFF bipolar cells respond with opposite polarities because, in darkness, glutamate release from photoreceptors hyperpolarizes ON bipolar cells by activating mGluR6Rs that close cation channels and depolarizes OFF bipolar cells by activating AMPA/KA receptors that open cation channels. The axon terminals of ON and OFF bipolar cells stratify in two distinct sublaminae of the IPL, contacting the dendrites of ON and OFF ganglion cells and amacrine cells. The outer half of the IPL encodes responses to light OFF and the inner half encodes responses to light ON.[3,4] The ON and OFF sublaminae are further divided into ten bipolar cell types that have axon terminals in different morphological strata (Fig. 23.2), encoding different aspects of the visual scene.[5]

An additional major functional subdivision of both the ON and OFF sublaminae is a stratification of the processes of temporally distinct responses.[5-7] Retinal neurons generate either transient or sustained responses to continuous visual stimulation that encode the temporal and spatial features of the visual scene, respectively. A transient response can signal the offset and/or onset of a light stimulus, while a sustained response can signal the duration of the light stimulus. Similar to the ON and OFF response separation, sustained and transient responsiveness is first observed at the bipolar cells. Transient and sustained ON and OFF signals recorded in bipolar cells have been attributed to variations in the types of mGluR6 and AMPA/KARs expressed on ON and OFF bipolar cell dendrites, respectively.[2,6] Transient bipolar cell terminals contact transient amacrine cell and ganglion cell processes in the mid IPL and sustained bipolar cell terminals contact the sustained amacrine cell and ganglion cell processes near the inner and outer margins of the IPL.[6-8]

These functional divisions in the IPL illustrate a common organizational property of the retina. Distinct retinal neuron types have different morphological classifications. The morphology of retinal neurons determines their functional properties by constraining the IPL strata at which they receive inputs and make outputs. Because neuron morphology is so tightly correlated with neuron physiology in the retina, many studies make use of the morphological properties to help elucidate neuronal function. However, the organization of the IPL may be disrupted by degenerative photoreceptor diseases (Box 23.1).

Synaptic mechanisms shape excitatory signals in the IPL

When ON and OFF bipolar cells are depolarized by increments or decrements of illumination, respectively, they release glutamate, which excites ganglion cells and amacrine cells. Bipolar cells do not spike or fire action potentials, but instead use slow-graded depolarizations to signal.[9,10] Their transmitter release machinery is optimized for sustained glutamate release (Fig. 23.3A). L-type calcium channels in bipolar cell terminals mediate a sustained calcium influx

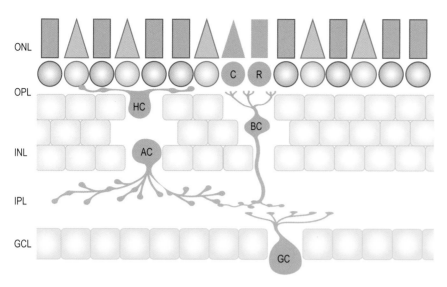

Figure 23.1 Schematic cross-sectional view of retinal layers, showing the vertical and lateral signaling pathways. Vertical signaling pathways are composed of photoreceptors, bipolar cells and ganglion cells. Two lateral pathways, comprised of horizontal cells in the OPL and amacrine cells in the IPL, modulate the flow of information along the vertical pathway. The somata of neurons are located in three cellular layers. Rod (R) and cone (C) photoreceptor somas are located in the outer nuclear layer (ONL); horizontal cell (HC), bipolar cell (BC) and amacrine cell (AC) somas are located in the inner nuclear layer (INL); ganglion cell somas are located in the ganglion cell layer (GCL). Synaptic connections occur in two layers. Synapses between photoreceptors, bipolar cells and horizontal cells are made in the outer plexiform layer (OPL). Synapses between bipolar cells, amacrine cells and ganglion cells are made in the inner plexiform layer (IPL).

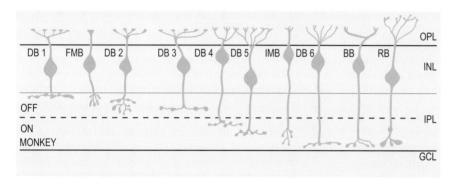

Figure 23.2 Bipolar cells form parallel retinal signaling pathways. Vertical view drawings of Golgi stained bipolar cells from macaque monkey retina illustrate distinct morphological subtypes. The axon terminals stratify at different depths of the IPL. ON bipolar cells have axons that terminate in the inner half of the IPL (ON sublamina) and OFF bipolar cells have axons that terminate in the outer half of the IPL (OFF sublamina). Rod bipolar cell (RB) dendrites exclusively contact rods. The remaining cone bipolar cells (CB) only contact cones. The OFF (FMB) and ON (IMB) midget bipolar cells and the blue bipolar cell (BB) transmit color information. (Reprinted from Ghosh KK, Bujan S, Haverkamp S, Feigenspan A, Wässle H. Types of bipolar cells in the mouse retina. J Comp Neurol 2004; 469:70–82; with permission from Wiley-Blackwell.)

Box 23.1 Degenerative Photoreceptor Diseases Disrupt the Organization of the Inner Plexiform Layer

Many retinal diseases, such as macular degeneration and retinitis pigmentosa, result in the degeneration of photoreceptors. It was long thought that the inner retina was spared in photoreceptor degeneration. Recent studies[71,72] show that this is clearly not accurate. The inner nuclear layer and IPL of animal models of photoreceptor degeneration show aberrant organization and the formation of inappropriate synaptic contacts, strongly suggesting that the elimination of photoreceptors providing the sensory input to inner retina leads to the profound disorganization of the inner retina. This large-scale structural and presumably functional reorganization of the inner retina points out that caution must be exercised when using approaches to restore vision with photoreceptor transplantation or prosthetic devices.

because they do not desensitize to prolonged depolarization, unlike other subtypes of calcium channels found at conventional CNS synapses. Similar to photoreceptor and hair cell synapses, bipolar cells possess specialized ribbons that are located at release zones, containing large numbers of tethered, glutamate-filled vesicles that mediate sustained signaling.

The excitatory output of each bipolar cell class is shaped by both pre- and postsynaptic processes (Fig. 23.3B). Response termination at bipolar cell to ganglion cell synapses[11,12] is attributed to the clearance of glutamate by presynaptic transporters, which limit the amplitude and time course of ganglion cell excitation.[11] Blocking glutamate transporters, located in the IPL, increased the amplitude and duration of the excitatory signal from bipolar cells to ganglion cells.[11,12] This role of transporters is unique to the retina[11–13] and some other specialized synapses in other parts of the CNS.[14–16] Elsewhere in the CNS, responses are typically terminated by either the chemical degradation of transmitter or rapid diffusion of transmitter from the synapse.[17,18] Glutamate transporters may regulate excitatory responses in ganglion cells by limiting signaling attributable to glutamate spillover. Spillover signaling results from released glutamate activating postsynaptic receptors that are distant from the release sites.

Glutamate transporters may affect signaling in the IPL in several ways. Transporters may improve the temporal response properties of some classes of ganglion cells by

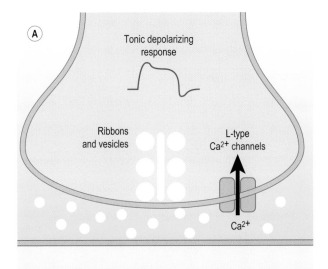

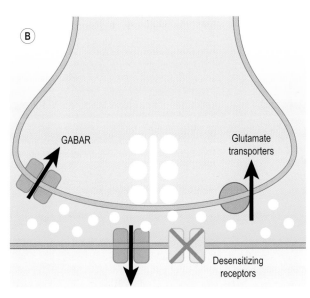

Figure 23.3 (A) Schematic of a bipolar cell terminal showing the processes that contribute to sustained glutamate release. Sustained, graded depolarizations, prolonged calcium influx through L-type calcium channels and a large pool of ribbon-associated vesicles contribute to tonic glutamate release. **(B)** Schematic illustrating the synaptic mechanisms that shape excitatory signaling from bipolar cells to ganglion cells. Transporters remove glutamate from the synapse, limiting excitatory signaling. Desensitizing glutamate receptors (X) attenuate responses to sustained glutamate release. Glutamate release may be reduced by the activation of presynaptic $GABA_A$ and $GABA_C$ receptors.

speeding the decay of ganglion cell light responses. Another possible role of transporters in the IPL is to limit crosstalk between different sublaminae in the IPL to preserve the functional segregation of distinct layers.[5,7] The findings that glutamate transporters regulate the spillover activation of glutamate receptors[11,19,20] suggests that a potential site for controlling IPL signaling is the regulation of glutamate transporters, which control the excitatory signal to ganglion cells.

The desensitizing AMPA receptor (AMPAR), a subtype of postsynaptic glutamate receptor on ganglion cell dendrites,

also shapes ganglion cell responses to glutamate released from bipolar cells. Reducing AMPAR desensitization enhances excitatory light responses in ganglion cells, suggesting that AMPAR desensitization shapes ganglion cell responses.[11,21] Additionally, glutamate and protons that are released from bipolar cells can restrict glutamate release.[22-24] Both glutamate and protons are contained in synaptic vesicles and act presynaptically upon metabotropic glutamate receptors and calcium channels, respectively, on bipolar cell terminals to reduce vesicular release.

Amacrine cells mediate inhibition in the IPL

Excitatory signaling in the IPL is also controlled by inhibitory amacrine cell inputs. Amacrine cells receive excitation from bipolar cells and mediate inhibition in the IPL by releasing either GABA or glycine onto ganglion cell dendrites, bipolar cell axon terminals and other amacrine cells (Fig. 23.4). Approximately 50 percent of amacrine cells are GABAergic and 50 percent are glycinergic. Ganglion cells receive mainly amacrine cell input,[25] underscoring the importance of inhibition in shaping the retinal output. Bipolar cell terminals are inhibited by presynaptic amacrine cell inputs, limiting excitatory signaling. Amacrine cells also inhibit other amacrine cells, resulting in serial, inhibitory synaptic interactions[26-28] that shape the timing and spatial sensitivity of inhibition. Using ERG recordings, some of the synaptic processing in the IPL can be assessed non-invasively (Box 23.2).

Presynaptic GABAergic inhibition occurs at the terminals of all classes of bipolar cells and is mediated by GABA receptors (GABARs).[29,30] The retina is unique because it possesses two types of ionotropic GABA receptors, the $GABA_A$ and $GABA_C$ receptors. Both receptor subtypes gate chloride channels, but they are composed of distinct molecular subunits, have distinct pharmacological profiles and have different biophysical properties. $GABA_C$ receptors are more sensitive to GABA than $GABA_A$ receptors and as a result have much slower activation and decay kinetics in response to GABA application compared to $GABA_A$Rs. These distinct receptor properties shape GABA-mediated synaptic responses in bipolar cells. $GABA_A$R-mediated light-evoked inhibitory postsynaptic currents (L-IPSCs) rise and decay rapidly and determine the rise time and peak amplitude of bipolar cell inhibition[31] (Fig. 23.5B). In contrast, $GABA_C$R-mediated L-IPSCs rise and decay slowly and determine the duration of bipolar cell inhibition (Fig. 23.5B). These two forms of presynaptic inhibition distinctly limit bipolar cell outputs. Slow $GABA_C$Rs limit the extent of glutamate release and the duration of excitatory responses in amacrine and ganglion cells, while fast $GABA_A$Rs limit the initial glutamate release and the initial postsynaptic excitatory responses, as detailed below.

In different bipolar cell classes, the L-IPSC time course depended on distinct contributions of $GABA_A$Rs and $GABA_C$Rs. GABAergic L-IPSCs in rod bipolar cells decay slowly because they are dominated by $GABA_C$Rs, while in OFF cone bipolar cells L-IPSCs decayed rapidly, reflecting a larger $GABA_A$R contribution (Fig. 23.5C). These different

Amacrine cells

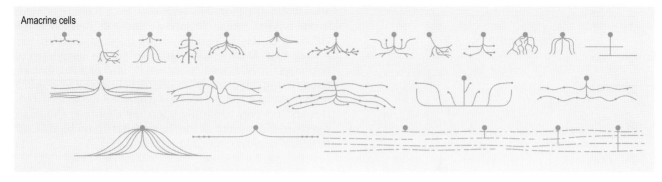

Figure 23.4 The major types of amacrine cells in mammalian retina. Amacrine cells comprise the most diverse class of retinal neurons. This illustrates amacrine cells that were morphologically identified in rabbit retina. Narrow field amacrine cells are shown in the top row and wide field amacrine cells are shown in the lower two rows. (Adapted from Masland RH. Neuronal diversity in the retina. Curr Opin Neurobiol 2001; 11:431–436; with permission from Elsevier. Copyright Elsevier 2001.)

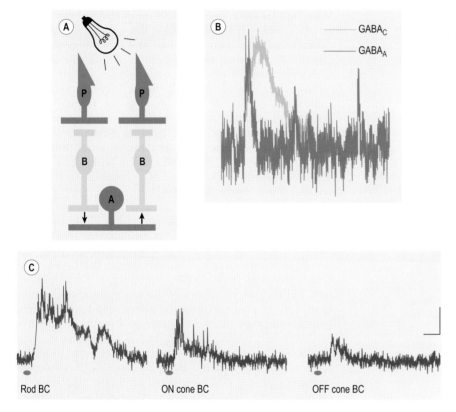

Figure 23.5 GABA$_A$ and GABA$_C$ receptors shape light-evoked inhibition at bipolar cell axon terminals. (A) Diagram illustrating light stimulation and recording procedures for measuring GABAergic L-IPSCs in bipolar cells. **(B)** Pharmacologically isolated GABA$_A$- and GABA$_C$-receptor-mediated L-IPSCs recorded from a bipolar cell, scaled to the same peak amplitude to compare response kinetics. The GABA$_A$-receptor-mediated L-IPSC exhibited fast rise and decay times, while the GABA$_C$-receptor-mediated L-IPSC exhibited slow rise and decay times. GABA-mediated L-IPSCs were isolated by pharmacologically blocking glycine receptors. The scale bar is 200 ms. (Reprinted with permission from Eggers ED, Lukasiewicz PD. Receptor and transmitter release properties set the time course of retinal inhibition. J Neurosci 2006; 26:9413–25.) **(C)** A gradient of GABA$_C$-receptor-mediated inputs differently shape bipolar cell responses. GABAergic L-IPSCs (isolated with glycine antagonist strychnine) were recorded from the bipolar cell classes. Rod bipolar cells had a large slow L-IPSC, consistent with large GABA$_C$ receptor-mediated input. ON cone bipolar cells had a faster L-IPSC and less GABA$_C$ receptor input. OFF cone bipolar cells had an even faster GABAergic L-IPSC and primarily fast GABA$_A$ receptor input. The scale bar is 5 pA and 200 ms. (Reprinted with permission from Eggers ED, McCall MA, Lukasiewicz PD. Presynaptic inhibition differentially shapes transmission in distinct circuits in the mouse retina. J Physiol 2007; 582:569–82.)

Box 23.2 Assessing IPL Function using ERGs

ERG recordings are routinely used clinically and in the laboratory for assessing photoreceptor and bipolar cell function. Retinal function attributed to synaptic interactions in the IPL can also be assessed with ERG recordings (see Chapter 24 for details). Oscillatory potentials (OPs) that are superimposed on the ERG B-wave (attributed to ON bipolar cell responses) reflect synaptic activity in the IPL.[73] Photopic negative responses and scotopic negative responses, ERG response components observed when cone and rod pathways, respectively, are activated, is thought to reflect ganglion cell responses to light-elicited inputs.[74] These more subtle components of the ERG may be helpful in assessing IPL function in the healthy and diseased retina.

contributions influence bipolar cell outputs. Removing GABA$_C$R-mediated presynaptic inhibition in bipolar cells enhances ON, but not OFF pathway signaling in the IPL,[19] suggesting that presynaptic inhibition is asymmetric (Fig. 23.6). Different complements of GABA$_A$Rs and GABA$_C$Rs appear to match the time course of inhibition with the time course of rod and cone photoreceptor input to different bipolar cells.[32]

There are two forms of presynaptic inhibition of bipolar cell terminals by amacrine cells. Local presynaptic or feedback inhibition shapes the time course[33] and extent of glutamate release.[19] Lateral inhibition, mediated by long-distance presynaptic inhibitory signaling, contributes to the antagonistic receptive field surround of ganglion cells.[34–36]

In vivo light responses

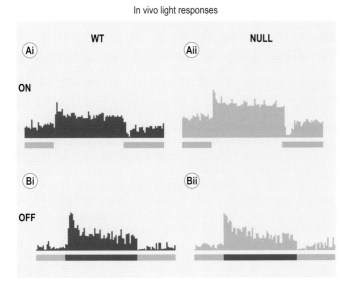

Figure 23.6 Asymmetric presynaptic inhibition differentially affects ON and OFF pathway signaling. Peristimulus time histograms (PSTHs) illustrating spontaneous and light-evoked firing in WT (Ai) and GABA$_C$R null (Null) (Aii) ON-center ganglion cells (GCs); and WT (Bi) and Null (Bii) OFF GCs. The lower traces in (A) and (B) indicate the duration of the stimuli, a bright, centered spot for ON GCs and a dark, centered spot for OFF GCs, presented on an adapting background. Spontaneous and light-evoked firing rates were significantly increased only in ON GCs in Null mice, compared to WT mice. No difference in spontaneous or light-evoked firing rates was seen between the WT and Null OFF GCs. (Modified from Sagdullaev BT, McCall MA, Lukasiewicz PD. Presynaptic inhibition modulates spillover, creating distinct dynamic response ranges of sensory output. Neuron 2006; 50:923–35; with permission from Elsevier. Copyright Elsevier 2006.)

GABAergic feedback inhibition changes the timecourse of bipolar cell signaling

GABA receptors on bipolar cell axon terminals mediate inhibitory feedback from amacrine cells that reduces and shortens bipolar cell transmitter release.[33] In the mouse, these distinct GABA receptors temporally tune inhibitory input to bipolar cells.[28] Recordings of light-evoked inhibitory inputs from rod bipolar cells show that slowly responding GABA$_C$ receptors prolong the response decay (Fig. 23.5B) and rapidly activating and decaying GABA$_A$ receptors (Fig. 23.5B) determine the rise time and peak amplitude of the response.[31] Because bipolar cell light-evoked excitatory responses are slow, GABA$_C$ receptors are thought to be better suited than GABA$_A$ receptors for mediating inhibitory feedback signals. Consistent with this, feedback inhibition to salamander bipolar cells[33,37] is more sensitive to GABA$_C$ receptor blockade than GABA$_A$ receptor blockade.

The GABA$_C$ receptor-mediated feedback inhibition of bipolar cell release also determines the nature of the excitatory signal to amacrine and ganglion cells.[19,38] When feedback to bipolar cell terminals is reduced with GABA$_C$ receptor blockers, glutamate release is increased and the NMDA receptor-mediated component of the ganglion cell light responses are enhanced, attributable to the spillover activation of NMDA receptors.[19,38] So GABA$_C$ receptor-mediated

feedback can control the excitatory ganglion cell response to light by limiting the spillover activation of NMDA receptors.

Both ON and OFF bipolar cells receive presynaptic input at their axon terminals from amacrine cells, suggesting that signaling to ON and OFF ganglion cells is shaped by presynaptic inhibition. However, presynaptic inhibition differentially shapes ON and OFF pathway signaling in the IPL.[19] When GABA$_C$ receptor-mediated presynaptic inhibition was eliminated, ON, but not OFF, ganglion cell responses were augmented, consistent with asymmetric presynaptic inhibition (Fig. 23.6). Electrophysiological measurements showed that presynaptic inhibition alters the response range in ON ganglion cells by limiting glutamate release from ON, but not OFF bipolar cells.

GABAergic inputs to the bipolar cell axon terminals contribute to surround signaling in the retina

While the receptive field surround of ganglion cells is often attributed to lateral interactions in the outer retina,[39,40] subsequent work demonstrated that inhibitory pathways in the IPL also contribute to surround inhibition measured in amacrine and ganglion cells.[34,35,41,42] Recordings from amacrine and ganglion cells suggest that GABA$_A$ and GABA$_C$ receptors on bipolar cell terminals, and GABA$_A$ receptors on the dendrites of third-order cells, mediate surround signaling.[35,41,43] The details of the lateral inhibitory signals that modulate distinct bipolar cell signaling pathways are starting to emerge. Since each bipolar cell class has different proportions of GABA$_A$ and GABA$_C$ receptors, lateral, inhibitory signaling for different bipolar cell pathways may vary.[30,32]

We do not know the relative roles of lateral inhibition feeding back to bipolar cell terminals versus lateral inhibition feeding forward onto third-order neuron dendrites. Experiments in salamander retina show that dim surround illumination activated GABA$_C$ receptors on bipolar terminals, but did not activate GABA$_A$ receptors on ganglion cell dendrites. GABA$_A$ receptor-mediated IPSCs were recorded in ganglion cells, but only in response to bright surround illumination.[36] These differences may be attributable to either the different complements of GABA receptors on bipolar and ganglion cells, or different inhibitory pathways contacting the two cell types.

The contributions of the inner and outer retina to ganglion cell receptive field surround organization

Although there is evidence for lateral inhibition in both the OPL and the IPL, the relative roles of these two pathways in the formation of the ganglion cell surround are not known. The roles of these inner and outer retinal pathways were investigated by measuring the effects of surround illumination on ganglion cells responses after either IPL or OPL signaling was blocked.[36] Dim surround illumination reduced the ganglion cell response to center illumination. The dim surround effect was diminished upon GABA$_C$ receptor blockade, consistent with an IPL signaling role. Brighter surround illumination further reduced ganglion cell center responses, but this effect was insensitive to GABA or glycine receptor blockade, suggesting that it was not mediated by the inner retina. Carbenoxolone or HEPES, both of which decrease horizontal cell feedback to cones,[44,45] reduced the bright surround effects,[36,46,47] suggesting they were mediated in the OPL. Thus lateral, inhibitory pathways in the OPL and IPL may affect ganglion cell responses to center illumination in different ways. The effects of bright surround illumination were mediated by horizontal cells in the outer retina, while the effects of dim surround illumination were mediated by GABAergic amacrine cells in the inner retina.

Glycinergic inhibition plays several different roles in the IPL

Presynaptic glycinergic inhibition has different roles than GABAergic inhibition. Rod and OFF cone (but not ON cone) bipolar cells receive input from glycinergic amacrine cells.[28,32] Glycine signaling plays a prominent role in the rod signaling pathway. In mammals, rod signals flow to ganglion cells by a unique pathway, consisting of rod bipolar cells and AII amacrine cells that both depolarize in response to light. AII amacrine cells then split the rod signal into the ON and OFF pathways within the IPL by contacting ON and OFF bipolar cell terminals via electrical and inhibitory chemical synapses, respectively. Slow rod signals are transmitted to OFF bipolar cell terminals by AII amacrine cells, which in turn contact OFF ganglion cells that send rod signals to the brain. Light-evoked presynaptic inhibition is slow and mediated by glycine, matching the time course of rod signaling. In contrast to GABAergic inhibition, the slow time course of glycinergic inhibition is determined primarily by transmitter release and not receptor properties.[48]

The two inhibitory transmitters GABA and glycine are utilized in the IPL in distinct signaling pathways. In mammalian retina, GABAergic amacrine cells have wide field processes, signal laterally and are confined to the same sublaminae,[49] where they play a critical role in the processing of spatial information and contribute to the receptive field surround of ganglion cells.[34,35] Wide field GABAergic amacrine cells synapse onto bipolar cell terminals and ganglion cell dendrites to form two IPL surround signaling pathways. In contrast to wide field GABAergic amacrine cells, glycinergic amacrine cells have narrow field processes that extend vertically within the IPL, mediating signaling between different sublaminae.[5,50] Glycinergic inhibition, which is reduced by light, acts in a push–pull manner with increased bipolar cell excitation to enhance light responses in amacrine, bipolar and ganglion cells.[51-53]

Neuromodulators in the IPL

In addition to inhibitory synaptic transmission that shapes signaling in the IPL, neuromodulators are also prevalent in the retina that shape inner retinal signaling. Dopamine, the best-studied neuromodulator in the retina, is found in a small subset of amacrine cells (reviewed by Witkovsky[54]). Although the population of dopaminergic amacrine cells is small in number, their processes tile or cover the entire retina. GABA is co-localized with dopamine in this class of amacrine cells. Dopamine can act as a paracrine substance that is released into the extracellular space, affecting both local and distant targets. Dopamine is a mediator of light adaptation in the retina that acts in both the inner and outer retina. Dopamine reduces signaling through rod pathways and enhances signaling through cone pathways. At the IPL, dopamine reduces rod-mediated signaling by uncoupling the electrical synapses between AII amacrine cells. Dopamine may also modulate fast synaptic signaling by $GABA_A$ and $GABA_C$ receptors in the IPL.

Numerous neuroactive peptides are localized to specific populations of amacrine cells with immunocytochemistry.[55] Throughout the CNS, neuropeptides modulate the actions of fast excitatory and inhibitory transmitters or the activity of voltage-gated channels. The precise mechanisms by which neuropeptides modify visual information processing in the retina remain unknown. Different amacrine cells contain neuroactive peptides, such as neuropeptide Y, somatostatin, vasoactive intestinal polypeptide and enkephalins. Imaging experiments suggest that both neuropeptide Y[56] and somatostatin[57] inhibit calcium currents in rod bipolar cell terminals, presumably limiting glutamate release.

Parallel ganglion cell output pathways

Another critical processing mechanism in the inner retina is the separation of visual information into bipolar cell pathways. Parallel bipolar cell signaling pathways relay different aspects of the visual signal to 10–15 morphological types of ganglion cells.[49,58] Ganglion cells are also divided into two major ON and OFF classes, both possessing antagonistic surrounds[59] (Fig. 23.7). ON and OFF types can be further divided into X- and Y-cells that respond to illumination in either a sustained (X) or transient (Y) fashion, similar to bipolar cells[60] and correspond to distinct morphological types, alpha and beta ganglion cells, respectively.[61] The X and Y ganglion cells are similar to the parvo (P) and magno (M) classes of primate ganglion cells, corresponding to the morphological midget and parasol classes, respectively.[62] P-ganglion cells transmit form and color information, while M-ganglion cells transmit motion and spatial relationship information to the lateral geniculate nucleus.[63] Retinal diseases such as diabetic retinopathy and glaucoma can result in structural and functional pathologies in the IPL and in ganglion cells (Boxes 23.3 and 23.4).

Ganglion cells encode color information

Midget ganglion cells transmit color information to the brain – they receive input from midget bipolar cells. Their central, excitatory, response is determined by a strong single red or green cone input, while the inhibitory surround is determined by weak mixed cone inputs, resulting in red–green color opponency. Another ganglion cell that processes color is a bistratified ganglion cell that is excited by blue light, via blue bipolar cells that make contacts at the inner IPL, and inhibited by yellow light, via red and green signaling bipolar cells that make contacts at the outer IPL to process blue–yellow color opponency. These two ganglion

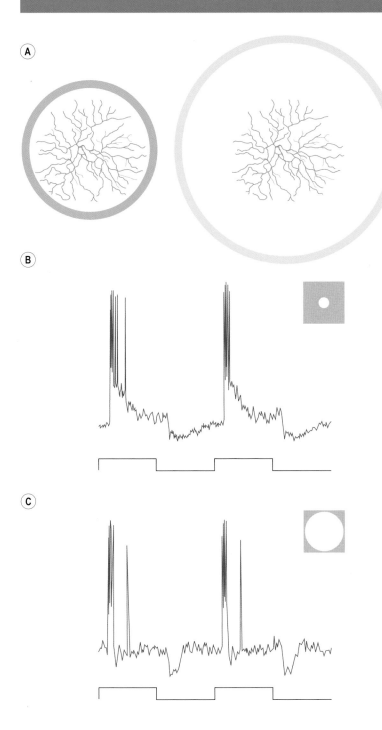

Figure 23.7 (**A**) Drawing of a parasol ganglion cell dendritic arbor. A stimulus the diameter of the solid dark green line would activate the receptive field center, and a stimulus the diameter of the solid light green line would also activate the receptive field surround. (Reprinted with permission from Diller L, Packer OS, Verweij J et al. L and M cone contributions to the midget and parasol ganglion cell receptive fields of macaque monkey retina. J Neurosci 2004; 24:1079–88.) (**B**) Responses of an ON parasol ganglion cell to a 300-μm-diameter spot (stimulus trace below) square wave stimulus that would activate the receptive field center. (**C**) Responses of a parasol ganglion cell to a 2000-μm-diameter spot (stimulus trace below) square wave stimulus that would activate both the center and surround. The response is smaller because of the surround antagonism. (Reprinted with permission Davenport CM, Detwiler PB, Dacey DM. Effects of pH buffering on horizontal and ganglion cell light responses in primate retina: evidence for the proton hypothesis of surround formation. J Neurosci 2008; 28:456–64.)

cell pathways are the main carriers of color information into the visual cortex.[64]

Directional selective ganglion cells

An additional ganglion cell type – directionally selective (D-S) cells – responds to the movement of images, but only in preferred directions. Image motion in the preferred direction excites D-S cells, but image motion in the null direction elicits no response because inhibition reduces the excitatory input. Although D-S ganglion cells were first described over 40 years ago,[65] the precise synaptic basis for their behavior remains unknown. One point of agreement between most studies is that a type of amacrine cell called the starburst amacrine cell is critical for D-S responses in ganglion cells. The genetic ablation of starburst cells, which provide both cholinergic and GABAergic input to D-S cells, results in the elimination D-S responses.[66]

Intrinsically photosensitive ganglion cells

Surprisingly, about 1–3 percent of ganglion cells can directly sense light.[67] This phenomenon has been attributed to presence of the photopigment melanopsin.[68] Disrupting rod and cone signaling to these ganglion cells does not eliminate their light responsiveness, consistent with their intrinsic photosensitivity.[67] Light responses from these ganglion cells are distinct from conventional ganglion cells. These responses

Box 23.3 Ganglion Cell and IPL Pathologies Occur in Diabetic Retinopathy

While diabetic retinopathy is viewed clinically as a disease affecting retinal vasculature, a large body of work shows that it also has neurodegenerative effects in the IPL and in ganglion cells. Diabetic patients and experimental animals have abnormal oscillatory potentials in their ERG responses, consistent with signaling abnormalities in the IPL.[75] Morphometric measurements show that ganglion cell dendritic trees are smaller and less elaborate in diabetic retinas.[76] The functional and morphological abnormalities observed in diabetic retinas are often observed before vascular-related pathologies,[75] suggesting a primary retinal dysfunction before vascular problems arise. These early functional deficits also suggest an avenue of detection of early diabetic retinopathy.

Box 23.4 Glaucoma, a Leading Cause of Blindness, is Attributed to Ganglion Cell Loss

Animal models of glaucoma show that ganglion cell morphology, synaptic connectivity and function is altered in early stages of the disease.[77–79] These changes are manifested in the IPL and suggest that synaptic inputs to ganglion cells and their ability to process visual information is perturbed. The mechanism of damage to ganglion cells in glaucoma is unclear, but these early changes provide an opportunity for ganglion cell protection, and possibly prevention of their eventual death.

are long-lasting and activated by moderate to bright light intensities. Intrinsically photosensitive ganglion cells also receive input from higher-sensitivity rod and cone synaptic pathways, which elicit lower light intensity responses. Although these ganglion cells are relatively few in number, they have extensive dendritic fields that completely tile the retina, leading to responsivity across the retina. These ganglion cells project to brain areas that control pupillary reflexes and the entrainment of circadian rhythms.

Ganglion cell types form a retinal mosaic

The subtypes of neurons that form the bipolar cell and ganglion cell parallel pathways each sample the entire retina, forming a mosaic of cells through a developmental process called tiling. Each individual cell type is distributed across the surface of the retina, with very little overlap between cells of the same type.[69] Since each pathway is sampling a different kind of visual information, this allows each cell type to reassemble the entire visual scene and contribute unique information to the downstream visual pathways.[70] This tiling mechanism occurs through developmental processes that are not yet well understood.

Conclusions

The IPL is a critical stage in visual processing. Distinct bipolar cell types comprise parallel input streams to the IPL. This information is then shaped by synaptic interaction between inhibitory amacrine cells with the bipolar cell inputs and

ganglion cell outputs of the IPL. The ability to detect the direction of movement and certain colors occurs because of processing in the IPL.

References

1. Slaughter MM, Miller RF. 2-Amino-4-phosphonobutyric acid: a new pharmacological tool for retina research. Science 1981; 211:182–185.
2. DeVries SH. Bipolar cells use kainate and AMPA receptors to filter visual information into separate channels. Neuron 2000; 28(3):847–856.
3. Nelson R, Famiglietti JEV, Kolb H. Intracellular staining reveals different levels of stratification for on- and off-center ganglion cells in cat retina. J Neurophys 1978; 41(2):472–483.
4. Famiglietti J, Edward V., Kaneko A, Tachibana M. Neuronal architecture of on and off pathways to ganglion cells in carp retina. Science 1977; 198:1267–1269.
5. Roska B, Werblin F. Vertical interactions across ten parallel, stacked representations in the mammalian retina. Nature 2001; 410(6828):583–587.
6. Awatramani GB, Slaughter MM. Origin of transient and sustained responses in ganglion cells of the retina. J Neurosci 2000; 20:7087–7095.
7. Wu SM, Gao F, Maple BR. Functional architecture of synapses in the inner retina: segregation of visual signals by stratification of bipolar cell axon terminals. J Neurosci 2000; 20:4462–4470.
8. Ichinose T, Shields CR, Lukasiewicz PD. Sodium channels in bipolar cells enhance the light sensitivity of ganglion cells. Invest Ophthalmol Vis Sci 2003; 44:E-Abstract 1007.
9. Lagnado L, Gomis A, Job C. Continuous vesicle cycling in the synaptic terminal of retinal bipolar cells. Neuron 1996; 17:957–967.
10. von Gersdorff H, Sakaba T, Berglund K, Tachibana M. Submillisecond kinetics of glutamate release from a sensory synapse. Neuron 1998; 21:1177–1188.
11. Higgs MH, Lukasiewicz PD. Glutamate uptake limits synaptic excitation of retinal ganglion cells. J Neurosci 1999; 19:3691–3700.
12. Matsui K, Hosoi N, Tachibana M. Active role of glutamate uptake in the synaptic transmission from retinal nonspiking neurons. J Neurosci 1999;19:6755–6766.
13. Eliasof S, Werblin F. Characterization of the glutamate transporter in retinal cones of the tiger salamander. J Neurosci 1993; 13:402–411.
14. Otis TS, Wu YC, Trussell LO. Delayed clearance of transmitter and the role of glutamate transporters at synapses with multiple release sites. J Neurosci 1996; 16:1634–1644.
15. Kinney GA, Overstreet LS, Slater NT. Prolonged physiological entrapment of glutamate in the synaptic cleft of cerebellar unipolar brush cells. J Neurophys 1997; 78:1320–1333.
16. Isaacson JS. Glutamate spillover mediates excitatory transmission in the rat olfactory bulb. Neuron 1999; 23:377–384.
17. Isaacson JS, Nicoll RA. The uptake inhibitor L-trans-PDC enhances responses to glutamate but fails to alter the kinetics of excitatory synaptic currents in the hippocampus. J Neurophysiol 1993; 70(5):2187–2191.
18. Sarantis M, Ballerini L, Miller B, Silver RA, Edwards M, Attwell D. Glutamate uptake from the synaptic cleft does not shape the decay of the non-NMDA component of the synaptic current. Neuron 1993; 11:541–549.
19. Sagdullaev BT, McCall MA, Lukasiewicz PD. Presynaptic inhibition modulates spillover, creating distinct dynamic response ranges of sensory output. Neuron 2006; 50(6):923–935.
20. Chen S, Diamond JS. Synaptically released glutamate activates extrasynaptic NMDA receptors on cells in the ganglion cell layer of the rat retina. J Neurosci 2002; 22:2165–2173.
21. Lukasiewicz PD, Lawrence JE, Valentino TL. Desensitizing glutamate receptors shape excitatory synaptic inputs to tiger salamander retinal ganglion cells. J Neurosci 1995; 15:6189–6199.
22. Higgs MH, Tran MN, Romano C, Lukasiewicz PD. Group III metabotropic glutamate receptors inhibit glutamate release from retinal bipolar cells. Invest Ophthalmol Vis Sci 2000; 41:S621.
23. Awatramani GB, Slaughter MM. Intensity-dependent, rapid activation of presynaptic metabotropic glutamate receptors at a central synapse. J Neurosci 2001; 15:741–749.
24. Palmer MJ, Hull C, Vigh J, von Gersdorff H. Synaptic cleft acidification and modulation of short-term depression by exocytosed protons in retinal bipolar cells. J Neurosci 2003; 23(36):11332–11341.
25. Masland RH. The fundamental plan of the retina. Nat Neurosci 2001; 4(9):877–886.
26. Zhang J, Chang-Sub J, Slaughter MM. Serial inhibitory synapses in retina. Vis Neurosci 1997; 14:553–563.
27. Roska B, Nemeth E, Werblin FS. Response to change Is facilitated by a three-neuron disinhibitory pathway in the tiger salamander retina. J Neurosci 1998; 18:3451–3459.
28. Eggers ED, Lukasiewicz PD. GABAA, GABAC and glycine receptor-mediated inhibition differentially affects light-evoked signaling from mouse retinal rod bipolar cells. J Physiol 2006; 572:215–225.
29. Euler T, Wassle H. Different contributions of GABAA and GABAC receptors to rod and cone bipolar cells in a rat retinal slice preparation. J Neurophys 1998; 79:1384–1395.
30. Shields CR, Tran MN, Wong RO, Lukasiewicz PD. Distinct ionotropic GABA receptors mediate presynaptic and postsynaptic inhibition in retinal bipolar cells. J Neurosci 2000; 20(7):2673–2682.
31. Eggers ED, Lukasiewicz PD. Temporal shaping of IPSCs by GABAA, GABAC and glycine receptors. Soc Neurosci Abstr 2005; 955.3.
32. Eggers ED, McCall MA, Lukasiewicz PD. Presynaptic inhibition differentially shapes transmission in distinct circuits in the mouse retina. J Physiol 2007; 582(Pt 2):569–582.
33. Dong C, Werblin FS. Temporal contrast enhancement via GABAC feedback at bipolar terminals in the tiger salamander retina. J Neurophys 1998; 79:2171–2180.
34. Cook PB, McReynolds JS. Lateral inhibition in the inner retina is important for spatial tuning of ganglion cells. Nat Neurosci 1998; 1:714–719.
35. Flores-Herr N, Protti DA, Wassle H. Synaptic currents generating the inhibitory surround of ganglion cells in the mammalian retina. J Neurosci 2001; 21:4852–4863.
36. Ichinose T, Lukasiewicz PD. Inner and outer retinal pathways both contribute to surround inhibition of salamander ganglion cells. J Physiol 2005; 565(Pt 2):517–535.

37. Freed MA, Smith RG, Sterling P. Timing of quantal release from the retinal bipolar terminal is regulated by a feedback circuit. Neuron 2003; 38(1):89–101.

38. Matsui K, Hasegawa J, Tachibana M. Modulation of excitatory synaptic transmission by GABA(C) receptor-mediated feedback in the mouse inner retina. J Neurophysiol 2001; 86(5):2285–2298.

39. Werblin FS. The control of sensitivity in the retina. Sci Am 1973; 228:70–79.

40. Mangel SC. Analysis of the horizontal cell contributions to the receptive field surround of ganglion cells in the rabbit retina. J Physiol (Lond) 1991; 442:211–234.

41. Volgyi B, Xin D, Bloomfield SA. Feedback inhibition in the inner plexiform layer underlies the surround-mediated responses of AII amacrine cells in the mammalian retina. J Physiol (Lond) 2002; 539(Pt 2):603–614.

42. Taylor WR. TTX attenuates surround inhibition in rabbit retinal ganglion cells. Vis Neurosci 1999; 16(2):285–290.

43. Bloomfield SA, Xin D. Surround inhibition of mammalian AII amacrine cells is generated in the proximal retina. J Physiol (Lond) 2000; 523(Pt 3):771–783.

44. Kamermans M. Hemichannel mediated inhibition in the outer retina. Science 2001; 292:1178–1180.

45. Verweij J, Hornstein EP, Schnapf JL. Surround antagonism in macaque cone photoreceptors. J Neurosci 2003; 23(32):10249–10257.

46. Davenport CM, Detwiler PB, Dacey DM. The mechanism of blue-yellow opponency in the primate small bistratified cell. FASEB Summer Research Conferences Abstract 2006.

47. McMahon MJ, Packer OS, Dacey DM. The classical receptive field surround of primate parasol ganglion cells is mediated primarily by a non-GABAergic pathway. J Neurosci 2004; 24(15):3736–3745.

48. Eggers ED, Lukasiewicz PD. Receptor and transmitter release properties set the time course of retinal inhibition. J Neurosci 2006; 26(37):9413–9425.

49. Wassle H. Parallel processing in the mammalian retina. Nat Rev Neurosci 2004; 5(10):747–757.

50. Werblin F, Roska B, Balya D. Parallel processing in the mammalian retina: lateral and vertical interactions across stacked representations. Prog Brain Res 2001; 131:229–238.

51. Manookin MB, Beaudoin DL, Ernst ZR, Flagel LJ, Demb JB. Disinhibition combines with excitation to extend the operating range of the OFF visual pathway in daylight. J Neurosci 2008; 28(16):4136–4150.

52. Molnar A, Werblin F. Inhibitory feedback shapes bipolar cell responses in the rabbit retina. J Neurophysiol 2007; 98(6):3423–3435.

53. Hsueh HA, Molnar A, Werblin FS. Amacrine to amacrine cell inhibition in the rabbit retina. J Neurophysiol 2008.

54. Witkovsky P. Dopamine and retinal function. Doc Ophthalmol 2004; 108(1):17–40.

55. Karten HJ, Brecha N. Localization of neuroactive substances in the vertebrate retina: evidence for lamination in the inner plexiform layer. Vision Res 1983; 23(10):1197–1205.

56. D'Angelo I, Brecha NC. Y2 receptor expression and inhibition of voltage-dependent Ca²⁺ influx into rod bipolar cell terminals. Neuroscience 2004; 125(4):1039–1049.

57. Johnson J, Caravelli ML, Brecha NC. Somatostatin inhibits calcium influx into rat rod bipolar cell axonal terminals. Vis Neurosci 2001; 18(1):101–108.

58. Rockhill RL, Daly FJ, MacNeil MA, Brown SP, Masland RH. The diversity of ganglion cells in a mammalian retina. J Neurosci 2002; 22(9):3831–3843.

59. Kuffler SW. Discharge patterns and functional organization of mammalian retina. J Neurophys 1953; 16:37–68.

60. Enroth-Cugell C, Robson JG. The contrast sensitivity of retinal ganglion cells of the cat. J Physiol (Lond) 1966; 187:517–552.

61. Boycott BB, Wassle H. The morphological types of ganglion cells of the domestic cat's retina. J Physiol (Lond) 1974; 240:397–419.

62. Callaway EM. Structure and function of parallel pathways in the primate early visual system. J Physiol 2005; 566(Pt 1):13–19.

63. Van Essen DC, Gallant JL. Neural mechanisms of form and motion processing in the primate visual system. Neuron 1994; 13(1):1–10.

64. Dacey DM. Parallel pathways for spectral coding in primate retina. Annu Rev Neurosci 2000; 23:743–775.

65. Barlow HB, Hill RM, Levick WR. Retinal ganglion cells responding selectively to direction and speed of image motion in the rabbit. J Physiol (Lond) 1964; 173:377–407.

66. Yoshida K, Watanabe D, Ishikane H, Tachibana M, Pastan I, Nakanishi S. A key role of starburst amacrine cells in originating retinal directional selectivity and optokinetic eye movement. Neuron 2001; 30(3):771–780.

67. Berson DM, Dunn FA, Takao M. Phototransduction by retinal ganglion cells that set the circadian clock. Science 2002; 295(5557):1070–1073.

68. Panda S, Provencio I, Tu DC et al. Melanopsin is required for non-image-forming photic responses in blind mice. Science 2003; 301(5632):525–527.

69. Wässle H. Parallel processing in the mammalian retina. Nat Rev Neurosci 2004; 5(10):747–757.

70. Field GD, Chichilnisky EJ. Information processing in the primate retina: circuitry and coding. Annu Rev Neurosci 2007; 30:1–30.

71. Jones BW, Watt CB, Frederick JM et al. Retinal remodeling triggered by photoreceptor degenerations. J Comp Neurol 2003; 464(1):1–16.

72. Strettoi E, Porciatti V, Falsini B, Pignatelli V, Rossi C. Morphological and functional abnormalities in the inner retina of the rd/rd mouse. J Neurosci 2002; 22(13):5492–5504.

73. Wachmeister L. Oscillatory potentials in the retina: what do they reveal. Prog Retin Eye Res 1998; 17:485–521.

74. Bui BV, Fortune B. Ganglion cell contributions to the rat full-field electroretinogram. J Physiol 2004; 555(Pt 1):153–173.

75. Barber AJ. A new view of diabetic retinopathy: a neurodegenerative disease of the eye. Prog Neuropsychopharmacol Biol Psychiatry 2003; 27(2):283–290.

76. Gastinger MJ, Kunselman AR, Conboy EE, Bronson SK, Barber AJ. Dendrite remodeling and other abnormalities in the retinal ganglion cells of Ins2 Akita diabetic mice. Invest Ophthalmol Vis Sci 2008;49(6): 2635–2642.

77. Fortune B, Bui BV, Morrison JC et al. Selective ganglion cell functional loss in rats with experimental glaucoma. Invest Ophthalmol Vis Sci 2004; 45(6):1854–1862.

78. Weber AJ, Harman CD. Structure-function relations of parasol cells in the normal and glaucomatous primate retina. Invest Ophthalmol Vis Sci 2005; 46(9):3197–3207.

79. Stevens B, Allen NJ, Vazquez LE et al. The classical complement cascade mediates CNS synapse elimination. Cell 2007; 131(6):1164–1178.

CHAPTER 24

Electroretinogram of Human, Monkey and Mouse

Laura J. Frishman & Minhua H. Wang

Introduction

The electroretinogram (ERG) is a useful tool for objective, non-invasive assessment of retinal function both in the clinic and the laboratory. It is a mass electrical potential that represents the summed response of all the cells in the retina to a change in illumination. Recordings can be made in vivo under physiological or nearly physiological conditions using electrodes placed on the corneal surface. For standard recordings in the clinic,[1] ERGs such as those illustrated in Figure 24.1 are recorded from alert subjects who are asked not to blink or to move their eyes. Anesthesia, selected for having minimal effects on retinal function, may be required for recordings from some very young human subjects, and is generally used for recordings in animals. The positive and negative waves of the ERG reflect the summed activity of overlapping positive and negative component potentials that originate from different stages of retinal processing. The choice of stimulus conditions and method of analysis will determine which of the various retinal cells and circuits are generating the response. Information about retinal function provided by the ERG is useful for diagnosing and characterizing retinal diseases, as well as for monitoring disease progression, and evaluating the effectiveness of therapeutic interventions.

Much of the basic research on the origins, pathophysiology and treatment of retinal diseases that occur in humans is carried out in animal models. The ERG provides a simple objective approach for assessing retinal function in animals. In recent years it has been of particular benefit in studies of mouse (rat, and other species) models of retinal disease, and it has been valuable for evaluating the retinal drug toxicity.[2] The ERG also has been useful for characterizing changes in retinal function that occur as a consequence of genetic alterations in mice and other models, which affect the transmission and processing of visual signals in the retina.[3,4]

This chapter provides information on retinal origins and interpretation of the ERG, with emphasis on advances in our understanding of the ERG that have occurred through pharmacological dissection studies in macaque monkeys whose retinas are similar to those of humans. The similarity of the waveforms of flash ERGs of humans and macaques can be seen in Figure 24.2. The chapter also will examine origins of the mouse ERG, and consider similarities and differences

between mouse and primate ERG. Although the focus of this chapter is on primate and rodent retinas, it is important to note that the ERG is a valuable tool for assessing retinal function of all classes of vertebrates. This includes amphibians and fish in which classical studies of the retina were carried out (for a review, see Dowling[5]) and in which current work continues to improve our understanding of retinal function under normal, genetically altered[6] and pathological conditions.

This chapter provides useful background information for interpreting the ERG, but does not provide a comprehensive review of the characteristics of ERGs associated with retinal disorders encountered in the clinic. Recent texts by Fishman et al,[7] Heckenlively & Arden[8] and Lam[9] are recommended resources for learning more about clinical applications.

Generation of the ERG

Radial current flow

The ERG is an extracellular potential that arises from currents that flow through the retina as a result of neuronal signaling, and for some slower ERG waves, potassium (K^+) currents in glial cells. Local changes in the membrane conductance of activated cells give rise to inward or outward ion currents, and cause currents to flow in the extracellular space (ECS) around the cells, creating extracellular potentials. Although all retinal cell types can contribute to ERGs recorded at the cornea, the contribution of a particular type may be large, or hardly noticeable in the recorded waveform, depending upon several factors detailed below.

The orientation of a cell type in the retina is a major factor in determining the extent to which its activity will contribute to the ERG. A schematic drawing of the mammalian retina, with the various cell types labeled, can be found in Chapter 21 (Figs 21.1 & 21.2). When activated synchronously by a change in illumination, retinal neurons that are radially oriented with respect to the cornea, i.e. the photoreceptors, and bipolar cells, make larger contributions to the major waves of the ERG than the more laterally oriented cells and their processes, i.e. the horizontal and amacrine cells. The major waves at light onset are, as marked in Figures 24.1 and 24.2, an initial negative-going a-wave (mainly from photoreceptors),

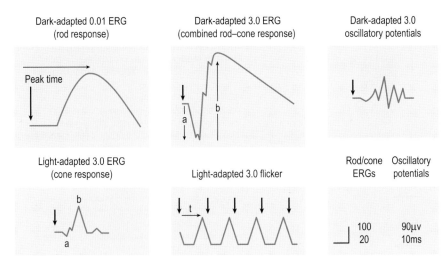

Dark-adapted 0.01 ERG
(rod response)

Peak time

Dark-adapted 3.0 ERG
(combined rod–cone response)

a

b

Dark-adapted 3.0
oscillatory potentials

Light-adapted 3.0 ERG
(cone response)

b

a

Light-adapted 3.0 flicker

t

Rod/cone   Oscillatory
ERGs   potentials

100   90μv
20   10ms

Figure 24.1 Standard tests for full field clinical electrophysiology. The Society for the Clinical Electrophysiology of Vision (ISCEV) has recommended five standard ERG tests with full-field stimulation for use worldwide in clinical electrodiagnostic facilities. This figure, from the 2008 update of "ISCEV Standard for Full-Field Clinical Electroretinography"[1] shows examples, but not norms, of ERGs elicited using the recommended tests in normal humans. Light calibrations in candela seconds per meter squared (cd.s.m^{-2}) for each test appear above the ERGs. Large arrowheads indicate the time at which the stimulus flash occurred. Dotted arrows show common ways to measure time-to-peak ("t", implicit time), and a- and b-wave amplitudes. For some purposes, e.g. quantitative analyses to separate various ERG components, amplitude measurements at a fixed time after stimulus onset are more appropriate (see Fig. 24.12). (From Marmor MF, Fulton AB, Holder GE, Miyake Y, Brigell M, Bach M. ISCEV Standard for full-field clinical electroretinography (2008 update). Documenta Ophthalmologica 2009;118(1): pp. 69–77, used with permission.)

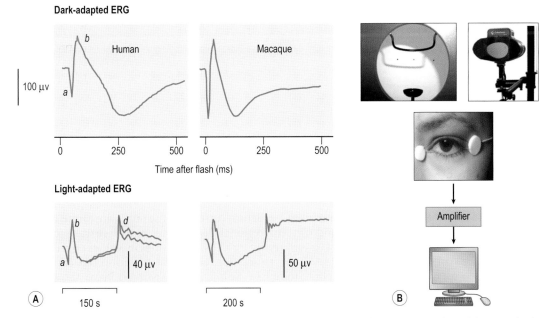

Dark-adapted ERG

100 μv

b

a

Human

Macaque

0   250   500   0   250   500

Time after flash (ms)

Light-adapted ERG

b

a

d

40 μv

50 μv

(A) 150 s   200 s

Amplifier

(B)

Figure 24.2 Full-field flash electroretinograms (ERGs) of human subjects and macaque monkeys. (A) Top: Dark-adapted ERG responses to brief high energy flashes of ~400 scotopic troland seconds (sc td.s) from darkness that occurred at time zero for a normal alert human subject (left) and an anesthetized macaque (right). For macaque ERGs, and ERGs from all animals in subsequent figures, subjects were anesthetized. (Adapted from Robson & Frishman 1998,[16] used with permission.) Bottom: Light-adapted ERGs in response to longer duration flashes (150 or 200 ms). For both subjects, white Ganzfeld flashes of 4.0 log photopic (ph) td were presented on a steady rod-saturating background of 3.3 log sc td. (Adapted from Sieving et al 1994,[110] used with permission.) **(B)** ERG recording setup: Recordings made using a traditional Ganzfeld bowl (left) or more modern LED-based full-field stimulator. Use of DTL fiber electrodes[32] is illustrated. ERG recordings are amplified and sent to a computer for averaging, display and analysis.

which is truncated by a positive-going b-wave, mainly generated by ON (depolarizing) bipolar cells. For longer duration flashes (Fig. 24.2, bottom row), the light-adapted ERG includes another positive-going wave, the d-wave, at light offset, with major contribution from OFF (hyperpolarizing) bipolar cells. Currents that leave retinal cells and enter the ECS at one retinal depth (the current source), will leave the

ECS to re-enter the cells at another (the current sink), creating a current dipole. These retinal currents also travel through the vitreous humor to the cornea, where the ERG can be recorded non-invasively, as well as through the extraocular tissue, sclera, choroid and high resistance of the retinal pigment epithelium (RPE) before returning to the retina. Local ERGs can be recorded near the retinal generators using

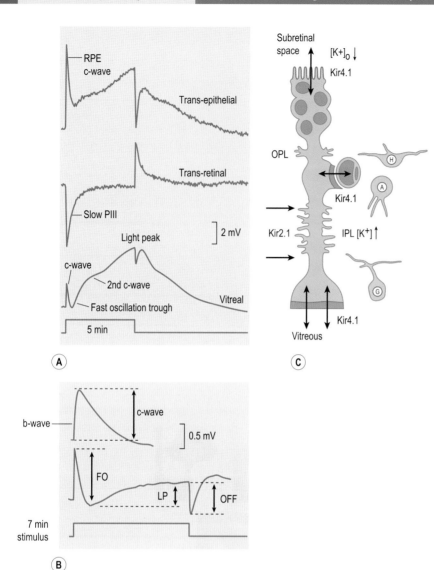

Figure 24.3 Distal retinal components of the direct current (dc)-ERG, and distribution of inward rectifying potassium channels, Kir2.1 and Kir4.1 in Müller cells. (**A**) Simultaneous intraretinal (local) and vitreal (global) ERG recordings from the intact eye of the cat. Top: Trans-retinal, and trans-epithelial (TEP) potentials were recorded in response to a 5-minute period of illumination. The a- and b-waves cannot be seen due to the compressed time scale. The intraretinal recordings show the two (sub) components of the c-wave: (1) the retinal pigment epithelium (RPE) c-wave, which is the TEP recorded between a microelectrode in the subretinal space (SRS) and a retrobulbar reference, and (2) slow PIII, the cornea-negative trans-retinal component generated by Müller cell currents (see text), recorded between the same microelectrode and the vitreal electrode. The vitreal ERG (bottom), recorded between the vitreal electrode and retrobulbar reference, is the sum of the RPE c-wave, and slow PIII. The c-wave is followed by the fast oscillation trough (FOT) and then the light peak (LP), both of which are exclusively RPE responses. (From Steinberg et al 1985,[161] used with permission.) (**B**) dc-ERG recorded from a C57BL/6 mouse in response to a 7-minute period of illumination. FO, LP, and OFF-response amplitudes are marked by arrows. Upper record shows the early portion of the response on an expanded (×5) time scale. Amplitude calibration: 0.5 mV. (From Wu et al 2004,[35] used with permission.) (**C**) Strongly rectifying Kir2.1 channels are widely distributed in the Müller cell membrane. K+ in synaptic regions enters Müller cells through strongly rectifying Kir2.1 channels, and exits through weakly rectifying (bidirectional) Kir4.1 channels concentrated in the ILM, OLL and Müller cell processes around blood vessels. (Modified from Kofuji et al 2002,[14] used with permission.)

intraretinal microelectrodes in animals, while simultaneously recording the global ERG elsewhere in the current path, e.g. from the corneal surface, or using an electrode in the vitreous humor, with a reference electrode behind the eye.[10,11] Such recordings have provided useful information about the origins of the various waves of the ERG.

Glial currents

The ERG waveform also will be affected by glial cell currents. Retinal glial cells include Müller cells, RPE cells, and radial astrocytes in the optic nerve head. One crucial function of glia is to regulate extracellular K+ concentration, $[K^+]_o$, to maintain the electrochemical gradients across cell membranes that are necessary for normal neuronal function. Membrane depolarization and spiking in retinal neurons that occur in response to changes in illumination lead to leak of K+ from the neurons and to K+ accumulation in the ECS. Membrane hyperpolarization, in contrast, leads to lower $[K^+]_o$ as the membrane leak conductance is reduced, but the Na+ K+ ATPase in the membrane continues to transport K+ into the cell.

K+ currents in Müller cells move excess K+ from areas of high $[K^+]_o$ to areas of lower $[K^+]_o$ by a process called spatial

buffering.[12,13] Return currents in the retina are formed by Na+ and Cl−. The regional distribution and electrical properties of inward rectifying K+ (Kir) channels in Müller cells (see Fig. 24.3C) are critical for the spatial buffering capacity of the cells. Studies of Kofuji and co-workers have shown that strongly rectifying Kir2.1 channels in Müller cells are located in 'source' areas, particularly in synaptic regions where $[K^+]_o$ is elevated due to local neuronal activity.[14,15] In contrast, Kir4.1 channels (weakly rectifying) are densest in Müller cell endfeet, in inner and outer limiting membranes and in processes around blood vessels. K+ enters Müller cells through Kir2.1 channels that minimize K+ outflow and K+ exits the Müller cells via the more bidirectional Kir4.1 channels to enter the extracellular 'sink' areas with low $[K^+]_o$: the vitreous humor, subretinal space (SRS), and blood vessels.[14]

ERG waves associated with glial K+ currents have a slower timecourse than waves related to currents around the neurons whose activity causes the changes in $[K^+]_o$. As $[K^+]_o$ increases or decreases with neuronal activity, the resulting glial K+ currents will be related to the integral of the K+ flow rate, e.g. the glial contribution to the b-wave modeled in Figure 24.11.[16] Other waves generated by glial K+ currents in Müller cells, or other retina cells with glial function, such as the RPE, include the c-wave and slow PIII (Fig. 24.3A,B), both related to the reduction in $[K^+]_o$ in the SRS that occurs when photoreceptors hyperpolarize in response to a strong flash of light.[15,17-21] The negative scotopic threshold response (nSTR) and photopic negative response (PhNR), both originating from inner retinal activity, are also thought to be mediated by glial K^+ currents in retina or optic nerve head.[22-29]

Stimulus conditions

Aside from structural and functional aspects of the retina, stimulus conditions are of great importance in determining the extent to which particular retinal cell or circuits contribute to the ERG. Signals will be generated in rod pathways, in cone pathways or both depending upon the stimulus energy, wavelength and temporal characteristics, as well as upon the extent of background illumination, with rods responding to, and being desensitized by lower light levels than cones. Fully dark-adapted, ERGs driven by rods only (i.e. scotopic ERGs) are thus useful for assessing rod pathway function (Figs 24.1 and 24.2, top), and light-adapted ERGs driven only by cones (i.e. photopic ERGs), for assessing cone pathway function (Figs 24.1 and 24.2, bottom). Figure 24.1, bottom right, also shows responses to 30 Hz flicker which isolates cone-driven responses because rod circuits do not resolve well such high frequencies. Very bright light flashes elicit small wavelets superimposed on the b-wave called oscillatory potentials (OPs) that are generated by circuits proximal to bipolar cells.[30,31] OPs can be isolated by band-pass filtering: 75–300 Hz in Figure 24.1, top right.

The spatial extent of the stimulus is an important factor in ERG testing. For standard clinical tests, as illustrated in Figure 24.1, and listed in Box 24.1A, as well as for most testing of animal models, a full-field (Ganzfeld) flash of light is used (Fig. 24.2B). A full-field stimulus generally elicits the largest responses because more retinal cells are activated and the extracellular current is larger than for focal stimuli. It also has the advantage that all regions of retina are evenly illuminated, and with respect to background illumination, evenly adapted. Pupils are generally dilated for full-field stimulation. More spatially localized (focal) stimuli are useful for analysis of function of particular retinal regions, e.g. foveal vs peripheral regions in primates. Multifocal stimulation allows assessment of many small regions simultaneously.

ERG responses illustrated in Figure 24.1 are to a minimum set of stimuli selected by the International Society for the Clinical Electrophysiology of Vision (ISCEV) to efficiently acquire standard, comparable data on rod and cone pathway function from clinics and laboratories around the world.[1] Names of the standard tests are listed in Box 24.1A. Tests using stimuli presented over a fuller range of stimulus conditions to allow more complete or specific evaluation of retinal function, are listed in Box 24.1B, and a few of these tests will be described later in this chapter.

Non-invasive recording of the ERG

ERGs can be recorded from the corneal surface using various types of electrodes. A commonly used electrode, with good

---

Box 24.1 Standard and More Specialized ERG Tests

A. Standard ERG tests

Described by ISCEV Standard for full-field clinical Electroretinography (2008 update).[1] All numbers are stimulus calibrations in $cd.s.m^{-2}$

- Dark-adapted 0.01 ERG ("rod response")
- Dark-adapted 3.0 ERG ("maximal or standard combined rod–cone response")
- Dark-adapted 3.0 oscillatory potentials ("oscillatory potentials")
- Light-adapted 3.0 ERG ("single-flash cone response")
- Light-adapted 3.0 flicker ERG ("30 Hz flicker")
- Recommended additional response: either dark-adapted 10.0 ERG or dark-adapted 30.0 ERG

B. Specialized types of ERG and recording procedures

- Macular or focal ERG
- Multifocal ERG (see published guidelines[162])
- Pattern ERG (see published standard[91])
- Early receptor potential (ERP)
- Scotopic threshold response (STR), negative and positive
- Photopic negative response (PhNR)
- Direct-current (dc) ERG
- Electro-oculogram (see published standard[39])
- Long-duration light-adapted ERG (ON–OFF responses)
- Paired-flash ERG
- Chromatic stimulus ERG (including S-cone ERG)
- Dark and light adaptation of the ERG
- Dark-adapted and light-adapted luminance-response analyses
- Saturated a-wave slope analysis
- Specialized procedures for young and premature infants[163]

---

signal-to-noise characteristics, is a contact lens with a conductive metal electrode set into it (Burian Allen electrode). It has a lid speculum to reduce effects of blinking and eye closure. In the bipolar form of the electrode, the outer surface of the lid speculum is coated with conductive material that serves as the reference. This type of electrode is best tolerated (in alert subjects) when a topical anesthetic is used. Other types of contact lens electrodes have been used as well, e.g. the jet electrode which is disposable. Some clinicians and researchers use thin mylar fibers impregnated with silver particles, called DTL electrodes,[32] as illustrated in Figure 24.2B, or gold foil, or wire loop electrodes (H-K loop) that hook over the lower eyelid. For rodents, metal wires, in loops or other configurations placed in contact with the corneal surface are often used. Some labs use cotton wick electrodes and some use DTL fibers under contact lenses, or another form of contact lens electrode.[26,33] Corneas are kept hydrated with a lubricating conductive solution in all cases. The reference electrode can be placed under the ey lid, e.g. the speculum of a contact lens electrode, as described above, or remotely, for example on the temple, the forehead, or the cornea of the fellow eye. ERG signals ranging from > microvolt to a millivolt or more, peak to peak, for responses to strong stimuli, are amplified, and digitized for computer averaging and analysis. Filtering is done to remove signals outside the frequency range of retinal responses to stimulation (<1 and >300 Hz), and to remove line frequency noise (e.g. 50 or 60 Hz).

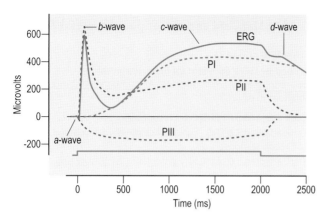

Figure 24.4 Dark-adapted ERG of the cat and its three component processes revealed by induction of ether anesthesia. The solid black line indicates the ERG of the intact cat eye in response to a 2 s flash (14 Lamberts) from darkness. Dashed lines indicate three processes (PI, PII, and PIII), numbered in the order in which they disappeared from the ERG during induction of ether anesthesia. (Modified from Granit 1933,[36] used with permission.)

Classical definition of components of the ERG

The origins of the various waves of the ERG have been of longstanding interest to clinicians and researchers. Our current understanding of the cellular origins of the ERG profits from extensive knowledge, as described in previous chapters, of the functional microcircuitry of the retina, and particularly of the physiology and cell biology of the retinal cell types, and the identity and action of retinal neurotransmitters, their receptors, transporters and release mechanisms. However, a classical study, using ether anesthesia, provided the first pharmacological separation of ERG components.

Granit's[36] classical pharmacological dissection of the ERG, illustrated in Figure 24.4, provided valuable insights on origins of ERG waves as well as a nomenclature for waves based on their distinct retinal origins. Component "processes" were found to disappear from the ERG during the induction of ether anesthesia in the following order: process (P)I, the slow c-wave response that follows the b-wave; generated predominantly by the RPE; PII, the b-wave, generated by bipolar cells; and eventually PIII, the photoreceptor-related responses that remained the longest. PII and PIII are still commonly used terms for ERG components generated by ON bipolar cells and photoreceptors respectively.

Slow PIII, the c-wave and other slow components of the direct current (dc)-ERG

PIII of the ERG can be separated into a fast and a slow portion: fast PIII is the a-wave, which reflects photoreceptor current (see below), and slow PIII results from Müller cell currents induced by photoreceptor-dependent reduction in $[K^+]_o$ in the SRS (Fig. 24.3C). Negative-going slow PIII and the positive-going pigment epithelial response to the same

reduction in subretinal $[K^+]_o$ add together to form the c-wave,[15,17-19,34] which, as shown in Figures 24.3 and 24.4, is positive-going in the dark-adapted ERG of the cat. This is because, as illustrated by the intraretinal recordings from intact cat eye in Figure 24.3A, the positive-going RPE contribution, is larger than the negative-going Müller cell contribution.[35] In mice, the c-wave also is positive-going,[35] as shown in Figure 24.3B. In humans and monkeys, slow PIII and the RPE c-wave are more equal in amplitude, and the corneal c-wave is less positive. Two slower potentials that arise from the RPE, the fast oscillation potential (FO) and light peak (LP), are also present in cat and mouse dc-ERG recordings (Fig. 24.3). The mouse LP is much smaller in amplitude than that in the cat. The cellular mechanisms that generate these slow waves were reviewed in more detail in previous reviews of ERG origins.[37,38]

In alert human subjects it is not possible to obtain stable dc-ERG recordings necessary for recording slow events, e.g. those arising from the RPE, that occur over seconds or minutes, because the eyes move too frequently. Therefore, to measure slow potentials, the electro-oculogram (EOG), an eye-movement-dependent voltage, is recorded. The EOG is a corneo-fundal potential that originates largely from the RPE; its amplitude changes with illumination, being maximal at the peak of the light peak. Use of EOGs to evaluate retinal/RPE function is described in an ISCEV Standard publication on clinical EOG.[39]

Full-field dark-adapted (Ganzfeld) flash ERG

Figure 24.5 shows full-field dark-adapted ERG responses to a range of stimulus strengths for a human subject (left), a macaque monkey, whose retina and ERG are similar to that of human (middle), and a C57BL/6 mouse (right). The mouse ERG is similar to the primate ERG, but larger in amplitude (see calibrations). For higher stimulus strengths than shown in the figure, the mouse ERG develops larger oscillatory potentials than those generally seen in the primate ERGs (e.g. Fig. 24.6). The ERGs in Figure 24.5 were generated almost entirely, except for responses to the strongest stimuli, by the most sensitive, primary rod circuit, This circuit is described more fully in previous chapters. For all three subjects, the strongest stimuli evoked an a-wave, followed by a b-wave. For stimuli more than two log units weaker than the strongest one, b-waves were still present, but a-waves were no longer visible. B-waves can be seen in responses to weaker stimuli than a-waves partly because of the convergence of many rods (20–40) onto the rod bipolar cells generating the response, which increases their sensitivity,[40–42] and partly because of the large radial extent of ON bipolar cells in the retina. The slow negative wave in the ERGs of the three subjects in response to the weakest stimuli, called the (negative) scotopic threshold response (nSTR), and the equally sensitive positive (p)STR are related to amacrine and/or ganglion cell activity, as described more fully in a later section.[22,25-27,43] The high sensitivity of the STRs relative to b-waves (and a-waves) reflects the additional convergence of rod signals in the primary rod circuit in the inner retina proximal to the rod bipolar cells.[40,44,45]

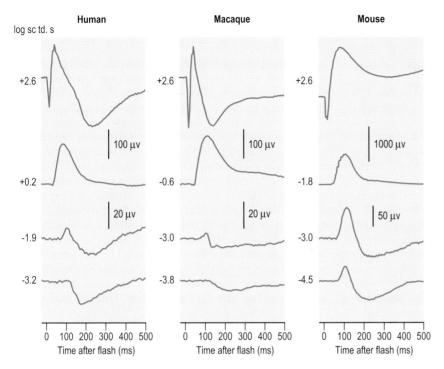

Figure 24.5 Dark-adapted full-field ERG of a human subject, a macaque monkey and a mouse. ERGs of a human, macaque and C57BL/6 mouse, in response to brief (<5 ms) flashes from darkness generated by computer-controlled LEDs. Responses to the weakest stimuli were rod-driven (scotopic), whereas for the strongest stimuli they were driven by both rods and cones. (Adapted from Robson & Frishman 1998,[16] used with permission.)

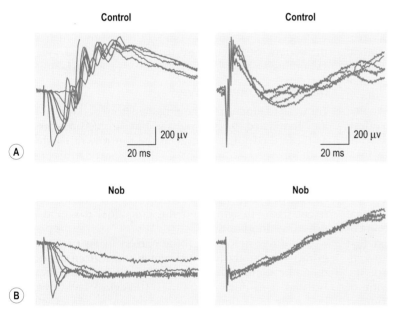

Figure 24.6 Negative ERG of mouse lacking a b-wave. Single-flash responses on short (left) and long (right) time scales from a control C57BL/6 mouse (**A**) and a *Nob1* mouse (**B**). ERGs on the left are responses to flashes of 0.11, 0.98, 2.57, 4.37, 17.4, 27.5, and 348 sc cd.s.m^{-2}, on the right, to flashes of 40.0, 68.4, 102, 170, and 348 sc cd.s.m^{-2}. (From Kang Derwent et al 2007,[58] used with permission.)

Dark-adapted a-wave

It has long been appreciated that the dark-adapted a-wave primarily reflects the rod receptor photocurrent. The a-wave generator was localized to photoreceptors in classical intra-retinal recording studies in mammalian retinas, some of which included current source density (CSD), or source-sink, analyses.[10,11,46–48] The most direct demonstration of the a-wave's cellular origin was provided in such experiments by Penn and Hagins[47,48] in isolated rat retina. These experiments produced evidence that light suppressed the circulating (dark) current of the photoreceptors, and the investigators proposed that this suppression is seen in the ERG as the a-wave.

Negative ERGs

The receptoral origin of the a-wave also was demonstrated in early studies in amphibians using compounds that blocked synaptic transmission, Mg^{2+}, Co^{2+}, and Na$^+$-aspartate, and isolated photoreceptor signals in the ERG, while abolishing responses of postreceptoral neurons.[49,50] These manipulations also caused the b-wave to disappear, indicating its postreceptoral origin. As our understanding of synaptic pharmacology has improved, it has become more common to use glutamate agonists and antagonists to block transfer of signals from photoreceptors to specific second-order neurons. For example, blocking metabotropic transmission to depolarizing (ON) bipolar cells, with

L-2-amino-4-phosphonobutyric acid (APB or AP4) an mGluR6 receptor agonist,[51] eliminates the b-wave, and produces a negative ERG.[52,53] Much of the remaining ERG is the rod photoreceptor response, although late negative signals also arise from OFF pathway neurons.[54,55]

An essentially identical negative ERG to that after APB occurs in mice in which the mGluR6 receptor is genetically deleted,[56] or when there are mutations in the mGluR6 receptor or other proteins whose function is necessary for normal signal transduction in ON bipolar cells.[57] For example, Figure 24.6 shows the typical negative ERG of the dark-adapted *Nob1*, i.e. no b-wave, mouse.[58,59] The *Nob1* mouse has a mutation in the *Nyx* gene that encodes nyctalopin, a protein found in ON bipolar cell dendrites.[60,61] Mutation of this protein produces a negative ERG in human patients who are diagnosed with X-linked complete congenital stationary night blindness (CNSB-1).[62,63] Negative ERGs also occur for other forms of CSNB caused by mutations in mGlur6 receptors, and in *Nob3* and *Nob4* mice with such mutations,[57,64] as well as in *Nob2* mice in which glutamatergic transmission from photoreceptors is compromised.[65] Although the ERG of the *Nob1*, and other mice lacking b-waves, is almost entirely negative-going, it rises from its trough at the timecourse of the c-wave. Photoreceptor-dependent slow responses such as c-wave and slow PIII are not affected by blockade of postreceptoral responses in neural retina.

Retinal ischemia due to compromised inner retina circulation also isolates the a-wave and eliminates postreceptoral ERG components. This was demonstrated in early experiments in monkeys by occlusion of the central retinal artery.[66,67] A "negative ERG" in which b-waves are reduced or missing is a common clinical readout of central retinal artery and vein occlusions, as well as other disorders affecting postreceptoral retina such as melanoma-associated retinopathy, X-linked retinoschisis, muscular dystrophy or toxic conditions.[7,9]

Modeling

The utility of the a-wave in studies of normal and abnormal photoreceptor function was advanced by the development of quantitative models based on single cell physiology that could predict the behavior both of the isolated photoreceptor cells and ERG a-wave in the same or similar species. Hood and Birch[68,69] demonstrated that the behavior of the leading edge of the dark-adapted a-wave in the human ERG can be predicted by a model of photoreceptor function derived to describe in vitro suction electrode recordings of currents around the outer segments of single primate rod photoreceptors.[70] Lamb and Pugh[71,72] followed a simplified kinetic model of the leading edge of the photoreceptor response (in vitro current recordings initially in amphibians) that took account of the stages of the biochemical phototransduction cascade in vertebrate rods (see Chapter 21). This model was subsequently shown to predict the leading edge of the human dark-adapted a-wave generated by strong stimuli,[73] and has been used extensively in clinical studies of retinal disease, and in analyses of photoreceptor function in animal models. A simplified formulation presented by Hood and Birch[74-76] often is used (see legend to Fig. 24.7), and can be adjusted to application to cone signals as well. Figure 24.7 shows fits of

Hood and Birch's model to the dark-adapted a-wave of a normal human subject and a patient with retinitis pigmentosa (RP).

Photoreceptor models of Hood and Birch,[74-76] Lamb and Pugh,[72] and more recent ones, with improved fits,[55] all provide parameters to represent the maximum amplitude of the a-wave, which is R_{max} in equations of Hood and Birch, and the sensitivity (S) of the response. R_{max} and S vary depending upon the pathology and stimulus conditions (e.g. adaptation level). For example, S is thought to be more affected than R_{max} in eyes whose photoreceptors are hypoxic and more generally for abnormalities in the transduction cascade or increases in retinal illumination, whereas R_{max} is more affected by photoreceptor loss. Both parameters may be affected in RP, but in the case illustrated in Figure 24.7, R_{max} was affected more than S.[77]

Although models of the leading edge of the a-wave yield useful parameters for describing the health of the photoreceptors, simpler approaches, using stronger flashes than those advised by the ISCEV Standard (Fig. 24.1 and Box 24.1)[1] are helpful. Hood and Birch,[77,78] and other investigators,[79] have noted that a pair of strong flashes, or even a single flash that nearly or just saturates the rod response can be analyzed without fitting a model, to make a rough estimate of R_{max}, and to measure a peak time that is related to S. Such measurements for a single flash[77] are illustrated in the bottom row of Figure 24.7. The flash strength used was 4.0 log sc td.s which is 63 times (1.8 log units) higher than the current ISCEV standard flash[1] for mixed rod-cone ERG (assuming an 8 mm pupil).

Mixed rod-cone a-wave

For weak to moderate stimuli flashed from darkness, a-waves are dominated by rod signals, but stronger flashes such as those used for responses in the top row of Figure 24.5 elicit mixed rod–cone ERGs. To investigate relative contributions of rod and cone-driven responses to the ERG, it is necessary to separate them. Figure 24.8 shows the ERG (red circles) of a dark-adapted macaque in response to two different flash strengths, and the individual rod- and cone-driven contributions.

Rod-driven responses (blue circles) in Figure 24.8 were extracted by subtracting the isolated cone-driven response to the same stimulus from the full mixed rod–cone response. Isolated cone-driven responses (triangles) were obtained by briefly (1 s) suppressing, rod-driven responses with an adapting flash, and then measuring the response to the original test stimulus, presented 300 ms after offset of the rod-suppressing flash. Cone-driven responses in primates recover to full amplitude within about 300 ms, whereas rod-driven responses take at least a second, making it possible to isolate the cone-driven responses. The cone photoreceptor-driven portion of the leading edge of the a-wave represented about 20 percent of the saturated response in Figure 24.8. Model lines for the rod (blue) and cone (purple) photoreceptor responses are modifications of Lamb and Pugh's model.[55,72] The entire cone-driven a-wave (green line) is larger than the modeled photoreceptor contribution (purple line) because it includes additional negative-going signals from postreceptoral OFF pathway that can be eliminated with ionotropic glutamate receptor antagonists, as described in a later section on light-adapted ERG.

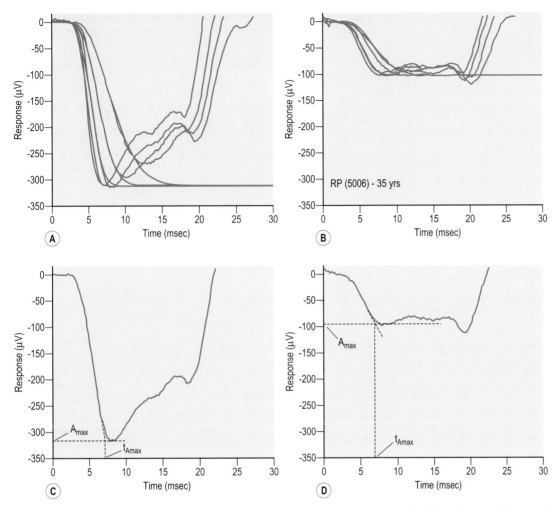

Figure 24.7 Modeling the leading edge of the dark-adapted a-wave. (A) and **(B)** Rod-isolated ERG of a normal human subject (A), and a retinitis pigmentosa (RP) patient (B) in response to flashes of increasing strengths near a-wave saturation. Responses are fit with the Hood & Birch[75] adaptation of the Lamb & Pugh[72] model, expressed in the following equation:

$$R(I,t) = \left(1 - \exp\left[-I * S * (t - td)^2\right]\right) * R_{max} \text{ For } t > t_d$$

where response amplitude, R, is a function of flash strength, I, and the time, t, after the occurrence of brief (impulse) flash of light. S is the sensitivity that scales the stimulus strength, R_{max} is the maximum amplitude, and t_d, a brief delay commonly between 2.4 to 4 ms. **(C)** and **(D)** Simple approach for estimating A_{max}, and t_{max}, indicators for model parameters, R_{max} and S for the normal human subject (C) and RP patient. (D). (From Hood & Birch 2006,[77] used with permission.)

Timecourse of the rod photoreceptor response

The leading edge of the a-wave is the only portion of the photoreceptor response that normally is visible in the brief flash ERG. As shown for the *Nob1* mouse in Figure 24.6, elimination of the major postreceptoral contributions to the dark-adapted ERG leaves the c-wave present in the record. For studying photoreceptor responses in isolation from all other ERG components, a paired flash technique was developed by Pepperberg[80] and colleagues to derive, in vivo, the photoreceptor response. The derived photoreceptor responses for a macaque monkey are shown for three different stimulus strengths, all of which saturated the b-wave in Figure 24.9 (top). The time course of the derived response is similar to that of current recordings from individual rod outer segments in the monkey,[70] although the derived response tends to reach its peak a little earlier in vivo in the ERG than in vitro. This approach has been used in studies in humans, macaques, and other animals, to study responses of rods and cones. Note that the (derived) photoreceptor response is quite large at its peak time for the ERG responses to weaker stimuli, when the leading edge of the a-wave is too small to be seen. The nicely illustrates the overlapping nature of ERG components arising from different stages of retinal processing.

Dark-adapted b-wave (PII)

It is well accepted that the dark-adapted b-wave arises primarily from ON bipolar cells. Results of intraretinal recording and CSD analyses were consistent with its originating from bipolar (or Müller cells),[10,11,66] and as described above, pharmacologic blockade of postreceptoral responses and specifically those of ON bipolar cells was found to eliminate the b-wave,[53] as did mutations that prevent signaling by ON bipolar cells (Fig. 24.6).[57,59]

There also is good evidence that for most of its dynamic range, the dark-adapted b-wave is generated by rod bipolar

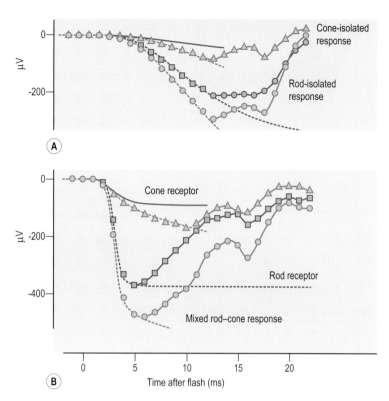

Figure 24.8 Mixed rod–cone dark-adapted ERG of macaque monkey showing separate components. (**A**) ERG responses of a macaque to a brief blue LED flash of 188 sc td.s (57 ph td.s,) and (**B**) to a xenon-white flash of 59,000 sc td.s (34,000 ph td.s). In both plots, the largest response, (red circles) is the mixed rod-cone a-wave; the red dashed line shows the modeled mixed rod/cone response. The second largest response is the (isolated) rod-driven response (filled blue squares and circles), and the dashed blue line is the modeled rod photoreceptor contribution. The second to smallest response (filled triangles) is the isolated cone-driven response, and the dashed green line shows the modeled response (including postreceptoral portion). The smallest response (purple line) is the modeled cone photoreceptor contribution, based on findings after intravitreal injection of *cis*-piperidine-2,3-dicarboxylic acid (PDA, 5 mM intravitreal concentration, assuming a vitreal volume of 2.1 ml in macaque eye in this and subsequent figures). The stimulus for part A (for 8.5 mm pupil) was 1 ph cd.s.m^{-2}, i.e. about 3 times weaker (for cones) than the ISCEV standard flash of 3 cd.s.m^{-2} whereas for part B it was 200 times stronger than the standard flash. (From Robson et al 2003,[55] used with permission.)

cells in the primary rod circuit. Only for very strong stimuli, will rod signals pass via gap junctions to cones and then to cone bipolar cells, which also will then contribute to the dark-adapted flash response.[41,44,81,82] For the dark-adapted, scotopic ERGs it has been possible to isolate Granit's PII from other ERG components and to compare its characteristics with those of rod bipolar cells recorded in retinal slices. One approach is to pharmacologically isolate PII.

Figure 24.10A shows the dark-adapted ERG of a C57BL/6 mouse before and after intravitreal injection of the inhibitory neurotransmitter γ-aminobutyric acid (GABA) to suppress inner retinal activity. GABA receptors are present in bipolar cell terminals, as well as in amacrine and ganglion cells of inner retina, as described in Chapter 23. The figure shows that the most sensitive positive and negative going waves (p- and n-STR) were removed by GABA,[26,83] leaving a pharmacologically isolated PII, likely generated by rod bipolar cells. Loss of the sensitive STRs was not due to a general loss of retinal sensitivity; a-wave amplitude and kinetics, reflecting photoreceptor function, did not change after GABA injection (Fig. 24.10C).

Figure 24.11B shows PII, isolated by intravitreal GABA from ERGs of four C57BL/6 mice in response to stimuli producing about 1 Rh* per rod. Also included in Figure 24.11A is PII isolated in humans using a weak adapting light to suppress the very sensitive STRs.[84] Isolated PII has been superimposed on an average of several patch electrode current recordings from rod bipolar cells in a mouse retinal slice preparation, in response to similarly weak stimuli.[85] The timecourse of PII isolated from the ERG and of single cell currents is remarkably similar, supporting the view that isolated PII reflects rod bipolar cell activity. A primary role for an alternative generator for PII, the Müller cells, is

unlikely in mice because mice with genetically inactivated Kir4.1 channels in their retinas still produce b-waves.[15]

Figure 24.11B plots ERG amplitude, measured at a fixed time (110 ms) near the peak the b-wave (and pSTR) before and after GABA injection. After GABA injection, with the pSTR removed, the response rises in proportion to stimulus energy and then saturates in a manner that is adequately described by a simple hyperbolic function (see Fig. 24.10 legend). These findings are consistent with a single mechanism, i.e. one retinal cell type, producing the isolated PII component of the ERG.[83,86]

Although bipolar cells play a major role in generating b-waves, Müller cells also contribute. The PII component, pharmacologically isolated from macaque ERG (Fig. 24.11B), is similar in its rising phase to PII of the other species illustrated in Figure 24.11A However PII in macaque (and cat) recovers more slowly to baseline. Macaque PII in Figure 24.11B has been analyzed into a fast component, and a slow component that is a low-pass filtered version of the fast component. In cat, intravitreal injection of Ba^{2+} to block inward rectifying K$^+$ channels in Müller cells did not eliminate the b-wave,[25,87] but did remove a slow portion of isolated PII that is similar in timecourse to a low pass filtered version of the response.[38] Although the slow component (in response to a brief flash) is lower in amplitude than the fast component, the area under the two curves is similar in both cat and the macaque. With longer-duration stimuli such as those used in early studies in amphibians,[88] contributions of neuronal and glial generators of the b-wave could be equal.

Scotopic threshold response (STR)

As illustrated in Figures 24.5 and 24.10, for very weak flashes from darkness, near psychophysical threshold in humans,[27,89]

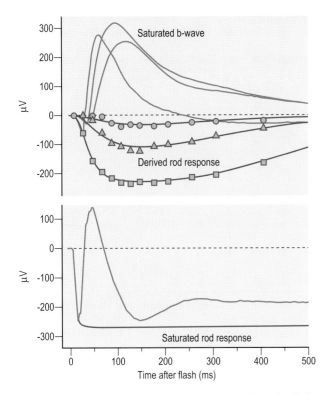

Figure 24.9 Derived rod photoreceptor response from the dark-adapted ERG of a macaque monkey. Top: ERG responses of a macaque are shown to three test stimuli (0.28, 1.24, and 6.86 sc td.s). The photoreceptor response was derived using the paired flash technique of Pepperberg et al:[80] a rod saturating probe flash followed a test flash at fixed intervals after its onset, and the residual response to the probe was subtracted from the probe alone response to derive the rod receptor response at each time point (data points). **Bottom**: ERG response to a stimulus of 2.6 log sc td.s for which the rod photoreceptor response was saturated. Model lines in this and the upper plot were generated using equations modified from Robson et al.[55]

the nSTR and pSTR dominate the ERG of most mammals studied. The nSTR was shown in intraretinal analyses in cat to be generated more proximally in the retina than PII.[27] In non-invasive recordings, the p and nSTR can be suppressed and thus separated from PII, using a weak adapting background, inhibitory neurotransmitters (GABA, as illustrated in Figs 24.10 and 24.11, or glycine).[90] Blockade of inontropic glutamate receptor antagonists (see legend to Fig. 24.11), or use of an agonist for N-methyl-D-aspartate (NMDA) receptors present on inner retinal neurons, also remove the STRs.[86,91] The effects of GABA on the pSTR in mouse (measured at 110 ms after the flash), and the nSTR (measured at 200 ms) are shown in Figure 24.10B and D respectively. In both cases, after GABA injection the remaining responses can be modeled as the PII component alone.

The plots in Figure 24.10B and D include a linear model of the dark-adapted ERG that takes into account contributions not only from PII, but also from the pSTR, nSTR (and PIII) components of the mouse ERG. These components add together to form the ERG response (blue solid line) at a particular time after the stimulus flash. The model assumes that each ERG component (dashed and red lines) initially rises in proportion to stimulus strength, and then saturates in a

characteristic manner, as has been demonstrated in single cell recordings in mammalian retinas[70,92] as well as for a- and b-waves in numerous other studies. To isolate PII at higher stimulus strengths when the photoreceptor contribution to the ERG is significant (see Fig. 24.9), the PIII (photoreceptor) component must be removed, A model line for PIII is shown in Figure 24.10B and D for the higher stimulus strengths.[26]

The neuronal origins of the nSTR and pSTR are species-dependent. In macaques, the nSTR may originate predominantly from ganglion cells; it was eliminated in severe experimental glaucoma with selective ganglion cell death, whereas the pSTR remained.[43] In contrast, in rodents the pSTR requires the healthy function of ganglion cells. The pSTR was eliminated in mice and rats by removing ganglion cells (optic nerve crush (ONC) or transection and subsequent degeneration) and in mice in which ganglion cells were deleted genetically, whereas the nSTR was only partially removed in rat, and of normal amplitude in mice.[22,93–95] In cat (and human) the nSTR also survives ganglion cell loss,[96] and in cat, it has been demonstrated that nSTR generation involves K^+ currents in Müller cells.[24]

Mice lacking connexin 36 gap junction proteins (Cx36) provide additional insights to the neuronal origins of the STR. Cx36 is expressed in the electrical synapses (gap junctions) between AII amacrine cells, AII amacrine cells and ON cone bipolar cells as well as between the rod and cone terminals.[82,97] In the scotopic ERG of mice with Cx36 genetically deleted, pSTR remains, whereas the nSTR is absent.[44] Because Cx36 is essential for transmission of ON pathway signals to ganglion cells in the primary rod pathway,[82] preservation of the pSTR in Cx36(−/−) mice suggests that the pSTR, shown to be ganglion cell related in rodents, must depend upon activity of OFF, rather than ON ganglion cells. The nSTR, in contrast, depends on the syncytium of AII amacrine cells and their electrical synapses with ON cone bipolar cells, no longer functional in Cx36-deficient mice. Figure 24.11A includes an example of PII isolated from the scotopic ERG of a Cx36(−/−) mouse (already lacking the nSTR) but also lacking the pSTR, due to a ganglion cell lesion subsequent to ONC. This isolated PII is very similar to PII isolated pharmacologically or by light adaptation, as well as to the rod bipolar cell recordings from the retinal slice.

Because of their specific inner retinal origins, the n- and pSTRs are of particular interest in animal models of diseases that affect the inner retina, such as glaucoma. They provide non-invasive readouts of pathologic changes in function, and can be used to document effects of neuroprotective agents.[22,43,98]

Light-adapted, cone-driven ERGs

Isolating cone-driven responses

To study the cone-driven (photopic) flash ERG, rods must be suppressed, or unable to respond due to characteristics of the stimuli selected. Typically for recording photopic ERGs, rods are rendered unresponsive by imposing a steady background light sufficient to saturate their responses (i.e. >25 cd.m^{-2}). Another approach is to briefly saturate rod responses, and to wait for cones to recover. Cones recover more quickly than rods: in primates, in about 300 ms

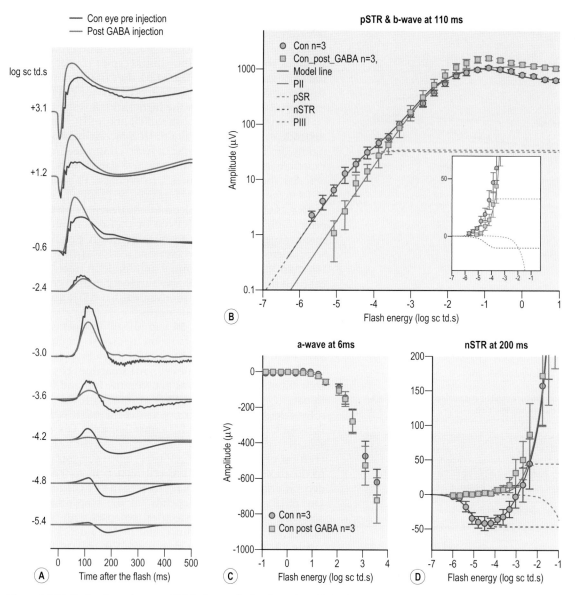

Figure 24.10 Dark-adapted mouse ERG before and after intravitreal injection of GABA: analysis of ERG components. (A) ERG of a C57BL/6 mouse in response to brief flashes of increasing stimulus strength recorded before and after intravitreal injection of GABA (30 mM intravitreal concentration assuming for this and subsequent panels, a 20 μl vitreal volume in mouse). **(B)** ERGs of three mice measured at 110 ms (mean ± SEM) after a brief flash, at the peak of the pSTR and b-wave. The modeled full ERG (blue line) includes four components: PII, pSTR, nSTR and PIII. All components are assumed to increase in proportion to stimulus strength before saturating. Explicitly, the exponential saturation of the pSTR, nSTR and PIII is defined by:

$$V = V_{max}(1 - \exp(-I/I_o))$$

where V_{max} is the maximum saturated amplitude, and I_o, the stimulus strength for an amplitude of $(1 - 1/e)V_{max}$.

The hyperbolic relation used for PII is described by the function:

$$V = V_{max}I/(I + I_o)$$

where V_{max} has the same meaning as for the exponential function but I_o is the stimulus strength at which the amplitude is $V_{max}/2$.[26]

(C) ERGs measured at 6 ms after a brief flash, to measure a-wave amplitude on its leading edge. **(D)** ERGs measured at 200 ms to measure nSTR amplitude. Model lines for the full response and for component parts are as described for panel B.

as described above for results in Figure 24.8.[55] In mice and rats, the recovery time of cone signals is longer, about a second.[93,99]

Rod and cone signals also can be separated by selection of appropriate stimulus wavelength. In trichromatic primates, L-cones can be isolated using red light, at wavelengths greater than 630 nm, and excluding wavelengths much lower. However mice and rats do not have L-cones, and the

spectral sensitivity of their medium wave-length M-cones peaks near 515 and 510 nm respectively[99,100] and overlaps with that of rods, which peaks around 500 nm. Scotopic stimulus calibrations are thus adequate, and more appropriate than photopic calibrations (based on human spectral sensitivity) for the rodent M-cone responses. However, for any subject, spectral separation of responses of individual cone types is possible using ERG flicker photometry.[101] In

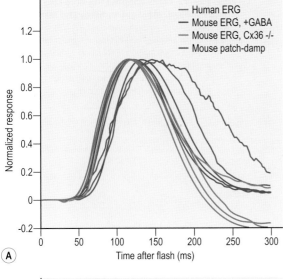

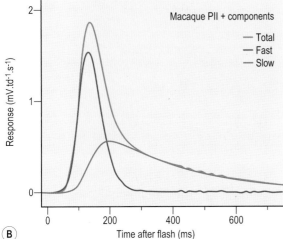

Figure 24.11 Rod bipolar cell component, PII, of the dark-adapted ERG of a human subject, macaque monkey and mouse.
(**A**) Comparison of rod bipolar cell current from patch recordings in mouse retinal slice (Field & Rieke, 2002) with isolated PII (by weak light adaptation) from humans, isolated PII from ERGs of six C57BL/6 mice by intravitreal injection of GABA (32–46 mM) and from a Cx36(–/–) mouse lacking ganglion cells. (From Cameron et al 2006,[84] which was modified from Robson et al 2004;[83] used with permission.) (**B**) PII of macaque, isolated by pharmacologic blockade [6,7-dinitroquinoxaline-2,3-dione (DNQX), 0.1 mM; N-methyl-D-aspartic acid (NMDA, 3 mM)] of inner retinal responses. PII has been analyzed into a fast component, proposed to be a direct reflection of the post-synaptic current, and a slow component that is a low-pass filtered version of the faster component, believed to be the Müller-cell contribution. (From Robson & Frishman 1998,[16] and unpublished observations.)

mice and rats such studies have isolated short-wavelength UV (S-) cones that peak around 358 or 359 nm.[100,102]

Separation of rod and cone signals in mice can also be achieved using genetically manipulated models in which either rods or cones have been inactivated. Examples of such models include: Rho(–/–) (rod opsin knockout) mice which have cone function only, but also eventual cone degeneration[103], Tra(–/–) (transducin α knockout) with cone function only[104], Cnga3(–/–) (cone cyclic nucleotide channel deficient) mice[4] or *GNAT 2*[cpfl3] (G$_T$ protein) mutant mice with rod function only.[105] Mice with inactivated or degenerated

rods and cones, i.e. rodless, coneless mice,[106] can be used to study function of a more recently identified photoreceptor pigment, melanopsin, found in a small population of retinal ganglion cells. Contributions to ERGs from melanopsin-driven responses are thought to be minimal.[107]

Figure 24.12 shows photopic ERG responses to brief stimulus flashes on a rod-saturating background for human, macaque monkey, and mouse. As shown above for the dark-adapted ERG, the responses to brief flashes are similar in man and macaque. In mice, however, the b-wave rises more slowly to its peak, does not return to baseline as early, and OPs are more prominent. Very strong stimuli produce more prolongation of responses in mice, and produce larger OPs (inset on the right of Fig. 24.12) than in primates. For macaques, a second peak after the b-wave, probably an OFF pathway response (i-wave),[108] emerges, and in some cases (not shown) more of the response is below the baseline level.

Light-adapted a-wave

The a-waves in the photopic ERG are smaller than those in the scotopic ERG, reflecting the large difference in retinal densities of cones vs rods. In human and macaque about 5 percent of the photoreceptors are cones, and only up to 3 percent in mice, and fewer in rats are cones. In addition, use of glutamate analogs has shown that much of cone-driven a-wave is postreceptoral in origin. Results of using glutamate analogs in macaques[109,110] are illustrated in Figure 24.13. The a-wave was reduced in amplitude by the ionotropic glutamate receptor antagonists PDA (*cis*-2,3-piperidine-dicarboxylic acid) (and kynurenic acid, KYN, not shown) that block kainate and AMPA receptors of OFF bipolar and horizontal cells as well as cells of inner retina. In contrast, APB, which eliminates the b-wave, did not reduce the amplitude of the leading edge of the a-wave. Figure 24.13 also shows that PDA had a similar effect to that of aspartate, used to block all glutamatergic transmission, or to cobalt (Co^{2+}), used to block voltage-gated Ca^{2+} channels that are essential for vesicular release of the neurotransmitter glutamate, for signal transmission to postreceptoral neurons.

PDA-sensitive postreceptoral neurons, rather than cones, or APB-sensitive ON bipolar cells (and more proximal cells receiving ON signals), were found to generate the leading edge of the a-wave for the first 1.5 log units of increasing flash strength that elicited a cone-driven a-wave in macaque, reaching an amplitude of 10–15 μV in experiments of the studies illustrated in Figure 24.13.[109] Postreceptoral cells continued to contribute 25–50 percent of the leading edge of the a-wave for higher stimulus strengths when cone photoreceptor contributions also were present. The relative sizes of the postreceptor and receptor-driven portions of the photopic a-wave were modeled in Figure 24.8 for cone-isolated responses in the macaque dark-adapted ERG-based results after PDA. Responses up to ~5 ms after the flash in Figure 24.8 were cone photoreceptor responses, as indicated by the blue model line.[55]

Experiments in mouse to investigate the origins of the photopic a-wave yielded similar results to those in macaque. PDA-sensitive responses dominated the leading edge of the a-wave for the initial 1.5 log units, and at least half of the response to stronger stimuli. Photoreceptor

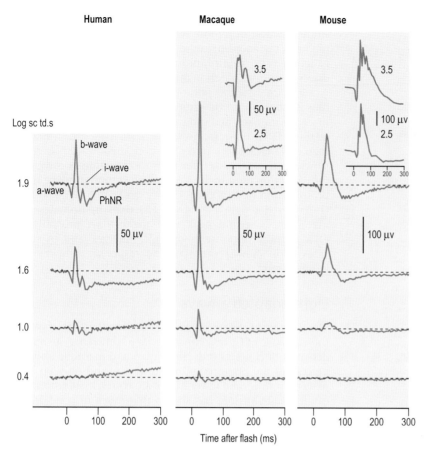

Figure 24.12 Light-adapted full-field ERG of a human subject, macaque monkey and mouse.
ERGs measured in response to brief (<5 ms) white LED flashes for primates, green for mouse, except for insets for the macaque and mouse showing responses from different subjects to white xenon- flashes. Steady background was 100 sc cd.m^{-2} for primates, 63 sc cd.m^{-2} for mice. (Unpublished observations.)

signals were present in response to strong flashes up to 8 ms after the flash. A significant portion of the a-wave also was removed by NMDA, which suppresses responses of inner retinal neurons, indicating contributions to the leading edge of the a-wave from more proximal neurons in the OFF pathway (unpublished observations, Maeda, Kaneko, and Frishman).

Light-adapted b-wave

Intraretinal recording studies described above for studying origins of the dark-adapted ERG, indicated a bipolar cell origin for light-adapted b-waves, but ON cone bipolar cells, rather than rod bipolar cells were implicated.[46,111] Origins of the photopic b-wave also were studied in the same series of experiments in macaques as those illustrated in Figure 24.13. Figure 24.14A, top right, shows the macaque photopic ERG response to long-duration flashes, before and after intravitreal injection of APB (eye 1, left).[110] APB removed the transient b-wave, supporting a role for ON bipolar cells in generating the response (although the possibility of glial mediation was not eliminated in these studies). In contrast, when PDA was injected first (Eye 2, right), to remove OFF pathway and inner retinal influences (and horizontal cell inhibitory feedback), the b-wave was larger in amplitude and the maximum amplitude was sustained for the duration of the stimulus. These findings indicate that cone-driven b-waves in primates are normally transient because they are truncated by (PDA-sensitive) OFF pathway contribution of opposite polarity to the ON bipolar cell contribution (a

push–pull effect[109,110]), or by inhibitory feedback via cones from horizontal cells.[110,112] Only an isolated photoreceptor response of slow onset, but more rapid offset, remained after all postreceptoral activity was eliminated, either by APB followed by PDA, or PDA followed by APB.

The outcome in mouse of parallel experiment to that in Figure 24.14A to determine origins of the photopic b-wave is shown in Figure 24.16B. The figure shows that the control response to a 200 ms stimulus in a C57BL/6 mouse is different from that in primates, and in fact, the b-wave looks similar to the response in macaque after PDA. APB removed the b-wave in the mouse, but did not reveal a d-wave, only a tiny positive intrusion was present.[113,114] Similar results have been observed in rat.[115] PDA injection in mouse produced only a small increase in b-wave amplitude; but it removed OPs, and the small positive intrusion. APB combined with PDA left a negative ERG of slower onset than in the normal ERG due to removal of the leading edge of the a-wave. The slow recovery of the isolated photoreceptor-related response, compared to that of the macaque, was due to the presence, in mouse, of slow PIII in the ERG.[116]

Recent studies in mice and rats found that TTX significantly reduced the amplitude of the light-adapted b-wave.[93,116] This finding is consistent with presence of TTX-sensitive voltage-gated sodium channels in cone bipolar cells (in rats).[117] In macaques, in contrast, the light-adapted b-wave amplitude increased, rather than decreased after TTX, probably due to removal of the negative-going photopic negative response (PhNR) which originates from spiking neurons in inner retina.[28,108,112]

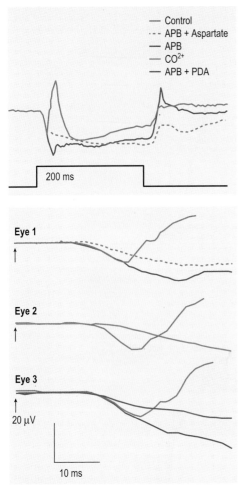

Figure 24.13 Postreceptoral contributions to the a-wave of the macaque monkey ERG. Comparison of effects of intravitreally injected L-2-amino-4-phosphonobutyric acid (APB, 1mM) and APB+PDA (5 mM) with effects of aspartate (ASP, 0.05 M) and cobalt (Co^{2+}, 1 mM); on photopic a-waves of three different eyes of two macaques. The inset at the top shows the control response of eye #1 to the 200 ms stimulus of 3.76 log td (100 cd.m^{-2}) on a steady background of 3.3 log td (35.5 cd.m^{-2}). The a-waves are shown below on an expanded time scale. For the stimulus used, the a-wave leading edge (10 μV) was mostly postreceptoral in origin.

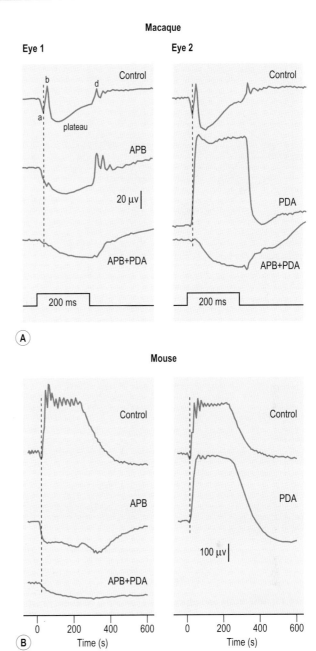

Figure 24.14 Effects of APB and PDA on the light-adapted ERG of macaque monkey and mouse. (A) Macaque: Intravitreal injections were given in two different eyes sequentially: APB followed by PDA for eye 1, and PDA followed by APB for eye 2. The vertical line shows the time of the a-wave trough in the control response. The 200 ms stimulus and drug concentrations were the same as reported for Fig. 24.15. (From Bush & Sieving 1994,[109] used with permission from Association for Research in Vision and Ophthalmology.) **(B)** Mouse: Control ERG and response after intravitreal injection of APB (1 mM) and PDA (0.5 mM)+ APB in a C57BL/6 mouse. The 200 ms stimulus was 4.6 log sc td (3.9 log cd.m^{-2}) on a steady background of 2.6 log sc td (63 cd.m^{-2}). (From Shirato & Frishman, unpublished.)

Light-adapted d-wave

The d-wave is a positive-going wave at light offset that is a characteristic of the photopic ERG. The d-wave is prominent in all-cone retinas, but also can be seen in the photopic ERG of the mixed rod–cone retina of man and macaque monkey for stimulus durations of 150 or 200 ms, as shown in Figures 24.2A and 24.14A. As described above, the mouse (and rat) photopic ERG does not include a d-wave in response to offset of light flashes of only a few hundred ms in duration.[114-116]

Intraretinal analysis in monkey retina initially indicated that d-waves represent a combination of a rapid positive-going offset of cone receptors followed by the negative-going offset of the b-wave.[66,67,118] However intravitreal injection of glutamate analogs in macaques revealed a prominent role, at light offset, for OFF cone bipolar cells as well, in generating d-waves (see Fig. 24.14).[110,119] The relatively slower offset of cone photoreceptor responses in rodents may explain, in part, the lack of d-waves for stimuli that elicit them in primates. The lack of d-waves in rodents is not due to differences in relative numbers of ON and OFF cone bipolar cells in rodents and primates, the proportions are similar.[120]

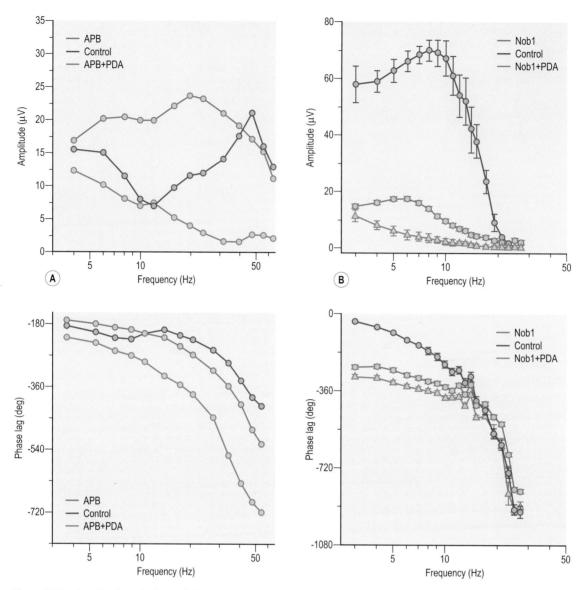

Figure 24.15 Amplitude and phase of the temporal frequency response of the light-adapted ERG in macaque monkey and mouse.
(**A**) Flicker ERG response of three macaques (mean ± SEM) to sinusoidal stimulation before and after pharmacological separation of ON and OFF pathway components of the fundamental response. Flicker mean luminance was 457 cd.m^{-2}; 80 percent contrast, on a 40 cd.m^{-2} white background. Blue circles: control; red circles: after APB (1mM); green circles: after APB + PDA, 5mM). (From Kondo 2001,[123] used with permission.) (**B**) Flicker ERG responses of three C57BL/6 mice (mean ± SEM) and three *Nob1* mice to sinusoidal stimulation. Mean luminance was 63 cd.m^{-2}, 100 percent contrast. Blue circles: fundamental response of control; red circles: *Nob1* mice; triangles: *Nob1* mice after intravitreal injection of PDA (5.2–6.2 mM). (From Shirato et al 2008,[113] used with permission from Association for Research in Vision and Ophthalmology.)

Flicker ERG

The fast flicker ERG (nominally 30 Hz flicker) is used to assess the health of cone-driven responses in humans because rod-driven responses do not respond to such fast flicker, resolving up to about 20 Hz. It is a particularly valuable test for monitoring residual cone function in degenerative diseases, such as retinitis pigmentosa. Studies using APB and PDA in macaques have shown that most of the fast flicker response (e.g. Fig. 24.1) is generated by postreceptoral cells in ON and OFF pathways.[121,122] Intravitreal injection of APB, to block the b-wave, left a delayed peak that was practically eliminated by addition of PDA to block OFF pathway responses.

The interactions of ON and OFF pathways in flicker ERGs have been examined in macaque over a wide range of temporal frequencies.[123] The typical temporal frequency response function of the photopic ERG of the macaque, illustrated in Figure 24.15A, is similar to that seen in humans.[124] Response amplitude dips at frequencies around 10–12 Hz before rising to a maximum amplitude around 50 Hz. Injections of APB and PDA revealed that the dip is due to cancellation of contributions to the response at different phases from ON and OFF pathways. Other experiments demonstrated small more proximal retinal contributions to the flicker responses, at the fundamental frequency of stimulation, and prominent TTX-sensitive contributions to the second harmonic component of the response that peaked around 8 Hz.[122]

Flicker ERGs in mice differ from those in primates in several ways. The frequency response for the primary rod pathway in mice extends up to 30 Hz and secondary rod pathway responses up to 50 Hz.[81] These ranges overlap substantially with the range for cone pathway responses shown in Figure 24.15B for a control C57BL/6 mouse.[81,113,124] The overlap of rod- and cone-driven frequency responses means that flicker cannot easily be used in rodents to selectively stimulate cone-driven responses. The lower photopic frequency response of mice relative to primates is due at least in part to the slower recovery of cone photoreceptors in mice.[113] Figure 24.17B also shows the frequency response of *Nob1* mice that lack ON pathway responses. The low amplitude of responses in *Nob1* mice indicates that OFF pathway contributions to the light-adapted frequency response are smaller in mice than in primates. The figure also shows that ON and OFF pathway responses do not interact to cancel responses at midrange frequencies in mice. Similar to macaques, however, the photoreceptors, whose contribution was isolated by injection of PDA, contributed only a small signal.[113] Second harmonic responses in C57BL/6 mice, that peak around 17 Hz, were found to be greatly reduced either by loss of ON pathway responses,[113] or injection of TTX.

Oscillatory potentials

Oscillatory potentials (OPs) in flash ERG responses to strong stimuli consist of a series of high-frequency, low-amplitude wavelets superimposed on the b-wave. OPs are present under dark- and light-adapted conditions. The mixed rod–cone flash ERG of a human (Fig. 24.1 top): includes at least four OPs that can be extracted by filtering the response to remove frequencies lower than ~75 Hz signals. The number of wavelets varies between four and ten depending upon stimulus conditions; the temporal characteristics of OPs vary as well. As an example, a recent study described dark-adapted OPs in humans as ranging in frequency from 100 Hz or less to more than 200 Hz, with a peak around 150 Hz for moderate stimulus strengths, but closer to 100 Hz for the strongest stimuli.[125] Light-adapted OPs also can include a high-frequency band peaking around 150 Hz, but have a lower-frequency band peaking that peaks around 75 Hz and extends to frequencies of about 50 Hz, below the low-frequency cutoff (75 Hz) for isolating OPs recommended by ISCEV.

There is consensus, from experiments in amphibians and mammals, that OPs are generated by inner retinal neurons,[30] making them useful for evaluating inner retinal function. However the discrete origins of different OPs, numbered in order from the first to occur, and mechanisms of their generation, are not well understood. Intraretinal studies using stimuli that elicited responses from both rod and cone systems in macaques localized OPs, as a group, to inner retina.[31] For a brief light flash, the major OPs in the photopic ERG were found to be APB-sensitive, indicating an origin in the ON pathway, but later OPs in the brief flash response, and at light offset for longer duration stimuli were found to originate from OFF pathway.[126,127] TTX reduced amplitudes of later OPs more than early ones in rabbits, but this finding has not been consistent across species.[127,128] Inhibitory neurotransmitters glycine and GABA suppress OPs in monkeys and other mammals, as do iontotropic glutamate receptor antagonists that block transmission of signals from bipolar

cells to amacrine and ganglion cells. Consistent with this, PDA removed OPs from the mouse light-adapted ERG illustrated in Figure 24.16B. GABAergic involvement is limited to specific receptor types; genetic deletion of GABA_C receptors in mice enhanced amplitudes of the OPs,[129] whereas blockade of GABA_A receptors removed OPs in C57BL/6 mice (unpublished observations).

Although the observations described above indicate involvement of amacrine or retinal ganglion cells in generating OPs, the role of ganglion cells has been controversial. Loss of OPs after ganglion cell death has not been observed consistently across species, making amacrine cells more likely generators. Reductions in OPs have been reported in conditions such as diabetic retinopathy that compromise inner retinal circulation, but not ganglion cell function selectively. However, in Ogden's studies in macaques, optic nerve section and subsequent ganglion cell degeneration lead to disappearance of the OPs.[130] Further, in the macaque photopic "flash" ERG studies of macular regions (using a multifocal stimulus), both severe experimental glaucoma and TTX removed a high-frequency band of OPs (centered at 143 Hz), while a lower-frequency band (at 77 Hz) remained intact.[128,131] In contrast, OPs were still present in full-field dark-adapted ERGs.[43]

The mechanisms for generating OPs are likely to involve both intrinsic membrane properties of cells and neuronal interactions/feedback circuits. Oscillatory activity has been observed most frequently in amacrine cells (ACs). For example, GABAergic wide-field ACs (WFAC) isolated from white bass retina generated high-frequency (>100 Hz) oscillatory membrane potentials (OMPs) in response to extrinsic depolarization.[132] These OMPs were shown to arise from "a complex interplay between voltage-dependent Ca^{2+} currents and voltage- and Ca^{2+}-dependent K^+ currents".[132]

A feedback mechanism is consistent with effects on OPs of GABA and glycine, both of which participate widely in feedback circuits in inner retina, e.g. between amacrine cells and bipolar cells or other amacrine cells (see Chapter 23). A feedback model has been proposed to account for high-frequency oscillatory (or "rhythmic") activity seen in some mammalian ganglion cell recordings from whole retina that involves electrical synapses, indicated by tracer coupling patterns, as well as inhibitory feedback circuits between amacrine and ganglion cells.[133]

OPs are more prominent in mice than in primates or rats. Peak temporal frequencies of OPs, in dark-adapted C57BL/6J mice range from 100 to 120 Hz or higher, as in humans, whereas they are about 70–85 Hz in light-adapted mice.[134] In rats where OPs are much smaller overall, two frequency bands of OPs have been reported for dark-adapted ERG, one peaking around 70 Hz and the other, more like other species, around 120–130 Hz. In the light-adapted ERG, only the lower-frequency band is present, similar to the findings in mice.[135] In mice and rats, OPs under dark- and light-adapted conditions are present at frequencies lower than common in primates, i.e. down to 50 Hz, and in mice small OPs can occur at frequencies as low as 30–40 Hz.[134]

Photopic negative response

A slow negative wave after the b-wave, called the photopic negative response (PhNR) can be seen in the light-adapted

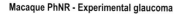

Macaque PhNR - Experimental glaucoma

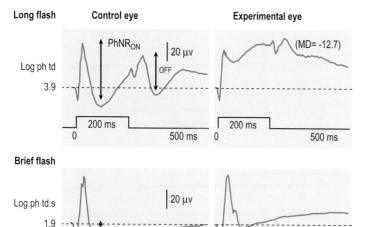

Figure 24.16 Photopic negative response (PhNR) in macaque monkey and human in normal and glaucomatous eyes.
(**A**) Full-field flash ERG showing the PhNR of a macaque in response to long (top) and brief (middle) red LED flashes on a rod saturating blue background (3.7 log sc td) from the control (left) and "experimental" (middle) fellow eye with laser-induced glaucoma, and the difference between control and experimental records (right). Arrows mark the amplitude of the PhNR. (Adapted from Viswanathan et al 1999,[28] used with permission from Association for Research in Vision and Ophthalmology.)
(**B**) Full-field flash ERGs of an age-matched 63-year-old normal human subject and a patient with POAG under similar stimulus conditions to those used above for monkeys. (Adapted from Viswanathan et al 2001,[29] used with permission from Association for Research in Vision and Ophthalmology.)

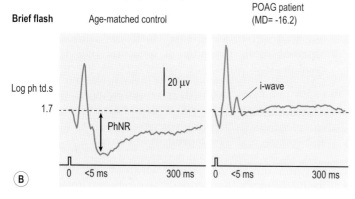

ERG, more prominently in primates than in mice and rats. In humans and monkeys the PhNR is thought to reflect spiking activity of retinal ganglion cells.[28,29,122] As shown in Figure 24.16, the PhNR is reduced in macaque eyes with experimental glaucoma (or after TTX injection),[28] and in humans with primary open angle glaucoma (POAG).[29] It also is reduced in several other disorders affecting the optic nerve head and inner retina. The slow timecourse of the PhNR suggests glial involvement, perhaps via K^+ currents in astrocytes in the optic nerve head set up by increased $[K^+]_o$ due to spiking of ganglion cells. PhNRs can be evoked using white flashes on a white background, if the flash is very bright. However a red LED flash on a blue background, as was used for responses in Figure 24.16, elicits PhNRs over a wider range of stimulus strengths. The red flash may minimize spectral opponency that would reduce ganglion cell responses, and use of a blue background suppresses rods while minimizing light adaptation of L-cone signals.[112] The PhNR can be exaggerated relative to other major ERG components using slowed focal stimuli confined to the macula.[136]

In rodents, with relatively fewer ganglion cells than in primates, the PhNR is small, and probably originates mainly

from amacrine cells. It is reduced in amplitude by TTX, PDA blockade of block transmission to inner retina in rats and mice, NMDA to suppress inner retinal activity,[137] but not by loss of ganglion cells in mice, as shown in the inset to Fig. 24.17,[116] or rat.[93] Glial involvement in generation of the PhNR in rodents is likely. In Royal College of Surgeons rats, a large photopic ERG negative response develops in the degenerating retina,[137] due at least in part to increased density of Kir4.1 channels in Müller cell endfeet. These channels are critical for producing glial currents that contribute to the ERG.[14,15]

Pattern ERG

A common technique for assessing ganglion cell function is to record a pattern ERG (PERG). The stimulus is usually a contrast-reversing checkerboard or grating pattern for which changes in local luminance occur with each reversal, but the mean luminance remains constant. This causes the linear signals that produce a- and b-waves to cancel, leaving only the non-linear signals in the ERG. Non-linear signals that compose the PERG are known to depend upon

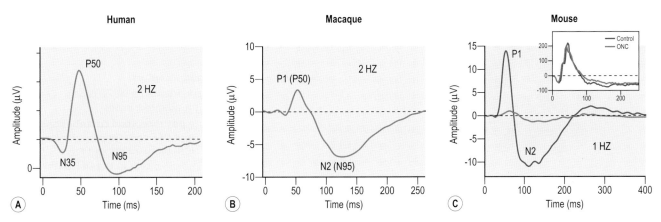

Figure 24.17 Transient pattern electroretinogram (PERG) of human, macaque monkey and mouse. (A) Representative transient PERG of a normal human subject, elicited with a contrast-reversing checkerboard pattern (0.8 deg checks), mean luminance >80 cd.m^{-2}, modulated at 2 Hz, 100 percent contrast. Amplitude for normal humans ranges from 2 to 8 μV, according to the "ISCEV standard for clinical pattern Electroretinography 2007 update" from which the figure was adapted. (From Holder et al 2007,[91] used with permission.) **(B)** Transient PERG of a macaque, elicited with a checkerboard (0.5 deg checks), mean luminance of 55 cd.m^{-2} modulated at 2 Hz, 84 percent contrast. (Unpublished, Luo & Frishman). **(C)** PERG of a C57BL/6 mouse before (blue line) and 40 days after (red line) unilateral optic nerve crush (ONC) when ganglion cells had degenerated. The pattern was a 0.05 c/deg horizontal bar grating, mean luminance 45 sc cd.m^{-2}, contrast-reversed at 1 Hz, 90 percent contrast. Inset shows brief flash ERGs for the same eye: 57 cd.s.m^{-2} flash on a 63 sc cd.m^{-2} background before (black line) and 40 days after (red line) ONC. (From Miura et al 2009,[116] used with permission.)

the functional integrity of retinal ganglion cells. PERGs are eliminated, while a- and b-waves of the flash ERG are still present (e.g. see Fig. 24.17), following optic nerve section or crush that causes degeneration of ganglion cell axons and subsequently, their cell bodies in several mammals. Initial studies done in cat[138], were extended to monkey[139], rodents[116,140–143] and a human with accidental nerve transection[144]

The PERG has been used widely in clinical studies assessing ganglion cell function in eyes with glaucoma and other diseases that affect inner retina (see reviews by Holder[145] and Bach and Hoffman).[146] The PERG can be a sensitive test in early glaucoma when visual field deficits are minimal.[146–148] It shares this attribute with the PhNR which has similar retinal origins to the PERG in macaque.[122]

The PERG is also useful for study of rodent models of glaucoma. The DBA/2J mouse model of inherited glaucoma has normal ganglion cell numbers at 2 months of age, and progresses to massive retinal ganglion cell degeneration by 12–14 months.[149] This is reflected in the PERG which is of normal amplitude in young DBA/2J mice, whereas it is practically eliminated in older mice.[143] Because the PERG is a non-invasive measure it can be used in these mice and other rodent models of glaucoma to track progression of ganglion cell loss as well as to document effects of therapies to slow progression and protect neurons.

PERGs can be recorded as transient responses to low reversal frequencies, 1–2 Hz, or as steady state responses to higher frequencies, e.g. 8 Hz. The transient PERG has prominent early positive and later negative waves, named respectively, P50 and N95 (see Fig. 24.17) in humans, denoting the timing of the peak and trough following each pattern reversal.[91] Both waves reflect ganglion cell activity, although P50 may include some non-spiking input.[150] The early positive (P1) and later negative (N2) component (Fig. 24.17) in the mouse transient PERG, different in exact timing, but have similar general appearance to P50 and N95 in primates.[116,143]

Multifocal ERG

The multifocal ERG (mfERG) provides a technique for simultaneously recording local ERG responses from many small retinal regions, typically 60 or 103, over 35–40 degrees of the visual field. Focal ERGs from individual regions are recorded in some clinics and labs,[151] but sampling more than a couple of regions is time consuming. Figure 24.18A illustrates the stimulus and mfERG records from a normal human subject, The individual hexagons of the stimulus were reversed in contrast following a pseudorandom (m-) sequence which is shifted in time uniquely for each sector to enable extraction via correlation of individual responses.[152,153] The reversal rate was locked to the frame rate of the visual stimulator, 75 Hz for CRTs, 60 Hz for LCD. Each hexagon had a 50 percent chance of reversing contrast on each frame change. The test is usually done under light-adapted conditions, for which the foveal response is large, allowing changes in function, such as occur in Stargardt's disease, or other macular dystrophies to be easily seen. The scotopic mfERG, rarely attempted, was found to be more susceptible than light-adapted recordings, to effects of scattered light.[154] Dark-adapted recordings in rodents required use of large focal regions,[155] and although difficult, some light-adapted recordings in rodents have been possible.[156]

Studies using glutamate analogs have shown that despite different methods of generation and response timing of the mfERG, the cellular origins of the major positive and negative waves are essentially the same as the light-adapted flash ERG; see Figure 24.18B.[157] This was the case for the standard fast mfERG, as well as when the stimulus presentation was slowed by interleaving blank frames between m-steps, to allow the full ERG to form.[158]

A small optic nerve head component from axons of retinal ganglion cells may also be observed in the mfERG, especially when stimuli are arranged to optimize the response.[159,160]

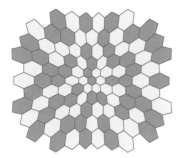

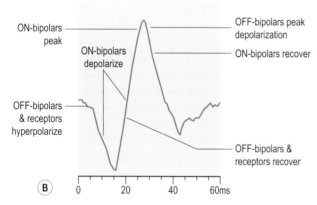

500nv

0 100ms

Ⓐ

ON-bipolars peak

ON-bipolars depolarize

OFF-bipolars & receptors hyperpolarize

OFF-bipolars peak depolarization

ON-bipolars recover

OFF-bipolars & receptors recover

Ⓑ   0   20   40   60ms

Figure 24.18 Multifocal electroretinogram (mfERG) and its major components. (A) mfERG trace array (field view) using 103 hexagons scaled with eccentricity and covering about 35 deg of visual angle, as illustrated above the traces. (From Hood et al 2008,[162] used with permission.) **(B)** Model for the retinal contributions to the human mfERG based on results from the macular region of a macaque after pharmacological separation of components. (From Hood et al 2002,[157] used with permission from Association for Research in Vision and Ophthalmology.)

The optic nerve head component is the likely generator of high-frequency oscillatory potentials in the slow sequence mfERG that are reduced in experimental glaucoma in macaques.[128,131]

Closing comments

The ERG provides a means for non-invasive evaluation of normal and abnormal retinal function in humans and in animal models. Its value as a test has increased as we have better understood the retinal circuits involved and mechanisms of generation for each wave. Our current understanding of the origins of the ERGs recording in standard testing, and in more specialized testing, as reviewed in this chapter, are summarized in Boxes 24.2 and 24.3 respectively.

Box 24.2 Retinal Cells Contributing to the Flash and Flicker ERG in Standard Clinical Testing

a-wave	Photoreceptors
	Dark-adapted a-wave: rods Light-adapted: cones
	Late postreceptoral contributions from OFF cone (hyperpolarizing: HCB) bipolar cells and more proximal cells in the OFF pathway
b-wave	Bipolar cells
	Dark-adapted b-wave, rod bipolar cells (RBC)
	Light-adapted b-wave, ON cone (depolarizing: DCB), OFF cone bipolar cells, and horizontal (Hz) cell feedback via the cones
d-wave	Bipolar cells
	Mainly a light-adapted response: OFF cone bipolar cells, offset of cone photoreceptors, and offset of ON cone bipolar cells
	The d-wave is not present in mice and rats
Oscillatory potentials (OPs)	Amacrine and ganglion cells, bipolar cell terminals?
'30 Hz' fast flicker	Bipolar cells
	ON and OFF cone bipolar cells, and a small cone photoreceptor contribution

---

Box 24.3 Retinal Cells Contributing to the ERG in Specialized Testing

Scotopic threshold response	Inner retinal neurons
pSTR	Amacrine cells (monkey)
	Retinal ganglion cells (rodents)
nSTR	Retinal ganglion cells (monkey)
	Partially retinal ganglion cells (rats, human?)
	Amacrine cells (AII) (mice)
	Partially amacrine cells (rats, human)
	Glial currents
Photopic negative response (PhNR)	Retinal ganglion cells (human, monkey)
	Amacrine cells (rodents)
	Glial currents
Pattern ERG (PERG)	Retinal ganglion cells (human, monkey, rodent)
	Glial currents (transient PERG: N95, N2?)
Multifocal ERG	Initial negative and positive waves have similar origins to a- and b-waves in photopic ERG (monkey, and presumably human)

References

1. Marmor MF, Fulton AB, Holder GE, Miyake Y, Brigell M, Bach M. ISCEV Standard for full-field clinical electroretinography (2008 update). Doc Ophthalmol 2009; 118(1):69–77.
2. Perlman I. Testing retinal toxicity of drugs in animal models using electrophysiological and morphological techniques. Doc Ophthalmol 2009; 118(1):3–28.
3. Peachey NS, Ball SL. Electrophysiological analysis of visual function in mutant mice. Doc Ophthalmol 2003; 107(1):13–36.
4. Tanimoto N, Muehlfriedel RL, Fischer MD, et al. Vision tests in the mouse: Functional phenotyping with electroretinography. Frontiers Biosci 2009; 14:2730–2737.
5. Dowling JE. The retina: an approachable part of the brain. Cambridge, MA: The Belknap Press, 1987.
6. Fadool JM, Dowling JE. Zebrafish: a model system for the study of eye genetics. Progr Retin Eye Res 2008; 27(1):89–110.
7. Fishman GA, Birch DG, Holder GE, Brigell MG. Electrophysiological testing in disorders of the retina, optic nerve, and visual pathways, 2nd edn. Washington: American Academy of Ophthalmology, 2001.
8. Heckenlively J, Arden GB. Principles and practice of clinical electrophysiology of vision, 2nd edn. Cambridge, MA: MIT Press, 2006.
9. Lam BL. Electrophysiology of vision; Clinical testing and applications. Boca Raton, FL: Taylor and Francis, 2005.
10. Brown KT, Wiesel TN. Localization of origins of electroretinogram components by intraretinal recording in the intact cat eye. J Physiol 1961; 158:257–280.
11. Brown KT, Wiesel TN. Analysis of the intraretinal electroretinogram in the intact cat eye. J Physiol 1961; 158:229–256.
12. Newman E, Reichenbach A. The Muller cell: a functional element of the retina. Trends Neurosci 1996; 19(8):307–312.
13. Newman EA, Frambach DA, Odette LL. Control of extracellular potassium levels by retinal glial cell K, siphoning. Science 1984; 225(4667):1174–1175.
14. Kofuji P, Biedermann B, Siddharthan V, et al. Kir potassium channel subunit expression in retinal glial cells: implications for spatial potassium buffering. Glia 2002; 39(3):292–303.
15. Kofuji P, Ceelen P, Zahs KR, Surbeck LW, Lester HA, Newman EA. Genetic inactivation of an inwardly rectifying potassium channel (Kir4.1 subunit) in mice: phenotypic impact in retina. J Neurosci 2000; 20(15):5733–5740.
16. Robson JG, Frishman LJ. Dissecting the dark-adapted electroretinogram. Doc Ophthalmol 1998; 95(3–4):187–215.
17. Steinberg RH, Linsenmeier RA, Griff ER. Three light-evoked responses of the retinal pigment epithelium. Vision Res 1983; 23(11):1315–1323.
18. Steinberg RH, Oakley B, 2nd, Niemeyer G. Light-evoked changes in [K+]0 in retina of intact cat eye. J Neurophysiol 1980; 44(5):897–921.
19. Witkovsky P, Dudek FE, Ripps H. Slow PIII component of the carp electroretinogram. J Gen Physiol 1975; 65(2):119–134.
20. Oakley B, 2nd. Potassium and the photoreceptor-dependent pigment epithelial hyperpolarization. J Gen Physiol 1977; 70(4):405–425.
21. Oakley B, 2nd, Green DG. Correlation of light-induced changes in retinal extracellular potassium concentration with c-wave of the electroretinogram. J Neurophysiol 1976; 39(5):1117–1133.
22. Bui BV, Fortune B. Ganglion cell contributions to the rat full-field electroretinogram. J Physiol 2004; 555(Pt 1):153–173.
23. Frishman LJ, Sieving PA, Steinberg RH. Contributions to the electroretinogram of currents originating in proximal retina. Vis Neurosci 1988; 1(3):307–315.
24. Frishman LJ, Steinberg RH. Light-evoked increases in [K$^+$]o in proximal portion of the dark-adapted cat retina. J Neurophysiol 1989; 61(6):1233–1243.
25. Frishman LJ, Steinberg RH. Intraretinal analysis of the threshold dark-adapted ERG of cat retina. J Neurophysiol 1989; 61(6):1221–1232.
26. Saszik SM, Robson JG, Frishman LJ. The scotopic threshold response of the dark-adapted electroretinogram of the mouse. J Physiol 2002; 543(Pt 3):899–916.
27. Sieving PA, Frishman LJ, Steinberg RH. Scotopic threshold response of proximal retina in cat. J Neurophysiol 1986; 56(4):1049–1061.
28. Viswanathan S, Frishman LJ, Robson JG, Harwerth RS, Smith EL, 3rd. The photopic negative response of the macaque electroretinogram: reduction by experimental glaucoma. Invest Ophthalmol Vis Sci 1999; 40(6):1124–1136.
29. Viswanathan S, Frishman LJ, Robson JG, Walters JW. The photopic negative response of the flash electroretinogram in primary open angle glaucoma. Invest Ophthalmol Vis Sci 2001; 42(2):514–522.
30. Wachtmeister L. Oscillatory potentials in the retina: what do they reveal. Progr Retin Eye Res 1998; 17(4):485–521.
31. Heynen H, Wachtmeister L, van Norren D. Origin of the oscillatory potentials in the primate retina. Vision Res 1985; 25(10):1365–1373.
32. Dawson WW, Trick GL, Litzkow CA. Improved electrode for electroretinography. Invest Ophthalmol Vis Sci 1979; 18(9):988–991.
33. Bayer AU, Cook P, Brodie SE, Maag KP, Mittag T. Evaluation of different recording parameters to establish a standard for flash electroretinography in rodents. Vision Res 2001; 41(17):2173–2185.
34. Linsenmeier RA, Steinberg RH. Origin and sensitivity of the light peak in the intact cat eye. J Physiol 1982; 331:653–673.
35. Wu J, Peachey NS, Marmorstein AD. Light-evoked responses of the mouse retinal pigment epithelium. J Neurophysiol 2004; 91(3):1134–1142.
36. Granit R. The components of the retinal action potential in mammals and their relation to the discharge in the optic nerve. J Physiol 1933; 77(3):207–239.
37. Frishman LJ. Electrogenesis of the ERG. In: Ryan SJ, ed. Retina. St. Louis, MO: Elsevier/Mosby, 2005:103–135.
38. Frishman LJ. Origins of the ERG. In: Heckenlively J, Arden GB, eds. Principles and practice of clinical electrophysiology of vision. Cambridge, Mass: MIT Press, 2006:139–183.
39. Brown M, Marmor M, Vaegan, Zrenner E, Brigell M, Bach M. ISCEV Standard for Clinical Electro-oculography (EOG) 2006. Doc Ophthalmol 2006; 113(3):205–212.
40. Freed MA, Smith RG, Sterling P. Rod bipolar array in the cat retina: pattern of input from rods and GABA-accumulating amacrine cells. J Comp Neurol 1987; 266(3):445–455.
41. Smith RG, Freed MA, Sterling P. Microcircuitry of the dark-adapted cat retina: functional architecture of the rod-cone network. J Neurosci 1986; 6(12):3505–3517.
42. Tsukamoto Y, Morigiwa K, Ueda M, Sterling P. Microcircuits for night vision in mouse retina. J Neurosci 2001; 21(21):8616–8623.
43. Frishman LJ, Shen FF, Du L, et al. The scotopic electroretinogram of macaque after retinal ganglion cell loss from experimental glaucoma. Invest Ophthalmol Vis Sci 1996; 37(1):125–141.
44. Abd-El-Barr MM, Pennesi ME, Saszik SM, et al. Genetic dissection of rod and cone pathways in the dark-adapted mouse retina. J Neurophysiol 2009; 102:1945–1955.
45. Sterling P, Freed MA, Smith RG. Architecture of rod and cone circuits to the on-beta ganglion cell. J Neurosci 1988; 8(2):623–642.
46. Heynen H, van Norren D. Origin of the electroretinogram in the intact macaque eye – II. Current source-density analysis. Vision Res 1985; 25(5):709–715.
47. Penn RD, Hagins WA. Signal transmission along retinal rods and the origin of the electroretinographic a-wave. Nature 1969; 223(5202):201–204.
48. Penn RD, Hagins WA. Kinetics of the photocurrent of retinal rods. Biophys J 1972; 12(8):1073–1094.
49. Furukawa T, Hanawa I. Effects of some common cations on electroretinogram of the toad. Japn J Physiol 1955; 5(4):289–300.
50. Sillman AJ, Ito H, Tomita T. Studies on the mass receptor potential of the isolated frog retina. II. On the basis of the ionic mechanism. Vision Res 1969; 9(12):1443–1451.
51. Slaughter MM, Miller RF. 2-amino-4-phosphonobutyric acid: a new pharmacological tool for retina research. Science 1981; 211(4478):182–185.
52. Knapp AG, Schiller PH. The contribution of on-bipolar cells to the electroretinogram of rabbits and monkeys. A study using 2-amino-4-phosphonobutyrate (APB). Vision Res 1984; 24(12):1841–1846.
53. Stockton RA, Slaughter MM. B-wave of the electroretinogram. A reflection of ON bipolar cell activity. J Gen Physiol 1989; 93(1):101–122.
54. Robson JG, Frishman LJ. Photoreceptor and bipolar cell contributions to the cat electroretinogram: a kinetic model for the early part of the flash response. J Optical Soc Am A Optics Image Sci Vis 1996; 13(3):613–622.
55. Robson JG, Saszik SM, Ahmed J, Frishman LJ. Rod and cone contributions to the a-wave of the electroretinogram of the macaque. J Physiol 2003; 547(Pt 2):509–530.
56. Masu M, Iwakabe H, Tagawa Y, et al. Specific deficit of the ON response in visual transmission by targeted disruption of the mGluR6 gene. Cell 1995; 80(5):757–765.
57. McCall MA, Gregg RG. Comparisons of structural and functional abnormalities in mouse b-wave mutants. J Physiol 2008; 586(Pt 18):4385–4392.
58. Kang Derwent JJ, Saszik SM, Maeda H, et al. Test of the paired-flash electroretinographic method in mice lacking b-waves. Vis Neurosci 2007; 24(2):141–149.
59. Pardue MT, McCall MA, LaVail MM, Gregg RG, Peachey NS. A naturally occurring mouse model of X-linked congenital stationary night blindness. Invest Ophthalmol Vis Sci 1998; 39(12):2443–2449.
60. Gregg RG, Mukhopadhyay S, Candille SI, et al. Identification of the gene and the mutation responsible for the mouse nob phenotype. Invest Ophthalmol Vis Sci 2003; 44(1):378–384.
61. Pesch K, Zeitz C, Fries JE, et al. Isolation of the mouse nyctalopin gene nyx and expression studies in mouse and rat retina. Invest Ophthalmol Vis Sci 2003; 44(5):2260–2266.
62. Bech-Hansen NT, Naylor MJ, Maybaum TA, et al. Mutations in NYX, encoding the leucine-rich proteoglycan nyctalopin, cause X-linked complete congenital stationary night blindness. Nature Genet 2000; 26(3):319–323.
63. Boycott KM, Pearce WG, Musarella MA, et al. Evidence for genetic heterogeneity in X-linked congenital stationary night blindness. Am J Hum Genet 1998; 62(4): 865–875.
64. Dryja TP, McGee TL, Berson EL, et al. Night blindness and abnormal cone electroretinogram ON responses in patients with mutations in the GRM6 gene encoding mGluR6. Proc Natl Acad Sci USA 2005; 102(13):4884–4889.
65. Chang B, Heckenlively JR, Bayley PR, et al. The nob2 mouse, a null mutation in Cacna1f: anatomical and functional abnormalities in the outer retina and their consequences on ganglion cell visual responses. Vis Neurosci 2006; 23(1):11–24.
66. Brown KT. The electroretinogram: its components and their origins. Vision Res 1968; 8(6):633–677.
67. Brown KT, Watanabe K, Murakami M. The early and late receptor potentials of monkey cones and rods. Cold Spring Harbor Symp Quant Biol 1965; 30:457–482.
68. Hood DC, Birch DG. A quantitative measure of the electrical activity of human rod photoreceptors using electroretinography. Vis Neurosci 1990; 5(4):379–387.
69. Hood DC, Birch DG. The A-wave of the human electroretinogram and rod receptor function. Invest Ophthalmol Vis Sci 1990; 31(10):2070–2081.
70. Baylor DA, Nunn BJ, Schnapf JL. The photocurrent, noise and spectral sensitivity of rods of the monkey Macaca fascicularis. J Physiol 1984; 357:575–607.
71. Lamb TD, Pugh EN, Jr. G-protein cascades: gain and kinetics. Trends Neurosci 1992; 15(8):291–298.
72. Lamb TD, Pugh EN, Jr. A quantitative account of the activation steps involved in phototransduction in amphibian photoreceptors. J Physiol 1992; 449:719–758.
73. Breton ME, Schueller AW, Lamb TD, Pugh EN, Jr. Analysis of ERG a-wave amplification and kinetics in terms of the G-protein cascade of phototransduction. Invest Ophthalmol Vis Sci 1994; 35(1):295–309.
74. Hood DC, Birch DG. Human cone receptor activity: the leading edge of the a-wave and models of receptor activity. Vis Neurosci 1993; 10(5):857–871.
75. Hood DC, Birch DG. Light adaptation of human rod receptors: the leading edge of the human a-wave and models of rod receptor activity. Vision Res 1993; 33(12):1605–1618.
76. Hood DC, Birch DG. Phototransduction in human cones measured using the alpha-wave of the ERG. Vision Res 1995; 35(20):2801–2810.
77. Hood DC, Birch DG. Measuring the health of the human photoreceptors with the leading edge of the a-wave. In: Heckenlively J, Arden GB, eds. Principles and practice of clinical electrophysiology of vision. Chicago: CV Mosby, 2006:487–502.

78. Hood DC, Birch DG. Assessing abnormal rod photoreceptor activity with the a-wave of the electroretinogram: applications and methods. Doc Ophthalmol 1996; 92(4):253–267.

79. Cideciyan AV, Jacobson SG. An alternative phototransduction model for human rod and cone ERG a-waves: normal parameters and variation with age. Vision Res 1996; 36(16):2609–2621.

80. Pepperberg DR, Birch DG, Hood DC. Photoresponses of human rods in vivo derived from paired-flash electroretinograms. Vis Neurosci 1997; 14(1):73–82.

81. Nusinowitz S, Ridder WH, 3rd, Ramirez J. Temporal response properties of the primary and secondary rod-signaling pathways in normal and Gnat2 mutant mice. Exp Eye Res 2007; 84(6):1104–1114.

82. Deans MR, Volgyi B, Goodenough DA, Bloomfield SA, Paul DL. Connexin36 is essential for transmission of rod-mediated visual signals in the mammalian retina. Neuron 2002; 36(4):703–712.

83. Robson JG, Maeda H, Saszik SM, Frishman LJ. In vivo studies of signaling in rod pathways of the mouse using the electroretinogram. Vision Res 2004; 44(28):3253–3268.

84. Cameron AM, Mahroo OA, Lamb TD. Dark adaptation of human rod bipolar cells measured from the b-wave of the scotopic electroretinogram. J Physiol 2006; 575(Pt 2):507–526.

85. Field GD, Rieke F. Non-linear signal transfer from mouse rods to bipolar cells and implications for visual sensitivity. Neuron 2002; 34(5):773–785.

86. Robson JG, Frishman LJ. Response linearity and kinetics of the cat retina: the bipolar cell component of the dark-adapted electroretinogram. Vis Neurosci 1995; 12(5):837–850.

87. Frishman LJ, Yamamoto F, Bogucka J, Steinberg RH. Light-evoked changes in [K$^+$]o in proximal portion of light-adapted cat retina. J Neurophysiol 1992; 67(5):1201–1212.

88. Miller RF, Dowling JE. Intracellular responses of the Muller (glial) cells of mudpuppy retina: their relation to b-wave of the electroretinogram. J Neurophysiol 1970; 33(3):323–341.

89. Frishman LJ, Reddy MG, Robson JG. Effects of background light on the human dark-adapted electroretinogram and psychophysical threshold. J Optical Soc Am A Optics Image Sci Vis 1996; 13(3):601–612.

90. Naarendorp F, Sieving PA. The scotopic threshold response of the cat ERG is suppressed selectively by GABA and glycine. Vision Res 1991; 31(1):1–15.

91. Holder GE, Brigell MG, Hawlina M, Meigen T, Vaegan, Bach M. ISCEV standard for clinical pattern electroretinography – 2007 update. Doc Ophthalmol 2007; 114(3):111–116.

92. Barlow HB, Levick WR, Yoon M. Responses to single quanta of light in retinal ganglion cells of the cat. Vision Res 1971; Suppl 3:87–101.

93. Mojumder DK, Sherry DM, Frishman LJ. Contribution of voltage-gated sodium channels to the b-wave of the mammalian flash electroretinogram. J Physiol 2008; 586(10):2551–2580.

94. Moshiri A, Gonzalez E, Tagawa K, et al. Near complete loss of retinal ganglion cells in the math5/brn3b double knockout elicits severe reductions of other cell types during retinal development. Dev Biol 2008; 316(2):214–227.

95. Saszik SM. The dark-adapted electroretinogram (ERG) of the mouse: inner retinal contributions and the effects of light adaptation. College of Optometry. Houston: University of Houston, 2003.

96. Sieving PA. Retinal ganglion cell loss does not abolish the scotopic threshold response (STR) of the cat and human ERG Clin Vis Sci 1991; 2:149–158.

97. Guldenagel M, Ammermuller J, Feigenspan A, et al. Visual transmission deficits in mice with targeted disruption of the gap junction gene connexin36. J Neurosci 2001; 21(16):6036–6044.

98. Bui BV, Edmunds B, Cioffi GA, Fortune B. The gradient of retinal functional changes during acute intraocular pressure elevation. Invest Ophthalmol Vis Sci 2005; 46(1):202–213.

99. Lyubarsky AL, Falsini B, Pennesi ME, Valentini P, Pugh EN, Jr. UV- and midwave-sensitive cone-driven retinal responses of the mouse: a possible phenotype for coexpression of cone photopigments. J Neurosci 1999; 19(1):442–455.

100. Jacobs GH, Fenwick JA, Williams GA. Cone-based vision of rats for ultraviolet and visible lights. J Exp Biol 2001; 204(Pt 14):2439–2446.

101. Jacobs GH, Neitz J, Krogh K. Electroretinogram flicker photometry and its applications. J Optical Soc Am A Optics Image Sci Vis 1996; 13(3):641–648.

102. Jacobs GH, Neitz J, Deegan JF, 2nd. Retinal receptors in rodents maximally sensitive to ultraviolet light. Nature 1991; 353(6345):655–656.

103. Toda K, Bush RA, Humphries P, Sieving PA. The electroretinogram of the rhodopsin knockout mouse. Vis Neurosci 1999; 16(2):391–398.

104. Calvert PD, Krasnoperova NV, Lyubarsky AL, et al. Phototransduction in transgenic mice after targeted deletion of the rod transducin alpha-subunit. Proc Natl Acad Sci USA 2000; 97(25):13913–13918.

105. Chang B, Dacey MS, Hawes NL, et al. Cone photoreceptor function loss-3, a novel mouse model of achromatopsia due to a mutation in Gnat2. Invest Ophthalmol Vis Sci 2006; 47(11):5017–5021.

106. Lucas RJ, Freedman MS, Lupi D, Munoz M, David-Gray ZK, Foster RG. Identifying the photoreceptive inputs to the mammalian circadian system using transgenic and retinally degenerate mice. Behav Brain Res 2001; 125(1–2):97–102.

107. Fu Y, Zhong H, Wang MH, et al. Intrinsically photosensitive retinal ganglion cells detect light with a vitamin A-based photopigment, melanopsin. Proc Natl Acad Sci USA 2005; 102(29):10339–10344.

108. Rangaswamy NV, Frishman LJ, Dorotheo EU, Schiffman JS, Bahrani HM, Tang RA. Photopic ERGs in patients with optic neuropathies: comparison with primate ERGs after pharmacologic blockade of inner retina. Invest Ophthalmol Vis Sci 2004; 45(10):3827–3837.

109. Bush RA, Sieving PA. A proximal retinal component in the primate photopic ERG a-wave. Invest Ophthalmol Vis Sci 1994; 35(2):635–645.

110. Sieving PA, Murayama K, Naarendorp F. Push-pull model of the primate photopic electroretinogram: a role for hyperpolarizing neurons in shaping the b-wave. Vis Neurosci 1994; 11(3):519–532.

111. Heynen H, van Norren D. Origin of the electroretinogram in the intact macaque eye – I. Principal component analysis. Vision Res 1985; 25(5):697–707.

112. Rangaswamy NV, Shirato S, Kaneko M, Digby BI, Robson JG, Frishman LJ. Effects of spectral characteristics of Ganzfeld stimuli on the photopic negative response (PhNR) of the ERG. Invest Ophthalmol Vis Sci 2007; 48(10):4818–4828.

113. Shirato S, Maeda H, Miura G, Frishman LJ. Postreceptoral contributions to the light-adapted ERG of mice lacking b-waves. Exp Eye Res 2008; 86(6):914–928.

114. Sharma S, Ball SL, Peachey NS. Pharmacological studies of the mouse cone electroretinogram. Vis Neurosci 2005; 22(5):631–636.

115. Xu L, Ball SL, Alexander KR, Peachey NS. Pharmacological analysis of the rat cone electroretinogram. Vis Neurosci 2003; 20(3):297–306.

116. Miura G, Wang MH, Ivers KM, Frishman LJ. Retinal pathway origins of the pattern ERG of the mouse. Exp Eye Res 2009; 89(1):49–62.

117. Cui J, Pan ZH. Two types of cone bipolar cells express voltage-gated Na$^+$ channels in the rat retina. Vis Neurosci 2008; 25(5–6):635–645.

118. Whitten DN, Brown KT. The time courses of late receptor potentials from monkey cones and rods. Vision Res 1973; 13(1):107–135.

119. Ueno S, Kondo M, Ueno M, Miyata K, Terasaki H, Miyake Y. Contribution of retinal neurons to d-wave of primate photopic electroretinograms. Vision Res 2006; 46(5):658–664.

120. Strettoi E, Volpini M. Retinal organization in the bcl-2-overexpressing transgenic mouse. J Comp Neurol 2002; 446(1):1–10.

121. Bush RA, Sieving PA. Inner retinal contributions to the primate photopic fast flicker electroretinogram. J Optical Soc Am A Optics Image Sci Vis 1996; 13(3):557–565.

122. Viswanathan S, Frishman LJ, Robson JG. Inner-retinal contributions to the photopic sinusoidal flicker electroretinogram of macaques. Macaque photopic sinusoidal flicker ERG. Doc Ophthalmol 2002; 105(2):223–242.

123. Kondo M, Sieving PA. Primate photopic sine-wave flicker ERG: vector modeling analysis of component origins using glutamate analogs. Invest Ophthalmol Vis Sci 2001; 42(1):305–312.

124. Krishna VR, Alexander KR, Peachey NS. Temporal properties of the mouse cone electroretinogram. J Neurophysiol 2002; 87(1):42–48.

125. Hancock HA, Kraft TW. Human oscillatory potentials: intensity-dependence of timing and amplitude. Doc Ophthalmol 2008; 117(3):215–222.

126. Rangaswamy NV, Hood DC, Frishman LJ. Regional variations in local contributions to the primate photopic flash ERG: revealed using the slow-sequence mfERG. Invest Ophthalmol Vis Sci 2003; 44(7):3233–3247.

127. Dong CJ, Agey P, Hare WA. Origins of the electroretinogram oscillatory potentials in the rabbit retina. Vis Neurosci 2004; 21(4):533–543.

128. Zhou W, Rangaswamy N, Ktonas P, Frishman LJ. Oscillatory potentials of the slow-sequence multifocal ERG in primates extracted using the Matching Pursuit method. Vision Res 2007; 47(15):2021–2036.

129. McCall MA, Lukasiewicz PD, Gregg RG, Peachey NS. Elimination of the rho1 subunit abolishes GABA(C) receptor expression and alters visual processing in the mouse retina. J Neurosci 2002; 22(10):4163–4174.

130. Ogden TE. The oscillatory waves of the primate electroretinogram. Vision Res 1973; 13(6):1059–1074.

131. Rangaswamy NV, Zhou W, Harwerth RS, Frishman LJ. Effect of experimental glaucoma in primates on oscillatory potentials of the slow-sequence mfERG. Invest Ophthalmol Vis Sci 2006; 47(2):753–767.

132. Vigh J, Solessio E, Morgans CW, Lasater EM. Ionic mechanisms mediating oscillatory membrane potentials in wide-field retinal amacrine cells. J Neurophysiol 2003; 90(1):431–443.

133. Kenyon GT, Moore B, Jeffs J, et al. A model of high-frequency oscillatory potentials in retinal ganglion cells. Vis Neurosci 2003; 20(5):465–480.

134. Lei B, Yao G, Zhang K, Hofeldt KJ, Chang B. Study of rod- and cone-driven oscillatory potentials in mice. Invest Ophthalmol Vis Sci 2006; 47(6):2732–2738.

135. Forte JD, Bui BV, Vingrys AJ. Wavelet analysis reveals dynamics of rat oscillatory potentials. J Neurosci Methods 2008; 169(1):191–200.

136. Kondo M, Kurimoto Y, Sakai T, et al. Recording focal macular photopic negative response (PhNR) from monkeys. Invest Ophthalmol Vis Sci 2008; 49(8):3544–3550.

137. Machida S, Raz-Prag D, Fariss RN, Sieving PA, Bush RA. Photopic ERG negative response from amacrine cell signaling in RCS rat retinal degeneration. Invest Ophthalmol Vis Sci 2008; 49(1):442–452.

138. Maffei L, Fiorentini A. Electroretinographic responses to alternating gratings in the cat. Exp Brain Res 1982; 48(3):327–334.

139. Maffei L, Fiorentini A, Bisti S, Hollander H. Pattern ERG in the monkey after section of the optic nerve. Exp Brain Res 1985; 59(2):423–425.

140. Berardi N, Domenici L, Gravina A, Maffei L. Pattern ERG in rats following section of the optic nerve. Exp Brain Res 1990; 79(3):539–546.

141. Porciatti V. The mouse pattern electroretinogram. Doc Ophthalmol 2007; 115(3):145–153.

142. Porciatti V, Pizzorusso T, Cenni MC, Maffei L. The visual response of retinal ganglion cells is not altered by optic nerve transection in transgenic mice overexpressing Bcl-2. Proc Natl Acad Sci USA 1996; 93(25):14955–14959.

143. Porciatti V, Saleh M, Nagaraju M. The pattern electroretinogram as a tool to monitor progressive retinal ganglion cell dysfunction in the DBA/2J mouse model of glaucoma. Invest Ophthalmol Vis Sci 2007; 48(2):745–751.

144. Harrison JM, O'Connor PS, Young RS, Kincaid M, Bentley R. The pattern ERG in man following surgical resection of the optic nerve. Invest Ophthalmol Vis Sci 1987; 28(3):492–499.

145. Holder GE. Pattern electroretinography (PERG) and an integrated approach to visual pathway diagnosis. Progr Retin Eye Res 2001; 20(4):531–561.

146. Bach M, Hoffmann MB. Update on the pattern electroretinogram in glaucoma. Optom Vis Sci 2008; 85(6):386–395.

147. Hood DC, Xu L, Thienprasiddhi P, et al. The pattern electroretinogram in glaucoma patients with confirmed visual field deficits. Invest Ophthalmol Vis Sci 2005; 46(7):2411–2418.

148. Ventura LM, Porciatti V, Ishida K, Feuer WJ, Parrish RK, 2nd. Pattern electroretinogram abnormality and glaucoma. Ophthalmology 2005; 112(1):10–19.

149. John SW, Smith RS, Savinova OV, et al. Essential iris atrophy, pigment dispersion, and glaucoma in DBA/2J mice. Invest Ophthalmol Vis Sci 1998; 39(6):951–962.

150. Viswanathan S, Frishman LJ, Robson JG. The uniform field and pattern ERG in macaques with experimental glaucoma: removal of spiking activity. Invest Ophthalmol Vis Sci 2000; 41(9):2797–2810.

151. Miyake Y. Focal macular electroretinography. Nagoya J Med Sci 1998; 61(3–4):79–84.

152. Sutter EE. The fast m-transform : a fast computation of cross-correlations with binary m-sequences. SIAM J Computing 1991; 20(4):686–694.

153. Sutter EE, Tran D. The field topography of ERG components in man – I. The photopic luminance response. Vision Res 1992; 32(3):433–446.

154. Hood DC, Wladis EJ, Shady S, Holopigian K, Li J, Seiple W. Multifocal rod electroretinograms. Invest Ophthalmol Vis Sci 1998; 39(7):1152–1162.

155. Nusinowitz S, Ridder WH, 3rd, Heckenlively JR. Rod multifocal electroretinograms in mice. Invest Ophthalmol Vis Sci 1999; 40(12):2848–2858.

156. Paskowitz DM, Nune G, Yasumura D, et al. BDNF reduces the retinal toxicity of verteporfin photodynamic therapy. Invest Ophthalmol Vis Sci 2004; 45(11):4190–4196.

157. Hood DC, Frishman LJ, Saszik S, Viswanathan S. Retinal origins of the primate multifocal ERG: implications for the human response. Invest Ophthalmol Vis Sci 2002; 43(5):1673–1685.

158. Hood DC, Seiple W, Holopigian K, Greenstein V. A comparison of the components of the multifocal and full-field ERGs. Vis Neurosci 1997; 14(3):533–544.

159. Sutter EE, Shimada Y, Li Y, Bearse MA. Mapping inner retinal function through enhancement of adaptive components in the m-ERG. Vision Sci Appl OSA Tech Dg Ser 1999; 1:52–55.

160. Sutter EE, Bearse MA, Jr. The optic nerve head component of the human ERG. Vision Res 1999; 39(3):419–436.

161. Steinberg RH, Linsenmeier RA, Griff ER. Retinal pigment epithelial cell contributions to the electroretinogram and electrooculogram. Prog Retin Res 1985; 4:33–66.

162. Hood DC, Bach M, Brigell M, et al. ISCEV guidelines for clinical multifocal electroretinography (2007 edition). Doc Ophthalmol 2008; 116(1):1–11.

163. Fulton AB, Brecelj J, Lorenz B, Moskowitz A, Thompson D, Westall CA. Pediatric clinical visual electrophysiology: a survey of actual practice. Doc Ophthalmol 2006; 113(3):193–204.

Regulation of Light through the Pupil

Randy Kardon

The pupillary opening appears to occupy a central location, but if carefully measured, it is actually situated slightly inferior and nasal to the center of the cornea. Thus, the pupil center may not correspond exactly to the optical axis defined by a target viewed in the distance and the fovea. The major functions of the pupil are outlined in Figure 25.1 and are summarized as follows.

First, pupil movement in response to changing light intensity aids in optimizing retinal illumination to maximize visual perception. In dim light, dilation of the pupil provides an immediate means for maximizing the number of photons reaching the retina, which in turn supplements the slower dark adaptive mechanisms involving retinal gain control at the photoreceptor and bipolar cell level. With exposure to bright light, pupil constriction can reduce retinal illumination by up to 1.5 log units within 0.5 seconds. Although this reduction in retinal illumination is only a portion of the 12 log unit range of light sensitivity of the retina, it provides an important and immediate means for light adaptation.[47] Patients with a fixed immobile pupil are usually symptomatic during an abrupt change in illumination; they may be photophobic when they are subjected to sudden increases in light, and they may not be able to discern objects in their environment when they first enter dim lighting conditions. These symptoms are described by patients with an immobile pupil because compensatory retinal photoreceptor adaptation is not fast enough. This emphasizes the important role of the pupil in optimizing visual perception in a timely fashion over a wide range of lighting conditions of the environment.

Second, the diameter of the pupil can also contribute to improving (up to a point) the image quality of the retina when the steady-state pupil diameter is small. A small pupil reduces the degree of chromatic and spherical aberration.[10,57] Part of the reason is that a smaller aperture size limits the light rays entering the optical system to the central cornea and lens, avoiding more peripheral portions of the cornea and lens, where aberrations are greater (Box 25.1).

Third, a small pupil increases the depth of focus of the eye's optical system, similar to the known pinhole effect of camera lenses used for photography.[9] When a subject views objects at near, not only the accommodation power of the eye changes but the near response of pupil contraction helps bring objects into better focus by increasing the depth of focus afforded by the smaller aperture size.

Besides the physiologic functions of the pupil explained earlier and outlined in Figure 25.1, the pupil diameter and its movement under different conditions also provide an important indicator used for clinical assessment of a patient. The clinical aspects of pupil function (Fig. 25.2) (Box 25.2) consist of (1) pupil movement as an objective indicator of afferent input, (2) pupil diameter as an indicator of wakefulness, (3) pupil inequality as a reflection of autonomic nerve output to each iris, (4) the influence of pupil diameter on the optical properties of the eye, and (5) the pupil response to drugs as a means of monitoring pharmacologic effects.

Inequality of the pupils, termed *anisocoria*, is another important clinical state of the pupils because it may represent autonomic nerve interruption to the iris from the sympathetic or parasympathetic nervous system, direct damage to the iris sphincter or dilator muscle, or pharmacologic exposure of the iris to mydriatic or miotic drugs. The clinical significance of anisocoria and the various causes of it are covered in more detail later in this chapter.

Large pupils may produce clinically significant visual symptoms in the form of aberrations in form and color of images, particularly at night or after dilating drops, when pupil diameter is largest. A large immobile pupil resulting from scarring, dilating drops, or damage to the iris sphincter muscle or its nerve supply may produce extreme sensitivity to light. This is because the normal function of the pupil in controlling retinal illumination is impaired.

The pupil may be used clinically as a pharmacologic indicator of peripheral or central drug effects. With topical delivery of drugs into each eye with eye drops, the sensitivity of the iris sphincter or dilator muscle can be compared between the two eyes because, under normal conditions, the two eyes should be matched in their response. Dilute concentrations of cholinergic or adrenergic drops, which normally cause little, if any, pupil response, may produce asymmetric exaggerated mydriasis (adrenergic, sympathomimetic drugs) or constriction (cholinergic, parasympathetic drugs) if the iris has been deprived of its nerve innervation for as little as a few days. In cases of unilateral oculosympathetic nerve damage, topical drugs such as cocaine or apraclonidine may be used to confirm the diagnosis. Once the diagnosis of an

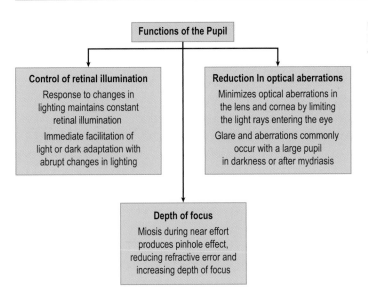

Figure 25.1 Functions of the pupil include control of retinal illumination, reduction in optical aberrations, and improved depth of focus.

Box 25.1 Optics and the pupil

After refractive surgery, younger patients, who usually have larger pupils in dim light compared with older individuals, often experience bothersome symptoms of glare and image degradation, especially at night, as a result of optical aberrations.

The area of the pupil is large enough in these patients to exceed the corneal optical zone of refractive surgery. Most refractive surgeons attempt to address this problem by carefully measuring the pupil diameter in dim lighting conditions preoperatively and then adjusting the optical zone of corneal refractive surgery according to pupil diameter.

Topical drops that produce miosis may reduce pupil diameter enough in dim lighting conditions (without affecting accommodation) to alleviate the symptoms of aberration.

There is a limit to the beneficial optical effects of a small pupil because constriction beyond a certain diameter results in image degradation as a result of increased diffraction and reduction of retinal illumination below an optimal level.[11] Therefore there exists an optimal range of pupil diameter for vision, and this size may vary somewhat, depending on the individual optical characteristics of a person's eye.

---

oculosympathetic lesion is confirmed, hydroxyamphetamine drops may be used a few days later to localize the lesion to the preganglionic or the postganglionic site along the sympathetic nervous system chain (covered in more detail later in this chapter). The pupil response to topical inhibitors of narcotics (naloxone) has also been used to study the effect of narcotic tolerance in addicted individuals. Pupillary responses to psychosensory stimuli has also been used in a laboratory setting to aid in the diagnosis and monitoring of treatment in psychiatric disorders such as depression and schizophrenia and as an objective indicator of cognitive function.

In this chapter the physiology of the normal pupil is discussed, and examples of abnormalities of pupil function are also shown to apply knowledge of normal pupil physiology to understanding various pathologic states. The reader who is interested in pursuing these subjects in much greater detail is advised to consult the excellent book by Loewenfeld.[42]

The neuronal pathway of the pupil light reflex and near pupil response

To understand the major factors that can affect the diameter and movement of the pupil to various stimuli, it is important to know the basic neuronal pathway for the pupil light reflex and near response. This is schematically depicted in Figure 25.3. The pupil light reflex consists of three major divisions of neurons that integrate the light stimulus to produce a pupil contraction: (1) an afferent division, (2) an interneuron division, and (3) an efferent division.

The afferent division consists of retinal input from photoreceptors, bipolar neurons, and ganglion cells. Axons of retinal ganglion cells from each eye provide light input information that is conveyed by synapses to interneurons located in the pretectal olivary nucleus of the midbrain. In turn, these interneurons distribute pupil light input to neurons in the right and left Edinger–Westphal nuclei through crossed (decussating) and uncrossed (non-decussating) connections. From here, the neurons of the Edinger–Westphal nucleus send their preganglionic parasympathetic axons along the oculomotor nerve to synapse at the ciliary ganglion in each orbit. The neurons in the ciliary ganglion give rise to postganglionic parasympathetic axons that travel in the short ciliary nerves to the globe (suprachoroidal space), where they synapse with the iris sphincter muscle.

The pupil constriction to a near stimulus involves activation of neurons in the rostral brainstem that relay their signal to the same Edinger–Westphal neurons that are activated in the light reflex. Therefore the efferent pathway for the near pupil constriction is the same as for the light reflex, but the input pathway to the Edinger–Westphal nucleus differs.

The integration of the pupil light reflex and pupil near response, including the anatomy of the involved neurons, their receptive field properties, and their response to various attributes of light stimuli, has been recently reviewed.[28]

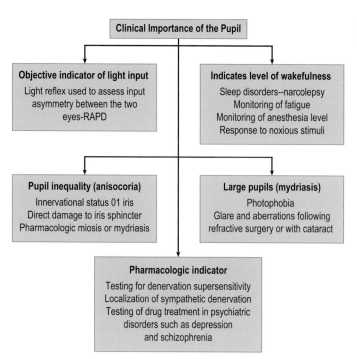

Figure 25.2 Clinical importance of the pupil. RAPD, Relative afferent pupillary defect.

Box 25.2 Diagnosis and the relative afferent pupillary defect

The amount of transient pupil contraction to a light stimulus or the steady-state diameter of the pupil under constant illumination can reflect the health of the retina and optic nerve and may be used to detect disease.

The most common clinical test for assessing input asymmetry between the two eyes is the alternating light test, commonly referred to as the swinging flashlight test. As a light is alternated back and forth between the right and left eyes, the clinician observes the pupil movements in response to the light.

If the two eyes are matched with respect to retinal and optic nerve input, the pupil movements appear similar when either eye is stimulated. However, if one eye's input has been diminished because of disease affecting the retina or optic nerve, the pupil responses to light shown in that eye become noticeably less during the alternating light test.

When this input asymmetry is observed during the alternating light test, it is called a relative afferent pupillary defect (RAPD): RAPDs are discussed in greater length in a subsequent portion of this chapter.

Pupil diameter can also be used to determine the extent of midbrain supranuclear inhibition, which is also related to an individual's state of wakefulness.

An excited, aroused person will have larger-diameter pupils because of the increase in central inhibition of the parasympathetic nerves innervating the iris sphincter, which originate in the midbrain, and the increase in sympathetic tone to the dilator muscle.

Conversely, a sleepy individual, a fatigued individual, or one under the influence of general anesthesia or narcotics will have smaller pupils as a result of central disinhibition at the level of the midbrain. Careful monitoring of the diameter of the pupil in this setting may be clinically useful for ascertaining the presence of sleep disorders such as narcolepsy, the level of anesthesia, or the presence of narcotics.

The extent of pupil dilation to sensory stimuli such as pain or sound may also serve as an objective indicator of how intact the sensory input is.

In the following sections, these neuronal pathways are summarized.

Afferent arm of the pupil light reflex

The neural integration of the pupil light reflex begins with the afferent pathway in the retina, consisting of the photoreceptors, bipolar cells, and ganglion cells. For many years, it was disputed whether it was the rods or cones that contribute to the pupil light reflex and whether these were the same photoreceptors as those contributing to visual perception. Extensive experimental and psychophysical work has shown that the neuronal pathways mediating the pupil light reflex and visual perception share the same photoreceptors. Until

recently, it was thought that all photoreceptor input was shared by both conscious perception of light and the pupil light reflex. In previous studies, almost all changes in light input producing a change in visual perception also produced a comparable change in pupil size. In fact, in almost every way measured, the pupil responses to light parallel those of visual perception. For example, the wavelength-sensitivity profile of pupil threshold for a small transient contraction as the color of light is changed from blue to red exactly parallels the same wavelength-sensitivity profile of visual perception. The shift in sensitivity is also the same as the eye is changed from a condition of light adaptation to dark adaptation (Purkinje shift), providing further evidence that the same photoreceptors are used for both pupil and vision.

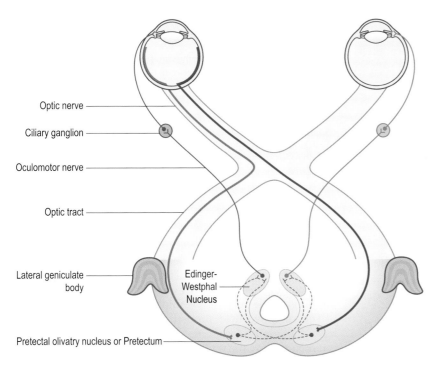

Optic nerve

Ciliary ganglion

Oculomotor nerve

Optic tract

Lateral geniculate body

Edinger-Westphal Nucleus

Pretectal olivatry nucleus or Pretectum

Figure 25.3 Diagram of the nerve pathways involved in the pupil light reflex. Afferent input from the nasal retina crosses to the contralateral side, and the pupil input from retinal ganglion cell axons exits the optic tract in the brachium of the superior colliculus to synapse at the pretectal olivary nucleus. The temporal retinal input from the same eye follows a similar course on the ipsilateral side. The neurons in the pretectal olivary nucleus send crossed and uncrossed fibers by way of the posterior commissure to the Edinger–Westphal nucleus on each side. From here, the preganglionic parasympathetic fibers travel with the oculomotor nerve and then synapse at the ciliary ganglion. The postganglionic parasympathetic neurons pass from the ciliary ganglion by way of the short ciliary nerves to the iris sphincter muscle.

Patients with various abnormalities of rods and cones can be shown to have the same deficits in color vision or lack of appropriate sensitivity change during dark adaptation when the results of visual threshold are compared with pupil threshold for small contractions to light stimuli.[50] Both rods and cones contribute to the pupil light reflex, but to a different extent depending on the lighting conditions.

Under conditions of dark adaptation and in response to low-intensity lights, the pupil light reflex becomes a sensitive light meter and is mediated primarily by rods, giving rise to low-amplitude pupil contractions. However, with brighter light stimuli and under conditions of greater light adaptation, the cones dominate most of the transient pupil light contractions. Therefore the rods are primarily responsible for the pupil's ability to give rise to small contractions in response to low-intensity lights under conditions of dark adaptation – they provide a high sensitivity to the pupil light reflex at low light levels, just as they do for visual perception. The cones provide the input responsible for the larger transient pupil contractions that are easily observed under direct clinical observation, occurring at suprathreshold levels of light, mainly under photopic conditions. Loewenfeld[37] has summarized the extensive literature on this subject and her text should be consulted by the reader desiring a more complete discussion of this topic. It is also believed that the bipolar cells, which receive input from the photoreceptors, are the same neurons providing input to the pupil light reflex as those mediating visual perception.

Although it appears that the pupil and visual systems share rod and cone photoreceptor input, a number of interesting observations in recent years has revealed a previously unrecognized and important new aspect of light transduction through the pupil pathway.

New evidence in the last few years has shown that the rod and cone input to the pupil light reflex is mediated by a special class of retinal ganglion cells containing the primitive visual pigment melanopsin found in the retina of lower animals.[60-71] Besides being activated by rod and cone input causing a transient pupil response, the melanopsin retinal ganglion cell is also directly sensitive to light, providing a sustained steady-state pupil constriction to light. This intrinsic, direct activation pathway of the melanopsin-containing retinal ganglion cells causes the cell to discharge in a sustained way and is directly proportional to steady-state light input, similar to a DC light meter, which does not show classical light adaptation properties. In genetically altered mice that completely lack functional rods and cones, it was discovered that a rather robust pupil light reflex was still present. This unexpected finding was followed by a series of studies to identify what retinal element could be contributing to the pupil light reflex in the absence of rod and cone input. Through clever labeling experiments, a specific ganglion cell was identified containing melanopsin, which was itself photosensitive, with a broad spectral peak centering on about 490 nm, which is blue light. These melanopsin ganglion cells have been found to project both to the suprachiasmatic nucleus in the hypothalamus and also to the pretectal nucleus (the site of the first midbrain interneuron synapse for the pupil light reflex pathway (Box 25.3)).

Elegant electrophysiologic recordings coupled with the study of response properties of these ganglion cells have revealed that the melanopsin-containing retinal ganglion cells provide the midbrain pathway for the pupil light reflex and also provide light sensing information for the diurnal regulating areas of the hypothalamus that modulate the circadian rhythm. These melanopsin ganglion cells also receive rod and cone input to the pupil light reflex, but are also capable of transduction of light directly (under photopic conditions), without photoreceptor input, and may be responsible for providing more of steady-state light input to the brain. This helps to explain why some patients blind from photoreceptor loss still exhibit both a pupil light reaction to bright blue light[72] and also maintain a circadian rhythm, while patients blind from optic nerve lesions (loss of melanopsin ganglion cell input) often lack a normal circadian rhythm.

As stated earlier, the melanopsin-containing retinal ganglion cells appear to mediate the classic midbrain pathway of the pupil light reflex. However, there is also evidence that ganglion cells conveying visual information to the occipital cortex may also play a role in modulation of pupil movement in response to different types of visual stimuli. For example, patients with isolated occipital infarcts have homonymous visual field defects that show corresponding pupil defects to small (2 degrees in diameter) focal lights presented to the same cortically blind visual field area. This phenomenon has been reported previously, but only recently has the correspondence between the shape characteristic of the homonymous pupil and visual field defect been fully appreciated.[1,3,4,12,20–23,26,30,44,46] This correspondence provides compelling evidence for a role of cortical mediation of the pupil light reflex when small, focal light stimuli are used. In addition, the pupil has also been shown to respond to changes in complex stimuli such as spatial frequency, motion, and contrast, providing additional evidence for a higher-level cortical process that is capable of mediating pupil contractions to visual stimuli.[5,6,13,49,54,58,59] Such evidence implies that other types of ganglion cells, in addition to the classic luminance responding cells that project to the midbrain, may also participate in the pupil reflex, perhaps the same ones that also mediate visual perception.

The interneuron arm of the pupil light reflex

The ganglion cell axons conveying light input to the classic pupil light reflex pathway segregate from the rest of the ganglion cell axons at the distal portion of the optic tract before the lateral geniculate nucleus. As in the visual input pathway, the pupil ganglion cell axons from the nasal retina (temporal field) decussate at the chiasm to the opposite side, and the axons from the temporal retina (nasal field) stay on the same side. Therefore pupil ganglion cell axons from homonymous areas of the visual field (temporal field from the contralateral eye and nasal field from the ipsilateral eye) distribute within the optic tract. From there, they travel in the brachium of the superior colliculus and synapse in the

midbrain with the next neurons in the light reflex located at the olivary pretectal nucleus. These neurons represent interneurons because they serve to integrate the afferent input coming from the retina with the efferent output of the pupil light reflex exiting the midbrain.

The receptive field properties of the pretectal neurons have been elucidated in the awake primate.[18,19] These interneurons are the way-station for the converging receptive field impulses of the retinal ganglion cells from the retina and are fewer in number, summating the ganglion cell input at this location. The receptive field of each pretectal neuron has been found to receive input from ganglion cells over a large area of retina (up to 20 degrees). Some of these neurons exhibit a "flat" response, firing equally well from input over its entire receptive field. However, another subset exhibits a "foveal-weighted" response, discharging at a higher frequency when a stimulus is placed near the center of the visual field (and receptive field of the neuron). This may partly explain why the pupil light reflex appears to be more sensitive to light coming from the center of the visual field.

Patients with a relatively small area of damage to their central visual field have also been found to show an obvious decrease in the pupil light reflex in the affected eye compared with the fellow eye, which may relate to the receptive field properties of the pretectal interneurons. The pretectal neurons discharge at a frequency that is linear to the log of intensity of the light stimulus given. However, not all of these neurons respond in the same range; it appears that some are more sensitive in different ranges of intensities, so together, there is an interneuron response covering at least a 4 log unit range of input. Neurons in the pretectum send crossed and uncrossed fibers, through the posterior commissure, to the small population of neurons comprising the paired Edinger–Westphal nuclei. This allows afferent input from the pretectal nucleus on each side to be distributed almost equally to the pupil efferent pathway originating in the Edinger–Westphal nucleus (Box 25.4).

The uncrossed pathway appears to have evolved during development of binocularity and stereovision. Animals with eyes located more to the side of the head (e.g. birds, rabbits) have almost completely crossed pathways with no significant uncrossed component. That is why shining a light in one eye of these animals produces an almost totally crossed input to the pretectum, which then sends an almost completely crossed output to the Edinger–Westphal nucleus. The result is that only the pupil of the eye being stimulated will contract. Cats are between birds and primates in this

Box 25.5 Pupil-involving third nerve palsy

The pupillary preganglionic fibers are located on the superior aspect of the oculomotor nerve as it exits the midbrain, but soon come to lie on the medial aspect.[34]

This is the reason aneurysms of the circle of Willis that lie in this area, such as aneurysms of the posterior communicating artery, often cause pupillary efferent deficits early on, because of the medial location of the artery with respect to the oculomotor nerve.

evolutionary respect, with approximately 70 percent of their pupil pathway crossed. Placing a pet cat so that one of its eyes points more toward a light produces a greater reaction of the pupil in the eye facing the light, causing an anisocoria. In humans, in whom the crossed and uncrossed pathways are almost equal, the direct and consensual pupil light reflex is equal. This is why illuminating one eye normally does not result in pupil inequality (anisocoria). Similarly, input deficits to one eye caused by damage to the retina or optic nerve should not normally produce an anisocoria in humans.

In some individuals the crossed pathway slightly exceeds the uncrossed pathway in both the retina and midbrain, leading to a slightly greater pupil response in the eye stimulated compared with the pupil contraction of the fellow eye, similar to cats, but not to the same extent. This consensual deficit, termed *contraction anisocoria*, is small and can usually be recognized only with the aid of pupillographic recordings.

The efferent arm of the pupil light reflex

The efferent arm of the pupil light reflex is diagrammatically summarized in Figure 25.4, which also shows the site of common lesions along this pathway that may be encountered in clinical practice. The neurons in the Edinger–Westphal nucleus send preganglionic axons into the right and left fascicle of the oculomotor nerve (third nerve) to join the motor axons destined for the eye muscles, as well as the preganglionic accommodative fibers that originate in nearby nuclei. The right and left fascicles of the third nerve exit the midbrain through the subarachnoid space, where each continues as the third nerve toward the orbital apex (Box 25.5).

After passing through the cavernous sinus to the orbital apex, the preganglionic pupillary fibers and accommodative fibers synapse in the ciliary ganglion (parasympathetic ganglion). A lesion at this site may produce an Adie's pupil. From here, the last neurons in the chain, the postganglionic neurons, pass into the eye by way of the short ciliary nerves, located in the suprachoroidal space, where they distribute to the anterior segment of the eye to innervate the iris sphincter muscle. The postganglionic accommodative fibers, which outnumber the pupil fibers 30:1, supply the ciliary muscle within the ciliary body of the eye. The postganglionic pupillary neurons appear to innervate the iris sphincter muscle in a segmental distribution over approximately 20 clock-hour sections. This is why lesions of the ciliary ganglion, such as in Adie syndrome, usually cause a number of sectors of the iris sphincter to become acutely denervated, with loss of the pupil contraction only in these segments (Box 25.6).

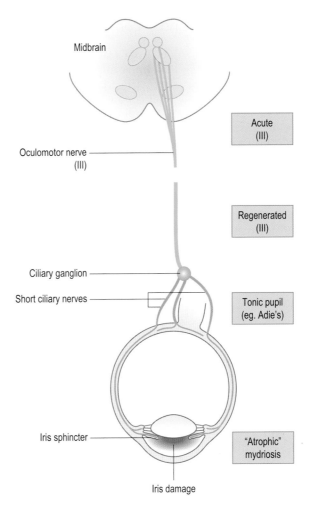

Figure 25.4 Innervation of the iris sphincter, from the Edinger–Westphal nucleus by way of oculomotor nerve, ciliary ganglion, and short ciliary nerves, with some of the causes of a fixed dilated pupil.

Box 25.6 Light-near dissociation

The postganglionic parasympathetic accommodative axons, which innervate the smooth muscle of the ciliary body, outnumber the postganglionic light reflex axons, which innervate the iris sphincter muscle in a ratio of 30:1.

Damage to the postganglionic parasympathetic axons can occur as a result of an Adie's pupil or trauma or after orbital surgery. After acute injury, the surviving nerve cell bodies within the ciliary ganglion sprout axons and grow toward the ciliary body muscle and the iris sphincter after about 8 to 12 weeks.

Because the accommodative cell bodies outnumber the pupil light reflex cell bodies, almost all of the axonal sprouts reaching the iris sphincter muscle originate from the accommodative cell bodies. This reinnervation of the iris sphincter is therefore aberrant because the pupil sectors that were denervated still do not respond to light, but they now contract in response to activation of the accommodative neurons, hence producing a light-near dissociation of the pupil reflex.

The pupil near reflex and accommodation

When fixation of the eye is shifted from a far to a near object of interest, the eyes converge, the intraocular lenses accommodate, and both pupils constrict. This triad of ocular vergence, accommodation, and pupil contraction is called the *near response*. Despite many contentions in the literature claiming that this pupil constriction is exclusively dependent on either convergence or accommodation, clinical and experimental data indicate that any one of the three functions can be selectively abolished or elicited without affecting the others. These experimental and clinical observations have resulted in general agreement that the impulses that cause accommodation, convergence, and pupil constriction must arise from different cell groups within the oculomotor nucleus and travel by way of separate fibers to their effector muscles. Accommodation, convergence, and pupillary constriction are associated movements and are not tied to one another in the manner usually referred to by the term *reflex*. They are controlled, synchronized, and associated by supranuclear connections, but they are not caused by one another. The components of the near pupil reflex were recently summarized.[28,34]

The miosis of the near response and the pupillary constriction to light have a single final common efferent pathway from the Edinger–Westphal nucleus to the iris sphincter by way of the ciliary ganglion. They primarily differ in the origin of the supranuclear pathways that are elicited by light and near that both converge on the Edinger–Westphal nucleus. With a near stimulus, such as accommodation, the pupil normally constricts even with no change in retinal luminance. It is important to realize that this near reflex of the pupil is mediated by the same efferent nerve pathway originating from the same neurons in the Edinger–Westphal nucleus that mediate the pupil light reflex. There does not appear to be a separate neuronal efferent pathway that mediates the near pupil constriction.

However, the supranuclear control over this response is different from the one mediating the light reflex. In the case of the light reflex, the supranuclear input comes from the pretectal nucleus, as described in the previous section. In the case of the near reflex involving accommodation, convergence, and miosis, the supranuclear input is thought to originate from cortical areas surrounding visual cortex and from cortical areas within the frontal eye fields. The cortical neurons providing input for the near reflex are thought to synapse at least once, before passing ventral toward the visceral neurons overlying the oculomotor complex in the midbrain. This is because a cortical lesion in this area does not produce atrophy within the oculomotor nuclear complex (it is at least one synapse removed). It is also important to realize that the near reflex consists of convergence of the eyes, accommodation, and pupil contraction, all of which should be thought of as co-movements and, as stated earlier, are not strictly dependent on one another. Any one of the three co-movements may occur in the absence of the others, as discussed by Loewenfeld.[37] Because the supranuclear pathway for the near reflex passes ventral in the midbrain and the supranuclear pathway for the light reflex passes dorsal in the midbrain, the two systems may be differentially affected by disease processes (Box 25.4).

The supranuclear neuronal input from a near visual task stimulates the pupil constrictor neurons located in the visceral part of the Edinger–Westphal nuclei. The same supranuclear neuronal input also stimulates the more numerous accommodative neurons, located nearby in the remaining visceral portion of the Edinger–Westphal nucleus. These preganglionic neurons give rise to accommodative axons that travel together with the pupil preganglionic light reflex neurons within the oculomotor nerve to synapse at the ciliary ganglion in the orbit (see previous section).

In summary, with a near stimulus, both the accommodative neurons (which mediate ciliary muscle contraction) and light reflex neurons (which mediate iris sphincter contraction) in the Edinger–Westphal visceral motor nuclei are stimulated from a supranuclear level. This gives rise to a separate neuronal output of accommodative and light reflex preganglionic neurons by way of the oculomotor nerve to the ciliary ganglion, which in turn gives off separate postganglionic innervation to the ciliary body and iris sphincter muscles. The preganglionic and postganglionic light reflex pathways make use of the same neurons to mediate pupil contraction to either near or light stimuli.[26]

Pupil reflex dilation: central and peripheral nervous system integration

Normally, when the pupil dilates, two integrated processes take place: the iris sphincter relaxes, and the iris dilator contracts, helping actively pull the pupil open.[38,39] Because the iris sphincter is stronger than the dilator muscle, pupil dilation does not readily occur until the sphincter muscle relaxes. Relaxation of the iris sphincter is accomplished by supranuclear inhibition of the Edinger–Westphal nucleus at a central nervous system level, most notably from the reticular activating formation in the brainstem. It appears from animal studies that this neuronal inhibitory pathway involves the central nervous system's sympathetic class of neurons. These sympathetic neurons pass through the periaqueductal gray area and innervate the pupil efferent neurons at the Edinger–Westphal nucleus and at the synapse there is an α_2-adrenergic receptor activation.[35] When this central inhibition is active, the preganglionic parasympathetic output from the Edinger–Westphal nucleus is suppressed, resulting in a relative relaxation of the iris sphincter and pupil dilation. When this inhibition is inactive, such as during sleep, with anesthesia, or with narcotics, the preganglionic neurons fire at a high rate, causing miosis. The neurons of the Edinger–Westphal nucleus are unique in this respect because their baseline discharge frequency, without any input, is high. If all input to these neurons is disconnected, they fire at a high rate, which results in a sustained pupil contraction and miosis. This is why deep sleep or anesthesia, which reduces almost all inhibitory supranuclear input to the Edinger–Westphal nucleus, results in small pupils.

Alternatively, during a state of wakefulness, the supranuclear inhibition is active and the neurons of the Edinger–Westphal nucleus are suppressed, causing the pupils to become larger again. If a light stimulus is given at this point, a train of neuronal impulses from the retina and then the pretectal interneuron will arrive at the Edinger–Westphal nucleus, which momentarily overcomes this inhibition, causing the pupil to constrict. If the light is turned off or the retina begins to become light adapted, the supranuclear inhibition again dominates, causing a reflex dilation of the pupil.

Almost all of the conditions mentioned previously cause changes in pupil diameter resulting from the modulation of the neuronal output from the Edinger–Westphal nucleus. In addition, the same factors causing a reflex dilation of the pupil also result in an increase in output to the peripheral sympathetic nervous system innervating the iris dilator muscle. The sympathetic nerve activity can be thought of as a "turbo-charge" for pupil dilation. Peripheral sympathetic nerve activation is not a requirement for pupil dilation to occur (parasympathetic inhibition alone can accomplish that to some extent), but it greatly enhances the dynamics of pupil dilation in terms of speed and maximal pupil diameter attained.

The sympathetic outflow to the iris dilator muscle can be thought of as a paired three-neuron chain (Fig. 25.5) on both the right and left side of the central and peripheral nervous system without decussations. The first neuron originates in the hypothalamus and descends through the brainstem on each side into the lateral column of the spinal cord, where it synapses at the cervicothoracic level of C7–T2. The second preganglionic neuron leaves this level of the spinal cord and travels over the apical pleura of the lung and into the spinal rami to synapse at the superior cervical ganglion at the level of the carotid artery bifurcation on the right and

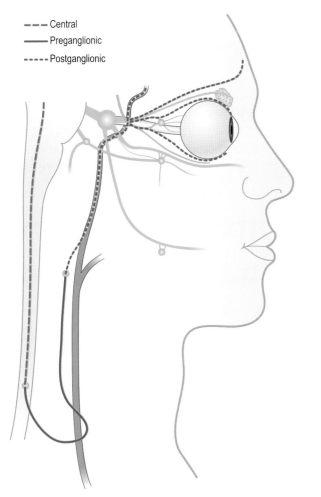

Central
Preganglionic
Postganglionic

Figure 25.5 Sympathetic innervation to the eye, showing the three-neuron chain of central, preganglionic, and postganglionic fibers. (Modified from Maloney WF et al: Am J Ophthalmol 1980; 90:394.)

left side of the neck. The third neuron, the postganglionic neuron, follows a long course along the internal carotid artery into the head and orbit. As these neurons pass through the cavernous sinus, they are associated with the abducens and then the trigeminal nerve before entering the orbit and distributing to the iris dilator muscle via the long ciliary nerves.[25]

In addition to the neuronal mechanisms involved in pupil dilation, humoral mechanisms also may contribute to pupil diameter. Circulating catecholamines in the blood (e.g. a bolus released from the adrenal glands) may act directly on the iris dilator muscle either through the blood-stream or, potentially, indirectly through the tears, resulting in mydriasis. Clinical conditions that influence the integration of the parasympathetic inhibition, sympathetic stimulation, and humoral release of neurotransmitters may take various forms and may affect the dynamics of reflex dilation in a characteristic manner that may be diagnostic of clinical conditions. This is revisited later in this chapter when pupil inequality and conditions that impede pupil dilation are discussed.

Other neuronal input to the iris

In addition to the autonomic nerves supplying the iris, sensory innervation to the iris is provided by the ophthalmic division of the trigeminal nerve.[24,48] However, these sensory nerves may play an additional role in modulating pupil diameter. It is well known to cataract surgeons that mechanical and chemical irritation of the eye can cause a strong miotic response that is non-cholinergic and fails to reverse with autonomically acting drugs. In rabbits and cats the response seems to be caused by the release of substance P or closely related peptides from the sensory nerve endings, but in monkeys and humans, substance P has little or no miotic effect. Cholecystokinin (in nanomolar amounts) caused contraction of isolated iris sphincter from monkeys and humans.[2] Intracameral injections in monkeys caused miosis that was not prevented by tetrodotoxin or indomethacin, indicating that the miosis was not caused by either stimulation of nerve endings or release of prostaglandins but by direct action on sphincter receptors. The cholecystokinin antagonist lorglumide caused competitive inhibition of the response.

Structure of the iris

Iris sphincter, iris dilator, and iris color

It is important to understand the structure of the iris and its histology to understand how the iris tissue accommodates changes in pupil diameter during contraction and dilation and how disorders of the iris tissue affect pupil movement. The iris can be divided into two main layers: the posterior leaf and the anterior leaf (Fig. 25.6). The posterior iris leaf contains the dilator muscle, the sphincter muscle, and the posterior pigmented epithelium. From a front view of the iris, the dilator muscle is located circumferentially, in the midperiphery of the iris.

The sphincter muscle is located just inside the pupillary border; its circumference is made up of approximately 20

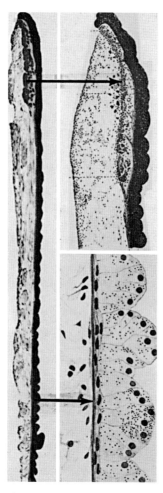

Figure 25.6 Histology of the iris in cross section. Upper arrow points to sphincter muscle drawn in higher magnification; lower arrow points to dilator muscle of bleached preparation drawn in higher magnification. (From Saltzmann M: Anatomy and histology of the human eye-ball, Chicago: University of Chicago Press, 1912.)

motor segments connected together but innervated individually by postganglionic branches of the ciliary nerve. In the normal iris these segments receive nerve excitation in a roughly simultaneous fashion, and the entire circumference contracts in concert. Both the dilator and sphincter muscles are derived embryologically from the anterior layer of the two layers of posterior pigmented epithelium.

The more superficial, anterior iris leaf consists of connective tissue stroma with cells, blood vessels, and nerves supplying the sphincter and dilator, but there is no epithelial layer in primate species. The different components of the posterior and anterior iris undergo structural alterations to accommodate changes in pupil diameter during contraction and dilation.[45] These alterations in structure confer mechanical non-linearities influencing how much the pupil can constrict or dilate in response to changes in light or pharmacologic stimulus and was extensively studied by Loewenfeld.[40]

The mechanical non-linearities are important because they impose limitations on the range of pupil diameter over which the extent of pupil movement can be used for assessing neuronal reflexes to light stimuli or near stimuli or for pharmacologic testing of the pupil. For example, a person

with small, 3-mm diameter pupils in dim light would obviously not show as large a pupil contraction to a standard light stimulus compared with a person with 5-mm diameter pupils, but the retina and optic nerves of both persons may be completely normal. A similar situation would occur if one attempts to quantify the response to a topical miotic or mydriatic agent. Therefore, the structure of the iris can pose physical constraints on pupil movement in response to sensory stimuli or pharmacologic agents, and this should be considered carefully when comparing the response in different eyes.

The color of the iris is determined by its mesodermal and ectodermal components. In Caucasians, the stroma is relatively free of pigment at birth. The stroma absorbs the long wavelengths of light, allowing the shorter (blue) wavelengths to pass through to the pigmented epithelium, where they are reflected back, causing the iris to appear blue. If pigmentation does not develop in the anterior stromal layers, the iris remains blue throughout life. If the stroma becomes denser and contains significant numbers of melanosomes, the blue color gives way to gray. The accumulation of pigment in the iris melanocytes of individuals destined to have non-blue irides occurs during the first year of life and is dependent on sympathetic innervation of the melanocytes (derived from neural crest cells). Interruption of the oculosympathetic nerve supply to one eye during this time period usually results in heterochromia, with the denervated iris remaining blue. In a heavily pigmented iris, the fine pattern of iris vessels is hidden by pigment and the surface of the iris looks brown and velvety.

Properties of light and their effect on pupil movement

Properties of light stimulating the retina that affect the pupil response include intensity, duration, temporal frequency, area, perimetric location, state of retinal adaptation, wavelength, and spatial frequency. There is a wealth of information on how these properties of light stimuli affect the pupillary response with regard to latency and amplitude of movement. Loewenfeld[41,42] has presented the most complete review of this topic in her book on the pupil, which should be consulted for a detailed literature review and for examples illustrating these different light effects. In general, the amplitude of pupil movement increases in proportion to the log light intensity of the stimulus, whereas the latency time of the pupil light reflex (time from stimulus onset to beginning of pupil contraction) becomes shorter (Fig. 25.7). With increasing duration of light stimulus, the contraction amplitudes become greater and more prolonged. With long-duration light stimuli, after an initial contraction, the pupil may undergo oscillations (hippus) and undergo slow dilation, or 'pupil escape,' because of light adaptation (Fig. 25.7). Table 25.1 summarizes the different light effects.

Most investigations of the pupillary light reflex have focused on the response of the pupil to changes in light level because the neuronal pathway for this reflex was thought to respond only to stepwise changes in light intensity. With the advent of computer graphics and sophisticated software programs, more complex stimuli can be presented to allow properties of spatial frequency, color, motion, and

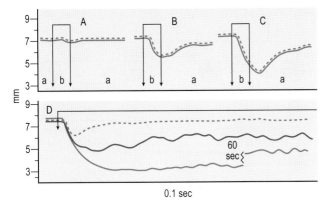

Figure 25.7 (**A**), (**B**), and (**C**): Dark-adapted normal subject. Light flashes, b, of increasing intensity were given to the right eye to produce increasing pupillary constriction. Latent period decreases with intensity of flash. Both pupils were recorded simultaneously using an infrared pupillography device. The right pupil tracing (solid line) and the left pupil tracing (broken line) move in synchrony. (**D**) Reactions of the pupil to prolonged light of different intensities. At the dimmest intensity there was a short pupil light constriction and the pupil dilated (escaped) during the light stimulus. At the brighter intensities, the contractions were larger and more sustained, also exhibiting oscillations (hippus).
(From: Lowenstein O, Loewenfeld IE. In: Davso H, ed. The eye, vol 3. New York: Academic Press, 1962.)

luminance to be controlled more carefully. A number of investigators have taken advantage of this technology to investigate whether the pupil is capable of responding to visual stimuli that change in color or spatial frequency when the average luminance does not change.[21–25,28,31]

The results of these studies provide evidence that the pupil contracts to either an onset or offset of spatial frequency or color exchanges. From a practical standpoint, these responses allow the pupil response to be used as an objective indicator of visual acuity and color discrimination. From a theoretical standpoint, the pupil response to isoluminant stimuli provides a means to explore how and where different signals are processed in the visual system. From these studies it is becoming more apparent that visual cortex plays a role in modulating the pupil response to these complex stimuli, providing evidence that small pupil reflexes elicited by properties of visual stimuli besides luminance involve more than just midbrain processing.

Relative afferent pupillary defects

Clinical observation of the pupil light reflex

One of the most important clinical uses of the pupil has to do with its role in assessing afferent input from the retina, optic nerve, and anterior visual pathways (chiasm, optic

Table 25.1 Effect of properties of light stimuli on the pupillary light reflex

Stimulus property	Effect on pupillary light reflex
Light intensity	Amplitude of contraction increases linearly over at least a 3 log unit range of stimulus intensity (stimulus under photopic conditions). The entire stimulus-response function resembles an 'S'-shaped curve. Latency time, the time from stimulus onset to the start of pupil contraction (200–450 msec), becomes more prolonged with dimmer light stimuli (in the range of 20–40 msec further delay/log unit decrement of light intensity).
State of light adaptation	In the dark-adapted state, the threshold light intensity needed to produce a pupil contraction becomes less as rods are brought into play. However, rods in the dark-adapted state do not produce as much increase in pupil contraction in response to increases in stimulus intensity, compared with cones in the mesopic and photopic states.
Duration	When stimulus duration is shorter than 70 msec, there is a reciprocal relationship between the duration and intensity, which is required to produce a given pupil contraction amplitude. With longer-duration stimuli, the pupil contracts more, there is a shorter latency time (up to a point), and the pupillary contraction is more sustained; however, pupil escape (relative dilation) may occur as a result of light adaptation.
Area	The pupillary light reflex shows much greater area summation properties than visual perception (for visual threshold of perception, summation is minimal with stimuli greater than 1 degree). With full-field Ganzfeld stimuli, the pupil threshold can be equal to visual threshold (or even smaller); with stimuli smaller than 1–2 degrees, visual threshold is usually more sensitive (by 0.5–1.0 log units).
Perimetric location	Under dark adaptation, the fovea shows a decreased sensitivity compared with surrounding retinal areas because of the lack of rods here. In mesopic and photopic adaptation, the pupil responds greatest in the central field; the temporal field response is usually greater than the nasal field response.
Spectral sensitivity	The wavelength sensitivity of the pupillary light reflex follows that of visual perception with a blue shift under dark adaptation and a peak sensitivity at green under photopic conditions.
Temporal frequency	The normal pupil cannot move much faster than 4 Hz because of the relatively slow contraction of smooth muscle. Animals with striated iris muscle (pigeons) can easily follow a 10-Hz stimulus. At frequencies of 9–25 Hz, the steady-state pupil diameter increases, indicating loss of sensitivity in neuronal integration of light within this frequency range.
Spatial frequency	When the change in average luminance across a stimulus patch is kept constant, the pupil undergoes small contractions when a sinusoidal grating is presented or when the grating bars are alternated between dark and light. The mechanism is thought to be independent of a luminance response. The greater the spatial frequency, the less the pupil contracts to the stimulus and this has been correlated with visual acuity.
Motion	Recent evidence suggests that the pupil may respond to a motion stimulus even under isoluminant conditions.

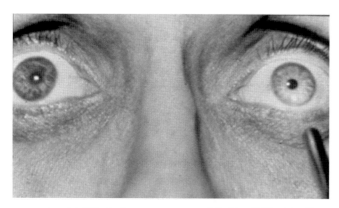

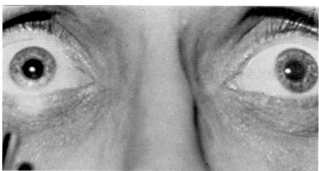

Figure 25.8 A patient is shown with a large relative afferent pupil defect of the right eye. In the top photograph a light is shined in the normal left eye and both pupils constrict to a small diameter. When the light is alternated to stimulate the right eye (bottom photograph), the pupils hardly constrict at all.

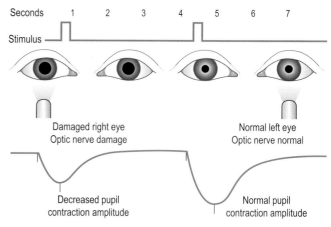

Figure 25.9 Pupillographic demonstration of a right relative afferent pupil defect. With short light pulses the pupil light reflex (bottom tracings) are of lesser amplitude when the stimulus was given to the right eye compared with stimulus to the left eye.

tract, and midbrain pathways). The pupil light reflex sums the entire neuronal input from the photoreceptors, bipolar cells, ganglion cells, and axons of ganglion cells. Therefore, damage at any one of these sites along the visual pathway will reduce the amplitude of pupil movement in response to a light stimulus.[42,43] The pupil light reflex can show considerable variation even among normal individuals as a result of supranuclear influences on the midbrain pupillomotor center that are not related to afferent input of the retina and optic nerve. However, the pupil light reflex is symmetric between the two eyes of a normal individual. The normal symmetry of input between the two eyes allows the clinician to pick up any asymmetric damage between the two eyes by simply comparing how well the pupil contracts to a standard light shined into one eye compared with alternating the light over to the other eye.[36] Observation of the pupil movement in response to alternating the light back and forth between the two eyes (Figs 25.8 and 25.9) is the basis for the alternating light test or swinging flashlight test for assessing the RAPD.[51,52] The RAPD, or input asymmetry, can be quantified in log units using neutral density filters of increasing strength placed in front of the better responding eye. A log density filter that neutralizes or balances the asymmetry of pupil movement between the two eyes, until they are matched, is chosen and is taken as the log unit RAPD.

Another important property of the pupil light reflex is that when a diffuse light stimulus enters the eye, the entire area of the visual field is summated into the pupil response, with some increased weight given to the central 10 degrees.[42] The area summating properties of the pupil light reflex make the amplitude of its movement roughly proportional to the

amount of working visual field. Therefore, damage to peripheral portions of the retina and visual field defects outside of the central field can also reduce the amplitude of the pupil light reflex. This is in contrast to other objective tests of visual function such as the Ganzfeld electroretinogram (which measures diffuse damage to the retina and not local damage) and visual evoked potentials (which are center weighted and therefore affected mainly by central field loss). The pupil light reflex is one of the few objective reflexes that can be used as a clinical test for detecting and quantifying abnormalities in the retina, optic nerve, optic chiasm, or optic tract, and is also able to detect regional visual field damage in the center or periphery.

Estimating the amount of the RAPD in log units is important to confirm and validate how much visual field damage (asymmetry between the two eyes) is present and whether it is consistent with the results of the visual field test. For example, a patient with a small amount of macular degeneration in one eye and not the other might be expected to have only a 0.3 log unit RAPD. However, if that patient had a 1.0 log unit RAPD, some other cause of visual loss, such as a previous branch retinal artery occlusion or optic neuropathy, would have to be considered. In addition, the area and extent of visual field loss would be expected to be more than that caused by a small amount of macular degeneration. The importance of quantifying the RAPD cannot be overemphasized. In general, in the case of unilateral visual damage, loss of the central 5 degrees of the visual field results in an RAPD of approximately 0.3 log units. Loss of the entire central area of field (10 degrees) causes about a 0.6 to 0.9 log unit RAPD. Each visual field quadrant outside of the macula is worth about 0.3 log units (Fig. 25.10), but the temporal field loss seems to result in more loss of pupil input compared with the nasal field quadrants. Examples of the log unit magnitude of the RAPD expected for common clinical disorders are given in Table 25.2. The correlation between the relative afferent defect and the area and extent of visual field loss, however, is only approximate. Differences between the two may be important clues as to the cause and extent of damage to the anterior visual system (Box 25.7).

Recent studies using computerized pupillography to more precisely quantify the log unit RAPD have also revealed that

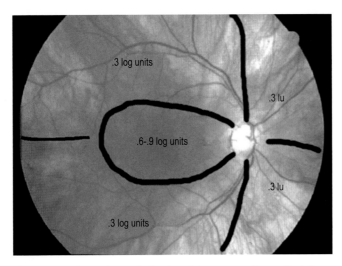

Figure 25.10 Approximate distribution of the amount of log unit relative afferent pupil defect expected for corresponding loss of input in the regions of the retina shown, assuming the other eye is normal.

some normal subjects with normal visual fields and examination can have a small 0.3 log unit RAPD.[25,33,55] Therefore small RAPDs discovered incidentally in a patient without ocular complaints and a normal examination can probably be dismissed.

Estimating the amount of the pupillomotor input asymmetry (the RAPD) can also be done more subjectively, without using neutral density filters, by grading the asymmetry of pupil response as +1, +2, +3, or +4. This subjective grading can also be classified according to the amount of pupil escape or dilation of the pupils as the light is alternated to the other eye.[7] However, most subjective grading of the RAPD has serious limitations; it is subject to some large errors resulting from age variations in pupil diameter and pupil mobility. For example, a patient with small pupils and small pupil contractions to light may have a large RAPD that may appear deceptively small on the basis of the small differences in pupil excursion observed as the light is alternated between the two eyes. However, the amount of neutral density filter needed to dim the better eye until the small

Table 25.2 Common diseases producing relative afferent pupillary defects (RAPD) and expected magnitude of defect

Condition	Site	Log unit RAPD	Influencing factors
Intraocular hemorrhage	Anterior chamber or vitreous (dense)	0.6–1.2	Density of hemorrhage
Intraocular hemorrhage	Anterior chamber (diffuse)	0.0–0.3	Density of hemorrhage
Intraocular hemorrhage	Preretinal (central vein occlusion or diabetic)	0.0	Preretinal location does not significantly reduce light
Diffusing media opacity	Cataract or corneal scar	0.0–0.3 (in opposite eye)	Dispersion of light producing increase in light input
Unilateral functional visual field loss	None	0	
Central serous retinopathy (CSR) or cystoid macular edema (CME)	Retina (fovea)	0.3 log units	Area of retina involved
Central or branch retinal vein occlusion (CRVO, BRVO)	Inner retina	0.3–0.6 (non-ischemic) 0.9 (ischemic)	Area of visual field defect and degree of ischemia
Central or branch retinal artery occlusion (CRAO, BRAO)	Inner retina	0.3–3.0	Area and location of retina involved
Retinal detachment	Outer retina	0.3–2.1	Area and location of detached retina (e.g. 0.6 log units for macula + 0.3 log units for each quadrant)
Anterior ischemic optic neuropathy	Optic nerve head	0.6–2.7	Extent and location of visual field defect
Optic neuritis (acute)	Optic nerve	0.6–3.0	Extent and location of visual field defect
Optic neuritis (recovered)	Optic nerve	0.0–0.6	No visual field defect, residual RAPD
Glaucoma	Optic nerve	Usually none, if symmetric damage to both eyes	Degree of visual field asymmetry between the two eyes correlates with the log unit RAPD
Compressive optic neuropathy	Optic nerve	0.3–3.0	Extent and location of visual field defect
Chiasmal compression	Optic chiasm	0.0–1.2	Asymmetry of visual field loss, unilateral central field involvement
Optic tract lesion	Optic tract	0.3–1.2 (in the eye with temporal field loss)	Incongruity of homonymous field defect, hemifield pupillomotor input asymmetry
Postgeniculate damage	Visual radiations Visual cortex	0.0	Stimulus light area (no RAPD but definite pupil perimetry defects)
Midbrain tectal damage	Olivary pretectal area of pupil light input region of midbrain	0.3–1.0	Similar to optic tract lesions, but no visual field defect

Box 25.7 Relative afferent pupillary defect and management

The amount of the relative afferent papillary defect (RAPD) is correlated largely with the amount of asymmetry of visual field deficit between the two eyes, and helps substantiate abnormal results of visual field testing.[8,25,31,53] This can often help the clinician determine whether a patient's visual field defects are reliable and reflect the true pathologic state.

The correlation between the visual field asymmetry and the RAPD is also useful in following the course of disease to determine whether there is a worsening or improvement in function over time.

The RAPD is a relative measure of the input of one eye compared with the other. Bilateral symmetric damage should not produce an RAPD.

A patient who exhibited a definite RAPD in one eye on the first visit may show no RAPD at all on the follow-up visit. This may represent improvement in the previously damaged eye. However, it may also indicate that there is now damage in what was previously the better eye, matching the damage to the other eye, so that there are now symmetric visual field defects and no RAPD.

Figure 25.11 Computerized infrared pupillographic instrument, which records the bright image of both pupils simultaneously, is shown. This instrument can be used to give full-field light stimuli to either eye produced on a monochrome monitor inside the box, or it may also be used to present focal light stimuli to provide a form of objective pupil perimetry.

contractions are equal could easily approach 0.9 to 1.2 log units, representing substantial input damage. Estimating the size of an RAPD without using filters is much like estimating an ocular deviation "by Hirschberg" without doing a prism cover test. More accurate quantification of the RAPD is accomplished by determining the log unit difference needed to balance the pupil reaction between the two eyes using photographic neutral density filters, as described previously. If one pupil does not move well because of weakness of the sphincter or pharmacologic immobility, one can still check for a RAPD by observing the pupil that still works – and comparing its direct reaction with its consensual reaction.

Computerized pupillometry

Various computerized, infrared-sensitive pupillometers are commercially available. Most of these elegant instruments can precisely record the dynamics of pupillary movement in the light or in the dark (Fig. 25.11). Once recorded, the pupillary information can be analyzed by sophisticated software (see pupil tracings in the alternating light test, shown in Fig. 25.9). This allows quantitative information about the pupillary light reflex to be assessed. In the near future, this may help automate the clinical determination of pupil input deficits caused by retinal and optic nerve diseases.[14,15,17,31–33] Such instrumentation is also useful for detecting and diagnosing causes of anisocoria (unequal pupils) when both pupils are recorded simultaneously (refer to the section on oculosympathetic defects).

Pupil perimetry

The pupil light reflex may also be used to obtain objective information about the sensitivity of local areas of the visual field by recording small pupil contractions in response to focal light stimuli placed in different perimetric locations. An automated perimeter can be modified to record pupil responses to each focal light stimulus to produce a form of

objective pupil perimetry.[26,30] A video camera is pointed at the pupil, and the amplitude of each light reaction is measured and stored in the computer. This has turned out to be helpful as an objective form of perimetry and as a way to localize lesions in the pupillary pathways (Fig. 25.12). Pupil perimetry is also useful in cases of non-organic, functional visual loss to show objectively that messages are indeed going normally into the brain from parts of the visual field in which the patient claims to see nothing.

Efferent pupillary defects

Anisocoria

When a pupil inequality is seen, it usually means that damage has occurred to either the iris sphincter or dilator muscle, their innervation has been interrupted, or there are external pharmacologic factors influencing pupil movement (Fig. 25.13). To sort out which muscle is not working right, it helps to know how the anisocoria is influenced by light. It is worth noting that an anisocoria always increases in the direction of action of the paretic iris muscle, just as an esotropia increases when gaze is in the direction of action of a weak lateral rectus muscle. If the iris sphincter is paretic, the lighting condition that normally brings into action that muscle accentuates the weakness; adding bright light tends to increase the anisocoria. Alternatively, if the iris dilator is paretic, the anisocoria is expected to increase as light is taken away because reflex dilation to darkness is impaired. Table 25.3 provides a summary of the common causes of anisocoria that are discussed in the following section.

Visual Threshold

Visual testing time = 18 minutes

db

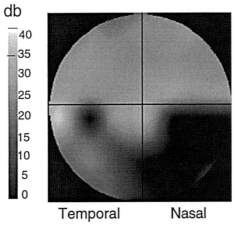

Temporal　　　Nasal

Pupil Contraction

Pupil testing time = 8 minutes

mm

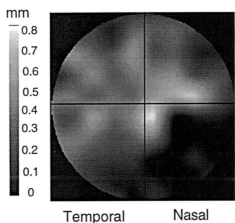

Temporal　　　Nasal

Figure 25.12 Example of a patient with anterior ischemic optic neuropathy showing the correspondence between the visual field and pupil field defect.

Figure 25.13 The cause of anisocoria in room light (top panel) is determined by first adding bright light to determine whether the anisocoria increases (middle panel) or decreases (bottom panel).

Pupil inequality that increases in the dark

In patients who have pupil inequality that increases in the dark, the problem is distinguishing Horner syndrome from a simple anisocoria (or physiologic anisocoria). In both of these conditions, the pupil inequality becomes greater in dim light but the dynamics of pupil dilation is impaired in oculosympathetic defects, but not in simple anisocoria. Other features also characterize simple anisocoria from Horner syndrome.

A simple anisocoria may vary from day to day, or even from hour to hour, and it is visible in about one-fifth of the normal population. In some people it may be present most of the time, and the larger pupil may always be in the same eye. In other people it may come and go, and the larger pupil may often switch to the other eye. Physiologic anisocoria is not related to refractive error. The cause of physiologic anisocoria is not known with certainty, but current evidence favors a transient, asymmetric, supranuclear inhibition of the Edinger–Westphal nucleus. This would cause the pupil on the more inhibited side to be larger. If this mechanism is correct, it would also explain why any stimulus that transiently overcomes supranuclear inhibition, such as bright light, near stimulus, sleep, or anesthesia, causes the physiologic anisocoria to decrease. Simple anisocoria is considered benign.

The characteristic dilation lag of the Horner's pupil can easily be seen in the office with a hand light shining from below. At the time the room lights are switched off, the reflex dilation of the two pupils should be simultaneously observed and the smaller pupil assessed to see whether it dilates slower than the other pupil does (Box 25.8). Pupil dilation is normally a combination of sphincter relaxation and dilator contraction. This combination produces a prompt dilation in a normal pupil when the illumination is abruptly decreased.

Table 25.3 Common causes of anisocoria and associated features

Condition	Cause	Anisocoria	Light response	Near response	Slit lamp	Pharmacologic testing
Acute Adie's pupil	Denervation of parasympathetic postganglionic nerves to pupillary sphincter (segmental)	Anisocoria increases in bright light	Segmental loss of light reaction in some sphincter areas around the circumference	Same areas where light response is lost also show loss of near constriction	Remaining innervated sphincter areas pucker with light and pull denervated segments	Supersensitivity to 0.1% pilocarpine; check response of both pupils in darkness after 30 minutes
Chronic Adie's pupil (greater than 8 weeks after event)	Reinnervation of denervated sphincter segments by postganglionic accommodative nerves	May be no anisocoria in room light or affected pupil may be the smaller pupil	Poor response to light; poor dilation in darkness as a result of tonically contracted segments that have reinnervation	Light-near dissociation is present with good near effort from the patient	Similar appearance as in the acute state, reinnervated segments show diffuse contraction to near	As segments become reinnervated, cholinergic supersensitivity is lost; small, tonic pupil dilates normally to anticholinergics
Pharmacologic mydriasis (anticholinergic)	Scopolamine patch, eyedrops, plants (jimson weed)	Anisocoria increases in bright light	Loss of response to light; residual small reaction may occur with submaximal exposure or after sufficient time has elapsed	Same degree of loss of near response as loss of light response; near point of accommodation is more remote	Any residual light response of the sphincter is diffuse and not segmental	Subsensitivity to pilocarpine of all concentrations compared with the opposite unaffected eye (observed in dim light or darkness)
Pharmacologic mydriasis (adrenergic)	Low concentration of adrenergics found in over-the-counter eyedrops for red eyes, cocaine, Neo-Synephrine	Anisocoria increases in bright light, but not as much as in anticholinergic mydriasis	Diminished response to light, but dilator muscle can be overcome by strong sphincter constriction to bright light	Same diminished response as light reaction; near point of accommodation is unaffected and is normal	Besides diminished reaction, the pupil movement looks normal and is not segmental	Reversal of anisocoria with adrenergic blockade (dipyridamole or thymoxamine); may be overcome with pilocarpine
Damage to iris sphincter	Ischemia, angle-closure glaucoma, herpes zoster iritis, trauma, after anterior segment surgery	Anisocoria increases in bright light	Loss of response to light, some sphincter segments may be more affected than others	Usually affected to the same degree as the light reflex; may have normal accommodative amplitude	Transillumination defects may be present	Lack of response to 1% pilocarpine in damaged areas of the iris sphincter
Iron or copper mydriasis	Intraocular foreign body	Anisocoria increases in bright light	Loss of response to light, not usually segmental	Usually affected to the same degree as the light reflex	Usually heterochromia is present, with the iris being darker	May show cholinergic supersensitivity
Third nerve palsy	Trauma, compression, rarely ischemia	Anisocoria increases in bright light	Loss of response to light, not usually segmental unless aberrant regeneration is present	Usually affected to the same degree as the light reflex; accommodative amplitude is decreased	Usually symmetric weakness along the circumference of the sphincter to light reaction	May show cholinergic supersensitivity
Mechanically scarred iris	Trauma, iritis	Anisocoria increases in bright light	Loss of response to light when viewed without slit lamp	Usually affected to the same degree as the light reflex; accommodative amplitude is normal	Small movement of the scarred down iris can usually be observed; synechiae can often be seen after dilation	Lack of response to 1% pilocarpine
Physiologic anisocoria	Asymmetric inhibition at the Edinger–Westphal nucleus	Anisocoria increases in bright light	Normal; normal dilation in response to darkness and auditory stimulation	Normal; anisocoria lessens with a good near response	Normal	Normal response to topical agents; cocaine lessens the anisocoria
Oculosympathetic defect (Horner syndrome)	Sympathetic nerve palsy; ptosis is usually present of the upper and lower lid; anhidrosis may be present	Anisocoria increases in bright light	Normal; slow dilation in response to darkness and auditory stimulation	Normal; anisocoria lessens with a good near response	Normal; heterochromia is common if palsy occurred early in life	Cocaine increases the anisocoria; supersensitivity is usually demonstrated to adrenergic agents
Congenital pseudo-Horner syndrome	Unknown; anisocoria is often present in old photographs from infancy	Anisocoria increases in bright light	Appears normal, but the smaller pupil does not dilate well to darkness or auditory stimulation	Normal; anisocoria lessens with a good near response	Normal	May give a false-positive cocaine test; when direct-acting agents or anticholinergic drops are used, the pupil still fails to dilate as well as the fellow eye

Box 25.8 Horner syndrome

Clinically, Horner syndrome is recognized by looking for the associated signs, such as ptosis, "upside-down ptosis" of the lower lid, and in a fresh case, conjunctival injection, decreased sweating on the involved side, and in some cases, lowered intraocular pressure on the affected side.

Pain on the side of the face with the smaller pupil (jaw, ear, and cheek pain) is often an important sign pointing to a carotid artery dissection as a cause of the Horner syndrome (Fig. 25.14).

The patient with Horner syndrome has a weak dilator muscle in one iris, and as a result, that pupil dilates more slowly than the normal pupil (Fig. 25.15). If the sympathetic lesion is complete, the affected pupil will dilate only by sphincter relaxation; this process takes longer than with an intact sympathetic innervation to the dilator muscle. This asymmetry of pupil dilation produces an anisocoria that becomes largest 4 to 5 seconds after the lights are turned out. Dilation of the eye with the oculosympathetic defect is a much slower process than most people imagine. After the lights have been out for

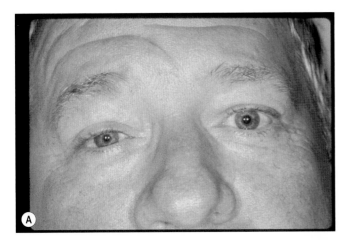

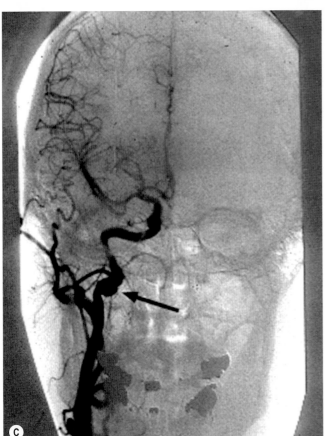

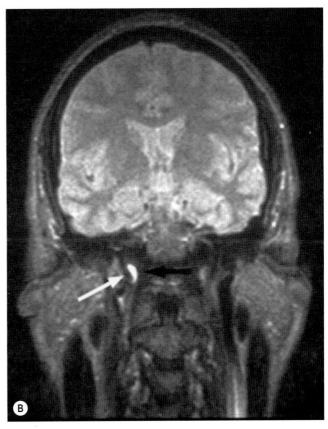

Figure 25.14 (**A**) Example of a patient with acute Horner syndrome on the right with accompanying facial pain. Note the miosis and ptosis on the right eye. This patient had a dissection of the internal carotid artery demonstrated by the magnetic resonance imaging scan (**B**) showing a bright signal in the wall of the carotid artery resulting from blood (white "comma" adjacent to carotid lumen in lower left; arrow). The dissection created a false lumen that appears as a focal enlargement of the lumen of the carotid artery demonstrated in the carotid angiogram on the right (arrow) (**C**).

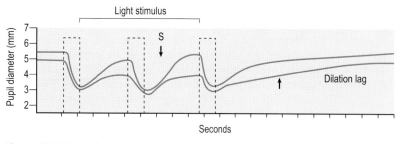

Figure 25.15 Example of a pupillographic demonstration of dilation lag of the pupil in the eye with Horner syndrome. The smaller pupil is slow to dilate, causing an increase in anisocoria during the early phase of dilation and accentuated by a loud sound (S). In the last part of the tracing, no further stimulus was given to allow the dilation lag to be observed in the tracing. The dilation lag is a delayed dilation of the smaller pupil; after 5 seconds in darkness, the smaller pupil slowly dilates as a result of inhibition of the sphincter muscle, despite loss of sympathetic nerve contribution to pupil dilation.

10 to 20 seconds, the anisocoria lessens as the sympathectomized pupil gradually catches up because of continual relaxation of the iris sphincter due to central inhibition of the parasympathetic pathway at the Edinger–Westphal nucleus. The delayed dilation of the involved eye due to a deficit of sympathetic innervation to the dilator muscle is a process referred to as *dilation lag* (Fig. 25.15).

Often, the initial increase in anisocoria during the first few seconds of darkness can be accentuated by adding an auditory stimulus just after the lights are turned out. This causes the normal pupil to dilate forcefully in response to the extra sympathetic stimulation evoked by the loud noise, but in the eye with the oculosympathetic defect, this maneuver has little effect. Comparison of the dynamics of dilation of the two pupils is a quick and simple way of distinguishing Horner syndrome from simple anisocoria, and it is a test that does not require pupillary drug testing. It works well most of the time, especially in young people with mobile pupils, but if the dilation lag is inconclusive, cocaine eye drops should be used to confirm the diagnosis of Horner syndrome.

Pharmacologic diagnosis of Horner syndrome with cocaine or apraclonidine

Cocaine's action is to block the reuptake of the norepinephrine that is normally released from the nerve endings. If, because of an interruption in the sympathetic pathway, norepinephrine is not being released, cocaine has no adrenergic effect. A Horner's pupil will dilate less to cocaine than the normal pupil, regardless of the location of the lesion. Forty-five minutes after cocaine drops have been placed in both eyes, the anisocoria should have clearly increased because the normal pupil has dilated more than the Horner's pupil.

A solution of 4 percent cocaine HCl drops in both eyes is recommended (never more than 2 drops) to be sure that the iris gets a full mydriatic dose; this concentration of cocaine does not normally produce a corneal epithelial defect. It should be said that 2 percent, 4 percent, and 5 percent cocaine have also been used as a diagnostic test for Horner syndrome and work well. The pH of the solution usually causes significant burning after topical application. Pre-application of topical anesthetic helps counteract the discomfort. After 40 to 60 minutes have elapsed, the anisocoria is measured in room light. The patient should stay active during that time so that adequate sympathetic discharge is occurring (sleeping during this time might result in neither pupil dilating to cocaine). If there is little dilation of the eye with a suspected oculosympathetic defect and this pupil never dilated well in darkness before the drop was given, even after 30 seconds of darkness, a false-positive cocaine test also must be considered. This can occur if the iris is in some way held in a miotic state by either scarring or aberrant reinnervation of the iris sphincter. In such cases, adding a direct-acting sympathomimetic agent to both eyes (i.e. 2.5 percent phenylephrine) at the conclusion of a positive cocaine test should easily dilate the iris of the suspected eye if there really is an oculosympathetic defect, and the cocaine-induced anisocoria should almost be eliminated. In some cases the 2.5 percent phenylephrine may even cause the eye with the oculosympathetic defect to have the larger pupil as a result of supersensitivity. Pseudo-Horner syndrome, caused by the reasons stated earlier, results in inadequate dilation to direct-acting sympathetic agents.

The likelihood of a diagnosis of Horner syndrome with cocaine testing increases in proportion to the amount of anisocoria (measured 50 to 60 minutes after the instillation of cocaine). Unlike the hydroxyamphetamine test (used for localizing the lesion to either the preganglionic or postganglionic neuron), calculation of the change in the anisocoria from before to after cocaine application is unnecessary. It has been found that if there is at least 0.8 mm of pupillary inequality after cocaine, the presence of Horner syndrome is highly likely.[29]

Recently, it has been proposed that a new pharmacologic test using 0.5 percent apraclonidine be used for the diagnosis of Horner's syndrome in place of cocaine.[74–78] Typically, 30 minutes following topical apraclonidine administered to both eyes, the miotic eye with the oculosympathetic defect dilates and if anything, the normal pupil becomes smaller (in dim light), causing the anisocoria to reverse, or at least lessen. In patients with anisocoria of other causes, such as physiologic anisocoria, no mydriasis occurs. Apraclonidine has an advantage over cocaine in that it will actively dilate the affected eye and not the normal eye, making its action a positive (mydriatic) one in the affected eye rather than a negative one in the affected eye and a positive (mydriatic) one in the unaffected eye, as with cocaine. Unlike phenylephrine, whose corneal penetration varies widely among individuals, apraclonidine readily penetrates the cornea and gains access to the iris, so the limiting factor to its mydriatic effect is whether α_1 supersensitivity is present in the iris dilator muscle. Apraclonidine has potential advantages over phenylephrine, not only in its ease of corneal penetration, but also in the fact that it does not need to be diluted. Further studies will clarify the diagnostic efficacy of apraclonidine for the diagnosis of oculosympathetic defects.

Pharmacologic localization of the denervation in Horner syndrome

The site of the oculosympathetic lesion in Horner syndrome is a question of considerable clinical importance because many postganglionic defects are caused by benign vascular headache syndromes or more serious carotid dissections and a preganglionic lesion is sometimes the result of the spread of a malignant neoplasm.

Hydroxyamphetamine eye drops help localize the site of the lesion in Horner syndrome. The clinician would like to know where the lesion is because that knowledge directs the radiographic workup (e.g. to the internal carotid artery rather than to the pulmonary apex). Horner syndrome sometimes presents itself in such a characteristic setting that further efforts at localizing the lesion are not needed. This is true of patients with cluster headaches or after surgical procedures that interrupt the oculosympathetic chain in a known location.

Hydroxyamphetamine acts by releasing norepinephrine from storage in the sympathetic nerve endings. When the lesion is postganglionic, most, if not all, of the nerves are dead, and no norepinephrine stores are available for release. When the lesion is complete, a pupil like this will not dilate at all in response to hydroxyamphetamine. However, a period of almost 1 week from the onset of damage may be needed before the dying neurons and their stores of norepinephrine are gone. Therefore a hydroxyamphetamine test given within a week of a postganglionic lesion may give a false preganglionic localization if the norepinephrine stores have not yet disappeared. Horner's pupils resulting from preganglionic or central lesions dilate *at least* normally because the postganglionic neuron with its stores of norepinephrine, although disconnected, is still intact. In fact, when the lesion is in the preganglionic neuron, the Horner's pupil often becomes larger than the normal pupil – apparently resulting from "decentralization supersensitivity."

The hydroxyamphetamine test is simple: The pupils are measured *before* and 40 to 60 minutes *after* hydroxyamphetamine drops have been put in both eyes, and the change in anisocoria from baseline state to the after drops state in room light is noted. If the Horner's pupil – the smaller one – dilates less than the normal pupil, the anisocoria increases and it can be concluded that the lesion is in the postganglionic neuron. If the smaller pupil dilates so well that it becomes the larger pupil, the lesion is preganglionic and the postganglionic neuron with its norepinephrine stores is still intact. The examiner should wait at least 2 to 3 days after using cocaine drops before using hydroxyamphetamine; cocaine seems to block hydroxyamphetamine's effectiveness because it blocks the uptake of hydroxyamphetamine into the postganglionic nerve terminal, thus inhibiting its action.

To understand how to better interpret the hydroxyamphetamine test, we studied hydroxyamphetamine mydriasis in patients with a known lesion location and in those in whom the location of the lesion was unknown. It appears that postganglionic lesions (along the carotid artery) can be separated from the non-postganglionic lesions (in the brainstem, spinal cord, upper lung, and lower neck) with a degree of certainty that varies with the amount of change in anisocoria induced when the drops are put in both eyes.[16] An increase in the anisocoria of greater than 0.5 mm (change in anisocoria from prehydroxyamphetamine to posthydroxyamphetamine) makes a postganglionic lesion highly likely.

Congenital and childhood Horner syndrome

When a child has a unilateral ptosis and miosis, the first question is whether it is really Horner syndrome. The ptosis of Horner syndrome is moderate, never complete. Sometimes, the elevation of the lower lid (upside-down ptosis) is persuasive. A child with a congenital Horner syndrome and naturally curly hair has, on the affected side of the head, hair that will seem limp and lank. The shape of the hair follicles appears to depend on intact sympathetic innervation, as does the iris pigment. In children, iris color usually occurs after the first 9 to 12 months of life because of accumulation of melanosomes in iris melanocytes that are innervated by the sympathetic nerves. Therefore, if the iris in the eye with Horner syndrome does not develop pigmentation during this period, i.e. there is heterochromia, then an oculosympathetic defect was present early in life. However, the presence of heterochromia does not indicate the cause or whether it was congenital versus acquired during the first year of life.

Cocaine eye drops may be of some help in diagnosing Horner syndrome in children. If there is no significant dilation of the smaller pupil, a drop of 2.5 percent phenylephrine should then be administered to each eye to substantiate that the pupil can dilate to a direct-acting sympathomimetic agent and that this is not pseudo-Horner syndrome (see previous section on the cocaine test). Other signs may also be helpful for diagnosing Horner syndrome in children. The most telling sign is a hemifacial flush that can occur on the normally innervated side in contrast to the blanch that occurs on the side with the oculosympathetic defect when the infant is nursing or crying. In an air-conditioned office it may be hard to decide whether there is an asymmetry of sweating. A cycloplegic refraction can sometimes produce an atropinic flush everywhere except on the affected face and forehead and thus can unexpectedly solve the diagnostic problem.

In infants the hydroxyamphetamine drop test is not very helpful in localizing the lesion because orthograde transsynaptic dysgenesis takes place at the superior cervical ganglion after an early interruption of the preganglionic oculosympathetic neuron. This results in fewer postganglionic neurons, even though there was no postganglionic injury, and this, in turn, produces weak mydriasis and ambiguous answers from the hydroxyamphetamine test in children.[56] Therefore, if a hydroxyamphetamine drop test is used, the interpretation needs to take this possibility into consideration. If, in response to hydroxyamphetamine, the affected eye with relative miosis dilates as well as or better than the fellow eye, it is probably a preganglionic lesion. However, any result showing less mydriasis in the affected eye compared with the fellow eye cannot be localized with certainty because of the reasons listed earlier. Horner syndrome that has been clearly acquired in infancy should be evaluated for neuroblastoma, a treatable tumor, using a combination of imaging and testing the urine for metabolites of adrenergic compounds secreted from the tumor.

Pupil inequality that is increased in bright light

When the anisocoria increases with bright light, this implies that the larger pupil is abnormal and may not contract well to light because of direct damage to the iris sphincter muscle from trauma or surgery or because of ischemic atrophy, scarring (synechia) of the iris to the lens from previous inflammation, pharmacologic mydriasis, or denervation of the parasympathetic nerve supply to the iris sphincter (see Fig. 25.4). The following sections outline the most important clinical observations and tests that may be needed to sort out the cause of the efferent pupil defect (see Table 25.2).

Examination of the iris with high magnification using the slit-lamp biomicroscope

Trauma to the globe usually results in a torn sphincter or an iris that shows transillumination defects at the slit lamp. The pupil is often not round, and there may be other evidence of ocular injury. Naturally, such a pupil does not constrict well to light. The residual reaction is often associated with the remaining normal sectors of the iris sphincter because the traumatic tears are usually segmental. An atrophic sphincter resulting from previous herpes zoster iritis may also reveal large geographic areas of transillumination defects seen with the slit lamp from previous ischemic vasculitic insults to the iris during the uveitis.

However, if the iris tissue looks normal, the examination is focused on whether any part of the iris sphincter is contracting with light. If some contraction occurs, it should be determined whether the residual contraction is diffuse over the whole circumference or segmental. If there is no segmental movement of the iris, the possibility of atropinic mydriasis should be considered.[27] However, a completely blocked light reaction can sometimes be seen when the sphincter is totally denervated by a preganglionic lesion (third nerve palsy) or a postganglionic lesion (fresh tonic pupil), in acute angle closure (iris ischemia), or in the presence of an intraocular iron foreign body (iron mydriasis). If the dilated pupil still has some response to light, the dilation could be caused by partial denervation of the sphincter or by incomplete atropinization or adrenergic mydriasis. When the light reaction is poor because the dilator muscle is in spasm (because of adrenergic mydriatics like phenylephrine), the pupil is large, the conjunctiva blanched, and the lid retracted. In such a case, the near point of accommodation is usually normal but may be slightly decreased as a result of spherical aberration and a shallow depth of field caused by the dilated pupil. However, in the case of adrenergic mydriasis, there usually is some light reaction to very bright lights because the stronger iris sphincter can usually overcome pulling by the adrenergic effects on the dilator muscle.

If there is some residual light reaction, the next step is to look with the slit lamp for sector palsy of the iris sphincter. When the dilator is in a drug-induced adrenergic spasm or when the cholinergic receptors in the iris sphincter are blocked by an atropine-like drug, the entire 360 degrees of the sphincter muscle is less effective. This is not the case when the postganglionic nerve fibers have been interrupted: Adie's pupils with a residual light reaction (about 90 percent of them) have segmental contractions of the remaining normal segments of the sphincter. Some preganglionic partial third nerve palsies also have regional sphincter palsies, but these can usually be attributed to an associated preexisting diabetic autonomic neuropathy or aberrant regeneration. This means that when the examiner sees a pupil with a weak light reaction but no segmental palsy, he or she should consider a drug-induced mydriasis and then perhaps look again for lid and motility signs of a third nerve paresis.

Pharmacologic response of the iris sphincter to cholinergic drugs

Cholinergic supersensitivity

If weak pilocarpine (0.0625 percent, 0.1 percent, or 0.125 percent) is applied to both eyes and the affected (dilated) pupil constricts more than the normal pupil (actually becoming the smaller pupil in darkness), the iris sphincter has probably lost some of its innervation and has become supersensitive (Fig. 25.16). Cholinergic supersensitivity can occur within 5 to 7 days. The conclusion that cholinergic supersensitivity exists assumes that corneal penetration of the drug is the same in each eye (i.e. both corneas are healthy and untouched, tear function is normal, and the eyelids are working properly in both eyes). It seems likely that with postganglionic denervation (damage at the ciliary ganglion or distal to it) the sphincter will show more supersensitivity than in the preganglionic case (third nerve palsy). However, it appears that the differences are not great. For all of these reasons, cholinergic supersensitivity of the iris sphincter is now considered only a confirmatory sign of Adie syndrome. In fact, the results of supersensitivity testing may be ambiguous in Adie syndrome, depending on the chronicity of the condition. As reinnervation takes place over time in Adie syndrome (with accommodative cholinergic nerves growing into the iris sphincter), the reinnervated sphincter segments may lose their cholinergic supersensitivity.[59] If a patient shows the presence of cholinergic supersensitivity in one eye without any segmental palsies, other causes besides Adie's pupil should be reconsidered. Subtle signs of ptosis or diplopia should be looked for once more to leave no doubt that the oculomotor nerve is not affected. It is rare for an ambulatory patient to have an isolated sphincter palsy without other signs of oculomotor nerve palsy as a result of compressive damage to the intracranial third nerve caused by a tumor or aneurysm.

Testing of iris cholinergic sensitivity is best performed when comparing an affected eye with its fellow, normal eye. Testing for whether cholinergic supersensitivity is present in both eyes, without comparing the response with the normal fellow eye in the same subject, is problematic because there is considerable variation in the cholinergic response among different individuals. Even some normal eyes can respond to 0.05 percent pilocarpine, so for supersensitivity testing to be the most meaningful, the test is best used in the setting of unilateral causes.

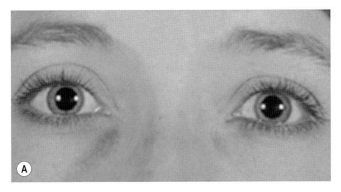

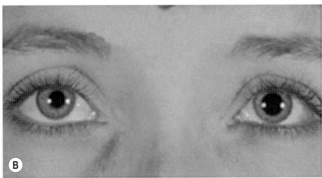

Near effort

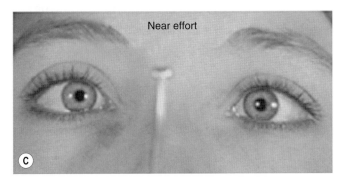

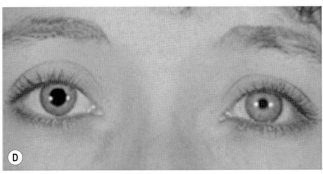

Figure 25.16 Example of Adie's pupil (**A**) showing little anisocoria in darkness (**B**) and an increasing anisocoria in room light (**C**). A light-near dissociation with a greater contraction of the left pupil to a near stimulus compared with light (**D**), indicating a chronic state resulting from reinnervation of the iris sphincter by accommodative neurons. The patient still exhibited signs of cholinergic supersensitivity (lower right), with the involved pupil contracting more to 0.1 percent pilocarpine than the fellow pupil.

Subsensitivity of the iris sphincter to cholinergic testing

If the normal pupil constricts only a small amount to dilute cholinergic agents and the dilated pupil not at all, the mydriasis may be caused by the presence of an anticholinergic drug like atropine, which inhibits the receptors on the iris sphincter muscle. A stronger concentration of pilocarpine is then needed to settle this point. If, on application of 1 percent pilocarpine in each eye, the affected pupil does little or nothing and the unaffected pupil constricts normally, the pupil is not larger because of nerve denervation, but rather, because of a problem in the sphincter muscle itself. The following are different non-neuronal causes of mydriasis:

- anticholinergic mydriasis (e.g. scopolamine, cyclopentolate, atropine)
- traumatic iridoplegia (look for sphincter tears, appearing as divots at the pupil border; pigment dispersion on the corneal endothelium of lens; and angle recession)
- angle-closure glaucoma (ischemia of the iris sphincter that occurs during the time when the intraocular pressure is high)
- previous herpes zoster iritis causing direct damage to the iris sphincter
- synechiae causing a bound-down iris that is mechanically immobile
- fixed pupil following anterior segment surgery
- ocular ischemia to the anterior segment, damaging the iris sphincter.

The cause for a loss of function of the iris muscle following anterior segment surgery is unknown. In some cases with a postoperative rise in intraocular pressure, the cause may be an ischemic insult to the iris sphincter. An autoimmune process may be responsible, but it has not been proven. It may be related to the same process as Urrets–Zavalia syndrome, in which a dilated, fixed pupil may occur following penetrating keratoplasty.

Adie's tonic pupil: postganglionic parasympathetic denervation

Young adults (more commonly women than men) may suddenly find that one pupil is large or that they cannot focus as well with one eye at near. Slit-lamp examination usually shows segmental denervation of the iris sphincter, with some remaining normal segments still reacting to light. Within the first week, supersensitivity of the iris and ciliary muscle to cholinergic substances can be demonstrated. After about 2 months, nerve re-growth is active and fibers originally bound for the ciliary muscle (they outnumber the sphincter fibers by 30 : 1) start arriving (aberrantly) at the iris sphincter and the ciliary muscle. The light reaction of the denervated segments does not return, but the reinnervated segments now show contraction to a near stimulus. This produces the characteristic "light-near dissociation" of Adie syndrome (Fig. 25.16), as well as a return of some accommodative amplitude. Therefore the presence of light-near dissociation in this setting is a sign of a chronic Adie's pupil with reinnervation and is not a sign of an acute Adie's pupil. The pupil contraction to a near stimulus is often tonic, being slow to dilate when gaze is shifted to a distant target. Although there is also some return of accommodative amplitude, the dynamics of focusing in the affected eye are also slowed and not normal. Patients often complain of difficulties when trying to refocus from near to far because the relaxation of accommodation is slower in the affected eye. Eventually, the affected pupil becomes the smaller of the two pupils, especially in dim light, as a result of the amount of reinnervation by cholinergic accommodative neurons, which keep the sphincter in a contracted state. It turns out that the segmental palsy of the iris sphincter in Adie syndrome and the individual sphincter segment responses to light, near and pilocarpine, can be seen especially well by infrared video recording of transillumination of the iris (Fig. 25.17). Many of these patients also lack normal motor jerk reflexes and may also have decreased vibratory sensation, indicating a similar process occurring in the spinal cord neurons. However, Adie syndrome is not associated with any major neurologic disorder or significant dysfunction. The cause of Adie's pupil is not well understood, but it has been hypothesized that an immune reaction may mediate the damage to ciliary ganglion and spinal neurons. Younger children may get an Adie's pupil after having chickenpox. After about 10 years, almost 50 percent

of patients with Adie's pupil show evidence of a similar process occurring in the other eye.

Pupil involvement in third nerve palsy

There is an old clinical rule of thumb stating that if the pupillary light reaction is spared in the setting of a third nerve palsy, the palsy is probably not caused by compression or injury, but more likely, it is caused by small vessel disease, such as might be seen in diabetes. It is still a fairly good rule, provided one bears in mind that a small but definite number of pupil-sparing third nerve palsies are caused by midbrain infarcts and should have neuroimaging studies. Because the preganglionic parasympathetic nerves for the pupil light reflex are located on the medial side of the intracranial portion of the third nerve as it exits the midbrain, compression of the third nerve in this location results in some element of iris sphincter palsy (Box 25.5). The most common cause of this would be an aneurysm (i.e. of the posterior communicating artery) or pituitary apoplexy (sudden lateral expansion of a pituitary adenoma pressing on the medial aspect of the third nerve). Pupil involvement is often incomplete, so it is important to look for iris sphincter weakness by observing for anisocoria in bright light. In the absence of

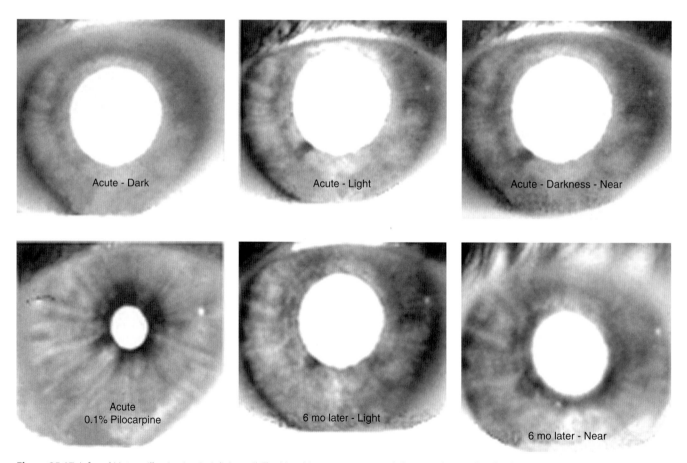

Figure 25.17 Infrared iris transillumination in Adie's pupil. The iris sphincter appears as a dark ring at the pupillary border when it contracts, as observed with infrared iris transillumination. In the patient shown, almost all of the iris sphincter was denervated, except for the segment at the 7 o'clock meridian, which darkened when made to contract with light or near (top row, center and right panels). With low-dose pilocarpine, the rest of the sphincter muscle that was denervated is the area where supersensitivity was present (bottom row, left panel). Every area darkened after the pilocarpine, except the normal segment at the 7 o'clock position, which remains lighter (not contracted), as shown in the upper right panel. After 6 months, the pupil started to become smaller and contracted to near (bottom right panel), but it remained unresponsive to light, except for the 7 o'clock segment (center right panel).

aberrant regeneration (from chronic compression), we have yet to observe a case of a pupil involving third nerve palsy from an aneurysm that showed segmental palsies; all of the cases so far have shown symmetric involvement of the iris sphincter over its circumference. Some patients with ischemic third nerve palsy and diabetes have shown mild pupil involvement with elements of segmental palsies. It appears that these patients may have had preexisting pupil involvement from diabetic autonomic neuropathy.

Aberrant regeneration in the third nerve

The third cranial nerve carries bundles of nerves supplying different extraocular muscles (medial rectus, inferior rectus, inferior oblique, superior rectus, and levator palpebrae muscles), as well as preganglionic parasympathetic nerves to the iris sphincter and ciliary body. Injury to the third nerve and glial scaffolding, through which individual nerve bundles pass, causes the nerve fibers to regrow, and they often end up in the wrong place. For example, the eye may inappropriately turn in when the patient is trying to look down, or the pupil may inappropriately constrict with depression, adduction, or supraduction of the globe (Fig. 25.18). With eyelid involvement, the lid fissure may widen with infraduction, adduction, or supraduction of the eye. Aberrant regeneration of the oculomotor nerve may be primary or secondary. In secondary aberrant regeneration, a third nerve palsy precedes the aberrant regeneration by at least 8 weeks. In primary aberrant regeneration, there is no preceding nerve palsy; the damage to the nerve slowly progresses simultaneous with the process of aberrant regeneration. Primary aberrant regeneration is clinically important to recognize because it is almost always caused by a slow compression of the intracranial third nerve by a tumor or aneurysm.

Light-near dissociation: evaluation of the near response

The pupil response to a near effort should be observed as a standard part of the pupil evaluation. Any time the pupil light reaction seems weak, it is important to check to see whether the pupils constrict better to near than they do to light. If they do, this is called a *light-near dissociation*. Causes of light-near dissociation are summarized in Table 25.4. These are categorized by three major mechanisms:

1. Loss of light input resulting from severe damage to the afferent visual system (retina or optic nerve pathways)

2. Interruption of the light input pathways to the Edinger–Westphal nucleus from the pretectum (Argyll Robertson pupils, dorsal midbrain syndrome)

3. Aberrant regeneration of the pupillary sphincter from accommodative fibers (Adie syndrome) or extraocular muscle neurons from the oculomotor nerve (medial rectus fibers or accommodative fibers from third nerve aberrant regeneration).

When the pupil fails to dilate

When one or both pupils stay small and miotic, even in darkness, a number of reasons may be responsible (Table 25.5). To better understand the different possible

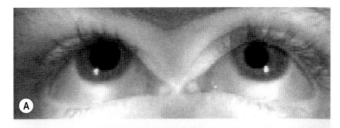
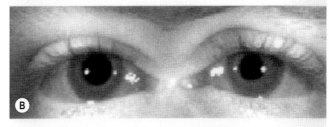
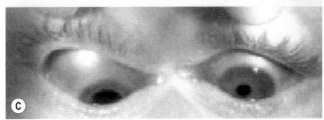

Figure 25.18 Primary aberrant regeneration of the left third nerve following chronic compression of the oculomotor nerve by a meningioma. In darkness (top and center panels) the left pupil is the smaller of the two pupils as a result of innervation by motor nerves. Nerves that normally would have supplied the inferior rectus muscle now are innervating the iris sphincter muscle, causing pupil contraction on downgaze (lower panel).

Table 25.4 Causes of light-near dissociation of the pupil

Cause	Location	Mechanism
Severe loss of afferent light input to both eyes	Anterior visual pathway (retina, optic nerves, chiasm)	Damage to the retina or optic nerve pathways
Loss of pretectal light input to Edinger–Westphal nucleus	Tectum of the midbrain	Infectious (Argyll Robertson pupils) or compression (pinealoma) or ischemia (stroke)
Adie syndrome	Iris sphincter	Aberrant reinnervation of sphincter by accommodative neurons
Third nerve aberrant reinnervation	Iris sphincter	Aberrant reinnervation of sphincter by accommodative neurons or medial rectus neurons

Table 25.5 Causes of poor pupil dilation in darkness

Cause	Location	Mechanism
Past inflammation or surgical trauma	Posterior iris surface or sphincter	Scarring or synechiae of the iris resulting from past iritis
Acute trauma	Sphincter	Prostaglandin release causing sphincter spasm
Adie tonic pupil Third nerve aberrant reinnervation	Sphincter	Aberrant reinnervation of iris sphincter by accommodative or extraocular motor neurons that are not inhibited in darkness
Pharmacologic miosis	Iris sphincter	Cholinergic influence
Unilateral episodic spasm of miosis	Postganglionic parasympathetic neuron	Uninhibited episodic activation of postganglionic neurons
Congenital miosis (bilateral)	Sphincter	Developmental abnormality
Fatigue, sleepiness	Edinger–Westphal nucleus	Loss of inhibition at midbrain from reticular activating formation
Lymphoma, inflammation, infection	Periaqueductal gray matter	Interruption of inhibitory fibers to the Edinger–Westphal nucleus
Central acting drugs	Reticular activating formation, midbrain	Narcotics, general anesthetics
Old age (bilateral miosis)	Reticular activating formation, midbrain	Loss of inhibition at midbrain from reticular activating formation
Oculosympathetic defect	Sympathetic neuron interruption	Horner syndrome

mechanisms it is important to understand what normally happens in darkness to allow the pupil to dilate. When a light stimulus is terminated, two mechanisms cause the pupil to dilate. The majority of pupil dilation comes about from inhibition to the Edinger–Westphal nucleus in the midbrain. This reduces the firing of the preganglionic parasympathetic neurons in the Edinger–Westphal nucleus, causing relaxation of the iris sphincter. Within a few seconds, sympathetic nerve firing increases, serving to augment the pupil dilation by active contraction of the dilator muscle. The combined inhibition of the iris sphincter and stimulation of the iris dilator is a carefully integrated neuronal reflex. The inability of the pupil to dilate in darkness may result from the following causes:

1. Mechanical limitations of the pupil (scarring)
2. Pharmacologic miosis
3. Aberrant reinnervation of cholinergic neurons to the iris sphincter that are not normally inhibited in darkness (accommodative or extraocular motor neurons)
4. Lack of inhibitory input signal getting to the Edinger–Westphal nucleus
5. Lack of sympathetic input to the dilator muscle.

References

1. Alexandridis E, Krastel H, Reuther R. Disturbances of the pupil reflex associated with lesions of the upper visual pathway. Albrecht von Graefes Arch Klin Exp Ophthalmol 1979; 209:199.
2. Almegrad B, Stjernschantz J, Bill A. Cholecystokinin contracts isolated human and monkey iris sphincters: a study with CCK receptor antagonists. Eur J Pharmacol 1992; 211:183.
3. Barbur JL, Forsyth PM. Can the pupil response be used as a measure of visual input associated with the geniculo-striate pathway? Clin Vis Sci 1986; 1:107.
4. Barbur JL, Harlow AJ, Sahraie A. Pupillary responses to stimulus structure, colour, and movement. Ophthal Physiol Opt 1992; 12:137.
5. Barbur JL, Keenleyside MS, Thomson WD. Investigation of central visual processing by means of pupillometry. In Kulikowski JJ, Dickinson CM, Murray IJ, eds. Seeing colour and contour. Oxford: Pergamon Press, 1989.
6. Barbur JL, Thomson WD. Pupil response as an objective measure of visual acuity. Ophthal Physiol Opt 1987; 7:425.
7. Bell RA et al. Clinical grading of relative afferent pupillary defects. Arch Ophthalmol 1993; 111:938.
8. Brown RH et al. The afferent pupillary defect in asymmetric glaucoma. Arch Ophthalmol 1987; 105:1540.
9. Campbell FW. The depth of field of the human eye. Optica Acta 1957; 4:157.
10. Campbell FW, Green DG. Optical and retinal factors affecting visual resolution. J Physiol 1965; 181:576.
11. Charman WN, Jenning JAM, Whitefoot H. The refraction of the eye in relation to spherical aberration and pupil diameter. Vis Res 1978; 17:737.
12. Cibis G, Campos E, Aulhorn E. Pupillary hemiakinesia in supragniculate lesions. Arch Ophthalmol 1975; 93:1252.
13. Cocker KD, Moseley MJ. Visual acuity and the pupil grating response. Clin Vis Sci 1992; 7:143.
14. Cox TA. Pupillography of a relative afferent pupillary defect. Am J Ophthalmol 1986; 101:250.
15. Cox TA. Pupillographic characteristics of simulated relative afferent pupillary defects. Invest Ophthalmol Vis Sci 1989; 30:1127.
16. Cremer SA et al. Hydroxyamphetamine mydriasis in Horner's syndrome. Am J Ophthalmol 1990; 110:71.
17. Fison PN, Garlick DJ, Smith SE. Assessment of unilateral afferent pupillary defects by pupillography. Br J Ophthalmol 1979; 63:195.
18. Gamlin PDR, Clarke RJ. The pupillary light reflex pathway of the primate. J Am Optom Assoc 1995; 66:415.
19. Gamlin PDR, Zhang H, Clarke RJ. Luminance neurons in the pretectal olivary nucleus mediate the pupillary light reflex in the rhesus monkey. Exp Brain Res 1995; 106:177.
20. Hamann K et al. Videopupillographic and VER investigations in patients with congenital and acquired lesions of the optic radiation. Ophthalmologica 1979; 178:348.
21. Harms H. Grundlagen, Methodik und Bedeutung der Pupillenperimetrie fur die Physiologie und Pathologie des Schorgans. Albrecht von Graefes Arch Klin Exp Ophthalmol 1949; 149:1.
22. Hellner KA, Jensen W, Muller A. Video processing pupillographic perimetry in hemianopsia. Klin Mbl Augenheik 1978; 172:731.
23. Hellner K, Jensen W, Muller-Jensen A. Video-processing pupillography as a method for objective perimetry in pupillary hemiakinesia. In: Greve EL, ed. The proceedings of the second international visual field symposium, Tubingen, 1976. Doc Ophthalmol Proc Series, vol 14. The Hague: Dr W Junk Publishers, 1977.
24. Huhtala A. Origin of myelinated nerves in the rat iris. Exp Eye Res 1976; 22:259.
25. Johnson LN, Hill RA, Bartholomew MJ. Correlation of afferent pupillary defect with visual field loss on automated perimetry. Ophthalmology 1988; 95:1649.
26. Kardon RH. Pupil perimetry. In Current opinion in ophthalmology, vol 3. Philadelphia: Current Science, 1992.
27. Kardon RH. Anatomy and physiology of the pupil. Section III. The autonomic nervous system: pupillary function, accommodation, and lacrimation. In: Miller NM, Newman NJ, eds. Walsh and Hoyt's clinical neuro-ophthalmology, vol 1, 5th edn. Baltimore: Williams & Wilkins, 1998.
28. Kardon RH, Corbett JJ, Thompson HS. Segmental denervation and reinnervation of the iris sphincter as shown by infrared videographic transillumination. Ophthalmology 1998; 105:313.
29. Kardon RH, Haupert C, Thompson HS. The relationship between static perimetry and the relative afferent pupillary defect. Am J Ophthalmol 1993;115:351.
30. Kardon RH, Kirkali PA, Thompson HS. Automated pupil perimetry. Ophthalmology 1991; 98:485.
31. Kardon RH et al. Critical evaluation of the cocaine test in the diagnosis of Horner's syndrome. Arch Ophthalmol 1990; 108:384.
32. Kawasaki A, Moore P, Kardon RH. Variability of the relative afferent pupillary defect. Am J Ophthalmol 1995; 120:622.
33. Kawasaki A, Moore P, Kardon RH. Long-term fluctuation of relative afferent pupillary defect in subjects with normal visual function. Am J Ophthalmol 1996; 122:875.
34. Kerr FWL, Hollowell OW. Location of pupillomotor and accommodation fibres in the oculomotor nerve: experimental observations on paralytic mydriasis. J Neurol Neurosurg Psychiatr 1964; 27:473.
35. Koss MC. Pupillary dilation as an index of central nervous system alpha2-adrenoceptor activation. J Pharmacol Methods 1986; 15:1.

36. Levatin P. Pupillary escape in disease of the retina or optic nerve. Arch Ophthalmol 1959; 62:768.

37. Loewenfeld IE. The light reflex. In: The pupil: anatomy, physiology and clinical applications, vol 1, Ames, IO & Detroit, MI: Iowa State University Press and Wayne State University Press, 1993.

38. Loewenfeld IE. Methods of pupil testing. In: The pupil: anatomy, physiology and clinical applications, vol 1, Ames, IO & Detroit, MI: Iowa State University Press and Wayne State University Press, 1993.

39. Loewenfeld IE. Reactions to darkness. In: The pupil: anatomy, physiology and clinical applications, vol 1, Ames, IO & Detroit, MI: Iowa State University Press and Wayne State University Press, 1993.

40. Loewenfeld IE. The reaction to near vision. In: The pupil: anatomy, physiology and clinical applications, vol 1, Ames, IO & Detroit, MI: Iowa State University Press and Wayne State University Press, 1993.

41. Loewenfeld IE. Reflex dilation. In: The pupil: anatomy, physiology and clinical applications, vol 1, Ames, IO & Detroit, MI: Iowa State University Press and Wayne State University Press, 1993.

42. Loewenfeld IE, Newsome DA. Iris mechanics: I. Influence of pupil diameter on dynamics of pupillary movements. Am J Ophthalmol 1971; 71:347.

43. Lowenstein O, Kawabata H, Loewenfeld I. The pupil as indicator of retinal activity. Am J Ophthalmol 1964; 57:569.

44. Narasaki S et al. Videopupillographic perimetry and its clinical application. Jpn J Ophthalmol 1974; 18:253.

45. Newsome DA, Loewenfeld IE. Iris mechanics: II. Influence of pupil diameter on details of iris structure. Am J Ophthal 1971; 71:553.

46. Reuther R, Alexandridis E, Krastel H. Disturbances of the pupil reflex associated with cerebral infraction in the posterior cerebral artery territory. Arch Psychiatr Nervenkr 1981; 229:249.

47. Rushton WAW. Visual adaptation: the Ferrier lecture. Proc R Soc Biol Lond 1965; 162:20.

48. Saari M et al. Wallerian degeneration of the myelinated nerves of cat iris after denervation of the ophthalmic division of the trigeminal nerve: an electron microscopic study. Exp Eye Res 1973; 17:281.

49. Slooter JH, van Noren D. Visual acuity measured with pupil responses to checkerboard stimuli. Invest Ophthalmol Vis Sci 1980; 19:105.

50. ten Doesschate J, Alpern M. Response of the pupil to steady state retinal illumination: contribution by cones. Science 1965; 149:989.

51. Thompson HS, Corbett JJ. Asymmetry of pupillomotor input. Eye 1991; 5:36.

52. Thompson HS, Corbett JJ, Cox TA. How to measure the relative afferent pupillary defect. Surv Ophthalmol 1981; 26:39.

53. Thompson HS et al. The relationship between visual acuity, pupillary defect, and visual field loss. Am J Ophthalmol 1982; 93:681.

54. Ukai K. Spatial pattern as a stimulus to the pupillary system. J Opt Soc Am 1985; 1094.

55. Volpe NJ et al. Portable pupillography of the swinging flashlight test to detect afferent pupillary defects. Ophthalmology 2000; 107:1913.

56. Weinstein JM, Zweifel TJ, Thompson HS. Congenital Horner's syndrome. Arch Ophthalmol 1980; 98:1074.

57. Westheimer G. Pupil diameter and visual resolution. Vis Res 1964; 4:39.

58. Young RSL, Han B, Wu P. Transient and sustained components of the pupillary responses evoked by luminance and color. Vis Res 1993; 33:437.

59. Young RSL, Kennish J. Transient and sustained components of the pupil response evoked by achromatic spatial patterns. Vis Res 1993; 33:2239.

60. Hannibal J, Hindersson P, Knudson SM, Georg B, Fahrenkrug J. The photopigment melanopsin is exclusively present in pituitary adenylate cyclase-activating polypeptide-containing retinal ganglion cells of the retinohypothalamic tract. J Neurosci 2002; 22.

61. Hattar S, Liao HW, Takao M, Berson DM, Yau KW. Melanopsin containing retinal ganglion cells: architecture, projections, and intrinsic photosensitivity. Science 2002; 295:1065–1070.

62. Fu Y, Liao HW, Do MTH, Yau KW. Non-image-forming ocular photoreception in vertebrates. Curr Opin Neurobiol 2005; 15:415–422.

63. Berson DM. Strange vision: ganglion cells as circadian photoreceptors. Trends Neurosci 2003; 26:314–320

64. Lucas RJ, Freedman MS, Lupi D, et al. Identifying the photoreceptive inputs to the mammalian circadian system using transgenic and retinally degenerate mice. Behav Brain Res 2001; 125:97–102.

65. Gamlin PDR, McDougal DH, Pokorny J, et al. Human and macaque pupil responses driven by melanopsin-containing retinal ganglion cells. Vision Res 2007; 47:946–954.

66. Dacey DM, Liao HW, Peterson BB, et al. Melanopsin-expressing ganglion cells in primate retina signal colour and irradiance and project to the LGN. Nature 2005; 433:749–754.

67. Van Gelder RN. Non-visual ocular photoreception. Ophthalm Genet 2001; 195–205.

68. Peirson S, Foster RG. Melanopsin: another way of signaling light. Neuron 2006; 49:331–339.

69. Gooley JJ, Lu J, Fischer D, Saper CB. A broad role for melanopsin in non-visual photoreception. J Neurosci 2003; 23:7093–7106.

70. Provencio I, Rollag MD, Castrucci AM. Photoreceptive net in the mammalian retina. Nature 2002; 415:493.

71. Hattar S, Lucas RJ, Mrosovsky N, et al. Melanopsin and rod–cone photoreceptive systems account for all major accessory visual functions in mice. Nature 2003; 424:76–81.

72. Kawasaki A, Kardon RH. Intrinsically photosensitive retinal ganglion cells. J Neuro-ophthalmol 2007; 27(3):195–204.

73. Kardon R, Anderson SC, Damarjian TG, Grace EM, Stone E, Kawasaki A. Chromatic pupil responses: preferential activation of the melanopsin-mediated versus outer photoreceptor-mediated pupil light reflex. Ophthalmology 2009; 116(8):1564–1573.

74. Kardon RH. Are we ready to replace cocaine with apraclonidine in the pharmacologic diagnosis of Horner syndrome? J Neuro-ophthalmol 2005; 25:69–70.

75. Morales J, Brown S, Abdul-Rahim AS, Crosson C. Ocular effects of apraclonidine in Horner's syndrome. Arch Ophthalmol 2000; 118:951–954.

76. Brown SM, Aouchiche R, Freedman KA. The utility of 0.5 percent apraclonidine in the diagnosis of Horner syndrome. Arch Ophthalmol 2003; 121:1201–1203.

77. Chen PL, Chen JT, Lu DW, Chen YC, Hsiao CH. Comparing efficacies of 0.5 percent apraclonidine with 4 percent cocaine in the diagnosis of Horner syndrome in pediatric patients. J Ocul Pharmacol Ther 2006; 22:182–187.

78. Koc F, Kavuncu S, Kansu T, Acaroglu G, Firat E. The sensitivity and specificity of 0.5 percent apraclonidine in the diagnosis of oculosympathetic paresis. Br J Ophthalmol 2005; 89:1442–1444.

CHAPTER 26

Ganglion-Cell Photoreceptors and Non-Image-Forming Vision

Kwoon Y. Wong & David M. Berson

Overview

In the past decade, compelling evidence has emerged for a novel class of photoreceptors in the mammalian retina. These neurons are ganglion cells that express the photopigment melanopsin and respond autonomously to bright light with a sustained depolarization and increase in spike frequency. These intrinsically photosensitive retinal ganglion cells (ipRGCs) differ radically in form and function from the classical rod and cone photoreceptors. In this chapter, we discuss the origins of this discovery, the idiosyncratic physiology of these cells and roles they play in retinal processing and visual behavior in health and disease.

Historical roots

For most contemporary retinal scientists, the recent discovery of photoreceptors in the inner retina may have come as a shock, but the roots of the idea are actually traceable to the very beginnings of retinal science (see reviews[1-3]). Beginning with Descartes and through the first half of the 19th century, it was assumed that the photoreceptive elements of the retina lined its inner surface, nearest the incoming light.[3] For example, Treviranus,[4] the first to conduct systematic microscopic studies of the retina, suggested that optic fibers passed through all retinal layers to end as photosensitive papillae at the vitreal surface. Bidder[5] accurately described the outer retinal location of the rods, but hypothesized that they served a reflective function, like a tapetum, and continued to view optic fibers in the inner retina as the locus of phototransduction. The lack of photosensitivity of the fibers of the optic disk, as indicated by the blind spot, eventually undermined this view, and prompted Bowman[6] and Helmholtz[7] to propose ganglion cells as the true photoreceptors. It was not until the latter half of the 19th century that photoreception was convincingly localized to the outer retina. The key insights were made by Heinrich Müller, who inferred the retinal depth of the photoreceptors by a clever geometric analysis of the parallactic displacement of shadows of retinal vasculature (the "Purkinje tree"). He also detected "visual purple" or rhodopsin in the rods, which Franz Boll soon thereafter demonstrated was a photosensitive pigment.

Fifty years would pass before interest in the possibility of inner-retinal photoreception would re-emerge. The catalyst was the groundbreaking work of Clyde Keeler, who identified the first animal model of inherited retinal degeneration. This strain of mice – which Keeler termed "*rodless*" – is allelic with the *rd1* strain (formerly *rd*); both carry the *Pde6b^rd1* mutation and phenotypically resemble human autosomal recessive retinitis pigmentosa (OMIM 180072).[8,9] Keeler eventually achieved wide acclaim for this seminal contribution to the genetics of retinal disease, but a key observation he made in these mice received little notice at the time. Though his *rodless* mice were apparently blind when tested behaviorally or electroretinographically, they unexpectedly retained robust pupillary responses to light.[10] This led him to suggest that "in mammals the iris may function independent of vision" either through intrinsic photosensitivity of the smooth muscle itself or by "direct stimulation of the internal nuclear or ganglionic cells" by light.[10,11]

Over the next 70 years, a series of studies in retinally degenerate mammals confirmed the persistence of the pupillary response and suggested that the functional blindness might not be as complete as Keeler had believed. Various photic effects on behavior or physiology were noted, despite the absence of an outer retina in histological material or a detectable electroretinogram (see reviews[12,13]). These included avoidance of a visual cliff, photic suppression of activity in open field test, visually guided avoidance of shock in a shuttlebox, chromatic discrimination, entrainment of circadian rhythms, suppression of pineal melatonin synthesis, and suppression of spontaneous firing in the superior colliculus.[12,14-19] A number of these studies demonstrated that eye removal abolished these residual photoresponses and overtly echoed Keeler's speculation about the possible existence of inner retinal photoreceptors, including ganglion cells.[15,17,18] These suggestions gained only limited currency, probably because improved anatomical studies had begun to cast doubt on the completeness of the loss of outer retinal photoreceptors, especially in the peripheral retina. These studies showed that the mouse retina, once believed to possess only rod photoreceptors, actually had a modest population of cones as well, and that these degenerated far more slowly than rods in the *rd1* model. Perhaps 20 percent survived beyond 80 days of age, albeit without intact outer segments, and a few lasted at least a year.[20] This clouded the interpretation of the behavioral studies, virtually all of which had been conducted in mice young enough to have had many surviving cones. With the benefit of hindsight, however, it seems plausible that at least some of the residual light-driven effects reported in these early studies were indeed mediated by inner retinal photoreceptors. Even in *rd1* animals as young as 6–10 weeks old, surviving rods or cones seem unable to support visual function: silencing ganglion-cell phototransduction in such animals (as described below) abolishes the photic influences on circadian rhythms, the pupil, locomotor activity, and melatonin synthesis that these mice would otherwise exhibit.[21]

In the 1990s, Russell Foster and colleagues took up the work on retinally degenerate mice, applying more rigorous methods to provide compelling evidence for inner retinal photoreception.[13,22–28] Their innovations included appropriate controls for genetic background and the use of very old *rd1* animals and several genetically modified mouse strains in which they confirmed nearly complete loss of cones as well as rods. They also showed that there was no loss of sensitivity of circadian photoreception as rod and cone degeneration progressed and that the residual visually evoked behaviors extended beyond circadian regulation to other outputs of what they called the non-image-forming (NIF) visual centers of the brain. These included pupillary constriction, acute suppression of locomotor activity and acute suppression of melatonin release by light. The case for novel inner photoreceptors was bolstered by parallel observations in human patients with advanced outer retinal disease[29–31] and by assessment of the spectral tuning of residual photoresponses in retinally degenerate animals. Yoshimura & Ebihara showed that the photic influence on circadian rhythms in retinally degenerate mice was most effective at 480 nm, clearly distinct from the optimal wavelength for the known rod and cone photopigments.[32,33] Remarkably similar spectral data were obtained for residual pupillary responses.[26]

Discovery of melanopsin and ganglion-cell photoreceptors

The identities of the mysterious inner-retinal photoreceptors and their photopigment were established in a flurry of studies in the years 2000 to 2005. These developments have been thoroughly reviewed elsewhere,[28,34–37] so we provide only an abbreviated summary here. A pivotal contribution was the discovery of melanopsin.[38,39] This novel opsin (coded by the *opn4* gene) derives its name from the dermal melanophores of frogs, the photosensitive cells of the skin in which the gene was first identified.[38] Ignacio Provencio and colleagues showed that in mice and primates, including humans, the protein was expressed solely in a small minority of cells in the inner retina, mainly in the ganglion cell layer. They speculated that melanopsin could be the photopigment of the postulated inner-retinal photoreceptors and that the neurons expressing it might be the cells of origin of the retinohypothalamic tract. These ideas were at odds with an alternative suggestion, introduced a few years earlier, that the relevant photopigment might not be an opsin but, rather, a cryptochrome, a blue-light-absorbing flavoprotein (see

reviews[35,40]). However, overwhelming evidence for melanopsin's central role soon emerged from studies in rodents and in heterologous expression systems.

Two key studies in this cohort focused on rat retinal ganglion cells shown by retrograde axonal tracing to innervate the suprachiasmatic nucleus (SCN) of the hypothalamus, the brain's circadian pacemaker.[41–44] The first showed that these ganglion cells expressed melanopsin,[45–47] while the second demonstrated that they generated robust electrical responses to light even when pharmacologically or mechanically isolated from other retinal neurons.[48,49] Because this capacity for autonomous phototransduction distinguishes these cells from all other RGCs, they are termed intrinsically photosensitive retinal ganglion cells (ipRGCs) or ganglion-cell photoreceptors. Both mice and primates were also soon shown to possess ipRGCs with structural and functional properties much like those in the rat (see Box 26.1).[46,50–53]

The presence of melanopsin within physiologically identified ipRGCs was first demonstrated directly by Hattar et al.[46,50,53] The opsin was found not only in the cell body, but also in their dendrites[46,54] which, like the soma, are directly photosensitive.[48] Further support for the view that melanopsin was the sensory photopigment in these cells came from the opsin-like action spectrum of the light response in ipRGCs[48,51,53] and from the observation that the intrinsic photosensitivity of these cells was abolished in melanopsin knock-out mice.[51,55] The capacity of melanopsin to function as a sensory photopigment was first

---

Box 26.1 An Unrecognized Drawback of Enucleation or Evisceration?

Ocular enucleation and evisceration are common surgical therapies for managing chronic debilitating ocular pain and improving cosmesis of blind eyes.[157,158] Such surgery is typically indicated only in eyes lacking useful vision. However, the recent discovery of inner retinal photoreceptors should prompt a reassessment of best practices in such cases. When the blindness is attributable to outer retinal diseases such as retinitis pigmentosa or Leber's congenital amaurosis, or to chronic retinal detachment, ganglion-cell photoreceptors may continue to provide functionally useful signals to the brain.[29,30,31,66,159] These signals may not be consciously perceived by the patient and may escape detection by most standard clinical tests of visual function. Such residual functions can nonetheless have significant impact on the quality of life for blind patients, since they may entrain their circadian rhythms.[30] Loss of ipRGCs through enucleation or evisceration would be expected to abolish such synchronization of circadian rhythms, with associated disruption of normal sleep patterns, daytime sleepiness, and other complications that plague many blind patients.[160] It seems likely that photic effects on physiology, including regulation of neuroendocrine physiology, alertness and mood, may also persist in such patients and be sacrificed when the eyes are removed. There is thus a pressing need to develop both a more inclusive concept of "useful vision" in ocularly blind patients and a more comprehensive panel of assessments of photosensory capacity when enucleation or evisceration is being weighed as a treatment option. Although tests of circadian or neuroendocrine effects would in principle allow residual inner retinal photoreceptor function to be evaluated, these would require highly specialized testing. Assessments of residual pupillary responses may be a promising avenue to explore.[133,159,161]

demonstrated biochemically,[56] and subsequently confirmed by electrophysiology and calcium imaging in heterologous expression systems.[57–59] There is remarkably good concordance in the spectral sensitivity in this system, as assessed from absorbance of heterologously expressed or purified melanopsin, from the action spectrum of ipRGCs, and from behaviors mediated by inner retinal photoreceptors; all closely adhere to a retinaldehyde template function with a best wavelength near 480 nm[26,32,33,48,51,53,58–63] (but see refs 56 & 57).

Distinctive functional properties of ipRGCs

Ganglion-cell photoreceptors have physiological properties that are optimized for their roles in NIF vision and contrast markedly with those of the rod and cone photoreceptors that feed the cortical circuits mediating the perception of form, color and motion.

Melanopsin chromophore and pigment bistability

In both vertebrate and invertebrate photopigments, the opsin apoprotein is covalently linked to a retinaldehyde molecule derived from vitamin A that serves as the light-absorbing moiety, or chromophore. In darkness, the retinaldehyde is in the 11-*cis* form. Absorption of a photon converts it into all-*trans* retinaldehyde. This triggers a conformational change in the opsin which in turn activates a G-protein, initiating the transduction cascade. Before the photopigment can undergo another cycle of photoexcitation, the all-*trans* retinaldehyde must be reisomerized to 11-*cis*. For rod and cone photoreceptors, such reisomerization is carried out through multiple enzymatic steps. Several of these occur in cells of the retinal pigment epithelium (RPE), which are essential for the visual cycle in rods and important, if not obligatory, for cones (see Chapter 13 for more information). The visual cycle in invertebrate photoreceptors appears to be radically different, in part because their photopigments are bistable: after photoexcitation, all-*trans* retinaldehyde remains covalently bound to the opsin and is reisomerized to 11-*cis* by subsequent absorption of light in a process known as photoreversal.[64]

The visual cycle of melanopsin in ipRGCs has not been well characterized. The chromophore of melanopsin in situ is unknown, but is doubtless a retinoid, and is most likely to be 11-*cis* retinaldehyde.[56–61,65] Sequence homology indicates that melanopsin resembles invertebrate opsins more than it does vertebrate ones,[38,39] implying that it might be bistable. There is growing evidence in support of this view[57,61,65–68] (but see ref. 69). If melanopsin is indeed bistable, it would be expected to be less reliant than rods or cones on the enzymatic machinery of the RPE for the regeneration of its chromophore, and might be utterly independent. Indeed, ipRGC photoresponses can be recorded for many hours in vitro in retinas isolated from the RPE.[48] Furthermore, neither acute pharmacological disruption of the retinoid cycle nor genetic deletion of the essential isomero-hydrolase of the RPE (RPE65) abolishes the melanopsin-based photoresponse of ipRGCs or their ability to

photoentrain circadian rhythms[66,70] (but see ref. 65). The evidence implies that to a remarkable degree, the melanopsin in ipRGCs can be loaded with its chromophore and can recover from photoactivation independent of retinoid processing by RPE. Melanopsin might acquire its retinoid mainly in the form of all-*trans* retinal which it then photo-isomerizes to a *cis* isomer to form a photoexcitable pigment. This autonomy is important for the intrinsic photosensitivity of ipRGCs, because these cells are located far from the RPE (see Box 26.2).

Spectral tuning

As noted above, melanopsin has peak sensitivity in the blue region at around 480 nm, different from the spectral sensitivities of the rod and cone visual pigments (Fig. 26.1). This wavelength roughly coincides with the peak of solar emission, perhaps the result of evolutionary pressures to optimize the sensitivity of this system to daylight. Though melanopsin is often described as a blue-light-sensitive pigment, it is important to recognize that, like other opsin-based photopigments, its spectral tuning is rather broad. Throughout the spectral range from 340 to 590 nm, sensitivity is within 2 log units of that seen with the optimal wavelength (Fig. 26.1). Though there is evidence that ipRGCs project to the lateral geniculate nucleus (LGN) and could thus contribute to cortical function,[53,71] we are unaware of any evidence that the unique spectral tuning of melanopsin is exploited in the conscious appreciation of color, which is well accounted for by the three cone photopigments and trichromatic theory (see Box 26.3).

The spectral behavior of the ipRGC output signal is much more complex than predicted simply from the action spectrum of the intrinsic light response as measured in dark-adapted ipRGCs. This is in part because melanopsin appears to be a bistable photopigment (see the previous section). The absorption spectrum of the activated form of the pigment, and thus of photoreversal, is shifted to longer wavelengths with respect to that of melanopsin in the dark

Box 26.2 Sparing of ipRGC Function in Conditions Involving the RPE: Retinal Detachment and Leber's Congenital Amaurosis

There are at least two clinical conditions in which disruption of normal interactions between RPE and neural retina compromise rod and cone function but may permit continued phototransduction by ipRGCs. Retinal detachment disrupts the close contact between rod and cone outer segments and the RPE. This compromises the bidirectional exchange of retinoids across the subretinal space and thus the photosensitivity of rods and cones. Leber's congenital amaurosis (LCA), an autosomal recessive, early-onset form of retinitis pigmentosa, is caused in some cases by mutations in the *rpe65* gene. This disrupts the function of RPE65, a retinoid isomerase essential for the regeneration of 11-*cis* chromophore for rod and cone photopigments. Both retinal detachment and LCA have devastating consequences for photosensitivity of rods and cones. Evidence from animal studies (see main text) predicts that photosensitivity of ipRGCs should be substantially preserved under these conditions and should thus provide some useful photic information to the brain.

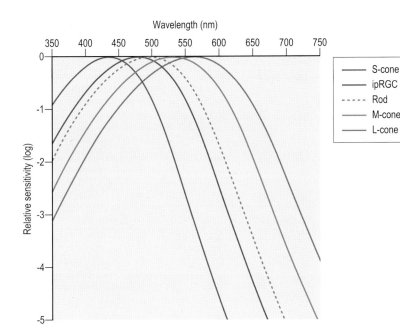

Box 26.3 Spectral Issues in the Design of Intraocular Lens Implants

Designers of intraocular lens implants must grapple with the fact that there are both harmful and salutary effects of short-wavelength visible light. The goal must be to strike the appropriate balance between photoprotection and photoreception. High-pass filters that block blue as well as ultraviolet wavelengths offer better retinal photoprotection than those that block only ultraviolet light, while leaving rod and cone photoresponses (which peak near 500 nm and 555 nm respectively) relatively unaffected. However, because melanopsin is most sensitive to blue light with peak sensitivity at 480 nm, lenses that block blue wavelengths can compromise NIF photoreception.[162] On the other hand, reducing NIF photoreception in this manner might have the benefit of reducing the harmful effects of light exposure at night on the NIF visual system, as discussed in the main text below. These factors need to be considered in the design and selection of lens implants. A recent study showed that for the typical person, long-pass filtering with a sharp filter cutoff near 445 nm may provide the best compromise between photoprotection and light reception.[163]

state (e.g. refs 61 & 67). How this shapes the ipRGC output signal under physiologically relevant lighting conditions is still in dispute,[67,69,72] but it suggests the possibility of a form of spectral opponency under some conditions. A second source of complexity and potential spectral opponency is the functional input to the ipRGCs from classical rod and cone photoreceptors, discussed in detail below. Because the threshold for melanopsin activation is above that for rods and cones, the spectral tuning of ipRGCs can be expected to be intensity dependent. Furthermore, there is evidence in primates that the cone input to ipRGCs is itself spectrally opponent, with activation of short-wavelength cones driving OFF responses in ipRGCs, and activation of mid- and long-wavelength cones driving ON responses.[53]

Invertebrate-like phototransduction cascade

In a photoreceptor cell, the light-absorbing pigment signals through an intracellular biochemical pathway to transduce light into electrical activity across the cell membrane. Interestingly, in ipRGCs this phototransduction cascade is more similar to those in invertebrate photoreceptors than to those in rods and cones. We have already noted that melanopsin has greater sequence homology with invertebrate photopigments than with the rod and cone opsins.[38,39] Upon stimulation by light, melanopsin activates a G-protein belonging to the $G_{q/11}$ family, which in turn signals to the effector enzyme phospholipase C (PLC).[58,59,61,76-77] PLC then somehow causes the opening of a cation channel, which has been proposed to belong to the family of transient receptor potential (TRP) channels.[77-79] While the signaling intermediates coupling PLC to the cation channel remain to be elucidated, the critical components are likely to be within or closely associated with the plasma membrane because patches of cell membrane isolated from ipRGCs continue to exhibit light-evoked electrical responses.[73] Thus, the entire ipRGC phototransduction cascade is very similar to that in *Drosophila* photoreceptors[80] but differs drastically from the rod/cone cascade[81] (see also Chapter 18) (Fig. 26.2). This invertebrate-like phototransduction cascade has prompted speculation that ipRGCs and invertebrate photoreceptors share a common evolutionary origin.[82] It also confers upon ipRGCs properties well suited to NIF vision, as discussed below.

Depolarizing photoresponse with action potentials

The endpoint of the melanopsin phototransduction cascade is the opening of ion channels in the plasma membrane. These behave like cation-selective channels.[48,49] The interior of an ipRGC is negatively charged relative to the extracellular space so the opening of light-gated channels results in net influx of cations and membrane depolarization. In this

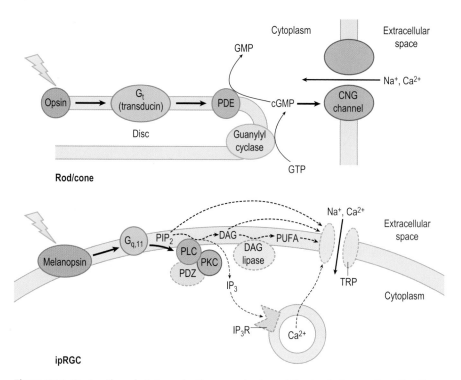

Figure 26.2 Contrasting phototransduction cascades in vertebrate photoreceptors. *Top:* Phototransduction cascade in rod and cone photoreceptors. In darkness, guanylyl cyclase converts GTP into cGMP, which binds to and opens cyclic nucleotide-gated (CNG) channels in the plasma membrane. The resulting continuous influx of cations maintains the cell in a depolarized state. Light absorption excites the opsin molecule within the disk membrane. This activates the G-protein transducin (G_t), which in turn stimulates phosphodiesterase ("PDE"). PDE cleaves cGMP to GMP, closing the CNG channels and hyperpolarizing the cell. *Bottom:* Phototransduction cascade in ipRGCs. Solid lines indicate established components and pathways; dotted lines indicate those that are possible but not proven. When melanopsin is excited by light, it signals through a G-protein of the $G_{q/11}$-family to activate phospholipase C ("PLC"). This ultimately opens a cation channel that probably belongs to the TRP superfamily, but the identity of the channel and the gating mechanism remain unknown. The signaling pathway is closely linked to the membrane and may involve depletion of the PLC substrate phosphatidyl 4,5-bisphosphate (PIP_2), its metabolite diacylglycerol ("DAG"), or polyunsaturated fatty acids ("PUFA") generated from DAG by DAG lipase. An isoform of protein kinase C (PKC) also appears to play an important role and, by analogy with *Drosophila* phototransduction, may be linked to PLC by an INAD-like scaffolding protein that contains PDZ domains. The cytosolic product of PIP_2 hydrolysis, inositol 1,4,5-trisphosphate (IP_3), acts at its cognate receptor (IP_3R) to liberate calcium from intracellular stores. Though such calcium mobilization is not necessary for phototransduction, it appears to play a significant modulatory role.

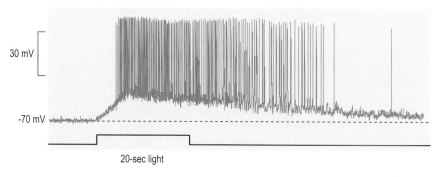

Figure 26.3 Melanopsin-driven light response of an ipRGC. Ganglion-cell photoreceptors can respond to light through not only their melanopsin phototransduction but also rod/cone-driven synaptic input. Here, the light-evoked response of a rat ipRGC was recorded intracellularly in the presence of pharmacological agents that blocked synaptic input, thus isolating the intrinsic photosensitivity of this cell. The dotted black line marks the prestimulus level of the membrane potential. Notice that the light-evoked action potentials begin long after stimulus onset and that they persist throughout and even long after the stimulus (K. Y. Wong, unpublished).

ipRGCs once again resemble invertebrate rhabdomeric receptors, which likewise depolarize when illuminated, and differ from the vertebrate rods and cones, which are hyperpolarized by light. If the depolarizing ipRGC light response is large enough, it brings the membrane potential above the threshold for activating voltage-gated sodium channels, triggering action potentials (Fig. 26.3). This represents yet another divergence from the rod/cone photoreceptors, which do not spike. Action potentials are necessary for ipRGCs, as for other RGCs, to propagate electrical information faithfully along the optic nerve and tract to relatively distant central visual centers. By contrast, rod and cone axons are very short, and passive spread of the electrical signals generated in the outer segment is sufficient for appropriate voltage modulation of the axon terminal.

Sensitivity

A salient feature of ipRGCs is the relative insensitivity of their melanopsin-mediated responses to light. The threshold intensity for the ipRGC intrinsic photoresponse is 1 to 2 orders of magnitude (i.e. 10–100 times) higher than that for cones, and up to 5 log units higher than that for rods.[48,53,83] Such insensitivity is apparently due mainly to the low abundance of melanopsin molecules in ipRGCs, and the resulting low probability of photon absorption.[84] This in turn is related to the fact that ipRGCs lack any structural specialization for increasing the packing of pigment-laden membrane in the cell, such as the disks in the outer segments of rods and cones or the rhabdomeric microvilli of invertebrate photoreceptors. The insensitivity of intrinsic, melanopsin-based responses in ipRGCs means that such responses are triggered mainly by bright sunlight during the day, which is the most effective stimulus for circadian photoentrainment and full pupillary constriction. Because the dynamic range of these responses complements those of the rod and cone photoreceptors, and because all three of these signals converge on the ipRGCs (see below), the ipRGCs and at least some of the NIF visual responses they drive are able to function over the entire intensity range of naturally encountered light stimuli.[53,55]

Kinetics

The temporal characteristics of the melanopsin-based ipRGC photoresponse are drastically different from those of the rod and cone photoreceptors, with remarkably slower onset and termination, and great stability in the face of sustained illumination. Rod and cone photoresponses begin within several milliseconds of light onset and peak within tens of milliseconds. Even the fastest ipRGC intrinsic responses, evoked by very high light intensities ($>10^{14}$ photons/cm^2/sec at 480 nm), are much slower, with the first action potential appearing no earlier than a few hundreds of milliseconds after light onset and peak firing rate achieved only after several seconds. Latencies rise dramatically for less intense flashes, and for near-threshold light stimuli ($\sim10^{11}$ photons/cm^2/sec at 480 nm) can be as long as tens of seconds to the first spike and several minutes before the peak discharge is reached.[48] The basis for such sluggish onset is unknown, but it is of interest that when melanopsin was heterologously introduced into three different cell culture systems, the photoresponses recorded had very slow onset in all three cases.[57–59] Such sluggishness appears unlikely to be an intrinsic property of melanopsin, because single-photon responses in ipRGCs reportedly have relatively brisk onsets.[84] The long latencies, especially near threshold, seem more likely to result from the temporal integration of many low-amplitude, long-lasting, single photon events.

In response to a prolonged light flash, the melanopsin-driven response gradually decays from an early peak to a lower steady-state level. This reflects light adaptation, which prevents saturation of the phototransduction cascade.[85] After light adaptation is complete (which typically requires several minutes), membrane potential and thus spike frequency remain elevated for as long as the light stimulus persists.

Analysis of activity-dependent induction of the immediate early gene *cfos* in putative ipRGCs implies that these photoreceptors can respond continuously to light for at least 19 hours.[86] Thus, the ipRGC intrinsic photoresponse is far more sustained (Fig. 26.3) than the cone response, which adapts much faster and more completely, making the response very transient. The ability of ipRGCs to continuously signal the presence of bright light for many hours is a key feature of the NIF visual system, which features tonic responses to steady diffuse illumination and the ability to integrate photon flux over periods of more than an hour, as explained in further detail below.

The termination of the melanopsin light response is extraordinarily slow. After the cessation of light stimulation, cone and rod responses terminate within hundreds of milliseconds and several seconds, respectively. But termination of the intrinsic photoresponse of ipRGCs takes up to several minutes, especially for very high light intensities.[48,51,53] The reason for this slow recovery is not established, but the similarity of melanopsin to invertebrate rhabdomeric opsins suggests that it results in part from the thermal stability of the light-activated (metarhodopsin) form of the photopigment.[61,68] If so, the slowly decaying post-stimulus voltage response is directly analogous to the persistent depolarizing afterpotential of invertebrate photoreceptors.[64] This is supported by the observation that the poststimulus depolarization can be suppressed by exposure to long-wavelength light, presumably by triggering photoreversal of the pigment.[68] However, light is not required for response termination, because this can occur in complete darkness. The light-independent mechanisms for response termination are unknown. By analogy to other better characterized phototransduction cascades, however, they seem likely to involve phosphorylation of melanopsin followed by arrestin binding.

Morphology, retinal distribution and receptive field

In all mammals studied to date, melanopsin-expressing retinal ganglion cells represent a small minority of all RGCs. These are now recognized as comprising multiple cell types. The best studied type has a very large, sparse dendritic arbor stratifying in outermost inner plexiform layer (IPL), abutting the inner nuclear layer (INL).[46,48,49,53,54,87] In mouse and primate retinas, there is an additional wide-field melanopsin ganglion cell type stratifying in inner IPL[53,54,71,87–89] (Fig. 26.4). In humans, there are roughly equal numbers of these two cell types, and most of the outer-stratifying type have somas displaced to the INL. Bistratified cells have also been described,[49,88,89] although it is not clear if these define a distinct cell type. Recent evidence in the mouse suggests there are several additional types with very low levels of melanopsin expression.[90] Differences among these cell types in functional properties or projections to the brain are just beginning to be explored.[90,91] A very few human cones of the far retinal periphery have been reported to express melanopsin, but their functional significance is unknown.[92]

The morphology and distribution of human ipRGCs have been described in detail.[52,53] Of the ~1.2 million RGCs in humans, only about 3000 (about 0.2 percent) express melanopsin. The somas of these cells are relatively large in

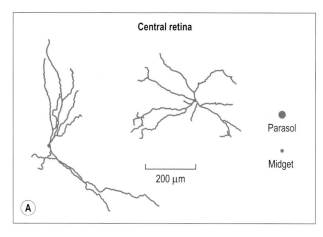

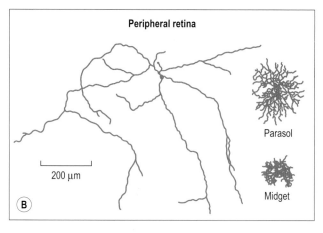

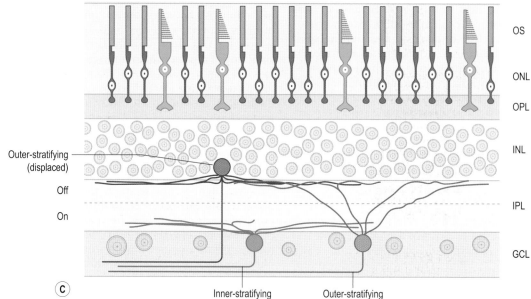

Figure 26.4 Morphology of ipRGCs. *Top:* Morphology of primate ipRGCs as seen in wholemounted retina. The dendritic arbors of these cells cover much less area in the central retina (*left*) than they do in the periphery (*right*), but at all eccentricities they have among the largest fields of any primate ganglion cells. For comparison, the much smaller and more densely branching dendritic fields of midget and parasol cells (two ganglion cell types mediating image-forming vision) are also shown. (Reproduced with permission from Dacey DM, Liao HW, Peterson BB, Robinson FR, Smith VC, Pokorny J, et al. Melanopsin-expressing ganglion cells in primate retina signal colour and irradiance and project to the LGN. Nature. 2005 Feb 17;433(7027):749-54. By permission from Macmillan Publishers Ltd.[53]) *Bottom:* Dendritic stratification of ipRGCs as seen in a schematic cross-section of the retina. There are two main types of ipRGCs, differing in the level at which their dendrites arborize within the inner plexiform layer ("IPL"). Outer-stratifying ipRGCs have dendrites that terminate in the OFF sublamina of the IPL and their cell bodies are located either in the ganglion cell layer ("GCL"; *red cell*) or displaced to the inner nuclear layer ("INL"; *blue cell*). The inner-stratifying ipRGCs (*green cell*), by contrast, have dendrites restricted to the ON sublamina of the IPL and have somas in the GCL. In primates, including humans, inner-stratifying cells and displaced outer-stratifying cells apparently predominate. OS, outer segments; ONL, outer nuclear layer; OPL, outer plexiform layer.

primates, typically 15–25 μm in diameter. Their dendritic fields are the largest of any ganglion-cell type, ranging from 300 μm in the central retina to 1200 μm in the periphery (Fig. 26.4). The dendrites of neighboring cells overlap extensively so that the entire retina, except for the fovea, is covered by a network of melanopsin-containing processes (Fig. 26.5). Because both the cell bodies and the dendrites of these cells are intrinsically photosensitive,[48,53] each ipRGC integrates photons over a region comparable to its dendritic field size. These large receptive fields and low spatial densities make ipRGCs ill-suited for fine spatial discriminations, but ideal for the pronounced spatial integration that characterizes NIF functions. They contrast sharply with the very small receptive fields of rods and cones (<10 μm in diameter), the foundation for the fine-grained retinal

representations at the cortical level that permit high acuity spatial vision.

Resistance to pathological states

Several studies have suggested that ipRGCs are more resistant to certain types of retinal damage than are other ganglion-cell types. For instance, after axotomy, melanopsin-expressing RGCs have a higher survival rate.[93] In addition, melanopsin RGCs are less susceptible to artificially induced chronic ocular hypertension in Sprague-Dawley rats,[94] although their number are certainly reduced.[95] Furthermore, ganglion cells exhibiting cFos induction by light (presumably including ipRGCs) were much less vulnerable to systemically administered monosodium glutamate,[96] suggesting that ipRGCs

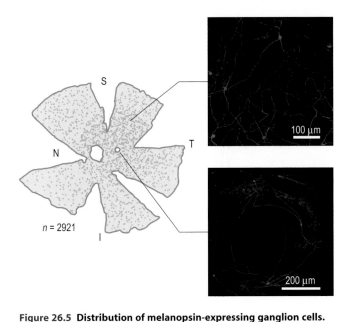

Figure 26.5 Distribution of melanopsin-expressing ganglion cells.
In this experiment, ipRGCs in a macaque retina were revealed by anti-melanopsin immunofluorescence. *Left:* Plot of a flattened whole retina, with each dot representing the cell body of a melanopsin-immunopositive cell. In this retina, a total of 2921 melanopsin-expressing ganglion cells were found. S, superior; I, inferior; N, nasal; T, temporal. *Right:* The network of melanopsin-immunoreactive dendrites in the peripheral retina (*top*) and around the fovea (*bottom*). Notice the absence of melanopsin staining in the fovea. (Reproduced with permission from Dacey DM, Liao HW, Peterson BB, Robinson FR, Smith VC, Pokorny J, et al. Melanopsin-expressing ganglion cells in primate retina signal colour and irradiance and project to the LGN. Nature. 2005 Feb 17;433(7027):749–54. By permission from Macmillan Publishers Ltd.[53])

are relatively resistant to excitotoxicity. The molecular basis for these kinds of resistance is largely unknown, although the PI3 K/Akt signaling pathway appears to be involved.[97] Further investigation into the molecular basis of ipRGC survival after retinal injuries may lead to novel strategies for preventing and/or treating various retinal diseases (see also Box 26.4).

Synaptic input

Unlike conventional RGCs, ipRGCs do not require synaptic input to respond to light because they are directly photosensitive. Nonetheless, both primate and rodent ipRGCs receive synaptic inputs from amacrine and bipolar cells, and these have consequences for the timing, spectral behavior and sensitivity of the light-evoked discharges of these cells. A schematic diagram summarizing the intraretinal synaptic inputs to ipRGCs appears in Figure 26.7.

Bipolar cell input

There is strong structural and physiological evidence for synaptic contacts from bipolar cells onto ipRGC dendrites. The earliest anatomical evidence came from electron-microscopic immunohistochemical data in mice showing ribbon synaptic (bipolar) contacts onto melanopsin-expressing dendrites[98] and this has been supported by subsequent structural observations.[87,88,99] Bipolar-cell input to ipRGCs has also been confirmed electrophysiologically. Rodent ipRGCs are excited by applied glutamate, the transmitter released by

bipolar-cell terminals. Spontaneous glutamate-mediated excitatory postsynaptic currents are detectable in ipRGCs, and these almost certainly derive from bipolar cell synapses.[83,100] Under appropriate recording conditions, white light elicits two excitatory ON response components in ipRGCs. The first has a high threshold, is sluggish in its onset and termination, and is insensitive to synaptic blockers; these properties testify to its origin in intrinsic melanopsin-based phototransduction. The other component occurs much more rapidly after light onset, can be abolished by pharmacological blockade of synaptic transmission, and is attributable to excitatory synaptic input from ON bipolar cells.[53,83,100] Both rods and cones contribute to these extrinsic light responses, making them about five orders of magnitude more sensitive to light than the melanopsin photoresponse[53,83] (Fig. 26.8).

It is surprising that the ON channel provides the dominant bipolar input to all ipRGCs. ON bipolar cell axons are thought to contact ganglion cell dendrites only in the inner

Human embryonic kidney cells

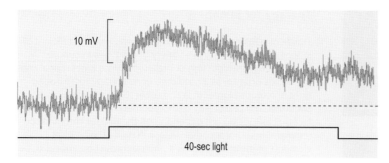

rd/rd mouse retinal ganglion cell

Figure 26.6 Induction of photosensitivity by ectopic melanopsin expression. Artificially expressing the gene for melanopsin in cells that normally lack photosensitivity can render them light-responsive. *Top:* Intracellular recording of the light-evoked depolarization in a human embryonic kidney cell transiently transfected with the mouse melanopsin gene (K. Y. Wong, unpublished). *Bottom:* An intracellular recording of light response in a ganglion cell in a retinal degeneration mouse retina virally transfected with the melanopsin gene. (Reproduced with permission from Lin B, Koizumi A, Tanaka N, Panda S, Masland RH. Restoration of visual function in retinal degeneration mice by ectopic expression of melanopsin. Proc Natl Acad Sci U S A. 2008 Oct 14;105(41):16009–14.[156]) This light sensitivity was induced by the genetic manipulation because this morphological type of ganglion cell is not intrinsically photosensitive in control retinas. In both cases, the artificially induced light responses resemble the intrinsic light response of ipRGCs (compare with Fig. 26.3).

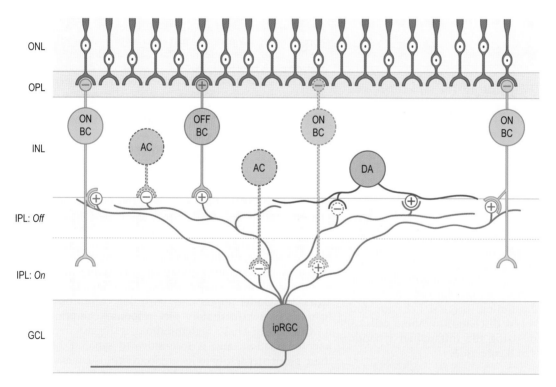

Figure 26.7 Synaptic circuits involving outer-stratifying ipRGCs. In rat and mouse retinas, outer-stratifying ganglion-cell photoreceptors are known to receive direct synaptic inputs from ON bipolar cells ("ON BC"), OFF bipolar cells ("OFF BC"), and amacrine cells ("AC"). ON bipolar cells normally innervate ganglion cells in the proximal, ON sublamina of the IPL, but because the dendrites of outer-stratifying ipRGCs are mainly in the distal, OFF sublamina, ON bipolar cells use ectopic synapses in the OFF sublamina to signal to these ipRGCs (*left and right ON BCs*). However, since the proximal dendrites of these ipRGCs are in the ON sublamina, it is conceivable that some ON bipolar cells synapse onto these proximal dendrites (*middle ON BC*). Weak input from OFF bipolar cells is presumably mediated by synaptic contacts in the OFF sublamina. Amacrine cells provide a strong inhibitory input to outer-stratifying ipRGCs, although it remains to be determined whether this takes place in the OFF sublamina, the ON sublamina, or both. In addition, outer-stratifying ipRGCs exert synaptic output to some dopaminergic amacrine cells ("DA"), whose processes costratify in the OFF sublamina with the ipRGC dendrites. It is possible that dopaminergic amacrine cells signal to outer-stratifying ipRGCs through the inhibitory transmitter GABA and there is evidence that their release of dopamine modulates phototransduction in these ganglion cells.

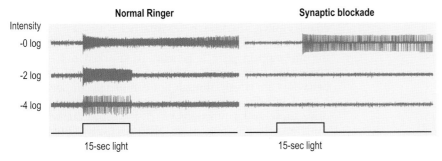

Figure 26.8 A rat ipRGC generating rod/cone-driven and melanopsin-based light responses. Extracellular recordings of spiking in a rat ipRGC evoked by light stimuli at three intensities (dimmest at bottom, brightest at top); the vertical lines in the traces are action potentials. *Left:* Under normal physiological conditions, this cell is capable of generating both rod/cone-driven and melanopsin-based light responses. At the two dimmest light intensities, only rod/cone-mediated responses are induced, and they have fast onset and termination. At the highest intensity, the melanopsin response is also activated, resulting in prolonged poststimulus spiking. *Right:* In the presence of drugs that block synaptic transmission, only the intrinsic melanopsin light response is induced. Notice the slow response onset and prolonged poststimulus persistence, which are similar to those in the intracellular recording shown in Figure 26.3. (K. Y. Wong, unpublished)

half of the IPL, that is, in the ON sublamina.[101,102] While this arrangement is congruent with the ON input to the inner-stratifying variety of ipRGCs, it seems at odds with the observation that the ON channel also dominates the synaptic inputs to the outer-stratifying type, which arborizes largely or exclusively within the OFF sublamina.[48,53,54,87–90] Because many of the outer-stratifying cells have somas in the ganglion cell layer, their dendrites must traverse the ON sublayer en route to the OFF sublayer, so some of the ON channel input may occur in the ON sublayer on proximal dendrites (Fig. 26.7). Indeed, Belenky et al[98] reported that virtually all bipolar cell contacts onto melanopsin dendrites were found in the inner half of the IPL, although they did not indicate whether these dendrites derived from the outer- or inner-stratifying subtype of melanopsin RGCs. Østergaard et al[99] have also suggested that rod bipolar cells (a subtype of ON bipolar cells) make direct contact onto the somas of melanopsin-expressing RGCs in the ganglion cell layer. However, contacts solely on the soma or proximal dendrites would severely restrict the size of the synaptically driven receptive fields of these cells, when in fact they are large and coextensive with the dendritic arbor.[53,83] This suggests that there must be additional ON bipolar input to distal dendrites in the OFF sublayer. This inference is further supported by the observation that displaced ipRGCs, which have somas in the INL and dendrites restricted to the OFF sublamina, nonetheless exhibit synaptically mediated ON responses.[53,103] There is emerging evidence that these paradoxical ON bipolar inputs are made by ectopic ON bipolar cell axon terminals in the OFF sublamina[103] (Fig. 26.7).

Pharmacological blockade of ON bipolar cells and amacrine cells reveals in some outer-stratifying ipRGCs a very small depolarizing ipRGC response at light offset, indicating a relatively weak input from OFF bipolar cells.[83] The anatomical basis for this influence may be the sparse bipolar cell contacts onto melanopsin dendrites in the OFF sublamina reported in the marmoset retina,[87] although these contacts could also be ectopic ON bipolar terminals.

Amacrine cell input

Ganglion-cell photoreceptors also receive substantial synaptic input from amacrine cells. Initial evidence came from an electron microscopic study in mouse documenting amacrine cell synaptic release sites in close apposition to melanopsin-containing dendrites in the IPL.[98] The dominant effect of these inputs is almost certainly inhibitory because virtually all amacrine cells contain one of two major inhibitory transmitters – γ-aminobutyric acid (GABA) or glycine. However, excitatory amacrine cell influences cannot be excluded because a wide variety of neurotransmitters or modulators have been found to be co-expressed with GABA or glycine in amacrine cells, including dopamine, acetylcholine and various neuropeptides, and some of these can exert excitatory effects. To date, however, only inhibitory modulation on ipRGCs has been documented. Studies in rodent retina have demonstrated that exogenously applied or endogenously released GABA or glycine trigger inhibitory chloride currents in ipRGCs and that such currents can be activated by light, primarily at light onset.[83,88,100] In primate ipRGCs, amacrine inputs have not been investigated electrophysiologically but there is anatomical evidence for them in the marmoset, where they contact both outer- and inner-stratifying melanopsin-expressing dendrites and outnumber those from bipolar cells,[87] consistent with electrophysiological evidence in rat ipRGCs that inhibitory synaptic conductances can dominate excitatory ones.[83] However, under normal physiological conditions, the net effect of light-driven synaptic input to ipRGCs is depolarizing (Fig. 26.8). This is presumably because the resting potential of ipRGCs is close to the reversal potential for chloride, so that increased chloride conductance triggered by amacrine-cell synaptic input induces little change in membrane potential.

There are at least two dozen subtypes of amacrine cells in mammalian retinas, differing in dendritic stratification and field size as well as in their neurochemical signature.[104–106] Work to identify which of these amacrine types innervate the ipRGCs has barely begun, but several have already been identified in the mouse.[88] One of these is the dopaminergic amacrine cell,[88,99,107] which has processes stratifying almost exclusively in the most distal sublayer of the IPL,[108] the same stratum in which dendrites of outer-stratifying ipRGCs arborize.[46,48,53,54] Dopamine is a key retinal neuromodulator, and its primary role is to help retinal cells and circuits adapt to different background lighting conditions.[104,109] Thus, ipRGCs can potentially adapt to light not only through

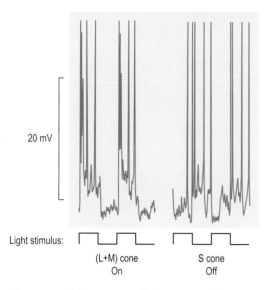

20 mV

Light stimulus:

(L+M) cone
On

S cone
Off

Figure 26.9 Color-opponent light responses in macaque ipRGCs.
Light stimuli selectively modulating L-cone and M-cone influences on
ipRGCs drive depolarizing or "ON" responses to light (*left*), whereas
S-cone-isolating stimuli trigger hyperpolarizing or "OFF" responses (*right*).
(Reproduced with permission from Dacey DM, Liao HW, Peterson BB, Robinson FR,
Smith VC, Pokorny J, et al. Melanopsin-expressing ganglion cells in primate retina signal
colour and irradiance and project to the LGN. Nature. 2005 Feb 17;433(7027):749-54. By
permission from Macmillan Publishers Ltd.[53])

intrinsic mechanisms ("photoreceptor adaptation"[85]), but
also through synaptic inputs ("network adaptation"). Supporting this possibility, dopamine has been shown to regulate the transcription of melanopsin in rat over a time course
of several hours,[110] and preliminary studies have demonstrated additional, more acute effects of dopamine on the
intrinsic photoresponse of ipRGCs (Van Hook, Wong &
Berson, unpublished observations). There is indirect evidence that cholecystokinin-containing amacrine cells might
also be presynaptic to the ipRGCs.[111]

Color coding of synaptic inputs

In primate ipRGCs, the intrinsic melanopsin-based responses
are invariably depolarizing, but the polarity of the extrinsic,
synaptically mediated light response is wavelength-
dependent.[53] Blue light elicits hyperpolarizing extrinsic
responses, whereas longer wavelengths evoke depolarizing
ones (Fig. 26.9). While the depolarizing responses are almost
certainly driven by ON bipolar cells, the circuit responsible
for the hyperpolarizing response to short-wavelength light
remains to be determined. Inhibitory input from amacrine
cells may be involved. The overall spectral behavior of
ipRGCs is unusually complex, because it is shaped not only
by their synaptic inputs but also by the two spectrally distinct states of their melanopsin photopigment which exert
opposing photic effects on the phototransduction cascade
(see above).

Synaptic output and physiological functions

Ganglion-cell photoreceptors send axons to various brain
regions involved in NIF visual functions, such as the pupillary light reflex, circadian photoentrainment, and masking

of circadian rhythms. Melanopsin phototransduction
enables ipRGCs to drive these visual behaviors in the absence
of functional rods and cones (see above). Conversely, when
the melanopsin gene is genetically knocked out in mice,
abolishing the intrinsic photosensitivity of ipRGCs, these
behaviors are largely, though not completely, preserved (see
below), indicating that the rod/cone photoreceptors can also
drive these visual functions.[21,33,55,112,113] However, when the
ganglion-cell photoreceptors are selectively killed either
genetically or pharmacologically, leaving the rest of the
retina including the rods and cones intact, these NIF behavioral responses to light were largely abolished.[114–116] Thus,
ipRGCs are essential for channeling rod/cone-driven light
signals to the various NIF visual brain centers, presumably
by relaying their synaptic inputs from bipolar and amacrine
cells, as described earlier. The ipRGCs also make synaptic
outputs within the retina and to at least some brain regions
involved in image-forming vision. This section surveys both
the intraretinal and extraretinal outputs of these ganglion
cells and their possible functional roles.

Intraretinal output

The first suggestion of a possible intraretinal output of
ganglion-cell photoreceptors emerged almost simultaneously with the definitive proof of their existence. Hankins &
Lucas[117] reported that light alters the latency of the human
cone electroretinogram (ERG) through an intraretinal mechanism driven by a non-classical opsin (Fig. 26.10). While
their study did not identify that opsin, subsequent work
leaves little doubt that it is melanopsin. The spectral tuning,
high threshold, and extensive temporal integration of the
ERG modulation they reported closely parallel those of the
melanopsin-dependent responses of ipRGCs and the behavioral responses they mediate. Related findings in the mouse
support a role for melanopsin in the regulation of the ERG,
although the nature of that regulation appears somewhat
different than in the human eye.[118]

Another early indication of possible intraretinal outputs
of ipRGCs came from the imaging study of Sekaran et al.[50]
In retinas of retinally degenerate mice, they detected light-
evoked calcium signals in many more cells of the ganglion
cell layer than were melanopsin immunoreactive. Similar
findings were reported in a study using photic induction of
the immediate early gene *cfos* to identify retinal neurons
activated directly or indirectly by ipRGCs in a functionally
rodless/coneless mouse.[119] The interpretation of these results
has been complicated somewhat by the recent evidence that
immunohistochemical staining underestimates the number
and diversity of ganglion cells expressing melanopsin.[90]
However, the prevalence of Fos immunopositive cells in the
inner nuclear layer[119] seems greater than can be accounted
for by the rare displaced ipRGCs and seems to suggest that
some amacrine and/or bipolar cells might receive excitatory
input from ipRGCs.

Evidence strongly indicating just such an input has
recently emerged from studies of dopaminergic amacrine
(DA) cells. As mentioned previously, ipRGC dendrites costratify with those of DA cells and seem to receive synaptic
inputs from them. Studies of the light responses of the
dopamine cells suggest that this synaptic input is recipro-
cated.[120] Most DA cells have brisk, relatively transient
ON responses to light. These are abolished by the ON

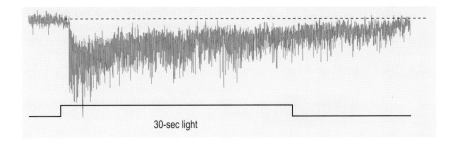

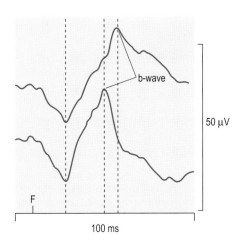

Figure 26.10 Intraretinal synaptic outputs and modulatory influences of ipRGCs. *Top:* Intracellular recording of a dopaminergic amacrine cell made under conditions blocking rod/cone-driven inputs, and revealing excitation from ipRGCs. The signatures of melanopsin photoresponses – sustained excitation, slow onset, and poststimulus persistence – are evident. Unlike the intracellular recordings shown in the previous figures, this is a "voltage clamp" recording, which monitors the ionic currents across the cell membrane. Downward deflection indicates that light induces an inward flow of cations into the dopaminergic amacrine cell. The black dotted line marks the prestimulus baseline level (K. Y. Wong, unpublished). *Bottom:* ipRGC signaling to dopaminergic amacrine cells might modulate the image-forming visual system in human. These electroretinograms were recorded from a human subject at night under dark-adapted conditions. The top recording was obtained at 9:00 p.m., showing a b-wave (which arises from ON bipolar cell light responses) with a relatively long latency to peak. A 15-min bright light pulse delivered at 11:00 p.m. accelerated the b-wave recorded at 3:00 a.m. (*bottom*). A plausible explanation for this phenomenon is that the bright light stimulated ipRGCs and consequently dopaminergic amacrine cells, elevating the release of dopamine which then modulated the kinetics of the ON bipolar cell photoresponse. "F" = flash. (From Hankins MW, Lucas RJ. The primary visual pathway in humans is regulated according to long-term light exposure through the action of a nonclassical photopigment. Curr Biol. 2002 Feb 5;12(3):191–8. Copyright Elsevier 2002.[117])

bipolar cell blocker L-(+)-2-amino-4-phosphonobutyric acid (L-AP4),[121] indicating that they are driven exclusively by ON bipolar cell input. However, a minority of DA cells continue to be excited by light in the presence of L-AP4. These residual responses have all the hallmarks of mediation by melanopsin-expressing ipRGCs: they are sluggish, very sustained (Fig. 26.10), and exhibit the action spectrum of melanopsin-based phototransduction, peaking in sensitivity near 480 nm. The inference that ipRGCs mediate this excitatory influence is further supported by the finding that it can be recorded from about 25 percent of the DA cells in *rd/rd* mice.[120] This centrifugal transmission of ipRGC signals can potentially regulate the intraretinal release of dopamine. In fact, two studies have shown significant light-evoked retinal dopamine release in retinally degenerate rats lacking virtually all rod and cone photoreceptors[122,123] (but see ref. 124). Because retinal dopamine plays a key role in modulation of retinal physiology by light adaptation and circadian phase, the influence of ipRGCs on DA cells may mediate the centrifugal influences of melanopsin on the ERG discussed earlier.

This centrifugal influence is likely to involve dendrodendritic synapses between presynaptic ipRGCs and postsynaptic DA cells in the most distal sublayer of the IPL, where these two cell types costratify (Fig. 26.7). An ionotropic

glutamatergic synapse is probably involved, because the sustained, L-AP4-resistant light responses in DA cells could be abolished by blockade of AMPA/kainate-type receptors.[120,125] This is the first known instance in the mammalian retina of a ganglion cell's dendrites as the presynaptic element at a chemical synapse, though there is a precedent in catfish.[126] Gap junctional contacts constitute an alternative route by which ipRGCs might influence other retinal neurons. Sekaran et al[50] argued for just such communication based on their observation that the gap junctional blocker carbenoxylone abolished the light-evoked calcium signal in some neurons in *rd* mice. This intriguing suggestion needs to be followed up with more definitive tests for the presence of gap junctions, heterotypic tracer coupling and functional electrical synapses between ipRGCs and other retinal neurons. In any case, gap junctions do not seem likely to be the conduit for ipRGC signals to DA cells because these are blocked by glutamate receptor antagonists, as noted above.[120,125]

Central projections

The axonal projections of melanopsin RGCs have been characterized in greatest detail in rodents[45,46,71,90,127–130] (Fig. 26.11), though some data are available in primates.[53] Most of the available information concerns the projections of the

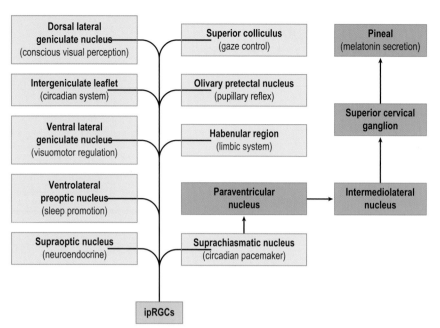

Figure 26.11 Central projections and possible functional roles of ipRGCs. The major brain nuclei receiving axonal innervations from melanopsin RGCs are listed in this diagram, with the monosynaptic projections in light green and the indirect pathway leading to the pineal in dark green. The visual behaviors served by these nuclei (summarized in parentheses) are mainly reflexive, non-image-forming functions.

outer stratifying variety, and characterizing the outputs of other subtypes remains an important near-term goal. Ganglion-cell photoreceptors are believed to influence their postsynaptic targets by releasing both L-glutamate and pituitary adenylate cyclase-activating polypeptide (PACAP).[129,131,132]

The pupillary light reflex

One of the three main targets of melanopsin RGCs is the olivary pretectal nucleus (OPN). This small component of the pretectal region, which lies just rostral to the superior colliculus, is an essential link in the polysynaptic circuit driving the pupillary light reflex (PLR). The direct input from ganglion-cell photoreceptors to the OPN is thought to account for the persistence of the PLR when rod and cone contributions are lost either through inherited degeneration, or by genetic or pharmacological manipulation.[10,11,26,55,133] The properties of this persistent PLR are strikingly concordant with those of the direct light responses of ipRGCs, including a very high threshold, an action spectrum peaking near 480 nm, and a complete loss of the response when the melanopsin gene is deleted.[26,31,55,133] Rods and cones also clearly drive the PLR. Thus, in melanopsin-knockout mice with intact rods and cones, the PLR is normal at low or moderate light intensities that are subthreshold for melanopsin activation; it is only at high intensity ($>10^{11}$ photons/cm^2/sec) that the PLR is incomplete.[55] The outer and inner retinal photoreceptor contributions to the PLR differ in terms of kinetics, as revealed by experiments involving humans and macaques. The time course of the PLR closely mirrors that of the light response in ipRGCs, with rod/cone photoreceptors mediating the fast component and the endogenous phototransduction generating the slower component which outlasts the light stimulus[133,134] (Fig. 26.12). When pharmacological agents were injected into the eye of a macaque to block rod/cone signaling to the inner retina but leaving melanopsin phototransduction intact, only the sluggish component of the ipRGC light response and of the PLR remained.[133] Thus, the classical photoreceptors mediate brisk pupillary responses to dim and moderate intensities,

whereas melanopsin-based phototransduction by ipRGCs drives sluggish pupillary responses to bright light. As noted earlier, ipRGCs receive substantial synaptic inputs from rod and cone circuits, and are thus a likely conduit for the influence of classical photoreceptors on the PLR. Indeed, when most ipRGCs are killed by genetic modification, the rod-component of the PLR is effectively abolished.[114] Available evidence leaves no doubt that the outer-stratifying type of ipRGCs contributes to the pupillomotor circuit, but leaves open the possibility that inner-stratifying types may contribute as well.[130] Further information about the PLR is available in Chapter 25.

Circadian photoentrainment and photic modulation of the pineal

Other than the OPN, the main brain targets of ipRGC axons are the SCN of the hypothalamus and the intergeniculate leaflet (IGL), which is a part of the lateral geniculate nucleus of the thalamus. Both of these are components of the circadian visual system. In all living organisms, many physiological processes (e.g. sleep–wake cycles, body temperature, food intake) display daily rhythms that help the organisms cope with predictable environmental change. In mammals, the central pacemaker for these rhythms is the SCN. Clock cells in the SCN have neuronal activities that exhibit intrinsic rhythms with a period of about 24 hours. These rhythms are known as circadian rhythms (*circa* = approximately; *dies* = day). During the "subjective night," the SCN coordinates hormonal, physiological and behavioral adaptations to the nocturnal environment, and during the "subjective day," it adapts the brain and body to the daytime world. Because the period of the SCN rhythms is not exactly 24 hours, these rhythms must be reset periodically so that the internal clock remains in phase with the external light–dark cycles. In other words, subjective day (or night) needs to be synchronized, or "entrained," with the environmental light cycle (Fig. 26.13). Such entrainment is accomplished mainly by the phase-shifting effects of light on the SCN, mediated by its direct input from the retina. If the SCN clock is too fast,

Pupillary light reflex

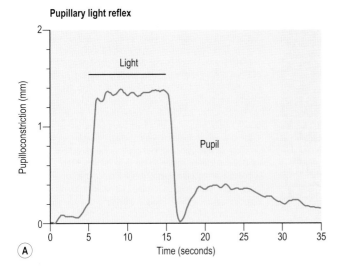

(A)

ipRGC light response

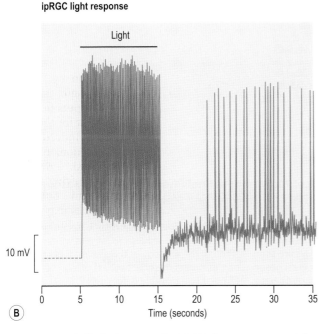

(B)

Figure 26.12 Similarity between the ipRGC photoresponse and the pupillary light reflex in macaque. The time course of the macaque pupillary light reflex closely follows that of the ipRGC light response. In both cases, a fast component that is mediated mainly by rods and cones is detected during the light stimulus, whereas a more sluggish, melanopsin-based component is evident after light off. (From Gamlin PD, McDougal DH, Pokorny J, Smith VC, Yau KW, Dacey DM. Human and macaque pupil responses driven by melanopsin-containing retinal ganglion cells. Vision Res. 2007 Mar;47(7):946–54. Copyright Elsevier 2007.[133]) This temporal separation of rod/cone- and melanopsin-driven components may allow the pupillary light reflex to serve as a diagnostic test to reveal rod/cone versus ganglion cell defects in patients (See Box 26.5).

subjective night starts too early, so the organism is exposed to environmental light during early subjective night, which delays the clock. On the other hand, if the clock is too slow, subjective night starts and ends too late, so the SCN gets photic input during late subjective night, which advances the clock.[41,43,44] The same mechanism is responsible for the recovery from jet lag when traveling across time zones. While both classical and ganglion-cell photoreceptors can mediate photoentrainment, all light signals are channeled through

Box 26.5 The Pupillary Light Reflex as a Diagnostic Test

The PLR has potential utility as a diagnostic tool to differentiate between outer retinal disorders affecting rod or cone circuits and inner retinal diseases affecting ganglion cells.[161] As illustrated in Fig. 26.12, components of the PLR mediated by the rod and cone photoreceptors are temporally separable from those mediated by melanopsin. The earliest component of the light-driven constriction is mediated almost exclusively by rods and/or cones. By contrast, the slowly decaying post-stimulus constriction is driven by melanopsin; this presumably reflects the persistent depolarizing after-potential in ipRGCs that, in turn, is thought to be supported by a slowly depleting pool of photoactivated (meta) melanopsin (see "Melanopsin chromophore and pigment bistability" in the main text above). It should also be possible to tease out specific photoreceptor contributions by exploiting differences among them in threshold and saturating intensities and in spectral sensitivity. Red light may be presented under rod-saturating photopic conditions to selectively evoke a cone-driven PLR, whereas bright blue light is most effective for inducing a melanopsin-mediated response that persists beyond stimulus termination. To elicit a response driven mainly by the rods, a dim green light presented under a dark-adapted condition is ideal.

ipRGCs. These reach the circadian pacemaker through a monosynaptic pathway called the retinohypothalamic tract, and also through an indirect pathway involving the IGL.[135] Because a diurnal organism is typically exposed to daylight for 8–16 hours per day but light can phase-shift a misaligned clock only at the transitions between the light and dark phases, it is crucial that the retinal signal to SCN is sustained throughout those 8–16 hours so that it is still active near the end of the light phase. Therefore, the sustainedness of the melanopsin photoresponse is important for entraining diurnal animals like humans. Indeed, melanopsin-driven light responses recorded from the SCN retain the sustained nature of the ipRGC intrinsic photoresponse.[136,137] Melanopsin is sufficient for photoentrainment not only in rodents, but also in primates including humans.[30,31,138] Rods and cones, on the other hand, generate relatively transient light responses, and are therefore somewhat less effective in mediating entrainment.[21]

The SCN sends several output signals to drive circadian rhythms throughout the organism. One of the primary outputs is a polysynaptic circuit that leads to the pineal gland, which releases the hormone melatonin. Under the influence of the SCN clock, melatonin release is lowest at subjective day and highest at subjective night. This hormone induces a range of physiological responses throughout the body that are suitable to nighttime, such as sleepiness for diurnal animals.[139] In addition to entraining the circadian rhythm of melatonin secretion, light can acutely suppress secretion of this molecule during the subjective night. This may be associated with significant adverse health consequences. For example, there is a high correlation between the amount of light exposure at night and the incidence of breast cancer.[140-142] This may be attributable in part to the photic suppression of nocturnal melatonin, which has been shown to exert various oncostatic and antitoxic effects.[143] Melatonin suppression by light at night may therefore be one risk factor responsible for the higher incidence of breast

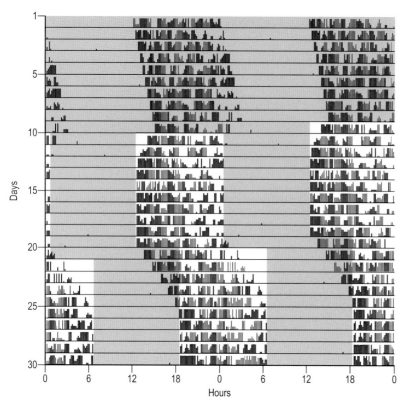

Figure 26.13 Photoentrainment of a diurnal animal.
In this simulated actogram, the locomotor activity of a diurnal animal is plotted as vertical black bars for about a month. This diagram is double-plotted, with each row displaying two consecutive 24-hour periods and each 24-hour period plotted twice. During the first 10 days, the animal is kept in constant darkness, allowing its circadian pacemaker to "free run". Since this animal's circadian clock has a period that is slightly longer than 24 hours, its locomotor activity starts slightly later each day, as reflected in the gradual rightward shift of the period of activity. On days 11 through 20, on the other hand, a 12-hour light:12-hour dark lighting schedule is presented, and the animal's active period becomes synchronized or "entrained" to the light–dark cycle. On day 21, the light–dark cycle is abruptly delayed by 6 hours, mimicking the consequences of a westward flight that crosses 6 time zones. It takes the circadian clock several days to re-entrain the animal to this new cycle.

cancer in industrialized countries. Thus, it is important to minimize light exposure at subjective night, especially for night-shift workers whose subjective night is during the daylight hours. It has been shown that in humans, melatonin release is most effectively suppressed by blue light close to the best wavelength for melanopsin phototransduction.[60,144] Therefore, if exposure to light at subjective night is unavoidable, red light would presumably be less harmful than broad-spectrum light (See also Box 26.6.).

Acute regulation of activity and sleep

Melanopsin RGCs project to the ventrolateral preoptic nucleus (VLPO), a brain region implicated in the regulation of sleep.[71,127] Indeed, besides influencing the sleep–wake cycle through entrainment of the circadian system, melanopsin-mediated photoreception has been shown to acutely modulate sleep and alertness. The polarity of this effect is apparently species-dependent and related to diurnal or nocturnal lifestyle. In humans, these ipRGC outputs increase alertness and vigilance,[31,145] whereas in nocturnal rodents they induce sleep.[146,147]

A related acute effect of light, termed "negative masking," has emerged from studies of circadian activity rhythms in nocturnal rodents.[148] Bright light presented at subjective night rapidly suppresses wheel-running and other locomotor activity that would otherwise occur at this circadian time (Fig. 26.14). Both classical and ganglion-cell photoreceptors contribute to this photic response, but in different ways. In rodless–coneless mice (as in wild-type animals), steady illumination by bright light continuously suppresses activity for at least 3 hours. In contrast, in melanopsin-knockout animals, in which rods and cones are the only functional

Box 26.6 Inner-retinal Photoreception, Seasonal Affective Disorder and Sleep in the Blind

In seasonal affective disorder (SAD), patients suffer recurring bouts of depression, fatigue, and lethargy in fall and winter, when days are short.[167] Though the etiology of the disease is unknown, several observations, including the effectiveness of phototherapy,[168,169] point to a possible defect in NIF photoreception and prolonged periods of nocturnal melatonin release in winter.[170] The efficacy of phototherapy is probably at least partly attributable to its ability to suppress melatonin release. While single-wavelength studies have shown that blue light is more effective than longer wavelengths due to the peak sensitivity of melanopsin at ~480 nm,[60] certain polychromatic stimuli have been reported to be even more potent by strongly activating not only melanopsin but also the rod and cone photopigments.[171] A recent study found that a specific missense variant of the melanopsin gene occurred at higher frequency in SAD patients than in healthy control subjects.[172] The functional consequences of this variation for melanopsin photochemistry or the intrinsic photosensitivity of ipRGCs are unknown.

Another melatonin-related clinical application concerns the treatment of sleep disorders in blind people. While some blind people retain inner retinal photosensitivity and hence NIF visual functions despite the total loss of functional rods and cones,[29–31] others suffer more serious retinal degenerations and do not have any capacity to support NIF vision. One of the major problems experienced by this latter group is their complete lack of photoentrainment, causing their sleep–wake cycle to be uncoupled from the solar cycle. An often effective treatment is the daily administration of melatonin, to mimic the SCN output signal that normally occurs once every 24 hours in properly entrained sighted individuals.[160,173]

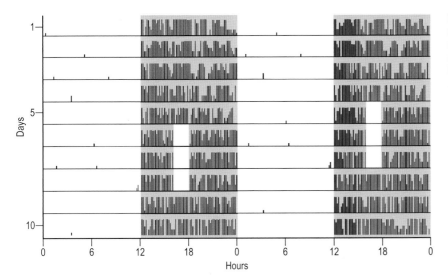

Figure 26.14 Negative masking of a nocturnal animal. This simulated actogram of an entrained nocturnal animal shows that a 2-hour light pulse delivered during the night phase acutely suppresses locomotor activity. This non-image-forming visual response is known as "negative masking".

photoreceptors, the suppression of activity by light is less effective, decays gradually and has largely dissipated by 1.5 hours.[149] This indicates that the sustained melanopsin photoresponse of the ipRGCs is necessary for normal negative masking, especially when triggered by prolonged light pulses. Furthermore, ipRGCs appear to be an essential relay for rod and cone influence over the masking mechanism, because killing these ganglion cells eliminates the acute photic suppression of wheel running activity.[114–116] It remains unclear which brain regions are involved in negative masking. Lesions of image-forming visual centers such as the visual cortex, lateral geniculate nucleus and superior colliculus do not abolish negative masking, and there is conflicting evidence on a possible role for the SCN.[135] Interestingly, negative masking appears to be driven by neuronal pathways distinct from those mediating acute photic regulation of sleep.[146]

Lateral geniculate nucleus and conscious light perception

In primates and rodents, melanopsin RGCs make substantial projections to the dorsal lateral geniculate nucleus (dLGN), the thalamic relay of retinal signals in the image-forming visual pathway.[53,90] In a recent study, a blind woman evidently lacking functional rods and cones reported conscious perception of light, with peak sensitivity near 480 nm, as expected for a melanopsin-based mechanism.[31] Thus, whereas the sluggish light responses and large dendritic fields of ipRGCs make them ill-suited for the analysis of fine details or motion in the retinal image, they may nevertheless contribute to conscious appreciation of ambient illumination, presumably through their projection to the dLGN. As mentioned above, synaptic input to ipRGCs in primates is color-opponent.[53] Thus, it is conceivable that ipRGCs contribute in some way to conscious discrimination of color, although there is presently no evidence for this. There are also weak projections from ipRGCs to the superior colliculus,[71] a midbrain center involved in redirecting gaze to visual targets. Photic effects on collicular activity have been detected in mice with very advanced outer retinal degeneration.[12] Thus, ipRGCs may play some role in spatial vision.

Development

Studies in rodents clearly demonstrate that ipRGCs are the first photoreceptors to become fully functional in the developing retina.[51,150,151] This is likely to be true also in the human eye. Melanopsin is expressed in human fetuses by 9 weeks post-conception,[152] many weeks before the appearance of rod and cone opsins.[153,154] Though outgrowing retinofugal axons reach their brain targets at around the onset of melanopsin expression,[155] it is unlikely that any environmental light can reach the fetal eye at intensities sufficient to drive melanopsin phototransduction. Thus, the functional significance of such early emergence of melanopsin expression is unknown. Neither deletion of melanopsin nor killing of ipRGCs early in development result in any gross alteration in retinal structure.[113–115]

References

1. Polyak SL. The vertebrate visual system; its origin, structure, and function and its manifestations in disease with an analysis of its role in the life of animals and in the origin of man, preceded by a historical review of investigations of the eye, and of the visual pathways and centers of the brain. Chicago: University of Chicago Press, 1957.
2. Finger S. Origins of neuroscience: a history of explorations into brain function. New York: Oxford University Press, 2001.
3. Wade NJ. Visual neuroscience before the neuron. Perception 2004; 33(7):869–889.
4. Treviranus GR. Beiträge zur Aufklärung der Erscheinungen und Gesetze des organischen Lebens. Volume 1, issue 1. Ueber die blättige Textur der Crystallinse des Auges als Grund des Vermögens, einerlei Gegenstand in verschiedener Entfernung deutlich zu sehen, und über den innern Bau der Retina. Bremen: Heyse, 1835.
5. Bidder F. Zur Anatomie der Retina, insbesondere zur Würdigung der stabförmigen Körper in derselben. Archiv für Anatomie, Physiologie und wissenschaftliche Medicin 1839:371–384.
6. Bowman W. Lectures on the Parts concerned in the Operations on the Eye, and on the structure of the retina, delivered at the Royal London Ophthalmic Hospital, Moorfields, June 1847. To which are added, a paper on the vitreous humor; and also a few cases of ophthalmic disease. London: Brown, Green and Longmans, 1849.
7. Helmholtz H. Beschreibung eines Augenspiegels zur Untersuchung der Netz haut im lebenden Auge. Arch Physiol Heilk 1851;11:827–830.
8. Pittler SJ, Keeler CE, Sidman RL, Baehr W. PCR analysis of DNA from 70-year-old sections of rodless retina demonstrates identity with the mouse rd defect. Proc Natl Acad Sci USA 1993; 90(20):9616–9619.
9. Chang B, Hawes NL, Hurd RE, Davisson MT, Nusinowitz S, Heckenlively JR. Retinal degeneration mutants in the mouse. Vision Res 2002; 42(4):517–525.
10. Keeler CE. Iris movements in blind mice. Am J Physiol 1927; 81:107–112.
11. Keeler CE. Blind mice. J Exp Zool 1928; 51(4):495–508.
12. Dräger UC, Hubel DH. Studies of visual function and its decay in mice with hereditary retinal degeneration. J Comp Neurol 1978;180(1):85–114.

13. Foster RG, Provencio I, Hudson D, Fiske S, De Grip W, Menaker M. Circadian photoreception in the retinally degenerate mouse (rd/rd). J Comp Physiol [A] 1991; 169(1):39–50.

14. Hopkins AE. Vision and Retinal Structure in Mice. Proc Natl Acad Sci USA 1927; 13(7):488–492.

15. Karli P. Étude de la valeur fonctionnelle d'une rétine dépourvue de cellules visuelles photo-réceptrices. Arch Sci Physiol (Paris). 1954; 8:305–327.

16. Bovet D, Bovet-Nitti F, Oliverio A. Genetic aspects of learning and memory in mice. Science 1969; 163(863):139–149.

17. Dunn J, Dryer R, Bennett M. Diurnal variation in plasma corticosterone following long term exposure to continuous illumination. Endocrinology 1972; 90(6):1660–1663.

18. Ebihara S, Tsuji K. Entrainment of the circadian activity rhythm to the light cycle: effective light intensity for a Zeitgeber in the retinal degenerate C3H mouse and the normal C57BL mouse. Physiol Behav 1980; 24(3):523–527.

19. Goto M, Ebihara S. The influence of different light intensities on pineal melatonin content in the retinal degenerate C3H mouse and the normal CBA mouse. Neurosci Lett 1990; 108(3):267–272.

20. Carter-Dawson LD, LaVail MM, Sidman RL. Differential effect of the rd mutation on rods and cones in the mouse retina. Invest Ophthalmol Vis Sci 1978; 17(6):489–498.

21. Panda S, Provencio I, Tu DC et al. Melanopsin is required for non-image-forming photic responses in blind mice. Science 2003; 301(5632):525–527.

22. Colwell CS, Foster RG. Photic regulation of Fos-like immunoreactivity in the suprachiasmatic nucleus of the mouse. J Comp Neurol 1992; 324(2):135–142.

23. Provencio I, Wong S, Lederman AB, Argamaso SM, Foster RG. Visual and circadian responses to light in aged retinally degenerate mice. Vision Res 1994; 34(14):1799–1806.

24. Freedman MS, Lucas RJ, Soni B et al. Regulation of mammalian circadian behavior by non-rod, non-cone, ocular photoreceptors. Science 1999; 284(5413):502–504.

25. Lucas RJ, Freedman MS, Munoz M, Garcia-Fernandez JM, Foster RG. Regulation of the mammalian pineal by non-rod, non-cone, ocular photoreceptors. Science 1999; 284(5413):505–507.

26. Lucas RJ, Douglas RH, Foster RG. Characterization of an ocular photopigment capable of driving pupillary constriction in mice. Nat Neurosci 2001; 4(6):621–626.

27. Mrosovsky N, Lucas RJ, Foster RG. Persistence of masking responses to light in mice lacking rods and cones. J Biol Rhythms 2001; 16(6):585–588.

28. Foster RG. Keeping an eye on the time: the Cogan Lecture. Invest Ophthalmol Vis Sci 2002; 43(5):1286–1298.

29. Czeisler CA, Shanahan TL, Klerman EB et al. Suppression of melatonin secretion in some blind patients by exposure to bright light. N Engl J Med 1995; 332(1):6–11.

30. Klerman EB, Shanahan TL, Brotman DJ et al. Photic resetting of the human circadian pacemaker in the absence of conscious vision. J Biol Rhythms 2002; 17(6):548–555.

31. Zaidi FH, Hull JT, Peirson SN et al. Short-wavelength light sensitivity of circadian, pupillary, and visual awareness in humans lacking an outer retina. Curr Biol 2007; 17(24):2122–2128.

32. Yoshimura T, Ebihara S. Spectral sensitivity of photoreceptors mediating phase-shifts of circadian rhythms in retinally degenerate CBA/J (rd/rd) and normal CBA/N (+/+)mice. J Comp Physiol [A] 1996; 178(6):797–802.

33. Hattar S, Lucas RJ, Mrosovsky N et al. Melanopsin and rod-cone photoreceptive systems account for all major accessory visual functions in mice. Nature 2003; 424(6944):76–81.

34. Berson DM. Strange vision: ganglion cells as circadian photoreceptors. Trends Neurosci 2003; 26(6):314–320.

35. Van Gelder RN. Making (a) sense of non-visual ocular photoreception. Trends Neurosci 2003; 26(9):458–461.

36. Fu Y, Liao HW, Do MT, Yau KW. Non-image-forming ocular photoreception in vertebrates. Curr Opin Neurobiol 2005; 15(4):415–422.

37. Provencio I. Melanopsin cells. In: Masland RH, Albright T, eds. The senses: a comprehensive reference - vision. Amsterdam: Elsevier, 2008.

38. Provencio I, Jiang G, De Grip WJ, Hayes WP, Rollag MD. Melanopsin: an opsin in melanophores, brain, and eye. Proc Natl Acad Sci USA 1998; 95(1):340–345.

39. Provencio I, Rodriguez IR, Jiang G, Hayes WP, Moreira EF, Rollag MD. A novel human opsin in the inner retina. J Neurosci 2000; 20(2):600–605.

40. Sancar A. Regulation of the mammalian circadian clock by cryptochrome. J Biol Chem 2004; 279(33):34079–34082.

41. Klein DC, Moore RY, Reppert SM. Suprachiasmatic nucleus: the mind's clock. New York: Oxford University Press, 1991.

42. van Esseveldt KE, Lehman MN, Boer GJ. The suprachiasmatic nucleus and the circadian time-keeping system revisited. Brain Res Brain Res Rev 2000; 33(1):34–77.

43. DeCoursey PJ. Functional organization of circadian systems in multicellular animals. In: Dunlap JC, Loros JL, DeCoursey PJ, eds. Chronobiology. Sunderland, MA: Sinauer Assoc., Inc; 2004:145–178.

44. Hastings MH, Herzog ED. Clock genes, oscillators, and cellular networks in the suprachiasmatic nuclei. J Biol Rhythms 2004; 19(5):400–413.

45. Gooley JJ, Lu J, Chou TC, Scammell TE, Saper CB. Melanopsin in cells of origin of the retinohypothalamic tract. Nat Neurosci 2001; 4(12):1165.

46. Hattar S, Liao HW, Takao M, Berson DM, Yau KW. Melanopsin-containing retinal ganglion cells: architecture, projections, and intrinsic photosensitivity. Science 2002; 295(5557):1065–1070.

47. Hannibal J, Hindersson P, Knudsen SM, Georg B, Fahrenkrug J. The photopigment melanopsin is exclusively present in pituitary adenylate cyclase-activating polypeptide-containing retinal ganglion cells of the retinohypothalamic tract. J Neurosci 2002; 22(1):RC191.

48. Berson DM, Dunn FA, Takao M. Phototransduction by retinal ganglion cells that set the circadian clock. Science 2002; 295(5557):1070–1073.

49. Warren EJ, Allen CN, Brown RL, Robinson DW. Intrinsic light responses of retinal ganglion cells projecting to the circadian system. Eur J Neurosci 2003; 17(9):1727–1735.

50. Sekaran S, Foster RG, Lucas RJ, Hankins MW. Calcium imaging reveals a network of intrinsically light-sensitive inner-retinal neurons. Curr Biol 2003; 13(15):1290–1298.

51. Tu DC, Zhang D, Demas J et al. Physiologic diversity and development of intrinsically photosensitive retinal ganglion cells. Neuron 2005; 48(6):987–999.

52. Hannibal J, Hindersson P, Ostergaard J et al. Melanopsin is expressed in PACAP-containing retinal ganglion cells of the human retinohypothalamic tract. Invest Ophthalmol Vis Sci 2004; 45(11):4202–4209.

53. Dacey DM, Liao HW, Peterson BB et al. Melanopsin-expressing ganglion cells in primate retina signal colour and irradiance and project to the LGN. Nature 2005; 433(7027):749–754.

54. Provencio I, Rollag MD, Castrucci AM. Photoreceptive net in the mammalian retina. This mesh of cells may explain how some blind mice can still tell day from night. Nature 2002; 415(6871):493.

55. Lucas RJ, Hattar S, Takao M, Berson DM, Foster RG, Yau KW. Diminished pupillary light reflex at high irradiances in melanopsin-knockout mice. Science 2003; 299(5604):245–247.

56. Newman LA, Walker MT, Brown RL, Cronin TW, Robinson PR. Melanopsin forms a functional short-wavelength photopigment. Biochemistry 2003; 42(44):12734–12738.

57. Melyan Z, Tarttelin EE, Bellingham J, Lucas RJ, Hankins MW. Addition of human melanopsin renders mammalian cells photoresponsive. Nature 2005; 433(7027):741–745.

58. Panda S, Nayak SK, Campo B, Walker JR, Hogenesch JB, Jegla T. Illumination of the melanopsin signaling pathway. Science 2005; 307(5709):600–604.

59. Qiu X, Kumbalasiri T, Carlson SM et al. Induction of photosensitivity by heterologous expression of melanopsin. Nature 2005; 433(7027):745–749.

60. Brainard GC, Hanifin JP, Greeson JM et al. Action spectrum for melatonin regulation in humans: evidence for a novel circadian photoreceptor. J Neurosci 2001; 21(16):6405–6412.

61. Koyanagi M, Kubokawa K, Tsukamoto H, Shichida Y, Terakita A. Cephalochordate melanopsin: evolutionary linkage between invertebrate visual cells and vertebrate photosensitive retinal ganglion cells. Curr Biol 2005; 15(11):1065–1069.

62. Torii M, Kojima D, Okano T et al. Two isoforms of chicken melanopsins show blue light sensitivity. FEBS Lett 2007; 581(27):5327–5331.

63. Walker MT, Brown RL, Cronin TW, Robinson PR. Photochemistry of retinal chromophore in mouse melanopsin. Proc Natl Acad Sci USA 2008; 105(26):8861–8865.

64. Hillman P, Hochstein S, Minke B. Transduction in invertebrate photoreceptors: role of pigment bistability. Physiol Rev 1983; 63(2):668–772.

65. Fu Y, Zhong H, Wang MH et al. Intrinsically photosensitive retinal ganglion cells detect light with a vitamin A-based photopigment, melanopsin. Proc Natl Acad Sci USA 2005; 102(29):10339–10344.

66. Tu DC, Owens LA, Anderson L et al. Inner retinal photoreception independent of the visual retinoid cycle. Proc Natl Acad Sci USA 2006; 103(27):10426–10431.

67. Mure LS, Rieux C, Hattar S, Cooper HM. Melanopsin-dependent nonvisual responses: evidence for photopigment bistability in vivo. J Biol Rhythms 2007; 22(5): 411–424.

68. Qiu X, Berson DM. Melanopsin bistability in ganglion-cell photoreceptors. Invest Ophthalmol Vis Sci 2007; 48:E-Abstract 612.

69. Mawad K, Van Gelder RN. Absence of long-wavelength photic potentiation of murine intrinsically photosensitive retinal ganglion cell firing in vitro. J Biol Rhythms 2008; 23(5):387–391.

70. Doyle SE, Castrucci AM, McCall M, Provencio I, Menaker M. Nonvisual light responses in the Rpe65 knockout mouse: rod loss restores sensitivity to the melanopsin system. Proc Natl Acad Sci USA 2006; 103(27):10432–10437.

71. Hattar S, Kumar M, Park A et al. Central projections of melanopsin-expressing retinal ganglion cells in the mouse. J Comp Neurol 2006; 497(3):326–349.

72. Rollag MD. Does melanopsin bistability have physiological consequences? J Biol Rhythms 2008; 23(5):396–399.

73. Graham DM, Wong KY, Shapiro P, Frederick C, Pattabiraman K, Berson DM. Melanopsin ganglion cells use a membrane-associated rhabdomeric phototransduction cascade. J Neurophysiol 2008; 99(5):2522–2532.

74. Isoldi MC, Rollag MD, Castrucci AM, Provencio I. Rhabdomeric phototransduction initiated by the vertebrate photopigment melanopsin. Proc Natl Acad Sci USA 2005; 102(4):1217–1221.

75. Contin MA, Verra DM, Guido ME. An invertebrate-like phototransduction cascade mediates light detection in the chicken retinal ganglion cells. FASEB J 2006; 20(14):2648–2650.

76. Peirson SN, Oster H, Jones SL, Leitges M, Hankins MW, Foster RG. Microarray analysis and functional genomics identify novel components of melanopsin signaling. Curr Biol 2007; 17(16):1363–1372.

77. Warren EJ, Allen CN, Brown RL, Robinson DW. The light-activated signaling pathway in SCN-projecting rat retinal ganglion cells. Eur J Neurosci 2006; 23(9):2477–2487.

78. Sekaran S, Lall GS, Ralphs KL et al. 2-Aminoethoxydiphenylborane is an acute inhibitor of directly photosensitive retinal ganglion cell activity in vitro and in vivo. J Neurosci 2007; 27(15):3981–3986.

79. Hartwick AT, Bramley JR, Yu J et al. Light-evoked calcium responses of isolated melanopsin-expressing retinal ganglion cells. J Neurosci 2007; 27(49):13468–13480.

80. Hardie RC, Raghu P. Visual transduction in Drosophila. Nature 2001; 413(6852):186–193.

81. Pepe IM. Recent advances in our understanding of rhodopsin and phototransduction. Prog Retin Eye Res 2001; 20(6):733–759.

82. Arendt D. Evolution of eyes and photoreceptor cell types. Int J Dev Biol 2003; 47(7–8):563–571.

83. Wong KY, Dunn FA, Graham DM, Berson DM. Synaptic influences on rat ganglion-cell photoreceptors. J Physiol 2007; 582(Pt 1):279–296.

84. Do MT, Kang SH, Xue T et al. Photon capture and signalling by melanopsin retinal ganglion cells. Nature 2009; 457(7227):281–287.

85. Wong KY, Dunn FA, Berson DM. Photoreceptor adaptation in intrinsically photosensitive retinal ganglion cells. Neuron 2005; 48(6):1001–1010.

86. Hannibal J, Vrang N, Card JP, Fahrenkrug J. Light-dependent induction of cFos during subjective day and night in PACAP-containing ganglion cells of the retinohypothalamic tract. J Biol Rhythms 2001; 16(5):457–470.

87. Jusuf PR, Lee SC, Hannibal J, Grunert U. Characterization and synaptic connectivity of melanopsin-containing ganglion cells in the primate retina. Eur J Neurosci. 2007; 26(10):2906–2921.

88. Viney TJ, Balint K, Hillier D et al. Local retinal circuits of melanopsin-containing ganglion cells identified by transsynaptic viral tracing. Curr Biol 2007; 17(11): 981–988.
89. Schmidt TM, Taniguchi K, Kofuji P. Intrinsic and extrinsic light responses in melanopsin-expressing ganglion cells during mouse development. J Neurophysiol 2008; 100(1):371–384.
90. Ecker JL, Dumitrescu ON, Wong KY, et al. Melanopsin-expressing retinal ganglion-cell photoreceptors: cellular diversity and role in pattern vision. Neuron 2010; 67(1): 49–60.
91. Schmidt TM, Kofuji P. Functional and morphological differences among intrinsically photosensitive retinal ganglion cells. J Neurosci 2009; 29(2):476–482.
92. Dkhissi-Benyahya O, Rieux C, Hut RA, Cooper HM. Immunohistochemical evidence of a melanopsin cone in human retina. Invest Ophthalmol Vis Sci 2006; 47(4):1636–1641.
93. Robinson GA, Madison RD. Axotomized mouse retinal ganglion cells containing melanopsin show enhanced survival, but not enhanced axon regrowth into a peripheral nerve graft. Vision Res 2004; 44(23):2667–2674.
94. Li RS, Chen BY, Tay DK, Chan HH, Pu ML, So KF. Melanopsin-expressing retinal ganglion cells are more injury-resistant in a chronic ocular hypertension model. Invest Ophthalmol Vis Sci 2006; 47(7):2951–2958.
95. Wang HZ, Lu QJ, Wang NL, Liu H, Zhang L, Zhan GL. Loss of melanopsin-containing retinal ganglion cells in a rat glaucoma model. Chin Med J (Engl) 2008; 121(11):1015–1019.
96. Chambille I. Retinal ganglion cells expressing the FOS protein after light stimulation in the Syrian hamster are relatively insensitive to neonatal treatment with monosodium glutamate. J Comp Neurol 1998; 392(4):458–467.
97. Li SY, Yau SY, Chen BY et al. Enhanced survival of melanopsin-expressing retinal ganglion cells after injury is associated with the PI3 K/Akt pathway. Cell Mol Neurobiol 2008; 28(8):1095–1107.
98. Belenky MA, Smeraski CA, Provencio I, Sollars PJ, Pickard GE. Melanopsin retinal ganglion cells receive bipolar and amacrine cell synapses. J Comp Neurol 2003; 460(3):380–393.
99. Østergaard J, Hannibal J, Fahrenkrug J. Synaptic contact between melanopsin-containing retinal ganglion cells and rod bipolar cells. Invest Ophthalmol Vis Sci 2007; 48(8):3812–3820.
100. Perez-Leon JA, Warren EJ, Allen CN, Robinson DW, Lane Brown R. Synaptic inputs to retinal ganglion cells that set the circadian clock. Eur J Neurosci 2006; 24(4): 1117–1123.
101. Famiglietti EV, Jr., Kolb H. Structural basis for ON-and OFF-center responses in retinal ganglion cells. Science 1976; 194(4261):193–195.
102. Nelson R, Famiglietti EV, Jr., Kolb H. Intracellular staining reveals different levels of stratification for on- and off-center ganglion cells in cat retina. J Neurophysiol 1978; 41(2):472–483.
103. Dumitrescu ON, Pucci FG, Wong KY, Berson DM. Ectopic retinal ON bipolar cell synapses in the OFF inner plexiform layer: contacts with dopaminergic amacrine cells and melanopsin ganglion cells. J Comp Neurol 2009; 517(2):226–244.
104. Dowling JE. The retina: an approachable part of the brain. Cambridge, MA: The Belknap Press of Harvard University Press, 1987.
105. Lin B, Masland RH. Populations of wide-field amacrine cells in the mouse retina. J Comp Neurol 2006; 499(5):797–809.
106. MacNeil MA, Masland RH. Extreme diversity among amacrine cells: implications for function. Neuron 1998; 20(5):971–982.
107. Vugler AA, Redgrave P, Semo M, Lawrence J, Greenwood J, Coffey PJ. Dopamine neurones form a discrete plexus with melanopsin cells in normal and degenerating retina. Exp Neurol 2007; 205(1):26–35.
108. Zhang DQ, Stone JF, Zhou T, Ohta H, McMahon DG. Characterization of genetically labeled catecholamine neurons in the mouse retina. Neuroreport 2004; 15(11):1761–1765.
109. Witkovsky P. Dopamine and retinal function. Doc Ophthalmol 2004; 108(1): 17–40.
110. Sakamoto K, Liu C, Kasamatsu M, Pozdeyev NV, Iuvone PM, Tosini G. Dopamine regulates melanopsin mRNA expression in intrinsically photosensitive retinal ganglion cells. Eur J Neurosci 2005; 22(12):3129–3136.
111. Shimazoe T, Morita M, Ogiwara S et al. Cholecystokinin-A receptors regulate photic input pathways to the circadian clock. Faseb J 2008; 22(5):1479–1490.
112. Ruby NF, Brennan TJ, Xie X et al. Role of melanopsin in circadian responses to light. Science 2002; 298(5601):2211–2213.
113. Panda S, Sato TK, Castrucci AM, Rollag MD, DeGrip WJ, Hogenesch JB, et al. Melanopsin (Opn4) requirement for normal light-induced circadian phase shifting. Science 2002; 298(5601):2213–2216.
114. Güler AD, Ecker JL, Lall GS et al. Melanopsin cells are the principal conduits for rod-cone input to non-image-forming vision. Nature 2008; 453(7191):102–105.
115. Hatori M, Le H, Vollmers C et al. Inducible ablation of melanopsin-expressing retinal ganglion cells reveals their central role in non-image forming visual responses. PLoS ONE 2008; 3(6):e2451.
116. Göz D, Studholme K, Lappi DA, Rollag MD, Provencio I, Morin LP. Targeted destruction of photosensitive retinal ganglion cells with a saporin conjugate alters the effects of light on mouse circadian rhythms. PLoS ONE 2008; 3(9):e3153.
117. Hankins MW, Lucas RJ. The primary visual pathway in humans is regulated according to long-term light exposure through the action of a nonclassical photopigment. Curr Biol 2002; 12(3):191–198.
118. Barnard AR, Hattar S, Hankins MW, Lucas RJ. Melanopsin regulates visual processing in the mouse retina. Curr Biol 2006; 16(4):389–395.
119. Barnard AR, Appleford JM, Sekaran S et al. Residual photosensitivity in mice lacking both rod opsin and cone photoreceptor cyclic nucleotide gated channel 3 alpha subunit. Vis Neurosci 2004; 21(5):675–683.
120. Zhang DQ, Wong KY, Sollars PJ, Berson DM, Pickard GE, McMahon DG. Intraretinal signaling by ganglion cell photoreceptors to dopaminergic amacrine neurons. Proc Natl Acad Sci USA 2008; 105(37):14181–14186.
121. Slaughter MM, Miller RF. 2-amino-4-phosphonobutyric acid: a new pharmacological tool for retina research. Science 1981; 211(4478):182–185.
122. Morgan WW, Kamp CW. Dopaminergic amacrine neurons of rat retinas with photoreceptor degeneration continue to respond to light. Life Sci 1980; 26(19):1619–1626.
123. Vugler AA, Redgrave P, Hewson-Stoate NJ, Greenwood J, Coffey PJ. Constant illumination causes spatially discrete dopamine depletion in the normal and degenerate retina. J Chem Neuroanat 2007; 33(1):9–22.
124. Boelen MK, Boelen MG, Marshak DW. Light-stimulated release of dopamine from the primate retina is blocked by 1-2-amino-4-phosphonobutyric acid (APB). Vis Neurosci 1998; 15(1):97–103.
125. Zhang DQ, Zhou TR, McMahon DG. Functional heterogeneity of retinal dopaminergic neurons underlying their multiple roles in vision. J Neurosci 2007; 27(3):692–699.
126. Sakai HM, Naka K, Dowling JE. Ganglion cell dendrites are presynaptic in catfish retina. Nature 1986; 319(6053):495–497.
127. Gooley JJ, Lu J, Fischer D, Saper CB. A broad role for melanopsin in nonvisual photoreception. J Neurosci 2003; 23(18):7093–7106.
128. Morin LP, Blanchard JH, Provencio I. Retinal ganglion cell projections to the hamster suprachiasmatic nucleus, intergeniculate leaflet, and visual midbrain: bifurcation and melanopsin immunoreactivity. J Comp Neurol 2003; 465(3):401–416.
129. Hannibal J, Fahrenkrug J. Target areas innervated by PACAP-immunoreactive retinal ganglion cells. Cell Tissue Res 2004; 316(1):99–113.
130. Baver SB, Pickard GE, Sollars PJ. Two types of melanopsin retinal ganglion cell differentially innervate the hypothalamic suprachiasmatic nucleus and the olivary pretectal nucleus. Eur J Neurosci 2008; 27(7):1763–1770.
131. Hannibal J, Moller M, Ottersen OP, Fahrenkrug J. PACAP and glutamate are co-stored in the retinohypothalamic tract. J Comp Neurol 2000; 418(2):147–155.
132. Hannibal J. Neurotransmitters of the retino-hypothalamic tract. Cell Tissue Res 2002; 309(1):73–88.
133. Gamlin PD, McDougal DH, Pokorny J, Smith VC, Yau KW, Dacey DM. Human and macaque pupil responses driven by melanopsin-containing retinal ganglion cells. Vision Res 2007; 47(7):946–954.
134. Zhu Y, Tu DC, Denner D, Shane T, Fitzgerald CM, Van Gelder RN. Melanopsin-dependent persistence and photopotentiation of murine pupillary light responses. Invest Ophthalmol Vis Sci 2007; 48(3):1268–1275.
135. Morin LP, Allen CN. The circadian visual system, 2005. Brain Res Rev 2006; 51(1):1–60.
136. Drouyer E, Rieux C, Hut RA, Cooper HM. Responses of suprachiasmatic nucleus neurons to light and dark adaptation: relative contributions of melanopsin and rod-cone inputs. J Neurosci 2007; 27(36):9623–9631.
137. Wong KY, Graham DM, Berson DM. The retina-attached SCN slice preparation: an in vitro mammalian circadian visual system. J Biol Rhythms 2007; 22(5):400–410.
138. Silva MM, Albuquerque AM, Araujo JF. Light-dark cycle synchronization of circadian rhythm in blind primates. J Circadian Rhythms 2005; 3:10.
139. Pandi-Perumal SR, Srinivasan V, Spence DW, Cardinali DP. Role of the melatonin system in the control of sleep: therapeutic implications. CNS Drugs 2007; 21(12):995–1018.
140. Jasser SA, Blask DE, Brainard GC. Light during darkness and cancer: relationships in circadian photoreception and tumor biology. Cancer Causes Control 2006; 17(4):515–523.
141. Stevens RG. Artificial lighting in the industrialized world: circadian disruption and breast cancer. Cancer Causes Control 2006; 17(4):501–507.
142. Reiter RJ, Tan DX, Korkmaz A et al. Light at night, chronodisruption, melatonin suppression, and cancer risk: a review. Crit Rev Oncol 2007; 13(4):303–328.
143. Hoang BX, Shaw DG, Pham PT, Levine SA. Neurobiological effects of melatonin as related to cancer. Eur J Cancer Prev 2007; 16(6):511–516.
144. Thapan K, Arendt J, Skene DJ. An action spectrum for melatonin suppression: evidence for a novel non-rod, non-cone photoreceptor system in humans. J Physiol 2001; 535(Pt 1):261–267.
145. Lockley SW, Evans EE, Scheer FA, Brainard GC, Czeisler CA, Aeschbach D. Short-wavelength sensitivity for the direct effects of light on alertness, vigilance, and the waking electroencephalogram in humans. Sleep 2006; 29(2):161–168.
146. Lupi D, Oster H, Thompson S, Foster RG. The acute light-induction of sleep is mediated by OPN4-based photoreception. Nat Neurosci 2008; 11(9):1068–1073.
147. Altimus CM, Guler AD, Villa KL, McNeill DS, Legates TA, Hattar S. Rods-cones and melanopsin detect light and dark to modulate sleep independent of image formation. Proc Natl Acad Sci USA 2008; 105(50):19998–20003.
148. Mrosovsky N. Masking: history, definitions, and measurement. Chronobiol Int 1999; 16(4):415–429.
149. Mrosovsky N, Hattar S. Impaired masking responses to light in melanopsin-knockout mice. Chronobiol Int 2003; 20(6):989–999.
150. Hannibal J, Fahrenkrug J. Melanopsin containing retinal ganglion cells are light responsive from birth. Neuroreport 2004; 15(15):2317–2320.
151. Sekaran S, Lupi D, Jones SL et al. Melanopsin-dependent photoreception provides earliest light detection in the mammalian retina. Curr Biol 2005; 15(12):1099–1107.
152. Tarttelin EE, Bellingham J, Bibb LC et al. Expression of opsin genes early in ocular development of humans and mice. Exp Eye Res 2003; 76(3):393–396.
153. Bibb LC, Holt JK, Tarttelin EE et al. Temporal and spatial expression patterns of the CRX transcription factor and its downstream targets. Critical differences during human and mouse eye development. Hum Mol Genet 2001; 10(15):1571–1579.
154. Hendrickson A, Bumsted-O'Brien K, Natoli R, Ramamurthy V, Possin D, Provis J. Rod photoreceptor differentiation in fetal and infant human retina. Exp Eye Res 2008; 87(5):415–426.
155. Finlay BL. The developing and evolving retina: using time to organize form. Brain Res 2008; 1192:5–16.
156. Lin B, Koizumi A, Tanaka N, Panda S, Masland RH. Restoration of visual function in retinal degeneration mice by ectopic expression of melanopsin. Proc Natl Acad Sci USA 2008; 105(41):16009–16014.
157. Shah-Desai SD, Tyers AG, Manners RM. Painful blind eye: efficacy of enucleation and evisceration in resolving ocular pain. Br J Ophthalmol 2000; 84(4):437–438.
158. Merbs SL. Management of a blind painful eye. Ophthalmol Clin North Am 2006; 19(2):287–292.
159. Wilhelm H. The pupil. Curr Opin Neurol 2008; 21(1):36–42.

160. Skene DJ, Arendt J. Circadian rhythm sleep disorders in the blind and their treatment with melatonin. Sleep Med 2007; 8(6):651–655.

161. Kawasaki A, Kardon RH. Intrinsically photosensitive retinal ganglion cells. J Neuroophthalmol 2007; 27(3):195–204.

162. Mainster MA. Violet and blue light blocking intraocular lenses: photoprotection versus photoreception. Br J Ophthalmol 2006; 90(6):784–792.

163. van de Kraats J, van Norren D. Sharp cutoff filters in intraocular lenses optimize the balance between light reception and light protection. J Cataract Refract Surg 2007; 33(5):879–887.

164. Zhang Y, Meister M, Pawlyk BS, Bulgakov OV, Li T, Sandberg MA. Responses of intrinsically-photosensitive retinal ganglion cells after melanopsin-gene transfection. Invest Ophthalmol Vis Sci 2007; 48:E-Abstract 4604.

165. Bi A, Cui J, Ma YP, Olshevskaya E et al. Ectopic expression of a microbial-type rhodopsin restores visual responses in mice with photoreceptor degeneration. Neuron 2006; 50(1):23–33.

166. Lagali PS, Balya D, Awatramani GB et al. Light-activated channels targeted to ON bipolar cells restore visual function in retinal degeneration. Nat Neurosci 2008; 11(6):667–675.

167. Rosenthal NE, Sack DA, Gillin JC et al. Seasonal affective disorder. A description of the syndrome and preliminary findings with light therapy. Arch Gen Psychiat 1984; 41(1):72–80.

168. Macchi MM, Bruce JN. Human pineal physiology and functional significance of melatonin. Front Neuroendocrinol 2004; 25(3–4):177–195.

169. Golden RN, Gaynes BN, Ekstrom RD et al. The efficacy of light therapy in the treatment of mood disorders: a review and meta-analysis of the evidence. Am J Psychiatry 2005; 162(4):656–662.

170. Wehr TA, Duncan WC Jr, Sher L et al. A circadian signal of change of season in patients with seasonal affective disorder. Arch Gen Psychiat 2001; 58(12):1108–1114.

171. Revell VL, Skene DJ. Light-induced melatonin suppression in humans with polychromatic and monochromatic light. Chronobiol Int 2007; 24(6): 1125–1137.

172. Roecklein KA, Rohan KJ, Duncan WC et al. A missense variant (P10L) of the melanopsin (OPN4) gene in seasonal affective disorder. J Affect Disord 2009; 114(1-3):279–285.

173. Sack RL, Brandes RW, Kendall AR, Lewy AJ. Entrainment of free-running circadian rhythms by melatonin in blind people. N Engl J Med 2000; 343(15):1070–1077.

Overview of the Central Visual Pathways

Joanne A. Matsubara & Jamie D. Boyd

The role of the central visual pathways is to process and integrate visual information that travels to the brain by means of the optic nerves. Although the eye is responsible for transducing patterns of light energy into neuronal signals, it is the brain that is ultimately responsible for visual perception and cognition. Vision at the level of the central nervous system is perhaps the best understood of all the sensory systems, partly because there is a wealth of information on its neuroanatomic and functional organization. In this chapter, an overview of the central visual pathways is given, beginning with a brief description of the retinal projections and their putative functional roles and ending with an overview of the consequences of lesions along the visual pathway. Schematic views of the central nuclei that receive a retinal projection are shown in Figures 27.1 and 27.2. The most significant, in terms of the number of optic nerve fibers, is the projection to the lateral geniculate nucleus (LGN) in the thalamus. The LGN, described in detail in Chapter 29, is an important relay in the visual pathway involved with form vision. The precision and specificity of the retinogeniculate connections are crucial to form vision. From the LGN, the visual pathway proceeds to the primary visual cortex, V1. It is at the level of V1 that visual signals undergo complex processing, which is detailed in Chapter 30. Visual processing of a combination of stimulus features such as form, color, contrast, motion, texture, depth and is further enhanced by over 30 extrastriate cortical areas, some of which are described in Chapter 31. Finally, we discuss in detail the role of experience and visual deprivation on the development of the central visual pathways in Section II.

Targets of the retinal projections

The major targets of the retinal ganglion cell axons are the LGN, superior colliculus (SC), pretectum, and pulvinar. There is a weaker projection to several small hypothalamic nuclei (including suprachiasmatic, supraoptic, paraventricular nuclei) and the accessory optic system (AOS) (including the nucleus of the optic tract (NOT) and the dorsal, medial, and terminal nuclei) (Box 27.1, Fig. 27.2 and see Fig. 9.8).

The LGN of the thalamus is the major termination site of the retinal ganglion cells, and plays an important role in the visual pathway leading to the primary visual cortex.

Approximately 90 percent of all retinal ganglion cells project to the LGN, which is laminated and displays retinotopic organization. Each LGN layer receives input from a specific eye and class of ganglion cell. While electrophysiological studies have determined that the neuronal signals coming into and leaving the LGN are not appreciably different, the processing that takes place in the LGN appears to be involved in regulating the flow of information to the primary visual cortex (V1), the major projection target of the LGN (see Chapter 29).

The SC is a midbrain structure which, in conjunction with the cortical frontal eye fields and the brainstem reticular formation, is involved in the generation of visually guided saccadic eye movements[1,2] (see Chapter 9). The SC is a laminated, retinotopically organized nucleus, and as seen in the LGN, the retinal projection retains eye segregation with alternating columns of left and right eye terminals forming a banded pattern throughout the superficial layers. Approximately 10 percent of all retinal ganglion cells project to the SC. The majority of retinal axons that terminate in the SC are small caliber, originate from ganglion cells with small dendritic fields, and do not project to other retinal targets.[3]

The pretectal complex, a group of small midbrain nuclei, is just rostral to the SC. It receives signals from a group of small-diameter retinal ganglion cells with large receptive fields and is involved with the control of the pupillary light reflex (see Chapter 25) by means of a projection to the Edinger–Westphal nucleus of the oculomotor complex. The pupillary light reflex demonstrates a consensual response primarily resulting from crossed and uncrossed optic nerve fibers that enter each pretectal complex which in turn sends a bilateral projection to the Edinger–Westphal nucleus (see Chapter 9).

Retinal ganglion cells also project to three of four major subdivisions of the pulvinar nucleus of the thalamus.[4,5] The pulvinar is the largest nuclear mass in the primate thalamus and receives a projection from the small-caliber fibers from the optic nerve and the SC. It projects to several visual cortical areas, including V1, and extrastriate, parietal areas. Thus, the pulvinar represents a pathway that can bypass the LGN to get to V1 and may play a role in processing form vision. More recent studies point to a role for the pulvinar in the coding of the "importance" of visual stimuli (i.e. visual salience or "attention").[6] It has been shown that the

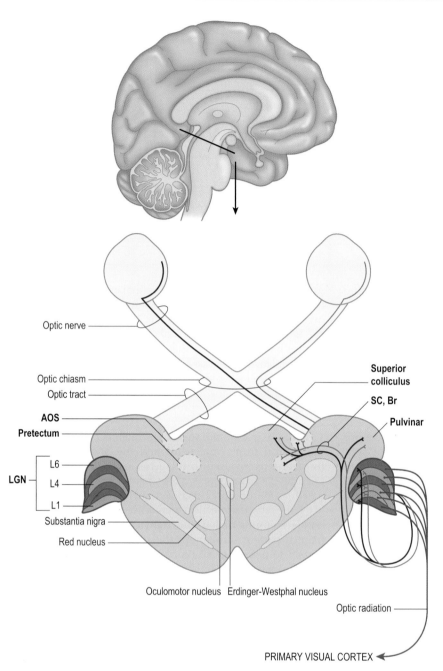

Figure 27.1 Schematic drawing of a section through the human brain at the level of the midbrain-diencephalon transition. Several of the target sites of the retinal projection are present at this level, and labeled in bold on the right: the superior colliculus, pretectum, accessory optic system (AOS), pulvinar and lateral geniculate nucleus (LGN). Optic nerve fibers originating from ganglion cells of the contralateral nasal retina cross at the optic chiasm and join the uncrossed optic nerve fibers from the ipsilateral temporal retina to form the optic tract. The layers of the LGN, the major termination site of the optic nerve fibers, receive contralateral eye (layers 1,4,6; blue) or ipsilateral eye (layers 2,3,5; red) inputs. Cells in all six layers of the LGN project to the primary visual cortex via the optic radiation. SC, Br, Brachium of the superior colliculus.

Optic nerve

Optic chiasm

Optic tract

Superior colliculus

SC, Br

AOS

Pretectum

Pulvinar

LGN — L6, L4, L1

Substantia nigra

Red nucleus

Oculomotor nucleus   Erdinger-Westphal nucleus

Optic radiation

PRIMARY VISUAL CORTEX

Box 27.1 Central targets of the optic nerves

- LGN
- Superior colliculus
- Pretectum
- Pulvinar
- Hypothalamic nuclei
- AOS

pulvinar integrates neural signals associated with eye and hand and arm movements and may receive signals associated with saccadic eye movements, which suggests its role is also one of formulating reference frames for hand-eye coordination.[7–9]

Several small hypothalamic nuclei receive a direct retinal projection. The suprachiasmatic nucleus receives a sparse projection from fibers that leave the dorsal surface of the optic chiasma and has been implicated in the synchronization of circadian rhythms.[10] The paraventricular and supraoptic nuclei are likely also involved with the regulation of the light–dark cycle for neuroendocrine functions.

The AOS consists of several small nuclei, the lateral terminal nucleus (LTN), the medial terminal nucleus (MTN), and the dorsal terminal nucleus (DTN), as well as the NOT in the midbrain.[11] The AOS plays an important role in optokinetic nystagmus (OKN) in which slow compensatory and pursuit-type eye movements alternate with fast saccadic-type eye movements in response to viewing prolonged large field motion (see Chapter 9). In primates, lesions of the NOT and

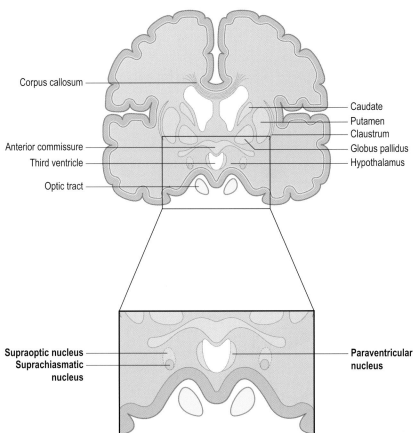

Corpus callosum

Anterior commissure
Third ventricle

Optic tract

Caudate
Putamen
Claustrum
Globus pallidus
Hypothalamus

Supraoptic nucleus
Suprachiasmatic
nucleus

Paraventricular
nucleus

Figure 27.2 Schematic drawing of a section through the human brain at the level of the anterior commissure. Retinal target in the hypothalamic area are labeled in bold (bottom enlargement). A small number of fibers branch off the optic tract, forming the retinohypothalamic tract, and terminate in the supraoptic, suprachiasmatic and paraventricular nuclei of the hypothalamus. Visual input to the hypothalamus drives the light–dark entrainment of neuroendocrine function and other circadian rhythms.

the DTN have been shown to modify the OKN and reduce or abolish optokinetic after-nystagmus (OKAN).[12]

Visual field lesions

Much of our knowledge of the retinotopic organization of the central visual nuclei arises from the visual field defects associated with lesions along the major pathway leading to form vision, the retinogeniculocortical pathway (Box 27.2). Figure 27.3 illustrates several known anatomical lesions, from the retina to the occipital lobes, and their subsequent effect on the visual fields. In the example shown in Figure 27.3A complete interruption of one optic nerve, which may occur with severe degenerative disease or injury, results in permanent blindness in the affected eye. Partial interruption of the nerve fibers results in a partial loss of the visual field and can occur with glaucoma, optic disc drusen, pits, infarcts, or optic neuritis.

Slightly further along the visual pathway, interruption of the decussating optic nerve fibers in the optic chiasm (Fig. 27.3B) results in loss of vision in the temporal visual hemifields of both eyes called bitemporal hemianopia. Damage of this type commonly occurs with pituitary tumors as they grow and compress the overlying optic chiasm. Rarely, pressure is exerted on the lateral edge of the optic chiasm (Fig. 27.3C) – damage arises primarily in the uncrossed fibers,

---

Box 27.2 Visual field lesions

A. Monocular blindness is caused by a lesion of the optic nerve or nerve fiber layer and may vary from complete blindness to various sizes of scotoma, depending on the extent of the lesion.

B. Bitemporal hemianopia results from a lesion at the level of the optic chiasm affecting the decussating optic nerve fibers.

C. Nasal hemianopia results from a lesion of the uncrossed fibers in the optic nerve at the level of the optic chiasm. Nasal visual field defects not necessarily respecting the vertical meridian are common with optic nerve head disease.

D. Homonymous hemianopia with an afferent pupillary defect in the contralateral eye results from a complete interruption at the level of the optic tract.

E. Normal visual fields may accompany an afferent pupillary defect caused by a lesion that is specific to the brachium of the superior colliculus, sparing the retinogeniculate fibers.

F. Homonymous hemianopia may also result from a lesion at the level of the LGN, which can be distinguished from (D) by the absence of an afferent pupillary defect.

G. Homonymous hemianopia with sparing of the macular field results from a lesion at the level of the optic radiation and is usually seen after vascular lesions of the occipital lobes. Several hypotheses exist for the macular sparing, including overlapping vascular fields of the middle and posterior cerebral arteries in the occipital lobes where the macula is represented.

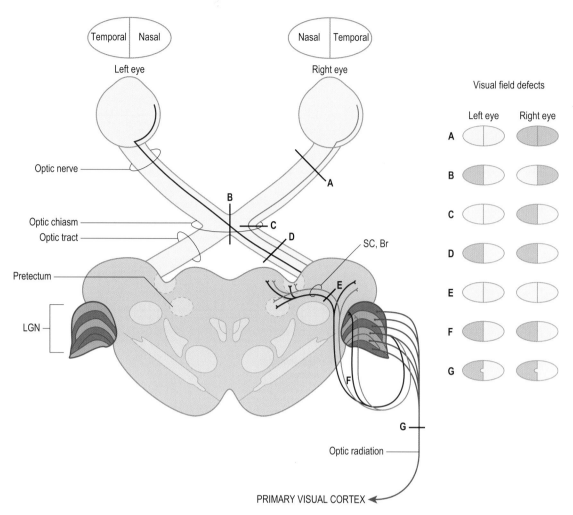

Figure 27.3 Visual field defects associated with lesions of the visual pathway. The bars (A–G) depict the location of lesions, and the associated visual field defect is shown at the right (shaded area represents visual field loss). A: Lesion of the right optic nerve causes blindness in the right eye. B: Optic chiasm lesion results in a bitemporal hemianopia, a loss of vision in the temporal visual field of both eyes. C: Lesion affecting only the uncrossed fibers of the optic chiasm results in a loss of vision in the nasal field of the ipsilateral eye. D: Lesion of the optic tract causes homonymous hemianopia, or complete loss of vision in the contralateral visual field. An afferent pupillary defect also arises from a lesion at this level. E: Lesion affecting the fibers in the brachium of the superior colliculus (SC, Br) results in an afferent pupillary defect, but with intact visual fields because the projection to the lateral geniculate nucleus (LGN) remains intact. F: Lesions at the level of the LGN are similar to lesions at D and result in homonymous hemianopia, but with an intact, normal pupillary reflex. G, Lesions at the level of the optic radiation also result in a homonymous hemianopia, but with sparing of macular vision.

resulting in loss of vision in the nasal hemifield, or nasal hemianopia, ipsilateral to the compression.

Lesions that occur after the chiasm are characterized by visual field defects that involve the temporal hemifield of the contralateral eye and the nasal hemifield of the ipsilateral eye; this is because of the partial decussation of optic nerve fibers at the optic chiasm. In this type of lesion, visual field loss occurs on the side contralateral to the lesion. Complete interruption at the level of the optic tract (Fig. 27.3D) or beyond results in loss of vision in the contralateral visual field, i.e. homonymous hemianopia. It is usually difficult to determine whether the site of the lesion is at the level of the optic tract, LGN, or visual cortex for most homonymous hemianopias. In cases in which the lesion is in the optic tract, the homonymous hemianopia may be accompanied by an afferent pupillary defect in the contralateral eye. The reason for this is likely because ganglion cells in the nasal retina outnumber those in the temporal retina, thus causing

a greater loss of fibers for the pupillary response of the contralateral eye[14] (however, also see reference 3) Lesions at the level of the brachium of the SC (Fig. 27.3E), may result in an afferent pupillary defect in the contralateral eye, but with intact, normal visual fields because lesions at this level spare the retinogeniculate fibers.

Lesions involving the LGN (Fig. 27.3F) are often difficult to distinguish from other optic tract lesions, but usually present with a visual field defect and an intact contralateral afferent pupillary reflex because lesions at this level normally spare the retinal fibers terminating in the pretectum responsible for the afferent pupillary reflex.

Lesions involving the optic radiation (Fig. 27.3G) or the visual cortex result in homonymous hemianopia (Box 27.3), but with sparing of the central, macular field. Several hypotheses have been put forward to explain macular sparing, but in many patients, macular sparing is likely more apparent than real because it is likely an artifact of visual

Box 27.3 Blindsight

Blindsight is a phenomenon in which people with lesions of primary visual cortex and hemifield defects are unconsciously aware of visual stimuli presented in their blind hemifield. Methodology used to obtain these results may be responsible for the apparent visual awareness and thus the concept of blindsight is heavily surrounded by controversy. However, when methodology can be ruled out, blindsight may be explained by the existence of retinal inputs to central targets other than the LGN, specifically the superior colliculus and pulvinar, and their projections to extrastriate cortical areas.[15]

field testing caused by poor foveal fixation or eye movements. In other patients in whom macular sparing is not artifactual, the most likely explanation is the immense areal dimension of the cortical representation of the retinal fovea. The posterior cerebral artery supplies the occipital poles, but in some individuals they are supplied by both the posterior and middle cerebral arteries. Thus infarcts of the posterior cerebral artery would affect only part of the occipital poles, sparing portions of the foveal representation and sparing macular vision.

References

1. Munoz DP et al. On your mark, get set: brainstem circuitry underlying saccadic initiation. Can J Physiol Pharmacol 2000; 78:934.
2. Schall JD. Neural basis of saccade target selection. Rev Neurosci 1995; 6:63.
3. Rodieck RW, Watanabe M. Survey of the morphology of macaque retinal ganglion cells that project to the pretectum, superior colliculus, and parvicellular laminae of the lateral geniculate nucleus. J Comp Neurol 1993; 338:289.
4. Grieve KL, Acuna C, Cudeiro J. The primate pulvinar nuclei: vision and action. Trends Neurosci 2000; 23:35.
5. O'Brien BJ, Abel PL, Olabvarria JF. The retinal input to calbindin-D28k-defined subdivisions in macaque inferior pulvinar. Neurosci Lett 2001; 312:145–148.
6. Robinson DL, Petersen SE. The pulvinar and visual salience. Trends Neurosci 1992; 15:127.
7. Acuna C, Gonzalez F, Dominguez R. Sensorimotor unit activity related to intention in the pulvinar of behaving Cebus apella monkeys. Exp Brain Res 1983; 52:411.
8. Cudeiro J et al. Does the pulvinar–LP complex contribute to motor programming? Brain Res 1989; 484:367.
9. Robinson DL, McClurkin JW, Kertzman C. Orbital position and eye movement influences on visual responses in the pulvinar nuclei of the behaving macaque. Exp Brain Res 1990; 82:235.
10. Moore RY. Organization of the primate circadian system. J Biol Rhythms 1993; 8:S3.
11. Fredericks CA et al. The human accessory optic system. Brain Res 1988; 454:116.
12. Schiff D, Cohen B, Buttner-Ennever J, Matsuo V. Effects of lesions of the nucleus of the optic tract on optokinetic nystagmus and after-nystagmus in the monkey. Exp Brain Res 1990; 79:225.
13. Curcio CA, Allen KA. Topography of ganglion cells in human retina. J Comp Neurol 1990; 300:5.
14. Schmid R, Wilhelm B, Wilhelm H. Naso-temporal asymmetry and contraction anisocoria in the pupillomotor system. Grafes arch Clin Exp Ophthalmol 2000; 238:123.
15. Stoerig P. Blindsight, conscious vision, and the role of primary visual cortex. In: Martinex-Conde S, Macknik S, Martinez L, Alonso J-M, Tse PU, eds. Fundamentals of awareness, multi-sensory integration and high-order perception. Visual Perception Part 2. Progress in Brain Research, vol. 155. Amsterdam: Elsevier, 2006:217–234.

CHAPTER 28

Optic Nerve

Jeffrey L. Goldberg

Introduction

The two optic nerves carry the axons of retinal ganglion cells (RGCs), and along these axons transmit all of the visual information from the inner retina (Chapter 23) to the brain (Chapters 29–31). Thus, diseases that affect the optic nerve commonly cause vision loss. The retina and optic nerve are developmentally an outgrowth of the forebrain, and, like other white matter tracts of the central nervous system (CNS), the optic nerve does not repair itself after most types of injury. Thus, blindness from optic nerve injury or degenerative disease is typically irreversible. This chapter reviews the principal aspects of optic nerve anatomy, development, and physiology, and discusses the pathologic changes at the molecular and cellular levels in the context of clinical disease.

Optic nerve anatomy

Retinal ganglion cell axons within the nerve fiber layer

RGC axons begin at the RGC bodies in the inner layer of the retina. Although most neurons in this layer are RGCs and the layer is commonly referred to as the *ganglion cell layer*, in humans approximately 3 percent of cells in the central retina and up to 80 percent in the peripheral retina may be other cell types, primarily displaced amacrine cells.[1] In addition, studies in mammals have demonstrated a few displaced RGCs located in the inner nuclear layer.[2] As discussed in Chapter 23, the RGCs receive input from bipolar cells and amacrine cells, and project their axon towards the vitreous, whereupon they turn approximately 90° and project towards the optic nerve head in the *nerve fiber layer*. The nerve fiber layer is not quite radially arranged around the optic nerve head (optic disk), as axons near the center of our visual field course away from the fovea, and then towards the optic disk, entering in the superior and inferior portions of the disk (Fig. 28.1). This interesting anatomy prevents axons from crossing the high-sensitivity fovea, where they might otherwise scatter light and degrade visual acuity. The axons from more peripheral RGCs are more superficial (vitread) to those arising from less peripheral ganglion cells[3] (Fig. 28.2). In addition, there is strict segregation of those fibers arising from RGCs located superior to the temporal horizontal meridian (raphe) and those fibers arising from RGCs located inferior to the horizontal raphe. Because of this segregation, visual field defects corresponding to injury to RGC axons typically have stereotyped patterns, e.g. superior or inferior nasal steps, temporal wedges, or arcuate scotomas. These are called nerve fiber bundle defects, and with other visual field defects are covered in more detail in Chapter 35. The nerve fiber layer can be measured using various modalities in human eyes (Box 28.1).

Intrascleral optic nerve

The optic nerve itself is considered to begin at the optic nerve head which, when viewed through the front of the eye, is observed as the *optic disk*. There is an approximately 1-mm component of optic nerve within the intrascleral part of the globe, which includes the lamina cribrosa (Fig. 28.3). Once at the optic disk, ganglion cell axons turn away from the vitreous and dive into the optic disk towards the brain. Axons arising from more peripheral RGCs are peripheral within the optic nerve head.[4] The optic disk contains the nerve fibers around its edge, the *neuroretinal rim*, and the central *cup* which does not contain RGC axons. The *cup-to-disk ratio* may range from 0 to 1.0, depending on natural variation in the size of the disk, and whether there is less than a full complement of axon fibers, for example as a result of damage from glaucomatous optic neuropathy (see below).

Intraorbital optic nerve

The optic nerve extends approximately 30 mm from the globe to the optic canal.[5] The straight-line distance from the back of the globe to the optic canal is much less (the exact amount depending on individual orbital depth), with the excess optic nerve laxity allowing for free movement of the globe during eye movements. Excessive proptosis stretches the optic nerve tautly, resulting in some cases in direct injury to or even avulsion of the nerve itself. Beginning behind the globe, the nerve is ensheathed by the layered meninges extending from the brain, bathing the optic nerve in cerebrospinal fluid and providing vascular support along the length of the nerve (see below).

The retinotopic segregation of RGC axons (topographic correspondence of optic nerve fibers to retinal location),

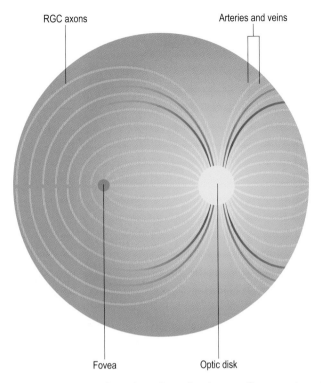

Figure 28.1 Axons of retinal ganglion cells take a specific course through the retina, such that fibers more temporal to the fovea skirt around the edge of the macula and enter the optic nerve head closer to the superior or inferior poles. Neither the axons nor the blood vessels typically cross the horizontal meridian of the retina.

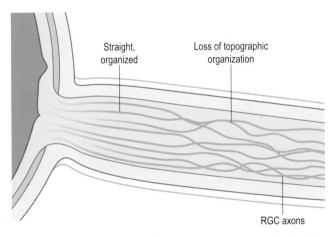

Figure 28.2 Retinal ganglion cell axons enter the optic nerve head with specific retinotopic organization (as in Fig. 28.1). With increasing distance from the optic nerve head, however, the axons become increasingly jumbled, losing their topographic organization. Thus focal damage to nerve at the nerve head will produce specific retinotopic defects, while focal damage to axons further posterior to the eye will produce scotomas that lack retinotopic organization.

Box 28.1

The number of RGC axons in the optic nerve is, in most cases, probably well-approximated by the number of RGC axons in the nerve fiber layer of the retina immediately adjacent to the optic nerve. The peripapillary nerve fiber layer can be measured quantitatively using a variety of modalities including coherence tomography (OCT), which measures layer thickness based on reflectance of light from the interfaces between layers, and by scanning laser polarimetry (SLP), which estimates nerve fiber layer thickness based on the interaction of polarized light with organized bundles of microtubules inside axons. Each of these technologies may face limitations. OCT estimation of RGC axons may be complicated by the presence of retinal astrocytes, which accompany RGC axons through the nerve fiber layer and contribute to nerve fiber layer thickness. SLP estimation of axon integrity may be limited by degradation or disorganization of RGC microtubules that may precede axon loss. All such technologies are limited by their inability to measure optic nerve axons specifically, and rather substitute peripapillary axon measurements.

The loss of retinotopy is not absolute, because the fibers from the nasal (but not temporal) macula continue to be located centrally within the nerve over a considerable distance.[10] Fibers from large RGCs are less retinotopically organized than those from smaller ganglion cells.[10] Although studies in humans are difficult to do ante-mortem, post-mortem studies of nerves with specific visual field defects demonstrate similar findings, while post-mortem studies of developing fetuses and adult eyes also show loss of retinotopy within the transition from the optic nerve head to the nerve.[6,11] Within the orbit, the optic nerve travels within the muscle cone formed by the superior rectus, lateral rectus, inferior rectus, and medial rectus muscles. Tumors within the cone are common sources of compression of the optic nerve, or *compressive optic neuropathy*. Examples of these tumors include cavernous hemangioma, hemangiopericytoma, fibrous histiocytoma, lymphoma, and schwannoma. In addition, enlargement of the muscles themselves, particularly the inferior rectus and/or the medial rectus in Grave's ophthalmopathy, may also compress the optic nerve.

The optic canal

The optic nerve enters the cranium via the optic canal, a 5–12 mm passage which lies immediately superonasal to the superior orbital fissure.[12] The optic canal contains some axons of sympathetic neurons destined for the orbit, as well as the ophthalmic artery. The latter lies immediately inferolateral to the optic nerve itself, covered in dura. At the distal end of the canal, there is a half-moon-shaped segment of dura which overhangs the optic nerve superiorly, and thereby lengthens the canal by a few millimeters. As in the intraorbital portion of the optic nerve, within the canal, and immediately posterior to the optic canal, meningeal tissue ensheathes the optic nerve. Benign tumors of the meninges, or meningiomas, are frequent causes of compressive optic neuropathies in these locations. Small tumors within the canal itself, where there is very little free space, may lead to compressive optic neuropathy without a radiographically visible tumor.

particularly the segregation between axons arising from the superior and inferior retina, is gradually lost as the axons course through the nerve (Fig. 28.3). There is only moderate retinotopy in the initial segment of the optic nerve.[6] The retinotopy decreases distally,[7,8] and then becomes ordered vertically for eventual nasal decussation near the chiasm.[9]

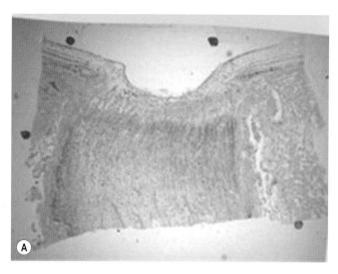

Figure 28.3 (**A**) Histochemical staining of the human optic nerve head. (**B**) The optic nerve head is at various points adjacent to retina (R), choroid (C), and sclera (S), and demonstrates a varied cellular morphology along this distance. (1) Superficial nerve fiber layer; (2) Anterior prelaminar region; (3) Posterior prelaminar region; (4) Laminar region; (5) Retrolaminar region. (A from Figure 1 in: Sigal et al, Exp Eye Res 2009; 90(1):70–80; B from Scheme 1 in: Trivino et al, Vision Res 1996; 36(14):2015–28.)

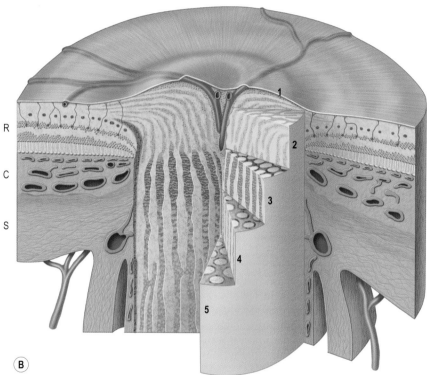

Intracranial optic nerve and the optic chiasm

Once the nerve has entered the cranium, there is a highly variable (8–19 mm, mean of 12 mm) length of nerve until the chiasm is reached.[13] The length of the chiasm itself is approximately 8 mm. The intracranial optic nerve and chiasm are immediately above the planum sphenoidale and sella turcica, the latter of which contains pituitary gland. There is approximately 10 mm between the inferior part of the nerve and the superior part of the pituitary. Tumors of the pituitary which increase in size enough to compress the chiasm may therefore cause compressive optic neuropathy.

At the optic chiasm, RGC axons from the temporal retina remain ipsilateral, and those from the nasal retina cross the chiasm and course towards the contralateral brain. The ratio of fibers which cross versus those which do not cross is anatomically 53:47[14] and functionally 52:48.[15] This small difference between the number of crossing and non-crossing fibers is commonly reported to be responsible for the relative afferent pupillary defect seen in disorders of the optic tract, in which an afferent pupillary defect is seen contralateral to the injured tract (see Chapter 25), but could also reflect the fact that some fibers from specialized cells within the retinal responsible for the pupillary reflex may cross from the temporal retina into the contralateral optic tract.[15]

The optic tract and lateral geniculate nucleus

Although the optic nerve anatomically ends at the chiasm, RGC axons continue within the optic tract until the lateral geniculate nucleus, superior colliculus, pretectal nuclei, or hypothalamus (see below). Circuitry and processing by these targets are discussed in Chapters 25 and 29.

Optic nerve axon counts and dimensions

In the normal adult human optic nerve, manual techniques have demonstrated an estimated 1,200,000 RGC axons per nerve;[16] automated counting algorithms give figures between 700,000[16] and 1,400,000[17,18] axons. There is a strong correlation between the size of the neuroretinal rim and the number of axons,[18,19] and between the number of axons and the size of the scleral canal in primates,[20] although the degree of correlation is controversial.[18,21]

Many factors affect the number of axons within the optic nerve, from inherited differences to damage from disease, i.e. optic neuropathy (see below). In addition, there is a gradual loss of RGCs during normal human aging, with an approximate 5000 axon loss per year of life.[16,17,22,23] For unclear reasons, there is a smaller degree of loss of macular RGCs with age, compared to peripheral RGCs, and this may reflect contraction of the macula with time.[24]

Although the number of RGC axons entering the optic nerve is fairly constant, the diameter of the optic nerve varies widely. At the disk itself, where the fibers are completely unmyelinated, the mean vertical diameter of the disk is 1.9 mm (range 1.0–3.0 mm) and the mean horizontal diameter is 1.7–1.8 mm (range 0.9–2.6 mm).[19,25] The mean area of the disk is 2.7 mm² (range 0.8–5.5 mm²). The mean area of the neuroretinal rim (not including the cup) is 2.0 mm² (range 0.8–4.7 mm²). Because the axonal tissue entering the optic disk varies much less than the size of the optic disk itself, the optic cup in the center of the disk can vary greatly without necessarily reflecting any underlying deficit in the number of ganglion cell axons. The diameter of the optic nerve approximately doubles posterior to the globe as a result of myelination of the axons.

Microscopic anatomy and cytology

Axons

There are no other neuronal cell bodies within the optic nerve, making it a pure white matter tract of the central nervous system. Although the optic nerve itself may contain other small nerves, particularly tiny peripheral nerves (branches of the trigeminal system) which carry pain sensation or control vascular tone, the vast majority of the optic nerve is composed of the approximately 1,200,000 axons of the retinal ganglion cells. Optic nerve axons are collected in fascicles, which are separated by pia-derived septa. The number of fascicles ranges from approximately 50 to 300, being maximal immediately retrobulbar and at the optic canal.[26] The mean axon diameter is slightly less than 1 µm, with a unimodal distribution skewed to the left. The mean diameter may decrease during aging, either through loss of large-diameter axons,[27] or from a general, aging-induced

atrophy. Compared with the optic tract, there is relatively little segregation of axons by size within the optic nerve, except for a tendency for finer axons to be located inferocentrotemporally.[7,28]

There is variability of axonal diameter and myelin thickness from the retina to the brain. Studies in the ferret show that the diameters of the largest axons increase as the distance from the retina increases.[29] The diameter of individual axons is regulated by multiple factors, including oligodendrocytes[30] and activity,[31] increases during development, and correlates with local accumulation of specifically phosphorylated neurofilament proteins.[32]

Oligodendrocytes and myelin

Conduction of nerve impulses down the axons in the optic nerve depends on the presence of myelin, a fatty multilaminated structure which insulates each axon and greatly increases the speed and efficiency of conduction (discussed below). The retrolaminar optic nerve is completely myelinated under non-pathological circumstances (Fig. 28.4). There are non-myelinated axons of the peripheral nervous systems within the adventitia of the central retinal artery[33] and non-myelinated and myelinated Schwann cells around peripheral axons of the outer dura.[34] Each axon is myelinated with several lamellae of myelin bilayers, with the number of lamellae varying from axon to axon[35] but in strict proportion to the diameter of the axon. Individual oligodendrocytes elaborate an average of 20–30 processes per cell, each of which myelinates 150–200 µm of axon length.[36]

Oligodendrocytes and axons regulate each other during development and throughout adulthood. Oligodendrocytes depend on the presence of axons for their survival,[37] and for production of myelin proteins.[38] Axons regulate oligodendrocyte survival and proliferation, and thus their number, through expression of specific signaling proteins[39] and through axonal electrical activity,[40] controlling oligodendrocyte number to match the number of axons.[41] Conversely, axons are signaled by oligodendrocytes to regulate axon number and diameter, and to prevent axon sprouting or branching in the optic nerve.[30,42] Later, in the adult, oligodendrocytes and myelin are partially responsible for inhibition of axon regeneration after optic nerve damage or degeneration (see below).[43,44]

Astrocytes

Astrocytes, named for their stellate appearance, are ubiquitous glial components of the central nervous system, and function in white matter tracts like the optic nerve to regulate ionic and energy homeostasis. Astrocytes are highly efficient at transporting potassium, and increases in the level of extracellular potassium as a result of repolarization are buffered by astrocytes. Their ability to accumulate glycogen may allow them to serve as an energy source for the optic nerve in the absence of glucose (e.g. ischemia), by shuttling lactate to adjacent axons.[45] Astrocytes form the glial-limiting membrane,[46] and their processes are concentrated at the nodes of Ranvier and in contact with nearby capillaries. This positions the astrocyte to play a role in transportation of substances between the local circulation and the axons; in inducing

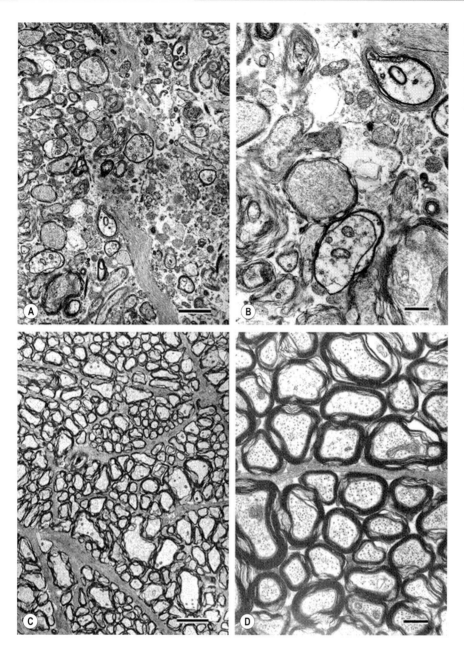

Figure 28.4 Transverse sections through the adult rat optic nerve demonstrate typical profiles of myelinated axons and the fibrous glial septae that separate bundles of axons. Scale bars are 4 mm in the left panel, 500 nm in the right. (From Figure 3 in, Gong & Amemiya, Exp Eye Res 2001; 72:363–9.)

endothelial cells to form the blood–brain barrier;[47] and perhaps in signaling blood vessels to dilate or constrict according to local metabolic needs. Astrocytes also mediate connectivity between optic nerve axons and the adjacent connective tissues, such as the pial septa, the adventitia of the central retinal artery and vein, and the pia in a layer named the glial mantle of Fuchs.[33]

Interestingly, the most common intrinsic tumor of the optic nerve is an astrocytoma, or optic nerve glioma. This is usually a low-grade tumor of well-differentiated astrocytic cells that appear hair-like, or "pilocytic." Pilocytic astrocytomas are usually seen in childhood, and have a favorable prognosis. They are also commonly seen in association with neurofibromatosis. Rarely, more malignant neoplasms of astrocytic origin develop in adults. These resemble the higher-grade astrocytic neoplasms found elsewhere in the central nervous system, and are usually fatal.

Microglia

Microglia, a type of resident macrophage, are an important cellular component of the optic nerve. Although their origin was debated for decades, they have been shown experimentally to be of peripheral, bone marrow origin, and are not derived from the neuroectoderm that yields neurons, astrocytes, and oligodendrocytes.[48,49] Microglia share several markers with macrophages, with both having Fc receptors (for immunoglobulin), C3 receptors (for complement), binding of Griffonia isolectin B4, and antigenicity for F4/80 and ED1 monoclonal antibodies.[50] In the human optic nerve, microglia can be seen at 8 weeks after conception, when they are relatively undifferentiated. Microglia become more differentiated during fetal development, going from tuberous to amoeboid to a ramified morphology.[51] They are associated with axon bundles, but not necessarily with blood vessels, and are found in both the nerve parenchyma and its meninges. Nerve and meningeal microglia are similar ultrastructurally except for vacuoles and endoplasmic reticulum in the former. This may be due to their phagocytosis of dying axons during development.[52] Microglia share several characteristics of immune capacities of macrophages. By phagocytosing extracellular material, degradation within intracellular compartments can occur. This may be followed by antigen presentation on the cell surface. In combination with certain histocompatibility antigens, this presented antigen can cause stimulation of T lymphocytes and subsequent immune system activation (see below).

Meninges and meningothelial cells

The optic nerve is covered with three layers of meninges, dura, arachnoid, and pia. The meninges can also be divided up into *pachymeninges* (dura) and *leptomeninges* (arachnoid and pia). The outermost dura is a thick fibrovascular tissue, which is in immediate contiguity with the sclera, periorbita, and the dural layer of the lining of the cranial contents. The middle arachnoid layer is a loose, thin, fibrovascular tissue. The innermost pia is a thin, tightly adherent layer with extensions into the nerve itself forming the pial septae, through which the fascicles of ganglion cell axons course.[26] In the optic canal, there are numerous trabeculae connecting the dura through the arachnoid to the pia, which reduce the free space of the nerve sheath in this area.[53]

The space between the dura and arachnoid is the subdural space, while the space between the arachnoid and pia is the subarachnoid space. The subdural space around the optic nerve is small, and is not in communication with the intracranial subdural space. The subarachnoid space, on the other hand, is in communication with the intracranial subarachnoid space. The optic nerve subarachnoid space ends anteriorly within a blind pouch just before the optic disk. This extension of the subarachnoid space from the brain into the orbit explains why elevated intracranial pressure may cause compression of the optic nerve via elevating the hydraulic pressure. There is an appreciable pressure within the subarachnoid space, measuring from 4 to 14 mmHg.[54]

Meningothelial cells may have several functions,[55] including wound repair and scarring, phagocytosis, and collagen production. The meningeal vessels include pericytes and endothelial cells, which form tight junctions.[34] Macrophages[56] and mast cells[57] are also found in the meninges. The role of these and other resident cells in meningeal inflammations is only speculative.

Blood supply

Optic nerve head

Studies of histological sections and of corrosion casting of the vessels themselves have added greatly to our understanding of the optic nerve blood supply[58] (Fig. 28.5). The ophthalmic artery provides the major vascular supply to the inner retina and optic nerve. The central retinal artery branches off of the intraorbital ophthalmic artery, and enters the optic nerve during approximately 12 mm behind the globe. In the retina, RGC bodies and the nerve fiber layer are supplied by capillary branches derived primarily from the

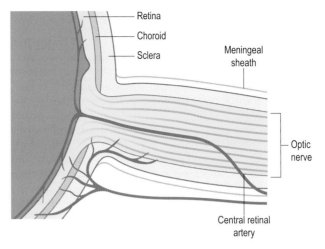

Figure 28.5 Vascular anatomy of the optic nerve head. The optic nerve is supplied by branches from the central retinal artery, from penetrating meningeal arterioles, from recurrent arterioles off of the inner retinal vasculature, and from diffusion from the fenestrated blood supply of the choroid.

central retinal artery arising out of the optic nerve head. As it branches on the optic disk into the retinal arterioles, the central retinal artery provides partial perfusion of the superficial optic disk via small capillaries. Along the optic nerve itself, however, the central retinal artery provides minimal perfusion of the nerve through which it courses.[59]

In contrast, branches of the medial and lateral short posterior ciliary arteries, originating from the ophthalmic artery, provide the major blood supply to the optic nerve head, as well as the choroid. Their branches perfuse the optic nerve, both anteriorly via direct branches and posteriorly via a retrograde arteriolar investiture of the optic nerve (see below). Anastomoses of posterior ciliary artery branches[60,61] form the circle of Zinn-Haller, which contributes significant perfusion to the optic nerve head. In addition, there are contributions from recurrent choroidal arterioles to the prelaminar and laminar optic nerve head, and contributions from recurrent pial arterioles to the laminar and retrolaminar optic nerve head. A clinical implication of the common posterior ciliary arterial source of the choroid and deep optic nerve head is that they will fluoresce simultaneously during the earliest phase of fluorescein angiography, before the retinal arterioles transit dye.

Unlike the choroidal vessels, which have fenestrated endothelial cells, optic nerve vessels have non-fenestrated endothelial cells with tight junctions, surrounded by pericytes. Optic nerve vessels therefore share the same blood–nerve barrier characteristics as the blood–brain barrier. Only a restricted number of molecules can cross the blood–nerve barrier. For example, gadolinium enhancement on magnetic resonance imaging is not seen unless there is some pathological process within the optic nerve that would disrupt the blood–nerve barrier, e.g. inflammation.[62,63] Another major difference between choroidal and optic nerve head vessels is that only the latter can autoregulate, i.e. maintain approximately constant blood flow despite most changes in intravascular or intraocular pressure.[64–66] Thus there is compensation of perfusion for a wide range of intraocular pressures in normal individuals. A dysregulation in optic nerve head autoregulation, and any ischemia that would follow, may contribute to the pathophysiology of glaucoma (see below).

Intraorbital optic nerve and optic canal

The intraorbital optic nerve is perfused primarily by the pial circulation, which branches off of the ophthalmic artery either directly or indirectly via recurrent branches of the short posterior ciliary arteries. The pia sends penetrating vessels into the intraorbital optic nerve along the fibrovascular pial septae, from which a capillary network extends into neural tissue (axons and glia). Similarly, the intracanalicular optic nerve is perfused by up to three branches of the ophthalmic artery, namely a medial collateral, lateral collateral, and ventral branch,[67,68] which perfuse the pial surface and then penetrate the nerve. An important clinical implication of the pial supply relates to optic nerve sheath meningiomas. If a surgeon strips the meningioma away from the nerve, then the pia will be removed as well, resulting in loss of the blood supply to the nerve, and possible infarction and blindness. Another clinical correlation is related to the small amount of free space within the canal, when occupied by blood from shearing of the delicate vessels there, this may result in a sight-threatening compressive hematoma[68] (and see below).

Intracranial optic nerve, chiasm, and optic tract

The intracranial optic nerve and chiasm are perfused by the internal carotid artery and its branches, primarily the anterior cerebral, anterior communicating, and the superior hypophyseal artery. The posterior chiasm may also be perfused by branches of the posterior communicating artery. The optic tract is predominantly perfused by branches of the posterior communicating and anterior choroidal arteries. Similar to the intraorbital and intracanalicular optic nerve, the blood supply occurs via small pial penetrating vessels. However, there is no dura or arachnoid surrounding the optic nerve posterior to the optic canal, and thus the risk of infarction is not from stripping a tumor off the nerve, but from inadvertent detachment of the fine vessels during surgical manipulation.

Vascular biology

Vessels of the human optic nerve contain endothelial cells separated from pericytes by a basement membrane, with endfeet of astrocytes surrounding the two.[69] The tight junctions between endothelial cells contribute to the blood–nerve barrier, which is present in normal optic nerve throughout its length. The anterior part of the optic nerve, however, shares a distinctive feature with the area postrema, choroid plexus, and median eminence, in that there is local disturbance of this blood–brain barrier. Although the optic nerve capillary endothelial cells in the anterior optic nerve have tight junctions, leakage can occur from adjacent choriocapillaris through the border tissue of Elschnig at the level of the lamina choroidalis.[70,71] Further leakage from this area into the subretinal space may be prevented by tight junctions between the glia making up the intermediary tissue of Kuhnt.[70,72] The blood–nerve barrier may be incompetent in focal inflammatory disorders, such as optic neuritis.[73]

The optic nerve differs from almost all other areas of the CNS in that there is connective tissue around vessels.[33] The vessels are closely surrounded by morphologically specialized collagen[74] and pial fibroblasts, beyond which are astrocytic foot processes.[33,69,75] These pial septae are not a constant feature of mammalian optic nerves; rats, for example, do not have septae.

The vessels within the nerve have their own innervation. These nervi vasorum contain multiple neurotransmitters[76] including adrenergic, cholinergic, and calcitonin gene-related peptide; neuropeptide Y; substance P; and vasoactive intestinal peptide.[77] Human posterior ciliary arteries respond to the potent vasoconstrictor angiotensin II.[78]

Optic nerve development

Generation of optic nerve oligodendrocytes and myelination

Oligodendrocytes develop from oligodendrocyte precursor cells (OPCs). OPCs migrate from the brain into the optic nerve, perhaps from the base of the preoptic recess, and this migration may continue during adult life.[79] The OPC has a complex regulatory mechanism, with in vitro and animal studies implicating platelet-derived growth factor (PDGF), basic fibroblast growth factor (bFGF), ciliary neurotrophic factor (CNTF), and other factors in the control of differentiation, division, and renewal of the progenitor cell population.[80,81] An intrinsic developmental clock controls the differentiation of OPCs into oligodendrocytes,[82] after which the cell is mitotically unresponsive to PDGF.[83–87] In normal development, some OPCs persist into the adult and may participate in remyelination after injury or in diseases like optic neuritis.[88]

After oligodendrocyte differentiation from OPCs, there is typically a short delay before myelination starts, which may be regulated by inhibitory signals from axons activating the notch signaling pathway.[89] The developmental regulation of myelination of the optic nerve and tract occurs earliest at the brain end of the optic nerve, then travels distally toward the eye, i.e. proximally with respect to the ganglion cell axon. This is opposite to myelination elsewhere in the CNS, which usually occurs from the neuron cell body outward along the axon. The brain-to-eye direction of myelination correlates with animal studies demonstrating OPC migration and then differentiation outward along the optic nerve.[90,91] In humans, oligodendrocytes are not seen in the optic nerve before 18 weeks of gestation.[90] Later, optic tract and intracranial nerve axons begin myelination at 32 weeks of gestation, with the process virtually complete at birth. The axons adjacent to the globe begin myelination only at birth; all are myelinated by about 7 months of age.[92] Further myelination continues, however, and thickening of axonal myelin may continue past 2 years, or longer.

In longitudinal studies of rat optic nerve development,[93] axons that become myelinated increase in diameter. The signal for change in caliber is generated by the oligodendrocytes.[30,94] Although unmyelinated and not yet myelinated axons can conduct electrical impulses in the developing optic nerve, there is a surprising increase in conduction velocity during development, independent of diameter or myelination status, the reasons for which are unclear.[95]

Interestingly, intraretinal RGC axons are not normally myelinated in most species; indeed, there are normally no oligodendrocytes in the retina. Studies of rat optic nerve head and anterior optic nerve demonstrate an abnormal transition zone from unmyelinated axons to myelinated axons, with varying lengths and thicknesses of myelin, unusual nodal region morphology, and few normal oligodendrocytes.[96] Studies of rat optic nerve suggest that a signal found at the lamina cribrosa blocks the migration of OPCs into the retina.[97] This barrier is possibly due to the presence of specialized optic nerve head astrocytes,[79] serum factors from a focally deficient blood–nerve barrier[98,99] or the adhesion substrate tenascin.[100] In rabbits, which have no lamina cribrosa, there are axons myelinated by oligodendrocytes in the retina, and intraretinal OPCs can be identified.[97] Further, the retinal portion of the ganglion cell axon has no constitutive barrier to myelination because ganglion cell axons of retina transplanted into rat midbrain can be myelinated,[99] and OPCs injected intraretinally can differentiate and myelinate RGC axons.[101] Myelination may also be artifactually induced by injuring the retina from the scleral side. This trauma leads to migration of Schwann cells and subsequent intraretinal myelination.[102] In human eyes, myelination of part of the nerve fiber layer is occasionally seen; it is assumed that this is secondary to ectopic oligodendrocytes, although there are reports of Schwann cells myelinating intraretinal axons in the cat[103] and rat.[104] Although oligodendrogliomas within other parts of the central nervous systems are sometimes seen, neoplasms of oligodendrocytes within the optic nerve are very rare.

Generation of optic nerve astrocytes

Astrocyte precursor cells have been considerably less well studied. Like oligodendrocytes, developmental regulation of astrocyte number is determined by the RGC.[105] Astrocyte precursor cells may depend on CNTF or LIF for differentiation into astrocytes,[106] at which point they upregulate the expression of glial fibrillary acidic protein (GFAP), an intermediate filament protein. GFAP itself is important to optic nerve development, as transgenic mice lacking GFAP demonstrate abnormal myelination and blood–brain barriers.[107] Astrocytes may also be critical for optic nerve development in that they may serve as the substrate over which extending ganglion cell axons grow.[108]

Development of optic nerve meninges

The anatomy and development of the optic nerve meninges have been well described.[34,55] All three meningeal layers are formed from fibroblast-like cells, which vary in ultrastructural appearance from typical process-bearing fibroblasts to mesothelial cells lining the subdural and subarachnoid space. The meningeal layers are particularly rich in collagen and elastin. Trabeculae between the arachnoid and pia are formed from collagen and are lined with the same mesothelial cells that line the subarachnoid space. The most rapid proliferation of human meningeal cells during development occurs within 8–18 weeks of conception, with centrally located fibroblasts and peripherally located cells containing glycogen that increase during this time.[34,56] By 14 weeks, the three layers of the meninges can be distinguished. These glycogen-rich cells go on to produce large amounts of collagen, which is deposited in the dura after 14 or 15 weeks.

Axon number

Studies in animals suggest that approximately 50–100 percent excess RGCs are produced in the retina. During early development, the number of RGCs decreases by programmed cell death called apoptosis (discussed in the context of axon injury, below), in many cases because not all of the axons properly connect to their target regions within the brain.[109,110] This form of programmed cell death may be adaptive if it does not release toxic intracellular components into the extracellular environment and thereby does not affect neighboring cells that are already integrated to survive. Likewise, in humans, up to two-thirds of developing RGCs die, predominantly during the second trimester.[111]

Axon growth

Considerable progress has been made in recent years in our understanding of the development of the optic nerve, particularly at the level of molecular and cellular regulation. Axon growth requires specific extracellular signals – merely inhibiting cell death is not sufficient for inducing axon growth in developing RGCs.[112,113] The most potent signals for axon growth are the same peptide trophic factors that most strongly promote RGC survival, and multiple trophic factors can often combine to induce even greater axon growth.[113–116] The best-studied peptide trophic factors for RGC axon growth are brain derived neurotrophic factor (BDNF), ciliary neurotrophic factor (CNTF), insulin-like growth factor (IGF), basic fibroblast growth factor (bFGF), and glial cell line-derived neurotrophic factor (GDNF), although it is not clear how critical any of these are for axon growth during normal development.[112,117,118] Other than peptide trophic factors, extracellular matrix molecules like laminin and heparin sulfate proteoglycans, and cell adhesion molecules like L1 and N-cadherin are expressed in the visual pathway[119,120] and provide a critical substrate along which axons extend. Eliminating N-cadherin expression, for example, clearly decreases RGC axon growth in vivo.[121] Finally, other extracellular signals as simple as purine nucleotides may be critical to induce axon growth, although the mechanism of this activity is still mysterious.[122]

Electrical activity (discussed below) is not required for axon growth, but likely plays a role in sculpting axonal morphology. Thus, after injecting tetrodotoxin into the eye to block action potentials,[123] or in experimental mice in which synaptic release of neurotransmitters is essentially eliminated,[124] RGCs and other CNS neurons successfully elongate their axons to their targets in the absence of electrical activity. RGC activity may increase the rate of axon arborization in their target fields by stabilizing growing branches,[125–127] and may influence the specificity of local connectivity and targeting of axons.[128–130] Considerable progress has been made in understanding these molecular mechanisms for axon growth, and these have been reviewed elsewhere.[131]

Axon guidance

Optic nerve axon guidance is heavily influenced by the earliest pathfinding axons,[132] which serve as a substrate and guidance cue for the growth of the axons that follow. During development, RGC axons are guided by neuronal and glial molecules that attract and repel axonal growth cones.[133,134]

For example, embryonic RGC axons are repelled by Müller glia away from deeper retinal layers into the nerve fiber layer.[135,136] Within the retina, the interesting course of retinal axons in the macular nerve fiber layer are likely due to a foveal repellant combined with an optic disk attractant.[137] RGC axons are directed centrally in the retina away from chondroitin sulfate proteoglycans (CSPGs)[138] and are attracted into the optic nerve head by netrins.[139,140] A semaphorin, sema5A, may contribute to keeping the developing axons within the substance of the developing optic nerve.[141] Ipsilateral-projecting axons are repelled at the optic chiasm by slit1 and slit2,[142] ephrin-B,[143,144] and CSPGs.[145] The developmental ordering of the axons as they pass through the chiasm and go on to innervate the lateral geniculate primarily reflects their time of arrival.[146,147]

RGC axons reach four main targets: (1) the lateral geniculate nucleus of the thalamus; (2) the superior colliculus; (3) the pretectal nucleus; and (4) the suprachiasmatic nucleus. Most axons course unbranched to the lateral geniculate nucleus, but studies in many animals,[148] including primates,[149] have demonstrated the existence of axon collaterals to the other targets, suggesting that the same may be true for human optic nerves.

The lateral geniculate nucleus is the primary synaptic target for visual information processing on the way to the visual cortex (discussed at length in Chapter 29). The primate lateral geniculate nucleus has six layers: layers 1 and 2 consist of large (magnocellular) cells, and layers 3, 4, 5, and 6 consist of small (parvocellular) cells. The magnocellular layers receive input from the midget RGCs, while the parvocellular layers receive input from the parasol RGCs. Layers 2, 3, and 5 receive input from the ipsilateral retina, while layers 1, 4, and 6 receive input from the contralateral retina. In addition, there are interlaminar cells forming the koniocellular layers. These cells are smaller than those of the parvocellular layers, and receive synaptic input from blue-yellow RGCs.[150]

A second set of optic nerve axons project to the superior colliculus, a paired structure of the dorsal midbrain. In lower animals, e.g. rodents, the majority of fibers of the optic nerve project to the superior colliculus. In primates, including humans, only a minority does so, with the majority of axons destined instead for the lateral geniculate nucleus. These fibers arise from most of the retina, except for the central 10 degrees.[151] The superior colliculus has a laminar organization, and is retinotopically mapped.[152] Most studies of superior colliculus architecture have been done in animals, although the human superior colliculus appears to have similar anatomy.[153] The function of the superior colliculus is to coordinate retinal and cortical control of saccades and fixation.[152] Fibers branching to midbrain terminal nuclei stabilize the retinal image during movement of the eye, head, or body.[154] The role of the superior colliculus in extraocular movements is covered in Chapter 9.

The pretectal nucleus in the midbrain receives RGC fibers important for the pupillary response.[155,156] The pretectal projection is bilateral, and in addition fibers from each pretectal nucleus project both ipsilaterally and contralaterally to the Edinger–Westphal subnuclei of the third nerve nucleus. The latter then projects parasympathetic axons along the course of the third nerve to the ciliary ganglion within the orbit, where a synapse is made. Postsynaptic parasympathetic fibers continue to the pupilloconstrictor muscle, constriction of which is visible as the pupillary light reflex. The combined ipsilateral and contralateral projections from the retina to the pretectal nuclei, and from the pretectal nuclei to the Edinger–Westphal subnuclei necessitate that a light stimulus causing impulses along either optic nerve will cause equal constriction of both pupils. This is the physiological basis for detecting an afferent pupillary defect (also called a Marcus Gunn pupil). If there is a functional difference in the number of optic nerve axons carrying impulses from each retina, then light shined into the eye with fewer axons will cause a lesser amount of (bilaterally symmetric) pupillary constriction than light shined into the other eye. By alternating the light between the two eyes and observing the difference in pupil size dependent on which eye the light is shined into, one can detect which optic nerve is functionally conducting less than the other. This measure is highly correlated with differences in the number of surviving RGCs or RGC axons.[157] It has been proposed that there are differences between the myelin surrounding axons destined for the lateral geniculate nucleus and those responsible for the pupillomotor reflex because of the existence, albeit rare, of patients with demyelinative optic neuropathies and monocular blindness with preserved pupillary reflexes.[158] These patients do not have apparent conduction toward the lateral geniculate nucleus but do have conduction toward the pretectal areas responsible for the pupillary reflex. Detection of the afferent pupillary defect is discussed at length in Chapter 25.

A fourth projection of optic nerve axons is via the retinohypothalamic tract to the suprachiasmatic nucleus and the paraventricular nucleus[159–161] within the hypothalamus. These fibers are responsible for circadian control of the sleep–wake cycle, temperature, and other systemic functions.

Retrograde labeling studies in pigeons[162] and rats[163] indicate that only a few percent of RGC axons project to the pretectal and suprachiasmatic nuclei. Recently, a population of intrinsically photosensitive RGCs has been identified,[164] that express the photopigment melanopsin,[165–167] and project to the subthalamic nuclei to mediate circadian rhythm and pupillary light response.[168,169] Thus, even if all rod and cone photoreceptors are lost (as in retinitis pigmentosa, for example), these intrinsically photosensitive RGCs can continue to collect light for circadian rhythm entrainment, for example.

Optic nerve physiology

Retinal ganglion cell electrophysiology and synaptic transmission

In the retina, RGCs receive synaptic input from bipolar and amacrine cells, which primarily use glutamate as the major excitatory neurotransmitter. RGCs have both NMDA and non-NMDA ionotropic glutamate receptors, as well as metabotropic receptors.[170–175] Other neurotransmitters modulating RGC activity include GABA,[174] acetylcholine,[176,177] and aspartate (acting on NMDA receptors), as well as serotonin and dopamine.[178] Within the retina itself the levels of glutamate are modulated by Müller cells, which have glutamate transporters, and which contain the enzyme glutamine synthetase, converting glutamate to the amino acid glutamine. Glutamate reuptake within the retina is necessary for maintaining appropriate RGC function.[179-182] The electrophysiology of RGCs is complex,[183-188] and is discussed at length in Chapter 23 and elsewhere.[189] RGCs themselves are glutamatergic, but they also use other neurotransmitters, including substance P and serotonin.[190-194]

Axonal conduction

Action potentials

RGC axons transmit information via action potentials, which are all-or-nothing spikes of electrical activity. This is in contrast to most intercellular communication within the retina, where *graded potentials* are used to transmit information. With action potentials, the actual amount of voltage change, i.e. depolarization, is the same, while the number of impulses per second and the distribution of impulses within various axons is the mechanism by which visual information is carried down the optic nerve. Conduction down individual axons occurs via the same biophysical mechanisms which occur in any myelinated axon (see section on myelin, below). Variations in conduction velocity of individual axons may help coordinate conduction time between RGCs located at differing distances between their retinal location and the optic nerve.[195]

The mechanism of axonal conduction is straightforward. At rest, the inside of an axon is at a negative voltage with respect to the outside of the axon. This negative *resting potential* primarily results from the higher concentration of potassium within the axon compared to its extracellular concentration. As a small number of potassium ions flow down their concentration gradient from the inside of the axon to the outside, this *leak current* out of the axon results in a negative potential within the axon. Outward flux of potassium continues until the charge separation becomes too great, and can no longer be driven by the concentration difference between inside and outside the axon. The point at which the gradient for potassium concentration is balanced by the gradient for separation of charge results in an equilibrium potential for potassium. While the resting potential is primarily determined by potassium, there is also a smaller contribution from sodium flowing down its concentration gradient. In the case of sodium, the concentration in the extracellular space is high, compared with a low concentration within the axon. This concentration gradient would induce sodium to enter the axon, and movement of a small number of sodium ions results in a positive sodium equilibrium potential. However, in a resting (non-depolarizing) axon the conductance for sodium is much smaller than that for potassium, and therefore the final resting potential is weighted much more by the equilibrium potential for potassium than for sodium.

The situation changes during axonal conduction. Partial axon depolarization (resulting from an adjacent section of membrane that is already depolarized) induces opening of voltage-sensitive sodium channels located within the membrane. These allow much greater amounts of sodium to enter the axon, and the positive sodium ions cause the axon to become more positive, i.e. *depolarized*. In this case, the potential across the membrane is now weighted far more by the sodium equilibrium potential than by the potassium equilibrium potential. Why do the sodium channels open? These channels are voltage-sensitive, and a relatively small depolarization of the membrane will cause them to open transiently. Increased potential within the axon rapidly affects adjacent sections of the axon, and causes sodium channels in these adjacent areas of axonal membrane to likewise become partially depolarized. These then completely open, and cause a compete depolarization. This chain of events continues down the axon, resulting in a transmission of action potential.

Repolarization is the return of the axonal membrane potential to the original (negative resting potential) state, and is necessary for more than a single action potential to be transmitted down the axon. Repolarization is due to the closure of the voltage-sensitive sodium channels and a transient opening of voltage-sensitive potassium channels. Once the latter occurs, the membrane resting potential is weighted more by the potassium equilibrium potential than by the sodium equilibrium potential. This results in a restoration of membrane potential to the resting state.

In the process of transmission of an action potential there is movement of sodium and potassium ions along their concentration gradients. If action potentials continued indefinitely, the sodium and potassium concentrations would reach equilibrium across the axonal membrane, resulting in loss of their concentration gradients and conduction block. In order to re-establish these concentration gradients, there is a relatively slow redistribution process of these ions via the Na^+-K^+-ATPase. This is a highly energy-dependent process, and will fail to happen if axonal metabolism is disturbed.

Role of oligodendrocytes and myelin

Myelin confers two electrophysiological functions. First, it decreases capacitance, meaning less sodium ion-positive charge needs to enter the axon in order to depolarize the membrane. Second, it increases resistance, meaning that there is less leakage of charge across the membrane. Together, these properties decrease the amount of ionic flux needed to achieve changes in voltage across the membrane, saving on the Na^+-K^+-ATPase activity and thus energy needed to maintain ionic homeostasis after conduction.

Oligodendrocytes and myelin confer other critical properties for axonal conduction. Neonatal unmyelinated axons have a low density of sodium channels spread along the length of the axon, which allows a measure of axon conduction. In the adult, however, ion channels are not distributed uniformly along a myelinated axon. Instead, the channels are segregated into patches located within the small areas where the axon is unmyelinated, called nodes of Ranvier.[196] Conduction in a myelinated axon, called *saltatory* conduction, becomes much faster, as depolarization "jumps" from one node to the next. The clustering of sodium channels into nodes, as well as an important developmental switch in sodium channel isoforms, are also induced by oligodendrocytes.[197-199]

Role of astrocytes

Although neurons are classically considered the only electrically active cell population in the CNS, astrocytes also have ion channels. For example, multiple voltage-sensitive sodium channels have been demonstrated in rat optic nerve astrocytes.[196,200,201] Both sodium and potassium channels have been studied electrophysiologically in cultured astrocytes.[202,203] Within the intact optic nerve, non-vesicularly released glutamate from the axon may inducing glial cell spiking[204] and possibly other effects.[205] Whether ion channels or neurotransmitter receptors on optic nerve glia are physiologically important in vivo has yet to be demonstrated, but their presence suggests that optic nerve astrocytes may be more functionally specialized than previously recognized.

Axonal transport

The entire length of the RGC axon must be maintained by transporting proteins and other subcellular constituents many centimeters from the cell body, where the nucleus and most of the protein synthetic machinery resides. Axonal transport occurs in two directions, orthograde (away from the cell body and towards the brain), and retrograde (towards the cell body and away from the brain). By injecting radioactive, fluorescent, or enzymatically active macromolecular tracers into either the eye or the terminal fields of the axons, the course and timing of orthograde and retrograde transport can be determined. Studies have generally been performed in experimental mammals, with findings generally consistent across species.[206] The rates of transport of specific moieties may vary as a function of stage of optic nerve development, suggesting a developmental regulation of axonal transport.[207] Different rates of axonal transport may be related to differences in the motor proteins interacting with microtubules, e.g. kinesin[208] for fast transport and dynein[209] for slow transport.

Fast axonal transport, at 90–350 mm/day,[210,211] carries several subcellular organelles, such as the vesicles of neurotransmitter used in synaptic transmission, towards the axon terminal.[212] This class of transport continues despite transection of the axon distal to the cell body, suggesting that intraaxonal components are sufficient for the process to occur. Slow axonal transport is divided into two classes, a slower class at 0.2–1 mm/day that carries cytoskeletal proteins, such as neurofilament proteins and the tubulins that make up microtubules,[213] and a more rapid class at 2–8 mm/day, which carries proteins such as actin and myosin, metabolic enzymes, and mitochondria.[214-216] Slow axonal transport does not continue after transection of the axon distal to the cell body, unlike fast axonal transport.[217]

Retrograde transport occurs at about half the velocity of fast orthograde transport. Retrograde transport of neurotrophic molecules during development may signal the cell body that the axon is reaching its targets in the brain. How do growth signals at the axon get passed along to the distant cell body, where gene transcription and translation are induced to support survival and axon growth? Neurotrophin-mediated activation of Trk receptors leads to endocytosis of activated, ligand-bound receptors into clathrin-coated vesicles. These signaling endosomes carry with them the machinery of the ras-raf-MAP kinase signaling cascades, and can continue to activate these pathways once inside the cell.[218] Neurotrophin signaling endosomes may then be retrogradely transported back to the cell body along microtubules, where they can activate transcriptional regulators and induce new gene expression.[219-221] The relationship between vesicle localization and signaling effects remains a largely unanswered question, but is clearly important for the maintenance of normal axon physiology.

Optic nerve injury

Clinical implications

The optic nerve is commonly involved in disease, resulting in optic neuropathy and visual loss. Glaucoma is the most common optic neuropathy; others include inflammatory optic neuropathies such as optic neuritis associated with the demyelinating disease multiple sclerosis; ischemic optic neuropathy usually affecting older adults; and compressive optic neuropathy, commonly associated with a tumor or aneurysm. Although not proven, some researchers believe that Leber's hereditary optic neuropathy and some of the toxic optic neuropathies may reflect damage to the RGC body within the retina. Except for these, most optic neuropathies are due to damage directly to the optic nerve. Therefore, the vast majority of optic neuropathies involve axonal injury. Importantly, in all forms of axon injury, there is typically no capacity of RGC axons to regrow in the adult optic nerve, and thus little endogenous repair.

Types of optic nerve injury

Traumatic optic neuropathy

Although traumatic optic neuropathy, particularly optic nerve transection, is one of the least common human scenarios in which RGC axons are injured, it has been the best studied animal model by far. In humans, concussive injury can indirectly traumatize the optic nerve and injure axons, as might happen after a car accident or other blunt injury. Direct transection, though more rare, might result from a bullet, or fragment of broken orbital bone, or by optic nerve avulsion. In animal models, typically, the optic nerve is surgically cut or crushed within the orbit behind the eye, although experiments crushing the nerve intracranially have also been studied. Regenerative response is assayed by labeling RGC axons and determining whether any extend past the lesion site.[222] Interestingly, although injuring the optic nerve at increasing distances from the eye delays the onset of RGC death,[223] there may be less regenerative response with more distal injury.[224]

Ischemic optic neuropathy

Because the vascular supply to the optic nerve is complex and varies qualitatively along its course, a variety of clinical syndromes may result from ischemia or infarction at each location. Arteritic and non-arteritic anterior ischemic optic neuropathies are associated with histopathologically verifiable occlusion of posterior ciliary arteries. An infarction of the central retinal artery or one of its branches will result in RGC axon loss, with a pale disk and loss of axons within the optic nerve itself. Ischemia may also result from hypotension or severe blood loss, although the resulting pathophysiology may vary.[225,226] Finally, there is a tremendous controversy as to whether the effects of intraocular pressure causing axonal damage in glaucomatous optic neuropathy is an ischemic or compressive process.

The pathophysiology of myelinated axonal ischemia is complex.[227,228] The presence of myelination affects the metabolism of the axon. Because less current is necessary for depolarization to occur, less ATP is needed for transmission of action potentials. At the same time, the myelinated axon is more sensitive to anoxic damage. Although neonatal optic nerve, which is not myelinated, does not easily suffer irreversible damage from anoxia, more mature nerves from animals that have just undergone myelination are highly sensitive to anoxia. Similarly, optic nerves from *md* rats, a myelin-deficient rat strain, are sensitive to anoxia.[229] Finally, myelinated axons (white matter) are relatively less sensitive to ischemia than neuronal cell bodies (gray matter).[230]

Ischemic optic nerve injury can be modeled in experimental animals via a variety of approaches. A powerful method is to occlude the posterior ciliary arteries, best done in primates, but which has the disadvantage of also affecting the choroidal circulation.[231,232] Infusion of the vasoconstrictor endothelin-1 into the perineural space is a good model of subacute to chronic ischemia, and can be performed in a variety of species.[233–239] More recently, direct vascular damage, tissue edema, and vascular compromise have been modeled at the optic disk[240,241] and at the retrobulbar optic nerve.[242]

Optic neuritis and inflammation

Inflammation of the optic nerve, or optic neuritis, is the most common form of acute optic neuropathy in young and middle-aged adults, and is a frequent harbinger of multiple sclerosis.[243,244] Demyelination is the most common pathological accompaniment of optic nerve inflammation, and conduction block due to this or other effects of inflammation[245] is responsible for the usually temporary loss of visual function seen in patients with optic neuritis. Demyelination itself would not necessarily cause loss of RGCs, and as in multiple sclerosis, multiple rounds of demyelination and associated inflammation may be required for axonal loss.[246–248] Axonal loss in optic neuritis has been long appreciated clinically as optic atrophy and loss of the nerve fiber layer,[249] and experimental models of optic nerve inflammation can be used to elucidate how this axonal damage occurs.[250–252] It is likely that one major effect of inflammation is on the axonal cytoskeleton, as witnessed by changes in microtubule and neurofilament organization,[252] and by changes in axonal transport.[253–255]

In cases of acquired loss of myelin, for example in idiopathic optic neuritis, there is abnormal conduction because of the changes in resistive and capacitive properties of the membrane brought about by demyelination. This can happen via loss of the compact morphology of the myelin alone, i.e. without frank demyelination.[256] A low density of internodal sodium channels in adult optic nerve may allow some conduction after demyelination, as in optic neuritis.[257] Axonal conduction becomes slowed, however, and in some cases may be blocked ("conduction block"), both resulting in decreased visual function. Even in a completely demyelinated axon, a low density of internodal sodium channels in demyelinated optic nerve may still allow conduction.[257] Another phenomenon peculiar to demyelination is worsening of vision with heat or exercise, or *Uhthoff's phenomenon.* Increased temperature and exercise are thought to decrease the sodium channel open-time during depolarization, resulting in less charge entering the axon, and a decreased

likelihood that an adjacent section of demyelinated axon will be able to depolarize enough to cause opening of its own voltage-sensitive sodium channels. This leads to temperature-sensitive conduction block.

Optic neuritis has been modeled in rat and mouse models of experimental autoimmune encephalomyelitis, in which rodents' immune systems are stimulated to react against myelin-associated proteins. Immunization against myelin-oligodendrocyte glycoprotein (MOG) induces multiple sclerosis-like demyelination throughout the brain including in the optic nerve,[258] but creating transgenic mice with T cells directed against MOG, or passively transferring anti-MOG T cells in rats, creates demyelinating disease largely confined to the optic nerve, mimicking optic neuritis.[259,260] In such models, RGC axons are incidentally severed and fail to regenerate, and as after optic nerve trauma, RGCs die with a 1–2 week delay.[261]

Compressive optic neuropathy

Compression of the optic nerve is a common clinical correlate of a large number of optic neuropathies. Besides the obvious causes such as neoplasms and aneurysms, the optic nerve can also be compressed by enlarged extraocular muscles (as in Grave's ophthalmopathy), edema (as seen with indirect traumatic optic neuropathy, where the nerve is contused within the optic canal), or optic disk drusen. Compression intrinsic to the nerve itself may occur in some forms of glaucomatous optic neuropathy, where increased intraocular pressure may cause bowing out and shifting of the lamina cribrosa, constricting the bundles of optic nerve axons within the cribrosal pores.[262] Although all three layers of meninges are present in the intraorbital and intracanalicular optic nerve, only the pia continues along the intracranial optic nerve. The clinical relevance of this is that a tumor arising from the optic nerve meninges themselves, i.e. an optic nerve sheath meningioma, will normally continue through the optic canal, but then extend along the sphenoid bone, and not along the course of the intracranial optic nerve or optic chiasm. This means that it is extremely rare for an optic nerve sheath tumor to extend towards the chiasm, and thereby affect the other side. The more common way for optic nerve sheath meningiomas to become bilateral is by directly extending along the sphenoid bone meninges.

The effects of chronic experimental compression of the intraorbital optic nerve have been delineated at the microscopic and ultrastructural level.[263,264] Demyelination occurs initially, followed by remyelination, even while the axons are still compressed, with relatively little axonal loss. Demyelinated axons or the direct effects of pressure might be expected to lead to conduction block, which would be reversible. This could therefore explain the remarkable return of visual function after removal of tumors compressing the optic nerve.[265]

Glaucoma

The most common optic neuropathy is glaucomatous optic neuropathy, distinguished by a distinct morphology of progressive excavation of the nerve head without significant pallor of the remaining neuroretinal rim. Within the retina, there are decreased numbers of RGC bodies in glaucoma,[266–268] and this likely reflects death by apoptosis.[269–272]

The number of RGCs lost correlates with the visual field deficit.[273] In addition to the RGC cell body loss, there is loss of the ganglion cell axons, manifested by segmental loss of the nerve fiber layer,[274–278] increased cup-to-disk ratio, thinning of the optic nerve[279] and chiasm,[280] changes in postsynaptic cell counts within the lateral geniculate nucleus[281–284] (the main target of RGC axons in higher animals), and in the cerebral cortex.[285,286]

That the site of primary injury is likely at the optic disk is consistent with a wide variety of evidence demonstrating pathology at that level,[268,287] particularly at the lamina cribrosa.[262,288–299] The focal areas of optic nerve axonal injury correlate with the focal areas of cell body loss in the retina and visual field defects.[22,23] Thus the most typical glaucomatous visual field defects spread in an arcuate pattern but stop at the horizontal meridian in the nasal field. This site in the visual field corresponds to the temporal raphe of the nerve fibers within the retina. Although there is a small distance between adjacent RGCs across the temporal raphe, the axons that correspond to these RGCs are separated by a large distance at the optic disk (Fig. 28.1). In those cases when defects are seen on both sides of the horizontal meridian, it can be seen that spread has occurred in both the superior and inferior visual fields, but not directly across the meridian. Thus the spread of visual field loss in glaucoma reflects the spread of pathology at the optic nerve head, followed later by RGC cell body loss in the retina.

What mechanisms induce RGC death in glaucoma? Glaucoma has been extensively modeled in animals (see reviews[300,301]). Studies of tissue from human patients with glaucoma and non-human primates and other mammals with experimental glaucoma confirm changes at the optic nerve head, such as with respect to bowing out of the lamina cribrosa, intra-axonal accumulation of organelles (consistent with blocked axonal transport; Fig. 28.6), and Wallerian degeneration distal to the lamina cribrosa.[262,302,303] Whether due to mechanical trauma of axons,[223,304–306] ischemia,[238] generation of nitric oxide,[307,308] or other causes, axonal injury causes changes in RGCs, eventually resulting in death. Increased intraocular pressure perturbs rapid anterograde and retrograde axonal transport at the lamina cribrosa.[309–312] This may cause RGCs to be deprived of neurotrophic factors or other survival signals produced by brain targets. Studies in experimental animals have shown that injury to the optic nerve, for example from increased intraocular pressure in experimental glaucoma, blocks the retrograde transport of the neurotrophic factor brain-derived neurotrophic factor (BDNF) along with its associated receptor, TrkB.[313–315] Interestingly, one of the genes associated with a hereditary form of glaucoma, optineurin,[316] may in its mutated form contribute to RGC sensitivity to optic nerve insults by failing to adequately participate in retrograde transport of neurotrophin signaling complexes, or by oxidative dysregulation.[317–320]

Papilledema

Papilledema is defined as optic disk edema secondary to elevated intracranial pressure, and should not be used nonspecifically to denote disk edema. Pathological examination of optic nerve heads with papilledema reveals intra-axonal edema, consistent with abnormalities of axonal transport. However, the pathogenesis of papilledema is controversial. While primate experiments show definite abnormalities of

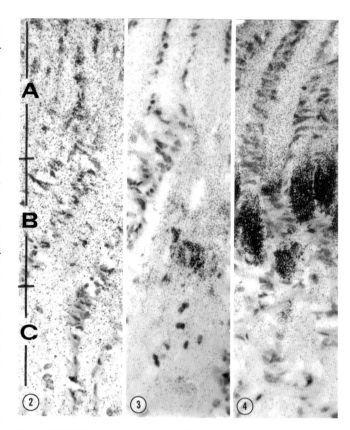

Figure 28.6 Failure of axoplasmic flow in the optic nerve injury associated with elevated intraocular pressure. The radiographs capture the optic nerve through the prelaminar region (**A**), the lamina cribrosa (**B**), and the post-laminar orbital optic nerve (**C**). Optic nerves were harvested after tritiated leucine was injected into the eyes of normal monkeys (left), or eyes of monkeys with moderately (center) or more severely (right) experimentally elevated intraocular pressure. With increasing IOP, increasing accumulation of radioactive label is appreciated at the lamina cribrosa, with an associated decrease in grains transported posterior to this point. (From Figures 2–4 of Anderson & Hendrickson, Invest Ophthalmol 1974; 13(10):771. Reproduced with permission from Association of Research in Vision and Ophthalmology.)

slow axonal transport,[321,322] evidence for abnormalities of fast axonal transport (which would include retrograde transport) is less clear,[322,323] and in part depends on experimental paradigms where papilledema is simulated by ocular hypotony.[324,325] It is also unclear whether the visual loss associated with chronic papilledema results from disturbances of axonal transport, or from ischemia due to congestion of the optic nerve head.

Retinal ganglion cell death after optic nerve injury

Death of RGCs is the final common pathway underlying virtually all diseases of the optic nerve, including glaucomatous optic neuropathy, anterior ischemic optic neuropathy, optic neuritis, and compressive optic neuropathy. In many diseases affecting the RGC axons (e.g. glaucoma or arteritic ischemia), visual loss is permanent, reflecting the fact that RGC loss is irreversible. In some cases (e.g. chronic compressive optic neuropathy, acute optic neuritis, or papilledema) visual loss can be reversed when the axonal damage is relieved, presumably because RGC death has not yet occurred. Even in glaucoma, there is now evidence that there is some period of RGC electrophysiological dysfunction and visual field deficit, that precedes frank cell death.[326–328] However, if allowed to continue, these disorders eventually can result in permanent RGC death.

Time course

The rate and timing of RGC death after optic nerve injury depends on the species, age, RGC cell body size, and the distance from the site of injury to the RGC. For example, after optic nerve injury in goldfish and frogs, ganglion cells do not die, but hypertrophy and regenerate axons within 1–2 months.[329–331] Conversely, in adult rodents, optic nerve injury leads to initiation of cell death within 5–7 days, and 50–90 percent cell death taking weeks to sometimes months.[304–306,332–335] Similarly, ganglion cells from squirrel and owl monkeys die approximately 4–6 weeks after axotomy.[336,337] In humans with chiasmal compression for at least 6 months, there is extensive loss of ganglion cells from the nasal hemiretinas; yet even 35 days after transection of the ipsilateral optic tract there is sparing of some temporal ganglion cells.[338] In contrast, neonatal RGCs are far more sensitive to axotomy than adults,[304,339,340] perhaps because naturally occurring developmental cell death is taking place at the same time. As an example, in postnatal rodents, 90 percent of RGCs die within 48 hours of optic nerve crush.

It is controversial whether the response of ganglion cells to axotomy depends on the distance between the lesion and the cell body. Some studies have shown that the shorter that distance, the more rapid the degeneration[332,333] and the more severe the effect.[341–344] This could be due to lack of a trophic signaling along a length of residual axon, or more rapid transmission of a signal mediating cell death, either mechanism possibly mediated through retrograde axonal transport.[306,337,345] Others have suggested that the timing of ganglion cell degeneration is not correlated with the distance to the lesion.[336,337] Axotomized fibers from the macula degenerate later than those more peripheral, suggesting that the distance from the site of axotomy to the retina does not affect timing of RGC death.[337] Differences in ganglion cell survival also depend on the site of optic nerve transection in relation to the vascular supply of the retina. If the ophthalmic or central retinal arteries are transected, then the inner retina, including the ganglion cell layer, will be infarcted,[346] rapidly speeding RGC death.

Apoptosis

A variety of techniques have been used to show that the death of RGCs after axotomy occurs by apoptosis.[223,269,347–350] Pathological studies show condensation of the nucleus and cytoplasm, budding of the cytoplasmic membrane, cleavage of DNA into 180–200-bp fragments, fragmentation of the cell into membrane-bound bodies, and heterophagy of these bodies.[351,352] The apoptosis program is an intricate and sequential activation of multiple intracellular enzymes, including cysteine-dependent aspartate-directed proteases (caspases), and endonucleases, regulated from the mitochondrial membrane by the Bcl-2 family of proteins. Proapoptotic family members such as Bax, Bak, and Bok are held in check by antiapoptotic members such as Bcl-2 and Bcl-xL. A third group in this family, including Bim and Bad,

cannot activate apoptosis by themselves, but can facilitate apoptosis by interfering with Bcl-2 or Bcl-xL's ability to regulate Bax, for example. If anti-apoptotic proteins are unable to balance the proapoptotic family members, either because they are sequestered by Bim-like proteins, or are in short supply due to inadequate production, the anti-apoptotic proteins will increase mitochondrial membrane permeability, and enhance release of cytochrome c into the cytoplasm. Cytochrome c binds to an adaptor protein Apaf-1, forming a complex that can initiate a cascade of enzymatic degradation of cellular components by a family of enzymes called caspases, beginning with the activation of procaspase-9. Also released from the mitochondria, a protein called Smac/Diablo binds to and inactivates an anti-apoptotic protein called Inhibitor of Apoptosis (IAP), which normally represses activation of procaspases. Thus by at least two mechanisms, mitochondrial permeabilization by Bax inclines the balance towards apoptosis. Once the *apoptosome* is formed from the factors released from the mitochondria, the activation of caspases takes place. These in turn activate other caspases, and this cascade of events results in the activation of other enzymes within the cell which results in the breakdown of DNA and other macromolecules, membrane-bound compartmentalization of cell components, and then phagocytosis by adjacent cells.

Besides axonal injury, several other modes of cell death may induce apoptosis. For example, the binding of tumor necrosis factor to tumor necrosis factor receptors may induce apoptosis via death domain signaling and subsequent activation of caspase-8, initially bypassing the caspase-9 activation by the mitochondria. Glutamate excitotoxicity may also induce apoptosis,[353] and RGCs are particularly susceptible to high levels of glutamate in the extracellular space.[175,354–356] It has been proposed that excitotoxic RGC death may explain pathological conditions, such as glaucoma.[357] The evidence supporting this is that antagonists of glutamate excitotoxicity mediated by NMDA receptors generally decrease RGC death after optic nerve injury,[358,359] and that there may be elevations of glutamate levels within the eye after optic nerve injury,[360] or experimental or human glaucoma.[361] Changes in NMDA receptor subunit expression follow axonal injury,[362] and may underlie changes in susceptibility to excitotoxicity. Variability in the sensitivity of RGCs to excitotoxicity in animal models,[363] and the failure of clinical efficacy of NMDA receptor inhibition for prevention of progression in glaucoma, however, raises significant questions about this mechanism.[364]

Signaling of axonal injury

How does axonal injury result in induction of apoptosis? RGC death after optic nerve injury or target removal appears to take place through several mechanisms, including lack of neurotrophic factors from the target tissue, excitotoxicity from physiological or pathological levels of glutamate, free radical formation, leakage of cellular constituents out the end of the axon, proliferation of macroglia, activation of microglia, build-up of excess retrogradely transported macromolecules, and breakdown of the blood–brain barrier.[117,332,341,343,359,365–367] Even the contralateral retina may be affected by a unilateral optic nerve injury, suggesting the involvement of intrachiasmal signaling or axon fibers that run towards the retina.[368]

RGCs are highly dependent on neurotrophic factors during development and in the adult for their survival. Support for the role of neurotrophin dependence comes from experiments using identified neurotrophic factors to rescue injured neurons, including RGCs after axotomy in adult animals.[117,118,335,369-375] The survival of ganglion cells in adult animals is facilitated by BDNF, FGF, GDNF, CNTF, and conditioned medium from transected sciatic nerves.[118,376-380] Purified neonatal RGCs (which are axotomized during dissociation) can be kept alive for significant periods beyond the normal time of developmental death with a cocktail of factors, including BDNF, CNTF, forskolin, and insulin.[381] Supplying neurotrophin alone is insufficient, as responsiveness to neurotrophin also appears to require cAMP elevation[382,383] or physiological levels of electrical activity.[113] Microglia may play a role in ganglion cell death, based on experiments using macrophage inhibitory factor (MIF) to inhibit microglial activation and enhance ganglion cell survival after axotomy.[384]

Our understanding of how neurotrophic factor deprivation causes neuronal cell death derives from understanding how these factors normally maintain neuronal survival. For example, TrkB functions as a receptor for BDNF, and both TrkB and BDNF are expressed in RGCs during postnatal development.[385] TrkB itself is regulated by activity or cAMP levels.[383] TrkB binding by BDNF causes autophosphorylation, inducing binding of specific adapter proteins, and activation of a cascade of GTP-binding proteins and kinases called the Ras-Raf-MAP kinase cascade.[386] Ras activity also leads to activation of PI3-kinase, which transduces a separate set of signals through activation of Akt (protein kinase B).[387,388]

If the optic nerve is injured, BDNF or other neurotrophins introduced into the vitreous or retina increases survival of RGCs.[117,118,335,369-371] However, there is no direct evidence that adult RGCs are normally dependent on BDNF or other neurotrophic factors, and it is possible that the increased survival seen after neurotrophin administration is not a *specific* rescue from neurotrophin deprivation, but rather a generalized pro-survival effect. Furthermore, retrograde axonal transport is rapid, and the subacute time-course by which RGCs die after axonal injury does not reflect the time-course of interrupted retrograde axonal transport. This probably explains why damage to the RGC axon in several different places results in similar rates of death after axonal injury.

Phagocytosis and immune activation

Optic nerve injury also leads to activation of local glial cells and recruitment of immune cells. For example, during Wallerian degeneration, resident microglia as well as circulating monocytes phagocytose myelin debris,[72] and there may also be a contribution from oligodendrocytes and astrocytes.[77,104,389-391] Probably most phagocytes in the degenerating optic nerve are derived from the circulation, although not all studies are in agreement.[392,393] A partial answer to the contradictory reports of phagocytic cells after optic nerve injury may come from a study in which several zones of the monkey nerve were examined after transection behind the globe.[394] Under these conditions, phagocytosis by macrophages of presumed hematogenous origin were seen focally, whereas endogenous cells, perhaps oligodendrocytes, phagocytosed at distant sites of Wallerian degeneration. This is supported by the findings of ED1-positive macrophages in regions of GFAP-negative crushed optic nerve.[395] Microglia can be observed to invade the myelin sheath and disrupt the myelin lamellae, the fragments of which are then phagocytosed.[396] Deficiencies in the activation of microglia may lead to decreased clearance of myelin debris, and subsequent inhibition of axonal regeneration;[397] conversely, increasing macrophage-mediated clearance of debris can increase regeneration.[398]

Gliosis

The role of optic nerve astrocytes in the pathologic processes of gliosis is complex.[399,400] As a reaction either to local injury or to remote death of adjacent axons, astrocytes hypertrophy and extend processes, partly as a type of scar formation, but likely also as part of the process of recruiting or modulating other cell types. Astrocyte activation after injury may explain in part the failure of ganglion cell axons to regenerate after injury[44] (see below).

Failure of axon regeneration

In adult mammals, there is essentially no axon regenerative response after optic nerve injury. This contrasts with lower vertebrates – for example, not only do goldfish RGCs survive after their axons are transected,[401] but they are also able to extend neurites and ultimately establish correct connections with their targets.[402] Research in the early 20th century showed that mammalian RGCs, like other centrally projecting neurons, make only abortive attempts to regenerate their axons. Even when conditions permit axonal extension, only a small minority of RGCs eventually extend their axons. For example, severed rodent RGC axons approach the optic nerve head but then reverse and meander in another direction.[403] Analysis of the proximal stump of the transected optic nerve just posterior to the optic nerve head shows early sprouting termed *abortive regeneration*, which then fails to continue and dies back.[404]

Thus in most cases, optic nerve injury and axon loss leads to permanent vision loss. Our understanding of the mechanisms preventing axon regeneration in adult mammals continues to advance, particularly with respect to the role of inflammation, discussed above, of glial-associated inhibitory signals, and of neuron-intrinsic limitations to axon repair, discussed below.

Glial inhibition of neurite extension

Since the early experiments of Aguayo and colleagues, it has long been known that RGCs and other central nervous system (CNS) neurons regenerate axons into peripheral nervous system (PNS) grafts (e.g. sciatic nerve), but not into CNS tissue.[222,405] Since RGCs do not extend axons in the absence of specific extracellular signals (discussed above), regenerative failure might be explained in part by a relative inability of mature CNS astrocytes and oligodendrocytes to secrete trophic signals after injury. Evidence has supported this hypothesis, but it turns out that astrocytes and oligodendrocytes also actively inhibit axons from regenerating. The non-permissive nature of the CNS (but not PNS) substrate for axonal elongation is therefore of great interest, and has resulted in a large number of candidate molecules. The most studied inhibitory molecules are components of oligodendrocyte myelin and proteoglycans expressed by astrocytes, and these molecules and their signaling mechanisms have recently been reviewed in detail.[406]

Resident microglia and/or macrophages recruited to the site of injury may also inhibit or fail to support axon growth. Although macrophages/microglia do increase in number at the site of injury,[395,407] they may not be appropriately activated for support of axonal extension,[398,408] and may differ from other macrophages in fundamental ways.[409] If activated macrophages/microglia are poorly able to phagocytose degraded myelin, then the inhibitory signals found in myelin may prevent axonal regeneration.[397]

Neuron-intrinsic limitations to axon regeneration

Interest has also focused on the possibility that failure of regeneration reflects a specific inability of an *adult* RGC to extend its axon, since embryonic axons are able to do so when transplanted into adult tissue. Recent research has focused on determining whether the rate and extent of axon growth depends purely on extracellular signals and substrates, as discussed above in regards to optic nerve development, or also on the intrinsic state of the neuron. Elsewhere in the CNS, embryonic neurons can regenerate their axons quite readily, but they lose their capacity to regenerate with age.[410–412] This developmental loss of regenerative ability has generally been attributed to the maturation of astrocytes and oligodendrocytes, and to the production of myelin, all of which strongly inhibit axon growth after injury. In experiments in which the inhibitory environment is removed or molecularly blocked, however, only a few percent of axons regenerate, and functional recovery typically proceeds remarkably slowly (discussed further below). For example, RGCs may take two months to regenerate through a peripheral nerve graft.[413,414]

These experiments indicate that the neurons themselves are partly responsible for regenerative failure. For example, axons from postnatal day 2 or older hamster retinas lose the ability to reinnervate even embryonic tectal explants.[415] When cultured in the complete absence of CNS glia, embryonic RGCs extend their axons 10 times faster than postnatal or adult RGCs.[416] Interestingly, the decrease in axon growth ability of RGCs occurs sharply at birth during the period of target innervation, but target contact may not be responsible for this change. Rather, retinal maturation, and possibly a membrane-associated signal from retinal amacrine cells, may be sufficient to induce the developmental decrease in RGC axon growth ability.[416] Once this signal is sent, the loss is permanent – removal of the amacrine cells does not allow the RGCs to speed up again. This may explain the failure of RGC regeneration, at least in part – even in a perfect environment, RGCs may lack the intrinsic capacity to rapidly grow their axons again, after they have done it once during development.

What molecular changes underlie the developmental loss in rapid axon growth ability? RGCs could gradually increase expression of genes that limit axon growth or decrease expression of genes necessary for faster axonal elongation, or both. For instance, in the developing chick, RGCs lose axon growth responsiveness to laminin either by down-regulating specific integrins (laminin receptors),[417–419] or by down-regulating the activation of such integrins,[420] although they continue to respond to laminin-2 (merosin).[421] Enhancing integrin signaling enhances regeneration in animal models.[422] Similarly, a difference in trophic receptor levels or responsiveness to trophic signals could explain these differences. For example, depolarization rapidly elevates TrkB receptors on the surface of CNS neurons.[381,423]

Indeed, almost any molecule in the cell could be crucial to determining axon growth.[131] It is attractive to first consider transcription factors that could activate the expression of whole programs of axon growth, and recent data have pointed to a family of transcription factors that may regulate axon regenerative ability in RGCs.[424,425] Transcription factors are not the only master regulators in a cell, however. Levels of specific complements of proteins can be regulated post-translationally by ubiquitination and degradation; the ubiquitin ligase anaphase-promoting complex (APC) has a significant role in regulating the axon growth rate of cerebellar granule neurons.[426] Cytoplasmic kinases or phosphatases such as Ca^{2+}-dependent signaling kinases, calmodulin kinases (CaMKs), may provide a downstream mechanism for axon growth control as well.[427] Second messengers such as cAMP and cGMP have also been implicated in the RGC's intrinsic axon regenerative control, perhaps by modulating responsiveness to positive neurotrophic factors[113] or to myelin-associated axon growth inhibitors.[428] Finding the molecules that, when modified, truly enhance functional regenerative repair remains a major ongoing goal in optic nerve research today.

Optic nerve repair

The major goal of any optic nerve repair strategy must be to protect or regenerate the connections of RGC axons to their targets in the brain, as these connections are ultimately what serve vision.

Optic nerve remyelination

In the case of optic neuritis, where the primary defect in RGC function is attributable to oligodendrocyte demyelination, strategies to enhance remyelination are paramount. Currently, the major approach involves treating patients with steroids, which presumably interferes with an ongoing inflammatory insult and allows a faster re-wrapping of optic nerve axons and return to baseline vision. In the presence of demyelination, oligodendrocytes may reconstitute themselves and remyelinate axons,[429] although the nature of the inducing chemical signals has not been well characterized.[430] Similarly, chronic treatment with immune modulators decreases the risk of subsequent demyelinating events.

Neuroprotection and retinal ganglion cell survival

A first step to protecting optic nerve axons from a degenerative process is preventing RGC death. Neuroprotection, first proposed for other central nervous system diseases, has been considered in recent years as a possible treatment for optic neuropathies.[431,432] Neuroprotective strategies include blocking retinal excitotoxicity mediated by glutamate (which binds to N-methyl-D-aspartate [NMDA] receptors and induces RGC death);[433] activation of small molecule receptors, which may enhance neuronal resistance to insult;[434,435] inhibition of nitric oxide synthases, which may prevent axonal injury at the lamina cribrosa;[436] and immunization with certain synthetic polypeptides.[437] Furthermore, as was seen for RGCs in vitro,[113] elevation of cAMP levels in RGCs enhances their regenerative ability after optic nerve injury in vivo.[438-440] None of these strategies has yet been successfully proven in randomized controlled clinical trials in human patients with glaucoma. However, there is a variety of laboratory evidence using cell culture or animal models of optic nerve disease to suggest that one or more neuroprotective methodologies eventually may be successful in clinical disease.

Most promising may be administration of neurotrophic factors which maintain RGC survival.[117,335,370] Indeed, ample evidence exists to support a role for neurotrophins in stimulating RGC survival in vivo. The same peptide trophic factors that stimulate RGC axon growth in vitro enhance survival and axon regeneration in vivo. GDNF,[441-443] BDNF,[444-446] and CNTF[447] strongly promote RGC survival. Generally, these neurotrophic factors tend to work on any optic neuropathy model. For example, CNTF has a potent effect in enhancing RGC survival after optic nerve transection in the adult rat;[118,438,448-450] in the adult mouse;[451,452] and in the adult cat.[440,453] Sustained delivery with viral vectors may be as or more effective.[452,454,455] Similarly, peptide trophic factors have proven neuroprotective in multiple animal models for ocular hypertension or glaucoma.[447,456,457] Thus neurotrophic factors provide a potent approach for RGC neuroprotection in multiple optic neuropathies.

Similarly, the convergence of many cell death signals onto the common pathways of apoptosis (discussed above) provides another avenue for therapeutic intervention and neuroprotection. Interfering with the apoptosis cascade at any of these points may provide a mechanism to protect RGCs from death after optic nerve injury. Inhibitors of cytochrome c, mitochondrial membrane depolarization, caspase activation, and IAP signaling are all in development for neuroprotection in the eye, brain, or elsewhere in the body.

Regeneration of RGC axons

The same neurotrophic factors that strongly promote retinal ganglion cell survival also tend to strongly promote RGC axon regeneration in vitro[113] and in vivo.[452] Recent evidence has extended the list of neurotrophic signaling molecules beyond these traditional families. For example, a small, macrophage-derived, calcium-binding protein named oncomodulin binds RGCs and strongly promotes RGC axon growth in vitro and in vivo.[458]

Furthermore, our understanding of the mechanisms of axon growth has generated much progress in understanding regenerative failure. For example, understanding that the regulation of both repulsion and collapse of growth cones by inhibitory molecules rely on the same intracellular signaling pathways suggests that manipulating these common regulators can increase regeneration. Thus, blocking the growth cone collapsing activity of the small GTPase rho increases RGC regeneration in the inhibitory environment of the optic nerve in vivo.[459] Such manipulations may be

combined with cAMP elevation and trophic factor delivery to enhance regeneration even further.[460]

Does the loss of intrinsic axon growth ability by RGCs[416] contribute to their failure to regenerate after injury in the adult? As mentioned above, a loss of intrinsic axon growth ability could explain why in many animal models of optic nerve injury, regeneration proceeds remarkably slowly, even when glial inhibitory cues have been removed. For example, most RGCs take 2-3 months to regenerate through peripheral nerve grafts to the superior colliculus.[413,414] This is far slower than the 10 days they would take if they elongated their axons at an embryonic growth rate of 10 mm/day. It is not clear whether this developmental switch will be reversible, either: soluble signals from optic nerve glia or peripheral nerve glia, or from retinal or superior collicular cells were not able to reverse the loss of rapid axon growth ability in RGCs in vitro, suggesting that the developmental switch may normally be permanent.[416] There remains the possibility, however, that other signals, or the discovery and manipulation of genes involved in this transition, may revert the postnatal neurons to their embryonic axon growth ability, and that this may be critical to increase regeneration in the CNS. Thus a combinatorial approach may be required – both the intrinsic neuronal growth state and the extrinsic environment may have to be optimized for successful regeneration after injury.

"Neuroenhancement" of retinal ganglion cell function

Before RGCs have died, and even before their axons are fully damaged in the optic nerve and would require a proregenerative therapy, RGCs may be partially inhibited in the optic nerve from their full functional capacity. For example, glaucoma causes dysfunction and then death of RGCs, but the window between dysfunction and death in humans is not currently known. Acute studies of IOP raising and lowering in humans, and chronic studies of IOP raising and lowering in animal models, both suggest that a window between dysfunction and death exists in this degenerative disease.[326-328] Thus, factors known to enhance RGC health and, ultimately, survival, could be applied to intervene before death and potentially reverse dysfunction, an approach termed "neuroenhancement". These remain major goals for the biomedical community for patients with glaucoma.

Neuroenhancement has been used in the literature either generically to refer to improvement of axon regeneration, or more specifically to improvement of cognitive function in degenerative brain disease.[461-463] Compared to the many years it would take a glaucoma trial to demonstrate neuroprotection (e.g. one would ideally demonstrate that progression of visual field losses is slowed or halted), it may be possible to measure an enhancement of RGC structure or function acutely. A potential "neuroenhancement" therapy would target RGCs that are dysfunctional and/or atrophic. With the same neurotrophic factors that stimulate neuroprotection in the long term, such dysfunctional RGCs could hypertrophy to their normal size, or resume normal electrical responsiveness or optic nerve transmission of visual information, or both. It is likely that the same neurotrophic factors that stimulate RGC survival, such as GDNF,[441-443]

BDNF,[444-446] or CNTF,[447] will also stimulate measurable neuroenhancement in these cells.

Conclusions

The optic nerve is the critical conduit from the eye to the brain. The past decade has seen dramatic increases in our fundamental understanding of optic nerve physiology and pathophysiology, particularly at the cellular and molecular levels, but much remains to address the many optic neuropathies that lead to vision loss and blindness. The next decade will likely prove even more exciting, as advances in regenerative medicine begin to transition from the bench to the clinic.

Acknowledgments

Supported by NIH EY020297, EY017971 and NS061348 (JLG); the Glaucoma Foundation (JLG); and unrestricted departmental grants from Research to Prevent Blindness, Inc. I am indebted to Leonard Levin for the foundation of this chapter from his writing in the prior edition, and I gratefully acknowledge Raul Corredor for sharing preliminary writing. Portions of this chapter were adapted from:

(1) Goldberg JL. Signals regulating growth and regeneration of retinal ganglion cells. In: Chalupa L ed. Eye, retina and visual system of the mouse. Boston: MIT Press; and

(2) Levin LA. Optic nerve. In: Kaufman PL, Alm A, eds. Adler's Physiology of the eye. St Louis: CV Mosby, 2002.

References

1. Curcio CA, Allen KA. Topography of ganglion cells in human retina. J Comp Neurol 1990; 300(1):5–25.
2. Linden R. Displaced ganglion cells in the retina of the rat. J Comp Neurol 1987; 258(1):138–143.
3. Ogden TE. Nerve fiber layer of the macaque retina: retinotopic organization. Invest Ophthalmol Vis Sci 1983; 24(1):85–98.
4. Minckler DS. The organization of nerve fiber bundles in the primate optic nerve head. Arch Ophthalmol 1980; 98(9):1630–1636.
5. Wolff E. The anatomy of the eye and orbit. Philadelphia: Blakiston, 1948:263.
6. Fitzgibbon T, Taylor SF. Retinotopy of the human retinal nerve fibre layer and optic nerve head. J Comp Neurol 1996; 375(2):238–251.
7. Reese BE, Ho KY. Axon diameter distributions across the monkey's optic nerve. Neuroscience 1988; 27:205.
8. Jeffery G. Distribution and trajectory of uncrossed axons in the optic nerves of pigmented and albino rats. J Comp Neurol 1989; 289:462.
9. Chan SO, Guillery RW. Changes in fiber order in the optic nerve and tract of rat embryos. J Comp Neurol 1994; 344(1):20–32.
10. Naito J. Retinogeniculate projection fibers in the monkey optic nerve: a demonstration of the fiber pathways by retrograde axonal transport of WGA-HRP. J Comp Neurol 1989; 284(2):174–186.
11. FitzGibbon T. The human fetal retinal nerve fiber layer and optic nerve head: a DiI and DiA tracing study. Vis Neurosci 1997; 14(3):433–447.
12. Maniscalco JE, Habal HB. Microanatomy of the optic canal. J Neurosurg 1978; 48(3):402–406.
13. Renn WH, Rhoton AL Jr. Microsurgical anatomy of the sellar region. J Neurosurg 1975; 43(3):288–298.
14. Kupfer C, Chumbley L, Downer JC. Quantitative histology of optic nerve, optic tract and lateral geniculate nucleus of man. J Anat 1967; 101(3):393–401.
15. Schmid R, Wilhelm B, Wilhelm H. Naso-temporal asymmetry and contraction anisocoria in the pupillomotor system. Graefes Arch Clin Exp Ophthalmol 2000; 238(2):123–128.
16. Balazsi AG et al. The effect of age on the nerve fiber population of the human optic nerve. Am J Ophthalmol 1984; 97:760.
17. Mikelberg FS et al. The normal human optic nerve: axon count and axon diameter distribution. Ophthalmology 1989; 96:1325.
18. Mikelberg FS et al. Relation between optic nerve axon number and axon diameter to scleral canal area. Ophthalmology 1991; 98(1):60–63.
19. Quigley HA et al. The size and shape of the optic disc in normal human eyes. Arch Ophthalmol 1990; 108(1):51–57.
20. Quigley HA, Coleman AL, Dorman-Pease ME. Larger optic nerve heads have more nerve fibers in normal monkey eyes. Arch Ophthalmol 1991; 109(10):1441–1443.
21. Falck FY, Klein TB, Higginbotham EJ. Larger optic nerve heads have more nerve fibers in normal monkey eyes. Arch Ophthalmol 1992; 110(8):1042–1043.
22. Harwerth RS, Quigley HA. Visual field defects and retinal ganglion cell losses in patients with glaucoma. Arch Ophthalmol 2006; 124(6):853–859.
23. Kerrigan-Baumrind LA et al. Number of ganglion cells in glaucoma eyes compared with threshold visual field tests in the same persons. Invest Ophthalmol Vis Sci 2000; 41(3):741–748.
24. Harman A et al. Neuronal density in the human retinal ganglion cell layer from 16–77 years. Anat Rec 2000; 260(2):124–131.
25. Jonas JB, Gusek GC, Naumann GO. Optic disc, cup and neuroretinal rim size, configuration and correlations in normal eyes. Invest Ophthalmol Vis Sci 1988; 29(7):1151–1158.
26. Jeffery G et al. The human optic nerve: fascicular organisation and connective tissue types along the extra-fascicular matrix. Anat Embryol 1995; 191(6):491–502.
27. Repka MX, Quigley HA. The effect of age on normal human optic nerve fiber number and diameter. Ophthalmology 1989; 96:26.
28. Sanchez RM, Dunkelberger GR, Quigley HA. The number and diameter distribution of axons in the monkey optic nerve. Invest Ophthalmol Vis Sci 1986; 27:1342.
29. Baker GE, Stryker MP. Retinofugal fibres change conduction velocity and diameter between the optic nerve and tract in ferrets. Nature 1990; 344:342.
30. Colello RJ, Pott U, Schwab ME. The role of oligodendrocytes and myelin on axon maturation in the developing rat retinofugal pathway. J Neurosci 1994; 14(5 Pt1):2594–2605.
31. Fernandez E et al. Visual experience during postnatal development determines the size of optic nerve axons. Neuroreport 1993; 5(3):365–367.
32. Nixon RA et al. Phosphorylation on carboxyl terminus domains of neurofilament proteins in retinal ganglion cell neurons in vivo: influences on regional neurofilament accumulation, interneurofilament spacing, and axon caliber. J Cell Biol 1994; 126(4):1031–1046.
33. Anderson DR, Hoyt WF. Ultrastructure of intraorbital portion of human and monkey optic nerve. Arch Ophthalmol 1969; 82:506.
34. Sturrock RR. Development of the meninges of the human embryonic optic nerve. J Hirnforsch 1987; 28:603.
35. Friedrich VL, Mugnaini E. Myelin sheath thickness in the CNS is regulated near the axon. Brain Res 1983; 274:329.
36. McLoon SC et al. Transient expression of laminin in the optic nerve of the developing rat. J Neurosci 1988; 8:1981.
37. Ludwin SK. Oligodendrocyte survival in Wallerian degeneration. Acta Neuropathol (Berl) 1990; 80:184.
38. Kidd GJ, Hauer PE, Trapp BD. Axons modulate myelin protein messenger RNA levels during central nervous system myelination in vivo. J Neurosci Res 1990; 26:409.
39. Barres BA et al. Does oligodendrocyte survival depend on axons? Curr Biol 1993; 3(8):489–497.
40. Barres BA, Raff MC. Proliferation of oligodendrocyte precursor cells depends on electrical activity in axons. Nature 1993; 361(6409):258–260.
41. Burne JF, Staple JK, Raff MC. Glial cells are increased proportionally in transgenic optic nerves with increased numbers of axons. J Neurosci 1996; 16(6):2064–2073.
42. Colello RJ, Schwab ME. A role for oligodendrocytes in the stabilization of optic axon numbers. J Neurosci 1994; 14(11 Pt 1):6446–6452.
43. Savio T, Schwab ME. Rat CNS white matter, but not gray matter, is non-permissive for neuronal cell adhesion and fiber outgrowth. J Neurosci 1989; 9:1126.
44. Schwab ME. Myelin-associated inhibitors of neurite growth. Exp Neurol 1990; 109:2.
45. Wender R et al. Astrocytic glycogen influences axon function and survival during glucose deprivation in central white matter. J Neurosci 2000; 20(18):6804–6810.
46. Ffrench-Constant C, Raff MC. The oligodendrocyte-type-2 astrocyte cell lineage is specialized for myelination. Nature 1986; 323:335.
47. Janzer RC, Raff MC. Astrocytes induce blood-brain barrier properties in endothelial cells. Nature 1987; 325:253.
48. Chan WY, Kohsaka S, Rezaie P. The origin and cell lineage of microglia: new concepts. Brain Res Rev 2007; 53(2):344–354.
49. Hickey WF, Kimura H. Perivascular microglial cells of the CNS are bone marrow-derived and present antigen in vivo. Science 1988; 239:290.
50. Stoll G, Trapp BD, Griffin JW. Macrophage function during Wallerian degeneration of rat optic nerve: Clearance of degenerating myelin and Ia expression. J Neurosci 1989; 9:2327.
51. Sturrock RR. Microglia in the human embryonic optic nerve. J Anat 1984; 139:81.
52. Sturrock RR. An electron microscopic study of macrophages in the meninges of the human embryonic optic nerve. J Anat 1988; 157:145.
53. Hayreh SS The sheath of the optic nerve. Ophthalmologica 1984; 189:54.
54. Liu D, Michon J. Measurement of the subarachnoid pressure of the optic nerve in human subjects. Am J Ophthalmol 1995; 119(1):81–85.
55. Anderson DR. Ultrastructure of meningeal sheaths: Normal human and monkey optic nerves. Arch Ophthalmol 1969; 82:659.
56. Sturrock RR. A quantitative histological study of cell division and changes in cell number in the meningeal sheath of the embryonic human optic nerve. J Anat 1987; 155:133.
57. Levin LA, Albert DM, Johnson D. Mast cells in human optic nerve. Invest Ophthalmol Vis Sci 1993; 34(11):3147–3153.
58. Anderson DR, Braverman S. Reevaluation of the optic disk vasculature. Am J Ophthalmol 1976; 82(2):165–174.
59. Lieberman MF, Maumenee AE, Green WR. Histologic studies of the vasculature of the anterior optic nerve. Am J Ophthalmol 1976; 82(3):405–423.
60. Olver JM, Spalton DJ, McCartney AC. Microvascular study of the retrolaminar optic nerve in man: the possible significance in anterior ischaemic optic neuropathy. Eye 1990; 4(Pt 1):7–24.

61. Olver JM. Functional anatomy of the choroidal circulation: methyl methacrylate casting of human choroid. Eye 1990; 4(Pt 2):262–272.

62. Guy J et al. Disruption of the blood-brain barrier in experimental optic neuritis: immunocytochemical co-localization of H2O2 and extravasated serum albumin. Invest Ophthalmol Vis Sci 1994; 35(3):1114–1123.

63. Guy J et al. Intraorbital optic nerve and experimental optic neuritis. Correlation of fat suppression magnetic resonance imaging and electron microscopy. Ophthalmology 1992; 99(5):720–725.

64. Geijer C, Bill A. Effects of raised intraocular pressure on retinal, prelaminar, laminar, and retrolaminar optic nerve blood flow in monkeys. Invest Ophthalmol Vis Sci 1979; 18(10):1030–1042.

65. Weinstein JM et al. Regional optic nerve blood flow and its autoregulation. Invest Ophthalmol Vis Sci 1983; 24(12):1559–1565.

66. Riva CE, Grunwald JE, Petrig BL. Autoregulation of human retinal blood flow. An investigation with laser Doppler velocimetry. Invest Ophthalmol Vis Sci 1986; 27(12):1706–1712.

67. Francois J, Fryczkowski A. The blood supply of the optic nerve. Adv Ophthalmol 1978; 36:164–173.

68. Chou PI, Sadun AA, Lee H. Vasculature and morphometry of the optic canal and intracanalicular optic nerve. J Neuroophthalmol 1995; 15(3):186–190.

69. Sturrock RR. Vascularization of the human embryonic optic nerve. J Hirnforsch 1987; 28:615.

70. Rao K, Lund RD. Degeneration of optic axons induces the expression of major histocompatibility antigens. Brain Res 1989; 488:332.

71. Flage T. Permeability properties of the tissues in the optic nerve head region in the rabbit and the monkey: An ultrastructural study. Acta Ophthalmol 1977; 55:652.

72. Tsukahara I, Yamashita H. An electron microscopic study on the blood-optic nerve and fluid-optic nerve barrier. Graefes Arch Clin Exp Ophthalmol 1975; 196:239.

73. Guy J, Rao NA. Acute and chronic experimental optic neuritis: Alteration in the blood-optic nerve barrier. Arch Ophthalmol 1984; 102:450.

74. Sawaguchi S et al. The collagen fibrillar network in the human pial septa. Curr Eye Res 1994; 13(11):819–824.

75. Cohen AI. Ultrastructural aspects of the human optic nerve. Invest Ophthalmol 1967; 6:294.

76. Lincoln J et al. Innervation of normal human sural and optic nerves by noradrenaline- and peptide-containing nervi vasorum and nervorum: effect of diabetes and alcoholism. Brain Res 1993; 632(1–2):48–56.

77. Ye XD, Laties AM, Stone RA. Peptidergic innervation of the retinal vasculature and optic nerve head. Invest Ophthalmol Vis Sci 1990; 31:1731.

78. Nyborg NC, Nielsen PJ. Angiotensin-II contracts isolated human posterior ciliary arteries. Invest Ophthalmol Vis Sci 1990; 31:2471.

79. Ling TL, Mitrofanis J, Stone J. Origin of retinal astrocytes in the rat: evidence of migration from the optic nerve. J Comp Neurol 1989; 286:345.

80. David S. Neurite outgrowth from mammalian CNS neurons on astrocytes in vitro may not be mediated primarily by laminin. J Neurocytol 1988; 17:131.

81. Dreyer EB et al. An astrocytic binding site for neuronal Thy-1 and its effect on neurite outgrowth. Proc Natl Acad Sci USA 1995; 92(24):11195–11199.

82. Tang DG, Tokumoto YM, Raff MC. Long-term culture of purified postnatal oligodendrocyte precursor cells. Evidence for an intrinsic maturation program that plays out over months. J Cell Biol 2000; 148(5):971–984.

83. Leifer D et al. Monoclonal antibody to Thy-1 enhances regeneration of processes by rat retinal ganglion cells in culture. Science 1984; 224:303–306.

84. Bartsch U, Kirchhoff F, Schachner M. Immunohistological localization of the adhesion molecules L1, N-CAM, and MAG in the developing and adult optic nerve of mice. J Comp Neurol 1989; 284:451.

85. Skoff RP. The fine structure of pulse labeled (3H-thymidine cells) in degenerating rat optic nerve. J Comp Neurol 1975; 161:595.

86. Bogler O, Noble M. Measurement of time in oligodendrocyte-type-2 astrocyte (O-2A) progenitors is a cellular process distinct from differentiation or division. Dev Biol 1994; 162(2):525–538.

87. Gao FB, Durand B, Raff M. Oligodendrocyte precursor cells count time but not cell divisions before differentiation. Curr Biol 1997; 7(2):152–155.

88. Shi J, Marinovich A, Barres BA. Purification and characterization of adult oligodendrocyte precursor cells from the rat optic nerve. J Neurosci 1998; 18(12):4627–4636.

89. Givogri MI et al. Central nervous system myelination in mice with deficient expression of Notch1 receptor. J Neurosci Res 2002; 67(3):309–320.

90. Politis MJ. Exogenous laminin induces regenerative changes in traumatized sciatic and optic nerve. Plast Reconstr Surg 1989; 83:228.

91. Colello RJ et al. The chronology of oligodendrocyte differentiation in the rat optic nerve: evidence for a signaling step initiating myelination in the CNS. J Neurosci 1995; 15(11):7665–7672.

92. Magoon EH, Robb RM. Development of myelin in human optic nerve and tract: A light and electron microscopic study. Arch Ophthalmol 1981; 99:655.

93. Lev-Ram V, Grinvald A. Ca2+- and K+-dependent communication between central nervous system myelinated axons and oligodendrocytes revealed by voltage-sensitive dyes. Proc Natl Acad Sci USA 1986; 83:6651.

94. Sanchez I et al. Oligodendroglia regulate the regional expansion of axon caliber and local accumulation of neurofilaments during development independently of myelin formation. J Neuroscience 1996; 16(16):5095–5105.

95. Foster RE, Connors BW, Waxman SG. Rat optic nerve: electrophysiological, pharmacological and anatomical studies during development. Brain Res 1982; 255:371.

96. Hildebrand C, Remahl S, Waxman SG. Axo-glial relations in the retina-optic nerve junction of the adult rat: electron-microscopic observations. J Neurocytol 1985; 14:597.

97. Ffrench-Constant C et al. Evidence that migratory oligodendrocyte-type-2 astrocyte (O-2A) progenitor cells are kept out of the rat retina by a barrier at the eye-end of the optic nerve. J Neurocytol 1988; 17:13.

98. Tso MO, Shih CV, McLean IW. Is there a blood brain barrier at the optic nerve head? Arch Ophthalmol 1975; 93:815.

99. Perry VH, Lund RD. Evidence that the lamina cribrosa prevents intraretinal myelination of retinal ganglion cell axons. J Neurocytol 1990; 19:265.

100. Bartsch U et al. Tenascin demarcates the boundary between the myelinated and non-myelinated part of retinal ganglion cell axons in the developing and adult mouse. J Neurosci 1994; 14(8):4756–4768.

101. Laeng P et al. Transplantation of oligodendrocyte progenitor cells into the rat retina: extensive myelination of retinal ganglion cell axons. Glia 1996; 18(3):200–210.

102. Perry VH, Hayes L. Lesion-induced myelin formation in the retina. J Neurocytol 1985; 14:297.

103. Bussow H. Schwann cell myelin ensheathing CNS axons in the nerve fibre layer of the cat retina. J Neurocytol 1978; 7:207.

104. Jung HJ, Raine CS, Suzuki K. Schwann cells and peripheral nervous system myelin in the rat retina. Acta Neuropathol (Berl) 1978; 44:245.

105. Burne JF, Raff MC. Retinal ganglion cell axons drive the proliferation of astrocytes in the developing rodent optic nerve. Neuron 1997; 18(2):223–230.

106. Mi H, Barres BA. Purification and characterization of astrocyte precursor cells in the developing rat optic nerve. J Neurosci 1999; 19(3):1049–1061.

107. Liedtke W et al. GFAP is necessary for the integrity of CNS white matter architecture and long-term maintenance of myelination. Neuron 1996; 17(4):607–615.

108. Lucius R et al. Growth stimulation and chemotropic attraction of rat retinal ganglion cell axons in vitro by co-cultured optic nerves, astrocytes and astrocyte conditioned medium. Int J Develop Neurosci 1996; 14(4):387–398.

109. Penfold PL, Provis JM. Cell death in the development of the human retina: phagocytosis of pyknotic and apoptotic bodies by retinal cells. Graefes Arch Clin Exp Ophthalmol 1986; 224(6):549–553.

110. Ilschner SU, Waring W. Fragmentation of DNA in the retina of chicken embryos coincides with retinal ganglion cell death. Biochem Biophys Res Comm 1992; 183(3):1056–1061.

111. Provis JM et al. Human fetal optic nerve: overproduction and elimination of retinal axons during development. J Comp Neurol 1985; 238:92.

112. Goldberg JL, Barres BA. The relationship between neuronal survival and regeneration. Annu Rev Neurosci 2000; 23:579–612.

113. Goldberg JL et al. Retinal ganglion cells do not extend axons by default: promotion by neurotrophic signaling and electrical activity. Neuron 2002; 33(5):689–702.

114. Jo SA, Wang E, Benowitz LI. Ciliary neurotrophic factor is and axogenesis factor for retinal ganglion cells. Neuroscience 1999; 89(2):579–591.

115. Logan A et al. Neurotrophic factor synergy is required for neuronal survival and disinhibited axon regeneration after CNS injury. Brain 2006; 129(Pt 2):490–502.

116. Loh NK et al. The regrowth of axons within tissue defects in the CNS is promoted by implanted hydrogel matrices that contain BDNF and CNTF producing fibroblasts. Exp Neurol 2001; 170(1):72–84.

117. Mansour-Robaey S et al. Effects of ocular injury and administration of brain-derived neurotrophic factor on survival and regrowth of axotomized retinal ganglion cells. Proc Natl Acad Sci USA 1994; 91(5):1632–1636.

118. Mey J, Thanos S. Intravitreal injections of neurotrophic factors support the survival of axotomized retinal ganglion cells in adult rats in vivo. Brain Res 1993; 602(2):304–317.

119. Lafont F et al. In vitro control of neuronal polarity by glycosaminoglycans. Development 1992; 114(1):17–29.

120. Reichardt LF, Tomaselli KJ. Extracellular matrix molecules and their receptors: functions in neural development. Annu Rev Neurosci 1991; 14:531–570.

121. Riehl R et al. Cadherin function is required for axon outgrowth in retinal ganglion cells in vivo. Neuron 1996; 17(5):837–848.

122. Benowitz LI et al. Axon outgrowth is regulated by an intracellular purine-sensitive mechanism in retinal ganglion cells. J Biol Chem 1998; 273(45):29626–29634.

123. Shatz CJ, Stryker MP. Prenatal tetrodotoxin infusion blocks segregation of retinogeniculate afferents. Science 1988; 242(4875):87–89.

124. Verhage M et al. Synaptic assembly of the brain in the absence of neurotransmitter secretion. Science 2000; 287(5454):864–869.

125. Cantallops I, Haas K, Cline HT. Postsynaptic CPG15 promotes synaptic maturation and presynaptic axon arbor elaboration in vivo. Nat Neurosci 2000; 3(10):1004–1011.

126. Cohen-Cory S. BDNF modulates, but does not mediate, activity-dependent branching and remodeling of optic axon arbors in vivo. J Neurosci 1999; 19(22):9996–10003.

127. Rashid NA, Cambray-Deakin MA. N-methyl-D-aspartate effects on the growth, morphology and cytoskeleton of individual neurons in vitro. Brain Res Dev Brain Res 1992; 67(2):301–308.

128. Catalano SM, Shatz CJ. Activity-dependent cortical target selection by thalamic axons. Science 1998; 281(5376):559–562.

129. Kalil RE et al. Elimination of action potentials blocks the structural development of retinogeniculate synapses. Nature 1986; 323(6084):156–158.

130. Katz LC, Shatz CJ. Synaptic activity and the construction of cortical circuits. Science 1996; 274(5290):1133–1138.

131. Goldberg JL. How does an axon grow? Genes Dev 2003; 17(8):941–958.

132. Sretavan DW et al. Disruption of retinal axon ingrowth by ablation of embryonic mouse optic chiasm neurons. Science 1995; 269(5220):98–101.

133. Goodman CS. Mechanisms and molecules that control growth cone guidance. Annu Rev Neurosci 1996; 19:341–377.

134. Haupt C, Huber AB. How axons see their way – axonal guidance in the visual system. Front Biosci 2008; 13:3136–3149.

135. Bauch H, Stier H, Schlosshauer B. Axonal versus dendritic outgrowth is differentially affected by radial glia in discrete layers of the retina. J Neurosci 1998; 18(5):1774–1785.

136. Stier H, Schlosshauer B. Axonal guidance in the chicken retina. Development 1995; 121(5):1443–1454.

137. Airaksinen PJ, Doro S, Veijola J. Conformal geometry of the retinal nerve fiber layer. Proc Natl Acad Sci USA 2008; 105(50):19690–19695.

138. Brittis PA, Silver J. Multiple factors govern intraretinal axon guidance: a time-lapse study. Mol Cell Neurosci 1995; 6(5):413–432.

139. de la Torre JR et al. Turning of retinal growth cones in a netrin-1 gradient mediated by the netrin receptor DCC. Neuron 1997; 19(6):1211–1224.

140. Deiner MS et al. Netrin-1 and DCC mediate axon guidance locally at the optic disc: loss of function leads to optic nerve hypoplasia. Neuron 1997; 19(3):575–589.

141. Oster SF et al. Invariant Sema5A inhibition serves an ensheathing function during optic nerve development. Development 2003; 130(4):775–784.
142. Plump AS et al. Slit1 and Slit2 cooperate to prevent premature midline crossing of retinal axons in the mouse visual system. Neuron 2002; 33(2):219–232.
143. Nakagawa S et al. Ephrin-B regulates the ipsilateral routing of retinal axons at the optic chiasm. Neuron 2000; 25(3):599–610.
144. Williams SE et al. Ephrin-B2 and EphB1 mediate retinal axon divergence at the optic chiasm. Neuron 2003; 39(6):919–935.
145. Chung KY et al. Axon routing at the optic chiasm after enzymatic removal of chondroitin sulfate in mouse embryos. Development 2000; 127(12):2673–2683.
146. Reese BE. The chronotopic reordering of optic axons. Perspect Dev Biol 1996; 3(3):233–242.
147. Chalupa LM, Meissirel C, Lia B. Specificity of retinal ganglion cell projections in the embryonic rhesus monkey. Perspect Dev Biol 1996; 3(3):223–231.
148. Bowling DB, Michael CR. Projection patterns of single physiologically characterized optic tract fibres in cat. Nature 1980; 286(5776):899–902.
149. Kondo Y et al. Single retinal ganglion cells sending axon collaterals to the bilateral superior colliculi: a fluorescent retrograde double-labeling study in the Japanese monkey (Macaca fuscata). Brain Res 1992; 597(1):155–161.
150. Hendry SH, Reid RC. The koniocellular pathway in primate vision. Annu Rev Neurosci 2000; 23:127–153.
151. Bunt AH et al. Monkey retinal ganglion cells: morphometric analysis and tracing of axonal projections, with a consideration of the peroxidase technique. J Comp Neurol 1975; 164(3):265–285.
152. Wurtz RH. Vision for the control of movement. The Friedenwald Lecture. Invest Ophthalmol Vis Sci 1996; 37(11):2130–2145.
153. Hilbig H et al. Neuronal and glial structures of the superficial layers of the human superior colliculus. Anat Embryol (Berl) 1999; 200(1):103–115.
154. Fredericks CA et al. The human accessory optic system. Brain Res 1988; 454:116.
155. Itoh K et al. A pretectofacial projection in the cat: a possible link in the visually-triggered blink reflex pathways. Brain Res 1983; 275:332.
156. Baleydier C, Magnin M, Cooper HM. Macaque accessory optic system: II: Connections with the pretectum. J Comp Neurol 1990; 302:405.
157. Lagreze WA, Kardon RH. Correlation of relative afferent pupillary defect and estimated retinal ganglion cell loss. Graefes Arch Clin Exp Ophthalmol 1998; 236(6):401–404.
158. Lhermitte F, Guillaumat L, Lyon CO. Monocular blindness with preserved direct and consensual pupillary reflex in multiple sclerosis. Arch Neurol 1984; 41:993.
159. Sadun AA, Schaechter JD, Smith LE. A retinohypothalamic pathway in man: light mediation of circadian rhythms. Brain Res 1984; 302:371.
160. Schaecter JD, Sadun AA. A second hypothalamic nucleus receiving retinal input in man: the paraventricular nucleus. Brain Res 1985; 340:243.
161. Johnson RF, Morin LP, Moore RY. Retinohypothalamic projections in the hamster and rat demonstrated using cholera toxin. Brain Res 1988; 462:301.
162. Gamlin PD et al. The neural substrate for the pupillary light reflex in the pigeon (Columba livia). J Comp Neurol 1984; 226(4):523–543.
163. Young MJ, Lund RD. The retinal ganglion cells that drive the pupilloconstrictor response in rats. Brain Res 1998; 787(2):191–202.
164. Berson DM, Dunn FA, Takao M. Phototransduction by retinal ganglion cells that set the circadian clock. Science 2002; 295(5557):1070–1073.
165. Gooley JJ et al. Melanopsin in cells of origin of the retinohypothalamic tract. Nat Neurosci 2001; 4(12):1165.
166. Provencio I et al. A novel human opsin in the inner retina. J Neurosci 2000; 20(2):600–605.
167. Provencio I, Rollag MD, Castrucci AM. Photoreceptive net in the mammalian retina. This mesh of cells may explain how some blind mice can still tell day from night. Nature 2002; 415(6871):493.
168. Hattar S et al. Central projections of melanopsin-expressing retinal ganglion cells in the mouse. J Comp Neurol 2006; 497(3):326–349.
169. Morin LP, Blanchard JH, Provencio I. Retinal ganglion cell projections to the hamster suprachiasmatic nucleus, intergeniculate leaflet, and visual midbrain: bifurcation and melanopsin immunoreactivity. J Comp Neurol 2003; 465(3):401–416.
170. Ohishi H et al. Distribution of the messenger RNA for a metabotropic glutamate receptor, mGluR2, in the central nervous system of the rat. Neuroscience 1993; 53(4):1009–1018.
171. Rothe T, Bigl V, Grantyn R. Potentiating and depressant effects of metabotropic glutamate receptor agonists on high-voltage-activated calcium currents in cultured retinal ganglion neurons from postnatal mice. Pflugers Arch 1994; 426(1–2):161–170.
172. Li X et al. Studies on the identity of the rat optic nerve transmitter. Brain Res 1996; 706(1):89–96.
173. Matsui K, Hosoi N, Tachibana M. Excitatory synaptic transmission in the inner retina: paired recordings of bipolar cells and neurons of the ganglion cell layer. J Neurosci 1998; 18(12):4500–4510.
174. Rorig B, Grantyn R. Rat retinal ganglion cells express Ca^{2+}-permeable non-NMDA glutamate receptors during the period of histogenetic cell death. Neurosci Lett 1993; 153(1):32–36.
175. Siliprandi R et al. N-methyl-D-aspartate-induced neurotoxicity in the adult rat retina. Vis Neurosci 1992; 8(6):567–573.
176. Feller MB et al. Requirement for cholinergic synaptic transmission in the propagation of spontaneous retinal waves. Science 1996; 272(5265):1182–1187.
177. Keyser KT et al. Amacrine, ganglion, and displaced amacrine cells in the rabbit retina express nicotinic acetylcholine receptors. Vis Neurosci 2000; 17(5):743–752.
178. Kubrusly RC, de Mello MC, de Mello FG. Aspartate as a selective NMDA receptor agonist in cultured cells from the avian retina. Neurochem Int 1998; 32(1):47–52.
179. Matsui K, Hosoi N, Tachibana M. Active role of glutamate uptake in the synaptic transmission from retinal non-spiking neurons. J Neurosci 1999; 19(16):6755–6766.
180. Higgs MH, Lukasiewicz PD. Glutamate uptake limits synaptic excitation of retinal ganglion cells. J Neurosci 1999; 19(10):3691–3700.
181. Pow DV, Barnett NL, Penfold P. Are neuronal transporters relevant in retinal glutamate homeostasis? Neurochem Int 2000; 37(2–3):191–198.
182. Barnett NL, Pow DV. Antisense knockdown of GLAST, a glial glutamate transporter, compromises retinal function. Invest Ophthalmol Vis Sci 2000; 41(2):585–591.
183. Taschenberger H, Grantyn R. Several types of Ca^{2+} channels mediate glutamatergic synaptic responses to activation of single Thy-1-immunolabeled rat retinal ganglion neurons. J Neurosci 1995; 15(3 Pt 2):2240–2254.
184. Schmid S, Guenther E. Developmental regulation of voltage-activated Na^{+} and Ca^{2+} currents in rat retinal ganglion cells. Neuroreport 1996; 7(2):677–681.
185. Schmid S, Guenther E. Alterations in channel density and kinetic properties of the sodium current in retinal ganglion cells of the rat during in vivo differentiation. Neuroscience 1998; 85(1):249–258.
186. Guenther E et al. Maturation of intrinsic membrane properties in rat retinal ganglion cells. Vision Res 1999; 39(15):2477–2484.
187. Schmid S, Guenther E. Voltage-activated calcium currents in rat retinal ganglion cells in situ: changes during prenatal and postnatal development. J Neurosci 1999; 19(9):3486–3494.
188. Taschenberger H, Juttner R, Grantyn R. Ca^{2+}-permeable P2X receptor channels in cultured rat retinal ganglion cells. J Neurosci 1999; 19(9):3353–3366.
189. Velte TJ, Masland RH. Action potentials in the dendrites of retinal ganglion cells. J Neurophysiol 1999; 81(3):1412–1417.
190. Caruso DM, Owczarzak MT, Pourcho RG. Colocalization of substance P and GABA in retinal ganglion cells: a computer-assisted visualization. Vis Neurosci 1990; 5(4):389–394.
191. Caruso DM et al. GABA-immunoreactivity in ganglion cells of the rat retina. Brain Res 1989; 476(1):129–134.
192. Ehrlich D et al. Differential effects of axotomy on substance P-containing and nicotinic acetylcholine receptor-containing retinal ganglion cells: time course of degeneration and effects of nerve growth factor. Neuroscience 1990; 36(3):699–723.
193. Ehrlich D, Keyser KT, Karten HJ. Distribution of substance P-like immunoreactive retinal ganglion cells and their pattern of termination in the optic tectum of chick (Gallus gallus). J Comp Neurol 1987; 266(2):220–233.
194. Pickard GE, Rea MA. Serotonergic innervation of the hypothalamic suprachiasmatic nucleus and photic regulation of circadian rhythms. Biol Cell 1997; 89(8):513–523.
195. Stanford LR. Conduction velocity variations minimize conduction time differences among retinal ganglion cell axons. Science 1987; 238(4825):358–360.
196. Black JA et al. Immuno-ultrastructural localization of sodium channels at nodes of Ranvier and perinodal astrocytes in rat optic nerve. Proc R Soc Lond Biol 1989; 238:39.
197. Kaplan MR et al. Differential control of clustering of the sodium channels Na(v)1.2 and Na(v)1.6 at developing CNS nodes of Ranvier. Neuron 2001; 30(1):105–119.
198. Kaplan MR et al. Induction of sodium channel clustering by oligodendrocytes. Nature 1997; 386(6626):724–728.
199. Van Wart, A, Matthews G. Impaired firing and cell-specific compensation in neurons lacking nav1.6 sodium channels. J Neurosci 2006; 26(27):7172–7180.
200. Black JA et al. Sodium channels in astrocytes of rat optic nerve in situ: immuno-electron microscopic studies. Glia 1989; 2:353.
201. Oh Y, Black J, Waxman SG. The expression of rat brain voltage-sensitive Na^{+} channel mRNAs in astrocytes. Molec Brain Res 1994; 23(1–2):57–65.
202. Bevan S et al. Voltage gated ionic channels in rat cultured astrocytes, reactive astrocytes and an astrocyte-oligodendrocyte progenitor cell. J Physiol (Paris) 1987; 82:327.
203. Barres BA, Chun LLY, Corey DP. Ion channels in vertebrate glia. Annu Rev Neurosci 1990; 13:441.
204. Kriegler S, Chiu SY. Calcium signaling of glial cells along mammalian axons. J Neurosci 1993; 13(10):4229–4245.
205. Jeffery G et al. Cellular localisation of metabotropic glutamate receptors in the mammalian optic nerve: a mechanism for axon-glia communication. Brain Res 1996; 741(1–2):75–81.
206. Levine J, Willard M. The composition and organization of axonally transported proteins in the retinal ganglion cells of the guinea pig. Brain Res 1980; 194:137.
207. Willard M, Simon C. Modulations of neurofilament axonal transport during the development of rabbit retinal ganglion cells. Cell 1983; 35:551.
208. Amaratunga A et al. Inhibition of kinesin synthesis and rapid anterograde axonal transport in vivo by an antisense oligonucleotide. J Biol Chem 1993; 268(23):17427–17430.
209. Dillman JFI, Dabney LP, Pfister KK. Cytoplasmic dynein is associated with slow axonal transport. Proc Natl Acad Sci USA 1996; 93(1):141–144.
210. Aschner M, Rodier PM, Finkelstein JN. Increased axonal transport in the rat optic system after systemic exposure to methylmercury: differential effects in local vs systemic exposure conditions. Brain Res 1987; 401(1):132–141.
211. Crossland WJ. Fast axonal transport in the visual pathway of the chick and rat. Brain Res 1985; 340:373.
212. Morin PJ et al. Isolation and characterization of rapid transport vesicle subtypes from rabbit optic nerve. J Neurochem 1991; 56:415.
213. Mercken M et al. Three distinct axonal transport rates for tau, tubulin, and other microtubule-associated proteins: evidence for dynamic interactions of tau with microtubules in vivo. J Neuroscience 1995; 15(12):8259–8267.
214. Black MM, Lasek RJ. Axonal transport of actin: Slow component b is the principal source of actin for the axon. Brain Res 1979; 171:401.
215. Willard M et al. Axonal transport of actin in rabbit retinal ganglion cells. J Cell Biol 1979; 81:581.
216. Giorgi PP, DuBois H. Labelling by axonal transport of myelin-associated proteins in the rabbit visual pathway. Biochem J 1981; 196:537.
217. Frizell M, McLean WG, Sjostrand J. Slow axonal transport of proteins: blockade by interruption of contact between cell body and axon. Brain Res 1975; 86:67.
218. Howe CL et al. NGF signaling from clathrin-coated vesicles: evidence that signaling endosomes serve as a platform for the Ras-MAPK pathway. Neuron 2001; 32(5):801–814.
219. MacInnis BL, Campenot RB. Retrograde support of neuronal survival without retrograde transport of nerve growth factor. Science 2002; 295(5559):1536–1539.
220. Riccio A et al. An NGF-TrkA-mediated retrograde signal to transcription factor CREB in sympathetic neurons. Science 1997; 277(5329):1097–1100.
221. Watson FL et al. Rapid nuclear responses to target-derived neurotrophins require retrograde transport of ligand-receptor complex. J Neurosci 1999; 19(18):7889–7900.

222. So KF, Aguayo AJ. Lengthy regrowth of cut axons from ganglion cells after peripheral nerve transplantation into the retina of adult rats. Brain Res 1985; 328(2):349–354.
223. Berkelaar M et al. Axotomy results in delayed death and apoptosis of retinal ganglion cells in adult rats. J Neurosci 1994; 14(7):4368–4374.
224. You SW, So KF, Yip HK. Axonal regeneration of retinal ganglion cells depending on the distance of axotomy in adult hamsters. Invest Ophthalmol Vis Sci 2000; 41(10):3165–3170.
225. Johnson MW, Kincaid MC, Trobe JD. Bilateral retrobulbar optic nerve infarctions after blood loss and hypotension. A clinicopathologic case study. Ophthalmology 1987; 94(12):1577–1584.
226. Connolly SE, Gordon KB, Horton JC. Salvage of vision after hypotension-induced ischemic optic neuropathy. Am J Ophthalmol 1994; 117(2):235–242.
227. Stys PK. Anoxic and ischemic injury of myelinated axons in CNS white matter: from mechanistic concepts to therapeutics. J Cereb Blood Flow Metab 1998; 18(1):2–25.
228. Petty MA, Wettstein JG. White matter ischaemia. Brain Res Brain Res Rev 1999; 31(1):58–64.
229. Waxman SG et al. Anoxic injury of mammalian central white matter: decreased susceptibility in myelin-deficient optic nerve. Ann Neurol 1990; 28:335.
230. Marcoux FW et al. Differential regional vulnerability in transient focal cerebral ischemia. Stroke 1982; 13(3):339–346.
231. Hayreh SS, Baines JA. Occlusion of the posterior ciliary artery. 3. Effects on the optic nerve head. Br J Ophthalmol 1972; 56(10):754–764.
232. Hayreh SS, Baines JA. Occlusion of the posterior ciliary artery. 1. Effects on choroidal circulation. Br J Ophthalmol 1972; 56(10):719–735.
233. Cioffi GA, Van Buskirk EM. Microvasculature of the anterior optic nerve. Surv Ophthalmol 1994; 38(Suppl):S107-S116.
234. Cioffi GA et al. An in vivo model of chronic optic nerve ischemia: the dose-dependent effects of endothelin-1 on the optic nerve microvasculature. Curr Eye Res 1995; 14(12):1147–1153.
235. Orgul S et al. An endothelin-1-induced model of chronic optic nerve ischemia in rhesus monkeys. J Glaucoma 1996; 5(2):135–138.
236. Nishimura K et al. Effects of endothelin-1 on optic nerve head blood flow in cats. J Ocul Pharmacol Ther 1996; 12(1):75–83.
237. Orgul S et al. An endothelin-1 induced model of optic nerve ischemia in the rabbit. Invest Ophthalmol Vis Sci 1996; 37(9):1860–1869.
238. Cioffi GA, Sullivan P. The effect of chronic ischemia on the primate optic nerve. Eur J Ophthalmol 1999; 9(Suppl 1):S34-S36.
239. Oku H et al. Experimental optic cup enlargement caused by endothelin-1-induced chronic optic nerve head ischemia. Surv Ophthalmol 1999; 44(Suppl 1):S74-S84.
240. Bernstein SL et al. Functional and cellular responses in a novel rodent model of anterior ischemic optic neuropathy. Invest Ophthalmol Vis Sci 2003; 44(10):4153–4162.
241. Chen CS et al. A primate model of non-arteritic anterior ischemic optic neuropathy. Invest Ophthalmol Vis Sci 2008; 49(7):2985–2992.
242. Duan Y et al. Retinal ganglion cell survival and optic nerve glial response after rat optic nerve ischemic injury. In: American Heart Association Investigators' Meeting. New York: AHA, 2008.
243. Rizzo JFD, Lessell S. Risk of developing multiple sclerosis after uncomplicated optic neuritis: a long-term prospective study. Neurology 1988; 38(2):185–190.
244. Optic Neuritis Study Group. The 5-year risk of MS after optic neuritis. Experience of the optic neuritis treatment trial. Neurology 1997; 49(5):1404–1413.
245. Yarom Y et al. Immunospecific inhibition of nerve conduction by T lymphocytes reactive to basic protein of myelin. Nature 1983; 303(5914):246–247.
246. Trapp BD et al. Axonal transection in the lesions of multiple sclerosis. N Engl J Med 1998; 338(5):278–285.
247. Perry VH, Anthony DC. Axon damage and repair in multiple sclerosis. Phil Trans R Soc Lond B Biol Sci 1999; 354(1390):1641–1647.
248. Evangelou N et al. Quantitative pathological evidence for axonal loss in normal appearing white matter in multiple sclerosis. Ann Neurol 2000; 47(3):391–395.
249. MacFadyen DJ et al. The retinal nerve fiber layer, neuroretinal rim area, and visual evoked potentials in MS. Neurology 1988; 38(9):1353–1358.
250. Hayreh SS et al. Experimental allergic encephalomyelitis. I. Optic nerve and central nervous system manifestations. Invest Ophthalmol Vis Sci 1981; 21(2):256–269.
251. Sergott RC et al. Antigalactocerebroside serum demyelinates optic nerve in vivo. J Neurol Sci 1984; 64(3):297–303.
252. Zhu B et al. Axonal cytoskeleton changes in experimental optic neuritis. Brain Res 1999; 824(2):204–217.
253. Guy J et al. Axonal transport reductions in acute experimental allergic encephalomyelitis: qualitative analysis of the optic nerve. Curr Eye Res 1989; 8(3):261–269.
254. Guy J et al. Quantitative analysis of labelled inner retinal proteins in experimental optic neuritis. Curr Eye Res 1989; 8(3):253–260.
255. Rao NA, Guy J, Sheffield PS. Effects of chronic demyelination on axonal transport in experimental allergic optic neuritis. Invest Ophthalmol Vis Sci 1981; 21(4):606–611.
256. Gutierrez R et al. Decompaction of CNS myelin leads to a reduction of the conduction velocity of action potentials in optic nerve. Neurosci Lett 1995; 195(2):93–96.
257. Waxman SG et al. Low density of sodium channels supports action potential conduction in axons of neonatal rat optic nerve. Proc Natl Acad Sci USA 1989; 86:1406.
258. Storch MK et al. Autoimmunity to myelin oligodendrocyte glycoprotein in rats mimics the spectrum of multiple sclerosis pathology. Brain Pathol 1998; 8(4):681–694.
259. Bettelli E et al. Myelin oligodendrocyte glycoprotein-specific T cell receptor transgenic mice develop spontaneous autoimmune optic neuritis. J Exp Med 2003; 197(9):1073–1081.
260. Shao H et al. Myelin/oligodendrocyte glycoprotein-specific T-cells induce severe optic neuritis in the C57BL/6 mouse. Invest Ophthalmol Vis Sci 2004; 45(11):4060–4065.
261. Guan Y et al. Retinal ganglion cell damage induced by spontaneous autoimmune optic neuritis in MOG-specific TCR transgenic mice. J Neuroimmunol 2006; 178(1–2):40–48.
262. Fontana L et al. In vivo morphometry of the lamina cribrosa and its relation to visual field loss in glaucoma. Curr Eye Res 1998; 17(4):363–369.
263. Clifford-Jones RE, McDonald WI, Landon DN. Chronic optic nerve compression. An experimental study. Brain 1985; 108(Pt 1):241–262.
264. Clifford-Jones RE, Landon DN, McDonald WI. Remyelination during optic nerve compression. J Neurol Sci 1980; 46(2):239–243.
265. Guyer DR et al. Visual function following optic canal decompression via craniotomy. J Neurosurg 1985; 62(5):631–638.
266. Giles CL, Soble AR. Intracranial hypertension and tetracycline therapy. Am J Ophthalmol 1971; 72(5):981–982.
267. Minckler DS. Histology of optic nerve damage in ocular hypertension and early glaucoma. Surv Ophthalmol 1989; 33(Suppl):401–411.
268. Quigley HA. Ganglion cell death in glaucoma: pathology recapitulates ontogeny. Aust N Z J Ophthalmol 1995; 23(2):85–91.
269. Quigley HA et al. Retinal ganglion cell death in experimental glaucoma and after axotomy occurs by apoptosis. Invest Ophthalmol Vis Sci 1995; 36(5):774–786.
270. Kerrigan LA et al. TUNEL-positive ganglion cells in human primary open-angle glaucoma. Arch Ophthalmol 1997; 115(8):1031–1035.
271. Okisaka S et al. Apoptosis in retinal ganglion cell decrease in human glaucomatous eyes. Jpn J Ophthalmol 1997; 41(2):84–88.
272. Dkhissi O et al. Retinal TUNEL-positive cells and high glutamate levels in vitreous humor of mutant quail with a glaucoma-like disorder. Invest Ophthalmol Vis Sci 1999; 40(5):990–995.
273. Quigley HA, Dunkelberger GR, Green WR. Retinal ganglion cell atrophy correlated with automated perimetry in human eyes with glaucoma. Am J Ophthalmol 1989; 107(5):453–464.
274. Hoyt WF, Frisen L, Newman NM. Fundoscopy of nerve fiber layer defects in glaucoma. Invest Ophthalmol 1973; 12(11):814–829.
275. Quigley HA, Miller NR, George T. Clinical evaluation of nerve fiber layer atrophy as an indicator of glaucomatous optic nerve damage. Arch Ophthalmol 1980; 98(9):1564–1571.
276. Airaksinen PJ et al. Diffuse and localized nerve fiber loss in glaucoma. Am J Ophthalmol 1984; 98(5):566–571.
277. Iwata K, Kurosawa A, Sawaguchi S. Wedge-shaped retinal nerve fiber layer defects in experimental glaucoma preliminary report. Graefes Arch Clin Exp Ophthalmol 1985; 223(4):184–189.
278. Drance SM. The early structural and functional disturbances of chronic open-angle glaucoma. Robert N. Shaffer lecture. Ophthalmology 1985; 92(7):853–857.
279. Stroman GA et al. Magnetic resonance imaging in patients with low-tension glaucoma. Arch Ophthalmol 1995; 113(2):168–172.
280. Iwata F et al. Association of visual field, cup-disc ratio, and magnetic resonance imaging of optic chiasm. Arch Ophthalmol 1997; 115(6):729–732.
281. Weber AJ et al. Experimental glaucoma and cell size, density, and number in the primate lateral geniculate nucleus. Invest Ophthalmol Vis Sci 2000; 41(6):1370–1379.
282. Yücel YH et al. Loss of neurons in magnocellular and parvocellular layers of the lateral geniculate nucleus in glaucoma. Arch Ophthalmol 2000; 118(3):378–384.
283. Vickers JC et al. Magnocellular and parvocellular visual pathways are both affected in a macaque monkey model of glaucoma. Aust NZ J Ophthalmol 1997; 25(3):239–243.
284. Chaturvedi N, Hedley-Whyte ET, Dreyer EB. Lateral geniculate nucleus in glaucoma. Am J Ophthalmol 1993; 116(2):182–188.
285. Crawford ML et al. Experimental glaucoma in primates: changes in cytochrome oxidase blobs in V1 cortex. Invest Ophthalmol Vis Sci 2001; 42(2):358–364.
286. Crawford ML et al. Glaucoma in primates: cytochrome oxidase reactivity in parvo- and magnocellular pathways. Invest Ophthalmol Vis Sci 2000; 41(7):1791–1802.
287. Hernandez MR, Pena JD. The optic nerve head in glaucomatous optic neuropathy. Arch Ophthalmol 1997; 115(3):389–395.
288. Pena JD et al. Elastosis of the lamina cribrosa in glaucomatous optic neuropathy. Exp Eye Res 1998; 67(5):517–524.
289. Thale A, Tillmann B, Rochels R. SEM studies of the collagen architecture of the human lamina cribrosa: normal and pathological findings. Ophthalmologica 1996; 210(3):142–147.
290. Hollander H et al. Evidence of constriction of optic nerve axons at the lamina cribrosa in the normotensive eye in humans and other mammals. Ophthalmic Res 1995; 27(5):296–309.
291. Quigley H, Pease ME, Thibault D. Change in the appearance of elastin in the lamina cribrosa of glaucomatous optic nerve heads. Graefes Arch Clin Exp Ophthalmol 1994; 232(5):257–261.
292. Fukuchi T et al. Sulfated proteoglycans in the lamina cribrosa of normal monkey eyes and monkey eyes with laser-induced glaucoma. Exp Eye Res 1994; 58(2):231–243.
293. Hernandez MR. Ultrastructural immunocytochemical analysis of elastin in the human lamina cribrosa. Changes in elastic fibers in primary open-angle glaucoma. Invest Ophthalmol Vis Sci 1992; 33(10):2891–2903.
294. Fukuchi T et al. Extracellular matrix changes of the optic nerve lamina cribrosa in monkey eyes with experimentally chronic glaucoma. Graefes Arch Clin Exp Ophthalmol 1992; 230(5):421–427.
295. Spileers W, Goethals M. Structural changes of the lamina cribrosa and of the trabeculum in primary open angle glaucoma (POAG). Bull Soc Belge Ophtalmol 1992; 244:27–35.
296. Quigley HA et al. Morphologic changes in the lamina cribrosa correlated with neural loss in open-angle glaucoma. Am J Ophthalmol 1983; 95(5):673–691.
297. Susanna R Jr. The lamina cribrosa and visual field defects in open-angle glaucoma. Can J Ophthalmol 1983; 18(3):124–126.
298. Quigley HA, Addicks EM. Regional differences in the structure of the lamina cribrosa and their relation to glaucomatous optic nerve damage. Arch Ophthalmol 1981; 99(1):137–143.
299. Emery JM et al. The lamina cribrosa in normal and glaucomatous human eyes. Trans Am Acad Ophthalmol Otolaryngol 1974; 78(2):OP290–7.
300. Morrison JC et al. Understanding mechanisms of pressure-induced optic nerve damage. Prog Retin Eye Res 2005; 24(2):217–240.
301. Whitmore AV, Libby RT, John SW. Glaucoma: thinking in new ways-a role for autonomous axonal self-destruction and other compartmentalised processes? Prog Retin Eye Res 2005; 24(6):639–662.
302. Quigley HA, Addicks EM. Chronic experimental glaucoma in primates. II. Effect of extended intraocular pressure elevation on optic nerve head and axonal transport. Invest Ophthalmol Vis Sci 1980; 19(2):137–152.

303. Quigley HA et al. Optic nerve damage in human glaucoma. II. The site of injury and susceptibility to damage. Arch Ophthalmol 1981; 99(4):635–649.

304. Allcutt D, Berry M, Sievers J. A qualitative comparison of the reactions of retinal ganglion cell axons to optic nerve crush in neonatal and adult mice. Brain Res 1984; 318(2):231–240.

305. Allcutt D, Berry M, Sievers J. A quantitative comparison of the reactions of retinal ganglion cells to optic nerve crush in neonatal and adult mice. Brain Res 1984; 318(2):219–230.

306. Barron KD et al. Qualitative and quantitative ultrastructural observations on retinal ganglion cell layer of rat after intraorbital optic nerve crush. J Neurocytol 1986; 15(3):345–362.

307. Neufeld AH, Hernandez MR, Gonzalez M. Nitric oxide synthase in the human glaucomatous optic nerve head. Arch Ophthalmol 1997; 115(4):497–503.

308. Neufeld AH. Nitric oxide: a potential mediator of retinal ganglion cell damage in glaucoma. Surv Ophthalmol 1999; 43(Suppl 1):S129-S135.

309. Anderson DR, Hendrickson A. Effect of intraocular pressure on rapid axoplasmic transport in monkey optic nerve. Invest Ophthalmol 1974; 13:771.

310. Quigley HA, Guy J, Anderson DR. Blockade of rapid axonal transport. Effect of intraocular pressure elevation in primate optic nerve. Arch Ophthalmol 1979; 97(3):525–531.

311. Radius RL. Pressure-induced fast axonal transport abnormalities and the anatomy at the lamina cribrosa in primate eyes. Invest Ophthalmol Vis Sci 1983; 24:343.

312. Minckler DS, Bunt AH, Johanson GW. Orthograde and retrograde axoplasmic transport during acute ocular hypertension in the monkey. Invest Ophthalmol Vis Sci 1977; 16(5):426–441.

313. Johnson EC et al. Chronology of optic nerve head and retinal responses to elevated intraocular pressure. Invest Ophthalmol Vis Sci 2000; 41(2):431–442.

314. Pease ME et al. Obstructed axonal transport of BDNF and its receptor TrkB in experimental glaucoma. Invest Ophthalmol Vis Sci 2000; 41:764–774.

315. Quigley HA et al. Retrograde axonal transport of BDNF in retinal ganglion cells is blocked by acute IOP elevation in rats. Invest Ophthalmol Vis Sci 2000; 41(11):3460–3466.

316. Rezaie T et al. Adult-onset primary open-angle glaucoma caused by mutations in optineurin. Science 2002; 295(5557):1077–1079.

317. Anborgh PH et al. Inhibition of metabotropic glutamate receptor signaling by the huntingtin-binding protein optineurin. J Biol Chem 2005; 280(41):34840–34848.

318. Chalasani ML et al. A glaucoma-associated mutant of optineurin selectively induces death of retinal ganglion cells which is inhibited by antioxidants. Invest Ophthalmol Vis Sci 2007; 48(4):1607–1614.

319. De Marco N et al. Optineurin increases cell survival and translocates to the nucleus in a Rab8-dependent manner upon an apoptotic stimulus. J Biol Chem 2006; 281(23):16147–16156.

320. Park BC et al. Studies of optineurin, a glaucoma gene: Golgi fragmentation and cell death from overexpression of wild-type and mutant optineurin in two ocular cell types. Am J Pathol 2006; 169(6):1976–1989.

321. Tso MO, Hayreh SS. Optic disc edema in raised intracranial pressure. IV. Axoplasmic transport in experimental papilledema. Arch Ophthalmol 1977; 95(8):1458–1462.

322. Anderson DR, Hendrickson AE. Failure of increased intracranial pressure to affect rapid axonal transport at the optic nerve head. Invest Ophthalmol Vis Sci 1977; 16(5):423–426.

323. Radius RL, Anderson DR. Fast axonal transport in early experimental disc edema. Invest Ophthalmol Vis Sci 1980; 19(2):158–168.

324. Minckler DS, Tso MO, Zimmerman LE. A light microscopic, autoradiographic study of axoplasmic transport in the optic nerve head during ocular hypotony, increased intraocular pressure, and papilledema. Am J Ophthalmol 1976; 82(5):741–757.

325. Minckler DS, Bunt AH. Axoplasmic transport in ocular hypotony and papilledema in the monkey. Arch Ophthalmol 1977; 95(8):1430–1436.

326. Ventura LM, Porciatti V. Restoration of retinal ganglion cell function in early glaucoma after intraocular pressure reduction: a pilot study. Ophthalmology 2005; 112(1):20–27.

327. Ventura LM et al. The relationship between retinal ganglion cell function and retinal nerve fiber thickness in early glaucoma. Invest Ophthalmol Vis Sci 2006; 47(9):3904–3911.

328. Ventura LM, Venzara FX 3rd, Porciatti V. Reversible dysfunction of retinal ganglion cells in non-secreting pituitary tumors. Doc Ophthalmol 2009; 118:155–162.

329. Murray M, Grafstein B. Changes in the morphology and amino acid incorporation of regenerating goldfish optic neurons. Exp Neurol 1969; 23:544–560.

330. Humphrey MF, Beazley LD. Retinal ganglion cell death during optic nerve regeneration in the frog Hyla moorei. J Comp Neurol 1985; 263:382–402.

331. Humphrey MF. A morphometric study of the retinal ganglion cell response to optic nerve severance in the frog Rana pipiens. J Neurocytol 1988; 17:293–304.

332. Grafstein B, Ingoglia NA. Intracranial transection of the optic nerve in adult mice: Preliminary observations. Exp Neurol 1982; 76:318–330.

333. Misantone LJ, Gershenbaum M, Murray M. Viability of retinal ganglion cells after optic nerve crush in adult rats. J Neurocytol 1984; 13(3):449–465.

334. Villegas-Perez MP et al. Rapid and protracted phases of retinal ganglion cell loss follow axotomy in the optic nerve of adult rats. J Neurobiol 1993; 24(1):23–36.

335. Peinado-Ramon P et al. Effects of axotomy and intraocular administration of NT-4, NT-3, and brain-derived neurotrophic factor on the survival of adult rat retinal ganglion cells. A quantitative in vivo study. Invest Ophthalmol Vis Sci 1996; 37(4):489–500.

336. Radius RL, Anderson DR. Retinal ganglion cell degeneration in experimental optic atrophy. Am J Ophthalmol 1978; 86(5):673–679.

337. Quigley HA, Anderson DR. Descending optic nerve degeneration in primates. Invest Ophthalmol Vis Sci 1977; 16:841–849.

338. Kupfer C. Retinal ganglion cell degeneration following chiasmal lesions in man. Arch Ophthalmol 1963; 70:256–260.

339. Goldberg S, Frank B. Do young axons regenerate better than old axons? Exp Neurol 1981; 74:245–259.

340. Stone J. The naso-temporal division of the cat's retina. J Comp Neurol 1966; 126:585–600.

341. Thanos S et al. Specific transcellular staining of microglia in the adult rat after traumatic degeneration of carbocyanine-filled retinal ganglion cells. Exp Eye Res 1992; 55(1):101–117.

342. Leinfelder PJ. Retrograde degeneration in the optic nerves and tracts. Am J Ophthalmol 1940; 23:796–802.

343. Watson WE. Cellular responses to axotomy and to related procedures. Br Med Bull 1974; 30:112–115.

344. Lieberman AR. A review of the principal features of perikaryal responses to axon injury. Int Rev Neurobiol 1971; 14:49–124.

345. Anderson DR. Ascending and descending optic atrophy produced experimentally in squirrel monkeys. Am J Ophthalmol 1973; 76:693–711.

346. Madison R, Moore MR, Sidman RL. Retinal ganglion cells and axons survive optic nerve transection. Int J Neurosci 1984; 23(1):15–32.

347. Garcia-Valenzuela E et al. Apoptosis in adult retinal ganglion cells after axotomy. J Neurobiol 1994; 25(4):431–438.

348. Rehen SK, Linden R. Apoptosis in the developing retina: paradoxical effects of protein synthesis inhibition. Braz J Med Biol Res 1994; 27(7):1647–1651.

349. Levin LA, Louhab A. Apoptosis of retinal ganglion cells in anterior ischemic optic neuropathy. Arch Ophthalmol 1996; 114(4):488–491.

350. Cellerino A, Galli-Resta L, Colombaioni L. The dynamics of neuronal death: a time-lapse study in the retina. J Neurosci 2000; 20(16):RC92.

351. Kerr JF, Wyllie AH, Currie AR. Apoptosis: a basic biological phenomenon with wide-ranging implications in tissue kinetics. Br J Cancer 1972; 26(4):239–257.

352. Gavrieli Y, Sherman Y, Ben-Sasson SA. Identification of programmed cell death in situ via specific labeling of nuclear DNA fragmentation. J Cell Biol 1992; 119(3):493–501.

353. Ankarcrona M et al. Glutamate-induced neuronal death: a succession of necrosis or apoptosis depending on mitochondrial function. Neuron 1995; 15(4):961–973.

354. Lucas M, Solano F, Sanz A. Induction of programmed cell death (apoptosis) in mature lymphocytes. FEBS Lett 1991; 279(1):19–20.

355. Olney JW. The toxic effects of glutamate and related compounds in the retina and the brain. Retina 1982; 2(4):341–359.

356. Sisk DR, Kuwabara T. Histologic changes in the inner retina of albino rats following intravitreal injection of monosodium L-glutamate. Graefes Arch Clin Exp Ophthalmol 1985; 223(5):250–258.

357. Schumer RA, Podos SM. The nerve of glaucoma! Arch Ophthalmol 1994; 112(1):37–44.

358. Russelakis-Carneiro M, Silveira LC, Perry VH. Factors affecting the survival of cat retinal ganglion cells after optic nerve injury. J Neurocytol 1996; 25(6):393–402.

359. Yoles E, Muller S, Schwartz M. NMDA-receptor antagonist protects neurons from secondary degeneration after partial optic nerve crush. J Neurotrauma 1997; 14:665–675.

360. Yoles E, Schwartz M. Elevation of intraocular glutamate levels in rats with partial lesion of the optic nerve. Arch Ophthalmol 1998; 116(7):906–910.

361. Dreyer EB et al. Elevated glutamate levels in the vitreous body of humans and monkeys with glaucoma. Arch Ophthalmol 1996; 114(3):299–305.

362. Kreutz MR et al. Axonal injury alters alternative splicing of the retinal NR1 receptor: the preferential expression of the NR1b isoforms is crucial for retinal ganglion cell survival. J Neurosci 1998; 18(20):8278–8291.

363. Ullian EM et al. Invulnerability of retinal ganglion cells to NMDA excitotoxicity. Mol Cell Neurosci 2004; 26(4):544–557.

364. Levin LA, Peeples P. History of neuroprotection and rationale as a therapy for glaucoma. Am J Manag Care 2008; 14(Suppl 1):S11-S14.

365. Cragg BG. What is the signal for chromatolysis? Brain Res 1970; 23:1–21.

366. Schnitzer J, Scherer J. Microglial cell responses in the rabbit retina following transection of the optic nerve. J Comp Neurol 1990; 302:779–791.

367. Cui Q, Harvey AR. At least two mechanisms are involved in the death of retinal ganglion cells following target ablation in neonatal rats. J Neurosci 1995; 15:8143–8155.

368. Bodeutsch N et al. Unilateral injury to the adult rat optic nerve causes multiple cellular responses in the contralateral site. J Neurobiol 1999; 38(1):116–128.

369. Maffei L et al. Schwann cells promote the survival of rat retinal ganglion cells after optic nerve section. Proc Natl Acad Sci USA 1990; 87(5):1855–1859.

370. Castillo BJ et al. Retinal ganglion cell survival is promoted by genetically modified astrocytes designed to secrete brain-derived neurotrophic factor (BDNF). Brain Res 1994; 647(1):30–36.

371. Weibel D, Kreutzberg GW, Schwab ME. Brain-derived neurotrophic factor (BDNF) prevents lesion-induced axonal die-back in young rat optic nerve. Brain Res 1995; 679(2):249–254.

372. Klocker N et al. Brain-derived neurotrophic factor-mediated neuroprotection of adult rat retinal ganglion cells in vivo does not exclusively depend on phosphatidyl-inositol-3'-kinase/protein kinase B signaling. J Neurosci 2000; 20(18):6962–6967.

373. Klocker N et al. Both the neuronal and inducible isoforms contribute to upregulation of retinal nitric oxide synthase activity by brain-derived neurotrophic factor. J Neurosci 1999; 19(19):8517–8527.

374. Klocker N, Cellerino A, Bahr M. Free radical scavenging and inhibition of nitric oxide synthase potentiates the neurotrophic effects of brain-derived neurotrophic factor on axotomized retinal ganglion cells in vivo. J Neurosci 1998; 18(3):1038–1046.

375. Klocker N et al. In vivo neurotrophic effects of GDNF on axotomized retinal ganglion cells. Neuroreport 1997; 8(16):3439–3442.

376. Carmignoto G et al. Effect of NGF on the survival of rat retinal ganglion cells following optic nerve section. J Neurosci 1989; 9(4):1263–1272.

377. Thanos S et al. Survival and axonal elongation of adult rat retinal ganglion cells: in vitro effects of lesioned sciatic nerve and brain derived neurotrophic factor. Eur J Neurosci 1989; 1:19–26.

378. Sievers J et al. Fibroblast growth factors promote the survival of adult rat retinal ganglion cells after transection of the optic nerve. Neurosci Lett 1987; 76:157–162.

379. Bahr M, Vanselow J, Thanos S. Ability of adult rat ganglion cells to regrow axons in vitro can be confluenced by fibroblast growth factor and gangliosides. Neurosci Lett 1989; 96:197–201.

380. Clarke DB et al. Effect of neurotrophin-4 administration on the survival of axotomized retinal ganglion cells in adult rats. Soc Neurosci Abstr 1993; 19:1104.

381. Meyer-Franke A et al. Characterization of the signaling interactions that promote the survival and growth of developing retinal ganglion cells in culture. Neuron 1995; 15:805–819.
382. Shen S et al. Retinal ganglion cells lose trophic responsiveness after axotomy. Neuron 1999; 23(2):285–295.
383. Meyer-Franke A et al. Depolarization and cAMP elevation rapidly recruit TrkB to the plasma membrane of CNS neurons. Neuron 1998; 21(4):681–693.
384. Thanos S, Mey J, Wild M. Treatment of the adult retina with microglia-suppressing factors retards axotomy-induced neuronal degradation and enhances axonal regeneration in vivo and in vitro. J Neurosci 1993; 13:455–466.
385. Perez MT, Caminos E. Expression of brain-derived neurotrophic factor and of its functional receptor in neonatal and adult rat retina. Neurosci Lett 1995; 183(1–2):96–99.
386. Segal RA, Greenberg ME. Intracellular signaling pathways activated by neurotrophic factors. Ann Rev Neurosci 1996; 19:463–489.
387. Dudek H et al. Regulation of neuronal survival by the serine-threonine protein kinase Akt. Science 1997; 275(5300):661–665.
388. Hemmings BA. Akt signaling: linking membrane events to life and death decisions. Science 1997; 275(5300):628–630.
389. Cook RD, Wisniewski HM. The role of oligodendroglia and astroglia in Wallerian degeneration of the optic nerve. Brain Res 1973; 61:191.
390. Ludwin SK. Phagocytosis in the rat optic nerve following Wallerian degeneration. Acta Neuropathol (Berl) 1990; 80:266.
391. Stoll G, Mueller HW. Lesion-induced changes of astrocyte morphology and protein expression in rat optic nerve. Ann NY Acad Sci 1988; 540:461.
392. Ling EA. Electron microscopic studies of macrophages in Wallerian degeneration of rat optic nerve after intravenous injection of colloidal carbon. J Anat 1978; 126:111.
393. Stevens A, Bahr M. Origin of macrophages in central nervous tissue. A study using intraperitoneal transplants contained in Millipore diffusion chambers. J Neurol Sci 1993; 118(2):117–122.
394. Cook RD, Wisniewski HM. The spatio-temporal pattern of Wallerian degeneration in the rhesus monkey optic nerve. Acta Neuropathol (Berl) 1987; 72:261.
395. Podhajsky RJ et al. A quantitative immunohistochemical study of the cellular response to crush injury in optic nerve. Exp Neurol 1997; 143(1):153–161.
396. Liu KM, Shen CL. Ultrastructural sequence of myelin breakdown during Wallerian degeneration in the rat optic nerve. Cell Tissue Res 1985; 242:245.
397. Reichert F, Rotshenker S. Deficient activation of microglia during optic nerve degeneration. J Neuroimmunol 1996; 70(2):153–161.
398. Lazarov-Spiegler O et al. Transplantation of activated macrophages overcomes central nervous system regrowth failure. FASEB J 1996; 10(11):1296–1302.
399. Butt AM, Colquhoun K. Glial cells in transected optic nerves of immature rats. I. An analysis of individual cells by intracellular dye-injection. J Neurocytol 1996; 25(6): 365–380.
400. Butt AM, Kirvell S. Glial cells in transected optic nerves of immature rats. II. An immunohistochemical study. J Neurocytol 1996; 25(6):381–392.
401. Wanner M et al. Re-evaluation of the growth-permissive substrate properties of goldfish optic nerve myelin and myelin proteins. J Neurosci 1995; 15(11):7500–7508.
402. Matsumoto N, Kometani M, Nagano K. Regenerating retinal fibers of the goldfish make temporary and unspecific but functional synapses before forming the final retinotopic projection. Neuroscience 1987; 22(3):1103–1110.
403. Sawai H et al. Brain-derived neurotrophic factor and neurotrophin-4/5 stimulate growth of axonal branches from regenerating retinal ganglion cells. J Neurosci 1996; 16(12): 3887–3894.
404. Zeng BY et al. Regenerative and other responses to injury in the retinal stump of the optic nerve in adult albino rats: transection of the intraorbital optic nerve. J Anat 1994; 185(Pt 3):643–661.
405. Aguayo AJ, David S, Bray GM. Influences of the glial environment on the elongation of axons after injury: transplantation studies in adult rodents. J Exp Biol 1981; 95: 231–240.
406. Yiu G, He Z. Glial inhibition of CNS axon regeneration. Nat Rev Neurosci 2006; 7(8): 617–627.
407. Frank M, Wolburg H. Cellular reactions at the lesion site after crushing of the rat optic nerve. Glia 1996; 16(3):227–240.
408. Zeev-Brann AB et al. Differential effects of central and peripheral nerves on macrophages and microglia. Glia 1998; 23(3):181–190.
409. Castano A, Bell MD, Perry VH. Unusual aspects of inflammation in the nervous system: Wallerian degeneration. Neurobiol Aging 1996; 17(5):745–751.
410. Kalil K, Reh T. A light and electron microscopic study of regrowing pyramidal tract fibers. J Comp Neurol 1982; 211(3):265–275.
411. Reh T, Kalil K. Functional role of regrowing pyramidal tract fibers. J Comp Neurol 1982; 211(3):276–283.
412. Saunders NR et al. Growth of axons through a lesion in the intact CNS of fetal rat maintained in long-term culture. Proc R Soc Lond B Biol Sci 1992; 250(1329): 171–180.
413. Aguayo AJ et al. Growth and connectivity of axotomized retinal neurons in adult rats with optic nerves substituted by PNS grafts linking the eye and the midbrain. Ann N Y Acad Sci 1987; 495:1–9.
414. Bray GM et al. The use of peripheral nerve grafts to enhance neuronal survival, promote growth and permit terminal reconnections in the central nervous system of adult rats. J Exp Biol 1987; 132:5–19.
415. Chen DF, Jhaveri S, Schneider GE. Intrinsic changes in developing retinal neurons result in regenerative failure of their axons. Proc Natl Acad Sci USA 1995; 92(16): 7287–7291.
416. Goldberg JL et al. Amacrine-signaled loss of intrinsic axon growth ability by retinal ganglion cells. Science 2002; 296(5574):1860–1864.
417. Cohen J et al. Developmental loss of functional laminin receptors on retinal ganglion cells is regulated by their target tissue, the optic tectum. Development 1989; 107(2): 381–387.
418. de Curtis I et al. Laminin receptors in the retina: sequence analysis of the chick integrin alpha 6 subunit. Evidence for transcriptional and posttranslational regulation. J Cell Biol 1991; 113(2):405–416.
419. de Curtis I, Reichardt LF. Function and spatial distribution in developing chick retina of the laminin receptor alpha 6 beta 1 and its isoforms. Development 1993; 118(2):377–388.
420. Ivins JK, Yurchenco PD, Lander AD. Regulation of neurite outgrowth by integrin activation. J Neurosci 2000; 20(17):6551–6560.
421. Cohen J, Johnson AR. Differential effects of laminin and merosin on neurite outgrowth by developing retinal ganglion cells. J Cell Sci Suppl 1991; 15:1–7.
422. Condic ML. Adult neuronal regeneration induced by transgenic integrin expression. J Neurosci 2001; 21(13):4782–4788.
423. Du J et al. Activity- and Ca^{2+}-dependent modulation of surface expression of brain-derived neurotrophic factor receptors in hippocampal neurons. J Cell Biol 2000; 150(6):1423–1434.
424. Blackmore MG et al. A developmentally regulated family of transcription factors controls axon growth in CNS neurons. Washington: Society for Neuroscience, 2008:725.23.
425. Moore DL, Blackmore MG, Goldberg JL. Transcriptional control of intrinsic axon growth ability in retinal ganglion cells. Washington: Society for Neuroscience, 2008:725.24.
426. Konishi Y et al. Cdh1-APC controls axonal growth and patterning in the mammalian brain. Science 2004; 303(5660):1026–1030.
427. Wayman GA et al. Regulation of axonal extension and growth cone motility by calmodulin-dependent protein kinase I. J Neurosci 2004; 24(15):3786–3794.
428. Cai D et al. Neuronal cyclic AMP controls the developmental loss in ability of axons to regenerate. J Neurosci 2001; 21(13):4731–4739.
429. Carroll WM, Jennings AR, Mastaglia FL. The origin of remyelinating oligodendrocytes in antiserum-mediated demyelinative optic neuropathy. Brain 1990; 113:953.
430. Stanhope GB, Billings-Gagliardi S, Wolf MK. Myelination requirements of central nervous system glia in vitro: statistical validation of differences between glia from adult and immature mice. Glia 1990; 3:125.
431. Schwartz M et al. Potential treatment modalities for glaucomatous neuropathy: neuroprotection and neuroregeneration. J Glaucoma 1996; 5(6):427–432.
432. Weinreb RN, Levin LA. Is neuroprotection a viable therapy for glaucoma? Arch Ophthalmol 1999; 117(11):1540–1544.
433. Dreyer EB. A proposed role for excitotoxicity in glaucoma. J Glaucoma 1998; 7(1): 62–67.
434. Yoles E, Wheeler LA, Schwartz M. Alpha2-adrenoreceptor agonists are neuroprotective in a rat model of optic nerve degeneration. Invest Ophthalmol Vis Sci 1999; 40(1): 65–73.
435. Wen R et al. Alpha 2-adrenergic agonists induce basic fibroblast growth factor expression in photoreceptors in vivo and ameliorate light damage. J Neurosci 1996; 16(19):5986–5992.
436. Neufeld AH, Sawada A, Becker B. Inhibition of nitric-oxide synthase 2 by aminoguanidine provides neuroprotection of retinal ganglion cells in a rat model of chronic glaucoma. Proc Natl Acad Sci USA 1999; 96(17):9944–9948.
437. Schori H et al. Vaccination for protection of retinal ganglion cells against death from glutamate cytotoxicity and ocular hypertension: Implications for glaucoma. Proc Natl Acad Sci USA 2001; 98(6):3398–3403.
438. Cui Q et al. Intraocular elevation of cyclic AMP potentiates ciliary neurotrophic factor-induced regeneration of adult rat retinal ganglion cell axons. Mol Cell Neurosci 2003; 22(1):49–61.
439. Monsul NT et al. Intraocular injection of dibutyryl cyclic AMP promotes axon regeneration in rat optic nerve. Exp Neurol 2004; 186(2):124–133.
440. Watanabe M et al. Intravitreal injections of neurotrophic factors and forskolin enhance survival and axonal regeneration of axotomized beta ganglion cells in cat retina. Neuroscience 2003; 116(3):733–742.
441. Jiang C et al. Intravitreal injections of GDNF-loaded biodegradable microspheres are neuroprotective in a rat model of glaucoma. Mol Vis 2007; 13:1783–1792.
442. Thanos C, Emerich D. Delivery of neurotrophic factors and therapeutic proteins for retinal diseases. Expert Opin Biol Ther 2005; 5(11):1443–1452.
443. Ward MS et al. Neuroprotection of retinal ganglion cells in DBA/2J mice with GDNF-loaded biodegradable microspheres. J Pharm Sci 2007; 96(3):558–568.
444. Ko ML et al. The combined effect of brain-derived neurotrophic factor and a free radical scavenger in experimental glaucoma. Invest Ophthalmol Vis Sci 2000; 41(10): 2967–2971.
445. Ko ML et al. Patterns of retinal ganglion cell survival after brain-derived neurotrophic factor administration in hypertensive eyes of rats. Neurosci Lett 2001; 305(2):139–142.
446. Martin KR et al. Gene therapy with brain-derived neurotrophic factor as a protection: retinal ganglion cells in a rat glaucoma model. Invest Ophthalmol Vis Sci 2003; 44(10):4357–4365.
447. Pease ME et al. CNTF over-expression leads to increased retinal ganglion cell survival in experimental glaucoma. Invest Ophthalmol Vis Sci, 2008.
448. Hu Y, Cui Q, Harvey AR. Interactive effects of C3, cyclic AMP and ciliary neurotrophic factor on adult retinal ganglion cell survival and axonal regeneration. Mol Cell Neurosci 2007; 34(1):88–98.
449. Lingor P et al. ROCK inhibition and CNTF interact on intrinsic signalling pathways and differentially regulate survival and regeneration in retinal ganglion cells. Brain 2008; 131(Pt 1):250–263.
450. Thoenen H et al. Neurotrophic factors and neuronal death. Ciba Found Symp 1987; 126:82–95.
451. Cui Q, Harvey AR. CNTF promotes the regrowth of retinal ganglion cell axons into murine peripheral nerve grafts. Neuroreport 2000; 11(18):3999–4002.
452. Leaver SG et al. AAV-mediated expression of CNTF promotes long-term survival and regeneration of adult rat retinal ganglion cells. Gene Ther 2006; 13(18):1328–1341.
453. Watanabe M, Fukuda Y. Survival and axonal regeneration of retinal ganglion cells in adult cats. Prog Retin Eye Res 2002; 21(6):529–553.
454. van Adel BA et al. Delivery of ciliary neurotrophic factor via lentiviral-mediated transfer protects axotomized retinal ganglion cells for an extended period of time. Hum Gene Ther 2003; 14(2):103–115.
455. Weise J et al. Adenovirus-mediated expression of ciliary neurotrophic factor (CNTF) rescues axotomized rat retinal ganglion cells but does not support axonal regeneration in vivo. Neurobiol Dis 2000; 7(3):212–223.

456. Ji JZ et al. CNTF promotes survival of retinal ganglion cells after induction of ocular hypertension in rats: the possible involvement of STAT3 pathway. Eur J Neurosci 2004; 19(2):265–272.

457. Unoki K, LaVail MM. Protection of the rat retina from ischemic injury by brain-derived neurotrophic factor, ciliary neurotrophic factor, and basic fibroblast growth factor. Invest Ophthalmol Vis Sci 1994; 35(3):907–915.

458. Yin Y et al. Oncomodulin is a macrophage-derived signal for axon regeneration in retinal ganglion cells. Nat Neurosci 2006; 9(6):843–852.

459. Lehmann M et al. Inactivation of Rho signaling pathway promotes CNS axon regeneration. J Neurosci 1999; 19(17):7537–7547.

460. Hu Y, Cui Q, Harvey AR. Interactive effects of C3, cyclic AMP and ciliary neurotrophic factor on adult retinal ganglion cell survival and axonal regeneration. Mol Cell Neurosci 2007; 34(1):88–98.

461. Brenner MJ et al. Delayed nerve repair is associated with diminished neuroenhancement by FK506. Laryngoscope 2004; 114(3):570–576.

462. Normann C, Berger M. Neuroenhancement: status quo and perspectives. Eur Arch Psychiatry Clin Neurosci 2008; 258(Suppl 5):110–114.

463. Schloesser RJ, Chen G, Manji HK. Neurogenesis and neuroenhancement in the pathophysiology and treatment of bipolar disorder. Int Rev Neurobiol 2007; 77: 143–178.

CHAPTER 29

Processing in the Lateral Geniculate Nucleus (LGN)

Vivien Casagrande & Jennifer Ichida

The lateral geniculate nucleus: the gateway to conscious visual perception

Conscious perception requires the visual information that passes through the dorsal lateral geniculate nucleus (LGN) in primates. Although the retina sends axons to many subcortical nuclei, only the pathway from the retina to LGN to cortex is critical to visual awareness.

The LGN is a distinctively layered structure and is located at the posterior lateral margin of the dorsal thalamus (Fig. 29.1). Although there is agreement that the LGN provides the key visual gateway to cortex, there is less agreement over LGN function.[1,2] The main reason is that receptive field properties of LGN cells are so similar to those of their retinal ganglion cell inputs. Given that visual signals are transformed in other visual areas it is puzzling that a similar transformation can not be identified in the LGN. The most reasonable explanation is that the main role of the LGN is to regulate the flow and strength of visual signals sent to V1. Evidence indicates that this regulation is much more complex than simply opening and closing a single gate. The LGN and visual cortex, working together, form part of a dynamic system designed to rapidly and efficiently process the most useful visual signals for survival. The parallel processing of visual information, elaborate circuitry and many extraretinal inputs, including the massive visual cortical feedback to the LGN, all suggest that visual signals are regulated in a variety of specific and general ways within the LGN.

The purpose of this chapter is to illuminate the structure and function of the LGN by first reviewing LGN anatomy and physiology and then current controversies over signal processing. The field has moved from static descriptions of LGN cell receptive field properties to dynamic descriptions that take into account eye movements, motor planning, arousal level, attention and possibly even more complex information concerning visual memories. In the following two sections, we describe the basic architecture, connections and neurochemistry of the LGN. The next four sections consider the LGN in a functional context beginning with the basics of signal processing in section four. Then we consider the potential functional impact of attention, motor planning, and binocular rivalry. The final section summarizes key points.

Overview of lateral geniculate anatomy

Layers and maps

The LGN is an elongate layered structure with each layer representing the opposite visual hemifield (Fig. 29.2). Within each layer the opposite hemifield is represented such that the superior and inferior visual fields are located toward the lateral and medial zones of the layer, respectively, and the central (toward the fovea) and peripheral visual fields are located, respectively, at the posterior and anterior zones of each layer. Each layer receives monocular input from the retina and only those layers receiving input from the contralateral eye (nasal retina) represent the entire contralateral visual hemifield since LGN layers receiving input from the ipsilateral eye (temporal retina) cannot represent the monocular portion of the hemifield. This means that the ipsilateral layers are always shorter in all dimensions than the corresponding contralateral layers (Fig. 29.2). The maps in each layer are retinotopically aligned such that each point in visual space is represented along a line perpendicular to the layers. The alignment of the maps is so precise that a cell-free gap, representing the optic disk (see arrows in Figs 29.1B and 29.2), exists in all contralaterally innervated layers in order to maintain retinotopic alignment with the ipsilaterally innervated layers. This is because the optic disk contains no receptors and is located in the nasal retina.

There are three types of cell layers, those with large cells (the magnocellular or M layers), those with medium size cells (the parvocellular or P layers) and those with very small cells (the koniocellular or K layers). All primates have at least two P and two M layers with K layers lying below each M and P layer. In humans and some other primates, each P layer can split into two or more layers in the portion of the

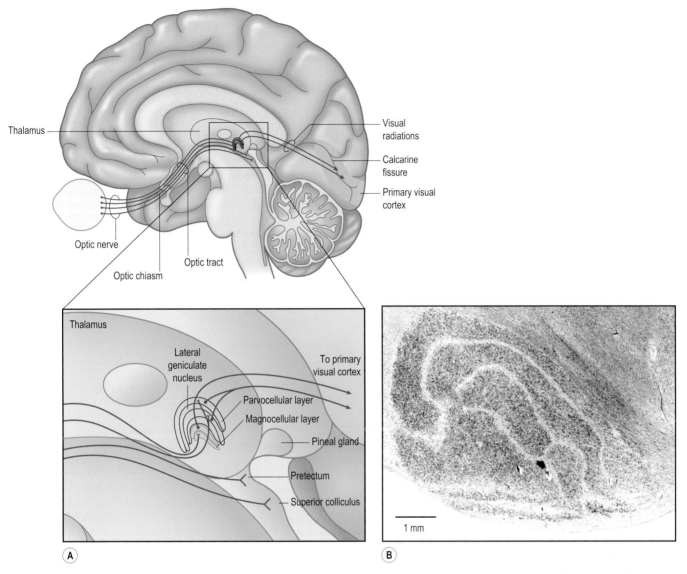

Figure 29.1 (**A**) Schematic diagram showing the location of the human LGN in relationship to the rest of the brain, with a magnified diagram of the LGN. (Modified from Kandel et al.[48]) (**B**) Nissl (cell) stained coronal section through the human LGN showing the laminar organization. Arrow in (B) indicates the optic disc representation which appears as a cell free gap in contralaterally innervated layer P4. (Reproduced with permission from Hickey & Guillery.[3])

LGN that represents central vision (~2–17 degrees). The normal human LGN can contain as few as two or as many as six P layers.[3]

Among primates, the laminar organization and topography of the LGN has been studied in the most detail in the macaque monkey. Malpeli and colleagues[4,5] performed careful reconstructions of the entire LGN documenting the retinotopic organization and number of M and P cells in each LGN layer (K cells were not counted). These studies have shown that in macaque monkeys the representation of the central visual field is magnified in the LGN (Fig. 29.2 and see Fig. 29.5 below). The magnified representation of the central few degrees in the LGN is easy to understand given that each LGN cell receives input from approximately one to three retinal ganglion cells and ganglion cell density is much higher in central retina.

Only within the portion of the nucleus representing eccentricities from 2 to 17 degrees do the two P layers split to represent four or more P layers, an arrangement which may reflect an increase in the diversity of retinal ganglion cells representing this portion of visual space. Classical textbook sections of the LGN typically are taken from this region of the LGN which shows four P layers and two M layers; as mentioned, we now know that K layers lie below each P and M layer as depicted for the macaque monkey in Figure 29.3.

Cell classes

All LGN cells can be grouped into two principal cell classes, *relay* cells that send an axon to visual cortex and *interneurons* whose axons remain within the LGN. LGN relay cells use

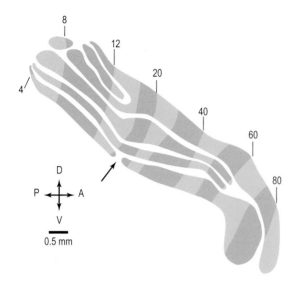

Figure 29.2 A reconstructed parasagittal section through the LGN of a macaque monkey. Light and dark bands represent eccentricity hemizones that have been mapped onto the LGN layers. Numbers refer to the eccentricity at the border between two adjacent eccentric hemizones. The gap in magnocellular layer 1 (M1) represents the optic disk (arrow). The cluster of arrows indicates anatomical orientation (D, dorsal; V, ventral; P, posterior; A, anterior). (Reproduced with permission from Malpeli et al.[5])

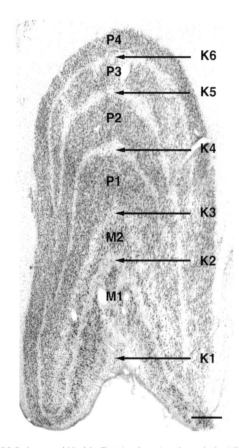

Figure 29.3 A coronal Nissl (cell) stained section through the LGN of a macaque monkey showing the parvocellular (P), magnocellular (M), and koniocellular (K) layers. At this cross-sectional level of the nucleus there are four P layers, two M layers, and six K layers. Scale bar = 500 μm.

the transmitter glutamic acid (glutamate) and interneurons use the transmitter gamma-amino butyric acid (GABA). Relay cells and interneurons occur in a ratio of approximately 4:1 throughout the LGN and exhibit distinct dendritic morphologies. The LGN contains several classes of interneurons[6] whose general role is to regulate signals that will ultimately pass to cortex from the retina via relay cells. The M, P, and K cells mentioned previously are all classes of relay cells and, in addition to size, can be distinguished based on dendritic morphology, calcium binding protein content, physiological properties, and axonal projection patterns to cortex.[1,7–9] P and M LGN cells play a major role in setting up the properties of V1 cells, but it remains controversial as to whether P and M cells belong to just two classes. Evidence from retinal inputs, axon projection patterns and physiology indicate that there are subclasses of P and M cells that perform somewhat different functions in V1[10] (see also below). It is even more likely that K cells are not a single LGN cell class given the diversity of their connections to cortex[8] (see also below).

Inputs: the retina

Classically it was argued that there were two pathways from the retina to the LGN, the P and the M pathways; all other retinal ganglion cells were presumed to project to other subcortical nuclei.[11] Evidence from more recent studies has shown that 10 or more different types of retinal ganglion cells send axons to the LGN based primarily on morphology[12] (Fig. 29.4). This finding complicates the picture because it becomes unclear whether K, M, and P LGN cells, as described above, receive unique inputs from only one class of ganglion cell or more than one class. It is evident from projection patterns of LGN axons in V1 that at least 10 axon classes have been identified that could support the argument that each retinal ganglion cell class has a unique conduit to cortex via a labeled line through the LGN[8] (see also Fig. 29.6 below).

Inputs: extraretinal sources and cortical feedback

In addition to retinal input, LGN cells receive input from a variety of both visual and non-visual extraretinal sources. These extraretinal afferent sources provide a much larger input to the LGN, in terms of synapse number, than does the retina (see review[2]). These extraretinal inputs also regulate the flow of information in the LGN in a variety of ways given the different transmitters they contain and the number of transmitter receptors that exist on LGN neurons[13] (Fig. 29.5). Input from extraretinal visual sources maintains retinotopic fidelity, meaning, regions representing a common point in visual space are connected. Extraretinal visual input to the LGN has been documented from the following areas in primates with their transmitter type in parentheses: primary visual cortex (glutamate), some extrastriate areas (likely glutamate), superior colliculus (glutamate), pretectum (GABA), parabigeminal nucleus (acetylcholine or ACh), and the visual sector of the thalamic

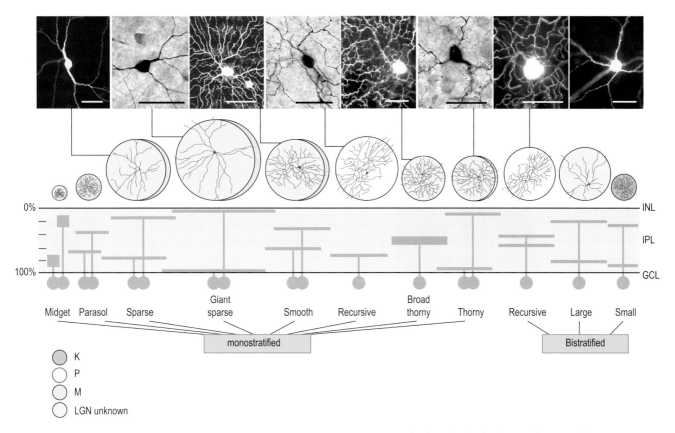

Figure 29.4 Many different types of retinal ganglion cells project to the LGN. Photomicrographs show details of cell morphology (scale bars = 50 µm) and the cell drawings and disks show the relative dendritic field sizes. The midget (red circle) and parasol (gray) cells are known to project to the P and M layers, respectively. The small bistratified cells (blue) are known to project to the K layers. Several other types of retinal ganglion cells (yellow) are also known to project to the LGN, but the layer specificity of their projection is currently unknown. The other cell types shown do not project to the LGN. (Modified from Dacey.[12] © 2004 Massachusetts Institute of Technology, by permission of the MIT Press.)

reticular nucleus (GABA). Of these extraretinal visual sources the two major sources, in terms of synapse number, come from the thalamic reticular nucleus and the primary visual cortex. The input from the thalamic reticular nucleus has been studied in great detail, and together with the primary visual cortex, provide retinotopically localized regulation of LGN signals to cortex. The massive cortical feedback pathway is highly specific in that cells in different subdivisions of layer VI of V1 send axons to either the P or M LGN layers but not both. This feedback also targets the K LGN layers and the thalamic reticular nucleus but only via axon collaterals destined for either the P or M LGN layers at least in owl monkeys.[14]

Extraretinal input to the LGN also arrives from several non-visual sources. The most well studied of these afferents are the cholinergic afferents from the pons that innervate all LGN layers and also contain the transmitter nitric oxide at least in the cat.[15] Other afferent sources include a serotonergic input from the dorsal raphé nucleus and a histaminergic input from the hypothalamus.[16–19]

Outputs: projections to V1 and beyond

In primates the bulk of the efferent axonal output from the LGN terminates within the primary visual cortex and the visual sector of the thalamic reticular nucleus (TRN). Although not documented in primates, in cats the projection to the equivalent of the visual sector of the TRN comes from collateral branches of LGN relay cell axons not from a separate projection that goes solely to this thalamic nucleus. A smaller efferent projection from the LGN also has been reported to terminate in several extrastriate visual areas. The most well documented of these projections is to the middle temporal area (MT). This projection appears to originate from K LGN cells, based upon double labeling studies using retrograde tracers and immunocytochemistry for calbindin, and arises from K cells that do not send collateral projections to V1.[20] Although this projection involves only about 1 percent of LGN cells, it has been implicated in the residual vision referred to as "blindsight" in patients that have lost their primary visual cortex[21] (Box 29.1).

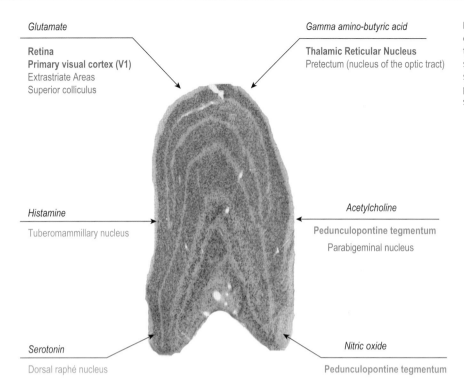

Glutamate

Retina
Primary visual cortex (V1)
Extrastriate Areas
Superior colliculus

Gamma amino-butyric acid

Thalamic Reticular Nucleus
Pretectum (nucleus of the optic tract)

Histamine

Tuberomammillary nucleus

Acetylcholine

Pedunculopontine tegmentum
Parabigeminal nucleus

Serotonin

Dorsal raphé nucleus

Nitric oxide

Pedunculopontine tegmentum

Figure 29.5 Diagram illustrating the brain areas connected directly with the macaque LGN and their chemical messages. Input from visual sources is indicated in red, input from non-visual sources is in blue. Bold text indicates areas that provide the heaviest input to the LGN in terms of synapse number. (Modified from Casagrande et al.[38])

In primates, studies have shown that P cells send axons principally to the middle and lower tier of layer IVC of V1 (using the nomenclature of Brodmann[22]). In addition, some P cells in macaque monkeys may send axons to IVA. The projection to layer IVA appears to be a specialization of only certain primates since it is absent in chimpanzees and likely also humans. M cells send their efferent axons to the middle and upper tier of layer IVC of primary visual cortex. Finally, K cells send their axons to cortical layers IVA (in macaque monkeys) and IIIB/IIIA where they terminate in patches of cells whose mitochondria contain high amounts of cytochrome oxidase (CO), the "CO blobs" and to layer I (see review[23]). Some M and P cells also have sparse projections to layer VI in V1. The pathways to V1 are summarized in Figure 29.6.

LGN circuits: How are visual signals regulated?

Feedback and feedforward pathways

Visual signals within the LGN are regulated by circuits involving two groups of inhibitory neurons, local inhibitory neurons and TRN inhibitory neurons. These inhibitory neurons are connected in feedforward and feedback circuits to LGN relay cells involving, respectively, retinal input to local inhibitory cells and relay cell input to TRN neurons as shown schematically in Figure 29.7A. The feedforward pathway involves a unique synaptic arrangement referred to as a triad in which a retinal terminal synapses upon a relay cell dendrite and the dendrite of an inhibitory interneuron

Box 29.1 The "blindsight" controversy

Cats (A; courtesy of Julie Mavity-Hudson) show only a slight loss in visual acuity following bilateral removal of primary visual cortex[49] and tree shrews (B; courtesy of Robert Strawn) (the closest living relatives of primates) show no deficit in acuity following similar lesions.[50] In both of these species it has been argued that vision is preserved via alternative subcortical routes to extrastriate cortex. Humans, however, appear to be completely blind following similar damage to V1. Nevertheless, it has been argued by some investigators that residual vision within the "blind" field (so called *"blindsight"*) exists even in patients with damage or complete loss of V1. Controversy persists, however, over whether such vision simply reflects small remaining pieces of V1 that are functional or is really the result of a separate functional pathway to extrastriate cortex as has been demonstrated in cats and tree shrews. Even among those who believe that a functional retinal route exists in humans and other primates that bypasses V1 there is debate as to whether this route passes through the LGN via direct projections from koniocellular (K) cells to extrastriate cortex[21] or involves a pathway from the retina to the superior colliculus to the pulvinar to extrastriate cortex as has been demonstrated in tree shrews.[51]

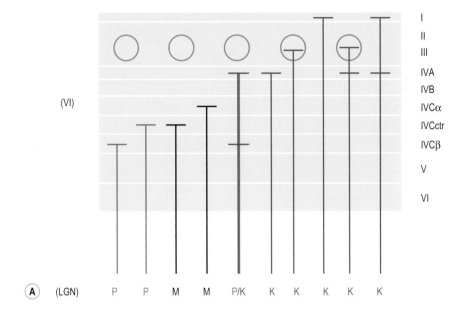

(LGN)

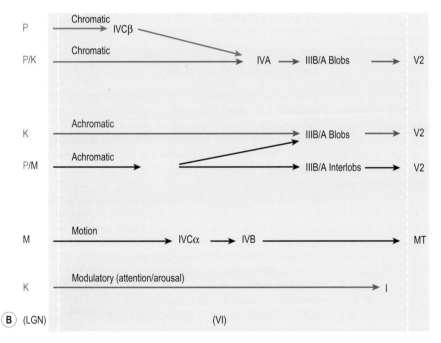

Figure 29.6 (**A**) Simplified schematic diagram of the efferent projections of LGN cells. Parvocellular (P, red), magnocellular (M, black), and koniocellular (K, blue) LGN cells have different projections to primary visual cortex (V1). P cells mainly terminate in lower and central layer IVC with some projections to layer IVA. M cells have the bulk of their terminations in upper and central layer IVC. K cells in the cytochrome oxidase blobs in layer III, layer IVA and layer I. K cell projections to layer I and the blobs in layer III usually arise from populations of K cells that lie in different LGN layers. (Modified from Casagrande & Xu.[23]) (**B**) Connections and possible functions of the eight parallel pathways within V1. In this diagram the ventral to dorsal sequence of layers in V1 is laid out from left to right. V1 projects directly to different compartments within the secondary visual area (V2) and to the middle temporal visual area (MT). Not shown: V1 also sends axons to the dorsal medial visual area (DM) also called V3a which exit from CO blob columns, to the third area (V3), and very sparsely to the fourth visual area (V4). See text for details.

(Fig. 29.7B). The dendrite of the inhibitory interneuron, in turn, makes a dendro-dendritic synapse on the same relay cell dendrite. In contrast, in feedback inhibition the relay cell synapses on the dendrite of a TRN neuron that then sends an inhibitory connection back to terminate on the dendrite of the same relay cell (Fig. 29.7A). Given that feedforward and feedback circuits can influence how retinal input is modulated, further documentation of ultrastructural differences in the circuits within the different layers of the primate LGN will become critical to the development of accurate physiological models.

Circuit neurochemistry

Although the feedback and feedforward circuits to LGN relay cells form the basic architecture for controlling visual signals, the final information that is sent on to the visual cortex is dependent in complex ways on all of the inputs that can impinge upon this circuitry and the variety of transmitter receptors that are activated by these inputs. Given space limitations we can only give a few examples here. Many of the same inputs that contact LGN relay cells are known to also contact inhibitory interneurons within

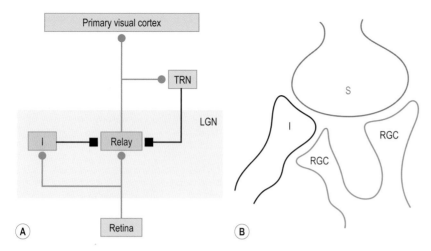

Figure 29.7 Simplified diagram of visual thalamic circuitry. (**A**) Diagram of feedforward and feedback inhibitory pathways that influence LGN relay cells. Excitatory inputs are indicated in green. Inhibitory inputs are indicated in black. TRN: thalamic reticular nucleus; LGN: lateral geniculate nucleus; I: LGN interneuron. (Modified from Casagrande & Xu[23] with permission of publisher.) (**B**) Schematic diagram showing the spine triad circuit within the LGN. Within a triad, a retinal afferent (RGC) synapses on a relay cell dendritic spine (S) and an interneuron dendrite (I), which, in turn, synapses onto the same relay cell dendrite.

the LGN or within the TRN although the transmitter receptors in some cases differ. The visual cortex is known to provide the most massive extraretinal input to the LGN. This glutamatergic input arriving from cells in layer VI of cortex activates both ionotropic and a specific class (group 1) of metabotropic glutamate receptors, both of which can directly depolarize relay cells although (importantly) with very different time courses and potentially different effects on the firing pattern of LGN relay cells[24] (see also[2]). Since cortical axons also synapse with inhibitory interneurons and TRN neurons that themselves have both types of receptors, the net output from the relay cell can not easily be predicted even from a simple circuit involving feedback from visual cortex.

A second major input to the LGN comes from cholinergic neurons in the pons (the parabrachial region in the cat). These cholinergic fibers innervating the interneurons and relay cells of the LGN and the TRN also are capable of releasing a second transmitter, nitric oxide. The impact of this input on the output of visual signals to cortex is complicated by the fact that ACh can activate both fast ionotropic nicotinic receptors and slower metabotropic M1 receptors on relay cells; both result in relay cell depolarization. These same ACh-containing pontine axons activate M2 receptors on interneurons and TRN cells resulting in hyperpolarization of these cells. The net impact of this ACh input on LGN relay cells, therefore, is generally excitatory since relay cells are directly depolarized and feedback and feedforward inhibition blocked. Under this condition one would predict an increased transfer ratio of retinal signals to cortex. In contrast, activation of the inhibitory circuitry would lower the transfer ratio and, under some circumstances, could change the temporal structure of signals arriving from the retina. The cortical and pontine inputs and their receptors in the LGN just described are only two examples that have been worked out for the cat LGN. The number of axonal inputs to the LGN and the variety of effects that each can have depending upon which receptor is activated argues strongly that LGN gating is not equivalent to simply opening or closing a connection between the retina to the cortex.

Signal processing in the LGN

Receptive field properties and parallel processing

With some exceptions the receptive field properties of LGN cells are inherited from retinal ganglion cells even though the synaptic inputs from other sources outnumber those of retinal axons within the LGN. It has been argued that retinal axons, therefore, are "drivers" for LGN cells and the other inputs are "modulators" based upon the impact that each source has on LGN relay cell output. The impact of retinal axons is mirrored in their anatomy given that these axons terminate closer than other excitatory inputs to the axon hillock of relay cells, have much larger terminals, and signal through fast ionotropic glutamate channels located at these synaptic sites on LGN relay cell dendrites.[25]

As in the retina, the majority of primate LGN cells have receptive fields defined as ON or OFF center with opposing surrounds. Some K LGN relay cells, however, appear to have non-standard visual receptive fields that have been difficult to characterize. Included in this category are cells with large diffuse fields, cells with no clear surround, and cells with more complex properties such as direction selectivity. Since some non-standard cells have also been identified in the primate retina, these unusual LGN cells also may reflect their retinal inputs (see reviews[12,26]).

When the input from the retina is measured within an LGN relay cell in the form of synaptic or S-potentials and these S-potentials are compared to the output of that same cell recorded as action potentials, the *ratio* of S potentials that result in action potentials (the *transfer ratio*) is less than 1.0, typically around 0.3–0.5 in an anesthetized preparation[27]. This would seem, ostensibly, to be a wasteful loss of important visual information. Kaplan et al[27] however, argued that such a decrease in the transfer ratio of signals between retina and LGN is non-linear and is largest when higher stimulus contrasts are used. They argue that this non-linear transfer of signals from retina to LGN serves, in part, to protect the cortex from response saturation under conditions of high stimulus contrast and thus preserves the dynamic range of cortical cells. Others have argued that relevant visual signals, such as information about stimulus contrast, also may be enhanced at the level of the LGN. For example, Hubel & Wiesel[28] suggested that LGN cells recorded in the cat exhibit a stronger surround mechanism than their retinal inputs and concluded from these data that the role of the LGN was to sharpen contrast differences and reduce responses to diffuse illumination (see also[29]). Although other interpretations of Hubel & Wiesel's original data exist,[1] it is evident that the LGN can act as both a spatial and temporal filter for the visual signals that it receives from the retina. We will have more to say about these filtering mechanisms below, but in order to appreciate how the LGN can filter or gate retinal signals, it is first necessary to describe more about the basic LGN signaling properties.

Physiology of M, P, and K cells

Numerous studies have investigated the detailed physiology of LGN P and M cells in different primate species (see reviews[1,2]). Typically, studies of M, P, or K cells have assumed that information is coded in the rate of production of action potentials over some period of time in relationship to the stimulus. There also is evidence that information is contained in the pattern of action potentials. For example, in sleeping cats, cells within the LGN can exhibit a regular bursting pattern of action potentials. Bursting has been observed in awake monkeys, but whether it is capable of transmitting information in awake animals has been the subject of some debate.[30] Sherman and his colleagues have argued that bursting performs a non-linear amplification of the visual signal, sacrificing signal discrimination for signal detectability in the awake animal (see review[2]). According to this hypothesis bursting is more efficient at alerting an animal to novel stimuli while tonic firing is more efficient for transferring information about stimulus quality. Thus far, this hypothesis has not been confirmed.[30]

In general, using firing rate as a criterion, P LGN cells can be distinguished from other cell classes, as a group, by their sensitivity to higher spatial and lower temporal frequencies measured using high contrast gratings to stimulate their receptive field centers. M cells, as a group, have the highest temporal frequency sensitivity and have the lowest thresholds for contrast measured in their receptive field centers at their preferred spatial and temporal frequency. M cells excel at temporal resolution, in part, because they signal the entrance or withdrawal of a stimulus in their receptive fields whereas P cells give a sustained response to the presence of a stimulus. Many K cells fall between the two other classes in terms of spatial and temporal resolution suggesting that the visual resolution of the whole system may reflect a combined contribution of all of these cell classes under some conditions. In terms of wavelength selectivity, there is evidence that both P and K cells contribute in some diurnal primates. In primates that have trichromatic vision, such as humans and macaque monkeys, there is evidence that many P cells transmit information about red and green in the same manner as their ganglion cell inputs. Similarly, some LGN K cells transmit information about blue/yellow based on their retinal inputs. Since the majority of primates are actually dichromats (meaning that they only

have two cones), however, indicates that the P pathway is color blind in most primates. This is because primates with two cone types always have one S cone and one other longer wavelength cone for chromatic comparison. If S cone information is transmitted exclusively by K cells, as most investigators believe, and not P cells, then this condition renders P cells color blind since P cells would only get input from a single cone.[31] Although it has been popular to link the physiological properties of LGN cell classes to complex behaviors such as P = form, M = motion, and K = color there is unlikely to be a direct link between the behavior of single cells and the behavior of the whole organism (Box 29.2).

The influence of the "extra-classical" surround

Although the receptive fields of most primate LGN cells are pictured classically as concentric ON center/OFF surround fields or the reverse, it is known that stimuli presented well beyond the classical receptive fields of LGN cells could influence the behavior of these cells.[32,33] It is currently accepted that LGN cells do respond to direct stimulation of their extra-classical surrounds and that these surrounds overlap the classical receptive field and extend beyond its borders. This suppressive surround is often referred to as the non-linear surround or suppressive surround, as stimuli of either sign (ON or OFF) reduce the driven responsiveness of neurons.[34,35] This form of suppression likely plays an important role in adjusting the gain of LGN responses to visual stimuli. Some investigators have suggested that this cryptic surround could contribute to the phenomenon known as visual pop-out where highly salient stimuli, such as a red ball against a field of blue balls, pops out from the background. Studies designed to manipulate the extraclassical surround of LGN cells also indicate that M cells show greater extra-classical surround suppression than P cells and that this suppression likely originates in the retina, as suggested by much earlier studies of the periphery effect.[36]

The impact of feedback

As mentioned above, cortical feedback from V1 to the LGN forms the largest modulatory pathway impacting the responses of LGN cells. Feedback from V1 influences circuits in LGN both directly and indirectly via the TRN. Cells in layer VI of V1 send axons to both the TRN and LGN (as described above) and extend their axons to a larger area of the LGN since V1 receptive fields view a larger area of visual space. Additionally, V1 cells get feedback of their own from higher visual cortical areas such as the area MT. Given that feedback pathways in cortex are quite fast this means that slower messages arriving from the LGN will intersect continually with feedback messages. This dynamic interplay means that every message from the retina can be modified in the LGN based on information received from higher-order visual areas. Such a system can be highly efficient because messages of key interest can be allotted a higher transfer ratio than messages of lower relevance. For example, Sillito and colleagues[34] demonstrated that manipulations of the activity in primate area MT can significantly change the activity in the LGN. Sillito and colleagues increased the activity in a localized region of MT by injecting a small amount of a GABA antagonist. Simultaneous to this injection in MT, they recorded from cells in the LGN and found that the increase in activity in

Box 29.2 Linking physiology to behavior: lesions in M and P layers

It has been popular to link the properties of cells in specific visual pathways to complex behaviors. For example, it is common to hear that P cells are for form and red/green color discrimination, K cells are for blue/yellow color discrimination and M cells are for discriminating motion. We know, however, from physiology that the majority of cells in all LGN classes respond to high contrast moving gratings of the correct temporal and spatial frequency. So, of course, they "see" motion and form. We also know that we can discriminate a huge variety of colors. Clearly if retinal ganglion cells were already dedicated to a particular complex visual behavior the brain would not waste synaptic time sending the information first to the LGN, then to V1, etc.

Direct tests of the contributions of P and M cells to visual behavior have been attempted by examining visual losses after making lesions within either the P layers or the M layers of the LGN in macaque monkeys. The monkeys were then trained to look at a fixation spot with one eye only (the other eye was covered). Stimuli to be discriminated were presented in various locations including the part of the visual field corresponding to the damaged portion of the M or P layers. Each lesion extended into adjacent

layers and therefore included some of the K layers. The behavioral tests demonstrated that monkeys *can discriminate form and motion with either pathway* assuming that the stimuli are high enough contrast and are presented in the spatial and temporal frequency range of the remaining pathway. Therefore, it is incorrect to say that P cells are necessary for form discrimination while M cells are necessary for motion discrimination. Instead, the behavioral tests revealed that after P layer lesions (including the associated K layers 4–6) monkeys lost the ability to discriminate high spatial frequencies (**B**, below) and any forms made up solely of these frequencies and colors, whereas after M and associated K layer lesions the monkeys lost the ability to discriminate high temporal frequencies (and any movements that were very fast) (**C**). The monkeys also suffered a major unpredicted deficit in contrast sensitivity following a loss of P cells (**B**). Taken together, these results suggest that the properties that distinguish cell classes at the retina and LGN provide basic information about contrast, size, wavelength, and speed and that higher level analysis, such as that necessary for the discrimination of complex shapes and the movement of these shapes in space, is performed in cortex.

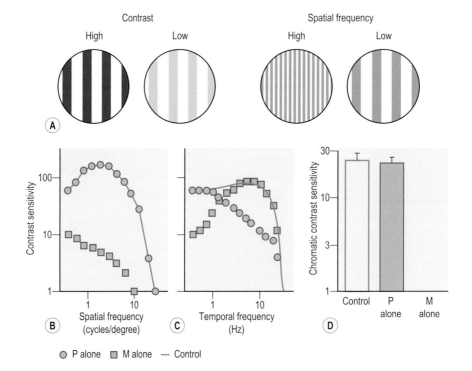

Visual losses after selective ablation of the M, P, and associated K layers of the LGN in macaque monkeys. Contrast sensitivity for all spatial frequencies is reduced when only the M and K layers 1 through 3 remain intact after lesioning the P layers and the K layers 4 through 6. The solid green line in (**B**) and (**C**) shows sensitivity of the normal monkey; circles show the contribution of the P pathway (after M lesions), and squares the contribution of the M pathway (after P lesions). (**C**) Contrast sensitivity to a grating with low spatial frequency is reduced at lower temporal frequencies when only P cells remain. (**D**) Color contrast sensitivities are lost when only the M cells remain. (Reproduced with permission from Kandel et al.[48])

MT can both increase and decrease responses of LGN cells. This example shows that the flow of information through the LGN can be modulated by high-level cortical feedback.

The LGN and arousal, attention and conscious vision

Given the diversity of non-retinal LGN connections and the diversity of connections that each of these input sources has with the rest of the brain, it stands to reason that LGN cells must be modulated by many different sensory, motor, and cognitive messages. Closer inspection of the types of circuits that impact the LGN suggests that signals leaving this nucleus are mainly influenced by four types of information. The most well-accepted impact of these non-retinal sources on the LGN concerns changes with the state of an animal. Numerous investigations, beginning more than 50 years ago, have shown that LGN cells change their firing patterns depending on whether animals are awake or asleep. During sleep LGN cells exhibit burst firing which has been interpreted as a disconnection between the thalamus and cortex.[37] During the waking state the LGN is dominated by tonic firing which modulates in relationship to visual stimulation. Whether the infrequent bursting of LGN cells during wakefulness conveys special visual meaning (perhaps alerting animals to something unusual) remains a subject of debate[2] (see above and also[30]). Regardless, there is little doubt that the LGN changes its message to cortex depending on the level of arousal.

The second source of information that likely impacts the LGN is visual attention. The influence of attention on the LGN, however, is controversial for two reasons:

(1) there exists an historic bias against the idea that sensory thalamic relay nuclei should do more than stably relay their messages in a feedforward manner to cortex and
(2) there has been difficulty demonstrating an impact of attention on individual LGN cells (see review[38]).

Demonstrating the impact of attention in LGN poses a problem mainly because the classical receptive fields of individual LGN cells in primates are quite small. Additionally, recording from single LGN cells in awake, behaving monkeys is challenging since monkeys cannot hold their eyes completely stationary. This means that stimuli must be presented over areas larger than an individual LGN cell's receptive field center in order to avoid artifactual stimulation of the edges of a receptive field during experiments. The result is that LGN cells respond poorly because the inhibitory surrounds of their receptive fields are being stimulated at the same time as their excitatory centers are being stimulated. In extrastriate cortex (see Chapter 31) receptive fields are quite large and, as a result, it has been much easier to demonstrate that changes in visual attention impact the responses of individual cells because stimulation can be confined to individual receptive fields. The other issue is that many LGN cells drive one V1 cell so that effects of attention on any one individual LGN cell could be minute but add up to a large impact when one considers the number of LGN cells that synapse with any one V1 cell. Therefore, it is not surprising that one of the first demonstrations of the impact of visual attention on the LGN used functional imaging (fMRI) in human subjects and compared activation patterns seen in each LGN when subjects were asked to shift spatial attention to different locations.

Looking at measures of neural activity at the population level with fMRI, O'Connor et al,[39] using flickering checkerboard stimuli of high and low contrast, showed that responses in LGN were reduced to unattended stimuli and enhanced to attended stimuli and that the shift between the baseline activation of the left and right LGN could be seen in anticipation of where the stimulus would come on even without retinal input. Surprisingly, O'Connor et al[39] also showed that the changes in activation relative to attention were stronger in LGN than in primary visual cortex (V1). This result is surprising because one might predict that the attentional shifts in LGN are a result of feedback from V1. Alternatively, it may be that both arousal and more specific spatial attentional effects are conveyed to the LGN through direct or indirect pathways involving either the large cholinergic pathway from the pons (pedunculopontine tegmental nucleus, PPT) or the GABAergic pathway from the thalamic reticular nucleus (TRN) or both. The latter pathway is of particular interest given the original proposal by Crick[40] that the TRN acts as an "attentional searchlight". In support of this proposal, McAlonan et al[41] showed that increases in visual attention in monkeys reduced the responses of individual visual TRN neurons (whose main output is to the LGN). In the LGN both McAlonan et al[41] and Royal et al[42] were able to show, recording from single P and M LGN neurons in monkeys, that attention modulates their responses rapidly at latencies that occur before signals can reach cortex (Fig. 29.8).

The LGN and motor planning

The third modulatory input to LGN responses comes from eye movements. Visual signals from the retina can only be processed usefully in the context of action. In other words, as human beings we use vision to interact with the world so our visual world is in constant motion. To maintain stability we need to know where our head, eyes, limbs, etc. are in relationship to the objects around us. It is known that auditory and visual maps in the superior colliculus of the midbrain are converted to common motor maps so that congruence between head-centered auditory space and eye-centered visual space exists in motor output neurons from this structure.[43] The question is, do LGN cells modulate their activity in relationship to motor activity? Again, because of technical challenges it has been difficult to test this idea. General support for the idea that LGN cells modulate in relationship to motor commands comes from studies of saccadic suppression. The brain suppresses signals related to our eye movements presumably to remove potential retinal blur and allow efficient processing between fixations.

There is now good evidence that such suppression occurs at the level of the LGN (see reviews[38,44]). Recording from single LGN cells in awake, behaving monkeys Royal et al[44] were able to show that all LGN cell classes (K, M, and P) modulate their activity in relationship to saccadic eye movements. What was especially interesting in the latter study was that LGN cells modulated their activity in complete darkness in anticipation of the pending eye movements. Activity in

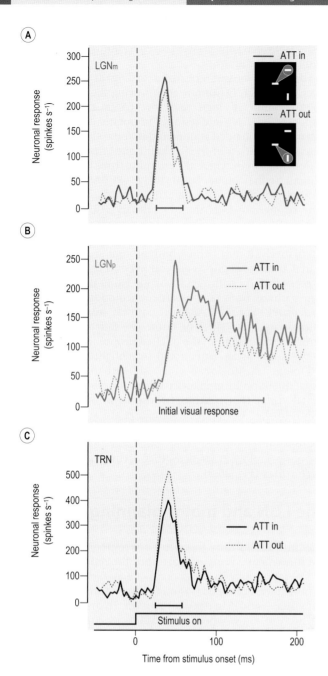

Figure 29.8 Attentional modulation of cells in the LGN. Spike density plots of a cell's response to an attended stimulus (ATTin, solid traces) and an unattended stimulus (ATTout, dashed traces). The inset illustrates the "spotlight" of attention that is directed to the circle representing the receptive field. Responses are aligned to stimulus onset (the dashed vertical line). (**A**) Responses of sample magnocellular LGN cell (LGNm). (**B**) Responses of sample parvocellular LGN cell (LGNp). (**C**) Responses of sample TRN cell. (Reproduced with permission from McAlonan et al.[41] Reproduced by kind permission of Springer Science + Business Media.)

LGN cells was suppressed as much as 100 ms prior to a "motor plan" to move the eyes to a new location. Once the eyes landed at their new location, activity significantly increased (Fig. 29.9). This pattern of changes in LGN activity could account for the perceptual phenomenon of saccadic suppression and help facilitate efficient transmission of relevant signals during fixation. It is noteworthy, that many investigators have argued that attention and planning saccadic eye movements are intimately related. Perhaps the cholinergic PPT–LGN pathway conveys information about both shifts in attention and the anticipated saccadic eye movements that accompany these attentional shifts.[38]

The LGN and binocular rivalry and visual awareness

The fourth source of information that can modulate LGN activity is interocular. Although LGN cells are activated only by one eye, in the 1970s Sanderson et al[33] showed that cat LGN cell responses can be suppressed by stimulation of the other eye. These results were subsequently replicated in primate LGN. A number of investigators have argued that binocular interactions in the LGN might account for binocular rivalry. When two stimuli are presented that cannot be fused into a coherent picture by the two eyes (say a face presented to one eye and a house presented to the other eye), rivalry is induced such that one sees first the face and then the house and so forth, alternately. This means that we are only "aware" or conscious of one image at a time. Is the LGN modulated by such changes in awareness? Recently, at least two groups have demonstrated positive results using functional imaging with fMRI.[45,46] These groups clearly showed that signals in both LGNs and V1 fluctuated while subjects viewed rivalrous stimuli. These results argue that the LGN is perceptually "aware" or conscious. Where do these signals come from? There is anatomical evidence that they could exist within the circuitry of the LGN itself[47] or arrive via feedback connections from V1, either directly or indirectly via the TRN. Since fMRI signals have poor temporal resolution it will take further study with single electrodes to dissect these pathways in more detail.

Conclusions

The primate LGN relays all retinal signals necessary for conscious visual perception to cortex. Each LGN cell receives monocular input from only a few ganglion cells of a given class. Since the LGN receives input from as many as 10 different morphological classes of ganglion cells, however, it follows that multiple relay cell classes exist in each morphologically defined layer. K LGN cells are the most diverse relay cell class based on morphology and physiology. Some K cells transmit yellow/blue chromatic signals but the function of the majority of K LGN cells remains unknown. M cells likely transmit signals important to low resolution form and motion, while P cells transmit detail and some P cells transmit red/green chromatic information. The LGN also contains several classes of interneurons that regulate the flow of information from the retina through feedforward and feedback inhibitory circuitry. In addition to the retina, the LGN receives input from a variety of other sources including a major feedback projection from V1. These inputs dynamically regulate the signals that the LGN receives from the retina, depending on planned action, eye movements, and attention. Thus, visual signals arriving from the retina immediately become part of a dynamic circuit in the LGN that is designed to process that information for the most efficient use.

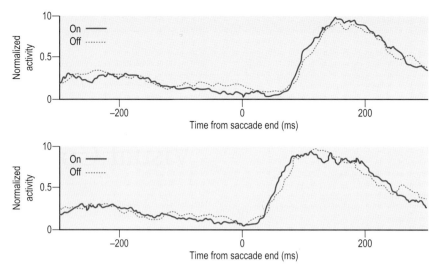

Figure 29.9 LGN cell activity during saccades. Activity of LGN ON-center and OFF-center cells that demonstrate significant perisaccadic modulations during spontaneous saccades. (**A**) Spike density functions for 25 LGN cells (10 ON-center, 15 OFF-center; 5M, 19P, 1K) that demonstrated significant saccadic suppression and postsaccadic facilitation during spontaneous saccades. Trials are aligned on saccade start (0 ms). (**B**) Spike density functions for the same 25 LGN cells with trials aligned on saccade end (0 ms) to demonstrate the time course of the postsaccadic facilitation. Spike density functions were normalized to the highest peak-firing rate observed in the two curves. (Reproduced with permission from Royal et al.[44] "Reproduced by kind permission of Springer Science + Business Media")

References

1. Casagrande VA, Norton TT. Lateral geniculate nucleus: a review of its physiology and function. In: Leventhal AG, ed. The neural basis of visual function. London: Macmillan Press, 1991:41–84.
2. Sherman SM, Guillery RW. Exploring the thalamus and its role in cortical function, 2nd edn. Cambridge, MA: MIT Press, 2005.
3. Hickey TL, Guillery RW. Variability of laminar patterns in the human lateral geniculate nucleus. J Comp Neurol 1979; 183(2):221–246.
4. Malpeli JG, Baker FH. The representation of the visual field in the lateral geniculate nucleus of Macaca mulatta. J Comp Neurol 1975; 161(4):569–594.
5. Malpeli JG, Lee D, Baker FH. Laminar and retinotopic organization of the macaque lateral geniculate nucleus: magnocellular and parvocellular magnification functions. J Comp Neurol 1996; 375(3):363–377.
6. Bickford ME, Carden WB, Patel NC. Two types of interneurons in the cat visual thalamus are distinguished by morphology, synaptic connections, and nitric oxide synthase content. J Comp Neurol 1999; 413(1):83–100.
7. Casagrande VA. A third parallel visual pathway to primate area V1. Trends Neurosci 1994; 17(7):305–310.
8. Casagrande VA, Yazar F, Jones KD, Ding Y. The morphology of the koniocellular axon pathway in the macaque monkey. Cereb Cortex 2007; 17(10):2334–2345.
9. Hendry SH, Reid RC. The koniocellular pathway in primate vision. Annu Rev Neurosci 2000; 23:127–153.
10. Boyd JD, Mavity-Hudson JA, Casagrande VA. The connections of layer 4 subdivisions in the primary visual cortex (V1) of the owl monkey. Cereb Cortex 2000; 10(7):644–662.
11. Perry VH, Cowey A. Retinal ganglion cells that project to the superior colliculus and pretectum in the macaque monkey. Neuroscience 1984; 12(4):1125–1137.
12. Dacey DM. Origins of perception: Retinal ganglion cell diversity and the creation of parallel visual pathways. In: Gazzaniga MS, ed. The cognitive neurosciences III. Cambridge, MA: MIT Press, 2004:281–301.
13. Casagrande VA, Royal DW, Sary G. Extraretinal inputs and feedback mechanisms to the lateral geniculate nucleus (LGN). In: Bowers D, Kremers J, House A, eds. The primate visual system: a comparative approach. Chichester, UK: John Wiley and Sons, 2005:191–211.
14. Ichida JM, Casagrande VA. Organization of the feedback pathway from striate cortex (V1) to the lateral geniculate nucleus (LGN) in the owl monkey (Aotus trivirgatus). J Comp Neurol 2002; 454(3):272–283.
15. Bickford ME, Gunluk AE, Guido W, Sherman SM. Evidence that cholinergic axons from the parabrachial region of the brainstem are the exclusive source of nitric oxide in the lateral geniculate nucleus of the cat. J Comp Neurol 1993; 334(3):410–430.
16. De Lima AD, Singer W. The brainstem projection to the lateral geniculate nucleus in the cat: identification of cholinergic and monoaminergic elements. J Comp Neurol 1987; 259(1):92–121.
17. Morrison JH, Foote SL. Noradrenergic and serotoninergic innervation of cortical, thalamic and tectal structures in Old and New World monkeys. J Compar Neurol 1986; 8:117–138.
18. Kayama Y, Shimada S, Hishikawa Y, Ogawa T. Effects of stimulating the dorsal raphe nucleus of the rat on neuronal activity in the dorsal lateral geniculate nucleus. Brain Res 1989; 489(1):1–11.
19. Uhlrich DJ, Manning KA, Pienkowski TP. The histaminergic innervation of the lateral geniculate complex in the cat. Vis Neurosci 1993; 10(2):225–235.
20. Nassi JJ, Callaway EM. Multiple circuits relaying primate parallel visual pathways to the middle temporal area. J Neurosci 2006; 26(49):12789–12798.
21. Vakalopoulos C. A hand in the blindsight paradox: a subcortical pathway? Neuropsychologia 2008; 46(14):3239–3240.
22. Brodmann K. Vergleichende Lokalisationslehre der Grosshirnrinde in ihren Prinzipien dargestellt auf Grund des Zellenbaues. Leipzig: J. A. Barth, 1909.
23. Casagrande VA, Xu X. Parallel visual pathways: a comparative perspective. In: Chalupa L, Werner JS, eds. The visual neurosciences. Cambridge, MA: MIT Press, 2004:494–506.
24. Godwin DW, Vaughan JW, Sherman SM. Metabotropic glutamate receptors switch visual response mode of lateral geniculate nucleus cells from burst to tonic. J Neurophysiol 1996; 76(3):1800–1816.
25. Sherman SM. The thalamus is more than just a relay. Curr Opin Neurobiol 2007; 17(4):417–422.
26. Xu X, Ichida JM, Allison JD, Boyd JD, Bonds AB, Casagrande VA. A comparison of koniocellular, magnocellular and parvocellular receptive field properties in the lateral geniculate nucleus of the owl monkey (Aotus trivirgatus). J Physiol 2001; 531(Pt 1):203–218.
27. Kaplan E, Purpura K, Shapley RM. Contrast affects the transmission of visual information through the mammalian lateral geniculate nucleus. J Physiol 1987; 391:267–288.
28. Hubel DH, Wiesel TN. Integrative action in the cat's lateral geniculate body. J Physiol 1961; 155:385–398.
29. Wiesel TN, Hubel DH. Spatial and chromatic interactions in the lateral geniculate body of the rhesus monkey. J Neurophysiol 1966; 29(6):1115–1156.
30. Ruiz O, Royal D, Sary G, Chen X, Schall JD, Casagrande VA. Low-threshold Ca²⁺-associated bursts are rare events in the LGN of the awake behaving monkey. J Neurophysiol 2006; 95(6):3401–3413.
31. Casagrande VA, Khaytin I, Boyd JD. Evolution of the visual system: mammals – color vision and the function of parallel visual pathways in primates. In: Butler A, ed. Encyclopedia of neuroscience. Berlin & Heidelberg: Springer, 2008:1472–1475.
32. McIlwain JT. Some evidence concerning the physiological basis of the periphery effect in the cat's retina. Exp Brain Res 1966; 1(3):265–271.
33. Sanderson KJ, Bishop PO, Darian-Smith I. The properties of the binocular receptive fields of lateral geniculate neurons. Exp Brain Res 1971; 13(2):178–207.
34. Sillito AM, Jones HE. Corticothalamic interactions in the transfer of visual information. Phil Trans R Soc Lond B Biol Sci 2002; 357(1428):1739–1752.
35. Solomon SG, White AJ, Martin PR. Extraclassical receptive field properties of parvocellular, magnocellular, and koniocellular cells in the primate lateral geniculate nucleus. J Neurosci 2002; 22(1):338–349.
36. Alitto HJ, Usrey WM. Origin and dynamics of extraclassical suppression in the lateral geniculate nucleus of the macaque monkey. Neuron 2008; 57(1):135–146.
37. Steriade M. Arousal: revisiting the reticular activating system. Science 1996; 272(5259):225–226.
38. Casagrande VA, Sary G, Royal D, Ruiz O. On the impact of attention and motor planning on the lateral geniculate nucleus. Prog Brain Res 2005; 149:11–29.
39. O'Connor DH, Fukui MM, Pinsk MA, Kastner S. Attention modulates responses in the human lateral geniculate nucleus. Nat Neurosci 2002; 5(11):1203–1209.
40. Crick F. Function of the thalamic reticular complex: the searchlight hypothesis. Proc Natl Acad Sci USA 1984; 81(14):4586–4590.
41. McAlonan K, Cavanaugh J, Wurtz RH. Guarding the gateway to cortex with attention in visual thalamus. Nature 2008; 456(7220):391–394.
42. Royal D, Sary G, Schall JD, Casagrande VA. Does the lateral geniculate nucleus (LGN) pay attention? J Vision 2004; 4(8):622a.
43. Sparks DL. Conceptual issues related to the role of the superior colliculus in the control of gaze. Curr Opin Neurobiol 1999; 9(6):698–707.
44. Royal DW, Sary G, Schall JD, Casagrande VA. Correlates of motor planning and postsaccadic fixation in the macaque monkey lateral geniculate nucleus. Exp Brain Res 2006; 168(1–2):62–75.
45. Haynes JD, Deichmann R, Rees G. Eye-specific effects of binocular rivalry in the human lateral geniculate nucleus. Nature 2005; 438(7067):496–499.
46. Wunderlich K, Schneider KA, Kastner S. Neural correlates of binocular rivalry in the human lateral geniculate nucleus. Nat Neurosci 2005; 8(11):1595–1602.
47. Bickford ME, Wei H, Eisenback MA, Chomsung RD, Slusarczyk AS, Dankowsi AB. Synaptic organization of thalamocortical axon collaterals in the perigeniculate nucleus and dorsal lateral geniculate nucleus. J Comp Neurol 2008; 508(2):264–285.
48. Kandel ER, Schwartz JH, Jessell TM. Principles of neural science. Principles of neural science. New York: McGraw Hill, 2000.
49. Berkley MA, Sprague JM. Striate cortex and visual acuity functions in the cat. J Comp Neurol 1979; 187(4):679–702.
50. Ware CB, Casagrande VA, Diamond IT. Does the acuity of the tree shrew suffer from removal of striate cortex? A commentary on the paper by ward and Masterton. Brain Behav Evol 1972; 5(1):18–29.
51. Brown LE, Goodale MA. Koniocellular projections and hand-assisted blindsight. Neuropsychologia 2008; 46(14):3241–3242.

CHAPTER 30

Processing in the Primary Visual Cortex

Vivien Casagrande & Roan Marion

Overview: The primary visual cortex constructs local image features

The visual system provides a description of the location and identification of objects that have survival value to the species. This description must be made in a dynamic world in which our gaze is constantly shifting and in which objects move. To be useful, the description must select items of importance depending on context and memory; in essence the visual system is designed to make useful predictions about visual reality, not to actually represent reality. Implicit in the architecture of the visual system reside mechanisms tuned by evolution and development to optimize useful vision. Beginning in the retina, the visual system selects what is needed to accomplish this goal. The retina contributes to this selection process by eliminating information about absolute light intensity, emphasizing local image contours, and compressing the visual signal information into a manageable size during a manageable time window to be transmitted to the lateral geniculate nucleus (LGN, see Chapter 29). The LGN contributes by regulating the flow of visual signals. Primary visual cortex (V1), as we will learn, contributes by coding aspects of local image features including their size, orientation, local direction of movement, and binocular disparity. All of these local descriptions of stimulus quality are critical for the more global and complex identification of objects ("what") and spatial relations ("where").

Classically, V1 is described as an area where information provided by the separate LGN channels is combined in different ways before it is sent to extrastriate visual areas for further processing. In order to do its job, V1 must solve the geometry puzzle of representing all stimulus qualities necessary for the subsequent steps of analyses within tissue that maps the outside world, i.e. responds to different parts of the visual field. V1 accomplishes this goal by a division of labor between different layers and different iterated modules within each layer. In this classic view, V1 is pictured as the beginning stage for several sets of serial processes that take place in feedforward hierarchies of visual areas; areas specialized to perform a variety of higher-order tasks. Currently, however, there is growing recognition that V1 does more. V1 does not represent stimulus qualities statically. Via dense and intricate horizontal and feedback pathways, V1 dynamically interacts with higher visual and non-visual areas, as well as the thalamus, and alters its activity and message depending on attention, memory, context, and reward. Therefore, for efficiency, V1 cells bias their messages to emphasize moment to moment relevance. In this chapter we initially provide the reader with a road map to the organization of V1, we show how the anatomical architecture of the V1 sets limits on how visual signals can be processed, and we emphasize the laminar architecture, connections, and cell types. Next we review the traditional view of how receptive field properties are put together in V1 and review the hypothesis that the parallel inputs from the LGN relate in functionally specific ways to output channels of V1 (see also Chapter 31). The second to last section then briefly reviews more controversial ideas concerning the functions of V1, including whether V1 is conscious and participates in memory or is modified by reward and attention. The final section of this chapter summarizes key points.

Overview of cortical organization: a general road map

Human V1 ranges in size from 1400 to 3400 mm^2, is about 2 mm thick, and is located within the occipital lobe extending from the posterior pole along the medial wall of the hemisphere[1] (Fig. 30.1). Like the rest of the cerebral cortex, V1 contains 6 principal layers. V1 is often called striate cortex in recognition of its original identification by an Italian medical student (Francesco Gennari) more than 200 years ago. Gennari observed that if he sliced into the posterior pole of an unfixed human brain and observed the cortical layers of the slice a striking white stripe was visible, known today as the *stria of Gennari*. The stria of Gennari actually marks one heavily myelinated layer in the middle of the gray matter of V1. V1 also is referred to by several other names including, most commonly, area 17 of Brodmann[2] or primary visual cortex. For convenience we will use the term V1 in this chapter.

Damage to V1 results in a hole (scotoma) or blind spot in one visual hemifield. The location of damage to V1 can be predicted based upon the topographic map of the

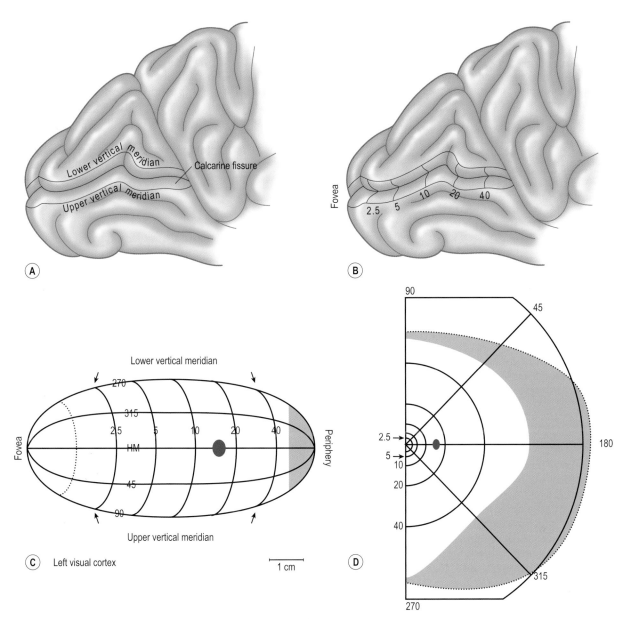

Figure 30.1 (**A**) Left occipital lobe showing the location of V1 within the calcarine fissure. (**B**) View of calcarine fissure of V1. The lines indicate the coordinates of the visual field map. The representation of the horizontal meridian runs approximately along the base of the calcarine fissure. The vertical lines mark the iso-eccentricity contours from 2.5° to 40°. V1 wraps around the occipital pole to extend about 1 cm onto the lateral convexity where the fovea is represented. (**C**) Schematic map showing the projection of the right visual hemifield upon the left visual cortex by transposing the map illustrated in (B) onto a flat surface. The row of dots indicates approximately where V1 folds around the occipital tip. The blue oval marks the region of V1 corresponding to the contralateral eye's blind spot. It is important to note that considerable variation occurs among individuals in the exact dimensions and location of V1. HM = horizontal meridian. (**D**) Right visual hemifield plotted with a Goldmann perimeter. The purple region corresponds to the monocular temporal crescent which is mapped within the most anterior 8 to 10 percent of V1. (Reproduced from Horton & Hoyt[6] with permission of the publisher.)

opposite visual hemifield that is known to exist in this region in all mammalian species. In humans, detailed knowledge about the manner in which the visual field is mapped onto V1 has been obtained from a variety of sources including clinical assessments of damage, results of electrical stimulation, and transcranial magnetic stimulation (TMS). The latter is a non-invasive method used to excite neurons by using rapidly changing magnetic fields to produce weak electrical currents.[3] Both direct invasive electrical stimulation and non-invasive TMS can give rise to the experience of

localized phosphenes (flashes of light) experienced by the subject. Other methods of developing functional maps of V1 have used positron emission tomography (PET) and, most commonly, functional magnetic resonance imaging (fMRI).[4–6] In the latter case the subject is presented with topographically restricted stimuli in different locations in space and the relative activation of V1 measured as a change in blood flow. Knowledge about the retinotopic organization of V1 in humans is relevant not only to the clinical evaluation of damage, but also to research efforts designed to develop visual prosthetic devices involving visual cortical electrical stimulation for individuals with incurable retinal diseases or loss of eyes (Box 30.1). All of the methods used to understand the map of the visual field in V1 are in general agreement showing, as illustrated in Figure 30.1, that the fovea is represented in the occipital pole and the far periphery is represented in the anterior margin of the calcarine fissure with the upper and lower visual fields being mapped onto its lower (lingual gyrus) and upper (cuneus gyrus) banks, respectively.[6] As in the LGN, the visual field map in human cortex is distorted such that the representation of central vision occupies much more tissue than does peripheral vision. Whether the foveal representation in V1 is expanded over what would be predicted simply by assigning each retinal ganglion cell or LGN cell the same amount of cortical tissue has been the subject of considerable debate. One explanation for this controversy is that the proportion of cortex devoted to central vision has been shown to be highly individually variable, at least in macaque monkeys.[7] The latter finding indicates that the relative amount of tissue devoted to the fovea could be magnified at the cortical level relative to the retina in some individuals but not others.

Layers, connections, and cells of V1: The inputs, outputs, and general wiring

As in other cortical areas, V1 has six main layers that can be identified in a cell stain as shown in Figure 30.2. These layers contain two main types of cells, pyramidal and stellate, in an 80/20 proportion. The layers and sublayers of V1 have been named in different ways depending on investigator interpretation. The most common laminar scheme is the one adopted by Brodmann.[2] Although Brodmann likely made a major mistake in his interpretation of V1 layers in all primates[8] his scheme is the most accepted so we use it here (Fig. 30.2). As in other areas of neocortex, input from the thalamus ends primarily in the middle layers (IVC of V1), with the superficial layers (above IVC) sending information to other cortical areas and the layers below IVC sending signals to subcortical targets (Fig. 30.2).

LGN inputs

Studies done in anesthetized monkeys have demonstrated that activation of V1 neurons depends completely on input from the LGN, because when the LGN is inactivated, visually evoked potentials in V1 are blocked.[9] LGN axons carrying signals from the left and right eyes and from K, M, and P layers remain segregated at the first synapse in V1 in non-human primates and presumably also humans (Fig. 30.3). The left and right eye segregation of axons from the M and P LGN layers ending in layer IVC of V1 is referred to as *ocular dominance columns*. These "columns" have been directly demonstrated to exist in humans as shown in the complete reconstruction of the pattern of left and right eye input to V1 in Figure 30.4.[10] These zebra-stripe-like patterns of monocular segregation from LGN can be revealed in tangential sections of layer IVC of V1 in humans postmortem following long-standing loss of one eye. As shown in Figure 30.4, the patterns are revealed by a histochemical stain for the mitochondrial enzyme, cytochrome oxidase (CO) which becomes down-regulated in the zones of layer IVC innervated by LGN axons from the missing eye. As mentioned and shown schematically in Figure 30.3, K, M, and P LGN axons in macaque monkeys also terminate in separate V1 layers or sublayers. M and P axons synapse in sublayers of IVC, and K LGN axons synapse in layers IVA, III, and I. Also, K LGN axons are likely made up of several classes since those that end in different V1 layers appear to come from different populations of LGN cells in macaques.[11] It is noteworthy, however, that the LGN does not project to layer IVA in apes and humans,[12] reminding us that care must be taken in translating data directly from other primates to humans.

The fact that input from the left and right eyes remains segregated at the first synapse in V1 raises an interesting question concerning the retinotopic map. Recall that in the LGN each layer contains a continuous map of the opposite visual hemifield. This means that in cortex there must be two topographic maps, one for each eye, within the *same* layer, at least at the first synapse from the LGN. This is exactly what was found in layer IV of the macaque monkey using detailed electrophysiological recordings.[13] Tangential recordings made within layer IVC revealed that the segments of the visual field mapped in one ocular dominance column for one eye were also represented in the adjacent column that received input from the other eye. As we will see below, information from the two eyes and from K, M, and P cells is combined in different ways within other layers of V1, hence, most cells in V1 receive a combination of many extrinsic and intrinsic inputs.

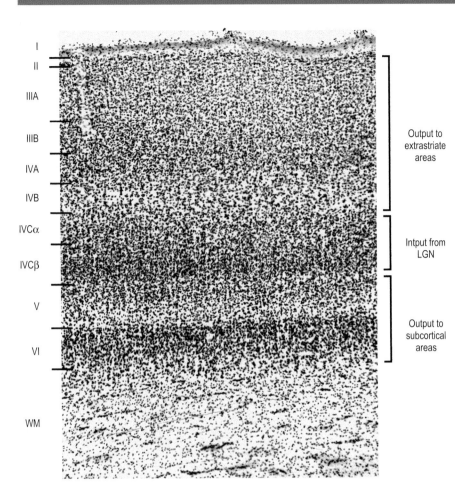

I
II
IIIA
IIIB
IVA
IVB
IVCα
IVCβ
V
VI
WM

Output to extrastriate areas

Intput from LGN

Output to subcortical areas

Figure 30.2 Nissl-stained section through macaque monkey V1. As indicated by the brackets, layer IVC receives its main input from the lateral geniculate nucleus. The layers above IVC send projections to other cortical areas, and the layers below IVC send projections to subcortical areas (see text for details). Roman numerals and letters refer to the layers of cortex. WM = white matter.

Other inputs to V1

Besides the LGN, V1 receives a variety of other modulatory inputs both from subcortical and cortical areas. These inputs include serotonergic, noradrenergic, and cholinergic inputs from the brainstem and basal forebrain nuclei, respectively.[14] The latter inputs show differences in density in the V1 layers, but show a much less specific pattern of innervation than do LGN inputs. Other input sources include the intralaminar nuclei of the thalamus and the pulvinar, both of which send broad projections most heavily to layers I and II of V1. Additionally, there are retinotopically more specific sources of input to V1, including from the claustrum and visual (V) areas 2, 3, 4, & 5 (V4 and V5 are also referred to as the dorsal lateral (DL) and middle temporal (MT) visual areas, respectively).[8,15] As a general rule any area to which V1 projects also sends feedback to V1. However, some higher-order visual areas in the temporal and parietal lobes that do not receive direct projections from V1 send axons to V1.[16] With the exception of the claustrum, whose axons overlap with M and P axons in layer IVC, all of the other extrastriate visual inputs to V1 terminate outside of layer IVC.

So why are there so many other inputs to V1 if the main drive comes from the LGN? As with the LGN described earlier, the numerous non-geniculate inputs to V1 regulate which visual signals will be transmitted to higher-order visual areas. An example of the impact that these non-LGN connections to V1 can have has been demonstrated using fMRI neural imaging methods. Using fMRI in humans, it has

been shown that topographic regions of V1 can be activated simply by asking normal subjects to imagine (with eyes closed) visual objects within particular areas of the visual field (i.e. in the *absence* of any direct stimulus to the retina).[17] Additionally, evidence indicates that, in congenitally blind subjects, major rearrangements of connections can take place such that somatosensory inputs can now activate V1,[18] or changes in receptor distribution at the retina can cause rewiring of V1[18] (Box 30.2). These findings argue that non-LGN inputs can have a strong impact on activity in V1 and that V1 is highly plastic during development.

Cell classes and connections within V1

In V1, cells containing glutamate account for the vast majority (~80 percent) with the remaining cells containing GABA.[19] These two main cell classes are morphologically distinct. There are two types of glutamate-containing cells: the spiny stellate cells that occur mainly in layer IV, and the pyramidal cells that occur in all of the layers. Both of these glutamate-containing neurons have a high density of dendritic spines. In contrast to the pyramidal cells, the inhibitory GABAergic cells have few to no spines on their dendrites. The latter are multipolar neurons whose dendritic arbors come in a variety of shapes. Many subclasses of GABAergic interneurons have been identified based also on physiology and the presence of different types of proteins such as different calcium-binding proteins, or various peptides. Although many models have been proposed to explain the role(s) of

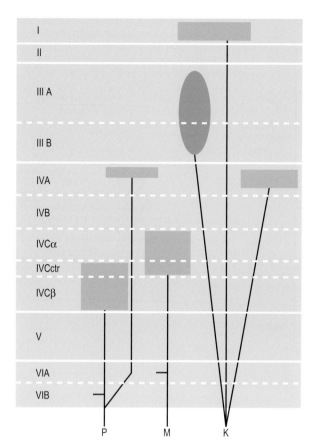

Figure 30.3 A schematic of the main LGN projection patterns to visual cortex in macaque monkeys. M and P LGN cells project, respectively, to layer IVCα and IVCβ while K LGN cells project to IVA, the CO blobs in layer III and to layer I. Note, no LGN pathway to IVA exists in apes and humans. Roman numerals refer to V1 layers; P, parvocellular; M, magnocellular; K, koniocellular. (Reproduced from Casagrande & Kass[8] by kind permission of Springer Science + Business Media.)

Right striate cortex

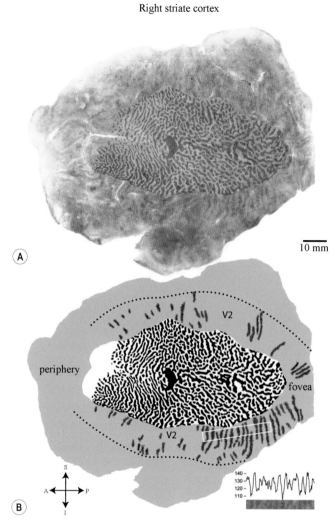

Figure 30.4 The complete pattern of ocular dominance columns in the human brain. (**A**) The surface area of the flat mount of the right hemisphere is 9413 mm². The V2/V3 border can be identified, based on a subtle change in the pattern of CO activity. Alternating dark and light stripes are visible in some portions of V2. Their density (see inset in B) fluctuated with an average period of 2.3 mm. (**B**) Thresholded columns, Stripes were drawn where visible in V2; the white box shows where the CO density in (A) was sampled to measure V2 stripe periodicity by counting the number of peaks. Dotted line indicates V2/V3 border. S, Superior; A, anterior; I, inferior; P, posterior. Striate cortex = V1. (Reproduced from Adams et al[10] with permission of the publisher.)

different types of inhibitory interneurons in V1 there is no general agreement on function for any class.

Connections between layers can be made by both excitatory and inhibitory neurons (see reviews[20,21]). Efforts to trace the overall flow of information in V1 have been hampered by the multiplicity of connections and diversity of cell classes. Using pharmacological manipulations it was demonstrated that layer IVC becomes active first, then the upper layers, followed by the lower layers.[22] Studies of cell responses in slices of visual cortex in monkeys (where photic uncaging of glutamate has been used to stimulate small groups of neurons) have shown that, although connections can be quite precise, there exists tremendous diversity of responses even among cells thought to have common inputs in the same layer.[23] If one attempts to trace anatomical pathways beyond the main input layer IVC, it also is difficult to decipher how separate output pathways are constructed in V1 due to the tremendous opportunity of intermixing; nevertheless, V1 does appear to give rise to distinct output pathways that emerge from somewhat different populations of input cells (Fig. 30.5).

Most of the connections between V1 cells are local, either within a layer, or within a vertically defined column of cortex approximately 350–500 μm wide. There are, however, connections of up to 3 mm in macaque monkeys that occur typically between cells with similar properties (e.g. selectivity for the same orientation or ocular preference). These long tangential connections are generally found in layers I, III, and V.[24] These long horizontal connections have been implicated in the changes seen in V1 cell responses when areas outside their classical receptive fields are stimulated. We discuss this issue in more detail below (see also Box 30.3).

Output pathways from V1

Some cells in all V1 layers except IVC send axons to other brain areas (see reviews[8,25]). The lower layers V and VI send axons to the thalamus and to the midbrain and pons. Layer VI is unique in that cells in this layer provide direct feedback

Box 30.2 Inherited retinal abnormalities can cause major reorganization of visual cortex

Cones are silent under dim light conditions. Since the normal human fovea is made up exclusively of cones, humans are blind in the area representing the fovea in dim light when only rods are operating. Using functional magnetic resonance imaging (fMRI) Baseler and colleagues (2002) were able to confirm that the area of V1 representing the fovea is inactive in normal humans under these dim light conditions. As the schematic below shows the fovea is represented over a substantial piece of tissue in cortex (several square centimeters). In other words, a large portion of V1 is presumably idle while you look for your seat in a dark theater or take a romantic moonlight walk.

Interestingly, these same investigators, using the fMRI were able to demonstrate a large-scale developmental reorganization in the human rod monochromats. Rod monochromats are congenitally color-blind individuals that almost completely lack retinal cones. In rod monochromats the foveal representation in V1 becomes reorganized so that it now responds strongly to visual stimuli presented under rod viewing conditions. These findings reinforce earlier work in both humans and animals suggesting that V1 is plastic and capable of large-scale reorganization when inputs are abnormal during development.

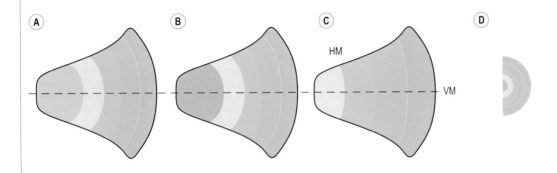

Schematic of the activation of V1 for normals (**A** and **B**) or for rod monochromats (**C**) under scotopic (**B** and **C**) and photopic (A) conditions. (**D**) represents the central 16 degrees of one visual hemi-field as described in the text. There is a mapping between visual space and cortical space, represented here by the correspondence between the colors in the schematic diagram of the right visual field (**D**) and the colors in the schematic diagrams of the left V1 unfolded and flattened (**A**, **B**, and **C**). Under normal lighting conditions the entire visual cortex of normal viewers is activated (**A**). Under scotopic conditions in normals the foveal representation is inactive represented in gray (**B**). In rod monochromats no fovea exists and the cortex is reorganized to represent the rest of visual space (**C**). In (**A**), (**B**), and (**C**) the vertical meridian (VM) runs along the upper and lower edges of the figure while the horizontal meridian (HM) is represented by the dashed horizontal line.

to the LGN, and indirect inhibitory feedback to the LGN via axon collaterals to the thalamic reticular nucleus (see also Chapter 29). This feedback is precise such that M and P LGN cells receive axons from distinct cell groups in layer VI.[26] Cells in layer V provide the major driving input to many cells in the pulvinar nucleus of the thalamus in monkeys; the pulvinar, in turn, provides input to a number of extrastriate areas that also feedback signals to V1. In addition, cells in layer V send a major projection to the superficial layers of the superior colliculus and other midbrain areas such as the pretectum, as well as nuclei in the pons that are concerned with eye movements. Thus, V1 is in a position to inform these structures of its activities.

The superficial cortical layers of V1 provide output connections to a number of extrastriate cortical areas. These connections emerge from different layers or modules within layers suggesting that they carry different messages. Currently there is ongoing debate about how many pathways actually leave V1 and how separate they are in terms of their physiological signatures (Fig. 30.6). Regardless, most investigators agree that in macaque monkeys the largest output connection is to visual area 2 (V2). Connections to V2 emerge from three or four populations of cells. Cells within the CO-rich blobs of layers IIIA and IIIB send a major output to thin CO-rich bands in V2 (see Chapter 31 for details). Cells located in the CO pale interblobs and cells in layer IVB send output to different locations within the pale and CO thick bands of V2. In addition to these connections there are direct connections from V1 layer IIIB and IVB to visual areas V3, V3A, V4, and V5.[8,25] As the technology improves, the story of how many V1 to V2 channels exist will continue to be revised.[25,27,28]

Receptive field properties: How is V1 different from the LGN?

Examination of the receptive field properties of V1 neurons suggests that visual signals are transformed from those seen

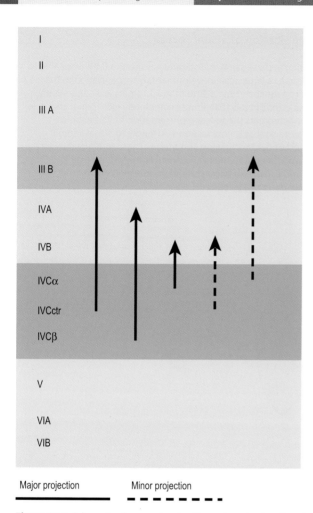

Major projection ——————   Minor projection ▪ ▪ ▪ ▪ ▪ ▪ ▪ ▪

Figure 30.5 Schematic diagram showing the main projections from layer 4 to layer 3 in the owl monkey. IVCα projects primarily to IVB. IVCβ projects primarily to IVA. IVCctr, which receives overlapping input from the M and P LGN axons, projects to IIIB where K axons from LGN also terminate. IIIA, the major output layer to visual area V2, receives signals from layer IVC only indirectly via other subdivisions of cortical layers III and layer V. (Modified from Boyd et al[50] with permission of the publisher.)

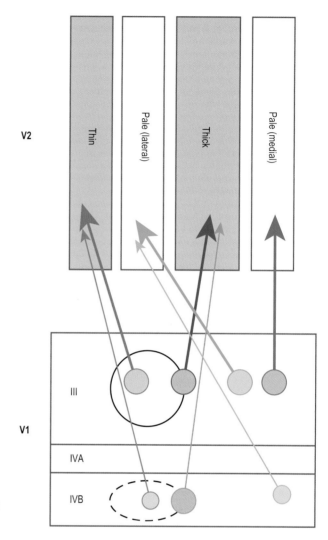

Figure 30.6 Recent data suggest that there may be four output pathways from V1 to V2 based on anatomy. Each pathway is represented by arrows. One pathway (red) projects from the CO blobs in layer III (indicated by a circle) to the CO thin dark stripes in V2, indicated in blue. A smaller number of cells (orange) from layer IVB in V1 also project to the CO thin stripes in V2. A second pathway (dark green) emanates from cells in the interblobs of layer III and projects to the CO pale medial stripes in V2. The third pathway (light green) projects to the pale lateral stripes of V1 that also receive input from the interblobs in layer III but additionally receive projections (yellow-green) from layer IVB. The fourth pathway (blue) originates from both cells at the edges of blobs in layer III as well as cells in layer IVB that align with the edge of blob cells in layer III. (This description of the pathways from V1 to V2 is taken from the work of Federer et al[27] and Angellucci (personal communication 2009) with permission of the authors.)

in the retina and LGN. In other words, new properties emerge in V1 such as binocularity and sensitivity to stimulus orientation and movement direction. At the same time, V1 cells retain the retinotopic selectivity of their LGN cell inputs, although receptive fields are a bit larger.

In the 1950s and 1960s Hubel & Wiesel[29,30] began to characterize the properties of V1 receptive fields in cats and monkeys using a variety of patterns including line segments and spots of light displayed on a screen. In these seminal studies they showed that V1 cells could be subdivided based on their responses to such patterns. Hubel & Wiesel[31] proposed that the cell types in V1 were arranged in serial order of complexity beginning with those that receive input directly from the LGN which they termed **simple cells**. They originally proposed that simple cell responses could best be explained by assuming that the receptive fields of a number of LGN cells were aligned as shown in Figure 30.7.[32] Simple cells differ from LGN and retinal ganglion cells, most of which have more or less circular receptive fields. Simple cells have elongated receptive fields with adjacent excitatory and

inhibitory regions. Hubel & Wiesel called these cells simple because it appeared that the responses of these cells to complex shapes could be predicted by linear summation of their responses to individual spots of light. As can be appreciated by examining Figure 30.7, simple cells can give different responses depending upon the spatial arrangement of their inhibitory and excitatory regions. For example, although all the cells shown in Figure 30.7 respond to the same orientation, cells with longer receptive fields such as shown in panel C will respond to a narrower range of orientations than those with shorter receptive fields as shown in panel A. As this figure also illustrates, the receptive fields of simple

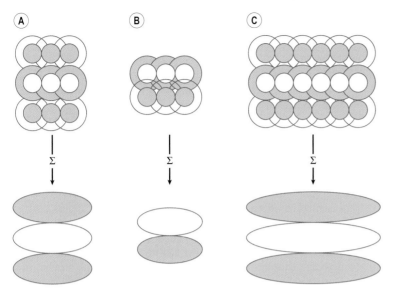

Figure 30.7 Orientation-selective receptive fields can be created by summing the responses of neurons with non-oriented, circularly symmetric receptive fields. The receptive fields of three hypothetical neurons are shown. Each hypothetical receptive field has adjacent excitatory and inhibitory regions. A comparison of (**A**) and (**C**) illustrates that the degree of orientation selectivity can vary depending on the number of neurons combined along the main axis. (Reproduced from Wandell[32] with permission of the publisher.)

cells require that LGN ON and OFF center cells are aligned, since it is the center responses of these cells that dominate the subfield responses of simple cells within V1.

Hubel & Wiesel also identified other cell classes with more complex response properties. These cells, generally called **complex cells**, are different from simple cells in that their responses to stimuli cannot be predicted based upon linear addition of the cell's response to spots of light presented in different parts of the receptive field; complex cells do not have discrete regions of excitation and inhibition. Instead, complex cells respond to preferred orientations like simple cells but complex cells respond equally well to a preferred stimulus anywhere in their receptive field. A special type of V1 cell referred to as an **end-stopped cell** responds only if the correctly oriented stimulus is of appropriate length. Extending the length of a bar beyond the field into an inhibitory zone of an end-stopped cell diminishes the cell's responses suggesting that these cells may signal more complex shapes. Hubel & Wiesel originally proposed that the receptive fields of each cell class (namely, LGN, simple, complex, and end-stopped cells) built upon the properties of their predecessors in serial order. We know now that such a straightforward model can not account for all of the properties observed, that many complex cells can receive input directly from the LGN, and that end-stopped cells can either be simple or complex cells.

Other receptive field properties that emerge within area V1 are direction selectivity and binocularity. Although in some mammals such as rabbits, **direction selectivity** is a characteristic of many ganglion cells, in primates there are very few retinal ganglion and LGN cells that exhibit this property.[33] In V1 there are many cells that respond best to one direction of motion of a stimulus. Explanations for how direction selectivity emerges generally argue for a time delay in inputs either from different classes of LGN cells or between cortical cells with the same orientation selectivity such that one cell either enhances or suppresses the response of the other.

In addition to orientation and direction selectivity, many cortical neurons in cats and monkeys are **binocular** (i.e. receive signals from both eyes). V1 is the first place where feedforward visual information from the two eyes is fused. In binocular V1 neurons, there exists a range of cells with many responding somewhat more to one eye than the other. The bias in ocular response is such that ocular preference within a column extending from layer I to layer VI tends to reflect the preference of cells in layer IV within that column. This is because cortical cells are connected preferentially in vertical columns. Binocular cells with slightly displaced monocular fields (i.e. cells with a disparity between the receptive fields of the left and right eye) have been proposed as the possible initial substrate for stereoscopic vision, although many researchers argue that the ability to see in 3D is actually the function of visual areas beyond V1.

Columns and modules: Outlining the functional architecture of V1

As we have seen, V1, like the LGN, is arranged in layers containing cells of different types. Unlike the LGN, new receptive field properties such as selectivity for stimulus orientation, movement direction, and binocularity are constructed in V1. Additionally, information about spatial frequency, temporal frequency, brightness, and color contrast, sent by cells of the LGN, must be either preserved or incorporated into the coding of V1 cells. Given the precision of the visuotopic map in V1, this means that critical stimulus attributes must be coded in an iterated manner to cover each location in visual space so that stimulus features that provide information relevant to form and movement can be appreciated equally well without holes or gaps at different locations in the visual field.

Hubel & Wiesel[31] were cognizant of the problem that local stimulus features would need to be represented repeatedly at each locale in the V1 map. What they noticed early on in their studies was that orientation preference in cat and monkey V1 changed regularly as the electrode was moved tangentially within any layer (Fig. 30.8). An advance of 1–2 mm was usually found to be sufficient to rotate twice through 180 degrees of orientation preference. This distance also was found to be sufficient to include at least one left

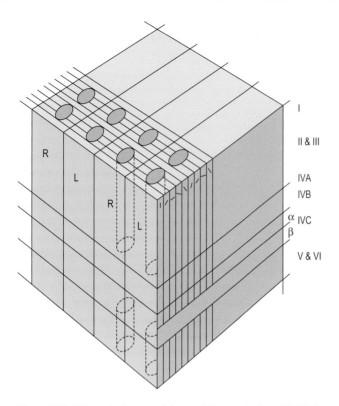

Figure 30.8 Schematic diagram of the modular organization of V1. Each module (or hypercolumn; see text for details) consists of two ocular dominance columns (representing right and left eyes), a series of orientation columns (representing 180 degrees of rotation) and cytochrome oxidase blobs (representing color information). (Reproduced from Livingstone & Hubel[28] with permission of the publisher.)

and one right eye ocular dominance column. From this information Hubel & Wiesel constructed a model in which they proposed that the cortex is organized as a series of repeating modules called **hypercolumns**. They argued that each hypercolumn should contain all of the machinery necessary to analyze one portion of visual space. Subsequently, Livingstone & Hubel[28] argued that CO blobs should be added to this modular organization as zones uniquely equipped to transmit color signals to the next cortical level. Although there is debate as to whether CO blobs are actually uniquely designed for color processing,[8] the fact that these modules are the targets of LGN input from a separate class of cells, the K cells, suggests that CO blobs do process some important attributes, perhaps color contrast and brightness contrast. Moreover, there appear to be enough CO blobs so that whatever is processed within these modules can be represented across the V1 topographic map without gaps. Since CO blobs are positioned in the centers of ocular dominance columns in macaque monkeys, they were added as another dimension to be included within a V1 hypercolumn (Fig. 30.8).

The geometric problem is not so difficult for the cortex to solve when only three stimulus properties (i.e. orientation, ocular dominance, and color) must be represented, but when more properties such as spatial frequency, temporal frequency, direction selectivity, texture and binocular disparity (among others) are added, the task becomes more challenging. Currently, a number of investigators are engaged in

modeling how different feature maps may be represented in V1.[34,35]

Optical imaging of signals in V1 has been used to further elucidate the relationship between maps of different stimulus properties in single animals. The technique of optical imaging of intrinsic signals makes use of tiny differences in the reflected light from cortex based upon dynamic differences in oxygenated and deoxygenated blood that occur as a result of changes in the activity of cells. Using this invasive technique in non-human primates, it has been found that changes in orientation selectivity are represented mainly in pinwheel formation, with some regions also showing more gradual linear or abrupt fractures in the orientation map. The structure of orientation maps in different primates and in other species shows a great deal of similarity suggesting that orientation selective cells are organized the same way in humans. Maps of different stimulus qualities also suggest that, although not organized exactly as originally envisioned in the hypercolumn model of Hubel & Wiesel, maps of stimulus attributes are nevertheless iterated in such a manner that there are no "holes" in the map across space. Figure 30.9 shows an optically imaged map of orientation preference in owl monkey V1, V2, and V3.[36] The functional compartments in V2 are also apparent in this surface image.

How do parallel inputs relate to parallel outputs?

It has been popular to suggest that there is a direct link between the input and output pathways of V1. There is considerable support for the idea that two hierarchies of extrastriate visual areas exist: one devoted to object vision, or *what* something is, and one to support spatial vision and complex tasks related to *where* items are in space relative to ourselves.[28,37] The "what" and "where" pathways, also called the ventral and dorsal streams, consist of two hierarchies of visual areas that receive input from separate cell groups in V1 as described earlier. The ventral stream consists of a pathway from V2 through V4 to the temporal cortex and the dorsal stream consists of a pathway from V2 through MT to the parietal cortex[25] (see review[38]). It is less clear whether the dorsal and ventral extrastriate pathways are directly linked to the K, M, and P LGN input pathways to V1. The best evidence for such a direct link comes from studies in which input from the M and P pathways and associated K cells were blocked in the LGN of macaque monkeys with micro injections of GABA.[39] This study demonstrated that the majority of input to visual area MT comes from M LGN cells or M and neighboring K cells since the two could not be inactivated separately in these studies. In spite of this finding, some MT cells could still be driven by the remaining P and/or K LGN cells.

As described in Chapter 29, we now know that some K cells send axons directly to area MT so this is not surprising. A fairly direct pathway for signals from M LGN cells to area MT has also been demonstrated anatomically, given that cells in layer IVCα, the target layer for LGN M cells, send axons directly to cells in layer IVB which, in turn, can send signals to area MT. Nevertheless, cells in layer IVB that project to MT do not reflect the receptive field properties of M cells; instead most are complex direction-selective cells

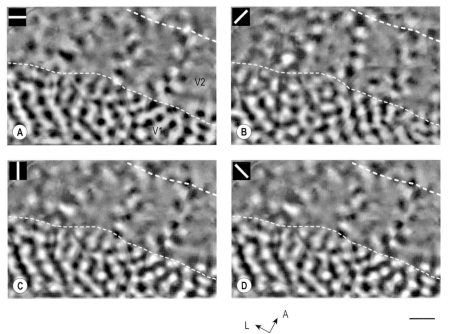

Figure 30.9 Orientation difference maps of V1 and V2 in the owl monkey obtained from optical imaging of the cortical surface. Four orientation difference maps are presented as follows: (**A**) 0/90°; (**B**) 45/135°; (**C**) 90/0°; (**D**) 135/45°. The data were obtained by presenting the monkey with full field rectangular gratings (duty cycle, 0.2; spatial frequency, 0.5 cycles/degree; temporal frequency, 2 Hz). The thin and thick dashed lines delineate the V1/V2 and V2/V3 borders, respectively. Note that orientation preference is represented continuously throughout V1 but in a series of alternating bands of high and low selectivity within V2 and V3. High selectivity for the stimulus is indicated by black or white while low selectivity is indicated by gray. A, Anterior; L, lateral. Scale bar, 1 mm. (Reproduced from Xu et al[36] with permission of the publishers.)

whose receptive fields are presumably constructed through complex cortical circuits.[40] Even more opportunity for integration between pathways seems to exist before signals enter the ventral stream ("what" pathway). Blockade of the P layers and surrounding K layers does not silence cells within output layers IIIA and IIIB, both of which respond well with either M or P layers blocked.[41] Moreover, anatomically, much of the output to the ventral stream leaves from layer IIIA which gets no direct input from layer IVC, but receives signals only after they have passed to other layers. Thus, both the wiring and physiology suggest that considerable integration of signals takes place in V1 before they are transmitted into the ventral stream for further analysis of object identity. Finally, as discussed in Chapter 29, the fact that lesions of either M or P layers in the LGN (together with associated K layers) do not completely eliminate either form or motion vision reinforces the view that it is inappropriate to equate complex visual behavior with the threshold properties of retinal and LGN cells.

Does V1 Do More?

In the classical view of V1, the visual field is analyzed by small receptive fields within separate modules and V1 layers. These receptive fields are depicted as static, isolated, and stable with respect to attributes being tested including orientation, direction, spatial frequency, etc. The emphasis in this classical model is one of serial processing where information constructed within these small localized receptive fields in V1 is sent, via feedforward connections, to higher visual areas where it is combined in different ways until complex objects or motions are acted upon by the brain areas concerned with executive function. In this model, V1 becomes a staging area to prepare signals for more complex perceptually relevant tasks performed only at the highest cortical levels. This classical view, however, does not take into account that the responses in V1 cells can be affected by time, the visual context of a stimuli and previous experience. Yet the anatomy described above alerts us to the fact that more information may be returning from higher visual areas back to V1 than is traveling the other way, and that local V1 areas are heavily interconnected both vertically across layers and tangentially via horizontal axon pathways, and that many other input pathways to V1 from the brainstem and thalamus exist. Returning to our introductory remarks, we remind the reader that the visual system is not designed to represent all physical constants, but to represent what is important for survival. In this section we give examples to show that V1 may do much more than classical models would predict.

The importance of time

What is difficult to appreciate from descriptions of wiring alone is that the visual system is highly dynamic. The problem is that the visual system must maintain stability while we move and objects move relative to us. Therefore, the receptive fields of V1 neurons can only provide useful snapshots within short windows of time. When researchers measure receptive field properties, they generally average the firing response of cells over long periods of time (a half second to several seconds). We can discriminate objects in as little as 250 ms; so the question is, does a cell prefer the same orientation or spatial frequency if you sampled at different windows of time? The answer is no. Using peak firing rate as a measure of preference, investigators have shown that cells shift their orientation and spatial frequency preferences over time significantly. Ringach and colleagues[42,43] have demonstrated that many V1 cells shift both their orientation tuning and spatial frequency tuning quite dramatically over time with some cells actually changing from being excited by to being inhibited by the same stimulus attribute in as little as 20 ms. As techniques for sampling from many neurons over time in awake monkeys become more sophisticated, it is clear from these examples that our concept of how V1 cells contribute to vision will have to move from static depictions of single simple cell receptive fields to plots that also represent changes in response over time across a network of cells.

The importance of context

As described earlier, the connections of V1 are complex and include horizontal connections that integrate information across separate parts of the visual field, as well as massive, fast-conducting feedback connections.[44,45] Although these feedback connections are retinotopically precise,[44] they arise from higher visual area neurons that have very large, complex receptive fields. Therefore, this heavy feedback connects individual higher-order visual cells with large numbers of V1 cells. When the visual field is analyzed using the small receptive fields of V1 cells, one forgets that these limited views of visual space will need to be combined with the rest of the field, both in space and in time. Clearly, both horizontal and feedback connections could provide the substrate for such information. The difficulty, however, of determining the role of connections that do not provide the main drive to a cell is that their impact can only be measured against the baseline or driven activity that the V1 cell exhibits.

Within the past few years a number of investigators have successfully demonstrated that the activity of a V1 cell can be modulated by the activity of its neighbors even when these neighbors lie well outside of the classically defined receptive field of a V1 cell. One method used to demonstrate the impact of stimulating neighboring cells is illustrated in Figure 30.10, which shows an example of a V1 cell that prefers (based on stimulation of its classical receptive field) a vertical bar of light as shown in (A). When, however, bars of similar orientation are presented in the receptive fields of neighboring cells simultaneously, the response of the V1 cell is significantly reduced. Note that stimulating the neighbors without simultaneously stimulating the receptive field of the V1 cell under study has *no effect*. Interestingly, as shown in (C) the V1 cell's response is actually enhanced if the neighbors are stimulated with a different orientation. Finally in (D) if the neighbors are stimulated with orientations that include a mix of matched and unmatched orientations, the V1 cell response is unaffected. In other words context matters. These effects at the single V1 cell level are surprisingly similar to the effects one sees perceptually. It is known that the context of a line segment changes its perceptual salience. A line segment by itself is quite salient but when it is embedded among like neighbors it becomes much less obvious. When the neighbors are different the salience is again high as the "different" orientation of the target "pops out" from the background. If the background and target are similar the target is less visible. There are many more examples.[46]

Other forms of context also appear to matter for V1 cells. Although there currently exists debate as to the timing of effects in V1, a number of laboratories have shown that visual spatial attention to a target can modulate the activity of V1 such that areas representing the focus of attention show enhanced activity at the expense of areas of V1 that are not the focus of attention.[47] As discussed in Chapter 29, such

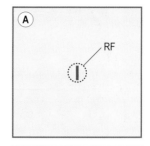

Classical RF response

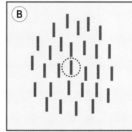

Response reduced by similar stimuli in surround

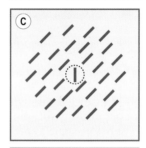

Response enhanced by stimuli of non-optimal orientation

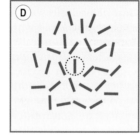

Response same assuming all elements have the same contrast

Figure 30.10 The context of a line segment changes both the local firing rate of cells and its perceptual salience. (**A**) A lone line segment is perceptually most salient and cells with a preference for this orientation will produce a high firing rate. (**B**) The same line segment, now embedded in similar ones, is much less conspicuous and the cells representing that line will show significantly lower firing rate to the same stimulus as in "A". (**C**) The line segment may pop out when its orientation differs from that of the surrounding line segments, restoring its perceptual salience to some extent and the cells representing that line segment will again show a high firing rate. (**D**) Global stimulus aspects are taken into consideration, since a similar local orientation difference does not produce popout when surrounding elements all have different orientations. Thus, when a V1 neuron is stimulated by presenting these displays, so that its RF only covers the same center line segment in all cases, contextual modulation signals the perceptual salience of the center line segment. (Modified from Lamme[46] with permission of the publisher.)

attention effects even have been reported at the single cell level in the LGN.

Even more surprising are results that show that reward and memory can impact the responses of V1 cells. Recently Shuler & Bear[48] reported that when adult rats experience an association between visual stimuli and subsequent rewards, the responses of a substantial fraction of cells in V1

Box 30.3 Does V1 remember?

How do we remember details of our visual world when higher-order visual areas do not appear to have the selectivity of early sensory areas such as V1? This problem was recently addressed using fMRI of human subjects in an elegant set of experiments designed to test if V1 showed sustained activity during a working memory task when no visual stimulus was present.[49] As illustrated in (**A**) subjects were presented with two different orientations (1 and 2) and then shown a number indicating which orientation they should remember during the blank working memory period. At the end of this working memory period they were shown a test pattern and asked to judge if the test pattern was rotated clockwise or counterclockwise relative to the remembered pattern. Although the task was difficult the subjects were able to perform the task with ~75 percent accuracy. Imaging results clearly showed that orientations held in working memory could be decoded by the relative change in activity patterns in V1. This surprising result does indeed support the conclusion that cells in V1 are capable of remembering visual patterns in the absence of stimulation at least for several seconds.

Orientation decoding results for areas V1–V4 (**B**). The accuracy of orientation decoding for remembered gratings in the working memory experiment (green circles), unattended presentations of low-contrast gratings (red triangles), and generalization performance across the two experiments (blue squares). Error bars indicate s.e.m. Decoding was applied to the 120 most visually responsive voxels in each of V1, V2, V3, and V3A–V4 (480 voxels for V1–V4 pooled), as determined by their responses to a localizer stimulus (1–4 u eccentricity). Individual areas V3A and V4 showed similar decoding performance but had fewer available voxels, so these regions were combined. (Figure reprinted from Harrison & Tong[49] with permission from the publisher.)

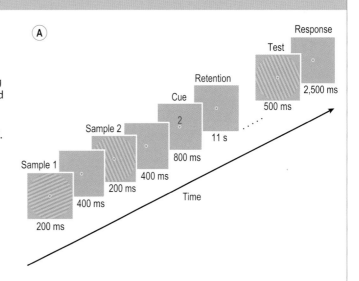

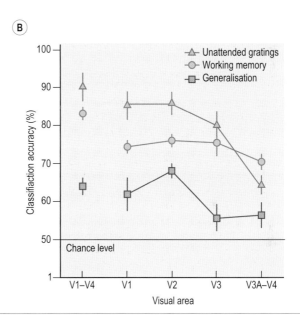

---

gradually begin to predict the timing of reward. In addition to revealing a remarkable type of response plasticity in V1, these findings demonstrate that reward-timing activity, which is considered a higher brain function, can actually occur very early in the sensory pathway. These results, and others showing how memory can affect V1 activity in humans[49] (Box 30.3), challenge our traditional interpretation of the role played by V1 neurons in conscious visual perception.

Conclusion

V1 is perhaps the most thoroughly researched area in the primate brain. Motivated by the fact that V1 is necessary for conscious visual perception in primates, researchers have produced volumes of work on the firing patterns, projections, and internal anatomy of this area. Some of the most enduring themes of these writings are that V1 cell firing encodes for local features (especially orientation) and that these cells are organized modularly in retinotopic maps. However remarkable our knowledge of V1 is, it is more remarkable that this tremendous amount of knowledge has only begun to explain the function of V1. We are only now beginning to understand V1 as part of a dynamic network that transmits signals to higher visual areas while simultaneously being transformed by feedback signals from these areas. This work argues V1 is not a mere conduit from the retina to higher-order brain areas but alters its message dynamically based on context so that only the essentials are transmitted for efficient processing.

References

1. Andrews TJ, Halpern SD, Purves D. Correlated size variations in human visual cortex, lateral geniculate nucleus, and optic tract. J Neurosci 1997; 17(8):2859–2868.

2. Brodmann K. Vergleichende lokalisationlehre der grosshirnrinde in ihren prinzipien dargestelt auf des zellenbaues. Leipzig: J. A. Barth, 1909.

3. Meyer BU, Diehl R, Steinmetz H, Britton TC, Benecke R. Magnetic stimuli applied over motor and visual cortex: influence of coil position and field polarity on motor responses, phosphenes, and eye movements. Electroencephalogr Clin Neurophysiol Suppl 1991; 43:121–134.

4. Engel SA, Rumelhart DE, Wandell BA et al. fMRI of human visual cortex. Nature 1994; 369(6481):525.

5. Fox PT, Miezin FM, Allman JM, Van Essen DC, Raichle ME. Retinotopic organization of human visual cortex mapped with positron-emission tomography. J Neurosci 1987; 7(3):913–922.

6. Horton JC, Hoyt WF. Quadrantic visual field defects. A hallmark of lesions in extrastriate (V2/V3) cortex. Brain 1991; 114(Pt 4):1703–1718.

7. Van Essen DC, Newsome WT, Maunsell JH. The visual field representation in striate cortex of the macaque monkey: asymmetries, anisotropies, and individual variability. Vision Res 1984; 24(5):429–448.

8. Casagrande VA, Kaas JH. The afferent, intrinsic, and efferent connections of primary visual cortex in primates. Primary visual cortex of primates. New York: Plenum, 1994:201–259.

9. Malpeli JG, Schiller PH, Colby CL. Response properties of single cells in monkey striate cortex during reversible inactivation of individual lateral geniculate laminae. J Neurophysiol 1981; 46(5):1102–1119.

10. Adams DL, Sincich LC, Horton JC. Complete pattern of ocular dominance columns in human primary visual cortex. J Neurosci 2007; 27(39):10391–10403.

11. Casagrande VA, Yazar F, Jones KD, Ding Y. The morphology of the koniocellular axon pathway in the macaque monkey. Cereb Cortex 2007; 17(10):2334–2345.

12. Casagrande VA, Khaytin I, Boyd JD. Evolution of the visual system: mammals – color vision and the function of parallel visual pathways in primates. In: Butler A, ed. Encyclopedia of neuroscience. Berlin & Heidelberg: Springer, 2008:1472–1475.

13. Hubel DH, Wiesel TN. Laminar and columnar distribution of geniculo-cortical fibers in the macaque monkey. J Comp Neurol 1972; 146(4):421–450.

14. Jones EG. The thalamus of primates. In: Björklund A, Hökfelt T, eds. Handbook of chemical neuroanatomy. New York: Elsevier, 1998.

15. Lyon DC, Kaas JH. Connectional and architectonic evidence for dorsal and ventral V3, and dorsomedial area in marmoset monkeys. J Neurosci 2001; 21(1):249–261.

16. Rockland KS, Ojima H. Multisensory convergence in calcarine visual areas in macaque monkey. Int J Psychophysiol 2003; 50(1–2):19–26.

17. Chen W, Kato T, Zhu XH, Ogawa S, Tank DW, Ugurbil K. Human primary visual cortex and lateral geniculate nucleus activation during visual imagery. Neuroreport 1998; 9(16):3669–3674.

18. Burton H, Snyder AZ, Conturo TE, Akbudak E, Ollinger JM, Raichle ME. Adaptive changes in early and late blind: a fMRI study of Braille reading. J Neurophysiol 2002; 87(1):589–607.

19. Fitzpatrick D, Lund JS, Schmechel DE, Towles AC. Distribution of GABAergic neurons and axon terminals in the macaque striate cortex. J Comp Neurol 1987; 264(1):73–91.

20. Callaway EM. Local circuits in primary visual cortex of the macaque monkey. Annu Rev Neurosci 1998; 21:47–74.

21. Xu X, Callaway EM. Laminar specificity of functional input to distinct types of inhibitory cortical neurons. J Neurosci 2009; 29(1):70–85.

22. Bolz J, Gilbert CD, Wiesel TN. Pharmacological analysis of cortical circuitry. Trends Neurosci 1989; 12(8):292–296.

23. Callaway EM, Katz LC. Photostimulation using caged glutamate reveals functional circuitry in living brain slices. Proc Natl Acad Sci USA 1993; 90(16):7661–7665.

24. Rockland KS, Lund JS. Widespread periodic intrinsic connections in the tree shrew visual cortex. Science 1982; 215(4539):1532–1534.

25. Sincich LC, Horton JC. The circuitry of V1 and V2: integration of color, form, and motion. Annu Rev Neurosci 2005; 28:303–326.

26. Ichida JM, Casagrande VA. Organization of the feedback pathway from striate cortex (V1) to the lateral geniculate nucleus (LGN) in the owl monkey (Aotus trivirgatus). J Comp Neurol 2002; 454(3):272–283.

27. Federer FC, Ichida JM, Jeffs J, Angelucci A, Angelucci A. Three streams of anatomical projections from primary visual cortex to the cytochrome oxidase stripes of marmoset area V2. New York: Society for Neuroscience, 2007.

28. Livingstone MS, Hubel DH. Anatomy and physiology of a color system in the primate visual cortex. J Neurosci 1984; 4(1):309–356.

29. Hubel DH, Wiesel TN. Receptive fields, binocular interaction and functional architecture in the cat's visual cortex. J Physiol 1962; 160:106–154.

30. Hubel DH, Wiesel TN. Receptive fields and functional architecture of monkey striate cortex. J Physiol 1968; 195(1):215–243.

31. Hubel DH, Wiesel TN. Ferrier lecture. Functional architecture of macaque monkey visual cortex. Proc R Soc Lond B Biol Sci 1977; 198(1130):1–59.

32. Wandell BA. Foundations of vision. Sunderland, Mass: Sinauer Associates, Inc., 1995.

33. Amthor FR, Grzywacz NM. Directional selectivity in vertebrate retinal ganglion cells. Rev Oculomot Res 1993; 5:79–100.

34. Khaytin I, Chen X, Royal DW et al. Functional organization of temporal frequency selectivity in primate visual cortex. Cereb Cortex 2008; 18(8):1828–1842.

35. Swindale NV. How different feature spaces may be represented in cortical maps. Network 2004; 15(4):217–242.

36. Xu X, Bosking W, Sary G, Stefansic J, Shima D, Casagrande V. Functional organization of visual cortex in the owl monkey. J Neurosci 2004; 24(28):6237–6247.

37. Milner AD, Goodale MA. The visual brain in action. Oxford: Oxford University Press, 1995.

38. Merigan WH, Maunsell JH. How parallel are the primate visual pathways? Annu Rev Neurosci 1993; 16:369–402.

39. Nealey TA, Maunsell JH. Magnocellular and parvocellular contributions to the responses of neurons in macaque striate cortex. J Neurosci 1994; 14(4):2069–2079.

40. Recanzone GH, Wurtz RH, Schwarz U. Responses of MT and MST neurons to one and two moving objects in the receptive field. J Neurophysiol 1997; 78(6):2904–2915.

41. Allison JD, Melzer P, Ding Y, Bonds AB, Casagrande VA. Differential contributions of magnocellular and parvocellular pathways to the contrast response of neurons in bush baby primary visual cortex (V1). Vis Neurosci 2000; 17(1):71–76.

42. Bredfeldt CE, Ringach DL. Dynamics of spatial frequency tuning in macaque V1. J Neurosci 2002; 22(5):1976–1984.

43. Ringach DL, Hawken MJ, Shapley R. Dynamics of orientation tuning in macaque V1: the role of global and tuned suppression. J Neurophysiol 2003; 90(1):342–352.

44. Angelucci A, Bullier J. Reaching beyond the classical receptive field of V1 neurons: horizontal or feedback axons? J Physiol Paris 2003; 97(2–3):141–154.

45. Salin PA, Bullier J. Corticocortical connections in the visual system: structure and function. Physiol Rev 1995; 75(1):107–154.

46. Lamme VAF. Beyond the classical receptive field: contextual modulation of V1 responses. In: Chalupa L, Werner JS, eds. The visual neurosciences. Cambridge, MA: MIT Press, 2004:720–732.

47. Tong F. Primary visual cortex and visual awareness. Nat Rev Neurosci 2003; 4(3):219–229.

48. Shuler MG, Bear MF. Reward timing in the primary visual cortex. Science 2006; 311(5767):1606–1609.

49. Harrison SA, Tong F. Decoding reveals the contents of visual working memory in early visual areas. Nature 2009; 458(Issue 7238):632–635.

50. Boyd JD, Mavity-Hudson JA, Casagrande VA. The connections of layer 4 subdivisions in the primary visual cortex (V1) of the owl monkey. Cereb Cortex 2000; 10(7):644–662.

CHAPTER 31

Extrastriate Visual Cortex

Jamie D. Boyd & Joanne A. Matsubara

What is extrastriate visual cortex?

The term "extrastriate" refers to all visually responsive cortex other than primary visual (striate) cortex, and does not receive strong direct projections from the lateral geniculate nucleus (LGN) (see Chapter 30). A central hypothesis in studying extrastriate cortex is that it is composed of discrete cortical areas which can be identified by histology, retinotopic mapping, patterns of connections with other areas, and the unique contribution each area makes to visual processing. Extrastriate areas that receive strong inputs directly from primary visual cortex can be thought of as a lower tier within a processing hierarchy, compared to a higher tier of areas that receive their main visual inputs from relays through other extrastriate areas.[54] The more relays between an extrastriate area and primary visual cortex, the higher its position in the processing hierarchy and, generally speaking, the more complex is its visual processing. Superimposed on this hierarchy is the concept of parallel streams of visual areas that are more strongly interconnected with each other than with areas of different streams, and that are all concerned with some common aspect of visual processing, such as object recognition (ventral stream) or object location (dorsal stream).[45] At least 25 cortical areas predominately or exclusively engaged in vision have been described in monkeys;[54] the number of visual areas identified in humans is rapidly approaching, and will probably surpass, that total (Fig. 31.1, Box 31.1 and see also Box 31.6).[133,160]

Methods used to identify extrastriate areas in monkeys and humans

Identifying cortical areas and their borders is neither straightforward nor completely resolved, and different schemata for parceling extrastriate cortex into areas exist, depending upon how the criteria of histology, retinotopy, connectivity, and physiology are weighted. There is most consensus regarding areas lower in the hierarchy, nearer to V1, and most divergence of opinion for higher-level areas in the temporal and parietal lobes.[71,118,133,160,177]

Histology

Cortical areas can be expected to have a uniform and distinct histology, such that anatomical borders can be recognized by changes in the laminar organization, density, or size of neurons, myelinated fibers, enzyme activity, neurotransmitter levels and their receptors, and so forth. However, not all borders can be recognized with all stains. Early parcellation schemes based on classical histological techniques are still in wide use,[162] although containing many known inaccuracies. Based on Nissl staining of neuronal cell bodies (cytoarchitectonics), Brodmann[23] divided most of extrastriate occipital cortex, in both human and monkeys, into only two areas, 18 and 19, each of which includes multiple areas defined by other criteria such as retinotopy and connectivity. Other histologic stains not available to Brodmann, such as cytochrome oxidase (CO; a metabolic enzyme that serves as a marker for long-term activity levels of neurons) histochemistry, reveal the borders of the many visual areas within the cortical territory occupied by Brodmann's areas 18 and 19.[76,91,159,161] Even with improved staining techniques, however, it is likely that some areal boundaries may escape detection by any histologic methods.

Retinotopic mapping

Each visual area comprises a single map of receptive field locations, with V1 demonstrating the most complete map of visual space (see Chapter 30). These maps can be studied with electrophysiologic recordings in monkeys and with functional magnetic resonance imaging (fMRI) or positron emission tomography (PET) in humans. However, none of the extrastriate maps are as regular as in striate cortex (V1). Extrastriate areas have larger receptive fields and more scatter in the position of each receptive field, resulting in "cruder" maps. Between extrastriate areas, there is generally a smooth transition of receptive field locations. Contiguous visual areas can have "mirror-image" representations, where the topography of the visual field map is reversed at the border of the two contiguous areas. When a mirror-image area like V1 borders a non-mirror-image representation like V2, the vertical meridian representation forms the border of the two

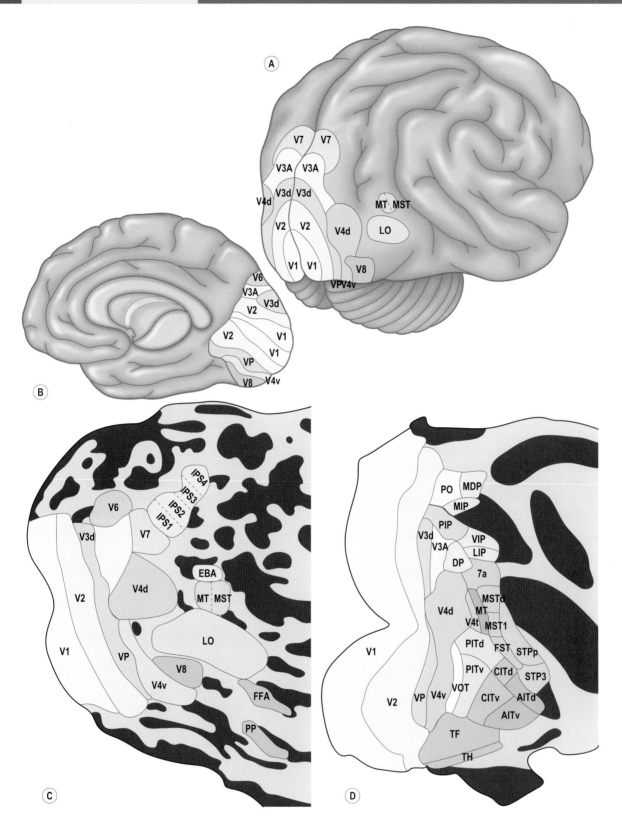

Figure 31.1 Human (**A**, **B**, **C**) and monkey (**D**) visual cortical areas. Visual cortical areas are shown on a schematic diagram of the human brain from the posterior aspect (A) and mid-sagittal plane (B). Many of the areas cannot be appreciated on the surface view of the brain, and a flat map of the human (C) visual cortical areas is also shown. Areas involved in vision are colored and labeled. The depths of the sulci are shown in black and gyri in white to give a perspective to location relative to the sulcal patterns. For comparison, a flat map of visual areas in the macaque monkey is shown in (D). Homologous areas, insofar as they can be identified, are given the same color and nomenclature, but many more areas have been studied in monkey brain than in human. In the occipital cortex of humans, there are: the second visual area (V2); the third visual area, broken into a dorsal (V3d) and a posterior half (VP); V3 anterior (V3A); the ventral (V4v) and dorsal (V4d) subdivisions of V4; and the sixth, seventh and eighth visual areas (V6, V7 and V8). Visual responsive areas in the parietal lobes include: the middle temporal area (MT); the medial superior temporal area (MST); the lateral occipital area (LO); and the extrastriate body area (EBA). Several visual areas in the intraparietal sulcus have been recently studied and named (IPS1, IPS2, IPS3, IPS4). In the occipitotemporal cortex are found the fusiform face area (FFA) and the parahippocampal place area (PPA). Monkey extrastriate areas shown in (D) include, in the temporal lobe: the posterior inferotemporal area, with dorsal (PITd) and ventral (PITv) subdivisions; the central inferotemporal area, with dorsal (CITd) and ventral (CITv) subdivisions; the anterior inferotemporal area, with dorsal (AITd) and ventral (AITv) subdivisions; the superior temporal polysensory area, with anterior (STPa) and posterior (STPp) subdivisions; the floor of the superior temporal sulcus (FST); and temporal areas F (TF) and H (TH). In the parietal lobe are found: the medial and superior temporal area, with dorsal (MSTd) and lateral (MSTl) subdivisions; the parieto-occipital area (PO); the posterior intraparietal area (PIP); the lateral intraparietal area (LIP); the ventral intraparietal area (VIP); the medial intraparietal area (MIP); the medial dorsal parietal area (MDP); the dorsal prelunate area (DP); and Brodmann's area 7a (7a). (Adapted from Swisher et al 2007,[149] Larsson & Heeger 2006,[88] Sereno & Tootell 2005[133] and Tootell et al 2003.[160]) (Extrastriate areas have been variously named by the order in which they were studied, by their position relative to sulci and gyri, and by the classical histological areas to which they most closely correspond. A single area can be named by all three methods, as for the Middle Temporal area (MT), known also as V5 and as Brodmann's area 37, and similar sounding names can refer to unrelated areas, as for V7, named in order of its discovery, and 7a, a completely unrelated visual area located in Brodmann's cytoarchitectonic area 7.)

Box 31.1 Organization of extrastriate visual areas

"Lower tier" extrastriate visual areas receive predominantly from primary visual cortex.

"Higher tier" extrastriate areas receive less from primary visual cortex, and more from both lower and higher tier areas and generally have more complex visual processing.

Lower Tier Areas
- V2
- V3
- MT
- PO

Higher Tier Ventral Stream Areas (Object Recognition)
- V4
- PIT
- CIT
- AIT
- TH
- TF

Higher Tier Dorsal Stream Areas (Motion and Location of Objects)
- MST
- LIP
- 7a
- VIP
- AIP

Connection patterns

Each visual area is interconnected with a subset of the other visual areas and subcortical visual structures, giving it a unique connectional signature. Neuroanatomic tracing methods in monkeys involve injecting tracers into one area either to label cells projecting to the injected area or to label axons projecting from the injected area to other areas. Tracing connectivity in human visual cortex was classically limited to postmortem staining for degenerating axons after cortical lesions.[30] Two new techniques suitable for studying connectivity in human extrastriate areas have recently been developed. Connections can be traced between cortical areas with an MRI technique that analyses the anisotropy of diffusion of water within oriented bundles of axons (diffusion tensor analysis; DTI).[58,104] Currently, DTI is not, and may never be, sensitive enough to demonstrate all connections among visual areas,[47,83] but it has proved useful in demonstrating pathological changes.[135,155] Connectivity can also be inferred by imaging a rise in activity of target areas following noninvasive activation of a projecting cortical area with transcranial magnetic stimulation (TCMS).[14,56,122] TCMS induces a brief electrical current inside cortical neurons by external application of a large, rapidly fluctuating magnetic field, and, with proper stimulation paradigms, can either activate or inactivate cortical areas.[35,102,154] It is important to note that connections are between retinotopically corresponding points in the visual field representation of the interconnected areas. Thus connectivity information has been used to infer receptive field maps. This is especially true for interhemispheric connections, which, especially for areas in the lower tier of the processing hierarchy, are concentrated in representations of the vertical meridian.

Functional specificity

The functional properties of neurons are often characteristic of each extrastriate area. A range of visual stimuli, selected to emphasize, for example, motion versus form and color information, can be used to selectively activate specific areas. These studies, like those detailed above on retinotopy, can be performed in humans using functional imaging techniques (fMRI, PET) or in monkeys using imaging or

areas and allows continuity of receptive field locations across the border. Visual field sign (mirror image or non-mirror image) can be computed from the local gradient of receptive field position, and can be used to infer borders of extrastriate areas.[132] Extrastriate maps also tend to show breaks in the retinotopy (split representations) along the horizontal meridian representation (Fig. 31.2). These breaks often form the borders of visual areas, allowing smooth transitions of receptive field locations from one area to the next, at the expense, in this case, of splitting the receptive field map within a single area. The splitting of maps within a single area and the continuity of receptive field positions at the borders of different areas make it difficult to identify areal boundaries on the basis of receptive field maps alone.

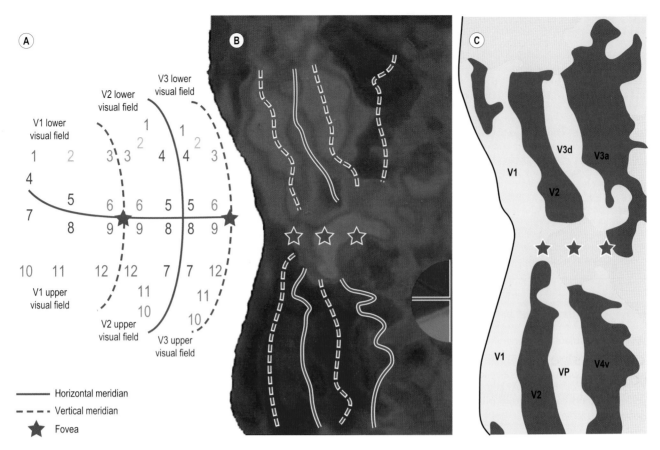

Figure 31.2 Maps of human visual areas, generated by functional imaging using visual stimuli that vary in retinotopic locations. (**A**) Schematic view of the retinotopy in V1, V2, and V3 (including upper and lower visual fields). The vertical (dashed line) and horizontal (solid line) meridians and foveal representations (stars) are shown. The numbers represent corresponding retinotopic locations in visual space in the three areas. The maps in (B) and (C) are flattened views of a right hemisphere, as if the cortex had been removed from the brainstem and flattened. Dorsal is to the top of each map and ventral is towards the bottom. In making the reconstruction, a cut (visible to the left in B and C) was introduced along the horizontal meridian representation of V1 to relieve the curvature of the cortical mantle and reduce distortion. (**B**) A map of cortical response plotted as a function of the polar angle of an annular stimulus (see text). The dashed and solid lines show the representations of the vertical and horizontal meridians, which form the boundaries of several cortical areas. The asterisks mark the location of the foveal representation. (**C**) A map of the local visual field sign, where yellow is a mirror image representation, like V1, and blue is a non-mirror image representation like V2. The locations and extent of V1, V2, V3d/VP, V3A, and V4v can be identified in this fashion. (From Tootell et al 1997.[156])

electrophysiological techniques. In vivo imaging in humans can reveal functional deficits due to cortical lesions caused by stroke, and localize the site of the damage to particular cortical areas. Repetitive TCMS can be used to reversibly inactivate extrastriate cortical areas in humans, allowing deficits in function to be associated with the inactivated area.[6,35]

Comparing visual areas in monkeys and humans

Related to the problem of identifying different areas in a single species is that of identifying homologous visual areas in different species. The functional organization of cortical areas has been studied extensively in monkeys, but it has been only with the recent advent of new non-invasive methods, that comparable studies have been done in humans. Most non-human data are from a genus of Old World monkeys (*Macaca* sp.). Although these species are more closely related to apes (including humans) than other primate groups (e.g. New World monkeys and prosimians), at least 25 million years of evolutionary history separate humans and macaque monkeys, with many new cortical

areas being added in humans.[81] Lower-level areas (V1, V2, V3d/VP, and MT) appear to be present in all primates, including humans, but there are important species differences in cortical organization for higher-level visual areas.[118,133,160] In addition to similarities in histology, retinotopy, connectivity, and physiology, relative position on the cortical surface and with respect to common sulcal and gyral landmarks is used to compare human visual areas with those of monkeys. This is not straightforward, however, as borders of cortical areas relative to sulcal landmarks vary among individuals, even more so in humans than in monkeys.[79] Moreover, there is no clear human analog of the macaque monkey's lunate sulcus, in which lie visual areas V2, V3, and V4.[5] Comparisons between human and monkey extrastriate cortex are also made more difficult by the fact that human V1 does not extend onto the external aspect of the hemisphere as much as in monkeys; accordingly, all of the human extrastriate areas appear displaced posteriorly relative to their positions in monkeys. The best positional clues to correspondence between monkey and human extrastriate areas are from each area's location relative to other cortical areas,[118] which does seem to be relatively well conserved.

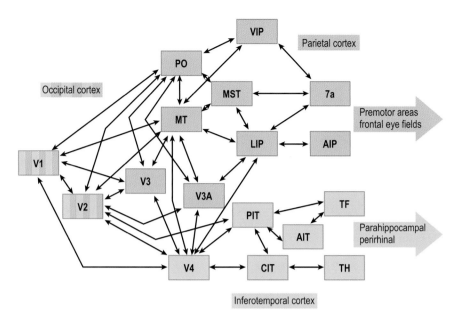

Figure 31.3 Summary diagram depicting the hierarchy of visual areas associated with the ventral and dorsal processing streams in monkey and the major interconnections between them. All connections are shown with bidirectional arrows to emphasize that each projection to a higher level visual area is matched by a feedback projection. Areas V1 and V2 are colored both red and green to depict their contributions to processing both "what" and "where" information for further, more segregated processing in extrastriate cortex. The dorsal stream areas, colored in red, process information about object location (where stream) and project to parietal cortical zones, eventually ending in premotor and frontal eye fields. The ventral stream areas, colored in green, process details of the form, color, shape of objects (what stream) and project to inferotemporal cortical zones and parahippocampal and perirhinal areas.

Processing streams in extrastriate cortex

As nearly all of the projections from the LGN terminate solely in striate cortex, the extrastriate areas receive their visual inputs through area 17, either directly or indirectly. (Some extrastriate areas do receive minor thalamic inputs from the lateral geniculate body[144,182] and the pulvinar, which may contribute to blindsight.[37]) The main extrastriate projections of V1 are to V2, V3, V4, and MT. As mentioned earlier (see Chapter 30), these projections arise from different subsets of neurons in V1 with different laminar and columnar distributions and likely different functional properties. Zeki[186] first pointed out that striate cortex thus acts to initially segregate and then recombine the visual inputs of the three retinogeniculocortical pathways, magnocellular, parvocellular, and koniocellular, such that different visual information is parceled out to different extrastriate cortical areas for further analysis. As described in Chapter 30 and shown in Figure 31.3, projections from V1 give rise to two main streams of extrastriate cortical processing, a dorsal stream devoted to spatial vision ("where" stream) and a ventral stream devoted to object vision ("what" stream).[45] As in V1, there is often a tangential organization of the inputs and outputs from different streams, forming a mosaic pattern when viewed from the cortical surface. This columnar and tangential organization of connectivity is often marked by differences in histochemical, particularly CO, staining.

V2

In nearly all mammals, the second visual area, V2, borders V1 laterally and dorsally, extending around V1 nearly completely in the case of primates.[127] V2 has a mirror-image visual field map, with the representation of the vertical meridian forming the boundary with V1 and a split representation of the horizontal meridian forming the boundary

Box 31.2 Area V2

- Part of both the "what" and "where" streams
- Composed of thick, thin, and pale CO stripe compartments
- Thin stripes receive from CO blobs of V1 and project to V4; unoriented and color-selective receptive fields predominate
- Thick stripes receive from the interblobs of V1 and project to MT; orientation and retinal disparity-sensitive receptive fields predominate
- Pale stripes receive from the interblob regions of V1 and project to V4; orientation-selective receptive fields predominate
- Local connections in V2 interconnect all CO stripe compartments
- In humans, V2 is split into four quadrants, with an upper and lower visual field representation in each of the two hemispheres
- Lesions in V2 give rise to selective visual loss depending on the CO stripe compartment affected

with V3d dorsally and VP ventrally.[136] V2 functions closely with V1 to organize information for output to higher visual areas such as V4 and MT (Box 31.2).[143]

The segregation of inputs, outputs, and physiological properties within area V2 is as dramatic as within V1 and, as in V1, is associated with a special pattern of CO staining. CO staining in sections of monkey V2 shows a series of parallel stripes of alternating dark and light staining that are oriented perpendicularly to the V1/V2 border.[77,158,] The V2 stripes, which are larger than the blobs in V1, extend across the whole width of the area, allowing its borders to be demarcated histologically using the CO stain.[115] The darkly staining stripes differ in thickness, with thin and thick stripes alternating, thus defining three compartments (thick, thin, and the pale interstripes) within V2. Sections through human V2 also reveal a striped organization when stained for cytochrome oxidase and other histological stains.[25,31,78,179] The

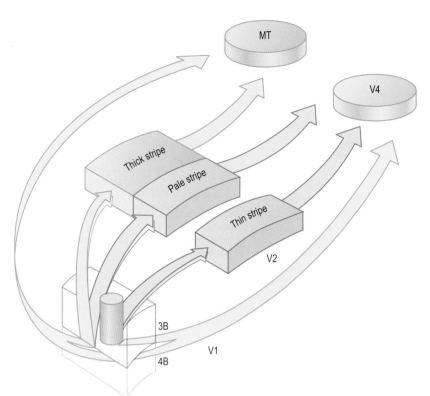

Figure 31.4 Schematic diagram depicting the parcellation of information from V1 into CO stripe compartments of V2 and onwards to MT and V4. MT, a dorsal stream area associated with motion processing, receives a direct projection from layer 4B of V1, from both blob and interblob columns. Interblob areas in layers 3A, 3B, and 4B of V1 project to the thick stripes of V2, and thence into MT. Interblob areas in layers 3A, 3B, and 4B of V1 also project to the pale stripes of V2, which then project into V4, a ventral stream area associated with color and form processing. V4 also receives from the blob compartments of V1 via the thin stripes of V2 as well as a direct projection from the blobs and interblobs of V1. Although not shown in the diagram, blob and interblob streams through V2 terminate in different columns within V4. (Adapted from Sincich & Horton, 2005.[143])

stripes are not as regular as in monkeys, and differentiation into thick and thin stripes is not clear based on CO staining. However, histologic stains known to differentiate thick and thin stripes in monkeys demonstrate a clearer stripe organization[76,159] with a periodicity of about 0.7–1 cm for a complete set of stripes, several times larger than in monkeys. As in monkeys, the extent of the stripes shows the extent of V2.[2,159] The presence of stripes in human V2 suggests that it may have the same intricate functional and connectional organization as shown in other primates (see below).

Each of the CO stripes has a different set of connections (Fig. 31.4). The thin stripes receive input from CO blob columns within V1, whereas the thick stripes and the pale-staining interstripes receive input from inter-blob columns.[141] In each case, the projection originates from magnocellular-dominated layer 4B (layer 3C of Hassler; see Chapter 30 for the equation of Hassler's and Brodmann's layering schemata), layer 3B, which combines inputs from magno- and parvocellular channels (plus koniocellular input in CO blob columns), and layer 3A, which has little or no direct input from LGN recipient layers. The corticocortically and subcortically projecting cells of V2 are also segregated into CO compartments. The thin stripes and interstripes project to V4,[44,137,187] a ventral stream area involved in form processing, whereas the thick stripes project to MT,[44,138] a dorsal stream area associated with motion processing. Cells projecting subcortically to the superior colliculus are also preferentially clustered in the thick stripes.[1] However, not all connections are segregated by stripe type. Feedback connections from higher-order visual areas tend not to be stripe specific and the local pathways within V2 interconnect stripes of all types.[89,98] Thus there is a substrate for binding different attributes of a visual stimulus into a coherent perceptual whole, even at relatively early levels of visual processing.

There are differences in the receptive field properties of cells in V2 that correlate with CO stripe type,[44,94,123,139] although the absolute degree of segregation of receptive field types within and between CO compartments is debated.[66,144] Studies reporting segregation of single cell receptive field properties generally agree that thin stripes contain proportionately more cells with receptive fields that are unoriented and color opponent, whereas the interstripes contain more oriented cells and fewer color-coded units. Clustered in the thick stripes are cells that are oriented and sensitive for retinal disparity, which is important for generating depth perception. The clustering of properties within stripe types is thus consistent with the physiology of the areas to which they project.

The segregation of selectivity for orientation, color, and disparity within compartments of monkey V2 is supported by imaging experiments.[28,82,95,126,164,178] There is one report of imaging experiments in human using high-resolution techniques to show that, within V2, stripes activated more strongly by slowly changing chromatic stimuli interleave with those activated by rapidly changing luminance stimuli, with a periodicity roughly equal to that of human V2 stripes.[92] Within an individual CO compartment, there is a columnar organization according to preferred orientation[95,163] (thick and pale stripes), disparity[28] (thick stripes), luminance[178] (thin stripes), or color[95,180] (thin stripes), similar to maps of orientation preference in V1 (see Chapter 30).

The visual topography of V2 is interesting in that the CO stripes are oriented perpendicular to lines of isopolarity in V2, but parallel to the isoeccentricity lines. If V2 had a fine-grained continuous map of the visual field, sub-modalities of visual function associated with each stripe type would be restricted to parts of the visual field mapped onto that stripe. In fact, the visual map in V2 consists of three distinct, interleaved maps.[125,140] Each region of visual space is represented three times, once for each stripe type. The regular progression of receptive field positions seen when traversing across a single stripe are interrupted by jumps in receptive field location at stripe borders. This ensures that visual space is represented completely in all of the stripe types of V2.

Areas of the dorsal stream

The dorsal stream of visual processing deals with the general problem of where objects are in visual space and how to manipulate them. The dorsal stream solves two main problems, processing motion, and decoding position and 3-D features from 2-D retinal images. Dorsal stream-bound outputs from V1 and V2 funnel through MT and V3 to reach areas of the parietal cortex.

MT/V5 and related areas

A group of visual areas exists in the superior temporal sulcus (STS) of the macaque monkey, with likely homologs in other primate species, including humans. These areas are recognized by their interconnections and by the fact that many of their neurons are sensitive to the direction of motion of moving stimuli.[118] The largest of these areas, V5, also known as MT (for Middle Temporal area), is probably the most easily identified visual area outside of V1. In addition to direction selectivity, MT is marked by heavy myelination and dark, patchy CO staining, a complete representation of the contralateral visual field, and strong direct input from V1 (Box 31.3). In humans, MT corresponds to an area near the occipitotemporal junction, in the dorsal part of the inferior temporal sulcus, showing the dense myelination and dark, patchy CO staining seen in other primates.[31,159] Motion selectivity in human MT/V5 has been shown with imaging and evoked potential studies comparing activation with moving versus stationary stimuli; such studies also show human MT to have higher contrast sensitivity than surrounding cortex and to respond poorly to contours demarcated solely by wavelength.[8,93,159,188]

MT receives projections from direction-selective cells in layers 4B (Hassler's 3C) and 6 of both blob and interblob columns in V1[103,142] and from the thick stripes of area V2.[53,138] As discussed previously (Chapter 30), layer 4B is associated with the magnocellular pathway, while V2 thick stripes receive a mixture of both magnocellular and parvocellular inputs. Nevertheless, blocking the magnocellular layers of

the LGN greatly affects the responses in MT, whereas blocking the parvocellular layers has much smaller effects on responses in MT, suggesting a dominance of the magnocellular pathway to MT.[100] MT thus serves as a gateway for magnocellular information into other areas important for motion processing in the STS, the parietal lobe, and the frontal eye fields. Not surprisingly, lesions of MT in monkeys lead to deficits in motion processing, assessed behaviorally. Monkeys with lesions in MT have elevated detection thresholds of a motion signal in the midst of masking motion noise.[112] The detection threshold for contrast of a visual stimulus was not affected by the MT lesion, showing that the MT lesion effect was selective for motion perception. Deficits for eye movements made to moving targets are also impaired by MT lesions (Box 31.4), whereas eye movements made to stationary targets are unaffected by the MT lesions,

---

Box 31.4 Cortical control of smooth pursuit and saccadic eye movements

- Smooth pursuit involves motion processing in dorsal stream areas, MT, and MST
- Other cortical areas such as the frontal eye fields (FEF) and supplementary eye fields (SEF) contribute to the control of smooth pursuit
- Asymmetric smooth pursuit function indicates focal lesions in the lateral occipitotemporal (MT/MST) or dorsomedial frontal (FEF and SEF) regions (lesions limited to V1 do not impair smooth pursuit)

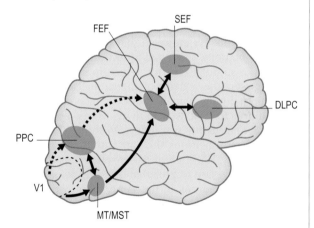

- Saccadic eye movements involve posterior parietal areas (PPC), the frontal eye fields (FEF), and the dorsolateral prefrontal cortex (DLPC)
- Parallel descending pathway from the FEF and SEF to the caudate nucleus, midbrain, and brainstem structures are also involved with saccadic eye movements
- Unilateral, focal cerebral lesions throughout frontal and parietal lobes may cause abnormalities of the in latency and accuracy of saccades
- Bilateral frontal or frontoparietal lesions may abolish saccades, sparing the quick phase of the vestibular ocular reflex (VOR), leading to acquired ocular motor apraxia
- Complete loss of all saccades, including reflexive saccades, quick phases, and smooth pursuit is called acquired supranuclear ocular motor apraxis and has been reported after small bilateral infarcts of both the posterior parietal cortex and FEF[110]

Box 31.3 Area V5/MT

- Extrastriate area of the dorsal (where) stream
- Displays motion, direction, and disparity sensitivity
- Strong direct inputs from layers 4B and 6 of V1
- Patchy CO staining and heavy myelination pattern
- Located in the occipitotemporal junction near the inferior temporal sulcus in humans
- Clinical condition of akinetopsia (loss of motion perception) develops after bilateral lesions to the lateral occipitotemporal cortex; unilateral lesions lead to a contralateral deficit that may go unnoticed by patients

suggesting that monkeys with MT lesions have more diffi-culty responding to moving than to stationary stimuli.[113] Human patients with lesions that include MT have difficulty with motion perception, a condition known as akinetop-sia.[185] One such patient complained that she could not cross the street because of her inability to judge the speed of a car, although she could identify the car itself without difficulty, showing the specificity of the deficit.[147,193] Temporary inacti-vation of MT with TCMS causes a similar akinetopsia.[13]

Several forms of columnar organization are found within MT. These include organizations for direction tuning, dispar-ity tuning, and selectivity for wide-field versus local motion contrast.[4,15,40] The tangential organization of direction selec-tivity in MT resembles the organization of orientation selec-tivity in V1 (see Chapter 30); regions of gradual change in preference are interrupted by sudden shifts of approximately 180 degrees. Approximately 180 degrees of axis of motion are represented in 400 to 500 μm of cortex, similar to the size of orientation columns in V1. This organization of direction-selective neurons made possible an elegant experi-ment using cortical microstimulation in awake behaving monkeys, showing how the stimulus selectivity of neurons in visual cortex underlies the perception of visual stimuli.[129] A microelectrode in MT in the middle of a cluster of neurons that shared a common preferred direction of motion was used to stimulate those neurons while the monkey performed a motion-discrimination task. Monkeys were required to discriminate between motion shown either in the direction preferred by the neurons or in the opposite direction, and noise was added to make the difficulty of discrimination near the monkeys' threshold for being able to perform the task. When electrical micro-stimulation was applied, the monkeys indicated that the motion was in the stimulated neurons' preferred direction more frequently than without stimulation. Thus a functional link was estab-lished between the activity of direction-selective neurons and perceptual judgments of motion direction.[32]

In addition to motion selectivity, MT is also important for stereoscopic depth perception. Within MT, patches of disparity-selective neurons alternate with neurons not selec-tive for disparity. Within a disparity-selective patch, there is a smooth progression of preferred disparity across the corti-cal surface and neurons in the same column have similar disparity tuning.[40] The presence of disparity-tuned neurons in MT is not surprising, given the disparity tuning found in the thick stripes of V2 and the connections between the thick stripes and MT. Microstimulation experiments on disparity-tuned columns in MT similar to the previously described experiment for the direction columns show that electrical stimulation of clusters of disparity-selective MT neurons bias perceptual judgments of depth in a way that is predictable from the disparity preference of neurons at the stimulation site.[39] Selectivity for stereoscopic depth has also been shown in human MT using fMRI.[21] Thus MT is also important in the perception of depth, as well as motion. Moreover, the two certainly interact, as image movement is an important cue for depth perception.[105]

V5 is strongly interconnected with three nearby motion-selective areas in the superior temporal sulcus, the lateral and dorsal divisions of the medial superior temporal areas (MSTl and MSTd), and area FST in the Fundus of the Supe-rior Temporal sulcus.[17,43,99] FST has a lower proportion of direction-selective cells than MT or MST, while MST has cells with larger receptive fields than in MT. MST can be divided into a lateral division preferring smaller stimuli and a dorsal division with cells responding to "optic flow" stimuli, i.e. rotation and expansion of objects, as would occur during movement through the environment. In humans, only a single visual area has been identified in this region, and it has been tentatively identified as MST, based on its large receptive fields and responses to optic flow stimuli.[12,50,176] An area dorsal to MT in humans has been identified as being specialized for processing images of the human body,[49] perhaps being involved in the analysis of biological motion.[9]

V3

V3 consists of a narrow belt of cortex immediately adjacent to V2, with the representation of the horizontal meridian forming its border with V2, and the vertical meridian repre-sentation forming its outer border.[189] The dorsal flank of V3 contains the lower visual field representation, and the ventral flank contains the upper field representation. Because the two parts of V3 appeared discontinuous across the represen-tation of the horizontal meridian, being interrupted by V4, and because some studies also suggested functional and con-nectional differences between the two parts, the ventral flank was originally referred to as the ventral posterior area (VP) as distinguished from dorsal V3 (V3d), or V3 proper.[26,52]

V3 has neurons selective for the basic stimulus parameters including orientation, color, direction, and binocular dispar-ity. In both dorsal and ventral halves, cells preferring similar orientations are grouped into columns.[3,175] As in MT, cells with similar preferred disparity are also arranged in columns.[3] About half the cells are strongly selective for direction of motion, more than any cortical area other than MT.[3,67] VP (ventral V3) has been reported to have more cells selective for color and fewer cells selective for direction.[52]

V3 receives projections from V1 and V2 and is strongly interconnected with dorsal stream areas MT (and surround-ing areas), and areas in the parietal lobe (PIP and VIP), although it also is connected with V4.[52] Although it was originally thought that only dorsal V3 received projections from V1,[52] connections between V1 and VP have been demonstrated.[96] The V1 projections originate mostly from magnocellular-dominated layer 4B; the connections from V2 are discontinuous, suggesting they arise from a subset of V2 CO compartments, but it is not yet known which subset.[52] V3 itself does not show a modular organization with CO staining, and it is not known how or if different classes of inputs from V1 and V2 are segregated within it.

V3 has been identified in human functional imaging studies, by its position relative to V2 and its similar visual field organization,[136,156] though it is not as narrow relative to V2 as it is in monkeys.[48] Functional imaging studies show similar properties compared to monkey V3. Human V3 is sensitive to direction of motion and stereoscopic depth, but not as strongly as V3a, MT, and parietal areas.[21,145,146,168] With regards to differences between the dorsal and ventral halves, these studies show little difference between V3d and VP in humans. Interestingly, VP did show, compared to dorsal V3, more activation during a color discrimination task relative to activation during a brightness discrimination task,[29] con-sistent with earlier results in monkey V3. However, the same study showed a similar, though not as extensive, difference in relative activation of upper and lower visual field representations of V2 and V1. The consensus is thus emerging that V3d and VP are two parts of a single area, and that the differences between V3d and VP in monkeys are either not large enough to warrant division into two areas, or are due to mistakenly including adjoining areas in the analysis of VP due to its extreme narrowness in macaque monkeys.[96,118,160]

V3A

V3A corresponds to a complete representation of the visual field located anterior to dorsal V3.[19,59] Although originally thought to function as an "accessory" to V3,[192] this area is not especially tied to V3 functionally. In monkeys, the neurons of V3A differ from those of neighboring dorsal V3 in being less selective to the speed and direction of stimulus motion.[64,174] In humans, however, V3A is more motion sensitive than V3.[156] Repetitive TCMS to V3A disrupts judgment of stimulus velocity as much as TCMS to MT.[101] In both humans and monkeys, V3A is activated by disparity stimuli[168] and responds better to achromatic stimuli than to color-defined stimuli,[34,93] placing it in the dorsal stream of visual processing. V3A receives ascending projections from V1, V2, and V3 and projects most strongly to other dorsal stream areas, including MT and MST[17] and to parietal area LIP, an important area for coordinating stereoscopic vision with prehensile hand movements.[108] Projections to temporal lobe areas exist, but are relatively minor.[46]

PO/V6

PO and V6 were defined independently in macaque monkeys, but were later concluded to correspond, at least in part, to the same area in the parietal-occipital sulcus.[62] V6 contains a complete representation of the visual field, with the representation of the horizontal meridian forming the posterior border, and the vertical meridian forming the anterior border.[61] Peripheral visual field representation is enhanced in V6, relative to other areas. V6 receives an extensive, topographically organized, projection from V1,[63] arising mostly from layer 4B, but with a component from layer 3B as well. As a dorsally located projection zone of V1, V6 resembles the medial area (M) and/or dorsal medial area (DM) described previously in New World monkeys[81] (see Chapter 30). V6 is connected with other dorsal stream areas, and provides input to parietal cortex, especially cortex immediately anterior to it, know variously as V6A or MIP.[63] An area claimed to correspond to V6 in humans was discovered in the posterior and superior part of the parieto-occipital sulcus, at the medial border of V3 and V3A.[124] Large field stimuli were needed to generate responses in this area, as, similar to V6 in macaque monkeys, the map emphasizes the periphery of the visual field.

Parietal lobe areas

The dorsal stream of visual processing continues in a constellation of areas in the parietal lobe, particularly in the intraparietal sulcus, which form the interface between perceiving visual stimuli and planning motor actions for controlling arm and eye movements in space.[68] In the macaque monkey, the most important of these are the lateral intraparietal area (LIP) and surrounding areas and area 7a. LIP receives input from MT and MST and, in turn, projects to area 7a.[90] Area 7a sends ascending projections to frontal cortex, including the frontal eye fields, which are important in programming eye movements (Box 31.4), as well as cingulate and parahippocampal cortices.[27] Area 7a is also interconnected with subcortical structures related to ocular and vestibulo-ocular function.[51]

Receptive fields of intraparietal areas are large, as much as 60 degrees across, and often extend across the midline. Different classes of intraparietal neurons have been identified with sensitivities for: depth relative to the fixation plane, linear movement in depth, rotation in depth, 3D orientation for object manipulation, and 3D axis or surface orientation.[128] Neurons in area 7a also receive extra-retinal oculomotor inputs, and their firing may be modulated by the horizontal and vertical position of the eyes in the orbits.[7] The combined eye position and visual signal in area 7a is sufficient to determine the spatial location of an object. Accordingly, lesions in this area lead to deficits in reaching for an object in three dimensions using visual guidance.[87]

As in monkeys, the intraparietal sulcus in humans is important for visuomotor tasks comprising target selections for arm and eye movements, object manipulation, and visuospatial attention. At least four intraparietal visual areas have been identified in humans,[134,149] although the homology of each of these areas with individual areas in macaque monkeys is unresolved, probably reflecting true species differences.[68,117] Lesions of parietal visual areas in humans lead to apraxia, neglect, and Balint's syndrome. Similarly, TCMS of human posterior parietal cortex degrades accuracy of saccades to remembered locations,[119, 18] disrupts visually guided reaching behavior,[171] and reduces detection and exploration of contralateral stimuli.[106,114,170]

Areas of the ventral stream

The ventral stream of visual processing deals with the general problem of object recognition. Ventral stream-bound outputs from V1 and V2 funnel through V4 to reach areas of the temporal lobe.

V4

V4 was first described in the macaque monkey as a target field of projections from V2 and V3 in the lunate sulcus (Box 31.5).[191] V4 is sandwiched between ventral V3 and MT, with the lower visual quadrant represented dorsally and the upper visual quadrant represented ventrally.[65] The upper visual quadrant apparently extends into the temporal lobe, although the border of V4 here is unclear. Retinotopic mapping with functional imaging suggests that the temporal extension, V4v, is actually a separate area distinct from V4d.[59] The borders of V4, especially with regard to the issue of extension into the temporal lobe, are not architectonically distinct.[91] A transitional area, V4t is more closely related to MT than V4, and separates V4 proper from MT.[43] In humans, as in monkeys, the boundaries of V4 have been difficult to resolve. It has been proposed that a dorsal V4d and a ventral V4v exist as in monkeys.[75] It has also been proposed that, instead of a single V4d, two complete maps of the visual

Box 31.5 Area V4

- Extrastriate area of the ventral (what) stream
- Displays orientation and color sensitivity as well as binocular disparity
- Strong inputs from blob and interblobs of V1 and thin stripes of V2 as well as V3
- Projects to PIT and CIT in the inferotemporal areas
- Located in the lingual and fusiform gyri in humans
- Bilateral lesions may cause complete cerebral achromatopsia (loss of color perception) in which patients perceive the world in different shades of gray; unilateral lesions may go unnoticed by patients

field, given the names LO1 and LO2, occupy the space between V3 and MT, and that V4v is a completely separate area.[88,149] Monkey V4d has also been proposed to be composed of rostral and caudal subdivisions,[148] although they may not correspond to the proposed human subdivisions.

V4 receives ascending projections from V1, V2, and V3.[169] The neurons of origin of the V1 projection are concentrated in the foveal part of V1 and arise from neurons in layers 3A and 3B, from both CO blobs and interblobs.[109,183] Inactivation studies show that V4 depends on both the parvocellular and magnocellular LGN channels.[57] From V2, projections arise from cells in thin stripes and interstripes, so neurons projecting from V2 to V4 interdigitate with those projecting to MT.[137,187] Although no mosaic organization is present in CO staining in V4, the projections from thin stripes and interstripes appear to be segregated within V4, suggesting that the separation of blob/interblob-related information may continue in V4, although not seen in the histology.[55,181] The size and arrangement of these columns are not well characterized, although they are larger than the modules in V2.

The main outputs of V4 are to areas PIT and CIT in the temporal lobe, involved in detailed form analysis for object vision. These temporal lobe projections also demonstrate segregation related to the CO-defined streams, as injection sites in V4 labeling thin stripes tended to receive more restricted feedback from temporal cortex than injection sites labeling interstripes.[181] Peripheral, but not central, field representations of V4 are connected with parietal areas (LIP, PIP, VIP) which could involve V4 in circuits serving spatial vision and spatial attention.[169]

V4 contributes to many aspects of visual processing, including form, color, visual attention, and stereopsis.[120] Cells in V4 show selectivity for stimulus hue, orientation, size, and binocular disparity.[41,42,131,190] Only a small proportion of cells in V4 show direction selectivity. Cells selective for color are concentrated in ventral V4,[190] and this area may best be considered an area separate from V4d. Cells in ventral V4, but not in V1 or V2, display shifts in preferred hue with changes in illumination. This is necessary for color constancy, the ability to perceive an object as being the same color when illuminated with light of different spectral composition.[86] Computation of color constancy, which entails taking the spectral information from an object and subtracting the spectral information from a large area surrounding it, is thought to be accomplished in V4 neurons by their large, spectrally sensitive, suppressive surrounds.[131]

Many V4 neurons show strong systematic tuning for the orientation and acuteness of angles and curves.[121] Unlike the responses of neurons in V2 to complex shapes,[80] the tuning in V4 can not be accounted for by selectivity for the orientation of a single edge of the angle, but is determined by the curvature at a specific boundary location within the shape. The computation of curvature selectivity in V4 from more basic orientation and spatial frequency-selective inputs from V1 and V2 is an intermediate step towards the complex shape selectivity found in inferotemporal cortex.[116]

Inferotemporal cortex

The areas of the inferotemporal cortex form the final stage of the ventral visual pathway for the recognition of visual objects.[60] The inferotemporal cortex receives projections from V2 and V4, and is interconnected with many multimodal brain sites including the hippocampal formation, prefrontal cortex, amygdala, and the striatum. Although inferotemporal cortex definitely contains several different visual areas, the borders of these areas are unclear, and several different naming schemes are used.[172] Nevertheless, it is generally agreed that IT cortex can be divided into more posterior areas that receive dense inputs from V4 and V2 (subdivisions of PIT; roughly corresponding to the classical cytoarchitectonic temporal occipital area, TEO), and more anterior and ventral areas (CIT and AIT; within cytoarchitectonic temporal area TE) that are interconnected strongly with PIT/TEO but only weakly with V4 and not at all with V2.[17,46,107,169]

PIT/TEO

Within PIT, retinotopic mapping shows a complete hemifield representation ventrally (PITv) and some evidence for a dorsal (PITd) subdivision.[16,20,59,173] Receptive fields are larger and shape selectivity is more complex for neurons in PIT/TEO than in V4, requiring particular arrangements of multiple contour fragments,[22] but PIT neurons are not selective for complex objects as are neurons in TE.

PIT receives most of its input from V4, with some inputs from V2; the V2 inputs arise from thin and interstripes, the same compartments that project to V4.[107] Individual columns in PIT receive projections from multiple, separate patches in V4. Adjacent columns in PIT receiving inputs from separate, interleaved sets of V4 patches, showing that the segregation of connectivity (and likely function) that is found in V2 continues through V4 and into PIT.[55] PIT projects to visual areas in TE.

Not surprisingly, given its inputs from areas known to contain color-responsive cells, PIT contains a class of cells that respond selectively to colors.[34,85,150] Color-selective cells in PIT resemble those in V1 and V4 in their tuning properties, responding selectively to a similarly narrow range of hues and saturation, but with larger receptive field sizes.[74] Divergent feedback projections from these cells to V4 is thought to create the large, color-selective surrounds of V4 neurons. Color-selective cells are found in separated patches several millimeters wide throughout a large extent of occipitotemporal cortex including ventral V4, PIT/TEO, and posterior portions of CIT/TE, suggesting that this region could best be described as comprising a single cortical area with a large-scale modular organization.[33,157] Monkeys with lesions of medial occipitotemporal cortex that include PIT and surrounding cortex show deficits in color discrimination.[24,36]

In humans, a color-selective area (known variously as V8, V0, and hV4) has been reported adjoining ventral V4, in the region of the lingual and fusiform gyri.[10,19,73] Although its receptive field organization and homology with visual areas in monkeys is still unclear,[133,177] it appears to correspond to the area damaged in clinical cases of cerebral achromatopsia.[153,184]

TE

With further progression along the ventral pathway into cytoarchitectonic area TE, retinotopic organization of visual areas is no longer apparent. Neurons have large receptive fields that nearly always include the fovea but also include most of the contralateral visual field, and many receptive fields extend bilaterally.[72] Neurons in TE show stimulus selectivity for images of complex objects or shapes such as hands and faces, and show translation invariance, i.e. a neuron responds similarly to its preferred stimuli no matter where they are placed in the receptive field. TE neurons may also show invariance for stimulus size and rotation.[151]

The images of complex objects that stimulate TE neurons can be systematically simplified to find the minimal critical features for activation.[151] A cell that responded to a picture of a water bottle, for example, also responded to a simpler stimulus composed of a vertical ellipse with a downward projection, but it did not respond to the ellipse alone. There is a columnar organization to the selectivity for these critical features in inferotemporal cortex, such that cells within the same cortical column respond to similar, although not identical, features. Furthermore, adjacent columns have partially overlapping feature selectivity, as, for example, profile and frontal views of faces.[152] Feature selectivity in inferotemporal cortex partly depends on experience, such that training monkeys to discriminate a specific set of shapes increases the proportion of TE cells responding to the shapes of the training set.[84]

In humans, as in monkeys, non-retinotopic, object-selective visual areas ventral and anterior to V4 can be recognized.[69,71] Due to the expansion of the human brain, however, these areas are in lateral occipital (LO) cortex rather than extending into the temporal lobe. Not containing an orderly map of the visual field, they are defined functionally as areas that show a greater fMRI response to images of objects compared to non-object control images.[69] Adaptation, the lessening of fMRI responses to repeated presentation of similar visual stimuli, was used to show that neurons in LO generalize responses for the same objects across differences in size, shape, and viewpoint,[70,130] i.e. LO shows constancy similarly to monkey inferotemporal cortex.

Face-selective responses were first shown for single neurons in monkey inferotemporal cortex.[72] Several concentrated patches of face-selective cells in this region have been identified by fMRI and verified by single unit recording within the identified patches.[165–167] Within a patch, 97% of the cells were face selective, showing, on average, an almost 20-fold larger response to faces than to non-face stimuli. In humans, there are at least two extrastriate areas that respond more to pictures of faces than objects, one in the middle fusiform gyrus (FFA), and one more posteriorly, in the

---

Box 31.6 Summary of major points

- Extrastriate visual cortex contains a constellation of richly interconnected areas comprising large portions of the occipital, parietal, and temporal lobes and contributes to a diverse array of visual information-processing tasks
- Extrastriate visual areas are identified by a combination of features including receptive field mapping, histology, patterns of connections with other brain regions, and functional specificity
- In both monkeys and humans a dorsal visual pathway dealing with object location extends through V2, V3, MT, PO/V6, and into parietal cortex
- In both monkeys and humans a ventral pathway dealing with object recognition extends through V2, V4, and into temporal cortex

---

inferior occipital cortex OFA, although it should be stressed that these areas are not homologous with face-preferring regions in monkeys.[71] Patients with lesions of these areas display difficulty in face processing, including prosopagnosia, a specific inability to recognize individual faces.[11,38,111]

Visual cortical areas in TE project to more ventral and rostral parts of the temporal lobe which integrate polymodal sensory inputs and are critically involved in the formation of memory, including visual memory.[60] Functional imaging in humans suggests an area on the parahippocampal gyrus (parahippocampal place areas; PPA) which is preferentially activated by images of places, i.e. buildings and landscapes, and appears to be involved in the encoding of places in memory (Box 31.6).[97]

References

1. Abel PA, O'Brien BJ, Lia B, Olavarria JF. Distribution of neurons projecting to the superior colliculus correlates with thick cytochrome oxidase stripes in macaque visual area 2. J Comp Neurol 1997; 377:313–323.
2. Adams DL, Sincich LC, Horton JC. Complete pattern of ocular dominance columns in human primary visual cortex. J Neurosci 2007; 27:10391–10403.
3. Adams DL, Zeki S. Functional organization of macaque v3 for stereoscopic depth. J Neurophysiol 2001; 86:2195–2203.
4. Albright TA, Desimone R, Gross CG. Columnar organization of directionally selective cells in visual area mt of the macaque. J Neurophysiol 1984; 51:16–31.
5. Allen JS, Bruss J, Damasio H. Looking for the lunate sulcus: A magnetic resonance imaging study in modern humans. Anat Rec A Discov Mol Cell Evol Biol 2006; 288:867–876.
6. Amassian VE, Cracco RQ, Maccabee PJ, Cracco JB, Rudell AP, Eberle L. Transcranial magnetic stimulation in study of the visual pathway. J Clin Neurophysiol 1998; 15:288–304.
7. Andersen RA, Essick GK, Siegel RM. Neurons of area 7 activated by both visual stimuli and oculomotor behavior. Exp Brain Res 1987; 67:316–322.
8. Anderson SJ, Holliday IE, Singh KD, Harding GF. Localization and functional analysis of human cortical area v5 using magneto-encephalography. Proc Biol Sci 1996; 263:423–431.
9. Astafiev SV, Stanley CM, Shulman GL, Corbetta M. Extrastriate body area in human occipital cortex responds to the performance of motor actions. Nat Neurosci 2004; 7:542–548.
10. Bartels A, Zeki S. The architecture of the colour centre in the human visual brain: New results and a review. Eur J Neurosci 2000; 12:172–193.
11. Barton JJ. Disorders of face perception and recognition. Neurol Clin 2003; 21:521–548.
12. Beauchamp MS, Yasar NE, Kishan N, Ro T. Human mst but not mt responds to tactile stimulation. J Neurosci 2007; 27:8261–8267.
13. Beckers G, Homberg V. Cerebral visual motion blindness: Transitory akinetopsia induced by transcranial magnetic stimulation of human area v5. Proc Biol Sci 1992; 249:173–178.
14. Bestmann S, Baudewig J, Siebner HR, Rothwell JC, Frahm J. Bold MRI responses to repetitive TMS over human dorsal premotor cortex. Neuroimage 2005; 28:22–29.
15. Born RT, Bradley DC. Structure and function of visual area mt. Annu Rev Neurosci 2005; 28:157–189.
16. Boussaoud D, Desimone R, Ungerleider LG. Visual topography of area teo in the macaque. J Comp Neurol 1991; 306:554–575.
17. Boussaoud D, Ungerleider LG, Desimone R. Pathways for motion analysis: Cortical connections of the medial superior temporal and fundus of the superior temporal visual areas in the macaque. J Comp Neurol 1990; 296:462–495.

18. Brandt SA, Ploner CJ, Meyer BU, Leistner S, Villringer A. Effects of repetitive transcranial magnetic stimulation over dorsolateral prefrontal and posterior parietal cortex on memory-guided saccades. Exp Brain Res 1998; 118:197–204.

19. Brewer AA, Liu J, Wade AR, Wandell BA. Visual field maps and stimulus selectivity in human ventral occipital cortex. Nat Neurosci 2005; 8:1102–1109.

20. Brewer AA, Press WA, Logothetis NK, Wandell BA. Visual areas in macaque cortex measured using functional magnetic resonance imaging. J Neurosci 2002; 22:10416–10426.

21. Bridge H, Parker AJ. Topographical representation of binocular depth in the human visual cortex using fMRI. J Vis 2007; 7:1511–1514.

22. Brincat SL, Connor CE. Underlying principles of visual shape selectivity in posterior inferotemporal cortex. Nat Neurosci 2004; 7:880–886.

23. Brodmann K. Localisation in the cerebral cortex: the principles of comparative localisation in the cerebral cortex based on cytoarchitectonics. New York: Springer, 2006.

24. Buckley MJ, Gaffan D, Murray EA. Functional double dissociation between two inferior temporal cortical areas: Perirhinal cortex versus middle temporal gyrus. J Neurophysiol 1997; 77:587–598.

25. Burkhalter A, Bernardo KL. Organization of corticocortical connections in human visual cortex. Proc Natl Acad Sci USA 1989; 86:1071–1075.

26. Burkhalter A, Felleman DJ, Newsome WT, Van Essen DC. Anatomical and physiological asymmetries related to visual areas v3 and vp in macaque extrastriate cortex. Vision Res 1986; 26:63–80.

27. Cavada C. The visual parietal areas in the macaque monkey: Current structural knowledge and ignorance. Neuroimage 2001; 14:S21–S26.

28. Chen G, Lu HD, Roe AW. A map for horizontal disparity in monkey V2. Neuron 2008; 58:442–450.

29. Claeys KG, Dupont P, Cornette L et al. Color discrimination involves ventral and dorsal stream visual areas. Cereb Cortex 2004; 14:803–822.

30. Clarke S. Association and intrinsic connections of human extrastriate visual cortex. Proc R Soc Lond B 1994; 257:87–92.

31. Clarke S. Modular organization of human extrastriate visual cortex: evidence from cytochrome oxidase pattern in normal and macular degeneration cases. Eur J Neurosci 1994; 6:725–736.

32. Cohen MR, Newsome WT. What electrical microstimulation has revealed about the neural basis of cognition. Curr Opin Neurobiol 2004; 14:169–177.

33. Conway BR, Moeller S, Tsao DY. Specialized color modules in macaque extrastriate cortex. Neuron 2007; 56:560–573.

34. Conway BR, Tsao DY. Color architecture in alert macaque cortex revealed by fMRI. Cereb Cortex 2006; 16:1604–1613.

35. Cowey A. The ferrier lecture 2004 what can transcranial magnetic stimulation tell us about how the brain works? Phil Trans R Soc Lond B Biol Sci 2005; 360:1185–1205.

36. Cowey A, Heywood CA, Irving-Bell L. The regional cortical basis of achromatopsia: A study on macaque monkeys and an achromatopsic patient. Eur J Neurosci 2001; 14:1555–1566.

37. Cowey A, Stoerig P. The neurobiology of blindsight. Trends NeuroSci 1991; 14:140–145.

38. Damasio AR, Damasio H, Van Hoesen GW. Prosopagnosia: anatomic basis and behavioral mechanisms. Neurology 1982; 32:331–341.

39. DeAngelis GC, Cumming BG, Newsome WT. Cortical area mt and the perception of stereoscopic depth. Nature 1998; 394:77–680.

40. DeAngelis GC, Newsome WT. Organization of disparity-selective neurons in macaque area mt. J Neurosci 1999; 19:1398–1415.

41. Desimone R, Schein SJ. Visual properties of neurons in area v4 of the macaque: sensitivity to stimulus form. J Neurophysiol 1987; 57:835–868.

42. Desimone R, Schein SJ, Moran J, Ungerleider LG. Contour, color and shape analysis beyond the striate cortex. Vision Res 1985; 25:441–452.

43. Desimone R, Ungerleider LG. Multiple visual areas in the caudal superior temporal sulcus of the macaque. J Comp Neurol 1986; 248:164–189.

44. DeYoe EA, Van Essen DC. Segregation of efferent connections and receptive field properties in visual area V2 of the macaque. Nature 1985; 317:5861.

45. DeYoe EA, Van Essen DC. Concurrent processing streams in monkey visual cortex. Trends Neurosci 1988; 11:219–226.

46. Distler C, Boussaoud D, Desimone R, Ungerleider LG. Cortical connections of inferior temporal area teo in macaque monkeys. J Comp Neurol 1993; 334:125–150.

47. Dougherty RF, Ben-Shachar M, Bammer R, Brewer AA, Wandell BA. Functional organization of human occipital-callosal fiber tracts. Proc Natl Acad Sci USA 2005; 102:7350–7355.

48. Dougherty RF, Koch VM, Brewer AA, Fischer B, Modersitzki J, Wandell BA. Visual field representations and locations of visual areas V1/2/3 in human visual cortex. J Vis 2003; 3:586–598.

49. Downing PE, Jiang Y, Shuman M, Kanwisher N. A cortical area selective for visual processing of the human body. Science 2001; 293:2470–2473.

50. Dukelow SP, DeSouza JF, Culham JC, van den Berg AV, Menon RS, Vilis T. Distinguishing subregions of the human mt+ complex using visual fields and pursuit eye movements. J Neurophysiol 2001; 86:1991–2000.

51. Faugier-Grimaud S, Ventre J. Anatomic connections of inferior parietal cortex (area 7) with subcortical structures related to vestibulo-ocular function in a monkey (Macaca fascicularis). J Comp Neurol 1989; 280:1–14.

52. Felleman DJ, Burkhalter A, Van Essen DC. Cortical connections of areas v3 and vp of macaque monkey extrastriate visual cortex. J Comp Neurol 1997; 379:21–47.

53. Felleman DJ, Kaas JH. Receptive-field properties of neurons in middle temporal visual area (MT) of owl monkeys. J Neurophysiol 1984; 52:488–513.

54. Felleman DJ, Van Essen DC. Distributed hierarchical processing in the primate cerebral cortex. Cerebr Cortex 1991; 1:1–47.

55. Felleman DJ, Xiao Y, McClendon E. Modular organization of occipito-temporal pathways: Cortical connections between visual area 4 and visual area 2 and posterior inferotemporal ventral area in macaque monkeys. J Neurosci 1997; 17:3185–3200.

56. Ferrarelli F, Haraldsson HM, Barnhart TE et al. A [17f]-fluoromethane pet/tms study of effective connectivity. Brain Res Bull 2004; 64:103–113.

57. Ferrera VP, Nealey TA, Maunsell JH. Responses in macaque visual area v4 following inactivation of the parvocellular and magnocellular lgn pathways. J Neurosci 1994; 14:2080–2088.

58. Ffytche DH, Catani M. Beyond localization: from hodology to function. Phil Trans R Soc Lond B Biol Sci 2005; 360:767–779.

59. Fize D, Vanduffel W, Nelissen K et al. The retinotopic organization of primate dorsal v4 and surrounding areas: A functional magnetic resonance imaging study in awake monkeys. J Neurosci 2003; 23:7395–7406.

60. Fujita I. The inferior temporal cortex: Architecture, computation, and representation. J Neurocytol 2002; 31:359–371.

61. Galletti C, Fattori P, Gamberini M, Kutz DF. The cortical visual area v6: Brain location and visual topography. Eur J Neurosci 1999; 11:3922–3936.

62. Galletti C, Gamberini M, Kutz DF, Baldinotti I, Fattori P. The relationship between v6 and po in macaque extrastriate cortex. Eur J Neurosci 2005; 21:959–970.

63. Galletti C, Gamberini M, Kutz DF, Fattori P, Luppino G, Matelli M. The cortical connections of area v6: An occipito-parietal network processing visual information. Eur J Neurosci 2001; 13:1572–1588.

64. Gaska JP, Jacobson LD, Pollen DA. Spatial and temporal frequency selectivity of neurons in visual cortical area v3a of the macaque monkey. Vision Res 1988; 28:1179–1191.

65. Gattass R, Sousa AP, Gross CG. Visuotopic organization and extent of v3 and v4 of the macaque. J Neurosci 1988; 8:1831–1845.

66. Gegenfurtner KR. Cortical mechanisms of colour vision. Nat Rev Neurosci 2003; 4:563–572.

67. Gegenfurtner KR, Kiper DC, Levitt JB. Functional properties of neurons in macaque area v3. J Neurophysiol 1997; 77:1906–1923.

68. Grefkes C, Fink GR. The functional organization of the intraparietal sulcus in humans and monkeys. J Anat 2005; 207:3–17.

69. Grill-Spector K, Kourtzi Z, Kanwisher N. The lateral occipital complex and its role in object recognition. Vision Res 2001; 41:1409–1422.

70. Grill-Spector K, Kushnir T, Edelman S, Avidan G, Itzchak Y, Malach R. Differential processing of objects under various viewing conditions in the human lateral occipital complex. Neuron 1999; 24:187–203.

71. Grill-Spector K, Malach R. The human visual cortex. Annu Rev Neurosci 2004; 27:649–677.

72. Gross CG. Single neuron studies of inferior temporal cortex. Neuropsychologia 2008; 46:841–852.

73. Hadjikhani N, Liu AK, Dale AM, Cavanagh P, Tootell RB. Retinotopy and color sensitivity in human visual cortical area v8. Nat Neurosci 1998; 1:235–241.

74. Hanazawa A, Komatsu H, Murakami I. Neural selectivity for hue and saturation of colour in the primary visual cortex of the monkey. Eur J Neurosci 2000; 12:1753–1763.

75. Hansen KA, Kay KN, Gallant JL. Topographic organization in and near human visual area v4. J Neurosci 2007; 27:11896–11911.

76. Hockfield S, Tootell RB, Zaremba S. Molecular differences among neurons reveal an organization of human visual cortex. Proc Natl Acad Sci USA 1990; 87:3027–3031.

77. Horton JC. Cytochrome oxidase patches: A new cytoarchitectonic feature of monkey visual cortex. Phil Trans R Soc Lond B 1984; 304:199–253.

78. Horton JC, Hedley-White ET. Mapping of cytochrome oxidase patches and ocular dominance columns in human visual cortex. Phil Trans R Soc Lond B 1984; 304:255–272.

79. Iaria G, Petrides M. Occipital sulci of the human brain: variability and probability maps. J Comp Neurol 2007; 501:243–259.

80. Ito M, Komatsu H. Representation of angles embedded within contour stimuli in area V2 of macaque monkeys. J Neurosci 2004; 24:3313–3324.

81. Kaas JH. The evolution of the visual system in primates. In: Chalupa L, Werner JS, eds. The visual neurosciences. Cambridge, MA: HUP, 2004.

82. Kaskan PM, Lu HD, Dillenburger BC, Kaas JH, Roe AW. The organization of orientation-selective, luminance-change and binocular-preference domains in the second (V2) and third (v3) visual areas of new world owl monkeys as revealed by intrinsic signal optical imaging. Cereb Cortex 2009; 19:1394–1407.

83. Kim M, Ducros M, Carlson T et al. Anatomical correlates of the functional organization in the human occipitotemporal cortex. Magn Reson Imaging 2006; 24:583–590.

84. Kobatake E, Wang G, Tanaka K. Effects of shape-discrimination training on the selectivity of inferotemporal cells in adult monkeys. J Neurophysiol 1998; 80:324–330.

85. Komatsu H, Ideura Y, Kaji S, Yamane S. Color selectivity of neurons in the inferior temporal cortex of the awake macaque monkey. J Neurosci 1992; 12:408–424.

86. Kusunoki M, Moutoussis K, Zeki S. Effect of background colors on the tuning of color-selective cells in monkey area v4. J Neurophysiol 2006; 95:3047–3059.

87. Lamotte RH, Acuna C. Defects in accuracy of reaching after removal of posterior parietal cortex in monkeys. Brain Res 1978; 139:309–326.

88. Larsson J, Heeger DJ. Two retinotopic visual areas in human lateral occipital cortex. J Neurosci 2006; 26:13128–13142.

89. Levitt JB, Yoshioka T, Lund JS. Intrinsic cortical connections in macaque visual area V2: Evidence for interaction between different functional streams. J Comp Neurol 1994; 342:551–570.

90. Lewis JW, Van Essen DC. Corticocortical connections of visual, sensorimotor, and multimodal processing areas in the parietal lobe of the macaque monkey. J Comp Neurol 2000; 428:112–137.

91. Lewis JW, Van Essen DC. Mapping of architectonic subdivisions in the macaque monkey, with emphasis on parieto-occipital cortex. J Comp Neurol 2000; 428:79–111.

92. Liu J. Neural circuits for chromatic and temporal signals in human visual cortex, Thesis, Applied Physics, Stanford University, 2006; p.152.

93. Liu J, Wandell BA. Specializations for chromatic and temporal signals in human visual cortex. J Neurosci 2005; 25:3459–3468.

94. Livingstone M, Hubel D. Segregation of form, colour, movement, and depth: Anatomy, physiology, and perception. Science 1988; 240:740–749.

95. Lu HD, Roe AW. Functional organization of color domains in V1 and V2 of macaque monkey revealed by optical imaging. Cereb Cortex 2008; 18:516–533.

96. Lyon DC, Kaas JH. Evidence for a modified v3 with dorsal and ventral halves in macaque monkeys. Neuron 2002; 33:453–461.

97. Maguire EA, Frith CD, Cipolotti L. Distinct neural systems for the encoding and recognition of topography and faces. Neuroimage 2001; 13:743–750.

98. Malach R, Tootell RBH, Malonek D. Relationship between orientation domains, cytochrome oxidase stripes, and intrinsic horizontal connections in squirrel monkey V2. Cereb Cortex 1994; 4:151–165.

99. Maunsell JH, van Essen DC. The connections of the middle temporal visual area (mt) and their relationship to a cortical hierarchy in the macaque monkey. J Neurosci 1983; 3:2563–2586.

100. Maunsell JHR, Neally TA, DePriest DD. Magnocellular and parvocellular contributions to responses in the middle temporal visual area (mt) of the macaque monkey. J Neurosci 1990; 10:3323–3334.

101. McKeefry DJ, Burton MP, Vakrou C, Barrett BT, Morland AB. Induced deficits in speed perception by transcranial magnetic stimulation of human cortical areas v5/mt+ and v3a. J Neurosci 2008; 28:6848–6857.

102. Merabet LB, Theoret H, Pascual-Leone A. Transcranial magnetic stimulation as an investigative tool in the study of visual function. Optom Vis Sci 2003; 80:356–368.

103. Movshon JA, Newsome WT. Visual response properties of striate cortical neurons projecting to area mt in macaque monkeys. J Neurosci 1996; 16:7733–7741.

104. Mukherjee P, Berman JI, Chung SW, Hess CP, Henry RG. Diffusion tensor mr imaging and fiber tractography: theoretic underpinnings. AJNR Am J Neuroradiol 2008; 29:632–641.

105. Nadler JW, Angelaki DE, DeAngelis GC. A neural representation of depth from motion parallax in macaque visual cortex. Nature 2008; 452:642–645.

106. Nager W, Wolters C, Munte TF, Johannes S. Transcranial magnetic stimulation to the parietal lobes reduces detection of contralateral somatosensory stimuli. Acta Neurol Scand 2004; 109:146–150.

107. Nakamura H, Gattass R, Desimone R, Ungerleider LG. The modular organization of projections from areas V1 and V2 to areas v4 and teo in macaques. J Neurosci 1993; 13:3681–3691.

108. Nakamura H, Kuroda T, Wakita M et al. From three-dimensional space vision to prehensile hand movements: the lateral intraparietal area links the area v3a and the anterior intraparietal area in macaques. J Neurosci 2001; 21:8174–8187.

109. Nakamura K, Colby CL. Visual, saccade-related, and cognitive activation of single neurons in monkey extrastriate area v3a. J Neurophysiol 2000; 84:677–692.

110. Newman NJ, Galetta S, Biousse V, Barton JJS, Fletcher WA, Jacobson DM. Disorders of ocular motility. AAN Continuum: Lifelong Learning in Neurology 2003; 9:79–148.

111. Newman NJ, Galetta S, Biousse V, Barton JJS, Fletcher WA, Jacobson DM. Disorders of vision. AAN Continuum: Lifelong Learning in Neurology 2003;9:11–78.

112. Newsome WT, Pare EB. A selective impairment of motion perception following lesions of the middle temporal visual area (mt). J Neurosci 1988; 8:2201–2211.

113. Newsome WT, Wurtz RH, Dursteler MR, Mikami A. Deficits in visual motion processing following ibotenic acid lesions of the middle temporal visual area of the macaque monkey. J Neurosci 1985; 5:825–840.

114. Nyffeler T, Cazzoli D, Wurtz P et al. Neglect-like visual exploration behaviour after theta burst transcranial magnetic stimulation of the right posterior parietal cortex. Eur J Neurosci 2008; 27:1809–1813.

115. Olavarria JF, Van Essen DC. The global pattern of cytochrome oxidase stripes in visual area V2 of the macaque monkey. Cereb Cortex 1997; 7:395–404.

116. Orban GA. Higher order visual processing in macaque extrastriate cortex. Physiol Rev 2008; 88:59–89.

117. Orban GA, Claeys K, Nelissen K et al. Mapping the parietal cortex of human and non-human primates. Neuropsychologia 2006; 44:2647–2667.

118. Orban GA, Van Essen D, Vanduffel W. Comparative mapping of higher visual areas in monkeys and humans. Trends Cogn Sci 2004; 8:315–324.

119. Oyachi H, Ohtsuka K. Transcranial magnetic stimulation of the posterior parietal cortex degrades accuracy of memory-guided saccades in humans. Invest Ophthalmol Vis Sci 1995; 36:1441–1449.

120. Pasupathy A. Neural basis of shape representation in the primate brain. Prog Brain Res 2006; 154:293–313.

121. Pasupathy A, Connor CE. Shape representation in area v4: Position-specific tuning for boundary conformation. J Neurophysiol 2001; 86:2505–2519.

122. Paus T, Jech R, Thompson CJ, Comeau R, Peters T, Evans AC. Transcranial magnetic stimulation during positron emission tomography: A new method for studying connectivity of the human cerebral cortex. J Neurosci 1997; 17:3178–3184.

123. Peterhans E, von der Heydt R. Functional organization of area V2 in the alert macaque. Eur J Neurosci 1993; 5:509–524.

124. Pitzalis S, Galletti C, Huang RS et al. Wide-field retinotopy defines human cortical visual area v6. J Neurosci 2006; 26:7962–7973.

125. Roe AW, Ts'o DY. Visual topography in primate V2: multiple representation across functional stripes. J Neurosci 1995; 15:3689–3715.

126. Roe AW, Ts'o DY. Specificity of color connectivity between primate V1 and V2. J Neurophysiol 1999; 82:2719–2730.

127. Rosa MG, Krubitzer LA. The evolution of visual cortex: Where is V2? Trends Neurosci 1999; 22:242–248.

128. Sakata H, Taira M, Kusunoki M, Murata A, Tanaka Y. The tins lecture. The parietal association cortex in depth perception and visual control of hand action. Trends Neurosci 1997; 20:350–357.

129. Salzman CD, Britten KH, Newsome WT. Cortical microstimulation influences perceptual judgements of motion direction. Nature 1990; 346:174–177.

130. Sawamura H, Georgieva S, Vogels R, Vanduffel W, Orban GA. Using functional magnetic resonance imaging to assess adaptation and size invariance of shape processing by humans and monkeys. J Neurosci 2005; 25:4294–4306.

131. Schein SJ, Desimone R. Spectral properties of v4 neurons in the macaque. J Neurosci 1990; 10:3369–3389.

132. Sereno MI, McDonald CT, Allman JM. Analysis of retinotopic maps in extrastriate cortex. Cereb Cortex 1994; 4:601–620.

133. Sereno MI, Tootell RB. From monkeys to humans: what do we now know about brain homologies? Curr Opin Neurobiol 2005; 15:135–144.

134. Shikata E, McNamara A, Sprenger A et al. Localization of human intraparietal areas aip, cip, and lip using surface orientation and saccadic eye movement tasks. Hum Brain Mapp 2008; 29:411–421.

135. Shimony JS, Burton H, Epstein AA, McLaren DG, Sun SW, Snyder AZ. Diffusion tensor imaging reveals white matter reorganization in early blind humans. Cereb Cortex 2006; 16:1653–1661.

136. Shipp S, Watson JD, Frackowiak RS, Zeki S. Retinotopic maps in human prestriate visual cortex: The demarcation of areas V2 and v3. Neuroimage 1995; 2:125–132.

137. Shipp S, Zeki S. Segregation of pathways leading from area V2 to areas v4 and v5 of macaque monkey visual cortex. Nature 1985; 315:322–325.

138. Shipp S, Zeki S. The organization of connections between areas v5 and V2 in macaque monkey visual cortex. Eur J Neurosci 1:333–353.

139. Shipp S, Zeki S. The functional organization of area V2, i: Specialization across stripes and layers. Vis Neurosci 2002; 19:187–210.

140. Shipp S, Zeki S. The functional organization of area V2, ii: The impact of stripes on visual topography. Vis Neurosci 2002; 19:211–231.

141. Sincich LC, Horton JC. Divided by cytochrome oxidase: A map of the projections from V1 to V2 in macaques. Science 2002; 295:1734–1737.

142. Sincich LC, Horton JC. Independent projection streams from macaque striate cortex to the second visual area and middle temporal area. J Neurosci 2003; 23:5684–5692.

143. Sincich LC, Horton JC. The circuitry of V1 and V2: Integration of color, form, and motion. Annu Rev Neurosci 2005; 28:303–326.

144. Sincich LC, Park KF, Wohlgemuth MJ, Horton JC. Bypassing V1: A direct geniculate input to area mt. Nat Neurosci 2004; 7:1123–1128.

145. Singh KD, Smith AT, Greenlee MW. Spatiotemporal frequency and direction sensitivities of human visual areas measured using fMRI. Neuroimage 2000; 12:550–564.

146. Smith AT, Greenlee MW, Singh KD, Kraemer FM, Hennig J. The processing of first- and second-order motion in human visual cortex assessed by functional magnetic resonance imaging (fmri). J Neurosci 1998; 18:3816–3830.

147. Stasheff SF, Barton JJS. Deficits in cortical visual function. Ophthalmol Clin North Am 2001; 14:217–242.

148. Stepniewska I, Collins CE, Kaas JH. Reappraisal of dl/v4 boundaries based on connectivity patterns of dorsolateral visual cortex in macaques. Cereb Cortex 2005; 15:809–822.

149. Swisher JD, Halko MA, Merabet LB, McMains SA, Somers DC. Visual topography of human intraparietal sulcus. J Neurosci 2007; 27:5326–5337.

150. Takechi H, Onoe H, Shizuno H et al. Mapping of cortical areas involved in color vision in non-human primates. Neurosci Lett 1997; 230:17–20.

151. Tanaka K. Mechanisms of visual object recognition studied in monkeys. Spat Vis 2000; 13:147–163.

152. Tanaka K. Columns for complex visual object features in the inferotemporal cortex: Clustering of cells with similar but slightly different stimulus selectivities. Cereb Cortex 2003; 13:90–99.

153. Tanaka Y, Kitahara K, Nakadomari S, Kumegawa K, Umahara T. Trans: Analysis with magnetic resonance imaging of lesions in cerebral achromatopsia. Nippon Ganka Gakkai Zasshi 2002; 106:154–161.

154. Terao Y, Ugawa Y. Studying higher cerebral functions by transcranial magnetic stimulation. Clin Neurophysiol 2006; 59(suppl):9–17.

155. Thomas C, Moya L, Avidan G et al. Reduction in white matter connectivity, revealed by diffusion tensor imaging, may account for age-related changes in face perception. J Cogn Neurosci 2008; 20:268–284.

156. Tootell RB, Mendola JD, Hadjikhani NK et al. Functional analysis of v3a and related areas in human visual cortex. J Neurosci 1997; 17:7060–7078.

157. Tootell RB, Nelissen K, Vanduffel W, Orban GA. Search for color 'center(s)' in macaque visual cortex. Cereb Cortex 14:353–363.

158. Tootell RB, Silverman MS, De Valois RL, Jacobs GH. Functional organization of the second cortical visual areas in primates. Science 1983; 220:737–739.

159. Tootell RB, Taylor JB. Anatomical evidence for mt and additional cortical visual areas in humans. Cereb Cortex 1995; 5:39–55.

160. Tootell RB, Tsao D, Vanduffel W. Neuroimaging weighs in: humans meet macaques in "primate" visual cortex. J Neurosci 2003; 23:3981–3989.

161. Tootell RBH, Hamilton SL, Silverman MS. Topography of cytochrome oxidase activity in owl monkey cortex. J Neurosci 1985; 5:2786–2800.

162. Triarhou LC. A proposed number system for the 107 cortical areas of economo and koskinas, and brodmann area correlations. Stereotact Funct Neurosurg 2007; 85:204–215.

163. Ts'o DY, Frostig RD, Lieke EE, Grinvald A. Functional organization of primate visual cortex revealed by high resolution optical imaging. Science 1990; 249:417–420.

164. Ts'o DY, Roe AW, Gilbert CD. A hierarchy of the functional organization for color, form and disparity in primate visual area V2. Vision Res 2001; 41:1333–1349.

165. Tsao DY, Freiwald WA, Knutsen TA, Mandeville JB, Tootell RB. Faces and objects in macaque cerebral cortex. Nat Neurosci 2003; 6:989–995.

166. Tsao DY, Freiwald WA, Tootell RB, Livingstone MS. A cortical region consisting entirely of face-selective cells. Science 2006; 311:670–674.

167. Tsao DY, Livingstone MS. Mechanisms of face perception. Annu Rev Neurosci 2008; 31:411–437.

168. Tsao DY, Vanduffel W, Sasaki Y et al. Stereopsis activates v3a and caudal intraparietal areas in macaques and humans. Neuron 2003; 39:555–568.

169. Ungerleider LG, Galkin TW, Desimone R, Gattass R. Cortical connections of area v4 in the macaque. Cereb Cortex 2008; 18:477–499.

170. Valero-Cabre A, Rushmore RJ, Payne BR. Low frequency transcranial magnetic stimulation on the posterior parietal cortex induces visuotopically specific neglect-like syndrome. Exp Brain Res 2006; 172:14–21.

171. Van Donkelaar P, Lee JH, Drew AS. Transcranial magnetic stimulation disrupts eye-hand interactions in the posterior parietal cortex. J Neurophysiol 2000; 84:1677–1680.

172. Van Essen DC. Organization of visual areas in macaque and human cerebral cortex. In: Chalupa L, Werner JS, eds. The visual neurosciences. Cambridge, MA: HUP, 2004.

173. Van Essen DC, Felleman DJ, DeYoe EA, Olavarria J, Knierim J. Modular and hierarchical organization of extrastriate visual cortex in the macaque monkey. Cold Spring Harbor Symp Quant Biol 1990; 55:679–696.

174. Vanduffel W, Fize D, Mandeville JB et al. Visual motion processing investigated using contrast agent-enhanced fmri in awake behaving monkeys. Neuron 2001; 32:565–577.

175. Vanduffel W, Tootell RB, Schoups AA, Orban GA. The organization of orientation selectivity throughout macaque visual cortex. Cereb Cortex 2002; 12:647–662.

176. Wall MB, Lingnau A, Ashida H, Smith AT. Selective visual responses to expansion and rotation in the human mt complex revealed by functional magnetic resonance imaging adaptation. Eur J Neurosci 2008; 27:2747–2757.

177. Wandell BA, Dumoulin SO, Brewer AA. Visual field maps in human cortex. Neuron 2007; 56:366–383.

178. Wang Y, Xiao Y, Felleman DJ. V2 thin stripes contain spatially organized representations of achromatic luminance change. Cereb Cortex 2007; 17:116–129.

179. Wong-Riley MT, Hevner RF, Cutlan R et al. Cytochrome oxidase in the human visual cortex: Distribution in the developing and the adult brain. Vis Neurosci 1993; 10:41–58.

180. Xiao Y, Wang Y, Felleman DJ. A spatially organized representation of colour in macaque cortical area V2. Nature 2003; 421:535–539.

181. Xiao Y, Zych A, Felleman DJ. Segregation and convergence of functionally defined V2 thin stripe and interstripe compartment projections to area v4 of macaques. Cereb Cortex 1999; 9:792–804.

182. Yoshida K, Benevento LA. The projection from the dorsal lateral geniculate nucleus of the thalamus to extrastriate visual association cortex in the macaque monkey. Neurosci Lett 1981; 22:103–108.

183. Yukie M, Eiichi I. Laminar origin of direct projection from area V1 to v4 in the rhesus monkey. Brain Res 1985; 346:383–386.

184. Zeki S. A century of cerebral achromatopsia. Brain 1990; 113:1721–1777.

185. Zeki S. Cerebral akinetopsia (visual motion blindness). A review. Brain 1991; 114:811–824.

186. Zeki S, Shipp S. The functional logic of cortical connections. Nature 1988; 335:311–317.

187. Zeki S, Shipp S. Modular connections between areas V2 and v4 of macaque monkey visual cortex. Eur J Neurosci 1989; 1:494–506.

188. Zeki S, Watson JD, Lueck CJ, Friston KJ, Kennard C, Frackowiak RS. A direct demonstration of functional specialization in human visual cortex. J Neurosci 1991; 11:641–649.

189. Zeki SM. The secondary visual areas of the monkey. Brain Res 1969: 13:197–226.

190. Zeki SM. Colour coding in rhesus monkey prestriate cortex. Brain Res 1973; 53:422–427.

191. Zeki SM. The cortical projections of foveal striate cortex in the rhesus monkey. J Physiol 1978; 277:227–244.

192. Zeki SM. The third visual complex of rhesus monkey prestriate cortex. J Physiol 1978; 277:245–272.

193. Zihl J, von Cramon D, Mai N. Selective disturbance of movement vision after bilateral brain damage. Brain 1983; 106:31–340.

Early Processing of Spatial Form

Robert F. Hess

Introduction

Vision is our most developed sense and unsurprisingly a substantial amount of brain processing is devoted to it, with over half the primate brain involved in vision-related processing.[1] A first step in understanding the nature of this processing involves an appreciation of the capabilities and specialization that define human vision. Such an understanding of our overall visual strengths and weaknesses provides a framework within which the emerging neurophysiology and neuroanatomy of different parts of the pathway can be best understood.

Early visual processing of spatial form has to contend with the fact that information that first impinges on receptors in the eye must be transmitted to the main processing sites in the brain. There are transmission bandwidth considerations to ensure the optic nerve/optic track remains of manageable size. What this means is that of all the information contained in a retinal image, only a fraction of it can be efficiently transmitted and processed; hard choices have to be made. This leads us into the concept of sensory filtering and information compression. The fact that individual cells have limited dynamic ranges also leads to a consideration of distributed parallel processing. These notions, namely filtering, compression, and parallel processing, are important principles in sensory analysis in general and visual analysis in particular.

Foveal window of visibility

It has only been relatively recently that quantitative methods have been developed to help quantify human vision. These developed from the realization of the importance of contrast and object size (or spatial frequency). That contrast is fundamental to the visibility of objects had been known for a long time. Bouguer[2] developed the first quantitative measurements of contrast sensitivity, using the visibility of the shadow of a rod produced by a candle at a specified distance away. That the size of objects (more correctly their retinal image size) also determines their visibility has also been appreciated for a long time, being the basis of acuity testing in the last century. The realization that the relationship

between these two variables could provide a more complete description of human vision was a logical consequence of the emerging work on optical transfer functions and was accelerated by the need for calibrated photographic surveillance methods during World War II.[3] Amid great controversy, particularly from certain optical physicists of the time, Campbell and co-workers[4,5] used this approach to great effect to provide a quantitative description of the capability of the human visual system, which in turn provided a starting point for considerations of optical quality of the human eye and its neural limitations. This function, often referred to as the spatial contrast sensitivity function, provided the first quantitative description of our "window of visibility" and makes explicit the information that our visual system processes best.

Figure 32.1 displays a sinusoidal grating stimulus that was first developed by Fergus Campbell and John Robson at Cambridge in which the spatial frequency (number of cycles subtended in one degree at the eye) is modulated along the abscissa (low spatial frequencies on the left, high spatial frequencies on the right) and contrast is modulated along the ordinate (low contrasts at the top, high contrasts at the bottom). The reader can observe an inverted U-shaped region in which the stripes are visible – this is a demonstration of our window of visibility.

In Figure 32.2 laboratory measurement of the contrast sensitivity function is displayed for a normal observer; its overall shape mirrors that seen in Figure 32.1. Here, contrast sensitivity, the reciprocal of the contrast needed for threshold detection, is plotted against the spatial frequency of a luminance-defined sinusoidal grating (filled circles), in cycles per degree of angle subtended at the eye. Our window of visibility is defined by an inverted U-shaped function that shows we are best at detecting objects subtending about a half to a third of a degree (approximately half the width of one's finger nail at arms length). Our sensitivity for detecting smaller objects (i.e. of higher spatial frequency) progressively declines, as it also does for detecting objects larger (i.e. of lower spatial frequency) than one degree. Information in the very low contrast range, particularly if it is of very low or very high spatial frequency, is not processed at all by our visual system; human vision is specialized in the intermediate spatial frequency range and within this limited range our contrast sensitivity is good to one part in 500 (i.e. 0.2

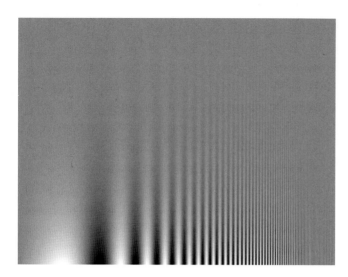

Figure 32.1 A demonstration of the window of visibility. A sinusoidal grating is increased in spatial frequency, going from left to right along the abscissa and its contrast is reduced, going from bottom to top along the ordinate. The inverted U-shaped region over which the stripes are seen is a facet of our visual systems (developed by Campbell & Robson).

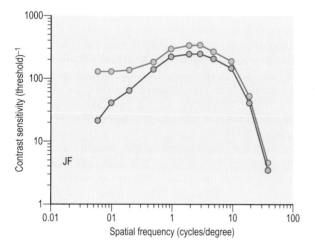

Figure 32.2 The contrast sensitivity function for sinusoidal (blue circles) and squarewave (red circles) gratings. Contrast sensitivity (reciprocal of the contrast threshold) is plotted against the spatial frequency for vertically oriented gratings. At each spatial frequency, five cycles of the grating were present and the length of the bars was greater than the critical length.[66] The exceptions were the data at 30 and 40 c/d where 120 cycles were used. When measured under comparable conditions of areal summation (length and height), the contrast sensitivity function has a broader peak centered at around 3 c/d and a low spatial frequency decline beginning at around 0.5 c/d (data from Hess & Howell[10]).

percent) under monocular conditions. Other animals have visual capabilities adapted to match their particular needs; the cat is better than the human at detecting low spatial frequencies and the falcon is better than the human at detecting high spatial frequencies. These benefits, however, come with associated costs. The cat is much worse than us at detecting high spatial frequencies and the falcon much worse than us at detecting low spatial frequencies. In other words the size of the window of visibility is approximately the same for different animals but shifts along the abscissa (i.e. the spatial frequency axis) to best suit the needs of the

animal. For the cat, this means things within pouncing distance (resulting in a shift of the function to lower spatial frequencies) and for the falcon, for detecting ground prey when hovering high in the sky (resulting in a shift of the function to higher spatial frequencies).

What limits our contrast sensitivity?

Why are we poor at detecting objects of very high and very low spatial frequency? In terms of the former, an obvious possibility is that the retinal image may not contain this information in the first place due to optical (i.e. cornea, lens) losses. Campbell & Green,[4] using a laser interferometric technique to by-pass the optics, showed that the fall-off of contrast sensitivity at high spatial frequencies only had a small (about one-third) optical component. They argued that two-thirds of the sensitivity loss at high spatial frequencies was neural in nature, due to limitations at the retina/brain (Box 32.1). More recently it has been suggested that the quantal fluctuations in light itself may account for the loss at high spatial frequency.[6] Although not yet resolved, in light of the discussion below in the section on luminance effects, the original suggestion that this fall-off is neural in nature seems more likely. The sensitivity loss at low spatial frequencies only occurs for sinusoidal stimuli (i.e. as opposed to squared-edged stimuli, as shown in Figure 32.1) and many different explanations have been put forward including, relative image stabilization,[7] a consequence of reduced cycles[8] or lateral inhibition.[9] None of these hold up to scrutiny; providing a comparable temporal stimulation and having the same number of spatial cycles does not eliminate the low spatial frequency decline in sensitivity.[10] The lateral inhibition explanation can also be rejected for reasons that are outlined in the section below on the relation to single cells. Its explanation has proved elusive.

What is the relationship between the contrast sensitivity function and the response of single cortical cells?

Neurons at various levels of the visual pathway exhibit spatially overlapping excitatory and inhibitory receptive field properties[11] that endow them with size or spatial frequency dependence,[12] as illustrated in Figure 32.5. This size selectivity becomes greater as one progresses from retina to cortex owing to the increased strength of the antagonistic surround. These neurons also exhibit contrast thresholds, quasi-linear contrast response regions and contrast saturation responses.[13] Indeed, efforts to outline the optimum spatial frequency response of a sample of contrast-sensitive neurons in the monkey have provided a distribution similar to that of the human behavioral contrast sensitivity function.[14] This can be seen in Figure 32.3 from the histogram of the distribution of cellular contrast sensitivities in monkey V1 relative to human behavioral sensitivities. The contrast sensitivity function represents the envelope of all the contrast-responding cells with the most sensitive ones defining the threshold limit.

Visual cortical neurons have sufficiently strong surrounds to their classical receptive fields to make them selective over a limited part of the spatial frequency range;[15–17] an example is shown in Figure 32.4. Thus different populations

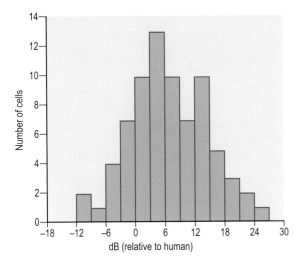

Figure 32.3 Comparison of single cell and behavior. Histogram of the distribution of a sample of 75 cortical V1 cells in the monkey where each threshold is given relative to the human psychophysical threshold (from Hawken & Parker[14]).

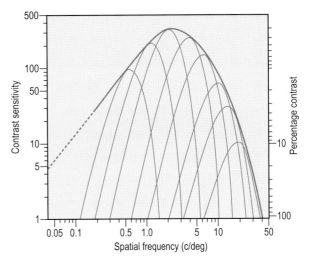

Figure 32.5 Illustration of the relationship of the spatial frequency selectivity of single cortical cells, measured neurophysiologically, to that of the overall contrast sensitivity function of the visual system, measured psychophysically.

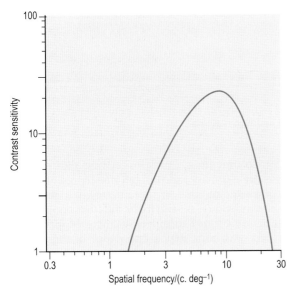

Figure 32.4 Typical spatial tuning function of a cortical cell in area V1. Cellular response in spikes/sec is plotted against the spatial frequency of an optimally oriented grating stimulus (data from Hawken & Parker[67]).

of cortical cells underlie the contrast sensitivity curve at different locations along the spatial frequency axis; the overall contrast sensitivity curve being the envelope of the sensitivity of the most sensitive cortical cells. This is diagrammatically illustrated in Figure 32.5.

Psychophysical supporting evidence comes from the finding that adaptation or prolonged viewing of stimuli of one particular spatial frequency desensitizes responses for only stimuli of a similar spatial frequency, also suggesting that the overall curve is composed of a number of more discrete "channels".[8,18] At one time it was thought that the low spatial frequency fall-off in sensitivity was a reflection of the surround inhibition of the largest (or lowest spatial frequency-tuned) mechanism whose peak was located around 1 c/deg.[9] There is now ample evidence to doubt this;

individual spatial frequency selective mechanisms have been reported down to as low as 0.2 c/deg.[10,19,20] Thus individual cortical mechanisms extend across the full extent of the visible spatial frequency range. The reason why the lower spatial frequency mechanisms are less sensitive is unknown but one possibility is that it may be due to a purely retinal cause; larger receptive fields must inevitably contain more inactive rods (contributing no signal but possibly contributing noise) under photopic conditions.

The contribution of M & P pathways to contrast sensitivity

In the retina there are morphologically and functionally different populations of ganglion cells[21] two of which, the parasol (magno; 10%) and midget (parvo; 80%) retinal ganglion cells send their afferents along the retino-geniculate pathway. These projections are kept separate in the geniculate terminating in the magno- and parvo-cellular layers respectively (so called M-cells and P-cells). The afferents of these two geniculate cell types are sent to different layers of 4C of the visual cortex (i.e. 4C alpha and 4C beta) and ultimately make a dominant (though not exclusive) contribution to the dorsal and ventral extrastriate streams respectively.

Do these two parallel systems carry the same or different contrast sensitivity information?

The combination of single cell recording[22] and lesion studies[23–25] has revealed that these two systems carry different though overlapping visual information to the cortex. The M-cell information is biased to lower spatial and midtemporal frequencies. P-cell information is biased to midhigher spatial and lower temporal frequencies. A particular subdivision of the M-cell system, the Y-like M cell, exhibits properties that suggest it may have a special role[26] in perception. The results shown in Figure 32.6 are from the lesion studies of Merigan and colleagues[23–25] and represent how the spatial and temporal contrast sensitivity function is affected

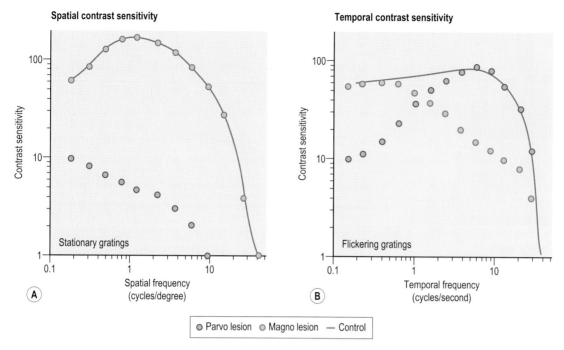

Figure 32.6 The relative contribution of M- and P- systems to spatial and temporal contrast sensitivity from the LGN lesion work of Merigan and colleagues.[23–25]

by lesions to parvo- and magno-cellular regions of the LGN of the monkey. Parvo-cellular lesions have a large effect on spatial contrast sensitivity at all spatial frequencies for static stimuli. Magno-cellular lesions reduce temporal contrast sensitivity in the mid range for spatial stimuli of low spatial frequency. Parvo-cellular lesions also produce a loss of color vision since red/green chromatic information is conveyed by this system.

The contribution of different cortical areas to contrast sensitivity

A number of different regions of the visual cortex contain a retinotopic map and are therefore given the status of being "distinct" visual areas.[27] V1 receives the majority of the input from the geniculate although there is evidence that V2[28] and MT[29] also receive a direct input in primates, although its role may be more modulatory in nature.[30] Although V1 contains cells with both small and large receptive field sizes at each eccentricity (approximately a factor of 4 in receptive field size at any one eccentricity from the data of Hubel & Wiesel[31]), cells with smaller receptive fields progressively drop out as eccentricity increases, resulting in a shift of the mean receptive field size with eccentricity to larger values.[31] Other visual areas have cells with larger receptive field size and consequently diminished retinotopy. Area V2 has cells that are tuned to lower spatial frequencies than those found in V1,[32] however there is no compelling evidence from lesion studies that cells other than in V1 make a major contribution to spatial contrast sensitivity.[33]

The effect of disease on contrast sensitivity

Having a more quantitative approach for specifying the quality of normal vision also provides a more sensitive

Box 32.1

A contrast sensitivity loss at high spatial frequencies may be optical or neural in nature.

method for documenting visual loss. Given that contrast sensitivity is not a unitary function but one representing the envelope of a large number of more narrowly tuned functions (i.e. the individual sensitivity of different populations of cortical neurons tuned to different ranges of spatial frequency), there was hope that a knowledge of the relative susceptibility of these neural populations might lead to a better understanding of a number of retinal and cortical diseases affecting the visual system. This expectation has been only partially realized because although the specification of the contrast sensitivity function provided more information than before it also has inherent limitations. Without an understanding of these limitations any simple interpretation of contrast sensitivity in disease becomes hazardous. The strengths and weaknesses of the approach can be best appreciated by one or two examples, one involving a developmental condition called amblyopia, the other, an acquired condition called optic neuritis. The reader can imagine how comparable issues could concern the interpretation of anomalous contrast sensitivity in related conditions, for example glaucoma and age-related maculopathy.

One of the first conditions to be studied with the contrast sensitivity approach was amblyopia, a common developmental disorder in which the vision in one eye is compromised in early life. It was shown (Fig. 32.7) that high spatial frequencies were selectively affected.[34] Although in some strabismic amblyopes it was subsequently shown[35] that additional and unrelated losses can occur at low spatial frequencies, the picture today is not substantially different from that originally outlined by Gstalder & Green[34] over 30 years ago. Indeed its application to the study of amblyopia has highlighted some potential pitfalls of using this approach. The first concerns the fact that it is limited to threshold. Amblyopes need much more contrast in order to

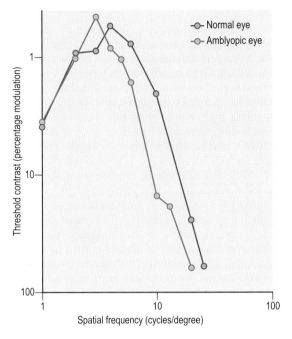

Figure 32.7 The first contrast sensitivity function measured in amblyopia. The reciprocal of the contrast threshold is plotted against the stimulus spatial frequency. The higher the spatial frequency, the greater the loss of sensitivity of the amblyopic (red symbols) compared with the fellow fixing eye (blue symbols) (data from Gastalder & Green[34]).

detect higher spatial frequencies; however at and above their raised thresholds they perceive contrast normally with their amblyopic eye.[36] This is illustrated in Figure 32.8 in which results are shown for contrast matches made between the eyes of the two most common types of amblyopia, strabismic (A) (i.e. turned eye) and (B) non-strabismic (i.e. anisometropia without a turned eye). Thresholds are raised (unfilled symbols) but as the contrast is raised to suprathreshold levels, it is perceived normally (solid diagonal line indicates matches that are veridical). The difference between the raised thresholds and normal perception of suprathreshold contrast is abrupt in the case of strabismic amblyopes (Fig. 32.8A) but more gradual in the case of non-strabismic, anisometropic amblyopes (Fig. 32.8B). In other words, thresholds are not representative of contrast vision above threshold; the contrast sensitivity function is highly nonlinear when it comes to contrast.

In optic neuritis, which commonly occurs as part of multiple sclerosis, similar high spatial frequency loss is observed in the contrast sensitivity function but it is not just limited to threshold. Sufferers of optic neuritis who need, for example, a factor of 10 more contrast to detect a given spatial frequency will perceive that spatial frequency to be a factor of 10 reduced in contrast, no matter what its absolute contrast is.[37] With respect to the results of Figure 32.8 for strabismic and non-strabismic amblyopia, the equivalent contrast-matching function for a person with optic neuritis would run parallel but below the normal matching line (solid diagonal). Amblyopia and optic neuritis can exhibit similar threshold losses but their suprathreshold contrast losses are very different.

Another limitation is that contrast sensitivity may be normal and yet spatial vision may be severely degraded.

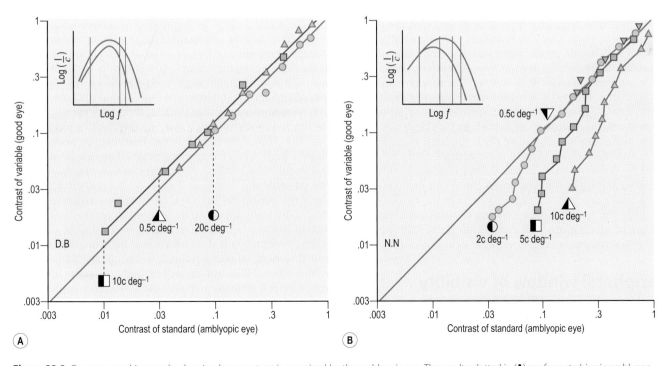

Figure 32.8 Contrast matching results showing how contrast is perceived by the amblyopic eye. The results plotted in (**A**) are for a strabismic amblyope while those in (**B**) are for a non-strabismic anisometrope. The amblyopic eye sees a number of fixed contrasts and a variable contrast stimulus seen by the fellow normal eye is used to match it. The half-filled black/white symbols are thresholds, the solid diagonal line represents the prediction for a normal observer who sees the contrasts comparably in the two eyes (from Hess & Bradley[36]).

Box 32.2

Measurement of constant sensitivity has minimal value for differential diagnosis as similar losses occur in different diseases but is of value in monitoring sensitivity changes.

Box 32.3

Contrast sensitivity supplies information about threshold detection not suprathreshold perceptions.

Box 32.4

To delineate the regional distribution of the contrast loss, sensitivity has to be measured for localized stimuli at multiple field positions.

Some amblyopes perceive severe spatial distortions that would render a 1-D sinusoidal grating unrecognizable (i.e. spatially scrambled), yet the contrast thresholds for detecting such a scrambled stimulus may be normal.[38] Finally the fact that the assessment involves large repetitive patterns means that it supplies no information about how the dysfunction is distributed across the visual field corresponding to the test stimulus. For example, the contrast sensitivity loss of strabismic and anisometropic amblyopes when measured with a relatively large (i.e. 10 degree) centrally fixed stimulus is very similar, yet the visual field is affected very differently in these two conditions; the central field is selectively affected in strabismics and the whole field is evenly affected in anisometropes.[39,40]

In optic neuritis, initial reports suggested that the loss of contrast sensitivity could be restricted to just a small subset of spatial channels producing a very localized loss of contrast sensitivity within the contrast sensitivity function.[41] Later studies showed that contrast sensitivity could be reduced in a variety of different ways; high-frequency loss alone, low-frequency loss alone, similar loss at all spatial frequencies, or a regional frequency loss (the latter was quite rare). More importantly, all of these different signatures of the contrast sensitivity loss in optic neuritis could be simulated by different types of visual field loss, suggesting that the exact nature of the loss across the visual field could be the primary determinant of the contrast sensitivity loss.[42] To decipher the regional nature of any visual loss necessitates the use of much more localized stimuli (i.e. a Gabor patch) and measurements made at multiple field positions.

Peripheral window of visibility

One of the more important ways the bandwidth of the visual system has been kept within limits is due to the fact that our ability to process information depends upon where it is imaged on our retinae. The density of retinal ganglion cells is greatest in the fovea[11] and much more of the cortical area is devoted to the processing of foveal information than to comparable regions of the periphery.[43] As a consequence of

this, our best spatial vision is obtained when fixating with the fovea. Visual acuity is best at the fovea and is progressively reduced away from the fovea which is one of the reasons why we can't read print when we look away from the text. The way in which our visual spatial capacity varies with eccentricity has been the subject of many investigations, but the work of Robson & Graham[44] has been particularly insightful. They measured the rate at which contrast sensitivity falls off with eccentricity and showed that the higher the spatial frequency the faster the decline in sensitivity with eccentricity and that, when considered in terms of a relative eccentricity metric (i.e. eccentricity not in absolute degrees but in number of cycles at a particular spatial frequency), all spatial frequencies higher than the peak (i.e. > 1 c/d) followed the same rule; sensitivity was lost at the rate of 60 periods/decade. A later study by Pointer & Hess[45] extended this to lower spatial frequencies (i.e. 0.8–0.2 c/d) and showed that a different rule was in operation for these frequencies, namely sensitivity when considered in units of relative eccentricity (i.e. number of equivalent spatial cycles) fell off at a much more gradual rate (i.e. 30 periods/decade). We still don't know why there are two different rules, one for mid-high spatial frequencies and another for low spatial frequencies, though one possibility is that spatial frequencies in these two ranges are processed in different cortical areas, each with their characteristic eccentricity dependence. For example V1 is likely to process the mid-high spatial frequency information, while other higher areas, say V2, process lower spatial frequency information.[32]

Whatever the reason, the consequence when we consider eccentricity in terms of degrees is that our window of visibility narrows in a very characteristic way for progressively more eccentric locations. Our visual system selectively loses sensitivity at high spatial frequencies; at low spatial frequencies our sensitivity is relatively constant (Fig. 32.9). It's not that the fovea is best at high spatial frequencies and the periphery is best at low spatial frequencies, as is sometimes thought, humans are just as good at detecting low spatial frequencies across the visual field but simply better at detecting high spatial frequencies with the fovea. The fovea is specialized for the detection of high spatial frequencies.

It is worth stressing that contrast sensitivity measurements are limited by their very nature to threshold and may not give the complete picture. For example, a peripherally imaged high spatial frequency target that can be detected may require say 10 times more contrast than the same target viewed with the fovea, as Figure 32.9 suggests. However, when the peripheral high spatial frequency target is detectable, it does not appear to be reduced in contrast compared with its foveal counterpart. In fact, if the contrast of a peripherally viewed target is slowly increased from below the detection threshold, at some point it is first detected but at this point its perceived contrast will be high, in fact as high as it would be if it were viewed with the fovea. This is illustrated by matching the perceived contrast of a foveal and peripheral spatial target as shown in Figure 32.10. A range of different fixed contrasts for a peripherally viewed target is matched with the same target of variable contrast viewed by the fovea. If foveal and peripheral targets are perceived to be of the same contrast, the results would fall along the solid diagonal line. In fact they do, apart from the fact that a higher contrast is needed to first detect the peripheral target

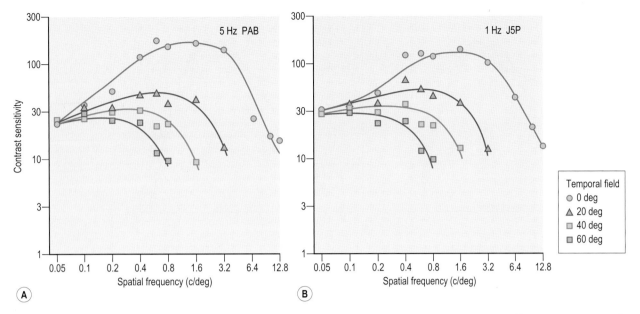

Figure 32.9 The characteristic loss of high spatial frequency sensitivity with eccentric viewing (data from Pointer & Hess[45]). Note that low spatial frequency sensitivity shows little or no decline with eccentricity.

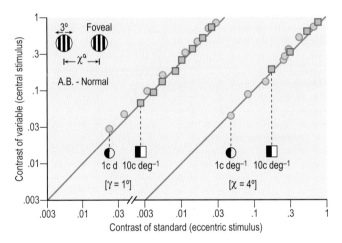

Figure 32.10 Contrast matches made between a peripheral and foveal target. The fixed contrasts were presented peripherally and a foveally viewed stimulus of variable contrast was used to obtain a perceptual match. The half-filled symbols refer to the contrast thresholds and the solid diagonal line represents the prediction that contrast is seen veridically in the periphery (data from Hess & Bradley[36]).

compared with that of its fovea counterpart (half-filled symbols). It is as if there is a switch that simply truncates the contrast range of the mechanism processing the peripheral stimulus.

Is the periphery specialized for detecting anything?

The fact that we can detect a rapidly flickering (> 60 Hz) pattern-less field in the periphery and not in the fovea[46] does suggest that for some forms of purely temporal stimulation, the periphery does show a specialization. This is shown in Figure 32.11 where the temporal contrast sensitivity of the fovea and periphery (60 degrees) are compared for a stimulus of low spatial frequency (0.25 c/d). Notice that the

Box 32.5

Reduced temporal acuity for a large pattern-less field suggests a peripheral loss of function.

periphery exhibits slightly better temporal contrast sensitivity (and a higher temporal acuity) in the high temporal frequency range. However this does not apply to the temporal sensitivity of patterned stimuli (i.e. > 1 c/d) whose sensitivity in the periphery is marginally lower than their foveal counterparts.[45]

Why does contrast sensitivity for high spatial frequencies decline with eccentricity?

First, as previously explained, the visual system allocates fewer neural resources to peripheral vision; peripherally, there is a reduced density of retinal ganglion cells and proportionally less allocated cortical area. Also the size of retinal receptive fields increases with eccentricity[47] and there is a progressive loss of smaller receptive fields with eccentricity in the cortex.[31] This inevitably means that the spatial range over which we can see restricts with eccentricity. This translates to a foveal specialization. Foveal analysis can be undertaken at a large number of different spatial scales and this provides optimum performance on a number of tasks that rely on combining information across different scales. These include the so-called "spatial hyperacuities",[48] which have erroneously been considered "acuity" measures but are more readily understood in terms of sensitivity,[49] often involving multi-scale analyses.[50]

Luminance

One of the truly impressive things about visual processing is the fact that it can operate with such high sensitivity over

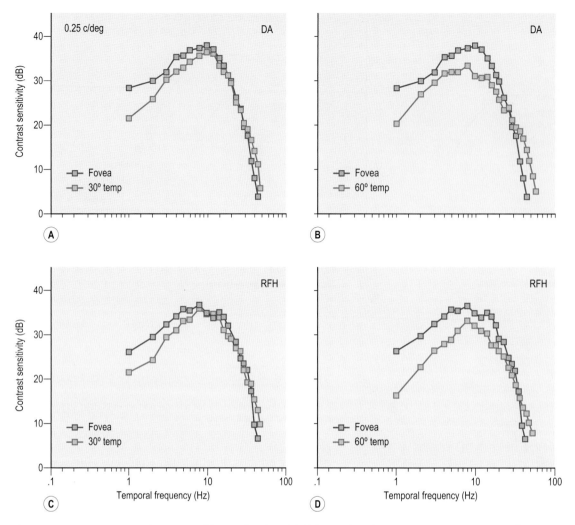

Figure 32.11 Superior temporal sensitivity of the peripheral visual field for a uniform field stimulus (data from Allen & Hess[46]).

such a large luminance operating range. Our luminance operating range is over a million, million to one (i.e. 15 log units) – from a moonlit night to direct sunlight. The fact that we have two receptor systems, one designed to operate optimally during the day, the cones, and another designed to operate optimally during the night, the rods, certainly helps. However it still means that each receptor system has a 7 log unit operating range over which to maintain optimal sensitivity. To achieve this feat, cells in the retina undergo light adaptation. They change their response gain depending on the prevailing luminance. This allows individual cells to be just as sensitive, that is able to respond to just as small luminance increments at very different mean luminances, often many log units apart. The combined effects of the receptoral duplicity and postreceptoral luminance-dependent response gain changes, allowing our visual processing to operate largely independent of the prevailing luminance conditions. Light adaptation is accomplished in the retina;[51] however the cortex also makes an important contribution to our visual stability when the ambient light level is reduced. In the primary visual cortex the information contained in the retinal image has been decomposed into its spatial components, and cells with different-sized receptive fields that carry

pieces of this spatial puzzle vary in a differential way with luminance. The spatial tuning of these cells remains constant with reducing luminance (unlike their retinal counterparts[12]), only their sensitivity changes and it does so in a way that is dependent on the spatial tuning of the cell.[52] This is illustrated in Figure 32.12 where the spatial tuning of a cortical simple (A) and complex (B) cell from the cat is shown at a number of different mean luminance conditions.

The cell's response, in terms of contrast sensitivity using an auditory criterion, is plotted against the stimulus spatial frequency. Notice the peak response of the cell remains constant as the ambient light level is reduced by 4 log units (300 cd/m² to 0.03 cd/m²). The same is true for the temporal tuning properties as luminance is reduced.[52] The lower the peak spatial tuning, the more resistant it is to reductions in the ambient luminance. This means that as luminance is reduced, we lose the contribution of the cells tuned to higher spatial frequencies conveying fine scale information but we retain the information processed by cells tuned to lower spatial frequencies that convey coarse scale information. This results is a relatively stable vision as luminance is reduced but at the expense of spatial detail. The spatial scale dependency of the sensitivity change as luminance is reduced

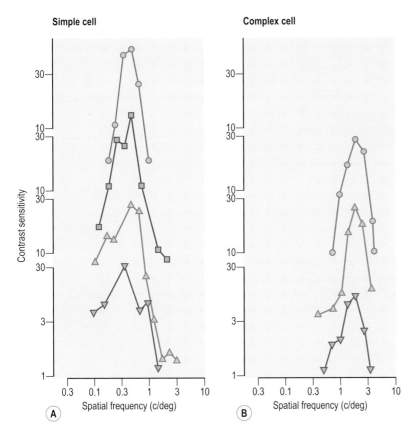

Simple cell   **Complex cell**

Figure 32.12 Spatial frequency tuning curves from a simple (**A**) and complex (**B**) cell in cat cortex (using an auditory criterion) at a range of different mean luminance (circles 300 cd/m²; squares 30 cd/m²; triangles 3 cd/m²; inverted triangles 0.03 cd/m²). Notice that the peak sensitivity is maintained and only sensitivity declines with mean luminance (data from Hess[52]).

is best seen in psychophysical measures of sensitivity. This was first shown by Van Nes & Bouman[53] and their results are reproduced in Figure 32.13.

Contrast sensitivity is plotted against the mean luminance for a range of different spatial frequencies. Notice that there is a similar form to the response (Fig. 32.13C) at each spatial scale; a sloping region (termed the Rose–de Vries region) where sensitivity depends on the reducing luminance (varies with the square root of the luminance) and a flat region (termed the Weber region) where sensitivity is invariant with mean luminance. The luminance that corresponds to the transition from one behavior to the other also depends on the spatial frequency of the stimuli; the lower the spatial frequency, the lower the luminance needs to be before it begins to affect sensitivity. This is also seen in the responses of individual cortical cells tuned to different spatial scales.[52] A limit is reached for human vision[10] at a spatial frequency of 0.2 c/d, thought to be the largest summation area available (corresponding to the lowest spatial frequency tuned cell).

Chromatic sensitivity

Primates are unique amongst mammals in having well-developed color vision. The beginnings of this are seen in the retina where information is distributed between different sub-populations of ganglion cells coexisting in the same retinal regions.[47] The majority of ganglion cells (80%) are of the midget variety (parvo) and these convey one type of chromatic information reflecting their long/medium

wavelength receptoral input. Another retinal circuit comprising the short wavelength cones connected to small bi-stratified ganglion cells conveys another form of chromatic information reflecting the short/long+medium wavelength opponency of their receptoral input.[54] The parasol retinal ganglion cells (magno) are not so numerous (10%) and carry the achromatic signals, reflecting a receptoral input comprising the addition of all three photoreceptor signals. These three different signals are kept separate in the lateral geniculate nucleus in the parvo- (chromatic long/medium wavelength opponent system), konio- (short/long+medium wavelength opponent system) and the magno-cellular layers (the achromatic system).[55] The short/long+medium wavelength opponent system that underlies our perception of blue/yellow has a relatively sparse representation as only 7% of cones are short wavelength.[56] For this reason, the representation of the konio-cellular layers in the geniculate is weak, however there is evidence from both neuro-imaging in humans[57] and single cell physiology in animals[58] that this signal is amplified in the transmission from geniculate to cortex such that there is a comparable representation for the two types of chromatic systems in the cortex.[59]

Our chromatic windows of visibility can be obtained by making similar measurements to those already described for luminance-defined sinusoidal gratings by using stimuli whose periodicity is defined solely by either a long/medium wavelength modulation (perceived as red/green bars) or short/long+medium wavelength modulation (perceived as blue/yellow bars). A demonstration is shown in Figure 32.14. The first detailed measurements[60] of our chromatic sensitivities are shown in Figure 32.15 where modulation

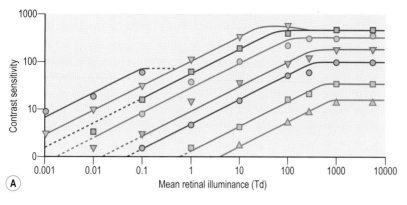

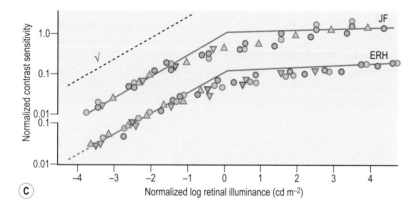

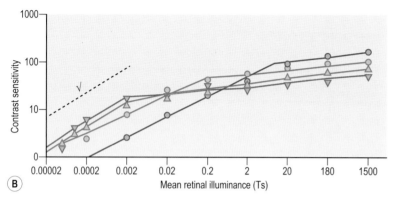

Figure 32.13 The effect of mean luminance on contrast sensitivity. In (**A**), the results of van Nes & Bouman[53] for spatial frequencies in the mid to high range. In (**B**), the results of Hess & Howell[10] for low spatial frequencies. In (**C**), normalized sensitivity at different spatial frequencies showing the characteristic features of the dependence of sensitivity on mean luminance (data from Hess & Howell[10]).

sensitivity is plotted against the spatial frequency of the stimulus as described above (see Fig. 32.1) for luminance-defined stimuli. The difference now is that the stimuli are only visible by virtue of their chromatic content (red/green or blue/yellow symbols), being invisible to the achromatic system. Notice that the sensitivity characteristics of the two chromatic systems are quite similar. Two differences are immediately obvious between these chromatic contrast sensitivity functions (red/green or blue/yellow symbols) and their achromatic counterparts (black/green or black/yellow symbols), namely the chromatic function is lowpass in its overall shape (as opposed to the bandpass shape of the achromatic contrast sensitivity function) and secondly sensitivity is reduced at higher spatial frequencies. Our chromatic sensitivity is specialized for the low spatial frequency range where its sensitivity is superior to that of the achromatic system. Therefore, color vision does not provide fine detail information but does help to detect boundaries

between large regions that have subtly different chromatic content.

These overall contrast sensitivity functions represent the envelope of numerous much more selectively tuned mechanisms, just as has been shown for the achromatic contrast sensitivity function.[61,62] Their bandpass shape suggests that they represent the responses of chromatically tuned cortical neurons that are optimally responsive to different spatial frequency ranges, similar to their achromatic counterparts. A comparison of how chromatic sensitivity varies across the visual field highlights changes that occur to the peripheral chromatic window of visibility.[63] "Blue/yellow" sensitivity shows a gradual decline with eccentricity, not unlike that found for achromatic stimuli. However, sensitivity for detecting "red/green" stimuli falls off much more sharply with eccentricity. Therefore, the fovea is specialized for red/green chromatic processing. This is illustrated in Figure 32.16 where, on the right, weighting functions are shown that

Figure 32.14 Demonstration of the chromatic window of visibility for the human visual system (from Mark E McCourt, NDSU). The sinusoidal bars are chromatically defined (by red/green chromatic contrast in the left frame (**A**) and by blue/yellow chromatic contrast in the right frame (**B**)) and vary in spatial frequency along the abscissa and chromatic contrast along the ordinate. Notice that the bars are now seen in an inverted-L shaped area (i.e. lowpass shaped function). Reproduced with the permission of Mark McCourt.

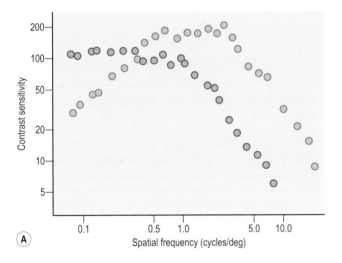

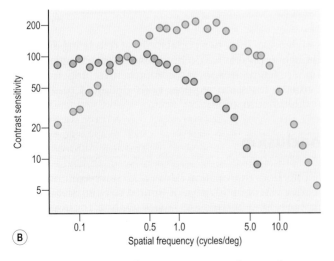

Figure 32.15 Comparison of the contrast sensitivity functions for achromatic and chromatic stimuli. In each case the chromatic (red/green on left and blue/yellow on right) and achromatic (black/green on left and yellow/black on right) contrast sensitivities are compared from the work of Mullen.[60]

Box 32.6

A selective loss of red/green chromatic sensitivity may simply reflect a central field deficit not a selective chromatic deficit per se

approximate the sensitivity fall-off with eccentricity for red/green chromatic stimuli (top) and achromatic stimuli (bottom) from the work of Mullen et al.[63] On the left, the perceptual effects of these different eccentricity dependences are illustrated. Red/green chromatic stimuli are much more visible in the fovea compared with the periphery. Achromatic (and blue/yellow chromatic) stimuli are only slightly less visible in the periphery.

Suprathreshold sensitivity

The use of contrast sensitivity to define a window of visibility for the human visual system is an important step in appreciating the quality of our visual perception, but it by no means represents the complete picture. The visual system is highly non-linear, comprising cells with different thresholds and contrast gain control, specialized to encode contrast in a distributed fashion.[22] One consequence of this is that threshold and suprathreshold sensitivity may be quite different. For example, Georgeson & Sullivan[64] showed that the typical inverted U-shaped contrast sensitivity function measured using thresholds as described above, is not present for stimuli above threshold. In other words, more contrast may be needed to detect a stimulus of higher spatial frequency at threshold without the stimuli looking different in their contrast above threshold. The suprathreshold contrast sensitivity function is relatively flat compared with its threshold counterpart. This is shown in Figure 32.17 where the matched contrast is plotted against the stimulus spatial frequency for a range of different suprathreshold contrasts. The same is true for the periphery;[36] more contrast may be needed for

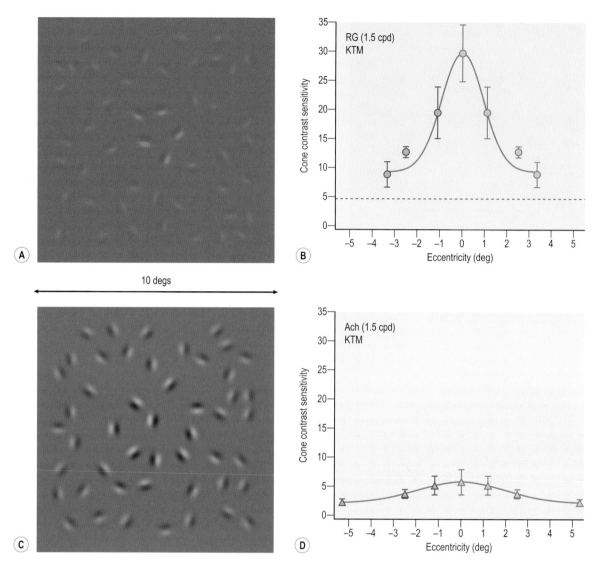

Figure 32.16 On the right, the fall-off in sensitivity for achromatic and red/green chromatic stimuli is depicted. The sensitivity fall-off for blue/yellow is similar to that for the achromatic stimulus. On the left, the simulated visibility effect this would have on the perception of an array of spatially narrowband stimuli. Notice the foveal specialization for red/green chromatic stimuli (data from Mullen et al[63]).

peripheral stimuli to be detected without their perception at threshold being any different in terms of contrast (see Fig. 32.10).

There are a number of different ways of thinking about this. First, one could assume that the visual thresholds are noise-limited, where the signal is reduced to such levels that any intrinsic noise begins to play an important part. The shape of the contrast sensitivity function reflects the spatial frequency dependence of the noise whose influence is progressively diminished as the signal (i.e. contrast) is increased. Alternatively, the lack of spatial frequency dependency reflected in contrast constancy at suprathreshold contrast levels could be a side-effect of contrast gain control,[65] a network of interactions that regulates the contrast responsiveness of individual cells. Finally one could argue that thresholds reflect the properties of cells with the best sensitivity whereas suprathreshold measures reflect the properties of different cells, those with optimal contrast coding properties in the higher contrast range; the spatial frequency

dependence of these two populations of cells being different. Although at present we do not know the exact explanation, from a practical viewpoint a comprehensive understanding of visual processing requires both threshold and suprathreshold information.

Conclusion

Not only does information have to be conveyed from retina to cortex via an optic nerve of limited size but also sensory cells have limited dynamic ranges. Only a fraction of the information that impinges on the retina can be transmitted to cortex; to maintain relatively high sensitivity, it needs to be parceled out among a population of cells such that all cells carry only a part of the information. Parallel processing is a fundamental part of visual processing. We have different windows of visibility for achromatic and chromatic processes, for photopic and scotopic processes, for central and

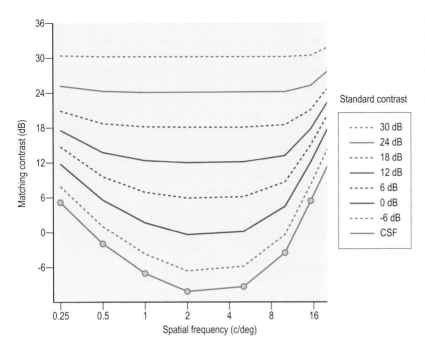

Figure 32.17 The shape of the contrast sensitivity function at threshold (lower red filled circles) compared with contrast matching functions above threshold (colored curves. The flattening of the function was first shown by Georgeson & Sullivan.[64] These results are simulations from a contrast gain control model (Hess et al[65]).

peripheral processes and for threshold and suprathreshold processes. The fovea exhibits a specialization in terms of spatial scale and in terms of red/green chromatic processing. The periphery exhibits a specialization for detecting rapid temporal change for low spatial frequency stimuli. Without the need to maintain comparable spatial and temporal sensitivity throughout the visual field, substantial saving can be made in the processing bandwidth. A more complete understanding of the early visual processing of spatial form involves an appreciation of the neural substrate of each of these different aspects of the parallel processing of visual information.

Acknowledgments

I acknowledge the grant support provided by NSERC and CIHR.

References

1. Van Essen DC, Anderson CH, Felleman DJ. Information processing in the primate visual system: an integrated systems perspective. Science 1992; 255(5043):419–423.
2. Bouguer P. Traite d'optique sur la gradation de la lumiere. Paris; 1760.
3. Selwyn EWH. The photographic and visual resolving power of lenses. Photographic J 1948; 88(B):6–12 and 46–57.
4. Campbell FW, Green DG. Optical and retinal factors affecting visual resolution. J Physiol Lond 1965; 181:576–593.
5. Campbell FW, Gubisch RW. Optical quality of the human eye. J Physiol 1966; 186(3):558–578.
6. Banks MS, Geisler WS, Bennett PJ. The physical limits of grating visibility. Vision Res 1987; 27(11):1915–1924.
7. Arend LE, Jr. Temporal determinants of the form of the spatial contrast threshold MTF. Vision Res 1976; 16(10):1035–1042.
8. Blakemore C, Campbell FW. On the existence of neurones in the human visual system selectively sensitive to the orientation and size of retinal images. J Physiol Lond 1969; 203:237–260.
9. Campbell FW, Johnstone JR, Ross J. An explanation for the visibility of low frequency gratings. Vision Res 1981; 21(5):723–730.
10. Hess RF, Howell ER. Detection of low spatial frequencies: A single filter or multiple filters? Ophthal Physiol Opt 1988; 8:378–385.
11. Rodieck RW. The vertebrate retina. San Francisco: WH Freeman and company; 1973.
12. Enroth-Cugell C, Robson JG. The contrast sensitivity of retinal ganglion cells of the cat. J Physiol Lond 1966; 187:517–552.
13. Tolhurst DJ. The amount of information transmitted about contrast by neurones in the cat's visual cortex. Vis Neurosc 1989; 2:409–416.
14. Hawken MJ, Parker AJ. Detection and discrimination mechanisms in the striate cortex of the old-world monkey. In: Blakemore C, ed. Vision: coding and efficiency. Cambridge: Cambridge University Press, 1990:103–116.
15. Campbell FW, Cooper GF, Enroth-Cugell C. The spatial selectivity of the visual cells of the cat. J Physiol 1969; 203(1):223–235.
16. Maffei L, Fiorentini A. The Visual Cortex as a spatial frequency analyser. Vision Res 1973; 13:1255–1267.
17. De Valois RL, Albrecht DG, Thorell LG. Spatial frequency selectivity of cells in macaque visual cortex. Vision Res 1982; 22:545–559.
18. Pantle A, Sekular R. Size detecting mechanisms in human vision. Science 1968; 62:1146–1148.
19. Kranda K, Kulikowski JJ. Adaptation of coarse grating under scotopic and photopic conditions. J Physiol Lond 1976; 257:35P.
20. Stromeyer CF, Klein S, Dawson BM, Spillmann L. Spatial frequency masking in vision; critical bands and spread of masking. J Opt Soc Am 1982; 62:1221–1232.
21. Levick WR. Sampling of information space by retinal ganglion cells. In: Pettigrew JD, Sawrl KJ, eds. Visual neuroscience. Cambridge: Cambridge University Press, 1986:33–43.
22. Kaplan E, Shapley RM. The primate retina contains two types of ganglion cells, with high and low contrast sensitivity. Proc Natl Acad Sci USA 1986; 83(8):2755–2757.
23. Merigan WH. Chromatic and achromatic vision of macaques: role of the P pathway. J Neurosci 1989; 9(3):776–783.
24. Merigan WH, Byrne CE, Maunsell JH. Does primate motion perception depend on the magnocellular pathway? J Neurosci 1991; 11(11):3422–3429.
25. Merigan WH, Katz LM, Maunsell JH. The effects of parvocellular lateral geniculate lesions on the acuity and contrast sensitivity of macaque monkeys. J Neurosci 1991; 11(4):994–1001.
26. Rosenberg A, Husson RT, Mallik A, Issa NP. Frequency-doubling in early visual system underlies sensitivity to second-order stimuli. J Vision 2008; 8(6):281a.
27. Zeki S. Functional specialization in the visual cortex of the rhesus monkey. Nature 1978; 274:423–428.
28. Girard P, Bullier J. Visual activity in area V2 during reversible inactivation of area 17 in the macaque monkey. J Neurophysiol 1989; 62:1287–1302.
29. Sincich LC, Park KF, Wohlgemuth MJ, Horton JC. Bypassing V1: a direct geniculate input to area MT. Nat Neurosci 2004; 7(10):1123–1128.
30. Sherman SM, Guillery RW. Exploring the thalamus and its role in cortical function. Cambridge, MA: The MIT Press, 2006.
31. Hubel DH, Wiesel TN. Functional architecture of the macaque monkey visual cortex. Ferrier Lecture. Proc R Soc Lond Biol 1977; 198:1–59.
32. Foster KH, Gaska JP, Nagler M, Pollen DA. Spatial and temporal frequency selectivity of neurones in visual cortical areas V1 and V2 of the macaque monkey. J Physiol 1985; 365:331–363.
33. Merigan WH, Nealey TA, Maunsell JH. Visual effects of lesions of cortical area V2 in macaques. J Neurosci 1993; 13(7):3180–3191.
34. Gstalder RJ, Green DG. Laser interferometric acuity in amblyopia. J Pediatr Ophthalmol 1971; 8:251–256.
35. Hess RF, Howell ER. The threshold contrast sensitivity function in strabismic amblyopia: Evidence for a two type classification. Vision Res 1977; 17(9):1049–1055.
36. Hess RF, Bradley A. Contrast coding in amblyopia is only minimally impaired above threshold. Nature. 1980; 287:463–464.
37. Hess RF. Contrast vision and optic neuritis: neural blurring. J Neurol, Neurosurg Psychiatr 1983; 46:1023–1030.

38. Hess RF, Campbell FW, Greenhalgh T. On the nature of the neural abnormality in human amblyopia; neural aberrations and neural sensitivity loss. Pflugers Archiv – Eur J Physiol 1978; 377(3):201–207.

39. Hess RF, Campbell FW, Zimmern R. Differences in the neural basis of human amblyopias: effect of mean luminance. Vision Res 1980; 20:295–305.

40. Hess RF, Pointer JS. Differences in the neural basis of human amblyopias: the distribution of the anomaly across the visual field. Vision Res 1985; 25:1577–1594.

41. Regan D, Silver R, Murray TJ. Visual acuity and contrast sensitivity in multiple sclerosis – hidden visual loss: an auxiliary diagnostic test. Brain 1977; 100(3):563–579.

42. Hess RF, Plant GT. Recent advances in optic neuritis. Cambridge: Cambridge University Press, 1986.

43. Daniel PM, Whitteridge D. The representation of the visual field on the cerebral cortex in monkeys. J Physiol 1961; 159:203–221.

44. Robson JG, Graham N. Probability summation and regional variation in contrast sensitivity across the visual field. Vision Res 1981; 21:409–418.

45. Pointer JS, Hess RF. The contrast sensitivity gradient across the human visual field: emphasis on the low spatial frequency range. Vision Res 1989; 29:1133–1151.

46. Allen D, Hess RF. Is the visual field temporally homogeneous? Vision Res 1992; 32(6):1075–1084.

47. Peichl L, Wassle H. Size, scatter and coverage of ganglion cell receptive field centres in the cat retina. J Physiol 1979; 291:117–141.

48. Westheimer G. The spatial grain of the perifoveal visual field. Vision Res 1982; 22:157–162.

49. Carney T, Klein SA. Optimum spatial localization is limited by contrast sensitivity. Vision Res 1999; 39(3):503–511.

50. Hess RF, Watt RJ. Regional distribution of the mechanisms that underlie spatial localization. Vision Res 1990; 30:1021–1031.

51. Dowling JE. The Retina. Cambridge: The Belknap Press of Harvard University Press, 1987.

52. Hess RF. Vision at low light levels: role of spatial, temporal and contrast filters. Ophthal Physiol Opt 1990; 10:351–359.

53. Van Nes FL, Bouman MA. Spatial modulation transfer in the human eye. J Opththalmol Soc Am 1967; 57:401–406.

54. Dacey DM. Parallel pathways for spectral coding in primate retina. Annu Rev Neurosci 2000; 23:743–775.

55. Martin PR, White AJ, Goodchild AK, Wilder HD, Sefton AE. Evidence that blue-on cells are part of the third geniculocortical pathway in primates. Eur J Neurosci 1997; 9(7):1536–1541.

56. Curcio CA, Allen KA, Sloan KR et al. Distribution and morphology of human cone photoreceptors stained with anti-blue opsin. J Comp Neurol 1991; 22; 312(4):610–624.

57. Mullen KT, Dumoulin SO, Hess RF. Color responses of the human lateral geniculate nucleus: Selective amplification of S-cone signals between the lateral geniculate nucleus and primary visual cortex measured with high-field fMRI. Eur J Neurosci 2008; 28:1911–1923.

58. Cottaris NP, De Valois RL. Temporal dynamics of chromatic tuning in macaque primary visual cortex. Nature 1998; 395(6705):896–900.

59. Mullen KT, Dumoulin SO, McMahon KL, Bryant M, de Zubicaray GI, Hess RF. A comparison of the BOLD fMRI response to achromatic, L/M opponent and S-cone opponent cardinal stimuli in human visual cortex. I perceptually-matched vs contrast-matched stimuli. Eur J Neurosci 2007; 25:491–502.

60. Mullen KT. The contrast sensitivity of human colour vision to red-green and blue-yellow chromatic gratings. J Physiol Lond 1985; 359:381–400.

61. Losada MA, Mullen KT. The spatial tuning of chromatic mechanisms identified by simultaneous masking. Vision Res 1994; 34(3):331–341.

62. Losada MA, Mullen KT. Color and luminance spatial tuning estimated by noise masking in the absence of off-frequency looking. J Ophthalmol Soc Am A 1995;12:250–260.

63. Mullen KT, Sakurai M, Chu W. Does L/M cone opponency disappear in human periphery? Perception 2005; 34(8):951–959.

64. Georgeson MA, Sullivan GD. Contrast constancy: deblurring in human vision by spatial frequency channels. J Physiol 1975; 252(3):627–656.

65. Hess RF, Baker DH, May KA, Wang J. On the decline of 1st and 2nd order sensitivity with eccentricity. J Vis 2008; 8(1):19, 1–2.

66. Howell ER, Hess RF. The functional area for summation to threshold for sinusoidal gratings. Vision Res 1978; 18:369–374.

67. Hawken MJ, Parker AJ. Spatial properties of neurons in the monkey striate cortex. Proc R Soc Lond B Biol Sci 1987; 231(1263):251–288.

CHAPTER 33

Visual Acuity

Dennis M. Levi

Visual acuity is a measure of the keenness of sight. The Egyptians used the ability to distinguish double stars as a measure of visual acuity more than 5000 years ago.[1] Over the centuries visual acuity has been studied, measured and analyzed because it represents a fundamental limit in our ability to see. Consequently, visual acuity has been used as a criterion for military service and various other occupations, driving, and for receiving social security benefits.

Visual acuity is limited primarily by the optics of the eye and by the anatomy and physiology of the visual system. As such, visual acuity is perhaps the key clinical measure of the integrity of the optical and physiological state of the eye and visual pathways (Box 33.1).

Defining and specifying visual acuity

How do we define the keenness of sight? Visual acuity is used to specify a spatial limit, i.e. a threshold in the spatial dimension.[2] Over the centuries there have been a large number of different ideas about how to define, measure and specify visual acuity, and these can be distilled down to four widely accepted criteria:

- Minimum visible acuity – detection of a feature
- Minimum resolvable acuity – resolution of two features
- Minimum recognizable acuity – identification of a feature
- Minimum discriminable acuity – discrimination of a change in a feature (e.g. a change in size, position or orientation).

These different criteria actually represent different limits and may be determined by different aspects of the visual pathway (Table 33.1).

Minimum visible acuity

- Minimum visible acuity refers to the smallest object that one can detect.

As early as the 17th century, de Valdez measured the distance at which a row of mustard seeds could no longer be counted, and early astronomers like Robert Hooke were interested in the size of stars that could be detected and their relation to retinal anatomy.[1] In this context, the minimum visible acuity refers to the smallest target that can be detected.

Under ideal conditions, humans can detect a long, dark wire (like the cables of the Golden Gate bridge) against a very bright background (like the sky on a bright sunny day) when they subtend an angle of just 0.5 arc seconds ($\approx$ 0.00014 degrees). It is widely accepted that the minimum visible acuity is so small because the optics of the eye (described below) spread the image of the thin line, so that on the retina it is much wider, and the fuzzy retinal image of the wire casts a shadow which reduces the light on a row of cones to a level which is just detectably less than the light on the row of cones on either side. In other words, although we specify the minimum visible acuity in terms of the angular size of the target at the retina, it is actually limited by our ability to discriminate the intensity of the target relative to its background.

Increasing the target size, up to a point, is equivalent to increasing its relative intensity. Hecht & Mintz[3] measured the visual resolution of a black line against a background over a large range of background brightness. They found that the minimum visible acuity varied from about 10 minutes at the lowest background levels, to about 0.5 arc seconds at the highest. The limiting factor in the case of minimum visible acuity is the observers' sensitivity to small variations in the stimulus intensity ($\Delta I/I$), i.e. their contrast sensitivity (discussed below). Indeed, Hecht & Mintz[3] state that the retinal image produced by the 0.5 arc second line represents a "fine fuzz of a shadow extending over several cones", and they calculated that at the highest intensity levels tested, the foveal cones occupying the center of the shadow suffers a drop in intensity (relative to the neighboring cones) of $\approx$ 1 percent – just at the limit of intensity discrimination.

Although the minimum visible acuity represents one limit to spatial vision, it is in fact a limit in the ability to discern small changes in contrast, rather than a spatial limit per se, and minimum visible acuity is not used clinically.

Minimum resolvable acuity

- Minimum resolvable acuity refers to the smallest angular separation between neighboring objects that one can resolve.

More than 5000 years ago the Egyptians assessed visual acuity by the ability of an observer to resolve double stars.[1] There is currently still debate about how best to define and measure resolution. However, today, the minimum resolvable acuity is much more likely to be assessed by determining the finest black and white stripes that can be resolved. Under ideal conditions (e.g. high contrast and luminance), humans with very good vision can resolve black and white stripes when one cycle subtends an angle of approximately 1 minute of arc (0.017 degrees). This minimum resolvable acuity represents one of the fundamental limits of spatial vision: it is the finest high-contrast detail that can be resolved. In foveal vision the limit is determined primarily by the spacing of photoreceptors in the retina. The visual system "samples" the stripes discretely, through the array of receptors at the back of the retina (Fig. 33.1). If the receptors are spaced such that the whitest and blackest parts of the grating fall on separate cones (Fig. 33.1B), we should be able to make out the grating. But if the entire cycle falls on a single cone (Fig. 33.1C), we will see nothing but a gray field (or

we may experience a phenomenon called **aliasing**, in which we misperceive the width or orientation of the stripes). Cones in the fovea have a center-to-center separation of about 0.5 minutes of arc (0.008 degrees), which fits nicely with the observed acuity limit of 1 minute of arc (0.017 degrees – since we need two cones per cycle to be able to perceive it accurately), and each foveal cone has a "private" line to a ganglion cell. Rods and cones in the retinal periphery are less tightly packed together, and many receptors converge on each ganglion cell. As a result, visual acuity is much poorer in the periphery than in the fovea (more about that later).

Box 33.1 Visual Acuity

Visual acuity is a measure of the keenness of sight. Visual acuity is limited primarily by the optics of the eye and by the anatomy and physiology of this visual system. As such, visual acuity is perhaps the key clinical measure of the integrity of the optical and physiological state of the eye and visual pathways.

Table 33.1 Summary of the different forms of acuity and their limits

Type of acuity	Measured	Acuity (degrees)
Minimum visible	Detection of a feature	0.00014
Minimum resolvable	Resolution of two features	0.017
Minimum recognizable	Identification of a feature	0.017
Minimum discriminable	Discrimination of a change in a feature	0.00024

Spatial Frequency lower than sampling limit

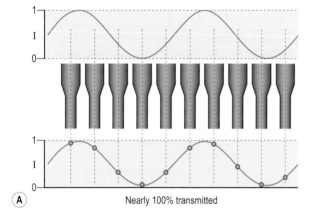

(A) Nearly 100% transmitted

Spatial Frequency equal to sampling limit

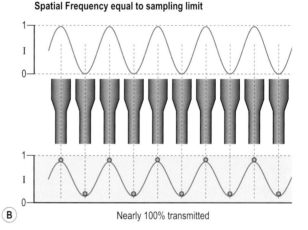

(B) Nearly 100% transmitted

Spatial Frequency higher than sampling limit

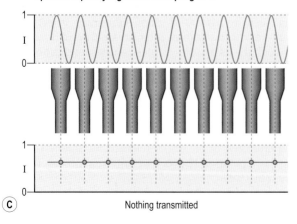

(C) Nothing transmitted

Figure 33.1 Spatial sampling of a sinusoid whose spatial frequency is (**A**) lower than the sampling limit set by the spacing of the cones, (**B**) equal to the sampling limit set by the spacing of the cones, (**C**) higher than the sampling limit set by the spacing of the cones.

Minimum recognizable acuity

- Minimum recognizable acuity refers to the angular size of the smallest feature that one can recognize or identify.

Although this method has been used since the 17th century, the approach still used by eye doctors today was introduced more than a century ago by Herman Snellen and his colleagues. Snellen constructed a set of block letters for which the letter as a whole was five times as large as the strokes that formed the letter (Fig. 33.2). The distance of the patient was varied until they could no longer accurately read the letters. In later adaptations of the Snellen test, the viewer was positioned at a constant distance (typically 20 feet [6 meters]), and the size of the letters, rather than the position of the viewer, was altered and visual acuity was defined as:

The distance at which the patient can just identify the letters / The distance at which a person with "normal" vision can just identify the letters

Thus, "normal" vision came to be defined as 20/20 (6/6 in metric units). To relate this back to visual angle, a 20/20 letter is designed to subtend an angle of 5 arc minutes (0.083 degrees) at the eye, and each stroke of a 20/20 letter subtends an angle of 0.017 degrees (our old familiar 1 arc minute). Thus, if you can read a 20/20 letter, you can discern detail that subtends 1 minute of arc. If you have to be at 20 feet to read a letter that someone with normal vision can read at 40 feet, you have 20/40 vision (worse than normal). Although 20/20 is often considered the gold standard, most healthy young adults have an acuity level closer to 20/15.

Illiterate E and Landolt C charts are based on the same principles as the Snellen chart.

Although Snellen's notation for visual acuity is commonly used, there are other schemes for specifying acuity (Table 33.2). For example, the Minimum Angle of Resolution (MAR) is the angular size of detail of the visual acuity letter in minutes of arc – it is the Snellen denominator (the number below the line, e.g. 40) over the Snellen numerator (the number above the line, e.g. 20). Thus, for example a Snellen acuity of 20/20 is equivalent to an MAR of 1'; 20/40 to an MAR of 2' and 20/100, an MAR of 5'. Another method for specifying acuity is the Snellen fraction (the Snellen numerator divided by the denominator or 1/MAR). Acuity is sometimes specified as LogMAR (the logarithm of the minimum angle of resolution) and sometimes as Log Visual Acuity (the logarithm of the Snellen fraction). Snell & Sterling[4] developed a metric for quantifying visual loss due to injury or disease. The Snell–Sterling efficiency scale sets 20/20 (MAR = 1') as 100% efficiency, and reduces the efficiency by a fixed percentage ($\approx$ 84%) for every 1 minute loss of acuity.

There are many stimulus and subject variables that influence the minimum recognizable visual acuity (discussed below).

Minimum discriminable acuity

- Minimum discriminable acuity refers to the angular size of the smallest *change* in a feature (e.g. a change in size, position or orientation) that one can discriminate.

Perhaps the most studied example of minimum discriminable acuity is our ability to discern a difference in the relative positions of two features. Our visual systems are very good at telling where things are. Consider two abutting horizontal lines, one slightly higher than the other (Fig 33.3A). It is very easy to discern that, for example the right line is higher than the left (i.e. discriminate the relative positions of the two lines) even from a long way away. This is an example of a class of visual tasks that have been given the label "*hyperacuity*" by Gerald Westheimer.[5] These tasks all have in common that they involve judging the relative position of objects, and Westheimer coined the term hyperacuity, because, under ideal conditions, humans can make these judgments with a precision that is finer than the size or spacing of foveal cones.

The smallest misalignment that we can reliably discern is known as Vernier acuity – named after the Frenchman, Pierre Vernier, whose scale, developed in the 17th century, was widely used to aid ship's navigators. The success of the Vernier scale was based on the fact that humans are very adept at judging whether nearby lines are lined up or not. Thus, Vernier alignment is still widely used in precision machines, and even in the dial switches in modern ovens.

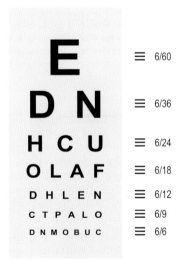

E	≡ 6/60
D N	≡ 6/36
H C U	≡ 6/24
O L A F	≡ 6/18
D H L E N	≡ 6/12
C T P A L O	≡ 6/9
D N M O B U C	≡ 6/6

Figure 33.2 Snellen acuity chart.

Table 33.2 Different acuity notations

Snellen (Imperial)	Notation (Metric)	MAR	logMAR	Decimal	SF (c/deg)
20/200	6/60	10	1	0.1	3
20/20	6/6	1	0.0	1.00	30
20/10	6/3	0.5	−0.3	2	60

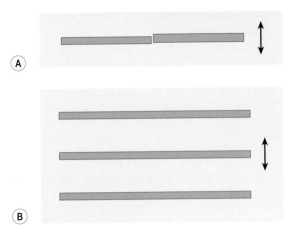

Figure 33.3 Two "hyperacuity" configurations: Vernier acuity (**A**) and bisection acuity (**B**). In both cases, the subject's task is to judge the relative position of the features.

Under ideal conditions, Vernier acuity may be just three arc seconds (≈ 0.0008 degrees)! This performance is even more remarkable when you consider that it is about ten times smaller than even the smallest foveal cones. Note that the optics of the eye spread the image of a thin line over a number of retinal cones, and that the eyes are in constant motion, and this performance appears even more remarkable.

Vernier acuity is not the most remarkable form of hyperacuity. Guinness World Records 2005 describes the "highest hyperacuity" as follows: In April 1984, Dr. Dennis M. Levi (yes – that's yours truly) "… repeatedly identified the relative position of a thin, bright green line within 0.8 seconds of arc (0.00024 deg). This is equivalent to a displacement of some 0.25 inches (6 mm) at a distance of 1 mile (1.6 km)." This "remarkable" position acuity was accomplished with a bisection task (Fig. 33.3B), but can actually be understood based on a straightforward model of human spatial vision.[6]

As remarkable as the other hyperacuities (sometimes also called "position" acuities) might seem, they do not defy the laws of physics. Geisler[7] calculated that if you placed a machine (known as an ideal discriminator) at the retina, and this machine knew precisely the pattern of photons absorbed by the retinal photoreceptors when the stripes were aligned, and the pattern of photons absorbed when they were misaligned, that this machine could actually perform an order of magnitude better than even the best humans. So, the information about the Vernier offset is present in the pattern of photons absorbed by the photoreceptors; however, humans must be able to interpret the information despite the constant motion of the eyes. Thus hyperacuity must ultimately be limited by neurons in the visual cortex that are able to interpolate positional information with high resolution.

There is a good deal of evidence that different mechanisms limit position judgments for closely spaced (or abutting) targets, and for widely separated targets. In the nearby case, both contrast and contrast polarity are important. For example, in a two-line Vernier target Vernier acuity is better when the lines are both either bright or dark than if one line is bright and the other dark. For longer-range position judgments where the targets are well separated, neither contrast nor contrast polarity nor the local stimulus details matter

very much. For long-range position judgments the visual system must localize each of the features, and then compare the position labels of separate cortical mechanisms. The idea that cortical receptive fields have position labels (in addition to labels for other stimulus dimensions) is consistent with the topographical mapping of visual space in the brain (i.e. each point in space is systematically mapped on to the visual cortex). More than a century ago, Hermann Lotze[8] wrote:

"So long as the opinion is maintained that the space relations of impressions pass as such into the soul, it must of course, in the interest of the soul, be further held that each impression is conveyed to it by a distinct fibre, and that the fibres reach the seat of the soul with their relative situation wholly undisturbed".

Although we now know much more about the anatomy and physiology of the visual system than was known in the 1880s, Lotze clearly recognized that there must be a topographical representation of the world in the visual nervous system (if not the soul!), and that each "fibre" must carry a label about the position of the "impression" that it carried. For this reason, position labels were called "Lotze's local signs". Lotze also concluded that local signs played an important role in directing eye movements toward stimuli in the peripheral field of vision. So how accurate is "local sign" information? It turns out that humans can localize the position of a single peripheral feature to within about 1–2 percent of the eccentricity of the target, a little more precisely than we can make saccadic eye movements to peripheral targets.[9]

Limiting factors in visual acuity

In this section we explore the optical, anatomical and physiological factors that limit visual acuity. In the fovea, optics anatomy and neural mechanisms conspire to limit our visual acuity.

Optical quality of the eye

The optics of the eye spread the retinal image. Consider the retinal image of a distant star. Because stars are really far away, they are considered to be a "point source", i.e. if they were in perfect focus and the eye's optics were perfect, they would subtend an infinitely small angle at the eye. Point sources are useful if you want to learn about the quality of the eye's optics. As it turns out, the eye's optics are far from perfect; in fact your camera might have better optical quality. The eye's optics spread the image, so that a point in space forms a distribution (like a Gaussian) on the retina, as illustrated in Figure 33.4A. This distribution, like the office football pool, is known as the **point spread** function (if the source is a line it's called the line spread function). Figure 33.4B–D shows the retinal light distribution for a pair of nearby stars. As the stars become closer and closer together, their images overlap to a greater and greater degree, so that when they are very close, they look like a single distribution. When the separation between the stars on the retina is less than half the spread of each image, they will appear as a single star. This is known as the Rayleigh limit (Fig. 33.4D). The Rayleigh limit is determined mainly by the wavelength

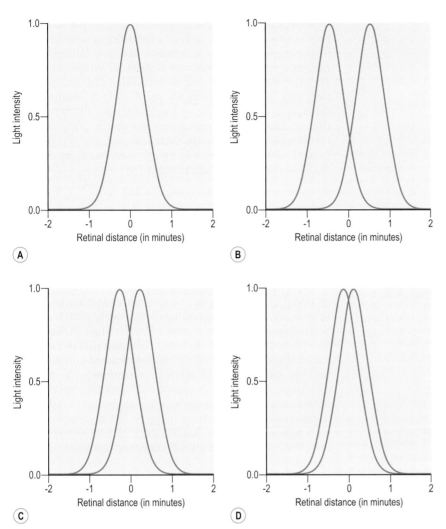

Figure 33.4 (**A**) Retinal image of a single point. Although the point is infinitesimally small, its image is spread on the retina. (**B–D**) Retinal image of a pair of nearby points, separated by distances equal to two (**B**), one (**C**) and half (**D**) the spread of each image. The latter case is known as the Rayleigh limit.

of the light and the size of the pupil (we'll soon discuss why). As noted earlier, the ability to distinguish double stars was one of the earliest measures of visual acuity, used by the Egyptians more than 5000 years ago.[1]

Another way to determine the quality of an optical system is to measure its **modulation transfer function**. This is typically done by passing test patterns of sinusoidal gratings (Fig. 33.5) of known contrast through the optical system, and measuring the contrast in the image. Sinewaves are characterized by their **spatial frequency** (i.e. the number of stripes in a given distance, usually specified in cycles per degree), their **contrast** (i.e. the difference in the luminance of the peaks and troughs divided by the sum of the luminance of the peaks and troughs), and their **phase** (the position of the peaks). Sinewaves are especially useful because even after they are degraded by an optical system, they maintain their characteristic shape (i.e. once a sinewave always a sinewave) they just become smaller in amplitude (and they may shift in phase). The ratio of image contrast to object contrast is a measure of image quality. By measuring the ratio of image contrast to object contrast for a range of object spatial frequencies, we can measure the modulation transfer of any optical system. The red line in Figure 33.6A shows the average modulation transfer functions of a large group of

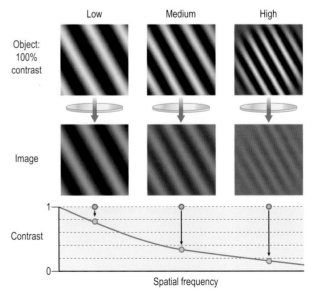

Figure 33.5 The modulation transfer function.

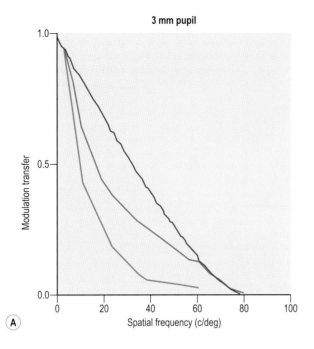

(A)

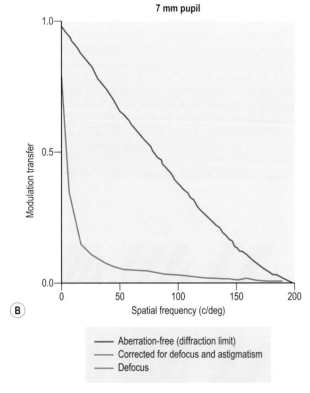

(B)

- Aberration-free (diffraction limit)
- Corrected for defocus and astigmatism
- Defocus

Figure 33.6 The modulation transfer function of the human eye. (**A**) The red line is the mean MTF of 12 eyes with a 3-mm pupil with refractive error corrected. The blue line is the MTF of an aberration-free eye (i.e. limited by diffraction only). The green line shows schematically the effect of uncorrected refractive error. (Adapted from Liang J. & Williams, D.R. (1997). Aberrations and retinal image quality of the normal human eye. J Opt Soc Am A Opt Image Sci Vis., 14, 2873–83.[10]) (**B**) The red line is the mean MTF of 14 eyes with a 7.3-mm pupil with refractive error corrected). The blue line is the MTF of an aberration-free eye (i.e. limited by diffraction only). (Adapted from Campbell, F. W., and Green, D. G. (1965). Optical and retinal factors affecting visual resolution. J Physiol 181: 576–593.[55])

observers with 3-mm pupils (about the average size of the pupil in a well-lighted room). As the object spatial frequency increases the modulation transfer declines, falling to near zero at a spatial frequency of around 80 cycles per degree. The spatial frequency at which the modulation transfer function falls to zero is called the **cutoff spatial frequency**. At this spatial frequency the optics do not transmit the sinusoidal variations in the luminance of the object. The theoretical cutoff spatial frequency (in cycles/degree), like the Rayleigh limit, is determined mainly by the wavelength of the light (λ), and the diameter of the pupil (d):

$$\text{Cutoff SF}(\text{c/deg}) = \pi/180 * \text{d}/\lambda$$

where d and λ are both specified in millimeters (mm). For red light (630 nm or 0.000630 mm as used by Liang & Williams[10]) and a 3-mm pupil (about as small as they get) the predicted cutoff spatial frequency is 83 c/deg.

What limits optical image formation of the eye?

As noted above, the optics of the human eye are imperfect. Why do the eye's optics not match up to a fine camera? One limitation is **diffraction** produced by the pupil of the eye (or the lens aperture of a camera). For example, when light passes through the aperture, instead of staying in a straight line, the light from each ray will be scattered into different directions. This scatter is known as diffraction, and its consequence is to spread or defocus the image and to reduce the transfer of high spatial frequencies in the modulation transfer function. The blue lines in Figure 33.6 show the diffraction limit (i.e. the modulation transfer of an aberration free optical system). As noted above, the cutoff spatial frequency in the modulation transfer function depends on the pupil diameter. The red line in Figure 33.6B shows the modulation transfer function of a large group of human eyes for a large (7-mm) pupil size. Note that for the large pupil diameter the modulation transfer function is much poorer than predicted by the theoretical limit set by diffraction.

In both panels of Figure 33.6, the eye's optics are always worse than the theoretical limit, especially when the pupil is large. One important reason is that the eye exhibits **aberrations**. One form of aberrations in the eye is known as **spherical** aberrations, in which light rays passing through different parts of the eye's optics are focussed at slightly different points in the image plane. Larger pupils will result in more aberrations because more peripheral rays of light will enter the eye. There are also **chromatic** aberrations. If the light source (like starlight) is a mixture of light of different wavelengths, then not all of the wavelengths will be in focus on the retina (at any given time), so the eye will have different point spreads for different wavelengths.

Refractive error and defocus results in a marked loss of image quality

The refractive error of the eye is determined by the refractive power of the optical components of the eye (i.e. the power of the cornea and lens) and the length of the eyeball. To focus a distant point source on the retina, the refractive power of the optical components of the eye must be perfectly matched to the length of the eyeball. This perfect match is known as emmetropia. If there is an uncorrected refractive error or the optics of the eye are defocused, the image of a distant object on the retina will be spread out (i.e. the point spread function will widen) and the modulation transfer function will be low (green line in Fig. 33.6A).

For myopes, the power of the optical components is too strong for the length of the eyeball, causing the light rays to converge too much to focus on the retina. Myopia cannot be remedied through accommodation because accommodation will increase the convergence of the light. The amount of image spread depends on the amount of defocus, and the pupil size. Reducing pupil size increases the depth of focus, so that defocus has less effect with small than with large pupils. As a rule of thumb, 1 diopter of uncorrected simple myopia will result in a decrease of Snellen acuity, on average, to $\approx$ 20/60.

Myopia can be corrected with negative (minus) lenses, which diverge the rays. On the other hand, hyperopia (farsightedness) occurs when the power of the optical components is too weak for the length of the eyeball, causing the light rays to not converge enough to focus on the retina (the image plane lies behind the retina in an unaccommodated hyperopic eye). If the hyperopia is not too severe, a young hyperope can compensate by accommodating, and thereby increasing the power of the eye.

The most powerful refracting surface in the eye is the cornea, which contributes about two-thirds of the eye's refracting power. When the cornea is not spherical the result is astigmatism. With astigmatism, a point source would not have a single point focus on the retina (i.e. the pointspread function would be asymmetric), and lines of different orientations would be focussed in different planes. Although there are other causes for astigmatism, it is usually caused by asymmetry of the front surface of the cornea. Cylindrical lenses that have two focal points can correct astigmatism (e.g. they have different power in the horizontal and vertical meridians).

Photoreceptor size and spacing; aperture size; the "Nyquist" limit; aliasing

Retinal anatomy plays a very important role in setting the limits for foveal resolution acuity. Foveal cones are arranged in a regular arrangement (sometimes called a triangular array) and are densely packed (see Fig. 33.7). This dense packing is critical to fine resolution. The reason is simple. The visual world is represented by continuous variations in light intensity. However, our visual system "samples" the world discretely, i.e. by looking at the light distribution through many small individual photoreceptors. The more closely packed the photoreceptors, the better the light distribution can be represented to the visual nervous system. As illustrated in Figure 33.1, there is a sampling limit. In order

to properly represent the peaks and troughs in the intensity profile of a sinewave, there must be at least two cones for each cycle of the grating. This is known as the **Nyquist limit** and is represented in Figure 33.1B. More formally, the Nyquist limit occurs when the sinewave has a spatial period (i.e. the peak to peak distance) that is two times the cone spacing. When the spatial period is smaller than twice the cone spacing (i.e. when the sinewave spatial frequency is higher than the Nyquist frequency) the phenomenon known as "aliasing" may occur, i.e. rather than the cones signaling a perfect replica of the sinewave on the retina, the signal is distorted – it appears to be a sinewave with a lower spatial frequency than the original (Fig. 33.8). Since cones in the fovea have a center to center separation of around 0.5 arc minutes, the cone sampling (or Nyquist) limit is a grating spatial period of about 1 minute (i.e. a grating spatial frequency of 60 cycles per degree. Note that because the cones have triangular packing, geometry dictates that the Nyquist frequency will actually be about 15 percent higher or roughly 69 c/deg). This represents a fundamental limitation set by the spacing of the foveal cones, and it is not far off the limitations set by the eye's optics. Thus, in the fovea, the cone spacing is nicely matched to the eye's optics. As we'll see later in this chapter, in peripheral vision cone spacing changes dramatically, while the optics change only a little.

The notion that there is a "highest" spatial frequency that can be accurately represented by the visual system (the Nyquist frequency) raises the interesting question of whether aliasing actually occurs in human vision. Fortunately for us, our optics "protect" us from the effects of aliasing in the fovea by reducing the contrast of high spatial frequencies. Interestingly, it is possible to create a very high spatial frequency pattern on the retina bypassing the optics of the eye (this can be accomplished using lasers to create interference patterns). Under these conditions, humans report the appearance of aliasing – fine gratings appear coarse (and often appear to change their orientation). Figure 33.8 illustrate how viewing a grating through too coarse a grid produces the kind of wavy, coarse appearance produced by aliasing.

Cone to ganglion cell convergence

If many cones converged on a single ganglion cell, the advantage of having closely spaced cones would be lost. Recall however that in the fovea each cone effectively has a "private line" to the brain via the midget bipolar and ganglion cells. So, the receptive field centers of midget ganglion cells are effectively one cone wide. In peripheral vision resolution is limited by the retinal ganglion cells that are widely

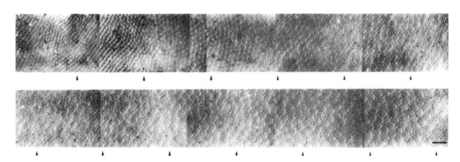

Figure 33.7 Inner segments of a human cone mosaic in a strip extending from the foveal center (upper left arrow) along the temporal horizontal meridian. The large cells are cones and the small cells are rods. The horizontal scale bar indicates 10 μm. (From Hirsch, J, Curcio, C. The spatial resolution capacity of human foveal retina. Vision Research, 29, 1095–1101; 1989.[94])

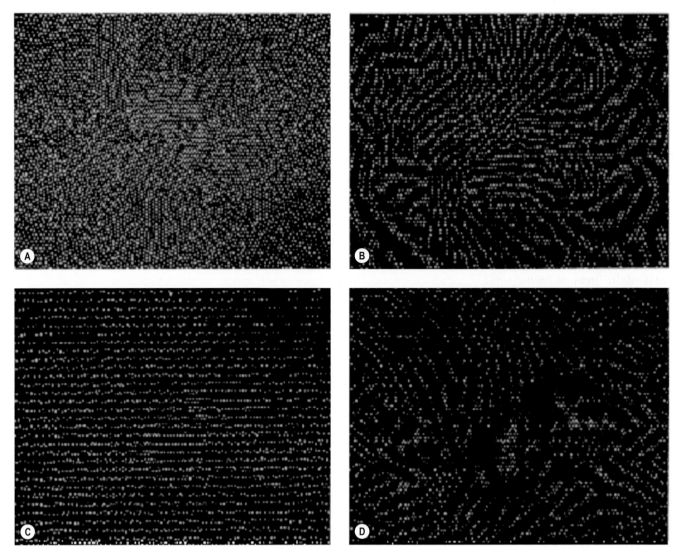

Figure 33.8 (**A**) The foveal cone mosaic is represented as sample points (bright dots). In the other panels it has been sandwiched with gratings of 40 cpd (**C**), 80 cpd (**B**) and 110 cpd (**D**). (From Williams DR. Topography of the foveal cone mosaic in the living human eye. Vision Research, 28, 433–54; 1988. Copyright Elsevier 1988.[95])

spaced (sparsely sampled) and that pool information from many cones.

Eccentricity

In contrast to the tight packing of cones in the fovea, cone density decreases dramatically with retinal eccentricity, and there is substantial convergence of cones to ganglion cells (Chapter 32). Humans have many more rods (about 90 million) than cones (about 4–5 million), and they have very different geographical distributions on the retina (Fig. 33.9). The cones are most concentrated in the center of the fovea, and their density drops off dramatically with retinal eccentricity. Rods are missing from the center of the fovea, and their density increases to a peak at about 20 degrees, and then declines again. As can be seen in Figure 33.7, the cones in the fovea are much more densely packed than in the peripheral retina, and there are no rods in the section from the foveal center. This "rod-free" area (about 300 micrometers on the retina) subtends a visual angle of

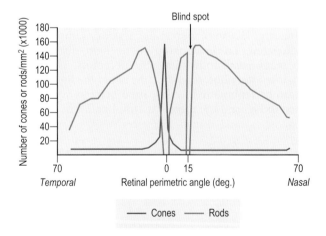

Figure 33.9 Distribution of cones (blue) and rods (red) across the retina. (Adapted from Osterberg, G. (1935). Topography of the layer of rods and cones in the human retina. Acta. Ophthal., Suppl. 6, 1–103.[96])

about 1 degree. Moreover, unlike the fovea, in the retinal periphery there is considerable convergence of cones onto ganglion cells.

Almost 140 years ago, Aubert & Forster[11] demonstrated that visual acuity declines in an orderly fashion with eccentricity – an important observation that has been often repeated and extended to a multiplicity of tasks over the years. About 100 years later, Weymouth[12] showed that many visual functions degrade approximately **linearly** with eccentricity, and we can characterize the rate of fall-off of many visual functions (as well as the anatomical structures that limit performance) by the gradient, which can be represented by a single number called E_2. E_2 represents the eccentricity at which the foveal value has doubled.[13,14]

Daniel & Whitteridge[15] coined the term "**cortical magnification factor**" or CMF, to describe the cortical distance (usually specified in mm) devoted to one degree in the visual field. Although there are still uncertainties about the precise CMF of the fovea, in humans, the best estimates (based on functional imaging) suggest that it is about 20 mm/deg (i.e. 1 degree in visual space is represented over a distance of about 20 mm in the cortical representation of the fovea). In contrast, at an eccentricity of 10 degrees, 1 degree in visual space is represented over a distance of only about 1.5 mm in the cortex.

The CMF is generally specified in mm/deg. A convenient simplification is that the inverse CMF (i.e. the number of degrees per millimeter) varies approximately linearly with eccentricity (Fig. 33.10A). Thus, in the human fovea, approximately 0.05 degrees of visual space (3') is represented in 1 mm of striate cortex, while at 10 degrees eccentricity about 0.67 degrees is represented in 1 mm. The red line in Figure 33.10B shows the gradient (i.e. the rate of variation of inverse magnification with eccentricity), by plotting the ratio of peripheral to foveal inverse magnification at each eccentricity. Essentially, this normalizes inverse magnification at each eccentricity by the foveal value. An advantage of using the gradient is that it makes it simple to compare changes in anatomy or performance with eccentricity for different structures and functions.

For inverse cortical magnification (the red line in Figure 33.10B) the gradient is steep and E_2 is $\approx$ 0.77 degrees.[14] In contrast, the gradient for midget retinal ganglion cell size (the blue line in Figure 33.10B – from Peterson & Dacey[16]) is shallow, and E_2 is about 3.7 degrees. The importance of the gradients of performance is that they may provide clues as to the structures, at various levels of the visual pathway, which limit performance. Indeed, one of the reasons Polyak[17] gave for undertaking his classical study of the retina was to obtain a better understanding of the "striking difference between central and peripheral acuity". Different psychophysical tasks also have different gradients. For example Vernier acuity has a steep gradient, while contrast sensitivity (discussed below) has a shallow gradient.

Figure 33.11 shows a visual acuity chart in which the letters are "scaled" in size, so that each letter covers an approximately equal cortical distance, and interestingly, the letters should be approximately equally visible!

Why is the foveal representation in the cortex so highly magnified? The visual system must make a trade off. High resolution (like our foveas) requires a great deal of resources – a dense array of photoreceptors; a one-to-one line from

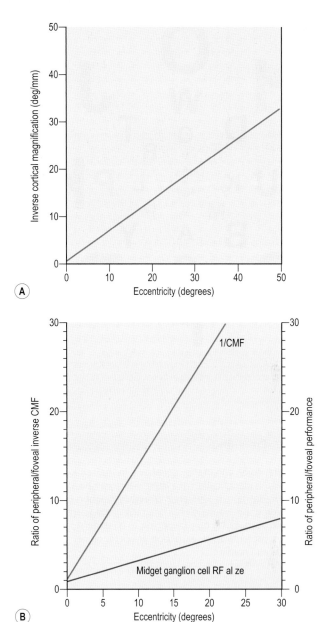

Figure 33.10 (**A**) The inverse cortical magnification factor (CMF in degrees per millimeter) versus eccentricity. (**B**) The red line shows the gradient (i.e. the rate of variation) of inverse magnification with eccentricity, by plotting the ratio of peripheral to foveal inverse magnification at each eccentricity. For inverse cortical magnification the gradient is steep. In contrast, the gradient for midget retinal ganglion cell size (the blue line) is shallow.

photoreceptors to retinal ganglion cells, and a large chunk of cortex. If we could see the entire visual field with such high resolution, we might need to have eyes and brains too large to fit in our heads! Thus, we have evolved a visual system that provides high resolution in the center, and lower resolution in the periphery. In large part, the high foveal magnification in the cortex reflects the topography of the retina[18] (see Fig. 33.7); however, there appears to be additional magnification of the foveal representation in the cortex,[19] and the striate cortex contains more than 100 times more cells than LGN.

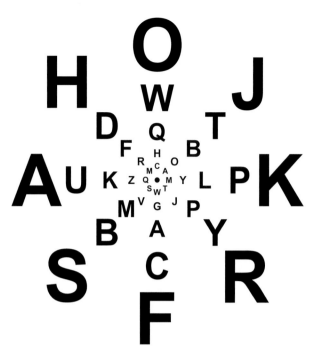

Figure 33.11 A visual acuity chart in which the letter size has been "scaled" according to the cortical magnification factor. (From Anstis SM. A chart demonstrating variations in acuity with retinal position. Vision Research, 14, 589–92; 1974. Copyright Elsevier 1974.[97])

Box 33.2 Crowding

In peripheral vision and in the central field in strabismic amblyopia, the identification of a letter, which can be easily identified in isolation, is severely impaired by neighboring letters. This "crowding" phenomenon has been discussed scientifically for some 60 years, but is only beginning to be understood. Crowding represents an essential bottleneck for object recognition and reading.

Crowding in peripheral vision

In peripheral vision, the identification of a letter is severely impaired by neighboring letters (Box 33.2). This "crowding" phenomenon[20] has been discussed scientifically for some 60 years, but is only beginning to be understood (see recent reviews[21,22]). The reader can experience this phenomenon by viewing Figure 33.12. When looking at the fixation point, the letter R is clearly legible in the top panel, where it is presented in isolation. However, it is much more difficult to read in the lower panel, when flanked by other letters.

Inspection of Figure 33.12 makes it obvious that crowding does not result in reduced apparent contrast – rather, crowded letters are high contrast but indistinct or jumbled together. Tyler & Likova[23] also note their strong subjective impression of a "gray, or inchoate, smudge between the two outer letters, including the inner parts of those letters."

In peripheral vision, the spatial extent of crowding depends on eccentricity, and can be as large as ≈ 0.5 times the target eccentricity. Several recent studies have varied both target size and eccentricity, and show that the extent of peripheral crowding is more or less invariant to target size.[24–27]

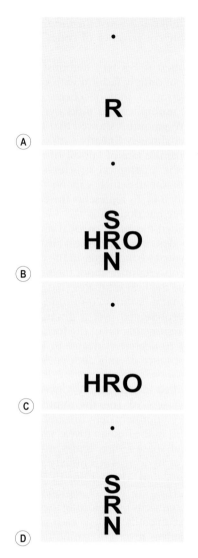

Figure 33.12 Crowding. The reader can experience crowding by fixating the dot, and trying to identify one letter: in isolation (**A**), surrounded by 4 random flanking letters (**B**), surrounded by 2 horizontally placed random flanking letters (**C**), surrounded by 2 vertically placed random flanking letters (**D**). (From Levi DM. Crowding—An essential bottleneck for object recognition: A mini-review. Vision Res., 48, 635–54; 2008. Copyright Elsevier 2008.[21])

The strength and extent of peripheral crowding are much greater than the strength and extent of masking,[24,28] so that in peripheral vision, the suppressive spatial interactions due to nearby flanks are not likely to be a consequence of simple contrast masking.

A number of studies, using very different stimuli and tasks, have shown convincingly that crowding is indifferent to whether the target and flanks are presented to the same eye or to different eyes (target to one eye, flanks to the other[14,29–31]). The fact that crowding occurs when target and flanks are presented to separate eyes immediately places the site of the interaction at or beyond the site of combination of information of the two eyes. Remarkably, this dichoptic interaction even occurs when the flankers are presented around the blindspot of one eye, and the target in the "monocular" region corresponding to the blindspot of the other eye.[32] This is both surprising and interesting because

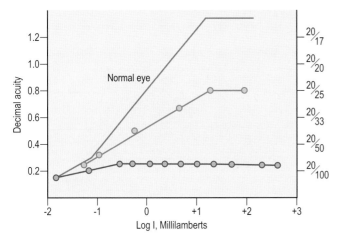

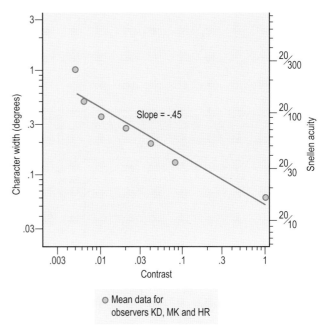

Figure 33.13 Acuity vs log luminance for normal fovea (red) and for patients with retinal lesions due to a disease of Bruch's membrane (green) and macular toxoplasmosis (blue). (Modified from Doc Ophthalmol.[40])

Figure 33.14 The effect of contrast on Snellen acuity. (From Legge GE, Rubin GS, Luebker A, et al. Psychophysics of reading—V. The role of contrast in normal vision. Vision Res., 27, 1165–77; 1987. Copyright Elsevier 1987.[47])

there is a complete absence of direct retinal afferents from one eye to this region of cortex.

Note that there is no evidence of crowding when looking directly at the letters in Figure 33.12. Except near the limit of resolution[33] in the normal fovea the extent of "crowding" is proportional to stimulus size and cannot easily be distinguished from ordinary masking.[34,35] At the limit, foveal crowding extends over a tiny distance just 4–5 minutes of arc.[33,36–38]

We will return to the issue of crowding in the section on amblyopia.

Luminance

Visual acuity is usually measured under moderate photopic luminance conditions (80 to 320 cd/m^2), and under those conditions (and indeed luminance levels ranging from full moonlight to a bright sunny sky) visual acuity remains fairly constant.[2,39,40] At very low luminance levels, visual acuity is markedly reduced. Under very low (scotopic) luminance conditions visual acuity is mediated by rods, and rod acuity asymptotes at an acuity level between 20/100 and 20/200. Figure 33.13 shows the acuity vs log luminance function for normal fovea (red) and for patients with retinal lesions due to a disease of Bruch's membrane (green) and macular toxoplasmosis (blue). Retinal disease lowers both the upper asymptote and the slope of the acuity vs luminance function.

Similarly, Vernier acuity for abutting lines is markedly reduced at low luminance levels.[42] Reducing luminance alters the visibility of a stimulus. For example, the contrast (ΔI/I) necessary to detect a thin line increases as retinal illuminance is reduced. Specifically, both line detection thresholds and (abutting) Vernier thresholds decrease with the square root of retinal illuminance, suggesting that photon statistics may be the main source of information loss.[7,41,42] Interestingly, when the effect of luminance on target visibility is taken into account (by making the target contrast a fixed multiple of the detection threshold at each luminance level) Vernier acuity becomes independent of target luminance.[42]

Contrast

Visual acuity is generally tested with high contrast (≈ 100% contrast) optotypes, and reducing the contrast results in a systematic reduction of acuity.[43–46] Figure 33.14 (from Legge et al[47]) shows that one's visual acuity for Sloan letters varies linearly on log–log coordinates, with a log–log slope of ≈ −0.5. In other words, visual acuity improves in proportion to the square root of the target contrast. Similarly, other types of acuity (e.g. abutting Vernier acuity) depend strongly on target contrast. Interestingly, relative position judgments with widely separated targets show little dependence on contrast, luminance, or polarity, as might be expected if these judgments are limited by a local sign mechanism (discussed earlier).

Time

For many visual tasks including visual acuity and abutting Vernier acuity, performance is degraded when the stimulus presentation is brief.[48,49] This effect can largely be understood on the basis of stimulus energy or visibility, i.e. reducing the duration below a "critical" duration, reduces the number of quanta and the visibility of the stimulus. This effect of duration can be cancelled by increasing the illumination or contrast of the stimulus to maintain a constant visibility (or a constant number of absorbed quanta[2,49,50]).

Motion

Even when you try to hold your eyes completely stationary, they continue to execute small but important movements. These involuntary eye drifts and small jerks keep the retinal image "fresh"; if the eye muscles are temporarily paralyzed – say, as a result of taking curare[51] – the entire visual world gradually fades from view. Using retinal image stabilization, Keesey[52] showed that these small involuntary eye

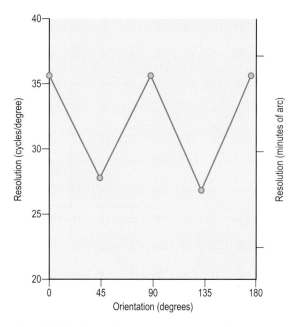

Figure 33.15 Oblique effect. (Modified from J Physiol.[98])

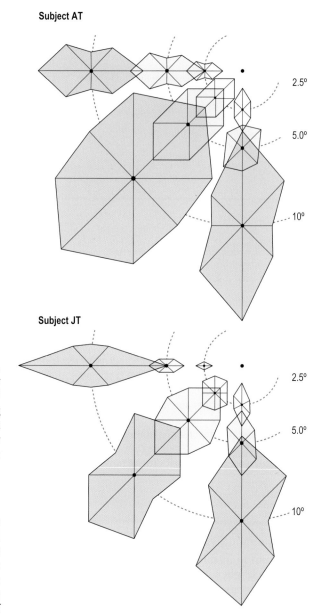

Figure 33.16 The 2-dimensional shape of "crowding" in foveal (the small dot in the center) and peripheral vision (at 2.5. 5 and 10 degrees). (From Toet A, Levi DM. The two-dimensional shape of spatial interaction zones in the parafovea. Vision Research, 32, 1349–57; 1992. Copyright Elsevier 1992.[38])

movements have little or no effect on visual acuity. Indeed, visual acuity measured with Landolt C's and Vernier acuity are unaffected by retinal image motion of up to ≈ 2.5 degrees/second.[53] Beyond that, Vernier acuity for targets consisting of closely spaced dots or lines is strongly degraded by fast motion. For very large blurred stimuli (low spatial frequency sinusoids), Vernier acuity is quite robust to even very high velocities.[54]

Anisotropies

There are anisotropies in visual acuity. For example, in the fovea visual acuity and contrast sensitivity at high spatial frequencies are better for horizontal and vertical gratings than for oblique (Fig. 33.15). The W-shaped function shown in Figure 33.15 is known as the oblique effect.[55] It is not an optical effect since it occurs when interference fringes are imaged directly on the retina, thus bypassing the optics of the eye.[56] In peripheral vision the oblique effect gives way to a radial organization, with acuity being better for gratings oriented along a line connecting the fovea to the peripheral location (radial orientation) and worse for the orthogonal orientation. The most profound anisotropy is found under conditions of crowding. Crowding in peripheral vision is much more extensive when the flankers are radially arranged than when they are tangentially arranged (Fig. 33.16).[27,38] The reader can see this effect for themselves in the lower two panels of Figure 33.12. In panel (D), the flanks are arranged radially, in panel (C), tangentially.

Visual acuity and reading

The ability to read quickly and accurately is crucial in our society, and difficulty in reading is one of the most frequent reasons that patients visit their eye doctor. Clearly, if acuity is substantially reduced, reading will be impaired. However, recent work shows that it is text spacing not text size that limits reading speed. When the text spacing is closer than a critical spacing, reading is slowed. The critical spacing for reading is equal to the critical spacing for crowding[22,27,57] (Fig. 33.17). The notion that crowding limits reading is surprising and perhaps controversial, because reading rate has generally been thought to be linked to letter size rather than spacing. Changing the viewing distance for normal reading material changes both size and spacing; however, experiments measuring reading performance versus letter size with normal and doubled letter spacing (i.e. 1.1 and 2.2 times the letter size) reveal that it is spacing that matters.[57] Thus crowding appears to represent a limit for reading in normal and amblyopic central vision.[57] This is consistent with the conclusions of Legge and his co-workers that the "visual span" (i.e. the number of letters in a line of text that can be recognized reliably without moving the eyes) represents a sensory bottleneck for reading.[58–60]

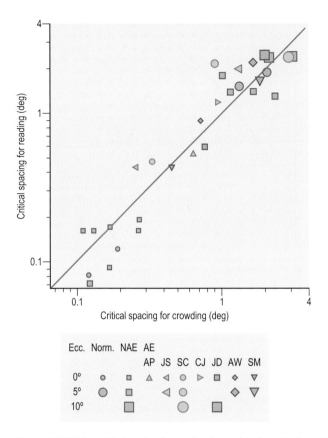

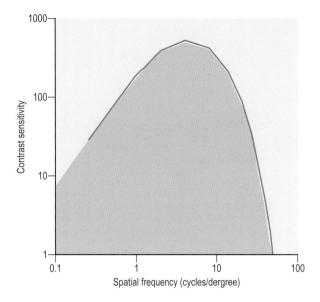

Figure 33.18 The contrast sensitivity function: our window of visibility.

Figure 33.17 The critical spacing for reading is equal to the critical spacing for crowding for normal foveal, peripheral and amblyopic vision. (Reproduced with permission from Association for Research in Vision and Ophthalmology[57])

While the critical spacing for crowding and reading are also identical in peripheral vision (suggesting that crowding represents one limit to fast reading in the periphery), reading slows with increasing eccentricity even when crowding is controlled for (by scaling with eccentricity), for reasons that remain mysterious.

Spatial vision with low contrast

Up until now we've been discussing the smallest, high-contrast details that we can resolve; however, most of the objects in our world are larger than the resolution limit, and have lower contrast.

How can we measure and quantify how well we see objects that are larger than the resolution limit? Recall how the performance of an optical system was measured earlier in this chapter. Test patterns consisting of sinusoidal gratings (see Fig. 33.5) of known contrast and a range of sizes (spatial frequencies) were passed through the optical system, and the ratio of image contrast to object contrast was used as a measure of modulation transfer of the optical system. Sinusoidal gratings have some important advantages for testing optical systems, and therefore should be useful in characterizing human pattern vision. There is, however, a problem. In an optical system, it is relatively easy to measure the contrast of the image; however, that's not so easily done in

human vision, because the image is on the retina. More importantly, how well we see sinusoidal gratings will be determined by both the optical components of the eye, and by any neural processing in the visual pathway. Thus, what we really want to know is about the quality of the image that reaches the brain and gives rise to perception. Schade[61] first addressed this by showing people sinusoidal gratings and asking them to adjust the contrast of the gratings until they could just be detected. By repeating these measurements for a wide range of different spatial frequencies (stripe widths), Schade was able to describe the "contrast sensitivity function" – i.e. contrast sensitivity for gratings of different spatial frequencies. Almost 10 years later, Campbell & Green[55] published what is widely regarded as a classical paper, describing a relatively simple method for measuring the contrast sensitivity function, and showed that both optical and neural processes influence the function.

The contrast sensitivity function represents our window of visibility

The red line in Figure 33.18 shows a contrast sensitivity function (CSF). The vertical axis shows the observer's contrast sensitivity plotted as a function of the spatial frequency of the sinusoidal gratings (with low spatial frequencies, or broad stripes represented on the left, and high spatial frequencies or fine stripes, on the right). Surprisingly, the function is shaped like an upside-down U. The modulation transfer of the eye (or any other optical system) reduces the contrast of high spatial frequencies, so the fall-off of sensitivity at high spatial frequencies is to be expected based on the optics of the eye. However, the fall-off of sensitivity at low spatial frequencies is unexpected. Poor optics will not reduce sensitivity at low spatial frequencies, so this low spatial frequency fall-off must be due to neural factors (we'll discuss these later).

Contrast sensitivity is obtained by measuring the smallest amount of contrast needed to detect the target (the contrast threshold), and sensitivity is defined as the reciprocal of the threshold contrast. Thus, if the contrast at threshold is 0.01,

Figure 33.19 Contrast and SF modulated grating (Robson & Campbell). (From Robson, J., and Campbell, F. (1997). A quick demonstration of your own contrast sensitivity function. In D. G. Pelli and A. M. Torres, Thresholds: Limits of Perception. New York: NY Arts Magazine.[99])

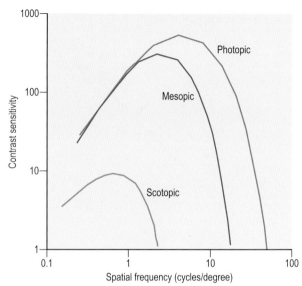

Figure 33.20 The CSF at photopic, mesopic and scotopic luminance levels.

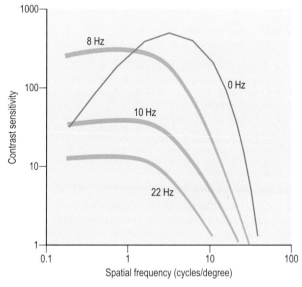

Figure 33.21 The CSF at 3 temporal frequencies. Note that it changes its shape at higher temporal frequencies.

contrast sensitivity will be 100, and a contrast threshold of 0.1 will give a contrast sensitivity of 10. Thus, the red line in Figure 33.18 corresponds to the lowest contrast one can see, and the pink region under the curve represents the window of visibility. Any pattern whose spatial frequency and contrast fall within the pink region will be visible, while any combination of spatial frequency and contrast in the region above the curve will be invisible. For example, a spatial frequency of 1 c/degree will be invisible if its contrast is less than 0.01 and will be visible with any contrast greater than 0.01. For this spatial frequency, the contrast sensitivity is 100. Note that contrast sensitivity equal to 1 corresponds to a contrast of 100 percent. This is the rightmost point on the CSF (at around 50 c/degree), and corresponds to the observer's resolution limit. It is the finest grating the observer can detect with 100 percent contrast.

Figure 33.19 allows you to visualize your own CSF. This figure shows a sinusoidal grating whose contrast increases continuously from the top of the figure to the bottom, and whose spatial frequency increases continuously from the left side of the graph to the right side. If you hold the figure at a distance of about 50 cm, you will notice the inverted U shape where the grating fades from visibility to invisibility. This is your own CSF – because it tells us how your sensitivity varies over a wide range of target sizes, it is more informative than visual acuity.

Figure 33.18 illustrates a typical CSF, obtained with foveal viewing of stationary gratings at high levels of illumination (i.e. using cone vision). The CSF changes its shape at low luminance (Fig. 33.20). For example, at very low (scotopic) luminance levels, sensitivity is markedly reduced at all spatial frequencies, and the low spatial frequency fall off is

less pronounced. At intermediate (mesopic) luminance levels, CSF is reduced mainly at high spatial frequencies. The CSF also changes its shape when the gratings are moved, flickered on and off or temporally modulated in other ways (Fig. 33.21). These changes in the shape of the CSF reflect neural changes.

The CSF in peripheral vision

Visual resolution declines with eccentricity. How the CSF changes with eccentricity depends in large part on how it is measured. In early studies, peripheral CSFs were measured with a patch of gratings that was fixed in size (say 1 degree). As illustrated in Figure 33.22A, when a fixed-size patch is viewed in the periphery, contrast sensitivity decreases dramatically at all spatial frequencies – with the most marked

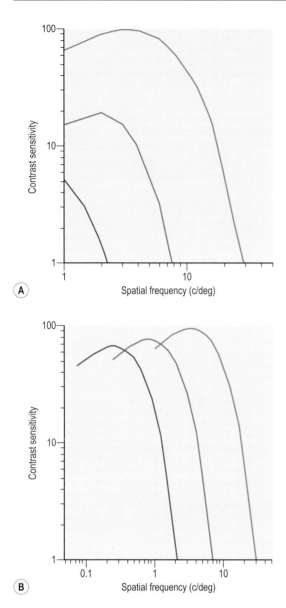

(A)

(B)

Figure 33.22 Contrast sensitivity functions at 3 eccentricities: fovea (green), 7.5 degrees (red) and 30 degrees (blue). (**A**) With fixed field size (2 degree half circle). (**B**) With field size scaled in proportion to ganglion cell density at each eccentricity. (Modified from Rovamo, J., Virsu, V. & Näsänen, R. (1978). Cortical magnification factor predicts the photopic contrast sensitivity of peripheral vision. Nature, 271, 54–6. By permission from Macmillan Publishers Ltd.[62])

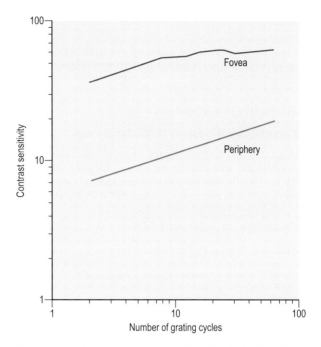

Figure 33.23 Contrast sensitivity vs number of cycles for fovea (blue) and periphery (red). (Modified from Robson JG, Graham N. Probability summation and regional variation in contrast sensitivity across the visual field. Vision Res., 21, 409–18; 1981. Copyright Elsevier 1981.[63])

effect at high spatial frequencies. Based on these types of measurements, it was commonly assumed that the periphery was basically able to function to detect light and movement, but not to be capable of making useful judgments about patterns or form. However, Rovamo et al[62] showed that the peak contrast sensitivity of the periphery could be almost (but not quite) as high as that of the fovea, if the stimulus size was magnified in order to compensate for the reduced representation of the periphery in the visual nervous system. This basic idea, now widely known as **magnification theory**, has been shown to apply to a wide variety of visual stimuli and tasks. Rovamo et al[62] magnified the stimulus size by a scale factor that was actually based on retinal ganglion cell density at each eccentricity, so in essence, they equated the number of ganglion cells that were stimulated by the patch

of grating at each eccentricity. (This "scaling" was accomplished by viewing the same patch on the screen from successively closer viewing distances as the eccentricity increased, thus increasing the angular size of the stimulus patch.) The result is that peak contrast sensitivity is almost the same at each eccentricity (Fig. 33.22B). Note, however, that even with the larger field size, both the peak of the CSF and the cutoff spatial frequency shift toward lower spatial frequencies as eccentricity increases.

How is it that peripheral contrast sensitivity improves so much at low spatial frequencies when the patch size is increased? Robson & Graham[63] showed that the improvement could largely be explained by a process known as **probability summation**. They measured contrast sensitivity for patches with different numbers of grating cycles. They found that contrast sensitivity improved as they added cycles, and the improvement followed a power function (i.e. a straight line on log–log axes). In the fovea, the function saturated when there were around 8 cycles, so that adding yet more cycles did not improve sensitivity much (Fig. 33.23 – blue line). However, in peripheral vision sensitivity improved up to 64 cycles (the maximum number they tested; see Fig. 33.23 – red line). Recall that the scaling method of Rovamo et al increased the field size when testing the periphery. Thus, the periphery would get the benefit of having more cycles of the grating than the fovea, and as Robson & Graham showed the periphery uses every cycle it gets! Robson & Graham argued that the improvement in sensitivity as more cycles are presented, accrues from having more opportunities to detect the pattern. Their notion of probability summation across space was that the grating would be detected if it landed on a mechanism (receptive field) that was sensitive to the target, and signaled its presence, and they further assumed that each receptive field acted independently. Based on these simple assumptions, they were able to predict the effect of the number of cycles by simply pooling the responses of independent mechanisms sensitive to the target.

Clinical testing of visual acuity

Visual acuity chart design considerations

The original Snellen chart consisted of seven letter sizes, with a single (20/200) letter at the top, two (20/100) letters on the next line, with the number of characters (letters and numbers) increasing on each subsequent line, and the inter-letter spacing varying from line to line see (Fig. 33.2). Moreover, the change in letter size between the 20/200 and

20/100 lines is very coarse (50%), while the change from 20/50 to 20/40 is much finer (20 percent), so that testing accuracy varies depending on the acuity level.

Most contemporary charts incorporate some or all of the design principles (Box 33.3) proposed by Bailey & Lovie;[64] the same number of letters on each line; a constant ratio from one letter size to the next (i.e. each line is a fixed percentage smaller than the previous line); spacing between letters and lines proportional to letter size, and nearly equal legibility to the optotypes (or nearly equal average legibility at each letter size). A distinguishing feature of charts that incorporate these principles is the V-shaped appearance, since each line is proportionally smaller than the preceding line. Figure 33.24 shows two such charts, the Bailey–Lovie chart and the Early Treatment of Diabetic Retinopathy Study

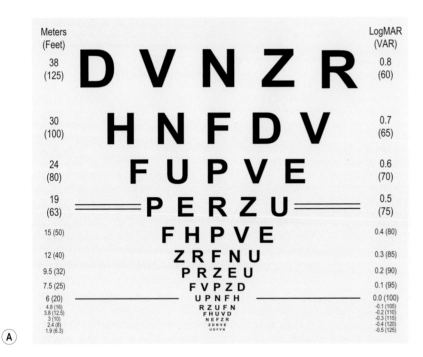

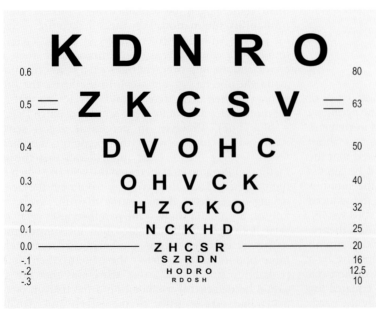

Figure 33.24 The Bailey–Lovie chart (**A**) and the Early Treatment of Diabetic Retinopathy Study (ETDRS) charts (**B**). Courtesy of Dr. Ian Bailey.

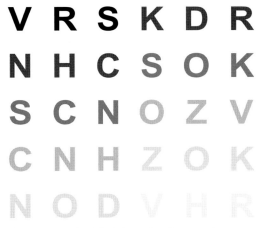

Figure 33.25 The Pelli–Robson chart for measuring contrast sensitivity. (From Pelli, D. G., Robson, J. G., & Wilkins, A. J. (1988) The design of a new letter chart for measuring contrast sensitivity. Clinical Vision Sciences 2, 187–199. Copyright Elsevier 1988.[68])

(ETDRS) charts. The constant ratio (logarithmic) size scaling is based on the psychophysical principle that just-noticeable differences and their variability scale logarithmically.[65]

One impetus for using proportional spacing is the idea that letter spacing should be proportional to letter size in order to keep the effect of contour interaction consistent across acuity levels.[36] However, in peripheral vision[24] and in the central field of strabismic amblyopes crowding[66] is a fixed spatial distance, not a fixed multiple of the target size.

Clinically, acuity can be measured with a variety of optotypes and charts. These include illiterate E's; Landolt C's and letters. One of the great advantages of using letters for measuring acuity is that the likelihood of randomly guessing the correct letter is low ($\approx$ 4% or 1 out of 26 if all the letters are used). Many charts used a reduced letter set, and for these the guess rate will be higher. For example, the ETDRS chart uses the Sloan letter set (C,D,H,K,N, O,R,S,V,Z), while the Bailey–Lovie chart uses the British letter set (D,E,F,H,N,P,R,U,V,Z). For both of these, the guess rate is 10 percent (1 out of 10). Operationally, visual acuity charts come in several formats: printed, projected and computer-generated.

There is a wide range of special methods for testing infants, toddlers and others who have limited capacity to respond to standard acuity charts. These run the gamut from objective tests such as visual-evoked potentials (VEP), optokinetic nystagmus (OKN) and preferential looking (PL), often used for infants and discussed in Chapter 38, to symbol and picture charts.

Clinical tests for CSF

Contrast sensitivity is not routinely measured in the clinical setting; however, there are now several methods available for doing so. These include the Functional Acuity Contrast Test (FACT CS Test – Stereo-Optical Co) that replaced the original Vistech chart.[67] This chart has 5 different sinusoidal grating frequencies, each presented at 9 contrast levels, allowing determination of the full CSF. A more popular approach is the Pelli–Robson CS chart[68] (Fig. 33.25) which consists of 16 "triplets" of letters. The letters in each triplet have the same contrast, with the contrast between successive triplets decreasing by 0.15 log units. Since the patient's task is to simply read the letters as far down the chart as they can, it is essentially identical to other visual acuity measures. What is different is that letter size is constant (0.5 degrees at 3 meters or 20/120) and contrast varies. There are also several conventional charts (varying letter size) that are available in both high- and low-contrast

versions (e.g. the Bailey–Lovie chart and the Regan low contrast chart). It should be noted that different clinical contrast sensitivity tests measure different aspects of vision. For example, the Pelli–Robson test (fixed large letter size) provides information about the sensitivity near the peak of the CSF, while the low-contrast variable letter size tests (e.g. Bailey–Lovie and Regan charts) provide information about sensitivity along the high-frequency limb of the CSF (i.e. the smallest letter size that can be resolved at a fixed low contrast). Although contrast sensitivity testing provides useful information for clinical trials, assessing treatment outcomes, etc., it is not clear how widely used it is in clinical practice. For example, fewer than 20% of eye-care practitioners who responded in a survey, frequently or always measured contrast sensitivity in pre-operative cataract patients.[69]

Glare

Glare – the effect of a bright or dazzling light source can be uncomfortable and can lead to a loss of visual function (known as disability glare). Disability glare is a consequence of excessive light scatter which reduces the contrast of the retinal image. Clinically, disability glare can be estimated by measuring the reduction in visual acuity or contrast sensitivity in the presence of a peripheral glare source, and there are several clinical tests available (e.g. the Miller–Nadler Glare tester; the Brightness Acuity Tester; the Optec 3000 and Vistech MCT8000). Disability glare tests have proved useful in clinical trials (e.g. ref. 70) and research on aging (discussed below) but the utility of glare testing in routine practice is not clear.

Development of spatial vision

We used to think of infant vision as "a buzzing, blooming, confusion".[71] However, studies over the past 20 years or so have shown that the visual system is much more developed at birth than we used to think (see Chapter 38).

Development of visual acuity and CSF

The emerging picture suggests that peak contrast sensitivity increases, and the peak of the CSF shifts toward higher spatial frequencies, with increasing age. Low spatial frequency sensitivity develops much more rapidly than high spatial frequency sensitivity. Thus, peak contrast sensitivity may reach adult levels as early as about 9 weeks of age, whereas sensitivity at higher spatial frequencies continues to develop dramatically, and there remains a roughly 20-fold difference in the contrast sensitivity of adults and 8-month-olds.[72] Visual acuity is only adult-like around 3–5 years post-natal.[73]

Development of hyperacuity

The early development of Vernier actually appears to be somewhat delayed, compared to resolution. Shimojo and co-workers[74,75] suggested that Vernier acuity is initially worse than grating acuity, and then shows a dramatic improvement (relative to grating acuity) between about 2 and 8 months. At 8 months, Vernier acuity was about twice as good as grating acuity. The Vernier acuity of very young infants remains controversial. However, the important point is that Vernier acuity and grating acuity reach adult levels of performance at different ages. The attainment of adult levels of hyperacuity – which is 6 to 10 times better than grating acuity is delayed compared to resolution.

It is interesting to speculate why Vernier acuity has a slow late development, reaching adult levels considerably later than resolution. If we consider that Vernier acuity is limited by the precision with which the brain "knows" the position of each afferent, then it is perhaps not so surprising that intrinsic positional uncertainty can only reach adult levels after retinal development is fully complete. The primary postnatal changes in the retina concern differentiation of the macular region.[73] After birth, foveal receptor density and cone outer segment length both increase, as foveal cones become thinner and more elongated. There is a dramatic migration of ganglion cells and inner nuclear layers from the foveal region, as the foveal pit develops during the first 4 months of life, and it is not until about 4 years of age that the fovea is fully adult-like.[76] From birth to beyond 4 years of age, cone density increases in the central region, due to both migration of receptors, and decreases in their dimensions. Both of these factors result in finer cone sampling (by decreasing the distance between neighboring cones). It is likely that alterations in cone spacing and the light-gathering properties of the cones during early development contribute a great deal toward the improvements in acuity and contrast sensitivity during the first months of life. The massive migration of retinal cells, and the alterations in the size of retina and eyeball (along with changes in interpupillary distance), may necessitate the plasticity of cortical connections early in life. It seems reasonable to speculate that the brain cannot "know" the positions of foveal cones with high levels of precision, until after these changes in the retina are complete. Interestingly, the peripheral retina appears to develop much more rapidly than the fovea.[76]

Visual acuity through the lifespan

Visual acuity reaches adult levels around 3–5 years of age, and remains rather constant until around 55–60 years of

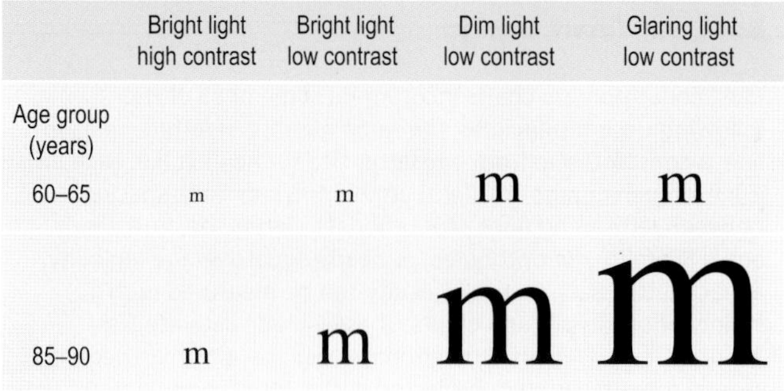

Figure 33.26 The letter m sized in proportion to the median acuity for 4 measures of visual acuity in 60–65-year-olds and in 85–90-year-olds.[78]

age, after which it begins to decline.[77] The decline is important because the graying of the population of the US. According to US Census 2000, 35 million Americans are 65 years and over (12.4 percent of the population). Those over 85 years are growing at roughly three times the rate of the general population – 50,000 Americans are over 100 years old.

There are a number of reasons why one might expect visual function to decline with age. These include:

- decreased pupil size particularly in dim light
- decreased lens transmission particularly for short-wavelength (blue) light
- increased intraocular light scattering.

All of these result in reduced retinal image contrast (particularly for small targets) and reduced retinal illumination. Thus, one might expect that visual performance is impaired most under low-contrast and low-light conditions NOT the high-luminance and high-contrast conditions normally used to test visual acuity.

Until ~10 years ago, only high-contrast acuity was known in elders. However, the SKI study, which has followed some 900 elderly persons from Marin county, CA over a period of about 7 years, suggests that high-contrast acuity greatly underestimates the loss of visual function in the elderly.[78] Figure 33.26 shows this graphically by presenting the letter m sized in proportion to the median acuity for 4 measures of visual acuity in 60–65-year-olds and in 85–90-year-olds.

For high-contrast, high-illumination targets, acuity is reduced by about a factor of 1.8 in the older group (i.e. the "m" must be about 1.8 times larger). Low-contrast acuity with glare is reduced by about a factor of 3 in the younger (60–65) group, but by almost a factor of 7 in the older group. Although the SKI study population is probably not representative of the population at large (it is primarily white, well educated with high socioeconomic status) it has important implications for understanding the visual problems of the elderly. Specifically, it shows that high-contrast visual acuity greatly underestimates the true age-related decline in vision, and that older persons with good standard visual acuity may be visually impaired under conditions of reduced contrast, reduced lighting, changing light level, or reduced contrast in the presence of glare.

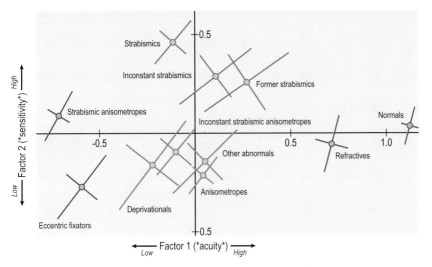

Figure 33.27 Mean locations of the 11 clinically defined categories in the two-factor space. The diagonal bars show 1 SEM measured along the principal axes of the elliptical distributions, which we use only as descriptive indicators of dispersion. (Reproduced with permission from Association for Research in Vision and Ophthalmology[85])

Box 33.4 Amblyopia

Amblyopia is a developmental disorder of spatial vision usually associated with the presence of strabismus, anisometropia or form deprivation early in life.[79] Reduced visual acuity in an apparently healthy eye, despite appropriate optical correction is the sine qua non of amblyopia. However, there is a great deal of evidence showing that amblyopia results in a broad range of neural, perceptual, and clinical abnormalities.

Amblyopia

Amblyopia is a developmental disorder of spatial vision usually associated with the presence of strabismus, anisometropia or form deprivation early in life (Box 33.4).[79] Amblyopia is clinically important because, aside from refractive error, it is the most frequent cause of vision loss in infants and young children,[80] and it is of basic interest because it reflects the neural impairment which can occur when normal visual development is disrupted. The damage produced by amblyopia is generally expressed in the clinical setting as a loss of visual acuity in an apparently healthy eye, despite appropriate optical correction; however, there is a great deal of evidence showing that amblyopia results in a broad range of neural, perceptual, and clinical abnormalities.[81–84] Currently there is no positive diagnostic test for amblyopia. Instead, amblyopia is diagnosed by exclusion: in patients with conditions such as strabismus and anisometropia, a diagnosis of amblyopia is made through exclusion of uncorrected refractive error and underlying ocular pathology.

In humans, amblyopia occurs naturally in about 2–4 percent of the population[79] and the presence of amblyopia is almost always associated with an early history of abnormal visual experience: binocular misregistration (strabismus), image degradation (high refractive error and astigmatism, anisometropia), or form deprivation (congenital cataract, ptosis). The severity of the amblyopia appears to be associated with the degree of imbalance between the two eyes (e.g. dense unilateral cataract results in severe loss), and to the age at which the amblyogenic factor occurred. Precisely how these factors interact is as yet unknown, but it is evident that different early visual experiences result in different functional losses in acuity and contrast sensitivity amblyopia[85] (Fig. 33.27), and a significant factor that distinguishes performance amongst amblyopes is the presence or absence of binocular function. Amblyopes who lack binocularity (primarily strabismic amblyopes) show much greater losses in Vernier and Snellen acuity than in grating resolution.[86,87]

Crowding and amblyopia

Extensive crowding occurs in the central field of strabismic amblyopes. As first reported by Irvine:[88] "single letters or direction of ... an E could be identified by the amblyopic eye, if viewed one letter at a time, but when placed in conjunction with other letters in a line, confusion affected the interpretation." The ophthalmologist Hermann Burian noted that his orthoptist recorded better visual acuities in amblyopic patients than he did. The orthoptist used isolated letters to measure acuity, whereas he used letters arranged in a row.[89] The phenomenon of crowding in strabismic amblyopia has been repeatedly confirmed.[20,35,36,66,86,87,90–93]

While the extent of crowding in the central field of amblyopic observers was originally thought to be proportional to the amblyope's uncrowded acuity, and similar to normal observers when the resolution deficit was taken into account,[36] the story seems to be a bit more complicated. Recent studies suggest that crowding is qualitatively different in strabismic and anisometropic amblyopes.[88,89] In anisometropic amblyopes crowded acuity is proportional to the amblyope's uncrowded acuity, similar to normal vision with blur, whereas in strabismic amblyopes it is worse, often very much so, similar to normal peripheral vision.

References

1. Wade NJ. Image, eye and retina (invited review). J Opt Soc Am 2007; 24:1229–1249.
2. Westheimer, G. Visual acuity. In: Kaufman PL, Alm A, Adler FH, eds. Adler's physiology of the eye: clinical application. St Louis, MO: CV Mosby, 2003:453–469.
3. Hecht S, Mintz EU. The visibility of single lines at various illuminations and the retinal basis of visual resolution. J Gen Physiol 1939: 22:593–612.
4. Snell AC, Sterling S. The percentage evaluation of macular vision. Trans Am Ophthalmol Soc 1925: 23:204–227.
5. Westheimer G. Visual hyperacuity. Progress in Sensory Physiology, vol. 1. New York: Springer, 1981:1–30.
6. Klein SA, Levi DM. Hyperacuity thresholds of 1 second: theoretical predictions and empirical validation. J Opt Soc Am A 1985; 2:1170–1190.
7. Geisler WS. Physical limits of acuity and hyperacuity. J Opt Soc Am A 1984; 1:775–782.
8. Lotze H. Microcosmus: an essay concerning man and his relation to the world. Edinburgh, Scotland: TT Clark, 1885.
9. Levi DM, Tripathy SP. Localization of a peripheral patch: the role of blur and spatial frequency. Vision Res 1996; 36:3785–3803.
10. Liang J, Williams DR. Aberrations and retinal image quality of the normal human eye. J Opt Soc Am A Opt Image Sci Vis 1997; 14:2873–2883.
11. Aubert H, Foerster K. Beitraege zur Kenntnisse der indirecten Sehens. Graefes Archiv fur Ophthalmologie 1857; 3:1–37.
12. Weymouth FW. Visual sensory units and the minimal angle of resolution. Am J Ophthalmol 1958; 46:102.
13. Levi DM, Klein SA, Aitsebaomo P. Detection and discrimination of the direction of motion in central and peripheral vision of normal and amblyopic observers. Vision Res 1984; 24:789–800.
14. Levi DM, Klein SA, Aitsebaomo AP. Vernier acuity, crowding and cortical magnification. Vision Res 1985; 25:963–977.
15. Daniel PM, Whitteridge D. The representation of the visual field on the cerebral cortex in monkeys. J Physiol 1961; 159:203–221.
16. Peterson BB, Dacey DM. Morphology of human retinal ganglion cells with intraretinal axon collaterals. Vis Neurosci 1998; 15:377–387.
17. Polyak S. The vertebrate visual system. Chicago: University of Chicago Press, 1957.
18. Wässle H, Grünert U, Röhrenbeck J, Boycott BB. Cortical magnification factor and the ganglion cell density of the primate retina. Nature 1989; 341:643–646.
19. Azzopardi P, Cowey A. The overrepresentation of the fovea and adjacent retina in the striate cortex and dorsal lateral geniculate nucleus of the macaque monkey. Neuroscience 1996; 72:627–639.
20. Stuart JA, Burian HM. A study of separation difficulty. Its relationship to visual acuity in normal and amblyopic eyes. Am J Ophthalmol 1962; 53:471–477.
21. Levi DM. Crowding – an essential bottleneck for object recognition: a mini-review. Vision Res 2008; 48:635–654.
22. Pelli DG, Tillman KA. The uncrowded window of object recognition. Nat Neurosci 2008; 11:1129–1135.
23. Tyler CW, Likova LT. Crowding: a neuroanalytic approach. J Vision 2007; 7:1–9.
24. Levi DM, Hariharan S, Klein SA. Suppressive and facilitatory spatial interactions in peripheral vision: peripheral crowding is neither size invariant nor simple contrast masking. J Vision 2002; 2:167–177.
25. Tripathy SP, Cavanagh P. The extent of crowding in peripheral vision does not scale with target size. Vision Res 2002; 42:2357–2369.
26. Pelli DG, Palomares M, Majaj NJ. Crowding is unlike ordinary masking: distinguishing feature integration from detection. J Vision 2004; 4:1136–1169.
27. Pelli DG, Tillman KA, Freeman J, Su M, Berger T, Majaj NJ. Crowding and eccentricity determine reading rate. J Vision 2007; 7(2):1–36.
28. Andriessen JJ, Bouma H. Eccentric vision: Adverse interactions between line segments. Vision Res 1976; 16:71–78.
29. Flom MC, Heath GG, Takahashi E. Contour interaction and visual resolution: contralateral effect. Science 1963; 142:979–980.
30. Westheimer G, Hauske G. Temporal and spatial interference with Vernier acuity. Vision Res 1975; 15:1137–1141.
31. Kooi FL, Toet A, Tripathy SP, Levi DM. The effect of similarity and duration on spatial interaction in peripheral vision. Spatial Vision 1994; 8:255–279.
32. Tripathy SP, Levi DM. Long-range dichoptic interactions in the human visual cortex in the region corresponding to visual field. Vision Res 1994; 34:1127–1138.
33. Danilova MV, Bondarko VM. Foveal contour interactions and crowding effects at the resolution limit of the visual system. J Vision 2007; 7(2):25:1–18.
34. Levi DM, Klein SA, Hariharan S. Suppressive and facilitatory spatial interactions in foveal vision: foveal crowding is simple contrast masking. J Vision 2002; 2:140–166.
35. Hariharan S, Levi DM, Klein SA. "Crowding" in normal and amblyopic vision assessed with Gaussian and Gabor C's. Vision Res 2005; 45:617–633.
36. Flom MC, Weymouth FW, Kahneman D. Visual resolution and contour interaction. J Opt Soc Am 1963; 53:1026–1032.
37. Bouma H. Interaction effects in parafoveal letter recognition. Nature 1970; 226:177–178.
38. Toet A, Levi DM. The two-dimensional shape of spatial interaction zones in the parafovea. Vision Res 1992; 32:1349–1357.
39. Shlaer S. The relation between visual acuity and illumination. J Gen Physiol 1937; 21:165–188.
40. Sloan LL. Variation of acuity with luminance in ocular diseases and anomalies. Doc Ophthalmol 1969; 26:384–393.
41. Geisler WS, Davila KD. Ideal discriminators in spatial vision: two-point stimuli. J Opt Soc of Am A 1985; 2:1483–1497.
42. Waugh SJ, Levi DM. Visibility, luminance and Vernier acuity. Vision Res 1993; 33:527–538.
43. Ludvigh EJ. Effect of reduced contrast on visual acuity as measured with Snellen test letters. Arch Ophthalmol 1941; 25:469–474.
44. Herse PR, Bedell HE. Contrast sensitivity for letter and grating targets under various stimulus conditions. Optom Vis Sci 1989; 66:774–781.
45. Rabin J, Wicks J. Measuring resolution in the contrast domain: the small letter contrast test. Optom Vis Sci 1996; 73:398–403.
46. Johnson CA, Casson EJ. Effects of luminance, contrast, and blur on visual acuity. Optom Vis Sci 1995; 72:864–869.
47. Legge GE, Rubin GS, Luebker A. Psychophysics of reading – V. The role of contrast in normal vision. Vision Res 1987; 27:1165–1177.
48. Baron WS, Westheimer, G. Visual acuity as a function of exposure duration. J Opt Soc Am 1973; 63:212–219.
49. Waugh SJ, Levi DM. Visibility, timing and Vernier acuity. Vision Res 1993; 33:505–526.
50. Hadani I, Meiri AZ, Guri M. The effects of exposure duration and luminance on the 3-dot hyperacuity task. Vision Res 1984; 24:871–874.
51. Matin L, Picoult E, Stevens JK, Edwards MW Jr, Young D, MacArthur R. Oculoparalytic illusion: visual-field dependent spatial mislocalizations by humans partially paralyzed with curare. Science 1982; 216:198–201.
52. Keesey UT. Effects of involuntary eye movements on visual acuity. J Opt Soc Am 1960; 50:769–774.
53. Westheimer G, McKee SP. Visual acuity in the presence of retinal-image motion. J Opt Soc Am 1975; 65:847–850.
54. Levi DM. Pattern perception at high velocities. Current Biol 1996; 6:1020–1024.
55. Campbell FW, Green DG. Optical and retinal factors affecting visual resolution. J Physiol 1965; 181:576–593.
56. Mitchell DE, Freeman RD, Westheimer G. Effect of orientation on the modulation sensitivity for interference fringes on the retina. J Opt Soc Am 1967; 57:246–249.
57. Levi DM, Song S, Pelli DG. Amblyopic reading is crowded. J Vision 2007; 7(2):1–17.
58. Legge, GE, Mansfield JS, Chung STL. Psychophysics of reading. XX Linking letter recognition to reading speed in central and peripheral vision. Vision Res 2001; 41(6):725–743.
59. Yu D, Cheung S-H, Legge GE, Chung STL. Effect of letter spacing on visual span and reading speed. J Vision 2007; 7(2):1–10.
60. Legge GE, Cheung S-H, Yu D, Chung STL, Lee H-W, Owens DP. The case for the visual span as a sensory bottleneck in reading. J Vision 2007; 7(2):1–15.
61. Schade OH Sr. Optical and photoelectric analog of the eye. J Opt Soc Am 1956; 46:721–739.
62. Rovamo J, Virsu V, Näsänen R. Cortical magnification factor predicts the photopic contrast sensitivity of peripheral vision. Nature 1978; 271:54–56.
63. Robson JG, Graham N. Probability summation and regional variation in contrast sensitivity across the visual field. Vision Res 1981; 21:409–418.
64. Bailey IL, Lovie JE. New design principles for visual acuity letter charts. Am J Optom Physiol Opt 1976; 53:740–745.
65. Westheimer G. Scaling of visual acuity measurements. Arch Ophthalmol 1979; 97:327–330.
66. Levi DM, Hariharan S, Klein SA. Suppressive and facilitatory interactions in amblyopic vision. Vision Res 2002; 42:1379–1394.
67. Ginsburg AP. A new contrast sensitivity vision test chart. Am J Optom Physiol Opt 1984; 61:403–407.
68. Pelli DG, Robson JG, Wilkins AJ. The design of a new letter chart for measuring contrast sensitivity. Clin Vis Sci 1988; 2:187–199.
69. Bass EB, Steinberg EP, Luthra R et al. Variation in ophthalmic testing prior to cataract surgery. Results of a national survey of optometrists. Cataract Patient Outcome Research Team. Arch Ophthalmol 1995; 113:27–31.
70. Chylack LT Jr, Wolfe JK, Friend J et al. Validation of methods for the assessment of cataract progression in the Roche European-American Anticataract Trial (REACT). Ophthalmic Epidemiol 1995; 2:59–75.
71. James W. Principles of Psychology. Cambridge, MA: Harvard University Press, 1981 (originally published in 1890).
72. Norcia AM, Tyler CW, Hamer RD. Development of contrast sensitivity in the human infant. Vision Res 1990; 30:1475–1486.
73. Boothe RG, Dobson V, Teller DY. Postnatal development of vision in human and non-human primates. Ann Rev Neurosci 1985; 8:495–545.
74. Shimojo S, Birch EE, Gwiazda J, Held R. Development of Vernier acuity in infants. Vision Res 1984; 24:721–728.
75. Shimojo S, Held R. Vernier acuity is less than grating acuity in 2- and 3-month-olds. Vision Res 1987; 27:77–86.
76. Yuodelis C, Hendrickson A. A qualitative and quantitative analysis of the human fovea during development. Vision Res 1986; 26:847–855.
77. Hirsch MJ, Wick RE, eds. Vision of the aging patient; an optometric symposium. Philadelphia: Chilton Co., 1960.
78. Haegerstrom-Portnoy G. The Glenn A Fry Award Lecture 2003: Vision in elders – summary of findings of the SKI study. Optom Vis Sci 2005; 82:87–93.
79. Ciuffreda KJ, Levi DM, Selenow A. Amblyopia: basic and clinical aspects. Boston: Butterworth-Heinemann, 1991.
80. Sachsenweger R. Problems of organic lesions in functional amblyopia. In: Arruga H, ed. International strabismus symposium. Basel: Karger, 1968.
81. Kiorpes L. Visual processing in amblyopia: animal studies. Strabismus 2006; 14:3–10.
82. Levi DM. Visual processing in amblyopia: human studies. Strabismus 2006; 14:11–19.
83. Barrett BT, Bradley A, McGraw PV. Understanding the neural basis of amblyopia. Neuroscientist 2004; 10:106–117.
84. Hess RF. Amblyopia: site unseen. Clin Exp Optom 2001; 84:21–36.
85. Mckee SP, Levi DM, Movshon JA. The pattern of visual deficits in amblyopia. J Vis 2003; 3:380–405.
86. Levi, DM, Klein, SA. Hyperacuity and amblyopia. Nature 1982; 298:268–270.
87. Levi DM, Klein SA. Vernier acuity, crowding and amblyopia. Vision Res 1985; 25:975–991.
88. Irvine RS. Amblyopia Ex Anopsia. Observations on retinal inhibition, scotoma, projection, light difference discrimination and visual acuity. Trans Am Ophthalmol Soc 1945; 66:527–575.

89. Burian HM, von Noorden GK. Binocular vision and ocular motility. Theory and management of strabismus. St Louis, Missouri: CV Mosby, 1974.

90. Bonneh YS, Sagi D, Polat U. Local and non-local deficits in amblyopia: acuity and spatial interactions. Vision Res 2004; 44:3099–3110.

91. Song S, Pelli DG, Levi DM. Three limits on letter identification by normal and amblyopic observers. In preparation.

92. Hess RF, Jacobs RJ. A preliminary report of acuity and contour interactions across the amblyope's visual field. Vision Res 1979; 19:1403–1408.

93. Hess RF, Dakin SC, Tewfik M, Brown B. Contour interaction in amblyopia: scale selection. Vision Res 2001; 41:2285–2296.

94. Hirsch J, Curcio CA. The spatial resolution capacity of human foveal retina. Vision Res 1989; 29:1095–1101.

95. Williams DR. Topography of the foveal cone mosaic in the living human eye. Vision Res 1988; 28:434–454.

96. Osterberg G. Topography of the layer of rods and cones in the human retina. Acta Ophthalmol 1935; 6(suppl):1–103.

97. Anstis SM. Letter: a chart demonstrating variations in acuity with retinal position. Vision Res 1974; 14:589–592.

98. Campbell FW, Kulikowski JJ, Levinson J. The effect of orientation on the visual resolution of gratings. J Physiol 1966; 187:427–436.

99. Robson J, Campbell F. A quick demonstration of your own contrast sensitivity function. In: Pelli DG, Torres AM, eds. Thresholds: Limits of Perception. New York: NY Arts Magazine, 1997.

CHAPTER 34

Color Vision

Jay Neitz, Katherine Mancuso, James A. Kuchenbecker & Maureen Neitz

The ability to perceive color is a highly valued sensory capacity, and it has been a subject of experimental inquiry for over 200 years. Prior to the development of modern biological techniques, breakthroughs in color vision research stemmed from careful consideration of perceptual experiences. From color-mixing and matching experiments it was deduced that three different receptors in the eye, each maximally sensitive to a different region of the visible spectrum, were required to explain normal color vision. Thus, the three-component theories of Young and Helmholtz, as well as Hering's conflicting hypothesis of three paired, opponent color processes were developed in the 1800s, long before the three types of retinal cone photopigment were isolated and characterized. It is now understood that normal, or trichromatic, color vision is mediated by three types of cone photoreceptor – designated short- (S), middle- (M), and long- (L) wavelength-sensitive – the activities of which are, indeed, combined in later opponent organization to provide color perception. The first direct measurements of the absorption properties of primate cone photopigments occurred in the 1960s using a technique called microspectrophotometry. Since then, additional techniques including the electroretinogram and molecular genetics have allowed for refinement of the spectral sensitivity curves for the S, M, and L pigments, shown in Figure 34.1.

The fact that most humans have trichromatic color vision explains why computer monitors and televisions mix just three primary colors, red, green, and blue, in order to produce colors that match almost all those seen in the real world. Likewise, if humans were tetrachromatic, having four cone photopigments, televisions with four primaries would be required. The presence of three classes of cone photoreceptor also explains color blindness. That is, loss of function of each individual cone class is associated with an inherited form of color vision deficiency: protan defects involve loss of L cone function, deutan defects with loss of M cone function, and tritan defects with loss of S cone function (Box 34.1). Each type of color vision defect results from rearrangements, deletions, or mutations in the genes encoding the corresponding L, M, or S photopigment, and color vision tests have been developed which can detect changes in the cone complement. Observations that cannot be explained by the presence of three types of cone photoreceptor, however, involve the appearance of color. The neural

circuitry for color vision extracts signals from only three cone types, and yet gives rise to six hue percepts arranged in opponent pairs: blue-yellow, red-green, and black-white. These topics, including the molecular genetics of color vision deficiencies, tests of color vision, and the neural circuitry of color perception, will be discussed in detail in the following sections.

Molecular genetics of color vision and color deficiencies

Although the tendency for color vision defects to run in families and to be more prevalent among men had long been recognized, a major advancement in understanding the biological basis of color vision and its defects came in 1986 when Jeremy Nathans and colleagues cloned and sequenced the genes encoding the S, M, and L cone opsins and verified that the L and M opsin genes are in adjacent positions on the X-chromosome.[1-3] Further breakthroughs in understanding differences in the quality of color vision among individuals with anomalous trichromacy followed, with identification of the amino acids involved in spectral tuning, which are responsible for the ~30 nm difference in peak sensitivity of the L and M pigments underlying normal color vision.[4,5]

The majority of inherited red-green color vision defects are caused by rearrangement and deletions of the L and M opsin genes on the X-chromosome that come about by meiotic recombination during oogenesis in females. A dichromatic phenotype results in males who inherit an X-chromosome in which all but one of the opsin genes have been deleted. A protanopic type defect is characterized by the absence of L opsin genes, and a deuteranopic type defect is characterized by the absence of M opsin genes (Box 34.2). Although humans often have more than just two opsin genes on the X-chromosome, only two are typically expressed, and these determine color vision phenotype. If meiotic recombination creates an array in which the first two genes encode pigments with identical spectral properties, a male who inherits the array will be dichromatic, either protanopic or deuteranopic. However, a male with an array in which the first two genes encode opsins of the same class (M or L), but differ slightly in spectral sensitivities, will be an anomalous trichromat. He will be protanomalous if the genes encode

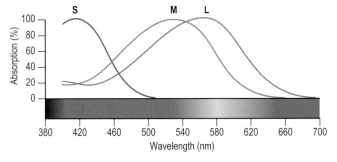

Figure 34.1 Absorption spectra of the three types of cone photopigment, short- (S), middle- (M), and long- (L) wavelength sensitive. The S photopigment is maximally sensitive to wavelengths near 415 nm, the M pigment to wavelengths near 530 nm, and the L pigment to wavelengths near 560 nm. Neural comparison of the rates of quantal absorption by the three cone types gives rise to four unique hue percepts – blue, yellow, red, and green – in addition to hundreds of intermediate colors in the visible spectrum.

Box 34.1 Nomenclature for inherited color vision deficiencies

Protan defects are the least common form of red-green color vision defects affecting about 2 percent of males in the US. There are two categories: dichromatic (protanopia), and anomalous trichromatic (protanomaly), each affecting about 1 percent of males. The severity of protan defects varies over a relatively narrow range. Protan comes from the Greek root for 'the first.' Protan color vision is mediated by S cones and M cones.

Deutan defects are the most common form of red-green color vision defects, affecting about 6 percent of males in the US. There are two categories: dichromatic (deuteranopia) and anomalous trichromatic (deuteranomaly). Deuteranomaly and deuteranopia affect ~5 percent and ~1 percent of males, respectively. The severity of deutan defects varies over a relatively broad range, from nearly normal to nearly dichromatic. Deutan comes from the Greek root for 'the second.' Deutan color vision is mediated by S cones and L cones.

Tritan defects are relatively rare and are characterized by impaired blue-yellow color vision. There is only one category, dichromatic (tritanopia). Tritan defects display incomplete penetrance, meaning there is variability in the degree to which color vision is impaired among individuals with the same underlying gene defect, even within a family. Tritan comes from the Greek root for 'the third.' Tritan color vision is mediated by M and L cones.

Box 34.2 Visual pigments, cone opsins, and official gene designations

Visual pigments, also known as photopigments, are composed of an apoprotein and an 11-*cis* retinal chromophore. The genes, OPN1LW, OPN1MW, and OPN1SW, each encode an apoprotein (termed opsin). The chromophore is a vitamin A derivative that absorbs ultraviolet light; when covalently bound to an opsin, the chromophore absorption spectrum is shifted to longer wavelengths. Amino acid differences among the opsins are responsible for the differences in the absorption spectra of the three cone classes (Fig. 34.1). Within the L and M classes, variations or polymorphisms in the amino acid sequences produce relatively small spectral shifts giving rise to variants of L and M photopigments.

- OPN1MW: middle-wavelength sensitive cone opsin, expressed in M cones, commonly referred to as 'green' cones. OPN1MW is on the X-chromosome at location Xq28. Deutan defects are most commonly associated with the absence or lack of expression of the OPN1MW gene(s).
- OPN1LW: long-wavelength sensitive cone opsin, expressed in L cones, commonly referred to as 'red' cones. OPN1LW is on the X-chromosome at location Xq28. Protan defects are most commonly associated with the absence or lack of expression of the OPN1LW gene(s).
- OPN1SW: short-wavelength sensitive cone opsin, expressed in S cones, commonly referred to as 'blue' cones. OPN1SW is on chromosome 7 at 7q32.1. Tritan defects are caused by a missense mutation in one copy of the OPN1SW gene. A missense mutation is a change in the amino acid coding sequence of a gene that substitutes one amino acid for another in the encoded protein.

---

color discrimination that could be reliably classified as just outside the normal range until the L/M separation was reduced to about 12 nm or less.[9] Another surprising finding occurred with the development of adaptive optics, a technique that allows resolution of the cone photoreceptor mosaic in living humans. Selective bleaching allowed identification of the different cone types and revealed that individuals with normal color vision have roughly the same number and arrangement of S cones, but there is tremendous variation in the ratios of L and M cones.[10,11] This was startling in light of the widely held belief that the quality of red-green color vision would be affected by extreme cone ratios, and it is a testament to the fact that the visual system is tuned for detecting very small differences.

Inherited blue-yellow color vision, or tritan, defects are considerably more rare than red-green defects. They are caused by mutations in the S opsin gene,[12–15] and are inherited in an autosomal dominant fashion, meaning that an individual who is heterozygous for an S opsin gene mutation will often exhibit the phenotype. Recent evidence from adaptive optics imaging of tritan subjects suggests that the defect is an S-cone dystrophy, similar to rod dystrophy caused by heterozygous mutations in the rod pigment rhodopsin in retinitis pigmentosa.[15] Tritan defects are said to be incompletely penetrant, meaning not everyone with an S opsin gene defect exhibits the phenotype, but this is perhaps due to a progressive loss of S cones which must reach a critical threshold before the phenotype is manifested.

---

two M opsins, or deuteranomalous if they encode two L opsins. The extent of color vision loss in anomalous trichromats is determined by the degree of similarity in the spectral peaks of the two pigments.[6–8]

It has been assumed that the 30 nm spectral separation between the L and M pigments was optimized during evolution, and that there is also an optimal ratio of L and M cones. However, recent studies have shown that much smaller spectral separations between, and skewed proportions of, L and M cones still provide robust red-green color vision. For example, red-green chromatic sensitivity was recently demonstrated to exhibit a non-linear relationship with photopigment proximity, and observers did not show a reduction in

Tests of color vision

As described above, color vision defects are the result of genetic changes that are responsible for either the alteration or loss of cone photopigments. Since color vision is based on neural comparisons between different classes of cone photoreceptor, both the loss of a photopigment type and changes in the peak sensitivity of a photopigment such that there is a reduction in the differential responses of the two cone classes, will result in a reduced ability to distinguish colors.

Pseudoisochromatic plate tests are perhaps the most familiar tool for diagnosing color vision deficiencies. Popular examples include Ishihara's test of color vision and the Hardy, Rand, and Rittler (HRR) pseudoisochromatic plates. The plates consist of printed pages with dots of various colors and shades of gray, each containing a colored symbol(s) that is visible to individuals with normal color vision, but invisible to those with certain color vision defects. Subjects are asked to identify the symbols, and the specific pattern of errors on a series of plates in a test allows for diagnosis of the likely presence or absence of a color vision defect, and for some tests, an indication of the type and severity.

A fundamentally different testing strategy is used with arrangement tests, in which subjects are asked to arrange a series of colored discs 'in order' so that each disk is placed adjacent to the discs that are most similar in color appearance. Misordering of the disks allows a diagnosis of the presence or absence of a color vision defect. Popular examples of arrangement tests are the Farnsworth-Munsell 100 Hue Test and its abridged version, the Farnsworth-Munsell Dichotomous D-15 Test.

The Rayleigh color match is often referred to as the 'gold standard' for color vision testing. It is performed on the anomaloscope, an instrument which contains an optical system that produces two side-by-side lighted fields. One field, the 'test light,' is a monochromatic amber color, and the other is a mixture of red and green light. The person being tested adjusts the ratio of red-to-green light in the mixture until it exactly matches the amber test light. Setting a match with a higher or lower ratio of red-to-green light compared to the match made by a person with normal color vision is diagnostic of an inherited color vision abnormality caused by a shift in the spectral peak of either the L or M pigment. This test is extraordinarily sensitive and can detect genetic alterations in the spectral sensitivities of the photopigments, even in subjects in whom the alteration in the photopigment has little or no effect on the person's ability to discriminate between different colors. Its extreme sensitivity in detecting the presence of anomalous photopigments is why the anomaloscope has been widely adopted as the 'gold standard.' However, it must be emphasized that mild photopigment abnormalities may be associated with little or no loss in color discrimination ability. Even with shifted pigments, the quality of color vision may still be quite good. Combining genetic analysis with the Rayleigh match can provide a more precise assessment of the likely quality of color vision.

Each of the widely used color vision tests has its separate advantages and is considered to be best under different circumstances. Because of this, for general diagnosis of color vision defects, the results from a battery of tests can provide the most complete picture from which to make a differential diagnosis. Ideally, for diagnosis of color vision defects, one should obtain information on the 'type' of problem. Is it a red-green or blue-yellow deficiency, or mixed? If it is red-green, is it a 'protan' or 'deutan' defect? Is it a genetic disorder or is it acquired? The test should also provide information about severity. Does the person have a severe dichromatic form, or a milder 'anomalous' form of deficiency? If it is an anomalous form, is it very mild, mild, moderate, or more severe? Under some conditions it is impractical to administer a battery of tests. Cole et al[16] recently recommended the latest edition of the Richmond HRR Pseudoisochromatic test as the 'one of choice for clinicians who wish to use a single test for color vision.' It has plates for detecting protan, deutan, and tritan defects, and its classification of mild, medium, and strong categories for deutan and protan defects is useful. To absolutely distinguish dichromats from anomalous trichromats, the anomaloscope, the D15, or both should be used in conjunction with a pseudoisochromatic plate test.

Genetic testing is the most reliable means of distinguishing acquired color vision defects from inherited color vision defects. Presently, genetic tests are not commercially available; however, the technology exists to perform such tests because the molecular genetics of inherited color vision defects is well understood, as described in the previous section.

Color appearance

The neural circuitry for color vision receives input from only three cone types (Fig. 34.2) and yet gives rise to six hue percepts arranged in opponent pairs: blue-yellow, red-green, and black-white. The dogma has been that for blue-yellow opponency, signals from S cones are combined antagonistically with an additive signal from M and L cones, abbreviated as S-(M+L); and for red-green opponency, signals from L cones are combined antagonistically with those from M cones (abbreviated as L-M). However, color vision testing in humans has established that the spectral signatures of hue perception actually involve contributions from all three cone types in the following manner: blue = (S+M)-L; yellow = L-(S+M); red = (S+L)-M; and green = M-(S+L). Therefore, while much progress has been made in identifying neurons in the visual system with color opponent responses, those matching the spectral signatures of human perception remain to be found. Results from molecular biology have shed new light on the evolution of primate color vision, imposing constraints on the possibilities for the underlying circuitry. Evolutionary constraints coupled with recent observations from human subjects with a retinal circuitry deficit are leading toward a new understanding of the color vision circuits.

Blue-yellow circuitry

In order to understand the neural operations responsible for transforming the cone signals into perception, it is helpful to first consider reduced color vision systems with fewer cone types and fewer fundamental sensations.

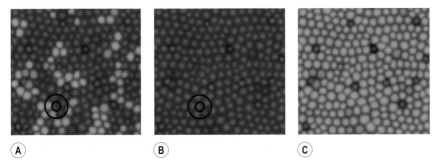

Figure 34.2 (**A**) Illustration of the arrangement of the three cone classes in a trichromat. S cones (colored blue) represent a minority, ~5 percent of the total, and are roughly evenly spaced across the retinal mosaic. The arrangement of L cones (colored red) and M cones (colored green) is very close to random for trichromats. Midget ganglion cells have center-surround receptive fields with the center derived from a single cone (for example, within the inner ring of the two concentric circles). The six adjacent cones are the most important contributors to the surround (enclosed by the outer ring). (**B**) Arrangement of cones in a deuteranope, a dichromat with only two different cone classes, L and S. In a dichromat, usually all of the cones in the surround are of the same class as the center input to a midget ganglion cell. (**C**) Arrangement of cones in a protanope, a dichromat with only M and S cones.

Short-wavelength cones are an evolutionarily ancient photoreceptor type and blue-yellow color vision appears to be the ancestor from which all other color vision evolved. Our dichromatic primate ancestors had two types of cone, short- and long-wavelength-sensitive and, presumably, they also had similar circuits underlying their vision as modern-day dichromats, including a high-acuity spatial vision circuit served by the midget ganglion cell system (introduced in Chapter 21), in addition to the blue-yellow color vision circuit.

It is not completely clear which ganglion cell subtypes are specialized for carrying information corresponding to blue-yellow color vision. In primates, the S-cones have only two significant output pathways that have been described thus far (Fig. 34.3, left). One pathway is the straight-through output to the receptive field center of S-cone-specific ON bipolar cells which, in turn, output to the *small bistratified ganglion cells* providing a 'blue-on' signal.[17] The second pathway from S-cones occurs through H2 horizontal cells to the terminals of adjacent L/M cones,[18] which then output to *midget ganglion cells* via L/M specific bipolar cells; however, the physiological role of this anatomical connection has remained unclear. Historically, the small bistratified ganglion cells have been viewed as the major candidate to serve as a neural substrate for blue-yellow color vision. However, they all respond to blue and inhibit to yellow, leaving us without a physiological basis for the yellow half of this opponent system. Furthermore, recent assessment of human subjects with a genetic mutation that renders the ON bipolar pathways inactive seems to cast doubt onto the role of small bistratified cells in blue perception. Specifically, the mutation occurs in the metabotropic glutamate receptor expressed on the ON-bipolar cells, and clinically, these subjects present with congenital stationary night blindness[19] (Box 34.3). If small bistratified cells were, indeed, the physiological substrate for blue perception, with no input from the S-cone ON bipolars, CSNB patients would be expected to have a tritan color vision defect. However, standard tests indicate that they have normal, trichromatic color vision. While the visual mechanisms associated with this ON-pathway deficit require further inquiry, these results suggest that blue perceptions may be mediated by a separate pathway, leading us now to consider the possible role of the second anatomical S-cone

> **Box 34.3 A clue for blue: Congenital Stationary Night Blindness (CSNB)**
>
> The two types of photoreceptor in the human eye, rods and cones, serve different functions. Rods provide vision only under very low light levels, such as at night when little light is available. In contrast, cones serve vision at relatively higher light levels, such as daylight and standard indoor lighting.
>
> While L/M cones synapse with both ON and OFF bipolar cells, the S cones and rods synapse with only ON-type bipolar cells. The dendrites of ON bipolar cells express a metabotropic glutamate receptor, mGluR6, and mutations in this receptor result in ON-pathway deficit. Curiously, while rod-mediated vision is compromised in CSNB patients, producing loss of nocturnal vision, S-cone mediated (blue) color vision appears to be unaffected.

output, via H2 horizontal cells, to the midget ganglion cell pathway.

In the region of primate retina that serves highest visual acuity, the fovea, more than 90 percent of ganglion cells are of the 'midget' variety, so called because of their small dendritic arbors, which in the central 7–10 degrees connect to a single cone via a 'midget' bipolar cell. There are two types of midget ganglion cells, ON- and OFF-center, which receive input through a corresponding pair of ON- and OFF-midget bipolar cells, which have opposite responses to the neurotransmitter, glutamate, released by cones (Chapters 22, 23). Midget ganglion cells do not receive direct center input from S-cones, only from M/L cones. Thus, in a dichromatic ancestor to humans that had only S and L cones, the midget ganglion cell system's major function was to compare absorptions of S vs. L cones to provide achromatic luminance signals (Fig. 34.2b); presumably giving rise to white percepts through the ON-pathway and black percepts through the OFF-pathway (Fig. 34.3, top right). Because S cones are relatively sparse and roughly evenly spaced across the retina, and the signal strength to neighboring cones falls off exponentially with distance, only a small number of midget ganglion cells would receive significant input from S-cones in their receptive field surrounds, providing a

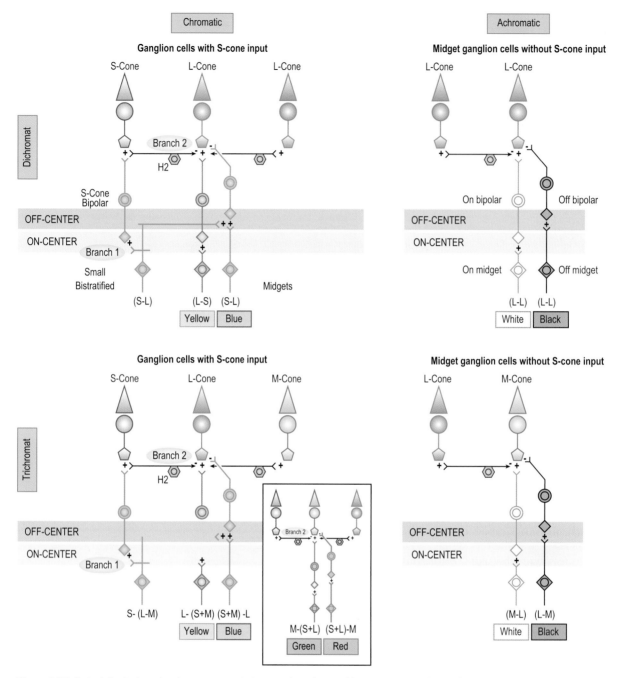

Figure 34.3 Retinal circuits based on known anatomical connections that would give rise to signals matching the spectral signatures of human perception following addition of a third cone type. 'Branch 1' indicates the straight-through S-cone output to the *small bistratified ganglion cell pathway*; 'branch 2' is a second known anatomical connection that may provide S-cone output to the *midget ganglion cell pathway* via H2 horizontal cells (see text). The addition of M cones to a dichromat only changes the 'surrounds' of small bistratified ganglion cell receptive fields, and they would continue to have similar spectral response properties, making them an unlikely substrate for hue sensations of red, green, blue, and yellow. In contrast, both the 'center' and 'surround' of midget ganglion cell receptive fields become altered by the addition of M cones, splitting the pre-existing blue-yellow circuit into two organizations with distinct spectral response properties, one for blue-yellow and one for red-green (inset). Even though midget ganglion cells in trichromats respond both to diffuse red or green lights and dark-light edges, ultimately the two percepts must be separated from each other at a higher level of visual processing; we have illustrated the idea that only the 'black-white' signals from purely L/M opponent circuits are propagated to the cortex (bottom right).

possible substrate for L vs. S cone comparisons required for blue-yellow opponency (Fig. 34.3, top left).

Red-green circuitry

The idea that midget ganglion cells with S-cone input in their receptive field surround may serve as a substrate for

blue-yellow perception seems plausible when considering the effects produced when a third type of cone was added during evolution. Results from molecular genetics indicate that trichromatic color vision arose relatively recently from a gene duplication event that added M cones. In order for this low-probability genetic event to get passed-on and eventually confer routine trichromacy to Old World primates, including humans, it must have produced an immediate advantage for the primate ancestor by adapting a pre-existing visual circuit for a new dimension of red-green color vision. In ancestral primates with only S and L cones, the small bistratified ganglion cell circuitry would have been little affected in its spectral response characteristics by the addition of M cones (Fig. 34.3, bottom left); that is, the center would remain unaltered due to the S-cone specific connections made by small bistratified cells, and the surround would be transformed from 'L' to 'L+M.' Thus, the simple addition of a third cone type would not 'split' this ganglion cell class into two spectral types, one for blue-yellow and one for red-green color vision, as required by the evolutionary constraint that an immediate advantage was produced. In contrast, when M cones were randomly added to the pre-existing midget system, the small subset of spectrally opponent midget ganglion cells would have automatically become segregated into two different classes. The center of the receptive field would have become either 'L' or 'M;' accordingly, the receptive field surrounds would be transformed to either 'M + S' or 'L + S,' automatically producing midget cells with two different spectral response characteristics – one corresponding to red-green and the other to blue-yellow color vision (Fig. 34.3, inset).

In summary, it appears plausible that midget ganglion cells receiving S-cone input in their receptive field surround may provide the physiological substrate giving rise to hue circuits in the cortex. The cone forming the receptive field center can be either L or M, and the ganglion cells can be either ON or OFF center. The resulting four possible combinations would correspond to the known spectral signatures of the four main hue percepts of trichromats. That is, an ON-center ganglion cell receiving input from an M cone center with S and L cones in the surround would provide M-(S+L)–green; this same receptive field through an OFF-center ganglion cell produces (S+L)-M–red. An L cone center with an M and S surround would result in L-(S+M)–yellow; this same receptive field through an OFF-center ganglion cell produces (S+M)-L–blue.

Black-white circuitry

A remaining unresolved mystery in color vision stems from the fact that addition of a randomly distributed long-to-middle wavelength sensitive cone type causes almost every midget ganglion cell in the retina, as well as the cells they project to in lateral geniculate nucleus (LGN), to respond to either spatially uniform red or green lights in addition to dark-light edges. Earlier color vision models thus attributed the red-green system to outputs of the L and M cones, about 95 percent of the cone population, while the whole blue-yellow system was considered to be centered on just the remaining 5 percent of cones, the S cones. Because an imbalance seems inherently implausible based on the similar perceptual acuities of both color systems, it has been proposed that the preponderance of L/M cones indicates that these cone types alone are used for luminance (dark-light) detection, suggesting that only a fraction of the spectrally opponent information from L/M cones actually contributes to color vision. While this idea seemed heretical at first, it has become more plausible in light of recordings from neurons in primary visual cortex (VI) which receive direct input from the axons of color opponent neurons in the LGN.

The receptive field properties of VI neurons are quite different from those of the LGN, in that few cortical neurons respond to spatially uniform stimulation and most are selective for the orientation of edges. Furthermore, most VI neurons respond well to achromatic modulation and less well or not at all to chromatic modulation, which, in contrast, is a powerful stimulus to most LGN cells. Altogether, only about 5–10 percent of neurons in V1 respond robustly to purely chromatic modulation and little, if at all, to achromatic modulation.[20] This lends credence to the idea that even though LGN cells receiving input from purely L/M opponent receptive fields respond to both spatially uniform colors and black-white edges, most of the chromatic signals become subtracted-out and disregarded, perhaps by a second round of lateral antagonism at the level of the cortex, analogous to horizontal cell interactions at the level of the retina. A second round of cortical antagonism would have been serendipitous in preserving the neural outputs of the pre-existing achromatic luminance circuitry in primate ancestors when a third cone type was added, leaving this system dedicated to the original purpose for which it evolved, responding to edges (Fig. 34.3, bottom right). Accordingly, only the small subset of midget ganglion cells receiving input from all three cone types would be able to pass color signals to higher cortical levels (Fig. 34.3, bottom left), thereby rectifying the apparent imbalance between the inputs to the red-green and blue-yellow systems.

Future directions

Much has been learned about the neural types and their interconnections in the retina responsible for luminance and color, and there is a growing body of information about the higher visual centers; yet, how these ultimately operate to provide vision remains a fascinating puzzle. Many clues have come from information about the evolution of the visual system and about its anatomy and physiology, and continued progress will undoubtedly arise from rapidly emerging technologies. One such example has resulted from the development of viral vector-mediated gene therapy to address human visual disorders. Using recombinant adeno-associated virus containing cone-specific promoter elements, it is possible to target transgene expression to subclasses of mammalian cone,[21,22] including in primates.[23] Introducing new cone photopigments into animal models with reduced color vision provides an experimental tool for probing the neural circuitry for color vision and understanding the evolutionary requirements for developing a new dimension of red-green color vision.

There is currently no cure for red-green color vision deficiency. However, in the majority of cases the cone photoreceptors are healthy and viable, and viral-mediated gene therapy is a feasible option for delivering the missing cone opsin gene to photoreceptors to rescue the deficit. If all of the neural circuitry required for taking advantage of a third cone type is present in dichromatic individuals, as was also possibly the case in dichromatic primate ancestors, gene therapy holds promise for providing red-green color blind adults with full trichromatic color vision.

References

1. Nathans J, Piantanida TP, Eddy RL, Shows TB, Hogness DS. Molecular genetics of inherited variation in human color vision. Science 1986; 232:203–210.

2. Nathans J, Thomas D, Hogness DS. Molecular genetics of human color vision: the genes encoding blue, green, and red pigments. Science 1986; 232:193–202.

3. Vollrath D, Nathans J, Davis RW. Tandem array of human visual pigment genes at Xq28. Science 1988; 240:1669–1672.

4. Neitz M, Neitz J, Jacobs GH. Spectral tuning of pigments underlying red-green color vision. Science 1991; 252:971–974.

5. Asenjo AB, Rim J, Oprian DD. Molecular determinants of human red/green color discrimination. Neuron 1994; 12:1131–1138.

6. Regan BC, Reffin JP, Mollon JD. Luminance noise and the rapid determination of discrimination ellipses in colour deficiency. Vision Res 1994; 34:1279–1299.

7. Neitz J, Neitz M, Kainz PM. Visual pigment gene structure and the severity of human color vision defects. Science 1996; 274:801–804.

8. Sharpe LT, Stockman A, Jägle H, et al. Red, green, and red-green hybrid pigments in the human retina: correlations between deduced protein sequences and psychophysically measured spectral sensitivities. J Neurosci 1998; 18:10053–10069.

9. Barbur JL, Rodriguez-Carmona M, Harlow JA, Mancuso K, Neitz J, Neitz M. A study of unusual Rayleigh matches in deutan deficiency. Vis Neurosci 2008; 25:507–516.

10. Roorda A, Williams DR. The arrangement of the three cone classes in the living human eye. Nature 1999; 397:520–522.

11. Roorda A, Neitz J. Chromatic topography of the retina. J Opt Soc Am A 2000; 17(3): 495–496.

12. Weitz CJ, Miyake Y, Shinzato K, et al. Human tritanopia associated with two amino acid substitutions in the blue sensitive opsin. Am J Hum Genet 1992; 50:498–507.

13. Weitz CJ, Went LN, Nathans J. Human tritanopia associated with a third amino acid substitution in the blue sensitive visual pigment. Am J Hum Genet 1992; 51: 444–446.

14. Gunther KL, Neitz J, Neitz M. A novel mutation in the short-wavelength sensitive cone pigment gene associated with a tritan color vision defect. Vis Neurosci 2006; 23: 403–409.

15. Baraas RC, Carroll J, Gunther KL, et al. Adaptive optics retinal imaging reveals S-cone dystrophy in tritan color vision deficiency. J Opt Soc Am A 2007; 24: 1438–1447.

16. Cole BL, Lian KY, Lakkis C. The new Richmond HRR pseuodisochromatic test of colour vision is better than the Ishihara test. Clin Exp Optom 2006; 89:73–80.

17. Dacey DM, Lee BB. The blue-ON opponent pathway in primate retina originates from a distinct bistratified ganglion cell type. Nature 1994; 367:731–735.

18. Dacey DM, Lee BB, Stafford DK, Pokorny J, Smith VC. Horizontal cells of the primate retina: cone specificity without spectral opponency. Science 1996; 271: 656–659.

19. Dryja TP, McGee TL, Berson EL, et al. Night blindness and abnormal cone electroretinogram ON responses in patients with mutations in the GRM6 gene encoding mGluR6. Proc Natl Acad Sci USA 2005; 102(13):4884–4889.

20. Solomon SG, Lennie P. The machinery of colour vision. Nature Rev Neurosci 2007; 8(4):276–286.

21. Li Q, Timmers AM, Guy J, Pang J, Hauswirth WW. Cone-specific expression using a human red opsin promoter in recombinant AAV. Vision Res 2007;48.

22. Mauck MC, Maunuso K, Kuchenbecher J, et al. Longitudinal evaluation of expression of virally delivered transgenes in gerbil cone photoreceptors. Vis Neurosci 2008; 25: 273–282.

23. Mancuso K, Hendrickson AE, Connor TB, Jr, et al. Recombinant adeno-associated virus targets passenger gene expression to cones in primate retina. J Opt Soc Am A 2007; 24: 1411–1416.

CHAPTER 35

The Visual Field

Chris A. Johnson & Michael Wall

Introduction

Perimetry and visual field testing have been used as diagnostic procedures for evaluation of visual function for nearly 200 years, and a historical review of these techniques may be found in several publications.[1-5] During this period of time, a variety of new techniques have been developed, although the standard clinical procedure of detecting a small white target superimposed on a uniform background at different peripheral locations in the field of view is still the method that is most commonly used by eye care specialists. Perimetry is used clinically to detect functional losses produced by pathology to the visual pathways, for differential diagnostic purposes to identify the location of a visual deficit, to monitor the status of acute and chronic disease processes, and to evaluate the efficacy of treatment. In spite of the similarity of current test procedures to early techniques, significant progress has recently been made in the presentation, analysis and interpretation of visual field findings. This has resulted in the automation of procedures, standardization of methods, the immediate statistical evaluation of test results and enhanced efficiency of testing. In this chapter, we will provide an overview of the psychophysical basis for perimetry, a discussion of the types of perimetric test strategies, the data representation formats that are used for visual field testing, important factors for interpreting visual field results, recognition of patterns of visual field loss, strategies for monitoring progression of visual field changes, and guidelines for assessment of visual field information in a clinical setting. Additionally, we will briefly describe some of the new visual field test procedures that have been shown to provide salient information concerning the functional status of the visual system and have been reported by many investigators to be clinically useful.

The psychophysical basis for perimetry

For conventional perimetric testing, the patient's task is to detect a small white target superimposed on a uniform background luminance at different visual field locations. The detection task is based on the increment or differential light threshold, which determines the minimum amount of light that must be added to a stimulus (ΔL) in order to make it distinguishable from the background (L) and is known as Weber's Law, or $\Delta L/L = C$.[6] In order for Weber's Law to be valid, the background adapting luminance of the perimeter must be in the photopic range (typically 31.5 apostilbs, or 10/candelas per meter squared), pupil size must be sufficiently large (greater than 2 mm), ocular media must be relatively clear, and other test conditions (stimulus size, stimulus duration, etc.) must be within appropriate levels. Under these conditions of photopic adaptation in a normal eye and visual system, the fovea is the visual field location with the highest sensitivity. As eccentricity from the fovea increases, there is a significant drop in sensitivity out to about 3 degrees, then a more gradual sensitivity decline out to about 30 degrees radius, followed by a more rapid decrease beyond 30 degrees. The temporal visual field (away from the nose) extends farther than the nasal visual field, and the inferior visual field extends farther than the superior visual field. The three-dimensional shape of the sensitivity profile of the visual field has therefore often been referred to as the "hill of vision" or the "island of vision in a sea of blindness", as shown in Figure 35.1. Note that there is a "hole" in the island of vision that represents the blind spot, a location where the optic nerve exits the eye where no photoreceptors or other retinal neural elements are present. The blind spot is represented as an area of non-seeing by a dark solid oval located approximately 15 degrees from the center of view (fixation) in the temporal visual field extending one-third above and two-thirds below the horizontal meridian. From a clinical standpoint, one of the primary purposes of perimetry and visual field testing is to search for departures from this normal sensitivity profile (sensitivity losses) and determine the shape and location of the deficit.

As with all psychophysical assessments, there is variability in the values obtained for repeated measures in the sensitivity to light increments. Several laboratories have indicated that under clinical testing conditions, the standard deviation of repeated measures is approximately 3 dB for individual visual field locations with normal sensitivity.[7-15] As sensitivity declines in damaged visual field areas, there is an increase in variability of about 300–400 percent that extends the 95 percent confidence interval to nearly the entire operating range of the perimeter.[15] One method of quantitatively evaluating this variability is by obtaining a frequency of seeing (FOS) curve, which plots the percentage of stimuli seen as a function of the luminance of the stimulus. The FOS curve is typically quite steep in areas with normal sensitivity, and becomes flatter at greater eccentricities and in damaged visual field areas. From a clinical perspective, it is important to minimize response variability so that subtle changes in sensitivity that are due to pathology can be more readily detected. This can be accomplished by using larger stimuli, modifying the type of stimulus used, adjusting the response estimate algorithm and several other procedures.[7–15]

The physiologic basis for perimetry

Proper interpretation of visual field results requires knowledge of the anatomical arrangement of the visual pathways. As shown in Figure 35.2, a panoramic image of the visual field of view is formed on the retina by the optical

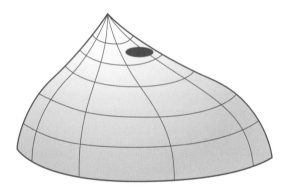

Figure 35.1 Three-dimensional representation of the eye's sensitivity to light throughout the visual field. An illustration for a right eye (OD) is presented.

components (cornea, lens, pupil) of the eye. The light image is converted into electrical impulses that are transmitted through more than 100 million photoreceptors to other retinal neural elements and eventually to approximately 1 million retinal ganglion cells. The retinal ganglion cell axons enter the optic nerve and travel back to eventually reach primary visual cortex. As the optic nerve fibers leave the eye and travel towards the brain, there is a separation of nerve fibers subserving the nasal and temporal visual field along the vertical meridian. This separation occurs at the optic chiasm, whereby the nasal visual field from one eye combines with the temporal visual field of the other eye, so that the left half of vision is represented in the right hemisphere and the right half of vision is represented in the left hemisphere. At this point in the visual pathways, there is not a point-for-point retinotopic mapping of locations between the two eyes. The nerve fibers continue on to a portion of the thalamus (the lateral geniculate nucleus) where they synapse with nerve fibers that convey information to primary visual cortex known as area 17 or V1. In primary visual cortex, there is a point-for-point mapping of retinotopic information from both eyes (under normal development) as a consequence of rearrangement of fibers traveling from the optic chiasm. Beyond primary visual cortex, there are complex interactions with other areas of the brain that are beyond the scope of this chapter.

Types of perimetric testing

Currently, there are three different forms of visual field testing that are performed clinically: kinetic perimetry, static perimetry, and suprathreshold static perimetry. Each of these procedures has particular advantages and disadvantages, as described below, and each of them serves distinct purposes

Figure 35.2 Schematic illustration of the human visual pathways.

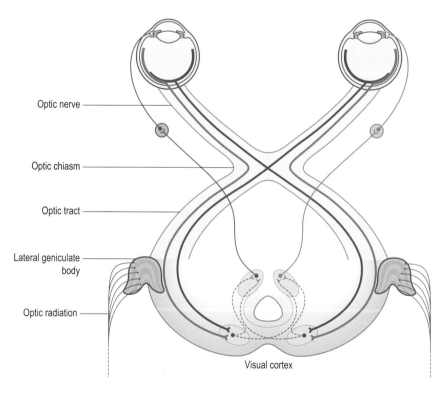

Optic nerve

Optic chiasm

Optic tract

Lateral geniculate body

Optic radiation

Visual cortex

for clinical diagnostic testing. In this view, it should be noted that these techniques should be regarded as complementary, with only moderate amounts of overlap in clinical performance.

Kinetic perimetry

As implied by its name, kinetic perimetry involves the movement of a small target on a uniform background consisting of a 1- or 2-meter dark tangent screen or a light hemispherical bowl perimeter (Goldmann or Goldmann-like perimeter). A skilled kinetic perimetrist usually requires a considerable amount of apprenticeship time (6–12 months) to become proficient in these techniques. A thorough description of kinetic perimetry strategies is beyond the scope of this chapter, although an excellent description of kinetic perimetric techniques and strategies may be found in the book by Anderson.[4]

Kinetic perimetry consists of mapping the hill of vision (boundary between seeing and non-seeing) by moving a target of predetermined size and luminance from the periphery towards the fixation point along successive meridians separated by approximately 15 degrees of angular extent. The rate of smooth movement should be 4–5 degrees per second for greater eccentricities, and 1–2 degrees per second for locations near fixation. For a particular target, the boundary locations are connected by a line of equal sensitivity known as an isopter. Multiple isopters can be generated by changing the size or luminance of the target. For the far periphery, target size is usually varied, whereas luminance is altered for evaluation of nearer eccentricities. A normal isopter is relatively egg-shaped, with the long axis of the egg extending to the temporal visual field and the short portion of the axis of the egg projecting to the nasal visual field. Similarly, the inferior visual field extends farther peripherally than the superior visual field. If an isopter corresponds to this shape, then no further evaluation is needed. However, if local indentations of an isopter are present, further moving target scans should be directed from the periphery in a direction perpendicular to an imaginary line drawn between the two points. If further definition of the isopter contour is needed, this process should be repeated. Typically, between 3 and 6 isopters are required to provide a thorough characterization of the visual field.

Within the central 30–40 degrees radius of the visual field, a series of arcuate "spot checks" are usually performed between isopters and in locations where one or more isopters show an indentation. Static spot checks are performed with the kinetic isopter stimulus, inside the isopter and consist of a brief flash of the target. A location where a spot check is missed should be checked again to confirm non-detection. In these missed spot check areas or regions with local recession or depression of isopters, scotoma mapping should be performed, beginning at the presumed center of the defect and moving outwards in 8 radial directions (horizontal, vertical and oblique). Additional intermediate scans can also be performed to better characterize the shape of the deficit, and multiple targets should be used to indicate the depth of the deficit and the slope of its borders. The final result is a topographic mapping of the hill of vision to provide a two-dimensional representation of the three-dimensional structure of the visual field.

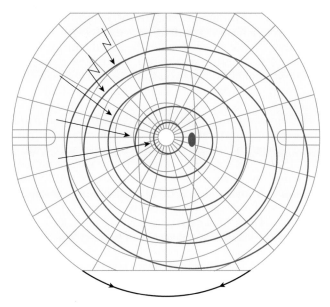

Figure 35.3 Representation of a normal visual field for kinetic visual field testing using a Goldmann-type perimeter.

Advantages of kinetic perimetry include the ability to characterize the entire central and peripheral visual field, enhanced flexibility and interaction between the examiner and the participant, the ability to accurately characterize the shape of visual field deficits, and a highly efficient method of evaluating the far peripheral visual field. Disadvantages consist of higher variability than other forms of perimetric testing, large individual differences in clinical performance among examiners, poorly standardized procedures, limited normative population characteristics, and difficulty in characterizing generalized or widespread visual field sensitivity loss for different age groups.

Although there are no specific quantitative features that can be easily described, Figure 35.3 presents an example of a normal visual field. Most kinetic perimetric testing is performed manually, except that several automated perimeters have recently incorporated kinetic testing procedures.[16–20]

Static perimetry

To date, static perimetry is the most common method of clinical visual field testing, particularly for automated perimeters. Static perimetry involves the detection of a small stationary target that is superimposed on a uniform white background. At each visual field location that is tested, the measurement consists of the minimum amount of light increment (differential light threshold) that can be just distinguished from the background. Different visual field locations are tested along various radial meridians between the central point of fixation and eccentric locations, or according to a Cartesian grid of locations that usually bracket the horizontal and vertical meridians. A variety of test strategies have been developed for static perimetry, including the ascending method of limits,[2] bracketing or staircase procedures,[21–23] forecasting (Bayesian) threshold estimation procedures,[24–29] and other efficient and adaptive testing strategies.[30–33] Most eye care specialists today use some form of adaptive forecasting procedure such as the Swedish Interactive Threshold

Algorithm (SITA), Zippy Estimation of Sequential Thresholds (ZEST), Tendency Oriented Perimetry (TOP) or a similar procedure.[24-33] Additionally, methods of evaluating test performance (false positive responses, false negative responses, and blind spot fixation checks) and alignment (eye and head position) tracking have also been developed for static perimetry.[34-39]

The representation of visual field results from automated static perimetry is typically displayed as numerical sensitivity values, general and pattern deviations from average expected normal values for a specific age group, gray scale sensitivity representations of visual field sensitivities, probability plots (deviations from average normal), and in some instances three-dimensional sensitivity density plots. Also, for the Humphrey Field Analyzer, summary statistics known as visual field indices are also determined to provide an indication of generalized or widespread sensitivity loss (mean deviation [MD]), localized sensitivity loss (pattern standard deviation [PSD]), visual field loss produced by asymmetric superior and inferior visual hemifield loss characteristic of glaucoma (glaucoma hemifield test [GHT]), and visual field progression (glaucoma progression analysis [GPA]) and rate of progression (glaucoma progression index [GPI]). Similar types of analyses are available for other automated perimeters.

Advantages of automated static perimetry include the use of a standardized test procedure, the ability to exchange information from one device and office to another, availability of an age-corrected normative database, immediate comparison of individual results to normal population characteristics, and methods for monitoring response reliability, attention and cooperation, proper alignment and greater reproducibility. Disadvantages of automated static perimetry consist of time-consuming and limited methods of evaluating the entire peripheral visual field, less flexible test procedures, a more demanding burden on the patient's attention and performance, and very high test–retest variability when sensitivity is less than 20dB for a size III target.

Figure 35.4 presents examples of gray scale plots of static perimetry conducted for (a) a normal visual field, (b) a generalized or widespread visual field loss and (c) a localized visual field deficit.

Suprathreshold static perimetry

Suprathreshold static perimetry is generally used as a rapid screening procedure to detect visual field deficits. However, there are several drawbacks to suprathreshold static visual field testing using standard automated perimetry: (1) with a few exceptions,[40-46] the vast majority of basic research in this area has been directed towards threshold estimation strategies rather than suprathreshold screening procedures, (2) there are a limited number of clinical studies that have been performed to evaluate the performance characteristics of these procedures, (3) threshold estimation strategies have become considerably more efficient and can actually take less time than some suprathreshold screening procedures, and (4) there is no consensus as to which suprathreshold visual field screening procedures are most suitable for routine use.

There are a wide variety of forms of suprathreshold static perimetric methods that are currently in use. The stimulus

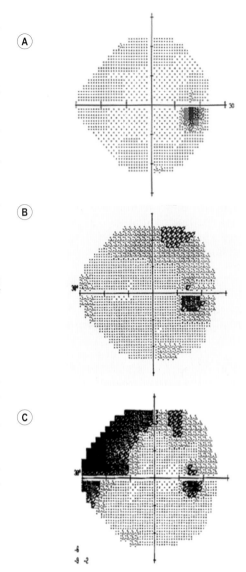

Figure 35.4 A gray scale representation of static perimetry test results obtained for (**A**) a normal visual field, (**B**) generalized or widespread visual field loss, and (**C**) localized visual field loss.

presentation patterns range from a small number of targets (less than 30) to nearly 200 locations. Some strategies present all targets at a fixed stimulus luminance, others use multiple stimulus luminance levels that are varied according to eccentricity, some perform threshold estimates for locations that have sensitivity below normal limits, some present multiple stimulus luminances in damaged locations to estimate the magnitude of sensitivity loss, some provide a retest of missed locations to confirm suspected sensitivity losses, and some use an adaptive strategy that provides greater spatial resolution of test points for suspicious visual field locations. Only a few studies have compared results obtained from different procedures or have evaluated screening results in comparison to threshold visual field techniques.[40-46] This has created an opportunity for future investigations to determine methods that are appropriate for visual field screening in a clinical setting as well as population-based screening.

Suprathreshold static perimetry has the advantage of providing simple and efficient visual field screening techniques

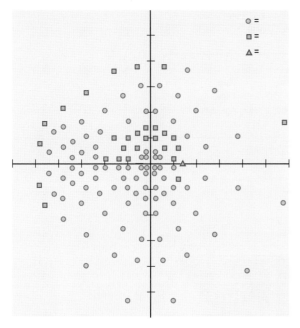

○ =
□ =
△ =

Figure 35.5 An example of a two-zone, threshold-related 120 point visual field screening test on the Humphrey Field Analyzer for the right eye of a primary open angle glaucoma patient with a superior arcuate defect.

for the patient and the examiner, standardized procedures and the ability to test the entire peripheral visual field. Disadvantages include limited quantitative information, lower sensitivity and specificity in comparison to threshold techniques, a lack of appropriate decision rules and scoring systems that have been validated, and limitations in the ability to monitor progressive changes.

Figure 35.5 presents visual field screening results for the right eye of a patient with superior arcuate glaucomatous visual field loss using a multiple stimulus luminance test strategy for the Humphrey Field Analyzer.

Detection of perimetric sensitivity loss and interpretation of results

Perhaps the best method of providing instruction about interpretation of visual field results is to conduct a systematic overview of the findings that are presented on the printed output. Figure 35.6 presents a single visual field analysis of the results for the left eye of a patient with an inferior arcuate glaucomatous visual field deficit that was obtained using the 24-2 SITA Standard test procedure on the Humphrey Field Analyzer II.

The top portion of the visual field printout shown in Figure 35.6 provides (from left to right and top to bottom) an indication of the patient's name, the eye that was tested, the patient's ID code and date of birth. Below this is information pertaining to the test conditions used (background luminance, stimulus size, fixation target, fixation monitoring method), the test strategy (SITA, Full Threshold, Suprathreshold screening), performance characteristics (false positive errors, false negative errors, fixation losses), the patient's age, and ocular conditions (distance refraction, pupil size,

visual acuity). False negative errors that exceed 33 percent are outside normal limits, although this occurs frequently in eyes with visual field loss.[47] Excessive false positives are infrequent, and are often associated with excessive fixation losses because patients are "trigger happy".[48] For the Full Threshold procedure a criterion of 33 percent is used, but for SITA a new algorithm is being used, and 15 percent is the appropriate criterion for excessive false positives.[49] Fixation losses that exceed 33 percent are considered to be excessive[34,35] but methods can be employed to dramatically reduce excessive fixation losses, and head tilts can falsely elevate this percentage.[50]

Below this information are the patient's foveal threshold (if it is enabled for the test), an XY graphical presentation of sensitivity values, and a gray scale representation of visual sensitivity. The sensitivity values are indicated in decibels (dB), which for standard automated perimetry is defined as the contrast (increment threshold or Weber fraction, $\Delta L/L$) needed to just detect a stimulus superimposed on the uniform background and reflects visual field sensitivity on a logarithmic scale. A value of 0 dB represents the brightest stimulus that can be produced by the instrument and superimposed on the background. Each 1 dB increment represents a 0.1 log unit increase in stimulus luminance. Thus, a 10 dB sensitivity represents an ability to detect a target that is 10 times dimmer than the maximum stimulus luminance, a 20 dB stimulus represents an ability to detect a target that is 100 times dimmer than the maximum stimulus luminance, and so forth. The gray scale representation of visual field sensitivity provides an immediate indication of the sensitivity properties of the hill of vision, and is able to convey areas of reduced sensitivity in the form of darkened portions of the graph. This provides a very rapid method of depicting the location, size, shape and depth of area of reduced visual field sensitivity.

Immediately below the gray scale representation are the values for the Mean Deviation (MD), Pattern Standard Deviation (PSD) and Glaucoma Hemifield Test (GHT) calculations. MD indicates the deviation of the patient's visual field from the average normal results for an individual of the same age. PSD is the departure (root mean square [RMS] deviation) from the average decline in sensitivity from the fovea to the periphery out to 30 degrees radius (0.6 dB per degree for a Size III target). In some ways, PSD may be regarded as a quantitative indicator of visual field perturbation (irregularities in visual field surface with eccentricity), particularly because the edges of scotomas provide a distinct transition from a smooth sensitivity profile. The Glaucoma Hemifield Test (GHT) compares the sensitivity of clustered points for mirror image regions of the superior and inferior hemifields, as shown in Figure 35.7. The five regions in the superior and inferior hemifields are designed to be representative of arcuate nerve fiber bundle areas that are most frequently involved in glaucomatous visual field losses (nasal steps, paracentral losses, partial arcuate and arcuate defects). In normal eyes, there are only minimal differences in sensitivity between these regions in the superior and inferior hemifields. However, in glaucoma and some other optic neuropathies, the damage to retinal locations corresponding to the superior and inferior hemifields is often asymmetric, resulting in large sensitivity differences between regions in the superior and inferior hemifields.

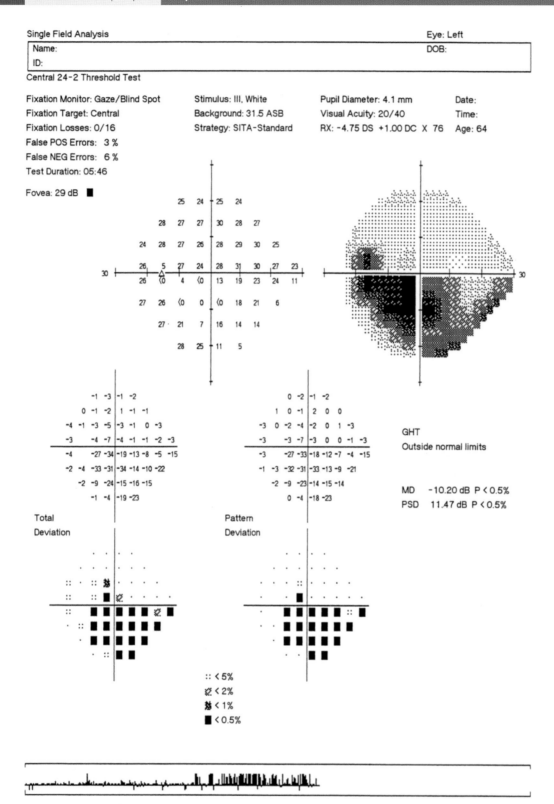

Single Field Analysis   Eye: Left

Name: DOB:
ID:

Central 24-2 Threshold Test

Fixation Monitor: Gaze/Blind Spot   Stimulus: III, White   Pupil Diameter: 4.1 mm   Date:
Fixation Target: Central   Background: 31.5 ASB   Visual Acuity: 20/40   Time:
Fixation Losses: 0/16   Strategy: SITA-Standard RX: -4.75 DS +1.00 DC X 76   Age: 64
False POS Errors: 3 %
False NEG Errors: 6 %
Test Duration: 05:46

Fovea: 29 dB ■

GHT
Outside normal limits

MD   -10.20 dB P < 0.5%
PSD   11.47 dB P < 0.5%

Total
Deviation

Pattern
Deviation

:: < 5%
▨ < 2%
▩ < 1%
■ < 0.5%

Figure 35.6 Standard Automated Perimetry (SAP) results using the SITA Standard 24-2 test procedure on the Humphrey Field Analyzer for the left eye of a patient with primary open angle glaucoma, depicting an inferior arcuate visual field defect. Note that eye alignment, as indicated by the gaze tracking presentation, became variable towards the end of the test procedure.

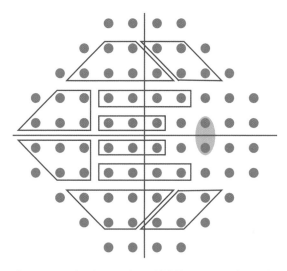

Figure 35.7 The clusters of visual field locations used to evaluate superior and inferior hemifield asymmetry as part of the glaucoma hemifield test. The clusters were selected to generally correspond to the pattern of nerve fiber bundles entering the optic nerve head.

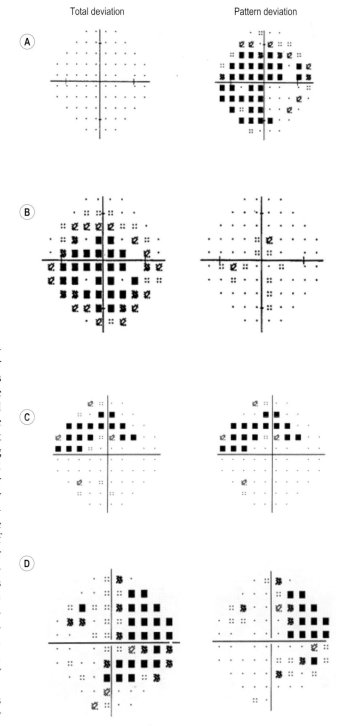

Figure 35.8 Examples of the results for Total and Pattern Deviation probability plots for (**A**) a "trigger happy" patient with abnormally high sensitivity related to pressing the response button too often, (**B**) generalized or widespread visual field loss, (**C**) localized visual field loss, and (**D**) mixed localized and widespread visual field loss. Note that these data representations provide highly useful clinical information.

To the left of the visual field indices are the Total Deviation and Pattern Deviation graphs. The upper left graph for Total Deviation indicates the deviation of the patient's threshold sensitivity from the average normal observer of the same age (plus values indicate higher than average normal sensitivity and minus values reflect lower than average normal sensitivity). The Pattern Deviation values to the right are derived by adjusting the patient's visual field according to the 85th percentile (7th most sensitive point for the 24-2 stimulus presentation pattern) to account for widespread or generalized sensitivity differences (i.e. higher sensitivity values for the patient are adjusted down and lower sensitivity values are adjusted up). Once this is accomplished, the deviation from average normal sensitivity for a person of the same age is calculated. Below the numerical values for the Total and Pattern Deviation calculations are the probability plot graphs. For each visual field location, these plots display single dots for values that are within the normal 95 percent confidence limits, and dithered symbols of increasing darkness and density to indicate locations with sensitivity below the average age-adjusted normal 5 percent, 2 percent, 1 percent and 0.5 percent probability levels, respectively. Comparison of the Total and Pattern Deviation probability plots can be very useful for interpreting the visual field.

Figure 35.8 presents four examples of visual field results represented by the Total and Pattern Deviation probability plots. Figure 35.8A displays a normal Total Deviation plot, with many abnormal results for the Pattern Deviation plot. This indicates that the patient's eye had sensitivity values that were above the 95 percent normal age-adjusted values, which is commonly caused by a "trigger happy" patient who inappropriately responds to targets that are not usually detected by normal individuals. Fixation losses and false positive errors are often outside normal limits in these cases as well. Figure 35.8B shows abnormalities that appear for the Total Deviation probability plot, but are not present for the Pattern Deviation probability plots. This is indicative of widespread or diffuse visual field loss. Figure 35.8C shows Total and Pattern Deviation probability plots that appear

to be nearly identical, which is indicative of localized visual field sensitivity loss. Finally, Figure 35.8D shows abnormal visual field locations for both the Total and Pattern Deviation probability plots, with some locations indicated as abnormal for both graphs and other locations

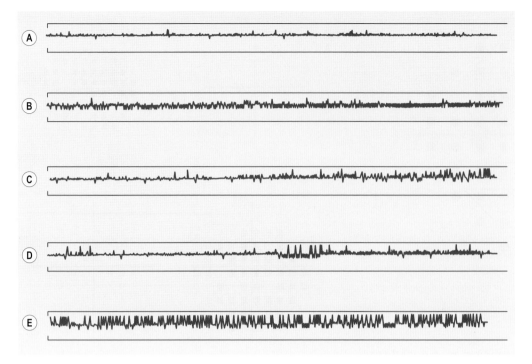

Figure 35.9 Examples of gaze tracking results for the Humphrey Field Analyzer. Upward deflections indicate alignment difficulties (eye and head movements) and downward deflections indicate blinking or droopy eyelid: (**A**) good gaze tracking, (**B**) excessive blinking, (**C**) droopy eyelid, (**D**) fatigue, and (**E**) alignment problems.

demonstrating an abnormality on the Total Deviation plot but not on the Pattern Deviation plot. This is a common presentation, which indicates a mixture of localized and widespread or diffuse visual field sensitivity loss.

Below this is the name and address of the eye clinic in the lower right, and a printout of gaze tracking from left to right. Gaze tracking is a valuable component of the visual field printout. During the test, a video camera is monitoring the location of the first Purkinje image (corneal reflex) relative to the edges of the pupil (retinal red reflex) from an infrared light source that does not interfere with the test procedure. This method can then be used to monitor eye and head alignment during the test, as well as eye blinks and droopy lids during the visual field test procedure. The visual field printout has a time line from left to right that presents eye blinks and upper eyelid drooping (a contiguous series of downward deflections) as well as eye and head alignment (upward tick mark deflections) throughout the test procedure. As shown in Figure 35.9, this information can reveal much about the attention and cooperation of the patient during the test procedure. Figure 35.9A shows the results for a patient with good fixation and blinking behavior throughout the test. Figure 35.9B shows the results for a patient who develops excessive blinking and tear film breakup that may be related to the tested eye drying out from exposure to the infrared light (a rest break can assist in minimizing this problem). Figures 35.9C–E present results for a patient who becomes sleepy (droopy upper eyelid) throughout the test, a patient that becomes fatigued as the test progresses, and a patient with head and eye alignment problems during the test, respectively.

It is important to keep in mind that all of the information provided on this printout is essential, and that proper interpretation of visual fields requires taking all of these results into account. Often, there is a temptation to concentrate on portions of the printout and ignore or discount other information that is presented. This is not a good habit to establish, however, because it can lead to misinterpretation of the visual field findings and artifactual results due to specific testing circumstances that can be missed. It is also critical to remember that visual field testing is only one part of the vision examination, and that additional information from the patient's exam, ocular and medical history, family medical history, and other factors need to be included in the final assessment.

Patterns of visual field loss associated with different pathologic conditions

Proper evaluation of patterns of visual field loss is probably the most vital aspect of visual field interpretation. It is important to keep in mind the three-dimensional nature of visual field properties so that X (width), Y (height) and Z (depth) coordinates are all taken into account. The shape and extent of a visual field defect, whether it has steep or sloping boundaries, and whether it "respects" (terminates abruptly) certain anatomical borders are all important features to consider. Figure 35.10 presents some schematic representations of fibers at different levels in the visual pathways and the patterns of visual field loss that correspond to damage at these locations.

In this section, we will outline some key features and patterns of visual field loss that will be of assistance in visual field interpretation.

- *Generalized or widespread visual field sensitivity loss*. This can be produced be a variety of conditions, ranging

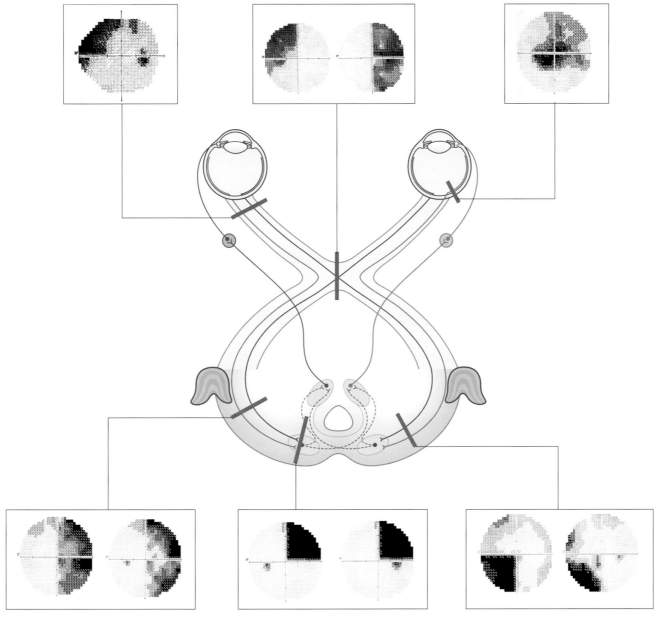

Figure 35.10 Examples of visual field loss produced by damage to the visual pathways at various locations from the retina to primary visual cortex.

from changes in the ocular media to pathology to various portions of the visual pathways to attentional and cognitive factors. It is therefore a rather non-specific diagnostic sign.

- *Ring scotomas.* These defects (either a total or a partial ring of sensitivity loss in the mid periphery surrounded by normal visual field locations) are typically indicative of retinitis pigmentosa or some other type of retinal degeneration.
- *Central scotomas.* Generally, these types of defects are associated with diseases of the macular region of the retina, although optic neuropathies can also produce central scotomas.
- *Defects that "respect" the nasal horizontal meridian.* Generally these types of defects are associated with the

retinal ganglion cell nerve fiber bundles, and are indicative of glaucoma or some other type of optic neuropathy. However, compromises of the retinal vasculature (e.g. a branch artery occlusion) can also produce this type of defect.

- *Arcuate scotomas.* These types of defects again typically affect either the arcuate nerve fiber bundles (glaucoma or other optic neuropathies) or the retinal vasculature.
- *Nasal steps.* These defects are primarily associated with glaucoma or other optic neuropathies.
- *Centrocecal scotomas.* These defects represent pathology to the papillomacular bundle of nerve fibers that subserve the region between the blind spot and fixation. Usually, the deficit is produced by optic neuropathies, but they can also be created by retinal

disease, particularly those that involve edema (e.g. central serous retinopathy).

- *Visual field constriction.* This is a deficit that can be produced by a variety of causes, ranging from retinal disease to optic neuropathies.
- *Defects that "respect" the vertical meridian.* Because of the separation of nerve fibers along the vertical border at fixation, these defects are found for damage to the visual pathways at the optic chiasm or after the optic chiasm.
- *Bitemporal defects.* In nearly all cases, a bitemporal visual field defect that respects the vertical meridian indicates damage to the visual pathways at the optic chiasm.
- *Defects that resemble a horizontal tongue.* This is an uncommon visual field deficit that is only infrequently seen. The deficit can be for the "tongue", or the tongue can be the only remaining visual field areas of seeing. It is produced by damage to visual pathway fibers at the level of the lateral geniculate body.
- *Homonymous visual field defects.* Damage to the visual pathways that is on the same side (right or left) of vision for both eyes is homonymous, and indicates a deficit that has occurred after the optic chiasm. A total homonymous hemianopic visual field defect (the entire right or left side of vision is missing) is indicative of a defect after the optic chiasm, but can be at any location within the optic radiations.
- *Congruous and incongruous defects.* In general, the greater the amount of incongruity of homonymous defects between eyes, the closer the site of damage is to the optic chiasm. Deficits for primary visual cortex (occipital lobe) are highly congruous, parietal lobe and temporal lobe deficits are semi-congruous.
- *Pie in the sky defects.* A homonymous visual field deficit that resembles a piece of pie missing from the superior visual field of both eyes is usually associated with a temporal lobe lesion.
- *Pie on the floor defects.* A homonymous visual field deficit that resembles a missing piece of pie that has fallen to the floor is more often the result of a parietal lobe lesion.
- *Cookie-cutter punched out lesions.* Visual field deficits that look as though one used a hole punch or cookie cutter to represent loss in both eyes is highly indicative of an occipital lobe (primary visual cortex) lesion.

We hope that these guidelines will assist you in the interpretation of visual field information.

Determination of visual field progression

The ability to accurately determine visual field progression and the rate of visual field progression is a vital component in following the functional visual status of patients over time. However, there is currently no general agreement or consensus concerning the best method of following visual field progression. This is best depicted by the fact that every multicenter clinical trial that has employed visual field

results as an outcome measure has used a different method of defining progression.[14,51-55] A review of the various methods used to determine visual field progression has been reported by several sources.[14,51-55] Basically, visual field progression techniques can be divided into the categories of clinical judgment methods, classification systems, event analyses and trend analyses.[51] Each procedure has its advantages and disadvantages. It should also be noted that most of the procedures for determining visual field progression have been directed towards glaucoma.

Clinical judgment consists of a practitioner using his/her knowledge and experience to make a determination of whether or not the visual field has undergone systematic changes over time. The main advantage of this procedure is that it is an easy, efficient method of evaluating visual field progression. The primary disadvantage is that there is a large amount of variability among eye care specialists for interpreting visual field progression. Previous investigations have demonstrated that even highly experienced, competent eye care specialists have a tendency to "overcall" visual field progression.

Classification systems consist of staging the amount of visual field damage into a series of categories. A historical review of visual field staging and classification systems for glaucoma has recently been published.[56] In general, most classification systems use between 5 and 7 different levels of visual field damage to characterize the spectrum of loss. Exceptions to this include the systems derived by the Advanced Glaucoma Intervention Study (AGIS) and the Collaborative Initial Glaucoma Treatment Study (CIGTS), which use 20 stages.[57,58] Advantages of classification systems include their use by several multicenter clinical trials and the ability to assign a quantitative value to visual field damage, whereas disadvantages consist of the use of a discrete outcome rather than a continuous function, and the lack of knowing the measurement scale (linear, non-linear, etc.) that the system is based upon.

Event analysis is most commonly performed by comparing the most recent visual field to those obtained at baseline visits. For this type of analysis, the intent is to determine whether visual field changes from baseline (progression or improvement) are within or outside expected limits. This technique has been used by several multicenter trials and is available on commercially available automated perimeters.[57-60] Advantages of this procedure are that it is able to monitor visual field changes over time, and the application of more rigorous statistical procedures to analyze results. Included for disadvantages are the absence of evaluating interim visual fields between the current and baseline results, and the large influence of spurious results (meaningfully higher or lower sensitivity values for a single visit).

Trend analysis has also been used for many multicenter clinical trials, most commonly as a linear least-squares regression.[61-65] This procedure analyzes all visual fields that have been obtained over time and determines the best fitting equation (usually a linear function) to describe the data. The slope of the line can then be used to provide an estimate of visual field change. Advantages of this procedure include the use of all the visual field data to obtain an estimate of change, based on a long-standing formal statistical model. Limitations of this procedure are related to a requirement for the slope to be statistically significantly different from zero, and a minimum of 6 to 8 visual fields that are necessary to achieve appropriate high clinical performance. Recently, there have been several enhancements that have been introduced to improve the performance of trend analysis, but it still requires a minimum of 5 to 6 visual fields before satisfactory performance can be achieved (i.e. high sensitivity and specificity).[66-68] Note that the requirement of at least 5 to 6 visual fields holds for all of the current procedures for analyzing visual fields, which means that an accurate, precise and reliable estimate of visual field status may require several years of follow-up before an informed decision can be made with existing techniques. Clearly from a clinical perspective, there is a strong need for new, innovative approaches to this perplexing problem.

How do the various methods of determining visual field progression compare? Several laboratories have made such comparisons using the same visual field data set to evaluate according to the various methods. In summary, these studies indicate that there are considerable differences in the amount of time to detect progression, the number of reliable visual fields that are determined to be progressing, the number of progressing visual fields that are determined to be unchanged, and the influence of variability (long and short term) on final outcomes.[14,52-55] Moreover, there is agreement among the various methods only about 50–60 percent of the time.[14,52-55]

Given these difficulties, what can be done to assist the eye care specialist in determining visual field progression at the present time? Here are a few helpful hints:

- Examine the complete visual field history of the patient, not just the current and the most recent previous results. Otherwise, false positives (inappropriate evaluation of change) and false negatives (inappropriate evaluation of stability) will frequently occur.
- Use all of the available methods of assessing longitudinal visual field status (e.g. Overview Analysis, Change Analysis, Glaucoma Change Probability, Glaucoma Progression Analysis, Glaucoma Progression Index, Progressor, etc.).[63-65,68-71]
- Compare visual field results with other clinical examination findings, the patient's ocular and general medical history, and other pertinent information (visual fields alone do not provide all of the necessary information).
- When in doubt, perform an additional visual field test, use another type of visual field testing, and perform assessments of structural retinal, optic disc or neural (imaging) characteristics to confirm suspected visual field findings.[72]
- Consult with colleagues to determine their analysis and interpretation of results.

In the future, what can be done to improve our ability to determine visual field changes over time? We believe that there are four factors that can assist us in this endeavor:

(1) The development of visual field techniques that provide larger changes that are related to pathology. As described in a later section, several techniques have been developed that appear to detect changes earlier than standard automated perimetry.

(2) The development of procedures to reduce variability in visual field measures, both within a single session and from one session to another. There have been some methods that have been introduced that appear to be of assistance in this regard.[28,29,66-68]

(3) Development of better, more sophisticated statistical analysis techniques. Again, there have been some advances in statistical, neural network and related analytical approaches that have been made in recent times.[73-76] However, clinical demands on optimizing patient care would be tremendously advanced if more reliable, accurate, precise and efficient methods could be developed for evaluating a patient's visual field status over time.

(4) Development of methods to analyze "noise" and variability to search for meaningful signals of pathology. Traditionally, visual field variability has been viewed as something that needs to be eliminated, and there has been little research directed towards a systematic analysis of this information to look for markers of pathology. There are many areas of medicine and ophthalmology that have shown value in seeking out such markers. It is likely that there are "signatures" of specific patterns of "noise" or variability that are indicative of pathology of a dysfunctional visual system.

In summary, each of these suggestions for the future is directed towards improving the ability to extract meaningful signals and markers of visual system pathology and changes in visual status that can be revealed through visual field testing.

Finally, one of the most important factors that eye care specialists must consider in the management of patients is the rate of visual field progression. This is a crucial issue in determining the management and treatment regimen for patients, and in assessing how their visual impairment may affect their quality of life and activities of daily living. In this regard, Progressor and the Glaucoma Progression Index provide a means of forecasting the future visual field characteristics for the patient.[63,64,71] However, it should be kept in mind that these procedures are based on a linear regression model, and it is not clear whether this or non-linear methods would be most appropriate for predicting future visual field outcome for the patient. This topic has only recently received attention by the research community, and it is clear that much additional work is needed.

A guide for interpretation of visual field information

It is important to have a systematic approach for interpreting visual field information (Box 35.1). We recommend using a relatively simple analysis sequence, but emphasize that it is critical to maintain the proper order of this sequence to minimize errors of interpretation (false positives, false negatives, improper localization of the defect, and other difficulties).

(1) Decide if the visual fields are reliable and if artifactual test results are present.

Box 35.1 Visual field interpretation guidelines

- Detection – Is the visual field of each eye within normal limits or does it have abnormalities?
- Identification – What is the pattern of visual field loss? Is the visual field loss in one eye or both eyes? What is the location or locations of damage to the visual pathways that could account for this loss (differential diagnosis)?
  - For each eye, is the visual field normal or abnormal?
  - If abnormal, is it in one or both eyes?
  - What is the general location of the defect (superior, inferior, nasal, temporal)?
  - What is the shape and specific visual field location of the deficit?
  - Does the deficit correlate with anatomical features and boundaries?
- Staging of visual field loss – What is the severity of visual field loss? Is the loss abrupt or gradual?
- Artifactual visual loss – Does the deficit resemble a lens rim holder, a droopy eyelid, or some other non-pathologic or physiologic result?
- Functional visual loss – Does the pattern and severity of visual field loss exceed that expected on the basis of other clinical findings?
- Longitudinal follow-up of visual field loss – Is the visual field defect remaining stable, improving or getting worse over time? If the visual field loss is changing over time, what is the rate of change over time?
- Practical visual field issues – How does the visual field loss affect the patient's quality of life and activities of daily living? Does it impair the ability to drive, navigate the environment, reduce employment effectiveness?

(2) Place the visual fields for both eyes in front of you, with the left eye on the left and the right eye on the right. For each eye separately, determine whether the visual field is within or outside of normal limits. Automated visual field testing provides assistance in this analysis by presenting a comparison of individual results to those of average findings for a person of the same age. If both visual fields are within normal limits, the analysis is complete.

(3) Determine if the visual field loss (outside normal limits) is in one eye or both eyes. If the loss is in one eye, then the source of the visual damage is before the optic chiasm. Visual field loss in both eyes can be due to insults affecting the visual pathways at the optic chiasm and beyond, or could be due to diseases influencing the optics, retinas or optic nerves of both eyes.

(4) For each eye, determine the general location of the defect (superior, inferior, nasal or temporal), and if the visual field loss is extensive, assess the predominant location of the deficit. A deficit that is in the temporal visual fields of each eye (bitemporal) is often due to factors affecting the optic chiasm, deficits affecting the nasal visual fields of each eye are usually due to retinal or optic nerve disorders, and losses occurring in the nasal visual field of one eye and the temporal visual field of the other eye are usually related to damage occurring at locations after the optic chiasm. In this view, it is now possible to localize the most likely areas of damage to specific regions of the visual pathways.

(5) Determine the shape and specific location of the defect. This particular aspect of visual field interpretation is probably the most difficult, but also most important part of the assessment. There are many physiologic and pathologic factors that can produce generalized or widespread visual field sensitivity loss (small pupils, lack of attention, fatigue, corneal opacities, lens opacities, ocular and neurologic disorders, etc.), making this a rather non-specific finding for localization of the etiology of visual field sensitivity loss.

Some retinal diseases can produce specific patterns of visual field loss: (a) scalloped edges of isopters in the far periphery, (b) ring or partial ring scotomas in the mid periphery produced by retinitis pigmentosa, (c) arcuate defects produced by branch artery occlusions, (d) central scotomas related to macular diseases, (e) multiple foci of visual field loss at scattered points the visual field from diabetic retinopathy or other retinal degenerative disease processes, and (f) irregular deficits that do not correspond to other anatomical landmarks throughout the visual pathways.

Diseases affecting the optic nerve can produce: (a) centro-cecal scotomas produced by damage to the papillomacular bundle of retinal ganglion cells subserving the region between the macula and the blind spot, (b) arcuate defects created by damage to the superior and inferior arcuate nerve fiber bundles entering the top and bottom of the optic nerve head (often these defects will "respect" [terminate at] the nasal horizontal meridian), (c) central scotomas, (d) constriction of the visual field, (e) enlargement of the blind spot, and (f) temporal wedge defects produced by the radial fibers from the periphery that enter the nasal portion of the optic nerve head (corresponding to the temporal visual field).

Chiasmal deficits most commonly produce bitemporal visual field losses that "respect" (terminate at) the vertical meridian for both eyes. Sometimes, a deficit can be located at the anterior portion of the chiasm and produce a central scotoma in one eye (damage to the optic nerve of that eye) and a superior temporal defect in the other eye. Binasal defects that "respect" the vertical midline can also occur, but they are very rare because they involve compression of the chiasm from both the left and right sides. Beyond the chiasm, visual field defects are homonymous (occurring on the left side of vision for both eyes or the right side of vision for both eyes). If the deficit is incongruous between the two eyes (greater loss in one eye than in the other, and only moderate overlap between the defective areas in the two eyes), the damage has occurred prior to primary visual cortex (area 17 or V1). The greater the congruity between deficits for the two eyes, the farther the insult is from the chiasm and the closer it is to primary visual cortex.

(6) If there are multiple visual fields that have been obtained for different visits, determine whether the results indicate improvement, stability or progression of visual loss. It is important to review all of the visual field results and not just rely on comparisons of the current results with those obtained for the most recent previous visit. Additionally, comparison of the visual field interpretation of multiple tests with other findings for the clinical examination is also essential.

(7) Finally, if there is uncertainty or doubt concerning the visual field interpretation, it is always helpful to get the opinion and assessment of colleagues. Different

practitioners do not always observe results in the same manner and may have various features that are regarded with greater or lesser importance.

New perimetric test procedures

Many new visual field test procedures have been introduced, particularly within the past 20 years (Box 35.2). Some of these techniques have been successful, while others have not been found to have sufficient clinical or basic research

value and have therefore been dropped from further consideration. We will provide a brief summary of the methods that have been shown to be important for clinical diagnostic purposes and for understanding the basis of underlying physiologic and pathologic mechanisms.

Short wavelength automated perimetry (SWAP)

Stiles reported the two-color increment threshold technique for psychophysically isolating and measuring the sensitivity of individual color vision mechanisms.[77] This procedure was adapted for clinical application by Kitahara[78] and Kranda and King-Smith,[79] although the procedures were too cumbersome and time consuming for routine clinical use. Subsequent studies by Sample and colleagues[80-89] and Johnson and associates[90-99] reported a modification of an existing automated perimeter (the Humphrey Field Analyzer) to permit the evaluation of short-wavelength-sensitive (blue) color vision mechanisms, which is referred to as Short Wavelength Automated Perimetry (SWAP) within a time frame that was comparable to standard automated perimetry (SAP). A collaborative study identified optimal test conditions for SWAP, which now serves as the basis for SWAP tests that are available on several commercially available perimeters.[84]

Investigations now indicate that SWAP may detect glaucomatous visual field deficits an average of 3–5 years earlier than conventional SAP, and in some cases can occur up to 10 years earlier than SAP visual field deficits.[80-100] SWAP can sometimes characterize glaucomatous patterns of visual field loss and predict the onset and location of future SAP deficits. Progression of visual field loss appears to be more rapid than for SAP deficits. Isolation of short wavelength mechanisms is maintained throughout the range of glaucomatous deficits, including regions of advanced visual field loss. Additionally, SWAP has been shown to be useful for characterizing visual field losses in other ocular and neurologic diseases affecting the visual pathways,[101] and Bayesian forecasting procedures have been developed for more efficient testing and expansion of the dynamic range.[102-105] Figure 35.11 displays an example of SWAP visual field loss for the right eye of a patient with inferior glaucomatous visual field loss.

Frequency doubling technology (FDT) perimetry

A low spatial frequency sinusoidal grating (less than 2 cycles/degree) that is rapidly counterphase (square wave) flickered at greater than 15 Hz produces a percept of twice as many light and dark bars than are physically present, known as "frequency doubling".[106-108] This phenomenon has been incorporated into a clinical diagnostic test procedure known as Frequency Doubling Technology (FDT) perimetry, which has now undergone two generations of testing devices.[109-110] However, the observer's task is not to evaluate the frequency doubled nature of the percept, but rather to determine the minimum contrast at which the stimulus can be detected. The first generation (FDT) perimeter presented 17 or 19 large (10-degree square) targets to various visual field locations, whereas the second generation device (the Humphrey Matrix) can provide a larger number of smaller targets

---

Box 35.2 Visual field test procedures

Contrast sensitivity and incremental light detection

- Standard Automated Perimetry (SAP) – A method of determining the eye's sensitivity to light at different locations throughout the field of view.

Spatial visual field tests

- High Pass Resolution Perimetry (HPRP) – A procedure that determines the minimum size of a low-contrast stimulus necessary for it to be detected.
- Rarebit Perimetry – A method of providing fine detail mapping of visual field locations by determining the detectability of tiny light dots.

Temporal visual field tests

- Flicker perimetry – A visual field procedure for determining the highest rate of flicker that can be detected (critical flicker fusion or CFF perimetry), the minimum amplitude (contrast) of flicker that can be detected (temporal modulation perimetry), or the minimum light increment needed to detect an illuminated target as flickering on a uniform background (luminance pedestal flicker perimetry).
- Motion perimetry – A method for determining sensitivity to motion throughout the visual field by determining the minimum displacement needed to detect motion (displacement threshold perimetry), the amount of directional coherence needed to detect motion among a collection of random dots (motion coherence perimetry) or the size of visual field area needed to detect motion of a subset of random dots.

Spatio-temporal visual field tests

- Frequency Doubling Technology (FDT) perimetry and Matrix perimetry – A method for presenting a low spatial frequency sinusoidal grating flickering at a high temporal frequency to determine the amount of contrast needed for detection.

Color perimetry

- Short Wavelength Automated Perimetry (SWAP) – A visual field technique that isolates and measures the sensitivity of short wavelength mechanisms.

Electrophysiological perimetry

- Multifocal Electroretinogram (mfERG) – A method of measuring electrical retinal signals for localized visual field regions.
- Multifocal visual evoked potentials (mfVEP) – A technique that measures the electrical activity from primary visual cortex for alternating stimuli presented to small regions of the visual field.

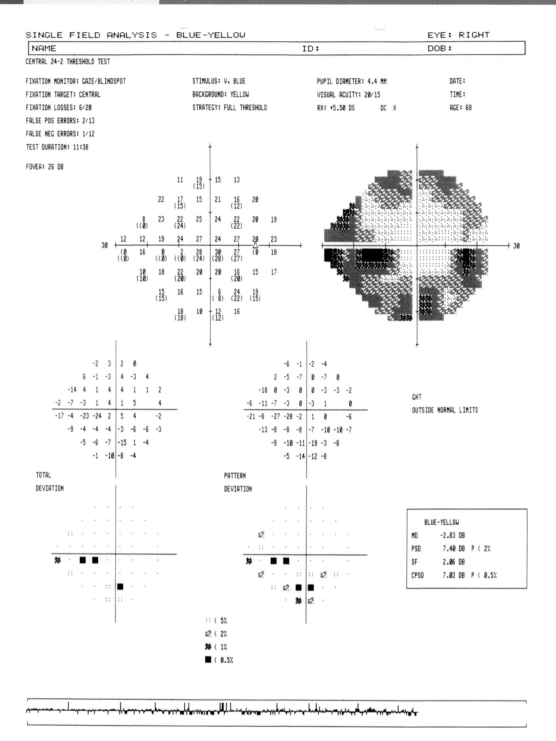

Figure 35.11 Short Wavelength Automated Perimetry (SWAP) results for the right eye (OD) of a patient with primary open angle glaucoma, revealing a partial inferior arcuate visual field defect.

(5-degree or 2-degree squares) to produce the 30-2, 24-2, 10-2 and macula presentation patterns. Details concerning these procedures and methods may be found in other publications.[111-128]

FDT perimetry has been found to be clinically effective in determining visual field loss that is produced by glaucoma and other ocular and neurologic diseases affecting the visual pathways.[111-128] Additionally, some reports have indicated that FDT perimetry deficits can precede losses detected by

SAP, and that they may be predictive of future visual field outcomes.[129,130] The recent advent of the Humphrey Matrix has provided a number of additional features for this procedure, and has also been reported to have uniform variability over the entire response range of the instrument, a distinct benefit for monitoring the status of moderate to advanced visual field loss. Figure 35.12 presents FDT results for the left eye of a patient with superior glaucomatous visual field loss, and Figure 35.13 shows the results for the same eye using

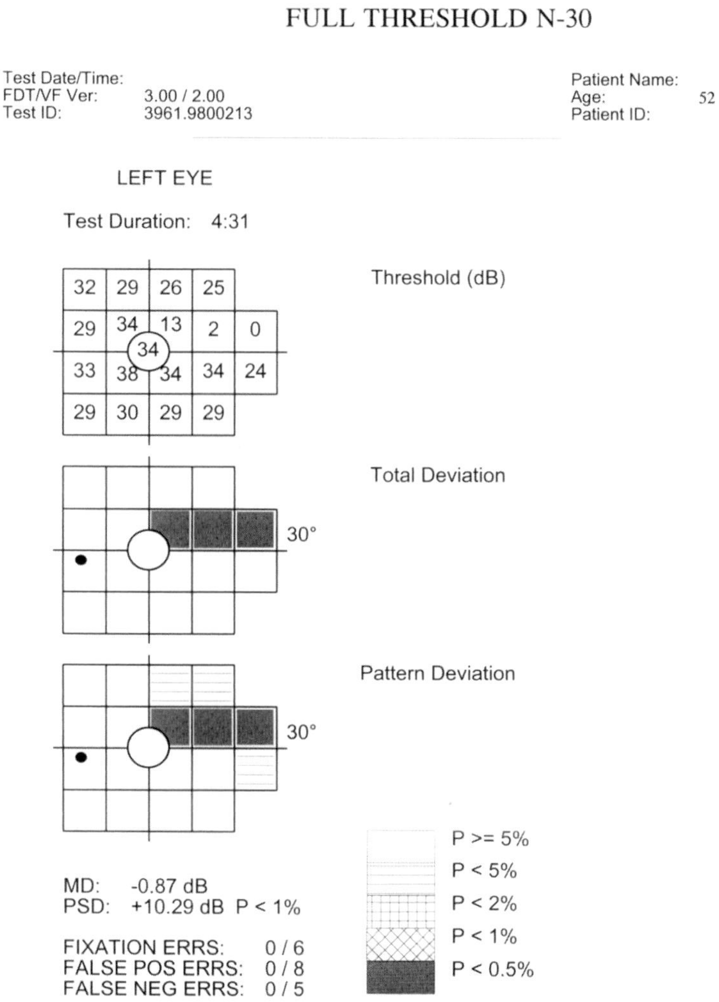

FULL THRESHOLD N-30

Test Date/Time:
FDT/VF Ver: 3.00 / 2.00
Test ID: 3961.9800213

Patient Name:
Age: 52
Patient ID:

Figure 35.12 Frequency Doubling Technology results for the left (OS) eye of a patient with primary open angle glaucoma that displays a superior nasal step.

LEFT EYE

Test Duration: 4:31

Threshold (dB)

32	29	26	25	
29	34	13	2	0
	(34)			
33	38	34	34	24
29	30	29	29	

Total Deviation

30°

Pattern Deviation

30°

P >= 5%
P < 5%
P < 2%
P < 1%
P < 0.5%

MD: -0.87 dB
PSD: +10.29 dB P < 1%

FIXATION ERRS: 0 / 6
FALSE POS ERRS: 0 / 8
FALSE NEG ERRS: 0 / 5

the Humphrey Matrix 24-2 test procedure obtained approximately one year after the initial FDT testing session.

Flicker and temporal modulation perimetry

Four forms of flicker perimetry have been developed in recent years: (1) determining the highest temporal flicker frequency that can be distinguished from a steady light (critical flicker fusion [CFF] perimetry), (2) evaluating the minimum amplitude of temporal alternation needed to detect flicker (temporal modulation perimetry [TMP]), (3) detection of a letter or shape of a series of rapidly alternating dots that are flickering in counterphase to other dots (flicker-defined form [FDF] perimetry) and (4) determining the minimum amount of flicker superimposed on a luminance increment needed to detect flicker from a steady light (luminance pedestal flicker [LPF] perimetry).[131-144]

Each of these procedures has certain advantages and disadvantages, but no study to date has compared all four tests in normal controls and patients with ocular disease to determine their clinical merits and deficits. However, a comparison between CFF and TMP indicated that in glaucoma, TMP was more sensitive for separation of findings from normal controls and glaucoma patients.[131] No systematic investigations of FDF or LPF perimetry in ocular and neurologic diseases have been reported to date.

Early investigations reported that high temporal frequencies were optimal for detection of glaucomatous visual field loss,[133] but subsequent studies indicated that a high temporal frequency sensitivity loss was characteristic of older normal controls, and that middle temporal frequencies (about 8 Hz) provided the greatest dynamic range, the lowest variability, and the best ability to detect early visual field losses.[136] Flicker perimetry also appears to be quite resistant to the influence of blur, and other test variables, but can sometimes be influenced by changes in pupil size and other stimulus presentation, physiologic and pathologic conditions.[134-139] Several studies have reported that flicker perimetry is able to detect glaucomatous visual field loss prior to SAP and has clinical value in predicting the onset and location of subsequent SAP visual field loss.[134-144] Figure 35.14 presents temporal modulation flicker perimetry results for the left eye of a patient with superior glaucomatous visual field loss.

Motion perimetry

As with flicker perimetry, motion perimetry is conducted according to several different methods: (1) detection of rapid movement of a single stimulus presented at various visual field locations (displacement threshold perimetry), (2) determination of a minimum proportion of dots moving

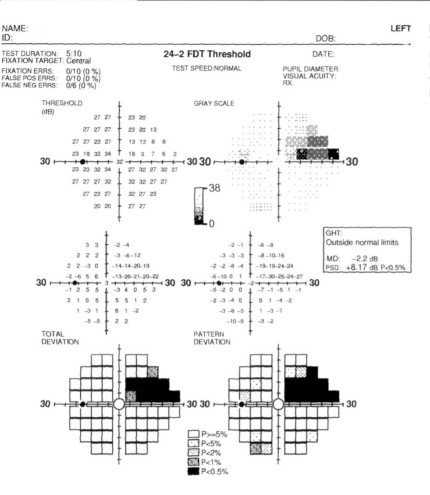

Figure 35.13 Humphrey Matrix Frequency Doubling Technology results for the same left eye (OS) of the patient with primary open angle glaucoma shown in Figure 35.12, exhibiting a superior nasal step. This test was performed one year after the results that were presented in Figure 35.12.

in the same direction that are embedded with a group of stationary dots or a group of dots that are moving in random directions (motion coherence perimetry), and (3) motion of a cluster of dots within a group of stationary dots.[145-158] In these instances, the measurement variables are minimum position displacement, minimum proportion of coherently moving dots, and minimum size of clustered moving dots, respectively.

Similar to flicker perimetry, motion detection perimetry is a robust measurement, being resistant to the influence of varying degrees of blur, contrast variations and background luminances. In this view, motion detection is a very salient feature that can be easily processed by peripheral visual field mechanisms. Additionally, motion perimetry appears to have uniform variability across its entire dynamic response range, and has been reported to be quite effective in detecting early visual field loss in a variety of diseases affecting the visual pathways.[145-158] Figure 35.15 presents the results for the left eye of a patient with inferior glaucomatous visual field loss.

High pass resolution perimetry

High pass resolution perimetry (HPRP) presents a series of ring stimuli consisting of a dark ring surrounding a circular central white stimulus to different visual field locations.[159-169] The observer's task is to indicate when they are able to detect the ring stimulus, and the size of the ring is varied to determine a threshold measurement. The HPRP stimuli are presented at 85 percent contrast so that detection (distinguishing

a target from the uniform background) and resolution (distinguishing one target from another) thresholds are essentially equivalent. The test is presented on a computer-controlled video monitor, and is an adaptive, interactive procedure.

Studies performed in glaucoma and other ocular and neurologic disorders indicate that HPRP is a useful diagnostic test procedure for detection and characterization of visual field loss,[159-169] has good test–retest reproducibility[165] and can detect progression of glaucomatous visual field loss an average of approximately 18 months earlier than SAP.[166] Although HPRP appears to have many advantages as a clinical diagnostic test procedure for characterizing visual field properties, it has limited commercial availability at the present time. Figure 35.16 presents results for the right eye of a patient with superior visual field loss due to glaucoma (left), with results from the Humphrey Field Analyzer also presented (right) for comparison.

Rarebit perimetry

Rarebit perimetry is a relatively new test procedure that uses a computer-driven video display to present one or two small white dots (pixels) on a uniform background.[170-174] The observer is asked to respond whether they saw one or two small targets during each stimulus presentation. By using a high-density pixel map, Rarebit perimetry evaluates correct and incorrect responses to construct regions of normal and abnormal detection of stimuli in an effort to locate gaps and discontinuities in performance throughout the visual field.

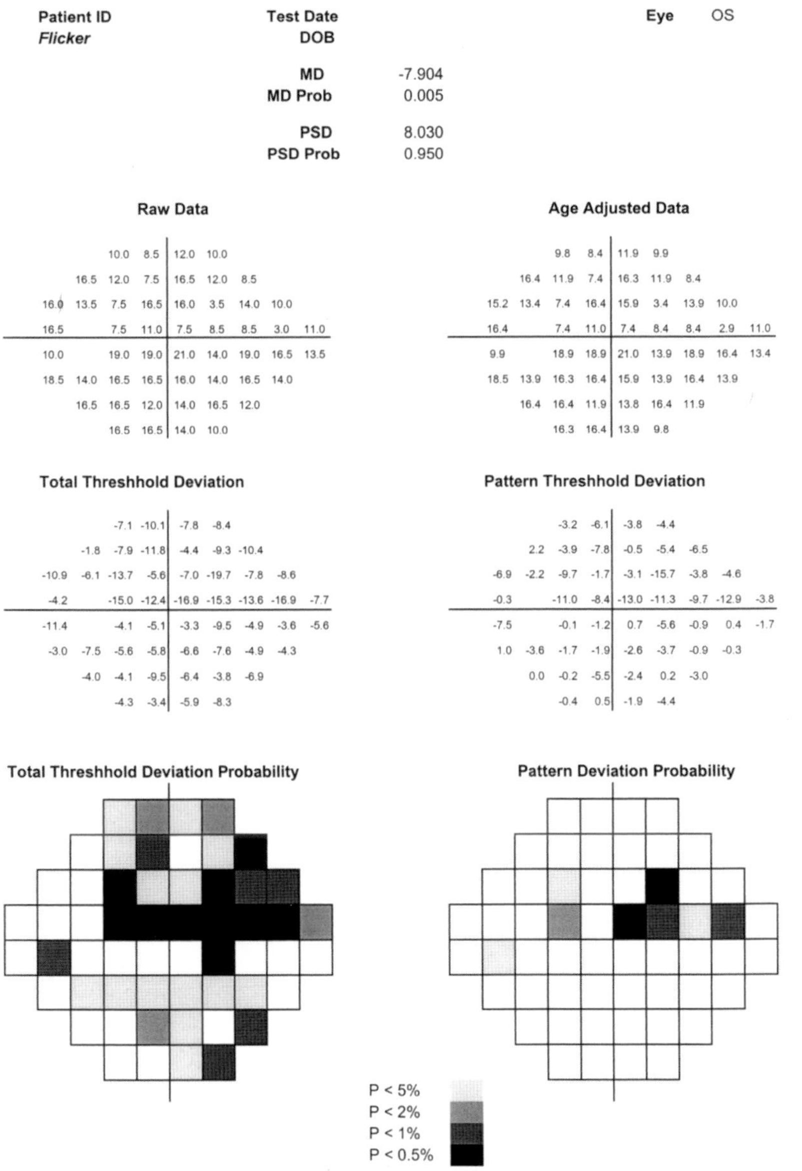

Patient ID *Flicker* **Test Date** DOB **Eye** OS

MD	-7.904
MD Prob	0.005
PSD	8.030
PSD Prob	0.950

Figure 35.14 Temporal modulation perimetry (flicker perimetry for the left eye (OS) of a patient with primary open angle glaucoma, revealing a superior arcuate visual field defect and an inferior partial arcuate defect.

Raw Data

```
          10.0  8.5 | 12.0 10.0
     16.5 12.0  7.5 | 16.5 12.0  8.5
16.0 13.5  7.5 16.5 | 16.0  3.5 14.0 10.0
16.5       7.5 11.0 |  7.5  8.5  8.5  3.0 11.0
10.0      19.0 19.0  21.0 14.0 19.0 16.5 13.5
18.5 14.0 16.5 16.5  16.0 14.0 16.5 14.0
     16.5 16.5 12.0  14.0 16.5 12.0
          16.5 16.5  14.0 10.0
```

Age Adjusted Data

```
           9.8  8.4 | 11.9  9.9
     16.4 11.9  7.4 | 16.3 11.9  8.4
15.2 13.4  7.4 16.4 | 15.9  3.4 13.9 10.0
16.4       7.4 11.0 |  7.4  8.4  8.4  2.9 11.0
9.9       18.9 18.9  21.0 13.9 18.9 16.4 13.4
18.5 13.9 16.3 16.4  15.9 13.9 16.4 13.9
     16.4 16.4 11.9  13.8 16.4 11.9
          16.3 16.4  13.9  9.8
```

Total Threshhold Deviation

```
           -7.1 -10.1 | -7.8  -8.4
      -1.8 -7.9 -11.8 | -4.4  -9.3 -10.4
-10.9 -6.1 -13.7 -5.6 | -7.0 -19.7 -7.8 -8.6
 -4.2      -15.0 -12.4| -16.9 -15.3 -13.6 -16.9 -7.7
-11.4       -4.1  -5.1  -3.3  -9.5 -4.9 -3.6 -5.6
 -3.0 -7.5  -5.6  -5.8  -6.6  -7.6 -4.9 -4.3
      -4.0  -4.1  -9.5  -6.4  -3.8 -6.9
            -4.3  -3.4 | -5.9  -8.3
```

Pattern Threshhold Deviation

```
           -3.2  -6.1 | -3.8  -4.4
       2.2 -3.9  -7.8 | -0.5  -5.4 -6.5
-6.9  -2.2 -9.7  -1.7 | -3.1 -15.7 -3.8 -4.6
-0.3      -11.0  -8.4 |-13.0 -11.3 -9.7 -12.9 -3.8
-7.5       -0.1  -1.2   0.7  -5.6 -0.9  0.4 -1.7
 1.0  -3.6 -1.7  -1.9  -2.6  -3.7 -0.9 -0.3
       0.0 -0.2  -5.5  -2.4   0.2 -3.0
           -0.4   0.5 | -1.9  -4.4
```

Total Threshhold Deviation Probability

Pattern Deviation Probability

P < 5%
P < 2%
P < 1%
P < 0.5%

In this manner, the intent of Rarebit perimetry is to identify areas of patchy sensitivity loss produced by glaucoma and other ocular and neurologic disorders that might be overlooked by SAP in the evaluation of early disease onset.

Information from clinical evaluation of Rarebit perimetry is somewhat limited to date, although these preliminary findings have been shown to be quite promising.[170-174] Although many of the features of Rarebit perimetry are quite appealing, empirical findings from clinical studies will ultimately determine its utility for clinical evaluations. The sensitivity, specificity, reproducibility and ability to monitor visual field progression with Rarebit perimetry need to be assessed carefully. Figure 35.17 presents the results of Rarebit perimetry testing in the right eye of a patient with superior and inferior glaucomatous visual field loss.

Multifocal visual evoked potentials (mfVEP)

Until recently, electrophysiological evaluation of visual field deficits produced by ocular and neurological diseases have had a rather limited, narrow application to clinical diagnostic procedures. The recent advent of multifocal techniques to measure the electroretinogram (mfERG) and visual evoked potential (mfVEP) has made it possible to expand these capabilities.[175-186] These techniques present rapidly alternating checkerboard patterns to various locations throughout the visual field according to a modified binary M-sequence, whereby the onset–offset characteristics of the checkerboard are the same for each location, but are subject to a delay in the initiation of the sequence at each of the visual field locations. An autocorrelation function is then used to extract localized signals from the complex waveform to obtain individual electrophysiologic recordings from a variety of visual field locations.[175-186]

Although the mfERG is useful for evaluation of many retinal disorders,[175-179] it has not been particularly beneficial for evaluation of glaucoma, and does not correlate well with traditional SAP visual field findings. On the other hand, mfVEP results have been shown to correlate quite well with SAP results, and have been reported to have signal to noise

Motion Detection Perimetry

Left

Figure 35.15 Visual field results for motion perimetry conducted on the right eye (OD) of a patient with an inferior arcuate visual field defect produced by primary open angle glaucoma.

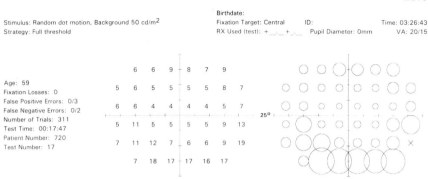

Stimulus: Random dot motion, Background 50 cd/m²
Strategy: Full threshold

Birthdate:
Fixation Target: Central   ID: Time: 03:26:43
RX Used (test): +__.__ +_.___ Pupil Diameter: 0mm   VA: 20/15

Age: 59
Fixation Losses: 0
False Positive Errors: 0/3
False Negative Errors: 0/2
Number of Trials: 311
Test Time: 00:17:47
Patient Number: 720
Test Number: 17

6	6	9	8	7	9		
5	6	5	5	5	5	8	7
6	6	4	4	4	4	5	7
5	11	5	5	5	5	9	13
7	11	12	7	6	6	9	19
7	18	17	17	16	17		

25°

20°   19 = Not Seen

20°   X = Not Seen

Number of Congratulations: 9
Number of Pauses: 4
Number of Mistakes: 0
Number of Errors: 1
Number of Mis-localizations: 1
Number of Outliers Retested: 9

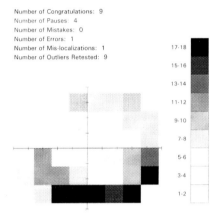

17-18	
15-16	
13-14	
11-12	
9-10	
7-8	
5-6	
3-4	
1-2	

Total Deviation Probability Plot

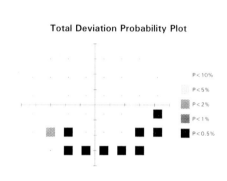

P < 10%
P < 5%
P < 2%
P < 1%
P < 0.5%

Suprathreshold Statistics

Mean Target Size:	36.77 ± 38.67
Mean Reaction Time:	427.79 ± 70.27
Mean Localization Error:	18.91 ± 13.34
Mean Relative Localization Error:	-14.05 ± 14.95

Threshold Statistics

Mean Target Size:	26.95 ± 38.56
Mean Reaction Time:	453.88 ± 71.66
Mean Localization Error:	19.77 ± 14.55
Mean Relative Localization Error:	-14.44 ± 16.99
Mean Deviation:	-1.89

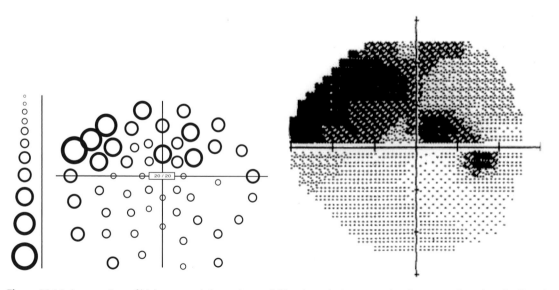

Figure 35.16 A comparison of high-pass resolution perimetry (left) and standard automated perimetry conducted on the Humphrey Field Analyzer (right) for the right eye (OD) of a patient with primary open angle glaucoma. (Reproduced with permission from Birt CM et al. J Glaucoma 1998; 7:111.)

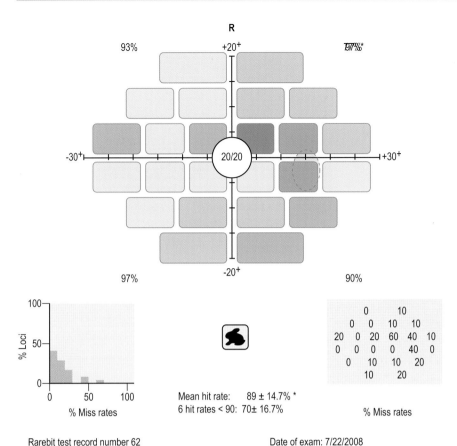

Figure 35.17 Superior and inferior arcuate visual field loss for the right eye (OD) of a patient with primary open angle glaucoma as determined with Rarebit automated perimetry.

Rarebit test record number 62
Pesantubee Michelene DOB 9/13/1953 Age 54
Dx nl tension glauc c pigment disp synd
*

RIGHT eye   Acuity: 2-/20   Correction:

Date of exam: 7/22/2008
Performance: + Tests: 130/5 Errors: 0
Test time: 4:30   MRT: 0.49   Examiner: JMV
Resolution 1280 * 960   Rarebit evrsion 4

ratios that remain relatively constant throughout its dynamic range, and to utilize the asymmetry between eyes to indicate subtle losses in glaucoma.[180–186] Although this procedure has undergone tremendous advances in recent years, there are also a number of disadvantages that remain:

(1) the response range is rather limited, so evaluation of patients with significant moderate to advanced glaucomatous damage cannot be performed,

(2) testing and setup times are long when compared to SAP,

(3) analysis of results is presently cumbersome and time-consuming, and

(4) the procedure requires careful monitoring to minimize the influence of confounding variables (attention loss, daydreaming, presence of alpha waves, etc.).

Figure 35.18 presents the results of multifocal VEP testing in a patient with inferior glaucomatous visual field loss in both eyes.

Conclusions

Perimetry and visual field testing are important diagnostic procedures that can provide a wealth of useful clinical information concerning a patient's functional vision status, ability to detect objects in the periphery, and effectively navigate through the environment. Additionally, visual field evaluation provides a useful guide concerning the location of damage to the visual pathways and the type of disease entity that is producing the visual field loss. The pattern of sensitivity loss and the relationship of visual field deficits between the two eyes can provide valuable insights about the locus of involvement for pathology that is affecting the visual pathways. In this view, it is useful to compare the diagnostic power of visual fields to other clinical test procedures. Measurement of visual acuity is undoubtedly the most common assessment of visual function. However, reduced visual acuity (e.g. 20/100) can be due to any factor influencing the visual pathways from the cornea to primary visual cortex and beyond. In this sense, visual acuity may be a sensitive measure of form vision that is important for reading, object identification and other related tasks, but it is not very specific in terms of where the anomaly is occurring in the visual pathways. Visual field testing, on the other hand, is able to provide specific information about the location of pathology by evaluating the shape, location and density of visual field sensitivity loss. Prior to the development and use of neurologic imaging procedures (CT, MRI, fMRI, PET, etc.), visual fields were used as a primary indicator of the location of ocular or neurological pathology.

It should be kept in mind that no single test procedure, including visual field testing, can provide the clinician with complete information about a patient's functional visual

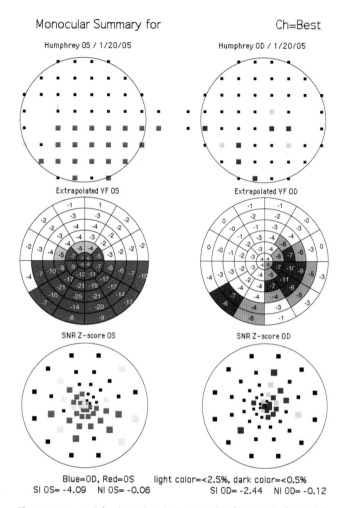

Figure 35.18 Multifocal visual evoked potential (mfVEP) results for the left and right eyes of a patient with primary open angle glaucoma resulting in an inferior arcuate to altitudinal visual field defect in the left eye (OS) and an inferior partial arcuate visual field defect in the right eye (OD).

status (although if this could be accomplished, clinical examinations could be performed much more rapidly). Therefore, it is helpful to keep in mind that visual field results should be interpreted within the context of information obtained from other portions of the eye examination, the patient's history (and family history), their primary complaints, and other pertinent information.

The intent of this chapter is to provide some guidelines for conducting and interpreting visual field information, and we hope that it has provided this service and will continue to be a resource for the future. However, as with most clinical ophthalmic procedures, there is no substitute for obtaining first-hand clinical experience with visual field testing. We hope that this chapter will provide you with the appropriate tools to achieve a high degree of expertise in the analysis and interpretation of perimetry.

References

1. Duke-Elder S, Scott GI. System of ophthalmology. Vol VII – The foundations of ophthalmology, visual fields. London: Henry Plimpton, 1971:393–425.
2. Aulhorn E, Harms H. Visual perimetry. In: Jameson D, Hurvich L, eds. Visual psychophysics: handbook of sensory physiology, Vol VII/4. New York: Springer-Verlag, 1972.
3. Harrington DO, Drake MV. The visual fields: text and atlas of clinical perimetry. St Louis: CV Mosby, 1990.
4. Anderson DR. Perimetry with and without automation. St Louis: CV Mosby, 1987.
5. Greve EL. Single and multiple stimulus static perimetry: The two phases of perimetry. The Hague: Dr W Junk Publishers, 1973..
6. Weber EH. In: Boring EG, ed. A history of experimental psychology. New York: Appleton-Century-Crofts, 1950.
7. Weber J, Rau S. The properties of perimetric thresholds in normal and glaucomatous eyes. Ger J Ophthalmol 1992; 1:79–85.
8. Chauhan BC, Tompkins JD, LeBlanc RP, McCormick TA. Characteristics of frequency-of-seeing curves in normal subjects, patients with suspected glaucoma and patients with glaucoma. Invest Ophthalmol Vis Sci 1993; 34:3534–3540.
9. Spenceley SE, Henson DB. Visual field test simulation and error in threshold estimation. Br J Ophthalmol 1996; 80:304–308.
10. Wall M, Maw RJ, Stanek KE, Chauhan BC. The psychometric function and reaction times of automated perimetry in normal and abnormal areas of the visual field in patients with glaucoma. Invest Ophthalmol Vis Sci 1996; 37:878–885.
11. Wall M, Kutzko KE, Chauhan BC. Variability in patients with glaucomatous visual field damage is reduced using size V stimuli. Invest Ophthalmol Vis Sci 1997; 38: 426–435.
12. Henson DB, Chaudry S, Artes PH, Faragher EB, Ansons A. Response variability in the visual field: comparison of optic neuritis, glaucoma, ocular hypertension, and normal eyes. Invest Ophthalmol Vis Sci 2000; 41:417–421.
13. Spry PGD, Johnson CA, McKendrick AM, Turpin A. Variability components of standard automated perimetry and frequency doubling technology perimetry. Invest Ophthalmol Vis Sci 2001; 42:1404–1410.
14. Chauhan BC, Garway-Heath DF, Goñi FJ, et al. Practical recommendations for measuring rates of visual field change in glaucoma. Br J Ophthalmol 2008; 92:569–573.
15. Chauhan BC, Johnson CA. Test-retest variability characteristics of Frequency Doubling Perimetry and conventional perimetry in glaucoma patients and normal controls. Invest Ophthalmol Vis Sci 1999; 40:648–656.
16. Johnson CA, Keltner JL, Jacob MP: New test procedures for the SQUID automated perimeter. Doc Ophthalmol Proc Series 1985; 42:91–94.
17. Johnson CA, Keltner JL: Optimal rates of movement for kinetic perimetry. Arch Ophthalmol 1987; 105:73–75.
18. Miller KN, Shields MB, Ollie AR. Automated kinetic perimetry with two peripheral isopters in glaucoma. Arch Ophthalmol 1989; 107:1316–1320.
19. Dolderer J, Vonthein R, Johnson CA, Schiefer U, Hart W. Scotoma mapping by semi-automated kinetic perimetry: the effects of stimulus properties and the speed of subjects' responses. Acta Ophthalmol Scand 2006; 84:338–344.
20. Nevalainen J, Paetzold J, Krapp E, Vonthein R, Johnson CA, Schiefer U. The use of semi-automated kinetic perimetry (SKP) to monitor advanced glaucomatous visual field loss. Graefes Arch Clin Exp Ophthalmol 2008; 246:1331–1339.
21. Bebie H, Fankhauser F, Spahr J. Static perimetry: strategies. Acta Ophthalmol 1976; 54:325–338.
22. Koch P, Roulier A, Fankhauser F. Perimetry – the information theoretical basis for its automation. Vision Res 1972; 12:1619–1630.
23. Johnson CA, Chauhan BC, Shapiro LR. Properties of staircase procedures for estimating thresholds in automated perimetry. Invest Ophthalmol Vis Sci 1992; 33:2966–2974.
24. Bengtsson B, Olsson J, Heijl A, Rootzen H. A new generation of algorithms for computerized threshold perimetry, SITA. Acta Ophthalmol Scand 1997;75(4): 368–375.
25. Bengtsson B, Heijl A, Olsson J. Evaluation of a new threshold visual field strategy, SITA, in normal subjects. Swedish Interactive Thresholding Algorithm. Acta Ophthalmol Scand 1998; 76(2):165–169.
26. Bengtsson B, Heijl A. Comparing significance and magnitude of glaucomatous visual field defects using the SITA and Full Threshold strategies. Acta Ophthalmol Scand 1999; 77(2):143–146.
27. Vingrys AJ, Pianta MJ. A new look at threshold estimation algorithms for automated static perimetry. Optom Vis Sci 1999; 76:588–595.
28. Turpin A, McKendrick AM, Johnson CA, Vingrys AJ. Properties of perimetric threshold estimates from full threshold, ZEST, and SITA-like strategies, as determined by computer simulation. Invest Ophthalmol Vis Sci 2003; 44:4787–4795.
29. Anderson AJ, Johnson CA. Comparison of the ASA, MOBS, and ZEST threshold methods. Vision Res 2006; 46:2403–2411.
30. Weber J, Klimaschka T. Test time and efficiency of the dynamic strategy in glaucoma perimetry. Ger J Ophthalmol 1995; 4:25–31.
31. Morales J, Weitzman ML, González de la Rosa M. Comparison between Tendency-Oriented Perimetry (TOP) and octopus threshold perimetry. Ophthalmology 2000; 107:134–142.
32. Anderson AJ. Spatial resolution of the tendency-oriented perimetry algorithm. Invest Ophthalmol Vis Sci 2003; 44:1962–1968.
33. Capris P, Autuori S, Capris E, Papadia M. Evaluation of threshold estimation and learning effect of two perimetric strategies, SITA Fast and CLIP, in damaged visual fields. Eur J Ophthalmol 2008; 18:182–190.
34. Nelson-Quigg JM, Twelker JD, Johnson CA. Response properties of normal observers and patients during automated perimetry. Arch Ophthalmol 1989; 107:1612–1615.
35. Bickler-Bluth M, Trick GL, Kolker AE, Cooper DG. Assessing the utility of reliability indices for automated visual fields. Testing ocular hypertensives. Ophthalmology 1989; 96:616–619.
36. Johnson CA, Nelson-Quigg JM. A prospective three year study of response properties of normals and patients during automated perimetry. Ophthalmology 1993; 100:269–274.
37. Bengtsson B. Reliability of computerized perimetric threshold tests as assessed by reliability indices and threshold reproducibility in patients with suspect and manifest glaucoma. Acta Ophthalmol Scand 2000; 78:519–522.
38. Keltner JL, Johnson CA, Cello KE, et al. Ocular Hypertension Treatment Study Group. Visual field quality control in the Ocular Hypertension Treatment Study (OHTS). J Glaucoma 2007; 16:665–669.
39. Hong S, Yeom HY, Kim CY, Seong GJ. Comparison between indices of Humphrey matrix and Humphrey perimetry in early glaucoma patients and normal subjects. Ann Ophthalmol 2007; 39:318–320.

40. Johnson CA, Keltner JL, Balestrery FG. Suprathreshold static perimetry in glaucoma and other optic nerve disease. Ophthalmology 1979; 86:1278–1286.

41. Johnson CA, Keltner JL. Automated suprathreshold static perimetry. Am J Ophthalmol 1980; 89:731–741.

42. Henson DB. Visual field screening and the development of a new screening program. J Am Optom Assoc 1989; 60:893–898.

43. Langerhorst CT, Bakker D, Raakman MA. Usefulness of the Henson Central Field Screener for the detection of visual field defects, especially in glaucoma. Doc Ophthalmol 1989; 72:279–285.

44. Siatkowski RM, Lam BL, Anderson DR, Feuer WJ, Halikman AM. Automated suprathreshold static perimetry screening for detecting neuro-ophthalmologic disease. Ophthalmology 1996; 103:907–917.

45. Artes PH, Henson DB, Harper R, McLeod D. Multisampling suprathreshold perimetry: a comparison with conventional suprathreshold and full-threshold strategies by computer simulation. Invest Ophthalmol Vis Sci 2003; 44:2582–2587.

46. Artes PH, McLeod D, Henson DB. Response time as a discriminator between true- and false-positive responses in suprathreshold perimetry. Invest Ophthalmol Vis Sci 2002; 43:129–132.

47. Bengtsson B, Heijl A. False-negative responses in glaucoma perimetry: indicators of patient performance or test reliability? Invest Ophthalmol Vis Sci 2000; 41:2201–2204.

48. Reynolds M, Stewart WC, Sutherland S. Factors that influence the prevalence of positive catch trials in glaucoma patients. Graefes Arch Clin Exp Ophthalmol 1990; 228:338–341.

49. Olsson J, Bengtsson B, Heijl A, Rootzén H. An improved method to estimate frequency of false positive answers in computerized perimetry. Acta Ophthalmol Scand 1997; 75:181–183.

50. Sanabria O, Feuer WJ, Anderson DR. Pseudo-loss of fixation in automated perimetry. Ophthalmology 1991; 98:76–78.

51. Spry PGD, Johnson CA. Identification of progressive glaucomatous visual field loss. Surv Ophthalmol 2002; 47:158–173.

52. Vesti E, Johnson CA, Chauhan BC. Comparison of different methods for detecting glaucomatous visual field progression. Invest Ophthalmol Vis Sci 2003; 44:3873–3879.

53. Smith SD, Katz J, Quigley HA. Analysis of progressive change in automated visual fields in glaucoma. Invest Ophthalmol Vis Sci 1996; 37:1419–1428

54. Nouri-Mahdavi K, Hoffman D, Ralli M, Caprioli J. Comparison of methods to predict visual field progression in glaucoma. Arch Ophthalmol 2007; 125:1176–1181.

55. Heijl A, Bengtsson B, Chauhan BC, et al. A comparison of visual field progression criteria of 3 major glaucoma trials in early manifest glaucoma trial patients. Ophthalmology 2008; 115:1557–1565.

56. Brusini P, Johnson CA: Staging functional damage in glaucoma: review of different classification methods. Sur Ophthalmol 2007; 52:156–179.

57. Advanced Glaucoma Intervention Study. 2. Visual field test scoring and reliability. Ophthalmology 1994; 101:1445–1455.

58. Musch DC, Lichter PR, Guire KE, Standardi CL. The Collaborative Initial Glaucoma Treatment Study: study design, methods, and baseline characteristics of enrolled patients. Ophthalmology 1999; 106:653–662.

59. Leske MC, Heijl A, Hyman L, Bengtsson B. Early Manifest Glaucoma Trial: design and baseline data. Ophthalmology 1999; 106:2144–2153.

60. Heijl A, Leske MC, Bengtsson B, Bengtsson B, Hussein M. Early Manifest Glaucoma Trial Group. Measuring visual field progression in the Early Manifest Glaucoma Trial. Acta Ophthalmol Scand 2003; 81:286–293.

61. Hitchings RA, Migdal CS, Wormald R, Poinooswamy D, Fitzke F. The primary treatment trial: changes in the visual field analysis by computer-assisted perimetry. Eye 1994; 8:117–120.

62. Wild JM, Hussey MK, Flanagan JG, Trope GE. Pointwise topographical and longitudinal modeling of the visual field in glaucoma. Invest Ophthalmol Vis Sci 1993; 34:1907–1916.

63. Fitzke FW, Hitchings RA, Poinoosawmy D, McNaught AI, Crabb DP. Analysis of visual field progression in glaucoma. Br J Ophthalmol 1996; 80:40–48.

64. McNaught AI, Crabb DP, Fitzke FW, Hitchings RA. Visual field progression: comparison of Humphrey Statpac2 and pointwise linear regression analysis. Graefes Arch Clin Exp Ophthalmol 1996; 234:411–418.

65. Wilkins MR, Fitzke FW, Khaw PT. Pointwise linear progression criteria and the detection of visual field change in a glaucoma trial. Eye 2006; 20:98–106.

66. Crabb DP, Fitzke FW, McNaught AI, Edgar DF, Hitchings RA. Improving the prediction of visual field progression in glaucoma using spatial processing. Ophthalmology 1997; 104:517–524.

67. Strouthidis NG, Scott A, Viswanathan AC, Crabb DP, Garway-Heath DF. Monitoring glaucomatous visual field progression: the effect of a novel spatial filter. Invest Ophthalmol Vis Sci 2007; 48:251–257.

68. Gardiner SK, Crabb DP. Examination of different pointwise linear regression methods for determining visual field progression. Invest Ophthalmol Vis Sci 2002; 43:1400–1407.

69. Anderson DR, Patella VM. Automated static perimetry. St Louis: CV Mosby, 1990.

70. Viswanathan AC, Fitzke FW, Hitchings RA. Early detection of visual field progression in glaucoma: a comparison of PROGRESSOR and STATPAC 2. Br J Ophthalmol 1997; 81:1037–1042.

71. Bengtsson B, Heijl A. A visual field index for calculation of glaucoma rate of progression. Am J Ophthalmol 2008; 145:343–353.

72. Johnson CA, Cioffi GA, Liebmann JR, Sample PA, Zangwill L, Weinreb RN. The relationship between structural and functional alterations in glaucoma: a review. Semin Ophthalmol 2000; 15:221–233.

73. Racette L, Medeiros FA, Bowd C, Zangwill LM, Weinreb RN, Sample PA. The impact of the perimetric measurement scale, sample composition, and statistical method on the structure-function relationship in glaucoma. J Glaucoma 2007; 16:676–684.

74. Boden C, Chan K, Sample PA, et al. Assessing visual field clustering schemes using machine learning classifiers in standard perimetry. Invest Ophthalmol Vis Sci 2007; 48:5582–5590.

75. American Academy of Ophthalmology. Automated perimetry. Ophthalmology 1996; 103:1144–1151.

76. Zeyen T. Interpretation of automated perimetry. Bull Soc Belge Ophtalmol 1997; 267:191–197.

77. Stiles WS. Increment thresholds and the mechanisms of colour vision. Doc Ophthalmol 1949; 3:138–165.

78. Kitahara K, Tamaki R, Noji J, Kandatsu A, Matsuzaki H. Extrafoveal Stiles pi mechanisms. Doc Ophthalmol 1982; 35:397–404.

79. Kranda K, King-Smith, PE. What can color thresholds tell us about the nature of underlying detection mechanisms? Ophthalmic Physiol Opt 1984; 4:83–87.

80. Sample PA, Weinreb RN, Boynton RM. Isolating color vision loss of primary open angle glaucoma. Am J Ophthalmol 1988; 106:686–691.

81. Sample PA, Weinreb RN. Color perimetry for assessment of primary open angle glaucoma. Invest Ophthalmol Vis Sci 1990; 31:1869–1875.

82. Sample RA, Weinreb RN. Progressive color visual field loss in glaucoma. Invest Ophthalmol Vis Sci 1992; 33:2068–2071.

83. Sample PA, Martinezz GA, Weinreb RN. Short-wavelength automated perimetry without lens density testing. Am J Ophthalmol 1994; 118:632–641.

84. Sample PA, Johnson CA, Haegerstrom-Portnoy G, Adams AJ. Optimum parameters for short-wavelength automated perimetry. J Glaucoma 1996; 5:375–383.

85. Sample PA, Martinez GA, Weinreb RN. Color visual fields: A 5 year prospective study in eyes with primary open angle glaucoma. In: Mills RP, ed. Perimetry Update 1992/93. New York: Kugler Publications 1993:473–476.

86. Sample PA, Taylor JD, Martinez GA, Lusky M, Weinreb RN. Short wavelength color visual fields in glaucoma suspects at risk. Am J Ophthalmol 1993; 115:225–233.

87. Sample PA, Medieros FA, Racette L, et al. Identifying glaucomatous vision loss with visual-function-specific perimetry in the diagnostic innovations in glaucoma study. Invest Ophthalmol Vis Sci 2006; 47:3381–3389.

88. Racette L, Sample PA. Short-wavelength automated perimetry. Ophthalmol Clinics North Am 2003; 16:227–236.

89. Sample PA. Short-wavelength automated perimetry: its role in the clinic and for understanding ganglion cell function. Prog Ret Eye Res 2000; 19:369–383.

90. Johnson CA, Adams AJ, Casson EJ, Brandt JD. Blue-on-Yellow perimetry can predict the development of glaucomatous visual field loss. Arch Ophthalmol 1993; 111:645–650.

91. Johnson CA, Adams AJ, Casson EJ, Brandt JD. Progression of early glaucomatous visual field loss for Blue-on-Yellow and standard White-on-White automated perimetry. Arch Ophthalmol 1993; 111:651–656.

92. Johnson CA, Brandt JD, Khong AM, Adams AJ. Short wavelength automated perimetry (SWAP) in low, medium and high risk ocular hypertensives: Initial baseline findings. Arch Ophthalmol 1995; 113:70–76.

93. Johnson CA, Adams AJ, Casson EJ. Blue-on-yellow perimetry: A five year overview. In: Mills RP, ed. Perimetry Update 1992/93. New York: Kugler Publications 1993: 459–466.

94. Johnson CA. Selective vs non-selective losses in glaucoma. J Glaucoma 1994; 3:S32–S44 (Feature Issue – Journal Supplement).

95. Demirel S, Johnson CA. Incidence and prevalence of Short Wavelength Automated Perimetry (SWAP) deficits in ocular hypertensive patients. Am J Ophthalmol 2001; 131:709–715.

96. Demirel S, Johnson CA. Isolation of short wavelength sensitive mechanisms in normal and glaucomatous visual field regions. J Glaucoma 2000; 9:63–73.

97. Demirel S, Johnson CA. Short Wavelength Automated Perimetry (SWAP) in ophthalmic practice. J Am Optom Assn 1996; 67:451–456.

98. Casson EJ, Johnson CA, Shapiro LR. A longitudinal comparison of Temporal Modulation Perimetry to White-on-White and Blue-on-Yellow Perimetry in ocular hypertension and early glaucoma. J Opt Soc Am 1993; 10:1792–1806.

99. Lewis RA, Johnson CA, Adams AJ. Automated static visual field testing and perimetry of short-wavelength-sensitive (SWS) mechanisms in patients with asymmetric intraocular pressures. Graefe's Arch Clin Exp Ophthalmology 1993; 231:274–278.

100. Sit AJ, Medieros FA, Weinreb RN. Short-wavelength automated perimetry can predict glaucomatous visual field loss by ten years. Semin Ophthalmol 2004; 19:122–124.

101. Keltner JL, Johnson CA. Short Wavelength Automated Perimetry (SWAP) in neuro-ophthalmologic disorders. Arch Ophthalmol 1995; 113:475–481.

102. Turpin A, Johnson CA, Spry PGD. Development of a maximum likelihood procedure for Short Wavelength Automated Perimetry (SWAP). In: Wall M, Mills RP, eds. Perimetry update 2000/2001. The Hague: Kugler Publications, 2001:139–147.

103. Bengtsson B. A new rapid threshold algorithm for short-wavelength automated perimetry. Invest Ophthalmol Vis Sci 2003; 44:1388–1394.

104. Bengtsson B, Heijl A. Normal intersubject threshold variability and normal limits of the SITA SWAP and full threshold SWAP perimetric programs. Invest Ophthalmol Vis Sci 2003; 44:5029–5034.

105. Bengtsson B, Heijl A. Diagnostic sensitivity of fast blue-yellow and standard automated perimetry in early glaucoma: a comparison between different test programs. Ophthalmology 2006; 113:1092–1097.

106. Kelly DH. Frequency doubling in visual responses. J Opt Soc Am A 1966: 56:1628–1633.

107. Kelly DH. Non-linear visual responses to flickering sinusoidal gratings. J Opt Soc Am 1981; 71:1051–1055.

108. Maddess T, Henry H. Performance of non-linear visual units in ocular hypertension and glaucoma. Clin Vis Science 1992; 7:371–383.

109. Johnson CA, Wall M, Fingeret M, Lalle P. A primer for frequency doubling technology perimetry. Skaneateles, New York: Welch Allyn, 1998.

110. Spry PGD, Johnson CA, Anderson AJ, et al. A primer for frequency doubling technology (FDT) perimetry using the Humphrey matrix. Skaneateles, New York: Welch Allyn, 2008.

111. Johnson CA, Samuels SJ. Screening for glaucomatous visual field loss using the frequency-doubling contrast test. Invest Ophthalmol Vis Sci 1997; 38:413–425.

112. Fujimoto N, Adachi-Usami E. Frequency doubling perimetry in resolved optic neuritis. Invest Ophthalmol Vis Sci 2000; 41:2558–2560.

113. Wall M, Neahring RK, Woodward KR. Sensitivity and specificity of frequency doubling perimetry in neuro-ophthalmic disorders: a comparison with conventional automated perimetry. Invest Ophthalmol Vis Sci 2002; 43:1277–1283.

114. Girkin CA, McGwin G, DeLeon-Ortega J. Frequency doubling technology perimetry in non-arteritic ischaemic optic neuropathy with altitudinal defects. Br J Ophthalmol 2004; 88:1274–1279.

115. Sheu SJ, Chen YY, Lin HC, Chen HL, Lee IY, Wu TT. Frequency doubling technology perimetry in retinal disease – preliminary report. Kaohsiung J Med Sci 2001; 17:25–28.
116. Parikh R, Naik M, Mathai A, Kuriakose T, Muliyil J, Thomas R. Role of frequency doubling technology perimetry in screening of diabetic retinopathy. Indian J Ophthalmol 2006; 54:17–22.
117. White AJ, Sun H, Swanson WH, Lee BB. An examination of physiological mechanisms underlying the frequency-doubling illusion. Invest Ophthalmol Vis Sci 2002; 43:3590–3599.
118. Zeppieri M, Demirel S, Kent K, Johnson CA. Perceived spatial frequency of sinusoidal gratings. Optom Vis Sci 2008; 85:318–329.
119. Anderson AJ, Johnson CA. Frequency doubling technology perimetry. Ophthalmol Clin North Am 2003; 16:213–225.
120. Anderson AJ, Johnson CA, Fingeret M, et al. Characteristics of the normative database for the Humphrey Matrix perimeter. Invest Ophthalmol Vis Sci 2005; 46:1540–1548.
121. Clement CI, Goldberg I, Graham S, Healey PR. Humphrey matrix frequency doubling perimetry for detection of visual field defects in open-angle glaucoma. Br J Ophthalmol 2009; 93:582–588.
122. Brusini P, Salvatet ML, Zeppieri M, Parisi L. Frequency doubling technology perimetry with the Humphrey Matrix 30–2 test. J Glaucoma 2006; 15:77–83.
123. Spry PG, Hussin HM, Sparrow JM. Clinical evaluation of frequency doubling perimetry using the Humphrey Matrix 24–2 threshold strategy. Br J Ophthalmol 2005; 89:1031–1035.
124. Taravati P, Woodward KR, Keltner JL, et al. Sensitivity and specificity of the Humphrey Matrix to detect homonymous hemianopias. Invest Ophthalmol Vis Sci 2008; 49:924.
125. Huang CQ, Carolan J, Redline D, et al. Humphrey Matrix perimetry in optic nerve and chiasmal disorders: comparison with Humphrey SITA standard 24-2. Invest Ophthalmol Vis Sci 2008; 49:927.
126. Artes PH, Hutchison DM, Nicolela MT, LeBlanc RP, Chauhan BC. Threshold and variability properties of matrix frequency-doubling technology and standard automated perimetry in glaucoma. Invest Ophthalmol Vis Sci 2005; 46:2451–2457.
127. Johnson CA, Cioffi GA, Van Buskirk EM. Evaluation of two screening tests for frequency doubling technology perimetry. Perimetry Update 1998/1999. Amsterdam: Kugler Publications, 1999:103–109.
128. Spry PG, Hussin HM, Sparrow JM. Performance of the 24–2-5 frequency doubling technology screening test: a prospective case study. Br J Ophthalmol 2007; 91: 1345–1349.
129. Medeiros FA, Sample PA, Weinreb RN. Frequency doubling technology perimetry abnormalities as predictors of glaucomatous visual field loss. Am J Ophthalmol 2004; 137:863–871.
130. Bayer AU, Erb C. Short wavelength automated perimetry, frequency doubling technology perimetry, and pattern electroretinography for prediction of progressive glaucomatous standard visual field defects. Ophthalmology 2002; 109:1009–1017.
131. Yoshiyama KK, Johnson CA. Which method of flicker perimetry is most effective for detection of glaucomatous visual field loss? Invest Ophthalmol Vis Sci 1997; 38:2270–2277.
132. McKendrick AM, Johnson CA. Temporal properties of vision. In: Alm A, Kaufmann P, eds. Adler's Physiology of the Eye, 10th edn. St. Louis: CV Mosby, 2002:511–530.
133. Tyler CW. Specific deficits of flicker sensitivity in glaucoma and ocular hypertension. Invest Ophthalmol Vis Sci 1981; 100:135–146.
134. Lachenmayr BJ, Drance SM, Douglas GR, Mikelberg FS. Light-sense, flicker and resolution perimetry in glaucoma: a comparative study. Graefes Arch Clin Exp Ophthalmol 1991; 229:246–251.
135. Lachenmayr BJ, Drance SM, Chauhan BC, House PH, Lalani S. Diffuse and localized glaucomatous field loss in light-sense, flicker and resolution perimetry. Graefes Arch Clin Exp Ophthalmol 1991; 229:267–273.
136. Casson EJ, Johnson CA, Shapiro LR. A longitudinal comparison of Temporal Modulation Perimetry to White-on-White and Blue-on-Yellow Perimetry in ocular hypertension and early glaucoma. J Opt Soc Am 1993; 10:1792–1806
137. Casson EJ, Johnson CA. Temporal modulation perimetry in glaucoma and ocular hypertension. In: Mills RP, ed. Perimetry update 1992/93. New York: Kugler Publications, 1993:443–450.
138. Matsumoto C, Takada S, Okuyama S, Arimura E, Hashimoto S, Shimomura Y. Automated flicker perimetry in glaucoma using Octopus 311: a comparative study with the Humphrey Matrix. Acta Ophthalmol Scand 2006; 84:866–872.
139. Swanson WH, Ueno T, Smith VC, Pokorny J. Temporal modulation sensitivity and pulse-detection thresholds for chromatic and luminance perturbations. J Opt Soc Am A 1987; 4:1992–2005.
140. Anderson AJ, Vingrys AJ. Interactions between flicker thresholds and luminance pedestals. Vision Res 2000; 40:2579–2588.
141. Anderson AJ, Vingrys AJ. Effect of eccentricity on luminance-pedestal flicker thresholds. Vision Res 2002; 42:1149–1156.
142. Anderson AJ, Vingrys AJ. Multiple processes mediate flicker sensitivity. Vision Res 2001, 41:2449–2455.
143. Quaid PT, Flanagan JG. Defining the limits of flicker defined form: effect of stimulus size, eccentricity and number of random dots. Vision Res 2005; 45:1075–1084.
144. Goren D, Flanagan JG. Is flicker-defined form (FDF) dependent on the contour? J Vis 2008; 22(8):15.1–15.11.
145. Ruben S, Fitzke F. Correlation of peripheral displacement thresholds and optic disc parameters in ocular hypertension. Br J Ophthalmol 1994; 78:291–294.
146. Johnson CA, Marshall D, Eng K. Displacement threshold perimetry in glaucoma using a Macintosh computer system and a 21 inch monitor. In: Mills RP, Wall M, eds. Perimetry update 1994/95. Amsterdam: Kugler Publications, 1995:103–110.
147. Westcott MC, Fitzke FW, Hitchings RA. Abnormal motion displacement thresholds are associated with fine scale luminance sensitivity loss in glaucoma. Vision Res 1998; 38:3171–3180.
148. Wall M, Ketoff KM. Random dot motion perimetry in patients with glaucoma and in normal subjects. Am J Ophthalmol 1995; 120:587–596.
149. Wall M, Jennisch CS, Munden PM. Motion perimetry identifies nerve fiber bundle like defects in ocular hypertension. Arch Ophthalmol 1997; 115:26–33.
150. Wall M, Jennisch CS. Random dot motion stimuli are more sensitive than light stimuli for detection of visual field loss in ocular hypertension patients. Optom Vis Sci 1999; 76:550–557.
151. Joffe KM, Raymond JE, Chrichton A. Motion coherence perimetry in glaucoma and suspected glaucoma. Vision Res 1997; 37:955–964.
152. Bosworth CF, Sample PA, Weinreb RN. Perimetric motion thresholds are elevated in glaucoma suspects and glaucoma patients. Vision Res 1997; 37:1989–1997.
153. Bosworth CF, Sample PA, Gupta N, Bathija R, Weinreb RN. Motion automated perimetry identifies early glaucomatous field defects. Arch Ophthalmol 1998; 116:1153–1158.
154. Silverman SE, Trick GL, Hart WM. Motion perception is abnormal in primary open angle glaucoma and ocular hypertension. Invest Ophthalmol Vis Sci 1990; 31:722–729.
155. Bullimore MA, Wood JA, Swenson K. Motion perception in glaucoma. Invest Ophthalmol Vis Sci 1993; 34:3526–3533.
156. Bosworth CF, Sample PA, Williams JM, Zangwill L, Kee B, Weinreb RN. Spatial relationships of motion automated perimetry and optic disc topography in patients with glaucomatous optic neuropathy. J Glaucoma 1999; 8:281–289.
157. Sample PA, Bosworth CF, Blumenthal EZ, Girkin C, Weinreb RN. Visual function-specific perimetry for indirect comparison of different ganglion cell populations in glaucoma. Invest Ophthalmol Vis Sci 2000; 41:1783–1790.
158. Shabana N, Cornilleau PV, Carkeet A, Chew PT. Motion perception in glaucoma patients: a review. Surv Ophthalmol 2003; 48:92–106.
159. Frisen L. Acuity perimetry: estimation of neural channels. Int Ophthalmol 1988; 12:169–174.
160. Wall M, Lefante J, Conway M. Variability of high-pass resolution perimetry in normals and patients with idiopathic intracranial hypertension. Invest Ophthalmol Vis Sci 1991; 32:3091–3095.
161. Wall M, Conway MD, House PH, Allely R. Evaluation of sensitivity and specificity of spatial resolution and Humphrey automated perimetry in pseudotumor cerebri patients and normal subjects. Invest Ophthalmol Vis Sci 1991; 32:3306–3312.
162. Sample PA, Ahn DS, Lee PC, Weinreb RN. High-pass resolution perimetry in eyes with ocular hypertension and primary open-angle glaucoma. Am J Ophthalmol 1992; 113:309–316.
163. Frisen L. High-pass resolution perimetry: A clinical review. Doc Ophthalmol 1993; 83:1–25.
164. Chauhan BC, LeBlanc RP, McCormick TA, Mohandas RN, Wijsman K. Correlation between the optic disc and results obtained with conventional, high-pass resolution and pattern discrimination perimetry in glaucoma. Can J Ophthalmol 1993; 28:312–316.
165. Chauhan BC, House PH. Intratest variability in conventional and high-pass resolution perimetry. Ophthalmology 1991; 98:79–83.
166. Chauhan BC, House PH, McCormick TA, LeBlanc RP. Comparison of conventional and high-pass resolution perimetry in a prospective study of patients with glaucoma and healthy controls. Arch Ophthalmol 1999; 117:24–33.
167. Chauhan BC. The value of high-pass resolution perimetry in glaucoma. Curr Opin Ophthalmol 2000; 11:85–89.
168. Ennis FA, Johnson CA. Are high-pass resolution perimetry thresholds sampling limited or optically limited? Optom Vis Sci 2002; 79:506–511.
169. Wall M, Chauhan B, Frisen L, House PH, Brito C. Visual field of high-pass resolution perimetry in normal subjects. J Glaucoma 2004; 13:15–21.
170. Frisen L. New, sensitive window on abnormal spatial vision: rarebit probing. Vision Res 2002; 42:1931–1939.
171. Martin L, Wanger P. New perimetric techniques: a comparison between rarebit and frequency doubling technology perimetry in normal subjects and glaucoma patients. J Glaucoma 2004; 13:268–272.
172. Brusini P, Salvatet ML, Parisi L, Zeppieri M. Probing glaucoma visual damage by rarebit perimetry. Br J Ophthalmol 2005; 89:180–184.
173. Salvatet ML, Zeppieri M, Parisi L, Brusini P. Rarebit perimetry in normal subjects: test-retest variability, learning effect, normative range, influence of optical defocus, and cataract extraction. Invest Ophthalmol Vis Sci 2007; 48:5320–5331.
174. Yavas GF, Kusbeci T, Eser O, Ermis SS, Cosar M, Ozturk F. A new visual field test in empty sella syndrome: rarebit perimetry. Eur J Ophthalmol 2008; 18:628–632.
175. Bearse MA Jr, Sutter EE. Imaging localized retinal dysfunction with the multifocal electroretinogram. J Opt Soc Am A 1996; 13:634–640.
176. Chan HL, Brown B. Multifocal ERG changes in glaucoma. Ophthalmic Physiol Opt 1999; 19:306–316.
177. Hood DC, Zhang X. Multifocal ERG and VEP responses and visual fields: comparing disease-related changes. Doc Ophthalmol 2000; 100:115–137.
178. Fortune B, Bearse MA, Cioffi GA, Johnson CA. Selective loss of an oscillatory component from temporal retinal multifocal ERG responses in glaucoma. Invest Ophthalmol Vis Sci 2002; 43:2638–2647.
179. Chan HH. Detection of glaucomatous damage using multifocal ERG. Clin Exp Optom 2005; 88:410–414.
180. Graham SL, Klistorner AL, Grigg JR, Billson FA. Objective VEP perimetry in glaucoma: asymmetry analysis to identify early deficits. J Glaucoma 2000; 9:10–19.
181. Klistorner A, Graham SL. Objective perimetry in glaucoma. Ophthalmology 2000; 107:2283–2299.
182. Hood DC, Greenstein VC. Multifocal VEP and ganglion cell damage: applications and limitations for the study of glaucoma. Prog Retin Eye Res 2003; 22:201–251.
183. Graham SL, Klistorner AL, Goldberg. Clinical application of objective perimetry using multifocal visual evoked potentials in glaucoma practice. Arch Ophthalmol 2005; 123:729–739.
184. Grippo TM, Hood DC, Kandani FN, Greenstein VC, Liebmann JM, Ritch R. A comparison between multifocal and conventional VEP latency changes secondary to glaucomatous damage. Invest Ophthalmol Vis Sci 2006; 47:5331–5336.
185. Fortune B, Demirel S, Zhang X, et al. Comparing multifocal VEP and standard automated perimetry in high-risk ocular hypertensives and early glaucoma. Invest Ophthalmol Vis Sci 2007; 48:1173–1180.
186. Klistorner A, Graham SL, Martins A, et al. Multifocal blue-on-yellow visual evoked potentials in early glaucoma. Ophthalmology 2007; 114:1613–1621.

CHAPTER 36

Binocular Vision

Ronald S. Harwerth & Clifton M. Schor

Introduction

Humans and other animals with frontally located eyes attain binocular vision from the two retinal images through a series of sensory and motor processes that culminate in the perception of singleness and stereoscopic depth. Keen stereopsis is considered the paramount consequence of binocular vision because optimal performance depends on the normal functioning of all of the underlying vision processes, including central fixation with normal visual acuity in each eye, accurate oculomotor control to obtain bifoveal fixation, normal inter-retinal correspondence of visual directions, sensory mechanisms to produce haplopia (single vision), and neural mechanisms to extract selective depth signals from objects that are nearer or farther than the plane of fixation. Consequently, ordinary binocular vision represents a highly coordinated organization of motor and sensory processes and if any component in the organization fails, then binocular vision and stereopsis must be compromised to some extent. Clinically, the causes and significance of abnormal binocular vision are the primary emphasis, usually with a goal to alleviate any interference with clear, comfortable, single binocular vision. However, clinical applications must be established from an understanding of normal binocular vision. The topics included in this chapter were selected to provide the basic information on normal and abnormal binocular vision that forms the foundation for the clinical science of binocular vision.

Two eyes are better than one

It is obvious that an organism should gain an advantage by having two eyes rather than one. For example, it may be assumed that binocular vision is an important attribute because 75–80 percent of the neurons in primary visual cortex have inputs from both eyes.[6,84] There are also several functional advantages to having two frontal eyes.

1. The binocular field of view is larger than either monocular field.[65] The normal visual field for each eye extends to approximately 60 deg superiorly, 60 deg nasally, 75 deg inferiorly, and 100 deg temporally from fixation. Thus, with binocular viewing, the horizontal extent of the visual field is increased from 160 deg to 200 deg, with a central 120 deg of overlap, or superimposition, of the monocular visual fields.

2. Any image distortion due to optical or pathological defects in one eye can be masked by a normal image in the other eye. In general, the degree of dysfunction from a unilateral disorder or bilateral eye disease can be predicted from the performance of the un- or less-impaired eye.[33,110,111,108]

3. Pragmatically, having two eyes provides a resource to retain functional vision if an accident or disease causes the loss of sight of one eye.

4. Normal binocular vision improves functional vision by binocular summation and stereopsis. Binocular summation, which is the improvement in visual sensitivity with binocular viewing compared to monocular viewing,[26] provides only a small increase in sensitivity when measured by threshold responses,[17,72] but may be a larger factor for performance measured with suprathreshold stimuli.[74] Stereopsis, on the other hand, is the quantitative and qualitative improvement in relative depth perception that occurs with binocular viewing. At its most elementary level, keen stereopsis benefits both the predator, at the point of attack, and the prey, by unmasking the camouflage of predators.[14]

Visual direction

Ordinarily, people are not aware that they are using two eyes, but rather, they have an impression of viewing with one eye located about midway between the two eyes (the egocenter, or cyclopean eye). Thus, the basic concept of visual space involves the relationship between the locations of objects in physical space and their perceived (subjective) spatial locations.[113] These concepts are illustrated diagrammatically in Figure 36.1, which presents representations of physical space in panel A and subjective space in panel B. Perceived directions of objects are based on several factors, including the retinal locations of their images. Each retinal image location is associated with a specific visual direction, called its local sign or oculocentric direction.[83,113] The primary visual direction is the oculocentric direction of an object that is fixated along the primary line of sight and is imaged on the center of the fovea (represented by the

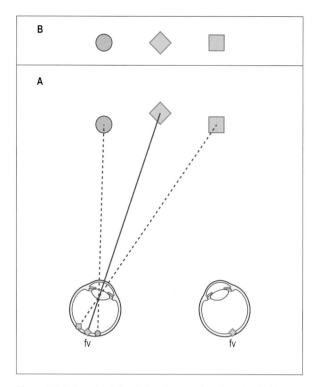

Figure 36.1 Perceived visual direction as a function of retinal image location. (**A**) The relationships between objects in physical space and their retinal images. The square represents a fixated object (primary line of sight) and the non-fixated objects (circle and square) are secondary lines of sight. (**B**) The relationships between retinal image locations and their visual directions (oculocentric directions). The perceived visual directions of the non-fixated objects (secondary visual directions of the circle and square) are relative to the primary visual direction of the fixated diamond.

diamond in Fig. 36.1). Secondary lines of sight are the oculocentric directions of other retinal image locations relative to the primary visual direction. As illustrated in Figure 36.1, the perceived visual directions of non-fixated objects (represented by the circle and square) are relative to the visual direction of a fixated object (represented by the diamond). In the example, an image located to the left side of the fovea is perceived to the right of the fixated object, e.g. the visual direction of the square. The normal ability to distinguish differences in oculocentric directions is extremely accurate, as demonstrated by Vernier alignment thresholds, which are on the order of 5 arcsec for foveal viewing.[98,175]

In a natural, dynamic world, objects are localized with respect to the head and body rather than to the line of sight and, therefore, retinal image information is not sufficient to define valid relationships between the physical and perceived locations of real objects. The perceived direction of an object in space relative to the body, its egocentric direction, is derived from the combination of its oculocentric visual direction with information about eye orientation in the head and head position relative to the trunk of the body. Egocentric localization is referenced to an egocenter midway between the two eyes and the combined information allows an individual to distinguish between retinal image motion caused by eye movements and movement of the object.[73]

From a clinical point, egocentric direction is an important concept for understanding the alterations of normal subjective visual space of patients with a misalignment of the two eyes from strabismus. The general principles of egocentric direction are illustrated in Figure 36.2. In the first examples (Fig. 36.2A&B), with normal binocular vision the retinal oculocentric directions from the two retinal images of an object at (panel A) or close to (panel B, circle) the plane of fixation are combined to produce a single direction

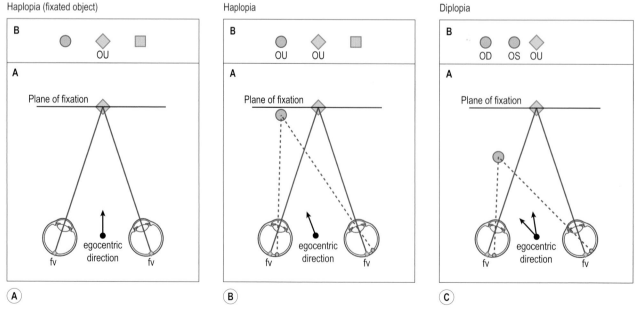

Figure 36.2 Egocentric directions for haplopia and diplopia. (**A**) The oculocentric directions of a fixated object are combined to produce a single egocentric direction. (**B**) The retinal images of non-fixated objects that are near the plane of fixation also will be combined to produce a single egocentric direction. (**C**) When the disparity of retinal images is larger than the normal fusion range (Panum's fusional area), the oculocentric directions are not combined and the object has two egocentric directions (physiological diplopia).

(haplopia) relative to the observer's egocenter. In contrast, for objects that are outside the range of single vision, either nearer or farther than the plane of fixation, a single object has two separate egocentric directions that correspond to the oculocentric direction of each of the retinal images relative to the perspective of the egocenter, i.e. diplopia, illustrated in Figure 36.2C.

Binocular eye movements are defined by combinations of version and vergence ocular movements (see Chapter 9). Versions are yoked or conjugate changes in the visual axes of the two eyes in the same direction and amount; vergence movements involve rotations of the visual axes in opposite directions. Only the version component of binocular eye position influences the perception of egocentric direction, which is sensed as the average position of the two eyes. The vergence component is nullified because vergence movements of the two eyes have opposite signs and their correlates cancel one another in the computation of the average of right and left eye positions. Thus, when the two eyes fixate objects to the left or right of the midline in asymmetric convergence, only the version, or conjugate, component of eye position contributes to perceived direction.[82]

In addition to averaged eye positions, the subjective computation of egocentric directions of non-fixated objects involves a reconciliation of differences in oculocentric visual directions between the two eyes. For example, objects that are small distances in front or behind the plane of fixation are imaged at retinal locations with different angular distances from the visual axis and, therefore, have different oculocentric directions in each of the two eyes. When the retinal image disparity is small, the target still appears single, and the binocular egocentric direction is based upon the average oculocentric direction for the two eyes.[153] In other words, the binocular egocentric direction deviates from either of the monocularly perceived directions by one-half of the angular disparity (Fig. 36.2B). However when the retinal images have unequal contrast, the average is biased or weighted toward the direction of the higher-contrast image.[7]

As object distances from the plane of fixation increase, retinal image disparities become large, and an object appears diplopic with two separate egocentric directions. Egocentric directions of diplopic images are perceived as though both monocular components had paired images on corresponding points in the contralateral eye (Fig. 36.2C). The relative egocentric direction of each of the diplopic images depends on whether the object is nearer or farther than the plane of fixation. The example (Fig. 36.2C) illustrates that, for an object closer than the plane of fixation, the egocentric direction associated with the retinal image of the right eye is to the left of the egocentric direction produced by the image of the left eye. This form of diplopia, produced by near objects, is called "crossed" diplopia due to the right–left/left–right relationship between the eyes and perceived locations of the object. Conversely, objects farther than the plane of fixation will be perceived with uncrossed diplopia, where the right eye sees an object located on the right side and the left eye sees an object located on the left side.

In normal binocular vision, diplopia of non-fixated objects, classified as physiological diplopia, occurs as a natural consequence of the lateral separation of frontally located eyes and the topographical organization of visual directions across the retina. In contrast, patients with strabismus experience pathological diplopia, wherein objects in the plane of fixation are perceived as doubled.[170] In pathological diplopia, the retinal image location caused by misalignment of the visual axis of the deviating eye causes a fixated target to be seen in different egocentric directions during binocular viewing, and yet, under monocular conditions each eye veridically perceives its direction. Under binocular viewing conditions, only one of the two visual directions seen in pathological diplopia can be veridical at any given moment. In addition to diplopia, a form of visual confusion confounds the space perception of strabismic patients. An object imaged on the fovea of the deviating eye has the same egocentric direction as another object imaged on the fovea of the fixating eye, even though their physical locations are much different. Diplopia and visual confusion create an intolerable visual environment that is typically alleviated during binocular vision by the development of suppression, which is a suspension of the vision in the deviating eye. Another solution to diplopia is anomalous retinal correspondence, which is an alteration of the normal topography of oculocentric directions.[18,171,172]

Normal retinal correspondence

The perceptions of haplopia and diplopia arise from the sense of oculocentric visual directions of corresponding or non-corresponding retinal areas that are stimulated during binocular vision. Binocular retinal correspondence is defined by the set of retinal image locations that produces perceptions of identical visual directions when viewing with one eye, or with the other, or with both eyes simultaneously.[113] Identical visual directions associated with corresponding points depend primarily on retinal image locations and their associated oculocentric directions, rather than on eye position and egocentric directions. This fundamental principle was demonstrated by Hering's famous "window experiment"[113,172] that illustrated the relationship between retinal image locations and perceived visual directions showed that objects imaged on the foveas have the same perceived visual direction regardless of their actual positions relative to the observer. Similarly, afterimages placed on the fovea of each eye always appear superimposed irrespective of the vergence or version positions of the two eyes.[10] The identical visual directions of the afterimages show that normally corresponding retinal points are combined physiologically before eye position information is used to compute perceived egocentric direction.

These relationships for binocular visual directions are illustrated in Figure 36.3, in which the two foveal images of a fixated object (diamond – panel A) share a common visual direction and are fused to produce the perception of a single haplopic object in subjective space (panel B). The specific pairs of retinal image locations in the two eyes that are perceived to have identical visual directions are called corresponding retinal points or areas.[83,113,114] The foveas represent an important pair of corresponding retinal points, but there are many other pairs associated with secondary lines of sight and visual directions. For example, in Figure 36.3 the square represents the location of an object in the right visual field that falls on paired retinal areas with a common

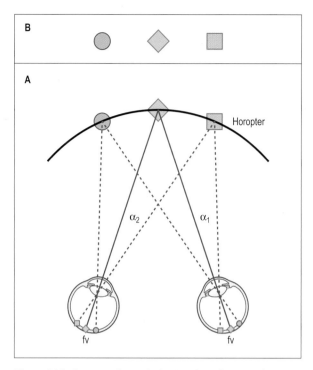

Figure 36.3 Corresponding retinal points. Retinal areas in the two eyes with identical oculocentric directions (panel B) are corresponding retinal points. The locations of objects along the horizontal meridian that are imaged on corresponding points (panel A) define the longitudinal horopter. The physical location of an object that is on the horopter is quantified by the longitudinal visual angles (α_1 & α_2).

visual direction for the two eyes, thus defining another set of corresponding retinal areas. Similarly, the circle in the left visual field represents a location of an object in the left visual field that is imaged on corresponding retinal points. For an infinite fixation distance, a large number of potential object locations are imaged on corresponding retinal points. These object locations in space that are imaged onto corresponding retinal points can be imagined to define a surface that resembles a cylinder with an infinite radius of curvature. This surface of points stimulates the perception of identical visual directions for the two eyes and is called the horopter.[113,114,154] At finite viewing distances, the surface is reduced to a horizontal line and a vertical line that pass through the intersections of the lines of sight when the eyes fixate in symmetrical convergence.

Points in tertiary locations (elevated and displaced laterally from the lines of sight) are closer to one eye than to the other and subtend larger visual angles at the more proximal eye. As a result, all physical points in tertiary locations at finite viewing distances are imaged with vertical disparities and, by definition; their images do not fall on corresponding points.[149] At these finite viewing distances, only points imaged along the horizontal retinal meridian or along the vertical retinal meridian in the midsagittal plane can be formed on corresponding points without vertical disparities. The locations of object points that are imaged on corresponding points along horizontal meridians through the foveas define the longitudinal horopter,[83,113] shown diagrammatically in Figure 36.3 by the arc passing through the fixated object and the two peripheral objects that stimulate

corresponding retinal points. A vertical horopter is defined by the location of object points in the midsagittal plane that are imaged on corresponding points along the vertical meridian.[149,151,169]

The relationship between object and image space for the longitudinal horopter is quantified by longitudinal visual angles, defined as the horizontal visual angles subtended by the primary and secondary lines of sight for each eye. The longitudinal visual angles for an object in the right visual field are illustrated by angles α_1 and α_2 in Figure 36.3. Similarly, longitudinal visual angles can be defined for the object in the left visual field. The lateral separation of the eyes in the head results in each eye having a different perspective of objects not on the horopter, and these differences in perspective are used perceptually to reconstruct a three-dimensional space from the two-dimensional retinal images. The objects that are closer or farther than the horopter stimulate non-corresponding points, which is defined as a binocular disparity, retinal disparity, or simply, disparity[177] of the retinal images. The horizontal component of binocular disparity is the unique stimulus for stereoscopic depth perception and is quantified by the difference in the longitudinal visual angles, also called horizontal disparity angles.[83,113,114,149] The horopter serves as a zero disparity reference from which the magnitudes of non-zero disparities are computed. In the absence of other depth cues, the curvature and slant of the horopter will influence the perceived shape and orientation of surfaces seen in depth. Optical devices such as spectacle lenses that magnify and distort one of the two eyes' retinal images will alter the horopter and produce predictable errors of space perception.[114]

The empirical longitudinal horopter illustrated in Figure 36.3 is a smooth surface that is symmetrical about the fixation point, but the actual shape and curvature of the empirical horopter will depend on physiological and optical factors that affect the retinal images and their cortical representations. If the locations of corresponding retinal areas simply were determined by equal angular distances from the primary line of sight, then the horopter would be a circle passing through the fixation point and the entrance pupil of each eye[14,113] (geometric or theoretical horopter or Vieth-Muller circle) (Fig. 36.4). An analogy of "cover points" has been used to describe corresponding retinal locations determined by retinal distances.[113] If the retina of one eye were to be translated and carefully aligned to cover the retina of the other eye, then a pin stuck into the retina would penetrate cover points, or corresponding retinal areas. Cover points, by simple geometry, would become corresponding points because the angles included by lines drawn from two points on the circumference of a circle to any other two points on the circle are equal. Consequently, the longitudinal visual angles will be equal ($\alpha_1 = \alpha_2$), and the theoretical horopter will be a section of a circle, specified as the Vieth-Muller circle (VMO), as illustrated in Figure 36.4.

Except for rare cases, the empirically defined horopter does not actually coincide with the Vieth-Muller circle, but the theoretical longitudinal horopter is an important reference for zero disparity; in fact, clinical tests of stereopsis use the Vieth-Muller circle as the zero disparity reference. Another important reference in physical space is the objective fronto-parallel plane (OFPP), which is the plane at the fixation distance that is parallel to a line passing through the entrance

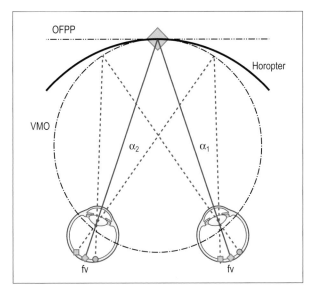

Figure 36.4 The empirical horopter (solid line) in relation to the Vieth-Muller circle (VMO; dot-dash line), defined by the locations of objects with equal longitudinal visual angles (α_1 & α_2), and the objective fronto-parallel plane (OFPP; dot-dot-dash line), defined by a plane at the fixation distance that is parallel to a line passing through the entrance pupils of the two eyes.

pupils of the two eyes (Fig. 36.4). Although, both the Vieth-Muller circle and the OFPP are used as reference surfaces to describe the empirical horopter, they are predicated on different perceptual correlates. The Vieth-Muller circle is based on a physical geometric assumption that identical visual directions are associated with retinal points that are equidistant from their respective foveas. The empirical longitudinal horopter, which is based primarily on identical visual directions, is the only pure measure of pairs of retinal corresponding points.[154] On the other hand, the OFPP, which is a measure of perceived surface shape and orientation, depends on egocentric directions derived by combining oculocentric directions and information about absolute distance and direction. The OFPP can be used as a frame of reference for adjusting the depth of objects at different retinal eccentricities so that the objects appear to lie in a flat plane parallel to the face (apparent fronto-parallel plane, or AFPP). The location, shape, and slant of the AFPP is based on disparity, azimuth cues, and absolute distance cues.[3] The AFPP is used to evaluate how disparity is scaled by absolute distance and how slant is influenced by azimuth estimates to enable subjects to estimate the depth of surface shapes and orientations. Both the empirical longitudinal horopter and the AFPP are used to evaluate the influence of optical distortions, such as magnification, on perceived space, since they usually yield similar results.

The empirical longitudinal horopter and the AFPP can be measured in several ways, based on different properties of visual directions and stereoscopic depth.[113] The empirical longitudinal horopter can be determined by identical visual directions for the two eyes; by the highest stereoacuity for points at various eccentricities from fixation; by the center of the sensory fusion range known as Panum's fusional area (see Fig. 36.7 below); and by object locations eccentric to fixation which result in a zero stimulus for disparity vergence. The vergence criterion assumes that horizontal version

movements guide the intersection of the visual axes along the empirical horopter, even if this surface is curved in depth. The AFPP horopter can be measured by positioning eccentric objects so that they all appear to lie in the same plane, or by positioning eccentric objects so that they appear to be equidistant from the observer (i.e. equidistant from the egocenter).

Abnormal retinal correspondence

The properties of corresponding retinal areas underlying normal binocular vision provide a stable sensory organization of the two eyes consistent with the functional properties of neurons in primary visual cortex[123,157] that are present in infancy[27] (see Chapter 41). However, the normal sensory organization of binocular vision can be altered substantively in infantile strabismus by the development of anomalous retinal correspondence and/or suppression.[172] Clinically, the type and extent of sensory adaptation is an important factor to be considered in treatments that are intended to reestablish functional binocular vision for a strabismic patient.[87,122,166,172]

Sensory adaptations in strabismus are necessary to eliminate the constant diplopia (two egocentric directions for all objects) and common visual directions for fixated and nonfixated objects (visual confusion) that would otherwise prevail[172] (Fig. 36.5A). Most strabismic patients do not experience diplopia and visual confusion, but their single vision is a result of suppression[53,86] and/or anomalous retinal correspondence.[2,4,18,25] Suppression causes the perception of objects normally visible to the deviating eye to be eliminated during simultaneous binocular viewing to provide single vision.[173] Suppression in some exotropes may involve an entire hemi-field,[58] but for esotropes the suppression is limited to a specific area (sometimes called a "suppression scotoma"), bounded by the fovea of the deviating eye and the area stimulated by the object that is fixated by the non-deviating eye[86] (the diplopia point; Fig. 36.5B).

Anomalous retinal correspondence (Fig. 36.5C) is an adapted shift in the visual directions of the deviated eye relative to the normal visual directions of the fixating eye. The result is that, during binocular viewing, a peripheral retinal area of the deviating eye acquires a common visual direction to that of the fovea of the fixating eye.[4,25,138,172] This shift in retinal correspondence produces a difference in the visual directions between the two eyes for foveally imaged objects, which can be sometimes demonstrated clinically with the Hering-Bielschowsky test for anomalous correspondence.[172] For this test, after-images are formed on the foveas of both eyes. Even with an interocular deviation from strabismus, a patient with normal correspondence will perceive the after-images in a common direction, but a patient with deep anomalous retinal correspondence will perceive the after-images in different visual directions. The subjective separation of the monocular after-images is referred to as the angle of anomaly, which represents the magnitude of the perceptual shift from the normal corresponding relationship between the two foveas. When the magnitude of the angle of anomaly is equal to the magnitude of the oculomotor misalignment (harmonious anomalous retinal correspondence), then objects in space can appear single even though

Normal retinal correspondence

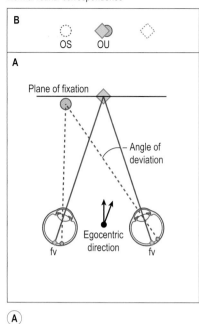

Suppression

Anomalous retinal correspondence

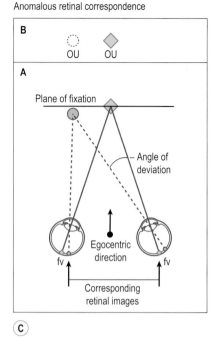

Figure 36.5 Retinal correspondence and suppression in strabismus. (**A**) A strabismic patient with normal retinal correspondence and without suppression would have diplopia, two egocentric directions of single objects (represented by solid and dashed images in panel B) and visual confusion, a common visual direction for two separate objects (represented by the superimposition of the images of the fixated diamond and the circle, which is imaged on the fovea of the deviating eye in panel B). (**B**) The elimination of diplopia and confusion by suppression of the retinal image of the deviating eye. (**C**) The elimination of diplopia and confusion by anomalous retinal correspondence, an adaptation of visual directions of the deviated eye.

one of the eyes is deviated. On the other hand, when the angle of anomaly is not equal to the angle of strabismic deviation (unharmonious anomalous retinal correspondence), the residual disparity will cause diplopia unless suppression is also present. The angular separation of the diplopic images is referred to as the subjective angle and in unharmonious anomalous retinal correspondence the subjective angle is usually smaller than the angle of strabismus (objective angle).

As illustrated by the data for a patient with infantile esotropia (Box 36.1) tests for the sensory adaptations of a strabismic patient are variable[68,87,121,166,170] and generally the diagnosis of anomalous retinal correspondence for a given strabismic patient is test dependent.[48] Tests for anomalous retinal correspondence either locate the extra-foveal point in the deviating eye that has the same visual direction as fovea in the fixating eye, or quantify the angular separation of the visual directions of the foveas of the two eyes.[172] Tests in the former category, e.g. Bagolini striated lenses or the Worth four-dot test, have a higher probability of being positive for anomalous retinal correspondence than tests in the latter category, such as the Hering-Beilschowsky after-image procedure.[48] The dependence of the state of a patient's retinal correspondence on the clinical procedure may be the result of variation in the angle of anomaly across the retina[42,138] or some other property of the mechanisms underlying anomalous retinal correspondence.[91,138,166,184] However, the practical consequence is illustrated by the patient with infantile esotropia (Box 36.1) who demonstrated harmonious anomalous retinal correspondence with the Bagolini striated lens test and normal retinal correspondence with the major amblyoscope and after-image tests.

> **Box 36.1 Infantile esotropia with anomalous retinal correspondence**
>
> - College student, 26 years old
> - Infantile strabismus
> - Prior treatment
>   Surgery at age 2 years
>   Patching ages 2–4 years
>   Surgery at age 8 years
> - Cover test: Constant, unilateral, right eye esotropia
>   Distance: 20^Δ ET
>   Near: 18^Δ ET
> - Refractive errors and BVAs
>   OD: +2.00 −2.00 × 180 20/20
>   OS: +1.00 −0.5 × 180 20/20
> - Lateral heterophoria: constant suppression of OD
> - Bagolini striated lens test: Lines intersect at the fixation point – harmonious anomalous retinal correspondence.
> - Amblyoscope: Equal objective and subjective angles – normal retinal correspondence.
> - Hering-Bielschowsky afterimage test: Images intersect at the fixation point – normal retinal correspondence.
> - Stereopsis with two-pencil test: Better accuracy with both eyes viewing.
>
> *Comment*: The patient has a constant, unilateral esotropia without amblyopia. He demonstrates harmonious anomalous retinal correspondence under the fovea-to-extrafovea testing conditions of the Bagolini test and normal retinal correspondence with the more dissociating tests (amblyoscope) or the test that evaluates fovea-to-fovea correspondence (after-image test).

Three main theories on anomalous retinal correspondence have been proposed.[138] The first theory states that, with anomalous correspondence the two eyes perceive space independently, much like the two hands explore space. Each eye computes egocentric visual direction from its own sense of eye position and oculocentric retinal image location. In effect, there is a lack of correspondence.[23,168] The observation that the separation between foveal afterimages changes during spontaneous changes in the magnitude of the angle of strabismus supports this theory. This co-variation between the angle of anomaly and objective angle of strabismus results in a stable percept of space, even though the magnitude of the strabismus is unstable and varies in time.[66] The second theory proposes that there is a shift in retinal correspondence[18,34,172] so that retinal points that are normally disparate are shifted to acquire an anomalous coupling in the striate cortex,[91,184] e.g. the fovea of the fixating eye is anomalously coupled with a peripheral area of the deviating eye. Measures of the identical visual directions horopter in strabismics reveal that correspondence is mainly shifted in the periphery, and the angle of anomaly decreases near the fovea in the region of space that lies between the visual axes that is projected to opposite cortical hemispheres.[20,47,68] The third theory proposes that strabismics who have no suppression are able to perceive space singly due to enlarged Panum's fusional areas.[2,5,77] These three theories are not mutually exclusive and all of them can be expressed to some degree in the same strabismic patient, or different mechanisms may be present in different forms of strabismus, e.g. primary versus secondary microstrabismus.

Binocular (retinal) disparity

The empirical longitudinal horopter represents the specific locations of objects in physical space that have zero retinal image disparity. Objects not located on the horopter will have some amount of lateral binocular (retinal) disparity. With horizontal retinal disparity, an object is imaged on laterally separated non-corresponding retinal areas and, as the case for the empirical horopter, the longitudinal visual angles are not equal. Horizontal retinal disparity is a unique binocular stimulus for stereoscopic depth perception, horizontal disparity vergence eye movements, and diplopia, either physiological or pathological. In each case, the perceptual or motor response is a consequence of the relationship between the object's disparate images and the horopter.

The following relationships concerning binocular disparities are important to the understanding of binocular vision.

1. Binocular disparities are classified as uncrossed (distal) or crossed (proximal), based on the relationship between the disparate object and the horopter,[83,177] as illustrated in Figure 36.6. For example, the square in the right visual field represents an object that is farther than the fixation point (represented by the diamond), and the secondary lines of sight for the square intersect behind the horopter. The binocular disparity produced by the square is classified as uncrossed and quantified by the difference between the longitudinal visual angles α_1 and α_2. Perceptually, uncrossed disparities give rise to a sense of relative "far" stereoscopic depth or, if the disparity is large, to uncrossed diplopia (i.e. double vision with the right eye's image seen to the left side

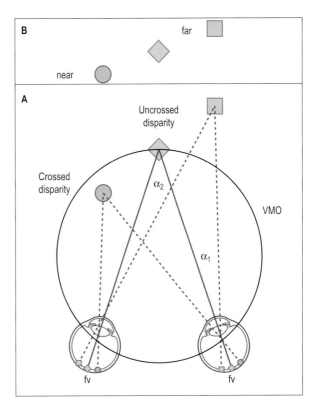

Figure 36.6 Binocular disparity and the perception of stereoscopic depth. Objects that are not located on the longitudinal horopter have binocular disparity, which may be crossed disparity (circle in panel A) and be perceived as nearer than the fixated object (the relative depth of the circle with respect to the diamond in panel B) or uncrossed disparity (square in panel A) and be perceived as farther than the fixated reference object (panel B).

of the left eye's image). As oculomotor stimuli, uncrossed disparities evoke divergence eye movements. By a parallel classification, crossed disparities (illustrated by the circle in the left field, Fig. 36.6) give rise to a perception of "near" stereoscopic depth or crossed diplopia and elicit convergence eye movements.

2. Binocular disparities are classified as either relative or absolute.[31,177] The absolute disparity of a target is the difference between the angle subtended by the target at the two entrance pupils of the eyes and the angle of convergence. A relative disparity is defined as the difference between two absolute disparities. Stereoscopic depth normally arises from discriminating relative disparities, wherein judgments are based on the perception of the depth of one object relative to another (e.g. the relative depth of the square with respect to the diamond, Fig. 36.6). Stereopsis with relative disparities, as measured by clinical testing, is quite fine compared to the perception of depth based solely on absolute disparities.[172] In contrast, an absolute disparity is the optimal stimulus for motor fusion, such as occurs with prism-induced disparity vergence or with changes in fixation distances, which change disparity of all objects in the visual field by a constant amount.[89]

3. For a person with normal binocular vision, there is a finite range of binocular disparities that provides a clear perception of relative depth with normal fusion (haplopia), and larger disparities produce diplopia.[109,113] The disparity range for single vision, called Panum's fusional area (PFA),

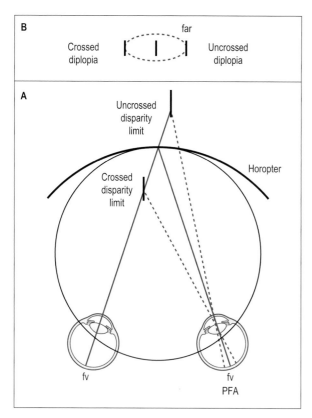

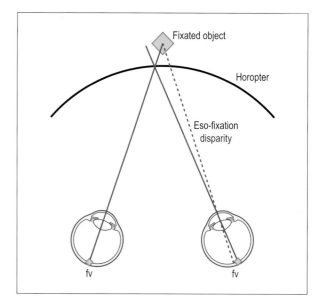

Figure 36.7 The binocular disparity range for single binocular vision (Panum's fusional area). With constant fixation, an object is moved along the primary line of sight of the left eye until the binocular disparity of the right eye's image exceeds the fusion limit, i.e. appears diplopic, first with crossed disparity and then with uncrossed disparity (A). The dimensions of Panum's fusional area (PFA) are illustrated in panel B.

Figure 36.8 The displacement of the horopter with an eso-fixation disparity. Vergence errors cause a displacement of the horopter from the fixation point and retinal disparity of the fixated object. Single binocular vision will be maintained if the vergence error is smaller than Panum's fusional area (fixation disparity).

is illustrated in Figure 36.7. Panel A of the figure illustrates the measurement of Panum's area. A test object is moved along the primary line of sight of the left eye while the two eyes maintain binocular fixation at one point in space. The crossed disparity limit for fusion is the greatest distance that the test-object can be moved toward the subject before it appears diplopic. Similarly, the object can be moved away from the subject to establish an uncrossed disparity limit for fusion. Thus, in subjective space (panel B) Panum's fusional area (illustrated by the dashed oval) defines the disparity range for fused binocular vision, and objects subtending disparities outside this range will be diplopic unless one ocular image is suppressed. When the target is intersected by the primary line of sight of the left eye, the disparate image of the near target seen by the right eye is projected to the left visual field in crossed diplopia and the image of the far target is projected to the right visual field in uncrossed diplopia. In addition, there is also a vertical disparity range that is compatible with binocular sensory fusion, although it is generally smaller than the horizontal disparity range. Although the size and shape of PFAs depend on retinal eccentricity and spatio-temporal stimulus properties,[146,165] the classical PFA for foveal vision is an oval with approximate horizontal and vertical dimensions of 15 × 6 arcmin.[113]

4. Most individuals with normal binocular vision will have small vergence errors known as fixation disparities.[76,116,137] Because the retinal images only need to fall within Panum's fusional areas for single binocular vision to

occur, a small residual misalignment of the visual axes (vergence error) can occur, resulting in a retinal disparity of a fixated object, i.e. a fixation disparity, without causing diplopia. This fixation disparity between the fixated object and the intersection of the primary lines of sight displaces the longitudinal horopter from the fixated object. An example illustrated in Figure 36.8 shows a small convergence error (eso-fixation disparity) for the object of fixation (diamond). The convergence error causes an inward shift of the horopter, (the locations of normal corresponding points), i.e. in normal binocular vision, identical primary visual directions result from bifoveal stimulation and, therefore, the horopter always passes through the intersection of the visual axes. Consequently, the fixated object lies beyond the horopter and, by definition, has an uncrossed retinal disparity, described clinically as an eso-fixation disparity. Conversely, an exo-fixation disparity causes a displacement of the horopter beyond the fixation distance, and crossed disparity of the fixated object. In any case, the magnitude of a fixation disparity cannot be larger than the width of Panum's fusional area if the patient is to retain single binocular vision.

5. Clinically, fixation disparities have been considered a sign of stress on fusional vergence mechanisms.[116,117,120] The clinical tests are based on the coincidence of the primary line of sight and the principal visual direction with normal retinal correspondence (see Fig. 36.8) and, therefore, the angular separation between dichoptic objects with a common visual direction is a measure of the convergence error (fixation disparity) at the given fixation distance. The assumption that clinically significant fixation disparity is caused by oculomotor stress is generally based on Ogle's extensive investigations of fixation disparity – forced vergence functions, which demonstrated systematic changes in the direction and magnitude of fixation disparity with variation in the magnitude of the stimulus for disparity vergence (see Chapter 9).[116,117] Three examples of fixation disparity –

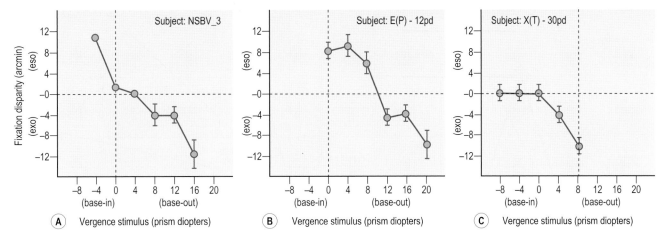

Figure 36.9 Fixation disparity as a function of prism-induced disparity vergence. The fixation disparity (defined in Fig. 36.8) changes systematically with the magnitude of the disparity vergence stimulus. The prismatic power that eliminates the fixation disparity, i.e. where the function crosses the abscissa, is defined as the associated phoria. The functions represent data for a patient with normal binocular vision (**A**), and patients with esophoria (**B**) or exophoria (**C**).

forced vergence functions are presented in Figure 36.9. These data were obtained using subjective alignment of dichoptic nonius lines, i.e., alignment of two stimuli, one seen by each eye, to define corresponding retinal locations while the subject viewed a binocular fusion stimulus. The separate functions represent data for a patient with clinically normal binocular vision (Fig. 36.9A), a patient with esophoria (Fig. 36.9B), and a patient with an intermittent exotropia (Fig. 36.9C). The functions show that fixation disparities occur when a large fusional vergence component is required to correct a vergence error such as a heterophoria. The amount of base-out prism needed to eliminate the vergence effort is determined from the prismatic power at which the fixation disparity was zero. This occurs when the forced vergence function crosses the abscissa (i.e. the "associated" phoria).[116,120] An example of the clinical application of these concepts is illustrated by the patient with asthenopia and esophoria with a base-out associated phoria for near-vision tasks that is presented in Box 36.2. The fixation disparity can be a useful error with significant heterophorias because it provides a steady state stimulus to overcome the heterophoria and maintain ocular alignment with fusion.[137] When large vergence responses are needed to overcome phoria they may be associated with asthenopia.[120]

In summary, broad ranges of binocular disparities are present under all conditions of normal seeing and the utilization of disparity information by the visual sensory and motor systems is the most fundamental process of normal binocular vision. The clinical evaluation of a patient's binocular vision includes tests of several aspects of retinal disparity, including relative disparities by measures of stereoacuity, absolute retinal disparities by measures of disparity-induced vergence eye movements, and fixation disparities.[137,123]

Stereopsis

Stereopsis is the perception of relative distance, or the depth separation, between objects that occurs as a result of neural

---

Box 36.2 Asthenopia with uncorrected astigmatism and oculomotor imbalance

- 19-year-old college student
- Visual complaints: Eye strain (burning and a "pulling" sensation) that she associates with reading and her other near work. Near-sighted, has worn glasses for 5 years and her vision is OK with a spectacle correction.
- Current spectacle correction & Vas
  OD: −2.00 −0.50 × 075; 20/25
  OS: −2.00 −0.25 × 090; 20/30
- Refractive errors & VAs
  OD: −2.25 −0.75 × 070; 20/15
  OS: −2.25 −0.75 × 105; 20/15
- Phorometric findings
  Distance heterohoria: 1^Δ E(P)
  Near phoria: 4^Δ E(P)
- Ductions at near: Base-in $5^\Delta/12^\Delta/4^\Delta$, Base out $8^\Delta/14^\Delta/7^\Delta$
  Near associated phoria: 2^Δ Base out.

Comment: The patient's asthenopia could be the result of astigmatism or oculomotor imbalance from an eso-fixation disparity caused by esophoria.

---

processing of the relative horizontal binocular disparities between the monocular retinal images. Binocular disparities are present because the lateral separation of the eyes in the head provides each eye with a slightly disparate view of a given object. Thus, as demonstrated by the seminal research of Wheatstone[180] and Julesz,[90] these horizontal (lateral) retinal image disparities produce stereoscopic depth. By themselves, vertical disparities do not result in depth perception; however, they are used to estimate azimuth and absolute distance in order to scale horizontal disparities into stereo-depth magnitudes.[3,53,57,99,132,178]

Clinical measures of the smallest detectable stereoscopic depth (stereoacuities) are quantified by a stereo-angle (η) that is equal to the difference in the longitudinal visual angles (α_1 and α_2), or the parallax (convergence) angles (δ_1

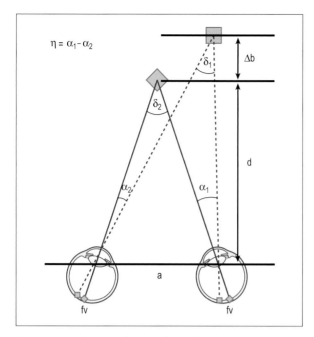

Figure 36.10 The quantification of binocular disparity. The binocular disparity (η) of an object (represented by the square), with respect to a fixated object (diamond) is equal to the difference in the longitudinal visual angles (α_1 & α_2) or the parallax angles (δ_1 & δ_2). The magnitude of binocular disparity is dependent on the fixation distance (d), the depth interval (Δb), and the interpupillary distance (a), see text for details.

and δ_2), as illustrated in Figure 36.10. For a conventional stereogram (see Fig. 36.14), η can be derived directly from the angular differences in positions of common objects in the two half-views of the stereogram. The retinal disparity associated with the physical depth of an object in normal vision (illustrated in Fig. 36.10) can be determined by the relationship:

$$\eta = [(a * \Delta b)/d^2] * c$$

where a is the interpupillary distance, Δb is the depth interval that is equal to the distance between the disparate object and the viewing distance, d is the viewing distance, and c is a constant to obtain the stereo-angle in dimensions of arcmin ($c = 3,438$) or arcsec ($c = 206,264$). The function shows that, for any given fixation distance, the relationship between depth interval and retinal disparity is linear, but the relationship varies across fixation distances by the square of the distance. For example, if an individual with a 60 mm PD can perceive stereoscopic depth from a retinal disparity of 10 arcsec, a depth difference of 0.12 mm can be discriminated at a 40 cm viewing distance. However, the same stereo-angle requires a 3 cm depth difference at 6 m, or an 800 m depth difference at a 1000 m viewing distance. The relationships between binocular disparities, linear depth, and observation distance for larger, suprathreshold disparities, are illustrated in Figure 36.11.

To maintain a valid perceptual constancy between depth and disparity, the relationship between stereo-angles and viewing distances requires an observer to scale the horizontal disparity with viewing distance. Perceptual information about viewing distance may be interpreted from the vertical disparity gradients that occur naturally in tertiary directions

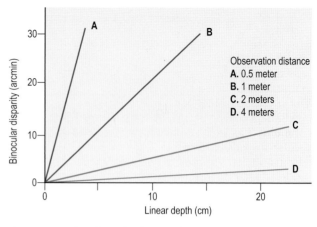

Figure 36.11 The relationship between binocular disparity and linear depth for four fixation distances. The disparity–depth relationship is linear for any constant fixation distance, but non-linear across fixation distances.

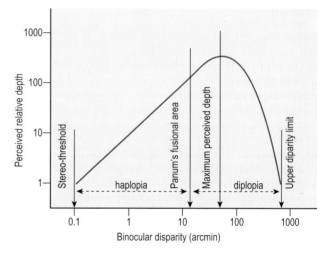

Figure 36.12 The relationship between perceived depth and binocular disparity for disparities from the lower to the upper limits of stereoscopic depth perception (modified from Tyler 1983[165]).

of gaze, in which a nearby target subtends a larger visual angle vertically in one eye than in the other.[53,57,99,132,149,178] If the target is elevated above the visual plane, it has a vertical disparity that increases with horizontal eccentricity and inversely with viewing distance. Other sources of perceptual information about viewing distance include the convergence angle,[24,49] retinal cues, such as oblique disparities,[178] and non-stereoscopic or perspective cues.[60]

Quantitative and qualitative stereopsis

The perception of stereoscopic depth at a constant viewing distance increases with disparity over a relatively large range.[28,80,112,143,148,165] The relationship between perceived depth and disparity from the lower (D_{min}) to upper (D_{max}) limits of stereoscopic vision is illustrated by Figure 36.12. A perception of relative depth cannot be detected from disparities below the stereo-threshold (D_{min}), but the relationship becomes proportional (quantitative stereopsis) for disparities larger than the stereothreshold.[28,165] Objects with larger disparities appear in increasing depth even beyond the limit of fusion (Panum's fusional limit), and the maximum perceived stereoscopic depth occurs when the stimulus is clearly diplopic. With larger binocular disparities (qualitative stereopsis) the perception of depth decreases to the upper limit of stereopsis where the relative depth of diplopic objects can no longer be resolved. Qualitative depth has a direction, or sign, but a vague amplitude.[112]

The specific parameters of the depth–disparity relationship will depend on the spatial and temporal properties of the stimulus,[28] as well as on other non-stereoscopic factors. Binocular disparities above the stereothreshold produce a vivid quantitative sense of depth and a strong impression of solid three-dimensional structure, but the disparity threshold for the perception of depth is altered by various changes in stimulus parameters, such as contrast,[67,96,142] spatial frequency,[73,80,96,147,148] and viewing duration.[71,118,150] For example, while depth in a stereogram may be perceived with just a spark of illumination,[118] the stereo-depth detection threshold decreases with exposure duration for both line[118,150] and random dot patterns.[69,71] Quantitative stereopsis responds to stimuli that appear fused, the magnitude of perceived depth increases with disparity magnitude up to about 1 degree,[28] and the depth percept improves with exposure duration up to a critical duration where the stereothreshold is independent of duration.[50,69] A qualitative sense of stereo-depth can be seen with even larger disparities, but the magnitude of its range is difficult to specify, i.e. a range of qualitative stereopsis with disparities as large as 10 degrees.[50,112]

In addition to the disparity range, qualitative stereopsis and quantitative stereopsis differ in two important ways. Quantitative depth perception improves with exposure duration,[118] while qualitative depth perception is optimal with briefly presented stimuli and fades with exposures longer than 0.5 sec.[95] Thus, these two forms of stereopsis can also be classified as sustained-quantitative and transient-qualitative. Additionally, the two forms of stereopsis can be distinguished by the spatial aspects of their stimuli,[39,140] especially by the similarity between the stimuli for both eyes that is required for quantitative, but not for qualitative, stereopsis.[124] Qualitative stereopsis arises from briefly presented stimuli even when a difference in size, shape, and/or contrast polarity exists between the two eyes.[39,79,124,140] Transient-qualitative stereopsis is a useful alerting percept for the approximate location of objects that appear suddenly, and it is also a stimulus to initiate vergence alignment of the eyes to transient changes in both small and large retinal disparities.[39] Quantitative stereopsis, on the other hand, is keenly useful for judging the shape, depth, and orientation of continuously viewed features near the plane of fixation.

The stereoscopic appreciation of either quantitative or qualitative depth is not merely a function of relative retinal disparity. As indicated by the equation for computing retinal image disparity (η), the linear depth interval (Δb) depends on both disparity and viewing distance (d). The equation illustrates that retinal image disparity is scaled by absolute distance to yield stereo-depth magnitude in linear units. The magnitude of relative depth and the degree of surface curvature (depth ordering) stimulated by retinal disparity both depend on viewing distance and a zero disparity surface can have many different shapes. Because the empirical longitudinal horopter is flatter than the theoretical Vieth-Muller circle, its curvature decreases (becomes less concave or more convex than the Vieth-Muller circle) with increasing viewing distance[113] (Fig. 36.3). Even a horopter that follows the Vieth-Muller circle changes curvature with viewing distance and approaches a flat surface at an infinite viewing distance. Nevertheless, flat surfaces appear flat at all viewing distances and, thus, without viewing distance information, the pattern of retinal image disparities across the visual field is insufficient to sense either depth ordering (surface curvature) or depth magnitude.[53] Similarly, a given pattern of horizontal disparities can correspond to different slants about a vertical axis presented at various horizontal gaze eccentricities (i.e. azimuth).[113] Convergence distance and direction of gaze are important pieces of information used to interpret slant from the disparity fields associated with slanting surfaces.[3]

Clinical anomalies of both quantitative and qualitative stereopsis have been identified. Patients with strabismus or binocular dysfunction demonstrate elevated thresholds for quantitative stereopsis, even when their eyes are aligned by optical or surgical procedures.[12,13,121,166] However, if any residual stereopsis is present, once disparity exceeds the elevated threshold, it stimulates depth perception to the normal upper disparity limit for quantitative stereopsis[139,174] with normal qualitative stereopsis at large disparities.[37] Interestingly, a normal range for sensory fusion, or Panum's fusional areas, can also accompany a total loss of stereopsis[146] but, for strabismic patients with a loss of binocular sensory fusion, stereopsis invariably is lost.[172] For these individuals, binocular images appear to wash over one another, with no binocular binding of aligned images. Their vergence is unstable, and their eyes continually shift their fixation disparity.

Non-strabismic individuals may have sub-clinical anomalies of stereopsis, as demonstrated by directional biases in their ability to sense transient-qualitative depth from large (>1 deg) crossed and uncrossed disparities. Subjects with these sub-clinical anomalies generally discriminate the direction of depth better for flashed near objects than for flashed far objects. In subjects with severe stereoanomalies, transient-qualitative stereo-depth perception with one or both signs of disparity may be absent.[128,129] Non-strabismic individuals who have a stereoanomaly can have normal quantitative stereopsis.[28,88] The selective loss of transient disparity in one direction suggests that the disparity-detecting neurons are grouped into different pools tuned to broad ranges of crossed and uncrossed disparity, and sensitivity to one direction of disparity may be reduced or lost without affecting the other. A related motor anomaly of transient vergence has been found in patients with a stereoanomaly, who exhibit a marked reduction or absence of either convergence or divergence responses to large, briefly presented disparities.[39,88]

Stereoacuity

Stereoacuity is one of the qualitative aspects of binocular vision commonly evaluated during clinical vision testing. It is the smallest relative binocular disparity stimulus for depth that can be detected (Fig. 36.13). The binocular disparity angles for depth detection normally are smaller than the resolution angles for visual acuity, even for unpracticed clinical patients, and are typically less than 30 arcsec. In practice, the terminology for visual acuity is maintained for

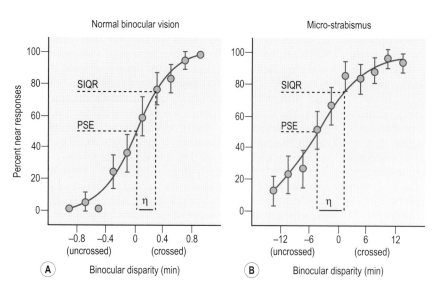

Figure 36.13 Psychometric functions for depth discrimination for: (**A**) an observer with normal binocular vision and (**B**) an observer with deficient stereopsis from micro-strabismus. The percentages of responses that a test stimulus appeared to be nearer than the reference stimulus are plotted as a function of the sign (negative values for uncrossed disparities and positive values for crossed disparities) and magnitude of disparity. The subjects' stereoacuities (η) are defined by the disparity range between the point of subject equality (PSE) and the semi-intraquartile range (SIQR) (adapted from Harwerth et al 2003[69]).

stereopsis. A patient's stereo performance is specified as a stereoacuity (disparity sensitivity), rather than stereothreshold. While the true definition of stereoacuity is the reciprocal of the stereothreshold, stereoacuities are usually specified by arcsec of disparity at threshold and, by usage, the terms stereothreshold and stereoacuity have become clinically interchangeable.

Normal stereoscopic vision is experience-dependent and requires normal binocular visual experience over a period of early childhood (see Chapter 38). However, even after the early period of development, clinical testing of stereopsis is important to detect binocular abnormalities in children or adults, and most of the clinical tests are designed as screening tools for distinguishing normal from abnormal (e.g. stereoblindness, microstrabismus, or monofixation) binocular vision. The stereo threshold of a normal observer[177] is usually lower than the smallest disparity presented by most clinical tests, which employ a limited number of stimuli with relatively large disparity increments. Improving the accuracy of clinical stereopsis tests would require a greater number of stimuli with smaller disparity increments, without necessarily improving the detection rate for abnormal binocular vision. Nevertheless, it is an important concept that for any given patient, the relationship between stereoscopic depth perception and binocular disparity is probabilistic over a range of disparity values, as shown by the psychometric function (Fig. 36.13) for the detection or discrimination of stereoscopic depth.

Consider a stereopsis test in which a test stimulus, with variable disparity direction and magnitude, is located below a zero-disparity reference stimulus.[75] The test stimulus is presented to the observer many times, changing the direction and magnitude of the disparity between presentations, or trials. For each trial, the observer judges the relative direction of depth of the test stimulus with respect to the reference stimulus. When the trials are completed, the percentages of near responses are plotted as a function of disparity magnitude and sign (direction) of the test stimulus (Fig. 36.13). In the range of probabilistic performance, the discrimination of near stereoscopic depth with respect to a reference stimulus of zero disparity systematically improves as the sign of

the disparity becomes appropriate for near (i.e. crossed) and as the magnitude of the binocular disparity increases. The disparity between the test and reference stimuli for which the percent "near" responses equals 50 percent denotes the point of subjective equality (PSE) in the depth of the test and reference stimuli. The observer's PSE represents a disparity bias, or constant error, at the specific retinal eccentricity tested. Since this disparity (the PSE) is perceived as having zero depth difference from the reference stimulus, the PSE is, by definition, a point on the longitudinal horopter. The slope of the psychometric function represents the variability of the stereo depth percept and the corresponding stereothreshold. The binocular disparity required to achieve depth discrimination with a given level of accuracy (typically 75 percent) defines the observer's stereothreshold. Operationally, the stereothreshold (η) is quantified by the semi-intraquartile range (SIQR) of the psychometric function, i.e. by the range of disparities from the PSE to the disparity that gives a "near" response rate of 75 percent.

The fundamental differences between normal and reduced stereoacuities are illustrated in Figure 36.13 by comparison of the psychometric functions for a subject with normal binocular vision and a patient with primary micro-strabismus.[68] The subject with normal binocular vision (Fig. 36.13A) had nearly perfect (100 percent) depth discrimination for binocular disparities of 0.6 arcmin, either uncrossed or crossed. Chance performance (PSE) occurred at zero disparity, indicating that the horopter was located at the fixation point, and the stereothreshold (η) determined from the SIQR was 12 arcsec. The performance of the microstrabismic subject, whose clinical data are presented in Box 36.3, is illustrated by Figure 36.13B. The psychometric function was similar in showing a systematic relationship between discrimination and disparity, although the range of binocular disparities required was fifteen times larger. The PSE for the microstrabismic subject was 4.5 arcmin of uncrossed disparity and the stereothreshold was greater than 320 arcsec. This example shows that accurate measurements of stereoacuites for patients with abnormal binocular vision, such as micro-strabismus (Box 36.3) require the use of a different range and larger increments of binocular disparities than needed

Box 36.3 Reduced stereoacuity associated with primary microstrabismus

- College student, diagnosed at age 22 years of age
- No surgery or orthoptic treatment
- Cover test: Constant, esotropia with alternating fixation
  Distance: 8^Δ ET
  Near: 6^Δ ET
- Refractive errors and VAs:
  OD: Non-significant 20/20
  OS: Non-Significant 20/20
- Lateral heterophoria at distance with Maddox Rod – normal retinal correspondence.
- Bagolini striated lens test: at near. Lines intersect at the fixation point – harmonious anomalous retinal correspondence.
- No stereopsis with the StereoOptical circles test.
- Reduced stereoacuity (304 arcsec) with psychophysical testing (see Fig. 36.13B).

Comment: The patient has primary microstrabismus with central harmonious retinal correspondence and reduced stereoacuity, but with normal peripheral retinal correspondence and disparity vergence oculomotor responses (see Harwerth et al 2003[68] for details).

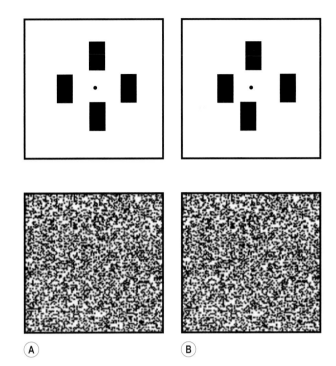

Figure 36.14 Examples of stereograms with contour-defined objects (local stereopsis) or disparity-defined objects (global stereopsis or cyclopean perception). The disparity patterns in the contour and random-dot stereograms are identical in size and disparity magnitude. Most individuals with normal binocular vision can learn to free-fuse the stereograms by either over-convergence or under-convergence and observe the stereoscopic depth in each example. However, it should be noted that the two viewing strategies produce opposite directions of relative depth.

for normal stereoacuity, but otherwise the psychophysical data are similar. While most clinical tests cover a sufficient range of disparities for evaluating both normal and abnormal subjects, the disparity increments are too large for accurately measuring normal stereothresholds and, thus, an individual's clinical performance is compared to expected suprathreshold performance levels for patients with normal binocular vision. For the primary purpose of screening for abnormal binocular vision, determining that a patient's stereoacuity is at an expected level of performance is sufficient to demonstrate that their sensory and motor processes are compatible with normal binocular vision.

Stereopsis is not usually evaluated with real-depth stimuli, but rather most clinical tests of stereopsis utilize stereograms with either contour-defined or disparity-defined forms (examples of the two forms of stereograms are shown in Figure 36.14) and most individuals with normal binocular vision can observe the relative depth in the stereograms with "free fusion" either by over-convergence (crossed-eye viewing) or under-convergence (straight-eye viewing), but it should be noted that the two viewing strategies produce opposite directions of relative depth. These two configurations of stereograms are often distinguished as being tests of local vs. global stereopsis,[165] and each has some desirable attributes for clinical evaluations. For example, stereograms that test local stereopsis use high-contrast test figures having unambiguous relative disparities that can be detected by all patients with normal binocular vision, and even by some patients with micro-strabismus or monofixation.[13,29,121,172] Thus, contour-defined stereograms provide valuable information for differentiating between normal and subnormal binocular vision. However, this form of stereogram unavoidably contains monocular cues that are discernable in targets having large disparities. These monocular cues could allow discrimination of disparate stimuli based on non-stereoscopic

information. Such non-stereoscopic cues are illustrated by the stereogram in Figure 36.14A. When properly fused, the contours in this stereogram provide clear stereoscopic depth appreciation from both crossed and uncrossed binocular disparities, but the relative differences in the positions of the stimuli are apparent without fusion or stereopsis.

The design of the random dot stereogram, which tests global stereopsis, or cyclopean perception,[90] eliminates almost all non-stereoscopic cues. Figure 36.14B is a random-dot stereogram of the same stimulus pattern as the contour stereogram in Figure 36.14A. Even though the locations, directions, and disparity magnitudes of the stimuli of the random-dot and contour stereograms are identical, the pattern is more difficult to see in the random-dot stereogram.[21] Because of the inherent ambiguity of which element in one half-view is the correct "match" for a given element in the other half-view of random dot stereograms, global stereopsis is a more complex second-order process than local stereopsis.[90,104]

The correct identification of a disparity-defined form embedded into the random-dot stereogram is definitive evidence that the patient's stereoacuity is higher than the disparity of the form and, therefore, clinical testing using random dot stereograms may be preferred. This property is especially useful for children because the identification of a familiar form is cognitively simpler than the discrimination of relative depth. However, clinical testing of global stereopsis requires a relatively long viewing duration to identify the stereoscopic form[21,71,133] and the relative direction of stereoscopic depth in random-dot stereograms can be detected with smaller disparities than needed for identification of the disparity-defined form.[70] In addition, an accurate registration of the random-dot patterns on corresponding retinal points is essential for form identification, for that reason, they may not be ideal for assessing the degree of binocularity in patients with subnormal binocular vision from microstrabismus or the monofixation syndrome,[12,172] although patient's with primary microstrabismus can perceive the depth in some forms of random-dot stereograms.[68]

Stereoacuity with refractive defocus

Occasionally a patient can be stereo-deficient or stereoblind even though they have normal eye alignment and visual acuity.[51] More often, stereoacuities are reduced by some factor that has affected the retinal image quality of one or both eyes. For example, optical defocus of the retinal images, as occurs with uncorrected refractive errors, degrades stereoacuity in proportion to the magnitude of defocus.[101,134,176] The effect of blurring the retinal images is greater for stereoacuity than for other resolution tasks, such as visual acuity or instantaneous displacement thresholds.[176]

Unilateral optical defocus produces a greater reduction of stereoacuity than does symmetrical bilateral defocus,[54] which is consistent with other paradoxical effects on stereoacuity caused by interocular differences in stimulus parameters.[32,67,96,142] For example, improving the focus in one eye will, paradoxically, degrade stereopsis more than equal defocus in both eyes. With unilateral defocus, stereoacuity decreases as the amount of defocus increases until stereopsis is suspended by interocular differences greater than about two diopters.[54,119] The filtering of high spatial frequencies that occurs in defocused images cannot fully explain the resulting reduction in stereoacuity[179] and, hence, foveal suppression may also play a role.[163]

Because of the effect of optical defocus on stereoacuity, usually a clinician's primary goal is to provide an accurate refractive correction for each eye in order to optimize binocular vision and stereoacuity. However, for one group of patients – presbyopes – interocular refractive error corrections are often purposely unbalanced to correct one eye for distance vision and the other eye for near vision.[42] The scheme of providing an ocular correction for distance vision with one eye and a near vision correction for the other eye, either with single vision contact lenses,[40,41,42,144] refractive surgery,[22,85] or intraocular lens implants,[59,61,62] is called monovision.[41,145,161] The monovision procedure is successful for a majority of patients with normal binocular vision and usually the dominant (sighting) eye is corrected for distance vision because less disorientation is experienced during perceptual adaptation to the anisometropic refractive condition.[40,107] After adaptation, most patients do not experience symptoms of asthenopia, nor are they aware of blurred vision at far or near viewing distances.[145] Although their binocular field of view and motor fusion ranges are generally normal, many monovision patients will experience diplopia at night or in dim illumination (when their pupils are dilated), or when they view bright targets of high contrast.[145] Some visual functions do not adapt to unilateral optical defocus, resulting in a loss of binocular summation for high spatial frequencies, reduced binocular visual acuity, and degraded stereoacuity with foveal suppression.[38,40,45,107]

Spatial frequency and contrast effects on stereopsis

Stereoacuity is influenced both by spatial frequency and contrast. Studies of stereopsis, using narrowly tuned, spatially filtered stimuli, have provided evidence that retinal disparities are processed by independent channels that are tuned to different spatial frequencies and different types of disparity.[46,67,80,93,96,142,147,148,185] Disparity can be described either as a positional offset of the two retinal images (measured in angular units) or a phase offset, where phase describes the disparity as a proportion of the luminance spatial period to which the disparity detector is tuned.[73,147] The neural mechanisms tuned to high spatial frequencies (>2.5 c/deg) are sensitive to positional disparity and have higher sensitivities than the mechanisms tuned to lower spatial frequencies, which are sensitive to phase disparity.[147,148,174]

Results from early experiments, using small, broadband stimuli, suggested that stereoacuity is relatively unaffected by lowering contrast and luminance unless the reduction diminishes the monocular visibility of the stereoscopic stimuli.[115] However, more recent investigations, using narrowband stimuli that are more specific for the early spatial filters in the binocular vision pathway, have found that stereoacuity is reduced with decreased binocular contrast, and the reduction is greater for high than low spatial frequency stimuli.[67,96,147] The effects of interocular differences are similar for contrast and optical defocus and stereoacuity is reduced more with unilateral or asymmetric reductions of contrast than with binocular or symmetric contrast reductions.[67,96,147] Additionally, the relative effects of contrast asymmetries are greater for low than for high spatial frequencies.[32] The improvement in stereoacuity with increasing bilateral contrast can be explained by an increased stimulus strength, which either intensifies signals from disparity-selective neurons or recruits signals from more disparity-selective neurons. However, the paradoxical decrease in stereoacuity with unilaterally increased contrast is not consistent with the improved signal strength explanation, but requires suppression, or masking, interactions.[163]

For disparities above the stereothreshold, the contrast sensitivity function for depth perception has a unimodal peak when measured with a binocular phase disparity of approximately 90 deg.[73,155,156] Using small or narrowband Gabor or Difference-of-Gaussian stimuli having low- or mid-range spatial frequencies, contrast sensitivity falls off as disparity increases, up to the upper disparity limit of stereopsis (D_{max}), but the upper limit of stereoscopic depth perception is also dependent on stimulus bandwidth (size).[174,181]

Spatial distortions from aniseikonia

The two retinal images need to be similar in size and shape to support normal binocular vision; nonetheless, the visual system can tolerate small interocular differences without complete disruption of binocular fusion. An inequality in the size and/or shape of the two ocular images (aniseikonia) can occur from differences in the optics of the eyes, in the retinal receptor mosaics, or in cortical magnification.[113] However, the most common cause of aniseikonia is a

Box 36.4 Spatial distortions caused by aniseikonia

- Electronics technician, 58 years old
- Cataract surgery for left eye with an IOL implant
- Refractive errors before surgery

    OD: +3.25 −1.25 × 010

    OS: +2.25 −0.50 × 180
- Refractive errors after surgery

    OD: +3.25 −1.25 × 010

    OS: +0.50 −2.75 × 005
- Symptoms after IOL surgery: Clear vision with each eye, but while reading a newspaper binocularly, the pages looked farther away on the left side and they appeared trapezoidal. He also complained of problems with depth perception, such as reaching and pressing the floor button in the elevator or soldering on a circuit board.

Comment: Using the rough rule-of-thumb of 1 percent magnification per diopter of anisometropia, the patient has an overall magnification of 2.75 percent with the right eye image larger, plus an additional 1.5 percent magnification in the vertical meridian of the right-eye image. Thus, the anisometropia following the cataract surgery has both an overall and meridional component, but the meridional component is the more likely cause of spatial distortions. The patient's complaints are consistent with an induced effect from the aniseikonia, and although a 1.5 percent image size difference may not cause symptoms in all patients, this electronics technician was quite disturbed by his altered perception.

difference in the refractive errors of the two eyes (anisometropia).[1,14] In refractive anisometropia, the optical components of the two eyes differ in power. Spherical optical differences produce a relative overall size discrepancy in the two retinal images, and astigmatic optical differences produce a relative meridional size difference in the two retinal images. Anisometropia as a result of the natural development of refractive errors is a common cause of aniseikonia, although nowadays aniseikonia is more often associated with anisometropia as a result of intraocular lens implantation (Box 36.4), refractive surgery, or penetrating keratoplasty.[11,55,59,81]

There are large individual differences in symptoms and toleration of aniseikonia. Aniseikonia is usually considered clinically significant when the image size difference is greater than 4 percent, but many patients experience distortions in spatial perception and/or uncomfortable binocular vision, headaches, and eyestrain with differences as small as 2 percent.[19,113] The symptoms of eyestrain (asthenopia) are not specific[1,19] and the diagnosis of aniseikonia in a patient with fusion and stereopsis must be made from measurements of retinal image size differences[94,102,162] or from assessment of the aniseikonic spatial distortions.[19,113] The size differences of the retinal images pose problems both for stereopsis and for control of eye alignment. The size discrepancy generated by an anisometropic spectacle correction produces disparities that vary with eye position, causing a non-comitant variation of heterophoria (anisophoria).[141,145] The oculomotor system can adapt tonic vergence to compensate for anisophoria. However, when a person is unable to adapt to the correction, the symptoms of eyestrain occur with attempts to adjust eye alignment as gaze shifts from one direction to another. These motor symptoms can be confused with

sensory disorders normally assumed to result from aniseikonia.[52] Anisometropic contact lens corrections do not produce anisophoria because the optical center of the lens remains fixed with respect to the pupil center during eye movements.

The types of spatial distortions perceived from the relative magnification of one retinal image can be understood by the altered perception of the orientation of the fronto-parallel plane that is associated with aniseikonia.[113] Unequal retinal image sizes cause the apparent fronto-parallel plane to be slanted around an axis at the fixation point by an amount and direction determined by three components of aniseikonia: the geometric effect, the induced effect, and declination errors. The three components of aniseikonic perceptual distortions are illustrated by stereograms in Figure 36.15, which were constructed for crossed-eye free fusion, and when viewed in a stereoscope or with straight-eye free fusion, the effects will be opposite to the descriptions in the text.

The geometric effect, simulated in Figure 36.15B, occurs with retinal image size differences in the horizontal meridian and is a consequence of the horizontal disparity caused by lateral magnification. The fronto-parallel plane appears slanted toward the eye with the larger horizontal magnification. The physiological basis for the apparent slant in the geometric effect is illustrated in Figure 36.16. To simulate the effect, an observer views an objectively flat plane (plane PFQ) through a ×090 size lens or afocal magnifier. A size lens is a small Galilean telescope without refractive power for objects at infinity. In this example, it produces magnification in the horizontal meridian of the right eye (×090). When viewed through the size lens, the plane does not appear parallel to the face, but rotated about the fixation point to the plane P′FQ′; i.e. such that point P is perceived as nearer and point Q as farther. The amount of perceived rotation is proportional to the magnification difference between the two retinal images produced by the size lens. The direction and magnitude of rotation result from the change in the horizontal disparity gradient produced by the uniocular horizontal magnification. The disparities across the gradient are in opposing directions in the left and right visual fields. In this example the gradient produces crossed disparities for objects in the left visual field and uncrossed disparities for objects in the right visual field. The perceptually slanted plane illustrated in panel B of Figure 36.15 incorporates a perceptual surface rotation and an apparent change in perceived size. Objects in the left visual field appear smaller than in the right. The size distortion is consistent with learned relationships between size and distance, e.g. if one object appears farther than another, but without a correlated difference in retinal image sizes, the cue conflict causes the farther object to be perceived as larger.

The second component of aniseikonia, the induced effect, is caused by a relative magnification difference in the vertical meridians of the two eyes, which can be simulated by a ×180 or vertical meridional magnifier. The induced effect is interesting because, as can be observed by the stereogram (Fig. 36.15C), vertical binocular disparities, by themselves, do not result in the perception of stereoscopic depth, but vertical meridional aniseikonia in a real-world visual environment causes a perceived rotation of the fronto-parallel plane. The induced effect is similar to the geometric effect in that it is perceived as a rotation of the fronto-parallel plane about a vertical axis, but the induced effect is in the opposite

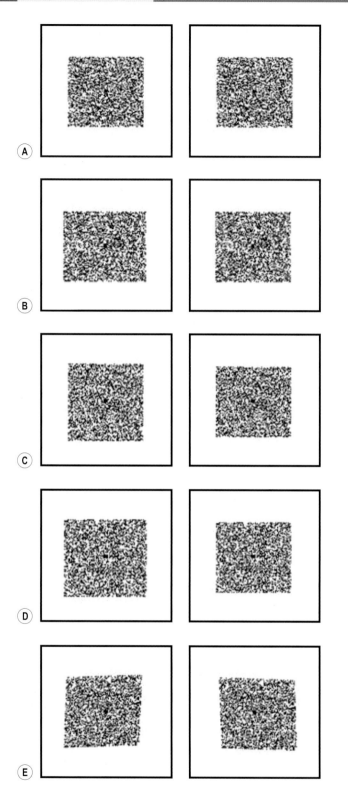

Figure 36.15 A series of stereograms to demonstrate the components of relative magnification from aniseikonia to illustrate subjective spatial distortions. The random-dot patterns represent fronto-parallel plane stereograms that have been constructed to simulate various components of aniseikonia during binocular fusion with crossed-eye viewing (other viewing strategies will produce perceptual effects that are opposite to those described). (**A**) Normal spatial perception with equal-sized images for each eye. For an observer without aniseikonia, the perceived fronto-parallel plane should not be distorted. (**B**) A stereogram with a relative magnification of 10 percent in the horizontal meridian to demonstrate the geometric (×090) effect. The observer should perceive a tilt of the fronto-parallel plane about a vertical axis with the right edge appearing more remote than the left edge of the plane. (**C**) A magnification of 10 percent in the vertical meridian to simulate the magnification of the induced (×180) effect. For most observers, vertical disparities in a stereogram will not produce the apparent distortion of the fronto-parallel plane that can be seen in a natural visual environment. (**D**) A combination of horizontal and vertical magnification to produce an overall magnification of 10 percent. (**E**) Relative magnification at oblique meridians (right eye: ×045 and left eye: ×135 for crossed-eye fusion) to simulate the declination effect. The fused perception should be a slanted surface with the upper edge appearing more remote that the lower edge.

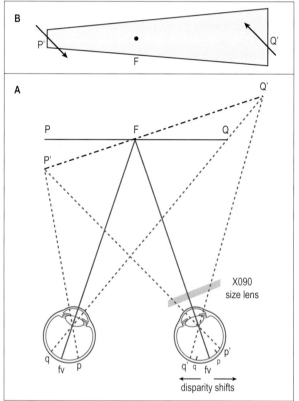

Figure 36.16 The physiological basis of spatial distortions caused by the aniseikonia from meridional magnification in the horizontal meridian of the right eye. The fronto-parallel plane PFQ appears to be rotated to a plane P'FQ' as a result of the binocular disparity produced by meridional magnification. See the text for details (adapted from Ogle 1964[113]).

direction. The OFPP appears tilted away from the eye viewing through the vertical magnifier. Because the phenomenon is similar in form, but opposite in direction to the geometric effect, it is called the induced effect to convey the notion that the distortion is the same as if it was "induced" by a horizontal meridional magnifier over the fellow eye.

The induced effect results from the magnification of the vertical disparities, which are inherently present at relatively near viewing distances due to differences in ocular perspective of the target. Naturally occurring vertical disparities increase as the eccentricity of the target increases and as the viewing distance decreases.[167] If the viewing distance is sensed from convergence, then target azimuth can be derived from the rate of increase of vertical disparity with eccentricity (i.e. from the vertical disparity gradient). Vertical disparity gradients are used to scale disparity in order to quantify depth and to interpret slant information.[3,56] Horizontal disparity alone is an ambiguous stimulus for slant perception. Identical horizontal disparity patterns can correspond to different slant patterns for surfaces seen in symmetrical and asymmetrical convergence. For example, while a zero disparity pattern describes the Vieth-Muller circle, this constant disparity pattern also describes a surface slant that increases with horizontal gaze eccentricity.[114] In order to interpret slant from horizontal disparity correctly, knowledge of target distance and gaze eccentricity are needed. The combination of vertical disparity gradients and convergence provides this information. When vertical magnifiers distort vertical disparity gradients, the surface slant is changed to conform to the distorted vertical disparity gradient.[83] This induced effect is weak, or absent, for targets located at an infinite viewing distance because retinal images of remote targets lack a difference in perspective significant enough to produce a vertical disparity gradient. For most observers, induced and geometric effects are about equal in magnitude and opposite in direction[113] thus, overall magnification differences, as opposed to meridional magnification differences, have little effect on the perception of space (see fronto-parallel plane stimulus D, Fig. 36.15).

The third component of aniseikonia is a declination error (perceived tilt about a horizontal axis) that arises from retinal image shear, or magnifications in oblique meridians (see stereogram E, Fig. 36.15). The direction of rotation (inclination or declination) of the oblique magnifier is predicted from the combined vertical and horizontal disparity components of the magnified retinal images. For example, the stereogram was constructed to simulate ×045 magnification for the right eye and ×135 magnification for the left eye with crossed-eye viewing, which creates a relative crossed disparity in the lower visual field compared to the upper (Fig. 36.15E). The resulting perception is a slanted surface with the bottom part of the surface closer. Oblique magnification can also cause horizontal tilt about a vertical axis if the horizontal and vertical components are unequal. Magnifiers at ×45 and ×135 degrees have equal horizontal and vertical magnification components, so that the resulting geometric and induced effects tend to cancel one another. However, oblique magnification along other meridians produces unequal amounts of geometric and induced distortion, and the stronger distortion usually dominates the horizontal tilt percept.

In the final analysis, the perceptual distortions from aniseikonia are a combination of the three components, geometric effect, induced effect, and declination error, with individual variability in the degree of tolerance and ability to adapt to stereoscopic misrepresentation of relative depth. These principles are illustrated by the case described in Box 36.4, where a relatively small induced effect caused disturbing spatial distortions for an electronics technician.

Motion-in-depth

The properties of visual direction and stereoscopic depth perception have been much more extensively studied with static stimuli at a constant fixation distance than under dynamic stimulus conditions. However, providing valid judgments of the trajectories and distances of moving objects, to capture moving objects or avoid collision with them, is a vital function of binocular vision.[60] The mechanisms underlying perception of the distance and the direction of moving objects are not as well understood, but there are at least two possible mechanisms for the perception of motion-in-depth.[36,126] First, the binocular disparity of an object moving in depth will change with time and, thus, the trajectory and distance could be interpolated from the responses of static disparity-sensitive mechanisms over time. The second type of mechanism for the perception of motion-in-depth is one that is sensitive to differential velocities of the retinal images. An object that is moving in depth will generate movements in the two retinal images that differ in speed and direction.

In natural scenes, as illustrated in Figure 36.7, changes in binocular disparity and differential interocular velocities are correlated. For this example, an object is moved along the visual axis of the left eye until the retinal image of the right eye is displaced outside the retinal region corresponding to Panum's fusional limit. During this measurement, if the test object had been moved from the uncrossed to the crossed disparity limits, it would have appeared to be moving in a depth trajectory toward the observer's left eye. The apparent trajectory is a property of the relative directions and velocities of the retinal images[125] and, if in another case, the two retinal images of an object move in the same direction and velocity, the object would not appear to move toward the head; rather, it would be perceived as moving in the fronto-parallel plane. If the retinal images move in opposite directions (i.e. temporal-ward), the object would appear to move towards the observer's head. In each example, the object's motion-in-depth may have been detected by mechanisms sensitive to the sign and magnitude of binocular disparity, which have changed with time. On the other hand, for any given time, the direction and velocity of movement of each of the object's retinal images are different; thus, the object's motion-in-depth could have been based on differential velocity information.

Some investigations of stereomotion have provided psychophysical and physiological evidence for velocity-dependent mechanisms,[8,9,127] while other studies have shown comparable perception of motion-in-depth in the absence of coherent velocity information.[36,126] Threshold and adaptation experiments have provided evidence for neural channels that are sensitive to the direction of motion in depth, with different channels selectively sensitive to approaching and receding motion. However, stereo-motion can also be detected in dynamic random-dot stereograms that produce changes in disparity over time without consistent signals for interocular velocity differences. Regardless of the mechanism, the normal sensitivity to stereomovement requires about 5–10 times greater disparity magnitudes than stereo-acuity measures with static stimuli.[36,177] Motion-in-depth may be present in patients with stereodeficiencies from early strabismus or amblyopia as a residual form of stereopsis.[103] But on the other hand, many individuals with normal visual acuity and stereopsis have areas of the binocular field that are insensitive, or even blind, to motion-in-depth.[130] Currently, clinical tests of stereomotion have not been developed.

Suppression in normal binocular vision

In complex three-dimensional scenes, there are several types of binocular information that must be combined to produce an integrated percept of depth and distance. First, some of the visual information is coherent and unambiguous, such as images formed within Panum's fusional areas and the coherent information contributes to normal fusion and stereoscopic depth perception. However, some forms of binocular information do not stimulate binocular fusion. For example, regions of space may be clearly visible to only one eye (i.e. partially occluded to other eye), and this fragmented binocular information also contributes to the perceptions of distance and depth. In addition, information from binocular images that are uncorrelated, either because the matches are ambiguous or would conflict with other cues, influences depth and distance perception. Visual confusion, in which there is superimposition of separate diplopic images arising from objects seen behind or in front of the plane of fixation, is an example of uncorrelation of binocular images that can be interpreted to suggest depth or distance. The binocular visual system attempts to preserve as much of the depth information as possible from all three sources to make inferences about objective space without introducing ambiguity or confusing space perception. At times, however, conflicts between the two eyes are so great that the conflicting information must be resolved by suppressing or suspending the perception arising from one of the retinal images.

Four classes of stimuli have been identified that appear to evoke different interocular suppression mechanisms in normal binocular vision.

1. *Interocular blur suppression* occurs when the retinal images have significant differences in defocus or contrast.[152] A wide variety of natural conditions produce unequal contrast of the two ocular images, including naturally occurring anisometropia, unequal amplitudes of accommodation, and asymmetric convergence on targets that are closer to one eye than to the other. The relative differences in defocus can be partially eliminated by differential accommodation of the two eyes[105] and by interocular suppression of the blur.[145] The latter mechanism is requisite for adaptation to a monovision correction of presbyopia. Monovision suppression allows clear, non-blurred, binocular percepts and retention of stereopsis albeit with the stereo-threshold elevated by approximately two-fold.[38,145]

2. *Suspension* is the mechanism that eliminates one of the two ocular percepts that arise during physiological diplopia, in which targets lying in front of or behind the singleness horopter would be seen as doubled.[30] Binocular eye alignment optimizes depth information from binocular disparities near the horopter, but at the same time it produces large disparities for the objects that are at some distance in front of or behind the plane of fixation. Even with binocular disparities that are well outside the limits of Panum's fusional areas, the perception of diplopia under normal casual viewing conditions is rare because of the suppression of one image. The suppression of physiological diplopia is called suspension because this form of suppression does not alternate between the two images. Instead, only one image is continually suppressed, favoring visibility

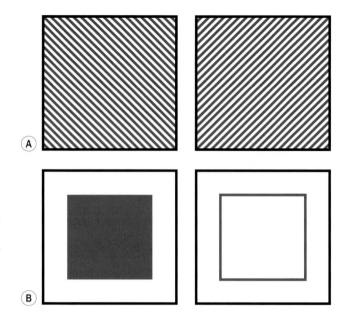

Figure 36.17 Stereograms to demonstrate the perceptual consequences of binocular viewing of non-fusible objects or surfaces. (**A**) Binocular rivalry. (**B**) Binocular luster.

of the target imaged on the nasal hemi-retina.[35,43,92] However, suspension is not obligatory, and calling attention to the disparate target can evoke physiological diplopia. It has also been suggested that suspension may be involved in the permanent suppression of pathological diplopia associated with strabismus.[44,63,135,136]

3. *Binocular (retinal) rivalry suppression* is a rhythmic alternating perception of each of the retinal images when substantial differences in their sizes and/or shapes preclude their fusion.[15] The classic demonstration of binocular rivalry arising from non-fusible images formed on or near corresponding retinal points is shown in Figure 36.17A, which presents orthogonal-oriented gratings to the two eyes. In this example, rivalry suppression can take the form of either exclusive domination or mosaic domination. During exclusive or wholesale domination, each of the two images alternates between dominance and suppression: one set of lines will be seen and, after several seconds, the image of the other set of lines appears to wash over the first. In mosaic domination, the two monocular images become fragmented, and small interwoven retinal patches from each eye alternate independently of one another. In this case, piecemeal suppression is regional, localized to the vicinity of the contour intersections.

The perceived rivalrous patches alternate between the two ocular images approximately once every 4 seconds, but the rate of oscillation and its duty cycle vary with the degree of difference between the two ocular images and with the stimulus strength. The rate of rivalry increases as the orientation difference between the targets increases beyond 22 degrees,[43,135] indicating that rivalry is not likely to occur within the tuned range for cortical orientation columns. Levelt[97] formulated a series of rules describing how the dominance and suppression phases of binocular rivalry vary with the strength of stimulus variables such as brightness, contrast, and motion. He concluded that the duration of the suppression phase of one of the retinal images decreases when its stimulus visibility or strength is increased relative to the retinal image of the fellow eye. If the strengths of the stimuli for both eyes are increased, then the suppression phase for each eye decreases, and the oscillation rate of rivalry increases. Reducing the stimulus contrast of both eyes also reduces the oscillation rate, until at a very low level of contrast, rivalry is replaced with a binocular summation of non-fusible targets.[100]

The perception of rivalry has a latency of approximately 200 msec so that briefly presented non-fusible patterns appear superimposed.[78] However, rivalry occurs between dichoptic patterns that are alternated rapidly at 7 Hz or faster, indicating an integration time of at least 150 msec.[183] When rivalrous and fusible stimuli are presented simultaneously, fusion takes precedence over rivalry,[15] and the onset of a fusible target can terminate suppression, although the fusion mechanism takes time (150–200 msec) to become fully operational.[64,184] Suppression and stereo-fusion appear to be mutually exclusive outcomes of binocular rivalry stimuli presented at a given retinal location,[16,164] but stereoscopic depth and rivalry can be observed simultaneously when fusible contours are superimposed on rivalrous backgrounds.

Because both binocular rivalry suppression and strabismic suppression result in a functional loss of visual information from one eye during binocular viewing, binocular rivalry suppression and strabismic suppression have been hypothesized to be related phenomena.[172] However, the patterns of suppression[159] are different for normal and for esotropic subjects suggesting that strabismic observers do not demonstrate normal binocular rivalry and that strabismic suppression and normal binocular rivalry suppression are mediated by different neural mechanisms.[158,160]

Another form of binocular rivalry occurs when un-fusible surface brightnesses (e.g. black and white) or surface colors (e.g. red and green) are dichoptically superimposed. In this form, the phenomenon is known as binocular luster or as luminance or color rivalry.[83,114] Figure 36.17B presents a stereogram that evokes binocular luster, which is usually described as a shimmering, unstable surface percept, as though one surface is being seen through another. The perception of luster requires a certain level of binocular vision, but does not require fusion ability and, therefore, testing for binocular luster has been advocated for evaluating binocularity and anomalous retinal correspondence in strabismic patients.

4. *Permanent-occlusion suppression* is a more stable form of suppression that occurs when one eye views a contoured stimulus while the other eye views a spatially homogeneous field.[106] Under these conditions, the image of the eye viewing the contoured field dominates, while the image of the eye viewing the homogeneous field is almost continually suppressed. Because of the stability of the dominance/suppression percept under these viewing conditions, the term "permanent suppression" is used to distinguish this type of suppression from binocular rivalry suppression.

There are many occasions in which the two eyes see dissimilar forms as a result of partial occlusion in the near visual field. Natural partial occlusion occurs when one eye's view of a distal target is partially obstructed by a nearby object, such as the nose, or when distal objects are viewed through a narrow aperture. Under these conditions, the occluding object is normally suppressed to retain a constant view of the background. This phenomenon can be demonstrated by holding a cylindrical tube in front of one's right eye and placing the palm of one's left hand in front of the left eye near the distal end of the tube. The combined stable percept is that of a hole in the left hand. The hand is seen as the region surrounding the aperture through which the background is viewed. This ecologically valid example of occlusion gives priority to the background seen through the aperture. The mechanisms underlying stable, permanent suppression are unlike those underlying rivalry suppression because different changes in the increment-threshold spectral sensitivity function are produced by these two phenomena.[131]

Summary

Binocular vision is an intricate organization of biological and psychological components to provide haplopia and an accurate subjective representation of the depth and distance of every object in the environment. The present chapter has presented some of the basic principles of the normal and clinically abnormal sensory processes, but they are just one part of the system. Equally important are the oculomotor components (Chapter 9), which in normal binocular vision interact with sensory mechanisms to place the ocular images on appropriate retinal areas to support fusion and stereopsis, or with abnormal oculomotor control as in strabismus, to enlist adaptive sensory mechanisms of suppression and anomalous retinal correspondence. Further, whether an individual has normal or abnormal binocular vision is a product of environmental influences in infancy (Chapters 38 & 40). During an early sensitive period, normal visual experience is required to develop the full capacity of the visual system, otherwise, abnormal visual experience will produce abnormal monocular and/or binocular visual processes. Clinical application of the science of binocular vision involves all of these components with the goal to alleviate any interference with clear, single binocular vision.

References

1. Achiron LR, Witkin NS, Ervin AM, Broocker G. The effect of relative spectacle magnification on aniseikonia, J Am Optom Assoc 1998; 69:591–599.
2. Awaya S, von Noorden GK, Romano PE. Sensory adaptations in strabismus. Anomalous retinal correspondence in different positions of gaze. Am Orthopt J 1970; 20:28–35.
3. Backus BT, Banks MS, van Ee R, Crowell JA. Horizontal and vertical disparity, eye position, and stereoscopic slant perception. Vision Res 1999; 39:1143–1170.
4. Bagolini B. Anomalous correspondence: definition and diagnostic methods. Doc Ophthalmol 1967; 23:346–398.
5. Bagolini B, Capobianco NM. Subjective space in comitant squint. Am J Ophthalmol 1965; 59:430–442.
6. Baker FH, Grigg P, von Noorden GK. Effects of visual deprivation and strabismus on the response of neurons in the visual cortex of the monkey, including studies on the striate and prestriate cortex in the normal animal. Brain Res 1974; 66:165–208.
7. Banks MS, van Ee R, Backus, BT. The computation of binocular visual direction. A re-examination of Manfield and Legge (1996). Vision Res 1997; 37:1605–1613.
8. Beverly KI, Regan D. Selective adaptation in stereoscopic depth perception, J Physiol Lond 1973; 232:40–41P.
9. Beverly KI, Regan D. Evidence for the existence of neural mechanisms selectively sensitive to the direction of movement in space. J Physiol Lond 1973; 235:17–29.
10. Bielschowsky A. Application of the afterimage test in the investigation of squint. Am J Ophthalmol 1937; 20:408–412.
11. Binder PS. The effect of suture removal on keratoplasty astigmatism. Am J Ophthalmol 1988; 105:637–645.
12. Birch EE, Stager DR. Random dot stereoacuity following surgical correction of infantile esotropia. J Ped Ophthalmol Strab 1995; 32:231–235.

13. Birch EE, Stager DR, Berry P, Everett ME. Prospective assessment of acuity and stereopsis in amblyopic esotropes following early surgery. Invest Ophthalmol Vis Sci 1990; 31:758–765.

14. Bishop PO. Binocular vision. In: Moses RA, Hart WM, eds. Adler's Physiology of the Eye, 8th edn. St Louis: CV Mosby, 1987:619–689.

15. Blake R, Logothetis NK. Visual competition. Nat Rev Neurosci 2002; 3:13–21.

16. Blake R, O'Shea RP. "Abnormal fusion" of stereopsis and binocular rivalry. Psychol Rev 1988; 95:151–154.

17. Blake R, Sloane M, Fox R. Further developments in binocular summation. Percept Psychophys 1981; 30:266–276.

18. Boeder P. The response shift. Doc Ophthalmol 1967; 23:88–100.

19. Borish IM. Anisometropia and aniseikonia. In: Clinical refraction, 3rd edn. Chicago, Professional Press, 1975:257–306.

20. Boucher JA. Common visual direction horopters in exotropes with anomalous correspondence. Am J Optom Arch Am Acad Optom 1967; 44:547–572.

21. Bradshaw MF, Rogers BJ, De Bruyn B. Perceptual latency and complex random-dot stereograms. Perception 1995; 24:749–759.

22. Braun EHP, Lee J, Steinert RF. Monovision in LASIK. Ophthalmology 2008; 115:1196–1202.

23. Brock FW. Projection habits in alternate squints. Am J Optom 1940; 17:193–207.

24. Brown JP, Ogle KN, Reiher L. Stereoscopic acuity and observation distance. Invest Ophthalmol 1965; 4:894–900.

25. Burian HM. Anomalous retinal correspondence: its essence and its significance in diagnosis and treatment. Am J Ophthalmol 1951; 34:237–253.

26. Campbell FW, Green DG. Monocular versus binocular visual acuity. Nature 1965; 208:191–192.

27. Chino YM, Smith EL, Hatta, S, Cheng H. Postnatal development of binocular disparity sensitivity in neurons of the primate visual cortex. J Neurosci 1997;17:296–307.

28. Cisarik PM, Harwerth RS. Stereoscopic depth magnitude estimation. Effects of stimulus spatial frequency and eccentricity. Behav Brain Res 2005;160:88–98.

29. Clarke WN, Noel LP. Stereoacuity testing in the monofixation syndrome. J Ped Ophthalmol Strab 1990; 27:161–163.

30. Cline D, Hofstetter HW, Griffin JR. Dictionary of Visual Science, 4th edn. Radnor PA: Chilton Trade Book Publishing, 1989.

31. Collewijn H, Steinman RM, Erkelens CF, Regan D. Binocular fusion, stereopsis and stereoacuity with a moving head. In: Regan D, ed. Vision and Visual Dysfunction, vol 9. Boca Raton: CRC Press, 1991:121–136.

32. Cormack LK, Stevenson SB, Landers DD. Interactions of spatial frequency and unequal monocular contrasts in stereopsis. Perception 1997; 26:1121–1136.

33. Crabb DP, Viswanathan AC. Integrated visual fields: a new approach to measuring the binocular field of view and visual disability. Graefes Arch Clin Exp Ophthalmol 2005; 243:210–216.

34. Crone RA. Diplopia. Amsterdam: Excerpta Medica, 1973.

35. Crovitz HF, Lipscomb DB. Dominance of the temporal visual fields at a short duration of stimulation. Am J Psychol 1963; 76:631–637.

36. Cumming BG, Parker AJ. Binocular mechanisms for detecting motion-in-depth. Vision Res 1994; 34:483–495.

37. Demanins R, Wang YZ, Hess RF. The neural deficit in strabismic amblyopia: sampling considerations. Vision Res 1999; 39:3575–3585.

38. Du Toit R, Ferreira JT, Nel ZJ. Visual and non-visual variables implicated in monovision wear. Optom Vis Sci 1998; 75:119–125.

39. Edwards M, Pope DR, Schor CM. Luminance contrast and spatial-frequency tuning of the transient-vergence system. Vision Res 1998; 38:705–717.

40. Erickson P, McGill EC. Role of visual acuity, stereoacuity, and ocular dominance in monovision patient success. Optom Vis Sci 1992;69:761–764.

41. Erickson P, Schor CM. Visual function with presbyopic contact lens correction. Optom Vis Sci 1990; 67:22–28.

42. Evans BJW. Monovision: a review. Ophthal Physiol Opt 2007; 27:417–439.

43. Fahle M. Binocular rivalry. Suppression depends on orientation and spatial frequency. Vision Res 1982; 22:787–800.

44. Fahle M. Naso-temporal asymmetry of binocular inhibition. Invest Ophthalmol Vis Sci 1987; 28:1016–1017.

45. Fawcett SL, Herman WK, Alferi CD, Castleberry KA, Parks MM, Birch EE. Stereoacuity and foveal fusion in adults with long-standing surgical monovision. J Pediatr Ophthalmol Strabismus 2001;5:342–347.

46. Felton TB, Richards W, Smith RA. Disparity processing of spatial frequencies in man. J Physiol Lond 1972; 225:349–362.

47. Flom MC. Corresponding and disparate retinal points in normal and anomalous correspondence. Am J Optom Physiol Opt 1980; 57:656–665.

48. Flom MC, Kerr KE. Determination of retinal correspondence. Multiple-testing results and the depth of anomaly concept. Arch Ophthalmol 1967; 77:200–213.

49. Foley JM. Binocular distance perception. Psychol Rev 1980; 87:411–434

50. Foley JM, Applebaum TH, Richards WA. Stereopsis with large disparities. Discrimination and depth magnitude. Vision Res 1975; 15:417–421.

51. Fredenburg P, Harwerth RS. The relative sensitivities of sensory and motor fusion to small disparities. Vision Res 2001;41:1969–1979.

52. Friedenwald JS. Diagnosis and treatment of anisophoria. Arch Ophthalmol 1936; 15:283–307.

53. Garding J, Porrill J, Mayhew J E W, Frisby J P. Stereopsis, vertical disparity and relief transformations. Vision Res 1995; 35:703–722.

54. Geib T, Baumann C. Effect of luminance and contrast on stereoscopic acuity. Graefe's Arch Clin Exp Ophthalmol 1990; 228:310–315.

55. Genvert GI, Cohen EJ, Arentsen JJ, Laibson PR. Fitting gas-permeable contact lenses after penetrating keratoplasty. Am J Ophthalmol 1985; 99:511–514.

56. Gillam BJ, Blackburn SG. Surface separation decreases stereoscopic slant but a monocular aperture increases it. Perception 1998; 27:1267–1286.

57. Gillam B, Lawergren B. The induced effect, vertical disparity, and stereoscopic theory. Percept Psychol 1983; 34:121–130.

58. Gobin MH. The limitation of suppression to one half of the visual field in the pathogenesis of strabismus. Br Orthop J 1968; 25:42–49.

59. Gobin L, Rozema JJ, Tassingnon MJ. Predicting refractive aniseikonia after cataract surgery in anisometropia. J Cataract Refract Surg 2008;34:1353–1361.

60. Graham CH. Visual space perception. In Graham CH, ed. Vision and Visual Perception. New York: Wiley, 1965:504–547.

61. Greenbaum S. Monovision pseudophakia. J Cataract Refract Surg 2002; 28: 1439–1443.

62. Handa T, Mukuno K, Uozato N, et al. Ocular dominance and patient satisfaction after monovision by intraocular lens implantation. J Cataract Refract Surg 2004; 30:769–774.

63. Harrad R. Psychophysics of suppression. Eye 1996; 10:270–273.

64. Harrad RA, McKee SP, Blake R, Yang Y. Binocular rivalry disrupts stereopsis. Perception 1994; 23:15–28.

65. Harrington DO. The visual fields. St Louis: CV Mosby, 1964:114–119.

66. Hallden U. Fusional phenomena in anomalous correspondence. Acta Ophthalmol 1952; 37:1–93.

67. Halpern DL, Blake R. How contrast affects stereoacuity. Perception 1988; 17:3–13.

68. Harwerth RS, Fredenberg PM. Binocular vision with primary microstrabismus. Invest Ophthalmol Vis Sci 2003; 44:4293–4306.

69. Harwerth RS, Fredenberg PM, Smith EL. Temporal integration for stereopsis. Vision Res 2003; 43:505–517.

70. Harwerth RS, Rawlings SC. Pattern and depth discrimination from random dot stereograms. Am J Optom Physiol Optics 1975; 52:248–257.

71. Harwerth RS, Rawlings SC. Viewing time and stereoscopic threshold with random-dot stereograms. Am J Optom Physiol Optics 1977; 54:452–457.

72. Harwerth RS, Smith EL. Binocular summation in man and monkey. Am J Optom Physiol Optics 1985; 62:439–446.

73. Harwerth RS, Smith EL, Crawford MLJ. Motor and sensory fusion in monkeys: psychophysical measurements. Eye 1996; 10:209–216.

74. Harwerth RS, Smith EL. Levi DM. Suprathreshold binocular interactions for grating patterns. Percept Psychophys 1980; 27:43–50.

75. Harwerth RS, Smith EL, Siderov J. Behavioral studies of local stereopsis and disparity vergence in monkeys. Vision Res 1995; 35:1755–1770.

76. Hebbard FW. Comparison of subjective and objective measurements of fixation disparity. J Opt Soc Am 1962; 52:706–712.

77. Helveston EM, von Noorden GK, Williams F. Sensory adaptations in strabismus. Retinal correspondence in the "A" or "V" pattern. Am Orthopt J 1970; 20:22–27.

78. Hering E. Beitrage zur Physiologie, Vol 5. Leipzig: Engelmann, 1861.

79. Hess RF, Baker CL, Wilcox LM. Comparison of motion and stereopsis: linear and non-linear performance. J Opt Soc Am A Opt Image Sci Vis 1999; 16:987–994.

80. Hess RF, Wilcox LM. Linear and non-linear filtering in stereopsis. Vision Res 1994; 34:2431–2438.

81. Holladay JT, Prager TC, Ruiz RS, Lewis JW, Rosenthal H. Improving the predictability of intraocular lens power calculations. Arch Ophthalmol 1986; 104:539–541.

82. Howard IP. Human Visual Orientation. Chichester, Wiley, 1982.

83. Howard IP, Rogers BJ. Seeing in Depth. Depth Perception, vol 2. Ontario, Canada: Porteuous Thornhill, 2002.

84. Hubel DH, Wiesel TN. Ferrier lecture. Functional architecture of macaque monkey visual cortex. Proc R Soc Lond B Biol Sci 1977; 198:1–59.

85. Jain S, Ou R, Azar DT. Monovision outcomes in presbyopic individuals after refractive surgery. Ophthalmology 2001; 108:1430–1433.

86. Jampolsky A. Characteristics of suppression in strabismus. Arch Ophthalmol 1955; 54:683–696.

87. Jampolsky A. Esotropia and convergent fixation disparity of small degree: differential diagnosis and management. Am J Ophthalmol 1956; 41:825–833.

88. Jones R. Anomalies of disparity detection in the human visual system. J Physiol Lond 1977; 264:621–640.

89. Jones R. Horizontal disparity vergence. In Schor CM, Ciuffreda KJ, eds. Vergence Eye Movements. Basic and Clinical Aspects. Boston: Butterworths, 1983:297–316.

90. Julesz B. Foundations of Cyclopean Perception. Chicago: University of Chicago Press, 1971.

91. Kerr KE. Anomalous correspondence – the cause or consequence of strabismus? Optom Vis Sci 1998; 75:17–22.

92. Kollner H. Das funktionelle Uberwiegen der nasalen Netzhauthalften im gemeinschaftlichen Sehfeld. Archiv Augenheilkunde 1914; 76:153–164.

93. Kontesevich LL, Tyler CW. Analysis of stereothresholds for stimuli below 2.5 c/deg. Vision Res 1994; 34:2317–2329.

94. Kramer P, Shippman S, Bennett G, Meininger D, Lubkin V. A study of aniseikonia and Knapp's law using a projection space eikenometer. Binocul Vis Strabismus Q 1999; 14:197–201.

95. Landers D, Cormack LK. Stereoscopic depth fading is disparity and spatial frequency dependent. Invest Opthalmol Vis Sci 1999; 40s:416.

96. Legge GE, Gu Y. Stereopsis and contrast. Vision Res 1989; 29:989–1004.

97. Levelt W. Psychological Studies On Binocular Rivalry. The Hague: Mouton, 1968.

98. Levi DM, Klein SA, Aitsebaomo P. Vernier acuity, crowding and cortical magnification. Vision Res 1985; 25:963–977.

99. Liu L, Stevenson SB, Schor CM. A polar coordinate system for describing binocular disparity. Vision Res 1994; 34:1205–1222.

100. Liu L, Tyler CW, Schor CM. Failure of rivalry at low contrast: evidence of a suprathreshold binocular summation process. Vision Res 1992; 32:1471–1479.

101. Lovasik J, Szymkin M. Effects of aniseikonia, retinal illuminance and pupil size on stereopsis. Invest Ophthalmol Vis Sci 1985; 26:741–750.

102. Lubkin LV, Shippman S, Bennett G, Meininger D, Kramer P, Poppinga P. Aniseikonia quantification: error rate of rule of thumb estimation. Binocul Vis Strabismus Q 1999; 14:191–196.

103. Maeda M, Sato M, Ohmura T, Miyazaki Y, Wang AH, Awaya S. Binocular depth-from-motion in infantile and late-onset esotropia patients with poor stereopsis. Invest Ophthalmol Vis Sci 1999; 40:3031–3036.

104. Marr D, Poggio T. A computational theory of human stereo vision. Proc Roy Soc Lond 1979; 204:301–328.

105. Marran L, Schor CM. Lens induced aniso-accommodation Vision Res 1997; 38:3601–3619.

106. Mauk D, Francis EL, Fox R. The selectivity of permanent suppression. Invest Ophthalmol Vis Sci 1984; 25s:294.

107. McGill EC, Erickson P. Sighting dominance and monovision distance binocular fusional ranges. J Am Optom Assoc 1991; 62:738–742.

108. Mills RP. Correlation of quality of life with clinical symptoms and signs at the time of diagnosis. Trans Am Ophthalmol Soc 1998; 96:753–812.

109. Mitchell DE. A review of the concept of "Panum's fusional areas." Am J Optom Physiol Optics 1966; 43:387–401.

110. Musch DC, Farjo AA, Meyer RF, Waldo MN, Janz NK. Assessment of health-related quality of life after corneal transplantation. Am J Ophthalmol 1997; 124:1–8.

111. Nelson-Quigg JM, Cello K, Johnson CA. Predicting binocular field sensitivity from monocular visual field results. Invest Ophthalmol Vis Sci 2000; 41:2212–2221.

112. Ogle KN. Disparity limits of stereopsis. Arch Ophthalmol 1952; 48:50–60.

113. Ogle KN. Researches in Binocular Vision. New York: Hafner Publishing Co, 1964.

114. Ogle KN. The optical space sense. In: Davson H, ed. The Eye, vol 4. New York: Academic Press, 1962.

115. Ogle KN, Groch J. Stereopsis and unequal luminosities of the images in the two eyes. Arch Ophthalmol 1956; 56:878–895.

116. Ogle KN, Martens TG, Dyer JA. Oculomotor imbalance in binocular vision and fixation disparity. Philadelphia: Lea & Febiger, 1967.

117. Ogle KN, Mussey F, Prangen AD. Fixation disparity and fusional processes in binocular single vision. Am J Ophthalmol 1949; 32:1069–1087.

118. Ogle KN, Weil MP. Stereoscopic vision and the duration of the stimulus. Arch Ophthalmol 1958; 59:4–17.

119. Ong J, Burley WS. Effect of induced anisometropia on depth perception. Am J Optom Arch Am Acad Optom 1972; 49:333–335.

120. Otto JMN, Kromeier M, Bach M, Kommerell G. Do dissociated or associated phorias predict the comfortable prism? Graefes Arch Clin Exp Ophthalmol 2008; 246:631–639.

121. Parks MM. Monofixation syndrome. Trans Am Ophthalmol Soc 1969; 67:609–657.

122. Parks MM. Sensorial adaptations in strabismus. In Duane TD, ed. Clinical Ophthalmology Vol 1. Philadelphia: Harper & Row; 1987:1–14.

123. Poggio GF, Gonzalez F, Krause F. Stereoscopic mechanisms in monkey visual cortex: binocular correlation and disparity selectivity. J Neurosci 1988; 8:4531–4550.

124. Pope DR, Edwards M, Schor CM. Extraction of depth from opposite-contrast stimuli. Transient system can, sustained system can't. Vision Res 1999; 39:4010–4017.

125. Regan D. Depth from motion and motion-in-depth. In Regan D, ed. Vision and Visual Dysfunction, vol 9. Boca Raton: CRC Press, 1991:137–169.

126. Regan D. Binocular correlates of the direction of motion in depth. Vision Res 1993; 33:2359–2360.

127. Regan D, Beverly KI. Electrophysiological evidence for the existence of neurons sensitive to the direction of depth movement. Nature 1973; 246:504–506.

128. Richards W. Stereopsis and stereoblindness. Exp Brain Res 1970; 10:380–388.

129. Richards W. Anomalous stereoscopic depth perception. J Opt Soc Am 1971; 61:410–414.

130. Richards W, Regan D. A stereo field map with implications for disparity processing. Invest Ophthalmol 1973; 12:904–909.

131. Ridder WH, Smith EL, Manny RE, Harwerth RS, Kato K. Effects of interocular suppression on spectral sensitivity. Optom Vis Sci 1992; 69:171–256.

132. Rogers BJ, Bradshaw MF. Vertical disparities, differential perspective and binocular stereopsis. Nature 1993; 361:253–255.

133. Saye A, Frisby JP. The role of monocularly conspicuous features in facilitating stereopsis from random-dot stereograms. Perception 1975; 4:159–171.

134. Schmidt PP. Sensitivity of random-dot stereoacuity and Snellen acuity to optical blur. Optom Vis Sci 1994; 71:466–471.

135. Schor CM. Visual stimuli for strabismic suppression. Perception 1977; 6:583–593.

136. Schor CM. Zero retinal image disparity: a stimulus for suppression in small angle strabismus. Doc Ophthalmol 1978; 46:149–160.

137. Schor CM. Fixation disparity and vergence adaptation. In: Schor CM, Ciuffreda KJ, eds. Vergence Eye Movements. Basic and Clinical Aspects. Boston: Butterworths, 1983:465–516.

138. Schor CM. Binocular sensory disorders. In: Regan D, ed. Vision and Visual Dysfunction, vol 9. Boca Raton: CRC Press, 1991:179–223.

139. Schor CM, Bridgeman B, Tyler CW. Spatial characteristics of static and dynamic stereoacuity in strabismus. Invest Ophthalmol Vis Sci 1983; 24:1572–1579.

140. Schor CM, Edwards M, Pope D. Spatial-frequency tuning of the transient-stereopsis system. Vision Res 1998; 38:3057–3068.

141. Schor CM, Gleason G, Lunn R. Interactions between short-term vertical phoria adaptation and non-conjugate adaptation of vertical pursuits. Vision Res 1993; 33:55–63.

142. Schor CM, Heckman T. Interocular differences in contrast and spatial frequency. Effects on stereopsis and fusion. Vision Res 1989; 29:837–847.

143. Schor CM, Heckman T, Tyler C W. Binocular fusion limits are independent of contrast, luminance gradient and component phases. Vision Res 1989; 29:821–835.

144. Schor CM, Landsman L, Erickson, P. Ocular dominance and interocular suppression of blur in monovision. Am J Optom Physiol Optics 1987; 64:723–730.

145. Schor CM, McCandless JW. An adaptable association between vertical and horizontal vergence. Vision Res 1995; 35:3519–3527.

146. Schor CM, Tyler CW. Spatio-temporal properties of Panum's fusional area. Vision Res 1981; 21:683–692.

147. Schor CM, Wood I. Disparity range for local stereopsis as a function of luminance spatial frequency. Vision Res 1983; 23:1649–1654.

148. Schor CM, Wood I C, Ogawa J. Spatial tuning of static and dynamic local stereopsis. Vision Res 1984; 24:573–578.

149. Schreiber KM, Hillis JM, Filippini HR, Schor CM, Banks MS. The surface of the empirical horopter. J Vis 2008; 8(3)7:1–20.

150. Sheedy JE, Fry GA. The perceived direction of the binocular image. Vision Res 1979; 19:201–211.

151. Shipley T, Rawlings SC. The nonius horopter – I. History and theory. Vision Res 1970; 10:1255–1262.

152. Shortess GK, Krauskopf, J. Role of involuntary eye movements in stereoscopic acuity. J Opt Soc Am, 1961; 51:555–559.

153. Siderov J, Harwerth RS, Bedell HE Stereopsis, cyclovergence and the backwards tilt of the vertical horopter. Vision Res 1999; 39:1347–1357.

154. Simmons DR, Kingdom FAA. Contrast thresholds for stereoscopic depth identification with isoluminant and isochromatic stimuli. Vision Res 1994; 34:2971–2982.

155. Simpson TL, Barbeito R, Bedell HE. The effect of optical blur on visual acuity for targets of different luminances. Ophthalmic Physiol Opt 1986; 6:279–281.

156. Smallman HS, MacLeod DIA. Size-disparity correlation in stereopsis at contrast threshold. J Opt Soc Am A 1994; 11:2169–2183.

157. Smith EL, Chino YM, Ni J, Ridder WH, Crawford MLJ. Binocular spatial phase tuning characteristics of neurons in the macaque striate cortex. Neurophysiol 1997; 78:351–365.

158. Smith EL, Fern K, Manny RE, Harwerth RS Interocular suppression produced by rivalry stimuli. A comparison of normal and abnormal binocular vision. Optom Vis Sci 1994; 71:479–491.

159. Smith EL, Levi DM, Harwerth RS, White JM. Color vision is altered during binocular rivalry. Science 1982; 218:802–804.

160. Smith EL, Levi DM, Manny RE, Harwerth RS, White JM. The relationship between binocular rivalry and strabismic suppression. Invest Ophthalmol Vis Sci 1985; 26:80–87.

161. Snyder C. Monovision: a clinical review. Spectrum 1989; 4:30–36.

162. Stephens GL, Polasky M. New options for aniseikonia correction: the use of high index materials. Optom Vis Sci 1991; 68:899–906.

163. Stevenson SB, Cormack LK. A contrast paradox in stereopsis, motion detection, and vernier acuity. Vision Res 2000; 40:2881–2884.

164. Timney B, Wilcox LM, St John R. On the evidence for a 'prue' binocular process in human vision. Spatial Vis 1989; 4:1–15.

165. Tyler CW. Sensory processing of binocular disparity. In Schor CM, Ciuffreda KJ, eds. Vergence Eye Movements. Basic and Clinical Aspects. Boston: Butterworths, 1983:199–295.

166. Tychsen L. Causing and curing infantile esotropia in primates: the role of decorrelated binocular input. Trans Am Ophthalmol Soc 2007; 105:564–593.

167. van Ee R, Schor CM. Unconstrained stereoscopic matching of lines. Vision Res 2000; 40:151–162.

168. Verhoeff FH. Anomalous projection and other visual phenomena associated with strabismus. Arch Ophthalmol 1938; 19:663–699.

169. von Helmholtz H. Treatise on physiological optics, vol III (Translated from the 3rd German edition and edited by JPC Southhall). New York: Dover, 1962.

170. von Noorden GK. Infantile esotropia. A continuing riddle. Am Orthop J 1984; 34:52–62.

171. von Noorden GK. Amblyopia: a multidisciplinary approach. Proctor lecture. Invest Ophthalmol Vis Sci 1985; 26:1704–1716.

172. von Noorden GK. Binocular Vision and Ocular Motility. St Louis: CV Mosby, 1990.

173. Wensveen JM, Harwerth RS, Smith EL. Clinical suppression in monkeys reared with abnormal visual experience. Vision Res 2001; 41:1593–1609.

174. Wensveen JM, Harwerth RS, Smith EL. Binocular vision deficits associated with early anisometropia: behavioral observations. J Neurophysiol 2003; 90:3001–3011.

175. Westheimer G. Visual hyperacuity. Prog Sensory Physiol 1981; 1:1–30.

176. Westheimer G. The spatial sense of the eye. Invest Ophthalmol Vis Sci 1979; 18:893–912.

177. Westheimer G. The Ferrier Lecture, 1992. Seeing depth with two eyes: stereopsis. Proc R Soc Lond B Biol Sci 1994; 257:205–214.

178. Westheimer G, McKee SP. Stereoscopic acuity with defocused and spatially filtered images. J Opt Soc Am 1980; 70:772–778.

179. Westheimer G, Pettet M W. Detection and processing of vertical disparity by the human observer. Proc R Soc Lond B Biol Sci 1992; 250:243–247.

180. Wheatstone C. Contributions to the physiology of vision. I. On some remarkable, and hitherto unobserved, phenomena of binocular vision. Phil Trans R Soc Lond 1938; 128:371–394.

181. Wilcox LM, Hess RF. Dmax for stereopsis depends on size, not spatial frequency content. Vision Res 1995; 35:1061–1069.

182. Wolfe JM. Afterimages, binocular rivalry, and the temporal properties of dominance and suppression. Perception 1983; 12:439–445.

183. Wolfe JM. Stereopsis and binocular rivalry. Psychol Rev 1986; 93:269–282.

184. Wong AM, Lueder GT, Burkhalter A, Tychsen L. Anomalous retinal correspondence: neuroanatomic mechanism in strabismic monkeys and clinical findings in strabismic children. J AAPOS 2000; 4:168–174.

185. Yang Y, Blake R. Spatial frequency tuning of human stereopsis. Vision Res 1991; 31:1177–1189.

CHAPTER 37

Temporal Properties of Vision

Allison M. McKendrick & Chris A. Johnson

Our visual perception arises from the interpretation of light information, which varies in space, wavelength, and time. It is the latter of these attributes that is explored in this chapter. Subjectively, the world appears to be stable despite continuous changes in the visual scene. How does the visual system respond to and interpret light variations that occur as a function of time?

The duration of a light affects both the ease with which we can see it, as well as its subjective appearance. This chapter emphasizes the first of these issues, that is, how sensitive we are to temporal variations in light and which factors influence our sensitivity. Temporal sensitivity cannot be studied in isolation because other stimulus attributes, such as spatial properties, chromaticity, background features and surround characteristics, all influence our ability to detect temporal variations. In the natural world, most temporal variation occurs through image motion. This may arise from motion of the observer, the eyes, or the object itself. Motion is a special form of temporal variation in which the change with time is associated with a change in spatial position.

This chapter summarizes a number of basic phenomena that describe the sensitivity of our visual system to temporal information. The application of these phenomena to the clinical study of abnormalities of visual processing, such as in disease, is also discussed.

Temporal summation and the critical duration

To detect the presence of something in the visual world, it must be present for a finite period of time. Although a single quantum of light may be sufficient to generate a neural response, multiple quanta are generally required during a short period before the light is reliably seen, a property known as *temporal summation*. In the human visual system, temporal summation occurs for durations of approximately 40 to 100 milliseconds, depending on the spatial and temporal properties of the object and its background, the adaptation level, and the eccentricity of the stimulus.[1-5] The maximum time over which temporal summation can occur is the *critical duration*.

Let's say we wish to determine how long a light needs to be presented on a dark background to be visible. In general

terms, a more intense light does not need to be presented for as long as a less intense one to reach threshold visibility. The relationship between the luminance of the light and the duration of its exposure to reach visibility is linear over a limited range. Provided that the light pulse is shorter than the critical duration, it will be at threshold when the product of its duration and its intensity equals a constant. The formula that describes this time–intensity reciprocity is *Bloch's law*:[6]

$$Bt = K$$

where: B = Luminance of the light, t = Duration, K = A constant value.

Bloch's law is shown schematically in Figure 37.1A. When stimulus intensity and duration are plotted on log-log coordinates, as in Figure 37.1A, Bloch's law describes a line with a slope of −1. When the critical duration is reached, the threshold intensity versus duration function is described by a horizontal line; that is, a constant intensity is required to reach threshold. Bloch's law has been shown to be generally valid for a wide range of stimulus and background conditions, including both foveal and peripheral viewing. Once the duration of the stimulus exceeds the critical duration, the luminance required for it to reach visibility is classically considered to be constant.

The preceding discussion assumes that whenever the observer's threshold is exceeded, he or she will respond accurately to the stimulus. This predicts an abrupt and idealized transition between the two curves, as depicted in Figure 37.1A. In the real world, both visual stimuli and the physiologic mechanisms that we use to detect them are subject to random fluctuations in response. We may consider the length of the stimulus presentation to be divided into a number of discrete time intervals. The signal is detected when the response exceeds threshold in at least one interval and the probability of detection in each interval is considered independent. This description of the probabilistic nature of visual detection is known as *probability summation over time*.[7] The concept of probability summation is included in many models of temporal visual processing and is thought to be at least partly responsible for the less-than-abrupt transition between the region of temporal summation and constant intensity that occurs under some experimental conditions.[3] One experimental situation in which this arises is

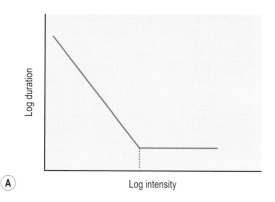

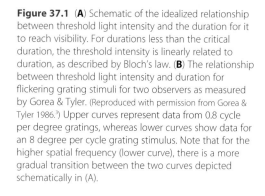

Figure 37.1 (**A**) Schematic of the idealized relationship between threshold light intensity and the duration for it to reach visibility. For durations less than the critical duration, the threshold intensity is linearly related to duration, as described by Bloch's law. (**B**) The relationship between threshold light intensity and duration for flickering grating stimuli for two observers as measured by Gorea & Tyler. (Reproduced with permission from Gorea & Tyler 1986.[3]) Upper curves represent data from 0.8 cycle per degree gratings, whereas lower curves show data for an 8 degree per cycle grating stimulus. Note that for the higher spatial frequency (lower curve), there is a more gradual transition between the two curves depicted schematically in (A).

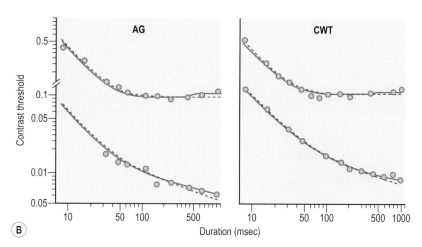

when threshold contrast is measured as a function of signal duration for sinusoidal grating stimuli. This is illustrated in Figure 37.1B.[3] In Figure 37.1B the upper curve shows threshold duration data for a 0.8 cycle per degree grating, and the lower curve shows those for an 8 cycle per degree grating. The upper curve conforms well to the schematic diagram that is illustrated in Figure 37.1A. For the lower curve depicting results for the 8 cycle per degree grating, however, a gradual transition is observed. In this latter case the actual critical duration (traditionally the point of intercept of the two slopes in Fig. 37.1A) is somewhat difficult to define. Data similar to those shown in Figure 37.1B, have been interpreted to indicate that the critical duration increases with increasing spatial frequency.[1,4] Gorea & Tyler[3] use an alternative form of analysis that includes the effects of probability summation; they conclude that the critical duration is minimally affected by spatial frequency.

Factors affecting the critical duration

The time interval defined by the critical duration depends on properties of both the stimulus and the background. The critical duration has been shown to vary with light adaptation level; that is, with brighter backgrounds, the critical duration decreases.[8–12] Conversely, with dark adaptation, the critical duration increases.[13,14] Unless dark adapted, the size of the stimulus also affects the critical duration, with larger stimuli having a decreased critical duration.[9,12,15] Retinal eccentricity also influences the critical duration, as does the visual task.[16,17] Temporal summation is also affected by the

spectral composition (wavelength or color) of the light stimuli, with isolated chromatic stimuli having longer temporal integration than achromatic (luminance) stimuli.[18–20] For colored lights, the critical duration decreases with increased chromatic saturation of the background, similar to the decrease in critical duration with increased luminance for achromatic stimuli.[21] Figure 37.2 demonstrates how chromaticity and retinal eccentricity markedly alter the sensitivity to temporal pulses and the critical duration.[20]

Temporal sensitivity to periodic stimuli

The preceding section has considered how the human visual system responds to aperiodic stimuli (for example, a single pulse of light). We will now consider how the visual system responds to periodic stimuli (repeatedly flickering stimuli). Most research in this area has been directed to explore the following questions: (1) what is the fastest flicker rate that can be detected by the human visual system (the critical flicker fusion frequency); and (2) what factors influence sensitivity to flicker slower than this critical rate?

Critical flicker fusion frequency

When a light is turned on and off repeatedly in rapid succession, the light appears to flicker, provided the on and off intervals are greater than some finite time interval. If the lights are flickered fast enough, we perceive the flashes as a

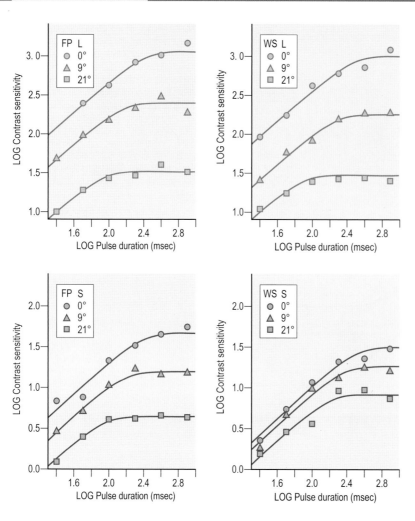

Figure 37.2 Log contrast sensitivity versus pulse duration for equiluminant chromatic pulses. Data are shown for two observers (left and right panels). The upper panels show performance for "red-green" modulation (long-wavelength sensitive [L] and middle-wavelength sensitive [M] cone modulation); whereas the lower panels show data for short wavelength modulation. (Reproduced with permission from Swanson et al 2008.[20])

single fused light rather than a series of flashes. In simple terms, when the perception of fusion occurs, we have reached the limit of the temporal-resolving ability of our visual system. The transition from the perception of flicker to that of fusion occurs over a range of temporal frequencies; the boundary between the two is called the *critical flicker fusion (CFF) frequency.* The value of the CFF varies, depending on a large number of both stimulus and observer characteristics. Some of the important factors that influence the CFF (Box 37.1) are discussed in the following sections.

Effect of stimulus luminance on CFF

In general, the CFF increases as the luminance of the flashing stimulus increases. This relationship is known as the *Ferry–Porter law,* which states that CFF increases as a linear function of log luminance.[22,23] The Ferry–Porter law is valid for a wide range of stimulus conditions and is illustrated in Figure 37.3.[24] The lower curves (*solid lines* and *symbols*) show data collected in the fovea, and the upper curves show data collected at 35 degrees eccentricity. For both locations the upper curves are for smaller targets (0.05 degree foveally and 0.5 degree eccentrically), and the lower curves are for larger targets (0.5 degree foveally and 5.7 degrees eccentrically). Figure 37.3 demonstrates several interesting observations about the relationship between CFF and luminance. First, the Ferry–Porter law is upheld despite changes in stimulus size. Second, the linear relationship between log luminance

> ### Box 37.1 Critical flicker frequency (CFF)
>
> - The critical flicker frequency (CFF) describes the fastest rate that a stimulus can flicker and just be perceived as a flickering rather than stable.
> - While the CFF is dependent on the temporal resolution of visual neurons, it is also considered to be a measure of conscious visual awareness because, at CFF threshold, an identical flickering stimulus varies in percept from flickering to stable. Functional magnetic resonance imaging demonstrates involvement of the frontal and parietal cortex in the conscious perception of flicker.[173]
> - CFF has been utilized as a measure of conscious visual awareness in a wide range of pharmacological and psychological research studies.
> - Perimetric tests of CFF have also been developed with most research directed towards visual field assessment in glaucoma.

and CFF is present for both central and peripheral viewing, although the slope of this relationship increases in the periphery, implying faster processing.[24] The Ferry–Porter law holds not only for spot targets but also for grating stimuli.[24] For scotopic luminance levels, at which rods mediate detection, CFF decreases substantially to approximately 20 Hz and no longer obeys the Ferry–Porter law.[23,25,26]

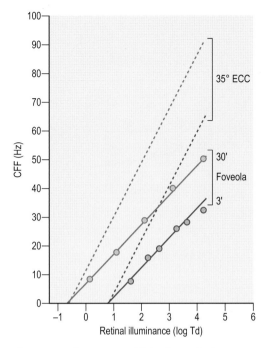

Figure 37.3 Photopic critical flicker fusion (CFF) versus intensity functions measured in the fovea (solid symbols and lines) and at 35 degrees eccentricity (dotted and dashed lines) from Tyler & Hamer. (Reproduced with permission from Tyler & Hamer 1990.[24]). The test stimulus was 660 nm presented on a equiluminant white surround. Foveal test stimuli were 0.5 degree (large filled circles) and 0.05 degree (small filled circles). Eccentric stimuli were 5.7 degrees (dashed lines) and 0.5 degree (dotted lines). Note the adherence to the Ferry–Porter law for all eccentricities tested. ECC, Eccentricity.

Effect of stimulus chromaticity on CFF

The linear relationship between CFF and log luminance, as described by the Ferry–Porter law, is also valid for purely chromatic stimuli.[27–30] However, the slope of the relationship has been shown to vary with stimulus wavelength.[28] This relationship is demonstrated in Figure 37.4, which shows CFF versus illuminance functions derived from four separate studies.[27–30] In all four studies the foveal CFF illuminance functions are well fit by Ferry–Porter lines, and in all cases the functions for green (middle wavelength) lights were found to be steeper than those for red (long wavelength) lights. The steeper slope for green stimuli has been interpreted as evidence supporting the green cone pathways being inherently faster than the red cone pathways for the transmission of information near the CFF.[28] The CFF is lowest for blue stimuli detected by the short-wavelength pathways.[31,32]

Effect of eccentricity on CFF

The CFF varies as a function of eccentricity in the visual field. If the stimulus size and luminance are kept constant, the CFF increases with eccentricity over the central 50 degrees or so of the visual field and then decreases with further increases in eccentricity.[33,34] This is illustrated in Figure 37.5, which plots the CFF as a function of eccentricity in the temporal visual field.

Effect of stimulus size on CFF: the Granit–Harper law

As shown in Figure 37.3, the CFF increases with stimulus size. For a wide range of luminances, CFF increases linearly with the logarithm of the stimulus area. This relationship is known as the *Granit–Harper law*, named after the

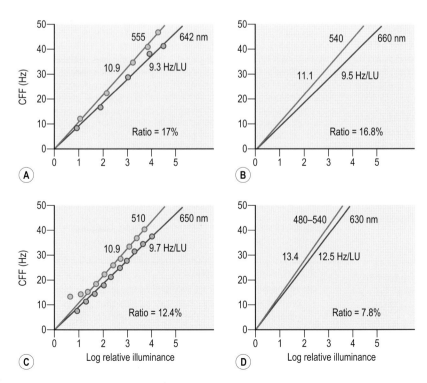

Figure 37.4 Critical flicker fusion (CFF) versus illuminance functions for red and green flicker measured by Hamer & Tyler. (Reproduced with permission from Hamer & Tyler 1992.[28]). Note that all data are well fit by the Ferry–Porter law, with green lights having a steeper slope than red. (**A**) Data for one subject from Hamer & Tyler.[28] The green light function is steeper by 17 percent. (**B**) CFF data from Ives[29] for red (650 nm) and green (510 nm) flicker stimuli viewed foveally. The green light function is steeper by 12.4 percent. (**C**) CFF data for red (660 nm) and green (540 nm) flickering stimuli for a single observer tested by Pokorny & Smith.[30] The green light function is steeper by 16.8 percent. (**D**) CFF data for green light (mean of 480 and 540 nm data) and red light (630 nm) flicker from Giorgi.[27] The green light function is steeper than the red by 7.8 percent.

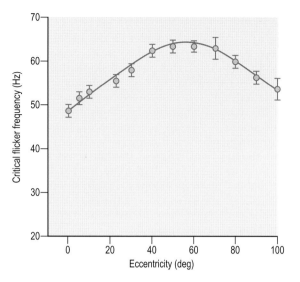

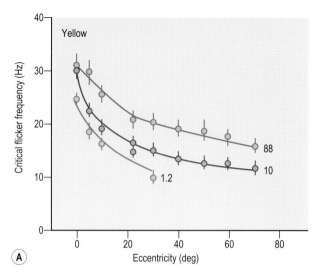

Figure 37.5 Critical flicker frequency (mean ± standard deviation) as a function of eccentricity in the temporal visual field of the right eye of one observer. (Reproduced with permission from Rovamo & Raninen 1984.[34]) Stimulus area was 88.4 degrees[2], pupil diameter was 8 mm, and retinal illuminance was 2510 photopic trolands.

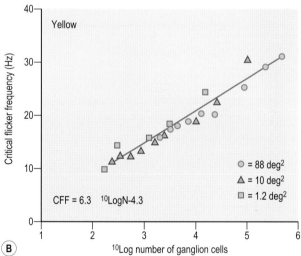

Figure 37.6 (**A**) Critical flicker frequency (mean ± standard deviation) for targets viewed eccentrically for one observer reported by Rovamo & Raninen. (Reproduced with permission from Rovamo & Raninen 1988.[38]) Similar target sizes were used at all eccentricities, but retinal illuminance was F scaled by reducing the average stimulus luminance in inverse proportion to photopic Ricco's area.[144] The numbers on the right of the curves refer to the stimulus area in degrees[2]. When eccentricity increased from 0 to 70 degrees, the stimulus luminance nominally decreased from 50 to 0.80 photopic cd/m[2]. Note that critical flicker frequency (CFF) decreases with increasing eccentricity irrespective of stimulus area. (**B**) CFF data from (A) plotted as a function of the number of retinal ganglion cells stimulated.

investigators who first reported it.[35] The Granit–Harper law holds for a wide range of luminances, retinal eccentricities out to 10 degrees, and stimuli as large as almost 50 degrees. However, subsequent investigators have determined that it is not the overall area of the stimulus that is critical, but rather the local retinal area with the best temporal resolution. This was demonstrated by Roehrig,[36,37] who measured the same value for the CFF for a complete 49.6-degree field in comparison to an annulus of the same diameter, with its central 66 percent not illuminated. Because the midperiphery has better temporal resolution than central vision, an eccentric annulus produced the same CFF as the full 49.6-degree stimulus. The Granit–Harper law does not hold under dim light conditions in which rods mediate performance.

Beyond the fovea, a revised form of the Granit–Harper law is required to fit the changes in CFF with stimulus area. Rovamo & Raninen[38] have shown that the Granit–Harper law can be generalized across the visual field by replacing the retinal stimulus area with the number of ganglion cells stimulated. In this more general case, the CFF increases linearly with the logarithm of the number of ganglion cells stimulated. This is illustrated in Figure 37.6. Figure 37.6A plots CFF against eccentricity for three different stimulus areas. The CFF decreases with increasing eccentricity irrespective of the stimulus area. Figure 37.6B plots the same CFF data as a function of the number of retinal ganglion cells stimulated at each eccentricity and results in a linear relationship between these two parameters.

Temporal contrast sensitivity

The CFF defines an upper limit for temporal sensitivity (Box 37.2), beyond which we can no longer detect that a light is flickering. How sensitive are we to flicker below the CFF, and how does the visual system respond to more complex temporal variations of light than simple flashes or trains of flashes? Classical work on human temporal contrast sensitivity was performed by De Lange in the 1950s.[39–41] De Lange

---

Box 37.2 Temporal contrast sensitivity

There are several versions of clinical perimetry that measure variants of temporal contrast sensitivity. These include:

- flicker on a pedestal temporal modulation perimetry, such as in the Medmont perimeter
- mean modulated contrast sensitivity perimetry
  - both of these display small spot targets that vary in contrast either about the mean luminance of the perimeter background or above and below a luminance pedestal
- spatio-temporal contrast sensitivity is measured in Frequency Doubling Technology Perimetry.

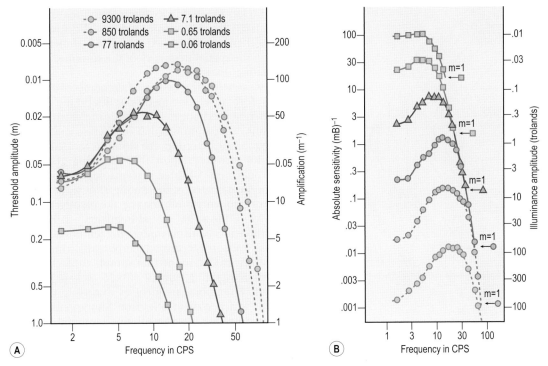

Figure 37.7 Temporal sensitivity data collected by Kelly. (Reproduced with permission from Kelly 1961.[42]) Visual attenuation characteristics, as measured with sinusoidal variation about different levels of retinal illuminance, are shown. (**A**) Threshold modulation ratios for just detectable flicker plotted as a function of temporal frequency. (**B**) Absolute amplitudes of retinal illuminance variation for just noticeable flicker plotted as a function of temporal frequency. Absolute sensitivity is the reciprocal of threshold absolute amplitude at each frequency.

evaluated temporal sensitivity to flicker using mathematical analysis of temporal waveforms and linear filter theory.

Figure 37.7 shows the results from flicker-sensitivity experiments by Kelly.[42] The vertical axis plots flicker sensitivity, which is the reciprocal of the flicker threshold, measured as a "threshold modulation ratio," which is the extent that the sinusoidally modulated light deviates from its average direct current (DC) component. The modulation ratio is calculated as a fraction of its DC value because the amplitude cannot be greater than the DC component (negative values of luminance are not physically possible). The modulation ratio expresses the amplitude as a percentage deviation of the stimulus from its average value.

The left panel of Figure 37.7 plots the modulation ratio as a function of flicker frequency. A series of curves are shown, each obtained at a different average retinal illumination or adaptation level. The curves define the flicker detection boundary for the particular level of adaptation. For a given level of adaptation, any combination of frequency and modulation amplitude below the curve is seen as flickering, whereas any combination above the curve is perceived as a steady light. The point at which the curve intersects the abscissa (x-axis) corresponds to the CFF. Note that at low adaptation levels, the shape of the curve is low-pass, meaning that modulation sensitivity is similar for low temporal frequencies and then falls off systematically for higher temporal frequencies. At high adaptation levels, the shape of the curve is band-pass, meaning that modulation sensitivity is greatest for a band of intermediate temporal frequencies and systematically falls off for lower and higher temporal frequencies.

With increasing retinal illuminance, more conditions are seen as flickering, consistent with the simpler Ferry–Porter law. At low frequencies the curves are similar, indicating that low-frequency flicker reaches threshold at a similar value of modulation ratio for all levels of photopic adaptation. This is not the case for high frequencies, at which the threshold is determined by both adaptation level and frequency. At higher luminance levels the sensitivity for flicker detection peaks at frequencies of approximately 15 to 20 Hz. This is similar to the Brücke brightness enhancement effect, which is discussed later.

The right panel of Figure 37.7 replots the data of Figure 37.7A as a function of the absolute amplitude of the high-frequency flicker. This has the effect of reversing the curves so that the amplitude sensitivity is greatest at the lowest adaptation level. It can be seen that the curves of Figure 37.7B approach a common asymptote at high frequencies. This convergence of curves measured at many different luminance levels implies that at high temporal frequencies, sensitivity is predicted by the absolute amplitude of the signal, independent of adaptation level. This behavior is not apparent at the low temporal frequency range.

Chromatic temporal sensitivity

Chromatic temporal contrast sensitivity measured with sinusoidal grating stimuli also varies with retinal illuminance,[43] however there are two key differences when compared to performance measured with achromatic stimuli. Firstly, chromatic functions are generally low-pass, that is, sensitivity is similar for a range of low temporal frequencies before falling off at higher temporal frequencies. Secondly, the

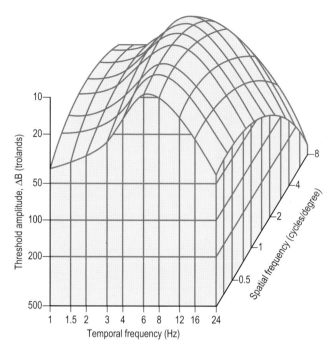

Figure 37.8 Human spatiotemporal amplitude threshold surface obtained with circular gratings of 16 degrees viewed binocularly with natural pupils. (Reproduced with permission from Kelly 1972.[46])

high-frequency cut-off is at lower temporal frequencies, that is, a lower CFF is found for chromatic in comparison to achromatic stimuli for a given retinal illuminance.[43]

Spatial effects on temporal sensitivity

The work of De Lange[40] and Kelly[42] describes our temporal contrast sensitivity characteristics, that is, how sensitive we are to temporal sinusoidal variations in stimulus contrast. This does not consider the effect that the spatial characteristics of the light source have on sensitivity. Human contrast sensitivity depends on both the spatial and temporal properties of the stimulus.[44,45] Figure 37.8 shows a surface plot of the human *spatiotemporal contrast sensitivity* function, as derived by Kelly, from a large number of psychophysical measurements.[44,46,47] One axis of the graph shows the temporal frequency of the stimulus; the other shows the spatial frequency. The height represents the observer's contrast sensitivity for the particular spatial and temporal conditions.[44,47,48] Paths running through the curve parallel to the temporal frequency axis represent the temporal contrast sensitivity functions, and those running parallel to the spatial frequency axis represent the spatial contrast sensitivity functions. The temporal contrast sensitivity function is band-pass at low spatial frequencies, becoming low-pass at high spatial frequencies.

Mechanisms underlying temporal sensitivity

The shape of the human spatial contrast sensitivity function has been explained in terms of a multiple-channel model in which the total curve is the window of a number of channels, each tuned to a different peak spatial frequency.[49–51] Similar to spatial processing, there is evidence for a discrete number of channels for temporal processing, each tuned to a different peak temporal frequency.[52–55] There appear to be fewer

independent temporal frequency filters than there are filters conveying spatial frequency information. The number of temporal mechanisms identified has been shown to be dependent on the spatial frequency of the stimulus, in addition to the retinal location.

How can we determine whether different mechanisms are governing performance? One method relies on the assumption that different mechanisms result in different subjective representations of the stimulus; that is, when separate mechanisms are governing detection, the stimulus looks different. Kelly[44] and Kulikowski[56] report that flickering gratings at threshold produce one of three percepts depending on the spatiotemporal parameters of the grating: slow flicker, apparent motion, or frequency doubling (i.e. the perceived spatial frequency is twice the actual spatial frequency). This is consistent with three mechanisms underlying temporal processing. Watson & Robson[54] measured temporal discrimination, that is, the ability to determine that stimuli are flickering at different rates. For grating patches of 0.25 cycle per degree, there were three temporal frequencies that were uniquely discriminable, which Watson & Robson[54] interpreted as evidence for three unique filters mediating temporal processing. Mandler & Makous[53] modeled temporal discrimination performance and also found evidence for three mechanisms underlying temporal processing.

An alternative method for identifying mechanisms underlying visual processing is that of *selective adaptation*. If a stimulus that activates one mechanism to a greater extent than another mechanism is presented, the sensitivity of the activated mechanism will be selectively depressed if the stimulus is viewed for a prolonged period. This should allow other, less sensitive mechanisms to detect the stimulus and hence allow their profile to be explored. Selective adaptation has been widely used to isolate the mechanisms underlying color vision processing.[57] Hess & Snowden[52] used selective adaptation with temporal stimuli to uncover temporal mechanisms; an example of their data is displayed in Figure 37.9. The first stage of these experiments (not shown in the figure) involved measuring foveal contrast detection thresholds for a wide range of spatial and temporal conditions. These measures were used to set the contrast of the *probe* stimulus, which was set to be just detectable (4 dB above threshold). The detectability of this probe stimulus was then measured in the presence of masking stimuli of the same spatial frequency but different temporal frequency. The contrast of the masking stimulus was varied until the probe was just visible.

Figure 37.9A presents data for three probes presented foveally with no spatial content (0 cycles per degree) but flickering at either 1, 8, or 32 Hz. The figure shows the contrast sensitivity of the mask plotted against the mask temporal frequency. Three mechanisms are revealed: one low-pass (*red circles*) and two band-pass (*blue circles, green triangles*). Figure 37.9B shows data for a probe of 3 cycles per degree, and only two mechanisms are revealed. The band-pass mechanism, centered on higher temporal frequencies in Figure 37.9A disappears when the spatial frequency content of the stimulus increases. Using a similar method, Snowden & Hess[58] have shown that for retinal eccentricities of 10 degrees, only two mechanisms are found, reducing to a single mechanism at eccentricities of greater than 30 degrees.

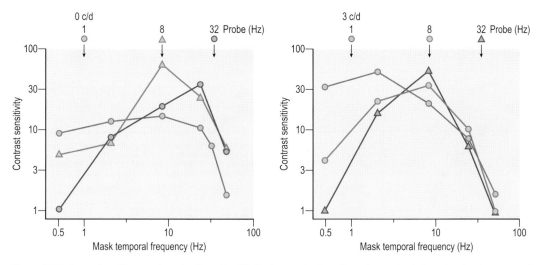

Figure 37.9 Properties of temporal mechanisms identified using a selective adaptation method. (Reproduced with permission from Hess & Snowden 1992.[52]) Mask contrast sensitivity is plotted against mask temporal frequency for detection of a just suprathreshold probe. Data are for probes of 1, 8, and 32 Hz. The left panel shows data for a probe of 0 cycles per degree, for which three mechanisms are revealed: one low-pass and two band-pass. The right panel shows data for a 3 cycle per degree probe, and only two mechanisms are revealed.

More recent evidence suggests that these channels are not as independent as initially thought. Using a masking paradigm, Cass & Alais[59] demonstrated two channels underlying human temporal vision. High-frequency masks were able to suppress low-frequency targets suggesting interaction between temporal channels. The high-frequency channel was orientation invariant, which suggests a precortical origin (as significant orientation selectively of visual neurons is not apparent until V1). In contrast, the low-frequency filter demonstrated orientation dependence, suggesting a cortical origin.

So what is the neural basis of these psychophysically measured temporal channels? Visual information from the retina is carried to the visual cortex, via the lateral geniculate nucleus, by several major neural pathways (magnocellular, parvocellular, and koniocellular). These pathways have been shown to carry largely independent, but sometimes overlapping, visual information. Magnocellular neurons are capable of processing achromatic fast flicker.[60,61] Parvocellular retinal ganglion cells are well established to be the physiological substrate for red–green chromatic modulation.[61] It is important to note, however, that retinal ganglion cell responses persist at considerably higher temporal frequencies than the human psychophysical CFF.[62] This is particularly the case for chromatic modulation where the human psychophysical function limit is approximately 10–15 Hz, yet parvocellular retinal ganglion cells will respond up to 30–40 Hz. These differences are explained by models where responses from multiple single retinal cells converge on cortical detection mechanisms. The exact manner in which this convergence occurs and how the site of convergence is represented in the visual cortex is an active area of current research.[20,63,64]

Surround effects on temporal sensitivity

Our sensitivity to temporal variations depends not only on the properties of the flickering light but also on those of the background surrounding the light. In the dark (scotopic conditions), detection of light increments is mediated by rods, and in the light (photopic conditions), detection is mediated by cones. For flicker sensitivity, it has been shown that when photopic flickering lights are presented on dark backgrounds, interactions between rods and cones (rod–cone interactions) act to decrease sensitivity.[65-70] The effects of rod–cone interactions on flicker sensitivity are most significant at high temporal frequencies (Fig. 37.10) and in the retinal periphery. Suppressive effects between cone mechanisms of flicker thresholds (e.g. long-wavelength-sensitive cones and medium-wavelength-sensitive cones) have also been demonstrated.[71,72]

Within the photopic range, luminance differences between a flickering target and its surround are particularly important in determining temporal contrast sensitivity for low temporal frequencies.[73-76] This is illustrated in Figure 37.11. The *crosses* represent the temporal contrast sensitivity for a surround equal to the average luminance of the flickering light (25 cd/m²), and sensitivity at low temporal frequencies is constant under this condition. For either a dark surround (0 cd/m², *squares*) or a light surround (50 cd/m², *circles*), the sensitivity drops off at low temporal frequencies.

The presence of a surround differing in luminance from that of the flickering target creates a contrast border at the edge of the target. Spehar & Zaidi[76] have demonstrated that the presence of such a border decreases temporal contrast sensitivity. Whether the background is brighter or dimmer is not important; rather the decrease in temporal contrast sensitivity depends on the magnitude of the contrast at the border, with larger border contrasts resulting in larger decreases in temporal contrast sensitivity at low temporal frequencies. In short, we are most sensitive to temporal changes when the luminance of the background is matched to the average of the flickering target. One experimental paradigm in which such surround effects become important is the case of luminance-pedestal flicker, which is discussed in the following section.

Differences between mean-modulated and luminance-pedestal flicker

Historically, studies of temporal sensitivity have generated flicker using one of two methodologies, which are illustrated

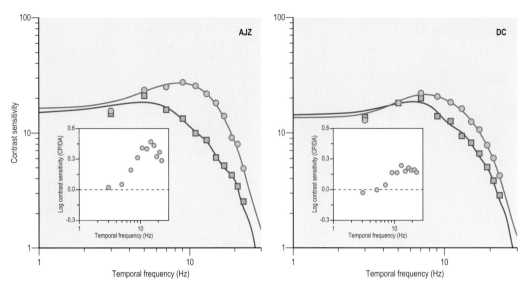

Figure 37.10 The effect of rod–cone interactions on temporal contrast sensitivity. Temporal contrast sensitivity to a 80 photopic troland sinusoidal stimulus measured after (1) 30 minutes of dark adaptation (closed squares) where the rods are fully sensitive; and (2) during the first 5 minutes following the termination of a 2-minute exposure to a 10,000 troland broad band light (partial rod bleach). After dark adaptation, temporal contrast sensitivity is attenuated for temporal frequencies above 6–8 Hz. Panel insets show the difference in log contrast sensitivity between the two conditions for each observer. (Reproduced with permission from Zele et al 2008.[65])

schematically in Figure 37.12.[77] The upper panel of Figure 37.12 illustrates mean-modulated flicker, in which flicker is modulated about a mean luminance. Flicker generated in this fashion results in no change in the time-averaged luminance (examples of studies using mean-modulated flicker include those by De Lange[40] and Roufs[75]). The lower panel of Figure 37.12 illustrates luminance-pedestal flicker, in which flicker is generated by modulating a luminance increment over time. This results in both a flickering component and an increase in time-averaged luminance above the background level (examples of studies using luminance-pedestal flicker include those of Alexander & Fishman;[69] Eisner;[78] Eisner, Shapiro & Middleton;[79] and Anderson & Vingrys[77]). Because both of these methods have been used to assess temporal sensitivity, it is important to understand whether they yield equivalent or differing results.

Luminance-pedestal flicker yields equivalent results to mean-modulated flicker only if the luminance pedestal has no effect on thresholds, which seems unlikely. In the previous section, we discussed how flicker sensitivity changes with light adaptation (see Fig. 37.10); as such, it may be expected that the local increase in luminance created by the luminance pedestal may act to increase thresholds. This is the case and has been reported by Anderson & Vingrys.[77]

Furthermore, the presence of the pedestal causes a difference between the surround and the flickering target. The surround affects temporal sensitivity (see Fig. 37.11), so it is likely that the presence of a non-luminance-matched surround, created by the very nature of the luminance-pedestal flicker stimulus, will result in interactions that alter our sensitivity to this type of target.[76] It has been shown that both local adaptation mechanisms and surround interactions are important to explaining the differences in thresholds between mean-modulated and luminance-pedestal flicker thresholds.[80,81]

The effects of flicker on perception

The presence of flicker can alter both the apparent color and brightness of a light.[82–88] When a light is flickered, its apparent brightness varies depending on the frequency of the flicker, with maximum apparent brightness occurring between 5 and 20 Hz. This phenomenon is known as the *Brücke brightness enhancement effect*.[89] Such brightness enhancement can be demonstrated experimentally by matching the brightness of a flickering light to that of a non-flickering standard.[85] When the light is flickered at rates faster than those that result in brightness enhancement, the apparent brightness of the flickering light decreases. Eventually, the apparent brightness plateaus, and at rates beyond the CFF, the apparent brightness is the same as the apparent brightness of a steady light equal to the time-averaged luminance of the flickering light. This observation was made by Talbot[90] and Plateau[91] and is known as the *Talbot–Plateau law*. Lights flickering above the CFF, at which point flicker is invisible, appear identical to steady lights of the same chromaticity and time-averaged luminance. Under most conditions of adaptation, the Talbot–Plateau law appears to hold. One exception is the case of blue targets presented on bright yellow backgrounds, in which case the target appears more yellow than a steady light, even when it is not visibly flickering.[88]

Temporal phase segmentation

An important task of the visual system is to distinguish objects from their background, a task known as *image segmentation* (see Chapter 32). Useful information for the process of image segmentation includes differences in luminance and color between the image and its background. Differences in temporal phase are also sufficient for image segmentation.[92–95] For example, if a field of identical dot elements is rapidly flickered from black to white in phase

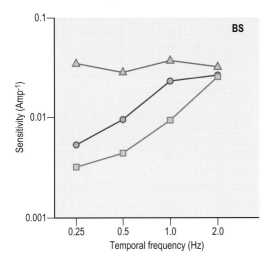

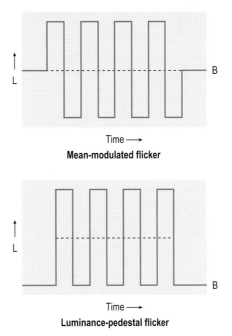

Time ⟶
Mean-modulated flicker

Time ⟶
Luminance-pedestal flicker

Figure 37.12 Schematic of mean-modulated (upper panel) and luminance-pedestal (lower panel) flicker from Anderson & Vingrys. (Reproduced with permissions from Anderson & Vingrys 2000.[77]) The time-averaged luminance during the presentation of flicker is represented by the dashed lines, and B represents the background luminance. Note that for luminance-pedestal flicker, the time-averaged luminance is greater than the background by an amount that is the luminance pedestal.

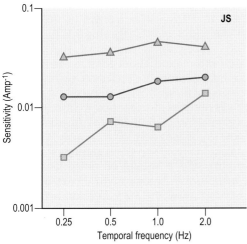

Figure 37.11 Surround effects on temporal sensitivity. (Reproduced with permission from Spehar & Zaidi 1977.[76]) Amplitude sensitivity is plotted as a function of temporal frequency for different surround conditions: squares, 0 cd/m² (dark surround); triangles, 25.0 cd/m² (equiluminant surround); and circles, 50 cd/m² (light surround).

(i.e. all dots appear white simultaneously, then black simultaneously), the field appears to be uniformly flickering. If, however, elements in a subregion flicker in counterphase to the rest (i.e. within the subregion the dots are white when those outside are black), a contour between the subregion and the rest of the display is observed. At high temporal frequencies (greater than approximately 10 Hz), the difference in phase of individual elements cannot be detected; however, the contour between regions is easily visualized provided the gap between regions is less than about 0.4 degree.[93,95] Temporal image segmentation is possible only at low temporal frequencies for chromatic stimuli.[95]

Clinical applications of temporal sensitivity measurements

Measurements of temporal processing have been used extensively for clinical testing. Both measures of CFF and temporal contrast sensitivity have been used to explore visual function in disease. Often, these measures are performed across the visual field, with use of perimetric strategies (see Chapter 35). There are a number of different perimetric techniques for exploring temporal sensitivity across the visual field: CFF perimetry (which measures the CFF for small spot targets at various field locations), temporal modulation perimetry (which measures temporal contrast sensitivity for small spot targets either about a mean luminance or displayed on a luminance pedestal), and frequency doubling perimetry. Frequency doubling perimetry measures contrast sensitivity for patches of sinusoidal grating of low spatial and high temporal frequency, parameters that result in the grating appearing to have twice its actual spatial frequency.

The development of perimetric methods for measuring temporal sensitivity has been largely directed to the detection of glaucoma or identification of glaucomatous visual field progression.[96–103] Abnormalities in temporal processing have also been identified in a range of other diseases, using perimetric and other strategies, including: Parkinson's disease, dyslexia, age-related macular degeneration, multiple sclerosis, optic neuritis, high-risk drusen, central serous chorioretinopathy, and migraine.[104–115]

Temporal segmentation stimuli have also been used in a clinical setting to assess the patency of the magnocellular system. Letters created from small rapidly flickering dots are presented, and the subject must then identify the letters. The dots within the letters flicker in opposite phase to the dots external to the letter. Temporal segmentation stimuli have been used to identify abnormalities in temporal sensitivity in dyslexia and glaucoma.[116–119] A new commercially available perimeter (the Heidelberg Edge Perimeter) utilizes the principle of temporal segmentation.

Motion processing

The detection of motion is an extension of temporal processing that forms a far-reaching separate area of study. *Motion* is a change in spatial position over time. Indeed, practically everything of visual interest in the world is moving, if not as a result of movement of the object itself, then by movement of the observer's head or eyes, causing the image on the retina to move. As we have already discussed, the ability of the visual system to detect temporal changes in luminance is exquisite. The visual system is also extremely sensitive to changes in spatial position. Hence, it may seem that to extract motion information, all that is needed is a comparison of separate measurements of an object's location at various times. Nevertheless, special visual mechanisms are specifically used for the processing of the linked spatiotemporal information that constitutes motion. This section briefly describes some perceptual and psychophysical data on human motion perception. A brief description of the neurophysiologic substrates underlying motion perception is included because this provides a basis for understanding the applications of motion tasks to the study of functional visual loss in disease. There are a number of detailed reviews of motion processing elsewhere (for example: Derrington et al,[120] Vaina & Soloviev[121]).

Psychophysical and perceptual evidence for unique motion processing

A number of perceptual phenomena provide evidence for the existence of mechanisms designed specifically to process motion information. One of these is known as the *motion aftereffect*, which occurs after prolonged viewing of a moving stimulus (this is also known as *motion adaptation*).[122–125] After prolonged staring at a moving stimulus, stationary objects appear to move in the opposite direction and the apparent speed of moving objects is distorted. Importantly, the apparent position of such stationary objects does not change, suggesting that motion and position are encoded separately.

Another example demonstrating that the perception of motion is not merely a simple representation of physical motion is that of *apparent motion*. Apparent motion occurs when spatially separate lights are flashed in sequence, giving the perception of movement despite each light remaining

stationary. This phenomenon is sometimes used in shop fronts, where the motion induced by sequential flashing of lights is designed to draw attention. Apparent motion can be induced by a single pair of lights flashed in succession. This movement is also known as *phi* and was first described in detail by Wertheimer.[126]

The neural encoding of motion

Further evidence for motion as a primary sensory dimension arises from the existence of distinct neuronal architecture capable of supporting the encoding of motion information. Hubel & Wiesel[127] observed that some neurons in cortical area V1 have directionally selective receptive fields. Cells with this property respond well when a stimulus moves in one direction, but they respond in a limited fashion, or not at all, when the stimulus moves in the opposite direction. Direction-selective neurons are not evenly distributed throughout area V1; they are predominantly located in layers 4A, 4B, 4Cα, and 6.[128] Layer 4Cα receives its main input from magnocellular pathways,[129] suggesting that the magnocellular pathway is part of a functional specialization designed to carry information about motion. As discussed earlier, magnocellular neurons are preferentially involved in the visual processing of flicker.[60,130–132] Signals from the magnocellular pathway are passed from 4Cα to layer 4B, from which neurons connect to cortical area MT (medial temporal, otherwise known as area V5).[133–136] In area MT a high proportion of directionally selective cells are present (approximately 80 percent).[137,138] It should be noted that not all magnocellular neurons follow the path to V5/MT; a number of neurons converge with the parvocellular neurons in area V1.[139,140]

The importance of area V5/MT to motion perception has been demonstrated by behavioral studies of primates with selective V5MT lesions. A V5/MT lesion causes performance deficits in motion tasks, without corresponding losses to visual acuity or color perception.[141,142] Using a dynamic random-dot motion task, Newsome & Paré[141] investigated motion perception in the presence of MT lesions in rhesus monkeys. This task requires the observer to identify the predominant direction of motion within a field of randomly moving dots (Fig. 37.13). The proportion of dots moving in a common direction (coherence ratio) is varied to determine the direction identification threshold. Key to this task is that the moving dots be selected randomly for each frame of

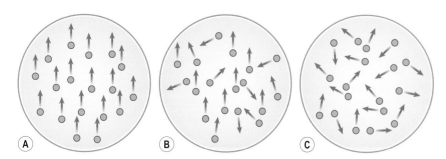

Figure 37.13 Representation of a global-dot motion stimulus. (Reproduced with permission from Edwards & Badcock 1994.[144]) The subject is required to identify the global direction of motion of the dots (up/down). The signal strength is varied by changing the number of signal dots moving in that direction in successive frames. The remaining dots, the noise dots, moved in random directions. The dots that comprise the signal dots are chosen at the start of each frame. (**A**) One hundred percent signal strength condition. (**B**) Fifty percent signal strength condition. (**C**) Zero percent signal strength condition; there is no global motion of the dots, only random local motion.

Box 37.3 Motion perception

- Normal motion perception is a precursor for many tasks of daily living such as driving, avoiding obstacles, and navigating through the environment.
- Motion perception is tested widely throughout the clinical and behavioral sciences.
- Deficits in the ability to detect the direction of motion are considered cortical as the earliest stage of the visual processing hierarchy capable of discriminating direction of motion is V1.
- Deficits in the ability to integrate local motion signals into coherent global percepts imply extrastriate dysfunction.

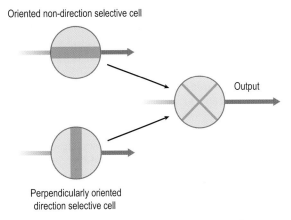

Oriented non-direction selective cell

Output

Perpendicularly oriented direction selective cell

Figure 37.14 Schematic of the motion streak model of motion processing described by Geisler.[150] The output of a direction selective neuron is combined with that of a neuron, or population of neurons, with oriented receptive fields but no directional selectivity that respond to the motion streak created during the temporal integration period. This combination of outputs significantly enhances the selectivity of the motion detection unit.

presentation so that the direction of motion cannot be determined by tracking the location of a single dot. Rather, the local motion signals must be combined over a large area of the visual field to determine the global motion percept. Newsome & Paré[141] found striking elevations in such motion thresholds to be present when selective lesions to area MT were made. Newsome et al[143] further demonstrated that the activity in a single MT cell correlated directly with the monkey's trial by trial performance in a motion discrimination task. They also demonstrated that microstimulation of tiny regions of MT can bias a monkey's motion perception in a manner predictable from the response characteristics of cells adjacent to the stimulating electrode.

Zihl et al,[145] Hess et al,[146] and Baker et al[147] provide a series of reports investigating the motion perception of a human patient with a bilateral loss of the superior temporal cortical region. This patient was found to have a specific deficit to motion processing while other measured forms of visual processing were normal or relatively intact. In particular, the patient performed poorly on a random-dot motion task similar to that used by Newsome & Paré.[141] This pattern of deficit is consistent with that found when area V5/MT is lesioned in primates, providing evidence for a similar specialized area for motion processing in humans.

In summary, motion perception (Box 37.3) appears to be performed largely by the visual pathway leading from the magnocellular layers of the lateral geniculate nucleus through area V1 to area MT. This pathway is not the only carrier of motion information in humans, however; area MT also receives input from the superior colliculus via the pulvinar.[148] In addition, Nawrot & Rizzo[149] have also demonstrated that the cerebellum may also play an important role in human motion processing. In their study, motion direction discrimination deficits were found in a group of patients with acute midline cerebellar lesions. These patients were unable to discern the direction of motion in a global motion task, similar to the deficit reported following MT lesions in primates.[141,142] Area V3 also appears important to human motion perception, particularly when spatial judgments are required. Brain imaging studies support the importance of areas V5/MT and V3 to human visual motion perception.[150,151]

While the magnocellular pathways possess the temporal resolution capacity to subserve motion perception there is clear evidence for additional input of information from orientation and form processing to motion perception. In 1999 Geisler[152] proposed a new model for motion perception that involves the combination of responses from detectors oriented orthogonal to the direction of motion and those oriented along the direction of motion (Fig. 37.14). The premise for this model was the observation that the visual system integrates information over time, hence that moving objects should leave behind 'motion streaks' along the direction of motion. Such streaks potentially provide information about the direction of image motion that is available to orientation selective neurons of the primary visual cortex. There is psychophysical evidence[152–154] for visual motion perception being influenced by motion streaks as well as neurophysiological evidence for responsivity to motion streak signals in primary visual cortex.[155] It has therefore become apparent that motion and form processing are far more interlinked than previously thought. Such interlinking needs to be considered in the interpretation of clinical deficits of motion perception.

Clinical applications of motion processing

Motion perception has been explored extensively in a range of clinical conditions because motion is a unique sensory dimension, processed by a relatively well-understood neural pathway. As discussed in the preceding section, the detection of stimulus motion occurs early in the motion pathway (retina to V1), the ability to discriminate the direction of motion is first present in cortical neurons within V1, and the ability to extract global motion information occurs in the extrastriate visual areas (MT or higher). The clinical study of motion processing often uses tasks chosen to localize deficits of motion processing to these sequential stages. For example, if the ability to detect stimulus motion and discriminate its direction is intact yet global motion perception impaired, this suggests damage at extrastriate areas.

The ability to both detect motion and discriminate its direction declines with age.[156,157] Motion processing abnormalities have been documented in Parkinson's disease.[158,159] Global motion processing, as measured using random-dot global motion stimuli (similar to that illustrated in Fig. 37.12), has been shown to be affected in people with migraines[160] and Alzheimer's disease.[161]

Visual motion processing has also been extensively studied in individuals with glaucoma. Perimetric approaches to testing motion processing have been developed because the visual dysfunction associated with glaucoma is more readily measured in the peripheral visual field (see Chapter 35). For example, random-dot motion perimetry assesses global motion perception at discrete locations across the visual field. A number of studies have found random dot motion perimetric deficits in glaucoma.[162–169] The magnitude of such motion deficits in glaucoma has been shown to be correlated with the extent of loss for contrast discrimination tasks designed to assess magnocellular processing, consistent with the source of the motion perception loss arising prior to cortical processing.[165] Peripheral motion displacement thresholds (the minimum distance for which a line or random-dot pattern that is laterally displaced appears to move) are also elevated in glaucoma.[167,170–172]

The temporal properties of vision are critical to the detection and perception of visual stimuli. The visual processing of motion, flicker, and other temporal characteristics is vital for us to effectively interact with and navigate through our environment. Because these visual functions are highly susceptible to pathophysiologic insults to the visual pathways, tests of motion sensitivity, flicker sensitivity, and other temporal visual functions are extensively used as sensitive clinical diagnostic procedures.

References

1. Breitmeyer BG, Ganz L. Temporal studies with flashed gratings: inferences about human transient and sustained channels. Vision Res 1977; 17:861–865.
2. Burr DC. Temporal summation of moving images by the human visual system. Proc R Soc Lond B Biol Sci 1981; 211(1184):321–339.
3. Gorea A, Tyler CW. New look at Bloch's law for contrast. J Opt Soc Am A 1986; 3(1):52–61.
4. Legge GE. Sustained and transient mechanisms in human vision: temporal and spatial properties. Vision Res 1978; 18:69–81.
5. Snowden RJ, Braddick OJ. The temporal integration and resolution of velocity signals. Vision Res 1991; 31(5):907–914.
6. Bloch A. Experience sur la vision. Comptes Rendus de la Societe de Biologie (Paris) 1885; 37:493–495.
7. Watson AB. Probability summation over time. Vision Res 1979; 19:515–522.
8. Sperling HG, Jolliffe CL. Intensity-time relationship at threshold for spectral stimuli in human vision. J Opt Soc Am 1965; 55:191–199.
9. Saunders RM. The critical duration of temporal summation in the human central fovea. Vision Res 1975; 15:699–703.
10. Mitsuboshi M, Kawabata Y, Aiba TS. Color-opponent characteristics revealed in temporal integration time. Vision Res 1987; 27:1197–1206.
11. Krauskopf J, Mollon JD. The independence of the temporal integration properties of the individual chromatic mechanisms. J Physiol Lond 1971; 219:611–623.
12. Barlow HB. Temporal and spatial summation in human vision at different background intensities. J Physiol Lond 1958; 141:337–350.
13. Stewart BR. Temporal summation during dark adaptation. J Opt Soc Am 1972; 62:449–457.
14. Montellese S, Sharpe LT, Brown JL. Changes in critical duration during dark-adaptation. Vision Res 1979; 19:1147–1153.
15. Graham CH, Margaria R. Area and intensity-time relation in the peripheral retina. Am J Physiol 1935; 113:299–305.
16. Baumgardt E, Hillman B. Duration and size as determinants of peripheral retinal response. J Opt Soc Am 1961; 51:340–344.
17. Connors MM. Luminance requirements for hue perception and identification, for a range of exposure durations. J Opt Soc Am 1970; 60:958–965.
18. Dain SJ, King-Smith PE. Visual thresholds in dichromats and normals: the importance of post-receptoral processes. Vision Res 1981; 21:573–580.
19. Smith VC, Bowen RW, Pokorny J. Threshold temporal integration of chromatic stimuli. Vision Res 1984; 24(7):653–660.
20. Swanson WH, Pan F, Lee BB. Chromatic temporal integration and retinal eccentricity: psychophysics, neurometric analysis and cortical pooling. Vision Res 2008; 48:2657–2662.
21. Kawabata Y. Temporal integration at equiluminance and chromatic adaptation. Vision Res 1994; 34(8):1007–1018.
22. Ferry E. Persistence of vision. Am J Sci 1892; 44:192–207.
23. Porter T. Contributions to the study of flicker. Proc R Soc A 1902; 62:313–329.
24. Tyler CW, Hamer RD. Analysis of visual modulation sensitivity. IV. Validity of the Ferry–Porter law. J Opt Soc Am A 1990; 7:743–758.
25. Hecht S, Verrijp C. Intermittent stimulation by light. III. The relation between intensity and critical flicker fusion frequency for different retinal locations. J Gen Physiol 1933; 17:251–265.
26. Ives HE. A theory of intermittent vision. J Opt Soc Am Rev Sci Instrum 1922; 6:343–361.
27. Giorgi A. Effect of wavelength on the relationship between critical flicker frequency and intensity in foveal vision. J Opt Soc Am 1963; 53:480–486.
28. Hamer RD, Tyler CW. Analysis of visual modulation sensitivity. V. Faster visual response for G- than R-cone pathway. J Opt Soc Am A 1992; 9(11):1889–1904.
29. Ives HE. Studies in the photometry of lights of different colours. II. Spectral luminosity curves by the method of critical frequency. Philos Mag 1912; 24:352–370.
30. Pokorny J, Smith VC. Luminosity and CFF in deuteranopes and protanopes. J Opt Soc Am 1972; 62:111–117.
31. Brindley GS, Du Croz JJ, Rushton WAH. The flicker fusion frequency of the blue-sensitive mechanism of colour vision. J Physiol (Lond) 1966; 183:497–500.
32. Hess RF, Mullen KT, Zrenner E. Human photopic vision with only short wavelength cones: post-receptoral properties. J Physiol 1989; 417:151–172.
33. Hartmann E, Lachenmayr B, Brettel H. The peripheral critical flicker frequency. Vision Res 1979; 19:1019–1023.
34. Rovamo J, Raninen A. Critical flicker frequency and M-scaling of stimulus size and retinal illuminance. Vision Res 1984; 24:1127–1131.
35. Granit R, Harper P. Comparative studies on the peripheral and central retina. II Synaptic reactions in the eye. Am J Physiol 1930; 95:211–227.
36. Roehrig W. The influence of area on the critical flicker fusion threshold. J Psychol 1959; 47:317–330.
37. Roehrig W. The influence of the portion of the retina stimulated on the critical flicker-fusion threshold. J Psychol 1959; 48:57–63.
38. Rovamo J, Raninen A. Critical flicker frequency as a function of stimulus area and luminance at various eccentricities in human cone vision: a revision of Granit–Harper and Ferry–Porter laws. Vision Res 1988; 28(7):785–790.
39. De Lange H. Relationship between critical flicker frequency and a set of low-frequency characteristics of the eye. J Opt Soc Am 1954; 44:380.
40. De Lange H. Research into the dynamic nature of the human fovea-cortex systems with intermittent and modulated light. I. Attenuation characteristics with white and colored light. J Opt Soc Am 1958; 48:777.
41. De Lange H. Research into the dynamic nature of the human fovea-cortex systems with intermittent and modulated light. II. Phase shift in brightness and delay in color perception. J Opt Soc Am 1958; 48:784.
42. Kelly D. Visual responses to time-dependent stimuli, 1. Amplitude sensitivity measurements. J Opt Soc Am 1961; 51:422.
43. Swanson WH et al. Temporal modulation sensitivity and pulse-detection thresholds for chromatic and luminance perturbations. J Opt Soc Am A 1987; 4(10):1992–2005.
44. Kelly DH. Frequency doubling in visual responses. J Opt Soc Am 1966; 56:1628–1633.
45. Robson JG. Spatial and temporal contrast-sensitivity functions of the visual system. J Opt Soc Am 1966; 56:1141–1142.
46. Kelly D. Adaptation effects on spatio-temporal sine-wave thresholds. Vision Res 1972; 12:89–101.
47. Kelly DH. Motion and vision. II. Stabilized spatio-temporal threshold surface. J Opt Soc Am 1979; 69(10):1340–1349.
48. Kelly DH. Motion and vision. I. Stabilized images of stationary gratings. J Opt Soc Am 1979; 69:1266–1274.
49. Blakemore C, Campbell FW. On the existence of neurones in the human visual system selectively sensitive to the orientation and size of retinal images. J Physiol 1969; 203:237–261.
50. Campbell FW, Robson JG. Applications of Fourier Analysis to the visibility of gratings. J Physiol 1968; 197:551–566.
51. Graham N, Nachmias J. Detection of grating patterns containing two spatial frequencies: a comparison of single channel and multiple-channels models. Vision Res 1971; 11(3):251–259.
52. Hess RF, Snowden RJ. Temporal properties of human visual filters: number, shapes and spatial covariation. Vision Res 1992; 32:47–59.
53. Mandler MB, Makous W. A three channel model of temporal frequency perception. Vision Res 1984; 24:1881–1887.
54. Watson AB, Robson JA. Discrimination at thresholds: labelled detectors in human vision. Vision Res 1981; 21:1115–1122.
55. Yo C, Wilson HR. Peripheral temporal frequency channels code frequency and speed inaccurately but allow accurate discrimination. Vision Res 1993; 33:33–45.
56. Kulikowski JJ. Effect of eye movements on the contrast sensitivity of spatio-temporal patterns. Vision Res 1971; 11:261–273.
57. Stiles WS. Separation of the "blue" and "green" mechanisms of foveal vision by measurements of increment thresholds. In: Mechanisms of colour vision. London: Academic Press, 1978:418–434.
58. Snowden RJ, Hess RF. Temporal frequency filters in the human peripheral visual field. Vision Res 1992; 32(1):61–72.
59. Cass J, Alais D. Evidence for two interacting temporal channels in human visual processing. Vision Res 2006; 46:2859–2868.
60. Kaplan E, Shapley R. The primate retina contains two types of ganglion cells, with high and low contrast sensitivity. Proc Natl Acad Sci USA 1986; 83:2755–2757.
61. Lee BB. Receptive field structure in the primate retina. Vision Res 1996; 36:631–644.
62. Lee BB et al. Luminance and chromatic modulation sensitivity of macaque ganglion cells and human observers. J Opt Soc Am A 1990; 7:2223–2236.
63. Lee BB, Sun H, Zucchini W. The temporal properties of the response of macaque ganglion cells and central mechanisms of flicker detection. J Vision 2007; 7(14):1–6.
64. Smith VC. et al. Sequential processing in vision: The interaction of sensitivity regulation and temporal dynamics. Vision Res 2008; 48:2649–2656.
65. Zele AJ, Cao D, Pokorny J. Rod-cone interactions and the temporal impulse response of the cone pathway. Vision Res 2008; 48:2593–2598.
66. Lange G, Denny N, Frumkes TE. Suppressive rod-cone interactions: evidence for separate retinal (temporal) and extraretinal (spatial) mechanisms in achromatic vision. J Opt Soc Am A 1997; 14:2487–2498.

67. Goldberg SH, Frumkes TE, Nygaard RW. Inhibitory influence of unstimulated rods in the human retina: evidence provided by examining cone flicker. Science 1983; 221:180–182.
68. Coletta NJ, Adams AJ. Rod-cone interaction in flicker detection. Vision Res 1984; 24:1333–1340.
69. Alexander KR, Fishman GA. Rod-cone interaction in flicker perimetry. Br J Ophthalmol 1984; 68:303–309.
70. Arden GB, Hogg CR. Rod-cone interactions and analysis of retinal disease. Br J Ophthalmol 1985; 69:404–415.
71. Coletta NJ, Adams AJ. Spatial extent of rod-cone and cone-cone interaction for flicker detection. Vision Res 1986; 26:917–925.
72. Eisner A. Non-monotonic effect of test illuminance on flicker detection: a study of foveal light adaptation with annular surrounds. J Opt Soc Am A 1994; 11:33–47.
73. Harvey LO. Flicker sensitivity and apparent brightness as a function of surround luminance. J Opt Soc Am 1970; 60:860–864.
74. Keesey UT. Variables determining flicker sensitivity in small fields. J Opt Soc Am 1970; 60:390–398.
75. Roufs JAJ. Dynamic properties of vision. I. Experimental relationships between flicker and flash thresholds. Vision Res 1972; 12:261–278.
76. Spehar B, Zaidi Q. Surround effects on the shape of the temporal contrast-sensitivity function. J Opt Soc Am A 1997; 14(9):2517–2525.
77. Anderson AJ, Vingrys AJ. Interactions between flicker thresholds and luminance pedestals. Vision Res 2000; 40:2579–2588.
78. Eisner A. Suppression of flicker response with increasing test illuminance: roles of temporal waveform, modulation depth and frequency. J Opt Soc Am A 1995; 12:214–224.
79. Eisner A, Shapiro AG, Middleton J. Equivalence between temporal frequency and modulation depth for flicker response suppression: analysis of a three-process model of visual adaptation. J Opt Soc Am A 1998; 15:1987–2002.
80. Zele AJ, Vingrys AJ. Defining the detection mechanisms for symmetric and rectified flicker stimuli. Vision Res 2007; 47(21):2700–2713.
81. Anderson AJ, Vingrys AJ. Multiple processes mediate flicker sensitivity. Vision Res 2001; 41:2449.
82. Ball RJ. An investigation of chromatic brightness enhancement tendencies. Am J Optom Physiol Opt 1964; 41:333–361.
83. Ball RJ, Bartley SH. Changes in brightness index, saturation, and hue produced by luminance-wavelength-temporal interactions. J Opt Soc Am 1966; 66:695–699.
84. Bartley SH. Some effects of intermittent photic stimulation. J Exp Psychol 1939; 25:462–480.
85. Bartley SH. Brightness comparisons when one eye is stimulated intermittently and the other steadily. J Psychol 1951; 34:165–167.
86. Bartley SH. Brightness enhancement in relation to target intensity. J Psychol 1951; 32:57–62.
87. Bartley SH, Nelson TM. Certain chromatic and brightness changes associated with rate of intermittency of photic stimulation. J Psychol 1960; 50:323–332.
88. Stockman A, Plummer DJ. Color from invisible flicker: a failure of the Talbot–Plateau law caused by an early 'hard' saturating non-linearity used to partition the human short-wave cone pathway. Vision Res 1998; 38:3703–3728.
89. Brucke E. Uber die Nutzeffect intermitterender Netzhautreizungen. Sitzungsberichte der Mathematisch-Naturwissenschaftlichen. Classe der Kaiserlichen Akademie der Wissenschaften 1848; 49:128–153.
90. Talbot HF. Experiments on light. Philos Mag Ser 3, 1834; 5:321–334.
91. Plateau J. Sur un principle de photometrie. Bulletins de L'Academie Royale des Sciences et Belles-lettres de Bruxelles 1835; 2:52–59.
92. Fahle M. Figure-ground discrimination from temporal information. Proc R Soc Lond B Biol Sci 1993; 254:199–203.
93. Forte J, Hogben JH, Ross J. Spatial limitations of temporal segmentation. Vision Res 1999; 39:4052–4061.
94. Leonards U, Singer W, Fahle M. The influence of temporal phase differences on texture segmentation. Vision Res 1996; 36:2689–2697.
95. Rogers-Ramachandran DC, Ramachandran VS. Psychophysical evidence for boundary and surface systems in human vision. Vision Res 1998; 38:71–77.
96. Lachenmayr BJ, Drance SM, Douglas GR. Light-sense, flicker and resolution perimetry in glaucoma: a comparative study. Graefe's Arch Clin Exp Ophthalmol 1991; 229:246–251.
97. Cello KE, Nelson-Quigg JM, Johnson CA. Frequency doubling technology perimetry for detection of glaucomatous visual field loss. Am J Ophthalmol 2000; 129:314–322.
98. Johnson CA, Samuels SJ. Screening for glaucomatous visual field loss with frequency-doubling perimetry. Invest Ophthalmol Vis Sci 1997; 38:413.
99. Maddess T, Henry GH. Performance of non-linear visual units in ocular hypertension and glaucoma. Clin Vis Sci 1992; 7(5):371–383.
100. Yoshiyama KK, Johnson CA. Which method of flicker perimetry is most effective for detection of glaucomatous visual field loss? Invest Ophthalmol Vis Sci 1997; 38:2270–2277.
101. Casson EJ, Johnson CA, Shapiro LR. A longitudinal comparison of Temporal Modulation perimetry to White-on-White and Blue-on-Yellow perimetry in ocular hypertension and early glaucoma. J Opt Soc Am 1993; 10:1792–1806.
102. Casson EJ, Johnson CA. Temporal modulation perimetry in glaucoma and ocular hypertension. In: Perimetry Update 1992/3. New York: Kugler Publications, 1993.
103. Casson EJ, Johnson CA, Nelson-Quigg JM. Temporal modulation perimetry: the effects of aging and eccentricity on sensitivity in normals. Invest Ophthalmol Vis Sci 1993; 34:3096–3102.
104. Bodis-Wollner I. Visual deficits related to dopamine deficiency in experimental animals and Parkinson's disease patients. Trends Neurosci 1990; 13(7):296–302.
105. Coleston DM et al. Precortical dysfunction of spatial and temporal visual processing in migraine. J Neurol Neurosurg Psychiat 1994; 57:1208–1211.
106. Coleston DM, Kennard C. Responses to temporal visual stimuli in migraine, the critical flicker fusion test. Cephalalgia 1995; 15:396–398.
107. Evans BJ, Drasdo N, Richards IL. An investigation of some sensory and refractive visual factors in dyslexia. Vision Res 1994; 34(14):1913–1926.
108. Fujimoto N, Adachi-Usami E. Frequency doubling perimetry in resolved optic neuritis. Invest Ophthalmol Vis Sci 2000; 41:2558–2560.
109. Grigsby SS et al. Correlation of chromatic, spatial, and temporal sensitivity in optic nerve disease. Invest Ophthalmol Vis Sci 1991; 32:3252–3262.
110. Mason RJ et al. Abnormalities of chromatic and luminance critical flicker frequency in multiple sclerosis. Invest Ophthalmol Vis Sci 1982; 23:246–252.
111. Mayer MJ et al. Foveal flicker sensitivity discriminates ARM-risk from healthy eyes. Invest Ophthalmol Vis Sci 1992; 33(11):3143–3149.
112. McKendrick AM et al. Visual field losses in subjects with migraine headaches. Invest Ophthalmol Vis Sci 2000; 41(5):1239–1247.
113. McKendrick AM et al. Visual dysfunction between migraine events. Invest Ophthalmol Vis Sci 2001; 42:626.
114. Phipps JA, Guymer RH, Vingrys AJ. Temporal sensitivity deficits in patients with high-risk drusen. Aust NZ J Ophthalmol 1999; 27:265–267.
115. Vingrys AJ, Pesudovs K. Localised scotomata detected with temporal modulation perimetry in central serous chorioretinopathy. Aust NZ J Ophthalmol 1999; 27(2):109–116.
116. Barnard N, Crewther SG, Crewther DP. Development of a magnocellular function in good and poor primary school-age readers. Optom Vis Sci 1998; 75(1):62–68.
117. Flanagan JG et al. The phantom contour illusion letter test: a new psychophysical test for glaucoma? In: Mills RP, Wall M, eds. Perimetry Update 1994/1995. Amsterdam/New York: Kugler Publications, 1995
118. Quaid PT, Flanagan JG. Defining the limits of flicker defined form: effect of stimulus size, eccentricity and number of random dots. Vision Res 2005; 45:1075–1084.
119. Goren D, Flanagan JG. Is flicker-defined form (FDF) dependent on the contour? J Vis 2008; Apr 22:8(4)(15):1–11.
120. Derrington AM, Allen HA, Delicato LS. Visual mechanisms of motion analysis and motion perception. Ann Rev Psychol 2004; 55:181–205.
121. Vaina LM, Soloviev S. First order and second-order motion: neurological evidence for neuroanatomically distinct systems. Prog Brain Res 2004; 144: 197–212.
122. Hiris E, Blake R. A new perspective on the visual motion aftereffect. Proc Natl Acad Sci USA 1992; 89:9025–9028.
123. Nishida S, Sato T. Positive motion after-effect induced by bandpass-filtered random-dot kinematograms. Vision Res 1992; 32(9):1635–1646.
124. Sekuler RW, Ganz L. Aftereffect of seen motion with a stabilized retinal image. Science 1963; 139:419–420.
125. Mather G et al. The motion aftereffect reloaded. Trends Cogn Sci 2008; 12:481–487.
126. Wertheimer M. Experimentelle Studien uber das Sehen von Bewegung. Zietschrift fur Psychologie 1912; 61:161–265.
127. Hubel D, Weisel T. Receptive fields and functional architecture of monkey striate cortex. J Physiol 1968; 195:215–243.
128. Hawken MJ, Parker AJ, Lund JS. Laminar organisation and contrast sensitivity of direction-selective cells in the striate cortex of the Old World monkey. J Neurosci 1988; 10:3541–3548.
129. Hendrickson AE, Wilson JR, Ogren MP. The neuroanatomical organisation of pathways between the dorsal lateral geniculate nucleus and visual cortex in Old World and New World primates. J Comp Neurol 1978; 182:123–136.
130. Merigan W, Maunsell J. Macaque vision after magnocellular lateral geniculate lesions. Vis Neurosci 1990; 5:347–352.
131. Schiller P, Malpeli J. Functional specificity of lateral geniculate nucleus laminae of the rhesus monkey. J Neurophysiol 1978; 41:788–797.
132. Callaway EM. Structure and function of parallel pathways in the primate early visual system. J Physiol 2005; 566:13–19.
133. DeYoe EA, Van Essen DC. Segregation of efferent connections and receptive field properties in visual area V2 of the macaque. Nature 1985; 317:58–61.
134. Livingstone MS, Hubel DH. Connections between layer 4B of area 17 and the thick cytochrome oxidase stripes of area 18 in the squirrel monkey. J Neurosci 1987; 7:3371–3377.
135. Shipp S, Zeki S. The organisation of connections between areas V5 and V1 in macaque monkey visual cortex. Eur J Neurosci 1989; 1:309–332.
136. Shipp S, Zeki S. The organisation of connections between areas V5 and V2 in macaque monkey visual cortex. Eur J Neurosci 1989; 1:333–354.
137. Albright TD. Direction and orientation selectivity of neurons in visual area MT of the macaque. J Neurophysiol 1984; 52(6):1106–1130.
138. Maunsell JH, Van Essen DC. Functional properties of neurons in middle temporal visual area of the macaque monkey. I. Selectivity for stimulus direction, speed, and orientation. J Neurophysiol 1983; 49:1127–1147.
139. Malpeli JG, Schiller PH, Colby CL. Response properties of single cells in monkey striate cortex during reversible inactivation of individual lateral geniculate laminae. J Neurophysiol 1981; 46:1102–1119.
140. Nealey TA, Maunsell JHR. Magnocellular and parvocellular contributions to the responses of neurons in macaque striate cortex. J Neurosci 1994; 14:2069–2079.
141. Newsome WT, Paré EB. A selective impairment of motion perception following lesions of the middle temporal visual area (MT). J Neurosci 1988; 8:2201–2211.
142. Schiller PH. The effects of V4 and middle temporal (MT) area lesions on visual performance in the rhesus monkey. J Neurosci 1993; 10:717–746.
143. Newsome WT, Britten KH, Movshon JA. Neuronal correlates of a perceptual decision. Nature 1989; 341:52–54.
144. Edwards M, Badcock DR. Global Motion Perception: Interaction of on and off pathways. Vision Res 1994; 34:2849–2858.
145. Zihl J, von Cramon D, Mai N. Selective disturbance of movement vision after bilateral brain damage. Brain 1983; 106 (Pt 2):313–340.
146. Hess RH, Baker CL, Zihl J. The "motion-blind" patient: Low-level spatial and temporal filters. J Neurosci 1989; 9:1628–1640.
147. Baker CL, Hess RF, Zihl J. Residual motion perception in a "motion-blind" patient, assessed with limited-lifetime random dot stimuli. J Neurosci 1991; 11:454–461.
148. Berman RA, Wurtz RH. Exploring the pulvinar path to visual cortex. Prog Brain Res 2008; 171:467–473.
149. Nawrot M, Rizzo M. Motion perception deficits from midline cerebellar lesions in human. Vision Res 1995; 35(5):723–731.

150. Greenlee MW. Human cortical areas underlying the perception of optic flow: brain imaging studies. Int Rev Neurobiol 2000; 44:269–292.

151. Bartels A, Logothetis NK, Moutoussis K. fMRI and its interpretations: an illustration on direction selectivity in area V5/MT. Trends Neurosci 2008; 31(9):444–453.

152. Geisler WS. Motion streaks provide a spatial code for motion direction. Nature 1999; 400(6739):65–69.

153. Ross J, Badcock DR. Coherent global motion in the absence of coherent velocity signals. Curr Biol 2000; 10(11):679–682.

154. Burr DC, Ross J. Direct evidence that "speedlines" influence motion mechanisms. J Neurosci 2002; 22(19):8661–8664.

155. Geisler WS et al. Motion direction signals in the primary visual cortex of cat and monkey. Vis Neurosci 2001; 18(4):501–516.

156. Bennett PJ, Sekuler R, Sekuler AB. The effects of aging on motion detection and direction identification. Vision Res 2007; 47(6):799–809.

157. Willis A, Anderson SJ. Effects of glaucoma and aging on photopic and scotopic motion perception. Invest Ophthalmol Vis Sci 2000; 41(1):325–335.

158. Castelo-Branco M et al. Motion integration deficits are independent of magnocellular impairment in Parkinson's disease. Neuropsychologia 2009; 47:314–320.

159. Trick GL, Kaskie B, Steinman SB. Visual impairment in Parkinson's disease: deficits in orientation and motion discrimination. Optom Vis Sci 1994; 71(4):242–245.

160. McKendrick AM, Badcock DR, Motion processing deficits in migraine. Cephalalgia 2004; 24(5):363–372.

161. Rizzo M, Nawrot M. Perception of movement and shape in Alzheimer's disease. Brain 1998; 121:2259–2270.

162. Bosworth CF et al. Motion automated perimetry identifies early glaucomatous field defects. Arch Ophthalmol 1998; 116(9):1153–1158.

163. Joffe KM, Raymond JE, Chrichton A. Motion coherence perimetry in glaucoma and suspected glaucoma. Vision Res 1997; 37(7):955–964.

164. Wall M, Ketoff KM. Random dot motion perimetry in patients with glaucoma and in normal subjects. Am J Ophthalmol 1995; 120(5):587–596.

165. McKendrick AM, Badcock DR, Morgan WH. The detection of both global motion and global form is disrupted in glaucoma. Invest Ophthalmol Vis Sci 2005; 46(10):3693–3701.

166. Wall M, Jennisch CS, Munden PM. Motion perimetry identifies nerve fiber bundle-like defects in ocular hypertension. Arch Ophthalmol 1997; 115:26–33.

167. Bosworth CF, Sample PA, Weinreb RN. Perimetric motion thresholds are elevated in glaucoma suspects and glaucoma patients. Vision Res 1997; 37:1989–1997.

168. Karwatsky P et al. Defining the nature of motion perception deficits in glaucoma using simple and complex motion stimuli. Optom Vis Sci 2006; 83:466–472.

169. Wall M et al. Repeatability of automated perimetry: a comparison between standard automated perimetry with stimulus size II and V, Matrix and motion perimetry. Invest Ophthalmol Vis Sci 2009; 50:974–979.

170. Bullimore MA, Wood JM, Swenson K. Motion perception in glaucoma. Invest Ophthalmol Vis Sci 1993; 34:3526–3533.

171. Verdon-Roe GM et al. Exploration of the psychophysics of a motion displacement hyperacuity stimulus. Invest Ophthalmol Vis Sci 2006; 47:4847–4855.

172. Westcott MC, Fitzke FW, Hitchings RA. Abnormal motion displacement thresholds are associated with fine scale luminance sensitivity loss in glaucoma. Vision Res 1998; 38:3171–3180.

173. Carmel D, Lavie N, Rees G. Conscious awareness of flicker in humans involves frontal and parietal cortex. Curr Biol 2006: 16:907–911.

Development of Vision in Infancy

Anthony M. Norcia

The limited behavioral repertoire of the infant and the impossibility of instructing the test subject have made it necessary for vision scientists interested in human visual development to adapt the classical methods of psychophysics and electrophysiology for use with infants and pre-verbal children. First these methodological adaptations and their interpretation in the context of a hierarchical model of visual processing are considered. Within this framework, the criteria for selecting material for review are discussed.

Methodologies for assessing infant vision and their interpretation

Preferential looking

Infants' spontaneous visual fixation is attracted to certain stimuli more readily than to others.[1,2] In particular, infants prefer to look at patterned stimuli rather than regions of uniform brightness. This spontaneous behavior has served as the basis for a quantitative measure of stimulus visibility known as *forced-choice preferential looking* (FPL).[3] In the FPL task, the infant is confronted with a randomized series of patterns of varying visibility, presented either on the left or the right of a test screen. An observer judges whether the infant's fixation behavior is biased to the left or the right on a trial-by-trial basis. If the observer's judgments agree (or disagree) systematically with the actual position of the stimulus, it can be said that the infant's behavior is under the control of the stimulus. Distributions of the observer's judgments for a series of stimulus values are used to plot a psychometric function that relates the observer's percent correct to the stimulus values presented to the infant. Thresholds are estimated by curve fitting and interpolation to a criterion value of percent correct.

Visual evoked potentials

Visual evoked potentials (VEPs) are electrical brain responses that are triggered by the presentation of a visual stimulus. VEPs are distinguished from the spontaneous electroencephalogram (EEG) due to their consistent time of occurrence after the presentation of the stimulus (time-locking).

For example, the abrupt contrast reversal of a checkerboard pattern consistently produces a positive potential at the surface of the scalp at a latency of approximately 100 msec in adults.[4] Time-locked responses to abrupt presentations are referred to as *transient VEPs*. A second method of recording VEPs, the steady-state method, uses temporally periodic stimuli. For commonly used pattern reversal stimuli, the frequency of the repetition is often specified as the pattern reversal rate in reversals per sec. This rate is twice the stimulus fundamental frequency (in Hz), which is more commonly used to describe the temporal frequency of pattern onset–offset stimuli. As the stimulus repetition rate increases, the responses to successive stimuli begin to overlap. At high stimulation rates, the response is comprised of only a small number of components that occur at exact integer multiples of the stimulus frequency. Activity at each of the frequency components of the steady-state response is characterized by its amplitude and phase, where phase represents the temporal delay between the stimulus and the evoked response.

The surface-recorded VEP reflects the activity of cortical visual areas, with contributions from subcortical generators being apparent only under highly specialized recording conditions.[5–7] The primary adaptations of adult VEP recording techniques for infants involve the control of fixation through the use of fixation toys or super-imposed video images and the rejection of trials when the infant's fixation was not centered on the stimulus.

Ocular following movements

Both infants and adults make reflexive eye movements following the presentation of a moving target. Optokinetic nystagmus (OKN) is characterized by a repetitive saw-tooth waveform. Rapid displacement of large fields also elicits short latency ocular following movements.[8] Ocular following can also take the form of slower, pursuit-like movements.[9] Reflexive eye movements are controlled by a combination of cortical and subcortical mechanisms.[10] Infrared tracking, electro-oculography, and naked-eye observation of the preponderant direction of eye motion (DEM) are the primary assays used in infants and pre-verbal children.[9,11]

Hierarchy of visual processing

Figure 38.1 presents the schematic framework of visual processing that is used to focus the discussion of empirical studies of visual function in infants and young children. The visual processing hierarchy is divided into three stages: early, middle, and late. The progression from early to late correlates roughly with an ascent from the retina to the cortex and with a functional hierarchy corresponding to the complexity of the information extracted at each level. In this view, early vision begins in the retina and continues through the lateral geniculate and on into primary visual cortex. By the level of primary visual cortex, stimulus attributes such as orientation, direction of motion, and disparity have been extracted from the retinal images.[12–14] Middle vision – the process by which local measurements of image features such as line orientation are integrated across space – begins no sooner than primary visual cortex and no doubt extends through a number of first- and second-tier extrastriate visual areas. The content of the representation at the level of middle vision includes information regarding the shape of extended contours, figure/ground relationships, the symmetry of objects, surface depths, but not the identity of the objects in the scene. The identification of objects (object recognition), which involves not only visual perception, but memory, is conceptualized as occurring in higher-order visual and visual association areas functionally associated with "late vision".

Each of the different methods for assessing visual function in the pre-verbal child has a different relationship to the visual processing hierarchy. The FPL technique depends on the integrity of the early visual system as well as additional mechanisms responsible for the spontaneous preference for pattern (labeled "preference generator" in Figure 38.1). Whether or not middle or late mechanisms are invoked may depend on the discrimination the infant is called on to perform. Orienting behaviors could be driven from many levels of the cortical hierarchy or from subcortical structures. In any case, the output of the preference generator must produce robust fixation behavior that can be detected reliably by the FPL observer. Information regarding the location of the stimulus can be lost at the level of early vision, or at the level of the preference generator or by the observer of the infant's behavior. Given the additional sites for potential information loss after early vision, FPL is a conservative estimator of the function of the early part of the visual pathway.

Like FPL, the VEP depends on the integrity of the retina and an unknown amount of cortical processing. Fixation, in the sense that the stimulus must fall on central retina is required, but spontaneous orientation to a preferred stimulus is not. Electrical activity in the visual pathway is obscured by non-stimulus-related electrical activity associated with the EEG and muscle activity, as well as electrode-motion artifacts. The obscuring experimental noise can be reduced effectively, either through time-locked averaging or spectral analysis. At this point relatively little is known about the contribution of extrastriate cortical areas to the VEP. Given this, the VEP is quite likely to reflect the capabilities of early vision, but caution must be used in inferring the integrity of later stages of processing – especially if simple stimuli are used.

Ocular following movements require the integrity of the retina, certainly, but given the substantial role of subcortical mechanisms in the control of eye movements,[10] it is difficult to specifically relate eye movement data to the hierarchy of cortical mechanisms in Figure 38.1.

This review emphasizes developmental studies that have used the VEP. The rationale for this choice is several-fold. First, there is now sufficient evidence to indicate that the infant VEP is generated distal to the site of orientation selectivity,[15,16] direction selectivity,[17–19] Vernier offset detection,[20] and binocular correlation detection.[21,22] All these features are considered to be the outputs of early vision. In adults, it has been found that the VEP reflects both rivalry and suppression[23–27] as well as several aspects of middle vision, including figure-ground segmentation based on either texture or motion.[28] Second, the VEP does not require visual preference or transfer of information through the observation of spontaneous behavior and is thus less likely to underestimate the capabilities of early vision. Third, the VEP provides a rich source of information regarding the temporal dynamics of the visual response. Finally, there are already

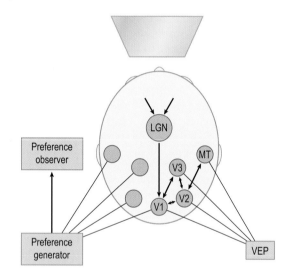

Visual hierarchies

	Early	**Middle**	**Late**
Attributes	Contrast Color Disparity Direction Speed Orientation	Collinearity Symmetry Figure/ground Occlusion Texture Depth Shape	Object categories recognition
Where	Retina, LGN, and V1	V1-Vn	Face area, Place area Temporal lobe
When	Up to 100 msec Sensation	100 to 200 msec Perception	After 200 msec Cognition

Figure 38.1 Visual processing hierarchy. Top panel: schematic diagram of early visual pathways. VEP method records activity directly from several early visual areas. Preference-based behavioral measures require an additional "stage" at which preference and spontaneous fixation behavior is generated as well as a behavioral observation stage.

excellent reviews that have emphasized FPL and OKN measures of developing visual function.[9,29–31] In deciding which studies to include, emphasis has been placed on those results that have been replicated by more than one research group, wherever possible. Data from the other methods are selectively discussed when these data can help to fill in gaps or when they illustrate particularly sharp contrasts.

Spatio-temporal vision

The retinal images contain a precise spatio-temporal mapping of the visual scene onto two-dimensional surfaces. At the most basic level of processing, the visual system must extract the contrast of the retinal images as a function of time and spatial scale. Visual sensitivity is limited by both spatial and temporal factors. Infant developmental studies have tended to focus on sensitivity along one dimension at a time – by measuring contrast sensitivity as a function of spatial frequencies for a fixed temporal frequency or vice versa. Sensitivity depends strongly on both parameters. While the FPL technique can be used at any combination of spatio-temporal frequencies, the eye movement and VEP measures each require temporally modulated stimuli. Given the fundamental importance of contrast sensitivity for subsequent visual processing, contrast sensitivity and the related function, grating acuity are among the few visual functions to have been studied extensively with each of the major methods discussed above.

Figure 38.2 plots peak contrast sensitivity as a function of age as determined by the steady-state VEP,[32] directional eye movements,[33–35] and FPL[34,36] methods. Each of these studies obtained peak sensitivity measures at a mid-range of temporal frequencies (around 5–10 Hz). There is considerable development of contrast sensitivity in each of the techniques, but the absolute contrast sensitivity is higher with the VEP. By 10 weeks of age, infant peak contrast sensitivity over the 0.25–1 cycle per degree (cpd) range is within about a factor of 2–4 of adult levels measured on the same apparatus.[32,37] Skoczenski & Norcia[38] found a factor of 4 difference between infant and adult sensitivities at 1 cpd. Shannon and co-workers[39] found sensitivities at 1.2 cpd that were a factor of 11 lower than adults at 2 months, with the difference decreasing to a factor of 4 at 3 months. Contrast sensitivity measured with the steady-state reversal VEP develops over progressively longer intervals as spatial frequency increases[32] (Fig. 38.3).

In contrast to the VEP, several behavioral measurements of contrast sensitivity in this age range are much lower than adult levels. Rasengane and colleagues[36] reported that low spatial frequency flicker sensitivity of 2-month-olds was a factor of 45 lower than adults, with 3- and 4-month-olds being a little less than 20 times less sensitive with FPL. Brown and colleagues[35] used a directional eye movement measure (0.31 cpd/ 15.5 deg/sec drift) and found that 3-month-olds were a factor of 100 less sensitive than adults on the same measure. Dobkins & Teller[34] measured both FPL and directional eye movement thresholds in 3-month-olds. They found that infants were almost 30 times less sensitive on the directional eye movement measure and about 60 times less sensitive when FPL and adults' forced choice thresholds were compared. FPL and directional eye movement thresholds were within 20% of each other in the infants. In adults DEM thresholds were higher than psychophysical thresholds by a factor of 2–3, depending on whether

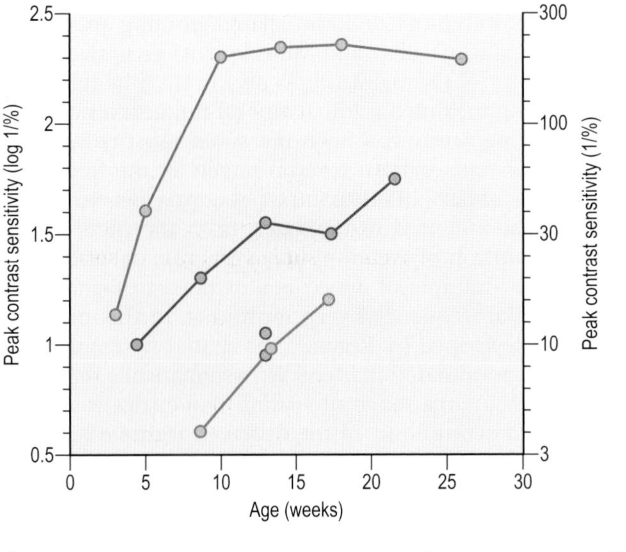

Figure 38.2 Peak contrast sensitivity measured with the steady-state (VEP), directional eye movement (DEM), and FPL methods. The VEP study[32] (red circles) used grating patterns that were reversed in contrast at 6 Hz (mean luminance of 220 cd/m²). Peak sensitivity was derived from recordings over the 0.25 to 1 cpd range. The DEM studies (blue circles,[33] half-filled circle[34]) used 0.07 to 2.4 cpd gratings drifting at a constant velocity of 7 deg/sec or 0.25 cpd gratings drifting at 6 Hz. Dobkins & Teller[34] measured FPL thresholds for 0.25 cpd/6Hz, as well (purple circle). Rasengane et al[36] measured contrast sensitivity for 10 deg luminance fields over the 1 to 25 Hz range (green circles). Peak sensitivity at any temporal frequency is plotted. Contrast sensitivity improves rapidly within each method.

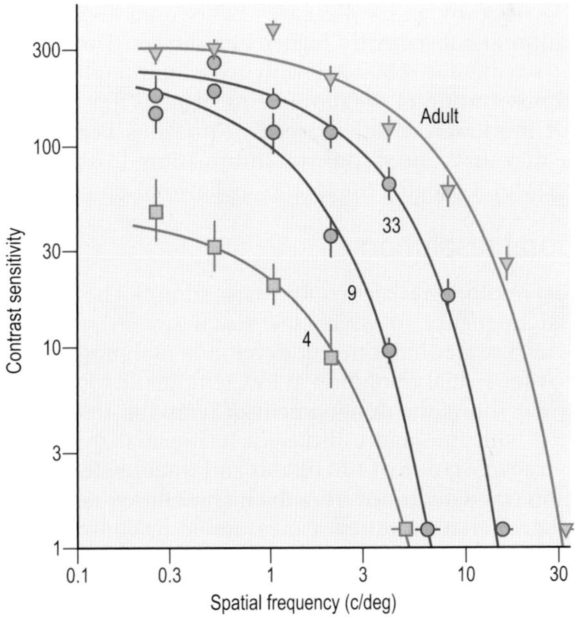

Figure 38.3 VEP contrast sensitivity is re-plotted from Norcia et al[32] as a function of spatial frequency and age for 6 Hz pattern reversal. Sensitivity development is progressively delayed at higher spatial frequencies.

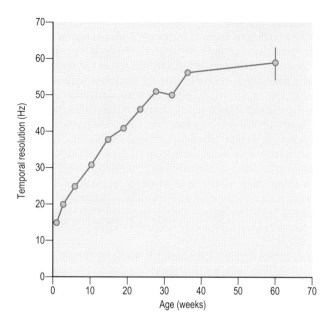

Figure 38.4 Flicker resolution frequency, measured with the steady-state VEP.[152] Resolution frequency increased linearly up to age 40 weeks. Adult values are plotted at 60 weeks. Infants in the 20–30-week age range have resolution acuities that are nearly adult.

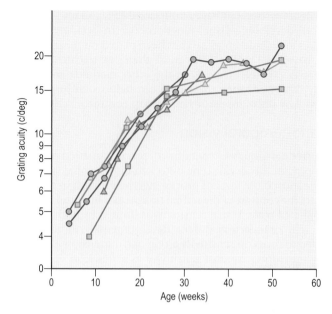

Figure 38.5 Grating acuity as a function of age for 5–10 Hz pattern reversal stimuli. Each study employed the swept spatial frequency technique. Acuity growth functions are similar across studies, with acuity increasing from 4–6 cpd in one month olds to around 15–20 cpd around 8 months of age. Data are re-plotted as follows: blue circles,[153] purple circles,[32] tan triangles,[154] blue triangles,[155] red squares,[156] green squares.[89]

the subject's task was detection of the direction of motion or simple contrast detection.

Hainline & Abramov[33] used directional eye movements recorded by an infrared eye tracker to measure contrast sensitivity. The observer made a forced choice judgment on the output of the tracker (noise level 0.5 deg) rather than on naked eye observation. Contrast sensitivity with this method develops to adult levels by 5 months of age (see Fig. 38.2). Absolute thresholds are lower than those measured with the VEP by a factor of about 4. Hainline & Abramov's contrast sensitivities are higher than those observed by Dobkins & Teller[34] or Brown and colleagues[35] who used naked eye observation at substantially higher luminances. The difference in sensitivities obtained with naked eye and instrumented observation of eye movements suggests that at least some of the lower sensitivity seen in previous behavioral studies may have been due to information loss in the observer who is judging the infants' behavioral output.

Temporal resolution

Temporal resolution is highest for coarse stimuli. The highest temporal frequency to which the visual system responds undergoes a somewhat different developmental progression than that for spatial resolution at low temporal frequencies. Figure 38.4 shows the development of temporal resolution measured with luminance flicker (a low spatial frequency target) in a large group of 130 infants and 6 adults. Temporal resolution was determined from the highest flicker rate from a large series that produced a measurable response. Adult temporal resolutions are approached by 20–30 weeks of age, a time at which grating acuity is still significantly lower than in adults (see Figs 38.3 & 38.5). A psychophysical study in children 4–7 years of age found temporal resolution to be fully adult at age 4, but that grating acuity continued to improve until about 6 years of age.[40]

Grating acuity

At the limit of the high spatial frequency limb of the contrast sensitivity function lies the observer's grating acuity. Grating acuity is limited by the optical quality of the eye, the spacing of the photoreceptors and the spatial pooling properties of the ganglion cells and subsequent receptive field mechanisms.[41-44] Grating acuity is also limited by temporal factors, being maximal at low temporal frequencies.

VEP grating acuity has been measured most commonly using steady-state, pattern reversal targets in the frequency range of 5–10 Hz (10–20 contrast reversals per sec). The acuity measurement is extrapolated from the high spatial frequency portion of the amplitude versus spatial frequency function, an example of which is illustrated in Figure 38.5 for the spatial frequency sweep technique. In this method, originally developed by Regan,[45] the spatial frequency of a temporally modulated pattern is systematically changed (swept) over a large range of spatial frequencies that span the expected acuity limit of the observer. Figure 38.5 plots grating acuity as a function of age for such pattern reversal stimuli. Each study employed the swept spatial frequency technique. Acuity growth functions are similar across studies, with acuity increasing from 4–6 cpd in 1-month-olds to around 15–20 cpd around 8 months of age.

VEP acuity has also been measured with pattern onset–offset stimuli, in both transient and steady-state paradigms. Two studies of the transient on–off acuity growth function found that acuity improved from approximately 2 cpd at 1 month to 30 cpd by 5 months.[46,47] A third study which used checks rather than gratings[48] found an acuity of 2.3 cpd (corrected for Fourier fundamental spatial frequency of the checks) at 8 weeks, with an increase to 8 cpd at 24 weeks.

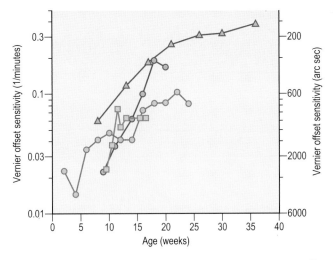

Figure 38.6 Behaviorally determined Vernier sensitivity from: triangles,[55] blue circles,[57] red circles,[59] squares.[60]

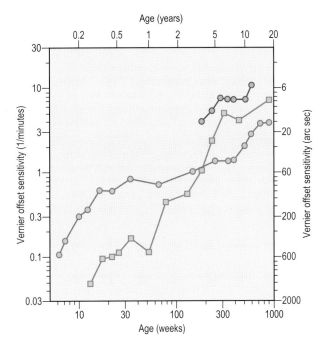

Figure 38.7 Vernier acuity determined by the visually evoked potential (red circles).[74] Shown for comparison are behavioral data from Carkeet and co-workers[62] (blue circles) and from Zanker and colleagues (squares).[58]

When both transient onset–offset and 6 Hz contrast reversal stimuli were used to measure acuity in the same infants, two different rates of growth were found: transient onset–offset acuity increased 0.63 octaves per month versus 0.28 octaves per month with 6 Hz pattern reversal.[46] These observations that the rate at which acuity increases depends on the response component being measured suggest that different post-receptor visual mechanisms have different rates of development.

Vernier acuity

Vernier acuity refers to a collection of spatial localization tasks requiring the detection of a misalignment relative to a reference. Adult Vernier thresholds are significantly better than would be predicted based on either the optical or anatomical properties of the eye. Therefore, Vernier acuity is considered one of the hyperacuities.[49-51] Since cortical processing is believed to be a critical factor limiting the hyperacuities,[52-54] the time course for the development of Vernier acuity has been investigated with great interest.

Most investigations of Vernier acuity during the first year of life have been cross-sectional, employing behavioral responses to moving stimuli.[55-58] However, two studies used stationary stimuli.[59,60] The results of several of these behavioral studies, plotted over the first 6 months of life, are summarized in Figure 38.6. On average, Vernier acuity improves by about a factor of 6–8 over the first 6 months of life, with the best thresholds recorded to be around 200 seconds, 1.3 to 1.8 log units poorer than adults.

While one group reported that by 5 years of age performance on their Vernier acuity task became comparable to that of adults,[58] others report that Vernier development is incomplete at age 5.[61-63] Based on data from preschool children, Vernier acuity is 2 times the adult threshold at 5.6 years of age (confidence interval 3.5–6.5 years).[62] Thus, there is some agreement in the literature that the development of Vernier acuity is incomplete during the early school years. However it is not clear if development is complete before 18 to 20 years of age.[64-69] Typical adult thresholds are in the 3–8 second of arc range.[69-71]

As noted above, the majority of the behavioral paradigms used to investigate the development of Vernier acuity contained motion. Skoczenski & Aslin[72] have demonstrated that temporally modulated stimuli improve the Vernier thresholds of 3-month-old infants and suggest that Vernier thresholds obtained with moving stimuli may be governed by a local motion mechanism rather than a position-sensitive mechanism. Therefore, our understanding of the developmental time course of Vernier acuity and its relationship to the development of other visual functions is potentially confounded by some of the stimuli that have been used in behavioral paradigms to gain the infant's attention and interest in the task.

The visual evoked potential (VEP) offers a unique solution to this problem. Norcia and colleagues[73] have suggested that by analyzing the separate Fourier components in the steady state visually evoked potential, Vernier response components may be isolated from those arising from stimulus motion. The VEP response to the introduction of a Vernier offset (alignment/misalignment) differs from the response to the return to alignment (misalignment/alignment). This asymmetric response to the introduction and then removal of a Vernier offset is reflected in the odd harmonics of the response to the stimulus modulation. The even harmonics in the evoked potential are produced in response to the symmetrical spatial aspects of the stimulus modulation (e.g. local motion of the offset grating) and may be used to examine *motion* acuity. The results obtained with a sweep VEP technique[74] are shown in Figure 38.7. There is a rapid improvement in Vernier acuity over the first 4 months of life. By 6 months of age, Vernier acuity has improved by about a factor of 7 reaching a threshold of nearly 70 seconds, approximately one log unit poorer than the normal adult values obtained psychophysically. Between 6 months of age

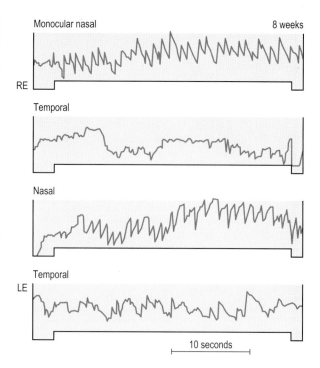

Figure 38.8 Directional asymmetry of monocular optokinetic nystagmus adapted from Naegele & Held.[99] Monocular OKN is robust for nasalward stimulus motion (left-ward motion in the right eye and right-ward motion in the left eye) but is weak for temporal-ward motion. This directional asymmetry declines during the first few months of life.

and 7.5 years, improvement occurs at a slower rate reaching about 40 seconds, a 1.75 times improvement over the course of about 7 years. After the age of 7.5 years there is another more rapid improvement in Vernier acuity. Although quantitatively better thresholds were obtained psychophysically by Carkeet and colleagues with stationary stimuli, there is a similar period during the early school years where no significant change in threshold was found.

The developmental time course for Vernier acuity appears to occur in at least 3 phases. There is a rapid development over the first 4–6 months of life, then a more gradual improvement up to about 7–10 years of age, followed by a more rapid improvement to reach adult levels no later than 18–20 years of age. Box 38.1 describes clinical applications of acuity measures and their interpretation.

Optotype acuity

Optotype recognition acuity as measured with standard eye charts shows a similar prolonged developmental time course. Improvement has been reported to continue until 25[75] to 29[76] years of age, reaching 0.56 MAR (minimum angle of resolution expressed in minutes of arc[75]) to 0.67–0.69 MAR.[76–78] At age 3–5 years, chart acuity is 1.25 MAR,[79,80] about a factor of 1.8 poorer than the most conservative estimate of the adult acuity noted above. At 6.8 years of age, chart acuity has improved to 1.09 MAR,[80] a factor of 1.6 poorer than the most conservative estimate of normal adult recognition acuity.

Motion

The detection of motion involves the determination of speed and direction. The presence of direction-selective mechanisms early in development has been demonstrated using each of the three major methods, FPL, VEP and OKN. Directionally appropriate eye movements can be seen at term or even before.[9,95] Uncertainty remains as to whether these early ocular following responses are controlled by cortical or subcortical pathways. Direction selectivity has been demonstrated behaviorally using FPL and looking time-habituation methods by 6–8 weeks.[96,97] VEP responses associated with changes of direction have been recorded by 10 weeks of age.[17,19]

Motion direction asymmetries

On any measure, the adult visual system shows roughly equal sensitivity for all directions of motion.[98] Developing infants, on the other hand, show large, systematic biases in their monocular oculomotor and VEP responses. Monocular OKN is robust for nasal-ward motion, but is weak for temporal-ward motion during the first 3–6 months of life[99–101] as shown in Figure 38.8. The time to attain a symmetric monocular OKN response may depend on the stimulus velocity[101] with time to maturity being later for higher image velocities.

Young infants also show monocular VEPs response asymmetries suggestive of a nasal-ward/temporal-ward bias in cortical responses.[18,19,102–105] These response asymmetries manifest themselves in the monocular steady-state VEP made in response to rapidly oscillating gratings. In adults, oscillating gratings produce responses primarily at twice the stimulus frequency (at the rate that stimulus direction changes, F2). In young infants, an additional response component is present at the stimulus frequency (first harmonic response, F1; see Figure 38.9). The first harmonic is 180 degrees out of phase in the two eyes. This pattern of response – a significant first harmonic that is of opposite phase in the two eyes – is consistent with a response bias that is in opposite directions in the two eyes. The absolute direction of the bias, nasal-ward or temporal-ward, cannot be directly determined from the steady-state response. It is not known at present whether the cortical motion asymmetry tapped by the VEP causes the oculomotor asymmetry or whether the two phenomena represent immaturities

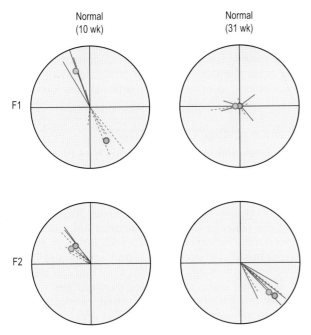

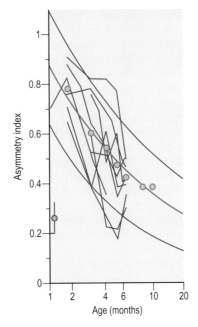

Figure 38.9 VEP motion asymmetry, adapted from Norcia et al.[18] Steady-state VEPs to monocular oscillatory motion contain significant odd-harmonics in young infants (F1). The phase of these components is shifted by 180 degrees between the two eyes in an 8-week-old infant (top left plot, individual trials are indicated by separate lines radiating from zero). The presence of odd harmonics suggests that the response to left and right motion is asymmetric and that the dominant direction is opposite in the two eyes. The odd harmonic components decline in amplitude over the first 5–6 months of age as indicated by the data from a normal 31-week-old (right panels). Second harmonic response (F2) are present at both ages.

Figure 38.10 Developmental sequence for symmetric motion VEPs.[19] The degree of motion response asymmetry can be quantified by calculating the fraction of the total response (first plus second harmonic) that is contributed by the first harmonic (asymmetry index). The monocular oscillatory motion VEP is dominated by the first harmonic in early infancy (asymmetry index greater than 0.5), but the degree of asymmetry declines rapidly over the first 6 months for 6 Hz, 1 cpd targets. Red circles indicate mean responses for infants between 1.5 and 10 months. Longitudinal recordings are indicated by the thin lines. The smooth curves indicate a fit to the mean growth function ± 1 standard deviation. The blue circle indicates the average asymmetry index for five infants between 0.5 and 1 month of age.

in independent mechanisms with similar developmental sequences.

Symmetric cortical motion responses develop during the first 6 months of life (Fig. 38.10) for 6 Hz oscillatory motion of low spatial frequency gratings. Figure 38.10 plots the ratio of amplitudes at the first harmonic to the sum of amplitudes at the first and second harmonics. This index runs from 1.0 for a completely asymmetric response to 0 for a completely symmetric response. Infants reach adult levels by about 5 months of age for low spatial frequency targets oscillating at 6 Hz.

The VEP motion asymmetry has been recorded in infants as young as 2 months of age,[19] suggesting that cortical direction selectivity is present at this time. Interestingly, the motion asymmetry was undetectable in infants younger than 8 weeks at 6 Hz, 1 c/deg. Development of motion-specific VEPs over the neonatal period has also been observed[17] in a different stimulation paradigm. Wattam-Bell[17] also found that the age of first direction-specific responses was earlier for lower stimulus velocities. Box 38.2 describes applications of the motion VEP to patients with strabismus.

Binocular vision

This section explores the development of the sensory aspects of binocular vision, sensory fusion (the neurological combination of the image from each eye into a single percept) and stereopsis (the perception of depth based on a horizontal

---

Box 38.2 Motion asymmetries in strabismus

The monocular motion VEP asymmetry persists in patients with early-onset esotropia,[18,19,104,106–108] but is not present in patients with late-onset esotropia.[109,110] Levels of VEP motion asymmetry can be reduced by alternate occlusion therapy[103] or by early, successful surgical alignment.[102,104,111,112] The development of motion-processing mechanisms is thus dependent on the presence of normal binocular interactions during early visual development and can be disrupted by abnormal binocular interactions that are secondary to strabismus. Early and accurate correction of strabismus appears to restore a degree of binocular interaction that is sufficient to promote more normal development of motion-processing mechanisms.

---

disparity between the retinal images in each eye). As with all visual processes, the parameters of the stimulus and the method of measurement may limit the infant's performance as well as our understanding of the developing visual system. Dichoptic stimulus presentations and the need to eliminate confounding monocular cues present unique challenges to studies of the development of fusion and stereopsis and have guided the selection of the studies reviewed.

Fusion

The most direct approach in determining the onset of sensory fusion in human infants has been to record a "cyclopean" VEP using one of two different stimuli. The first approach popularized by Baitch & Levi[113] and applied to infants by France & Ver Hoeve[114] presents a uniform field to each eye that is modulated at two different temporal frequencies (e.g. 8 Hz right eye, 6 Hz left eye). A cortical response recorded at either the sum (14 Hz) or the difference frequency (2 Hz, also termed the beat frequency) suggests the presence of binocular neurons that integrate and respond to the input from both eyes. Using this approach, France & Ver Hoeve[114] reported that a binocular response was present in the majority of normal infants by 2 months of age. The two-frequency approach has also been used to study the development of fusion and rivalry by Brown et al.[115] In that study, beat frequency responses were recordable only for stimuli that had the same orientation in the two eyes.

The second approach to identify the integration of information from the two eyes employs dynamic random dot correlograms. Correlograms consist of separate fields of random dots presented in a dichoptic manner to the two eyes. Correlograms alternate between two different phases – correlated and anticorrelated. In the correlated phase, the dot patterns are identical in the two eyes. In the anticorrelated phase, a dark dot in the pattern presented to one eye corresponds to a bright dot in the pattern presented to the fellow eye. The cortical response to the alternation between the two phases is recorded. The "cyclopean VEP" response to these correlated/anticorrelated stimuli has been interpreted as fusion without stereopsis since horizontal disparities are not induced by the stimuli.[116-118] Using this approach, the onset of sensory fusion has been reported to occur between 3 and 5 months of age.[21,22,119]

Stereopsis and disparity sensitivity

Adult observers are able to associate the sign of disparity, crossed or uncrossed, with a percept of whether a stereoscopic target appears to be nearer or farther than fixation. These "signed" perceptual reports are definitive evidence for perceptual stereopsis. Unfortunately, perceptual reports of this nature cannot be obtained from pre-verbal subjects, and other less definitive criteria must be used to determine whether stereopsis, per se, is likely to be present. Most studies of the development of stereopsis in pre-verbal subjects have taken sensitivity to horizontal disparity as being synonymous with the presence of adult-like stereopsis. However, recent studies in macaque have suggested that cells that can support a signed percept of depth are found in higher-level visual areas, rather than in V1. Cells in V1 and MT respond in a disparity tuned fashion to stimuli (reverse contrast images in the two eyes) that do not support a percept of depth.[120,121] Disparity tuning per se, while necessary for stereopsis is thus not entirely sufficient to support depth from disparity percepts (stereopsis). Higher-level cortical areas do distinguish these two stimuli and are also tuned for relative disparity (differences in the disparity/distance of different surfaces in the visual field), whereas cells in V1 and MT are not.[122-124] Relative disparity sensitivity underlies form perception based on disparity, while absolute disparity processing may underlie the control of vergence eye movements (see Neri[125] and Parker[122] for reviews of disparity processing in different cortical areas). Additional criteria beyond simple sensitivity to disparity that can be used in non-invasive studies in humans will be needed to fully

understand the developmental sequence for stereopsis in humans. In the meantime, the existing studies of the development of stereopsis will be interpreted conservatively – as being indicative of the development of disparity sensitivity – leaving open the possibility that development of more complex disparity processing mechanisms may occur at later ages.

Age of onset of disparity sensitivity

Due to the limited number of reports that have investigated disparity sensitivity in the developing visual system with the visual evoked potential, this section will emphasize studies that have applied the behavioral techniques of preferential looking and ocular following movements. Behaviorally, studies of the onset of disparity sensitivity have used real depth tests, local or linear stereograms and global or random dot stereograms.[126] However, real depth targets and local tests of stereopsis contain monocular cues that can yield a positive response and confuse the interpretation of the results. Since many of the early reports that investigated the onset of disparity sensitivity relied on real depth tests with multiple cues to depth, these will not be reviewed (see Daw[31] for review). Studies employing random dot stereograms that rely on the detection of a horizontal displacement in the image presented to each eye without monocular cues will be emphasized. Results obtained from linear stereograms with adequate controls will also be summarized.

There is remarkable consistency among the various laboratories and techniques concerning the onset of disparity sensitivity in random dot patterns that have no monocular cues. Sensitivity emerges between 3 and 5 months of age whether determined with the VEP[22,119,127,128] or by behavioral methods.[119,127,129-131]

Several reports from the same laboratory using disparities ranging from 58 to 32 minutes of arc, have also reported the onset of local or linear stereopsis to be between 3 and 5 months of age.[132-135]

The age at which at least 75% of infants demonstrate sensory fusion or disparity sensitivity is summarized across studies in Figure 38.11. Details concerning the stimulus and the measurement method for each study are provided in the figure legend.

Development of disparity sensitivity

Following the emergence of measurable disparity sensitivity between 3 and 5 months of age, the improvement in sensitivity for line stereograms during the first year of life has been reported to be quite rapid. For a small sample of infants followed longitudinally (n = 16), sensitivity for crossed disparities improved from 58 min at 14.8 weeks of age to 1 min of arc by 20 weeks of age.[132] A similar time course was noted for uncrossed disparities that emerged slightly later at 16.8 weeks and improved from 58 min of arc to 1 minute by 23.2 weeks. Since the five adults tested on the same apparatus averaged 97.6% correct for the 1 min crossed disparity stimulus and 80.2% for the 0.5 min crossed disparity stimulus but only 62.6% correct for the 0.5 min uncrossed disparity target, the authors suggested that the performance of the 5–6-month-old infants was adult-like.[132]

However, it is important to recall that stereoacuity belongs to a family of localization tasks known as hyperacuities.

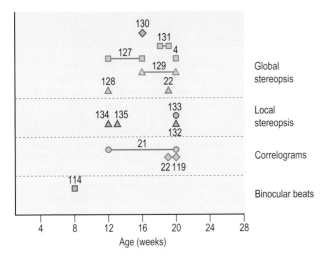

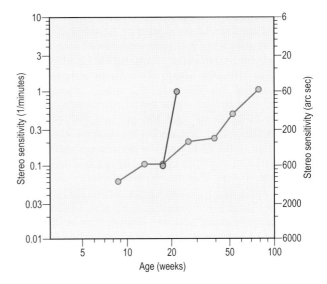

Figure 38.11 Age at which at least 75% of those tested demonstrated fusion or stereopsis. Filled symbols represent studies using the visually evoked potential (VEP), open symbols show studies using behavioral paradigms (FPL), gray symbols indicate both FPL and VEP. 1.[114] 6 & 8 Hz dichoptic presentation, 2.[119] correlograms, 20 min disparity (FPL), 30 min disparity (VEP), 3.[22] correlograms, 40 min disparity, 4.[21] correlograms, 5.[134] 45 min crossed disparity, 6.[135] 32 min crossed disparity, 7.[132] 58 min crossed and uncrossed disparity, 8.[133] 58 min disparity, 9.[128] alternation of 100 min crossed to 100 min uncrossed disparity, 10.[127] < 1735 disparity, 11.[129] 30 min crossed disparity, 12.[130] 60 min disparity, 13 random dot stereogram.[131]

Figure 38.12 The development of disparity sensitivity (seconds of arc) as a function of age (months). Global disparity sensitivity with Polaroid random dot stereograms (red circles[127]). The smallest linear disparity that 50% or more of the infants tested reached criterion (blue circles[132]).

Hyperacuities provide finer localization than predicted based on either the optical or anatomical[49,51] properties of the eye. Adult stereothresholds obtained with real depth stimuli are extremely fine and range from about 2 to 5 seconds with the Howard Dolman apparatus.[70,136,137] Similar stereothresholds of about 5 seconds have been reported with the diastereo test[138–140] another real depth test of stereopsis. Adult stereothresholds with random dot stereograms have been reported to range from 5.37 sec,[137] to about 30 sec[141] with Simons[142] reporting 15.1 sec. Thus, a closer look at the time course for the development of stereoacuity appears warranted. Based on the number of behavioral studies available from a variety of laboratories covering a large age range, the development of global stereopsis was selected for review. As mentioned previously, global stereopsis has the additional advantage of being based solely on the detection of the horizontal image displacement between the two eyes since the stimulus contains no monocular cues.

Figure 38.12 shows the development of disparity sensitivity during the first 18 months of life. The red symbols indicate the mean stereoacuity measured behaviorally in response to random dot stereograms.[127] The blue symbols are taken from Birch et al[132] and represent the smallest linear disparity at which 50% or more of the infants tested reached criterion. Over the first year of life, global stereoacuity improves by a factor of 8, to about 120 seconds of arc, but has not yet reached adult levels. This improvement is greater than the factor of 4 or 5 reported for the development of spatial resolution over the same time period.

Figure 38.13 illustrates the continued development of global disparity sensitivity during the preschool years. All show continued improvement during the pre-school years. Two studies also report thresholds from normal adults under similar test conditions. Ciner et al[143] (cyan triangles) indicate

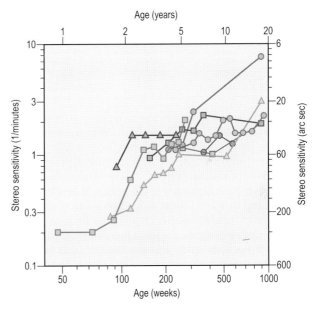

Figure 38.13 The development of global stereopsis (seconds of arc) as a function of age (years). All studies are based on a behavioral response to a random dot stereogram during the preschool years. Cyan triangles,[143] green squares,[157] tan circles,[142] blue triangles,[158] purple squares,[145] grey squares,[141] olive circles,[159] red circles.[144]

that at age 5 disparity sensitivity is still about a factor of 3 poorer than the adult. Simons[142] (tan circles) also finds sensitivity is a factor of 3 poorer for 6-year-olds compared to adults.

Four studies that measured disparity sensitivity with the commercially available TNO test in school-aged children are illustrated in Figure 38.13. Walraven & Janzen[144] (red circles) report that the ability to detect disparity in random dot stereograms improves by a factor of 2 between the ages of 4 and

12 years. The data also suggest additional improvement up to 18 years of age, the oldest reported in the study. When the small number (n = 4) of 6–10-year-olds tested by Sloper & Collins[141] (purple squares) are compared to 12 adults with normal vision, disparity sensitivity remains a factor of 2 poorer than that measured in the adults. Heron and co-workers[145] (red circles) report that disparity sensitivity as measured by the TNO was adult-like at 5 years of age. These investigators also noted that among the 4 stereotests evaluated in the report, the TNO gave the poorest thresholds and the largest variability in their sample. Using a real depth test, the Frisby Stereo Test, they found that disparity sensitivity was not yet adult-like at 7 years of age. Fox et al[136] reported that with the Howard–Dolman real depth test, children 3 to 5 years of age were about 2.5 times less sensitive to disparity than adults (12.6 versus 4.9 seconds of arc). The Howard–Dolman task involves a depth alignment judgment where the presence of stereopsis is indicated when the binocular threshold is better than the monocular threshold.

In summary, sensory fusion has been demonstrated as early as 2 months of age while disparity sensitivity emerges no later than 3 and 5 months of age. Although there is about a factor of 8 improvement in disparity sensitivity for globally defined targets over the first year of life, disparity sensitivity does not reach adult levels until some time after age 5. At age 5 or 6, disparity sensitivity is still about 3 times poorer than that of the adult. Box 38.3 discusses the clinical implications of multiple developmental sequences.

Summary

Maturation of visual function in humans occurs over very different time scales, depending on the particular aspect of function that is being measured. Within a particular visual function, development may proceed at quite different rates during different post-natal periods. In several of the functions reviewed above, there is an early, rapid period of development that is followed by a long, slow second phase that lasts into childhood or even adolescence. Low spatial frequency contrast sensitivity appears to be the most precocious function yet measured. Flicker sensitivity, another function that is measured with low-spatial-frequency targets also matures relatively quickly, reaching adult levels during the first 6 months. Grating acuity measured with mid-frequency contrast reversal has its initial rapid period of development during the first 8 months, while Vernier acuity appears to develop at a different rate over this period and is far from adult levels.

Variations in the developmental sequences across visual functions presumably reflect limitations imposed at different levels of the visual hierarchy shown in Figure 38.1. The critical immaturity limiting contrast sensitivity, grating acuity and temporal resolution may lie in the early part of the pathway comprised of the retina and early cortical visual areas. Functions such as Vernier and stereo acuity are known to be quite sensitive to contrast in adults and thus some of the immaturity in these functions is likely to be secondary to the reduced contrast sensitivity of the infant. However, other factors must also limit infant Vernier and stereoacuity, since their developmental sequences are longer than those underlying contrast sensitivity and grating acuity.

Box 38.3 Sensitive and critical periods in visual development

Periods of rapid maturation in many sensory systems are associated with high levels of sensitivity to the quality and structure of the inputs from the environment[146]. Periods during development in which the effects of visual experience on development are particularly strong are referred to as "sensitive periods". If visual experience during a sensitive period is essential for development and permanently alters performance, this type of sensitive period is referred to as a "critical period"[147]. Sensitive and critical periods differ in the extent to which experience has a permanent effect. If normal experience is not present during a critical period, restoration of normal input cannot rehabilitate normal function. Abnormal visual experience either before or after this period is not effective in disrupting development.

The presence and timing of a sensitive/critical period for the development of binocular vision in humans has been inferred from studies of explicitly binocular functions such as the inter-ocular transfer of the tilt-after effect[148] or stereoacuity[149] in patients who had strabismus with onset at different times after birth and in whom the strabismus persisted for different durations. Banks and co-workers[148] developed a simple descriptive mathematical model that they used to infer the time during development that the binocular visual system is most sensitive to disruption by strabismus. Using interocular transfer of the tilt after-effect as the endpoint measure of binocular development they found that abnormal visual experience around 1.7 years of age was maximally deleterious to final development for patients with infantile esotropia, but that the peak occurred at 1.0 year in patients with late-onset esotropia. The same model was used to locate the sensitive period for the development of stereopsis[149]. Patients with infantile esotropia showed peak sensitivity to abnormal experience around 0.36 years, but that patients with accommodative esotropia had maximal sensitivity around 1.7 years. The fact that different outcome measures yield different estimates of the sensitive period, as do different kinds of strabismus suggests that there are multiple binocular sub-systems within the visual system, as a whole, and that these each have different developmental sequences. The existence of different sensitive periods for different visual functions is well established in both animal[150] and human models[151].

References

1. Fantz R. A method for studying early visual development. Percept Motor Skills 1956; 6:13.
2. Fantz R. Pattern vision in young infants. Psychol Rec 1958; 8:43–47.
3. Teller DY. The forced-choice preferential looking procedure: a psychophysical technique for use with human infants. Infant Behav Dev 1979; 2:135–153.
4. Regan D. Human brain electrophysiology: evoked potentials and evoked magnetic fields in science and medicine. New York: Elsevier, 1989.
5. Ducati A, Fava E, Motti ED. Neuronal generators of the visual evoked potentials: intracerebral recording in awake humans. Electroenceph Clin Neurophysiol 1988; 71(2):89–99.
6. Noachtar S, Hashimoto T, Luders H. Pattern visual evoked potentials recorded from human occipital cortex with chronic subdural electrodes. Electroenceph Clin Neurophysiol 1993; 88(6):435–446.
7. Arroyo S, Lesser RP, Poon WT, Webber WR, Gordon B. Neuronal generators of visual evoked potentials in humans: visual processing in the human cortex. Epilepsia 1997; 38(5):600–610.
8. Miles FA. Short-latency visual stabilization mechanisms that help to compensate for translational disturbances of gaze. Ann NY Acad Sci 1999; 871:260–271.
9. Hainline L. Conjugate eye movements of infants. In: Early visual development, Normal and abnormal. New York: Oxford University Press, 1993:47–79.
10. Leigh R, Zee D. The neurology of eye movements, 3rd edn. Oxford: Oxford University Press, 1999.
11. Shupert C, Fuchs AF. Development of conjugate human eye movements. Vision Res 1988; 28(5):585–596.
12. Hubel DH, Wiesel TN. Receptive fields, binocular interaction and functional architecture in the cat's visual cortex. J Physiol 1962; 160:106–154.

13. Barlow HB, Blakemore C, Pettigrew JD. The neural mechanism of binocular depth discrimination. J Physiol 1967; 193(2):327–342.

14. Schiller PH, Finlay BL, Volman SF. Quantitative studies of single-cell properties in monkey striate cortex. II. Orientation specificity and ocular dominance. J Neurophysiol 1976; 39(6):1320–1333.

15. Braddick OJ, Wattam-Bell J, Atkinson J. Orientation-specific cortical responses develop in early infancy. Nature 1986; 320(6063):617–619.

16. Manny RE. Orientation selectivity of 3-month-old infants. Vision Res 1992; 32(10):1817–1828.

17. Wattam-Bell J. Development of motion-specific cortical responses in infancy. Vision Res 1991; 31(2):287–297.

18. Norcia AM, Garcia H, Humphry R, Holmes A, Hamer RD, Orel-Bixler D. Anomalous motion VEPs in infants and in infantile esotropia. Invest Ophthalmol Vis Sci 1991; 32(2):436–439.

19. Birch EE, Fawcett S, Stager D. Co-development of VEP motion response and binocular vision in normal infants and infantile esotropes. Invest Ophthalmol Vis Sci 2000; 41(7):1719–1723.

20. Skoczenski AM, Norcia AM. Development of VEP Vernier acuity and grating acuity in human infants. Invest Ophthalmol Vis Sci 1999; 40(10):2411–2417.

21. Braddick O, Atkinson J, Julesz B, Kropfl W, Bodis-Wollner I, Raab E. Cortical binocularity in infants. Nature 1980; 288(5789):363–365.

22. Petrig B, Julesz B, Kropfl W, Baumgartner G, Anliker M. Development of stereopsis and cortical binocularity in human infants: electrophysiological evidence. Science (New York) 1981; 213(4514):1402–1405.

23. Cobb W, Morton H, Ettlinger G. Cerebral potentials evoked by pattern reversal and their suppression by visual rivalry. Nature 1967; 216(5120):1123.

24. Lansing RW. Electroencephalographic correlates of binocular rivalry in man. Science (New York) 1964; 146:1325–1327.

25. Brown RJ, Norcia AM. A method for investigating binocular rivalry in real-time with the steady-state VEP. Vision Res 1997; 37(17):2401–2408.

26. Wright KW, Fox BE, Eriksen KJ. PVEP evidence of true suppression in adult onset strabismus. J Pediatr Ophthalmol Strabismus 1990; 27(4):196–201.

27. Norcia AM, Harrad RA, Brown RJ. Changes in cortical activity during suppression in stereoblindness. Neuroreport 2000; 11(5):1007–1012.

28. Bach M, Meigen T. Similar electrophysiological correlates of texture segregation induced by luminance, orientation, motion and stereo. Vision Res 1997; 37(11):1409–1414.

29. Dobson V. The behavioral assessment of visual acuity in human infants. In: Berkeley MA, Stebbins WC, eds. Comparative perception. Pittsburgh: Wiley, 1990.

30. Hamer R, Mayer L. The development of spatial vision. In: Albert DM, Jakobiec FA, eds. Principles and practice of ophthalmology: basic sciences. Philadelphia: WB Saunders Co, 1994:578–608.

31. Daw N. Visual development, 2nd edn. Berlin: Springer, 1995.

32. Norcia AM, Tyler CW, Hamer RD. Development of contrast sensitivity in the human infant. Vision Res 1990; 30(10):1475–1486.

33. Hainline L, Abramov I. Eye movement-based measures of development of spatial contrast sensitivity in infants. Optom Vis Sci 1997; 74(10):790–799.

34. Dobkins KR, Teller DY. Infant contrast detectors are selective for direction of motion. Vision Res 1996; 36(2):281–294.

35. Brown AM, Lindsey DT, McSweeney EM, Walters MM. Infant luminance and chromatic contrast sensitivity: optokinetic nystagmus data on 3-month-olds. Vision Res 1995; 35(22):3145–3160.

36. Rasengane TA, Allen D, Manny RE. Development of temporal contrast sensitivity in human infants. Vision Res 1997; 37(13):1747–1754.

37. Kelly JP, Borchert K, Teller DY. The development of chromatic and achromatic contrast sensitivity in infancy as tested with the sweep VEP. Vision Res 1997; 37(15):2057–2072.

38. Skoczenski AM, Norcia AM. Neural noise limitations on infant visual sensitivity. Nature 1998; 391(6668):697–700.

39. Shannon E, Skoczenski AM, Banks MS. Retinal illuminance and contrast sensitivity in human infants. Vision Res 1996; 36(1):67–76.

40. Ellemberg D, Lewis TL, Liu CH, Maurer D. Development of spatial and temporal vision during childhood. Vision Res 1999; 39(14):2325–2333.

41. Banks MS, Bennett PJ. Optical and photoreceptor immaturities limit the spatial and chromatic vision of human neonates. J Opt Soc Am A 1988; 5(12):2059–2079.

42. Wilson HR. Development of spatiotemporal mechanisms in infant vision. Vision Res 1988; 28(5):611–628.

43. Banks MS, Crowell JA. Front-end limitations to infant spatial vision: examination of two analyses. In: Simons K, ed. Early visual development: normal and abnormal. New York: Oxford University Press, 1993.

44. Wilson HR. Theories of infant visual development. In: Simons K, ed. Early visual development: normal and abnormal. New York: Oxford University Press, 1993.

45. Regan D. Speedy assessment of visual acuity in amblyopia by the evoked potential method. Ophthalmologica 1977; 175(3):159–164.

46. Orel-Bixler D, Norcia AM. Differential growth of acuity for steady-state pattern reversal and transient pattern VEPs. Clin Vis Sci 1987; 2:1–9.

47. Marg E, Freeman DN, Peltzman P, Goldstein PJ. Visual acuity development in infants: Evoked potential measurements. Invest Ophthalmol 1976; 15:150–153.

48. DeVries-Khoe LH, Spekreijse H. Maturation of luminance and pattern EPs in man. Doc Ophthalmol Proc Ser 1982; 31:461–475.

49. Westheimer G. Editorial: Visual acuity and hyperacuity. Invest Ophthalmol 1975; 14(8):570–572.

50. Westheimer G. The spatial sense of the eye. Proctor lecture. Invest Ophthalmol Vis Sci 1979; 18(9):893–912.

51. Morgan M. Hyperacuity. In: Regan D, ed. Spatial vision, vol 10: Vision and visual dysfunction. Boca Raton, FL: CRC Press, 1991.

52. Barlow HB. Reconstructing the visual image in space and time. Nature 1979; 279(5710):189–190.

53. Westheimer G. The spatial grain of the perifoveal visual field. Vision Res 1982; 22(1):157–162.

54. Stanley OH. Cortical development and visual function. Eye (London) 1991; 5(Pt1):27–30.

55. Shimojo S, Birch EE, Gwiazda J, Held R. Development of Vernier acuity in infants. Vision Res 1984; 24(7):721–728.

56. Manny RE, Klein SA. A three alternative tracking paradigm to measure Vernier acuity of older infants. Vision Res 1985; 25(9):1245–1252.

57. Shimojo S, Held R. Vernier acuity is less than grating acuity in 2- and 3-month-olds. Vision Res 1987; 27(1):77–86.

58. Zanker J, Mohn G, Weber U, Zeitler-Driess K, Fahle M. The development of Vernier acuity in human infants. Vision Res 1992; 32(8):1557–1564.

59. Manny RE, Klein SA. The development of Vernier acuity in infants. Curr Eye Res 1984; 3(3):453–462.

60. Brown AM. Vernier acuity in human infants: rapid emergence shown in a longitudinal study. Optom Vis Sci 1997; 74(9):732–740.

61. Gonzalez EG, Steinbach MJ, Ono H, Rush-Smith N. Vernier acuity in monocular and binocular children. Clin Vis Sci 1992; 7:257–261.

62. Carkeet A, Levi DM, Manny RE. Development of Vernier acuity in childhood. Optom Vis Sci 1997; 74(9):741–750.

63. Kim E, Enoch JM, Fang MS et al. Performance on the three-point Vernier alignment or acuity test as a function of age: measurement extended to ages 5 to 9 years. Optom Vis Sci 2000; 77(9):492–495.

64. Whitaker D, Elliot D, MacVeigh D. Variations in hyperacuity performance with age. Ophthalmic Physiol Opt 1992; 12:29–32.

65. Reich L, Lahshminarayanan V, Enoch JM. Analysis of the method of adjustment of testing potential acuity with the hyperacuity gap test: a preliminary report. Clin Vis Sci 1991; 6:451–456.

66. Odom JV, Vasquez RJ, Schwartz TL, Linberg JV. Adult Vernier thresholds do not increase with age; Vernier bias does. Invest Ophthalmol Vis Sci 1989; 30(5):1004–1008.

67. Lakshminarayanan V, Aziz S, Enoch JM. Variation of the hyperacuity gap function with age. Optom Vis Sci 1992; 69(6):423–426.

68. Vilar E, Giraldez-Fernandez MJ, Enoch JM, Lakshminarayanan V, Knowles R, Srinivasan R. Performance on three-point Vernier acuity targets as a function of age. J Opt Soc Am 1995; 12(10):2293–2304.

69. Li RW, Edwards MH, Brown B. Variation in Vernier acuity with age. Vision Res 2000; 40(27):3775–3781.

70. Berry RN. Quantitative relations among Vernier, real depth, and stereoscopic depth acuities. J Exp Psych 1948; 38:708–721.

71. Westheimer G, McKee SP. Spatial configurations for visual hyperacuity. Vision Res 1977; 17(8):941–947.

72. Skoczenski AM, Aslin RN. Spatiotemporal factors in infant position sensitivity: single bar stimuli. Vision Res 1992; 32(9):1761–1769.

73. Norcia AM, Wesemann W, Manny RE. Electrophysiological correlates of Vernier and relative motion mechanisms in human visual cortex. Vis Neurosci 1999; 16(6):1123–1131.

74. Skoczenski AM, Norcia AM. Late maturation of visual hyperacuity. Psychol Sci 2002; 13(6):537–541.

75. Frisen L, Frisen M. How good is normal visual acuity? A study of letter acuity thresholds as a function of age. Albrecht von Graefes Archiv fur klinische und experimentelle Ophthalmologie 1981; 215(3):149–157.

76. Elliott DB, Yang KC, Whitaker D. Visual acuity changes throughout adulthood in normal, healthy eyes: seeing beyond 6/6. Optom Vis Sci 1995; 72(3):186–191.

77. Brown B, Lovie-Kitchin J. Repeated visual acuity measurement: establishing the patient's own criterion for change. Optom Vis Sci 1993; 70(1):45–53.

78. Owsley C, Sekuler R, Siemsen D. Contrast sensitivity throughout adulthood. Vision Res 1983; 23(7):689–699.

79. Sprague JB, Stock LA, Connett J, Bromberg J. Study of chart designs and optotypes for preschool vision screening – I. Comparability of chart designs. J Pediatr Ophthalmol Strabismus 1989; 26(4):189–197.

80. Bowman RJ, Williamson TH, Andrews RG, Aitchison TC, Dutton GN. An inner city preschool visual screening programme: long-term visual results. Br J Ophthalmol 1998; 82(5):543–548.

81. Ridder WH, 3rd, Rouse MW. Predicting potential acuities in amblyopes: predicting post-therapy acuity in amblyopes. Doc Ophthalmol 2007; 114(3):135–145.

82. Simon JW, Siegfried JB, Mills MD, Calhoun JH, Gurland JE. A new visual evoked potential system for vision screening in infants and young children. J AAPOS 2004; 8(6):549–554.

83. Lim M, Soul JS, Hansen RM, Mayer DL, Moskowitz A, Fulton AB. Development of visual acuity in children with cerebral visual impairment. Arch Ophthalmol 2005; 123(9):1215–1220.

84. Skoczenski AM, Good WV. Vernier acuity is selectively affected in infants and children with cortical visual impairment. Dev Med Child Neurol 2004; 46(8):526–532.

85. John FM, Bromham NR, Woodhouse JM, Candy TR. Spatial vision deficits in infants and children with Down syndrome. Invest Ophthalmol Vis Sci 2004; 45(5):1566–1572.

86. da Costa MF, Salomao SR, Berezovsky A, de Haro FM, Ventura DF. Relationship between vision and motor impairment in children with spastic cerebral palsy: new evidence from electrophysiology. Behav Brain Res 2004; 149(2):145–150.

87. Bradfield YS, France TD, Verhoeve J, Gangnon RE. Sweep visual evoked potential testing as a predictor of recognition acuity in albinism. Arch Ophthalmol 2007; 125(5):628–633.

88. Till C, Westall CA, Koren G, Nulman I, Rovet JF. Vision abnormalities in young children exposed prenatally to organic solvents. Neurotoxicology 2005; 26(4):599–613.

89. Birch EE, Hoffman DR, Uauy R, Birch DG, Prestidge C. Visual acuity and the essentiality of docosahexaenoic acid and arachidonic acid in the diet of term infants. Pediatric Res 1998; 44(2):201–209.

90. McKee SP, Levi DM, Movshon JA. The pattern of visual deficits in amblyopia. J Vis Electronic Resource 2003; 3(5):380–405.

91. Birch EE, Swanson WH. Hyperacuity deficits in anisometropic and strabismic amblyopes with known ages of onset. Vision Res 2000; 40(9):1035–1040.

92. Bradley A, Freeman RD. Is reduced Vernier acuity in amblyopia due to position, contrast or fixation deficits? Vision Res 1985; 25(1):55–66.

93. Levi DM, Klein S. Hyperacuity and amblyopia. Nature 1982; 298(5871):268–270.

94. Hou C, Good WV, Norcia AM. Validation study of VEP Vernier acuity in normal-vision and amblyopic adults. Invest Ophthalmol Vis Sci. 2007; 48(9):4070–4078.

95. Dubowitz LM, Dubowitz V, Morante A, Verghote M. Visual function in the preterm and fullterm newborn infant. Dev Med Child Neurol 1980; 22(4):465–475.

96. Wattam-Bell J. Visual motion processing in one-month-old infants: habituation experiments. Vision Res 1996; 36(11):1679–1685.

97. Wattam-Bell J. Visual motion processing in one-month-old infants: preferential looking experiments. Vision Res 1996; 36(11):1671–1677.

98. Gros BL, Blake R, Hiris E. Anisotropies in visual motion perception: a fresh look. J Opt Soc Am 1998; 15(8):2003–2011.

99. Naegele JR, Held R. The postnatal development of monocular optokinetic nystagmus in infants. Vision Res 1982; 22(3):341–346.

100. Lewis TL, Maurer D, Chung JY, Holmes-Shannon R, Van Schaik CS. The development of symmetrical OKN in infants: quantification based on OKN acuity for nasalward versus temporalward motion. Vision Res 2000; 40(4):445–453.

101. Roy N, Lachapelle P, Lepore F. Maturation of the optokinetic nystagmus as a function of the speed of stimulation in fullterm and preterm infants. Clin Visual Science 1989; 4:357–366.

102. Norcia AM, Hamer RD, Jampolsky A, Orel-Bixler D. Plasticity of human motion processing mechanisms following surgery for infantile esotropia. Vision Res 1995; 35(23–24):3279–3296.

103. Jampolsky A, Norcia AM, Hamer RD. Preoperative alternate occlusion decreases motion processing abnormalities in infantile esotropia. J Pediatr Ophthalmol Strabismus 1994; 31(1):6–17.

104. Bosworth RG, Birch EE. Direction-of-motion detection and motion VEP asymmetries in normal children and children with infantile esotropia. Invest Ophthalmol Vis Sci 2007; 48(12):5523–5531.

105. Mason AJ, Braddick OJ, Wattam-Bell J, Atkinson J. Directional motion asymmetry in infant VEPs – which direction? Vision Res 2001; 41(2):201–211.

106. Shea SJ, Chandna A, Norcia AM. Oscillatory motion but not pattern reversal elicits monocular motion VEP biases in infantile esotropia. Vision Res 1999; 39(10):1803–1811.

107. Kommerell G, Ullrich D, Gilles U, Bach M. Asymmetry of motion VEP in infantile strabismus and in central vestibular nystagmus. Doc Ophthalmol 1995; 89(4):373–381.

108. Anteby I, Zhai HF, Tychsen L. Asymmetric motion visually evoked potentials in infantile strabismus are not an artifact of latent nystagmus. J AAPOS 1998; 2(3):153–158.

109. Hamer RD, Norcia AM, Orel-Bixler D, Hoyt CS. Motion VEPs in late-onset esotropia. Clin Vis Sci 1993; 8(1):55–62.

110. Brosnahan D, Norcia AM, Schor CM, Taylor DG. OKN, perceptual and VEP direction biases in strabismus. Vision Res 1998; 38(18):2833–2840.

111. Gerth C, Mirabella G, Li X et al. Timing of surgery for infantile esotropia in humans: effects on cortical motion visual evoked responses. Invest Ophthalmol Vis Sci 2008; 49(8):3432–3437.

112. Tychsen L, Wong AM, Foeller P, Bradley D. Early versus delayed repair of infantile strabismus in macaque monkeys: II. Effects on motion visually evoked responses. Invest Ophthalmol Vis Sci 2004; 45(3):821–827.

113. Baitch LW, Levi DM. Evidence for nonlinear binocular interactions in human visual cortex. Vision Res 1988; 28(10):1139–1143.

114. France TD, Ver Hoeve JN. VECP evidence for binocular function in infantile esotropia. J Pediatr Ophthalmol Strabismus 1994; 31(4):225–231.

115. Brown RJ, Candy TR, Norcia AM. Development of rivalry and dichoptic masking in human infants. Invest Ophthalmol Vis Sci 1999; 40(13):3324–3333.

116. Eizenman M, Westall CA, Geer I et al. Electrophysiological evidence of cortical fusion in children with early-onset esotropia. Invest Ophthalmol Vis Sci 1999; 40(2):354–362.

117. Livingstone MS. Differences between stereopsis, interocular correlation and binocularity. Vision Res 1996; 36(8):1127–1140.

118. Poggio GF, Gonzalez F, Krause F. Stereoscopic mechanisms in monkey visual cortex: binocular correlation and disparity selectivity. J Neurosci 1988; 8(12):4531–4550.

119. Birch E, Petrig B. FPL and VEP measures of fusion, stereopsis and stereoacuity in normal infants. Vision Res 1996; 36(9):1321–1327.

120. Cumming BG, Parker AJ. Responses of primary visual cortical neurons to binocular disparity without depth perception. Nature 1997; 389(6648):280–283.

121. Krug K, Cumming BG, Parker AJ. Comparing perceptual signals of single V5/MT neurons in two binocular depth tasks. J Neurophysiol 2004; 92(3):1586–1596.

122. Parker AJ. Binocular depth perception and the cerebral cortex. Nature reviews. 2007; 8(5):379–391.

123. Uka T, Tanabe S, Watanabe M, Fujita I. Neural correlates of fine depth discrimination in monkey inferior temporal cortex. J Neurosci. 2005; 25(46):10796–10802.

124. Umeda K, Tanabe S, Fujita I. Representation of stereoscopic depth based on relative disparity in macaque area V4. J Neurophysiol. 2007; 98(1):241–252.

125. Neri P. A stereoscopic look at visual cortex. J Neurophysiol 2005; 93:1823–1826.

126. Julesz B. Binocular depth perception without familiarity cues. Science (New York) 1964; 145:356–362.

127. Birch EE, Salomao S. Infant random dot stereoacuity cards. J Pediatr Ophthalmol Strabismus 1998; 35(2):86–90.

128. Skarf B, Eizenman M, Katz LM, Bachynski B, Klein R. A new VEP system for studying binocular single vision in human infants. J Pediatr Ophthalmol Strabismus 1993; 30(4):237–242.

129. Archer SM, Helveston EM, Miller KK, Ellis FD. Stereopsis in normal infants and infants with congenital esotropia. Am J Ophthalmol 1986; 101(5):591–596.

130. Shea SL, Fox R, Aslin RN, Dumais ST. Assessment of stereopsis in human infants. Invest Ophthalmol Vis Sci 1980; 19(11):1400–1404.

131. Fox R. Stereopsis in animals and human infants: a review of behavioral investigations. In: Aslin RN, Alberts J, Petersen M, eds. Development of perception: The visual system. New York: Academic Press, 1981.

132. Birch EE, Gwiazda J, Held R. Stereoacuity development for crossed and uncrossed disparities in human infants. Vision Res 1982; 22(5):507–513.

133. Birch EE, Gwiazda J, Held R. The development of vergence does not account for the onset of stereopsis. Perception 1983; 12(3):331–336.

134. Birch EE, Shimojo S, Held R. Preferential-looking assessment of fusion and stereopsis in infants aged 1–6 months. Invest Ophthalmol Vis Sci 1985; 26(3):366–370.

135. Gwiazda J, Bauer J, Held R. Binocular function in human infants: correlation of stereoptic and fusion-rivalry discriminations. J Pediatr Ophthalmol Strabismus 1989; 26(3):128–132.

136. Fox R, Patterson R, Francis EL. Stereoacuity in young children. Invest Ophthalmol Vis Sci 1986; 27(4):598–600.

137. Reading RW, Tanlami T. Finely graded binocular disparities from random-dot stereograms. Ophthalmic Physiol Opt 1982; 2:47–56.

138. Hofstetter H, Bertsch J. Does stereopsis change with age? Am J Optom Physiol Optics 1976; 53:664–667.

139. Hofstetter H. Absolute threshold measurements with the diastereo test. Arch Soc Am Ophthalmol Optom 1968; 6:327.

140. Woo GC, Sillanpaa V. Absolute stereoscopic thresholds as measured by crossed and uncrossed disparities. Am J Optom Physiol Optics 1979; 56(6):350–355.

141. Sloper JJ, Collins AD. Reduction in binocular enhancement of the visual-evoked potential during development accompanies increasing stereoacuity. J Pediatr Ophthalmol Strabismus 1998; 35(3):154–158.

142. Simons K. Stereoacuity norms in young children. Arch Ophthalmol 1981; 99(3):439–445.

143. Ciner EB, Schanel-Klitsch E, Scheiman M. Stereoacuity development in young children. Optom Vis Sci 1991; 68(7):533–536.

144. Walraven J, Janzen P. TNO stereopsis test as an aid to the prevention of amblyopia. Ophthalmic Physiol Opt 1993; 13(4):350–356.

145. Heron G, Dholakia S, Collins DE, McLaughlan H. Stereoscopic threshold in children and adults. Am J Optom Physiol Optics 1985; 62(8):505–515.

146. Berardi N, Pizzorusso T, Maffei L. Critical periods during sensory development. Curr Opin Neurobiol 2000 Feb; 10(1):138–145.

147. Knudsen EI. Sensitive periods in the development of the brain and behavior. J Cog Neurosci 2004; 16(8):1412–1425.

148. Banks MS, Aslin RN, Letson RD. Sensitive period for the development of human binocular vision. Science (New York) 1975; 190(4215):675–677.

149. Fawcett SL, Wang YZ, Birch EE. The critical period for susceptibility of human stereopsis. Invest Ophthalmol Vis Sci 2005; 46(2):521–525.

150. Harwerth RS, Smith EL, 3rd, Duncan GC, Crawford ML, von Noorden GK. Multiple sensitive periods in the development of the primate visual system. Science (New York) 1986; 232(4747):235–238.

151. Lewis TL, Maurer D. Multiple sensitive periods in human visual development: evidence from visually deprived children. Dev Psychobiol 2005; 46(3):163–183.

152. Apkarian P. Temporal frequency responsivity shows multiple maturational phases: state-dependent visual evoked potential luminance flicker fusion from birth to 9 months. Vis Neurosci 1993; 10(6):1007–1018.

153. Norcia AM, Tyler CW. Spatial frequency sweep VEP: visual acuity during the first year of life. Vision Res 1985; 25(10):1399–1408.

154. Allen D, Norcia AM, Tyler CW. Comparative study of electrophysiological and psychophysical measurement of the contrast sensitivity function in humans. Am J Optom Physiol Optics 1986; 63(6):442–449.

155. Sokol S, Moskowitz A, McCormack G. Infant VEP and preferential looking acuity measured with phase alternating gratings. Invest Ophthalmol Vis Sci 1992; 33(11):3156–3161.

156. Auestad N, Montalto MB, Hall RT et al. Visual acuity, erythrocyte fatty acid composition, and growth in term infants fed formulas with long chain polyunsaturated fatty acids for one year. Ross Pediatric Lipid Study. Pediatr Res 1997; 41(1):1–10.

157. Ciner EB, Schanel-Klitsch E, Herzberg C. Stereoacuity development: 6 months to 5 years. A new tool for testing and screening. Optom Vis Sci 1996; 73(1):43–48.

158. Birch EE, Hale LA. Operant assessment of stereoacuity. Clin Vis Sci 1989; 4:295–300.

159. Williams S, Simpson A, Silva PA. Stereoacuity levels and vision problems in children from 7 to 11 years. Ophthalmic Physiol Opt 1988; 8(4):386–389.

CHAPTER 39

Development of Retinogeniculate Projections

S. M. Koch & E. M. Ullian

The dorsal lateral geniculate nucleus of the thalamus (dLGN) is the gateway through which visual information is transmitted from the retina to the cortex; therefore, visual perception relies on the ability of dLGN neurons to faithfully relay the specific features of the visual world that are encoded by the retina. Accordingly, the connections between retinal ganglion cells (RGCs) and dLGN neurons are highly precise and organized.[1,2] For instance, RGC projections are arranged such that neighboring cells in the retina innervate neighboring cells in the dLGN creating a retinotopic map in the thalamus that is capable of maintaining the spatial features of the visual scene.[3] Furthermore, the resolution of the retinal map is also maintained in the dLGN since at maturity few RGCs converge onto dLGN neurons; consequently, dLGN receptive fields are similar to those of RGCs in their size and structure.[4] Another key feature of retinogeniculate organization, which is important for binocular vision, is that the axons arising from each eye are segregated within the dLGN.[1] As a result, mature dLGN neurons are monocularly driven.[1,2] Finally, some cell-type-specific organization also exists in the dLGN since different types of dLGN neurons reside in distinct laminae and receive inputs from different subtypes of functionally distinct RGCs.[2,5–8]

How does precise retinogeniculate circuitry get established during development? Numerous experiments have demonstrated that many aspects of retinogeniculate connectivity are initially imprecise, and that the immature circuit must subsequently undergo refinement in order to achieve its finely tuned mature form.[9–16] Much of this refinement occurs before the onset of vision,[10,11] and it is now well established that this early refinement requires both spontaneous retinal activity and molecular cues that are present over development.[11,17–19]

Retinogeniculate projections are refined during development

Anatomical studies have shown that when the axons from the two eyes initially invade the dLGN their arbors are extensively overlapped,[13,20] and functional studies have demonstrated that these overlapping projections give rise to binocularly innervated dLGN neurons.[14,21] After this initial connectivity is established there is a specific developmental window during which the axons from each eye segregate into non-overlapping territories.[15,16] This "eye-specific segregation" results in a pattern of ipsilateral and contralateral projections that is highly stereotyped both in size and placement within the dLGN. Eye-specific inputs also segregate functionally. As RGC axons become restricted to the appropriate region of the dLGN their functional synaptic connections in the inappropriate region are lost, resulting in monocularly driven dLGN neurons.[14,21] In mouse, electrophysiological recordings have demonstrated that dLGN neurons are initially weakly innervated by 12–30 RGCs which get refined down to 4–6 inputs from each eye as segregation proceeds, and finally to just 1–3 strong inputs from a single eye at maturity.[14,21,22] While the convergence of RGC inputs onto individual dLGN neurons gets reduced, the nearest neighbor relationships among RGCs are maintained, and thusly, the retinotopic map is sharpened.[23]

In addition to eye-specific segregation, RGC axons undergo further restriction of their arbors into functionally distinct sublaminae. For example, each major class of ganglion cell consists of two subtypes: cells that are depolarized (ON ganglion cells) or hyperpolarized (OFF ganglion cells) by light onset.[24] At maturity these parallel ON and OFF pathways are segregated in the dLGN at the level of single neurons; individual dLGN neurons generally respond to increments or decrements in light but not both. In some species, such as the ferret, dLGN neurons receiving ON input and OFF input reside in two distinct sublaminae within each eye-specific layer,[25] and these sublaminae develop just after eye-specific segregation and before opening.[26] The emergence of On and Off sublaminae before the onset of vision precludes a physiological assessment of whether dLGN neurons transiently receive converging ON and OFF synaptic inputs. However, this is likely because before these On and Off sublaminae form RGC axons initially arborize over both the inner and outer half of each eye-specific layer.[26]

Among the classes of RGCs that project to the dLGN are Y cells which process movement, X cells which are responsible for image acuity, and a heterogeneous population of W cells.[2,10,27] Whether individual dLGN neurons initially receive synaptic input from these multiple classes of ganglion cells is unclear.[10] In the primate, however, magnocellular and parvocellular pathways project to distinct regions of the dLGN from early stages as their axons innervate targets.[28,29] This suggests that during primate development single dLGN neurons may receive inputs from just one class of RGC. This specificity could be due, at least in part, to the generation of different retinal ganglion cell classes at different times during development,[30] and to the fact that the axons of parvo cells reach the dLGN before those of magno cells.[28] In addition, in primates the axons from the two eyes may initially innervate the dLGN in a somewhat eye-specific manner; thus, the primate retinogeniculate connection may require less refinement than that of other mammals.[31]

Interestingly, studies also suggest that once the mature pattern of anatomical and functional connections has been established they must be actively maintained.[32–35] Taken together, these data indicate that the mature retinogeniculate connection must be sculpted out of an initially over-connected circuit, and that the extent of this refinement may vary from species to species. These data also suggest that retinogeniculate development occurs over several stages beginning with the segregation of eye-specific inputs, followed by a more fine-scale refinement phase that sharpens the retinotopic map, and a final maintenance phase during which this connection remains malleable.[36]

Activity-dependent refinement of retinogeniculate projections

Once it was appreciated that much of the initial development and refinement of retinogeniculate projections occurs before the onset of vision, these findings raised the question of whether synaptic transmission and/or neuronal activity are required for refinement. The first demonstrations that action potentials are required for eye-specific segregation came from studies in which tetrodotoxin (TTX) was used to block sodium channels in fetal cat brain. This activity blockade prevented the formation of eye-specific domains.[37,38] Further experiments found that spontaneous activity is generated in the retina during eye-specific segregation and when spontaneous retinal activity is disrupted eye-specific layers do not form.[39] In addition, activity disruptions that prevent axon refinement also lead to larger than normal receptive fields.[40–42]

After eye-specific segregation, neural activity also contributes to the refinement of ON and OFF inputs in the dLGN. In the ferret visual system either blockade of retinal activity or postsynaptic N-methyl-D-aspartate (NMDA) receptor activity disrupts the formation of ON and OFF sublaminae.[43,44] This suggests that synaptic transmission from the retina to the dLGN is necessary for at least this one aspect of retinogeniculate refinement; although the specific role of synaptic transmission and plasticity in this and other aspects of retinogeniculate refinement warrants closer investigation.

After retinal inputs are segregated within the dLGN neural activity continues to be required for maintaining the precision established during development. For example, blockade of retinal activity after eye-specific layers have formed in the ferret causes the silenced projections to lose the territory that they had previously occupied.[32] Similarly, blockade of late stage activity in mice with TTX disrupts synaptic refinement, preventing the elimination of weak inputs and the strengthening of remaining inputs.[35] In addition, analysis of a mutant mouse that develops excessive retinal activity after

eye-specific segregation found that this increased activity is capable of desegregating retinal inputs. This mutant mouse, termed nob1 (no b wave), lacks a protein necessary for photoreceptor to bipolar cell synaptic transmission and at the time when vision through photoreceptors would normally ensue it instead develops abnormally high rates of spontaneous retinal activity. Since spontaneous retinal activity is unaltered in these mice during the period when eye-specific segregation normally occurs, their axons initially segregate normally.[33,45] However, with the onset of the abnormally high-frequency activity the eye-specific inputs desegregate, leading to increased axonal overlap. In addition to this late requirement for normal spontaneous activity, visual input also appears to play an important role in the maintenance of retinogeniculate refinement. If animals are deprived of visual input when the late stages of within-eye refinement are nearly complete, dLGN neurons will become multiply innervated again, essentially reverting to a more immature innervation state.[34,35]

Together these experiments indicate that spontaneous activity is required for driving both eye-specific and On–Off segregation and that both spontaneous and visually evoked retinal activity are necessary for maintaining segregation.

What parameters of activity drive refinement?

Retinal activity is necessary during each stage of retinogeniculate development, but what aspects of activity are required? Theoretical studies propose that patterned activity that preferentially correlates the spiking of nearby cells but not distant cells is likely to be essential. In the visual system, patterned retinal activity is known to be present before vision,[46,47] and developmental mechanisms within the retina produce a pattern of activity that meets these theoretical criteria for segregation. For instance, both multielectrode array recordings and calcium imaging experiments have revealed that bursts of activity are correlated among neighboring RGCs and these bursts propagate in a wave-like fashion across the retina.[18,48–50] These "retinal waves" have now been observed in many vertebrate species suggesting that this activity pattern may be a conserved mechanism governing vertebrate visual system development (Fig. 39.1).

Three stages of patterned wave activity have been described in the developing retina. Interestingly, at each stage the activity is mediated by distinct mechanisms and displays unique properties, such as the frequency of the activity, the area of the retina involved in the correlated activity, and the speed of the propagating wave.[51–53] These studies suggest that the unique features of these stages might contribute to the specific aspects of refinement that occur during each stage. The first stage of retinal activity, stage I, consists of infrequent bursts of patterned activity which appear embryonically and are largely mediated by communication through gap junctions.[54] The bulk of retinogeniculate refinement occurs during stage II waves. Stage II waves are mediated by acetylcholine release from starburst amacrine cells[48,55] and occur embryonically in mammals born with their eyes open and postnatally in mammals born with their eyes closed.[54,56] The final stage of patterned spontaneous activity, stage III, is mediated by glutamatergic transmission from bipolar cells, spans the period of eye opening, and is propagated, at least in part, due to glutamate spillover.[18,33,54,56–59] Interestingly,

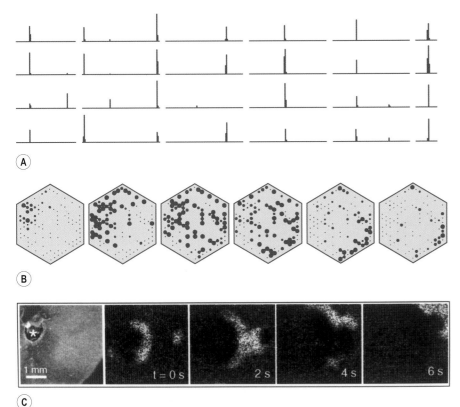

Figure 39.1 The immature retina generates a pattern of synchronized bursting activity before vision. (**A**) Multielectrode recording of the spike activity from many cells in the isolated but intact ferret retina reveals the presence of rhythmic bursting in individual cells (*each line*). The bursts of four neighboring cells (shown by the *vertical histograms*) are correlated in time. (**B**) During a burst of activity, retinal ganglion cells are activated sequentially as a wave propagates across the recording electrodes. Shown here is an example of such a wave propagating across the hexagonal multielectrode array. *Left to right*: Sequential "snapshots" of the activity recorded every 0.5 second. Each dot is a cell, and the size of the dot is proportional to the spike rate. (**C**) Retinal waves are easily observed using calcium imaging. Changes in fluorescence intensity over time detected using a low-light camera show two waves that collide in a neonatal mouse retina. First panel shows the fluorescence labeling by fura-2, and white regions in the subsequent images show elevations in intracellular calcium. * = Optic nerve head.

during stage III waves, ON and OFF RGCs fire at different rates and are anti-correlated[60-62] as do other subtypes of RGCs such as X, Y, and W, which could play a role in the segregation of these distinct types of inputs.[63]

By what mechanisms do retinal waves drive changes in retinogeniculate connectivity? A common mechanism that appears to underlie the development of connection specificity throughout the nervous system is activity-dependent competition among presynaptic inputs for postsynaptic space.[64,65] This competition is driven by neural activity, but how? Evidence that eye-specific segregation involves competition among RGC inputs comes from studies in which activity was either blocked or enhanced in one eye resulting in alterations to the pattern of ipsilateral and contralateral projections. In these studies when epibatidine was used to block cholinergic transmission in just one eye, the projections from the blocked eye lost territory in the dLGN while the projections from the active eye expanded their coverage.[39] The converse was also true; that is, when activity was increased in one eye by intraocular administration of either forskolin or the cyclic AMP analog, CPT-cAMP (which increase the size, frequency, and speed of waves), the projections from the more active eye gained territory in the dLGN at the expense of the non-enhanced eye's projections.[66,67]

Both experimental and theoretic studies suggest that in addition to overall activity levels, specific aspects of patterned spontaneous activity may drive segregation through Hebbian mechanisms that strengthen and eliminate synaptic inputs over time.[68] This Hebbian model of synaptic competition predicts that ganglion cells that "fire together, wire together".[69] For Hebbian mechanisms to drive segregation, neighboring RGCs must fire in a correlated manner while more distant RGCs or RGCs from different eyes must exhibit less correlated spiking; thereby allowing for the strengthening of the temporally correlated neighboring inputs and the weakening of the uncorrelated inputs. Indeed, these requirements are met as a consequence of wave propagation. For example, during a wave neighboring RGCs are co-activated causing their firing to become more synchronized than that of non-neighboring RGCs. In addition, the random nature of wave initiation and the relatively long refractory period between waves results in a low probability of coincident firing of more distant ganglion cells or ganglion cells in separate eyes. This favors the segregation of inputs from different eyes, and could simultaneously remodel both the connectivity between the two eyes and refine the retinotopic map within each eye.[70,71]

Anatomical studies in mice have shown that stage II waves are of particular importance for eye-specific segregation since this segregation occurs during stage II waves and disruption of stage II waves through binocular injections of epibatidine prevents eye-specific segregation during this period[39] (Fig. 39.2). These data show that stage II waves are required for eye-specific segregation – but are they permissive or instructive for segregation? The data regarding this question are somewhat conflicting. Studies of mice lacking the nicotinic acetylcholine receptor subunit β2 have suggested that retinal wave structure is critical for driving segregation since in these mice, which retain retinal activity but have severely altered wave structure,[54,72,73] eye-specific refinement does not occur during stage II.[54,73,74] On the other hand, the first study to abolish retinal waves without disrupting the overall level of retinal activity did not find any deficit in eye-specific segregation.[75] In addition, studies of other

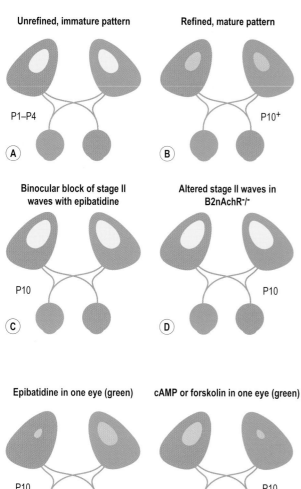

Unrefined, immature pattern

Refined, mature pattern

P1–P4

(A)

P10+

(B)

Binocular block of stage II waves with epibatidine

Altered stage II waves in B2nAchR-/-

P10

(C)

P10

(D)

Epibatidine in one eye (green)

cAMP or forskolin in one eye (green)

P10

(E)

P10

(F)

Ephrin A2 / 3 / 5 -/-

Ephrin A2 / 3 / 5 -/- and binocular epibatidine

P10

(G)

P10

(H)

Figure 39.2 Both spontaneous retinal activity and ephrin signaling are required for the normal segregation of eye-specific inputs in the dLGN. Diagram illustrates projection patterns in the rodent dLGN. (**A**) Initially the projections from both eyes overlap in the dLGN. In rodents this pattern can be seen between P1–P4. (**B**) During stage II retinal waves the inputs from each eye segregate into non-overlapping regions with their mature pattern emerging around P10. (**C**) Pharmacological ablation of stage II waves prevents segregation during this period. (**D**) In the β2nAchR knockout mouse, waves with altered correlation structure and frequency also prevent segregation during stage II. (**E**), (**F**) Eye-specific segregation involves activity-dependent competition. (E) When stage II waves are blocked in one eye with epibatidine the projections from the blocked eye lose territory in the dLGN while the projections from the normally active eye expand. (F) When stage II waves are increased in size and frequency in one eye, the projections from the enhanced eye gain territory in the dLGN at the expense of the other eye's projections. (**G**) Ephrin As are required for the normal size, position, and pattern of eye-specific layers. Mice lacking ephrin A2, A3, and A5 undergo eye-specific segregation but do not form normal eye-specific laminae. (**H**) When activity is blocked in both eyes of ephrin A2, A3, and A5 knockout mice eye-specific segregation does not occur, and ipsilateral and contralateral projects are overlapped throughout most of the dLGN.

for driving eye-specific segregation, such as the amount of correlated firing that occurred in bursts with frequencies greater than 10 Hz. In addition, they found the relative difference in the correlation strengths between neighboring and distant RGCs to be critical, rather than the overall activity of the RGCs or the precise time window of RGC firing. This study suggests that retinal waves encode information that instructs retinogeniculate refinement, though this topic remains controversial. As mentioned above, in the nob1 mouse the onset of altered late-stage waves causes eye-specific inputs to desegregate. Interestingly, the altered waves in the nob1 mice display correlations that are stronger between neighboring RGCs than distant RGCs, similar to the correlation structure found by Feller and colleagues to promote eye-specific segregation. This suggests that, at least for stage III waves, the levels of activity can also affect retinogeniculate development. In the nob1 mouse it is likely that very high frequency of waves leads to increased co-incident firing of RGCs in different eyes, thus leading to the re-growth of new inputs and/or the strengthening of residual inputs.[33,36]

Synaptic inputs change strength with segregation

Electrophysiological recordings have shown that the convergence of functional synaptic inputs onto individual dLGN neurons dramatically decreases over development.[14,21,22,78] As discussed above, evidence suggests that specific aspects of retinal wave structure likely play an instructive role in retinogeniculate refinement by driving Hebbian learning rules at retinogeniculate synapses. Presumably, the correlated firing of neighboring RGCs leads to increased synaptic strength which in turn stabilizes the axon arbors containing the strengthened synapses. Conversely, uncorrelated firing leads to decreased synaptic strength which ultimately leads to the elimination of the weak synapses and the axon collaterals on which they reside. So, what are the synaptic learning rules that operate at retinogeniculate synapses, and what molecular mechanisms translate changes in synaptic strength into

mice with disrupted waves indicated that although these animals display defects in eye-specific segregation,[74] when certain aspects of the waves were rescued this did not lead to the rescue eye-segregation.[76]

In an effort to determine the specific parameters of waves that might be necessary for eye-specific segregation the Feller laboratory analyzed a number of mutant mice with different retinal wave disruptions and varying degrees of eye-specific segregation.[77] Based on these comparisons they concluded that several aspects of retinal waves are likely to be essential

altered morphology? The answers to these questions are still largely unknown; however, some insight has been gained and is discussed below.

The first study to demonstrate that retinogeniculate synapses can undergo synaptic plasticity stimulated RGC afferents in a retinogeniculate slice preparation with high-frequency trains and showed that this strengthened retinogeniculate synapses in an NMDA receptor-dependent fashion. However, one caveat to this study was that the high-frequency trains were faster than those typically observed during retinal waves.[79] Using a preparation containing intact retinal inputs to the dLGN combined with patch-clamp recordings Mooney et al demonstrated that, during segregation, retinal inputs are capable of driving postsynaptic dLGN spiking, a prerequisite for Hebbian-based learning.[80] More recently, Butts and colleagues, using stimulation that more accurately reflects endogenous activity patterns, described a burst-timing-dependent plasticity protocol that was capable of both strengthening or weakening retinogeniculate synapses depending on whether presynaptic and postsynaptic bursts (10–20 Hz) were coincident or non-overlapping.[81] In addition to this homosynaptic learning rule, synaptic weakening may also occur via heterosynaptic depression. In support of this idea, trains of synaptic stimuli from one eye were shown to lead to depression of the opposite eye inputs.[82] Undoubtedly, future studies will further elucidate the types of plasticity mechanisms that influence the development of this connection. In particular, since the distinct stages of retinal activity differ in their wave parameters it will be interesting to find out whether the synaptic plasticity rules operating in the dLGN change over these stages. For instance, in ferrets it was reported that NMDAR activation is not required for eye-specific segregation, but is required for On–Off segregation.[43] Interestingly, toward the end of the period of stage III waves, both spontaneous activity and vision were shown to preferentially drive the glutamate receptor subunit 1 into retinogeniculate synapses.[83]

While, as previously mentioned, the number of RGCs innervating individual dLGN neurons decreases dramatically over development,[21,22,78] the total number of synapses made onto individual dLGN neurons seems to remain fairly constant.[22] This finding is analogous to the refinement of motor neuron inputs at the neuromuscular junction where a single motor neuron takes over the synaptic territory of multiple motor neurons.[65] The mechanisms controlling total synapse number are not known. In addition, some mutant mice retain excess retinal inputs into adulthood, yet these extra inputs do not drive aberrantly large synaptic currents.[35,84] These studies suggest that the total synaptic drive received by dLGN neurons is under homeostatic regulation.

Molecular mechanisms involved in activity-dependent axonal segregation

A major area of research that remains is to elucidate the molecular mechanisms that translate changes in synaptic drive into altered synapse and axon morphology. The use of transgenic mouse lines has led to tremendous advancement in our understanding of these molecular mechanisms. Screening of mutant animals that display fairly normal retinal waves yet have severely disrupted eye-specific segregation has demonstrated that several types of immune-related

molecules are likely to be involved in this process. The first such example was the class I major immunohistocampatibility complex (MHC). This protein is upregulated by activity in the dLGN during the time of eye-specific segregation, and MHC I knockout mice display overlapping eye inputs.[85–87] The neuronal pentraxins (NPs) are another example of activity-regulated genes shown to be involved in eye-specific segregation, and interestingly, NPs are also immune-related molecules that share some sequence similarity to the short pentraxins, C-reactive protein, and serum amyloid P. Like MHC, NPs are highly expressed during retinogeniculate refinement, and animals lacking two NPs, NP1 and NP2, display overlapping eye inputs past the normal period of segregation.[88] Both MHC and NPs may directly affect synaptic function.[85,89,90] Another example of molecular mechanisms that may couple functional and structural changes in the dLGN comes from studies of the complement pathway. Several strains of mice with deletions of protein in the classical complement pathway have overlapping eye-specific input in the dLGN. The complement pathway is well known for mediating phagocytosis of dead or infected cells. Because complement appears to co-localize with a subset of synapses in the dLGN, one possible role for the complement pathway is that it may tag weak inputs that are slated for removal during the period of eye-specific segregation.[84]

Molecular mechanisms guiding the formation of eye-specific axonal territories

In addition to the molecular mechanisms discussed above, activity-independent molecular mechanisms also play a role in shaping the mature pattern of retinogeniculate connections. Because the size and position of the eye-specific regions in the dLGN are highly stereotyped from animal to animal and symmetrical on both sides of the thalamus, it has long been speculated that molecular cues must determine where inputs from each finally coalesce.[11] What cues might guide the inputs from the two eyes into their distinct territories? Interestingly, in achiasmatic sheep dogs with uncrossed eye projections, inputs from the normally crossed nasal and the uncrossed temporal retina still segregate in the dLGN,[91] strongly indicating that molecular cues direct eye-specific segregation by marking the temporal and nasal retina. Work on retinotectal mapping has shown that the nasal and temporal retina can be distinguished by their differing expression of EphAs[92] and both the EphAs and their binding partners, ephrin-As, play a critical role in eye-specific segregation in the dLGN. Mice lacking three of the ephrins, A2, A3, and A5 still display activity-dependent segregation of eye-specific inputs, but the eye-specific regions are fractured and do not form distinct layers in the dLGN.[93] If these triple knockout mice are subjected to activity blockade during the period of eye-specific segregation their retinal inputs form a permanent salt and pepper pattern of intermingled inputs. In addition, retinal overexpression of EphA3 or EphA5 can lead to mistargeting of retinal inputs to the incorrect eye-specific layer despite normal retinal patterned activity.[94] These data indicate that EphA–EphrinA interactions shape the final pattern of eye-specific inputs. These also indicate that activity and ephrin signaling act in parallel and are both required for the normal organization of retinogeniculate inputs. In addition, activity manipulations do not seem to affect ephrin expression, nor does the loss or over-expression of ephrins impact activity levels, further suggesting that these pathways act largely independently from one another.[19,94]

Summary

The mature retinogeniculate pathway is highly organized in structure and function, and a clearer understanding of how this precise organization develops continues to emerge. It is now well established that both activity-dependent and -independent mechanisms shape this final organization. Specific features of spontaneous retinal activity appear to be critical, suggesting that activity may work in an instructive fashion. An important task for future studies will be to further elucidate the synaptic and molecular mechanisms that turn activity into structural changes in this pathway. The use of transgenic animals and additional genetic and molecular screens will allow for greater molecular insight into the mechanisms that underlie both the activity-dependent and -independent aspects of this process. Each stage of refinement, between eye, within eye, ON vs OFF, may employ some similar and unique mechanisms to mediate refinement. In summary, the retinogeniculate projection has proven to be a fascinating example of central nervous system development, and further studies in this system will continue to advance our understanding of the processes leading to developmental refinement throughout the nervous system.

Acknowledgments

The authors would like to thank Dr. Andrew D. Huberman for critical comments on this chapter. This work is supported by the March of Dimes, Basil O'Connor Award.

References

1. Kaas JH, Guillery RW, Allman JM. Some principles of organization in the dorsal lateral geniculate nucleus. Brain Behav Evol 1972; 6(1):253–299.
2. Rodieck RW. Visual pathways. Annu Rev Neurosci 1979; 2:193–225.
3. Roskies A, Friedman GC, O'Leary DD. Mechanisms and molecules controlling the development of retinal maps. Perspect Dev Neurobiol 1995; 3(1):63–75.
4. Bullier J, Norton TT. Comparison of receptive-field properties of X and Y ganglion cells with X and Y lateral geniculate cells in the cat. J Neurophysiol 1979; 42(1, Pt 1):274–291.
5. Leventhal AG, Rodieck RW, Dreher B. Central projections of cat retinal ganglion cells. J Comp Neurol 1985; 237(2):216–226.
6. Leventhal AG, Rodieck RW, Dreher B. Retinal ganglion cell classes in the Old World monkey: morphology and central projections. Science 1981; 213(4512):1139–1142.
7. Huberman AD et al. Architecture and activity-mediated refinement of axonal projections from a mosaic of genetically identified retinal ganglion cells. Neuron 2008; 59(3):425–438.
8. Huberman AD et al. Genetic identification of an On-Off direction-selective retinal ganglion cell subtype reveals a layer-specific subcortical map of posterior motion. Neuron 2009; 62(3):327–334.
9. Constantine-Paton M, Cline HT, Debski E. Patterned activity, synaptic convergence, and the NMDA receptor in developing visual pathways. Annu Rev Neurosci 1990; 13:129–154.
10. Garraghty PE, Sur M. Competitive interactions influencing the development of retinal axonal arbors in cat lateral geniculate nucleus. Physiol Rev 1993; 73(3):529–545.
11. Goodman CS, Shatz CJ. Developmental mechanisms that generate precise patterns of neuronal connectivity. Cell 1993; 72(Suppl):77–98.
12. Shatz CJ, Sretavan DW. Interactions between retinal ganglion cells during the development of the mammalian visual system. Annu Rev Neurosci 1986; 9:171–207.
13. Sretavan DW, Shatz CJ. Prenatal development of retinal ganglion cell axons: segregation into eye-specific layers within the cat's lateral geniculate nucleus. J Neurosci 1986; 6(1):234–251.
14. Shatz CJ, Kirkwood PA. Prenatal development of functional connections in the cat's retinogeniculate pathway. J Neurosci 1984; 4(5):1378–1397.
15. Sretavan DW, Shatz CJ. Prenatal development of cat retinogeniculate axon arbors in the absence of binocular interactions. J Neurosci 1986; 6(4):990–1003.
16. Sretavan DW, Shatz CJ. Axon trajectories and pattern of terminal arborization during the prenatal development of the cat's retinogeniculate pathway. J Comp Neurol 1987; 255(3):386–400.
17. Penn AA, Shatz CJ. Brain waves and brain wiring: the role of endogenous and sensory-driven neural activity in development. Pediatr Res 1999; 45(4 Pt 1):447–458.
18. Wong RO, Meister M, Shatz CJ. Transient period of correlated bursting activity during development of the mammalian retina. Neuron 1993; 11(5):923–938.
19. Pfeiffenberger C et al. Ephrin-As and neural activity are required for eye-specific patterning during retinogeniculate mapping. Nat Neurosci 2005; 8(8):1022–1027.
20. Sretavan D, Shatz CJ. Prenatal development of individual retinogeniculate axons during the period of segregation. Nature 1984; 308(5962):845–848.
21. Ziburkus J, Guido W. Loss of binocular responses and reduced retinal convergence during the period of retinogeniculate axon segregation. J Neurophysiol 2006; 96(5):2775–2784.
22. Chen C, Regehr WG. Developmental remodeling of the retinogeniculate synapse. Neuron 2000; 28(3):955–966.
23. Tavazoie SF, Reid RC. Diverse receptive fields in the lateral geniculate nucleus during thalamocortical development. Nat Neurosci 2000; 3(6):608–616.
24. Nelson R, Famiglietti E Jr, Kolb H. Intracellular staining reveals different levels of stratification for on- and off-center ganglion cells in cat retina. J Neurophysiol 1978; 41(2):472–483.
25. Stryker MP, Zahs KR. On and off sublaminae in the lateral geniculate nucleus of the ferret. J Neurosci 1983; 3(10):1943–1951.
26. Cucchiaro J, Guillery RW. The development of the retinogeniculate pathways in normal and albino ferrets. Proc R Soc Lond B Biol Sci 1984; 223(1231):141–164.
27. Wingate RJ, Fitzgibbon T, Thompson ID. Lucifer yellow, retrograde tracers, and fractal analysis characterise adult ferret retinal ganglion cells. J Comp Neurol 1992; 323(4):449–474.
28. Meissirel C et al. Early divergence of magnocellular and parvocellular functional subsystems in the embryonic primate visual system. Proc Natl Acad Sci USA 1997; 94(11):5900–5905.
29. Snider CJ et al. Prenatal development of retinogeniculate axons in the macaque monkey during segregation of binocular inputs. J Neurosci 1999; 19(1):220–228.
30. Rapaport DH et al. Genesis of neurons in the retinal ganglion cell layer of the monkey. J Comp Neurol 1992; 322(4):577–588.
31. Huberman AD et al. Early and rapid targeting of eye-specific axonal projections to the dorsal lateral geniculate nucleus in the fetal macaque. J Neurosci 2005; 25(16):4014–4023.
32. Chapman B. Necessity for afferent activity to maintain eye-specific segregation in ferret lateral geniculate nucleus. Science 2000; 287(5462):2479–2482.
33. Demas J et al. Failure to maintain eye-specific segregation in nob, a mutant with abnormally patterned retinal activity. Neuron 2006; 50(2):247–259.
34. Hooks BM, Chen C. Vision triggers an experience-dependent sensitive period at the retinogeniculate synapse. J Neurosci 2008; 28(18):4807–4817.
35. Hooks BM, Chen C. Distinct roles for spontaneous and visual activity in remodeling of the retinogeniculate synapse. Neuron 2006; 52(2):281–291.
36. Huberman AD. Mechanisms of eye-specific visual circuit development. Curr Opin Neurobiol 2007; 17(1):73–80.
37. Sretavan DW, Shatz CJ, Stryker MP. Modification of retinal ganglion cell axon morphology by prenatal infusion of tetrodotoxin. Nature 1988; 336(6198):468–471.
38. Shatz CJ, Stryker MP. Prenatal tetrodotoxin infusion blocks segregation of retinogeniculate afferents. Science 1988; 242(4875):87–89.
39. Penn AA et al. Competition in retinogeniculate patterning driven by spontaneous activity. Science 1998; 279(5359):2108–2112.
40. Grubb MS et al. Abnormal functional organization in the dorsal lateral geniculate nucleus of mice lacking the beta 2 subunit of the nicotinic acetylcholine receptor. Neuron 2003; 40(6):1161–1172.
41. Chandrasekaran AR et al. Evidence for an instructive role of retinal activity in retinotopic map refinement in the superior colliculus of the mouse. J Neurosci 2005; 25(29):6929–6938.
42. Mrsic-Flogel TD et al. Altered map of visual space in the superior colliculus of mice lacking early retinal waves. J Neurosci 2005; 25(29):6921–6928.
43. Cramer KS, Sur M. Blockade of afferent impulse activity disrupts on/off sublamination in the ferret lateral geniculate nucleus. Brain Res Dev Brain Res 1997; 98(2):287–290.
44. Dubin MW, Stark LA, Archer SM. A role for action-potential activity in the development of neuronal connections in the kitten retinogeniculate pathway. J Neurosci 1986; 6(4):1021–1036.
45. Gregg RG et al. Identification of the gene and the mutation responsible for the mouse nob phenotype. Invest Ophthalmol Vis Sci 2003; 44(1):378–384.
46. Galli L, Maffei L. Spontaneous impulse activity of rat retinal ganglion cells in prenatal life. Science 1988; 242(4875):90–91.
47. Maffei L, Galli-Resta L. Correlation in the discharges of neighboring rat retinal ganglion cells during prenatal life. Proc Natl Acad Sci USA 1990; 87(7):2861–2864.
48. Feller MB et al. Requirement for cholinergic synaptic transmission in the propagation of spontaneous retinal waves. Science 1996; 272(5265):1182–1187.
49. Meister M et al. Synchronous bursts of action potentials in ganglion cells of the developing mammalian retina. Science 1991; 252(5008):939–943.
50. Wong RO. Early functional neural networks in the developing retina. Nature 1995; 374(6524):716–718.
51. Wong RO. Retinal waves and visual system development. Annu Rev Neurosci 1999; 22:29–47.
52. Sernagor E, Mehta V. The role of early neural activity in the maturation of turtle retinal function. J Anat 2001; 199(Pt 4):375–383.
53. Firth SI, Wang CT, Feller MB. Retinal waves: mechanisms and function in visual system development. Cell Calcium 2005; 37(5):425–432.
54. Bansal A et al. Mice lacking specific nicotinic acetylcholine receptor subunits exhibit dramatically altered spontaneous activity patterns and reveal a limited role for retinal waves in forming ON and OFF circuits in the inner retina. J Neurosci 2000; 20(20):7672–7681.
55. Zheng JJ, Lee S, Zhou ZJ. A developmental switch in the excitability and function of the starburst network in the mammalian retina. Neuron 2004; 44(5):851–864.

56. Syed MM et al. Stage-dependent dynamics and modulation of spontaneous waves in the developing rabbit retina. J Physiol 2004; 560(Pt 2):533–549.

57. Wong WT et al. Developmental changes in the neurotransmitter regulation of correlated spontaneous retinal activity. J Neurosci 2000; 20(1):351–360.

58. Zhou ZJ, Zhao D. Coordinated transitions in neurotransmitter systems for the initiation and propagation of spontaneous retinal waves. J Neurosci 2000; 20(17):6570–6577.

59. Blankenship AG et al. Synaptic and extrasynaptic factors governing glutamatergic retinal waves. Neuron 2009; 62(2):230–241.

60. Wong RO, Oakley DM. Changing patterns of spontaneous bursting activity of on and off retinal ganglion cells during development. Neuron 1996; 16(6):1087–1095.

61. Myhr KL, Lukasiewicz PD, Wong RO. Mechanisms underlying developmental changes in the firing patterns of ON and OFF retinal ganglion cells during refinement of their central projections. J Neurosci 2001; 21(21):8664–8671.

62. Kerschensteiner D, Wong RO. A precisely timed asynchronous pattern of ON and OFF retinal ganglion cell activity during propagation of retinal waves. Neuron 2008; 58(6):851–858.

63. Liets LC et al. Spontaneous activity of morphologically identified ganglion cells in the developing ferret retina. J Neurosci 2003; 23(19):7343–7350.

64. Lichtman JW, Colman H. Synapse elimination and indelible memory. Neuron 2000; 25(2):269–278.

65. Sanes JR, Lichtman JW. Development of the vertebrate neuromuscular junction. Annu Rev Neurosci 1999; 22:389–442.

66. Stellwagen D, Shatz CJ. An instructive role for retinal waves in the development of retinogeniculate connectivity. Neuron 2002; 33(3):357–367.

67. Stellwagen D, Shatz CJ, Feller MB. Dynamics of retinal waves are controlled by cyclic AMP. Neuron 1999; 24(3):673–685.

68. Shatz CJ. Emergence of order in visual system development. Proc Natl Acad Sci USA 1996; 93(2):602–608.

69. Hebb DO. The organization of behavior. New York: Wiley, 1949.

70. Eglen SJ The role of retinal waves and synaptic normalization in retinogeniculate development. Phil Trans R Soc Lond B Biol Sci 1999; 354(1382):497–506.

71. Miller KD. Synaptic economics: competition and cooperation in synaptic plasticity. Neuron 1996; 17(3):371–374.

72. Sun C et al. Retinal waves in mice lacking the beta2 subunit of the nicotinic acetylcholine receptor. Proc Natl Acad Sci USA 2008; 105(36):13638–13643.

73. Muir-Robinson G, Hwang BJ, Feller MB. Retinogeniculate axons undergo eye-specific segregation in the absence of eye-specific layers. J Neurosci 2002; 22(13): 5259–5264.

74. Rossi FM et al. Requirement of the nicotinic acetylcholine receptor beta 2 subunit for the anatomical and functional development of the visual system. Proc Natl Acad Sci USA 2001; 98(11):6453–6458.

75. Huberman AD et al. Eye-specific retinogeniculate segregation independent of normal neuronal activity. Science 2003; 300(5621):994–998.

76. Torborg C et al. L-type calcium channel agonist induces correlated depolarizations in mice lacking the beta2 subunit nAChRs. Vision Res 2004; 44(28):3347–3355.

77. Torborg CL, Hansen KA, Feller MB. High frequency, synchronized bursting drives eye-specific segregation of retinogeniculate projections. Nat Neurosci 2005; 8(1):72–78.

78. Jaubert-Miazza L et al. Structural and functional composition of the developing retinogeniculate pathway in the mouse. Vis Neurosci 2005; 22(5):661–676.

79. Mooney R, Madison DV, Shatz CJ. Enhancement of transmission at the developing retinogeniculate synapse. Neuron 1993; 10(5):815–825.

80. Mooney R et al. Thalamic relay of spontaneous retinal activity prior to vision. Neuron 1996; 17(5):863–874.

81. Butts DA, Kanold PO, Shatz CJ. A burst-based "Hebbian" learning rule at retinogeniculate synapses links retinal waves to activity-dependent refinement. PLoS Biol 2007; 5(3):e61.

82. Guido W. Refinement of the retinogeniculate pathway. J Physiol 2008; 586(Pt 18):4357–4362.

83. Kielland A et al. Activity patterns govern synapse-specific AMPA receptor trafficking between deliverable and synaptic pools. Neuron 2009; 62(1):84–101.

84. Stevens B et al. The classical complement cascade mediates CNS synapse elimination. Cell 2007; 131(6):1164–1178.

85. Goddard CA, Butts DA, Shatz CJ. Regulation of CNS synapses by neuronal MHC class I. Proc Natl Acad Sci USA 2007; 104(16):6828–6833.

86. Huh GS et al. Functional requirement for class I MHC in CNS development and plasticity. Science 2000; 290(5499):2155–2159.

87. Corriveau RA, Huh GS, Shatz CJ. Regulation of class I MHC gene expression in the developing and mature CNS by neural activity. Neuron 1998; 21(3):505–520.

88. Bjartmar L et al. Neuronal pentraxins mediate synaptic refinement in the developing visual system. J Neurosci 2006; 26(23):6269–6281.

89. Xu D et al. Narp and NP1 form heterocomplexes that function in developmental and activity-dependent synaptic plasticity. Neuron 2003; 39(3):513–528.

90. O'Brien RJ et al. Synaptic clustering of AMPA receptors by the extracellular immediate-early gene product Narp. Neuron 1999; 23(2):309–323.

91. Williams RW, Hogan D, Garraghty PE. Target recognition and visual maps in the thalamus of achiasmatic dogs. Nature 1994; 367(6464):637–639.

92. McLaughlin T, O'Leary DD. Molecular gradients and development of retinotopic maps. Annu Rev Neurosci 2005; 28:327–355.

93. Pfeiffenberger C, Yamada J, Feldheim DA. Ephrin-As and patterned retinal activity act together in the development of topographic maps in the primary visual system. J Neurosci 2006; 26(50):12873–12884.

94. Huberman AD et al. Ephrin-As mediate targeting of eye-specific projections to the lateral geniculate nucleus. Nat Neurosci 2005; 8(8):1013–1021.

CHAPTER 40

Developmental Visual Deprivation

Yuzo M. Chino

The vision of newborn infants is crude. As infants experience the normal visual environment, their vision rapidly improves with different visual capabilities emerging at different ages. Determining the exact timing for the behavioral onset of specific visual functions and identifying the critical factors that limit their development have been the primary focus of perceptual and physiological studies on vision development. Although immaturities of the physiological optics and ocular motility are known to affect the infant's vision soon after birth, the maturation of the retina and, to a greater extent, the visual cortex largely set a limit on their normal perceptual development.[1,2]

The neuronal connections of the visual cortex are malleable for a considerable time after birth, the "critical" or "plastic" period of development. The postnatal development of the visual cortex, therefore, requires normal visual experience and precise matching of the images in the two eyes. Experiencing binocularly discordant images early in life, a binocular imbalance, has devastating effects on the development of the visual system because after eye opening, the neurons in the highly plastic visual cortex receive robust signals from the two eyes that do not match. Prevalent abnormal visual conditions that can cause binocular imbalance are early monocular form deprivation, unilateral defocus, and ocular misalignment. Experiencing binocular imbalance during early infancy causes binocular vision disorders, and if untreated, amblyopia is likely to develop.

Topics in this chapter cover what we currently know about the perceptual consequences of early abnormal visual experience, the neural basis of altered vision, and the synaptic and molecular mechanisms of cortical plasticity.

Macaque monkeys are ideal animal models for exploring the neural mechanisms underlying developmental vision disorders in humans. The anatomical and physiological organizations of their visual system are nearly identical to humans. Perceptual studies in normal mature monkeys have extensively documented the striking similarities in monocular and binocular visual capabilities between macaque monkeys and humans.[3,4] The relative (scaled) time course of normal visual development in macaque monkeys parallels that of humans.[5-8] The primate visual cortex is structurally[9] and functionally[10,11] more developed at or near birth than the visual cortex in lower species.

Many important discoveries on vision development have been made on sub-primate species (e.g. cats, ferrets, rats, and mice).[12-16] However, in lower animals it is not always possible to establish a link between neural and perceptual deficits resulting from early abnormal visual experience. This chapter, therefore, will primarily review studies on non-human primates. Studies with human infants are mentioned when appropriate, and research in sub-primate species is described in detail mostly where the neural and molecular basis of cortical plasticity is discussed.

Effects of early monocular form deprivation

Constant monocular form deprivation

Monocular form deprivation can result from an occlusion of the image in one eye or from a severely degraded image in the affected eye. Congenital dense cataracts and ptosis are the common causes of monocular form deprivation in human infants. To create primate models of monocular form deprivation, the eyelids of infant monkeys are surgically closed,[17] or more recently, by wearing diffuser lenses in front of one eye.[18]

Perceptual deficits

All binocular functions including local/global stereopsis and binocular summation of contrast sensitivity are severely compromised or lost following early monocular form deprivation.[17,19] The visual sensitivity of the deprived eye is dramatically reduced or virtually lost, *form deprivation amblyopia* (Fig. 40.1A).[17,19,20] Importantly, the severity of the contrast sensitivity loss resulting from monocular form deprivation is directly related to the degree of retinal image degradation during early infancy (Fig. 40.1B).[18]

Monocular deprivation also leads to an abnormal elongation of the eyes, hence, the development of myopic refractive errors.[18,21,22] The deleterious effects of early monocular deprivation on spatial vision development are generally far more severe than the anomalies resulting from form deprivation in both eyes, *bilateral form deprivation* (Fig. 40.2).[23,24] As in monkeys with monocular form deprivation,

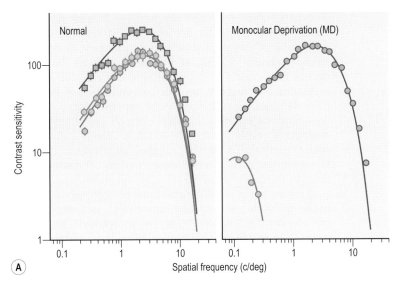

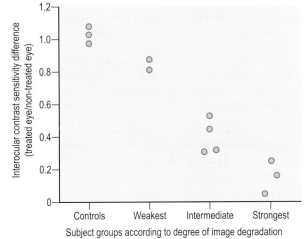

Figure 40.1 Effects of early monocular form deprivation on contrast sensitivity in macaque monkeys. (**A**) Spatial contrast sensitivity functions from normal (left) and monocularly form-deprived (right) monkeys. Red circles indicate contrast sensitivity for the left eye in the normal and the affected eye in the deprived monkey. Green circles in Normal and purple circles in MD indicate contrast sensitivity for the fellow eyes. Blue squares show binocular contrast sensitivity. (Modified from Harwerth et al 1981.[17]) (**B**) Effects of the degree of image degradation on contrast sensitivity loss. (Redrawn from Smith et al 2000.[18])

bilateral form deprivation leads to a significant loss of binocular functions including the detection of stereoscopic cues and binocular summation of contrast sensitivity (Fig. 40.2). In these binocularly deprived monkeys, the binocular contrast sensitivity function (square symbols) overlaps with the better monocular sensitivity function (open circles).[23]

Neural changes

The most consistent effect of early monocular form deprivation on development is anomalous changes in the ocular dominance distribution of V1 neurons (*ocular dominance plasticity*). During the critical period of development, the afferent fibers from the LGN representing the two eyes compete for consolidation of functional connections in V1 (*binocular competition*).[25,26] This early binocular competition is activity dependent; hence, depriving normal signals from one eye puts the affected eye into a competitive disadvantage and leads to a severe loss of functional connections in V1 from the deprived eye. The ocular dominance columns of the input layer in V1 (layer IVC) representing the deprived eye exhibit a substantial shrinkage.[26–28] The axon arbors of the afferent fibers from the LGN in the deprived columns show abnormal structural changes[29] and

the intrinsic long-range horizontal connections extending over multiple ocular dominance columns reorganize their wiring pattern in the cat primary visual cortex.[30,31]

Electrophysiological studies consistently report the severe loss of binocularly driven cells, i.e. neurons that can be activated by stimulation of either eye. In addition, there is a clear shift in the ocular dominance distribution of cortical neurons away from the deprived eye. Specifically, the percentage of V1 neurons that can be activated or dominated by stimulation of the deprived eye is significantly decreased (Fig. 40.3).[26,27,32] The reduced functional innervation from the deprived eye is, at least in part, the neural basis for "undersampling" of visual scenes by the affected eye in form deprivation amblyopia.[33]

For subcortical structures, there is a mild shrinkage of cell bodies of LGN neurons that receive input signals from the deprived eye.[25,26,34] However, the response properties of these primate LGN neurons are largely unaffected by early monocular form deprivation.[26,35,36] Interestingly, there is considerable evidence for functional alterations in the cat LGN due to early abnormal visual experience.[12,37–39] For the primate retina there are no significant structural or functional abnormalities due to early monocular form deprivation. Together, major neural changes resulting from early

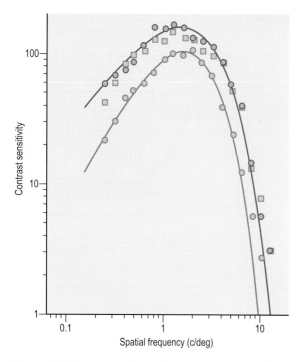

Figure 40.2 Contrast sensitivity and binocular summation after binocular form deprivation. Spatial contrast sensitivity function of the right (red circle) and left (blue circle) eye after 16 weeks of deprivation beginning at 3 weeks of age. Green square symbols indicate contrast sensitivity under binocular viewing conditions. (Modified from Harwerth et al 1991.[23] Reproduced from Association for Research in Vision and Ophthalmology.)

monocular form deprivations *in primates* occur beyond the LGN, i.e. begins in the primary visual cortex.

Intermittent monocular deprivation

Experiencing "normal vision" during early monocular form deprivation (*intermittent monocular form deprivation*) reduces some of the deleterious effects of early *constant* monocular form deprivation.[32,40-42] The effects of early intermittent deprivations have been studied using different rearing regimens including daily alternating monocular deprivation, reverse occlusion, and monocular form deprivation with daily brief periods of unrestricted vision.

Alternating monocular deprivation

Daily alternations of form deprivation between the two eyes have very little impact on the perceptual development of either eye in cats.[41,42] Consistent with this observation, the spatial receptive-field properties such as orientation selectivity are normal. However, the same daily alternating deprivation devastates the development of binocular vision. Local stereopsis is lost and the proportion of binocularly driven neurons in area 17 is severely reduced. In monkeys, daily alternating monocular occlusion beginning *at birth* leads to a variety of abnormal eye positions and eye movements including strabismus and/or saccadic disconjugacy, i.e. the amplitudes of saccades in the occluded eye are less than that in the viewing eye.[43,44]

Reverse occlusion

The effects of constant form deprivation in one eye, including spatial contrast sensitivity loss and ocular dominance

shift in V1 away from the deprived eye, can be reversed if vision of the originally deprived eye is restored early in development and the fellow non-deprived eye is occluded, *reverse occlusion*[45,46] (Fig. 40.4). The timing of the reverse occlusion is critical in determining the effectiveness of this procedure because the "recovery" of functions in the originally deprived eye may occur at the expense of the originally non-deprived eye. For example, contrast sensitivity can be restored if reverse occlusion occurs relatively early in the critical period, i.e. if the original deprivation is short. However, this early reversal leads to a loss of contrast sensitivity in the newly deprived (or originally non-deprived) eye (red circles in Fig. 40.4A)[45] and causes a corresponding shift in the ocular dominance distribution of V1 neurons favoring the initially deprived eye.[46] There is an optimal time for the reversal of monocular occlusion in order to achieve near normal contrast sensitivity for both eyes (Fig. 40.4B).[13,45] In all cases, the binocular functions are diminished.[45] Similar effects of reverse occlusion have been extensively studied in cats, and the results have contributed in advancing our understanding of the neural mechanisms underlying the breakdown and recovery of visual functions from early monocular form deprivation.[13,14,47]

The clinical significance of these findings is that this kind of animal study can provide key information for developing an effective clinical strategy for treating amblyopia with various *patching regimens*.

Brief unrestricted vision during monocular deprivation

Providing brief daily periods of normal vision to the deprived eye during early monocular deprivation prevents or reduces the severity of form deprivation amblyopia in monkeys (Fig. 40.5A).[40] Constant form deprivation (0 hour of unrestricted vision), as previously described, causes severe amblyopia of the deprived eye and a large shift in the ocular dominance of V1 neurons away from the deprived eye (Fig. 40.5B). However, only *one hour* of unrestricted (normal) vision every day during the deprivation period (12 hours/day) dramatically improves the contrast sensitivity of the deprived eye, hence reducing the severity of form deprivation amblyopia.[40] In these monkeys the extent of abnormal ocular dominance shift in V1 is significantly reduced (Fig. 40.5B).[32] In stark contrast, the same "preventive" measure, even with 4 hours of daily unrestricted vision during the 12-hour deprivation period, does not prevent a severe loss of disparity sensitive neurons, highlighting the extremely fragile nature of developing binocular connections in V1. However, one hour of unrestricted vision is effective in reducing binocular suppression in V1 (i.e. the binocular responses of V1 neurons are less than monocular responses) that is prevalent in monkeys reared with constant form deprivation (Fig. 40.5C).[32]

Constant monocular form deprivation leads to an elongation of the eye, and hence, the development of myopic refractive errors in the deprived eye.[18,48,49] However, a brief period of unrestricted vision during the deprivation period reduces the degree of myopic refractive error.[50]

The clinical relevance of these studies is that the timely removal of the conditions that produce degradation of images or image occlusion (e.g. severe hyperopic anisometropia, cataract, or ptosis) is critically important for the prevention of amblyopia. If that is not immediately possible,

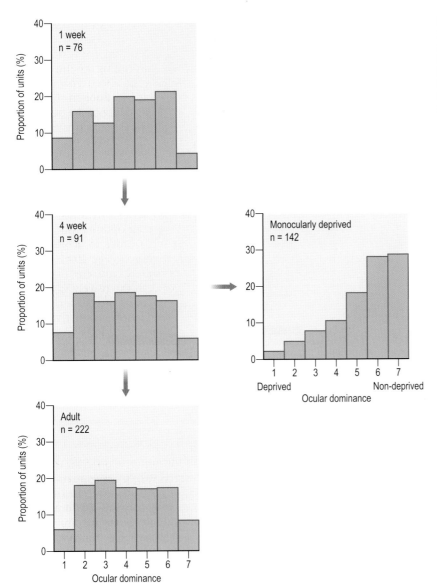

Figure 40.3 Ocular dominance (OD) distributions of V1 neurons in normal infant (1 week & 4 weeks), normal adult, and monocularly form-deprived monkeys. Cells in OD 1 and 7 are exclusively driven by the contralateral or ipsilateral eye, respectively. Cells in OD 4 are binocularly balanced and neurons in OD 2 and 3 or OD 5 and 6 are dominated by the contralateral or ipsilateral eye, respectively. (Data from Chino et al 1997,[10] Sakai et al 2007.[32])

stop-gap manipulations such as lifting a drooping eyelid or keeping corrective lenses even for a short period of time every day are likely to have preventive effects against form deprivation amblyopia and development of myopic refractive errors.[40] Taken together, studies of this kind and those on the effects of temporal variations in image quality, e.g. alternating or reverse occlusion, can provide key information in developing effective clinical strategies for the management of amblyopia in human infants.[40,47]

Critical period

The critical (plastic) period of vision development is traditionally defined as the postnatal period during which visual deprivation leads to long-term or permanent structural and/or functional changes of the visual system. The critical period differs substantially between species, the visual functions affected by deprivation, sites of neural alterations, and the nature of the visual deprivation, e.g. dark rearing, monocular form deprivation, monocular defocus, or ocular misalignment.[14] For example, the critical period for primates, unlike

sub-primate species, begins at or near birth (Fig. 40.6).[9,19,51] Binocular functions are generally more readily disrupted by early visual deprivations than monocular spatial vision. The critical period for experience-dependent changes differs between cortical sites, e.g. V1, V2, V4 or MT, and cortical layers within a given cortical site. The higher stages of processing, e.g. supra- and infra-granular layers compared to input layer within V1 or cortical sites later in the hierarchy of extrastriate visual areas, appear to have longer periods of plasticity.[14]

Critical period for monocular form deprivation

There are multiple "plastic" periods for different visual functions in macaque monkeys (Fig. 40.6).[19,51] Spectral sensitivity functions have relatively short critical periods that begin soon after birth and last for 3 months for scotopic spectral sensitivity and 6 months for photopic spectral sensitivity. The critical period for visual acuity loss is much longer, lasting over 24 months. Binocular vision development can be disrupted by monocular deprivation starting as late as 25

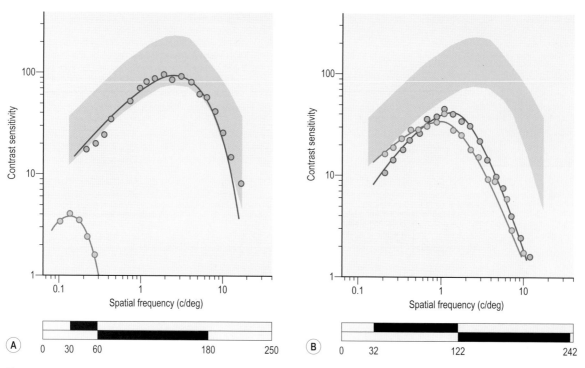

Figure 40.4 Spatial contrast sensitivity functions after reverse monocular occlusion. The functions for the originally non-deprived (deprived after reversal) eye are shown with blue circles. The initial deprivation began at 21 days of age in all groups. (**A**) Reversal after 4 weeks of monocular deprivation. (**B**) After 3 months. Shaded area shows the normal range of contrast sensitivity in normal monkeys. (Modified from Harwerth et al 1989.[45] Reproduced by kind permission of Springer Science + Business Media.)

months of age (roughly equivalent to 8 years in humans).[19] However, the "end" of the critical period for binocular functions has not been determined for monkeys.

Because the sensitivity of the visual cortex to deprivations varies substantially during the critical periods, the timing of deprivation, i.e. *onset and duration*, has significant effects on the severity of perceptual and neural deficits. At what age is monocular form deprivation likely to have the most damaging effects in monkeys? The perceptual development of contrast sensitivity and visual acuity in monkeys is most vulnerable to monocular form deprivation during the first 5 postnatal months. A sharp drop of sensitivity to deprivation occurs after this initial period of heightened sensitivity, followed by a gradual decline over an extended period of time, i.e. > 24 months (Fig. 40.6).[19,51]

For ocular dominance plasticity in V1 of monkeys, the most severe shrinkage of ocular dominance columns for the deprived eye occurs with the *earliest* onset age, e.g. 1 week of age.[28] The degree of the shrinkage becomes progressively smaller as the onset of deprivation is delayed and there is no obvious shrinkage if the onset is set at the 12th postnatal week. Thus, contrary to a classic observation,[27] the ocular dominance columns in layer IVC of monkey V1 are most sensitive to monocular deprivation right after birth.

These behavioral and anatomical studies reinforce the *clinical view* that removal of dense congenital cataract combined with good optical quality lenses or correction of ptosis at the earliest possible postnatal time is essential to minimize the negative impact of monocular form deprivation in humans.[52–54]

The critical period for ocular dominance in cats begins about 3–4 weeks of age when the optics of their eyes becomes relatively clear, peaks around 6–8 weeks, and gradually decreases during the next 12–14 weeks.[25,55,56] A similar timing of the critical period for monocular deprivation has been reported for rats and ferrets with minor variations.[57–59] The critical period is longer for monkeys than that in lower species and appears to be generally correlated with animal's life expectancy.[15,60] It is difficult to determine the precise critical period of vision development for humans in part because of the difficulties associated with conducting experiments on human infants and dependence on clinical observations for data collection. Although the critical period for humans varies considerably for specific visual tasks as evidenced in animal studies, the critical period for experience-dependent changes in humans is thought to begin soon after birth (within 6 months or earlier), peak around 1–3 years of age, and decline slowly until 8–10 years of age or later.[14]

Molecular mechanisms of ocular dominance plasticity

The molecular mechanisms of experience-dependent ocular dominance plasticity have been extensively studied in rodents because the visual cortex of rodents is in general organized similarly to that in higher mammals[61,62] and a wide range of genetic manipulations are readily accessible in rodents.[14–16,47,63,64] Synaptic events following monocular deprivation (*binocular competition*) consist of an initial reversible reduction in functional connections to the deprived eye[58] and rewiring of the upper layer long-range connections.[30,31] These are followed by extended periods of strengthening of the responses to the non-deprived eye and an eventual structural reorganization (*consolidation*)[15,16,25,29] (Fig. 40.7A). The initial rapid changes occur as a result of imbalance in the strength of the input signals between the deprived and

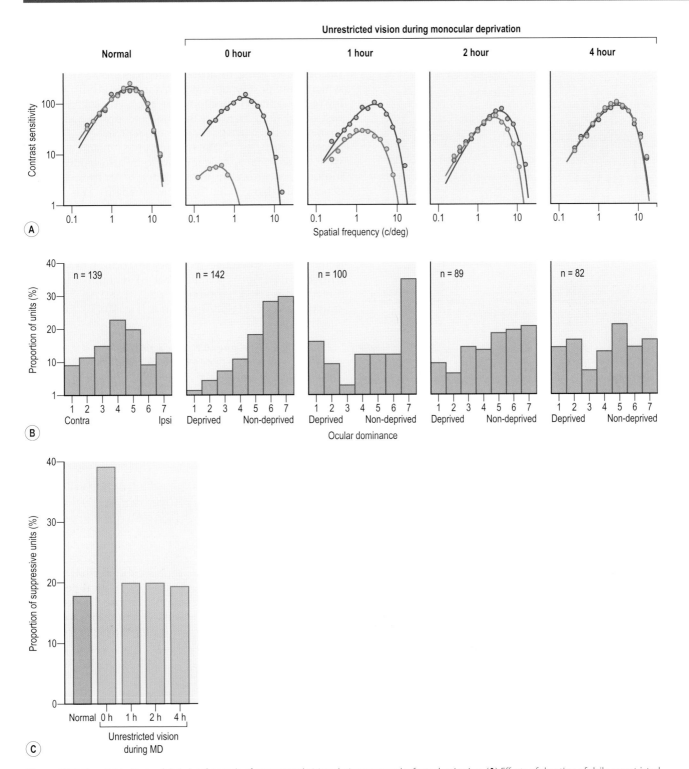

Figure 40.5 Beneficial effects of daily brief periods of unrestricted vision during monocular form deprivation. (**A**) Effects of duration of daily unrestricted vision on contrast sensitivity for representative monkeys. Red circles in experimental monkeys indicate contrast sensitivity for the deprived eye. (Data from Wensveen et al 2006.[40]) (**B**) Ocular dominance distributions of V1 neurons. (**C**) The proportion of binocularly suppressive V1 neurons. (Data from Sakai et al 2006.[32])

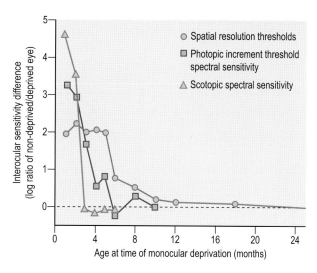

Figure 40.6 The critical periods of development in macaque monkeys. Scotopic spectral sensitivity (triangles). Photopic spectral sensitivity (squares). Acuity (circles). (Modified from Harwerth et al 1986.[51])

the non-deprived eyes. This imbalance disrupts the timing of the firing of action potentials between the pre-synaptic (LGN) and post-synaptic (V1) neurons for the deprived eye. Uncorrelated firing of action potentials between the pre-synaptic and post-synaptic neurons in V1 leads to a weakening of synaptic connections for the deprived eye while well-timed firing between the pre- and post-synaptic cells strengthens the synapses for the non-deprived eye ("*Hebbian synaptic plasticity*").[65–68] The timing of pre-synaptic and post-synaptic neuronal spiking appears to be also modulated by inhibitory neurons in the circuitry.[15,16,69] These activity-dependent changes in synaptic strength also involve long-term potentiation (LTP) of input signals from the non-deprived eye and long-term depression (LTD) of signals from the deprived eye.[66,70,71]

Closely associated with LTP and LTD synaptic plasticity is the role of glutamate receptors in cortical neurons, in particular, *n-methyl-D-aspartate* (NMDA), α-amino-3-hydroxy-5-methyl-4-isoxazolepropionic acid (AMPA), and γ-aminobutyric-acid (GABA) receptors (Fig. 40.7B). Excitatory

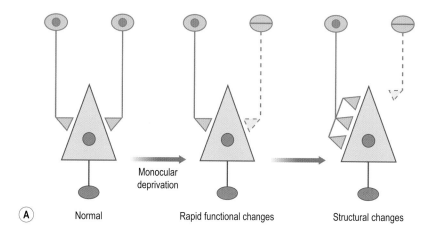

Figure 40.7 Cortical mechanisms of ocular dominance plasticity. (**A**) Classic view of "binocular competition" in the primary visual cortex following early monocular form deprivation. (**B**) Molecular mechanisms of experience-dependent cortical plasticity. (See reviews by Hensch 2005[15] and Tropia et al 2009.[16])

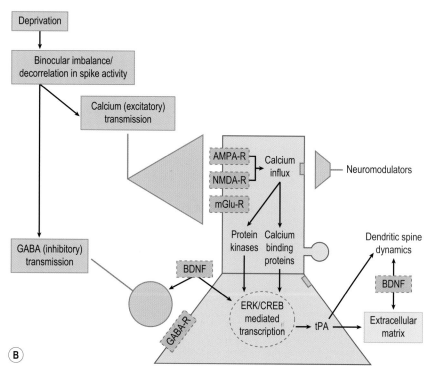

synaptic transmission is mediated by NMDA and AMPA receptors while inhibitory synaptic transmission is regulated by GABA-Aα receptors. Each of these glutamate receptors has a broad range of critical roles in regulating the balance between excitation and inhibition and ocular dominance plasticity.[15,16]

NMDA receptors are made up of three types of subunits (NR1, NR2A and NR2B). The normal postnatal changes in the expression of NR1 and NR2A or NR2B subunits, e.g. the NR2A/NR2B ratio is low at birth and gradually increases during postnatal development, are also experience-dependent. As a result, the developmental changes of the NMDA subunit expression due to monocular deprivation, i.e. the activity-dependent regulation of the NR2A/NR2B ratios, are intimately involved in regulating the plasticity of the visual cortex. For example, an increase in the NR2A or NR2A/2B ratios leads to the induction of LTP and to a heightened sensitivity of synapses to modification.[72–74]

AMPA receptors, composed of GluR2 and either GluR1 or GluR3 subunits, are also involved in synaptic plasticity.[16,75–77] Synaptic strength is determined by the AMPA receptor density and calcium permeability. Repetitive activation of synapses leads to increased insertion of AMPA subunits into postsynaptic neuronal membrane resulting in LTP whereas reductions in synaptic activation, e.g. by monocular deprivation, remove AMPA receptors leading to LTD.[14,16,78]

Synaptic depolarization via activation of NMDA and AMPA receptors induces *calcium influx* and activates an *intracellular signaling cascade*. The second messenger molecules that are directly involved in synaptic strength and ocular dominance plasticity include protein kinase (PKA), calcium/calmodulin-dependent protein kinase II (CaMKII), extracellular signal-regulated kinase 1,2 (ERK), cyclic AMP responsive element binding protein (CREB), and protein synthesis machinery. PKA, CaMKII, and ERK rapidly promote ocular dominance plasticity by modulating synaptic strength by phosphorylating plasticity-regulating molecules in glutamate or GABA receptors. This kinase signaling leads to the activation of CREB.[79–81] The changes initiated by activation of intracellular second messenger molecules along with the action of brain-derived neurotrophic factor (BDNF) leads to the enhanced expression of molecules that act on tissue plasminogen activation (tPA). BDNF and tPA can initiate the changes of dendritic spine motility,[82] spine density,[83] and extracellular matrix[16] that ultimately result in rewiring of cortical circuits favoring the non-deprived eye.

GABA-mediated inhibition is important in maintaining the balance between cortical excitation and inhibition, hence in regulating the timing of ocular dominance plasticity.[15,69,83,84] Manipulation of the normal level of GABAergic transmission in the developing brain can delay or advance the onset of the critical period. Preventing the maturation of GABA mediated transmission or dark rearing delays the critical period. Enhancing GABA transmission by infusing GABA agonist, e.g. benzodiazepines,[84] or facilitating the growth of GABAergic interneurons by BDNF can advance the onset of the critical period.[85,86]

The clinical significance of these manipulations to "reset" the excitatory and inhibitory balance in V1 by various pharmacological methods is that the results may provide new insights into the mechanisms underlying critical periods,

and thus, reveal potential means to promote functional recovery in adults with developmental vision disorders.[63,64]

Another class of molecules involved in synaptic plasticity is the *neuromodulators* including norepinephrine (noradrenaline), serotonin, and acetylcholine. These molecules are abundantly present throughout the cortex. Enhancing adrenergic, serotoninergic or cholinergic modulator systems facilitates the onset of ocular dominance plasticity.[87–90] Changes in the expression of neuromodulators influence the level of LTP and LTD induction by modifying the intracellular calcium concentration by second messenger pathways and results in the associated structural reorganizations of local connections to the visual cortex.[91]

Effects of early monocular defocus

Constant monocular defocus

A less severe form of monocular image degradation results from large differences in refractive errors between the two eyes (*anisometropia*). Normal primates begin their life with modest but binocularly balanced hyperopic refractive errors that decline to a normal refractive state during early infancy (*emmetropization*).[92,93] If infants have large differences in refractive state between the two eyes, they are unable to focus with both eyes. To avoid experiencing binocularly discordant images, infants begin to fixate with one eye (typically with the eye having a less severe refractive error) and as a result, the non-fixating eye experiences a defocused image. Monkey models of anisometropia are simulated by rearing infant monkeys with monocular defocusing lenses[40,92–96] or by monocular atropinization.[97]

Perceptual deficits

The perceptual consequences of untreated early anisometropia are generally similar to the anomalies found after early monocular form deprivation; impoverished binocular vision, reduced contrast sensitivity for high-spatial-frequency stimuli, and lower optotype acuity in the affected eye (*anisometropic amblyopia*).[95–98] However, the perceptual deficits in anisometropes are generally less severe than in monocular form deprivation and are spatial frequency dependent (Fig. 40.8A). Anomalies vary substantially between individuals depending on the etiological factors and rearing histories, e.g. the degree of defocus.

Neural changes

Abnormal alterations in cortical physiology that result from early unilateral defocus are also similar to, but milder than cortical deficits in monocularly form-deprived monkeys. The ocular dominance distribution of V1 neurons is marked by a substantial loss of binocularly balanced cells (ocular dominance between 3 and 5) and by a milder but significant shift away from the affected eye[95,96,99] (Fig. 40.8B). The sensitivity of V1 neurons to binocular disparity is substantially reduced and complex cells are more severely affected than simple cells[95] (Fig. 40.8C). The spatial resolution and contrast sensitivity of V1 neurons for the affected eye of a *severely* anisometropic monkey are slightly lower than that of the fellow eye.[96,99] However, the magnitude of deficits in the spatial receptive-field properties of V1 neurons is too small to

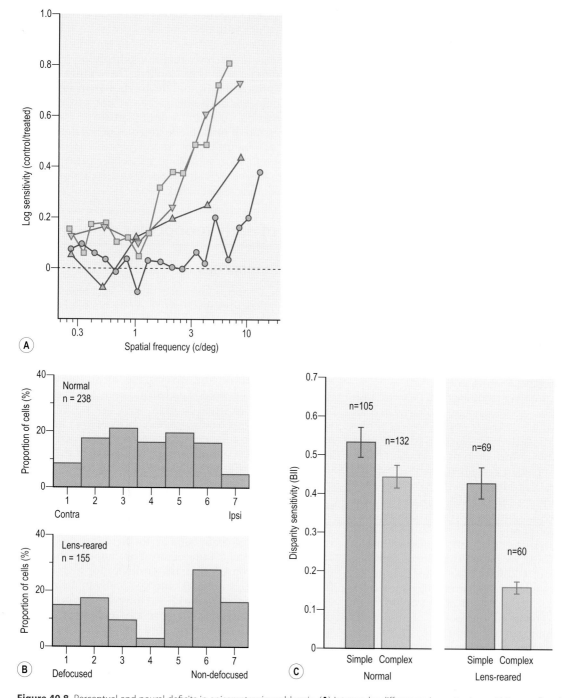

Figure 40.8 Perceptual and neural deficits in anisometropic amblyopia. (**A**) Interocular differences in contrast sensitivity as a function of stimulus spatial frequency in 4 different monkeys reared with unilateral lens-defocus. (**B**) Ocular dominance distribution of V1 neurons in normal (top) and lens-reared (bottom) monkeys. (**C**) Disparity sensitivity loss in V1 of anisometropic monkeys. Note that complex cells had more severe deficits. (Modified from Smith et al 1985,[98] 1997.[95])

account for their perceptual loss in acuity and contrast sensitivity.[1,96]

Alternating defocus

Early monocular defocus can be alternated daily between the two eyes to prevent the development of amblyopia.[94,100] If infant monkeys experience alternating defocus, the monocular response properties of V1 neurons, such as orientation selectivity, spatial frequency tuning, and/or spatial

resolution, do not show response alterations that favor one eye over the other because each eye receives uninterrupted vision on alternate days during early infancy.[100] However, daily alternating defocus leads to a spatial-frequency-dependent loss of local stereopsis (elevated disparity threshold) (Fig 40.9B). This reduction of local stereopsis is exaggerated for high-spatial-frequency stimuli because larger defocus generates greater conflicts between signals coming from the two eyes.[94] Moreover, the disparity sensitivities of V1 neurons are significantly reduced in these

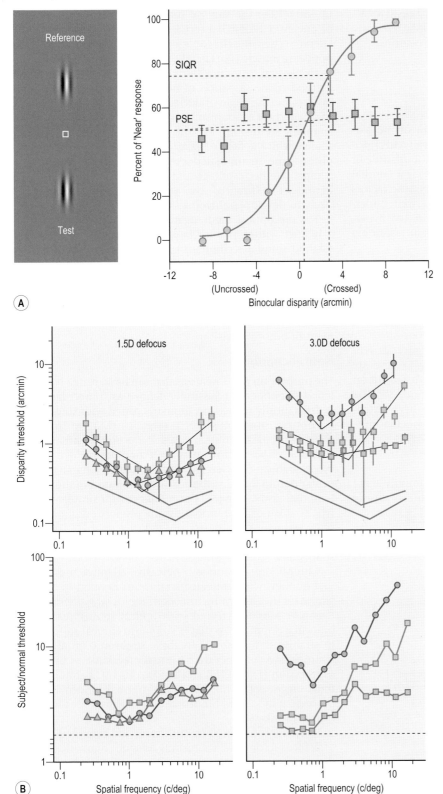

Figure 40.9 Effects of early alternating defocus on disparity sensitivity. (**A**) Stimulus used to test local stereopsis in monkeys (left), and a psychometric function for binocular disparity discrimination (right). Stereoacuity is defined by disparity differences between the point of subject equality (PSE) and the semi-interquartile range (SIOR). Monocular control data are illustrated with small squares. (**B**) Disparity sensitivity loss as a function of spatial frequency in monkeys reared with 1.5 diopter alternating defocus (left top) and interocular differences (left bottom), and comparable values for monkeys reared with 3.0 diopter alternating defocus (right column). Thick lines signify the range of threshold values for normal monkeys. (Modified from Wensveen et al, 2003.[94])

monkeys and this reduction is also spatial frequency dependent.[100]

These observations support the traditional view that local stereopsis is spatial frequency dependent, and that binocular disparity information is processed by independent channels tuned to different spatial frequencies.[101,102] It is also evident that the local disparity processing mechanisms in V1 can be independently compromised by early abnormal visual experience depending on their spatial frequency tuning properties. Finally, the effects of early alternating defocus on binocular vision development underscore the importance of having the normal presence of disparity-sensitive neurons in

V1 for local stereopsis although disparity-sensitive V1 neurons alone are not sufficient to support fine stereopsis, i.e. it requires further processing by extrastriate neurons.[103,104]

Effects of early strabismus

Strabismus is a chronic deviation of the visual axes that emerges shortly after birth.[105,106] The direction of the axis deviation can be convergent (esotropia), divergent (exotropia), or vertical (hypertropia). The etiology of infantile or congenital strabismus is not known. There is a clear familial tendency of developing strabismus, but the genetic factors responsible for infantile strabismus are not well understood.[106] In primates, a high degree of uncorrected hypermetropia soon after birth is known to result in esotropia (accommodative esotropia).[107-110]

Perceptual deficits

Strabismic infants experience double vision (diplopia) immediately after the onset of misalignment. If "normal" alignment is not achieved in a timely manner, binocular vision anomalies, such as deficient stereoscopic vision[95,111,112] (Fig. 40.10), reduced binocular summation of contrast sensitivity,[95] and clinical suppression[113] are likely to develop. Amblyopia may develop if strabismus is not treated for an extended period of time during the critical period.[1,33,96] The effects of experiencing early strabismus have been extensively studied in monkeys by artificially creating ocular misalignment shortly after birth.

Animal models of strabismus

In monkey models, human strabismus is commonly *simulated* by surgical[96,112] or optical[95,114-120] methods shortly after birth. As mentioned previously, alternating monocular occlusion from birth leads to strabismus and these monkey models have served primarily for studies on ocular

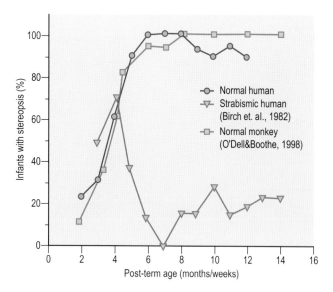

Figure 40.10 Perceptual deficits in strabismic amblyopes. Development of stereopsis in normal human (blue circles) and monkey (green squares) infants, and human infants with infantile esotropia (red triangles). (Data from Birch 2003[105] and O'Dell & Boothe 1998.[8])

motility.[43] For the surgical method, the extraocular muscle of one eye (the medial rectus muscle for exotropia and the lateral rectus muscle for esotropia) is sectioned and the opposing muscle (the lateral or medial rectus muscle, respectively) is tied to induce the misalignment.[96,112] The surgical method creates a period of non-comitant or paralytic strabismus, i.e. the angle of deviations changes with the field of gaze. This type of strabismus is less common in humans than comitant strabismus, i.e. the angle of deviation does not change with the field of gaze. In monkeys with surgical strabismus, the manipulated eye is immediately placed at a competitive disadvantage and the non-deviating fellow eye becomes the "fixating" eye. As a result, the deviating eye is likely to develop amblyopia.[96,121]

For the optical method, infant monkeys are fitted with a helmet with a pair of base-in or base-out prisms. This method simulates comitant strabismus in humans. Neither eye is disadvantaged because their fixation frequently alternates between the two eyes, and thus, the prism-reared monkeys are less likely to develop amblyopia.[95,114,121] However, the effects of early surgical or optical strabismus are quite devastating on the perceptual and neural development of binocular vision.

Neural changes

The proportion of V1 neurons that can be activated by stimulation of either eye (classically defined as "binocular cells") is dramatically reduced in monkeys reared with surgical[96,122] or optical strabismus[95,114,123] (Fig. 40.11A). In monkeys reared with paralytic strabismus, the ocular dominance distribution of V1 neurons may shift away from the deviating eye if strabismus is severe.[96] Similar but smaller changes in ocular dominance distribution are also found in the middle temporal visual area (MT) of monkeys with simulated paralytic strabismus.[124] Importantly, the proportion of disparity-sensitive neurons is drastically reduced in V1[95,115-117] (Figs. 40.11B, C) of stereo-deficient monkeys reared with optical strabismus.

Clinical suppression is a pervasive adaptive mechanism that humans with untreated strabismus develop in order to eliminate the disturbing impact of diplopia and visual confusion.[106] Monkeys reared with optical strabismus exhibit binocular suppression that is similar to clinical suppression in strabismic humans.[113] In these strabismic monkeys, V1 neurons have a higher than normal prevalence of binocular suppression (Fig 40.11B and see Fig. 40.13C below),[95,115-117] thus, suggesting a link between abnormal binocular signal processing in V1 and binocular vision deficits for these strabismic monkeys.

Unless accompanied by consistent and extended periods of defocus in one eye, strabismus does not dramatically alter the monocular spatial response properties of V1 neurons.[1,2,96] Even if V1 neurons exhibit mild reductions in contrast sensitivity and/or spatial resolution in the deviating eye of severely amblyopic monkeys, such deficits in V1 are too small to account for their sensitivity losses.[1,96]

Effects of onset age and duration of strabismus

A critical issue for the management of infantile esotropia is at what age corrective measures should be taken to preserve

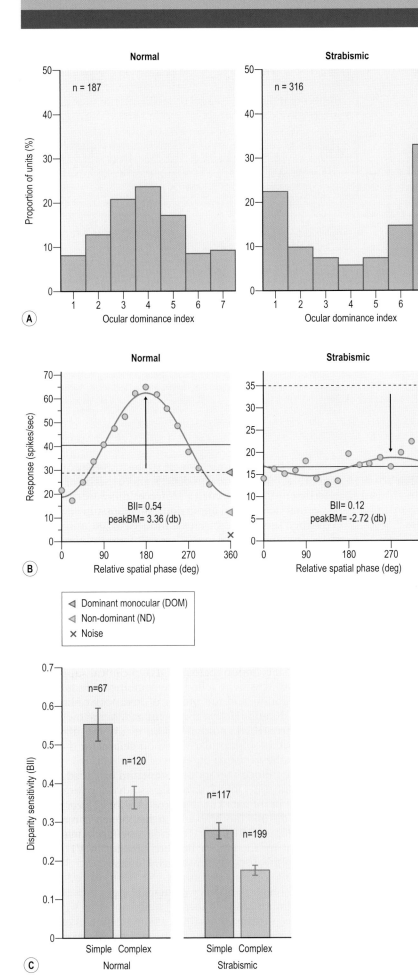

Figure 40.11 Neural deficits in monkeys reared with optically induced strabismus. (**A**) Ocular dominance distribution of V1 neurons in normal (left) and strabismic (right) monkeys. (**B**) Representative disparity tuning functions of V1 neurons for normal (left) and strabismic (right) monkeys. Note that the neuron from a strabismic monkey exhibited a severe loss of disparity tuning and robust binocular suppression (binocular responses < dominant monocular response). (**C**) The average disparity sensitivity (BII) of V1 neurons in normal (left) and strabismic (right) monkeys. (Modified from Smith et al 1997.[95])

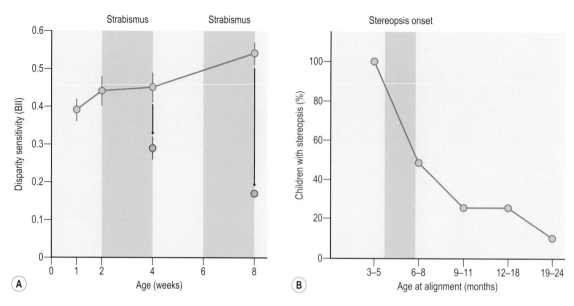

Figure 40.12 Effects of onset age of strabismus and timing of alignment on disparity sensitivity. (**A**) Effects of onset age of strabismus on disparity sensitivity of V1 neurons in infant monkeys. Duration of optical strabismus was kept for 2 weeks while the onset age of strabismus was set either at 2 weeks or 6 weeks of age. Normal development of disparity sensitivity in V1 is also illustrated (red circles). Note that the reduction in disparity sensitivity is far greater for the late-onset group than for the early-onset group. (Data from Chino et al 1997[10] and Kumagami et al 2000.[115]) (**B**) Effects of alignment age on stereopsis in human strabismic infants. Blue shading indicates the known onset age of stereopsis for normal human infants (i.e. between 4 and 6 months of age). (Modified from Birch et al 2000.[127])

stereoscopic vision and establish alignment. Recent animal studies support the *clinical view* that surgical correction before the known onset age of stereopsis (4–6 months of age) should be beneficial for maintaining binocular sensory functions[109,125–128] and reducing eye movement anomalies.[118–120,129]

Onset age

Experiencing a brief period (2 weeks, equivalent to 2 months in humans) of optical strabismus beginning at 4 weeks of age (equivalent to 4 months in humans) drastically reduces the proportion of disparity-sensitive neurons and increases the prevalence of binocular suppression in monkey V1.[115] These binocular response deficits are *more severe* when infant monkeys experience optical strabismus *after* 4 weeks of age (6 to 8 weeks) rather than *before* 4 weeks of age (2 to 4 weeks)[115,116] (Fig. 40.12A). Optical strabismus creates well-focused images that do not match between the two eyes. Consequently, unlike monocular deprivation, the most sensitive segment of the critical period for binocular disparity sensitivity is not immediately after birth when spatial vision is still very crude. Instead, it is soon after the known onset age of stereopsis around 4 to 8 weeks of age (equivalent to 4 to 8 months in humans). By this time of normal development, V1 neurons attain the qualitatively adult-like monocular spatial tuning properties and responsiveness.[10,11,130–132] Consequently, V1 neurons become more sensitive to small discrepancies in the images between the two eyes and as a result, their binocular response properties are more readily disrupted.

The clinical relevance of these findings is evident. As mentioned, the earlier the surgical or optical correction of infantile esotropia, the better the preservation of fine binocular vision and later alignment of the visual axes[109,125–128] (Fig. 40.12B).

Finally, the latest onset age of strabismus that can cause substantial binocular vision deficits and/or amblyopia is yet to be determined in primates. However, there are substantial perceptual deficits and anomalous alterations in V1 and V2 even when surgical strabismus is induced as late as 6 months of age (roughly equivalent to 2 years of age in humans).

Duration

In human infants, longer durations of strabismus result in greater deficits of binocular vision (stereopsis).[127] For infant monkeys reared with prisms, a similar relationship holds for disparity sensitivity deficits in V1. However the duration of optical strabismus has little impact on the disparity sensitivity loss of V1 neurons if it is longer than 2 weeks (Fig. 40.13B).[116] It is important to note that the effects of duration on disparity sensitivity reflect the combined effects of strabismus duration and the timing of interocular alignment, i.e. longer duration delays alignment and, hence the restoration of unrestricted "normal" visual experience during the critical period for binocular vision development.

What is the *minimum* duration of misalignment required to alter the binocular response properties of V1 neurons in infant monkeys? Only 7 days of prism-rearing (equivalent to 4 weeks in humans) beginning at 4 weeks of age are sufficient to disrupt the binocular disparity sensitivity of V1 neurons (Fig. 40.13B).[115–117] Three days of optical strabismus (equivalent to 12 days in human infants) do not significantly alter the disparity sensitivity. However, after 3 days of experiencing optical strabismus, V1 neurons exhibit an abnormally high prevalence of binocular suppression (Fig. 40.13C). Importantly, binocular suppression in V1 after 3 days of prism-rearing is as strong as suppression in monkeys that experience much longer durations (months or years) of strabismus. Thus, the very first neural alteration in V1

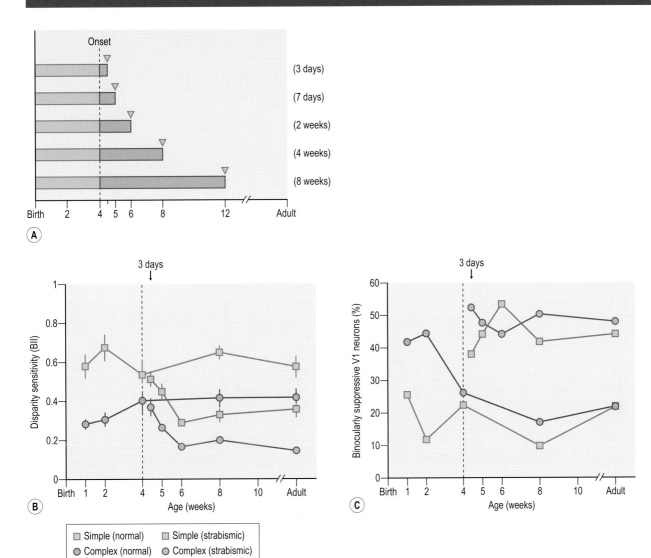

Figure 40.13 Effects of duration of optical strabismus on binocular responses of V1 neurons. (**A**) Rearing regimen. The onset was kept at 4 weeks of age. The duration of prism-rearing was varied between 3 days and 8 weeks. Green triangles indicate the time of V1 recording experiments. (**B**) Effects of duration on disparity sensitivity of V1 neurons. Normal data are also illustrated (blue and green symbols). Data for strabismic adults come from monkeys reared with surgically induced strabismus. (**C**) Effects of strabismus duration on the prevalence of V1 neurons exhibiting binocular suppression. (Data from Chino et al 1997,[10] Mori et al 2002,[116] Zhang et al 2005.[117])

following the onset of strabismus is the emergence of robust binocular suppression.[117]

Eye movement anomalies

Strabismic monkeys and humans exhibit temporal-to-nasal asymmetries in pursuit eye movement and visual tracking under monocular viewing,[44,106,124,133] similar to those found in normal infants.[133,134] Also, strabismic humans often exhibit latent nystagmus.[120,135] These eye movement anomalies result from abnormal control of cortical neurons over the subcortical nuclei that support eye movements and visual tracking.[44,119,124,133,134,136]

The timing of corrective measures for infantile esotropia has significant impact on preventing eye movement anomalies.[118,120,129] In a series of recent studies by Tychsen and his colleagues, newborn monkeys were reared from birth with a goggle containing a 20-diopter prism in front of each eye

until either 3 weeks of age ("*early repair*") or 3 to 6 months of age ("*late repair*"). Monkeys in the "late repair group" exhibited temporal-to-nasal asymmetries of smooth pursuit and OKN under monocular viewing, latent nystagmus, and persistent ocular misalignment. However, monkeys in the "early repair group" did not exhibit any of these motor deficits.[118,120,129] The motor deficits in the late repair group appeared to be linked to their abnormal visual cortical physiology and anatomy.[95,115,116,119,120,133,137]

The clinical implications of these studies are that the earlier the corrective measures for infantile esotropia the better the normal maturation of oculomotor functions (see review[109]).

Amblyopia

Amblyopia is a developmental disorder of spatial vision. Amblyopic primates show reduced contrast sensitivity that

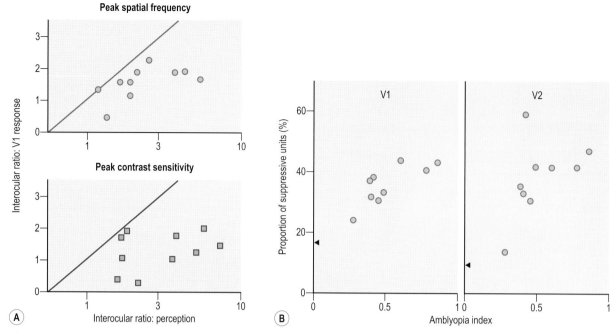

Figure 40.14 Amblyopia and cortical physiology in monkeys. (**A**) The average peak spatial frequency (top) and average peak contrast sensitivity (bottom) of V1 neurons as a function of perceptual differences in contrast sensitivity between the amblyopic and fellow eye of strabismic and anisometropic monkeys. Note that behavioral deficits are much greater than cortical deficits. (Modified from Kiorpes & Movshon 2003.[1]) (**B**) Relationship between the degree of amblyopia and the prevalence of V1 and V2 neurons exhibiting binocular suppression. Triangles indicate normal control data.

is restricted to or exaggerated for mid- to high-spatial-frequency ranges of stimuli and exhibit lower optotype acuity for the affected eye.[96] The nature and extent of visual anomalies in amblyopes differ widely depending on the onset age, the status of binocular functions,[138] and the type of amblyopia, i.e. strabismic, anisometropic, or form deprivation amblyopia.[33,138,139]

Perceptual deficits

In addition to acuity and contrast sensitivity losses in all types of amblyopia, the visual system of strabismic and anisometropic amblyopes is known to be intrinsically "noisy" and their visual performance is readily disrupted by introduction of noise in the stimuli.[33,140] The sensitivity of amblyopic humans to second-order stimuli is also reduced.[141] Strabismic and anisometropic amblyopes exhibit difficulties in global perceptual tasks that require precise pooling of neighboring local feature information over extended ranges of space. These perceptual difficulties emerge in the form of abnormal contour integration, motion integration, and/or object segmentation.[142–145] As mentioned above, fixation instability is frequently diagnosed in the affected eye of amblyopes under monocular viewing. Without adequate control procedures, fixation instability could make testing sensory capacities of an amblyopic eye and the interpretation of results difficult.[135]

Neural changes

Studies on the neural basis of *form deprivation amblyopia*, as we have reviewed, have given us a relatively clear view of how the form deprived eye loses its functional connections with the primary visual cortex (V1) and how the visual sensitivity of the affected eye is altogether lost even to the most basic, robust visual stimuli[56] (see also recent reviews[13–16]).

For all types of amblyopic monkeys, lower sampling in V1 for the affected eye (ocular dominance shift away from the deviating eye) and/or reduced spatial resolution and contrast sensitivity of V1 neurons are commonly invoked to explain the mechanisms underlying "low-level" perceptual deficits, i.e. reduction in contrast sensitivity and acuity.[1,2,33,139] Recent fMRI studies of amblyopic humans report substantial reductions in BOLD signals for stimulation of the amblyopic eye.[146,147]

Unlike monocular form deprived monkeys, a relatively small percentage of V1 neurons in strabismic or anisometropic monkeys exhibit reduced contrast sensitivity at high spatial frequencies and/or lower spatial resolutions and *only in severely amblyopic monkeys*.[1,96] Moreover, a shift in ocular dominance distribution of V1 neurons away from the affected eye is relatively mild in anisometropic amblyopes and often absent in strabismic amblyopes.[1,96] Obviously the magnitude of the V1 anomalies is too small to account for the severity of perceptual losses for these amblyopic monkeys (Fig 40.14A).[1,96]

An emerging view on the neural basis of strabismic or anisometropic amblyopia, therefore, is that unlike in monocular form deprivation, the major cortical alterations that may limit a wide range of monocular performance of anisometropic and strabismic amblyopes are more likely to occur beyond V1 (in extrastriate visual areas), while relatively mild changes in monocular response properties may begin in V1 of severely amblyopic monkeys.[1,2,33,139] Another emerging view is that interocular suppression highly prevalent in V1 and V2 of strabismic or anisometropic monkeys may play a critical role in the emergence of amblyopia (Fig. 40.14B).[2,33,95] Finally, the neural basis of more complex anomalies in amblyopic monkeys, such as excessive crowding and position uncertainty,[145] deficits in contour integration, abnormal topographic mapping of visual space or topographic "jitter"[148,149] and higher-order cognitive disorders,[150] has not been systematically investigated for monkeys using electrophysiological or brain imaging methods. All of these higher-order perceptual deficits in amblyopes are more likely to involve abnormal signal processing in extrastriate visual cortex.

Improved visual performance in adult subjects with developmental disorders

Although the impact of early deprivations on the visual system development becomes increasingly small with age, the mature visual brain is not completely hard wired and exhibits limited plasticity.[151-154] The nature and extent of adult plasticity in normal primate visual brain is still a matter of considerable debate.[155,156] Another form of "adult plasticity" that is receiving increasing attention is *perceptual learning* where visual performance of primates improves with repeated practice. Moreover, such learning appears to occur as a result of alterations of specific functional connections in the extrastriate visual cortex.[157,158] These discoveries raise the possibility that reduced visual performance of adults who had experienced early deprivations could be improved by training and practice. In human amblyopes, repeated practice significantly improves performance of adults and children for a variety of visual tasks including visual acuity (see review[159]). In stereo-deficient adult monkeys reared with early optical strabismus, stereoacuity improves by several factors after extended periods of training and testing, and this improvement appears to be associated with the preservation of disparity sensitivity in extrastriate cortical neurons.[160]

The clinical significance of these observations is that although the earliest intervention of abnormal visual conditions during infancy is most critical in preventing or minimizing the impact of binocular vision disorders, adults with certain developmental vision anomalies may benefit from well-thought-out treatment regimens (but see Helveston 2005[161] and Horton 2005[162]).

Summary

The primate visual brain is highly malleable during early infancy and the maturation of visual functions depends on normal visual experience during the critical period of development. Normal visual experience requires clear unrestricted images in each eye and precise matching of the images of the two eyes. If one eye is deprived of clear vision during the critical period, causing binocular imbalance, the infant's visual brain is abnormally reorganized. Binocular imbalance commonly results from monocular form deprivation, monocular defocus, and/or misalignment of visual axes. Anomalous reorganization of the wiring in the visual brain largely begins in the primary visual cortex (V1), but not in the retina. Neural deficits resulting from early abnormal visual experience are likely to be more extensive in the extrastriate visual areas. Binocular visual capacities, such as stereopsis and binocular summation of contrast sensitivity, are compromised or completely lost. If the visual conditions responsible for binocular imbalance are untreated, amblyopia is likely to develop in the affected eye. As reviewed, animal studies are designed to reveal the neural basis of developmental vision anomalies and provide key information that is beneficial for establishing effective treatment strategies for binocular vision disorders in humans.

References

1. Kiorpes L, Movshon JA. Neural limitations on visual development in primates. In: Chalupa LM, Werner JS, eds. The visual neurosciences, MIT Press: Cambridge, MA, 2003:159–173.

2. Chino YM, Bi H, Zhang B. The postnatal development of the neuronal response properties in primate visual cortex. In: Kaas J, Collins C, eds. Primate vision. Boca Raton: CRC Press, 2004:81–108.

3. Harwerth RS, Smith EL 3rd. Binocular summation in man and monkey. Am J Optom Physiol Opt 1985; 62(7):439–446.

4. Harwerth RS, Smith EL 3rd. Rhesus monkey as a model for normal vision of humans. Am J Optom Physiol Opt 1985; 62(9):633–641.

5. Boothe RG, Kiorpes L, Williams RA, Teller DY. Operant measurements of contrast sensitivity in infant macaque monkeys during normal development. Vision Res 1988; 28(3):387–396.

6. Kiorpes L. Development of Vernier acuity and grating acuity in normally reared monkeys. Vis Neurosci 1992; 9(3–4):243–251.

7. Birch EE, Gwiazda J, Held R. Stereoacuity development for crossed and uncrossed disparities in human infants. Vision Res 1982; 22(5):507–513.

8. O'Dell C, Boothe RG. The development of stereoacuity in infant rhesus monkeys. Vision Res 1997; 37(19):2675–2684.

9. Horton JC, Hocking DR. An adult-like pattern of ocular dominance columns in striate cortex of newborn monkeys prior to visual experience. J Neurosci 1996; 16(5):1791–1807.

10. Chino YM, Smith EL 3rd, Hatta S, Cheng H. Postnatal development of binocular disparity sensitivity in neurons of the primate visual cortex. J Neurosci 1997; 17(1):296–307.

11. Maruko I, Zhang B, Tao X, Tong J, Smith EL 3rd, Chino YM. Postnatal development of disparity sensitivity in visual area 2 (V2) of macaque monkeys. J Neurophysiol 2008; 100(5):2486–2495.

12. Sherman SM, Spear PD. Organization of visual pathways in normal and visually deprived cats. Physiol Rev 1982; 62(2):738–855.

13. Mitchell DC. The effects of selected forms of early visual deprivation on perception. In: Chalupa LM, Werner JS, eds. The visual neurosciences, MIT Press: Cambridge, MA, 2003:189.

14. Daw NW. Visual development, 3rd edn. New York: Springer, 2006.

15. Hensch TK. Critical period plasticity in local cortical circuits. Nat Rev Neurosci 2005; 6(11):877–888.

16. Tropea D, Van Wart A, Sur M. Molecular mechanisms of experience-dependent plasticity in visual cortex. Phil Trans R Soc Lond B Biol Sci 2009; 364(1515):341–355.

17. Harwerth RS, Crawford ML, Smith EL 3rd, Boltz RL. Behavioral studies of stimulus deprivation amblyopia in monkeys. Vision Res 1981; 21(6):779–789.

18. Smith EL 3rd, Hung LF. Form deprivation myopia in monkeys is a graded phenomenon. Vision Res 2000; 40(4):371–381.

19. Harwerth RS, Smith EL 3rd, Crawford ML, von Noorden GK. Behavioral studies of the sensitive periods of development of visual functions in monkeys. Behav Brain Res 1990; 41(3):179–198.

20. Boothe RG, Dobson V, Teller DY. Postnatal development of vision in human and nonhuman primates. Annu Rev Neurosci 1985; 8:495–545.

21. Wiesel TN, Raviola E. Myopia and eye enlargement after neonatal lid fusion in monkeys. Nature 1977; 266(5597):66–68.

22. Smith EL 3rd, Harwerth RS, Crawford ML, von Noorden GK. Observations on the effects of form deprivation on the refractive status of the monkey. Invest Ophthalmol Vis Sci 1987; 28(8):1236–1245.

23. Harwerth RS, Smith EL 3rd, Paul AD, Crawford ML, von Noorden GK. Functional effects of bilateral form deprivation in monkeys. Invest Ophthalmol Vis Sci 1991; 32(8):2311–2327.

24. Crawford ML, Blake R, Cool SJ, von Noorden GK. Physiological consequences of unilateral and bilateral eye closure in macaque monkeys: some further observations. Brain Res 1975; 84(1):150–154.

25. Hubel DH, Wiesel TN. The period of susceptibility to the physiological effects of unilateral eye closure in kittens. J Physiol 1970; 206(2):419–436.

26. Hubel DH, Wiesel TN, LeVay S. Plasticity of ocular dominance columns in monkey striate cortex. Phil Trans R Soc Lond B Biol Sci 1977; 278(961):377–409.

27. LeVay S, Wiesel TN, Hubel DH. The development of ocular dominance columns in normal and visually deprived monkeys. J Comp Neurol 1980; 191(1):1–51.

28. Horton JC, Hocking DR. Timing of the critical period for plasticity of ocular dominance columns in macaque striate cortex. J Neurosci 1997; 17(10):3684–3709.

29. Antonini A, Stryker MP. Plasticity of geniculocortical afferents following brief or prolonged monocular occlusion in the cat. J Comp Neurol 1996; 369(1):64–82.

30. Trachtenberg JT, Trepel C, Stryker MP. Rapid extragranular plasticity in the absence of thalamocortical plasticity in the developing primary visual cortex. Science 2000; 287(5460):2029–2032.

31. Trachtenberg JT, Stryker MP. Rapid anatomical plasticity of horizontal connections in the developing visual cortex. J Neurosci 2001; 21(10):3476–3482.

32. Sakai E, Bi H, Maruko I et al. Cortical effects of brief daily periods of unrestricted vision during early monocular form deprivation. J Neurophysiol 2006; 95(5):2856–2865.

33. Levi DM. Visual processing in amblyopia: human studies. Strabismus 2006; 14(1):11–19

34. Vital-Durand F, Garey LJ, Blakemore C. Monocular and binocular deprivation in the monkey: morphological effects and reversibility. Brain Res 1978; 158(1):45–64.

35. Blakemore C, Vital-Durand F. Effects of visual deprivation on the development of the monkey's lateral geniculate nucleus. J Physiol 1986; 380:493–511.

36. Levitt JB, Schumer RA, Sherman SM, Spear PD, Movshon JA. Visual response properties of neurons in the LGN of normally reared and visually deprived macaque monkeys. J Neurophysiol 2001; 85(5):2111–2129.

37. Chino YM, Cheng H, Smith EL 3rd, Garraghty PE, Roe AW, Sur M. Early discordant binocular vision disrupts signal transfer in the lateral geniculate nucleus. Proc Natl Acad Sci USA 1994; 91(15):6938–6942.

38. Cheng H, Chino YM, Smith EL 3rd, Hamamoto J, Yoshida K. Transfer characteristics of X LGN neurons in cats reared with early discordant binocular vision. J Neurophysiol 1995; 74(6):2558–2572.

39. Chino YM, Kaplan E. Abnormal orientation bias of LGN neurons in strabismic cats. Invest Ophthalmol Vis Sci 1988; 29(4):644–648.

40. Wensveen JM, Harwerth RS, Hung LF, Ramamirtham R, Kee CS, Smith EL 3rd. Brief daily periods of unrestricted vision can prevent form deprivation amblyopia. Invest Ophthalmol Vis Sci 2006; 47(6):2468–2477.

41. Blake R, Hirsch HV. Deficits in binocular depth perception in cats after alternating monocular deprivation. Science 1975; 190(4219):1114–1116.

42. Tieman DG, McCall MA, Hirsch HV. Physiological effects of unequal alternating monocular exposure. J Neurophysiol 1983; 49(3):804–818.

43. Das VE, Mustari MJ. Correlation of cross-axis eye movements and motoneuron activity in non-human primates with "A" pattern strabismus. Invest Ophthalmol Vis Sci 2007; 48(2):665–674.

44. Fu L, Tusa RJ, Mustari MJ, Das VE. Horizontal saccade disconjugacy in strabismic monkeys. Invest Ophthalmol Vis Sci 2007; 48(7):3107–3114.

45. Harwerth RS, Smith EL 3rd, Crawford ML, von Noorden GK. The effects of reverse monocular deprivation in monkeys. I. Psychophysical experiments. Exp Brain Res 1989; 74(2):327–347.

46. Crawford ML, de Faber JT, Harwerth RS, Smith EL 3rd, von Noorden GK. The effects of reverse monocular deprivation in monkeys. II. Electrophysiological and anatomical studies. Exp Brain Res 1989; 74(2):338–347.

47. Mitchell DE, Sengpiel F. Neural mechanisms of recovery following early visual deprivation. Phil Trans R Soc Lond B Biol Sci 2009; 364(1515):383–398.

48. Smith EL 3rd, Hung LF. The role of optical defocus in regulating refractive development in infant monkeys. Vision Res 1999; 39(8):1415–1435.

49. Smith EL 3rd, Hung LF, Harwerth RS. The degree of image degradation and the depth of amblyopia. Invest Ophthalmol Vis Sci 2000; 41(12):3775–3781.

50. Smith EL, Hung LF, Kee CS, Qiao Y. Effects of brief periods of unrestricted vision on the development of form deprivation myopia in monkeys. Invest Ophthalmol Vis Sci 2002; 43(2):291–299.

51. Harwerth RS, Smith EL 3rd, Duncan GC, Crawford ML, von Noorden GK. Multiple sensitive periods in the development of the primate visual system. Science 1986; 232(4747):235–238.

52. Birch EE, Stager DR. Prevalence of good visual acuity following surgery for congenital unilateral cataract. Arch Ophthalmol 1988; 106(1):40–43.

53. Drummond GT, Scott WE, Keech RV. Management of monocular congenital cataracts. Arch Ophthalmol 1989; 107(1):45–51.

54. Wright KW, Matsumoto E, Edelman PM. Binocular fusion and stereopsis associated with early surgery for monocular congenital cataracts. Arch Ophthalmol 1992; 110(11):1607–1609.

55. Olson CR, Freeman RD. Profile of the sensitive period for monocular deprivation in kittens. Exp Brain Res 1980; 39(1):17–21.

56. Wiesel TN, Hubel DH. Single-cell responses in striate cortex of kittens deprived of vision in one eye. J Neurophysiol 1963; 26:1003–1017.

57. Fagiolini M, Pizzorusso T, Berardi N, Domenici L, Maffei L. Functional postnatal development of the rat primary visual cortex and the role of visual experience: dark rearing and monocular deprivation. Vision Res 1994; 34(6):709–720.

58. Gordon JA, Stryker MP. Experience-dependent plasticity of binocular responses in the primary visual cortex of the mouse. J Neurosci 1996; 16(10):3274–3286.

59. Issa NP, Trachtenberg JT, Chapman B, Zahs KR, Stryker MP. The critical period for ocular dominance plasticity in the ferret's visual cortex. J Neurosci 1999; 19(16):6965–6978.

60. Berardi N, Pizzorusso T, Maffei L. Critical periods during sensory development. Curr Opin Neurobiol 2000; 10(1):138–145.

61. Niell CM, Stryker MP. Highly selective receptive fields in mouse visual cortex. J Neurosci 2008; 28(30):7520–7536.

62. Wang Q, Burkhalter A. Area map of mouse visual cortex. J Comp Neurol 2007; 502(3):339–357.

63. Hooks BM, Chen C. Critical periods in the visual system: changing views for a model of experience-dependent plasticity. Neuron 2007; 56(2):312–326.

64. Morishita H, Hensch TK. Critical period revisited: impact on vision. Curr Opin Neurobiol 2008; 18(1):101–107.

65. Antonini A, Stryker MP. Rapid remodeling of axonal arbors in the visual cortex. Science 1993; 260(5115):1819–1821.

66. Frenkel MY, Bear MF. How monocular deprivation shifts ocular dominance in visual cortex of young mice. Neuron 2004; 44(6):917–923.

67. Bi G, Poo M. Synaptic modification by correlated activity: Hebb's postulate revisited. Annu Rev Neurosci 2001; 24:139–166.

68. Song S, Miller KD, Abbott LF. Competitive Hebbian learning through spike-timing-dependent synaptic plasticity. Nat Neurosci 2000; 3(9):919–926.

69. Heinen K, Baker RE, Spijker S, Rosahl T, van Pelt J, Brussaard AB. Impaired dendritic spine maturation in GABAA receptor alpha1 subunit knock out mice. Neuroscience 2003; 122(3):699–705.

70. Artola A, Brocher S, Singer W. Different voltage-dependent thresholds for inducing long-term depression and long-term potentiation in slices of rat visual cortex. Nature 1990; 347(6288):69–72.

71. Heynen AJ, Yoon BJ, Liu CH, Chung HJ, Huganir RL, Bear MF. Molecular mechanism for loss of visual cortical responsiveness following brief monocular deprivation. Nat Neurosci 2003; 6(8):854–862.

72. Liu L, Wong TP, Pozza MF et al. Role of NMDA receptor subtypes in governing the direction of hippocampal synaptic plasticity. Science 2004; 304(5673):1021–1024.

73. Massey PV, Johnson BE, Moult PR et al. Differential roles of NR2A and NR2B-containing NMDA receptors in cortical long-term potentiation and long-term depression. J Neurosci 2004; 24(36):7821–7828.

74. Cho KK, Khibnik L, Philpot BD, Bear MF. The ratio of NR2A/B NMDA receptor subunits determines the qualities of ocular dominance plasticity in visual cortex. Proc Natl Acad Sci USA 2009; 106(13):5377–5382.

75. Daw NW, Reid SN, Beaver CJ. Development and function of metabotropic glutamate receptors in cat visual cortex. J Neurobiol 1999; 41(1):102–107.

76. Wang XF, Daw NW. Long term potentiation varies with layer in rat visual cortex. Brain Res 2003; 989(1):26–34.

77. Dolen G, Bear MF. Role for metabotropic glutamate receptor 5 (mGluR5) in the pathogenesis of fragile X syndrome. J Physiol 2008; 586(6):1503–1508.

78. Jiang B, Trevino M, Kirkwood A. Sequential development of long-term potentiation and depression in different layers of the mouse visual cortex. J Neurosci 2007; 27(36):9648–9652.

79. Cancedda L, Putignano E, Impey S, Maffei L, Ratto GM, Pizzorusso T. Patterned vision causes CRE-mediated gene expression in the visual cortex through PKA and ERK. J Neurosci 2003; 23(18):7012–7020.

80. Pham TA, Impey S, Storm DR, Stryker MP. CRE-mediated gene transcription in neocortical neuronal plasticity during the developmental critical period. Neuron 1999; 22(1):63–72.

81. Mower AF, Liao DS, Nestler EJ, Neve RL, Ramoa AS. cAMP/Ca^{2+} response element-binding protein function is essential for ocular dominance plasticity. J Neurosci 2002; 22(6):2237–2245.

82. Majewska A, Sur M. Motility of dendritic spines in visual cortex in vivo: changes during the critical period and effects of visual deprivation. Proc Natl Acad Sci USA 2003; 100(26):16024–16029.

83. Mataga N, Mizuguchi Y, Hensch TK. Experience-dependent pruning of dendritic spines in visual cortex by tissue plasminogen activator. Neuron 2004; 44(6):1031–1041.

84. Fagiolini M, Fritschy JM, Low K, Mohler H, Rudolph U, Hensch TK. Specific GABAA circuits for visual cortical plasticity. Science 2004; 303(5664):1681–1683.

85. Huang ZJ, Kirkwood A, Pizzorusso T et al. BDNF regulates the maturation of inhibition and the critical period of plasticity in mouse visual cortex. Cell 1999; 98(6):739–755.

86. Hanover JL, Huang ZJ, Tonegawa S, Stryker MP. Brain-derived neurotrophic factor overexpression induces precocious critical period in mouse visual cortex. J Neurosci 1999; 19(22):RC40.

87. Kasamatsu T, Pettigrew JD. Depletion of brain catecholamines: failure of ocular dominance shift after monocular occlusion in kittens. Science 1976; 194(4261):206–209.

88. Bear MF, Singer W. Modulation of visual cortical plasticity by acetylcholine and noradrenaline. Nature 1986; 320(6058):172–176.

89. Gu Q, Singer W. Involvement of serotonin in developmental plasticity of kitten visual cortex. Eur J Neurosci 1995; 7(6):1146–1153.

90. Gu Q. Contribution of acetylcholine to visual cortex plasticity. Neurobiol Learn Mem 2003; 80(3):291–301.

91. Kirkwood A, Rozas C, Kirkwood J, Perez F, Bear MF. Modulation of long-term synaptic depression in visual cortex by acetylcholine and norepinephrine. J Neurosci 1999; 19(5):1599–1609.

92. Smith EL 3rd, Hung LF, Harwerth RS. Effects of optically induced blur on the refractive status of young monkeys. Vision Res 1994; 34(3):293–301.

93. Smith EL 3rd. Spectacle lenses and emmetropization: the role of optical defocus in regulating ocular development. Optom Vis Sci 1998; 75(6):388–398.

94. Wensveen JM, Harwerth RS, Smith EL 3rd. Binocular deficits associated with early alternating monocular defocus. I. Behavioral observations. J Neurophysiol 2003; 90(5):3001–3011.

95. Smith EL 3rd, Chino YM, Ni J, Cheng H, Crawford ML, Harwerth RS. Residual binocular interactions in the striate cortex of monkeys reared with abnormal binocular vision. J Neurophysiol 1997; 78(3):1353–1362.

96. Kiorpes L, Kiper DC, O'Keefe LP, Cavanaugh JR, Movshon JA. Neuronal correlates of amblyopia in the visual cortex of macaque monkeys with experimental strabismus and anisometropia. J Neurosci 1998; 18(16):6411–6424.

97. Kiorpes L, Boothe RG, Hendrickson AE, Movshon JA, Eggers HM, Gizzi MS. Effects of early unilateral blur on the macaque's visual system. I. Behavioral observations. J Neurosci 1987; 7(5):1318–1326.

98. Smith EL 3rd, Harwerth RS, Crawford ML. Spatial contrast sensitivity deficits in monkeys produced by optically induced anisometropia. Invest Ophthalmol Vis Sci 1985; 26(3):330–342.

99. Movshon JA, Eggers HM, Gizzi MS, Hendrickson AE, Kiorpes L, Boothe RG. Effects of early unilateral blur on the macaque's visual system. III. Physiological observations. J Neurosci 1987; 7(5):1340–1351.

100. Zhang B, Matsuura K, Mori T et al. Binocular deficits associated with early alternating monocular defocus. II. Neurophysiological observations. J Neurophysiol 2003; 90(5):3012–3023.

101. Schor CM, Wood IC, Ogawa J. Spatial tuning of static and dynamic local stereopsis. Vision Res 1984; 24(6):573–578.

102. Yang Y, Blake R. Spatial frequency tuning of human stereopsis. Vision Res 1991; 31(7–8):1177–1189.

103. Cumming BG, Parker AJ. Responses of primary visual cortical neurons to binocular disparity without depth perception. Nature 1997; 389(6648):280–283.

104. Cumming BG, Parker AJ. Local disparity not perceived depth is signaled by binocular neurons in cortical area V1 of the Macaque. J Neurosci 2000; 20(12):4758–4767.

105. Birch EE. Stereopsis in infants and its developmental relation to visual acuity. In: Simons K, ed. Early development: normal and abnormal. Oxford: Oxford University Press, 1993.

106. von Noorden GK. Binocular vision and ocular motility: theory and management, 5th edn. St Louis: Mosby, 1996.

107. Fawcett SL, Birch EE. Risk factors for abnormal binocular vision after successful alignment of accommodative esotropia. J AAPOS 2003; 7(4):256–262.

108. Birch EE, Fawcett SL, Morale SE, Weakley DR Jr., Wheaton, DH. Risk factors for accommodative esotropia among hypermetropic children. Invest Ophthalmol Vis Sci 2005; 46(2):526–529.

109. Birch EE, Wang J. Stereoacuity outcomes after treatment of infantile and accommodative esotropia. Optom Vis Sci 2009; 86:647–652.

110. Quick MW, Eggers HM, Boothe RG. Natural strabismus in monkeys. Convergence errors assessed by cover test and photographic methods. Invest Ophthalmol Vis Sci 1992; 33(10):2986–3004.

111. Crawford ML, Harwerth RS, Smith EL, von Noorden GK. Loss of stereopsis in monkeys following prismatic binocular dissociation during infancy. Behav Brain Res 1996; 79(1–2):207–218.

112. Harwerth RS, Smith EL 3rd, Crawford ML, von Noorden GK. Stereopsis and disparity vergence in monkeys with subnormal binocular vision. Vision Res 1997; 37(4):483–493.

113. Wensveen JM, Harwerth RS, Smith EL 3rd. Clinical suppression in monkeys reared with abnormal binocular visual experience. Vision Res 2001; 41(12):1593–1608.

114. Crawford ML, von Noorden GK. Optically induced concomitant strabismus in monkeys. Invest Ophthalmol Vis Sci 1980; 19(9):1105–1109.

115. Kumagami T, Zhang B, Smith EL 3rd, Chino YM. Effect of onset age of strabismus on the binocular responses of neurons in the monkey visual cortex. Invest Ophthalmol Vis Sci 2000; 41(3):948–954.

116. Mori T, Matsuura K, Zhang B, Smith EL 3rd, Chino YM. Effects of the duration of early strabismus on the binocular responses of neurons in the monkey visual cortex (V1). Invest Ophthalmol Vis Sci 2002; 43(4):1262–1269.

117. Zhang B, Bi H, Sakai E et al. Rapid plasticity of binocular connections in developing monkey visual cortex (V1). Proc Natl Acad Sci USA 2005; 102(25):9026–9031.

118. Wong AM, Foeller P, Bradley D, Burkhalter A, Tychsen L. Early versus delayed repair of infantile strabismus in macaque monkeys: I. ocular motor effects. J AAPOS 2003; 7(3):200–209.

119. Tychsen L, Wong AM, Foeller P, Bradley D. Early versus delayed repair of infantile strabismus in macaque monkeys: II. Effects on motion visually evoked responses. Invest Ophthalmol Vis Sci 2004; 45(3):821–827.

120. Richards M, Wong A, Foeller P, Bradley D, Tychsen L. Duration of binocular decorrelation predicts the severity of latent (fusion maldevelopment) nystagmus in strabismic macaque monkeys. Invest Ophthalmol Vis Sci 2008; 49(5):1872–1878.

121. Harwerth RS, Smith EL 3rd, Boltz RL, Crawford ML, von Noorden GK. Behavioral studies on the effect of abnormal early visual experience in monkeys: spatial modulation sensitivity. Vision Res 1983; 23(12):1501–1510.

122. Crawford ML, von Noorden GK. The effects of short-term experimental strabismus on the visual system in Macaca mulatta. Invest Ophthalmol Vis Sci 1979; 18(5):496–505.

123. Crawford ML, von Noorden GK, Meharg LS et al. Binocular neurons and binocular function in monkeys and children. Invest Ophthalmol Vis Sci 1983; 24(4):491–495.

124. Kiorpes L, Walton PJ, O'Keefe LP, Movshon, JA, Lisberger, SG. Effects of early-onset artificial strabismus on pursuit eye movements and on neuronal responses in area MT of macaque monkeys. J Neurosci 1996; 16(20):6537–6553.

125. Wright KW. Strabismus management. Curr Opin Ophthalmol 1994; 5(5):25–29.

126. Birch EE, Stager DR, Everett ME. Random dot stereoacuity following surgical correction of infantile esotropia. J Pediatr Ophthalmol Strabismus 1995; 32(4):231–235.

127. Birch EE, Fawcett S, Stager DR. Why does early surgical alignment improve stereoacuity outcomes in infantile esotropia? J AAPOS 2000; 4(1):10–14.

128. Fawcett S, Leffler J, Birch EE. Factors influencing stereoacuity in accommodative esotropia. J AAPOS 2000; 4(1):15–20.

129. Hasany A, Wong A, Foeller P, Bradley D, Tychsen L. Duration of binocular decorrelation in infancy predicts the severity of nasotemporal pursuit asymmetries in strabismic macaque monkeys. Neuroscience 2008; 156(2):403–411.

130. Hatta S, Kumagami T, Qian J, Thornton M, Smith EL 3rd, Chino Y. Nasotemporal directional bias of V1 neurons in young infant monkeys. Invest Ophthalmol Vis Sci 1998; 39(12):2259–2267.

131. Endo M, Kaas JH, Jain N, Smith EL 3rd, Chino YM. Binocular cross-orientation suppression in the primary visual cortex (V1) of infant rhesus monkeys. Invest Ophthalmol Vis Sci 2000; 41(12):4022–4031.

132. Zhang B, Zheng J, Watanabe I et al. Delayed maturation of receptive field center/surround mechanisms in V2. Proc Natl Acad Sci USA 2005; 102(16):5862–5867.

133. Bosworth RG, Birch EE. Direction-of-motion detection and motion VEP asymmetries in normal children and children with infantile esotropia. Invest Ophthalmol Vis Sci 2007; 48(12):5523–5531.

134. Brown RJ, Wilson JR, Norcia AM, Boothe RG. Development of directional motion symmetry in the monocular visually evoked potential of infant monkeys. Vision Res 1998; 38(9):1253–1263.

135. Zhang B, Stevenson SS, Cheng H et al. Effects of fixation instability on multifocal VEP (mfVEP) responses in amblyopes. J Vis 2008; 8(3):16 1–14.

136. Watanabe I, Bi H, Zhang B et al. Directional bias of neurons in V1 and V2 of strabismic monkeys: temporal-to-nasal asymmetry? Invest Ophthalmol Vis Sci 2005; 46(10):3899–3905.

137. Tychsen L. Causing and curing infantile esotropia in primates: the role of decorrelated binocular input (an American Ophthalmological Society thesis). Trans Am Ophthalmol Soc 2005; 105:564–593.

138. McKee SP, Levi DM, Movshon JA. The pattern of visual deficits in amblyopia. J Vis 2003; 3(5):380–405.

139. Kiorpes L. Visual processing in amblyopia: animal studies. Strabismus 2006; 14(1):3–10

140. Pelli DG, Levi DM, Chung ST. Using visual noise to characterize amblyopic letter identification. J Vis 2004; 4(10):904–920.

141. Wong EH, Levi DM, McGraw PV. Is second-order spatial loss in amblyopia explained by the loss of first-order spatial input? Vision Res 2001; 41(23):2951–2960.

142. Chandna A, Pennefather PM, Kovacs I, Norcia AM. Contour integration deficits in anisometropic amblyopia. Invest Ophthalmol Vis Sci 2001; 42(3):875–878.

143. Kozma P, Kiorpes L. Contour integration in amblyopic monkeys. Vis Neurosci 2003; 20(5):577–588.

144. Mussap AJ, Levi DM. Orientation-based texture segmentation in strabismic amblyopia. Vision Res 1999; 39(3):411–418.

145. Levi DM. Crowding – an essential bottleneck for object recognition: a mini-review. Vision Res 2008; 48(5):635–654.

146. Barnes GR, Hess RF, Dumoulin SO, Achtman RL, Pike GB. The cortical deficit in humans with strabismic amblyopia. J Physiol 2001; 533(Pt 1):281–297.

147. Muckli L, Kiess S, Tonhausen N, Singer W, Goebel R, Sireteanu R. Cerebral correlates of impaired grating perception in individual, psychophysically assessed human amblyopes. Vision Res 2006; 46(4):506–526.

148. Hess RF, Field DJ. Is the spatial deficit in strabismic amblyopia due to loss of cells or an uncalibrated disarray of cells? Vision Res 1994; 34(24):3397–3406.

149. Levi DM, Klein SA, Sharma V. Position jitter and undersampling in pattern perception. Vision Res 1999; 39(3):445–465.

150. Sharma V, Levi DM, Klein SA. Undercounting features and missing features: evidence for a high-level deficit in strabismic amblyopia. Nat Neurosci 2000; 3(5):496–501.

151. Kaas JH, Krubitzer LA, Chino YM, Langston AL, Polley EH, Blair N. Reorganization of retinotopic cortical maps in adult mammals after lesions of the retina. Science 1990; 248(4952):229–231.

152. Chino YM, Smith EL 3rd, Kaas JH, Sasaki Y, Cheng H. Receptive-field properties of deafferentated visual cortical neurons after topographic map reorganization in adult cats. J Neurosci 1995; 15(3 Pt 2):2417–2433.

153. Giannikopoulos DV, Eysel UT. Dynamics and specificity of cortical map reorganization after retinal lesions. Proc Natl Acad Sci USA 2006; 103(28):10805–10810.

154. Darian-Smith C, Gilbert CD. Topographic reorganization in the striate cortex of the adult cat and monkey is cortically mediated. J Neurosci 1995; 15(3 Pt 1):1631–1647.

155. Smirnakis SM, Brewer AA, Schmid MC et al. Lack of long-term cortical reorganization after macaque retinal lesions. Nature 2005; 435(7040):300–307.

156. Calford MB, Chino YM, Das A et al. Neuroscience: rewiring the adult brain. Nature 2005; 438(7065):E3; discussion E3–4.

157. Yang T, Maunsell JH. The effect of perceptual learning on neuronal responses in monkey visual area V4. J Neurosci 2004; 24(7):1617–1626.

158. Chowdhury SA, DeAngelis GC. Fine discrimination training alters the causal contribution of macaque area MT to depth perception. Neuron 2008; 60(2):367–377.

159. Levi DM, Li RW. Improving the performance of the amblyopic visual system. Phil Trans R Soc Lond B Biol Sci 2009; 364(1515):399–407.

160. Nakatsuka C, Zhang B, Watanabe I et al. Effects of perceptual learning on local stereopsis and neuronal responses of V1 and V2 in prism-reared monkeys. J Neurophysiol 2007; 97(4):2612–2626.

161. Helveston EM. Visual training: current status in ophthalmology. Am J Ophthalmol 2005; 140(5):903–910.

162. Horton JC. Vision restoration therapy: confounded by eye movements. Br J Ophthalmol 2005; 89(7):792–794.

CHAPTER 41

The Effects of Visual Deprivation After Infancy

Lindsay B. Lewis & Ione Fine

Introduction

Blind individuals tend to show better performance than sighted individuals on a variety of auditory and tactile tasks. This is thought to be due both to *compensatory hypertrophy* – enhanced neuronal processing within auditory and somatosensory cortices, and to *cross-modal plasticity* – whereby regions of occipital cortex that normally only process visual information begin to process auditory and tactile stimulation. While there is an extensive animal literature examining the long-term effects of early visual deprivation, we still do not have a full understanding of the neurophysiological changes that underlie cross-modal plasticity within the human brain. Here we discuss what is currently known about the behavioral and neurophysiological effects of long-term visual deprivation, and discuss the potential of new imaging techniques to reveal new insights.

Over the last two decades it has been well documented that those who suffer from visual deprivation, especially early in life, show an enhanced ability to use tactile and auditory cues that seems to be accompanied by remarkable changes in the response properties of occipital cortex.

While there is a growing literature demonstrating behavioral differences and differential neural responses to auditory and tactile stimuli in occipital cortex, the neurophysiological changes that occur as a result of blindness in humans are still fairly unclear. There is of course an extensive literature examining the anatomical and neural consequences of visual deprivation using both primate and non-primate animal models. However the literature on the effects of blindness on humans and in animal models has perhaps not been as mutually informative as might be hoped. One reason for this is that it is possible to examine the effects of visual deprivation with high temporal resolution within individual or small groups of cells within animal models, but the tools for assessing the effects of visual deprivation on humans are limited to techniques with much lower spatial or temporal resolution (Fig. 41.1). This limitation in the tools available to study the effects on long-term visual deprivation in humans is of particular concern because it is not clear how easily the animal literature can be generalized to humans. Humans rely more heavily on vision than almost all other animals and have a far larger number of visual sensory areas.[1-3] Fortunately, as illustrated in Box 41.1,

techniques such as functional magnetic resonance imaging, high-resolution structural imaging, transcranial magnetic stimulation, magnetic resonance spectroscopy, and functional connectivity analyses are now beginning to offer new ways of non-invasively studying the neurophysiological changes that occur as a result of blindness in humans. Table 41.1 summarizes research discussed in this article examining performance on and measuring brain responses to a wide range of auditory and tactile tasks.

The neuronal effects of visual deprivation

Cross-modal processing in visually normal development

During development, while much of sensory cortical architecture and connectivity is determined by genetically controlled factors, further refinement of connections, both within and between brain areas, appears to be dependent on early sensory input during a "critical" period.[4-8] These experience-dependent effects are believed to be mediated by adjustments in the weighting/gain of pre-existing connectivity,[9] the guidance of new projections to target areas,[4] and the elimination of exuberant projections.[10] A common underlying theme behind all these processes is that competition, based on experience, mediates connectivity; more active projections are strengthened at the expense of those that are less active.[11]

In infant animals, there is substantial physiological evidence for exuberant connectivity within the layers of visual cortex itself (primates,[12] kittens[13,14]); as well as between visual cortex and other cortical areas (primates,[15] cats[16]) (Fig 41.2). In kittens, there have been reports of projections from auditory/temporal areas to visual cortex,[10,17-19] from somatosensory and fronto-parietal cortex to visual cortex.[20] as well as from the subcortical visual thalamus (lateral geniculate nucleus, LGN) to auditory cortex.[4,21] These findings suggest that during early development the anatomical connections required for multisensory input to occipital cortex exist, even between primary sensory cortices.[22] A variety of animal models of normal visual development have

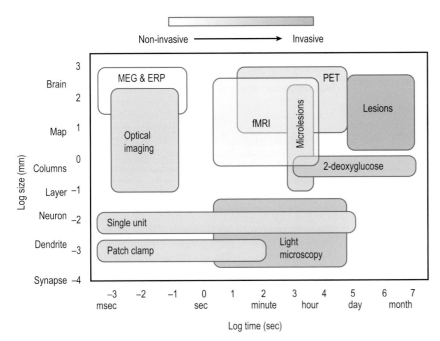

Figure 41.1 Schematic illustration of the ranges of spatial and temporal resolution of a subset of the various experimental techniques currently used for studying neural activity. The vertical axis represents the spatial extent of the technique, with the boundaries indicating the largest and smallest sizes of the region from which the technique can provide useful information. Thus, single-unit recording can only provide information from a small region of space, typically 10 to 50 μm on a side. The horizontal axis represents the minimum and maximum time intervals over which information can be collected with the technique. (Adapted from Churchland & Sejnowski.[3])

demonstrated "pruning" (or re-allocation) of a large proportion of these cross-modal connections during the course of maturation.[10,15]

A similar "pruning" is thought to occur in normal human infancy. As a result, it has been suggested that cross-modal connectivity may be more pronounced in infancy than in adulthood. A common technique for measuring brain responses in infants is by measuring event-related potentials (ERP) – the electrical activity of the brain through the skull and scalp. While auditory event-related potential (ERP) responses are usually small or absent in occipital (visual) regions of the human adult brain, large auditory ERP responses are found in occipital regions of 6-month-old infants.[23] These responses decrease as a function of age, implying a reduction in the strength of projections of auditory information to visual cortex.

As shown in Figure 41.2, and as will be described further below, many animal models suggest that, in the absence of visual input, the normal pruning process fails to occur. According to this model, cross-modal responses in early blind subjects might be mediated through the retention of an "infantile" pattern of anatomical connectivity. However, in contradiction to these results, recent examination of the volume of white and gray matter of early blind human subjects using voxel-based morphometry[24] suggests a reduction in connectivity between cortical areas.

Cross-modal processing in visually normal adults

Over the last three decades, improvements in neuroimaging techniques have made it increasingly easy to study neural processing within humans. Studies before the early 1990s tended to rely on positron emission tomography (PET), where a short-lived biologically active radioactive tracer isotope (such as FDG, an analog of glucose) is injected into the living subject and local neural activity is inferred based on resulting tissue concentrations within the area of interest. Since the early 1990s, functional magnetic resonance imaging (fMRI) has become the technique of choice due to its low invasiveness, lack of radiation exposure, and relatively wide availability (Fig. 41.3). fMRI measures the hemodynamic (blood oxygen level-dependent, BOLD) response, which is proportional to blood flow and the relative concentrations of deoxygenated to oxygenated blood in the brain area of interest.

However, it should be noted that the precise relationship between neuronal signals and the BOLD response is still not clear. The relationship between blood flow, metabolic rate, and neuronal responses remains a matter of active research, and it is even possible that these relationships are altered within the occipital cortex of blind subjects. BOLD responses should not therefore be considered synonymous with neuronal activity.

While primary visual cortex has not been shown to exhibit cross-modal responses to purely tactile or auditory stimuli in visually normal adults, BOLD responses in primary visual cortex can be modulated by information from other senses.[25] Furthermore, many extrastriate visual areas do show cross-modal responses in sighted individuals. Area MT+ shows BOLD responses to tactile motion stimuli,[26,27] and the lateral occipital complex (LOC)/inferior temporal (IT) gyrus shows BOLD responses during tactile object recognition tasks.[28–30] Extrastriate visual areas also show BOLD activation during tactile embossed letter[31] and tactile orientation discrimination[32] tasks.

Recently, transcranial magnetic stimulation (TMS) has been used to examine whether responses to tactile stimuli within occipital cortex play a functional role in sighted subjects. Weak electric currents are induced in brain tissue by rapidly changing magnetic fields (electromagnetic induction), thereby triggering neuronal activity over a relatively well-localized region of the cortical surface (Fig. 41.4). TMS applied over regions of the occipital cortex interferes with

---

Box 41.1 Non-invasive techniques for studying neurophysiological changes that occur as a result of blindness

Diffusion tensor imaging (DTI). A magnetic resonance (MR) method in which image intensities are proportional to the strength of diffusion along the direction of the magnetic diffusion gradient. Currently, DTI is principally used for the imaging of white matter, since axons in white matter tend to have strong diffusion gradients. More recently tractography (also known as fiber tracking) algorithms have been applied to DTI data in order to estimate the pathways of white matter in the brain.

Electroencephalography (EEG) is the measurement of summed activity of post-synaptic currents produced by the brain as recorded from electrodes placed on the scalp. A surface EEG reading is the summation of the synchronous activity of thousands of neurons that have similar spatial orientation, radial to the scalp. Currents tangential to the scalp are not picked up by EEG, and activity from deep sources is more difficult to detect than currents near the skull. EEG has very high temporal resolution (~1 ms), but poor spatial resolution. Indeed, it is theoretically impossible (the "inverse problem") to calculate the 3D spatial sources of EEG signals from the 2D electrode array. With assumptions, some estimates of spatial sources can be made but these estimates have spatial resolution on the order of a few centimeters. Recently EEG has been used in conjunction with MEG or fMRI in an attempt to achieve better spatial resolution.

Functional magnetic resonance imaging (fMRI). Hemoglobin is diamagnetic when oxygenated but paramagnetic when deoxygenated. The magnetic resonance signal of blood is therefore slightly different depending on the level of oxygenation. Hemodynamic or blood oxygen level dependent (BOLD) responses are proportional to blood flow and the relative concentrations of deoxygenated to oxygenated blood in the brain area of interest. It should be noted that the precise relationship between neural signals and BOLD remains a matter of active research. With sophisticated research paradigms fMRI can provide spatial resolution on the order of 1 millimeter and temporal resolution of less than a second, but generally has a spatial resolution of a few millimeters and temporal resolution of a few seconds. Since the early 1990s, fMRI has come to dominate the brain mapping field

due to its low invasiveness, lack of radiation exposure, and relatively wide availability.

Magnetoencephalography (MEG) is the measurement of summed magnetic fields produced by synchronous electrical activity in the brain as recorded from superconducting quantum interference devices (SQUIDS) placed on the scalp. It is only those magnetic fields generated by bundles of neurons that are perpendicular (as compared to radial in the case of EEG) to the cortical surface that can be measured. Like EEG, MEG has very high temporal resolution, but also suffers from the "inverse" problem. However magnetic fields are less distorted by factors like the skull, allowing somewhat better spatial resolution than EEG.

Magnetic resonance spectroscopy (MRS). An MR method in which levels of metabolites (such as myo-inositol, choline, creatine, N-acetylaspartate, choline, lactate, glutamate, glutamine, and GABA) associated with development and neuronal functioning are measured. Currently, because of the small concentrations of these metabolites, MRS is limited to a spatial resolution on the order of several millimeters and a temporal resolution on the order of minutes.

Positron emission tomography (PET). A short-lived biologically active radioactive tracer isotope is injected into the living subject and local neural activity is inferred based on resulting tissue concentrations within the area of interest. Provides spatial resolution on the order of millimeters, temporal resolution on the order of 10 seconds. Limitations to the use of PET include the high cost and the need to limit radiation exposure.

Transcranial magnetic stimulation (TMS). Weak electric currents are induced in brain tissue by electromagnetic induction due to rapidly changing magnetic fields. These electric currents can be used to trigger neuronal activity over a relatively well-localized region of the cortical surface. TMS is often used to demonstrate causality. If performance for a given task is suppressed by TMS stimulation to a particular brain area, this provides evidence that the region is used in performing that task.

---

the ability of sighted subjects to perform tactile orientation[32] and tactile spatial distance[33] tasks, suggesting functional involvement of these areas in tactile processing. However, it should be noted that other studies have failed to find visual cortical activation for tactile tasks in sighted subjects using fMRI[32,34-37] or show interference with tactile tasks using occipital TMS.[33,38] It therefore remains unclear how important a role visual cortex plays in the performance of tactile tasks in sighted subjects.

The anatomical connections that underlie these responses in normally sighted individuals are still being determined. Until fairly recently cross-modal modulation within primary sensory areas was assumed to be mediated by feedback from multimodal association areas outside visual cortex, such as intraparietal and superior temporal sulcus cortex, insular cortex, and perhaps even frontal and prefrontal cortex.[22,39,40] However, recent anatomical studies have found direct cross-modal projections between sensory cortices in adult primates. Direct projections have been found from auditory to visual cortex,[41-43] from visual to auditory cortex,[44,45] and from visual to somatosensory cortex.[46]

Cross-modal processing in early blind individuals

Measurements of cross-modal responses within early and late blind subjects have tended to either focus on "basic stimuli", such as simple tactile discrimination or auditory frequency tasks, or have focused on tasks which have strong functional significance, such as Braille reading, auditory localization, or auditory language. While cross-modal plasticity seems to be stronger for functionally relevant tasks, it should be noted that these tasks/stimuli also tend to be more complex, and are therefore more likely to evoke large brain responses.

The accumulation of evidence suggests that, in early blind subjects, primary as well as extrastriate visual areas are extensively recruited for processing of other senses. In contrast, for late blind subjects, there is only strong evidence for activation of extrastriate visual areas – and activation within these areas may not necessarily play a functional role.

Tactile performance

It is often suggested that blind subjects have enhanced tactile abilities. However it is still unclear to what extent differences

Table 41.1 Cross-modal plasticity in visual cortex: a summary of evidence for cross-modal plasticity discussed in this chapter

		Study type	Early blind	Late blind	Sighted
Tactile activation	Braille words (Sadato et al. 1996, 1998)	PET	P & E	–	–
	Braille words (Cohen et al. 1999)	PET	P & E	E	–
	Braille words (Burton et al. 2002)	fMRI	P & E	P & E	–
	Braille words (Gizewski et al. 2003)	fMRI	P & E	–	N
	Angle, width, character discrimination (Sadato et al. 1996, 1998)	PET	P & E	–	N
	Sweep (no discrimination) (Sadato et al. 1996, 1998)	PET	N	–	N
	Braille letters (Sadato et al. 2002)	fMRI	P & E	E	N
	Braille letters (naïve to Braille) (Sadato et al. 2004)	fMRI	–	E	N
	Embossed letters (Burton et al. 2006)	fMRI	P & E	P & E	E
	Tactile motion (Hagen et al. 2002; Blake et al. 2004)	fMRI/PET	–	–	E
	Tactile object recognition (Amedi et al. 2001, 2002; James et al. 2002; Pietrini et al. 2004)	fMRI	–	–	E
	Tactile orientation (Sathian et al. 2002)	PET	–	–	E
	Tactile spatial frequency (Sathian et al. 2002)	PET	–	–	N
Auditory activation	Frequency discrimination (Alho et al. 1993; Kujala et al. 1995)	ERP/MEG	Y	–	N
	Frequency discrimination (Kujala et al. 1997)	ERP	Y	Y	N
	Frequency discrimination (Kujala et al. 2005)	fMRI	P & E	–	N
	Sound localization (Kujala et al. 1992)	ERP	Y	–	N
	Sound localization (Leclerc et al. 2000; Gougoux et al. 2005)	PET/ERP	P & E	–	N
	Sound localization (De Volder et al. 1999)	PET	P & E	–	N
	Sound localization (Weeks et al. 2000)	PET	E – RH only	–	N
	Sound localization (Voss et al. 2006)	PET	–	E – RH dom	N
	Auditory motion (Poirier et al. 2006)	fMRI	MT+	–	Y
	Auditory motion (Saenz et al. 2008)	fMRI	MT+	–	N
	Auditory language (Burton et al. 2003)	fMRI-PER	P & E	–	N
	Auditory language (Amedi et al. 2003)	fMRI	P & E	P & E	N
	Auditory language (Burton et al. 2006)	fMRI	–	P & E	N
Tactile functional role	Braille words (Cohen et al. 1999)	TMS	Y	N	–
	Braille words (Hamilton et al. 2000)	Stroke	Y	–	–
	Braille words (Kupers et al. 2007)	TMS	Y	–	–
	Braille experience (Liu et al. 2007)	fMRI-PER	Y	–	–
	Braille letters (Cohen et al. 1997)	TMS	Y	–	N
	Embossed letters (Cohen et al. 1997)	TMS	Y	–	N
	Spatial distance (Merabet et al. 2004)	TMS/Stroke	Y	–	Y
	Roughness (Merabet et al. 2004)	TMS/Stroke	N	–	N
	Tactile sensations (Ptito et al., 2008)	TMS	Y	–	N
	Tactile orientation (Zangaladze et al. 1999; Sathian et al. 2002)	TMS	–	–	Y
Auditory functional role	Frequency discrimination (Stevens et al. 2007)	fMRI-PER	Y	–	N
	Auditory language (Amedi et al. 2003)	fMRI-PER	P	–	N
	Sound localization (Leclerc et al. 2000; Gougoux et al. 2005)	PET/ERP	P & E	–	–
	Sound localization (Collignon et al. 2007)	TMS	Y	–	N
	Auditory language (Amedi et al. 2004)	TMS	Y – RH only	–	N
	Pitch (Collignon et al. 2007)	TMS	N	–	N
	Intensity (Collignon et al. 2007)	TMS	N	–	N

P = cross-modal responses found in primary visual cortex; E = cross-modal responses found in extrastriate cortex; Y = cross-modal responses found in visual cortex; N = cross-modal responses not found in visual cortex; – = cross-modal responses not studied; TMS = functional significance established using TMS; fMRI-PER = functional significance established by correlating cross-modal responses as measured using fMRI with task performance, across subjects; RH = right hemisphere; dom = dominance.

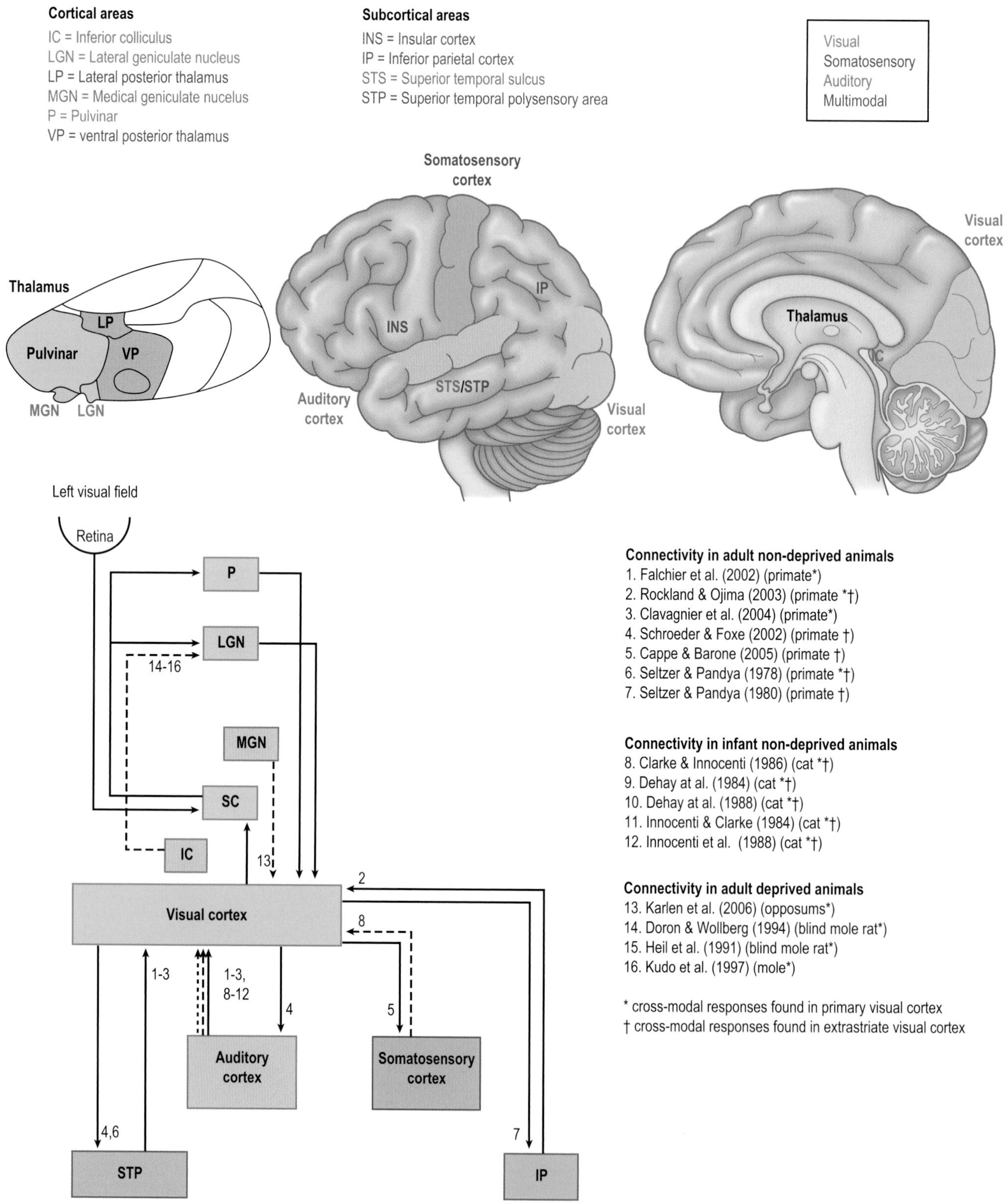
Cortical areas
IC = Inferior colliculus
LGN = Lateral geniculate nucleus
LP = Lateral posterior thalamus
MGN = Medical geniculate nucelus
P = Pulvinar
VP = ventral posterior thalamus
Subcortical areas
INS = Insular cortex
IP = Inferior parietal cortex
STS = Superior temporal sulcus
STP = Superior temporal polysensory area
Visual
Somatosensory
Auditory
Multimodal
Somatosensory cortex
Thalamus
LP
Pulvinar
VP
MGN
LGN
INS
IP
STS/STP
Auditory cortex
Visual cortex
Thalamus
IC
Visual cortex
Left visual field
Retina
P
LGN
14-16
MGN
SC
IC
13
2
Visual cortex
8
1-3
1-3, 8-12
4
5
Auditory cortex
Somatosensory cortex
4,6
7
STP
IP
Connectivity in adult non-deprived animals
1. Falchier et al. (2002) (primate*)
2. Rockland & Ojima (2003) (primate *†)
3. Clavagnier et al. (2004) (primate*)
4. Schroeder & Foxe (2002) (primate †)
5. Cappe & Barone (2005) (primate †)
6. Seltzer & Pandya (1978) (primate *†)
7. Seltzer & Pandya (1980) (primate †)
Connectivity in infant non-deprived animals
8. Clarke & Innocenti (1986) (cat *†)
9. Dehay at al. (1984) (cat *†)
10. Dehay at al. (1988) (cat *†)
11. Innocenti & Clarke (1984) (cat *†)
12. Innocenti et al. (1988) (cat *†)
Connectivity in adult deprived animals
13. Karlen et al. (2006) (opposums*)
14. Doron & Wollberg (1994) (blind mole rat*)
15. Heil et al. (1991) (blind mole rat*)
16. Kudo et al. (1997) (mole*)
* cross-modal responses found in primary visual cortex
† cross-modal responses found in extrastriate visual cortex

Figure 41.2 A map of major visual and cross-modal connections to visual cortex, including connections known to exist in infant (dashed lines) and visually deprived (dotted lines) animals. Contralateral projections from the left visual field are shown. Green represents visual, blue represents somatosensory, red represents auditory and purple represents multimodal areas. References are provided for cross-modal connections to visual cortex.

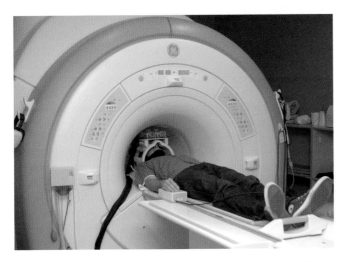

Figure 41.3 A subject about to undergo a magnetic resonance imaging procedure. The table the subject is lying on is moved into the bore of the scanner. (Courtesy of the Van de Veer Institute.)

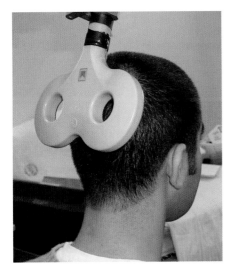

Figure 41.4 A subject undergoing transcranial magnetic stimulation (TMS) using a butterfly coil. (Courtesy of the Kastner laboratory.)

in tactile performance between sighted and blind subjects should be attributed to blindness per se as compared to practice effects.

For obvious reasons, blind subjects tend to have more extensive experience with Braille than even well-trained sighted subjects (indeed, most sighted teachers of Braille read using vision rather than touch). As a result, non-Braille tactile tasks are typically considered a better measure of the direct effects of blindness (as compared to experience) on tactile abilities. Interestingly, it is not entirely clear whether blind subjects do exhibit enhanced acuity for non-Braille tactile discrimination tasks. In one study, no significant differences in sensory, touch, or two-point discrimination thresholds were found between early blind and sighted subjects.[47] However in several more recent studies, tactile acuity has been found to be significantly higher in early blind subjects than sighted subjects for embossed Roman letters[38] and grating orientation discrimination[48,49] tasks. Taken together

these data suggest that visual deprivation in early blind subjects may only provide a weak or selective (for some types of task, but not all) advantage for tactile processing.[50]

Most of the benefits in tactile discrimination abilities in blind as compared to sighted subjects are likely to be due to an interaction between blindness and the effects of increased experience and reliance on visual cues. In several studies, although blind subjects initially exhibit superior performance to sighted subjects on a range of tactile tasks, when practice is accounted for between blind and sighted subjects, these differences disappear.[47,51] Experiments such as these support the notion that the superiority in tactile tasks found in the blind is driven at least in part by greater tactile experience.

However, in another recent study it has been shown that sighted subjects perform better at Braille letter discrimination when they are blindfolded during a preceding five-day period than when they are allowed normal vision during that five-day period, irrespective of the amount of training in Braille. Indeed, blindfolded subjects who have not been trained at all in the task perform better than non-blindfolded subjects trained in the task over a five-day period.[52,53] These findings suggest that visual deprivation does play a role in the enhanced tactile abilities of blind subjects.[50] Furthermore, the fact that sighted Braille teachers read Braille visually (from the shadows of the letters) rather than by touch, suggests that loss of visual input is important for tactile fluency (Pascual-Leone, personal communication).

Given that both visual deprivation and tactile experience play significant roles in subserving the tactile performance benefits observed in blind subjects, an understanding of how deprivation and experience interact may be important in developing effective Braille instruction programs, especially in the partially sighted.

Braille tactile processing

Numerous studies have examined cross-modal responses to Braille reading in the visual cortex of blind subjects. In an early ERP study, early blind subjects were reported to show a more posterior/occipital distribution of response during Braille reading than sighted subjects who read embossed Roman letters.[54] Since then, several neuroimaging studies have demonstrated that early blind subjects consistently show greater responses to Braille words than to non-words (or other appropriate control stimuli) in primary as well as extrastriate visual cortex. In contrast, late blind subjects show greater responses than sighted subjects in extrastriate but not primary visual cortex[34–36,55] (see reviews,[50,56] though see[57,58]). In early blind subjects the magnitude of BOLD responses to Braille in primary visual cortex is highly correlated on an individual basis with verbal memory ability, suggesting a relationship between the extent of cross-modal responses and Braille abilities,[59] at least in those who are visually deprived at an early age.

Several TMS studies provide clear evidence for the functional role of visual cortex in Braille reading in early blind subjects. TMS delivered to visual cortex while discriminating Braille letters (note that this is a somewhat simpler task than reading Braille text) induces significantly more errors than sham stimulation in early but not late blind subjects (Fig. 41.5).[55] This disruption of Braille performance suggests that the visual cortex does play a functional role in Braille reading

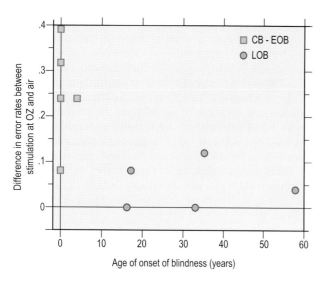

Figure 41.5 Errors in Braille letter discrimination for individual subjects during TMS stimulation over the mid-occipital region (OZ) and control (Air) as a function of age of onset of blindness. CB-EOB = congenitally blind and early onset blind, LOB = late onset blind. Subjects identified 25 Braille letters presented in five strings of five letters (all non-words) for each scalp position stimulated. Errors were identified as wrong identification or an inability to identify letters. (From Cohen et al.[55])

at least in early blind subjects[38,55] (see reviews[50,56,60]). Additionally, there is the interesting case study of an early blind woman who, following an occipital stroke, lost the ability to read Braille without loss of her ability to detect Braille letters or loss of her other somatosensory abilities.[61]

Similarly, while TMS delivered to the visual cortex of early blind subjects does not seem to interfere with the detection of Braille tactile stimulation, it does interfere with the reading of Braille.[50,62,63] Collectively these studies provide strong evidence that the responses found to Braille reading in visual cortex are not epiphenomena, and in fact are functionally necessary for the comprehension of Braille. Furthermore, it seems that these responses in visual cortex belong to a later part of an extended network of areas subserving Braille processing. TMS applied to occipital cortex interferes with Braille processing only after a delay of 50–80 ms, as compared to lesser time periods of 20–40 ms when applied to somatosensory cortex[62] (see reviews[58,63,64]).

Overall, in early blind subjects the preponderance of evidence suggests that extrastriate and primary areas of visual cortex are recruited for Braille reading, and that cross-modal responses in visual cortex are functionally relevant. In late blind subjects, it is probable that extrastriate (though not primary) visual cortex is activated during Braille processing; however, it remains likely that activation in visual cortex during Braille reading in late blind subjects does not have a functional role.

Non-Braille tactile processing

Cross-modal responses in visual cortex of early as well as late blind subjects have also been shown for a variety of non-Braille tactile discrimination tasks. In an early ERP study, early blind subjects showed a more posterior/occipital distribution of response than sighted subjects during even a simple passive tactile task where subjects passed their finger over a random-dot pattern.[54] Furthermore, when performing

same/different judgments of grooves or embossed letters, early blind subjects showed an increase in visual cortex activation (as measured using PET), whereas sighted subjects showed a decrease in visual cortex activation. In this task, neither blind nor sighted subjects showed changes in visual cortex activation to these tasks when no response was required, suggesting that attention to the tactile stimulation may be necessary.

As is the case for Braille reading, it seems fairly certain that both primary and extrastriate visual areas respond to non-Braille tactile discrimination in early blind subjects, and extrastriate but not primary visual areas respond to non-Braille tactile discrimination in late blind subjects. Only early blind subjects show responses in primary visual cortex to a same/different Braille letter tactile discrimination task (letters were discriminated rather than read), whereas both early and late blind subjects show responses in extrastriate visual cortex. The recruitment of primary visual cortex for non-reading tactile discrimination may be limited to early blind subjects.[65] Interestingly, recently late blind subjects who are naïve to Braille also show BOLD responses to tactile discrimination in extrastriate visual cortex, suggesting that cross-modal responses to tactile input in extrastriate cortex may not be dependent on experience with Braille.[37,56]

The tactile recruitment of visual cortex in early blind subjects seems to have a functional role. TMS delivered to the visual cortex of early blind subjects while performing an embossed letter discrimination task induces significantly more errors in early blind than in sighted subjects, suggesting a functional role of visual cortex even in non-Braille reading tactile tasks.[38] However, visual cortex may only play a functional role for certain types of tactile processing; in early blind subjects, TMS delivered to occipital areas interferes with performance during spatial distance, but not roughness judgments for raised dot stimuli.[33]

There is also evidence for cross-modal responses in visual cortex to complex tactile processing. During tactile imagery, early blind subjects (none of whom ever had vision) exhibited ERP responses in extrastriate visual cortex.[66] ERP responses over occipital cortex were also shown during encoding of tactile stimuli, and mental rotation of those stimuli.[67] Moreover, early blind subjects show category-related patterns of BOLD response for tactile manmade objects in inferotemporal (IT) and fusiform areas of visual cortex.[30]

Responses in visual cortex to tactile stimulation have been evoked in sighted as well as early and late blind subjects for certain stimuli. However one difficulty in studying the effects of tactile and auditory stimulation on occipital cortex is the possibility that occipital responses within sighted and late blind subjects may be due to visual imagery. Sighted, early and late blind subjects all showed BOLD responses to embossed Roman letter identification (a task likely to have elicited visual imagery in both sighted and late blind subjects) in visual cortex.[31] In late, but not early blind subjects, tactile motion and tactile face perception evoke BOLD responses in area MT+ and fusiform face area (FFA), respectively – a finding, again, thought likely to be due to visual imagery.[68]

Auditory processing

Most studies have found little evidence for overall improved auditory sensitivity in blind subjects.[69] However, there are

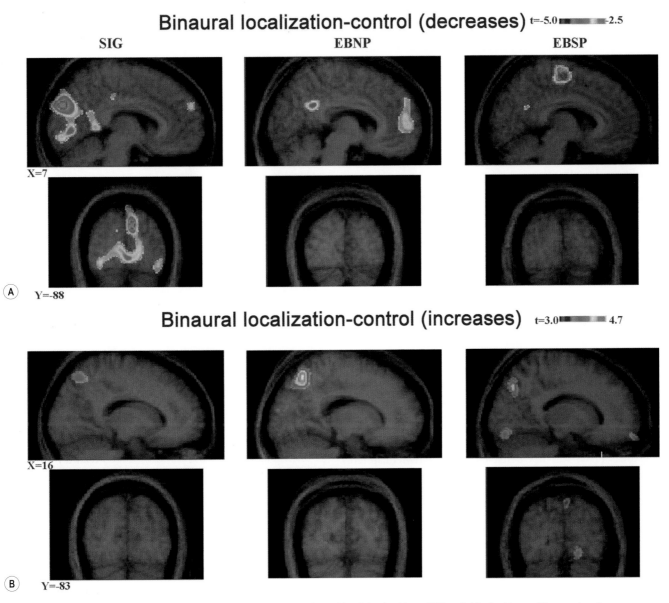

Figure 41.6 PET responses during a binaural sound localization task performed by sighted subjects (SIG), early blind subjects with normal performance on a monaural sound localization task (EPNP) and early blind subjects with superior performance (EBSP). Subjects were asked to localize a pair of sounds while they were lying within the scanner by pointing with a joystick, placed at the subject's side, toward the apparent source of stimulation. The control task consisted of pointing in alternation to the left and right after hearing a stimulus pair presented in the frontal position. (**A**) Decreased cerebral blood flow (CBF) is observed in the visual striate and extrastriate cortices. (**B**) Increased CBF is seen in the right ventral extrastriate cortex for the EBSP group, but not for the other two groups. X and Y coordinates refer to standardized stereotaxic space. The significance criterion was calculated as a t-value of 3–3.5. (From Gougoux et al.[74])

some selective benefits of blindness on auditory perception, as described below, and visual cortex seems to be recruited for at least certain aspects of auditory processing.

In early blind subjects there is evidence for generally superior localization abilities,[70,71] particularly in peripheral space.[72,73] Auditory BOLD responses in visual cortex may also be correlated with sound localization accuracy[74,75] (Fig. 41.6).

In an early study, ERPs evoked by a dichotic listening task (where subjects were asked to listen to different auditory tones in each ear) were reported to have shown a more posterior scalp distribution in early blind than in sighted subjects. This was true for auditory input from both the attended as well as the unattended ear.[76] This result suggests that visual areas may subserve auditory processing regardless of attention. However, more recently, auditory responses in visual cortex for auditory frequency discrimination have been shown in both early and late blind subjects, but only when attention is directed to the auditory stimulus (MEG,[77] ERP[78] and fMRI[79]) (see review[60]). In another attentional auditory ERP study, early blind subjects have been reported to exhibit larger and more posteriorly distributed N2 responses than sighted subjects, as well as faster auditory target detection times.[80]

Recently, it has been shown in sight recovery subjects who were blinded at an early age (as well as in early blind subjects) that visual extrastriate area MT+ responds selectively to auditory motion but not other stimuli,[81] suggesting that cross-modal plasticity can be influenced by the normal functional specialization of a cortical region. Evidence for functionally relevant auditory processing in visual cortex of early blind subjects comes from studies of auditory localization and auditory language, as described below.

Visual cortex also seems to be recruited in blind subjects trained to use auditory sensory substitution devices (Box 41.2, Fig. 41.7).

Auditory localization

Auditory ERP responses in primary and extrastriate visual cortex have been shown in early blind subjects while performing sound localization tasks,[83] and these responses seem to be larger in those early blind subjects who behaviorally demonstrate superior sound localization accuracy. The extent of activation in early blind subjects is correlated on an individual basis with sound-source localization accuracy (PET,[74] see Fig. 41.6; ERP[75]). This suggests that cross-modal responses to sound localization in visual cortex may have functional importance.

These cross-modal responses within visual cortex may also be lateralized; in another PET study, auditory responses in extrastriate visual cortex in early blind subjects to sound localization are reported in the right, but not the left, visual cortex.[84] Responses to sound-source localization have additionally been shown with PET in extrastriate visual cortex of late blind subjects, again with a seemingly right hemisphere dominance.[85] Moreover, TMS delivered to right occipital areas (TMS was not delivered to the left hemisphere) interferes with an auditory localization task, but does not significantly interfere with an auditory pitch or intensity discrimination task.[86] This finding suggests that in early blind subjects right visual cortex plays a functional role in auditory localization, and is perhaps specifically involved

Box 41.2 Sensory substitution: replacing vision with audition

The enhanced abilities of blind individuals to make use of auditory and tactile cues have inspired the development of a variety of sensory substitution devices. While few are currently in common use, it seems likely that as miniaturization and processing capacities improve, such devices will quickly become more commonplace. Interestingly, a few studies suggest that use of these sensory substitution devices may elicit occipital responses within blind subjects, and that these responses may be highly dependent on training.

In one early study, early blind and blindfolded sighted subjects were trained with an ultrasonic echolocation device which transduces ultrasonic echoes into auditory signals.[1] During a sound localization task without the device, larger PET activations were found in extrastriate (and probably primary) visual cortex of early blind subjects than were found within blindfolded sighted subjects. Moreover, while training with the device had no effect on visual cortex activation for blindfolded sighted subjects, in the case of early blind subjects training with the device resulted in an additional increase in extrastriate (and probably primary) visual cortex activation during sound localization.

In a second study, early blind and blindfolded sighted subjects were trained to use a visual-to-auditory sensory substitution device (prosthesis substituting vision with audition, PSVA) for spatial pattern recognition. After training with the device larger PET activations were again found in the extrastriate visual cortex of early blind than blindfolded sighted subjects (see Fig. 41.7). These findings suggest that differences between early blind and sighted subjects might be mediated by a pre-existing ability in early blind subjects to recruit visual cortex for non-visual spatial processing that extended relatively easily to the sensory substitution device.

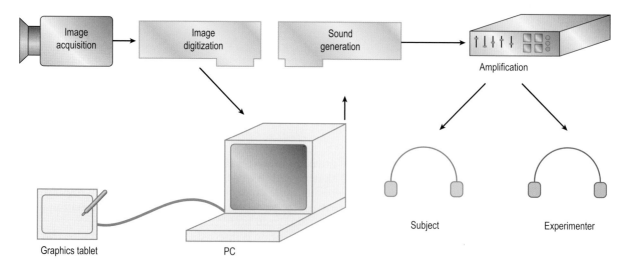

Figure 41.7 Schematic representation of the sensory substitution equipment used by Arno et al.[82] Early blind and blindfolded sighted control subjects were trained to use a PSVA (prosthesis substituting vision with audition) device over the course of 12 one-hour training sessions, distributed over 8–10 weeks. Visual patterns were generated using a graphics tablet connected to a PC. The audio translation of these visual patterns (each pixel was associated with a specific frequency) was provided to the subject (and experimenter) via headphones. During each training session, subjects learned to move the pen on the graphics tablet to scan each pattern, and during each test session (before the first training session, and after the 6th and 12th training sessions), subjects were asked to reproduce each pattern via arranging metallic bars and squares in a frame. (Adapted from Arno et al.[82])

in the analysis of spatial components necessary for such tasks.

This right hemisphere bias for sound location in early and late blind subjects is consistent with laterality effects shown in auditory cortex of sighted subjects. In sighted subjects sound localization tasks result in right hemisphere engagement, based on measurements using magnetoencephalography (MEG; a method of measuring the magnetic fields produced by electrical activity in the brain).[87] Analogous laterality effects have been found in sighted subjects. For example, lesions of the right hemisphere can result in deficits in visuospatial processing[88] (see also[86]).

Auditory language

Several studies have shown that both primary and extrastriate visual cortex in early blind subjects are recruited for auditory language, with a greater extent of BOLD activation for semantic tasks compared to phonological processing tasks or passive indecipherable sounds[89–91] (see review[89]). Similarly, BOLD responses in primary and extrastriate visual cortex have been shown for auditory language processing in late blind subjects who are naïve to Braille.[31,89] Cross-modal responses to auditory verbal content in primary and extrastriate visual cortex seem not to be heavily dependent on the onset of blindness occurring early in life.

TMS delivered to visual cortex (particularly the left hemisphere) of early blind subjects during verb generation in response to auditory-presented nouns results in verb-generation errors, that seem to be primarily semantic,[92]

consistent with left hemisphere lateralization for language processing (Fig. 41.8). It should be noted that the late timing of the TMS pulses (660 ms after word presentation onset), as well as subject reports indicating clear auditory perception of the stimulus, make it highly unlikely that the TMS affected the auditory processing of the stimulus per se, as compared to the processing of verbal content. It seems that visual cortex is involved in high-level verbal processing in early blind subjects.

Cross-modal connectivity within occipital cortex of early blind individuals

Most evidence for changes in connectivity in humans as a result of blindness has surfaced relatively recently, using a variety of techniques including fMRI (functional magnetic resonance imaging), TMS (transcranial magnetic stimulation), high-resolution MRI, and diffusion tensor imaging (DTI). Despite the growing interest in the connectivity underlying cross-modal plasticity, the routes by which auditory and tactile information reaches striate and extrastriate visual cortex in early blind individuals are still unclear. One possibility is that early visual deprivation results in a failure to prune early exuberant corticocortical and/or thalamocortical cross-modal connections. Alternatively, early deprivation may permit top–down connections from multimodal areas to extend into primary and secondary visual areas.

Functional evidence about the source of cross-modal plasticity is somewhat contradictory. Some functional evidence

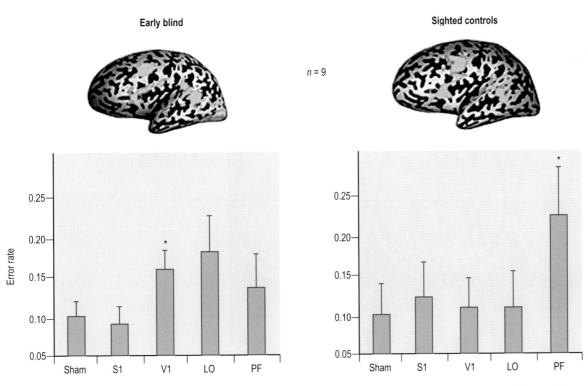

Figure 41.8 Verb-generation error rates in blind and sighted groups as a function of repetitive TMS (rTMS) site. In blind subjects, rTMS over left V1 increased error rates relative to sham and right S1 stimulation; a similar trend was observed with left prefrontal (PF) and lateral-occipital (LO) stimulation. In sighted control subjects only left PF stimulation increased error rates relative to sham and S1 stimulation. Insets on each graph show the group-average pattern of fMRI activation in the left hemisphere during verb generation versus rest.[59] Note that V1 only shows significant activation in the blind group, consistent with the larger error rates produced by rTMS over this region. Statistical parametric maps were thresholded at $p < 0.05$ and corrected for multiple comparisons with use of a random-effect general linear model. (From Amedi et al.[92])

suggests that the primary source of tactile cross-modal plasticity in visual cortex of blind subjects is corticocortical, not thalamocortical. Stimulation of primary somatosensory cortex in early blind subjects by TMS yields activation in primary visual cortex.[93] Tactile stimulation of one hand of blind subjects leads to bilateral activation of visual cortex.[34,35,65] Because only the contralateral side of the body is represented at the thalamic level, the bilateral nature of these visual responses implies that tactile information must be delivered to visual cortex by cortical mechanisms beyond the thalamus. However, measurement of resting state functional connectivity (correlations between BOLD activity across the brain while a subject is at rest) suggests that early blind subjects show decreased functional connectivities compared to sighted subjects within the occipital visual cortices as well as between the occipital visual cortices and the parietal somatosensory, frontal motor and temporal multisensory cortices. Interestingly the greater the blind subjects' use of Braille, the stronger the functional connectivities between these brain areas, suggesting that increased functional connectivities are not dependent on visual deprivation per se, but rather caused by increased reliance on auditory and tactile modalities enforced by loss of vision.[94]

Anatomical data seem to consistently show a decrease in corticocortical connectivity within early blind human subjects, though there are still a relatively small number of studies in the literature. A recent study using voxel-based morphometry based on high-resolution MR imaging[24] found that early blind subjects show significant atrophy of the geniculostriate system (Fig. 41.9). Loss of gray matter occurred within the dorsal lateral geniculate nucleus, primary and extrastriate cortex. White matter reductions were observed in the optic nerves, optic radiations, and optic tracts as well as within occipitotemporal projections and the posterior part of the corpus callosum (which links the visual areas of both hemispheres), suggesting large-scale atrophy of afferent projections to the visual cortex in early blind individuals. There was no evidence of any regions of increased white matter volume as might be expected if early blind subjects retained an "infantile" pattern of connectivity.

DTI is a magnetic resonance imaging technique which measures the restricted diffusion of water in tissue, thereby allowing the imaging of white matter tracts. In recent years the inclusion of fiber tracking algorithms in DTI analysis has been used to estimate tractography, the pathways formed

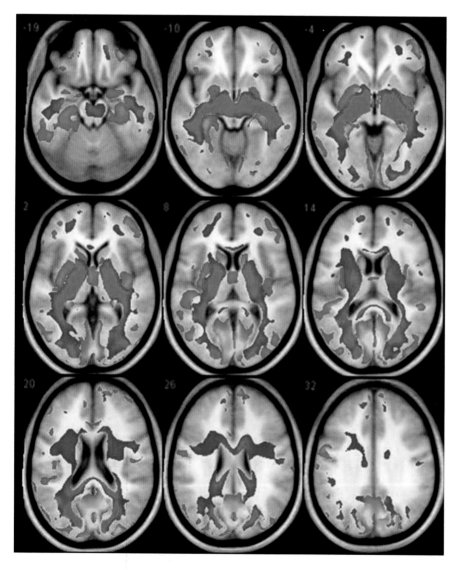

Figure 41.9 Regions showing a reduction in gray (*red*) and white (*blue*) matter in congenitally blind (CB) as compared to normally sighted (NS) individuals. No areas were found that showed a reduction in gray or white matter in NS as compared to CB (areas showing NS < CB). Volume differences are presented on horizontal sections, with the left side of the figure referring to the left side of the brain. Statistical parametric maps were thresholded at $p < 0.05$ corrected for multiple comparisons with the false discovery rate (FDR) algorithm implemented in SPM2. (From Ptito et al.[95] Reproduced by kind permission of Springer Science + Business Media.)

by white matter connections.[96] Currently there are only a few studies examining changes in tractography as a result of blindness. To date no differences in tractography have been found between late blind and sighted subjects, suggesting that optic pathways in late blind subjects remain intact.[97] However DTI, like high-resolution MRI, has found some evidence of optic pathway atrophy in early blind subjects.[98,99]

In contrast, a large number of animal models suggest that visual deprivation leads to increased cross-modal connectivity between subcortical thalamic nuclei and visual cortex, as well as from other sensory cortical areas to visual cortex. In adult opossums early bilateral eye enucleation results in abnormal thalamocortical and corticocortical projections. Abnormal connections projecting to visual cortex have been noted from both auditory (medial geniculate nucleus, MGN) and somatosensory (ventral posterior nucleus, VP) thalamus.[7,100] These same authors also noted abnormal corticocortical projections from auditory and somatosensory cortex to visual cortex.[7,100] Similarly, in early bilaterally deprived cats, portions of the anterior ectosylvian area that are typically driven purely by visual input have been found to be almost completely taken over by auditory and somatosensory inputs,[101,102] and in early bilaterally deprived monkeys, some primary and extrastriate visual cortical areas are found to be recruited by tactile input.[103,104]

So, while functional and animal data support the notion that responses with the occipital cortex are driven by increased corticocortical connectivity, there is currently no evidence for any such increases in connectivity within early blind humans. One possibility is that early blind have similar (or weaker) thalamo-cortical or corticocortical anatomical connections than sighted individuals, but that these connections play a much stronger role in modulating functional responses in blind subjects.

Neuronal structure within occipital cortex of early blind individuals

In early blind subjects, primary and extrastriate visual cortical areas are generally shown to exhibit greater than normal glucose metabolism and regional cerebral blood flow.[105,106] These findings support the notion that the visual cortex of early blind individuals remains "active" as a result of cross-modal plasticity.

However, in animal deprivation models the effects of deprivation seem to include a variety of neuronal changes including a retardation in the development of glial cells,[107] disruption of the inhibitory GABA pathway,[108,109] disruptions in synaptogenesis and pruning,[110–112] and losses of functional tuning.[113–116] It should also be noted that many of these changes observed in animal models would not necessarily have an observable effect using techniques that have currently been used to examine neural deterioration in humans (MR imaging of macroanatomy, and PET measurements of glucose metabolism or blood flow). One interesting potential tool is magnetic resonance spectroscopy (MRS), a noninvasive technique capable of measuring changes in the levels of metabolites (such as myo-inositol, choline, creatine, N-acetylaspartate, choline, lactate, glutamate, glutamine, and GABA) associated with development and neuronal functioning.

Differences in cross-modal processing between the periphery and the fovea

Cross-modal connectivity seems to differ between peripheral and foveal representations of space. In primate models it has been shown that adult (non-deprived) anatomical connections between auditory and multimodal temporal areas and primary visual cortex map preferentially to peripheral rather than foveal representations.[41,42] Consistent with this, many cross-modal interactions in visually normal individuals are stronger for peripherally presented stimuli.[117,118]

Analogously, cross-modal responses seem to be stronger within regions of visual cortex that in sighted subjects subserve the peripheral visual field. Braille and auditory word responses within primary visual cortex are found within peripheral rather than foveal areas of primary visual cortex.[58] Moreover, in blind subjects cross-modal responses to sound source localization tend to be stronger for sounds originating from peripheral space.[72]

One possible explanation for stronger cross-modal responses in the periphery is that cortical areas that normally represent peripheral representations have a greater potential for plasticity. For many aspects of visual processing, processing in the periphery is qualitatively different from processing in the fovea, even after scaling for acuity. For example, in sighted subjects visual perceptual learning effects tend to be larger in the periphery than in the fovea.[119]

Blindfolding studies

It has recently been shown that cross-modal responses to Braille and auditory stimuli can be elicited within the visual cortex of sighted individuals after relatively short periods of blindfolding. In a group of these experiments[53,120] (see also reviews[50,63]), sighted subjects were visually deprived via blindfolding for a period of five days, and BOLD responses in visual cortex were measured while they performed auditory and tactile tasks. In the auditory task, subjects listened to a series of tones and performed a same/different task on each tone compared to the previous tone. In the tactile task, subjects performed a same/different task on pairs of Braille symbols. On the first day of blindfolding, there was no response in visual cortex to either the auditory or tactile tasks. However, by the fifth day of blindfolding activation was seen in an anatomical region likely to include primary visual cortex during both the auditory and tactile tasks (Fig. 41.10).

In parallel to the findings of the effect of TMS on Braille letter discrimination abilities in early blind subjects,[38,55] TMS applied to occipital cortex in blindfolded sighted subjects (but not sighted control subjects) on the fifth day also significantly interfered with performance on the tactile Braille discrimination task,[53] suggesting that BOLD responses in visual cortex during this task are functionally relevant. Furthermore, on the sixth day, just 24 hours after removal of the blindfold (the blindfold was used during scanning), the activation of visual cortex in response to tactile or auditory stimulation was no longer apparent, and TMS applied to occipital cortex no longer interfered with the tactile Braille task. These latter findings indicate that exposure to vision for a single day is sufficient to eliminate the visual cortical activation induced during the blindfolding period.

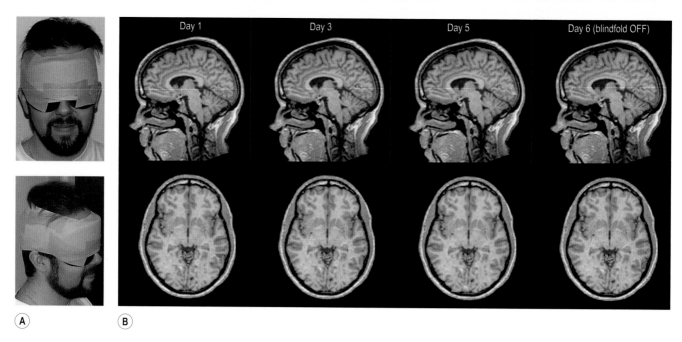

Figure 41.10 Normally sighted subjects were randomized to a blindfolded or non-blindfolded group. Both groups were immersed in an intensive tactile stimulation program that included formal Braille instruction over five days. Group difference tactile activation maps are shown (overlaid on the anatomy of a single representative subject) contrasting the blindfolded and non-blindfolded groups for each fMRI session of the study. Activation maps were corrected for multiple comparisons by a joint voxelwise significance/cluster size threshold of ($p < 0.05$; estimated through Monte Carlo simulation using Brain Voyager) in a procedure designed to allow detection of weak but spatially extensive activations. (From Merabet et al.[53])

Additional evidence for effects of short-term visual deprivation on cross-modal responses in visual cortex has been provided by a second research group. In this study, sighted subjects were blindfolded for two hours, and then cortical activation was assessed with fMRI as they performed several tactile discrimination tasks. A comparison of blindfolded subjects vs control subjects revealed a task-specific increase in activation in lateral occipital complex (LOC) for a tactile global form task (but not a gap detection task), as well as an unexpected decrease in activation in intermediate visual areas for both tactile tasks.[121]

These data imply that activation of visual cortex during processing of other senses is not necessarily an exclusive feature of blindness; some mechanisms that can mediate cross-modal plasticity are present even in the normal brain.[50] The rapid nature of these changes suggests that some connections between different sensory areas may be pre-existing, and simply latent until "unmasked" as a result of specific experiences such as temporary deprivation.[64,120] In other words, when a sensory cortical area is deprived of its normal input, existent cross-modal connections allow this area to respond to input from other modalities.

However, it should be noted that the mechanisms uncovered by blindfolding paradigms do not necessarily represent the same mechanisms that mediate cross-modal plasticity in blind subjects. In blindfolded sighted subjects responses in visual cortex to input from other modalities could be a consequence of "unmasking" or release from inhibition (a fast, and easily reversed, effect), whereas in early blind subjects the same finding could be a consequence of the establishment of a different type of connectivity (not necessarily one involving an increase in white matter) – a slow effect that would be practically impossible to study in blindfolded sighted subjects.[64] This possibility seems particularly likely

given that the evidence for functional significance of cross-modal activations in late blind subjects is not very strong – indeed it is curious that TMS studies using Braille stimuli show functional significance of cross-modal activation in blindfolded sighted subjects, but not in late blind subjects.

One possibility is that during early stages of vision loss there may be a reinforcement of pre-existing connections (similar to what has been demonstrated in the short term in blindfolded sighted subjects) which may in the long-term lead to permanent structural modifications in pathways between visual cortex and other sensory areas.[63] Long-term connectivity patterns may fundamentally differ between early and late blind subjects – perhaps accounting for the generally weaker and less extensive cross-modal findings in late blind subjects, especially within primary occipital cortex.

Restoration of vision

The existence of cross-modal plasticity in visual cortex raises interesting questions regarding the effects of cross-modal plasticity on restoration of visual function. One possibility is that cross-modal responses to other forms of sensory input within visual cortex might be beneficial in maintaining neuronal functioning and preventing atrophy. On the other hand if allocation of visual cortex to other senses is irreversible, then cross-modal processing could reduce the ability of visual cortical neurons to process restored visual input.

TMS has been used in blind subjects to investigate the relationship between induced phosphenes and neuronal function in visual cortex.[122] The likelihood of phosphene elicitation is correlated with the amount of residual vision; subjects with severe blindness (no light perception, NLP) only rarely perceive phosphenes.

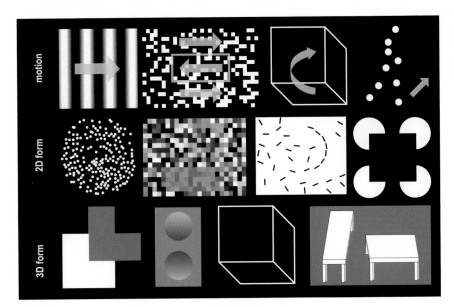

Figure 41.11 Stimuli used for tests of MM's form, depth, and motion processing. Stimuli shown to visually normal controls were always blurred using a low pass filter that was designed to match MM's acuity losses. (Adapted from Fine et al.[123])

These data are supported by the small number of studies examining restored vision in those who were blind since childhood. All these studies suggest severe losses in acuity and difficulties interpreting the visual world. The extent of visual processing deficits seems to depend in a complex way on the age and severity of blindness, and the age at which sight is restored, as illustrated by comparing two recent case studies.

For example, sight recovery subject MM was born with normal vision, but lost vision (leaving LP vision) at age 3 due to a chemical accident. At 43 he underwent corneal and limbal stem-cell transplant in one eye. Post-operatively MM's visual acuity was substantially impaired (due to neural losses) as were fMRI responses in primary visual cortex. When neural acuity losses were compensated for, his performance on 2D form, color, and motion tasks was remarkably good, as shown in Figure 41.11, and he showed strong MT+ responses to moving stimuli. However 3D form, depth, object, and face identification were extremely poor, and he showed no extrastriate responses to images of faces or objects.[123]

SRD was blind from birth due to dense congenital cataracts which were removed at age 12.[124] When tested at age 34 this subject also showed acuity losses, and a lack of impairment for simple form processing. However SRD showed greater proficiency than MM at higher-level visual tasks such as 3D form, depth, and face vs non-face discrimination. Like MM, motion processing appeared relatively unimpaired.

The retention of visual motion processing in these subjects is surprising (while formal tests of SRD's motion processing are not reported, no deficits have been noted), especially given recent evidence that after early blindness MT+ selectively processes auditory motion information.[81]

One possibility is that MM had time to establish motion processing before becoming blind at the age of 3. While SRD's cataracts were congenital, her pre-operative acuity was not measured (it was LP or above) and she may have retained sufficient visual acuity while blind to subserve basic motion processing. One possibility is that to a certain degree

motion processing may be "hard-wired" and require relatively little visual experience; another (as discussed below) is that the ability to process motion was maintained due to the existence of auditory motion-selective responses with that brain area.[81]

There are several possible reasons for the difference in performance between these two subjects on 3D form tasks. Perhaps the residual visual input received by SRD while blind provided enough stimulation for extrastriate cortex to retain the potential to regain complex visual function once the cataracts were removed at age 12. Alternatively, SRD's better performance might be the result of recovering sight at a relatively early age. Finally, given that extrastriate cortex probably remains relatively plastic even into adulthood, it is possible that her ability to perform these tasks gradually improved in the decades that followed sight restoration.

More recently, a study of MM and another early blind subject who regained some vision in adulthood demonstrated robust activations in area MT+ to both auditory and visual stimulation.[81] In these subjects MT+ responded to auditory motion, while in visually normal subjects MT+ did not show similar auditory responses. These auditory responses in MT+ were selective for motion as compared to other complex auditory stimuli, suggesting that cross-modal plasticity can be influenced by the normal functional specialization of a cortical region. These results further demonstrate that in the case of sight recovery, cross-modal responses can coexist with regained visual responses within the visual cortex (Fig. 41.12). The ability to process auditory input within MT+ as a result of visual deprivation does not seem to usurp the ability of MT+ to process visual input, and the restoration of visual responses does not seem to block auditory motion responses within MT+. These data therefore suggest that changes in neuronal function that occur as a result of cross-modal plasticity in visual cortex need not be detrimental to restored visual function. Indeed, the presence of functionally appropriate cross-modal responses may even help preserve the ability of an area to process visual information.

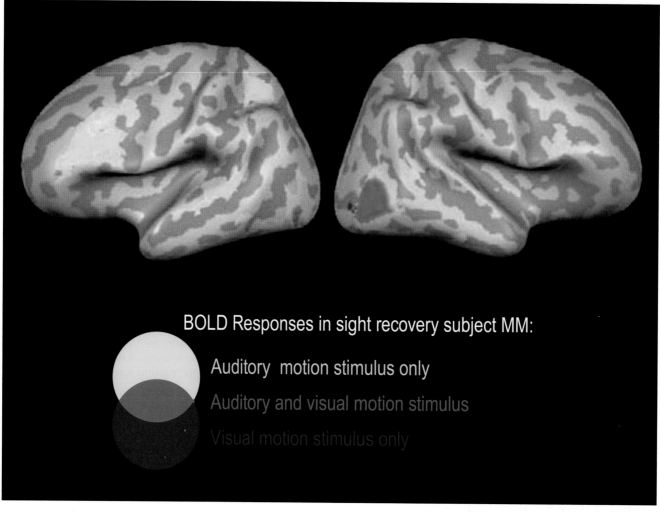

Figure 41.12 Responses to auditory motion stimuli and visual motion stimuli within sight recovery subject MM. Note the near-complete overlap between visual and auditory responses, indicating co-localization of auditory and visual motion responses. (Adapted from Saenz et al.[81])

Concluding remarks

Over the last decade a wide variety of studies have established that blindness results in remarkable changes in the response properties of human occipital cortex, and have further shown that these changes in response properties underlie a variety of skill differences between blind and sighted subjects.

However at this present date there is still much to be learned about the neurophysiological substrates of human cross-modal plasticity. Potential sources of cross-modal responses in occipital cortex include thalamic input, direct input from primary auditory or somatosensory areas, and feedback connections from polysensory extrastriate cortex. These connections may be mediated by pre-existing anatomical pathways, deviations in developmental "pruning", or entirely novel anatomical pathways. Patterns of connectivity are, of course, also likely to differ between early and late blind subjects, between foveal and peripheral cortex, and as a function of individual experience, such as the age of learning Braille. Fortunately, over the next decade the

increasing use of powerful non-invasive methods such as MRS, DTI, and TMS to examine the effects of blindness is likely to provide increasing insight into the biological foundations that underlie this remarkable example of plasticity within the human brain.

It is, of course, also likely that the factors that affect cross-modal plasticity interact. Early blind subjects tend to be much more fluent at skills such as Braille, use of a cane, or sign language than later blind subjects. One reason for this may be that these skills are easier to learn at an early age, when cortex is more plastic, and occipital cortex shows larger cross-modal plasticity. However the greater skills of the early deprived may also be partially due to behavioral and cultural factors – early blind subjects tend to learn Braille at school, and as a result learn to rely on Braille more heavily than those deprived at a later age. What is most likely is that these two factors may reinforce each other: Early blind subjects may learn sensory substitution skills more easily than those deprived later in life, this makes them rely on these skills more heavily, giving them practice and thereby further improving those skills, and this, in turn, increases the amount of cross-modal plasticity that occurs. In those deprived later in life, this "circle of competency" may be much more difficult to establish.

A deeper understanding of how neural deterioration as a result of blindness and cross-modal plasticity might interact will also prove increasingly important as new sight restoration procedures (such as retinal implants, epithelial stem cell replacements, gene therapies, and retinal transplants) become available. The cross-modal plasticity that occurs as a result of sensory deprivation is of benefit to the blind individual, playing an important role in allowing them to make better use of his/her remaining senses. It is therefore important to be aware that restoring vision may have the unintended consequence of interfering with cross-modal abilities. This is of particular concern, because, as described earlier, the neural deterioration that occurs as a result of early blindness does seem to severely limit an individual's ability to make use of restored vision.

Take, for example, the classic description of sight recovery subject SB's inability to deal with crossing the road after sight restoration:[125]

"He found the traffic frightening, and would not attempt to cross even a comparatively small street by himself. This was in marked contrast to his former behavior, as described to us by his wife, when he would cross any street in his own town by himself. In London, and later in his home town, he would show evident fear, even when led by a companion whom he trusted, and it was many months before he would venture alone."

Clearly, new visual information interfered with SB's ability to use cross-modal skills that he had relied on preoperatively. The decision to implant a prosthetic device or carry out a restorative surgery needs to consider not only the potential loss to any residual vision or hearing but also the potential of the device to interfere with the development of (in the case of children) or the use of cross-modal processing. Given that the ability to make use of the restored sense will often be limited, any potential deterioration in a patient's ability to navigate with a cane or read Braille, must be treated as a serious concern.

Acknowledgments

Many thanks to Geoffrey Boynton, Vivian Ciaramitaro, Karen Dobkins, and Melissa Saenz for comments on this manuscript.

References

1. De Volder AG, Catalan-Ahumada M, Robert A et al. Changes in occipital cortex activity in early blind humans using a sensory substitution device. Brain Res 1999; 826(1):128–134.
2. van Essen DC. Corticocortical and thalamocortical information flow in the primate visual system. Prog Brain Res 2005; 149:173–185.
3. Churchland PS, Sejnowski TJ. Perspectives on cognitive neuroscience. Science 1988; 242(4879):741–745.
4. Catalano SM, Shatz CJ. Activity-dependent cortical target selection by thalamic axons. Science 1998; 281(5376):559–562.
5. Crair MC, Gillespie DC, Stryker MP. The role of visual experience in the development of columns in cat visual cortex. Science 1998; 279(5350):566–570.
6. Crowley JC, Katz LC. Early development of ocular dominance columns. Science 2000; 290(5495):1321–1324.
7. Kahn DM, Krubitzer L. Massive cross-modal cortical plasticity and the emergence of a new cortical area in developmentally blind mammals. Proc Natl Acad Sci USA 2002; 99(17):11429–11434.
8. Sur M, Angelucci A, Sharma J. Rewiring cortex: the role of patterned activity in development and plasticity of neocortical circuits. J Neurobiol 1999; 41(1):33–43.
9. Ruthazer ES. You're perfect, now change – redefining the role of developmental plasticity. Neuron 2005; 45(6):825–828.
10. Innocenti GM, Clarke S. Bilateral transitory projection to visual areas from auditory cortex in kittens. Brain Res 1984; 316(1):143–148.
11. Maurer D, Lewis TL, Mondloch CJ. Missing sights: consequences for visual cognitive development. Trends Cogn Sci 2005; 9(3):144–151.
12. Callaway EM. Prenatal development of layer-specific local circuits in primary visual cortex of the macaque monkey. J Neurosci 1998; 18(4):1505–1527.
13. Price DJ, Blakemore C. Regressive events in the postnatal development of association projections in the visual cortex. Nature 1985; 316:721–724.
14. Price DJ, Blakemore C. The postnatal development of the association projection from visual cortical area 17 to area 18 in the cat. J Neurosci 1985; 5(9):2443–2452.
15. Rodman HR, Consuelos MJ. Cortical projections to anterior inferior temporal cortex in infant macaque monkeys. Vis Neurosci 1994; 11(1):119–133.
16. Innocenti GM. Exuberant development of connections, and its possible permissive role in cortical evolution. Trends Neurosci 1995; 18(9):397–402.
17. Dehay C, Bullier J, Kennedy H. Transient projections from the fronto-parietal and temporal cortex to areas 17, 18 and 19 in the kitten. Exp Brain Res 1984; 57(1):208–212.
18. Dehay C, Kennedy H, Bullier J. Characterization of transient cortical projections from auditory, somatosensory, and motor cortices to visual areas 17, 18, and 19 in the kitten. Behav Brain Res 1988; 29(3):237–244.
19. Innocenti GM, Berbel P, Clarke S. Development of projections from auditory to visual areas in the cat. J Comp Neurol 1988; 272(2):242–259.
20. Clarke S, Innocenti GM. Organization of immature intrahemispheric connections. J Comp Neurol 1986; 251(1):1–22.
21. Ghosh A, Shatz CJ. Pathfinding and target selection by developing geniculocortical axons. J Neurosci 1992; 12(1):39–55.
22. Ghazanfar AA, Schroeder CE. Is neocortex essentially multisensory? Trends Cogn Sci 2006; 10(6):278–285.
23. Neville HJ. Developmental specificity in neurocognitive development in humans. In: Gazzaniga M, ed. The cognitive neurosciences. Cambridge, MA: MIT Press, 1995:219–231.
24. Ptito M, Schneider FC, Paulson OB, Kupers R. Alterations of the visual pathways in congenital blindness. Exp Brain Res 2008; 187(1):41–49.
25. Watkins S, Shams L, Tanaka S, Haynes JD, Rees G. Sound alters activity in human V1 in association with illusory visual perception. Neuroimage 2006; 31(3):1247–1256.
26. Blake R, Sobel KV, James TW. Neural synergy between kinetic vision and touch. Psychol Sci 2004; 15(6):397–402.
27. Hagen MC, Franzen O, McGlone F, Essick G, Dancer C, Pardo JV. Tactile motion activates the human middle temporal/V5 (MT/V5) complex. Eur J Neurosci 2002; 16(5):957–964.
28. Amedi A, Malach R, Hendler T, Peled E, Zohary E. Visuo-haptic object-related activation in the ventral visual pathway. Nat Neurosci 2001; 4(3):324–330.
29. James TW, Humphrey GK, Gati JS, Servos P, Menon RS, Goodale MA. Haptic study of three-dimensional objects activates extrastriate visual areas. Neuropsychologia 2002; 40(1):1706–1714.
30. Pietrini P, Furey ML, Ricciardi E et al. Beyond sensory images: Object-based representation in the human ventral pathway. Proc Natl Acad Sci USA 2004; 101(15):5658–5663.
31. Burton H, McLaren DG, Sinclair RJ. Reading embossed capital letters: an fMRI study in blind and sighted individuals. Hum Brain Mapping 2006; 27(4):325–339.
32. Sathian K, Zangaladze A. Feeling with the mind's eye: contribution of visual cortex to tactile perception. Behav Brain Res 2002; 135(1–2):127–132.
33. Merabet L, Thut G, Murray B, Andrews J, Hsiao S, Pascual-Leone A. Feeling by sight or seeing by touch? Neuron 2004; 42(1):173–179.
34. Sadato N, Pascual-Leone A, Grafman J et al. Activation of the primary visual cortex by Braille reading in blind subjects. Nature 1996; 380(6574):526–528.
35. Sadato N, Pascual-Leone A, Grafman J, Deiber MP, Ibanez V, Hallett M. Neural networks for Braille reading by the blind. Brain 1998; 121(Pt 7):1213–1229.
36. Gizewski ER, Gasser T, de Greiff A, Boehm A, Forsting M. Cross-modal plasticity for sensory and motor activation patterns in blind subjects. Neuroimage 2003; 19(3):968–975.
37. Sadato N, Okada T, Kubota K, Yonekura Y. Tactile discrimination activates the visual cortex of the recently blind naive to Braille: a functional magnetic resonance imaging study in humans. Neurosci Lett 2004; 359(1–2):49–52.
38. Cohen LG, Celnik P, Pascual-Leone A et al. Functional relevance of crossmodal plasticity in blind humans. Nature 1997; 389:180–183.
39. Lewis JW, Beauchamp MS, DeYoe EA. A comparison of visual and auditory motion processing in human cerebral cortex. Cereb Cortex 2000; 10(9):873–888.
40. Calvert GA. Crossmodal processing in the human brain: insights from functional neuroimaging studies. Cereb Cortex 2001; 11(12):1110–1123.
41. Falchier A, Clavagnier S, Barone P, Kennedy H. Anatomical evidence of multimodal integration in primate striate cortex. J Neurosci 2002; 22(13):5749–5759.
42. Rockland KS, Ojima H. Multisensory convergence in calcarine visual areas in macaque monkey. Int J Psychophysiol 2003; 50(1–2):19–26.
43. Clavagnier S, Falchier A, Kennedy H. Long-distance feedback projections to area V1: implications for multisensory integration, spatial awareness, and visual consciousness. Cogn Affect Behav Neurosci 2004; 4(2):117–126.
44. Schroeder CE, Foxe JJ. The timing and laminar profile of converging inputs to multisensory areas of the macaque neocortex. Brain Res Cogn Brain Res 2002; 14(1):187–198.
45. Schroeder CE, Smiley J, Fu KG, McGinnis T, O'Connell MN, Hackett TA. Anatomical mechanisms and functional implications of multisensory convergence in early cortical processing. Int J Psychophysiol 2003; 50(1–2):5–17.
46. Cappe C, Barone P. Heteromodal connections supporting multisensory integration at low levels of cortical processing in the monkey. Eur J Neurosci 2005; 22(11):2886–2902.
47. Pascual-Leone A, Torres F. Plasticity of the sensorimotor cortex representation of the reading finger in Braille readers. Brain 1993; 116(Pt 1):39–52.

48. Goldreich D, Kanics IM. Tactile acuity is enhanced in blindness. J Neurosci 2003; 23(8):3439–3445.

49. Van Boven RW, Hamilton RH, Kauffman T, Keenan JP, Pascual-Leone A. Tactile spatial resolution in blind Braille readers. Neurology 2000; 54:2230–2236.

50. Theoret H, Merabet L, Pascual-Leone A. Behavioral and neuroplastic changes in the blind: evidence for functionally relevant cross-modal interactions. J Physiol Paris 2004; 98(1–3):221–233.

51. Grant AC, Thiagarajah MC, Sathian K. Tactile perception in blind Braille readers: a psychophysical study of acuity and hyperacuity using gratings and dot patterns. Percept Psychophys 2000; 62(2):301–312.

52. Kauffman T, Theoret H, Pascual-Leone A. Braille character discrimination in blindfolded human subjects. Neuroreport 2002; 13(5):571–574.

53. Merabet LB, Hamilton R, Schlaug G et al. Rapid and reversible recruitment of early visual cortex for touch. PLoS ONE 2008; 3(8):e3046.

54. Uhl F, Franzen P, Lindinger G, Lang W, Deecke L. On the functionality of the visually deprived occipital cortex in early blind persons. Neurosci Lett 1991; 124(2):256–259.

55. Cohen LG, Weeks RA, Sadato N, Celnik P, Ishii K, Hallett M. Period of susceptibility for cross-modal plasticity in the blind. Ann Neurol 1999; 45(4):451–460.

56. Sadato N. How the blind "see" Braille: lessons from functional magnetic resonance imaging. Neuroscientist 2005; 11(6):577–582.

57. Burton H, Snyder AZ, Conturo TE, Akbudak E, Ollinger JM, Raichle ME. Adaptive changes in early and late blind: a fMRI study of Braille reading. J Neurophysiol 2002; 87(1):589–607.

58. Burton H. Visual cortex activity in early and late blind people. J Neurosci 2003; 23(10):4005–4011.

59. Amedi A, Raz N, Pianka P, Malach R, Zohary E. Early "visual" cortex activation correlates with superior verbal memory performance in the blind. Nat Neurosci 2003; 6(7):758–766.

60. Kujala T, Alho K, Naatanen R. Cross-modal reorganization of human cortical functions. Trends Neurosci 2000; 23(3):115–120.

61. Hamilton R, Keenan JP, Catala M, Pascual-Leone A. Alexia for Braille following bilateral occipital stroke in an early blind woman. Neuroreport 2000; 11:237–240.

62. Hamilton RH, Pascual-Leone A. Cortical plasticity associated with Braille learning. Trends Cogn Sci 1998; 2(5):168–174.

63. Pascual-Leone A, Amedi A, Fregni F, Merabet LB. The plastic human brain cortex. Annu Rev Neurosci 2005; 28:377–401.

64. Amedi A, Merabet LB, Bermpohl F, Pascual-Leone A. The occipital cortex in the blind: Lessons about plasticity and vision. Curr Directions Psychol Sci 2005; 16:306–311.

65. Sadato N, Okada T, Honda M, Yonekura Y. Critical period for cross-modal plasticity in blind humans: a functional MRI study. Neuroimage 2002; 16(2):389–400.

66. Uhl F, Kretschmer T, Lindinger G et al. Tactile mental imagery in sighted persons and in patients suffering from peripheral blindness early in life. Electroencephalogr Clin Neurophysiol 1994; 91(4):249–255.

67. Rosler F, Roder B, Heil M, Hennighausen E. Topographic differences of slow event-related brain potentials in blind and sighted adult human subjects during haptic mental rotation. Brain Res Cogn Brain Res 1993; 1(3):145–159.

68. Goyal MS, Hansen PJ, Blakemore CB. Tactile perception recruits functionally related visual areas in the late-blind. Neuroreport 2006; 17(13):1381–1384.

69. Starlinger I, Niemeyer W. Do the blind hear better? Investigations on auditory processing in congenital or early acquired blindness. I. Peripheral functions. Audiology 1981; 20(6):503–509.

70. Lessard N, Pare M, Lepore F, Lassonde M. Early-blind human subjects localize sound sources better than sighted subjects. Nature 1998; 395(6699):278–280.

71. Muchnik C, Efrati M, Nemeth E, Malin M, Hildesheimer M. Central auditory skills in blind and sighted subjects. Scandinavian audiology 1991; 20(1):19–23.

72. Roder B. Improved auditory spatial tuning in blind humans. Nature 1999; 400:162–166.

73. Zwiers MP, Van Opstal AJ, Cruysberg JR. A spatial hearing deficit in early-blind humans. J Neurosci 2001; 21(9):RC142: 1–5.

74. Gougoux F, Zatorre RJ, Lassonde M, Voss P, Lepore F. A functional neuroimaging study of sound localization: visual cortex activity predicts performance in early-blind individuals. PLoS Biol 2005; 3(2):e27.

75. Leclerc C, Saint-Amour D, Lavoie ME, Lassonde M, Lepore F. Brain functional reorganization in early blind humans revealed by auditory event-related potentials. Neuroreport 2000; 11(3):545–550.

76. Alho K, Kujala T, Paavilainen P, Summala H, Naatanen R. Auditory processing in visual brain areas of the early blind: evidence from event-related potentials. Electroencephalogr Clin Neurophysiol 1993; 86(6):418–427.

77. Kujala T, Huotilainen M, Sinkkonen J et al. Visual cortex activation in blind humans during sound discrimination. Neurosci Lett 1995; 183(1–2):143–146.

78. Kujala T, Alho K, Huotilainen M et al. Electrophysiological evidence for cross-modal plasticity in humans with early- and late-onset blindness. Psychophysiology 1997; 34(2):213–216.

79. Kujala T, Palva MJ, Salonen O et al. The role of blind humans' visual cortex in auditory change detection. Neurosci Lett 2005; 379(2):127–131.

80. Roder B, Rosler F, Neville HJ. Effects of interstimulus interval on auditory event-related potentials in congenitally blind and normally sighted humans. Neurosci Lett 1999; 264(1–3):53–56.

81. Saenz M, Lewis LB, Huth AG, Fine I, Koch C. Visual Motion Area MT+/V5 Responds to Auditory Motion in Human Sight-Recovery Subjects. J Neurosci 2008; 28(20):5141–5148.

82. Arno P, De Volder AG, Vanlierde A et al. Occipital activation by pattern recognition in the early blind using auditory substitution for vision. Neuroimage 2001; 13(4):632–645.

83. Kujala T, Alho K, Paavilainen P, Summala H, Naatanen R. Neural plasticity in processing of sound location by the early blind: an event-related potential study. Electroencephalogr Clin Neurophysiol 1992; 84(5):469–472.

84. Weeks R, Horwitz B, Aziz-Sultan A et al. A positron emission tomographic study of auditory localization in the congenitally blind. J Neurosci 2000; 20(7):2664–2672.

85. Voss P, Gougoux F, Lassonde M, Zatorre RJ, Lepore F. A positron emission tomography study during auditory localization by late-onset blind individuals. Neuroreport 2006; 17(4):383–388.

86. Collignon O, Lassonde M, Lepore F, Bastien D, Veraart C. Functional cerebral reorganization for auditory spatial processing and auditory substitution of vision in early blind subjects. Cereb Cortex 2007; 17(2):457–465.

87. Kaiser J, Lutzenberger W, Preissl H, Ackermann H, Birbaumer N. Right-hemisphere dominance for the processing of sound-source lateralization. J Neurosci 2000; 20(17):6631–6639.

88. Heilman KM, Bowers D, Valenstein E, Watson RT. The right hemisphere: neuropsychological functions. J Neurosurg 1986; 64(5):693–704.

89. Burton H, Diamond JB, McDermott KB. Dissociating cortical regions activated by semantic and phonological tasks: a fMRI study in blind and sighted people. J Neurophysiol 2003; 90(3):1965–1982.

90. Burton H, Snyder AZ, Diamond JB, Raichle ME. Adaptive changes in early and late blind: a fMRI study of verb generation to heard nouns. J Neurophysiol 2002; 88(6):3359–3371.

91. Roder B, Stock O, Roesler F, Bien S, Neville H. Plasticity of language functions in blind humans: an fMRI study. Cognitive Neuroscience Society Abstracts 2001.

92. Amedi A, Floel A, Knecht S, Zohary E, Cohen LG. Transcranial magnetic stimulation of the occipital pole interferes with verbal processing in blind subjects. Nat Neurosci 2004; 7(11):1266–1270.

93. Wittenberg GF, Werhahn KJ, Wassermann EM, Herscovitch P, Cohen LG. Functional connectivity between somatosensory and visual cortex in early blind humans. Eur J Neurosci 2004; 20(7):1923–1927.

94. Liu Y, Yu C, Liang M et al. Whole brain functional connectivity in the early blind. Brain 2007; 130(Pt 8):2085–2096.

95. Ptito M, Fumal A, de Noordhout AM, Schoenen J, Gjedde A, Kupers R. TMS of the occipital cortex induces tactile sensations in the fingers of blind Braille readers. Exp Brain Res 2008; 184(2):193–200.

96. Minati L, Aquino D. Probing neural connectivity through Diffusion Tensor Imaging (DTI). In: R Trappl, ed. Cybernetics and systems, Vienna: ASCS 2006:263–268.

97. Schoth F, Burgel U, Dorsch R, Reinges MH, Krings T. Diffusion tensor imaging in acquired blind humans. Neurosci Lett 2006; 398(3):178–182.

98. Park HJ, Jeong SO, Kim EY et al. Reorganization of neural circuits in the blind on diffusion direction analysis. Neuroreport 2007; 18(17):1757–1760.

99. Shimony JS, Burton H, Epstein AA, McLaren DG, Sun SW, Snyder AZ. Diffusion tensor imaging reveals white matter reorganization in early blind humans. Cereb Cortex 2006; 16(11):1653–1661.

100. Karlen SJ, Kahn DM, Krubitzer L. Early blindness results in abnormal corticocortical and thalamocortical connections. Neuroscience 2006; 142(3):843–858.

101. Rauschecker JP. Compensatory plasticity and sensory substitution in the cerebral cortex. Trends Neurosci 1995; 18:36–43.

102. Rauschecker JP, Korte M. Auditory compensation for early blindness in cat cerebral cortex. J Neurosci 1993; 13(10):4538–4548.

103. Hyvarinen J, Carlson S, Hyvarinen L. Early visual deprivation alters modality of neuronal responses in area 19 of monkey cortex. Neurosci Lett 1981; 26(3):239–243.

104. Toldi J, Rojik I, Feher O. Neonatal monocular enucleation-induced cross-modal effects observed in the cortex of adult rat. Neuroscience 1994; 62(1):105–114.

105. Wanet-Defalque MC, Veraart C, De Volder A et al. High metabolic activity in the visual cortex of early blind human subjects. Brain Res 1988; 446(2):369–373.

106. De Volder AG, Bol A, Blin J et al. Brain energy metabolism in early blind subjects: neural activity in the visual cortex. Brain Res 1997; 750:235–244.

107. Muller CM. Dark-rearing retards the maturation of astrocytes in restricted layers of cat visual cortex. Glia 1990; 3(6):487–494.

108. Chen L, Yang C, Mower GD. Developmental changes in the expression of GABA(A) receptor subunits (alpha(1), alpha(2), alpha(3)) in the cat visual cortex and the effects of dark rearing. Brain Res Mol Brain Res 2001; 88(1–2):135–143.

109. Morales B, Choi SY, Kirkwood A. Dark rearing alters the development of GABAergic transmission in visual cortex. J Neurosci 2002; 22(18):8084–8090.

110. Winfield DA. The postnatal development of synapses in the different laminae of the visual cortex in the normal kitten and in kittens with eyelid suture. Brain Res 1983; 285(2):155–169.

111. O'Kusky JR. Synapse elimination in the developing visual cortex: a morphometric analysis in normal and dark-reared cats. Brain Res 1985; 354(1):81–91.

112. Huttenlocher PR, de Courten C. The development of synapses in striate cortex of man. Hum Neurobiol 1987; 6(1):1–9.

113. Wiesel TN, Hubel DH. Effects of visual deprivation on morphology and physiology of cells in the cat's lateral geniculate body. J Neurophysiol 1963; 26:978–993.

114. Wiesel TN, Hubel DH. Single cell responses in striate cortex of kittens deprived of vision in one eye. J Neurophysiol 1963; 26:1003–1017.

115. Wiesel TN, Hubel DH. Comparison of the effects of unilateral and bilateral eye closure on cortical unit responses in kittens. J Neurophysiol 1965; 28(6):1029–1040.

116. Wiesel TN, Hubel DH. Extent of recovery from the effects of visual deprivation in kittens. J Neurophysiol 1965; 28(6):1060–1072.

117. Shams L, Kamitani Y, Thompson S, Shimojo S. Sound alters visual evoked potentials in humans. Neuroreport 2001; 12(17):3849–3852.

118. Zhang N, Chen W. A dynamic fMRI study of illusory double-flash effect on human visual cortex. Exp Brain Res 2006; 172(1):57–66.

119. Fine I, Jacobs RA. Comparing perceptual learning across tasks: A review. J Vision 2002; 2(5):190–203.

120. Pascual-Leone A, Hamilton R. The metamodal organization of the brain. In: Casanova C, Ptito M, eds. Progress in Brain Research; 2001.

121. Weisser V, Stilla R, Peltier S, Hu X, Sathian K. Short-term visual deprivation alters neural processing of tactile form. Exp Brain Res 2005; 166(3–4):572–582.

122. Gothe J, Brandt SA, Irlbacher K, Roricht S, Sabel BA, Meyer BU. Changes in visual cortex excitability in blind subjects as demonstrated by transcranial magnetic stimulation. Brain 2002; 125(Pt 3):479–490.

123. Fine I, Wade A, Boynton GMB et al. The neural and functional effects of long-term visual deprivation on human cortex. Nature Neurosci 2003; 6(9).

124. Ostrovsky Y, Andalman A, Sinha P. Vision following extended congenital blindness. Psychol Sci 2006; 17(12):1009–1014.

125. Gregory RL, Wallace JG. Recovery from early blindness: a case study. Exp Psychological Soc Monograph 2. Cambridge: Heffer and Sons, 1963.

Index

T